The Bees of the World

The Bees

of the World
SECOND EDITION

Charles D. Michener

Entomology Division
University of Kansas Natural History Museum
and Biodiversity Research Center
and
Entomology Program, Department of Ecology
and Evolutionary Biology
University of Kansas

The Johns Hopkins University Press
Baltimore

© 2000, 2007 The Johns Hopkins University Press
All rights reserved. Published 2007
Printed in the United States of America on acid-free paper
9 8 7 6 5 4 3 2 1

The Johns Hopkins University Press
2715 North Charles Street
Baltimore, Maryland 21218-4363
www.press.jhu.edu

Library of Congress Cataloging-in Publication Data

Michener, Charles Duncan, 1918–
   The bees of the world / Charles D. Michener.—2nd ed.
     p.   cm.
   Includes bibliographical references.
   ISBN-13: 978-0-8018-8573-0 (hardcover : alk. paper)
   ISBN-10: 0-8018-8573-6 (hardcover : alk. paper)
   1. Bees—Classification. I. Title.
   QL566.M53   2007
   595.79'9—dc22   2006023201

A catalog record for this book is available from the British Library.

Title page illustration from H. Goulet and J. T. Huber (1993). Used with permission.

To my many students, now scattered over the world,
from whom I have learned much

and to my family, who lovingly tolerate an obsession with bees

Contents

Preface to the Second Edition ix
Preface to the First Edition xi
Abbreviations xvi

1. About Bees and This Book 1
2. What Are Bees? 3
3. The Importance of Bees 4
4. Development and Reproduction 6
5. Solitary versus Social Life 12
6. Floral Relationships of Bees 16
7. Nests and Food Storage 23
8. Parasitic and Robber Bees 30
9. Body Form, Tagmata, and Sex Differences 42
10. Structures and Anatomical Terminology of Adults 44
11. Structures and Terminology of Immature Stages 57
12. Bees and Sphecoid Wasps as a Clade 59
13. Bees as a Monophyletic Group 60
14. The Origin of Bees from Wasps 63
15. Classification of the Bee-Sphecoid Clade 65
16. Bee Taxa and Categories 66
17. Methods of Classification 76
18. The History of Bee Classifications 77
19. Short-Tongued versus Long-Tongued Bees 83
20. Family-Level Phylogeny and the Proto-Bee 88
21. The Higher Classification of Bees 93
22. Fossil Bees 98
23. The Geological History of Bees 100
24. Diversity and Abundance 102
25. Dispersal 105
26. Biogeography 106
27. Reduction or Loss of Structures 110
28. New and Modified Structures 112
29. Family-Group Names 117
30. Explanation of Taxonomic Accounts in Sections 36 to 121 118
31. Some Problematic Taxa 120
32. The Identification of Bees 121
33. Key to the Families, Based on Adults 122
34. Notes on Certain Couplets in the Key to Families (Section 33) 126
35. Practical Key to Family-Group Taxa, Based on Females 127
36. Family Stenotritidae 129
37. Family Colletidae 132
    38. Subfamily Colletinae 136
        39. Tribe Paracolletini 138
        40. Tribe Colletini 167
        41. Tribe Scraptrini 171
    42. Subfamily Diphaglossinae 173
        43. Tribe Caupolicanini 174
        44. Tribe Diphaglossini 177
        45. Tribe Dissoglottini 179
    46. Subfamily Xeromelissinae 180
    47. Subfamily Hylaeinae 187
    48. Subfamily Euryglossinae 220
49. Family Andrenidae 235
    50. Subfamily Alocandreninae 238
    51. Subfamily Andreninae 239

52. Subfamily Panurginae 270
   53. Tribe Protandrenini 273
   54. Tribe Panurgini 285
   55. Tribe Nolanomelissini 290
   56. Tribe Melitturgini 291
   57. Tribe Protomelitturgini 295
   58. Tribe Perditini 296
   59. Tribe Calliopsini 306
60. Subfamily Oxaeinae 316
61. Family Halictidae 319
   62. Subfamily Rophitinae 322
   63. Subfamily Nomiinae 332
   64. Subfamily Nomioidinae 345
   65. Subfamily Halictinae 348
      66. Tribe Halictini 354
      67. Tribe Augochlorini 393
68. Family Melittidae 413
   69. Subfamily Dasypodainae 416
      70. Tribe Dasypodaini 417
      71. Tribe Promelittini 422
      72. Tribe Sambini 423
   73. Subfamily Meganomiinae 426
   74. Subfamily Melittinae 429
75. Family Megachilidae 434
   76. Subfamily Fideliinae 436
      77. Tribe Pararhophitini 437
      78. Tribe Fideliini 438
   79. Subfamily Megachilinae 441
      80. Tribe Lithurgini 444
      81. Tribe Osmiini 448
      82. Tribe Anthidiini 491
      83. Tribe Dioxyini 538
      84. Tribe Megachilini 543
85. Family Apidae 587
   86. Subfamily Xylocopinae 592
      87. Tribe Manueliini 595
      88. Tribe Xylocopini 596
      89. Tribe Ceratinini 611
   90. Tribe Allodapini 619
   91. Subfamily Nomadinae 633
      92. Tribe Hexepeolini 637
      93. Tribe Brachynomadini 639
      94. Tribe Nomadini 643
      95. Tribe Epeolini 646
      96. Tribe Ammobatoidini 653
      97. Tribe Biastini 656
      98. Tribe Townsendiellini 659
      99. Tribe Neolarrini 660
      100. Tribe Ammobatini 661
      101. Tribe Caenoprosopidini 666
102. Subfamily Apinae 667
   103. Tribe Isepeolini 672
   104. Tribe Osirini 674
   105. Tribe Protepeolini 678
   106. Tribe Exomalopsini 680
   107. Tribe Ancylini 685
   108. Tribe Tapinotaspidini 687
   109. Tribe Tetrapediini 694
   110. Tribe Ctenoplectrini 697
   111. Tribe Emphorini 700
   112. Tribe Eucerini 707
   113. Tribe Anthophorini 742
   114. Tribe Centridini 753
   115. Tribe Rhathymini 762
   116. Tribe Ericrocidini 763
   117. Tribe Melectini 770
   118. Tribe Euglossini 778
   119. Tribe Bombini 785
   120. Tribe Meliponini 803
   121. Tribe Apini 830

Literature Cited 833
Addenda 905
Index of Terms 907
Index of Taxa 913

Color plates follow pages 46 and 78.

Preface to the Second Edition

Of course I was pleased when Johns Hopkins University Press indicated an interest in a revised edition of *The Bees of the World.*

A large review or revisional work like the original *The Bees of the World* inevitably goes out of date as new findings or interpretations are made, and also as errors or omissions in the original book are discovered.

For the original edition (usually referred to below as Michener, 2000) relevant publications were surveyed through 1997, with some additional material being included as addenda or otherwise as it came to my notice into 1999. Some publications that appeared in 1998 and 1999 were not cited or were inadequately utilized and are now properly incorporated, as are the items in the Addenda of the original edition. For the second edition, I have tried to cover literature through 2005, with additional material for 2006.

As in the original edition, arbitrary decisions about rank or recognition of taxa were often needed. Some recently proposed genera and subgenera are synonymized below, even though they constitute recognizable and even useful groups, because I am following so far as possible the practices involved in writing the first edition. The main point is that the classification should represent relationships, or similarities when phylogenetic relationships are in doubt. A classification that emphasizes differences can result in an unnecessary multiplication of taxa that (1) often can be distinguished only with difficulty or (2) represent odd derivatives whose relationships are better represented by inclusion within the recognized groups. There is no doubt, however, that some of the taxa here synonymized will be resurrected when new classifications are proposed, based on phylogenetic hypotheses that are yet to be developed.

Noteworthy developments in recent years are the number of phylogenetic analyses based on molecular characters, morphological characters, or both, prepared for groups of bees, and the classificatory changes based on these analyses. When the hypotheses are robust, I have modified the text in response. When it seems that changes in taxa studied or in the characters included in the analyses would change the outcome considerably, I usually report the study but, for the sake of stability, I do not change the classification. Changes in classifications as a result of cladograms subject to major change are not justified, because stability is an important feature for classifications. We should change a classification when we know that the change is justified, but otherwise we should not.

The acknowledgments for the first addition still stand, of course. Some of those listed have generously provided additional help. Other persons who have helped for this edition, as follows:

John S. Ascher, New York City, New York, USA;

Michael S. Engel, Lawrence, Kansas, USA (the color photos and plates of fossil bees, etc.);

Molly G. Rightmyer, Lawrence, Kansas, USA (Epeolini);

Allan H. Smith-Pardo, Lawrence, Kansas, USA (Augochlorini);

Michael Terzo, Mons, Belgium (Ceratina).

I especially appreciate the contribution of "Key to the Subgenera of Centris" by Ricardo Ayala, Estacion Chamela, Instituto de Biologia, Universidad Nacional Autonama de México, Ciudad de México.

I also appreciate the careful typing and editing by Anna J. Michener.

Lawrence, Kansas
July 18, 2006

Preface to the First Edition

In some ways this may seem the wrong time to write on the systematics of the bees of the world, the core topic of this book. Morphological information on adults and larvae of various groups has not been fully developed or exploited, and molecular data have been sought for only a few groups. The future will therefore see new phylogenetic hypotheses and improvement of old ones; work in these areas continues, and it has been tempting to defer completion of the book, in order that some of the new information might be included. But no time is optimal for a systematic treatment of a group as large as the bees; there is always significant research under way. Some genera or tribes will be well studied, while others lag behind, but when fresh results are in hand, the latter may well overtake the former. I conclude, then, that in spite of dynamic current activity in the field, now is as good a time as any to go to press.

This book constitutes a summary of what I have been able to learn about bee systematics, from the bees themselves and from the vast body of literature, over the many years since I started to study bees, publishing my first paper in 1935. Bee ecology and behavior, which I find fully as fascinating as systematics, are touched upon in this book, but have been treated in greater depth and detail in other works cited herein.

After periods when at least half of my research time was devoted to other matters (the systematics of Lepidoptera, especially saturniid moths; the biology of chigger mites; the nesting and especially social behavior of bees), I have returned, for this book, to my old preoccupation with bee systematics. There are those who say I am finally finishing my Ph.D. thesis!

My productive activity in biology (as distinguished from merely looking and being fascinated) began as a young kid, when I painted all the native plants that I could find in flower in the large flora of Southern California. When, after a few years, finding additional species became difficult, I expanded my activities to drawings of insects. With help from my mother, who was a trained zoologist, I was usually able to identify them to family. How I ultimately settled on Hymenoptera and more specifically on bees is not very clear to me, but I believe it had in part to do with *Perdita rhois* Cockerell, a beautiful, minute, yellow-and-black insect that appeared in small numbers on Shasta daisies in our yard each summer. The male in particular is so unbeelike that I did not identify it as a bee for several years; it was a puzzle and a frustration and through it I became more proficient in running small Hymenoptera, including bees, through the keys in Comstock's *Introduction to Entomology*.

Southern California has a rich bee fauna, and as I collected more species from the different flowers, of course I wanted to identify them to the genus or species level. Somehow I learned that T. D. A. Cockerell at the University of Colorado was the principal bee specialist active at the time. Probably at about age 14 I wrote to him, asking about how to identify bees. He responded with interest, saying that Viereck's *Hymenoptera of Connecticut* (1916) (which I obtained for $2.00) was not very useful in the West. Cresson's *Synopsis* (1887) was ancient even in the 1930s, but was available for $10.00. With these inadequate works I identified to genus a cigar box full of bees, pinned and labeled, and sent them to Cockerell for checking. He returned them, with identifications corrected as needed, and some specimens even identified to species.

Moreover, Cockerell wrote supporting comments about work on bees and invited me to meet him and P. H. Timberlake at Riverside, California, where the Cockerells would be visiting. Timberlake was interested in my catches because, although I lived only 60 miles from Riverside, I had collected several species of bees that he had never seen. Later, he invited me to accompany him on collecting trips to the Mojave and Colorado deserts and elsewhere.

Professor and Mrs. Cockerell later invited me to spend the next summer (before my last year in high school) in Boulder with them, where I could work with him and learn about bees. Cockerell was an especially charming man who, lacking a university degree, was in some ways a second-class citizen among the university faculty members. He had never had many students who became seriously interested in bees, in spite of his long career (his publications on bees span the years from 1895 to 1949) as the principal bee taxonomist in North America if not the world. Probably for this reason he was especially enthusiastic about my interest and encouraged the preparation and publication of my first taxonomic papers. Thus I was clearly hooked on bees well before beginning my undergraduate work at the University of California at Berkeley.

As a prospective entomologist I was welcomed in Berkeley and given space to work among graduate students. During my undergraduate and graduate career, interacting with faculty and other students, I became a comparative morphologist and systematist of bees, and prepared a dissertation (1942) on these topics, published with some additions in 1944. The published version included a key to the North American bee genera, the lack of which had sent me to Professor Cockerell for help a few years before. Especially important to me during my student years at Berkeley were E. Gorton Linsley and the late Robert L. Usinger.

There followed several years when, because of a job as lepidopterist at the American Museum of Natural History, in New York, and a commission in the Army, my research efforts were taken up largely with Lepidoptera and with mosquitos and chigger mites, but I continued to do limited systematic work on bees. It was while in the Army, studying the biology of chigger mites, that I had my first tropical experience, in Panama, and encountered, for the first time, living tropical stingless honey bees like *Trigona* and *Melipona* and orchid bees like *Euglossa* at orchid flowers. In 1948 I moved to the University of Kansas, and since about 1950 almost all of my research has been on bees.

Until 1950, I had gained little knowledge of bee behavior and nesting biology, having devoted myself to systematics, comparative morphology, and floral relationships, the last mostly because the flowers help you find the bees. In 1950, however, I began a study of leafcutter bee biology, and a few years later I began a long series of studies of nesting biology and social organization of bees, with emphasis on primitively social forms and on the origin and evolution of social behavior. With many talented graduate students to assist, this went on until 1990, and involved the publication in 1974 of *The Social Behavior of the Bees*. Concurrently, of course, my systematic studies continued; behavior contributes to systematics and vice versa, and the two go very well together.

Across the years, I have had the good fortune to be able to study both behavior and systematics of bees in many parts of the world. In addition to shorter trips of weeks or months, I spent a year in Brazil, a year in Australia, and a year in Africa. The specimens collected and ideas developed on these trips have been invaluable building blocks for this book.

Without the help of many others, preparing this book in its present form would have been impossible. A series of grants from the National Science Foundation was essential. The University of Kansas accorded me freedom to build up a major collection of bees as part of the Snow Entomological Division of the Natural History Museum, and provided excellent space and facilities for years after my official retirement. Students and other faculty members of the Department of Entomology also contributed in many ways. The editorial and bibliographic expertise of Jinny Ashlock, and her manuscript preparation along with that of Joetta Weaver, made the job possible. Without Jinny's generous help, the book manuscript would not have been completed. And her work as well as Joetta's continued into the long editorial process.

It is a pleasure to acknowledge, as well, the helpful arrangements made by the Johns Hopkins University Press and particularly the energy and enthusiasm of its science editor, Ginger Berman. For marvelously detailed and careful editing, I thank William W. Carver of Mountain View, California.

The help of numerous bee specialists is acknowledged at appropriate places in the text. I mention them and certain others here with an indication in some cases of areas in which they helped: the late Byron A. Alexander, Lawrence, Kansas, USA (phylogeny, *Nomada*); Ricardo Ayala, Chamela, Jalisco, Mexico (Centridini); Donald B. Baker, Ewell, Surrey, England, UK; Robert W. Brooks, Lawrence, Kansas, USA (Anthophorini, Augochlorini); J. M. F. de Camargo, Ribeirão Preto, São Paulo, Brazil (Meliponini); James W. Cane, Logan, Utah, USA (Secs. 1-32 of the text); Bryan N. Danforth, Ithaca, New York, USA (Perditini, Halictini); H. H. Dathe, Eberswalde, Germany (palearctic Hylaeinae); Connal D. Eardley, Pretoria, Transvaal, South Africa (Ammobatini); the late George C. Eickwort, Ithaca, New York, USA (Halictinae); Michael S. Engel, Ithaca, New York, USA (Augochlorini, fossil bees); Elizabeth M. Exley, Brisbane, Queensland, Australia (Euryglossinae); Terry L. Griswold, Logan, Utah, USA (Osmiini, Anthidiini); Terry F. Houston, Perth, Western Australia (Hylaeinae, *Leioproctus*); Wallace E. LaBerge, Champaign, Illinois, USA *(Andrena,* Eucerini); G. V. Maynard, Canberra, ACT, Australia (*Leioproctus*); Ronald J. McGinley, Washington

D.C., USA (Halictini); Gabriel A. R. Melo, Ribeirão Preto, São Paulo, Brazil (who read much of the manuscript); Robert L. Minckley, Auburn, Alabama, USA (Xylocopini); Jesus S. Moure, Curitiba, Paraná, Brazil; Christopher O'Toole, Oxford, England, UK; Laurence Packer, North York, Ontario, Canada (Halictini); Alain Pauly, Gembloux, Belgium (Malagasy bees, African Halictidae); Yuri A. Pesenko, Leningrad, Russia; Stephen G. Reyes, Los Baños, Philippines (Allodapini); Arturo Roig-Alsina, Buenos Aires, Argentina (phylogeny, Emphorini, Tapinotaspidini, Nomadinae); David W. Roubik, Balboa, Panama (Meliponini); Jerome G. Rozen, Jr., New York, N.Y., USA (Rophitini, nests and larvae of bees, and ultimately the whole manuscript); Luisa Ruz, Valparaíso, Chile (Panurginae); the late S. F. Sakagami, Sapporo, Japan (Halictinae, Allodapini, Meliponini); Maximilian Schwarz, Ansfelden, Austria (*Coelioxys*); Roy R. Snelling, Los Angeles, California, USA (Hylaeinae); Osamu Tadauchi, Fukuoka, Japan (*Andrena*); Harold Toro, Valparaíso, Chile (Chilicolini, Colletini); Danuncia Urban, Curitiba, Paraná, Brazil (Anthidiini, Eucerini); Kenneth L. Walker, Melbourne, Victoria, Australia (Halictini); V. B. Whitehead, Cape Town, South Africa (*Rediviva*); Paul H. Williams, London, England, UK (*Bombus*); Wu Yan-ru, Beijing, China; Douglas Yanega, Belo Horizonte, Minas Gerais, Brazil, and Riverside, California.

The persons listed above contributed toward preparation or completion of the book manuscript, or the papers that preceded it, and also in some cases gave or lent specimens for study; the following additional persons or institutions lent types or other specimens at my request: Josephine E. Cardale, Canberra, ACT, Australia; Mario Comba, Cecchina, Italy (*Tetralonia*); George Else and Laraine Ficken, London, England, UK; Yoshihiro Hirashima, Miyazaki City, Japan; Frank Koch, Berlin, Germany; Yasuo Maeta, Matsue, Japan; the Mavromoustakis Collection, Department of Agriculture, Nicosia, Cyprus (Megachilinae).

The illlustrations in this book are designed to show the diversity (or, in certain cases, similarity or lack of diversity) among bees. It was entirely impractical to illustrate each couplet in the keys—there are thousands of them—and I made no effort to do so, although references to relevant text illustrations are inserted frequently into the keys. Drs. R. J. McGinley and B. N. Danforth, who made or supervised the making of the many illustrations in Michener, McGinley, and Danforth (1994), have permitted reuse here of many of those illustrations. The other line drawings are partly original, but many of them are from works of others, reproduced here with permission. I am greatly indebted to the many authors whose works I have used as sources of illustrations; specific acknowledgments accompany the legends. In particular I am indebted to J. M. F. de Camargo for the use of two of his wonderful drawings of meliponine nests, and to Elaine R. S. Hodges for several previously published habitus drawings of bees. Modifications of some drawings, additional lettering as needed, and a few original drawings, as acknowledged in the legends, are the work of Sara L. Taliaferro; I much appreciate her careful work.

The colored plates reproduce photographs from the two sources indicated in the legends: Dr. E. S. Ross, California Academy of Sciences, San Francisco, California, USA, and Dr. Paul Westrich, Maienfeldstr. 9, Tübingen, Germany. I am particularly indebted to Drs. Ross and Westrich for making available their excellent photographs. It is worth noting here that many other superb photographs by Westrich were published in his two-volume work on the bees of Baden-Württemberg (Westrich, 1989).

Svetlana Novikova and Dr. Bu Wenjun provided English translations of certain materials from Russian and Chinese, respectively. Their help is much appreciated.

The text has been prepared with the help of the bees themselves, publications about them, and unpublished help from the persons listed above. I have not included here the names of all the persons responsible for publications that I have used and from which I have, in many cases, derived ideas, illustrations, bases for keys, and other items. They are acknowledged in the text. Several persons, however, have contributed previously unpublished keys that appear under their authorship in this book. Such contributions are listed below, with the authors' affiliations.

"Key to the Palearctic Subgenera of *Hylaeus*" by H.H. Dathe, Deutsches Entomologisches Institut, Postfach 10 02 38, D-16202 Eberswalde, Germany.

"Key to the New World Subgenera of *Hylaeus*" by Roy R. Snelling, Los Angeles County Museum of Natural History, 900 Exposition Boulevard, Los Angeles, California 90007, USA.

"Key to the Genera of Osmiini of the Eastern Hemisphere," "Key to the Subgenera of *Othinosmia*," and "Key to the Subgenera of *Protosmia*" by Terry L. Griswold, Bee Biology and Systematics Laboratory, UMC 53, Utah State University, Logan, Utah 84322-5310, USA.

"Key to the Genera of the Tapinotaspidini" by Arturo Roig-Alsina, Museo Argentino de Ciencias Naturales, Av. A. Gallardo 470, 1405 Buenos Aires, Argentina.

I have modified the terminology employed in these keys, as necessary, to correspond with that in use in other parts of this book (see Sec. 10). Several contributions became so modified by me that the original authors would scarcely recognize them. I have identified them by expressions such as "modified from manuscript key by . . ."

Names of authors of species are not integral parts of the names of the organisms. In behavioral or other nontaxonomic works I omit them except when required by editors. But in this book, which is largely a systematic account, I have decided to include them throughout for the sake of consistency.

A measure of the success of this book will be the need for revision as new work is completed and published. Not only does this book contain a great deal of information about bees, but, by inference or explicitly, it indicates myriad topics about which more information is needed. I hope that it points the way for the numerous researchers who will take our knowledge beyond what is here included, and beyond what is to be found in the nearly 2,500 items in the Literature Cited.

Lawrence, Kansas
1999

ABBREVIATIONS

The following are used in the text:

BP = before the present time

Code = International Code of Zoological Nomenclature

Commission = International Commission on Zoological Nomenclature

L-T = long-tongued (see Sec. 19)

myBP = million years before the present

s. str. (sensu stricto) = in the strict sense

s. l. (sensu lato) = in the broad sense

S-T = short-tongued (see Sec. 19)

S1, S2, etc. = first, second, etc., metasomal sterna

scutellum = mesoscutellum

scutum = mesoscutum

stigma = pterostigma of forewing

T1, T2, etc. = first, second, etc., metasomal terga

The terminology of wing veins and cells also involves abbreviations; see Section 10.

The Bees of the World

1. About Bees and This Book

Since ancient times, people have been drawn to the study of bees. Bees are spritely creatures that move about on pleasant bright days and visit pretty flowers. Anyone studying their behavior should find them attractive, partly because they work in warm sunny places, during pleasant seasons and times of day. The sights and odors of the fieldwork ambience contribute to the well-being of any researcher. Moreover, bees are important pollinators of both natural vegetation and crops, and certain kinds of bees make useful products, especially honey and wax. But quite apart from their practical importance, at least since the time of Aristotle people have been interested in bees because they are fascinating creatures. We are social animals; some bees are also social. Their interactions and communications, which make their colonial life function, have long been matters of interest; we wonder how a tiny brain can react appropriately to societal problems similar to those faced by other social animals, such as humans. For a biologist or natural historian, bees are also fascinating because of their many adaptations to diverse flowers; their ability to find food and nesting materials and carry them over great distances back to a nest; their ability to remember where resources were found and return to them; their architectural devices, which permit food storage, for example, in warm, moist soil full of bacteria and fungi; and their ability to rob the nests of others, some species having become obligate robbers and others cuckoolike parasites. These are only a few of the interesting things that bees do.

I consider myself fortunate to work with such a biologically diverse group of insects, one of which is the common honey bee, *Apis mellifera* Linnaeus. In terms of physiology and behavior, it is the best-known insect. Educated guesses about what happens in another bee species are often possible because we know so much about *Apis mellifera*. In this book, however, *Apis* is treated briefly, like all other bee taxa, its text supplemented by references to books on *Apis* biology; the greater part of this book concerns bees (the great majority) that are not social.

Sections 2 to 28, and what follows here, are intended to provide introductory materials important to an understanding of all bees and aspects of their study. Some topics are outlined only briefly to provide background information; others are omitted entirely; still others are dealt with at length and with new or little-known insights when appropriate.

This book is largely an account of bee classification and of phylogeny, so far as it has been pieced together, i.e., the systematics of all bees of the world. All families, subfamilies, tribes, genera, and subgenera are characterized by means of keys and (usually brief) text comments to facilitate identification. I include many references to such revisional papers or keys as exist, so that users can know where to go to identify species. About 17,500 species have been placed as to genus and subgenus (see Sec. 16); no attempt has been made even to list species here, although the approximate number known for each genus and subgenus is given in Table 16-1, as well as under each genus or subgenus in Sections 36 to 121. Aspects of bee biology, especially social and parasitic behavior, nest architecture, and ecology, including floral associations, are indicated. Major papers on bee nesting biology and floral relationships are also cited. The reader can thus use this book as a guide to the extensive literature on bee biology. Because the male genitalia and associated sterna of bees provide characters useful at all levels, from species to family, and because they are often complex and difficult to describe, numerous illustrations are included, as well as references to publications in which others are illustrated.

Besides entomologists, this book should be useful to ecologists, pollination biologists, botanists, and other naturalists who wish to know about the diversity and habits of bees. Such users may not be greatly concerned with details of descriptive material and keys, but should be able to gain a sense of the taxonomic, morphological, and behavioral diversity of the bee faunas with which they work. As major pollinators, bees are especially important to pollination biologists. I hope that by providing information on the diversity of bees and their classification and identification, this book will in some mostly indirect ways contribute to pollination biology.

To a significant degree, studies of bees, especially nontaxonomic studies, have developed in Russia more or less independently from those in the rest of the world. It is fortunate, therefore, that a huge list of publications by Russian authors and others in the former U.S.S.R., with summaries in English, has recently been published (Pesenko and Astafurova, 2003). This large work, covering the period from 1771 to 2002 provides information on 3,027 publications by 1,126 authors. Michener (2000) and many earlier works utilized the publications in Russian on bee systematics and the like; it is especially in pollination biology, ecology, nesting behavior, and related fields that much Russian work has been little known in the West.

The title of this book can be read to indicate that the book should deal, to at least some degree, with all aspects of bee studies. It does not. All aspects of **apiculture**, the study and practice of honey bee culture, based on managed colonies of *Apis mellifera* Linnaeus and *A. cerana* Fabricius, are excluded. The findings about sensory physiology as well as behavioral interactions, including communication, foraging behavior, and caste control are virtually omitted, although they constitute some of the most fascinating aspects of biology and in the hands of Karl von Frisch led to a Nobel prize. A major work, principally about communication, is Frisch (1967).

Whether the scientific study of communication in *Apis* is part of apiculture is debatable, but the study of all the other species of bees is not; such studies are subsumed under the term **melittology**. Persons studying bees other than *Apis* and concerned about the negative and awkward expression "non-*Apis* bees" would do well to call themselves melittologists and their field of study melittology.

I would include under the term "melittology" the taxonomic, comparative, and life history studies of species of the genus *Apis,* especially in their natural habitats. This book is about melittology.

Users of this book may wonder about the lack of a glossary. Definitions and explanations of structures, given mostly in Section 10, are already brief and would be largely repeated in a glossary. The terms, including many that are explained only by illustrations, are therefore included in the Index of Terms, with references to pages where they are defined, illustrated, or explained. Some terminology, e.g., that relevant only to certain groups of bees, is explained in other sections, and indexed accordingly.

2. What Are Bees?

A major group of the order Hymenoptera is the Section Aculeata, i.e., Hymenoptera whose females have stings—modifications of the ovipositors of ancestral groups of Hymenoptera. The Aculeata include the wasps, ants, and bees. Bees are similar to one group of wasps, the sphecoid wasps, but are quite unlike other Aculeata. Bees are usually more robust and hairy than wasps (see Pls. 3-15), but some bees (e.g., *Hylaeus,* Pl. 1; *Nomada,* Pl. 2) are slender, sparsely haired, and sometimes wasplike even in coloration. Bees differ from nearly all wasps in their dependence on pollen collected from flowers as a protein source to feed their larvae and probably also for ovarian development by egg-laying females. (An exception is a small clade of meliponine bees of the genus *Trigona,* which use carrion instead of pollen.) Unlike the sphecoid wasps, bees do not capture spiders or insects to provide food for their offspring. Thus nearly all bees are plant feeders; they have abandoned the ancestral carnivorous behavior of sphecoid wasp larvae. (Adult wasps, like bees, often visit flowers for nectar; adult sphecoid wasps do not collect or eat pollen.)

Bees and the sphecoid wasps together constitute the superfamily Apoidea (formerly called Sphecoidea, but see Michener, 1986a). The Apoidea as a whole can be recognized by a number of characters, of which two are the most conspicuous: (1) the posterior pronotal lobe is distinct but rather small, usually well separated from and below the tegula; and (2) the pronotum extends ventrally as a pair of processes, one on each side, that encircle or nearly encircle the thorax behind the front coxae. See Section 10 for explanations of morphological terms and Section 12 for more details about the Apoidea as a whole.

As indicated above, the Apoidea are divisible into two groups: the sphecoid (or apoid) wasps, or Spheciformes, and the bees, or Apiformes (Brothers, 1975). Older authors also used the term Anthophila for the bees and the name was resurrected by Engel (2005); Apiformes is a junior synonym of Anthophila. No priority rules govern such names. I prefer Apiformes because it contrasts well with Spheciformes and because the word itself makes its meaning immediately clear. Structural characters of bees that help to distinguish them from sphecoid wasps are (1) the presence of branched, often plumose, hairs, and (2) the hind basitarsi, which are broader than the succeeding tarsal segments. The proboscis is in general longer than that of most sphecoid wasps. The details, and other characteristics of bees, are explained in Section 12.

A conveniently visible character that easily distinguishes nearly all bees from most sphecoid wasps is the golden or silvery hairs on the lower face of most such wasps, causing the face to glitter in the light. Bees almost never exhibit this characteristic, because their facial hairs are duller, often erect, often plumose, or largely absent. This feature is especially useful in distinguishing small, wasplike bees such as *Hylaeus* from similar-looking sphecoid wasps such as the Pemphredoninae.

The monophyletic Apiformes is believed to have arisen from the paraphyletic Spheciformes. **Monophyletic** is used here in the strict sense sometimes called **Holophyletic.** Such a group (1) arose from a single ancestor that would be considered a member of the group, and (2) includes all taxa derived from that ancestor. Groups termed **Paraphyletic** also arose from such an ancestor but do not include all of the derived taxa. Brief explanations of other terms used by systematists are appended to Section 12.

3. The Importance of Bees

Probably the most important activity of bees, in terms of benefits to humans, is their pollination of natural vegetation, something that is rarely observed by nonspecialists and is almost never appreciated; see Section 6. Of course the products of honey bees—i.e., wax and honey plus small quantities of royal jelly—are of obvious bernefit, but are of trivial value compared to the profoundly important role of bees as pollinators. Most of the tree species of tropical forests are insect-pollinated, and that usually means bee-pollinated. A major study of tropical forest pollination was summarized by Frankie et al. (1990); see also Jones and Little (1983), Roubik (1989), and Bawa (1990). In temperate climates, most forest trees (pines, oaks, etc.) are wind-pollinated, but many kinds of bushes, small trees, and herbaceous plants, including many wild flowers, are bee-pollinated. Desertic and xeric scrub areas are extremely rich in bee-pollinated plants whose preservation and reproduction may be essential in preventing erosion and other problems, and in providing food and cover for wildlife. Conservation of many habitats thus depends upon preservation of bee populations, for if the bees disappear, reproduction of major elements of the flora may be severely limited.

Closer to our immediate needs, many cultivated plants are also bee-pollinated, or they are horticultural varieties of bee-pollinated plants. Maintenance of the wild, bee-pollinated populations is thus important for the genetic diversity needed to improve the cultivated strains. Garden flowers, most fruits, most vegetables, many fiber crops like flax and cotton, and major forage crops such as alfalfa and clover are bee-pollinated.

Some plants require bee pollination in order to produce fruit. Others, commonly bee-pollinated, can self-pollinate if no bees arrive; but inbreeding depression is a frequent result. Thus crops produced by such plants are usually better if bee-pollinated than if not; that is, the numbers of seeds or sizes of fruits are enhanced by pollination. Estimates made in the late 1980s of the value of insect-pollinated crops (mostly by bees) in the USA ranged from $4.6 to $18.9 billion, depending on various assumptions on what should be included and how the estimate should be calculated. Also doubtful is the estimate that 80 percent of the crop pollination by bees is by honey bees, the rest mostly by wild bees. But whatever estimates one prefers, bee pollination is crucially important (see O'Toole, 1993, for review), and the acreages and values of insect-pollinated crops are increasing year by year.

Wild bees may now become even more important as pollinators than in the past, because of the dramatic decrease in feral honey bee populations in north-temperate climates due to the introduction into Europe and the Americas of mites such as *Varroa* and tracheal mites, which are parasites of honey bees. Moreover, there are various crops for which honey bees are poor pollinators compared to wild bees. Examples of wild bees already commercially used are *Osmia cornifrons* (Radoszkowski), which pollinates fruit trees in Japan; *Megachile rotundata* (Fabricius), which pollinates alfalfa in many areas; *Bombus terrestris* (Linnaeus), which pollinates tomatoes in European greenhouses, and other *Bombus* species that do the same job elsewhere. O'Toole (1993) has given an account of wild bee species that are important in agriculture, and the topic was further considered by Parker, Batra, and Tepedino (1987), Torchio (1991), and Richards (1993). Since honey bees do not sonicate tubular anthers to obtain pollen (i.e., they do not buzz-pollinate; see Sec. 6), they are not effective pollinators of Ericaceae, such as blueberries and cranberries, or Solanaceae such as eggplants, chilis, and tomatoes.

Many bees are pollen specialists on particular kinds of flowers, and even among generalists, different kinds of bees have different but often strong preferences. Therefore, anyone investigating the importance of wild bees as pollinators needs to know about kinds of bees. The classification presented by this book can suggest species to consider; for example, if one bee is a good legume pollinator, a related one is likely to have similar behavior. Proboscis length is an important factor in these considerations, for a bee with a short proboscis usually cannot reach nectar in a deep flower, and probably will not take pollen there either, so is unlikely to be a significant pollinator of such a plant.

Although some bees, e.g., Euglossini and many Meliponini, inhabit undisturbed forests, especially in the tropics, many and probably most bee species inhabit savannas and forest margins and thrive in moderately disturbed areas. Temperate forests were presumably never good places for most bees since they consist largely of trees that do not produce flowers visited by bees. There were probably fewer bees in primeval temperate forests than now inhabit the same areas. Now, with pastures, waste lands, road and field margins, and forests that have been opened by cutting, once scarce species of bees have become abundant and presumably play important roles as pollinators of the vegetation. Even unbroken prairie appears to have fewer bees than more or less abandoned disturbed areas. Thus Laroca (1983) found more species and individuals in a little used, often disturbed university campus area, not planted with lawn or other vegetation, than in nearby prairie in Kansas. Of course the bees in such areas are the species that may pollinate agricultural crops or ornamental plants and, therefore, may be selected for practical use as pollinators.

Although disturbed areas often have more bees than undisturbed areas, it does not follow, of course, that the more disturbance, the better for the bees. When a whole area becomes a monoculture of corn or any other crop, most bees become dependant on small waste areas or road verges, and with small isolated populations, the number of species and individuals will soon be reduced.

In many countries the populations of wild bees have been seriously reduced by human activity. Destruction of habitats supporting host flowers, destruction of nesting sites (most often in soil) by agriculture, roadways, etc., and overuse of insecticides, among other things, appear to be major factors adversely affecting wild bee populations. Introduction or augmentation of a major competitor for food, the honey bee, has probably also affected some species of wild bees. Recent accounts of such problems and some possible solutions were published by Banaszak (1995) and Matheson et al. (1996); see also O'Toole (1993).

National and international organizations are now seriously considering and publicizing the need to conserve native pollinators (mostly bees) to maintain agricultural production as well as survival and well-being of native vegetation. Recent accounts of pollination problems worldwide have been included in numerous reports; see Kevan and Imperatriz-Fonseca (2002) and Freitas and Pereira (2004).

A single example analyzing the economic impact of bees (*Apis* and Meliponini) on a 1,065-hectare coffee plantation (this crop does not even require insect pollination) is telling. Bees from two forest fragments (46 and 111 hectares) translated into ~US$60,000 per year added income for the farm (Ricketts et al., 2004). Another relevant work is Imperatriz-Fonesca, Saraiva, and De Jong (2006). A useful bibliography for the neotropics is Anonymous (2006).

One of the problems in verifying recent declines of pollinating insects has always been lack of firm quantitative data on abundance in times past. Collectors' recollections and museum specimens suffice for presence (and even absence) information, but not for changes in abundance. A recent study (Biesmeijer et al., 2006) in The Netherlands and United Kingdom, however, compares older (before 1980) versus recent (after 1980) data and shows substantial declines in local bee diversity, in particular, declines of oligolectic and other specialist species, and parallel declines in plant species dependent for pollination services on such bees.

4. Development and Reproduction

As in all insects that undergo complete metamorphosis, each bee passes through egg, larval, pupal, and adult stages (Fig. 4-1).

The haplodiploid system of sex determination has had a major influence on the evolution of the Hymenoptera. As in most Hymenoptera, eggs of bees that have been fertilized develop into females; those that are unfertilized develop into males. Sex is controlled by alleles at one or a few loci; heterozygosity at the sex-determining locus (or loci) produces females. Development without fertilization, i.e., with the haploid number of chromosomes, produces males, since heterozygosity is impossible. Inbreeding results in some diploid eggs that are homozygous at the sex-determining loci; diploid males are thus produced. Such males are ordinarily reproductively useless, for they tend to be short-lived (those of *Apis* are killed as larvae) and to have few sperm cells; moreover, they may produce triploid offspring that have no reproductive potential. Thus for practical purposes the sex-determining mechanism is haplodiploid.

When she mates, a female stores sperm cells in her spermatheca; she usually receives a lifetime supply. She can then control the sex of each egg by liberating or not liberating sperm cells from the spermatheca as the egg passes through the oviduct.

Because of this arrangement, the female (of species whose females are larger than males) is able to place female-producing eggs in large cells with more provisions, male-producing eggs in small cells. In *Apis*, the males of which are larger than the workers, male-producing cells are larger than worker-producing cells and presumably it is the cell size that stimulates the queen to fertilize or not to fertilize each egg. Moreover, among bees that construct cells in series in burrows, the female can place male-producing eggs in cells near the entrance, from which the resultant adults can escape without disturbing the slower-developing females. The number of eggs laid during her lifetime by a female bee varies from eight or fewer for some solitary species to more than a million for queens of some highly social species. Females of solitary bees give care and attention to their few offspring by nest-site selection, nest construction, brood-cell construction and provisioning, and determination of the appropriate sex of the individual offspring. Of course, it is such atttention to the well-being of offspring that makes possible the low reproductive potential of many solitary bees.

The eggs of nearly all bees are elongate and gently curved, whitish with a soft, membranous chorion ("shell") (Fig. 4-1a), usually laid on (or rarely, as in *Lithurgus*, within) the food mass provided for larval consumption. In bees that feed the larvae progressively (*Apis*, *Bombus*, and most Allodapini), however, the eggs are laid with little or no associated food. Eggs are commonly of moderate size, but are much smaller in highly social bees, which lay many eggs per unit time, and in *Allodapula* (Allodapini), which lays eggs in batches, thus several eggs at about the same time. Eggs are also small in many cleptoparasitic bees (see Sec. 8) that hide their eggs in the brood cells of their hosts, often inserted into the walls of the cells; such eggs are often quite specialized in shape and may have an operculum through which the larva emerges (see Sec. 8). Conversely, eggs are very large in some subsocial or primitively eusocial bees like *Braunsapis* (Allodapini) and *Xylocopa* (Xylocopini). Indeed, the largest of all insect eggs are probably those of large species of *Xylocopa*, which may attain a length of 16.5 mm, about half the length of the bee's body. Iwata and Sakagami (1966) gave a comprehensive account of bee egg size relative to body size.

The late-embryonic development and hatching of eggs

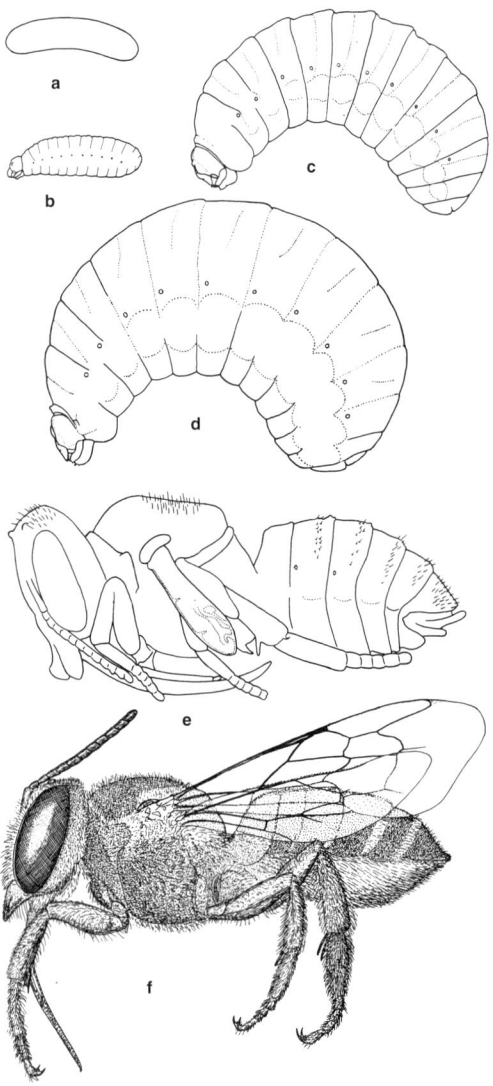

Figure 4-1. Stages in the life cycle of a leafcutter bee, *Megachile brevis* Cresson. **a**, Egg; **b-d,** First stage, half-grown, and mature larvae; **e,** Pupa; **f,** Adult. From Michener, 1953b.

has proved to be variable among bees and probably relevant to bee phylogeny. Torchio, in various papers (e.g., 1986), has studied eggs of several different bee taxa immersed in paraffin oil to render the chorion transparent. Before hatching, the embryo rotates on its long axis, either 90° or 180°. In some bees (e.g., Nomadinae) the chorion at hatching is dissolved around the spiracles, then lengthwise between the spiracles; eventually, most of the chorion disappears. In others the chorion is split but otherwise remains intact.

Larvae of bees are soft, whitish, legless grubs (Fig. 4-1b-d). In mass-provisioning bees, larvae typically lie on the upper surface of the food mass and eat what is below and in front of them, until the food is gone. They commonly grow rapidly, molting about four times as they do so. The shed skins are so insubstantial and hard to observe that for the great majority of bees the number of molts is uncertain. For the honey bee *(Apis)* there are five larval instars (four molts before molting into the pupal stage); and five is probably the most common number in published reports such as that of Lucas de Olivera (1960) for *Melipona*. In some bees, e.g., most nonparasitic Apinae other than the corbiculate tribes (i.e., in the old Anthophoridae), the first stage remains largely within the chorion, leaving only four subsequent stages (Rozen, 1991b); such development is also prevalent in the Megachilidae. In the same population of *Megachile rotundata* (Fabricius) studied by Whitfield, Richards, and Kveder (1987), some individuals had four instars and others five. The first to third instars were almost alike in size in the two groups, but the terminal fourth instar was intermediate in size between the last two instars of five-stage larvae.

Markedly different young larvae are found in most cuckoo bees, i.e., cleptoparasitic bees. These are bees whose larvae feed on food stored for others; details are presented in Section 8. Young larvae of many such parasites have large sclerotized heads and long, curved, pointed jaws with which they kill the egg or larva of the host (Figs. 84-5, 91-6, 105-3). They then feed on the stored food and, after molting, attain the usual grublike form of bee larvae.

Other atypical larvae are those of allodapine bees, which live in a common space, rather than as a single larva per cell, and are mostly fed progressively. Especially in the last instar, they have diverse projections, tubercles, large hairs, and sometimes long antennae that probably serve for sensing the movements of one another and of adults, and obviously function for holding masses of food and retaining the larval positions in often vertical nest burrows (Fig. 90-6). Many of the projections are partly retracted when the insect is quiet, but when touched with a probe or otherwise disturbed, they are everted, probably by blood pressure.

It has been traditional to illustrate accounts of bee larvae (unfortunately, this is largely not true for adults). The works of Grandi (culminating in Grandi, 1961), Michener (1953a), McGinley (1981), and numerous papers by Rozen provide drawings of mature larvae of many species. Various other authors have illustrated one or a few larvae each. Comments on larval structures appear as needed later in the phylogenetic and systematic parts of this book. Unless otherwise specified, such statements always concern mature larvae or prepupae. Accounts of larvae are listed in a very useful catalogue by McGinley (1989), organized by family, subfamily, and tribe. It is therefore unnecessary except for particular cases to cite references to papers on larvae in this book, and such citations are mostly omitted to save space.

As in other aculeate Hymenoptera, the young larvae of bees have no connection between the midgut and the hindgut, so cannot defecate. This arrangement probably arose in internal parasitoid ancestors of aculeate Hymenoptera, which would have killed their hosts prematurely if they had defecated into the host's body cavity. In some bees defecation does not begin until about the time that the food is gone; in others, probably as a derived condition, feces begin to be voided well before the food supply is exhausted. After defecation is complete the larva is smaller and often assumes either a straighter or a more curled form than earlier and becomes firmer; its skin is less delicate, and any projections or lobes it may have are commonly more conspicuous (Fig. 4-2). This last part of the last larval stage is called the **prepupa** or **defecated larva**; this stage is not shown in Figure 4-1. Most studies of larvae, e.g., those by Michener (1953a) and numerous studies by Rozen, are based on such larvae, because they are often available and have a rather standard form for each species; feeding larvae are so soft that their form frequently varies when preserved. Prepupae are often the stage that passes unfavorable seasons, or that survives in the cell for one to several years before development resumes. Houston (1991b), in Western Australia, recorded living although flaccid prepupae of *Amegilla dawsoni* (Rayment) up to ten years old; his attempts to break their diapause were not successful. Such long periods of developmental stasis probably serve as a risk-spreading strategy so that at least some individuals survive through long periods of dearth, the emergence of adults being somehow synchronized with the periodic blooming of vegetation. Even in nondesertic climates, individuals of some species remain in their cells as prepupae or sometimes as adults for long periods. Thus Fye (1965) reported that in a single population and even in a single nest of *Osmia atriventris* Cresson in Ontario, Canada, some individuals emerge in about one year, others in two years.

An account of delayed emergence as risk-spreading behavior in a desert bee, *Macrotera portalis* (Timberlake), was given by Danforth (1999b); see also Minckley, Cane, and Kervin (2000). An account of possible risk-spreading behavior in *Hylaeus grossus* Cresson, by construction of most cells in different places rather than assembled in one or few nests as do related forms, is by Michener and Brooks (2003).

Mature larvae of many bees spin **cocoons**, usually at about the time of larval defecation, much as is the case in sphecoid wasps. The cocoons are made of a framework of silk fibers in a matrix that is produced as a liquid and then solidifies around the fibers; the cocoon commonly consists of two to several separable layers. Various groups of bees, including most short-tongued bees, have lost cocoon-spinning behavior and often are protected instead by the cell lining secreted by the mother bee. Cocoon spinning sometimes varies with the generation. Thus in *Microthurge corumbae* (Cockerell), even in the mild cli-

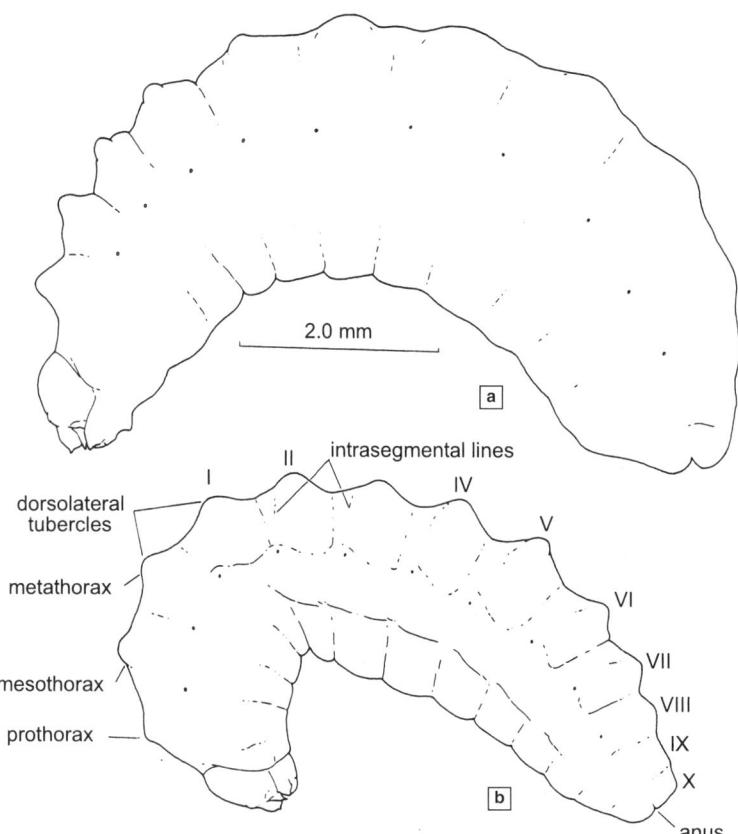

Figure 4-2. Change of a mature larva to a prepupa shown by last larval stadium of *Neffapis longilongua* Ruz. **a**, Predefecating larva; **b**, Postdefecating larva or prepupa. (The abdominal segments are numbered.) From Rozen and Ruz, 1995.

mate of the state of São Paulo, Brazil, the cocoons of the overwintering generation are firm and two-layered but those of the other generation consist of a single layer of silk (Mello and Garófalo, 1986). Similar observations were made in California by Rozen (1993a) on *Sphecodosoma dicksoni* (Timberlake), in which larvae in one-layered cocoons pupated without diapausing, whereas those in two-layered cocoons overwintered as prepupae. In other cases, in a single population, some individuals make cocoons and others do not. Thus in *Exomalopsis nitens* Cockerell, those that do not make cocoons pupate and eclose promptly, but those that make cocoons diapause and overwinter (Rozen and Snelling, 1986).

When conditions are appropriate, pupation occurs; for all eusocial species and many others this means soon after larval feeding, defecation, and prepupal formation are completed. In other species pupation occurs only after a long prepupal stage. **Pupae** are relatively delicate, and their development proceeds rapidly; among bees the pupa is never the stage that survives long unfavorable periods. Because they are delicate and usually available for short seasons only, fewer pupae than larvae have been preserved and described. Pupal characters are partly those of the adults, but pupae do have some distinctive and useful characters of their own (see Michener, 1954a). Most conspicuous are various spines, completely absent in adults, that provide spaces in which the long hairs of the adults develop. Probably as a secondary development, long spines of adults, like the front coxal spines of various bees, arise within pupal spines.

Adults finally appear, leave their nests, fly to flowers and mate, and, if females, according to species, either return to their nests or construct new nests elsewhere. Many bees have rather short adult lives of only a few weeks. Some, however, pass unfavorable seasons as adults; if such periods are included, the adult life becomes rather long. For example, in most species of *Andrena*, pupation and adult maturation occur in the late summer or fall, but the resulting adults remain in their cells throughout the winter, leaving their cells and coming out of the ground in the spring or summer to mate and construct new nests. In most Halictinae, however, although pupation of reproductives likewise occurs in late summer or autumn, the resulting adults emerge, leave the nest, visit autumn flowers for nectar, and mate. The males soon die, but the females dig hibernaculae (blind burrows), de novo or inside the old nest, for the winter. A few bees live long, relatively active adult lives. These include the queens of eusocial species and probably most females of the Xylocopinae and some solitary Halictinae. Among the Xylocopinae, a female Japanese *Ceratina* in captivity is known to have laid eggs in three different summer seasons, although only one was laid in the last summer (for summary, see Michener, 1985b, 1990d). Females of some solitary *Lasioglossum* (Halictinae), especially in unfavorable climates (only a few sunny days per summer month, as in Dartmoor,

England) provision a few cells, stop by midsummer, and provision a few more cells the following year (Field, 1996). Like the variably long inactivity of prepupae described above, this may be a risk-spreading strategy.

The male-female interactions among bees are diverse; they must have evolved to maximize access of males to receptive females and of females to available males. The mating system clearly plays a major role in evolution. Reviews are by Alcock et al. (1978) and Eickwort and Ginsberg (1980); the following account lists only a few examples selected from a considerable literature. Many male bees course over and around flowers or nesting sites, pouncing on females. In other species females go to particular types of vegetation having nothing to do with food or nests and males course over the leaves, pouncing on females when they have a chance. In these cases mating occurs quickly, lasting from a few seconds to a minute or two, and one's impression is that the female has no choice; the male grasps her with legs and often mandibles and mates in spite of apparent struggles. The male, however, may be quite choosy. In *Lasioglossum zephyrum* (Smith), to judge largely by laboratory results, males over the nesting area pounce on small dark objects including females of their own species, in the presence of the odor of such females, but do so primarily when stimulated by unfamiliar female odor, thus presumably discriminating against female nestmates, close relatives of nestmates, and perhaps females with whom they have already mated (Michener and Smith, 1987). Such behavior should promote outbreeding. Conversely, it would seem, males are believed to fly usually over the part of the nesting area where they were reared; they do not course over the whole nesting aggregation (Michener, 1990c). Such behavior should promote frequent inbreeding, since males would often encounter relatives, yet they appear to discriminate against their sisters. The result should be some optimum level of inbreeding.

In communal nests of *Andrena scotia* Perkins (as *A. jacobi* Perkins) studied in Sweden, over 70 percent of the females mated within the nests with male nestmates (Paxton and Tengo, 1996; Paxton et al., 2000). Such behavior, with its potential for inbreeding, may be common in communal bees. Given the rarity with which one sees mating in most species of bees, it may be that mating in nests is also common in some solitary species. In species that have several sex-determining loci, inbreeding may not be particularly disadvantageous, because deleterious genes tend to be eliminated by the haploid-male system.

In some bees, females tend to mate only once. Males in such species attempt to mate with freshly emerged young females, even digging into the ground to meet them, as in *Centris pallida* Fox (Alcock, 1989) or *Colletes cunicularius* (Linnaeus) (Cane and Tengö, 1981). In other species females mate repeatedly. The behavior of males suggests that there is sperm precedence such that sperm received from the last mating preferentially fertilize the next egg. Males either (1) mate again and again with whatever females they can capture, as in *Dianthidium curvatum* (Smith) (Michener and Michener, 1999), or (2) remain in copula for long periods with females as they go about their foraging and other activities, thus preventing the females from mating with other males (many Panurginae, personal observation).

In *Colletes cunicularius* (Linnaeus), *Lasioglossum zephyrum* (Smith), *Centris pallida* Fox, and many others, female-produced pheromones seem to stimulate or attract males, but in *Xylocopa varipuncta* Patton a male-produced pheromone attracts females to mating sites (Alcock and Smith, 1987). Some male *Bombus* scent-mark a path that they then visit repeatedly for females (Haas, 1949). In other species of *Bombus,* those with large-eyed males, the males wait on high perches and dash out to passing objects including *Bombus* females (Alcock and Alcock, 1983). Although playing a role in all cases, vision is no doubt especially important also in other bees with large-eyed males, such as *Apis mellifera* Linnaeus, the males of which fly in certain congregating areas and mate with females that come to those areas; see also the comments on mating swarms of large-eyed males in Section 28.

Most male bees can mate more than once, but in Meliponini and Apini the male genitalia or at least the endophallus is torn away in mating, so that after the male mates he soon dies.

Males in many species of bees in diverse families have enlarged and modified legs, especially the hind legs (see Sec. 28), or broad heads with long, widely separated mandibles. These are features that help in holding females for mating, and may be best developed in large males. Many males of *Megachile* have elaborately enlarged, flattened, pale, fringed front tarsi (Fig. 84-19). Wittmann and Blochtein (1995) found epidermal glands in the front basitarsi; at mating these tarsi hold the female's antennae, or cover her eyes. This behavior and gland product are presumably associated with successful mating or mate choice.

Large-headed males occur especially in some Andrenidae—both Andreninae and Panurginae—and in some Halictinae. Large heads appear to be characteristic of the largest individuals of certain species, no doubt as an allometric phenomenon. In two remarkable examples, one an American *Macrotera* (Panurginae) (Danforth, 1991b) and the other an Australian *Lasioglossum (Chilalictus)* (Halictinae) (Kukuk and Schwarz, 1988; Kukuk, 1997), the large-headed males (Figs. 4-3, 58-3, 58-4) have relatively short wings and are flightless nest inhabitants in communal colonies. The large-headed males also have large mandibles and fight to the death when more than one is present in a nest. Smaller males of each species have normal-sized wings and fly. Great size variation among males and macrocephaly may be most frequent in, or even limited to, communal species. Unlike most male bees that leave the nest permanently and mate elsewhere, short-winged males mate with females of their own colony. Thus such a male is often the last to mate with a female before she lays an egg.

In some other bees the male mating strategy also varies greatly with body size. Large males usually fly about the nesting sites, finding young females as they emerge from the ground or even digging them out of the ground, presumably guided by odor. Small males seek females on flowers or in vegetation near the nesting area. Such dual behavior is documented for *Centris pallida* Fox (Alcock, 1989) in the Centridini, and for *Habropoda depressa* (Fowler) (Barthell and Daly, 1995) and *Amegilla dawsoni* (Rayment) (Alcock, 1996), both in the Anthophorini.

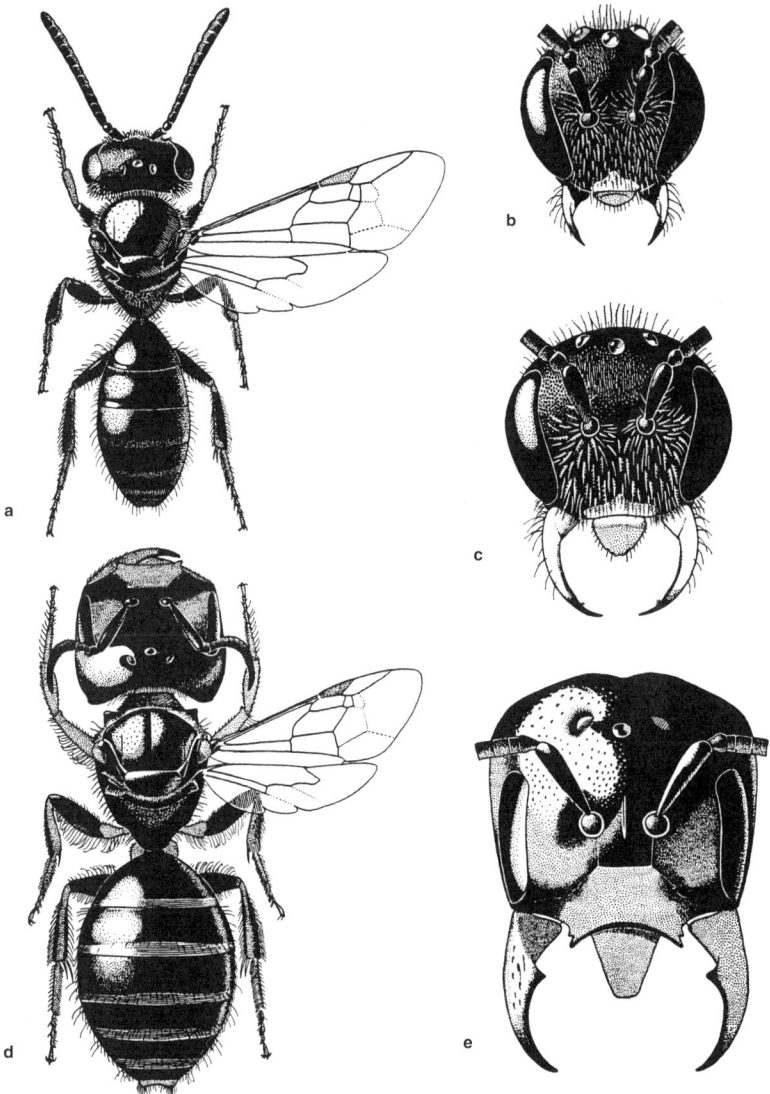

Figure 4-3. Male morphs of *Lasioglossum (Chilalictus) hemichalceum* (Cockerell) from Australia. **a,** Ordinary male; **b, c,** Heads of same; **d,** Large, flightless male; **e,** Head of same. From Houston, 1970.

Such behavior seems akin to that of *Anthidium manicatum* (Linnaeus), in which large males have mating territories that include flowers visited by females (Severinghaus, Kurtak, and Eickwort, 1981), whereas small ones are not territorial, and to that of certain *Hylaeus* (Alcock and Houston, 1996), in which large males with a strong ridge or tubercle on S3 are territorial whereas small ones with reduced ventral armature or none are not territorial. The ventral armature is apparently used to grasp an adversary against the thoracic venter by curling the metasoma.

An interesting and widespread feature in Hymenoptera is the prevalence of yellow (or white) coloration on the faces of males. If a black species has any pale coloration at all, it will be on the face (usually the clypeus) of males. Species with other yellow markings almost always have more yellow on the face of the male than on that of the female, although on the rest of the body yellow markings often do not differ greatly between the sexes. Groups like *Megachile* that lack yellow integumental markings frequently have dense yellow or white hairs on the face of the male, but not on that of the female. In mating attempts males usually approach females from above or behind, so that neither sex has good views of the face of the other. Therefore I do not suppose that the male's yellow face markings have to do with male-female recognition or mating. Rather, I suppose that they are involved in male-male interactions, when males face one another in disputes of various sorts. Sometimes, males of closely related species, such as *Xylocopa virginica* (Linnaeus) and *californica* Cresson, differ in that one (in this case *virginica*) has yellow on the face but the other does not. Someone should study the male-male interactions in such species pairs. Presumably, male behavior linked to yellow male faces is found in thousands of species of Hymenoptera.

Obviously, the variety of mating systems in bees deserves further study, both because of its interest for bee evolution and for evolutionary theory. Moreover, because of the frequency of morphological or chromatic correlates, mating systems and such correlates are important for systematists.

5. Solitary versus Social Life

Many works treat aspects of behavior of diverse kinds of bees. Specialized papers are cited throughout this book; some more general treatments are the books by Friese (1923), with its interesting colored plates of nests of European bees; Iwata (1976), with its review of previous work on the behavior of bees and other Hymenoptera; and O'Toole and Raw (1991), which offers readable accounts and fine illustrations of bees worldwide. A major aspect of behavior involves intraspecific interactions, i.e., social behavior in a broad sense. Courtship and mating are treated briefly in Section 4. Here we consider colonial behavior—its origin as well as its loss.

Some female bees are solitary; others live in colonies. A **solitary** bee constructs her own nest and provides food for her offspring; she has no help from other bees and usually dies or leaves before the maturation of her offspring. Sometimes such a female feeds and cares for her offspring rather than merely storing food for them; such a relationship is called **subsocial**.

A **colony** consists of two or more adult females, irrespective of their social relationships, living in a single nest. Frequently the females constituting a colony can be divided into (1) one to many **workers**, which do most or all of the foraging, brood care, guarding, etc., and are often unmated; and (2) one **queen**, who does most or all of the egg laying and is usually mated. The queen is often, and in some species always, larger than her workers, but sometimes the difference is only in mean size. In some social halictines, the largest females have extraordinarily large heads, often with toothed genae or other cephalic modifications probably resulting from allometry.

For many people, bees are thought of as honey-producing social insects living in perennial colonies, each of which consists of a queen and her many daughter workers. This is indeed the way of life for the honey bees (genus *Apis*) and the stingless honey bees (*Trigona, Melipona*, etc.) of the tropics. Queens and workers in these cases are morphologically very different, and the queen is unable to live alone (e.g., she never forages); nor do workers alone form viable colonies (they cannot mate and therefore cannot produce female offspring). These are the **highly eusocial** bees. Such bees always live in colonies, and new colonies are established socially, by groups or swarms. Only two tribes, the Apini and the Meliponini (family Apidae), consist of such bees.

Most bumble bees (Bombini) and many sweat bees (Halictinae) and carpenter bees and their relatives (Xylocopinae) may live in small colonies, mostly started by single females working as solitary individuals performing all necessary functions of nest construction, foraging, provisioning cells or feeding larvae progressively, and laying eggs. Later, on the emergence of daughters, colonial life may arise, including division of labor between the nest foundress (queen) and workers. These are **primitively eusocial** colonies. Queens and workers are essentially alike morphologically, although often differing in size; they differ more distinctly in physiology and behavior. Such colonies usually break down with production of reproductives; thus the colonies are obligately temporary rather than potentially permanent like those of highly eusocial bees.

Since in primitively eusocial bees the individual that becomes the queen cannot always be recogized until she has workers, the word **gyne** has been introduced for both potential queens and functional queens. The word is most frequently used for females that will or may become queens (Michener, 1974a), but have not yet done so. Thus it is proper to say that a gyne establishes her nest by herself in the spring, and becomes a queen when the colony develops.

Both permanent honey bee colonies and temporary bumble bee or halictine colonies are called **eusocial**, meaning that they have division of labor (egg-layer vs. foragers) among cooperating adult females of two generations, mothers and daughters. Such a definition is adequate for most bees; there is currently much discussion of modifying the definition of "eusocial" and relevant terms to make them useful across the board for all groups of social animals or, alternatively, eliminating them in favor of a system of terms that addresses the social levels among diverse animal species as well as the variability within species (see Crespi and Yanaga, 1995; Gadagkar, 1995; Sherman et al., 1995; Costa and Fitzgerald, 1996; and Wcislo, 1997a).

Not all bees that live in colonies are eusocial. Sometimes a small colony consists of females of the same generation, probably often sisters, that show division of labor, with a principal egg-layer or queen and one or more principal foragers or workers. Such colonies, called **semisocial**, may not be worth distinguishing from primitively eusocial colonies. As noted below, they often arise when the queen of a primitively eusocial colony dies and her daughters carry on, one of them commonly mating and becoming the principal egg-layer or replacement queen.

Some bee colonies lack division of labor or castes: all colony members behave similarly. Some such colonies are **communal**; two or more females use the same nest, but each makes and provisions her own cells and lays an egg in each of them. In most or all species that have communal colonies, other individuals in the same populations nest alone, and are truly solitary. Thus colonial life is facultative. A possible precursor of communal behavior arises when a nest burrow, abandoned by its original occupant, is then occupied by another bee of the same species (Neff and Rozen, 1995). Such behavior is rarely reported, because without marked bees, one does not know of it. A condition that appears to promote communal behavior is very hard soil or other substrate, because it is much easier to join other bees in a preexisting nest than to excavate a new nest starting at the surface (Michener and Rettenmeyer, 1956; Bennett and Breed, 1985).

A little-used additional term is **quasisocial**. It applies to the relatively rare case in which a few females occupying a nest cooperate in building and provisioning cells, but different individuals (as opposed to a single queen) lay eggs in cells as they are completed. That is, all the females have functional ovaries, mate, and can lay eggs. This may not be the terminal or most developed social state for any species of bees, but at times some colonies exhibit this condition.

When one opens a nest containing a small colony of bees, it is often impossible to recognize the relationships among the adult female inhabitants. The colony might be communal, quasisocial, or semisocial. Only observations and dissections will clarify the situation. Such colonies can be called **parasocial**, a noncommittal umbrella term used for a colony whose members are of a single generation and interact in any of the three ways indicated or in some as yet unrecognized way. At first, primitively eusocial colonies may look like parasocial colonies, but one individual, the queen (mother), is older, more worn, and sometimes larger than the others, which are workers (daughters). The queen commonly has enlarged ovaries and sperm cells in the spermatheca; workers usually do not.

Because many species pass through ontogenetic stages of sociality or are extremely variable in this regard, terms like "eusocial" should be applied to colonies, not species, except when dealing with permanently highly social forms like *Apis* and the Meliponini. For example, a nest may contain a single female, a gyne who has provided for and is protecting her immature progeny in a subsocial relationship. After emergence of the first adult workers, however, the nest contains a eusocial colony. There are species of Halictinae that have eusocial colonies in warmer climates but are solitary in cold climates (Eickwort et al., 1996). A single population may consist of some individuals functioning like solitary bees while others are eusocial, as observed in a New York population of *Halictus rubicundus* (Christ) (Yanega, 1988, 1989). In most Allodapini and some Ceratinini (in the Xylocopinae), although nests harboring colonies of two or more cooperating adult females exist, most nests contain a lone adult, rearing her young subsocially without benefit of a worker or other adult associate (Michener, 1990b). In the nests containing two or more adults, one is often the principal layer, thus a queen, and the others (or one of them), principal foragers and often unmated, and thus workers. Later, if the queen dies, one of the workers may become a queen; the result is a semisocial colony of sisters. But if several or all of the sisters become reproductive, the result is a quasisocial colony. Of course there are sometimes intergradations or mixtures. In such bees eusocial and other social relationships have arisen even though most individuals of the species never experience cooperative behavior among adult females.

The terminology summarized above is not always helpful; I introduce it here because some of the terms are often found in the literature and are used later in this book. A case in which the terminology ("communal," "semisocial," etc.) is not useful is found in the autumnal colonies of *Exoneura bicolor* Smith in Australia (Melna and Schwarz, 1994). The bees in such colonies can be divided into four classes, according to their activities. Yet there is no reproductive activity at this time; thus there is no queenlike or workerlike division of labor, but rather division along other lines. It may be that in the Allodapini, whenever two or more adults nest together, some sort of division of labor ensues.

Many kinds of bees that nest in the ground construct numerous nests in limited areas; a patch of earth, a path, or an earthen bank may be peppered with their holes (Fig. 5-1). Such groupings of individual nests are called **aggregations**. Each burrow may be made and inhabited by one female or may contain some sort of small colony (Fig. 5-2). Some aggregations doubtless result from the availability of local patches of suitable soil, but often the bees choose to aggregate in only part of an extensive area that appears uniform. Sometimes, gregarious behavior seems to be a response to the presence of other bees or bee nests—thus a social phenomenon; see Michener (1974a). In other cases bees may be returning to the site of their own emergence or "birth." In the literature, aggregations are sometimes called "colonies." I think it is best to avoid this usage and to limit the word "colony" as indicated above.

The above is a brief account of a large topic, the social diversity found among bees. Additional information and sources can be found in Michener, 1974a, 1985b, 1990c, d. The great abundance of the highly social forms (honey bees, stingless bees) almost wherever they occur suggests that such sociality itself is an enormous advantage in the presumed competition with other bees. The great body of literature on the theory of eusocial behavior of insects mostly addresses in one way or another the problem of how it is possible for attributes like those of workers to evolve and be passed on from generation to generation, even though they decrease the probability of their bearer's leaving progeny. Briefly expressed, an individual's overall

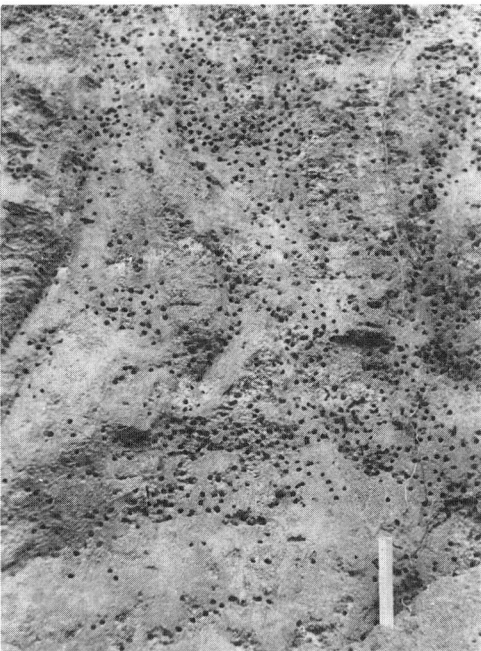

Figure 5-1. Part of an aggregation of nests made by females of *Trigonopedia oligotricha* Moure. The holes were in a vertical bank near Rio de Janeiro, Brazil. From Michener and Lange, 1958c.

Figure 5-2. Part of an aggregation of nests, each containing a eusocial colony of *Halictus hesperus* Smith, in Panama. The tumuli at the nest entrances make the site conspicuous. Photo by R. W. Brooks.

or inclusive fitness consists of (1) its direct fitness (its number of offspring and their contributions to subsequent generations; i.e., the fitness resulting from its own actions) and (2) an indirect effect resulting from its influence on the fitness of other individuals, weighted by its coefficient of relatedness to those individuals. Association of two individuals (x and y) should be favored by selection if x experiences no decrease in direct individual fitness that is not more than offset by an increased fitness received indirectly through the actions of y. A worker bee, a daughter of a queen, is closely related to the queen; and the queen's other offspring are genetically similar to the worker. The worker's individual fitness is zero if she produces no offspring, but the proliferation of genes like those of the worker is promoted by the benefits for the queen and her colony provided by the worker. And because of the haplodiploid sex-determination system in Hymenoptera, relationships between full sisters are closer than are mother-daughter relationships. Therefore a group of sisters (workers) may increase their inclusive fitness more by caring for their sisters, younger offspring of their mother, than by producing their own offspring. They thus gain in fitness by staying with their mother (the queen). This situation, resulting from haplodiploidy, is presumably a partial explanation of the frequency of evolution of eusociality in the Hymenoptera, compared to its rarity in other animals.

One must also observe, however, that associates in colonies are not always closely enough related to satisfy such thinking, perhaps because of multiple mating by gynes, or the formation of colonies by not necessarily related individuals from the general population. There must also be, then, additional factors that can promote colony formation. These are ecological factors, namely, mutualism, including such behavior as defense against natural enemies (Lin and Michener, 1972), cooperative nest construction, and continued protection of a mother's young offspring in spite of her death. Although behavioral studies (e.g., nest switching among communal nests) long ago suggested low coefficients of relationship among communal colony members, DNA fingerprinting makes such investigations easier and far more decisive. A recent study, containing relevant references to earlier works, is that of *Macrotera texana* (Cresson) (Panurginae). It showed that in this commonly communal bee, relationships among colony members did not differ significantly from relationships among non-nestmates of the same population (Danforth, Neff, and Barretto-Ko, 1996).

The terms explained above for various social levels among bees were often thought of as reflecting a possible evolutionary sequence of species from solitary to eusocial. Thus a parasocial sequence consisted of solitary, communal, semisocial, and eusocial species and a subsocial sequence consisted of solitary, subsocial, and eusocial species. It now seems probable that eusociality has often arisen directly from solitary antecedents (Michener, 1985b). Communal behavior is an alternative way of living together that does not usually lead to eusociality, according to Danforth, Neff, and Barretto-Ko (1996).

One caution is important in considering these matters: in haplodiploid insects like the Hymenoptera, the conditions for the *origin* of social behavior may differ from the requirements for the survival, maintenance, and subsequent *evolution* of social behavior. The expression "the evolution of social behavior" can include both, a fact that in the past has resulted in substantial confusion.

A review of the literature on the origins and evolution of sociality is beyond the scope of this book. Starr (1979) and Andersson (1984) provided comprehensive reviews. Radchenko (1993) gave a useful list of the many publications on social behavior in the Halictinae, a group that is particularly critical for evaluating theories about social behavior because of the many origins and losses of eusociality that have occurred in this subfamily. Packer (1991) and Richards (1994) have examined the distribution of sociality on phylogenies of the halictines *Lasioglossum (Evylaeus)* and *Halictus*, respectively; see also Packer, 1997. Valuable recent papers on the social evolution of the augochlorine bee *Augochlorella* are those of U. G. Mueller (see Mueller, 1997). See also Wcislo and Gonzalez (2006).

Clearly, there is no ready answer to the often-asked question about the number of times that eusocial behavior has arisen in the course of bee evolution. If each population of many species can become either more or less social, ranging from always social at the season of maximal activity to never social, the number of origins becomes both unknowable and useless. It is the wrong question. Nonetheless, there are interesting phylogenetic aspects to the occurrence of eusociality. So far as I know, it is never found in most bee families, although communal behavior occurs at least occasionally in nearly all families. Evidently, the Halictidae (especially Halictinae) and the Xylocopinae (especially Allodapini) have special potentials for repeated evolution of eusocial behavior. But even within these groups, there is much variation in the frequency of eusocial colonies, as shown by the efforts (cited above) to plot sociality on phylogenies. An interesting example is in the halictine subgenus *Lasioglossum* s. str., most species of which are consistently solitary. Packer (1997), however, cites meager evidence that one species, *L. aegyptellum* (Strand), can be eusocial; if this interpretation is correct, eusociality in this species is believed to be a recent evolutionary development in a clade of basically solitary species. Phylogenetic analysis of certain *Lasioglossum* species by Danforth, Conway, and Ji (2003), based on nucleotide sequences of three genes (one mitochondrial, two nuclear), indicated a single origin of eusociality within the genus but six reversals, i.e., changes from eusocial to solitary. Thus losses of sociality may be much more frequent than origins of sociality.

Questions about the origins of eusocial behavior are usually asked with the assumption that evolution is from solitary to eusocial. Certainly among bees as a whole this has been true. But there may be many cases in which species of primitively eusocial clades like those of many of the Halictinae have evolved to become solitary. Packer (1997) and his associates, when plotting behavior on cladograms, have discovered diverse cases of this kind; see also Wcislo and Danforth (1997).

6. Floral Relationships of Bees

Wind and bees are the world's most important pollinating agents. Bees are either beneficial or actually essential for the pollination, and therefore for the sexual reproduction, of much of the natural vegetation of the world, as well as for many agricultural crops (see Sec. 3). The pollinators are primarily female bees, which collect pollen as the principal protein source in their own food and especially to feed their larvae. Flowers produce not only nectar and sometimes oil but also excess pollen as bait or reward. The pollen that may fertilize ovules is that which bees lose inadvertently on floral stigmata as they go about collecting nectar, pollen, or other material. Male bees of nearly all species, as well as the females of parasitic species, take nectar from flowers but carry only the pollen that happens to stick to them. They thus play a role in pollination, but a less important one than that of the females, which actively collect pollen and (as workers) in eusocial groups are vastly more numerous than males. Parasitic bees are often not very hairy and thus probably play a less significant role in pollination than do males of hairy bees, which are likely to carry abundant pollen. Wcislo and Cane (1996) and Westerkamp (1996) gave excellent recent reviews of floral resource utilization by bees. Barth (1991) provided an excellent account of flowers and their insect pollinators, and referred to older books on the subject.

There must have been a sort of general or diffuse coevolution, diverse species of plants influencing and being influenced by a diverse fauna of bees. Many bees take nectar from the same flowers that provide them with pollen. Thus features of a flower that facilitate nectar collecting and the corresponding features of bees that also facilitate nectar collecting are important to the flower because of associated pollen transfer. Short-tongued or minute bees take nectar from shallow flowers like those of Apiaceae. Longer tongues are needed to remove nectar from deeper flowers. Most kinds of bees are generalists in kinds of nectar utilized, although they may exhibit preferences and may be unable to reach nectar in some kinds of flowers. A few bee species, however, have morphological adaptations, such as palpi that fit together to form a sucking tube, that are associated with apparent specialization for gathering nectar from particular kinds of flowers. Examples are described and illustrated by Houston (1983c) and Laroca, Michener, and Hofmeister (1989); see also Figure 19-6.

One group perhaps involved in population- or species-level coevolution with plant hosts consists of the bee *Rediviva* (Melittidae) and its principal floral host, *Diascia* (Scrophulariaceae), in South Africa. The front tarsi of females are equipped with fine, dense hairs that sop up oil from inside the floral spurs. The spurs vary in length in various populations or species of *Diascia*, and the forelegs of female *Rediviva* vary in length accordingly (Fig. 6-1); some have forelegs longer than the entire body (Fig. 6-2). Details and alternatives are discussed by Steiner and Whitehead (1990, 1991).

Complex interactions frequently characterize relationships between even ordinary nectar-collecting and pollen-collecting bees and their floral food sources. Just because a species of bee visits a flower species does not necessarily mean that the bee is a pollinator of that flower. Small bees on large flowers may collect pollen, nectar, or both without going near the stigmata. In this case there is no pollination; the bee is merely a thief. An example is *Perdita kiowi* Griswold, a whitish bee of the North American high plains that is a specialist harvester of pollen from the long stamens of the large, cream-colored flowers of *Mentzelia decapetela* that open in the late afternoon. It rarely goes near the pistil; presumably, pollination is ordinarily by moths.

Thievery, such as that described above, merely reduces the amount of pollen available for food or distribution by actual pollinating insects. Some bees, however, damage flowers while at the same time robbing. Various kinds of large bees, especially *Bombus* and *Xylocopa*, cut open the sides of tubular flowers and extract nectar without contacting the anthers. Thus not only is the corolla damaged but the amount of nectar reward for legitimate pollinators is greatly reduced. Meliponine bees may chew into closed flowers or anthers, removing nectar or pollen and causing the flower major damage if not its destruction.

The effectiveness of a bee as a pollinator depends on many factors, unfortunately not always studied by people investigating pollination. A bee that has come from other flowers on the same plant or the same clone is unlikely to cross-pollinate. A bee that combs pollen off most of its body and appendages for transport in the scopa (pollen-transporting brushes or areas) is probably less likely to pollinate the next flower than a bee that leaves the pollen where it lodges on its body as it seeks more. A bee that moistens pollen with nectar or oil for transport is presumably less likely to pollinate than a bee that carries pollen dry and loose. And the location where pollen is deposited on the body of the bee can be critical for later

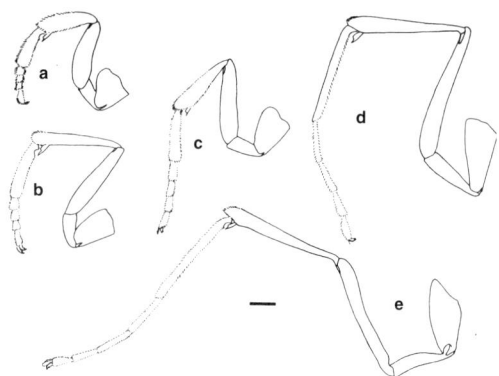

Figure 6-1. Front legs of females of *Rediviva*, showing elongation for oil collecting. **a,** *R. rufocincta* (Cockerell); **b,** *R. colorata* Michener; **c,** *R. peringueyi* (Friese); **d,** *R. longimanus* Michener; **e,** *R. emdeorum* Vogel and Michener. (Scale line = 1 mm.) From Vogel and Michener, 1985.

Figure 6-2. *Rediviva emdeorum* Vogel and Michener, female, showing the long front legs, which function to withdraw oil from spurs of *Diascia* flowers. From Vogel and Michener, 1985.

pickup by a floral stigma. Such factors depend not only on the floral structure but also on the movement patterns of bees, which may differ among different individual bees because they are partly learned, and will be different for different kinds of bees because they are partly species-specific. Students of pollination biology need to pay attention to these and many related matters. Too frequently, the assumption is made that because a particular bee species visits a flower species, that bee is a pollinator of that flower. Another unfortunate assumption is that bees of a common size (usually small) can be lumped as a single functional pollinating unit.

Nearly all eusocial bees and many solitary bees are floral generalists, whereas some solitary bees are floral specialists. Social bees are usually active for long seasons, so that, for them, floral specialization is impractical, because few flower species are in bloom for so long a period. *Bombus consobrinus* Dahlbom of Northern Europe, however, is a specialist on *Aconitum* and eusocial after an initial subsocial phase, like all nonparasitic *Bombus* (Mjelde, 1983). Eusocial bees often do show distinct "preferences," such that at a given time and place, the bee species visiting one flower species may be different from those visiting another. Such preferences are especially obvious in the American tropics, where numerous species of Meliponini are commonly active in the same vicinity, some of them segregated onto particular flowers. Unlike social bees, many solitary bees have short seasons of adult flight activity, and can therefore be specialists even if their favorite plant is in bloom for only a few weeks each year.

One would expect plants to evolve in ways that would promote floral specialization by bees, because a specialist is more likely to carry pollen to another plant of the same species than is a generalist that may next visit an entirely different kind of flower. This may not be as important a consideration as one might think, however, because of bee behavior that is called floral **constancy:** On any one trip, or during a longer period of time, individual bees tend to visit flowers of the same species. Whereas floral specialization by bees is presumably a result of inherent neural or morphological constraints, constancy is learned by each individual bee and may change with new opportunities, or may differ among individuals of the same species at the same time and place. Foraging generalist bees probably exhibit constancy because they can forage more efficiently (i.e., realize more gain per unit time) on one familiar floral type than on a diversity of types, each of which must be manipulated differently. Such aspects of bee behavior may be as important for pollination biology as is the bee's level of oligolecty or polylecty (see the definitions below).

Nectar and oil. Sugars in nectar are the principal source of carbohydrates in bees' diets. Nectar is eaten by adults as an energy source and mixed with pollen to make larval food. Nectar also contains some amino acids, and thus may also contribute toward a bee's nitrogen metabolism. Nectar for regurgitation into brood cells or for storage is carried to the nest in the crop.

Ingestion of nectar, of course, is by way of the proboscis. The gross structure of certain bee proboscides is shown in Figures 19-1 to 19-5. The actual mouth opening is on the anterior surface of the proboscis, near its base (Fig. 19-1c). The details of how nectar moves up the proboscis to the mouth are not fully understood and must vary in different kinds of bees. They involve a sheath consisting of the maxillary galeae, supplemented in long-tongued bees by the labial palpi, which surround the glossa. The flow by capillarity and labial, especially glossal, movement takes nectar toward the base of the proboscis. Some details for *Andrena* are provided by Harder (1983) and for *Apis* by Snodgrass (1956). The glossa is elaborately hairy and a significant part of the process may involve variations in the volume of nectar held among these hairs as they are alternately erected and depressed with protraction and retraction of the glossa. As shown in Figures 61-3, 86-1, and 116-2, the hairs, although sometimes simple, may be flattened and lanceolate, capitate, or branched in various ways.

As in most of biology there are exceptions to the generalities. Thus some plants in diverse families (Cucurbitaceae, Iridaceae, Krameriaceae, Malpighiaceae, Orchidaceae, Primulaceae, Scrophulariaceae, Solanaceae) secrete, instead of nectar, floral oils, which certain specialist bees collect and carry to the nest externally, i.e., in the scopal hairs, to mix with pollen and sometimes nectar to make larval food. The oils are believed to replace sugars in nectar as the larval energy source, but at least in *Centris vittata* Lepeletier both nectar and oils are included in larval food (Pereira and Garófalo, 1996). A review was by Buchmann (1987). Bees of the genus *Macropis* (Melittidae) collect floral oil from *Lysimachia* (Primulaceae) and use it in part to line (presumably to

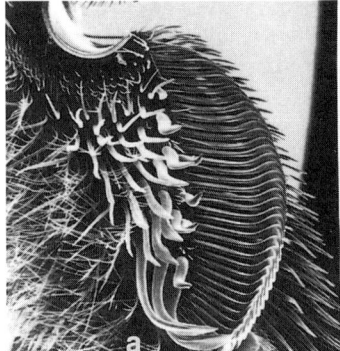 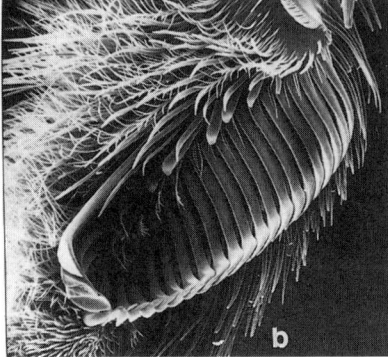 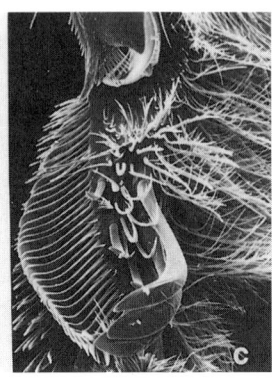

Figure 6-3. Ventral views of basitarsi of females. **a,** *Centris (Ptilotopus)* sp.; **b,** *C. (Paracentris)* near *tricolor* Friese; **c,** *C. (Heterocentris) trigonoides* Lepeletier. Combs of setae are for oil collecting. From Neff and Simpson, 1981.

waterproof) their brood cells (Cane et al., 1983). Adult bees rarely if ever ingest the oils and thus, like other bees, are dependent on nectar for their own energy sources. Since oil flowers do not produce nectar, oil-collecting bees must get needed sugars from other flowers. Oil-collecting bees have a striking array of pads, brushes, or combs of flattened setae (Figs. 6-3, 110-3a) with which to absorb or scoop the oil and to transport it back to the nest, sometimes mixed with pollen. The morphological details are discussed and illustrated by Vogel (1966 to 1990), Neff and Simpson (1981), and Cocucci, Sérsic, and Roig-Alsina (2000). Bees with such structures are found in the Centridini, Ctenoplectrini, Tapinotaspidini, and Tetrapediini in the Apidae and in the Melittinae in the Melittidae. Obviously, oil utilization has arisen independently, probably in each of these groups, just as the production of oil for reward has evolved independently in different families of plants (Vogel, 1988). The holarctic and Old World oil-collecting bees are specific to particular genera of plants, i.e., *Ctenoplectra* to *Momordica* and *Thladiantha* (Cucurbitaceae), *Macropis* to *Lysimachia* (Primulaceae), and *Rediviva* species mostly to *Diascia* (Scrophulariaceae). In the neotropics, however, oil-using bee genera and often species are not always specific to particular oil-producing genera or even families of plants.

Pollen. For most bees, pollen is the principal protein source; it is collected and carried to the nest as food for larvae and is also eaten by adults, especially females producing eggs. After dissecting for other purposes a thousand or more females of social halictine bees (mostly *Lasioglossum,* subgenus *Dialictus*), my impression is that large quantities of pollen in the crop were frequent in young adults, whose ovaries might enlarge, and in egg layers with large ovaries, but were virtually absent in old bees with slender ovaries, i.e., workers. Even among workers of highly social bees, whose ovaries will not enlarge greatly, it is the young ones that eat the most pollen, perhaps promoting development of their exocrine glands (Cruz-Landim and Serrao, 1994).

Pollen may initially stick to the bee's legs and body because it is spiny or sticky, or because of electrostatic charges. Some bees carry it back to the nest dry. Others (many Panurginae, Stenotritidae, Melittidae, and the corbiculate Apinae) moisten it with nectar to form a firm mass that can be carried with relatively few hairs to hold it in place. Oil-collecting bees moisten it with floral oils and possibly also nectar, thus sticking it to the oil-carrying scopal hairs. Finally, although pollen in bees' crops is partly used for their own nutrition, some is carried to the nests and regurgitated. All of the pollen used by bees of the subfamilies Hylaeinae and Euryglossinae to provision cells is carried in the crop, for these bees lack scopae for carrying pollen externally.

Thorp (1979) provided an excellent review of adaptations of bees for collecting and carrying pollen. These adaptations are both structural and behavioral. The details of hair structure associated with pollen gathering, manipulation, and transport have received considerable attention (Braue, 1913; Roberts and Vallespir, 1978; Thorp, 1979; Müller, 1996d; and papers by Pasteels and Pasteels cited in Pasteels, Pasteels, and Vos, 1983). Some aspects of these structures are characters of taxa described in the parts of this book on systematics. Figure 6-4 shows the scopa on the hind tibia and basitarsus of a eucerine bee, and Figures 102-2 and 120-11 show scopae reduced to form pollen baskets or corbiculae on the hind tibiae of corbiculate Apidae. Modified grooming movements are used by female bees for pollen handling. Pollen is commonly removed from anthers by the front tarsi or is dusted onto the body of the bee by its movement among floral parts. The forelegs may be pulled through the mouthparts if the bee eats the pollen, or they are pulled through the flexed middle legs whose opposable midfemoral and midtibial brushes remove the pollen. The pollen is then transferred to the hind legs, where it may be either held in the leg scopa for transport or, in megachilines among others, passed on to the metasomal scopa. Pollen dusted onto the bee's body is groomed off by the legs and transferred backward to the scopa. Details of these movements, and their many variations among taxa of bees, are described by Jander (1976) and Thorp (1979). Some of the best-known variations are in the remarkable ways in which pollen is loaded into the tibial corbicula by corbiculate Apidae, i.e., Apini, Bombini,

Figure 6-4. Hind leg of a female of *Svastra obliqua* (Say), a eucerine (L-T) bee, showing the scopa for transporting dry pollen on the tibia and basitarsus. The bare area on the lower outer surface of the femur constitutes the femoral corbicula in many S-T bees. Drawing by D. J. Brothers.

Euglossini, and Meliponini (Michener, Winston, and Jander, 1978).

Bees such as *Apis mellifera* Linnaeus are extreme generalists, and many others take pollen from various unrelated kinds of flowers. Such bees are called **polylectic**. Bee species or genera that specialize on a particular pollen taxon are called **oligolectic**. Some will collect pollen from a number of plant species of the same or related or even superficially similar families. These can be called **broadly oligolectic**. Others collect pollen from a few closely related species and are called **narrowly oligolectic**. The boundaries are indefinite, for there seems to be a continuum from the most broadly polylectic to the most narrowly oligolectic. Some authors have quantified this terminology. For example, Müller (1996b) suggested the following: *oligolectic,* at least 95 percent of the pollen grains from the scopa belong to one family, subfamily, or tribe; *polylectic with strong preference for one plant family,* 70 to 94 percent of the pollen grains, etc.; *polylectic,* 69 percent or less of the pollen grains, etc. Variations in the abundance of various plants, however, are likely to render such a system ineffective. Although these terms relate to pollen collecting, nectar or oil specialists are mostly also pollen specialists and thus oligolectic.

Frequently in any one area, or throughout its range, a bee species is restricted in its pollen collecting to a particular species of plant that has no close relatives in the vicinity. It is the usual view, based on considerable experience, that if a related plant species were present, the bees would utilize its pollen also, and that in other regions where related plants do exist, they will be visited by the same species of bee. For these reasons, the term **monolectic** is almost unused. Use of the term is appropriate, however, if it is clear that close relatives of the plant host are absent or not flowering in the area and season under study. For example, about 22 species of bees collect pollen only from *Larrea divaricata* in the southwestern United States (Hurd and Linsley, 1975). They are monolectic; there are no closely related plants in North America. Nonetheless, we usually call these bees oligolectic, thereby predicting that if other species of *Larrea* were present, they also would be utilized. The word "monolectic" is especially appropriate for a species of bee that collects pollen only from one species of flower, even in the presence of closely related flowers. *Anthemurgus passiflorae* (Robertson), a specialist on flowers of *Passiflora lutea* in the eastern United States, may be such a bee, for it does not visit other species of *Passiflora* so far as is known; but the size and color of flowers of the other regional species are entirely different from those of *P. lutea*.

Many oligolectic bee taxa (e.g., subgenera or genera) consist of related species specializing on the same or related plants. Examples are *Systropha* (Rophitinae), all species of which, so far as I know, use *Convolvulus* pollen more or less exclusively; *Macropis* (Melittinae), all species of which use pollen of *Lysimachia*; and the *Proteriades* group of *Hoplitis* (Osmiini), most members of which use *Cryptantha* pollen more or less exclusively, although visits to other plants for nectar result in taking some pollen.

Although many oligolectic bees appear to be dependent on their particular flowers, and do not occur outside of the ranges of those flowers, the plants are generally not dependent for pollination on their oligoleges. Plants often occur and reproduce outside the ranges of their oligoleges; pollination by polylectic bees or other insects is adequate for the plants' needs. Examples are given by Michener (1979a). As noted above, one can rarely recognize the coevolution of particular species of plants and bees; rather, the bees appear to have adapted to plant floral structure and chemistry, while the plant has commonly not adapted to any one oligolectic bee species or genus. In fact, readily accessible pollen characterizes some plants, such as willows *(Salix)*, that host numerous oligolectic species of bees. Often the bee's adaptation appears to be only behavioral, but there are many cases of probable morphological adaptation of a bee to a particular kind of flower. A common example is the sparse and often coarsely branched scopal hairs of bees such as *Tetralonia malvae* (Rossi) (Eucerini) and most *Diadasia* (Emphorini) that use coarse pollen like that of Malvaceae and Cactaceae. Hooked hairs on the mouthparts or front tarsi of females, which pull pollen away from anthers located deep in a small corolla, are other examples that occur in various unrelated bees. North American examples are *Andrena osmioides* Cockerell (Andreninae) and the above-mentioned *Proteriades* group of *Hoplitis* (Osmiini) on *Cryptantha* (Boraginaceae) and *Calliopsis (Verbenapis)* (Panurginae) on *Verbena* (Verbenaceae). European examples include *Colletes nasutus* Smith (Colletinae), *Andrena nasuta* Giraud (Andreninae), and *Cubitalia parvicornis* (Mocsáry) (Eucerini), all oligolectic on Boraginaceae (Müller, 1995). Some narrowly polylectic bees that frequently collect pollen from Boraginaceae have similar hooked hairs, as shown by the same author. A scopa consisting of simple sparse bristles is characteristic of bees that specialize on pollen of Onagraceae, plants whose pollen grains are webbed together by viscin threads. Examples are *Svastra (Anthedonia)* (Eucerini) and *Lasioglossum (Sphecodogastra)* (Halictini); for others, see Thorp (1979).

A morphological feature that has arisen independently in various groups of bees appears to be adaptive for collecting pollen from Lamiaceae and Scrophulariaceae, particularly from *Salvia* and its relatives. The facial vestiture consists of erect, rather short hairs having stiff, thickened bases tapering to slender tails that are usually hooked, bent to one side, or wavy. Such hairs are usually on the clypeus but are on the frons in *Rophites* s. str. In the best-developed cases, the face is flatter than in related species lacking such hairs. Müller (1996a) reviewed such bees in Europe and found them to be mostly oligolectic on Lamiaceae or narrowly polylectic on that family, Fabaceae, and Scrophulariaceae. He observed that such bees rub the anthers with their faces and remove the pollen from their faces with the front basitarsi; obviously, they then transfer the pollen to the scopa. Certain species in each of the following genera have such facial modifications: *Caupolicana* (Colletidae); *Andrena* (Andrenidae); *Rophites* (Halictidae); *Anthidium, Trachusa, Osmia,* and *Megachile* (Megachilidae); *Anthophora, Amegilla, Habropoda,* and *Tetraloniella* (Apidae).

A type of pollen presentation that has received considerable attention is in tubular anthers that perhaps protect pollen from damage by rain. Instead of dehiscing in usual ways, such anthers, found in diverse families, are poricidal, i.e., tubular with one or two holes in the distal ends through which pollen must escape. Such plants usually produce no nectar, but depend on pollen as a reward for bees. Many kinds of bees, both oligolectic and polylectic, obtain pollen from such flowers by vibrating (sonicating) them, the anther aperture usually directed toward the bee. Pollen shoots out and some of it clings to the bee, after which it can be handled in the usual way. The vibrating, caused by the wing muscles, results in bursts of audible sound, hence "buzz-pollination." A review is by Buchmann (*in* Jones and Little, 1983). Müller (1996a) records buzzing during pollen collecting from Lamiaceae by bees with bristles on the frons *(Rophites)* or clypeus. Such flowers do not have tubular anthers, but perhaps vibrations help to release the pollen from the anthers. An interesting aspect of sonication by bees is that not all kinds of bees do it. Conspicuous among bees that do not is *Apis mellifera* Linnaeus. Moreover, minute bees usually do not do it.

Vibrating behavior is widespread among bees and wasps, and is usually used by individuals finding it difficult to push through a small space or to loosen a pebble in nest construction. This is probably the ancestral function of such vibrating. Minute bees probably do not have the mass and energy to liberate pollen or pebbles in this way. *Apis* may have lost the tendency to sonicate anthers because it nests in the open or in large cavities and builds with malleable wax rather than hard soil, pebbles, etc., and therefore rarely needs sonication in nest construction. *Melipona* does sonicate flowers; this may seem to negate the argument based on *Apis*. The intricacies of its nests and the small nest entrances may have promoted retention of the behavior.

An unsolved question is whether oligolecty or polylecty is the ancestral condition for bees. No doubt evolution can go in both directions, but it seems reasonable to suppose that oligolecty is a specialized condition and therefore derived. Probable evidence for this supposition comes from unusual oligolectic species of generally polylectic groups. An example is *Lasioglossum (Hemihalictus) lustrans* (Cockerell), an oligolege on *Pyrrhopappus* (Asteraceae), in the midst of the huge and generally polylectic tribe Halictini. There is no reason to believe that *L. lustrans* is exhibiting a plesiomorphic condition; on the basis of morphology, it seems to be a derived species, although this is an impression, not based on a phylogenetic analysis.

Conversely, in diverse groups of bees all species are oligolectic, but on plants of different and often unrelated families. Examples are the tribe Perditini in the Panurginae and the tribe Emphorini in the Apinae. In such cases it seems clear that oligoleges have given rise to other oligoleges dependent on different host flowers. We have no evidence concerning how they became oligolectic in the first place. There are, however, various archaic bee taxa, i.e., basal branches of clades, that consist entirely or largely of oligolectic species. Examples are the Fideliinae, Lithurgini, Rophitinae, and Melittidae. Their phyletic positions suggest that more derived taxa containing many polylectic species may have arisen from taxa consisting of oligolectic species. In their study of the phylogeny of bees at the family level, Danforth et al. (2006) reached the same conclusion. One can support this idea with the notion that a specialist need be adapted to only a limited environment, e.g., the chemicals in its pollen food, or the floral structure of its host plant, to which it must adjust. A generalist, on the contrary, must be able to deal with environmental diversity, e.g., different chemicals in pollens and diverse floral structures, in different plants. Much evolution, therefore may have been from the simpler requirements of a specialist to the complex requirements of a generalist. The obvious advantage would be access to the much increased resources available to the generalist. As species-level phylogenies are worked out in genera like *Andrena, Colletes, Leioproctus,* and *Megachile* that contain both polylectic and oligolectic species, better understanding of this topic will develop. Müller (1996b) made such a study for western palearctic Anthidiini. He found evidence for transitions from oligolecty to polylecty and for transitions of oligoleges from one floral host to another, but he found no transitions from polylecty to oligolecty.

An interesting observation is that some species of plants have many oligolectic visitors while others have none. For example, in North America there are many oligoleges on *Helianthus* (Hurd, LaBerge, and Linsley, 1980). Some of them occasionally take pollen from other large Asteraceae, but most are almost exclusively dependent for pollen on *Helianthus,* to judge by my observations and collecting records. But no oligolege is known for the similar flowers of another large Asteraceae, *Silphium,* even though *Helianthus* and *Silphium* often flower in the same vicinity. I have no explanation for this rather common phenomenon. A combined botanical and entomological study would probably be worthwhile.

The frequency of oligolecty among bees also varies regionally. Michener (1954b) observed that oligoleges form a smaller percentage of the bee fauna in the moist tropics than in temperate regions, and that the maximum percentage of oligolectic species seems to be in xeric warm-temperate areas, at least in the Western Hemisphere. Good data are difficult to obtain, partly because of problems with the definition of the terms, but I believe that this regional pattern in percentage of oligolectic species is real, and occurs more or less worldwide. This pattern could be accentuated in the Western Hemisphere by the abundance of the largely oligolectic Panurginae in the xeric regions of both North and South America, but Pesenko (*in* Banaszak, 1995) wrote that in the former U.S.S.R. nearly half of the nonparasitic bee species of steppes and deserts are oligolectic, the percentage apparently being much less in more humid regions and in boreal regions. The same pattern is subjectively recognizable in Africa in spite of the scarcity of Panurginae there.

Even within broad areas, such as the Sonoran deserts of the United States and Mexico, the prevalence of oligolecty varies among districts. Minckley, Cane, and Kervin (2000) showed that bees oligolectic on *Larrea* are concentrated in and presumably most frequently arise in areas of least predictable flowering. *Larrea* flowers after rains of 12 mm or more; the driest deserts, where the plants may fail to flower for years, appear to be most prone to produce oligoleges. Relatives of most of them are oligoleges on other flowers.

Other substances collected by bees. Aside from materials for nest construction collected by many megachiline bees, many bees assiduously collect certain other substances. These include water, employed for temperature control in colonies, as in *Apis,* and for softening hard soil while excavating, as in *Ptilothrix* (Emphorini). Sweat bees (Halictinae) and some Meliponini take perspiration, probably for its water and salts, and can be quite bothersome to people in the process. These same groups of bees sometimes take salts from other sources, e.g., soil moistened by urine. Roubik (1989) lists various other bees that appear to be attracted to, and to take, inorganic salts.

Male euglossine bees collect aromatic fragrances from orchid flowers as well as from flowers of certain Araceae and a few other plant families. (Even larger quantities of the same and similar chemicals may come from rotting logs, fungi, and perhaps other objects in tropical forests, as noted by Whitten, Long, and Stern, 1993.) The functions of these compounds in euglossine bee biology are not clear (see Sec. 118, on Euglossini), but they are the bait or reward that attracts the bees to the flowers. Male euglossine bees are the sole pollinators of many species of neotropical orchids (see Dressler, 1968), but that function does not depend on pollen collected by the bees or dusted onto the bees' bodies. Orchid pollen is in fact useless for bees, because it is produced in saclike pollinia. The often complex orchid floral structures stick pollinia to bees' bodies at sites that later will come in contact with the stigmatic surfaces of other orchid flowers as the bees seek more of the fragrant compounds.

Just as some bees collect oil in place of the usual nectar for larval food, some species of *Trigona* (Meliponini) have another source, meat, to fill part or all of their protein needs. Some species not only collect pollen, but frequently visit carcasses of dead animals, where they collect bits of tissue, perhaps for nest construction but probably in some cases also for larval food. Three neotropical species of the same genus are obligately necrophagous; they do not collect pollen but take tissue from animal carcasses instead (Roubik, 1982; Baumgartner and Roubik, 1989). These bees, which can rather quickly skeletonize the carcass of a small animal, do not even collect nectar from flowers but use fruits and extrafloral nectaries as sugar sources (Noll et al., 1997). At least one of the carrion-feeding *Trigona* species is sometimes predaceous on soft protein sources—living wasp larvae in nests abandoned by the adult wasps, and eggs of toads (*Bufo* sp.) stranded by lowering water levels (Mateus and Noll, 2004; D. Roubik, personal comm., 2004).

Worker honey bees sometimes collect such strange materials as coal dust, brick dust, and flour. Presumably, such substances have no function in the hive; probably they are discarded.

7. Nests and Food Storage

The nests of bees are the places where their young are reared. They are always to some degree made by the mother, or, in social bees, by the workers. Nests and especially cells, their provisions, and larval behavior are full of meaningful details of importance not only for bee survival but also for our understanding of adaptations and of phylogeny. Malyshev's many papers were among the most important and detailed early studies of these matters, culminating in his summary work (Malyshev, 1935). Some of the best and most critical recent accounts are in Rozen's papers. References to many papers that include information on nest architecture are in the accounts of taxa in later parts of this book. General accounts were by Michener (1961a), Iwata (1976), Radchenko and Pesenko (1994a, b), and Radchenko (1995), and detailed summary accounts of certain taxa are those by Wille and Michener (1973) for the Meliponini and by Sakagami and Michener (1962) for the Halictinae.

Bee nests ordinarily contain or consist of brood cells (Fig. 7-1). A **cell** serves to protect the delicate immature stages, and in most cases the food, of the growing larva. It is the space in which a single immature bee grows, although in most species of *Bombus* a cluster of eggs is placed together in a small wax cell, and the cell is enlarged as the resulting larvae grow. Except in *Bombus*, cells are big enough initially to contain one mature bee each.

Most bee nests consist of more than cells, being burrows in the soil, in wood, or in pith. Typically, and probably ancestrally (because the pattern is common in sphecoid wasps), the nests are in the soil and the main burrow gives rise to lateral burrows, each of which ends in a single cell (Michener, 1964b; Radchenko and Pesenko, 1994a, b; Radchenko, 1995). As to nests of ancestral bees, note that Budrys (2001) believed that nests of ancestral

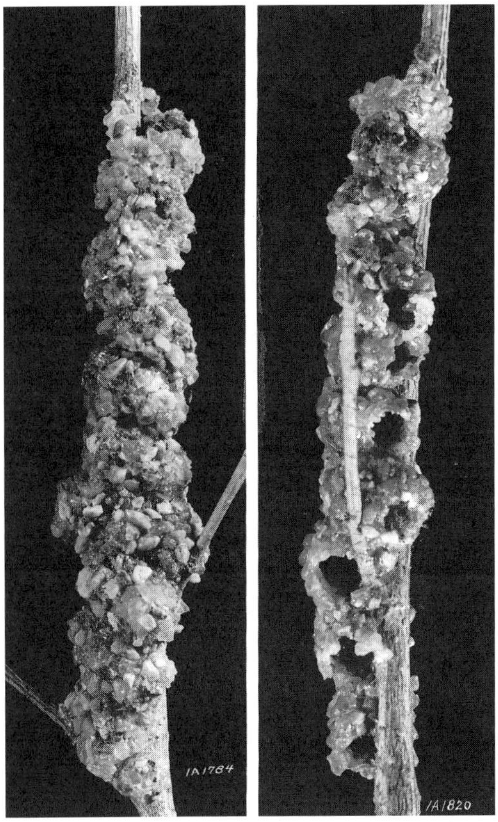

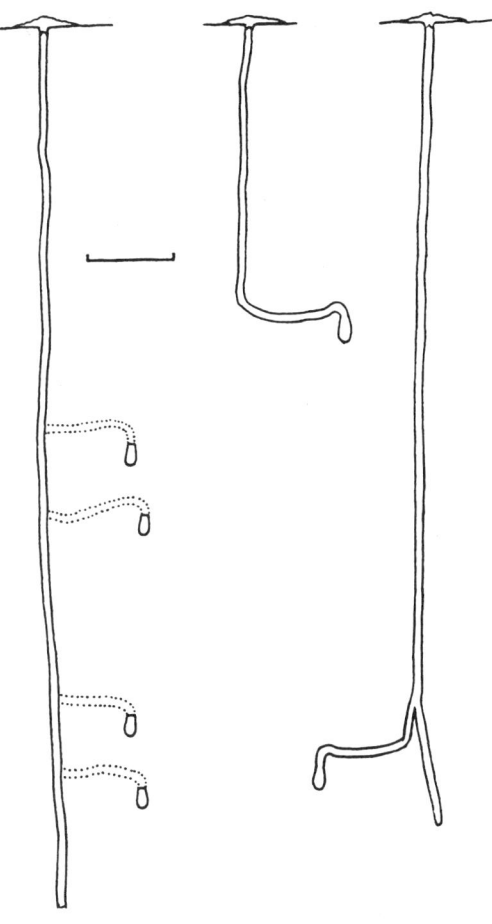

Figure 7-1. Cells of an anthidiine bee, *Dianthidium concinnum* (Cresson), made of pebbles and resin, constructed on an elm twig in Kansas. Emergence openings are shown on righthand photograph. From Fischer, 1951.

Figure 7-2. Diagrams of nests of a eucerine bee, *Peponapis fervens* (Smith), excavated in soil in Brazil. At left is a mature nest with the lateral burrows filled with earth and unrecognizable, their exact positions not determined. The other nests are relatively new, each with one newly constructed lateral burrow and an unprovisioned cell. (Scale line = 6 cm.) From Michener and Lange, 1958c.

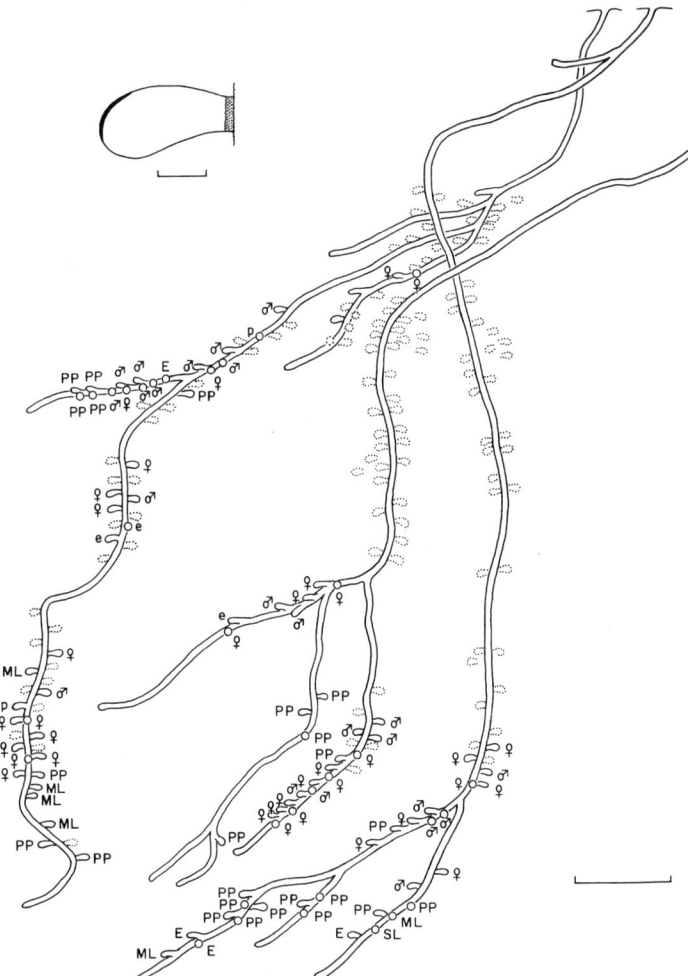

Figure 7-3. Diagram of three nests of a colonial halictine bee, *Halictus ligatus* Say, excavated in soil in Trinidad, showing sessile cells. Cells shown by dots were abandoned, earth-filled; the contents of other cells are indicated as follows: e, empty; E, egg; SL, small larva; ML, medium-sized larva; PP, large larva, usually prepupa; the sex symbols identify pupae of the sexes indicated. At upper left is a sectional view of a cell showing shape, earth closure (dotted), and feces of larva (black). (Scale lines = 5 mm for the cell; 10 cm for the nests.) From Michener and Bennett, 1977.

sphecoid wasps were in wood cavities. If this is true, ancestral bees may have occupied similar holes instead of burrowing in the ground. However, since sphecoid wasps are paraphyletic relative to bees and related lineages are ground-nesting, it is more likely that ground-nesting is ancestral for bees, as also concluded by Engel (2001b).

The cells are lined or unlined; the burrows themselves are unlined. Typically, each lateral is filled as a new lateral is excavated, saving the bees the trouble of pushing the excavated earth to the surface. Laterals may be well separated up and down the nest, as shown by the mature nest in Figure 7-2, or may all radiate from one level. Among the many other modifications of such architecture are horizontal cells, instead of vertical cells as in Figure 7-2; two or more cells per lateral; shortening of the laterals until the cells are sessile, arising directly from the main burrow (Fig. 7-3); and rotation until the main burrow is horizontal, entering a vertical bank instead of flat ground.

In some bees the cells, like the burrows, are unlined, mere excavations into the soil, usually broader than the burrows leading to them. Such bees include many Melittidae, most Xylocopinae, Fideliinae, and the genus *Perdita* in the Panurginae. If this cell type, resembling that of most sphecoid wasps, is ancestral for bees, it supports the Perkins-McGinley hypothesis of the proto-bee as a form with a pointed glossa like a melittid; see Section 20. The alternative, that the proto-bee had a broad glossa like that of a colletid, would favor the proto-bee's lining each cell with a secreted film, applied with the broad glossa as do colletids.

Unlike the taxa listed above, most bees excavate cells in a substrate (usually soil), line them with a smooth earthen layer, often made of fine clay from elsewhere in the burrow, tamp the cell surface smooth with the pygidial plate, and apply to this surface a secreted film of cellophane-like or waxlike material (Fig. 7-5). The "waxlike" material is a mixture that may not include wax; J. Rozen (in litt.) prefers simply to call it a shining secretion. For additional information, see Section 111. Two views of such cells are shown in Figure 7-4; see also Plate 16. Both the earthen layer and the secreted lining are derived features relative to those of sphecoid wasps; when sphecoid wasps make

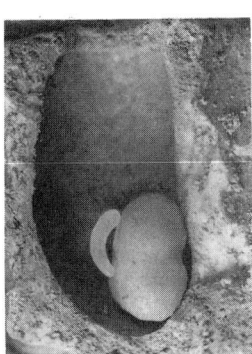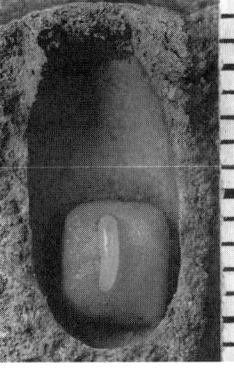

Figure 7-4. Cells of *Augochloropsis sparsilis* (Vachal) excavated into soil, showing the characteristic cell shape (saggital section at left, frontal section at right) as well as a pollen mass and egg. (The scale at the right is in millimeters.) From Michener and Lange, 1959.

similar structures, such as the lined cells made by some Pemphredoninae, the lining is not homologous to that constructed by bees. The earthen layer and secreted lining may also be derived features relative to those of the proto-bee. Such cells can be isolated or grouped in clusters made by excavating them close together. Some halictids, however, construct clusters of similar cells in cavities that the bees have excavated in the soil (Fig. 7-5); the cells are made from the homologues of the earthen cell linings. The view of Radchenko and Pesenko (1994a) that such construction of cells does not occur is incorrect.

Cells made by subdividing a burrow with transverse partitions, as is done by many Megachilinae, Hylaeinae, Ceratinini, and others, are usually not identical in shape, i.e., they are **heteromorphic**. Cells excavated or constructed in the soil or other substrate, such as rotting wood, are usually alike, i.e., **homomorphic**, for any one species. In the homomorphic cells of some common taxa like Andrenidae and Halictidae, one surface (the lower surface if the cells are horizontal) is flatter than the other surfaces, each cell thus being bilaterally symmetrical about a sagittal plane (Figs. 7-3, 7-4).

Megachiline bees usually make cells [sometimes only by means of partitions in an unlined burrow (Fig. 7-6; Pl. 16), but usually with whole cell walls] using foreign materials carried to the nest. Such materials can be cut pieces of leaves, chewed leaf pulp, plant hairs (sometimes supplemented with sticky material from stem or foliar trichomes, Müller, 1996c), resin, pebbles (Fig. 7-1), mud, etc. A secreted lining seems to be absent. In a few Megachilidae [*Heriades spiniscutis* (Cameron) (Fig. 7-6), Michener, 1968b; *Osmia (Metallinella)*, Radchenko, 1978; *Megachile (Sayapis) policaris* Say, Krombein, 1967; *Fidelia*, Rozen, 1977c; Lithurgini, Malyshev, 1930b], partitions between cells are sometimes or always omitted, so that larvae are reared in a common space with separate or contiguous food masses. Some megachilid bee nests are so constructed that they consist only of one or several cells made of resin, resin and pebbles, leaf pulp, or mud on the surfaces of rocks, walls, stems, or leaves. Examples are *Anthidiellum* s. str., whose nests usually consist of a single resinous cell exposed on a leaf, stem, or rock surface (Pl. 8), and most species of *Dianthidium* s. str. and *Megachile*

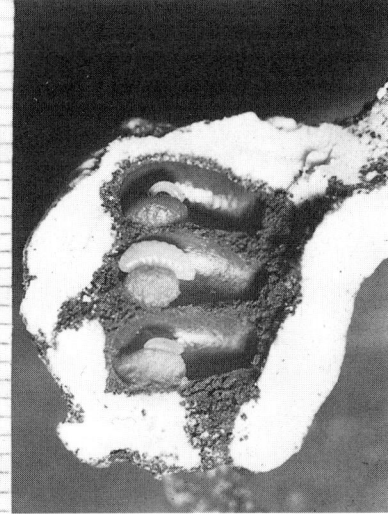

Figure 7-5. Nests of *Augochlorella striata* (Provancher) excavated into soil. At left, a cell cluster exposed by digging. At center, a nest poured full of plaster of Paris, then exposed by digging. At right, the same cell cluster opened to show three cells (the oldest in the center) in saggital section, the very thin earthen cell walls, and the earthen pillars supporting the cell cluster in the space here filled with plaster. (The scales are in millimeters; that at the right center relates only to the righthand photograph.) Photos by E. Ordway (left) and C. Rettenmeyer.

Figure 7-6. Parts of three nests of *Heriades spiniscutis* (Cameron) in dead, dry stems. The nest at the right had thin partitions made of pith fragments between the cells; the partitions are marked by horizontal lines. The other nests lack partitions. All the nests contain eggs or very young larvae in the upper ends of masses of provisions. To show the eggs, loose pollen was blown away from the nest at the left before photographing. From Michener, 1968b.

(Chalicodoma), whose nests consist of clusters of cells similarly exposed (Fig. 7-1). Those of *Dianthidium* are made of pebbles in a matrix of resin, whereas those of *Chalicodoma* are made of mud or sand impregnated with a secretion (probably of the labial glands, since these glands are enlarged) that renders the nest hydrophobic and able to withstand rain (Kronenberg and Hefetz, 1984b).

The simplest but architecturally derived bee nests are those of the allodapines. Such a nest is a cavity in a hollow stem, a burrow in a stem made by some other insect, or an unbranched burrow made by the female bee in a pithy stem. If necessary, such a tubular hollow is cleaned out by the bee, the bottom rounded out by tamped particles. A collar of pith or wood particles cemented together, probably by salivary materials, is constructed in such a manner that it narrows the entrance, permitting more efficient guarding. When there is a threat, the entrance is plugged by the somewhat flattened metasoma of the female. Immature stages are reared together, usually fed progressively, in the nest burrow. For illustrations, see Michener (1971a, 1990d) and Figures 90-4 and 90-5. Such nests, though simple, are not the ancestral type of bee nest; no doubt they are derived from nests like those of *Ceratina*, which are burrows in stems, subdivided into mass-provisioned cells by partitions of pith particles (Fig. 90-5a). The allodapine subgenus *Compsomelissa (Halterapis),* unlike most of its relatives, makes mass-provisioned nests somewhat like those of *Ceratina* but with the partitions omitted.

In contrast, in the corbiculate tribes Apini, Bombini, and Meliponini, cells are built of wax secreted by the metasomal wax glands, and except in Apini, mixed with other materials such as resin or pollen. The cells are in clusters or in **combs** (i.e., regular layers), usually in a cavity in a tree or in the ground, or in a cavity in a larger nest. Rarely, as in the groups of *Apis dorsata* Fabricius and *A. florea* Fabricius, the combs of cells are exposed, but protected by layers of bees. Details of cells of corbiculate Apidae and the nests in which they are found are explained by Michener (1961a, 1974a), Wille and Michener (1973), and numerous other works.

The most elaborate bee nests are those of the Meliponini (Figs. 7-7, 7-8, and 120-3), in which the clusters or combs of wax brood cells are surrounded by one or multiple layers of resin or wax involucrum. These layers, and masses of food-storage pots, are usually surrounded by batumen consisting of one or multiple layers of wax mixed with either resin or mud, sometimes forming an enormous, exposed nest, more often a nest hidden in a hollow tree or in the ground. For clarification of terminology, see Figure 7-8. The mixture of wax and resin is called **cerumen**. The multiple layers of cerumen around

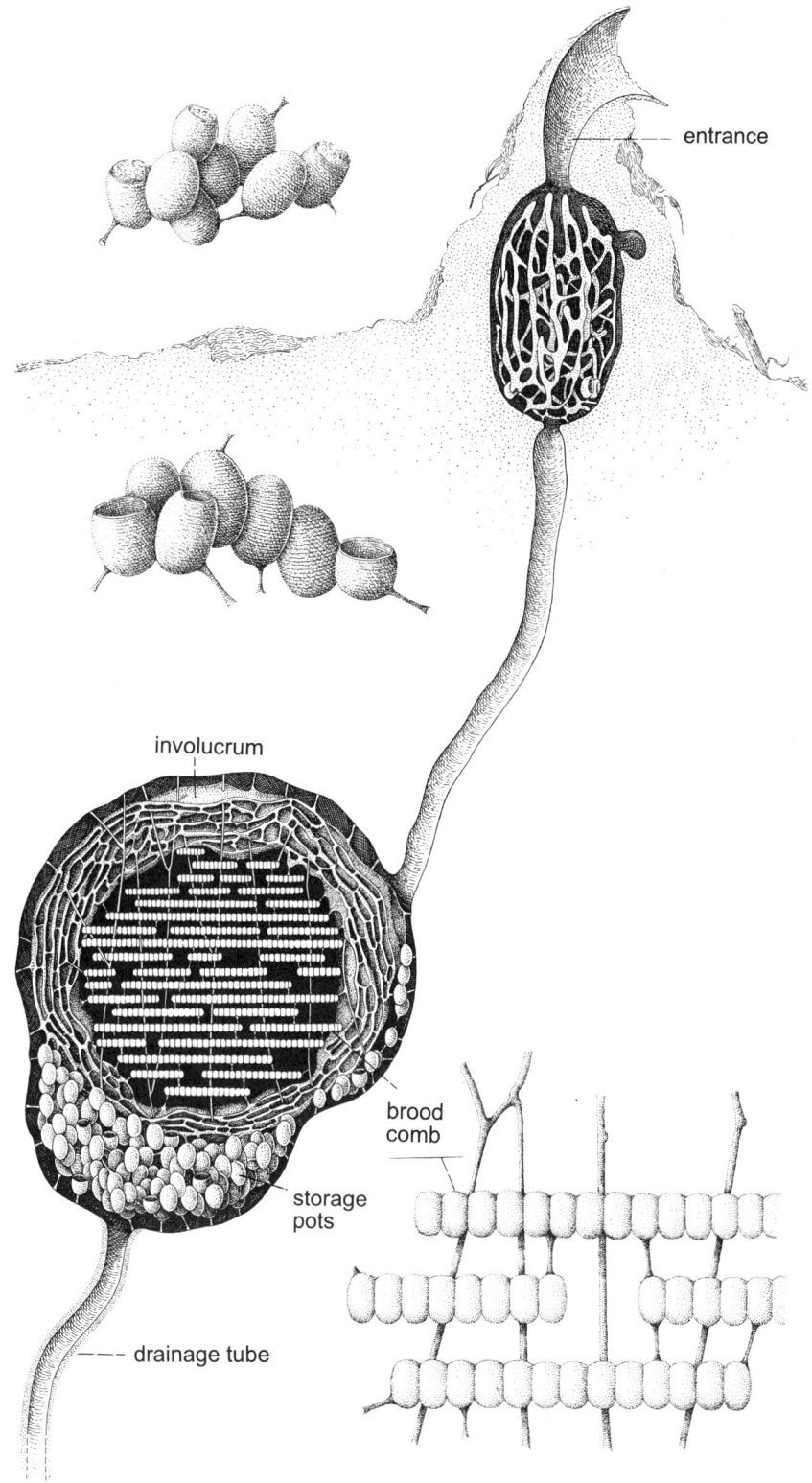

Figure 7-7. The subterranean nest of a colony of stingless bees (Meliponini), *Partamona testacea* (Klug). The horizontal combs of brood cells are supported by slender vertical pillars; the brood chamber is surrounded by multiple layers of cerumen, constituting the involucrum. The batumen is reduced to the thin lining of the cavity in the soil. (Enlarged drawings of the storage pots are at the upper left, of brood comb at lower right.) Drawing by C. M. F. de Camargo, from Michener, 1974a.

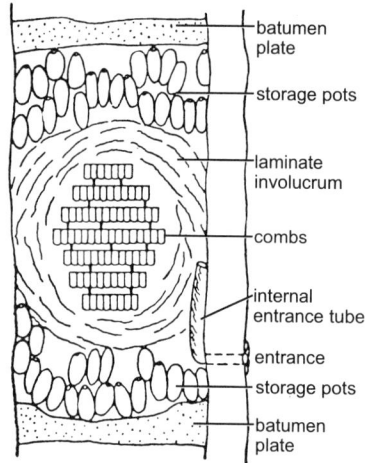

Figure 7-8. Diagram of a meliponine nest in a hollow tree, with parts labeled. The batumen typically surrounds the entire nest; here it consists of two plates that limit the nest area and a thin lining of batumen (not illustrated) that lines the entire cavity between the plates. From Wille and Michener, 1973.

the brood chamber are called the **involucrum**, and the plates or layers enclosing the whole nest are called **batumen**. Many works describe and illustrate such nests; some are Michener (1961a), Wille and Michener (1973), and especially the beautifully illustrated works of J. M. F. de Camargo, e.g., Camargo (1970) and Kerr et al. (1967).

Terms that relate to brood cell usage are **mass provisioning** vs. **progressive feeding**. Most bees provision each cell with enough food to suffice for all of larval growth; then after an egg is placed in a cell, the cell is closed. Usually, it is not opened again by the mother. These bees are mass provisioners. Some bees, however, feed the growing larvae at intervals. These are the progressive feeders. The only progressive-feeding bees are *Apis* (honey bees), most Allodapini (which, in fact, do not make cells; see above), and *Bombus* (bumble bees, in which the extent and nature of progressive feeding varies with the season, caste, and species). In certain Halictini (see the review in Michener, 1974a) and many *Ceratina* (Ceratinini) (see the review in Michener, 1990d), although mass provisioning occurs, cells are opened, feces are removed, and the cells are reclosed. This activity has led to apparently incorrect reports of progressive feeding (see Michener, 1974a).

In mass-provisioning species, cells serve to hold the provisions. The walls are often lined with a secreted layer that is impervious to water, so that if the provisions are liquid, they do not seep away. The cell walls must also be important in protecting the larva and pupa both from desiccation and from drowning, and in preventing the hygroscopic provisions (when not already liquid) from liquefying because of excess water. These rather obvious functions of cells have been dealt with by various authors such as Radchenko and Pesenko (1994) and Stephen, Bohart, and Torchio (1969). Less well-known functions include bactericidal and fungicidal activity that reduces invasion of cells by microorganisms and consequent spoilage of the provisions (Cane, Gerdin, and Wife, 1983). (Immature bees, at least the prepupae of *Nomia melanderi* Cockerell, also have bacteriostatic materials in or on the integument; Bienvenu, Atchison, and Cross, 1968). Water relations inside cells deserve further study. In some Halictinae the mature larva weighs over 60 percent more than the combined weight of the pollen mass and egg, because of water absorption by the larva from the atmosphere (May, 1972). Another suggestive observation is that in some ground-nesting bees like *Perdita* (Panurginae) that do not secrete a cell lining, there is instead a secreted covering protecting the food mass.

The secreted cell lining is probably mostly derived from products of **Dufour's gland,** which opens at the base of the sting (Cane, 1981, 1983b). It tends to be large in bees that construct cells in the ground, and it is commonly small in bees that construct cells elsewhere and that do not line their cells (see Pesotskaya, 1929, and papers by Lello cited in Lello, 1976). Salivary-gland products probably cause polymerization and solidification of the Dufour's gland product, which is initially a liquid (Albans et al., 1980). In *Anthophora* it seems that the Dufour's gland secretion consists largely of liquid triglycerides that are transformed into solid diglycerides (Batra and Norden, 1996). In this case the salivary secretion is evidently added to the food mass, and solidification of the liquid cell lining occurs on contact with the food mass. Dufour's gland products may also be found in the food mass in the cells of some bees but, contrary to old literature, in which it was called the alkaline gland, Dufour's gland does not contribute to the sting venom.

The food stored for larval consumption in brood cells of mass-provisioning bees takes various forms. Sometimes, as in *Hylaeus* and *Colletes* (both Colletidae), it is liquid, consisting primarily of nectar with some pollen admixture. In many others, e.g., *Anthophora, Megachile,* and *Trigona,* it is more viscous, as a result of containing more pollen, but nonetheless fills and takes the form of the part of the cell in which it is placed. In still other bees, such as *Leioproctus* (Colletidae), Halictidae, Andrenidae, Melittidae, and Xylocopini, the food mass is firm and carefully shaped, commonly a spheroidal or a flattened sphere, but sometimes, as in *Exomalopsis* (Apinae), *Dasypoda,* and *Macropis* (Melittidae), with basal projections that support the rest of the food mass. Such firm food masses are considered the ancestral condition by Radchenko and Pesenko (1994a, b). The spherical form and especially the supporting projections, as well as the loaf-shaped provisions of *Ceratina* and *Xylocopa,* minimize areas of contact with the cell surface, and possibly thus minimize moist sites where destructive molds may gain a foothold. A series of illustrations of various cell and provision types was presented by Stephen, Bohart, and Torchio (1969). In addition to nectar and pollen, the larval provisions sometimes, or perhaps regularly, contain glandular secretions. Salivary glands clearly contribute to the larval food, probably especially in the corbiculate Apidae but also in *Anthophora* (Batra and Norden, 1996). In addition to providing nutrition, the salivary secretions no doubt contribute to the preservation of larval food as well as of food stored in pots by the necrophagous species of *Trigona* (Meliponini) and probably others.

Food storage (both pollen and nectar or honey) for adult consumption or transferral to larvae occurs in the nests of relatively few bees. In *Apis* such storage is in cells much like worker brood cells; in Meliponini and Bombini it is in pots, quite different from brood cells, those of *Bombus* usually made from abandoned cocoons. In the Allodapini, pollen is stored on the walls of the nest burrows; nectar or honey is stored in large drops on the bodies of the larvae, where it will not be absorbed into the pith that usually forms the walls of the nest burrows.

8. Parasitic and Robber Bees

In many groups of organisms that store food for themselves or their young, parasitic or robber individuals, species, or genera can be found. Such forms steal or feed upon the stored food, often starving or more directly killing the hosts. Bees well illustrate such tendencies; reviews were by Bischoff (1927: 397-401), Grütte (1935), Bohart (1970), and Iwata (1976).

This section consists of four parts, as follows: first, on nest usurpation and robbing; second, on social parasites that live in the nests of social host bees; third, on cleptoparasites that leave their eggs, cuckoo/cowbird fashion, in the cells of their host bees; and fourth, on some common attributes of social parasites and cleptoparasites.

Usurpation and robbing. Intraspecific usurpation is probably frequent. It has been little studied because, unless bees have been marked for individual recognition, it is likely to go unnoticed. When a solitary bee comes out of a nest hole and is the same species that was seen there yesterday, one naturally assumes that it is the same individual. Usually, this assumption is correct, but various studies of wasps and bees made with marked individuals show that intruders are often present, that they enter nests, often fight with nest owners, and sometimes win and take over nests made by other individuals. Barthell and Thorp (1995) showed that in *Megachile apicalis* Spinola the usurpers average larger than their victims, at least in highly competitive situations. Presumably, larger individuals are better able to win contests.

Wcislo (1987) listed 17 species of bees in which usurpation or intraspecific robbing has been reported. Perhaps to facilitate nest recognition by the owner as well as, among social species, to reduce admission of foreign individuals while allowing access by nestmates, many bees and wasps appear to have distinctive individual or nest odors. Nest guarding, too, is common, especially in social species. It is my clear impression that the commonest function of such guards is to reject conspecific individuals that attempt to enter, although of course guards also react strongly to other insects, including parasites and predators. Summary accounts of individual and nest recognition in bees are by Michener and Smith (1987) and Breed and Bennett (1987).

As might be expected, the defenses implied in the preceding paragraph do not always succeed. In the Meliponini and Apini, robbing is frequent. Weak colonies may not be able to defend themselves against intraspecific or interspecific robbing, in spite of guards and constricted entrances. Two genera of Meliponini, the neotropical *Lestrimelitta* and the African *Cleptotrigona*, are specialist robbers. They have nests of their own but obtain food, not from flowers, but from nests of other species of Meliponini. Accounts of their robbing activity are by Portugal-Araújo (1958), Wittmann et al. (1990), and Sakagami, Roubik, and Zucchi (1993). They lack tibial corbiculae for carrying pollen and have other common features, as listed in Section 120 on the Meliponini, but these features appear to be convergent; the two genera evolved from different nonrobbing ancestors.

Social parasites. Parasitic bees, as distinguished from robbers, can be divided into two groups, social parasites and cleptoparasites. A female **social parasite** enters a nest of the social host and in some way replaces the queen, so that the workers of the host thereafter rear offspring of the parasite rather than of their own species. The host must be social but sometimes (as in Allodapini) is marginally so. The female parasites are found living in colonies of the hosts. Although derived from eusocial ancestors (see Sec. 5), the parasitic species lack a worker caste.

There are relatively few social parasites among bees (Table 8-1). In the genus *Bombus*, the subgenus *Psithyrus* is entirely parasitic and has lost the corbicula for carrying pollen. The subgenera *Alpinobombus* and *Thoracobombus* each have a parasitic species; they have normal or nearly normal but presumably functionless corbiculae. Thus in *Bombus* there have been three origins of social parasitism.

Allodapini is the only group other than Bombini that contains indisputably socially parasitic species. There are eight or nine such species, and four others considered as probable social parasites because of reduction in the pollen-carrying scopa (Table 8-1). (For an account of the parasites, see Michener, 1970.) Except for the two species of *Eucondylops*, two species of *Exoneura* subgenus *Inquilina*, and the two similar species of the *Braunsapis breviceps* group, each parasitic species appears to have arisen independently from nonparasitic relatives. There is one or more parasitic or probably parasitic species derived from or included in each of the following genera: *Allodape, Allodapula, Braunsapis, Exoneura,* and *Macrogalea*. Eleven independent origins of parasitic allodapine species are indicated, but in no case has a parasitic allodapine line undergone as much speciation to produce a group of parasitic species as has *Bombus (Psithyrus);* the largest such groups in the Allodapini contain only two known species.

In both *Bombus (Psithyrus)* and *Braunsapis kaliago* Reyes and Sakagami, females functionally replace queens of the host; the latter are sometimes killed but often remain alive. The female of the parasitic *Braunsapis kaliago* becomes unable to fly but participates in many nest activities such as the feeding of larvae (Batra, Sakagami, and Maeta, 1993).

Because of a scarcity of information on the often rather common parasitic halictids, the nature of their parasitism is often in doubt. Those that parasitize solitary hosts are of necessity cleptoparasitic (see below). Those that attack social halictines may also leave the nest promptly after oviposition, like other cleptoparasites. But some of those species that attack social halictines are more or less social parasites, staying in the host nest and possibly taking on qualities of the host queen. Field collecting and nest excavations suggest that in a species of *Microsphecodes* the females remain in the host (*Lasioglossum* subgenus *Dialictus*) nests and may be social parasites (Eickwort and Eickwort, 1972b). Knerer (1980) reported on two species of *Sphecodes* whose females were found in nests of hosts (*Halictus maculatus* Smith). Wcislo (1997) found similar

8. Parasitic and Robber Bees

Table 8-1. Social Parasites.

The notations are as follows: **(p)** Probable social parasite, recognized by the reduced scopa but not known from host nests; **(n)** Found in host nests. Numbers in parentheses after *subfamily* names indicate the probable numbers of origins of social parasitism. An asterisk (*) in front of a generic or subgeneric name indicates that all species of that taxon are parasitic. Specific names linked by an **and** represent two species that probably diverged from a common parasitic ancestor, and therefore are considered to represent a single origin of parasitism. Note that bees known to be cleptoparasites are not included in this list; see Table 8.2.

Higher taxa and parasitic taxa	Host taxa
Family Halictidae	
Subfamily Halictinae (3)	
Tribe Halictini[a]	
Sphecodes, n	*Halictus*
Microsphecodes, n	*Lasioglossum (Dialictus)*
Lasioglossum (Dialictus)[b], n	*Lasioglossum (Dialictus)*
Family Apidae	
Subfamily Xylocopinae (11)	
Tribe Allodapini	
Allodape greatheadi Michener, p	
Allodapula guillarmodi Michener, p	
Braunsapis bislensis Michener and Borges, n	*Braunsapis*
Braunsapis breviceps (Cockerell), n	*Braunsapis*
and *B. kaliago* Reyes and Sakagami, n	*Braunsapis*
Braunsapis natalica Michener, n?	*Braunsapis*
Braunsapis pallida Michener[c], p	
**Eucondylops konowi* Brauns, n, and	*Allodapula*
E. reducta Michener, n	*Allodapula*
**Effractapis furax* Michener, p	
**Inquilina excavata* (Cockerell), n,	*Exoneura*
and *I. schwarzi* Michener, n	*Exoneura*
Macrogalea mombasae (Cockerell), n	*Macrogalea*
**Nasutapis straussorum* Michener, n	*Braunsapis*
Subfamily Apinae (3)	
Tribe Bombini	
Bombus (**Psithyrus*), n	*Bombus*
B. (Alpinobombus) arcticus (Quenzel), n	*Bombus*
B. (Thoracobombus) inexspectatus (Tkalcŭ), n	

[a] As indicated in the text, some species of *Sphecodes* and its relatives may be more like social parasites than cleptoparasites. They are included in this table as reminders that they may be social parasites; their behavior is too little known for assurance on either count.

[b] The parasitic species of *Dialictus* were formerly segregated as a parasitic genus *Paralictus*.

[c] Some supposedly parasitic Australian *Braunsapis* species are now believed to be probably not parasitic.

behavior in a parasitic species of *Lasioglossum (Dialictus)* (the former *Paralictus*). There may be intermediates between social parasites and cleptoparasites among the parasitic halictines.

Cleptoparasites. The remaining parasitic bees, the great majority, are **cleptoparasites**. A cleptoparasite enters the nest of a host and lays an egg in a cell. In most cases the adult parasite then leaves, although sometimes (e.g., in *Hoplostelis* s. str.) it ejects the host and stays in the nest. The parasite larva feeds on the food that had been provided for a host larva. Such bees are often appropriately called "cuckoo bees." Rarely recorded is intraspecific cleptoparasitism (Field, 1992), in which a bee opens a cell of another individual of its own species and replaces the egg. Most cleptoparasites belong to their own obligately parasitic species, genera, tribes, or subfamilies. The host is commonly solitary, although some social Halictinae are hosts of cleptoparasites. Table 8-2 lists the cleptoparasitic taxa. Grütte (1935) published a valuable account of such parasites. Some genera are recognized as cleptoparasites only by reduction or lack of pollen-manipulating and pollen-carrying structures, especially the scopa of females, and their probable association with solitary hosts. In Table 8-2, such forms are marked "p" for probable.

Females of some mostly cleptoparasitic genera such as *Sphecodes* (Halictini) destroy the egg of the host and replace it with their own. In such cases one never finds a cell with two or more eggs; the *Sphecodes* egg is not noticeably different from that of other halictine bees. It is failure to find cells with two or more eggs, combined with the ordinary (i.e., hostlike) structure of the young parasitic larvae, that leads to the conclusion that the host egg is de-

Table 8-2. Cleptoparasites.

The notations are as follows: **(p)** Probable cleptoparasite; see Section 8. **(n)** Found in or reared from host nests, but the method of killing the host egg or larva is not known. **(ad)** Host egg or larva killed by adult female parasite; this is assumed if among many cells studied, none contained both host and parasite eggs. **(lo)** Host egg or young larva killed by active but otherwise rather ordinary parasite larva. **(lm)** Host egg or young larva killed by active young parasite larva with sclerotized head and sickle-shaped mandibles. Numbers in parentheses after *subfamily* names represent the probable numbers of origins of cleptoparasitism. Generic names are omitted when *all* species of a subfamily or tribe are cleptoparasitic. An asterisk (*) in front of a name indicates that all species of that taxon are parasitic. Note that bees known to be social parasites are not included in this list; see Table 8.1.

Higher taxa and parasitic taxa	Host taxa
Family Colletidae	
Subfamily Hylaeinae (1)	
Hylaeus (Nesoprosopis). in part, n	*Hylaeus (Nesoprosopis)*
Family Halictidae	
Subfamily Halictinae (9)	
Tribe Halictini[a]	
*Echthralictus, p (Derived from *Homalictus*)	
*Eupetersia, Sphecodes clade, p	
Halictus (*Paraseladonia), p	
Lasioglossum	
Dialictus[b], in part, ad	*Lasioglossum (Dialictus)*
*Paradialictus, p	
*Microsphecodes, Sphecodes clade, ad	Halictini
*Nesosphecodes, Sphecodes clade, p	
*Parathrincostoma, p	
*Ptilocleptis, Sphecodes clade, p	
*Sphecodes[c], ad	Halictinae and others[d]
Tribe Augochlorini	
Megalopta (*Noctoraptor), p	
Megommation (*Cleptommation), p	
*Temnosoma, p	
Family Megachilidae	
Subfamily Megachilinae (10)	
Tribe Osmiini	
*Bekilia, position doubtful, p	
Hoplitis (*Bytinskia), n	*Hoplitis*
Tribe Megachilini	
*Coelioxys, lm	*Megachile,* less frequently various Apidae (see Sec. 84)
*Radoszkowskiana, lm	*Megachile*
Tribe Anthidiini	
*Afrostelis, p	
*Euaspis, ad	*Megachile*
*Hoplostelis, ad	Euglossini
*Larinostelis, p	

(*continues*)

stroyed by the adult parasite. Sick et al. (1994) saw a female *Sphecodes* in an observation nest of *Lasioglossum (Evylaeus) malachurum* (Kirby) enlarge the opening of a host cell, enter the cell head first, for two minutes probably destroying the host egg, then back out, turn, and back into the cell, remaining there for five minutes probably laying her egg. She then came out and closed the cell with soil, and the next day, on leaving the nest, she closed the nest entrance. Genera known to parasitize host cells in this way are marked "ad" (for adult) in Table 8-2.

Females of most cleptoparasites, however, lay eggs in host cells without destroying the host egg or larva. The egg of the parasite may be (1) inserted into and hidden in the cell wall of an as yet unclosed cell while the host is out of the nest, or (2) laid in a finished and closed host cell by the parasitic mother through a hole that she makes and later seals in the cell wall or closure.

Rozen and Özbek (2003) and Rozen (2003a) described the mature oocytes or eggs of numerous cleptoparasitic megachilid and apid taxa. Relative to body size, eggs of cleptoparasites are smaller than those of related nonparasitic solitary bees. Moreover, larger num-

Table 8-2. Cleptoparasites *(continued)*

Higher taxa and parasitic taxa	Host taxa
Stelis, lo	Megachilinae
Dolichostelis, ad	*Megachile*
Xenostelis, p	
*Tribe Dioxyini, lo	Megachilinae
Family Apidae	
*Subfamily Nomadinae (1), lm	many groups, see Section 91
Subfamily Apinae (10)	
Tribe Ctenoplectrini	
Ctenoplectrina, p (Derived from *Ctenoplectra*)	
*Tribe Rhathymini, lm	*Epicharis*
*Tribe Ericrocidini, lm	Centridini
*Tribe Melectini, lm	Anthophorini
*Tribe Isepeolini, lm	*Colletes, Canephorula*
*Tribe Protepeolini, lm	*Diadasia*
*Tribe Osirini, lo	*Epeoloides* on *Macropis*, others on Tapinotaspidini
Tribe Tetrapediini	
Coelioxoides, lm	*Tetrapedia*
Tribe Euglossini	
Exaerete, ad, lo	*Eulaema, Eufriesea*
Aglae, n	*Eulaema*

[a] As indicated in the text, some species of parasitic halictids may be more like social parasites than cleptoparasites.

[b] A few North American species of *Dialictus* formerly placed in *Paralictus* are parasitic on other *Dialictus*.

[c] *Eupetersia, Microsphecodes,* and *Ptilocleptis* are probably members of the *Sphecodes* phyletic line, although no analysis has been made.

bers are ready to be laid than in nonparasitic solitary bees; sometimes there are more ovarioles than the plesiomorphic number for the family. Compared with eggs of solitary bees, those of cleptoparasites often show thickening and elaboration of ornamentation of the chorion and elaboration of the micropyle. Rozen (2003a) provided a table showing, for many cleptoparasitic Apidae, that the eggs are significantly smaller relative to body size in those that hide their eggs in open host brood cells, often still being provisioned, than in those that open completed and closed host cells for oviposition and do not need to hide their eggs because the host is no longer about.

Eggs hidden in cell walls are unusually small compared to those of other bees of the same size (see Sec. 4), and are quite diverse in structure, often differing widely from the usual bee egg with its soft chorion. Specialized eggs are laid by all Nomadinae and also by Protepeolini in the Apinae and *Coelioxys* in the Megachilini. Their eggs are inserted into the inside wall of the host cell or otherwise hidden before the cell is closed by the host, often before the provisioning is completed. Eggs inserted by cleptoparasitic bees into finished, closed cells are of ordinary size and form, like those of *Sphecodes*. Such eggs are those of Melectini, Ericrocidini, and others.

Interesting diversity exists in the Nomadinae not only in egg form but also in the manner in which the eggs are inserted in the hosts' cell walls (Rozen, 1991a, 1992a).

Some (e.g., *Doeringiella*) are completely buried and at right angles to the cell wall; some are thrust into the wall only partway and left with one end projecting (*Nomada*) (see Radchenko, 1981); some are doubled over in the cell wall (*Oreopasites*); and some are placed in the wall almost parallel to its surface with one side exposed, that side hardened and roughened, in contrast to the usually soft chorion (Biastini, illustrated by Rozen, Roig-Alsina, and Alexander, 1997). *Epeolus,* which lays eggs in *Colletes* cells composed of two cellophane-like layers, places its egg between the two layers, with the anterior end exposed. Females of nomadine genera have distinctive structures, especially of S6, presumably for their particular methods of egg laying (Figs. 8-10f, 91-2, 95-3c, d).

Unlike nonparasitic bees that ordinarily lay one egg per cell, cleptoparasitic forms as different as *Nomada* and *Coelioxys,* i.e., forms in different families, frequently put two to several eggs into parasitized cells. The resultant larvae then kill not only the host egg or larva but also their conspecific competitors until there is only one left alive. Some cleptoparasitic larvae are active and able to kill the host egg or larva with ordinary-sized but sharp mandibles [e.g., *Stelis,* "lo" (for larva ordinary) in Table 8-2]. It is usually young larvae that do this, but Rozen (1987a) indicated that even last-stage larvae of *Stelis* may have modifications for killing hosts. In contrast, the young larva of *Coelioxys* and parasitic Apidae has a large, usually more or less prognathous, sclerotized head and sickle-shaped mandibles with which to kill the host egg or young larva [e.g., *Coelioxys* (Fig. 84-5a, d), "lm" (for larval mandibles) in Table 8-2]. "Prognathous" means that the head and mouthparts are directed forward, rather than more or less downward as in other larval bees. In the cleptoparasitic Apidae the first-stage larvae are those specialized for killing the host or their conspecific competitors, whereas in *Coelioxys* (Megachilidae) it is the still small second- or third-stage larvae that have the largest mandibles (Fig. 84-5c, d). Rozen (1991a) described and illustrated the known first-stage larvae of parasitic tribes of Apinae (except for the parasites in the Euglossini) and compared them with first-stage larvae of Nomadinae. Only in the Rhathymini, among first-stage parasitic Apinae, is the larval head incompletely sclerotized, more or less spherical, and hypognathous. Young larval morphology and presumably behavior are thus convergent in various groups independently derived from nonparasitic ancestors.

Rozen (2001) has given a key to the mature larvae of cleptoparasitic bees, along with a summary of host relationships.

The numbers after subfamily names in Table 8-2 indicate the numbers of independent origins of cleptoparasitism within those subfamilies. These estimates may be high, since phylogenetic analyses remain to be done or are problematic. Thus *Radoszkowskiana* and *Coelioxys* may not represent separate origins of parasitism, and *Afrostelis*, *Larinostelis*, *Stelis*, and possibly *Euaspis* could have evolved from a single parasitic ancestor. The tribes Protepeolini and Isepeolini, even perhaps Osirini, might be basal branches of Nomadinae and thus have the same parasitic ancestor as that subfamily. Melectini, Ericrocidini, and Rhathymini might have a common parasitic ancestor. Thus the total number of origins of cleptoparasitism could be less than the 31 indicated in Table 8-2. Alexander (1990), in a conservative list that omitted the then doubtful entries such as the Hylaeinae and Osmiini, nonetheless enumerated 17 origins of cleptoparasitism.

The parasitism by members of the Colletidae is based on five Hawaiian species of *Hylaeus* (*Nesoprosopis*) reported to be parasites of other species of the same subgenus by Perkins (1899). He evidently found parasites in host nests, and recognized parasites by the reduced pollen-gathering hairs on the front tarsi of females, but gave no data to clarify or verify his conclusions. His work, however, was usually dependable and is supported by Daly and Magnacca (2003).

In spite of the taxonomically diverse groups of bees that have evolved cleptoparasites, many large and sometimes old and widespread groups have not done so. Except as noted above, the Colletidae is such a group. Parasites of any sort are unknown in the Andrenidae, the Nomiinae and Rophitinae in the Halictidae, the Melittidae, and various large groups of Apidae such as the Ceratinini, Xylocopini, and Eucerini.

The hosts of cleptoparasitic bees are always other bees. Emery's rule (see Wilson, 1971) for parasitic aculeate Hymenoptera indicates that parasites usually attack their close relatives. Cleptoparasitic bees that are similar to their nonparasitic relatives, so that both are in the same genus, tribe, or subfamily, are usually parasitic on members of that genus, tribe, or subfamily. Thus parasitic Halictini mostly parasitize other Halictini, although as noted in the discussion of *Sphecodes* (Sec. 66), a few *Sphecodes* species parasitize other halictids and even bees in other families. Parasitic Euglossini parasitize other Euglossini. Parasitic Megachilinae are parasitic on other megachilines except that a few of the many *Coelioxys* species attack *Anthophora, Centris, Euglossa,* or, reportedly, *Tetralonia* (Bischoff, 1927: 398), and *Hoplostelis* s. str. attacks Euglossini. The parasitic tribes of Apinae are, so far as is known, all parasitic on other tribes of Apinae, except for Isepeolini, which parasitizes Colletinae and *Epeoloides*, which parasitizes *Macropis* (Melittidae). As noted elsewhere, the Isepeolini could be a basal nomadine tribe.

A common observation is that species of cleptoparasitic bees vary greatly in size. Sometimes two size classes are evident; that they parasitize host species of different sizes is a common assumption, rarely verified. Such size classes of parasites can be explained (1) as probable cryptic species, each specializing on a host species of a certain size, (2) as "races" specializing on such hosts, or (3) as direct effects of food quantity on the growth and maturation of the larvae of the parasitic species. A study of *Coe-*

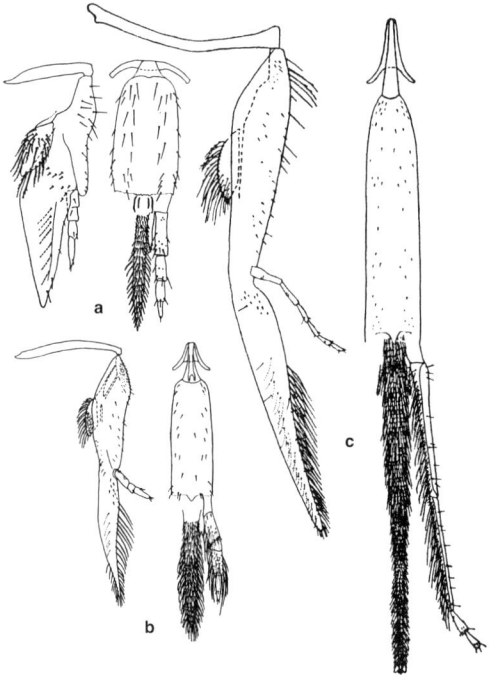

Figure 8-1. Proboscidial structures of parasitic and nonparasitic allodapine bees, maxilla at left, labium at right. **a, b,** The social parasites *Eucondylops reducta* Michener and *Nasutapis straussorum* Michener; **c,** The nonparasitic species *Allodapula melanopus* (Cameron). The bodies of these bees are roughly the same size; these drawings are to the same scale, thus showing the reduction of the proboscis in size as well as in palpal segmentation in the social parasites. Drawings modified from Michener, 1970.

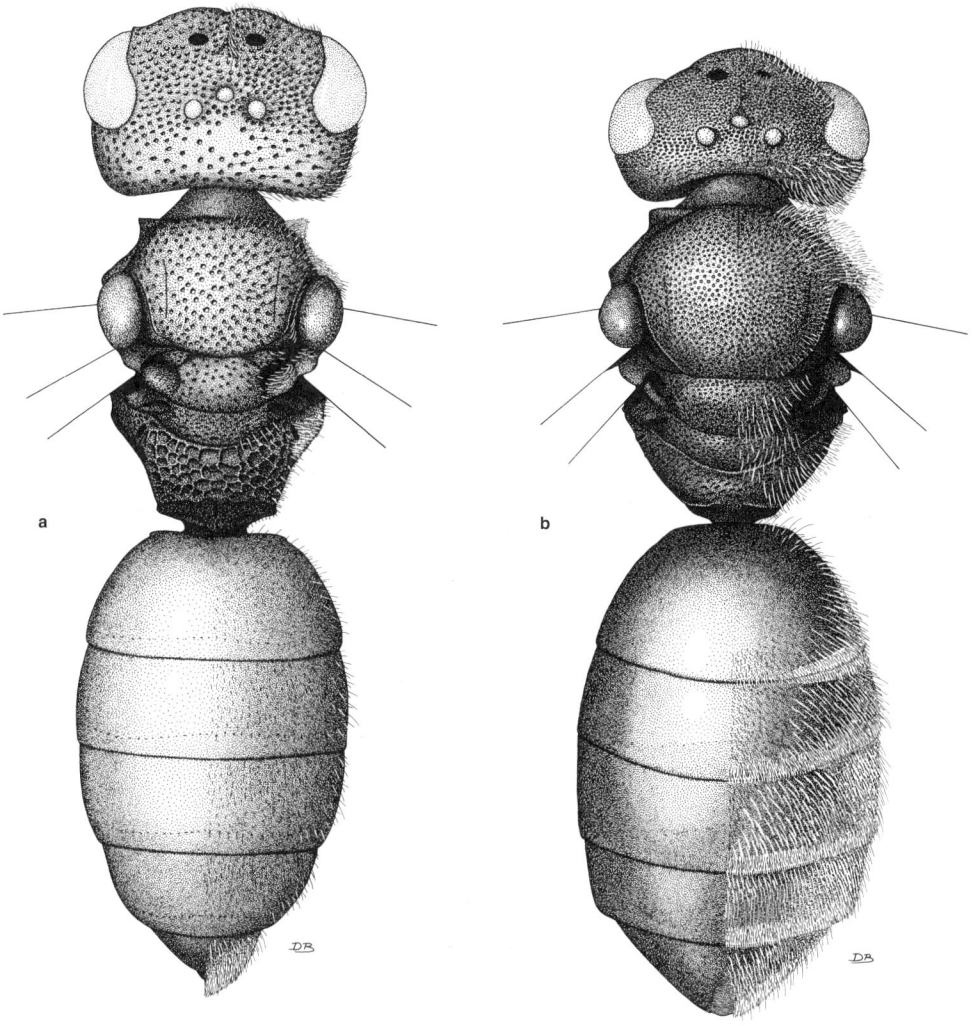

lioxys funeraria Smith parasitizing two different-sized species of *Megachile* in Michigan supports the third explanation, for on the basis of 41 loci, the genetic difference between samples of large and small *Coelioxys* from the two *Megachile* hosts could be explained entirely by sampling error (Packer et al., 1995). Presumably, there was one panmictic population of *Coelioxys* parasitizing the two species of *Megachile*.

The largest group of cleptoparasites is the Nomadinae, which, although assuredly an apid subfamily, is so different from its nonparasitic relatives that its closest relatives are unrecognized. Genera of Nomadinae are often rather host-specific (e.g., *Epeolus* on *Colletes*, *Triepeolus* mostly on Eucerini), but as a whole the subfamily parasitizes a wide range of bees, including Colletidae (Colletinae, Diphaglossinae), Andrenidae (all major subfamilies), Halictidae (all subfamilies), Melittidae (Melittinae, Dasypodainae), and Apidae (Anthophorini, Eucerini, Exomalopsini, Tapinotaspidini).

A curious finding is that in various species of *Nomada* the cephalic secretions of the males are chemically simi-

Figure 8-2. Bodies of females of Halictini. **a,** The cleptoparasitic *Sphecodes monilicornis* (Kirby); **b,** The nonparasitic *Lasioglossum malachurum* (Kirby); hairs are omitted on the left half of each. Note the coarse head and thoracic punctation and propodeal sculpturing and the strong dorsolateral pronotal angles of the cleptoparasite. Drawing by D. J. Brothers, from Michener, 1978b.

lar to Dufour's gland volatiles of the females of the host species, *Andrena* or *Melitta* (Tengö and Bergström, 1976, 1977). The species of *Nomada* tend to be rather host-specific, and for each host-parasite pair studied, the chemical similarities mentioned are evident. The Dufour's gland product is used to line brood cells; its odor presumably characterizes the nests. Cephalic secretions of males are similar to those of females in most species of *Andrena*, but the secretions of males and females of *Nomada* species are quite different; it seems likely that a mimetic relationship has evolved between host or host nest odor and male parasite odor. Attempted explanations for why it is the male parasite's odor that resembles the host female or nest odor are not yet convincing. Does the female *No-*

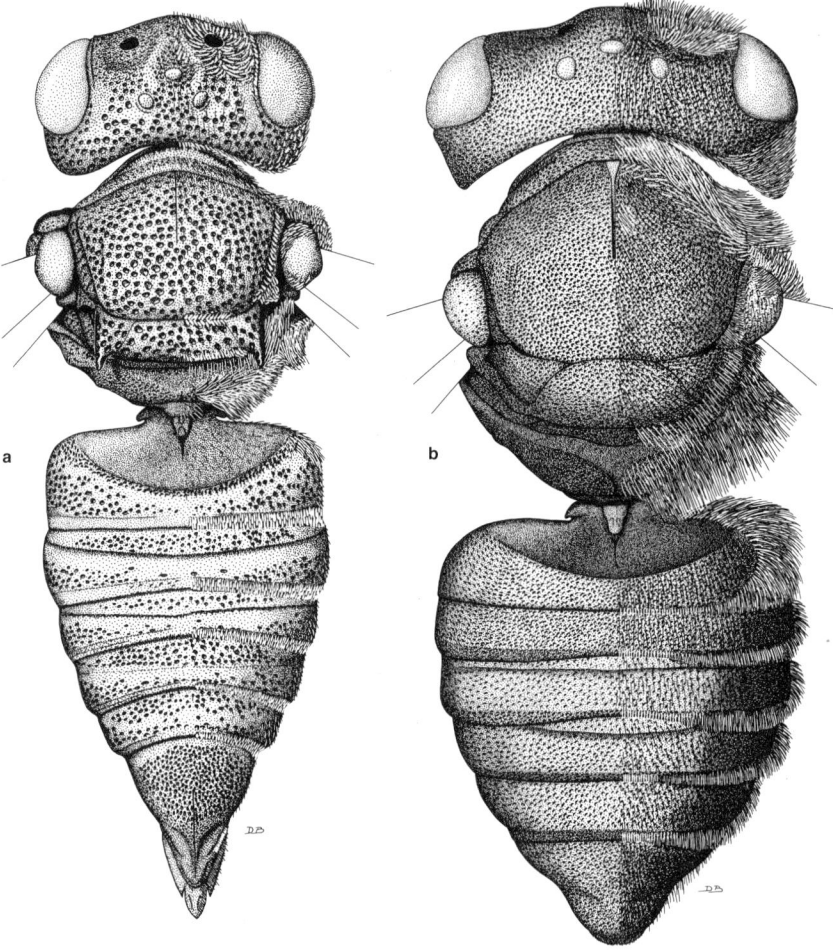

Figure 8-3. Bodies of females of Megachilini. **a,** The cleptoparasitic *Coelioxys octodentata* Say; **b,** Its host, *Megachile brevis* Say; hairs are omitted on the left half of each. Note the coarse punctation and pointed axillae (on each side of the scutellum) of the cleptoparasite. Drawing by D. J. Brothers.

mada learn from the male with whom she mated the odor needed to find an oviposition site? Or, if the female needs the host nest odor, perhaps to facilitate her entrance past defending owners into host nests, might she acquire the needed odor from the male at the time of copulation? Neither of these explanations seems likely!

Social parasites and cleptoparasites. The following paragraphs take up topics that relate to both types of parasitic bees.

Parasitic bees have many morphological features not or rarely found in nonparasitic bees; see the remainder of this section and see Sections 27 and 28. Social parasites have reduced scopae (Fig. 8-8). In the case of parasitic species of *Bombus*, this means that the corbiculae are reduced. Michener (1970) listed convergent features in socially parasitic Allodapini. Among such features is the reduction in the proboscis (Fig. 8-1); these parasites feed in the host nests rather than on flowers.

A surprising number of the cleptoparasitic bees are wasplike in appearance, partly because of reduced hairiness and loss of the scopa, features frequently enhanced by slender form, red coloration (especially of the metasoma) or yellow-and-black wasplike coloration, as in many species of *Nomada* (Pl. 2). Figures 8-2, 8-3, and 8-4 show in a general way the reduction of hairiness in parasites, as contrasted with their hosts, which in the case of Figures 8-2 and 8-3 are members of the same tribes as the parasites. The loss of the pollen-carrying scopa, the most decisive morphological characteristic of parasitic bees, is illustrated in Figure 8-5, which compares the hind leg of a cleptoparasitic *Sphecodes* with that of an ordinary, nonparasitic bee of the same tribe, the Halictini. As is often seen among cleptoparasites, the outer surface of the tibia has coarse setae or spines that probably help the parasite to push through a burrow, against an opposing host bee, and the basitarsus has lost not only scopal hairs but also the apical process and brush (penicillus) used by nest-making halictids in spreading secreted material on cell walls. Among cleptoparasites, various degrees of scopal reduction can be found. For comparison with Figure 8-5, Figure 8-6 shows degrees of reduction of the femoral and tibial scopa of other cleptoparasitic Halictini. In the

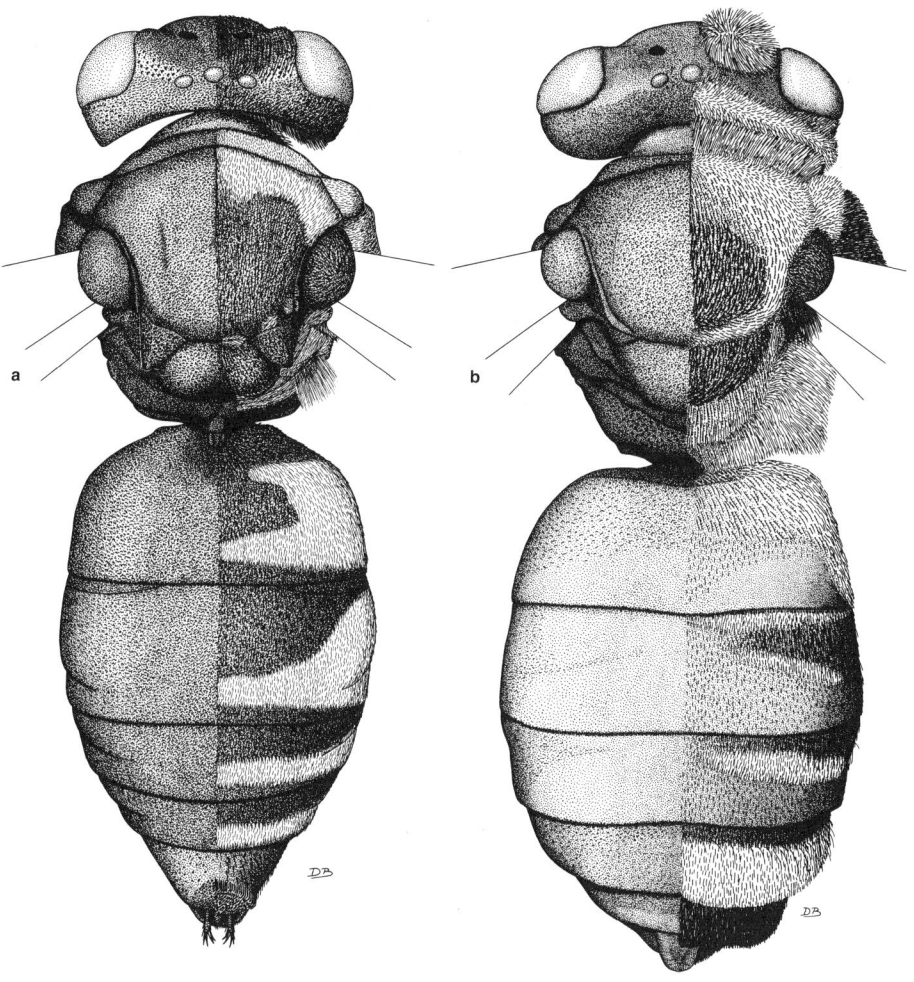

Figure 8-4. Bodies of females of Apidae. **a,** A cleptoparasite in the Nomadinae, *Triepeolus concavus* (Cresson); **b,** Its host in the Apinae, *Svastra obliqua* (Say). Hairs are omitted on the left half of each. Note the large, angular axillae at the sides of the scutellum of the cleptoparasite. Drawings by D. J. Brothers.

cleptoparasitic Megachilidae and Apidae, scopal reduction similar to that of Halictini is found; in the Nomadinae and some Megachilidae (Fig. 8-7), the reduction is complete. Partial reduction in the parasitic Allodapini (Apidae) is shown in Figure 8-8.

Other structures of female bees that are commonly reduced in cleptoparasites include the pygidial plate. Figure 8-9 shows that of the female of an ordinary species of Halictini, and the reduced plates of two cleptoparasitic species. The pygidial plate's usual function is probably the tamping of cell surfaces; cleptoparasites do not make cells and have reduced plates. Similarly, the basitibial plate, which probably helps nest-making bees to brace themselves while digging or tamping, is reduced in cleptoparasites (Figs. 8-9, 66-17d). In the Halictinae, the labrum of females ordinarily has an apical process with a strong median keel of unknown function. The process becomes broad and flat in cleptoparasites, thus more like the labrum of males (Fig. 8-9), or at least the keel is lost (Fig. 66-17c).

The apex of the female metasoma of cleptoparasitic forms is frequently modified for the placement of the eggs, but this is not the case for cleptoparasites whose adults destroy and replace the host egg (ad in Table 8-2), as shown by Figure 8-10a, b, which illustrates nearly identical apical sterna of nonparasitic and cleptoparasitic Halictini. Cleptoparasites that insert their eggs into cells or cell walls instead of replacing the host egg (lo and lm in Table 8-2) commonly have apical modifications, as shown in Figures 8-3, 8-4, 8-10d, f, 84-6b-e, 91-1, 101-2a-c, and 104-4. Even in social parasites, whose egg-laying should differ little if at all from that of the host, the apex of the metasoma is more pointed in the subgenus *Psithyrus* than in other *Bombus*, and the last tergum is perhaps more often scoop-shaped in parasitic Allodapini than in nonparasitic species.

Parasitic taxa, especially those in the Apidae, are structurally very different from one another and from their probable nonparasitic antecedents. The result, in the No-

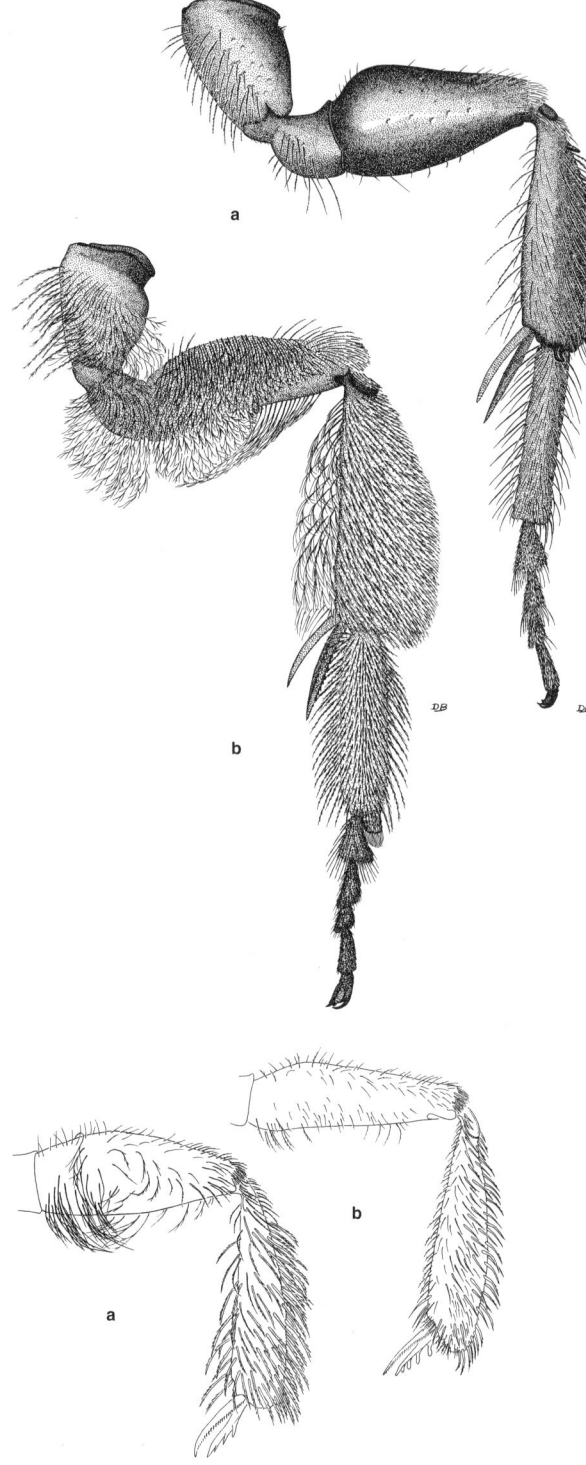

Figure 8-5. Hind legs of female halictine bees. **a,** The cleptoparasitic *Sphecodes monilicornis* (Kirby), showing the lack of a scopa and penicillus and the presence of large tibial spicules; **b,** *Lasioglossum malachurum* (Kirby), showing the strong scopa from the trochanter to the basitarsus and the distal process and penicillus on the basitarsus. Drawing by D. J. Brothers, from Michener, 1978b.

Figure 8-6. Posterior femora and tibiae of females of cleptoparasitic halictids, showing scopal reduction as compared to *Lasioglossum malachurum* (Fig. 8-5b). **a,** *Lasioglossum (Dialictus) asteris* (Mitchell), a parasitic species; **b,** *Echthralictus extraordinarius* (Kohl), a parasitic relative of *Homalictus*. Drawings by M. McCoy, from Michener, 1978b.

8. Parasitic and Robber Bees

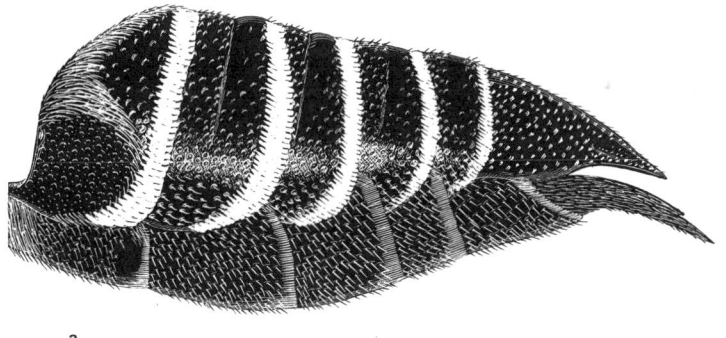

Figure 8-7. Side views of metasoma of female Megachilinae. **a,** The cleptoparasite *Coelioxys octodentata* Say; **b,** Its host, *Megachile brevis* Say. Note the hairiness and especially the ventral scopa of the latter. Drawings by D. J. Brothers.

Figure 8-8. Scopal reduction in parasitic Apidae. **a,** Hind leg of a cleptoparasitic species of Nomadinae, *Triepeolus concavus* (Cresson), showing complete lack of scopal hairs; **b, c,** Hind tibiae of the nonparasitic species of Allodapini, *Braunsapis simillima* (Smith), with sparse but functional scopal hairs, and of the social parasite *B. breviceps* (Cockerell), with the scopa reduced. a, Drawing by D. J. Brothers; b, c, from Reyes, 1991b.

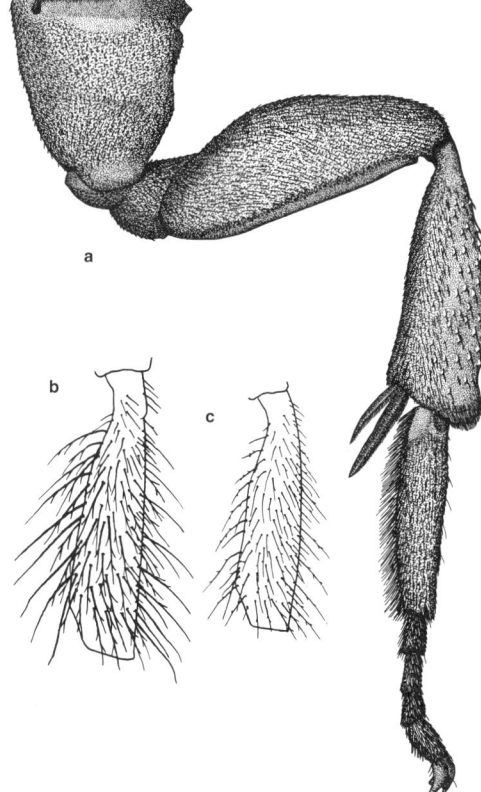

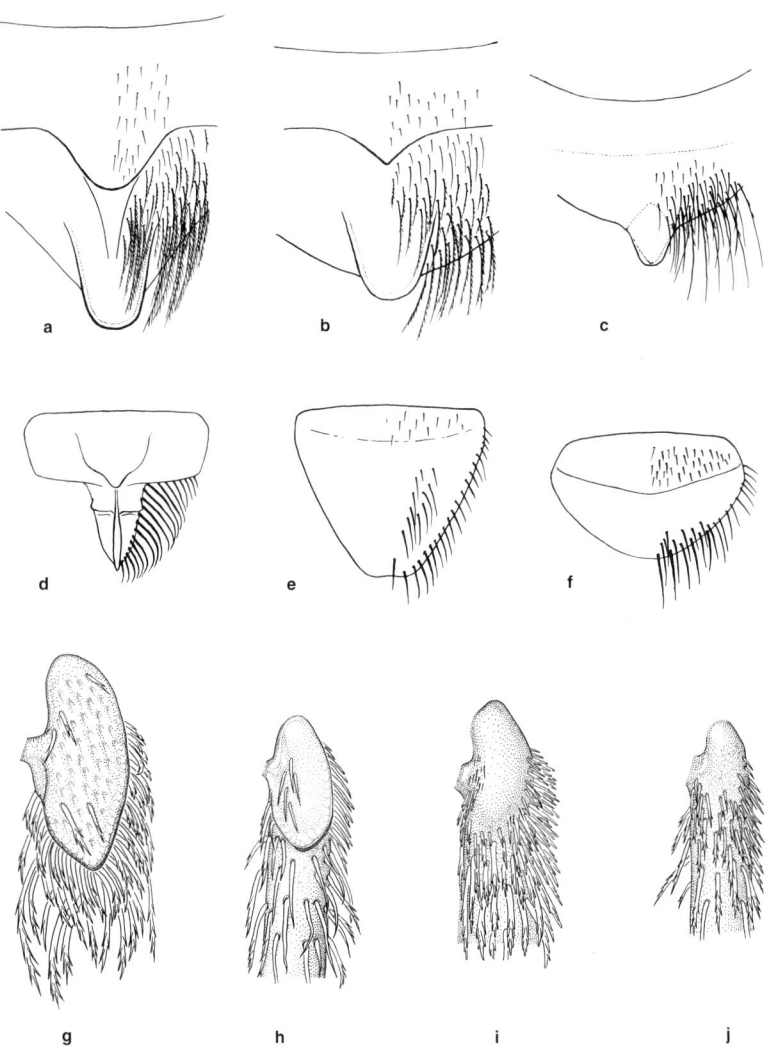

Figure 8-9. Structures of female halictine bees, nonparasitic species in lefthand column, otherwise cleptoparasitic forms. a-c, Pygidial plates. a, *Lasioglossum (Evylaeus) malachurum* (Kirby); b, *Lasioglossum (Dialictus) asteris* (Mitchell), a member of the cleptoparasitic group formerly called *Paralictus*; c, *Ptilocleptis polybioides* Michener. d-f, Labra. d, *L. malachurum* (Kirby); e, *Sphecodes monilicornis* (Kirby); f, *Ptilocleptis tomentosa* Michener. g-j, Basitibial plates. g, *L. malachurum* (Kirby); h, *Echthralictus extraordinarius* (Kohl); i, *S. monilicornis* (Kirby); j, *S. chilensis* Spinola. Drawings by M. McCoy, from Michener, 1978b.

madinae and Apinae, is recognition of numerous parasitic tribes. Their morphological diversity relative to probable ancestral taxa (when recognized) leads to the theory that the morphological features of parasitic taxa evolve relatively rapidly during and after the acquisition of obligatory parasitic behavior. Support for this idea comes from *Echthralictus*, a parasitic derivative of *Homalictus* (Halictinae). *Echthralictus* is found only in Samoa, where it must have evolved from its presumed hosts, the local *Homalictus*, during the relatively short history of Samoa, probably less than 2.6 million years. Yet *Echthralictus* differs in numerous morphological features from related halictids (see Sec. 66). If molecular characteristics evolved at a more uniform rate than morphological characteristics, molecular studies should indicate relationships of the parasitic taxa more clearly.

In females of nonparasitic bees there must be strong selection for the maintenance of such nest-making and food-collecting structures as characteristic mandibles, other mouthparts, basitibial plate, pygidial fimbria and plate, pollen-manipulating brushes, and the scopa. But once parasitism is established, such selection pressure vanishes and these structures are reduced or eliminated, just as eyes are reduced or lost in cave animals (see Sec. 27). Advantages probably accrue when nutrients and genetic machinery are not devoted to unneeded structures and associated behavior. At the same time, novel structures characteristic of parasites frequently develop, e.g., a strong cuticle and spines, lamellae, or carinae that probably protect the neck and petiolar regions (see Sec. 28). Thus there is a tendency for parasitic bees to be well defended against the jaws and perhaps stings of irate hosts. A possibly alternative strategy is seen in the cleptoparasite *Osiris*, which has relatively thin but smooth and shiny cuticle and no protective spines or carinae, but an enormous sting. Parasites commonly have stronger stings than their nonparasitic relatives, although, curiously, the cleptoparasitic megachilid tribe Dioxyini has the most reduced sting of any bee. The tendency for structures of female parasitic bees to resemble structures of males was emphasized by Wcislo (1999b).

The number of mature oocytes (i.e., eggs nearly ready

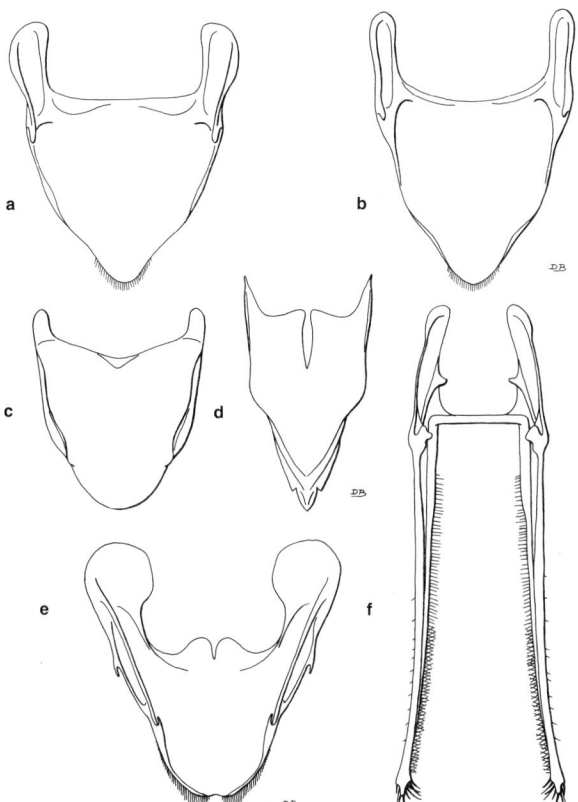

Figure 8-10. Sixth sterna of certain female bees, showing modifications related to egg placement by cleptoparasites that insert their eggs into cell walls. **a,** *Lasioglossum malachurum* (Kirby), nonparasitic; **b,** *Sphecodes monilicornis* (Kirby), a cleptoparasite that does not insert eggs but merely replaces the host egg; **c,** *Megachile brevis* Say, nonparasitic; **d,** *Coelioxys octodentata* Say, cleptoparasitic; **e,** *Svastra obliqua* (Say), nonparasitic; **f,** *Triepeolus concavus* (Cresson), cleptoparasitic (see also Fig. 8-4a). Species shown in e and f are in different subfamilies; nonetheless, e probably represents the type of sternum from which f evolved through various less-extreme steps found in the Nomadinae. Drawings by D. J. Brothers.

to be laid) in the ovaries in parasitic bees is frequently greater than the number in nonparasitic relatives, a convergent feature among parasitic bees. Presumably, a female solitary bee, which has to construct a cell and provision it before laying an egg, needs to have only one egg ready to be laid at a time. A parasite, by contrast, might find several cells in succession ready to receive its eggs, and therefore needs to have several eggs ready at any time. Some parasites simply have more than one mature oocyte per ovariole, whereas others have more than the usual number of ovarioles per ovary, the usual number being three for most bees, four for Apidae. In the whole subfamily Nomadinae, the number of ovarioles per ovary is variable but greater than four, most often five but up to ten (Alexander, 1996) and 17 in *Rhopalolemma* (Rozen, Roig-Alsina, and Alexander, 1997). In *Ericrocis* but not in other Ericrocidini dissected, the number is five instead of four. In the social parasite *Bombus (Psithyrus)* the number ranges from 6 to 18. In other parasitic bees that have been dissected, the number of ovarioles is the same as in the related nonparasites, i.e., three or four (Alexander and Rozen, 1987).

It is reasonably clear that in the Northern Hemisphere the percentage of parasitic species in a bee fauna tends to increase with distance from the Equator (Wcislo, 1987; Petanidou, Ellis, and Ellis-Adam, 1995). In northern Ellesmere Island there are two species of bees *(Bombus)*, one of which is a social parasite in nests of the other, but the trend is more reliably indicated in the larger faunas of middle and tropical latitudes. Possible explanations for this trend in percentages of parasitic species include (1) the synchronization caused by cool seasonal climates, such that many host nests are in the right condition for attack at the same time, (2) the competition for nest sites when synchronization is intense, some individuals thus being unable to find their own good sites, and (3) the short summer at high latitudes, leaving an individual that is delayed in nesting unable to produce offspring ready to overwinter, either because of the seasonal lack of the right flowers to provide food or because of autumnal cold weather. A delayed individual thus might profit from laying eggs in an already established nest. Obviously, these ideas are not mutually exclusive. Petanidou, Ellis, and Ellis-Adam (1995) believed that the unpredictability of the xeric warm-temperate Mediterranean climates contributed to their reduced percentage of parasitic species relative to percentages in cooler and less xeric areas. Another possibility is that the whole pattern is simply a result of the abundance of *Andrena* and its cleptoparasites in the genus *Nomada,* and of *Bombus* and its congeneric social parasites, in the cooler latitudes of the Northern Hemisphere, and that general ecological explanations such as are enumerated above are only marginally relevant, or not relevant at all. It remains to be determined whether percentages of parasitic species increase with latitude in bee faunas of southern Africa and South America. In Australia there are but few parasitic forms at any latitude.

9. Body Form, Tagmata, and Sex Differences

Bees of many genera can be identified to genus at a glance, or at least very promptly, by a person familiar with the bees of the relevant region. This is possible in part because of the diverse body shapes of bees. I have indicated general body shape in the text by a series of terms such that with a single word a person who knows a few common genera of bees can get an idea of what bees of an otherwise unknown genus look like. These terms, following Michener, McGinley, and Danforth (1994), are listed below, arranged in a general way from slender and relatively hairless to robust and hairy. References to the colored plates, as well as to the habitus drawings and photographs, are offered so that the reader might gain a better idea of the meanings of these terms. The terms are subjective, however, and can be used only to give a general idea of body form and often hairiness. (Terms marked by asterisks apply primarily to megachilids and therefore indicate the large-headed megachilid body form.)

Hylaeiform. Body form of *Hylaeus*, also suggestive of a pemphredonine wasp. Slender, the hairs inconspicuous without magnification; scopa inconspicuous or absent (Pl. 1; Figs. 90-1, 90-10b).

Nomadiform. Body form of *Nomada*. Slender, wasplike, not noticeably hairy, often with yellow or red markings; scopa absent (Pl. 2; Figs. 93-2, 94-1, 102-2).

Epeoliform. Body form of *Epeolus, Triepeolus,* or *Doeringiella.* Somewhat more robust than *Nomada* but nonetheless wasplike parasitic bees; scopa absent. Body often with areas of short, pale pubescence forming a conspicuous pattern (Pl. 2; Figs. 93-4, 95-1, 96-1, 97-1, 98-1, 100-1, 105-1).

Andreniform. Body form of *Andrena, Halictus,* or *Colletes.* Male often slender, its metasoma more parallel-sided than that of female (Pls. 3-5; Figs. 40-1, 49-1, 49-2, 58-3, 58-9, 59-4, 59-6, 60-1, 64-7).

*****Heriadiform.** See hoplitiform (Figs. 81-6, 81-7).

*****Hoplitiform.** Body form of *Hoplitis (Alcidamea), Heriades,* or more slender species of *Megachile (Callomegachile).* Similar to megachiliform but more slender, metasoma parallel-sided. The term **heriadiform** has been used for this body form, but implies greater slenderness (Pl. 6; Figs. 81-8, 84-9).

*****Chalicodomiform.** Between hoplitiform and megachiliform (Pls. 7, 8; Figs. 81-5, 82-9, 82-11, 84-15).

*****Megachiliform.** Body form of *Megachile (Megachile)* or *Anthidium.* Body heavy, head thick, metasoma rather wide, not parallel-sided (Pls. 7, 8; Figs. 81-10, 82-15, 84-8).

Trigoniform. Body form of *Trigona* and its relatives, e.g., of the genus *Partamona.* Metasoma small and robust to slender and parallel-sided; body not conspicuously hairy, i.e., hairs short, and metasoma usually shiny (Pl. 9; Fig. 108-1).

Apiform. Body form of workers of *Apis mellifera,* i.e., more robust than andreniform and more slender than euceriform (Pls. 9, 10).

Euceriform. Body form of *Eucera* or *Melissodes.* Similar to anthophoriform but somewhat less robust (Pls. 11, 15; Figs. 111-1, 112-1, 112-2, 117-1, 117-2).

Anthophoriform. Body form of *Bombus.* Robust, head, thorax, and sometimes metasoma with abundant hair, thus enhancing the aspect of robustness (Pls. 10, 12, 13, 15; Figs. 88-1, 106-1). The term **bombiform** may be used especially for those with a hairy metasoma.

Bombiform. See Anthophoriform (Pl. 14; Figs. 116-1, 116-7).

Many bees, of course, do not fall unequivocally into one or another of the above categories. For dry specimens, much depends on how full the crop was when the specimen was killed, how much the metasoma has telescoped in drying, and so forth. Nonetheless, these terms may be useful in suggesting the characteristic aspects of groups of bees. Some genera, such as *Coelioxys* because of its tapering conical metasoma, do not fall readily into any of the above categories.

At the outset, a question arises about the names for the three **tagmata**, or main parts of the body. Logically, they should be "head, thorax, and abdomen" or "prosoma, mesosoma, and metasoma." For simplicity I prefer the first series. But because the first true abdominal segment is incorporated into the thorax as the **propodeum**, the numbering of segments in the remainder of the abdomen should begin with 2 (as was done by Michener, 1944, 1954b). In bees, many taxonomically important structures are on one or another of the segments. But because of confusion as to an author's terminology, "first abdominal segment" could mean either the propodeum (if the reference is to be morphologically correct, as in Michener, 1944), or the segment next behind the propodeum, i.e., the first segment of the metasoma, as in most other works. To make it clear that I am following the customary system of numbering, I always speak of metasomal rather than abdominal terga and sterna. Thus the **first metasomal segment** is the segment behind the propodeal-metasomal constriction. I use the names **head, thorax** (including the propodeum), and **metasoma** in order to combine familiar, unequivocal terms with a term that is not confusing for segmental numbering. Abbreviations such as **T1** (first metasomal tergum), **S1** (first metasomal sternum), and so forth, are regularly used to save space.

The word "gaster" has been recommended by some authors as an alternative to "metasoma." The problem is that, especially for taxa other than bees, such as wasps and ants, it is morphologically deceptive. It really refers to the enlarged or swollen part. For bees it would not be confusing, for it would be entirely equivalent to "metasoma," i.e., the first and following metasomal segments. For an ant or wasp with a one-segmented petiole, gaster comprises the second and following metasomal segments, and for an ant with a two-segmented petiole, the gaster is the third and following metasomal segments. Thus if one hopes for a uniform terminology throughout the Aculeata or the Hymenoptera, in which homologous structures have the same names, "gaster" is unsatisfactory, because a particular numbered gastral segment may be numbered differently in related taxa.

The sexes in bees are often quite different from one another, and in the keys the sexes are sometimes treated separately. Most males have 13 antennal segments or antennomeres, as they are often called, but there are only 12 in some Euryglossinae (some species of *Euryglossina*), some Ammobatini *(Pasites, Melanempis,* and *Parammobatodes)*, some Ammobatoidini *(Holcopasites)*, some Biastini (some species of *Biastes*), some Halictini (some species of *Thrinchostoma*), and some Augochlorini *(Chlerogas)*; only 11 or 12 in *Systropha* (Rophitinae); and 14 in *Uromonia* s. str. (Meganomiinae). Females have 12 antennal segments, although there are only 11 in some species of *Euryglossina* (Euryglossinae). Males usually have seven exposed metasomal terga; females have six. Sometimes the apical terga are retracted beneath the preapical ones, so that female Halictinae, for example, usually appear to have only five terga, and in some male bees, such as *Protosmia*, T7 can be seen only with difficulty, because it is largely or wholly retracted and must be dissected out if one wishes to examine it. Females have stings, and males have sclerotized genitalia, but both are usually retracted, and in some females the sting is rudimentary. The sting and associated structures are reduced and not (or weakly) functional in many Andrenidae; more reduced, not at all useful for stinging, in the Meliponini; and maximally reduced, compared to all other bees, in the Dioxyini. For sting reduction see Figure 27-1.

10. Structures and Anatomical Terminology of Adults

The illustrations and text in this section provide the names for the external parts of adult bees. Many structures are simply labeled in the illustrations. Those that do not appear in the figures, or for which discussion or explanation is needed, are treated in the text, with the preferred terms appearing in boldface type at the places where they are defined or explained. The hope is that the illustrations plus the text will encourage uniformity of usage in future work.

The emphasis is on features that vary among kinds of bees; very few terms are explained or illustrated for structures that do not provide useful diagnostic characters in one or more bee taxa.

The morphology of the honey bee (*Apis mellifera* Linnaeus), which has been much studied, serves as a background for persons interested in the Apiformes as a whole. Snodgrass (1956) provided a valuable account of *Apis* morphology as well as a terminology for structures (derived from his earlier studies) that has served as a basis for subsequent works. Several authors studying non-*Apis* bees have given, in varying degrees of detail, their own accounts of external morphology, using the terminology of Snodgrass except where there were reasons for deviating, and establishing additional terminologies as needed. Such studies are those by Michener (1944) on *Anthophora*; Urban (1967a) on *Thygater*; Camargo, Kerr, and Lopes (1967) on *Melipona*; Eickwort (1969a) on *Pseudaugochlora*; Gerber and Akre (1969) on *Megachile*; Pesenko (1983) on *Nomioides*; and Brooks (1988) on *Anthophora*. There is no need to repeat here the details of morphology presented in some of these works, but the terminology to be used in subsequent parts of this book requires some explanation. It is that of Michener (1944), modified in various ways.

Many of the structures referred to in the systematic parts of this book (Sections 33 to 121) are labeled in Figures 10-1 to 10-15. These figures are mostly diagrammatic, intended to illustrate a maximum number of bee structures; only a few are based on particular bee species or genera. Terms that seem obvious from a perusal of these figures are not further described here. For fuller accounts, see Michener (1944) and the other morphological studies listed above.

The head. Figures 10-1 and 10-2 illustrate the major structures of the head. For names of mandibular structures I usually follow Michener and Fraser (1978); see Figure 10-2. For simplicity I often refer to **preapical teeth** on the upper margin instead of teeth of the pollex. The **pollex** is the upper margin of the mandible, above the acetabular groove, commonly ending in the preapical tooth or teeth; the **rutellum** is the rest of the distal part of the mandible, below the pollex, usually forming the major or lower apical mandibular tooth. A few bees (*Lithurgus*, many Xylocopinae) have a lower preapical tooth, i.e., a rutellar tooth, below the main apex (Figs. 78-3a, 90-9a, b). Other terms used herein are the **condylar ridge**, which arises near the mandibular condyle (at the lower or posterior basal angle of the mandible) and extends toward the

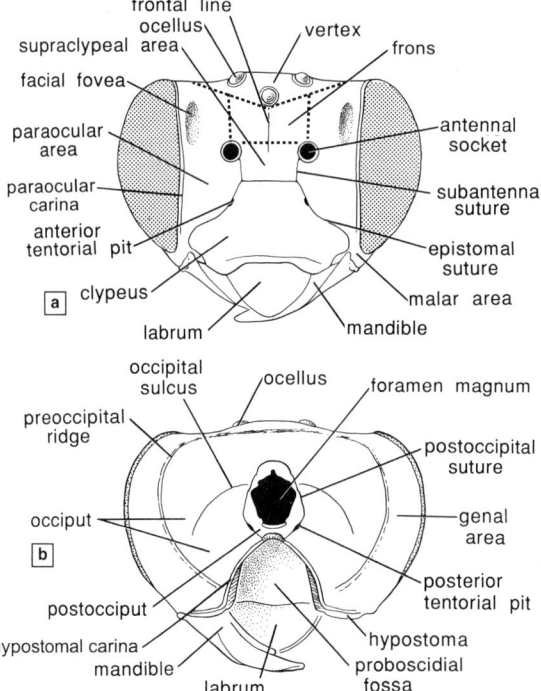

Figure 10-1. Diagrams of a bee's head, showing major structures.

a, Anterior view; **b,** Posterior view. From Michener, McGinley, and Danforth, 1994.

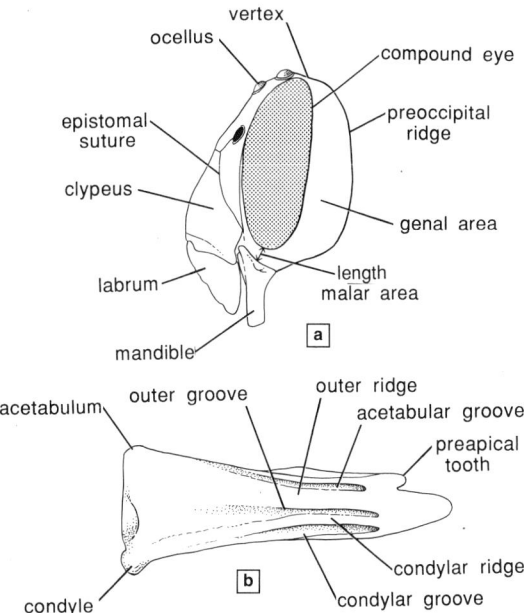

Figure 10-2. a, Diagrammatic lateral view of a bee's head; b, Outer surface of mandible. From Michener, McGinley, and Danforth, 1994.

apex of the mandible, and the **outer ridge,** which is the next ridge above the condylar ridge on the outer surface of the mandible.

The terminology for parts of the labrum is often confused by the use, in the genus *Andrena,* of the term process for the basal elevated plate. The term process is misleading because this plate does not project, as one expects of a process. In other bees, e.g., the Panurginae (see Ruz, 1986), the same structure is called the **basal area** of the labrum. Use of the word "process" in the sense of basal area is further confusing because in some bees, especially the Halictidae, there is an entirely different process on the apex of the labrum, here called the **apical process** of the labrum.

For descriptive purposes the face is often divided into ill-defined areas, as indicated by dotted lines in Figure 10-1. These are the two **paraocular areas,** the **supraclypeal area,** the **frons** or **supra-antennal area,** and the **vertex.**

The **malar area** (or malar space) is between the eye and the mandible; its length is the shortest distance from the eye to the mandible (Fig. 10-2); the width of this area is the width of the base of the mandible.

The **foveae** of the face (Fig. 10-1) (and of the sides of T2, Fig. 10-12) are depressions, usually black in color and therefore conspicuous when the ground color is pale. Those of the face are paired, one in each paraocular area, and lie largely or entirely above the level of the antennal bases, sometimes extending up onto the vertex between the ocelli and the upper ends of the compound eyes. The foveae, better developed in females than in males, may be punctiform or narrow, hairless grooves (e.g., in many Hylaeinae and Euryglossinae) or broad, slightly depressed areas, those in most Andreninae finely hairy. Schuberth and Schönitzer (1993) have given an anatomical account of the facial foveae of various taxa, showing that the epidermis beneath foveal cuticle consists of secretory cells. Presumably, the foveae are evaporative surfaces, but the products and their functions are unknown. In many bees that lack distinct facial foveae there are nonetheless areas of differentiated cuticle, presumably homologous to foveae. Such areas are usually more sparsely and less coarsely punctate than adjacent areas, and frequently differ in surface microstructure and in color (usually black). Thus every intergradation exists between complete lack of foveae (as in Megachilidae and most Halictinae) and distinct, depressed foveae. In keys and descriptions I do not indicate the presence of foveae unless the area is sufficiently depressed that at least one margin is distinct and the other indicated by a sharp change in texture or color.

The **antennae** arise from **antennal sockets,** sometimes called **alveoli.** For simplicity, the terms antennal or flagellar segments are used instead of flagellomeres. Nearly all bees have a **subantennal suture** extending from each antennal socket down to the epistomal suture. If the antennae arise close to the clypeus, there may be no subantennal sutures, and in some bees (most Andrenidae) there are two subantennal sutures below each antenna, defining a **subantennal area.** These sutures are easily seen as dark lines if the integument is pale but are often hard to see if it is black, and may be invisible if it is both black and coarsely punctate. The **epistomal suture** defines the upper limits of the clypeus. A longitudinal carina immediately mesal to the antennal base occurs in some bees. Typically, it is most elevated just above the antennal base and often forms a lamella that partially overlaps the antennal base (Fig. 10-3a). Such carinae are often called interantennal or interalveolar carinae. Such terms, however, suggest carinae extending from one antennal base to the other. The term **juxtantennal carinae,** proposed by Michener and Griswold (1994a), is therefore preferable for these structures.

In some bees (e.g., *Augochlora*) the anterior tentorial arms seem to have migrated downward, carrying with them the epistomal suture, which therefore becomes an-

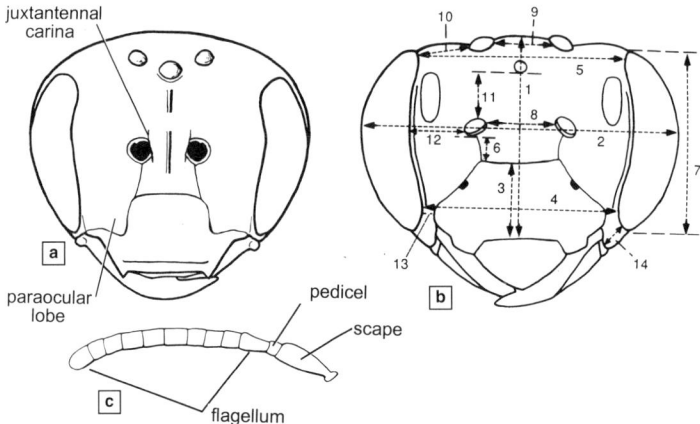

Figure 10-3. Diagrams of heads and antenna. **a**, Diagrammatic frontal view of a bee's head, showing structures not illustrated in **Figure 10-1**. (No known bee has both paraocular lobes and juxtantennal carinae.) **b**, Diagrammatic frontal view of a bee's head, showing how the following measurements are made: (1) length of head (or face); (2) width of head; (3) length of clypeus; (4) lower interocular distance; (5) upper interocular distance; (6) clypeoantennal distance (or length of subantennal suture if it is straight); (7), length of eye; (8) interantennal or interalveolar distance; (9) interocellar distance; (10) ocellocular distance; (11) antennocellar or alveolocellar distance; (12) antennocular or alveolocular distance; (13) clypeocular distance; (14) length of malar area. **c**, Antenna of female bee. From Michener, McGinley, and Danforth, 1994.

gled or lobed down into the clypeus on each side. The resultant lobe of the paraocular area into the clypeus is called the **paraocular lobe** (Fig. 10-3a).

The term **orbit** is often used for the eye margin, **inner orbit** for the frontal or facial margin, and **outer orbit** for the genal margin. An expression like "eyes converging below" is ordinarily exactly equivalent to "inner orbits converging below." Imaginary lines tangent to the upper or lower extremities of both eyes, as seen in a frontal view of the head, are sometimes useful in indicating the positions of the ocelli or of the clypeal apex. Such lines are called the **upper** and **lower ocular tangents**.

Descriptions of bees often include measurements or, more often, statements of relative dimensions of various structures, especially on the head. Figure 10-3b indicates how certain measurements should be made. Looking down on the vertex, one can use the **postocular tangent**, the imaginary line tangent to the posterior convexities of both eyes. The **ocelloccipital distance** is between a posterior ocellus and the point where the vertex curves or angles down onto the posterior surface of the head, i.e., usually the preoccipital ridge. Several of these terms can be reversed, e.g., *ocellocular* has the same meaning as *oculocellar*.

The **genal area** is the region behind the eye and in front of the preoccipital ridge. The ridge surrounding the concave posterior surface of the head above and laterally is called the **preoccipital ridge**. A carina sometimes found on this ridge is the **preoccipital carina**. It can be dorsal (behind the vertex only), lateral (behind the eye only), or complete (both dorsal and lateral). According to Silveira (1995a), the dorsal part of the preoccipital ridge in some Exomalopsini is actually on the posterior surface of the head, and a new transverse ridge (or carina), the **postocellar ridge** (or carina), is present just behind the ocelli.

The **proboscidial fossa** (Fig. 10-1) is the large, deep groove on the underside of the head into which the proboscis folds. The fossa is margined laterally by the **hypostomal carina**, the anterior end of which bends laterad behind the mandibular base. The underside of the head, lateral to the hypostomal carina and behind the mandibular base, is the **hypostomal area**, or, according to Eickwort (1969b), the postgena; this area is not the entire hypostoma because the latter includes also the walls and roof of the proboscidial fossa. The **paramandibular process** is an anterior projection of the hypostoma that approaches, butts against, or is fused with the lateral part of the clypeus, and in the latter case provides sclerotic closure of the mandibular socket. Bristles or hairs arising from a ridge on the paramandibular process and sometimes continuing laterally on the lateral extremity of the hypostomal carina constitute the **subgenal coronet** of some *Andrena* species.

A disproportionate number of characters used in the higher classification of bees are based on proboscidial features. Most sclerites of the **proboscis** are clear from Figure 10-4, but some explanation is needed. Sometimes, the length of the proboscis is expressed as the point that it reaches under the body of the bee when retracted, which means when folded, at rest, or, as often expressed, in repose. The proboscis consists of three "segments." The basal one, containing the maxillary **cardines** (sing. **cardo**), extends backward in repose from their articulations to the cranium. The middle one, containing the maxillary **stipites** (sing. **stipes**), prepalpal parts of the **galeae**, and the labial **prementum**, extends forward in repose from the apices of the cardines. The distal segment, containing the **galeal blades** or postpalpal parts of the galeae, the **labial palpi**, and the **paraglossae** and **glossa**, extends again backward from the apex of the middle segment. In repose the apices of these third-segment parts may not reach out of the proboscidial fossa, or may ex-

PLATE 1

Slender and sparsely haired (hylaeiform) bees commonly nest in narrow burrows in pithy stems. Note here superficially similar representatives of two families: Colletidae (Hylaeinae) and Apidae (Xylocopinae).

Above, *Hylaeus variegatus* (Fabricius), female, from Germany, on flower of *Tanacetum vulgare,* photo by P. Westrich (a relatively robust example of a hylaeiform bee, family Colletidae); **Middle,** *Hylaeus angustatus* (Schenck), male, from Germany, on inflorescence of *Allium,* photo by P. Westrich (hylaeiform); **Below,** *Ceratina* sp., female, from California, on flower of *Diplacus aurantiacum,* photo by E. S. Ross (hylaeiform, Apidae).

PLATE 2

Cleptoparasitic bees tend to be sparsely haired and wasplike in coloration, or as in the lower figures, with a strong pattern produced by short, pale, dense hairs. These three are all females. Note the complete absence of pollen-carrying scopal hairs.

Above, *Nomada armata* Herrich-Schäffer, female, from Germany, photo by P. Westrich (nomadiform, Apidae); **Middle,** *Nomada* sp., female, from California, photo by E. S. Ross (nomadiform); **Below,** *Epeolus* sp., female, from California, photo by E. S. Ross (epeoliform, Apidae).

PLATE 3

One of the commonest genera of bees in North America and Eurasia is *Andrena*. Note the scopal hairs of the two females, well loaded with pollen in the middle figure.

Above, *Andrena tscheki* Morawitz, female, from Germany, photo by P. Westrich (andreniform, Andrenidae); **Middle,** *Andrena* sp., female, from Oregon, photo by E. S. Ross (andreniform); **Below,** *Andrena* sp., male, from California, photo by E. S. Ross (andreniform).

PLATE 4

Andreniform bees are among the most common in the world. Here are representatives of two families, Colletidae and Halictidae, and of two subfamilies (Nomiinae and Halictinae) of the latter.

Above, *Scrapter heterodoxus* (Cockerell), male, from South Africa, photo by E. S. Ross (andreniform, Colletidae); **Middle,** *Nomia melanderi* Cockerell, female, from Wyoming, photo by E. S. Ross (andreniform, Halictidae); **Below,** *Halictus farinosus* Smith, female, from California, photo by E. S. Ross (andreniform, Halictidae).

PLATE 5

Additional andreniform bees exhibit diverse coloration, bright green and dark brown forms both in the halictid tribe Augochlorini, and yellow and black in the andrenid subfamily Panurginae. Minute yellow and black bees like this North American *Perdita* occur in most of the world, but belong to other groups of bees: the Nomioidinae (Halictidae) in the Old World, *Habralictus* (Halictidae) in the American tropics, and various Euryglossinae (Colletidae) in Australia.

Above, *Augochlorella pomoniella* (Cockerell), female, from California, photo by E. S. Ross (andreniform, Halictidae); **Middle,** *Megalopta* sp., female, a nocturnal halictid from Brazil, photo by E. S. Ross (andreniform, Halictidae); **Below,** *Perdita (Perdita)* sp., female, from California, photo by E. S. Ross (andreniform, Andrenidae).

PLATE 6

Rather slender-bodied Megachilidae. This family is often called "Leafcutting bees," but none of these three genera make nest cells with pieces of leaves. These species, all of the tribe Osmiimi, use resins, mud, or chewed leaf pulp in making brood cells.

Above, *Heriades crenulata* Nylander, male, from Germany, on flower of *Anthemis tinctoria,* photo by P. Westrich (a relatively robust heriadiform bee, family Megachilidae); **Middle,** *Hoplitis (Hoplitis) adunca* (Panzer), female, from Germany, photo by P. Westrich (hoplitiform, Megachilidae); **Below,** *Osmia* sp., female, from California, photo by E. S. Ross (hoplitiform, an unusually slender-bodied *Osmia,* family Megachilidae).

PLATE 7

Three members of the megachilid genus *Megachile*. The robust ones, middle and lower figures, are those whose females cut leaf pieces for making brood cells; they are the true leaf-cutting bees.

Above, *Megachile (Sayapis) fidelis* Cresson, female, from California, photo by E. S. Ross (chalicodomiform, Megachilidae); **Middle,** *Megachile* sp., female, from Arizona, photo by E. S. Ross (megachiliform, Megachilidae); **Below,** *Megachile* sp., male mating, with expanded front tarsi covering female's eyes, from California, photo by E. S. Ross (megachiliform).

PLATE 8

Most members of the megachilid tribe Anthidiini have striking yellow and black patterns, as shown here for three genera.

Above, *Trachusa (Heteranthidium) cordaticeps* (Michener), male, from Arizona, photo by E. S. Ross (megachiliform, Megachilidae); **Middle,** *Rhodanthidium siculum* (Spinola), female, from Tunisia, photo by E. S. Ross (chalicodomiform, Megachilidae); **Below,** *Anthidiellum notatum robertsoni* (Cockerell), female, from California, on resin cell attached to stem, photo by E. S. Ross (anthophoriform, Megachilidae).

PLATE 9

Among the highly social bees, diverse only in the tropics, are two major groups: the Meliponini here illustrated by two species of *Trigona,* and the true honey bees, illustrated here and in Plate 10 (upper figure) by two tropical Asiatic species of *Apis.* Plate 10 (middle) illustrates a more robust member of the Meliponini.

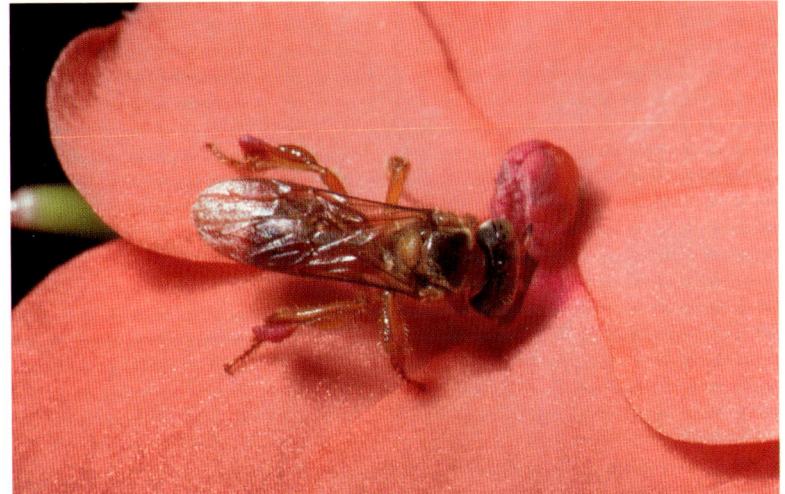

Above, *Trigona (Trigona) ferricauda* Cockerell, worker, from Panama, on flower of *Impatiens,* photo by E. S. Ross (trigoniform, Apidae); **Middle,** *Trigona (Trigona) williana* Friese, worker, from Peru, photo by E. S. Ross (trigoniform); **Below,** *Apis florea* Fabricius, workers, from India, photo by E. S. Ross (apiform, Apidae).

PLATE 10

Two species of highly social Apinae, and below, a carpenter bee (Apinae, Xylocopini). Females of the latter excavate large burrows, often in solid wood, hence the vernacular name.

Above, *Apis cerana* Fabricius, worker, from Thailand, photo by E. S. Ross (apiform, Apidae); **Middle,** *Meliponula (Meliponula) bocandei* (Spinola), worker, from Congo, transferring pollen from middle leg to hind leg while hovering in front of flower, photo by E. S. Ross (robust apiform or short-haired anthophoriform, Apidae); **Below,** *Xylocopa (Koptortosoma) caffra* (Linnaeus), female, from Namibia, photo by E. S. Ross (anthophoriform, Apidae).

PLATE 11

Rather large, hairy Apinae, tribe Eucerini. Most males of this tribe have long antennae, as shown in the middle figure.

Above, *Melissodes* sp., female, from California, photo by E. S. Ross (euceriform, Apidae); **Middle,** *Melissodes* sp., male, from California, on flower of *Bidens laevis,* photo by E. S. Ross (euceriform); **Below,** *Xenoglossa angustior* Cockerell, female, from Arizona, photo by E. S. Ross (euceriform, Apidae).

PLATE 12

Many of the robust (anthophoriform) Apinae have striking color patterns, often made by area of short, dense pubescence. *Thyreus* is a cleptoparasitic genus (tribe Melectini) that lacks pollen-carrying scopal hairs; *Amegilla* (lower figure) has a well-developed tibial (white) and basitarsal (black) scopa.

Above, *Thyreus* sp., from Congo, photo by E. S. Ross (anthophoriform, Apidae); **Middle,** *Thyreus histrionicus* (Illiger), female, from Germany, photo by P. Westrich (anthophoriform); **Below,** *Amegilla quadrifasciata* (Villers), female, from Germany, photo by P. Westrich (anthophoriform, Apidae).

PLATE 13

Three species of robust, hairy, anthophoriform bees, all in the Apidae, subfamily Apinae. The middle figure shows a form that appears to mimic species of *Bombus* in coloration.

Above, *Anthophora fulvitarsis* Brullé, female, from Germany, photo by P. Westrich (anthophoriform, Apidae); **Middle,** *Anthophora bomboides stanfordiana* Cockerell, female, from California, taking water, photo by E. S. Ross (anthophoriform); **Below,** *Amegilla (Zonamegilla)* sp., female, from Sabah, photo by E. S. Ross (anthophoriform, Apidae).

PLATE 14

Bumblebees *(Bombus)* exhibit diversity among species in the patterns of coloration of their long pubescence.

Above, *Bombus sylvarum* (Linnaeus), queen, from Germany, photo by P. Westrich (bombiform, Apidae); **Middle,** *Bombus* sp., worker on *Hypericum* flower, from California, photo by E. S. Ross (bombiform); **Below,** *Bombus hypnorum* (Linnaeus), queen, from Germany, photo by P. Westrich (bombiform).

PLATE 15

The top two photographs are of orchid bees (Euglossini), each with an attached orchid pollinium. The lower figure of *Colletes* cells illustrates the membranous brood-cell lining, characteristic of the Colletidae.

Above, *Eulaema* sp., male, from Ecuador, with orchid pollinium on dorsum, photo by E. S. Ross (euceriform to anthophoriform, Apidae); **Middle,** *Euglossa* sp., male, from Brazil, with orchid pollinium on dorsum, photo by E. S. Ross (anthophoriform, Apidae); **Below,** nest of *Colletes similis* Schenck, from Germany, artificially exposed in a cliff, showing the cellophane-like cell lining surrounding two provisioned cells, photo by P. Westrich (Colletidae).

PLATE 16

Bees construct nests in diverse situations, mostly well hidden, although Plate 8 (below) shows an exposed nest. (Top) Cluster of halictid cells in the ground, opened to show the provisions (yellow) and immature stages. (Middle) Three cells of a megachilid in a hollow stem, each cell nearly filled with provisions and with a young larva. (Below) Similar megachilid cells in a dead snail shell.

Above, cell cluster of spring nest of *Lasioglossum (Evylaeus) malachurum* (Kirby), in Germany, showing pollen balls, eggs (extreme right and left), and small larvae, photo by P. Westrich (Halictidae); **Middle,** nest of *Osmia (Osmia) rufa* (Linnaeus), in Germany, in dead bamboo stem, showing mud partitions between cells and small larvae feeding on stored pollen, photo by P. Westrich (Megachilidae); **Below,** nest of *Osmia (Helicosmia) aurulenta* (Panzer), in Germany, in shell of dead snail, opened to show two cells, photo by P. Westrich.

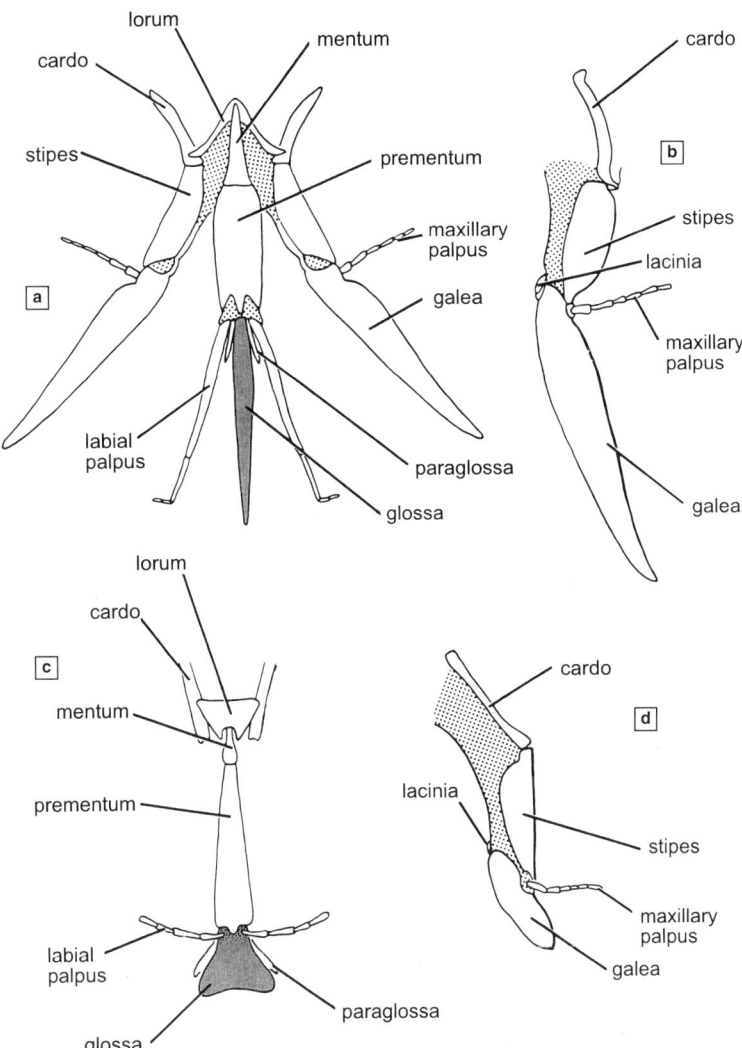

Figure 10-4. Diagrams of proboscides of bees. **a,** Spread proboscis of a long-tongued bee; **b,** Maxilla of the same; **c,** Labium of a short-tongued, in this case colletid, bee, showing portions of maxillary cardines at the base; **d,** Maxilla of the same. From Michener, McGinley, and Danforth, 1994.

tend back on the ventral surface of the body to or even beyond the apex of the metasoma. The proboscis is extended, for example to probe a flower for nectar, by unfolding these segments. It can project downward from the bee's head, or forward. For descriptive purposes I consider it to project downward, but some other authors make the other decision. What I consider the anterior surface of the proboscis, they call the dorsal surface.

Proboscidial structures are illustrated and labeled not only in Figure 10-4 but also in Figures 19-1 and 19-2. The terminology of the parts of the glossa is extensive and is indicated in Figures 19-3 and 19-4.

The galeal blade, i.e., the postpalpal part of the galea, is divided by a longitudinal **galeal rib** into (1) a thin, anterior, hairless part, its sclerotized surfaces compressed together (Roig-Alsina and Michener, 1993, fig. 15), called by J. Plant (manuscript, 1991) the **galeal velum;** and (2) a posterior part that supports hairs, and whose inner and outer surfaces are largely separated. In S-T bees the apices of these parts are commonly separated by a notch near the apex of the galeal blade. The galeal rib, which often bears a series of hairs, appears as a strengthening element, principally in L-T bees. The prepalpal part of the galea is the **subgalea.**

There has been confusion about the naming of the basal sclerites of the labium. The main labial sclerite, which is in the middle "segment" of the labium and to which the labial palpi are attached, is regularly called the prementum. Basal to the prementum, and supported in the membrane between the apical parts of the maxillary cardines, are one or two sclerites that can be called the **postmentum.** In most Hymenoptera there is only one such sclerite. In many bees there is a sclerite immediately basal to the prementum that tapers basally and is thus more or less triangular. This is what I call the **mentum** (Michener, 1944, 1985a), following Snodgrass, who, however, in 1956 called it the postmentum. Plant and Paulus (1987) consider it the distal part of the postmentum, the basal part being the lorum. In some bees the mentum is partly (as in some *Andrena* and *Panurgus*) or wholly (as in *Ctenocolletes,* some *Andrena*) membranous (Michener, 1985a), but can be identified by its position and shape. In other bees, such as the Halictidae, the membranous area is smaller, not so distinctively shaped, and may represent either the reduced mentum or merely a membrane between the prementum and the postmentum. Basal to the mentum, in most bees, is the **lorum**. It was called submentum by Michener (1944), but Winston (1979) and others pointed out that because there is only one basal labial sclerite in other Hymenoptera, the lorum should be considered a new structure, special to the bees, not homologous to the submentum of some other insect orders. Plant and Paulus (1987), however, as indicated above, regarded it as the basal part of the postmentum. For convenience, I term the labial sclerites, starting basally, to be **lorum, mentum,** and **prementum.** When the lorum expands as a weak sclerotization occupying space between the cardines, I call it the **loral apron.** When, as in most Halictidae, the mentum is partly or wholly membranous and doubtfully recognizable, I nonetheless tentatively regard the single basal labial sclerite as the lorum and loral apron, not as a fusion product of lorum plus mentum. These sclerites are fused in some L-T bees, but this is not as common or so evident as was indicated by Plant and Paulus (1987), who illustrated quite distinct sclerites as fused in some cases.

The basal part of the proboscis, i.e., the part that attaches to the head, is called the **labiomaxillary tube** (stippled in Fig. 21-2). Its skeletal parts are the cardines, which are strong rods in the wall of the tube, and it is further strengthened by flexible strips called the **conjunctival thickenings.** (Additional longitudinal thickenings on the anterior surface are of maxillary lacinial origin in the Halictidae.) For illustrations, see Figures 21-2b and 59-1. In many bees the lower end of the conjunctival thickening is separated as a small sclerite, the **suspensorium of the prementum,** that connects the thickening to a lateral notch in the prementum. In the posterior surface of the labiomaxillary tube, the lorum varies greatly, as described above and by Michener (1985a). It may be a rather weak, flat loral apron, i.e., a thickening or sclerotization of the posterior wall of the labiomaxillary tube occupying most of the space between the cardines. It may be more limited in area but elevated around the connection to the mentum, or reduced to a strongly sclerotized, V-shaped structure connected medially to the mentum (Fig. 10-4a). In the last two alternatives, when the labium is retracted, the base of the mentum and median (apical) part of the lorum together commonly project, forming a lobe extending posteriorly from the labiomaxillary tube. This lobe is the **proboscidial lobe** (Fig. 21-2a). In some bees, when the proboscis is retracted, this lobe projects upward and into the **postoccipital pouch,** a large pit below the foramen magnum (Roig-Alsina and Michener, 1993).

The thorax. The bee thorax (Fig. 10-5) is a compact structure consisting of sclerites of the pro-, meso- and metathoracic segments, which bear the legs and wings, and the first true abdominal segment, termed the **propodeum.** The prothorax is represented primarily by the large **pronotum,** which extends ventrally at each side as a process that meets or nearly meets its fellow behind the fore coxae. The **propleura** and **prosternum** are in front of this lateroventral extension of the pronotum. The **pronotal lobe** is a useful landmark. The **dorsolateral angle** of the pronotum is in front of and somewhat mesad from the pronotal lobe; often there is a ridge, carina, or lamella connecting the lobe to the dorsolateral angle. Sometimes a ridge, carina, or elevated zone extends between the two dorsolateral angles along the posterior margin of the pronotum; this is the **pronotal collar.** Another ridge or carina may extend directly downward from the dorsolateral angle toward the front coxa.

In dorsal view, the mesothorax can be divided into four distinct sclerites: the **scutum,** the **scutellum,** and the paired **axillae.** The suture between each axilla and the scutellum is the **axillar suture;** other sutures in this area are easily recognized by name, e.g., scutoscutellar. Laterally, the mesothorax is represented by the **mesepisternum,** sometimes referred to as the mesopleuron. The mesepisternum is sometimes divided by the nearly vertical **episternal groove,** formerly called the pre-episternal groove. The episternal groove may extend down, after meeting the anterior end of the horizontal **scrobal groove,** onto the lower anterior part of the mesepisternum, as in most Colletinae and Halictinae (Fig. 20-5b); it may curve posteriorly in an arc that merges indistinguishably with the scrobal groove as in *Andrena* and many Apinae (Fig. 20-5c); or it may be absent (Fig. 20-5a) or so short as not to reach the scrobal groove, as in the Megachilinae. The **scrobe** is a small pit on the scrobal groove in front of the meso-metepisternal suture. The area above the scrobal groove and behind the episternal groove is often more convex and shiny than adjacent areas; it is the **hypoepimeral area** (not labeled in Fig. 10-5). The depressed and largely hidden anterior margin of the mesepisternum is the **prepectus,** according to Brothers (1975).

The lateral (as distinguished from ventral) part of the mesepisternum is divisible into an anterior-facing surface and a lateral-facing surface. The angle between these surfaces varies from gradually rounded (Fig. 20-5a) through sharply angular to carinate and even lamellate. To avoid expressions like "angle between anterior and lateral surfaces of mesepisternum," Michener and Griswold (1994a) introduced the term **omaulus** for this angle. For sphecoid wasps, "omaulus" is used only if the angle is carinate; for bees the term is broadened and used even if the angle is rounded, so that one can record "omaulus rounded" or "omaulus carinate," or even "omaulus lamellate." The omaulus, which is not shown in Figure 10-5, is anterior to the episternal groove when the groove is present, and the **preomaular area** is anterior to the omaulus, i.e., it is the anterior, forward-facing surface of the mesepisternum.

Dorsally, the metathorax consists of a sclerite, the **metanotum,** which is obliquely divided at each side by the **transmetanotal suture.** The **metepisternum** (or metapleuron) forms the lateral surface of the metathorax. The **wing bases** are located above the upper margins of the mesepisternum and the metepisternum.

The middle and hind coxae of bees seem superficially to be shifted posteriorly so that the middle leg appears to arise from the lower end of the metepisternum and the hind leg from the propodeum (Fig. 10-5a). Of course this

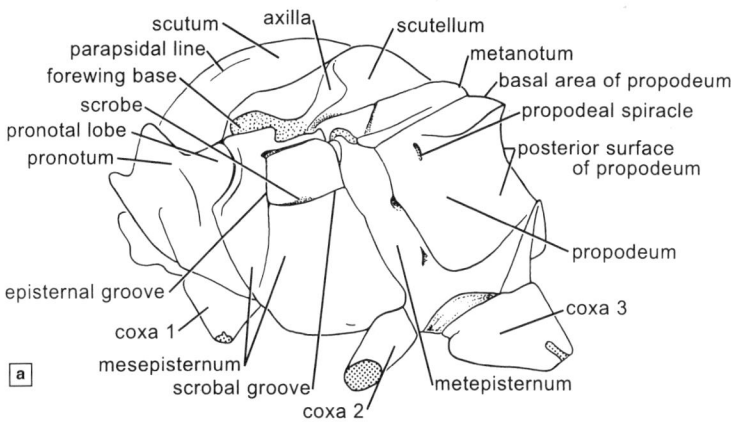

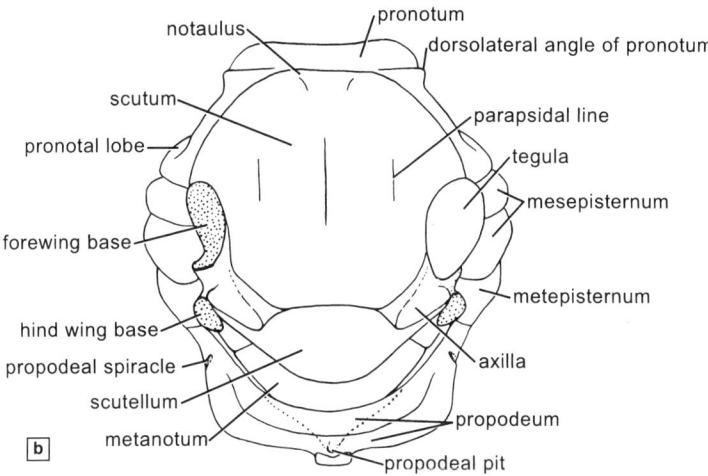

Figure 10-5. Diagrams of a bee's thorax. **a**, Lateral view; **b**, Dorsal view. (The tegula is omitted in a and the left side of b. The propodeal triangle is indicated by dotted lines in b.) From Michener, McGinley, and Danforth, 1994.

is not true, as careful examination shows; nevertheless the middle coxa, which in many bees is expanded upward to form a vertically elongated cylinder (see Michener, 1981b), displaces or is partly hidden beneath the lower part of the metepisternum.

The form and subdivisions of the propodeum are important systematically. Many bees have a pair of impressed lines on the propodeum (dotted in Fig. 10-5b), beginning near its anterior dorsolateral parts and extending downward and posteromedially and nearly meeting in or above the **propodeal pit,** a median depression of the lower posterior surface. These lines, together with the anterior dorsal margin of the propodeum, enclose the **propodeal triangle.** Morphologically, this triangle is the **metapostnotum** (Brothers, 1976). The shape of the propodeum as seen in profile is quite independent of the triangle. The whole propodeum may be vertical or nearly so, dropping from the posterior margin of the metanotum (Fig. 20-5a, c). In this case it is termed **declivous**. But there is frequently a more or less horizontal or sloping basal region (Fig. 20-5b), sometimes separated by a sharp line or carina from the declivous posterior surface, as shown in Figure 10-5a. The horizontal part is called the **basal zone** or **basal area** of the propodeum, sometimes called the **enclosure** when it is set off or enclosed by carinae; usually its sculpturing is distinctive, e.g., with striae radiating from its base. The basal area may be part of the propodeal triangle, or may extend beyond the triangle, at least laterally. Sometimes the basal area, as recognized by its sculpturing, is vertical like the rest of the posterior propodeal surface. The term "basal area" is applicable even if no sharp line separates it from the vertical surface and even if it is slanting or vertical rather than horizontal. In some bees the two surfaces are continuously rounded, one onto the other in a broad, curving surface; in that case the term "basal area" is not definable unless there is distinctive surface sculpturing. In a few bees the triangle is reduced in size, its lateral margins meeting and continuing posteriorly as a single line (sometimes not recognizable) to the propodeal pit. In other cases (e.g., in some *Xylocopa*) the reduction seems to have continued until there is no triangle, but only a median longitudinal line extending to the pit. In other bees, when the triangle is not recognizable, it is because the lines that demarcate it are weak or absent.

The wings. Wings are illustrated, and the veins la-

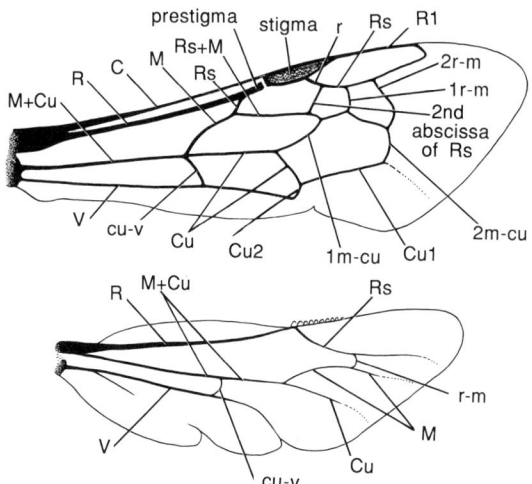

Figure 10-6. Diagram of the wings of a bee, showing the vein terminology of Michener (1944). From Michener, McGinley, and Danforth, 1994.

beled, in Figure 10-6, using a modified Comstock and Needham system. Wings are described as though spread, so that the direction toward the costal margin (where the stigma is in the forewing) is called **anterior**; toward the wing apex, **distal**. To save space, the word **stigma** is used in place of pterostigma. Because the homologies of the veins are not very certain, as well as because of some comparable-looking veins that often are referred to as a group, yet have very different morphological names, it has seemed best to continue the use of terms that are morphologically noncommittal for certain cells and veins much used in taxonomy. The names of cells and certain noncommittal names for veins are shown in Figure 10-7. Table 10-1 gives the equivalents, in Comstock-Needham terms, of these names.

Of special importance are three veins that all look like crossveins: the second abscissa of Rs, first r-m, and second r-m, to use the Comstock and Needham system. These veins help to define the submarginal cells, which are usually either three or two in number. When there are only two submarginal cells, one sometimes does not know whether the missing vein is the second abscissa of Rs or the first r-m; both losses can apparently occur, and both result in two submarginal cells, as illustrated by Peters (1969). *Hyleoides* (Colletidae) (Sec. 47) illustrates the impossibility of knowing which vein is lost. Hylaeinae have two submarginal cells; in most genera the first is much longer than the second, suggesting that the first is really the fusion product of the first and second, but in *Hyleoides* the reverse seems to be true. Expression is greatly simplified by using similar terminology for all three veins. In the past they have been called first, second, and third transverse cubital veins. These veins, however, have nothing to do with the cubitus; in fact the cubitus is in a very different part of the wing. I prefer to call them **first, second,** and **third submarginal crossveins** (1st, 2nd, 3rd in Fig. 10-7). This terminology agrees approximately with that of Diniz (1963), which was proposed for the same reason.

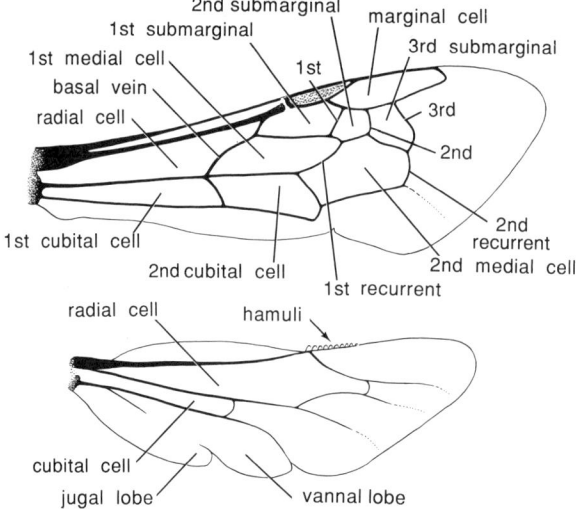

Figure 10-7. Diagram of the wings of a bee, showing the terminology of cells and morphologically noncommital terms for certain veins. (The notations 1st, 2nd, and 3rd refer to the submarginal crossveins.) From Michener, McGinley, and Danforth, 1994.

Table 10-1. Morphologically Noncommittal Terms for Certain Forewing Cells and Veins and Their Equivalents in Comstock-Needham Terminology.

Noncommittal terms	Comstock-Needham terms
Forewing Cells	
marginal	2nd R1
first submarginal	1st R1
second submarginal	1st Rs
third submarginal	2nd Rs
Forewing Veins	
basal	M
first recurrent	1st m-cu
second recurrent	2nd m-cu
first submarginal crossvein	2nd abscissa of Rs
second submarginal crossvein	1st r-m
third submarginal crossvein	2nd r-m
prestigma	1st abscissa of R1
anterior margin of marginal cell	2nd abscissa of R1
posterior margin of marginal cell	r and Rs
posterior margins of submarginal cells	Rs+M, 2nd and following abscissae of M

It has the drawback that the first is not technically a crossvein, but is thought to be a transverse section of a longitudinal vein, Rs.

Louis (1973) reviewed prior alar terminologies and proposed a new nomenclature for veins, attempting to avoid considerations of homology and phylogeny. He called the submarginal crossveins the first to third RM, or radiomedial veins. I believe that the less technical expression, submarginal crossveins, gives a clearer indication of what these veins are really like and of their relation to the submarginal cells.

Figure 10-8 shows the stigma, marginal cell, and nearby structures, with lines to show how measurements that are used in the keys and by various authors are to be made. The width of the prestigma is measured to the costal margin of the wing, and the expression is thus a misnomer, for it is more than the width of the prestigma

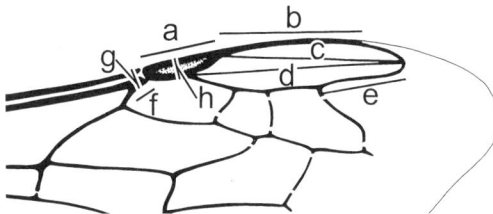

Figure 10-8. Forewing of *Ceratina rupestris* Holmberg, showing how certain measurements are made, as follows: (a) length of stigma; (b) length of costal edge of marginal cell or of margin of cell on the costa or wing margin (not a useful measurement in a wing like this, in which the cell diverges gradually from the wing margin); (c) length of marginal cell beyond stigma; (d) length of marginal cell; (e) length of free part of marginal cell; (f) length of prestigma; (g) width of prestigma (to wing margin); (h) width of stigma. The breaks in the wing veins are the alar fenestrae.

proper. The length of the **costal edge** (or **margin**) **of the marginal cell** or of the marginal cell on the costa is repeatedly used. This is the measurement from the apex of the stigma distad to the apex of the cell, or, if the cell is truncate, to the point where the cell diverges abruptly from the wing margin. As suggested by Figure 10-8, it is a poor measurement to use for wings in which the marginal cell bends gradually away from the wing margin. A common usage is **basal vein** for the first abscissa of vein M in the forewing (Fig. 10-7).

The **jugal lobe** and **vannal lobe** of the hind wing are both measured from the wing base to the apices of the lobes. Thus, on Figures 10-6 and 10-7 one might say that the jugal lobe is about two-thirds as long as the vannal lobe. The terminology of the veins and cells of the posterior parts of the wings varies; the word "vannal" is sometimes replaced by "anal," and the abbreviation V by A. Thus the vannal lobe is sometimes called the anal lobe.

The **alar fenestrae** are small, clear areas occupying specific sites in various veins (Fig. 10-8). **Flexion lines,** often faintly visible in the wing membrane, cross veins at these fenestrae.

The legs. Some authorities advocate a system for identifying parts of legs that assumes that all legs are pulled out laterally at right angles to the long axis of the body. Although I appreciate the logic of that system, I here follow the more traditional system in which the legs are considered to be in their normal positions. Thus, the corbicula of corbiculate Apidae is on the outer, not the anterior, surface of the hind tibia, and the two hind tibial spurs are outer and inner, not anterior and posterior. Some additional positional terminology is indicated by numbers in Figure 10-9.

The **tibial spurs** are the movable inferior apical spurs on the tibiae; there is one spur (part of the **strigilis**) on the front tibia, one on the middle tibia, and in nearly all bees two on the hind tibia. The **tibial spines** (Fig. 10-9) are immovable, sharp, superior apical projections, usually small in size, often blunt or minute, found in some bees. There are none, one, two, or rarely three spines per tibia; often they are mere angles. The tibial spur of the front leg consists of a main axis or **malus**, and a thin platelike scraper or **velum** directed toward the main axis of the leg; the velum usually does not extend to the apex of the malus. In some Apinae a prong or projecting ridge on the anterior side of the malus is termed the **anterior velum** by Schönitzer (1986) and Schönitzer and Renner (1980). The **inner hind tibial spur** of the hind leg is especially important taxonomically. This spur usually has two toothed margins. It is the inner one that is commonly elaborated in various ways. Following custom, I have described this margin as ciliate if it has slender, almost hairlike projections (usually numerous), although in many cases the appearance is like that of a fine comb. Because the finely serrate or ciliate (or intermediate) condition is common in Hymenoptera (Gennerich, 1922) and frequently the same as that of the outer hind tibial spur and of the middle tibial spur, such spurs are often described as simple, meaning, I suppose, unmodified. It may be, however, either coarsely serrate or pectinate. Again following custom, I have described a spur as pectinate if its inner margin is produced into several long, coarse, often blunt projections, even though the number of such projections is in some cases reduced to only one or two. The sockets of the hind tibial spurs and their relation to the tibia vary among taxa and were the subject of a study by Cane (1979).

The **basitibial plate** (Figs. 8-9, 10-9) is on the upper or outer side of the base of the hind tibia of many bees. It is best developed in females and presumably is important for support as bees move up or down their burrows in the soil or tamp the cell surfaces with the pygidial plate. The importance of the latter function is suggested by the observation that most female bees either have both pygidial and basitibial plates or lack both. Commonly, the basitibial plate is surrounded by a carina or a sharp line of some sort and its vestiture (if any) differs from that of adjacent regions, but it may be indicated only by a series of tubercles, as in many Euryglossinae, or even by a single tubercle that indicates its apex; in some cases (as in many *Xylocopa*) its apex is represented by a structure near the middle of the tibia instead of more or less near the base. In some bees (e.g., certain groups of *Centris*) the basal or central part of the basitibial plate is sharply elevated above the rest of the plate surface. Such an area is called the **secondary basitibial plate**.

Most S-T bees possess a pair of brushes or combs, best developed in females, on the middle legs. One is on the underside of the tibia, sometimes also including a brush on the basitarsus. The other is on the basal part of the femur, sometimes extending onto the trochanter. These structures are opposable and are used in cleaning or transferring pollen from the ipsilateral (same side) foreleg (Jander, 1976). They are termed the **midtibial brush** or **comb** and the **midfemoral brush** or **comb** (see Sec. 13). A comb is a single row of bristles, whereas a brush is less organized.

On the inner surface of the hind tibia of most bees is an area of variable size covered with hairs of uniform length, usually blunt, truncate, or briefly bifid. These hairs, the **keirotrichia** (Fig. 10-11), appear to serve for cleaning the wings. In some bees they are replaced by longer, more ordinary hairs that may function as part of the scopa in females.

On the hind basitarsus of many female bees is a distal process that extends beyond the base of the second tarsal segment. (For simplicity, I use the expression "tarsal segment" instead of tarsomere.) Sometimes this process bears on its apex a small brush, the **penicillus** (Fig. 10-9). (This brush bears no relation to the tibial tuft known as the penicillum in the Meliponini.)

Between the **tarsal claws** there is often a protruding, padlike **arolium** (Fig. 10-10). Its lower distal surfaces are almost always dark, often black, a fact that helps to distinguish it from associated pale structures. See Michener (1944) and Figure 10-10 for details of the structures between the claws. It is likely that comparative study of these structures would yield new characters of value for bee phylogeny or systematics, although loss of the arolium has occurred repeatedly among bees.

The **scopae**. Female bees have scopae for holding and transporting pollen; males do not. Exceptions are the Hylaeinae and Euryglossinae in the Colletidae, parasitic and robber bees in various families, and queens of highly eusocial bees (Meliponinae and Apinae), all of which lack scopae. In the keys and descriptive comments I often refer to the scopa without reminding the user that scopae are found only on females. The **scopa** consists of pollen-carrying hairs. These are not usually the hairs and brushes with which pollen is removed from flowers, but are the brushes on which pollen is carried back to the nest. Some pollen may be carried on various parts of the body, but scopae occur principally on the hind legs (Figs. 6-4, 8-5) or on the metasomal sterna (Fig. 8-7). In most bees the scopal hairs are on the hind legs, but in nonparasitic Megachilidae they are on the metasomal sterna; in some colletids and halictids they are on both the underside of the metasoma and on the hind legs. If fringes of scopal hairs surround a space in which pollen is carried, they are said to form a **corbicula**. The best-known corbicula, on the outer side of the hind tibia of the corbiculate Apidae (Euglossini, Bombini, Meliponini, Apini), consists not only of the fringes of hairs but also of the enclosed concave or flat surface (Fig. 10-11). Other corbiculae are on the undersides of the hind femora of Andreninae, Halictidae, Colletidae, and others, and on the sides of the propodeum of many species of *Andrena*, some halictids, and some colletids.

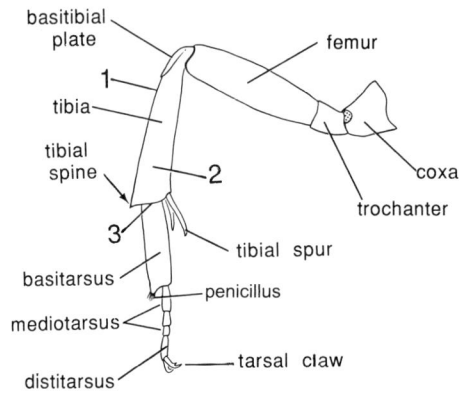

Figure 10-9. Hind leg of a female bee, hairs omitted except those that form the penicillus. The numeral 1 indicates the posterior or upper margin of the tibia; 2, the outer surface; and 3, the distal or apical margin. Modified from Michener, McGinley, and Danforth, 1994.

The **metasoma**. For simplicity and to save space, as noted in Section 9, metasomal terga and sterna are referred to as T1, T2, etc., and S1, S2, etc., T1 and S1 constituting the basal segment of the metasoma (Fig. 10-12). Each metasomal tergum or sternum (except for the anteriormost and the reduced apical ones) consists of a plate commonly marked by some transverse lines, as follows. First, across the anterior margin, always completely hidden in the intact metasoma, is the **antecostal suture**. The equivalent internal ridge is the site of attachment of longitudinal intersegmental muscles. The very narrow rims of the tergum and sternum anterior to the antecostal suture are the acrotergite and acrosternite, plus the apodemal margin that expands to form apodemes laterally. Second, nearer to the middle of each plate is another transverse line, the **gradulus**. Typically, the surface basal

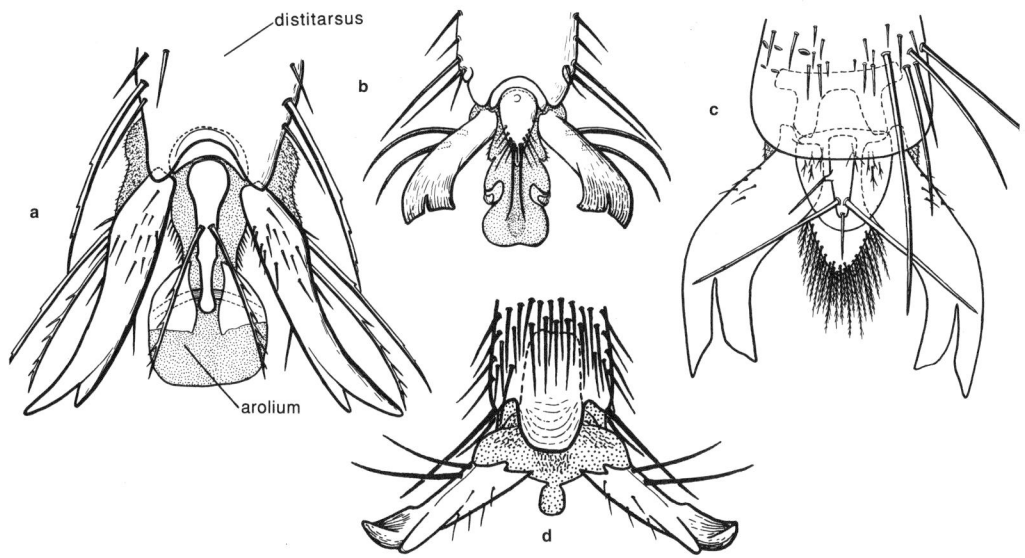

Figure 10-10. Dorsal views of apices of tarsi. **a**, *Andrena mimetica* Cockerell; **b**, *Anthophora edwardsii* Cresson; **c**, *Xylocopa orpifex* Smith; **d**, *Anthidium atripes* Cresson. Arolia are well developed in a and b, absent or greatly reduced in c and d. The median, arolium-like structure in d is membranous, not obvious in dry specimens, and *Anthidium* is considered to lack arolia. From Michener, 1944.

to the gradulus, i.e., the **pregradular area** or **disc**, is slightly elevated compared to that behind it, rendering the gradulus a minute step, as shown in Figure 10-12b. The ends of the tergal graduli, unless bent strongly to the rear, are usually near the spiracles. If bent strongly to the rear, the resultant longitudinal lines are called **lateral parts** or **lateral arms of the graduli** or, if carinate, **lateral gradular carinae**, sometimes elevated to form **lateral gradular lamellae** or **lateral gradular spines**. The graduli—except for their lateral arms, when present—are often concealed by the preceding terga or sterna on the intact metasoma but, especially on T2 and S2, are sometimes exposed or can be exposed easily by slightly extending the metasoma artificially. Third, near the posterior margin of each tergum and sternum is usually another transverse line, the **premarginal line**, separating the **marginal zone** (posterior marginal area of Michener, 1944; apical depression of Timberlake, 1980b) from the rest of the sclerite (Fig. 10-12). This zone is often depressed but in other cases differs only in sculpturing from the area basal to it; sometimes the marginal zone is not differentiated at all. The region between the gradulus and the premarginal line can be called the **disc** when a name is needed for it. Sometimes, e.g., on T2 of *Exomalopsis*, the premarginal line is arched far forward, so that the marginal zone is broad and the disc reduced to a transverse zone. The dorsolateral parts of the tergal discs (between the graduli and the premarginal lines), especially on T2 to T4, are often somewhat elevated, convex, and frequently shiny. These **dorsolateral convexities** frequently accentuate the premarginal lines, which limit the convexities posteriorly. The narrow posterior margin of the marginal zone is often recognizably different from the rest of the zone in sculpture or is elevated; this is termed simply the **margin**. Packer (2004a) provided an account of sternal variation among bees.

T1 differs from other terga because its base is constricted for the narrow connection with the thorax. Its dorsal, horizontal surface is nonetheless similar to that of succeeding terga, often having a marginal zone, premarginal line, and disc, each easily recognizable. Toward the base, T1 is strongly declivous. Often there is a transverse line or carina near the summit of the declivous surface and more or less separating it from the horizontal surface. This line or carina may be the gradulus of T1, although I do not use that term for it. It is the line or carina delimiting the anterior surface or anterior concavity of T1, and it is well developed in *Heriades* and certain other small

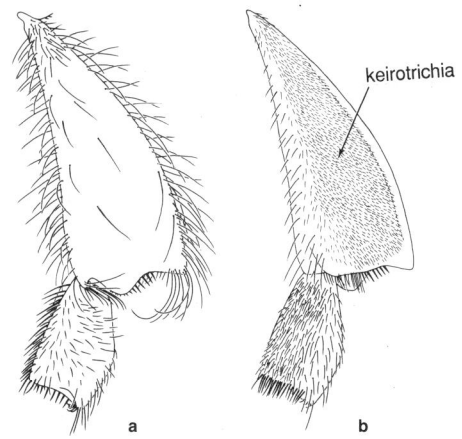

Figure 10-11. Hind tibia and basitarsus of a worker of *Plebeia frontalis* (Friese). **a**, Outer surfaces, showing the scopa reduced to fringes around the smooth and largely hairless tibial corbicula; **b**, Inner surfaces. From Michener, McGinley, and Danforth, 1994.

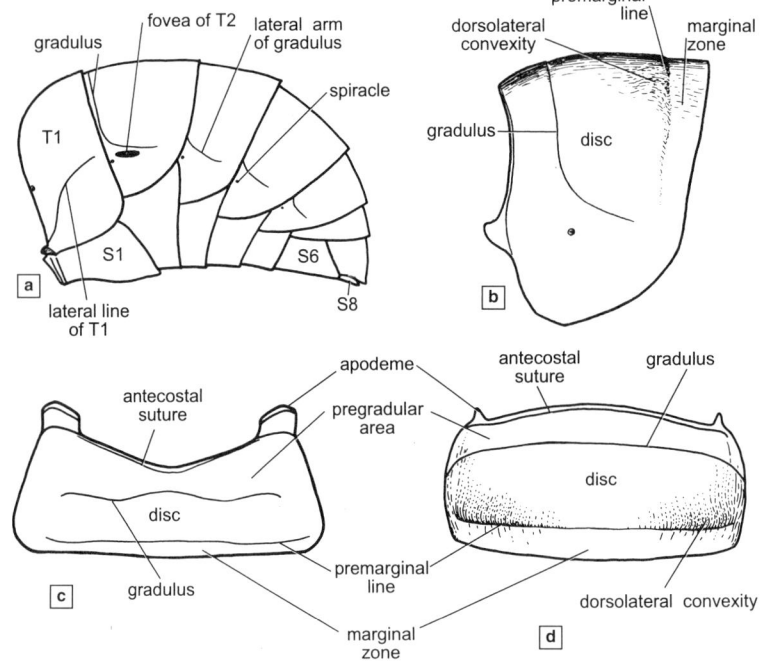

Figure 10-12. Metasomal structures of a male bee. **a,** Diagrammatic lateral view of the metasoma; **b,** Lateral view of T3; **c,** Ventral view of S3; **d,** Dorsal view of T3. a, b, and c, modified from Michener, 1944.

Osmiini. In addition, the **lateral line of T1** (Fig. 10-12a) is usually present.

In various colletids and andrenids, a lateral fovea of T2 (Fig. 10-12a) is usually developed if the facial foveae are developed. The lateral fovea of T2, like the facial fovea, is lined with apparently secretory cells (Ramos et al., 2004). It is likely that whatever the function of facial foveae, that of the lateral T2 foveae is the same. Both foveae are little evident in males, but often distinct in females.

The terga are often provided with transverse bands of pale hair. These are often nonhomologous, being on different parts of the terga. The terms **metasomal bands** and **fasciae** are applied indiscriminately, apical bands if on or overlapping the marginal zones, basal bands if on the discs.

The **pygidial plate** (Fig. 10-13) is a usually flat plate, commonly surrounded laterally and posteriorly by a carina or a line and in some cases produced as an apical projection, on T6 of females and T7 of males. The line demarcating the pygidial plate may be a median elaboration of the gradulus of T6 (females) or T7 (males), as is well shown, for example, in female Eucerini. In many bees the gradulus is absent laterally, so that only the part demarcating the plate is present. In other cases, the gradulus is entirely transverse, i.e., it extends across the tergum basal to the pygidial plate, which nonetheless is margined. (This observation makes one wonder if pygidial plates could have two, nonhomologous origins.) Sometimes the pygidial plate is reduced to a flat-topped spine or is completely absent. It is more often absent or rudimentary in males than in females. The **prepygidial fimbria** is a band of dense hairs across the apex of T5 of females. It is conspicuously different from, usually denser than, the apical hair bands or fasciae that may be present on preceding terga (Fig. 10-13). The prepygidial fimbria is considered absent if the hair band on T5 is like that on T4 and more anterior terga. In females of Nomadinae the prepygidial fimbria is modified, consisting of uniform short hairs on the often sloping surface of the apical median part of T5, forming an area suggestive of a pygidial plate, sometimes distinctly outlined by a ridge or change of surface slope and texture. This area is called the **pseudopygidial area**. Dense hairs on T6 of females, on each side of the pygidial plate, constitute the **pygidial fimbria** (Fig. 10-13), which is divided into two parts by the plate.

T7 of the female, always completely hidden, consists of two weakly sclerotized plates called **T7 hemitergites**; these are part of the sting apparatus. Each contains a spiracle, which thus readily identifies the T7 hemitergites (Fig. 10-14). The T8 hemitergites are similar-sized plates lacking spiracles. The second valvifers or female gonocoxites which are also similar-sized plates that lack spiracles, give rise basally to the rami of the second valvulae and apically to the **third valvulae** or **female gonostyli**, also called sting sheaths. Other terminology includes **stylet** for the fused **second valvulae**, **lancets** for the **first valvulae**, and **valve** for the dorsal flap near the base of each

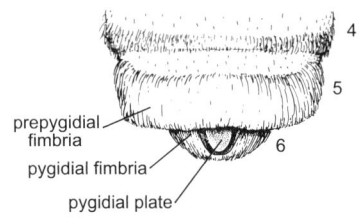

Figure 10-13. Diagram of apex of metasoma of a female bee, such as a eucerine. The tergal numbers are indicated on the righthand side.

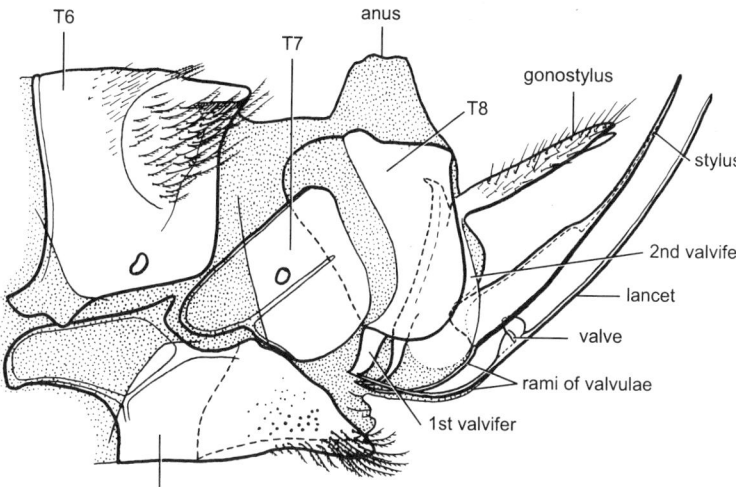

Figure 10-14. Apex of metasoma of a female *Halictus farinosus* Smith, the sting apparatus artificially extruded. From Michener, 1944.

lancet. The small **first valvifer** is called the triangular plate in literature on honey bees.

There is now good evidence that the sting apparatus varies among taxa and should be routinely extracted and studied in comparative examinations of genus-level relationships among bees. The principal paper on this topic is by Packer (2003). Rightmyer (2004) introduced to melittology a more recent terminology based on homologies recognized by Scudder (1971), which do not agree with the homologies accepted by Michener (1944, 2000). The principal changes in terminology are shown in Table 10-2. With the currently increased use of characters in the female terminalia for systematic studies (Packer, 2003, 2004a; see also Fig. 27-1), the terminology becomes important. As with wing venation, the noncommittal terminology will not change with differing opinions about homologies and may sometimes be more useful than the morphologically correct terminology.

The genitalia and S7 and S8 of male bees exhibit many interesting characters and may be dissected out for study, although they are almost always retracted in killed specimens. In some groups, e.g., *Heriades* in the Megachilinae, more anterior sterna such as S6 and S5, and rarely even S4 and S3, are also hidden and modified, but the expression "hidden sterna" is commonly used for S7 and S8. On a fresh or freshly relaxed specimen it is usually possible to reach between the apical exposed tergum and sternum and, with a hooked needle, pull out the genitalia and hidden sterna. In most cases, such dissection is not too difficult, but in the Megachilini the numerous hidden sterna are firmly connected to one another and laterally to the terga; moreover, they are often delicate and easily torn apart medially, so that successful dissection may be difficult. Beginners should start with other groups. Packer (2004a) has treated the variation in metasomal sterna of females for a broad selection of the Apoidea, thus providing a basis for recognition of previously little used or unrecognized characters.

S8 of males usually has a median basal point or angle for muscle attachment that is absent on other sterna. It is called the **spiculum** (Fig. 10-15b). Sterna and terga of both sexes, except for T1 and S1, have a basolateral projection or **apodeme** on each side (Fig. 10-12).

The male genitalia (Fig. 10-15a) have on each side, distal to the **gonobase**, a **gonocoxite**, to the distal end of which is usually attached the **gonostylus**. Although the gonocoxite may have some hairs, the gonostylus is frequently quite hairy and thus easily recognized. In most cases the gonocoxite and gonostylus are partly fused, often showing their articulation only on one side of the area of union. Frequently the fusion is complete or the gonostylus is lost, so that instead of a two-segmented appendage, one finds an unsegmented appendage; in this case one may not know whether the gonostylus is absent or fully fused to the gonocoxite. The structure in this case is called the **gonoforceps**. Commonly the distal part is hairy, in which case that part probably represents the gonostylus. In various groups one can find related forms of both sorts, some with distinct gonostyli and gonocoxites, others with the fusion or the loss of gonostyli complete.

In many groups of bees the male gonostylus is divided to its base, so that there appear to be two gonostyli on each side, arising from adjacent parts of the gonocoxite (Fig. 10-15c). These are called the **upper** or **dorsal gonostylus** and the **lower** or **ventral gonostylus**. The latter is often absent, in which case the upper gonostylus is simply called the gonostylus. The upper gonostylus is the only

Table 10-2. Terminology of Sting and Associated Parts.

Noncommittal terms (e.g., Snodgrass, 1956)	Terms accepted by Michener (1944, 2000)	Terms used by Scudder (1971) and Rightmyer (2004)
Triangular plate	1s valvifer	gonangulum
Oblong plate	2nd valvifer	2nd gonocoxa
Sting sheath	3rd valvula or gonostylus	gonoplac
Lancet	1st valvula	1st gonapophysis
Stylet	fused 2nd valvulae	fused 2nd gonapophyses

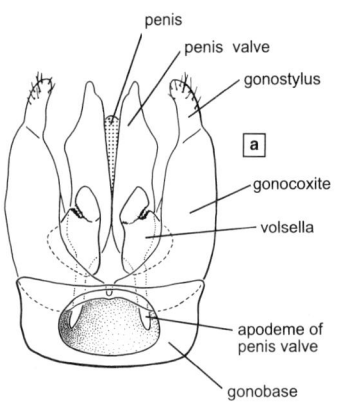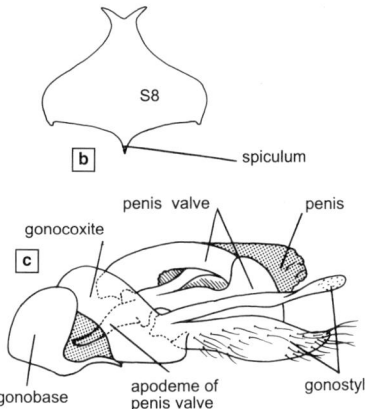

Figure 10-15. Terminal structures of male bees. **a,** Diagram of ventral view of genitalia; **b,** Diagram of S8; **c,** Lateral view of male genitalia of *Coelioxoides exulans* (Holmberg), showing upper and lower gonostyli. a, b, modified from Michener, McGinley, and Danforth, 1994; c, from Roig-Alsina, 1990.

one called a gonostylus in most literature. The lower gonostylus takes on various shapes and aspects in different bees and may not be homologous in all cases. It looks quite like a gonostylus, distally directed and hairy, in such bees as *Epicharis* and *Eufriesea*, and in *Coelioxoides* (Fig. 10-15c) (Apidae). Commonly, the base is in contact with or even in common with the base of the upper gonostylus. In some halictines the lower gonostylus is similar to the upper, with minute hairs, but in many other halictines the lower gonostylus is directed basad and in such cases has usually been called the **retrorse lobe**. In allodapine bees the ventroapical plate, which bears peglike setae, may be the lower gonostylus.

The **volsella** (Fig. 10-15a) is often most easily identified by the heavily sclerotized dark teeth on the opposable surfaces of the **digitus** and **cuspis**. In other Hymenoptera these parts clearly function as pincers, but in bees the volsellae are reduced (often wanting) and the digitus becomes fused to the body of the volsella and thus immovable. Nonetheless, Snodgrass (1941) showed muscles that move the digitus in *Andrena* and *Macropis*. In Figure 10-15a the distal mesal volsellar structure is the digitus; lateral to it is the cuspis.

The **penis valves** (Fig. 10-15a) are connected on the dorsal surface, near their bases, by a bridge. In many Apinae this bridge is expanded posteriorly to form a dorsal plate called the **spatha**. Some bees have an often large and complex, largely membranous **endophallus** that is usually invaginated within what is here called the penis (see Roig-Alsina, 1993). The genus *Apis* is unique among all Hymenoptera in its enormous endophallus and the reduction of all other external parts of the male genitalia (Fig. 121-3).

Some special terms used only for the male genitalia of Meliponini, and defined in Section 120, are **amphigonal**, **rectigonal**, and **schizogonal**. Special terminologies used in the past for genitalic structures of Bombini are explained in Section 119, especially Table 119-1.

Genitalia and hidden sterna of many bees have been illustrated by Snodgrass (1941), and in the multitude of taxonomic works cited in the accounts of bee taxa in subsequent sections of this book. Particularly large collections of illustrations of these features in diverse taxa are to be found in Saunders (1882, 1884), Strohl (1908), Mitchell (1960, 1962), and Michener (1954b, 1965b). The various works of Radoszkowski are also rich in genitalic illustrations.

11. Structures and Terminology of Immature Stages

Eggs of bees are described in Sections 4 and 8. They are not rich in external characters, although size varies as described in those sections, and the chorion of some cleptoparasitic bees is variously modified (Sec. 8). Rozen (2003a) and Rozen and Ozbek (2003) have investigated eggs of numerous, especially parasitic, bees and in addition to characters of the chorion in general, have found interesting differences in the structure of the micropyle.

Larvae, as described in Section 4, are generally soft and grublike and superficially seem rather devoid of differentiating characters. However, as shown by Michener (1953a) and especially by the many works of J. G. Rozen, they have numerous characters that, for phylogenetic studies, are well worth investigating because they appear to be independent of the characters of adults usually used in such work. Special features of larvae of various bee taxa (e.g., Allodapini and young larvae of various cleptoparasites) are described in the accounts of those taxa. What follows here concerns the morphological features and terminology of ordinary larvae, primarily mature larvae or prepupae (see Sec. 4). Most comparative work has been based on prepupae, partly because, as compared with earlier stages, they are larger, tougher, and found more often because they are commonly longer-lived.

Larvae or prepupae of bees are legless and grublike, most of them shaped about as in Figure 11-1, although a

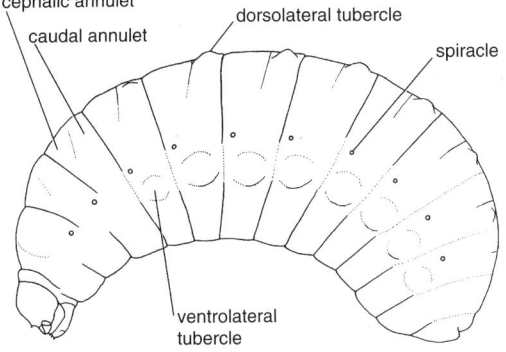

Figure 11-1. Diagrammatic lateral view of a bee larva.

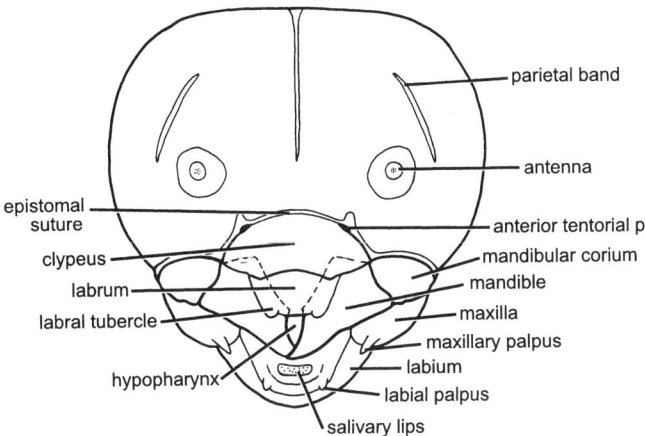

Figure 11-2. Diagrammatic frontal and lateral views of the head of a bee larva.

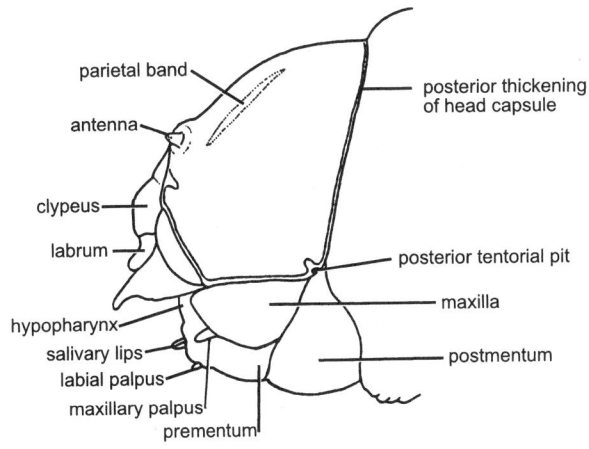

few are slender and almost wormlike (Emphorini) and others are extremely fat, only about twice as long as broad (predefecating larvae of *Holcopasites,* Ammobatoidini). Some have very strong tubercles (Panurginae, especially Perditini; also many Allodapini) and some have conspicuous hairs or spicules. Most segments are divided by an intrasegmental line (Fig. 4-2), usually weaker than the intersegmental lines, into **cephalic** and **caudal annulets.** Spiracles, which have some useful characteristics, can be seen by clearing in a 10-percent aqueous solution of KOH and then examining them with a compound microscope. Longitudinal series of tubercles, if present, may be dorsolateral, ventrolateral, or sometimes ventral; rarely, there are also mid-dorsal or midventral tubercles but usually not in series.

The terminology of the cephalic structures is shown in Figure 11-2. In forms that do not spin cocoons, the labium and maxillae are reduced in size and by partial fusion.

Pupae (Fig. 4-1e) have received little study compared with larvae (see Michener, 1954a; Michener and Scheiring, 1976; Rozen, 2000). Because they are short-lived, they are found less often than mature larvae. Also, most of their features are essentially those of the adults and hence of little importance for phylogenetic studies. They do, however, often have large, robust but soft (like the rest of the integument) spines on the legs or body that seem quite different from structures of adults. Even these spines, however, may usually relate directly to adult features, for they seem to be places in which the long hairs of adults can develop. In particular, at the ends of leg segments, long hairs directed apically would have no place to develop were it not for pupal spines.

Early pupae are yellowish white, like larvae, but as they get older the integument of the adult aquires its coloration, which shows clearly through the pupal integument.

12. Bees and Sphecoid Wasps as a Clade

The bees and the sphecoid wasps have long been regarded as allied groups (Comstock, 1924) and are united as the superfamily Apoidea. As indicated in Section 2, a character traditionally used to demonstrate this relationship is the pronotal lobe, which in these groups is differentiated but rather small and usually well separated from the tegula, whereas in other aculeate Hymenoptera the posterolateral part of the pronotum reaches the tegula.

Brothers (1975) emphasized this character (Brothers' character 21.2), and Lanham (1981) stated it as enlargement of the anterolateral parts of the mesoscutum, which results in the prothoracic lobe feature (see also Brothers' character 27.1). Brothers also listed two other strong synapomorphies that unite sphecoids and bees: (1) a ventrolateral extension of the pronotum to encircle or nearly encircle the thorax behind the front coxae (Brothers' 23.2), and (2) an enlargement of the metapostnotum (propodeal triangle), carrying the third phragma posteriorly (Brothers' 35.3; see also Brothers, 1976). Other, less impressive characters uniting bees and sphecids are the shortened pronotum, its posterior margin commonly broadly concave (Brothers' 18.2 and 22.l); and the fusion of the mesopleural suture with the intersegmental suture (Brothers' 33.l.l). Most of these characters also exist in other aculeate groups; they are therefore weaker than the strong characters listed above. The manner of cleaning the thoracic dorsum is also an interesting character (Jander, 1976). Most Hymenoptera use forward scraping by the front tarsi for this purpose; many sphecoids and most bees use the middle tarsi. The noncorbiculate Apidae appear to have reverted to wasplike cleaning behavior; otherwise the use of the middle tarsi seems to be a feature showing the relationship of sphecoid wasps and bees.

To me the evidence for close relationship between sphecoids and bees is highly convincing, in spite of Lanham's (1981) contrary views suggesting that the characters listed by Brothers (1975), Lomholdt (1982), and others as uniting sphecoid wasps and bees are convergent rather than synapomorphous.

Additional material on the origin of bees will be found in Sections 14, 22, and 23, on fossil bees and their antiquity.

This is as appropriate a place as any to explain briefly some terms used here and in subsequent sections that are in common use by systematists, yet are perhaps little understood by others. The terms **monophyletic** (**holophyletic**) and **paraphyletic** are explained at the end of Section 2. A **clade** is a monophyletic group, i.e., the organisms subsumed by a branch of a cladogram.

A **cladogram** is a treelike diagram representing a hypothesis of phylogenetic relationships. It is in contrast to a **phenogram**, which is a tree based on numbers of differences or similarities rather than on phylogeny. The term **dendrogram** is used for any type of treelike diagram, or one made without a clear indication of the methodology used in its preparation. The term **cladistic** is said of an analysis that seeks to infer the branching sequence of a phylogeny, or of a classification based on such an analysis. A cladistic analysis or classification is the same as a phylogenetic analysis or classification. The term **phenetic** is said of a classification or analysis based on degrees of difference among taxa, as distinguished from a phylogenetic or cladistic classification or analysis based on genealogy, i.e., the sequence of branching in a cladogram. An **apomorphy** is a character state that is derived (not ancestral) relative to other states of the same character; a **synapomorphy** is an apomorphy shared by two or more taxa and inferred to have been present in their common ancestor. A **plesiomorphy** is a character state that is ancestral relative to all other states of the same character. A **symplesiomorphy** is a plesiomorphy shared by two or more taxa because of ancestral relationships. The word **polyphyletic** (or **diphyletic**) is said of a taxon whose distinctive features arose independently (nonhomologously) in several (or two) clades. Such a taxon must be divided into several (or two) taxa that have shared, but nonhomologous, features. In a dichotomous cladogram, any two branches arising from a single point are sisters; each is the **sister group** to the other. The word **taxon** (pl. **taxa**) is used for any named systematic unit at any classificatory level.

13. Bees as a Monophyletic Group

The bees have long appeared to constitute a monophyletic or holophyletic unit (Michener, 1944; Brothers, 1975; Lomholdt, 1982). This view was strongly supported by the more recent phylogenetic studies of Alexander (1992), Brothers and Carpenter (1993), Alexander and Michener (1995), Brothers (1999), Melo (1999), and others. The following is an annotated list of some syna-pomorphies that demonstrate the monophyly of the bees:

Character a. Some of the hairs of bees are plumose or at least branched (Fig. 13-1). In most Hymenoptera they are simple, although plumose hairs are found in some other groups, e.g., some Mutillidae. Contrary to the usual opinion, I doubt that plumose hairs of bees arose as pollen-collecting and pollen-carrying structures, although of course some bees take advantage of plumosity to enhance these functions. In many bees the scopal hairs are simple, yet nevertheless carry pollen, showing that plumosity is not necessary for a pollen carrier. Moreover, plumose hairs are often found in locations where pollen is never carried, e.g., around the anterior thoracic spiracles and on the male genitalia and hidden sterna. Further, hairs are branched (plumose) in many different ways, some of them not at all suitable for pollen-collecting or pollen-carrying, indicating that the degree or type of plumosity may be under various selective pressures having nothing to do with pollen.

Possibly plumosity first arose as one way (an alternative to a great number of hairs) in which forms in a xeric environment could decrease air flow near the integumental surface, and thus reduce water loss. Simultaneously, since hairs are often pale, plumosity could have been one way to increase the pale coloration often characteristic of insects in xeric environments. Presumably, pale coloration both reflects heat, helping to prevent overheating, and serves for protective coloration on the pale soils and pale vegetation characteristic of deserts.

Character b. With few exceptions, bee larvae eat pollen mixed with nectar or floral oil, or glandular secretions of adults that eat pollen and nectar. The larvae are carnivorous in related Hymenoptera except for *Krombeinictus nordenae* Leclercq, a Sri Lankan crabronine wasp that feeds pollen to its larvae (Krombein and Norden, 1997a,b). There are three carrion-eating species of *Trigona*, as noted in Sections 6 and 120. Adults of other social bees sometimes eat eggs of their competitors, queens eat trophic eggs laid by workers. Larvae of some and adults of other cleptoparasitic bees kill and possibly eat eggs or young larvae of their hosts. Otherwise bees are entirely phytophagous (= plant feeders). Since I do not think that plumose hairs arose for pollen-collecting, I do not think that feeding on pollen is a character that duplicates character **a**.

Eating of pollen by adults (especially females) is another synapomorphy of bees, unknown in sphecoid wasps, but it is probably a correlate of character **b**. Adult wasps use prey fluids as a protein source. Bees have no prey and hence eat pollen.

Character c. The hind basitarsus is broader than the subsequent tarsal segments (Fig. 10-9). In other ac-

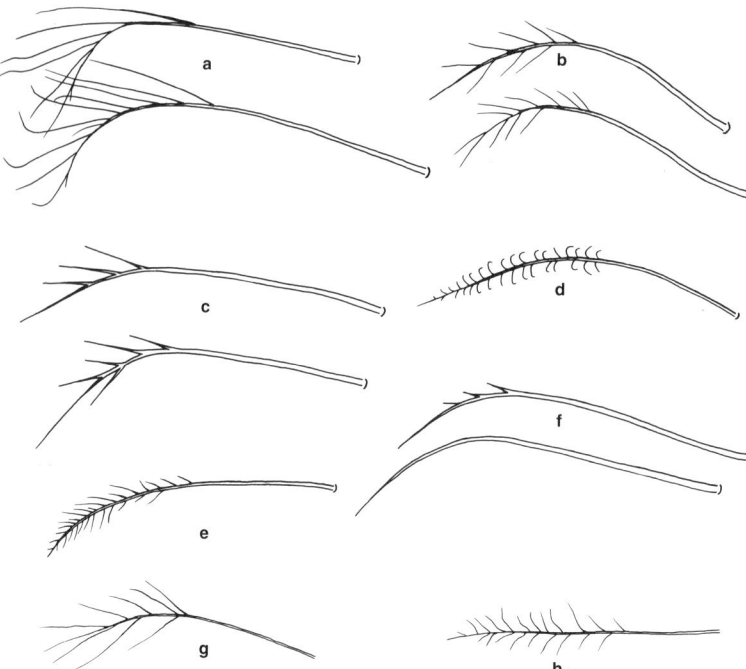

Figure 13-1. Hairs from tibial scopas (a-g) and sternal scopa (h) of bees of the genus *Leioproctus* (Colletinae). **a,** *L. (Perditomorpha) erithrogaster* Toro and Rojas; **b,** *L. (Perditomorpha) brunerii* (Ashmead); **c,** *L. (Leioproctus) fulvoniger* Michener; **d,** *L. (Glossopasiphae) plaumanni* Michener; **e,** *L. (Protodiscelis) palpalis* (Ducke)?; **f,** *L. (Pygopasiphae) wagneri* (Vachal); **g,** *L. (Reedapis) bathycyaneus* Toro. **h,** *L. (Reedapis) bathycyaneus* Toro. From Michener, 1989.

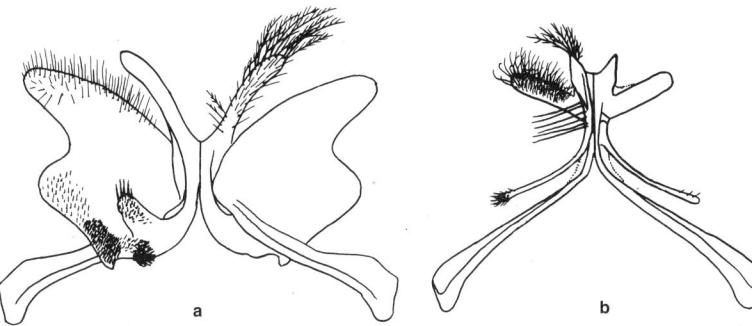

Figure 13-2. Seventh sterna of male bees of the family Colletidae. **a,** *Leioproctus (Perditomorpha) inconspicuus* Michener; **b,** *Diphaglossa gayi* Spinola. (Dorsal views are at the left.) From Michener, 1986b, 1989.

uleates the hind basitarsus is of about the same width as tarsal segments 2 through 5. The broad hind basitarsus may be related to pollen manipulation and transport. It occurs also, however, in forms that do not collect pollen, such as males and parasitic females, and in forms that do not carry pollen externally on a scopa, such as the Hylaeinae, although it is not as well developed as in female pollen collectors. In minute Euryglossinae, such as *Euryglossula*, and in males of some other bees, the hind basitarsus is only about 1.3 times as wide as the second tarsal segment.

Character d. The larval maxilla has one apical papilla, the palpus (Fig. 11-2). In related aculeates there are two papillae, the palpus and the galea. Lack of the galea was therefore considered a larval synapomorphy of bees by Lomholdt (1982). A galea or galea-like projection is found, however, in various bee larvae (some melittids and apines), but this galea is not a fully distinct papilla like the palpus.

Character e. The seventh metasomal tergum (T7) of the female is membranous mid-dorsally. In related aculeates this tergum forms a sclerotized arch. In bees the sclerotized parts are reduced to two lateral hemitergites, one on each side, that form part of the sting apparatus (Fig. 10-14).

Character f. A posterior strigil, consisting of hind tibial spurs opposed by a brush of short hairs on a shallow basal concavity of the hind basitarsus, is absent. Such a strigil is present in related aculeates, and its loss must be a synapomorphy for bees (see Lanham, 1960). Lomholdt's (1982) remark that some bees possess the posterior strigil seems to be an error. Wasps use this strigil to clean the hind legs; each leg is pulled through the contralateral strigilis, one after the other. Bees clean hind legs by rubbing the two legs against one another. This behavioral character, although synapomorphic for bees, cannot be considered independent from the strigilar loss.

Character g. Only one sperm cell develops from each spermatocyte. Wasps produce four sperm cells from each spermatocyte. This character, cited by Lomholdt (1982), has been examined in so few taxa that no certainty exists as to its distribution.

Character h. The foreleg is cleaned (or pollen is transferred from it to the middle leg) by drawing it through the flexed (at femorotibial joint) middle leg (Jander, 1976). This movement is not seen in other Hymenoptera. It is associated in females and some males with the midfemoral brush, a brush on the lower side of the middle femur-trochanter, and the midtibial brush, a brush of often ordinary hairs on the lower side of the middle tibia and sometimes basitarsus. When the leg is flexed, these brushes are opposable and both surfaces of the foreleg are cleaned at one stroke (see Secs. 6 [pollen], 10). In the Hylaeinae and Euryglossinae these brushes are weak and limited to the distal part of the tibia and the base of the femur (and sometimes the trochanter). These bees carry pollen to the nest in the crop, thus eating it rather than transferring it to a scopa. The midleg apparatus therefore probably serves only for grooming. The same is true of many male bees. In bees the ancestral cleaning movements of the foreleg by the mouthparts serve for transfer of pollen to the mouthparts for eating.

Character i. S7 and S8 of the male are modified and concealed by S6, or only the apical process of S8 (and sometimes that of S7) is exposed. In bees, S7 is commonly greatly elaborated, with lobes and hairy surfaces that may have tactile or evaporative functions (Fig. 13-2). These are common features in the Colletidae, Andrenidae, Rophitinae, Fideliini, etc. Conversely, S7 may be reduced to little more than a transverse ribbon, as in the Xylocopini. In sphecoids S7 and S8 are relatively unmodified, S7 being usually exposed.

Character j. The bristles on the outer surfaces of the tibiae are usually absent or weak, although strong, presumably secondarily, in some parasitic forms like the Nomadinae. Such bristles are very common in sphecoid wasps.

Character k. The basitibial plate (Fig. 10-9) is present, especially in females. It is well developed in almost all bees that excavate nests and shape cells in the soil, and is no doubt an ancestral bee feature, although it has been lost in many bees that do not make their own cells in the soil. It is absent in sphecoid wasps or it is present as a convergent development in a ground-nesting species of Pemphredoninae, a subfamily whose members commonly nest in holes in wood or stems (McCorquodale and Naumann, 1988).

Character l. The mandible of the larva is simple to minutely denticulate or ends in two teeth. In related aculeates the mandible ends in three or more teeth. Such reduction in bee larvae is not surprising, considering that they mostly feed on pollen and nectar and do not eat arthropod prey.

Character m. G. Melo has pointed out to me that the cleft claws (Fig. 10-10) found in most bees may be a synapomorphy for bees, although they revert to simple claws in some bees, principally females. Claws are not cleft in sphecoid wasps.

Character n. A character used by Gauld and Bolton (1988) to distinguish bees from sphecoid wasps is the course of flexion lines in the forewing. In bees a line cuts across the first recurrent vein and no line cuts across vein M, i.e., the posterior margin of the submarginal cells. Presumably this pattern is synapomorphic for bees. In sphecoid wasps the pattern is reversed. Unfortunately, these lines and the fenestrae where they cross veins are often difficult or impossible to see, and the universality of this character is not verified.

Lomholdt (1982) used the presence of a single spur on the middle tibia as a synapomorphy of bees separating them from the Larridae, properly called Crabronidae (Menke, 1993). In Lomholdt's sense, the Larridae consist of the Sphecidae s. l. minus Sphecinae and Ampulicinae. Many sphecoids, however, including those in the Larrinae, Philanthinae, and Pemphredoninae, also have only one middle tibial spur (Bohart and Menke, 1976). Perhaps Lomholdt derived his statement on middle tibial spurs from Brothers (1975); it has to be remembered that Brothers' lists of characters for each taxon contain ancestral states, not states that arose within the taxon, so that when he says sphecids have two middle tibial spurs, he is saying that the presence of two spurs is an ancestral character for sphecids. He is not excluding the fact that many of them, like all bees, have lost one of the spurs.

The above list is substantial enough to assure monophyly for bees. Several of the characters, however, may not be independent. The broad hind basitarsus (character c) may have evolved as an enhancement of pollen-manipulation or pollen-carrying capacity. In pollen-collecting females the breadth and flatness of the basitarsus are striking; the characteristic may have been carried over in a less noticeable condition to males, parasitic females, and the Hylaeinae and Euryglossinae (which carry pollen in the crop rather than externally). The broadening and this novel function of the hind basitarsus may have led to the loss of the posterior strigil (character **f**). If character **c** has to do with pollen, then obviously it is also related to **b**, the larval food, which of course is related to **l**. Character **h** is also associated with the use of pollen, and hence related to character **c**. Finally, if (as I doubt) the plumose hairs (character **a**) arose as pollen holders, and thereafter spread to males and were retained in the evolution of parasitic bees, then character **a** also is related to **c**.

Nonetheless, the monophyly of the bees seems clear. There is no group that could have evolved from the bees, and they therefore cannot be paraphyletic. Robertson (1904) suggested that the bees may be diphyletic, that is, that those with pygidial plates and those without could have arisen from sphecoid wasps with and without such plates. He even gave names (Pygidialia and Apygidialia) to the two groups of bees. This idea is understandable for the limited fauna known to him (eastern North America) but is not tenable when the world fauna is examined, as stated by Michener (1944). For example, colletids in eastern North America are apygidialate, but numerous related colletids, mostly in the Southern Hemisphere, have pygidial plates. As will be shown later, pygidial plates are plesiomorphic for the Apoidea. They have been lost in diverse groups of bees and sphecoid wasps.

14. The Origin of Bees from Wasps

It seems reasonably certain that bees arose from forms which, were they alive today, would be considered as Spheciformes—Sphecidae in the broad sense. I know of no synapomorphy of all sphecoid wasps that would justify regarding them as a monophyletic sister group of the bees. The Spheciformes are morphologically diversified to such an extent that various authors have commented on the incongruity of the custom of dividing the bees into several families while regarding their closest relatives as constituting a single family, the Sphecidae.

Many times in the past (as long ago as Latreille in 1802), authors have regarded some sphecoid groups as families, e.g., Larridae, Crabronidae, Nyssonidae, etc. Lomholdt (1982) addressed the problem in a cladistic study in which he divided the Spheciformes into the Sphecidae s. str. (for Sphecinae and Ampulicinae) and the Larridae, which should be called Crabronidae. More recent studies such as that of Alexander (1992) showed that the Sphecinae and Ampulicinae do not form a single clade.

The Crabronidae are held together by one strong synapomorphy, the double salivary opening of the larva, which is shared neither with the bees nor with the Sphecidae s. str. In the crabronid subfamily Astatinae, the larval salivary opening has been described as single. This appears to have been an observational error; whatever the explanation, the Astatinae do have a double opening like that of other Crabronidae (Evans, 1958). It is thus quite likely that Lomholdt was right in considering this family as monophyletic and the sister group to the bees.

Malyshev (1968) and others have speculated that it is in the Pemphredoninae that one finds the closest sphecid relatives of the bees. This conjecture is based in part on the small, slender body and stem-nesting habits of many of these wasps, superficially suggesting the bee genus *Hylaeus* (Colletidae). Malyshev emphasized the idea that the pemphredonines' numerous small homopterous prey, sweet because of their filter chambers and content of honeydew, might be a preadaptation to the evolution of dependence upon nectar and pollen by bees. Many other sphecoids provision their cells with only one or a few larger prey items that are not sweet. More important, some pemphredonines, unlike other sphecoid wasps, line their nest cells with a secretion, as do most bees, but the secretion is not of the same origin as that of bees (McGinley, 1980), and the cell linings are therefore not homologous in the two taxa. Adult females of some Pemphredoninae secrete silk or silklike material from epidermal glands opening on apical terga or sterna (G. Melo, personal communication, 1995) and this material is the source of the cell linings of at least some species. In bees, on the other hand, the cell linings are secreted by Dufour's gland and salivary glands.

To me and evidently also to Radchenko and Pesenko (1994), none of this is convincing evidence of a relationship between Pemphredoninae and bees, but there are morphological indications, probably convergent, of a relationship of the bees to the crabronine branch of the Crabronidae—the branch that leads to Larrinae, Philanthinae, and Crabroninae in Bohart and Menke's (1976: 32) dendrogram. For example, in most Hymenoptera, including most sphecoids, the middle coxa is not greatly different in structure from the hind coxa (Fig. 14-1a), there being a short basicoxite separated by a groove from the large disticoxite (Michener, 1981b). In the Crabroninae (including Oxybelini), Trypoxylonini, and some Larrini (but not in Philanthini) the disticoxite of the middle coxa is reduced and the basicoxite enlarged, as in all bees (Fig. 14-1b, c). Moreover, R. McGinley (personal communication, 1981) found similarities in the maxillary (galeal and stipital) structure between bees and certain Philanthini that suggested a possible relationship. Fi-

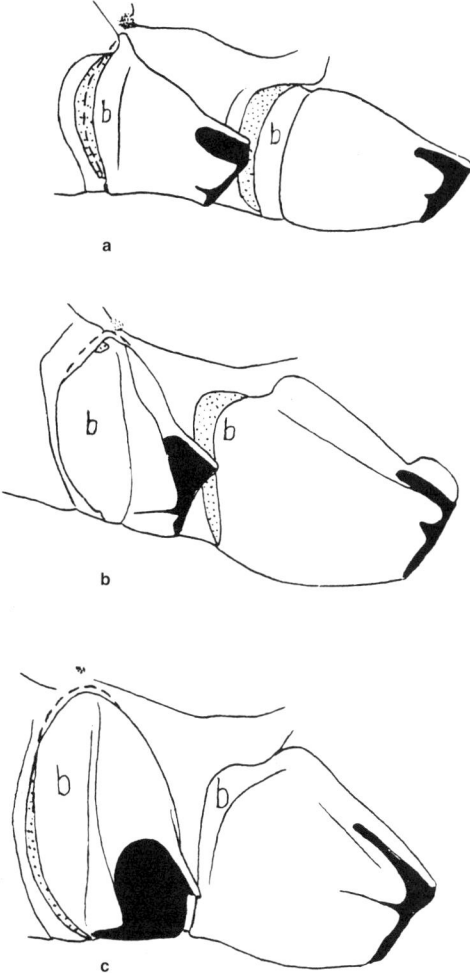

Figure 14-1. Middle (left) and hind (right) coxae. **a**, *Philanthus gibbosus* (Fabricius) (Crabronidae); **b**, *Tachytes* sp. (Crabronidae); **c**, *Anthidium illustre* Cresson (Megachilidae). The basicoxite, marked "b," is enlarged at the expense of the disticoxite of the middle coxa of bees and some Crabronidae. From Michener, 1981b.

nally, in Alexander and Michener's (1995) phylogenetic study of S-T bees, the bees arose in all analyses from among the few Crabronidae included in the study, which therefore formed a paraphyletic group. These wasps, however, were included as an outgroup, and the study should not be viewed as informative about the relationships among the wasps. Although none of the adult crabronid synapomorphies is strong, the paired larval salivary openings of Crabronidae are a unique synapomorphy, as noted above. The bees have various synapomorphies, as listed in Section 13. Therefore, it is probable that bees and Crabronidae (Larridae *sensu* Lomholdt, 1982) are sister groups. Of course the Series Spheciformes, the group called sphecoid wasps, including the Sphecidae and Ampulicidae as well as the Crabronidae, is indeed paraphyletic, for the Series Apiformes arose from within it.

15. Classification of the Bee-Sphecoid Clade

Lomholdt (1982), Brothers (1975), Michener (1944, 1979a), Gauld and Bolton (1988), and others, including older authors such as Comstock (1924), have advocated placing the sphecoid wasps and the bees in the same superfamily. The custom of separating them as the superfamilies Sphecoidea and Apoidea obscures their close relationship to one another, as compared to other superfamilies of aculeate Hymenoptera. It has been common to recognize only one family of sphecoid wasps in spite of their considerable diversity, but to recognize several families of bees.

The placement of sphecoid wasps and bees in the same superfamily has led authors such as Gauld and Bolton (1988) and Melo and Gonçalves (2005) to recognize only two major families in the superfamily, Sphecidae and Apidae s. l., thus correcting the inconsistent classificatory treatment of the sphecoid wasps vs. the bees. Both the Sphecidae and the Apidae in this sense seem diverse; subjectively, they seem to contain groups at least as different from one another as the families of Chalcidoidea. Moreover, the Sphecidae—in the sense of sphecoid wasps or Spheciformes—is paraphyletic (see Sec. 14). A better idea, therefore, is to divide both sphecoid wasps and bees into several families. Lomholdt (1982) divided the sphecoid wasps into Sphecidae and Larridae (= Crabronidae), as noted above. This eliminates the paraphyly of the Sphecidae s. l. Several families of bees are long established and already have subordinate taxa considered as subfamilies, tribes, etc. It seems desirable to maintain them. Melo and Gonçalves (2005), however, have in general reduced bee families to subfamilies of Apidae, many subfamilies to tribes, and various tribes to subtribes.

Because a family-group name based on the generic name *Apis* antedates a name based on *Sphex* (Michener, 1986a), the name of the bee-sphecoid superfamily should be Apoidea, not Sphecoidea as most previous works have had it. This is not unreasonable, since the bees are a much larger group than the sphecoid wasps. The Apoidea in this sense is divided into families, some of which (Sphecidae, Crabronidae, etc.) are wasps and constitute the informal paraphyletic group or Series Spheciformes of Brothers (1975), while the others (Colletidae, Apidae, etc.) are bees, the Series Apiformes of Brothers and the Anthophila of some older authors. This classification is summarized in Table 15-1.

I have used Crabronidae in the discussion above without bias concerning recognition of several additional families, for example as was done by Krombein (1979). I suspect that recognition of such families is desirable. The following sections will consider what bee families should be recognized.

Table 15-1. Classification of the Superfamily Apoidea.

Families	Series
Family Ampulicidae	
Family Sphecidae	sphecoid wasps, or Spheciformes
Family Crabronidae (could be subdivided)	
Family Stenotritidae	
Family Colletidae	
Family Andrenidae	
Family Halictidae	bees, or Apiformes
Family Melittidae	
Family Megachilidae	
Family Apidae	

16. Bee Taxa and Categories

Classifications, of course, are based in large part on phylogeny. Some specialists (cladists) base classification entirely on phylogeny; others consider also information from diverse sources in developing a classification. No one, however, should presume to make a classification without having all available phylogenetic information. For practical purposes, I present here some information on bee classification, prior to the section on phylogeny, because use of the family-group names makes explanation and understanding easier. Recent phylogenetic studies have not overturned this classification in a major way; it is therefore possible to discuss phylogeny using for the most part the taxa that have been accepted in the past by many specialists. To provide ready reference to the classification that will be elaborated later, Table 16-1 lists the taxa (family to subgenus) here accepted.

In dealing with a large group such as the bees, it is inevitable that the classification will be unsatisfactory in some areas even though quite satisfying in others. This situation arises partly because of intrinsic differences among living bees. In some groups the taxa are well differentiated and easily organized hierarchically, whereas in other groups characters occur in diverse combinations among taxa difficult to differentiate and resistant to unambiguous classification. Another reason for differences in the usefulness of taxonomic constructs is the amount and kind of study to which each group (e.g., family) has been subjected. Some taxa have been analyzed phylogenetically, others have not, and some such analyses are convincing while others are not. Thus the parts of the classification differ in their usefulness. It has not been practical to make phylogenetic studies of all groups as part of the preparation of this book. Some such studies (Roig-Alsina and Michener, 1993; Alexander and Michener, 1995), published separately partly in order to present details not appropriate here, were made specifically in the hope of clarifying the family-level classification for this book.

The tradition has been to recognize large genera of bees, like *Andrena, Lasioglossum,* and *Megachile.* I believe this is desirable, for it allows many biologists to recognize the genera and know what is meant by the names. In the same way, I find names like *Culex, Aedes,* and *Drosophila* more useful to me, and I think to biologists in general, than the many generic names that would result if the many subgenera they comprise were all raised to generic rank.

A frequent result of maintaining large genera is the development of multiple subgenera. Melittologists are sometimes criticized for extensive use of subgenera. Species-groups, named informally with specific names, could be used instead of subgenera; such a procedure would avoid burdening the literature with numerous subgeneric names and their associated formalities such as type species and problems of homonymy. For large genera, however, with hundreds or thousands of species, the choice is not between recognizing a genus with subgenera vs. a genus with species groups. It is between recognizing a genus with subgenera vs. several or many genera, because many of the current subgenera are quite different from one another, sometimes recognizable on flowers or even in flight, and some specialists already prefer to recognize them as genera. I prefer, nonetheless, to retain inclusive genera for the reason indicated above. In the end, however, these decisions are subjective regardless of one's views on systematic methodology.

One of the advantages of subgenera is that they need not be cited. Unlike generic names, which are required, subgeneric names are optional parts of the scientific nomenclature of organisms. Thus, *Megachile (Eutricharaea) rotundata* can also be written *Megachile rotundata.* Anyone can simply ignore subgeneric names. It is fortunate that much of the activity of splitters therefore can be ignored by those who wish to do so; this tactic is more difficult when the taxa are called genera.

The appropriate genus-level classification varies among groups. In *Andrena,* in spite of its numerous subgenera, almost no modern author proposes elevating the subgenera to generic rank. In *Lasioglossum,* as explained in the account of that genus, certain authors do recognize subdivisions such as *Dialictus, Evylaeus,* and *Lasioglossum* s. str. as genera, although I consider them to be subgenera. Since there is no useful definition of genus or subgenus, such differences of opinion are matters of judgment about which there is no right or wrong interpretation. The Anthidiini, a group prone to striking morphological variables like carinae and lamellae on various parts of the body, have been broken up into numerous genera; a few recent authors (Warncke, Westrich) place most of its species in *Anthidium.* Finally, perhaps largely for historical reasons, the Eucerini are divided into many genera rather than subgenera of one huge genus. In the systematic part of this book I have tried to reduce the diversity of treatments both by uniting some genera in much "split" taxa and by breaking up a few taxa that have usually seemed "lumped," such as the old genus *Nomia.* The problem, of course, is that such efforts are largely subjective, because there is no objective basis for deciding whether two taxa are distinct from one another subgenerically, generically, tribally, or whatever. This is true, moreover, whether the taxa are recognized by phenetic differences or by phylogenetic positions. See Section 30.

No bee specialist will be satisfied with all aspects of Table 16-1. It does indicate the classification that will be used in the sections of this book concerned with systematics, i.e., Sections 36 to 121.

In Table 16-1 the authors of family-group names (superfamily to subtribe, or for our purposes, family to tribe) are given. Authors' names do not appear elsewhere. Traditionally less attention is given to them than for genus-group and species-group names, but details can be found in Michener (1986a), and repeated and updated in Engel (2005).

The total number of species placed to genus and subgenus, as indicated in Table 16-1, is about 17,500. As indicated elsewhere, there remain named species not placed to subgenus and not enumerated in the table. Thus in genera like *Megachile, Coelioxys* (especially of South America), and *Eucera* s. l., many species remain unplaced as to subgenus and therefore are not included in Table 16-1. This situation is presumably worst in the little-studied parts of the world, where many species were named in past centuries, but modern treatments of such faunas scarcely exist. The number of described species of bees (not including those relegated to synonymy) may well be over 18,000, and new species yet to be found and named will probably exceed the number of new synonyms recognized. A guess as to the total number of bee species in the world is therefore near or above the often-mentioned figure of 20,000. Given the number of cryptic species that will probably be found with the advent of molecular methods and more careful morphological analyses, the total number of species could be even higher. I am indebted to Dr. John S. Ascher for data and discussion of these matters.

Table 16-1. The Recent Bee Taxa (Family to Subgenus).

This table gives the authors' names for the higher taxa, and estimates of the approximate numbers of included species for genera and subgenera. Depending on the status of the systematics of the particular group, these numbers are based on revisions, estimates, or merely number of specific names. They relate to named species but are often uncertain because revisors were not certain about the status of some names or because of unpublished synonymies or other findings. Isolated descriptions of new species were commonly ignored. The numbers in parentheses after family-group names are those of the corresponding text sections. Subgeneric names are in italics. At the end of the table are the totals. Fossil taxa are excluded; see Section 22.

Taxon	#	Taxon	#	Taxon	#
Family **Stenotritidae** Cockerell(36)		*Perditomorpha*	45	*Ptiloglossa*	40
		Protodiscelis	5	Tribe Diphaglossini Vachal (44)	
Ctenocolletes	10	*Protomorpha*	9	*Cadeguala*	2
Stenotritus	11	*Pygopasiphae*	2	*Cadegualina*	2
		Reedapis	3	*Diphaglossa*	1
Family **Colletidae** Lepeletier (37)		*Sarocolletes*	6	Tribe Dissoglottini Moure (45)	
		Spinolapis	3	*Mydrosoma*	9
Subfamily Colletinae Lepeletier (38)		*Tetraglossula*	5	*Mydrosomella*	1
Tribe Paracolletini (39)		*Torocolletes*	2	*Ptiloglossidia*	1
Brachyglossula	4	*Urocolletes*	1	Subfamily Xeromelissinae Cockerell (46)	
Callomelitta	11	Lonchopria		Chilicola	
Chrysocolletes	5	*Biglossa*	9	*Anoediscelis*	19
Eulonchopria		*Ctenosibyne*	1	*Chilicola* s. str.	4
Ethalonchopria	2	*Lonchoprella*	1	*Chilioediscelis*	3
Eulonchopria s. str.	3	*Lonchopria* s. str.	3	*Hylaeosoma*	10
Glossurocolletes	2	*Porterapis*	1	*Oediscelis*	20
Hesperocolletes	1	Lonchorhyncha	1	*Oroediscelis*	7
Leioproctus		Neopasiphae	3	*Prosopoides*	2
Actenosigynes	1	Niltonia	1	*Pseudiscelis*	2
Albinapis	1	Paracolletes		Chilimelissa	18
Andrenopsis	4	*Anthoglossa*	8	Geodiscelis	2
Baeocolletes	3	*Paracolletes* s. str.	8	Xenochilicola	3
Cephalocolletes	5	Phenacolletes	1	Xeromelissa	1
Ceratocolletes	2	Trichocolletes		Subfamily Hylaeinae Viereck (47)	
Chilicolletes	1	*Callocolletes*	1	Amphylaeus	
Cladocerapis	9	*Trichoclletes* s. str.	22	*Agogenohylaeus*	3
Colletellus	1	Tribe Colletini Lepeletier (40)		*Amphylaeus* s. str.	1
Colletopsis	1	Colletes	330	Calloprosopis	1
Euryglossidia	22	Mourecotelles		Hemirhiza	1
Excolletes	1	*Hemicotelles*	2	Hylaeus	
Filiglossa	4	*Mourecotelles* s. str.	8	*Abrupta*	1
Glossopasiphae	1	*Xanthocotelles*	11	*Alfkenylaeus*	5
Goniocolletes	21	Tribe Scraptrini Melo and		*Analastoroides*	1
Halictanthrena	1	Gonçalves (41)		*Cephalylaeus*	2
Hexantheda	2	Scrapter	31	*Cephylaeus*	1
Holmbergeria	2	Subfamily Diphaglossinae Vachal (42)		*Cornylaeus*	2
Hoplocolletes	1	Tribe Caupolicanini Michener (43)		*Dentigera*	20
Kylopasiphae	1	Caupolicana		*Deranchylaeus*	49
Lamprocolletes	18	*Alayoapis*	3	*Edriohylaeus*	1
Leioproctus s. str.	125	*Caupolicana* s. str.	31	*Euprosopellus*	4
Nesocolletes	5	*Willinkapis*	1	*Euprosopis*	5
Nomiocolletes	5	*Zikanapis*	11	*Euprosopoides*	10
Odontocolletes	8	Crawfordapis	1	*Gephyrohylaeus*	3

(*continues*)

Table 16-1. The Bee Taxa *(continued)*

Subfamily Hylaeinae *(continued)*		Brachyhesma s. str.	22	Chrysandrena	14
Gnathoprosopis	7	Henicohesma	2	Cnemidandrena	45
Gnathoprosopoides	2	Microhesma	16	Conandrena	2
Gnathylaeus	1	Callohesma	34	Cordandrena	7
Gongyloprosopis	5	Dasyhesma	21	Cremnandrena	1
Heterapoides	8	Euhesma		Cryptandrena	5
Hoploprosopis	1	Euhesma s. str.	65	Cubiandrena	2
Hylaeana	9	Parahesma	1	Dactylandrena	4
Hylaeopsis	25	Euryglossa	36	Dasyandrena	3
Hylaeorhiza	1	Euryglossina		Derandrena	10
Hylaeteron	5	Euryglossella	8	Diandrena	25
Hylaeus s. str.	92	Euryglossina s. str.	54	Didonia	7
Koptogaster	2	Microdontura	1	Distandrena	11
Laccohylaeus	1	Pachyprosopina	1	Erandrena	1
Lambdopsis	18	Quasihesma	10	Euandrena	74
Macrohylaeus	1	Euryglossula	7	Fumandrena	11
Meghylaeus	1	Heterohesma	2	Fuscandrena	1
Mehelyana	1	Hyphesma	7	Geissandrena	1
Metylaeus	6	Melittosmithia	4	Genyandrena	2
Metziella	1	Pachyprosopis		Gonandrena	6
Nesoprosopis	68	Pachyprosopis s. str.	7	Graecandrena	20
Nesylaeus	1	Pachyprosopula	7	Habromelissa	1
Nothylaeus	34	Parapachyprosopis	9	Hesperandrena	9
Orohylaeus	1	Sericogaster	1	Holandrena	16
Paraprosopis	47	Stenohesma	1	Hoplandrena	23
Planihylaeus	5	Tumidihesma	2	Hyperandrena	2
Prosopella	1	Xanthesma		Iomelissa	1
Prosopis	46	Argohesma	8	Larandrena	7
Prosopisteroides	4	Chaetohesma	10	Leimelissa	4
Prosopisteron	76	Xanthesma s. str.	13	Lepidandrena	18
Pseudhylaeus	5	Xenohesma	17	Leucandrena	16
Rhodohylaeus	21			Longandrena	3
Spatulariella	18			Malayapis	1
Sphaerhylaeus	2	Family **Andrenidae** Latreille (49)		Margandrena	7
Xenohylaeus	4			Melanapis	4
Hyleoides	8	Subfamily Alocandreninae Michener (50)		Melandrena	64
Meroglossa	20	Alocandrena	1	Melittoides	4
Palaeorhiza		Subfamily Andreninae Latreille (51)		Micrandrena	103
Anchirhiza	2	Ancylandrena	5	Nemandrena	3
Callorhiza	40	Andrena		Nobandrena	13
Ceratorhiza	2	Aciandrena	26	Notandrena	16
Cercorhiza	13	Aenandrena	7	Oligandrena	2
Cheesmania	3	Agandrena	3	Onagrandrena	24
Cnemidorhiza	20	Anchandrena	2	Orandrena	24
Eupalaeorhiza	3	Andrena s. str.	83	Oreomelissa	13
Eusphecogastra	3	Aporandrena	2	Osychnyukandrena	2
Gressittapis	2	Archiandrena	3	Oxyandrena	1
Hadrorhiza	3	Augandrena	3	Pallandrena	4
Heterorhiza	12	Avandrena	7	Parandrena	13
Michenerapis	1	Belandrena	5	Parandrenella	8
Noonadania	2	Biareolina	1	Pelicandrena	1
Palaeorhiza s. str.	15	Brachyandrena	4	Planiandrena	4
Paraheterorhiza	2	Callandrena	79	Plastandrena	33
Trachyrhiza	1	Calomelissa	6	Poecilandrena	29
Zarhiopalea	5	Campylogaster	14	Poliandrena	33
Pharohylaeus	2	Carandrena	39	Psammandrena	2
Xenorhiza	5	Carinandrena	1	Ptilandrena	13
Subfamily Euryglossinae Michener (48)		Celetandrena	1	Rhacandrena	4
Brachyhesma		Charitandrena	2	Rhaphandrena	3
Anomalohesma	1	Chlorandrena	49	Rufandrena	2

(continues)

Table 16-1. The Bee Taxa *(continued)*

Taxon	Count	Taxon	Count	Taxon	Count
Subfamily Andreninae *(continued)*		*Panurgus* s. str.	30	Oxaea	8
Scaphandrena	53	*Simpanurgus*	1	*Protoxaea*	3
Scitandrena	1	Tribe Nolanomelissini (55)			
Scoliandrena	2	*Nolanomelissa*	1		
Scrapteropsis	18	Tribe Melitturgini Newman (56)		Family **Halictidae** Thomson (61)	
Simandrena	41	*Borgatomelissa*	2		
Stenomelissa	3	*Flavomeliturgula*	6	Subfamily Rophitinae Schenck (62)	
Suandrena	11	*Gasparinahla*	1	*Ceblurgus*	1
Taeniandrena	23	*Melitturga*	13	*Conanthalictus*	
Tarsandrena	6	*Meliturgula*	11	*Conanthalictus* s. str.	2
Thysandrena	21	*Mermiglossa*	1	*Phaceliapis*	11
Trachandrena	30	*Plesiopanurgus*	4	*Dufourea*	130
Troandrena	5	Tribe Protomelitturgini Ruz (57)		*Goeletapis*	1
Tylandrena	14	*Protomelitturga*	1	*Micralictoides*	8
Ulandrena	31	Tribe Perditini Robertson (58)		*Morawitzella*	1
Xiphandrena	1	*Macrotera*		*Morawitzia*	3
Zonandrena	17	*Cockerellula*	13	*Penapis*	3
Euherbstia	1	*Macrotera* s. str.	6	*Protodufourea*	5
Megandrena		*Macroterella*	6	*Rophites*	
Erythrandrena	1	*Macroteropsis*	6	*Flavodufourea*	2
Megandrena s. str.	1	*Perdita*		*Rophitoides*	4
Orphana	2	*Allomacrotera*	2	*Rophites* s. str.	13
Subfamily Panurginae Leach (52)		*Alloperdita*	6	*Sphecodosoma*	
Tribe Protandrenini Robertson (53)		*Callomacrotera*	2	*Michenerula*	1
Anthemurgus	1	*Cockerellia*	25	*Sphecodosoma* s. str.	2
Anthrenoides	30	*Epimacrotera*	18	*Systropha*	25
Chaeturginus	2	*Glossoperdita*	4	*Xeralictus*	2
Liphanthus		*Hesperoperdita*	3	Subfamily Nomiinae Robertson (63)	
Leptophanthus	7	*Heteroperdita*	13	*Dieunomia*	
Liphanthus s. str.	4	*Hexaperdita*	29	*Dieunomia* s. str.	5
Melaliphanthus	2	*Pentaperdita*	13	*Epinomia*	4
Neoliphanthus	1	*Perdita* s. str.	441	*Halictonomia*	10
Pseudoliphanthus	4	*Perditella*	7	*Lipotriches*	
Tricholiphanthus	3	*Procockerellia*	5	*Afronomia*	7
Xenoliphanthus	4	*Pseudomacrotera*	1	*Austronomia*	102
Neffapis	1	*Pygoperdita*	43	*Clavinomia*	1
Parapsaenythia	2	*Xeromacrotera*	1	*Lipotriches* s. str.	99
Protandrena		*Xerophasma*	2	*Macronomia*	45
Austropanurgus	1	Tribe Calliopsini Robertson (59)		*Maynenomia*	1
Heterosarus	59	*Acamptopoeum*	11	*Melanomia*	2
Metapsaenythia	2	*Arhysosage*	6	*Nubenomia*	15
Parasarus	1	*Calliopsis*		*Trinomia*	6
Protandrena s. str.	50	*Calliopsima*	15	*Mellitidia*	19
Pterosarus	40	*Calliopsis* s. str.	12	*Nomia*	
Psaenythia	80	*Ceroliopoeum*	1	*Acunomia*	33
Pseudopanurgus	32	*Hypomacrotera*	3	*Crocisaspidia*	11
Rhophitulus		*Liopoeodes*	1	*Hoplonomia*	20
Cephalurgus	5	*Liopoeum*	4	*Leuconomia*	39
Panurgillus	21	*Micronomadopsis*	20	*Nomia* s. str.	6
Rhophitulus s. str.	3	*Nomadopsis*	13	*Paulynomia*	2
Incertae Sedis		*Perissander*	7	*Pseudapis*	
Stenocolletes	1	*Verbenapis*	4	*Pachynomia*	4
Tribe Panurgini Leach (54)		*Callonychium*		*Pseudapis* s. str.	69
Avpanurgus	1	*Callonychium* s. str.	6	*Ptilonomia*	3
Camptopoeum		*Paranychium*	5	*Reepenia*	3
Camptopoeum s. str.	13	*Litocalliopsis*	1	*Spatunomia*	2
Epimethea	12	*Spinoliella*	6	*Sphegocephala*	6
Panurginus	49	Subfamily Oxaeinae Ashmead (60)		*Steganomus*	7
Panurgus		*Mesoxaea*	7	Subfamily Nomioidinae Börner (64)	
Flavipanurgus	5	*Notoxaea*	1	*Cellariella*	2

(continues)

Table 16-1. The Bee Taxa (continued)

Subfamily Nomioidinae *(continued)*		Mexalictus	5	Megommation	
Ceylalictus		Microsphecodes	7	*Cleptommation*	1
Atronomioides	11	Nesosphecodes	3	*Megaloptina*	2
Ceylalictus s. str.	13	Paragapostemon	1	*Megommation* s. str.	1
Meganomioides	2	Parathrincostoma	2	*Stilbochlora*	1
Nomioides	60	Patellapis		Micrommation	1
Subfamily Halictinae Thomson (65)		*Archihalictus*	16	Neocorynura	65
Tribe Halictini Thomson (66)		*Chaetalictus*	35	Paroxystoglossa	9
Agapostemon		*Dictyohalictus*	12	Pseudaugochlora	7
Agapostemon s. str.	43	*Lomatalictus*	4	Rhectomia	4
Agapostemonoides	1	*Pachyhalictus*	30	Rhinocorynura	5
Caenohalictus	55	*Patellapis* s. str.	5	Temnosoma	7
Dinagapostemon	8	*Zonalictus*	68	Thectochlora	1
Echthralictus	2	Pseudagapostemon		Xenochlora	4
Eupetersia		*Brasilagapostemon*	3		
Eupetersia s. str.	21	*Neagapostemon*	6		
Nesoeupetersia	8	*Pseudagapostemon* s. str.	16	Family **Melittidae** Schenck (68)	
Glossodialictus	1	Ptilocleptis	3		
Habralictus		Rhinetula	1	Subfamily Dasypodainae Börner (69)	
Habralictus s. str.	21	Ruizantheda	4	Tribe Dasypodaini Börner (70)	
Zikaniella	1	Sphecodes	285	Dasypoda	35
Halictus		Thrincohalictus	1	Eremaphanta	
Argalictus	8	Thrinchostoma		*Eremaphanta* s. str.	6
Halictus s. str.	4	*Diagonozus*	5	*Popovapis*	2
Hexataenites	11	*Eothrincostoma*	7	Hesperapis	
Lampralictus	1	*Thrinchostoma* s. str.	44	*Ambylapis*	6
Monilapis	31	Urohalictus	1	*Capicola*	6
Nealictus	2	Tribe Augochlorini Beebe (67)		*Capicoloides*	1
Odontalictus	2	Andinaugochlora		*Carinapis*	7
Pachyceble	22	*Andinaugochlora* s. str.	2	*Disparapis*	1
Paraseladonia	1	*Neocorynurella*	2	*Hesperapis* s. str.	1
Platyhalictus	14	Ariphanarthra	1	*Panurgomia*	6
Protohalictus	13	Augochlora		*Xeralictoides*	1
Ramalictus	1	*Augochlora* s. str.	86	*Zacesta*	1
Seladonia	36	*Oxystoglossella*	27	Tribe Promelittini Michener (71)	
Tytthalictus	4	Augochlorella		Afrodasypoda	1
Vestitohalictus	35	*Augochlorella* s. str.	15	Promelitta	1
Homalictus		*Ceratalictus*	5	Tribe Sambini Michener (72)	
Homalictus s. str.	94	*Pereirapis*	6	Haplomelitta	
Papualictus	6	Augochlorodes	1	*Atrosamba*	1
Quasilictus	1	Augochloropsis		*Haplomelitta* s. str.	1
Lasioglossum		*Augochloropsis* s. str.	46	*Haplosamba*	1
Acanthalictus	1	*Paraugochloropsis*	92	*Metasamba*	1
Australictus	11	Caenaugochlora		*Prosamba*	1
Austrevylaeus	19	*Caenaugochlora* s. str.	13	Samba	1
Callalictus	8	*Ctenaugochlora*	4	Subfamily Meganomiinae Michener (73)	
Chilalictus	134	Chlerogas	9	Ceratomonia	1
Ctenonomia	196	Chlerogella		Meganomia	4
Dialictus	465	*Chlerogella* s. str.	15	Pseudophilanthus	
Eickwortia	2	*Ischnomelissa*	7	*Dicromonia*	1
Evylaeus	60	Chlerogelloides	2	*Pseudophilanthus* s. str.	3
Glossalictus	1	Corynura		Uromonia	
Hemihalictus	1	*Callistochlora*	3	*Nesomonia*	1
Lasioglossum s. str.	162	*Corynura* s. str.	18	*Uromonia* s. str.	1
Paradialictus	1	Halictillus	2	Subfamily Melittinae Schenk (74)	
Parasphecodes	99	Megalopta		Macropis	
Pseudochilalictus	1	*Megalopta* s. str.	27	*Macropis* s. str.	10
Sellalictus	35	*Noctoraptor*	3	*Paramacropis*	1
Sphecodogastra	8	Megaloptidia	3	*Sinomacropis*	5
Sudila	6	Megaloptilla	3		

(continues)

Table 16-1. The Bee Taxa *(continued)*

Taxon	Count	Taxon	Count	Taxon	Count
Subfamily Melittinae *(continued)*		Alcidamea	72	*Protosmia* s. str.	19
Melitta		Annosmia	31	Pseudoheriades	7
Dolichochile	1	Anthocopa	74	Stenoheriades	10
Melitta s. str.	26	Bytinskia	4	Stenosmia	11
Rediviva	21	Chlidoplitis	2	Wainia	
Redivivoides	1	Coloplitis	2	*Caposmia*	4
		Cyrtosmia	1	*Wainia* s. str.	3
		Dasyosmia	2	*Wainiella*	2
Family **Megachilidae** Latreille (75)		Eurypariella	1	Xeroheriades	1
		Exanthocopa	1	Tribe Anthidiini Ashmead (82)	
Subfamily Fideliinae Cockerell (76)		Formicapis	1	Acedanthidium	1
Tribe Pararhophitini Popov (77)		Hoplitina	6	Afranthidium	
Pararhophites	3	*Hoplitis* s. str.	43	*Afranthidium* s. str.	9
Tribe Fideliini Cockerell (78)		Jaxartinula	2	*Branthidium*	10
Fidelia		Kumobia	4	*Capanthidium*	12
Fidelia s. str.	3	Megahoplitis	1	*Domanthidium*	1
Fideliana	2	Megalosmia	4	*Immanthidium*	5
Fideliopsis	5	Microhoplitis	1	*Mesanthidiellum*	3
Parafidelia	2	Monumetha	6	*Mesanthidium*	8
Neofidelia	2	Nasutosmia	2	*Nigranthidium*	2
Subfamily Megachilinae Latreille (79)		Pentadentosmia	24	*Oranthidium*	3
Tribe Lithurgini Newman (80)		Penteriades	2	*Xenanthidium*	1
Lithurgus		Platosmia	8	*Zosteranthidium*	1
Lithurgopsis	11	Prionohoplitis	6	Afrostelis	5
Lithurgus s. str.	15	Proteriades	22	Anthidiellum	
Microthurge	4	Robertsonella	3	*Ananthidiellum*	1
Trichothurgus	13	Hoplosmia		*Anthidiellum* s. str.	7
Tribe Osmiini Newman (81)		*Hoplosmia* s. str.	3	*Chloranthidiellum*	1
Afroheriades	5	*Odontanthocopa*	9	*Clypanthidium*	3
Ashmeadiella		*Paranthocopa*	1	*Loyolanthidium*	8
Arogochila	18	Noteriades	9	*Pycnanthidium*	22
Ashmeadiella s. str.	33	Ochreriades	2	*Ranthidiellum*	2
Chilosima	2	Osmia		Anthidioma	1
Cubitognatha	1	*Acanthosmioides*	22	Anthidium	
Isosmia	2	*Allosmia*	3	*Anthidium* s. str.	75
Atoposmia		*Cephalosmia*	5	*Callanthidium*	2
Atoposmia s. str.	12	*Diceratosmia*	5	*Gulanthidium*	1
Eremosmia	14	*Erythrosmia*	13	*Nivanthidium*	1
Hexosmia	2	*Euthosmia*	1	*Proanthidium*	8
Bekilia	1	*Helicosmia*	81	*Severanthidium*	10
Chelostoma		*Hemiosmia*	6	*Turkanthidium*	5
Ceraheriades	1	*Melanosmia*	108	Anthodioctes	
Chelostoma s. str.	27	*Metallinella*	1	*Anthodioctes* s. str.	36
Eochelostoma	1	*Monosmia*	1	*Bothranthidium*	1
Foveosmia	19	*Mystacosmia*	1	Apianthidium	1
Gyrodromella	6	*Neosmia*	8	Aspidosmia	2
Prochelostoma	1	*Orientosmia*	1	Austrostelis	8
Haetosmia	3	*Osmia* s. str.	22	Aztecanthidium	3
Heriades		*Ozbekosmia*	1	Bathanthidium	
Amboheriades	11	*Pyrosmia*	33	*Bathanthidium* s. str.	1
Heriades s. str.	46	*Tergosmia*	6	*Manthidium*	1
Michenerella	32	*Trichinosmia*	1	*Stenanthidiellum*	2
Neotrypetes	13	Othinosmia		Benanthis	2
Pachyheriades	5	*Afrosmia*	1	Cyphanthidium	2
Rhopaloheriades	1	*Megaloheriades*	7	Dianthidium	
Toxeriades	1	*Othinosmia* s. str.	5	*Adanthidium*	4
Tyttheriades	1	Protosmia		*Deranchanthidium*	2
Hofferia	2	*Chelostomopsis*	4	*Dianthidium* s. str.	20
Hoplitis		*Dolichosmia*	1	*Mecanthidium*	2
Acrosmia	5	*Nanosmia*	6	Duckeanthidium	5

(continues)

Table 16-1. The Bee Taxa *(continued)*

Taxon	Count
Subfamily Megachilinae *(continued)*	
Eoanthidium	
Clistanthidium	5
Eoanthidium s. str.	6
Hemidiellum	1
Salemanthidium	2
Epanthidium	
Ananthidium	2
Carloticola	3
Epanthidium s. str.	18
Euaspis	12
Gnathanthidium	1
Hoplostelis	
Hoplostelis s. str.	3
Rhynostelis	1
Hypanthidioides	
Anthidulum	4
Ctenanthidium	4
Dichanthidium	1
Dicranthidium	6
Hypanthidioides s. str.	1
Larocanthidium	10
Michanthidium	2
Mielkeanthidium	2
Moureanthidium	5
Saranthidium	7
Hypanthidium	
Hypanthidium s. str.	16
Tylanthidium	1
Icteranthidium	25
Indanthidium	1
Larinostelis	1
Neanthidium	1
Notanthidium	
Allanthidium	6
Chrisanthidium	3
Notanthidium s. str.	1
Pachyanthidium	
Ausanthidium	1
Pachyanthidium s. str.	11
Trichanthidiodes	1
Trichanthidium	3
Paranthidium	
Paranthidium s. str.	4
Rapanthidium	1
Plesianthidium	
Carinanthidium	1
Plesianthidium s. str.	1
Spinanthidiellum	2
Spinanthidium	5
Pseudoanthidium	
Exanthidium	4
Micranthidium	3
Pseudoanthidium s. str.	18
Royanthidium	6
Semicarinella	1
Tuberanthidium	4
Rhodanthidium	
Asianthidium	7
Meganthidium	1
Rhodanthidium s. str.	5
Serapista	4
Stelis	
Dolichostelis	6
Heterostelis	9
Malanthidium	1
Protostelis	1
Pseudostelis	3
Stelidomorpha	3
Stelis s. str.	75
Trachusa	
Archianthidium	7
Congotrachusa	1
Heteranthidium	13
Legnanthidium	1
Massanthidium	3
Metatrachusa	2
Orthanthidium	1
Paraanthidium	7
Trachusa s. str.	1
Trachusomimus	2
Ulanthidium	6
Trachusoides	1
Xenostelis	1
Tribe Dioxyini Cockerell (83)	
Aglaoapis	3
Allodioxys	4
Dioxys	15
Ensliniana	3
Eudioxys	2
Metadioxys	3
Paradioxys	2
Prodioxys	3
Tribe Megachilini Latreille (84)	
Coelioxys	
Acrocoelioxys	25
Allocoelioxys	45
Boreocoelioxys	17
Coelioxys s. str.	52
Cyrtocoelioxys	39
Glyptocoelioxys	50
Haplocoelioxys	5
Liothyrapis	35
Mesocoelioxys	1
Neocoelioxys	7
Platycoelioxys	1
Rhinocoelioxys	5
Synocoelioxys	5
Torridapis	14
Xerocoelioxys	10
Megachile	
Acentron	11
Amegachile	30
Argyropile	7
Austrochile	10
Austromegachile	25
Callomegachile	91
Cestella	1
Chalicodoma	31
Chalicodomoides	2
Chelostomoda	14
Chelostomoides	31
Chrysosarus	25
Creightonella	50
Cressoniella	12
Cuspidella	1
Dasymegachile	20
Eumegachile	1
Eutricharaea	236
Gronoceras	10
Grosapis	1
Hackeriapis	90
Heriadopsis	1
Largella	3
Leptorachis	30
Litomegachile	7
Matangapis	1
Maximegachile	2
Megachile s. str.	9
Megachiloides	60
Megella	3
Melanosarus	8
Mitchellapis	6
Moureapis	8
Neochelynia	5
Neocressoniella	2
Paracella	39
Parachalicodoma	1
Platysta	2
Pseudocentron	55
Pseudomegachile	80
Ptilosaroides	1
Ptilosarus	9
Rhodomegachile	3
Rhyssomegachile	1
Sayapis	18
Schizomegachile	1
Schrottkyapis	1
Stelodides	1
Stenomegachile	4
Thaumatosoma	2
Trichurochile	1
Tylomegachile	2
Xanthosarus	26
Zonomegachile	2
Radoszkowskiana	4
Incertae Sedis	
Neochalicodoma	2
Stellenigris	1
Family **Apidae** Latreille (85)	
Subfamily Xylocopinae Latreille (86)	
Tribe Manueliini Sakagami & Michener (87)	
Manuelia	3
Tribe Xylocopini Latreille (88)	

(continues)

Table 16-1. The Bee Taxa *(continued)*

Taxon	Count
Subfamily Xylocopinae *(continued)*	
Xylocopa	
Alloxylocopa	6
Biluna	5
Bomboixylocopa	5
Cirroxylocopa	1
Copoxyla	4
Ctenoxylocopa	6
Dasyxylocopa	1
Diaxylocopa	1
Gnathoxylocopa	1
Koptortosoma	196
Lestis	2
Maaiana	6
Mesotrichia	23
Monoxylocopa	1
Nanoxylocopa	1
Neoxylocopa	49
Nodula	7
Notoxylocopa	2
Nyctomelitta	3
Prosopoxylocopa	1
Proxylocopa	16
Rhysoxylocopa	8
Schonnherria	29
Stenoxylocopa	6
Xenoxylocopa	3
Xylocopa s. str.	8
Xylocopoda	2
Xylocopoides	6
Xylocopsis	1
Xylomelissa	65
Zonohirsuta	4
Tribe Ceratinini Latreille (89)	
Ceratina	
Calloceratina	10
Catoceratina	1
Ceratina s. str.	20
Ceratinidia	26
Ceratinula	30
Chloroceratina	2
Copoceratina	2
Crewella	12
Ctenoceratina	10
Euceratina	16
Hirashima	4
Lioceratina	7
Malgatina	1
Megaceratina	1
Neoceratina	8
Pithitis	9
Protopithitis	1
Rhysoceratina	2
Simiceratina	3
Xanthoceratina	9
Zadontomerus	25
Tribe Allodapini Cockerell (90)	
Allodape	30
Allodapula	
Allodapula s. str.	9
Allodapulodes	4
Dalloapula	2
Braunsapis	87
Compsomelissa	
Compsomelissa s. str.	6
Halterapis	22
Effractapis	1
Eucondylops	2
Exoneura	
Brevineura	26
Exoneura s. str.	40
Inquilina	2
Exoneurella	4
Exoneuridia	
Alboneuridia	1
Exoneuridia s. str.	2
Macrogalea	11
Nasutapis	1
Subfamily Nomadinae Latreille (91)	
Tribe Hexepeolini Roig-Alsina & Michener (92)	
Hexepeolus	1
Tribe Brachynomadini Roig-Alsina & Michener (93)	
Brachynomada	
Brachynomada s. str.	8
Melanomada	7
Kelita	
Kelita s. str.	4
Spinokelita	1
Paranomada	3
Trichonomada	1
Triopasites	2
Tribe Nomadini Latreille (94)	
Nomada	795
Tribe Epeolini Robertson (95)	
Doeringiella	35
Epeolus	109
Odyneropsis	
Odyneropsis s. str.	10
Parammobates	4
Pseudepeolus	5
Rhinepeolus	1
Rhogepeolus	5
Thalestria	1
Triepeolus	141
Tribe Ammobatoidini Michener (96)	
Aethammobates	1
Ammobatoides	6
Holcopasites	16
Schmiedeknechtia	5
Tribe Biastini Linsley & Michener (97)	
Biastes	4
Neopasites	
Micropasites	3
Neopasites s. str.	2
Rhopalolemma	2
Tribe Townsendiellini Michener (98)	
Townsendiella	3
Tribe Neolarrini Fox (99)	
Neolarra	
Neolarra s. str.	11
Phileremulus	3
Tribe Ammobatini Handlirsch (100)	
Ammobates	
Ammobates s. str.	30
Euphileremus	7
Xerammobates	3
Melanempis	5
Oreopasites	11
Parammobatodes	7
Pasites	21
Sphecodopsis	
Pseudodichroa	2
Sphecodopsis s. str.	8
Spinopasites	1
Tribe Caenoprosopidini Michener (101)	
Caenoprosopina	1
Caenoprosopis	1
Subfamily Apinae Latreille (102)	
Tribe Isepeolini Rozen, Eickwort, & Eickwort (103)	
Isepeolus	11
Melectoides	10
Tribe Osirini Handlirsch (104)	
Epeoloides	2
Osirinus	7
Osiris	21
Parepeolus	
Ecclitodes	1
Parepeolus s. str.	4
Protosiris	4
Tribe Protepeolini Linsley & Michener (105)	
Leiopodus	5
Tribe Exomalopsini Vachal (106)	
Anthophorula	
Anthophorisca	29
Anthophorula s. str.	29
Isomalopsis	2
Chilimalopsis	2
Eremapis	1
Exomalopsis	
Diomalopsis	2
Exomalopsis s. str.	55
Phanomalopsis	15
Stilbomalopsis	13
Teratognatha	1
Tribe Ancylini Michener (107)	
Ancyla	10
Tarsalia	7
Tribe Tapinotaspidini Roig-Alsina & Michener (108)	
Arhysoceble	5
Caenonomada	3
Chalepogenus	
Chalepogenus s. str.	21
Lanthanomelissa	5

(continues)

Table 16-1. The Bee Taxa *(continued)*

Subfamily Apinae *(continued)*		Hamatothrix	1	*Dasymegilla*	6
Monoeca	6	Lophothygater	1	*Heliophila*	91
Paratetrapedia		Martinapis		*Lophanthophora*	33
Amphipedia	1	*Martinapis* s. str.	2	*Melea*	9
Lophopedia	7	*Svastropsis*	1	*Mystacanthophora*	19
Paratetrapedia s. str.	14	Melissodes		*Paramegilla*	66
Tropidopedia	2	*Apomelissodes*	4	*Petalosternon*	21
Xanthopedia	5	*Callimelissodes*	14	*Pyganthophora*	66
Tapinotaspis	3	*Ecplectica*	8	*Rhinomegilla*	4
Tapinotaspoides	4	*Eumelissodes*	72	Deltoptila	10
Trigonopedia	4	*Heliomelissodes*	2	Elaphropoda	6
Tribe Tetrapediini Michener & Moure (109)		*Melissodes* s. str.	23	Habrophorula	3
		Psilomelissodes	1	Habropoda	50
Coelioxoides	3	*Tachymelissodes*	3	Pachymelus	
Tetrapedia	13	Melissoptila	60	*Pachymelopsis*	5
Tribe Ctenoplectrini Cockerell (110)		Micronychapis	1	*Pachymelus* s. str.	15
Ctenoplectra	24	Mirnapis	1	Tribe Centridini Cockerell & Cockerell (114)	
Ctenoplectrina	2	Notolonia	1		
Tribe Emphorini Robertson (111)		Pachysvastra	1	Centris	
Alepidoscelis	6	Peponapis	13	*Acritocentris*	4
Ancyloscelis	25	Platysvastra	1	*Aphemisia*	3
Diadasia	45	Santiago	2	*Centris* s. str.	35
Diadasina		Simanthedon	1	*Exallocentris*	1
Diadasina s. str.	4	Svastra		*Heterocentris*	17
Leptometriella	3	*Anthedonia*	2	*Melacentris*	18
Meliphilopsis	2	*Brachymelissodes*	2	*Paracentris*	25
Melitoma	10	*Epimelissodes*	13	*Ptilocentris*	1
Melitomella	3	*Idiomelissodes*	1	*Ptilotopus*	12
Ptilothrix	13	*Svastra* s. str.	3	*Schisthemisia*	2
Toromelissa	1	Svastrides	4	*Trachina*	15
Tribe Eucerini Latreille (112)		Svastrina	1	*Wagenknechtia*	5
Agapanthinus	1	Syntrichalonia	2	*Xanthemisia*	4
Alloscirtetica		Tetralonia		*Xerocentris*	8
Alloscirtetica s. str.	36	*Eucara*	7	Epicharis	
Megascirtetica	1	*Tetralonia* s. str.	1	*Anepicharis*	3
Canephorula	1	*Thygatina*	9	*Cyphepicharis*	1
Cemolobus	1	Tetraloniella		*Epicharana*	6
Cubitalia		*Glazunovia*	1	*Epicharis* s. str.	3
Cubitalia s. str.	4	*Loxoptilus*	2	*Epicharitides*	7
Opacula	1	*Pectinapis*	4	*Epicharoides*	4
Pseudeucera	1	*Tetraloniella* s. str.	115	*Hoplepicharis*	4
Eucera		Thygater		*Parepicharis*	2
Eucera s. str.	50	*Nectarodiaeta*	2	*Triepicharis*	2
Hetereucera	60	*Thygater* s. str.	23	Tribe Rhathymini Lepeletier (115)	
Oligeucera	1	Trichocerapis		Nanorhathymus	2
Pteneucera	8	*Dithygater*	1	Rhathymus	8
Synhalonia	104	*Trichocerapis* s. str.	5	Tribe Ericrocidini Cockerell & Atkins (116)	
Eucerinoda	1	Ulugombakia	1		
Florilegus		Xenoglossa		Acanthopus	2
Euflorilegus	5	*Eoxenoglossa*	2	Aglaomelissa	1
Florilegus s. str.	5	*Xenoglossa* s. str.	5	Ctenioschelus	2
Floriraptor	1	Tribe Anthophorini Dahlbom (113)		Epiclopus	3
Gaesischia				Ericrocis	2
Dasyhalonia	2	Amegilla	253	Hopliphora	7
Gaesischia s. str.	19	Anthophora		Mesocheira	1
Gaesischiana	3	*Anthomegilla*	8	Mesonychium	12
Gaesischiopsis	7	*Anthophora* s. str.	11	Mesoplia	
Pachyhalonia	3	*Anthophoroides*	6	*Eumelissa*	5
Prodasyhalonia	1	*Caranthophora*	6	*Mesoplia* s. str.	18
Gaesochira	1	*Clisodon*	2	Tribe Melectini Westwood (117)	

(continues)

Table 16-1. The Bee Taxa *(continued)*

Subfamily Apinae *(continued)*		*Dasybombus*	2	Melipona	40
Afromelecta		*Diversobombus*	4	Meliponula	
Acanthomelecta	1	*Eversmannibombus*	1	*Axestotrigona*	12
Afromelecta s. str.	2	*Exilobombus*	1	*Meliplebeia*	12
Brachymelecta	1	*Fervidobombus*	20	*Meliponula* s. str.	1
Melecta		*Festivobombus*	1	Meliwillea	1
Eupavlovskia	2	*Fraternobombus*	1	Nannotrigona	9
Melecta s. str.	48	*Funebribombus*	2	Nogueirapis	3
Melectomimus	1	*Kallobombus*	1	Oxytrigona	8
Paracrocisa	3	*Laesobombus*	1	Paratrigona	28
Pseudomelecta	5	*Megabombus*	14	Pariotrigona	1
Sinomelecta	1	*Melanobombus*	14	Paratrigonoides	1
Tetralonioidella	10	*Mendacibombus*	12	Partamona	
Thyreomelecta	7	*Mucidobombus*	1	*Parapartamona*	7
Thyreus	123	*Orientalibombus*	3	*Partamona* s. str.	34
Xeromelecta		*Pressibombus*	1	Plebeia	
Melectomorpha	2	*Psithyrus*	29	*Plebeia* s. str.	30
Nesomelecta	3	*Pyrobombus*	43	*Scaura*	4
Xeromelecta s. str.	1	*Rhodobombus*	3	*Schwarziana*	2
Zacosmia	1	*Robustobombus*	5	Plebeina	1
Tribe Euglossini Latreille (118)		*Rubicundobombus*	1	Scaptotrigona	24
Aglae	1	*Rufipedibombus*	2	Trichotrigona	1
Eufriesea	52	*Senexibombus*	4	Trigona	
Euglossa	103	*Separatobombus*	2	*Duckeola*	2
Eulaema	25	*Sibericobombus*	7	*Frieseomelitta*	10
Exaerete	6	*Subterraneobombus*	9	*Geotrigona*	16
Tribe Bombini Latreille (119)		*Thoracobombus*	19	*Heterotrigona*	37
Bombus		*Tricornibombus*	3	*Homotrigona*	1
Alpigenobombus	6	Tribe Meliponini Lepeletier (120)		*Lepidotrigona*	4
Alpinobombus	5	Austroplebeia	9	*Papuatrigona*	1
Bombias	2	Cephalotrigona	3	*Tetragona*	17
Bombus s. str.	10	Cleptotrigona	2	*Tetragonisca*	30
Brachycephalobombus	2	Dactylurina	2	*Trigona* s. str.	30
Coccineobombus	2	Hypotrigona	6	Trigonisca	23
Confusibombus	1	Lestrimelitta	8	Tribe Apini Latreille (121)	
Crotchiibombus	1	Liotrigona	8	Apis	11
Cullumanobombus	4	Lisotrigona	3		

Total genera	443
Total genera and subgenera	1,234[a]
Total described species placed as to genus and subgenus	17,533[b]

[a]This is the total number of genus-group taxa that are not subdivided in this classification. The number was obtained by counting all genera and subgenera, except that, for genera in which subgenera are recognized, the typical subgenera (labeled s. str.) were not counted.

[b]See Section 16 and the legend for this table for explanations of the criteria for counting species included in this total.

17. Methods of Classification

It will always be the case, in a study of such a large group as the bees, that some parts of the study are based on relatively recent investigations made with modern methods, whereas other parts, by necessity, are based on antiquated data and methods of analysis. The family-level taxa (including subfamilies and tribes) of some families have been analyzed using cladistic methods (Roig-Alsina and Michener, 1993; Alexander and Michener, 1995). Likewise, the genus- and species-level taxa of certain groups have been analyzed using morphology, molecular characters (usually DNA), or both. In some cases one's confidence in the phylogenetic results is limited, because the number of characters employed was small, as was the case in a cladistic analysis of the Meliponini (Michener, 1990a), or the methods were outdated, as in Michener's analysis of the Melittidae (Michener, 1981a), or the number of species analyzed was too small to represent the diversity of the group. More seriously, most groups have not yet received such treatment.

When a relevant phylogenetic analysis has been made, I have usually based classificatory decisions on that analysis. One must remember, however, that the results of such an analysis are hypotheses, not facts. Moreover, an analysis commonly yields many different cladograms. Authors usually offer reasons for preferring some of their cladograms and rejecting others, but decisions necessary for making a classification are often arbitrary in spite of phylogenetic analysis and various consensus and character-weighting methods. The use of additional characters, e.g., larval, molecular, or behavioral, or the introduction of new internal or external morphological characters, may clarify these matters. Remarkable changes in phylogenetic results can arise simply from including additional taxa. But the classification one settles on at a given time must be made on the basis of the information then available. It is therefore often tentative.

I do not, in fact, believe that classifications should always be based exclusively on phylogenetic hypotheses. The proper functions of classifications include those that facilitate data storage and retrieval. Decisions based on the less reliable parts of a cladogram are likely to be corrected by later work, and thus run counter to these functions. I see nothing wrong with paraphyletic taxa as practical or temporary expedients, provided they are labeled as such and are justified by distinct differentiating characters, even if plesiomorphic. The Melittidae is such a "taxon of convenience," probably not monophyletic.

Users of phylogenetic information should always base their work on cladograms and associated information on the strength of each clade, not on often imperfect and usually incomplete summaries of phylogenies in the form of classifications. It is a disservice to users of systematic information to offer them classifications instead of properly documented cladograms. In nearly all cladograms some parts are supported by numerous strong characters and other parts are less certain; users should have such information at hand, so that they know which parts of a cladogram are reliable and which are not. They cannot glean such information from classifications.

In preparing and trying to improve the classification of bees, I have found the information in cladograms very useful. An example concerns bumble bees. The genus *Bombus* consists of phenetically similar species; it has been broken up into many subgenera, but they are notoriously similar and difficult to separate. A preliminary cladistic analysis (Williams, 1985) indicated that the parasitic genus *Psithyrus* arose from within a paraphyletic *Bombus*, not from an ancestor of all *Bombus*. Williams therefore divided the Bombini into three genera showing the following phylogenetic relationships: *Mendacibombus (Psithyrus, Bombus)*. Williams' decision was based on few characters; and *Mendacibombus* is similar to certain subgenera included in *Bombus*. If *Psithyrus* did not exist, one would not think of giving *Mendacibombus* generic rank. A derived group, *Psithyrus*, would dictate the classification of the group from which it arose. Recognition or identification of *Bombus* would be made difficult because one would have to separate it from *Mendacibombus*.

This particular problem became moot when more comprehensive phylogenetic analysis (Williams, 1994) showed that *Mendacibombus* was itself paraphyletic, and that *Psithyrus* is closest to other subgenera (Fig. 119-7). It seemed best, then, to place the whole clade, including *Psithyrus*, in the genus *Bombus*. I follow Williams (1994) in this decision because of morphological similarity and the apparent behavioral similarity to *Psithyrus* of parasitic species in two other mostly nonparasitic subgenera.

For the many groups that have never received any cladistic analysis, I have usually used the traditional taxa, largely phenetically based, and modified as seemed appropriate. Often it is possible to make an informed judgment about phylogeny, and such judgments influenced my taxonomic decisions. Of course the classifications will be improved as more analyses are made.

18. The History of Bee Classifications

The species of bees have been classified in many different ways. An exhaustive historical treatment of the classifications proposed would occupy a great deal of space. The following pages briefly summarize the classifications found in some of the major publications on the subject. To facilitate comparison of these classifications, I have sometimes rearranged the groups so that the sequence does not inhibit comparisons. The genera I list are not necessarily all those listed by the authors, but are representatives of the various currently accepted tribes, subfamilies, or families. Thus a reader can determine what groups were intended by the authors. For example, where for brevity I have written *Megachile* after Megachilidae, most authors also included *Osmia, Chalicodoma, Chelostoma, Heriades*, etc. Of course, when authors placed these genera in different tribes or higher categories, this fact has been indicated. The genera listed vary, largely for geographical reasons; for example, a classification of European bees cannot be expected to include taxa not found in Europe. Misplaced genera that are not types of higher taxa are often ignored; in some cases the authors had not seen authentic specimens and had therefore placed the genera on the basis of descriptions. In other cases they were simply careless. Modern generic names and accepted spellings have been introduced as necessary to permit easy understanding.

Because the small Nomadinae have been included as a block in most classifications, but the genera and tribes cited often differ among authors, I have used the word "pasitines" to indicate this group instead of listing particular genera or tribes. The word "pasitines" has the advantage that it is familiar, yet no currently recognized tribe is called Pasitini. Therefore the word "pasitines" can reasonably stand for the entire group of tribes, Ammobatoidini to Caenoprosopidini in Table 16-1.

Kirby (1802), in the first major account of the bees of any area (Britain), placed all bees in two genera, *Melitta* for the S-T bees and *Apis* for the L-T bees. In the same year Latreille (1802b) recognized the same two groups as families, with certain supergeneric subdivisions as shown in Table 18-1. The idea of recognizing numerous genera was quickly established; Klug (1807b), only five years after Kirby's monograph, gave a summary listing 32 bee genera.

Lepeletier (1835, 1841) not only gave a later classification by Latreille that is similar to the above, but also presented his own classification. He separated the social Apinae *(Bombus, Melipona, Apis)* in a separate account (1835). Other bees (1841) were divided into two major groups, the solitary nesters and the parasites, the latter including some that we now know to be nonparasitic: *Ceratina, Allodape,* and *Hylaeus*. Each of the two major groups was divided into three families (Table 18-2). The recognition of parasitic bees as families separate from nonparasitic bees affected bee classifications for years, parasites like *Psithyrus* being placed in families different from those of closely related nonparasites, in this case

Table 18-1. Classification of Bees Based on Latreille (1802b).

Family Andrenetes (Andrenetae)
    Division I. Tongue blunt (*Colletes, Hylaeus*)
    Division II. Tongue pointed (*Andrena, Dasypoda*)
Family Apiares (Apiariae)
    Megachiles (*Megachile*)
    Nomades (*Epeolus, Melecta, Nomada*)
    Euceres (*Eucera*)
    Podaliries (*Centris, Podalirius = Anthophora*)
    Claviceres (*Clavicera = Ceratina*)
    Xilocopes (*Xylocopa*)
    Euglosses (*Euglossa*)
    Bourdons (*Bombus*)
    Apiares domestiques (*Apis*)

other *Bombus*. It was not until the time of Robertson (1904) that such parasitic forms were regularly placed with their nonparasitic relatives in bee classifications, although Thomson (1872) had also correctly placed them. A century later, Tkalcū (1972), as noted below, again postulated that in their phylogenetic history the parasitic bees never possessed a scopa but evolved from wasps, independently from other bees.

Some features of Lepeletier's classification that seem extraordinary to modern melittologists are the placement of *Hylaeus* and *Ceratina* among the parasitic bees, of *Rhathymus* in the same tribe with *Sphecodes*, of *Melitta* in the Xylocopites, and of pasitines and parasitic megachilids in the same tribe.

Schenck (1861, 1869), in accounts of the bees of Nassau, Germany, provided the classification summarized in Table 18-3, using the 1869 version with interpretation as necessary from that of 1861. All subfamily names (with family name endings by modern standards) were based on generic roots. The placement of *Ceratina* in the Anthophoridae removed it from the parasitic groups, where it does not belong, and *Melitta* was removed from association with *Xylocopa*. Segregation of the parasitic megachilids from other parasites was an important step. A noteworthy problem, which extended through many later classifications, concerned "short-tongued" bees that have a somewhat elongate glossa, e.g., *Dasypoda, Dufourea, Melitturga, Panurginus, Panurgus,* and *Systropha*. The classification contained enough taxa (subfamilies), but these genera were scattered in an almost random way; for example, the three of these genera placed in the Panurgidae fall in three families by current standards, but *Dufourea, Halictoides, Rophites,* and *Systropha*, scattered among three subfamilies by Schenck, belong in one.

Thomson (1872) made a classification that in various ways is more modern than those of earlier writers, although he called the main named divisions tribes. His classification is summarized in Table 18-4. Except for the association of *Epeolus, Nomada,* and the pasitines with

Table 18-2. Classification of Bees Based on Lepeletier (1835, 1841).

Solitary-nesting Bees
  Family Podilegides
    Tribe Eulmites (*Euglossa, Eulaema*)
    Tribe Anthophorites (*Anthophora, Eucera, Melitturga, Systropha*)
    Tribe Xylocopites (*Centris, Epicharis, Melitta, Xylocopa*)
  Family Gastrilegides (*Anthidium, Chelostoma, Lithurgus, Megachile*)
  Family Merilegides
    Tribe Andrenites (*Andrena, Halictus, Nomia*)
    Tribe Panurgites (*Dasypoda, Dufourea, Panurgus*)
    Tribe Colletides (*Colletes*)

Social Bees
  Family Apiarides
    Tribe Apiarites (*Apis*)
    Tribe Meliponites (*Melipona*)
  Family Bombides (*Bombus*)

Parasitic Bees
  Family Psithyrides (*Psithyrus*)
  Family Dimorphides
    Tribe Melectites (*Aglae, Ceratina, Epeolus, Melecta, Mesoplia, Nomada*)
    Tribe Phileremides (*Coelioxys, Dioxys,* pasitines, *Stelis*)
  Family Monomorphides
    Tribe Prosopites (*Hylaeus*)
    Tribe Rhathymites (*Rhathymus, Sphecodes*)

Table 18-3. Classification of Bees Based on Schenck (1861, 1869).

Contrary to current practice, Schenck used the -idae ending for subfamilies.

Subfamily Andrenidae (*Andrena, Colletes, Hylaeus* = *Halictus, Nomia*)
Subfamily Prosopidae (*Prosopis* = *Hylaeus*)
Subfamily Sphecodidae (= Rhathymidae in 1869) (*Sphecodes*)
Subfamily Panurgidae (*Dasypoda, Dufourea, Panurgus*)
Subfamily Rophitidae (*Halictoides, Rophites*) (In 1869 these were included in the Panurgidae.)
Subfamily Melittidae (*Macropis, Melitta, Panurginus*)
Subfamily Megachilidae (*Anthidium, Lithurgus, Megachile*)
Subfamily Anthophoridae (*Anthophora, Ceratina, Eucera, Melitturga, Systropha*)
Subfamily Xylocopidae (*Xylocopa*)
Subfamily Apidae (*Apis, Bombus*)
Subfamily Psithyridae (*Psithyrus*)
Subfamily Melectidae (*Epeolus, Melecta, Nomada,* pasitines)
  (In 1869 the pasitine bees were put in a separate subfamily, the Phileremidae.)
Subfamily Stelidae (*Coelioxys, Dioxys, Stelis*)

Table 18-4. Classification of Bees Based on Thomson (1872).

Solitary Bees
  Tribe Halictina [*Colletes, Halictus, Hylaeus, Rophites* (including *Dufourea* and *Halictoides*), *Sphecodes*]
  Tribe Andrenina (*Andrena, Panurgus*)
  Tribe Megachilina (*Anthidium, Coelioxys, Dioxys, Megachile, Stelis*)
  Tribe Megillina (*Ceratina, Cilissa* = *Melitta, Dasypoda, Eucera, Macropis, Megilla* = *Anthophora*)
  Tribe Nomadina (*Epeolus, Melecta, Nomada,* pasitines)

Social Bees
  Tribe Bombina (*Apathus* = *Psithyrus, Bombus*)
  Tribe Apina (*Apis*)

Table 18-5. Classification of Bees Based on Schmiedeknecht (1882).

I. Solitary bees
  A. Podilegidae (Scopulipedes, leg collectors)
    a. Femorilegidae (femur collectors)
      Andrenidae (*Andrena, Colletes, Halictus, Nomia*)
      Panurgidae (*Biareolina, Dasypoda, Panurgus, Rophites*)
      Xylocopidae (*Ceratina, Xylocopa*)
    b. Crurilegidae (tibia collectors)
      Melittidae (*Macropis, Melitta*)
      Anthophoridae (*Ancyla, Anthophora, Eucera, Systropha*)
  B. Gastrilegidae (Dasygastrae, belly collectors)
    Megachilidae (*Anthidium, Lithurgus, Megachile, Osmia*)
  C. Pseudoparasitae (nonparasitic bees without a scopa)
    Prosopidae = Hylaeidae
    Sphecodidae (of course this is now known to be parasitic)

II. Social bees
  Apidae (*Apis*)
  Bombidae (*Bombus*)

III. Parasitic bees
  Psithyridae (*Psithyrus*)
  Melectidae (*Melecta, Nomada,* pasitines)
  Stelidae (*Coelioxys, Dioxys, Stelis*)

Melecta, Thomson put the parasitic bees where they belong—*Sphecodes* in the halictids, the megachilid parasites in the Megachilidae, and *Psithyrus* with *Bombus.* For the first time, *Halictus* appeared in a major taxon different from that of *Andrena; Colletes* and *Hylaeus* were in the same tribe, along with *Rophites,* and the melittids were among what are frequently called the anthophorines, i.e., the noncorbiculate Apidae.

In spite of Thomson's finding that most parasitic taxa do not belong in their own separate families but can be associated with their nonparasitic relatives, Schmiedeknecht (1882), Friese (1895), and subsequent works as late as Schmiedeknecht (1930) reverted to a system similar to that of Lepeletier. They recognized three sections, the solitary nest-making bees, the social bees, and the parasitic bees. In the first section these authors provided considerable classificatory structure, as seen in Schmiedeknecht's 1882 version, which is summarized in Table 18-5. Schmiedeknecht's families agreed with Schenck's (1861) in uniting Panurgidae and Rophitidae under the former name. *Ceratina* was included in the Xylocopidae. Bombidae was recognized as separate from the Apidae.

PLATE 17

Fossil bees (photographs courtesy of Michael S. Engel).

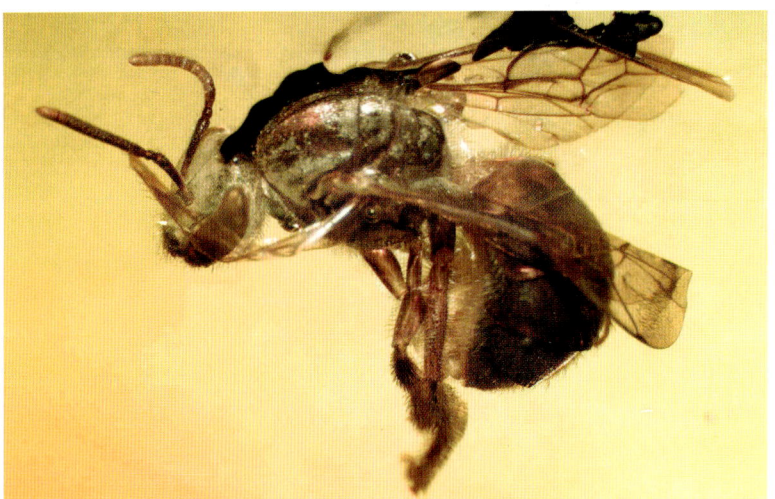

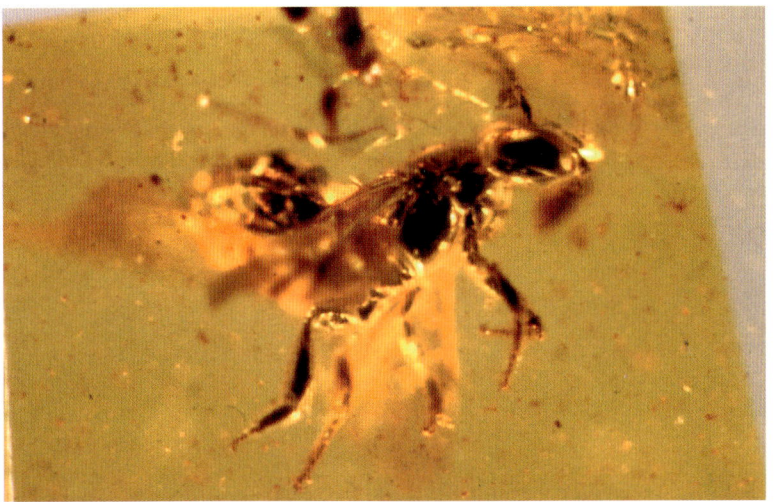

Above, *Halictus petrefactus* Engel & Peñalve [Halictidae, Halictini] from the early Miocene of Rubielos de Mora Basin, Spain. **Middle**, *Oligochlora eickworti* Engel [Halictidae, Augochlorini] in early Miocene amber from the Dominican Republic. **Below**, *Boreallodape mollyae* Engel [Apidae, Xylocopinae] in middle Eocene amber from the Baltic region.

PLATE 18

Fossil L-T bees (photographs courtesy of Michael S. Engel).

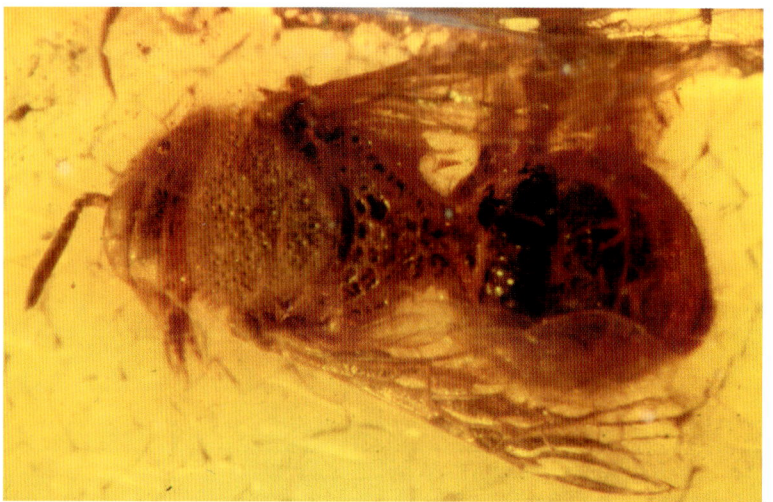

Above, *Ctenoplectrella grimaldii* Engel [Megachilidae, Osmiini] in middle Eocene Baltic amber. **Middle**, *Glyptapis mirabilis* Cockerell [Megachilidae, Osmiini] in middle Eocene Baltic amber. **Below**, *Boreallodape striebichi* Engel [Apidae, Xylocopininae] in middle Eocene Baltic amber.

PLATE 19

Fossil corbiculate Apinae (photographs courtesy of Michael S. Engel).

Above, *Protobombus indecisus* Cockerell [Apidae, Electrapini] in middle Eocene amber from the Baltic region. **Middle**, *Melissites trigona* Engel [Apidae, Melikertini] in middle Eocene Baltic amber. **Below**, *Succinapis goeleti* Engel [Apidae, Melikertini] in middle Eocene Baltic amber.

PLATE 20

Fossil corbiculate Apinae (photographs courtesy of Michael S. Engel).

Above, *Nogueirapis silacea* (Wille) [Apidae, Meliponini] in latest Oligocene amber from Chiapas, Mexico. **Middle**, *Cretotrigona prisca* (Michener & Grimaldi) [Apidae, Meliponini] in late Cretaceous amber from New Jersey. **Below**, *Apis henshawi* Cockerell [Apidae, Apini] a fossil honey bee, from the Oligocene of Germany.

Table 18-6. Classification of Bees Based on Ashmead (1899a).

Family Colletidae (*Colletes, Diphaglossa, Paracolletes*)
Family Prosopidae (*Euryglossa, Prosopis* = *Hylaeus*)
Family Andrenidae
    Subfamily Andreninae (*Ancyla, Andrena, Melitta, Nomia, Stenotritus*)
    Subfamily Halictinae (*Augochlora, Halictus, Systropha*)
    Subfamily Sphecodinae (*Sphecodes, Temnosoma*)
Family Panurgidae (*Dasypoda, Dufourea, Hylaeosoma* = *Chilicola, Macropis, Panurgus, Rophites*)
Family Megachilidae
    Subfamily Osmiinae (*Heriades, Osmia*)
    Subfamily Megachilinae (*Ctenoplectra, Lithurgus, Megachile*)
    Subfamily Anthidiinae (*Anthidium*)
Family Stelidae
    Subfamily Stelidinae (*Stelis*)
    Subfamily Coelioxinae (*Allodape, Chilicola, Coelioxys, Dioxys,* pasitines)
Family Nomadidae (*Aglae, Epeolus, Exaerete, Melecta, Nomada, Osiris*)
Family Anthophoridae (*Anthophora, Centris, Emphor* = *Ptilothrix, Eucera, Exomalopsis, Melitturga*)
Family Ceratinidae (*Ceratina*)
Family Xylocopidae (*Oxaea, Xylocopa*)
Family Euglossidae (*Euglossa*)
Family Bombidae (*Bombus*)
Family Psithyridae (*Psithyrus*)
Family Apidae
    Subfamily Meliponinae (*Melipona*)
    Subfamily Apinae (*Apis*)

Table 18-7. Classification of Bees Based on Robertson (1904).

Apygidialia
  Colletoidea
    Family Colletidae (*Colletes*)
    Family Prosopididae (*Prosopis* = *Hylaeus*)
  Trypetoidea
    Family Megachilidae
      Subfamily Osmiinae
        Tribe Osmiini (*Osmia*)
        Tribe Trypetini (*Trypetes* = *Heriades*)
      Subfamily Megachilinae
        Tribe Megachilini (*Megachile*)
        Tribe Coelioxyini (*Coelioxys*)
    Family Stelidae
      Subfamily Trachusinae (*Trachusa*)
      Subfamily Anthidiinae
        Tribe Stelidiini (*Stelis*)
        Tribe Anthidiini (*Anthidium*)
  Ceratinoidea
    Family Ceratinidae (*Ceratina*)
    Family Exoneuridae (*Allodape, Exoneura*)
    Family Xylocopidae (*Xylocopa*)
  Apoidea
    Family Apidae (*Apis, Bombus, Psithyrus*)
Pygidialia
  Andrenoidea
    Family Andrenidae (*Andrena*)
    Family Panurgidae
      Subfamily Panurginae (*Panurgus*)
      Subfamily Protandreninae (*Protandrena*)
    Family Halictidae (*Augochlora, Halictus, Sphecodes*)
    Family Nomiidae (*Paranomia* = *Nomia*)
    Family Dufoureidae (*Dufourea, Halictoides, Rophites*)
    Family Macropididae (*Macropis*)
  Anthophoroidea
    Family Anthophoridae (*Anthophora*)
    Family Euceridae (*Eucera*)
    Family Emphoridae (*Emphor* = *Ptilothrix, Melitoma*)
    Family Melectidae (*Melecta,* pasitines, and presumably *Nomada* and *Epeolus*)

Ashmead (1899a) greatly modified Schmiedeknecht's (1882) system and included genera from all parts of the world. His classification, summarized in Table 18-6, placed all parasitic bees in families of their own, except for the Sphecodinae; he presumably did not realize that his Sphecodinae consisted of parasitic forms. Anomalies, in view of our present knowledge, were the placement of parasitic euglossines in the Nomadidae, the placement of the melittids and dufoureines in the Panurgidae, and the positions of such genera as *Chilicola* (two places), *Oxaea, Ctenoplectra,* and *Allodape. Melitturga* was equally out of place; its position in the Anthophoridae, although traditional, was incorrect. I have ignored some of Ashmead's careless placements of genera that were little known to him.

Robertson (1904) thoughtfully developed a new classification for bees; his families were widely accepted by North American hymenopterists such as Viereck (1916) and by American textbook writers. Table 18-7 summarizes it, with some interpretation based on Robertson's 1903 papers. A noteworthy feature of Robertson's classification is recognition of the two large groups, Pygidialia and Apygidialia. As stated elsewhere, the pygidial plate has been lost repeatedly and independently. Robertson was not familiar with the numerous pygidialate colletids or the remnants of such plates in many of his Ceratinoidea and in the megachilid tribe Lithurgini. Like Thomson, but probably independently, Robertson placed the parasitic forms appropriately except for the association of *Nomada* and its relatives with *Melecta*. He was the first to properly recognize the limits of his Dufoureidae (= Rophitinae); Schenck had named the family but had somehow put *Dufourea* itself in the Panurgidae. Unfortunately, it is not clear where Robertson would have placed *Melitta* (he omitted it because it does not occur in his area), although he said that it would not be near *Macropis*.

Börner (1919) constructed a classification that, for S-T bees, foreshadowed some later classifications. The parasitic bees were appropriately placed, as shown by Grütte (1935), who followed Börner's system. Some taxa, such as *Macropis, Melitta,* and the Ceratinini, were misplaced, and the Nomiinae and Halictinae were mixed, as were the Osmiini and Megachilini. Börner did not mention the pasitine bees, but probably he intended them to be in

Table 18-8. Classification of Bees Based on Börner (1919).

Family Colletidae
    Subfamily Prosopinae (*Hylaeus*)
    Subfamily Colletinae (*Caupolicana, Colletes*)
Family Andrenidae
    Subfamily Andreninae (*Andrena*)
    Subfamily Panurginae (*Macropis, Melitta, Panurgus*)
Family Halictidae
    Subfamily Halictinae
        Tribe Nomiini (*Agapostemon, Augochlora, Nomia*)
        Tribe Halictini (*Halictus, Paragapostemon, Sphecodes*)
        Tribe Nomioidini (*Nomioides*)
    Subfamily Halictoidini (*Dufourea, Rophites*)
Family Megachilidae
    Subfamily Osmiinae (*Osmia, Stelis*)
    Subfamily Megachilinae (*Anthidium, Coelioxys, Megachile*)
Family Nomadidae
    Subfamily Ceratininae (*Allodape, Ceratina*)
    Subfamily Nomadinae (*Nomada*)
Family Apidae
    Subfamily Anthophorinae
        Tribe Eucerini (*Centris, Eucera, Exomalopsis, Melissodes, Tetrapedia*)
        Tribe Anthophorini (*Anthophora*)
        Tribe Xylocopini (*Xylocopa*)
    Subfamily Apinae
        Tribe Bombini (*Bombus, Euglossa, Psithyrus*)
        Tribe Apini (*Apis*)
        Tribe Meliponini (*Melipona*)

Table 18-9. Classification of Bees Based on Michener (1944).

Family Colletidae
    Subfamily Euryglossinae (*Euryglossa*)
    Subfamily Hylaeinae (*Hylaeus*)
    Subfamily Chilicolinae (*Chilicola, Xeromelissa*)
    Subfamily Colletinae
        Tribe Paracolletini (*Paracolletes*)
        Tribe Colletini (*Colletes*)
        Tribe Caupolicanini (*Caupolicana*)
    Subfamily Stenotritinae (*Stenotritus*)
    Subfamily Diphaglossinae (*Diphaglossa*)
Family Andrenidae
    Subfamily Andreninae (*Andrena*)
    Subfamily Panurginae
        Tribe Panurgini (*Panurgus, Protandrena*)
        Tribe Melitturgini (*Melitturga*)
    Subfamily Oxaeinae (*Oxaea*)
Family Halictidae
    Subfamily Dufoureinae (*Dufourea, Rophites, Systropha*)
    Subfamily Nomiinae (*Nomia*)
    Subfamily Halictinae (*Augochlora, Halictus, Sphecodes, Temnosoma*)
Family Melittidae
    Subfamily Melittinae (*Melitta*)
    Subfamily Macropidinae (*Macropis*)
    Subfamily Dasypodinae (*Dasypoda*)
    Subfamily Ctenoplectrinae (*Ctenoplectra*)
Family Megachilidae
    Subfamily Lithurginae (*Lithurgus*)
    Subfamily Megachilinae
        Tribe Megachilini (*Coelioxys, Heriades, Megachile, Osmia*)
        Tribe Anthidiini (*Anthidium, Dioxys, Stelis*)
Family Apidae
    Subfamily Fideliinae (*Fidelia*)
    Subfamily Anthophorinae
        Tribe Exomalopsini (*Exomalopsis*)
        Tribe Ancylini (*Ancyla*)
        Tribe Nomadini (*Nomada*)
        Tribe Epeolini (*Epeolus*)
        Tribe Osirini (*Osiris*)
        Tribe Protepeolini (*Protepeolus* = *Leiopodus*)
        Tribe Epeoloidini (*Epeoloides*)
        Seven tribes of pasitine bees
        Tribe Emphorini (*Melitoma, Ptilothrix*)
        Tribe Eucerini (*Eucera*)
        Tribe Anthophorini (*Anthophora*)
        Tribe Hemisiini (*Centris, Epicharis*)
        Tribe Melectini (*Melecta*)
        Tribe Rhathymini (*Rhathymus*)
        Tribe Ericrocini (*Ctenioschelus, Ericrocis, Mesoplia*)
    Subfamily Xylocopinae
        Tribe Ceratinini (*Allodape, Ceratina, Exoneura*)
        Tribe Xylocopini (*Xylocopa*)
    Subfamily Apinae
        Tribe Euglossini (*Aglae, Euglossa, Eulaema, Exaerete*)
        Tribe Bombini (*Bombus, Psithyrus*)
        Tribe Meliponini (*Melipona*)
        Tribe Apini (*Apis*)

the Nomadinae. Börner's classification is summarized in Table 18-8.

The older classifications of bees were based largely on various characters of mouthparts, wings, legs, and scopa. Bischoff (1934) was among the first to call attention to various little-used characters of the body, such as subantennal sutures and the episternal groove, as well as to the jugal lobe of the hind wing. Grütte (1935) made use of Bischoff's findings in a study of parasitic bees, and subsequent classifications, such as that of Michener (1944), utilized the same characters.

A classification for the bees of the world developed by Michener (1944) is summarized in Table 18-9. Some principal features of this classification were the placement of *Melitturga* in the Panurginae, of the Nomiinae and Dufoureinae in the Halictidae, of *Chilicola* in the Colletidae, and of *Oxaea* in the Andrenidae. Note also the enlarged Melittidae and the recognition of the Lithurginae as a distinct subfamily of the Megachilidae. The broad Apidae, including all L-T bees except the Megachilidae, was also novel. The enormous number of tribes in the Anthophorinae had not been anticipated by previous general bee classifications, although most of the tribes had been named before 1944.

Michener (1965b) summarized the bee classification and modified that of 1944 as follows: The tribe Caupolicanini was transferred to the Diphaglossinae. The subfamily Euherbstiinae was added to the Andrenidae.

Table 18-10. Classification of Bees Based on Warncke (1977a).

Family Andrenidae
    Subfamily Colletinae
        a. (*Colletes, Hylaeus*)
        b. (*Caupolicana*)
    Subfamily Andreninae
        a. (*Andrena*)
        b. (*Melitturga, Oxaea, Panurgus*)
    Subfamily Halictinae
        a. (*Rophites, Systropha*)
        b. (*Halictus, Nomia, Nomioides, Sphecodes*)
Family Apidae
    Subfamily Melittinae
        a. (*Dasypoda, Pararhophites*)
        b. (*Ctenoplectra, Macropis, Melitta*)
    Subfamily Megachilinae
        a. (*Lithurgus*)
        b. (*Anthidium, Stelis, Dioxys, Osmia, Coelioxys, Megachile*)
    Subfamily Ceratinae (sic)
        a. (*Exomalopsis, Fidelia*)
        b. (*Allodape, Ceratina*)
    Subfamily Anthophorinae
        a. (*Ancyla, Manuelia, Xylocopa*)
        b. (*Dasiapis* = *Diadasia, Eucera, Lanthamelissa* (sic), *Tapinotaspis, Tetrapedia*)
        c. (*Ancyloscelis, Anthophora, Caenonomada, Epeoloides, Melecta*)
    Subfamily Nomadinae
        a. (*Biastes, Epeolus*)
        b. (*Nomada*, most pasitines)
    Subfamily Apinae
        a. *Melipona*
        b. *Apis, Bombus*
        c. *Euglossa*

Fideliinae was raised to family rank. Likewise, Anthophorinae was raised to family rank with the following subfamilies: Exomalopsinae (for Exomalopsini and Ancylini), the Nomadinae (for tribes Nomadini through the pasitine bees in the 1944 classification), the Anthophorinae (for the Emphorini to Ericrocini), and the Xylocopinae (for Ceratinini and Xylocopini). Of course, the Apinae of 1944 were also raised to family rank, and two subfamilies were recognized, Bombinae for the tribes Euglossini and Bombini and Apinae for the tribes Meliponini and Apini. Name changes were Emphorini to Melitomini, Hemisiini to Centridini, and Ericrocini to Ctenioschelini.

Some of these changes resulted from the tradition of recognizing numerous families and the hesitation of others to accept a broad family Apidae. Recognition of families like Bombidae, Meliponidae, and Xylocopidae was widespread, and the 1965 classification was a compromise. I now consider that raising the Anthophorinae to family level, and the related changes, were mistakes.

A modified world classification can be extracted from Michener (1979a: 297-323). It incorporated the changes that appeared in 1965 except that the Euherbstiinae was incorporated into the Andreninae. Additionally, *Oxaea* and its relatives were placed in the Oxaeidae, Halictinae was divided into three tribes (Augochlorini, Halictini, Nomioidini), *Dioxys* was placed in a separate megachilid tribe, *Ancyla* was tentatively put in the Exomalopsini, additional tribes of Anthophoridae (Eucerinodini, Tetrapediini, Canephorulini, Pararhophitini, Isepeolini) were recognized, and the Meliponini was raised to subfamily rank.

In general, based on the classification above but with more categories are the classifications of Engel (2001b, 2005). Another version by Melo and Gonçalves (2005) is similar but with all the Apiformes included in a single family, Apidae, which therefore equals the Apiformes.

Table 16-1 shows the classification accepted for the present book.

Sustera (1958), reacting to many of the same findings that led to the development of my classification of 1979, had proposed a similar classification 21 years earlier. It placed the Ceratininae and Xylocopinae in a separate family, the Xylocopidae. Its most unusual feature was the placement of the Nomadinae (Nomadini and Ammobatini) in the Andrenidae while Epeolini, Epeoloidini, and the remaining pasitines were placed in a subfamily Biastinae in the family Anthophoridae. This division and placement of Nomadinae had been indicated earlier in a diagram by Pittioni and Schmidt (1942, pl. I).

Although the classifications by Michener outlined above have been widely used, other authors have proposed very different classifications. The samples discussed below demonstrate the persistent lack of agreement about bee classification and phylogeny. A principal reason for the preparation of the present work is to present cladistic patterns and classifications that best represent our current knowledge of these insects.

Table 18-10 summarizes a classification, limited largely to European genera, by Warncke (1977a); tribal names were not used, but the genera were nonetheless grouped into units (here lettered a, b, or c) falling below the subfamily level. Some unusual features of this classification include the placement of *Pararhophites* in the Melittinae, of the Melittinae in the Apidae (L-T bees), and of *Fidelia* and *Exomalopsis* close together and in the Ceratininae, the wide separation of *Ceratina* from *Manuelia* and *Xylocopa,* and the placement of *Ancyla* with the latter. The genera included by Warncke in each group of Anthophorinae are so diverse that I have listed them all rather than merely presenting a representative of each group.

Finally, to further demonstrate the lack of general agreement on bee evolution and classification, I note that Tkalců (1972, 1974a) proposed that all parasitic bees arose, not from pollen-collecting ancestors, but from nonpollen-collecting ancestors of the pollen-collecting groups. If true, this would greatly change accepted classifications, for it implies different nonpollen-collecting wasp ancestors for major groups of bees. I believe, on the contrary, that there is abundant morphological evidence for the relations of parasitic bees to different nonparasitic taxa; see Section 8.

The status of the corbiculate Apidae (Euglossini,

Bombini, Meliponini, and Apini) is of special historical interest, partly because honey bees and bumble bees are included, but also because the group has often been recognized, in the above classifications and by Michener (1990a), as a taxon of family rank. The four tribes consistently came out as a single clade in analyses by Roig-Alsina and Michener (1993), who included them as part of a large subfamily Apinae but called them the "apine clade" (the apine line of Silveira, 1993b). To recognize the corbiculate Apidae as a family would require the recognition of numerous other families, a procedure that does not seem helpful or appropriate. Yet because of their relationship to one another, as well as because of history, it is convenient to have a term for the four tribes considered collectively. The name "*corbiculate Apidae*" is appropriate. Additional characters showing their monophyletic relationship are enumerated in Section 102. The difficulty of giving them a technical name within our system is somewhat troubling.

19. Short-Tongued versus Long-Tongued Bees

At least since Kirby (1802), it has been the custom of specialists on bees, unlike other hymenopterists, to devote a great deal of attention to mouthparts. Continuing this practice, recent phylogenetic studies (Roig-Alsina and Michener, 1993; Alexander and Michener, 1995) have been heavily weighted by numerous characters of mouthparts. An old atlas of proboscides of bees, that of Saunders (1891), shows many of the characters discussed below. From at least as early as Kirby's monograph (1802) to the present, it has been common to recognize two groups, the short-tongued (S-T) bees and the long-tongued (L-T) bees. Some older classifications gave these two groups formal status, as indicated in Section 18 for the classifications of Kirby (1802) and Latreille (1802b). Later classifications mostly divided bees into more families, a series of S-T families followed by a series of L-T families. Thus the distinction between S-T and L-T families tended to be preserved.

S-T bees as a group are paraphyletic, having given rise to the L-T bees (see Sec. 20). Indeed, as shown below, the family Melittidae is probably the paraphyletic source of the L-T bees. But since S-T and L-T bees differ in many features, distinguishing these groups remains useful and they are differentiated below.

In L-T bees the first two segments of the labial palpi are ordinarily elongate, flattened and sheathlike (Fig. 19-1b), forming, along with the maxillary galeae, a tube or channel in which the glossa can move back and forth. In S-T bees these palpal segments are unmodified (Fig. 19-2b, d) or the first one is sometimes elongate, the first two being elongate only in a few genera, such as *Morawitzia* and *Rophites* s. str. (Rophitinae), *Protomeliturga* (Panurginae), and *Andrena (Callandrena) micheneriana* LaBerge (Andreninae). L-T bees also lack the galeal comb, or it is extremely reduced; commonly, however, they have a concavity and comb on the posterior distal margin of the maxillary stipes (Fig. 19-1a), although the comb is absent in most Megachilidae and is probably lost in most Nomadinae, and in some of those forms the concavity is also absent. S-T bees commonly have a comb (Fig. 19-5a) on the inner surface of the galea (not visible in outer view, Fig. 19-2a; but see Figs. 21-1, 21-2a). This comb is absent in some S-T groups, e.g., the Halictinae (Fig. 21-2b) except *Corynura* and its relatives. Except for *Eremaphanta* in the Melittidae, S-T bees lack a stipital comb and concavity. In L-T bees the glossa is nearly always elongate (Fig. 19-1b) with a deep, longitudinal groove on its posterior surface (Fig. 19-3b-d), the lips of the groove margined by small hairs and nearly meeting to enclose a channel (Fig. 19-3c), the glossal canal (Michener and Brooks, 1984). On the anterior surface of the glossal canal there is almost always a longitudinal thickening, the glossal rod, lateral to which are minute, simple seriate hairs directed mesodistally. Clearly, these are synapomorphies; in nearly all other bees the glossa has at most a broad, shallow concavity running along all or part of the posterior surface and margined by hairs similar to other glossal hairs. The seriate hairs are exposed, relatively large, often branched, and directed laterodistally (Fig. 19-4e, f). As shown by Michener and Brooks (1984), however, in *Melitturga* (Panurginae) the channel is almost as well formed as in an L-T bee.

Other characters that distinguish most or all L-T bees from most or all S-T bees are indicated by Michener and Greenberg (1980), Roig-Alsina and Michener (1993), and Alexander and Michener (1995). As noted by these authors as well as by Laroca, Michener, and Hofmeister (1989), the expressions L-T and S-T are not always appropriate, for there are L-T bees with short glossae (Fig. 8-1) and S-T bees with long glossae (Fig. 19-5c). Among

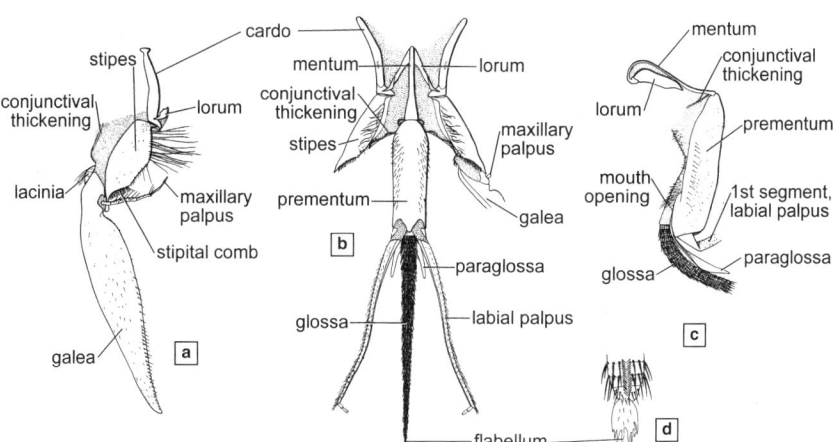

Figure 19-1. Proboscis of an L-T bee, *Anthophora edwardsii* Cresson. **a,** Outer view of maxilla; **b,** Posterior view of labium and basal parts of maxilla; **c,** Lateral view of labium, the distal parts of the glossa and labial palpus omitted; **d,** Flabellum. From Michener, 1944.

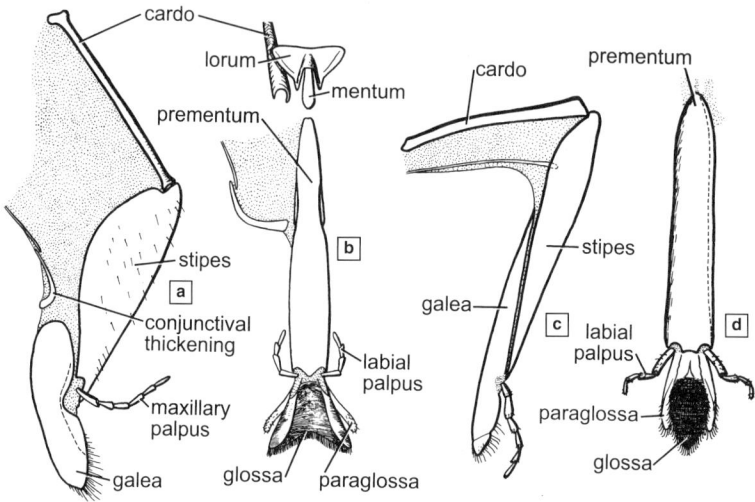

parasitic Allodapini there exist species obviously related to the nonparasitic L-T allodapines but with the basal segments of the labial palpi and the glossa relatively short (Fig. 8-1). This trend reaches its extreme in the South African parasitic genus *Eucondylops* (Michener, 1970). The parasitic allodapines are mostly not known to visit flowers; they must feed in the nests of their host bees, other allodapines. Thus they do not need equipment for extracting nectar from flowers and appear to have lost it. Likewise, as emphasized by Silveira (1993a), the genus *Ancyla* (Ancylini), which visits shallow-flowered plants such as the Apiaceae (Popov, 1949b), has no long flat seg-

Figure 19-2. Proboscides of S-T bees. **a, b,** *Colletes fulgidus* Swenk; **c, d,** *Halictus farinosus* Smith. (a and c are outer views of maxillae; b and d are posterior views of labia.) The not or weakly sclerotized basal connections of the prementum are not shown in d. From Michener, 1944.

ments of the labial palpi (Fig. 107-2), and yet it seems to be a relative of *Tarsalia,* an obvious L-T bee (see Silveira, 1993b). Warncke (1979c) separated *Ancyla* and *Tarsalia* only subgenerically. Finally, *Ctenoplectra,* often given familial status because of its combination of characteristics

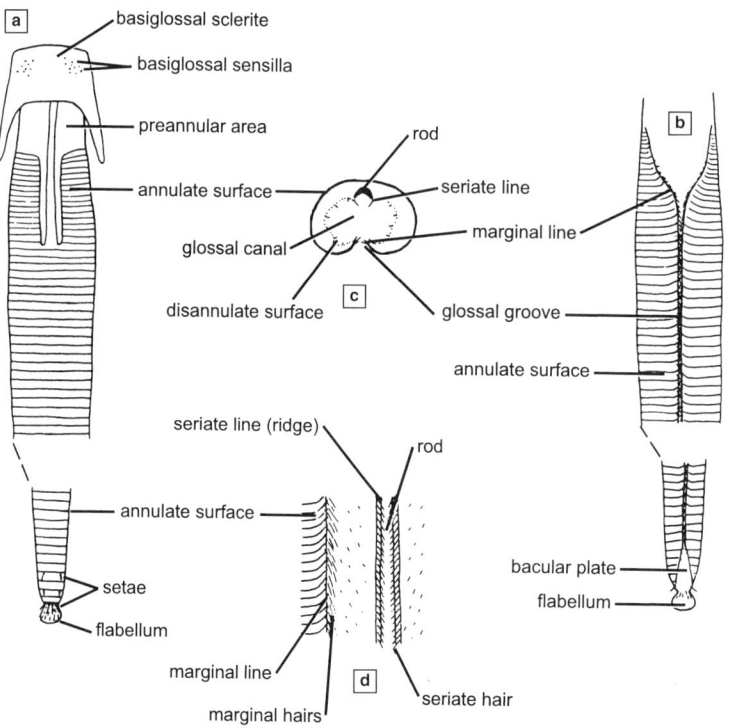

Figure 19-3. Diagrams of the glossa of an L-T bee, with structures labeled. **a, b,** Anterior and posterior surfaces; **c,** Cross section of same, the anterior surface above; **d,** Inner surface of a portion of the glossal canal and adjacent edge of the annulate surface, flattened out. From Michener and Brooks, 1984.

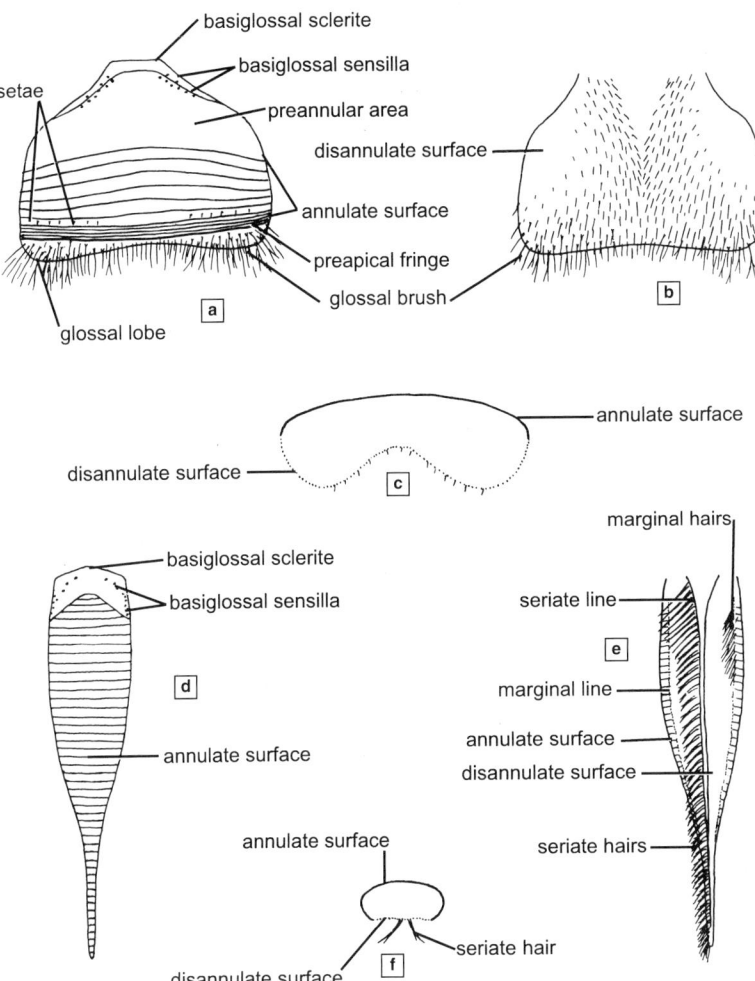

Figure 19-4. Diagrams of colletid and halictid glossae, with structures labeled. **a, b,** Anterior and posterior surfaces of the glossa of a colletid; **c,** Cross section of same, the anterior surface above; **d, e,** Anterior and posterior surfaces of the glossa of a halictid; **f,** Cross section of same, the anterior surface above. From Michener and Brooks, 1984.

of L-T bees with the labial palpi of S-T bees (Fig. 110-1c, d; Michener and Greenberg, 1980), is a member of the L-T bee clade, according to Roig-Alsina and Michener (1993); it probably lost the palpal characteristics of that clade.

Conversely, there are many S-T bees with long mouthparts. In some groups, such as the Halictinae and Xeromelissinae, the basal parts of the proboscis, i.e., the cardines, stipites, and prementum, are elongated (Fig. 19-2c), unlike those of L-T bees. In others, like some Panurginae, Nomiinae, Halictinae, and Rophitinae, the glossa is elongate (Fig. 19-5c), and in some ways, e.g., the development of a flabellum in some Panurginae (see Michener and Brooks, 1984), resembles that of L-T bees.

As described above, most S-T bees with long mouthparts have elongations of major structures (cardines, stipites, prementum, glossa). A few S-T bees, however, have unusual elongations of other structures that presumably help in imbibing nectar from deep or protected sources. Examples, all in the Colletidae, are (1) maxillary palpi that fit together to form a tube (a species of *Euhesma* in the Euryglossinae; Houston, 1983c), (2) labial palpi that form a tube (*Niltonia* in the Colletinae; Laroca, Michener, and Hofmeister, 1989), and (3) a pencil, formed by enormous galeal setae and the labial palpi, that probably draws nectar by capillary action (*Leioproctus* subgenus *Filiglossa,* in the Colletinae; Fig. 19-6).

In spite of the problems indicated in the preceding paragraphs, the division of all bees into two units, the S-T and L-T bees, is distinct; it is derived rather than ancestral species or genera in each unit that blur the distinction. Phylogenetic studies show that the L-T bees constitute a holophyletic unit. The S-T bees, however, are paraphyletic; they gave rise to the L-T bees and may have no common synapomorphies; their common characters mentioned above are plesiomorphies as judged by comparisons with sphecoid wasps, as are other characters enumerated in the papers cited, with one possible exception. The galeal comb, widespread among S-T bees, is believed to be not homologous to the galeal comb found in most sphecoid wasps because of its position on the galea (see characters 26 and 27, Alexander and Michener, 1995). It could therefore be a synapomorphy for S-T bees, although lost in some of them. More likely it is a synapo-

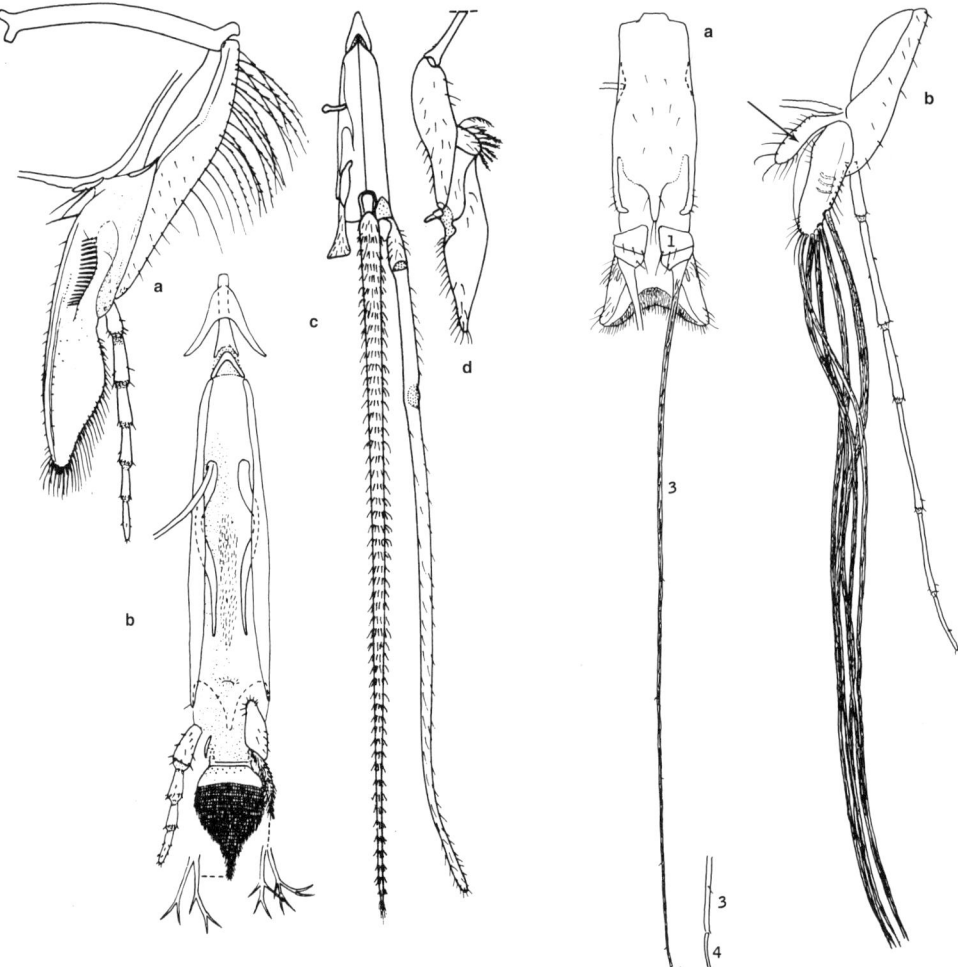

Figure 19-5. Proboscides of S-T bees. **a, b,** Inner surface of maxilla and anterior surface of labium (with enlargements of apical hairs of glossa and paraglossa) of *Macropis europaea* Warncke; **c, d,** Labium (with truncate paraglossa at left and labial palpus at right of glossa) and maxilla (base of cardo omitted) of *Neffapis longilingua* Ruz. *Neffapis* is a long-tongued S-T bee. From Michener, 1981a, and Rozen and Ruz, 1995.

Figure 19-6. Proboscis of *Leioproctus (Filiglossa) filamentosus* (Rayment). **a,** Labium, with mentum, lorum, and most of one palpus omitted; enlargement of apex of palpus below. **b,** Outer view of maxilla, with cardo omitted. (Labial palpal segments are numbered; an arrow indicates the unusually large lacinia.) Presumably, the pencil of filaments consisting of palpi and galeal setae draws nectar by capillarity.

morphy for bees as a whole, lost in some S-T bees and nearly all L-T bees, although indicated even in some L-T bees such as *Xeromelecta* (Melectini).

A few other character states of adults exhibit a similar distribution and, although variable, appear to be generally derived among S-T bees, while ancestral in L-T bees. Examples are the elevated basal area of the labrum; the short basistipital process; the large, diverging seriate hairs of the glossa (Fig. 19-4e), possibly lost in female and most male colletids; and the midtibial and midfemoral brushes or combs. These are all found in S-T bees and are seemingly apomorphic compared to sphecoid wasps and to the L-T families of bees, which lack these features.

A viewpoint that I have abandoned but which nonetheless requires discussion is that L-T bees were derived from the Panurginae or their antecedents. This possibility is suggested by certain panurgine characters similar to those of L-T bees. As indicated above, in certain panurgines the first segment of the labial palpus is long and (rarely) the first two segments are elongate, suggestive of L-T bees. Moreover, in those Panurginae with an elongate glossa the disannulate surface is usually somewhat invaginated (strongly so in *Melitturga;* see Michener and Brooks, 1984) and the apex has a well-formed flabellum.

This flabellum, especially in some species of *Perdita* (Fig. 28-2h, i) is incredibly similar to that of many L-T bees, the resemblance including not only its shape but its anterior curvature, the row of setae across its anterior surface, the hairs elsewhere on it, etc. (Michener and Brooks, 1984). It seems most unlikely that such an elaborate structure could have evolved independently in these panurgines and in L-T bees. In many doubtful cases elsewhere it seems hopeless to suppose that we will ever decide which similar characters of two organisms are homologous and which result from convergence. In this case, however, the answer is clear. Among S-T bees only certain panurgines have such characters, which are clearly apomorphic, suggesting L-T bees. If L-T bees arose from panurgines, then the L-T bees are a sister group to the rather long-tongued panurgines—*Calliopsis, Melitturga, Perdita,* etc. Yet panurgines have striking synapomorphies not shared by L-T bees. For example, they lack or nearly lack the male gonobase and they have two subantennal sutures under each antenna. To regain the gonobase after losing it seems most improbable. Furthermore, the relationship of L-T bees to the short-tongued Melittidae seems well established. The common characters of panurgines and L-T bees, then, appear to be a remarkable case of convergence (see Secs. 27, 28).

Michener's (1944: 231) reasons for deriving the melittid/L-T bees from the Panurginae or their immediate ancestors are also not tenable. The tapering mentum characteristic of L-T bees and panurgine bees, on which I based this conclusion, is also found in some Colletinae and even in some Andreninae (Michener, 1985a) and therefore does not point to a panurgine origin.

In a study of bee larvae, Michener (1953a) observed that many characters seemed to be derived in the S-T bees and ancestral (like those of more ancestral Hymenoptera) in the L-T bees. This is the reverse of the relationship that would be anticipated if derived larval characters were associated with the long tongue and other derived adult characters. For example, a rather broad mandibular apex, often with two large teeth; the presence of body setae and hair-like spicules; and well-developed, more or less cylindrical antennae are plesiomorphic for larval bees as judged by their presence in other Hymenoptera. These are widespread features in L-T bees. The corresponding apomorphies, slender mandibular apices usually lacking large teeth, the near absence of body setae, and the weak antennal papillae are features of S-T bees (and certain derived L-T bees like *Apis*). If the phylogeny with Colletidae near the root (Fig. 20-1a, b) is correct, then the L-T bee seemingly ancestral larval attributes must have reappeared in the melitted-L-T bee clade or they were retained from ancestral aculeates but independently lost in S-T bees. However, as indicated in subsequent sections, it seems likely that the Colletidae is not near the root of the bee phylogeny. In this case the larval features listed above for S-T bees may have arisen early in the evolution of an S-T clade (see Fig. 20-1d), whereas the melittid-L-T clade retained ancestral features. It must be remembered, however, that within both L-T and S-T bees, the larval characters mentioned above vary; the conclusions suggested are therefore not very strongly supported by larval characters.

20. Family-Level Phylogeny and the Proto-Bee

Phylogeny, presumably, was at least in the back of the minds of the proponents of some of the classifications (and their modifications) summarized in Section 18. Recent studies, directed toward understanding the relationships of major groups like families, subfamilies, and tribes, provide a more overt and detailed look at bee phylogeny than was available in the past. These studies (Roig-Alsina and Michener, 1993; Alexander and Michener, 1995, and Danforth et al., 2006), however, leave various aspects of bee phylogeny still in doubt. Aspects of the phylogeny among major groups of bees, mostly families, will be considered here; phylogeny within families will be considered when possible in the systematic treatment later in this book. The emphasis here is on which is the basal branch of bee phylogeny, i.e., what is the sister group to all the rest of the bees? Of course this involves consideration of what was the protobee, i.e., the most recent common ancestor of all the bees (see Michener, 2000; Radchenko and Pesenko, 1994a,b)?

There are two principal hypotheses in answer to this question. The traditional view is that the Colletidae is the sister to other bees, principally because all female and most male colletids have the glossa bilobed or truncate, resembling in shape that of most other Hymenoptera, e.g., the Crabronidae, the sister group to the bees (Sec. 14). The alternative view is that the truncate or emarginate glossa of female colletids is a derived feature that functions to paint the characteristic colletid cell lining onto the brood cells. Thus the ancestral bee had a pointed glossa, and the Melittidae + L-T bees is the sister group to a clade that includes all other S-T bees (Fig. 20-1d). Although I prefer the second alternative, both are elaborated below.

The phylogenetic studies of Roig-Alsina and Michener (1993) and Alexander and Michener (1995) were based primarily on adult characters of 124 species of bees, one or more representatives of nearly every tribe, including representatives of virtually all genera whose familial affinities were in doubt. The methods were parsimony analyses of various kinds, as explained in the papers cited above; the S-T and L-T bees were analyzed separately. Each analysis resulted in several to many equally parsimonious trees, a common result when many of the characters used are convergent. Outgroups for the study of L-

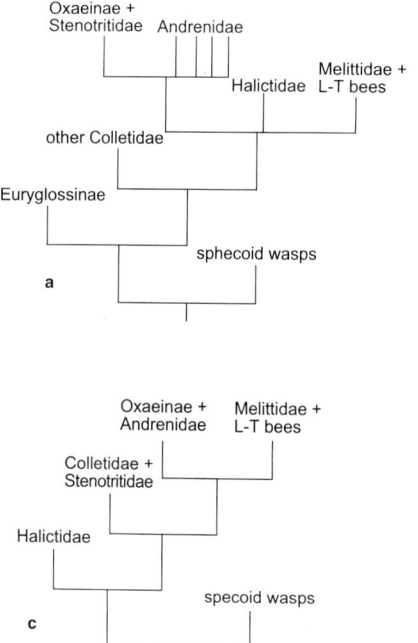

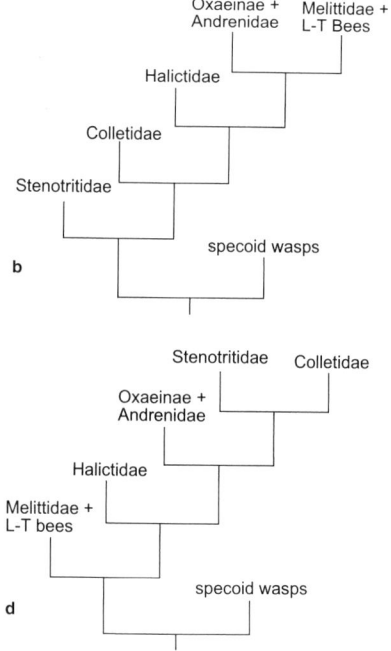

Figure 20-1. Representative alternative cladograms showing relations among families of S-T bees, with certain subfamilies of doubtful position also shown. The Euryglossinae are within the Colletidae except in cladogram a. The sphecoid wasps (Spheciformes) are shown here in a simplified way; as indicated in Sections 14 and 15, they are a paraphyletic group from which the bees arose. **a**, Tree for families, modified from figure 5 of Alexander and Michener (1995), which was a strict consensus tree based on 109 characters weighted equally for 57 bee taxa and 8 spheciform taxa, derived from minimum-length trees found by Goloboff's NONA; **b**, Same, but from figure 6 of Alexander and Michener (1995), using Goloboff's implied weights of characters; **c**, Same, from figure 10 of Alexander and Michener (1995), based on 114 characters of equal weight in an island of 226 trees; **d**, Same, from figure 12 of Alexander and Michener (1995), based on 114 characters of equal weight in an island of 336 trees.

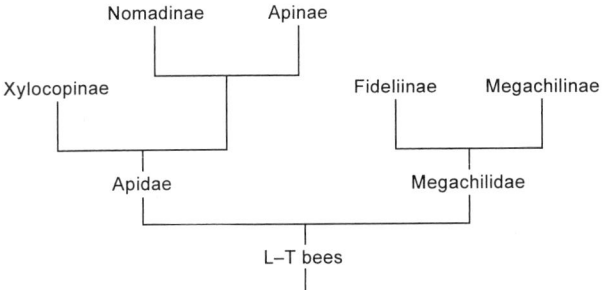

Figure 20-2. Cladogram showing relations among families and subfamilies of L-T bees, modified from figure 2b of Roig-Alsina and Michener (1993).

T bees were six melittid genera; for S-T bees, eight sphecoid wasps. The trees showed relationships among exemplars, i.e., the 124 species studied, but for practical purposes the relationships were considered to be among genera. For the most part these results were summarized to show relationships among subfamilies or families, since those taxa were in general supported by the phylogenetic analyses.

Figure 20-1 illustrates some of the results for S-T bees. The L-T bees were consistently found to have been derived from among the S-T melittid bees, verifying the conclusions of Michener and Greenberg (1980). Our 1995 study included only six melittids, however, and the relations of L-T bees to the different melittids are inconsistent; a new study with a larger sample of melittids should be made to determine, if possible, more about the origin of L-T bees from among the melittids. A problem with the Melittidae as the term is used in this book is that it is almost certainly paraphyletic. For this reason Alexander and Michener (1995) and Danforth et al. (2006) recognized three families, the Dasypodaidae, the Meganomiidae, and the Melittidae s. str. For this book, I have retained Melittidae in its usual sense because a phylogenetic analysis using more genera and more characters is needed. A phylogeny of the whole family Melittidae s. l. was given by Michener (1981a), but it did not clearly show what groups should have the family level. I think it is best to retain the Melittidae s. l. as a known paraphyletic family until it is clear how it should be divided; for example, are there three (or some other number) of groups to be regarded as families? The best supported hypothesis is that of Danforth et al. (2006); see below and Section 68.

Roig-Alsina and Michener (1993) determined that the L-T bees should be divided into only two families, Megachilidae and Apidae, as shown in Figure 20-2, but it would be easy to justify division of the Megachilidae into Fideliidae and Megachilidae s. str., particularly if the Pararhophitini and Fideliini are clearly sister groups. Our analyses indicated uncertainty on this point.

Returning now to the S-T bees, there are major differences among the cladograms of Alexander and Michener, as indicated in the sample shown in Figure 20-1. Note that the first or basal branch of bee phylogeny is depicted, in different trees, as being every S-T family except the Andrenidae! In the account below, factors supporting each of the principal hypotheses about the basal group of bees are considered.

The proto-bee was colletid-like: As noted above, phylogenetic trees for bees have ordinarily been envisioned as having the Colletidae at the base. The traditional reason for this view is the truncate or emarginate glossa of members of this family (Figs. 10-4c, 19-4a, 20-3a), which is shaped like that of the sphecoid wasps and other Hymenoptera, the assumption being that the pointed glossa of other bees is a derived character. Michener (1944), after listing 36 characters that he regarded as ancestral for bees, and thus characteristic of the hypothetical proto-bee, found that bees of the tribe Paracolletini possessed all of them. With some exceptions these characters were possible or probable plesiomorphies, shared with sphecoid wasps. Alternatively, as noted above, the suggestion has recently been made (Michener, 1981a: 17) that the Melittidae rather than the Colletidae might be near the root of the phylogenetic tree of bees. This possibility will be explored below, after further consideration of colletid characters.

The colletids are united by diverse plesiomorphies mostly shared with various other families of S-T bees. For example, except for the tribes Diphaglossini and Dissoglottini and the genus *Hesperocolletes* in the Colletinae, colletids have a well-formed episternal groove extending

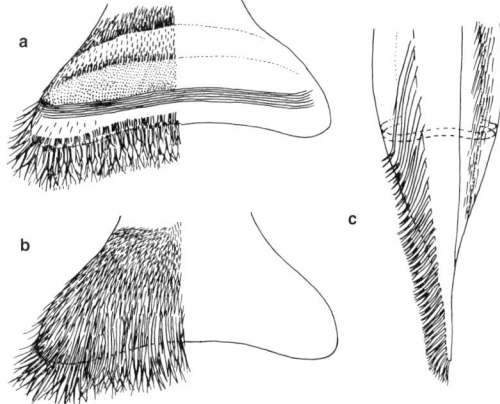

Figure 20-3. Diagrams of glossae of *Palaeorhiza parallela* (Cockerell). **a, b,** Anterior and posterior views of female; **c,** Posterior view of male, showing seriate hairs and, by broken lines, the course of one of the annuli (rows of hairs) on the anterior surface of the glossa. (The terminologies for these two glossal types are shown in Figure 19-4a, e.) From Michener and Brooks, 1984.

well below the scrobal groove (Fig. 20-5b). Moreover, the volsellae are free, with distinct digitus and cuspis (Fig. 10-15a). These are among the ancestral characters listed in 1944, as noted above. Another such character is that colletids (only five genera examined) are the only bees known to have seven pairs of ostia in the metasomal part of the dorsal vessel of males, six in females (Wille, 1958). Other bees have reduced numbers of ostia.

A supposed plesiomorphy of some colletids, the lack of scopal hairs for carrying pollen, has been considered evidence of a near-root position for the Colletidae. Jander (1976) suggested that internal (crop) pollen-carrying by Hylaeinae (a colletid subfamily) is ancestral relative to external (scopal) transport by other bees. He observed that in *Hylaeus* pollen-grooming movements that get pollen to the mouth only take pollen from the head and forelegs. Pollen on other parts of the body is groomed off and lost. I have verified these observations with another species of *Hylaeus*. Jander considered evolution from less to more efficient gathering, so that pollen landing on all parts of the body can be utilized, to be more probable than the reverse. The possession of plesiomorphic pollen transport by the colletid subfamilies Hylaeinae and Euryglossinae, but of apomorphic scopal transport in other bees, would support the near-root position for Colletidae.

As Jander noted, however, the slender, nearly hairless *Hylaeus* body is similar to that of *Ceratina* (see Pl. 1) and other small Xylocopinae that also nest in stems. In these xylocopines the scopa, although present, is reduced, and they presumably carry much of their pollen in the crop. Since these bees were apparently derived from hairy, fully scopate ancestors, I believe that the slender body, hairlessness, lack of scopa, and resultant loss of the ability to use pollen landing on the thorax and metasoma may be derived features of Hylaeinae associated with nesting in narrow burrows. I believe that the Hylaeinae, like the Ceratinini, arose from hairy bees that had a scopa. If this is true, Jander's observations do not indicate a near-root position for the Hylaeinae or the Colletidae, but rather a derived scopal loss.

Female and most male colletids have certain glossal features that are not found in other bees but that also occur in sphecoid wasps. Thus the glossal shape, the broad disannulate surface of the glossa, and the lack of differentiated seriate hairs (Fig. 20-3b) are as in sphecoid wasps and many other Hymenoptera, and the classical view is that these characters are plesiomorphic for bees, suggesting that Colletidae is the sister group to all other bees. Certain other glossal characters, the glossal lobes and brush and the preapical fringe (Fig. 19-4a) (the latter absent in some Euryglossinae), are found only in the Colletidae and are presumably apomorphic for this family. Along with the form of S7 of the male, these colletid synapomorphies indicate that Colletidae is a monophyletic group.

Extrinsic information that has been used to strengthen the classical hypothesis (Colletidae as the sister group to all other bees) is biogeographical: A major colletid clade, the Paracolletini, is greatly developed in Australia and temperate parts of South America, with the perhaps related Scraptrini in Southern Africa. Such a disjunct southern distribution is at least suggestive of antiquity.

The Euryglossinae is endemic to Australia and the Hylaeinae is most diverse there.

The protobee was melittid-like: Perkins (1912) and McGinley (1980), as explained by Michener (1981b, 1992c), Michener and Brooks (1984), Radchenko and Pesenko (1994a, b), and Alexander and Michener (1995), cast doubt on the classical hypothesis that an obtuse glossa is ancestral among bees, and suggested that the acute glossa of certain male Australian and New Guinea colletids (*Hemirhiza, Meroglossa,* and *Palaeorhiza*) may be ancestral, with the obtuse or bilobed glossa being a "special development" in the words of Perkins. According to this idea, here called the Perkins-McGinley hypothesis, the most ancestral bees or the wasps from which they evolved must have had a short, acute glossa like that of most S-T bees. It is not outlandish to believe that a crabronid-like wasp could have had such a glossa, for the crabronid genus *Pseudoscolia* today has exactly that, a remarkably beelike glossa (Michener, 2005), although its other structures are those of a crabronid wasp.

According to this hypothesis, the male glossa has no known special function different from that of females, but female colletids evolved a broad, obtuse glossa which serves to paint their distinctive cellophane-like lining material onto the cell and sometimes the burrow walls. Males initially would have retained an acute glossa, and in three genera they still do (Fig. 20-3c), but perhaps because the acute glossa had no special advantages for males and required maintenance of separate genetic machinery, it disappeared in most male colletids. The truncate or bilobed glossa and associated characters of all female and most male colletids would then be synapomorphies among bees, and reversions toward the ancestral sphecoid glossal shape.

But it is not legitimate to assume reversion in glossal

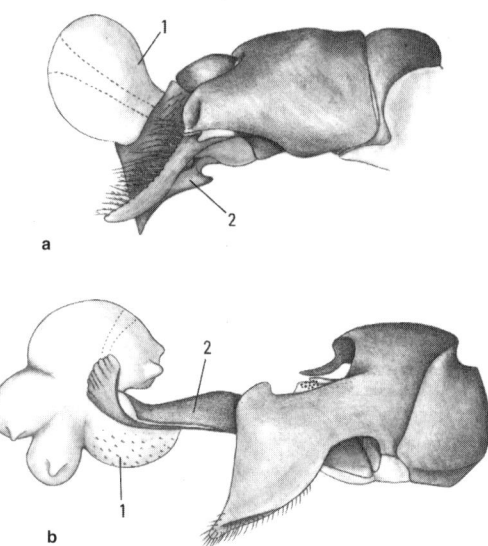

Figure 20-4. Lateral views of male genital capsule with endophallus everted. **a,** *Ptilothrix bombiformis* (Cresson); **b,** *Perdita albipennis* Cresson. (1, endophallus, 2, penis valve.)
From Roig-Alsina, 1993.

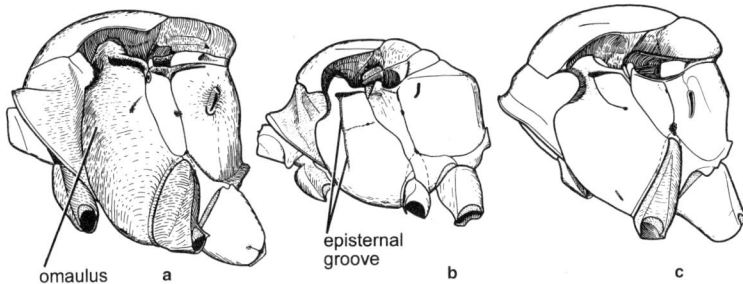

Figure 20-5. Lateral views of thoraces. **a,** *Anthidium atripes* Cresson; **b,** *Halictus farinosus* Smith; **c,** *Bombus* sp. (The upper internal extension of the middle coxa, when present, is indicated by broken lines.) From Michener, 1944.

shape to the sphecoid glossal form from an acute antecedent unless one also accepts the reversion in the associated features. I emphasize the gross difference in nearly every attribute between male and female glossae (Fig. 20-3) in *Hemirhiza, Meroglossa,* and *Palaeorhiza.* The male structures can easily be compared and the parts homologized in detail with those of most bees with a short, acute glossa like that of *Andrena,* whereas the female glossa is like that of other colletids (see Michener and Brooks, 1984). Other Hylaeinae in which males have somewhat intermediate glossae (*Amphylaeus* and *Hylaeus* subgenus *Hylaeorhiza*; Michener, 1992c and Fig. 46-1) do not help in locating the root of the phylogenetic tree, since they could have retained a glossa of an intermediate type irrespective of whether the evolutionary direction was from acute to obtuse or the reverse.

A character that was not used in the phylogenetic analyses but nonetheless seems to be of interest at the family level is the eversible endophallus of the male genitalia (Fig. 20-4). Unlike sphecoid wasps and other Hymenoptera, most bees have such a structure (Roig-Alsina, 1993). On the basis of 122 species of bees studied, it is absent in Colletidae and in the Oxaeinae and Andreninae, but is present in all other bees, including the Stenotritidae, Alocandreninae, and Panurginae. If its absence is plesiomorphic and derived from the wasps, its distribution supports the basal position of Colletidae, but as with the glossal characters, the endophallus might have arisen in the proto-bee and been lost in Colletidae and some Andreninae.

Another character worth special attention that was used in the analyses by Alexander and Michener (1995) is the middle coxa. In sphecoid wasps and melittid and L-T bees the coxa is fully exposed (Fig. 20-5a,c). In other bees, i.e., S-T bees except melittids, the coxa is hemicryptic, meaning that it is elongate with its upper end hidden under the pleura (Fig. 20-5b) (Michener, 1981b). No doubt the latter is a derived condition, a synapomorphy for S-T bees except melittids, and therefore a strong support for the phylogeny shown in Figure 20-1d, in which the melittid/L-T clade is the sister group to all the rest.

The form of S7 of the male is relevant to this discussion. In various families of bees, unlike sphecoid wasps, it has apical processes or lobes, often elaborate in shape and vestiture. It attains a sort of culmination of such features in most Colletidae, in which the disc is greatly reduced, and the one to three pairs of apical processes are often complex and large (Fig. 13-2). This not only is a synapomorphy for Colletidae but suggests the origin of Colletidae from among the other families of S-T bees that have less extreme modification of S7. Sphecoid wasps have nothing of this sort.

Further discussion: Returning now to family-level cladograms, Figures 20-1a and 1b suggest the classical hypothesis of glossal evolution, the glossal characters being polarized on the basis of a sphecoid wasp outgroup. They show the Colletidae at or near the base of the tree. Figure 20-1d, however, supports the Perkins-McGinley hypothesis of glossal evolution. For purposes of coding glossal characters, sexual dimorphism was considered an intermediate state between an ancestral state with a pointed glossa (and associated characters described above) in both sexes and a derived state in which both sexes have a broadly truncate or bilobed glossa. In Figure 20-1c Halictidae is the basal branch, sister to all other bees, an idea for which I see little support.

Extrinsic support for the antiquity of the clade S-T bees except Melittidae comes from the widely disjunct distribution of certain components. They are the *Hesperapis-Eremaphanta* clade of Dasypodainae, found in the western USA, southern Africa, and Central Asia (Michener, 1981a); the Fideliinae, found in Chile, South Africa, Morocco, and Central and southwestern Asia; and the Meliponini, found in tropical regions of the world. The Meliponini show little vagility; for example, among Recent meliponine taxa none or almost none reached the Antilles without human aid in spite of their abundance and diversity on the Caribbean continental margins. Their widely disjunct distribution, therefore, must indicate a long history.

The fossil record also suggests that the S-T families Colletidae, Andrenidae, and Halictidae may have arisen later than the melittid and L-T families; see Section 23 and Michener, 1979a; Zeuner and Manning, 1976; Michener and Grimaldi, 1988a, b; and Michener and Poinar, 1996. If this is true, it supports the topology of Figure 20-1d. The main points are these (they may well be significant, although negative paleontological data [absence of fossils] are always questionable): The oldest known fossil bee is an L-T bee (Meliponini), from the Cretaceous. Late Eocene fossils from the Baltic Amber, totaling about 36 species of bees, include diverse L-T taxa, various probable Melittidae, one halictid, and no andrenids, no colletids. Later, Oligomiocene bees from the Dominican amber include not only L-T bees, but also colletid, andrenid (panurgine), and numerous halictid taxa. For more details, see Section 23.

Analysis of glossal characters coded in the light of the

Perkins-McGinley hypothesis resulted in various cladograms (Alexander and Michener, 1995), of which Figure 20-1d summarizes one that showed the melittid/L-T clade as the sister group to all other bees, and colletids as a more recent clade. In no case, however, did *Meroglossa*, the only colletid in the cladistic study with a pointed glossa in the male, appear in a basal position within or outside of the Colletidae. It always appeared within the Hylaeinae, which was not a basal group of Colletidae in any of the analyses. (This position of the Hylaeinae is consistent with biogeographical information; see Sec. 23.) Thus the pointed glossa was shown as an apomorphy appearing within the Colletidae, convergent with the pointed glossa of noncolletid bees, and contrary to the Perkins-McGinley hypothesis. I believe one still must say that we do not know whether the colletid glossal shape is a plesiomorphy derived from sphecoid wasps or a synapomorphy of female and most male colletids, although to me the latter seems more likely.

I find it probable, as did Radchenko and Pesenko (1994a, b), that the proto-bee was not similar to the Hylaeinae, the colletid subfamily containing three genera with a pointed glossa in males. Assuming that the pointed glossa is ancestral for bees, it evolved into a colletid glossa once in females; males retained a pointed glossa that was converted to a female-type glossa on several occasions but is retained in three hylaeine genera (Michener, 1992c). The alternative, that the pointed glossa is derived, would require that it appear independently in some male hylaeinae and in an ancestor to Andrenidae, etc. It is a complex structure, as indicated above, far more so than is shown by shape alone, and unlikely to have evolved twice, but it clearly did evolve independently in *Pseudoscolia*.

Given, then, that the proto-bee was not *Hylaeus*-like, I agree with Radchenko and Pesenko (1994a, b) that it was a hairy bee that carried pollen externally, probably in a scopa. To judge by the behavior of andrenids, halictids, melittids, and many colletids, it stored doughlike rather than liquid provisions as larval food, made nests consisting of branching burrows in the soil (see Sec. 7), each leading to a horizontal cell (vertical cells often contain liquid provisions), tamped the cell walls with the pygidial plate, metamorphosed in cocoons made by mature larvae, and perhaps did not line its cells with secreted material (since cocoon-spinning forms sometimes do not secrete cell linings). These features could be correct whether its glossa was colletid-like or acute, i.e., whether it was colletid-like or melittid-like.

A reasonable speculation is that the proto-bee carried pollen among the hairs on its general surface, as *Pararhophites* seems to do, although it does have a small hind tibial scopa. Some bees evolved a sternal scopa and became megachilids. They may not have had ancestors with leg scopae. Others evolved a scopa on the hind leg, the scopa largely tibial in the Melittidae, Apidae, Stenotritidae, and Panurginae; in other taxa the scopae were not only tibial, but also femoral and trochanteral. In this last group, some bees added to their pollen-carrying capacity with scopal hairs on the metasomal sterna and on the sides of the terga, as in *Dieunomia, Systropha,* and *Homalictus*. Others added hairs and even a corbicula on the side of the propodeum, as in *Andrena*. Perhaps the scopa was reduced in Xeromelissinae and lost in Hylaeinae and Euryglossinae. Consideration of individual characters and how they might be interpreted in the study of the evolutionary history of bees is useful and informative. However, the best way to construct reliable phylogenetic patterns is through analyses involving simultaneous use of many characters. Figure 20-1 illustrates results of such analyses using morphological characters. Phylogenetic trees are hypotheses, not factual, and new studies often result in changes. No doubt the most reliable such tree for family groups of bees is that of Danforth et al. (2006), upon which Figure 20-6 is based. The data used were morphological (109 characters following Roig-Alsina and Michener, 1993, and Alexander and Michener, 1995) and molecular, using data from five genes (4,229 nucleotide sites). Analyses of both the DNA data and DNA + morphology, using both parsimony and Bayesian methods, gave well-supported similar trees (Fig. 20-6) that support the paraphyly of the Melittidae s. l., the basal position of this group relative to other bees, and the clade consisting of the other S-T families.

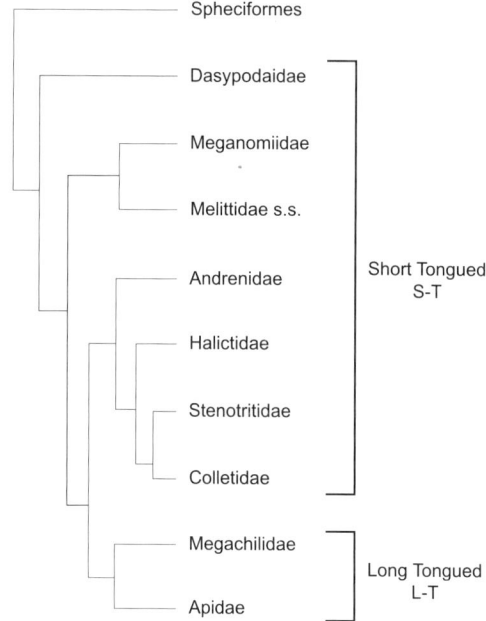

Figure 20-6. Relationships among families of bees. Data used were from 58 genera of bees (1 to 6 species each) and, for outgroups, 16 genera of spheciform wasps. Parsimony and Bayesian analyses of data from 109 morphological characters and five genes (4,299 nucleotide sites) gave similar topological results.

21. The Higher Classification of Bees

The following paragraphs serve to identify those groups recognized as families, as well as the relationships among the families. They concern only extant families; the extinct groups that have been given family-group status are ignored. They add detail to the information already provided by the dendrograms (Figs. 20-1, 20-2) and Table 16-1. They are not intended as descriptions of the families, but rather as indications of some of the reasons for decisions about the family-level classification. Keys to the families of bees will be found in Sections 33 to 35.

Stenotritidae. The two genera of this family, both Australian, are very similar to one another in most characters, although the male S7 is similar to that of most colletids in *Stenotritus*, much more simple in *Ctenocolletes*. In the phylogenetic studies by Alexander and Michener (1995), *Ctenocolletes* appeared in such diverse locations (see Fig. 20-1) that its phylogenetic position was uncertain. Stenotritidae therefore was given family status. In both cladistic and nearest-neighbor (phenetic) analyses of larval structures by McGinley (1981), stenotritrids fell among the Paracolletini in the Colletinae. These findings support the placement shown in Figure 20-1c and may indicate that stenotritids should be included in the Colletidae. In the stenotritids, however, the glossa is rounded and lacks the familial synapomorphies listed for Colletidae in Section 20 and below. Also in stenotritids, the scopa is on the hind tibia, not developed on the trochanter and femer as in scopate Colletidae.

Colletidae. There is so much diversity within the Colletidae that the subfamilies could easily be, and at various times have been, considered as separate families. No strong claim can be made for uniting them on the basis of general phenetic similarity; compare, for example, *Hylaeus* and *Caupolicana,* or even *Hylaeus* and *Colletes.* There are, however, significant familial synapomorphies. The glossa of most colletids is recognizable by synapomorphies at least of females, that are not found in any other groups of Hymenoptera, i.e., the glossal lobes and brush and the preapical fringe. Moreover, if the Perkins-McGinley hypothesis (see Sec. 20) is correct, there are other colletid glossal synapomorphies, e.g., shape, and there is at least one synapomorphy other than the glossal characters that unites all colletid subfamilies. This is the very reduced disc of S7 of the male, with long apodemes extending basolaterally and one to three pairs of usually hairy and elaborate apical processes or lobes (Fig. 13-2). Nothing of the sort is found in sphecoid wasps. In some other bees, such as some Stenotritidae, Melittidae, some panurgine Andrenidae, rophitine Halictidae, and even some Apinae, S7 of males has apical lobes or processes, but it is rarely so modified as those in Colletidae. A few colletids have probably lost the typical colletid S7; examples are the colletine *Glossurocolletes,* the euryglossine *Euryglossina (Euryglossella),* and the hylaeines *Hylaeus (Edriohylaeus* and *Metziella).* In each case, however, closely related genera or subgenera have the usual colletid structure.

Interestingly, the colletids whose males have pointed glossae *(Hemirhiza, Meroglossa, Palaeorhiza),* and those like *Amphylaeus,* with a slightly pointed glossa, do not constitute a subfamily or other major group of their own. Their females and other male characters are typical of the subfamily Hylaeinae.

Given the fact that males of a few colletids have glossae that are in all features like those of acute-tongued, non-colletid, S-T bees, one may expect to discover bees that are in most features colletids but that have acute glossae in both sexes, or at any rate glossae that are not of the usual colletid style. Possibly such bees should be included in the Colletidae, since obviously the colletid-style glossa is not an essential feature of the family, at least in males. Candidates for possible inclusion in the Colletidae were the Oxaeinae and Stenotritidae. This idea is attractive because of the similarity of these families, in various features, to the colletid subfamily Diphaglossinae. The species of Stenotritidae, indeed, resemble Colletidae in many ways; they share with colletids certain glossal probable synapomorphies (no recognizable seriate hairs; annulate surface not broader than the disannulate surface) as well as the male S7 structure in *Stenotritus* but not in *Ctenocolletes.* The reasons for retaining a separate family, Stenotritidae, are indicated above. The Oxaeinae, however, have distinctive mouthparts that are more similar in some ways to those of Halictidae, and larvae that do not share the usual colletid characteristics (see below). In most of the analyses by Alexander and Michener (1995), Oxaeinae fell with Andreninae; in other analyses Oxaeinae was the sister group to the Stenotritidae, but in no case did it appear among colletids. Apparently, there are no bees with a pointed glossa in both sexes that should be included in Colletidae. The monophyly of the Colletidae and exclusion of the Stenotritidae from the Colletidae are supported by Brady and Danforth (2004) who found a unique intron, not found elsewhere, in all colletid subfamilies in the study (Diphaglossinae was not included). Monophyly of the Colletidae was also shown by Danforth et al. (2006) on the basis of numerous molecular as well as morphological characters (see Fig. 20-6).

Colletidae as here constituted is paraphyletic if the Australian subfamily Euryglossinae is the basal clade of bees, sister group to all other bees, as indicated in Figure 20-1a. In other analyses, however, Euryglossinae appeared as the sister group to all other colletids or to the Hylaeinae. Although the basal position of Euryglossinae is attractive, I am impressed by similarities to Hylaeinae, the large sclerite supporting the curved galeal comb (Figs. 21-1, 48-1) in both subfamilies, among other characters, being probably synapomorphic.

No doubt one reason for the impression that Euryglossinae may be an ancestral group is the short proboscis and the very short, often almost rectangular, galeal blade, and frequently the reduction or absence of the galeal velum. The galea, however, is not at all like that of any sphecoid wasp known to me, although these wasps also lack a velum as distinctive as that of most bees. Moreover, euryglossines like *Pachyprosopis* have a well-formed galeal velum, as in other bees. The galeal structure is further discussed in Section 48 (Euryglossinae).

Andrenidae. Major maxillary types of Andrenidae and Halictidae are diagrammed in Figure 21-2. In most of the analyses by Alexander and Michener (1995), the Andrenidae appeared as a monophyletic unit when the Oxaeinae, frequently given family rank, was included. The Panurginae and the Andreninae have been associated in the past because of the presence of two subantennal sutures on each side of the face, a synapomorphy shared by few other bees. Less diagnostic characters are the presence, in most, of facial foveae (also found in some colletids and, in less well-defined forms, in various other bees) and of a basal premental fragmentum, also found in most Melittidae. Indeed the proboscis as a whole closely resembles that of melittids (see Figs. 21-2c, 72-1a-c, 74-1b, c) except for the form of the mentum and lorum. Two subantennal sutures also are found in Oxaeinae, as well as in Stenotritidae and a few colletids. The presence of two subantennal sutures is a character that may have arisen independently in different apoid groups, and appears to have been lost in some species of *Protandrena (Heterosarus)* in the subfamily Panurginae. The association of Andreninae and Panurginae is further supported by McGinley's (1981) phenetic study of larvae; all the andrenids except Oxaeinae clustered close together in a nearest-neighbor analysis as well as in a phenogram. (In the phenogram, however, the genus *Nomia* in the Halictidae fell in the same group.)

As indicated in Sections 50 and 51, the genus *Alocandrena*, hitherto placed in the Andreninae, cannot be retained in that subfamily and is placed in a separate sub-

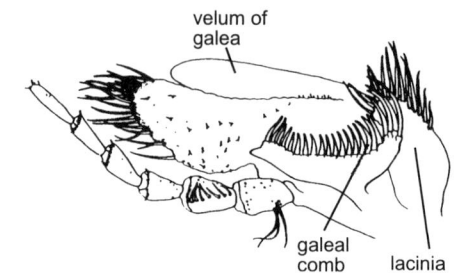

Figure 21-1. Inner view of maxilla of *Hylaeus basalis* Smith, showing the large curved sclerite supporting the galeal comb. (See also Figure 47-1.) From Alexander and Michener, 1995.

family, Alocandreninae. This at least reduces the heterogeneity of the Andreninae. Further study may well support the hypothesis that Andreninae is a paraphyletic group, even after removal of *Alocandrena*.

The inclusion of the Oxaeinae in the Andrenidae was supported by Michener (1944), but most recent works gave oxaeines familial status until Alexander and Michener (1995) reunited them with Andrenidae as a result of phylogenetic analyses. Rozen (1993b), in a careful study of andrenid larvae, found the Oxaeinae to be the sister group of the South American *Euherbstia* in the Andreninae, thus supporting a position in the Andrenidae. The Oxaeinae have many apomorphies, and phenetically they are very different from other Andrenidae (for maxillae, see Fig. 21-2). The possibility exists that a few characters like the two subantennal sutures overinfluenced the phylogenetic process, and that oxaeines are indeed an isolated

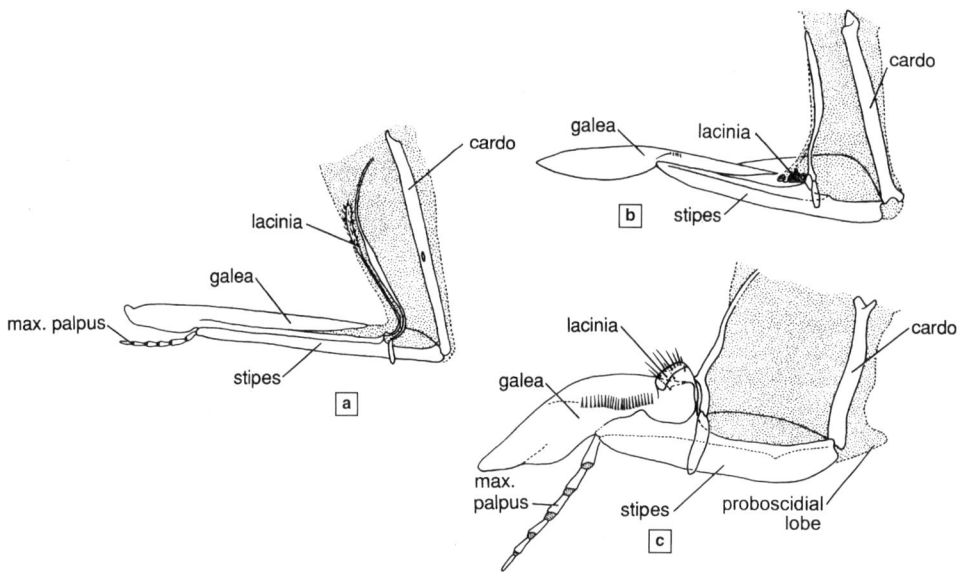

Figure 21-2. Diagrams of inner views of maxillae of three types of S-T bees. **a**, *Halictus quadricinctus* (Fabricius); **b**, *Oxaea flavescens* Klug; **c**, *Andrena erythrogaster* (Ashmead). (The membranous labiomaxillary tube is stippled.) From Michener, McGinley, and Danforth, 1994.

family-level group without close relation to the other Andrenidae. Cane (1983a, b) regarded the Colletidae, Oxaeinae, and Halictidae as related because of the apomorphic presence of macrocyclic lactones in their Dufour's glands. These lactones are rare chemicals, unique among bees, and they are lacking in the panurgines and andrenines. They could have evolved twice or been lost in andrenines and panurgines, but clearly they make the cladograms shown (Fig. 20-1) more questionable, particularly as to the position of the Oxaeinae.

The Oxaeinae are also unusual in that, as in the Halictidae, the first (cardines) and second (prementum and stipites) segments of the proboscis are long and slender and the lorum and mentum are much simplified (Michener, 1985a). Depending on one's interpretation of the structures, the mentum is either short and membranous, as in halictids, or sclerotized and indistinguishably fused to the lorum. I prefer the former interpretation, since such complete fusion is unknown in other bees. The loral apron is broad and sclerotized, as it is in some halictids, and the galea tapers gradually to a pointed base as in halictids (Fig. 21-2b) (Michener and Greenberg, 1985). The nearly hairless, sclerotized lacinia set against the inner anterior stipital margin, the broadly expanded, convex outer stipital margin, the stiff rather than flexible median region of the galea near the palpal base (the palpus is absent in *Oxaea*), the fused suspensoria of the prementum, and other unique synapomorphies emphasize the distinctness of the Oxaeinae.

In larval characters the Oxaeinae are not close to those of any other bees; they occupy an isolated position among the S-T bees in McGinley's (1981) nearest-neighbor (phenetic) study. Rozen (1964a, 1993b) also pointed out unique features of oxaeine larvae, the striking but probably convergent similarities to larvae of Nomadinae, and the sister-group relationship (based on larvae) to *Euherbstia* (Andreninae), as noted above.

Halictidae. The Halictidae form a monophyletic unit that is easily characterized. The position of the lacinia, drawn high up on the anterior surface of the labiomaxillary tube, away from other skeletal structures of the mouthparts, is unique among bees and characteristic of all halictids (Fig. 21-2a) (Michener and Greenberg, 1985). Almost as characteristic is the fusion of the hypostoma and tentorium almost to their anterior ends (Fig. 21-3j) (Michener, 1944, Alexander and Michener, 1995).

The subfamily Rophitinae has sometimes been given family rank (e.g., by Robertson, 1904, as Dufoureidae). Indeed, rophitine genera have traditionally been placed close to panurgine or melittid genera, and Cane (1983a) showed the similarity of *Dufourea* to those groups in the chemistry of the Dufour's gland secretions. Some rophitine features, such as cocoon spinning (except in *Conanthalictus*) and associated larval structures (unlike those of other halictids), and the moderately large and usually apically lobate S7 of males, are plesiomorphies relative to other halictids. The relatively large labrum and low position of the antennae are apomorphies that, because of variability and the consequent difficulty of concisely describing the character states, were not exploited by Alexander and Michener (1995). Nonetheless, they might have made the Rophitinae a monophyletic group instead of the paraphyletic one that their analysis found. The position and form of the lacinia and the broad covering of the upper part of the middle coxa (Fig. 20-5b) are halictid synapomorphies that unify the family and argue against familial status for the Rophitinae, as does the extensive fusion of the tentorium and hypostoma. The membranous mentum, largely membranous or uniformly lightly sclerotized lorum, and other characters of the family Halictidae are also, of course, found in Rophitinae and were among the features that led Michener (1944) to include rophitines in the Halictidae.

Melittidae. No unique synapomorphy is known for this group, but it is well characterized by a combination of characters (Michener and Greenberg, 1980). The following characters are like those of many other S-T bees and unlike the L-T families: labral base elevated, galeal comb present, stipital comb and marginal concavity absent (concavity and marginal hairs present in *Eremaphanta*), basistipital process short, seriate hairs of glossa large and diverging, and glossa without specializations characteristic of L-T bees. At the same time, in the following characters, which are synapomorphies of the whole melittid/L-T bee clade, melittids are like most L-T bees and generally unlike other S-T families: base of mentum uniformly tapering and curled to attach to lorum (Fig. 19-1b, c), loral apron reduced to slender basolateral arms so that the lorum is V-shaped, and lower ends of anterior conjunctival thickening of proboscis not separated as a distinct suspensorial sclerite of the prementum, as it is in all the families discussed previously and as shown beside the maxilla in Figure 21-2. Thus the members of the Melittidae are S-T bees with certain characters of the L-T bees.

Several characters have been proposed as synapomorphic for melittids. The almost complete loss of the episternal groove, both above and below the scrobal groove, is such a character. But being also found in a few other S-T bees and in megachilids among L-T bees, this is not a strong character. The paraglossae are unusually small, even absent or fused to the suspensoria in some Sambini. But they are larger in the Meganomiinae. Roig-Alsina and Michener (1993) considered a minute sclerite in the cardo-stipital articulation of the maxilla to be characteristic of melittids and absent in other bees. It is also absent in *Hesperapis* (Dasypodainae), and Alexander and Michener (1995) did not find it a useful characteristic of melittids. These authors, therefore, did not consider the melittids to have synapomorphies and therefore regarded them as a paraphyletic group from among the members of which the L-T bees arose. Michener (1981a), in a review of Melittidae, also found no synapomorphies for the entire group.

Alexander and Michener (1995) considered the paraphyly of Melittidae s. l. to be established and, like Danforth et al. (2006), recognized three families, as follows: Melittidae s. str., Dasypodaidae, and Meganomiidae. They were all characterized as subfamilies of Melittidae by Michener (1981a). I believe, however, that it is premature to recognize these families of melittoid bees in a general work such as this, because further changes in familial classification might well be made in the near future.

Melittidae is already a small family. Melittid phylogeny, using exemplars of all or at least more genera, should be investigated. Alexander and Michener's analysis was inconclusive about the relationship of *Macropis* and *Melitta* and did not include *Rediviva;* it may be that a cladistic classification would require more than three families (or union of melittids, megachilids, and apids into one family, Apidae). I think it is best to await further studies before breaking up the Melittidae s. l., in order to avoid more unnecessary vacillation in family-level classification.

Megachilidae. The phylogeny of the L-T bees was dealt with by Roig-Alsina and Michener (1993). A summary is presented in Figure 20-2. As a group, they are monophyletic but could all be put in one family, the Apidae. The group characters that unite the Megachilidae and Apidae are those of the L-T bees, enumerated in Section 19.

The common Megachilidae, i.e., the subfamily Megachilinae, constitute a cohesive and easily recognized unit. Its characters include the broad labral articulation, the labrum nearly as long as or longer than broad and more or less rectangular, the presence of a dististipital process, the presence of the metasomal scopa (Fig. 8-7b) in nonparasitic forms, the lack of scopa on the hind legs, the lack of basitibial and usually of pygidial plates, and the lack of pygidial and prepygidial fimbriae. All these are synapomorphies not shared with ancestral members of the Apidae, although the losses have arisen more than once and are shared with some derived Apidae such as the Apini.

Fideliini has been shown to be a basal megachilid group (Rozen, 1970a); it shares various characters (listed above) with Megachilinae, but has some plesiomorphies relative to both Megachilinae and Apidae, such as the presence of well-developed free volsellae and of apical lobes of S7 of the male arising from a slender base, i.e., from a small disc. S7 thus resembles that of colletids and some other S-T bees, not that of other megachilids. Compared to two in Megachilinae, the presence of three submarginal cells is a conspicuous plesiomorphy. Fideliini also has apomorphies, such as the short second abscissa of M-Cu of the hind wing, the presence of alar papillae, and the long hairs (not functioning as a scopa) on the hind tibiae of females. In larval characters the fideliines are similar to megachilines. McGinley's (1981) nearest-neighbor analysis showed larvae of Fideliini to be most similar to those of Lithurgini, a connection that accords with the latter's being the megachiline tribe with the most ancestral characters.

Similarities of Pararhophitini to Fideliini are apparent in such characters as the loss of basitibial and pygidial plates (or it may be that, in both tribes, the pygidial plate is extremely expanded) and loss of pygidial and prepygidial fimbriae. A remarkable feature of the Pararhophi-

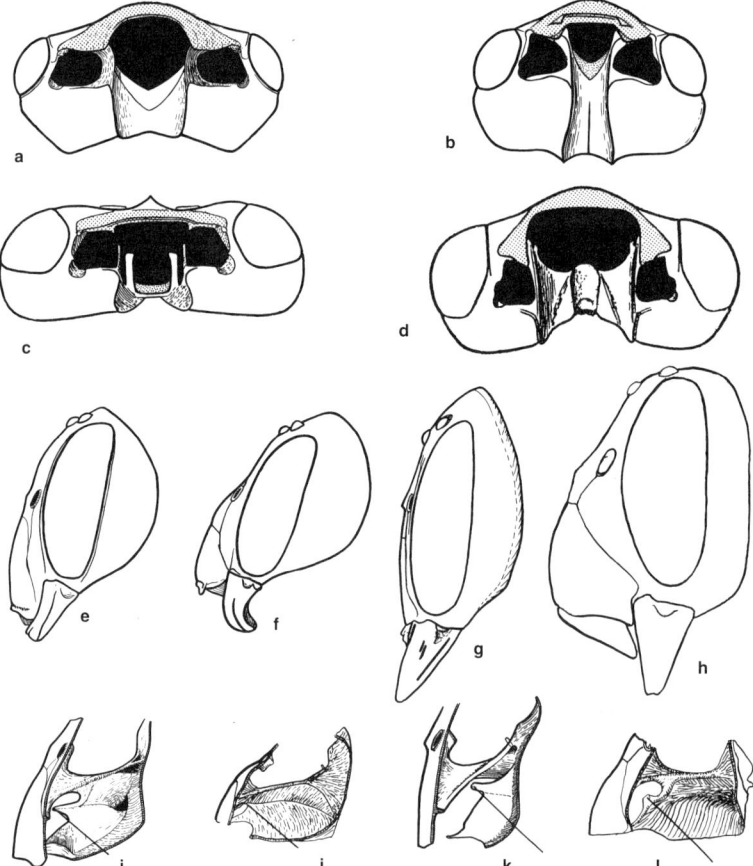

Figure 21-3. Head structures of certain bees. a-d, Ventral views of heads with labrum, mandibles, and proboscis removed; the lateral black areas are the mandibular sockets; the clypeus is shaded, and shows the lower lateral areas bent back in d, not at all in c, and slightly bent back in a and b. **a,** *Andrena mimetica* Cockerell; **b,** *Halictus farinosus* Smith; **c,** *Xylocopa tabaniformis orpifex* Smith; **d,** *Anthophora edwardsii* Cresson.

e-h, Side views of the heads of the same species in the same sequence, showing the protuberance of the clypeus associated with the degree to which the lower lateral areas are bent backward.

i-l, Dissections of the heads of the same species in the same sequence, showing lateral view of the tentorium and the extent of its fusion with the hypostoma of the proboscidial fossa. The location of the articulation of the maxillary cardo in each is indicated by a guide line. From Michener, 1944.

tini is the lack of a distinct scopa; to judge by dry specimens, pollen sticks to various parts of the body, although the hind tibiae do bear possible scopae. *Pararhophites* is the only nonparasitic megachilid without a sternal scopa. Indeed, inclusion of *Pararhophites* in the Megachilidae is a relatively new idea and still in need of reexamination (McGinley and Rozen, 1987).

Apidae. This family is here used in the same broad sense that I employed in 1944. Subsequently, and unfortunately, I agreed with various authors that the Anthophoridae warranted familial recognition as distinct from the Apidae s. str., i.e., the corbiculate Apidae. My decision published in 1944 was based on traditional systematic methods and intuitional phylogenetic trees. Roig-Alsina and Michener (1993) came to the same viewpoint after cladistic analysis. It is convenient, however, to have some shorthand terms for groups of tribes within the Apidae. Distinguishing the subfamilies Nomadinae, Xylocopinae, and Apinae provides for part of this need. Apinae, however, is an especially large and diverse group, containing groups that have elsewhere been given family status, resulting in such names as Anthophoridae, Bombidae, Emphoridae, Euceridae, Meliponidae, Xylocopidae, etc. It is therefore useful to divide the Apinae into (1) the **corbiculate Apinae** (usually actually called the corbiculate Apidae), for the Euglossini, Bombini, Meliponini, and Apini, i.e., Apidae in the sense of Michener (1990a), and (2) the **noncorbiculate Apinae**, for all the other tribes of Apinae. The noncorbiculate Apinae, which are equivalent to the Anthophoridae or Anthophorinae of some older classifications, are united by no synapomorphies, are very diverse morphologically, and consist of those L-T bees that do not fall into some other recognized group. These are the reasons why recognition of Anthophoridae is no longer justified.

Finding synapomorphies for the whole of Apidae is difficult. The characters listed in the following paragraphs are synapomorphies for at least the more primitive Apidae; most of them fail in some of the more derived tribes. They separate most Apidae from the Megachilidae except as noted above for the megachilid tribes Fideliini and Pararhophitini.

In most Apidae, the lower lateral parts of the clypeus and often also the lateral parts of the labrum are bent back (Fig. 21-3d). This is also true of the Fideliini. This feature, associated with the protuberance of the clypeus (Fig. 21-3e-h), making more space for the folded proboscis, is completely lost in some apids like *Xylocopa* that have a flat clypeus. There is usually a preapical concavity on the stipes containing a comb, except in most parasitic forms and some others. The episternal groove is absent below the scrobal groove, i.e., it is neither complete nor entirely absent, but forms an arc from the upper episternal groove to the posterior scrobal groove (Fig. 20-5c). Exceptionally, a few Nomadinae (e.g., some Ammobatini) have a long episternal groove as illustrated by Grütte (1935). The feature may not be homologous to the episternal groove of many S-T bees. (Also in the megachilid tribe Fideliini the episternal groove is short, as in most Apidae; it is entirely absent in Megachilinae and Pararhophitini.) In apids that have been examined there are only five pairs of ostia in the metasomal part of the dorsal vessel (Wille, 1958); in Megachilinae there are six, and the same is likely to be true of Fideliini and Pararhophitini.

A group that has often been given family status but is here included in the Apidae, subfamily Apinae, is the Ctenoplectrini. Michener and Greenberg (1980) considered it to be a family between the melittids and the L-T bees, i.e., as the sister group to L-T bees, for it has nearly all the features of L-T bees except that the labial palpi are like those of S-T bees. Roig-Alsina and Michener (1993), however, in their phylogenetic study of L-T bees, found that *Ctenoplectra* fell in the midst of the Apinae, as near to the Eucerini as to any other tribe. They therefore regarded Ctenoplectrini as a tribe of Apinae. Alexander and Michener (1995) showed it once more as the sister group to L-T bees, but their analysis (intended for S-T bees) is probably not meaningful in this respect, for it included few L-T bees and none of those found to be close to *Ctenoplectra* by Roig-Alsina and Michener.

22. Fossil Bees

Fossil bees are rare, and with some exceptions only those in amber are well enough preserved to shed light on phylogeny. The fossils so preserved are probably biased toward bees that used resin in nest construction, and therefore sometimes became mired and trapped as they collected the resin. Bees are prone to the development of convergent features in the wing venation, legs, and other parts often visible in fossils, whereas the mouthparts and male genitalia—so often important in phylogenetic studies—are almost never visible. Therefore the relationships, even the family, of a fossil are often not apparent. Nonetheless, some interesting results have emerged from studies of fossil bees, and more can be expected. The geological history of bees was reviewed by Engel (2001b, 2004b) and Grimaldi and Engel (2005); see also Section 23. The only comprehensive account of fossil bees (Zeuner and Manning, 1976) was written by people not familiar with Recent bees, and accordingly it contributes nothing to our understanding of apiform history beyond bringing together and summarizing relevant literature. For photographs of bee fossils, see plates 17 to 20.

I have not tried to include fossil genera in the systematic account, except where their relationships are clear. I list in Table 22-1 the generic or subgeneric names based on fossil species, in the same format used for the synonymies of Recent bees. When the higher taxon to which a genus belongs is clear, further information may be given in the account of that taxon. In the table the higher taxa are indicated in bold face. It must be admitted that such taxa may sometimes be in doubt. For example, in the genera of "Halictinae" described by Engel (2002a), the basal vein is shown in his illustrations as curved, a halictine character, but it appears curved throughout its length or medially, not specifically toward the basal end as in Recent Halictinae. So while placement in Halictinae based primarily on the curvature of the basal vein is probably correct, some question remains.

The late Jurassic or early Cretaceous hymenopteran described as *Palaeapis* and placed in the Apidae by Hong (1984) is not a bee. The long stigma suggests a trigonalid, but it may be an aculeate wasp. The slender hind basitarsus, the strong venation close to the wing apex, and the low position of the antennae, as well as the apparently almost globose scape, indicate that it is not one of the Apiformes.

See Addenda for additional material (including a new family) on fossil bees.

Table 22-1. Generic and Subgeneric Names of Bees Based on Fossils.

Anthophorites Heer, 1849: 97. Type species: *Anthophorites mellona* Heer, 1850, by designation of Cockerell, 1909d: 315. **Anthophorini?**

Apiaria Germar, 1839: 210. Type species: *Apiaria dubia* Germar, 1839, monobasic. [See also Germar, 1849: 66.] **Osmiini?**

Bombusoides Motshulsky, 1856: 28. Type species: *Bombusoides mengei* Motshulsky, 1856, monobasic.

Boreallodape Engel, 2001b: 78. Type species: *Boreallodape baltica* Engel, 2001, by original designation. **Boreallodapini.**

Calyptapis Cockerell, 1906a: 41. Type species: *Calyptapis florissantensis* Cockerell, 1906, monobasic. **Bombini.**

Apis (Cascapis) Engel, 1999b: 187. Type species: *Apis armbrusteri* Zeuner, 1931, by original designation. **Apini.**

Chalcobombus Cockerell, 1908d: 326. Type species: *Chalcobombus humilis* Cockerell, 1908, by designation of Cockerell, 1909e: 11 = *Protobombus* Cockerell, 1908. **Electrapini.**

Megachile (Chalicodomopsis) Engel 1999a: 3. Type species: *Megachile glaesaria* Engel, 1999, by original designation. **Megachilini.**

Cretotrigona Engel, 2000a: 3. Type species: *Trigona prisca* Michener and Grimaldi, 1988, by original designation. **Meliponini.**

Ctenoplectrella Cockerell, 1909d: 314; also described as new by Cockerell, 1909e: 19. Type species: *Ctenoplectrella viridiceps* Cockerell, 1909, monobasic. **Osmiini.**

Cyrtapis Cockerell, 1908c: 339. Type species: *Cyrtapis anomalus* Cockerell, 1908, monobasic. **Halictinae.**

Eckfeldapis Lutz, 1993: 180. Type species: *Eckfeldapis electrapoides* Lutz, 1993, by original designation = *Electrapis* Cockerell, 1908. **Electrapini.**

Eickwortapis Michener and Poinar, 1997: 354. Type species: *Eickwortapis dominicana* Michener and Poinar, 1997, by original designation. **Halictinae.**

Electrapis Cockerell, 1908d: 326; also described as new by Cockerell, 1909e: 7. Type species: *Apis meliponoides* Buttel-Reepen, 1906, by designation of Cockerell, 1909e: 8. **Electrapini.**

Augochlora (Electraugochlora) Engel, 2000b: 32. Type species: *Augochlora leptoloba* Engel, 2000, by original designation. **Augochlorini.**

Electrobombus Engel, 2001b: 88. Type species: *Electrobombus samlandensis* Engel, 2001, by original designation. **Electrobombini.**

Electrolictus Engel, 2001b: 38. Type species: *Electrolictus antiguus* Engel, 2001, by original designation. **Halictinae.**

Eomacropis Engel, 2001b: 46. Type species: *Eomacropis glaesaria* Engel, 2001, by original designation. **Eomacropidini.**

Glaesosmia Engel, 2001b: 75. Type species: *Glaesosmia genalis* Engel, 2001, by original designation. **Osmiini.**

Glyptapis Cockerell, 1909d: 314; also described as new by Cockerell, 1909e: 13. Type species: *Glyptapis mirabilis* Cockerell, 1909, monobasic and by designation of Cockerell, 1909e: 14. **Osmiini.**

(continues)

Table 22-1. Generic and Subgeneric Names of Bees Based on Fossils *(continued)*

Hauffapis Armbruster, 1938: 37. Type species: *Hauffapis scheuthlei* Armbruster, 1938 = *Apis armbrusteri* Zeuner, 1931, by designation of Zeuner and Manning, 1976: 243. [*Hauffapis* is not a valid name; see Section 121.] **Apini.**

Kelneriapis Sakagami, 1978 (June): 232. Type species: *Trigona eocenica* Kelner-Pillault, 1970, monobasic. **Meliponini.**

Kelnermelia Moure and Camargo, 1978 (Nov. 17): 565. Type species: *Trigona eocenica* Kelner-Pillault, 1970, by original designation = *Kelneriapis* Sakagami, 1978 (June). **Meliponini.**

Kronolictus Engel, 2002a: 256. Type species: *Kronolictus volcanus* Engel, 2002, by original designation. **Halictinae.**

Libellulapis Cockerell, 1906a: 42. Type species: *Libellulapis antiquorum* Cockerell, 1906, monobasic. **Panurginae?**

Liotrigonopsis Engel, 2001b: 135. Type species: *Liotrigonopsis rozeni* Engel, 2001, by original designation. **Meliponini.**

Lithandrena Cockerell, 1906a: 44. Type species: *Lithandrena saxorum* Cockerell, 1906, monobasic. **Andreninae?**

Lithanthidium Cockerell, 1911d: 225. Type species: *Lithanthidium pertriste* Cockerell, 1911, monobasic. **Anthidiini.**

Electrapis (Melikertes) Engel, 1998: 95. Type species: *Electrapis stilbonota* Engel, 1998. **Melikertini.**

Meliponorytes Tosi, 1896: 352. Type species: *Meliponorytes succini* Tosi, 1896, by designation of Sandhouse, 1943: 570. **Meliponini.**

Melissites Engel, 2001b: 129. Type species: *Melissites trigona* Engel, 2001, by original designation. **Melikertini.**

Ocymoromelitta Engel, 2002a: 261. Type species: *Ocymoromelitta sorella* Engel, 2002, by original designation. **Halictinae.**

Oligochlora Engel, 1997a: 336. Type species: *Oligochlora eickworti* Engel, 1997, by original designation. **Augochlorini.**

Paleoeuglossa Poinar, 1999: 29. Type species: *Paleoeuglossa melissiflora* Poinar, 1999, by original designation. **Euglossini.**

Paleomelitta Engel, 2001b: 42. Type species: *Paleomelitta nigripennis* Engel, 2001, by original designation. **Paleomelittidae.**

Pelandrena Cockerell, 1909f: 159. Type species: *Pelandrena reducta* Cockerell, 1909, monobasic. **Andreninae?**

Apis (Priorapsis) Engel, 1999b: 188. Type species: *Apis vestuta* Engel, 1998, by original designation. **Apini.**

Probombus Piton, 1940: 218. Type species: *Probombus hirsutus* Piton, 1940, monobasic. **Bombini.**

Prohalictus Armbruster, 1938: 48. Type species: *Prohalictus schemppi* Armbruster, 1938, monobasic. **Halictinae.**

Trigona (Proplebeia) Michener, 1982: 44. Type species: *Trigona dominicana* Wille and Chandler, 1964, by original designation. **Meliponini.**

Protobombus Cockerell, 1908d: 326; also described as new by Cockerell, 1909e: 9. Type species: *Protobombus indecisus* Cockerell, 1908, monobasic; also by designation of Cockerell, 1909e: 10. **Electrapini.**

Protolithurgus Engel, 2001b: 51. Type species: *Protolithurgus ditomeus* Engel, 2001, by original designation. **Protolithurgini.**

Protomelecta Cockerell, 1908: Type species: *Protomelecta brevipennis* Cockerell, 1908, monobasic. **Melectini.**

Pygomelissa Engel and Wappler, in Wappler and Engel, 2003: 909. Type species: *Pygomelissa lutetia* Engel and Wappler, 2003, by original designation.

Electrapis (Roussyana) Manning, 1961: 306. Type species: *Apis palmnickenesis* Roussy, 1937, by original designation. **Melikertini.**

Oligochlora (Soliapis) Engel, 2000b: 47. Type species: *Oligochlora rozeni* Engel, 2000, by original designation. **Augochlorini.** [Engel (2002a) considered this a synonym of *Oligochlora s. str.*]

Sophrobombus Cockerell, 1908d: 326; also described as new by Cockerell, 1909g: 21. Type species: *Sophrobombus fatalis* Cockerell, 1908, monobasic = *Electrapis* Cockerell, 1908. **Electrapini.**

Succinapis Engel, 2001b: 115. Type species: *Succinapis proboscidia* Engel, 2001, by original designation. **Melikertini.**

Apis (Synapis) Cockerell, 1907a: 229. Type species: *Apis henshawi* Cockerell, 1907, monobasic. **Apini.**

Thaumastombus Engel, 2001b: 109. Type species: *Thaumastombus andreniformis* Engel, 2001, by original designation. **Electrapini.**

23. The Geological History of Bees

The oldest known fossil Spheciformes date from the early Cretaceous. (For relevant geological terminology, see Table 23-1.) Lomholdt (1982) speculated that they might be pemphredonine wasps, although Rasnitsyn (1980) suggested that two of the genera should be in the Ampulicidae. Lomholdt also speculated that sphecoid wasps must have existed in the late Jurassic.

Bees as we know them are dependent on products of angiosperm flowers (nectar, pollen, sometimes oil) for food. The group therefore is usually believed to have arisen at the same time as, or after, the angiosperms. Various authors (e.g., Baker and Hurd, 1968) have noted that the more primitive groups of angiosperms (e.g., Magnoliaceae) are largely pollinated by beetles. Hence it seemed likely that bees arose or at least became common with the subsequent evolution of angiosperms. In their review of angiosperm biogeography, Raven and Axelrod (1974) indicated that angiosperm fossils first appear in middle Early Cretaceous, that by early Late Cretaceous angiosperm pollen (Muller, 1970) is becoming more abundant than spores of ferns and pollen of gymnosperms, and that by the end of the Cretaceous there was much diversity among angiosperms. Some plant families present at that time (e.g., Myrtaceae and Aquifoliaceae, genus *Ilex*) are now visited extensively by bees, and it is probable that bees were present and perhaps abundant at that time (the Maastrichtian). More recent studies (Crane and Herendeen, 1996; Crepet, 1996) established the abundance and diversity of angiosperms earlier in the Cretaceous with abundant floral remains of various angiosperm families in early to middle Albian floras (about 110 myBP) and with sparse angiosperm remains in early to middle Aptian.

Among the Albian floral structures recently recognized are some that suggest modern bee-pollinated flowers, according to Crepet (1996); it is likely that early bees utilized these flowers, and that the origin of bees was not delayed until the appearance of known bee-pollinated families of plants. Bee-pollinated plants tend to produce limited quantities of sticky pollen that does not blow extensively in the wind. It may be, therefore, that bee-pollinated plants became abundant, especially in dry areas, well before their pollens appeared abundantly in the fossil record. For these reasons one can postulate that bees arose before the middle Cretaceous. An important review of early bee and angiosperm evolution was by Grimaldi (1999); see also Engel (2000b), Grimaldi and Engel (2005), and Danforth et al. (2006). Engel (2004b) gave a necessarily speculative account of the geological history of bees. For example, he said that the end-of-Mesozoic extinction had little effect on bees; perhaps he was right, but only one fossil specimen (*Cretotrigona*) is known from before that event.

Although the use of gymnosperm pollen by bees is not now usual, the possibility exists of pollen sources exploited by bees even before the Cretaceous. Some members of the Triassic to Cretaceous gymnosperm group Bennettitales had bracts, quite possibly colored and petal-like, around their reproductive structures. Probably the bracts served, along with rewards such as pollen and perhaps nectar, to attract pollinators. Is there a possibility that ancestral bees could have been among such pollinators? If so, they radiated with the advent of abundant angiosperms, but bees did not have to arise at so late a time.

The possible importance of gymnosperms for early bees is also suggested by the observations of Ornduff (1991) on the male cones of an Australian cycad that produces abundant pollen as well as a distinct fruity odor. Workers of *Trigona carbonaria* Smith collect the pollen. Ornduff notes that cycads may have arisen in the Paleozoic.

The oldest fossil bee and the only known Mesozoic bee fossil is a worker of the late Cretaceous *Cretotrigona prisca* (Michener and Grimaldi) from New Jersey amber, believed to be nearly 80 million years old. A cloud hangs over the dating, for the piece of amber was collected long ago and its label could be wrong; moreover, the bee closely resembles a modern genus of highly derived bees and is

Table 23-1. Geological Time Scale and Approximate Age in Millions of Years to Base of Each Unit.

Geological units	myBP[a]	Notes relevant to text
RECENT AND TERTIARY		
Recent and Pleistocene	1.5	—
Pliocene	5	—
Miocene	23	—
Oligocene	30	Dominican amber, Florissant shale
Eocene		
Late	38	Baltic amber
Middle	50	—
Early	56	—
Paleocene	65	—
CRETACEOUS		
Late Cretaceous		*Uruguay* nests
Maastrichtian	74	—
Campanian	83	*Trigona prisca*
Santonian	86	—
Coriacian	88	—
Turonian	90	—
Cenomanian	97	—
Early Cretaceous		
Albian	112	diverse angiosperms
Aptian	124	sparse angiosperms
Barremian-Hauterivian	135	—
Valanginian-Berriasian	146	—
JURASSIC		
Late Jurassic (Maim)	157	—
Middle Jurassic (Dogger)	178	—
Early Jurassic (Lias)	208	—
TRIASSIC	245	S. Hasiotis' cells

[a] Million years before present; from Taylor and Taylor, 1993.

associated with certain insect fossils expected to be Tertiary. Analyses of the amber, however, show that it is similar to known Cretaceous amber from New Jersey. Relevant papers are those of Michener and Grimaldi (1988a, b), Grimaldi, Beck, and Boon (1989), and Rasnitsyn and Michener (1991). Rejection of the dating, as is done in the English summary of Radchenko and Pesenko (1994), is not appropriate. If the dating is correct, it strengthens the case for an origin of bees before the angiosperms became common, because the Meliponini to which it belongs contains extremely unwasplike insects. It was highly social; the fossil is a worker, to judge by its very small metasoma.

The next significant date for fossil bees is the late Eocene, because of fossils in the Baltic amber, some 35-40 myBP. For a review of this material, amounting to 36 species, none of which belongs to an extant genus, see Engel (2001b). The following remarks are largely based on that work. The few S-T bees in the Baltic amber are one halictid and two melittid-like species, a possible third being Salt's *"Andrena" wrisleyi* (see also Michener, 2000). (Other unplaced names, based on badly preserved fossils or inadequate descriptions, may represent other melittid-like forms.) The 34 remaining species were L-T bees, Megachilidae and Apidae. Thus by the late Eocene there had been substantial radiation of L-T bees, a great part of which is indicated by 21 species belonging to four tribes (three extinct) of corbiculate Apidae recognized in the Baltic amber. The scarcity of S-T bees except melittid-like forms, including complete absence of the Andrenidae and Colletidae, is in sharp contrast to later fossil material and to the Recent fauna. Thus the known fossil record is consistent with an early development of the melittid/L-T bee clade relative to other S-T bees. In further support of the antiquity of this clade is the widely disjunct distribution of some of its basal members among the dasypodaine melittids, as indicated in Sections 26 and 70 and Michener (1981a). Moreover, the African-Chilean desert distribution of the Fideliini and the pantropical distribution of the Meliponini, both L-T tribes, suggest great antiquity. However, Danforth et al. (2004), using nucleotide date sets from a nuclear gene, estimated the divergence of halictid subfamilies as in the Cretaceous. If accurate, this is not necessarily in conflict with an early Cretaceous origin of bees; the Cretaceous was a very long period (Table 23-1).

A possible explanation for the prevalence of L-T bees in the Baltic Amber is that such bees are the ones that often use resin in nest construction, and so sometimes become entrapped and thus preserved in amber. However, later amber fossils do not lack the major families of S-T bees. By Oligomiocene times, 20-30 myBP, the bee fauna had a more modern aspect. It appears that after the late Eocene there must have been extinction of many taxa including most tribes of the rich Eocene fauna of corbiculate Apidae, perhaps surviving since the Cretaceous. It is likely that Engel (2004b) was right in associating this extinction with the global cooling that characterized the Eocene-Oligocene border. There appears to have been a second radiation, including the major S-T families, to produce the Oligomiocene to Recent fauna. A sample of bees, from the Oligocene of Florissant, Colorado, shows S-T and L-T bees nearly equally abundant, with about 13 species of S-T bees plus one melittid, and 16 L-T bees. Another sample, possibly of middle Miocene age although commonly attributed to the Oligomiocene, unfortunately consists of few species but is more nearly comparable to the Baltic amber because the fossils are also in amber; it is from the Dominican Republic. It includes three meliponine species (L-T) and seven species of S-T bees: five Halictinae, one Colletidae (Xeromelissinae), and one Andrenidae (Panurginae) (Michener and Poinar, 1997; Engel, 1997b). Here, too, unlike the Eocene, S-T families are well represented, as they are in the Florissant and Recent faunas. It is reasonably clear that in the late Eocene, S-T families other than melittids were scarce or absent, whereas probable melittids and L-T bees were rather diverse. By the Oligomiocene, however, the principal S-T families were all present.

Evidence from fossil bee nests may conflict with that from fossil bees themselves. The South American ichnogenus *Uruguay* is based on clusters of bee cells in paleosoil, separated from the surrounding matrix by a space (see photographs by Genise and Bown, 1996). The spiral cell closures and form of the cells and clusters are extremely similar to those of certain Augochlorini such as *Pseudaugochlora* (Halictidae). They are dated as latest Cretaceous or early Tertiary, and thus should antedate the late Eocene Baltic amber, which contains only one recognizable halictid, as noted above. This may mean that halictids were present at that time, at least in South America, or that some other bees, probably now extinct, made cell clusters similar to those of some Recent halictids. Such an explanation is not improbable, since subterranean cell clusters of a more or less similar nature appear to have arisen in species of Halictini, Augochlorini, Nomiinae, and even one subgenus *(Proxylocopa)* in the Xylocopini.

S. Hasiotis and associates (for references, see Engel, 2001b) reported brood cells, some supposedly of social bees, in Triassic paleosoils and petrified wood. These ichnofossils were not made by bees. The order Hymonoptera does not appear in the fossil record until the Triassic, when Xyelidae in the suborder Symphyta are found. The earliest fossils of Aculeata are in the upper Jurassic; the earliest members of the Apoidea, in this case Spheciform wasps, in the Early Cretaceous. Bees should have arisen at that time or after that. But even if all these dates are meaningless, Hasiotis' cells lack features that identify them as made by bees. None have the spiral pattern in the cell closure that characterizes bee cells made of soil or wood particles.

24. Diversity and Abundance

Michener (1979a) published on the biogeography of bees, and the following material is in part derived from that paper. Bees appear to attain their greatest abundance, greatest numbers of species, and probably greatest numbers of genera and subgenera, not in the tropics, but in various warm-temperate, xeric regions of the world. It is easy to make statements like this, but to provide suitable documentation for them is difficult. The taxonomic literature is a poor guide because of the different levels of knowledge in different areas. One can better compare numbers of species taken in limited areas where bee specialists have worked for many years, because most of the species will be recognized, even if not always properly identified, in such a study. The problem is that such areas differ in size and local vegetational, edaphic, and topographic conditions and diversity. Thus differences in the faunas as published result in part from differences in such factors. Similar problems exist if one compares faunal lists for larger areas—states, countries, etc. I therefore must admit that some of my statements are simply impressions based on field experiences and examination of literature. The data given below and by Michener (1979a), however, provide some idea of the numbers of species in certain regions. In my own experience collecting bees in tropical areas I have been impressed by the relative scarcity of bees (both individuals and species), other than the highly social forms (Meliponini and Apini), compared to the abundance of bees in xeric, warm-temperate regions. The richest tropical areas are those of the Americas, and, as indicated below, they are perhaps exceptions to this statement in numbers of species. Michener (1979a) listed many surveys of bee faunas in diverse areas, and the number of species in each. Additional surveys published since 1979 or simply omitted from that work do not alter the general picture, as shown by Petanidou, Ellis, and Ellis-Adam (1995). The great numbers of bee species in regions of topographic and climatic diversity were illustrated by the recent lists of 536 species from Slovenia (Gogala, 1999) and 655 species from the Czech Republic and Slovakia (Pridal, 2004). The large bee fauna of Poland (469 species) listed by Banaszak (2000) illustrates the rich fauna of mesic Europe. A recent report (Schwenninger, 1999) lists 258 species from the vicinity of Stuttgart, Germany. As expected, more northern areas have fewer species, e.g., 284 species in Sweden (Nilsson, 2003).

The bee fauna is particularly rich in the Mediterranean basin and thence eastward to Central Asia, and in the Madrean region of North America (= Californian and the desertic regions of the southwestern United States and northern Mexico). The large fauna of Spain, 1,043 species recorded by Cebellos (1956) without very detailed collecting or intensive studies, suggests great richness, but Spain includes diverse areas, from high montane habitats to Mediterranean macchia, and some Spanish species are thus boreal or mid-European rather than Mediterranean. The same mixture of habitats characterizes francophone Europe, for which a recent list consists of 913 species (Rasmont et al., 1995). The large fauna of southwestern France, 491 species recorded by Pérez (1890), is indicative of the rich Mediterranean bee fauna. Local faunas in the Mediterranean area may contain over 300 species, and with careful collecting the number might in some places exceed 400; see Graeffe (1902) for the Trieste coastal area, 366 species.

For the equivalent nearctic region, the Madrean, the enormous bee fauna of California is impressive, but California includes even more climatic and floral zones than Spain. Part of the great size of the Californian fauna, 1,985 species according to a modification of Moldenke and Neff's (1974) enumeration (see Michener, 1979a), is due to the large area, the north-south extent, and the altitudinal and precipitational range found within the state. In the chaparral or macchia region, i.e., the area having a Mediterranean climate, the richness of the fauna is indicated by the 439 species taken from the vicinity (within 16.6 km = 10 miles) of Riverside, according to Timberlake (personal communication, 1950). Palm Springs, lying on the interface between such an area and the Sonoran Desert, has an even larger bee fauna (probably about 500 species; Timberlake, personal communication, 1950), thanks in part to local topographic and vegetational diversity.

The bee fauna of Mexico, recently analyzed (Ayala, Griswold, and Yanega, 1996), supports the idea that bees are more numerous and diverse in xeric temperate areas than in the tropics. Of 1,800 identified species in Mexico (certainly many are yet to be found and described), the largest numbers are from the xeric northern states—Chihuahua (396 species) and Sonora (359 species)—and the peninsula of Baja California (445 species).

Moldenke (1976a, b) compared the bee faunas of various vegetational areas in California and the Pacific Northwest. He found the chaparral or macchia areas and the sparse forests of the southern mountains (i.e., the Californian areas of Mediterranean climate) richest in bee species, the Californian deserts (the northwestern part of the Sonoran Desert) nearly as rich, and the boreal forests, the grasslands, and the coastal zone progressively poorer. The actual numbers of species varied greatly according to how he divided the area, but representative numbers are 676 for the southern chaparral areas, 668 for the deserts, 589 for the mountain forests of California, thence diminishing to 129 for the coastal strip.

Other xeric, warm-temperate areas such as central Chile and Argentina, much of Australia, and western parts of southern Africa also possess large bee faunas, although to judge by the taxonomic literature and my collecting impressions, these faunas are smaller than those cited in the preceding paragraphs. Unfortunately, there are few faunal data for these areas. Moldenke (1976b) compared the faunas of the temperate coastal areas of Chile and the western United States. For the climatically Mediterranean area of Chile and for the deserts, he reports 183 and 176 species, respectively, while comparable figures for California are 676 (for the southern part of the Californian Mediterranean area only) and 668 (for the Californian deserts). The full meaning of these numbers is not clear, in part because of a problem with the data (see Michener, 1979a: 283 footnote), and in part because of the less intense collecting in Chile than in California. Moldenke, however, believed the differences in numbers of species between comparable Californian and Chilean areas to be real, and I see no reason to disagree with him.

In mesic temperate areas such as the eastern United States or central Europe, the numbers of individuals and species are markedly lower than those in the xeric regions of the same continents. Local lists generally have fewer than 300 species, exceptions mostly being those for more southern localities or larger areas. One of the most carefully collected localities in the world must be Carlinville, Illinois, where Robertson (1928) found 297 species within a 10-mile (16.6-km) radius during a 12-year study. The 566 species reported for Germany (Stoeckhert, 1954) and 693 for Austria (Schwarz, Gusenleitner, and Kopf, 2005) might cause one to question the view that Central Europe has a smaller fauna than that of the Mediterranean, but it must be remembered that Germany and Austria are moderately large and diverse areas that have been meticulously collected in many localities by a large number of specialist collectors of bees. Such careful collecting has never been done in any large Mediterranean area.

Bee faunas become impoverished as one approaches and enters the Arctic, in spite of the abundance of flowers in arctic habitats. Obviously, the abundant arctic flowers are mostly not pollinated by bees.

The data from the tropics appear to support the view that tropical bee faunas are not as large as those of some temperate xeric and mesic areas. The fauna of Java (Friese, 1914b; Lieftinck, in litt., 1977), only about 193 species, should be viewed from the perspective that it has been studied by persons having special interests in bees and that Java is a large, altitudinally diverse island that was connected with the Asian mainland so recently that it has a rich oriental fauna, not a specialized insular fauna.

My impressions of the African tropics, in the absence of appropriate data, suggest that the bee fauna is richer than that of the oriental tropics. The American tropics are much richer still. Thus Panama is smaller in area than Java but has a much larger known bee fauna (353 species, Michener, 1954b), and many species are yet to be recorded. Indeed, the vicinity of Belém, Pará, Brazil, a low, flat region, has a much larger fauna (255 species, Ducke, 1906) than the whole of Java with all of its topographic diversity.

There are various possible explanations for the abundance of bees in some xeric areas and their scarcity, relative to what would be expected from experience with other organisms, in the tropics. Most bees store their highly perishable larval food (usually pollen mixed with nectar; see Sec. 7) in cells, excavated in the soil, that are only thinly lined with secreted waxy or cellophane-like material. In humid environments the loss from fungal attacks on such food and on immature bees is substantial and sometimes catastrophic. J. G. Rozen has suggested (in litt., 1979) that another problem for bees in humid areas may be hygroscopic liquification of the food provided for the larvae, which therefore drown. As noted below, the bee groups that are most successful in humid areas are mostly those that no longer nest in the soil, or that do not use simple cells excavated in it. However, an unusual case of larvae in excavated cells in the tropics surviving for months below the water table in sand is that of *Epicharis zonata* Smith, which lines and seals its cells with thick resinous material (Roubik and Michener, 1980). Cane (1996) found that in New Jersey, USA, populations of *Megachile addenda* Cresson and associated *Coelioxys* survive annual flooding to a depth of 1 meter for four months (Cane, Schiffhauer, and Kervin, 1996). *Megachile* cells do not appear to be waterproof, but the cocoons may be. Even larvae in the rather simple cells of Andrenidae and Halictidae may survive flooding for a few days (Michener, personal observations). Nonetheless, most bees do not live under such conditions.

In view of these and other reports of the success of some bees under not merely humid but very wet conditions, one wonders if the relative scarcity of solitary bees in the tropics might be a result of other factors. For example, predation on bee larvae by ants may be more intense in tropical than in temperate regions. Another such factor may be the success of a few kinds of highly social bees in the tropics, as suggested by D. W. Roubik (in litt., 1979). The genus *Apis* in southern Asia and in Africa (until recently absent in the Americas) and the other highly social bees, the Meliponini, are often the most abundant bees in terms of individuals in the tropics. Each such species must be, from the standpoint of floral resources, the ecological equivalent of a number of species of nonsocial bees, for the workers of highly social species are not only abundant but active all year. Competition for food from aggressive generalists could have an important influence on the tropical bee faunas.

That competition with highly social bees may actually be important for other bees is suggested by observations made by D. W. Roubik and me on Île Royal (Île du Diable), one of the three Îles du Salut, about 10 km off the coast of French Guiana. Large numbers of small bees, mostly *Hylaeus* and *Lasioglossum (Dialictus)*, were swept from small herbaceous flowers. The same species visited the same kinds of flowers on the mainland, but there these bees were scarce. On the mainland, small Meliponini (*Trigona, Plebeia*, etc.) were abundant on the same flowers; but Meliponini were absent on the island. Because of their poor ability to disperse over water, meliponines would rarely if ever reach the islands, and because each colony has only one queen, probably mated only once, the effective population size would be extremely small and extermination therefore probable even if they did reach the island. Similar situations should be investigated elsewhere.

Although Mediterranean climates (winter wet season, warm dry summers, macchia or chaparral vegetation) have rich bee faunas, as do warm-temperate desertic areas with more or less regular rainfall like those of the Sonoran and Chihuahuan deserts of North America, the deserts and semideserts of Argentina, and the desertic areas of the Middle East, mere aridity does not assure a large bee fauna. Tropical and subtropical dry areas usually have rather poor bee faunas, both in diversity and numbers of individuals. For example, there is no evidence of an enriched bee fauna along the southern edge of the Sahara, in spite of the proximity of a very rich Mediterranean fauna in North Africa, and in spite of climatically habitable areas that must have joined these zones in various places in recent, more humid times. And in several other dry regions in or near the tropical zone, there is no evidence of faunal richness compared to nearby areas. Thus northeastern Brazil, northern Australia, and northwestern India, so far as is known, have poor faunas compared to other parts of the same continents. Tropical savannas such as those of East Africa, Venezuela, Panama (Michener, 1954), and northern Queensland also have poor faunas, and the drier tropical areas are in effect depauperate savannas. Warm-temperate grasslands, like those of the southern Great Plains of North America, have moderate bee faunas that become rich where arid and interspersed with xeric vegetation.

These views are challenged by recent studies of bee communities in tropical America. New methods of trapping and year-long surveys, partly unpublished, are showing at least as many species as in rich temperate areas, at sites in both Brazil and Panama. Pedro (1996) listed 442 species (in 110 genera) taken in two areas of savanna forest (cerrado and cerradão, one of them much disturbed) in the state of São Paulo. She also summarized previously published smaller lists for several other Brazilian localities. These areas are near or south of the Tropic of Capricorn, and marginally tropical. Of the 442 species from the São Paulo sites, 33 were Meliponini, demonstrating a bee fauna of a strongly tropical character, but not like that of deep tropical forests. Further work is needed to determine whether American tropical bee faunas are indeed smaller than those of xeric warm-temperate areas, but recent discoveries throw doubt on this contention. The richness of the neotropical fauna was recently illustrated by Pedro and Camargo's (1999) list of 729 species from the state of São Paulo, Brazil.

There are groups of bees that have no affinity for Mediterranean and associated, more xeric climates. Meliponini especially are by far most richly developed in the moist tropics. To a lesser degree the same is true of the Apini, Augochlorini, Centridini, Ceratinini, Ctenoplectrini, Ericrocidini, Euglossini, Rhathymini, Tapinotaspidini, Tetrapediini, and Xylocopini. With the exceptions of many Augochlorini, Centridini, and Tapinotaspidini and the cleptoparasitic Ericrocidini and Rhathymini, these groups do not live in cells in the soil, as do most solitary bees, and therefore should be less subject than most bees to destruction of immature stages by moisture and fungi.

25. Dispersal

The present distribution of bee taxa depends on (1) the climatic and vegetational factors considered in Section 24 and (2) intercontinental and other barriers and the Tertiary and probably late Cretaceous continental movements considered in Section 26. But present distribution also depends on bees' ability to disperse and to reach suitable areas under their own power.

Because bees fly well, one might think that they would be rather successful at crossing barriers, such as water or areas that are climatically or vegetationally inhospitable. A female bee usually mates early in adult life and carries enough sperm cells in her spermatheca to last for part or all of her reproductive life. One can therefore assume that, except for those of the few highly social bees (Meliponini, Apini), individual females transported across a barrier might be able to nest, reproduce, and establish a population. The ability of a bee in a new area to fly about and perhaps find suitable nest sites and food sources also would seem to enhance the probability of establishment.

Nonetheless, distributional data suggest that most groups of bees are not particularly good at crossing major barriers. Most bees fly only in good weather; accordingly, they are likely to be in their nests during storm winds. Moreover, individuals of successive generations of solitary to primitively social species commonly return to the same nesting site, so that they tend to be quite sedentary (Michener, 1974a). Thus for the majority of kinds of bees, dispersal has been by slow spread across continents or to nearby land masses, or by transport on moving continents. The Antillean and eastern Indonesian faunas, however, show that scattered islands between continents can serve as stepping stones for many taxa. (Western Indonesian islands were so recently part of the continent of Asia that no over-water dispersal is needed to explain their bee faunas.)

The monumental work on bees of Madagascar (Pauly et al., 2001) reports 244 species from that island. Although there are certainly new species to be found, the fauna is far better known than previously. Nearly all species and several genera are restricted to Madagascar. Major groups that one might expect in Madagascar but that are absent include the Colletinae, Andrenidae except one species of Melitturgini, Melittidae except Meganomiinae, Osmiini except *Heriades* and *Stenoheriades,* and Nomadini.

Some evidence suggests that solitary to primitively social bees that nest in wood or stems are more likely to cross water barriers than are those that nest in the ground, presumably because wood and stems containing nests are sometimes carried above water in floating islands of vegetation. The bee faunas of oceanic islands, however, include minute forms that nest in soil in addition to the moderate-sized to large, wood-nesting species. For example, on the oceanic islands of the Pacific (Fiji, Samoa, Hawaii, Micronesia, etc.) there are a few moderate-sized, wood-nesting *Lithurgus* and *Megachile,* perhaps carried to some of the islands by nests in Polynesian boats. Otherwise, the fauna consists mostly of small, ground-nesting *Homalictus* (Halictini) or of *Hylaeus;* the latter nest in wood, in stems, or in holes in rocks or soil. It is only among the small forms, especially *Hylaeus* in Hawaii (Perkins, 1899) and *Homalictus-Echthralictus* in Samoa (Perkins and Cheesman, 1928), that there has been significant evolution in these isolated oceanic islands. Thus it seems that dispersal, presumably by wind at least for *Homalictus,* must have favored small forms, whereas the larger ones probably came later and perhaps in some cases with the help of humans. The Galápagos Islands lack small bees and have only one large form, the wood-nesting *Xylocopa darwini* Cockerell, which belongs to the tropical American subgenus *Neoxylocopa*.

Less-isolated island groups frequently have richer faunas, including small as well as larger bees, both those that nest in soil and those that nest in wood. For example, the rather large *Megachile (Creightonella) frontalis* (Fabricius) [and its subspecies or allied species *M. atrata* (Smith)] ranges across 6,400 km from Sumatra to the Solomon Islands; this is a ground-nesting species, and presumably the adults have flown or been blown across the water barriers. The Antillean fauna has probably all arrived across the water. Yet it includes not only bees that nest in wood and minute bees, but also moderate-sized to large, ground-nesting bees such as *Agapostemon, Anthophora, Caupolicana, Centris,* and *Melissodes.* They probably were blown from Yucatan or elsewhere by hurricanes or other storms. The New Zealand bee fauna consists largely of middle-sized ground-nesting *Leioproctus* rather closely related to those of Australia, about 1,400 km away. It is the smallest fauna of any substantial land area except for arctic and antarctic regions; rare over-water dispersal seems to be the only reasonable explanation for its fauna.

The highly social bees (Meliponini, Apini) present special biogeographical problems. They disperse by swarming or by absconding as colonies, not by the action of individuals, most of which are in any case nonreproductive workers. In Meliponini, a new colony is established by individuals from the parent colony that go back and forth provisioning the new nest before a young queen goes there. Thus dispersal by flight across even a few hundred meters of water would be impossible. The absence of Meliponini from the Greater Antilles (except for one species probably introduced by human agency), in spite of their abundance on the Caribbean mainland shores, supports this view, although there are fossil meliponines from Hispaniola. In view of the distribution of Meliponini in the East Indies, as far east as the Solomon Islands and south to Australia, these bees must occasionally be carried across substantial water barriers as colonies in natural rafts or perhaps even in hollow logs floating in the sea. They store food supplies, and those that inhabit hollow logs often close entrances with waterproof resin under unfavorable conditions, and therefore might survive weeks of drifting. In the Apini, dispersal is by swarms or migrating colonies that may fly for distances of perhaps dozens of kilometers or—across habitable country where they can stop—hundreds of kilometers. Yet traversal of a broad ocean by an organized swarm would be impossible. There are three *Apis* species in the Philippines, but the water gaps there were much narrower or absent when the sea level was lowered during the glacial periods.

26. Biogeography

The distributions of the various groups of bees (tribes to families) are indicated in Table 26-1, which is a modified version of a table presented by Michener (1979a). The following is an explanation of the columns.

1. **Aust.** Australia, including Tasmania, New Guinea, the Bismarck Archipelago, and nearby islands such as the Solomons.

2. **NZ.** New Zealand.

3. **Orient.** The oriental faunal region, i.e., tropical Asia from Sri Lanka, India and Pakistan below the Himalayas, across southeastern Asia to Vietnam and southeastern China, also Taiwan, the Philippines, and western Indonesia. (Most of China and Japan are in the eastern part of the palearctic region, not in the oriental region. Likewise, the mountainous parts of northern India and of Pakistan and its western area in Baluchistan are palearctic.)

4. **Madag.** Madagascar.

5. **Afr.** Sub-Saharan Africa (Africa north of the Sahara is palearctic).

6. **Palear.** The palearctic faunal region, including Europe, northern Africa, Turkey and the Middle East, northern India and Pakistan, most of China, and Japan. Although much of Pakistan is oriental, the western Baluchistan area is palearctic.

7. **Nearct.** The nearctic faunal region, including the Mexican plateau and surrounding mountains. Seemingly nearctic areas on the mountains of Chiapas, Mexico, and Guatemala are explained separately in the text but are not included under the term "nearctic" for present purposes.

8. **Neotr.** The neotropical faunal region from tropical Mexico southward through South America, excluding regions 9 and 10.

9. **Antill.** The Greater and Lesser Antilles, excluding Trinidad which is included in region 8.

10. **Arauc.** The Araucanian region, i.e., Chile and adjacent parts of western and southern Argentina.

Across the bottom of Table 26-1, the totals show the numbers of higher-category (tribe to family) bee taxa in each region. Because the areas represented in Table 26-1 often grade into one another, arbitrary decisions were often necessary. Michener (1979a) provided details not repeated here. Radchenko and Pesenko (1994) provided a similar table, except that their columns are for the usual six biotic areas; comparison with Table 26-1 may be useful.

The place of origin of bees remains obscure, but one can speculate as to both the place and its climate. The xeric interior of the old continent of Gondwanaland, particularly West Gondwanaland (Africa-South America), has been suggested as the area of origin of angiosperms (Raven and Axelrod, 1974). It presumably had a seasonal temperate climate. Xeric regions, especially those with sandy soils, are commonly areas of abundance for sphecoid wasps, most of which nest in the ground. It is not unlikely that bees arose from these wasps in such a place, or, if bees already existed, that they radiated in such a place. Most of them have retained their association with xeric areas and have been, compared to the angiosperms, relatively unsuccessful in adapting to humid climates.

The complete absence of unusual archaic bees in New Zealand may indicate that there were no bees in Gondwanaland when New Zealand became isolated by the splitting of that continent in the late Jurassic, some 157 myBP (Smith, Smith, and Funnell, 1994). New Zealand was one of the early fragments to separate from Gondwanaland. As noted in Section 25, the bees of New Zealand appear to be the few that arrived and established themselves after overwater dispersal from Australia; the New Zealand bees all belong to Australian genera. The distance now is about 1,400 km.

The distributions of most bee taxa are not disjunct, and where disjunctures do occur they are limited to neighboring continents, e.g., holarctic taxa whose distributions could have been attained during warmer times with the continents in their present positions. Similarly, many tropical groups occur in Africa and from Sri Lanka and India eastward across southern Asia, often with little differentiation between forms in the two areas. An earlier, more humid (not wet-tropical but savanna) climate across the Arabian peninsula, southern Iran, and western Pakistan would connect or nearly connect these areas for the bees concerned, even with the continents in their present positions. Since more humid conditions undoubtedly existed in this area in the not very distant past, this disjuncture, like that for cool-temperate forms across Bering Strait, is easy to understand. Some of the taxa involved are *Braunsapis, Ctenoplectra, Megachile* (three or more subgenera), *Pachyhalictus, Tetralonia (Thygatina), Thrinchostoma,* and *Xylocopa* (various subgenera).

Some taxa are found on most or all continents and even many islands, and obviously have considerable potential for crossing major water barriers. The dispersal of some such groups probably preceded the present arrangement of the continents. Numerous taxa, by occurring on both sides of a major physiogeographical or climatic barrier, provide some information on their antiquity or their dispersal ability.

Many organisms inhabit Mediterranean and desertic climates of both North and South America and are absent from the intervening tropics. These amphitropical distributions have long been a subject of interest (Raven, 1963; Raven and Axelrod, 1974). In North America most amphitropical bees are primarily Sonoran and Chihuahuan desertic elements, although *Caupolicana* and *Ptilothrix* each has an eastern North American species. In South America, amphitropical bees occur either in the Argentinan-southern Brazilian area, often in its more xeric parts, or in the Araucanian region, or both. Some amphitropical genera and subgenera are *Anthophorula, Caupolicana, Martinapis, Protoxaea, Ptilothrix,* and *Zikanapis.*

Since there is no evidence for an arid corridor through the neotropics during the Tertiary (or at any other time), long-distance dispersal probably accounts for amphitropical distributions, perhaps facilitated by local xeric areas, as suggested by Michener (1954). The eleva-

Table 26-1. Summary of Distribution of Families, Subfamilies, and Tribes of Bees.

Areas (column headings) are explained in the text. Plus signs indicate presence, neither diversity nor abundance necessarily implied; m indicates marginal presence, i.e., the taxon enters the area only marginally, or one or a few of its species extend well into the area but not halfway across it. The second symbols in some of the notations are subjective indications of relative diversity of the taxon in those areas, + indicating more diverse, − less diverse, than indicated for the taxon by a simple + without a second symbol. Introductions by human agency are ignored.

Taxon	1 Aust	2 NZ	3 Orient	4 Madag	5 Afr	6 Palear	7 Nearct	8 Neotr	9 Antill	10 Arauc
STENOTRITIDAE	+	−	−	−	−	−	−	−	−	−
COLLETIDAE										
Paracolletini	++	+	−	−	−	−	m	+	−	++
Colletini	−	−	m	−	+	+	+	+	+	+
Scraptrini	−	−	−	−	+	−	−	−	−	−
Caupolicanini	−	−	−	−	−	−	+	+	+	+
Diphaglossini	−	−	−	−	−	−	−	+	−	+
Dissoglottini	−	−	−	−	−	−	−	+	−	−
Xeromelissinae	−	−	−	−	−	−	m	+	m	+
Hylaeinae	++	+−	+	+	+	+	+	+	+	−
Euryglossinae	+	−	−	−	−	−	−	−	−	−
ANDRENIDAE										
Alocandreninae	−	−	−	−	−	−	−	+	−	−
Andreninae	−	−	m	−	+	++	++	+	−	+
Protandrenini	−	−	−	−	−	−	+	+	−	+
Panurgini	−	−	−	−	−	++	+	−	−	−
Nolanomelissini	−	−	−	−	−	−	−	−	−	+
Melitturgini	−	−	−	m	+	+	−	−	−	−
Protomeliturgini	−	−	−	−	−	−	−	+	−	−
Perditini	−	−	−	−	−	−	+	m	m	−
Calliopsini	−	−	−	−	−	−	+	+	−	+
Oxaeinae	−	−	−	−	−	−	m	+	−	−
HALICTIDAE										
Rophitinae	−	−	m	−	+	++	++	+−	−	+−
Nomiinae	+	−	++	+	++	+	+	m	m	−
Nomioidinae	+	−	+	+	+	+	−	−	−	−
Halictini	+	+−	+	+	+	+	+	+	+	+
Augochlorini	−	−	−	−	−	−	+	++	+	+
MELITTIDAE										
Dasypodaini	−	−	m	−	+	+	+	−	−	−
Promelittini	−	−	−	−	−	+	−	−	−	−
Sambini	−	−	−	−	+	−	−	−	−	−
Meganomiinae	−	−	−	+	++	m	−	−	−	−
Melittinae	−	−	+	−	+	+	+	−	−	−
MEGACHILIDAE										
Pararhophitini	−	−	−	−	−	+	−	−	−	−
Fideliini	−	−	−	−	+	m	−	−	−	+
Lithurgini	+	−	+	+	+	+	+	++	+	+
Osmini	−	−	+−	+−	+	++	+	m	m	−
Anthidiini	+−	−	+	+−	+	+	+	+	−	+−
Dioxyini	−	−	m	−	+−	++	+	−	−	−
Megachilini	+	−	+	+	+	+	+	+	+	+
APIDAE, Xylocopinae										
Manueliini	−	−	−	−	−	−	−	−	−	+
Xylocopini	+	−	++	+	++	+	+	++	+	−
Ceratinini	+	−	+	+−	+	+	+	+	+	−
Allodapini	+	−	+	+	++	m	−	−	−	−

(*continues*)

Table 26-1. (continued)

Taxon	1 Aust	2 NZ	3 Orient	4 Madag	5 Afr	6 Palear	7 Nearct	8 Neotr	9 Antill	10 Arauc
APIDAE, Nomadinae										
Hexepeolini	—	—	—	—	—	—	+	—	—	—
Brachynomadini	—	—	—	—	—	—	+	+	—	+
Nomadini	m	—	+	—	+	++	++	++	+	—
Epeolini	—	—	—	—	+	++	++	++	+	+
Ammobatoidini	—	—	—	—	+	+	+	—	—	+
Biastini	—	—	—	—	—	+	+	—	—	—
Townsendiellini	—	—	—	—	—	—	+	—	—	—
Ammobatini	—	—	—	+	++	+	+	—	—	—
Caenoprosopidini	—	—	—	—	—	—	—	+	—	—
APIDAE, Apinae										
Isepeolini	—	—	—	—	—	—	—	+	—	+
Osirini	—	—	—	—	—	+−	+−	+	—	+
Protepeolini	—	—	—	—	—	—	m	+	—	—
Exomalopsini	—	—	—	—	—	—	+	++	+	—
Ancylini	—	—	m	—	—	+	—	—	—	—
Tapinotaspidini	—	—	—	—	—	—	—	++	+	+
Tetrapediini	—	—	—	—	—	—	—	+	—	—
Ctenoplectrini	m	—	+	—	+	m	—	—	—	—
Emphorini	—	—	—	—	—	—	+	++	—	+
Eucerini	—	—	+−	+−	+	+	+	+	+−	+
Anthophorini	+	—	+	+	+	++	++	+	+	+
Centridini	—	—	—	—	—	—	+	++	+	+
Rhathymini	—	—	—	—	—	—	—	+	—	—
Ericrocidini	—	—	—	—	—	—	+	++	+	+
Melectini	+−	—	+	+−	+	++	+	m	+−	—
Euglossini	—	—	—	—	—	—	—	+	m	—
Bombini	—	—	+	—	—	++	++	+	—	+−
Meliponini	+	—	+	+−	+	—	—	++	m	—
Apini	—	—	++	+	+	+	—	—	—	—
Total higher taxa present	18	3	26	18	31	36	41	43	24	28

tion of the Andes progressively increased the possibility of intercontinental dispersal by creating cool or arid habitats near the Equator. There is an alternative to long-distance dispersal, however. From time to time in its history, a xeric-adapted genus may give rise to a species able to persist in mesic or humid habitats. For example, *Ashmeadiella*, a generally xeric-adapted genus, has one species that ranges eastward in North America as far as Indiana and Georgia, and another that occurs from North Carolina to Florida. Should such a species reach another xeric area, it might well speciate there, and if it then disappeared in the intervening mesic area, one could have separate clusters of species in the two xeric areas and a problem in explaining how the genus traversed the mesic area.

In Africa the situation is different, because there is now a savanna corridor (with arid areas here and there) through eastern Africa between the palearctic (Mediterranean) region and the temperate xeric Cape region. Various genera such as *Andrena* and *Nomada* are present in this corridor and in southern Africa, to which they appear to have dispersed from the north. Bees with amphitropical disjunct distributions in the Old World exist, however, although much less commonly than in the New World. Examples are *Fidelia* (in southern Africa and Moroccan deserts) and *Aglaoapis* (South Africa and Eurasia).

Other disjunct distributions of bees, described by Michener (1979a), include Asiatic groups that range through the islands to Australia, and Malagasy forms with relatives in Africa or in a few cases in Asia. Others are the Paracolletini, which are found principally in temperate parts of Australia and South America, and the Fideliini in desertic parts of Chile, southern Africa, and Morocco, with a relative, the Pararhophitini, in palearctic desert areas in Central and western Asia. Dispersal between Australia and South America through cool-temperate Antarctica might have been possible across moderate water gaps as recently as the beginning of the Oligocene (38 myBP), although land connection was broken in the late Cretaceous, about 80 myBP (Smith, Smith, and Funnell, 1994); see Section 23. This route could have been traversed by the paracolletines. Connections to Africa were disrupted earlier. By the end of the Cretaceous, while relatively narrow seas separated tropical Africa and South America, the temperate parts of these continents were already well separated. Thus, the Fideliini and perhaps the Paracolletini may have originated as long ago as the Late Cretaceous.

Dispersal among other xeric areas also presents interesting problems. The primarily Australian-North American distribution of the *Hackeriapis-Chelostomoides* group of *Megachile* has been discussed elsewhere. (*Hackeriapis* is Australian, north to New Guinea savannas; *Chelostomoides* is North American, especially Madrean, south to Colombia.) Long-distance dispersal over water seems extremely unlikely between Australia and North America. The Bering Straits would seem a likely route, but if it were the route used, why is the nearest relative of *Chemostomoides* in Australia instead of in Asia? Possibly the related Asiatic subgenus *Chelostomoda* has something to do with this problem. Alternatively, dispersal between Australia and South America across moderate water gaps and through cool-temperate Antarctica might have been possible in the Paleogene, but if this route were used, what happened to the group in most of South America, some of which is climatically similar to the desertic areas of North America and Australia where *Chelostomoides* and *Hackeriapis,* respectively, are abundant and diversified?

As noted elsewhere, *Hesperapis* (southern Africa and western North America) also occurs in deserts of both the Northern and Southern hemispheres; its closest relative is *Eremaphanta* from deserts of Central Asia. The unusual feature of the *Hesperapis* group's distribution in the Old World is its absence between Central Asia and South Africa. It probably became extinct between these areas; there is no need to postulate long-distance dispersal. The problem is, how did it get to the New World?

Anthophora of the subgenus *Heliophila* of North America and those of the Mediterranean-Central Asian area are presumably similar because of common ancestry. Detailed studies have not discovered group differences between them (Brooks, 1988). Some bees may have moved directly between the Old World and New World dry areas. Raven and Axelrod (1974) suggest that eastern North America and Western Europe were in latitudes suitable to warm, seasonally dry climates in the late Cretaceous and early Eocene, and were separated by only moderately broad seas. If so, bees of dry areas could have been exchanged more readily than under present conditions, but probably only by long-distance dispersal. *Hesperapis* might have moved between the continents at the same time as *Heliophila.*

The disjunctions noted previously for the Meliponini are unique among bees in that they involve a group that is characteristic of the moist tropics, and with minimal ability to cross water. Yet they occur in all tropical areas of the world except the oceanic islands. The genus *Trigona* occurs in the neotropical region and from southern Asia to Australia, but is absent from Africa. Speculations on how they might have attained their present distribution can be found in Michener (1979a, 1990a) and Kerr and Maule (1964); all that one can say with certainty is that Meliponini is an old group, probably older than most of the taxa discussed above that do not show transoceanic distributions.

The effectiveness of a barrier, as shown by a lack of disjunct distributions across it, can also tell us interesting things about the taxa concerned. An outstanding biogeographical observation to be made about bees is that, if one ignores cosmopolitan taxa and the Meliponini and Fideliini, the bees of Africa and South America are very different. As shown in Table 26-1, both continents have numerous taxa of bees not found in the other. Therefore, it is reasonable to believe that largely tropical taxa like the Augochlorini, Centridini, Diphaglossinae, Ericrocidini, Euglossini, Exomalopsini, and Oxaeinae, widespread in South America but absent from the Old World, including Africa, originated after the separation of Africa and South America. Likewise, taxa like the Nomiini, Nomioidini, Ctenoplectrini, and Allodapini, widespread in the Old World, including Africa, but absent from South America, originated or became widespread in Africa after the separation of these two continents. In the south, that separation occurred in the early Cretaceous (about 140 myBP), but the continents were joined in the equatorial region for a long time, i.e., until the middle Cretaceous, about 100 myBP (Smith, Smith, and Funnell, 1994), and must have been close together in that region long after that. Presumably, the tropical taxa listed above originated after the continental separation became too wide and perhaps after the Cretaceous.

Of course, different taxa frequently tell different stories about past continental connections. For example, the Paracolletini indicate past connections or proximities of the southern continents, whereas many Hylaeinae and the Euryglossinae are unique in Australia, with numerous genera being found nowhere else. Although the Euryglossinae occur only in Australia, the Hylaeinae are widespread; African and South American Hylaeinae are similar to those of the rest of the world but are not similar to the rich Australian fauna. Presumably, this means that the Paracolletini are an older group, and the Hylaeinae and Euryglossinae are enough younger that connections between Australia and other southern continents were fully broken before these bees became common. This idea is consistent with the nonbasal position of Hylaeinae and sometimes Euryglossinae in the phylogenetic analyses of Alexander and Michener (1995); see also Section 20.

27. Reduction or Loss of Structures

Systematists devote a great deal of time and energy to finding shared characters that have not evolved convergently and that therefore characterize taxa and are useful in developing phylogenetic hypotheses. Some of the most interesting characters that organisms have, however, are those that are convergent and that therefore suggest some common behavior or common environmental challenge perhaps working through behavior. The following are some examples of characters that have arisen two or more times in the course of bee evolution, assuming our phylogenetic and classificatory ideas to be reasonably correct. There are many more such characters; many were listed by Michener (1944), who recorded, in a section on comparative morphology, taxa that exhibit common features. Those selected for mention here are particularly prominent or functionally interesting.

Losses of structures are the commonest and often the most easily understood convergences. In phylogenetic studies, synapomorphies based on losses must always seem weak compared to those based on novel structures, because losses are frequent and often repetitive. Independent losses of the same structures usually have morphologically identical outcomes, but the losses are not homologous and therefore can be misleading for phylogenetic analyses requiring recognition of homologies. It is not legitimate in an analysis based on parsimony to code losses as different characters simply because, according to a phylogenetic hypothesis, they must have arisen independently in different clades. Because they might be homologous and the phylogenetic hypothesis thus wrong, one must tolerate a reduced consistency index and say that *if* the phylogeny is correct, the losses were independent, i.e., not homologous.

Throughout zoology, losses of unused structures or organs, like eyes of cave animals, are usual. Losses of the nest-making and pollen-collecting and pollen-manipulating structures of parasitic and robber bees are of a similar nature; see Section 8. Such bees commonly lack anterior basitarsal brushes, which are common in other bees and used in removing pollen from anthers. Scopal reduction or loss is almost universal in parasitic and robber bees (Figs. 8-5 to 8-8), regardless of the family to which they belong, for such bees do not transport pollen loads. Other losses associated with the parasitic way of life include loss or reduction of the stipital comb of L-T bees, of the basitibial plate (see next paragraph), and of the pygidial and prepygidial fimbriae (see next paragraph), and reduction in the size of or rarely (as in *Melanempis*) loss of the jugal lobe of the hind wing (Roig-Alsina and Michener, 1993).

Some of these same losses, i.e., of basitibial plates and pygidial and prepygidial fimbriae, also occur in some nonparasitic bees such as the corbiculate Apidae, and the jugal lobe is absent in Bombini and Euglossini. Information on the possible function of the jugal lobe and why it is sometimes reduced or absent is completely lacking. The pygidial plate and basitibial plates seem ancestral for bees, at least in females, because of the presence of these structures in basal clades, and the presence of pygidial plates in some sphecoid wasps. In parasitic bees these plates tend to disappear (Fig. 8-9). Associated with these plates, especially in females, are usually the pygidial and prepygidial fimbriae—dense fringes of hair different from any fringes that may be present on more anterior terga. All these structures are well developed in ground-nesting bees that excavate brood cells of uniform shape and with smooth walls. The basitibial plates are used to brace the bee while she tamps the cell wall with the apex of her metasoma—i.e., with the pygidial plate and associated fimbriae. Not surprisingly, since they do not construct nests, male bees frequently have the plates and fimbriae reduced or absent. In females of diverse taxa that do not construct cells in the soil (or sometimes in rotting wood), these plates and fimbriae are commonly reduced or lacking. Examples among nonparasitic bees include nearly all Hylaeinae, Megachilinae, and corbiculate Apidae, and the genus *Colletes*. The Xylocopinae mostly have reduced or much modified basitibial and pygidial plates. Females of cleptoparasitic Halictidae and Apidae mostly have pygidial plates but lack or nearly lack the fimbriae and the basitibial plates.

The following paragraphs describe a few other interesting losses that have occurred during bee evolution. The basic number of segments of the maxillary palpus is six in Hymenoptera in general and in many bees. This is anomalous, since for insects generally the basic number is five; almost certainly the so-called basal segment in Hymenoptera is in reality the palpifer. In diverse groups of bees the number of segments is reduced to five, four, three, two, perhaps one; and in a few cases the maxillary palpi are absent, as in *Oxaea, Rhathymus, Melanempis,* and *Pasites maculatus* Jurine. Reduction from six to five, four, three, or two segments is known in the Perditini (Panurginae), Megachilinae, Apinae, etc.; and to three segments within the tribe Eucerini. The functional significance of this reduction or loss is unknown. In the genus *Perdita*, different species of which have from six to two segments, there is no evident difference in associated mouthpart structures that might suggest how the palpi function. However, except for the Perditini, reduction in the number of segments of the maxillary palpi is very rare in S-T bees but common in L-T bees. One can therefore speculate that the small divergent apical segments of the labial palpi of L-T bees may have the same (tactile?) function as the maxillary palpi. Increases above the basic number of segments are very rare and are not loss characters. Worth noting are *Andrena grosella* Grünwaldt, extraordinary in having nine-segmented maxillary palpi (Grünwaldt, 1976), and *Xeromelissa wilmattae* Cockerell, variable but with up to eight-segmented maxillary palpi.

The basic number of segments of the labial palpus is four. This number is not often reduced as it is for the maxillary palpus, but the labial palpi of *Eulaema* (Euglossini) and *Hoplitis (Coloplitis)* (Osmiini) are two-segmented, and of *Effractapis* (Allodapini), *Neffapis* (Panurginae), and *Xeromelissa* (Xeromelissini), three-segmented. Increase above the basic number is extremely rare, but is

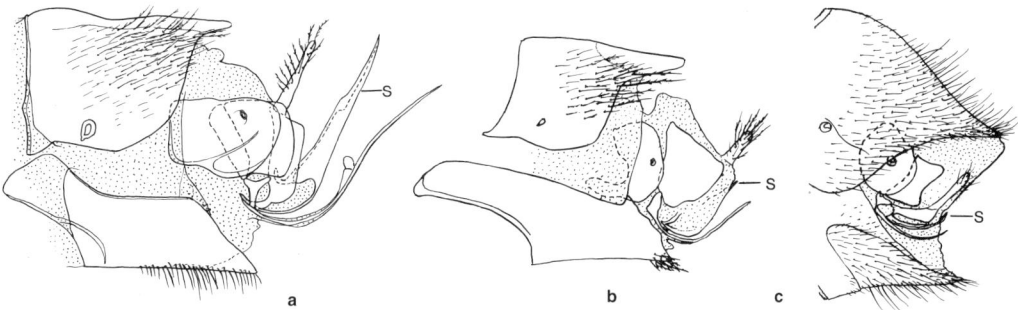

Figure 27-1. Reduction of the sting; the sting apparatus is artificially extruded. **a**, *Hesperapis* sp.? (Melittidae); **b**, *Perdita albipennis* Cresson (Panurginae); **c**, *Melipona* sp., worker (Meliponini). (s indicates the sting stylus, i.e., the fused second valvulae.) The large sclerites with setae are T6 and S6; the palplike structures with hairs are gonostyli or third valvulae. See also Figure 10-14.

known, for *Leioproctus (Hexantheda) missionicus* (Ogloblin) (Colletidae), six, seven, or nine segments; and for *Andrena grossella* Grünwaldt (Andrenidae), nine segments, the same number as for its maxillary palpi (Grünwaldt, 1976).

Most bees have an arolium between the claws of each leg. Some bees have lost arolia (Fig. 10-10); examples include *Leioproctus (Urocolletes)* in the Colletidae; Oxaeinae in the Andrenidae; some species of *Trachusa* and *Hypanthidiodes*, as well as all species of *Anthidium* and allied genera in the Anthidiini; and nearly all Megachilini. In addition, *Amegilla, Centris, Pachymelus (Pachymelopsis), Pachysvastra, Ptilothrix, Svastrina, Zacosmia,* and the Euglossini and Xylocopini in the Apidae lack arolia. Relatives in all cases have arolia. This is by no means a comprehensive list of bees that lack arolia, but it shows that their loss occurred many times and so now characterizes some species, some genera, and some tribes in higher taxa that otherwise possess arolia. No known characteristics of behavior or nest structure are associated with arolial loss.

Although aculeate Hymenoptera are especially known for the stings of females, which for bees are commonly used for defensive purposes (Fig. 10-14), the reduction or loss of stings is surprisingly common and occurs at least in Andrenidae, Megachilidae, and Apidae. In Andrenidae, stings are somewhat reduced, often lacking the valve of the first valvula (Fig. 27-1b) that pumps the venom into the wound (Ruz, 1986; Michener, 1986c). The sting is much more reduced in the Meliponini (Fig. 27-1c), quite incapable of puncturing anything, and is the most reduced of all in the Dioxyini. The last is a cleptoparasitic megachilid group. In most such parasites, however, the sting is well developed, presumably for defense against irate hosts.

Many additional repetitive, i.e., convergent, loss characters, for example of distal veins of the wings in small bees (Danforth, 1989a), are indicated elsewhere in this book as well as by Michener (1944). There is no need to repeat here an account that could fill many pages.

It has long been believed that, once lost, the same (homologous) structure cannot reappear at a later date. Although for a phylogenetic analysis this is probably a reasonable assumption, there is increasing evidence that the genetic mechanism may persist, inactivated, for long periods and may then be reactivated, causing the reappearance of the structure. When this happens, the clade may have as a synapomorphy the potential for a character state, but not all members demonstrate this potentiality. A possible example is the loss and possible resurrection of submarginal cross veins. Certainly in two subgenera of *Perdita*, a genus that normally has two submarginal cells, an intercalary cell exists; it is small and triangular, thus unlike the three-celled plesiomorphic condition, but it shows that veins can split or arise de novo. In some other bees with three ordinary submarginal cells, this condition may be derived from two-celled ancestors. Some of Roig-Alsina and Michener's (1993) analyses of L-T bees indicated such development; perhaps these authors were wrong to reject such hypotheses. The resurrection of a structure is easy to understand when it disappears in only one sex (or stage), for example the male, but is retained in the female, and later in phylogeny reappears in the male. Genes for the character would be continuously present and expressed in females, and could be reactivated in males to cause reappearance of the character in that sex.

28. New and Modified Structures

In contrast to loss characters, new or newly modified structures are more likely to be unique synapomorphies that can be used to recognize clades in phylogenetic studies. Nonetheless, remarkable cases of convergence exist, provided that our phylogenetic hypotheses are correct. If our hypotheses are considered to be incorrect, worse problems of understanding phylogenies usually arise. The following are some examples of apparently new structures that seem to have arisen independently in different phyletic lines of bees.

The flabellum of the glossa in some species of Panurginae, especially *Perdita* and various Calliopsini, resembles in detail the flabellum of many L-T bees (Figs. 28-1, 28-2, 86-1). The presumably ancestral bee glossa lacked a flabellum and had abundant long annular and seriate hairs, as in Melittidae, most Halictidae (Fig. 28-1a, b), most Andrenidae, and even a few colletid males. Distal enlargement of the glossa is found sporadically, weakly in Nomioidini (Fig. 28-1c), more strongly in Rophitinae (Fig. 28-1d-f). Even within the Panurginae, some genera lack such a flabellum-like structure, as shown by *Protandrena* (Fig. 28-2g). Others have a well-developed flabellum, as shown for *Calliopsis* in Figure 28-1i, j. Finally, in *Perdita* (Fig. 28-2h, i) the flabellum is essentially like that of L-T bees (Figs. 28-2a-f, 86-1). If the L-T bees arose from among the melittids, they had nothing to do with the Panurginae, and the flabellum as well as details of its structures arose independently. More information is given in Section 19. As also shown in Section 19, other aspects of glossal structure, for example large, divergent, seriate hairs, may have arisen independently, depending on the phylogeny that one accepts.

A remarkable feature that crops up occasionally in all major families is hairs on the eyes. It characterizes no tribe or higher-level taxon except Apini, but is usually characteristic of a few species, a subgenus, or sometimes a genus. Examples are as follows: a few species of *Leioproctus* (Colletinae); *Parapsaenythia* (Panurginae); *Agapostemon (Agapostemonoides), Caenohalictus, Rhinetula,* and some other Halictini; *Caenaugochlora* (Augochlorini); most *Coelioxys* (Megachilini); two subgenera of *Pachyanthidium* (Anthidiini); some species of *Holcopasites* (Ammobatoidini); *Trichonomada* (Brachynomadini); *Trichotrigona* (Meliponini); and *Apis* (Apini). As can be seen from this list, hairy eyes can be found in social as well as solitary bees, in cleptoparasites as well as nonparasitic forms—no functional explanation for the repeated origin of hairy eyes is evident. In *Apis* the eye-hairs are reported to monitor air flow, but why should a scattering of other bees have the same structure?

The ancestral thoracic shape for bees is cylindrical, with the scutellum, metanotum, and basal area of the propodeum horizontal or slanting (Figs. 20-5b, 28-3b). In a few taxa that live in narrow burrows in wood, this is also a derived state, sometimes accentuated, no doubt related to providing the elongate, slender body needed to use such nest burrows. Such cases include *Chelostoma, Heriades,* and *Osmia (Pyrosmia) cephalotes* Morawitz (Megachilidae); and Ceratinini and Allodapini. *Hylaeus (Heteraopoides)* (Hylaeinae) (Fig. 28-3a) is more elongate than other *Hylaeus* (Pl. 1), and is perhaps the most slender of all bees. In many bees, however, the thorax is shortened by the propodeum becoming entirely vertical, sometimes also by the metanotum and even the posterior half of the scutellum likewise becoming vertical (Fig. 20-5a, c). The result is a nearly spherical thorax (Fig. 28-3c). All intermediates exist. Examples of bees with quite spherical thoraces are mostly in the Megachilidae and Apidae, but also occur in the Diphaglossinae (Colletidae) and Oxaeinae (Andrenidae). As was suggested by Michener (1944), the spherical form is compatible with rapid and often hovering flight, and it is bees with such thoraces that exhibit such capability.

Danforth (1989a) has explained some of the convergent characters found in bee wings. Of course, there is a phylogenetic component as well; a moderately large stigma is clearly plesiomorphic relative to a minute stigma or a stigma that is essentially lost as in some *Xylocopa, Centris,* and Oxaeinae. Nonetheless, there is a strong size-related component to wing characters, and even congeneric species, if of quite different sizes, may have conspicuously different wings. In minute forms the stigma is relatively large, the distal wing veins are often reduced and withdrawn from the apical region of the wing, and other wing veins and cells tend to be more transverse in the relatively shorter and broader wings. The converse features characterize wings of large bees. All this, of course, is related to aspects of flight mechanics. Since both large and small species occur in most families of bees, features characteristic of each size have arisen (and probably been lost) repeatedly during bee evolution. Illustrations throughout Sections 36 to 121 should be examined in this connection, but Figure 28-4 illustrates the wing shape and venation for three species of a single genus, *Allodape,* that are only moderately different in size (see the legend). When the wings are drawn the same size (length), as in this figure, the size- related differences are particularly conspicuous. See also Figure 41-1 for wings of small and large species of the genus *Scrapter.*

Large, fast-flying bees frequently have papillae on the distal parts of the wings, beyond the veins (Fig. 85-2a). They are present in bees as distantly related as *Caupolicana* (Colletidae), *Centris* and *Eulaema* (Apidae), and large *Megachile* (Megachilidae). Their presence is frequently associated with bare or partly bare wings, whereas most bees that lack or have only small papillae have rather uniformly, minutely hairy wings (Fig. 85-2b).

A curious relationship is common between the lengths of the basal antennal segments. In most bees the scape is much longer than the first flagellar segment, and the second flagellar segment is somewhat shorter than the first. In bees with a long flagellum, such as most male Eucerini, the first segment is shortened, often broader than long. Thus the longer the flagellum, the relatively if not actually shorter is its first segment (Fig. 112-8c-e). The few male Eucerini with rather short antennae have long first

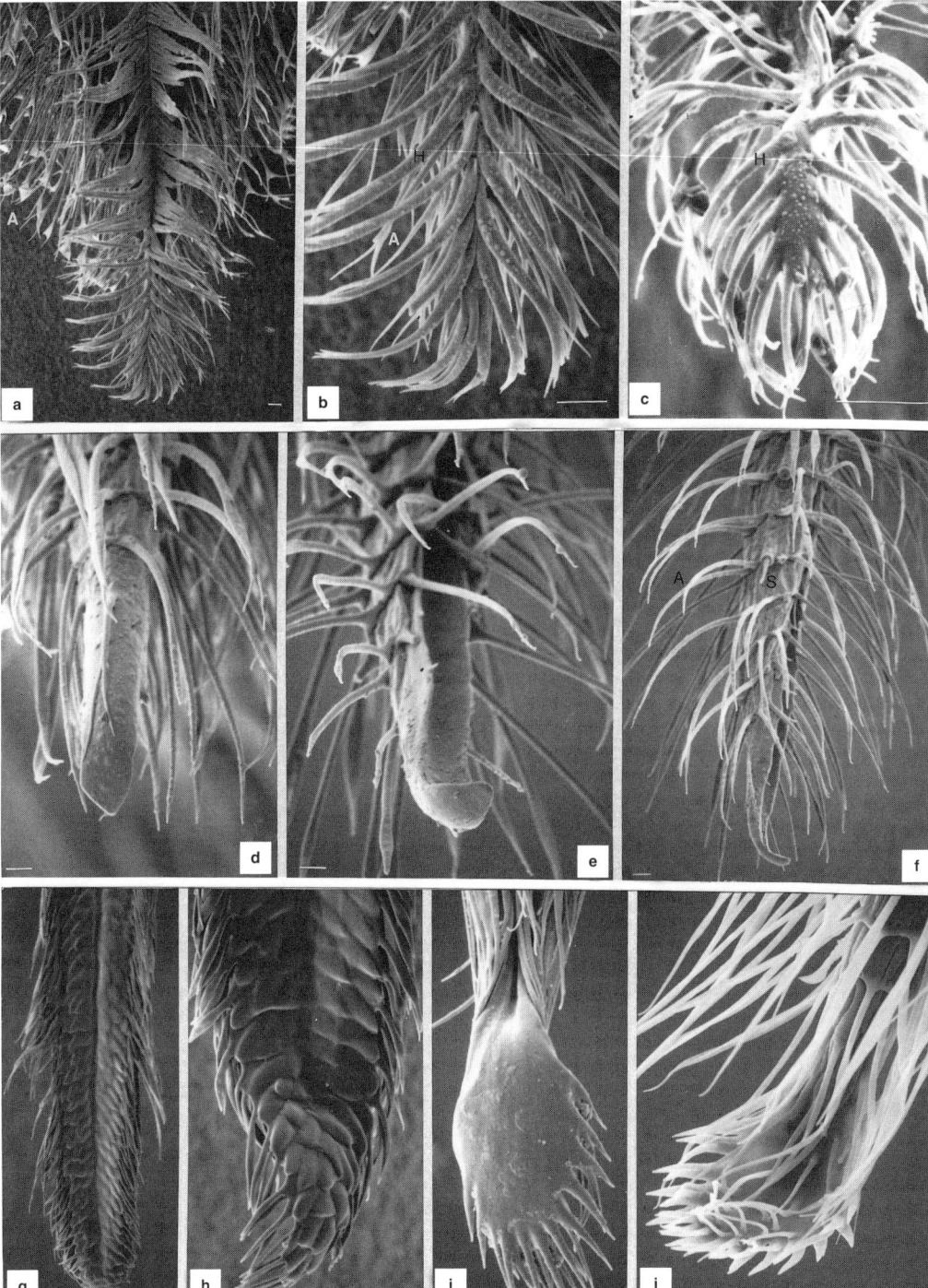

Figure 28-1. Apices of glossae. **a, b,** *Ruizantheda divaricatus* (Vachal), posterior views; **c,** *Ceylalictus divisus* (Cameron), posterior view, showing weak distal modification; **d-f,** *Rophites algirus trispinosus* Pérez, anterior, posterior, and lateral views, showing flabellum-like structure; **g, h,** *Panurginus calcaratus* (Scopoli), posterior views; **i,** *Calliopsis trifasciatum* (Spinola), posterior view; **j,** *Calliopsis australior* Cockerell, anterior view, showing flabellum. (A, annular hair; H, seriate hair. Scale lines = 0.01 mm.) See also Figure 84-1. From Michener and Brooks, 1984.

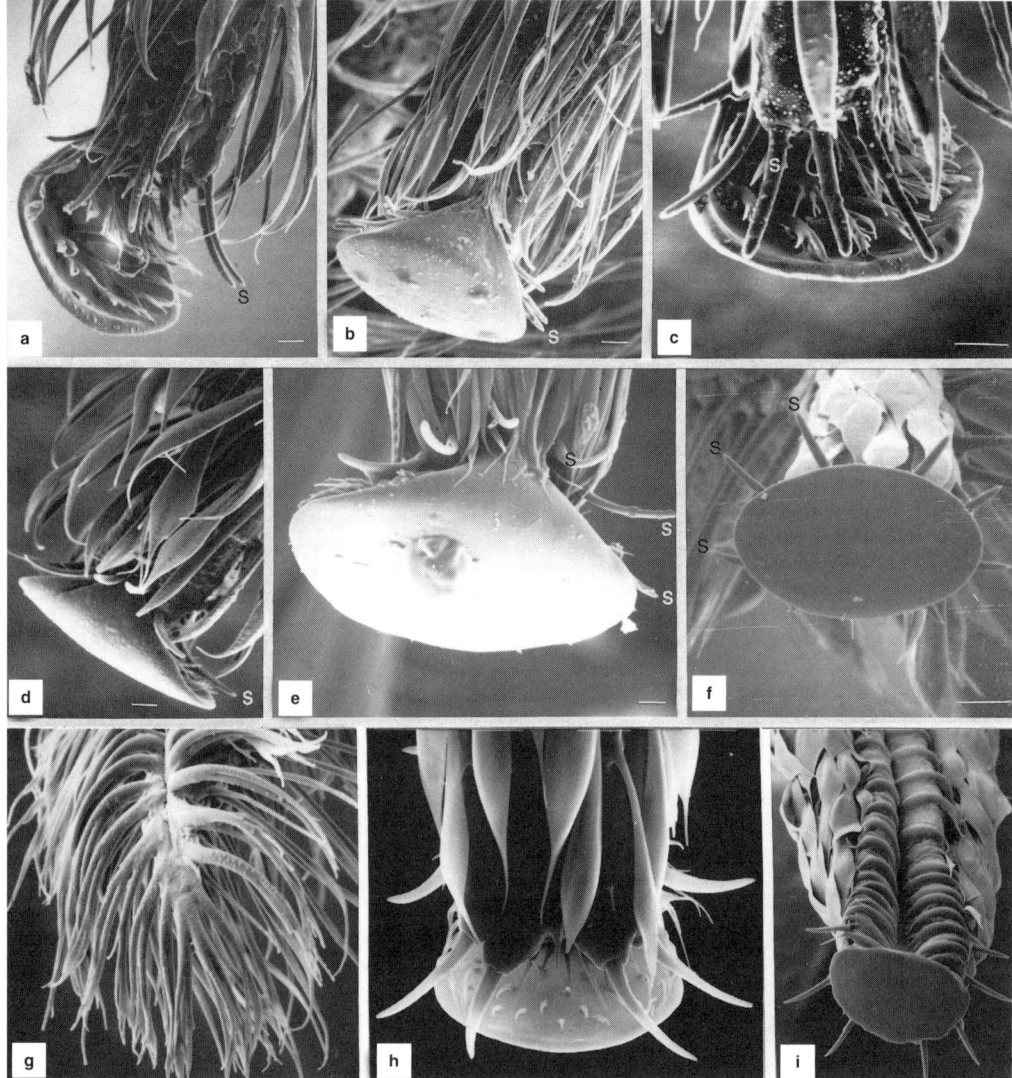

Figure 28-2. Apices of glossae. **a,** *Anthidiellum perplexum* (Smith), anterolateral view; **b, c,** *Spinanthidium volkmanni* (Friese), posterior and anterior views; **d,** *Euaspis carbonaria* (Smith), posterior view; **e,** *Megachile melanophaea* Smith, posterior view; **f,** *Ashmeadiella bigeloviae* (Cockerell), anterior view; **g,** *Protandrena bancrofti* Dunning, posterior view; **h,** *Perdita tridentata* Stevens, posterior view; **i,** *Perdita zebrata zebrata* Cresson, anterior view. (s indicates seta. Scale lines = 0.01 mm.) See also Figure 84-1. From Michener and Brooks, 1984.

flagellar segments; an example is *Xenoglossa*. There must be some mechanical reason for these regularly recurring relationships.

Males with unusually short antennae, although sometimes with the first flagellar segment long, are found among bees whose males form mating swarms or hover for long periods apparently in a mate-seeking context and have large eyes, convergent above. Presumably, the large eyes have to do with aerial pursuit of females. Such bees are *Melitturga* (Panurginae), Oxaeinae, some *Exoneura* (Allodapini), and *Xanthesma* subgenus *Xenohesma* (Euryglossinae). *Macrogalea* (Allodapini) probably belongs in this list, although its male behavior is unknown. Some bees whose males have enlarged eyes and exhibit the relevant behavior described above do not have short antennae; examples are *Apis* and some *Xylocopa*.

A set of carinae or lamellae is widespread among bees, almost always in the same positions. Perhaps they have to do with strengthening the integument, but their positions are such that they might also have to do with defense of the neck, the metasomal base, and the bases of the antennae. They are (1) the juxtantennal carina beside and sometimes overlapping the antennal base, (2) the preoccipital carina, over and/or lateral to the posterior concavity of the head, (3) the carina of the pronotal lobe and dorsolateral angle of the pronotum, (4) the omaular carina,

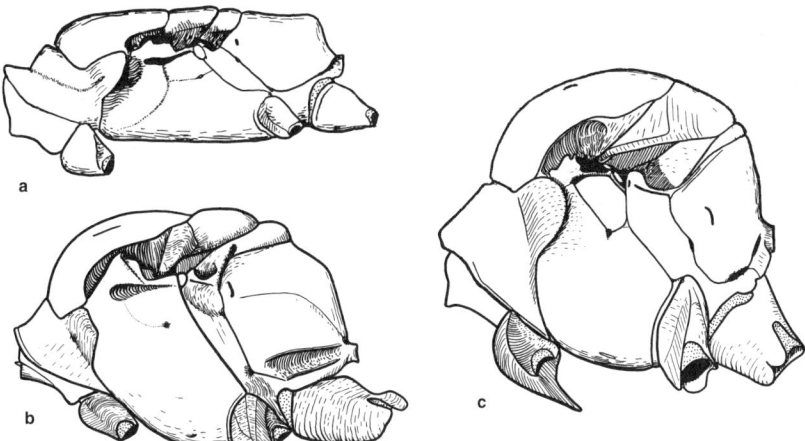

Figure 28-3. Side views of thorax. **a,** *Hylaeus (Heterapoides) delicatus* (Cockerell); **b,** *Andrena mimetica* Cockerell; **c,** *Xylococpa orpifex* Smith. b and c, from Michener, 1944.

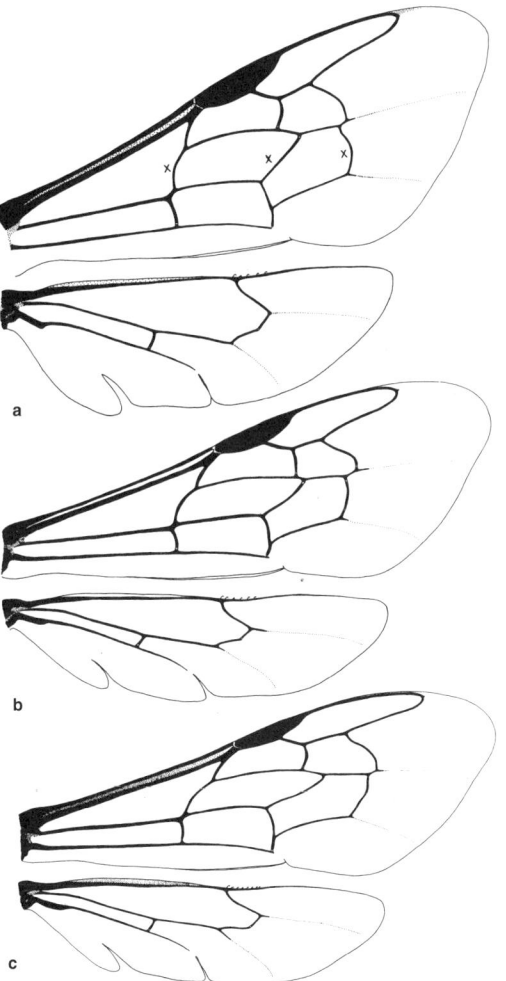

Figure 28-4. Wings of *Allodape*. **a,** *A. interrupta* Vachal; **b,** *A. exoloma* Strand; **c,** *A. mucronata* Smith. Wing lengths are 5.6, 6.0, and 8.0 mm, respectively. Note that, from smallest to largest body and wing length, the stigma becomes relatively smaller and narrower, the wings become more slender, the veins labeled x in the first figure become more longitudinal and associated cells more elongate. From Michener, 1975b.

(5) the transverse carina on the scutellum, marking its posterior extension above the metanotum, and (6) the transverse carina of T1. Nearly all these are sometimes elevated to form lamellae rather than carinae. Some, most, or all of them are found in some colletines, especially *Eulonchopria*, in some Hylaeinae, and in some Apidae, and are especially common in Megachilinae.

Other structures probably serving for defense of the neck and petiolar regions are spines or angular projections on the dorsolateral pronotal angles, pronotal lobes, and axillae, and less commonly on other sclerites. Spines directed posteriorly and possibly protecting the metasomal base are especially diverse in Dioxyini, occurring in different genera on the posterior angles of the scutum, the axillae, the scutellum, and the metanotum (Table 83-1; Michener, 1996b). Angularly produced axillae occur in diverse bee taxa, such as most *Coelioxys*, some *Heriades*, some *Stelis* (all Megachilidae); some *Callonychium* (Andrenidae); and some *Eulonchopria* (Colletidae). For comments on these features and parasitism, see Section 8, under "Social Parasites and Cleptoparasites."

The inner hind tibial spur of females, described in Section 10, is ancestrally finely serrate or ciliate, to judge by its condition in sphecoid wasps. In many groups of bees, however, its inner margin is coarsely serrate to coarsely pectinate, and sometimes it bears only one large tooth or is even toothless. Figures 39-8, 66-13, and 67-5e-n illustrate various types. Especially diverse spurs are found in the Halictini, Augochlorini, and Colletinae; less diversity exists in certain other groups. I suspect that this structure has to do with combing pollen or, in some bees, oils, off the metasoma and perhaps other areas having scopal hairs as part of the unloading process in brood cells. Spurs are ciliate or finely serrate, i.e., like the plesiomorphic condition, in most parasitic bees (e.g., *Temnosoma*, Fig. 67-5e) and in nearly all males, although pectinate in a few males, e.g., *Chlerogas* (Augochlorini). Pectinate spurs must have arisen independently and probably also been lost in many groups of *Leioproctus* and other Paracolletini, various Euryglossinae, diverse genera or subgenera of Halictidae, and some Apidae such as *Tetrapedia*. The especially finely pectinate spurs of *Ctenoplectra* (Apidae) presumably are used in manipulating floral oils; those of *Tetrapedia* may have the same function. Although not found in Megachilidae, the pectinate condition is approached in the coarse teeth of the outer margin of the same spur in *Ashmeadiella femorata* (Michener) (Osmiini) and its relatives. There are no data to suggest that the many female bees that do not have pectinate inner hind tibial spurs are less efficient in unloading pollen or oil than are those that have such spurs. It could be, however, that different kinds of pollen (sticky or dry, coarse or fine) are best manipulated with different kinds of spurs. Many species with pectinate spurs, e.g., most Halictini, however, are polylectic, not specialists on one or another type of pollen.

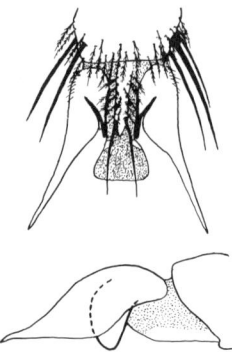

Figure 28-5. Hind tarsal claws of female of *Xeromelecta californica* (Cresson) (Melectini), dorsal and lateral views. The setae and the arolium are omitted from the lateral view.

The tarsal claws of various (but not all) cleptoparasitic Apidae have a distinctive form, the inner rami appearing as short, flattened blades whose apices are rounded or truncate, and the outer rami swollen basally and tapering, not much curved apically (Figs. 28-5 and 117-6; for comparison, see Fig. 10-10). This claw form has evolved in cleptoparasitic bees as dissimilar as some genera of Melectini, Protepeolini, and Isepeolini of the Apinae and in various tribes of Nomadinae; its function is entirely unknown.

Many male bees have enlarged and sometimes grotesquely modified hind legs, but no major group of bees consistently has such legs; rather, some species of many genera are so equipped. They are particularly frequent in the Nomiinae and Xeromelissinae, but also occur with varying degrees of enlargement in Centridini, Colletinae, Emphorini, Halictini, etc. Presumably, such males use the legs to hold females at copulation. Such behavior has been described for *Nomia* (Wcislo and Buchmann, 1995). Toro and Magunacelaya (1987), working with Xeromelissinae, described and illustrated the strong musculature occupying the swollen femora. Behavioral differences should be recognizable between related species with and without such legs. See also Section 4.

The preceding paragraphs describe only a sample of the new structures or arrangements that appear to have arisen independently among different bees. More examples can be gleaned from Michener (1944). An interesting case of specialized facial hairs of various species in diverse families, the hairs used for pollen collecting, is discussed in Section 6. The yellow facial marking of many male Hymenoptera is also probably a repeatedly derived feature; see Section 4.

29. Family-Group Names

Sections 29 through 31 contain practical information concerning the systematic sections of this book. They are intended not only to facilitate the use of the systematic sections but to indicate what weight should be placed on the various types of content.

The authorship and dating of family-group names of bees were dealt with by Michener (1986a, 1997b) with up-dating by Engel (2005). These names, being coordinate at all levels (subtribe to superfamily), are attributed to the same author and date regardless of their rank and termination [Code, ed. 3, art. 36; see also art. 11(f)ii]. Thus the Megachilini, Megachilinae, and Megachilidae all have the same author and date. It is not customary to cite authors and dates for these names, but for the accepted suprageneric taxa Table 16-1 provides authors' names. For dates and for synonymous or otherwise rejected family-group names, see the three papers cited above.

As noted by Michener (1986a), several of the best-known family-group names of bees would have to be changed if strict priority were to be observed. These names are Colletidae, Paracolletini, Halictidae, Anthidiini, and Anthophorini. Opinion 1713 of the Commission (1993) conserved the familiar family-group names, rendering changes unnecessary.

Family-group names based on *Dasypoda* (a bee) and *Dasypus* (a mammal) are identical, Dasypodidae. The Commission has ruled that the stem of *Dasypoda* should be Dasypoda-, so that the family name of the bee would be Dasypodaidae, thus avoiding homonymy with Dasypodidae, the mammal (Alexander, Michener, and Gardner, 1998).

An extremely unfortunate circumstance is the diversity of meanings for a single name with the same author that can result from our nomenclatural system. The worst example among bees is the various meanings that have been given by different authors to the family name Apidae, as follows:

1. Only the genus *Apis* = tribe Apini.
2. The corbiculate Apidae, i.e., tribes Apini, Bombini, Euglossini, and Meliponini.
3. The Apinae of the present work.
4. The Apidae of the present work.
5. All L-T bees.
6. All L-T bees plus Melittidae.
7. All bees.

Most of these meanings can be found in one or another recent work; one does not have to delve into ancient history to find supporters of diverse interpretations. Probably, someone will also use Apidae for all bees plus some or all sphecoid wasps.

Because some rules of nomenclature differ for names at different categorical levels, it is well to remember the three groups of names. Familiar **family-group names** (and their terminations) are for the categories of subtribe (-ina), tribe (-ini), subfamily (-inae), family (-idae), and superfamily (-oidea). **Genus-group names** are for the categories of genus and subgenus. **Species-group names** are for species and subspecies. A taxon can be transferred (up or down) to any level within its group, without a name change except for the endings of family-group names.

30. Explanation of Taxonomic Accounts in Sections 36 to 121

It is obviously impossible for one person to be familiar with all the taxa of bees. I have examined specimens of virtually all genera and most subgenera, but there are without doubt species that do not agree with the characters that I have chosen to use in the keys or descriptive comments. I have often depended upon publications by others for characters, even though I realize that this can be dangerous. I hope that the classification, keys, comments, and literature citations in Sections 36 to 121 will provide a useful summary of the world's bee taxa and an introduction to melittology, even though I am aware that keys will fail for some species and that frustrations will be numerous, e.g., when keys require both sexes or present other difficulties. When one has a single specimen, and the key requires characters from both sexes, what does one do? At least the key indicates some characters that need to be examined. Used along with other sources of information, such as the descriptive comments, even a very imperfect key can often be found useful.

The synonymy for each subgenus, or genus if it is not divided into subgenera, are thought to be complete, in the sense that all synonymized genus-group names are cited, with the type species indicated for each. For some names, the details of nomenclatural problems summarized here were presented more fully in a list of genus-group names (Michener, 1997b).

To satisfy taxonomic practices, indications of significant nomenclatural changes are appended to the synonymies. New synonyms are indicated as such, as are changes in status (subgenus to genus, genus to subgenus). It is presumptuous to label as new synonyms those subgeneric names that for purely subjective reasons are not accepted (not different enough, too few species, etc.). I have indicated such new synonymy merely as part of the necessary program of keeping track of names.

Sometimes, comments following a synonymy deal with nomenclatural matters not fully explained in the formal context of a synonymy.

For brevity, the descriptive comments offered for a taxon often do not duplicate characters indicated in the keys, except as their discussion may be needed. Thus to obtain all the descriptive material on the bees constituting a given subgenus, for example, one must read the comments (if any) in the subgenus text, the key to subgenera, the descriptive material on the genus, the key to genera, and so on to the higher taxa.

It is impossible and probably not even desirable in an account of a large group like all the bees to attempt to describe the same features for all taxa, as would normally be done in a revisional study. For all bees, such an approach would require impossibly long descriptions that would tend to obscure the crucial diagnostic characters. Instead, I have tried to note a few particularly distinctive characters in each case. In addition, because it is often useful, I indicate size (body length). Because male genitalia, S7 and S8 (the hidden sterna), and often other sterna are usually complex, difficult to describe, and highly diagnostic, I have included references to works where these structures are illustrated. For this purpose I frequently do not refer to papers containing one or a few species descriptions that are illustrated, but in other cases, especially if there are few good illustrations for a genus, I do cite such papers. I frequently mention variation among species, hoping to call characters to the attention of those who will study at the species level. In both keys and descriptive comments there are sometimes species mentioned as exceptions to a cited character, or as having a particular unusual feature. I have not always listed all such species; I usually mention just one or two to give an idea of variations among species.

The descriptive material for each terminal taxon (genus, or if it is divided into subgenera, then subgenus) is followed by a paragraph marked with a ■, starting with statement of range, then proceeding to number of species and references to revisions or keys to species. For the range I usually list the countries, provinces, or states that margin the range, not those that fall in the middle. When a city has the same name as a province or state, the meaning is the province or state. For simplicity, words like "palearctic" or "neotropical," representing faunal regions, are often used in place of or in addition to names of countries. Such regions are indicated in Section 26, in the text associated with Table 26-1. Important points are that *nearctic* includes the whole temperate part of North America, including the Mexican plateau, and that *oriental* is used for the tropical orient, whereas temperate parts of China and Japan are palearctic.

The number of species in a taxon is not always precise, because of variations in the state of the literature for different taxa. Sometimes I have merely used the number of species names proposed up to the time of a published listing. If there has been a recent revision, I use the number of species recognized in the revision. Among larger bees, and among all bees in some areas, the number of species names proposed is usually greater, sometimes much greater, than the number of species recognized by a reviser. But especially among small bees in some areas, a revision may include new species that greatly increase the number of species previously recognized. Examples are Toro and Moldenke's (1979) account of Chilean Xeromelissinae, which increased the number of known species from 10 to 49; and Timberlake's revisional papers (1954-1980) on North American *Perdita* (including *Macrotera*), which increased the number of species in the USA from 113 to 498. Whether a revisional study will increase or decrease the number of species recognized in a taxon cannot always be predicted. I believe that full knowledge of bee species worldwide would increase the number of species recognized, partly by the discovery of entirely new ones and partly by subdividing currently recognized species. But because much synonymy remains to be discovered, the increase in total number of species will be less than the number of new species. Given this unsatisfactory state of affairs, I have not felt it worthwhile to be meticulous about adding the numbers of new species described in the years following a revision or listing, except for small genera. My totals are good enough to give an idea of the numbers of named or recognized species. Palearctic genera revised by Warncke present a special problem because, at least in the few groups with which I am familiar, his "species" often included several similar but distinct species. The result is much false synonymy. The palearctic fauna is the most difficult in the world to study because of the lack of catalogues and continent-wide revisions combined with the great number of species, many of them named long ago with minimal descriptions in diverse languages and no illustrations. I am certain that my estimates of numbers of species often suffer from these problems.

Following the paragraph on range, number of species, and revisions, if any, are sometimes paragraphs on species groups within the genus or subgenus, or on floral biology and nesting biology if there is anything interesting to say. At least references to accounts of these matters are included when information is available. References to original papers are provided in most cases, but for taxa whose biology has been much studied, such as Halictinae and Meliponinae, I often refer to review papers.

The references to revisions or biology appear under the highest relevant taxon. For example, if a family has been revised, this fact is noted in the account of that family but is not necessarily repeated under each included subfamily, tribe, and genus. Likewise, the keys and other useful material in major faunal works and catalogues on bees as a whole (Table 32-1) are not referenced under each included taxon.

31. Some Problematic Taxa

Scattered through the families of bees are several large species complexes for which the current generic classification is arbitrary and will probably be revised in the near future. I am not referring here to the many differences of opinion on rank—should a given group of species be regarded as a species group or be given subgeneric status, should a taxon be a subgenus or a separate genus, or should a taxon be regarded as a tribe or a subfamily? Such problems are perpetual; they have no right or wrong solutions. I refer instead to cases where the taxa to be recognized are uncertain, either because of conflicting character complexes or the existence of intermediates between largely separable taxa. The following list includes the largest of such complexes.

Leioproctus and *Lonchopria* in the Colletinae (Sec. 39).

Evylaeus and *Dialictus* and other groups in the genus *Lasioglossum* in the Halictini (Sec. 66).

Hoplitis, Osmia, Atoposmia, Hoplosmia, etc., in the Osmiini (Sec. 81).

Megachile, including subgenera *Chalicodoma* and *Creightonella*, in the Megachilini (all placed in the genus *Megachile*) (Sec. 84).

Eucera (Synhalonia) and *Tetraloniella* in the Eucerini (Sec. 112).

Within each of these complexes, intermediates exist among the taxa. Sometimes they are intermediates only on the basis of key characters, which break down in probably derived taxa. In such cases the current classification may be suitable with some changes in key characters. Other intermediate forms, like *Lonchopria (Lonchoprella)*, which connects *Leioproctus* and *Lonchopria*, are probably basal taxa. Phylogenetic studies have not been made, or at least are not yet reliable, for any of these groups. Molecular studies should provide an infusion of many new characters. I believe that the development of sounder classifications should in each case await both morphological and molecular studies, although I doubt that they will solve all the problems.

Each of the problem groups listed above, and numerous other problems, are explained in some detail in the systematic part of this book, Sections 36 through 121. The enormous number of similar subgenera of *Andrena* contrast sharply with the quite divergent subgenera of some other genera, for example *Leioproctus*. Such inconsistencies are numerous and are not likely to be solved in the near future.

32. The Identification of Bees

For the practical problems of identification of bees to the genus level, it may be easier to use a regional key than to go through the worldwide treatment below, even though many of the keys herein are regional. Identification to species is of course more difficult. Throughout Sections 36 to 121, I have included references to works containing keys or revisions of species. For some taxa and some areas, no such treatments exist. Table 32-1 lists some works containing keys to bee genera of certain areas, keys to species, or catalogues of species. Many other lists of the bees found in particular areas have been published, without keys to facilitate recognition of the genera or species. Nonetheless, such lists are often useful. Many were cited by Michener (1979a).

McGinley's (1987) key to the major groups of bees, based on mature larvae, was part of a key to the families of Hymenoptera. Most bee families were split, with subfamilies and constituent tribes appearing in different places in the key, because in general there are no good larval familial characters. It is thus impossible to present here a key to families based on larval characters. Only the Megachilidae (in reality the Megachilinae) emerged at a single place in McGinley's key. Larvae of that family can usually be recognized by the relatively dense hairs (both setae and spicules) over much of the body, but even this character breaks down, for *Macrogalea* (of the Allodapini) also has such hairs, and *Fidelia* and *Pararhophites* (of the Megachilidae) lack them.

Table 32-1. Some Faunal Works That List or Contain Keys for the Identification of Bees of Certain Regions.
Asterisks (*) indicate works containing keys to genera but no lists or other treatment of species.

Amiet, 1996, 1999, and Amiet, Herrmann, Müller and Neumeyer, 2001, 2004, Swizerland (keys, descriptions, maps).
Ayala, 1990. Jalisco, Mexico.
Bingham, 1897. India, Sri Lanka, Burma.
*Chiappa, Rojas, and Toro, 1990. Chile.
Dalla Torre, 1896. World.
*Diniz, 1962. Portugal, Spain.
Frey-Gessner, 1899-1912. Europe.
Friese (see Schmiedeknecht, below)
Gogala, 1999, Slovenia (list of species)
Hedicke, 1930. Europe.
Hurd, 1979. North America.
Ler (ed.), 1993. Oriental Russia.
Michener, 1951a. North America.
Michener, 1954b. Panama.
Michener, 1965b. Australian region.
*Michener, McGinley, and Danforth, 1994. North and Central America.
Mitchell, 1960, 1962. Eastern North America.
Móczár, 1957-1967. Hungary.
Müller, Krebs, and Amiet, 1997. Middle Europe.
Ornosa and Ortiz-Sanchez, 2004, revision of Colletidae, Melittidae, and Apidae (s. str.) of Iberian Peninsula, other volumes anticipated.
Osychnyuk, Panfilov, and Ponomareva, 1978. European portion of former USSR.
Pauly et al., 2001. Madagascar (revision).
Rasmont, Ebmer, Banaszak, and Zanden, 1995. Francophone Europe.
Scheuchl, 1996, 1997, 2000, three of a planned six volumes on bees of Germany and Austria, well illustrated, with keys, etc.
Schmiedeknecht, 1882, 1884, plus Friese, 1895, 1896c, 1897a, 1898a, 1901. Europe.
Schmiedeknecht, 1930. North and Central Europe.
Schwarz, Gusenleitner, Westrich, and Dathe, 1996. Central Europe.
Schwenninger, 1999, species of Stuttgart, Germany.
Silveira, Melo, and Almeida, 2002. Brazil (keys to genera and subgenera with lists of species).
Smith-Pardo, 2003, Columbia.
*Stephen, Bohart, and Torchio, 1969. Northwestern America.
Stoeckhert, 1933, 1954. Germany.
Viereck, 1916. Connecticut.
Westrich, 1984, 1989. Germany.
Wu, 1965b. China.
Wu, 2000, treatment of Melittidae and Apidae of China.

33. Key to the Families, Based on Adults

The key to families below is intended to work for the great majority of bees, but exceptions to some of the characters of some couplets exist, usually in rare or geographically limited taxa. These problems are addressed in the notes in Section 34, each note keyed to the pertinent couplet number.

Some general attributes of the families are discussed in Section 21, and the bases for the recognition of families are discussed in Sections 18 to 21, as well as in the main systematic text, Sections 36 to 121. Many of the diagnostic familial characters are in the proboscis, which must be extended for study. Moreover, most of the characters that can be seen without extending the proboscis or dissecting the male genitalia and hidden sterna are variable within families and so not valuable in identifying families. Paradoxically, then, it is often easier to identify the subfamily or tribe, or even the genus, of a bee than to identify its family. The regional keys to genera found in the works listed in Table 32-1, above, often facilitate identification. Section 35, which deals with the practical problems of identification of female bees, should also be helpful.

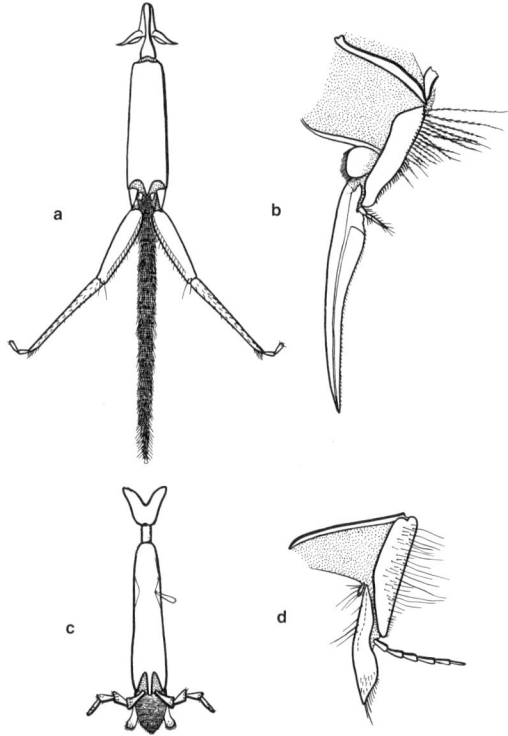

Figure 33-1. Proboscides. **a, b,** Labium and maxilla of *Anthidium atripes* Cresson (Megachilidae); **c, d,** Labium and maxilla of *Andrena mimetica* Cockerell (Andrenidae). From Michener, 1944.

Key to the Families of Bees, Based on Adults

1. Labial palpus with first two segments elongate (Fig. 33-1a), flattened, the last two segments small, usually diverging laterally from axis of first two, not flattened, rarely absent; galeal comb absent or rarely weakly indicated; stipital comb and concavity commonly present (Figs. 19-1a, 33-1b); galeal blade elongate, commonly as long as or longer than stipes (Fig. 33-1b); volsella frequently absent or difficult to recognize, rarely with distinct digitus and cuspis [L-T (long-tongued) bees] 2
—. Labial palpus with the four segments similar to one another (Fig. 33-1c), or first or rarely first two elongate but not much flattened; galeal comb commonly present; stipital comb and concavity absent (Fig. 33-1d); galeal blade usually shorter than stipes (Fig. 33-1d); volsella commonly well developed, usually with recognizable digitus and cuspis [S-T (short-tongued) bees] 3

2(1). Labrum with basolateral angles enlarged, base forming broad articulation with clypeus, labrum thus widest at base (Fig. 33-2a); labrum at least 0.8 times as long as broad and usually as long as broad or longer; forewing with two submarginal cells, usually about equal in length (except with three in Fideliini); scopa, when present, restricted to metasomal sterna .... Megachilidae (Sec. 75)
—. Labrum with basolateral angles little developed, articu-

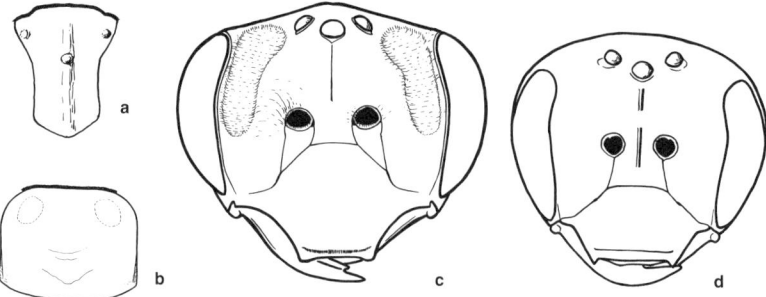

Figure 33-2. Labra and faces. **a,** Labrum of *Heriades apriculus* Griswold, female (Megachilidae); **b,** Labrum of *Anthophora edwardsii* Cresson, male (Apidae); **c,** Face of *Andrena mimetica* Cockerell, female (Andrenidae); **d,** Face of *Halictus farinosus* Smith, female (Halictidae). (The heavy lines across the tops of a and b represent diagrammatically the clypeal articulations. b, c, from Michener 1944.

lation with clypeus thus narrower than full width of labrum (Fig. 33-2b); labrum usually broader than long, but in some parasitic forms (where scopa is absent) labrum elongate; forewing with two or three submarginal cells, rarely only one; scopa, when present, on hind leg, particularly the tibia, and usually absent on metasomal sterna .... Apidae (Sec. 85)

3(1). Glossa pointed at apex, sometimes with flabellum...... 4
—. Glossa bluntly rounded, truncate, or bilobed at apex (except pointed in males of three hylaeine genera from Australia-New Guinea area); flabellum absent 7

4(3). Lacinia represented by scalelike lobe with hairs near base of galea (Fig. 33-1b, d); mentum and lorum forming proboscidial lobe (Figs. 33-3b-f, 33-4b), both at least partly sclerotized; lorum not flat 5
—. Lacinia inconspicuous or displaced, not a scalelike lobe at base of galea (Fig. 21-2a, b); mentum and lorum not forming proboscidial lobe (Figs. 33-3h, i, 34-4a), mentum sometimes membranous; lorum membranous or nearly flat sclerotized membrane (apron) between cardines (Figs. 33-3h, 33-4a) 6

5(4). Lorum more or less platelike but produced in middle for attachment to base of mentum; facial fovea present in females (Fig. 33-2c) and some males, fovea sometimes a groove rather than broad as in figure; subantennal area almost always defined by two subantennal sutures below each antennal socket (Fig. 33-2c) ....
.... Andreninae and Panurginae (Andrenidae) (Sec. 49)
—. Lorum slender, V-shaped or Y-shaped, as in L-T bees (Fig. 33-1a); facial fovea absent; a single subantennal suture below each antennal socket (as in Fig. 33-2d) ....
.... Melittidae (Sec. 68)

6(4). Lacinia a small, hairless sclerite hidden between expanded stipites; subantennal area defined by two subantennal sutures below each antennal socket (as in Fig. 33-2c); stigma nearly absent; first flagellar segment as long as scape or longer .... Oxaeinae (Andrenidae) (Sec. 60)
—. Lacinia represented by small, hairy lobe on anterior surface of labiomaxillary tube above rest of maxilla (Fig. 21-2a); a single subantennal suture below each antennal socket (Fig. 33-2d); stigma well developed; first flagellar segment much shorter than scape ...... Halictidae (Sec. 61)

7(3). Apex of glossa bluntly rounded, without preapical fringe or apical glossal lobes; episternal groove absent below scrobal groove; scopa present on hind tibia, but absent on femur .... Stenotritidae (Sec. 36)
—. Apex of glossa truncate to bilobed (except pointed in males of three genera in Australia-New Guinea region); episternal groove usually present below scrobal groove; scopa, when present, well developed on hind femur as well as tibia .... Colletidae (Sec. 37)

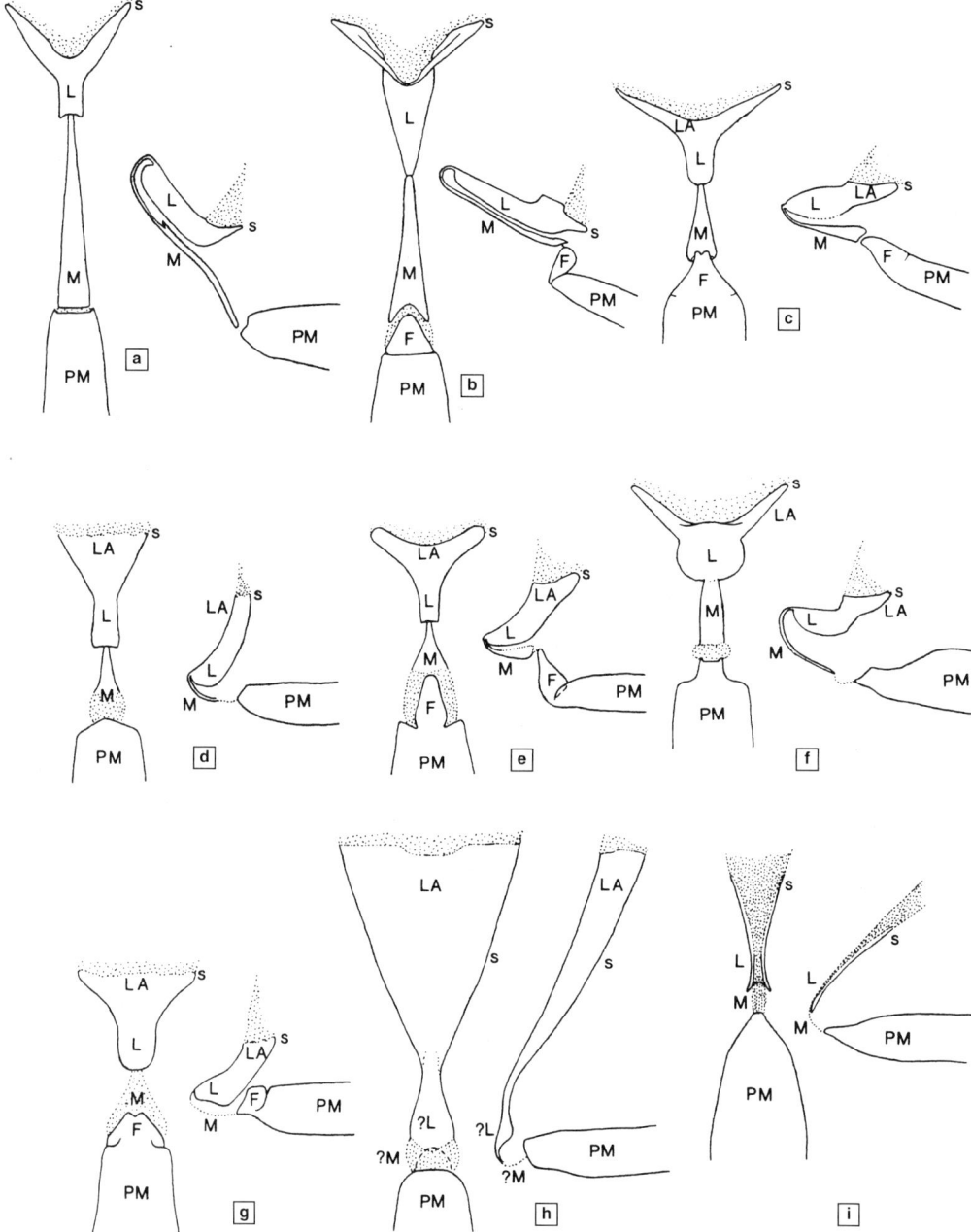

Figure 33-3. Diagrams of basal sclerites of labium, posterior views of lorum, mentum, and basal part of prementum, extended and arranged in a single plane, and lateral views of same sclerites in more natural position. **a,** *Anthophora occidentalis* Cresson (Apidae); **b,** *Melitta leporina* (Panzer) (Melittidae); **c,** *Melitturga clavicornis* (Latreille) (Andrenidae); **d,** *Panurgus calcaratus* (Scopoli) (Andrenidae); **e,** *Pseudopanurgus aethiops* (Cresson) (Andrenidae); **f,** *Megandrena enceliae* (Cockerell) (Andrenidae); **g,** *Andrena erythrogaster* (Ashmead) (Andrenidae); **h,** *Protoxaea gloriosa* (Fox) (Andrenidae); **i,** *Lasioglossum calceatum* (Scopoli) (Halictidae). Abbreviations used here and in Figure 33-4 are: LA, loral apron; L, lorum; M, mentum; PM, prementum; F, basal fragmentum of prementum; A, basal apodeme of prementum; S, area of lorum lying against shaft of cardo, or in L-T and melittid bees, against apex of cardo. (Only the profiles of unsclerotized mentums or portions of mentums are shown, as dotted lines. Dots represent membrane. Dotted areas above lorums represent the membranous posterior surface of the labiomaxillary tube, extending toward its attachment to the head.) From Michener, 1985a.

33. Key to the Families, Based on adults

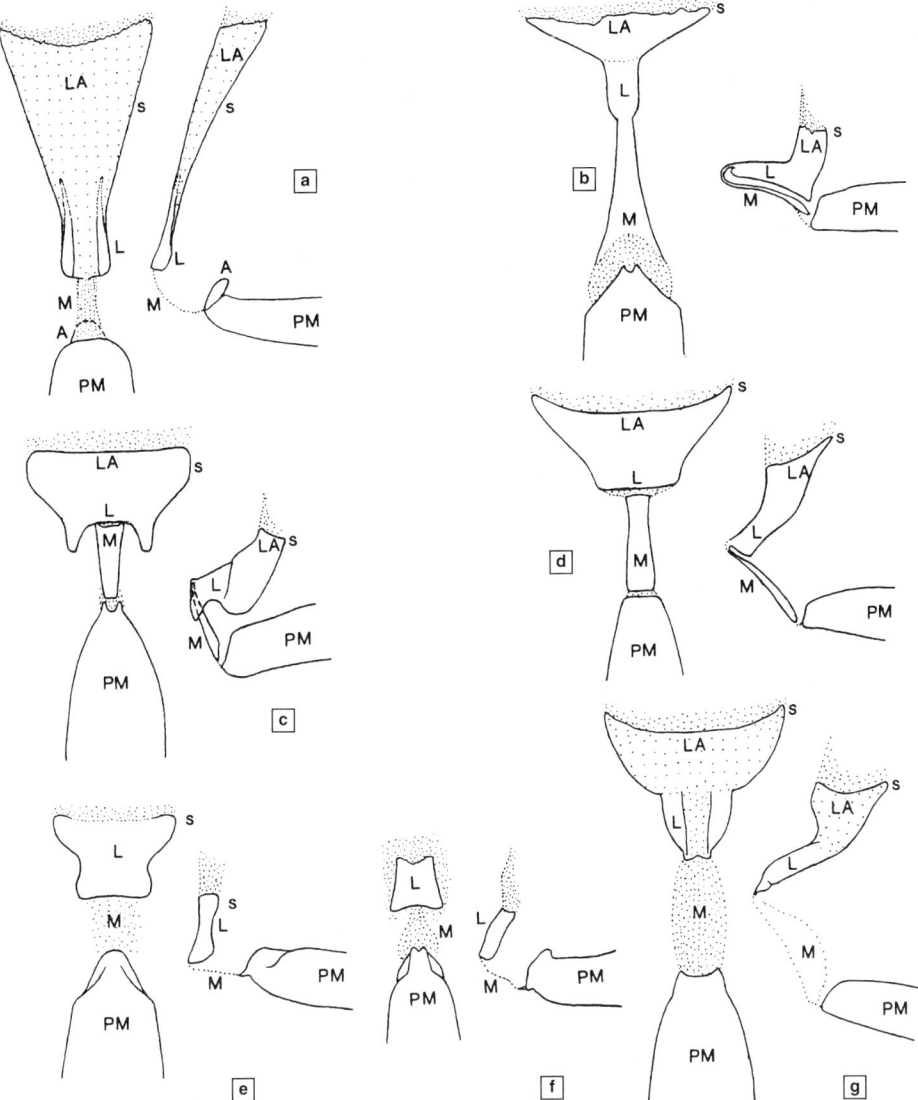

Figure 33-4. Diagrams of basal sclerites of labium, as explained for Figure 33-3. **a,** *Systropha curvicornis* (Scopoli) (Halictidae); **b,** *Lonchopria herbsti* Vachal (Colletidae); **c,** *Colletes inaequalis* Say (Colletidae); **d,** *Caupolicana hirsuta* Spinola (Colletidae); **e,** *Euryglossa subsericea* Cockerell (Colletidae); **f,** *Amphylaeus morosus* (Smith) (Colletidae); **g,** *Ctenocolletes albomarginatus* Michener (Stenotritidae). (For abbreviations, see legend, Fig. 33-3.) From Michener, 1985a.

34. Notes on Certain Couplets in the Key to Families (Section 33)

Couplet 1. See Section 19 for illustrations of variation in these characters. Some L-T bees do not agree with the statement on the labial palpus. In some subgenera of *Chelostoma* (a small, slender, holarctic megachilid), the third segment (as well as the first two) is broad; it is rather rigidly attached to the second, only one segment being nonflattened. In certain African and Malagasy social parasites in the Allodapini, the last two segments of the labial palpus do not contrast with the first two, as they do in most L-T bees; in one genus, *Effractapis,* the labial palpus has only three segments.

S-T bees rarely have more than the normal four segments of the labial palpus; extra segments are known in three palearctic species of *Andrena* (Andreninae) and three species of South American *Leioproctus* (Colletinae) (see Sec. 39). A stipital comb and associated concavity occur in the central Asian *Eremaphanta* (Dasypodainae), as in L-T bees. The few S-T bees in which the first two segments of the labial palpus are elongate include the Brazilian *Protomeliturga* (Panurginae) and the North American *Andrena (Callandrena) micheneriana* LaBerge (Andreninae).

Couplet 2. On the hind legs of female Fideliini (Africa, Chile), long hair suggests a scopa, but pollen is carried only on the metasomal scopa. And in the South African *Aspidosmia* (Anthidiini) the hind tibia bears long hairs not only suggestive of a scopa but carrying pollen in museum specimens.

In the Apidae the labrum is ordinarily little, if at all, longer than broad, but in some pasitine Nomadinae, parasitic bees without a scopa, mostly small, the labrum is much longer than broad.

Couplet 3. The hylaeine genera whose males have a pointed glossa are *Meroglossa, Palaeorhiza,* and *Hemirhiza,* all found in Australia and the New Guinea region. Like other Hylaeinae and unlike the families that run to 4 in this couplet, these three genera lack scopal hairs (the scopa is also absent in parasitic Halictidae) and have hairless, groovelike facial foveae.

Couplets 5 and 6. The only bees having two subantennal sutures below each antenna, such that the sutures are well separated at their lower ends, are in the Andrenidae (including Oxaeinae). A few other bees, e.g., the Stenotritidae, have two subantennal sutures on each side, but the sutures meet or nearly meet at their lower ends, producing a triangular subantennal area. In the Chilean *Euherbstia* (Andreninae) these sutures approach one another, leaving the margin of the subantennal area on the clypeus short, and in the Brazilian *Chaeturginus* (Panurginae) the subantennal sutures on each side nearly meet at the upper clypeal margin, but the subantennal area is long, over three times as long as wide, not a short triangle as in the stenotritids. A few Panurginae (Mexican and Arizona species of *Protandrena* s. l. and a Brazilian species of *Chaeturginus*) have only one subantennal suture on each side. Such forms differ from Melittidae in the yellow or white facial areas in the male, the truncate marginal cell, and the presence of facial foveae.

Unfortunately, subantennal sutures are easily seen only when the background is yellow or white. When the face is black, as in nearly all females and many males, these sutures are inconspicuous, often requiring removal of hairs if they are to be seen, and may be impossible to see if the surface is coarsely punctate.

35. Practical Key to Family-Group Taxa, Based on Females

Because the key to families (Sec. 33) depends heavily on characters that are difficult to see in dry specimens with the mouthparts in repose, a key based on more readily observable characters seems worthwhile. This key does not usually lead to families, but rather directs the user to tribes or subfamilies. It is based on females; for males, it is best to make the necessary examinations of mouthparts and use the key to families, Section 33.

The tibial hairs of *Pararhophites* (Megachilidae) look like a scopa but may not function for pollen carrying. For the purposes of this key, they are considered to be a scopa (see couplet 1).

Users of the key will find that both Xylocopinae and Apinae run to Apidae, couplet 11. See Section 85 for distinctions between the two.

Ancyla (Ancylini) and the Ctenoplectrini are apids that would run to Melittidae (couplet 11) on the basis of the palpal character. *Ancyla*, from xeric palearctic areas, is a genus of nondescript small anthophoriform bees hard to characterize without examination of the mouthparts. The Ctenoplectrini, from paleotropical and oriental areas, are easily recognized in the female by the broad, finely comblike inner hind tibial spur and the long oil-collecting hairs on the metasomal sterna, the hairs reduced but nonetheless evident in the parasitic genus *Ctenoplectrina*.

The specification "cleptoparasites and social parasites within Apinae" in couplet 23 means the tribes Ericrocidini, Isepeolini, Melectini, Osirini, Protepeolini, Rhathymini, and parts of Tetrapediini, Euglossini, and Bombini. See the key in Section 85.

Key to the Family-Group Taxa of Bees, Based on Adult Females

1. Scopa, consisting of hairs for carrying pollen, present (Figs. 6-4, 8-5b, 8-7b, 10-11a) 2
—. Scopa absent (Figs. 8-5a, 8-7a, 8-8a) 18
2(1). Scopa consisting of erect branched hairs, longest on S2, shorter on S1 and S3 (Fig. 46-1b), scopal hairs often present also on hind legs [body hylaeiform; submarginal cells two, second much smaller than first (Fig. 46-1)] neotropics) .... Xeromelissinae (Colletidae) (Sec. 46)
—. Scopa variable, but hairs not erect, not longest, and branched on S2 3
3(2). Scopa well developed on metasomal sterna (Fig. 8-7b) but absent on hind legs [submarginal cells two, usually about equal in length (Figs. 76-1a, 80-1, 81-1, 82-1, 83-2, 84-1), except three in Fideliini, which have long hairs on hind legs that are not used in carrying pollen] ....Megachilidae (Sec. 75)
—. Scopa on hind legs (Figs. 6-4, 8-5b, 10-11a), sometimes also on sterna 4
4(3). Scopa (sometimes as a tibial corbicula) on hind tibia and usually basitarsus, elsewhere not well developed, tibial scopa thus looking considerably larger than that of femur (Figs. 6-4, 10-11a) 5
—. Scopa on hind femur (Fig. 8-5b), where a ventral corbicula is usually evident, scopal hairs usually also present on trochanter, tibia, and basitarsus and sometimes on metasomal sterna 12
5(4). Facial fovea rather small but well defined (Fig. 59-1); two subantennal sutures well separated on clypeal margin below each antenna (Fig. 33-2c) [apex of marginal cell truncate or sometimes obliquely cut off (Figs. 50-1f, 53-1, 53-2, 54-1, 56-1, 58-1, 58-2, 59-2) and thus pointed, but apex well separated from wing margin] 6
—. Facial fovea absent or vaguely defined; one subantennal suture below each antenna (Fig. 33-2d) or *if* two, then the two nearly meeting on clypeal margin 7
6(5). Facial fovea deep, with conspicuous hairs (Fig. 50-1a, b) (Peru) .... Alocandreninae (Andrenidae) (Sec. 50)
—. Facial fovea shallow, hairless, shining .... Panurginae (Andrenidae) (Sec. 52)
7(5). Two subantennal sutures below each antenna, the two nearly meeting at clypeal margin (Chile) .... Andreninae (Andrenidae) (Sec. 51)
—. One subantennal suture below each antenna (Fig. 33-2d) 8
8(7). Body largely yellow; labrum with basolateral angles strongly developed, thus broadest at extreme base where articulated to clypeus (as in Fig. 33-2a); subantennal suture short, directed toward outer margin of antennal socket (pygidial and prepygidial fimbriae absent) (Palearctic deserts) .... Pararhophitini (Megachilidae) (Sec. 77)
—. Body usually exhibiting little or no yellow; labrum with basolateral angles little developed, thus not broadest at extreme base and articulation with clypeus shorter (as in Fig. 33-2b); subantennal suture usually directed toward middle or inner margins of antennal socket 9
9(8). Episternal groove extending below scrobal groove (as in Fig. 20-5b) although frequently shallow (antennae arising below middle of face) .... Rophitinae (Halictidae) (Sec. 62)
—. Episternal groove not extending below scrobal groove (Fig. 20-5a, c) 10
10(9). Glossa short, apex broadly rounded (inner hind tibial spur pectinate) (Australia)........ Stenotritidae (Sec. 36)
—. Glossa pointed, often with flabellum 11
11(10). L-T bees, first two segments of labial palpus elongate, flattened (Figs. 10-4a, 19-1b); episternal groove commonly present down to or curving into and joining scrobal groove (Fig. 20-5c) .... Apidae (Sec. 85)
—. S-T bees, first two segments of labial palpus similar in form to subsequent segments (Figs. 10-4c, 19-5b); episternal groove almost completely absent ....Melittidae (Sec. 68)
12(4). Facial fovea well developed, covered with short hairs (two subantennal sutures below each antenna, often difficult to see) (Fig. 33-2c) ....Andreninae (Andrenidae) (Sec. 51)
—. Facial fovea absent or not well defined, not bearing distinctive short hairs, but *if* defined, then bare 13

13(12). Stigma absent (Fig. 60-2a); two subantennal sutures below each antenna (as in Fig. 51-1a) (Western Hemisphere) .... Oxaeinae (Andrenidae) (Sec. 60)
—. Stigma present, although sometimes no wider than prestigma as measured to wing margin; ordinarily only one subantennal suture below each antenna (Fig. 33-2d)......14
14(13). Stigma almost always shorter than prestigma, vein r arising almost at its apex, margin of stigma in marginal cell concave or straight and not much longer than width of stigma (Fig. 43-1); large, robust, euceriform, hairy bees (Western Hemisphere) .... Diphaglossinae (Colletidae) (Sec. 42)
—. Stigma longer than prestigma, vein r arising near its middle or at least well before its apex, margin of stigma in marginal cell straight or convex, much longer than width of stigma; andreniform bees, much more slender than those of above alternative....................................... 15
15(14). Episternal groove extending little below scrobal groove .... Nomiinae (Halictidae) (Sec. 63)
—. Episternal groove extending far below scrobal groove (Fig. 20-5b), commonly onto venter of thorax 16
16(15). Basal vein only feebly arcuate (Fig. 39-5); glossa bilobed (Fig. 19-2a, b) ....Colletinae (Colletidae) (Sec. 39)
—. Basal vein strongly curved (Fig. 65-5); glossa acutely pointed (Figs. 19-2c, d, 28-1a-c) 17
17(16). T5 with prepygidial fimbria divided by medial longitudinal zone or triangle of short, dense hairs (Fig. 65-1j) and minute, dense punctations (the hairs sometimes absent) .... Halictinae (Halictidae) (Sec. 65)
—. T5 with prepygidial fimbria weak but continuous (Eastern Hemisphere) .... Nomioidinae (Halictidae) (Sec. 64)
18(1). Episternal groove extending far below scrobal groove (Fig. 20-5b) toward ventral surface of thorax (S6 exposed, not bifurcate) 19
—. Episternal groove absent or curving into scrobal groove (Fig. 20-5a, c), extending below scrobal groove only in Caenoprosopidini (in which S6 is retracted, only its bifurcate apex exposed) 21
19(18). Glossa pointed (Fig. 19-2c, d); basal vein strongly curved (Fig. 65-5); submarginal cells usually three .... Cleptoparasites in Halictinae, both tribes (Halictidae) (Sec. 65)
—. Glossa bilobed or broadly truncate (Fig. 19-2a, b); basal vein gently arcuate (Fig. 39-5); submarginal cells two, second usually much smaller than first (Figs. 47-2, 48-2, 48-3) 20
20(19). Supraclypeal area elevated abruptly above level of antennal sockets (Fig. 47-3a); pygidial plate usually absent, but *if* present, then broad, its margins converging posteriorly; anterior surface of T1 usually lacking longitudinal median groove .... Hylaeinae (Colletidae) (Sec. 47)
—. Supraclypeal area sloping up from level of antennal sockets; pygidial plate present, the apical part slender, parallel-sided or spatulate; anterior surface of T1 with longitudinal median groove .... Euryglossinae (Colletidae) (Sec. 48)
21(18). S6 retracted under S5 except for apex, metasomal venter thus appearing to be five-segmented; apex of S6 bilobed, bifurcate, or produced to median spine, frequently bearing rows or clumps of stiff setae (Fig. 91-2) .... Nomadinae (Apidae) (Sec. 91)
—. S6 more fully exposed, the metasomal venter thus recognizably six-segmented; apex of S6 not modified as above 22
22(21). Labrum with basolateral angles strongly developed, labrum thus broad at extreme base, where articulated to clypeus (Fig. 33-2a); labral shape more or less rectangular and usually longer than broad (forewing with two submarginal cells) ....Cleptoparasites in Megachilinae, all tribes (Megachilidae) (Sec. 79)
—. Labrum with basolateral angles weakly developed, labrum thus not broadest at extreme base, articulation with clypeus not extending full width of labrum (Fig. 33-2b); labral shape often less rectangular, often rounded apically, usually broader than long 23
23(22). Epistomal suture between lateral extremity and subantennal suture arcuate, upper part of clypeus thus almost parallel-sided (Fig. 90-2); submarginal cells two (Eastern Hemisphere) .... Social parasites within Allodapini (Apidae) (Sec. 90)
—. Epistomal suture not arcuate upward in such a way that upper part of clypeus is almost parallel-sided; submarginal cells usually three ....Cleptoparasites and social parasites within Apinae (Apidae) (Sec. 102)

36. Family Stenotritidae

This family comprises two Australian genera of moderate-sized to large, robust, euceriform, hairy bees (Fig. 36-2). Superficially, these bees closely resemble those of the American tribe Caupolicanini of the Diphaglossinae [although one stenotritid species, *Ctenocolletes smaragdinus* (Smith), is bright metallic green]. Unlike colletids, they have a well-developed tibial scopa and a reduced femoral scopa, and pollen is accordingly carried principally on the tibia, as shown by Houston (1984). The first flagellar segment is longer than the scape and petiolate (Fig. 36-1b). The glossa is short, thick, rounded, and lacks the preapical fringe, glossal lobes, and glossal brush, although the brush is perhaps represented by a few, sometimes bifid, apical hairs that are longer than the other glossal hairs. The prementum lacks a depression or fovea on its posterior (lower) surface. The two subantennal sutures below each antennal socket meet above the clypeus and delimit a small, triangular subantennal area under the antenna. The facial fovea, broad and not sharply defined, is absent in males. The ocelli are low on the face, nearer to the antennal bases than to the posterior margin of the vertex. The episternal groove is absent below the scrobal groove. There are three submarginal cells (Fig. 36-1a). The stigma is slender, slightly longer than the prestigma and of the same width (measured to the wing margin), and vein r arises near the apex, the margin within the marginal cell not convex; the wing membrane beyond the veins is papillate but also bears minute hairs. The prepygidial and pygidial fimbriae of the female are large and dense (Fig. 36-2).

The larva, characterized by Houston (1975b) and McGinley (1981), is not distinguishable at the family level from that of certain Colletinae.

Although stenotritids have often been placed in the Colletidae, McGinley's (1980) study emphasized their distinctness at the family level, and that conclusion was supported in a way by Alexander and Michener's (1995) phylogenetic investigation of S-T bees. As noted in Section 21, in this phylogenetic study the stenotritid (*Ctenocolletes*) fell in such diverse positions in different analyses that there was no basis for considering it closest to any one other bee group (see Fig. 20-1). My belief, notwithstanding, is that the stenotritids are either the basal branch, sister group to all other bees (Fig. 20-1b), or a group within or sister group to the Colletidae (Fig. 20-1c, d). Among characters in support of the colletid relationship are the lobes and reduced disc of S7 of the male *Stenotritus* (Fig. 36-1f) but not of *Ctenocolletes* (Fig. 36-1c). Similarity to the Melittidae, however, in the lack of a femoral or more basal scopa and lack of the episternal groove below the scrobal groove, should be considered. The lorum, as in various Andrenidae, especially *Megandrena* and *Alocandrena*, forms a stronger proboscidial lobe (Fig. 33-4g) than is found in many Colletidae, but it is even stronger in some Colletinae (Fig. 33-4b).

Some of the characters of Stenotritidae are common features of various large, fast-flying bees. These include the reduced stigma, papillate distal parts of the wings, the largely vertical propodeal profile, and perhaps the long, petiolate first flagellar segment. It is partly such charac-

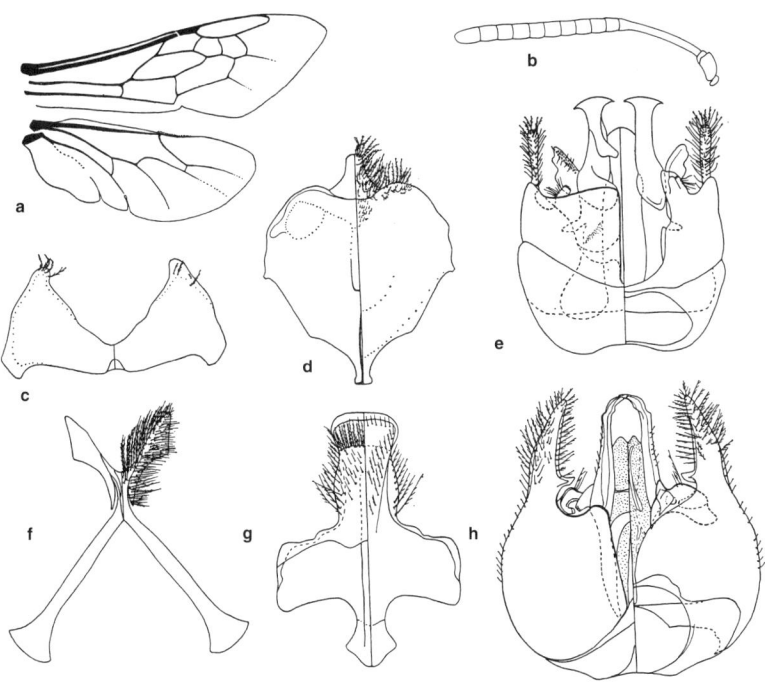

Figure 36-1. Structures of Stenotritidae. **a,** Wings of *Stenotritus pubescens* (Smith); **b,** Antenna of *Ctenocolletes nicholsoni* (Cockerell), male; **c-e,** S7, S8, and genitalia of male of *C. smaragdinus* (Smith); **f-h,** S7, S8, and genitalia of male of *S. pubescens* (Smith). (For the genitalia and sterna, dorsal views are at the left.) From Michener, 1965b.

ters that are responsible for the similarity of Stenotritidae to the Caupolicanini, and for the relationship between Stenotritidae and Oxaeinae indicated in some of the phylogenetic analyses of Alexander and Michener (1995).

The two genera of Stenotritidae are similar to one another in most characters, although strikingly different in the form of S7 and S8 of the males.

Key to the Genera of the Stenotritidae

1. T7 of male usually with well-developed pygidial plate; S7 of male a transverse band sometimes broadened and produced apically at each side, leaving large median emargination (Fig. 36-1c), with hairs on apical margin or projections; basal elevation of labrum of female undivided; inner hind tibial spur of female thickest at basal one-fourth to one-half, with long, coarse teeth ...... *Ctenocolletes*
—. T7 of male with bare area representing pygidial plate but not defined by carinae; S7 of male with disc greatly reduced, connecting long laterobasal apodemes and a pair of hairy apical lobes (Fig. 36-1f); basal elevated area of labrum of female binodulose; inner hind tibial spur of female tapering from base, with moderate-sized to coarse teeth .... *Stenotritus*

Genus *Ctenocolletes* Cockerell

Stenotritus (Ctenocolletes) Cockerell, 1929c: 358. Type species: *Stenotritus nicholsoni* Cockerell, 1929, monobasic.

Ctenocolletes is composed of large (14.0-20.5 mm long), euceriform, hairy, fast-flying bees (Fig. 36-2). They differ not only from *Stenotritus* but also from most other bees in (1) the broad, platelike or bandlike disc of S7 of the male, bearing short lateral apodemes and an entire to bilobed apical margin (Fig. 36-1c), and (2) the large S8 of the male, which has a short or slender apical process (Fig. 36-1d) and hairs on the apical one-third to one-half of the sternum, which is often exposed. On the basis of these sternal characters, *Ctenocolletes* could be regarded as the most primitive of bees, since S7 and S8 are more like the other sterna than like the totally different S7 and S8 found in most other bees. Supporting this view are the simple, articulated, hairy male gonostyli (Fig. 36-1e) and the well-formed male pygidial plate (except in *C. fulvescens* Houston). But *Ctenocolletes* is closely related to *Stenotritus*, which shares none of these features. It seems likely that regulatory factors lead S7 and S8 to develop more like ordinary sterna in *Ctenocolletes*. Clearly, bees

Figure 36-2. *Ctenocolletes nicholsoni* (Cockerell), female. From Goulet and Huber, 1993.

have genes for ordinary sterna; a regulatory change could lead to such sterna on segments 7 and 8 as well as on more anterior segments. Other bees with such ordinary-looking male S7 and S8 include the Oxaeinae (Andrenidae). In fact, the Oxaeinae have an even more platelike S8, without a spiculum. Other bees with a platelike S7 are *Melitta* (Melittidae) and *Euherbstia* and *Orphana* (Andreninae). The very different larval as well as adult characters argue against any close relationship between *Ctenocolletes* and these andrenid and melittid taxa. Some species of *Ctenocolletes* have an extraordinarily small propodeal triangle (incorrectly emphasized as a generic character by Michener, 1965b). Three species lack arolia in females (Houston, 1983b). Many other structures, including male genitalia, were illustrated by Michener (1965b) and Houston (1983a, b).

■ *Ctenocolletes,* found in Western Australia and the westernmost part of South Australia, comprises ten species, as revised by Houston (1983a, b; 1985).

Nesting, mating, and floral biology are described by Houston (1984, 1987). The nests resemble those of *Stenotritus* in major features; sometimes the cells are inclined, and only the distal parts are varnished with a secreted material.

Genus *Stenotritus* Smith

Stenotritus Smith, 1853: 119. Type species: *Stenotritus elegans* Smith, 1853, monobasic.

Oestropsis Smith, 1868a: 253, not Brauer, 1868. Type species: *Oestropsis pubescens* Smith, 1868, monobasic.

Gastropsis Smith, 1868b: xxxix, replacement for *Oestropsis* Smith, 1868. Type species: *Oestropsis pubescens* Smith, 1868, autobasic.

Melitribus Rayment, 1930a: 217. Type species: *Melitribus greavesi* Rayment, 1930, by original designation. [Rayment (1930b: 61) subsequently and invalidly designated *Gastropsis victoriae* Cockerell, 1906, as the type species.]

Stenotritus is composed of moderate-sized (body length 12-15 mm), euceriform, hairy, fast-flying bees. The characters of the hidden sterna suggest those of most colletids, S7 having a reduced disc, long basolateral apodemes, and two hairy apical processes (Fig. 36-1f); and S8 having a strong spiculum and a strong, subtruncate, hairy apical process (Fig. 36-1g). Presumably, these features are plesiomorphic, but the lack of a male pygidial plate and the immovable fusion of the gonostyli to the gonocoxites (Fig. 36-1h) appear to be synapomorphies of *Stenotritus*. Male genitalia and hidden sterna were illustrated by Michener (1965b); see also Figure 36-1f-h.

■ *Stenotritus* is known from the east to the west coasts of Australia and north to southern Queensland, but not from the tropical north. There are eleven species.

Important papers on nesting and floral biology are by Houston (1975b) and Houston and Thorp (1984). The nests are burrows in the soil, with more or less horizontal branches at the bottom, each leading to one or perhaps two bilaterally symmetrical (i.e., with flattened floor), horizontal cells lined with a very thin, water-repellant, secreted film. The provisions in each cell consist of a flattened ovoid mass, sometimes surrounded by liquid. Larvae do not spin cocoons.

37. Family Colletidae

As noted in Section 21, the Colletidae are morphologically diverse bees, such that one could easily justify recognizing several families among them, as some authors have done. These bees, however, have synapomorphies (as indicated below), and it seems reasonable to retain them as a single, large, worldwide family. It is most abundant and most diversified in temperate parts of Australia and South America. In the holarctic region there are only two common genera, *Colletes* and *Hylaeus;* neotropical genera enter the southern USA, especially the Southwest. By contrast, in Australia there are many genera. The family is relatively scarce in the moist tropics, especially so in the Indo-Malayan area.

Nearly all members of the family are easily characterized by glossal features not found in other bees. These features are found in all colletid females; as noted below, some of them are not found in certain males (see Secs. 20, 21). The glossa is short, commonly broader than long, truncate, bilobed (Figs. 19-2b; 20-3a, b; 37-1) or bifid, sometimes drawn out into two long, pointed processes (Fig. 39-12). The disannulate surface is as broad as the annulate surface, the former including an apical zone (beyond the preapical fringe) that is usually expanded into a pair of large apical glossal lobes that bear the conspicuous branched or simple hairs of the glossal brush (Fig. 37-1). The annuli, or those of one area, are fine and close, the annular hairs usually minute and blunt, capitate, or spatulate (Fig. 37-2). The distal end of the annulate surface is usually marked by the preapical fringe (Fig. 37-1). The disannulate surface is hairy but lacks seriate hairs (Fig. 20-3b). Some of these features—the apical zone or lobes derived from the disannulate surface, the glossal brush, the fine annuli and minute, blunt or spatulate annular hairs, and the preapical fringe—are unique synapomorphies of the Colletidae. The others may be synapomorphies or may be plesiomorphies derived from sphecoid wasps (see Secs. 20, 21; also Michener and Brooks, 1984, and Michener, 1992c).

The glossal features described above distinguish all female Colletidae from all other bees, and appear to show that the family is a monophyletic unit. Some of them are also found in most males. In males of *Hemirhiza, Meroglossa,* and *Palaeorhiza* (Hylaeinae), however, the glossa is pointed (Figs. 20-3c, 47-1i, j), much like that of *Andrena,* not only in shape but in having alternatives to all the characters listed in the preceding paragraph (Michener and Brooks, 1984). Michener (1992c) further elaborated this point and noted lesser sexual differences in glossal structure in most colletids, although not in certain species of *Colletes.* Many male colletids with the glossa shaped about as in their female conspecifics nonetheless lack the preapical fringe, apical zone, glossal lobes, and glossal brush, and have coarser annuli with pointed annular hairs. Males of *Amphylaeus* and *Hylaeus (Hylaeorhiza)* have an intermediate sort of glossal shape with a small to distinct median apical glossal point, but lack the other *Andrena*-like glossal features of males of *Hemirhiza,* etc. For details and relevant illustrations, see Section 47.

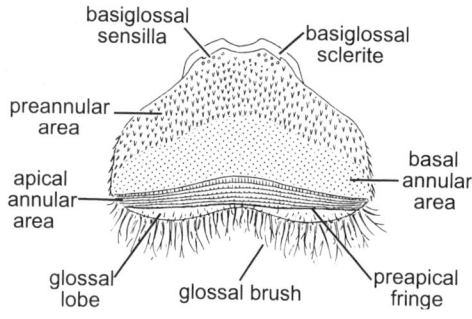

Figure 37-1. Diagram of glossa of a *Hyleoides* (Hylaeinae) female, showing typical structures of a colletid glossa. From Michener, 1992c, after McGinley, 1980. See also Figure 19-4a.

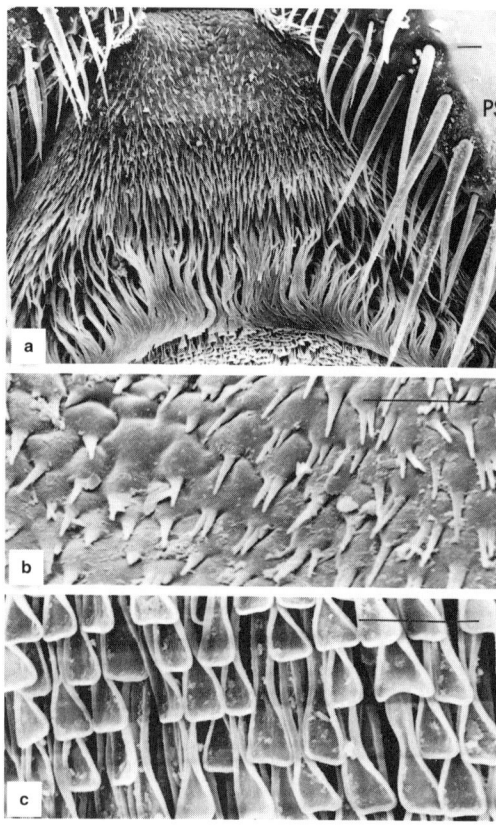

Figure 37-2. Glossa of *Niltonia virgilii* Moure (Colletinae), showing details of anterior surface of glossa. **a,** Basal part of glossa (PS = paraglossal suspensorium); **b,** Basal or preannular area; **c,** Apical annular area, showing transverse rows of spatulate annular hairs. (Scale lines = 0.01 mm.) From Laroca, Michener, and Hofmeister, 1989.

Less distinctive characters of the family, that is, characters shared with certain other families, can be described as follows: The labrum is usually much broader than long, and the apical margin in both sexes is fringed with bristles. Below each antenna is one subantennal suture, directed toward the inner margin of the antennal socket. [*Leioproctus semicyaneus* (Spinola) from Chile has two apparent subantennal sutures; in some Australian genera (*Xanthesma, Brachyhesma*) the antennal sclerites reach the clypeus, effectively eliminating the subantennal sutures.] The clypeus is usually relatively flat and its lower lateral parts are usually not much bent posteriorly on either side of the labrum. The mentum varies, from absent (i.e., membranous) to sclerotized and tapering from a broad apex to a narrow base. The base of the prementum is not a detached fragmentum. The lorum is either a flat plate (Fig. 33-4e, f) or elevated distally around the base of the mentum (Fig. 33-4b-d), in that case usually forming a strong proboscidial lobe (Fig. 33-4b). The galeal comb is present (Fig. 38-18a, b), but much reduced and consisting of only three or four bristles in *Scrapter*, and absent in *Leioproctus (Excolletes)*. The epispernal groove is usually fully developed (Figs. 28-3a, 37-3) and extends well below the scrobal groove, but in the Diphaglossini it is absent below the scrobal groove, and it is nearly absent in *Hesperocolletes* (Colletinae). The middle coxa appears shorter than the distance from its summit to the base of the hind wing (Figs. 28-3a, 37-3), because the upper quarter or more is hidden (Michener, 1981b). The propodeal triangle is hairless. The scopa is absent to well developed on the hind leg (trochanter to basitarsus) and sometimes also on the metasomal sterna. The disc of S7 of the male is usually reduced to almost nothing but supports long basolateral apodemes and one to three pairs of long, sometimes elaborate, haired apical lobes (Fig. 13-2); this feature is lost in a few groups of *Hylaeus* and *Leioproctus*, but is found in *Stenotritus* (Stenotritidae) and is approached in other families as well. S8 usually has a strong apical process. The volsella is present and free, with recognizable digitus and cuspis. The spatha is absent.

Larvae were described and illustrated in detail by McGinley (1981). Since no combination of characters distinguishes all colletid larvae from those of Andrenidae, Melittidae, etc., an enumeration of the larval characters here seems unwarranted (see McGinley, 1987). Characters of colletid pupae, with emphasis on differences among subfamilies, were tabulated by Torchio and Burwell (1987).

All colletid species are solitary, although some nest in aggregations. Species in the hairy subfamilies (Colletinae, Diphaglossinae) usually excavate burrows in the soil, although *Callomelitta* nests in rotting wood and certain *Colletes* nest in pithy stems. Nests of the species in sparsely haired subfamilies may be excavated in the soil (*Euryglossa*), but many are in preexisting holes in wood, stems, soil, volcanic rock, etc. Most species that excavate nests in the soil make cells, commonly only one or a few per lateral burrow, that are homomorphic. If more or less horizontal, as is common in Colletinae, the cells are usually bilaterally symmetrical (flatter on the lower side than elsewhere). If the cells are consistently vertical as in Diphaglossinae, then they are round in cross section and sym-

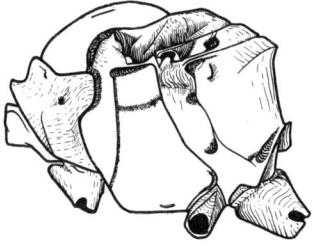

Figure 37-3. Lateral view of thorax of *Colletes fulgidus* Swenk, showing the long episternal groove and the small exposed part of the middle coxa. From Michener, 1944.

metrical if one ignores the region toward the cell entrance, which curves to one side. *Colletes*, however, often makes heteromorphous series of cells in subterranean burrows. Species of the sparsely haired subfamilies mostly make heteromorphous cells, often in series, although *Euryglossa* and allies make homomorphous cells.

A colletid synapomorphy is the cellophane-like cell lining (Pl. 15), a lining different from that of other bees. In spite of physical similarity, the cell linings were reported to be chemically different in *Hylaeus* and *Colletes*, in the former being silklike proteins presumably secreted by the salivary glands, in the latter, polyesters (specifically laminesters) from Dufour's gland (Hefetz, Fales, and Batra, 1979; Batra, 1980). The difference is probably not so great as this statement suggests (Espelie, Cane, and Himmelsbach, 1992); in both genera the materials are spread by the glossa before polymerization, and it is likely that in both cases salivary gland components and Dufour's gland secretions are mixed.

As noted by Torchio (1984), the provisions in colletid cells, except for those of *Leioproctus* and *Lonchopria* in the Paracolletini, are, so far as is known, liquid (but see *Scrapter*). The egg floats on the surface of the liquid, or in the case of *Colletes*, hangs from the top of the cell. The young larva curls on its side on the surface of the liquid. Torchio associated this with a 90° rotation of the late embryo, contrasted with 180° rotation in other Hymenoptera. The 90° rotation may be an apomorphy of the Colletidae. Unfortunately, the embryology is unknown for *Leioproctus* and *Lonchopria*, which have firm subspherical pollen masses and whose larvae may well lie on their ventral surfaces.

The Colletidae are divided into five subfamilies. At least superficially, these subfamilies are very different from one another. The Diphaglossinae (mostly neotropical) are medium-sized to very large, euceriform, densely hairy, fast-flying bees. The cosmopolitan, andreniform Colletinae are also moderately hairy. These subfamilies have large scopae on the hind legs (coxa to basitarsus) of the females and sometimes on the metasomal sterna as well. The remaining subfamilies contain mostly small, sometimes minute, usually slender, hylaeiform bees (Pl. 1), although many Euryglossinae are more andreniform and some are hoplitiform. The pubescence consists of short, generally sparse hairs. Species of the neotropical Xeromelissinae have small scopae on the hind legs and the first three metasomal sterna, but the cosmopolitan Hylaeinae and Australian Euryglossinae have no scopa and carry pollen internally, in the crop, instead of externally, on the scopa.

The phylogeny of 20 adult exemplars representing all colletid subfamilies was investigated by Alexander and Michener (1995) as part of a study of S-T bees. The Colletinae consistently appeared as paraphyletic, although the groups derived from within this subfamily varied. An analysis including many more colletid taxa is needed to settle on the most likely relationships. One group that was often derived from within the Colletinae in the analyses was the Diphaglossinae. I do not believe, however, that this is the correct position for Diphaglossinae, for the larvae of that subfamily spin cocoons. This behavior and the projecting labial region of the larva on which the salivary (silk) opening lies are ancestral features shared with sphecoid wasps. No other colletid spins cocoons or has such a projecting labium. It is unlikely that the Diphaglossinae arose from a group that does not spin cocoons, like the Colletinae, and re-evolved cocoon spinning and the structures necessary to do so. One of Alexander and Michener's consensus trees (their fig. 13) shows the Diphaglossinae as the sister group of all other colletids, a position supported by these larval and spinning characters. Nonetheless, the phylogenetic position of the Diphaglossinae is not firmly established. The restriction of the Diphaglossinae to the Western Hemisphere is not what one would expect of the basal colletid clade. The disjunct panaustral distribution of the Colletinae (ignoring the widespread *Colletes*) is much more suggestive of an ancient type.

Alexander and Michener's (1995) analyses usually showed the Hylaeinae and Xeromelissinae as sister groups, closely associated sometimes with the Euryglossinae and sometimes with the tribe *Scraptrini* of the Colletinae. In some analyses the Euryglossinae were part of this same clade. I tend to accept this relationship as likely, although in other analyses the Euryglossinae fell at the base of the Colletidae or the base of all bees. The analyses seem to establish the relation of Xeromelissinae to Hylaeinae, although the former has a small scopa and the latter lacks a scopa. They do not establish the position of Euryglossinae, but as noted in Section 21, the large, crescentic galeal comb on a curved sclerite in both Euryglossinae and Hylaeinae seems to be a unique synapomorphy indicating common ancestry.

The idea that the Euryglossinae might be the sister group to all other bees is supported by such observations as their restriction to Australia, the lack of a scopa, the unusually short proboscis, and, as has been emphasized by John Plant (manuscript, 1991), the lack of a galeal velum in most genera. Except for the association with Australia, these items are wasplike and therefore can be interpreted to support a basal position in bee phylogeny. I interpret these matters differently, however. The scopal loss I consider a probable synapomorphy in common with Hylaeinae, as is the crescentic sclerite bearing the fused bases of the bristles of the galeal comb, among other characters. In the euryglossine genus *Pachyprosopis* the galeal velum is present, just as it is in the Hylaeinae and most other bees. Its loss in several other genera could be a derived condition. The reverse hypothesis, that the galeal velum evolved in *Pachyprosopis* and in other bees, almost certainly would require it to arise twice, for *Pachyprosopis* is clearly a euryglossine, very different from other bees. The euryglossine galeal blade is not at all similar to that of sphecoid wasps, in spite of the lack of the velum in most genera of both taxa. Features of wasp galeal blades that differ from those of bees, including euryglossines, are the sclerotic plates on the inner galeal surface and the comb, which is not homologous to that of bees. The short proboscis of the Euryglossinae may be a special Australian development. Most native nectar sources in Australia are in the Myrtaceae, whose flowers are wide open like cups of nectar (Michener, 1965b). Long proboscides are thus not needed.

A recent catalog of neotropical species of Colletidae, published in five parts, will greatly facilitate studies, especially on *Colletes* and *Hylaeus* in South America. The parts are: 1, Paracolletini, Moure, Graf, and Urban, 1999; 2, Diphaglossinae, Urban and Moure, 2001; 3, Colletini, Moure and Urban, 2002a; 4, Hylaeinae, Urban and Moure, 2002; 5, Xeromelissinae, Moure and Urban, 2002b.

The subfamilies can be distinguished by reference to the following key.

Key to the Subfamilies of the Colletidae

1. Body usually hairy, female with well-formed scopa enclosing corbicula on underside of hind femur; prepygidial fimbria (apical hair band of T5) of female much stronger (hairs longer and denser) than hair bands (if any) of preceding terga, and T6 with abundant hair (pygidial fimbria) lateral to pygidial plate (except in genera that lack both fimbriae and pygidial plate); pygidial plate of female, if present, usually broad and tapering posteriorly; forewing with three submarginal cells or, *if* two, then second at least two-thirds as long as first, as though second submarginal crossvein is lost; distal submarginal crossvein sinuate, at acute angle to distal part of radial sector forming free part of marginal cell (Fig. 38-17) 2
—. Body with hairs short and relatively sparse, female lacking scopa or with a sparse or short scopa forming corbicula on underside of hind femur; prepygidial and pygidial fimbriae of female nearly always absent; pygidial plate of female absent or, *if* present, then usually narrow and parallel-sided posteriorly, or spinelike; forewing with two submarginal cells, second usually much shorter than first, as though first submarginal crossvein is lost; second submarginal crossvein usually not sinuate, usually at right or obtuse angle to distal part of radial sector (Fig. 47-2) 3
2(1). Stigma small, shorter than prestigma, as wide as prestigma measured to costal wing margin (Figs. 43-1, 44-1, 45-1); glossa deeply bifid with apical lobes commonly directed strongly apicolaterally (Western Hemisphere) ....
.... Diphaglossinae (Sec. 42)
—. Stigma usually large, at least longer than prestigma, usually wider than prestigma measured to wing margin (Figs. 39-2a, 39-5, 39-11, 41-1); glossa weakly bilobed to deeply bifid but, *if* the latter, then lobes commonly directed more apicad than laterally (except apicolaterally in *Leioproctus tomentosus* Houston from Western Australia)
.... Colletinae (Sec. 38)
3(1). Facial fovea absent or broad, at least one-third as wide

as long; female with sparse scopa on S1 to S3 and outlining ventral corbicula on hind femur (Fig. 43-1); longitudinal part of hypostomal carina usually longer than clypeus but, *if* not, then clypeus protuberant and its lower lateral extremities bent back around ends of labrum (neotropical) .... Xeromelissinae (Sec. 46)
—. Facial fovea usually a narrow groove, sometimes a broader area, wider than diameter of scape, absent in a few females and some males (Figs. 47-4, 47-6, 47-7); scopa absent; longitudinal part of hypostomal carina usually not longer than clypeus; clypeus usually not protuberant, not much bent back around ends of labrum 4

4(3). Supraclypeal area elevated abruptly above level of antennal socket (Fig 47-3a); pygidial and basitibial plates usually absent but, *if* present (as in a few Australian and New Guinea species), then pygidial plate of female broad, its margins converging posteriorly; anterior surface of T1 usually without longitudinal median groove; posterior (lower) surface of prementum with longitudinal, usually spiculate depression or fovea (Fig. 41-3a) (weak in a few males) margined by ridges that diverge on basal half of prementum and meet near base of subligular process .... Hylaeinae (Sec. 47)
—. Supraclypeal area sloping up from level of antennal socket; apical part of pygidial plate of female slender, sometimes a spine, its margins parallel or converging slightly toward apex or spatulate; basitibial plate usually indicated in female, sometimes only by one or more tubercles; anterior surface of T1 with longitudinal median groove; posterior (lower) surface of prementum lacking longitudinal medial fovea but with comparable spiculate area (Australia) .... Euryglossinae (Sec. 48)

38. Subfamily Colletinae

This subfamily is by far the largest and most abundant group of hairy colletids. It consists of small to rather large, generally hairy andreniform bees, most of them superficially resembling species of the genera *Andrena* or *Halictus*. A few small species are only sparsely hairy and are almost hylaeiform, especially in males. The first flagellar segment is shorter than the scape and not recognizably petiolate. The glossa usually has two short lobes (it is thus weakly bilobed, Fig. 19-2b, as in Fig. 37-1) but is sometimes deeply bifid (Fig. 39-12); the preapical fringe (at least in the female) and the glossal lobes (or apical glossal zone bearing the glossal brush) are well developed. The prementum lacks a spiculate depression on the posterior (lower) surface, or has a longitudinal median groove perhaps homologous to the depression; or, in the African genus *Scrapter*, the spiculate depression is well developed (Fig. 41-3b). The facial fovea is usually broad and ill-defined or absent, but sometimes is sharply defined, and sometimes (as in *Callomelitta*, some *Eulonchopria*, and some *Scrapter*) forms a groove. The episternal groove extends well below the scrobal suture (but forms only a broad, shallow depression in *Hesperocolletes*). The scopa on the hind leg of the female is large and dense to sparse, forming a corbicula on the underside of the femur; it is also well developed on the tibia and sometimes on the metasomal sterna and on the side of the propodeum. The scopa is almost absent, however, in *Leioproctus (Euryglossidia) cyanescens* (Cockerell), which seems to transport pollen in the crop instead of on the scopa (Houston, 1981b). There are two or three submarginal cells; if two, then the second is two-thirds the length of the first or longer, as if the second submarginal crossvein has been lost. The pygidial and prepygidial fimbriae of the female are usually present, often dense, but they are absent (margins thus similar to those of preceding terga) in *Colletes* and *Mourecotelles*.

The larva was characterized by McGinley (1981, 1987).

The distribution of the subfamily is worldwide except for the arctic and antarctic. Bees of this subfamily are uncommon (or in some areas absent) in moist tropics, and probably completely absent in much of tropical Asia and Indonesia. Except for the genus *Colletes,* the Colletinae are essentially austral, being most abundant and diversified in temperate areas of Australia and South America. One tribe, the *Scraptrini,* is found only in southern Africa.

Engel (2005) has discussed the features of *Scrapter* and concluded that they justify recognition of a subfamily, Scraptrinae. Variation within *Scrapter* and among the remaining genera here placed in the tribe Paracolletini is so great, however, that the only known constant character separating *Scrapter* from the other genera is the presence of a broad fovea on the posterior surface of the prementum (see Fig. 41-3). This character may be a plesiomorphy rather than a derived feature characterizing *Scrapter*. For the present I therefore recognize a tribe Scraptrini within the subfamily Colletinae. Other characters that might support recognition of the Scraptrini are those of larvae (McGinley, 1981); some analyses of these characters show *Scrapter* as the sister group to all other Colletinae. A relationship to the Euryglossinae rather than to other Colletinae is suggested not only by the premental fovea but by the nodulose margins of the basitibial plate of some species.

Differences in nesting biology between the Colletini and Paracolletini seem substantial. In the latter group, *Lonchopria* (Michener and Lange, 1957 and contained references) and *Leioproctus* (Michener and Lange, 1957; Michener, 1960) make burrows from near the ends of which laterals diverge, each lateral usually ending in a single, more or less horizontal cell, but sometimes, in both genera, there may be two or more cells in series. The cells are similar to those of *Halictus* or *Andrena*, homomorphic, bilaterally symmetrical about a vertical plane (because the lower surface is flatter than the upper), and lined with a thin secreted membrane. The larval food is a firm, subspherical pollen mass, the egg laid on top of it as in *Halictus* and *Andrena*. In the Colletini, nests of *Colletes* are well known; see the account of that genus. The burrow structure may be similar to that described above but more usually ends in burrows subdivided into series of cells, not shaped for particular cells. The cells are therefore heteromorphic. The partitions between cells are of the secreted cell-lining material as in Hylaeinae, not of soil. The provisions are semi-liquid, and the egg is attached by one end to the cellophane-like lining of the cell, above the provisions. In the Scraptrini (Rozen and Michener, 1968) the situation is intermediate, in the sense that the cell is merely the distal part of a lateral burrow, neither flattened on the lower side nor enlarged, but nonetheless probably homomorphic. The provisions fill the distal part of the cell, as they do in *Colletes,* although they are at first firm, only later, with absorption of water, becoming semi-liquid. The egg, however, is laid on top of the provisions, as in *Leioproctus*.

The phylogenetic study of families of S-T bees by Alexander and Michener (1995) indicated that the Colletinae are paraphyletic. Because of the very different phylogenetic hypotheses shown by different analyses for the eight exemplars of Colletinae, no acceptable phylogeny was evident, although *Colletes* and *Mourecotelles* consistently emerged as sisters. *Scrapter* sometimes appeared as the basal branch of the hylaeine clade, instead of as a colletine, in part because of its possession of a depression or fovea on the prementum. The one *Leioproctus* and one *Lonchopria* in the study were widely separated in spite of the existence of an intermediate (not in the analyses) between the two. It is clear that the Colletinae includes diverse elements. A needed step is a phylogenetic study of the forms here placed in Colletinae, using representatives of many more taxa than were used in the Alexander and Michener study.

Key to the Tribes of the Colletinae

1. Basitibial and pygidial plates absent; prepygidial and pygidial fimbriae lacking, in both sexes vestiture of T5 and T6 thus similar to that of preceding terga (Fig. 40-1); S7 of male with apicolateral lobes greatly enlarged, disc of sternum and apodemes reduced, slender and delicate, the lobes thus constituting the major part of S7 (Fig. 40-2b) .... *Colletini* (Sec. 40)
—. Basitibial and pygidial plates present, at least in females (pygidial plate absent in most males; basitibial plate absent in both sexes of a few Australian taxa); prepygidial and pygidial fimbriae of female present; S7 of male with apicolateral lobes of moderate size (Figs. 13-2a; 39-1, -3, -7, -9, -10, etc.) (sometimes greatly reduced or absent), disc of sternum and apodemes thus constituting the major part of S7 2
2(1). Posterior (lower) surface of prementum with a broad longitudinal depression or fovea margined by shiny ridges (Fig. 41-3b) that diverge on basal part of prementum and converge near base of subligular process; galeal comb reduced to three or four small bristles (Fig. 41-2b) (Africa) .... *Scraptrini* (Sec. 41)
—. Posterior (lower) surface of prementum lacking a fovea or with only a narrow medial groove; galeal comb usually well developed .... *Paracolletini* (Sec. 39)

39. Tribe Paracolletini

The Paracolletini, in a superficial way, replaces the holarctic genus *Andrena* in the Australian region and in temperate South America. It contains numerous andreniform bees, some of them with special features relevant to the flowers where they collect pollen. As noted in Section 20, these bees exhibit a series of characters listed as probably ancestral for bees in an old work (Michener, 1944) because of similarity to characters of wasps. Yet the paracolletines are not at all wasplike in appearance and, as indicated in Section 20, colletids are probably a derived group of S-T bees, not a basal group.

This is the most diverse of the tribes of Colletinae. Michener (2000) did not segregate it from the tribe Colletini, but the distinctive features of the latter now seem to justify tribal segregation. The distinctive features of the Paracolletini, however, are probable plesiomorphies and the tribe is likely to be paraphyletic. As suggested above, a phylogenetic study using numerous taxa is needed. *Callomelitta* is a very unusual genus and should perhaps be removed from the Paracolletini.

Paracolletini differ from Colletini in the presence of a pygidial plate and pygidial and prepygidial fimbriae, and almost always basitibial plates in females, present also in some males. S7 of the male has a disc to which apodemes and usually the two or four apicolateral lobes, often much reduced, are attached. The posterior surface of the prementum is smooth, convex, sometimes with a longitudinal median groove.

These bees are abundant in Australia and temperate southern South America, ranging north to Misoöl in Indonesia, to New Guinea, and in the Western Hemisphere, to Arizona, USA.

Stenocolletes Schrottky (1909c), which was originally placed among the colletids and would have to be a paracolletine, may be a protandrenine panurgine (see Sec. 53).

Key to the Genera of Paracolletini of the Western Hemisphere

1. Preoccipital carina strong, often lamella-like; pronotum dorsolaterally with strong transverse carina or lamella extending onto pronotal lobe; hind tibial hairs of female shorter than tibial diameter (neotropical to Arizona) .... *Eulonchopria*
—. Preoccipital and pronotal carinae (or lamellae) absent; many hind tibial hairs of female as long as or longer than tibial diameter 2
2(1). Malar space nearly as long as or longer than eye; S8 of male weakly sclerotized, lacking apical process (Fig. 39-13c) (Ecuador) .... *Lonchorhyncha*
—. Malar space little if any longer than flagellar width, usually virtually absent; S8 of male with strong median apical process 3
3(2). Labial palpi enormous, 8-9 mm long, in repose reaching S3 or S4; claws of both sexes deeply cleft, the two rami similar in shape and of almost equal length (Brazil) ......
.... *Niltonia*
—. Labial palpi unremarkable; claws with inner rami shorter than outer and differently shaped, at least in female, or, rarely, claws simple 4
4(3). Forewing with three submarginal cells, second usually about as long as third on posterior margin (but see subgenus *Lonchopria* s. str.); apical process of S8 of male lacking flat apical region resembling a pygidial plate; inner hind tibial spur of female coarsely palmate-pectinate, bases of teeth close together and diverging from thick part of spur (Fig. 39-8e); tibial scopa (except in subgenus *Lonchoprella*) extremely dense, hiding tibial surface; hind basitarsus of female weakly concave on outer surface near upper margin, this surface unlike that of tibia in appearance, the surface easily visible among hairs that are usually shorter than those of inner surface (South America) .... *Lonchopria*
—. Forewing with two or three submarginal cells, if with three, then second much shorter than third on posterior margin (except in *Leioproctus* subgenus *Cephalocolletes* and most specimens of subgenus *Reedapis*); apical process of S8 of male with flat, bare apical region on upper side, superficially resembling a pygidial plate and usually exposed at apex of metasoma; inner hind tibial spur of female ciliate to coarsely pectinate, not at all palmate and not thickened medially; scopa not hiding tibial surface; hind basitarsus of female flat or convex on outer surface, this surface superficially similar to that of tibia, its hairs longer than those of inner surface (ignoring hairs of upper margin) 5
5(4). Stigma small, vein r arising well beyond middle; costal margin of marginal cell 2.5-3.0 times as long as stigma; propodeum almost wholly declivous in profile; volsella of male large, vertically expanded, reaching dorsum of genital capsule, bifid (Fig. 39-1a); mandible of male tridentate (Fig. 39-1e) [middle and both hind tibial spurs strongly curved and coarsely pectinate (Fig. 39-8d) or outer hind spur of male sometimes dentate or almost simple; forewing with two submarginal cells] (South America) .... *Brachyglossula*
—. Stigma elongate, vein r arising at or slightly beyond middle (Fig. 39-5d-l); costal margin of marginal cell 1.5-2.0 times as long as stigma; propodeum usually with subhorizontal or sloping basal part curving onto steeply declivous posterior surface; volsella of male more or less horizontal, ventral, not attaining dorsum of genital capsule; mandible of male simple or bidentate (tibial spurs not curved and coarsely pectinate, or, if so, as in some species of subgenus *Reedapis*, then forewing with three submarginal cells) (South America) ....
.... *Leioproctus* (in part)

Key to the Genera of Paracolletini of the Australian Region

1. Marginal cell with apex on wing margin (Fig. 39-2a); facial fovea linear or nearly so (often very short or absent in male); mandible of female two to three times as long as basal width, ending in three equally conspicuous teeth; pygidial plate of female with lateral margins concave, the

apex very slender and parallel-sided (Australia)..............
.... *Callomelitta*
—. Marginal cell bent away from wing margin at apex (Figs. 39-5, 39-11), sometimes, as in *Leioproctus (Euryglossidia)*, only slightly so; facial fovea broad or absent; mandible of female usually four or more times as long as basal width, bidentate, lower tooth much longer than upper (upper tooth bilobed, giving a tridentate appearance, in *Paracolletes*); pygidial plate of female with lateral margins not strongly concave, apex neither slender nor parallel-sided (except in a few *Leioproctus*) 2

2(1). Stigma small, parallel-sided, truncated (in some cases obliquely) at base of vein r or rarely shortly beyond that point, not or scarcely tapering to apex within marginal cell; marginal cell on costal edge of wing 2.75 to 5.0 times as long as stigma; bare part of labrum uniformly convex, one to five times as broad as long 3

—. Stigma usually larger (Fig. 39-5) and usually not parallel-sided, apex usually tapering to a point on costal edge of marginal cell, vein r thus arising near middle of stigma; marginal cell on costal edge of wing 1.3 to 2.5 times as long as stigma; bare part of labrum usually with transverse ridge or otherwise not uniformly convex, usually more than five times as broad as long 4

3(2). Inner hind tibial spur of female finely serrate, rarely finely pectinate with slender teeth of approximately uniform length arising from a shaft that tapers rather uniformly toward apex (as in Fig. 39-8c); basitibial plate of female and sometimes of male fully defined, in some cases visible without removal of hairs, apex of plate rounded or blunt (angulate in *Paracolletes montanus* Rayment); mandible slender near base, usually expanded apically, female with upper apical tooth bidentate in unworn mandibles, mandibular apex of female thus tridentate; eyes parallel or converging below (Australia) ....
.... *Paracolletes*

—. Inner hind tibial spur of female coarsely pectinate, shaft thick near base and narrowing in region where most of teeth arise (Fig. 39-8f); basitibial plate of female defined only along posterior margin, or, at least apex not defined, plate never visible without removal of hairs; basitibial plate of male variable, acutely pointed if defined; mandible approximately parallel-sided, apex bidentate; eyes often diverging below (Australia) .... *Trichocolletes*

4(2). Metasoma with transverse, pale-yellow, integumental bands, broken or narrowed sublaterally, on subapical parts of terga; clypeus yellow in both sexes; scape of male greatly broadened (Australia) .... *Neopasiphae*

—. Metasoma without yellow integumental bands, yellowish-brown bands occasionally present but not emarginate or broken sublaterally; clypeus dark in female, rarely yellow in male; scape of male usually unmodified, sometimes thickened but not flat 5

5(4). Basal vein basal to cu-v of forewing; maxillary palpus about as long as width of galea, four-segmented; first recurrent vein received near basal one-third or one-fourth of second submarginal cell (Fig. 39-5b) (Australia) ....
.... *Phenacolletes*

—. Basal vein meeting or distal to cu-v of forewing (Fig. 39-5); maxillary palpus much longer than width of galea, six-segmented; first recurrent vein received beyond basal one-third of second submarginal cell (Fig. 39-5a, c-f)....
.... 15

6(5). S8 of male with two flat, delicate, apical processes, longer than body of sternum (Fig. 39-4b); supraclypeal area with longitudinal, strongly elevated, impunctate, shining carina or broad ridge extending from frontal carina down to upper margin of clypeus; distal three antennal segments of male modified (Fig. 39-4d) (Australia) .... *Glossurocolletes*

—. S8 of male ending in the usual single, commonly heavily sclerotized, apical process; supraclypeal area broadly convex, median part sometimes impunctate; apical antennal segments of male unmodified or rarely crenulate, or last one rarely broadened and flattened 7

7(6). Claws (at least of male, those of female unknown) each with broad, flat inner ramus arising near base (Fig. 39-4g); episternal groove below scrobal groove represented only by weak, short depression; strong carina just behind posterior orbit (Australia) .... *Hesperocolletes*

—. Claws cleft apically in male and usually in female, inner ramus pointed like outer ramus or sometimes in female reduced to a tooth or absent (in one undescribed species of *Leioproctus* somewhat like those of *Hesperocolletes*); episternal groove distinct below scrobal groove; no strong carina behind posterior orbit 8

8(7). S8 of male with median apical process slender and hairy at apex, pale, not exposed (Fig. 39-3b); volsella large, produced posteriorly, reaching beyond apex of articulated gonostylus (Fig. 39-3a); stigma less than one-half as long as marginal cell, measured on wing margin (Fig. 39-5a); apex of marginal cell bent gradually from wing margin for about one-sixth length of cell and bearing long appendage; profile of propodeum nearly vertical (Australia)... *Chrysocolletes*

—. S8 of male with median apical process robust, heavily sclerotized (Fig. 39-6c, g), its apex suggesting a pygidial plate and commonly exposed in repose, or (in some species of subgenus *Goniocolletes*) process broadened and appearing as extension of elongated disc (Fig. 39-7e); volsella not reaching gonostylus or reaching only its basal part, gonostylus fused to gonocoxite; stigma usually larger; apex of marginal cell only minutely bent away from wing margin; profile of propodeum usually with sloping or subhorizontal basal zone (Australian area) ....
.... *Leioproctus*

Genus *Brachyglossula* Hedicke

Brachyglossa Friese, 1922a: 577, not Boisduval, 1829. Type species: *Brachyglossa rufocaerulea* Friese, 1922, monobasic.
Brachyglossula Hedicke, 1922: 427, replacement for *Brachyglossa* Friese, 1922. Type species: *Brachyglossa rufocaerulea* Friese, 1922, autobasic.

This genus of large (body length 12-16 mm), dark-haired, unbanded bees is distinctive in appearance, except for the superficially similar *Leioproctus (Cephalocolletes) laticeps* (Friese). In some major features, such as the vestiture and form of the hind basitarsus of the female and the shape of the process of S8 of the male, *Brachyglossula* resembles *Leioproctus*. The other characters are sufficiently marked and unique, however, to support recognition at

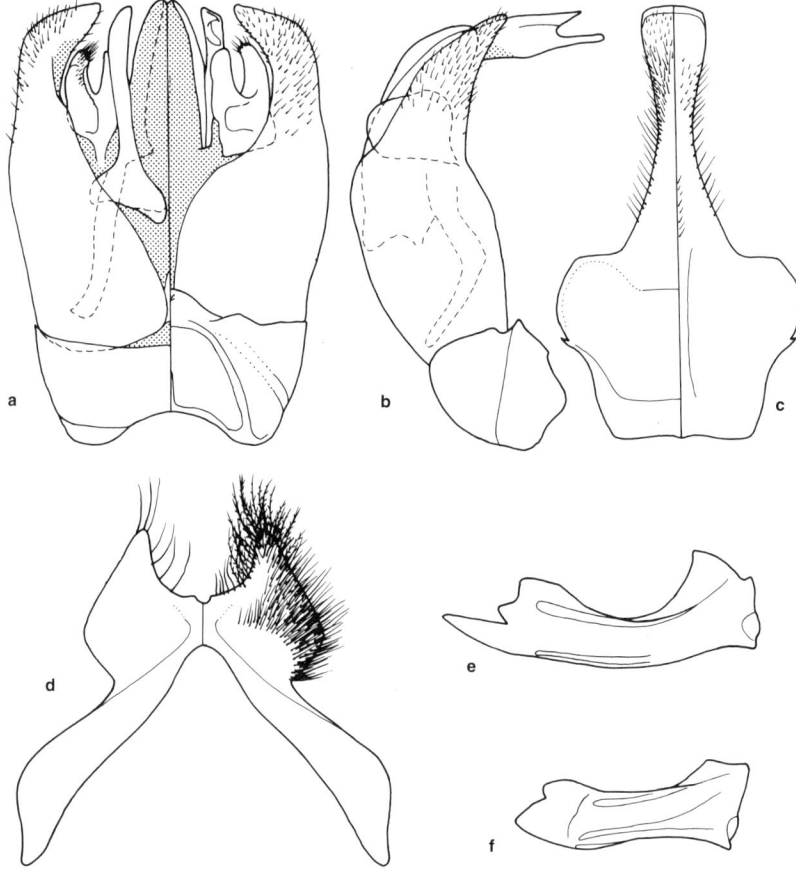

Figure 39-1. *Brachyglossula bouvieri* (Vachal). **a, b,** Dorsoventral and lateral views of male genitalia; **c, d,** Dorsoventral views of S8 and S7 of male; **e, f,** Mandibles of male and female. In the divided drawings, dorsal views are at the left. From Michener, 1989.

the genus level, even though the result, at least for the time being, is a probably paraphyletic genus *Leioproctus*. The integument is black or metallic blue, the metasoma in some species red. The inner orbits diverge below; the facial fovea is as described for *Lonchopria*. The preapical tooth of the mandible of the male is broad and weakly to strongly emarginate, the mandible thus at least weakly tridentate (Fig. 39-1e). All scopal hairs have strong axes and numerous fine side branches at more or less right angles to the axes. T1 is markedly narrower than T2. S2-S5 of the female support a well-developed, dense, plumose scopa. For illustrations of the male genitalia and sterna, see Figure 39-1a-d and Michener (1989).

■ *Brachyglossula* is found in Bolivia and Argentina (provinces of Misiones to Catamarca). There are four named species.

Brachyglossula may collect pollen principally from cactus flowers.

Genus *Callomelitta* Smith

Callomelitta Smith, 1853: 85. Type species: *Callomelitta picta* Smith, 1853, monobasic.

Binghamiella Cockerell, 1907c: 235. Type species: *Sphecodes antipodes* Smith, 1853, monobasic.

Binghamiella (Pachyodonta) Rayment, 1954: 48. Type species: *Binghamiella fulvicornis* Rayment, 1954, monobasic.

This is a genus of black or metallic blue-black bees with the head and thorax coarsely punctate, often with the metasoma and sometimes with parts of the thorax red. The body length is 6 to 10 mm. There are no conspicuous metasomal hair bands, and the prepygidial and pygidial fimbriae are of rather dense but short hairs. The jugal lobe of the hind wing is nearly five-sixths as long as the vannal lobe (Fig. 39-2) (one-fourth to three-fourths as long in most other Australian Colletinae). There are three submarginal cells. The inner hind tibial spur of the female is finely ciliate (Fig. 39-2) and the basitibial plate is one-third as long as the tibia or nearly so, the margins sometimes nodulose; in the male the basitibial plate is one-fifth to one-sixth as long as the tibia. For figures of genitalic and sternal characters as well as wing venation, see Michener (1965b).

■ *Callomelitta* ranges from Tasmania to northern Queensland, west to Western Australia. Most species are from eastern Australia. The 11 species were listed by Michener (1965b) and Cardale (1993).

The body form, coarse punctation, nonmetallic coloration, and red metasoma characteristic of *Callomelitta*

39. Tribe Paracolletini; *Calomellita* to *Chrysocelletes*

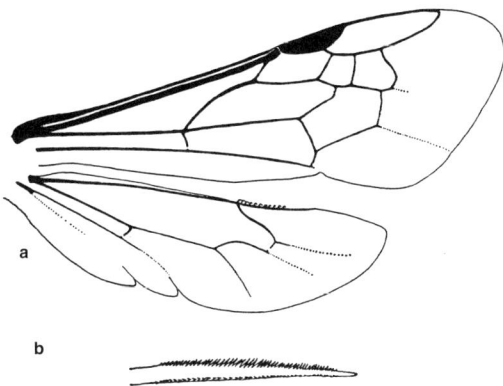

Figure 39-2. *Callomelitta picta* Smith. **a,** Wings; **b,** Inner hind tibial spur of female. From Michener, 1965b.

antipodes (Smith) produce a superficial similarity to a species of *Sphecodes* (Halictidae), and F. Smith accordingly described the species in that genus. The other ten species, however, because of their usually metallic coloration and more elongate body form, do not resemble *Sphecodes*.

Genus *Chrysocolletes* Michener

Leioproctus (Chrysocolletes) Michener, 1965b: 71. Type species: *Paracolletes moretonianus* Cockerell, 1905, by original designation.

This genus suggests *Phenacolletes* in its occipital development, the median ocellus being about midway between the antennal bases and the posterior edge of the vertex. The appressed, short pubescence of the metasoma also suggests *Phenacolletes*, but the rest of the body and the legs have the ordinary plumose hairs characteristic of *Leioproctus*. Body length is 9 to 13 mm. The marginal cell is bent away from the costa for a greater distance than in most *Leioproctus* (Fig. 39-5a). The male is unique in having the inner prongs of the hind claws much nearer the bases of the claws than is the case on the other legs. The facial foveae are absent. There are three submarginal cells; the stigma is subparallel-sided, less than half as long as the costal edge of the marginal cell (Fig. 39-5a). The inner hind tibial spur of the female is finely pectinate; that of the male, ciliate. S7 and S8 and the enormous volsellae are also distinctive; see Figure 39-3a-c and Michener (1965b) and Maynard (1996).

■ *Chrysocolletes*, which has been found in Queensland, New South Wales, Northern Territory, and Western Australia, comprises at least five species. The genus was revised by Maynard (1996).

Chrysocolletes moretonianus (Cockerell) is probably restricted in pollen collecting to species of Goodeniaceae.

Although originally described as a subgenus of *Leioproctus, Chrysocolletes* differs from *Leioproctus* in many characters, and its phylogenetic position is obscure. *Leioproctus crenulatus* Michener was placed in *Chrysocolletes* by Michener (1965b) and was thought to be intermediate between *Leioproctus* s. str. and *Chrysocolletes*. It now seems that the characters in which it resembles *Chryso-*

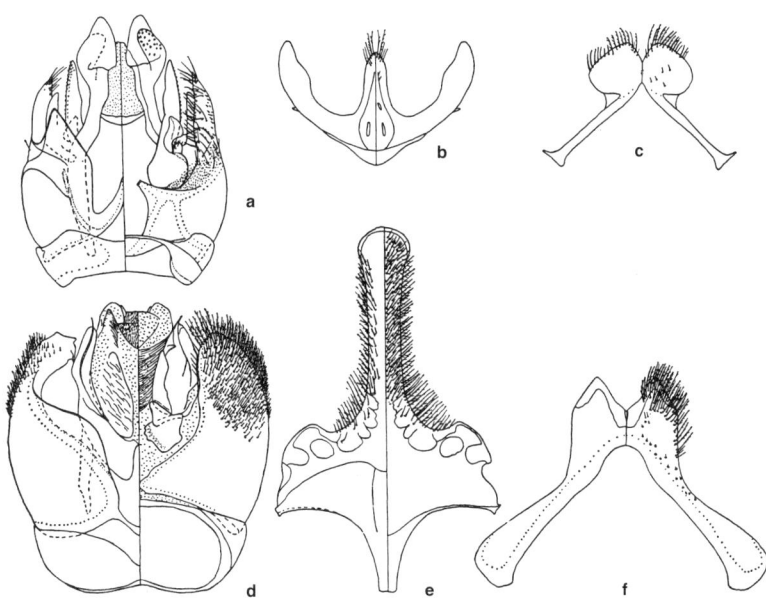

Figure 39-3. Male terminalia of Colletinae. **a-c,** Genitalia, S8, and S7 of *Chrysocolletes moretonianus* (Cockerell); **d-f,** Genitalia, S8, and S7 of *Phenacolletes mimus* Cockerell. (Dorsal views are at the left.) From Michener, 1965b.

colletes (e.g., the crenulate flagellum of the male) are probably convergent. *Leioproctus crenulatus* is transferred to the subgenus *Leioproctus* s. str., at least until that group is properly revised.

Genus *Eulonchopria* Brèthes

This is a genus of coarsely sculptured, nonmetallic bees 8 to 11 mm long with yellow apical integumental bands on at least some of the metasomal terga, but without hair fasciae on the terga, and often with plaited (longitudinally folded) forewings with darkened costal margins (Danforth and Michener, 1988), the whole thus yielding a superficial resemblance to eumenine wasps. The pubescence is short; in *Eulonchopria* s. str., at the anterior and posterior scutal angles and often elsewhere, the hairs are so short that each fits inside a puncture and is broadly plumose. Facial foveae are absent or deeply impressed and well defined in both sexes; when present, they are elongate and low on the face, that is, not reaching the summits of the eyes. The propodeal triangle has large, deep pits; some of the ridges margining the pits are produced, lamella-like or toothlike. The horizontal and vertical surfaces of the propodeum are separated by a sharp angle or lamella. The front basitarsus of the female in *Eulonchopria* s. str. ends with an outer apical process from which a comb extends basad on the outer edge of the basitarsus; in *Ethalonchopria* the comb is present but the apical process is absent. The inner hind tibial spur of the female is coarsely pectinate (three to five teeth); that of the male is coarsely toothed or ciliate, or the hind tibial spurs are completely absent. The hairs of the outer surface of the hind tibia of the female are short and not scopalike, especially on the distal half of the tibia. The basitibial plate of the male ends in a carina extending to the apex of the tibia. The stigma is nearly parallel-sided, vein r arising near the apex, and the margin within the marginal cell is oblique, not convex. There are three submarginal cells; the apex of the marginal cell is obliquely bent away from the wing margin or obliquely truncate. The genitalia and hidden sterna of males were illustrated by Michener (1963a), the wing venation by Danforth and Michener (1988).

Typical members of this genus possess various probable apomorphies that are unusual among bees and not shared by related groups (*Leioproctus* and other Colletinae). The deep, rather slender, bare, well-defined facial foveae (of one subgenus), however, are shared with the Australian genus *Callomelitta* and with some species of the African genus *Scrapter*, as well as with the Euryglossinae and Hylaeinae. This character is probably a plesiomorphy. Likewise, the distinct, slender male gonostyli of some species, unique for the Colletidae, are a possible plesiomorphy. The apparently disjunct distribution of *Eulonchopria* (Americas but absent in wet tropics), combined with its unusual characters and the morphological diversity of its species, suggests that *Eulonchopria* is an archaic group possessing many derived features.

Leioproctus simplicicrus Michener (1989) from Peru, originally incorrectly placed as an unusual species of *L. (Nomiocolletes)*, to which it runs in the keys to genera and subgenera, seems to connect *Eulonchopria* and *Leioproctus*, suggesting that *Eulonchopria* is a specialized derivative of the large paraphyletic genus *Leioproctus*. G. Melo pointed out to me that *L. simplicicrus* resembles *Eulonchopria* in its short pubescence, the obliquely truncate apex of the marginal cell, the apical yellow tergal bands, the carina along the hind tibia of the male, and (especially) its genitalia and hidden sterna (compare Michener, 1963a and 1989). (The female is unknown, and the characters of the scopa cannot be determined.) On the other hand, *L. simplicicrus* differs from *Eulonchopria* and resembles *Leioproctus* in the absence of a preoccipital carina, the absence of a lamella from the posterior pronotal lobe to the dorsum of the pronotum, and the relatively large stigma, which is broadest at the base of vein r and convex within the marginal cell. The propodeal triangle has a few weak rugae basally but lacks the sharp carinae or lamellae characteristic of *Eulonchopria* s. str. When both sexes are known, *L. simplicicrus* may well be placed as a distinct genus or subgenus; females of an unnamed species from Brazil may fall in the same group.

Key to the Subgenera of *Eulonchopria*

1. Facial fovea absent; omaulus not carinate .... *E. (Ethalonchopria)*
—. Facial fovea distinct; omaulus carinate .... *E. (Eulonchopria s. str.)*

Eulonchopria / Subgenus *Ethalonchopria* Michener

Eulonchopria (Ethalonchopria) Michener, 1989: 670. Type species: *Apista gaullei* Vachal, 1909, by original designation.

This subgenus differs in many features from the other subgenus. In nearly all of the subgeneric characters it is less strange than *Eulonchopria* s. str., that is, more like other colletines. Noteworthy are the punctate and only slightly concave foveal areas on the face, such that distinct foveae are absent; the simple axillae; and the jugal lobe of the hind wing, which extends little more than halfway from the wing base to the level of vein cu-v. Although most of the subgeneric characters are plesiomorphic relative to *Eulonchopria* s. str., the small jugal lobe and the small second submarginal cell are apomorphic. This subgenus is probably the sister group to *Eulonchopria* s. str.

■ The subgenus is known from Bolivia, southern Brazil, and eastern Colombia. The two species names, both dating from Vachal (1909), may represent only one species.

Eulonchopria / Subgenus *Eulonchopria* Brèthes s. str.

Eulonchopria Brèthes, 1909a: 247. Type species: *Eulonchopria psaenythioides* Brèthes, 1909, monobasic.

This subgenus contains the more ornate and extraordinary members of the genus. The carinate omaulus and produced, angulate axillae are especially unusual. The preoccipital ridge is expanded as a strong lamella. The carina on the upper margin of the hind tibia of the male is toothed. The jugal lobe of the hind wing nearly attains the level of cu-v. The apex of T7 of the male is bilobed or bidentate.

■ This subgenus ranges from Paraguay, Argentina (Salta province), and Brazil (Santa Catarina to Minas Gerais) northward to Venezuela, Colombia, Nicaragua, Mexico (Oaxaca to Sonora), and the USA (southern Arizona). The three named species, only one of them from South America, were revised by Michener (1963a); there are additional (undescribed) species in South America.

The species of *Eulonchopria* may be oligolectic visitors to flowers of *Acacia;* at least the North and Central American species appear to collect pollen regularly from that plant.

Eulonchopria s. str. contains two species groups. In the South American *E. psaenythioides* Brèthes and its undescribed relatives, the hind tibial spurs of the male are present, S8 of the male lacks an apical process, and there are no slender male gonostyli. In the two North and Central American species, *E. punctatissima* Michener and *oaxacana* Michener, the hind tibial spurs of the male are absent, S8 of the male has a small, apically expanded process, and there is a distinct, slender, hairy gonostylus.

Genus *Glossurocolletes* Michener

Leioproctus (Glossurocolletes) Michener, 1965b: 60. Type species: *Leioproctus bilobatus* Michener, 1965, by original designation.

This genus includes species that have hitherto been included in *Leioproctus,* but seem so different as to justify generic status. They agree in various features with *Leioproctus (Protomorpha),* although the species are larger (about 7.5 mm long) than the average of that group and more robust. In these respects, the lack of metasomal fasciae, and the reduced apical lobes of S7, this genus also resembles *Leioproctus (Odontocolletes)*. Noteworthy distinguishing features of the male of *Glossurocolletes* are the apically modified antenna (Fig. 39-4d) [differing in type of modification from those of *Leioproctus (Ceratocolletes)*], the swollen antennal scape, the broad mandible (in males three times as long as the basal width, less than four times as long as the minimum width), and the strong longitudinal ridge on the supraclypeal area. The really striking feature of the group, however, is the enormously elongate and biligulate apex of S8 of the male (Fig. 39-4b), the apices of the two lobes being exposed in undissected specimens. S7 of the male does not have a narrow neck where the apodemes join the small (almost insignificant) disc of the sternum, as is usual in colletids. Instead, the apodemes are broadly joined to one another, forming the transverse body of the sternum (Fig. 39-4c). Thus, the S7 of this genus does not look like the S7 of a colletid.

The following are additional characters of *Glossurocolletes.* The facial fovea is about three times as long as broad, impunctate but minutely roughened, and distinctly impressed. The eyes bear scattered hairs [very short and inconspicuous in *G. xenoceratus* (Michener)]. Of the three submarginal cells, the second is unusually small (Fig. 39-5c); the stigma is small, not parallel-sided, with vein r arising near its middle. The hind tibial spurs of both sexes are thick and strongly curved; both the outer margin of the outer spur and the inner margin of the inner spur are pectinate with coarse teeth. For illustrations of the genitalia, sterna, and antennae, see Figure 39-4 and Michener (1965b).

■ *Glossurocolletes* is found in Western Australia. The named species are *G. bilobatus* (Michener) and *xenoceratus* (Michener), the two distinguished by Michener (1965b).

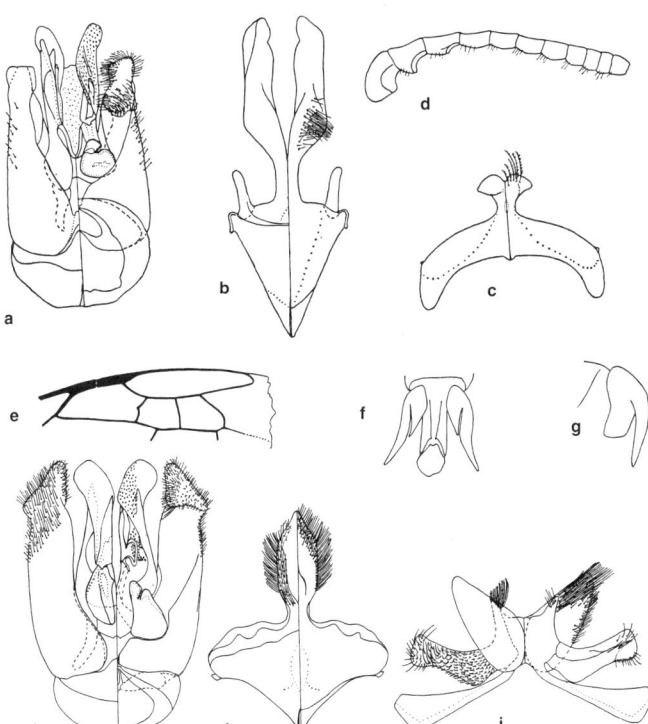

Figure 39-4. Details of paracolletine structures. **a-c,** Male genitalia, S8, and S7 of *Glossurocolletes bilobatus* (Michener); **d,** Male antennal flagellum of *G. xenoceratus* (Michener); **e-j,** *Hesperocolletes douglasi* Michener, male, portion of wing, claws and arolium, side view of claw, genitalia, S8, and S7. (In the divided drawings dorsal views are at the left. From Michener, 1965b.

Genus *Hesperocolletes* Michener

Hesperocolletes Michener, 1965b: 75. Type species: *Hesperocolletes douglasi* Michener, 1965, by original designation.

This genus, known only in the male, consists a moderate-sized (body length 12 mm), nonmetallic species similar in appearance to *Trichocolletes* and *Paracolletes* (especially *P. crassipes* Smith), as well as to *Leioproctus*. The stigma (Fig. 39-4e) is like that in those groups of *Leioproctus* having a slender stigma, such as the subgenus *Goniocolletes*. There are three submarginal cells, the second and third equal in length and together slightly longer than the first. The wholly vertical propodeum, as seen in profile, is almost matched in a few groups of *Leioproctus,* but the lack of a defined basitibial plate in the male separates *Hesperocolletes* from nearly all *Leioproctus*. The rather protuberant and yellowish clypeus of the male suggests that of *Paracolletes* and *Trichocolletes*. The most distinctive generic characters (male only) are (1) the strong carina around, and especially behind, each eye; (2) the near absence of the episternal groove (only a shallow concavity) below the scrobal groove; and (3) the deeply cleft claws, their inner prongs broad, flat, and directed more or less ventrally (Fig. 39-4g), suggesting the claws of many cleptoparasitic bees such as some female Melectini (Anthophorinae). For illustrations of structures including male genitalia and sterna, see Figure 39-4 and Michener (1965b).

■ This genus is known from a single specimen reportedly from Rottnest Island, Western Australia. Additional material of this genus has not been found at the type locality. The specimen had been handled by T. Rayment, who is known to have been careless about switching locality labels on specimens. The specimen certainly came from somewhere in Western Australia, however.

Genus *Leioproctus* Smith

This huge genus, with its many subgenera, is found both in the Australian area and in South America, principally in temperate parts of that continent. In the appearance, abundance, and diversity of its species it is equivalent to the genus *Andrena* of the holarctic region. The integument is black, metallic blue or green, or sometimes red on the metasoma; metasomal hair bands and colored integumental bands may be present or absent. The clypeus is usually rather uniformly punctured, with or without a weak, upper-median, depressed, densely punctate area, or if such an area is strongly evident, as in the subgenus *Kylopasiphae* and a few species of the subgenus *Perditomorpha*, then this area is often surrounded by extensive impunctate, usually convex areas, as in the genus *Lonchopria*. The facial foveae are usually not recognizable or are indicated by broad, weakly defined areas of slightly different texture, or (in a few Australian subgenera) the foveae are rather well defined, depressed, and rather broad (not groovelike). The mandible of females and most males has a preapical tooth on the upper margin; the expansion or tooth found on the lower margin in many male *Lonchopria* is absent. The labrum is usually more than three times as wide as long. The front basitarsus of the female usually lacks a well-formed comb of hairs on the outer margin, but such a row of hairs is present in the subgenera *Cephalocolletes, Nomiocolletes, Reedapis,* and *Spinolapis*. The inner hind tibial spur of the female is ciliate to coarsely pectinate, the bases of the teeth forming a uniform series, not crowded and diverging from one another as in females of *Lonchopria* and *Trichocolletes*. The tibial scopa, unlike that of *Lonchopria,* is not so dense as to completely obscure the tibial surface; the hairs of the inner surface of the hind tibia of the females are moderately long, simple (at least in a limited area), and do not form a zone of short keirotrichia. The hind basitarsus of the female tapers slightly, but the apex is more than half as wide as the maximum width near the base; the outer surface is flat or convex, its vesiture being superficially rather similar to that of the tibia, and its hairs are longer than those of the inner surface. The basitibial plate of the female is usually well defined, pointed or rounded, and sometimes hidden by hair, but it is sometimes absent in the subgenus *Excolletes*. The basitibial plate of the male is defined, but sometimes one margin is missing. There may be either two or three submarginal cells; if three, the second is usually much shorter than the third on the posterior margin. The stigma is moderate-sized, vein r arising at or slightly beyond its middle, and the margin within the marginal cell is usually convex (Fig. 39-5d-l). The sternal scopa is present or absent. The genitalia and sterna of many species and nearly all subgenera were illustrated by Michener (1965b, 1989) and in other works cited therein; see Figures 39-6 and 39-7.

A feature of this genus is that, although membership in the genus is indicated by a complex of characters, no one of these is invariable, and indeed characters that elsewhere among bees are often regarded as of generic importance vary within this genus. Thus arolia are absent from the subgenus *Urocolletes;* claws are simple instead of cleft in the females of several groups and all intermediate conditions are found in the *platycephalus* group of the subgenus *Leioproctus* s. str. and in the subgenus *Euryglossidia;* basitibial plates are partly gone in one species of the subgenus *Lamprocolletes* and wholly gone in females of some species of *Excolletes;* and the apical lobes of S7 of the male, characteristically present in this family, vary from four in number (subgenera *Andrenopsis, Ceratocolletes,* and *Euryglossidia* and the *advenus* group of *Leioproctus* s. str.) to two in most forms, these lobes very much reduced in the subgenera *Nesocolletes* and *Odontocolletes*.

A character that has often been considered as of generic or subgeneric importance, namely the inner hind tibial spur of the female (ciliate or pectinate, Fig. 39-8) is not always even a subgeneric character. It varies among species of the subgenera *Leioproctus* s. str., *Euryglossidia,* and *Perditomorpha*. Likewise, the number of submarginal cells, though usually a good subgeneric character (i.e., constant among species that closely resemble one another in other features), varies among species of the subgenera *Sarocolletes* and *Nomiocolletes* and even among individuals of an Australian species of *Leioproctus* s. str., *L. abnormis* (Cockerell).

The subgenera in South America are mostly well-defined taxa; those in Australia are less well understood. In Australia there is partial intergradation among some of the taxa that are here called subgenera; at least some of the group characters break down. The problem confronting

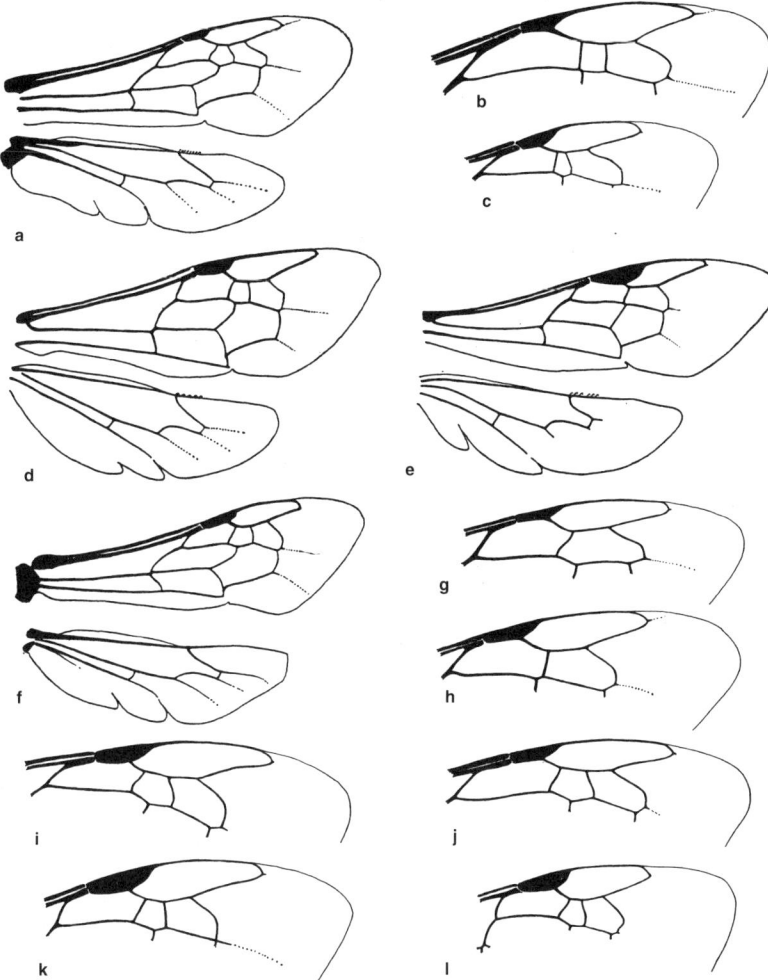

Figure 39-5 Wings and parts of forewings of Colletinae of the Australian region. **a,** *Chrysocolletes moretonianus* (Cockerell); **b,** *Phenacolletes mimus* Cockerell; **c,** *Glossurocolletes xenoceratus* Michener; **d,** *Leioproctus (Protomorpha) tarsalis* (Rayment); **e,** *L. (Filiglossa) filamentosus* (Rayment); **f,** *L. (Nodocolletes) dentiger* (Cockerell); **g,** *L. (Andrenopsis) flavorufus* (Cockerell); **h,** *L. (Baeocolletes) calcaratus* Michener; **i,** *L. (Nesocolletes) fulvescens* (Smith); **j,** *L. (Urocolletes) rhodurus* Michener; **k,** *L. (Leioproctus) unguidentatus* Michener; **l,** *L. (Excolletes) impatellatus* Michener. From Michener, 1989.

a comprehensive review of the situation at this time is that there are many unusual undescribed species in Australia, and many of the described species are known in only one sex. I believe that Australian workers, particularly G. V. Maynard, are rectifying this situation, and in fact her as yet unpublished thesis will help greatly, but I can present only the currently published information and indicate some of these problems in the discussions of various subgenera.

Leioproctus occurs in Australia (north to New Guinea and Misoöl, an island west of New Guinea), Tasmania, New Caledonia, New Zealand, and temperate South America. It is probably a paraphyletic taxon from which *Brachyglossula, Eulonchopria, Lonchopria,* and perhaps *Lonchorhyncha* and *Niltonia* were derived in South America and from which *Glossurocolletes, Neopasiphae, Paracolletes, Phenacolletes,* and *Trichocolletes* and perhaps *Chrysocolletes* and *Hesperocolletes* were derived in Australia. *Leioproctus* is nonetheless definable and generally useful taxonomically, in spite of intergrading through the one species of the subgenus *Lonchoprella* to the genus *Lonchopria*. Until more species are known and cladogeny is better studied, especially in the diverse Australian fauna, there is no point in attempting a new classification, for new combinations that would last only until the next revision would be the result. Table 39-1 lists the subgenera from the two areas where *Leioproctus* exists.

There is no known character that distinguishes the Australian and South American groups of the genus. In the absence of a phylogenetic analysis I would not use this fact in any biogeographical sense, for *Leioproctus* s. str. simply consists of *Leioproctus* species not readily at-

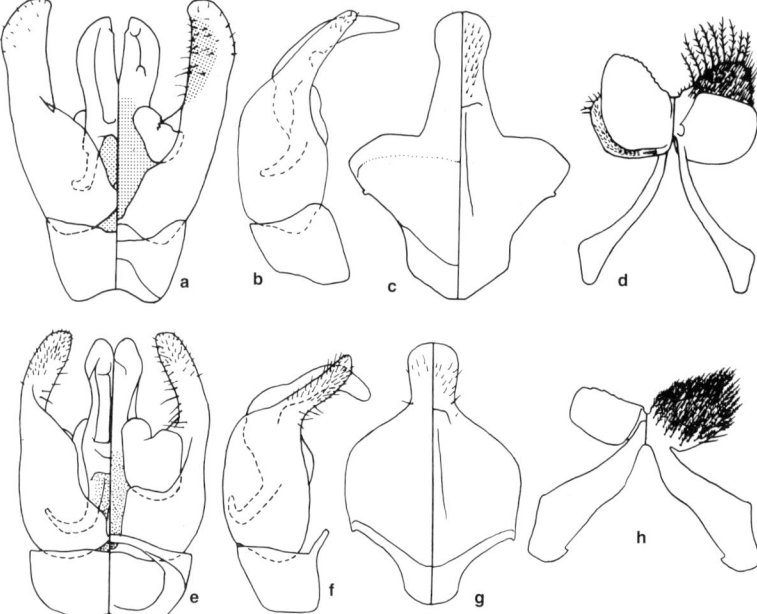

Figure 39-6. Male genitalia and hidden sterna of *Leioproctus*, showing similarity of certain subgenera. **a-d**, Dorsoventral and lateral views of genitalia, S8, and S7 of *L. (Perditomorpha) eulonchopriodes* Michener; **e-h**, Same structures of *L. (Halictanthrena) malpighiacearum* (Ducke). (Dorsal views are at the left.) From Michener, 1989.

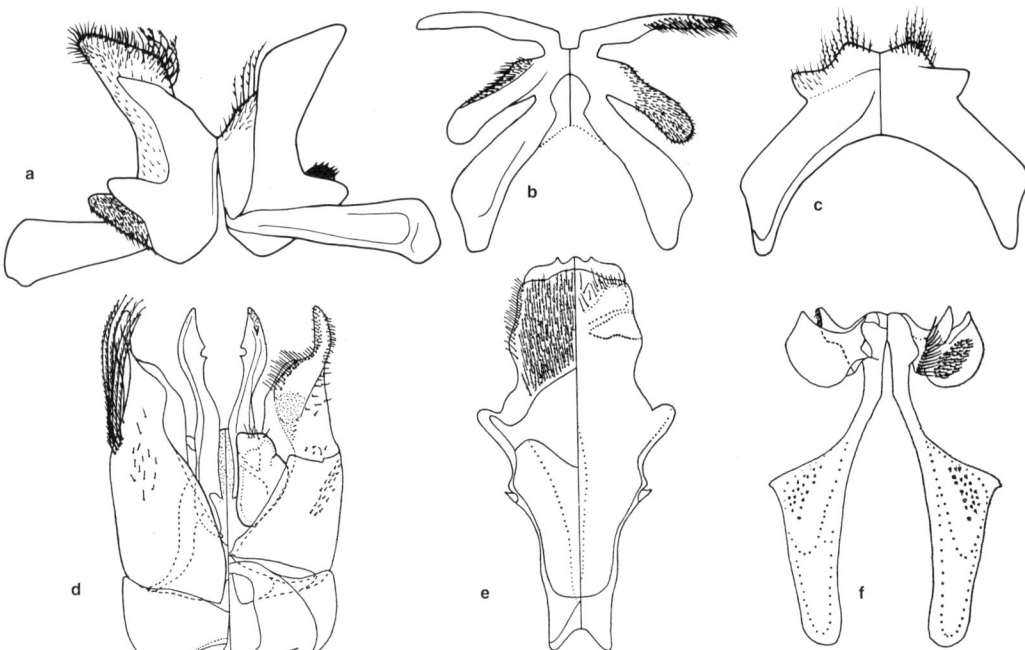

Figure 39-7. Male terminalia of *Leioproctus*, showing variation in S7 among diverse subgenera. **a,** *L. (Perditomorpha) brunerii* (Ashmead); **b,** *L. (Holmbergeria) rubriventris* (Friese); **c,** *L. (Kylopasiphae) pruinosus* Michener. **d-f,** Male genitalia, S8, and S7 of *L. (Goniocolletes) dolosus* Michener. a-c, from Michener, 1989; d-f, from Michener, 1965b.

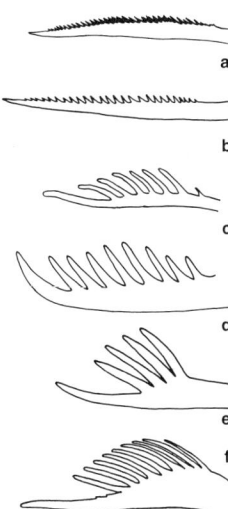

Figure 39-8. Variation in inner hind tibial spurs of females of Colletinae. **a**, *Leioproctus (Spinolapis) caerulescens* (Spinola); **b**, *L. (Perditomorpha) brunerii* (Ashmead); **c**, *L. (P.) inconspicuus* Michener; **d**, *Brachyglossula bouvieri* (Vachal); **e**, *Lonchopria (Lonchoprella) annectens* Michener; **f**, *Trichocolletes venustus* (Smith). Both a and b are considered ciliate, the others pectinate. a-e, from Michener, 1989; f, from Michner, 1965b.

tributable to any other subgenus. It is almost certainly paraphyletic. The lone South American species included in the subgenus by Michener (2000) and now placed in *Actenosigynes* shows no special affinity at the subgenus level with the Australian members of the genus.

Table 39-1. Subgenera of *Leioproctus* Segregated by Geographical Area.

South America	Australian Region
Actenosigynes	*Andrenopsis*
Albinapis	*Baeocolletes*
Cephalocolletes	*Ceratocolletes*
Chilicolletes	*Cladocerapis*
Glossopasiphae	*Colletellus*
Halictanthrena	*Colletopsis*
Hexantheda	*Euryglossidia*
Holmbergeria	*Excolletes*
Hoplocolletes	*Filiglossa*
Kylopasiphae	*Goniocolletes*
Nomiocolletes	*Lamprocolletes*
Perditomorpha	*Leioproctus* s. str.
Protodiscelis	*Nesocolletes*
Pygopasiphae	*Odontocolletes*
Reedapis	*Protomorpha*
Sarocolletes	*Urocolletes*
Spinolapis	
Tetraglossula	

Key to the Subgenera of *Leioproctus* of South America

1. T1-T4 in female and T1-T5 or T6 in male with enamel-like apical marginal zones of yellowish, green, bluish, or whitish, these zones usually at least partly impunctate and hairless .... *L. (Nomiocolletes)*
—. Terga without enamel-like apical marginal zones, with hairs and punctures near apical margins that are concolorous with other parts of terga, or translucent or brownish, or, rarely [in *L. (Perditomorpha) eulonchopriodes* Michener], with apical yellow bands, but the band of T2 absent 2
2(1). Submarginal cells three 3
—. Submarginal cells two 14
3(2). Dorsolateral angle of pronotum produced as small tooth projecting upward and outward (smallest in male); basitibial plate of female not easily seen because its hairs are erect, similar to those of adjacent parts of tibia, largely hiding marginal carinae .... *L. (Halictanthrena)*
—. Dorsolateral angle of pronotum low, rounded, scarcely evident; basitibial plate of female distinct, its hairs short, appressed, different from those of adjacent areas, its marginal carinae clearly exposed 4
4(3). Second submarginal cell on posterior margin usually at least three-fourths as long as third; second submarginal crossvein usually curved in a manner parallel to third, anterior margin of third submarginal cell thus at least two-thirds as long as posterior margin.................................... 5
—. Second submarginal cell on posterior margin much shorter than third (as in Fig. 39-5d, f); second submarginal crossvein usually straight, at least not curved parallel to third, anterior margin of third submarginal cell usually less than two-thirds as long as posterior margin 6
5(4). Mandible of male with preapical tooth; outer hind tibial spur of female pectinate, although more finely so than inner spur; metasoma with at least weak blue reflections .... *L. (Reedapis)*
—. Mandible of male simple; outer hind tibial spur of female coarsely ciliate; metasoma black............................
.... *L. (Cephalocolletes)*
6(4). Thorax dull, minutely roughened, almost lacking punctures; malar area as long as minimum diameter of flagellum; clypeus protuberant in lateral view by fully eye width... *L. (Torocolletes)*
—. Thorax with at least some areas of shining integument between strong punctures; malar area linear; clypeus not or little protuberant.. 7
7(6). Males (unknown in *Hoplocolletes*) 8
—. Females (unknown in *Holmbergeria*) 11
8(7). Subantennal suture little over half as long as diameter of antennal socket; supraclypeal and subantennal areas impunctate, shining, hairless, in conspicuous contrast to adjacent areas... *L. (Holmbergeria)*
—. Subantennal suture as long as diameter of antennal socket; supraclypeal and subantennal areas punctate, with hairs 9
9(8). Gonoforceps hairy to base; gonobase one-half as long as gonoforceps; apex of S6 with broad, shallow emargination; metasoma with pubescence all blackish ....
.... *L. (Leioproctus s. str.)*
—. Gonocoxite (or coxal part of gonoforceps) hairless;

gonobase one-third as long as gonoforceps or less; apex of S6 with the usual small, V-shaped (sometimes shallow) median emargination; metasoma with some or all hair pale, usually forming apical tergal bands 10

10(9). Labrum three times as wide as long, apical margin broadly emarginate .... *L. (Chilicolletes)*

—. Labrum little over twice as wide as long, apical margin convex or with small median emargination .... *L. (Sarocolletes)* (in part)

11(7). Inner hind tibial spur finely pectinate (almost ciliate) with over 25 teeth .... *L. (Actenosigynes)*

—. Inner hind tibial spur strongly pectinate with less than ten teeth 12

12(11). S2-S4 with apical bands of sparse, simple hairs not hiding surfaces of sterna .... *L. (Chilicolletes)*

—. S2-S4 with apical bands of long, dense hairs forming a ventral scopa that partially hides surfaces of sterna 13

13(12). Tibial and sternal scopal hairs with numerous short, fine branches projecting laterally from rachis (as in Fig. 13-1d) .... *L. (Sarocolletes)* (in part)

—. Tibial scopal hairs dividing to form few major branches; the sternal scopal hairs simple .... *L. (Hoplocolletes)*

14(2). Mandible of male simple; labrum about six times as wide as long, in female with apicolateral lobe bearing part of marginal fringe of bristles (Fig. 39-9a, b) 15

—. Mandible of male with preapical tooth on upper margin, as in female; labrum two to five times as wide as long, without apicolateral lobe (Fig. 39-9c) 16

15(14). Glossal lobes not much longer than basal width; scopal hairs of tibia and sterna with numerous short, fine branches (Fig. 13-1e); clypeal margin of male unmodified, truncate .... *L. (Protodiscelis)*

—. Glossa deeply divided, the lobes elongate, seven to ten times as long as basal width (as in Fig. 39-12); scopal hairs of tibia and sterna simple, or those of tibia with a few major branches; clypeal margin of male with short median lobe overhanging base of labrum .... *L. (Tetraglossula)*

16(14). Glossa deeply bifid, lobes about five times as long as basal width (Fig. 39-12a) .... *L. (Glossopasiphae)*

—. Glossal lobes short, not much if any longer than basal width 17

17(16). Labial palpus five- to nine-segmented, longer than maxillary palpus; hind tibia of male with strong carina from apex of basitibial plate to apex of tibia 27

—. Labial palpus four-segmented, usually shorter than maxillary palpus; hind tibia of male without longitudinal carina or, rarely, with weak carina arising behind apex of basitibial plate 18

18(17). S2-S5 of female covered with short, unbranched, erect hairs enlarged and curved posteriorly at tips and of uniform length except longer on S2; pygidial plate of male defined, at least posterior end limited by carina; hind tarsus of male elongate, segment 2 well over three times as long as broad .... *L. (Pygopasiphae)*

—. S2-S5 of female with broad apical bands of relatively long, simple or branched hairs; T7 of male with pygidial area indicated only by lack of hairs (but large and somewhat defined in *Kylopasiphae*); hind tarsus of male not especially elongate, segment 2 less than three times as long as greatest breadth 19

19(18). Females 20
—. Males 24

20(19). Tibial and sternal scopal hairs with numerous short, fine side branches projecting at right angles to rachis or curled basad (as in Fig. 13-1d) ......*L. (Sarocolletes)* (in part)

—. Tibial scopal hairs with long branches directed distad (Fig. 13-1a, b); sternum with hairs simple or their branches directed distad 21

21(20). Inner hind tibial spur coarsely pectinate with ten teeth or less 22

—. Inner hind tibial spur ciliate or finely pectinate with over a dozen teeth 23

22(21). Scopa of hind tibia formed around tibia without long, loose hairs extending above and below; basitibial plate with carinate margins not hidden by hair .... *L. (Perditomorpha)* (in part)

—. Hind tibia with a few long, loose hairs fully half as long as tibia on upper and lower margins; basitibial plate hidden by hair except sometimes at base ...... *L. (Kylopasiphae)*

23(21). Claws simple or with inner rami reduced to small teeth, shorter than basal diameters of outer rami; body metallic blue or greenish .... *L. (Spinolapis)*

—. Inner rami of claws strong, longer than basal diameters of outer rami, claws thus bifid; body lacking metallic coloration .... *L. (Perditomorpha)* (in part)

24(19). T7 with shiny, hairless, irregularly rough pygidial area, not narrowed posteriorly, defined across posterior border by weak carina, this area occupying much of dorsum of tergum; S7 with apical lobes much reduced, all in a single plane .... *L. (Kylopasiphae)*

—. T7 with dull or shiny, usually ill-defined pygidial area, sometimes a longitudinal strip, sometimes a broader area narrowed posteriorly; S7 with well-developed apical lobes, two to four on each side, usually at two levels 25

25(24). Body metallic bluish or greenish; margin of S6 produced midapically as rounded hairy lobe about one-third as wide as sternum, notched medially .... *L. (Spinolapis)*

—. Body almost always nonmetallic; margin of S6 broadly rounded, not produced midapically, with median notch, often broad and shallow 26

26(25). Metasoma rather broad and flattened, resembling that of female in shape .... *L. (Sarocolletes)* (in part)

—. Metasoma commonly rather slender, not flattened, usually distinctly different in shape from that of female .... *L. (Perditomorpha)*

27(17). Labial palpus 5-segmented; apicolateral lobes of S7 of male small, rounded .... *L. (Albinapis)*

—. Labial palpus 6- to 9-segmented; apicolateral lobes of S7 of male well-developed .... *L. (Hexantheda)*

Key to the Subgenera of *Leioproctus* of the Australian Region

1. Submarginal cells two 2
—. Submarginal cells three [except in some specimens of *L. (Leioproctus) abnormis* (Cockerell)] 6

2(1). First recurrent vein basal to first submarginal crossvein (Fig. 39-5h); clypeus and supraclypeal area usually flat, depressed, shining, largely impunctate; hind tibial spurs robust, curved apically, outer one nearly as coarsely toothed as inner .... *L. (Baeocolletes)*

—. First recurrent vein distal to or rarely meeting first submarginal crossvein; clypeus and supraclypeal area convex, the latter elevated above level of antennal sockets; hind tibial spurs slender, not strongly curved apically, outer one not coarsely toothed 3

3(2). Propodeum almost wholly vertical in profile; stigma small, nearly parallel-sided, little more than half as long as marginal cell on costal margin of wing (Fig. 39-5g) .... *L. (Andrenopsis)*

—. Propodeum with broad basal subhorizontal or horizontal zone, curving onto vertical posterior surface; stigma large, not parallel-sided, at least two-thirds as long as marginal cell on costal margin of wing (Fig. 39-5e) 4

4(3). Jugal lobe of hind wing extending well beyond level of cu-v; inner hind tibial spur of female ciliate .... *L. (Colletellus)*

—. Jugal lobe of hind wing short, not attaining level of cu-v (Fig. 39-5e); inner hind tibial spur of female usually pectinate 5

5(4). Galea with several very long apical hairs and labial palpus filamentose, about as long as face (Fig. 19-6) .... *L. (Filiglossa)*

—. Mouthparts unmodified .... *L. (Euryglossidia)*

6(1). Arolia absent .... *L. (Urocolletes)*

—. Arolia present 7

7(6). Basitibial plate of female and some males absent; pygidial plate of female with apical part slender, parallel-sided or slightly narrowed preapically; basal vein more or less transverse, slanting about 30° to costal margin of wing, and little if any longer than first abscissa of Rs (Fig. 39-5l) .... *L. (Excolletes)*

—. Basitibial plate present (but anterior side not defined in one species of *Lamprocolletes*); pygidial plate of female broad, sides converging posteriorly [except in *L. (Protomorpha) fallax* (Cockerell)]; basal vein slanting 45° or more to costal margin of wing, and much longer than first abscissa of Rs (Fig. 39-5d-f) 8

8(7). Clypeus and supraclypeal area flat, depressed, shining, at least partly impunctate, sometimes longitudinally striate, suture separating them weak; anterior basitarsus of female with long coarse bristles on outer surface .... *L. (Cladocerapis)*

—. Clypeus and supraclypeal area not flat, usually punctate, suture separating them distinct [weak or absent but whole area uniformly punctate in *L. (Protomorpha) tarsalis* Rayment]; anterior basitarsus of female with ordinary vestiture 9

9(8). Dorsolateral angles of pronotum much elevated above adjacent scutal surface so that dorsal pronotal margin between them is concave............................... *L. (Colletopsis)*

—. Dorsolateral angle of pronotum weak or absent, dorsal pronotal margin not concave 10

10(9). Malar area more than half as long as broad; S7 of male with apical lobes reduced (New Zealand) .... *L. (Nesocolletes)*

—. Malar area very short; S7 of male usually with two large apical lobes 11

11(10). Females... 12

—. Males 17

12(11). Scape not attaining anterior ocellus (small to middle-sized, robust, nonmetallic, commonly strongly punctate species; wings not reaching beyond apex of metasoma; flagellum short, middle segments mostly broader than long or scarcely longer than broad; propodeum, as seen in profile, with horizontal area, if present, usually shorter than metanotum) 13

—. Scape usually attaining anterior ocellus, but *if* shorter, then not agreeing with characters listed in parentheses above 14

13(12). Terga usually red, without apical hair bands; forewing length usually over 6 mm .... *L. (Odontocolletes)*

—. Terga usually black, usually with pale apical hair bands; forewing length commonly 5 mm or less .... *L. (Protomorpha)*

14(12). Metanotum with median tubercle (in some cases weak), projection, spine, or bifid process; propodeum as seen in profile vertical or nearly so, without subhorizontal basal area (nearly always metallic species)................. .... *L. (Lamprocolletes)*

—. Metanotum without median elevation or, if with small median tubercle, then propodeum with distinct horizontal basal area 15

15(14). Metasoma, including apical marginal zones of terga, densely punctate; clypeus with longitudinal, median, shining convexity, thus protuberant; claws simple .... *L. (Ceratocolletes)*

—. Metasoma not densely punctate or, if so, then marginal zones of terga not similarly punctate; clypeus not usually protuberant; claws usually cleft 16

16(15). Facial fovea commonly impressed; stigma small, slender, parallel-sided, little (if any) more than one-half as long as that part of marginal cell on wing margin (as in Fig. 39-5g); metasomal terga often with basal hair bands and often with very broad, translucent marginal zones .... *L. (Goniocolletes)*

—. Facial fovea not impressed; stigma usually not parallel-sided, more than one-half as long as that part of marginal cell on wing margin (Fig. 39-5k) [except in *L. subpunctatus* (Rayment)]; metasomal terga without basal hair bands, apical marginal zones rarely broadly translucent .... *L. (Leioproctus s. str.)*

17(11). Hind tibial spur one, minute, sometimes apparently absent 18

—. Hind tibial spurs two, of ordinary size 19

18(17). Hind tibia expanded, a row of coarse bristles on distal part of lower margin; hind basitarsus with broad, flat, median or basal projection from lower margin; mandible simple, sharply pointed, without preapical tooth .... *L. (Protomorpha)* (in part)

—. Hind tibia and basitarsus slightly modified but slender; mandible with preapical tooth .... *L. (Goniocolletes)* (in part)

19(17). Species small to middle-sized, robust, nonmetallic, commonly strongly punctate; wings not reaching beyond apex of metasoma; flagellum short, its middle segments mostly broader than long or scarcely longer than broad; propodeum, as seen in profile, with horizontal area, if present, usually shorter than metanotum 20

—. Species not agreeing with combination of characters listed above 21

20(19). S7 with apical lobes small, broadly attached to body of sternum; terga without apical hair bands .... *L. (Odontocolletes)*

—. S7 with two large apical lobes; terga often with apical hair bands .... *L. (Protomorpha)* (in part)

21(19). Metanotum with median tubercle (in some cases weak), projection, spine, or bifid process; propodeum as seen in profile vertical or nearly so, without subhorizontal basal area (nearly always metallic species).................. .... *L. (Lamprocolletes)*

—. Metanotum without median elevation or, *if* with small median tubercle, then propodeum with distinct horizontal basal area 22

22(21). Metasoma, including apical marginal zones of terga, densely punctate; last antennal segment sometimes enlarged and flattened .... *L. (Ceratocolletes)*

—. Metasoma not densely punctate or, if so, then marginal zones of terga not similarly punctate; last antennal segment unmodified 23

23(22). Facial fovea commonly impressed; stigma small, slender, parallel-sided, little (if any) more than one-half as long as that part of marginal cell on wing margin (as in Fig. 39-5g); metasomal terga often with basal hair bands and often with very broad translucent apical marginal zones; S7 produced posteriorly as slender neck to which apical lobes are attached (Fig. 39-7f) .... *L. (Goniocolletes)* (in part)

—. Facial fovea not impressed; stigma usually not parallel-sided, more than one-half as long as that part of marginal cell on wing margin (Fig. 39-5k) [except in *L. subpunctatus* (Rayment)]; metasomal terga without basal hair bands, apical marginal zones rarely broadly translucent; S7 with median portion not so slender and elongate .... *L. (Leioproctus* s. str.)

Leioproctus / Subgenus *Actenosigynes* Moure, Graf, and Urban, new status.

Actenosigynes Moure, Graf, and Urban, 1999: 2. Type species: *Leioproctus fulvoniger* Michener, 1989, by original designation.

This name was proposed as a genus for the only American species left in *Leioproctus* s. str. by Michener (1989, 2000). Moure, Graf, and Urban (1999) had both sexes and its separation from the Australian *Leioproctus* s. str. is appropriate, but like other similar forms, it is considered a subgenus of *Leioproctus*. It is not particularly similar to any one Australian species or species group. It differs from the great majority of Australian species in having four instead of two apicolateral lobes on S7 of the male. In this respect, however, it resembles the Australian *L. advena* (Smith) and its relatives, which Michener (1965b) noted might well be separated subgenerically from *Leioproctus* s. str. In other features, however, *L. fulvoniger* is quite different from *L. advena*. The presence of only two lobes on S7 is likely to be an apomorhic state, since four is a widespread condition in other colletid subfamilies.

■ The single species, *Leioproctus (Actenosigynes) fulvoniger* Michener, is found in southern Brazil.

Leioproctus / Subgenus *Albinapis* Urban and Graf, new status.

Albinapis Urban and Graf, 2000: 595. Type species: *Albinapis gracilis* Urban and Graf, 2000, by original designation.

Albinapis is a close relative of *Leioproctus (Hexantheda)*. *Leioproctus (Albinapis) gracilis* (Urban and Graf) **new combination** differs from *Hexantheda* in the 5-segmented labial palpi and especially in the small rounded apicolateral lobes of S7 of the male, as illustrated by Urban and Graf (2000a). For additional characters, see the subgenus *Hexantheda*.

■ The single species *Leioproctus (Albinapis) gracilis* (Urban and Graf) occurs in Rio Grande do Sul, Brazil.

Leioproctus / Subgenus *Andrenopsis* Cockerell

Andrenopsis Cockerell, 1905a: 363. Type species: *Andrenopsis flavorufus* Cockerell, 1905, monobasic.

This Australian subgenus contains rather robust, nonmetallic species, 7 to 8 mm long, the males nearly as robust as the females. The clypeus of the male is yellow in most species. The facial fovea of the female is broad; it is distinct because of its dull surface and sparser punctation than that on adjacent areas, but it is only slightly depressed. The metanotum is not tuberculate but broadly elevated to the level of the scutellum medially, and with a distinct, more or less vertical posterior declivity. The inner hind tibial spur is finely pectinate in the female; the inner teeth of the claws of the female are present but reduced. The inner hind tibial spur of the male is enlarged and curved in *Leioproctus (Andrenopsis) flavorufus* (Cockerell) but not in other species. The metasoma lacks hair bands but in some cases has orange-brown integumental bands. S7, S8 and the male genitalia were illustrated by Michener (1962b). In *L. (A.) douglasiellus* Michener and perhaps *L. (A.) nigrifrons* Michener there are a few very short hairs on the eyes.

■ *Andrenopsis* is known from Western Australia to Victoria and southern Queensland. The four species were listed by Michener (1965b) and Cardale (1993).

Leioproctus / Subgenus *Baeocolletes* Michener

Leioproctus (Baeocolletes) Michener, 1965b: 70. Type species: *Leioproctus calcaratus* Michener, 1965, by original designation.

This Australian subgenus consists of small (4.5-7.0 mm long), nonmetallic species; the red or partly red metasoma lacks hair bands. The species are more robust than those of the subgenus *Euryglossidia,* having much the form of *Andrenopsis*. The short scape and the form of the facial foveae suggest the subgenus *Protomorpha*. The tibial spurs and the position of the first recurrent vein are unique in the genus (see the key to subgenera). The flat, polished face of two of the species is suggestive of the subgenus *Cladocerapis,* but there is clearly no close affinity with that subgenus. The facial fovea is distinct in the female. The stigma is much more than one-half the length of the margin of the marginal cell on the costa (Fig. 28-8h). The propodeum has a distinct, roughened, subhorizontal basal zone, shorter than the metanotum, not separated from the vertical part by a carina or sharp differentiation. The claws in the female are simple. S7 of the male has two pairs of apical lobes having more or less the form of those of *Protomorpha*. The hidden sterna and male genitalia were illustrated by Michener (1965b).

■ The species of this subgenus are from Western Australia and New South Wales. The three described species were listed by Michener (1965b) and Cardale (1993).

Leioproctus / Subgenus *Cephalocolletes* Michener

Leioproctus (Cephalocolletes) Michener, 1989: 656. Type species: *Biglossa laticeps* Friese, 1906, by original designation.

The South American subgenus *Cephalocolletes* superficially resembles black species of the genus *Brachyglossula* or even the large, black-haired species of *Lonchopria (Biglossa)*; the body length is 11 to 15 mm. The relationship of *Cephalocolletes* to the subgenus *Reedapis* is shown by the male genitalia and sterna (Michener, 1989), the scopa, and the large second submarginal cell. *Cephalocolletes* and *Reedapis* are the only American subgenera of *Leioproctus* in which the second submarginal cell is usually rather long compared to the third. This is a *Lonchopria*-like feature, but in other respects these forms clearly belong to *Leioproctus*. *Cephalocolletes* differs from *Reedapis* in the very broad head, the interocular distance being much greater than the eye length; in the eyes, which scarcely converge below in the male and are parallel in the female; and in the vertex, which is much enlarged so that all the ocelli are nearer to the antennal bases than to the posterior margin of the vertex. The branches of the sternal scopal hairs of the female are somewhat longer and less numerous than in *Reedapis*.

■ This subgenus, known from the provinces of Tucumán to Mendoza, Argentina, was based on a single species, *Leioproctus laticeps* (Friese).

Four additional species placed here by Urban (1995d) are smaller, down to 8.75 mm in body length, and have less-developed heads with eyes converging below. Perhaps they constitute a group of more ordinary species from which the extreme, *Leioproctus laticeps*, was derived. With Urban's species included, the subgenus ranges east to Rio Grande do Sul and Santa Catarina, Brazil.

Leioproctus / Subgenus *Ceratocolletes* Michener

Leioproctus (Ceratocolletes) Michener, 1965b: 63. Type species: *Lamprocolletes antennatus* Smith, 1879, by original designation.

This Australian subgenus consists of moderate-sized (body length 10-12 mm), rather robust, almost eucericiform species, the body form of the male nearly as robust as that of the female. The metasomal terga in females and some males have apical white hair bands. The clypeus is medially protuberant and impunctate, except in the male of *Leioproctus antennatus* (Smith). The antenna of the male is rather long, and the apical segment is expanded on one side in some species. The hind legs in the male of some species are incrassate, the trochanters toothed, the tibiae bent, and the tibial spurs reduced in size. The stigma is rather long but slender, parallel-sided to vein r, more than half as long as the costal margin of the marginal cell. The propodeal profile slopes and gradually curves onto the vertical posterior surface, the sloping portion being shorter than the metanotum. The front coxa of both sexes has a hairy apical lobe; the inner hind tibial spur of the female is pectinate. The apex of S7 of the male has four laterally projecting lobes, two on each side, as in the *advena* group of *Leioproctus* s. str. S7, S8, and the male genitalia were illustrated by Michener (1965b) and Maynard (1993).

■ Specimens of the subgenus have been collected in Western Australia, Queensland, and New South Wales. The two species were revised by Maynard (1993).

Leioproctus / Subgenus *Chilicolletes* Michener

Leioproctus (Chilicolletes) Michener, 1989: 640. Type species: *Leioproctus delahozii* Toro, 1973, by original designation.

Bees of this Chilean subgenus, consisting of nondescript-looking species 7.0 to 10.5 mm in body length, closely resemble those of the subgenus *Perditomorpha* but have three submarginal cells. They differ from those of the subgenus *Sarocolletes* by the very different scopal hairs of the latter, and from American *Leioproctus* s. str. by the coarsely pectinate inner hind tibial spur of the female. My first inclination was to include this subgenus in *Perditomorpha* in spite of the three submarginal cells, but the presence of a comb of hairs on the anterior basitarsus of the female is also unlike any *Perditomorpha*. This weakly developed comb, however, is not like that found in the subgenera *Cephalocolletes*, *Nomiocolletes*, and *Reedapis*. Instead, it seems to be merely the abrupt line ending the vestiture of long hairs on the upper surface of the basitarsus, the lower surface being nearly bare. The other characters of *Chilicolletes* are within the range of variation found in the subgenus *Perditomorpha*. The presence of three submarginal cells suggests that this might be a surviving representative of the group from which *Perditomorpha* was derived, but it probably is not a specialized derivative of *Perditomorpha* (like the little groups here included in that subgenus, but placed by others under such names as *Belopria*, *Edwyniana*, and *Perditomorpha* s. str.), because reacquisition of a lost vein seems unlikely. Male genitalia and other structures of the type species were illustrated by Toro (1973b) and Michener (1989).

■ *Chilicolletes* is found in central Chile. It includes two species; only one of them, *Leioproctus delahozii* Toro, is described.

Leioproctus / Subgenus *Cladocerapis* Cockerell

Cladocerapis Cockerell, 1904b: 292. Type species: *Lamprocolletes cladocerus* Smith, 1862 = *Lamprocolletes bipectinatus* Smith, 1856, by original designation.

This Australian subgenus is closely related to *Leioproctus* s. str., with which it agrees in many characters and in appearance. The species of *Cladocerapis* are nonmetallic, 9 to 11 mm long. The facial fovea is absent. The most distinctive features are indicated in the key to subgenera above. The propodeum in profile has a horizontal surface, usually as long as the metanotum, and usually not sharply delimited but curving down to the vertical surface. The inner hind tibial spur of the female is pectinate. The metasoma lacks hair bands and is weakly punctured. Male genitalia and hidden sterna were illustrated by Michener (1965b) and Maynard (1992).

■ The subgenus ranges from Tasmania, Victoria, and eastern South Australia to southern Queensland, with a disjunct area in southwestern Australia. The nine species were revised by Maynard (1992) and listed by Cardale (1993). The subgenus is apparently largely restricted in pollen gathering to flowers of *Persoonia* (Proteaceae).

In one species, *Leioproctus bipectinatus* (Smith), the male flagellum is elaborately bipectinate, a character unique among bees. In all other species the flagellum is quite ordinary. As remarkable as is the male flagellum of *L. bipectinatus,* there are no other known features separating that species strongly from other species of the subgenus. A comparative study of the mating behavior of *L. bipectinatus* and an ordinary species of *Cladocerapis* should be of special interest. A possibly related bee is *L. (Leioproctus) macmillani* Houston, in which the male flagellar segments 2 to 10 each have a simple rather than branched process on each side. This species does not have the facial characteristics of *Cladocerapis,* but has instead an elongate head, with the malar area over twice as long as wide and more than one-half as long as the eye (Houston, 1991a).

Leioproctus / Subgenus *Colletellus* Michener

Leioproctus (Colletellus) Michener, 1965b: 70. Type species: *Andrenopsis velutinus* Cockerell, 1929 (not *Paracolletes velutinus* Cockerell, 1929, homonym in *Leioproctus*) = *Leioproctus velutinellus* Michener, 1965, by original designation.

This subgenus is unusual among Australian forms with two submarginal cells in having the hind tibial spur of the female (as well as the male) ciliate rather than pectinate, although this is also true of some species of the subgenus *Euryglossidia*. *Colletellus,* with a body length of 6 mm, is similar to the subgenus *Perditomorpha* from South America. It differs in the lobes of S7 of the male, the small inner ramus of the claw of the female, and the short scape. In fact, the scape does not approach the level of the anterior ocellus and the antenna as a whole is short, the median segments of the flagellum being much broader than long. The stigma is large, not parallel-sided, more than one-half as long as the costal edge of the marginal cell. The propodeum in profile has a long subhorizontal surface, curving onto the vertical surface. The metasoma lacks distinct hair bands, but has weak apical ones laterally. The male genitalia and hidden sterna were illustrated by Michener (1965b).

■ *Colletellus* is known from Western Australia. There is only one named species, *Leioproctus velutinellus* Michener.

Leioproctus / Subgenus *Colletopsis* Michener

Leioproctus (Colletopsis) Michener, 1965b: 58. Type species: *Leioproctus contrarius* Michener, 1965, by original designation.

This Australian subgenus contains a single slender, black species, 10 mm long, that lacks metasomal fasciae except at the extreme sides. The protuberant dorsolateral angle of the pronotum, a right-angular projection above the level of adjacent parts of the scutum, is unique in the genus. The large stigma is suggestive of *Leioproctus* s. str., but the essentially vertical propodeal profile, the curious vestiture of the frons and vertex, consisting of coarse, simple, scarcely tapering hairs, and other characters separate *Colletopsis* from that subgenus. The continuation of the frontal carina as a ridge extending the length of the supraclypeal area suggests the subgenus *Ceratocolletes,* but the large stigma and other characters of *Colletopsis* distinguish it. The facial fovea is twice as broad as an ocellar diameter, not impressed but recognizable by its dull, impunctate surface. The head, sides of the thorax, and metasoma are coarsely punctate. The inner hind tibial spur of the female is coarsely pectinate; the tibia is slender, its scopa consisting of sparse, very long, coarse hairs, each of which has only one or a few branches. The genitalia and hidden sterna were illustrated by Michener (1965b).

■ This subgenus is from Western Australia. The single species is *Leioproctus contrarius* Michener.

Leioproctus / Subgenus *Euryglossidia* Cockerell

Euryglossidia Cockerell, 1910b: 358. Type species: *Euryglossidia rectangulata* Cockerell, 1910, by original designation.
Notocolletes Cockerell, 1916b: 44. Type species: *Notocolletes heterodoxus* Cockerell, 1916, monobasic.
Paracolletes (Lysicolletes) Rayment, 1935: 208. Type species: *Paracolletes singularis* Rayment, 1935, by original designation.

The name of this subgenus is unfortunate, since *Euryglossidia* is neither related to nor particularly similar to *Euryglossa*. The name doubtless resulted in Rayment's renaming of the group as *Lysicolletes;* he usually employed the name *Euryglossidia* incorrectly for a group of *Euryglossa*. *Notocolletes* is based on a species with extraordinarily modified male legs (illustrated by Michener, 1965b); otherwise, it does not differ from *Euryglossidia*.

This Australian subgenus consists of slender species 4 to 10 mm long, usually feebly metallic green or blue, sometimes black, often with the metasoma partially or wholly red. The integument is frequently dull and minutely roughened, much of the body (e.g., the scutum) lacking punctures. Although in most of the species the inner hind tibial spur of the female is conspicuously pectinate, it is ciliate in *L. striatulus* (Rayment). Since this species is nonmetallic and shining-punctate rather than dull, I considered placing it in a separate subgenus, but in sternal and genital characters, the short jugal lobe (not reaching vein cu-v), and the sparse scopa it agrees with typical *Euryglossidia*. The stigma in *Euryglossidia* is large, much more than one-half as long as the costal edge of the marginal cell. The propodeum has a long horizontal surface, curving onto the subvertical surface without sharp differentiation. S7 of the male has four apical lobes, in some cases rather broadly connected basally. The hidden sterna and genitalia were illustrated by Michener (1965b).

■ Species of this subgenus are found in Australia, mostly south of the latitude of southern Queensland. There are 22 specific names listed by Michener (1965b); Cardale (1993) listed 20 species, not including *Leioproctus heterodoxus* (Cockerell).

Leioproctus (Euryglossidia) cyanescens (Cockerell) is unusual in the near absence of scopal hairs on the trochanter and femur of the female; the tibial scopa consists of sparse, simple hairs. Houston (1981b) found females with large quantities of pollen in the crop and concluded that they carry pollen in this way instead of externally, as do other Colletinae. He did not find their nests, however, and the possibility exists that the species is cleptoparasitic.

Leioproctus / Subgenus *Excolletes* Michener

Leioproctus (Excolletes) Michener, 1965b: 56. Type species: *Leioproctus impatellatus* Michener, 1965, by original designation.

This Australian subgenus, which is so distinctive that it could well be regarded as a separate genus, contains small (5.5-8.0 mm long), nonmetallic species superficially resembling members of the subgenus *Protomorpha*. It is remarkable for (1) the complete lack of a margined basitibial plate in the females and some males (the plate has distinctive short pubescence and is margined but small in other males), (2) the almost parallel-sided, almost spatulate, rear part of the pygidial plate in the female, (3) the nearly transverse basal vein of the forewing (see the key to subgenera and Fig. 8-8l), (4) the absence of the galeal comb, and (5) the low position of the antennal bases (in the female, almost twice as far from the vertex as from the anterior clypeal margin) and their proximity to the clypeus. Characters 1-4 are unique in *Leioproctus*, and 5 is approached by *L. microsomus* Michener and *finkei* Michener, species that were included in *Microcolletes* (= *Protomorpha*) by Michener (1965b) but were excluded and not placed subgenerically by Maynard (1991). The basal horizontal part of the propodeal profile is more than one-half the length of the vertical part and longer than the metanotum. The claws of the female are simple or cleft. The inner hind tibial spur of the female is ciliate. The hidden sterna and male genitalia were illustrated by Michener (1965b).

■ This subgenus has been found in New South Wales and Western Australia. Of several species, only one, *Leioproctus impatellatus* Michener, has been described.

Leioproctus / Subgenus *Filiglossa* Rayment

Filiglossa Rayment, 1959a: 324. Type species: *Filiglossa filamentosa* Rayment, 1959, by original designation.

Species of this subgenus are similar to the nonmetallic, shiny-punctate species of *Euryglossidia*, such as *Leioproctus striatulus* (Rayment). The mouthparts, however, are among the most extraordinary of any bee. Although the glossa and galea are short, as in related Colletinae, the apex of the galea gives rise to 4 to 12 huge setae (Fig. 19-6), each at least two-thirds as long as the labial palpus and frequently as long as or longer than the face. The first two segments of the labial palpus are wider than long, the third filamentous and about as long as the face, and the fourth short and even more slender. The body length is 4 to 6 mm. Male genitalia, hidden sterna, and other structures were illustrated by Maynard (1994).

■ *Filiglossa* occurs in eastern Australia from Victoria to southern Queensland. The four species were revised by Maynard (1994).

All species appear to be oligolectic visitors to flowers of *Persoonia* (Bernhardt and Walker, 1996). Unlike the species of subgenus *Cladocerapis,* these bees are probably too minute to pollinate *Persoonia* flowers.

Recognition of *Filiglossa* may make *Euryglossidia* paraphyletic.

Leioproctus / Subgenus *Glossopasiphae* Michener

Leioproctus (Glossopasiphae) Michener, 1989: 643. Type species: *Leioproctus plaumanni* Michener, 1989, by original designation.

The most distinctive character of this South American subgenus is the glossa (Fig. 39-12a), which has two long lobes like those of the subgenus *Tetraglossula*. The structure, however, must have evolved independently in *Glossopasiphae*, for *Tetraglossula* belongs with *Protodiscelis* in a group having simple mandibles in the male and an extremely short labrum with lateroapical lobes, at least in females. *Glossopasiphae*, however, like most *Leioproctus,* has bidentate male mandibles and a relatively long labrum (about twice as broad as long) without lateroapical lobes. The pubescence is largely black or dusky, but pale on the thoracic dorsum; tergal hair bands are absent. The body length is 12 to 14 mm. The inner orbits are very slightly diverging below in both sexes. The basal zone of the propodeum is steeply sloping in profile, shorter than the metanotum. The tibial scopa is dense and long, the hairs having many fine, short branches at more or less right angles to the rachis. The inner hind tibial spur of the female is coarsely pectinate. S2-S5 of the female bear a dense scopa of hairs similar to those of the tibial scopa. S7, S8, and the genitalia of the male were illustrated by Michener (1989).

■ *Glossopasiphae* is known from Santa Catarina, Brazil. The single species is *Leioproctus plaumanni* Michener.

Leioproctus / Subgenus *Goniocolletes* Cockerell

Goniocolletes Cockerell, 1907c: 231. Type species: *Goniocolletes morsus* Cockerell, 1907, by original designation.

This Australian subgenus includes some species that look like ordinary members of *Leioproctus* s. str., with body lengths of 10 to 15 mm, but most species have unusually short, clear wings with large, sparsely haired or hairless basal areas. Common tendencies are broad, translucent apices to the metasomal terga, basal as well as sometimes apical hair bands on these terga, and modified hind tibiae and basitarsi in the males. More decisive group characters are (1) the elongate disc or bases of apodemes of S7 of the male, with the apical lobes directed laterally, and (2) the broad hairy apical process of S8, which is at least half as wide as the disc of the sternum (Fig. 39-7e, f). The facial fovea is usually distinct and impressed in females but sometimes absent, usually weak in males but sometimes distinct, sometimes absent. The propodeum in profile usually has the basal zone subhorizontal, rarely almost vertical, usually separated from the rest of the triangle by a carina. The inner hind tibial spur of the female is pectinate. The pygidial plate of the male is distinctly defined laterally for some distance in front of the tergal apex (in most subgenera the plate is represented by an undefined bare area in the male). The disc of S7 of the male is produced apically as a slender stalk, to the extremity of which are attached two lobes that are directed laterally, their posterior margins thus not extending much behind the apex of the disc of the sternum (Fig. 39-7f). This as well as other male terminalia were illustrated by Michener (1965b).

■ This subgenus is found in temperate parts of Australia, north to central Queensland and Northern Territory; it is not known from Tasmania. The 21 named species were listed by Michener (1965b) and Cardale (1993).

Leioproctus / Subgenus *Halictanthrena* Ducke

Halictanthrena Ducke, 1907: 364. Type species: *Halictanthrena malpighiacearum* Ducke, 1907, monobasic.

Distinctive features of this South American subgenus are the sharp tooth on the dorsolateral angle of the pronotum (projecting dorsolaterally, shorter than nearby hairs, and in the male so small as to be difficult to see except in profile), the presence of erect hairs and the lack of distinctive short hairs on the basitibial plate of the female, and the large stigma, larger than in any other American colletine. The body length is 6.5 to 9.0 mm. The propodeum has a subhorizontal surface that is strongly sloping, about as long as the metanotum, gradually curving onto the declivous surface. The scopal hairs near the upper margin of the tibia, including those arising from the basitibial plate, are particularly coarse near their bases, shorter near the base of the tibia. S7, S8, and genitalia were illustrated by Michener (1989).

■ *Halictanthrena* is known from the state of Minas Gerais, Brazil. The single known species is *Leioproctus malpighiacearum* (Ducke).

Leioproctus / Subgenus *Hexantheda* Ogloblin

Hexantheda Ogloblin, 1948: 172. Type species: *Hexantheda missionica* Ogloblin, 1948, by original designation.

The long, six-, seven-, or nine-segmented labial palpi are unique, two-thirds as long as the prementum, much longer than the maxillary palpi. The metasoma of the male is robust, like that of a female. The body length is 8 to 12 mm. The inner orbits converge slightly below (except for the upper extremities) in the female, but are almost parallel in the male; the clypeus is protuberant in front of the eye by about an eye width, as seen in lateral view; and the labrum is over four times as wide as long. The vertex is considerably extended and convex behind the ocelli. The subhorizontal surface of the propodeum is a little shorter than the metanotum. The inner margin of the inner hind tibial spur of the female is pectinate with slender teeth. S7, S8, and genitalia of the male were illustrated by Ogloblin (1948) and Michener (1989).

■ *Hexantheda* occurs in the state of Paraná, Brazil, and the northern Argentine provinces of Misiones and Formosa.

Leioproctus (H.) missionica (Ogloblin) has 6- or 7-segmented labial palpi. Urban and Graf (2000) described the second known species of *Hexantheda*, *Leioproctus (H.) enneomera* **new combination,** which has 9-segmented labial palpi.

Leioproctus / Subgenus *Holmbergeria* Jörgensen

Holmbergeria Jörgensen, 1912: 100. Type species: *Holmbergeria cristariae* Jörgensen, 1912, monobasic.

This South American subgenus is known only from the males. When female characters are known, its relationships may be clarified. The body length is 9 to 11 mm. Pubescence forms strong apical bands of pale hair on the terga, which may be either red or black. The face is broad, the inner orbits parallel. Unlike that of all other subgenera, the supraclypeal area is small, convex, hairless, shining, and impunctate, and the small subantennal areas are flat, hairless, shining, and impunctate, in contrast to the rest of the face. The anterior margin of the median ocellus is midway between the antennal bases and the posterior margin of the vertex, or nearer to the former. The subhorizontal surface of the propodeum (in profile) is shorter than the metanotum, rounding onto the sloping posterior surface. S7, S8, and genitalia were illustrated by Michener (1989); see also Figure 39-7b.

■ *Holmbergeria* occurs from the provinces of Mendoza and Santiago del Estero, Argentina, to Paraguay. It includes two species (see Michener, 1989).

Leioproctus / Subgenus *Hoplocolletes* Michener

Leioproctus (Hoplocolletes) Michener, 1965b: 42. Type species: *Dasycolletes ventralis* Friese, 1924, by original designation.

This South American subgenus is recognized only on the basis of a combination of female characters most of which are duplicated in other groups of *Leioproctus;* the male is unknown. The combination of the characters, however, suggests a form quite dissimilar from other subgenera. The very coarsely punctate head and thorax of *Hoplocolletes* suggest *L. (Perditomorpha) iheringi* (Schrottky) and *eulonchopriodes* Michener, although the dorsum of the thorax is even more coarsely and less closely punctate. The metasoma, however, is nearly impunctate (T1, T2) or finely punctate (T3, etc.). *Hoplocolletes* is further differentiated by the subparallel inner orbits, the three submarginal cells, the very deeply impressed median line of the anterior half of the scutum, the deeply impressed parapsidal lines, the simple hind coxal and trochanteral hairs, the sparse, long, simple hairs of the femoral scopa, the long and mostly simple hairs of the tibial scopa, and the strong sternal scopa (see below). The pubescence is short, sparse, and blackish, but dense and pale on the metasomal sterna, where there are long, simple, but hooked scopal hairs. The body length is 12 mm. The ocelli are well forward, so that the anterior margin of the median ocellus is nearer to the antennal bases than to the posterior margin of the vertex.

■ The single known species, *Leioproctus ventralis* (Friese), was described from Sydney, Australia. The specimens must have been mislabeled, however, for the species is now known from the states of Rio de Janeiro, Espirito Santo, and Minas Gerais, Brazil (the last two based on specimens collected by G. Melo).

Leioproctus (Tetraglossula) anthracinus Michener has been misidentified as *L. ventralis* in some collections.

Leioproctus / Subgenus *Kylopasiphae* Michener

Leioproctus (Kylopasiphae) Michener, 1989: 641. Type species: *Leioproctus pruinosus* Michener, 1989, by original designation.

This South American subgenus is similar to *Perditomorpha*. The greatly reduced lobes of S7 of the male (Fig. 39-7c), not layered one above another but all on a plane, are the most distinctive feature; others include the hidden basitibial plate of the female and the tibial scopa of the female (see the key to subgenera above). The body length is 7.0 to 8.5 mm. The long antenna of the male (reaching beyond the tegula) also differentiates this subgenus from *Perditomorpha;* the short antenna of *Perditomorpha* is likely to be a derived feature uniting species of that subgenus. The clypeus is shiny and impunctate except for the lateral extremities, a longitudinal median punctate depression, and in the male a band of punctures across the upper margin. S7, S8, and the male genitalia were illustrated by Michener (1989).

■ *Kylopasiphae* occurs in desert areas in the Argentine provinces from Tucumán to Neuquén. The single species is *Leioproctus pruinosus* Michener.

Leioproctus / Subgenus *Lamprocolletes* Smith

Lamprocolletes Smith, 1853: 10. Type species: *Andrena chalybeata* Erichson, 1851, by designation of Cockerell, 1905a: 345.
Nodocolletes Rayment, 1931: 164. Type species: *Nodocolletes dentatus* Rayment, 1931 = *Andrena chalybeata* Erichson, 1851, by original designation.

This Australian subgenus went under the name *Nodocolletes* until G. Maynard (in litt., 1994) recognized the species-level synonymy indicated above, under *Nodocolletes*. Most species of *Lamprocolletes* are metallic, 10 to 14 mm in length, and resemble some species of *Leioproctus* s. str., for example, *L. carinatus* (Smith) and *plumosus* (Smith). *Lamprocolletes* differs in general from that subgenus, however, in the following characters: the stigma is slender, parallel-sided or nearly so, about one-half to nearly two-thirds as long as the costal margin of the marginal cell; the metanotum has a median tubercle, process, spine, or bifid projection; and the propodeum in profile is steeply sloping to essentially vertical. The inner hind tibial spur of the female is pectinate; the metasoma lacks hair bands. The subgenus *Lamprocolletes* as here understood is quite possibly an artificial unit, as is suggested by variation in the hidden sterna and genitalia of the male, shown by Michener (1965b).

■ This subgenus ranges from Queensland to Tasmania, and one species (*Leioproctus pacificus* Michener) is found on New Caledonia. There are 18 specific names; see the lists under *Nodocolletes* by Michener (1965b) and Cardale (1993).

Given the variability within *Lamprocolletes* and the occurrence of characteristics of that subgenus in various species of *Leioproctus* s. str., recognition of *Lamprocolletes* in its present sense is defensible only as a temporary expedient, awaiting a revision of Australian *Leioproctus*. For example, *L. (Leioproctus) insularis* (Cockerell) has a metanotal tubercle as in *Lamprocolletes*, but the stigma is large and the base of the propodeum is broadly subhorizontal. The stigma is very slender and parallel-sided in *L. (Leioproctus) megachalcoides* Michener, a species that is superficially like some *Lamprocolletes* but lacks the metanotal and propodeal peculiarities of that subgenus. Further, *L. tuberculatus* (Cockerell), tentatively placed in *Lamprocolletes* because of its appearance and metanotal and propodeal structure, has a relatively large stigma like that of some *Leioproctus* s. str. Finally, *L. sexmaculatus* (Cockerell) is hesitantly placed in *Leioproctus* s. str. because of its appearance and the white apical hair bands on the metasoma; nevertheless it has a slender stigma, a feeble median metanotal prominence, and a propodeum that is nearly vertical in profile.

Leioproctus / Subgenus *Leioproctus* Smith s. str.

Leioproctus Smith, 1853: 8. Type species: *Leioproctus imitatus* Smith, 1853, by designation of Cockerell, 1905a: 348.
Dasycolletes Smith, 1853: 14. Type species: *Dasycolletes metallicus* Smith, 1853, by designation of Cockerell, 1905a: 347.
Lioproctus Smith, 1879: 6, unjustified emendation of *Leioproctus* Smith, 1853.
Paracolletes (Heterocolletes) Rayment, 1935: 184. Type species: *Paracolletes capillatus* Rayment, 1935, by original designation.
Leioproctus (Anacolletes) Michener, 1965b: 59. Type species: *Lamprocolletes bimaculatus* Smith, 1879, by original designation.

This is by far the largest subgenus of *Leioproctus*. Among Australian forms, its species, along with those of *Cladocerapis,* are the ones most similar in appearance to species of *Andrena*. The body length varies from 6 to 17 mm. The facial fovea is absent or represented by a shining or smooth area, not impressed and usually not sharply defined. The stigma is usually not parallel-sided, usually more than one-half as long as the costal part of the marginal cell (some exceptions are listed below). The jugal lobe of the hind wing usually attains or surpasses vein cu-v and is about two-thirds as long as the vannal lobe, although in *L. advena* (Smith), *worsfoldi* (Cockerell), and a few others, it is only about one-half as long as the vannal lobe, and in *L. rudis* (Cockerell) and *opaculus* (Cockerell) it is little more than one-fourth as long as the vannal lobe. The subhorizontal basal zone of the propodeum is occasionally separated from the vertical part by a carina but usually not, in some cases considerably sloping and merging with the vertical surface. The inner hind tibial spur of the female is usually pectinate, but is ciliate in the *metallicus* group and in *L. crenulatus* Michener, *subpunctatus* (Rayment), and others. The metasoma usually lacks hair bands and is usually weakly punctured. S7 of the male usually has two apical lobes. S7, S8, and the male genitalia were illustrated by Michener (1965b, 1989) and Houston (1989, 1990, 1991a).

■ *Leioproctus* s. str. is abundant in Australia and occurs also in New Guinea, Misoöl (an island to the west), Lord Howe Island, Tasmania, and New Zealand. There are about 125 specific names in this subgenus; Cardale (1993) listed 113 from Australia.

Leioproctus s. str. is the colletine taxon with the maximum number of plesiomorphies, as determined by comparison with outgroups such as the Andreninae. It consists of the species of Colletinae that lack the apomorphies characteristic of other subgenera and genera. I am not philosophically opposed to recognition of a paraphyletic taxon, provided that it is monophyletic in the classical sense and readily distinguished from other taxa (see Sec. 16). If a group of similar organisms is a useful unit, the discovery that a distinctive taxon was derived from it does not make it less useful. But one cannot defend *Leioproctus* s. str. in this way. The best one can say is that, provisionally, recognition of *Leioproctus* s. str. is the practical solution to an uncomfortable problem. Several of the subgenera that were probably derived from *Leioproctus* s. str. in fact grade into it, as was pointed out for *Nodocolletes* (now *Lamprocolletes*) and *Goniocolletes* by Michener (1965b). As collecting in Australia makes more species known, and especially as it makes known both sexes of more species, it should be possible to discover the relationships among species of *Leioproctus* s. str. and to develop a more satisfactory classification. Meanwhile, I provisionally recognize the subgenus in the sense of Michener (1965b). Any other course at this point would result in many new combinations that would probably have to be changed when a proper study is made.

Leioproctus s. str. in the Australian region includes a great number of similar species, but divergence in striking characters is indicated in the following paragraphs:

1. In scattered, unrelated species the stigma is slender and nearly parallel-sided, as, for example, in *Leioproctus capillatus* (Rayment), *cinereus* (Smith), *crenulatus* Michener, *megachalcoides* Michener, *rhodopus* (Cockerell), *sexmaculatus* (Cockerell), *subpunctatus* (Rayment), and the group comprising *L. advena* (Smith), *ruficornis* (Smith), and *worsfoldi* (Cockerell).

2. Some species, such as *Leioproctus cinereus* (Smith), *insularis* (Cockerell), *rhodopus* (Cockerell), and *sexmaculatus* (Cockerell), having a median tubercle on the metanotum, may resemble the stock or stocks from which the subgenus *Lamprocolletes* arose.

3. In a small group of species the first recurrent vein is well beyond the middle of the second submarginal cell; the jugal lobe is short, not reaching cu-v; the basitibial plate is unusually large except in *Leioproctus rubellus* (Smith); and a carina often separates the horizontal from the dorsal surfaces of the propodeum. Each of these characters is found alone in other unrelated species, but their association may indicate a phyletic line. Within this line, reduction of the inner tooth of the tarsal claws of the female is noteworthy: in *L. maculatus* (Rayment) and *rubellus* (Smith) the claws are cleft as usual in the genus; in *L. unguidentatus* Michener the inner tooth is much reduced; and in *L. platycephalus* (Cockerell) and *truncatulus* (Cockerell) the claws are simple. Claws are simple also in the female of the distantly related *L. crenulatus* Michener.

4. In a few species the eyes are hairy. This is true of *Leioproctus nigriventris* (Friese), a very ordinary-appearing species. It is also true of *L. capillatus* (Rayment), which has been made the type of a subgenus, *Heterocolletes*, partly because of its hairy genitalia. Equally hairy genitalia occur in certain other species with bare eyes, and *L. capillatus* seems to fall within the range of variation of *Leioproctus* s. str.

5. The group of *Leioproctus conospermi* Houston consisting of three species has, among other unusual characters, reduced maxillary palpi, three- to five-segmented and shorter than the labial palpi. The scopa is sparse and consists of largely simple hairs, some bifid or trifid, and the hind basitarsus of the female is five to six times as long as broad. The two apical lobes of T7 of the male are much reduced. See Houston, 1989.

6. The 13 species of the group of *Leioproctus capito* Houston are unusual in having a relatively long proboscis and enlarged labial palpi. They are associated with flowers of *Eremophila*. See Houston, 1990.

7. *Leioproctus excubitor* Houston is unusual in having a much elongate first flagellar segment in both sexes, longer than segments 2 to 4 combined. *L. macmillani* Houston has a remarkably elongate head and pectinate male antennae; see the remarks under the subgenus *Cladocerapis* (Houston, 1991a). *L. apicalis* (Cockerell) is unusual in having a sparse tibial scopa consisting of simple rather than plumose hairs. As noted by Michener (1965b), *L. abnormis* (Cockerell) is in some ways intermediate between *Leioproctus* s. str. and the subgenus *Euryglossidia*.

8. *Leioproctus crenulatus* Michener was placed in *Chrysocolletes* by Michener (1965b), but I now believe that its resemblance to *Chrysocolletes* is convergent, and that until a detailed study of phylogeny is made, this species is best included in the catch-all subgenus *Leioproctus* s. str. Distinctive features include those cited under items (1) and (3) above and the rather closely punctate metasomal terga (except for the impunctate extreme margins), the position of the median ocellus midway between the antennal bases and the posterior edge of the vertex (as seen in profile), and the steeply sloping basal zone of the propodeum.

Anacolletes, synonymized above, is retained as a subgenus for a single species by Maynard (1997); see the account of the subgenus *Odontocolletes* for further explanation.

Leioproctus / Subgenus *Nesocolletes* Michener

Leioproctus (Nesocolletes) Michener, 1965b: 52. Type species: *Lamprocolletes fulvescens* Smith, 1876, by original designation.

This New Zealand subgenus is related to *Leioproctus* proper, specifically to the *metallicus* group, although in general it is more slender and has more long erect pubescence. The body length is 9 to 13 mm. Its most distinctive external feature is the long malar space, more than one-half as long as the width of the base of the mandible, as in the quite different American subgenus *Torocolletes*. The distinctness of *Nesocolletes* from *Leioproctus* s. str. is emphasized by S7 of the male, which, instead of having moderate-sized to large, apical lobes, has these lobes minute and broadly attached to the apex of the sternum, as illustrated by Michener (1965b). The stigma is slender, not parallel-sided, more than one-half as long as the costal side of the marginal cell. The jugal lobe of the hind wing does not reach vein cu-v and is only one-half as long as the vannal lobe. The basal zone of the propodeum is steeply sloping, shorter than the metanotum, and round-

ing onto the subvertical part, the basal part being so steep that the whole profile can be described as subvertical or strongly declivous. The inner hind tibial spur of the female is finely ciliate.
- This subgenus is known only fron New Zealand. Five specific names have been given to its members, as listed by Michener (1965b).

Leioproctus / Subgenus *Nomiocolletes* Brèthes

Nomiocolletes Brèthes, 1909c: 455. Type species: *Nomia joergenseni* Friese, 1908, by original designation.
Baptonedys Moure, Urban, and Graf, 1999: 12. Type species: *Lonchopria bicellularis* Ducke, 1910, by original designation. [New synonymy.]

In spite of its distinctive appearance, which is due to the enamel-like metasomal bands, much like those of *Nomia* (Halictidae), *Nomiocolletes* is similar to *Perditomorpha* but has three submarginal cells, except in *Leioproctus bicellularis* (Ducke), which has two. The modified hind legs of males, with the hind femur swollen except in *L. bicellularis* and the tibia broadened apically, suggest *Pygopasiphae*, but are quite different and no doubt independently evolved. As in *Cephalocolletes, Reedapis,* and some *Spinolapis,* there is a distinct row or comb of hairs on the front basitarsus of the female. The body length is 7.0 to 12.5 mm. The basal zone of the propodeum is sloping, much shorter than the metanotum, and curving rather abruptly to the declivity, so that the profile of the propodeum is mostly steeply declivous. In *L. bicellularis* the basal zone is narrowly horizontal but then curves gradually onto the declivity. The male S7, S8, and genitalia were illustrated by Michener (1989).
- *Nomiocolletes* occurs from Río Negro province, Argentina, and Bolivia to Ceará in northeastern Brazil, mostly in xeric areas. There are five species.

Leioproctus simplicicrus Michener (1989) was described in the subgenus *Nomiocolletes,* largely because of the metasomal color bands. It is known only from the male, and its characters are discussed under the genus *Eulonchopria,* to which it seems to be related. It now seems clear that it is not a *Nomiocolletes,* although it runs there in the keys to genera and subgenera. Its inclusion in *Eulonchopria* would eliminate some of the distinctions between that genus and *Leioproctus.* More specimens and the unknown female are needed before a decision about *L. simplicicrus* can be made.

That *Leioproctus bicellularis* (Ducke), placed in the subgenus *Perditomorpha* by Michener (1989), appears closer to *Nomiocolletes* was pointed out by G. Melo (personal communication, 1996). Its enamel-like apical yellow tergal bands (with fine punctures and sparse, short hairs, unlike those of other *Nomiocolletes*), row of hairs on the front basitarsus of the female, upwardly flexed apical lobes of the male S7 and apically modified gonoforceps, and the slightly expanded hind tibia of the male indicate placement in *Nomiocolletes.* It is perhaps justifiable to recognize *Baptomedys* as a subgenus of *Leioproctus,* differing from *Nomiocolletes* by having two instead of three submarginal cells, in addition to other characters. It may well make *Nomiocolletes* paraphyletic, however, and for the present I regard *Baptonedys* as a synonym of *Nomiocolletes.*

Leioproctus / Subgenus *Odontocolletes* Maynard

Leioproctus (Odontodolletes) Maynard, 1997: 140. Type species: *Paracolletes pachyodontus* Cockerell, 1915, by original designation.

This is the subgenus for which Michener (1965b) intended the name *Anacolletes.* The type specimen of the type species of *Anacolletes,* however, was found by Maynard (1997) to be a species here included in *Leioproctus* s. str.; she therefore proposed *Odontocolletes* to replace *Anacolletes* in its original sense.

This Australian subgenus agrees with most of the major external features of *Protomorpha,* although its species are larger (7.5-11.0 mm in length) than most species of that subgenus. The really distinctive feature is found in S7 of the male, in which the two broad apical lobes found in *Protomorpha* are reduced to mere hairy pads broadly connected to the body of the sternum, which therefore does not appear to be that of a colletid bee. The propodeal triangle is rugose, with large areolae lateromarginally; its profile is steeply sloping at the base but mostly subvertical. On the metanotum is a large, blunt, median tubercle. S7 and S8 and the male genitalia were illustrated by Michener (1965b) under the subgeneric name *Anacolletes* and by Maynard (1997).
- Although most diversified in Western Australia and perhaps Northern Territory, species of this subgenus are also found as far east as Queensland. Maynard (1997) revised the eight included species.

Leioproctus / Subgenus *Perditomorpha* Ashmead

Perditomorpha Ashmead, 1899a: 86. Type species: *Perditomorpha brunerii* Ashmead, 1899, monobasic.
Bicolletes Friese, 1908b: 341 (11 in reprint). Type species: *Bicolletes neotropica* Friese, 1908, by designation of Cockerell, 1915: 342.
Edwynia Moure, 1951c: 195, not Aldrich, 1930. Type species: *Pasiphae flavicornis* Spinola, 1851, by original designation.
Edwyniana Moure, 1954a: 165, replacement for *Edwynia* Moure, 1951. Type species: *Pasiphae flavicornis* Spinola, 1851, autobasic.
Belopria Moure, 1956: 305. Type species: *Belopria zonata* Moure, 1956, by original designation.

This is the largest South American subgenus of *Leioproctus.* Most species are small, but the body length ranges from 5 to 13 mm. The integument is nonmetallic or sometimes has a barely perceptible bluish tint, and the metasoma is sometimes red (blue in a rather large, undescribed Argentine species). The ocelli are near the summit of the vertex, the anterior margin of the median ocellus usually being well behind the midpoint between the antennal bases and the posterior edge of the vertex, but at the midpoint in *L. arnauellus* Michener and in the female of *L. brunerii* (Ashmead). The tibial spurs are straight, the inner margin of the inner hind spur of the female being pectinate or, less commonly, ciliate (Fig. 39-8b, c), as in *L. arnauellus* and *brunerii*. The second submarginal cell is subequal in length to the first. Sterna of the female usually lack a well-developed scopa but have apical bands of hair, the hair sometimes rather long and carrying pollen, thus functionally a scopa. These sternal hairs are usually simple or plumose on certain sclerites, but are sometimes coarsely branched. In *L. arnauellus, brunerii,* and *inconspicuus* Michener the sternal scopa is moderately developed, consisting of hairs with numerous rather short side branches. The apicolateral lobes of S7 of the male are two on each side, usually both broad and more or less rounded but sometimes narrower and elongate, as for example in *L. mourei* (Toro) and *zonatus* (Moure), or rather small and one of them somewhat elongate, as for example in *L. herrerae* (Toro), or broadened and extended basally, each bilobed, so that there are four apicolateral lobes on each side, as in *L. arnauellus* and *brunerii* (Fig. 39-7a). S7, S8, and the genitalia of the male were illustrated by Moure (1954a, 1956), Toro and Rojas (1970a), and Michener (1989); see also Figures 39-6a-d and 39-7a.

■ The subgenus *Perditomorpha* is abundant in temperate South America on both sides of the Andes from Bio Bio, Chile, and Neuquén Province, Argentina, north to Peru and the state of Ceará, Brazil. There are about 45 species, as listed by Michener (1989).

Nests of *Leioproctus zonatus* (Moure) were described by Michener and Lange (1957); see the discussion of nesting biology under the subfamily Colletinae (Sec. 38).

The following characters, which are often of subgeneric or generic importance in bees, vary within the subgenus. The facial foveae of females are usually completely absent or represented only by areas that are less densely punctate or slightly more shiny than surrounding regions. The foveae, however, are distinct, well defined, and impressed in females of *Leioproctus herrerae* Toro and *tristis* (Spinola). Certain species are very coarsely punctate, unlike most species, which are only moderately to finely punctate. One of the most coarsely punctate is *L. iheringi* (Schrottky). *L. eulonchopriodes* Michener is similarly coarsely punctate, with yellow apical integumental bands on T1, T3, and T4, and its apical tergal margins are upturned and carinate, as are those of *Eulonchopria psaenythioides* Brèthes. *L. chrysostomus* (Cockerell) is unusual in that the claws of each leg are asymmetrical, the principal ramus of the inner claw being enlarged and blunt.

In an old sense, the subgenus *Perditomorpha* comprised only two species, *Leioproctus arnauellus* Michener amd *brunerii* (Ashmead). They had long been separated from *Bicolletes* at the generic or subgeneric level (Moure, 1954a; Michener, 1965) by the characters noted in the descriptive comments above. Intermediate species make recognition of *Perditomorpha* in this old sense both impractical and undesirable, as shown by Michener (1989).

Leioproctus bicellularis (Ducke) was placed in *Perditomorpha* by Michener (1989); it seems nearer to *Nomiocolletes* (see above).

Leioproctus / Subgenus *Protodiscelis* Brèthes

Protodiscelis Brèthes, 1909a: 245. Type species: *Protodiscelis fiebrigi* Brèthes, 1909, monobasic.

Protodiscelis resembles the subgenus *Tetraglossula* in the simple mandibles of the male (a character found otherwise in American *Leioproctus* only in the unrelated subgenus *Cephalocolletes*) and in the presence of apicolateral lobes on the very short labrum of the female (Fig. 39-9a) (see the key to subgenera). These are both apomorphies that unite the two subgenera. *Protodiscelis* differs from *Tetraglossula*, however, in numerous features, such as the scarcely elongate glossa and the plumose scopal hairs (Fig. 13-1e) of both metasomal sterna and hind legs. The body length is 6 to 9 mm. In the female there is a shiny, depressed, usually ill-defined facial fovea extending onto the vertex and mesad toward the ocelli. The mandible of the female is unusually slender and its preapical tooth is small. The inner hind tibial spur of the female is finely pectinate. S7, S8, and genitalia of the male were illustrated by Michener (1989) and Melo (1996).

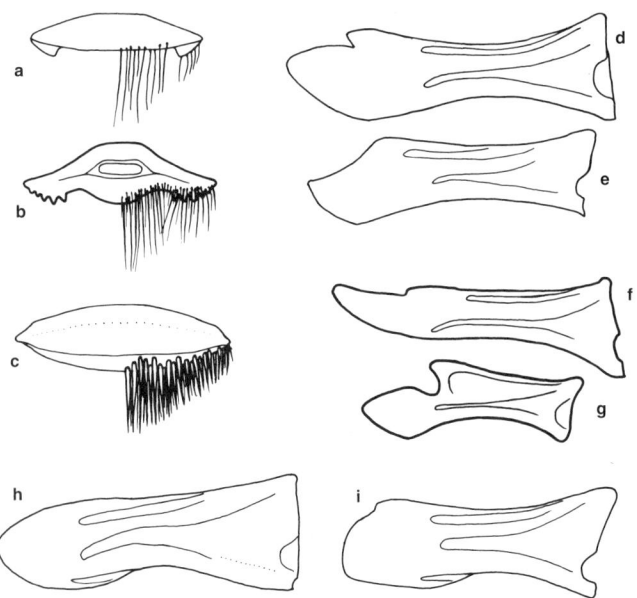

Figure 39-9. Structures of Paracolletini. a-c, Labra of *Leioproctus* females. **a**, *L. (Protodiscelis) spathigerus* Michener, the apicolateral lobes of which are larger than those in most other species of its subgenus; b, *L. (Tetraglossula) fucosus* Michener; **c**, *L. (Spinolapis) caerulescens* (Spinola), lacking apicolateral lobes. **d-i**, Mandibles of female and male of the following. **d, e**, *Lonchopria (Biglossa) robertsi* Michener; **f, g**, *L. (Ctenosibyne) cingulata* Moure; **h, i**, *L. (Porterapis) porteri* Ruiz. From Michener, 1989.

■ Species of this subgenus occur from the state of Paraíba in northeastern Brazil to the state of Paraná in southern Brazil and to Paraguay; G. Melo (in litt., 1995) reports the subgenus also from Argentina and Bolivia. There are five described and several undescribed species.

Some species of *Protodiscelis,* such as *Leioproctus fiebrigi* (Brèthes), have such short and sparse hair that the males superficially resemble *Chilicola* or other Xeromelissinae. Brèthes described a male, thinking that it was a female; of course it did not have a scopa or the other features of female Colletinae. The result was years of confusion about *Protodiscelis* (see Michener, 1989). *Leioproctus spathigerus* Michener is unusual for the broad, almost sheathlike hind and especially middle tibial spurs of the female; *L. echinodori* Melo has a similar but less extreme development.

Leioproctus / Subgenus *Protomorpha* Rayment

Protomorpha Rayment, 1959b: 334. Type species: *Protomorpha tarsalis* Rayment, 1959, by original designation.
Leioproctus (Microcolletes) Michener, 1965b: 55. Type species: *Paracolletes halictiformis* Cockerell, 1916 (not *Euryglossa halictiformis* Smith, 1879, homonym in *Leioproctus*) = *Leioproctus halictomimus* Michener, 1965 = *Paracolletes minutus* Cockerell, 1916, by original designation.

This Australian subgenus contains small (length 5-9 mm), completely nonmetallic forms, males of which have the body robust, like that of the females, the antennae short (described below), and the hind tibiae and basitarsi often modified. In a few species the metasoma is red. The subgenus grades into *Leioproctus* s. str. in most characters. In general, the small species of *Leioproctus* s. str. differ strikingly from *Protomorpha* in the long antennae of the males. As with the subgenus *Lamprocolletes,* most of the characters of *Protomorpha* are duplicated individually in one or another species of *Leioproctus* s. str., but the combination is diagnostic for *Protomorpha.* The facial fovea is impressed or represented by a smooth but not impressed area. The flagellum is short in both sexes, the second and third segments usually much wider than long, and the median segments, even in males, little if any longer than wide, but in *L. (P.) tarsalis* (Rayment) only the first two segments are broader than long. The stigma is of moderate size, not parallel-sided, more than one-half as long as the costal margin of the marginal cell. The propodeum in profile is entirely declivous or has a short horizontal basal area, shorter than the metanotum, and in some cases set off from the vertical surface by a carina or distinct angle. The inner hind tibial spur of the female is pectinate. In males of some species, the hind tibia and basitarsus are expanded, angulate. S7 of the male has a pair of large, posterolaterally directed lobes. S7, S8, and the male genitalia were illustrated by Michener (1965b) and Maynard (1991).

■ *Protomorpha* is widespread in Australia, being especially abundant in dry areas, and is known north to central Queensland. The five named species and four unnamed species were revised by Maynard (1991). She excluded as unplaced certain species included in *Microcolletes* by Michener (1965b). Among them was *Leioproctus tropicalis* (Cockerell) from Melville Island.

Leioproctus / Subgenus *Pygopasiphae* Michener

Leioproctus (Pygopasiphae) Michener, 1989: 647. Type species: *Leioproctus mourellus* Michener, 1989, by original designation.

The most distinctive characters of the South American subgenus *Pygopasiphae* are the capitate hairs of the metasomal sterna of the female, the distinct pygidial plate of the male, and the attenuate hind tarsi of the male (see the key to subgenera). The body length is 7 to 11 mm. The pubescence forms pale apical bands on T1 or T2 to T4. The inner hind tibial spur of the female is pectinate, with two to five large teeth. The pygidial plate of the female is unusually large, its apex rounded, not at all truncate. S7, S8, and the genitalia of the male were illustrated by Michener (1989).

■ *Pygopasiphae* occurs in the Argentine provinces of Catamarca, La Rioja, and Santiago del Estero, in xeric areas. The two named species were listed by Michener (1989); several undescribed species are recognized in collections.

Leioproctus / Subgenus *Reedapis* Michener

Leioproctus (Reedapis) Michener, 1989: 656. Type species: *Leioproctus bathycyaneus* Toro, 1973, by original designation.

Superficially, the rather large, robust species of *Reedapis* closely resemble those of the unrelated subgenus *Spinolapis,* which are found in the same area. *Reedapis* differs from *Spinolapis* in having three submarginal cells and bifid claws in the female. *Reedapis* is actually more similar to the subgenera *Perditomorpha* and *Chilicolletes.* It differs from *Perditomorpha* in having three submarginal cells, and from both of those subgenera not only in appearance but also in the distinct row or comb of hairs on the outer margin of the front basitarsus of the female and in the strongly pectinate condition, in females and some males, of both the inner and outer hind tibial spurs and of the midtibial spur. The closest relative of *Reedapis* is *Cephalocolletes;* these subgenera differ conspicuously in the characters indicated in the key. The integument in *Reedapis* is black, and at least the metasoma is weakly metallic blue. The body length is 9 to 15 mm. T1-T3 have apical bands of white hair, sometimes weak on T1 and weakly indicated on T4. The propodeum has a basal subhorizontal surface about as long as the metanotum. The second submarginal cell is usually subequal to or longer than the third on the posterior margin but in occasional individuals the second is much shorter than the third. S2-S5 of the female have a scopa of dusky or blackish, somewhat appressed, plumose hairs nearly as long as the exposed parts of the sterna, the branches mostly directed apicad, like those of the tibia (Fig. 13-1g, h). S7, S8, and genitalia of the male were illustrated by Toro (1973a) and Michener (1989).

■ *Reedapis* is found in Chile from Antofagasta to Maule. The three species were listed by Michener (1989) and revised as the "Grupo Semicyaneus" by Toro (1973a).

An unusual feature of *Leioproctus semicyaneus* (Spinola) is the presence of two subantennal "sutures" or lines below each antennal base. They converge toward the clypeus and meet above the upper margin of the clypeus, so that the subantennal area is triangular. This area is smooth, impunctate, unlike adjacent parts of the face. The inner line probably represents merely a change in surface sculpture, but it is fully as conspicuous as the subantennal sutures in many Andrenidae. *L. bathycyaneus* Toro has no such subantennal areas.

Leioproctus / Subgenus *Sarocolletes* Michener

Leioproctus (Sarocolletes) Michener, 1989: 643. Type species: *Lonchopria rufipennis* Cockerell, 1917, by original designation.

The scopal hairs, with numerous short, fine side branches (as in Fig. 13-1d), are a striking feature of this subgenus, shared only with the subgenera *Glossopasiphae* and *Protodiscelis*. Most other characters agree with those of *Chilicolletes*, or, in the case of species with two submarginal cells, with *Perditomorpha*. *Sarocolletes* differs further from those subgenera, however, in the well-developed ventral scopa of the female and in the relatively broad metasoma of the male, which therefore resembles a female. The body length is 7 to 11 mm. The pubescence forms apical bands of pale hair on the terga. There are usually three submarginal cells, but only two in *Leioproctus duplex* Michener. S7, S8, and the male genitalia were illustrated by Michener (1989).

■ *Sarocolletes* occurs from the provinces of Buenos Aires and Entre Ríos to Tucumán, Argentina, and to the state of Bahia, Brazil. Moure and Urban (1995) considered five named species. A sixth was published by Urban (1995d). Three others appear to be in collections.

Leioproctus / Subgenus *Spinolapis* Moure

Pasiphae Spinola, 1851: 226, not Latreille, 1819. Type species: *Pasiphae caerulescens* Spinola, 1851, designated by Sandhouse, 1943: 585.

Spinolapis Moure, 1951c: 193, replacement for *Pasiphae* Spinola, 1851. Type species: *Pasiphae caerulescens* Spinola, 1851, by original designation and autobasic.

From the superficially similar subgenus *Reedapis*, *Spinolapis* differs in having only two submarginal cells, in the reduced (or absent) inner tooth on the claws of the female, and in the finely pectinate or ciliate tibial spurs (Fig. 39-8a). The integument is largely metallic blue, or the head and thorax are black. The body length is 9 to 12 mm. The metasoma lacks hair bands. The anterior basitarsus of the female lacks or [in *Leioproctus cyaneus* (Cockerell)] has a comblike row of hairs on the outer margin. S2-S5 of the female bear rather short, pollen-carrying hairs, shorter than the exposed parts of the sterna, often somewhat erect, some simple, others with rather short, apically directed branches. The pygidial plate of the female is narrow and parallel-sided apically. S7, S8, and the genitalia of the male were illustrated by Michener (1989).

■ This subgenus occurs from Atacama, Chile, and Neuquén province, Argentina, south to Tierra del Fuego. The three species were listed by Michener (1989). A key to the species was given by Toro (2000).

Leioproctus / Subgenus *Tetraglossula* Ogloblin

Tetraglossula Ogloblin, 1948: 165. Type species: *Tetraglossula deltivaga* Ogloblin, 1948, by original designation.

The simple mandible of the male and the slender mandible of the female show the relationship of *Tetraglossula* to the subgenus *Protodiscelis*. The deeply bifid glossa of *Tetraglossula* suggests the subgenus *Glossopasiphae*, which, however, has a subapical mandibular tooth in the male and plumose scopal hairs. The body length is 6 to 12 mm. The metasomal hair bands are absent or weak. The labrum is about six times as wide as long, and has a transverse depression surrounded by carinae between the discal convexities; the lateral lobes are weak in the male, conspicuous and serrate or pectinate in the female (Fig. 39-9b). On the apical margin of the clypeus of the male is a small median lobe overhanging the labrum. The femoral scopa and that of the distal part of the trochanter, as well as that of S2-S5, consist of sparse or simple hairs. The inner margin of the inner hind tibial spur of the female is pectinate. T7 of the male has a triangular, bare pygidial plate, tapering to a narrow rounded apex. S7, S8, and the genitalia of the male were illustrated by Michener (1989).

■ This subgenus ranges from the state of Pará, Brazil, to the province of Buenos Aires, Argentina. The five species were listed by Michener (1989).

According to G. Melo (personal communication, 1996), *Tetraglossula* appears to be oligolectic on *Ludwigia* (Onagraceae).

Leioproctus / Subgenus *Torocolletes* Michener

Leioproctus (Torocolletes) Michener, 1989: 651. Type species: *Lonchopria fazii* Herbst, 1923, by original designation.

The largely dull, dark blue or black integument, lacking recognizable punctures except on the clypeus, is characteristic of this subgenus, as are the protuberant clypeus, distinct malar area, and reduced jugal lobe of the hind wing. In its distinct malar space, *Torocolletes* resembles the unrelated subgenus *Nesocolletes* from New Zealand as well as some undescribed Australian species of *Leioproctus* s. str. that also have elongate malar areas, but in other characters these Australian species are unlike both *Nesocolletes* and *Torocolletes*. The body length is 7 to 11 mm. The labrum of both sexes is convex, shining, and twice as broad as long or slightly more. The tibial spurs are straight, the inner margin of the inner hind spur of the female being finely pectinate to coarsely ciliate (illustrated by Toro, 1973a). The jugal lobe of the hind wing terminates well before cu-v and is scarcely over one-half as long as the vannal lobe. S7, S8, and male genitalia were illustrated by Toro (1973a) and Michener (1989).

■ This subgenus occurs in central Chile. The two species were listed by Michener (1989) and revised as the "Grupo Fazii" by Toro (1973a).

Leioproctus / Subgenus *Urocolletes* Michener

Leioproctus (Urocolletes) Michener, 1965b: 58. Type species: *Leioproctus rhodurus* Michener, 1965, by original designation.

This subgenus is unique among Australian colletids in the absence of arolia. Although this is a striking feature often characteristic of genera among bees, the similarity of *Urocolletes* in other features to various groups of *Leioproctus* indicates that it should be included in *Leioproctus* unless that genus is to be divided. The body length is nearly 12 mm. The finely pectinate or ciliate inner hind tibial spur is intermediate between that of the *metallicus* group of *Leioproctus* proper and that of the bulk of that subgenus. Facial fovea are not or scarcely recognizable. The forewing has three submarginal cells; the stigma is slender, parallel-sided before the base of vein r, and less than one-half as long as the costal side of the marginal cell. The propodeum is nearly vertical in profile. The metanotum has a large, median, rounded, produced portion, so that in profile it seems to have a large tooth. The claws of the female are cleft. S7 of the male has one major and one small apicolateral lobe; S8 is of the usual *Leioproctus* type with a strong apical process. These structures were illustrated by Michener (1965b).

■ *Urocolletes* is from Western Australia; only one species, *Leioproctus rhodurus* Michener, is known.

Genus *Lonchopria* Vachal

Lonchopria, found in temperate South America, contains diverse species ranging from rather small forms superficially not easily distinguished from small or middle-sized species of the genus *Leioproctus* to large and easily recognized forms like *Lonchopria* s. str. and the large species of the subgenus *Biglossa*. The very dense, strongly plumose tibial scopa of the female, contrasting with short, sparse, simple or restrictedly plumose hairs on the outer side of the hind basitarsus, is the most striking distinction of *Lonchopria* from other American colletines. The single species of *Lonchopria (Lonchoprella)* is an exception to this character, however. Males of *Lonchopria* differ from those of *Leioproctus* and related genera in the lack of a beveled apex of the process of S8 (Fig. 39-10). In *Leioproctus* the apex of that process is exposed and beveled, and superficially resembles a pygidial plate (Fig. 39-6). In *Lonchopria* the clypeus commonly has a depressed, median, closely punctate area, lateral and distal to which are more shining convex areas (not so in *Lonchopria* s. str.). The apex of the male mandible is commonly broadened by expansion of the lower margin, or it has a process on the lower margin. The labrum is usually less than three times as wide as long; it is usually shorter and wider in *Leioproctus*. The front basitarsus of the female *Lonchopria* has a comb of hairs on the outer margin. The inner hind tibial spur of the female is pectinate with four to eight very long teeth, their bases about as close as they can be, so that they diverge from a sometimes somewhat thickened part of the spur; this is also the case in the Australian genus *Trichocolletes* (Fig. 39-8e, f). The hind basitarsus of the female tapers toward the extreme apex, which is only about half as wide as the maximum width

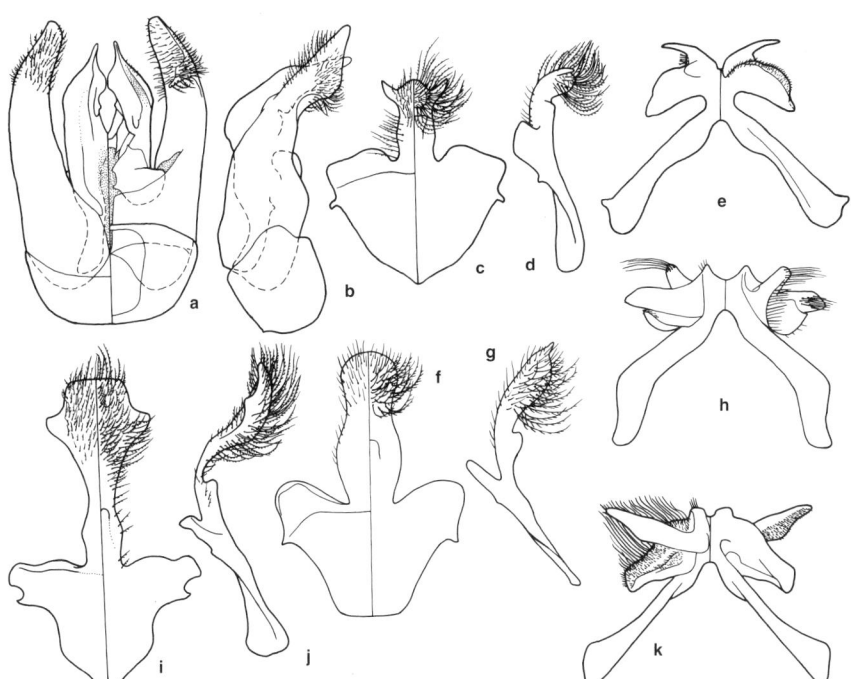

Figure 39-10. Male genitalia and hidden sterna of *Lonchopria*, dorsoventral and lateral views of genitalia and of S8, and dorsoventral views of S7. **a-e,** *L. (Biglossa) thoracica* (Friese); **f-h,** *L. (Lonchoprella) annectens* Michener; **i-k,** *L. (Biglossa) chalybaea* (Friese). (Dorsal views are at the left.) From Michener, 1989.

near the base. There are always three submarginal cells. The stigma is slender, not or little broader than the prestigma (measured to the wing margin), somewhat broader in the subgenus *Porterapis*. Vein r arises near the middle of the stigma (Fig. 39-11); the margin of the stigma within the marginal cell is convex, usually somewhat angulate. Male S7, S8, and genitalia were illustrated by Michener (1989) and by papers cited therein; see also Figure 39-10.

Key to the Subgenera of *Lonchopria*

1. Mandible of male with large tooth or process on lower margin; clypeus of female without or with weakly differentiated, closely punctate, upper median area; the lateral and apical clypeal areas, *if* differentiated, then merely less closely punctate than upper median area 2
—. Mandible of male often with obtuse preapical angle on lower margin but without large tooth or process; clypeus of female with flat or depressed, relatively closely punctate, upper median area contrasting with impunctate or sparsely punctate, convex, U-shaped lateral and apical region 3
2(1). Preapical tooth of male mandible enormous, separated from apical part of mandible (rutellum) by curved emargination (Fig. 39-9g); S8 of male with apical process downcurved, apex expanded and quadrangular; mandible of female unusually slender, preapical (pollex) tooth weakly developed (Fig. 39-9f) .... *L. (Ctenosibyne)*
—. Preapical tooth of male mandible absent; S8 of male with apical process short, pointed, scarcely downcurved; mandible of female more robust with well-developed preapical tooth .... *L. (Lonchopria s. str.)*
3(1). Glossa deeply bifid (Fig. 39-12b), each lobe longer than prementum; apex of mandible scoop-shaped, the preapical (pollex) tooth reduced in female, absent in male .... *L. (Porterapis)*

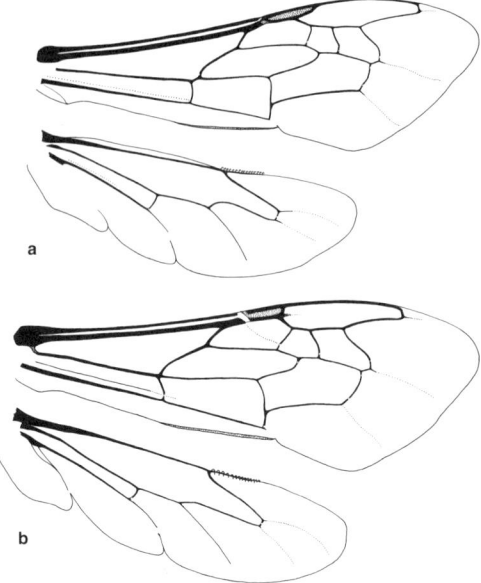

Figure 39-11. Wings of Paracolletini. **a,** *Lonchopria zonalis* (Reed); **b,** *Trichocolletes venustus* (Smith).

—. Glossa of the usual bilobed form (sometimes rather deeply so), each lobe less than one-third length of prementum; apex of mandible pointed, preapical tooth (which is sometimes double) usually strong 4
4(3). Tibial scopal hairs sparse (as in *Leioproctus*), not hiding tibial surface, which is therefore not greatly different from that of basitarsus; body length 7.5-9.0 mm; jugal lobe of hind wing reaching or surpassing vein cu-v, over two-thirds as long as vannal lobe measured from wing base .... *L. (Lonchoprella)*
—. Tibial scopal hairs extremely dense, hiding tibial surface, which therefore contrasts strongly with more sparsely haired basitarsus; body length variable but in most species over 9 mm; jugal lobe of hind wing usually not reaching vein cu-v and usually less than two-thirds as long as vannal lobe .... *L. (Biglossa)*

Lonchopria / Subgenus *Biglossa* Friese

Biglossa Friese, 1906c: 374. Type species: *Biglossa thoracica* Friese, 1906, by designation of Cockerell, 1914: 328.
Biglossidia Moure, 1948: 313. Type species: *Biglossa chalybaea* Friese, 1906, by original designation.
Aeganopria Moure, 1949a: 442. Type species: *Lonchopria nivosa* Vachal, 1909, by original designation.

This is the largest subgenus of *Lonchopria*. The metasoma is slightly blue or green to black, with or without pale hair bands, sometimes uniformly covered with pale hair or with broad mid-dorsal areas of pale hair. The body length is 8 to 14 mm. There is a flat or depressed, closely punctate upper median clypeal area (at least in females) surrounded laterally and below by large, convex, shining, and often hairless and impunctate areas. This feature is shared with certain other subgenera, but the ordinary bilobed glossa, distinct preapical tooth of the mandible, dense tibial scopal hairs, and strong, downcurved, hairy, and often ornate process of T8 of the male distinguish *Biglossa*. The mandible of the female is usually not expanded ventrad preapically, the lower mandibular margin therefore being uniformly curved (Fig. 39-9d). The lower mandibular margin of the male, preapically, is usually expanded ventrad, forming a preapical convexity (Fig. 39-9e). S8 of the male usually has a downcurved apical process, broadened and hairy distally, often with lateral or lateroapical projections from the enlarged apex. S7, S8, and the male genitalia were illustrated by Michener (1989); see also Figure 39-10.

■ *Biglossa* occurs in western Argentina (Mendoza to Jujuy) and north in the Andean uplift through Bolivia and Peru to Colombia. Some species occur in xeric lowlands while others occur at least as high as 3,874 m in the Peruvian Andes. There are nine named species, and others yet to be described. Moure (1949a) gave a key to the species that he placed in *Biglossidia,* and Michener (1989) listed the species.

Most of the species could be placed in a subgenus *Biglossidia* Moure, but if this were done, then *Biglossa* would stand as a monotypic subgenus for *L. thoracica* (Friese), derived from *Biglossidia* and differing from it in a few striking apomorphies. The terminalia of *L. thoracica* fall well within the range of variation among species placed in *Biglossidia*. It therefore does not seem that *L.*

thoracica differs enough from other species to justify putting it in a separate subgenus and thus making *Biglossidia* one more paraphyletic unit.

The male of *L. chalybaea* (Friese), the type species of Moure's *Biglossidia*, differs from the males of other species of the *Biglossidia* group in several characters, such as the large, right-angular, preapical tooth on the underside of the hind tibia and the swollen hind femur.

Lonchopria / Subgenus *Ctenosibyne* Moure

Lonchopria (Ctenosibyne) Moure, 1956: 311. Type species: *Lonchopria cingulata* Moure, 1956, by original designation.

This subgenus consists of a rather large (9.5-13.5 mm long), robust species similar to *Lonchopria* s. str. The tooth on the lower margin of the male mandible is a unique synapomorphy uniting *Lonchopria* s. str. and *Ctenosibyne*. T2-T4 have broad basal pale hair bands as well as apical bands. S8 of the male has its apical process large, downcurved, and apically abruptly broadened and truncate. S7, S8, and the male genitalia were illustrated by Moure (1956) and Michener (1989).

■ *Ctenosibyne* occurs on the southern Brazilian plateau, in the state of Paraná. The single known species is *Lonchopria cingulata* Moure.

The nests, deep burrows in a bank with lateral burrows each going to a single horizontal cell, were described by Michener and Lange (1957).

Lonchopria / Subgenus *Lonchoprella* Michener

Lonchopria (Lonchoprella) Michener, 1989: 673. Type species: *Lonchopria annectens* Michener, 1989, by original designation.

Lonchoprella is based on a single small species similar in appearance to *Lonchopria (Biglossa) robertsi* Michener. It differs from all other *Lonchopria* in having the tibial scopa similar to that of *Leioproctus*. It thus destroys the most conspicuous difference between the two genera, although other characters support placement of *Lonchoprella* in *Lonchopria*. The body is nonmetallic and lacks metasomal hair bands. The length is 7.5 to 9.0 mm. The mandible of the female is not expanded ventrad preapically, the lower margin being uniformly curved. The mandible of the male has a rounded preapical angle on the lower margin. S7, S8, and the male genitalia were illustrated by Michener (1989); S8 of the male has a hairy, downcurved apical process longer than the disc of the sternum, but the process is neither broadened nor ornate distally (Fig. 39-10f, g).

■ *Lonchoprella* is from the provinces of Santiago del Estero and Catamarca, Argentina. The one known species is *Lonchopria annectens* Michener.

The tibial scopal characters are probably plesiomorphic relative to those of other *Lonchopria*. I suppose *Lonchoprella* to be the sister group of all other *Lonchopria*; it may retain some characters derived from a *Leioproctus*-like ancestor.

Lonchopria / Subgenus *Lonchopria* Vachal s. str.

Lonchopria Vachal, 1905a: 204. Type species: *Lonchopria herbsti* Vachal, 1905 = *Colletes zonalis* Reed, 1892, monobasic.

This subgenus includes rather large, robust species having abundant pale hair usually forming apical bands on the metasomal terga. The most distinctive features are the lack of the preapical (pollex) tooth on the upper margin of the male mandible (although there is a large tooth on the lower margin) and the rather short, pointed, hairy, apical process of S8 of the male. The body length is 8.5 (for small males) to 15.0 mm. The clypeus of the female is punctate throughout; that of the male has the distal part broadly shining, impunctate, hairless or nearly so, and not strongly convex. The mandible of the female has the cap of the rutellum expanded ventrad so that the lower margin of the mandible is not a uniform curve. Male genitalia, sterna, and other structures were illustrated by Toro (1973a) and Michener (1989); see Figure 39-13a.

■ This subgenus is found from Atacama to Los Lagos, Chile. There are three or possibly four species, revised by Toro (1973a) and listed by Michener (1989).

Lonchopria s. str. is quite different from the other subgenera, and might have been regarded as a genus separate from *Biglossa* and *Porterapis* except for the intermediacy of *Ctenosibyne*. That subgenus has sparsely punctate areas on each side of the median depressed area of the clypeus in females, suggestive of the usually impunctate areas of *Biglossa* and *Porterapis*. Moreover, *Ctenosibyne* is intermediate between *Biglossa* and *Lonchopria* s. str. in gonobase, gonoforceps, and penis valve characters but it is more like *Biglossa* in S8.

Toro and de la Hoz (1976) have shown that the large tooth of the lower mandibular margin—and indeed the whole shape of the male mandible, in both *Lonchopria* s. str. and *Ctenosibyne*—fits the structure of the petiolar region of the female where the male holds the female while mating. As indicated by Toro and de la Hoz, these structures may function as isolating mechanisms among the sympatric species.

Lonchopria / Subgenus *Porterapis* Michener

Lonchopria (Porterapis) Michener, 1989: 678. Type species: *Lonchopria porteri* Ruiz, 1936, by original designation.

In size, body form, and the pale hair bands on the metasomal terga, this subgenus resembles *Lonchopria* s. str. In morphological details, however, *Porterapis* resembles *Biglossa*, from which it differs in its long glossal lobes (Fig. 39-12b), suggestive of *Leioproctus* subgenera *Glossopasiphae* and *Tetraglossula*, and its scoop-shaped mandibles (Fig. 39-9h, i). The body is nonmetallic, 12 to 13 mm long. The clypeus is flat, its lateral areas shining, nearly impunctate, and connected by a broad, similarly smooth zone across the lower end of the clypeus. The claws of the female are strongly curved, the inner ramus reduced to a small subbasal tooth. On S8 of the male the downcurved apical process is longer than the disc of the sternum, and abruptly broadened and hairy at the apex. S7, S8, and the male genitalia as well as mandibles were illustrated by Michener (1989).

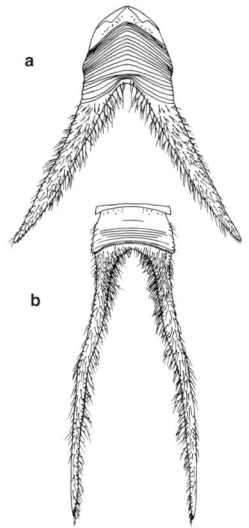

Figure 39-12. Unusually bifurcate glossae of Paracolletini. **a**, *Leioproctus (Glossopasiphae) plaumanni* Michener; **b**, *Lonchopria (Porterapis) porteri* Ruiz. From Michener, 1989.

bands and integumental color bands, and is 11.5 to 12.5 mm long. The inner hind tibial spur of the female is pectinate with about ten teeth, their bases well separated as in *Leioproctus*. There are three submarginal cells, the second and third subequal in length. S8 of the male is weakly sclerotized, almost quadrate, and lacks both an apical process and a spiculum (Fig. 39-13c). The volsella is largely membranous, weakly sclerotized only laterally, and lacks denticles. The two ventral prongs on the penis valve suggest *Lonchopria* s. str. (compare a, b, in Fig. 39-13), but other characters do not indicate a close relationship to that taxon. S7, S8, and the male genitalia were illustrated by Michener (1989).

■ *Lonchorhyncha* was known from only two Ecuadorian specimens of *L. ecuadoria* (Friese) (Michener, 1989); it is now known also from two males from Colombia that differ in minor ways from the Ecuadorian specimens (see Michener, 2000).

■ *Porterapis* is found in central Chile, north to Coquimbo. The single species is *Lonchopria porteri* Ruiz.

This subgenus is perhaps a derivative of *Biglossa*. It is so distinctive, however, that it seems worth recognizing, especially since the species of *Biglossa* have a synapomorphy (the preapical angle on the lower mandibular margin of males, absent in some probably derived species) suggesting that it and *Porterapis* might be sister groups.

Genus *Lonchorhyncha* Michener

Lonchorhyncha Michener, 1989: 667. Type species: *Diphaglossa ecuadoria* Friese, 1925, by original designation.

In no other American colletine does the head have an elongate clypeus, or malar areas about as long as an eye. The face is protuberant and extremely produced, the upper margin of the clypeus being below the lower ends of the eyes. The body is nonmetallic, lacks metasomal hair

Genus *Neopasiphae* Perkins

Neopasiphae Perkins, 1912: 114. Type species: *Neopasiphae mirabilis* Perkins, 1912, monobasic.

These are the only Australian hairy colletids having a pattern of yellow integumental markings on the metasoma. The yellow tergal bands are preapical, not on the marginal zones as in *Eulonchopria* and *Leioproctus (Nomiocolletes)*. The hind basitarsus and sometimes the tibia of the male are broad and flat, the basitarsus as wide as the tibia. The inner hind tibial spur of the female is briefly pectinate, the 9 to 12 teeth shorter in length than the diameter of the spur. An unusual feature is the absence of the gradulus of S2 in females; in males it is procurved medially rather than largely transverse as in *Leioproctus, Paracolletes,* and *Trichocolletes*. The body length is 7.0 to 10.5 mm. The male genitalia, sterna, and other characters were illustrated by Michener (1965b).

■ The species of this genus are known from Western Australia, with a doubtful record for Victoria. The three species were listed by Cardale (1993).

Genus *Niltonia* Moure

Niltonia Moure, 1964c: 52. Type species: *Niltonia virgilii* Moure, 1964, by original designation.

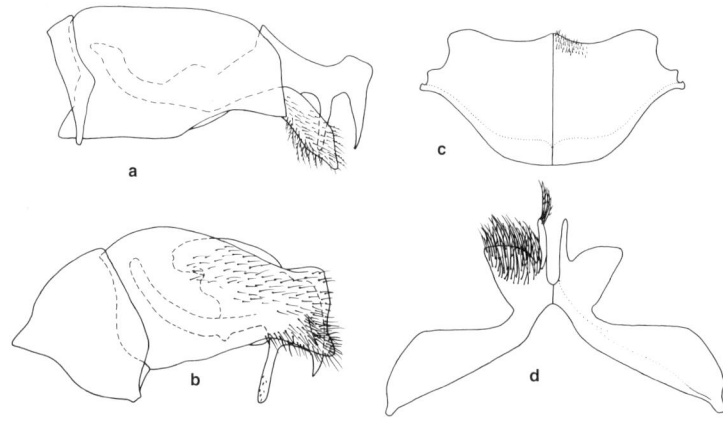

Figure 39-13. Male genitalia and sterna of Paracolletini. **a**, Lateral view of genitalia of *Lonchopria (Lonchopria) similis* Friese, showing ventral prongs of penis valve (compare with b); **b-d**, Genitalia, S8, and S7 of *Lonchorhyncha ecuadoria* (Friese). (Dorsal views of sterna are at the left.) From Michener, 1989.

The extremely long labial palpi are the outstanding feature of this genus, but this is by no means a *Lioproctus* with long palpi, as shown by a series of other unique features. *Niltonia* is not obviously allied to any group of *Leioproctus*. The integument is black and nonmetallic. Body length is 10.0 to 12.5 mm. Metasomal hair bands are completely absent. Although the proboscis is short, the labial palpus is enormous, 8 to 9 mm long, its fourth segment much longer than the first three together, tapering, and often extended beyond the apex of the metasoma (fig. 1 of Moure, 1964c; fig. 3 of Laroca and Almeida, 1985). The scopal hairs on the hind leg and on S2-S5 are branched. There are two submarginal cells; the stigma is rather small, vein r arising well beyond its middle and the margin within the marginal cell convex. T7 of the male has a triangular pygidial plate, sharply pointed apically, margined by carinae laterally. S7, S8, and the genitalia of the male were illustrated by Laroca and Almeida (1985) and by Michener (1989).

■ This genus is found in Brazil from Santa Catarina to Rio de Janeiro. The only species, *Niltonia virgilii* Moure, visits flowers of *Jacaranda*. For details of mouthparts and floral behavior, see Laroca, Michener, and Hofmeister (1989).

In certain features *Niltonia* resembles *Brachyglossula*. The large and dorsally expanded volsellae and the distal origin of vein r on the stigma are the most apparent such characters. Possibly both genera arose from a common ancestor within the *Leioproctus* group.

Genus *Paracolletes* Smith

This Australian genus embraces bees larger than the average *Leioproctus;* body length is 10 to 17 mm. The body is nonmetallic and black, or (in the male) the metasoma is partly or wholly red. The inner hind tibial spur of the female is usually finely serrate or ciliate, but in *P. plumatus* (Smith) the teeth are elongated, the spur thus finely pectinate, and in *P. cygni* (Cockerell) there are fewer and even longer teeth. Such pectinate spurs, however, differ strikingly from the almost palmate spurs of *Trichocolletes*. The propodeum is declivous, without differentiated horizontal and vertical (posterior) surfaces. In males the clypeus is sometimes yellow or partly so. The male antennae are often quite long, for which reason some species were described in *Tetralonia* (Eucerini). Genitalia and hidden sterna of both subgenera were illustrated by Michener (1965b).

It is possible to recognize a homogeneous *crassipes* group (*Paracolletes* proper) and a more diversified assembly of other species tentatively called the subgenus *Anthoglossa*. When males of more species become known, the situation should be clearer.

Key to the Subgenera of *Paracolletes*

1. Metasoma with apical tergal hair bands [except in *P. cygni* (Cockerell) and *montanus* Rayment]; basitibial plate of male completely defined, of female usually exposed, rounded apically (angulate in *montanus*), one-fifth as long as tibia or less (but one-fourth in *montanus*); jugal lobe of hind wing little more than half as long as vannal lobe (but nearly two-thirds in *cygni* and *montanus*); first flagellar segment of male longer than broad; apical lobes of S7 of male broad .... *P. (Anthoglossa)*

—. Metasoma without hair bands; basitibial plate of male not or scarcely defined anteriorly, of female hidden by hairs, bluntly rounded apically, nearly one-fourth as long as tibia; jugal lobe of hind wing about two-thirds as long as vannal lobe; first flagellar segment of male broader than long; apical lobes of S7 of male linear ....
.... *P. (Paracolletes s. str.)*

Paracolletes / Subgenus *Anthoglossa* Smith

Anthoglossa Smith, 1853: 16. Type species: *Anthoglossa plumata* Smith, 1853, monobasic.

The distinctive characters are indicated in the above key and discussion. Unfortunately, because the male of the type species, *Paracolletes plumatus* (Smith), is unknown, there may be instability in the application of the subgeneric name.

■ The species of *Anthoglossa* are mostly from Western Australia, but one is from Victoria. The eight species were listed by Michener (1965b) and Cardale (1993).

Paracolletes / Subgenus *Paracolletes* Smith s. str.

Paracolletes Smith, 1853: 6. Type species: *Paracolletes crassipes* Smith, 1853, monobasic.

The distinctive characters are indicated in the key to subgenera above.

■ This subgenus is found in southern Australia north at least to the latitude of central Queensland. The eight known species were listed by Michener (1965b) and Cardale (1993).

Genus *Phenacolletes* Cockerell

Phenacolletes Cockerell, 1905b: 301. Type species: *Phenacolletes mimus* Cockerell, 1905, monobasic.

Phenacolletes consists of large bees (body length 14 mm); the pubescence is short and appressed, not forming tergal bands, giving the bee the appearance of a larrine wasp such as *Tachysphex*. The antennae of the male are short, suggesting those of a female. There are three submarginal cells; the stigma is little more than one-half as long as the costal part of the marginal cell, this cell being strongly and gradually bent away from the costa and strongly appendiculate apically (Fig. 39-5b). The propodeum is vertical seen in profile. The keirotrichia of the female's hind tibia are short, forming a distinct band on the inner surface. This is a common feature of bees and wasps and probably plesiomorphic. Long keirotrichia, more like ordinary hairs, may be an apomorphy of *Leioproctus* and its derivatives. Thus *Phenacolletes* and *Leioproctus* could be sister groups. The apex of S7 of the male is bidentate, completely lacking the usual apical lobes (Fig. 39-3e). For S7, S8, and the male genitalia, see Figure 39-3d-f and Michener (1965b).

■ This genus is known from Western Australia and South Australia. The single species is *Phenacolletes mimus* Cockerell.

Genus *Trichocolletes* Cockerell

This Australian genus includes moderate-sized to large bees with body lengths 10 to 18 mm. Those that I have seen in the field differ from *Paracolletes* and *Leioproctus* in their exceedingly fast flight and frequent hovering. *Paracolletes* and *Leioproctus*, by contrast, fly like most species of *Andrena*. Whether the metasoma is red or black, the terga of most species have broad, translucent, testaceous to golden apical margins. In all species the terga have a sericeous texture, owing to very fine sculpturing and short, appressed pubescence. The inner hind tibial spurs of the female are pectinate, thickest basally or medially, the bases of the teeth crowded together (Fig. 39-8f), thus resembling the South American genus *Lonchopria* in this respect. The tarsal claws are fully cleft or the inner tooth is reduced or absent. The eyes of both sexes are commonly divergent below but sometimes parallel; in some species the eyes are hairy. The clypeus is usually protuberant. The second submarginal cell is usually rather large and quadrate but sometimes is small and narrowed toward the costal margin (Fig. 39-11b), as is usual in *Leioproctus*. Male genitalic and other structures were illustrated by Michener (1965b) and Houston (1990).

In view of the many combinations of the characters among species of the genus, clear-cut species groups separated by numerous characters are not apparent. It therefore seems best to divide the genus only as indicated below in spite of the morphological diversity of its extreme species. Noteworthy characters of single species are the greatly shortened labial palpi (second and third segments much broader than long) of *Trichocolletes hackeri* (Cockerell) and the swollen hind legs, lacking tibial spurs, in the male of *T. pulcherrimus* Michener.

Key to the Subgenera of *Trichocolletes*

1. Hind tibial spurs of male present; basitibial plate of female defined only posteriorly, apex not evident ....
   .... *T. (Trichocolletes s. str.)*
—. Hind tibial spurs of male absent; basitibial plate of female complete except for indefinite apex ....
   .... *T. (Callocolletes)*

Trichocolletes / Subgenus *Callocolletes* Michener

Trichocolletes (Callocolletes) Michener, 1965b: 80. Type species: *Trichocolletes pulcherrimus* Michener, 1965, by original designation.

Callocolletes consists of a single extraordinary species: in the male the legs are incrassate with a large tooth on the anterior trochanter and there are no hind tibial spurs. The basitibial plate of the female is complete except for the indefinite apex; the basitibial plate of the male is well defined. The stigma is more than 1.5 times as long as the prestigma.

■ *Callocolletes* is from Western Australia. The single species is *Trichocolletes pulcherrimus* Michener.

Callocolletes may well be merely the most elaborately modified of the species of *Trichocolletes*. I have not synonymized *Callocolletes*, however, because it has a character that appears to be plesiomorphic relative to other *Trichocolletes*—the retention of a carina on each side of the basitibial plate of the female. In other *Trichocolletes* the female has lost the carina along the anterior margin of the basitibial plate. This variable thus suggests that *Callocolletes* may be the sister group to all other *Trichocolletes*. In this case it is a matter of judgment whether *Callocolletes* should be synonymized; I have chosen to let the current classification stand.

Trichocolletes / Subgenus *Trichocolletes* Cockerell s. str.

Trichocolletes Cockerell, 1912: 176. Type species: *Lamprocolletes venustus* Smith, 1862, by original designation.

The basitibial plate of the female is defined only posteriorly, the apex not being evident; that of the male is variable, but when present is smaller than that of *Callocolletes*. The stigma is less than or about 1.5 times as long as the prestigma (Fig. 39-11b).

■ This subgenus is widespread in the temperate parts of Australia. Cardale (1993) lists 22 named species; see also Michener (1965b).

40. Tribe Colletini

The Colletini is best known for the genus *Colletes*, found in all continents except Australia, almost always recognizable by the sigmoid second recurrent vein and the eyes convergent below, so that the head, seen from the front, is heart shaped. The tribe also includes the genus *Mourecotelles* of temperate South America. Distinctive features are the greatly reduced disc of S7 of the male (Fig. 40-2b) and the enormous apical lobes that constitute most of the sternum. Further information on tribal characters is found in Section 38.

Although the genus *Colletes* is widespread and common in much of the world and is well known as a ground-nesting bee, this tribe and genus may have originated in South America, nesting in pithy stems. Long ago the lack of pygidial and basitibial plates in *Colletes* was noted; these lacks are unusual in ground-nesting bees, most of which use these structures in moving about in their burrows and in constructing cells in the soil. In the South American Andes, however, species of *Colletes* that nest in pithy stems are known (Benoist, 1942). It seems possible that *Colletes* arose as a stem-nesting bee, lost the pygidial and basitibial plates in a life style in which these structures are not useful, and then reverted to ground-nesting and spread over the world (except Australia) despite the lack of usual structures of ground-nesting bees. A South American origin of *Colletes* is supported by the great diversity among the South American species and by the presence, there, of the only closely related genus, *Mourecotelles*. The nesting habits of species of that genus are unknown, but they may also use pithy stems.

Key to the Genera of the Tribe Colletini

1. T1 with anterior surface not broadly concave, often with a longitudinal median concavity; dorsal surface of T1 about as long as exposed part of T2 and at least two-thirds as long as anterior surface of T1, as seen in profile; second recurrent vein usually sigmoid with posterior half arcuate distad (Fig. 40-3), rarely angulate distad at alar fenestra on flexion line; base of propodeum with short, subhorizontal to vertical basal zone usually defined posteriorly by a carina (Fig. 37-3) or exhibiting a sharp change in slope or sculpture and divided by longitudinal carinae that may delimit strong pits (worldwide except Indo-Australian region) .... *Colletes*
—. T1 with anterior surface broadly concave; dorsal surface of T1 much shorter than exposed part of T2 and less than two-thirds as long as anterior surface of T1, as seen in profile; second recurrent vein with posterior half straight or gently arcuate (as in Fig. 39-5), its anterior part not curved in the reverse direction, vein thus not sigmoid; base of propodeum with sloping basal zone nearly smooth, not traversed by longitudinal carinae (South America) .... *Mourecotelles*

Genus *Colletes* Latreille

Colletes Latreille, 1802a: 423. Type species: *Apis succincta* Linnaeus, 1758, monobasic. [For a later type designation, see Michener, 1997b.]

Evodia Panzer, 1806: 207. Type species: *Apis calendarum* Panzer, 1802 = *Apis succincta* Linnaeus, 1758, monobasic.

Monia Westwood, 1875: 221 (not Gray, 1850). Type species: *Monia grisea* Westwood, 1875, monobasic.

Monidia Cockerell, 1905c: 9, replacement for *Monia* Westwood, 1875. Type species: *Monia grisea* Westwood, 1875, autobasic.

Colletes (*Rhinocolletes*) Cockerell, 1910a: 242. Type species: *Colletes nasutus* Smith, 1853, monobasic.

Colletes (*Ptilopoda*) Friese, 1921b: 83. Type species: *Colletes maculipennis* Friese, 1921 = *Colletes spiloptera* Cockerell, 1917, monobasic.

Colletes (*Denticolletes*) Noskiewicz, 1936: 25, 486. Type species: *Colletes graeffei* Alfken, 1900, monobasic.

Colletes (*Puncticolletes*) Noskiewicz, 1936: 26, 490. [Not valid under the Code, ed. 3, art. 13(b), because no type species was designated. Warncke (1978) considered *Puncticolletes* a synonym of *Rhinocolletes*.]

Rhynchocolletes Moure, 1943b: 447. Type species: *Rhynchocolletes albicinctus* Moure, 1943, by original designation.

Colletes (*Pachycolletes*) Bischoff, 1954, in Stoeckhert, 1954: 73. Type species: *Apis cunicularia* Linnaeus, 1758, by original designation.

Colletes (*Albocolletes*) Warncke, 1978: 353. Type species: *Halictus albomaculatus* Lucas, 1849, by original designation.

Colletes (*Elecolletes*) Warncke, 1978: 330. Type species: *Colletes elegans* Noskiewicz, 1936, by original designation.

Colletes (*Nanocolletes*) Warncke, 1978: 341. Type species: *Colletes nanus* Friese, 1898, by original designation.

Colletes (*Simcolletes*) Warncke, 1978: 348. Type species: *Colletes similis* Schenck, 1853, by original designation.

A unique feature of *Colletes*, not found in any other bee (only weakly developed in a few species), is the outwardly arcuate posterior part of the second recurrent vein (Fig. 40-3). An unusual feature of females—shared, however, with *Mourecotelles*—is the lack of pygidial and prepygidial fimbriae (Fig. 40-1). In general form (Fig. 40-1) *Colletes* resembles *Andrena* and *Halictus*, from which it differs in the characters just listed and usually in the strongly convergent eyes (Fig. 40-2a). The body length is 7 to 16 mm. In the holarctic region, *Colletes* is the only common genus of hairy colletids. Male genitalia of many species were illustrated by Morice (1904); Swenk (1908); Metz (1910); Noskiewicz (1936); Stephen (1954); and Mitchell (1960); see also Figure 40-2.

The long synonymy above results from two tendencies. The first is the assigning of genus-group names to each unusual species. Thus *Monidia* contains a Mexican species in which the last antennal segment of the male is expanded and the hind tibiae of the male bear long hairs. *Denticolletes* contains a palearctic species in which the axilla is produced and angulate. *Rhinocolletes* (palearctic) and *Rhynchocolletes* (Brazilian) each contain a species in which the malar area is unusually long and the clypeus is produced (in *Rhynchocolletes*, the legs of the male are modified). *Ptilopoda* contains two species (Texas to Panama) with spotted wings and, in the male, somewhat

Figure 40-1. *Colletes cercidii* Timberlake, female. Drawing by E. R. S. Hodges, from Michener, McGinley, and Danforth, 1994.

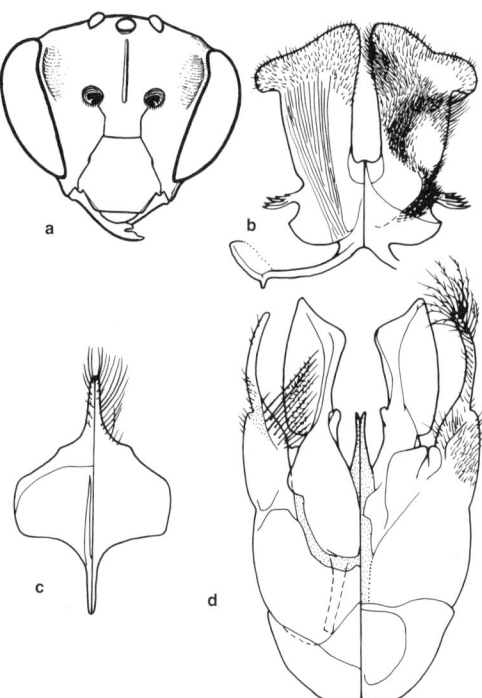

Figure 40-2. *Colletes.* **a,** Face of *C. fulgidus* Swenk, female; **b-d,** S7, S8, and genitalia of *C. everaertae* Michener, male. (Dorsal views are at the left.) a, from Michener, 1944.

modified hind legs. The second tendency is to give names to groups of the less exceptional species on the basis of forms found in limited geographical areas, in this case the western palearctic. These names should be ignored until the genus is reviewed worldwide and appropriate subgroups recognized.

Rhynchocolletes (in the above synonymy) is also unusual in the short mandible of the male (shorter than the malar area), the broad, deep concavity between the ocelli and the summit of the eye, and the course of the second recurrent vein, which is only gently arcuate outward in the posterior portion and straight in the anterior portion, thus not S-shaped, as in most *Colletes* (Fig. 40-3). *Rhynchocolletes* is known only in the male. The short mandible and the form of the second recurrent vein are suggestive of *Mourecotelles,* but neither character is decisive because intergradation occurs among more ordinary species of *Colletes.*

■ *Colletes* occurs in the temperate and tropical regions of all continents except for the Indo-Australian region. Its absence from Australia is noteworthy in view of the abundance of other Colletinae there. In North America the genus ranges as far north as Alaska. Particularly in temperate South America the fauna is diverse and beautiful; some species have metallic blue metasomas, some have bright-red thoracic hair, and various species have long malar areas suggesting *Rhinocolletes* and attaining the extreme in *Rhynchocolletes.* There are about 330 species, of which about 135 are from the palearctic region and nearly 90 from the nearctic region (including nearctic Mexico) and an estimated 90 are from the neotropical

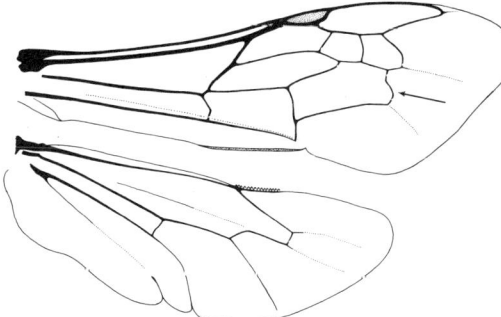

Figure 40-3. Wings of *Colletes* sp. The arrow indicates the sigmoid second recurrent vein characteristic of the genus.

region. *Colletes* has reached various islands, such as the Canary Islands and Cuba, but is not known from Madagascar in spite of its reaching southern Africa. The sub-Saharan fauna, however, is sparse, numbering about 15 species. The *Colletes* species of America north of Mexico were revised by Stephen (1954), those of Colorado were reviewed by Timberlake (1943b), those of the palearctic region by Noskiewicz (1936), of Britain by Richards (1937), of the western palearctic region by Warncke (1978), of the Ukraine by Osychnyuk (1970), and of Japan by Ikudome (1989).

The nesting biology of several species has been studied, for example, by Malyshev (1923b, 1927b), Michener and Lange (1957), Torchio (1965), Rozen and Favreau (1968), Scheloske (1974), and Torchio, Trostle, and Burdick (1988). A noteworthy feature of the burrow architecture of some species is that, instead of shaping each cell more or less identically, as do most other ground-nesting bees, they divide a burrow with transverse partitions made of the transparent cell-lining material. The result is a series of cells (Pl. 15) not identical in shape, the distal one rounded at one end, conforming with the rounded end of the burrow, the others truncate at both ends. In other *Colletes* species each lateral burrow ends in a single cell that is sometimes larger in diameter than the burrow; in this case, all cells are essentially alike in shape. The provisions of *Colletes* are liquid, as in *Hylaeus,* not at all like the firm ball of provisions characteristic of *Leioproctus* and *Lonchopria* as well as Andrenidae, Halictidae, etc. Unlike the eggs of most bees, those of *Colletes* are attached by one end to the upper wall of the cell, rather than being placed on the provisions. Even in *Hylaeus,* which makes similar cells containing liquid provisions, the egg floats on the surface of the provisions. Cell closure in *Colletes* is by a cellophane-like membrane, again as in *Hylaeus,* not by an earthen plug. Aspects of cell construction by various *Colletes* species were reviewed by Rozen and Michener (1968).

Colletes daviesanus Smith, which ordinarily nests in south-facing earth or sandstone banks, has become synanthropic in parts of Germany, boring into and damaging sandstone and mortar buildings (Scheloske, 1974). The cause of the damage is easily determined by the series of cellophane-like cells in holes in the deteriorating structures. The distribution of nests of this species in outcrops of certain sandstone strata was described in detail by Mader (1992). Mader (1999) gave a detailed compendium of literature on *Colletes* biology (with information also on other bank-nesting bees). Emphasis is on *C. daviesanus* Smith.

Although *Colletes* is ordinarily a ground-nesting genus, in the South American Andes species such as *C. rubicola* Benoist construct series of cells in dead, pithy stems (Benoist, 1942).

Genus *Mourecotelles* Toro and Cabezas

This genus, found only in temperate South America, appears to be the sister group to *Colletes*. Its apomorphies relative to *Colletes* include the shape of T1 (perhaps two independent characters) and the short, almost globose metasoma, T6 of the female being scarcely exserted. The apomorphies of *Colletes* include all the characters listed in the key to genera except those of T1. *Mourecotelles* has more characters like those of *Leioproctus* than does *Colletes*.

There are three subgenera of *Mourecotelles*. *Hemicotelles* and *Xanthocotelles* have simple claws in the female, an apomorphic character relative to Colletinae in general. The same two subgenera have what Toro and Cabezas (1977, 1978) regarded as a rudimentary pygidial plate, a character absent in *Mourecotelles* s. str. This plate, only a small, bare, elevated area at the apex of T6, may well be an apomorphic feature rather than a remnant of a pygidial plate, for it is narrowed to a point anteriorly, instead of broadened anteriorly like a typical pygidial plate. Even in *Mourecotelles* s. str. the same region is broadly elevated, quite unlike that of *Colletes,* although it is not clearly defined or bare as it is in the subgenus *Xanthocotelles*. Male genitalia and hidden sterna were illustrated by Toro and Cabezas (1977, 1978).

Key to the Subgenera of *Mourecotelles*

1. Basal zone of propodeum margined posteriorly by weak carina; middle flagellar segments of male about twice as long as wide (claws of female simple)...... *M. (Hemicotelles)*
—. Basal zone of propodeum not margined posteriorly by carina; middle flagellar segments of male less than twice as long as wide... 2
2(1). Claws of female simple; tibiae and tarsi reddish yellow
.... *M. (Xanthocotelles)*
—. Claws of female cleft; tibiae and tarsi largely blackish
.... *M. (Mourecotelles s. str.)*

Mourecotelles / Subgenus Hemicotelles Toro and Cabezas

Hemicotelles Toro and Cabezas, 1977: 46. Type species: *Lonchopria ruizii* Herbst, 1923, by original designation.

Hemicotelles, recognizable by the characters in the key to subgenera, includes species 12 to 14 mm long.

■ This subgenus ranges from Coquimbo to Aisén, Chile, and to Santa Cruz province, Argentina. The two species were revised by Toro and Cabezas (1977).

Mourecotelles / Subgenus Mourecotelles Toro and Cabezas s. str.

Mourecotelles Toro and Cabezas, 1977: 50. Type species: *Mourecotelles mixta* Toro and Cabezas, 1977, by original designation.

This subgenus, the only one with cleft claws in the female and also the only one in which the pygidial plate is clearly absent, consists of species 8 to 11 mm long.
■ *Mourecotelles* is found in Bolivia, Chile (Coquimbo to Aisén), and Argentina (Catamarca to Mendoza province). The eight described species were revised and illustrated by Toro and Cabezas (1977).

Mourecotelles / Subgenus Xanthocotelles Toro and Cabezas

Xanthocotelles Toro and Cabezas, 1978: 131. Type species: *Xanthocotelles adesmiae* Toro and Cabezas, 1978, by original designation.

Like *Mourecotelles* s. str., this subgenus consists of small species, 7 to 10 mm in body length.
■ This subgenus is found from Coquimbo to Aisén, Chile, and in the provinces of Catamarca and Mendoza, Argentina. The 11 known species were revised and illustrated by Toro and Cabezas (1978).

41. Tribe Scraptrini

The genus *Scrapter* has been recognized for some time as fitting poorly into the Paracolletini, with which it has been associated. Melo and Gonçalves (2005) placed it in its own tribe. It differs from other hairy colletids in its foveate prementum and from most in its reduced galeal comb (Fig. 41-2b); in appearance the premental fovea is like that of the Hylaeinae (Fig. 41-3) and Xeromelissinae. Probably because in many species of *Scrapter* the female basitibial plate is margined by a series of large tubercles or by broken carinae, the suggestion has been made that *Scrapter* might be related to the Euryglossinae instead of the Colletinae. However, in both *Scrapter* and Euryglossinae there are species in which the carinae are continuous, forming ordinary basitibial plates, no doubt a plesiomorphic character. The tuberculate margins must have arisen independently in the two groups. Also, in some species of *Scrapter* the facial foveae are narrow grooves, as in Hylaeinae and most Euryglossinae. There are other Colletinae (*Callomelitta*, some species of *Eulonchopria*), however, that share this character. Thus the similarities of *Scrapter* to other subfamilies may be convergent.

Genus *Scrapter* Lepeletier and Serville

Scrapter Lepeletier and Serville, 1828: 403 (not *Scrapter* Lepeletier, 1841). Type species: *Scrapter bicolor* Lepeletier and Serville, 1828, by designation of Vachal, 1897: 63. [For later type designations and confusion with *Scrapter* Lepeletier, see Michener, 1997b.]

Polyglossa Friese, 1909a: 123. Type species: *Polyglossa capensis* Friese, 1909, by designation of Cockerell, 1921a: 203. [For a later type designation by Sandhouse, see Michener, 1997b.]

Strandiella Friese, 1912a: 181. Type species: *Strandiella longula* Friese, 1912 = *Scrapter niger* Lepeletier and Serville, 1828, by designation of Cockerell, 1916a: 430.

Polyglossa (Parapolyglossa) Brauns, 1929: 134. Type species: *Polyglossa heterodoxa* Cockerell, 1921, by designation of Sandhouse, 1943: 584. [See Michener, 1997b.]

Aspects of the history of the name *Scrapter* were explained by Cockerell (1920a, 1930g, 1932a); these references all relate to *Scrapter* 1828, not 1841. Friese (e.g., 1909a) confused *Scrapter* with *Ctenoplectra* (Apinae) and placed various tropical African *Ctenoplectra* species in *Scrapter*.

Except for *Colletes*, this is the only hairy colletid genus (Pl. 4) found in Africa. It can be distinguished easily from *Colletes* by having only two submarginal cells (Fig. 41-1).

The body is nonmetallic, 8 to 12 mm long. Many species have apical metasomal tergal hair bands. The stigma is long, receiving vein r near the middle; it is usually broad (Fig. 41-1a) but in large species is more slender, especially in *Scrapter heterodoxus* (Cockerell), in which the margins basal to vein r are about parallel (Fig. 41-1b). The inner hind tibial spur of the female is unusually straight, tapering, and ciliate. As in *Lioproctus*, the keirotrichia are represented by long hairs. In the male, the form of S7 is often not very different from that of S6, the apex usually bilobed, but the large, complex apical lobes found in most other colletids are absent. In a few species such as *S. albitarsis* (Friese) and *calx* Eardley, S7 has small, hairless, laterally directed apical lobes. Hidden sterna, genitalia, and other structures of males were illustrated by Eardley (1996). T7 of the male has a pygidial plate.

■ *Scrapter* is found in South Africa, Namibia, and Zimbabwe. The 31 species were revised by Eardley (1996). Friese (1924c) gave a key to the species.

In a general way *Scrapter* is divisible into two major groups, as follows:

In the first: (a) facial fovea of female broad, mesal margin sometimes indistinct; (b) basitibial plate of female with marginal carinae (or at least lower one) tuberculate or lobed; (c) propodeal triangle finely roughened, separated from rest of propodeum by a fine line; (d) thoracic sculpturing not especially coarse; (e) claws of female cleft or simple; (f) body commonly larger and robust, with pale metasomal hair bands.

In the second: (a) facial fovea of female a narrow groove; (b) basitibial plate of female with simple marginal carinae; (c) propodeal triangle with striate dorsal surface separated from rest of propodeum by pitted lines; (d) thoracic sculpturing of extremely coarse punctures, midline and notauli deeply impressed on anterior end of scutum;

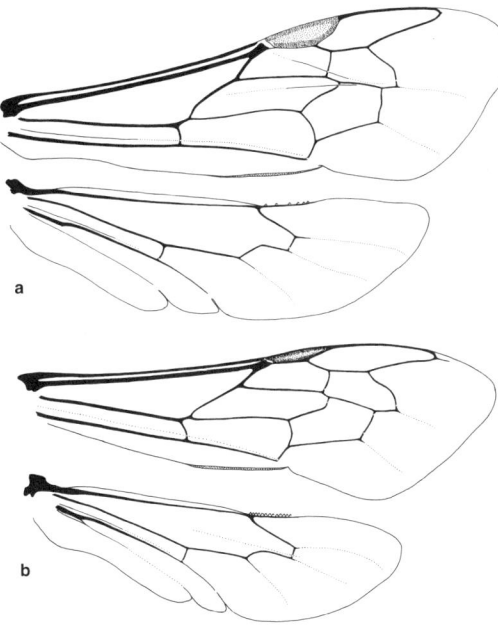

Figure 41-1. Wings of *Scrapter*. **a**, *S. nitidus* (Friese), forewing length 5.5 mm; **b**, *S. heterodoxus* (Cockerell), forewing length 8.0 mm. The more pointed forewing, more slender stigma, and more numerous hamuli characterize larger bees of many groups, in contrast to smaller relatives.

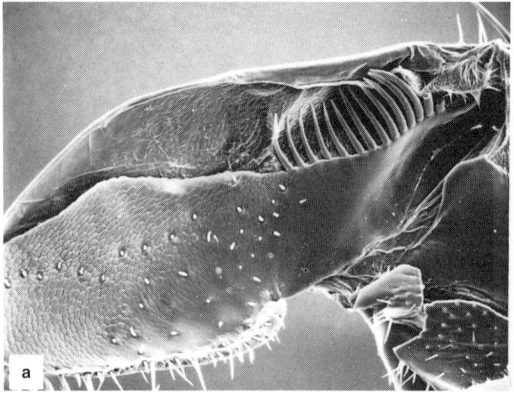
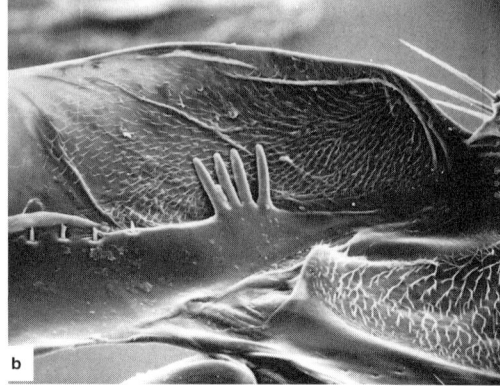

Figure 41-2. Inner surface of galea, showing galeal comb. **a,** *Lonchopria similis* Friese; **b,** *Scrapter* sp. SEM photos by R. W. Brooks.

(e) claws of female simple; (f) body small, slender, without metasomal hair bands.

All the genus-group names listed in the above synonymy pertain to the first group, but there are probably as many species in the second group as in the first. Although some of the group characters listed above are generic or subgeneric characters in other groups of bees, combinations of characters in a minority of the species of *Scrapter* break down the group differences.

Males of a few species have enlarged and modified hind legs, as seen in *Scrapter heterodoxus* (Cockerell) and *armatipes* (Friese), and even the middle legs (especially the basitarsi) are modified in the latter species. These features were illustrated by Brauns (1929) and Eardley (1996).

The nesting biology of two species was described by Rozen and Michener (1968). Nests are vertical burrows in the soil with laterals, each lateral leading to a slanting cell that is merely the end of the lateral, not enlarged; the cell is lined with a cellophane-like membrane, and the distal part of the cell is filled with firm to rather liquid provisions, not forming a ball as in *Leioproctus* and *Lonchopria* (Paracolletini).

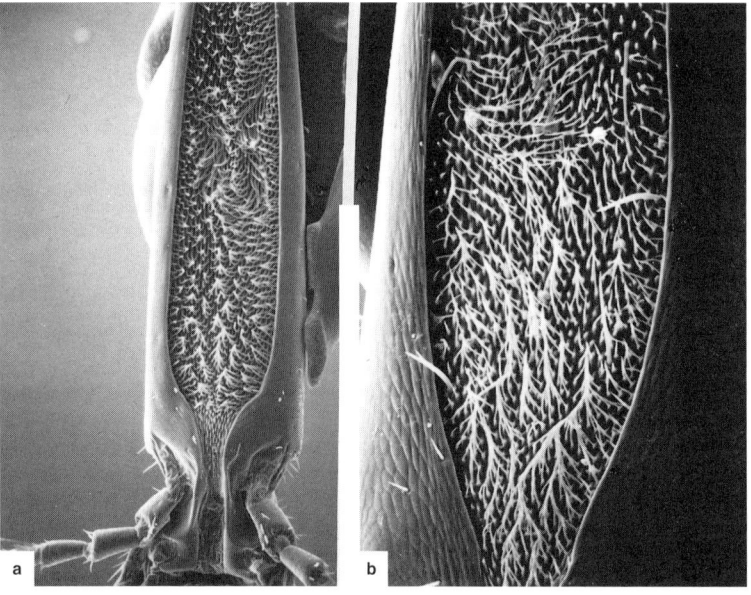

Figure 41-3. Fovea containing minute spicules or setae on posterior surface of prementum of Colletidae. **a,** *Hylaeus episcopalis* (Cockerell) (Hylaeinae); **b,** *Scrapter* sp. (Scraptrini). SEM photos by R. W. Brooks.

42. Subfamily Diphaglossinae

Most members of this American subfamily are large, robust, densely hairy, and euceriform, but some look like middle-sized andreniform Colletinae. The glossa is bifid, the two lobes usually pointed and extending apicolaterad; the preapical fringe is present in the female, absent or weakly developed in the male. The glossal brush is well developed. The prementum lacks a spiculate depression. The facial fovea is suggested by a broad, slightly depressed, impunctate area that extends up into the ocellocular region. The episternal groove is variable. The scopa is large and dense on the hind leg, forming a corbicula on the underside of the femur. The basitibial area of the female is covered with short, appressed hair but the basitibial plate is not indicated or only the posterior marginal carina is evident, except that the anterior carina also is evident in *Mydrosomella;* the basitibial plate of the male is absent. There are three submarginal cells. The stigma is shorter than the prestigma (Fig. 43-1), slender or almost absent, the sides parallel or converging toward vein r, which arises from the apex of the stigma. The margin of the stigma within the marginal cell is about as short as possible, that is, transverse to the long axis of the stigma, and straight or concave. The pygidial and prepygidial fimbriae of the female are strong; the pygidial plate is present in the female.

Larval characters are presented by McGinley (1981). The larvae can be distinguished from all other bee larvae by the elongate, spoutlike projection of the salivary lips, forming a circular or short transverse salivary opening. The Diphaglossinae are the only colletids that retain cocoon-spinning behavior and associated larval structures. Pupal characters were described by Torchio and Burwell (1987).

Nests consist of more or less vertical burrows with deep branches, each of which at its distal end usually bends up slightly, then curves sharply down to form a single vertical cell lined with a secreted film and holding the largely liquid provisions on which the egg floats. In various species, but not in *Crawfordapis,* the mature larva scrapes and breaks down the cell lining in the bottom end of the cell, then spins a cocoon that separates this end of the cell from the rest, forming in the bottom of the cell a fecal chamber from which liquid can escape. At the other end the cocoon consists of a strong operculum perforated by numerous round holes. A comparative study of cocoons and nest structures was made by Rozen (1984b). His treatment is so excellent and complete that references to older works on nesting biology cited by him are unnecessary here.

The most distinctive adult character is the reduction of the stigma. The other adult characters are within the range of variation found in the Colletinae. For this reason I long ago suspected that the Diphaglossinae were derived from the Colletinae, making the latter paraphyletic, a conclusion supported by most of the phylogenetic analyses by Alexander and Michener (1995). As noted in Section 37 on the Colletidae, however, the spoutlike salivary opening of diphaglossine larvae, on the one hand, and the reduced salivary lips and loss of cocoon spinning in all other colletids, on the other hand, are both synapomorphies relative to ancestral bees and wasps. Thus on the basis of these and associated larval characters, Diphaglossinae appears to be the sister group to all other Colletidae, as shown in a cladistic treatment of larvae by McGinley (1981) and as also indicated by Rozen (1984b) and by Michener (1986b) working with adult characters. (Larvae and cocoons are known for two tribes of Diphaglossinae, the Diphaglossini and Caupolicanini; nests of Dissoglottini have never been found.)

The Caupolicanini differ from other Diphaglossinae in such striking characters (the complete episternal groove, the long and petiolate first flagellar segment, the coarsely papillate distal parts of wings) that Michener (1944) separated Caupolicanini from other Diphaglossinae, placing the former in the Colletinae, the latter in its own subfamily. Moure (1945a) and later Michener (1954b, 1966a), however, placed the Caupolicanini within the Diphaglossinae, a view strongly supported by Alexander and Michener's (1995) phylogenetic study of adults and by the larval characters cited above.

The resemblance of members of this subfamily, and especially the Caupolicanini, to the Australian Stenotritidae and the American Oxaeinae is striking, and includes not only appearance but the reduced stigma, the papillate wings, and the long, petiolate first flagellar segment. Stenotritid and oxaeine larvae, however, show none of the synapomorphies of the Diphaglossinae, and the resemblance of the adults is almost certainly convergent.

Urban and Moure (2001) published a catalog of the species of Diphaglossinae.

Key to the Tribes of the Diphaglossinae
(Modified from Michener, 1986b)

1. Episternal groove complete; first flagellar segment nearly as long as, to longer than, scape, much longer than subsequent segments, petiolate (Fig. 42-1a).. Caupolicanini (Sec. 43)
—. Episternal groove absent below scrobal groove; first flagellar segment much shorter than scape, less than twice as long as middle flagellar segments, not or only moderately petiolate (Fig. 42-1b) 2
2(1). Notaulus represented by deep groove in anterior part of mesoscutum; malar space nearly one-third as long as eye or longer .... Diphaglossini (Sec. 44)
—. Notaulus weak or absent; malar space short or absent ....Dissoglottini (Sec. 45)

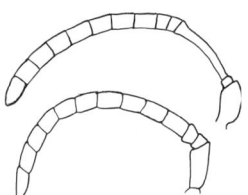

Figure 42-1. Antennae of males of Diphaglossinae. **a,** *Caupolicana yarrowi* (Cresson); **b,** *Mydrosoma brooksi* Michener.

43. Tribe Caupolicanini

In this tropical and subtropical American tribe of large, euceriform bees the lower part of the face is short, the malar space being short or absent. The notauli are strong. The jugal lobe of the hind wing is over three-fourths as long as the vannal lobe and extends beyond cu-v (Fig. 43-1). The second submarginal cell is much shorter than the first or third, and the first recurrent vein approximately meets the first submarginal crossvein (Fig. 43-1). The second recurrent vein more or less continues in the same direction as vein Cu_1 (Fig. 43-1). The distal parts of the wings are hairless but strongly papillate, the papillae often ending in slender hairlike points.

The Caupolicanini are divisible into two large genera and one small one, *Crawfordapis*, as shown below. Relationships among these genera are shown in a cladogram by Michener (1986b); *Caupolicana* appears there as the sister group to the other two genera together.

Key to the Genera of the Caupolicanini

1. Outer hind tibial spur of male immovably fused to tibia (Fig. 43-2a); hind basitarsus of female less than twice as long as broad, second hind tarsal segment broader than long; metasomal terga usually weakly metallic bluish or greenish .... *Ptiloglossa*
—. Outer hind tibial spur of male articulated at base; hind basitarsus of female more than twice as long as broad (Fig. 43-2c), second hind tarsal segment longer than broad; metasomal terga usually nonmetallic 2
2(1). S7 of male with no paired apical lobes; base of marginal cell prolonged as narrow sinus to apex of stigma (as in Fig. 43-1b) (Mesoamerica) .... *Crawfordapis*
—. S7 of male with paired apical lobes; base of marginal cell not prolonged as narrow sinus (Fig. 40-1a) .... *Caupolicana*

Genus *Caupolicana* Spinola

This genus is interpreted broadly to include *Zikanapis* and *Willinkapis*, taxa that could be given generic status. The principal characters are indicated in the key to genera (above).

The subgenera were reviewed by Michener (1966a), as were the species of North America and the West Indies.

Key to the Subgenera of *Caupolicana*

1. Metasomal terga rather distinctly metallic bluish; ventral apical lobe of S7 of male probably represented by broad, apically rounded, laterally directed, heavily sclerotized lateral apical projection that is hairless except mesally (South America) .... *C. (Willinkapis)*
—. Metasoma nonmetallic; ventral apical lobe of S7 not heavily sclerotized, not hairless, variable in size and shape but not as above, sometimes absent 2
2(1). S6 of male with apex rounded, apex rarely with broad, median, V-shaped notch but no produced region; ventrolateral extremities of T2-T4 lacking specialized regions; clypeus of male not over 0.76 times as long as wide 3
—. S6 of male with weak median apical projection that has a broad, median, V-shaped notch; ventrolateral extremities of T2-T4 of male with large areas of dense short hair of uniform length; clypeus of male about 0.85 times as long as wide .... *C. (Zikanapis)*
3(2). Inner orbits of male strongly converging above; ocellocular distance one-fourth of an ocellar diameter or less (Greater Antilles) .... *C. (Alayoapis)*
—. Inner orbits of male not or weakly converging above; ocellocular distance over one-third of an ocellar diameter and usually nearly equal to an ocellar diameter .... *C. (Caupolicana s. str.)*

Caupolicana / Subgenus *Alayoapis* Michener

Caupolicana (Alayoapis) Michener, 1966a: 728. Type species: *Megacilissa nigrescens* Cresson, 1869, by original designation.

In addition to the characters given in the key, this subgenus is distinctive in having a median area on S6 of the male that is nearly hairless or bears only short hairs, the posterior margin of S6 being rounded, the margin proper being a thin, hairless, translucent flange. The body length is about 15 mm. The male genitalia and hidden sterna were illustrated by Michener (1966a).

■ This subgenus is known from Cuba and Hispaniola. The three species were revised by Michener (1966a).

Caupolicana / Subgenus *Caupolicana* Spinola s. str.

Caupolicana Spinola, 1851: 212. Type species: *Caupolicana gayi* Spinola, 1851, designated by Sandhouse, 1943: 534.

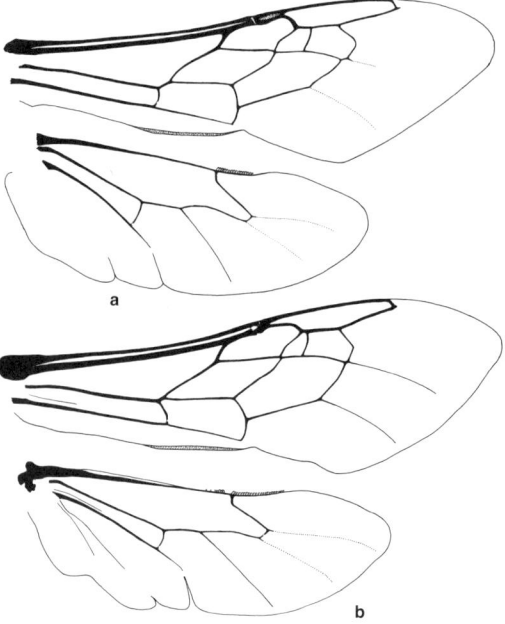

Figure 43-1. Wings of Caupolicanini. **a,** *Caupolicana hirsuta* (Spinola); **b,** *Ptiloglossa guinnae* Roberts.

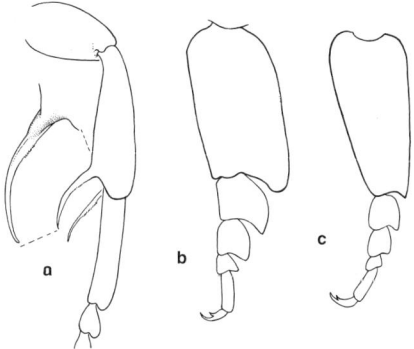

Figure 43-2. Legs of Caupolicanini. **a,** Hind leg of male of *Ptiloglossa mexicana* (Cresson), showing enlarged, immovable outer tibial spur; **b,** Hind tarsus of female of *Ptiloglossa* sp.; **c,** Hind tarsus of female of *Caupolicana yarrowi* (Cresson). From Michener, McGinley, and Danforth, 1994.

Megacilissa Smith, 1853: 123. Type species: *Megacilissa superba* Smith, 1853 = *Caupolicana fulvicollis* Spinola, 1851, monobasic.
Caupolicania Schulz, 1906: 238, unjustified emendation of *Caupolicana* Spinola, 1851.
Megalocilissa Schulz, 1906: 243, unjustified emendation of *Megacilissa* Smith, 1853.
Caupolicana (Caupolicanoides) Michener, 1966a: 725. Type species: *Caupolicana pubescens* Smith, 1879, by original designation.

This is the largest and most variable subgenus of *Caupolicana*. It is almost certainly paraphyletic, for the other subgenera all possess characters that could be derived from those of *Caupolicana* s. str. There is no compelling reason to maintain paraphyly in this case, but until the group is restudied in detail, I see no need to synonymize the other subgenera. The body length is 15.5 to 24.0 mm. Genitalia and hidden sterna were illustrated by Mitchell (1960) and Michener (1966a).
■ In South America, *Caupolicana* s. str. occurs from Valdivia, Chile, and San Luis, Argentina, and Uruguay north to São Paulo, Brazil, and in the west to Colombia. It is absent, so far as is known, in the moist tropics. In North America it ranges from the state of Oaxaca, Mexico, to Arizona, Kansas, Florida, and North Carolina, USA. There are over 25 species in South America (12 in Chile alone) and six more in North America. The North American species were revised by Michener (1966a); a new key is by Vergara and Michener (2004).

At least the North American species are most active early in the morning, for a few hours after first light, although they are sometimes taken later in the day. Works on nesting biology were cited by Rozen (1984b).

Caupolicanoides was based on a "typus" of *Caupolicana herbsti* Friese, a subjective synonym of *C. pubescens* Smith, in the Smithsonian Institution. It exhibits strikingly distinctive wing venation, as illustrated by Michener (1966a), with a broad marginal cell (about four times as long as broad), a stigma slightly wider at the apex than at the base, and the base of the marginal cell almost right-angular, not forming an acute angle at the apex of the stigma. There is no sign of abnormality in the specimen; all this is symmetrical in the two wings and differs in these respects from all other known Diphaglossinae. H. Toro and I have both reexamined the specimen concerned, however, and find it identical in other features to numerous specimens of *C. pubescens* having normal wing venation for *Caupolicana*. We therefore believe that *Caupolicanoides* was based on an abnormal individual.

Caupolicana / Subgenus *Willinkapis* Moure

Willinkapis Moure, 1953a: 66. Type species: *Ptiloglossa chalybaea* Friese, 1906, by original designation.

In this subgenus the metasomal terga are distinctly metallic bluish, more strongly metallic than in *Ptiloglossa*, and the dorsal apical lobe of S7 of the male is very slender, not spatulate, the ventral apical lobe as indicated in the key to subgenera. The body length is 18 to 19 mm.
■ This subgenus is from the Cordilleran region of Argentina (Mendoza, La Rioja) and Peru. As indicated by Michener (1966a) there are two species, only one of them described.

Caupolicana / Subgenus *Zikanapis* Moure

Zikanapis Moure, 1945a: 147. Type species: *Ptiloglossa zikani* Friese, 1925, by original designation.
Zikanapis (Foersterapis) Moure, 1964a: 441. Type species: *Zikanapis foersteri* Moure and Seabra, 1962, by original designation.

To elaborate on one of the key characters, a feature of male *Zikanapis* is that the ventrolateral extremities of T2 through T4 and sometimes T5 and T6 have large areas of short, dense, erect hair of uniform length; these areas lack longer hairs and appear dull and scarcely punctate in contrast to adjacent areas. This character suggests the genus *Ptiloglossa*. The body length is 14 to 20 mm. Male genitalia and hidden sterna were illustrated by Moure and Seabra (1962a), Moure (1964a), and Michener (1966a).
■ *Zikanapis* is known from the province of Santiago del Estero, Argentina, and Paraguay north to southern Arizona, USA. There are 11 species, of which two are North American. The North American species were reviewed by Michener (1966a), the South American species by Moure (1964a), and the Central American species by Michener, Engel, and Ayala (2003).

At least some of the species are matinal or largely so, even flying when it seems completely dark to human observers (observations by D. H. Janzen; Michener, 1966a).

Genus *Crawfordapis* Moure

Zikanapis (Crawfordapis) Moure, 1964a: 448. Type species: *Megacilissa luctuosa* Smith, 1861, by original designation.

The strongly elevated clypeus (above the level of adjacent parts of the face) and the short stigma (less than half as long as the prestigma, such that the base of the marginal cell is a slender sinus leading to the apex of the stigma) are *Ptiloglossa*-like features. The articulated outer hind tibial spur of the male, the widespread long hairs on the ventrolateral parts of the metasomal terga of the male, and certain other characters are plesiomorphic and in general *Caupolicana*-like. The simple, slender, hairy apical process, lacking lobes, of S7 of the male and the slender, styluslike apices of the gonoforceps, are unlike equivalent structures of other bees. The body length is 18 to 24 mm. Genitalia and hidden sterna were illustrated by Moure (1964a) and Michener (1966a).

■ *Crawfordapis* occurs in the mountains from western Panama to southern Mexico. The only recognized species is *C. luctuosa* (Smith), but the form, *C. crawfordi* (Cockerell), occurring in southern Central America may be specifically distinct.

As in other Diphaglossinae, the nests are more or less vertical burrows in the soil, with laterals each leading to a single vertical cell lined with a membrane containing the soupy larval food. An account of the nesting biology by Roubik and Michener (1985) contains references to earlier works on the same topic; a recent contribution concerning nest switching was by Jang, Wuellner, and Scott (1996). In their cold montane habitat, these bees are not so restricted to matinal hours of flight as are some of their relatives living at lower altitudes.

Genus *Ptiloglossa* Smith

Ptiloglossa Smith, 1853: 7. Type species: *Ptiloglossa ducalis* Smith, 1853, monobasic.
Ptiloglossa (Ptiloglossodes) Moure, 1945a: 153. Type species: *Megacilissa tarsata* Friese, 1900, by original designation.

This genus is easily recognized by the characters indicated in the key to genera. In addition, the clypeus is elevated above the level of adjacent parts of the face, the marginal cell [except in *P. tarsata* (Friese)] is prolonged at its base as a narrow sinus to the apex of the stigma, and, in the male, the ventrolateral extremities of T4 and T5 are broadly overlapped by the preceding sterna, dull, and densely covered with short, erect hair of uniform length. The body length is 15 to 20 mm. Male genitalia were illustrated by Michener (1954b) and Roberts (1971).

■ *Ptiloglossa* is found through tropical America from Santa Catarina, Brazil, and Cordoba province, Argentina, to southern Texas and Arizona, USA. It is absent from Chile. There are about 40 species. Moure (1945a) gave a key to some of them.

Ptiloglossodes was based on a clearly unusual species with much modified hind legs in the male.

Most of the species are matinal or almost nocturnal, becoming inactive in broad daylight. Roberts (1971) gave an account of a species that is active at low light intensities in morning and evening, but inactive during both the day and the night. He also described and illustrated the nests, which are vertical burrows in the soil with long laterals, each ending in a vertical cell lined with a cellophane-like membrane containing the liquid (watery) provisions on which the egg floats. He speculated that the principal protein source is yeasts that ferment the liquid, for there was little pollen in the provisions. Other material on the nesting biology of *Ptiloglossa* was provided by Rozen (1984b).

44. Tribe Diphaglossini

In this South American tribe the body of the larger species is euceriform whereas smaller species seem andreniform. The lower part of the face is elongate (see key to tribes). The first flagellar segment is not greatly longer than the others, not as long as the scape, and not petiolate. The notauli are strong. The jugal lobe of the hind wing is less than half as long as the vannal lobe and does not reach the level of cu-v. The submarginal cells decrease in length from first to third, or, rarely, the second and third are equal (Fig. 44-1). The first recurrent vein enters the second submarginal cell more or less medially, and the second recurrent vein is at a distinct angle to Cu_1. The distal parts of the wings are hairy, not strongly papillate. The male genitalia and hidden sterna of all genera were illustrated by Michener (1986b); see also Figures 44-2 and 13-2b.

The relationships among the genera are indicated in a cladogram by Michener (1986b); *Cadeguala* appears to be the sister to the other two genera. The same work provides a revision of the species of this tribe.

Key to the Genera of the Diphaglossini

1. Malar space about two-thirds as long as eye in female, three-fourths in male; S3 of male with broad, median, apical projection bearing stiff, basally directed setae (Chile) .... *Diphaglossa*
—. Malar space about one-third as long as eye; S3 of male simple 2
2(1). Third submarginal cell much smaller than second; apex of marginal cell obliquely, and conspicuously, truncate; S7 of male with a small lobe broadly attached on each side at apex (Fig. 44-2a) .... *Cadegualina*
—. Third submarginal cell about as large as second (Fig. 44-1a, b); apex of marginal cell narrow, the truncation inconspicuous or absent; S7 of male with three lobes, two of them large, on each side at apex (Fig. 44-2b, c) .... *Cadeguala*

Genus *Cadeguala* Reed

Cadeguala Reed, 1892: 234. Type species: *Colletes chilensis* Spinola, 1851 = *Colletes occidentalis* Haliday, 1836, by designation of Sandhouse, 1943: 532.

Policana Friese, 1910a: 651. Type species: *Colletes herbsti* Friese, 1910 = *Colletes albopilosus* Spinola, 1851, by designation of Sandhouse, 1943: 589.

Policana was long given generic status, but is so similar in adult morphology, although smaller in size, that it cannot be separated at the genus level. McGinley (1981) remarks that the differences between the larvae of *Cadeguala* and *Policana* do not support the recognition of separate genera.

Beyond the characters indicated in the key, *Cadeguala* differs from both of the other genera of Diphaglossini in the sparse and relatively short hairs of the sides of the female propodeum, beneath the longer, denser hairs of the upper lateral areas. The body length is 12 to 17 mm. The male genitalia and hidden sterna were illustrated by Michener (1986b).

■ This genus occurs from the Coquimbo region, Chile, and Bolivia south to Valdivia, Chile, and Río Negro, Argentina. It is particularly common in central Chile and Neuquén, Argentina. The two species were revised by Michener (1986b).

The nesting biology of *Cadeguala occidentalis* (Haliday) was described and illustrated by Claude Joseph (1926) and Torchio and Burwell (1987).

Genus *Cadegualina* Michener

Cadegualina Michener, 1986b: 187. Type species: *Bicornelia andina* Friese, 1925, by original designation.

Specimens of this genus (see key to genera) resemble fulvous-haired individuals of the common *Cadeguala occidentalis* (Haliday) but are probably more closely related to *Diphaglossa* (Michener, 1986b). The body length is 10 to 11 mm.

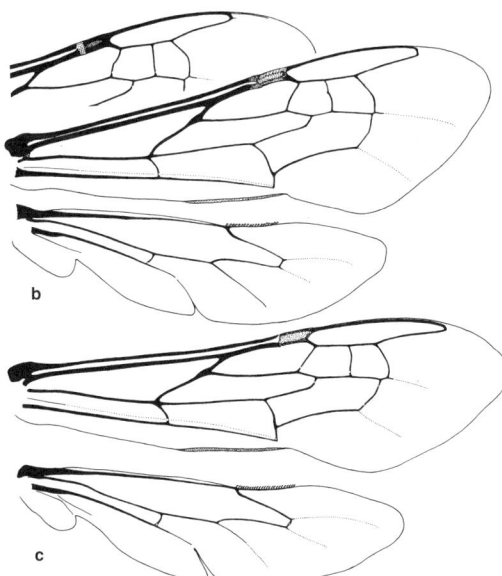

Figure 44-1. Wings of Diphaglossini. **a,** *Cadeguala occidentalis* (Haliday); **b,** *C. albopilosa* (Spinola); **c,** *Diphaglossa gayi* Spinola.

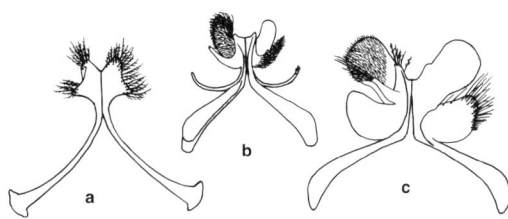

Figure 44-2. S7 of males of Diphaglossini. **a,** *Cadegualina andina* (Friese); **b,** *Cadeguala albopilosa* (Spinola); **c,** *Cadeguala occidentalis* (Haliday). (Dorsal views are at the left.) From Michener, 1986b.

■ *Cadegualina* ranges from Bolivia to Venezuela. There appear to be two species.

Genus *Diphaglossa* Spinola

Diphaglossa Spinola, 1851: 168. Type species: *Diphaglossa gayi* Spinola, 1851, monobasic.

Distinctive characters of this genus are the long lower part of the face, including the malar areas (see key to genera), with the inner orbits diverging below (slightly so in the male). In addition to illustrations in Michener (1986b), see Figure 13-2b. Because the large body (17-19 mm long) is covered with orange-red hair, these bees superficially resemble queens of *Bombus dahlbomii* Guérin-Méneville, found in the same area.

■ *Diphaglossa* is found in Chile, from Santiago to Valdivia. The only species is *D. gayi* Spinola.

45. Tribe Dissoglottini

Although *Ptiloglossidia* and larger species of *Mydrosoma* are euceriform, the smaller species are best called andreniform. The lower part of the face is short, as in the Caupolicanini. The first flagellar segment is about as long as the apical one, and less than one-half as long as the scape (female) or much shorter than any others and less than one-fourth as long as the scape (male), not petiolate. The jugal lobe of the hind wing is about one-half as long as the vannal lobe and does not reach the level of cu-v. Submarginal cells and recurrent veins are as described for the Diphaglossini, except that the first recurrent vein enters the second submarginal cell at the base or in the basal one-third (Fig. 45-1). The distal parts of the wing are hairy, not strongly papillate. The male genitalia and hidden sterna of all genera were illustrated by Michener (1986b).

This tribe was formerly called the Mydrosomini; the name Dissoglottini has priority, however, even though the genus *Dissoglotta* is considered a junior synonym of *Mydrosoma* (Michener, 1986b). The tribal name Ptiloglossidiini is also a synonym of Dissoglottini.

Key to the Genera of the Dissoglottini
1. Arolia absent; glossa moderately bifid, its apicolateral lobes longer than wide but not attenuate or pointed (Argentina) .... *Ptiloglossidia*
—. Arolia present; glossa strongly bifid, its apicolateral lobes attenuate, pointed 2
2(1). Second submarginal cell smaller than third; basitibial plate of female complete, margined (although hidden by hairs) (Argentina) .... *Mydrosomella*
—. Second submarginal cell larger than or rarely the same size as third (Fig. 45-1); basitibial plate of female at most a slightly elevated area with an elevated posterior margin .... *Mydrosoma*

Genus *Mydrosoma* Smith

Apista Smith, 1861: 148 (not Hübner, 1816). Type species: *Apista opalina* Smith, 1861, monobasic.
Mydrosoma Smith, 1879: 5. Type species: *Mydrosoma metallicum* Smith, 1879 = *Apista opalina* Smith, 1861, monobasic.
Bicornelia Friese, 1899a: 239. Type species: *Bicornelia serrata* Friese, 1899, monobasic.
Madrosoma Ashmead, 1899a: 94, *lapsus* for *Mydrosoma* Smith, 1879.
Egapista Cockerell, 1904a: 357, replacement for *Apista* Smith, 1861. Type species: *Apista opalina* Smith, 1861, autobasic.

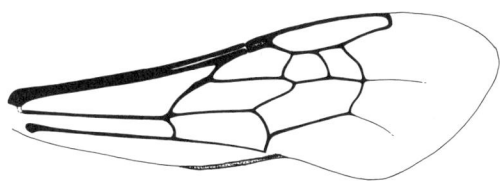

Figure 45-1. Forewing of *Mydrosoma bohartorum* Michener. From Michener, McGinley, and Danforth, 1994.

Dissoglotta Moure, 1945a: 144. Type species: *Dissoglotta stenoceratina* Moure, 1945, by original designation.

This genus of moderate-sized to large bees (body length 12-17 mm) is distinguished from its relatives by the small third submarginal cell (nearly always shorter than second, Fig. 45-1) and the reduced basitibial plate of the female (at most slightly elevated with a ridge along the posterior side, not around the apex). The hind legs of the male, especially the tibiae, are often enlarged and variously modified. Male genitalia and hidden sterna as well as hind legs were illustrated by Michener (1986b).

■ The genus ranges from Santa Catarina in southern Brazil through the tropics to Sinaloa, Mexico. The nine species are all rare in collections; the only one with known foraging habits visits flowers of *Triumfetta* (Tiliaceae) in Jalisco, Mexico, late in the afternoon (1730-1800 hrs). This genus was revised by Michener (1986b).

Genus *Mydrosomella* Michener

Mydrosomella Michener, 1986b: 194. Type species: *Diphaglossa* (?) *gaullei* Vachal, 1904, by original designation.

This genus contains a single species, one that looks superficially like a *Leioproctus* (Colletinae) because of its small size, 10.5 to 12.5 mm in length. (A few species of *Mydrosoma* are almost equally *Leioproctus*-like in appearance.) Some characters, such as that the third submarginal cell is longer than the second, and the almost complete marginal carina of the basitibial plate of the female, are as in many Colletinae and therefore must be plesiomorphic within the Diphaglossinae. Derived characters include a deep fossa for the scuto-scutellar suture. The genus was described and illustrated by Michener (1986b).

■ *Mydrosomella* is known from the provinces of Buenos Aires and Tucumán, Argentina. The only known species is *M. gaullei* (Vachal).

In view of the colletine-like features of *Mydrosomella*, it would be important to know whether the larva has the characteristics of the larvae of other Diphaglossinae.

Genus *Ptiloglossidia* Moure

Ptiloglossidia Moure, 1953a: 73. Type species: *Ptiloglossidia fallax* Moure, 1953, by original designation.

This genus, too, contains a single species, one easily distinguished from all other Diphaglossinae by its lack of arolia, but differing also in numerous other features, including the only moderately bifid glossa, which is thus intermediate between that of most Colletinae and the deeply bifid and attenuately produced glossa of other Diphaglossinae. The body length is 10.5 to 12.5 mm. I have not seen the female; hence all the information on the female is based on Moure's (1953a) description. A remarkable feature is that the head integument of the male is straw yellow, whereas the female has yellow on the labrum, mandible, and genal area only. Michener (1986b) described and illustrated the genus.

■ *Ptiloglossidia* is known only from the province of Salta, Argentina; it contains a single species, *P. fallax* Moure.

46. Subfamily Xeromelissinae

The bees of this Neotropical subfamily are small to minute, mostly slender, and hylaeiform. They are nonmetallic black, but sometimes show extensive white integumental bands on the terga and often white to yellow areas on the face, the metasoma rarely red. The first flagellar segment is much shorter than the scape or, in some males, nearly as long as the scape, cylindrical or tapering toward the base, not petiolate. The glossa of both sexes is broader than long, the apex emarginate. Females have a preapical glossal fringe; the annuli and annular hairs form a dense band basal to the preapical fringe, and basal to this band is an exceedingly fine pattern perhaps representing annuli. The annuli do not extend onto the posterior surface of the glossa, which in both sexes has numerous long hairs, grading into the large, long, branched hairs of the apical glossal brush. The male glossa lacks a recognizable preapical fringe; on its anterior surface it has well-separated annuli and pointed annular hairs. The prementum has an elongate, often narrow, depression or fovea on its posterior surface, margined by distinct raised lines. The lacinia, almost hairless and not easily recognized, stretches along the upper edge of the base of the galea or along the stipes. The galeal comb is represented by only a few to about ten rather weak bristles, thus differing markedly from that of the Hylaeinae. The stipes, prementum, and cardo are unusually long (as in the Halictinae). The facial fovea is not recognizable or is represented by a shining groove or depression above the middle of the eye, not extending up between the ocelli and the summit of the eye as is the case in most Hylaeinae. The ocular margin of the paraocular area in the vicinity of the upper one-fourth of the orbit is elevated, especially in those species having an emargination at this point. The episternal groove extends well below the scrobal suture, except in *Chilicola* subgenus *Chilioediscelis*. The basitibial plate is absent. There are two submarginal cells, the second much shorter than the first. The stigma is much longer than the prestigma, usually broad, but rather slender and parallel-sided basal to vein r in the Xeromelissini. The scopa on the hind legs is often not recognizable except by the presence of pollen. The hairs are relatively short and sparse (Fig. 46-1a), and on the femur are so arranged as to indicate the femoral corbicula of Colletinae and many other bees. The sternal scopa is recognizable on S1 to S3, the longest hairs being on S2 (Fig. 46-1b). The pygidial plate and prepygidial fimbria are absent. Male genitalia, sterna, and other structures were illustrated by Toro and Moldenke (1979) for all genera; see also Figure 46-4.

Larvae of *Chilicola* were illustrated and described by Eickwort (1967), and those of *Chilimelissa* were characterized (as *Chilicola*) by McGinley (1981); the identity of the latter was clarified by J. G. Rozen (in litt., 1995).

The Xeromelissinae occur in temperate and subtropical regions of southern South America (especially abundant in Chile) north mostly in arid zones or the Andean uplift to northeastern Brazil and to Colombia and Venezuela. They are present but rare in wet forest areas like Belém, Brazil, occurring also on the island of St. Vincent in the Lesser Antilles, in Central America (Panama to Guatemala), and northward to central Mexico (where, at the northern limit of the subfamily's range, species attain an altitude of 3,000 m).

Nests are in holes in hollow stems or beetle burrows in wood and consist of series of cells made with a cellophane-like membrane (Herbst, 1922; Claude-Joseph, 1926; Eickwort, 1967), or alternatively they may be in sandy soil where nests of *Geodiscelis* and *Chilimelissa* have been found (Michener and Rozen, 1999; Packer, 2004b). They do not differ conspicuously from those of *Hylaeus*.

The classification of the subfamily was reviewed by Michener (1995b), who recognized two tribes, Xeromelissini and Chilicolini; but Michener and Rozen (1999), on the basis of the genus *Geodiscelis*, concluded that the two tribes of Xeromelissinae merge and should not be recognized. Lawrence Packer has made a phylogenetic study of the Xeromelissinae, recognizing 18 genus-group taxa. Unfortunately this work is not yet published. The five genera of the subfamily here recognized can be distinguished by the following revision of Michener and Rozen's key:

Key to Genera of Xeromelissinae (modified from Michener and Rozen, 1999)

1. Stigma basal to vein r with margins diverging apically; beyond vein r, inner margin of stigma in marginal cell frequently convex, sometimes straight; thorax elongate; maxillary palpus not abruptly more slender at and beyond segment 4; paraocular area not invading clypeus as a strong lobe, epistomal sulcus continuing directly, straight or gently curved, from anterior tentorial pit to mandibular base (except in *Geodiscelis,* which has a strong paraocular lobe) 2

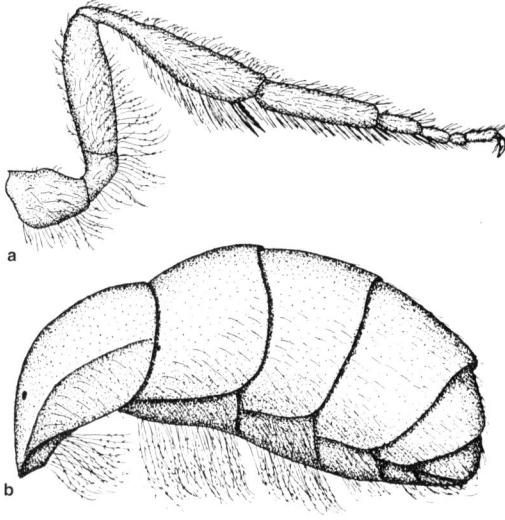

Figure 46-1. Female of *Chilicola ashmeadi* (Crawford) (Xeromelissinae). **a,** Hind leg; **b,** Lateral view of metasoma, showing the sternal scopa. From Eickwort, 1967.

—. Stigma basal to vein r nearly parallel-sided; beyond vein r margin of stigma in marginal cell straight; thorax not elongate; maxillary palpus abruptly more slender at and beyond segment 4; paraocular area produced downward into clypeus as an elongate lobe along mesal border of which anterior tentorial impression slants down into clypeus; epistomal sulcus (often weak or not recognizable) curving back from lower end of this impression to mandibular base 4

2(1). Paraocular area produced down as lobe (along with groove representing tentorial pit) invading clypeus; glossa with slender apicolateral lobes; metasoma with ivory integumental bands .... *Geodiscelis*

—. Paraocular area not produced downward as lobe into clypeus; glossa of usual short hylaeine or xeromelissine type; metasoma usually without ivory integumental bands 3

3(2). Basal sloping part of propodeum about as long as metanotum, less than half as long as declivitous vertical surface (as seen in profile); inner orbits of eyes nearly straight, not emarginate................................. *Xenochilicola*

—. Basal sloping or subhorizontal part of propodeum longer than metanotum, often as long as scutellum, half as long as vertical surface or frequently longer (as seen in profile); inner orbits of eyes emarginate or at least slightly concave at about upper third or fourth................. *Chilicola*

4(1). Anterior tentorial impression slanting down almost to apex of clypeus (Fig. 46-5a, b); labial palpus 4-segmented; maxillary palpus consisting of 6 easily seen segments, the first three markedly broader than the distal three .... *Chilimelissa*

—. Anterior tentorial impression not approaching apex of clypeus (Fig. 46-1c, d); labial palpus three-segmented; maxillary palpus consisting of three large segments followed, at least sometimes, by two to five extremely minute segments (these are easily broken off) .... *Xeromelissa*

Genus *Chilicola* Spinola

Chilicola is a rather large genus with considerable morphological diversity. Species range in length from 3 to 8 mm. The body is elongate; the propodeal upper surface is longer than that of *Xenochilicola* and the pronotum is usually longer than that of *Xenochilicola*, with a distinct dorsal surface on the same level as the scutum. The stigma is large and wide, the margin within the marginal cell distinctly convex (Fig. 46-2). S5 and S6 of the male are simple; the apical process of S8 is usually broad apically, truncate or bifid.

Toro and Moldenke (1979) revised the Chilean fauna; their work is the basis for subsequent taxonomic investigation of *Chilicola* and its relatives. Michener (2002b) showed that there is also a rich fauna of *Chilicola* in the Andes from Venezuela to Peru, consisting of species of the subgenera *Anoediscelis, Hylaeosoma,* and *Oroediscelis.* The *Chilicola* of Mexico and Central America were revised by Michener (1994) and the subgeneric classification of the genus was reviewed by Michener (1995b).

The subgenus *Anoediscelis* is quite possibly a paraphyletic group from which *Oediscelis* as well as *Chilicola* s. str. and *Chilioediscelis* were derived. The slender male hind leg and the proportions of its segments in *Anoediscelis* are like those of outgroups such as the Hylaeinae and

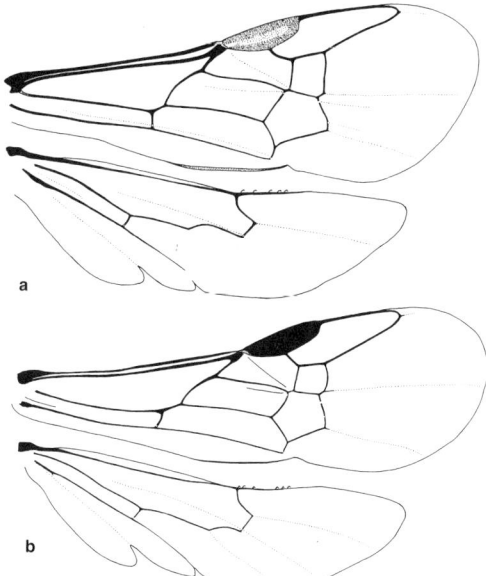

Figure 46-2. Wings of *Chilicola.* **a,** *C. (Anoediscelis) ashmeadi* (Crawford); **b,** *C. (Hylaeosoma) mexicana* Toro and Michener.

some *Chilimelissa; Anoediscelis* has no known apomorphies. On the other hand, present knowledge does not indicate any particular section of *Anoediscelis* from which the other subgenera may have arisen. The hind leg modifications of males that characterize *Chilicola* s. str., *Chilioediscelis,* and *Oediscelis* could have been reversed with a change in the mating system. That they could also arise more than once is suggested by the presence of similar hind legs in males of *Xeromelissa* (Toro, 1981).

The subgenera *Hylaeosoma, Prosopoides,* and *Pseudiscelis,* which have not proliferated in temperate or montane environments as have the other four subgenera, are not easily placed relative to *Anoediscelis. Prosopoides* and *Pseudiscelis* have common synapomorphies, such as the prolongation of the tentorial pits, and are presumably sister groups.

Key to the Subgenera of *Chilicola*

1. Hind tibial spurs strong and curved; clypeus and supraclypeal area rather flat and depressed, face thus seemingly slightly concave 2

—. Hind tibial spurs slender and almost straight; clypeus and supraclypeal area convex 3

2(1). Episternal groove extending down to lower part of thorax; claws cleft; hind tibia of male usually showing a transverse or slanting preapical cleft or depression on inner margin .... *C. (Chilicola* s. str.*)*

—. Episternal groove not extending below level of scrobe; claws of female simple; hind tibia of male somewhat thickened but ordinary in shape ....*C. (Chilioediscelis)*

3(1). Face with depression extending from antennal base toward area between ocelli and upper end of eye (Fig. 46-3d); second submarginal cell usually not extending beyond apex of stigma (Fig. 46-2b) ....*C. (Hylaeosoma)*

—. Face without depression slanting upward from antennal

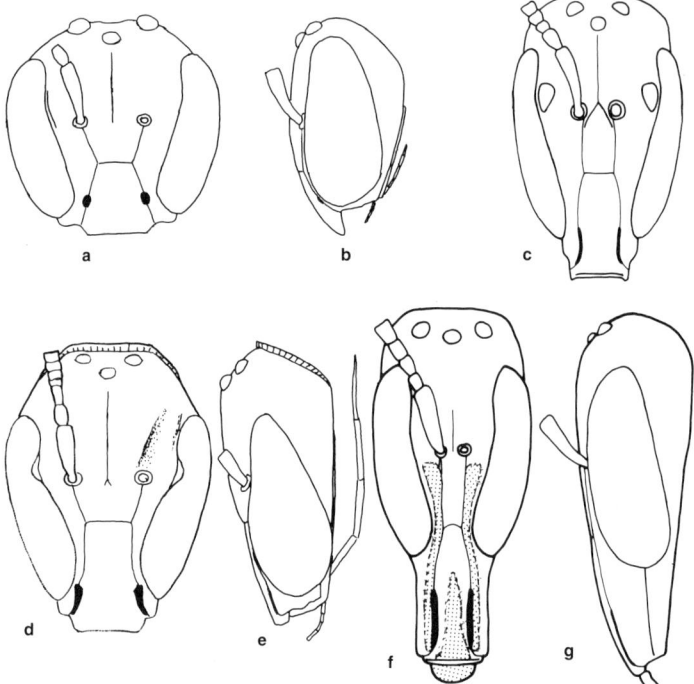

Figure 46-3. Facial and lateral views of heads of *Chilicola*, mandibles omitted. **a, b,** *C. (Anoediscelis) ashmeadi* (Crawford); **c,** *C. (Prosopoides) prosopoides* (Ducke); **d, e,** *C. (Hylaeosoma) polita* Michener; **f, g,** *C. (Pseudiscelis) rostrata* (Friese). (Females except f and g, which are male; stippled areas are yellowish.) In part from Michener, 1995b.

base; second submarginal cell ending distal to apex of stigma (Fig. 46-2a) 4

4(3). Anterior tentorial pit extended apicad as deep, shining groove along epistomal suture to area near clypeal apex, where suture bends laterad toward mandibular base (Fig. 46-3c, f); head over 1.5 times as long as wide 5

—. Anterior tentorial pit not greatly extended along epistomal suture; head wider than long to about 1.2 times as long as wide 6

5(4). A conspicuous, smooth, shining, well-defined facial fovea (less than three times as long as wide) near emargination in inner orbit (Fig. 46-3c); malar space broader than long .... *C. (Prosopoides)*

—. No clearly recognizable facial fovea (Fig. 46-3f); malar space about 1.5 to 3.0 times as long as broad (Fig. 46-3g) .... *C. (Pseudiscelis)*

6(4). Malar space 0.4 to 1.2 times as long as broad; line perpendicular to costal forewing margin at apex of stigma crossing submarginal cells near first transverse submarginal vein; S4 of male with two strong tubercles or projections. (hind femur of male not greatly enlarged) .... *C. (Oroediscelis)*

—. Malar space linear, nearly absent; line perpendicular to costal forewing margin at apex of stigma usually crossing submarginal cells at or beyond middle of second submarginal cell; S4 of male without tubercles or projections 7

7(6). Hind femur of male greatly swollen, as long as or longer than tibia, which is enlarged and modified distally .... *C. (Oediscelis)*

—. Hind femur of male not greatly swollen, shorter than tibia, which is also not or little modified ...... *C. (Anoediscelis)*

Chilicola / Subgenus *Anoediscelis* Toro and Moldenke

Anoediscelis Toro and Moldenke, 1979: 131. Type species: *Oediscelis herbsti* Friese, 1906, by original designation.
Stenoediscelis Toro and Moldenke, 1979: 135. Type species: *Oediscelis inermis* Friese, 1908, by original designation.

This is one of the two large subgenera of *Chilicola*. The species are mostly small (3-7 mm long). Even though some species are quite slender, the dorsal surface of the propodeum is equal to or shorter than the scutellum and usually shorter than the declivous posterior surface. S1 of the male is unmodified, as is the apex of the antenna, segment 13 being as long, or nearly as long, as 12. Male genitalic and other structures were illustrated by Toro and Moldenke (1979) and Michener (1994); see also Figure 46-4e-g. I know of no characters that reliably distinguish females from those of the subgenus *Oediscelis*.

■ *Anoediscelis* occurs from the provinces of Malleco, Chile, and Mendoza, Argentina, north in the Andean countries to Colombia; one species, *Chilicola ashmeadi* (Crawford), is found from Colombia to Mexico as far north as Nayarit and Puebla. The eight species found in Chile were reviewed by Toro and Moldenke (1979). One, as just mentioned, occurs in Mesoamerica. Ten Andean species were revised by Michener (2002b). Three of the Chilean species [*C. olmue* Toro and Moldenke, *orophila* Toro and Moldenke, and *minor* (Philippi)] were included by Toro and Moldenke (1979) in the subgenus *Heteroediscelis*.

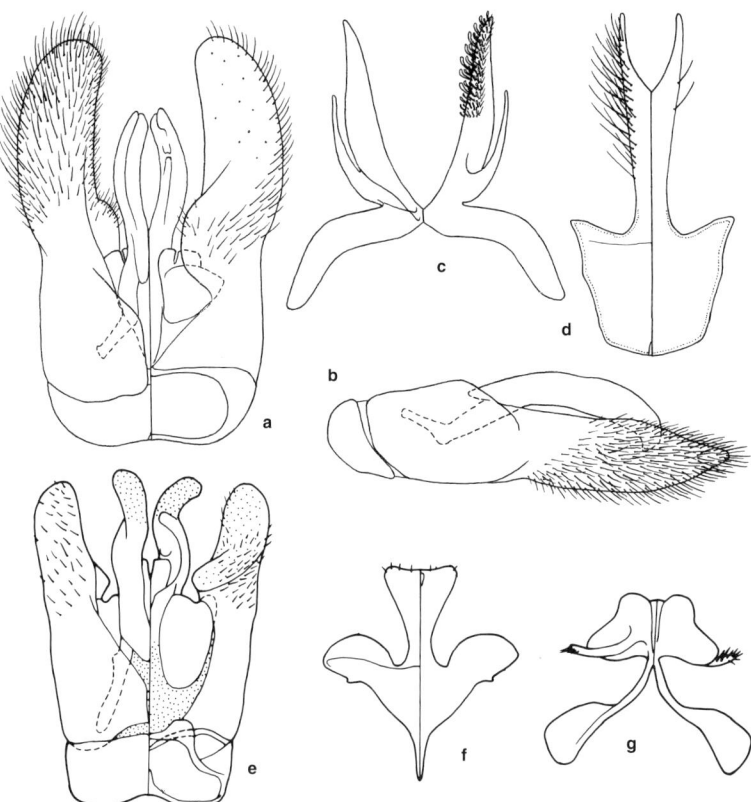

Figure 46-4. Male genitalia and hidden sterna of *Chilicola*. **a-d,** Genitalia, S7, and S8 of *C. (Hylaeosoma) mexicana* Toro and Michener; **e-g,** Genitalia, S8, and S7 of *C. (Anoediscelis) ashmeadi* (Crawford). From Michener, 1994.

Chilicola / Subgenus *Chilicola* Spinola s. str.

Chilicola Spinola, 1851: 210. Type species: *Chilicola rubriventris* Spinola, 1851, by designation of Sandhouse, 1943: 537.

This subgenus contains rather large species, 6 to 8 mm long. The males have enlarged hind legs (least so in *Chilicola colliguey* Toro and Moldenke) exhibiting the tibial modifications (weak in the same species) indicated in the key to subgenera. The other modifications of the hind legs of the males are similar to those of the subgenus *Oediscelis,* and it may be that *Chilicola* s. str. and *Oediscelis* should be united. The gently scoop-shaped face and curved hind tibial spurs are shared with the subgenus *Chilioediscelis.* Male genitalia and other structures were illustrated by Toro and Moldenke (1979). The best-known species, *C. rubriventris* Spinola, is unusual in having a red metasoma.

■ *Chilicola* s. str. is found in Chile, from Antofagasta to Aisén, and in Santa Cruz Province, Argentina. Four species were revised by Toro and Moldenke (1979).

Chilicola / Subgenus *Chilioediscelis* Toro and Moldenke

Chilicola (Chilioediscelis) Toro and Moldenke, 1979: 104. Type species: *Chilicola andina* Toro and Moldenke, 1979, by original designation.

Species of this subgenus are easily recognized by the characters indicated in the key to subgenera. They are moderately large for *Chilicola* (5-7 mm long), with moderately to considerably enlarged hind legs in the male (the tibia thickened, but not greatly modified in shape, and about as long as the enlarged femur). This subgenus is perhaps derived from *Chilicola* s. str., but because of the unusual characters listed in the key, especially the reduced episternal groove, I hesitate to unite it with *Chilicola* s. str. Male genitalia and other structures were illustrated by Toro and Moldenke (1979).

■ This rare subgenus is known from Coquimbo, Chile, to the province of Santa Cruz in Argentina. I have not examined its three species; my comments are based on Toro and Moldenke (1979), who revised the subgenus.

Chilicola / Subgenus *Hylaeosoma* Ashmead

Hylaeosoma Ashmead, 1898: 284. Type species: *Hylaeosoma longiceps* Ashmead, 1898, by original designation.
Hyloeosoma Ashmead, 1899b: 376, unjustified emendation of *Hylaeosoma* Ashmead, 1898.

The species of *Hylaeosoma* are unusually slender-bodied, 4.5 to 6.0 mm long, and either lack yellow markings or show yellow only on the clypeus of the males; the elongation of the thorax is indicated by the propodeum, whose dorsal surface is longer than not only the scutellum but also the declivous posterior surface of the propodeum. Moreover, the first metasomal segment is much longer than broad, in females as well as males. In this subgenus the stigma is usually large, about as long as the margin of the marginal cell on the costa, and a line through the apex of the stigma at right angles to the costa approximately meets the second submarginal crossvein (Fig. 46-2b). In other *Chilicola* such a line enters the second submarginal cell, which thus extends beyond the apex of the stigma (Fig. 46-2a). The head is much longer than broad (Fig. 46-3d). The emargination of the inner orbit is strong. A depression for the reception of the basal part of the antenna extends from the antennal base toward the ocellocular interval; sometimes this depression is rather weak, but even in such cases, near the antenna the mesal margin of the depression is shining and steeply sloping in contrast to the bulging frons. The preoccipital carina is present. The hind legs of both sexes are slender, those of the female bearing sparse, short hairs that can hardly be considered a scopa. Male genitalia and hidden sterna were illustrated by Toro and Michener (1975) and Michener (1994); see also Figure 46-4a-d.

■ This subgenus is known from São Paulo (Brazil), Bolivia, and the Ecuadorian Andes north to northeastern Mexico (Tamaulipas) and to St. Vincent in the Lesser Antilles. There are at least 10 species; those that have been named are mentioned by Michener (1994, 1995b, 2002b). Two additional fossil species are from Oligomiocene Dominican amber.

One group of species, including *Chilicola megalostigma* (Ducke) and *polita* Michener, is noteworthy for its largely impunctate and highly shining integument and the strong, reflexed preoccipital carina (Fig. 46-3d, e) that joins the hypostomal carina.

Chilicola / Subgenus *Oediscelis* Philippi

Oediscelis Philippi, 1866: 109. Type species: *Oediscelis vernalis* Philippi, 1866, by designation of Cockerell, 1919a: 185. [Toro and Moldenke (1979) erroneously list the type species as *O. plebeia* Philippi.]

Idioprosopis Meade-Waldo, 1914a: 451. Type species: *Idioprosopis chalcidiformis* Meade-Waldo, 1914, by original designation.

Oediscelisca Moure, 1946c: 243. Type species: *Oediscelis friesei* Ducke, 1907, by original designation.

Heteroediscelis Toro and Moldenke, 1979: 112. Type species: *Chilicola mantagua* Toro and Moldenke, 1979, by original designation (misspelled *Heteroesdiscelis* in heading but correct elsewhere).

As here understood, this subgenus contains species ranging in length from 4 to 8 mm. As in the subgenus *Anoediscelis,* the dorsal surface of the propodeum is equal to or shorter than the scutellum. In the two species placed in *Idioprosopis* by Toro and Moldenke (1979) the dorsal surface is much shorter and sloping. There is a projection or tooth on the hind trochanter of the male, except in *Chilicola gutierrezi* Moure. S1 of the male has a median spine or large truncate projection, except in the species placed in *Oediscelisca* by Moure (1946c). The last antennal segment of the male is often normal, as in *Anoediscelis,* but is much reduced in size in *C. hahni* Herbst and in the group called *Oediscelis* by Toro and Moldenke (1979). Finally, in their group called *Idioprosopis* the last antennal segment of the male is reduced to a mere nub. S7 of the male usually has four apical lobes, although there are only two in *C. gutierrezi* Moure and *friesei* (Ducke), and one pair is reduced in size in some others, as can be seen from the illustrations by Toro and Moldenke (1979).

■ *Oediscelis* occurs from Osorno, Chile, and Argentina north to Antofagasta, Chile, and Minas Gerais, Brazil, but is not known in the Andean region north of Chile. It comprises about 20 described species; the 17 of them from Chile were revised by Toro and Moldenke (1979).

Chilicola / Subgenus *Oroediscelis* Michener

Chilicola (Oroediscelis) Michener, 2002b: 26. Type species: *Oediscelis styliventris* Friese, 1908, by original designation.

Chilicola (Oreodiscelis) Michener, 2002b: 26. Incorrect original spelling, not available for use (see International Code of Zoological Nomenclature, ed. 4, Article 32.4, 32.5). The name was rendered correctly almost throughout the work.

This subgenus of moderate-sized to large (body length 5.5 to 8.5 mm) species, with or without a yellow mark on the clypeus of the male, is easily recognized by the characters listed in the key to subgenera.

■ *Oroediscelis* is known only from the Andean regions from Bolivia and Peru to Venezuela. Seven species have been described; see the revision by Michener (2002b).

Chilicola / Subgenus *Prosopoides* Friese

Prosopoides Friese, 1908b: 338 (reprint, p. 10). Type species: *Oediscelis paradoxus* "Ducke" Friese, 1908 = *Oediscelis prosopoides* Ducke, 1907, monobasic. [The name *Prosopoides* was published rather casually by Friese and attributed to Ducke. Ducke, however, had given no description, and I agree with Sandhouse (1943) in attributing the name to Friese.]

This subgenus contains small (3.5-5.0 mm long), unusually slender-bodied species without yellow markings, or with a diffuse median yellow streak on the clypeus. As in the subgenera *Hylaeosoma* and *Pseudiscelis,* the elongation of the thorax is indicated by the propodeum, the dorsal surface of which is longer than both the scutellum and the declivous posterior propodeal surface. The pronotum is unusually long, the dorsal surface (on a level with the scutum) being longer than the scutellum. Unlike *Hylaeosoma* but like *Pseudiscelis,* T1 of the female is about as long as broad. The head is about 1.6 times as long as broad, the upper part prolonged upward somewhat, as in *Hylaeosoma.* The most distinctive feature is the shining, well-defined, short, broad, facial fovea next to the very feeble (scarcely noticeable in the male) emargination of the inner orbit. There is a preoccipital carina. The hind legs of both sexes are slender, and both the femur and the tibia of the female have somewhat dense although short scopal hairs.

■ *Prosopoides* ranges from Pará to Santa Catarina, Brazil, and to Paraguay. There are two species, as listed by Michener (1995b).

Chilicola / Subgenus *Pseudiscelis* Friese

Oediscelis (Pseudiscelis) Friese, 1906a: 228. Type species: *Pseudiscelis rostrata* Friese, 1906, monobasic. [The type species was described in this combination, although the genus-group name was proposed as a subgenus of *Oediscelis*.]

This subgenus contains small (4.5-5.5 mm long), unusually slender-bodied species with a greatly elongated head, the malar space being about 1.5 to 3.0 times as long as the basal mandibular width, and the head being almost twice as long as wide or even longer. The upper part of the head is prolonged upward, as in the subgenus *Prosopoides*. Males have a long yellow stripe on the paraocular area margining the clypeus and supraclypeal area (Fig. 46-3f). The inner orbits are only gently emarginate, but more so than in *Prosopoides*. The characters of the mouthparts, preoccipital carina, hind legs, propodeum, and terminalia are as in *Prosopoides,* except that there is a membranous lobe at the apex of each penis valve.
■ *Pseudiscelis* occurs in the provinces of Salta and La Rioja in Argentina. There are two species.

Pseudiscelis is clearly related to *Prosopoides*. Each has its apomorphies; presumably they are sister groups. The relation to *Hylaeosoma* is less clear and the similarities are probably convergent, the very slender body perhaps relating to the use of small burrows for nesting.

Genus *Chilimelissa* Toro and Moldenke

Chilimelissa Toro and Moldenke, 1979: 149. Type species: *Chilimelissa luisa* Toro and Moldenke, 1979, by original designation.

Species of this genus range from 3 to 7 mm in length. They differ from other Xeromelissinae in their relatively long labrum, which is nearly as long as broad. The mouthparts are not very different in basic features from those of *Xeromelissa,* except that the last three segments of the maxillary palpus are of moderate size, although more slender than the first three. Most species have a noticeably long head; this feature reaches an extreme in *Chilimelissa rozeni* Toro and Moldenke, in which the clypeus is more than twice as long as wide and the malar area is considerably longer than the eye. Its genitalic, sternal, and most other characters, however, are typical of *Chilimelissa*. The opposite extreme is found in *C. brevimalaris* Toro, in which the malar space is about linear. *C. rozeni* is separated from other species by a considerable morphological gap. Not only is the lower part of its face greatly elongated, but the cardines, stipes, and mentum are very long. Most remarkable are the maxillary palpi of *C. rozeni:* segments 3 and 4 are each about as long as the stipes, and in repose the apices of the palpi lie between the hind coxae. Male genitalia and other structures were illustrated by Toro and Moldenke (1979) and Toro (1981).
■ This genus is known from Antofagasta to Nuble, Chile, and to Santa Cruz province, Argentina. A specimen of an undescribed species is from La Rioja Province, Argentina (Amer. Mus. Nat. Hist.). The 18 named species were revised and illustrated by Toro and Moldenke (1979) and Toro (1981, 1997). Nests in sandy soil were found by Packer (2004b).

Genus *Geodiscelis* Michener and Rozen

Geodiscelis Michener and Rozen, 1999: 2. Type species: *Geodiscelis megacephala* Michener and Rozen, 1999, by original designation.

Geodiscelis is characterized in the key to genera.
■ The two species of this genus are known from San Juan Province, Argentina.

As in *Chilimelissa,* the nests of *Geodiscelis* are in sandy soil.

Genus *Xenochilicola* Toro and Moldenke

Xenochilicola Toro and Moldenke, 1979: 145. Type species: *Xenochilicola mamigna* Toro and Moldenke, 1979, by original designation.

This genus contains minute bees, 2.5 to 3.0 mm long. The thorax is more robust than in *Chilicola,* as indicated by the largely vertical propodeal profile and the short pronotum, its dorsomedian part declivous in profile, lacking a surface at the level of the scutum. The stigma is smaller than usual in *Chilicola,* about two-thirds as long as the marginal cell on the costa, its margin within the marginal cell almost straight. The malar area is conspicuous, as long as or longer than the maximum flagellar width. S5 of the male is deeply and broadly emarginate, with a hairy apical process on each side of S6. S8 has a simple, slender apical process (broad and more or less bifid in *Chilicola*). The hind legs of the male are slender, and the penis valves lack membranous apical lobes. Toro and Moldenke (1979) illustrated the male genitalia and other structures.
■ This genus occurs in Chile from Tarapacá to Santiago. Its three species were revised by Toro and Moldenke (1979), and a key to the species was by Genaro and Parker (2005).

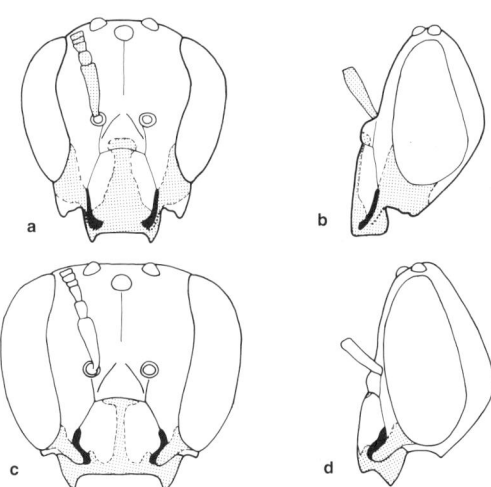

Figure 46-5. Facial and lateral views of heads of female Xeromelissini. **a, b,** *Chilimelissa luisa* Toro and Moldenke; **c, d,** *Xeromelissa wilmattae* Cockerell. (Stippled areas are ivory or yellow.) From Michener, 1995b.

Genus *Xeromelissa* Cockerell

Xeromelissa Cockerell, 1926a: 221. Type species: *Xeromelissa wilmattae* Cockerell, 1926, monobasic.

Although bees of this genus, which are about 6 mm in length, have a more ordinary (i.e., plesiomorphic) head than *Chilimelissa*, they also have derived features such as three-segmented labial palpi and minute, deciduous distal segments of the maxillary palpi, which therefore often appear three-segmented. The minute apical segments of the maxillary palpi vary in number from two to five, thus reaching a maximum of eight palpal segments, as compared to the normal maximum of six such segments in Hymenoptera. Some specimens appear to have no minute segments; I have called them deciduous, but the possibility exists that some individuals never had them. The hind legs of the male are strongly dilated, as in some species of *Chilicola;* in *Chilimelissa* they are more ordinary. The male genitalia and other structures were illustrated by Toro (1981).

■ *Xeromelissa* is found in desert areas of southern Peru (Arequipa) and northern Chile (Tarapacá). The only known species is *Xeromelissa wilmattae* Cockerell. Toro and Moldenke (1979) gave an account of the genus and its single species, based on female characters. The male was described and illustrated later (Toro, 1981).

47. Subfamily Hylaeinae

This is a group of minute to moderate-sized, mostly slender bees with short and generally sparse hair (Pl. 1), often superficially resembling small black sphecid wasps such as the Pemphredoninae. In Australia and New Guinea, rarely in Africa and the Philippines, some species are metallic blue or green, some have abundant yellow or white or even red markings, and some are rather large, up to 15 mm long. Most species have limited yellow or white marks on the face. The first flagellar segment is much shorter than the scape, cylindrical, not petiolate. The glossa of the female is broader than long, the apex truncate or concave (Figs. 47-1a-d; 20-3a), strongly so in *Hyleoides* so that the apex is deeply bilobed. The preapical fringe of the female glossa is distinct; the annuli are very close together, usually forming a narrow band basal to the preapical fringe, with short and blunt annular hairs. Basal to this band is a zone of exceedingly fine nodules probably representing greatly reduced annular hairs arranged in fine rows. (For the terminology of glossal parts, see Fig. 19-4.) The posterior surface of the glossa bears numerous fine, simple hairs, usually longest and densest laterally and apically and grading into the coarser and usually partly branched hairs of the apical glossal brush (Figs. 47-1b-d; 20-3b). The glossa of the male (Michener, 1992c) is usually shaped like that of the female but in the Australian region it is variable in shape (Fig. 47-1e-j). Usually, it lacks a preapical fringe or has an indication of it in the form of spatulate hairs. The annuli are well separated, covering most of the anterior surface of the glossa (Fig. 47-1e). The annular hairs are pointed. The posterior surface of the glossa is hairy, sometimes with a median bare area. The apex of the glossa often has

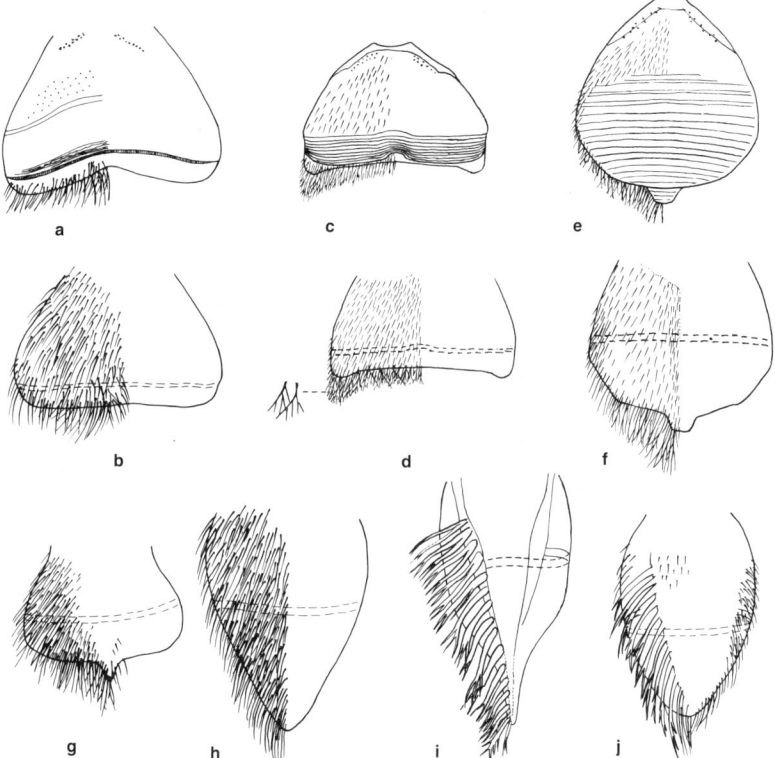

Figure 47-1. Diagrammatic sketches of glossae of Hylaeinae. **a, b,** Anterior and posterior views of female of *Hylaeus (Hylaeorhiza) nubilosus* (Smith); **c, d,** Same of *Amphylaeus (Amphylaeus) morosus* (Smith).

In most male Hylaeinae the glossae are shaped like those of these females; the following are exceptions: **e, f,** Anterior and posterior views of glossa of male *A. (A.) morosus* (Smith).

g-j, Posterior views of glossae of males (also exceptions): **g,** *Hylaeus (Hylaeorhiza) nubilosus* (Smith); **h,** *Amphylaeus (Agogenohylaeus) obscuriceps* (Friese); **i,** *Meroglossa torrida* (Smith); **j,** *Hemirhiza melliceps* (Cockerell). (Transverse broken lines on posterior views represent selected annuli seen through from the anterior surfaces; in figures i and j they curl around to the sides of the posterior surface; the coarse branched hairs in figures i and j are the seriate hairs.)

c-f, from Michener and Brooks, 1984; others from Michener, 1992c.

some branched hairs but lacks a distinct glossal brush or glossal lobes, such as are found in females. In males of *Hemirhiza, Meroglossa,* and *Palaeorhiza,* the glossa is relatively elongate, pointed like that of *Andrena* or a halictid (Figs. 47-1i, j; 19-4d, e). Its features in *Hemirhiza,* etc., as well as in *Amphylaeus* and *Hylaeus (Hylaeorhiza),* are explained below. The prementum in both sexes has a large, usually spiculate depression or fovea (Fig. 38-19a) on its posterior surface, margined by distinct, usually raised, lines. The facial fovea is a narrow, shining groove (e.g., Fig. 47-4b, h), in some males short (only twice as long as the ocellar diameter), very rarely only a pit. The episternal groove extends well below the scrobal suture. The scopa is absent, the hind femur lacking any indication of a femoral corbicula. The basitibial plate is usually absent but is rarely well developed. There are two submarginal cells, the second usually much shorter than the first, as in Figure 47-2a, so that the first submarginal crossvein seems to be lost, but in several taxa the second cell is about two-thirds as long as the first (Fig. 47-2b), and in *Hyleoides* it is almost as long as the first (Fig. 47-5a). The stigma is variable, as described for Colletinae. The pygidial plate is nearly always absent, but rarely is distinct. The prepygidial fimbria is absent, the the margin of T5 of the female resembling that of the preceding terga, or, in a few *Palaeorhiza,* the fimbria is present. The pygidial fimbria is absent, or is present such that the apical part of T6 of the female or T7 of the male is hairier than the preceding terga.

The subfamily Hylaeinae is found worldwide, and is most abundant and diversified in the Australian region.

Variation among hylaeine genera is much more extensive than I have indicated above. The apical lobes of S7, as in most colletids, are usually narrowly attached to the dorsal side of the small disc or body of the sternum. In several of the common taxa of the Hylaeinae found in the Australian region, the four apical lobes of the male S7 are of rather simple outline and have simple or plumose hairs. This condition, likely to be plesiomorphic, occurs in the genus *Hemirhiza* and in *Hylaeus (Hylaeteron, Macrohylaeus,* and *Prosopisteron).* It occurs also in a not greatly modified form in at least some species of *Hylaeus,* subgenera *Analastoroides, Gnathoprosopoides, Planihylaeus,* and *Sphaerhylaeus.* In the subgenus *Macrohylaeus,* and even in one of the species of the subgenus *Euprosopoides,* the lobes are not much modified from the *Hylaeus (Prosopisteron)* style. Such lobes are not found elsewhere in the world (except for a species of *Hylaeus (Prosopisteron)* introduced in South Africa), although similar-shaped but usually hairless lobes are found in the palearctic subgenus *Koptogaster.* In various taxa, the lobes are broadly fused to the disc of the sternum and on the same plane as the disc. This derived character has arisen repeatedly, not only in this subfamily but in others. (So much for the idea that convergence is unlikely in terminalia.) In the accounts of the taxa below, the word "fused" is used to describe such an arrangement. The most simplified lobes of S7 are those of the Australian *Hylaeus (Edriohylaeus),* in which the two lobes are directed posteriorly rather than laterally and are broadly fused to the disc of the sternum, which is at least twice as broad as usual. The modified lobes of S7 were illustrated by Metz

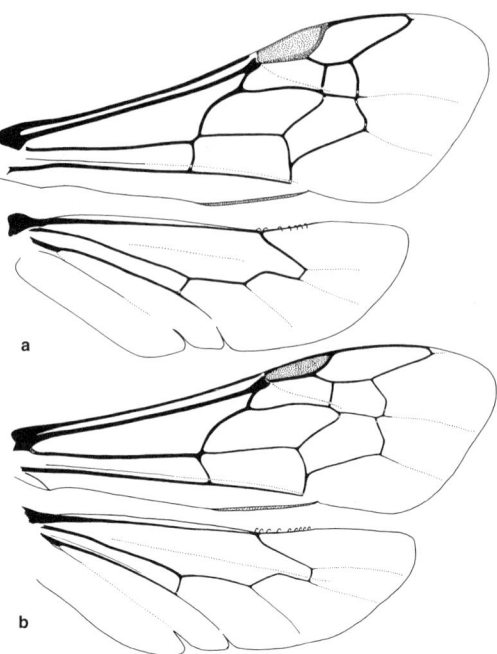

Figure 47-2. Wings of Hylaeinae. **a,** *Hylaeus nubilosus* (Smith); **b,** *Amphylaeus morosus* (Smith).

(1911), Méhelÿ (1935), Mitchell (1960), Michener (1965b), Houston (1975a, 1981a), Dathe (1979, 1986), Snelling (1985 and numerous papers cited below), Ikudome (1989), and others. In some, the lobes have grotesque shapes or pectinations, or rows of coarse setae.

Although all Hylaeinae have colletid-style glossae in females, the males of *Hemirhiza, Meroglossa,* and *Palaeorhiza* have a pointed glossa at least twice and usually several times as long as broad (see Secs. 20 and 37). On its posterior surface are two rows of coarse, distally branched, seriate hairs (Figs. 47-1i, j; 19-4e). Between these rows the surface is largely bare or has a few minute hairs. Because the annuli extend around the sides of the glossa, some hairs between the lateral margin of the glossa and the row of seriate hairs are annular hairs. The annuli are well separated and occupy most of the anterior surface of the glossa; the annular hairs are rather large and pointed. All the above are also features of glossae of both sexes of Andrenidae and Melittidae; most apply also to Halictidae. If the Perkins-McGinley hypothesis (Sec. 20) is correct, these are also ancestral features for Colletidae, now lost in all females.

The genus *Amphylaeus* retains, in the male, a somewhat pointed glossa. Like that of *Hylaeus (Hylaeorhiza)* (Fig. 47-1g), it is broader than long with a small apical protuberance in the subgenus *Amphylaeus* s. str. (Fig. 47-1e, f). It is markedly longer than broad and acute in the subgenus *Agogenohylaeus* (Fig. 47-1h), as also in the genus *Hemirhiza* (Fig. 47-1j). The posterior surface in *Amphylaeus* is densely hairy, and lacks seriate hairs. The annuli are well separated and occupy most of the anterior surface of the glossa, and the annular hairs are large and pointed. The annuli end at the sides of the glossa, not curving onto the posterior surface. Thus *Amphylaeus*, while retaining to some degree the pointed and perhaps plesiomorphic male glossal shape, lacks most of the other *Andrena*-like features of the posterior surface of the glossa. As in most other colletids, the annulate surface of the male glossa is quite different from that of the female. In *Amphylaeus* s. str. the preapical fringe of the glossa, a well-known female feature, may be indicated in the male by a row of coarse, spatulate hairs on the anterior side of the glossa basal to the midapical protuberance.

In the remaining genera of Hylaeinae the glossa is broader than long, similar in shape in the two sexes, and the apex is slightly rounded, truncate, or emarginate (deeply so in *Hyleoides*). The only exception is *Hylaeus (Hylaeorhiza)*, in which the male has a small median point arising from the otherwise broadly subtruncate glossal apex. In vestiture, however, male hylaeine glossae are like those of *Amphylaeus*, described above; that is, they differ from those of females in lacking a preapical fringe and clearly demarked glossal lobes and glossal brush, and in the nature of the annuli and annular hairs (Michener, 1992c).

No one has yet made a cladistic study of the Hylaeinae. Of the four Hylaeinae included in Alexander and Michener's (1995) phylogenetic study, which was based on adult characters of S-T (short-tongued) bees, one can say only that no consistent pattern was found. The only genus in that study with an *Andrena*-like glossa in the male, *Meroglossa*, did not appear as the basal hylaeine branch. As explained in Section 20, the Perkins-McGinley hypothesis was not supported by that study.

McGinley (1981) characterized the larvae of Hylaeinae, on the basis of *Amphylaeus*, *Hylaeus*, *Hyleoides*, and *Meroglossa*. He found numerous characters supporting the position of *Hylaeus* as sister group to the other three genera combined, but, obviously, larvae of many more genera should be studied before a reliable cladogram can be produced.

Most users of this work can ignore the key below because *Hylaeus* is the only genus found in most continents. A single rare species found on the mountains of central Africa is placed in the genus *Calloprosopis*. Otherwise, all genera other than *Hylaeus* are restricted to Australia, New Zealand, New Guinea, and nearby islands.

Key to the Genera of the Hylaeinae
(Based in part on Houston, 1975a, and Michener, 1965b)

1. Anterior tibial spine prolonged into long curved process, at least as long as basitarsal diameter (Fig. 47-3p); stigma with edge within marginal cell straight (Fig. 47-5a); posterior margin of T1 angulate near apex of lateral carina; apex of S1 transverse; S2 strongly produced downward at base (Fig. 47-3i) (Australian region).................... *Hyleoides*
—. Anterior tibial spine small or absent; stigma with edge within marginal cell usually convex (Figs. 47-2; 47-5b); posterior margin of T1 straight or with broadly rounded posterior lateral angle; apex of S1 with median cleft or slit; S2 not produced downward at base 2
2(1). T1-T3 enormous, enclosing apical segments (Fig. 47-3b); basal vein meeting cu-v or nearly so; second recurrent vein beyond second submarginal crossvein; gradulus of T2 absent, faintly indicated laterally, pregradular area densely hairy, especially laterodorsally (preoccipital carina present) (Australia, New Guinea) .... *Pharohylaeus*
—. T1-T3 of ordinary size, not hiding apical segments; basal vein distal to cu-v; second recurrent vein often meeting or basal to (although in some cases beyond) second submarginal crossvein (Figs. 47-2; 47-5); gradulus of T2 present, pregradular area not densely hairy 3
3(2). Males 4
—. Females.. 10
4(3). Glossa usually broader than long, gently rounded-truncate to weakly bilobed at apex (a small median point on otherwise subtruncate apex in *Hylaeus*, subgenus *Hylaeorhiza*, Fig. 47-1g).. 5
—. Glossa usually longer than broad, apex acute (Fig. 47-1h-j) or, at least with preapical margins meeting to form obtuse apical angle (in *Amphylaeus s. str.*, Fig. 47-1e, f).. 7
5(4). Bees brilliant metallic blue or green (with yellow markings); both recurrent veins outside limits of second submarginal cell or meeting submarginal crossveins; propodeal triangle largely dorsal, with strong carina separating dorsal from posterior surface (New Guinea) .... *Xenorhiza*
—. Bees nonmetallic or less brilliant blue or green; second recurrent vein and usually first received by second submarginal cell (except when submarginal and second medial cells are confluent, Fig. 47-5b); propodeal triangle with dorsal and posterior surfaces not separated by a carina, or, if so, then dorsal surface usually only a short zone...6
6(5). Gonobase reduced to narrow ring (metallic blue-green) (Africa) .... *Calloprosopis*
—. Gonobase large, forming a cuplike base to genital capsule .... *Hylaeus*
7(4). Fovea of T2 linear 8
—. Fovea of T2 punctiform or absent 9
8(7). Gena and scutum with yellow maculations; first metasomal segment appearing constricted in lateral view (Fig. 47-3j) (Australia) .... *Hemirhiza*
—. Gena and scutum without pale maculations; first metasomal segment not appearing constricted in lateral view (Australia) .... *Amphylaeus*
9(7). Face with large lateral depressions from sides of clypeus to above antennal sockets (Fig. 47-4i); gradulus of T2 exposed and arcuate posteriorly (i.e., recurved) medially; preoccipital carina absent; hind tibia with one or two spines on outer apical margin (Fig. 47-3o) (Australia) .... *Meroglossa*
—. Face without large lateral depressions; gradulus of T2 normally concealed and transverse; preoccipital carina usually present; hind tibia lacking spines on apical margin (New Guinea and nearby islands, Australia) ....*Palaeorhiza*
10(3). Outer apical margin of hind tibia with a pair of spines (Fig. 47-3m-o), the spines sometimes small or only one 11
—. Outer apical margin of hind tibia without spines (Fig. 47-3l) 12
11(10). Gradulus of T2 usually exposed and arcuate posteriorly (i.e., recurved) medially; fovea of T2 absent or punctiform (Australia) .... *Meroglossa*

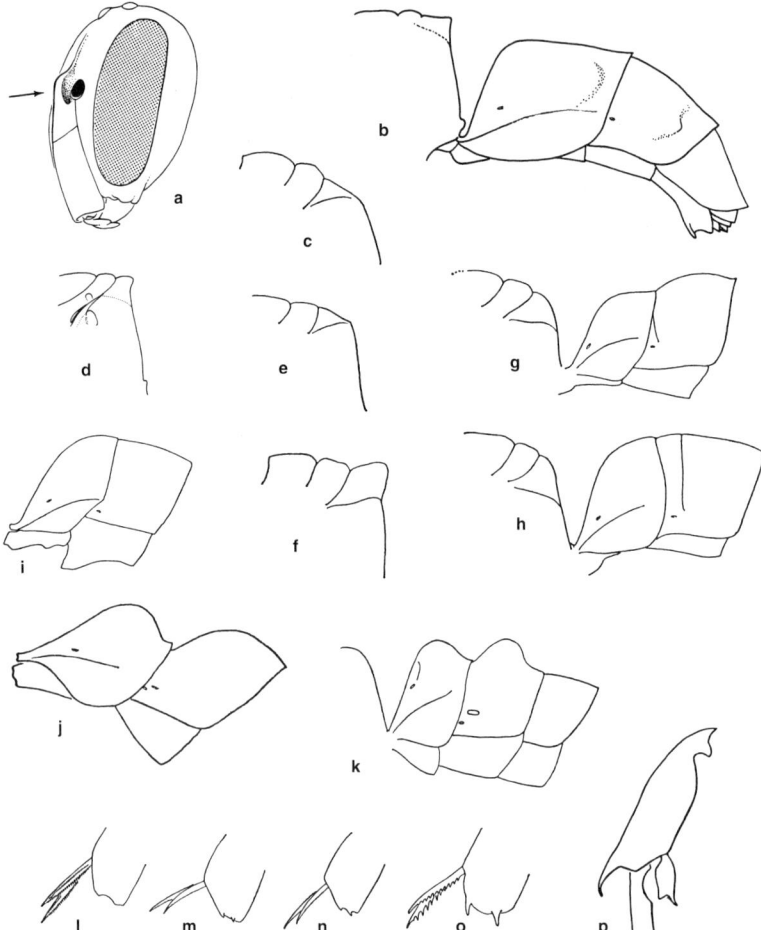

Figure 47-3. Structures of Hylaeinae. **a,** Elevated upper supraclypeal area, marked by arrow in laterofrontal view of head of *Hylaeus ellipticus* (Kirby); **b,** Propodeal profile and side view of metasoma of *Pharohylaeus lactiferus* (Cockerell); **c,** Propodeal profile of *Hylaeus (Macrohylaeus) alcyoneus* (Erichson); **d,** Same, of *Palaeorhiza (Callorhiza) stygica* Michener; **e,** Same, of *P. (Anchirhiza) mandibularis* Michener; **f,** Same, of *P. (Ceratorhiza) conica* Michener; **g,** Propodeal profile and base of metasoma of *Hylaeus (Analastoroides) foveatus* (Rayment), female; **h,** Same, of *Hylaeus (Hylaeorhiza) nubilosus* (Smith); **i,** Base of metasoma of *Hyleoides concinna* (Fabricius), showing projection of S2; **j,** Base of metasoma of *Hemirhiza melliceps* (Cockerell); **k,** Same, of *Hylaeus (Euprosopellus) dromedarius* (Cockerell); **l-o,** Apices of hind tibiae of females of *Hylaeus (Hylaeorhiza) nubilosus* (Smith), *Amphylaeus (Agogenohylaeus) nubilosellus* (Cockerell), *A. (Amphylaeus) morosus* (Smith), and *Meroglossa canaliculata* Smith; **p,** Front tibia of *Hyleoides concinna* (Fabricius). a, from Michener, McGinley, and Danforth, 1994; b, from Houston, 1975a; c, from Houston, 1981a; d-p, from Michener, 1965b.

—. Gradulus of T2 usually hidden and procurved medially; fovea of T2 linear (Australia) .... *Amphylaeus*
12(10). Preoccipital carina present (absent in the only known specimen of *Palaeorhiza bicolor* Hirashima and Lieftinck)...13
—. Preoccipital carina absent 15
13(12). Both recurrent veins outside limits of second submarginal cell or meeting submarginal crossveins; mesepisternum in front of middle coxa sometimes with strong spine or projection (New Guinea) .... *Xenorhiza*
—. Second recurrent vein and usually first received by second submarginal cell; mesepisternum in front of middle coxa simple or with ridge 14
14(13). Propodeal triangle nearly all on dorsal surface of propodeum except in species with protuberance on dorsum of propodeum; body usually bright metallic blue or green *or* with abundant white to yellow maculations on gena and mesothorax; vertical ridge in front of middle coxa usually strongly developed (New Guinea area, Australia) ....*Palaeorhiza*
—. Propodeal triangle having less than three-quarters of its length on dorsal surface of propodeum; body rarely brilliant metallic, gena and mesothorax usually without extensive white to yellow maculations; ridge in front of middle coxa weak or absent......................*Hylaeus* (in part)
15(12). Gena and scutum with yellow maculations; pro-

podeal triangle smooth, shiny, and evenly rounded in profile; T6 with a distinct pygidial plate; first metasomal segment slightly constricted in lateral view (Australia) .... *Hemirhiza*

—. Gena and scutum without yellow maculations or, if with them, then propodeal enclosure neither smooth, nor shiny, nor evenly rounded; pygidial plate usually absent; first metasomal segment not appearing constricted in lateral view 16

16(15). Base of hind tibia with elongate, glabrous ridge on outer side probably representing basitibial plate; propodeal triangle nearly all on dorsal surface (Africa) .... *Calloprosopis*

—. Base of hind tibia without glabrous ridge; propodeal triangle partly on declivous surface of propodeum *Hylaeus* .... (in part)

Genus *Amphylaeus* Michener

The species of this genus are moderate-sized to rather large, black, marked with yellow or white on the face, scutellum, metanotum, and legs; in general appearance they thus resemble some of the large species of Australian *Hylaeus*. The glossa of the male is bluntly to acutely pointed; the posterior surface lacks coarse seriate hairs, but has abundant slender hairs. The hind tibia of the female has two small spines on the outer apical margin. Among other Hylaeinae, only *Meroglossa* has similar, but usually larger, tibial spines. The lateral fovea of T2 is linear. Male genitalia and other structures were illustrated by Michener (1965b) and Houston (1975a); see also Figure 47-8a.

Key to the Subgenera of *Amphylaeus*

1. Male with clypeus and supraclypeal area indistinguishably fused, subantennal suture and upper lateral part of epistomal suture united as one strongly arcuate suture from tentorial pit to antennal base (Fig. 47-4a); interantennal distance of female equal to minimum clypeocular distance (Fig. 47-4b)........................ *A. (Amphylaeus s. str.)*

—. Male with complete epistomal suture of ordinary form joining subantennal sutures in usual way (Fig. 47-4c); female with interantennal distance nearly twice minimum clypeocular distance............................ *A. (Agogenohylaeus)*

Amphylaeus / Subgenus *Agogenohylaeus* Michener

Amphylaeus (Agogenohylaeus) Michener, 1965b: 148. Type species: *Prosopis nubilosellus* Cockerell, 1910, by original designation.

Agogenohylaeus contains moderate-sized species (body length 6.5-7.5 mm) that look like *Hylaeus*.

This subgenus occurs from Victoria to southern Queensland, Australia, in the dividing range and east of it. The three species were revised by Houston (1975a) and listed by Cardale (1993).

Amphylaeus / Subgenus *Amphylaeus* Michener s. str.

Amphylaeus Michener, 1965b: 147, 149. Type species: *Prosopis morosa* Smith, 1879, by original designation.

This subgenus consists of a single large species (body length 11-12 mm) having the remarkable male facial characters indicated in the key to subgenera above (Fig. 47-4a).

- *Amphylaeus* s. str., which has the same distribution as *Agogenohylaeus*, includes only one species, *A. morosus* (Smith).

The nests, which are similar to those of *Hylaeus*, were described (under the name *Meroglossa sculptifrons* Cockerell) by Michener (1960). Spessa, Schwarz, and Adams (2000) reported that about 23% of the nests of *Amphylaeus morosus* (Smith) contain two or three females, probably living communally.

Genus *Calloprosopis* Snelling

Calloprosopis Snelling, 1985: 27. Type species: *Hylaeus magnificus* Cockerell, 1942, by original designation.

This genus contains a single, rather large (8 mm long), dark-blue species—the only African hylaeine with metallic coloration. Since it may well be derived from *Hylaeus*, I have accepted its generic status with hesitation. In both sexes a ridge on the base of the hind tibia appears to represent the anterior side of the basitibial plate, the ridge ending abruptly in just the position where one would expect the basitibial plate to end. If this is a plesiomorphy, then *Hylaeus* exhibits the apomorphic condition (basitibial plate absent), and the two might be sister groups. A much more obvious difference from all other Hylaeinae, and no doubt an apomorphy, is the reduction of the male gonobase to a narrow ring around the genital foramen. Other characters of the male include the elongate volsellae and reduced S7 having a short body and two small apical lobes, each with a lateral and a posterior projection. Snelling (1985) has described and illustrated this genus.

- *Calloprosopis* is known only from rather high altitudes in Kenya. The single species is *C. magnifica* (Cockerell).

Genus *Hemirhiza* Michener

Hemirhiza Michener, 1965b: 147. Type species: *Palaeorhiza melliceps* Cockerell, 1918, by original designation.

This genus consists of one rather small (about 6 mm long) species, richly marked with yellow, including yellow on the gena and scutum (also on the face, Fig. 47-4e, f). The glossa of the male is pointed, nearly twice as long as wide, bearing strong seriate hairs on its posterior surface (Fig. 47-1j). The lack of a preoccipital carina distinguishes the species from nearly all *Palaeorhiza*; another distinguishing feature is the linear fovea on the side of T2. The male genitalia and hidden sterna were illustrated by Michener (1965b) and Houston (1975a). The apical lobes of S7, four in number, are rather broad (Fig. 47-8j), as in most species of *Hylaeus (Prosopisteron)*, and unlike those of the species of *Palaeorhiza* whose terminalia are known.

- *Hemirhiza* is found in southern Queensland and in New South Wales, Australia, in coastal and montane regions of high rainfall. The single species is *H. melliceps* (Cockerell).

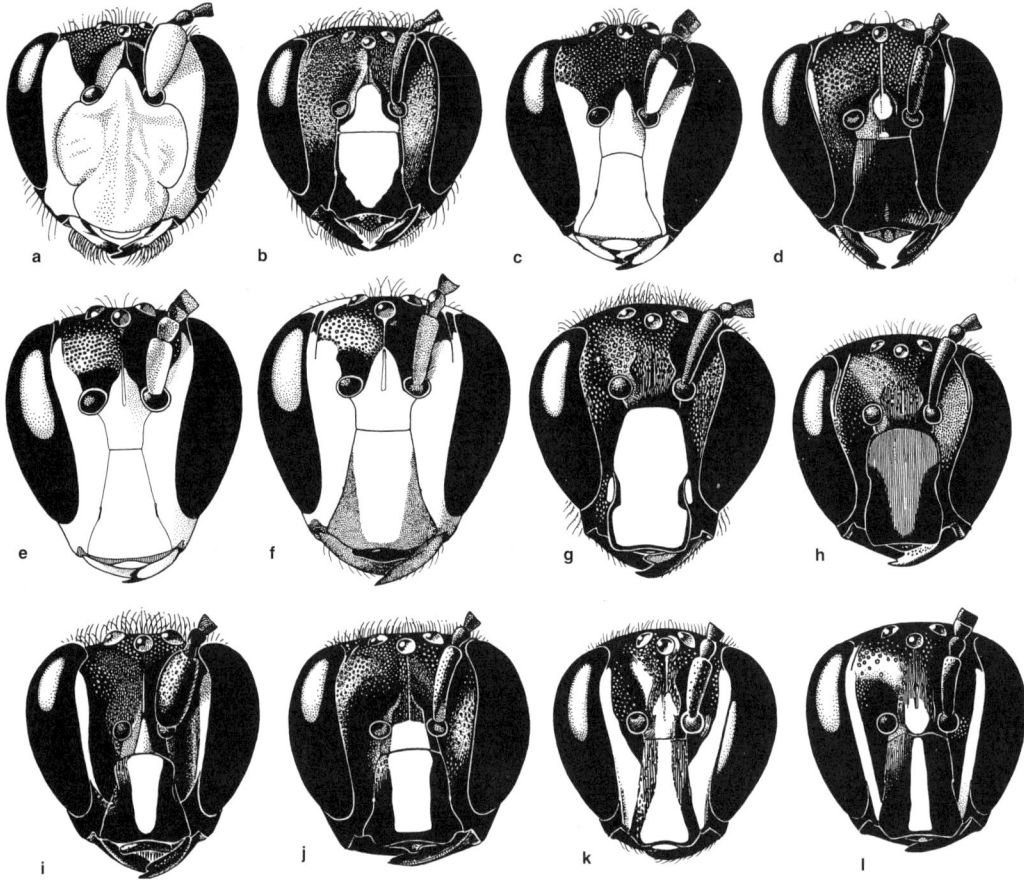

Figure 47-4. Faces of Australian genera of Hylaeinae, in each case male and female. **a, b,** *Amphylaeus (Amphylaeus) morosus* (Smith); **c, d,** *A. (Agogenohylaeus) nubilosellus* (Cockerell); **e, f,** *Hemirhiza melliceps* (Cockerell); **g, h,** *Hyleoides concinna* (Fabricius); **i, j,** *Meroglossa impressifrons penetrata* (Smith); **k, l,** *Pharohylaeus lactiferus* (Cockerell). From Houston, 1975a; illustrations by T. F. Houston.

Genus *Hylaeus* Fabricius

The genus *Hylaeus* has, at various times in the past, been known under the generic name *Prosopis*. The name *Hylaeus*, which has priority, has also been long and widely used; a proposal to suspend the rule of priority and place *Prosopis* on the list of *nomina conservanda* was not accepted. *Prosopis* is, however, the valid name of a holarctic subgenus of *Hylaeus* (see Popov, 1939a, and Michener, 1944). Dathe (1979) prepared an account of the nomenclatorial problem with facsimile copies of critical literature.

This is a worldwide genus of usually small bees, mostly with limited pale integumental markings on the head and thorax (Pl. 1; for facial marks, see Figs. 47-6, 47-7). The glossa of both sexes is short and subtruncate to weakly bilobed (Fig. 47-1a, b) (with a small median point in males of the subgenus *Hylaeorhiza*, Fig. 47-1g), and lacks seriate hairs but has fine hairs on the posterior surface. The mesepisternum normally lacks a strong ridge or spine in front of the middle coxa, although there is a distinct vertical ridge there in the subgenus *Hylaeorhiza*. The apex of the hind tibia, on the outer surface, lacks spines (Fig. 47-3l). So far as I know, *Hylaeus* differs from *Pharohylaeus* and *Calloprosopis* only or primarily in its lack of their special and no doubt derived features. *Hylaeus* is therefore probably paraphyletic, and these two genera could well be placed as subgenera within *Hylaeus*. They are abundantly different, however, and do not intergrade with *Hylaeus*; I have therefore chosen to retain them at the genus level. A detailed study should be made before they are added to *Hylaeus*. The case is clearer for the groups known as *Heterapoides* and *Gephyrohylaeus*, which are here relegated to subgeneric status. These taxa were given generic status in the past principally because of a single unusual venational characteristic, the fusion of the second submarginal cell with the second medial cell (Fig. 47-5b). Male genitalia and sterna were illustrated in each of the regional systematic works listed below, as well as by Michener (1954b) and Constantinescu (1973, 1974b).

Hylaeus occurs on all continents except Antarctica and on many islands. None of the Australian subgenera occurs outside of the Australia-New Guinea-New Zealand

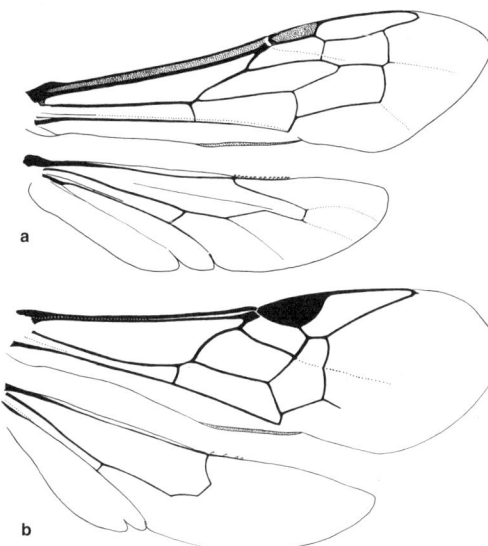

Figure 47-5. Wings of unusual Hylaeinae. **a**, *Hyleoides concinna* (Fabricius); **b**, *Hylaeus (Heterapoides) extensus* (Cockerell).

great majority, however, are small, nonmetallic, black or sometimes with red areas, usually showing restricted yellow or white areas on the face, thorax, and legs. Figures 47-6 and 47-7 give an idea of the variability in facial markings and structure. Rarely, as in *H. (Prosopisteron) albozebratus* Michener from Australia, there are much more extensive pale markings on all parts of the body. On the metasoma of *H. (Analastoroides) foveatus* Rayment from Australia are bright orange bands of tomentum. Although most *Hylaeus* are small, 4 to 7 mm in body length, some are smaller or larger, as indicated below and under certain subgenera. Some Australian subgenera *(Macrohylaeus, Meghylaeus)* contain species that are much larger than those of other continents, up to 15 mm long.

Many species of *Hylaeus* nest in dead stems, arranging cells made of cellophane-like material in series. Some species, however, make their cells alternatively or regularly in other small cavities—beetle burrows or nail holes in wood, cavities in volcanic rock, or even burrows made in earthern banks by other bees (Taylor, 1962b; Barrows, 1975; Torchio, 1984; Westrich, 1989). *Hylaeus pectoralis* Förster nests almost exclusively in abandoned galls of a reed gall fly, *Lipara* (Chloropidae) (Westrich, 1989). Laroca (1971b) described the nests of a species in small spherical galls, usually only one cell per gall. Torchio (1984) gave a particularly detailed account of the nesting biology of *Hylaeus leptocephalus* (Morawitz) (= *bisinuatus* Forster), which he induced to nest in glass tubes.

In view of the lack of a scopa, pollen is normally carried in the crop along with liquid, presumably nectar, and the provisions in cells are liquid, as in *Colletes*. In females of species of the subgenus *Prosopisteron* that I dissected, the crop was frequently full of pollen. Danks (1971), however, makes the remarkable statement concerning *Hylaeus brevicornis* Nylander that pollen "was initially brought in dry to the newly built cell-cup," and that dry pollen could be found in uncompleted cells, although after one or a few loads the larval food was in its normal semi-liquid form. How *H. brevicornis* carries dry pollen is not obvious; the finding should be verified.

One species, *Hylaeus (Hylaeopsis) tricolor* (Schrottky), nests in the cells of abandoned wasp nests *(Trypoxylon, Mischocyttarus)*. The *Hylaeus* cells are made by subdividing the wasp cells with the cellophane-like membrane that colletids use; there are one to three *Hylaeus* cells in series in each occupied wasp cell (Sakagami and Zucchi, 1978). One could regard the *Hylaeus* females in each wasp nest as simply aggregated and constructing clustered nests because of the clustering of suitable holes (wasp cells). The number of *Hylaeus* with enlarged ovaries associated with wasp nests, however, was over twice the number of cells being provisioned. Sakagami and Zucchi therefore assumed that there were communal or quasisocial relationships among the female bees. Elsewhere such relationships have not been postulated for Colletidae.

Most *Hylaeus* species visit and probably collect pollen from a variety of flowers, but some Australian forms are specialists, for example, the subgenus *Hylaeteron* on *Grevillea* and some *Xenohylaeus* on small yellow legumes. Recently, Scott (1997) found that three North American species appeared, in one locality, to be specialists on

area except (1) *Prosopisteron*, which has been introduced to South Africa and occurs (possibly introduced) in the Tuamotu Islands, (2) *Gephyrohylaeus*, which ranges from Australia into the oriental region as far as Borneo and the Philippines, and (3) *Euprosopoides*, which occurs in Micronesia. *Hylaeus* is scarce in New Guinea, in contrast to its abundance in temperate and subtropical Australia. It is also rare in the Sunda Islands and apparently in most of the Oriental faunal region. The sub-Saharan subgenera are restricted to Africa. The major North American subgenera are also abundant in the palearctic area, but the latter has a richer fauna (11 subgenera) than does the nearctic area (five subgenera, plus two neotropical subgenera that reach the southwestern USA).

Important regional systematic works on the genus include those of Metz (1911), Mitchell (1960), and Snelling (1966a) for North America; Méhelÿ (1935), Elfving (1951), Benoist (1959), Dathe (1980a), and Koster (1986) for Europe; Dathe (1993) for the Canary Islands; Snelling (1985) for sub-Saharan Africa; Houston (1975a, 1981a) for Australia; Ikudome (1989) for Japan; and Dathe (1980b, 1986) for Iran and Mongolia. These works contain illustrations of male genitalia and hidden sterna and sometimes other features. Moure (1960a) provided keys to numerous neotropical species, and Osychnyuk (1970) gave keys and descriptions for Ukrainian species. Meade-Waldo (1923) catalogued the species of this and other genera of Hylaeinae.

In most parts of the world there is relatively little diversity in aspect among the species of *Hylaeus*, and they constitute a rather small percentage of the total bee fauna. In Australia, however, *Hylaeus* is one of the major bee genera, and there is great diversity in aspect, as well as in structure (Figs. 47-3, 47-6, 47-7). Some species, mostly from Australia, are dark metallic blue or green, and a few in the subgenus *Prosopisteroides* from New Guinea are brilliantly metallic, like some species of *Palaeorhiza*. The

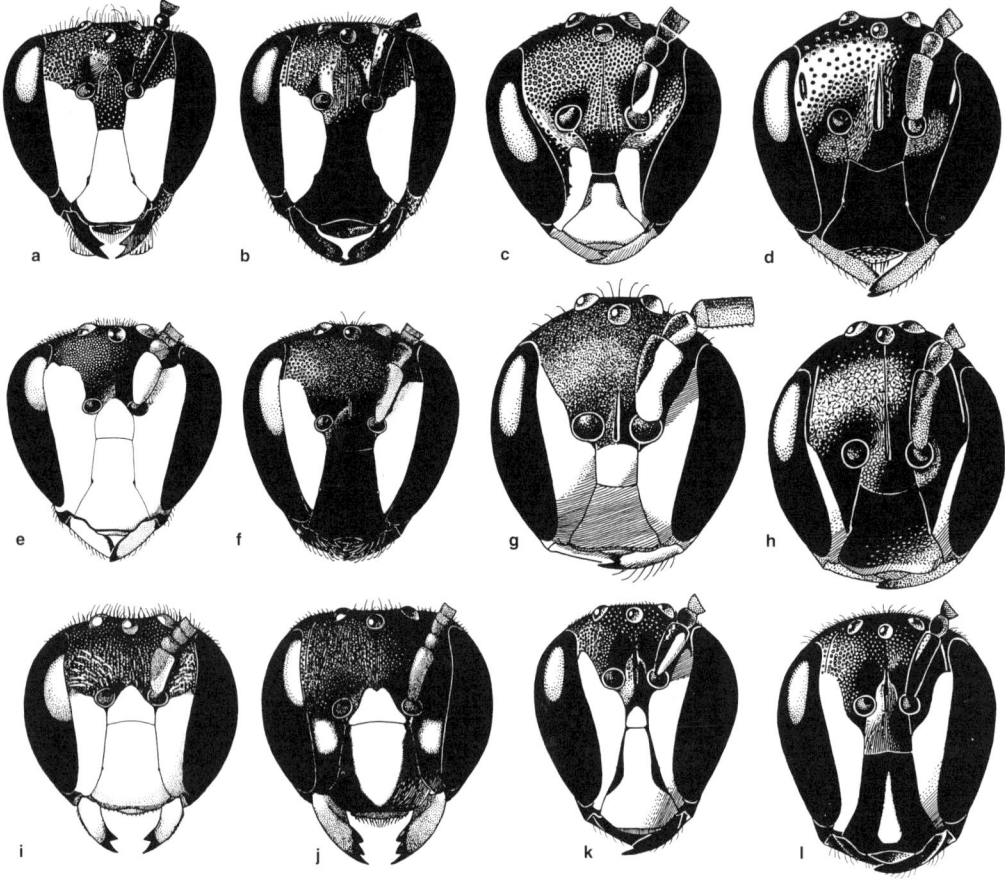

Figure 47-6. Faces of Australian *Hylaeus*, in each case male and female. **a, b,** *H. (Euprosopellus) chrysaspis* (Cockerell); **c, d,** *H. (Gephyrohylaeus) sculptus* (Cockerell); **e, f,** *H. (Gnathoprosopis) euxanthus* (Cockerell); **g, h,** *H. (Heterapoides) delicatus* (Cockerell); **i, j,** *H. (Hylaeteron) semirufus* (Cockerell); **k, l,** *H. (Hylaeorhiza) nubilosus* (Smith). From Houston, 1981.

pollen of Rosaceae. The view that most *Hylaeus* are polylectic may have arisen from the fact that pollen collecting is not visible because pollen is carried internally. Examination of pollen in nests is needed to determine the prevalence of pollen specialization in *Hylaeus*.

The keys to subgenera below are divided geographically, as follows: Australian region, sub-Saharan Africa, palearctic region, and Western Hemisphere. Because of the relationship of its *Hylaeus* fauna, Hawaii is included for this purpose with the palearctic fauna; if the Oriental fauna were better known, the Hawaiian connection might be just as close with the Oriental region. The subgenera of the Oriental region (meaning tropical Asia and nearby islands) are so little known that no key is provided; and, indeed, most species in that area are not placed in subgenera. The subgenera that are recorded from the Oriental region include *Lambdopsis, Nesoprosopis, Paraprosopis* (included among palearctic subgenera); *Gephyrohylaeus,* placed under the Australian region; and *Gnathylaeus, Hoploprosopis,* and *Nesylaeus,* which may be restricted to tropical Asia and associated islands. Snelling (1980) treated nine species from Sri Lanka and south and central India, providing descriptions and illustrations. Only one could be placed in a subgenus (*Paraprosopis*). Appropriate placement of Oriental species will require much larger collections and association of sexes; most species are known from one sex only. Table 47-1 shows the faunal areas in which the subgenera occur.

Since, in the holarctic and neotropical areas, there is only moderate size variation, that is, body length 3.5 to 9.0 mm, measurements are omitted for most subgenera from these areas.

Key to the Subgenera of *Hylaeus* of the Australia-New Guinea Area
(Modified from Houston, 1981a)

1. Second submarginal and second medial cells of forewing confluent (Fig. 47-5b); minute, slender bees with T1, at least in male, much longer than broad, as seen from above 2
—. Second submarginal and second medial cells of forewing separated, as in other bees (Fig. 47-2); size and form variable, but rarely so slender-bodied, or with T1 so slender 3

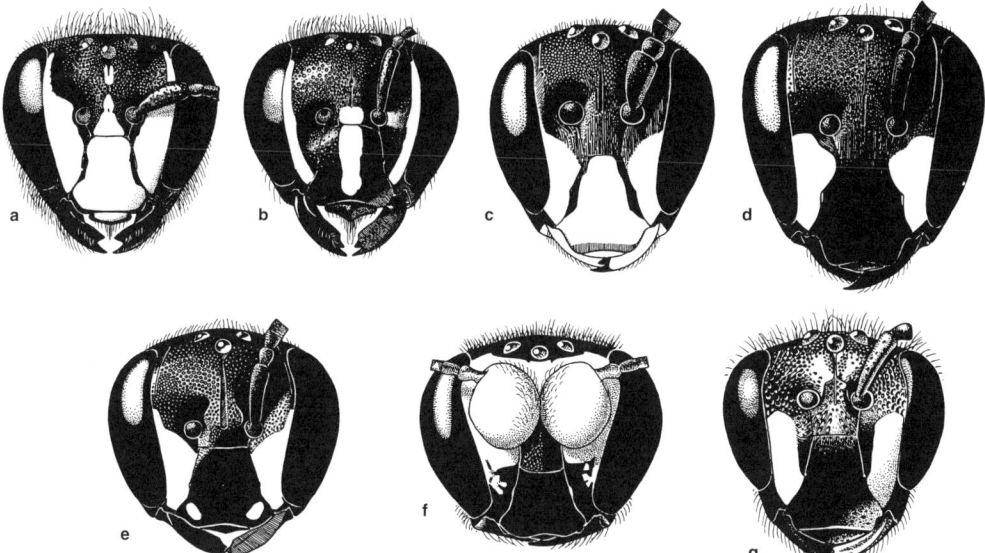

Fig. 47-7. Faces of Australian *Hylaeus*. **a, b,** *H. (Macrohylaeus) alcyoneus* (Erichson), male, female; **c, d,** *H. (Planihylaeus) daviesiae* Houston, male, female; **e,** *H. (Sphaerhylaeus) bicolorellus* Michener, female; **f,** *H. (S.) globuliferus* (Cockerell), male; **g,** *H. (Meghylaeus) fijiensis* (Cockerell), female. From Houston, 1981.

2(1). Mesepisternum broadly attaining or closely approaching propodeum, thus nearly eliminating metepisternum above coxa; preoccipital carina present; facial fovea of female short, not attaining summit of eye (Australia to Borneo and Philippines)........ *H. (Gephyrohylaeus)*
—. Mesepisternum separated from propodeum by metepisternum; preoccipital carina absent; facial fovea of female a long groove attaining summit of eye, as in most *Hylaeus* (Australia) ....*H. (Heterapoides)*
3(1). T1 of male, viewed laterally, constricted apically, and T2 strongly humped (Fig. 47-3k); T2 of female moderately humped (Fig. 47-3g) 4
—. T1 of male, viewed laterally, not constricted apically, and T2 not strongly humped; T2 of female not at all humped (Fig. 47-3h)... 5
4(3). Metasoma with bands of dense orange tomentum across apical margins of T1 and T3; terga with even, dense, fine pitting................................ *H. (Analastoroides)*
—. Metasoma without bands of orange tomentum; terga with irregular pitting, the pits of two sizes and partially confluent... *H. (Euprosopellus)*
5(3). Clypeus of female with fine longitudinal median carina (strongest apically); lower face longitudinally striate; face of male with very large depression on each side, from clypeus to well above antennal sockets; scape of male greatly expanded and flattened; subapical metasomal segments of male with large, expanded, pubescent, pregradular areas .... *H. (Xenohylaeus)*
—. Clypeus of female without median carina; lower face striate or not; face of male lacking depressions; scape of male not greatly expanded and flattened [except in *H. (Prosopisteron) semipersonatus* Cockerell, which lacks expanded, pubescent, pregradular areas] 6
6(5). Propodeum almost wholly vertical, dorsal surface shorter than metanotum and bearing a single row of regular pits bounded posteriorly by a carina 7
—. Propodeum variable, but dorsal surface never very short or bearing such a row of pits bounded posteriorly by a carina 9
7(6). Upper end of raised interantennal area about as wide as an antennal socket and merging with frons; posterior surface of scape of male without pits; female with distinct mesosternal brush-hairs, and outer apical spines of fore and middle tibiae unmodified .... *H. (Euprosopis)*
—. Upper end of raised interantennal area much narrower than an antennal socket and usually distinct from frons; posterior surface of scape of male with one or two distict pits; female without mesosternal brush-hairs, but outer apical spines of fore and middle tibiae modified into longitudinal carinae 8
8(7). Scutellum and metanotum usually with yellow areas; malar area of male no longer than one-fourth width; S8 of male with apex deeply bifid, hairy (Fig. 47-9g) ....
.... *H. (Euprosopoides)*
—. Scutellum and metanotum without pale areas; malar area of male about one-half as long as width; S8 of male with apex simple, usually hairless .... *H. (Laccohylaeus)*
9(6). Precoxal ridge of mesosternum distinct (propodeum almost entirely vertical, triangle smooth and shiny; scutellum and metanotum bright yellow).....................
.... *H. (Hylaeorhiza)*
—. Precoxal ridge of mesosternum indistinct or absent 10
10(9). Preoccipital carina present, at least medially; mandible sometimes broadest at blunt, edentate apex; scape of male sometimes globular 11
—. Preoccipital carina absent; mandible never broadest apically; scape of male never globular 14
11(10). Dorsal surface of propodeum as long as scutellum,

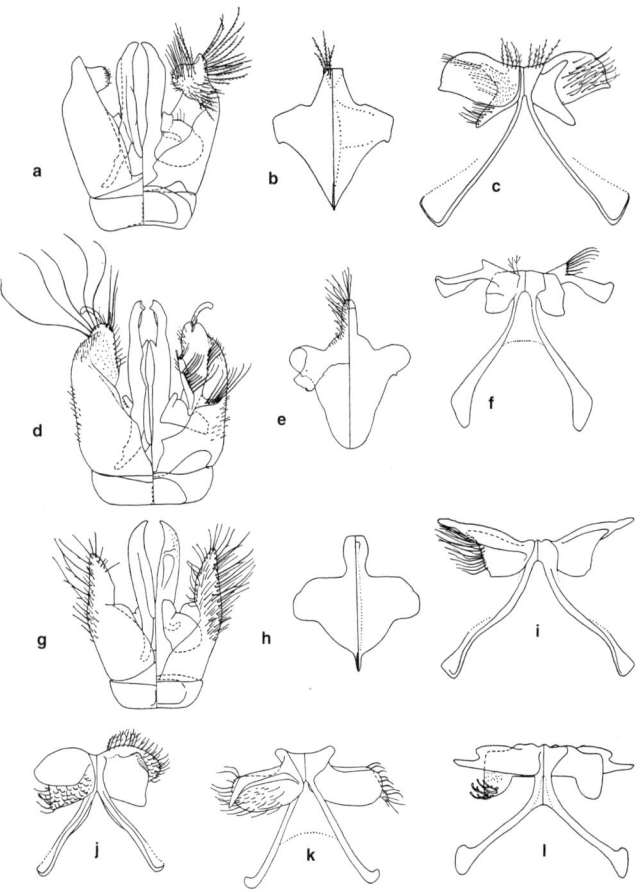

Figure 47-8. Male terminalia of Australian Hylaeinae. a-i, genitalia, S8, and S7 of the following: **a-c,** *Amphylaeus morosus* (Smith); **d-f,** *Meroglossa canaliculata* Smith; **g-i,** *Palaeorhiza stygica* Michener.

j-l, S7 of the following: **j,** *Hemirhiza melliceps* (Cockerell); **k,** *Hyleoides concinna* (Fabricius); **l,** *Pharohylaeus lactiferus* (Cockerell). (Dorsal views are at the left.) From Michener, 1965b.

triangle nearly all subhorizontal, smooth and shining to microtesellate and somewhat dull; maxillary palpus as long as thorax or longer (New Guinea).......................... .... *H. (Prosopisteroides)*
—. Dorsal surface of propodeum much shorter than scutellum, triangle thus with extensive area on posterior surface of propodeum, surface of triangle areolate, pitted, partly dull; maxillary palpus normal 12
12(11). Mandibles elongate, pollex forming the usual upper mandibular tooth; mandible broadest basally; scape of male almost hemispherical (Fig. 47-7f); frons of male with cavity above each antennal socket.......................... .... *H. (Sphaerhylaeus)*
—. Mandible short, pollex much broadened and rounded or subtruncate, not exceeded by mandibular apex, which is a small tooth; apical width of mandible equal to or greater than basal width [except in male of *H. (Gnathoprosopoides) philoleucus* (Cockerell), in which upper tooth (pollex) is broad and rounded]; scape of male slender to moderately swollen, seldom globular; frons of male without cavity above antennal socket 13
13(12). Dorsal surface of propodeal triangle entirely areolate and usually delimited laterally and posteriorly by carina [carina absent in *H. amiculus* (Smith)]; S7 of male with single pair of apical lobes............ *H. (Gnathoprosopis)*
—. Dorsal surface of propodeal triangle largely smooth and not at all delimited by carina; S7 of male with two pairs of apical lobes .... *H. (Gnathoprosopoides)*
14(10). Propodeum with short subhorizontal dorsal surface separated from long vertical posterior surface by distinct angle (Fig. 47-3c); body length 8.5-15.0 mm; metasoma metallic blue 15
—. Propodeum variable, its short subhorizontal dorsal surface not usually separated from vertical posterior surface by distinct line or angle; body length usually under 8 mm; metasoma rarely metallic blue 16
15(14). Prestigma about as long as stigma from its base to vein r; interantennal area strongly elevated and separated from frons by distinct edge (Fig. 47-7g); propodeum without a median prominence; mesosternum of female with brush-hairs .... *H. (Meghylaeus)*
—. Prestigma much shorter than stigma from its base to vein r; interantennal area convex but not delimited from frons (Fig. 47-7a, b); propodeal triangle with shiny median prominence; mesosternum of female without brush-hairs .... *H. (Macrohylaeus)*
16(14). Propodeal triangle areolate or coarsely roughened and delimited laterally and posteriorly by a carina; pronotal collar almost as high as scutum, contiguous laterally with the pronotal lobe through a carinate ridge; body at least partly yellow-brown to red-brown .... *H. (Rhodohylaeus)*
—. Propodeal triangle variable, not delimited by a carina or, *if* so, then pronotal collar much lower than scutum and

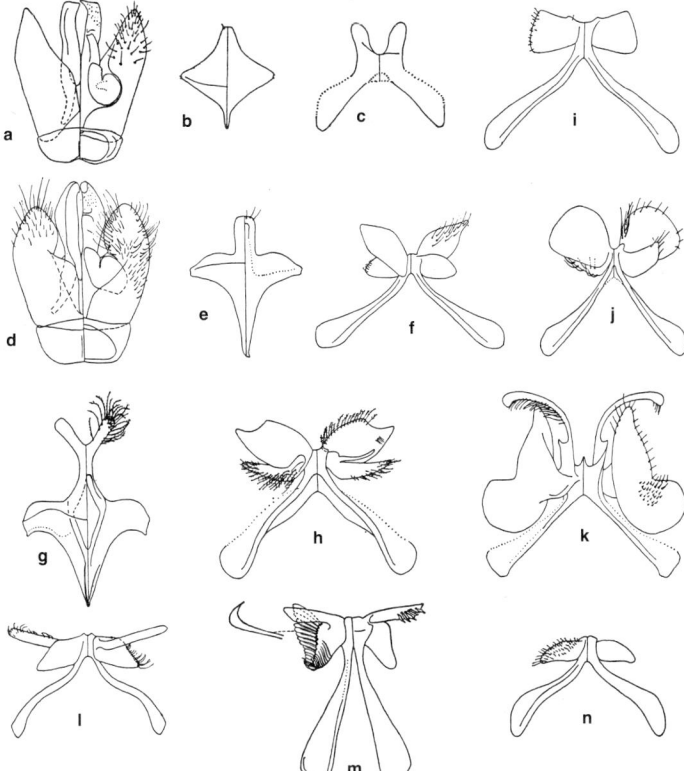

Figure 47-9. Male terminalia of Australian *Hylaeus*. a-f, genitalia, S8, and S7. **a-c,** *Hylaeus (Edriohylaeus) ofarrelli* Michener; **d-f,** *H. (Prosopisteron) perhumilis* (Cockerell). **g, h,** S8 and S7 of *H. (Euprosopoides) perconvergens* Michener. i-n, S7 of the following: **i,** *H. (Euprosopellus) dromedarius* (Cockerell); **j,** *H. (Prosopisteron) serotinellus* (Cockerell); **k,** *H. (Euprosopis) elegans* (Smith); **l,** *H. (Hylaeorhiza) nubilosus* (Smith); **m,** *H. (Xenohylaeus) rieki* Michener; **n,** *H. (Rhodohylaeus) ceniberus* (Cockerell). (Dorsal views of all are at the left.) From Michener, 1965b.

not contiguous with pronotal lobe through a carinate ridge; color variable 17

17(16). Mandible with outer ridge strong or carinate 18

—. Mandible with outer ridge inconspicuous 19

18(17). Mandible bidentate, the lower tooth usually longest by far; interantennal area not delimited from frons; hind margins of metasomal terga usually translucent and with laterally directed setae; outer margin of fore tarsus with long setae; apex of S8 of male bifid, dorso-apical lobes of S7 with acute apices .... *H. (Pseudhylaeus)*

—. Mandible of female tridentate, of male acutely bidentate; interantennal area elevated and delimited from frons by distinct edges; hind margins of metasomal terga rarely translucent and with laterally directed setae; outer margin of fore tarsus without long setae; apex of S8 of male not bifid, dorso-apical lobes of S7 rounded .... *H. (Hylaeteron)*

19(17). Lower face flat, longitudinally striate; propodeum with steeply sloping dorsal surface, triangle usually areolate anteriorly; lateral fovea of T2 large and rounded; subapical metasomal segments of male with large expanded pubescent pregradular areas; apex of S8 of male bifid and apex of dorso-apical lobes of S7 of male acute .... *H. (Planihylaeus)*

—. Without the combination of characters indicated above: lower face seldom both flat and striate; propodeum variable; lateral fovea of T2 rarely large and rounded; subapical metasomal segments of male with or without expanded pubescent pregradular areas; apex of S8 and dorso-apical lobes of S7 of male variable 20

20(19). Body very slender; T1 of male 1.5 times longer than broad, of female nearly as long as broad; propodeum with long, smooth, horizontal dorsal surface rounding evenly onto vertical posterior surface; S7 of male with only one pair of apical lobes broadly fused to broad body of sternum (Fig. 47-9c) .... *H. (Edriohylaeus)*

—. Body seldom very slender; T1 of male usually not longer than broad, of female usually much broader than long; propodeum variable, dorsal surface sometimes roughened anteriorly; S7 of male with two pairs of apical lobes narrowly connected to narrow body of sternum (Fig. 47-9f, j) ....*H. (Prosopisteron)*

Key to the Subgenera of *Hylaeus* of the Western Hemisphere (Males) (See account of *Orohylaeus*) (By R. R. Snelling)

For illustrations of male genitalia and hidden sterna of North American forms, see Metz (1911), Mitchell (1960), and the more specialized papers referred to under the subgenera.

1. Frons largely covered with conspicuous mat of short, dense, highly plumose pilosity, partly hidden behind swollen scape (neotropics)..................*H. (Gongyloprosopis)*

—. Frons without dense mat of short, plumose pilosity; scape swollen or not 2

2(1). Omaulus, at least below level of lower end of episternal groove, carinate 3

—. Omaulus rounded 5

3(2). Apical process of S8 flattened and broadly spatuliform, always visible in ventral view; side of propodeum never

Table 47-1. Distribution of the Subgenera of *Hylaeus*
i indicates introduction.

Subgenus	Neotropical	Nearctic	Palearctic	Oriental	Sub-Saharan	Australian
Abrupta	—	—	x	—	—	—
Alfkenylaeus	—	—	—	—	x	—
Analasteroides	—	—	—	—	—	x
Cephalylaeus	—	x	—	—	—	—
Cephylaeus	x	—	—	—	—	—
Cornylaeus	—	—	—	—	x	—
Dentigera	—	—	x	—	—	—
Deranchylaeus	—	—	—	—	x	—
Edriohylaeus	—	—	—	—	—	x
Euprosopellus	—	—	—	—	—	x
Euprosopis	—	—	—	—	—	x
Euprosopoides	—	—	—	x	—	x
Gephyrohylaeus	—	—	—	x	—	x
Gnathoprosopis	—	—	—	—	—	x
Gnathoprosopoides	—	—	—	—	—	x
Gnathylaeus	—	—	—	x	—	—
Gongyloprosopis	x	—	—	—	—	—
Heterapoides	—	—	—	—	—	x
Hoploprosopis	—	—	—	x	—	—
Hylaeana	x	x	—	—	—	—
Hylaeopsis	x	—	—	—	—	—
Hylaeorhiza	—	—	—	—	—	x
Hylaeteron	—	—	—	—	—	x
Hylaeus s. str.	—	x	x	—	—	—
Koptogaster	—	—	x	—	—	—
Laccohylaeus	—	—	—	—	—	x
Lambdopsis	—	—	x	x	—	—
Macrohylaeus	—	—	—	—	—	x
Meghylaeus	—	—	—	—	—	x
Mehelyana	—	—	x	—	—	—
Metylaeus	—	—	—	—	x	—
Metziella	—	x	—	—	—	—
Nesoprosopis	—	—	x	x	—	—
Nesylaeus	—	—	—	x	—	—
Nothylaeus	—	—	—	—	x	—
Orohylaeus	x	—	—	—	—	—
Paraprosopis	—	x	x	x	—	—
Planihylaeus	—	—	—	—	—	x
Prosopella	—	x	—	—	—	—
Prosopis	—	x	x	—	—	—
Prosopisteroides	—	—	—	—	—	x
Prosopisteron	—	—	—	—	i	x
Pseudhylaeus	—	—	—	—	—	x
Rhodohylaeus	—	—	—	—	—	x
Spatulariella	i	i	x	—	—	—
Sphaerhylaeus	—	—	—	—	—	x
Xenohylaeus	—	—	—	—	—	x

pruinose (introduced from Eurasia to southern California, central Chile) .... *H. (Spatulariella)*
—. Apical process of S8 not flattened and spatuliform; side of propodeum often densely pruinose 4
4(3). Spiracular area of propodeum enclosed by carina; propodeal triangle with coarse, more or less longitudinal rugae; thorax and/or metasoma usually coarsely punctate; omaulus with carina extending above lower end of episternal groove to pronotal lobe (neotropics) .... *H. (Hylaeopsis)*
—. Spiracular area open or, if enclosed, then propodeal triangle lacking coarse longitudinal rugae and thorax finely punctate; omaulus without carina above lower end of episternal groove (neotropics to southwestern USA) .... *H. (Hylaeana)*
5(2). Antennal scape much broader than long; S6 elevated

along midline; pronotum black (nearctic) ....
.... *H. (Cephalylaeus)*
—. Antennal scape usually about twice as long as broad or, if nearly as broad as long, then S6 flat and pronotal collar marked with yellow 6
6(5). Dorsolateral angle of pronotum in dorsal view slightly protuberant and sharply truncate; distal process of S8 bent downward at about 45° angle (southwestern USA, Mexico) .... *H. (Prosopella)*
—. Dorsolateral angle of pronotum in dorsal view obtuse or rounded; distal process of S8 straight or nearly so............ 7
7(6). Margins of interantennal elevation nearly parallel between antennal sockets, terminating on frons well above level of upper margins of antennal sockets; apical lobes of S7 laterally pectinate (Fig. 47-10g) (holarctic) ....
.... *H. (Hylaeus s. str.)*
—. Margins of interantennal elevation sharply convergent between antennal sockets and ending little, if any, above level of upper margins of antennal sockets; apical lobes of S7 not pectinate 8
8(7). Outer margin of front coxa with sharp laterobasal angle or tooth; apical process of S8 short, triangular, broad at base and tapering to blunt apex (first flagellar segment shorter than second) (nearctic).................... *H. (Metziella)*
—. Outer margin of front coxa without sharp laterobasal angle or tooth; apical process of S8 slender at least basally, usually parallel-sided for most of its length or bifid at apex 9
9(8). S7 with apical lobes rather small, flat, broadly attached to and on same plane as body of sternum; process of S8 with basal half slender, distal half robust and expanded distally to shallowly emarginate apex (Brazil) ....
.... *H. (Cephylaeus)*
—. S7 with apical lobes variable, large and elaborate or, if small, narrowly and flexibly attached to and on different plane from body of sternum; process of S8 slender, parallel-sided, but apex usually enlarged or deeply bifid 10
10(9). Apical process of S8 entire, more or less parallel-sided for most of its length and usually subapically broadened, apex rounded or truncate; first flagellar segment about as long as second (holarctic) .... *H. (Prosopis)*
—. Apical process of S8 deeply bifid at apex; first flagellar segment distinctly shorter than second (holarctic) ....
.... *H. (Paraprosopis)*

Key to the Subgenera of *Hylaeus* of the Western Hemisphere (Females) (See account of *Orohylaeus*) (By R. R. Snelling)

1. Omaulus, at least below lower end of episternal groove, carinate; spiracular area of propodeum usually enclosed by carina; anterior margin of pronotal collar often carinate or with distinct crest; metasomal sterna often very finely microstriate and weakly iridescent 2
—. Omaulus without carina; spiracular area of propodeum open or enclosed; anterior margin of pronotal collar rounded; metasomal sterna never finely microstriate and never iridescent 4
2(1). Labrum with single median tubercle; pronotal collar rounded in front; metasomal terga more or less polished and with sparse, fine, sharply defined punctures (introduced from Eurasia to southern California and Chile)
.... *H. (Spatulariella)*

—. Labrum with paired submedian longitudinal tubercles; pronotal collar with transverse carina or crest or, if rounded, metasomal terga microlineolate, satiny, without obvious punctures.................................. 3
3(2). Omaular carina absent above lower end of episternal groove; spiracular area of propodeum open or, *if* enclosed by carina, then basal area of propodeum not rugulose and metasomal terga without obvious punctures; pronotal collar not carinate or with weak anterior carina; thoracic punctures fine to moderate in size (neotropics to southwestern USA) .... *H. (Hylaeana)*
—. Omaular carina extending up to pronotal lobe; spiracular area of propodium enclosed by carina and basal area of propodeum usually coarsely rugose; pronotal collar usually anteriorly carinate or crested; thoracic (and often metasomal) punctures moderate to coarse (neotropics) .... *H. (Hylaeopsis)*
4(1). Dorsolateral angle of pronotum, in dorsal view, distinctly angulate or protuberant; anterior margin of pronotal collar sometimes with short, low, sublateral carina 5
—. Dorsolateral angle of pronotum rounded in dorsal view; anterior margin of pronotal collar rounded 6
5(4). Dorsolateral angle of pronotum slightly protuberant; propodeal triangle with transverse subbasal ridge, behind which surface is largely without rugulae; lateral carina of propodeum prominent along spiracular area (neotropics)... *H. (Gongyloprosopis)*
—. Dorsolateral angle of pronotum abruptly angulate (truncate) but not protuberant; entire propodeal triangle rugulose, rugulae primarily transverse; lateral carina of propodeum absent along spiracular area (southwestern USA, Mexico)... *H. (Prosopella)*
6(4). Gena as wide as eye or wider in lateral view, and mesepisternal punctures distinct; lateral propodeal carina absent or present only posteriorly; metasomal terga with only scattered, minute punctures 7
—. Gena usually conspicuously narrower than eye, but if as wide, then mesepisternum roughened and punctures weak and indistinct; lateral propodeal carina usually extending forward to spiracular area; metasomal terga variable, but often with conspicuous sparse to dense punctures 8
7(6). Front coxa with large, triangular lateral process or stout spine; upper end of facial fovea much nearer to inner eye margin than to lateral ocellus; side of face and pronotal lobe with pale markings (nearctic) ....*H. (Metziella)*
—. Front coxa without lateral process; facial fovea ending about midway between eye and lateral ocellus; side of face and pronotal lobe immaculate (nearctic) ....
.... *H. (Cephalylaeus)*
8(6). Outer margin of front coxa more or less distinctly angulate...9
—. Outer margin of front coxa evenly curved (not verified for *Cephylaeus*) 10
9(8). Facial fovea ending nearer to lateral ocellus than to inner eye margin; spiracular area of propodeum enclosed by carina (holarctic)..
.... *H. (Paraprosopis)* (in part)
—. Facial fovea ending nearer to inner eye margin than to lateral ocellus; spiracular area of propodeum open (holarctic)... *H. (Hylaeus s. str.)* (in part)

10(8). T1 without apicolateral patch of appressed, pale pubescence, but terga sometimes with narrow apical bands of white hairs laterally, and punctures on T1 and T2 usually fine, scattered, never dense ....(Brazil) *H. (Cephylaeus)*; (holarctic) *H. (Hylaeus s. str.)* (in part)
—. T1 with apicolateral patch of appressed, highly plumose white pubescence and/or punctures on T1 and T2 conspicuous, well-defined 11
11(10). T1 (and often T2) densely punctate, T1 almost always lacking apicolateral pubescent patch; spiracular area of propodeum often enclosed by carina, and mesepisternum finely punctate; facial fovea often ending at or mesal to midpoint between inner eye margin and lateral ocellus (Fig. 47-11l) (holarctic) ...... *H. (Paraprosopis)* (in part)
—. T1 sparsely to densely punctate, in latter case apicolateral pubescent patch present; if spiracular area of propodeum enclosed, then mesepisternum coarsely punctate; facial fovea always ending nearer to inner eye margin than to lateral ocellus (as in Fig. 47-11m) (holarctic) .... *H. (Prosopis)*

Key to Palearctic Subgenera of *Hylaeus* (Males)
(By H. H. Dathe)

For additional illustrations of male genitalia and other structures, see Méhelÿ (1935), Dathe (1980a), and other papers referred to under the subgenera. Illustrations accompanying this key (Figs. 47-10 and 47-11) are by H. Dathe.

1. Face entirely black, and frons broadly concave and shining between ocelli and antennal bases; middle basitarsus dilated at base; gonostylus truncate, the long, stiff bristles on lateral part of transverse margin nearly as long as gonoforceps (Fig. 47-10a) (scape conically swollen, white; S7 with apical lobes well developed, truncate at hairless tips; S8 with long spiculum and apical process, the latter bifurcate, hairy) .... *H. (Abrupta)*
—. Face with white or yellow areas, or, if entirely black, then frons convex and densely punctured medially; middle basitarsus normal; gonostylus apically rounded or pointed, its bristles shorter than gonoforceps 2
2(1). Gonostylus and gonocoxite distinctly separated by oblique constriction; outer margin of gonostylus convex, protruding; apical hair tufts of gonostylus dense and feathered (Fig. 47-10b), in normal position projecting from metasomal apex (genital capsule extraordinarily large; S7 with apical lobes finely pectinate; S8 with apical process short, both margins with hairs) .... *H. (Mehelyana)*
—. Gonostylus and gonocoxite not separated, their outline throughout convex or only weakly transversely constricted; apical hair tufts of genitalia not exposed in normal position 3
3(2). Gonoforceps conspicuously elongate, slender, distal third or thereabouts surpassing penis valves (Fig. 47-10c); apical process of S8 long, hairless, spoon-shaped, exposed at apex of metasoma (S7 with apical lobes small and simple, triangular, without hairs) .... *H. (Spatulariella)*
—. Gonoforceps of normal length and thickness, about as long as penis valves; S8 in normal position, concealed in metasoma 4
4(3). Penis valve with flat, rectangular membrane laterally, basally edge of membrane acutely angulate (Fig. 47-10d); gonostylus and gonocoxite separated by weak constriction (head in frontal view conspicuously longer than broad; pronotum thickened, dorsolateral angle square; propodeum steeply truncate; body robust, with coarse punctures; S7 with apical lobes consisting of two pairs of large membranes without hairs; S8 with apical process elongate, bilobate, with short hairs) .... *H. (Koptogaster)*
—. Penis valves and gonostylus/gonocoxite variable, but not as above 5
5(4). Dorsal carinae of the two penis valves in dorsal view parallel and in close contact to their apices (Fig. 47-10e), or, if separated, then only narrowly so, the inner ventral structures thus hidden from above (outline of the two penis valves together cuneiform to spindle-shaped) 6
—. Dorsal carinae of penis valves in dorsal view in contact basally but clearly diverging near bases or medially, often abruptly bent laterad, inner ventral structures clearly visible from above between separated penis valves 7
6(5). Apical lobes of S7 simple (four in number) with smooth margins (Fig. 47-10f), hairs absent or sparse and confined to lateral part of proximal lobe; scape usually slender; labrum and mandible frequently with yellow spots .... *H. (Paraprosopis)*
—. Apical lobes of S7 pectinate (Fig. 47-10g); scape commonly conically dilated or flattened, but if slender, then sternal callosity or lateral fringe of T1 absent; labrum and mandible black .... *H. (Hylaeus s. str.)*
7(5). S8 with prolonged basal part and apical process (Fig. 47-10i, j), the latter with short hairs or hairless and hooked downward; hook and incision in S6 reduced in some Japanese species, but the hook usually projecting from V-shaped incision in S6; scape broadened, shield-like [except in *H. melba* (Warncke) of northwestern Africa]; paraocular area lacking transverse flat impressions (S7 with apical lobes reduced, of various shapes, compact, with some short hairs; Fig. 47-10k) .... *H. (Lambdopsis)*
—. S8 more or less rhombiform, with short basal and apical parts (Fig. 47-11d, e, g), or apical process, if elongated, then with hairs and not hooked; S6 rounded or emarginate; scape slender or conically enlarged, or, if scutiform, then face with flat transverse impressions on paraocular area and below antennal sockets 8
8(7). Penis valves in dorsal view gently bent so that between them can be seen a pair of acute or truncate spines (Fig. 47-11a, b), these being ventral projections of penis valves (S7 with apical lobes reduced, triangular, with or without sparse hairs; S8 strongly reduced, rhombiform, hairless, apical process rarely somewhat elongate and filiform) .... *H. (Dentigera)*
—. Penis valves in dorsal view largely approximate, no spines thus visible from above between penis valves, ventral projections usually short and broad, not spinelike (Fig. 47-11c) 9
9(8). S8 with extremely elongated, curved, slender apical process, its apex with pair of hair tufts (Fig. 47-11d, e); S6 emarginate in middle; thorax, particularly mesepisterna, strikingly coarsely and strongly punctate; S7 with apical lobes reduced, slender, pointed, with sparse hairs (Fig. 47-11f) .... *H. (Nesoprosopis)*
—. S8 rhombic, with short, rounded or truncate, hairless apical process; S6 not emarginate; thorax usually finely

punctate; S7 with apical lobes reduced, compact, with hairs (Fig. 47-11h) that may be short and sparse............ .... *H. (Prosopis)*

Key to the Palearctic Subgenera of *Hylaeus* (Females) (By H. H. Dathe)

1. Mandible tridentate, inner tooth sometimes short (Fig. 47-11i) 2
—. Mandible with two teeth or apex bilobate (Fig. 47-11j) 5
2(1). Clypeus with broad, transverse, saddle-like depression below transverse supraclypeal projection and above two lower lateral clypeal projections; epistomal suture largely unrecognizable; face entirely black .... *H. (Abrupta)*
—. Clypeus slightly convex, without projections or teeth; supraclypeal projection absent; epistomal suture complete; face usually with yellow on paraocular area 3
3(2). Anterior coxa with obtuse process (Fig. 47-11k); clypeus rather rectangular, the middle evenly domed as seen from side .... *H. (Mehelyana)*
—. Anterior coxa without process; clypeal outline trapezoidal; clypeus flat or asymmetrically domed as seen from side 4
4(3). Paraocular spots often elongate, contiguous with inner orbits, or, if face entirely black, then T1 transversely obsoletely reticulated; T1 with lateral fringes, partly indistinct .... *H. (Dentigera)* (*brevicornis* group)
—. Paraocular spots usually rounded, contiguous with clypeal margin or, *if* face entirely black, then T1 integument smooth; T1 without lateral fringes ....*H. (Lambdopsis)* (in part)
5(1). Vertex swollen, in frontal view surpassing upper ocular margins by ocular width; head in frontal view nearly circular; inner margins of eyes not or only slightly convergent below; genal area broad, or, *if* head conspicuously rectangular and gena narrow, then thorax red (small species) .... *H. (Dentigera)* (*brachycephalus* group and *Hylaeus rubicola* Saunders)
—. Vertex convex as usual; head in frontal view rounded or trapezoidal, never rectangular; inner margins of eyes markedly convergent below; genal area narrow; thorax black, with red marks in only a few large species 6
6(5). Omaulus carinate or lamellate; malar area at least as long as basal flagellar diameter; thorax and clypeus mostly strongly punctate; propodeal triangle with coarse wrinkles .... *H. (Spatulariella)*
—. Omaulus rounded or merely angular; malar area shorter than basal flagellar diameter, rarely longer, in which case another character given above does not agree 7
7(6). Head conspicuously longer than broad in frontal view; pronotum thickened, dorsolateral angle square-truncate; mesoscutum coarsely and strongly wrinkled-punctate .... *H. (Koptogaster)*
—. Head shorter, circular or trapezoidal in frontal view; pronotum short, dorsolateral angle rounded or pointed; if mesoscutum coarsely punctate, then head always short 8
8(7). Facial fovea elongate, somewhat surpassing upper ocular margin, converging strongly toward ocelli, terminating closer to ocelli than to compound eye (Fig. 47-11l)... *H. (Paraprosopis)*
—. Facial fovea short and straight, barely reaching upper ocular margin and terminating closer to compound eye than to ocelli (Fig. 47-11m)... 9

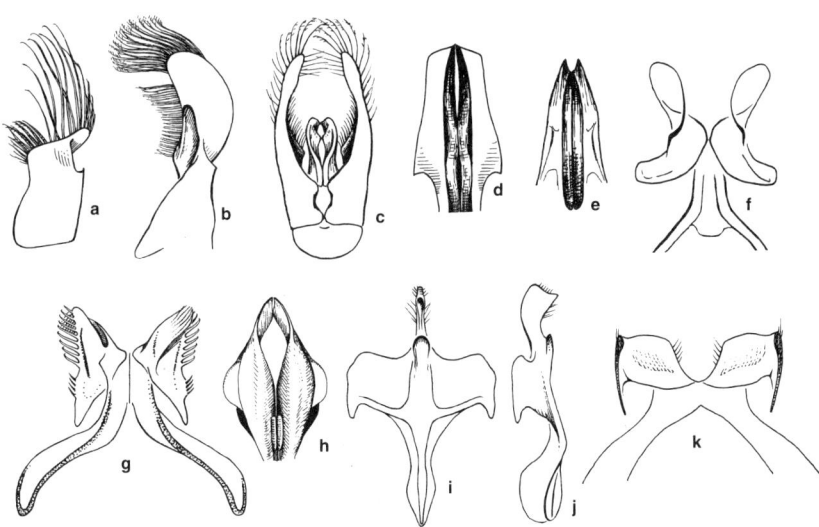

Figure 47-10. Structures of palearctic *Hylaeus*, males, dorsal views except g (dorsal at left, ventral at right) and i and j (which are ventral and lateral). **a,** *Hylaeus (Abrupta) cornutus* Curtis, left gonoforceps; **b,** *H. (Mehelyana) friesei* (Alfken), left gonoforceps; **c,** *H. (Spatulariella) hyalinatus* Smith, genital capsule; **d,** *H. (Koptogaster) punctulatissimus* Smith, penis valves; **e,** *H. (Hylaeus) angustatus* (Schenck), penis valves; **f,** *H. (Paraprosopis) ater* (Saunders), apical lobes of S7; **g,** *H. (Hylaeus) paulus* Bridwell, apical lobes of S7; **h,** *H. (Lambdopsis) annularis* (Kirby), penis valves; **i, j,** *H. (Lambdopsis) annularis* (Kirby), S8; **k,** *H. (Lambdopsis) annularis* (Kirby), S7. a-e, from Dathe, 1980a; f, from Dathe 1993; others are original by H. Dathe.

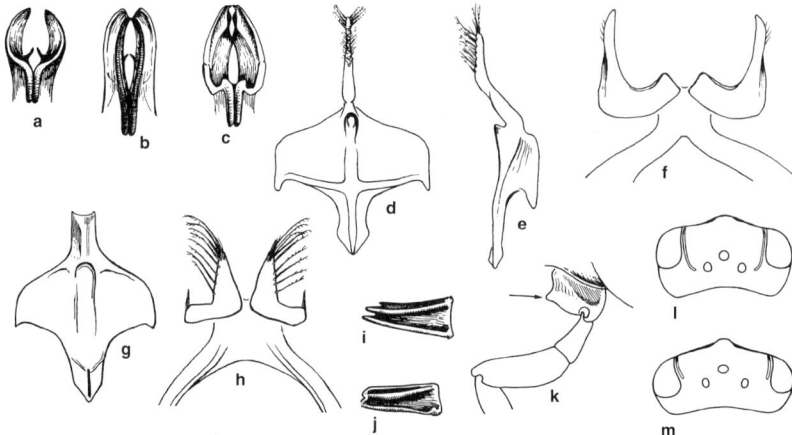

Figure 47-11. Structures of palearctic *Hylaeus*, a-h, males; i-m, females. **a,** *Hylaeus (Dentigera) brevicornis* Nylander, penis valves, dorsal view; **b,** *H. (Dentigera) pilosulus* (Pérez), penis valves, dorsal view; **c,** *H. (Prosopis) confusus* Nylander, penis valves, dorsal view; **d, e,** *H. (Nesoprosopis) pectoralis* Förster, S8, ventral and lateral views; **f,** *H. (Nesoprosopis) pectoralis* Förster, S7, dorsal view; **g,** *H. (Prosopis) hyrcanius* Dathe, S8, ventral view; **h,** *H. (Prosopis) hyrcanius* Dathe, S7, dorsal view; **i,** *H. (Lampdopsis) annularis* (Kirby), left mandible; **j,** *H. (Hylaeus) leptocephalus* (Morawitz), left mandible; **k,** *H. (Mehelyana) friesei* (Alfken), left anterior leg, coxal process marked by arrow; **l,** *H. (Paraprosopis) sinuatus* (Schenck), dorsal view of vertex with elongate facial foveae; **m,** *H. (Hylaeus) communis* Nylander, vertex with short facial foveae. From Dathe, 1980a, except g and h, from Dathe, 1980b, and d-f, original by H. Dathe.

9(8). Mesepisternum with strikingly coarse and strong but regular pitlike punctures; T1 without lateral fringe (T1 and T2 polished) .... *H. (Nesoprosopis)*
—. Mesepisternum with fine punctures or, *if* coarse and strong, then not regular and metasomal terga densely punctate; T1 often with lateral fringe........................... 10
10(9). Larger species (body length 5-9 mm) usually with short head; facial fovea usually with upper end well separated from eye; propodeum short, at least lateral and posterior surfaces mostly rounded, not delimited by carinae, with fine sculpture and covered with white feltlike hairs; T1 often with lateral fringe of white hairs, but *if* fringe absent, then metasomal base sometimes red ....
.... *H. (Prosopis)*
—. Smaller species (body length 3.5-8.0 mm) with elongated head; facial fovea shorter, upper end close to eye margin; propodeum usually sharp-edged or with carina around posterior surface; metasoma black, usually without lateral fringe on T1, but *if* T1 has fringe, then mesepisternum densely, finely punctate and propodeum rounded, with fine sculpture 11
11(10). T1 usually without lateral fringe and propodeum usually sharp-edged or carinate; *if* fringe on T1 present, then following terga without hair bands and propodeum rounded, with fine sculpture .... *H. (Hylaeus s. str.)*
—. T1 and T2 with lateral fringe, margins of T3 and T4 with bands of white hairs; propodeum rounded, partially with weak radial carinae.................................
.... *H. (Lambdopsis)* (in part, East Asian species)

Key to the Sub-Saharan Subgenera of *Hylaeus*
(Partly modified from Snelling, 1985)

The *Prosopisteron* species in the Australian region do not all agree with the characterization in couplet 1. The African species was introduced and is known only from the south coast of Cape Province.

1. Supraclypeal area gently sloping from midline to antennal sockets, not laterally margined; propodeum smooth, densely tessellate, without defined basal area; entire body densely tessellate, with conspicuous punctures; S7 of male with four similar, hairy apical lobes ....
.... *H. (Prosopisteron)*
—. Supraclypeal area elevated between antennal sockets and laterally margined; propodeum with defined basal area, usually coarsely rugose or roughened or sharply punctate, at least in part; S7 of male with two apical lobes or four that are dissimilar, at least one pair hairless 2
2(1). Apex of mandible acute, without distinct teeth; mandible elongate, slender, without grooves and ridges on outer surface ....*H. (Nothylaeus)*
—. Apex of mandible transverse or oblique, two- or three-toothed; mandible short and broad, with the usual grooves and ridges on outer surface 3
3(2). Integument very coarsely punctate; scutellum and metanotum each usually with a pair of spines (Fig. 47-12b); occipital carina present, sharp; omaulus sharply carinate... *H. (Metylaeus)*
—. Integument variously punctate; scutellum and metanotum without lateral spines; occipital carina often absent; omaulus not sharply carinate but sometimes with an obscure ridge.. 4
4(3). S7 with two apical lobes, these small, directed laterally or basolaterally, with only small setae; gonoforceps of male with distal one-fifth or more narrowed, attenuate, much exceeding apex of penis valve .... *H. (Alfkenylaeus)*
—. S7 with four apical lobes, proximal ones usually with coarse to very coarse setae; gonoforceps of male termi-

nating bluntly at about level of apex of penis valve (attenuate in some species of *Deranchylaeus*) 5
5(4). T1-T3 with abundant erect hairs on discs; proximal apical lobe of S7 of male without setae or with a median row of rather small, thickened setae .... *H. (Cornylaeus)*
—. T1-T3 with or without a few erect hairs on discs; proximal apical lobe of S7 of male with marginal row of large, very coarse setae .... *H. (Deranchylaeus)*

Hylaeus / Subgenus Abrupta Méhelÿ

Prosopis (Abrupta) Méhelÿ, 1935: 32, 137. Type species: *Hylaeus cornutus* Curtis, 1831, monobasic. [For date and authorship of this subgenus, see Michener, 1997b.]

The facial modifications as well as the male genitalia (Fig. 47-10a), as indicated in the key, are distinctive. The gonostylus of the male is perhaps distinctly separated from the gonocoxite (Fig. 47-10a), as in the subgenus *Mehelyana*, but is short, broad, and truncate. The male genitalia and other structures were illustrated by Méhelÿ (1935) and Dathe (1980a). This subgenus does not seem to be closely related to any other.

■ *Abrupta* occurs from Portugal and Morocco east through Europe, northern Africa, and southwestern Asia to Iran and Turkmenistan, north to 55°N in Denmark. The only species is *Hylaeus cornutus* Curtis.

Hylaeus / Subgenus Alfkenylaeus Snelling

Hylaeus (Alfkenylaeus) Snelling, 1985: 13. Type species: *Hylaeus namaquensis* Cockerell, 1942, by original designation.

Of the African subgenera with attenuate apices of the male gonoforceps (Fig. 47-13a), this is the one most similar to ordinary *Hylaeus*. Females are not separable by subgeneric characters from the subgenus *Deranchylaeus*. The body is coarsely punctate. The disc of S7 of the male is rather broad, with two small, hairless or nearly hairless lateroapical lobes directed laterobasally. The apical process of S8 of the male is slender, apically enlarged. The body length is 6 to 8 mm. Male genitalia and other structures were illustrated by Snelling (1985).

■ *Alfkenylaeus* occurs in Africa from Kenya, Upper Volta, and Senegal to the Transvaal in South Africa. The four species were revised by Snelling (1985). A fifth species, *Hylaeus arnoldi* (Friese), agrees with *Alfkenylaeus* in most features but the male has a less prolonged attenuation of the gonoforceps and S8 is of a totally different form, having two long, broad apical lobes instead of a median apical process.

Hylaeus / Subgenus Analastoroides Rayment

Analastoroides Rayment, 1950: 20. Type species: *Analastoroides foveata* Rayment, 1950, by original designation.

This subgenus (originally called a genus, but see Houston, 1981a) is unique among Hylaeinae in the bands of dense orange tomentum on T1 and T3; this feature, along with size (body length 8.0-12.5 mm) results in a remarkable superficial resemblance to some species of the genus *Hyleoides* and to some wasps of the genus *Alastor*. The strongly humped T2, especially in the male (Fig. 47-3g shows the female), and the constricted apex of T1 suggest the subgenus *Euprosopellus*, but these characters probably arose independently. In *Analastoroides* the male S7 has two pairs of hairy apical lobes, shaped and attached much as in *Prosopisteron* (Fig. 47-9f), whereas in *Euprosopellus* there is only one pair of lobes, directed laterally but broadly fused to the body of the sternum (Fig. 47-9i). Male genitalia and hidden sterna were illustrated by Houston (1981a).

■ *Analastoroides* is from the coastal zone of Victoria and New South Wales, Australia. The only known species is *Hylaeus foveatus* (Rayment).

Hylaeus / Subgenus Cephalylaeus Michener

Michener, 1942a: 273. Type species: *Prosopis basalis* Smith, 1853, by original designation.

Rather large (7-8 mm long) in comparison to other Nearctic *Hylaeus*, the female lacks pale markings and the male has a greatly enlarged scape. The two apical lobes of S7 of the male are broadly fused to an unusually wide sternal body. The apical process of S8 is reduced to a small, hairless point only about as long as its basal width. The male gonoforceps are widest apically, amd are provided with an apical crescentic zone of long hairs and a preapical rounded mesal projection on the ventral side. Metz (1911), Mitchell (1960), and Snelling (1968) illustrated the male genitalia and hidden sterna.

■ *Cephalylaeus* ranges across Canada (British Columbia to Newfoundland) and the northern USA, southward to California and in mountain ranges to Colorado. The two species were reviewed by Snelling (1968).

Hylaeus / Subgenus Cephylaeus Moure

Hylaeus (Cephylaeus) Moure, 1972: 280. Type species: *Hylaeus larocai* Moure, 1972, by original designation.

This subgenus consists of a small (length 5 mm) Brazilian species that lacks strong carinae. It may be related to the subgenus *Hylaeana;* specimens have not been available and should be reexamined. They might have an omaular carina below the lower end of the episternal groove, as does *Hylaeana;* this carina was not observed when *Hylaeana* was originally described and could have been missed when *Cephylaeus* was described. Male genitalia and other structures were illustrated by Moure (1972).

■ *Cephylaeus* is known from Paraná, Brazil. The only species is *Hylaeus larocai* Moure.

Hylaeus / Subgenus Cornylaeus Snelling

Hylaeus (Cornylaeus) Snelling, 1985: 8. Type species: *Prosopis aterrima* Friese, 1911, by original designation.

Cornylaeus seems closely related to another African subgenus, *Deranchylaeus*. Its main characters are its larger size (6.0-8.5 mm long), especially of the metasoma of the male, the abundant erect hair on the metasomal terga, and the presence, at least in the largest males, of a shining black prominence on each side of T3 and of S3. These same features occur in the largest males only of various species of *Gnathoprosopis* (Houston, 1981a), and in that case are not even specific characters. I have retained the name *Cornylaeus*, however, because in addition to the above mentioned characters, which could be allometric results of large body size, *Cornylaeus* differs from *Deranchylaeus* in characters of the male's hidden sterna, as follows: 1. S7 lacks setae on the proximal as well as the distal apical lobe, as in *Hylaeus aterrimus* (Friese), or has a row of small, thickened setae arising from the surface of each lobe. 2. The large marginal setae found in *Deranchylaeus* are absent. 3. The posterior margin of the disc of S8 is produced as a rounded shoulder lateral to the rather short, blunt median apical process, as in *Nothylaeus* (Fig. 47-13b). The male genitalia and other structures were illustrated by Snelling (1985).

■ This subgenus ranges from the Congo and Zimbabwe to the Transvaal, Natal, and the Transkei of South Africa. Its two species were revised by Snelling (1985).

Hylaeus / Subgenus *Dentigera* Popov

Prosopis (Dentigera) Méhelÿ, 1935: 45, 151. [Invalid because no type species was designated.]
Prosopis (Imperfecta) Méhelÿ, 1935: 48, 154 (part). [Invalid because no type species was designated; for a note on a later supposed designation, see Michener, 1997b.]
Prosopis (Dentigera) Popov, 1939a: 168. Type species: *Hylaeus brevicornis* Nylander, 1852, by original designation.

The distinctive male genitalic and sternal characters are indicated in the key and were illustrated by Méhelÿ (1935) and Dathe (1980a); see also Figure 47-11a, b. H. Dathe (in litt., 1990) remarks that *Dentigera* consists mainly of two species groups (*brachycephalus* group and *brevicornis* group) that are so different that they could be separated subgenerically. Females appear separately in the key to subgenera.

■ *Dentigera* occurs from Portugal to Iran and central Asia, north to 64° latitude, and south to the Mediterranean basin and India. It comprises 12 European species (Dathe, 1980a) and an estimated total of 20 species, the main concentration being in the Mediterranean area.

Hylaeus / Subgenus *Deranchylaeus* Bridwell

Hylaeus (Deranchylaeus) Bridwell, 1919: 136. Type species: *Prosopis curvicarinata* Cameron, 1905, by original designation.

This subgenus consists of small to middle-sized (4.0-6.5 mm long), coarsely punctate species. The most distinctive characters are in S7 of the male, which has four apical lobes, all somewhat attenuate, the distal ones without setae, the proximal ones with a largely marginal row of coarse setae; S8 has only a short, blunt apical process. The male genitalia and other structures were illustrated by Snelling (1985). Snelling (1985) indicated in his key to subgenera that in one or more species of *Deranchylaeus* the male gonoforceps are attenuate apically, as in the subgenera *Alfkenylaeus* and *Nothylaeus*, and he has verified this (in litt, 1990). Most species, however, have more or less blunt gonoforceps ending at about the level of the apices of the penis valves.

■ *Deranchylaeus* is widespread in sub-Saharan Africa. Although many species are South African, others are found in Kenya, the Congo basin, Nigeria, etc. Snelling (1985) listed 49 specific names for this subgenus, and Bridwell (1919) gave comments on numerous species.

Hylaeus / Subgenus *Edriohylaeus* Michener

Hylaeus (Edriohylaeus) Michener, 1965b: 124. Type species: *Hylaeus ofarrelli* Michener, 1965, by original designation.

This subgenus contains a single Australian species that is almost as small (body length 3.7-4.0 mm) and slender as species of the subgenera *Heterapoides* and *Gephyrohylaeus*. It also resembles minute species of the subgenus *Prosopisteron*. S7 of the male has only two, posteriorly directed, hairless apical lobes broadly fused to its rather broad disc (Fig. 47-9c) (this is unique in the Hylaeinae). The propodeum has a smooth horizontal dorsal surface rounding onto and as long as the vertical surface, the triangle being not or feebly defined and nearly all on the dorsal surface, as shown by the high position of the propodeal pit. Some species of *Prosopisteron* approach *Edriohylaeus* in the propodeal characters, the characters of S7 being the only thoroughly distinctive ones. *Hylaeus ofarrelli* Michener could be regarded as a derived species of *Prosopisteron*. Male genitalia and hidden sterna were illustrated by Michener (1965b) and Houston (1981a); see also Figure 47-9a-c.

■ *Edriohylaeus* is found in coastal and mountain regions of eastern Australia from Victoria to northern Queensland. The only species is *Hylaeus ofarrelli* Michener.

Hylaeus / Subgenus *Euprosopellus* Michener

Hylaeus (Euprosopellus) Michener, 1965b: 132. Type species: *Prosopis dromedaria* Cockerell, 1910, by original designation.

This is a subgenus of moderate-sized (7-12 mm long), nonmetallic species with yellow on the scutellum. One of the most distinctive features is the large dorsal hump on T2 (Fig. 47-3k) (weak in the female). The propodeum has a single row of pits across its base; behind this row the surface continues horizontally, gradually curving onto the vertical surface. S7 of the male has only two apical lobes, directed laterally and broadly fused to the narrow body of the sternum (Fig. 47-9i). Male genitalia and other structures were illustrated by Michener (1965b) and Houston (1981a).

■ This subgenus occurs across southern Australia and north on the Pacific coast to southern Queensland. The four species were revised by Houston (1981a) and listed by Cardale (1993).

Hylaeus / Subgenus *Euprosopis* Perkins

Euprosopis Perkins, 1912: 106. Type species: *Prosopis husela* Cockerell, 1910, by original designation.

This subgenus consists of moderate-sized (5-9 mm long), somewhat robust species, nonmetallic or dark metallic blue or green, often with reddish areas on the metasoma, and with extensive yellow areas on the face and sometimes on the thorax. The propodeum has a row of areoli across the base occupying the short, horizontal surface, which is margined posteriorly by a carina behind which the surface drops abruptly; the propodeal triangle is therefore mostly on the subvertical surface. S7 is unmistakable: its distal apical lobes are long, curved outward, and ribbonlike (Fig. 47-9k); and its proximal apical lobes are large and much expanded anteroposteriorly. Male genitalia and other structures were illustrated by Michener (1965b) and Houston (1981a).

■ Some of the commonest Australian *Hylaeus* are in this subgenus, which is found throughout Australia including Tasmania. The five species (and many synonyms) were revised by Houston (1981a) and listed by Cardale (1993).

Hylaeus / Subgenus *Euprosopoides* Michener

Hylaeus (Euprosopoides) Michener, 1965b: 131. Type species: *Prosopis fulvicornis* Smith, 1853 = *Prosopis ruficeps* Smith, 1853, by original designation.

Like the subgenus *Euprosopis,* this subgenus consists of medium-sized (6-13 mm long), rather robust bees, often with bluish or greenish tints, with the propodeum in profile mostly declivous, there being only a narrow basal or dorsal horizontal zone bearing a series of areoli and margined posteriorly by a carina. *Euprosopoides* differs most profoundly from *Euprosopis* in that S7 of the male has four hairy lobes (Fig. 47-9h), as in the subgenus *Prosopisteron,* except that the proximal lobes in most species of *Euprosopoides* are deeply divided, so that one might recognize six apical lobes. The apex of the apical process of S8 is deeply divided to form two dark lobes bearing long, plumose hairs (Fig. 47-9g). The malar area is broader than long; in *Euprosopis* it is ordinarily longer than broad. An unusual feature of *Euprosopoides* is the indication of a pygidial plate in the female. In some species it is rather well defined; in others the plate is indicated only by a carina across the apex of T6. Male genitalia and other structures were illustrated by Krombein (1950), Michener (1965b), and Houston (1981a).

■ *Euprosopoides* is widespread in Australia, including Tasmania as well as the northernmost parts of the continent, but may be absent in the central and northwestern parts of the continent. It also occurs in Micronesia. The seven Australian species were revised by Houston (1981a) and listed by Cardale (1993); the three Micronesian species were revised by Krombein (1950).

The occurrence of an Australian subgenus in Micronesia is unexpected, although Houston (1981a) called attention to the similarity in male genitalia and hidden sterna, as shown by Krombein's (1950) illustrations of Micronesian species. I have examined specimens of *Hylaeus guamensis* (Cockerell) and *H. yapensis* (Yasumatsu) and recognize no subgeneric characters that might differentiate these species from Australian *Euprosopoides.* They are of different species, however, and their occurrence in Micronesia is unlikely to be a result of introduction from Australia by human agency.

Hylaeus / Subgenus *Gephyrohylaeus* Michener

Gephyrohylaeus Michener, 1965b: 138. Type species: *Heterapis sandacanensis* Cockerell, 1919, by original designation.

The species of this subgenus are as small (3.0-3.5 mm long) and nearly as slender as those of the subgenus *Heterapoides.* The fusion of the second submarginal and second medial cells (Fig. 47-5b) is also as in *Heterapoides,* suggesting that these taxa are closely related. Indeed, most features of *Gephyrohylaeus* are derived relative to *Heterapoides,* and the latter may be paraphyletic, as the group from which *Gephyrohylaeus* arose. The mesepisternum attains the propodeum or nearly so, and the lower part of metepisternum is thus absent or merely a narrow strip. The preoccipital carina or sharp ridge is present. The facial fovea of the female is reduced to a short groove near the widest part of the face (Fig. 47-6d); the male has a pit in the same position. The distal lobes of S7 of the male are large, with thickened curved hairs at the apices; the proximal lobes are small. Male genitalia and other structures were illustrated by Michener (1965b) and Houston (1975a), the wing and thorax by Hirashima (1967b).

■ This rare subgenus is found in central and northern Queensland, Australia, and in New Guinea, Borneo, and the Philippines. It contains two or perhaps three species, *Hylaeus sculptus* (Cockerell) from Australia (Houston, 1975a), *H. sandacanensis* (Cockerell) from New Guinea and Borneo (Michener, 1965b), and perhaps a third species from Culion Island in the Philippines (Hirashima, 1967b).

Hylaeus / Subgenus *Gnathoprosopis* Perkins

Gnathoprosopis Perkins, 1912: 104. Type species: *Prosopis xanthopoda* Cockerell, 1910 = *P. euxantha* Cockerell, 1910, by original designation.

This subgenus contains small (3.5-7.0 mm) metallic blue or nonmetallic black species. The mandibles of both sexes, almost as broad as long, are rectangular because of the expansion of the pollex (probably) to form a gently convex apical margin much wider than the apex of the rutellum. The preoccipital carina is present. The horizontal base of the propodeum is areolate and separated from the declivous surface by a carina. S7 of the male has only two distinct apical lobes, hairless or nearly so, and directed posterolaterally; a straplike appendage may represent the proximal lobe. S8 of the male is produced posteriorly on each side of the median apical process. Male genitalia and other structures were illustrated by Michener (1965b) and Houston (1981a).

■ This subgenus is found throughout Australia. The seven species were revised by Houston (1981a) and listed by Cardale (1993). The Australian *Hylaeus (Gnathoprosopis) albonitens* (Cockerell) is now widespread in the Hawaiin Islands (Snelling, 2003).

Hylaeus / Subgenus *Gnathoprosopoides* Michener

Hylaeus (Gnathoprosopoides) Michener, 1965b: 127. Type species: *Prosopis eburniella* Cockerell, 1912 = *Prosopis philoleucus* Cockerell, 1910, by original designation.

Gnathoprosopoides is a close relative of *Gnathoprosopis* as shown by the short, broad mandibles (less extreme and, especially in males, narrower apically than basally) and by the presence of a preoccipital carina. The body length is 4.5 to 6.5 mm. *Gnathoprosopoides* differs from *Gnathoprosopis* in the relatively smooth, nonareolate propodeum, its basal horizontal surface curving gradually onto the declivous surface without an intervening carina. S7 of the male has four broad, hairy, apical lobes, the proximal ones expanded anteroposteriorly; *Gnathoprosopis* has two largely hairless lobes. The body of S8 lacks a posterior expansion on either side of the apical process. Male genitalia and associated structures were illustrated by Michener (1965b) and Houston (1981a). The separation of *Gnathoprosopis* and *Gnathoprosopoides* can be justified principally on the basis of S7 of the male; the two subgenera are probably sister groups and could be united, especially in view of the small number of species involved.

■ *Gnathoprosopoides* occurs from Tasmania and eastern South Australia to northern Queensland. The two species were revised by Houston (1981a) and listed by Cardale (1993).

Hylaeus / Subgenus *Gnathylaeus* Bridwell

Gnathylaeus Bridwell, 1919: 126, 133. Type species: *Gnathylaeus williamsi* Bridwell, 1919, by original designation.

This subgenus has short, broad mandibles, rounded at the apices, suggestive of those of the subgenus *Gnathoprosopis* but with channels on the outer surface.

■ *Gnathylaeus* is from the Philippine Islands. The single species, *Hylaeus williamsi* (Bridwell), is known only from females.

Hylaeus / Subgenus *Gongyloprosopis* Snelling

Hylaeus (Gongyloprosopis) Snelling, 1982: 16. Type species: *Prosopis cruenta* Vachal, 1910, by original designation.

In this subgenus the pronotal collar lacks a transverse crest or ridge; T1 and T2 are shiny between fine, scattered punctures; the scape of the male is swollen; and the frons of the male has a broad zone of dense, short, plumose hairs. S7 of the male has a rather broad, short body with two rather small, hairless or sparsely haired, variously shaped apical lobes. The apical process of S8 of the male is strong and hairless, pointed or narrowly rounded; the body of the sternum extends posteriorly as a rounded shoulder on either side of the apical process. Male genitalia and other structures were illustrated by Snelling (1982).

■ This subgenus is widespread in South America, at least from Bolivia and Argentina to Trinidad. The five species were reviewed by Snelling (1982).

The male of *Hylaeus cruentus* (Vachal) has a large process arising medially on the ventral mesal margin of the gonoforceps. Other species lack such a structure.

Hylaeus / Subgenus *Heterapoides* Sandhouse

Heterapis Cockerell, 1911a: 140 (not Linston, 1889). Type species: *Heterapis delicata* Cockerell, 1911, by original designation.

Heterapoides Sandhouse, 1943: 557, replacement for *Heterapis* Cockerell, 1911. Type species: *Heterapis delicata* Cockerell, 1911, autobasic.

The most slender and delicately built of all minute bees (3-4 mm long) belong to this subgenus; they probably nest in very narrow burrows. Figure 28-3a illustrates the slender thorax. The outstanding character of this subgenus and of *Gephyrohylaeus* is the confluence of the second submarginal and second medial cells of the forewing (Fig. 47-5b). This is a unique synapomorphy, found nowhere else among bees. It has been considered heretofore as a generic character, but the other characters are all duplicated or approximated in other subgenera of *Hylaeus*. For this reason, *Heterapoides* is here regarded as a subgenus of *Hylaeus*. Its male genitalia and other structures were illustrated by Michener (1965b) and Houston (1975a).

■ *Heterapoides* is found principally in and east of the Great Dividing Range in Australia, from South Australia and Victoria to Queensland. The eight species were revised by Houston (1975a) and listed by Cardale (1993).

Hylaeus / Subgenus *Hoploprosopis* Hedicke

Prosopis (Hoploprosopis) Hedicke, 1926: 415. Type species: *Prosopis quadricornis* Hedicke, 1926, by original designation.

Like the African subgenus *Metylaeus* and some *Nothylaeus*, *Hoploprosopis* (which is known only from males) has a large spine directed posteriorly from each side of the scutellum, and another from each side of the metanotum (illustrated in Fig. 47-12a and by Snelling, 1969). Moreover, the extremely coarse punctation is as in *Metylaeus*, as is the presence of a preoccipital carina and an omaular carina. As concluded by Snelling, the differences among forms having such spines are so great as to indicate that the possession of the spines is probably convergent. Other characters are unlike *Metylaeus*: the short scape of the male (shorter than the interantennal distance); the somewhat elongate, flattened (but not apically attenuate) gonoforceps of the male; the lack of the usual apical lobes of S7 of the male (only lateral projections with a few short hairs, broadly attached to the body of sternum); and the long, slender, apically expanded, and strongly hairy apical process of S8.

■ *Hoploprosopis* is known from the Philippine Islands. The single known species is *Hylaeus quadricornis* (Hedicke).

It should be remembered that by analogy with *Metylaeus*, females, or some of them, when found, are likely to lack the large thoracic spines.

Hylaeus / Subgenus *Hylaeana* Michener

Hylaeus (Hylaeana) Michener, 1954b: 28. Type species: *Hylaeus panamensis* Michener, 1954, by original designation.

In this subgenus of small species, 3.5 to 4.5 mm in length, the thorax is strongly punctate, but the discs of T1 and T2 are impunctate or have scattered minute punctures on a tessellate or lineolate, dull or weakly shining

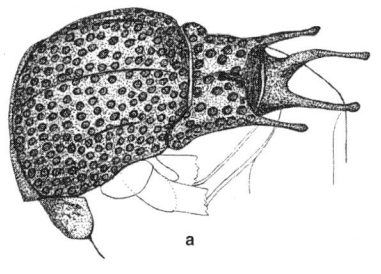 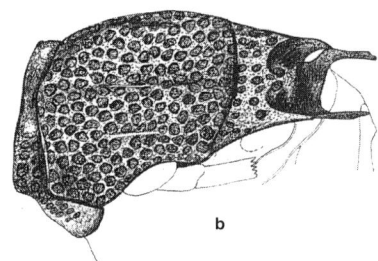

Figure 47-12. Dorsolateral views of thoraces of two male *Hylaeus* that appear to have developed, independently, large scutellar and metanotal spines and coarse punctation. **a,** *H. (Hoploprosopis) quadricornis* Hedicke from the Philippines; **b,** *H. (Metylaeus) cribratus* Bridwell from Africa. From Snelling, 1969.

surface. S7 of the male has two rather large, rounded apical lobes with hairs on the distal margins. S8 has a large apical process, somewhat expanded and truncate or weakly notched apically, with abundant plumose hairs laterally on the distal half of the process. The male genitalia and other structures were illustrated by Michener (1954b) and Snelling (1982).

■ This is a neotropical subgenus with one species, *Hylaeus panamensis* Michener, that reaches the southwestern USA from California to Texas (Snelling, 1966a, 1968). Other species occur in the Antilles (at least Jamaica) and in Central and South America at least as far south as Brazil. Snelling (1982) lists nine species, but there are probably at least ten more species in South America.

Hylaeus / Subgenus *Hylaeopsis* Michener

Hylaeus (Hylaeopsis) Michener, 1954b: 27. Type species: *Prosopis mexicana* Cresson, 1869, by original designation.

This subgenus differs from other American species of the genus *Hylaeus* in that the scutellum is commonly yellow. The preoccipital carina and the omaular carina are distinct, and the coarsely areolate basal area of the propodeum is separated from the posterior surface by a carina. S7 of the male has two apical lobes bearing small slender hairs, the lobes extending laterally, then basad, as in the subgenera *Prosopella* and *Lambdopsis*. In *Hylaeus (Hylaeopsis) cecidonastes* Moure from Brazil these lobes are greatly reduced. S8 of the male has a moderately long, hairless apical process, somewhat enlarged distally. The male gonostyli are somewhat attenuate apically, exceeding the penis valves; this and other structures were illustrated by Michener (1954b) and Moure (1972).

■ This large subgenus is widespread in South America and ranges into tropical Mexico at least as far north as Sonora. Snelling (1982) lists 13 described species for Central America and Mexico; there may well be 25 or more species in all.

Sakagami and Zucchi (1978) gave an account of the structure of, and behavior at, the nests of *Hylaeus (Hylaeopsis) tricolor* (Schrottky) in abandoned cells of old wasp nests (*Trypoxylon* and *Mischocyttarus*). These authors considered the bee communal or quasisocial; see the account under the genus *Hylaeus*.

Hylaeus / Subgenus *Hylaeorhiza* Michener

Hylaeorhiza Michener, 1965b: 141. Type species: *Prosopis nubilosa* Smith, 1853, by original designation.

This subgenus occupies a special position in the genus *Hylaeus*. The name was proposed at the genus level because of *Palaeorhiza*-like features; indeed, Meade-Waldo (1923) included the type species in the genus *Palaeorhiza*. Such features are the smoothly rounded propodeal profile, the ridge on the mesepisternum in front of the middle coxa, and the slender, almost straplike distal apical lobes of S7 of the male, in contrast to the basal lobes (Fig. 47-9l). *Hylaeorhiza* could be the sister group of *Palaeorhiza*. It is superficially very like *Amphylaeus (Agogenohylaeus)*. The male glossa is broad and short, as in most *Hylaeus*, but has a small median apical point (Fig. 47-1g), somewhat suggestive of *Amphylaeus*. The body is nonmetallic and moderate-sized (6-9 mm long), with yellow markings, including those on the scutellum and metanotum. The male genitalia and other structures were illustrated by Michener (1965b) and Houston (1981a).

■ *Hylaeorhiza* occurs in Tasmania and continental Australia, from Victoria and eastern South Australia to northern Queensland. The single species is *Hylaeus nubilosus* (Smith).

Hylaeus / Subgenus *Hylaeteron* Michener

Hylaeus (Hylaeteron) Michener, 1965b: 126. Type species: *Prosopis pulchricrus* Cockerell, 1915 = *Euryglossa semirufa* Cockerell, 1914, by original designation.

This subgenus, consisting of rather small (4-8 mm), robust species, resembles the subgenus *Pseudhylaeus* in having the outer ridge of the mandible, which is unusually strong and elevated as a carina. The propodeum is areolate at least basally. The female has tridentate mandibles and sparse capitate or spatulate hairs on the fore tarsi. The distal lobes of S7 of the male are broad and rounded, in contrast to the much smaller angulate or pointed proximal lobes. Male genitalia and other structures were illustrated by Michener (1965b) and Houston (1981a).

■ *Hylaeteron* is widespread in Australia, mostly in dry parts. The five species of this subgenus were revised by Houston (1981a) and listed by Cardale (1993).

Unlike most species of *Hylaeus*, which probably visit diverse flowers, species of *Hylaeteron* seem to be specialist feeders on *Grevillea* (Proteaceae).

Hylaeus / Subgenus *Hylaeus* Fabricius s. str.

Hylaeus Fabricius, 1793: 302. Type species: *Prosopis annulata* Fabricius, 1804 = *Apis annulata* Linnaeus, 1758, by designation of Latreille, 1810: 438. [For a later designation, see Michener, 1997b.]

Prosopis (Pectinata) Méhelÿ, 1935: 54, 161. Invalid because no type species was designated. [For a later supposed designation, see Michener, 1997b.]

Prosopis (Trichota) Méhelÿ, 1935: 63, 169. Invalid because no type species was designated. [For a later supposed designation, see Michener, 1997b.]

Prosopis (Nylaeus) Popov, 1939a: 169. Error for *Hylaeus* Fabricius, 1793.

Prosopis (Patagiata) Blüthgen, 1949: 77. Type species: *Prosopis difformis* Eversmann, 1852, by original designation.

Hylaeus (Nesohylaeus) Ikudome, 1989: 125. Type species: *Hylaeus niger* Bridwell, 1919, by original designation.

This is the Cressoni Division of Metz (1911). *Nesohylaeus* and *Patagiata* are probably derivitives of *Hylaeus* s. str. and are here considered synonyms of *Hylaeus* s. str.

In this subgenus, S7 of the male has four apical lobes. Those of one pair, directed laterad, are usually pointed, with slender hairs, but are reduced and difficult to see in *Hylaeus niger* Bridwell, i.e., *Nesohylaeus;* those of the distal pair are much broadened and their distal margins are straight or gently concave, and pectinate because of a series of flattened, apically pointed and usually hooked, hairlike processes (Fig. 47-10g). The apical process of S8 usually ends in two small lobes. Male genitalia and sterna were illustrated by Metz (1911), Méhelÿ (1935), Mitchell (1960), and Dathe (1980a, 1986); see Figure 47-10e, g. The supraclypeal mark of the male is slender, over one-half as long as the clypeus (absent in *H. niger*), and the antennal scape of many species is broadened, concave beneath. In addition to the characters listed above for *H. niger (Nesohylaeus)*, it is unusual in the wholly black face of the female and the lack of yellow on the head of the male except for the clypeus. The wholly black paraocular areas are most unusual. *H. difformis* Eversmann *(Patagiata)* is even more distinctive, because of its strong preoccipital carina and a translucent process at the apex of the male gonostylus.

■ This holarctic subgenus ranges in North America from coast to coast and from Alaska and Canada (British Columbia to Nova Scotia) south to Baja California and Coahuila, Mexico. In the palearctic region, *Hylaeus* s. str. occurs from Spain to eastern Siberia and Japan. Snelling (1970) listed 12 species in North America; one of them was adventive from Europe [*H. leptocephalus* (Morawitz) = *bisinuatus* Forster = *stevensi* (Crawford)]. *H. leptocephalus* (Morawitz) has also reached Hawaii, presumably from North America; in Hawaii it is known only from Oahu (Snelling, 2003). The subgenus is more richly represented in the palearctic region; there are 14 species from Europe but only three from Japan. The greatest known concentration of species is in the steppe and desertic areas of Asia, for Dathe (1980b, 1986) records 25 species from Mongolia and 14 from Iran; he believes there to be 80 palearctic species. Revisional or review papers are, for Europe, Dathe (1980a); for Japan, Ikudome (1989); and for North America, Metz (1911) and Snelling (1970).

Hylaeus / Subgenus *Koptogaster* Alfken

Prosopis (Koptogaster) Alfken, 1912: 23. Type species: *Prosopis bifasciata* Jurine, 1807, by designation of Meade-Waldo, 1923: 16.

Prosopis (Pseudobranchiata) Méhelÿ, 1935: 33, 139. Invalid because no type species was designated. [For a later supposed designation, see Michener, 1997b.]

Prosopis (Koptobaster) Popov, 1939a: 168, error for *Koptogaster* Alfken, 1912.

The coarse punctation of the scutum, scutellum, and T1, the elongate head (longer than broad), the sharply truncate dorsolateral angles of the pronotum, and the four broad, hairless apical lobes of S7 of the male are distinctive. Male genitalia and sterna were illustrated by Méhelÿ (1935) and Dathe (1980a); see also Figure 47-10d.

■ This subgenus occurs on the northern Mediterranean coast from France to the Middle East, north to Denmark, east through the Caucasus, Ukraine, and northern Iran. The two species were reviewed by Dathe (1980a).

Hylaeus / Subgenus *Laccohylaeus* Houston

Hylaeus (Laccohylaeus) Houston, 1981a: 88. Type species: *Prosopis cyanophila* Cockerell, 1910, by original designation.

This subgenus resembles the subgenus *Euprosopoides*, from which it differs in the simple, hairless apical process of S8 of the male, the presence of more than a single row of areoli across the base of the propodeum, and the lack of yellow on the scutellum and metanotum. The background color is black, faintly metallic blue on the metasoma; the body length is 5.0 to 6.5 mm. S7 of the male is like that of the subgenus *Prosopisteron* except that the hairs of the distal lobe are represented by a row of thickened bristles. The male genitalia and other structures were illustrated by Houston (1981a).

■ *Laccohylaeus* is known from coastal localities from southern to northern Queensland. The only species is *Hylaeus cyanophilus* (Cockerell).

Hylaeus / Subgenus *Lambdopsis* Popov

Prosopis (Lambdopsis) Méhelÿ, 1935: 65, 171. Invalid because no type species was designated.

Prosopis (Lambdopsis) Popov, 1939a: 169. Type species: *Melitta annularis* Kirby, 1802, by original designation.

Hylaeus (Boreopsis) Ikudome, 1991: 790. Type species: *Hylaeus macilentus* Ikudome, 1989, by original designation.

Hylaeus (Noteopsis) Ikudome, 1991: 791. Type species: *Hylaeus nanseiensis* Ikudome, 1989, by original designation.

The species placed in the subgenera *Boreopsis* and *Noteopsis* by Ikudome are included here with hesitation. They broaden the definition of *Lambdopsis*, as shown by comments on certain characters below, but the resemblance to typical *Lambdopsis* is shown in males by the small apical lobes of S7; the slender, bent down or hooked apex of S8 in *Boreopsis;* and the broad, bladelike penis valves of both *Boreopsis* and *Noteopsis*, although this character is variable, and in *Hylaeus ikedai* (Yasumatsu), a species placed in *Boreopsis*, the apices are attenuate. Typical conditions are shown in Figures 47-10h-k. Females of *Lambdopsis* typically lack the lateral face marks, or, if present, the marks are small, rounded, and attached to the clypeal margin, not to the orbits. In the Asiatic species placed in *Boreopsis* and *Noteopsis,* however, the lateral face marks of the female are well developed, but they are also in *Hylaeus (Lambdopsis) crassanus* (Warncke) from Europe.

This subgenus includes species in which the male scape is shieldlike [in *H. melba* (Warncke), from northwest Africa, it is not], and S7 has two small lobes, each somewhat contorted, attached to the short, usually broad body of the sternum. Each of the S7 lobes is more or less rectangular or triangular; each has a few short hairs in European species and in *Hylaeus (Lambdopsis) nipponicus* Bridwell and *nanseiensis* Ikudome from Japan, but the hairs are lacking in the Asiatic species placed in *Boreopsis* by Ikudome (1991). S8 of the male has a slender apical process, the pointed apex hooked downward in most species, but not in *H. (L.) nanseiensis.* Male genitalia and sterna were illustrated by Méhelÿ (1935), Dathe (1980a), and Ikudome (1989). The mandible of the female is three-toothed in European species (Fig. 47-11i) but two-toothed in the Asiatic species recently placed in *Boreopsis* and *Noteopsis.*

■ *Lambdopsis* ranges from Britain and Morocco to Mongolia, Siberia, and Japan, north to 63°N in Scandinavia, and south to Syria and the Caucasus. One species recently placed in *Noteopsis* is from Taiwan. There are six European species, about 18 in all (H. Dathe, in litt., 1996). European species were revised by Dathe (1980a).

Hylaeus / Subgenus *Macrohylaeus* Michener

Hylaeus (Macrohylaeus) Michener, 1965b: 133. Type species: *Prosopis vidua* Smith, 1853 = *Prosopis alcyonea* Erichson, 1842, by original designation.

Bees of this subgenus are rather large (8.5-11.0 mm) and robust, and have a metallic-blue metasoma. Noteworthy characters include the low pronotal collar; the virtual absence of the dorsolateral angles of the pronotum; the strong, carinate outer ridge of the mandible of the female, the ridge running parallel to the lower margin of the mandible almost to the base, then turning upward across the base of the mandible to the acetabulum; and the slender, parallel-sided stigma. S7 of the male is similar to that of males of the subgenus *Prosopisteron,* suggesting that *Macrohylaeus* may be a specialized derivative of that subgenus. The male genitalia and other structures were illustrated by Michener (1965b) and Houston (1981a).

■ *Macrohylaeus* ranges across the southern coast of Australia (including Tasmania, but is perhaps absent along the Australian Bight), north on the Pacific coast to southern Queensland. The only species is *Hylaeus alcyoneus* (Erichson).

Hylaeus / Subgenus *Meghylaeus* Cockerell

Hylaeus (Meghylaeus) Cockerell, 1929b: 314. Type species: *Palaeorhiza gigantea* Cockerell, 1926 = *Prosopis fijiensis* Cockerell, 1909, by original designation.

This subgenus is known only from females. The oldest specific name is based on an old, mislabeled specimen; the species almost certainly does not occur in Fiji in spite of the name. Another mislabeled specimen, the type of the synonymous *Prosopis chalybaea* Friese, was supposedly from New Zealand. This species, the largest of the hylaeine bees (12.5-15.0 mm long), is a robust, dark metallic blue. Unusual characters include the long prestigma, almost as long as the stigma from the base to vein r; the parallel-sided stigma; the nearly bare wings; the distinct malar space (Fig. 47-7g); and the tridentate mandible of the female, the teeth being subequal (in most other *Hylaeus* the lowest tooth is longest).

■ Aside from the mislabeled types, *Meghylaeus* is known from two localities on the coast of Victoria, Australia, and from a specimen labeled only "Tasmania." The only species is *Hylaeus fijiensis* (Cockerell).

Hylaeus / Subgenus *Mehelyana* Sandhouse

Prosopis (Barbata) Méhelÿ, 1935: 32, 138 (not Humphrey, 1797). Type species: *Prosopis friesei* Alfken, 1904, monobasic.
Prosopis (Mehelya) Popov, 1939a: 167 (not Csiki, 1903). Type species: *Prosopis friesei* Alfken, 1904, by original designation.
Mehelyana Sandhouse, 1943: 569, replacement for *Barbata* Méhelÿ, 1935, and *Mehelya* Popov, 1939. Type species: *Prosopis friesei* Alfken, 1904, autobasic.

The robust gonostylus of the *Mehelyana* male is distinctly separated from and about as long as the gonocoxite, and terminates in a luxuriant mass of plumose hairs (Fig. 47-10b). The separate gonostylus is supposedly a plesiomorphy, in *Hylaeus* shared only with the subgenus *Abrupta,* in which the separation is not well developed. The male genitalia and other structures were illustrated by Méhelÿ (1935) and Dathe (1980a).

■ *Mehelyana* occurs only in southeastern Europe, from Croatia to southern Romania and south to Greece. The only species is the rare *Hylaeus friesei* (Alfken).

Hylaeus / Subgenus *Metylaeus* Bridwell

Metylaeus Bridwell, 1919: 126, 131. Type species: *Metylaeus cribratus* Bridwell, 1919, by original designation.

A striking character of *Metylaeus* is that the known males and some females possess a pair of long, laterally compressed, posterolateral processes on the scutellum and a similar but usually smaller pair on the metanotum (Fig. 47-12b). The subgenera *Hoploprosopis* and some *Nothylaeus* have similar processes. Other characters are the presence of the preoccipital carina, the omaular carina (at least below), and a strong carina across the anterior margin of the pronotal collar and extending laterad to the pronotal lobe. The body length is 5.5 to 7.0 mm. The male genitalia and other structures were illustrated by Snelling (1985) and Hensen (1987).

■ *Metylaeus* is found in Africa from Nigeria to Namibia, east to Kenya and Mozambique, and also in Madagascar. The four African species were revised by Snelling (1985); two species are known in Madagascar, bringing the total to six species.

Hylaeus / Subgenus *Metziella* Michener

Hylaeus (Metziella) Michener, 1942a: 273. Type species: *Prosopis potens* Metz, 1911 = *Prosopis sparsa* Cresson, 1869, by original designation.

This subgenus resembles the subgenus *Prosopis* in its small size and form. Distinctive characters are the short, broad face, the apical process of S8 of the male, which is reduced to a pointed projection, and the two hairy apical lobes of S7, each attached to the body of the sternum near the middle of the lobe. The male genitalia and hidden sterna were illustrated by Metz (1911), Mitchell (1960), and Snelling (1968).

■ *Metziella* occurs from Quebec, Canada, to Georgia, USA, west to Michigan and Texas, USA. The only species is *Hylaeus sparsus* (Cresson). The subgenus was reviewed by Snelling (1968), but one of the two species then included was removed to *Paraprosopis* when its male was recognized (Snelling, 1970).

Hylaeus / Subgenus *Nesoprosopis* Perkins

Nesoprosopis Perkins, 1899: 75. Type species: *Prosopis facilis* Smith, 1879, by designation of Popov, 1939a: 168.

S7 in this subgenus has two moderate-sized apical lobes, both hairless or with a few short hairs (Fig. 47-11f). The apical process of S8 is long, its basal part directed more or less ventrad from the body of the sternum before bending apicad (Fig. 47-11e), and the apical part simple to strongly bifid, with few short hairs or with abundant long hairs. The ventrally directed basal section of the process of S8 is scarcely noticeable in *Hylaeus nippon* Hirashima, but is usually a distinctive feature of the subgenus. The only other subgenus sharing such a bent apical process of S8 is *Nesylaeus*, which has elongate male gonoforceps, exceeding the penis valves, as in *Nothylaeus* (Fig. 47-13a). The male genitalia and associated structures were illustrated by Perkins (1899), by Ikudome (1989), and (for *H. pectoralis* Förster) by Méhelÿ (1935).

■ Although proposed for the species swarm found in Hawaii, this subgenus ranges widely in the Oriental and palearctic regions, north to Finland (61°N) and west to France, and south to the Philippines. The Hawaiian species were revised by Perkins (1899), and a key was provided by Perkins (1910). Daly and Magnacca (2003) revised the 60 Hawaiian species, and supported the belief that five of them are cleptoparasites in nests of other species of the same subgenus; they are the only known cleptoparasitic Colletidae. The eight Japanese species, including one from the Bonin Islands, were revised by Hirashima (1977) and Ikudome (1989). The single European species, *Hylaeus pectoralis* Förster, which ranges from France to Japan, was included by Méhelÿ (1935) in his subgenus *Imperfecta* and by Dathe (1980a) in the subgenus *Prosopis*. Some additional tropical Oriental species, including one from southern China, probably belong in *Nesoprosopis*.

Hylaeus / Subgenus *Nesylaeus* Bridwell

Hylaeus (Nesylaeus) Bridwell, 1919: 147. Type species: *Hylaeus (Nesylaeus) nesoprosopoides* Bridwell, 1919, by original designation.
Neshylaeus Heider, 1935: 2245, unjustified emendation of *Nesylaeus* Bridwell, 1919.

In this subgenus the apical process of S8 of the male is basally bent, as in the subgenus *Nesoprosopis*, from which it differs by having the gonoforceps much attenuate apically, extending far beyond the apices of the penis valves, as in *Nothylaeus* (Fig. 47-13a). This is the only named Palearctic or Oriental subgenus exhibiting such attenuation, if one ignores the different kinds of attenuation found in the subgenera *Patagiata* and *Spatulariella;* the otherwise different African subgenera *Nothylaeus* and *Alfkenylaeus*, however, have apically attenuate gonoforceps similar to those of *Nesylaeus*..

■ This subgenus is known in tropical Asia north to the Philippines and southern China (Hong Kong area) (see Hirashima, 1977). The only named species is *Hylaeus nesoprosopoides* Bridwell.

Hylaeus / Subgenus *Nothylaeus* Bridwell

Nothylaeus Bridwell, 1919: 125, 126. Type species: *Prosopis heraldica* Smith, 1853, by original designation.
Nothylaeus (Anylaeus) Bridwell, 1919: 129. Type species: *Nothylaeus aberrans* Bridwell, 1919, by original designation.
Anhylaeus Heider, 1926: 184, unjustified emendation of *Anylaeus* Bridwell, 1919.

This group of often rather robust species, 5.5 to 7.0 mm long, was described as a genus and has frequently been recognized as generically distinct, for example, by Snelling (1985), because of the elongate, slender, almost needle-like mandibles of both sexes (Fig. 47-13d), a feature not found in other bees. The other characteristics, however, are those of *Hylaeus*. A single apomorphy, however striking, hardly seems to justify generic status. I have therefore placed *Nothylaeus* in *Hylaeus*. The name *Anylaeus* was proposed for species with the scutellum and metanotum produced posteriorly at the sides as in males and some females of the subgenus *Metylaeus*, and as in the Philippine *Hoploprosopis*. These conspicuous features seem to have arisen independently. The gonoforceps of the male are attenuate apically (Fig. 47-13a), as in the subgenera *Alfkenylaeus* and *Nesylaeus*. S7 of the male has a pair of apicolateral lobes, each partly divided into a small, basal part and a large, densely hairy, apical part (Fig. 47-13c). S8 of the male has a short, almost round, hairless apical process; the margins lateral to this process are strongly produced to form rounded shoulders (Fig. 47-13b). The male genitalia and hidden sterna were illustrated by Snelling (1985).

■ *Nothylaeus* is widespread in Africa from Liberia and Ethiopia to Cape Province, South Africa, and to Madagascar (R. Snelling, in litt., 1990). Snelling (1985) lists 34 specific names.

Hylaeus / Subgenus *Orohylaeus* Michener

Hylaeus (Orohylaeus) Michener, 2001: 2. Type species: *Hylaeus benoisti* Michener, 2001, by original designation.

I have hesitated to modify Snelling's keys to include *Orohylaeus;* instead I explain here its positions in those keys. Both sexes lack yellow facial markings. In the key to males of the subgenera of *Hylaeus* of the Western Hemisphere, *Orohylaeus* runs to couplet 6. There it agrees with *Prosopella* in the approximately right angular dorsolateral angle of the pronotum (although it is not as broadly trun-

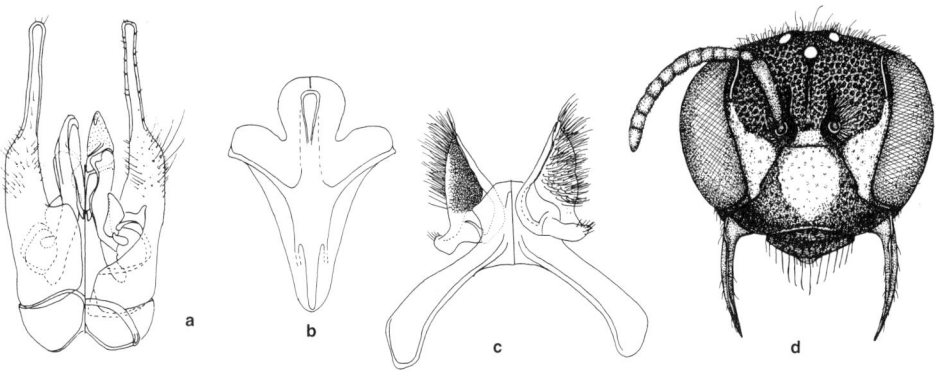

Figure 47-13. *Hylaeus (Nothylaeus) heraldicus* (Smith). **a-c,** Male genitalia, S8, and S7; **d,** Face of female showing needle-like mandibles. (In the divided figures, dorsal views are at the left.) From Snelling, 1985.

cate as in *Prosopella*), but it differs from that subgenus in the straight (although very short) distal process of S8 of the male. If one continues with the key, ignoring the pronotal character, no alternative of any couplet agrees with the characters of *Orohylaeus*. In the key to females, *Orohylaeus* does not run at all; in the first couplet, it agrees with the second alternative in lacking an omaular carina but with the first in the microstriate metasoma. If one ignores the metasomal character and goes to couplet 4, thence to 5, it runs best to *Prosopella* but does not agree in the propodeal rugae and carinae.

■ *Orohylaeus* is known only from Ecuador and contains one species, *H. (O.) benoisti* Michener.

Hylaeus / Subgenus *Paraprosopis* Popov

Prosopis (Campanularia) Méhelÿ, 1935: 50, 157, not Lamarck, 1816. Invalid because no type species was designated.

Prosopis (Paraprosopis) Popov, 1939a: 169. Type species: *Hylaeus pictipes* Nylander, 1852, by original designation.

This is the Asininus Division of Metz (1911). The subgenus includes small, slender species with fine sculpturing. The male antennal scape is unmodified. The apical process of S8 of the male is bifid at the apex, forming two elongate lobes, usually with short hairs. S7 of the male is variable and permits division of the subgenus into two or three species groups. In *Hylaeus pictipes* Nylander and most other palearctic species, and in certain nearctic species, e.g., *H. calvus* (Metz), S7 has four similar hairless apical lobes, those of each side broadly united basally (a single lobe?) but not situated on the same plane (Fig. 47-10f). In a few palearctic species, e.g., *H. lineolata* (Schenck), the lobes are similar but the proximal ones have hairs. In the majority of the nearctic species, such as *H. asininus* (Cockerell and Casad), the distal lobes of S7 are large, somewhat subdivided, and hairless, whereas the basal lobes are slender and hairy; see illustrations of Metz (1911), Méhelÿ (1935), Mitchell (1960), Snelling (1970), and Dathe (1980a).

■ In the New World, *Paraprosopis* ranges across North America from British Columbia, Canada, to Maine, USA, south to Baja California and Chihuahua, Mexico; in the Old World, from Madeira and the Canary Islands and Sweden to Siberia and the Bonin Islands in the Pacific, south to northern Africa, Yemen, and Sri Lanka. In the palearctic region, 27 species are known, in North America, 15 others, and in tropical Asia perhaps five others. A concentration of species is found in the Mediterranean area, and another four species are on the Canary Islands and Madeira alone. Revisions are by Snelling (1970) for North America, Dathe (1980a) for Europe, and Ikudome (1989) for Japan.

Hylaeus / Subgenus *Planihylaeus* Houston

Hylaeus (Planihylaeus) Houston, 1981a: 96. Type species: *Prosopis trilobata* Cockerell, 1910, by original designation.

The species of this subgenus known to me in 1965 were then included in the subgenus *Prosopisteron*, primarily because of the form of S7 of the male, but Houston (1981a) considers this group closer to the subgenera *Pseudhylaeus* and *Xenohylaeus*. It differs from the former in lacking a carinate outer ridge on the mandible, and from the latter in lacking a median clypeal carina in females and large paraocular depressions in males. *Planihylaeus* includes moderate-sized (5.0-7.5 mm long), nonmetallic species. The face below the antennae is flat and longitudinally striate (Fig. 47-7c, d). The propodeum is short, the sloping basal zone rounding evenly onto the declivous surface. Similarities of S7 of the male to that of the subgenus *Laccohylaeus* include a row of coarse bristles and the tapered apices of the distal lobes; the ventroapical lobes also may have coarse bristles. The apical process of S8 of the male is strongly bifid at its apex. The male genitalia and associated structures were illustrated by Houston (1981a).

■ This subgenus occurs in southwestern Australia and in coastal and montane areas from Tasmania through Victoria to southern Queensland. The five included species were revised by Houston (1981a) and listed by Cardale (1993).

Hylaeus / Subgenus *Prosopella* Snelling

Hylaeus (Prosopella) Snelling, 1966c: 139. Type species: *Hylaeus hurdi* Snelling, 1966, by original designation.

This subgenus is superficially similar to the subgenus *Paraprosopis,* from which it differs in the apically enlarged but nonbifid apex of the process of S8 of the male, the strong, truncate (in dorsal view) dorsolateral angles of the pronotum, and the coarse punctation of the head and thorax (suggesting the subgenus *Hylaeopsis*). S7 of the male has a single pair of small apical lobes, the apices of which are produced basad, like those of *Lambdopsis, Hylaeopsis,* and *H. (Prosopis) insolitus* Snelling. *Prosopella* may be related to *Hylaeopsis,* from which it differs in its rounded omaulus, lack of a preoccipital carina, the truncate dorsolateral pronotal angles, and the blunt rather than attenuate apices of the gonoforceps. It is not at all related to *Lambdopsis,* as indicated by the simple rather than notched S6 of the male, the slender, acute apices of the penis valves, and numerous other characters. The form of the lobes of S7 in *Prosopella* and *Lambdopsis* must be convergent. *H. (Prosopis) insolitus,* however, seems closely related to *Prosopella,* which it resembles in the form of S7 and S8 of the male, but it differs in punctation, the dorsolateral angles of the pronotum, etc. Male genitalia and hidden sterna were illustrated by Snelling (1966c).

■ *Prosopella* occurs in southern Arizona, USA, and in Chihuahua, Mexico. There are undescribed species farther south in Mexico, according to Snelling (1966a). The only described species is *Hylaeus hurdi* Snelling.

Hylaeus / Subgenus *Prosopis* Fabricius

Prosopis Jurine, 1801: 164. Suppressed by Commission Opinion 135 (1939).
Prosopis Fabricius, 1804: 293. Type species: *Mellinus bipunctatus* Fabricius, 1798 = *Sphex signata* Panzer, 1798, designated by Morice and Durrant, 1915: 416.
Prosapis Ashmead, 1894: 43, unjustified emendation of *Prosopis* Fabricius, 1804.
Prosopis (Cingulata) Méhelÿ, 1935: 43, 149. Type species: *Vespa pratensis* Geoffroy, 1785 (not Miller, 1759) = *Sphex signata* Panzer, 1798, monobasic.
Prosopis (Fasciata) Méhelÿ, 1935: 44, 150. Type species: *Prosopis facialis* Pérez, 1895 = *Prosopis trinotata* Pérez, 1895, monobasic.
Prosopis (Auricularia) Méhelÿ, 1935: 41, 147. Invalid because no type species was designated. [For a supposed subsequent designation, see Michener, 1997b.]
Prosopis (Navicularia) Méhelÿ, 1935: 34, 140. Invalid because no type species was designated.
Prosopis (Navicularia) Popov, 1939a: 168. Type species: *Mellinus variegatus* Fabricius, 1798 (as *P. variegata*), by original designation.
Prosopis (Fascista) Popov, 1939a: 168. Incorrect subsequent spelling for *Fasciata* Méhelÿ, 1935.

This subgenus, the Modestus Division of Metz (1911), includes the species placed in the subgenera *Auricularia, Cingulata, Fasciata,* and *Navicularia* by Méhelÿ (1935). S7 of the male has two apical lobes, narrowly attached to the disc of the sternum; the distolateral margins of the lobes are concave, nearly always margined by slender hairs, but the resulting two divisions are on the same plane (Fig. 47-11h), not one above the other and overlapping as when there are four lobes. The apical process of S8 of the male is simple or feebly emarginate at the apex (Fig. 47-11g). The pale supraclypeal mark of males is not over about half the length of the clypeus. The male antennal scape is commonly not modified. In some species the basal metasomal terga are red. The male genitalia and hidden sterna were illustrated in all the major works on the genus in the Holarctic region.

■ The subgenus is holarctic in distribution, ranging across the palearctic region from Britain to Japan and eastern Siberia, from 66°N in Scandinavia south to northern Africa, and in North America from Alaska, USA, to Quebec, Canada, south to Florida, USA, and Veracruz, Durango, and Baja California, Mexico. About 30 species are known in the palearctic region (4 are known from Britain, 12 from Europe, 2 from Japan, and 2 from Mongolia). (It appears that the subgenus is more richly represented in the western palearctic than in the eastern; the related subgenus *Nesoprosopis* shows the reverse pattern in Eurasia.) For the nearctic region, Snelling (1966a) lists 16 species. Snelling (1966b, 1975) treats the taxonomy of species of the western USA; species of the USA and parts of Eurasia are treated in works cited under the genus *Hylaeus,* especially Metz (1911), Mitchell (1960), Dathe (1980a), and Ikudome (1989).

H. Dathe (in litt., 1990) noted that the group of *Hylaeus variegatus* (Fabricius) consists of closely related species with coarse punctation and, commonly, red on the metasoma or yellow on the head and thorax.

Hylaeus / Subgenus *Prosopisteroides* Hirashima

Hylaeus (Prosopisteroides) Hirashima, 1967a: 134. Type species: *Hylaeus heteroclitus* Hirashima, 1967, by original designation.

This subgenus contains elongate species 5 to 7 mm in body length, some brilliantly metallic blue-green, others dark blue. A remarkable feature is the enormous maxillary palpus, as long as or longer than the thorax. Unfortunately, the genitalia and hidden sterna are unknown, but other structures were illustrated by Hirashima (1967a).

■ *Prosopisteroides* is known only from New Guinea. There are four species. Hirashima and Tadauchi (1984) gave a key and described the first known male of the subgenus.

Hylaeus / Subgenus *Prosopisteron* Cockerell

Prosopisteron Cockerell, 1906c: 17. Type species: *Prosopisteron serotinellum* Cockerell, 1906, monobasic.
Psilylaeus Snelling, 1985: 28. Type species: *Psilylaeus sagiops* Snelling, 1985 = *Prosopis perhumilis* Cockerell, 1914, by original designation.

Prosopisteron was proposed for a species with a most unusual character, i.e., an enormous stigma; *Hylaeus serotinellus* (Cockerell) is the only such species. The name *Psilylaeus* was proposed on the basis of African specimens of an introduced Australian species, as was pointed out to me by R. Snelling (in litt., 1988), who indicated the synonymy listed above. If *Prosopisteron* were divided, *Psilylaeus* would be the valid name of one of the groups.

Prosopisteron consists of small to moderate-sized

species (3.5-9.5 mm in body length), black or with dark-blue coloration usually restricted to the metasoma. The second submarginal cell is nearly always half as long as the first or less. The propodeum is almost always nonareolate but dull, the dorsal surface curving onto the posterior surface. S7 of the male has four apical lobes (Fig. 47-9d-f), all with slender hairs; the apical process of S8 is simple or has a shallow apical notch. The male genitalia and hidden sterna were illustrated by Michener (1965b); note that *Hylaeus douglasi* Michener was transferred to *Hylaeteron* by Houston (1981a). See also Figure 47-9d-f, j.

■ *Prosopisteron* is found throughout Australia, as well as in Tasmania, New Guinea, New Zealand, and the Chatham Islands. A species from the Tuamotu Islands should be restudied to see if it might be an introduced Australian species, for the genus is unknown from the islands between Australia and the Tuamotus. An Australian species, *Hylaeus (Prosopisteron) perhumilis* (Cockerell), has been found in southernmost South Africa, no doubt introduced. Thirteen collections in South Africa, ranging from Cape Town to Port Elizabeth (400 km), were made from 1930 to 1948. Lack of recent captures suggests that the species may have disappeared in Africa. I collected along that coast in 1966 and did not find it. The subgenus as presently constituted contains 76 named species; Cardale (1993) lists 55 from Australia. There will certainly be much synonymy among these names in due course, but there must also be many species yet undiscovered.

Hylaeus / Subgenus *Pseudhylaeus* Cockerell

Pseudhylaeus Cockerell, 1929b: 299. Type species: *Euryglossa albocuneata* Cockerell, 1913, by original designation.

Species in this subgenus exhibit considerable morphological diversity but are clearly related to one another. They are notable for their fine sculpturing and dull integument, the sericeous appearance of the metasoma due to fine appressed hairs, and the pallid apical margins of the terga, which carry abundant laterally directed hairs. The robust body form, 4.5 to 7.0 mm long, dull and nonmetallic cuticle, and pale (not bright yellow) markings probably led to two of the species being described in the genus *Euryglossa*. The face is usually flat, the clypeus sometimes being finely striate; the scape of the male is sometimes swollen. The maxillary palpus in some species is much enlarged, as long as or longer than the head, a feature suggesting the subgenus *Prosopisteroides* but one that probably evolved independently. The hidden sterna of the male are as in the subgenera *Xenohylaeus* and *Planihylaeus*. Male genitalia and other structures were illustrated by Michener (1965b).

■ *Pseudhylaeus* is found from Tasmania and southern Australia north to southern Queensland. There are five species names and about five undescribed species; those with names were listed by Cardale (1993).

Hylaeus / Subgenus *Rhodohylaeus* Michener

Hylaeus (Rhodohylaeus) Michener, 1965b: 124. Type species: *Prosopis cenibera* Cockerell, 1910, by original designation.

This is a subgenus of mostly small species (4.0 to 7.5 mm long), usually with red-brown coloration on the metasoma and often on the head and thorax as well. The sculpturing, usually strong, usually includes a carina or distinct angle on the omaulus. The propodeal triangle is largely coarsely areolate and delimited by a carina. The paraocular ridge is present, sometimes carinate. S7 of the male has a single pair of laterally directed apical lobes (Fig. 47-9n), their hairs slender. The male genitalia and hidden sterna were illustrated by Michener (1965b).

■ *Rhodohylaeus* is widespread in Australia but is not known in Tasmania. Twenty-one names have been placed in this subgenus (Michener, 1965b) and listed by Cardale (1993).

Hylaeus / Subgenus *Spatulariella* Popov

Prosopis (Spatularia) Méhelÿ, 1935: 69, 175 (not Van Deventer, 1904). Invalid because no type species was designated; also a junior homonym.
Prosopis (Spatulariella) Popov, 1939a: 169. Type species: *Hylaeus hyalinatus* Smith, 1842, by original designation.
Prosopis (Brachyspatulariella) Pittioni, 1950: 79. Type species: *Spatulariella helenae* Pittioni, 1950, by original designation.
Prosopis (Amblyspatulariella) Pittioni, 1950: 80. Type species: *Prosopis sulphuripes* Gribodo, 1894, by original designation.
Prosopis (Platyspatulariella) Pittioni, 1950: 80. Type species: *Prosopis punctata* Brullé, 1832, by original designation.

This subgenus is characterized by the elongate malar space (see the key to females), the sharp-edged to lamellate omaulus, and the elongate gonoforceps (Fig. 47-10c), which are at least nine times as long as their median widths in dorsal view, at least their distal thirds extending beyond the apices of the penis valves. Male genitalia and hidden sterna were illustrated by Méhelÿ (1935) and Dathe 1980a).

■ *Spatulariella* is found from Spain, Morocco, and the Canary Islands to Iran and central Russia and Kazakhstan. *Hylaeus (Spatulariella) punctatus* (Brullé) is adventive in southern California (Snelling, 1983a) and now also occurs in the San Francisco Bay area as well as in the vicinity of Washington, D.C. (J. Ascher, personal comm..). It also has appeared in central Chile (Toro, Frederick, and Henry, 1989). *H. (S.) hyalinatus* (Smith), another European species, is established in New York State (Ascher, 2001) and now occurs in New York City (J. Ascher, personal comm.). There are six European species (Dathe, 1980a) and a total of about 18 species. European species were revised by Pittioni (1950) and Dathe (1980a).

Hylaeus / Subgenus *Sphaerhylaeus* Cockerell

Gnathoprosopis (Sphaerhylaeus) Cockerell, 1929a: 217. Type species: *Gnathoprosopis globulifera* Cockerell, 1929, by original designation.

A remarkable feature of the subgenus *Sphaerhylaeus* is the extraordinary globose antennal scape of the male (Fig. 47-7f), hidden behind which is a cavity in the frons. This subgenus contains moderate-sized to rather large (7.0-12.5 mm long) nonmetallic species. The presence of a preoccipital carina suggests a relationship to the subgenera *Gnathoprosopis* and *Gnathoprosopoides* and the male sterna and genitalia are reasonably similar to those of the latter. They are perhaps even more similar to those of the subgenus *Prosopisteron,* from which both subgenera may have evolved. The preoccipital carina appears or disappears so many times in bee phylogeny that it cannot be considered a strong indication of relationship. The male genitalia and other structures were illustrated by Michener (1965b) and Houston (1981a).

■ This subgenus is known from the coast of southwestern Australia and from the coast of New South Wales. There are two known species.

Hylaeus / Subgenus *Xenohylaeus* Michener

Hylaeus (Xenohylaeus) Michener, 1965b: 136. Type species: *Hylaeus rieki* Michener, 1965, by original designation.

This subgenus includes medium-sized (6-9 mm long), black, nonmetallic species without yellow on the scutellum and metanotum. The male has a large concavity on each side of the face, occupying most of the paraocular area and thus suggesting the genus *Meroglossa*. The scape is much expanded and flat. The face of the female is rather flat, finely longitudinally striate below the antennae and the clypeus has a longitudinal median carina, strongest apically. The tergal margins have laterally directed hairs, often not as numerous as those of the subgenus *Pseudhylaeus*. S7 and S8 of the male are similar to those of the subgenus *Planihylaeus,* but the coarse bristles on the lobes of S7 are usually stronger, sometimes expanded and flattened (Fig. 47-9m). The male genitalia and associated structures were illustrated by Michener (1965b) and Houston (1981a).

■ *Xenohylaeus* is found in the southern half of Australia, north to southern Queensland; it is not known from Tasmania. Its four species were revised by Houston (1981a) and listed by Cardale (1993). Some of the species seem to be specialists on small yellow-flowered legumes.

Genus *Hyleoides* Smith

Hyleoides Smith, 1853: 32. Type species: *Vespa concinna* Fabricius, 1775, by designation of Taschenberg, 1883: 45.
Hylaeoides Dalla Torre, 1896: 51, unjustified emendation of *Hyleoides* Smith, 1853.

Members of this genus, the largest Hylaeinae except *Hylaeus (Meghylaeus),* are 10 to 14 mm long, rather robust, wasplike, and black with red or yellow integumental markings on all tagmata. [The similar markings of *Hylaeus (Analastoroides)* are due to orange tomentum.] The glossa is strongly bifid. The second submarginal cell is about as long as the first, the distal margin longer than the basal margin and perpendicular to the long axis of the wing (Fig. 47-5a). Male genitalia, hidden sterna, and other structures were illustrated by Michener (1965b) and Houston (1975a); see also Figure 47-8k. The long second submarginal cell suggests that, unlike what we find in other Hylaeinae, the second transverse submarginal vein is lost instead of the first. This conclusion is probably false, however, for *Hyleoides* seems to be a highly derived hylaeine. It is much more likely that the veins shifted during *Hyleoides'* evolution, indicating that one cannot know which submarginal crossvein is lost in many forms with two submarginal cells.

■ *Hyleoides* is found in Australia (no records in the northwestern quarter or the extreme north), including Tasmania. There is a record for Lord Howe Island (Michener, 1965b), presumably indicating an introduction by humans or possibly a mislabeled specimen. One species, *Hyleoides concinna* (Fabricius), has been introduced and established in New Zealand (Donovan, 1983b). The eight species were revised by Houston (1975a) and listed by Cardale (1993).

Genus *Meroglossa* Smith

Meroglossa Smith, 1853: 33. Type species: *Meroglossa canaliculata* Smith, 1853, monobasic.
Meroglossa (Meroglossula) Perkins, 1912: 99. Type species: *Meroglossa eucalypti* Cockerell, 1910, monobasic.

This is a genus of rather robust, nonmetallic, moderate-sized to rather large (7-12 mm long) Hylaeinae. Many species have rather extensive red-brown coloration with yellow markings, but a few are black with limited yellow markings and thus superficially resemble species of *Amphylaeus* or large species of *Hylaeus*. The male glossa is pointed and has seriate hairs on the posterior surface (Fig. 47-1i), suggesting the glossa of *Andrena*. The scape of the male is slightly to considerably enlarged (Fig. 47-4i). The male has a strong paraocular depression on each side extending from the upper clypeal area to above the antennal bases (Fig. 47-4i). These depressions are suggestive of those of *Hylaeus (Xenohylaeus),* but to judge by dissimilarity in other characters, the depressions are probably not homologous. The hind tibia has one or usually two spines on its outer apical margin (Fig. 47-3o). The gradulus of T2 is arcuate posteriorly, therefore usually exposed. The flagellum of the male is constricted and commonly bent at segment 5, and usually has one or more short setae arising from the bases of segments 4 and 5 on the underside. For further generic characterization and discussion of variation, see also Michener (1965b). Male genitalia, sternal, facial, and other characters (Fig. 47-8d-f) were illustrated by Michener (1965b) and Houston (1975a).

■ *Meroglossa* is found in all parts of continental Australia, and is especially abundant in the north, but is unknown from Tasmania. The 20 species were revised by Houston (1975a) and listed by Cardale (1993).

Nests are similar to those of *Hylaeus* (Michener, 1960).

Genus *Palaeorhiza* Perkins

This is a large and diverse but inadequately studied genus. Its species are larger than most Hylaeinae, 6 to 12 mm long, morphologically diversified, and often brilliantly metallic or with abundant yellow markings, although sometimes largely red or wholly black. The glossa of the male is pointed, two or more times as long as broad, with two rows of strong seriate hairs on the posterior surface (as in Fig. 19-4d, e). The preoccipital carina is usually present, at least medially. The large dorsal surface of the propodeum usually curves gently onto the posterior surface, the propodeal triangle being almost wholly on the dorsal surface (Fig. 47-3d, e) except in forms like *Palae-*

orhiza (Ceratorhiza) conica Michener with a modified propodeum (Fig. 47-3f). The vertical ridge in front of the mesocoxa is usually sharp. The apex of the hind tibia lacks spines on the outer margins. The male genitalia and hidden sterna of certain species were illustrated by Michener (1965b); see also Figure 47-8g.

The majority of species of *Palaeorhiza* occur in New Guinea, and all subgenera are represented there; this is the only bee group that has undergone major diversification on that island. The genus extends into other moist forested areas, west to the Moluccas and as far as Flores in the Lesser Sundas, south as far as southern Queensland in Australia, and east to the Bismarck Archipelago, the Solomon Islands, the Santa Cruz Islands, and the New Hebrides. An Australian species, *P. (Heterorhiza) flavomellea* Cockerell, or a close relative, occurs in New Caledonia. About 150 species have been described, many of them from single individuals; the total number of species must be much greater. The 25 Australian species were listed by Cardale (1993).

The species of *Palaeorhiza* are mostly easily distinguishable. Probably for this reason, it has not been the custom to examine or illustrate the male genitalia and hidden sterna, as is routinely done for other Hylaeinae and indeed for most bees. The subgeneric classification is therefore in a particularly primitive state. Although many of the subgenera differ sharply from one another, Hirashima and Lieftinck (1983) reported 19 species (16 of them new) that they could not assign to subgenus, and Hirashima (1988) did not place 12 new species in subgenera. In 1989, however, he named the subgenus *Callorhiza* to contain the unplaced species. The whole genus needs study in the light of genitalic and other characters that might permit better analysis. The following are some of the unusual characters of certain species or subgenera.

Palaeorhiza (Michenerapis) bicolor Hirashima and Lieftinck, known from a single male specimen, lacks a preoccipital carina. Hirashima and Lieftinck (1982) suggest that it may be generically distinct, but lack of a preoccipital carina by itself does not establish generic distinctness. *Hylaeus* and *Pharohylaeus* are other hylaeine genera in which this carina can be either present or absent.

Unlike those of most other Hylaeinae, females of the subgenus *Cercorhiza* have a well-developed pygidial plate, pygidial and prepygidial fimbriae, and a well-developed basitibial plate (Hirashima, 1975b, 1982b). Presumably these are ancestral features associated with nesting in the ground. At least two species of *Cercorhiza* are known to nest in burrows in the ground, unlike most Hylaeinae. The subgenus *Cnemidorhiza* also has a pygidial plate and fimbria in the female, but there is no basitibial plate. Instead, the basal part of the outer surface of the female hind tibia, where the basitibial plate should be, is broadened and coarsely roughened. This area probably functions as does a basitibial plate, or it may actually be a basitibial plate without marginal carinae. The only species of *Cnemidorhiza* whose nests are known makes burrows in the ground, like those of the subgenus *Cercorhiza* (Hirashima, 1981a). It is likely that *Cnemidorhiza* (with *Cercorhiza*) should form a genus separate from *Palaeorhiza,* but until further studies are made, particularly of males, such a change would be premature. This is written on the assumption that except for the two subgenera mentioned above, *Palaeorhiza* species, like most Hylaeinae, nest in holes in wood, stems, galls, etc. Unfortunately, *Palaeorhiza* nests are unknown, except for the two ground-nesting subgenera.

The genus *Xenorhiza* consists of a group of species that would easily fall within the range of variability found in *Palaeorhiza,* except that the male glossa is similar to that of the female, as in *Hylaeus* and most other colletids. One can easily imagine that this is a derived group of *Palaeorhiza* in which the female-type glossa was transferred to the male. On the other hand, it might be a relictual basal group from which forms with a pointed male glossa with seriate hairs arose. Pending further study, I am not modifying the current classification.

Key to the Subgenera of *Palaeorhiza*
(Modified from Hirashima and Lieftinck, 1982, with certain subgenera added on the basis of literature only)

1. Preoccipital carina absent; space between clypeus and compound eye narrower than width of middle ocellus; inner hind tibial spur of male strongly modified .... *P. (Michenerapis)*
—. Preoccipital carina present; space between clypeus and compound eye at least about as broad as middle ocellus; inner hind tibial spur of male slender and simple as usual 2
2(1). Surface of propodeal triangle strongly convex in middle, or with a conical projection 3
—. Surface of propodeal triangle not convex in the middle 4
3(2). Propodeal triangle convex in middle; T1 small, its basal portion distinctly constricted and subpetiolate; large, more or less slender and nonmetallic species .... *P. (Eusphecogastra)*
—. Propodeal triangle with rounded conical projection; T1 not constricted; large and robust species, with head and thorax black (with yellow markings) and metasoma blue-green .... *P. (Ceratorhiza)*
4(2). Propodeal triangle densely fluted longitudinally 5
—. Propodeal triangle not fluted 7
5(4). T2 and T3 each with band of short, dense, appressed white hair across base; anterior surface of T1 covered with dense, scale-like white hairs .... *P. (Gressittapis)*
—. T1 to T3 without short, dense pubescence as described above 6
6(5). Integument of head and thorax strongly sclerotized, with dense, usually strong punctures on the latter; inner hind tibial spur of female distinctly serrate; male T7 with a pair of long spines at apex, the spines broadly separated from each other; male mandible simple at apex .... *P. (Heterorhiza)*
—. Integument of head and thorax appearing softer; inner hind tibial spur of female simple; male T7 with or without a pair of projections, these not broadly separated when present; male mandible bidentate .... *P. (Paraheterorhiza)*
7(4). Posterior surface of propodeum hexagonal, surrounded by strong carina connected to longitudinal cari-

nae separating dorsal, dorsolateral, and lateral areas, dorsal area divided by longitudinal median carina; upper portion of mesepisternum flat, depressed .... *P. (Noonadania)*

—. Propodeum not so divided by strong carinae; upper part of mesepisternum convex 8

8(7). Anterior basitarsus slender, slightly or usually considerably longer than tarsal segments 2 to 5 taken together; clypeus and supraclypeal area flat (usually longitudinally rugoso-punctate)....................................... *P. (Cheesmania)*

—. Anterior basitarsus little if any longer than segments 2 to 5 taken together; clypeus and supraclypeal area convex 9

9(8). Thorax, especially mesepisternum, and propodeum extremely coarsely foveolate-punctate; propodeal triangle irregularly coarsely rugose, carinate on lateral margins .... *P. (Trachyrhiza)*

—. Thorax rather finely punctate; propodeal triangle not carinate on lateral margins 10

10(9). Female mandible large and edentate; scutellum, metanotum, and most of propodeal enclosure flat; second flagellar segment of male (in *Palaeorhiza mandibularis* Michener) distinctly longer than broad .... *P. (Anchirhiza)*

—. Female mandible as usual; scutellum, metanotum, and most of propodeal enclosure convex as usual; second flagellar segment of male at most as long as broad 11

11(10). Dorsal surface of propodeum with pair of small swellings apically, close to propodeal triangle, either depressed apically or densely and finely shagreened and dull 12

—. Propodeum without such swellings dorso-apically 13

12(11). Propodeal triangle longitudinally depressed in middle apically, smooth, shining; upper portion of supraclypeal area and median portion of frons longitudinally elevated, a broad longitudinal yellow stripe on this portion of face; inner hind tibial spur of female serrate .... *P. (Palaeorhiza s. str.)*

—. Propodeal triangle densely and finely shagreened and dull, transversely slightly concave; swelling of upper portion of supraclypeal area sharply defined from flat frons, no yellow stripe on frons; inner hind tibial spur of female normal .... *P. (Zarhiopalea)*

13(11). Glossa of male longer than head; male S5 concealed by S4; nonmetallic species, the posterior part of thorax, basal part of metasoma, and legs honey-colored (male S6 strongly convex in middle; propodeal triangle somewhat coarsely sculptured in middle) .... *P. (Eupalaeorhiza)*

—. Glossa of male much shorter; male S5 exposed as usual; species usually metallic, not honey-colored 14

14(13). Propodeal triangle punctate-roughened at least on apical portion, usually distinctly convex basally; inner hind tibial spur of female finely serrate (large, robust, strongly metallic species) .... *P. (Hadrorhiza)*

—. Propodeal triangle not punctate-roughened; inner hind tibial spur of female simple..15

15(14). Basal part of hind tibia of female without basitibial plate and not thickened; T6 of female lacking both fimbria and pygidial plate ....*P. (Callorhiza)*

—. Basal part of hind tibia of female either with basitibial plate or thickened and coarsely sculptured; T6 of female with fimbria of hairs differing from those of other terga and sometimes having pygidial plate 16

16(15). Female hind tibia thick basally, its dorsal surface usually broad and punctate-roughened or coarsely sculptured basally but without margined basitibial plate; female T6 with a pygidial fimbria of dense downy hairs in middle or a partly formed pygidial plate (female hind femur with apical tuft of black hairs, sometimes obscure) .... *P. (Cnemidorhiza)*

—. Basitibial plate present in female; female T6 with pygidial plate in middle, lateral to which is pygidial fimbria .... *P. (Cercorhiza)*

Palaeorhiza / Subgenus *Anchirhiza* Michener

Palaeorhiza (Anchirhiza) Michener, 1965b: 147. Type species: *Palaeorhiza mandibularis* Michener, 1965, by original designation.

In this subgenus the body is black, the metasoma is slightly metallic, and the head and thorax have yellow markings. The edentate mandible of the female and the small clypeus, broadly separated from the eye, are distinctive.

■ *Anchirhiza* occurs in Queensland, Australia, and in New Guinea. There are two species. Hirashima (1978a) characterized the subgenus.

Palaeorhiza / Subgenus *Callorhiza* Hirashima

Palaeorhiza (Callorhiza) Hirashima, 1989: 2. Type species: *Prosopis apicatus* Smith, 1863, by original designation.

This subgenus contains a diverse lot of species, that is, all the species that do not fall in any other subgenus. Some are fulvous, some black, some metallic, some have white or yellow markings, in some the metasoma is red; the malar space can be long or short, etc. Their common characters are largely or entirely plesiomorphic relative to related subgenera. Proper study will be possible only when both sexes are known for diverse species.

■ *Callorhiza* is known from Queensland, Australia, and from Misoöl in Indonesia, New Guinea, and the Solomon Islands. Hirashima (1989) listed 40 species.

Palaeorhiza / Subgenus *Ceratorhiza* Hirashima

Palaeorhiza (Ceratorhiza) Hirashima, 1978a: 81. Type species: *Palaeorhiza conica* Michener, 1965, by original designation.

Ceratorhiza consists of large black species with a greenish or purplish metallic metasoma and yellow markings on the head and thorax. The large projection of the propodeal triangle is unique.

■ This subgenus is found in New Guinea. The two species were revised by Hirashima (1978a).

Palaeorhiza / Subgenus *Cercorhiza* Hirashima

Palaeorhiza (Cercorhiza) Hirashima, 1982b: 88. Type species: *Palaeorhiza gressittorum* Hirashima, 1975, by original designation.

This is a subgenus of black, red, or metallic species having yellow marks limited to the faces of the males. As in the subgenus *Cnemidorhiza*, females have a pygidial plate and pygidial and prepygidial fimbriae. Unlike *Cnemidorhiza*, the basitibial plate is well developed, as illustrated by Hirashima (1975b).

■ *Cercorhiza* is found in New Guinea. The 11 species were revised by Hirashima (1982b); two more species were described later.

Most Hylaeinae and probably most *Palaeorhiza* nest in pithy stems or other above-ground situations and lack basitibial plates and pygidial plates and fimbriae. Species of *Cercorhiza* and *Cnemidorhiza* nest in the ground and probably dig their own burrows; they must have retained or redeveloped these plates and fimbriae. It has to be noted that nests of *Palaeorhiza* above ground have not been studied.

Palaeorhiza / Subgenus *Cheesmania* Hirashima

Palaeorhiza (Cheesmania) Hirashima, 1981b: 27. Type species: *Palaeorhiza (Cheesmania) amabilis* Hirashima, 1981, by original designation. This is not the same as *Palaeorhiza (Heterorhiza) amabilis* Hirashima and Lieftinck, 1982.

Cheesmania consists of slender, beautifully metallic bees without yellow markings or with yellow on the face of some males. The face differs from that of all other *Palaeorhiza* in that the clypeus and supraclypeal area are flat and usually densely, longitudinally rugoso-punctate.

■ This subgenus is found in New Guinea. The three species were revised by Hirashima (1981b).

Palaeorhiza / Subgenus *Cnemidorhiza* Hirashima

Palaeorhiza (Cnemidorhiza) Hirashima, 1981a: 1. Type species: *Prosopis elegans* Smith, 1864 = *Prosopis elegantissima* Dalla Torre, 1896, by original designation.

In this subgenus, as in the subgenus *Cercorhiza*, the female has a pygidial plate, although it varies among species in distinctness. The prepygidial and pygidial fimbriae are developed. The outer surface of the hind tibia is broadened basally and coarsely sculptured, but lacks a margined basitibial plate. At least *Palaeorhiza (Cnemidorhiza) gratiosa* Cheesman nests in the soil, like *Cercorhiza* (Hirashima, 1981a). The body is usually metallic, sometimes with the metasoma red, and at least the head and thorax have yellow markings.

■ *Cnemidorhiza* occurs in New Guinea, New Ireland, and Queensland, Australia. Of the 20 described species, 18 were reviewed by Hirashima (1981a).

Palaeorhiza / Subgenus *Eupalaeorhiza* Meade-Waldo

Eupalaeorhiza Meade-Waldo, 1914b: 403. Type species: *Eupalaeorhiza papuana* Meade-Waldo, 1914, by original designation.

This subgenus includes nonmetallic, partly testaceous species with an extremely long glossa in the male.

■ *Eupalaeorhiza* occurs in New Guinea and Misoöl, Indonesia. There are three species. The subgenus was fully characterized by Hirashima and Lieftinck (1983).

Palaeorhiza / Subgenus *Eusphecogastra* Hirashima

Palaeorhiza (Sphecogaster) Hirashima, 1978a: 72 (not Lacordaire, 1869). Type species: *Palaeorhiza paradisea* Hirashima, 1978, by original designation.

Palaeorhiza (Eusphecogastra) Hirashima, 1992: 395, replacement for *Sphecogaster* Hirashima, 1978. Type species: *Palaeorhiza paradisea* Hirashima, 1978, autobasic and by original designation.

This subgenus is unique in having T1 small and slightly petiolate. There is a distinct median swelling on the propodeal triangle, but one not nearly so high as in *Ceratorhiza*. The body is black with the metasoma largely red, sometimes with metallic tints. Yellow marks are absent, although the legs are sometimes yellow.

■ *Eusphecogastra* occurs in New Guinea. The three species were reviewed by Hirashima and Lieftinck (1982).

Palaeorhiza / Subgenus *Gressittapis* Hirashima

Palaeorhiza (Gressittapis) Hirashima, 1978a: 65. Type species: *Palaeorhiza miranda* Hirashima, 1978, by original designation.

In this subgenus the body is nonmetallic black with conspicuous yellow markings on the head and thorax. In addition to the areas of short white pubescence described in the key to subgenera, distinctive characters include the triangularly projecting dorsolateral angles of the pronotum and the low, broad, flat dorsal surface of the pronotal collar, sharply truncate anteriorly.

■ *Gressittapis* occurs in New Guinea. There are two species.

Palaeorhiza / Subgenus *Hadrorhiza* Hirashima

Palaeorhiza (Hadrorhiza) Hirashima, 1980: 108. Type species: *Prosopis imperialis* Smith, 1863, by original designation.

Hadrorhiza includes large, brilliantly green or blue-green species with yellow on the head and sometimes on the pronotal lobe.

■ This subgenus is found in Queensland, Australia, and in New Guinea. The three species were revised by Hirashima (1980).

Palaeorhiza / Subgenus *Heterorhiza* Cockerell

Palaeorhiza (Heterorhiza) Cockerell, 1929b: 316. Type species: *Palaeorhiza melanura* Cockerell, 1910, by original designation.

Heterorhiza consists of nonmetallic black species with extensive yellow markings. The morphological characters are distinctive, as indicated in the key to subgenera.

■ This subgenus occurs in New Guinea and islands to the west (Batjan, Misoöl, Obi, and Flores, all in Indonesia), as well as Queensland, Australia, and New Caledonia. There are 12 species. The subgenus was characterized, and a key to some of the species offered, by Hirashima and Lieftinck (1982).

Palaeorhiza / Subgenus *Michenerapis* Hirashima and Lieftinck

Palaeorhiza (Michenerapis) Hirashima and Lieftinck, 1982: 5. Type species: *Palaeorhiza bicolor* Hirashima and Lieftinck, 1982, by original designation.

This subgenus consists of a single nonmetallic black species with a largely red metasoma. The inner hind tibial spur of the male is short, thick, and strongly convex ventrally in the middle; the outer spur is also unusually short and thick.

■ *Michenerapis* occurs in New Guinea. It includes only *Palaeorhiza bicolor* Hirashima and Lieftinck.

Palaeorhiza / Subgenus *Noonadania* Hirashima

Palaeorhiza (Noonadania) Hirashima, 1978a: 68. Type species: *Palaeorhiza sculpturalis* Hirashima, 1978, by original designation.

The bees of this subgenus have the most carinate propodeum of any *Palaeorhiza*; unlike most subgenera, there is a sharp (and carinate) angle between the dorsal and posterior surfaces of the propodeum, and the propodeal triangle, instead of being almost entirely on the dorsal surface, extends onto the posterior surface. In the future, these are likely to be considered generic characters. The body is bright metallic green, purple, and coppery, with yellow markings restricted to the face and pronotum. The male of *Noonadania* is unknown.

■ *Noonadania* is found in New Guinea and the Bismarck Archipelago. There are two species.

Palaeorhiza / Subgenus *Palaeorhiza* Perkins s. str.

Palaeorhiza Perkins, 1908: 29. Type species: *Prosopis perviridis* Cockerell, 1905, by original designation.

The species of this subgenus are strongly metallic green, blue, or purple, or rarely black with the metasoma red; yellow markings are usually restricted but rarely entirely absent. Distinctive characters include the small prominence on each side, beside the posterior part of the propodeal triangle, and the coarsely pectinate inner hind tibial spur of the female. There is usually an incomplete pygidial plate in the female.

■ *Palaeorhiza* s. str. occurs in New Guinea and surrounding areas: northern Australia, the Bismarck Archipelago, the Solomon Islands, Kai, Ambon, and Timor. The 14 species were revised by Hirashima (1978b); another species was described later.

Palaeorhiza s. str., as used here, is far more restricted than *Palaeorhiza* s. str. of Michener (1965b). The latter included all *Palaeorhiza* not placed in any other subgenus. Now, with more named taxa, *Palaeorhiza* s. str. is well delimited and *Callorhiza* has taken over a reduced "wastebasket" function.

Palaeorhiza / Subgenus *Paraheterorhiza* Hirashima

Palaeorhiza (Paraheterorhiza) Hirashima, 1980: 104. Type species: *Palaeorhiza hilara* Cheesman, 1948, by original designation.

Like *Heterorhiza, Paraheterorhiza* consists of species with conspicuous pale markings and a longitudinally fluted propodeal triangle. The ground color of the thorax and basal metasomal terga, however, is fulvous rather than black as in *Heterorhiza*. The major morphological characters are indicated in the key to subgenera.

■ *Paraheterorhiza* occurs in New Guinea. The two species were revised by Hirashima (1980).

Palaeorhiza / Subgenus *Trachyrhiza* Hirashima

Palaeorhiza (Trachyrhiza) Hirashima, 1980: 100. Type species: *Palaeorhiza rugosa* Hirashima, 1980, by original designation.

This subgenus, known only from the male, consists of a strongly metallic species with yellow markings on the head and thorax. Its most distinctive character is the extremely strongly sculptured thorax, as indicated in the key to subgenera.

■ *Trachyrhiza* occurs in New Guinea. The only species is *Palaeorhiza rugosa* Hirashima.

Palaeorhiza / Subgenus *Zarhiopalea* Hirashima

Palaeorhiza (Zarhiopalea) Hirashima, 1982a: 57. Type species: *Palaeorhiza paradoxa* Hirashima, 1975, by original designation.

This subgenus consists of weakly to strongly metallic species, with or almost without yellow markings on the head and thorax. The propodeal triangle is slightly concave medially, minutely roughened, and dull; beside the distal part of the triangle, on each side, is a small prominence, as in *Palaeorhiza* s. str., from which it differs as indicated in the key to subgenera.

■ *Zarhiopalea* occurs in New Guinea. Five species have been described.

Genus *Pharohylaeus* Michener

Pharohylaeus Michener, 1965b: 141. Type species: *Meroglossa lactifera* Cockerell, 1910, by original designation.

This genus is probably a derivitive of *Hylaeus*, and, if so, could be relegated to subgeneric status. The most distinctive feature is that the first three metasomal segments are enlarged and enclose and largely hide the others (Fig. 47-3b). An approach to this condition can be seen in dried specimens of some other genera, particularly *Hyleoides*, in which the apical metasomal segments are sometimes largely telescoped into the first three. The preoccipital carina is present or absent. The scutellum, the metanotum, and the base of the propodeum tend to be on a single plane, the propodeal triangle nearly all on the subhorizontal basal area, which is separated rather abruptly from the vertical, declivous surface. The male genitalia, hidden sterna, and other structures were illustrated by Michener (1965b) and Houston (1975a); see also Figure 47-8l. The body length is 9 to 11 mm.

■ This genus occurs in central and northern Queensland, Australia, and in New Guinea. Of the two species, *Pharohylaeus lactiferus* (Cockerell) was illustrated and described by Houston (1975a) and *P. papuanus* by Hirashima and Roberts (1986).

Genus *Xenorhiza* Michener

Palaeorhiza (Xenorhiza) Michener, 1965b: 146. Type species: *Palaeorhiza hamada* Cheesman, 1948, by original designation.

Xenorhiza (Papuanorhiza) Hirashima, 1996: 80. Type species: *Xenorhiza krombeini* Hirashima, 1996, by original designation.

This name was originally proposed as a subgenus of *Palaeorhiza,* to which the included species are similar. Hirashima (1975a), however, found that the glossa of the male is short and bilobed, like that of the female, and therefore raised *Xenorhiza* to the generic rank. Details of the male glossal structure are not known. The body is brilliantly metallic green or blue with yellow markings like some species of *Palaeorhiza;* sometimes the metasoma is red. The body length is 6 to 8 mm. The second submarginal cell is short, little over one-third the length of the first, both the recurrent veins being outside the limits of that cell. The female in *Xenorhiza* s. str. usually has a large projection or spine on the mesepisternum in front of the middle coxa. The small segments of the front tarsus of the female are unusually short.

■ This genus is known only from New Guinea. Three species were revised by Hirashima (1975a), and a total of five by Hirashima (1996).

Except for the short, emarginate male glossa, *Xenorhiza* would fall within the broad limits of variation found in *Palaeorhiza.* I arbitrarily elect to follow Hirashima for the present in giving *Xenorhiza* generic rank, but predict that when other characters such as male genitalia and hidden sterna are known, they will show that *Xenorhiza* is a *Palaeorhiza* whose male has either retained a plesiomorphic glossal form or acquired the female form. Originally, the strong projection or spine in front of the middle coxa of females seemed to be a good generic character, but the two species separated by Hirashima (1996) as his subgenus *Papuanorhiza* lack this feature, thus reducing the distinction between *Xenorhiza* and *Palaeorhiza.*

48. Subfamily Euryglossinae

This Australian subfamily contains rather small to minute bees. They are mostly andreniform, more robust than many Hylaeinae; some are almost hoplitiform, a few nomadiform. They vary from wholly yellow to black, sometimes with yellow markings, sometimes with the metasoma or even most of the body red; rarely, they are metallic green or blue. In contrast to most Hylaeinae, the face is usually broad, and the clypeus does not extend high above the level of the tentorial pits (Fig. 48-2e). The first flagellar segment is much shorter than the scape, cylindrical or tapering toward the base, not petiolate. The glossa of both sexes is broader than long, its apex truncate or shallowly emarginate; there is a preapical fringe in females (weak or absent in some minute forms), but not in males. The prementum is short and robust, often only twice as long as wide, without a fovea on its posterior surface but with a field of spicules on the distal half or more, suggestive of the spiculate fovea of Hylaeinae and *Scrapter* (Colletinae). The lacinia is distinct, hirsute. The galeal comb is strong, crescentic (Fig. 48-1), the bases of the bristles fused to the sclerite from which they arise and thus to one another, as in the Hylaeinae. The galeal blade is extremely short, about as broad as long, trunctate or rounded apically but usually angular, often with an anterior appendage basally (Fig. 48-1), usually without a differentiated velum. There is one subantennal suture below each antenna, or none because the clypeus sometimes reaches the antennal sclerite (Fig. 48-2e; see discussion of this topic below). The facial fovea, often absent in males, is a distinct groove or sometimes a punctiform or a broad depression. The episternal groove extends well below the scrobal suture. The scopa is absent; females carry pollen to their nests in the crop rather than externally. There are two submarginal cells, the second much shorter than the first, or the second is open so that there is only one closed submarginal cell. The stigma is much longer than the prestigma, convex within the marginal cell, usually broad but rarely *(Dasyhesma)* nearly parallel-sided before the base of vein r. The wing membrane is not papillate. The anterior surface of T1 is usually broadly concave, with a longitudinal median groove. (The same is true of some Colletinae, but this character is useful for distinguishing Euryglossinae from Hylaeinae and Xeromelissinae.) Pygidial and prepygidial fimbriae of the female are present but often sparse, especially in minute species. The pygidial plate of the female is narrow, the distal part parallel-sided or even slightly spatulate, or the plate may be constricted basally; the plate is absent in most males, but in some species of *Callohesma* it is similar to that of the female.

Larvae were characterized by McGinley (1981), who studied those of two genera, *Euryglossa* and *Pachyprosopis*.

The subfamily Euryglossinae is found in Australia, including Tasmania. *Euryglossina (Euryglossina)* has been introduced in South Africa (one specimen collected; probably not established) and New Zealand (one species, widespread, Donovan, 1983a).

Most euryglossine nests are probably in the ground, as

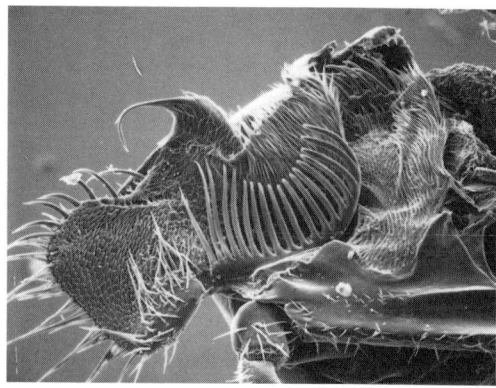

Figure 48-1. Inner surface of maxilla of *Euryglossa subsericea* Cockerell, showing the galeal comb. The base of the palpus is in the center of the lower margin. SEM photo by R. W. Brooks.

indicated by records of *Brachyhesma, Euryglossa, Euryglossula, Hyphesma,* and *Xanthesma,* but *Pachyprosopis* and *Euryglossina* apparently nest in holes (made by termites or beetles?) in wood. Michener (1965b) listed older references to nests; more recent papers are by Houston (1969) and Exley (1975b). The more or less horizontal cells in the soil, single or in short series, were described by Michener (1960) and Houston (1969). The cells, separated by earthen fill, are homomorphic, round in cross section, not flattened on the lower side, and lined with a secreted cellophane-like film; liquid provisions occupy one end of the cell, the egg floating on the surface. Nests in wood are less regular, and one nest of *Euryglossina pulchra* Exley had heteromorphic cells (the first cell rounded at the end, the others truncate) in series, separated from one another only by secreted membrane, the base of each later cell closing the one before.

The distribution of a number of characters within the subfamily is of interest. The Euryglossinae can be divided into two groups according to the size of the second submarginal cell. In the first group (Group A) this cell is much less than half as long as the first (or is absent), and the first recurrent vein joins the apical part or even the middle of the first submarginal cell, or meets the first submarginal crossvein (Fig. 48-4). Genera included in this group are *Brachyhesma, Euryglossina, Euryglossula,* and *Pachyprosopis.*

In the second group (Group B), the second submarginal cell is usually nearly half as long as the first to more than half as long as the first, and the first recurrent vein ordinarily joins the basal part of the second submarginal cell (Fig. 48-2), or meets the first submarginal crossvein (Fig. 48-3), although sometimes it enters the extreme apex of the first submarginal cell. This group includes all the remaining genera, except that many specimens of *Hyphesma* would fall better in the first group. The size of the second submarginal cell and the position of the first recurrent vein are sufficiently variable, even within species

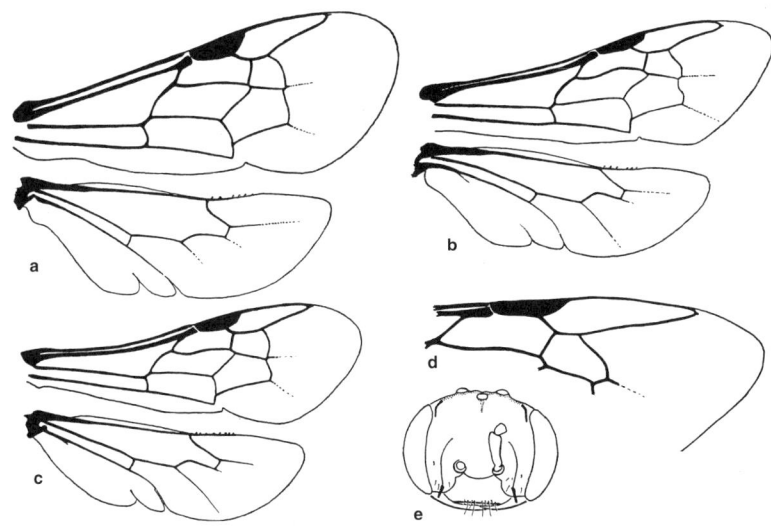

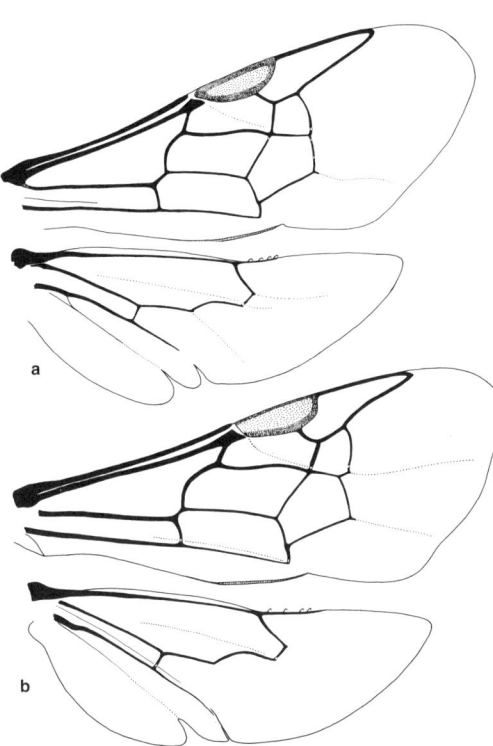

Figure 48-3. Wings of Euryglossinae, Group B. **a,** *Xanthesma furcifera* (Cockerell); **b,** *Hyphesma atromicans* (Cockerell).

Figure 48-2. Wings of Euryglossinae, Group B, and face of *Xanthesma*. **a,** *Euhesma goodeniae* (Cockerell); **b,** *Euryglossa subsericea* Cockerell; **c,** *Euhesma flavocuneata* Cockerell; **d,** Forewing fragment of *Dasyhesma abnormis* (Rayment); **e,** Face of *Xanthesma furcifera* (Cockerell), male. From Michener, 1965b.

but also among them, that these characters are unsatisfactory in keys, even though of apparent importance in grouping.

Another character of wing venation is in the posterior margin of the first submarginal cell. It is typically gently sinuate (Figs. 48-2, 48-3), but it is straight in Group A (Fig. 48-4). Moreover, it is only weakly sinuate in most *Xanthesma*, although more distinctly so in the subgenus *Xenohesma*. It is also straight in *Euhesma hemixantha* (Cockerell), a species whose generic position may be questioned.

A third wing character is indicated in the first couplet of the key to genera (below); in three of the genera of Group A (*Euryglossina, Euryglossula,* and *Pachyprosopis*), the first abscissa of vein Rs of the forewing is transverse (Fig. 48-4b), not oblique as in other bees, with resultant effects on the angles at the ends of this abscissa.

Members of genera of Group A are commonly smaller than many of those of Group B, although *Xanthesma* in Group B contains some minute bees. It is possible that the differences in wing structure mentioned above and in couplet 1 of the key are associated with size. Danforth (1989a) has shown an association between small size and more transverse rather than longitudinal orientation of certain vein segments. In view of this, probably the larger species in the first group of genera, i.e., some species of *Pachyprosopis*, evolved from minute ancestors. Moreover, all these characters might be regarded as a single tendency (reduction in size) for purposes of phylogenetic analysis. In that case, the first group of genera could be polyphyletic, derived from different ancestors and appearing similar in diverse characters because of small size. Because the minute species of *Brachyhesma* and *Xanthesma* have plesiomorphic characters associated with the base of the first submarginal cell (so that they run to 4 in the first couplet of the key), these genera appear to have become minute independently from the *Euryglossina-Euryglossula-Pachyprosopis* group.

A character not found in other bees is the prolongation of the lower end of the eye mesad above the mandibular base, the anterior mandibular articulation thus more or

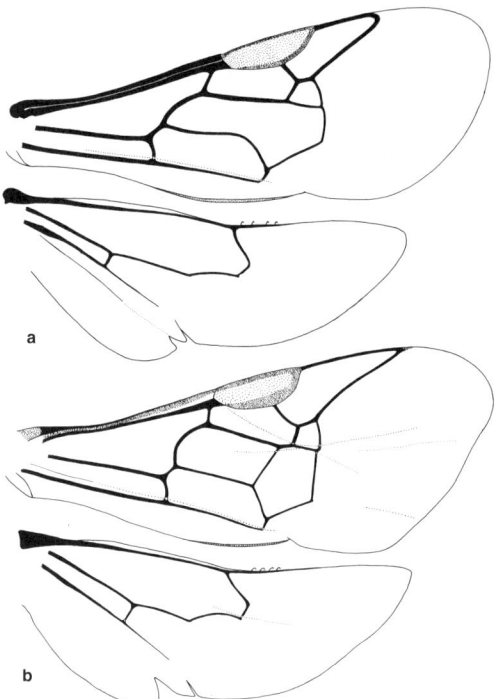

Figure 48-4. Wings of Euryglossinae, Group A. **a,** *Brachyhesma (Microhesma) incompleta* Michener; **b,** *Euryglossula chalcosoma* (Cockerell).

less on a line with the median axis of the eye, instead of on a line with the inner ocular orbit. This character occurs in females of *Pachyprosopis, Euryglossina, Euryglossula* (weakly), and *Hyphesma,* and in *Euhesma hyphesmoides* (Michener). It also occurs in males, but not females, of *Brachyhesma.*

The subantennal sutures of Euryglossinae are mostly ordinary. In *Brachyhesma, Hyphesma,* and *Xanthesma,* however, the subantennal sutures are absent and the clypeus broadly abuts against the antennal sclerites; in the last two genera the upper lateral parts of the clypeus are commonly drawn up to the antennal sclerites, well above the level of the arcuate median part of the epistomal suture. This is not necessarily very different from the condition in some *Euhesma,* e.g., in *E. wahlenbergiae* (Michener). In *Callohesma,* however, if Exley's (1974b) interpretation is correct, the apparent subantennal suture is double, with a slender clypeal ribbon extending up between two sutures from each upper clypeal angle to the antennal sclerite. This would be an elaboration of the condition found in *Hyphesma* and *Xanthesma.*

Key to the Genera of the Euryglossinae
(Modified from Michener, 1965b)

1. First abscissa of vein Rs transverse (Fig. 48-4b), so that posterior basal angle of first submarginal cell (often also apex of cell R1) is about 90°; lower end of eye of female protruding mesad above mandibular base [only slightly in *Euryglossula* and *Euryglossina (Microdontura)*], so that anterior mandibular articulation is usually on a line with median axis of eye; posterior margin of first submarginal cell straight (Fig. 48-4) (lateral fovea of T2 well defined, linear, rarely punctiform or absent in minute species; first recurrent vein joining first submarginal cell or rarely meeting first submarginal crossvein) 2
—. First abscissa of vein Rs oblique (Figs. 48-2, 48-3, 48-4a), so that posterior basal angle of first submarginal cell and apex of cell R1 are nearly always acute; lower end of eye (except in females of *Hyphesma* and *Tumidihesma* and males of *Brachyhesma*) not protruding mesad above mandible, anterior mandibular articulation therefore usually in line with inner orbit; posterior margin of first submarginal cell sinuate (Figs. 48-2, 48-3) [except in *Brachyhesma* (Fig. 48-4a) and *Euhesma hemixantha* (Cockerell)] 4
2(1). Second submarginal crossvein about one-third longer than first (as in Fig. 48-3b); costal margin of second submarginal cell sloping apically toward costa (as in Fig. 48-3b); labrum of female nearly always with strong apical spine (mandible of female bidentate, rarely simple) .... *Pachyprosopis*
—. Second submarginal crossvein usually little longer than first or absent; costal margin of second submarginal cell subparallel to costal margin of stigma; labrum usually without apical spine (minute species) 3
3(2). Basitibial plate of female defined (though in some cases very indistinctly and incompletely), one-fourth to one-sixth of length of tibia; eye of female protruding but little mesad over mandibular base; clypeus of female not strongly sloping inward but forming continuous arc with supraclypeal area, as seen in profile (Fig. 48-5a); marginal cell pointed on costa (Fig. 48-4b) .... *Euryglossula*
—. Basitibial plate of female not clearly defined, but margin indicated (often vaguely) by tubercles and ending near middle of tibia; eye of female strongly protruding mesad over mandibular base (except in subgenus *Microdontura*); clypeus of female sloping inward, at least below, usually at distinct angle to supraclypeal area (Fig. 48-5b); apex of marginal cell separated from costa, sometimes by less than width of a vein .... *Euryglossina*
4(1). Clypeus more than 3.5 times as broad as long, as seen from front; scape at least two-thirds as long as eye (Fig. 48-5c, d); antennal bases more than three times as far from median ocellus as from lower edge of clypeus; eye of male usually produced mesad above anterior mandibular articulation (minute, largely yellow forms, antennal sockets immediately above epistomal suture and subantennal sutures thus absent) .... *Brachyhesma*
—. Clypeus usually less than 3.5 times as broad as long; scape usually not much more than one-half as long as eye; antennal bases not much more than twice as far from median ocellus as from lower edge of clypeus; eye of male not produced mesad above anterior mandibular articulation 5
5(4). Costal margin of second submarginal cell distinctly sloping apically toward costa (Fig. 48-3b), the cell having more or less the same shape as that of *Pachyprosopis*; facial fovea of female with lower end curved mesad toward antennal base; eye of female strongly protruding mesad above anterior mandibular articulation (body black, without yellow markings) .... *Hyphesma*
—. Costal margin of second submarginal cell usually sub-

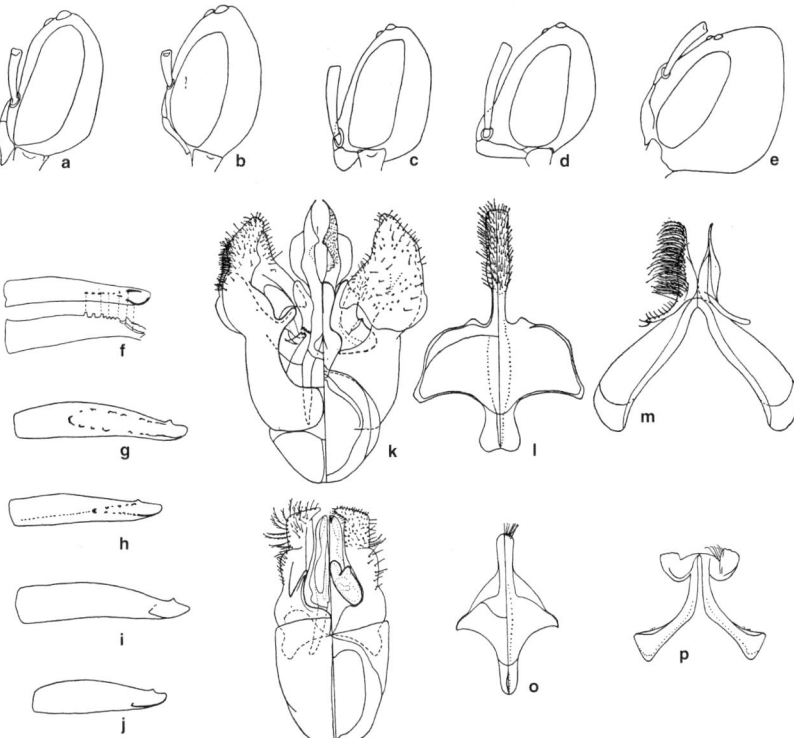

Figure 48-5. Structures of Euryglossinae. a-e, Side views of heads. **a,** *Euryglossula chalcosoma* (Cockerell), female; **b,** *Euryglossina (Euryglossina) hypochroma* Cockerell, female; **c,** *Brachyhesma (Microhesma) incompleta* Michener, male; **d,** *B. (Brachyhesma) sulphurella* (Cockerell), male; **e,** *Sericogaster fasciatus* Westwood, female.

f-j, Posterior views of hind tibiae of females, showing basitibial plates, mostly margined by tubercles as shown in outer view of f.

f, *Melittosmithia carinata* (Smith); **g,** *Sericogaster fasciata* Westwood; **h,** *Euryglossa laevigata* (Smith); **i,** *Xanthesma furcifera* (Cockerell); **j,** *Callohesma calliopsiformis* (Cockerell).

k-m, Male genitalia, S8, and S7 of *Callohesma calliopsiformis* (Cockerell); **n-p,** Same, of *Brachyhesma (Microhesma) incompleta* Michener. (Dorsal views are at the left.)

From Michener, 1965b.

parallel to costal margin of stigma (Fig. 48-2); facial fovea of female not curved mesad toward antennal base; eye of female not protruding mesad above anterior mandibular articulation [except in *Tumidihesma* and *Euhesma hyphesmoides* (Michener)] 6

6(5). T1 about as broad as long, as seen from above; body wholly black; distal part of pygidial plate of female narrower than last tarsal segment, apex upturned; inner hind tibial spur finely ciliate .... *Heterohesma*

—. T1 much broader than long *or* body largely yellow; *if* distal part of pygidial plate of female narrower than last tarsal segment, then inner hind tibial spur pectinate 7

7(6). Second submarginal cell strongly narrowed toward costa, about half as long on anterior side as on posterior side (Fig. 48-2d); second submarginal crossvein strongly curved or sinuate and at an angle of about 45° to first; head and thorax strongly and closely punctate, metasoma dull with dense, minute punctures .... *Dasyhesma*

—. Second submarginal cell little shorter on anterior side than on posterior side, second submarginal crossvein only gently curved and subparallel to first or at an angle of less than 40° to first (Fig. 48-2a-c); head and thorax with punctures fine or well separated, metasoma not dull with minute, dense punctures, although sometimes dulled by other sculpturing 8

8(7). Basitibial plate in both sexes indicated by two rows of large tubercles, the rows nearly meeting and terminating the "plate" well beyond middle of tibia (Fig. 48-5g); median ocellus closer to antennae than to posterior edge of vertex in female (Fig. 48-5e), and midway between these points in male .... *Sericogaster*

—. Basitibial plate not extending beyond middle of tibia, although a single row of tubercles may extend beyond middle; vertex less produced posteriorly so that median ocellus is at or behind midpoint between antennae and posterior edge of vertex in female and behind midpoint in male 9

9(8). Body slender, T1 seen from above little broader than long (extensive yellow pattern on body; mandible simple in both sexes) .... *Stenohesma*

—. Body of ordinary form, T1 seen from above much broader than long 10

10(9). Clypeus with strong longitudinal median carina (mandible simple) (male unknown) .... *Melittosmithia*

—. Clypeus without longitudinal carina 11

11(10). Anterior end of scutum, especially in female, nearly as broad as scutal width at anterior ends of tegulae; front of scutum curved down rather sharply and usually differently sculptured than rest of scutum; scutum coarsely puncured, in female with large, shining interspaces or smooth impunctate areas; basitibial area of female defined by large tubercles or basally by carinae, a particularly strong tubercle at apex (Fig. 48-5h) (inner hind tibial spur of female strongly pectinate; body usually without yellow markings) .... *Euryglossa*

—. Anterior end of scutum much narrower than width at anterior ends of tegulae; front of scutum curved down rather uniformly and usually not sculptured differently than rest of scutum; scutum shining and almost impunctate, as in *Tumidihesma,* or usually dull, minutely lineate or roughened, its punctation variable; basitibial area of female variable, usually without particularly strong tubercle at apex 12

12(11). Apex of marginal cell rounded or somewhat pointed, bent well away from wing margin; outer surface of hind tibia of female covered with simple bristles (integument yellow or with yellow markings) ...... *Callohesma*

—. Apex of marginal cell pointed on or almost on wing margin; outer surface of hind tibia of female usually with some plumose hairs in addition to simple bristles 13

13(12). Mandible of female (male unknown) tridentate; facial fovea linear, not bent toward ocelli; lower end of eye protruding mesad above anterior mandibular articulation .... *Tumidihesma*

—. Mandible simple or bidentate; facial fovea broader, or, *if* narrowly linear, then upper end bent toward ocelli; lower end of eye not protruding above anterior mandibular articulation (except in *Euhesma hyphesmoides* Michener) 14

14(13). Facial fovea of female slender, linear, upper fourth or more bent mesad toward ocelli (Fig. 48-2e); apical lobes of S7 of male usually directed laterad, not extending more than half a lobe width behind median apical margin of S7; mandible of female simple (with small preapical tooth in subgenus *Chaetohesma*); subantennal suture absent, upper lateral part of clypeus usually produced upward to antenna; claws of female simple (except with tooth in a few species of subgenus *Chaetohesma*); body commonly with extensive yellow markings .... *Xanthesma*

—. Facial fovea of female broad or broadly linear, upper end commonly not bent mesad; apical lobes of S7 of male commonly elongate, extending far behind median apical margin of S7; mandible of female with preapical tooth on upper margin; subantennal suture usually present, but upper lateral angle of clypeus sometimes attaining antennal base, eliminating subantennal suture; claws of female cleft or with inner tooth, rarely simple; body usually without extensive yellow markings .... *Euhesma*

Genus *Brachyhesma* Michener

This is a genus of minute (body length 2.7-4.0 mm), largely yellow-bodied bees. In spite of their small body size, the base of the first submarginal cell is acute (Fig. 48-4a), a plesiomorphy relative to *Euryglossina* and its relatives (see first couplet of the key to genera). As in those *Euryglossina* with more extensive wing venation, the second submarginal cell is complete, but less than half as long as the first, and the first recurrent vein (when present) ends near the apex of the first submarginal cell. The clypeus, when seen from the front, is 3.5 to 10 times as wide as long, and the antennae arise far down on the face, next to the clypeus, so that subantennal sutures do not exist (Fig. 48-5c, d). Thus the frons is large compared to that of other bees. The labrum of the female lacks the midapical spine found in most *Pachyprosopis* and some *Euryglossina,* but often has several coarse, spinelike apical setae. The facial foveae of females are linear, long, the upper ends curved mesad, usually almost to the lateral ocelli. In males they are long and slender but not curved toward the ocelli. The eyes of the male (not females as in *Euryglossina, Hyphesma,* and *Pachyprosopis*) protrude mesad over the mandibular bases, except in the subgenus *Henicohesma*. In males the hind tibial spurs appear to be absent, but are replaced by one to several large bristles, these sometimes curiously shaped. Unlike many other Euryglossinae, *Brachyhesma* has a small basitibial plate, about one-fifth as long as the tibia, weakly defined only on the posterior margin, or unrecognizable in some males. The claws of females are simple. Illustrations of various structures, including male genitalia, were given by Michener (1965b) and Exley (1968e, 1974c, 1975a, 1977); see also Figure 48-5n.

Brachyhesma, a rather large Australian genus, was revised by Exley (1968e, 1977) with keys to species in Exley (1968f, 1975a). On the basis of fragmentary evidence (Exley, 1968f) and excavation of a single nest (Houston, 1969), one can suppose that nests are regularly or always in the ground. All species visit flowers of Myrtaceae.

There are two major subgenera, *Brachyhesma* s. str. and *Microhesma,* and two small or monotypic subgenera, *Henicohesma* and *Anomalohesma*. The latter are not specialized derivatives of the large subgenera, for each has plesiomorphies not shared by the large subgenera. Presumably, the small subgenera are basal branches, each a sister group to one or both of the large subgenera. Subgeneric characters are largely unknown in females.

Key to the Subgenera of *Brachyhesma,* Based on Males
(Modified from Exley, 1977)

1. Supraclypeal area and upper part of clypeus protruding strongly forward between antennal bases (Fig. 48-5d); clypeus, in frontal view, a mere strip across lower end of face, its major area reflexed and exposed ventrally 2

—. Supraclypeal area and upper part of clypeus not protruding strongly forward between antennal bases (Fig. 48-5c); clypeus rather broadly exposed in frontal view, neither reflexed nor broadly exposed ventrally 3

2(1). Reflexed part of clypeus forming a flattened or concave triangular plate; scape nearly as long as to longer than eye (Fig. 48-5d) .... *B. (Brachyhesma* s. str.)

—. Reflexed part of clypeus convex, not forming triangular plate; scape much shorter than eye ..... *B. (Anomalohesma)*

3(1). Metasoma yellow with transverse brown tergal bands; profile of clypeus strongly convex (Fig. 48-5c); gonobase almost one-half length of genitalia (Fig. 48-5n) .... *B. (Microhesma)*

—. Metasoma completely yellow; profile of clypeus nearly flat; gonobase about one-third length of genitalia .... *B. (Henicohesma)*

Brachyhesma / Subgenus *Anomalohesma* Exley

Brachyhesma (Anomalohesma) Exley, 1977: 39. Type species: *Brachyhesma scapata* Exley, 1977, by original designation.

Although the male clypeus is a mere lower rim across the face as seen from the front, as in *Brachyhesma* s. str., and although it is flexed under as in that subgenus, it does not form a broad triangular plate as seen from below. Thus the male clypeal structure is intermediate between the more plesiomorphic form of *Microhesma* and the derived clypeal form of *Brachyhesma* s. str. In other characters, however, *Anomalohesma* is not simply an intermediate between these two subgenera; some features agree with one, others with the other. The short scape and hind basitarsus of the male and the small male gonobase are probable plesiomorphies, the first two in agreement with *Microhesma*, the last in agreement with *Brachyhesma* s. str. The simple hairs on S7 and the abundant long hairs on the apex of S8 of the male are as in *Microhesma*.
- This subgenus is known from south coastal Queensland, Australia. The single species is *Brachyhesma scapata* Exley.

Brachyhesma / Subgenus *Brachyhesma* Michener s. str.

Brachyhesma Michener, 1965b: 112. Type species: *Euryglossina sulphurella* Cockerell, 1913, by original designation.

As indicated in the key to subgenera, males of *Brachyhesma* s. str. have one of the most unusual clypeal structures found among bees. Other apparently apomorphic features are the long scape and the long hind basitarsus of males, longer than the hind tibia in most species. S7 of the male has branched hairs in nearly all species, and S8 has only a few very short hairs at the apex, as in the subgenus *Henicohesma*.
- This subgenus is found in all Australian states except Tasmania. It is particularly common in xeric regions and is unknown on the east coast or in the Great Dividing Range. There are 22 species, revised and keyed out by Exley (1968e, f; 1975a; 1977), and listed by Cardale (1993).

Brachyhesma / Subgenus *Henicohesma* Exley

Brachyhesma (Henicohesma) Exley, 1968e: 199. Type species: *Brachyhesma macdonaldensis* Exley, 1968, by original designation.

Henicohesma seems most closely related to the subgenus *Microhesma*. The strong differentiating characters are in the male: the ordinary-sized rather than enlarged gonobase, the branched hairs of S7 (as in *Brachyhesma* s. str.) and the few, short hairs on the apex of S8 (also as in *Brachyhesma* s. str.). Other characters are indicated in the key to subgenera and in the account of the generic characters.
- This subgenus is known from Northern Territory and Queensland, Australia. The two species were revised by Exley (1968e) and listed by Cardale (1993).

Brachyhesma / Subgenus *Microhesma* Michener

Brachyhesma (Microhesma) Michener, 1965b: 113. Type species: *Brachyhesma incompleta* Michener, 1965, by original designation.

Microhesma differs from all other subgenera in the enormous male gonobase, one-half as long as the whole genitalia, quite unlike that of related bees (Fig. 48-5n). Otherwise, its characters are either probably plesiomorphic, for example, the relatively short scape and male hind basitarsus and the forward-directed clypeus; or not readily polarizable, for example, the long hairs at the apex of male S8 and the simple hairs of S7.
- The species of *Microhesma* are recorded from all Australian states except Tasmania, but are most abundant in the north (Queensland, Northern Territory). The 16 described species were revised by Exley (1968e, f; 1975a; 1977) and listed by Cardale (1993).

The type species is remarkable for lacking the first recurrent vein (Fig. 48-4a).

Genus *Callohesma* Michener

Euryglossa (Callohesma) Michener, 1965b: 95. Type species: *Euryglossa calliopsiformis* Cockerell, 1905, by original designation.

This is a genus of small to moderate-sized bees (3.5-12.5 mm long) with extensive yellow markings on the body or the whole body sometimes yellow or greenish yellow; rarely, the metasoma or other areas are red or reddish brown. The apex of the marginal cell is bent away from the costal margin of the wing and usually rounded. This feature, although common in other groups of bees, is not found in other Euryglossinae and is doubtless a synapomorphy for the genus. The antennal bases are well above the clypeus. Either the subantennal sutures are distinctly present or there is a narrow ribbon that appears to be derived from the upper lateral angle of the clypeus, extending up to each antennal base. The result is two subantennal sutures on each side, very close together, sometimes fused (see Exley, 1974b). The male genitalia appear to have distinctive features not found in other Euryglossinae; they were illustrated along with other structures by Michener (1965b) and Exley (1974b).
- *Callohesma* is found in all Australian states except Tasmania. The 34 known species were revised by Exley (1974b) and listed by Cardale (1993). They visit flowers of Myrtaceae.

Genus *Dasyhesma* Michener

Dasyhesma Michener, 1965b: 102. Type species: *Dasyhesma robusta* Michener, 1965, by original designation.
Euryglossa (Dermatohesma) Michener, 1965b: 91. Type species: *Euryglossimorpha abnormis* Rayment, 1935, by original designation.

Two genus-group names were published simultaneously for species now placed in *Dasyhesma*, the name selected by Michener (2000).

This genus consists of unusually robust euryglossines about 8 mm long. The metasoma is made dull by dense, fine punctation. T1 has an obtusely angulate profile, and the anterior surface is thus more nearly vertical than in most other euryglossines, in which the profile of T1 is rounded. The propodeum, in profile, is nearly all subvertical, there being only a narrow, sloping upper zone. Other distinctive characters are the strongly converging eyes (the head shape is thus like that of many male *Colletes*), the strongly depressed, broadly linear facial fovea, the strong punctation of the head and thorax, and especially the shape of the second submarginal cell, as indicated in the key to genera and Figure 48-2d. Male genitalia and other structures were illustrated by Michener (1965b).

■ *Dasyhesma* is found in Western Australia. The genus was revised by Exley (2004) who recognized 21 species.

In 1965 it seemed that one species, now called *Dasyhesma abnormis* (Rayment), showed affinities to *Euhesma*. These affinities are indicated by euryglossine plesiomorphies, and the relation to *D. robusta* Michener was even then obvious.

Genus *Euhesma* Michener

Euhesma is here given generic rank because it seemed appropriate to give that rank to the former *Euryglossa* s. str., as well as to *Callohesma*, which was included in the genus *Euryglossa* by earlier authors, including Michener (1965b). *Euhesma* comprises the species left after removal of *Euryglossa* s. str. and *Callohesma;* it remains a diverse group without known synapomorphies and will probably be subdivided in the future. Probably it is paraphyletic. *Parahesma* is here included in *Euhesma* because its single species, unknown in the male, fits rather well into the broad range of variation found in *Euhesma*. Michener (1965b) discussed some of the characters and more unusual variants found in *Euhesma*, citing certain species as examples having the less common variations to be found in this genus. The following is a modification of that discussion: These are mostly small to moderate-sized bees (4.5-8.0 mm long). The dorsum of the thorax is minutely roughened and dull between small and widely separated punctures, except in *E. crabronica* (Cockerell) and *rufiventris* (Michener), in which it is coarsely punctate and shining. The body is black or greenish, the metasoma rarely red [*E. maculifera* (Rayment), *platyrhina* (Cockerell), *rainbowi* (Cockerell), *rufiventris*], in some cases with broken yellow bands or lateral spots. The clypeus and other face marks in some species are yellow, and rarely the thorax as well as the metasoma have extensive yellow markings [*E. australis* (Michener), *perditiformis* (Cockerell)]. The antennae are slightly above to slightly below the middle of the face (far above in *E. maculifera*). The subantennal sutures are longer than the diameter of the antennal socket to very short or essentially absent. The facial fovea is narrow and distinct [*E. crabronica, serrata* (Cockerell)] to broad, indistinct, or virtually absent. The basitibial plate is completely surrounded by a carina [*E. crabronica, hemichlora* (Cockerell), *hemixantha* (Cockerell)] or more often open distally and defined by carinae only laterally, the carinae uncommonly broken into tubercles as in *E. fasciatella* (Cockerell); the basitibial plate is one-fifth to nearly one-half as long as the tibia. The inner hind tibial spur of the female is usually coarsely serrate, in some [e.g., *E. neglectula* (Cockerell)] ciliate, in others [*E. perkinsi* (Michener), *wahlenbergiae* (Michener)] rather weakly pectinate. In some cases, each claw of the female has a large median tooth (as in *Euryglossa*), but more often the tooth is smaller and more nearly parallel to the main ramus of the claw; the tooth is very small in *E. hemichlora* and some others and absent, the claws simple, in females of *E. altitudinis* (Cockerell), *australis, malaris* (Michener), and *ridens* (Cockerell).

Some other noteworthy variations are as follows: Long labial palpi (as long as maxillary palpi) combined with a flattened clypeus occur in *Euhesma goodeniae* (Cockerell), *palpalis* (Michener), and especially *platyrhina* (Cockerell). *E. malaris* (Michener) has even longer labial palpi, but the clypeus is enormously protuberant and there is a long malar space unique in the genus. Extremely long maxillary palpi that fit together to form a tube are found in *E. tubulifera* (Houston).

In *Euhesma undulata* (Cockerell) the outer surfaces of the hind tibiae of the females lack or nearly lack plumose hairs, as in the genus *Callohesma*, which is similar in appearance to this species and may be related to it. In *E. crabronica* (Cockerell) the second recurrent vein is received two-thirds of the way from base to apex of the second submarginal cell, instead of near the base as in other *Euhesma*. *E. maculifera* (Rayment) is unusual in that the first recurrent vein is basal to the first submarginal crossvein.

Euhesma dolichocephala (Rayment) is unusual because of the very long head (length is to width as 10.0:7.5, the head much narrower than the thorax), the short flagellum (basal segments in both sexes more than twice as broad as long), and the heavy and strongly curved outer hind tibial spur in both sexes.

A group of small species like *Euhesma altitudinis* (Cockerell), *australis* Michener, *hemichlora* (Cockerell), and *ridens* (Cockerell) shows some yellow markings, at least in the male, and the dark areas are often more or less metallic green. The front basitarsus of the females is short and broad, the inner hind tibial spur serrate, the facial fovea absent, and the scape, especially in the male, short, no longer than the last flagellar segment. An interesting feature of this group is the reduction of the inner tooth of each claw; in males it is smaller than usual; in the females it is small or absent, as indicated above. This feature suggests the genus *Xanthesma*.

Euhesma fasciatella (Cockerell) and *neglectula* (Cockerell) are entirely black species; the inner hind tibial spur is ciliate, and there is a row of large tubercles along the outer margin of the hind tibia beyond the basitibial plate. In these respects the two species resemble the genus *Heterohesma*. The discovery of males of *Heterohesma* may shed light on this relationship.

In *Euhesma hyphesmoides* (Michener) the eyes are produced mesad above the mandibular bases, as in the genus *Hyphesma*. The species resembles *Hyphesma,* also, in general appearance (no yellow markings) but not in the generic characters of facial foveae (linear, but the lower ends not bent toward the antennal bases), wings, etc. The species may be related to *Tumidihesma*, which has similar eyes but linear foveae that are not bent toward either the antennal bases or the ocelli, and tridentate mandibles. Males of both *E. hyphesmoides* and *Tumidihesma* are unknown, and should contribute to our understanding of these bees.

Key to the Subgenera of *Euhesma*

1. Basitibial area of female margined by large tubercles and reaching middle of tibia, apex of area marked by strong tubercle; disc of scutum almost impunctate, smooth and shiny, in female; inner hind tibial spur of female coarsely pectinate as in *Callohesma* and *Euryglossa* (male unknown) .... *E. (Parahesma)*
—. Basitibial area of female shorter, apex usually not marked by large tubercle, margins usually indicated by carinae that are sometimes mostly absent, sometimes broken into large tubercles; disc of scutum punctate and usually minutely roughened and dull; inner hind tibial spur of female ciliate, serrate, or weakly pectinate, that is, with teeth short .... *E. (Euhesma s. str.)*

Euhesma / Subgenus *Euhesma* Michener s. str.

Euryglossa (Euhesma) Michener, 1965b: 88. Type species: *Euryglossa wahlenbergiae* Michener, 1965, by original designation.

The principal characters of this subgenus are indicated in the key. The wide variability among species is discussed under the genus *Euhesma*. The present classification is one of convenience. We need to know both sexes of many more species, after which a rational classification of the genus should be possible. Illustrations of male genitalia, sterna, and other characters were given by Michener (1965b) and Houston (1992b).

■ *Euhesma* s. str. is widespread in Australia, including Tasmania, but is not abundant in the north of the continent. Forty-five species were listed by Cardale (1993), and many new species remain to be described; E. Exley has described 20 new species taken on flowers of *Eremophila* (Myoporaceae). E. Exley has started a revisional study of *Euhesma* and revised the *walkeriana* species group (Exley, 2001) and the *crabronica* group (Exley, 2002).

Although some species are associated with Myrtaceae, this subgenus includes many species probably oligolectic on other flowers, such as *Euhesma wahlenbergiae* (Michener) on *Wahlenbergia* (Michener, 1965b), *E. tubulifera* (Houston) on *Calothamnus* (Houston, 1983c), and many species on *Eremophila*. Probably all the species with unusual mouthparts listed in the discussion of the genus are oligolectic on flowers other than Myrtaceae. Nests in the ground were described by Rayment (references in Michener, 1965b: 87).

Euhesma / Subgenus *Parahesma* Michener

Euryglossa (Parahesma) Michener, 1965b: 92. Type species: *Euryglossa tuberculipes* Michener, 1965, by original designation.

This subgenus is known from a single female specimen. It might have been left within the diverse subgenus *Euhesma*, although it deviates from all members of that subgenus in several characters; the principal ones are noted in the key. I list it here purely provisionally, because it has and may deserve a subgeneric name.

■ *Parahesma* is from the state of Victoria, Australia. *Euhesma tuberculipes* (Michener) is the only species.

Genus *Euryglossa* Smith

Euryglossa Smith, 1853: 17. Type species: *Euryglossa cupreochalybea* Smith, 1853, by designation of Meade-Waldo, 1923: 6.
Stilpnosoma Smith, 1879: 16. Type species: *Stilpnosoma laevigatum* Smith, 1879, monobasic.
Euryglossa (Euryglossimorpha) Strand, 1910: 40. Type species: *Euryglossa nigra* Smith, 1879, monobasic.

The name *Euryglossa* is used here in a different than usual sense, to include the subgenus *Euryglossa* s. str. of Michener (1965b) plus the genus *Stilpnosoma*. The latter is based on a single species, *Euryglossa laevigata* (Smith), that differs from the rest of the genus primarily in characters that can be seen as tendencies elsewhere in the genus. Thus *E. laevigata* is bright metallic green; some other species are metallic but less brightly so. *E. laevigata* has broad genal and vertex areas, so that the median ocellus of the female is about midway between the antennal bases and the posterior margin of the vertex; this condition is approached in some other species.

Euryglossa differs from nearly all other moderate-sized to large Euryglossinae in the rather cylindrical, hoplitiform appearance of the head and thorax, which results from the tendency toward a large and quadrate head and especially from the swollen anterior part of the scutum, which is parallel-sided in front of the tegulae rather than narrowing anteriorly (see the key to the genera). This feature is not well developed in males, many of which can be recognized immediately by their long antennae, sometimes with a flattened distal segment. The body length ranges from 5 to 15 mm, the males usually being much smaller than the females. Although some species are black, others have dark metallic blue or green coloration on the metasoma and thorax, or rather bright green on the whole body; some have a red metasoma or even much of the body may be red. One species, *E. limata* Exley, has extensive yellow markings suggesting species of *Callohesma*. Male genitalia and other structures were illustrated by Michener (1965b) and Exley (1976b).

■ *Euryglossa* is widespread in Australia, including Tasmania, but is not particularly common in xeric areas. The 36 described species were revised by Exley (1976b) and listed by Cardale (1993).

So far as is known, species of this genus make nests in the ground; references to relevant papers were given by Michener (1965b: 87). They visit flowers of Myrtaceae.

Genus *Euryglossina* Cockerell

This is a genus of minute bees (1.8-5.0 mm long) in which the first abscissa of Rs of the forewing is transverse, as described in the first couplet of the key to genera (Figs. 48-4b, 48-6). The body is nonmetallic black, and the clypeus, paraocular areas, supraclypeal area, parts of legs, and pronotal lobes are often yellow; the metasoma is often brownish, yellowish beneath; rarely, as in *Euryglossina (Euryglossina) aurantia* Exley and *E. (Microdontura) mellea* (Cockerell), the body is largely yellow. Usually, the whole clypeus of females slopes inward, at a distinct an-

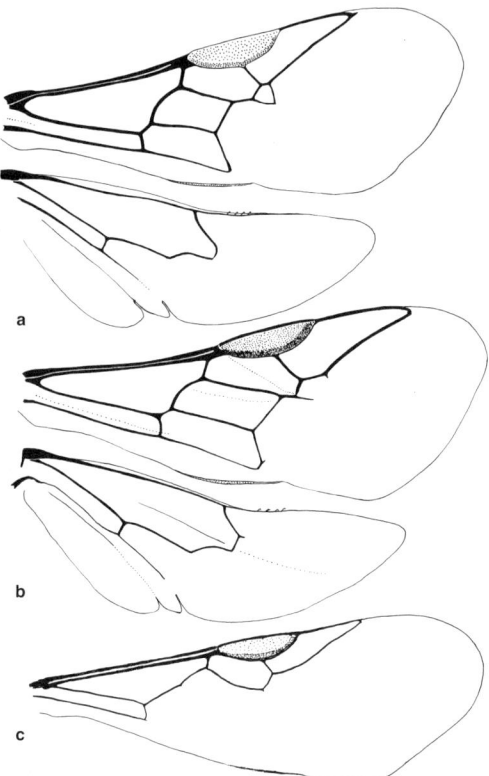

Figure 48-6. Wings of Euryglossinae, Group A. **a**, *Euryglossina (Microdontura) mellea* (Cockerell); **b**, *Euryglossina (Euryglossina) nothula* (Cockerell), a species having the *"Turnerella"* type wing venation; **c**, *Euryglossina (Quasihesma) moonbiensis* (Exley), diagram showing the most reduced wing venation known among bees.

gle to the supraclypeal area (Fig. 48-5b), thus distinguishing the genus from *Euryglossula*. Sometimes, however, only the apical part of the clypeus curves strongly inward, as seen in profile. The facial foveae are linear but extremely variable in length. The reduced wing venation was illustrated for several species by Michener (1965b); see also Figure 48-6.

I here interpret this genus more broadly than did Exley (1968d), to include two groups to which she gave generic status, i.e., *Euryglossella* and *Quasihesma*, as well as a species placed in *Pachyprosopis* by Michener (1965b), i.e., *Euryglossina (Pachyprosopina) paupercula* Cockerell, new combination. Various degrees of reduction of wing venation occur independently in different groups. For example, one species of the subgenus *Euryglossella* [*E. (Euryglossella) incompleta* (Exley)] lacks the first recurrent vein, thus having wing venation as in most *Quasihesma*. Likewise, reduction in the number of antennal segments to 12 in males (and to 11 in females of some species of *Quasihesma*) occurs in different groups, i.e., in most *Quasihesma* and some species of different species groups of *Euryglossina* s. str. Thus some characters once thought to have generic importance turn out to be variable. Even the labral spine of females, a principal character of *Euryglossella*, is shared by at least some *Quasihesma* (Exley, 1974a) and by *Pachyprosopina;* even in the related genus *Pachyprosopis*, which supposedly has such a spine, it is evidently missing (lost?) in one species, *P. (Pachyprosopis) cornuta* Exley (Exley, 1972).

Euryglossina is found throughout Australia, including Tasmania, but is especially abundant in the north and in xeric areas. So far as is known, all species visit principally flowers of Myrtaceae.

Key to the Subgenera of *Euryglossina*

1. Hind tibia (both sexes) with row of suberect, scalelike and spinelike setae on outer surface from near base to apex; apex of pygidial plate of female elongate, upturned .... *E. (Microdontura)*
—. Hind tibiae without such setae, usually with scattered small tubercles, largest one (near middle of tibia in females, slightly basal to middle in males) probably representing apex of otherwise undefined basitibial plate; apex of pygidial plate of female not upturned 2
2(1). Costal edge of marginal cell distinctly shorter than stigma (as in Fig. 48-6a); first recurrent vein, if present, entering first submarginal cell near middle; vein M+Cu between basal vein (M) and vein m-cu of forewing longer than basal vein; claws of female with small preapical tooth 3
—. Costal edge of marginal cell equal to or slightly longer than stigma; first recurrent vein, if present, entering first submarginal cell near apex or in distal one-fifth; vein M+Cu between basal vein (M) and vein m-cu of forewing about one-third as long as basal vein; claws of female simple 4
3(2). S7 of male with small to moderate-sized, laterally directed, apical lobes and broadly expanded, triangular basolateral apodemes; face of male with minute pit ("glandular opening") above and lateral to antennal base .... *E. (Quasihesma)*
—. S7 of males almost without small posteriorly directed apical lobes or with small ones, basolateral apodemes normal, almost straplike form; face of male lacking pit above and lateral to antennal base .... *E. (Euryglossella)*
4(2). Labrum of female without median apical spine; second submarginal cell usually not shaped as in *Pachyprosopis* .... *E. (Euryglossina s. str.)*
—. Labrum of female with median apical spine; second submarginal cell shaped like that of *Pachyprosopis* (as in Fig. 48-3b) .... *E. (Pachyprosopina)*

Euryglossina / Subgenus *Euryglossella* Cockerell

Euryglossella Cockerell, 1910c: 263. Type species: *Euryglossella minima* Cockerell, 1910, monobasic.

Zalygus Cockerell, 1929b: 321. Type species: *Zalygus cornutus* Cockerell, 1929, monobasic.

This taxon was given generic rank separate from *Euryglossina* by Exley (1968b) largely because of the strong spine on the labrum of females. The presence of such a spine in a bee that appears to be very like *Euryglossina* s. str., i.e., the subgenus *Pachyprosopina*, suggests that this character appears in rather different groups and may be independently lost or gained. Moreover, a species of the related genus *Pachyprosopis* seems to lack such a spine while others in the genus possess it. Couplet 2 of the key to subgenera, however, lists additional strong characters that distinguish *Euryglossella* (along with *Quasihesma*) from *Euryglossina* s. str. These characters include derived features, like those of the forewing, and plesiomorphies not shared with *Euryglossina,* like the cleft claws. I suspect that *Euryglossella-Quasihesma* is the sister group to *Euryglossina-Pachyprosopina*. *Euryglossella* averages smaller than other *Euryglossina* subgenera, ranging from 2.0 to 2.8 mm long; the other subgenera are not under 2.5 mm in length. Illustrations, including those of male genitalia, were presented by Michener (1965b) and Exley (1968b, 1974a). In all species of *Euryglossella* the second submarginal cell is absent, as in those species of *Euryglossina* s. str. formerly placed in *Turnerella* (Fig. 48-6b). In one species, *E. (E.) incompleta* (Exley), the second cubital cell of the forewing is also open, so that there are only four closed cells in the wing, as in *Quasihesma* and one species of *Euryglossina* s. str. Interesting as are these wing characters, they do not necessarily differentiate monophyletic groups.

■ The species of *Euryglossella* are from Queensland and Northern Territory, Australia. The eight known species were revised by Exley (1968b) and new keys were given by Exley (1974a, 1982); the species were listed by Cardale (1993).

Euryglossina / Subgenus *Euryglossina* Cockerell s. str.

Euryglossa (Euryglossina) Cockerell, 1910a (August): 211. Type species: *Euryglossa semipurpurea* Cockerell, 1910, monobasic.

Turnerella Cockerell, 1910c (October): 262. Type species: *Turnerella gilberti* Cockerell, 1910, monobasic.

Euryglossina and *Turnerella* have been distinguished on the basis of the more reduced wing venation of the latter, in which the second submarginal cell is incomplete and the second recurrent vein is absent (Fig. 48-6b). But in five species (Exley, 1969e) the second submarginal cell is complete, although the second recurrent vein is absent, as shown by both Michener (1965b) and Exley (1968d). There is also variation in the incompleteness of the second submarginal cell. Finally, there is a species, *E. proserpinensis* Exley, in which the venation is even more reduced, the second cubital cell being open, so that the forewing contains only four closed cells (first cubital, radial, first submarginal, and marginal), as is usual in the subgenus *Quasihesma* (Fig. 48-6c). Exley pointed out that there are some species in each "venational subgenus" that have 12-segmented antennae in the males. Both Exley (1968d) and Michener (1965b) suggested that venational reduction quite possibly occurred independently in different groups of *Euryglossina,* leaving *Turnerella* polyphyletic. Certainly if *Turnerella* (distinguished on the basis of only one or two venational characters) is recognized, then *Euryglossina* would be paraphyletic. The time has come to synonymize *Turnerella*! *Euryglossina,* in the sense used here, can be recognized by its lack of a labral spine in the female, the length of the costal margin of the marginal cell (about as long as the stigma), and the simple claws of the female. Male genitalia, hidden sterna, and other characters were illustrated by Michener (1965b) and Exley (1968d, 1969e, 1976d).

■ *Euryglossina* s. str. is found in all Australian states including Tasmania, and, as is indicated in the account of the subfamily, has been introduced into New Zealand and South Africa, although probably not established in the latter. There are about 54 described species and various undescribed species. The subgenus was revised by Exley (1968d), and sections of keys were extended by Exley (1976d); the species were listed by Cardale (1993).

Four species of the subgenus *Euryglossina* s. str. have been reported entering abandoned holes made by small beetles in dead wood (branches, log, posts, telegraph pole) (Exley, 1968d), and the nests of three of the species were found in such situations and described (Houston, 1969).

Euryglossina / Subgenus *Microdontura* Cockerell

Microdontura Cockerell, 1929b: 322. Type species: *Microdontura mellea* Cockerell, 1929, monobasic.

This subgenus consists of a species that differs markedly from the rest of *Euryglossina* not only in the characters indicated in the key to subgenera but in its unusually slender body. The costal edge of the marginal cell is shorter than the stigma, as in *Euryglossella* and *Quasihesma,* but the recurrent vein, unlike that in those subgenera, enters the first submarginal cell at the distal one-third or one-fourth (Fig. 48-6a), and the claws of the female are simple. The second submarginal cell is small but usually complete, rarely open.

■ *Microdontura* is found in Queensland and New South Wales, Australia. The single species, *Euryglossina mellea* (Cockerell), was treated by Exley (1968d).

Euryglossina / Subgenus *Pachyprosopina* Michener

Pachyprosopis (Pachyprosopina) Michener, 1965b: 108. Type species: *Euryglossa paupercula* Cockerell, 1915, by original designation.

This subgeneric name is included here tentatively. *Pachyprosopina* agrees with *Euryglossina* s. str. except for the spine on the apex of the labrum and the somewhat *Pachyprosopis*-like second submarginal cell, the latter found also in some species of *Euryglossina* s. str. *Pachyprosopina* is known from females of a single species, *Euryglossina paupercula* (Cockerell). It was removed from *Pachyprosopis* by Exley (1972) and placed only hesitantly in that genus by Michener (1965b). If males are found to resemble *Euryglossina* s. str. in genitalia, sterna, and other characters, presumably *Pachyprosopina* should be synonymized with *Euryglossina* s. str., since the only known difference then would be the labral spine.

■ *Pachyprosopina* is known from southwestern Australia. The only species is *Euryglossina paupercula* (Cockerell).

Euryglossina / Subgenus *Quasihesma* Exley

Quasihesma Exley, 1968c: 228. Type species: *Quasihesma moonbiensis* Exley, 1968, by original designation.

This taxon was proposed for a group of unusually minute species in which the second cubital cell of the forewing is open, so that there are only four closed cells, and in which the antennae are reduced to 12 segments in the male and 11 in the female. These characters fail, however, in *Euryglossina (Quasihesma) gigantica* (Exley), discovered later, which has five cells and normal antennal segmentation. Moreover, there are males with 12-segmented antennae in *Euryglossina* s. str., and wings with only four closed cells in that subgenus and in *Euryglossella*. Distinctive features of *Quasihesma* remain, however. To the characters indicated in the key to subgenera can be added the usually flat mandibles, which often show a striking color pattern, and the tuft or fringe of long hairs on the outer undersurface of the scape of males. The body length ranges from 1.8 to 3.5 mm. Various characters show the close relationship of *Quasihesma* to *Euryglossella*. The clypeus is usually very short and transverse, four to eight times as wide as long in both *Quasihesma* and *Euryglossella*. The labrum of the female has a spine (not verified for all species), and the claws of the female are cleft, as in *Euryglossella*. S8 of the male is transverse with a strong apical process in both subgenera, instead of being diamond-shaped as in *Euryglossina* s. str. Illustrations of male genitalia, hidden sterna, and other characters were provided by Exley (1968c, 1974a).

■ *Quasihesma* has been found in Queensland and Northern Territory, rarely in New South Wales, Australia. Keys to the ten known species were given by Exley (1974a, 1980), and the species were listed by Cardale (1993).

It would be easy to justify a united *Quasihesma-Euryglossella* as a genus distinct from *Euryglossina*. *Quasihesma* should probably be synonymized under *Euryglossella;* all its characters seem apomorphic with relation to *Euryglossella*, so that the latter is probably paraphyletic if *Quasihesma* is recognized. Until a cladistic analysis is done, however, it seems premature to synonymize *Quasihesma*.

Genus *Euryglossula* Michener

Euryglossula Michener, 1965b: 111. Type species: *Euryglossina chalcosoma* Cockerell, 1913, by original designation.

This is a genus of minute bees (body length 2.5-3.5 mm) at least superficially suggestive of *Euryglossina*. It differs from *Euryglossina* in its broader body and broader head, which is never quadrate as is often the case in *Euryglossina*. The clypeus is not curved backward and downward as in *Euryglossina;* thus the clypeus and supraclypeal area of the female, as seen in profile, form a continuous arc (Fig. 48-5a). The eyes are more similar in the two sexes than are those in *Euryglossina*, the lower inner angle protruding less over the base of the mandible in females, and more in males, than in that genus. In the type species the head and thorax have weak greenish or brassy metallic tints, unlike any *Euryglossina*, but this is not true of all *Euryglossula*. Unless the venational characters indicated in couplet 1 of the key to genera are convergent, *Euryglossula* may be the sister group of *Euryglossina-Pachyprosopis*. *Euryglossula* has plesiomorphic features, such as the head shape, combined with apomorphies relative to *Euryglossina-Pachyprosopis,* such as the fringe of long hairs on S5. The claws of the female are simple, as in some *Pachyprosopis (Pachyprosopula)* and some *Euryglossina*. Illustrations, including those of male genitalia, were provided by Michener (1965b) and Exley (1968a).

■ Species of *Euryglossula* occur in all Australian states except Tasmania. The seven species were listed by Cardale (1993). The genus was revised (without a key) by Exley (1968a), and a new species and key to species were presented by Exley (1969a).

Nests are found in earthen banks (Houston, 1969). Flowers visited are largely Myrtaceae.

Genus *Heterohesma* Michener

Heterohesma Michener, 1965b: 97. Type species: *Stilpnosoma clypeata* Rayment, 1954, by original designation.

Known only in the female, this is a genus of large (10 mm long), slender, almost nomadiform, black bees having a dull surface on the head and thorax and only scattered punctures. The posterior articulation of the mandible is far from the eye, the anterior articulation close to the eye. The facial fovea is very large and broad, scarcely recognizable. The clypeus is small, separated from the eye by more than the width of the scape, and the anterior margin has a median tooth. The basitibial plate, less than one-fourth the length of the tibia, is bounded by a row of tubercles, and a row of tubercles extends from the apex of the plate toward the apex of the tibia. Michener (1965b) and Exley (1983) provided illustrations.

■ The species of this genus are found in the mountains of eastern Australia from northern New South Wales to Tasmania. The two species were distinguished by Exley (1983).

Michener (1965b) suggested that *Heterohesma* may be related to the group of *Euhesma fasciatella* (Cockerell). If so, *Heterohesma* is probably a derived member of that group rather than a distinct genus. Males should be found and studied before such a decision is reached.

Genus *Hyphesma* Michener

Hyphesma Michener, 1965b: 103. Type species: *Pachyprosopis atromicans* Cockerell, 1913, by original designation.

This is a genus of small (3.5-5.0 mm long) black bees without pale markings, the males somewhat more hairy than most Euryglossinae and thus superficially suggesting small Colletinae or Halictinae. As indicated in the first couplet to the key to genera, the posterior basal angle of the first submarginal cell is acute. The posterior margin of that cell, unlike that in *Brachyhesma, Euryglossina, Euryglossula,* and *Pachyprosopis,* but like that in other Euryglossinae, is sinuate (Fig. 48-3b). The second submarginal cell, shaped as in *Pachyprosopis,* is less than half as long as the first. The eyes converge below, and especially in females they protrude mesad above the mandibular bases. The facial foveae of females are linear, their lower ends curved mesad toward the antennal bases; they are thus unlike those of all other bees. The upper margin of the clypeus extends upward as a narrow zone to each antennal base, there being no subantennal sutures. To a greater or lesser degree, the same clypeal structure is found in *Xanthesma,* no species of which is entirely black like *Hyphesma*. The tarsal claws of females are cleft or simple. Michener (1965b) and Exley (1975b) provided illustrations of male genitalia and hidden sterna.

■ *Hyphesma* occurs in all Australian states, including Tasmania. The seven known species were revised by Exley (1975b) and listed by Cardale (1993).

Most of the floral records are for Myrtaceae. Exley (1975b) described a nest in the soil.

Genus *Melittosmithia* Schulz

Smithia Vachal, 1897: 63 (not Milne-Edwards, 1851). Type species: *Scrapter carinata* Smith, 1862, by designation of Cockerell, 1910b: 358.

Melittosmithia Schulz, 1906: 244, replacement for *Smithia* Vachal, 1897. Type species: *Scrapter carinata* Smith, 1862, by designation of Cockerell, 1910b: 358.

The bees tentatively segregated from *Euhesma* and placed in *Melittosmithia* differ from *Euhesma* in lacking a subapical tooth on the upper mandibular margin of the female, the mandible thus simple (male unknown), and in having a thin, sharp, longitudinal, median clypeal carina. The inner hind tibial spur is ciliate or finely pectinate, an unusual feature in those *Euhesma* having the rather large size of *Melittosmithia*. Body length varies from 6.5 to 9.0 mm. The body is black without yellow markings, with the metasoma partly to wholly red. *Euhesma* could reasonably be synonymized into *Melittosmithia*, but until males are known, it seems best to retain *Melittosmithia* and *Euhesma* as genera. The type species of the two are very different.

■ *Melittosmithia* is found in New South Wales, Victoria, and South Australia. Four specific names, as listed by Cardale (1993), have been applied in this genus. Cockerell (1926b) gave a key to the species.

Genus *Pachyprosopis* Perkins

This genus is usually easily recognized by the venational characters indicated in the first two couplets of the key to genera. Exceptions exist, however, and are discussed below. The body is nonmetallic except for the females of *Pachyprosopis haematostoma* Cockerell, which are blue. The metasoma is sometimes red, and the head and body sometimes have yellow markings or are largely yellow. The head and thorax are finely roughened between small and often sparse punctures. The labrum has a strong apical spine except for *P. (Pachyprosopis) cornuta* Exley. The facial foveae are linear, short, and inconspicuous in some males. The flagellum is short, the middle segments being broader than long. Male genitalia and other structures were illustrated by Michener (1965b) and Exley (1972, 1976a).

Even in the unusual shape of the second submarginal cell, a character also found in *Hyphesma*, there are problems. In *Euryglossula fultoni* (Cockerell) the cell is sometimes shaped a little like that of *Pachyprosopis*, as it is in some specimens of *Euryglossina*. More specifically, in *Euryglossina narifera* (Cockerell) the second submarginal cell is as in *Pachyprosopis*. The same is true of some specimens of *E. hypochroma* Cockerell (see illustration in Exley, 1968d). Because they lack a labral spine, have very short antennae, and have eyes more sharply produced mesad above the mandibles in females, as well as because of their general form and maculation, such species are included in *Euryglossina*. Presumably, their *Pachyprosopis*-like feature is a result of convergence. The relationship of *Pachyprosopis* to *Euryglossina*, however, is close, as shown, for example, by the head shape. In both *Pachyprosopis* and *Euryglossina* (Fig. 48-5b) the clypeus (or at least the lower part of it, as seen in profile), is bent posteriorly, so that the face is strongly convex. A discussion of variability in *Pachyprosopis* and its relations to other genera was given by Michener (1965b).

Pachyprosopis visits primarily the flowers of Myrtaceae. Nests have been found in abandoned beetle burrows in wood [*P. (Pachyprosopis) haematostoma* Cockerell] and in "termite soil" at the bases of, or in, hollow trees [*P. (Parapachyprosopis) angophorae* Cockerell and *indicans* Cockerell] (Houston, 1969; Exley, 1972).

Pachyprosopis was revised by Exley (1972), with additions to the keys by Exley (1976a).

Key to the Subgenera of *Pachyprosopis* (Females)

1. Facial fovea with upper end on level of, or below, upper end of eye, nearer eye margin than to lateral ocellus, not curved mesad; margin of basitibial plate represented by tubercles that extend beyond middle of tibia; clypeus more than three times as wide as median length .... *P. (Pachyprosopis s. str.)*
—. Facial fovea with upper end above level of upper end of eye [except in *Pachyprosopis (Pachyprosopula) xanthodonta* (Cockerell)], nearer to lateral ocellus than to eye margin and curved mesad toward ocellus; margin of basitibial plate variable but not reaching middle of tibia; clypeus commonly less than three times as wide as median length 2
2(1). Thorax and metasoma lacking yellow areas; basitibial plate demarcated by tubercles in addition to carinae .... *P. (Parapachyprosopis)*
—. Thorax and metasoma with yellow areas; basitibial plate usually demarcated by carinae (with one tubercle in *Pachyprosopis flavicauda* Cockerell) ...... *P. (Pachyprosopula)*

Key to the Subgenera of *Pachyprosopis* (Males)

1. Facial fovea adjacent to concavity in inner orbit of eye, thus low on face and far from summit of eye .... *P. (Pachyprosopis s. str.)*
—. Facial fovea adjacent to upper one-third of eye, above concavity of inner orbit, and commonly approaching summit of eye 2
2(1). S8 with small, hairy apical lobe on each side of median apical process; S7 with long hairs on mesodistal margin of lateral apical lobe; scutellum not yellow .... *P. (Parapachyprosopis)*
—. S8 without lateroapical lobe; S7 usually without long hairs; scutellum yellow .... *P. (Pachyprosopula)*

Pachyprosopis / Subgenus *Pachyprosopis* Perkins s. str.

Pachyprosopis Perkins, 1908: 29. Type species: *Pachyprosopis mirabilis* Perkins, 1908, monobasic.

This subgenus is interpreted more narrowly here than by Michener (1965b), who included under *Pachyprosopis* s. str. the species here placed in *Parapachyprosopis*. Although ranging from 3.5 to 8.5 mm in length, species of this subgenus are in general larger than those of the other subgenera. The thorax and metasoma lack yellow markings. The short facial fovea of females is presumably an apomorphy in this genus, as is the long basitibial plate. The broad hairless zone on the outer side of the hind tibia of females is a synapomorphy shared with *Parapachyprosopis*, suggesting that the latter is the sister group of *Pachyprosopis* s. str.

■ *Pachyprosopis* s. str. is found in all Australian states except Tasmania. The seven species were revised by Exley (1972) and listed by Cardale (1993).

Pachyprosopis / Subgenus *Pachyprosopula* Michener

Pachyprosopis (Pachyprosopula) Michener, 1965b: 106. Type species: *Pachyprosopis kellyi* Cockerell, 1916, by original designation.

This subgenus contains minute or small species (3.5-5.5 mm long) with extensive yellow markings, at least on the males; the scutellum is yellow except in *Pachyprosopis flavicauda* Cockerell. The lack of a broad hairless band on the outer side of the hind tibia of females is a plesiomorphy suggesting that this subgenus may be the sister group to the other two subgenera taken together. The same interpretation is probably correct for the usually nontuberculate margins of the basitibial plate. Autapomorphies for the subgenus include a patch of minute spines on the dorsolateral part of the male gonocoxite [a character found also in *P. (Parapachyprosopis) indicans* Cockerell] and a reduction in the inner tooth of the claws, the claws thus simple in some females.

■ *Pachyprosopula* is known from all Australian states, including Tasmania. Its seven species were revised by Exley (1972), with a supplement to the key by Exley (1976a), and listed by Cardale (1993).

Pachyprosopis / Subgenus *Parapachyprosopis* Exley

Pachyprosopis (Parapachyprosopis) Exley, 1972: 17. Type species: *Pachyprosopis angophorae* Cockerell, 1912, by original designation.

This may be the sister group of *Pachyprosopis* s. str. Body length ranges from 4.5 to 7.0 mm. The thorax sometimes has yellow areas, although the scutellum is dark. Unusual apomorphic features are the hind basitarsi of males, which have a broad basal or median tooth or a pad of dense bristles (except unmodified in *P. indicans* Cockerell), and the two tufts or continuous band of long hairs with enlarged tips on S3 of males. The characters of male S7 and S8 given in the key to subgenera are also probably autapomorphies within *Pachyprosopis*.

■ This subgenus is widespread in middle and northern Australia but is not recorded from Victoria or South Australia. It consists of nine species (Cardale, 1993) and was revised by Exley (1972); a new key to species was published later by the same author (Exley, 1976a).

Genus *Sericogaster* Westwood

Sericogaster Westwood, 1835: 71. Type species: *Sericogaster fasciatus* Westwood, 1835, monobasic.
Holohesma Michener, 1965b: 102. Type species: *Stilpnosoma semisericea* Cockerell, 1905 = *Sericogaster fasciatus* Westwood, 1835, by original designation.

Westwood (1835) described this form as a wasp, and in fact it has a rather wasplike aspect. Westwood's name was not recognized as applying to a bee until Menke and Michener (1973) examined the type specimen. *Sericogaster* Dejean, 1835, has priority over Westwood's name but is a nomen nudum.

This genus is easily distinguished from all others by the enormous basitibial plate, demarcated by rows of large tubercles, in both sexes. The body is slender but large (7 to 11 mm long) for this subfamily, parallel-sided, black with yellowish-brown thoracic and metasomal markings. The head is developed posteriorly, the genal area being enlarged, so that it is much wider than the eye (Fig. 48-5e). The pygidial plate of the female is very slender, and upcurved apically as in *Euryglossa laevigata* (Smith) and *Heterohesma*.

■ This genus occurs in eastern Australia from New South Wales to central Queensland. It consists of a single species, *Sericogaster fasciatus* Westwood.

Genus *Stenohesma* Michener

Stenohesma Michener, 1965b: 99. Type species: *Stenohesma nomadiformis* Michener, 1965, by original designation.

This genus consists of slender, wasplike (more accurately, nomadiform) bees with extensive yellow markings on a black background. The body is 7.5 to 10.0 mm long. Indicative of its slender form is T1, which is nearly as long as broad, its median groove ending before the middle of the tergum instead of behind the middle as in other genera. The tooth on the upper mandibular margin of the female is weak and situated one-third of the mandibular length from the apex, so that the mandible appears simple; that of the male is likewise weak and situated one-fourth of the mandibular length from the apex. It is not certain that these teeth are homologous to those found near the apex of the mandible in *Callohesma*, *Euhesma*, *Euryglossa*, and *Sericogaster*. The male genitalia, hidden sterna, and various other structures were illustrated by Michener (1965b).

■ *Stenohesma* is known from northern Queensland, Australia. It contains a single species, *S. nomadiformis* Michener.

Genus *Tumidihesma* Exley

Tumidihesma Exley, 1996: 253. Type species: *Tumidihesma tridentata* Exley, by original designation.

Tumidihesma could be considered an odd *Euhesma*, since that genus as here understood includes a diversity of forms not united by any known synapomorphy. *Tumidihesma*, however, has distinctive features, and nothing useful would be accomplished by synonymizing it. Only the female is known; the relationships of the genus will be more evident when males are found. It consists of largely black, shiny species 6 mm in length with yellow facial marks in one species. The tridentate mandibular apices are especially distinctive. The lower ends of the eyes protrude mesad above the mandibular articulations, as in *Euhesma hyphesmoides* (Michener). The facial fovea is linear and nearly parallel with the eye margin, neither end being bent mesad.

■ This genus occurs in dry areas in South Australia, Northern Territory, and Western Australia. The two species were revised by Exley (1996). They visit flowers of Myrtaceae.

Genus *Xanthesma* Michener

Two names, *Xanthesma* and *Xenohesma*, were published simultaneously for elements now included in this genus. Michener (2000), as first revisor, selected *Xanthesma* as the generic name.

This is a genus of minute to moderate-sized (2.9-7.8 mm body length), often largely yellow bees. Michener (1965b) included some of the species in his then more restricted genus *Xanthesma*, while placing others in *Euryglossa sensu lato*. It now seems that the relationships are best shown by broadening the application of the name *Xanthesma* as indicated below. Of the characters listed in the key to genera, the reduced or absent preapical mandibular tooth, the lack of subantennal sutures, the usually simple female claws, and perhaps the others are probable synapomorphies that distinguish *Xanthesma* from *Euhesma*, the only part of the old (Michener, 1965b) genus *Euryglossa* to which *Xanthesma* may be closely related.

Xanthesma has been found in all Australian states except Tasmania, and is especially abundant in arid areas. So far as is known, all species visit flowers of Myrtaceae.

The subgenera listed below have not hitherto been assigned to a single genus. They are not always easy to separate, however, and one sometimes cannot place a species in its subgenus without having both sexes, as will become obvious with attempts to use the key.

Key to the Subgenera of *Xanthesma*

1. First recurrent vein entering apex of first submarginal cell; stigma as long as or longer than costal edge of marginal cell .... *X. (Argohesma)*
—. First recurrent vein entering base of second submarginal cell or meeting first submarginal crossvein; stigma shorter than costal edge of marginal cell 2
2(1). Eyes of male strongly converging above, upper end of eye less than ocellar diameter from lateral ocellus ....
.... *X. (Xenohesma)*
—. Eyes of male subparallel or converging below, upper end of eye more than ocellar diameter from lateral ocellus 3
3(2). Mandible of female with weak preapical tooth; male without vertical groove on paraocular area; front coxa of female with several distinctive bristles on apex mesal to base of trochanter .... *X. (Chaetohesma)*
—. Mandible of female simple; many males with vertical groove in paraocular area midway between antennal base and eye, and tuft of hair at lower end of groove (Fig. 47-2e); front coxa of female without such bristles ....
.... *X. (Xanthesma s. str.)*

Xanthesma / Subgenus *Argohesma* Exley

Argohesma Exley, 1969c: 528. Type species: *Argohesma eremica* Exley, 1969, by original designation.

Except for the weak characters of wing venation listed in the first couplet of the key to subgenera, this subgenus agrees in most features with *Xanthesma* s. str. An additional character is that the claws of males are simple, whereas those of *Xanthesma* s. str. are cleft. (Claws of females are simple in both subgenera.) The body is largely yellow to largely black; its length ranges from 2.4 to 3.7 mm. Illustrations of male genitalia and other structures were provided by Exley (1969c, 1974c).

■ The subgenus is recorded from all Australian states except New South Wales and Tasmania, but is abundant chiefly in Western Australia. The eight known species were listed by Cardale (1993). Exley (1969c) described and revised the genus; Exley (1974c) gave a new key.

Xanthesma / Subgenus *Chaetohesma* Exley

Chaetohesma Exley, 1978a: 373. Type species: *Chaetohesma tuberculata* Exley, 1978, by original designation.

In addition to the key characters, the following are of interest: The fore basitarsus of the female has a brush of dense setae, the setae short in *Xanthesma isae* (Exley). The claws are usually simple in females, cleft in males, but in *X. infuscata* (Exley) they are simple in both sexes, whereas in females of *X. baringa* (Exley), *foveolata* (Exley), *levis* (Exley), and *striolata* (Exley), the claws are toothed, as in many *Euhesma*. The size is small (body length 3.5-5.3 mm). Male genitalia and other structures were illustrated by Exley (1978a).

■ This subgenus is widespread in the xeric areas of Australia, mostly in the northern half of the continent; it has not been found in Victoria or South Australia. The ten described species were named and revised by Exley (1978a) and listed by Cardale (1993).

Xanthesma / Subgenus *Xanthesma* Michener s. str.

Xanthesma Michener, 1965b: 97. Type species: *Euryglossa furcifera* Cockerell, 1913, by original designation.

This is a subgenus of minute (body length 2.9-5.0 mm), largely yellow bees. Its characters are indicated in the key to subgenera. Its similarity to *Argohesma* is discussed under that subgenus. A relationship to *Xenohesma* is indicated by the presence in males of most species of a vertical groove in the space between the eye and the antenna, the lower end of the groove bearing a tuft of hairs (Fig. 48-2e). Illustrations of male genitalia and hidden sterna were published by Michener (1965b) and Exley (1969d, 1974c). Comments on the subgeneric characters were included in Exley (1978b).

■ Although known from all Australian states except Tasmania and Victoria, this subgenus seems to be particularly abundant in the north and west. The 13 named species were revised by Exley (1969d); a new key was provided by Exley (1974c). The species were listed by Cardale (1993).

Nests of the type species, *X. furcifera* (Cockerell), have been found in the ground (Exley, 1969d; Houston, 1969) and were illustrated by both authors.

Xanthesma / Subgenus *Xenohesma* Michener

Euryglossa (Xenohesma) Michener, 1965b: 96. Type species: *Euryglossa flavicauda* Michener, 1965, by original designation.

Males of this subgenus are easily recognized by the enlarged eyes, which converge above, their summits approaching the lateral ocelli as indicated in the key to subgenera. Some of the males are densely hairy, unlike those of other Euryglossinae. Some of the females associated with *Xenohesma* by Michener (1965b) and Exley (1969b) probably belong in the genus *Euhesma*. Characters of the true females of *Xenohesma* were clarified by Exley (1976c, 1978b); they do not seem to differ from those of *Xanthesma* s. str. Species of *Xenohesma*, however, are larger (4.5-7.8 mm long) than most *Xanthesma* s. str. and the body is largely black, often with yellow areas on the head and thorax, and the metasoma is often largely yellow. As in some other bees with large-eyed males, the antennal flagella are very short (see Sec. 28). Males of some species have a weak depression in the paraocular area between the eye and antennal base, with a tuft of long hairs at the lower end. This feature may be homologous to the groove and hair tuft found in males of most *Xanthesma* s. str. Illustrations of male genitalia, sterna, and other structures were provided by Michener (1965b) and Exley (1969b).

■ *Xenohesma* is widespread in xeric areas of Australia, being known from all states except Victoria and Tasmania. Seventeen species were included in the revision by Exley (1969b) and listed by Cardale (1993).

Xenohesma is probably a visitor to flowers of Myrtaceae; swarms of males hover around terminal twigs often away from flowers (Exley, 1976c).

49. Family Andrenidae

As is evident from Section 21, Andrenidae consists of four subfamilies, the Alocandreninae, Andreninae, Panurginae, and Oxaeinae, the last of which is so different from the others in a series of derived characters that it has often been given familial status. It not only differs greatly from other Andrenidae, but also, in certain characters (e.g., Dufour's gland lipids; Cane, 1983b), it resembles other bees, especially in the Colletidae. But in the phylogenetic study based on adult characters by Alexander and Michener (1995), Oxaeinae appeared as part of the Andrenidae, a family that usually was holophyletic in this study, although including Stenotritidae in some analyses. Rozen's (1993b, 1994a) studies of larvae showed the same thing; *Euherbstia* in the Andreninae appeared as the sister group to the Oxaeinae. Thus both larval and adult characters indicate that the Oxaeinae are much modified (i.e., with many derived characters) Andrenidae.

Andrenidae is a major family of andreniform or sometimes apiform S-T bees, some of the larger species almost euceriform. Diverse aspects are illustrated in Figures 49-1 and 49-2. The most distinctive character of the Andrenidae is the presence of two subantennal sutures below each antenna (Figs. 33-2c, 51-1a, 51-3e, etc.). These sutures are not strongly convergent below except in *Euherbstia* and *Chaeturginus;* they usually meet the epistomal suture at well separated locations, the subantennal area thus rectangular or quadrate, not triangular as in a few other bees that have two subantennal sutures that meet below. Rarely, in a few male *Protandrena sensu lato,* mostly from Mexico, one subantennal suture has probably been lost, so that only one remains. In species with a black face and coarse punctation, the subantennal sutures may be unrecognizable.

Other familial characters include the short to long, pointed glossa (Fig. 52-1), sometimes terminating in a flabellum (e.g., *Perdita,* Fig. 28-2h, i; *Calliopsis,* Fig. 28-

Figure 49-1. *Andrena,* males at left, females at right. Above, *A. wilkella* (Kirby); Below, *A. surda* Cockerell. From Michener, McGinley, and Danforth, 1994.

Figure 49-2. Andrenidae, Panurginae. Above, *Calliopsis (Micronomadopsis) scutellaris* Fowler, male; Center, *C. (Nomadopsis) edwardsii* Cresson, female; Below, *Protandrena (Metapsaenythia) abdominalis tricolor* (Cockerell), female. From Michener, McGinley, and Danforth, 1994.

1i, j). The disannulate surface is narrower than the annulate surface, the former rarely (*Melitturga*) invaginated to form a glossal groove suggestive of that of L-T bees. The serial hairs are large or moderately so and directed lateroapically, or are sometimes absent. The annular hairs are pointed, unbranched, and usually slender. Except for the Oxaeinae, there is a partly separated fragmentum at the base of the prementum. And except for some Panurginae, the episternal groove is absent below the level of the scrobal groove (Fig. 28-3b).

The classification within the Andrenidae presents problems, the chief of which is that the Andreninae may be paraphyletic. It is the group from which the Panurginae and Oxaeinae arise in the analyses based on adult characters by Alexander and Michener (1995) and those based on larval characters by Rozen (1993b, 1994a). I have recognized Alocandreninae because it is one of the diverse elements formerly in the Andreninae. Its removal leaves the Andreninae more cohesive, although still paraphyletic in the phylogenetic study of Alexander and Michener (1995). Moreover, *Alocandrena* has characters of the Panurginae as well of Andreninae; its placement in the latter subfamily was never very convincing.

The Alocandreninae and Panurginae, as well as the Oxaeinae, differ from the Andreninae by the truncate apex of the marginal cell (Figs. 50-1f, 53-1). This character is probably apomorphic, for the cell is pointed or narrowly rounded in related families such as the Melittidae and Colletidae. Sometimes in Panurginae the truncation is so oblique that the cell could be described as pointed. A second well-known character of Panurginae is the loss or near loss of the male gonobase. This is clearly an apomorphy, because almost all other Hymenoptera (except Meliponini and Apini) have a distinct, large gonobase. Within the Andreninae, however, it is small in *Andrena (Melittoides)* (Fig. 51-8), reduced in *Megandrena (Megandrena)* (Fig. 51-7a, b), and, for practical purposes, absent in the subgenus *Erythrandrena*. I believe that the loss of this structure is convergent in Panurginae and *Erythrandrena*, in view of the *Andrena*-like features of the latter. A third well-known panurgine feature is the loss of the scopa on the coxa, trochanter, and femur, pollen thus being carried primarily on the hind tibia. This too seems to be a panurgine and alocandrenine feature, perhaps derived, for in Colletinae and Halictidae, as well as most Andreninae, the scopa extends along the leg from the coxa to the tibia or basitarsus. But in certain Andreninae—*Euherbstia* and to a lesser extent *Orphana* and *Ancylandrena*—it is absent or weak except on the tibia, as in Panurginae and Alocandreninae.

Another panurgine synapomorphy is the virtual absence of the basal lateral lobes of the labrum, to which the tendonlike labral apodemes are attached. In Andreninae (except *Megandrena*), Alocandreninae, colletids, melittids, etc., the labrum has such a basal lateral lobe (visible only when the labrum is removed from the clypeus) on each side. Moreover, the labrum in Panurginae (and *Megandrena*) is nearly rectangular, whereas in the other groups listed it is ordinarily transverse, tapering laterally and rounded at each side. Another panurgine apomorphy is the long and often flattened first segment of the labial palpus, that segment usually at least as long as segments 2 to 4 taken together and commonly longer. Exceptions are *Anthemurgus* and *Camptopoeum (Epimethea)*, in which the first segment is short.

In the characters listed above, the probable plesiomorphic state is found in the Andreninae and the presumed apomorphic state in the Panurginae. There are, however, a few characters that may be synapomorphies of the Andreninae or a part of that group. The facial fovea of female Andreninae and Alocandreninae is covered with short, fine or velvety hairs, or, in the Chilean genera *Euherbstia* and *Orphana,* the fovea is absent. In female Panurginae the fovea is nearly always present (absent in *Melitturga*) and nearly hairless, as in colletids such as the Euryglossinae and Hylaeinae. This is presumed to be the plesiomorphic condition.

In most Panurginae as well as most other aculeate Hymenoptera of both sexes, on an area of the inner surface of the hind tibia there are many, usually short hairs of uniform length with blunt, capitate or bifid apices. These are the keirotrichia (Michener, 1981a); presumably their function is cleaning the wings. They are replaced in females of Alocandreninae and Andreninae by long hairs similar in aspect and, probably, function to the scopal hairs on the outer side of the hind tibia. In *Andrena (Melittoides)* (Andreninae) they are long but still have minutely bifid apices. Reanalysis of andrenid phylogeny with more taxa and appropriate outgroups might show the loss of short keirotrichia as a synapomorphy of Andreninae (together with Alocandreninae).

A character that supports the similarity of *Euherbstia* and *Orphana* to the Oxaeinae is the unmodified S7 of the male, its form in *Euherbstia* not very different from that of S6 (Fig. 51-6d, h). This could be a plesiomorphy, a significant synapomorphy, or convergence perhaps resulting from transfer of developmental control from more anterior sterna to S7.

Andrenidae occur on all continents except Australia. They are also almost absent from the tropical Asian region. In the north temperate areas, the genus *Andrena* is ubiquitous. In sub-Saharan Africa there are only a few genera and species. In the Western Hemisphere, however, especially in the temperate and xeric parts of both North and South America, there are many andrenid genera and species; in the moist tropics the numbers are small.

All species of Andrenidae nest in the soil, making their own burrows and cells, one or a short series at the end of each lateral burrow. Except in some Panurginae whose cells are unlined, the cells are lined with a shiny secretion. In the Oxaeinae the provisions are viscous and fill the lower ends of vertical cells. In other Andrenidae the provisions are firm and have the form of a sphere or flattened sphere in each cell. The lower surface of the usually horizontal cell (but slanting to nearly vertical in *Euherbstia,* Rozen, 1993b) is flatter than the upper surface in the Andreninae, the cell thus bilaterally symmetrical around a vertical plane, like the cells of Halictidae and Colletinae such as *Leioproctus.* In Panurginae and Oxaeinae this is not so, the cell being similarly shaped on all surfaces; or in some Panurginae the lower surface may be slightly flatter than the upper. The egg is laid on top of the food mass. No andrenid spins a cocoon. (Nests of Alocandreninae are unknown.)

Key to the Subfamilies of the Andrenidae

1. Stigma essentially absent; marginal cell over seven times as long as broad and only half as wide as widest submarginal cell (Fig. 60-2a); mentum absent or fused to flat lorum; proboscidial lobe absent; first flagellar segment as long as scape (Western Hemisphere) .... Oxaeinae (Sec. 60)
—. Stigma present; marginal cell shorter and broader, about as wide as widest submarginal cell; mentum and lorum both present, forming proboscidial lobe; first flagellar segment usually shorter than scape 2
2(1). Apex of marginal cell pointed or narrowly rounded on or near wing margin (Fig. 51-2) (facial fovea not or weakly evident in male, in female, when present, slightly depressed, covered with short hairs; gonobase of male distinct and large, reduced to narrow ring or almost absent in *Megandrena*) .... Andreninae (Sec. 51)
—. Apex of marginal cell truncate, sometimes obliquely so, cell thus pointed well away from wing margin (Figs. 50-1f, 53-1) 3
3(2). Facial fovea of both sexes a distinct deep pocket with short hairs (Fig. 50-1a, b); gonobase of male distinct and large (Fig. 50-1c-e) (Peru) .... Alocandreninae (Sec. 50)
—. Facial fovea hairless and shining, often absent in males, rarely also in females *(Melitturga)*; gonobase of male absent or nearly so .... Panurginae (Sec. 52)

50. Subfamily Alocandreninae

The type genus of this new subfamily is *Alocandrena*. The habitus is that of a moderate-sized (length 11 mm) *Andrena* with apical tergal hair bands. The truncate marginal cell (Fig. 50-1f); the small facial fovea (Fig. 50-1a, b); the deeply bilobed S7 of the male (Fig. 50-1e) with a small disc suggestive of the type of S7 common in Colletidae; the small, hairy volsella; and the undeveloped femoral and trochanteral scopa of the female are all suggestive of the Panurginae, although in *Euherbstia* and *Orphana*, provisionally included in Andreninae, the femoral and trochanteral scopa is also undeveloped. The hairy facial fovea, the well-developed male gonobase, and the long hairs (rather than short keirotrichia) on the inner surface of the hind tibia of the female, however, are indicative of the Andreninae. The Alocandreninae may be a derivative of a common ancestor of the two major subfamilies, perhaps the sister group to one or the other. Its male genitalia and hidden sterna were illustrated by Michener (1986c); see also Figure 50-1c-e.

The only genus of this subfamily is *Alocandrena*.

Genus *Alocandrena* Michener

Alocandrena Michener, 1986c: 68. Type species: *Alocandrena porteri* Michener, 1986, by original designation.

The characteristics of this genus are indicated above,

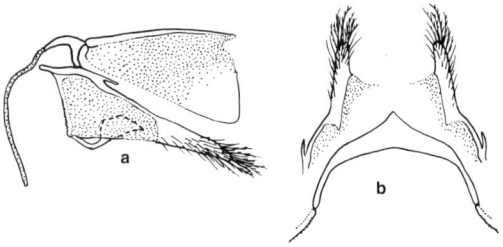

Figure 50-2. Reduced sting apparatus of female *Alocandrena porteri* Michener. **a,** Lateral view: the membranous strip at lower left represents the first valvula; the sting stylet is shown in broken lines behind the second valvifer; **b,** Dorsal view, the rudimentary sting stylet in the center. The hairy structures are the gonostyli. From Michener, 1986c.

in the discussion of the subfamily. An additional feature is the extremely reduced sting, the first valvula being merely a membranous ribbon, the sting a transverse sclerite with a midapical, only slightly acute projection (Fig. 49-2). Thus the sting is more reduced than in the Meliponini although less reduced than in Dioxyini.

■ *Alocandrena* occurs on the western slope of the Peruvian Andes. The only species is *A. porteri* Michener.

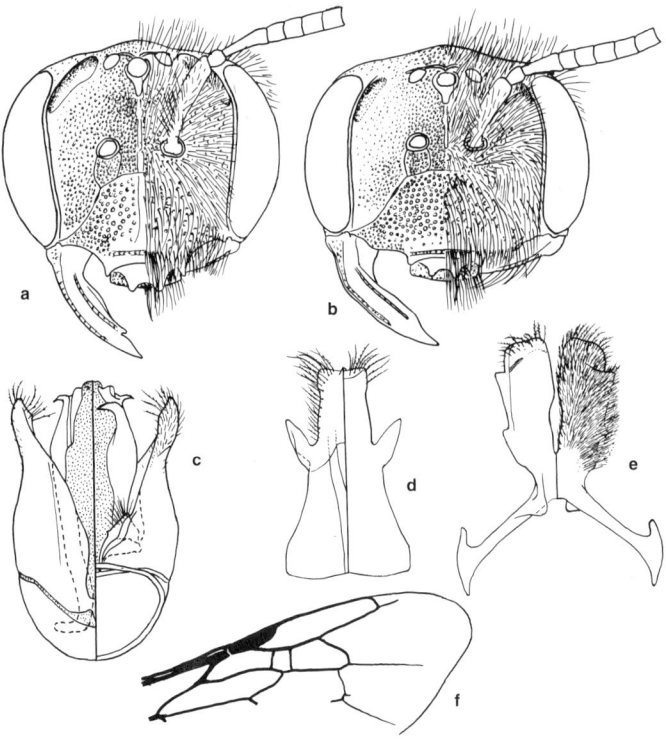

Figure 50-1. *Alocandrena porteri* Michener. **a, b,** Faces of female and male; **c-e,** Male genitalia, S8, and S7. (Dorsal views are at the left.) **f,** part of forewing. From Michener, 1986c.

51. Subfamily Andreninae

Some of the features of this subfamily are explained above in the discussion of the family Andrenidae. Andrenines are small to rather large, somewhat more hairy than most Panurginae (Fig. 49-1), and when yellow integumental markings are present, they are limited to the faces of males and, uncommonly, of females. Most have three submarginal cells (Fig. 51-2), but some species and even certain subgenera of *Andrena* have two, the second rather long, indicating that the second transverse submarginal vein has been lost.

The subfamily Andreninae could reasonably be divided into two tribes, thus: (1) The Andrenini for the genera *Andrena*, *Ancylandrena*, and *Megandrena*, characterized by the broad, velvety facial foveae of the females and found in the holarctic region, ranging south to South Africa and to Panama. (2) The Euherbstiini for the genera *Euherbstia* and *Orphana*, characterized by lack of facial foveae, and found in Chile. I have decided not to formalize these tribes for this work.

As indicated above, the Andreninae are primarily holarctic, but occur also in eastern and southern Africa, in mountains in the oriental region south as far as the Malay peninsula and southern India, and, in the Western Hemisphere, south to Panama, with disjunct forms in Chile. In most parts of the world the keys to genera are unnecessary because only one genus of the subfamily, *Andrena*, is found there. The other genera are found in xeric regions of the southwestern USA, northwestern Mexico, and Chile.

Key to the Genera of the Andreninae (Females)
1. Facial fovea depressed, velvety (Fig. 51-1a) 2
—. Facial fovea absent (or vaguely indicated by color or texture of cuticle) 4
2(1). Hairs of hind trochanter rather dense, short, the longer ones simple, not much curved; anterior surface of T1 broadly concave, much longer than dorsal surface of T1 (Fig. 51-1c, d) (southwestern North America) .... *Ancylandrena*
—. Some hairs of hind trochanter long, curved distad, plumose, forming a floccus (Fig. 51-1b) closing basal end of femoral corbicula; anterior surface of T1 with smaller concavity or groove, surface shorter than to slightly longer than dorsal surface of T1 (Fig. 51-1e, f) 3
3(2). Hind basitarsus more than half as long as hind tibia; stigma often broader than prestigma (measured to wing margin), margins usually converging basad from vein r (Fig. 51-2a, b) .... *Andrena*
—. Hind basitarsus about half as long as hind tibia; stigma about as wide as prestigma (measured to wing margin), margins parallel or nearly so from vein r to base of stigma (Fig. 51-2c) (southwestern USA) .... *Megandrena*
4(1). Claws each with the usual large inner tooth; inner subantennal suture much shorter than diameter of antennal socket and apparently not converging below toward outer suture (Chile) .... *Orphana*
—. Claws each with minute inner tooth; inner subantennal suture about as long as diameter of antennal socket and

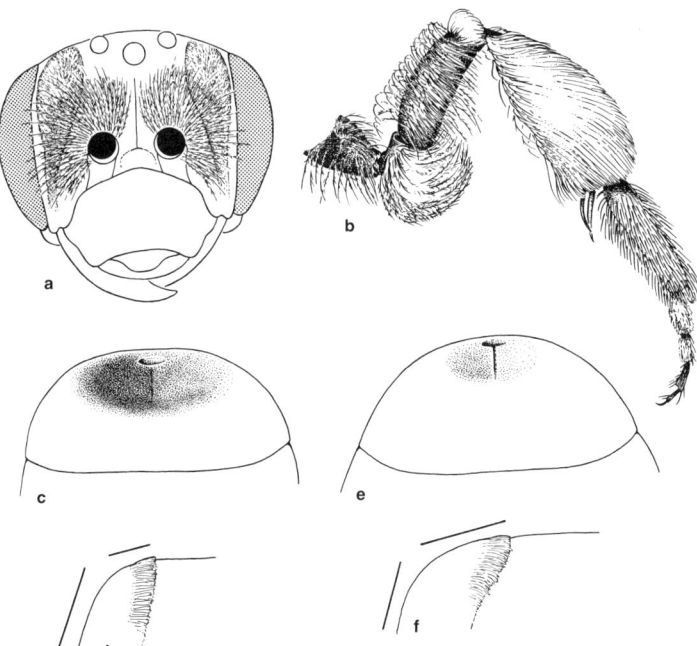

Figure 51-1. Andreninae. **a,** Face of *Andrena mariae* Robertson, female; **b,** Hind leg of *Andrena* sp., female, showing large floccus on trochanter and other scopal hairs on femur and tibia; **c, d,** T1 of *Ancylandrena larreae* Timberlake in dorsal and lateral views, with lines showing lengths of dorsal and anterior surfaces; **e, f,** Same, of *Megandrena enceliae* (Cockerell). From Michener, McGinley, and Danforth, 1994.

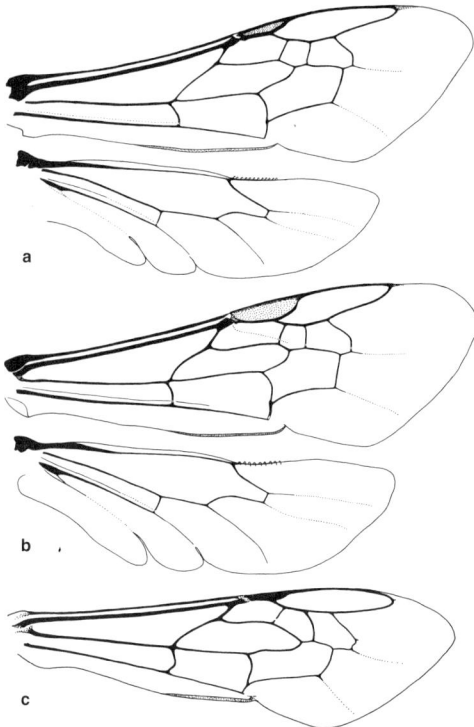

Figure 51-2. Wings of Andreninae. **a**, *Andrena (Callandrena) accepta* Viereck; **b**, *Andrena illinoiensis* Robertson; **c**, *Megandrena enceliae* (Cockerell). c, from Michener, McGinley, and Danforth, 1994.

converging below toward outer suture, subantennal area thus having only short margin on epistomal suture (Chile) .... *Euherbstia*

Key to the Genera of the Andreninae (Males)

1. Gonobase a narrow ring or essentially absent (Fig. 51-7a, b); mandible simple or with weak preapical shoulder on upper margin (southwestern USA) .... *Megandrena*
—. Gonobase with broad dorsal surface; mandible with preapical tooth on upper margin (except in *Euherbstia* from Chile) 2
2(1). Anterior surface of T1 largely convex, the concavity a median, longitudinal groove or depression, this surface shorter than to slightly longer than distance from its upper end to apex of T1; hind basitarsus five or more times as long as wide .... *Andrena*
—. Anterior surface of T1 largely flat (in *Euherbstia*) or concave, the concavity a median, longitudinal line or groove, this surface usually longer than distance from its upper end to apex of T1; hind basitarsus less than five times as long as wide 3
3(2). Mandible simple; inner subantennal suture converging below toward outer suture, so that subantennal area has only short margin on epistomal suture (Chile) .... *Euherbstia*
—. Mandible with preapical tooth on upper margin; inner subantennal suture not converging strongly below toward outer suture 4

4(3). S7 a broad plate with large median apical process (anterolateral processes partly hairy, not comparable to anterolateral apodemes, Fig. 51-6h); epistomal suture separated from antennal socket by much less than diameter of socket (Chile) .... *Orphana*
—. S7 with a relatively small body, large anterolateral apodemes, and two small apical lobes (Fig. 51-3d); epistomal suture separated from antennal socket by about diameter of socket (Fig. 51-3e) (southwestern North America) .... *Ancylandrena*

Genus *Ancylandrena* Cockerell

Andrena (Ancylandrena) Cockerell, 1930d: 5. Type species: *Andrena heterodoxa* Cockerell, 1930 (preoccupied) = *A. atoposoma* Cockerell, 1934, monobasic.

This genus consists of hairy, fast-flying *Andrena*-like bees 8 to 15 mm long. The integument is largely black, usually including the male clypeus, but the lower paraocular area of males is partly yellow (Fig. 51-3e). In both this genus and *Megandrena* the tibial and basitarsal scopa includes long hairs and therefore seems less compact than that of *Andrena;* the center of the basitibial plate of the female in both genera is covered by a patch of dense, short, erect black hairs. Male genitalia and other structures were illustrated by Zavortink (1974); see Figure 51-3.

■ *Ancylandrena* occurs in xeric parts of California, Nevada, and Arizona, USA, and in Baja California Norte and Sonora, Mexico. Four species were revised by Zavortink (1974); one additional species has been described.

The nest resembles that of *Andrena* with extremely long horizontal lateral burrows each leading to one cell. The cell wall is extremely thinly covered with a secreted film (Rozen, 1992a).

Genus *Andrena* Fabricius

This well-known genus (Fig. 49-1, Pl. 3) is characterized by the large, velvety facial foveae of the females (in which it resembles *Ancylandrena* and *Megandrena*; the strong femoral scopa, that of the trochanter long, curved distad (Fig. 51-1b), and closing the base of the femoral corbicula (thus resembling *Megandrena*); and the characters of T1 and the hind basitarsus listed in the key to genera. The male genitalia are usually recognizable by the dorsal preapical lobe of the gonocoxite emphasized in Figure 51-4a but see also Figure 51-5a, a lobe that is not found in similar form in other genera. S7 of the male is a simple, rather short plate with a short, bilobed apical process or with no apical process (Fig. 51-5e). The stigma is typically broader than that of the other genera of the subfamily (Fig. 51-2b), although it is slender in some groups of *Andrena* such as the subgenus *Callandrena* and especially *Melittoides* (Fig. 51-2a). Illustrations of many structures, including male genitalia and hidden sterna, were published by Pittioni (1948a, b), Popov (1949c, 1958b), Mitchell (1960), Hirashima (1962-1966a), Warncke (1965), Thorp (1969a), Hirashima and Tadauchi (1975), Svensson and Tengö (1976), Donovan (1977); Wu (1982b, d), Tadauchi and Hirashima (1983, 1984a, b, 1988), Tadauchi (1985), Dylewska (1987), and Tadauchi, Hirashima, and Matsumura (1987), and in numerous papers by LaBerge and Ribble and coauthors.

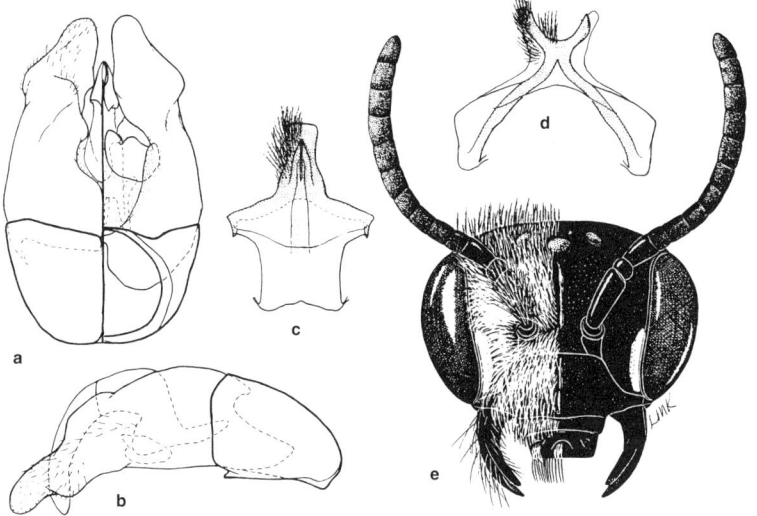

Figure 51-3. *Ancylandrena atoposoma* (Cockerell), male. **a-d,** Genitalia dorsoventral (dorsal to left) and lateral views, and ventral views of S8 and S7; **e,** Face, one side denuded to show pale lateral face mark. From Zavortink, 1974.

Andrena is found throughout the holarctic region, south in the Western Hemisphere to Panama, where one species occurs in the tropical lowlands, in Africa through the East African highlands and south to the Cape of Good Hope, and in Asia to the mountains of southern India and of Malaysia. Eastward, the genus occurs in Okinawa and Taiwan. The record for São Tomé in the Gulf of Guinea is an error based on a mislabeled specimen of a European species and the genus is probably absent in the lowland tropics of Africa; it is also absent from the Antilles.

Although species of *Andrena* are often distinctive in coloration (all black, gray-haired, red-haired, metallic blue or green, the metasoma sometimes red or amber-colored, sometimes with hair bands, etc.), they are similar morphologically and resistant to recognition of several easily distinguished genera or subgenera. Yet because there are so many species, and many recognizable groups within the genus, the genus has provided rich material for descriptions of subgenera; 96 subgenera are currently recognized!

Although in her work on Polish *Andrena*, Dylewska

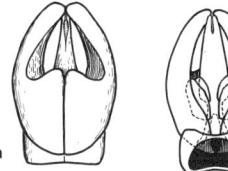

Figure 51-4. Diagrams (hairs omitted) of male genitalia of *Andrena mimetica* Cockerell. **a,** Dorsal view; **b,** Ventral view. From Michener, 1944.

(1974) placed the species in subgenera, in her later work on the species of north and middle Europe (Dylewska, 1987), she abandoned subgenera and placed the species in numbered groups and subgroups that she characterized. Her classification is indicated in Table 51-1; D. Baker (in litt., 1995) constructed a similar table with at least one subgeneric name for each of Dylewska's numbered groups, except for Groups 13 and 14. Such a reduced number of subgenera, if definable on a worldwide

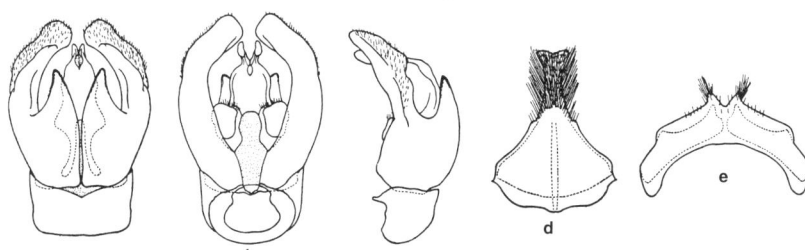

Figure 51-5. *Andrena (Callandrena) humeralis* LaBerge, male. **a-e,** Genitalia (dorsal, ventral, and lateral view), S8, and S7. (The dorsal lobe of the gonocoxite is indicated by heavy lines.) From LaBerge, 1967.

Table 51-1. Summary of Dylewska's (1987) Classification of *Andrena* of Northern and Central Europe.

Groups	Subgroups	Subgeneric names	Groups	Subgroups	Subgeneric names
1	labialis	*Holandrena*		ovatula	*Taeniandrena*
	agilissima	*Agandrena*		dorsata	*Simandrena* = *Platandrena* = *Stenandrena*
2	tibialis	*Plastandrena* = *Glyphandrena* = *Schizandrena* = *Mitsukuriapis*		gracella	*Graecandrena*
				longibarbis	*Distandrena*
	scita	*Scitandrena*		minutula	*Micrandrena* = *Andrenella*
3	lagopus	*Biareolina*		bisulcata	—
	haemorrhoa	*Trachandrena*	12	barbilabris	*Leucandrena*
4	colletiformis	*Brachyandrena*		aeneiventris	*Aenandrena*
5	hattorfiana	*Charitandrena*		coitana	*Oreomelissa*
	braunsiana	*Pallandrena*		dentiventris	*Parandrenella*
6	erberi	*Campylogaster*	13	lathyri	—
	cressonii	*Opandrena* (= *Holandrena*, according to LaBerge, 1986)		limbata	—
			14	ispida	—
			15	schulzi	*Ulandrena*
7	ventricosa	*Cryptandrena*		bucephala	—
	proxima	—	16	nigriceps	*Cnemidandrena*
8	suerinensis	*Suandrena*		hypopolia	—
	fuscosa	*Melanapis*		transitoria	—
9	curvungula	*Lepidandrena* = ?*Aporandrena*		bicolor	*Euandrena* = *Xanthandrena* = *Geandrena*
	florivaga	—		trimmerana	*Hoplandrena*
	oralis	*Orandrena*	17	helvola	*Andrena* s. str.
10	morio	*Melandrena* = *Gymnandrena* = *Bythandrena*		vaga	—
				fulvata	*Ptilandrena*
	flavipes	*Zonandrena*		nobilis	*Nobandrena*
	mucida	*Didonia* = *Solenopalpa* = *Conandrena*[a] = *Chaulandrena*		symphyti	—
			18	humilis	*Chlorandrena*
				truncatilabris	*Scaphandrena* = *Truncandrena*
	polita	*Poliandrena*			
	fulvago	*Chrysandrena*		aciculata	*Aciandrena*
	nitidiuscula	*Notandrena*	19	aerinifrons	*Carandrena*
	chrysosceles	—	20	sericata	*Parandrena* = ?*Larandrena*
	cordialis	*Cordandrena*		tarsata	—
	ensliniella	—		viridescens	—
	combinata	—	21	labiata	*Poecilandrena*
11	fulvida	—		marginata	*Margandrena*

[a] D. Baker (in litt., 1995) noted that *Gonandrena* erroneously replaced *Conandrena* as a synonym of *Didonia* in Dylewska's (1987) publication. The relation of *Conandrena* and *Chaulandrena* (= *Dactylandrena*) to *Didonia* is not here accepted.

basis, may be a step in the direction of a manageable classification for the enormous genus *Andrena*. For the present work, I have given keys and listed subgenera, but this exercise may have little value except to indicate the diversity in the genus and many of the characters that demonstrate it. The texts for individual subgenera are accordingly brief, and depend largely on the keys for morphological detail.

Gusenleitner and Schwarz (2002), in a list of species of *Andrena* with extensive annotations on Palearctic species, recognized 1443 species worldwide. Many more names were regarded as synonyms by these authors; additions and corrections were made by Gusenleitner, Schwarz, Ascher, and Scheuchl, 2005. All subgenera in the Western Hemisphere have been revised; 476 species are recognized. There are many new species, especially in the uplands of Mesoamerica.

Some regional works with keys for palearctic *Andrena* species and numbers of species reported from each region, are as follows: Perkins (1919), 61 British species; Vecht (1928), about 91 Netherlands species, including several from nearby countries not taken in the Netherlands; Kocourek (1966), 152 Czechoslovakian species; Hirashima (1966a), 61 Japanese species; Osychnyuk (1977), 153 species from the Ukraine; E. Stöckhert in Schmiedeknecht (1930) and Dylewska (1987), 156 species of north and central Europe. Other major keys to species are by Osychnyuk, Panfilov, and Ponomareva (1978) for European USSR and Osychnyuk (in Ler, 1993) for oriental Russia. Finally, a large work on the species of the central and eastern Palearctic region is in progress, with one of three projected volumes published (Osychnyuk et al., 2005).

Sub-Saharan species of *Andrena* are not numerous

(perhaps a dozen) but have never received careful, comparative study. They look much like palearctic species and subgenera, and there may be no distinctive sub-Saharan subgenera; at least none is now recognized.

All *Andrena* species nest in the ground. A few are communal, but most make individual nests, sometimes in large aggregations. Probably all communal species are sometimes also solitary, but it seems that most nests of such species are inhabited by several to many females. Most reports are of less than 40 females working in a nest, but Paxton and Tengö (1996) reported nests of *A. jacobi* Perkins in which up to 594 females shared a single nest entrance. Moreover, such nests may be perennial, that is, used for several years; Osgood (1989) reports one nest of *A. cratagei* Robertson with four entrances, its burrows probably enlarged by use, although most nests have only one entrance.

The cells are usually at the ends of lateral burrows radiating from the main burrow, but sometimes two or more cells are in series in a single lateral, as in the case of *A. labialis* (Kirby); see Radchenko (1981). The cells are lined with waxlike material and the pollen mass is smooth and more or less spherical (see photographs in Radchenko, 1981); an apparent exception is *A. viburnella* Graenicher, which, according to Stephen (1966), fills the bottom of a cell with pollen and places the egg in a large depression in the mass. These observations need to be verified, since they differ greatly from the usual *Andrena* behavior. Most species produce only one generation per year; they mature and then hibernate as adults in the cells, the two sexes emerging at more or less the same time the following year. It may be that autumnal species overwinter as prepupae. Such behavior has been observed for two species of the largely autumnal subgenus *Callandrena*, and *A. (C.) rudbeckiae* Robertson in Texas has a mixed strategy, some overwintering as adults, others as prepupae, according to Neff and Simpson (1997). A few species produce two generations per year. Michener and Rettenmeyer (1956), Youssef and Bohart (1968), and Schönitzer and Klinksik (1990) listed and summarized papers on the nesting biology of *Andrena*, but there are numerous other accounts of individual species, for example, Malyshev (1926), Hirashima (1962), Matsumura (1970), Davis and LaBerge (1975), Schrader and LaBerge (1978), Radchenko (1981), Gebhardt and Röhr (1987), Neff and Simpson (1997), and Maeta, Fujiwara, and Kitamura (2004).

Although many species of *Andrena* are polylectic, many others are strictly oligolectic. Some, such as *A. (Iomelissa) violae* Robertson, seem to be morphologically adapted to particular flowers, in this case employing a long glossa for the flowers of *Viola*. Others show little obvious morphological adaptation, but are nonetheless restricted in pollen collecting. An example is the whole subgenus *Onagrandrena* (see Linsley et al., 1973), which collects pollen from Onagraceae.

Some of the subgenera listed below are similar, and their validity is not evaluated here; the keys provided are essentially as constructed by others. Keys to subgenera have been provided by Hedicke (1933), Hirashima (1952), and Warncke (1968) for palearctic subgenera and by Lanham (1949) and LaBerge (1964, 1986) for nearctic subgenera.

Some special terminology used in *Andrena* is as follows:

The **facial quadrangle** is measured by the width between the eyes at the level of the lower margins of the antennal sockets, and by the length from the apex of the clypeus to the lower margin of the median ocellus. This term is used in the key to American subgenera.

The **basal area of the labrum** is the elevated part of the labrum, often called the "process of the labrum" in *Andrena* (see Sec. 10).

The **subgenal coronet** is a row of bristles or specialized hairs arising from a ridge on the paramandibular process of the hypostomal area, curving behind the mandibular base. The lateral part of this structure is concordant with the lateral extremity of the hypostomal carina (see Sec. 10).

The **floccus** or flocculus on the underside of the hind trochanter of females is a mass of long, branched hairs, curled distad, that closes the base of the femoral corbicula. When all the hairs on the underside of the trochanter are long and curled, the floccus is said to be **complete** (or perfect). When the hairs of the basal half of the trochanter are nearly straight and not particularly long, only the more distal hairs being long and curled, the floccus is considered **incomplete** (or imperfect).

Keys for the subgenera of three regions are provided below; the regions are North and Central America, the western palearctic region, and Japan.

Key to the Subgenera of *Andrena* of North and Central America (Females)
(Modified from LaBerge, 1986)

1. Inner hind tibial spur bent and broadened near base; tibial scopal hairs simple 2
—. Inner hind tibial spur neither bent nor broadened near base, or *if* so, then tibial scopal hairs plumose and spur only slightly broadened 4
2(1). Propodeal triangle coarsely sculptured .... *A. (Plastandrena)*
—. Propodeal triangle finely sculptured 3
3(2). Thoracic dorsum with hairs short, forming a dense mat .... *A. (Aporandrena)*
—. Thoracic dorsum with normal, long, plumose hairs not at all forming a mat obscuring surface .... *A. (Charitandrena)*
4(1). Middle and hind tibial spurs unusually thick, their apices abruptly hooked (terga without apical pale bands; facial fovea narrow) .... *A. (Euandrena)* (in part)
—. Middle and hind tibial spurs long and slender, their apices usually evenly curved.. 5
5(4). Inner hind tibial spur rather abruptly bent and slightly twisted at about one-third of its length from apex (T2-T4 with conspicuous, usually medially interrupted, white fasciae) .... *A. (Leucandrena)* (in part)
—. Inner hind tibial spur long, slender, not bent nor markedly twisted at apical third 6
6(5). Facial quadrangle broader than long; propodeal corbicula poorly formed, lacking long internal hairs, dorsal hairs extremely short, not flexed down over corbicular area (terga without pale apical fasciae) .... *A. (Oligandrena)*
—. Facial quadrangle usually at least as long as broad and often slightly longer; propodeal corbicula with at least long, flexed dorsal hairs 7

7(6). Galeal blade as long as clypeus or slightly longer; maxillary and/or labial palpi greatly elongate (T2-T4 with pale apical fasciae, often weak) 8
—. Galeal blade not as long as clypeus; maxillary and labial palpi not greatly elongate 10
8(7). Labial palpus greatly elongate, segments 1 plus 2 three to five times as long as segments 3 plus 4 and flattened; maxillary palpus of normal length, half as long as galeal blade; basal area of labrum not raised much above general level of surface .... *A. (Callandrena)* (in part)
—. Labial palpus elongate, but first two segments not unusually long nor much flattened; maxillary palpus elongate, more than half as long as galeal blade; basal area of labrum distinctly raised above general level of surface 9
9(8). Maxillary palpus with second segment at least 1.5 times as long as first; stipes distinctly narrowed medially and tapering apically, the apical third thus less than half as broad as basal third, with weak, plumose hairs .... *A. (Iomelissa)*
—. Maxillary palpus with second segment about as long as first; stipes linear, tapering from base to apex, with abundant, long, spinelike hairs .... *A. (Erandrena)*
10(7). Pronotum without dorsolateral angle, smoothly rounded posteriorly from one posterior pronotal lobe to the other, or *if* weak angle present, then without trace of lateral ridge or elevation 11
—. Pronotum with dorsolateral angle on posterior margin just above and in front of pronotal lobe, usually with a ridge or elevation extending down side of pronotum from dorsolateral angle, often with depressed or flattened area posterior to this ridge; angle and ridge occasionally quite weak and rarely only that part of ridge immediately above front coxa present 13
11(10). Propodeal triangle coarsely sculptured, often bounded posteriorly by a strong transverse carina, or *if* finely sculptured, then with longitudinal rugae at least basally and with transverse carina at apex 12
—. Propodeal triangle usually finely sculptured, usually tessellate, never with strong transverse posterior carina, or *if* coarsely sculptured, then without longitudinal basal rugae and/or without transverse apical carina 45
12(11). T2-T4 with depressed marginal zones less than half length of exposed part of each tergum medially .... *A. (Scrapteropsis)*
—. T2-T4 with depressed marginal zones half or more length of exposed part of each tergum medially .... *A. (Trachandrena)*
13(10). Pronotum laterally with deeply impressed groove cutting diagonally up and forward from pronotal spiracle to near midline anteriorly, this groove crossing and strongly depressing ridge extending down from dorsolateral angle; T2-T4 without apical pale fasciae or these weak and broadly interrupted; genal area broader than eye in side view; ocellocciptal distance about one and one-half to two and one-half ocellar diameters .... *A. (Tylandrena)*
—. Pronotum without lateral diagonal groove, or groove not deeply impressed and not depressing dorsoventral ridge, or *if* diagonal groove present and depressing ridge, then terga with distinct pale apical fasciae; genal area often narrow; ocellocciptal distance often one ocellar diameter or less 14

14(13). Submarginal cells two 15
—. Submarginal cells three 16
15(14). Propodeal triangle usually relatively coarsely sculptured; propodeal corbicula incomplete anteriorly, with internal hairs; integument usually metallic .... *A. (Diandrena)* (in part)
—. Propodeal triangle finely tessellate; propodeal corbicula complete anteriorly, with few or no internal hairs; integument black or in part red, not metallic .... *A. (Parandrena)*
16(14). Propodeum with dorsal surface poorly defined, sloping evenly from base to apex, with complete lateral carina separating dorsoposterior and lateral surfaces; clypeus usually very short, produced beyond lower ends of eyes by less than one-fourth median clypeal length .... *A. (Hesperandrena)* (in part)
—. Propodeum with well-defined dorsal surface and without complete lateral carina; clypeus normal, usually produced beyond lower ends of eyes by more than one-fourth median clypeal length 17
17(16). Tibial scopal hairs highly plumose throughout 18
—. Tibial scopal hairs simple or largely so, occasionally weakly plumose throughout 22
18(17). Subgenal coronet absent; vesture largely black or dark brown (Mexico) .... *A. (Celetandrena)*
—. Subgenal coronet present; vesture various but usually not all or mostly dark 19
19(18). Scopal hairs long and rather weak, with abundant short curved branches in outer half or more; maxillary palpal segments all exceptionally long .... *A. (Ptilandrena)* (in part)
—. Scopal hairs relatively short, with short stiff branches; maxillary palpal segments not all elongate 20
20(19). Propodeal corbicula complete anteriorly, with few or no internal hairs; first flagellar segment scarcely, if any, longer than second segment .... *A. (Archiandrena)* (in part)
—. Propodeal corbicula incomplete anteriorly, with abundant long, simple internal hairs; first flagellar segment about 1.5 times as long as second segment or more 21
21(20). Scopal hairs entirely or mostly dark brown to black; pronotal dorsolateral ridge close to pronotal lobe; first recurrent vein usually meeting second submarginal cell near middle of cell .... *A. (Ptilandrena)* (in part)
—. Scopal hairs pale ochraceous; pronotal ridge not distinct, or well separated from pronotal lobe; first recurrent vein meeting second submarginal cell at about two-thirds of its length from base of cell ...... *A. (Belandrena)* (in part)
22(17). Galeal blade with abundant short, hooked hairs; body usually 11 mm long or more (integument usually metallic blue or black) ....*A. (Scoliandrena)*
—. Galeal blade without hooked hairs; often small bees, less than 10 mm in length... 23
23(22). Frons below ocelli tessellate, without rugulae, or the rugulae sparse and extremely fine .... *A. (Derandrena)* (in part)
—. Frons below ocelli with parallel longitudinal rugulae, in terrugal spaces narrow, weakly tessellate or shagreened 24
24(23). Facial fovea extremely elongate, almost reaching vertex above, occupying most of paraocular space below;

ocelloccipital distance half an ocellar diameter or less; terga impunctate .... *A. (Oxyandrena)*
—. Facial fovea shorter than above; ocelloccipital distance often more than half an ocellar diameter; terga often punctate 25

25(24). Hypostomal carina with longitudinal part longer than lateral part, lamellate, as high as two-thirds length of last labial palpal segment (clypeus with free apical margin elongate, moderately upturned; T2-T4 with pale apical fasciae interrupted medially) .... *A. (Geissandrena)*
—. Hypostomal carina either with longitudinal part no longer than lateral part or carinate, not as high as half length of last labial palpal segment 26

26(25). Facial quadrangle considerably longer than broad 27
—. Facial quadrangle at least almost as broad as long 28

27(26). Malar area more than half as long as broad; pleural hairs black to dark brown ...... *A. (Dactylandrena)* (in part)
—. Malar area about half as long as broad; pleural hairs pale ochraceous to white .... *A. (Conandrena)*

28(26). Middle basitarsus expanded medially (T2-T4 and usually T1 with complete apical pale fasciae of hairs with close-set, short barbs, giving them a dull appearance; basal area of labrum strongly bidentate and reflexed) .... *A. (Cnemidandrena)*
—. Middle basitarsus not expanded medially, parallel-sided 29

29(28). Median third of clypeus impunctate, shiny; tibial scopal hairs weakly plumose throughout; scutellum shiny, unshagreened at least medially ...... *A. (Larandrena)*
—. Median third of clypeus punctate or dulled by dense shagreening or tessellation or both, sometimes with impunctate and shiny area occupying much less than one-third of clypeus; tibial scopal hairs usually simple, at least medially; scutellum often opaque, dulled by fine tessellation or shagreening 30

30(29). First flagellar segment only slightly longer than second, about as long as third segment; propodeal triangle with distinct longitudinal rugulae ending posteriorly in irregularly reticulate rugulae (propodeal corbicula incomplete anteriorly, with abundant internal hairs) .... *A. (Xiphandrena)*
—. First flagellar segment considerably longer than either second or third; propodeal triangle with or without longitudinal rugulae, often with finely reticular rugulae forming small areolae 31

31(30). Propodeal triangle with coarse longitudinal rugulae; metasomal terga distinctly punctate .... *A. (Rhaphandrena)*
—. Propodeal triangle finely tessellate or with irregular anastomosing rugulae; metasomal terga often impunctate or only weakly punctate 32

32(31). Scutum with extremely fine longitudinal rugulae visible at an angle to light source .... *A. (Nemandrena)* (in part)
—. Scutum variously shagreened, tessellate, or shiny, but without fine longitudinal rugulae 33

33(32). Genal area often with strong posterior carina; scutellum shiny, unshagreened; propodeal triangle with rather coarse irregular sculpturing; clypeus with apical margin usually broad, distinctly reflexed or turned upward .... *A. (Gonandrena)*

—. Genal area without posterior carina; scutellum usually shagreened or tessellate, but *if* shiny, then propodeal triangle finely granular or tessellate; clypeal apical margin often narrow, not reflexed upward 34

34(33). Malar area extremely short, linear, its length one-sixth to one-tenth of its width; dorsolateral angle of pronotum weak; basal area of labrum simple, as long as broad or longer, strap-shaped or U-shaped with sides diverging slightly basad; propodeal triangle roughened at least mediobasally .... *A. (Leucandrena)* (in part)
—. Malar area distinct, its length one-fourth to one-fifth of its width, but *if* linear, then dorsolateral angle of pronotum distinct and basal area of labrum distinctly bidentate or short and broad; propodeal triangle often finely tessellate, not roughened 35

35(34). Malar area extremely short, linear; terga without pale apical fasciae; vesiture entirely or largely black .... *A. (Onagrandrena)* (in part)
—. Malar area often distinct, its length one-fifth to one-fourth of its width or more, but *if* linear, then terga sometimes with pale apical fasciae or at least vesiture not largely black 36

36(35). Malar area at least half as long as broad; propodeal corbicula incomplete, male-like; vesiture black .... *A. (Dactylandrena)* (in part)
—. Malar area less than one-third as long as broad, usually one-fourth or one-fifth; propodeal corbicula usually complete anteriorly, but *if* incomplete, then not male-like; vesiture usually not entirely black 37

37(36). Scopal hairs long, simple, extremely sparse, scarcely obscuring surface of tibia; propodeal corbicula with long dorsal hairs but no anterior hairs, and with no or extremely few internal hairs .... *A. (Andrena s. str.)* (in part)
—. Scopal hairs abundant, usually long, surface of tibia effectively hidden; propodeal corbicula complete anteriorly, or *if* lacking long anterior hairs, then with abundant long, simple or barbed internal hairs 38

38(37). Terga without pale apical fasciae, or fasciae diffuse, consisting of hairs of about same length and color as more basal hairs; basal area of labrum almost always bidentate and slightly reflexed; propodeal corbicula usually complete .... *A. (Andrena s. str.)* (in part)
—. Terga with pale apical fasciae; basal area of labrum often simple; propodeal corbicula often incomplete anteriorly 39

39(38). Length 8-9 mm; stigma large, usually separated from vein r by two or three vein widths; pronotal dorsolateral angle distinct, dorsoventral ridge indistinct .... *A. (Notandrena)*
—. Length usually 10 mm or more, or, *if* smaller, then stigma narrow, *or* pronotum with distinct dorsoventral ridge, *or* both 40

40(39). Malar area distinct, one-sixth to one-fourth as long as broad 41
—. Malar area linear, less than one-sixth as long as broad 43

41(40). Basal area of labrum entire, short, three or more times as broad as long .... *A. (Anchandrena)*
—. Basal area of labrum bidentate or emarginate apicomedially, as long as broad at base or almost so 42

42(41). Pygidial plate large, as long as basal width; T1-T4 with pale apical fasciae .... *A. (Archiandrena)* (in part)

—. Pygidial plate narrower, longer than basal width, acute at apex unless worn; T2-T4 fasciate .... *A. (Leucandrena)* (in part)
43(40). Galeal blade sharply pointed, narrow, tapering directly from base to apex, spear-shaped .... *A. (Belandrena)* (in part)
—. Galeal blade broadened in basal half or more, and more or less obliquely tapered apically, not spear-shaped 44
44(43). Metasomal terga usually with weak, pale apical fasciae, weakly if at all punctate; pronotal dorsoventral ridge roughened below oblique groove; propodeal corbicula complete anteriorly .... *A. (Andrena s. str.)* (in part)
—. Metasomal T2-T4 with distinct, pale apical fasciae often interrupted medially, the terga distinctly punctate; pronotal dorsoventral ridge usually sharp, subcarinate, or *if* rounded below, then propodeal corbicula incomplete anteriorly .... *A. (Holandrena)*
45(11). Submarginal cells two; tibial scopal hairs largely simple (propodeal corbicula incomplete anteriorly, with abundant internal hairs) 46
—. Submarginal cells three, or *if* only two, then tibial scopal hairs largely plumose 47
46(45). Integument often metallic blue or blue-green; propodeum outside of triangle roughened by punctures .... *A. (Diandrena)* (in part)
—. Integument not metallic; propodeum outside of triangle with punctures sparse, not roughening surface .... *A. (Pelicandrena)*
47(45). Scutum between parapsidal lines with numerous, extremely fine, longitudinal rugulae .... *A. (Nemandrena)* (in part)
—. Scutum shagreened, tessellate, punctate or shiny, but without fine longitudinal rugulae 48
48(47). Mandible tridentate; subgenal coronet composed of several rows (5-8) of short, thick hairs forming a dense brush .... *A. (Dasyandrena)*
—. Mandible bidentate or simple; subgenal coronet composed of one or a few rows of short thick or plumose hairs 49
49(48). Propodeum with lateral surface completely set off from dorsal and posterior surfaces by a carina, profile declivous from base to apex; clypeus short .... *A. (Hesperandrena)* (in part)
—. Propodeum with lateral surface not set off by a carina, profile showing a distinct dorsal surface; clypeus longer 50
50(49). Maxillary palpus short, usually not exceeding galea or occasionally exceeding galea by length of terminal segment; stigma narrow; tibial scopal hairs almost always plumose throughout 51
—. Maxillary palpus exceeding galea by at least last two segments; stigma often broader than width of prestigma, as measured to wing margin; tibial scopal hairs usually simple, at least medially 53
51(50). Ocelloccipital distance about one-half ocellar diameter; integument usually slightly metallic blue-green; hind tibia cuneate with relatively short hairs; clypeal hairs plumose, all erect and only slightly curved near tips .... *A. (Augandrena)*
—. Ocelloccipital distance usually one ocellar diameter, but *if* shorter, then integument black or red, not metallic and/or hind tibia not cuneate, with long scopal hairs;

clypeal hairs plumose or largely so, not entirely erect, usually curving forward 52
52(51). Scopal hairs simple (ocelloccipital distance one ocellar diameter) .... *A. (Rhacandrena)* (in part)
—. Scopal hairs largely plumose .... *A. (Callandrena)* (in part)
53(50). Tibial scopal hairs highly plumose 54
—. Tibial scopal hairs simple or largely so 56
54(53). Propodeal corbicula complete; propodeum narrow at apex (at metasomal attachment); integument black, nonmetallic .... *A. (Simandrena)* (in part)
—. Propodeal corbicula incomplete, without long anterior hairs but with internal hairs; propodeum not unusually narrow at apex; integument often metallic 55
55(54). Scopal hairs of distal part of tibia short; hind tibia cuneate; clypeus distinctly punctate; integument black, nonmetallic .... *A. (Genyandrena)* (in part)
—. Scopal hairs longer; hind tibia not cuneate; clypeal punctures often obscured by dense tessellation or coarse shagreening; integument often metallic .... *A. (Euandrena)* (in part)
56(53). Length usually 9 mm or less; stigma large, first submarginal crossvein usually meeting marginal cell within two or three vein widths of stigma; propodeal corbicula usually incomplete anteriorly, often with internal hairs 57
—. Length usually more than 9 mm; stigma narrow or only moderately broad, first submarginal crossvein usually meeting marginal cell several vein widths from stigma; but *if* length 9 mm or less and stigma broad, then propodeal corbicula complete anteriorly, without internal hairs 59
57(56). Vesture of head, pleurae, and terga dark-brown; metasomal terga without pale apical fasciae; facial fovea narrow; facial quadrangle broader than long .... *A. (Cremnandrena)*
—. Vesture of head, pleurae, and terga not all dark brown; metasomal terga usually with pale apical fasciae; facial fovea usually broad; facial quadrangle usually as long as or longer than broad 58
58(57). Pronotum with weak dorsolateral angle, the dorsoventral ridge absent or extremely weak; face below ocelli tessellate, without distinct longitudinal rugulae or punctures .... *A. (Derandrena)* (in part)
—. Pronotum without both dorsolateral angle and ridge; face below ocelli rugulate and/or punctate .... *A. (Micrandrena)*
59(56). Hind tibia cuneate; tibial scopal hairs along posterior margin usually short, less than half as long as greatest tibial width, shortest subapically 60
—. Hind tibia not cuneate; tibial scopal hairs along posterior margin at least half as long as greatest tibial width 64
60(59). Subgenal coronet absent; ocelloccipital distance no more than half an ocellar width .... *A. (Genyandrena)* (in part)
—. Subgenal coronet present; ocelloccipital distance often longer than half an ocellar diameter 61
61(60). Integument metallic blue or blue-black; terga with abundant, coarse, round punctures; tibial scopal hairs long anteriorly and longest anteroapically, where each hair equals length of tibial spur .... *A. (Scaphandrena)* (in part)

—. Integument black; terga impunctate or with relatively small and/or sparse punctures; tibial scopal hairs short anteriorly as well as posteriorly, each anteroapical hair half length of tibial spur or less 62
62(61). Pleural hairs black; tergal vestiture black; facial fovea narrow, its upper end occupying about half ocellocular distance .... *A. (Scaphandrena)* (in part)
—. Pleural hairs white to fulvous and/or tergal vestiture not entirely black; facial fovea usually broad 63
63(62). Propodeal corbicula complete, bare internally .... *A. (Simandrena)* (in part)
—. Propodeal corbicula incomplete, with long, simple internal hairs .... *A. (Scaphandrena)* (in part)
64(59). Vestiture entirely black or thoracic dorsum with reddish-brown hairs; propodeal triangle relatively coarsely and irregularly scuptured, forming small areolae; tibial scopal hairs long, simple, sparse .... *A. (Onagrandrena)* (in part)
—. Vestiture not entirely black, or *if* so, then *either* propodeal triangle finely sculptured (tessellate or shagreened) *or* tibial scopal hairs simple but abundant 65
65(64). Labial palpal segments 2 and 3 each with next segment attached subapically, the third segment attached to second near middle of second segment; integument bright metallic blue-black; first recurrent vein meeting second submarginal cell about two-thirds of cell length from base .... *A. (Belandrena)* (in part)
—. Labial palpal segments 2 and 3 attached terminally, or, *if* subapically, then third attached to second well beyond middle of second; integument often black or red, occasionally metallic; first recurrent vein meeting second submarginal cell usually not much, if any, beyond middle of cell 66
66(65). T2-T4 with pale apical fasciae of hair 67
—. T2-T4 without pale apical fasciae of hair 73
67(66). T1 with pale apical fascia; basal parts of T2-T4 with abundant erect hairs arising from small punctures; wing membranes often infumate 68
—. T1 without fascia; basal areas of T2-T4 without erect hairs, or these not arising from distinct punctures; wing membranes usually not infumate 69
68(67). Galeal blade pointed, its outer apical margin straight; clypeus, dorsum of thorax, and metasomal terga shiny; propodeum outside of triangle impunctate .... *A. (Rhacandrena)* (in part)
—. Galeal blade with outer apical margin slightly convex, broader basally; at least clypeus and metasomal terga dulled or moderately dulled by coarse shagreening or tessellation; propodeum outside of triangle roughened by distinct punctures .... *A. (Euandrena)* (in part)
69(67). Terga distinctly punctate, their surfaces dulled by dense reticular shagreening; clypeus distinctly flattened posteromedially, with slightly elongate punctures, surface usually dulled by fine tessellation .. *A. (Taeniandrena)*
—. Terga often not distinctly punctate, their surfaces dulled by fine reticular shagreening or shiny; clypeus not flattened posteromedially, or, *if* so, then shiny, not tessellate, and punctures round, not elongate 70
70(69). Terga punctate; clypeus sparsely punctate, punctures mostly separated by two to three or more puncture widths; pronotum with oblique groove impressed .... *A. (Rhacandrena)* (in part)

—. Terga usually impunctate or weakly so; clypeal punctures more crowded; pronotum usually with oblique groove absent or scarcely impressed 71
71(70). Labrum below basal area with strong median crista, often lamellate, *and* with strong lateral crista and two shiny transverse sulci (metasomal terga tessellate, opaque) .... *A. (Psammandrena)*
—. Labrum below basal area with one median crista, not lamellate, *or* without cristae, without transverse sulci 72
72(71). Propodeum outside of triangle with punctures moderately roughening surface; dorsal femoral hairs often with abundant short, S-shaped branches bent away from rachis, appearing highly plumose; galeal blade generally narrow, its outer margin in apical half usually straight .... *A. (Euandrena)* (in part)
—. Propodeum outside of triangle without punctures, smoothly tessellate or shagreened; dorsal femoral hairs with relatively short, stiff barbs or simple; galeal blade with outer apical margin gently convex, broad at base .... *A. (Thysandrena)*
73(66). Vestiture entirely black; ocelloccipital distance one ocellar diameter or less .... *A. (Scaphandrena)* (in part)
—. Vestiture in part or largely pale; or, *if* mostly black, then ocelloccipital distance one and one-half to two ocellar diameters .... *A. (Melandrena)*

Key to the Subgenera of *Andrena* of North and Central America (Males)
(Modified from LaBerge, 1986)

1. Submarginal cells two 2
—. Submarginal cells three 6
2(1). First and often base of second flagellar segment excavated below... *A. (Pelicandrena)*
—. First and second flagellar segments terete or flattened and expanded laterally but not excavated below 3
3(2). Clypeus metallic blue, often with violaceous reflections; basal area of labrum usually bidentate; propodeal triangle usually moderately coarsely sculptured .... *A. (Diandrena)*
—. Clypeus yellow, white, or black; basal area of labrum bidentate or entire; propodeal triangle usually finely sculptured 4
4(3). Maxillary palpus short, when extended distad exceeding tip of galea by no more than length of last segment and usually less ....*A. (Callandrena)* (in part)
—. Maxillary palpus exceeding tip of galea by at least length of two apical segments 5
5(4). Clypeus and paraocular areas black; propodeum outside of triangle roughened by coarse punctures .... *A. (Onagrandrena)* (in part)
—. Clypeus and paraocular areas white to yellow; propodeum outside of triangle not roughened by coarse punctures .... *A. (Parandrena)*
6(1). Malar space one-third to one-half as long as broad, or longer, but *if* as short as one-third as long as broad, then anterior trochanter with spinelike process on anterior surface near apex; clypeus yellow; facial quadrangle usually as broad as long .... *A. (Dactylandrena)*
—. Malar space less than one-third as long as broad, or, *if* as long as one-third width, then *either* clypeus black *or* facial quadrangle longer than broad; anterior trochanter never with apical spine 7

7(6). Scutum with fine longitudinal rugulae, often reduced to short rugulae on posteromedian area; basal area of labrum with short median horn, often as long as rest of area (metasomal terga without pale fasciae) .... *A. (Nemandrena)*
—. Scutum without fine longitudinal rugulae; basal area of labrum without median horn 8
8(7). Mouthparts elongate, galeal blade as long as clypeus or slightly longer; maxillary palpus and/or labial palpus elongate; T2-T4 with pale, often weak, apical fasciae 9
—. Mouthparts short, galeal blade usually not as long as clypeus; neither maxillary nor labial palpus greatly elongate, but *if* galeal blade as long as clypeus, then metasomal terga lacking pale apical fasciae 11
9(8). Labial palpus greatly elongate, segments 1 and 2 three to five times as long as segments 3 and 4 taken together, first two segments greatly flattened; maxillary palpus normal in length, half as long as galeal blade .... *A. (Callandrena)* (in part)
—. Labial palpus elongate but normal, first two segments neither unusually long nor much flattened; maxillary palpus usually about as long as galeal blade or much longer 10
10(9). Maxillary palpus with second segment at least one and one-half times as long as first; stipes distinctly narrowed medially and tapering apically, the apical third less than half as broad as basal third, with relatively sparse, weak, plumose hairs .... *A. (Iomelissa)*
—. Maxillary palpus with second segment about as long as first; stipes linear, tapering extremely slightly toward apex, apical third almost as broad as basal third, with abundant long, stout, spinelike hairs .... *A. (Erandrena)*
11(8). Malar area about one-third as long as wide; facial quadrangle distinctly longer than broad (clypeus yellow or partially so) .... *A. (Conandrena)*
—. Malar area usually less than one-third as long as broad; facial quadrangle as long as broad or broader 12
12(11). Inner hind tibial spur slightly flattened basally with a membranous flange, spur often strongly curved, broader than outer spur near base 13
—. Inner hind tibial spur not broadened basally, not unusually curved near base, usually about as broad as outer spur near base 15
13(12). Propodeal triangle coarsely sculptured .... *A. (Plastandrena)* (in part)
—. Propodeal triangle finely sculptured, usually tessellate or shagreened 14
14(13). Flagellar segment 1 as long as segment 2 plus 3 or longer; maxillary palpus not reaching apex of galea .... *A. (Charitandrena)*
—. Flagellar segment 1 about as long as segment 3, longer than segment 2; maxillary palpus exceeding galea by at least last palpal segment .... *A. (Aporandrena)*
15(12). Middle and hind tibial spurs unusually thick, their apices rather abruptly hooked; terga without pale apical fasciae; clypeus black .... *A. (Euandrena)* (in part)
—. Middle and hind tibial spurs neither unusually thickened nor strongly hooked at apices; terga often with pale apical fasciae; clypeus usually black, occasionally yellow 16
16(15). Inner hind tibial spur rather abruptly bent and slightly twisted at about one-third length from apex; T3-T4 with pale apical fasciae and distinct punctures .... *A. (Leucandrena)* (in part)
—. Inner hind tibial spur long, slender, neither bent abruptly nor twisted in apical third; T3-T4 with or without pale apical fasciae, often impunctate 17
17(16). Propodeum declivous, its dorsal surface slanting from basal margin to apex; clypeus often short, scarcely if at all protruding below lower margins of compound eyes 18
—. Propodeum with distinct horizontal basal area; clypeus usually protruding below eyes, occasionally short.......... 19
18(17). Propodeum with lateral surface separated from lower posterior surface by short lateral carina; clypeal punctures obscured by dense, fine tessellation .... *A. (Hesperandrena)*
—. Propodeum with lateral surface not at all separated from lower posterior surface by a carina; clypeal punctures usually evident .... *A. (Celetandrena)*
19(17). Clypeus with lateral angle produced forward beyond base of mandible, to about same level as median apex of clypeus; head extremely broad, broader than thorax; dorsoventral pronotal ridges almost meeting middorsally, far from dorsolateral angles ....*A. (Cremnandrena)*
—. Clypeus with lateral parts not produced forward; head not much, if any, broader than thorax; dorsoventral pronotal ridge usually ending in dorsolateral angle, or pronotum lacking dorsolateral angle and ridge 20
20(19). Labial palpus with third segment arising near middle of second segment or slightly beyond, and fourth segment arising subapically from third; integument metallic blue (first recurrent vein meeting second submarginal cell two-thirds of length of cell from base; clypeus pale yellow) .... *A. (Belandrena)* (in part)
—. Labial palpus with second and third segments attached to succeeding segments apically or almost so; integument usually black or red 21
21(20). Pronotum with distinct dorsolateral angle along posterior margin, and with more or less distinct dorsoventral ridge extending down from angle to anterior coxa, area between ridge and pronotal lobe often shiny and impunctate, occasionally deep and narrow or rugulose 22
—. Pronotum without dorsolateral angle along posterior margin, without dorsoventral ridge 44
22(21). Maxillary palpus short, not exceeding galea (when extended distad) by more than length of apical segment of palpus; stigma usually narrower than prestigma, as measured to anterior margin of wing; clypeus yellow .... *A. (Callandrena)* (in part)
—. Maxillary palpus usually exceeding galea by at least length of last two palpal segments; stigma variable, often broader than or equal to prestigma, as measured to wing margin; but *if* palpus short *and* stigma narrow, then clypeus black 23
23(22). Genal area immediately below posterior mandibular articulation with a short, shiny, often fingerlike process 24
—. Genal area immediately below posterior mandibular articulation without a fingerlike process 25
24(23). Basal area of propodeum shorter than metanotum, triangle with fine but distinct, longitudinal rugulae;

metasomal terga distinctly punctate .... *A. (Ptilandrena)* (in part)
—. Basal area of propodeum longer than metanotum, triangle tessellate or finely areolate, without longitudinal rugulae; metasomal terga impunctate or indistinctly punctate .... *A. (Derandrena)* (in part)
25(23). Genal area with a large, blunt process extending downward and somewhat forward from ventral surface, which is otherwise concave (poorly developed in a few specimens) 26
—. Genal area without large process on ventral surface, which is flat, smoothly curved at sides 27
26(25). Clypeus yellow; propodeal triangle finely sculptured, shagreened or finely tessellate ...... *A. (Genyandrena)*
—. Clypeus black; propodeal triangle coarsely sculptured, irregularly rugulose .... *A. (Xiphandrena)*
27(25). Facial quadrangle distinctly broader than long; metasomal terga without pale apical fasciae; pronotum with dorsolateral angle indistinct, displaced toward midline (dorsoventral pronotal ridge distinct) .... *A. (Oligandrena)*
—. Facial quadrangle about as broad as long, or, *if* broader, then *either* metasomal terga with distinct pale apical fasciae *or* dorsolateral angle of pronotum distinct *or* both 28
28(27). Clypeus partially or wholly yellow or cream-colored 29
—. Clypeus entirely black 36
29(28). Pygidial plate well developed, narrow, V-shaped; first flagellar segment usually about two-thirds as long as second .... *A. (Archiandrena)*
—. Pygidial plate absent or vestigial (narrow and linear when present); first flagellar segment usually as long as or almost as long as second 30
30(29). Galeal blade sharply pointed and spear-shaped, evenly tapering from base to apex, narrow .... *A. (Belandrena)* (in part)
—. Galeal blade not sharply pointed, not spear-shaped, more or less sharply tapered apically, broad in basal one-third to one-half 31
31(30). Clypeus with apical margin broad, turned forward; propodeal triangle finely areolate mediobasally (T2-T5 with distinct pale apical fasciae) .... *A. (Geissandrena)*
—. Clypeus with apical margin short, not turned forward; propodeal triangle variously sculptured but usually tesselate 32
32(31). Genal area with posterior margin distinctly carinate; stigma large, first submarginal crossvein ending on marginal cell within two or three vein widths of stigma; first flagellar segment as long as or longer than second plus third .... *A. (Notandrena)*
—. Genal area with posterior margin rounded, or, *if* carinate, then stigma narrow and first submarginal crossvein ending on marginal cell several vein widths from stigma; first flagellar segment shorter than second plus third 33
33(32). Stigma large, broader than prestigma as measured to anterior wing margin; first submarginal crossvein meeting marginal cell within three or four vein widths of stigma or less; pronotum with dorsoventral ridge crossed by distinctly impressed oblique groove 34
—. Stigma narrower, or, *if* as broad as above, then first submarginal crossvein meeting marginal cell at more than four vein widths from stigma; pronotum with dorsoventral ridge relatively sharp, not depressed by oblique groove 35
34(33). Basal area of labrum trapezoidal, often with small but distinct median emargination; integument, except clypeus, black, nonmetallic .... *A. (Larandrena)*
—. Basal area of labrum entire, rounded, short (several times broader than long), *or*, if trapezoidal and slightly emarginate, then integument largely metallic blue .... *A. (Derandrena)* (in part)
35(33). Malar area distinct, six times as broad as long or longer; metasomal terga usually impunctate or indistinctly punctate; propodeal triangle finely sculptured .... *A. (Anchandrena)*
—. Malar area linear, six times as broad as long or shorter; metasomal terga more or less distinctly punctate; propodeal triangle moderately coarsely sculptured .... *A. (Holandrena)* (in part)
36(28). Pronotum with dorsoventral ridge extending down from dorsolateral angle interrupted by distinct, oblique, deeply impressed groove; T2-T4 usually without pale apical fasciae; clypeus short, not produced much beyond level of lower ends of compound eyes ...... *A. (Tylandrena)*
—. Pronotum with dorsoventral ridge extending down from dorsolateral angle not interrupted by distinctly impressed groove, or, *if* groove present, then a mere line; T2-T4 often with pale apical fasciae; clypeus often produced by one-fourth or more of median length beyond level of lower ends of compound eyes 37
37(36). Genal area with posterior margin distinctly carinate; clypeus with apical area long, distinctly turned forward; T2-T5 with more or less distinct pale apical fasciae .... *A. (Gonandrena)* (in part)
—. Genal area with posterior margin not carinate; clypeus with apical area usually short, not turned forward; T2-T5 often without pale apical fasciae 38
38(37). T2-T4 with distinct pale apical fasciae; metasomal terga impunctate or with fine, indistinct punctures; propodeal triangle often areolate to coarsely rugulose (malar area linear) .... *A. (Leucandrena)* (in part)
—. T2-T4 without pale apical fasciae; metasomal terga with small to coarse, distinct punctures; propodeal triangle smooth, tessellate to coarsely rugose 39
39(38). Malar area distinct, one-fourth to one-fifth as long as broad; terga without pale apical fasciae; mandible often with inferior basal tooth .... *A. (Andrena s. str.)* (in part)
—. Malar area linear, much less than one-fifth as long as broad; terga with pale apical fasciae; mandible lacking inferior tooth 40
40(39). Metasomal terga coarsely punctate; dorsal propodeal triangle moderately coarsely to coarsely sculptured 41
—. Metasomal terga impunctate or extremely finely punctate; dorsal propodeal triangle finely sculptured 42
41(40). Basal area of labrum short, four times as broad as long, entire to weakly emarginate .... *A. (Holandrena)* (in part)
—. Basal area of labrum long, bidentate, usually reflexed .... *A. (Rhaphandrena)*
42(40). T1-T5 or T2-T5 with pale apical fasciae composed of long, blunt hairs with abundant, extremely short barbs, giving hairs an opaque appearance in strong light;

basal area of labrum strongly bidentate and strongly reflexed, usually elevated well above margin of clypeus, its apical teeth turned under .... *A. (Cnemidandrena)*
—. T2–T5 with pale apical fasciae weak, composed of moderately long, pointed, weakly barbed white hairs; basal area of labrum entire, weakly to strongly bidentate, often not strongly reflexed 43

43(42). T1 and usually T2 medially with abundant long, erect to suberect hairs; basal area of labrum usually bidentate; maxillary palpal segments not all elongate ....
.... *A. (Andrena s. str.)* (in part)
—. T1 and T2 without long, erect to suberect hairs, T1 bare or with short, sparse, erect hairs; basal area of labrum entire or only weakly emarginate apically; maxillary palpal segments all moderately elongate ....
.... *A. (Ptilandrena)* (in part)

44(21). Maxillary palpus short, rarely exceeding galea when extended distad or exceeding galea by less than length of last two palpal segments 45
—. Maxillary palpus long, exceeding galea when extended distad by at least length of last two palpal segments 47

45(44). Basal area of labrum large, entire, subtriangular, about as long as broad; first flagellar segment twice as long as second; integument dark metallic blue or blue-black
.... *A. (Scoliandrena)*
—. Basal area of labrum smaller, usually bidentate or emarginate, broader than long; first flagellar segment often less than twice as long as second segment; integument occasionally metallic, usually black or in part red 46

46(45). Ocelloccipital distance half an ocellar diameter or less, rarely slightly more; integument dull metallic blue; clypeus and paraocular areas pale yellow ....
.... *A. (Augandrena)*
—. Ocelloccipital distance usually one ocellar diameter or more, but *if* as short as half an ocellar diameter, then *either* integument not metallic *or* paraocular areas without pale maculae *or* both .... *A. (Callandrena)* (in part)

47(44). Stigma large, first submarginal crossvein ending one to three vein widths from stigma; body 9 mm or less in length 48
—. Stigma narrower, first submarginal crossvein usually ending more than three vein widths from stigma; body usually more than 9 mm in length 49

48(47). Propodeal triangle margined by minutely carinate lateral bounding sutures; dorsal surface of propodeum not longer than scutellum .... *A. (Simandrena)* (in part)
—. Propodeal triangle margined by acarinate lateral bounding sutures, mere lines or slight depressions; dorsal surface of propodeum often longer than scutellum ....
.... *A. (Micrandrena)*

49(47). Pleurae and propodeal triangle coarsely or moderately coarsely sculptured, or, *if* triangle finely sculptured, then genal area narrow, about as wide as eye in side view
.... 50
—. Pleurae and propodeal triangle usually finely sculptured, often merely granular or tessellate, or, *if* moderately coarsely sculptured, then genal area much broader than eye in side view 56

50(49). Inner hind tibial spur slightly broadened basally by a membranous flange, often strongly curved, broader than outer spur; S6 bent downward and forward apicolaterally to form two reflexed blunt teeth; first recurrent vein meeting second submarginal cell near or only slightly beyond middle of cell ....
.... *A. (Plastandrena)* (in part)
—. Inner hind tibial spur neither broadened basally nor unusually curved, about as narrow as outer spur; S6 relatively flat apically, or, *if* margin reflexed, then not forming apicolateral teeth; first recurrent vein often meeting second submarginal cell two-thirds of length of cell or more from base 51

51(50). Propodeal triangle coarsely areolate or irregularly rugose, with transverse posterior carina separating dorsal from posterior surface 52
—. Propodeal triangle coarsely punctate or finely areolate, without transverse carina separating dorsal from posterior surface 53

52(51). Marginal zone of T2 one-third or more of median tergal length; third flagellar segment usually two-thirds as wide as long or longer, antennae in repose usually reaching beyond scutellum; first recurrent vein meeting second submarginal cell near middle of cell, rarely beyond .... *A. (Trachandrena)*
—. Marginal zone of T2 less than one-third of median tergal length; third flagellar segment distinctly more than two-thirds as wide as long, antennae in repose usually not reaching beyond middle of scutellum; first recurrent vein meeting second submarginal cell two-thirds or more of length of cell from base .... *A. (Scrapteropsis)*

53(51). Clypeus flattened mediobasally; metasomal terga distinctly punctate, surface and bottoms of punctures dulled by fine tessellation; T2–T5 with pale apical fasciae, but these often interrupted medially .... *A. (Taeniandrena)*
—. Clypeus usually not flattened mediobasally; metasomal terga punctate but shiny or moderately so, at most dulled by fine reticulate shagreening; T2–T5 without pale apical fasciae 54

54(53). Clypeus yellow; terga bright metallic blue; S2–S5 with distinct, white, subapical fimbriae ....
.... *A. (Scaphandrena)* (in part)
—. Clypeus black; terga dark metallic blue-black or black; sterna with or without subapical fimbriae, but these usually weak when present 55

55(54). First flagellar segment much shorter than second, second subequal in length to third; labial palpus short, second and third segments almost as broad as long; integument dark metallic blue-black with violaceous reflections .... *A. (Melandrena)* (in part)
—. First flagellar segment usually at least as long as second, rarely slightly shorter, second segment usually shorter than third; labial palpus normal, second and third segments slender, not nearly as broad as long; integument black or with extremely faint metallic reflections ....
.... *A. (Onagrandrena)* (in part)

56(49). Clypeus at least in part yellow or white 57
—. Clypeus black or metallic blue or green 61

57(56). Metasomal terga without pale apical fasciae; integument usually dark metallic blue-black ....
.... *A. (Scaphandrena)* (in part)
—. Metasomal terga with pale apical fasciae; integument usually dull black, nonmetallic 58

58(57). Clypeus dark except for small apicomedial spot or

subapical band of yellow occupying less than half of clypeal area; S2-S5 with exceptionally long subapical fimbriae .... *A. (Dasyandrena)* (in part)
—. Clypeus at least two-thirds yellow; metasomal sterna with subapical fimbriae absent or consisting of hairs of moderate length 59
59(58). Metasomal terga dull, opaque, tessellation obscuring punctures if present; S6 flat apically ....
.... *A. (Psammandrena)*
—. Metasomal terga strongly to moderately punctate, surface shiny or moderately shiny; S6 with apical margin often reflexed 60
60(59). Clypeus usually white or cream-colored; T2-T4, but rarely T1, with apical pale fasciae; S2-S5 with subapical fimbriae weak (short) or absent (*if* with weak sternal fimbriae, *then* without fascia on T1); hind tibia broad, cuneate .... *A. (Scaphandrena)* (in part)
—. Clypeus usually yellow; T1-T4 with apical pale fasciae; S2-S5 with well-formed pale subapical fimbriae; hind tibia not broadened apically, not cuneate ....
.... *A. (Rhacandrena)*
61(56). Genal area narrow, at most slightly broader than eye in lateral view; metasomal sterna usually with pale subapical fimbriae 62
—. Genal area conspicuously broader than eye in lateral view; metasomal sterna without pale subapical fimbriae 67
62(61). Propodeum with lateral sutures delimiting triangle slightly raised; clypeus impunctate or punctures obscured by dense regular tessellation or coarse shagreening; integument black, never metallic ....
.... *A. (Simandrena)* (in part)
—. Propodeum with lateral sutures demarcating triangle flat or slightly depressed, or, *if* raised and ridgelike, then integument at least slightly metallic; clypeus usually distinctly punctate, often shiny but occasionally impunctate and dull 63
63(62). T1-T5 with pale apical fasciae, bases of terga with abundant long, erect hairs; wing membranes moderately to deeply infumate .... *A. (Euandrena)* (in part)
—. T1-T5 or at least T1 without pale apical fasciae, bases of terga with or without erect hairs; wing membranes clear to moderately infumate 64
64(63). Metasomal sterna lacking pale subapical fimbriae; terga often without pale apical fasciae; propodeum outside of triangle distinctly punctate ....
.... *A. (Euandrena)* (in part)
—. Metasomal sterna with distinct pale subapical fimbriae; T2-T4 with pale apical fasciae; propodeum outside of triangle impunctate or nearly so 65
65(64). Sternal subapical fimbriae exceptionally long; space between clypeus and antennal sockets less than one socket diameter (western USA) .... *A. (Dasyandrena)* (in part)
—. Sternal subapical fimbriae moderately long; space between clypeus and antennal sockets one socket diameter or more 66
66(65). Galeal blade narrow, spear-shaped, outer margin of apical half straight; basal area of labrum simple ....
.... *A. (Oxyandrena)*
—. Galeal blade broad basally, outer margin of apical half or less gently concave; basal area of labrum usually emarginate apically .... *A. (Thysandrena)* (in part)

67(61). T2-T4 and often T5 with more or less distinct pale apical fasciae; ocelloccipital distance about one ocellar diameter, often less and rarely slightly more ....
.... *A. (Leucandrena)* (in part)
—. T2-T5 without pale apical fasciae, *or*, if present on some terga, then weak, and ocelloccipital distance at least one and one-half ocellar diameters, often more ....
.... *A. (Melandrena)* (in part)

Key to the Subgenera of *Andrena* of the Western Palearctic Region (Females)
(Modified from A. Molino-Pardo's translation (MS) of Warncke's 1968 key)

The subgenus *Oreomelissa,* containing the European *Andrena coitana* (Kirby), is omitted from the key to males, but see the key to Japanese subgenera. Warncke (1968) placed this species in *Stenomelissa* but neither described that subgenus nor placed it in his key. It presumably has nothing to do with the east Asian *Stenomelissa* Hirashima and LaBerge. Although *Carinandrena, Fuscandrena, Leimelissa, Longandrena, Osychnyukandrena,* and *Planiandrena* are palearctic taxa, they not included in this key. Four of them are monotypic. *Malayapis* is included, although it is oriental. For couplet 17, see also couplet 49; the subgenus *Holandrena* would seemingly run to couplet 18. In Warncke's key these couplet numbers are 17 and 48, respectively.

1. Inner side of hind femur with a row of small peglike or thornlike projections 2
—. Inner side of hind femur without such projections 7
2(1). Tibial scopa of simple hairs; labrum not conspicuously divided 3
—. Tibial scopa of small to large plumose hairs; labrum divided conspicuously into median and lateral parts 4
3(2). Facial fovea broad in upper part, narrow and deeply channeled in lower part .... *A. (Orandrena)*
—. Facial fovea uniformly broad, occupying one-third of paraocular area .... *A. (Cryptandrena)* (in part)
4(2). Facial fovea short, rectangular, almost twice as long as broad; hind femur slender and almost cylindrical ....
.... *A. (Avandrena)* (in part)
—. Facial fovea usually long and not rectangular; hind femur strong, its inner side flattened and limited above by a more or less conspicuously developed linear carina 5
5(4). Facial fovea short and drop-shaped; metasomal terga fairly strongly to strongly and densely punctate ....
.... *A. (Rufandrena)*
—. Facial fovea long, or, *if* drop-shaped, then metasomal terga with scattered deep punctures 6
6(5). Inner side of hind femur with weak carina and sparse, long bristles; lateral area of labrum small; metasomal terga usually with scattered deep punctures ....
.... *A. (Chlorandrena)*
—. Inner side of hind femur with strong carina and dense, short bristles; lateral area of labrum about as large as median area; metasomal terga strongly and densely punctate
.... *A. (Lepidandrena)*
7(1). Inner hind tibial spur usually distinctly and strongly broadened at base 8
—. Inner hind tibial spur not broadened at base, at most distal half convexly broadened 13

8(7). Mesepisternum and propodeal triangle strongly honeycomb-areolate 9
—. Mesepisternum and propodeal triangle finely roughened to smooth 11
9(8). Pronotum with dorsolateral angle elevated, abruptly cut off laterally .... *A. (Melanapis)*
—. Pronotum with more rounded dorsolateral angle 10
10(9). Propodeal corbicula only weakly rugose in posterior part; body black to red .... *A. (Plastandrena)*
—. Propodeal corbicula with strongly developed rugae, mostly lengthwise; body brilliant metallic blue .... *A. (Agandrena)*
11(8). Mesepisternum and propodeum glossy, coarsely and deeply punctate .... *A. (Scitandrena)*
—. Mesepisternum and propodeum roughened, moderately finely lined, with asymmetrical (one-sided) punctures 12
12(11). Propodeal triangle weakly rugosely areolate to posterior margin; inner side of hind femur conspicuously carinate; basal area of labrum small and usually triangular .... *A. (Suandrena)*
—. Propodeal triangle with middle area shagreened, covered with fine rugae; inner side of hind femur not carinate; basal area of labrum of medium size, deeply notched .... *A. (Pallandrena)*
13(7). Inner hind tibial spur distinctly convexly broadened a little beyond middle; inner side of hind femur almost always conspicuously carinate; pronotum not carinate 14
—. Inner hind tibial spur not much broadened; inner side of hind femur more or less rounded; pronotum carinate or not 15
14(13). Glossa greatly elongated, at least six times as long as broad .... *A. (Charitandrena)*
—. Glossa of normal length, about three times as long as broad .... *A. (Ulandrena)*
15(13). Propodeum declivous from the metanotum; facial fovea short and drop-shaped (head large) .... *A. (Cubiandrena)*
—. Propodeum with basal area more or less horizontal, or, if declivous, then facial fovea not as above 16
16(15). Facial fovea short and rectangular, at most twice as long as broad .... *A. (Avandrena)* (in part)
—. Facial fovea otherwise, usually conspicuously narrower below than above 17
17(16). Mesepisternum and propodeal triangle coarsely honeycomb-areolate 18
—. Mesepisternum and propodeal triangle at most weakly wrinkled 20
18(17). Facial fovea small, upper part drop-shaped, lower part greatly reduced and fading out .... *A. (Brachyandrena)*
—. Facial fovea normally long, deep, upper part broadly rounded, lower part almost reduced to a sulcus 19
19(18). Submarginal cells two .... *A. (Biareolina)*
—. Submarginal cells three .... *A. (Trachandrena)*
20(17). Mesepisternum with remarkably minute to fine punctures; head usually elongate and narrower 21
—. Mesepisternum strongly and deeply punctate; head short and broad 58
21(20). Dorsolateral angle of pronotum not strongly elevated as transverse ridge and without carina extending down from it, at most with hint of such carina 22
—. Dorsolateral angle of pronotum weakly to strongly elevated as transverse ridge, abruptly cut off laterally, and with carina extending down from it 47
22(21). Mesepisternum behind attachment of foreleg conspicuously compressed and flat to deeply grooved; hairs of thorax squamose to minutely branched .... *A. (Aenandrena)*
—. Mesepisternum behind attachment of foreleg normally rounded; thorax usually normally haired 23
23(22). Small to very small species; first submarginal crossvein reaching marginal cell near posterior margin of stigma 24
—. Usually not small, generally medium-sized to large species; first submarginal crossvein reaching marginal cell at least three vein widths from stigma 27
24(23). Propodeal triangle strongly rugose to rugosely areolated; clypeus not grooved; facial fovea long, usually broad, and occupying about one-third of paraocular area .... *A. (Micrandrena)* (in part) and *A. (Fumandrena)*
—. Propodeal triangle reticulated to shagreened, usually without rugae or weakly rugose at base, but if strongly rugose, then clypeus conspicuously grooved lengthwise; facial fovea usually not as above 25
25(24). Propodeal triangle near metanotum finely but conspicuously and evenly rugose; facial fovea moderately long, uniformly broad, and occupying about one-third of paraocular area .... *A. (Graecandrena)*
—. Propodeal triangle without rugae; facial fovea conspicuously narrowed below 26
26(25). Facial fovea short, not extending below upper margin of antennal socket .... *A. (Aciandrena)*
—. Facial fovea long, usually extending below upper margin of antennal socket .... *A. (Distandrena)*
27(23). Facial fovea long and narrow, deeply grooved at upper end, usually constricted somewhat at middle 28
—. Facial fovea short or broad, not deeply grooved 29
28(27). Body under 10 mm long .... *A. (Parandrenella)*
—. Body over 15 mm long .... *A. (Hyperandrena)*
29(27). Facial fovea short, comma- to drop-shaped, narrowed below 30
—. Facial fovea longer, somewhat narrowed below to occupying almost whole paraocular area 32
30(29). Glossa much more than four times as long as broad; clypeus more or less distinctly snout-shaped, convex, or basal area of labrum broad .... *A. (Didonia)* (in part)
—. Glossa not over about four times as long as broad; clypeus not to weakly protuberant; basal area of labrum usually normal, trapezoidal 31
31(30). Tibial scopa strongly plumose; glossa about four times as long as broad .... *A. (Chrysandrena)*
—. Tibial scopa not to very briefly plumose; glossa less than three times as long as broad .... *A. (Euandrena)*
32(29). Surface of propodeal corbicula impunctate and bare, its margin entirely densely hairy; metanotum with dense, light tuft of hairs directed forward .... *A. (Simandrena)*
—. Surface of propodeal corbicula either punctate and hairy or margin incomplete; metanotum usually not as above 33
33(32). Clypeus flattened and frequently more or less distinctly concave in middle; basal area of labrum short, as if strongly compressed in lengthwise direction; facial

fovea occupying almost whole paraocular area ....
.... *A. (Taeniandrena)* and *A. (Troandrena)*
—. Clypeus more or less strongly convex, or, *if* flattened, then basal area of labrum always of normal size and trapezoidal *or* facial fovea not so broad................................. 34
34(33). Mesepisternum and usually also propodeum finely punctate; propodeal triangle at most finely rugose like metanotum (metasoma not to indistinctly punctate) ....
....35
—. Mesepisternum and propodeum densely punctate; propodeum and especially triangle rugosely areolate 38
35(34). Facial fovea occupying almost whole paraocular area; clypeus strongly punctate; propodeal triangle finely reticulate, shagreened to glossy .... *A. (Hoplandrena)*
—. Facial fovea occupying scarcely half of paraocular area; clypeus usually finely punctate; propodeal triangle minutely roughened to reticulately shagreened 36
36(35). Metasomal terga almost bare except for complete or interrupted hair bands; facial fovea fading away upward; clypeus elongate .... *A. (Nobandrena)*
—. Metasomal terga usually distinctly hairy, without bands, or, *if* bands present, then propodeum densely punctate; facial fovea more distinctly defined above, but *if* vanishing, then clypeus not elongate 37
37(36). Very large species (body length 16-18 mm) with broad maxilla............................ *A. (Melittoides)*
—. Medium-sized to large species (less than 16 mm) with weakly developed maxilla .... *A. (Scaphandrena)*
38(34). Body under 10 mm long 39
—. Body over 10 mm long 43
39(38). Mesepisternum moderately finely punctate and shagreened 40
—. Mesepisternum rugosely areolate 41
40(39). Clypeus longer than broad, transversely convex; facial fovea long .... *A. (Poecilandrena)*
—. Clypeus broader than long, flattened; facial fovea short, not extending below level of antennal bases ....
.... *A. (Poliandrena)* (in part) and *A. (Tarsandrena)* (in part)
41(39). Clypeus distinctly broader than long; basal area of labrum very broad and short, rectangular ....
.... *A. (Cryptandrena)* (in part)
—. Clypeus as long as broad; basal area of labrum narrower.. 42
42(41). Terga uniformly densely punctate, apical marginal zones also punctate .... *A. (Cordandrena)*
—. Terga impunctate or punctate and apical marginal zones distinctly less punctate to impunctate ....
.... *A. (Micrandrena)* (in part)
43(38). T1 impunctate; metanotum with weak, forward-directed tuft of dense hairs .... *A. (Thysandrena)*
—. T1 strongly and usually densely punctate; or, *if* impunctate, then metasoma without bands and metanotum usually differently hairy 44
44(43). Clypeus transversely wrinkled; basal area of labrum narrow and elongated; pygidial plate with elevated area
....*A. (Leucandrena)* (in part)
—. Clypeus not transversely wrinkled; basal area of labrum usually trapezoidal; pygidial plate flat 45
45(44). Clypeus strongly convex; metasoma bare, without bands, at most with lateral, white hair patches (usually very large species)...................... *A. (Melandrena)*
—. Clypeus flattened; metasoma usually with distinct bands 46

46(45). Clypeus longitudinally grooved, appearing longer than broad .... *A. (Zonandrena)*
—. Clypeus smooth to weakly transversely grooved, appearing broader than long .... *A. (Poliandrena)* (in part)
47(21). Glossa greatly elongated; clypeus usually distinctly elongated 48
—. Glossa short; clypeus short, rarely elongated 49
48(47). Facial fovea occupying scarcely half of paraocular area; basal area of labrum trapezoidal; galea without hooked hairs .... *A. (Margandrena)*
—. Facial fovea occupying almost whole paraocular area; basal area of labrum very large; galea covered with long, hooked hairs .... *A. (Didonia)* (in part)
49(47). Mesepisternum strongly honeycomb-areolated; propodeal triangle strongly rugosely wrinkled; dorsal surface of propodeum short and declivous ....
.... *A. (Holandrena)*
—. Mesepisternum and dorsal area of propodeum remarkably finely sculptured, at most weakly rugosely wrinkled; dorsal surface of propodeum normally developed, more or less at right angle to posterior surface 50
50(49). Facial fovea narrow, short to scarcely elongate drop-shaped, or, *if* broadened over half of paraocular area, then propodeal corbicula with surface free of hairs, margin densely hairy all around, and clypeus usually elongated .... *A. (Ptilandrena)*
—. Facial fovea broader, but *if* narrow, then not drop-shaped but constricted in middle and propodeal corbicula not as above, *or* clypeus not elongated 51
51(50). Mandible falcate, without preapical tooth, but with large, rounded tooth on inner margin at distal end of basal fourth (Malaysia) .... *A. (Malayapis)*
—. Mandible with preapical tooth, but without tooth at end of basal fourth 52
52(51). Pygidial plate flat to weakly convex on disc and outer margin somewhat elevated; facial fovea narrow, constricted in middle, or, *if* broad, then metasoma strongly punctate 53
—. Pygidial plate with distinctly limited, raised central area; facial fovea broad, or, *if* narrow, then metasoma finely punctate 55
53(52). Facial fovea broad; metasoma and scutum strongly punctate; clypeus flattened..
....*A. (Poliandrena)* (in part) and *A. (Tarsandrena)* (in part)
—. Facial fovea narrow; metasoma and usually also scutum weakly punctate to impunctate; clypeus convex 54
54(53). Propodeal triangle shagreened, on base almost weakly rugose; scutum and especially metasoma not to finely punctate, rarely strongly so .... *A. (Carandrena)*
—. Propodeal triangle weakly rugosely wrinkled; scutum and especially metasoma strongly punctate ....
.... *A. (Notandrena)*
55(52). Metasomal terga usually glossy and more or less bare; clypeus short, usually transversely grooved; basal area of labrum distinctly elongated or short-triangular, not notched on distal margin.. 56
—. Metasomal terga usually shagreened and more or less densely, protrudingly haired; clypeus usually somewhat elongated, strongly punctate; basal area of labrum trapezoidal, distal margin usually notched 57
56(55). Basal area of labrum longer than broad.................
.... *A. (Leucandrena)* (in part)

—. Basal area of labrum short-triangular *A. (Larandrena)*

57(55). Maxillary palpus surpassing galea by about last three segments; active in spring .... *A. (Andrena s. str.)*

—. Maxillary palpus surpassing galea by about one segment; autumnal .... *A. (Cnemidandrena)*

58(20). Basal area of labrum large, notched; body length over 11 mm .... *A. (Campylogaster)*

—. Basal area of labrum small, although strongly elevated, entire; body length 11 mm or less .... *A. (Oreomelissa)*

Key to the Subgenera of *Andrena* of the Western Palearctic Region (Males)
(Modified from A. Molino-Pardo's translation (MS) of Warncke's 1968 key)

1. Apex of S8 broadened, and on each side a long, bent spine distinctly visible among metasomal hairs .... *A. (Rufandrena)*

—. Apex of S8 rarely strongly broadened, and never bearing visible spines 2

2(1). Mesepisternum and propodeal triangle strongly rugose-alveolar 3

—. Mesepisternum and propodeal triangle at most strongly rugosely wrinkled, commonly not alveolar, usually distinctly weaker in sculpturing 9

3(2). Dorsolateral angle of pronotum distinctly elevated as transverse ridge, with dorsoventral carina extending down from it 4

—. Pronotum without strong, elevated dorsolateral angle (ridge) and without well-developed dorsoventral carina below it 5

4(3). Clypeus yellow; propodeum largely declivous .... *A. (Holandrena)* (in part)

—. Clypeus dark; propodeum with distinct more or less horizontal and vertical surfaces .... *A. (Melanapis)*

5(3). Length under 7 mm; metasoma strongly and very densely punctate .... *A. (Brachyandrena)*

—. Length over 8 mm; metasoma strongly but usually only moderately densely punctate 6

6(5). Submarginal cells two; antenna of male short, like that of a female .... *A. (Biareolina)*

—. Submarginal cells three; antenna of male longer than that of female 7

7(6). Marginal zones of T2 to T4 occupying half of tergal lengths .... *A. (Trachandrena)*

—. Marginal zones of T2 to T4 normal, narrow 8

8(7). Metasoma shining like steel; body with whitish to yellowish-white hairs .... *A. (Agandrena)*

—. Metasoma not so shiny; body dark-haired .... *A. (Plastandrena)*

9(2). Pronotum laterally smooth or weakly or interruptedly carinate below weak dorsolateral angle 10

—. Pronotum laterally distinctly and sharply carinate below commonly strong dorsolateral angle 48

10(9). Mesepisternum and propodeum usually glossy, marked with deep and coarse, sievelike, dense punctures 11

—. Mesepisternum and propodeum usually shagreened, marked with distinctly finer, often elongate punctures with raised edges, *or* surface rugosely finely areolate 12

11(10). Metasoma finely and densely punctate; genitalia complex .... *A. (Scitandrena)*

—. Metasoma strongly to very strongly punctate; genitalia simple .... *A. (Campylogaster)* (in part)

12(10). Propodeum entirely declivous; ocelloccipital distance almost as long as four midflagellar segments taken together .... *A. (Cubiandrena)*

—. Propodeum with distinct more or less horizontal basal area; ocelloccipital distance seldom longer than length of one midflagellar segment 13

13(12). First submarginal crossvein meeting marginal cell close to stigma or at most three vein widths distant (body length not over 8 mm) 14

—. First submarginal crossvein meeting marginal cell more than three vein widths away from stigma 17

14(13). Propodeal triangle strongly to weakly rugose, usually areolate to posterior truncation of basal area (clypeus black) .... *A. (Micrandrena)* (in part) and *A. (Fumandrena)*

—. Propodeal triangle minutely reticulate to shagreened, at most with rugae along anterior edge 15

15(14). First flagellar segment at least as long as second and third taken together; clypeus usually longitudinally grooved and weakly elongated, always black; genitalia simple with elongated gonostylus, which is uniformly broad to apex .... *A. (Distandrena)*

—. First flagellar segment shorter than second and third taken together; clypeus never longitudinally grooved, black or yellow; genitalia usually shortened, gonostylus more or less shovel-shaped to strongly reduced 16

16(15). Clypeus yellow or rarely black; gonostylus slightly angled, shovel-shaped .... *A. (Aciandrena)*

—. Clypeus black or exceptionally yellow; gonostylus not as above .... *A. (Graecandrena)*

17(13). Broad central area of clypeus flattened, anterior edge truncate, punctation not strong; basal area of labrum shortened and transversely wrinkled basally .... *A. (Taeniandrena)* and *A. (Troandrena)*

—. Clypeus more or less distinctly convex, or, *if* flattened, then anterior edge convex and clypeus strongly shortened, *or* its punctation strong and deep; basal area of labrum not shortened 18

18(17). Glossa narrow, almost round in cross section and at least six times longer than broad 19

—. Glossa broader, flattened, at most three times longer than broad 24

19(18). Malar area distinctly, even if narrowly, developed; glossa very long 20

—. Malar area not developed; glossa not especially elongated 21

20(19). Clypeus yellow; mesepisternum with strongly inclined punctures having weak, downwardly elongated, raised margins .... *A. (Charitandrena)*

—. Clypeus black; mesepisternum with round punctures lacking distinctly raised margins .... *A. (Didonia)* (in part)

21(19). Mesepisternum strongly rugosely areolate; metasoma strongly punctate and glossy; clypeus usually flat (clypeus broader than long) .... *A. (Cryptandrena)*

—. Mesepisternum shagreened or weakly rugose because of raised edges of punctures; metasoma seldom strongly punctate; clypeus usually distinctly convex 22

22(21). T1 shagreened, with craterlike punctures (edges of punctures raised on all sides), or, *if* terga glossy and strongly punctate, then propodeal triangle toward metanotum at most weakly rugose .... *A. (Chlorandrena)*

—. T1 glossy, coarsely and usually densely punctate; propodeal triangle variable 23

23(22). First flagellar segment as long as or shorter than second, *or,* if longer, then scutal disc strongly glossy and scarcely punctate .... A. (Chrysandrena)
—. First flagellar segment longer than second; scutum strongly and densely punctate .... A. (Lepidandrena)
24(18). Clypeus black 25
—. Clypeus yellow (sometimes dark in stylopized individuals) 41
25(24). Propodeal triangle alveolarly areolate to posterior margin; mesepisternum not alveolar but punctures slightly sloped, with weakly raised margins (genitalia typically complex)............................ A. (Suandrena)
—. Propodeal triangle never alveolarly areolate; mesepisternum differently sculptured 26
26(25). Propodeal triangle more or less smooth and shining .... A. (Hoplandrena) (in part)
—. Propodeal triangle rugosely areolated 27
27(26). First flagellar segment very short, at most half as long as second 28
—. First flagellar segment more than half as long as second 30
28(27). Metasoma shining, strongly and densely punctate .... A. (Cordandrena)
—. Metasoma shagreened to weakly shining, not very densely punctate 29
29(28). Clypeus and thorax shining, densely and strongly punctate .... A. (Aenandrena)
—. Clypeus and thorax dull, sparsely and moderately finely punctate .... A. (Thysandrena)
30(27). First flagellar segment usually somewhat shorter than second, or at most as long as second 31
—. First flagellar segment distinctly longer than second 32
31(30). Smaller species, less than 10 mm long; posterior metasomal terga usually with weakly developed, narrow bands.................................... A. (Simandrena)
—. Larger species, over 10 mm long; posterior metasomal terga more or less bare .... A. (Melandrena) (in part)
32(30). Genitalia simple, gonostylus somewhat broadened toward apex and penis valve normally slender 33
—. Genitalia distinctly complex, with gonostylus broadened, shovel-like, and/or penis valve laterally broadened 37
33(32). Length under 10 mm; mesepisternum and propodeumn, including triangle, strongly rugosely wrinkled .... A. (Micrandrena) (in part)
—. Length over 10 mm, or, *if* smaller, then mesepisternum and propodeum only shagreened to finely rugosely wrinkled 34
34(33). Mandibles long and sickle-shaped, crossed; terga finely and densely punctate with broad, shining marginal zones .... A. (Hyperandrena)
—. Mandibles not long and crossed, or, *if* so, then terga with at most scattered punctures and marginal zones of normal width 35
35(34). Length less than 10 mm; mesepisternum shagreened and finely punctate .... A. (Euandrena)
—. Length over 10 mm; mesepisternum abundantly rugosely wrinkled 36
36(35). Clypeus densely and deeply punctate; metasoma usually strongly punctate; first flagellar segment almost as long as second and third taken together .... A. (Zonandrena)
—. Clypeus strongly convex, usually densely but shallowly punctate; metasoma usually finely and sparsely punctate; first flagellar segment usually only a little longer than second .... A. (Melandrena) (in part)
37(32). First recurrent vein meeting second submarginal cell usually distinctly before middle; marginal zones of terga usually strongly yellow to reddish yellow 38
—. First recurrent vein at or usually beyond middle of second submarginal cell; marginal zones of terga at most weakly yellowish 39
38(37). First recurrent vein ending in vicinity of base of second submarginal cell; clypeus normally long and distinctly convex .... A. (Pallandrena)
—. First recurrent vein ending little before middle of second submarginal cell; clypeus broader than long and flattened .... A. (Ulandrena) (in part)
39(37). Metasoma strongly and usually densely punctate; clypeus short and narrow, flattened .... A. (Poliandrena) (in part)
—. Metasoma with scarcely perceptible to fine punctures; clypeus of normal length and convex or very broad 40
40(39). Small species; propodeal triangle shagreened, not rugose .... A. (Avandrena)
—. Medium-sized species; propodeal triangle with conspicuous rugae near anterior margin ...... A. (Leucandrena)
41(24). Mesepisternum and propodeum finely reticulately shagreened, shallowly and finely punctate; propodeal triangle reticulate to granularly shagreened 42
—. Mesepisternum *or* propodeum *or* both rugosely areolate; propodeal triangle usually wrinkled to rugose 44
42(41). Malar area rather long .... A. (Melittoides)
—. Malar area linear or very little elongated..................... 43
43(42). Paraocular area in addition to clypeus yellow; scutum finely shagreened, finely, shallowly and sparsely punctate; metasoma almost bare .... A. (Nobandrena)
—. Paraocular area and clypeus dark, rarely with small spot of yellow; scutum usually coarsely shagreened, punctures usually not evident or coarser and indistinct; metasoma usually relatively densely haired .... A. (Scaphandrena)
44(41). Small species; S8 strongly thickened on outer side, with several dense hair tufts .... A. (Parandrenella)
—. Small to large species, *if* small, then S8 normal 45
45(44). Galeal blade short and narrow; labial palpus as long as or longer than glossa; paraocular area in addition to clypeus usually yellow 46
—. Galeal blade normal to somewhat elongate; labial palpus shorter than glossa; paraocular area dark 47
46(45). Penis valve narrow, ventral lamella broadened laterally; apex of gonostylus narrowly shovel-shaped .... A. (Poecilandrena)
—. Penis valve usually strongly bladder-shaped and enlarged; apex of gonostylus usually thickened and broadly shovel-shaped .... A. (Ulandrena) (in part)
47(45). Clypeus somewhat elongated; first submarginal crossvein meeting marginal cell only about three to four vein widths away from stigma .... A. (Orandrena)
—. Clypeus short and relatively broad; first submarginal crossvein meeting marginal cell almost 10 vein widths away from stigma .... A. (Poliandrena) (in part) and A. (Tarsandrena) (in part)
48(9). Glossa conspicuously elongated and almost cylindrical 49
—. Glossa short and more flattened................................. 50

49(48). Body large; malar area well developed; genal area normally developed .... *A. (Didonia)* (in part)
—. Body moderate-sized; malar area not unusual; posterior margin of genal area with tooth, or curved outward into carina .... *A. (Margandrena)*
50(48). Mesepisternum strongly and deeply punctate; head thick, with short clypeus; genitalia simple .... *A. (Campylogaster)* (in part)
—. Mesepisternum finely punctate or rugosely areolated; head not thickened; genitalia simple to complex 51
51(50). Mandibles of normal length, apices at most somewhat crossed 52
—. Mandibles long and sickle-shaped, apices more crossed over one another 54
52(51). Mesepisternum alveolarly areolate; propodeum strongly and densely punctate, dorsal area sloping, scarcely developed..................... *A. (Holandrena)* (in part)
—. Mesepisternum shagreened and at most weakly rugosely wrinkled; propodeum shagreened to weakly rugose, shallowly and sparsely punctate, dorsal area, including triangle, more or less normally developed 53
53(52). Propodeal triangle shagreened, with only a few flat and short rugae near anterior margin; metasoma shagreened and finely, inconspicuously punctate; clypeus dark; genitalia simple .... *A. (Cnemidandrena)*
—. Propodeal triangle rugosely areolate; metasoma usually shiny and strongly punctate; clypeus usually yellow; genitalia complex .... *A. (Poliandrena)* (in part) and *A. (Tarsandrena)* (in part)
54(51). Mesepisternum strongly rugose in longitudinal direction, almost undulate-areolate; clypeus and adjacent parts of face yellow; S8 with broad, isolated transverse carina on outer side before apex .... *A. (Holandrena)* (in part)
—. Mesepisternum finely sculptured, never longitudinally undulate-areolate; paraocular area seldom yellow; S8 normal, with at most weak thickening before apex 55
55(54). Posterior margin of genal area with outwardly curved carina, sometimes weak 56
—. Posterior margin of genal area angular, forming a point 57
56(55). Galeal blade of normal length; mesepisternum finely punctate; propodeal triangle shagreened with few short rugae near anterior margin; metasoma usually very finely punctate; gonostylar apex and penis valve scarcely broadened .... *A. (Carandrena)*
—. Galeal blade short; mesepisternum strongly punctate; propodeal triangle shagreened and weakly rugosely areolate; metasoma usually conspicuously punctate; gonostylar apex and penis valve distinctly broadened .... *A. (Notandrena)*
57(55). Clypeus yellow .... *A. (Larandrena)*
—. Clypeus dark.. 58
58(57). Clypeus flattened and broad, anterior margin usually distinctly indented; genitalia complex .... *A. (Andrena s. str.)*
—. Clypeus usually distinctly convex, anterior margin not or weakly indented; genitalia simple 59
59(58). Propodeal triangle rugosely wrinkled; first recurrent vein meeting second submarginal cell about in middle; first flagellar segment longer than second .... *A. (Ptilandrena)*
—. Propodeal triangle smooth and shiny to shagreened, without rugae; first recurrent vein meeting second submarginal cell far beyond middle; first flagellar segment usually very short, seldom longer than second .... *A. (Hoplandrena)* (in part)

Key to the Subgenera of *Andrena* of Japan (Females) (Modified from Hirashima, 1966a; I have not indicated here the doubts expressed by Hirashima concerning the placement of certain species of *Notandrena*, etc.)

1. Submarginal cells two .... *A. (Parandrena)*
—. Submarginal cells three 2
2(1). Body length about or less than 7 mm; T1 impunctate or finely and sparsely punctate .... *A. (Micrandrena)*
—. Body length about or more than 8 mm; if occasionally smaller, then T1 closely punctate 3
3(2). Propodeal triangle clearly indicated, coarsely roughened to strongly wrinkled, basal area bounded posteriorly by transverse carina or by strong wrinkles behind which surface is declivous 4
—. Propodeal triangle finely sculptured, or, *if* coarsely sculptured, then apex of triangle not bounded by a transverse carina nor by irregular rugae 5
4(3). Facial fovea separated from eye by wide, shiny space; scutum and scutellum either foveolate-punctate or very strongly rugoso-punctate; inner hind tibial spur neither widened nor curved near base; trochanteral floccus incomplete, scanty; body length less than 10 mm .... *A. (Trachandrena)*
—. Facial fovea and eye adjacent or separated by narrow space; scutum and scutellum rather strongly punctate but punctures not so close; inner hind tibial spur strongly widened and curved near base; trochanteral floccus complete or nearly so, dense; body length about or more than 12 mm .... *A. (Plastandrena)*
5(3). Tibial scopa well developed, composed of long, dense, branched hairs, and femoral scopal hairs also branched 6
—. Tibial scopa and femoral scopa largely composed of simple hairs 7
6(5). Head elongate; clypeus protuberant .... *A. (Stenomelissa)*
—. Head broader than long, more or less round in front view; clypeus not protuberant .... *A. (Chlorandrena)*
7(5). Third submarginal cell receiving second recurrent vein at end of cell or close to it...................... *A. (Habromelissa)*
—. Third submarginal cell receiving second recurrent vein well before end of cell 8
8(7). Middle basitarsus strongly expanded medially, subequal to or broader than hind basitarsus (the latter widened subbasally) ....*A. (Cnemidandrena)*
—. Middle basitarsus slender, not widened medially (or at most only slightly so) 9
9(8). T1 densely, finely punctate; T2 and T3 progressively more sparsely, finely punctate; metasoma sparsely hairy, without distinct hair fringes; posterior margins of metasomal terga yellowish-transparent; clypeus densely tessellate, dull, sparsely and more or less weakly punctate, with broad, median, longitudinal, impunctate space; antenna short, swollen toward apical segments, first flagellar segment one and one-half times as long as broad, about as long as second plus third segments; scutum tessellate, more or less weakly punctate, covered with rather short, pale fulvous hairs; propodeal triangle well indicated, distinctly

wrinkled all over; propodeum outside triangle densely rugulose or nearly shagreened; robust species, length about or less than 8 mm .... *A. (Poecilandrena)*

—. T1 impunctate or nearly so, or, *if* densely punctate, then larger and without combination of characters listed above 10

10(9). Metasoma with integument densely tessellate, therefore dull, impunctate or with indications of weak, sparse punctures; posterior margins of T3 and T4 each with complete, narrow band of short, dense, appressed, white hairs, a similar hair band usually on posterior margin of T2; interior of propodeal corbicula with sparse, usually coarse, simple hairs throughout .... *A. (Hoplandrena)*

—. Metasoma with integument either smooth or distinctly punctate, but *if* tessellate and impunctate (as in *Andrena* s. str. and some species of *Euandrena*), *then* appressed tergal hair bands lacking and/or propodeal corbicula without coarse hairs on interior surface................................. 11

11(10). Metasoma tessellate, impunctate, dull or slightly shiny; trochanteral floccus complete, well developed .... *A. (Andrena s. str.)*

—. Metasoma smooth and/or distinctly punctate, if tessellate and impunctate, then trochanteral floccus incomplete and scanty... 12

12(11). Inner hind tibial spur widened and curved near base (usually rather large and robust species) 13

—. Inner hind tibial spur at most gently curved, neither widened nor curved near base 14

13(12). Triangle of propodeum large, well indicated, sparsely wrinkled; propodeum outside triangle densely tessellate and densely rugose .... *A. (Plastandrena)*

—. Triangle of propodeum rather large to small, more or less well defined, rugose to coarsely sculptured; propodeum outside triangle densely punctate ....
.... *A. (Holandrena)* (in part)

14(12). Propodeal corbicula well developed, with dorsal fringe of long, dense, well-arranged, curled hairs, and with complete fringe of hairs anteriorly, interior of corbicula free of hairs medially; hind tibia rather short, dilated apically .... *A. (Simandrena)*

—. Propodeal corbicula poorly developed or at most moderately so, without fringe of hairs anteriorly, usually hairy on interior; hind tibia normal 15

15(14). Propodeal corbicula moderately well developed, with dorsal fringe of long, dense, rather well- to well-arranged hairs, interior of corbicula with coarse, simple hairs throughout, no complete fringe of hairs anteriorly, but frequently sparse branched hairs present on dorsal portion of anterior margin; trochanteral floccus incomplete, scanty to dense; tibial scopa compact, well developed; facial fovea separated from eye by a narrow punctate space; medium-sized to large, robust species...................... *A. (Melandrena)*

—. Propodeal corbicula poorly developed, with dorsal fringe of short, scanty hairs; or, *if* propodeal corbicula more or less well developed, then without combination of characters listed above 16

16(15). Propodeal corbicula with sparse, coarse, simple hairs nearly throughout interior; dorsal fringe of corbicula rather well indicated but not well developed, rather short and not especially dense; no fringe of hairs on anterior margin of corbicula; trochanteral floccus complete, dense; metasoma smooth and shiny, weakly to very weakly punctate; interocellar distance equal to or a little longer than ocelloccipital distance ....
.... *A. (Holandrena)* (in part)

—. Propodeal corbicula with soft, fine hairs in interior; *or,* if coarse hairs present in interior, then without combination of characters listed above 17

17(16). Metasoma short-oval, densely tessellate-punctate, nearly dull or weakly shiny, posterior margins of intermediate terga with narrow, appressed, pure-white hair bands, those of T3 and T4 complete; head broad and thin, nearly round in front view; clypeus hardly convex, tessellate, coarsely rugoso-punctate; basal area of labrum short, transverse, apex narrowly bilobed; scutum tessellate, rather densely punctate, covered with short, dull, pale, yellowish-brown hairs; propodeum roughened, triangle poorly defined, rugose basally, granulate apically; trochanteral floccus nearly complete, white; tibial scopa silver-white, narrowly brownish above basally, composed of rather coarse hairs .... *A. (Taeniandrena)*

—. Metasoma elliptical or elongate, smooth and shiny, or, *if* terga tessellate as in some species of *Euandrena*, then without combination of characters listed above 18

18(17). Metasomal terga tessellate, sometimes weakly so, therefore nearly dull or weakly shiny, impunctate or rarely with weak punctures; metasoma more or less hairy, posterior margins of intermediate terga with loose to more or less compact hair fringes; propodeal corbicula poorly developed, with dorsal fringe of loose or more or less well-arranged, rather long hairs, interior with sparse, fine hairs; trochanteral floccus incomplete, scanty; tibial scopa well developed; dorsal surface of propodeum shagreened to roughened; propodeal triangle finely to rather finely sculptured, usually less coarsely sculptured than rest of dorsal surface of propodeum; facial fovea rather narrow, upper end occupying about or less than one-half distance between eye and posterior ocellus; head and thorax with rather abundant, not especially long hairs; rather small species ....*A. (Euandrena)*

—. Metasomal terga, especially basal ones, smooth and shiny, or, *if* rarely very finely tessellate, then without combination of characters listed above; metasoma sparsely hairy, at least T1 bare dorsally 19

19(18). Mesopleura coarsely sculptured or at least rugose above, or with distinct punctures; dorsal surface of propodeum coarsely sculptured or at least shagreened; dorsal fringe of propodeal corbicula long, rather well to well arranged; trochanteral floccus usually complete......
.... *A. (Notandrena)*

—. Mesopleura finely tessellate, with or without weak, well-separated small punctures; dorsal surface of propodeum nearly smooth or at most densely tessellate, usually with weak, well-separated, small punctures; dorsal fringe of propodeal corbicula poor, composed of rather short to short, rather sparse to sparse hairs; trochanteral floccus nearly incomplete 20

20(19). Facial fovea very broad, with upper end occupying full space between eye and posterior ocellus; propodeal triangle large, with lateral margins convex outward; mesopleura weakly tessellate with distinct, well-separated punctures; tibial scopa compact, with hairs well arranged, not loose .... *A. (Calomelissa)*

—. Facial fovea much narrower; propodeal triangle large,

subtriangular, with sides not convex outward; mesopleura densely tessellate, impunctate or with an indication of weak punctures; tibial scopa large, composed of long, rather loose hairs .... *A. (Oreomelissa)*

Key to the Subgenera of *Andrena* of Japan (Males)
(Modified from Hirashima, 1966a)

1. Clypeus and frequently lower paraocular area ivory-white or yellow 2
—. Clypeus black 9
2(1). Malar area approximately as long as broad; apical segment of antenna uncinate .... *A. (Stenomelissa)*
—. Malar area linear; apical segment of antenna normal 3
3(2). Second recurrent vein at end of third submarginal cell, meeting third submarginal crossvein or nearly so .... *A. (Habromelissa)*
—. Second recurrent vein much basad of third submarginal crossvein 4
4(3). Clypeus strongly convex mediosubapically, ivory white medially .... *A. (Notandrena)* (in part)
—. Clypeus not strongly convex, largely yellow or ivory-white 5
5(4). Length about 6 mm, or occasionally slightly larger; metasomal terga, including T1, finely and densely punctate; propodeal triangle coarsely sculptured; dorsal surface of propodeum roughened; flagellum beneath yellowish brown; first flagellar segment approximately one and one-half times as long as broad, second much broader than long, about one-half as long as first, third as long as broad .... *A. (Poecilandrena)*
—. Length more than 7 mm; or, *if* occasionally smaller, then without combination of characters listed above 6
6(5). Head and thorax with abundant brown to blackish hairs, or at least with admixture of brown hairs 7
—. Head and thorax without admixture of brown hairs, primarily covered with yellowish hairs 8
7(6). Clypeus and lateral face marks ivory-white or only slightly yellowish; vertex strongly arched in front view .... *A. (Oreomelissa)*
—. Clypeus yellow, lateral face marks usually lacking; vertex not arched or only slightly convex in front view .... *A. (Calomelissa)*
8(6). Basal area of labrum rather small, not reflexed at apex, slightly emarginate; propodeal triangle rugulose to rugose basally, tessellate apically, shiny; body length about or less than 9 mm .... *A. (Chlorandrena)* (in part)
—. Basal area of labrum rather large, transverse, reflexed and deeply emarginate at apex; propodeal triangle rugulose or wrinkled all over; body length about 10 mm .... *A. (Holandrena)*
9(1). First flagellar segment at most as long as broad, about one-half as long as second; third and following segments about twice as long as broad, distinctly convex in front; malar area usually with sharp spine posteriorly in spring form (only) .... *A. (Hoplandrena)*
—. First flagellar segment longer than broad, or, *if* occasionally shorter, then without combination of characters listed above 10
10(9). Mandible long, slender, falciform, with triangular tooth or projection near base of inner margin .... *A. (Andrena s. str.)* (in part)
—. Mandible variable, without tooth basally 11

11(10). Mandible slender, curved, with sharp falciform apex; malar area evident, about one-third as long as broad; first flagellar segment slightly longer than broad, much shorter than elongate second segment (about 3:4.2); face covered with pale yellowish hairs; paraocular area, frons, and gena near eye with black hairs; length about 9 mm .... *A. (Andrena s. str.)* (in part)
—. Mandible normal, or, *if* falciform, then without combination of characters listed above 12
12(11). Propodeal triangle strongly and rather sparsely wrinkled, bounded by a transverse carina posteriorly 13
—. Propodeal triangle not bounded by a transverse carina posteriorly 15
13(12). Apical margins of metasomal terga reflexed .... *A. (Plastandrena)* (in part)
—. Apical margins of metasomal terga normal 14
14(13). Length about 10 mm; lower paraocular area with punctures more obscure than those on clypeus; propodeal triangle with wrinkles irregular .... *A. (Plastandrena)* (in part)
—. Length less than 9 mm; lower paraocular area with punctures distinct; propodeal triangle with wrinkles longitudinal .... *A. (Trachandrena)*
15(12). Length about or less than 7 mm; propodeal triangle large, weakly and rather coarsely sculptured; second submarginal crossvein ending close to stigma .... *A. (Micrandrena)*
—. Length greater than 7 mm; propodeal triangle finely sculptured, or, *if* coarsely sculptured, then second submarginal crossvein ending well away from stigma 16
16(15). Basal area of labrum protuberant .... *A. (Cnemidandrena)*
—. Basal area of labrum not protuberant 17
17(16). Gena well developed, carinate posteriorly............... .... *A. (Notandrena)*
—. Gena normal.. 18
18(17). First flagellar segment slightly longer than second plus third, these broader than long .... *A. (Chlorandrena)* (in part)
—. First flagellar segment shorter than next two segments taken together 19
19(18). Ocelloccipital distance about or more than twice as long as ocellar width................................. *A. (Melandrena)*
—. Ocelloccipital distance approximately equal to ocellar width 20
20(19). Propodeal triangle rather strongly wrinkled (metasoma very shiny, feebly tessellate, with an indication of sparse fine punctures; posterior margins of T2-T4 with lateral fringes of sparse white hairs; scutellum nearly smooth and shiny anteriorly; first flagellar segment much longer than second, second about as long as broad) .... *A. (Notandrena)* (in part)
—. Propodeal triangle less strongly wrinkled 21
21(20). Propodeal triangle large, densely, rather weakly wrinkled all over....................................... *A. (Simandrena)*
—. Propodeal triangle differently sculptured 22
22(21). Ocelloccipital distance longer than posterior ocellar diameter; metasomal terga tessellate, with distinct, close punctures .... *A. (Taeniandrena)*
—. Ocelloccipital distance at most as long as posterior ocellar diameter; metasomal terga tessellate, impunctate...... .... *A. (Euandrena)*

Andrena / Subgenus *Aciandrena* Warncke

Andrena (Aciandrena) Warncke, 1968: 62. Type species: *Andrena aciculata* Morawitz, 1886, by original designation.

■ This subgenus occurs in the southern palearctic region. Gusenleitner and Schwarz (2002) listed 26 names.

Andrena / Subgenus *Aenandrena* Warncke

Andrena (Aenandrena) Warncke, 1968: 64. Type species: *Andrena aeneiventris* Morawitz, 1872, by original designation.

■ This southern palearctic subgenus comprises seven species, according to Gusenleitner and Schwarz (2002).

Andrena / Subgenus *Agandrena* Warncke

Andrena (Agandrena) Warncke, 1968: 56. Type species: *Apis agilissima* Scopoli, 1770, by original designation.

■ This subgenus is from the palearctic region. Three west palearctic species are placed in this subgenus.

Andrena / Subgenus *Anchandrena* LaBerge

Andrena (Anchandrena) LaBerge, 1986: 496. Type species: *Andrena angustella* Cockerell, 1936, by original designation.

■ This subgenus occurs in California. The two species were revised by LaBerge (1986).

Andrena / Subgenus *Andrena* Fabricius s. str.

Andrena Fabricius, 1775: 376. Type species: *Apis helvola* Linnaeus, 1758, by designation of Viereck, 1912: 613. [Invalid designations of type species not originally included are listed by Michener (1997b).]
Anthrena Illiger, 1801: 127, unjustified emendation of *Andrena* Fabricius, 1775.
Anthocharessa Gistel, 1850: 82, unjustified replacement for *Andrena* Fabricius, 1775. Type species: *Apis helvola* Linnaeus, 1758, autobasic.

■ This is a holarctic subgenus, occurring as far north as subarctic Alaska, and as far south as Nuevo León, Mexico. LaBerge (1980) revised the 31 nearctic species, one of which, *Andrena clarkella* (Kirby), also occurs in the palearctic region. Warncke (1968) listed 15 west palearctic species. Tadauchi, Hirashima, and Matsumura (1987) revised the 16 Japanese species, two of which are considered subspecies of European species. Gusenleitner and Schwarz (2002) listed a total of 83 names.

Andrena / Subgenus *Aporandrena* Lanham

Andrena (Aporandrena) Lanham, 1949: 201. Type species: *Andrena coactipostica* Viereck, 1917, by original designation.

■ This Western Hemisphere subgenus includes only two species, one from California, the other from western Panama. Their close relationship remains uncertain, according to LaBerge (1969).

Andrena / Subgenus *Archiandrena* LaBerge

Andrena (Archiandrena) LaBerge, 1986: 482. Type species: *Andrena banksi* Malloch, 1917, by original designation.

■ This subgenus is found in the eastern and central United States. The three species were revised by LaBerge (1986).

Andrena / Subgenus *Augandrena* LaBerge

Andrena (Augandrena) LaBerge, 1986: 557. Type species: *Andrena plumiscopa* Timberlake, 1951, by original designation.

■ This subgenus occurs in Arizona and California. The three species were revised by LaBerge (1986).

Andrena / Subgenus *Avandrena* Warncke

Andrena (Avandrena) Warncke, 1968: 31. Type species: *Andrena avara* Warncke, 1967, by original designation.

■ This subgenus is found in the southern palearctic region. Warncke (1968) listed five west palearctic species; two were named later (see Gusenleitner and Schwarz, 2002).

Andrena / Subgenus *Belandrena* Ribble

Andrena (Belandrena) Ribble, 1968b: 221. Type species: *Andrena nemophilae* Ribble, 1968, by original designation.

■ This subgenus occurs in southwestern North America, including Texas and the Mexican plateau. The five species were revised by Ribble (1968b).

Andrena / Subgenus *Biareolina* Dours

Biareolina Dours, 1873: 288. Type species: *Andrena neglecta* Dours, 1873 = *Andrena lagopus* Latreille, 1809, designated by Sandhouse, 1943: 530.

■ This palearctic subgenus contains a single species, *Andrena lagopus* Latreille. Other species that were placed in *Biareolina* by Warncke (1968) are now included in *Scrapteropsis* (entirely nearctic) or *Trachandrena* (see LaBerge, 1971b, 1973).

Andrena / Subgenus *Brachyandrena* Pittioni

Andrena (Brachyandrena) Pittioni, 1948a: 54. Type species: *Andrena colletiformis* Morawitz, 1874, by original designation.

■ This subgenus is known from the southwest palearctic region to central Asia. Four names were listed by Gusenleitner and Schwarz (2002).

Andrena / Subgenus *Callandrena* Cockerell

Callandrena Cockerell, 1898b: 186. Type species: *Panurgus manifestus* Fox, 1894, monobasic.
Pterandrena Robertson, 1902a: 187, 193. Type species: *Andrena pulchella* Robertson, 1891 (not Jurine, 1807) = *Andrena accepta* Viereck, 1916, by original designation.

■ This primarily nearctic subgenus is especially well represented among autumnal species, and species of Mexico; moreover, it ranges farther into the tropics than any other *Andrena* in the New World, *A. (C.) vidalesi panamensis* Michener occurring in lowland central Panama. The 79 species were revised by LaBerge (1967).

Andrena / Subgenus *Calomelissa* Hirashima and LaBerge

Andrena (Calomelissa) Hirashima and LaBerge, 1963, *in* Hirashima, 1963: 241. Type species: *Andrena prostomias* Pérez, 1805, by original designation.

■ This subgenus is known from China and Japan. Five species were reviewed by Xu and Tadauchi (1995), and six were listed by Gusenleitner and Schwarz (2002).

Andrena / Subgenus *Campylogaster* Dours

Campylogaster Dours, 1873: 286. Type species: *Campylogaster fulvo-crustatus* Dours, 1873 = *Andrena erberi* Morawitz, 1871, designated by Sandhouse, 1943: 534.

■ This subgenus, from the southwest palearctic region, contains 14 species, according to Gusenleitner and Schwarz (2002).

Andrena / Subgenus *Carandrena* Warncke

Andrena (Carandrena) Warncke, 1968: 94. Type species: *Andrena aerinifrons* Dours, 1873, by original designation.

■ This largely Mediterranean subgenus includes 39 species (Gusenleitner and Schwarz, 2002). *Andrena bellidoides* LaBerge from the mountains of southern India probably also belongs to this subgenus. Osychnyuk (1984c) gave a key to species of the USSR.

Andrena / Subgenus *Carinandrena* Osychnyuk

Andrena (Carinandrena) Osychnyuk, 1993a: 18. Type species: *Andrena carinifrons* Morawitz, 1876, by original designation.

■ This central Asian subgenus contains a single species, *Andrena carinifrons* Morawitz.

Andrena / Subgenus *Celetandrena* LaBerge and Hurd

Andrena (Celetandrena) LaBerge and Hurd, 1965: 188. Type species: *Andrena vinnula* LaBerge and Hurd, 1965, by original designation.

■ This subgenus occurs on the Mexican plateau. The single species, *Andrena vinnula* LaBerge and Hurd, was described and illustrated by LaBerge and Hurd (1965).

Andrena / Subgenus *Charitandrena* Hedicke

Andrena (Charitandrena) Hedicke, 1933: 210. Type species: *Nomada hattorfiana* Fabricius, 1775, by original designation.

■ This subgenus contains a single palearctic species, *Andrena hattorfiana* (Fabricius), and a species from the Mexican plateau, *A. toluca* LaBerge. The subgenus and the Mexican species were described by LaBerge (1969).

Andrena / Subgenus *Chlorandrena* Pérez

Andrena (Chlorandrena) Pérez, 1890: 172. Type species: *Andrena humilis* Imhoff, 1832, by designation of Hedicke, 1933: 211.

■ This is a palearctic subgenus. Gusenleitner and Schwarz (2002) listed 49 names and Xu and Tadauchi (2002) revised the eight East-Asian species of *Chlorandrena*.

Andrena / Subgenus *Chrysandrena* Hedicke

Andrena (Chrysandrena) Hedicke, 1933: 211. Type species: *Apis fulvago* Christ, 1791, by original designation.

■ This subgenus contains 14 species in the palearctic region, according to Gusenleitner and Schwarz (2002). Eastern palearctic species formerly placed here have been transferred to *Chlorandrena* (O. Tadauchi, in litt., 1998).

Andrena / Subgenus *Cnemidandrena* Hedicke

Andrena (Cnemidandrena) Hedicke, 1933: 213. Type species: *Melitta nigriceps* Kirby, 1802, by original designation.

■ This is a holarctic subgenus, occurring as far north as subarctic Alaska, and south to Tlaxcala, Mexico. Donovan (1977) revised the 26 recognized nearctic species. Tadauchi and Xu (2002) revised the 14 East-Asian species of *Cnemidandrena*. Gusenleitner and Schwarz (2002) listed 19 palearctic species.

Andrena / Subgenus *Conandrena* Viereck

Andrena (Conandrena) Viereck, 1924a: 20. Type species: *Andrena bradleyi* Viereck, 1907, by original designation.

■ This is a nearctic subgenus. The two species were revised by LaBerge (1986). Warncke (1968) synonymized this subgenus with *Didonia,* but LaBerge (1986) did not accept this viewpoint.

Andrena / Subgenus *Cordandrena* Warncke

Andrena (Cordandrena) Warncke, 1968: 63. Type species: *Andrena cordialis* Morawitz, 1878, by original designation.

■ This is a subgenus of southern palearctic distribution. Four species were listed by Warncke (1968) for the west palearctic; one was described later; two species occur in China (Tadauchi and Xu, 2004).

Andrena / Subgenus *Cremnandrena* LaBerge

Andrena (Cremnandrena) LaBerge, 1986: 554. Type species: *Andrena anisochlora* Cockerell, 1936, by original designation.

■ This subgenus from California and Oregon contains a single species, *Andrena anisochlora* Cockerell. It was described and illustrated by LaBerge (1986).

Andrena / Subgenus *Cryptandrena* Pittioni

Andrena (Cryptandrena) Pittioni, 1948a: 49 (not Lanham, 1949). Type species: *Andrena ventricosa* Dours, 1873, by original designation.

■ This southern palearctic subgenus contains five species according to Warncke (1968).

Andrena / Subgenus *Cubiandrena* Warncke

Andrena (Cubiandrena) Warncke, 1968: 76. Type species: *Andrena cubiceps* Friese, 1914, by original designation.

■ This eastern Mediterranean subgenus contains two species.

Andrena / Subgenus *Dactylandrena* Viereck

Andrena (Dactylandrena) Viereck, 1924a: 20. Type species: *Andrena maura* Viereck, 1924 = *Andrena caliginosa* Viereck, 1916, by original designation.
Andrena (Chaulandrena) LaBerge, 1964: 314. Type species: *Andrena porterae* Cockerell, 1900, by original designation.

■ This subgenus is known only from the nearctic region. The four species were revised by LaBerge (1986). Warncke (1968) synonymized *Chaulandrena* with *Didonia*, but LaBerge did not accept that view.

Andrena / Subgenus *Dasyandrena* LaBerge

Andrena (Dasyandrena) LaBerge, 1977: 71. Type species: *Andrena obscuripostica* Viereck, 1916, by original designation.

■ This subgenus is found in western North America. The three species were revised by LaBerge (1977).

Andrena / Subgenus *Derandrena* Ribble

Andrena (Derandrena) Ribble, 1968a: 333. Type species: *Andrena vandykei* Cockerell, 1936, by original designation.

■ This is a nearctic subgenus; most species occur in the western United States, south to Baja California, Mexico. The ten species were revised by Ribble (1968a).

Andrena / Subgenus *Diandrena* Cockerell

Diandrena Cockerell, 1903a: 75. Type species: *Panurgus chalybaeus* Cresson, 1878, by original designation.

■ This is a subgenus of western North America, including Baja California, Mexico. The 25 species were revised by Thorp (1969a).

Andrena / Subgenus *Didonia* Gribodo

Didonia Gribodo, 1894: 106. Type species: *Didonia punica* Gribodo, 1894 = *Andrena mucida* Kriechbaumer, 1873, monobasic.
Solenopalpa Pérez, 1897a: lxvii, 1897b: 260. Type species: *Solenopalpa fertoni* Pérez, 1897 (not *Andrena fertoni* Pérez, 1895) = *Andrena solenopalpa* Benoist, 1945, monobasic.

■ This subgenus contains seven palearctic species, according to Gusenleitner and Schwarz (2002). Dylewska (1987) included in this subgenus the nearctic *Conandrena* and *Chaulandrena* (= *Dactylandrena*), which LaBerge (1986) did not appear to regard as closely related to one another or to *Didonia*.

Andrena / Subgenus *Distandrena* Warncke

Andrena (Distandrena) Warncke, 1968: 60. Type species: *Andrena longibarbis* Pérez, 1895, by original designation.

■ This is a southern palearctic subgenus. Gusenleitner and Schwarz (2002) listed eleven species.

Andrena / Subgenus *Erandrena* LaBerge

Andrena (Erandrena) LaBerge, 1986: 505. Type species: *Andrena principalis* LaBerge, 1986, by original designation.

■ A single species, *Andrena principalis* LaBerge from California, USA, and Baja California Norte, Mexico, constitutes this subgenus. It was described and illustrated by LaBerge (1986).

Andrena / Subgenus *Euandrena* Hedicke

Andrena (Euandrena) Hedicke, 1933: 213. Type species: *Andrena bicolor* Fabricius, 1775, by original designation.
Andrena (Xanthandrena) Lanham, 1949: 218. Type species: *Andrena auricoma* Smith, 1879, by original designation.
Andrena (Geandrena) LaBerge, 1964: 313. Type species: *Andrena caerulea* Smith, 1879, by original designation.

■ This is a holarctic subgenus, occurring as far north as subarctic Alaska and south to Baja California, Mexico. LaBerge and Ribble (1975) and LaBerge (1977) revised the 20 nearctic species. Gusenleitner and Schwarz (2002) listed 54 palearctic species. Tadauchi and Hirashima (1984a) gave a synopsis and key to the six Japanese species, one of which also occurs in Europe.

Andrena / Subgenus *Fumandrena* Warncke

Andrena (Fumandrena) Warncke, 1974: 98 (nomen nudum).
Andrena (Fumandrena) Warncke, 1975a: 57. Type species: *Andrena fumida* Pérez, 1895, by original designation.

■ This subgenus, found in the west palearctic region, was segregated from *Micrandrena* by Warncke (1975a). Eleven specific names were listed by Gusenleitner and Schwarz (2002).

Andrena / Subgenus *Fuscandrena* Osychnyuk

Andrena (Fuscandrena) Osychnyuk, 1994: 17. Type species: *Andrena fuscicollis* Morawitz, 1876, by original designation.

■ *Fuscandrena* contains a single central asiatic species, *Andrena fuscicollis* Morawitz.

Andrena / Subgenus *Geissandrena* LaBerge and Ribble

Andrena (Geissandrena) LaBerge and Ribble, 1972: 302. Type species: *Andrena trevoris* Cockerell, 1897, by original designation.

■ This subgenus occurs in northwestern North America. There is only one species, *Andrena trevoris* Cockerell; see LaBerge and Ribble (1972).

Andrena / Subgenus *Genyandrena* LaBerge

Andrena (Genyandrena) LaBerge, 1986: 544. Type species: *Andrena mackieae* Cockerell, 1937, by original designation.

■ This nearctic subgenus contains two species. It was revised by LaBerge (1986).

Andrena / Subgenus *Gonandrena* Viereck

Andrena (Gonandrena) Viereck, 1917a: 390. Type species: *Andrena persimulata* Viereck, 1917, monobasic.
Andrena (Tropandrena) Viereck, 1924a: 21. Type species: *Andrena fragilis* Smith, 1853, by original designation.

■ This nearctic subgenus is close to the palearctic *Notandrena*; see LaBerge and Ribble, 1972. The six species were revised by LaBerge and Ribble (1972).

Andrena / Subgenus *Graecandrena* Warncke

Andrena (Graecandrena) Warncke, 1968: 61. Type species: *Andrena graecella* Warncke, 1965, by original designation.

■ This is a subgenus of southern palearctic distribution. Twenty specific names were listed by Gusenleitner and Schwarz (2002).

Andrena / Subgenus *Habromelissa* Hirashima and LaBerge

Andrena (Habromelissa) Hirashima and LaBerge, 1964, *in* Hirashima, 1964b: 71. Type species: *Andrena omogensis* Hirashima, 1953, by original designation.

■ This subgenus occurs in Japan, where it is represented by a single species, and in China, where it is represented by two species.

Andrena / Subgenus *Hesperandrena* Timberlake

Andrena (Hesperandrena) Timberlake, 1949, *in* Lanham, 1949: 208. Type species: *Andrena escondida* Cockerell, 1938, by original designation.

■ This subgenus is found in California, USA, and Baja California Norte, Mexico. The nine species were revised by Thorp and LaBerge (2005).

Andrena / Subgenus *Holandrena* Pérez

Andrena (Holandrena) Pérez, 1890: 176. Type species: *Melitta labialis* Kirby, 1802, by designation of Hedicke, 1933: 214.
Opandrena Robertson, 1902a: 187, 193. Type species: *Andrena cressonii* Robertson, 1871, by original designation.

■ This is a holarctic subgenus. It occurs from British Columbia to Nova Scotia, Canada, south to Florida to California, USA, and to Zacatecas, Mexico. In the Eastern Hemisphere it occurs from westernmost Europe to Japan and Taiwan. There are four nearctic species and at least 11 additional species in the Eastern Hemisphere. LaBerge (1986) revised the nearctic species; Warncke (1968) and Schönitzer et al. (1995), the west palearctic species; Hirashima (1964b), the Japanese species; and Tadauchi and Xu (1998), the species of eastern Asia. Warncke (1968) retained *Opandrena* as a separate subgenus and placed one of the European species in it.

Andrena / Subgenus *Hoplandrena* Pérez

Andrena (Hoplandrena) Pérez, 1890: 170. Type species: *Melitta trimmerana* Kirby, 1802, by designation of Hedicke, 1933: 214.

■ This is a palearctic subgenus. Names for twelve Eurasian species were listed by Gusenleitner and Schwarz (2002). Tadauchi and Hirashima (1984b) gave a key to the six Japanese species (see also Hirashima, 1964b). Xu and Tadauchi (2005) revised the eleven East-Asian species.

Andrena / Subgenus *Hyperandrena* Pittioni

Andrena (Hyperandrena) Pittioni, 1948a: 58. Type species: *Apis bicolorata* Rossi, 1790, by original designation.

■ This Mediterranean subgenus includes two species, according to Warncke (1968).

Andrena / Subgenus *Iomelissa* Robertson

Iomelissa Robertson, 1900: 50. Type species: *Andrena violae* Robertson, 1891, monobasic.

■ This subgenus is found in the eastern and central USA. The single species, *Andrena violae* Robertson, has unusually elongate glossa, galea, and palpi, and is oligolectic on *Viola*. It was described and illustrated by LaBerge (1986).

Andrena / Subgenus *Larandrena* LaBerge

Andrena (Larandrena) LaBerge, 1964: 304. Type species: *Andrena miserabilis* Cresson, 1872, by original designation.

■ This holarctic subgenus contains one North American species (Ribble, 1967; LaBerge and Ribble, 1972) and five palearctic species, listed as in *Parandrena* by Warncke (1968). There are two Japanese species (Hirashima and Haneda, 1973).

Larandrena differs from *Parandrena* principally in having three submarginal cells.

Andrena / Subgenus *Leimelissa* Osychnyuk

Andrena (Leimelissa) Osychnyuk, 1984b: 20. Type species: *Andrena bairucumensis* Morawitz, 1876, by original designation.

■ This subgenus, which is not included in the keys (and I have not seen), occurs in Uzbekistan and central Asia. Osychnyuk (1984b) gave a key to four species. Gusenleitner and Schwarz (2002) listed six names.

Andrena / Subgenus *Lepidandrena* Hedicke

Andrena (Lepidandrena) Hedicke, 1933: 215. Type species: *Andrena curvungula* Thomson, 1870, by original designation.

■ This is a palearctic subgenus as here understood, but Warncke (1968) considered the New World *Aporandrena* to be a synonym. Eighteen palearctic species were listed by Gusenleitner and Schwarz (2002).

Andrena / Subgenus *Leucandrena* Hedicke

Andrena (Leucandrena) Hedicke, 1933: 215. Type species: *Apis sericea* Christ, 1791 (not Forster, 1771) = *Melitta barbilabris* Kirby, 1802, by original designation.

■ This is a holarctic subgenus, and one species, *Andrena barbilabris* (Kirby), is itself holarctic. In North America the subgenus ranges from Alaska to Labrador, and from California to Virginia; in Eurasia it occurs from Britain to Japan. LaBerge (1987) revised the nine nearctic species and Gusenleitner and Schwarz (2002) listed seven species from the palearctic region.

Andrena / Subgenus *Longandrena* Osychnyuk

Andrena (Longandrena) Osychnyuk, 1993a: 17. Type species: *Andrena longiceps* Morawitz, 1895, by original designation.

■ This central Asian subgenus contains three species.

Andrena / Subgenus *Malayapis* Baker

Andrena (Malayapis) Baker, 1995a: 67. Type species: *Andrena chrysochersonesus* Baker, 1995, by original designation.

The broad head, wider than the thorax, the falcate mandibles, the diverging inner orbits, and the projecting clypeal tooth on each side of the labrum are distinctive. The male is unknown.

■ *Malayapis* is known from the highlands of west Malaysia; it is the only *Andrena* known from Southeast Asia. The only species is *Andrena chrysochersonesus* Baker.

Andrena / Subgenus *Margandrena* Warncke

Andrena (Margandrena) Warncke, 1968: 91. Type species: *Andrena marginata* Fabricius, 1776, by original designation.

■ This south palearctic subgenus contains seven palearctic species listed by Gusenleitner and Schwarz (2002).

Andrena / Subgenus *Melanapis* Cameron

Melanapis Cameron, 1902a: 420. Type species: *Melanapis violaceipennis* Cameron, 1902 = *Andrena fuscosa* Erichson, 1835, monobasic.

■ Patiny (1997) made a preliminary revision of *Andrena (Melanapis)*, recognizing four species (eight taxa, counting subspecies). *Melanapis* ranges from Spain, the Canary Islands, and Mauritania east through Europe and North Africa to Iran and Central Asia.

Andrena / Subgenus *Melandrena* Pérez

Andrena (Melandrena) Pérez, 1890: 170. Type species: *Apis thoracica* Fabricius, 1775, by designation of Michener, 1997b: 36. [An earlier but invalid designation was listed by Michener (1997b).]
Andrena (Gymnandrena) Hedicke, 1933: 212. Type species: *Apis thoracica* Fabricius, 1775, by original designation.
Andrena (Cryptandrena) Lanham, 1949: 222 (not Pittioni, 1948). Type species: *Andrena carlini* Cockerell, 1901, by original designation.
Andrena (Bythandrena) Lanham, 1950: 140, replacement for *Cryptandrena* Lanham, 1949. Type species: *Andrena carlini* Cockerell, 1901, autobasic.

■ This is a holarctic subgenus. The 24 nearctic species were revised by Bouseman and LaBerge (1979) and by LaBerge (1987). Hirashima (1957, 1964a) recorded six species from Japan and gave a key to eight from the Far East. The total from the palearctic region is 40 (Gusenleitner and Schwarz 2002).

Andrena / Subgenus *Melittoides* Friese

Melittoides Friese, 1921a: 177. Type species: *Andrena melittoides* Friese, 1899, by absolute tautonymy.

Michener (2000) regarded *Melittoides* as a genus distinct from *Andrena* primarily because of the extraordinary male genitalia of *A. melittoides* Friese (Fig. 51-8). Other distinctive features are large size (body length 16-18 mm), the slender stigma (as in *Megandrena*, Fig. 51-2c), and the long first flagellar segment, in both sexes longer than the scape. Other species such as *A. (Melittoides) curiosa* (Morawitz), however, agree with *A. melittoides* in many features but have genitalia typical of *Andrena*.

An unpublished phylogenetic study of Andrenidae by John S. Ascher included *A. melittoides* and showed it in the midst of *Andrena* (Ascher, personal comm.), presumably indicating that its strange male genitalia are a derived feature not justifying exclusion of the species from the genus *Andrena*, a conclusion supported by the existence of species such as *A. curiosa*.

■ Four species are found from North Africa (Tunisia, Egypt) and southwestern Asia to the Caucasus.

Andrena / Subgenus *Micrandrena* Ashmead

Micrandrena Ashmead, 1899a: 89. Type species: *Micrandrena pacifica* Ashmead, 1899 = *Andrena melanochroa* Cockerell, 1898, by original designation.
Andrena (Andrenella) Hedicke, 1933: 210. Type species: *Melitta minutula* Kirby, 1802, by original designation.

■ This is a holarctic subgenus. The 19 nearctic species were revised by Ribble (1968a). Tadauchi (1985) revised the 11 Japanese species, two of which were considered the same as European species. Pittioni (1948b) revised a group of palearctic species. A total of 84 names for palearctic species was listed by Gusenleitner and Schwarz (2002).

Andrena / Subgenus *Nemandrena* LaBerge

Andrena (Nemandrena) LaBerge, 1971a: 48. Type species: *Andrena torulosa* LaBerge, 1971, by original designation.

■ This subgenus occurs in California and Oregon. The three species were revised by LaBerge (1971a).

Andrena / Subgenus *Nobandrena* Warncke

Andrena (Nobandrena) Warncke, 1968: 45. Type species: *Andrena nobilis* Morawitz, 1874, by original designation.

■ This is a southern palearctic subgenus. Thirteen specific names were listed by Gusenleitner and Schwarz (2002).

Andrena / Subgenus *Notandrena* Pérez

Andrena (Notandrena) Pérez, 1890: 173. Type species: *Andrena nitidiuscula* Schenck, 1853, designated by Hedicke, 1933: 216.

■ This is a primarily palearctic subgenus, but includes two species from the central and southern United States. The North American species were revised by LaBerge (1986). Fourteen palearctic species were listed by Gusenleitner and Schwarz (2002). Hirashima (1965a) included one Japanese species in this subgenus. Warncke regarded the nearctic *Gonandrena* as a synonym, but LaBerge (1986) did not accept this view (see also LaBerge and Ribble, 1972).

Andrena / Subgenus *Oligandrena* Lanham

Andrena (Oligandrena) Lanham, 1949: 207. Type species: *Andrena macrocephala* Cockerell, 1916, by original designation.

■ This subgenus, known only from California, contains two species and was revised by LaBerge (1986).

Andrena / Subgenus *Onagrandrena* Linsley and MacSwain

Andrena (Onagrandrena) Linsley and MacSwain, 1956: 111. Type species: *Andrena oenotherae* Timberlake, 1937, by original designation.

■ This subgenus is found in the southwestern USA and Baja California, Mexico. Hurd (1979) listed 24 species. All the species are specialist visitors to flowers of Onagraceae. The subgenus was revised by LaBerge and Thorp (2005).

Andrena / Subgenus *Orandrena* Warncke

Andrena (Orandrena) Warncke, 1968: 36. Type species: *Andrena oralis* Morawitz, 1876, by original designation.

■ This subgenus is found in the southern palearctic region and central Asia. Gusenleitner and Schwarz (2002) listed 24 names.

Andrena / Subgenus *Oreomelissa* Hirashima and Tadauchi

Andrena (Oreomelissa) Hirashima and Tadauchi, 1975: 176. Type species: *Andrena mitakensis* Hirashima, 1963, by original designation.

■ This is a palearctic subgenus ranging from Europe to Japan. The three Japanese species were revised by Hirashima and Tadauchi (1975); three species were reported from Tibet by Wu (1982b). The European species (which also reaches Japan), *Andrena coitana* (Kirby), was erroneously placed in *Stenomelissa* by Warncke (1968). A revision of eleven East Asian species of *Andrena* subgenus *Oreomelissa* was provided by Xu, Tadauchi, and Wu (2000). A total of 13 names were listed by Gusenleitner and Schwarz (2002).

Andrena / Subgenus *Osychnyukandrena* Michener

Andrena (Calcarina) Osychnyuk, 1993b: 60 (not d'Orbigny, 1826). Type species: *Andrena cochlearicalcar* Lebedev, 1933, by original designation.
Andrena (Osychnyukandrena) Michener, 2000: 254. Replacement for *Calcarina* Osychnyuk, 1993. Type species, *Andrena cochlearicalcar* Lebedev, autobasic.

■ This central Asian subgenus was erected for two species.

Andrena / Subgenus *Oxyandrena* LaBerge

Andrena (Oxyandrena) LaBerge, 1977: 135. Type species: *Andrena longifovea* LaBerge, 1977, by original designation.

■ This subgenus, known only from California, contains a single species, *Andrena longifovea* LaBerge.

Andrena / Subgenus *Pallandrena* Warncke

Andrena (Pallandrena) Warncke, 1968: 39. Type species: *Andrena pallidicincta* Brullé, 1832, by original designation.

■ This subgenus occurs in the southern palearctic region. Gusenleitner and Schwarz (2002) listed four species.

Andrena / Subgenus *Parandrena* Robertson

Parandrena Robertson, 1897: 337. Type species: *Panurgus andrenoides* Cresson, 1878, by designation of Cockerell, 1897a: 288.

■ *Parandrena* occurs from Ontario to British Columbia, Canada, south to California to Florida, USA, and in Tibet and Japan. Nine nearctic species were revised by LaBerge and Ribble (1972); one Japanese species was included by Hirashima (1965a) and one Tibetan species by Wu (1982b). Warncke (1968) regarded *Larandrena* as a synonym of *Parandrena,* and listed four west palearctic species, but these species would be included in *Larandrena* as understood by LaBerge and Ribble (1972). Hirashima and Haneda (1973) also discuss the close relationship of *Larandrena* and *Parandrena*. Four east-palearctic species listed by Gusenleitner and Schwarz (2002) are presumably real *Parandrena*.

Andrena / Subgenus *Parandrenella* Popov

Andrena (Parandrenella) Popov, 1958b: 112. Type species: *Andrena dentiventris* Morawitz, 1874, by original designation.

■ This is a southern palearctic subgenus. Popov (1958b) illustrated the male genitalia and hidden sterna and revised the subgenus, including four species. Eight names were listed by Gusenleitner and Schwarz (2002).

Andrena / Subgenus *Pelicandrena* LaBerge and Ribble

Andrena (Pelicandrena) LaBerge and Ribble, 1972: 346. Type species: *Parandrena atypica* Cockerell, 1941, by original designation.

■ This subgenus occurs from Oregon to Baja California Norte. The single species is *Andrena atypica* Cockerell; see the description by LaBerge and Ribble (1972).

Andrena / Subgenus *Planiandrena* Osychnyuk

Andrena (Planiandrena) Osychnyuk 1983: 794. Type species: *Andrena planirostris* Morawitz, 1876, by original designation.

■ This subgenus occurs in Kazakhstan and central Asia. Osychnyuk (1983) listed four species.

Andrena / Subgenus *Plastandrena* Hedicke

Andrena (Plastandrena) Hedicke, 1933: 217. Type species: *Melitta tibialis* Kirby, 1802, by original designation.
Andrena (Schizandrena) Hedicke, 1933: 218. Type species: *Andrena aulica* Morawitz, 1876 = *Melitta bimaculata* Kirby, 1802, by original designation.
Andrena (Glyphandrena) Hedicke, 1933: 212. Type species: *Apis carbonaria* Linnaeus, 1767, by original designation.
Andrena (Mitsukuriella) Hirashima and LaBerge, 1965, *in* Hirashima, 1965a: 472 (not Heding and Panning, 1954). Type species: *Nomia japonica* Smith, 1873, by original designation.
Andrena (Mitsukuriapis) Hirashima, LaBerge and Ikudome, *in* Ikudome, 1994: 6, replacement for *Mitsukuriella* Hirashima and LaBerge. Type species: *Nomia japonica* Smith, 1873, autobasic.

■ This is a holarctic subgenus ranging south as far as Oaxaca, Mexico. The five nearctic species were revised by LaBerge (1969). Popov (1958b) revised the subgenus in the palearctic region and recognized 29 species, illustrating the male genitalia and hidden sterna of various species. Warncke (1968) listed six western palearctic species and added one later, and Hirashima (1965a) records three Japanese species, including the two placed in *Mitsukuriapis*.

Andrena / Subgenus *Poecilandrena* Hedicke

Andrena (Poecilandrena) Hedicke, 1933: 218. Type species: *Andrena labiata* Fabricius, 1781, by original designation.

■ This subgenus occurs in the southern palearctic region. The species of eastern Asia were revised by Tadauchi and Xu (2000); in all, 21 species of the subgenus were recognized, of which only four occur in eastern Asia. Gusenleitner and Schwarz (2002) listed 29 names.

Andrena / Subgenus *Poliandrena* Warncke

Andrena (Poliandrena) Warncke, 1968: 71. Type species: *Andrena polita* Smith, 1847, by original designation.

■ This subgenus is found in the Mediterranean basin and eastward, and includes 33 names (Gusenleitner and Schwarz, 2002).

Andrena / Subgenus *Psammandrena* LaBerge

Andrena (Psammandrena) LaBerge, 1977: 83. Type species: *Andrena cercocarpi* Cockerell, 1936, by original designation.

■ This subgenus is found in Oregon and California. The two included species were revised by LaBerge (1977).

Andrena / Subgenus *Ptilandrena* Robertson

Ptilandrena Robertson, 1902a: 187, 192. Type species: *Andrena erigeniae* Robertson, 1891, by original designation.
Andrena (Eremandrena) LaBerge, 1964: 295. Type species: *Pterandrena pallidiscopa* Viereck, 1904, by original designation.

■ This is a holarctic subgenus. LaBerge (1987) revised the three nearctic species and synonymized the name *Eremandrena*. Ten palearctic specific names were included by Gusenleitner and Schwarz (2002).

Andrena / Subgenus *Rhacandrena* LaBerge

Andrena (Rhacandrena) LaBerge, 1977: 90. Type species: *Andrena brevipalpis* Cockerell, 1930, by original designation.

■ This is a nearctic subgenus. The four species were revised by LaBerge (1977).

Andrena / Subgenus *Rhaphandrena* LaBerge

Andrena (Rhaphandrena) LaBerge, 1971b: 507. Type species: *Andrena prima* Casad and Cockerell, 1896, by original designation.

■ This subgenus occurs in the southwestern USA. The two or probably three species were revised by LaBerge (1971b).

Andrena / Subgenus *Rufandrena* Warncke

Andrena (Rufandrena) Warncke, 1968: 33. Type species: *Andrena rufiventris* Lepeletier, 1841, by original designation.

■ This subgenus occurs in the western Mediterranean region. There are two species (Warncke, 1968).

Andrena / Subgenus *Scaphandrena* Lanham

Andrena (Scaphandrena) Lanham, 1949: 200. Type species: *Andrena montrosensis* Viereck and Cockerell, 1914, by original designation.
Andrena (Elandrena) Lanham, 1949: 203. Type species: *Andrena amplificata* Cockerell, 1910, by original designation.
Andrena (Truncandrena) Warncke, 1968: 46. Type species: *Andrena truncatilabris* Morawitz, 1878, by original designation.

■ This is a holarctic subgenus. The 22 nearctic species were revised by Ribble (1974). Gusenleitner and Schwarz (2002) listed 31 names for the palearctic region.

Andrena / Subgenus *Scitandrena* Warncke

Andrena (Scitandrena) Warncke, 1968: 51. Type species: *Andrena scita* Eversmann, 1852, by original designation.

■ This southern palearctic subgenus contains a single west palearctic species, *Andrena scita* Eversmann.

Andrena / Subgenus *Scoliandrena* Lanham

Andrena (Scoliandrena) Lanham, 1949: 223. Type species: *Andrena osmioides* Cockerell, 1916, by original designation.

■ This subgenus occurs in California, USA, and northern Baja California, Mexico. The two species were revised by LaBerge (1987).

Andrena / Subgenus *Scrapteropsis* Viereck

Andrena (Scrapteropsis) Viereck, 1922: 42. Type species: *Andrena fenningeri* Viereck, 1922, monobasic.
Andrena (Mimandrena) Lanham, 1949: 217. Type species: *Andrena imitatrix* Cresson, 1872, by original designation.

■ This is a nearctic subgenus found from southern Alberta, Canada, to Maine, USA, south to Florida and Texas, USA, and Baja California, Mexico; it is related to *Trachandrena* and *Biareolina*. LaBerge (1971b) revised the 18 species.

Andrena / Subgenus *Simandrena* Pérez

Andrena (Simandrena) Pérez, 1890: 174. Type species: *Andrena propinqua* Schenck, 1851 = *Melitta dorsata* Kirby, 1802, by designation of Hedicke, 1933: 218.
Andrena (Platandrena) Viereck, 1924a: 21. Type species: *Andrena nasonii* Robertson, 1895, by original designation.
Andrena (Stenandrena) Timberlake, 1949, *in* Lanham, 1949: 213. Type species: *Pterandrena pallidifovea* Viereck, 1904, by original designation.

■ This is a holarctic subgenus ranging south to Nuevo León and Baja California, Mexico. LaBerge (1989a) revised the nine nearctic species. Tadauchi and Xu (1995) gave a key to 18 palearctic species, indicating a total of 22 species for that region, while Gusenleitner and Schwarz listed 32 palearctic species.

Andrena / Subgenus *Stenomelissa* Hirashima and LaBerge

Andrena (Stenomelissa) Hirashima and LaBerge, 1965, *in* Hirashima, 1965b: 500. Type species: *Andrena halictoides* Smith, 1869, by original designation.

■ This subgenus occurs in Japan, China, and Korea. The three species were revised by Tadauchi and Hirashima (1988). See also Hirashima and Tadauchi (1975).

Andrena / Subgenus *Suandrena* Warncke

Andrena (Suandrena) Warncke, 1968: 52. Type species: *Andrena suerinensis* Friese, 1884, by original designation.

■ This is a southern palearctic subgenus. Gusenleitner and Schwarz (2002) listed 11 names.

Andrena / Subgenus *Taeniandrena* Hedicke

Andrena (Taeniandrena) Hedicke, 1933: 219. Type species: *Melitta ovatula* Kirby, 1802, by original designation.

■ This is a primarily palearctic subgenus for which Gusenleitner and Schwarz (2002) listed 23 names. *Andrena wilkella* (Kirby), a palearctic species, also occurs in eastern North America, where it was probably introduced. LaBerge (1989a) gave an account of the subgenus in North America, and Niemelä (1949) gave a key to northern European species. Tadauchi and Xu (2003) revised the ten species of *Taeniandrena* known from eastern Asia.

Andrena / Subgenus *Tarsandrena* Osychnyuk

Andrena (Tarsandrena) Osychnyuk, 1984a: 24. Type species: *Andrena tarsata* Nylander, 1848, by original designation.

■ The single west palearctic species of this subgenus was included in *Poliandrena* by Warncke (1968). The subgenus ranges from Europe to eastern Asia. Osychnyuk (1984a) gave a key to four species. Xu and Tadauchi (1999) revised the six species found in eastern Asia.

Andrena / Subgenus *Thysandrena* Lanham

Andrena (Thysandrena) Lanham, 1949: 213. Type species: *Andrena candida* Smith, 1879, by original designation.

■ This is a holarctic subgenus ranging south to Nuevo León and Sonora, Mexico. The 15 nearctic species were revised by LaBerge (1977). Gusenleitner and Schwarz (2002) listed six names for the palearctic forms.

Andrena / Subgenus *Trachandrena* Robertson

Trachandrena Robertson, 1902a: 187, 189. Type species: *Andrena rugosa* Robertson, 1891, by original designation.

■ This is a holarctic subgenus, occurring as far north as subarctic Alaska. It contains 24 nearctic species and one western palearctic species, *Andrena haemorrhoa* (Fabricius). Names of six palearctic species were listed by Gusenleitner and Schwarz (2002). The nearctic species were revised by LaBerge (1973).

Andrena / Subgenus *Troandrena* Warncke

Andrena (Troandrena) Warncke, 1974: 108. Nomen nudum.
Andrena (Troandrena) Warncke, 1975a: 81. Type species: *Andrena troodica* Warncke, 1975, by original designation.

■ This west palearctic subgenus was segregated from *Taeniandrena* by Warncke (1975a), who listed three species; two more were described later.

Andrena / Subgenus *Tylandrena* LaBerge

Andrena (Tylandrena) LaBerge, 1964: 312. Type species: *Cilissa erythrogaster* Ashmead, 1890, by original designation.

■ This is a nearctic subgenus ranging south to Nuevo León and Baja California, Mexico. The 14 species were revised by LaBerge and Bouseman (1970).

Andrena / Subgenus *Ulandrena* Warncke

Andrena (Ulandrena) Warncke, 1968: 43. Type species: *Andrena schulzi* Strand, 1921, by original designation.

■ This is a southern palearctic subgenus. Gusenleitner and Schwarz (2002) listed 31 names.

Andrena / Subgenus *Xiphandrena* La Berge

Andrena (Xiphandrena) LaBerge, 1971b: 504. Type species: *Andrena mendica* Mitchell, 1960, by original designation.

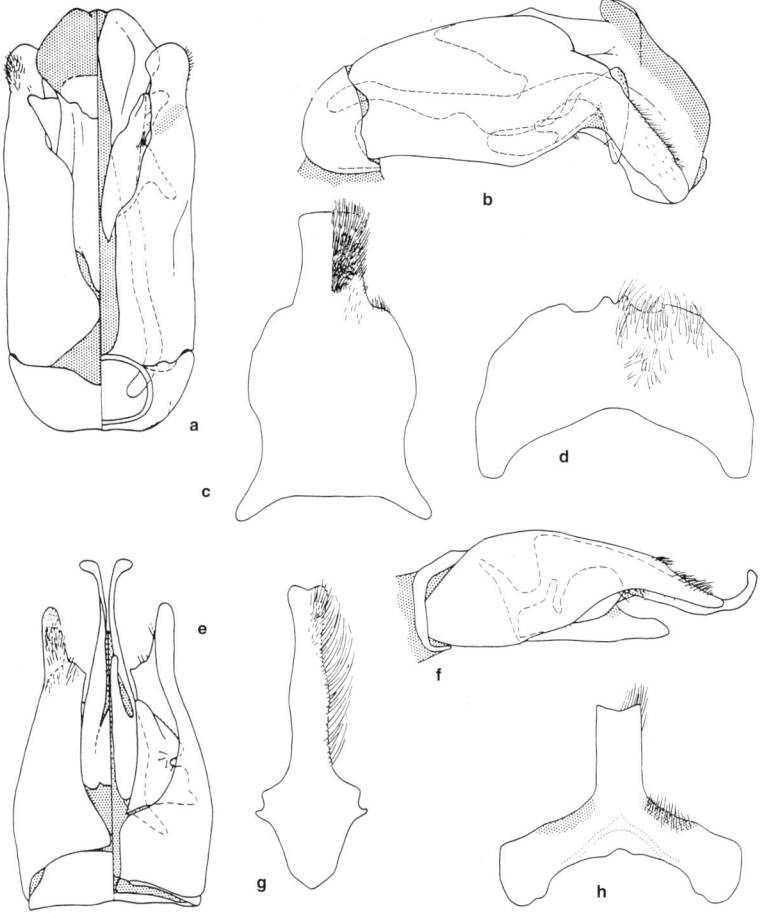

Figure 51-6. Chilean Andreninae, males. **a-d,** *Euherbstia excellens* (Friese), genitalia (dorsal side at the left), lateral view, S8, and S7 (ventral views); **e-h,** *Orphana inquirenda* Vachal, same structures in same sequence. From Rozen, 1971a.

■ This subgenus is from southeastern North America. The single species, *Andrena mendica* Mitchell, was considered by LaBerge (1971b).

Andrena / Subgenus *Zonandrena* Hedricke

Andrena (Zonandrena) Hedicke, 1933: 220. Type species: *Andrena flavipes* Panzer, 1799, by original designation.

■ This is a palearctic subgenus. Gusenleitner and Schwarz (2002) listed 17 names.

Genus *Euherbstia* Friese

Euherbstia Friese, 1925a: 8. Type species: *Euherbstia excellens* Friese, 1925, monobasic.

This genus contains a large (15-16 mm long), *Andrena*-like species with T2 and usually T3 largely orange, and with strong blue-green reflections on the thorax and metasomal terga. It is related to *Orphana* but differs in having S7 of the male consisting of a broad plate (Fig. 51-6d) with a small, bilobed median apical projection but without a strong apical process, and in the broad penis valves and other genitalic characters (see also the key to genera). This genus was described in detail by Moure (1950c) and illustrated (genitalia, sterna, etc.) by Rozen (1971a); see Figure 51-6a-d.

■ *Euherbstia* occurs in central Chile. The single species is *E. excellens* Friese.

Rozen (1993b) described the larva and nesting biology of *Euherbstia*. The nest burrows start in soil cracks; the cells are at the ends of short lateral burrows, and the cell axes are slanting to nearly vertical. Thus there is little in the nest architecture that would distinguish *Euherbstia* from one or another species of *Andrena*.

Genus *Megandrena* Cockerell

This genus is better differentiated from *Andrena* than is *Ancylandrena*, distinctive features being the greatly reduced male gonobase, the large, spiculate volsella, the dorsal projection of the fused penis valves (Fig. 51-7), and the reduced, parallel-sided stigma (see key to genera and Fig. 51-2c). The inner orbits of the eyes diverge below (only slightly in females).

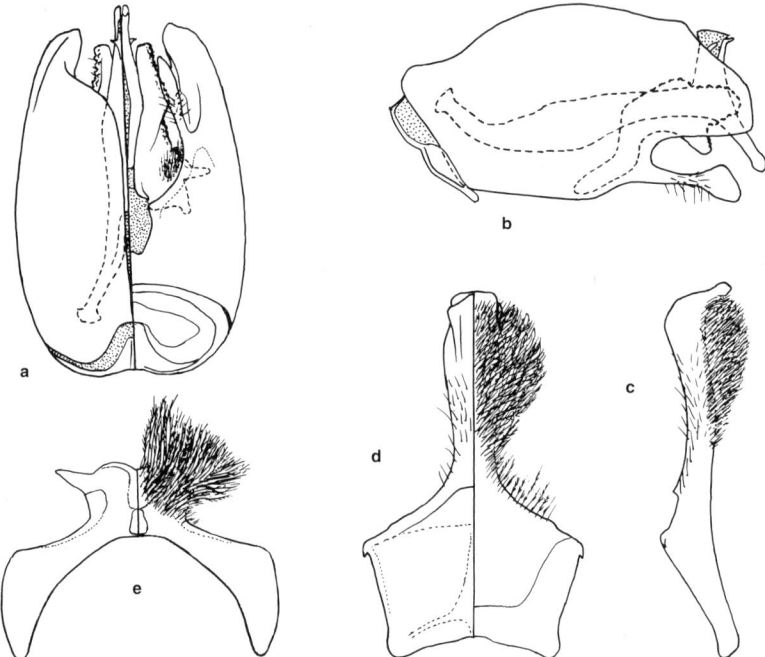

Figure 51-7. *Megandrena enceliae* (Cockerell). **a-e,** Male genitalia (dorsoventral and lateral views), S8 (dorsoventral and lateral views), and S7 (dorsoventral view). (Dorsal views are at the left.) Note the reduced gonobase and enormous volsella of the genitalia. From Michener, 1986c,

This genus contains two subgenera, both from the deserts of the southwestern United States. They were well differentiated by Zavortink (1972).

Key to the Subgenera of *Megandrena*

1. Long hairs of fore tarsus of female curved; gonobase of male absent or completely enclosed in bases of gonocoxites .... *M. (Erythrandrena)*
—. Long hairs of fore tarsus of female straight; gonobase of male present, a narrow ring (Fig. 51-7a, b) .... *M. (Megandrena s. str.)*

Megandrena / Subgenus *Erythrandrena* Zavortink

Megandrena (Erythrandrena) Zavortink, 1972: 61. Type species: *Megandrena mentzeliae* Zavortink, 1972, by original designation.

In this subgenus the metasoma is largely red with apical white tergal hair bands. The head of the male is particularly large and broad. The body length is 12 mm. Structures including male genitalia were illustrated by Zavortink (1972). The characteristics of the subgenus are so distinctive (e.g., virtual absence of a gonobase) that subgeneric recognition seems justified even though the genus *Megandrena* contains only two species.

■ *Erythrandrena* is known only from southern Nevada, USA. The single species is *Megandrena mentzeliae* Zavortink, a visitor to flowers of *Mentzelia tricuspis*.

Megandrena / Subgenus *Megandrena* Cockerell s. str.

Andrena (Megandrena) Cockerell, 1927a: 42. Type species: *Andrena enceliae* Cockerell, 1927, by original designation.

In this subgenus the metasoma is black with strong, white apical tergal hair bands. The body length is 13 to 16 mm. Noteworthy features of the male terminalia are the enormous volsellae, a distinct although membranous bridge between the volsellae, the apparently bifid gonocoxite (the slender, spatulate lower ramus probably represents the gonostylus, the upper one, a gonocoxal lobe), the reduced gonobase, and the divergent hairy lobes of S7. These structures were illustrated by Michener (1986c); see Figure 51-7.

■ This subgenus is found in the deserts of southern California, Nevada, and Arizona, USA. It, too, contains a single species, *Megandrena enceliae* (Cockerell), which visits, and may be oligolectic on, flowers of *Larrea*.

Genus *Orphana* Vachal

Orphana Vachal, 1909a: 35, 38. Type species: *Orphana inquirenda* Vachal, 1909, monobasic.
Leptoglossa Friese, 1925a: 9 (not Klug, 1839). Type species: *Leptoglossa paradoxa* Friese, 1925 = *Orphana inquirenda* Vachal, 1909, monobasic.
Ptoleglossa Friese, 1930: 127, replacement for *Leptoglossa* Friese, 1925. Type species: *Leptoglossa paradoxa* Friese, 1925 = *Orphana inquirenda* Vachal, 1909, autobasic.

This genus and *Euherbstia* differ from other Andreninae in their lack of depressed, hairy facial foveae in the

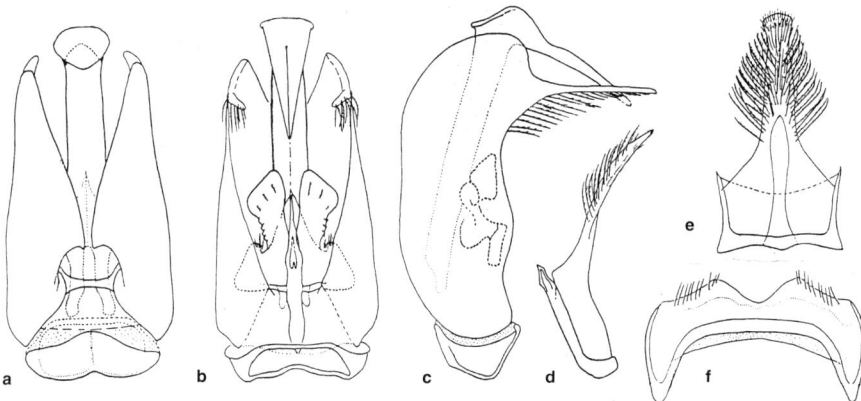

females. Distinctive characters are cited in the key to genera. In addition, S7 of the male is a transverse plate with a long median apical process (Fig. 51-6h). Such a process is absent in *Euherbstia*. The lateral extensions of this plate superficially suggest the lateral apodemes of S7 in bees such as *Ancylandrena* and *Megandrena* (Figs. 51-3d, 51-7e), but in *Orphana* they are partly hairy and not apodemal. In aspect, *Orphana* resembles a hairy *Andrena* or a large species of the colletid genus *Leioproctus;* the body length is 10 to 14 mm. Male genitalia, hidden sterna, and

Figure 51-8. Terminalia of another unusual andrenine, *Andrena* (*Melittoides*) *melittoides* Friese, male. **a-c,** Dorsal, ventral, and lateral views of genitalia; **d, e,** Lateral and ventral views of S8; **f,** Ventral view of S7. Drawings by C. O'Toole.

other structures were illustrated by Rozen (1971a); see also Figure 51-6e- h).

■ *Orphana* is found in central Chile from Coquimbo to Curico. The two species were described and illustrated by Rozen (1971a).

52. Subfamily Panurginae

The principal characters of this subfamily are discussed in the text for the family Andrenidae (Sec. 49). Some of the diversity in proboscidial structure is shown in Figure 52-1; in a few species in both Andreninae and Panurginae the first two segments of the labial palpus are elongate, as in L-T bees (Fig. 52-1c). The Panurginae are usually smaller, less hairy bees than the Andreninae, and sometimes bear yellow or cream-colored markings on all tagmata (in Andreninae such markings, when present, are limited to the face, commonly of males only). The Panurginae constitute a major part of the bee fauna in parts of North and South America, but are scarce in the tropics. In the Old World they are far less numerous, but do occur in the palearctic region and Africa. They are absent in Australia and in tropical Asia.

A major contribution to the phylogeny and classification of the Panurginae was by Ruz (1986). One part of this work, on Calliopsini and its relatives, has been revised and published (Ruz, 1991). The remainder, available as a thesis, includes keys, descriptions, a phylogeny, and 550 illustrations; all this has greatly facilitated my work on these bees. Other works on the subfamily include Warncke's (1972) treatment of west palearctic species and Eardley's (1991a) account of the species of southern Africa. Rozen (1966b) investigated larvae of various species; the larval characters, as indicated by his key, do not contradict the tribal classification used below.

Patiny (1999b) has divided the subfamily Panurginae of the Eastern Hemisphere into six tribes: Panurgini, Camptopoeumini (sic), Panurginini, Mermiglossini, Melitturgini, and Paramelittergini. This arrangement is for the forms placed below in two tribes, Panurgini and Melitturgini. Because the last of Patiny's tribal names is not based on any generic name, the tribal name is invalid; if the taxon is to be recognized, it would have to be called Meliturgulini, with *Meliturgula* as the type genus and distinguishing characters indicated by Patiny (1999b, p. 265, 272). The name Meliturgulini has already been proposed by Engel (2001b). Patiny's cladograms (1999b) show two major branches of Old World Panurginae, one for his first four tribes listed above, the other for the last two. In terms of the classification used here, the first three are in the Panurgini, the last three in the Melitturgini; if the fourth of Patiny's tribes were moved out of the Melit-

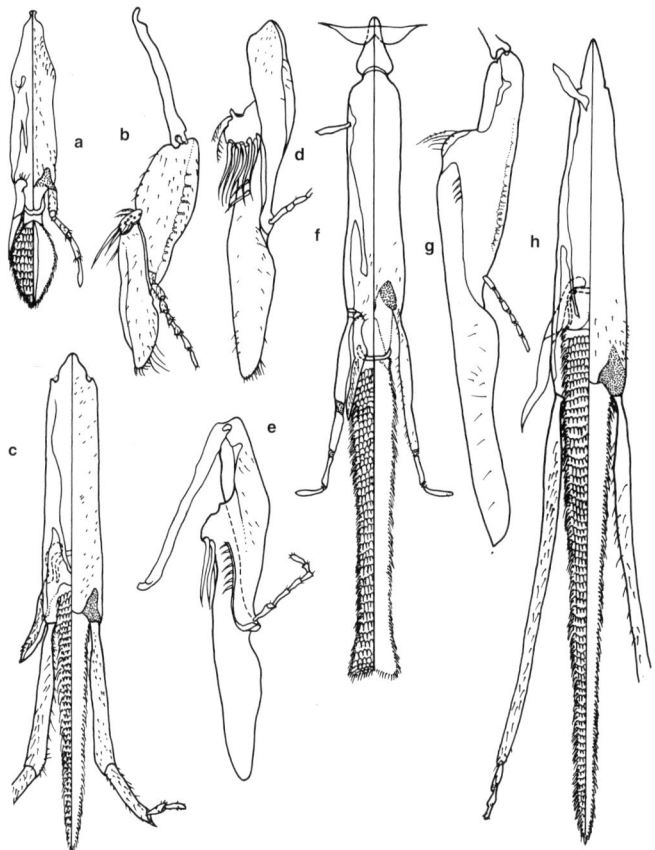

Figure 52-1. Proboscides of Panurginae. **a, b,** Labium (except basal parts) and maxilla (outer view) of *Anthemurgus passiflorae* Robertson; **c,** Labium (except basal parts) of *Protomeliturga turnerae* Ducke; **d, e,** Outer and inner views of maxilla of *Chaeturginus testaceus* (Ducke); **f, g,** Labium and maxilla (inner view) of *Flavomeliturgula tapana* (Warncke); **h,** Labium (except basal parts) of *Calliopsis (Nomadopsis) zonalis* Cresson. (In the divided figures, anterior views are at the left.) See also Figure 19-5c, d. From Ruz, 1986.

turgini and into the Panurgini, then the arrangements would be compatible. The distinctions indicated by Ruz (1986), and in the account below, between Melitturgini and Panurgini are not particularly impressive and Patiny's cladograms may be correct in associating *Mermiglossa* and *Plesiopanurgus* with the Panurgini rather than with the Melitturgini.

Since two major groups of Panurginae of the Eastern Hemisphere, Panurgini and Melitturgini, seem recognizable, I prefer to use them as tribes rather than recognizing six tribes, for the total number of genera (as recognized by Patiny, 1999b) is only nine; three of his tribes contain only one genus each. Recognition of subtribes might be appropriate.

A review of panurgine biology with many references to older literature was published by Rozen (1967a), and supplemented by Rozen (1989c) on the Protandrenini. The nests are burrows in the ground, with branches ending in isolated, usually subhorizontal cells or occasionally small series of cells. At least in the Perditini, Melitturgini, and Panurgini, many species are communal, and sometimes the nest burrows become many-branched systems. The cells are doubtfully flattened on their lower surfaces, or not flattened at all so that in cross section they appear to be circular. The walls of the cells are smooth, lined or partly lined with a secreted shiny "waxlike" membrane, except in the genus *Perdita*, in which the walls are unlined, or (in *P. graenicheri* Timberlake) are thinly lined, perhaps with nectar (Norden, Krombein, and Danforth, 1992). The larval food mass is spherical or somewhat flattened, the egg laid on top of it, and the food mass is covered with a waxy membrane in *Perdita, Calliopsis,* and at least some other Calliopsini (J. Rozen, personal communication, 1995). When the larva is large enough, it lies on its back, supported by large dorsal tubercles, with the partly eaten provisions on its venter, safeguarded from contact with the cell wall and possible moisture. When mature, the larva defecates onto the upper rear of the cell, except in the genus *Perdita*, in which the feces rest on the venter of the larva or prepupa. Prepupae pass any unfavorable season (e.g., winter) lying on their backs, supported by strong dorsal tubercles that reduce contact with the cell wall to several small points and thus perhaps reduce the probability of mold starting in extensive moist contacts with the cell wall.

Most Panurginae are oligolectic. Even within a single genus like *Perdita*, species are restricted to plants in diverse and unrelated families (see Sec. 6). Unlike Andreninae, most Panurginae transport pollen in a firm mass, apparently moistened with nectar, on each hind tibia, as do corbiculate Apidae. In *Panurgus* (Rozen, 1971b), however, and in certain subgenera of *Perdita*, pollen is carried dry on the scopa. Although for bees as a whole, dry transport is probably plesiomorphic, I suspect that it is derived in Panurginae. This behavior should be investigated more thoroughly in relation to phylogeny and scopal structure.

Key to the Tribes of Panurginae

1. S6 of female with curved (i.e., lateral parts oblique) marginal band of dense hairs (Fig. 59-3d), the band sometimes broken medially, elsewhere S6 with only widely scattered minute hairs; base of S6 of female with two broad membranous lobes (visible only on dissection) occupying space between apodemes (Fig. 59-3d); S5 of female with distal margin convex medially; gonostylus of male absent or reduced to small projection usually not reaching middle of penis valve, the latter enormous and complex (Fig. 59-3a, e, h); anterior tentorial pit in outer subantennal suture (Fig. 59-1) (Western Hemisphere) .... Calliopsini (Sec. 59)

—. S6 of female with marginal hair band undefined, poorly defined (Figs. 53-6g, 57-1d), or transverse (not curved), S6 commonly with moderate-sized hairs basal to band (in a few Melitturgini, especially *Plesiopanurgus*, very like Calliopsini but the band less curved); base of S6 of female without membranous lobes between apodemes (Figs. 53-6g, 57-1d); S5 of female with distal margin nearly straight to concave; gonostylus of male well developed, usually nearly reaching or exceeding apex of penis valve (or directed downward in *Neffapis*); penis valve variable in complexity; anterior tentorial pit at intersection of epistomal and outer subantennal sutures, or below, in epistomal suture, except in Nolanomelissini and some Melitturgini, in which it is in outer subantennal suture 2

2(1). Marginal cell usually much shorter than, usually about half as long as, distance from its apex to wing tip, broadly truncate, margin on costa little if any longer than stigma (Figs. 58-1, 58-2); submarginal cells two, or, rarely, a minute, petiolate intercalary cell between first and second; second submarginal cell (or third if there is a petiolate intercalary cell) less than two-thirds as long as first, rarely absent; integument often metallic and often with yellow markings; upper margin of hind tibia of male usually not toothed (North America) .... Perditini (Sec. 58)

—. Marginal cell slightly shorter to longer than distance from its apex to wing tip, moderately to narrowly and often obliquely truncate, margin on costa usually longer than stigma; submarginal cells two or three, and *if* two, then second submarginal cell at least two-thirds as long as first on posterior margin; integument usually nonmetallic, but *if* metallic, then with no yellow on metasoma; upper margin of hind tibia of male nearly always carinate or toothed (indistinctly carinate in Nolinomelissini) 3

3(2). Labial palpus with first two segments elongate (first over twice length of second), rather flattened; third segment arising preapically on second and directed laterally (as in L-T bees), third and fourth segments similar and small (Fig. 52-1c); basal vein strongly curved (uniformly curved, the curve not principally near base as in Halictinae); T7 of male strongly curled forward, with strong, blunt, apicolateral tooth (Brazil) .... Protomelitturgini (Sec. 57)

—. Labial palpus with all segments similar to one another *or* only first strongly elongate (as in Fig. 52-1h), first three elongate in *Nolanomelissa* and *Plesiopanurgus*, or first shortest; third segment arising apically on second, last segments thus not directed laterally; basal vein straight or gently curved; T7 of male usually not strongly curled, lacking apicolateral tooth 4

4(3). Stigma slender, almost parallel-sided, margin within marginal cell straight or nearly so; facial fovea of male absent (weakly evident in some species of *Melitturgula*);

valve of first valvula of sting reduced or absent (except in *Meliturgula*); inner orbits of male diverging below (except in *Plesiopanurgus*, which has the inner orbits sinuous, and the antenna thickened with the last segment pointed) 5

—. Stigma wider, broadest at level of vein r (Figs. 53-1, 53-2), margin within marginal cell convex except in *Liphanthus*; facial fovea of both sexes present (weakly defined in some males of *Panurgus*); valve of first valvula of sting usually well developed; inner orbits of male parallel or converging below [except somewhat diverging below in *Psaenythia, Rhophitulus (Cephalurgus)*, and *Panurgus (Flavipanurgus)*] 6

5(4). Hind tibia of male with dorsal carina distinct, toothed; inner orbits of female parallel or convergent below; labial palpus with segment 4 well developed, other segments variable but less sheathlike, second not the longest (palearctic region and Africa) .... *Melitturgini*

—. Hind tibia of male with dorsal carina inconspicuous, not toothed; inner orbits of female divergent below; labial palpus with segment 4 minute, segments 1-3 elongate, 2 and 3 sheathlike, fitting around glossa, segment 2 longest (Chile) .... *Nolanomelissini*

6(4). Episternal groove absent, short, or curving into and joining scrobal groove; T2-T5 of male with marginal zones usually hairy; foramen of male genitalia in deep sinus between bases of gonocoxites, its two halves facing one another (Fig. 54-3a, e, k) (holarctic) .... Panurgini (Sec. 54)

—. Episternal groove usually extending below scrobal groove, not curving to join scrobal groove; T2-T5 of male with marginal zones glabrous; genital foramen of male exposed at bases of gonocoxites (Fig. 53-6a, d) (Western Hemisphere) .... Protandrenini (Sec. 53)

53. Tribe Protandrenini

This is perhaps a paraphyletic tribe, one recognized principally by ancestral characters. Although Ruz (1986), in her taxonomic treatment, recognized it (as Anthemurgini) in its present sense for reasons of convenience, because the groups within it are morphologically similar and difficult to key out, in her phylogenetic treatment she divided it into four tribes. These were the Liphanthini (for *Liphanthus*), the Protandrenini (for *Protandrena*), the Austropanurgini (for *Austropanurgus*), and the Anthemurgini (for all other genera). The recognition of four tribes, three of them monotypic, that differ little from one another seems unnecessary and also unwise, considering the lack of robustness in this part of the cladogram (see the discussion under *Protandrena*).

The tribe Protandrenini, as here understood, consists of relatively slender, nearly all nonmetallic bees, almost always with yellow on the face of males, sometimes with yellow markings on all tagmata of both sexes, and occasionally with the metasoma red. Usually there are two submarginal cells, but in several genera there are three (Fig. 53-1). Apical hair bands on the metasomal terga are ordinarily absent. S7 of the male has two large apical lobes (small in *Liphanthus*, Fig. 53-4g-j) and the disc, to the extent that it is recognizable, is sometimes quite narrow (Figs. 53-3 to 53-6), suggesting the style common in Colletidae. The male gonostyli are rather large, at least half as long as the gonocoxites, and freely articulated to completely fused with the gonocoxites.

This tribe is abundant in North America and in temperate South America, and rare but present in the intervening moist tropics. It is not found in the Old World.

An Argentine bee of unknown relationship, but possibly a protandrenine, was named *Stenocolletes pictus* Schrottky; see the comments under that generic name below (it is not included in either key).

Key to the Genera of the Protandrenini (Males) (Modified from Ruz, 1986)

1. Forewing with three submarginal cells (Fig. 53-1) 2
—. Forewing with two submarginal cells (Fig. 53-2) 6
2(1). Eye pilose; pronotum with strong transverse dorsal lamella; punctation on parts of thorax strong, contiguous (South America) .... *Parapsaenythia*
—. Eye glabrous; pronotum with rounded ridge on dorsal margin; punctation well marked, usually not contiguous, to fine, weak 3
3(2). Stigma only slightly wider than prestigma, as measured to wing margin, sides subparallel or converging slightly basad from vein r, margin within marginal cell straight; T2 with narrow, deep postgradular depression, shallower on succeeding terga (South America) ....
.... *Liphanthus* (in part)
—. Stigma clearly wider than prestigma, sides converging basad from vein r, margin within marginal cell at least slightly convex (Fig. 53-1); T2 with postgradular depression usually shallow, always similar on succeeding terga
.... 4
4(3). Metasoma usually with yellow markings; T2 with lateral fovea well developed; mandible with preapical tooth (South America) .... *Psaenythia*
—. Metasoma without yellow markings (except in *Protandrena maculata* Timberlake); T2 with lateral fovea shallow; mandible simple 5
5(4). Paraocular area medially depressed, flat below; tentorial pit at intersection between outer subantennal and epistomal sutures; propodeal triangle pilose (South America) .... *Anthrenoides*
—. Paraocular area convex; tentorial pit in epistomal suture below intersection of outer subantennal suture; propodeal triangle usually glabrous (North and Central America) .... *Protandrena s. str.*
6(1). Glossa and labial palpus reaching middle of metasoma in repose, labial palpus three-segmented, third segment longer than first and second combined (Fig. 19-5c, d); maxillary palpus two-segmented; gonostylus directed downward at right angle to gonocoxite (Chile) .... *Neffapis*
—. Glossa and labial palpus not reaching beyond base of metasoma in repose, labial palpus four-segmented; maxillary palpus six-segmented; gonostylus directed posteriorly at side of penis valve 7
7(6). Stigma only slightly wider than prestigma, as measured to wing margin, sides subparallel or converging slightly basad from vein r, margin within marginal cell straight; T2 with narrow, deep postgradular depression, shallower on succeeding terga (South America) ....
.... *Liphanthus* (in part)
—. Stigma clearly wider than prestigma, sides converging basad from vein r, the margin within marginal cell at least slightly convex (Fig. 53-2); T2 with postgradular depression usually shallow, always similar on succeeding terga
.... 8
8(7). Head wider than thorax; orbits somewhat divergent below [Some wide-headed species of *Rhophitulus* s. str. might run here; for additional characters, see account of the genus.] (South America) ...... *Rhophitulus (Cephalurgus)*
—. Head usually narrower than thorax; orbits at least slightly convergent below 9
9(8). Propodeal triangle hairless 10
—. Propodeal triangle with hairs on at least lateral third .. 12
10(9). Punctation strong, punctures usually contiguous on some areas of thorax; omaulus sharp, at least dorsally; hind tibia with upper margin a strong, untoothed carina (Fig. 52-6j) (North and Central America) ....
.... *Pseudopanurgus*
—. Punctation well marked but punctures not contiguous to fine, weak; omaulus smoothly curved from lateral to anterior mesepisternal surfaces; hind tibia with upper margin at least at base carinate with teeth or serrate (as in Fig. 53-6i) [but untoothed in *Protandrena (Pseudosarus)*]
.... 11
11(10). Antennal flagellum about as long as head; propodeal triangle polished; tentorial pit far below intersection of outer subantennal and epistomal sutures (Brazil) ....
.... *Chaeturginus*
—. Antennal flagellum clearly longer than head; propodeal triangle striate basally; tentorial pit at intersection of

outer subantennal and epistomal sutures or just below it .... *Protandrena* (in part)
12(9). First flagellar segment about as long as second; genitalia with basal dorsal gonobase-like sclerotization; digitus (mesal lobe) of volsella prolonged, parallel-sided in ventral view, longer than rest of volsella (Fig. 53-3a) .... *Rhophitulus* s. str.
—. First flagellar segment longer than second; gonobase entirely absent; digitus of volsella not parallel-sided, shorter than rest of volsella 13
13(12). Yellow on lower half of face, areas of thorax, and legs; antennal flagellum longer than head; glossa somewhat shorter than to longer than prementum; metasoma red or partly red (nearctic) .... *Protandrena (Metapsaenythia)*
—. Yellow limited to small, faded spots on face; antennal flagellum shorter than head; glossa about half as long as prementum (Fig. 51-1a); metasoma black (nearctic) .... *Anthemurgus*

Key to the Genera of the Protandrenini (Females)
(Modified from Ruz, 1986)

1. Forewing with three submarginal cells (Fig. 53-1) 2
—. Forewing with two submarginal cells (Fig. 53-2) 6
2(1). Eye pilose; pronotum with dorsal margin a strong lamella (at least laterally); T2 to T5 with complete or interrupted median (not apical) hair bands (South America) .... *Parapsaenythia*
—. Eye glabrous; pronotum with dorsal margin a rounded ridge; metasomal terga without hair bands, except in some species of *Protandrena* s. str. 3
3(2). Metasomal terga usually with yellow markings; middle tibial spur with coarse teeth; hind femur on inner surface with longitudinal ridge (Fig. 53-6h); T1 to T5 densely punctate (South America) .... *Psaenythia*
—. Metasomal terga with no yellow markings (except in *Protandrena maculata* Timberlake from Mexico); middle tibial spur with fine teeth; hind femur on inner surface without longitudinal ridge; T1 to T5 with punctures separated by spaces as wide as punctures or wider 4
4(3). Stigma little wider than prestigma, as measured to wing margin, sides of stigma subparallel or slightly convergent basad from vein r, margin within marginal cell straight (lower paraocular area slightly convex) (South America) .... *Liphanthus* (in part)
—. Stigma about twice as wide as prestigma or wider, sides strongly convergent basad from vein r, margin within marginal cell at least slightly convex (Fig. 53-1) 5
5(4). Face black; paraocular area concave medially and below; tentorial pit at intersection of outer antennal and epistomal sutures; propodeal triangle pilose (inner orbits subparallel) (South America) .... *Anthrenoides*
—. Face often with yellow on lower half; paraocular area convex; tentorial pit in epistomal suture below intersection with outer subantennal suture; propodeal triangle glabrous (North and Central America)........................... .... *Protandrena* s. str.
6(1). Glossa and labial palpus reaching middle of metasoma in repose, labial palpus three-segmented, third segment longer than first and second combined (Fig. 19-5c, d); maxillary palpus two-segmented (Chile) .... *Neffapis*
—. Glossa and labial palpus not reaching base of metasoma in repose, labial palpus four-segmented; maxillary palpus six-segmented 7

7(6). Stigma little wider than prestigma, as measured to wing margin, the margin within marginal cell straight (South America) .... *Liphanthus* (in part)
—. Stigma about twice as broad as prestigma, as measured to wing margin, margin within marginal cell at least slightly convex (Fig. 53-2) 8
8(7). S2 to S5 or at least S4 each with irregular row of coarse setae on premarginal area (Fig. 53-3g) (face with conspicuous yellow marks; middle tibial spur finely toothed) (Brazil... *Chaeturginus*
—. S2 to S5 with normal hairs on premarginal areas 9
9(8). Fore coxa with well-developed, hairy, apical spine or process (Fig. 53-6k); middle tibial spur with small, fine, dense teeth (North and Central America) .... *Pseudopanurgus*
—. Fore coxa unmodified; middle tibial spur usually with distal teeth larger and better spaced than basal ones 10
10(9). Glossa about half as long as prementum; first segment of labial palpus less than half length of second to fourth segments taken together (Fig. 52-1a) (nearctic) .... *Anthemurgus*
—. Glossa somewhat shorter than to longer than prementum; first segment of labial palpus about as long as second to fourth segments taken together, or longer 11
11(10). Lateral third of propodeal triangle with hairs (South America) .... *Rhophitulus*
—. Propodeal triangle bare .... *Protandrena* (in part)

Genus *Anthemurgus* Robertson

Anthemurgus Robertson, 1902b: 321. Type species: *Anthemurgus passiflorae* Robertson, 1902, monobasic.

At least superficially, the bees of this genus resembles *Protandrena (Heterosarus),* although slightly more robust. The female is entirely black and the male has very restricted yellowish facial marks. The body length is 6 to 8 mm. *Anthemurgus* differs from related Protandrenini in its short mouthparts, the glossa being about half as long as the prementum and less than twice as long as broad,

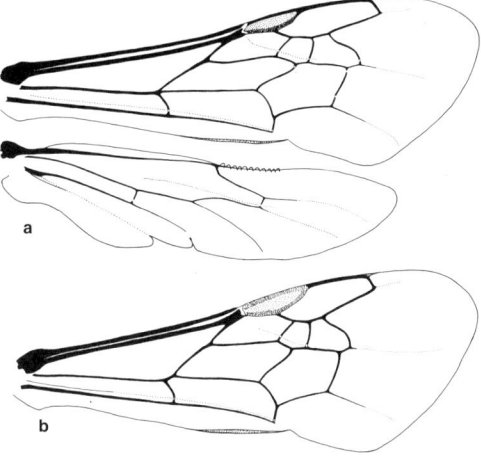

Fig. 53-1. Wings of Protandrenini with three submarginal cells. **a,** *Psaenythia bergi* Holmberg; **b,** *Anthrenoides meridionalis* (Schrottky).

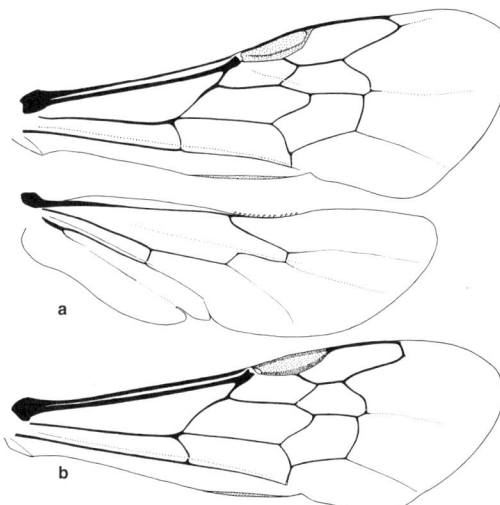

Fig. 53-2. Wings of Protandrenini with two submarginal cells. **a**, *Pseudopanurgus aethiops* (Cresson); **b**, *Protandrena (Heterosarus) neomexicana* (Cockerell), new combination.

the four segments of the labial palpus being almost equal in length (Fig. 52-1a), and the maxillary blade being not greatly longer than the prepalpal part of the galea. In this tribe, such short mouthparts occur otherwise only in a few species of *Protandrena* s. str. and are likely to be plesiomorphic features. Male genitalia and hidden sterna were illustrated by Mitchell (1960) and Ruz (1986); see also Figure 53-5h, i.

If the mouthparts were ignored, *Anthemurgus* would fall into *Protandrena,* but would not agree with any subgenus. It seems most similar to *Heterosarus,* from which it differs in its minutely hairy propodeal triangle, the lack of a V-shaped emargination in S6 of the male, and the strongly basally directed lateral arms of S8 of the male (Fig. 53-5h).

■ *Anthemurgus* ranges from central Texas, Kansas, and Illinois to North Carolina, USA. There is only one species, *A. passiflorae* Robertson; it collects pollen only from flowers of *Passiflora lutea.*

The nesting biology and immature stages of *Anthemurgus* were described by Neff and Rozen (1995).

Genus *Anthrenoides* Ducke

Anthrenoides Ducke, 1907: 368. Type species: *Anthrenoides alfkeni* Ducke, 1907 = *Protandrena meridionalis* Schrottky, 1906, monobasic.

This genus is perhaps related to *Psaenythia,* from which it differs in its small size (length 5-7 mm) and lack of yellow areas on the body, except usually on the face of the male. It therefore resembles superficially *Protandrena (Heterosarus)* and related groups. Resemblance to *Psaenythia* is indicated by the three submarginal cells (Fig. 52-1b), the laterally pilose propodeal triangle, the large although U-shaped apical emargination of S6 of the male, and the broad and membranous base of S8 of the male. Differences from *Psaenythia* include not only the size and coloration mentioned above, but the oblique impression on the side of the male gonocoxite (as in *Rhophitulus*), the longer apical lobes (constricted at their bases) on S7 of the male, and the finely serrate basal part of the middle tibial spur of the female. Male genitalia and other structures were illustrated by Ruz (1986).

■ This genus ranges from the state of Ceará, Brazil, to Paraguay and the province of Buenos Aires, Argentina, and perhaps Chile. There are 30 described species; Urban (2005) named 27 of these and gave a key to species.

Genus *Chaeturginus* Lucas de Oliveira and Moure

Chaeturginus Lucas de Oliveira and Moure, 1963: 575. Type species: *Rhophitulus testaceus* Ducke, 1907, by original designation.

This genus contains small species (length 5-7 mm) with shiny, smooth integument and sparse punctures. Most of the body is yellow or testaceous, although the head is partly black in the male. *Chaeturginus* is perhaps related to *Rhophitulus,* as indicated by Ruz (1986). It differs from that genus, however, in that the propodeal triangle is smooth, the apical lobes of S7 of the male are scarcely narrowed at the bases, the gonocoxite has neither a lateral impression nor a mid-dorsal lobe, and the middle tibial spur of the female is uniformly finely serrate or ciliate. As in *Rhophitulus,* there is a distinct dorsal remnant of the gonobase (Fig. 53-3d), a structure found in Panurginae only in these two genera. Unique features of

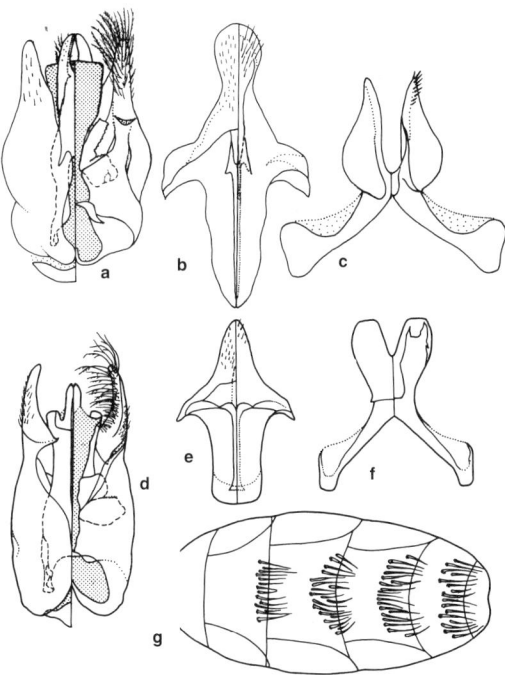

Fig. 53-3. Structures of protandrenine genera having a gonobase. **a-c,** Male genitalia, S8, and S7 of *Rhophitulus (Rhophitulus) frisei* Ducke; **d-f,** Same structures of *Chaeturginus testaceus* (Ducke); **g,** S2-S5 of female, *Chaeturginus testaceus* (Ducke). (In the divided figures, dorsal views are at the left.) From Ruz, 1986.

Chaeturginus are the irregular, partly double row of very coarse setae on each sternum, S2 to S5, of the female (Fig. 53-3g) or on S4 only, in which case (*C. alexanderi* Ruz and Melo) similar setae are on the ventral part of the mesepisternum; and the few very long, coarse, setae arising from the lacinia (Fig. 52-1d, e). Moreover, the lower ends of the inner and outer subantennal sutures meet or are close together at the epistomal suture, so that the elongate subantennal area scarcely or only narrowly reaches the clypeus; no other panurgine exhibits this character. These and other features were illustrated by Lucas de Oliveira and Moure (1963) and Ruz (1986).

■ This genus is known from the states of Amazonas and Pará south to São Paulo, Brazil. *Chaeturginus testaceus* (Ducke) is from Amazonas and Pará. Another species, *C. alexanderi* Ruz and Melo from São Paulo and Minas Gerais, has the outer subantennal suture scarcely indicated, an unusual feature for a panurgine.

Genus *Liphanthus* Reed

Liphanthus consists of mostly minute (3-7 mm long) black bees frequently bearing yellow markings, sometimes including metasomal areas, and the metasoma sometimes red. Most species have three submarginal cells, a few species, only two. Although most small bees have a relatively large stigma, that of *Liphanthus* is narrow, scarcely widened toward vein r, and the margin within the marginal cell is straight. In most species the male antennae are unusually long (Fig. 53-4a-c), the middle flagellar segments over twice as long as broad. S7 of the male is distinctive, with a small body but the usual large apodemal lobes, and without or with small apical lobes (Fig. 53-4g-j). The male gonostyli, much shorter than the gonocoxites, are articulated to the lower distal surfaces of the gonocoxites, and are thus quite different from those of other genera, in which the gonostyli are partly or fully fused to the gonocoxites or articulated to the extreme apices of the gonocoxites. Ruz and Toro (1983) illustrated male genitalia, sterna, and other structures; see also Figure 53-4.

Liphanthus is best known in Chile from Antofagasta to Chile Chico, but also occurs in Argentina from Jujuy to Santa Cruz Province. There are 26 described species. The genus was revised by Ruz and Toro (1983).

Key to the Subgenera of *Liphanthus* (Males)
(Based on Ruz and Toro, 1983)

1. Submarginal cells two; propodeal triangle laterally less than half length of metanotum (laterally) .... *L. (Neoliphanthus)*
—. Submarginal cells three; propodeal triangle laterally more than half length of metanotum (laterally) or almost the same length 2
2(1). Hind tibial spurs with apices curved like claws, subequal in length; vertex concave or almost straight in frontal view .... *L. (Xenoliphanthus)*

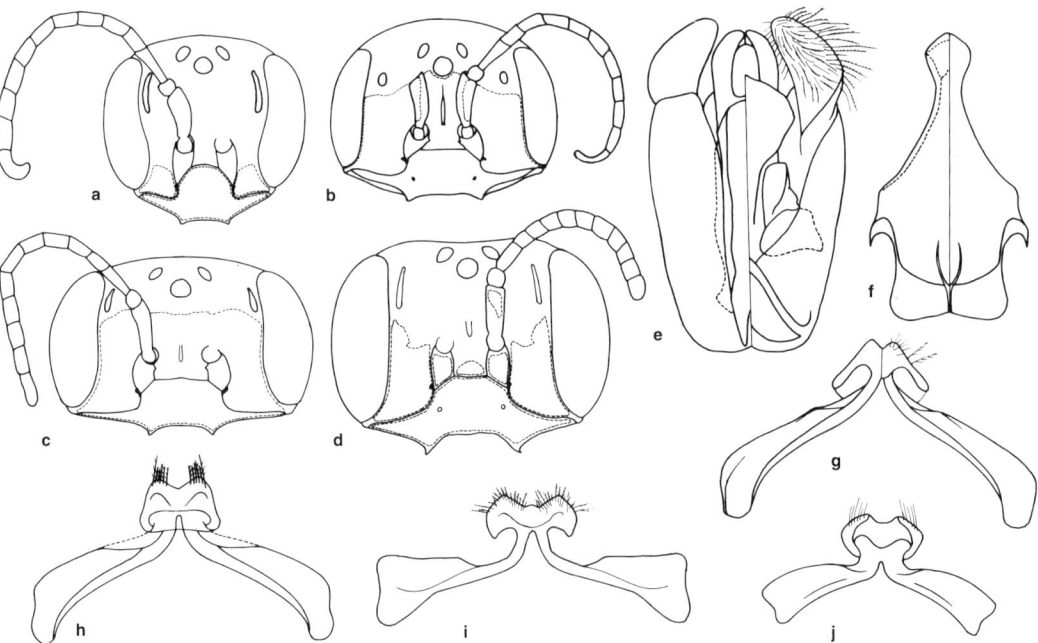

Fig. 53-4. Structures of males of *Liphanthus* (Protandreninae). a-d, Facial views; limits of yellow areas are indicated by dotted lines. **a,** *L. (Tricholiphanthus) leucostomus* Ruz and Toro; **b,** *L. (Liphanthus) sabulosus* Reed; **c,** *L. (Pseudoliphanthus) spiniventris* Ruz and Toro; **d,** *L. (Xenoliphanthus) parvulus* (Friese). **e-g,** Genitalia, S8, and S7 of *L. (L.) sabulosus* Reed. h-j, S7 of the following species: **h,** *L. (Pseudoliphanthus) spiniventris* Ruz and Toro; **i,** *L. (Xenoliphanthus) parvulus* (Friese); **j,** *L. (Tricholiphanthus) leucostomus* Ruz and Toro. (In the divided figures, dorsal views are at the left.) From Ruz and Toro, 1983.

—. Hind tibial spurs with apices only slightly curved, inner somewhat longer than outer; vertex convex in frontal view 3
3(2). Distance between antennal socket and inner orbit shorter than or similar to interantennal distance 4
—. Distance between antennal socket and inner orbit longer than interantennal distance 5
4(3). First flagellar segment about three times as long as broad; clypeus flattened basally .... *L. (Melaliphanthus)*
—. First flagellar segment less than twice as long as broad; clypeus convex .... *L. (Leptophanthus)*
5(3). S3 and S4 with posterior margins usually clearly concave (*if* only slightly concave, then with median, depressed, triangular, transparent area between apical thickenings) .... *L. (Tricholiphanthus)*
—. S3 and S4 with posterior margins convex or straight (without thickenings) 6
6(5). Pygidial plate present; head in ventral view with lateroventral area of clypeus widened mesally; frontal line in a very long and well-marked groove .... *L. (Liphanthus s. str.)*
—. Pygidial plate absent; head in ventral view with lateroventral area of clypeus widened laterally; frontal line scarcely marked, not in groove .... *L. (Pseudoliphanthus)*

Key to the Subgenera of *Liphanthus* (Females)
(Based on Ruz and Toro, 1983)
(Females of the subgenus *Tricholiphanthus* are not known.)

1. Submarginal cells two; propodeal triangle laterally about half length of metanotum (laterally) .... *L. (Neoliphanthus)*
—. Submarginal cells three; propodeal triangle laterally slightly more than half length of metanotum (laterally) or almost same length 2
2(1). Hind tibial spurs with apices curved like claws, subequal in length; vertex concave or almost straight in frontal view .... *L. (Xenoliphanthus)*
—. Hind tibial spurs with apices only slightly curved, inner somewhat longer than outer; vertex convex in frontal view 3
3(2). Lateral ocellus, in frontal view, above upper orbital tangent; inner orbits divergent dorsally .... *L. (Leptophanthus)*
—. Lateral ocellus, in frontal view, at same level as or slightly below upper orbital tangent; inner orbits convergent dorsally or almost subparallel 4
4(3). Outer subantennal suture almost straight; clypeus, inferior paraocular area, and base of mandible without yellow .... *L. (Melaliphanthus)*
—. Outer subantennal suture distinctly arcuate laterally; at least clypeus with yellow 5
5(4). Frontal line almost imperceptible; lower paraocular area and tegula without yellow .... *L. (Pseudoliphanthus)*
—. Frontal line distinct, in a groove; lower paraocular area and tegula with yellow .... *L. (Liphanthus s. str.)*

Liphanthus / Subgenus *Leptophanthus* Ruz and Toro

Liphanthus (Leptophanthus) Ruz and Toro, 1983: 277. Type species: *Psaenythia nigra* Friese, 1916 (not Friese, 1908) = *Liphanthus nitidus* Ruz and Toro, 1983, by original designation.

In this subgenus the metasoma lacks yellow, although the face of the male has yellow or white markings.

■ *Leptophanthus* is known from Antofagasta to Malleco, Chile, and Santa Cruz province, Argentina. The seven known species were included in the revision by Ruz and Toro (1983).

Liphanthus / Subgenus *Liphanthus* Reed s. str.

Liphanthus Reed, 1894: 645. Type species: *Liphanthus sabulosus* Reed, 1894, monobasic.

In *Liphanthus* s. str. the face of the male is largely yellow and the metasoma is dark with broad yellow bands on T1 and T2 only. In the female, pale facial marks are limited and the metasoma is red or black.

■ The known range of *Liphanthus* s. str. is Coquimbo to Cautín, Chile. The four species were included in the revision by Ruz and Toro (1983).

Liphanthus / Subgenus *Melaliphanthus* Ruz and Toro

Liphanthus (Melaliphanthus) Ruz and Toro, 1983: 271. Type species: *Liphanthus atratus* Ruz and Toro, 1983, by original designation.

In this subgenus of largely black species, there are limited yellow areas on the clypeus of the male.

■ The known range is Malleco and Arauco, Chile. The two species were included in the revision by Ruz and Toro (1983).

Liphanthus / Subgenus *Neoliphanthus* Ruz and Toro

Liphanthus (Neoliphanthus) Ruz and Toro, 1983: 274. Type species: *Liphanthus bicellularis* Ruz and Toro, 1983, by original designation.

Neoliphanthus is unique in the genus in having only two submarginal cells. It differs from other subgenera in other characters as well (see the key to subgenera); it is not merely a representative of another subgenus that has lost a submarginal crossvein.

■ This subgenus is known only from Linares, Chile. The only species is *Liphanthus bicellularis* Ruz and Toro.

Liphanthus / Subgenus *Pseudoliphanthus* Ruz and Toro

Liphanthus (Pseudoliphanthus) Ruz and Toro, 1983: 253. Type species: *Liphanthus rozeni* Ruz and Toro, 1983, by original designation.

In appearance, males of this subgenus resemble *Liphanthus* s. str., because of the largely yellow face, a yellow band on T2, and usually another on T1.

■ *Pseudoliphanthus* is known from Valparaíso to Malleco, Chile, and Neuquén, Argentina. The four species were included in the revision by Ruz and Toro (1983).

Liphanthus / Subgenus *Tricholiphanthus* Ruz and Toro

Liphanthus (Tricholiphanthus) Ruz and Toro, 1983: 267. Type species: *Liphanthus leucostomus* Ruz and Toro, 1983, by original designation.

Only males are known for this subgenus. They have yellow facial areas, but the metasoma is dark. The body is more densely hairy than most species in other subgenera.

■ *Tricholiphanthus* is known from Santiago to Malleco, Chile. The three species were included in the revision by Ruz and Toro (1983).

Liphanthus / Subgenus *Xenoliphanthus* Ruz and Toro

Liphanthus (Xenoliphanthus) Ruz and Toro, 1983: 260. Type species: *Psaenythia parvula* Friese, 1916, by original designation.

In this subgenus, yellow facial markings are rather extensive in the male, more restricted or absent in the female; the metasoma is at least partly red in both sexes.

■ *Xenoliphanthus* is found from Coquimbo to Ñuble, Chile. The four species were included in the revision by Ruz and Toro (1983).

Genus *Neffapis* Ruz

Neffapis Ruz, 1995, in Rozen and Ruz, 1995: 3. Type species: *Neffapis longilingua* Ruz, 1995, by original designation.

Neffapis consists of a single small (body length 6 mm), nonmetallic, finely punctate species with yellow face marks in both sexes. It has various characters that are unique. The long glossa and labial palpi (see the key to genera) are noteworthy; the glossa is more than three times as long as the prementum, and the labial palpi are about equally long, the first segment short, the second segment not greatly elongate, and the third (and last) extremely long (Fig. 19-5c, d). The pupa has a long proboscis to house these developing structures. Uniquely in the tribe, the maxillary palpi are minute and two-segmented. Male genitalia and other structures were illustrated by Ruz (*in* Rozen and Ruz, 1995).

Ascher (in Engel, 2005) placed *Neffapis* in a new tribe, Neffapini, apparently not near the Protandrenini, but associated with the Panugini and its mostly Old World relatives. Although relevant characters are cited, they do not seem very decisive. Ascher's work is not published; for the present I leave *Neffapis* in the Protandrenini as in Michener (2000).

■ This genus is known from Coquimbo, Chile. The only species is *Neffapis longilingua* Ruz.

Rozen (*in* Rozen and Ruz, 1995) investigated the nesting of *Neffapis* and described the immature stages. Cracks in the soil replace the usual main burrows of ground-nesting bees, and burrows, each to a cell, are equivalent to the laterals of many other bees. The cells are similar to those of other Protandrenini. The bees appear to be oligolectic on *Malesherbia* (Malesherbiaceae).

Genus *Parapsaenythia* Friese

Psaenythia (Parapsaenythia) Friese, 1908b: 42. Type species: *Psaenythia argentina* Friese, 1908 = *Caenohalictus serripes* Ducke, 1908, monobasic.

Parapsaenythia is a genus of black bees with yellow markings on the face and pronotum of the male. The body is very coarsely and closely punctate except for broad smooth marginal zones of T1 to T4 of females and T1 to T5 of males; the body length is 8 to 9 mm. This is the only panurgine with hairy eyes. Another unusual feature is the basal hair bands on the metasomal terga of females (weak in males), a feature found also in some *Protandrena* s. str. The propodeal triangle is pilose. S6 of the male has a quadrate midapical emargination, partly filled by a group of thickened hairs. The two apical lobes of S7 are constricted at their bases, as in *Protandrena* (Fig. 53-6c). The male gonocoxite and gonostylus are uniquely broadened, so that the genital capsule is much the widest at its apex; the gonostylus bears several very large plumose setae. Male genitalia and other structures were illustrated by Ruz (1986); see also Figure 53-6a-c.

■ This genus ranges from Minas Gerais, Brazil, southward through Paraguay to the province of Mendoza, Argentina. There are two species, *Parapsaenythia serripes* (Ducke) and *paspali* (Schrottky).

Genus *Protandrena* Cockerell

Protandrena consists of slender andreniform bees, the metasoma of the males commonly parallel-sided. They are black, rarely with metallic green or blue-green tints, sometimes with the metasoma red, commonly with yellow markings on the face, especially of males, and often with yellow on the pronotum, rarely with yellow also elsewhere on the thorax and on the metasomal terga [e.g., in the Mexican *P. (Protandrena) maculata* Timberlake]. The forewings have either two or three submarginal cells. The middle tibial spur of the female is finely toothed or ciliate basally, with coarser and more widely separated teeth distally, or even almost toothless distally, except that in the subgenus *Austropanurgus* the spur is finely toothed distally as well as basally. S7 of the male has a pair of large distal lobes (Fig. 53-5c, d), more or less constricted at their bases and not, as in *Pseudopanurgus*, mere extensions from the small disc or body of the sternum that unites the usual large apodemal arms. The gonostyli are over one-half as long as the gonocoxites, to the apices of which they are articulated or partly fused, except that in the subgenus *Parasarus* the gonostyli are less than one-third as long as the gonocoxites.

Protandrena is here recognized in a much broader sense than in the past, to include, among others, the species that have been placed in *Heterosarus, Metapsaenythia,* and *Pterosarus,* either as separate genera or as subgenera of *Pseudopanurgus. Protandrena* has hitherto been recognized as including only species with three submarginal cells, but in many groups of bees the second submarginal crossvein disappears, leaving only two submarginal cells in some species or individuals. In such common genera as *Andrena, Lasioglossum, Leioproctus, Nomada,* and *Sphecodes* there are species with only two submarginal cells, although most have three. It is therefore appropriate, whenever a taxon is recognized by the number of submarginal cells, to look at its relatives to see if this conveniently visible feature has been overemphasized in developing the classification. The three-celled *Protandrena* s. str. seems extremely similar to the two-celled *Metapsaenythia,* as was indicated by Timberlake (1975).

The generic or tribal recognition for what is here called *Protandrena* s. str. is supported by the cladogram of Ruz

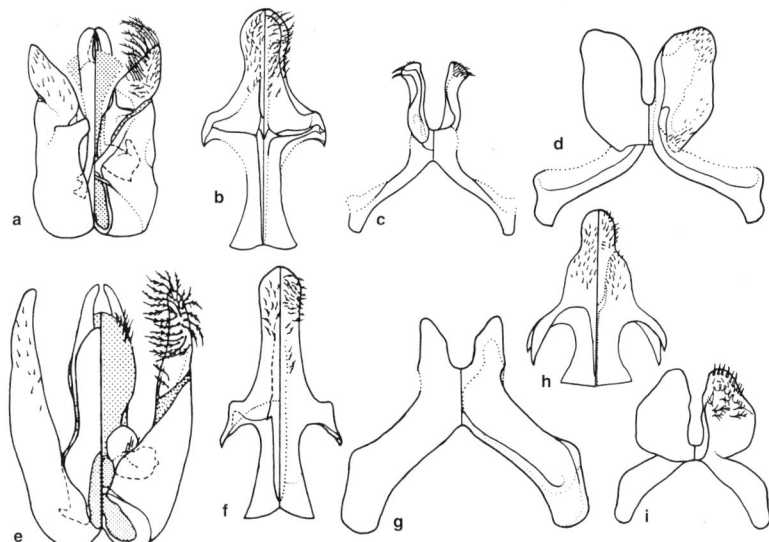

Figure 53-5. Structures of male Protandrenini. **a-c,** Genitalia, S8, and S7 of *Protandrena (Protandrena) mexicanorum* (Cockerell); **d,** S7 of *P. (Pterosarus) rudbeckiae* (Robertson); **e-g,** Genitalia, S8, and S7 of *Pseudopanurgus aethiops* (Cresson); **h, i,** S8 and S7 of *Anthemurgus passiflorae* Robertson. (In the divided figures dorsal views are at the left.) From Ruz, 1986.

(1986), but may not be justified because the relevant part of the cladogram appears to be far from robust. No apomorphies are shown for *Protandrena* s. str.; the apomorphies for the alternative line leading to nearly all other Panurginae are (1) a simple scopa, which reverses and appears elsewhere, probably in relation to the type of pollen being collected, and also is variable within *Protandrena* s. str. and (2) a relatively elongate glossa that also reverses and is probably related to floral type being utilized, and, moreover, is variable within *Protandrena* s. str.

The *Pseudopanurgus* to *Anthrenoides* clade of the Ruz cladogram involves the only polytomies in her cladogram (three and seven branches each). The synapomorphies are few and weak, unlike those in the rest of her cladogram, which is mostly strongly supported. As indicated above, I have chosen to incorporate several of these taxa into *Protandrena* on the basis that when the number of submarginal cells is ignored, the distinguishing features are not impressive. The taxa in this tribe that I retain at the genus level all differ in features with no known intermediate conditions closing the gaps between them and *Protandrena* s. l. Moreover, they have features that may be plesiomorphic relative to *Protandrena* s. l., such that a cladistic study of this group might show them to be branches basal to *Protandrena*.

Austropanurgus also has no obvious apomorphies, according to Ruz's cladogram, but does differ from the rest of *Protandrena* in the present sense in four characters, three of which also appear in one or more of the genera and subgenera of the *Pseudopanurgus* to *Anthrenoides* clade. Thus *Austropanurgus* also appears to be a member of the *Protandrena* group. For comments on *Pseudosarus* and *Xenopanurgus*, see the discussion of the subgenus *Heterosarus*. A species (*P. evansi* Ruz and Chiappa) that does not fall in any one of the subgenera as currently recognized was described by Ruz and Chiappa (2004).

I believe that *Protandrena* in the broad sense is comparable to *Calliopsis* in the degree of diversity among subgenera. It constitutes a much more useful taxon than the numerous genera sometimes recognized.

Key to the Subgenera of *Protandrena* (Males)

1. Submarginal cells three (occasional individuals have only two) (North and Central America) .... *P. (Protandrena s. str.)*
—. Submarginal cells two (Fig. 53-2b) 2
2(1). Propodeal triangle basally pilose (metasoma often red or largely so) (nearctic) .... *P. (Metapsaenythia)*
—. Propodeal triangle basally glabrous 3
3(2). Hind tibial spurs strongly curved at apices; first submarginal cell on posterior margin shorter than second (Chile) .... *P. (Austropanurgus)*
—. Hind tibial spurs or at least one of them slightly curved or almost straight; first submarginal cell on posterior margin about as long as or longer than second 4
4(3). Gonostylus less than one-third as long as gonocoxite; S6 scarcely notched apically (face black) (South America) .... *P. (Parasarus)*
—. Gonostylus over one-half as long as gonocoxite; S6 with deep midapical notch or slit.. 5
5(4). Scutum with punctures well marked, many of them separated by spaces larger than their diameters; S6 with midapical emargination narrow, deep (North and Central America) .... *P. (Pterosarus)*
—. Scutum with punctures very small, homogeneous, commonly dense; S6 midapical emargination V-shaped .... *P. (Heterosarus)*

Key to the Subgenera of *Protandrena* (Females)

1. Forewing with three submarginal cells (only two in occasional individuals) (North and Central America) .... *P. (Protandrena s. str.)*
—. Forewing with two submarginal cells (Fig. 53-2b) 2
2(1). Tibial scopa of rather long, abundant hairs with clearly visible branches (North and Central America) .... *P. (Pterosarus)*

—. Tibial scopa of sparser hairs that lack branches, or some of them with few, minute branches 3
3(2). Propodeal triangle basally pilose (metasoma often largely red) (nearctic) .... *P. (Metapsaenythia)*
—. Propodeal triangle basally glabrous 4
4(3). Hind tibial spurs not strongly curved at apices; anterior tentorial pit in epistomal suture slightly to distinctly below intersection between outer subantennal and epistomal sutures .... *P. (Heterosarus)*
—. Hind tibial spurs strongly curved at apices; anterior tentorial pit at intersection between outer subantennal and epistomal sutures 5
5(4). First submarginal cell on posterior margin shorter than second; face with yellow areas (Chile).................. *P. (Austropanurgus)*
—. First submarginal cell on posterior margin longer than second; face black (South America) .... *P. (Parasarus)*

Protandrena / Subgenus *Austropanurgus* Toro

Austropanurgus Toro, 1980: 209. Type species: *Austropanurgus punctatus* Toro, 1980, monobasic.

Austropanurgus consists of a small (6 mm long), slender, black species with yellow facial marks (these more extensive in the male); the metasoma is commonly reddish. Although *Austropanurgus* is widely separated from other taxa related to *Protandrena* by Ruz (1986), it falls in the weak part of her cladogram and seems to me to be best regarded as a subgenus of *Protandrena* s. l. Arguments in this connection are given in the discussion of the genus *Protandrena*. Distinctive features of *Austropanurgus* are the strongly curved apices of the subequal hind tibial spurs and the U-shaped emargination in the apex of S6 of the male. The latter appears in some of the related genera, such as *Rhophitulus (Cephalurgus)*, and in some species of *Protandrena* s. str., but not in other groups of *Protandrena*. Also unlike most *Protandrena*, the two apical lobes of S7 of the male are only somewhat constricted in uniting to the sternal disc at their bases; in other subgenera the disc or body of the sternum from which the lobes arise is usually conspicuously small. Male genitalia, sterna, and other structures were illustrated by Toro (1980) and Ruz (1986).

■ This subgenus, found in Coquimbo, Chile, consists of a single species, *Protandrena punctata* (Toro).

Protandrena / Subgenus *Heterosarus* Robertson

Heterosarus Robertson, 1918: 91. Type species: *Calliopsis parvus* Robertson, 1892, by original designation. [New status.]
Xenopanurgus Michener, 1952: 24. Type species: *Xenopanurgus readioi* Michener, 1952, by original designation.
Pseudosarus Ruz, 1980: 25. Type species: *Pseudosarus virescens* Ruz, 1980, monobasic.

This subgenus consists of small (4-7 mm long) species having the appearance of the subgenus *Pterosarus* but differing in the simple (or very shortly and sparsely branched) scopal hairs and the rather broadly V-shaped midapical emargination of S6 of the male. Male genitalia and hidden sterna were illustrated by Mitchell (1960), Timberlake (1975), and Ruz (1980, 1986, 1990).

■ This subgenus is widespread in North America but scarce in the Pacific states, diverse in the Rocky Mountain states and Arizona, and reasonably common east to the Atlantic Ocean; it ranges from southern Canada to Argentina and Chile but is rare in the moist tropics (e.g., only one species is known from Panama). Timberlake recognized 52 species in North and Central America; only a few (perhaps about seven) are known from South America. Timberlake (1964b) gave a key to North and Central American species, and later revised them (Timberlake, 1975).

One central Chilean species, *Protandrena (Heterosarus) virescens* (Ruz), and two Mexican and Arizona species, *P. (Heterosarus) readioi* (Michener) and *platycephala* (Ruz), are weakly to strongly metallic greenish or blue. The Chilean species, which was placed in a genus *Pseudosarus* Ruz, differs from most *Heterosarus* not only in being weakly metallic but in having the middle tibial spur of the female finely serrate or ciliate, without coarser teeth distally as in other *Protandrena*. The two Mexican and Arizona species were placed in a genus *Xenopanurgus*. It was first named on the basis of the highly aberrant *P. (H.) readioi*, whose unusual features are described by Shinn (1964). *P. (H.) platycephala*, however, bridges the gap between *Heterosarus* and *P. readioi* (Ruz, 1990). D. Yanega has concluded that these two species are related to the group of *P. (Heterosarus) bakeri* (Cockerell) (personal communication, 1995).

Some South American species sometimes confused with *Protandrena (Heterosarus)* in collections are here removed to *Rhophitulus (Panurgillus)*. They are so similar to *Heterosarus* that the male genitalic and sternal characters are the most convincing evidence that they belong in a different group. A relatively common South American species that remains in *Heterosarus* is *Protandrena (Heterosarus) nigra* (Spinola) (**new combination**); it is found in central Chile and in the lake district of Neuquén province, Argentina.

Protandrena / Subgenus *Metapsaenythia* Timberlake

Metapsaenythia Timberlake, 1969a: 89. Type species: *Calliopsis abdominalis* Cresson, 1878, by original designation.

This subgenus contains the only known species of *Protandrena* having two submarginal cells that also have hairs (inconspicuous) on the propodeal triangle. The thorax and metasoma are extraordinarily finely and closely punctate, the latter frequently red. Superficially, these bees are very like some of the species of *Protandrena* s. str. The V-shaped midapical notch of S6 of the male suggests *Heterosarus* and some *Protandrena* s. str. The long apical lobes of S7 with their retrorse apices also suggest *Protandrena* s. str. The scopa is of the simple type, but the hairs have a few short branches. Body length ranges from 4.5 to 9.0 mm. Male genitalia and hidden sterna were illustrated by Mitchell (1960) under the genus *Pseudopanurgus*, and by Ruz (1986).

■ This subgenus ranges from New Jersey to Georgia west to Kansas and Texas, USA, and south to Chihuahua and Sonora, Mexico. There are two species. The subgenus was revised by Timberlake (1969a).

At least the common species, *Protandrena abdominalis*

(Cresson), is oligolectic on *Monarda* (Lamiaceae). The other species, *P. sonorana* (Timberlake), is known only from the female and may not be a close relative of *P. abdominalis*.

Protandrena / Subgenus *Parasarus* Ruz

Parasarus Ruz, 1993, *in* Ruz and Rozen, 1993: 2. Type species: *Parasarus atacamensis* Ruz, 1993, by original designation.

This is a subgenus of small (4-5 mm long) *Heterosarus*-like bees without yellow face marks in either sex. The lobes of S7 of the male are hairless, and strongly constricted at their bases. S6 of the male, scarcely notched apically, lacks the deep slit or notch found in *Heterosarus* and *Pterosarus*. The hind tibial spurs of the female are strongly curved apically, as in *Austropanurgus*, but in the male only the outer spur is curved. Male genitalic and other characters were illustrated by Ruz (*in* Ruz and Rozen, 1993).

■ *Parasarus* is found from Atacama to Valparaíso, Chile, and in Catamarca Province, Argentina. Only one species, *Protandrena atacamensis* (Ruz), is described, but two others exist, according to Ruz and Rozen (1993).

Protandrena / Subgenus *Protandrena* Cockerell s. str.

Protandrena Cockerell, 1896: 91. Type species: *Andrena maurula* Cockerell, 1896, by designation of Sandhouse, 1943: 591.
Protandrena (*Austrandrena*) Cockerell, 1906d: 37. Type species: *Andrena modesta* Smith, 1879, by original designation.

Species of this subgenus vary from 4 to 12 mm long, the small ones tending to be particularly slender. All ordinarily have three submarginal cells, although occasional individuals have only two in one or both wings; in some, the metasoma is red, and the Mexican *Protandrena maculata* Timberlake has white or yellow thoracic and tergal markings, thus superficially resembling the South American genus *Psaenythia*. The glossa is shorter than the prementum, less than half that length in *P. (P.) mexicanorum* Cockerell. The first segment of the labial palpus is usually about as long as or longer than segments 2 to 4 together, but in *P. mexicanorum* it is considerably shorter than the combined lengths of those segments. The propodeal triangle is basally strigose or striate, either pilose or glabrous. The upper margin of the hind tibia of the male is toothed, as in *Psaenythia* (Fig. 53-6i). The tibial scopa is made up of either simple or branched hairs, or is sometimes intermediate with only a few hairs having branches. The claws of the female are bifid or simple. The two apical lobes of S7 of the male are rather slender and sometimes have retrorse extensions at the apices. Male genitalia and hidden sterna were illustrated by Mitchell (1960), Timberlake (1976), and Ruz (1986); see also Figure 53-5a-c.

■ This subgenus ranges from Wyoming, North Dakota, and New Jersey south to Texas, west to Arizona, USA, and south in Mexico to Oaxaca; it is also reported from Costa Rica. There are about 50 known species, most of them Mexican. A key to females was published by Timberlake (1955b), and the subgenus was revised by Timberlake (1976).

As indicated above, *Protandrena* s. str. varies in some of the features used to distinguish the two-celled subgenera. It is possible that it should be subdivided, but such a subdivision is not suggested by Timberlake's revision. Clearly, the subgenus warrants further study. It could be a paraphyletic group from which some of the two-celled subgenera arose. At one stage [e.g., Michener (1944)], *Protandrena* s. str. was included in *Psaenythia* because of the three submarginal cells. It now seems that these two genera are not especially close.

Protandrena / Subgenus *Pterosarus* Timberlake

Pseudopanurgus (*Pterosarus*) Timberlake, 1967: 10. Type species: *Calliopsis rudbeckiae* Robertson, 1895, by original designation.

This subgenus differs from all other "two-celled" subgenera in the plumose scopa of females, which resembles that of the genus *Pseudopanurgus*, and in the narrow, often deep, midapical emargination of S6 of the male, through which one can see a characteristic thickening or keel of S7. In appearance, species of this subgenus are small (5-9 mm long) and black, with yellow areas on the face and pronotum, principally of males. Male genitalia and other structures were illustrated by Mitchell (1960, under *Pseudopanurgus*) and Ruz (1986); see Figure 53-5d.

■ This subgenus is wide-ranging in North America, from Maine to Florida and westward to southern California, USA, and from southern Canada south to Guatemala. There are about 40 species. Timberlake (1967) gave a key to the species.

Genus *Psaenythia* Gerstaecker

Psaenythia Gerstaecker, 1868: 111. Type species: *Psaenythia philanthoides* Gerstaecker, 1868, by designation of Sandhouse, 1943: 592.
? *Stenocolletes* Schrottky, 1909c: 253. [See comments under the "genus" *Stenocolletes*.]

Psaenythia consists of moderate-sized species (7-14 mm long) nearly always having yellow markings on all tagmata, although a few females nearly lack the yellow. The size of these bees and their three submarginal cells suggest *Protandrena* s. str., and at one time the two taxa were considered congeneric. *Psaenythia* differs from *Protandrena*, however, in various features that led to its being given separate generic status by Ruz (1986) and me. A unique feature among Panurginae is the longitudinal ridge on the posterior side of the hind femur, carinate in females, and the depressed glabrous area below it (Fig. 53-6h). The spurs of both the middle and hind legs are coarsely serrate. S6 of the male has a large apical emargination, occupying most of the posterior margin of the sternum. The two apical lobes of S7 of the male are broadly connected to a moderate-sized sternal disc (Fig. 53-6f). S8 of the male lacks the large lateral apodemes found in related genera. The male genitalia are wide open at the base, the apodemes of the penis valve projecting basad through the genital foramen; the genitalia and

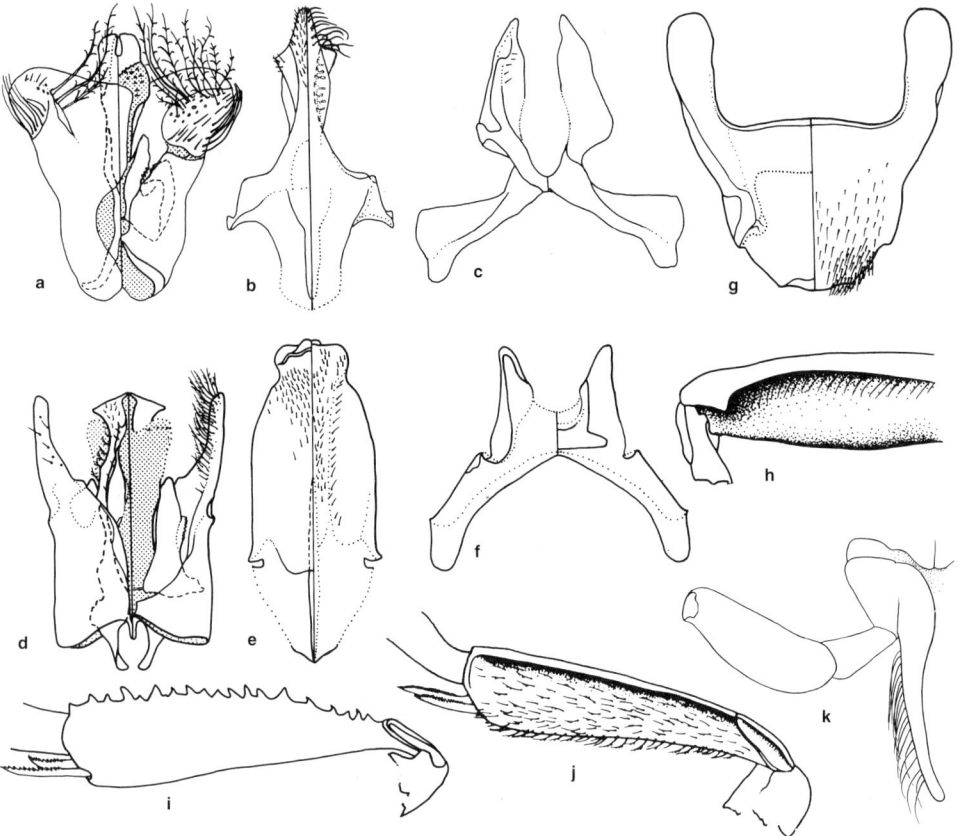

Figure 53-6. Structures of Protandrenini. **a-c,** Male genitalia, S8, and S7 of *Parapsaenythia paspali* (Ducke); **d-f,** Same structures, *Psaenythia philanthoides* Gerstaecker; **g,** S6 of female of *Psaenythia philanthoides* Gerstaecker; **h,** Inner surface of hind femur of same, showing ridge and row of hairs; **i,** Hind tibia of male of *Psaenythia philanthoides* Gerstaecker (showing serrate outer carina); **j,** Same of *Pseudopanurgus aethiops* (Cresson); **k,** Front coxa, trochanter, and femur of female of *Pseudopanurgus crenulatus* (Cockerell), showing hairy coxal spine. (In the divided figures, dorsal views are at the left.) From Ruz, 1986, except k from Michener, McGinley, and Danforth, 1994.

other structures were illustrated by Ruz (1986); see also Figure 53-6d-f.

■ *Psaenythia* occurs from Minas Gerais, Brazil, to central Argentina and Chile. It comprises about 80 species (Ruz, 1986). A revision of the Argentine species was published by Holmberg (1921).

Genus *Pseudopanurgus* Cockerell

Pseudopanurgus Cockerell, 1897a: 290. Type species: *Panurgus aethiops* Cresson, 1872, by original designation.
Protandrenopsis Crawford, 1903: 337. Type species: *Protandrenopsis fuscipennis* Crawford, 1903 = *Panurgus aethiops* Cresson, 1872, by original designation.
Friesina Moure, 1999a: 107. Type species: *Friesina carinulata* Moure, 1999, by original designation. [**New synonymy.**]

Moure (1999a) correctly indicated the relationship of *Friesina* to *Pseudopanurgus;* the only differentiating character that he cites is the distinct apical tergal, metasomal hair bands. However, as shown by Timberlake (1973), there are species of *Pseudopanurgus* such as *P. fasciatus* Timberlake, *P. fulvicornis* Timberlake, and *P. trifasciatus* Timberlake, with various degrees of metasomal hair banding. *P. trimaculatus* Timberlake has apical hair bands similar to those of *P. carinulatus* and, in addition, strong basal bands. Basal bands are feebly indicated in *P. carinulatus.*

Males of *Pseudopanurgus carinulatus* (Moure) **new combination** (females are unknown) run to *P. trimaculatus* Timberlake in Timberlake's (1973) key to species. The close relationship to *P. trimaculatus* is shown especially by the deeply bifurcate base of S8 in both species; see Timberlake's figure 50. The specimens of *P. carinulatus* were kindly lent by Dr. Frank Koch for study and examination of the genitalia and hidden sterna. The genitalia of *P. carinulatus* do not show differences from Timberlake's rather diagrammatic figure 49. Minor differences such as the more finely punctuate vertex and scutum and the weak or nearly absent basal tergal hair bands appear to differentiate *P. carinulatus* from *P. trimaculatus* at the specific level.

Most species of this genus are more robust than most

species of *Protandrena*. Length varies from 5 to 12 mm. They are usually more coarsely punctate, and males, in particular, usually have strong transverse postgradular depressions on T2 to T5. The body is black with yellow on the face of males and some females. All species have two submarginal cells. Distinctive generic characters are the large, hairy apical spine on the front coxa of females (Fig. 53-6k), not found in other Panurginae, and the strong but nondentate carina along the upper margin of the hind tibia of males (Fig. 53-6j). The scopa is strongly plumose, a feature seen among other Protandrenini only in *Protandrena (Pterosarus)* and some species of *P. (Protandrena)*. The pronotum has a transverse carina or lamella, not found in other Protandrenini except *Parapsaenythia* and the female of *Chaeturginus*. The large anterior surface of the mesepisternum, a feature contributing to the robustness of the thorax, is flat, and the omaulus is represented by a narrowly rounded angle; these features are not duplicated in other Protandrenini. Likewise contributing to the short, robust appearance of the thorax is the short dorsal surface of the propodeum, only about one-half as long as the metanotum. In other Protandrenini it is nearly as long as the metanotum or longer. S7 of the male has two apical lobes; unlike *Protandrena* they arise from a relatively large disc and are not constricted at their bases (Fig. 53-5g). A probably plesiomorphic feature is the uniformly serrate or ciliate middle tibial spur of the females; in most related taxa the distal part of the spur has coarser and more widely separated teeth. Mitchell (1960) and Timberlake (1973) illustrated male genitalia and eighth sterna of *Pseudopanurgus* species; these and other structures were illustrated by Ruz (1986). See also Figure 53-5e-g.

■ This genus ranges from Wyoming, Kansas, and Texas west to southernmost California, USA, south through Mexico and Central America to Costa Rica. The 32 species were reviewed by Timberlake (1973).

The name *Pseudopanurgus* was formerly used (e.g., by Michener, 1944) in a broader sense to include the subgenera with two submarginal cells here placed in *Protandrena*. Although I have elected to maintain *Pseudopanurgus* as a genus distinct from *Protandrena* on the basis of the characters listed above, one could justify adding it to *Protandrena* as a subgenus.

Genus *Rhophitulus* Ducke

Rhophitulus is a genus of small (4-7 mm long) bees resembling *Protandrena (Heterosarus)*. The clypeus of the male is partly or wholly yellow, whereas the female is wholly black or with yellow on the pronotum. The male resembles *Heterosarus* in the long apical lobes of S7, the lobes constricted at the bases and attached to a small discal area (Fig. 53-3c), and in the rather large V- or U-shaped midapical notch in S6. The male gonocoxite has an oblique lateral impression, and thus exhibits a lateral concavity (as in *Anthrenoides*; less developed in subgenus *Cephalurgus*) (Fig. 53-3a). The hind tibia of the female is commonly about twice as long as the basitarsus; in most other Protandrenini except *Liphanthus* it is less than twice the basitarsal length. Especially interesting is the distinct dorsal "remnant" of the gonobase, found among other Panurginae only in *Chaeturginus*.

As Ruz (1986) and Ruz and Melo (1999) correctly suggest, the dorsal "remnant" of the gonobase in *Rhophitulus* and *Chaeturginus* may be a new sclerotization of membrane rather than a remnant homologous to the gonobase found in most non-Panurgine bees.

Moure, in Schlindwein and Moure (1998), provided a new genus-group name for bees of this group, and a tentative key is provided below for subgenera of *Rhophitulus*. However, users should recognize that the classification is tentative, as was indicated by Ruz (1986) when she wrote "sometimes it is difficult to know exactly what *Rhophitulus* is." The close relationship of three subgenera is shown by the presence in the male of the dorsal sclerotization suggestive of a gonobase (shared only with *Chaeturginus*), the elongate digitus of the volsella, as well as the forms of S6, S7, and S8 of the male. Male genitalia and other structures were illustrated by Ruz (1986); see also Figure 53-3 a–c.

Key to the Subgenera of *Rhophitulus*

1. Entire lower face strongly convex, clypeus protuberant for full eye width in lateral view and produced apicad so that lower ocular tangent crosses face near base of clypeus; S6 of male with apical emargination V-shaped although rounded at anterior extremity; male gonocoxite with small dorsal lobe (Fig. 53-3a) .... (*R. Rhophitulus* s. str.)
—. Lower face not unusually convex, clypeus protuberant for two thirds eye width or less in lateral view and not produced apicad, so that lower ocular tangent crosses lower half of clypeus; S6 of male with apical emargination U-shaped; male gonocoxite without middorsal lobe 2
2. Head commonly wider than thorax, inner orbits of male diverging below, of female closest medially; clypeus rather flat, protuberant about one seventh to one half width of eye in lateral view; S6 of male with lobe on each side of emargination rounded or obtuse .... *R. (Cephalurgus)*
—. Head not wider than thorax, inner orbits converging below; clypeus protuberant one fourth to one half of width of eye in lateral view; S6 of male with lobe on each side of emargination acute (at least in some species) .... *R. (Panurgillus)*

Rhophitulus / Subgenus *Cephalurgus* Moure and Lucas de Oliveira

Cephalurgus Moure and Lucas de Oliveira, 1962: 2. Type species: *Cephalurgus anomalus* Moure and Lucas de Oliveira, 1962, by original designation.

This subgenus consists of more shiny species than the other subgenera, often with smooth integument between well-separated punctures. Unlike other species of *Rhophitulus*, some shiny species of *Cephalurgus* lack hairs on the propodeal triangle.

Ruz (1986) suggested that *Rhophitulus* and *Cephalurgus* might be synonymous; I suggest that subgeneric status would indicate their close relationship. Additional species may yet show that they grade one into the other.

■ This subgenus ranges from the state of Minas Gerais, Brazil, to Paraguay. Ruz (1986) listed the five described species.

Rhophitulus / Subgenus *Panurgillus* Moure, new status

Panurgillus Moure, 1998, in Schlindwein and Moure, 1998: 398. Type species: *Panurginus vagabundus* Cockerell, 1918, by original designation.

This subgenus contains species of very "ordinary" appearance, that is, they closely resemble species of *Protandrena (Heterosarus)* in form and punctation. In the absence of the male genitalic and sternal characters that differentiate the genus, those species would certainly be included in or near *Heterosarus*. It is likely that *Panurgillus* is a paraphyletic group from which *Cephalurgus* and *Rhophitulus* s. str. evolved.

■ *Panurgillus* is known from the state of Rio de Janeiro, Brazil, to the provinces of Salta and Buenos Aires, Argentina. Schlindwein and Moure (1998, 1999) recognized and provided keys to 21 species.

Rhophitulus / Subgenus *Rhophitulus* Ducke s. str.

Rhophitulus Ducke, 1907: 366. Type species: *Rhophitulus frisei* Ducke, 1907, by designation of Sandhouse, 1943: 597. [Sandhouse corrected the spelling of the species name to *friesei*, but since Ducke used *frisei* on pages 366 and 368, he must have intended that spelling, despite naming the species for H. Friese.]

The distinctive characters of the subgenus are indicated in the key above.

■ This subgenus occurs from Minas Gerais, Brazil, south to Argentina. Ruz (1986) listed three described species, but mentioned other, undescribed forms.

Genus *Stenocolletes* Schrottky [incertae sedis]

Stenocolletes Schrottky, 1909c: 253. Type species: *Stenocolletes pictus* Schrottky, 1909, monobasic.

Stenocolletes may be a name for a protandrenine bee from Argentina. As noted by Michener (1989), this generic name was included in the Colletidae by Schrottky and by subsequent authors, including Michener (1965b). The reason for that placement, however, is obscure, for Schrottky says that the tongue of his two specimens was hidden. Other family characters also were not indicated. *S. pictus* was based upon two male bees 7 mm long having three submarginal cells and yellow markings on all tagmata. No American colletid known to me has such integumental markings. The specimens appear to be lost, but most probably were panurgine bees of the genus *Psaenythia*. Schrottky (1913a) regarded it as the most distinctive of the hairy Colletidae, a fact in accordance with the idea that it was not a colletid.

54. Tribe Panurgini

Members of this holarctic tribe range from all black to those with extensive yellow maculation. The body form is similar to that of the Protandrenini. Most of the distinguishing tribal characters are indicated in the key to tribes, above. All Panurgini have two submarginal cells. The male gonocoxites are often more elongate than is usual in other tribes (Fig. 54-2a, e), and the inner margins of the penis valves usually have minute transverse ridges, otherwise known only in some Melitturgini. S7 of the male lacks the two large apical lobes arising from a small disc that are found in Protandrenini; the disc of the sternum is usually broad, but is narrow in *Camptopoeum* (Fig. 54-2g). The apodemal lobes in this same genus are long, as they are in Protandrenini, but they are much shorter in the other genera.

The Panurgini are found primarily in the palearctic region, but one genus, *Panurginus*, occurs also in North America, thus overlapping the distribution of the Protandrenini. *Panurginus* consists of largely black species, superficially resembling *Protandrena (Heterosarus)*, but it can be distinguished quickly from such similar-looking Protandrenini by the first recurrent vein, which meets (or is slightly basal or distal to) the first submarginal crossvein (Fig. 54-1c). In North American *Protandrena* and other Protandrenini the first recurrent vein is considerably distal to the first submarginal crossvein. This is not, however, a tribal character outside of temperate North America.

The form of S7 of the male is particularly informative in indicating major groups within the tribe. In *Camptopoeum* the disc of the sternum is small and parallel-sided (Fig. 54-2g), the apex somewhat produced and ending in a large, broad notch, and the apodemal processes are long, as noted above and as in the Protandrenini. In *Panurgus* and *Panurginus*, S7 is a rather broad plate with two, three, or four apical spines, angles, or lobes and relatively short, broad apodemal processes (Fig. 54-3c, g, i). In some species of *Panurginus*, such as *P. clavatus* (Warncke), S7 is medially produced, the four angles possibly indicated. *Avpanurgus* falls in a group by itself on the basis of S7, which has a short, broad sternal disc or body and a small, slender, bifid apical process (Fig. 54-2c). These characters are extensively used in the generic classification proposed below.

Key to the Genera of the Panurgini

1. S7 of male with disc small, about as broad as long, almost parallel-sided, with broad apical notch (Fig. 54-2g) and sometimes [e.g., in *Camptopoeum ruber* (Warncke)] with lateral lobe at each side, and with long somewhat pedunculate apodemal lobes; body usually with yellow markings (episternal groove punctate, extending slightly below scrobal level) (palearctic) .... *Camptopoeum*

—. S7 of male with disc rather large, usually apically with two, three, or four lobes, angles, or spines or a median process, disc much broader than long [except when drawn out as an apical process in *Panurginus* such as *P. clavatus* (Warncke)]; apodemal lobes of S7 not or little narrowed basally, relatively short (Figs. 54-2c, 54-3c, g, i); body often entirely black or only clypeus yellow, but body sometimes extensively yellow 2

2(1). S6 of male with posterior margin slightly to strongly produced medially, this projection with truncate to broadly concave apex margined by zone of short hairs (Fig. 54-3d); episternal groove entirely absent (body entirely black except for yellow clypeus of some males; first recurrent vein meeting or basal to first submarginal crossvein or nearly so, Fig. 54-1c) (holarctic) .... *Panurginus*

—. S6 of male with posterior margin thin and more or less straight or usually broadly bilobed (Figs. 54-2d, 54-3h), [thickened, and medially bilobed, with dense, diverging apical brushes of plumose hair in *Panurgus (Flavipanurgus)*] but usually without an apical zone of very short hairs; episternal groove present but usually short, not reaching level of scrobe, or rarely absent.......................... 3

3(2). S7 of male with slender midapical bifid process (Fig. 54-2c); gonostylus short, simple, about one-fourth as long as gonocoxite (Fig. 54-2a), articulated well before apex of gonocoxite; dorsal surface of propodeum twice as long as metanotum (body with yellow areas) (northern Africa) .... *Avpanurgus*

—. S7 of male without midapical process; gonostylus over one-fourth as long as gonocoxite, usually variously bidentate, bifid, or angulate, or with hair tufts or brushes (Fig. 54-3e, k); dorsal surface of propodeum little if any longer than metanotum (palearctic) .... *Panurgus*

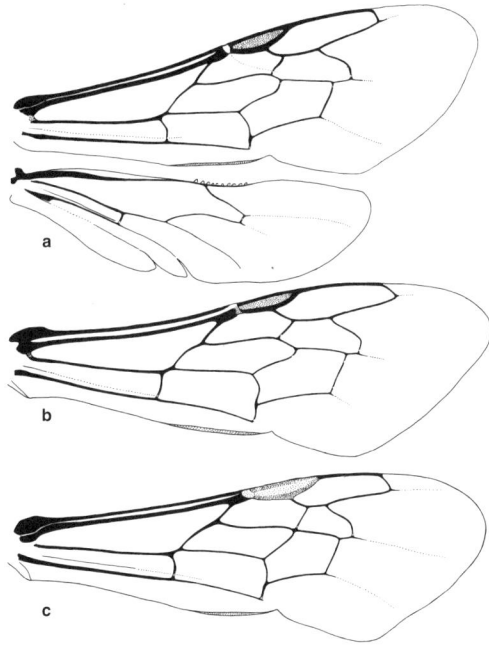

Figure 54-1. Wings of Panurgini. **a,** *Panurgus calcaratus* (Spinola); **b,** *Camptopoeum friesei* Mocsáry; **c,** *Panurginus occidentalis* (Crawford).

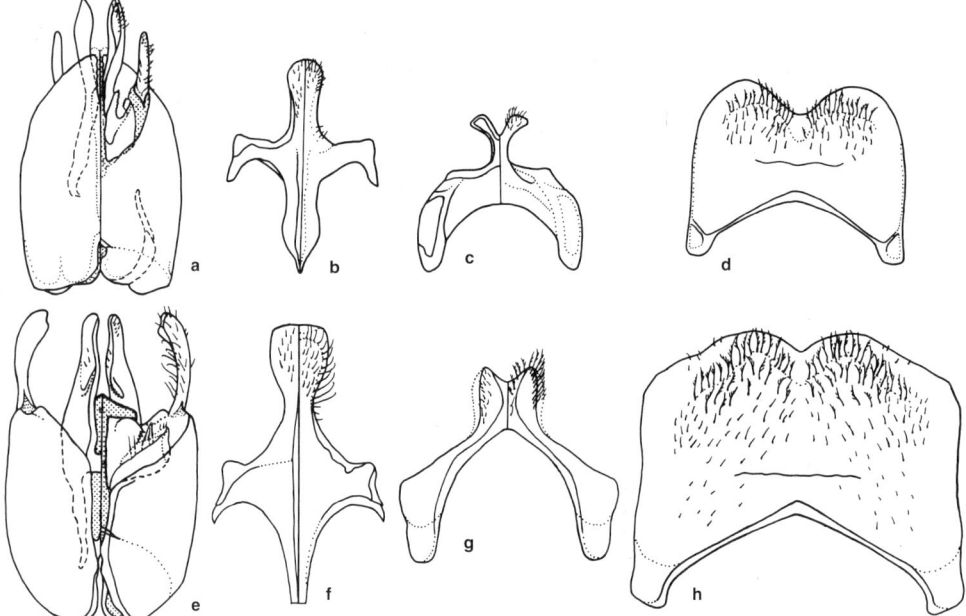

Figure 54-2. Structures of male Panurgini. **a-d**, Genitalia, S8, S7, and S6 of *Avpanurgus flavofasciatus* (Warncke); **e-h**, Same structures of *Camptopoeum (Epimethea) variegatum* (Morawitz). (In the divided figures, dorsal views are at the left.) From Ruz, 1986.

Genus *Avpanurgus* Warncke

Panurgus (Avpanurgus) Warncke, 1972: 70. Type species: *Panurgus flavofasciatus* Warncke, 1972, by original designation.

Avpanurgus consists of a species with extensive yellow markings on the head and thorax and with yellow tergal bands, narrowed medially. The body length is 7 to 8 mm. The male sterna and genitalia (Fig. 54-2a-d) are so different from those of other Panurgini that generic status seems warranted in spite of similarity in other features to *Panurgus*. The apex of S6 is broadly bilobed, with a rounded emargination between the lobes, not at all thickened or dentate. S7 has a short, transverse disc or body, extended basolaterally as the apodemal arms; at midapex is a slender, posteriorly directed process that is divergently bifid distally. The gonocoxites are elongate; the gonostylus is about one-fourth as long as the gonocoxite, simple, tapering, articulated well before the apex of the gonocoxite, and considerably exceeded by the slender penis valve. The episternal groove is distinct but short, not reaching down to the scrobal level. The first recurrent vein is distant from the first submarginal crossvein. The marginal cell is unusually narrowly truncate. The male genitalia and hidden sterna were sketched by Warncke (1972) and illustrated by Ruz (1986); see also Figure 54-2a-d.

■ This genus occurs in Algeria. It is known from a single species, *Avpanurgus flavofasciatus* (Warncke).

Genus *Camptopoeum* Spinola

In *Camptopoeum*, which consists of palearctic species, there are yellow maculations on the head and thorax and usually also the metasoma, the background color of which is occasionally red. The most distinctive feature is S7 of the male, which has a rather narrow, apically produced disc having a broad apical notch (Fig. 54-2g), with a lateral lobe in *C. ruber* (Warncke); the same sternum has long apodemal arms that are frequently narrowed near the bases. These long arms may well be plesiomorphic, but the reduced disc lacking apical lobes is an apomorphy.

Panurgus nadigi Warncke (1972) was placed in *Camptopoeum* (then treated as a subgenus) by Warncke, but clearly belongs elsewhere, as shown by the broad S7 of the male, which is suggestive of *Panurginus*.

Warncke (1987) gave a key to Turkish and Middle Eastern species. Patiny (1999c) listed the species of both subgenera of *Camptopoeum*, 13 for *Camptopoeum* s.str. and 12 for *Epimethea*.

Key to the Subgenera of *Camptopoeum*

1. Labrum with basal area triangular, delimited by converging carinae, those in females meeting and forming longitudinal median carina that ends near apex of labrum, and in males ending together near apex of labrum; clypeus protuberant one-fourth to one-half or more of eye width in lateral view; glossa longer than prementum; first segment of labial palpus about as long as second to fourth segments taken together .... *C. (Camptopoeum s. str.)*
—. Labrum with basal area delimited by transverse carina that curves basad laterally; clypeus not protuberant; glossa about one-half as long as prementum; first segment of labial palpus shorter than second to fourth segments taken together .... *C. (Epimethea)*

Camptopoeum / Subgenus *Camptopoeum* Spinola s. str.

Camptopoeum Spinola, 1843: 139. Type species: *Prosopis frontalis* Fabricius, 1804, by original designation.

Camptopaeum Spinola, 1851: 192, unjustified emendation of *Camptopoeum* Spinola, 1843.

Bees of this subgenus, often larger than *Epimethea* (5-10 mm long), have unbroken yellow bands on the terga. The galeal comb is absent in at least some species. The male genitalia and hidden sterna were illustrated by Ruz (1986) and Rozen (1988).

■ *Camptopoeum* ranges from Spain and Morocco through the Mediterranean region, north to Czechoslovakia and southern Russia, east to Central Asia, and south to Baluchistan in Pakistan. There are 13 species.

The nesting biology of one species was recorded by Rozen (1988). The nests are burrows in the ground; each of the few laterals ends in a slightly sloping cell, the cell smooth-walled with a thin, almost invisible, waterproof membrane covering part of the surface. The provisions, in the form of a flattened sphere, are not covered with a secreted membrane as they are in some Perditini and Calliopsini. At least in Perditini, the presence of such a membrane is correlated with the loss of the secreted cell lining.

Camptopoeum / Subgenus *Epimethea* Morawitz

Epimethea Morawitz, 1876: 61. Type species: *Epimethea variegata* Morawitz, 1876, by designation of Cockerell, 1922d: 1.

This subgenus consists of small species (4-6 mm long) with yellow metasomal bands either continuous [*Camptopoeum subflava* (Warncke) male], broken sublaterally [*C. subflava* female], usually represented only by lateral spots, or rarely entirely absent, the metasoma in the latter case being reddish [*C. ruber* (Warncke)]. The galeal comb is present. The genitalia and sterna were illustrated by Popov (1967) and Ruz (1986); see also Figure 54-2e-h.

■ This subgenus ranges from Romania, Greece, Morocco, and Algeria east to Siberia and China. There are 12 species.

Genus *Panurginus* Nylander

Scrapter Lepeletier, 1841: 260 (not *Scrapter* Lepeletier and Serville, 1828, which is a colletine). Type species: *Scrapter brullei* Lepeletier, 1841, by designation of Ashmead, 1899a: 84.
Panurginus Nylander, 1848: 223. Type species: *Panurginus niger* Nylander, 1848, monobasic.
Scrapteroides Gribodo, 1894: 112. Type species: *Scrapteroides difformis* Gribodo, 1894 = *Panurgus annulatus* Sichel, 1859, monobasic.
Greeleyella Cockerell, 1904d: 235. Type species: *Greeleyella beardsleyi* Cockerell, 1904, by original designation.
Birkmania Viereck, 1909: 50. Type species: *Birkmania andrenoides* Viereck, 1909 = *Panurginus polytrichus* Cockerell, 1909, by original designation.
Panurgus (Clavipanurgus) Warncke, 1972: 95. Type species: *Panurgus clavatus* Warncke, 1972, by original designation.

In this holarctic genus of small (5-9 mm long) black bees, the clypeus and parts of the legs of males are often yellow. The episternal groove, unlike that of *Panurgus*, is completely absent, the propodeal triangle has at least minute hairs laterally, and the first recurrent vein is close to (basad to distad from) the first submarginal crossvein. S6 of the male has a well-sclerotized, short, median apical area or projection (Fig. 54-3d), the distal margin of which is truncate or concave, often thickened, and provided with short hairs. S7 of the male may be a broad plate with four apical angles or teeth, as in many *Panurgus*, but the two median ones are approximate (Fig. 54-3c) and sometimes elongate, or, in the group of *Panurginus clavatus* (Warncke), the distal part of the plate is strongly produced posteriorly and the four teeth are thus less evident (Fig. 54-3j). The valve of the first valvula of the sting is rudimentary. Male genitalia and hidden sterna were illustrated by Richards (1932), Mitchell (1960), and Ruz (1986).

■ The distribution of this genus is from the Canary Islands east through Europe and North Africa to Siberia and Japan, and in North America from Alaska, USA, to northern Baja California, Mexico, east to New Jersey and Georgia, USA. Warncke (1972) recorded 12 species and numerous subspecies, some of them clearly deserving specific rank, in the western palearctic. Patiny (2003b) listed 20 species (not including *P. niger* Nylander!) in a catalog of Old World species. Ten additional species fall in the group called *Clavipanurgus* (Patiny, 2004a). In the nearctic region 18 species are recognized (Hurd, 1979). A revision of North American forms was provided by Crawford (1926) and a partial key was given by Michener (1936).

The nesting biology was described by Malyshev (1924b), Rozen (1971b), and Rust (1976). Nests of *Panurginus albopilosus* Lucas differ from those of most Panurginae in that cells are sessile, arising directly from the main burrow (Rozen, 1971b). Some species are facultatively communal, that is, nests occupied by single females as well as those occupied by several females are known. An additional important study of nesting biology of *Panurginus* is by Neff (2003), who investigated *P. polytrichus* Cockerell, a clearly polylectic species, in Texas.

Warncke (1972) segregated certain species as a subgenus *Clavipanurgus*. His descriptions of *Panurginus* and *Clavipanurgus* do not indicate any consistently differentiating characters, although yellow areas on the legs of males are often more extensive in *Clavipanurgus*, and the clypeus is more protuberant. *Panurgus anatolicus* (Warncke) and *clavatus* (Warncke) have long and intricately shaped male gonostyli much exceeding the penis valves (Fig. 54-3k) and may warrant subgeneric recognition as *Clavipanurgus*, but other species that Warncke placed under that name have much more ordinary male genitalia, as illustrated by Warncke (1972). Patiny (2003c) considered such groups, here and elsewhere (e.g., Patiny 2000), as genera and he regarded *Clavipanurgus* as a genus rather than a synonym of *Panurginus*.

Genus *Panurgus* Panzer

This palearctic genus consists of species with S7 of the male broad. Typically, its posterior margin (Fig. 54-3g) consists of four more or less equally spaced teeth (sometimes with a marginal convexity between the two median teeth), but some species have two long apical processes between which is a marginal convexity or median tooth. Such species are *Panurgus ovatulus* Warncke and *farinosus*

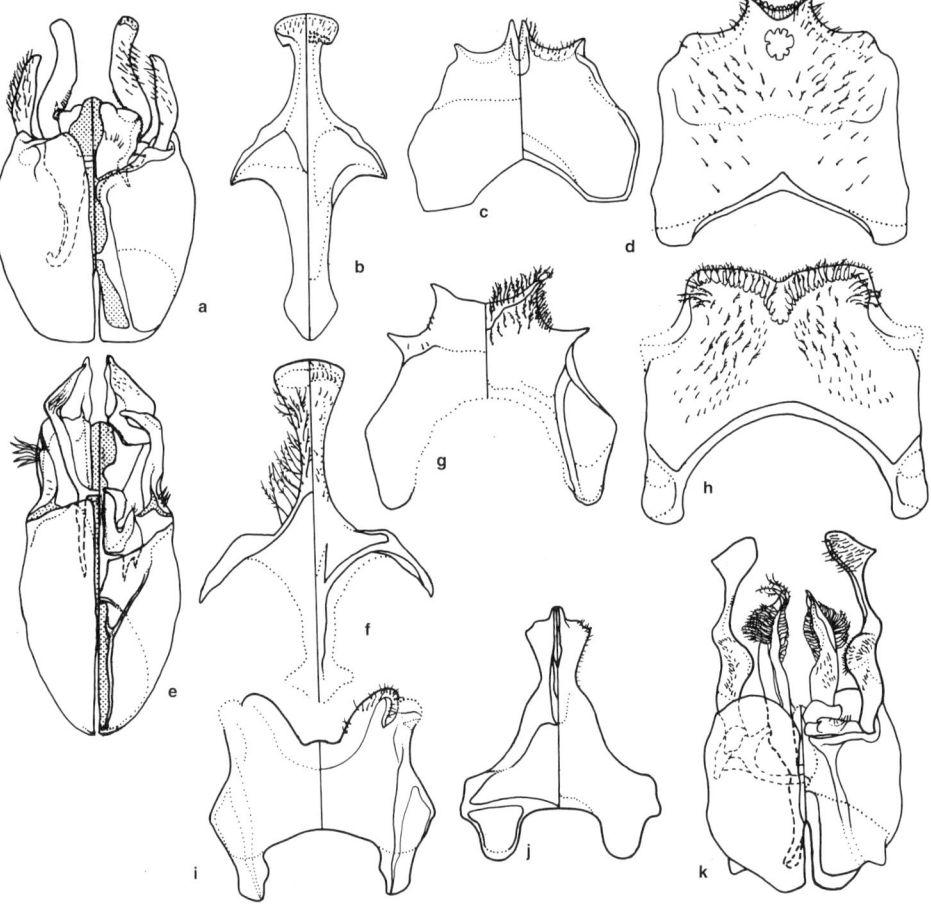

Figure 54-3. Structures of male Panurgini. **a-d,** Genitalia, S8, S7, and S6 of *Panurginus polytrichus* Cockerell; **e-h,** Same structures of *Panurgus (Panurgus) calcaratus* (Scopoli); **i,** S7 of *Panurgus (Flavipanurgus) venustus* Erichson; **j, k,** S7 and genitalia of *Panurginus clavatus* (Warncke). (In the divided figures, dorsal views are at the left.) From Ruz, 1986.

Warncke. Finally, in the subgenus *Flavipanurgus* S7 has apical lobes instead of teeth (Fig. 54-3i). The gonostylus is commonly deeply bifid, but sometimes bidentate or with a mere tooth instead of a long branch, or simple except for a tuft of coarse hairs, as in *P. (Panurgus) calcaratus* (Scopoli). Genitalia and hidden sterna were illustrated by Rozen (1971b) and Ruz (1986); see also Figure 54-3e-h. Segment 1 of the labial palpus is shorter than 2 to 4 taken together.

The short, wide S7 of the male is an apomorphy, to judge by other Panurginae, but is shared with *Panurginus*. It is possible that *Panurgus* in my sense is paraphyletic, and *Panurginus* an ex-group derived from it.

Key to the Subgenera of *Panurgus*

1. Body without yellow markings; scopal hairs minutely spiraled or zigzag; galeal comb absent; glossa longer than prementum (episternal groove weak, not reaching down to scrobal level) .... *P. (Panurgus s. str.)*
—. Body with yellow at least on face and metasoma; scopal hairs simple (female unknown in *Simpanurgus*); galeal comb present; glossa shorter than prementum 2
2(1). Episternal groove well marked, extending below scrobal level; inner orbits of male convex, closest to one another medially (Iberian peninsula) ...... *P. (Simpanurgus)*
—. Episternal groove weak, not reaching down to scrobal level; inner orbits of male diverging below (Iberian peninsula) .... *P. (Flavipanurgus)*

Panurgus / Subgenus *Flavipanurgus* Warncke

Panurgus (Flavipanurgus) Warncke, 1972: 69. Type species: *Panurgus flavus* Friese, 1897, by original designation.

Species of *Flavipanurgus* have yellow areas at least on the head and metasoma, and on the terga lateral spots sometimes uniting to form bands. The body length is 5 to 9 mm. The head is unusually large, and the inner orbits of the male diverge below, unlike those of other Panurgini. The most distinctive features are those of the male sterna and genitalia: S6 has four divergent projections, apicolateral teeth and midapical lobes, the latter well separated and covered with divergent, plumose hairs;

S7 is large and broad as in *Panurgus* s. str. but has a pair of apical lobes rather than teeth, these lobes (sometimes slender) being on either side of a broad emargination and mesal to the apicolateral angles (Fig. 54-3i). The gonostylus is large, deeply bifid from the base, with a basal lobe, or simple. Illustrations of male genitalia and hidden sterna were presented by Ruz (1986) and Warncke (1987).

■ *Flavipanurgus* is known from Spain and Portugal. Five species are recognized and a partial key was given by Warncke (1987).

Panurgus / Subgenus *Panurgus* Panzer s. str.

Panurgus Panzer, 1806: 209. Type species: *Andrena lobata* Panzer, 1799 = *Apis calcarata* Scopoli, 1763, by designation of Latreille, 1810: 439. [Later type designations were listed by Michener, 1997b.]

Eriops Klug, 1807b: 207, 227. Type species: *Andrena lobata* Panzer, 1799 = *Apis calcarata* Scopoli, 1763, monobasic.

Eryops Latreille, 1811: 716, unjustified emendation of *Eriops* Klug, 1807.

Panurgus (*Euryvalvus*) Patiny, 1999c: 316. Type species: *Apis banksiana* Kirby, 1802, by original designation. [**New synonymy.**]

Panurgus (*Pachycephalopanurgus*) Patiny, 1999c: 316. Type species: *Panurgus rungsii* Benoist, 1937, by original designation. [**New synonymy.**]

Panurgus (*Stenostylus*) Patiny, 1999c: 317. [not *Stenostylus* Pilsbury 1898]. Type species: *Panurgus ovatulus* Warncke, 1972, by original designation. [**New synonymy.**]

Panurgus (*Micropanurgus*) Patiny, in Ascher and Patiny, 2002: 140. Replacement name for *Stenostylus* Patiny. Type species: *Panurgus ovatulus* Warncke, 1972, autobasic and by original designation. [**New synonymy.**]

Patiny (1999b) regarded the subgenera *Simpanurgus* and *Flavipanurgus* as of uncertain position, but divided the rest of *Panurgus* (here called *Panurgus* s. str.) into four subgenera, *Euryvalvus*, *Pachycephalopanurgus*, *Panurgus* s. str., and *Stenostylus*, the last properly called *Micropanurgus*. Species to be included in each of these subgenera were indicated but since characters were not listed and type species were not designated, the three new subgeneric names were not valid. However, in a later paper (Patiny, 1999c), the subgenera were described and type species designated. Patiny (1999c) listed 17 species in *Panurgus* s.str., two in *Euryvalvus*, six in *Pachycephalopanurgus*, and two in *Stenostylus* (= *Micropanurgus*); a third *Micropanurgus* was described later (Patiny, 2002a).

This subgenus consists of completely black bees, often larger (5-14 mm long) than *Panurginus*. The first recurrent vein is sometimes rather close to the first transverse cubital, as in *Panurginus*, but is variable and usually well separated; in general, this character distinguishes *Panurgus* from *Panurginus*. An unusual character, otherwise known in Panurginae only in some species of *Perdita*, is the zigzag or spiral scopal hairs. The male genitalia and hidden sterna were illustrated by Ruz (1986) and sketched by Warncke (1972, 1987); see also Figure 54-3e-h.

■ This subgenus is known from the Canary Islands and the Mediterranean basin north to Sweden and east to China and Japan. Warncke (1972) lists 14 species as *Panurgus* s. str. in the western palearctic region, but there are many subspecies that apparently should have specific rank; probably there are at least 30 species.

The behavior of *Panurgus* s. str. was described by Münster-Swendsen (1968, 1970), Rozen (1971b), and Meyer-Holzapfel (1984). The nests may be in banks or flat ground, and may contain communal colonies (1 to 20 females in the case of Moroccan species; Rozen, 1971b). Cells are at the ends of lateral burrows, or sometimes constructed midway along such a burrow, and are similar to those of most Panurgini, that is, subhorizontal, smooth-walled, the lower surface not flattened, the walls coated with a secreted material, and the provisions a firm, somewhat flattened ball.

Panurgus / Subgenus *Simpanurgus* Warncke

Panurgus (*Simpanurgus*) Warncke, 1972: 67. Type species: *Panurgus phyllopodus* Warncke, 1972, by original designation.

Simpanurgus consists of *Panurgus*-like bees with yellow markings, including broad yellow metasomal bands. Body length is 8 to 9 mm. Features in which these bees resemble *Panurgus* s. str. include the broad S7 of the male, with its four widely spaced apical spines or angles, and other male sternal and genitalic characters. *Simpanurgus* differs from *Panurgus* s. str. in its yellow markings; the front tarsus of the male, which is broad and flat, the basitarsus about twice as long as broad and nearly as broad as the tibia, the other segments equally broad or nearly so but much shorter; the claws of the male, the inner rami of which are considerably shorter than the outer rami (an unusual feature); the better-developed episternal groove (see the key); the presence of a galeal comb; and the glossa, which is only about half as long as the prementum. The male antennal flagellum is clavate and flattened. The male genitalia and hidden sterna were sketched by Warncke (1972) and illustrated by Ruz (1986).

■ This subgenus is known only from Spain. The single species, *Panurgus phyllopodus* Warncke, is unknown in the female sex.

Simpanurgus was placed by Ruz (1986) as a member of the *Camptopoeum-Epimethea* clade. The characters that might support this position are the convex inner orbits of the male and the shape of the female hind basitarsus; the female, however, is unknown for *Simpanurgus*, and this character state was predicted, not observed. It seems appropriate to leave *Simpanurgus* as a subgenus of *Panurgus*.

55. Tribe Nolanomelissini

This tribe, proposed by Rozen (2003b), contains a Chilean species that is very distinctive, yet many of its features are duplicated in one or another genus of the Old World tribe Melitturgini. Distinctive features include the lack of pale markings; the long glossa and labial palpi (second segment longest, fourth minute); the three submarginal cells; the inner orbits that diverge below in both sexes; the lack of facial foveae (possibly vaguely indicated in females); the vestigial basitibial plates; the median slit of T5 of the female, surrounded by dense minute hairs (suggestive of Halictinae); and the large pygidial plate of the female with a median, ligulate, elevated area.

It might seem reasonable to include *Nolanomelissa* in the Melitturgini; however, the Melitturgini so expanded has no evident synapomorphies. Moreover, Ascher (2003) made a phylogenetic analysis of *Nolanomelissa* and related taxa based on molecular data; he concluded that *Nolanomelissa* is the sister group to all other Panurginae.

Genus *Nolanomelissa* Rozen

Nolanomelissa Rozen, 2003b: 94. Type species: *Nolanomelissa toroi* Rozen, 2003, by orignal designation.

The generic characters are those listed for the tribe.

■ The only known species of this genus and tribe is *Nolanomelissa toroi* Rozen, evidentially an oligolege on flowers of *Nolana rostrata* (Family Nolanaceae) in Huasco and Copiapó Provinces, Chile.

56. Tribe Melitturgini

The forms embraced by this tribe are more robust than most Protandrenini and Panurgini, the body being euceriform or sometimes almost anthophoriform. Unusual features are the lack of facial foveae in most males and some females and the downwardly divergent inner orbits of most males. Coloration ranges from black to largely yellow, the metasoma sometimes red. The propodeal triangle is largely pilose. Tribal characters are presented in the key to tribes.

The Melitturgini are found in the palearctic and Ethiopian faunal regions.

Key to the Genera of the Tribe Melitturgini

1. Body with few to extensive yellow areas; submarginal cells three (Fig. 56-1); eyes of male large, inner orbits conspicuously convergent above; antennal flagellum of male at least somewhat clavate (Fig. 56-2g); labrum partially pilose, the hairs short; episternal groove usually extending below scrobal level 2
—. Body without yellow markings; submarginal cells two; inner orbits of male not convergent above; antennal flagellum not clavate; labrum fully pilose, hairs long; episternal groove reaching but not extending below scrobal level 3
2(1). Posterior margins of submarginal cells subequal; tarsi of male with segments 2 to 4 expanded; stigma shorter than prestigma; T6 of male basally with conspicuous, broad, swollen hairy area (palearctic, Africa) .... *Melitturga*
—. Posterior margin of first submarginal cell nearly as long as that of second and third taken together; tarsi of male with segments 2 to 4 unmodified; stigma almost as long as to longer than prestigma; T6 of male unmodified (Africa, southwest Asia) 4
3(1). Scape of male robust; last antennal segment of male strongly tapered at apex (Fig. 56-2k); pronotum with strong dorsolateral lamella or protuberance; metasoma black; T1 (lateral view) with dorsal and anterior surfaces rounding smoothly; T2 and T3 with gradulus not strongly carinate, lateral part rather short; head slightly wider than long (palearctic) .... *Plesiopanurgus*
—. Scape and last antennal segment unmodified; pronotum with rounded and little-developed ridge on dorsal margin; metasoma red; T1 (lateral view) with dorsal and anterior surfaces meeting at almost 90 degrees; T2 and T3 with lateral part of gradulus strongly carinate, long; head conspicuously wider than long (Africa) ....*Mermiglossa*
4(2). First three segments of labial palpus elongate, fourth shorter; body usually partly or wholly yellow (glossa longer than face, at least in some species with apex broadly truncate (Fig. 52-1 f) .... *Flavomelitturgula*
—. First segment of labial palpus elongate, as long as or longer than second to fourth segments together; body variable in coloration but often mostly dark, never wholly yellow; glossa pointed 5
5(4). Maxillary palpus "nearly absent" (Patiny, 2001); glossa longer than face; first segment of labial palpus longer than second to fourth segments together .... *Gasparinahla*

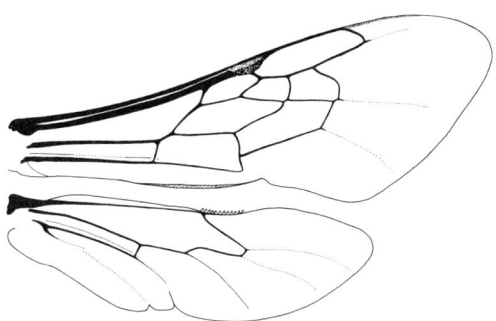

Figure 56-1. Wings of *Melitturga clavicornis* (Latreille).

—. Maxillary palpus 6-segmented; glossa shorter than face; first segment of labial palpus about as long as second to fourth together 6
6(5). Scutum of female densely covered with short, thickened, appressed hairs among which are scattered erect hairs; male gonostylus very slender, as long as gonocoxite; volsella spiculate, extending beyond gonocoxites; T7 of male with bifid apex; S6 and S7 of male with deep median apical emarginations .... *Borgatomelissa*
—. Scutum with scattered suberect hairs or in some species with appressed hairs, usually less dense than in *Borgatomelissa*; male gonostylus robust, half as long as gonocoxite or less (Fig. 56-2a); volsella not spiculate, extending little if at all beyond gonocoxites; T7 of male with apex simple; S6 and S7 with apical margins transverse or convex .... *Meliturgula*

Genus *Borgatomelissa* Patiny

Borgatomelissa Patiny, 2000: 101. Type species: *Andrena brevipennis* Walker, 1871, by original designation.
Sabulapis Patiny, 2002a: 43. This is a D. B. Baker manuscript name unnecessarily published by Patiny as a synonym of *Borgatomelissa*. Invalid because published with no type species designation and as a synonym. [**New synonymy.**]

This taxon could be considered as a subgenus of *Meliturgula* but Popov, Baker, etc., have thought it was generically distinct; see comments about the type species by Michener (2000) and as a *genus incertus* in Patiny (1999a,b). Distinctive characters, especially those of the male genitalia, are indicated in the key to the genera. The male genitalia were illustrated by Popov (1951a) as *Meliturgula arabica* Popov.

■ *Borgatomelissa* contains two species and is known from Saudi Arabia to Ethiopia and Mauritania.

Genus *Flavomelitturgula* Patiny

Flavomelitturgula Warncke, 1985: 229 *nomen nudum*.
Flavomelitturgula Patiny, 1999a: 251. Type species: *Poecilomelitta lacrymosa* Popov, 1967, by original designation.

Flavomelitturgula Warncke (1985) was tentatively listed under *Meliturgula* by Michener (2000). As originally published by Warncke it contained no species nor, of course, type species, and was a *nomen nudum*. Patiny (1999a), however, while attributing the name to Warncke, described it as a genus and designated a type species; the name should be attributed to Patiny.

The long glossa, as described in the key to the genera of Melitturgini, is one of the characteristic features of this genus. In at least some species it is not only long but broad and truncate (Fig. 52-1 f), a unique feature.

■ *Flavomelitturgula* contains six species found in southern Iran and Baluchistan, northwestern Pakistan.

Flavomelitturgula is a relative of *Meliturgula;* three of Patiny's (1999b) four cladograms show it as the sister goup to *Meliturgula*. If this cladistic pattern is accepted, it is a matter of judgment whether *Flavomelitturgula* is to be regarded as a separate genus or included (as a subgenus) in *Meliturgula*.

The species of *Flavomelitturgula* vary from nearly all yellow through patterned with conspicuous yellow markings (suggesting *Camptopoeum*) to almost completely black except for yellow areas on the legs and on the face of the male (Patiny, 2004a).

Genus *Gasparinahla* Patiny

Gasparinahla Patiny, 2001: 309. Type species: *Gasparinahla megapalpae* Patiny, 2001, by original designation.

Distinctive features of this genus are indicated in the key to genera. The body length is 6 to 7 mm.

■ This genus contains a single species, *Gasparinahla megapalpae* Patiny, from southern Iran.

Genus *Melitturga* Latreille

Melitturga Latreille, 1809: 176. Type species: *Eucera clavicornis* Latreille, 1806, monobasic.
Mellitturga Latreille, 1809: 177, *lapsus* for *Melitturga* Latreille, 1809.
Meliturga Lepeletier and Serville, 1828: 799, unjustified emendation of *Melitturga* Latreille, 1809.
Melitturga (Petrusianna) Patiny, 1998: 30. Type species: *Melitturga spinosa* Morawitz, 1892, by original designation. **[New synonymy.]**
Melitturga (Australomelitturga) Patiny, 1999a: 246. Type species: *Melitturga capensis* Brauns, 1912, by original designation. **[New synonymy.]**
Petrusia Patiny, 1999b: 255. Preoccupied and lacking both a description and a type designation; this name is a *nomen nudem,* but it was apparently intended for the same taxon as *Petrusianna*. **[New synonymy.]**

This genus consists of relatively large (12-13 mm long), robust, black species with yellow limited to the clypeus. The first segment of the labial palpus is longer than the second to fourth taken together. The pygidial plate of the male is well developed, bifurcate; T6 has a preapical prepygidial plate. S7 of the male consists largely of broad apodemal arms, narrowly connected apically with a very short, four-toothed to four-lobed disc; S8 has a long apical process. The gonostyli are long, as long as to much exceeding the slender penis valves, with hairy areas or hair tufts on the mesal surfaces. Male genitalia and other structures were illustrated by Tkalců (1974e, 1978a), Ruz (1986), and Eardley (1991a); see also Figure 56-2d-f. As the synonymy above indicates, Patiny recognized three subgenera of *Melitturga* including *Melitturga* s. str. They are distinguishable groups; for no strong reason I have chosen not to recognize them as subgenera. In *Melitturga* s. str. he included seven palearctic species. In *Australomelitturga* he included (Patiny, 2004b) six species (Angola to South Africa and Botswana as well as North Africa). In *Petrusianna* he included three species (Spain and Morocco east to Turkey and the Sinai).

■ This genus ranges from Portugal and Morocco north in Europe as far as Czechoslovakia, east through North Africa and Asia to China; it also occurs from Angola to South Africa and Botswana; there are 16 species. The species of southern Africa were revised by Eardley (1991a) and a new key was by Patiny (2004b); Warncke (1987) gave a key to Near Eastern species.

The nesting habits were described by Malyshev (1925b) and Rozen (1965c; 1971b). The nesting biology and larval characters of *Melitturga* are similar to those of other panurgines, which characters support the adult morphological evidence that *Melitturga* is one of the Panurginae. The cells are subhorizontal, the lower surface is not flattened, and the walls are lined with a secreted waterproof coating; the provisions form a flattened spheroid.

Similarity to the Oxaeinae was noted by Michener (1944), who pointed out that in male *Melitturga* and the Oxaeinae the antennae are very short, although the first flagellar segment is longer than the scape; the eyes are large and convergent above, and the ocelli are far down on the frons near the antennal bases. In both sexes the facial foveae are absent, the three submarginal cells are of more or less equal size, and the stigma is small (virtually absent in Oxaeinae). It now appears that these striking similarities are convergent and related to the similar courtship behavior of the two groups, in which males hover in the air and dart extremely rapidly toward passing females or other small objects, apparently being mediated by visual rather than olfactory cues (Rozen, 1965c).

Genus *Meliturgula* Friese

Meliturgula Friese, 1903b: 33. Type species: *Meliturgula braunsi* Friese, 1903, monobasic.
Poecilomelitta Friese, 1913a: 574 and 1913b: 585. Type species: *Poecilomelitta flavida* Friese, 1913, by designation of Sandhouse, 1943: 589. [*Poecilomelitta* was described as new twice.]
Meliturgula (Popovmeliturgula) Patiny, 1999a: 250. Type species: *Poecilomelitta ornata* Popov, 1951, by original designation.
Popovia Patiny, 1999b: 253, 1999d: 29. Preoccupied, lacked a type designation, and thus was a *nomen nudum,* but was intended for the same taxon as *Popovmeliturgula*. Since *Popovia* Patiny had no status, *Popovmeliturgula* should not have been considered by Patiny (1999a) as a new name; it was a new subgenus.

This genus varies from largely black forms superficially resembling *Melitturga* to forms with extensive yellow maculation including most of the metasoma. Body length ranges from 6 to 11 mm. The first segment of the labial palpus is about as long as the second to fourth taken together. The claws of the female vary from cleft to simple. The pygidial plate of the male is absent. The male

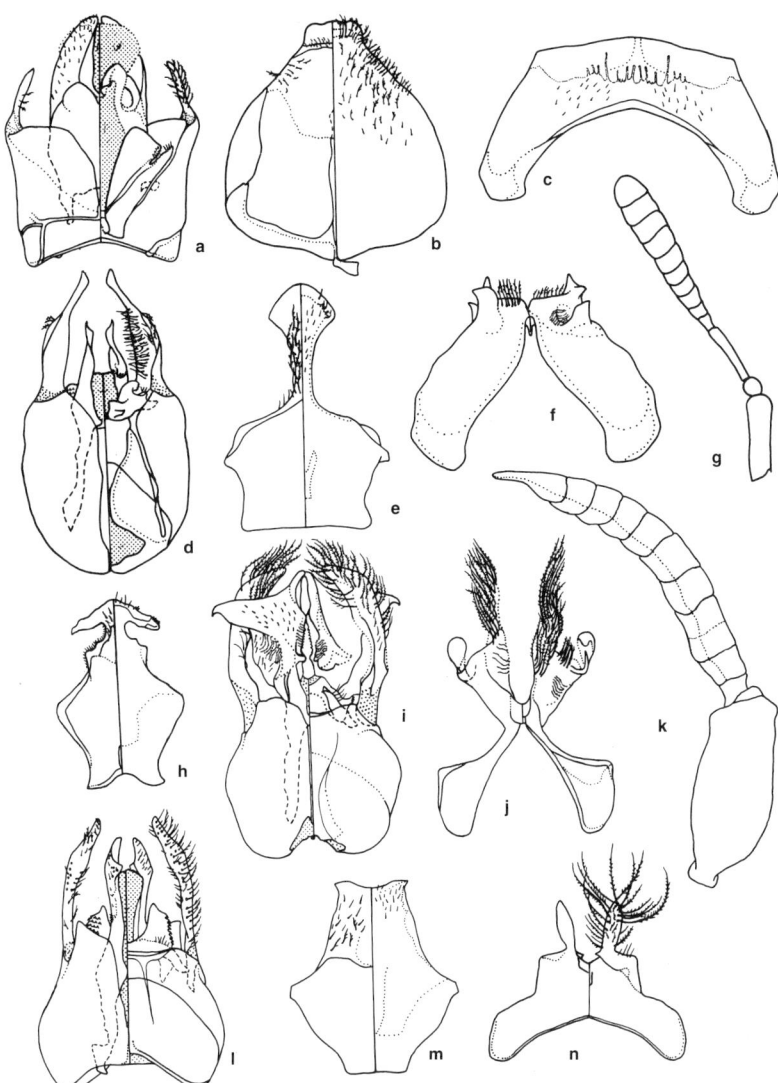

Figure 56-2. Structures of male Melitturgini. **a-c,** Genitalia, S8, and S7 of *Melitturgula braunsi* Friese; **d-f,** Same structures of *Melitturga clavicornis* Latreille; **g,** Antenna of same species; **h-j,** S8, genitalia, and S7 of *Plesiopanurgus cinerarius* Cameron; **k,** Antenna of same species; **l-n,** Genitalia, S8, and S7 of *Mermiglossa rufa* Friese. (In the divided figures, dorsal views are at the left.) From Ruz, 1986.

genitalia and hidden sterna (Fig. 56-2a-c) are remarkably different from those of other Melitturgini. S7 of the male is a simple, often very short sternum; S8 is ovoid, without a strong apical process. The simple gonostyli are much exceeded by the large penis valves. Male genitalia and other structures were illustrated by Popov (1951a, c), Ruz (1986), and Eardley (1991a); see also Figure 56-2a-c.

Patiny (1999a, b) recognized three subgenera of *Melitturgula*. *Melitturgula* s. str., with seven species, has the distribution indicated below for the genus; *Poecilomelitta*, with three species, is from Namibia and South Africa, and *Popovmelitturgula*, with one species, is from the Arabian Peninsula.

■ *Melitturgula* occurs from Cape Province, South Africa, north to Angola, Kenya, Egypt, Israel, and Saudi Arabia; also to Madagascar. There are seven species in southern Africa, one (undescribed) in Kenya, another in Egypt, at least two in Saudi Arabia, and one in Madagascar. The species of southern Africa were revised by Eardley (1991a) and a new key was by Patiny (2004b).

The genitalia and sterna of *Melitturgula ornata* (Popov) and *scriptifrons* (Walker) = *dzheddaensis* (Popov), as illustrated by Popov (1951a), seem to place these species easily within the genus *Melitturgula* as here understood.

The nesting biology of *Melitturgula* was described by Rozen (1968b). The nests are at least sometimes occupied by communal groups; one contained 36 females. The cell wall is only partly lined with a secreted film, but the structure is generally similar to that of most panurgines.

Genus *Mermiglossa* Friese

Mermiglossa Friese, 1912a: 188. Type species: *Mermiglossa rufa* Friese, 1912, monobasic.

This genus shares many strong apomorphies with *Plesiopanurgus*, from which it differs in the characters listed in the key to genera and the following characters: Body length is 8 to 9 mm; the metasoma is red and the pubescence is short. The glossa is three times as long as the prementum; the maxillary palpi are five-segmented. The antennae of the male are unmodified. S7 of the male has a large disc, broader and nonpedunculate apodemal lobes, and slender apical lobes, the last simple except for a few long, plumose hairs; S8 has a rather short, broad, truncate apical process; the penis valves are slender as seen from above, lacking the enormous lateral flanges found in *Plesiopanurgus,* but are vertically expanded apically. Male genitalia and other structures were illustrated by Ruz (1986) and Eardley (1991a); see also Figure 56-2l-n.

■ This genus is recorded only from Namibia but an undescribed species has been found in Kenya (J. S. Ascher, personal comm.). There is only one described species, *Mermiglossa rufa* Friese.

Genus *Plesiopanurgus* Cameron

Plesiopanurgus Cameron, 1907: 130. Type species: *Plesiopanurgus cinerarius* Cameron, 1907, monobasic.

Neopanurgus Schwammberger, 1971a: 2. Type species: *Neopanurgus richteri* Schwammberger, 1971, by original designation.

Plesiopanurgus (Zizopanurgus) Patiny, 1999b: 256. No designation of type species, the subgeneric name is therefore invalid.

Plesiopanurgus (Zizopanurgus) Patiny and Rasmont 1999: 78. Type species: *Panurgus cinerarius zizus* Warncke, 1987, by original designation. [**New synonymy.**]

Plesiopanurgus consists of robust black forms without yellow markings and with relatively long pale pubescence. Body length ranges from 7 to 11 mm. The clypeus is strongly protuberant and the glossa is long but less than twice as long as the prementum. Facial foveae are absent in both sexes, an unusual feature for Panurginae. The galeal comb is absent. The mandible has a strong, subbasal tooth on the upper margin. The outer subantennal suture is weak. The scape and flagellum in males are thickened, the last flagellar segment is tapered to a point, and flagellar segments 2 and 3 are fused or partly so (Fig. 56-2k). The claws of the female are simple. In males, S7 has a small disc, pedunculate apodemal lobes, and two complex, ornate apical lobes with areas of long hairs; S8 has a strong apical process with strong lateral projections at its apex; the male gonostyli are longer than the gonocoxites, hairy, and overlapped medially by the broad, pointed, lateral flanges of the penis valves. Male genitalia and sterna were illustrated by Schwammberger (1971a), Baker (1972a), and Ruz (1986); see also Figure 56-2h-j.

Zizopanurgus is a distinctive group, although in a genus of only four species its recognition is not necessary. It is found in Turkey and North Africa, while the other *Plesiopanurgus* species are from Iran and Baluchistan (Pakistan).

■ *Plesiopanurgus* occurs in Morocco and from eastern Turkey to Iran and Baluchistan in desert areas. There are four species. The genus was revised by Baker (1972a); Warncke (1987) discussed differences among forms that probably are specifically different. Probably all species visit flowers of *Convolvulus*.

57. Tribe Protomeliturgini

This tribe contains a single genus, *Protomeliturga*, that consists of a largely blackish species 5 to 8 mm long with limited cream-colored markings, such markings lacking on the metasoma. There are hairy apical tergal bands laterally, at least in the male. As indicated in the key to tribes, the labial palpi agree with the prime character of L-T bees (Fig. 52-1c). However, *Protomeliturga* must have evolved these key features independently, for its other characters are those of Panurginae, for example, the well-defined facial foveae, the presence of a galeal comb, the lack of a stipital comb, the two subantennal sutures on each side, the lack of a gonobase, etc.; see Figure 57-1. The marginal cell, however, curves gradually away from the costa, as in various nonpanurgine bees; it can hardly be called truncate. The labrum is hairy (bare medially and at base) and the propodeal triangle is hairy. The feeble episternal groove extends below the scrobal suture. Along the dorsal margin of the hind tibia of the male is an untoothed carina. The apex of the metasoma of the male is strongly curled downward and forward. The male gonostyli are much shorter than the gonocoxites and exceeded by the very large, fused penis valves and penis (Fig. 57-1e). The gonocoxites are broadly open basad as in the Melitturgini, the apodemes of the penis valves projecting through the opening as in many other bees but unlike those of most Panurginae. The genitalia, sterna, and other structures were illustrated by Ruz (1986, 1991); see also Figure 57-1e-g.

This tribe contains a single Brazilian genus and species.

Genus *Protomeliturga* Ducke

Protomeliturga Ducke, 1912: 63, 90. Type species: *Calliopsis turnerae* Ducke, 1907, monobasic.

The generic characters are those of the tribe.

■ *Protomeliturga* is known from the states of Ceará and Maranhão in northeastern Brazil. The only species is *Protomeliturga turnerae* (Ducke).

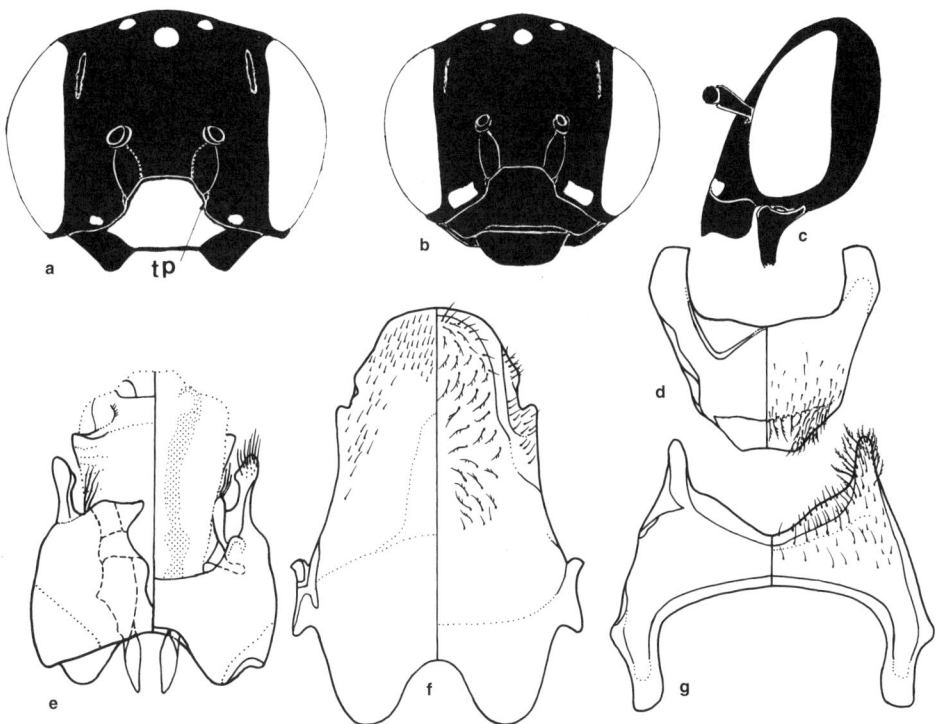

Figure 57-1. *Protomeliturga turnerae* (Ducke). **a,** Face of male; **b, c,** Face and side view of head of female; **d,** S6 of female; **e-g,** Male genitalia, S8, and S7. (Dorsal views at the left; tp is the tentorial pit; clear areas on the face are yellowish.) From Ruz, 1991.

58. Tribe Perditini

The tribe Perditini, all species of which have been placed, in the past, in the genus *Perdita,* consists of minute (2 mm long) to moderate-sized (10 mm) andreniform bees usually more robust than most Panurgini and Protandrenini, the metasoma, especially in females, flattened, and the terga sharply bent ventrad at each side as in *Homalictus* (Halictinae). Although an outstanding feature of the Perditini is the short, truncate marginal cell (Fig. 58-1), as shown in Figure 58-2 the cell is sometimes long enough that verbalizing the difference between it and that of some Panurgini and Protandrenini is difficult. Perditini tend to be sparsely haired, often with strong integumental color patterns. The integument may be black or metallic green or blue, commonly with yellow or white markings (Pl. 5), the metasoma or the whole body sometimes red, the pale markings variably expanded in some species until the whole body is yellow or white. The episternal groove is absent or curved, meeting the scrobe, and does not extend downward below the scrobal groove. The sting is extremely short, the first valvula lacking a valve and the first valvifer being linear rather than triangular. Interesting additional synapomorphic characters discovered by Danforth (1996) are (1) a longitudinal vertical septum inside the top of the head, behind the median ocellus, a feature unknown in other bees; (2) the lack of sternal graduli in females, and (3) the presence, in the anterolateral parts of terga, especially T5 of females and usually T6 of males, of a pale membranous line extending posteriad to or nearly to the spiracle. In other panurgines such a line is absent or does not reach the spiracle. Although in most Perditini the maxillary palpi are six-segmented, as they are in nearly all other Panurginae, in various Perditini these palpi are reduced, sometimes to only two segments; the whole range of variation occurs within the subgenus *Perdita* s. str., *P. (P.) halictoides* Smith having only two segments. The male genitalia and hidden sterna were illustrated in all of the revisional works of Timberlake (1954 to 1980a) as well as in Mitchell (1960), Ruz (1986), Snelling and Danforth (1992), and Danforth (1996). See Figures 58-7 and 58-8.

Species of this North American tribe are nearly all oligolectic, not on some one group of plants, but on diverse plant taxa as distant as Liliaceae, Asteraceae, and Fabaceae. This will be an ideal group for studies of floral host specificity, when the phylogeny of species and species groups is better understood.

Nests are burrows in the soil with branches, each branch ending in a single more or less horizontal cell. Many species are communal, several females using a single burrow. Torchio (1975), Danforth (1989b), Neff and Danforth (1992), and Norden, Krombein, and Danforth (1992) described nests of certain species and referred to papers on others; Rozen (1967a) provided an especially valuable review. One species, *Perdita maculigera maculipennis* Graenicher, nests shallowly in sand and appears to construct one cell at the end of each nest burrow instead of multicell nests (Michener and Ordway, 1963), but the observers found it difficult to follow the tiny burrows in sand and an observational error is possible. A de-

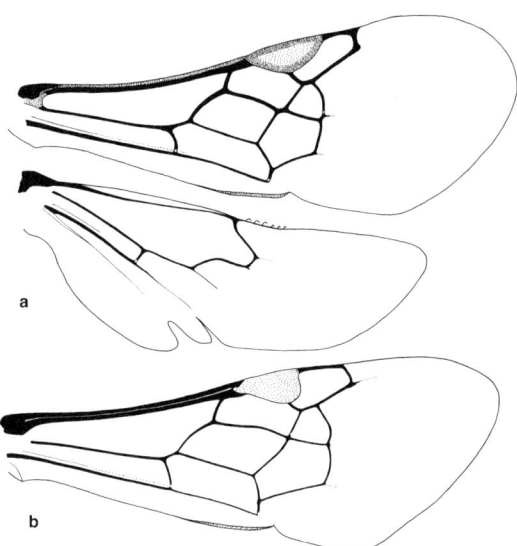

Figure 58-1. Wings of *Perdita*. **a,** *P. (Perdita) chihuahua* Timberlake; **b,** *P. (P.) maculigera maculipennis* Graenicher. The expanded stigma of *P. maculigera* is unique.

tailed study of the biology of *Perdita floridensis* Timberlake, some of whose nests are under water for five to six months of the year, was made by Norden et al. (2003).

A noteworthy tendency, especially in the genus *Macrotera* but also in some *Perdita*, is for males to have great variation in head size and shape, the larger males having relatively larger heads and longer mandibles (Figs. 58-3, 58-4). Such males appear to have an advantage in grasping females for mating, and a possible disadvantage

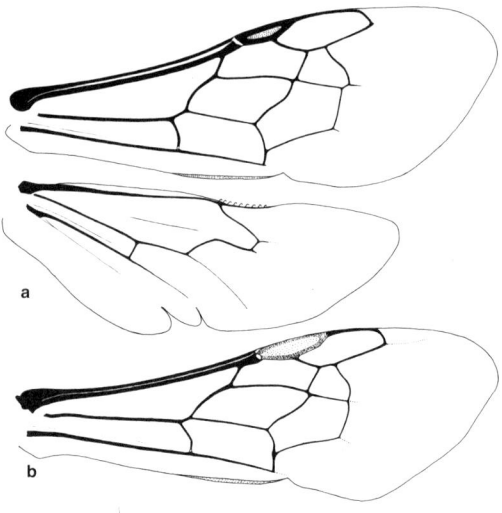

Figure 58-2. Wings of Perditini. **a,** *Macrotera (Macrotera) bicolor* Smith; **b,** *Perdita (Callomacrotera) acapulconis* Timberlake.

in the relatively small size of the wings. In *M. (Macroteropsis) portalis* (Timberlake) and *M. (Macroterella) mellea* (Timberlake) the males are dimorphic, the larger ones being flightless and unable to leave the nests (Danforth, 1991b; Norden, Krombein, and Danforth, 1992), as in some *Lasioglossum (Chilalictus)* in Australia (Halictinae). Such males, at least in *M. portalis,* mate with the females, presumably their close relatives, that are provisioning cells in the communal nest. Of course, a new nest will not have any flightless males, and females there must mate with "normal" males, presumably on flowers.

Key to the Genera of the Perditini

1. Outer groove of mandible broadened and fading away toward mandibular base (Fig. 58-5a); episternal groove usually curving posteriorly to scrobe (Fig. 58-6a) .... *Perdita*
—. Outer groove of mandible narrow basally (as it is apically) and bent upward, cutting diagonally across mandible to acetabulum (Fig. 58-5b); episternal groove absent or short, not curving posteriorly to scrobe (Fig. 58-6b) .... *Macrotera*

Genus *Macrotera* Smith

Macrotera consists of four subgenera commonly placed in the genus *Perdita*. Danforth (1991a) retained them in the genus *Perdita* but showed that they constitute the sister group of *Perdita* as it is here understood. In addition to the characters indicated in the key to subgenera, *Macrotera* has the following features. The stigma is slender, little if any broader than the prestigma, as measured to the wing margin, its margin within the marginal cell not or scarcely convex (Fig. 58-2a); the paraglossae of females are expanded into broad, brushlike apices; the body is nonmetallic, the yellow maculation usually absent except on the face, although the body may be honey-colored and the metasoma, at least of males, is commonly reddish; and the head of the male tends to be broad and quadrate, often highly variable in size within a population (Figs. 58-3, 58-4). Interestingly, the male genitalia and hidden sterna do not differ markedly between *Macrotera* and *Perdita;* as shown in Figure 58-7, the hidden sterna of *Macrotera* resemble those of some groups of *Perdita*.

Some interesting behavioral plesiomorphies retained by *Macrotera* and unknown in *Perdita* include the following. In species of *Macrotera* that have been studied, the cells are lined with a water-repellant secretion, as in most ground-nesting bees; in *Perdita* the cells are very incompletely lined or not lined at all. In *Macrotera,* however, the pollen ball in each cell is not covered by a water-repellent secretion, whereas in *Perdita*, which lacks the cell lining, the food mass is so covered. In *Macrotera* the feces of larvae are applied to the inner part of the cell, as in many other ground-nesting bees, including other Panurginae. In *Perdita* the feces of larvae are deposited on the venter of the larva, and thus kept away from the cell wall, possibly keeping them dry and reducing their potential as a source of fungal development. References to literature on these topics are listed under the tribe Perditini.

Key to the Subgenera of *Macrotera*

1. Facial fovea well developed and elongate; head of male usually enlarged, subquadrate, wider than long, inner orbits more or less divergent below; head of female as long as wide or slightly longer than wide; marginal cell about three times as long as wide, subequal in length to first submarginal cell 2
—. Facial fovea small, oval or punctiform; head in both sexes sometimes wider than long but less quadrate, usually not much enlarged in male and inner orbits not divergent downward; marginal cell usually about twice as long as wide, yet shorter than or subequal to first submarginal cell 3
2(1). Glossa reaching hind coxae in repose, two to three times as long as prementum; head of male much broader than long, widened anteriorly; mandible of male inserted below eye, only posterior articulation behind posterior eye margin; metasoma of male lacking special modifications except for broad, depressed form; tibial spurs briefly hooked at apices and not serrate on inner margins .... *M. (Macrotera* s. str.)
—. Glossa not over twice as long as prementum; head of male a little broader than long, only slightly widened anteriorly; mandible of male inserted partly behind posterior margin of eye; T7 and one or more metasomal sterna modified; tibial spurs strongly curved at apices and minutely serrate on inner margins ....*M. (Cockerellula)*
3(1). Head of male broader than long, rounded on sides and above, but truncate anteriorly, thus appearing almost semicircular in frontal view; mandible of male usually with half of base behind posterior margin of eye, either bidentate or simple at apex; stigma slender and tapering; maxillary palpus four- to six-segmented, apical segments short and more or less indistinct .... *M. (Macroteropsis)*
—. Head of male not enlarged, ordinary in shape although more or less wider than long; mandible of male rather short, simple, tapering, and inserted largely below eye, only posterior articulation behind posterior eye margin; stigma small and usually rounded posteriorly, but varying to very small and lanceolate; maxillary palpus with six distinct segments .... *M. (Macroterella)*

Macrotera / Subgenus *Cockerellula* Strand

Perdita (Lutziella) Cockerell, 1922e: 1 (not Enderlein, 1922). Type species: *Perdita opuntiae* Cockerell, 1922, by original designation.
Cockerellula Strand, 1932: 196, replacement for *Lutziella* Cockerell, 1922. Type species: *Perdita opuntiae* Cockerell, 1922, autobasic.

This subgenus is related to *Macrotera* s. str. but differs in its smaller size (3.5-7.0 mm long) and shorter proboscis, the glossa in repose only reaching the front coxae or slightly beyond, whereas in *Macrotera* s. str. it reaches the hind coxae.

■ *Cockerellula* occurs from North Dakota south and southwest through western Texas to Arizona, USA, and on to Baja California Sur and Puebla in Mexico. There are 13 species. Keys and revisions were by Timberlake (1953b, 1954, 1960, 1968) and Danforth (1996).

Figure 58-3. Small-headed male of *Macrotera portalis* (Timberlake). Drawing by N. Florenskya, from Danforth, 1996.

So far as is known, as for *Macrotera* s. str., all species use *Opuntia* as the pollen source. *M. (C.) opuntiae* (Cockerell) is well known as the bee that makes its burrows in sandstone (Custer, 1928; Bennett and Breed, 1985).

Macrotera / Subgenus *Macrotera* Smith s. str.

Macrotera Smith, 1853: 130. Type species: *Macrotera bicolor* Smith, 1853, monobasic.

Macrotera s. str. differs from other subgenera in its large size, the body length being 6 to 10 mm. The stigma is slender, about as broad as the prestigma, as measured to the wing margin (Fig. 58-2a).

- This subgenus is found from Oklahoma to southern Texas and New Mexico, USA, and south to the states of Jalisco, Michoacan, and Puebla in Mexico. The six species were revised by Timberlake (1958) and Snelling and Danforth (1992).

Neff and Danforth (1992) and Danforth and Neff (1992) gave accounts of the biology of *Macrotera texana* (Cresson). Its nests are communal. In spite of great variation in male head size, there is a continuum between the extremes, and, unlike those of *M. (Macroteropsis) portalis*, the largest males can fly. All species of *Macrotera* s. str. appear to depend on *Opuntia* for pollen.

Macrotera / Subgenus *Macroterella* Timberlake

Perdita (Macroterella) Timberlake, 1954: 360. Type species: *Perdita mortuaria* Timberlake, 1954, by original designation.

Macroterella includes the smallest species of the genus *Macrotera*, 3.0 to 4.5 mm long. The glossa is shorter than in other *Macrotera*, not longer than the prementum, and sometimes, e.g., in *M. (M.) mellea* (Timberlake), much shorter. The mandibles of the female are usually slender and simple, but in *M. mellea* are wider and have a preapical convexity.

- This subgenus ranges from New Mexico to southern California, USA, and the Mexican states of Chihuahua and Baja California. The six species were revised by Timberlake (1954, 1968) and Danforth (1996).

The males of *Macrotera mellea* (Timberlake) and perhaps other species are dimorphic, like those of *M. (Macroteropsis) portalis* (Timberlake) (Rozen, 1970c).

Macrotera / Subgenus *Macroteropsis* Ashmead

Macroteropsis Ashmead, 1899a: 85. Type species: *Perdita latior* Cockerell, 1896, by original designation.

This is a subgenus of rather small species, 3.5 to 6.0 mm long. Striking characters include the origin of the second segment of the labial palpus subapically on the first, which tapers to an acute point, and the reduction or loss of distal segments of the maxillary palpus. The frons and thorax have a weak metallic sheen, unlike those of other *Macrotera*. The stigma is slender for Perditini, only

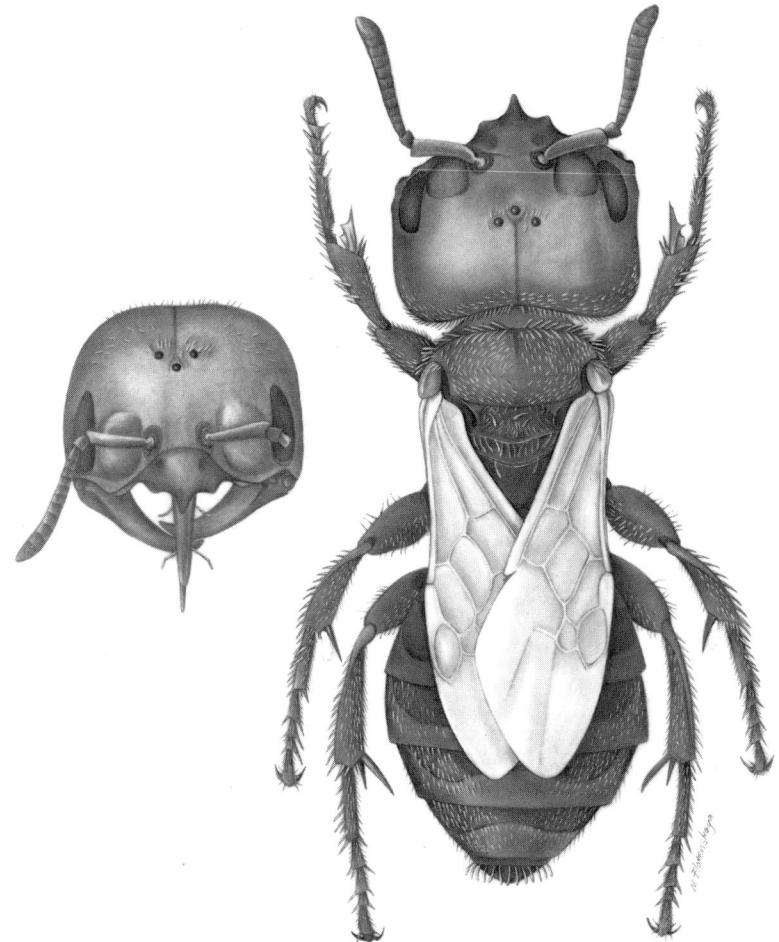

Figure 58-4. Large-headed male of *Macrotera portalis* (Timberlake). Compare with Figure 56-3. This form is flightless and does not leave the nest. Drawing by N. Florenskya, from Danforth, 1996.

slightly wider than the prestigma, as measured to the wing margin.

■ *Macroteropsis* occurs in desert areas from western Texas to Nevada and California, USA, south to Zacatecas in Mexico. The six species were revised by Timberlake (1954, 1962, 1980a) and by Danforth (1996).

The biology of *Macrotera portalis* Timberlake was discussed by Danforth (1991b, c); it is the best known of the species having dimorphic males, one morph being large-headed and flightless (see the discussion of the tribe Perditini).

Genus *Perdita* Smith

This genus is characterized in the key to genera and in the comments on *Macrotera*. In the great majority of species, the stigma is large, its margin within the marginal cell is convex, and an episternal groove curves posteriorly to the scrobe. Two of the most interesting synapomorphies of *Perdita* are the behavioral ones described under *Macrotera;* of course they have been verified for only a few species. Male genitalia and hidden sterna were illustrated by Mitchell (1960) and in the revisional works by Timberlake (1954 to 1980a); see also Figures 58-7 and 58-8. Figure 58-9 illustrates the general features of the genus.

Perdita, an enormous genus, is most abundant and diversified in the arid southwestern USA and northern Mexico, but limited numbers of species occur to the Atlantic and Pacific coasts, north to southern Canada, and one species occurs south to Guatemala. One undescribed species is reported from Hispaniola (G. Eickwort, personal communication, 1992), but otherwise the genus is unknown from the Antilles.

Although most species of Perditini, as well as all those in related tribes, have two submarginal cells, certain small groups of the genus *Perdita,* that is, the seemingly distantly related subgenera *Alloperdita* and *Xerophasma,* have a small, triangular, petiolate intercalary cell (Fig. 58-10) between the first and second submarginal cells. In one sense, such forms have three submarginal cells. I strongly suspect, however, that the intercalary cell is not homologous to the second submarginal cell of bees having three ordinary submarginal cells. Because the two subgenera have no close relatives with three ordinary cells, I believe that the intercalary cell is a derived feature, resulting from the splitting of a vein, rather than a plesiomorphic retention of a vein.

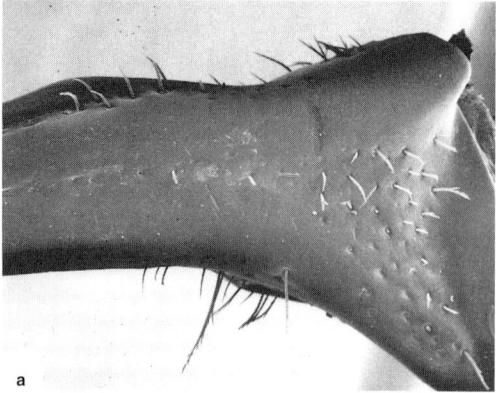

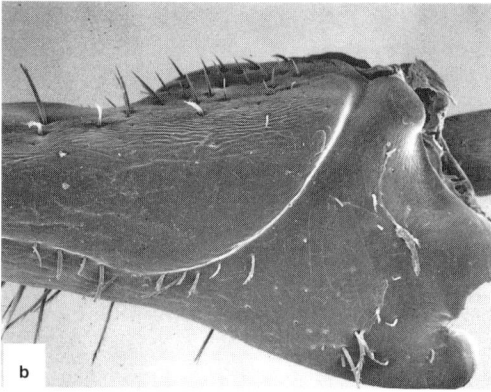

Figure 58-5. Outer surfaces of male mandibles (apices omitted) of Perditini. **a**, *Perdita octomaculata* (Say); **b**, *Macrotera texana* Cresson. From Danforth, 1996.

Timberlake (1954) explained that the subgenera *Callomacrotera, Cockerellia, Hexaperdita, Pentaperdita, Procockerellia,* and *Xeromacrotera* are closely related, as shown by their unusually copious scopa, the simple claws in the female, the bilobed base of S8 of the males (as in *Heteroperdita*), and other characters. In these subgenera, pollen is carried partly dry in the scopa, not moistened and stuck together (with nectar?) as is the case with most *Perdita*. Danforth (1991a), in a cladistic analysis, found that these same subgenera (except *Xeromacrotera,* which he did not have) constituted a monophyletic group. Future workers might well consider uniting these six subgenera under a single name.

Key to the Subgenera of *Perdita*
(Modified from Timberlake, 1954)

1. Ocelli large, ocellocular distance less than half ocellar diameter; body wholly pallid; intercalary cell present between first and second submarginal cells (Fig. 58-10) .... *P. (Xerophasma)*
—. Ocelli of ordinary size, ocellocular distance much greater than ocellar diameter; if body pallid, then intercalary cell absent 2
2(1). Stigma small and slender, width about or not much over half maximum width of marginal cell and less than width of apical truncation of that cell; basitibial plate more or less distinct, at least in the male 3
—. Stigma more or less large and broad, width ordinarily more than half width of marginal cell and as wide as apical truncation of that cell; basitibial plate usually absent or poorly developed, but sometimes well developed in both sexes, as in *Pygoperdita*.. 7
3(2). Metasoma broader than thorax and especially broad and depressed in male; hair of hind tibia and basitarsus of male unusually abundant, fine and appressed; basitibial plate developed in both sexes (head of male often greatly enlarged, facial fovea about three times as long as wide but weakly impressed; mandible of male with a broad biangulate or bidentate expansion on inner margin) .... *P. (Pseudomacrotera)*
—. Metasoma not broadened in female, and at most only moderately depressed and broadened in male; hair of hind tibia and basitarsus of male more or less long, curved, erect, and not much thickened by short, densely set branches; basitibial plate little developed in female 4
4(3). Marginal cell more or less shorter and smaller than first submarginal cell and squarely truncate at apex; stigma small and slender, but more or less rounded on posterior margin; facial fovea small and inconspicuous (very small bees with a satiny luster and appressed white pubescence; mandible simple and tapering in both sexes) .... *P. (Heteroperdita)*
—. Marginal cell subequal to first submarginal, apex a little obliquely truncate; stigma more or less narrowly lanceolate; facial fovea distinct................................. 5
5(4). Glossa elongate, reaching middle coxae in repose (facial fovea of female considerably widened at lower end; metasoma of male at most moderately broadened; mandible of male simple) ...... *P. (Glossoperdita)*
—. Glossa ordinary, not reaching much behind front coxae in repose 6
6(5). Stigma slender and tapering, hardly wide enough to show central pale streak; mandible of both sexes with a blunt inner tooth near apex; facial fovea of female slightly widened below; metasoma of male not broadened; front and middle femora of male subincrassate and apical segments of tarsi unusually large, with large claws .... *P. (Hesperoperdita)*
—. Stigma somewhat less slender; mandible of female with a small inner tooth, that of male usually simple and tapering; facial fovea of female very narrow; metasoma of male moderately depressed and broadened; legs of male ordinary .... *P. (Epimacrotera)*
7(2). Scopa of hind tibia copious, hairs long, often more or less crinkly or minutely branched, seemingly adapted for carrying pollen at least partly dry; claws of female simple; side of pronotum of male more or less deeply furrowed (except in *Allomacrotera*); S8 of male with bilobate base 8
—. Scopa of hind tibia composed of long, curved, simple, widely spaced hairs, seemingly adapted for carrying agglutinated pollen; claws of female more or less distinctly dentate within; side of pronotum of male usually not grooved; S8 of male with simple, median spiculum at base 14
8(7). Usually large species, 5 to 9 mm long; mandible of female dilated on inner margin and abruptly bent inward before apex, incurved part tapering and simple; metasoma of male no wider than thorax; claws of male hind

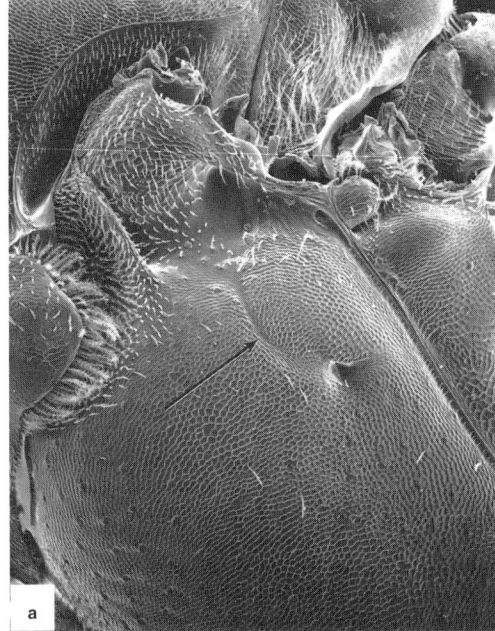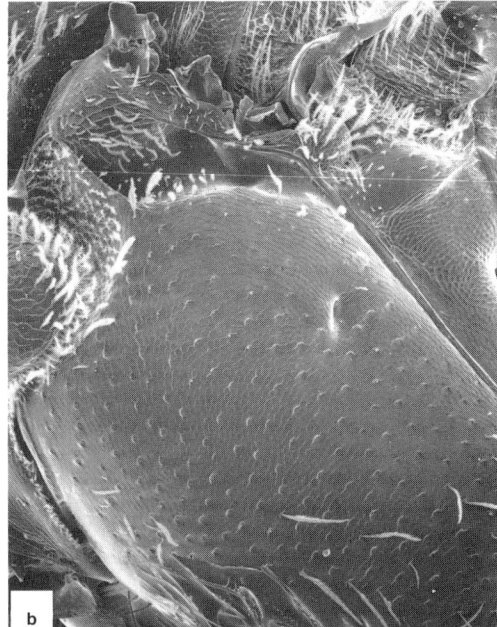

leg simple (except in *Allomacrotera* where cleft) 9
—. Smaller species, rarely exceeding 6 mm in length; mandible of female ordinary or only moderately expanded within, and but little incurved at apex; metasoma of male usually moderately broad and depressed; claws of male hind leg cleft 11
9(8). Claws of male hind leg cleft; side of pronotum not furrowed (maxillary palpus three-segmented) ....
.... *P. (Allomacrotera)*
—. Claws of male hind leg simple; side of pronotum more or less deeply furrowed 10
10(9). Maxillary palpus six-segmented; stigma moderately wide, sometimes less than half as wide as marginal cell; thorax green, mesonotum and scutellum never black but sometimes yellow; mandible of male long, slender, tapering, and curved .... *P. (Cockerellia)*
—. Maxillary palpus three- to five-segmented; stigma considerably larger and broader, more than half as wide to fully as wide as marginal cell; thorax green, disc of mesoscutum and scutellum black; mandible of male long, simple, abruptly bent inward before middle, slightly dilated beyond bend... *P. (Procockerellia)*
11(8). Maxillary palpus five-segmented (possibly indistinctly six-segmented in *Xeromacrotera*) 12
—. Maxillary palpus six-segmented 13
12(11). Body metallic blue or green, usually with white facial marks (tibial scopa of female unusually copious; metasoma of male depressed, considerably broader than thorax and usually without light markings) ....
.... *P. (Pentaperdita)*
—. Body yellow with black or nearly black markings on head and thorax [head of male very large, quadrate; malar area short but evident (almost obliterated in allies); disc of clypeus small, much narrower than lateral extensions of enormous paraocular areas] .... *P. (Xeromacrotera)*
13(11). Body large (length 6.5-9.0 mm), robust; metasoma with yellow bands and a definite pygidial plate in both

Figure 58-6. Lateral views of upper part of mesepisternum of Perditini. **a,** *Perdita octomaculata* (Say); **b,** *Macrotera portalis* (Timberlake). *Perdita* has the episternal and scrobal groove (arrow), lacking in *Macrotera*. From Danforth, 1996; in his Figure 7 the legend for these two photographs is reversed.

sexes; mandible of female having triangular process on inner margin about one-fourth of length from base; tibial scopa copious for a *Perdita* and distinctly plumose ....
.... *P. (Callomacrotera)*
—. Body small, of ordinary form; metasoma without distinct light markings, not subpygidiform at apex; mandible of female more or less dilated within; tibial scopa moderately copious, hairs simple and curved except on dorsal margin where they are mosslike with close-set, very short branches .... *P. (Hexaperdita)*
14(7). Marginal cell extremely short and oblique behind short, broad stigma; second submarginal cell triangular, first and second submarginal crossveins uniting at marginal cell (clypeus of male produced to sharp tooth or slender lobe on each side of base of labrum) ....
.... *P. (Perditella)*
—. Marginal cell never excessively shortened and oblique; second submarginal cell not triangular.......................... 15
15(14). Small intercalary cell usually present between first and second submarginal cells (as in Fig. 56-10); mandible of male with small subapical tooth on inner margin .... *P. (Alloperdita)*
—. Intercalary cell never present; mandible of male usually simple and acute at apex 16
16(15). T7 of male strongly emarginate on each side, setting off median pygidiform process or deeply bilobed process, or modifications thereof (female robust; head broader than long; margin of clypeus dentate on each side of labrum) .... *P. (Pygoperdita)*
—. T7 of male usually simple, tapering, or, *if* somewhat

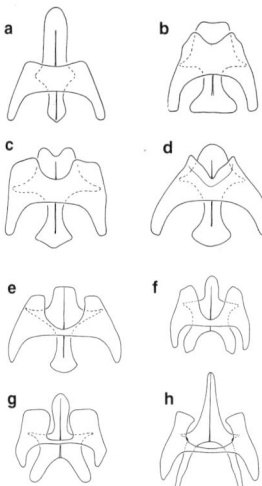

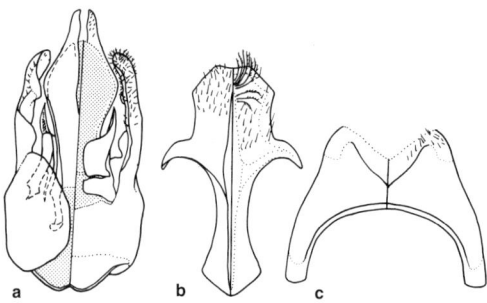

Figure 58-8. Male genitalia, S8, and S7 of *Perdita (Perdita) halictoides* Smith. (Dorsal views are at the left.) From Ruz, 1986.

Figure 58-7. Diagrams of ventral views of S7 and S8 of male Perditini; S8 is indicated in part by broken lines behind S7. **a**, *Macrotera (Macrotera) texana* Cresson; **b**, *Perdita (Epimacrotera) euphorbiae* Timberlake; **c**, *P. (Perdita) sphaeralceae* Cockerell; **d**, *P. (P.) octomaculata* (Say); **e**, *P. (P.) zonalis* Cresson; **f**, *P. (Hexaperdita) bishoppi* Cockerell; **g**, *P. (Cockerellia) coreopsidis* Cockerell; **h**, *P. (Procockerellia) albonotata* Timberlake. Note that in f, g, and h, S8 is basally emarginate, in contrast to the basal spiculum of the others. From Danforth, 1996.

modified, then never deeply emarginate on each side nor strongly bilobed at apex .... *P. (Perdita s. str.)*

Perdita / Subgenus *Allomacrotera* Timberlake

Perdita (Allomacrotera) Timberlake, 1960: 131. Type species: *Perdita stephanomeriae* Timberlake, 1954, by original designation.

This is a close relative of *Procockerellia*, from which it differs by the cleft hind claws of the male and the lack of lateral furrows on the pronotum. Body length is 5.5 to 6.0 mm.

■ *Allomacrotera* is found in southern California and Utah, USA. It includes two species. Timberlake (1971) gave a key to species of *Procockerellia* and *Allomacrotera*.

Perdita / Subgenus *Alloperdita* Viereck

Perdita (Alloperdita) Viereck, 1917b: 241. Type species: *Perdita novaeangliae* Viereck, 1907, monobasic.

Like those of *Xerophasma*, bees of this subgenus have an intercalary submarginal cell in the forewing. This character distinguishes *Alloperdita* from *Perdita* s. str., but the intercalary cell varies in size and is absent in an occasional individual. There may be no other character to separate *Alloperdita* from *Perdita* s. str. reliably, in which case *Alloperdita* should become a synonym. The size is like that of ordinary *Perdita*, 4 to 6 mm in body length.

■ This subgenus ranges from Massachusetts to Florida west to Texas, USA, and to Coahuila and Chihuahua, Mexico. All records in the USA are from coastal states. The six species were revised by Timberlake (1956).

Perdita / Subgenus *Callomacrotera* Timberlake

Perdita (Callomacrotera) Timberlake, 1954: 413. Type species: *Perdita maritima* Timberlake, 1954, by original designation.

This subgenus consists of large species, 6.5 to 9.0 mm long, the head and thorax of which is dark green or blue green; the metasoma is blackish with yellow bands in females, largely yellow in some males. The transverse head, broad metasoma, and unusually short maxillary palpi are characteristic.

■ *Callomacrotera* is known only from the Pacific coast of Mexico, from Guerrero to Nayarit. There are two species; see Timberlake (1954).

Perdita / Subgenus *Cockerellia* Ashmead

Cockerellia Ashmead, 1898: 284. Type species: *Perdita hyalina* Cresson, 1878 = *Perdita albipennis* Cresson, 1865, by original designation.
Philoxanthus Ashmead, 1898: 285. Type species: *Perdita beata* Cockerell, 1895, by original designation.

This subgenus consists of rather large species, 6 to 9 mm long, ranging from almost wholly yellow to the pattern of the commoner species, with broad yellow metasomal bands and metallic green head and thorax; rarely, the metasoma is entirely dark. As in *Procockerellia*, the claws of the female are simple, those of the male cleft on the front and middle legs, simple on the hind legs.

■ *Cockerellia* occurs in the southern prairie provinces of Canada, and from Idaho to New Jersey, USA, south to California and Georgia, USA, and to Veracruz and Zacatecas, Mexico. The 25 species were revised by Timberlake (1954), who also gave a separate key for the identification of females (Timberlake, 1953b).

All the species collect pollen from large-flowered Asteraceae, such as *Helianthus*.

Perdita / Subgenus *Epimacrotera* Timberlake

Perdita (Epimacrotera) Timberlake, 1954: 377. Type species: *Perdita ainsliei* Crawford, 1932, by original designation.

Epimacrotera consists of small species (2.5-5.0 mm long) closely related to *Glossoperdita* except for the short proboscis, the glossa in repose extending little behind the head. It seems likely that the two subgenera should be

united. *Epimacrotera* is much like typical *Perdita* but has a more slender stigma and well-defined basitibial plates.
■ This subgenus is known from southern California to western Texas, USA, and south to Baja California and Oaxaca, Mexico, with a possibly isolated species in Iowa, USA. There are 18 species. Timberlake (1954, 1958, 1962, 1968, 1980a) revised the subgenus.

Perdita / Subgenus *Glossoperdita* Cockerell

Glossoperdita Cockerell, 1916c: 43. Type species: *Glossoperdita pelargoides* Cockerell, 1916, monobasic.

This is a subgenus of small *Perdita* species, 3.5 to 5.0 mm long. The head and thorax are dark blue green, the metasoma orange. The head is somewhat longer than wide and the mouthparts are unusually long.
■ *Glossoperdita* is known in southern California and Arizona, USA, and in Sonora, Mexico. The four species were revised by Timberlake (1954, 1958).

All the species collect pollen from *Gilia* and related genera of Polemoniaceae.

Perdita / Subgenus *Hesperoperdita* Timberlake

Perdita (Hesperoperdita) Timberlake, 1954: 374. Type species: *Perdita ruficauda* Cockerell, 1916, by original designation.

This subgenus consists of species with a bluish-green sheen on the head and thorax and a red or orange metasoma, the metasoma sometimes with yellow bands. The body length is 4 to 6 mm. The wings are unusually small. Unlike several other subgenera with the stigma slender (and the genus *Macrotera*), male heads so far as is known are not enlarged, that is, no large-headed male morph is found on the flowers.
■ This subgenus is found in California, USA, and Baja California, Mexico, not in deserts. There are three species. Timberlake (1954) revised the group. So far as is known, the subgenus is dependent for pollen on the genus *Lotus* (Fabaceae).

Perdita / Subgenus *Heteroperdita* Timberlake

Perdita (Heteroperdita) Timberlake, 1954: 365. Type species: *Perdita rhodogastra* Timberlake, 1954, by original designation.

Heteroperdita contains minute species (length 2.5-4.0 mm) with extremely fine sculpturing, which produces a satiny sheen, and, usually, an appressed white pubescence. The head and thorax are dark metallic green, the metasoma whitish to orange or with white bands on an orange background. The mandibles are simple, acute, and the claws are almost simple, the inner tooth being rudimentary. The scopa consists of sparse, long hairs with short branches.
■ *Heteroperdita* ranges from New Mexico to southern California, USA, southeast to Coahuila and Durango, Mexico. The 13 species were revised by Timberlake (1954, 1968).

Probably all species collect pollen exclusively from *Coldenia* and *Heliotropium* (Boraginaceae).

Perdita / Subgenus *Hexaperdita* Timberlake

Perdita (Hexaperdita) Timberlake, 1954: 416. Type species: *Perdita ignota* Cockerell, 1896, by original designation.

This is a group of relatively small species (body length 3.5-7.0 mm) that are similar to *Pentaperdita* in many ways, but the maxillary palpi six-segmented and the mandibles of the females are weakly incurved at the apices and only weakly expanded on their inner margins.
■ *Hexaperdita* ranges from Florida to New Jersey west to Colorado and southern California, USA, and south to the states of Baja California, Jalisco, Zacatecas, and Veracruz, Mexico. The 29 species were revised by Timberlake (1956). All of them collect pollen from small-flowered Asteraceae such as *Heterotheca* and *Chrysopsis*.

Perdita / Subgenus *Pentaperdita* Cockerell and Porter

Perdita (Pentaperdita) Cockerell and Porter, 1899: 414. Type species: *Perdita albovittata* Cockerell, 1895, monobasic.

These relatives of *Cockerellia* differ from that subgenus in having five-segmented maxillary palpi, and, in the male, a broad metasoma and cleft claws on the posterior legs, like those of the other legs. They are smaller than most *Cockerellia;* the body length is 3.25 to 6.50 mm.
■ This subgenus occurs from Texas west to southern California, USA, and south to Coahuila, Zacatecas, and Durango, Mexico. The 13 species were revised by Timberlake (1954, 1956, 1958, 1868).

Perdita / Subgenus *Perdita* Smith s. str.

Perdita Smith, 1853: 128. Type species: *Perdita halictoides* Smith, 1853, monobasic.
Neoperdita Ashmead, 1899a: 85. Type species: *Perdita zebrata* Cresson, 1878, by original designation.
Perdita (Geoperdita) Cockerell and Porter, 1899: 415. Type species: *Perdita chamaesarachae* Cockerell, 1896, monobasic.
Perdita (Tetraperdita) Cockerell and Porter, 1899: 415. Type species: *Perdita sexmaculata* Cockerell, 1895, monobasic.
Zaperdita Robertson, 1918: 91. Type species: *Perdita maura* Cockerell, 1901 = *Perdita halictoides* Smith, 1853, by original designation.

This subgenus contains the majority of the species of the genus, and there is much variability among them. For example, the number of segments of the maxillary palpi varies from two to six. The head and thorax are blue to green, rarely almost without metallic tints, to wholly yellow. Body length varies from 3 to 8 mm. The general form of this and other subgenera is illustrated by Figure 58-9 and Plate 5.
■ This subgenus ranges from British Columbia to Ontario, Canada, south to Florida and California, USA, and on through Mexico to Guatemala. The 441 species were revised in various groups by Timberlake (1958, 1960, 1962, 1964a, 1968, 1971, 1980a).

The species of this subgenus are oligolectic on a very wide diversity of different flowers.

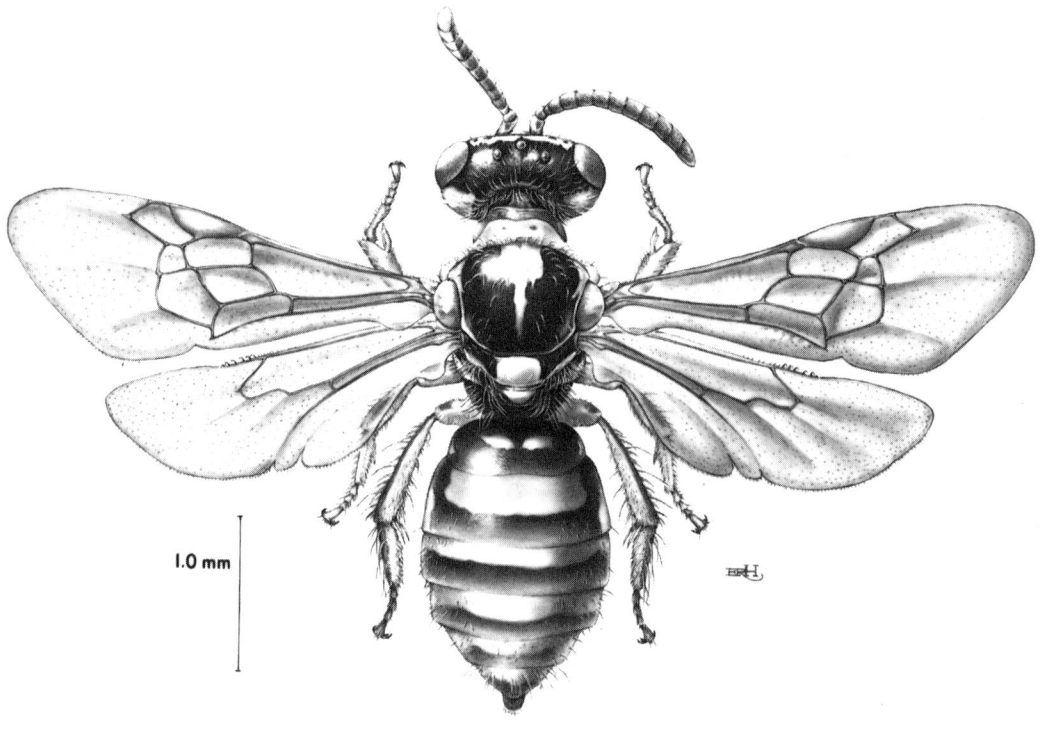

Figure 58-9. *Perdita (Perdita) stathamae* Timberlake, male. Drawing by E. R. S. Hodges, from Michener, McGinley, and Danforth, 1994.

Perdita / Subgenus *Perditella* Cockerell

Perdita (Perditella) Cockerell, 1899b: 312. Type species: *Perdita laneae* Cockerell, 1899, error for *P. larreae* Cockerell, 1896, by original designation.

This subgenus includes minute species, 2.0 to 3.5 mm long. The principal characters, those of the wing venation, may be convergent results of minute size rather than indications of close relationship (see Danforth, 1989a). The stigma is large, about as broad as the first submarginal cell; the marginal cell is extremely short and tilted at an angle to the axis of the submarginal cells.

■ This subgenus ranges in deserts from western Texas to southern California, USA, and south to Coahuila, Zacatecas, and Baja California, Mexico. There are seven species. Timberlake (1956) revised the species then known.

The species are oligolectic on plants of various families, such as Zygophyllaceae and Euphorbiaceae.

Perdita / Subgenus *Procockerellia* Timberlake

Perdita (Procockerellia) Timberlake, 1954: 402. Type species: *Perdita albonotata* Timberlake, 1954, by original designation.

In *Procockerellia* the apex of the mandible of the female is less strongly hooked than that of *Cockerellia*, but the principal differentiating characters are the longer glossa, reaching the apices of the front coxae in repose; the reduced maxillary palpi, three- to five-segmented instead of six-segmented; and the black disc of the scutum. Body length is 6 to 8 mm.

■ This subgenus is known from southern California, Arizona, and Utah, USA, and Sonora, Mexico. The five species were revised by Timberlake (1954, 1971).

These species may all be specialists on *Stephanomeria* (Asteraceae).

Perdita / Subgenus *Pseudomacrotera* Timberlake

Perdita (Pseudomacrotera) Timberlake, 1954: 349. Type species: *Perdita turgiceps* Timberlake, 1954, by original designation.

This subgenus has somewhat the aspect of a small (length 5.0-7.5 mm) *Macrotera* s. str., including great variability in male size and especially in male head size, but males with impressively large heads seem to be the common morph in collections. The broad metasoma of the male is also suggestive of *Macrotera*. The head and thorax are tinged with dark metallic green, unlike those of *Macrotera*. S6 of the male is greatly inflated and protruding.

■ *Pseudomacrotera* occurs in the deserts of Arizona and southern California, USA, and Baja California, Mexico. The only species is *Perdita turgiceps* Timberlake.

Although many specimens have been collected on diverse kinds of flowers in southern California, the pollen source remains unknown.

Perdita / Subgenus *Pygoperdita* Timberlake

Perdita (Pygoperdita) Timberlake, 1956: 275. Type species: *Perdita interrupta* Cresson, 1878, by original designation.

This subgenus closely resembles *Perdita* s. str. except for the characters indicated in the key to subgenera. The basitibial plate is usually more clearly defined than in *Perdita* s. str. Body length is 3 to 9 mm.

■ *Pygoperdita* is most abundant in California but ranges to Utah, Nebraska, Colorado, and Arizona, USA, British Colombia, Canada, and south to Baja California, Mexico. Timberlake (1956) revised the 43 species, supplementing the key in 1958.

Perdita / Subgenus *Xeromacrotera* Timberlake

Perdita (Xeromacrotera) Timberlake, 1954: 412. Type species: *Macrotera cephalotes* Cresson, 1878, by original designation.

The lone species of this subgenus shares most of the characters of *Pentaperdita* but is large (length about 6 mm) and largely yellow, without metallic coloration.

■ *Xeromacrotera* is known from Arizona, Nevada, and southern California, USA. There is only one species, *Perdita cephalotes* (Cresson).

Perdita / Subgenus *Xerophasma* Cockerell

Xerophasma Cockerell, 1923a: 1. Type species: *Xerophasma bequaerti* Cockerell, 1923 = *Perdita bequaertiana* Cockerell, 1951, by original designation.

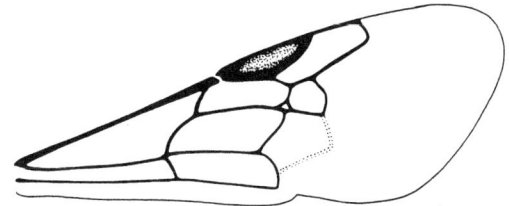

Figure 58-10. Forewing of *Perdita (Xerophasma) bequaertiana* Cockerell showing the intercalary cell between the two submarginal cells.

This subgenus contains large (8-9 mm long) pallid nocturnal species. Like species of *Alloperdita,* they have an intercalary cell in the forewing (Fig. 58-10).

■ The range is desert areas from western Texas to southern California. The two species were revised by Timberlake (1954).

Species of this subgenus, unlike all other *Perdita* species, and perhaps unlike all other Andrenidae, are nocturnal. They collect pollen from flowers of *Oenothera* and perhaps related genera of Onagraceae.

59. Tribe Calliopsini

The unity of this tribe is most clearly shown by the male genitalia, in which the gonostylus is reduced to a small hairy projection *(Acamptopoeum,* Fig. 59-3a; *Arhysosage)* or hairy area (in *Spinoliella,* Fig. 59-7d) or, more commonly, is absent (Fig. 59-3e, h); the gonocoxite is large and convex, often subglobose; and the penis valve is enormous and complex. The unity of the tribe, or actually of the genus *Calliopsis* as here understood, is further supported by the maculate faces of both sexes (Fig. 59-1) and the chemistry of the Dufour's gland products (Cane, 1983a). Other characters are the short episternal groove, curving into the scrobal suture and not extending below it and the two submarginal cells (Fig. 59-2). The body is somewhat robust, i.e., it has the form of *Andrena,* not more slender and parallel-sided as in many Protandrenini. The body is nonmetallic (rarely slightly bluish), and usually has yellow or cream-colored areas, at least on the face and often on other tagmata.

This is a Western Hemisphere tribe, well represented in both northern and southern temperate zones, especially xeric areas, but is nearly absent in the tropics. As for other andrenids, the nests are burrows in the soil, sometimes in aggregations. Each burrow is occupied by a single female. The more or less horizontal cells are at least partly lined with a thin, secreted, shiny "waxlike" lining. The food mass in each cell is spherical and covered with

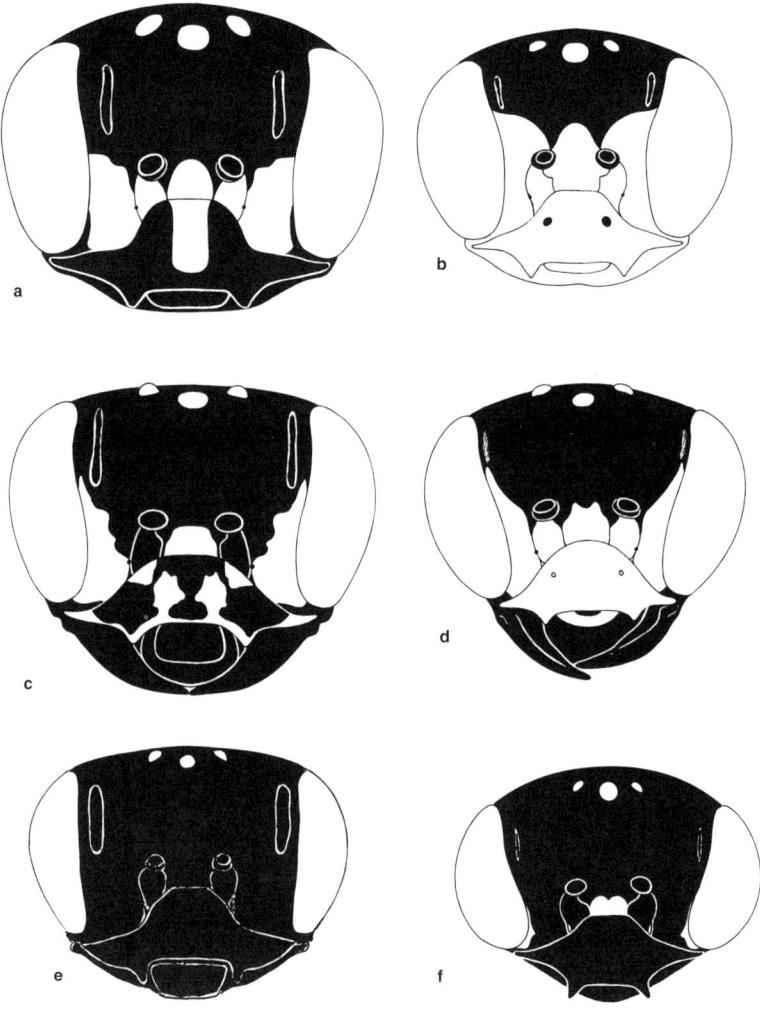

Figure 59-1. Faces of *Calliopsis*, in each case female and male. **a, b,** *C. (Calliopsis) andreniformis* Smith; **c, d,** *C. (Ceroliopoeum) laeta* (Vachal); **e, f,** *C. (Leiopoeodes) xenopous* Ruz. Clear areas are white or yellow. From Ruz, 1991.

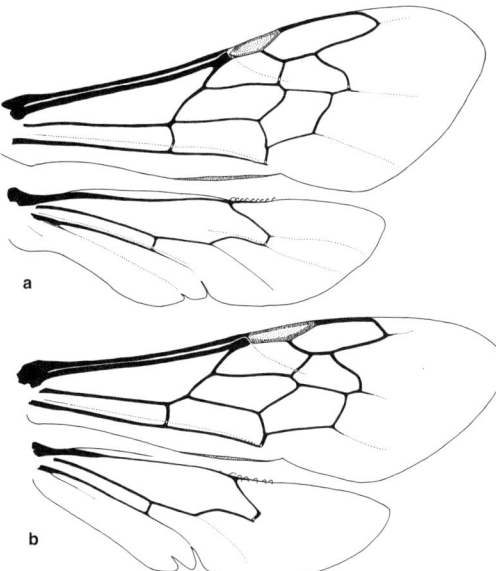

Figure 59-2. Wings of Calliopsini. **a,** *Calliopsis andreniformis* Smith; **b,** *Callonychium minutum* (Friese).

similar waterproofing, a feature duplicated, so far as is known, only in the Perditini. Details of nesting biology are provided by Rozen (1967a, 1970b), Shinn (1967), Danforth (1990), and Visscher and Danforth (1993).

The genera of this tribe were reviewed, and genitalic and other structures were illustrated, by Ruz (1986, 1991).

Key to the Genera of the Calliopsini (Males)
(Modified from Ruz, 1991)

1. Inner orbits convergent below; tentorial pit clearly below middle of outer subantennal suture (Fig. 59-1) 2
—. Inner orbits subparallel or divergent below; tentorial pit near middle of outer subantennal suture 4
2(1). T7 with median smooth area distally, area delimited by hairs and tapered at apex (S4 and S5 with distal margins straight or nearly so, without median projections) ....*Acamptopoeum*
—. T7 with pygidial plate rounded or truncate at apex, usually delimited by ridges or carinae laterally ....
....3
3(2). S4 with distal margin usually produced in middle, its surface broadly concave mesally; S5 with median projection (sometimes inconspicuous) on distal margin ....
....*Calliopsis*
—. S4 and S5 unmodified, transverse, without projections or concavity .... *Litocalliopsis*
4(1). Orbits strongly divergent below; lower paraocular area flat; most of body yellow; hind tibial spurs strongly curved; metasoma much wider than head; body length about 10 mm or nearly so (South America) ...... *Arhysosage*
—. Orbits usually subparallel, but if not, then lower paraocular area very swollen mesally; body with yellow markings but usually predominantly black; inner hind tibial spur almost straight, outer spur straight or somewhat curved; metasoma as broad as head; body length 4 to 7 mm.. 5
5(4). Metasoma at apex slightly curved or straight; paraocular area yellow only on lower part; clypeus (in ventral view) with projection or convexity of lateral, downward-directed part beside lateral margin of labrum; antennal socket (lower margin) usually at lower third of face (South America) .... *Spinoliella*
—. Metasoma at apex strongly curved downward and forward; paraocular area with yellow surpassing level of antennal socket, usually narrowly following orbit; clypeus (in ventral view) with inner margin of lateral area beside lateral margin of labrum usually almost straight; antennal socket (lower margin) usually at lower quarter of face (South America).. *Callonychium*

Key to the Genera of the Calliopsini (Females)
(Modified from Ruz, 1991)

1. Labrum with basal area excavated, distal part convex, protuberant in lateral view; inner orbits generally convergent below, or, *if* not, then lower paraocular area not swollen mesally 2
—. Labrum flat or with smooth, rounded, nearly transverse ridge, distal area flat; inner orbits subparallel or divergent below and lower paraocular area swollen mesally 4
2(1). Labrum with basal area pilose at least laterally, its distal margin strongly salient; hind tibia with keirotrichia widespread on most of inner surface (South America) 3
—. Labrum with basal area usually glabrous, but *if* pilose, then flat (without ridge); hind tibia with keirotrichia absent toward ventral margin to completely absent between basal and distal patches .... *Calliopsis*
3(2). Lower paraocular area mesally (immediately above sublateral concavity in upper clypeal margin) swollen and impunctate; apical hair bands of T1 to T4 with hairs short, scarcely surpassing tergal margins, and directed posterolaterally... *Litocalliopsis*
—. Lower paraocular area mesally not or weakly swollen, sometimes with small, impunctate area; apical hair bands of T1 to T4 with hairs long, surpassing tergal margins, directed posteriorly .... *Acamptopoeum*
4(1). Inner orbits divergent below; metasoma wider than thorax; middle and hind tibial spurs strongly curved at apices; lower paraocular area convex (South America) .... *Arhysosage*
—. Inner orbits subparallel; metasoma about as wide as thorax or narrower; middle and hind tibial spurs slightly curved; lower paraocular area strongly convex only on inner corner 5
5(4). Gena black; lower paraocular area with yellow not extending upward as narrow band along orbit; antennal sockets (lower margins) at lower one-third of face; claws bifurcate; facial fovea not linear (South America) .... *Spinoliella*
—. Gena with longitudinal yellow band to completely yellow; lower paraocular area with yellow spot, which is narrowed and extended upward along upper orbit; antennal sockets (lower margins) usually at lower one-fourth of face; claws simple; facial fovea linear (South America) .... *Callonychium*

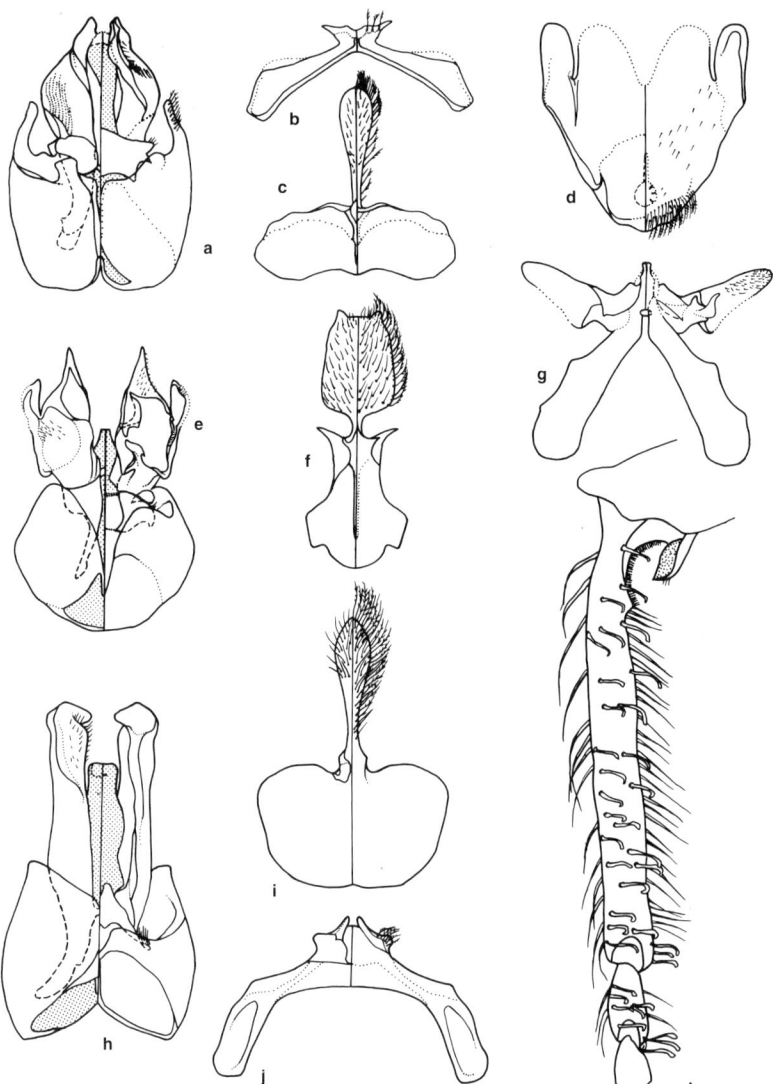

Figure 59-3. Structures of Calliopsini. **a-c,** Male genitalia, S7, and S8 of *Acamptopoeum submetallicum* (Spinola); **d,** S6 of female *Calliopsis (Calliopsis) andreniformis* Smith; **e-g,** Male genitalia, S8, and S7 of *C. (C.) andreniformis* Smith; **h-j,** Male genitalia, S8, and S7 of *C. (Verbenapis) verbenae* Cockerell and Porter; **k,** Front tarsal segments 1 and 2 of female of *C. (V.) verbenae* Cockerell and Porter, showing hooked hairs. (In the divided figures, dorsal views are at the left.) From Ruz, 1991.

Genus *Acamptopoeum* Cockerell

Friesea Schrottky, 1902a: 418, not Dalla Torre, 1895. Type species: *Friesea brasiliensis* Schrottky, 1902 = *Camptopoeum prinii* Holmberg, 1884, monobasic.

Acamptopoeum Cockerell, 1905a: 320. Type species: *Camptopoeum trifasciatum* Spinola, 1851, sensu Cockerell, 1905a = *Camptopoeum submetallicum* Spinola, 1851, monobasic, designated by Commission Opinion 1759 (1994).

Parafriesea Schrottky, 1906b: 118, replacement for *Friesea* Schrottky, 1902. Type species: *Friesea brasiliensis* Schrottky, 1902 = *Camptopoeum prinii* Holmberg, 1884, autobasic.

Species of this genus resemble rather large *Calliopsis* (8-11 mm long) without colored integumental bands on the terga, but with marginal hair bands as in *Calliopsis* s. str. Face marks and sometimes areas of the pronotum are yellow. Distinctive features not in the key to genera are the coarsely toothed middle tibial spur of the female, and the sting, which surpasses the stylus and is truncate at the apex. Shinn (1965) diagrammed the male genitalia and apical sterna; these and other structures were also illustrated by Ruz (1986); see Figure 59-3a-c.

■ *Acamptopoeum* is widespread in South America from Argentina and Chile north to Colombia. It is probably absent from the moist tropics. Ruz (1991) lists eight species. Shinn (1965) gave a key to species. Compagnucci (2004) described three more species and gave a key to the species of Argentina. The nesting biology and immature stages of *Acamptopoeum* were described by Rozen and Yanega (1999).

Genus *Arhysosage* Brèthes

Arhysosage Brèthes, 1922: 121. Type species: *Arhysosage johnsoni* Brèthes, 1922 = *Camptopoeum ochraceum* Friese, 1908, monobasic.
Ruiziella Timberlake, 1952a: 105 (April) [not Cortés, 1952 (March)]. Type species: *Camptopoeum ochraceum* Friese, 1908, by original designation.
Ruziapis Timberlake, 1952b: 528, *lapsus* for *Ruizapis* and replacement for *Ruiziella* Timberlake, 1952.
Ruizapis Timberlake, 1953a: 598, emendation of *Ruziapis* Timberlake, 1952.

This genus is related to *Spinoliella* and *Callonychium*, sharing with them the yellow and black to largely yellow body (length about 10 mm) and the low position of the antennal sockets. Its members can be differentiated by their inner orbits, which are distinctly divergent below, and the hind tibial spurs, which are distinctly curved toward the apices. In the male the body is almost completely yellow, and the mandible is strongly curved and elongate, with a preapical tooth and a projection on the upper margin. Genitalic and other structures were illustrated by Ruz (1986, 1991).

■ This genus is found in Argentina, from the provinces of Formosa and Catamarca south to Córdoba. There were three known species (Ruz, 1991); they were reviewed by Moure (1958b). A revision of the six species of *Arhysosage* now recognized was by Engel (2000d).

Genus *Calliopsis* Smith

Calliopsis is used here in the sense of Ruz (1991) to include forms traditionally placed in the genera *Calliopsis, Hypomacrotera, Liopoeum,* and *Nomadopsis*. To judge from the study by Ruz, the genus *Calliopsis* in the old sense was a paraphyletic taxon from which the others arose. I have not hesitated to recognize various other probably paraphyletic taxa when strongly supported by large morphological gaps from derived taxa, but there is no such situation in this case.

This genus is similar to *Acamptopoeum;* the two are sister groups (Ruz, 1991). *Calliopsis* differs from *Acamptopoeum* by the usually shorter and less dense thoracic hair, the presence in most cases of a transverse but not strongly salient ridge on the labrum, the median marginal projection of S4 of the male (sometimes inconspicuous), and, usually, such a projection also on S5. The sting of the female does not reach the stylus and is pointed at the apex. Male genitalia and sterna of various subgenera were illustrated by Mitchell (1960) and Ruz (1986, 1991). These structures of the North American subgenera having integumental metasomal color bands (old *Nomadopsis*) were illustrated by Rozen (1958), and structures of the subgenera with metasomal hair bands (the old *Calliopsis*) were illustrated by Shinn (1967); see also Figure 59-3e-j.

The species of *Calliopsis* are oligolectic or narrowly polylectic. The subgenus *Calliopsis* s. str. visits mostly small legumes, but at least some species, such as *C. (C.) andreniformis* Smith, are rather polylectic, utilizing pollen from other flowers as well (see Shinn, 1967). *Perissander* visits small Euphorbiaceae. *Calliopsima* collects pollen from yellow Compositae. *Verbenapis* collects pollen from, and seems to be morphologically adapted to, certain species of *Verbena*. *Hypomacrotera* includes one species that is associated with *Sphaeralcea* (Malvaceae) and others that are associated with small Solanaceae such as *Physalis*. The subgenera *Nomadopsis* and *Micronomadopsis* are not oligolectic as subgenera, but many of the included species are oligolectic.

Nests of *Calliopsis* are short burrows in the soil with laterals leading to cells, which are sometimes isolated, sometimes in short series. The most extensive account is by Shinn (1967) on *C. andreniformis* Smith. The cell is lined with a thin secreted layer, and the ball of larval food, when completed, is covered with a similar film (Rozen, 1958, 1970b). In some species the loose tumulus material closes the nest entrance; on leaving or entering, the bee has to dig through the loose earth of the tumulus. *C. (Micronomadopsis) larreae* (Timberlake) nests in areas where loose sand 2 to 8 cm thick covers the sandy clay where the cells are constructed. Bees entering or leaving their nests have to dig through the loose sand; those entering evidently keep their pollen loads in place and quickly find their burrows under the loose sand layer (Rust, 1988). The life cycle and behavior of a tropical species, *Calliopsis (Calliopsis) hondurasicus* Cockerell, were discussed by Wcislo (1999a). Even in the lowlands of Panama, this species has a short season of flight, limited to the early part of the dry season.

Key to the Subgenera of *Calliopsis* (Males)

1. Hind basitarsus with pubescence of dorsal margin at least partly about as long as basitarsus or longer; apical area of labrum clearly convex, almost glabrous except at margin; labrum with no defined ridge separating basal and apical areas (South America) .... *C. (Liopoeum)*
—. Hind basitarsus with pubescence of dorsal margin about half as long as basitarsus or shorter; apical area of labrum hairy, rather flattened; labrum with ridge or carina separating basal and apical areas [but entire labrum flat and pilose in *C. (Hypomacrotera) subalpina* (Cockerell)] 2
2(1). T2-T5 at least laterally with marginal hair bands 3
—. T2-T5 without marginal hair bands.............................. 6
3(2). Subantennal area black; metanotum laterally without velvety area; propodeal triangle basally smooth, medially concave (nearctic) .... *C. (Verbenapis)*
—. Subantennal area yellow; metanotum laterally with area (sometimes reduced to a small and narrow strip) of dense velvety hairs; propodeal triangle basally rugose or at least slightly roughened, medially flat 4
4(3). Clypeus (in lateral view) clearly protuberant; metanotum with conspicuous patch of velvet-brown hairs laterally (North and Central America) ....*C. (Calliopsis* s. str.*)*
—. Clypeus (in lateral view) almost flat; metanotum with small patch of velvet-white hairs laterally 5
5(4). Labrum with basal area delimited by strong carinate ridge; middle basitarsus somewhat shorter than hind; propodeal triangle basally with strong striae, striated part delimited posteriorly by strong transverse ridge (nearctic) .... *C. (Calliopsima)*

—. Labrum with basal area delimited by weak ridge; middle basitarsus much longer than hind; propodeal triangle basally with weak striae, not delimited posteriorly by strong transverse ridge (nearctic) .... *C. (Perissander)*

6(2). Metasoma without yellow marks 7

—. Metasomal terga almost always with interrupted or complete yellow bands 9

7(6). Antennal scape robust, less than three times as long as wide; stigma more than twice as broad as prestigma, as measured to wing margin, margin within marginal cell clearly convex (Argentina) .... *C. (Ceroliopoeum)*

—. Antennal scape normal, over four times as long as wide; stigma less than twice as broad as prestigma, margin within marginal cell straight or nearly so (Fig. 59-5) 8

8(7). Inner orbits subparallel; marginal cell about twice as long as second submarginal cell or longer (Fig. 59-5); hind tarsomeres 2 to 5 unmodified (nearctic) ....
.... *C. (Hypomacrotera)*

—. Inner orbits convergent below (Fig. 59-1f); marginal cell less than twice as long as second submarginal cell; hind basitarsus and tarsomeres 2 to 4 asymmetrical, and distitarsus widened medially (Argentina) ....
.... *C. (Liopoeodes)*

9(6). Front and middle tarsi slender, and tarsomeres 2 to 4 of hind tarsus often broadened laterally, asymmetrical (nearctic) .... *C. (Micronomadopsis)*

—. Tarsi with tarsomeres broadened distally, symmetrical (nearctic) .... *C. (Nomadopsis)*

Key to the Subgenera of *Calliopsis* (Females)

1. Stigma more than twice as broad as prestigma as measured to wing margin; stigmal margin within marginal cell clearly convex; midfemoral comb less than half length of femur (Argentina) .... *C. (Ceroliopoeum)*

—. Stigma less than twice as broad as prestigma as measured to wing margin; stigmal margin within marginal cell straight or nearly so; midfemoral comb about half length of femur 2

2(1). T2-T4 with marginal hair bands, at least laterally 3
—. T2-T4 without marginal hair bands................................ 8

3(2). Integument black (with some slight blue but no yellow); wing veins mostly dark; inner orbits subparallel; T1 and T2 polished, scarcely punctate (Argentina) ....
.... *C. (Liopoeodes)*

—. Integument with at least some yellow marks; wing veins mostly yellowish; inner orbits convergent below; terga with rather abundant punctation...................... 4

4(3). Middle tibial spur with about five coarse teeth, all on distal half; hind tibia with keirotrichia on inner surface at base and at apex only; S6 distally with fringe of hairs interrupted medially (South America) ....
.... *C. (Liopoeum)* (in part)

—. Middle tibial spur with somewhat coarse teeth, not limited to distal half; hind tibia with keirotrichia on most of inner surface or absent toward ventral border; S6 distally with hairs forming a continuous curved fringe 5

5(4). Front tarsus with hairs on inner surface sparse, mostly stiff, curved and not tapered at apices (Fig. 59-3k); metanotum laterally without patch of velvety hairs (nearctic) .... *C. (Verbenapis)*

—. Front tarsus with hairs on inner surface dense, unmodified; metanotum laterally with small (sometimes inconspicuous) patch of white, velvety hairs 6

6(5). Clypeus (in lateral view) protuberant, projecting forward one-third width of eye or more; paraocular area with lowest part noticeably wider than at level of antennal sockets, usually swollen on lower inner corner (North and Central America)......................... *C. (Calliopsis s. str.)*

—. Clypeus (in lateral view) less protuberant, projecting forward only about one-fourth width of eye or less; paraocular area with lowest part about as wide as at level of antennal sockets, only slightly convex 7

7(6). Propodeum mostly punctate, base with a conspicuously rugose triangle delimited by strong ridge or carina; metanotum laterally with clearly visible patch of white, velvety hairs (nearctic) .... *C. (Calliopsima)*

—. Propodeum posteriorly with extensive impunctate area, base usually little or not very strongly rugose, triangle delimited by rather rounded ridge; metanotum laterally with inconspicuous (difficult to see) patch of white, velvety hairs (nearctic) .... *C. (Perissander)*

8(2). Middle tibial spur with most teeth much longer than those of hind tibial spurs; S6 with apical marginal fringe of hairs broken medially (South America) ....
.... *C. (Liopoeum)* (in part)

—. Middle tibial spur with most teeth only slightly longer than those of hind tibial spurs; S6 with hairs forming a continuous curved fringe distally..................................... 9

9(8). Propodeal triangle polished; metasoma with no yellow markings (nearctic) .... *C. (Hypomacrotera)*

—. Propodeal triangle at least slightly rugose basally; metasoma almost always with complete or interrupted yellow integumental bands.. 10

10(9). Body length less than 10 mm; middle tibial spur with four teeth on distal half (nearctic) .... *C. (Micronomadopsis)*

—. Body length 10 mm or longer; middle tibial spur with more than four teeth on distal half (nearctic) ....
.... *C. (Nomadopsis)*

Calliopsis / Subgenus *Calliopsima* Shinn

Calliopsis (Calliopsima) Shinn, 1967: 834. Type species: *Calliopsis rozeni* Shinn, 1967, by original designation.

Most species of this subgenus are slightly more robust and larger (7-9 mm long) than most species of *Calliopsis* s. str. As in that subgenus, the body is black with yellow or cream-colored areas on the face and sometimes the pronotum. The glossa is longer than the prementum. The clypeus is flattened, and the distal margin of the basal area of the labrum is marked by a strong carina. The metanotum has a narrow patch of dense hairs basilaterally. The middle basitarsus is shorter than the hind basitarsus.

■ *Calliopsima* occurs from the southern prairie provinces of Canada to Chiapas in southern Mexico and across the continent from California to Florida, USA. It is unknown from the Northwest (Oregon, Washington, British Columbia) and from the Northeast (from Tennessee to the north and east). The 15 species were revised by Shinn (1967).

The biology of *Calliopsis pugionis* Cockerell was described in detail by Visscher and Danforth (1993).

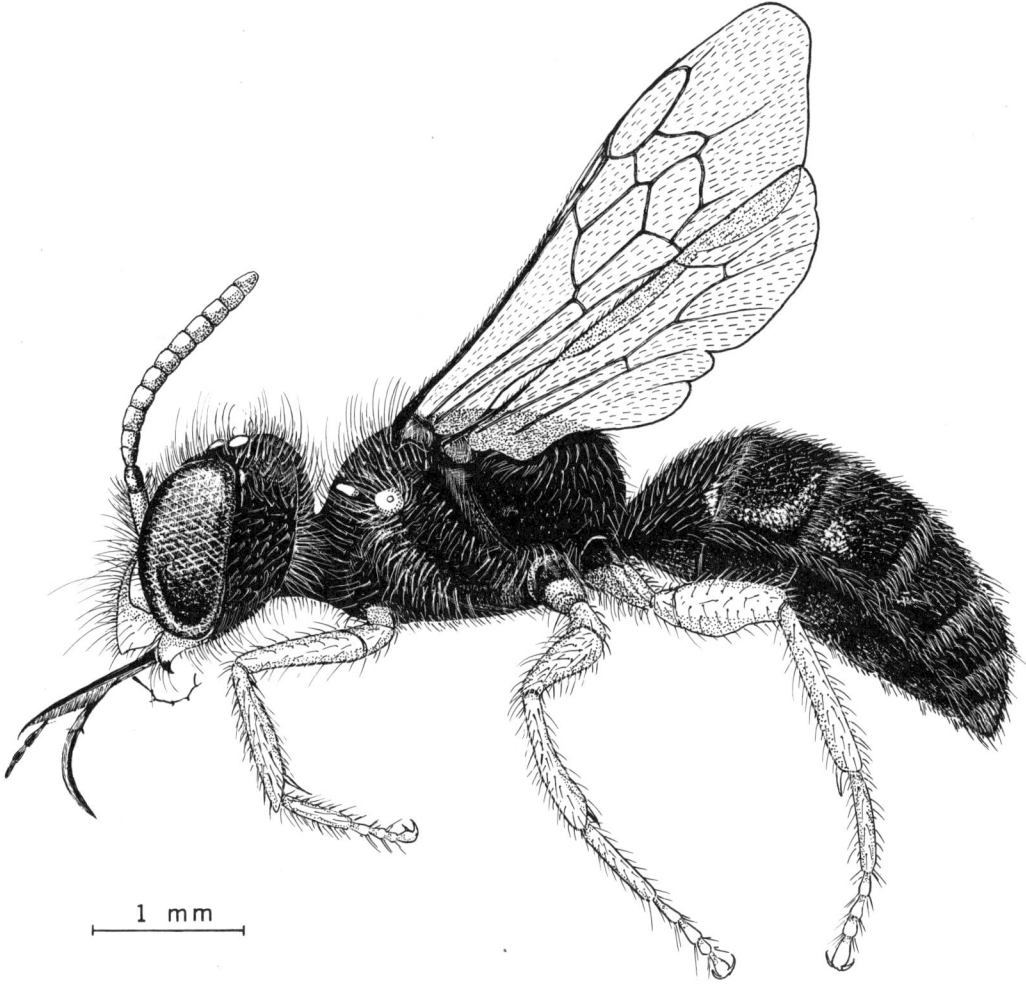

Figure 59-4. Side view of male of *Calliopsis (Calliopsis) andreniformis* Smith. The pale areas (legs and face) are bright yellow. Drawing by S. A. Earle, from Shinn, 1967.

Calliopsis / Subgenus *Calliopsis* Smith s. str.

Calliopsis Smith, 1853: 128. Type species: *Calliopsis andreniformis* Smith, 1853, by designation of Ashmead, 1899a: 85.

In this subgenus the face, legs, and pronotum have yellow or cream markings and are largely yellow in most males (Fig. 59-4). The body length is 5 to 8 mm. The glossa is somewhat shorter than the prementum. The metasomal terga are black, with distal hair bands. The axilla and metanotum laterally have dense patches of velvety hairs. In the male, the middle basitarsus is somewhat longer than the other basitarsi. Figure 59-3d-g shows metasomal structures of this subgenus.

- This subgenus ranges from southern Canada through the USA, except Oregon, Washington, and Idaho, south through Mexico and Central America to Panama, where it occurs in the lowlands. The 12 species were revised by Shinn (1967).

The biology of *Calliopsis andreniformis* Smith was described in detail by Shinn (1967); for a tropical species, see Wcislo (1999a).

Calliopsis / Subgenus *Ceroliopoeum* Ruz

Calliopsis (Ceroliopoeum) Ruz, 1991: 237. Type species: *Camptopoeum laetum* Vachal, 1909, by original designation.

In appearance, bees of this subgenus resemble species of *Liopoeum* that lack yellow tergal bands. They differ from bees of that subgenus in their short, sparse pubescence, the medially convex clypeus with convex areas lateral to the labrum, the rather robust male antennal flagellum, its last segment tapering and pointed, the slender hind tarsus of the male, and the middle tibial spur of the female, which is as long as the basitarsus, sinuate, and has well-separated teeth. The body length is 5 to 6 mm.

- *Ceroliopoeum* occurs in the provinces of Santiago del Estero and La Rioja, Argentina. There is only one species, *Calliopsis laeta* (Vachal).

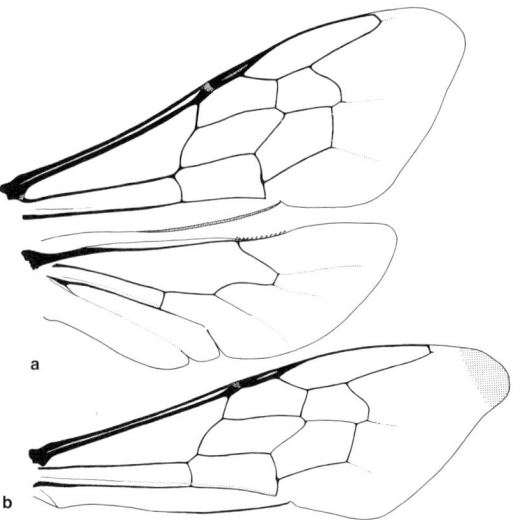

Figure 59-5. Wings of *Calliopsis (Hypomacrotera) subalpina* Cockerell, showing sexual dimorphism in shape and maculation. **a**, Female; **b**, Male.

Calliopsis / Subgenus *Hypomacrotera* Cockerell and Porter

Hypomacrotera Cockerell and Porter, 1899: 418. Type species: *Hypomacrotera callops* Cockerell and Porter, 1899, by original designation.

In this subgenus the black or red metasoma lacks both yellow markings and dense premarginal tergal bands. The body length is 5 to 8 mm. The marginal cell is narrowly and obliquely truncate and much longer than the distance from its apex to the wing tip (Fig. 59-5). The forewing of the male is longer and more pointed than that of the female. The stigma has parallel sides. The hind tarsus of the male is unmodified, and the apex of the forewing of the male is dark brown (Fig. 59-5b). Such a dark wing tip in males is found in the subgenus *Liopoeodes* and, less well developed, in some species of *Perissander* and *Calliopsis* s. str. S5 of the male has an elongate, tapered, distal, medial projection. In the female the middle tibial spur has fine, dense teeth.

■ This subgenus occurs in the southwestern USA (western Texas and Colorado to California) and in Mexico south to Durango and Baja California, in xeric areas. The three species were revised by Danforth (1994).

Calliopsis / Subgenus *Liopoeodes* Ruz

Calliopsis (Liopoeodes) Ruz, 1991: 236. Type species: *Calliopsis xenopous* Ruz, 1991, by original designation.

The species of this subgenus is similar to those *Liopoeum* species that lack yellow tergal markings, except that the apex of the forewing of the male is dark, as in *Hypomacrotera,* and the hind tarsus of the male is modified, segments 1 to 4 bearing apicolateral projections, segment 5 broadest medially. In the female the metasomal terga have apical white hair bands. Body length is 6 to 8 mm.

■ This subgenus is found in the provinces of Jujuy, Tucumán, and Salta, Argentina. The only species is *Calliopsis xenopous* Ruz.

Calliopsis / Subgenus *Liopoeum* Friese

Camptopoeum (Liopoeum) Friese, 1906b: 176. Type species: *Camptopaeum hirsutulum* Spinola, 1851, by designation of Sandhouse, 1943: 564.

Among the subgenera of *Calliopsis, Liopoeum* is one of the most distinctive. Some species lack yellow bands but have hair bands on the metasomal terga of females, as in *Calliopsima* and *Calliopsis* s. str., whereas others have yellow tergal marks and no hair bands, as in the *Nomadopsis* group of subgenera. The body length is 5 to 8 mm. *Liopoeum* is close to *Hypomacrotera* but differs in its long pubescence, especially on the legs. Hind tarsomeres 2 to 4 of the male are expanded laterally. S5 of the male has a long distal median projection, truncate or tapered at the apex. The middle tibial spur of the female has sparse and usually coarse teeth. Keirotrichia on the hind tibia are dense at the base and apex, but sparse or limited to the upper margin in the intermediate region. The apical fringe of S6 of the female is interrupted medially.

■ This subgenus is found in Chile and Argentina. There are four described species (Ruz, 1991).

Calliopsis / Subgenus *Micronomadopsis* Rozen

Nomadopsis (Micronomadopsis) Rozen, 1958: 107. Type species: *Nomadopsis fracta* Rozen, 1958, by original designation.

Species of this subgenus are similar to, but usually smaller (5-10 mm long) than, *Nomadopsis,* from which they differ in the slender segments 2 to 4 of the front tarsus of the male. The penis valve (in most species) is thick and elaborately contorted, thus differing from the more slender penis valves of *Nomadopsis.* The middle tibial spur of the female has only four teeth on its distal half.

■ *Micronomadopsis* is widespread in western North America from Washington to South Dakota, south to Texas, USA, and from Baja California to Durango, Mexico. The 20 species were revised by Rozen (1958).

Rozen (1958) recognized four species that he did not place subgenerically within the *Nomadopsis-Micronomadopsis* group, and Ruz (1991) did not clarify this situation. They may yet show that *Nomadopsis* and *Micronomadopsis* should be united.

Calliopsis / Subgenus *Nomadopsis* Ashmead

Nomadopsis Ashmead, 1898: 285. Type species: *Perdita zonalis* Cresson, 1879, *lapsus* for *Calliopsis zonalis* Cresson, 1879, by original designation.
Spinoliella (Claremontiella) Cockerell 1933c: 26. Type species: *Spinoliella euxantha* Cockerell, 1916 = *Calliopsis zonalis* Cresson, 1879, by original designation.
Nomadopsis (Macronomadopsis) Rozen, 1958: 93. Type species: *Nomadopsis micheneri* Rozen, 1958, by original designation.

This is a subgenus of relatively large (length 7-12 mm) species with yellow integumental tergal bands. Its distinctive features are indicated in the key to subgenera and by Ruz (1991). Although the two groups recognized by

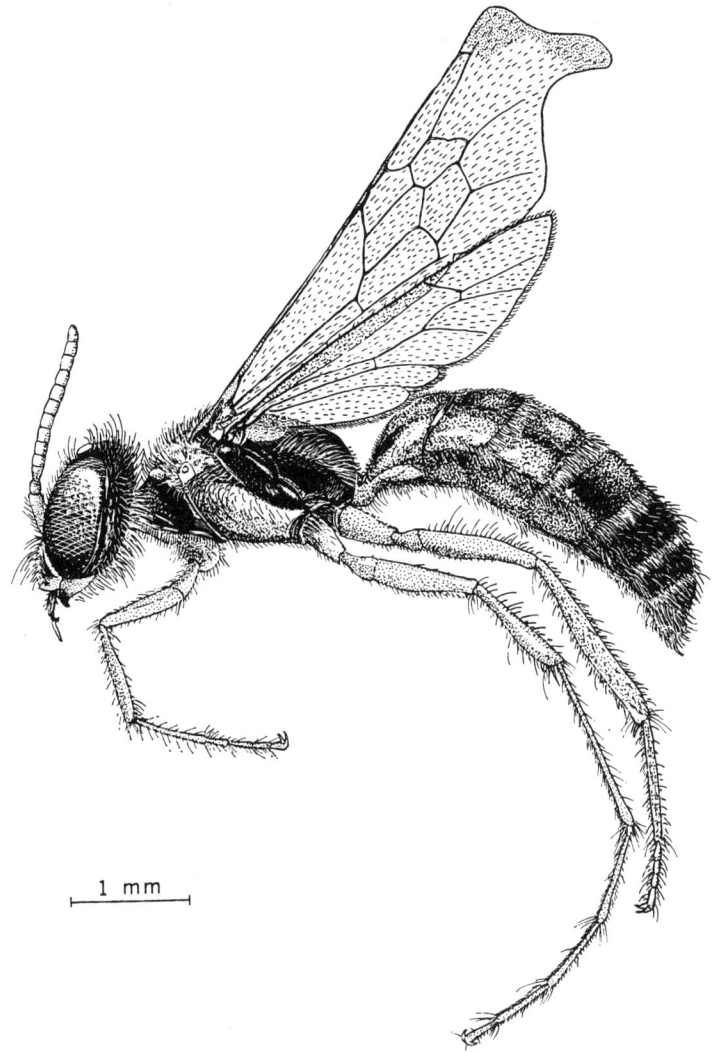

Figure 59-6. Male of *Calliopsis (Perissander) anomoptera* Michener. The falcate wing shape characterizes males of this species, but not other members of its subgenus. Note the long middle tarsus. Drawing by S. A. Earle, from Shinn, 1967.

Rozen (1958) and Ruz (1991) under the names *Nomadopsis* and *Macronomadopsis* can be distinguished in males (volsellae dentate in the former, not dentate in the latter), the differences seem trivial compared to those that separate other subgenera of *Calliopsis*.

■ This subgenus occurs in western North America from British Columbia, Canada, to western South Dakota and south to California and Texas, USA, and Baja California and Durango, Mexico. The 13 species were revised by Rozen (1958).

Calliopsis / Subgenus *Perissander* Michener

Calliopsis (Perissander) Michener, 1942a: 275. Type species: *Calliopsis anomoptera* Michener, 1942, by original designation.

This subgenus is similar to *Calliopsis* s. str. but differs in having the glossa much shorter than the prementum. Only the metanotum carries a narrow patch of dense hairs laterally, the patch sometimes difficult to see or absent. The basal area of the propodeal triangle is delimited posteriorly by a rounded ridge. The forewing of the male is sometimes brown at the apex, and the middle basitarsus is much longer than the other basitarsi.

■ *Perissander* is found from New Mexico to California, USA, south to Michoacan, Durango, and Baja California Sur, Mexico. The seven described species were revised by Shinn (1967).

The type species has a strangely shaped, falcate forewing with a dark apex in the male (Fig. 59-6); in other species the forewing is of the usual shape.

Calliopsis / Subgenus *Verbenapis* Cockerell and Atkins

Calliopsis (Verbenapis) Cockerell and Atkins, 1902: 44. Type species: *Calliopsis verbenae* Cockerell and Porter, 1899, monobasic.

Like *Calliopsima*, this subgenus consists of rather hairy bees (7-8 mm long) with yellow or cream-colored markings limited to the head and pronotum. The genitalia and associated sterna are more similar to those of *Calliopsima* than to those of *Calliopsis* s. str. and *Perissander. Verbenapis* differs from all these subgenera in that the metanotum laterally lacks velvety hairs and the subantennal and supraclypeal areas are black. The glossa is as in *Calliopsima*. The middle basitarsus is shorter than the hind basitarsus (as in *Calliopsima*). The median projection of S6 of the male has a minute median distal emargination. Male genitalia and hidden sterna are illustrated in Figure 59-3h-j. The apical process of S8 lacks the angles basad to the median constriction found in other subgenera. The inner surface of the front tarsus of the female bears rather short, rigid hairs, curved at the blunt apices (Fig. 59-3k).

■ This subgenus ranges from North Dakota to New Jersey, south to Arizona and Texas, USA, and the state of México; it is not known in the southeastern USA. There are four species, as revised by Shinn (1967).

Species of this subgenus are probably all oligolectic on flowers of the genus *Verbena*. The hooked hairs of the front tarsi of the female (Fig. 59-3k) serve to pull pollen from the slender corolla tube.

Genus *Callonychium* Brèthes

Callonychium is a South American genus of black-and-yellow species 3 to 7 mm long. They differ from *Spinoliella* in the yellow on the genal area, the antennal sockets (lower margins), which are usually at the lower fourth of the face, and the lateral parts of the clypeus, which are strongly bent posteriorly. In the male the basitibial plate is flat and delimited by a carina, the metasomal apex is strongly curled forward, and the sterna are considerably modified with ridges and projections. Male genitalia, sterna, and other structures were illustrated by Toro and Herrera (1980), Cure and Wittmann (1990), and Ruz (1991); see also Figure 57-7a-c.

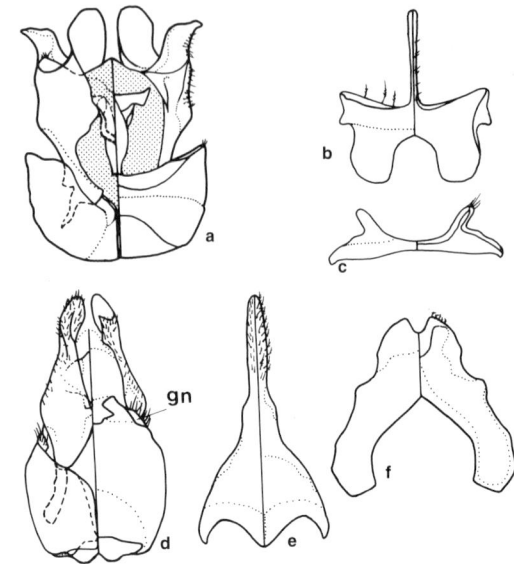

Figure 59-7. Genitalia, S8, and S7 of male Calliopsini. **a-c,** *Callonychium mandibulare* Brèthes; **d-f,** *Spinoliella nomadoides* (Spinola). (Dorsal views are are at the left; gn is the gonostylus.) From Ruz, 1991.

Key to the Subgenera of *Callonychium*

1. Axilla acute, yellow; S1 of male without premarginal process .... *C. (Callonychium* s. str.)
—. Axilla rounded, depressed, black; S1 of male with premarginal process .... *C. (Paranychium)*

Callonychium / Subgenus *Callonychium* Brèthes s. str.

Callonychium Brèthes, 1922: 120. Type species: *Callonychium argentinum* Brèthes, 1922, monobasic.

■ This subgenus is found in Brazil, Paraguay, and Argentina. There are six described species and additional undescribed forms (Ruz, 1991).

Callonychium / Subgenus *Paranychium* Toro

Callonychium (Paranychium) Toro and Herrera, 1980: 213. Not valid because no type species was designated.
Callonychium (Paranychium) Toro, 1989: 231. Type species: *Camptopoeum chilense* Friese, 1906, by original designation.

■ This subgenus is known in Chile and Argentina; J. Rozen (in litt., 1997) has found it in Peru. There are five named species (Ruz, 1991). The Chilean species were revised by Toro and Herrera (1980).

The male of *C. (P.) chilense* Friese has variously modified metasomal sterna. Toro (1985) has investigated the way in which these structures function during mating, when the male is curled around the apical part of the female's metasoma.

Genus *Litocalliopsis* Roig-Alsina and Compagnucci

Litocalliopsis Roig-Alsina and Compagnucci, 2003: 103.
Type species: *Litocalliopsis adesmiae* Roig-Alsina and Compagnucci, 2003, by original designation.

Litocalliopsis combines some characters of *Acamptopoeum* and *Calliopsis*. Thus the simple male S4 and S5, the pilose labrum of the female, and the keirotrichia widespread on the inner surface of the hind tibia of the female are as in *Acamptopoeum* and the presence of a male pygidial plate is as in *Calliopsis*.

The single known species of this genus has been found in Buenos Aires Province, Argentina. It appears to be oligolectic on flowers of *Adesmia* (Fabaceae).

Genus *Spinoliella* Ashmead

Spinoliella Ashmead, 1899a: 84. Type species: *Camptopoeum nomioides* Spinola, lapsus for *Camptopoeum nomadoides* Spinola, 1851, by original designation.
Spinoliella (Peniella) Toro and Ruz, 1972: 146. Type species: *Camptopoeum maculatum* Spinola, 1851, by original designation.

Spinoliella consists of yellow-and-black species 4 to 9 mm long, and is closer to *Callonychium* than to

Arhysosage. It differs from the former in its black genal areas, the position of the antennal sockets (near the lower third of the face), and the rather horizontal lateral part of the clypeus beside the mandibular articulation. The basitibial plate of the male is a shiny, convex area not delimited by a carina. Genitalia and other structures were illustrated by Toro and Ruz (1972) and Ruz (1991); see also Figure 59-7d-f.

■ This genus occurs in Chile, Argentina, and Peru (the last reported by J. Rozen, in litt., 1997). Six species are recognized (Ruz, 1991). The genus was revised by Toro and Ruz (1972).

Given the small number of species involved, the two subgenera recognized by Toro and Ruz (1972) seem unnecessary; they represent recognizable, but not very dissimilar, groups.

60. Subfamily Oxaeinae

This is a small American (USA to Argentina) group of large (13-26 mm long), robust, hairy bees (Fig. 60-1), apiform or euceriform. The glossa is short and acute, more or less as in *Andrena*, but in other features the proboscis is quite distinctive. The lacinia is unique; either it is absent or, if identified correctly, it consists of two small sclerites hidden between the expanded stipites (Fig. 21-2b), as are the suspensoria of the prementum (Michener and Greenberg, 1985). The mentum either is not recognizable or is fused to the large, flat lorum (Michener, 1985a). As in other Andrenidae, there are two subantennal sutures below each antenna (Fig. 60-2b). The scopa is dense from the hind coxa to the basitarsus, as it is in most Andreninae, but in Oxaeinae it seems denser and larger. Unlike other Andrenidae, the stigma is absent, the marginal cell is very long and slender (Fig. 60-2a), the clypeus is strongly protuberant, and the labrum is as long as or longer than broad. In the males the eyes converge strongly above, and in both sexes the ocelli are low on the face, so that the lateral ocelli and the antennal bases are commonly equidistant from the median ocellus (Fig. 60-2b). Further, the first flagellar segment is as long as or longer than the scape. These features are also found in the panurgine genus *Melitturga*, but almost certainly they are convergent and do not indicate close relationship. Male genitalia, hidden sterna, and other structures were illustrated by Popov (1941b), Moure and Seabra (1962b), Moure and Urban (1963), and Hurd and Linsley (1976); see also Figure 60-3. Unusual features shared with *Ctenocolletes* (Stenotritidae) are the large discs of S7 and S8 of the male, these quite unlike those of most other bees.

Both larvae and adults are so different from other Andrenidae that the Oxaeinae have often been given family status, but as noted in the account of the family Andrenidae (Sec. 49), the Oxaeinae fell within the Andrenidae in the phylogenetic analysis by Alexander and Michener (1995), as well as in the earlier study by Michener (1944). Rozen (1993b, 1994a), in studies of larvae, showed the Oxaeinae as the sister group of *Euherbstia*, here placed in the Andreninae. Graf (1972) strongly sup-

Figure 60-1. *Protoxaea gloriosa* (Fox) (Oxaeinae). Above, male; Below, female. These are large bees, the female 22 mm long. From Michener, McGinley, and Danforth, 1994.

ported placement of Oxaeinae in the Andrenidae, pointing out that Oxaeinae and Panurginae are the only bees studied by him in which the paramandibular process of the hypostoma is fused with the clypeus, closing the mandibular socket. In other bees the process approaches

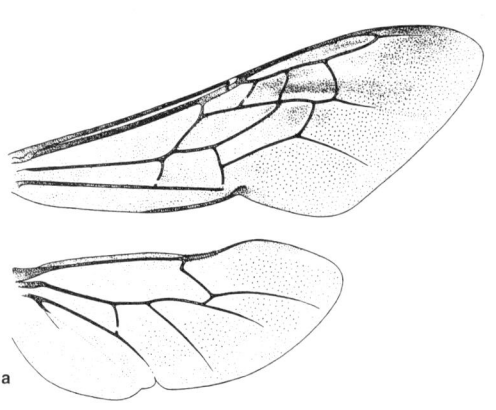

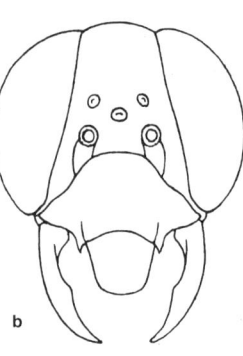

Figure 60-2. Structures of *Protoxaea gloriosa* (Fox), Oxaeinae.
a, Wings; **b**, Face of male, showing the position of the ocelli near the antennae. From Hurd and Linsley, 1976.

or even abuts against the inner surface of the clypeus without fusion (Michener, 1944: 206).

The larva is unusual, quite unlike other andrenid larvae, and in various characters is similar to that of the Nomadinae, an apparently unrelated taxon of L-T bees (Rozen, 1964a). The distinctive features of mature oxaeine larvae include the following: the apically cleft labrum, the much-reduced labiomaxillary region, the long and blade-like apex of the mandible, and the slitlike primary tracheal openings (Rozen, 1964a, 1993b).

The nests consist of deep (30-245 cm) vertical burrows in flat ground, and radiating, more or less horizontal laterals, each ending in a single, vertical cell lined with a "waxlike" film (Roberts, 1973). Provisions are somewhat liquid and fill the bottom of the cell. Thus the nests closely resemble those of Diphaglossinae (see Roberts, 1971) and are quite different from those of other Andrenidae. It may be significant that the cells of *Euherbstia,* here included in the Andreninae, are sloping to nearly vertical, although in other respects (e.g., firm spherical food mass) quite different from those of Oxaeinae.

The mutual similarity of all species of Oxaeinae is impressive. Hurd and Linsley (1976) revised the subfamily and recognized four genera.

Key to the Genera of the Oxaeinae

1. Maxillary palpus present, six-segmented; male gonostylus partly differentiated from gonocoxite and hairy (Fig. 60-3a) 2
—. Maxillary palpus absent; male gonostylus not recognizable, the apex of gonoforceps hairless (Fig. 60-3d) .... *Oxaea*
2(1). Mandible simple apically; clypeus and mandible of male without pale maculations 3
—. Mandible with preapical tooth on inner margin; clypeus and mandible of male with pale maculations (South America) .... *Notoxaea*
3(2). S8 of male with apex convex, not emarginate (Fig. 60-3b); T6 of male and T5 of female without conspicuous lateral tufts of long white hairs (North America) .... *Protoxaea*
—. S8 of male with median apical emargination (Fig. 60-3g); T6 of male and T5 of female each with lateral tuft of long white hairs (North America) .... *Mesoxaea*

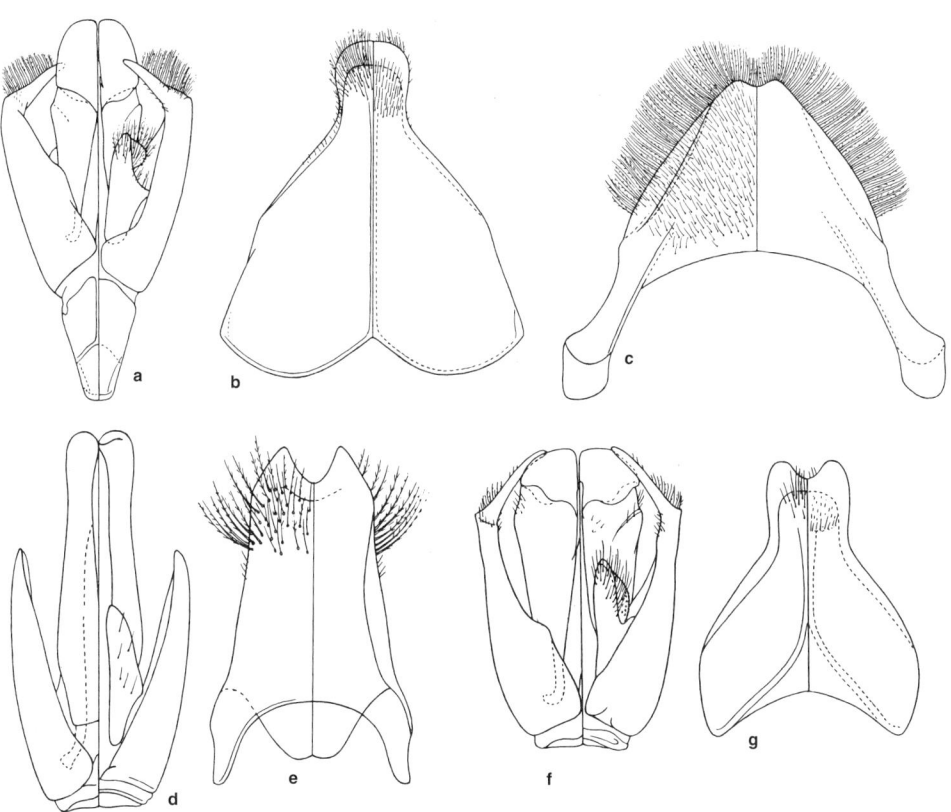

Figure 60-3. Structures of males of Oxaeinae. **a-c,** Genitalia, S8, and S7 of *Protoxaea (Protoxaea) gloriosa* (Fox); **d, e,** Genitalia and S8 of *Oxaea flavescens* Klug; **f, g,** Genitalia and S8 of *Protoxaea (Mesoxaea) nigerrima* (Friese). (Dorsal views are at the left.) From Hurd and Linsley, 1976.

The three genera, *Mesoxaea, Notoxaea* and *Protoxaea* differ from *Oxaea* in several plesiomorphies: the presence of maxillary palpi, the hairy and recognizable male gonostyli, and the relatively short penis valves (extending only slightly beyond the apices of the gonostyli). The subgenus *Notoxaea* may be a basal oxaeine, sister group to the rest of the subfamily, as suggested by its bidentate mandibles, a probable plesiomorphy. Michener (2000a) united these three genera under the name *Protoxaea* but no known apomorphies justify this action.

Genus *Mesoxaea* Hurd and Linsley

Mesoxaea Hurd and Linsley, 1976: 41. Type species: *Oxaea nigerrima* Friese, 1912, by original designation. [New status.]

In addition to the characters indicated in the key to genera, this genus differs from its close relative, *Protoxaea*, in its ringlike male gonobase (Fig. 60-3f).
■ This genus ranges from Chiapas, Mexico, to Louisiana, Texas, and southern Arizona, USA. It is largely absent from the Mexican plateau. The seven species were revised by Hurd and Linsley (1976).

Genus *Notoxaea* Hurd and Linsley

Notoxaea Hurd and Linsley, 1976: 21. Type species: *Oxaea ferruginea* Friese, 1898, by original designation.

In addition to the characters indicated in the key above, the lone species of this genus differs from others in the partly red or ferruginous metasomal tergal areas.
■ This genus occurs from Paraguay and Argentina (as far south as Mendoza) north to Mato Grosso and possibly to Piauí, Brazil. The only species is *Notoxaea ferruginea* (Friese).

Genus *Oxaea* Klug

Oxaea Klug, 1807a: 261. Type species: *Oxaea flavescens* Klug, monobasic.
Dasyglossa Illiger, 1807, in Klug, 1807b: 217, no valid species names; Klug, 1810: 44, included valid species. Type species: *Oxaea flavescens* Klug, 1807, by designation of Sandhouse, 1943: 544.

The names *Oxaea* and *Dasyglossa* appear to have been published at approximately the same time, late in 1807. *Dasyglossa* has not been used since, and in 1810 was discussed under the heading of *Oxaea* (in Klug, 1810). Even if *Dasyglossa* has priority, it should not be resurrected.

In the common species of this genus the metasomal terga in females and their posterior marginal zones in males are brilliantly metallic green. Other species, however, are scarcely metallic. Males usually have pale maculations on the clypeus, labrum, mandible, and antennal base.
■ This genus ranges from southern Brazil to Veracruz, Mexico, although it is rare in Mesoamerica. There are eight species (Hurd and Linsley, 1976).

Genus *Protoxaea* Cockerell and Porter

Protoxaea Cockerell and Porter, 1899: 410. Type species: *Megacilissa gloriosa* Fox, 1893, by original designation.

In this genus and *Mesoxaea,* the metasomal terga are black or brownish black. The male gonobase of *Protoxaea* differs from that of other Oxaeinae in being longer than wide (Fig. 60-3a). In other taxa it is a ring, much wider than long (Fig. 60-3d, f).
■ This genus occurs from Puebla, Morelos, and Guerrero, Mexico, north to Arizona, New Mexico, and Texas, USA. The three species were revised by Hurd and Linsley (1976).

61. Family Halictidae

The family Halictidae, the sweat bees, includes some of the commonest bees; in many temperate areas of the world, halictids dominate other bees in numbers of individuals, after *Apis* is ignored. They are mostly andreniform, although in the Nomiinae some Old World species are apiform, euceriform, or even more robust.

The Halictidae consistently appeared as a monophyletic unit in the phylogenetic analysis of S-T (short-tongued) bees by Alexander and Michener (1995). Features that are especially characteristic of the family are as follows: The lacinia is stretched up the anterior surface of the labiomaxillary tube and ends in a setose, often finger-shaped projection far above the rest of the maxilla (Figs. 21-2a, 61-1; Michener and Greenberg, 1985). In the Rophitinae this character is less developed, i.e., the lacinia is often not so far above the rest of the maxilla as it is in the other subfamilies, and may be pale and membranous. The wall of the proboscidial fossa of halictids is fused to the tentorium forward almost to the clypeus, and the articulatory process for the cardo is close behind the clypeus (Fig. 21-3j). Associated with the anterior position of this articulation is elongation of the cardines, and of the prementum as well, so that the proboscis is often quite long, because of its long basal segments. The base or prepalpal part of the galea tapers gradually to a point overlapping at least half the length of the stipes (Figs. 19-2c, 21-2a, 61-1). This character also occurs in *Chilicola* (Colletidae), as does the elongate cardo with forward articulation, and the extensive fusion of the wall of the proboscidial fossa with the tentorium. Presumably, these similarities of halictids and *Chilicola* are convergent.

Halictids agree with the characters of the S-T bees, as explained in Sections 18 and 19. The following are some other characters of the family: the body of the labrum, which is thick, elevated, and possibly homologous to the basal area of the labrum in Andrenidae, is much broader than long. In females an apical process, fringed with bristles, is often present, and sometimes is so long that the whole labrum is longer than broad. The one subantennal suture below each antenna is directed toward the lower edge of the socket or, in some Rophitinae (Fig. 61-2), mesad, thus ending near the inner margin of the socket. In this subfamily there is sometimes an outer subantennal suture, or more accurately a feeble line, so that there may appear to be two subantennal sutures, as in the Andrenidae. The facial foveae, as they are found in female Andrenidae and many Colletidae, are absent or slightly evident in some Rophitinae and Nomiinae (Schuberth and Schönitzer, 1993), but even in Halictinae there may be comparable areas. For example, brightly metallic *Agapostemon* females have a black strip along the anterior eye margin that is somewhat suggestive of a facial fovea. Unlike those of other families of bees, the annular hairs of the glossa, or some of them, are usually bifid or otherwise branched near their apices. In the Rophitinae and Nomiinae a minority of these hairs may be branched, but in Halictinae they are mostly bifid or trifid (Fig. 61-3).

Unlike the other families of S-T bees, the Halictidae have given rise to several cleptoparasitic groups, all in the subfamily Halictinae. *Sphecodes* is the largest and most familiar parasitic halictid genus. Some characteristic features of female halictids break down among the parasites. Thus basitibial and pygidial plates and a midapical process on the labrum, all of them usual halictid features, are not always present in parasitic forms (see Figs. 8-5, 8-6, 67-13d, g). Section 65 and Michener (1978b) describe other characteristics of parasitic Halictidae.

Although the monophyly of Halictidae as a family was

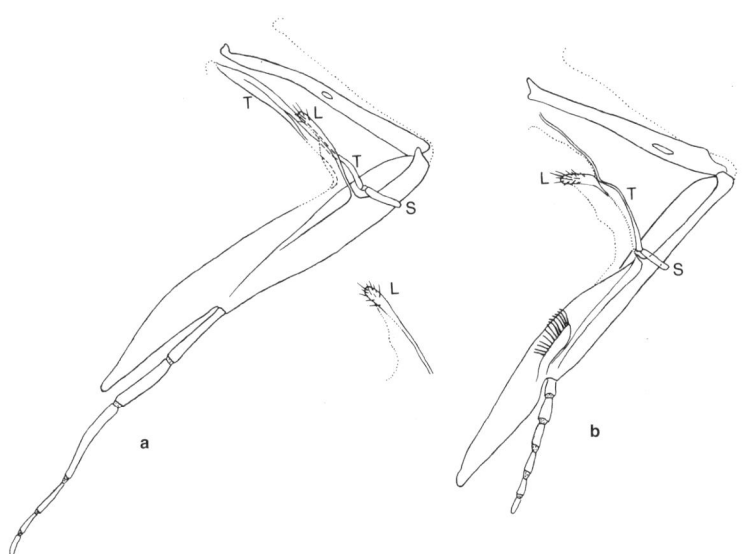

Figure 61-1. Inner views of maxillae of Halictidae. **a,** *Systropha curvicornis* (Scopoli), with lacinia drawn separately (Rophitinae); **b,** *Nomia melanderi* Cockerell (Nomiinae). An equivalent illustration of one of the Halictinae is Figure 21-2a. (L, lacinia; S, suspensorium of prementum; T, conjunctival thickening. Dotted lines indicate the profile of the labiomaxillary tube.) From Michener and Greenberg, 1985.

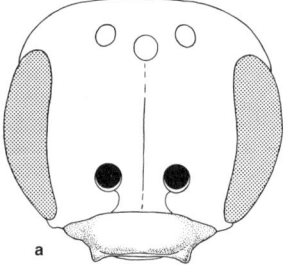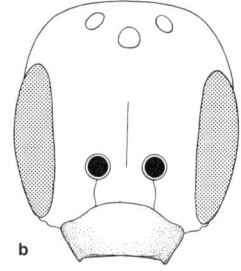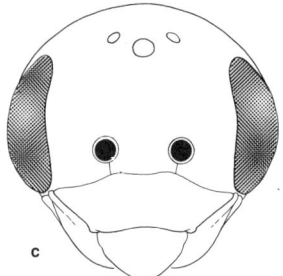

Figure 61-2. Faces of Rophitinae. **a,** *Conanthalictus caerulescens* Timberlake, male; **b,** *Sphecodosoma pratti* Crawford, female; **c,** *Dufourea calochorti* (Cockerell), female, with labrum flexed forward. From Michener, McGinley, and Danforth, 1994.

well supported in the study by Alexander and Michener (1995), relationships within the family were left in doubt. A further phylogenetic study of the Family Halictidae based on morphological characters was made by Pesenko (2000a). He showed that the Rophitinae as well as the other subfamilies and tribes are monophyletic. He placed the Nomioidinae as a tribe within the Halictinae, but as it was the basal branch of the latter, sister group to all the other Halictinae, its recognition as a subfamily is equally logical and I think desirable. He divided the tribe Halictini into three subtribes: the Halictina (which he considered paraphyletic), Sphecodina, and Gastrohalictina, the last containing the genera *Lasioglossum, Homalictus,* and *Urohalictus. Sphecodes* and its relatives probably constitute an ancient parasitic group and its recognition as a subtribe is justifiable.

Major features of the classification herein and of Pesenko's classification (above) were supported by studies of Danforth (2002) and Danforth et al. (2004) based on single-copy nuclear genes (2,234 aligned nucleotide sites in all). The phylogeny supported by Danforth et al. (2004) is summarized in Figure 61-4. Their classification differs from mine in that the four subtribes of Figure 61-4 are all given tribal status. I consider them subtribes to preserve the Tribe Halictini in its traditional sense. An important feature is the group of the tribe Halictini consisting of western hemisphere genera *Agapostemon* to *Caenohalictus.* This is the *Agapostemon* group of Roberts and Brooks (1987), properly the subtribe Caenohalictina. It has long been recognized informally although not defined, but see Pesenko (2004b). Danforth's tree also places the Mesoamerican *Mexalictus* in a clade with *Patellapis* of the Old World tropical and austral regions, a relationship suggested very tentatively by Michener (2000, p. 340, 370, left-hand columns).

Halictid nests are burrows in the soil or rarely in rotting wood. Details of architecture as well as behavior are variable, and are reported below under each subfamily.

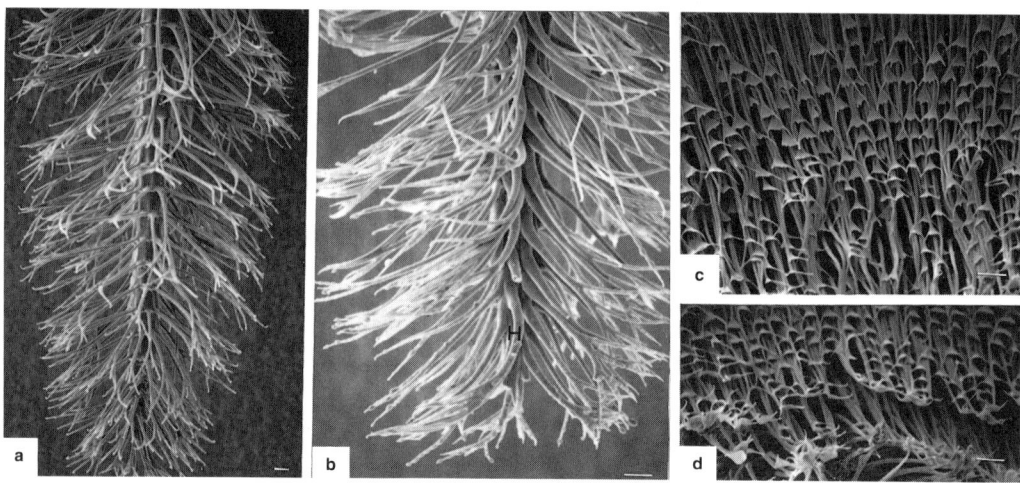

Figure 61-3. Details of glossa of Halictini, showing bifid or branched annular hairs. **a, b,** Apex of glossa of *Augochloropsis metallica* (Fabricius), anterior view and posterior view (H = seriate hair); **c, d,** Annular hairs on middle part of glossa of *Lasioglossum (Sphecodogastra) texanum* (Cresson) and *Sphecodes gibbus* (Linnaeus). (Scale lines = 0.01 mm.) SEM photos by R. W. Brooks, from Michener and Brooks, 1984.

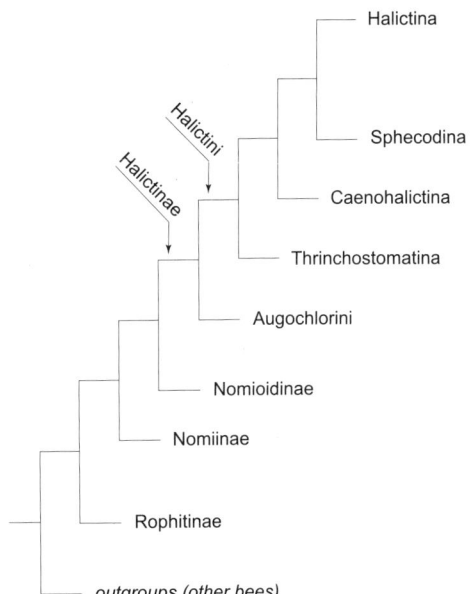

Figure 61-4. Phylogeny of Halictidae, modified from Danforth et al. (2004)

Key to the Subfamilies of the Halictidae

1. Antennae usually arising below midlength of eyes *and* antennal sockets separated from upper clypeal margin by little (if any) more than diameter of socket (Fig. 61-2); apex of labrum of female truncate or rounded (Fig. 61-2c), not produced to a process; hind basitarsus of female usually with apical angle or process but always without penicillus; trochanteral and femoral scopa reduced, longest hairs ordinarily shorter than those of tibial scopa .... Rophitinae (Sec. 62)
—. Antenna usually arising near midlength of eyes, or, *if* below that level, then separated from upper clypeal margin by much more than diameter of antennal socket; apex of labrum of female, except in some cleptoparasitic forms, produced to a process commonly bearing a strong keel; hind basitarsus of female, except in some parasitic forms, with apical process bearing penicillus; trochanteral and femoral scopa well developed except in parasitic forms, longest hairs ordinarily longer than those of tibial scopa and forming femoral corbicula [but femoral scopa poorly developed in some *Lipotriches* (Nomiinae) such as *L. tuleareasis* (Benoist)] 2

2(1). Epistemal groove below scrobal groove absent or a weak depression, sometimes directed strongly forward; when there are three submarginal cells, third usually about as long as first, or, *if* shorter, *then* usually twice as long as second (apex of marginal cell rather broadly rounded, Fig. 63-1 (prepygidial fimbria of female not divided medially; S7 of male a transverse plate with short apodemal arms and no midapical projection) ....
.... Nomiinae (Sec. 63)
—. Epistemal groove distinct and directed strongly downward below scrobal groove; when there are three submarginal cells, third shorter than first and less than twice as long as second 3

3(2). Apex of marginal cell pointed or minutely truncate; prepygidial fimbria of female divided by longitudinal median zone or triangle of very fine, dense hairs and punctation or sometimes nearly bare integument, the fimbria and its median zone absent in some cleptoparasitic forms; S7 of male consisting of a small discal region, long basolateral apodemes, and usually a midapical angle or process; S8 of male broader than long, variable in shape (S7 and S8 aberrant in *Urohalictus*) ....
.... Halictinae (Sec. 65)
—. Apex of marginal cell rounded or truncate; prepygidial fimbria of female not divided medially; S7 of male a broad and little modified sternum; S8 of male longer than broad, with broad spiculum and often long apical process (Eastern Hemisphere) ...... Nomioidinae (Sec. 64)

62. Subfamily Rophitinae

This is a group of andreniform bees that has been placed in various families in the past. The elongate and tapering prepalpal part of the galea; the slender, hairy lacinia high above the rest of the maxilla (Fig. 61-1a) (Michener and Greenberg, 1985), although not so isolated from the rest of the maxilla as in other halictids; and the usually largely membranous mentum and lorum (Fig. 33-4a; Michener, 1985a) are derived features characteristic of the family Halictidae. These features support inclusion of the Rophitinae in that family, although in *Systropha planidens* Giraud the mentum is darkened, not membranous, but, rather, rodlike, the proximal end partly enclosed between the lateral strips of the lorum illustrated by Michener (1985a). As in the Halictinae, the episternal groove is present below the level of the scrobal groove, but it is often a shallow valley, not a real groove. The scrobal groove, moreover, is often broad between the convex areas above and below it; the unusually convex lower mesepisternum contributes to this effect.

The most distinctive features of the Rophitinae are the following: The apex of the labrum of the female is broadly rounded (Fig. 61-2c), fringed with bristles, the apical labral process being either absent or represented by a depressed apical rim. The whole labrum is exposed between the closed mandibles. (Females of other halictids have an apical labral process that is hidden by the closed mandibles.) Many halictid males agree with the statements about the labrum of female rophitines, but the labrum in males of other halictids is usually shorter, less than half as long as the clypeus, whereas in rophitines it is usually more than half as long as the clypeus. The clypeus in rophitines is usually protuberant, its profile projecting at an angle to the nonprotuberant supraclypeal area, although the clypeal protuberance is not evident in *Xeralictus, Micralictoides, Sphecodosoma,* and some species of *Dufourea* and *Conanthalictus.* The clypeus is short, often no longer than the labrum, the epistomal suture between the anterior tentorial pits being only weakly arched (Fig. 61-2); in other halictids the clypeus occupies much more of the face, the epistomal suture between the subantennal sutures being arched high above the level of the tentorial pits (Fig. 33-2d). The antennal bases in Rophitinae are well below the middles of the eyes, and the subantennal sutures are not much if any longer than the diameter of an antennal socket (Fig. 61-2). Sometimes a faint line mesal to the subantennal suture suggests a second suture but is merely the margin of a smooth subantennal area that extends down to the small paraocular lobe. Genera having such an area are *Ceblurgus, Goeletapis, Penapis,* and *Systropha.* In the first three, the area is associated with the position of an anterior tentorial root, next to the antennal socket. This and other characters indicate that the South American rhophitines, *Ceblurgus, Goeletapis,* and *Penapis,* constitute a distinct clade not found on other continents, as first noted by Rozen (1997b). This group is sometimes given tribal status as the Penapini (Danforth et al., 2004). My initial reaction was to unite these three genera as subgenera of *Penapis.* Rozen, however, considers them to be as distinctive as are other genera of Rophitinae.

Except in *Xeralictus* the basal vein in Rophitinae is not as strongly curved as in Halictinae; its curvature is uniform (Fig. 62-1), not appearing principally in the basal region as in Halictinae. The scopa is well developed on the hind tibia, but weaker and consisting of simple hairs on the femur and trochanter. T5 of the female lacks the longitudinal specialized zone characteristic of most Halictinae, although the posterior margin of T5 usually is covered by short, dense hairs, and in some forms, especially *Xeralictus* but to a lesser extent *Systropha,* this covering extends forward medially to occupy a triangular area suggesting in its shape the broad median triangular area of such Halictinae as *Megalopta.* The pygidial plate of the female is rather ordinary, its margins being formed by the gradulus of T6; in other subfamilies the gradulus delimits a basal triangle that is extended posteriorly as a

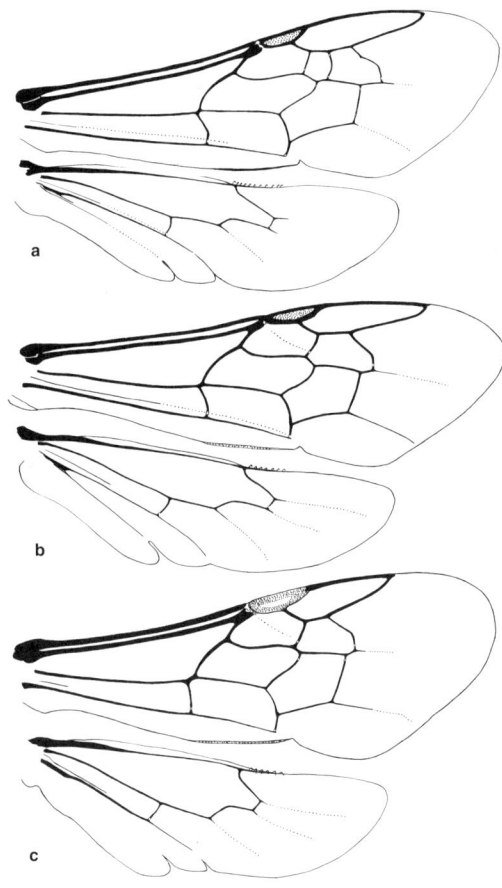

Figure 62-1. Wings of Rophitinae. **a,** *Systropha curvicornis* (Scopoli); **b,** *Dufourea marginata* (Cresson); **c,** *Micralictoides altadenae* (Michener).

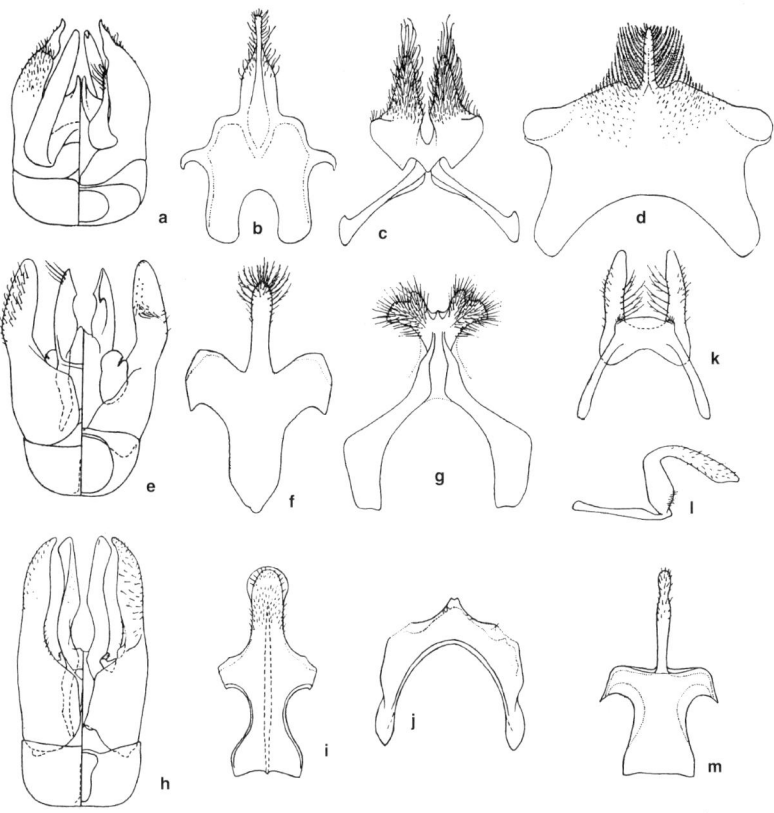

Figure 62-2. Structures of males of Rophitinae. **a-d,** Genitalia, S8, S7, and S6 of *Dufourea marginata* (Cresson); **e-g,** Genitalia, S8, and S7 of *Conanthalictus (Phaceliapis) nigricans* Timberlake; **h-j,** Same, of *Penapis penai* Michener; **k, l,** Ventral and lateral views of S7 of *Sphecodosoma pratti* Crawford; **m,** S8 of *S. pratti*. (Dorsal sides of genitalia are at the left.) From Michener, 1965c.

narrow strip or line, on each side of which, apically, is a more ordinary-looking pygidial plate depressed below the level of the triangle. S7 of the male (Fig. 62-2c, g, j) has two or rarely four, usually large, apical lobes arising from a small disc, except for *Ceblurgus, Penapis,* and *Xeralictus,* which have a large disc and either lack apical lobes or have small ones probably not homologous to the large lobes of other genera. S8 has a strong apical process and a subtruncate to bilobed base (Fig. 62-2b, f, i, m); i.e., it lacks a pointed spiculum, except for *Conanthalictus,* which has a strong spiculum. In *Rophites, Morawitzia,* and to varying degrees other Rophitini, like *Protomeliturga* in the Panurginae and one species of *Andrena* (Andreninae), the first two segments of the labial palpus are elongated and flattened in contrast to the last two segments (or last segment in *Rophites* s. str., which has three flattened long segments). This is a principal feature of L-T (long-tongued) bees; clearly, it must have arisen independently in a few genera of Halictidae and Andrenidae (see Sec. 19).

Although the Rhophitinae are easily distinguished from other subfamilies of Halictidae, most of the distinguishing characters of the Rophitinae are perhaps plesiomorphic. Alexander and Michener (1995), using a small sample of halictid genera in a parsimony analysis of adult characters, found the Rophitinae to be a paraphyletic group from which the Nomiinae and Halictinae were derived. *Nomioides* fell within the Rophitinae; but it is clearly an extrinsic element and is removed to a separate subfamily. The remaining Rophitinae do have probable apomorphic characters. The low antennal position and the short clypeus, with the epistomal suture not arched much above the level of the tentorial pits (Fig. 61-2), are probable synapomorphies of the Rophitinae. These characters were not included in the study by Alexander and Michener because they seemed variable and hard to present as a few discrete states, but in the context of the Halictidae they are important. The reduction of the scopa on the hind femur and trochanter may also be an apomorphy of Rophitinae, as compared to the strong scopa on these areas in Colletinae, Andreninae, and other Halictidae. Moreover, the larvae have distinctive characters that may be apomorphic for the subfamily (see below). The morphological study by Pesenko (2004b) and the molecular study by Danforth et al. (2004) found that the Rophitinae are monophyletic.

Danforth et al. (2004), on the basis of three single-copy nuclear genes, considers the Rophitinae to be divisible into two tribes, the Rophitini consisting of Old World and North American taxa, and the Penapini, consisting of the three South American genera.

Nests are burrows in the soil with lateral burrows leading to single cells or series of several cells. The cells are subhorizontal or slanting, short (only about 1.5 times as long as broad), and not or only slightly flattened on the lower surface; because the cells are short, observers are usually uncertain about any such flattening. The cell lining is a secreted membrane but it is dull and more pervious to water than that of the other subfamilies, and seems to be absent in *Protodufourea* (Rozen, Roig-Alsina, and Alexander, 1997); the larvae of Rophitinae may add a glistening lining during the feeding stage (Rozen, 1993a). The food mass in each cell is more or less spherical. Unlike those of all other Halictidae, most larvae spin cocoons, but those of one rophitine genus, *Conanthalictus,* do not do so (Rozen, 1993a). Overwintering is by prepupae, so far as is known.

Rozen (1993a) has described and compared various rophitine larvae. The larva of *Xeralictus* was described later by Snelling and Stage (1995a), although Rozen (1993a) had included it in a key to mature larvae. Rophitine larvae differ from those of other halictids in having paired, conical (not transverse) dorsolateral tubercles on most body segments, but those of the prothorax are either reduced, compared to those of other body segments, or absent. The larvae of *Conanthalictus* are very different from those of other known rophitine larvae, presumably because *Conanthalictus* does not spin a cocoon and has the recessed labiomaxillary region common to bee larvae that do not spin cocoons. Relevant details for *Conanthalictus* include fusion of the prementum and postmentum and loss of the lips of the salivary opening. *Xeralictus* may also not spin a cocoon; the one known postdefecating larva was not found in a cocoon, but it did not have the labiomaxillary reduction seen in *Conanthalictus.*

Warncke (1979b) revised the entire subfamily for the west palearctic region, placing all species except those of *Systropha* under the generic name *Rophites;* he sketched male genitalia and other structures. Below, the genera of the Western Hemisphere are keyed separately from those of the Eastern Hemisphere.

Key to the Genera of the Rophitinae of the Western Hemisphere

1. Forewing with two submarginal cells (Fig. 62-1b, c) 2
—. Forewing with three submarginal cells (Fig. 62-1a) 4
2(1). S8 of male without a spiculum but with a pair of basal lobes (Fig. 62-2b); penis valves large, elevated above bases of gonocoxites; volsella produced to a fingerlike hairy process (Fig. 62-2a); distance from apex of stigma to apex of marginal cell almost always at least as great as distance from apex of marginal cell to wing tip (Fig. 62-1b) (holarctic) .... *Dufourea*
—. S8 of male pointed to truncate at base (Fig. 62-2f, i), without pair of basal lobes; penis valves but little elevated above bases of gonocoxites; volsella relatively short, not produced to a fingerlike process (Fig. 62-2e, h); distance from apex of stigma to apex of marginal cell sometimes less than distance from apex of marginal cell to wing tip (Fig. 62-1c) 3
3(2). Dorsal surface of propodeum more than twice as long as metanotum; male with major apical lobes of S7 slender basally and strongly incurled apically; gonostylus slender and straplike (nearctic) .... *Micralictoides*
—. Dorsal surface of propodeum not or little longer than metanotum; male with apical lobes of S7 broad throughout and downcurved; gonostylus subtriangular and not well differentiated from gonocoxite (nearctic) ....
.... *Sphecodosoma* (in part)
4(1). Subantennal area sharply defined, smooth, impunctate, and hairless, extending down broadly to epistomal suture; anterior tentorial pit adjacent to lower outer edge of antennal socket; labial palpus longer than prementum, at least first three segments flattened (South America) ..
.... 5
—. Subantennal area often evident as small triangle, impunctate area below antenna, if present, not sharply defined, with punctures and hairs at least marginally; anterior tentorial pit on or just above epistomal suture; labial palpus equal to or shorter than prementum, not flattened (North America) 7
5(4). Clypeus four or five times as broad as long; mandible of both sexes with medial or basal tooth or projection on upper margin, in addition to preapical tooth; pygidial plate of male nearly linear (Brazil) .... *Ceblurgus*
—. Clypeus two or three times as broad as long; mandible with usual preapical tooth only; pygidial plate of male broad or absent 6
6(5). Maxillary palpus six-segmented; tibial scopal hairs simple; pygidial plate of male about as broad at base as long (Chile) .... *Penapis*
—. Maxillary palpus four-segmented; tibial scopal hairs plumose; pygidial plate of male absent (Peru) .... *Goeletapis*
7(4). Outer hind tibial spur and middle tibial spur coarsely serrate, some teeth long enough to make the spurs briefly pectinate (Fig. 62-4); S7 of male a broad plate without or with small apical lobe on each side, much smaller than disc of sternum; body length 8 mm or more (nearctic deserts) .... *Xeralictus*
—. Tibial spurs minutely ciliate, appearing simple; S7 of male with small disc to which are attached two or four large apical lobes, the lobes collectively much larger than disc; body length usually less than 7.5 mm 8
8(7). Dorsal surface of propodeum about two-thirds as long as scutellum and about one-half as long as posterior surface of propodeum, as seen from side (Fig. 62-3a); S7 of male with four apical lobes (nearctic) .... *Protodufourea*
—. Dorsal surface of propodeum about as long as scutellum, more than one-half as long as posterior surface of propodeum, as seen from side (Fig. 62-3b); S7 of male with two apical lobes (Fig. 62-2g, k) 9
9(8). Clypeal truncation more than twice as long as length of clypeus; body surface minutely roughened and usually more or less dull; S7 of male relatively flat; spiculum of S8 of male pointed (Fig. 62-2f) (nearctic) ....
.... *Conanthalictus*
—. Clypeal truncation less than twice as long as length of clypeus; body surface smooth and shining between punctures; S7 of male with disc at right angles to basal arms and apical lobes (Fig. 62-2l); base of S8 of male broadly truncate (Fig. 62-2m) (nearctic) .... *Sphecodosoma* (in part)

Key to the Genera of the Rophitinae of the Eastern Hemisphere

1. Margin of marginal cell on costa shorter than stigma and less than half as long as distance from apex of marginal cell to wing tip; two submarginal cells, second less than

half length of first; body with yellow integumental markings (palearctic) .... *Morawitzella*
—. Margin of marginal cell on costa as long as or longer than stigma, and as long as or longer than distance from apex of marginal cell to wing tip; two or three submarginal cells, but if two, then second more than half as long as first; body without yellow markings 2

2(1). Flagellum of male with first six segments more or less normal, remaining segments abruptly more slender than preceding segments, tightly curled, and sometimes reduced in number; metasoma of both sexes, at least laterally, with abundant, long, more or less erect hair; S8 of male with apical process very broad, projecting, thus suggesting in undissected specimens an eighth tergum (forewing with three submarginal cells) .... *Systropha*
—. Flagellum of male not so modified; metasoma without abundant long hair; S8 of male with apical process slender, capitate, often projecting but not suggesting a tergal plate 3

3(2). Male without pygidial plate, or sometimes with shiny bare area but this area not defined by carinae and not elevated; base of S8 of male deeply bilobed (Fig. 60-2b); labial palpus with first two segments, or at least second, slender and similar in width to third and fourth segments (metasomal hair bands absent or sparse, integument thus visible through bands of hair) (holarctic) .... *Dufourea*
—. Male with well defined, strongly elevated pygidial plate [not verified for *Rophites (Flavodufourea)*]; base of S8 of male rounded to weakly bilobed; labial palpus with first two segments broad and flattened in contrast to third and fourth segments (except that in *Rophites* s. str. third segment is on same axis as second and also rather broad basally, only fourth being freely articulated) 4

4(3). Metasomal terga with well-developed, dense, pale apical hair bands; forewing with two submarginal cells; pygidial plate of male margined by sharp carina (palearctic) .... *Rophites*
—. Metasomal terga without pale apical hair bands; forewing with three submarginal cells; pygidial plate of male not margined by sharp carina (Turkey) .... *Morawitzia*

Genus *Ceblurgus* Urban and Moure

Ceblurgus Urban and Moure, 1993: 102. Type species: *Ceblurgus longipalpis* Urban and Moure, 1993, by original designation.

Ceblurgus consists of a robust nonmetallic species about 7.5 mm long. The base of the mandible of the male has a large tuft of long white hairs, directed mesad when the mandibles are closed. The lacinia, if correctly identified, is membranous, with few hairs, and high up on the labiomaxillary tube, as in *Penapis*. And as in *Penapis* and *Goeletapis,* an anterior tentorial root is located against the lower lateral margin of the antennal socket. The short, four-segmented maxillary palpi are as in *Goeletapis,* but the three-toothed mandibles are unique in Rophitinae; all other rophitines have six-segmented maxillary palpi and mandibles with a single preapical tooth, the tooth sometimes extremely reduced. S7 of the male is a broad plate with broad basolateral arms; the apex is truncate and lacks lobes. In this respect *Ceblurgus* is similar to *Penapis* (Fig. 62-2j) and *Goeletapis*. S8 of the male is subtruncate, not bilobed, basally. The male gonostylus seems triangular, as is the case in many rophitines, but is indistinguishably fused to the gonocoxite; the volsella is elongate, without distinct digitus and cuspis, and is pointed apically (not a fingerlike, hairy projection as in *Dufourea*). The male genitalia and other structures were illustrated by Urban and Moure (1993).

■ *Ceblurgus* is known from the states of Bahia and Pernambuco, Brazil. It is the only rophitine from eastern South America. The single species is *C. longipalpis* Urban and Moure.

Genus *Conanthalictus* Cockerell

Conanthalictus consists of species with greenish or bluish coloration on at least some parts of the head and thorax, sometimes with a red metasoma; the body surface is minutely roughened. The basal area of the propodeum is as long as the scutellum (Fig. 62-3b). Plesiomorphic features include the small, chelate volsellae (Fig. 62-2e) (as usual in Halictidae) and the pointed spiculum of S8 of the male (Fig. 62-2f). S7 of the male is relatively flat, lacks the vertical disc of *Sphecodosoma,* and has two apical lobes (Fig. 62-2g). T6 of the male has a dense fringe that largely obscures T7. A flat-topped apical process on T7 of the male constitutes the pygidial plate, which is not elevated above the surface of the tergum as it is in *Rophites*. The male genitalia and hidden sterna of both subgenera were illustrated by Michener (1965c); see also Figure 62-2e-g.

All species of *Conanthalictus* appear to be oligolectic, mostly on certain genera of Hydrophyllaceae. The nesting biology has been described by Rozen and McGinley (1976) and Rozen (1993a). Except for their minute size, the nests are similar to those of *Dufourea*.

Key to the Subgenera of *Conanthalictus*

1. Head elongate oval, longer than broad; antennal bases twice as far from ocelli as from anterior clypeal margin .... *C. (Conanthalictus. s. str.)*
—. Head broader than long; antennal bases scarcely 1.5 times as far from ocelli as from anterior clypeal margin .... *C. (Phaceliapis)*

Conanthalictus / Subgenus *Conanthalictus* Cockerell s. str.

Halictus (Conanthalictus) Cockerell, 1901b: 209. Type species: *Halictus conanthi* Cockerell, 1901, monobasic.

Conanthalictus s. str. consists of minute, long-headed species, 3 to 4.5 mm long.

■ The range is western Texas to Arizona, USA. For the two species, see Timberlake (1961).

Conanthalictus / Subgenus *Phaceliapis* Michener

Conanthalictus (Phaceliapis) Michener, 1942a: 277. Type species: *Conanthalictus bakeri* Crawford, 1907, by original designation.

These are the broad-headed species of *Conanthalictus,* mostly larger than *Conanthalictus* s. str. (body length 3-7 mm).

■ The range is western Texas to southern California, USA, south at least to Chihuahua and Baja California,

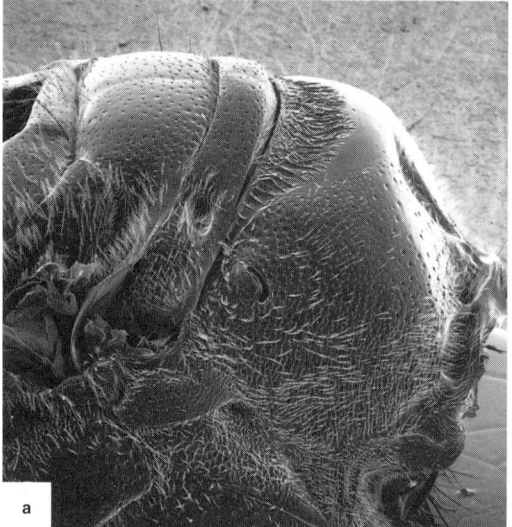 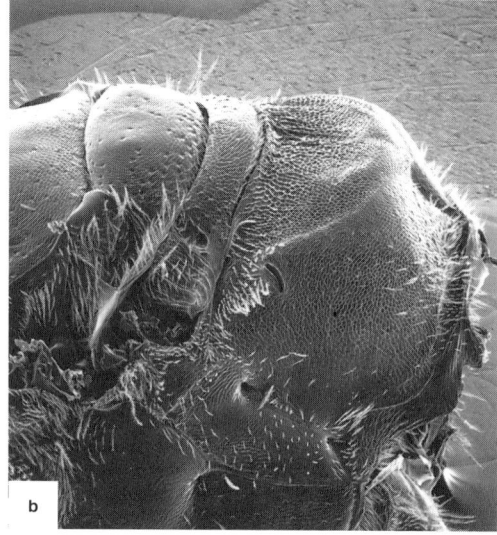

Figure 62-3. Dorsolateral views of posterior part of thorax of two rophitine genera. **a,** *Protodufourea parca* Timberlake; **b,** *Conanthalictus bakeri* Crawford. From Michener, McGinley, and Danforth, 1994.

Mexico. The 11 described species were reviewed by Timberlake (1961).

A specimen of an apparently undescribed western Texas species seems to be intermediate between *Conanthalictus* s. str. and *Phaceliapis*. Perhaps the two subgenera merge.

Genus *Dufourea* Lepeletier

Dufourea Lepeletier, 1841: 227. Type species: *Dufourea minuta* Lepeletier, 1841, by designation of Richards, 1935: 172. [See Michener, 1997b.]

Halictoides Nylander, 1848: 195. Type species: *Halictoides dentiventris* Nylander, 1848, by designation of Cockerell and Porter, 1899: 420.

Dufourea (Trilia) Vachal, 1900: 534. Type species: *Dufourea muoti* Vachal, 1899, monobasic.

Halictoides (Parahalictoides) Cockerell and Porter, 1899: 420. Type species: *Halictoides campanulae* Cockerell, 1897, by original designation.

Halictoides (Epihalictoides) Cockerell and Porter, 1899: 420. Type species: *Panurgus marginatus* Cresson, 1878, monobasic.

Conohalictoides Viereck, 1904a: 245 [also described as new in 1904b: 261]. Type species: *Conohalictoides lovelli* Viereck, 1904 = *Panurgus novaeangliae* Robertson, 1897, by original designation.

Neohalictoides Viereck, 1904b: 261. Type species: *Panurgus maurus* Cresson, 1878, by original designation.

Cryptohalictoides Viereck, 1904b: 261. Type species: *Cryptohalictoides spiniferus* Viereck, 1904, by original designation.

Mimulapis Bridwell, 1919: 162. Type species: *Mimulapis versatilis* Bridwell, 1919, by original designation.

Betheliella Cockerell, 1924b: 169. Type species: *Betheliella calocharti* Cockerell, 1924, monobasic.

Halictoides (Cephalictoides) Cockerell, 1924c: 244. Type species: *Halictoides paradoxus* Morawitz, 1867, by original designation.

Rophites (Dentirophites) Warncke, 1979b: 130. Type species: *Dufourea gaullei* Vachal, 1897, by original designation.

Rophites (Merrophites) Warncke, 1979b: 133. Type species: *Dufourea merceti* Vachal, 1907, by original designation.

Rophites (Microrophites) Warncke, 1979b: 133. Type species: *Rophites quadridentatus* Warncke, 1979, by original designation.

Rophites (Cyprirophites) Warncke, 1979b: 135. Type species: *Dufourea caeruleocephala cypria* Mavromoustakis, 1952, by original designation.

Rophites (Carinorophites) Warncke, 1979b: 136. Type species: *Dufourea rufiventris* Friese, 1898, by original designation.

Dufourea (Alpinodufourea) Ebmer, 1984: 360. Type species: *Dufourea alpina* Morawitz, 1865, by original designation.

Dufourea (Atrodufourea) Ebmer, 1984: 360. Type species: *Rophites atrata* Warncke, 1979, by original designation.

Dufourea (Minutodufourea) Ebmer, 1984: 361. Type species: *Dufourea minutissima* Ebmer, 1976, by original designation.

Dufourea (Afrodufourea) Ebmer, 1984: 362. Type species: *Dufourea punica* Ebmer, 1976, by original designation.

Dufourea (Glossadufourea) Ebmer, 1993b: 32. Type species: *Dufourea longiglossa* Ebmer, 1993, by original designation.

Dufourea is the major genus of the Rophitinae. It consists of andreniform bees 3.5 to 11.0 mm long, either nonmetallic black or dull greenish or bluish metallic, sometimes with the metasoma red. Metasomal pale hair bands are weak or absent. The clypeus is strongly protuberant at an angle to the unswollen supraclypeal area, as seen in lateral view. The male has no elevated pygidial area, although a bare, unelevated area that lacks marginal carinae may be recognizable. The volsella (probably its digitus) is elongate, fingerlike, and hairy distally (Fig. 62-2a), the cuspis being not or scarcely recognizable. The penis valve is strongly Z-shaped in lateral view; its middle section, vertical or sloping basad, elevates the base of the distal section well above the level of the gonocoxites; and the lower end of the middle section has a projection distad immediately mesal to the volsella. The base of S8 of the male is deeply bilobed (Fig. 62-2b), there being no median spiculum; the apical process is slender and usually capitate. Illustrations of the male genitalia, hidden sterna, and other structures are included in Popov (1957a, as *Trilia;* 1958c; 1959a), Mitchell (1960), Constantinescu (1974a), Wu (1982b, 1983f, 1987a), Ebmer (1984, 1987b, 1989, 1993b) and Pesenko (1998); see also Figure 62-2a-c.

■ *Dufourea* is a holarctic genus that ranges from the Canary Islands and Britain eastward to Japan and across the full width of North America. In the palearctic region it occurs north to Finland (64°N), and south to North African countries. In North America it occurs north to British Columbia and Alberta, Canada, and from Minnesota to Maine, USA, southward to the state of Oaxaca, Mexico. Species are far more numerous in western America, especially California, than in eastern America. In the palearctic region there are about 60 species, and in the nearctic region about 70 others. Revisional papers are by Ebmer (1984, 1987b, 1989, 1993b, for palearctic species; these exclude *Trilia*) and Warncke (1979b, west palearctic; this includes as *Rophites* all palearctic taxa in the subfamily except *Systropha*). Wu (1987a) gave a key to Chinese species of the *Halictoides* group.

Dufourea is diverse morphologically, especially in the males, with the result that many generic or subgeneric names have been proposed for particular species or species groups. Ebmer (1984) in particular surveyed the diversity of mouthparts. Mouthparts are particularly short in such forms as *D. gaullei* Vachal *(Dentirophites),* in which the glossa is only about half as long as the prementum. In contrast, in diverse species groups the glossa is longer than the prementum. In the several species (e.g., *D. cypria* Mavromoustakis) that are grouped under the name *Cyprirophites,* the labial palpi are half as long as the maxillary palpi or less; in all others they are over half as long as the maxillary palpi. Most species are oligolectic, and such proboscidial characters are no doubt related to floral associations.

A species with particularly long mouthparts is *Dufourea versatilis* (Bridwell), in which, presumably to provide space for such a long proboscis, the malar space is about as long as wide in males and half as long as wide in females. This is a unique feature in the Rophitinae, not duplicated in *D. longiglossa* Ebmer, which also has a very long glossa, over twice as long as the head, although its other mouthparts are only moderately long, the galeal blade being a little shorter than the head and covered with dense, short, erect hair.

Most of the characters that have been used to form subgenera of *Dufourea* are features of males. A widely used division by Timberlake (1939), Popov (1959a), and Wu (1987a) separated those in which S6 of the male is simple (*Dufourea* s. str.) from those in which S6 has modifications (all others, placed in *Halictoides*). Wu in particular lists other differences for palearctic species. The problem with this dichotomy is that the sternal modifications, such as those shown in Figure 62-2d, can take various forms, and there is thus no certainty that they are homologous; on the contrary, the structures of S6 are diverse and probably had several independent origins. Moreover, the modifications are sometimes so weak that one wonders whether certain species should be called *Dufourea* or *Halictoides.*

Structures of the legs, antennae, and sterna of the males have been used in the recognition of genera or subgenera. For the most part, one or very few species are segregated from the less modified forms by such characters. The following are a few examples: In *Dufourea versatilis* (Bridwell) *(Mimulapis)* the three basal flagellar segments are swollen, and on their upper surfaces are strongly concave and hairy; the middle tibia is greatly modified. In *Dufourea spinifera* (Viereck) *(Cryptohalictoides),* all the legs are grotesquely modified. In the species placed in *Cyprirophites,* S7 is particularly delicate, the parts all slender, and the two apical lobes are small and narrowly attached medially to the small disc of the sternum.

Most of the taxa with specialized male features are clearly derived from less modified sorts of *Dufourea.* In spite of their striking characters, it seems unnecessary to recognize them at the subgenus or genus level. To do so would leave a large, undifferentiated, paraphyletic group.

A few species of *Dufourea* have a paraocular lobe extending down into the clypeus, somewhat as in *Augochlora* (Fig. 67-6d) and certain other Halictinae; accompanying this lobe is a shining depression that may be associated with the articulation of the base of the maxillary cardo, which lies against the inner surface of the clypeus. If the species in which this occurs had other shared diagnostic features, they might constitute a useful group, but the epistomal suture is weak laterally and recognition of the lobe is not always easy. Species having such lobes include those of the groups called *Conohalictoides* [*D. monardae* (Viereck) and *novaeangliae* (Robertson)] and *Mimulapis* [*D. versatilis* (Bridwell)], as well as *D. pectinipes* Bohart.

The group called *Trilia* differs from other *Dufourea* in the presumably plesiomorphic character of having three submarginal cells. In other respects, however, these bees are *Dufourea,* as shown by Popov (1957a).

Most species of *Dufourea* are oligolectic, but some, such as the European *D. paradoxa* (Morawitz), appear to be polylectic. Nesting biology has been little studied for the genus *Dufourea.* Eickwort, Kukuk, and Wesley (1986) gave an account of that of *D. novaeangliae* (Robertson), an oligolege on *Pontederia.* Torchio et al. (1967) investigated the nesting biologies of several species. They found the burrows to be relatively shallow, with laterals each ending in one or rarely more subhorizontal or slanting cells thinly lined with a dull "waxlike" film and containing a spherical pollen mass. The larva feeds around the pollen mass, which therefore remains spherical as it is reduced in size.

Genus *Goeletapis* Rozen

Goeletapis Rozen, 1997b: 8. Type species: *Goeletapis peruensis* Rozen, 1997, by original designation.

Goeletapis is similar in appearance to *Penapis*, more slender than the remaining South American dufoureine genus, *Ceblurgus*. It resembles *Ceblurgus*, however, in the maxillary palpus, which is four-segmented; the labial palpus, the fourth segment of which is not directed away from the palpal axis; the female middle trochanter, which is ventrally carinate; and the tibial scopa, which is plumose.

■ This genus is known only from the arid western slope of Peru. The only species is *Goeletapis peruensis* Rozen.

Genus *Micralictoides* Timberlake

Dufourea (Micralictoides) Timberlake, 1939: 397. Type species: *Halictoides ruficaudus* Michener, 1937, by original designation.

Micralictoides consists of minute species (5.5-6.0 mm long), nonmetallic black, the metasoma sometimes red. Bees of this genus resemble small species of *Dufourea*, but the clypeus is not or little protuberant, as seen in profile; the volsellae are polished knobs, not long and hairy as in *Dufourea*; the male gonostyli are slender and straplike, about as long as the gonocoxites, from which they are clearly differentiated; S7 of the male has the two major apical lobes long, slender, and strongly curved mesad and sometimes also basad at their apices; and the base of the male S8 is truncate or subtruncate. As in some *Dufourea*, the legs and S6 of the male are unmodified and the male antennae are short, the middle flagellar segments not or scarcely longer than broad and without modified hairs.

■ This genus occurs in the southwestern United States (California, Nevada, and Arizona) and probably in Baja California, Mexico. The eight known species were revised by Bohart and Griswold (1987).

Genus *Morawitzella* Popov

Morawitzella Popov, 1957a: 916. Type species: *Epimethea nana* Morawitz, 1880, by original designation.

Morawitzella contains a minute species (length 3.5-4.0 mm) that differs from all other Rophitinae in its reduced wing venation (see the key to genera) and the yellow integumental markings on all tagmata. The head is much broader than long. The apical process of S8 of the male is not enlarged at its tip, and the base of S8 is truncate, not bilobed as in *Dufourea*. Ebmer (1984) has tabulated some of its characters, and Popov (1957a) illustrated various features, including the male genitalia and hidden sterna. The female is unknown.

■ This genus occurs in central Asia. The single species is *Morawitzella nana* (Morawitz).

It would be interesting to know whether Warncke's (1979b) illustration showing the maxillary palpus as seven-segmented is correct.

Genus *Morawitzia* Friese

Morawitzia Friese, 1902: 185. Type species: *Morawitzia panurgoides* Friese, 1902, by designation of Sandhouse, 1943: 574. [See Michener, 1997b.]

This genus differs from *Dufourea* not only in having three submarginal cells (like the *Dufourea* group called *Trilia*) but also in the following characters: the male pygidial plate is strongly elevated (not margined by a carina); the base of S8 is weakly emarginate; the two broad apical lobes of S7 are provided with long hairs; and there is a strong median angle on the dorsomedian margin of the male gonocoxite (such that the mesal margins of the gonocoxites are close together and parallel basal to this angle). The penis valves and volsellae are as described for *Rophites*. Male genitalia and hidden sterna were illustrated by Schwammberger (1975b). The body is robust, 11 to 13 mm long.

■ This genus is found in Turkey. The three species were reviewed by Schwammberger (1975b).

Genus *Penapis* Michener

Penapis Michener, 1965c: 324. Type species: *Penapis penai* Michener, 1965, by original designation.

This genus consists of rather slender, superficially *Dufourea*-like bees, 7 to 10 mm in length, lacking metallic coloration. Like the other South American genera, *Ceblurgus* and *Goeletapis*, it differs from most *Dufourea* in having three submarginal cells (as in the *Trilia* group of *Dufourea*); a clypeus that is not protuberant at an angle to the supraclypeal area, as seen in profile; small volsellae; a broad disc of the male S7, with long and rather broad basolateral arms but no apical lobes, the apex of the sternum being a median, truncate, or minutely bidentate projection; and a truncate rather than bilobed base of S8 of the male (Fig. 62-2h-j). The lacinia, as in *Ceblurgus*, is difficult to identify, but is probably a membranous lobe with a few hairs well above the rest of the maxilla. The pygidial plate of the male is broad, rounded at the apex, occupying most of the dorsum of T7, and delimited by a strong carina; such a plate is not found in other male American Rophitinae. Structures were illustrated by Michener (1965c) and Rozen (1997b).

■ *Penapis* is found in northern Chile, from Tarapacá to Coquimbo. The three described species were revised by Rozen (1997b).

Four genera, all in the Western Hemisphere, are unusual among Rophitinae in lacking the usual broad apical lobes on S7 of the male. These genera are *Ceblurgus*, *Goeletapis*, *Penapis*, and *Xeralictus*. The first three of these genera, i.e., those from South America, constitute the tribe Penapini of some authors.

Genus *Protodufourea* Timberlake

Protodufourea Timberlake, 1955a: 105. Type species: *Protodufourea wasbaueri* Timberlake, 1955, by original designation.

Protodufourea, which resembles *Dufourea* superficially, consists of nonmetallic species 5.5 to 7.0 mm long, the metasoma in some females red. These bees differ from all American *Dufourea* not only in having three submarginal cells but also in their robust form, the dorsum of the propodeum being but little longer than the metanotum (Fig. 62-3a); the small volsellae; the relatively low penis valves, not Z-shaped as seen in side view; and the subtruncate rather than bilobed base of S8 of the male. The short propodeum and robust form are as in *Sphecodosoma (Michenerula)*, a taxon from which *Protodufourea* differs in the number of submarginal cells and in the form of the four small apical lobes of the male S7, the disc of which is flat, not vertical as in *Sphecodosoma*. The genitalia and hidden sterna were illustrated by Michener (1965c) and Bohart and Griswold (1997).

■ This genus is found in California and Arizona. The five species were revised by Bohart and Griswold (1997).

Protodufourea species appear to be oligolectic visitors to flowers of *Phacelia* and *Emmenanthe* (Hydrophyllaceae). The nesting behavior was described by Rozen, Roig-Alsina, and Alexander (1997).

Genus *Rophites* Spinola

Rophites is a close relative of *Dufourea*, and these two taxa could be regarded as congeneric. The principal differences are as follows: In *Rophites* the metasomal terga have dense apical hair bands. In other Rophitinae the bands are absent or thin, not hiding the surfaces of the tergal margins. As in *Ceblurgus* and *Penapis*, T7 of the male *Rophites* has a strongly elevated pygidial plate margined by a sharp, raised carina (not verified for the subgenus *Flavodufourea*). In male *Morawitzia* there is a raised pygidial area, but it is not margined by a sharp carina; other rophitine male pygidial areas, if recognizable, are not defined, except for being hairless. In *Rophites* the base of S8 of the male is convex to subtruncate or feebly emarginate; in *Dufourea* it is deeply bilobed. The *Dufourea*-like characters of *Rophites* include (1) the hairy apical projection of the volsellar digitus, although it is shorter than in *Dufourea* and the cuspis is nearly as long as the digitus, the two pressed together (in *Dufourea* the cuspis is unrecognizable or much shorter than the long digitus); (2) the moderately Z-shaped penis valve, with an angle produced apicad near the volsella; and (3) the capitate apical process of S8 of the male.

Three subgenera are distinguishable.

Key to the Subgenera of *Rophites*

1. First segment of labial palpus strongly broadened in apical half; apical lobes of S7 of male broad basally and tapering to points.................................... *R. (Flavodufourea)*
—. First segment of labial palpus nearly parallel-sided; apical lobes of S7 of male slender basally, slightly enlarged and rounded apically...................................... 2
2(1). Labial palpal segments 1 to 3 flattened, fourth segment not flattened and sometimes divergent from axis of other segments; labial palpus longer than maxillary palpus; frons of female with several coarse bristles, these hooked or curved at tips (but bristles absent in *Rophites gusenleitneri* Schwammberger) .... *R. (Rophites s. str.)*
—. Labial palpus with at least segment 3 (as well as 4) not flattened, commonly diverging from axis of 1 and 2; labial palpus shorter than maxillary palpus; frons of female without coarse bristles .... *R. (Rhophitoides)*

Rophites / Subgenus *Flavodufourea* Ebmer

Dufourea (Flavodufourea) Ebmer, 1984: 373. Type species: *Dufourea flavicornis* Friese, 1913, by original designation.

The type species of this subgenus was originally placed in *Dufourea*, and subsequently in *Rophites* (Popov, 1946) and *Rhophitoides* (Schwammberger, 1975a). Its strong metasomal hair bands and the subtruncate or scarcely concave base of S8 of the male support its inclusion in *Rophites* in the sense of this work. The mouthparts are short; the maxillary palpus is shorter than the labial palpus, an unusual feature in Rophitinae but one that is found also in *Rophites* s. str. Mouthparts, genitalia, and hidden sterna were illustrated by Schwammberger (1975a) and Ebmer (1984).

■ *Flavodufourea* is known from Mongolia and Kasakhstan. Two species are recognized.

A. W. Ebmer (in litt., 1994) considers *Flavodufourea* to be closer to *Dufourea* than to *Rophites*. I have not seen specimens, and place it with *Rophites* because of the characters listed above, supported by Popov's (1946) judgment. *Flavodufourea* was returned to the genus *Dufourea* from its present placement, and a second species (from Kasakhstan) was described, by Patiny (2003a). The most distinctive feature of *Dufourea*, the deeply bilobed base of S8 of the male, was not reported by Patiny; therefore placement in *Dufourea* does not seem to be decisive.

Rophites / Subgenus *Rhophitoides* Schenck

Rhophitoides Schenck, 1861: 69 [for the date, see Michener, 1968a]. Type species: *Rhophitoides distinguendus* Schenck, 1861 = *Rophites cana* Eversmann, 1852, monobasic.

This subgenus consists of forms 6 to 8 mm in length. For distinguishing characters, see the key to the subgenera.

■ The range is from Morocco and France to Turkey and the Caucasus. The four species were revised (along with *Flavodufourea*) by Schwammberger (1975a).

Nesting biology has been reported by various authors, such as Enslin (1921), Malyshev (1925a), and Wilkaniec, Wójtowski, and Szyma (1985). The nests are as described above for the subfamily; there are one to four subhorizontal cells per nest. *Rophites canus* Eversmann is a significant pollinator of alfalfa in Eurasia and an oligolege on small legumes.

Rophites / Subgenus *Rophites* Spinola s. str.

Rophites Spinola, 1808: 8, 72. Type species: *Rophites quinquespinosus* Spinola, 1808, monobasic.
Rhophites Agassiz, 1846: 29, unnecessary emendation.

Members of this subgenus are usually larger than those of *Rhophitoides*, with a body length of 5 to 11 mm. The only species less than 8 mm long is *R. gusenleitneri* Schwammberger, which, in the absence of frontal bristles as well as in its size, resembles *Rhophitoides*. Male genitalia and other structures were illustrated by Schwammberger (1971b), Benedek (1973), Constantinescu (1974a), Ebmer and Schwammberger (1986), and Ebmer 1993a).

■ This subgenus ranges from Spain and Morocco to Asia Minor, southern Russia, and eastward to Mongolia, occurring north to 60°N in Europe. The 13 species were revised by Ebmer and Schwammberger (1986).

Müller (1996a) found that the bristles on the frons are used with buzzing to remove pollen from flowers of mints (Lamiaceae).

Genus *Sphecodosoma* Crawford

This genus consists of small to minute, nonmetallic, strongly punctate bees with the interspaces smooth, unlike those of *Conanthalictus;* the metasoma of the female is often red. Males differ from those of *Dufourea* in their short volsellae, without or with few hairs, the ordinary rather than elevated bases of the penis valves (so that the valves are not Z-shaped as seen in side view), the truncate base of S8 of the male (Fig. 62-2m), and the downcurved apical process of S8. Especially characteristic is the vertical disc of S7, lying more or less at right angles both to the basal arms and to the two apical lobes (Fig. 62-2l). The character of wing venation cited in the key to genera, couplet 2, is weak; it is therefore difficult to distinguish females from those of small *Dufourea*. The clypeus is not strongly protuberant as seen in side view, however, a helpful feature in distinguishing *Sphecodosoma* from *Dufourea*.

There are two subgenera of *Sphecodosoma*, each hitherto given generic status but actually closely related.

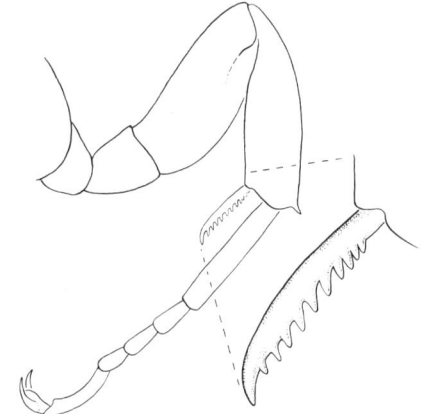

Figure 62-4. Middle leg of *Xeralictus timberlakei* Cockerell (Rophitinae), male, showing coarsely serrate tibial spur. That of the female has fewer but equally coarse teeth. From Michener, McGinley, and Danforth, 1994.

Key to the Subgenera of *Sphecodosoma*

1. Fore wing with two submarginal cells; dorsal surface of propodeum not or little longer than metanotum; body length about 6 mm .... *S. (Michenerula)*
—. Fore wing usually with three submarginal cells; dorsal surface of propodeum much longer than metanotum; body length 4.5 mm or less .... *S. (Sphecodosoma s. str.)*

Sphecodosoma / Subgenus *Michenerula* Bohart

Michenerula Bohart, 1965: 320. Type species: *Michenerula beameri* Bohart, 1965, by original designation.

This subgenus contains small (6 mm long) species distinguishable from similar small rophitines by the robust body form with the apex of the metasoma strongly curled under in the male. The thorax is robust in part because the propodeum is short, its dorsal surface little longer than the metanotum. Genitalic, sternal, and other structures were illustrated by Bohart (1965).

■ *Michenerula* is found in western Texas, USA, and south to Durango, Mexico. It contains one described species and probably a second, undescribed species.

Sphecodosoma / Subgenus *Sphecodosoma* Crawford s. str.

Sphecodosoma Crawford, 1907: 182. Type species: *Sphecodosoma pratti* Crawford, 1907, by original designation.

This subgenus consists of minute species (length 3.0-4.5 mm) that appear more slender than *Michenerula*, at least in part because of the longer dorsal surface of the propodeum. Genitalia and hidden sterna were illustrated by Michener (1965c).

■ *Sphecodosoma* s. str. ranges from southeastern Kansas, Oklahoma, and western Texas to southern California, USA, and south to the state of Oaxaca, Mexico. The two species were differentiated by Timberlake (1961).

Sphecodosoma s. str. seems to be oligolectic on *Nama* (Hydrophyllaceae). The nesting biology was described by Rozen and McGinley (1976, as *Conanthalictus*) and by Rozen (1993a).

Specimens with only two submarginal cells, because of the loss of one of the transverse cubital veins, are sometimes found, as well as specimens in which both veins are present but unusually close together.

Genus *Systropha* Illiger

Systropha Illiger, 1806: 145. Type species: *Andrena spiralis* Olivier, 1789 = *Eucera curvicornis* Scopoli, 1770, monobasic.

Systropha (Systrophidia) Cockerell, 1936b: 477. Type species: *Systropha ogilviei* Cockerell, 1936, monobasic.

Systropha is a genus of rather robust rophitines, 6 to 14 mm long. Except in the smaller species, the head is noticeably small, distinctly narrower than the thorax. The mouthparts are elongate but variable. The first three segments of the labial palpus are flattened and sheathlike, only the fourth segment being not flattened and sometimes diverging from the main axis of the palpus. The maxillary palpus is usually longer than the labial palpus, but in *S. ogilviei* Cockerell from southern Africa the former is shorter, the glossa and labial palpi being greatly elongated; these characters are the basis for the subgeneric name *Systrophidia*. The apically curled antennae of males are among the most remarkable among bees; see the key to the genera. The number of male antennal segments varies from 11 to 13. S7 of the male has two large apical lobes and a small disc. The distal part of the apical process of S8 is greatly expanded, the apical margin convex. The male genitalia, sterna, and other structures were illustrated by Ponomareva (1967), Popov (1967), Constantinescu (1974a), Warncke (1976b), Ebmer (1994), and Baker (1996c).

■ This genus is widespread in the palearctic region from Spain and Morocco east to Tadzhikistan, north as far as southern Germany. Southward, the genus is known in both East and West Africa, south to Namibia, and in Asia south to Sri Lanka and Thailand. There are about 25 species. The palearctic species were revised by Ponomareva (1967); palearctic and Oriental species were reviewed by Baker (1996c); and African and tropical Asian species were listed by Ebmer (1994).

At least some and perhaps all of the species are specialists on *Convolvulus* and carry large loads of its pollen not only in the tibial scopa but among the abundant plumose hairs of the venter, the sides, and even the posterior dorsum of the metasoma (see photo by Westrich, 1989: 878). The nesting biology was described by Malyshev (1925c) and Grozdanić and Vasić (1968). The laterals, each leading to a cell, are much shorter than those of most Rophitinae, mostly less than 1 cm long.

Genus *Xeralictus* Cockerell

Xeralictus Cockerell, 1927a: 41. Type species: *Xeralictus timberlakei* Cockerell, 1927, monobasic.

Xeralictus is a morphologically isolated genus of nonmetallic, rather large (length 11.0-12.5 mm) bees. The coarsely serrate middle and outer hind tibial spurs of both sexes (Fig. 62-4) are unique. The mandible of the male is broad, with a blunt dorsal angle before the middle. In the male the labrum has a median basal projection; there is a strong tooth behind the mandibular base and another on the posterior part of the hypostomal carina; S4 to S6 are highly modified; S7 is unique among Rophitinae, being a large plate, longer than broad, lacking long basolateral arms but having small, hairy lateroapical lobes; S8 has a rounded, not bilobed, base. T7 of the male is broad, hairy, and has no pygidial plate. Male genitalia and other structures were illustrated by Snelling and Stage (1995a).

■ This genus occurs in deserts of California, Nevada, and Arizona, USA, and Baja California, Mexico. The two species were revised by Snelling and Stage (1995a).

Xeralictus appears to be oligolectic on flowers of *Mentzelia* (Loasaceae).

63. Subfamily Nomiinae

Nomiinae is a major group of bees in the paleotropical and Old World austral regions (Africa, southern Asia, Australia) but is scarce in the holarctic region, with only two genera in North America, and is absent in South America. The species are andreniform or apiform, less commonly euceriform, or rarely almost anthophoriform. The marginal cell usually does not taper much toward the apex, which is rounded (Fig. 63-1). This character usually distinguishes the Nomiinae from the Halictinae, but it fails in some of the very small species of Nomiinae in which the marginal cell is almost as pointed as in Halictini. All the wing veins are strong, unlike those of *Lasioglossum* and *Homalictus* in the Halictinae. The Nomiinae are usually said to have the first and third submarginal cells long, the second much shorter, as in Figure 63-1a. Of course this distinction does not apply to *Steganomus*, which has only two submarginal cells. Also, in many of the smaller species, the third submarginal cell is much shorter than the first, although always much longer than the second; this condition is approached in Figure 63-1b. Thus, although usually distinctive, the wing venation is sometimes much like that of some Halictinae. Except in *Halictonomia*, the stigma is commonly more slender than is usual in the Halictinae, only about as wide as the prestigma (as measured to the wing margin), and the stigmal margin within the marginal cell is usually angularly convex, unlike that of the Halictinae. In small species of various genera this character fails, the stigma being larger and often lacking the angle within the marginal cell. As in nonparasitic Halictinae, the labrum of the female has a strong apical process with a median keel and a fringe of coarse bristles (Fig. 63-2m, n). The episternal groove extends only a short distance below the scrobal groove, or (in *Halictonomia*) both are largely or entirely absent. The basal vein of the forewing varies from slightly and uniformly curved to strongly curved, as in the Halictinae.

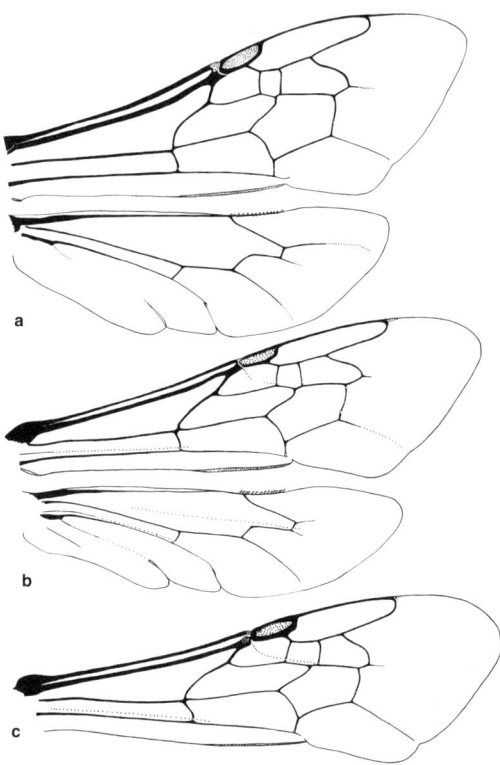

Figure 63-1. Wings of Nomiinae. **a,** *Lipotriches (Austronomia) australica* (Smith); **b,** *Nomia (Acunomia) melanderi* Cockerell; **c,** *Dieunomia nevadensis* (Cresson).

The prepygidial fimbria is not divided as it is in Halictinae. The structure of T6 of the female resembles that of the Halictinae; the gradulus is bent posterad in the middle to outline two sides of a small triangular area near the base of the tergum; from the apex of the triangle a narrow line extends posteriorly to or nearly to the apex of the tergum. This line may be the true pygidial plate, but on each side of it, in the distal part of the tergum, is a rather ordinary plate bisected by the line. The male genitalia usually have both upper and lower gonostyli, the latter not retrorse as in many Halictinae. The lower gonostyli are much smaller than the upper and rarely, as in *Reepenia*, absent. Pauly (1990, 1991) sketched male genitalia and sterna of most genera. S7 and S8 of the male are quite different from those of the Halictinae, S7 being transverse with two or four small apical projections that usually bear long hairs, S8 being compact and triangular or pentagonal, without long lateral extensions, and with a median apical peglike structure. *Paulynomia* is exceptional but not at all halictine-like in these sternal characters. T7 of the male is broad, often emarginate, without a pygidial plate.

Male Nomiinae are highly variable in the leg modifications that presumably help the male to hold its position on the female while mating (Wcislo and Buchmann, 1995). Similar features crop up in different genera and subgenera; Figure 63-2a-d illustrates hind tibial forms found in North American species of *Nomia (Acunomia)*. Figure 63-2e illustrates the similar modification in *Pseudapis,* some species of which have the strange spatulate hairs shown on the underside of the femur.

Nomiines nest in the ground; Iwata (1976) and Wcislo and Engel (1996) summarized the scant knowledge of nest architecture of this subfamily. In some species of *Lipotriches* s. str., each female makes a vertical burrow from which horizontal or somewhat slanting, sessile cells diverge (Hirashima, 1961; Michener, 1969a; Rayment, 1956), much as in nests of solitary *Lasioglossum (Dialictus)*. In most nomiines, however, the cells are vertical, either "hanging" from more or less horizontal branch burrows, as in *Dieunomia* (Cross and Bohart, 1960) and *Nomia (Leuconomia) candida* Smith (Michener, 1969a), or clustered. Rather loose cell clusters characterize nests of *Nomia (Acunomia) melanderi* Cockerell, but the closely related *N. (A.) nortoni* Cresson makes tight clusters (Ribble, 1965), as do *Lipotriches (Austronomia) australica*

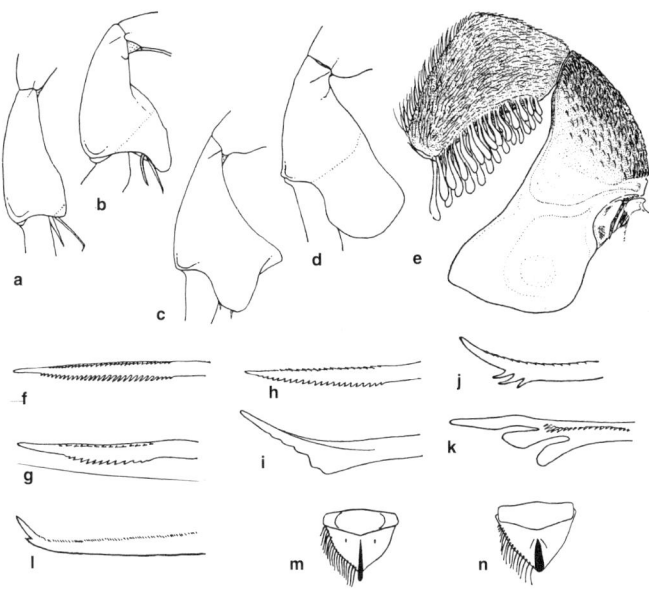

Figure 63-2. Structures of Nomiinae. a-e, Male tibial forms. **a**, *Nomia (Acunomia) angustitibialis* Ribble; **b**, *N. (A.) tetrazonata* Cockerell; **c**, *N. (A.) robinsoni* Cresson; **d**, *N. (A.) melanderi howardi* Crawford; **e**, *Pseudapis (Pseudapis) umesaoi* (Sakagami), including femur.

f-k, Female inner hind tibial spurs. **f**, *Reepenia bituberculata* (Smith); **g**, *Lipotriches (Austronomia) australica* Smith; **h**, *Nomia (Hoplonomia) lyonsiae* Cockerell; **i**, *Lipotriches (Lipotriches) halictella* (Cockerell); **j**, *L. (Austronomia) maai* (Michener); **k**, *Mellitidia gressitti* (Michener).

l, Outer hind tibial spur of female *Nomia (Acunomia) chlorosoma* Cockerell. **m, n**, Labra of females of *Mellitidia gressitti* (Michener) and *Ptilonomia plumosa* Michener.

a-d, from Ribble, 1965; e, based on Sakagami, 1961; f-n, from Michener, 1965b.

(Smith) (Rayment, 1956), *Pseudapis diversipes* (Latreille) (Rozen, 1986), and *P. bispinosa* (Brullé), the last published as *Nomia ruficornis* Spinola (Gutbier, 1916). Some species make cell clusters supported by pillars or rootlets in soil cavities. *Ptilonomia plumosa* (Michener) nests of this type were illustrated by Hirashima (1997). The same type of construction is found in *Nomia (Hoplonomia) punctulata* Dalla Torre, whose cell clusters are reused year after year, the nest being enlarged by presumed daughter groups of females (Masuda, 1943). The phylogeny and functions of nest structures would be particularly interesting to study in the Nomiinae in comparison to the better known Halictinae.

The cells are lined with shiny "waxlike" material, much as in the Halictinae. Those species that make more or less horizontal cells may construct them with the lower surface slightly flatter than the upper surface, although observations are doubtful on this point. Vertical cells are not flatter on one side, unlike the vertical cells of the Halictinae. The pollen masses of Nomiinae are much more diverse than those of the Halictinae, being sometimes subspherical (*Pseudapis*, see Rozen, 1986), at other times flat and lens-shaped (*Nomia*), and sometimes rectangular (*Lipotriches, Rhopalomelissa* group). In vertical cells the pollen ball may be on one side of the cell or in the bottom and not associated with any particular cell wall. A study of these features would be of much interest; they are known for too few species to be recognized as generic characters, but they may well have such significance, as is suggested by the marginal rim around the flattened pollen masses of three species of *Dieunomia* (Wcislo and Engel, 1996).

Many Nomiinae are solitary, but communal nesting has been reported repeatedly for different genera and subgenera; see above for *Nomia (Hoplonomia) punctulata* Dalla Torre by Masuda (1943). A detailed account for *Nomia (Acunomia) tetrazonata* Cockerell is by Wcislo (1993), who found nests containing from 1 to 20 working females and 23 to 191 cells. Vogel and Kukuk (1994) suggested quasisocial relationships in some nests of *Nomia (Austronomia) australica* Smith. Batra (1966b) found some evidence of division of labor among colony members in a species of *Lipotriches* and a species of *Pseudapis*. A survey of the literature by Wcislo and Engel (1997) indicated that nearly half of the nomiine species so far studied sometimes or regularly have more than one female per nest. There are no known cleptoparasitic Nomiinae.

The genera of Nomiinae have commonly been regarded as subgenera of *Nomia*. For this reason various regional revisional studies, aside from those in general accounts of bees, treated the whole subfamily under the generic name *Nomia*. Such works are by Hirashima (1961) for Japan; Warncke (1976a) for the west palearctic, with additions in 1980c; Strand (1913c) and Pauly (1990) for Africa; and Pauly (1991) for Madagascar. A key to the Malagasy genera of Nomiinae, which includes most of those found in Africa, was given by Pauly (in Pauly et al., 2001).

Key to the Genera of the Nomiinae of the Eastern Hemisphere

1. Malar space at least half as long as wide (apical tergal hair bands absent except sometimes at sides of T1 and T2; head only slightly, if any, wider than long) 2
—. Malar space nearly absent 3
2(1). Metanotum with median tubercle or projection (weak in *Mellitidia simplicinotum* Michener); mandible of female three-toothed; inner hind tibial spur of female pectinate with two large, broad teeth (Fig. 63-2k) (New Guinea region)... *Mellitidia*
—. Metanotum strongly convex, slightly depressed medially, without median tubercle; mandible of female two-toothed; inner hind tibial spur of female minutely ciliate-serrate (as in Fig. 63-2f) (New Guinea) .... *Ptilonomia*
3(1). Ocelli enlarged, ocellocular distance equal to or less than ocellar diameter; hind wing with dark area (due to dense minute hairs) along costal margin between hamuli and wing tip (body partly brown or testaceous; scutellum of male bituberculate) (New Guinea region)........ *Reepenia*
—. Ocelli not or only slightly enlarged, ocellocular distance greater than ocellar diameter; hind wing without dark area 4
4(3). Marginal zones of T2 to T4 (or, in males, to T5), and often also T1, except near bases, hairless, impunctate, usually opaque white, yellow, green, or blue, although sometimes concolorous with rest of metasoma, the frequently opalescent bands sometimes broken medially (basal area of propodeum subhorizontal to vertical, sometimes incompletely distinguished from rest of propodeal triangle; anterior end of scutum not or only medially bent sharply down to meet pronotum; pronotum not lamellate)................................... *Nomia* (in part)
—. Marginal zones of T1 to T4 variable, often hairy or punctate, and if impunctate, then often translucent, sometimes only that of T4 hairy or punctate, the zones never white, yellow, green, or blue 5
5(4). Tegula enormous, extending well behind level of scutoscutellar suture (Fig. 63-4o) *and* preoccipital carina present; anterior end of scutum medially bent sharply downward, the angle sharp and commonly carinate (pronotum with strong, translucent lamella from posterior lobe mesad) .... *Pseudapis*
—. Tegula not reaching level of scutoscutellar suture, or, *if* enlarged as above, then preoccipital carina absent; anterior end of scutum, if bent sharply downward, without sharp angle or carina along bend 6
6(5). Submarginal cells two; males and some females with translucent lamella on dorsolateral angle of pronotum, the lamella extending continuously mesad as carina all across pronotum; tegula large as in *Pseudapis* (Fig. 63-4o); mesepisternum of male, mesal to middle coxa, with tubercle or spine; antenna of male with last two segments enlarged and flattened, not pedunculate (African, oriental)... *Steganomus*
—. Submarginal cells three; pronotum frequently without lamella, although often with carina that may or may not extend across pronotum, and in some *Lipotriches* s. str. [e.g., *notabilis* (Schletterer)] that do have a lamella there is a break or angular notch between lamella of posterior pronotal lobe and that of dorsolateral angle of pronotum; tegula only rarely enlarged as in *Pseudapis*; mesepisternum of male without ventral tubercle or spine; antenna of male usually not enlarged and flattened at apex [except in *Spatunomia* and *Lipotriches (Clavinomia)*] 7
7(6). Propodeal triangle sloping, curving down onto posterior surface of propodeum; basal area of propodeum neither sculptured distinctively nor defined by strong line or carina 8
—. Propodeal triangle vertical or its basal part subhorizontal and usually bending rather abruptly onto declivous surface; basal area of propodeum recognizable by having coarse sculpturing such as radiating striae, *or* the sharp angle or carina delimiting it, *or* being subhorizontal in contrast to declivous surface behind it 9
8(7). Basitibial plate of female delimited by carina only along posterior side; lateral carina of T1 strong, reaching posterior margin of tergum; pronotum with strong transverse carina at and usually extending mesad from dorsolateral angle, but the carina depressed and weak or absent medially; T6 of male with median, flat, apical projection resembling a pygidial plate (Madagascar) .... *Sphegocephala*
—. Basitibial plate of female delimited by carina on both sides; lateral carina of T1 absent; pronotum without or with weak carina; T6 of male without apical projection (Madagascar).. *Halictonomia*
9(7). Mandible of female two- or three-toothed; last antennal segment of male not pedunculate, usually not modified (preoccipital carina absent) 10
—. Mandible of female simple; last antennal segment of male expanded, flattened, its base sometimes narrowly pedunculate 11
10(9). T2 to T4 with apical bands of hair or tomentum, at least on T4 and at least laterally, or *if* posterior marginal zones hairless as in some *Lipotriches* s. str., then pronotum with continuous transverse carina in front of posterior pronotal margin; male upper gonostylus directed mesoapically, not a broad plate (Fig. 63-3a) ....
 .... *Lipotriches* (in part)
—. T2 to T4 completely without apical bands of hair or tomentum, depressed marginal zones largely impunctate; pronotum without continuous transverse carina; male upper gonostylus directed apically and consisting of a broad, flat plate with ornate or incised margins (metanotum sometimes with broad lamelliform projections)....
 .... *Nomia* (in part)
11(9). Last antennal segment of male narrowly pedunculate; male with preoccipital carina; inner hind tibial spur of female coarsely serrate (Africa)............ *Spatunomia*
—. Last antennal segment of male fully as wide at base as preceding flagellar segments; male lacking preoccipital carina; inner hind tibial spur of female with margin a simple, untoothed lamella (Iran) ...... *Lipotriches (Clavinomia)*

Key to the Genera of Nomiinae of North America

1. Marginal zones of T2 to T4, except near base, hairless, impunctate, opaque green or blue; metasomal scopa absent; male gonostylus broad, flat, apex curved mesad and then basad (as in Fig. 63-3f) and usually ornamented with modified hairs .... *Nomia*
—. Marginal zones of T2 to T4 with hair bands (sometimes of dark hairs) and punctate except for margins, zones concolorous with adjacent areas; S2 to S5 and sides of T2

to T5 with well-developed scopa of long, straight hairs, those of medial region simple or nearly so, of lateral areas with branches; male gonostylus slender, directed apicad, with only a few simple hairs .... *Dieunomia*

Genus *Dieunomia* Cockerell

Dieunomia, the only strictly New World genus of Nomiinae, contains large (body length 7-23 mm), andreniform species not clearly related to any of the Old World genera. Except for a weak carina on the posterior lobe, the pronotum lacks a carina. The tibial spurs are unmodified. The basitibial plate of the female is completely margined by a carina that is often largely hidden by hairs; the plate is rounded apically. The hind legs of the males are usually moderately to greatly enlarged (Fig. 63-2a-d), and sometimes the middle legs are also modified. Features not found in other Nomiinae include (1) the dense scopa of S2 to S5 of the female, extending even onto the metasomal terga dorsolaterally, much as in *Systropha* (Rophitinae), (2) the form of T1 of both sexes, which has a longitudinal median depression extending to the summit of the anterior surface, the summit broadly V-shaped as seen from above and elevated above both the depressed zone behind it and the declivity in front of it, the profile of T1 thus angular, and (3) the thick head, such that the genal areas and vertex are broad and curve gradually onto its posterior surface. Male genitalia and hidden sterna were illustrated by Mitchell (1960).

Species of *Dieunomia* sometimes nest in great aggregations in sandy soil, even in land that is farmed, because the cells are below the level of plowing. The cells are vertical, "hanging" from rather long, horizontal laterals (Cross and Bohart, 1960). These bees are specialists on Asteraceae. Minckley et al. (1994) investigated the close seasonal relationship of *D. (Epinomia) triangulifera* (Vachal) to the flowering of *Helianthus annuus*.

Dieunomia has been divided into two subgenera.

Key to the Subgenera of *Dieunomia*

1. Labial palpus with first segment as long as or longer than segments 2 to 4 taken together; last antennal segment of male broader than preceding segments; tegula large, posterior end subtruncate and attaining level of scutoscutellar suture, lateral margin sinuate ....
.... *D. (Dieunomia s. str.)*
—. Labial palpus with first segment shorter than remaining segments taken together; last antennal segment of male unmodified; tegula as above or shorter, its lateral and posterior margins then convex .... *D. (Epinomia)*

Dieunomia / Subgenus *Dieunomia* Cockerell s. str.

Eunomia Cresson, 1875: 723, not Hübner, 1818. Type species: *Eunomia marginipennis* Cresson, 1875 = *Nomia kirbii* Smith, 1865, by designation of Cockerell, 1910f: 290.
Dieunomia Cockerell, 1899a: 14, replacement for *Eunomia* Cresson, 1875. Type species: *Eunomia marginipennis* Cresson, 1875 = *Nomia kirbii* Smith, 1865, autobasic and by designation of Cockerell, 1910f: 290.

This subgenus consists of very large species, body length 11 to 23 mm.

■ *Dieunomia* is found from Manitoba, Canada, and Colorado, Wisconsin, Ohio, and Maryland south to Texas and Florida, USA, and Tamaulipas, Mexico. The five species were revised by Blair (1935).

Dieunomia / Subgenus *Epinomia* Ashmead

Epinomia Ashmead, 1899a: 88. Type species: *Nomia persimilis* Cockerell, 1898 = *Nomia triangulifera* Vachal, 1897, by original designation.

The species of this subgenus are smaller than most *Dieunomia* s. str., 7 to 13 mm long.
■ *Epinomia* ranges from Oregon, Minnesota, Illinois, and Florida, USA, south to the states of Baja California, Jalisco, and Morelos, Mexico. The four species were revised by Cross (1958).

Genus *Halictonomia* Pauly

Nomia (Halictonomia) Pauly, 1980a: 123. Type species: *Halictus decemmaculatus* Friese, 1900, by original designation.

Halictonomia is a genus of small (length 4-9 mm), andreniform species with the whole body or the metasoma largely yellow-brown in some species, although others are entirely black. The hairs are sparse and do not form metasomal bands, or form weak apical bands laterally. The punctation is weak and fine. The hind legs and exposed sterna of the male are unmodified, as are all the tibial spurs, except that in *H. minuta* (Benoist) (see below) the hind tibial spurs of the male are absent. Two characters that are unique in the Nomiinae are the absence of the episternal and scrobal grooves (or the former indicated only near the upper extremity) and the large stigma, two or more times as broad as the prestigma (as measured to the wing margin). In all other Nomiinae the episternal groove is indicated from its upper end down to a short distance below the scrobal groove, and the stigma is little if any wider than the prestigma, or wider in some small species. Another unusual feature of *Halictonomia* is the slender, more or less equally long, upper and lower male gonostyli, the upper with a mesal basal lobe bearing modified setae, at least in *H. decemmaculata* (Fricse) and *minuta* (Benoist). Pauly (1990) diagramed the male genitalia and sterna.
■ This genus is found only in Madagascar. In his revision of Malagasy Nomiinae, Pauly (in Pauly et al., 2001) recognized 10 species of the genus *Halictonomia*.

Halictonomia minuta (Benoist) belongs here on the basis of its large stigma and lack of episternal and scrobal grooves. Unlike other species of *Halictonomia*, however, it lacks a strong longitudinal ridge on the supraclypeal area, and the scutellum is not biconvex. I consider its position tentative for the present.

Genus *Lipotriches* Gerstaecker

This is the largest genus of the Nomiinae, in its present sense including the species that Pauly (1990) segregated into several genera. Thus he segregated some African species related to the subgenus *Austronomia* into the genera *Afronomia, Macronomia,* and *Trinomia*. Asian material badly needs further study, but it either bridges the differences among these subgenera or could represent additional taxa in this complex. I prefer to indicate the similarity among these forms by placing all of them as subgenera of the one genus. This may turn out to be a paraphyletic unit, and exclusion of *Sphegocephala* from it is rather arbitrary; nonetheless, recognition of a large genus *Lipotriches* is convenient and in accordance with current practices elsewhere among bees. A subgenus that differs from the rest in striking morphological characters (see the key), for example, in its mandibular dentition, is *Nubenomia,* but except for those characters, *Nubenomia* is almost like *Melanomia,* which is more difficult to separate from other subgenera of *Lipotriches.* I have therefore included both *Nubenomia* and *Melanomia* within *Lipotriches,* as subgenera. Because of its simple female mandibles, *Clavinomia* is also distinctive; the inner hind tibial spur of the female, with its margin consisting of a smooth, untoothed lamella, is suggestive of *Lipotriches* s. str. The body of *Lipotriches* is usually markedly more slender than that of most other nomiines, a tendency that culminates in the slender and petiolate males of some species of *Lipotriches* s. str. formerly segregated as *Rhopalomelissa.* Male genitalia, sterna, and other structures were diagramed by Pauly (1990); see also Figure 63-3a-e.

As noted in the discussion of the Nomiinae above, *Lipotriches* is diverse in nest architecture, some species of *Lipotriches* s str. having sessile horizontal cells dispersed along vertical burrows while a species of the subgenus *Austronomia* builds its cells in clusters. Great diversity in nest structure is also known in the genus *Nomia.*

Key to the Subgenera of *Lipotriches*

1. Mandible of female tridentate, of male usually bidentate; basal area of propodeum reduced to transverse line or fine groove .... *L. (Nubenomia).*
—. Mandible of female bidentate, or simple in *Clavinomia,* of male simple (except bidentate in a few species of *Lipotriches* s. str.); basal area of propodeum larger, sometimes not defined medially or (rarely) entirely unrecognizable 2
2(1). Pronotum with continuous or medially or laterally notched transverse carina or lamella anterior to the scutum, at about level of scutum, the latter thus not bending down anteriorly (inner hind tibial spur of female commonly untoothed, margin a simple lamella, broadest near middle of spur, sometimes with a few teeth, Fig. 63-2i).. *L. (Lipotriches s. str.)*
—. Pronotum with transverse carina either absent or broadly interrupted medially, the anterior part of the scutum thus bending down medially to depressed pronotal margin 3
3(2). Mandible of female simple; inner hind tibial spur of female with simple, smooth-margined lamella; male with last two antennal segments curled to one side, the last segment much expanded, broader than long (Iran) ....
.... *L. (Clavinomia).*
—. Mandible of female with preapical tooth, thus bidentate; inner hind tibial spur of female pectinate, dentate, or minutely serrate (ciliate); male with apical antennal segments usually unmodified, but if modified, then not as above 4
4(3). Ocellocular distance less than twice ocellar diameter; glossa very slender, as long as face; basitibial plate of female delimited on both sides but apex open; scape of female short, not reaching median ocellus (Africa) ....
.... *L. (Maynenomia).*
—. Ocellocular distance usually twice ocellar diameter or more; glossa shorter than face; basitibial plate of female, if carinate on both sides, with apex almost always closed; scape of female reaching to or beyond median ocellus 5
5(4). Head about as long as broad; clypeus with well-developed lip below row of large apical hairs (Madagascar) ....
.... *L. (Melanomia).*
—. Head distinctly broader than long; clypeus with row of large apical hairs very near lower margin except in a few *Austronomia* and *Macronomia* species in which lip extends below hair bases 6
6(5). Basitibial plate of female margined by carina only on posterior side (Africa, oriental region) .... *L. (Macronomia).*
—. Basitibial plate of female with marginal carina complete except for base... 7
7(6). Propodeal basal area more or less horizontal, only posterior part of propodeal triangle declivous (Australia, oriental region, Africa)............................ *L. (Austronomia).*
—. Propodeal basal area more or less vertical, at least medially, so that whole propodeal triangle, or most of it, is vertical 8
8(7). Terga, at least most of T1, minutely tessellate, dull, with only fine, sparse punctation; hind femur of male with three strong teeth on undersurface (Africa, Arabia) .... *L. (Trinomia).*
—. Terga more shiny, often shagreened but not tessellate, usually more coarsely punctate; hind femur of male without or with one or two strong teeth (Africa, oriental region) .... *L. (Afronomia).*

Lipotriches / Subgenus *Afronomia* Pauly

Afronomia Pauly, 1990: 126. Type species: *Nomia picardi* Gribodo, 1894, by original designation.

Like *Trinomia,* this subgenus is similar to *Austronomia,* from which *Afronomia* differs in its more declivous propodeum (see the key to subgenera) and the presence of one or two teeth on the underside of the swollen hind femur of the male. *Afronomia* is larger (body length 10-12 mm) than African *Austronomia,* although some Australian *Austronomia* attain 12 mm. Males of *Afronomia* are not distinguishable by group characters from those of *Macronomia.*

■ This subgenus occurs from Ethiopia and Zaire south to Natal, Cape Province, and Namibia. The seven African species were revised by Pauly (1990).

At least three unidentified Asian species tentatively placed in this subgenus are smaller than African species (the smallest 7 mm long), and lack teeth on the hind femora of the males, although these femora are greatly swollen. Such species occur in India, Malaysia, and Java.

Lipotriches / Subgenus *Austronomia* Michener

Nomia (Austronomia) Michener, 1965b: 156. Type species: *Nomia australica* Smith, 1875, by original designation.

This is possibly a basal paraphyletic group from which arose various subgenera that have derived characters of various sorts. It is characterized by the complete basitibial plate of the female (either rounded or pointed), the lack of a continuous carina across the pronotum, the presence of apical hair bands on the terga (at least laterally on T4 and T5), the lack of teeth on the posterior femora of males, and the subhorizontal (at least laterally) basal area of the propodeum. The inner hind tibial spurs of the female vary from minutely serrate or ciliate (Fig. 63-2g) to pectinate (Fig. 63-2j). Body length varies from 6 to 12 mm. The male genitalia and sterna were illustrated by Michener (1965b) and Pauly (1990); see Figure 63-3a-e.

■ *Austronomia* is widespread in Australia (excluding Tasmania) and occurs also in New Caledonia, north through New Guinea, Indonesia, the Philippines, and Taiwan to Japan, the whole of southern Asia, Sri Lanka, and Madagascar, and in Africa from Ethiopia to Natal (South Africa) and west to Zaire. There are over 60 names for Australian species (Michener, 1965b), 14 African and 8 Malagasy species (Pauly, 1990), and perhaps about 20 in Asia and nearby insular areas. Hirashima (1978c) revised some of the species from Asia.

The tegulae in a few species are enormous, as in *Pseudapis* (Fig. 63-4o). Some Australian species are dark metallic green or blue instead of black; others have bluish reflections on the metasoma. *Lipotriches maai* (Michener) from New Guinea is bright metallic blue. One or more Indian species have, on T6 of the male, a projecting rounded median lobe, as in some species of *Macronomia*.

Lipotriches / Subgenus *Clavinomia* Warncke

Nomia (Clavinomia) Warncke, 1980c: 372. Type species: *Nomia clavicornis* Warncke, 1980, by original designation.

In addition to the characters indicated in the key to subgenera, *Clavinomia* has the following features: The basitibial plate of the female is complete, narrow, and pointed. The pronotal lobe is carinate, but the carina does not reach the dorsolateral angle of the pronotum, which is small; a carina extends down from this angle but there is no carina extending mesad toward the other dorsolateral angle. The pronotal margin is broadly depressed medially and the anterior part of the scutum is bent rather sharply downward medially. The terga have strong white apical hair bands, but on T6 of the male the apical margin is convex, bare, and shining white, suggesting the bands of some species of *Nomia*. The hind femur of the male is slightly enlarged, the undersurface bare except for a longitudinal series of long, curved hairs; the tibia is somewhat expanded preapically and apically, the lower surface nearly bare. The body length is 10 mm. The male genitalia, sterna, antenna, and other structures were illustrated by Warncke (1980c).

■ *Clavinomia* is known only from Iran. The only species is *Lipotriches clavicornis* (Warncke).

Lipotriches / Subgenus *Lipotriches* Gerstaecker s. str.

Lipotriches Gerstaecker, 1858: 460. Type species: *Lipotriches abdominalis* Gerstaecker, 1857 = *Sphecodes cribrosa* Spinola, 1843, monobasic.
Rhopalomelissa Alfken, 1926b: 267. Type species: *Rhopalomelissa xanthogaster* Alfken, 1926, by designation of Sandhouse, 1943: 596.
Nomia (Epinomia) Alfken, 1939: 113, not Ashmead, 1899.

Type species: *Nomia andrenoides* Vachal, 1903 = *Nomia andrei* Vachal, 1897, by original designation.
Alfkenomia Hirashima, 1956: 33, replacement for *Epinomia* Alfken, 1939. Type species: *Nomia andrenoides* Vachal, 1903 = *Nomia andrei* Vachal, 1897, autobasic.
Rhopalomelissa (Lepidorhopalomelissa) Wu, 1985: 58. Type species: *Nomia burmica* Cockerell, 1920, by original designation.
Rhopalomelissa (Trichorhopalomelissa) Wu, 1985: 58. Type species: *Rhopalomelissa hainanensis* Wu, 1985, by original designation.
Rhopalomelissa (Tropirhopalomelissa) Wu, 1985: 58. Type species: *Rhopalomelissa nigra* Wu, 1985, by original designation.

This subgenus is rather easily recognized by the characters indicated in the key to subgenera. Another useful character is the basitibial plate of the female, which, as in *Macronomia*, is defined by a carina only along the posterior margin, or if there is an indication of the anterior margin, it is weak. In males of some species the metasoma is slender, almost petiolate, and T1 is longer than broad. This is the group that was called *Rhopalomelissa*. In Africa are found all intermediates in metasomal form between that of *Rhopalomelissa* and the parallel-sided males of *Lipotriches* in the narrowest sense. The body length is 5 to 12 mm. In various species the metasoma is partly or wholly red. Metasomal hair bands are present on most species, at least at the sides of the posterior terga, but in some, such as *L. (L.) notabilis* (Schletterer), such bands are virtually absent. In the group of *L. (L.) panganina* Strand the tegulae are enlarged (as in Fig. 63-4o), exactly as in *Pseudapis*, although the body is slender, as in other *Lipotriches*. The male genitalia and sterna were illustrated by Michener (1965b) and Pauly (1990).

■ *Lipotriches* s. str. occurs from Senegal, Sudan, and Ethiopia, with a record for Libya, south through Africa to South Africa; east to Madagascar, Aden, India, and Sri Lanka and throughout southern Asia to China, north to northern Kyushu, Japan; also in the Philippines, Taiwan, and south through Indonesia, New Guinea, the Solomon Islands, and Australia as far south as southern Queensland. For Africa (including Madagascar) Pauly (1990) reported 58 species of *Lipotriches* s. str. There are 4 names for Australian species, at least 25 for species in southern Asia, Indonesia, Taiwan, New Guinea, etc., 10 for species in China (Wu, 1985), and 2 for species in Japan. The African species were reviewed by Pauly (1990). Many Asiatic nomiine species have not been placed as to genus and subgenus; some of them certainly belong to *Lipotriches* s. str.

Lipotriches / Subgenus *Macronomia* Cockerell

Nomia (Macronomia) Cockerell, 1917c: 468. Type species: *Nomia platycephala* Cockerell, 1917, by original designation.
Crinoglossa Friese, 1925b: 502. Type species: *Crinoglossa natalensis* Friese, 1925, monobasic.

Macronomia resembles *Lipotriches* s. str. in having the basitibial plate of the female margined only on the posterior side; sometimes even that side is scarcely defined. It differs from *Lipotriches* s. str. in the finely serrate to pectinate inner hind tibial spur of the female and the absence of a continuous carina or lamella across the pronotum. Males are not distinguishable by group characters from those of *Afronomia*. As in *Austronomia* the metasomal terga have apical hair bands. Body length is 8 to 13 mm.

- This subgenus occurs from Senegal to Ethiopia and Somalia, south throughout Africa to Namibia and Natal, South Africa, and Madagascar. Three unidentified species from India appear to fall in this subgenus. Forty-five species from Africa and Madagascar were listed by Pauly (1990).

The characters of wing venation that probably caused Cockerell to name this subgenus turn out to be variable and not particularly characteristic of the subgenus. In some species, T6 of the male has a median, rounded, apical projection, as in a few species of *Austronomia* and suggestive of the more truncate projection found in *Nomia (Nomia)* and *Sphegocephala*.

Lipotriches / Subgenus *Maynenomia* Pauly

Maynenomia Pauly, 1984d: 698. Type species: *Nomia maynei* Cockerell, 1937 = *Nomia testacea* Friese, 1914, by original designation.

Maynenomia consists of a single species that might well be included in the subgenus *Austronomia,* from which it differs by the characters indicated in the key to subgenera. Pauly (1990) emphasizes its "oval" head shape, a term that he uses generally for taxa with the head about as long as broad, but in *Maynenomia* it is markedly broader than long, although not as broad as in most nomiines. Body length is about 7 mm; the metasoma is red. T3 and T4 of the female have rather hairy posterior parts, although these do not form distinct bands. Pauly (1984d) illustrated the male genitalia and other structures.

- This subgenus occurs from Senegal and Kenya south to Transvaal, South Africa. There is only one species, *Lipotriches testacea* (Friese).

Lipotriches / Subgenus *Melanomia* Pauly

Melanomia Pauly, 1990: 149 [also described as new by Pauly, 1991: 310]. Type species: *Nomia melanosoma* Benoist, 1963, by original designation.

In spite of the mandibular differences indicated in the key to genera, *Melanomia* seems to be related to *Nubenomia,* as indicated by the extension of the clypeal lip beyond the bases of the apical row of large hairs, the head about as long as broad, the complete female basitibial plate, and the slender hind legs and relatively simple exposed sterna of males. The body is rather coarsely punctate, not very hairy; the hair on T4 and T5 of the female suggests hair bands but is dark in color. The body is 9 to 11 mm long, black or with large areas dark red.

- *Melanomia* is found in Madagascar. The two species were characterized by Pauly (1991).

Lipotriches / Subgenus *Nubenomia* Pauly

Nomia (Nubenomia) Pauly, 1980a: 122. Type species: *Nomia nubecula* Smith, 1875, by original designation.

Aside from the unusual mandibular characters indicated in the key to subgenera, *Nubenomia* is similar to *Melanomia*. As in that subgenus the clypeus extends as a lip below the row of large apical hairs, and the head is about as long as broad, unlike the great majority of *Lipotriches*. The basitibial plate of the female is complete and pointed, or, in *Lipotriches reichardia* (Strand), open at the apex. The hind legs and S1 to S6 of the males are not or little modified. The body is black with weak metasomal hair bands, the length 9 to 12 mm. The distal part of the forewing is strongly darkened, a feature also found in some species of the subgenus *Melanomia* and of *Nomia (Acunomia)*.

- This subgenus occurs in tropical Africa from Liberia to Kenya south to Zaire and Malawi, with one species in Madagascar. Fifteen species were listed by Pauly (1990) and the subgenus was revised by Pauly (2003), who recognized seven African and one Malagasy species.

Lipotriches / Subgenus *Trinomia* Pauly

Nomia (Trinomia) Pauly, 1980a: 122. Type species: *Nomia tridentata* Smith, 1875, by original designation. [New status.]

This subgenus is near *Austronomia,* differing in that the propodeal triangle is vertical, sometimes quite small, the basal area being unrecognizable or represented by a series of small pits. The apical hair bands on the terga are made up of curved golden hairs rather than plumose tomentum. The posterior leg of the male is swollen, and there are three teeth on the underside of the femur.

- This subgenus occurs from Gambia to Kenya, south to Cape Province and Natal, South Africa; also in Aden. The six species were listed by Pauly (1990) and revised by Pauly (1999a).

Genus *Mellitidia* Guérin-Méneville

Mellitidia Guérin-Méneville, 1831: 270. Type species: *Andrena australis* Guérin-Méneville, 1831, monobasic.
Melittidia Dalla Torre, 1896: 99, unjustified emendation of *Mellitidia* Guérin-Méneville, 1831.

Except for the subgenus *Nubenomia* in the genus *Lipotriches, Mellitidia* is the only nomiine with tridentate mandibles in the female. The median metanotal tubercle is usually large, but it is obtuse in *M. metallica* (Smith) and inconspicuous in *M. simplicinotum* (Michener). In other nomiines the metanotum either is simple or has broad, flat projections, as in two subgenera of *Nomia.* The inner hind tibial spur of the female *Mellitidia* is pectinate, with two large, flat teeth (Fig. 63-2k), a form not found elsewhere in the Nomiinae. The body is black (with abundant pale hairs), or weakly metallic greenish to bright green (with most hairs dark), 8.0 to 12.0 mm long. The genitalia and sterna were illustrated by Michener (1965b); see Figure 63-4a-e. Heads were illustrated by Hirashima (1967c).

- This genus is found in the New Guinea area, south to northern Queensland in Australia, north to the Bismarck Archipelago, and west to Ceram, Indonesia. There are 19 specific names; 12 were listed by Michener (1965b) and the others were introduced by Hirashima (1967c) in his revision of metallic species of the genus.

Genus *Nomia* Latreille

All of the Nomiinae with colored (white, yellow, blue, green) tergal bands (Pl. 4) belong to this genus. The colored parts of the marginal zones are largely or entirely impunctate and hairless, although the bases of these zones often have some punctures and hairs. In some species that clearly belong to *Nomia,* however, colored bands are absent, although the depressed black tergal marginal zones are hairless and impunctate except at their bases. Thus *Nomia (Hoplonomia) amboinensis* Cockerell and *N. (H.) flavipennis* Friese have the characteristic thoracic structure of their subgenus, but have black rather than colored metasomal bands. Likewise in the subgenus *Acunomia,* within the group that has been called *Maculonomia* on the basis of distinctive genitalic structure, *Nomia terminata* Smith and *fuscipennis* Smith lack colored bands, whereas *N. megasoma* Cockerell and *viridicinctula* Cockerell are banded. Male genitalia and hidden sterna were illustrated by Mitchell (1960), Ribble (1965), Michener (1965b), and Wu (1982a, 1983e); see also Figure 63-3f-n.

Key to the Subgenera of *Nomia*

1. Metanotum with two broad, lamelliform projections, the two often united to one another 2
—. Metanotum without projections 3
2(1). Lateral extremity of scutellum with lamelliform projection (Africa, oriental) .... *N. (Crocisaspidia)*
—. Lateral extremity of scutellum not produced (oriental, Australia, Madagascar) .... *N. (Hoplonomia)*
3(1). Outer hind tibial spur of female and some males bent near apex, usually with projection at the bend continuing direction of main shaft (Fig. 63-2l); middle tibial spur usually with a few coarse preapical teeth (basitibial plate of female completely margined by carinae except at base; posterior femur of male without tooth) 4
—. Outer hind tibial spur of female and most males tapering to unmodified apex (Fig. 63-2h); middle tibial spur with all teeth minute 5
4(3). S8 of male a small, compact sclerite with midapical peglike projection (as in Fig. 63-3g); S7 of male transverse, with two hirsute apical projections (as in Fig. 63-3h); male gonostylus flattened, commonly expanded, sometimes curved basad (as in Fig. 63-3f); S6 of male with apical margin emarginate, surface without oblique rows of setae; S5 of male with apex variously thickened, lobate, with hair tufts (as in Fig. 63-3j) (holarctic, Africa, oriental) .... *N. (Acunomia)*
—. S8 of male relatively large, broadened basally and preapically, lateral margins thus concave, apical peglike projection absent (Fig. 63-3l); S7 of male relatively large, basolateral apodemal lobes long, length of sclerite thus much more than half its width (Fig. 63-3m); male gonostylus rather slender, not flattened, much exceeding penis valves, directed posteriorly, and arising from distal end of elongated gonocoxite (Fig. 63-3k); S6 of male with apex rounded, surface with oblique rows of flattened setae (Fig. 63-3n); S5 of male unmodified (Australia) ....
.... *N. (Paulynomia)*
5(3). Male, in species with slender legs, with basitibial plate; T1 frequently without colored band and usually with pale hair band laterally; hind femur of male without tooth or with tooth near base on underside; body length 10 mm or less; colored bands white or nearly so (Africa, oriental) .... *N. (Leuconomia)*
—. Male without basitibial plate; T1 with colored band, without hair band; hind leg enlarged, femur with one or two preapical teeth on underside; body length 9 mm or more; bands more brightly colored (Africa, oriental) ....
.... *N. (Nomia s. str.)*

Nomia / Subgenus *Acunomia* Cockerell

Nomia (Paranomia) Friese, 1897b: 48, not Conrad, 1860. Type species: *Nomia chalybeata* Smith, 1875, by designation of Cockerell, 1910f: 290.

Nomia (Acunomia) Cockerell, 1930, in Cockerell and Blair, 1930: 11. Type species: *Nomia nortoni* Cresson, 1868, by original designation.

Nomia (Paranomina) Michener, 1944: 251, not Hendel, 1907, replacement for *Paranomia* Friese, 1897. Type species: *Nomia chalybeata* Smith, 1875, by original designation and autobasic.

Nomia (Curvinomia) Michener, 1944: 251. Type species: *Nomia californiensis* Michener, 1937 = *Nomia tetrazonata* Cockerell, 1910, by original designation.

Nomia (Maculonomia) Wu, 1982a: 275. Type species: *Nomia terminata* Smith, 1876, by original designation.

The principal subgeneric characters of *Acunomia* are given in the key to subgenera. The body length is 8 to 16 mm. In the North American group that has been called *Curvinomia,* as well as in certain Old World species, the middle tibial spur is finely toothed to the apex or bears a few teeth near the apex that are scarcely larger than the more basal teeth. I agree with Pauly (1990) that the three North American species with tapering male antennal flagella, for which the name *Acunomia* was originally proposed, need not be separated at present from the other North American and Old World species that lack this feature. *Nomia (A.) yunnanensis* Wu from China also has tapering flagella, although its sternal characters do not agree with those of *N. (A.) nortoni* and its relatives in North America.

■ *Acunomia* is found from Eritrea and Senegal south throughout Africa to eastern Cape Province, South Africa, east to Madagascar, India, southeast Asia, China, Taiwan, the Philippines, and Indonesia. In North America it occurs from Washington State east to North Dakota and New Jersey, south to Florida, USA, the Bahamas, Cuba, and the states of Baja California, Guerrero, and Veracruz, Mexico. There are about 33 species. Nine African species were revised by Pauly (1990), three Malagasy species by Pauly (1991), and nine American species by Ribble (1965). The Australian species listed by Michener (1965b) belong to the subgenus *Paulynomia,* but many of the Asiatic species of *Acunomia* were listed in that same paper and by Pauly (1990).

The subgenus *Acunomia* is highly variable and may be a paraphyletic group that should be subdivided. The arolia of some African species are reduced or, as in *Nomia (A.) speciosa* Friese and *theryi* Gribodo, absent. As in the subgenera *Crocisaspidia, Hoplonomia,* and *Paulynomia,* the outer hind tibial spur in *Acunomia* usually is sharply bent preapically, perhaps because of a large apical or scarcely preapical tooth projecting to one side (Fig. 63-2l). This is not true, however, in a few species, e.g., those placed in *Maculonomia* by Wu (1982a) and *N. (A.) thoracica* Smith, a species with a protuberant male clypeus and large male mandibles each with a median tooth on the upper margin. The tegulae are unusually large and broad posteriorly, reaching the level of the scutoscutellar suture, in *Acunomia* in the narrowest sense. They are almost as large in *Nomia (A.) senticosa* Vachal, but in the majority of species they taper posteriorly to a narrow point. Every intergradation occurs, and in some small groups (e.g., the group named *Maculonomia*) tegular shape varies widely among similar species. The gonostyli of the male are flat and sheetlike, sometimes almost parallel-sided and directed posteriorly, or broadly expanded and angular (as in the *Maculonomia* group), or, most frequently, expanded and curved medially and basad, as in *Hoplonomia* (Fig. 63-3f).

Nomia (Acunomia) melanderi Cockerell is widely known as the alkali bee, because it nests in great aggregations in moist alkali soil in the western United States. It was important in alfalfa pollination until an alternative pollinator, *Megachile rotundata* (Fabricius), came into use.

Nomia / Subgenus *Crocisaspidia* Ashmead

Crocisaspidia Ashmead, 1899a: 68. Type species: *Crocisaspidia chandleri* Ashmead, 1899, by original designation.

The scutellum is distinctive as compared to that of all other Nomiinae, for its lateral projections are flat with raised carinate edges and extend straight back, not upward, from the rest of the scutellum. The body is more robust than in most *Nomia* subgenera, although andreniform; length is 9 to 15 mm. This is the only group of *Nomia* in which the colored tergal bands are sometimes broken medially; the broken bands lead to a superficial similarity to *Thyreus* in the Melectinae.
■ This subgenus is primarily African, ranging from Senegal to Egypt and Somalia south throughout Africa to the Transvaal in South Africa; it ranges east, however, to Madagascar, Oman, Iran, Pakistan, and northern India. The 11 species were revised by Pauly (1990).

Nomia / Subgenus *Hoplonomia* Ashmead

Hoplonomia Ashmead, 1904b: 4. Type species: *Hoplonomia quadrifasciata* Ashmead, 1904, by designation of Cockerell, 1910f: 289.

In this subgenus the metanotum has two broad, basally fused, lamelliform projections, much as in *Crocisaspidia*, but the scutellum lacks projections or has only small ones in some males. The body is generally smaller and more slender than that of *Crocisaspidia*; its length is 6.5 to 11.5 mm. Male genitalia and sterna were illustrated by Michener (1965b).
■ *Hoplonomia* occurs in Japan, China, India, southeast Asia, the Philippines, Indonesia, New Guinea, the Bismarcks, the Solomon Islands, Australia south to southern Queensland, and Madagascar. It is not known from Africa. Over 20 names have been proposed for species of this group, some of which were listed by Michener (1965b).

Masuda (1943) gave an account of the biology of *Nomia (Hoplonomia) punctulata* Dalla Torre in Japan. Its nests are frequently occupied by several females, presumably a communal group.

Nomia / Subgenus *Leuconomia* Pauly

Nomia (Leuconomia) Pauly, 1980a: 124. Type species: *Nomia candida* Smith, 1875, by original designation.
Pronomia Pauly, 1997b: 102. Type species: *Pronomia pulawskii* Pauly, 1997, by original designation.

Leuconomia includes species smaller than most *Nomia* (body length 5-10 mm), with white or yellowish metasomal bands, and frequently with no colored band on T1. Some specimens of *Nomia (Leuconomia) lutea* (Warncke) have such a band, and some Oriental species regularly have a band on T1. T1 has a hair band laterally; it is sometimes very weak when there is a colored integumental band. One species from India has T1 and T2 red. The basitibial plate of the female is carinate only along the posterior margin or, in some Asiatic and Malagasy forms, the carina is complete. As in *Nomia* s. str., the middle and hind tibial spurs are simple, i.e., minutely serrate or ciliate. The tegulae are rounded, not attenuate posteriorly.
■ This subgenus is widespread in Africa, from Egypt to Liberia south to Cape Province, South Africa, and also occurs in Madagascar, Sri Lanka, India, Malaysia, and Java. Twenty African species were listed by Pauly (1990) and 25 were included in his revision of African species (Pauly, 2000). Two more are from Madagascar and at least ten from tropical Asia and nearby islands.

Because of their small size and the pale tergal bands, species of *Leuconomia* superficially resemble bees of the genus *Lipotriches* more than *Nomia*. The unmodified tibial spurs and tegulae and the relatively slender and simple male gonostyli support similarity to *Lipotriches*, but the subgenus *Leuconomia* does not resemble any particular subgenus of *Lipotriches*. *Leuconomia* resembles *Nomia*, however, in the characters indicated in the key to genera. The Asiatic species include some that are similar to the African species and others with enlarged hind legs lacking basitibial plates in the males but with completely defined basitibial plates in the females.

Three small (5.0-8.5 mm long) Malagasy species recently described as constituting the genus *Pronomia* fall in *Leuconomia* as here delimited. *Pronomia* is, however, unusual and similar to some Oriental species in the completely defined basitibial plate of both sexes and the colored band and lack of a lateral hair band on T1. The male genitalia of *Nomia pulawskii* Pauly, the type species, are also unusual, in having a row of long, blunt hairs on the attenuate gonostylus; the other two species do not have these features. Further study of Oriental and African members of this group should clarify the relationships and the groups that ought to be recognized.

Nomia / Subgenus *Nomia* Latreille s. str.

Nomia Latreille, 1804: 182. Type species: *Andrena curvipes* Fabricius, 1781, monobasic. *Nomia diversipes* Latreille, 1806, designated by Blanchard, 1849: pl. 125 (1847), was not originally included.
Nitocris Rafinesque, 1815: 123, unnecessary replacement for *Nomia* Latreille, 1804. Type species: *Andrena curvipes* Fabricius, 1781, autobasic.

This is a subgenus of moderate-sized (9-13 mm long), conspicuously banded species with tegulae rather narrowly rounded, not attenuate, posteriorly and the tibial spurs simple, i.e., without large teeth. It is the only subgenus of *Nomia* having one or two preapical teeth on the underside of the hind femur of the male, and also with T6 of the male produced medially to a truncate projection resembling a pygidial plate and hiding T7.

■ *Nomia* s. str. occurs in China, Thailand, India, Pakistan, Madagascar, and Africa from Gambia to Sudan south to Angola, the Transvaal, and Mozambique. The six species were reviewed by Pauly (1990).

Nomia / *Paulynomia* Michener

Nomia (Paulynomia) Michener, 2000: 326. Type species: *Nomia aurantifer* Cockerell, 1910, by original designation.

This group has been included in the subgenus *Curvinomia,* here included in *Acunomia.* The male differs from *Acunomia* in the very striking characters of the genitalia and sterna (see the key to subgenera); in these particulars *Acunomia* agrees with the subgenus *Hoplonomia.* The simple S5, relatively elongate S7, and slender, unflattened gonostyli (Fig. 63-3k-n) are likely to be plesiomorphic relative to *Acunomia, Hoplonomia,* and perhaps various other Nomiinae. The middle tibial spurs of the female are dark in color and armed with coarse, thornlike teeth along both margins of their distal three-fifths; some Oriental species of *Acunomia* approach this condition. The body length is 10 to 13 mm. Some individuals have a broken yellow band on T1; in *Acunomia* the band on T1 is usually complete or absent, not broken.

■ *Paulynomia* occurs from northern New South Wales to northern Queensland, Australia. The two species listed by Cardale (1993) are *Nomia (Paulynomia) aurantifer* Cockerell and *swainsoniae* Cockerell. Three names for Australian species were listed by Michener (1965b), *N. luteofasciata* Friese being a synonym of *N. aurantifer.*

Genus *Pseudapis* Kirby

Pseudapis is the major genus having greatly enlarged tegulae (Fig. 63-4o), these extending far back behind the level of the scutoscutellar suture. From other genera in which one or more species have precisely similar tegulae, presumably independently evolved, *Pseudapis* differs by the presence of a preoccipital carina behind the genal area and a carina across the anterior part of the scutum separating the small, vertical, anterior surface from the main part of the scutum. The lamella on the pronotal lobe (Fig. 63-4o) and extending mesad from it is translucent, larger than that of almost any other Nomiinae, and similar in the two sexes. The body form, especially of females, is short, markedly more robust than that of most other Nomiinae, euceriform or even anthophoriform rather than andreniform as is usual for the Nomiinae. In most species the metasomal terga (except T1), or at least the more posterior ones, have both basal and apical bands of pale hairs or tomentum. The body length ranges from 6 to 11 mm. Male genitalia and numerous other structures were illustrated by Sakagami (1961), Wu (1983e), and Pauly (1990).

Relationships among the genera recognized by Pauly (1990) seem best shown by regarding some of them as subgenera or synonyms of *Pseudapis.*

Key to the Subgenera of *Pseudapis*

1. Basitibial plate of female delimited by a carina only on posterior margin; hind leg of male not enlarged, tibia without apical process but bearing two apical spurs; male upper gonostylus broadly fused to gonocoxite, directed apicad, with mosslike hair covering dorsal surface (Africa) .... *P. (Pachynomia)*
—. Basitibial plate of female fully delimited, i.e., with a carina on both anterior and posterior margins; hind leg of male somewhat, often greatly, swollen, at least tibia with apical process, but without spurs or with only one spur; male upper gonostylus differentiated from gonocoxite, directed mesad across apex of genital capsule, and without mosslike hairs .... *P. (Pseudapis s. str.)*

Pseudapis / Subgenus *Pachynomia* Pauly

Nomia (Pachynomia) Pauly, 1980a: 124. Type species: *Nomia amoenula* Gerstaecker, 1870, by original designation.

Pachynomia looks superficially like *Pseudapis* s. str. but differs in the rather striking characters indicated in the key to subgenera. The male leg characters and part of the genitalic characters are probably plesiomorphic relative to *Pseudapis* s. str. As in *Pseudapis* s. str., the males of some species have a lateral, posteriorly directed tooth on the scutellum.

■ This subgenus is found from Senegal to Kenya south to Cape Province and Natal, South Africa. The four species were revised by Pauly (1990).

Pseudapis / Subgenus *Pseudapis* Kirby s. str.

Pseudapis W. F. Kirby, 1900: 15. Type species: *Pseudapis anomala* W. F. Kirby, 1900, monobasic.
Stictonomia Cameron, 1905: 192. Type species: *Stictonomia punctata* Cameron, 1905, monobasic.
Nomia (Nomiapis) Cockerell, 1919a: 208. Type species: *Nomia diversipes* Latreille, 1806, by original designation.
Nomia (Lobonomia) Warncke, 1976a: 99. Type species: *Nomia lobata* Olivier, 1811, by original designation.
Ruginomia Pauly, 1990: 103. Type species: *Nomia rugiventris* Friese, 1930, by original designation.

Pauly (1990) recognized all of the names listed above except *Lobonomia* as genera. The principal characters, however, appear in diverse combinations. In the females, the groups called *Pseudapis* and *Nomiapis* have the basitibial plate rounded, while in the groups called *Rugonomia* and *Stictonomia* it is pointed. No other characters appear to support this division into two groups. The lateral scutellar tooth of the male is present in *Stictonomia* but also in some species of both *Pseudapis s. strictissimo* and *Nomiapis* (and in some species of the subgenus *Pachynomia*). The lack of apical tergal hair bands is a character of *Nomiapis* and males of *Ruginomia,* but hair bands are sometimes absent on the basal half or more of the metasoma in *Pseudapis s. strictissimo.* There are no hind tibial spurs in males of *Ruginomia* and *Nomiapis,* but one spur is present in males of most but not all species of *Pseudapis s. strictissimo* and *Stictonomia.*

■ *Pseudapis* s. str. occurs in the Mediterranean basin (Morocco and Spain to Egypt), north in Europe to Aus-

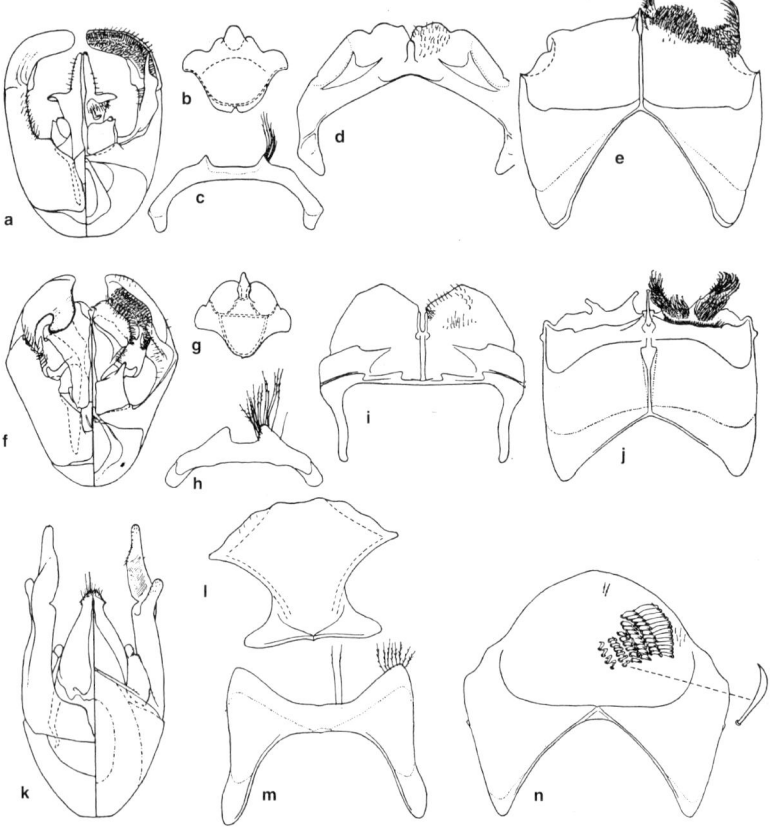

Figure 63-3. Structures of male Nomiinae. **a-e**, Genitalia, S8, S7, S6, and S5 of *Lipotriches (Austronomia) australica* (Smith); **f-j**, Same structures of *Nomia (Hoplonomia) lyonsiae* Cockerell; **k-n**, Genitalia, S8, S7, and S6 of *N. (Paulynomia) aurantifer* Cockerell. (Genitalia are divided, dorsal views on the left; sterna are ventral views, hairs omitted on the left.) From Michener, 1965b.

tria, Poland, and southern Russia, east to Turkey, central Asia, northern China, and Japan, south throughout Africa (especially the xeric areas, but a few records are for Zaire) to Cape Province and Natal, South Africa, and south in Asia to India and Thailand. The single Australian record is based on a mislabeled specimen (Michener, 1965b; Baker, 2002). *Pseudapis* does not occur in Madagascar in spite of its abundance in Africa. The total of about 69 species consists of about 37 sub-Saharan African species (Pauly, 1990), 22 west palearctic and Turkestan species (Warncke, 1976a), and perhaps not more than 10 additional species in tropical and east Asia. There appear to be very few Chinese species and only one in Japan. Revisions, using various generic or subgeneric names, are for the western palearctic region (Warncke, 1976a) and for Africa (Pauly, 1990). Baker (2002c) provided much information on the palearctic and oriental species of *Pseudapis,* and a key to 35 species from those regions. His phylogenetic treatment excluded African forms; for the palearctic and oriental forms *Pseudapis* and *Nomiapis* appear as sister groups which he regarded as genera. If the same pattern is evident when the large African fauna is included, I would recognize *Nomiapis,* perhaps as a subgenus rather than a genus to show, in the classification, its relationship to *Pseudapis* s. str.

Genus *Ptilonomia* Michener

Nomia (Ptilonomia) Michener, 1965b: 160. Type species: *Nomia plumosa* Michener, 1965, by original designation.

This genus resembles *Mellitidia* in having a distinct malar area, but is probably more closely related to *Reepenia.* The loosely plumose hairs that cover the head and thorax are longer than the similar hairs of *Reepenia. Ptilonomia* has the same form as a rather large (11-13 mm long), nonmetallic *Mellitidia*. The genitalia and sterna were illustrated by Michener (1965b), the head by Hirashima (1966b); see also Figure 63-4f-j.

■ *Ptilonomia* is known only from New Guinea. The three species were distinguished in a key by Hirashima (1966b).

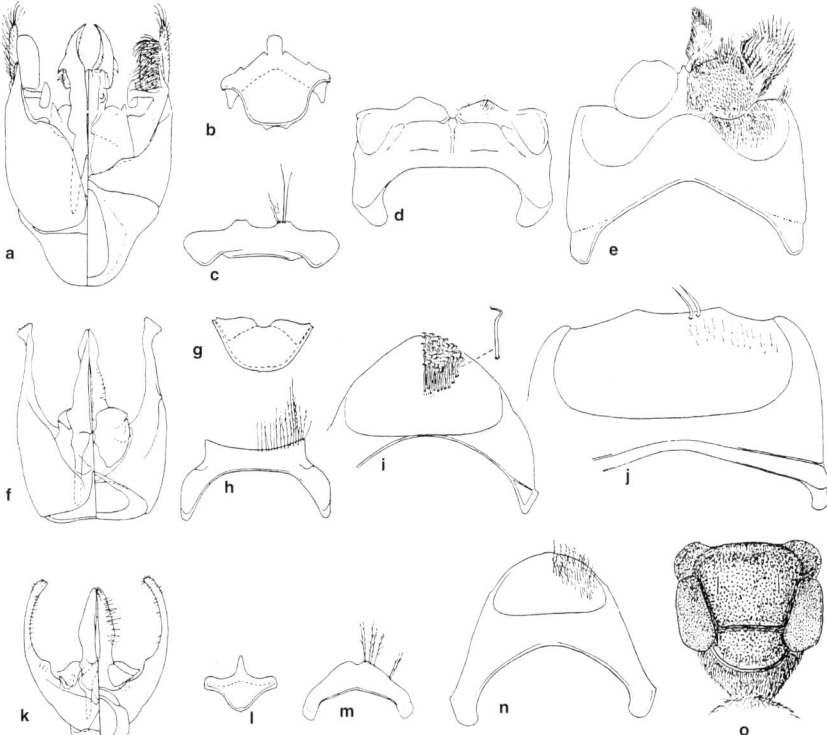

Figure 63-4. a-n, Structures of male Nomiinae. **a-e,** Genitalia, S8, S7, S6, and S5 of *Mellitidia gressitti* Michener; **f-j,** Same, of *Ptilonomia plumosa* Michener; **k-n,** Genitalia, S8, S7, and S6 of *Reepenia bituberculata* (Smith). (Genitalia are divided, dorsal views on the left; sterna are ventral views, hairs omitted on the left.) **o,** Dorsum of thorax of *Pseudapis (Pseudapis)* sp., showing large tegulae and lamellae on pronotal lobes. a-n, from Michener, 1965b.

Genus *Reepenia* Friese

Nomia (Reepenia) Friese, 1909b: 205. Type species: *Nomia variabilis* Friese, 1909, by designation of Sandhouse, 1943: 595; Michener, 1965b: 159; and Commission Opinion 788 (1966). [See Michener, 1997b.]

Megaloptodes Moure, 1959: 183. Type species: *Megalopta bituberculata* Smith, 1853, by original designation.

Reepenia is perhaps related to *Ptilonomia* and *Mellitidia*. In all three genera the head is but little broader than long as seen from the front, unlike the head of most other Nomiinae. The enlarged ocelli and brown or testaceous coloration of the body (which is 9 to 12 mm long) suggest that *Reepenia* is a nocturnal genus, although no behavioral observations yet support this view. Its appearance led to confusion with the well-known nocturnal neotropical augochlorine genus *Megalopta* by Smith (1853), Moure (1959), etc. Male genitalia and sterna were illustrated by Michener (1965b); see also Figure 63-4k-n.

■ This genus occurs in the New Guinea area, north to the Bismarck Archipelago, west to Misoöl, south to Kai and northern Queensland, Australia. The 11 species-group names, listed by Michener (1965b), probably involve only two or three species.

Genus *Spatunomia* Pauly

Nomia (Spatunomia) Pauly, 1980a: 121. Type species: *Nomia filifera* Cockerell, 1932, by original designation.

This is a genus of large bees (body length 13-15 mm) in which the metasoma is dark red and lacks hair bands. There is a strong preoccipital carina in the male, behind both the vertex and the genal areas. The pronotum lacks a carina (except on the pronotal lobe) but extends across in front of, and at the level of, the scutum, so that the scutum does not bend down to the pronotum. The basitibial plate of the female is complete, rounded distally; the inner hind tibial spur is very coarsely serrate, and the middle tibial spur has some enlarged distal teeth, as in many species of *Nomia (Acunomia)*. The hind legs of the male are moderately enlarged, with either one or two tibial spurs. The most remarkable features of the genus are the simple mandibles of the female (according to Pauly, 1990) and the pedunculate last antennal segment of the male, which has a slender base and a broad, flattened apical region. Male genitalia and other structures were illustrated by Pauly (1990).

■ *Spatunomia* occurs from Sudan to Namibia, and to Cape Province and Natal, South Africa. The two species were reviewed by Pauly (1990).

Genus *Sphegocephala* Saussure

Sphegocephala Saussure, 1890: 74. Type species: *Sphegocephala philanthoides* Saussure, 1890, monobasic.

Stegocephala Dalla Torre, 1896: 176, unjustified emendation of *Sphegocephala* Saussure, 1890.

Sphecophala Ashmead, 1899a: 89, error for *Sphegocephala* Saussure, 1890. Type species: *Sphegocephala philanthoides* Saussure, 1890, autobasic.

The body in *Sphegocephala* is coarsely punctate, the hairs inconspicuous and mostly dark, not forming metasomal bands. The gradulus of T3, however, often has a fringe of conspicuous white hairs. The head and parts of the thorax are often dark red. The heads of males of certain species are large, broader than the thorax. Body length is 7.5 to 11.0 mm. Characters other than those indicated in the key to genera include the defined basitibial plate of the male, the unmodified hind legs and exposed sterna of the male, the coarsely serrate to almost edentate inner hind tibial spur of the female (the outer hind and middle tibial spurs minutely serrate, i.e., unmodified). Genitalia and other structures were illustrated by Pauly (1991).

■ *Sphegocephala* is found only in Madagascar. The six species were characterized by Pauly (1991).

Genus *Steganomus* Ritsema

Steganomus Ritsema, 1873: 224. Type species: *Steganomus javanus* Ritsema, 1873, monobasic.

Cyathocera Smith, 1875: 47. Type species: *Cyathocera nodicornis* Smith, 1875, monobasic.

Nomia (Dinomia) Hirashima, 1956: 29. Type species: *Nomia taiwana* Hirashima, 1956, by original designation.

In its small (length 6-9 mm), robust body and enormous tegulae, this genus resembles *Pseudapis*. It differs from that genus and all other Nomiinae in having only two submarginal cells. It differs further from *Pseudapis* in having two hind tibial spurs in males [but see *P. (Pachynomia)*], a reduced prothoracic lamella in females, and genae without preoccipital carinae. The carina across the pronotum suggests *Lipotriches* s. str., not the lamella characteristic of *Pseudapis*. The male genitalia and other structures were illustrated by Pauly (1990).

■ *Steganomus* occurs from Senegal to Kenya south to Cape Province and Natal, South Africa, as well as in Sri Lanka, India, and east as far as Java and Taiwan. There are seven species. The four African species were revised by Pauly (1990).

64. Subfamily Nomioidinae

In spite of the name, *Nomioides* and its relatives do not resemble *Nomia*, either in appearance or in their combination of characters. They have usually been placed in the Halictinae, partly because of the curved basal vein of the forewing (Fig. 64-1a), but they differ from Halictinae in the unbroken prepygidial fimbria of the females and other characters listed below. In Alexander and Michener's (1995) study of the phylogeny of S-T bees, *Nomioides* usually appeared among the rophitine genera, as part of a paraphyletic group from which the Nomiinae and Halictinae arose. *Nomioides* differs from the Rophitinae in its most distinctive characters, however, and is here accorded subfamily rank, leaving the Rophitinae as monophyletic.

This Old World subfamily consists of minute (body length 2.5-6.5 mm), weakly sclerotized, andreniform bees, the head and thorax finely punctate and usually dull, usually metallic blue or green, and the metasoma flattened, especially in the female. All tagmata usually have yellow or cream markings (Fig. 64-3a-c), the metasoma sometimes being mostly yellow, although it may be wholly black. There are no pubescent fasciae on T1 to T4. Superficially, these bees closely resemble various groups of *Perdita* (Andrenidae, Panurginae) of North America, *Habralictus* (Halictinae) of the neotropics, *Eremaphanta* (Melittidae) of central Asia, and certain Euryglossinae (Colletidae) of Australia. The Nomioidinae differ from the Halictinae in the presence of a sparse fimbria on T5 of the female that is not divided medially by an area of specialized, fine vestiture and texture. Also, the labrum of the female has a small but well-defined apical process that lacks a keel, thus differing from all other halictid groups; in nonparasitic Halictinae, this process is larger and strongly keeled. The inner orbits are distinctly and somewhat angularly emarginate (Figs. 64-1b, 64-3a, b), as is usual in the Augochlorini. The anterior tentorial pit is at the apex of an acutely pointed projection of the paraocular area (Figs. 64-1b, 64-3a, b) into the clypeus. This paraocular lobe is not elevated above the surrounding areas and thus differs from the elevated rounded paraocular lobe that projects into the clypeus in many Augochlorini. T7 of the male has a truncate apical projection that doubtless represents the pygidial plate, but the plate is not indicated on the tergal surface and the margin of the tergum is thin, not reflexed as in the Halictini. S7 of the male is a simple plate, like the preceding sterna but smaller. The group was further characterized by Michener (1978a) and in detail by Pesenko (1983, 2000b), who tabulated differences between Nomioidinae and each tribe of Halictinae. Illustrations of the male genitalia, sterna, and other structures were provided by Ireland (1935) and Pesenko (1983, 2000b); see also Figure 64-2.

Pesenko (2000a) considered the nomioidines as a tribe of the subfamily Halictinae, but his phylogenetic analysis placed them as the basal branch of the Halictinae (sister-group to all other Halictinae) and thus they could also logically be considered as a separate subfamily, the rank that I prefer for them.

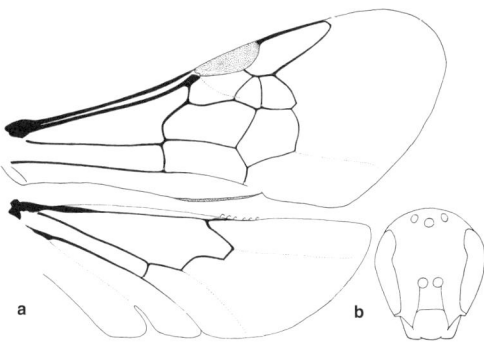

Figure 64-1. *Nomioides minutissimus* (Rossi) (Nomioidinae). **a,** Wings; **b,** Face, showing acute paraocular lobes extending down into the clypeus.

It has been usual to place all members of this subfamily in the genus *Nomioides*. Pesenko (1983, 1993), however, recognized three genera that seem to be as distinctive as genera elsewhere in the Halictidae.

The subfamily was revised by Blüthgen (1925) and, for the palearctic region, by Pesenko (1983), who also gave a good account of the morphology and a detailed statement of the differences among the genera. Pesenko (1996) revised the species of Madagascar. A large work by Pesenko and Pauly (2005) is on the nomioid species of Africa. Pesenko (2000b) gave an account of the phylogeny and classification of the subfamily.

So far as is known, the Nomioidinae are polylectic. Nests are slender vertical burrows in the soil with subhorizontal cells, one at the end of each lateral burrow, each lateral being filled with earth after the cell is completed. The cells and provisions are similar to those of the Halictinae. Some nests are occupied by presumably communal groups of females; there is no evidence of castes (Batra, 1966a; Radchenko, 1979). Unlike most Halictinae, both sexes overwinter as adults (Blüthgen, 1925).

Key to the Genera of the Nomioidinae

1. S8 of male with long apical process (Fig. 64-2e), the process much longer than broad, thus sternum over 2.5 times as long as wide; posterior depressed marginal zone of T2 and frequently other terga translucent yellowish, so that yellow base of T3 shows through .... *Nomioides*
—. S8 of male without or with short apical process (Fig. 64-2b), the process not or scarcely longer than broad, sternum thus less than 2.5 times as long as broad; posterior depressed marginal zone of T2 and frequently other terga brown or dusky, at least basally (margin often narrowly pallid translucent) so that base of T3 does not show through conspicuously 2
2(1). Second submarginal cell petiolate from costal side (Fig. 64-3d); median flagellar segments of male little longer than broad; inner hind tibial spur of female with one large tooth (Fig. 64-3e); S8 of male with apical

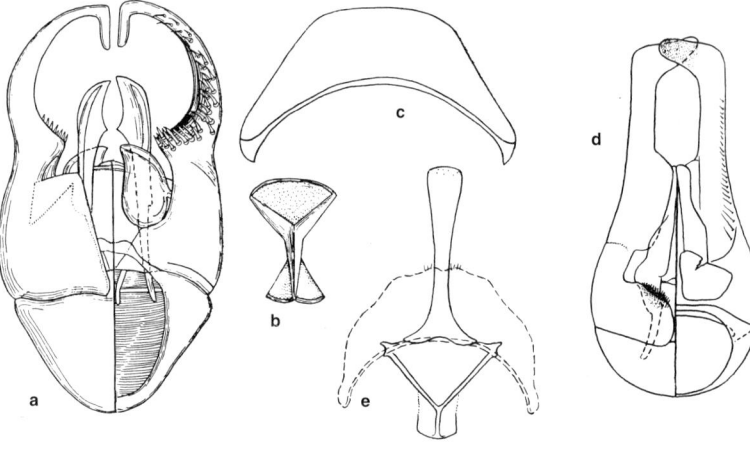

Figure 64-2. Structures of male Nomioidinae. **a-c,** Genitalia, S8, and S7 of *Ceylalictus variegatus* (Olivier); **d, e,** Genitalia, S8, and (in broken lines) S7 of *Nomioides minutissimus* (Rossi). (Divided drawings show dorsal views on the left.) a-c, based on Pesenko, 1983; d, e, from Michener, 1978a.

process short, broader than long or about as broad as long depending on how it is measured (Africa) .... *Cellariella*
—. Second submarginal cell not petiolate (as in Fig. 64-1a), although sometimes pointed on marginal cell; median flagellar segments of male about 1.5 times as long as broad; inner hind tibial spur of female with two large teeth; S8 of male completely without apical process, apex Y- shaped with membrane between arms of Y (Fig. 64-2b) .... *Ceylalictus*

Genus *Cellariella* Strand

Nomioides (Cellaria) Friese, 1913a: 575, not Ellis and Solander, 1786. Type species: *Nomioides arnoldi* Friese, 1913 = *Nomioides somalica* Magretti, 1899, monobasic.
Nomioides (Cellariella) Strand, 1926: 53, replacement for *Cellaria* Friese, 1913. Type species: *Nomioides arnoldi* Friese, 1913 = *Nomioides somalica* Magretti, 1899, autobasic.

■ *Cellariella* is widespread in Africa, from Somalia and Zaire south to Cape Province, South Africa. It also occurs in Madagascar. There are two species.

Genus *Ceylalictus* Strand

This genus includes fewer species than *Nomioides,* but there is more diversity among them than among species of *Nomioides,* as indicated by the three subgenera recognized by Pesenko (1983). Pesenko considered *Ceylalictus* a paraphyletic genus from which *Nomioides* arose. Another analysis by Pesenko (2000b) showed the same relationship, the subgenus *Atronomioides* being the sister group to all the rest of the Nomioidinae. If this relationship is verified, *Atronomioides* should doubtless be given generic rank.

The subgeneric positions of most species are unknown to me, although the principal subgenus is evidently *Ceylalictus* s. str.

Key to the Subgenera of *Ceylalictus* (Males)

1. Gonostylus of male slender, eight or more times as long as wide (Fig. 64-2a) .... *C. (Ceylalictus s. str.)*
—. Gonostylus of male broad, not over about four times as long as wide 2
2(1). Gonostylus of male three or four times as long as wide, the apex narrowly rounded; penis valve rather straight .... *C. (Meganomioides)*
—. Gonostylus of male about twice as long as broad, the apex broadly truncate; penis valve with strong bend near base .... *C. (Atronomioides)*

Ceylalictus / Subgenus *Atronomioides* Pesenko

Ceylalictus (Atronomioides) Pesenko, 1983: 186. Type species: *Ceylalictus warnckei* Pesenko, 1983, by original designation.

■ This subgenus is known from South Africa, equatorial Africa, and Madagascar to Iran, and southern China. There are eleven species.

Ceylalictus / Subgenus *Ceylalictus* Strand s. str.

Halictus (Ceylalictus) Strand, 1913a: 137. Type species: *Halictus horni* Strand, 1913, monobasic.
Nomioides (Eunomioides) Blüthgen, 1937: 3, nomen nudum. Type species: *Andrena variegata* Olivier, 1789, by original designation.

■ *Ceylalictus* s. str. ranges from Senegal, the Canary Islands, and Portugal to western China, mostly south of 50°N. To the south it occurs to Cape Province, South Africa, Madagascar, the Arabian peninsula, and eastward to Sri Lanka, India, southeast Asia, Indonesia, and the northern half of Australia. No Nomioidinae are known from New Guinea or islands to the eastward. One species, *C. (Ceylalictus) variegatus* (Olivier), is found from the western outposts of the genus in Europe and Africa (south to Namibia) all the way to China. The total number of known species is 13.

Ceylalictus / Subgenus *Meganomioides* Pesenko

Ceylalictus (Meganomioides) Pesenko, 1983: 183. Type species: *Nomioides karachensis* Cockerell, 1911, by original designation.

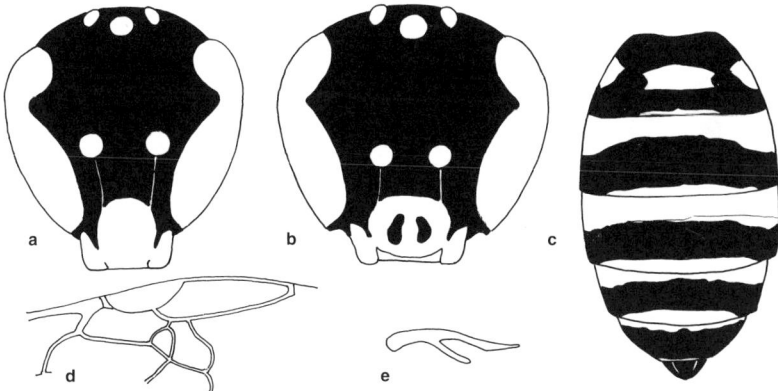

Figure 64-3. *Cellariella brooksi* Pesenko (Nomioidinae). **a, b,** Faces of male and female, showing yellow clypeus; **c,** Metasoma of female, showing yellow and blackish pattern; **d,** Portion of forewing, showing petiolate second submarginal cell; **e,** Inner hind tibial spur of female. From Pesenko, 1993.

■ This subgenus is known from western Pakistan and northern Africa, west as far as Algeria. There are two species.

Genus *Nomioides* Schenck

Nomioides Schenck, 1867: 333. Type species: *Apis minutissima* Rossi, 1790, designated by the Commission, Opinion 1319, 1985. See Michener, 1997b, and Pesenko and Kerzhner, 1981.

Nomioides (Erythronomioides) Pesenko, 1983: 176. Type species: *Nomioides socotranus* Blüthgen, 1925, by original designation.

Nomioides (Paranomioides) Pesenko, 1983: 175. Type species: *Nomioides steinbergi* Pesenko, 1983, by original designation.

Each of the two names proposed as a subgenus relates to a single unusual species. Subgeneric recognition seems unnecessary.

■ *Nomioides* ranges from Portugal and the Canary Islands eastward across Eurasia, north to about 51°N, to China (not to Japan). Farther south, it reaches southern Africa (Natal and Cape Province), India, Southeast Asia, Taiwan, and the Philippines. *Nomioides* is rare or absent in moist, forested areas, but often common in xeric regions, both temperate and tropical. There are about 60 species (Pesenko, 2000b). A key to the Asian species of the minutissimus group was provided by Pesenko (2004a).

65. Subfamily Halictinae

Halictinae is an enormous group of bees, the species of which are frequently common. All females have andreniform bodies; the same is true of males except that they are frequently much more slender, and have longer antennae, than females. Except in parasitic forms the labrum of females has a strong apical process, fringed with coarse bristles; the process usually has a strong dorsal keel (Fig. 65-1a, b, e, g) except in parasitic taxa such as *Sphecodes* and *Temnosoma* (Fig. 65-1i). The episternal groove is distinct and extends well below the level of the scrobe (Fig. 20-5b). This groove usually distinguishes the Halictinae from most species of the related subfamily Nomiinae; see the key to subfamilies (Sec. 61) and the account of Nomiinae (Sec. 63). The basal vein is strongly curved, especially so near its posterior or basal end (Fig. 65-5). Although sometimes duplicated in the Nomiinae and approached in the rophitine genus *Xeralictus,* the curvature of this vein usually separates the Halictinae from all other bees except the Nomioidini.

The Halictinae also differ from other halictid subfamilies in the form of S7 and S8 of the males. These sterna are small, weakly sclerotized, transverse, and seemingly fused together (Fig. 65-1n), although they can be separated (Fig. 65-1o, p). S7 usually lacks elaboration; S8 often has a small midapical process and a basal spiculum. A remarkable exception to all this is the uniquely large and heavily sclerotized S7 and S8 complex of *Urohalictus* (Fig. 66-23).

A longitudinal median area of extremely fine punctation and fine, short hairs, less commonly nearly bare and even shining (in *Thrinchostoma*), divides the prepygidial fimbria in most female Halictinae (Fig. 65-1j, k). This area is rather broad and triangular (with the base of the triangle near the tergal margin) in some Augochlorini, such as *Megalopta.* In the various parasitic taxa it is also triangular but weakly evident, or completely absent as in *Sphecodes*. Its disappearance is related to the weakening and disappearance of the prepygidial fimbria; in no case among Halictinae is there a well-developed fimbria on T5 not divided by this specialized area.

Nest architecture for the subfamily was described in detail by Sakagami and Michener (1962). Except for a few groups that nest in rotting wood [*Augochlora* s. str., *Megalopta,* and some species of *Neocorynura* in the Augochlorini, and a few species of *Lasioglossum (Dialictus)* in the Halictini], Halictinae nest in burrows in banks or flat soil. The nest entrances are commonly constricted and are guarded much of the time if more than one adult female lives in the nest, either by the head of a bee or, after disturbance, by the dorsal surface of the metasoma. The nest burrows probably primitively give rise to rather long lateral burrows, each of which ends in a single cell, or less commonly in a series of two or more cells end to end (Fig. 65-2). Serially arranged cells are found in laterals of unrelated forms such as *Agapostemon* and its relatives, *Lasioglossum (Ctenonomia),* and certain nonmetallic European species of *Lasioglossum (Dialictus)*. When the cell or cells of a lateral are complete, the lateral is usually filled with soil as other laterals are excavated. Sakagami, Matsumura, and Maeta (1985) discovered that the Japanese *Lasioglossum (Dialictus) allodalum* Ebmer and Sakagami constructs no laterals, the cells lying in series in the more or less horizontal main burrow. In the great majority of halictine nests, however, cells are not in series. In many cases they arise directly or by way of very short laterals from the main burrow (Figs. 7-3, 65-3). The cells are commonly scattered along the main burrow; in such nests the main burrows are often branched. In some cases the cells are or tend to be concentrated along a single short section of the main burrow, so that the cells are clustered. In some such cases the bees make burrows around the outside of the cell cluster, and such burrows may expand so that the cell cluster is isolated from the substrate except for a few pillars or supporting rootlets. In some species the cell cluster is built in an excavated cavity (Fig. 67-3) instead of being a remnant of the substrate that contains the cells. Cell clusters surrounded by air spaces are particularly frequent in the Augochlorini (Figs. 7-4, 7-5, 67-2, 67-3) but occur also in such diverse Halictini as *Halictus quadricinctus* (Fabricius) (Fig. 65-4), some species of *Caenohalictus,* and many species of *Lasioglossum (Evylaeus)* (Packer, 1991). The cell clusters of the European *H. quadricinctus* are remarkable for their large size and beautiful construction, and especially for the fact that most other species of *Halictus* make simple nests like those of *H. ligatus* Say (Fig. 7-3). Species that nest in rotting wood usually make rather irregular nests, very different superficially from those in soil, because suitable wood is often limited by intrusions of hard wood or bark; see Stockhammer (1966, 1967) and Sakagami (1964).

Considering the diversity in nest structure, the cells themselves are surprising uniform, bilaterally symmetrical about the saggital plane, the upper surface being more strongly convex than the lower surface. An exception is *Lasioglossum (Dialictus) allodalum* Ebmer and Sakagami, whose cells do not seem flatter on the lower surface, according to Sakagami, Matsumura, and Maeta (1985). Typically, the cells are more or less horizontal, but the axes are often somewhat irregularly oriented, and in various Augochlorini whose cells are in clusters, they are consistently vertical. Cells are lined with "waxlike" material that is at least partly of Dufour's gland origin and is rich in macrocyclic lactones (Cane, 1981).

Provisions are usually firm, subspherical (Pl. 16), less commonly rectangular, and the mass lies on the ventral surface of the cell near the distal end (Fig. 7-5). When the cells are vertical, the provisions are in exactly the same position, stuck to the now vertical flatter cell surface (Fig. 7-4). An egg is attached to the upper surface of the mass of provisions, or equivalent position if the cell is vertical, and the egg's major axis is parallel to that of the cell.

The majority of species of Halictinae are polylectic, although oligolectic species and even subgenera are known, for example *Lasioglossum (Sphecodogastra),* which is morphologically adapted to handling pollen of Onagraceae

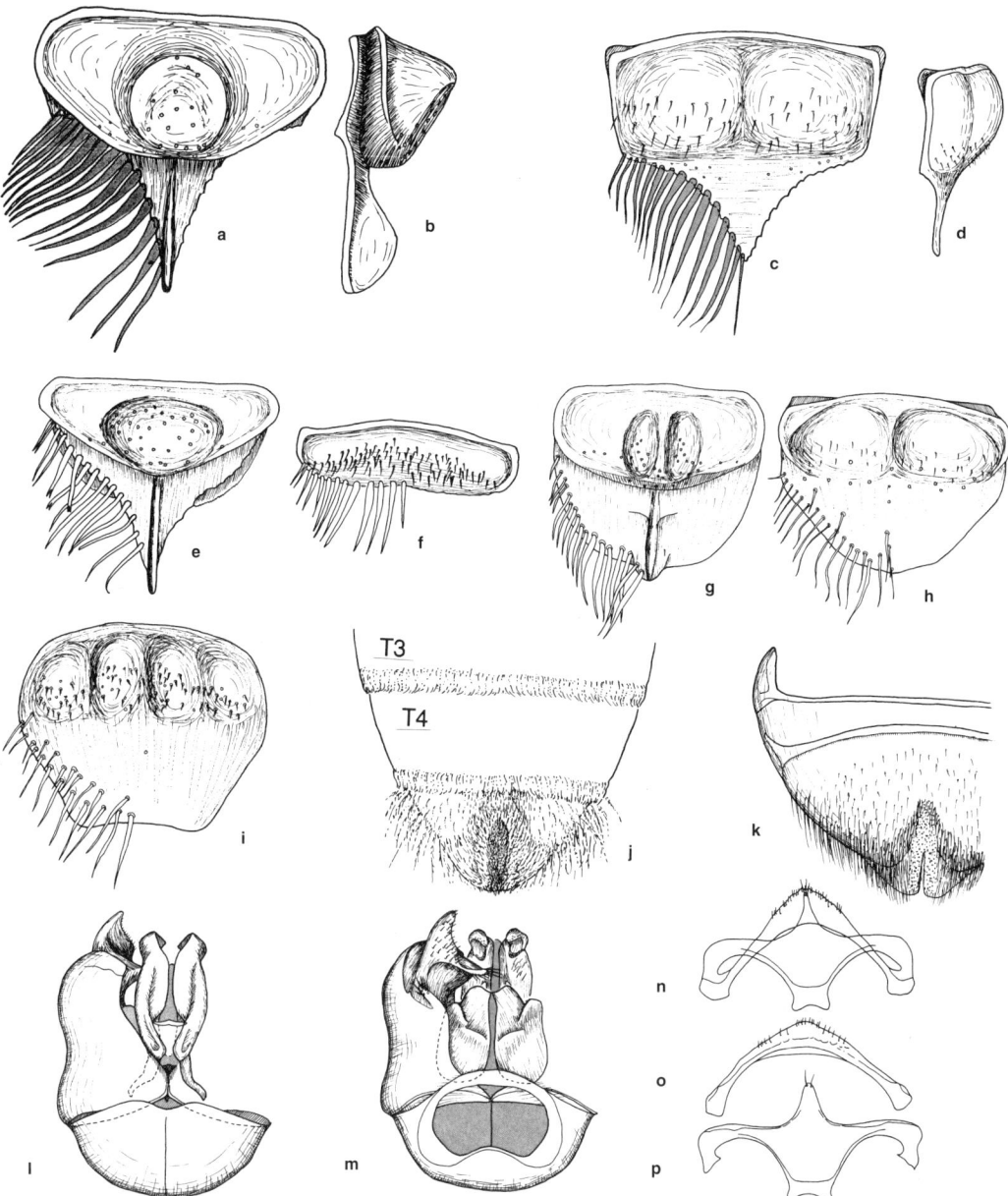

Figure 65-1. Structures of Halictinae. **a, b,** Labrum of *Pseudaugochlora graminea* (Fabricius) female, showing dorsal keel of apical process and greatly thickened basal area of labrum; **c, d,** Labrum of male of *P. graminea*, showing lack of keel on apical process; **e, f,** Labrum of female and of male of *Corynura chilensis* (Spinola); **g, h,** Labrum of female and of male of *Augochloropsis metallica* (Fabricius); **i,** Labrum of female of *Temnosoma smaragdinum* Smith, a cleptoparasite whose female lacks a keel on the apical process. **j,** Apex of metasoma of female *Halictus rubicundus* (Christ), showing median specialized area of T5; **k,** T5 of female of *Pseudaugochlora graminea* (Fabricius), showing slit in median specialized area. **l, m,** Dorsal and ventral views of male genitalia of *Corynura chilensis* (Spinola); **n,** S7 and S8 of same, fused; **o, p,** S7 and S8 of same, artificially separated. From Eickwort, 1969b, except j and k, which are modified from Michener, McGinley, and Danforth, 1994.

that is webbed together by viscin threads. There may be a causal relation between polylecty and the long season of flight of many Halictinae, as suggested in Section 6. Some of the solitary and communal species, such as most *Agapostemon* and *Lasioglossum (Dialictus) villosulum* (Kirby), have more than one generation per year and thus are active during a long season; the social species are necessarily active for a long season. A long season of activity must usually be associated with dependence on a diversity of flowers.

Figure 65-2. Nests of *Ruizantheda divaricata* (Vachal). **a, b,** Autumnal nests, showing hibernation chambers in which females will pass the winter; **c,** Spring nest occupied by 12 bees, with 12 cells being constructed or provisioned; **d,** A nest excavated at the same time as **c** but occupied by only one bee; **e,** Summer nest probably constructed by about 18 bees, only four of which survived when the nest was excavated; **f,** A lateral burrow, showing cells in series. (Earth-filled burrows are indicated by crosswise lines; the upper scale line represents 8 cm, the lower line, 1 cm.) From Michener and Lange, 1958d.

Most Halictinae pass unfavorable seasons like winter as adult females with sperm cells in their spermathecae, hibernating in hibernaculae that may or may not be extensions of their natal nests. Like bumblebees, they emerge with good weather in the spring and establish nests, usually solitarily but sometimes communally or even, perhaps, semisocially. They forage, provision cells, and rear offspring. Some species form eusocial colonies in which most of the early offspring are females, some or all of which will become workers. In such cases most of the males are produced in late summer; they mate with young females (gynes) that will hibernate and then be next year's prospective queens. The workers and males usually die before the onset of winter. Such a life cycle almost requires polylectic foraging, because of the season-long activity. In a few non-eusocial Halictinae, all of them species of *Lasioglossum* s. str., males as well as females survive temperate-climate winters; a review is included in Sakagami and Maeta (1990). In fully tropical climates, activity is sometimes continuous, as was indicated for four species studied by Michener, Breed, and Bell (1979) in Colombia.

Of all insects, the Halictinae exhibit the richest series of gradations between ordinary solitary ways of life and eusocial behavior with clearly recognizable queen and worker castes, i.e., castes that differ behaviorally and physiologically although scarcely morphologically, except sometimes for size (see Sec. 5). Social behavior has arisen in diverse clades, both in Augochlorini and Halictini, and has also been lost repeatedly. Danforth (2002) made a phylogenetic analysis of the family based on molecular data that indicated only three origins of eusociality, followed by various reversions to solitary life. A probable additional origin of eusocial behavior within the Augochlorini was discovered by Coelho (2002), who found *Augochloropsis iris* (Schrottky) to be eusocial in Brazil. No other species of *Augochloropsis* is known to be eusocial. In the same tribe, species of *Megalopta* are facultatively eusocial; when there is a colony, the dominant female tends to be larger, older, and with larger ovaries than the others (Wcislo and Gonzalez, 2006). Nest populations of Halictinae may consist of a single female and her eggs and larvae, or of multiple adult females having any of a range of diverse relationships. Sometimes, as in *Pseudagapostemon* and some species of *Agapostemon,* they are communal, each female presumably making and provisioning her own cells, the several individuals sharing only the entrance and main burrows. Sometimes, they are semisocial groups, probably usually of close relatives of the same generation, such as sisters, with one individual laying most of the eggs and the others acting more like workers. Such colonies regularly arise when the queen in a eusocial colony dies, and one of her several daughters becomes a replacement queen. On the contrary, an account of the relationships among colony members and among colonies of a regularly communal species of the *Lasioglossum* subgenus *Chilalictus* in Australia by Forbes

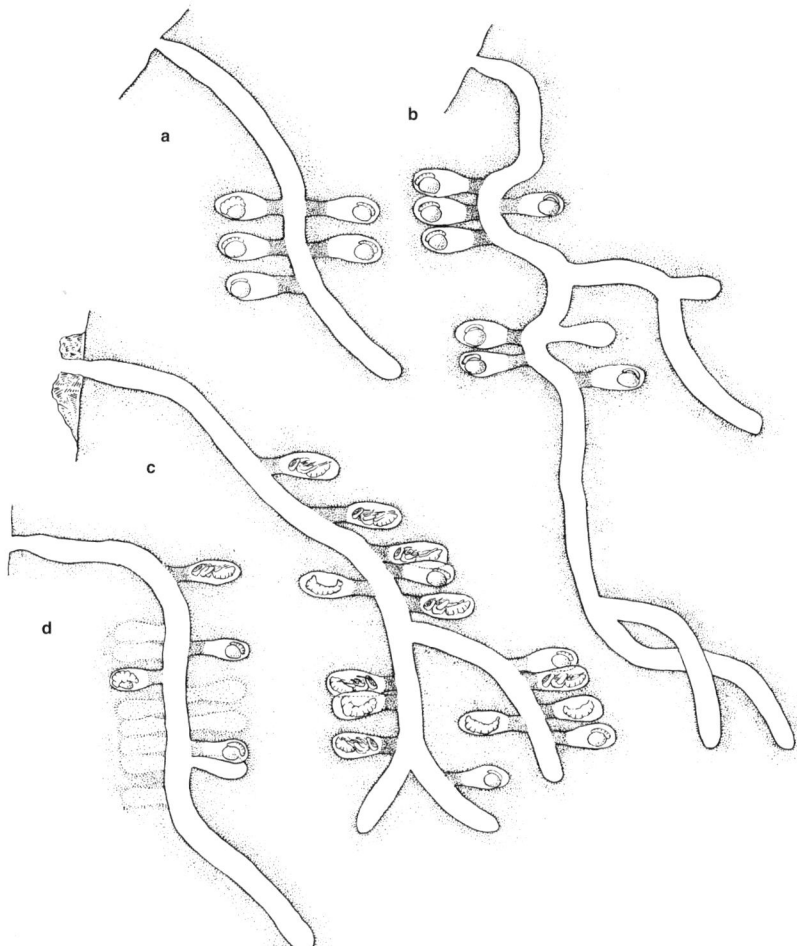

Figure 65-3. Nests of *Lasioglossum (Dialictus) rhytidophorum* (Moure) in earth bank. **a,** Spring nest made by single mother; **b-d,** Summer nests occupied by small eusocial colonies. From Sakagami and Michener, 1962.

et al. (2002) indicated otherwise. The study was based on DNA microsatellites and indicated that most colonies consisted of unrelated females. *Augochloropsis sparsilis* (Vachal) may be a species that regularly forms semisocial colonies without passing through a eusocial phase. In other cases the nest population of adults consists of eusocial mother-daughter groups, the mother commonly functioning as a queen, i.e., laying most of the eggs and not foraging much, while the daughters function like workers, often being unmated and having slender ovaries, capable of producing few if any eggs.

Even within single species, e.g., in *Halictus rubicundus* (Christ), the more montane and presumably the far northern populations are solitary and lack workers (Eickwort et al., 1996), while the lower altitude and more southern populations are social. Soucy (2002) compared social populations (elevation about 1,300 m in Utah) with solitary populations (elevation, 2,850 m in Colorado) of *Halictus rubicundus* (Christ) and provided additional information about this socially flexible species. Soucy and Danforth (2002), by analysis of mitochondrial DNA, found that North American populations showing social and solitary behavior belong to different lineages. Moreover, within a single population (New York City) some young females enter hibernation in June, like those of a solitary species that has one generation per year. Such females return to the same site the following spring to establish new nests. Other females of the same population, even from the same nests, remain in their mothers' nests and function as workers (Yanega, 1988, 1989).

Another such variable species is *Halictus sexcinctus* (Fabricius), which is solitary in central Europe but partly eusocial in southern Greece (Richards, 2001). The well-known *Lasioglossum malachurum* (Kirby) in western Europe makes nests founded by lone overwintered gynes whose early season eggs mostly develop into workers but whose last brood becomes reproductives. In Greece, however, probably because of the long season of activity, the colony often outlives its queen; many workers then mate, form semisocial colonies, and lay the eggs that will produce males and overwintering females (Richards, 2000). In halictines in general eusocial and communal colonies are characteristic of different genera or subgenera, but Richards, von Wettberg, and Rutgers (2003) reported a single population of *Halictus sexcinctus* in southern Greece in which some colonies were eusocial, others were communal. Morphological differences between eusocial and communal females suggested that two cryptic species

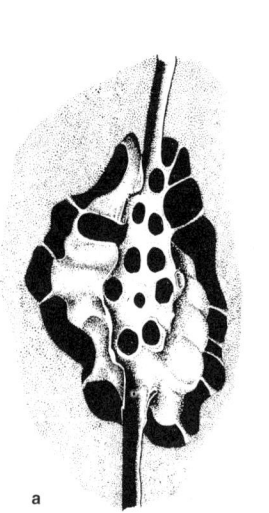
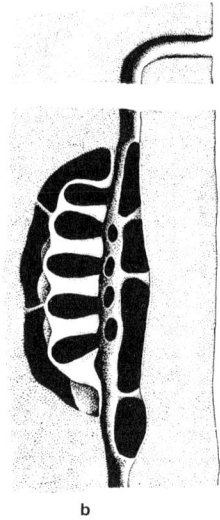
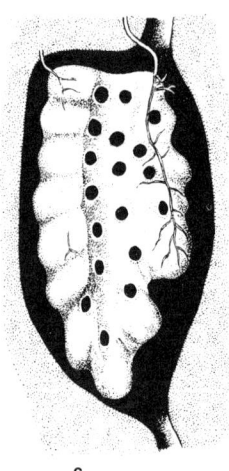

Figure 65-4. Cell clusters of *Halictus (Halictus) quadricinctus* (Fabricius). **a, b,** Clusters supported by earthen pillars; **c,** A cluster supported only by rootlets. From Sakagami and Michener, 1962.

might be involved. Mitochondrial DNA sequences, however, indicated that the two Greek forms were more similar to one another than to *H. sexcinctus* from elsewhere in Europe, strongly suggesting that the two Greek forms are conspecific and constitute a single population. More details on the Greek population were provided by Wyman and Richards (2003).

Even in the ordinarily solitary group *Lasioglossum* s. str., a laboratory nest of *L. mutilum* (Vachal) was inhabited by two females (Miyanaga, Maeta, and Mizuta, 1998). In such cases division of labor developed with one queenlike and one workerlike individual. In another species, *L. (L.) scitulum* (Smith), some spring females remain in nests with their overwintered mothers, forming colonies of two or rarely more individuals, both egg layers to judge by their ovarian development, the mother being the principal guard, the daughter the principal forager. Other young females disperse and establish their own nests. Similarly, in summer nearly 20 percent of the nests contain two females, in this case usually sisters, again both usually with enlarged ovaries, and with significant behavioral division of labor (Maeta, Yoshida, and Miyanaga, 2001; Miyanaga, Maeta, and Hoshikawa, 2000). See also the note on *L. aegyptiellum* (Strand) in the section on the genus *Lasioglossum*.

Much more extensive summaries of social behavior in the Halictinae can be found in Michener (1974a, 1990c); some of the numerous, more recent papers on this topic are the following: Packer, 1990, 1991, 1992, 1993b; Wcislo, 1993; Wille and Orozco, 1993; Yanega, 1993. Studies of individual and kin recognition in Halictinae were summarized by Michener and Smith (1987).

In addition to the nest-making Halictinae discussed above, there are various cleptoparasitic or socially parasitic taxa, as discussed by Michener (1978b); see also Section 8. These include *Megalopta (Noctoraptor)*, *Megommation (Cleptommation)*, and *Temnosoma* in the Augochlorini and *Sphecodes, Microsphecodes, Ptilocleptis, Eupetersia, Parathrinchostoma, Echthralictus, Halictus (Paraseladonia), Lasioglossum (Paradialictus)*, and a few species of *Lasioglossum (Dialictus)*, all in the Halictini. Of these, only *Sphecodes, Microsphecodes, Ptilocleptis,* and *Eupetersia* probably had a common parasitic ancestor; the others were derived from different nonparasitic ancestors. The parasites, nevertheless, have many common features, including strong cuticle and coarse sculpturing (Fig. 8-2); often large, falcate, single-pointed mandibles of the female (Fig. 67-13d, g); and the reduction or loss of the scopal hairs (Figs. 8-5, 8-6, 67-13e), the pygidial plate (Fig. 8-9a-c), the basitibial plates (Fig. 8-9g-j), the specialized median area of T5 of females, and the median keel of the labral process of females, as well as the eleva-

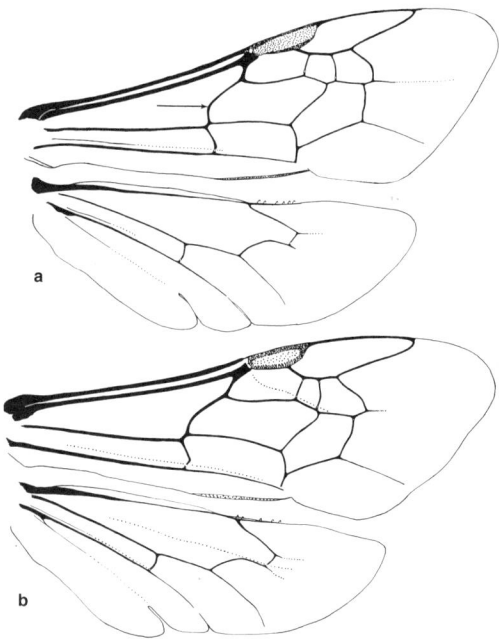

Figure 65-5. Wings of Halictinae. **a,** *Pseudagapostemon* sp.? (Halictini); **b,** *Augochlorella striata* (Provancher) (Augochlorini). (The arrow indicates the basal vein.)

tions of the basal part of the labrum (Fig. 8-9d-f, 67-13d, g). For details and other features of the parasites, see Section 8 and Michener (1978b).

It may seem inconsistent to place the species formerly included in *Paralictus* into *Dialictus,* thus with no parasitic genus or subgenus, while retaining such taxa for the other parasitic species. This is done because the former *Paralictus* species are so similar to other *Dialictus,* from which the *Paralictus* must have evolved. For the other parasites, their derivation is less obvious or less certain.

Melittologists have long looked among nonparasitic Halictinae to seek the taxon most closely related to *Sphecodes.* I now believe that it is futile to seek a nonparasitic genus like *Halictus* or *Lasioglossum* as sister to *Sphecodes.* Rather, I suspect that *Sphecodes* and its relatives constitute an ancient parasitic halictine lineage that has spread worldwide and is here assigned subtribal status as the Sphecodina. A possible plesiomorphy retained by this group is the form of S7 and S8 of the males, which are less reduced than those of most other Halictini, often more or less quadrate with basolateral apodemal arms and with an obtuse spiculum on S8. Included genera would be *Eupetersia, Microsphecodes, Nesosphecodes, Ptilocleptis,* and *Sphecodes.*

Most parasitic halictids parasitize their close relatives or at least bees of the same tribe, but some species of *Sphecodes* parasitize bees of other halictid tribes or subfamilies, the Augochlorini and Nomioidinae. Other species of *Sphecodes* parasitize quite unrelated bees, such as *Andrena, Calliopsis, Melitturga,* and *Perdita* in the Andrenidae and *Colletes* in the Colletidae. In most cases the parasitic female enters the nest of the host, destroys host eggs, and replaces them with her own; see Section 8 and Sick et al. (1994). This is quite unlike the method of parasitism used by most other groups of cleptoparasitic bees, in which it is the larva of the parasite that destroys the host egg or larva. In some species of *Sphecodes* and especially *Microsphecodes,* the parasitic female appears to remain in the nest with the host for a long period, suggesting social parasitism rather than cleptoparasitism; see Section 8.

Figure 61-4 is a phylogenetic tree for the Family Halictidae, and the basis for the classification. The Halictinae is divided into two tribes, the Augochlorini and Halictini. The principal tribal characters are indicated in the key to tribes, below. Presumably, these tribes are sister groups. An interesting plesiomorphic character is the galeal comb, retained among Halictinae only in certain South American Augochlorini such as *Corynura,* and lost in other Halictinae. Thus it is unlikely that the Augochlorini arose from within the Halictini, all of which lack the comb. On the other hand, it is the Halictini that have the plesiomorphic T5 of the female, lacking a cleft (Fig. 65-1j), and it is therefore unlikely that the Halictini arose from within the Augochlorini. My impression is that each tribe has both ancestral and derived characters relative to the other.

In addition to the tribal characters—often difficult to see—that are listed above and in the key to tribes below, there is a character of wing venation that frequently although not always distinguishes the two. In the Halictini the first recurrent vein usually meets the second submarginal cell, whereas in the Augochlorini it usually meets the second submarginal crossvein or the third submarginal cell. The difference is illustrated in Figure 65-5.

Fossil halictids have been described from late Eocene to Miocene deposits; see the review by Engel (1997a). The fossil augochlorine genus *Oligochlora* comprises three species from the Dominican amber (Engel, 1997b).

Key to the Tribes of the Halictinae

1. Longitudinal median specialized area of T5 of female of nonparasitic forms not divided by a notch or cleft (Fig. 65-1j); parasitic forms (recognized by loss or reduction of scopa and other pollen-carrying and -manipulating structures) nonmetallic or with dull greenish coloration; T7 of male with a transverse ridge, usually carinate, forming a false apex beneath which the tergum is strongly reflexed to the morphological apical margin, surface above the transverse ridge usually with a recognizable hairless pygidial plate .... Halictini (Sec. 66)
—. Longitudinal median specialized area of T5 of female of nonparasitic forms divided by a notch or cleft in tergal margin (Fig. 65-1k); parasitic forms mostly brilliant metallic green; T7 of male without pygidial plate and without transverse premarginal ridge or carina forming a false apex (Western Hemisphere) .... Augochlorini (Sec. 67)

66. Tribe Halictini

Although the major tribal characters of the Halictini are indicated in the key to tribes, some additional features that are easily seen can facilitate the recognition of certain Halictini because they are not found in other bees. In many species the distal veins of the forewing (Fig. 66-1), i.e., the third and sometimes the second submarginal crossvein and the second recurrent vein, are weaker than the neighboring veins, e.g., the first submarginal crossvein and vein Rs around the marginal cell. This character is especially developed in females of *Lasioglossum* and *Homalictus*, but is not found in many common genera such as *Halictus* and *Agapostemon* (Fig. 66-2). Likewise, the first and second hind tarsal segments in many male Halictini are either fused or broadly articulated to one another, as compared to the more distal articulations between, for example, the second and third segments. Even in forms like *Lasioglossum (Dialictus)*, in which the articulation between the first and second hind tarsal segments of the male appears normal, examination of cleared specimens shows some probable fusion. Associated with such fusion is the fact that the second hind tarsal segment in male Halictini is as long as or shorter than the third. By contrast, in the Augochlorini, the articulation is entirely unmodified and the second segment is longer than the third. The male genitalia often have a retrorse flap or lobe (Fig. 66-16), probably of gonostylar origin and here considered the ventral gonostylus, arising on the undersurface near the base of the dorsal gonostylus. In other taxa this structure is absent or directed downward or distad like the dorsal (principal) gonostylus.

Four subtribes are recognized, as shown in Figure 61-4. The genera in each are as follows:

Thrinchostomina: *Parathrincostoma, Thrinchostoma*.
Caenohalictina: *Agapostemon, Caenohalictus, Dinagapostemon, Habralictus, Paragapostemon, Pseudagapostemon, Rhinetula, Ruizantheda*.
Sphecodina: *Eupetersia, Microsphecodes, Nesosphecodes, Ptilocleptis, Sphecodes*.
Halictina: *Echthralictus, Glossodialictus, Halictus, Homalictus, Lasioglossum, Mexalictus, Patellapis, Thrinchohalictus, Urohalictus*.

In an important study, Pesenko (2000a, 2004b) examined the phylogeny for, and proposed a classification of, subtribes of the Halictini based on morphological data. This phylogeny is largely supported by studies by Danforth et al. (2004) based on coding regions of three nuclear genes, 2,234 aligned sites. Pesenko's subtribes are the tribes recognized by Danforth except that the Gastrohalictina (*Lasioglossum* and allies) is separated from the Halictina by Pesenko. I believe that the monophyly of these two groups (Halictina and Gastrohalictina) taken together is well shown by both morphology and DNA. Recognition of the Gastrohalictina would make the Halictina paraphyletic as shown by Pesenko (2000a).

In the Eastern Hemisphere most Halictini have black or dull greenish integument, but in the Western Hemisphere, although most species are black or greenish, some, especially in the Caenohalictina, are brilliant metallic

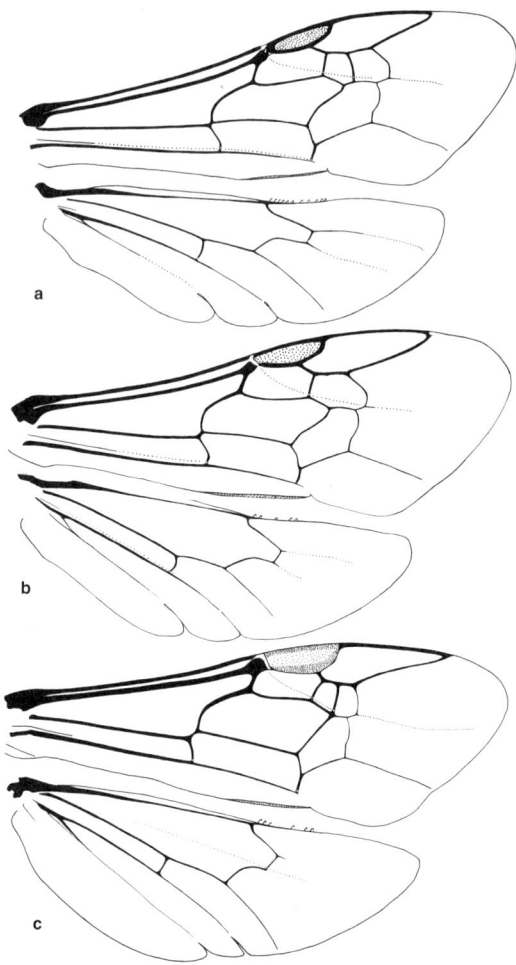

Figure 66-1. Wings of Halictini. **a,** *Lasioglossum (Lasioglossum) leucozonium* (Schrank); **b,** *L. (Hemihalictus) lustrans* (Cockerell); **c,** *Homalictus (Homalictus) dampieri* (Cockerell).

green or blue, superficially resembling many Augochlorini. Unlike those of most Augochlorini, the inner orbits of the eyes are usually only moderately and gently emarginate (Fig. 66-3).

This tribe contains bees that are abundant in all parts of the world. It tends to be morphologically monotonous, yet embraces very numerous species. Hence, as will be explained in more detail below, there exists much divergence among melittologists on the question of generic limits. The problem exists primarily in *Halictus, Lasioglossum,* and their allies. Some authors have taken the view that morphologically similar forms should be included in the same genus, regardless of the number of species involved. The result in this case is enormous genera. The extreme in this direction was Vachal, who included almost all nonparasitic Halictinae in the genus *Halictus*. His *Halicti genuini* and *Halicti intermedii*

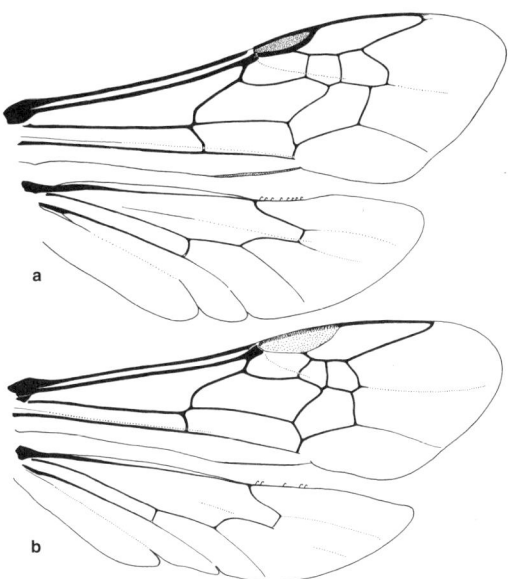

Figure 66-2. Wings of Halictini. **a,** *Agapostemon texanus* Cresson; **b,** *Habralictus trinax* (Vachal).

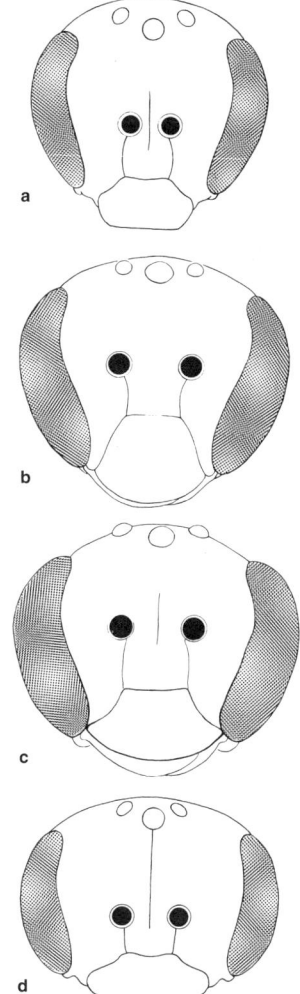

Figure 66-3. Faces of Halictini. **a,** *Lasioglossum (Evylaeus)* sp., male; **b,** *L. (Dialictus)* sp., male of a parasitic species formerly placed in *Paralictus*; **c,** *Ptilocleptis tomentosa* Michener, female; **d,** *Sphecodes carolinus* Mitchell, female. From Michener, McGinley, and Danforth, 1994.

(Vachal, 1911) are the Halictini as here understood. *Halicti genuini* included *Halictus, Lasioglossum,* and the like; *Halicti intermedii* included *Agapostemon* and its relatives. Warncke (1973b) and others also used *Halictus* in a broad sense. Various other authors, by contrast, such as Robertson (1902c), Mitchell (1960), and Moure and Hurd (1987), have recognized genera differing in seemingly trivial characters, such as the presence or absence of greenish reflections. Such splitting does result in smaller genera, although some nonetheless contain hundreds of species, and some grade into one another. Like Ebmer (1969-71, 1987a, 1988a), I have followed an intermediate course, dividing the old genus *Halictus* but not into very many generic units (Michener, 1944, 1993a; Michener, McGinley, and Danforth, 1994). It is clear that a worldwide generic study is needed before a stable, rational classification can be devised.

Nearly all Halictini have three submarginal cells. As in some other groups, however, a few unusual and sometimes rare species have two submarginal cells (Fig. 66-1b), and have therefore at some stage been given new subgeneric or generic names. Being unusual, the two-celled species sometimes received genus-group names before their three-celled relatives. One result is that *Hemihalictus,* which is based on a single species with two submarginal cells, has priority over certain other names based on more common species. In any event, number of submarginal cells is not necessarily a generic or subgeneric character.

The principal group of Halictini, insofar as number of species is concerned, is the *Halictus-Lasioglossum* group of the subtribe Halictina. This group is worldwide in distribution but is not represented by great diversity in South America, possibly because the Augochlorini are dominant there. *Lasioglossum* was long considered a synonym or subgenus of *Halictus,* and some authors still prefer that

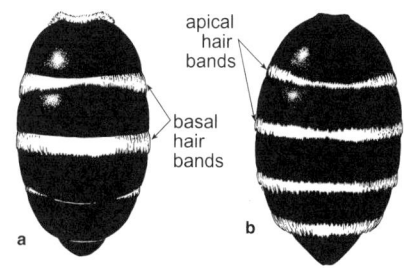

Figure 66-4. Metasomas of female Halictini, showing basal hair bands of *Lasioglossum* vs. apical hair bands of *Halictus*. **a,** *L. coriaceum* (Smith); **b,** *H. rubicundus* (Christ). From Michener, McGinley, and Danforth, 1994.

Table 66-1. Major Differences between *Halictus* and *Lasioglossum sensu lato*.

Feature	*Halictus*	*Lasioglossum s. l.*
Distal wing veins	Strong	Weak, at least in females
Apical tergal hair	Of densely plumose bands hairs[a]	Absent or rarely of weakly plumose hairs
Inferior basal process of penis valve	Inconspicuous, slender, rounded at apex	Broad, truncate or obliquely truncate
Ventral gonostylus or retrorse lobe of male gonostylus	Absent or, if present, directed apically, not retrorse	Commonly present, membranous, minutely hairy, usually retrorse

[a] Sometimes (e.g., subgenus *Vestitohalictus*) expanded over tergal surface, the apical bands thus not evident.

classification because of general similarity. Table 66-1 distinguishes these two genera. Superficial examination does not always distinguish immediately between the apical tergal hair bands characteristic of *Halictus* and the basal hair bands found in some *Lasioglossum* (Fig. 66-4). The difference in the inferior basal process of the penis valve (Table 66-1) is consistent, and supports the separation of the two genera. The differences of opinion concerning classification of the forms here included in *Lasioglossum* will be considered in the discussion of that genus.

The genera with the plesiomorphic character of strong venation, other than *Halictus*, do not constitute a group of related forms. Some fall in different subtribes. The South African *Patellapis* s. str. is quite similar to *Mexalictus*; they could both be archaic types surviving in limited areas. *Mexalictus* has paraocular lobes that are lacking in most *Patellapis*, but *P. braunsella* Michener has such lobes. *Mexalictus* is dull greenish, whereas *Patellapis* is not, but *M. polybioides* Packer lacks greenish reflections. In most *Patellapis* s. l. and some *Thrinchostoma*, S4 of the male is shortened medially, tends to be hidden under S3, and bears large apical or subapical bristles (Fig. 66-20). *Mexalictus polybioides* Packer approaches that condition, as does *Pseudagapostemon jenseni* (Friese) and, at least in sternal shape, *P. puelchanus* (Holmberg) (see Cure, 1989, fig. 4). Whether these similar and evidently derived features of African and American species result from convergence or from retention of archaic features lost in many species of most genera is not yet evident.

A large group of Old World tropical and austral Halictini is assembled here under the generic name *Patellapis*. As noted in the discussion of that genus, the genus *Homalictus* agrees with some *Patellapis* in the distinctive hind tibial vestiture of females, the strong sternal and ventrolateral tergal scopa, and the tendency for the metasoma of females to be flattened with the lateral extremities of the terga, especially T2, bent sharply ventromesally, forming a lateral angle along the metasoma. These are presumably derived characters relative to those of most Halictini. *Homalictus*, therefore, may be derived from or the sister group of *Patellapis*. If this is true, the weak distal wing veins of *Homalictus* and *Lasioglossum* probably arose independently. This would be contrary to the views of Danforth et al. (2004) who found *Homalictus* in the same clade as *Lasioglossum*.

Another group of Halictini comprises American genera related to *Agapostemon*, i.e., part of the subtribe Caenohalictina. The other taxa of this group are *Dinagapostemon*, *Paragapostemon*, *Rhinetula*, *Pseudagapostemon* and *Ruizantheda*. This group is more restricted than the *Agapostemon* group as understood by Eickwort (1969a, b) and Roberts and Brooks (1987) for whom it corresponded to the Caenohalictina as recognized (at whatever rank) by Pesenko (2004b) and Danforth et al. (2004). These authors included in it the genera *Caenohalictus* and *Habralictus*. The *Agapostemon* group in its narrow sense consists of rather large, robust species with a tendency toward enlarged and otherwise modified hind legs and toward yellow integumental metasomal bands in the male, a very unusual feature for Halictini. Some of the genera have long hair on the eyes. A probable synapomorphy for females of the *Agapostemon* group in its narrow sense (the first four

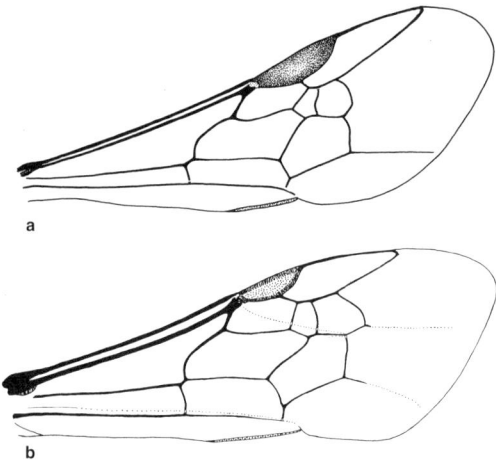

Figure 66-5. Forewings of Halictini. **a**, *Microsphecodes truncaticaudus* Michener; **b**, *Sphecodes gibbus* (Linnaeus). From Michener, McGinley, and Danforth, 1994.

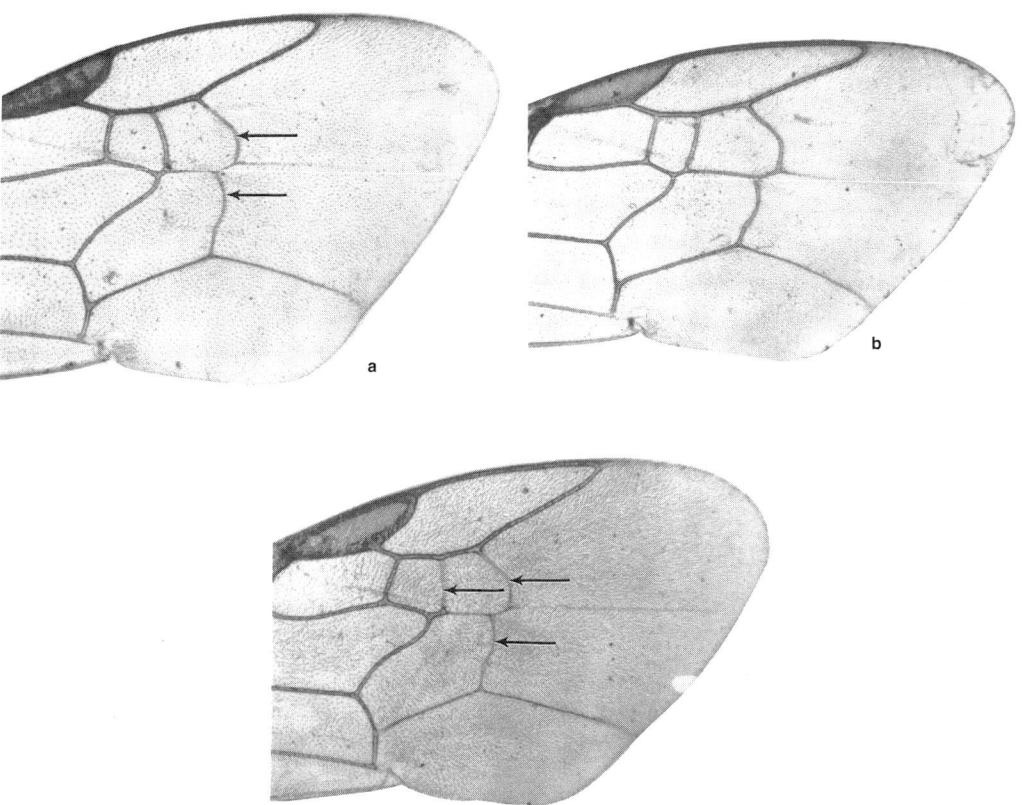

Figure 66-6. Distal venation in forewings of females of *Lasioglossum* and *Halictus*, showing weakening of veins in the former. **a**, Left, *L. (Lasioglossum) sisymbrii* (Cockerell); **b**, *H. rubicundus* (Christ); **c**, *L. (Evylaeus) quebecense* (Crawford). (Arrows indicate weakened veins.)

genera listed at the beginning of this paragraph) is a nonmetallic flat or concave strip along the inner margin of the eye, from the emargination downward. This strip is suggestive of a facial fovea. *Caenohalictus* and *Habralictus* have a nonmetallic area in the lower paraocular area next to the lower inner orbit, as in *Homalictus*. This area might be equivalent to the lowest part of the nonmetallic area of *Agapostemon*. No consistent apomorphy is known for the Caenohalictina. Engel (2001b), as noted above, found one quite ordinary looking fossil halictine species, *Electrolictus antiquus* Engel, perhaps near *Patellapis*, in the Eocene Baltic amber. Considering the worldwide abundance of halictines now and their diversity in later deposits such as the Dominican amber, it is of interest that more fossil halictines of Eocene age have not been found.

Aside from faunal works on bees as a whole, there are several important regional catalogues or reviews of the Halictini, or of certain groups that cut across subtribal or even modern generic limits, especially *Halictus* and *Lasioglossum*. Such works are listed here, as follows: For the Western Hemisphere, Moure and Hurd (1987); for the Eastern Hemisphere, Blüthgen (1920, 1921, 1923a, b, 1924a, c, 1926, 1928a, 1930, 1931, 1933, 1936, 1955); Perkins (1922); Ebmer (1969, 1987a, 1988a, b); Warncke (1975b); Pauly (1984b, 1999b); Pauly (in Pauly et al., 2001); and Pesenko et al. (2000). Cockerell (1937) provided notes on many African species.

In order to simplify the keys, separate keys to genera are given for the Western and Eastern hemispheres. In the first key, the temperate North American cleptoparasitic species sometimes placed in a subgenus or genus *Paralictus* run to *Lasioglossum (Dialictus)*; see the first option of the second couplet.

Key to the Genera of the Halictini of the Western Hemisphere (Females)

1. Scopa weak (Figs. 8-5a, 8-6) or absent; T5 with longitudinal median zone of fine punctation and short hairs weakly developed or absent; apical labral process without keel (as in Fig. 65-1i) or keel reduced to weak carina 2
—. Scopa present from hind trochanter to tibia (Fig. 8-5b), forming corbicula on underside of femur; T5 with well-developed longitudinal median zone of fine punctation and commonly short, dense hairs, this zone dividing prepygidial fimbria (Fig. 65-1j); apical labral process with strong longitudinal keel on anterior surface (Fig. 65-1a, b, e) 5
2(1). Second and third submarginal crossveins and second recurrent vein weaker than first submarginal crossvein

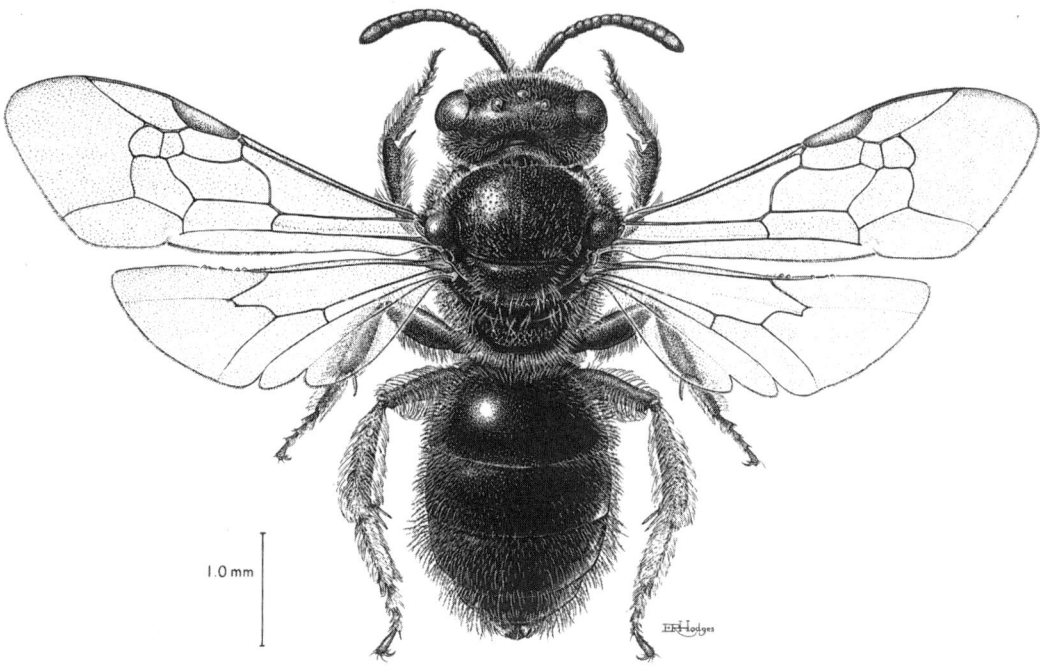

Figure 66-7. Female of *Lasioglossum (Dialictus) microlepoides* (Ellis). At this magnification the weakness of the distal veins of the forewing is not evident. Drawing by E. R. S. Hodges, from Michener, McGinley, and Danforth, 1994.

and first recurrent vein (as in Fig. 66-6c); head and thorax largely dull greenish; inner hind tibial spur pectinate with two to four teeth (nearctic) .... *Lasioglossum (Dialictus)* (in part)
—. Distal veins of forewing as strong as first submarginal crossvein (as in Fig. 66-6b); body black, metasoma often red; inner hind tibial spur ciliate or minutely serrate 3

3(2). Inner orbits of eyes strongly converging below (Fig. 66-3c); head little wider than long; clypeus about twice as broad as long (Fig. 66-3c); carina across pronotum between dorsolateral angles continuous; surface of S2 conspicuously convex in profile, its base strongly depressed, suggesting strong constriction between S1 and S2 (neotropical) .... *Ptilocleptis*
—. Inner orbits of eyes usually not strongly converging; head distinctly wider than long, as seen from front (Fig. 66-3d); clypeus three or more times as wide as long (Fig. 66-3d), rarely only twice as broad as long (Fig. 66-3d); carina between dorsolateral pronotal angles incomplete; surface of S2 usually not strongly convex in profile, apparent constriction between S1 and S2 being weak 4

4(3). Free part of marginal cell about or more than three times as long as part subtended by submarginal cells (Fig. 66-5a); T1 slightly longer than broad; T5 with apical margin bare, like that of preceding terga (neotropical) .... *Microsphecodes*
—. Free part of marginal cell about or less than twice as long as part subtended by submarginal cells (Fig. 66-5b); T1 usually broader than long; T5 with margin more hairy than that of preceding terga 16

5(1). Third and often second submarginal crossvein and second recurrent vein weaker than first submarginal crossvein (Fig. 66-6a, b) in females and some males; body not brilliantly metallic except in a few, mostly Antillean, species; metasomal terga without apical hair bands (Fig. 66-4a), basal hair bands present or absent .... *Lasioglossum* (in part)
—. Distal veins of forewing strong (Fig. 66-6b); coloration and hair bands variable 6

6(5). T1 to T4 with apical bands of posteriorly directed, plumose, tomental, pale hairs (Fig. 66-4b), sometimes limited to extreme sides of terga or to T5, terga sometimes also with basal bands; body not or weakly metallic 7
—. T1 to T4 without bands (Fig. 66-7) or with basal bands of hairs (Fig. 66-4a), and without apical bands (but bands sometimes suggested by the density of laterally directed hairs, as in *Mexalictus*); body coloration variable, sometimes bright metallic green or blue 8

7(6). Epistomal suture, at or lateral to tentorial pit, obtusely angulate .... *Halictus*
—. Epistomal suture, at or near tentorial pit, forming acute paraocular lobe protruding down into clypeus (as in Fig 67-6d) (South America) .... *Pseudagapostemon*

8(6). Posterior surface of propodeum enclosed by strong carina .... *Agapostemon*
—. Posterior surface of propodeum not or only partially surrounded by strong carina 9

9(8). T2 to T4 with basal bands or patches of white hair, bands or patches sometimes small and largely hidden under preceding terga (ignoring white hairs arising from gradulus) 10

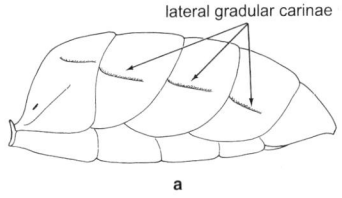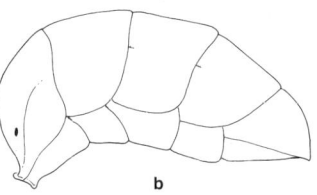

Figure 66-8. Side views of metasomas of female Halictini. **a,** *Rhinetula denticrus* Friese; **b,** *Dinagapostemon sicheli* (Vachal). From Michener, McGinley, and Danforth, 1994.

—. T2 to T4 without basal bands or patches of white hair except for series of pale hairs often arising from gradulus 11

10(9). Inner hind tibial spur with two or three teeth; hind tibia about as long as tarsal segments 1 to 4 taken together (South America) .... *Ruizantheda*

—. Inner hind tibial spur with five or more teeth; hind tibia about as long as tarsal segments 1 to 5 taken together (neotropical) .... *Dinagapostemon* (in part)

11(9). Inner hind tibial spur coarsely to minutely serrate; epistomal suture near tentorial pit forming paraocular lobe protruding into clypeus (as in Fig. 67-6d) *and* body nonmetallic or weakly metallic greenish blue (Mesoamerica to Arizona) .... *Mexalictus*

—. Inner hind tibial spur pectinate with two to eleven large teeth, the teeth longer than broad; epistomal suture near tentorial pit obtusely angulate or nearly straight (except forming paraocular lobe in *Paragapostemon*, which is brilliantly metallic blue or green) 12

12(11). Propodeal profile almost wholly steeply declivous, with only narrow sloping portion above, next to metanotum; T2 to T4 with lateral gradular carinae (Fig. 66-8a) *and* metasoma nonmetallic (neotropical) .... *Rhinetula*

—. Propodeal profile with subhorizontal or sloping upper portion about as long as or longer than metanotum; T2 to T4 without lateral gradular carina (Fig. 66-8b) *or* metasoma metallic blue or green *or* both 13

13(12). Eye bare or with scattered minute hairs; metasoma nonmetallic black or brown, rarely honey-colored, commonly with yellow markings (neotropical) ...... *Habralictus*

—. Eye with long hair, hair about as long as diameter of median ocellus or longer; metasoma metallic, without yellow markings 14

14(13). Inner hind tibial spur usually with three teeth; subhorizontal basal zone of propodeum over 1.5 times as long as metanotum (neotropical) .... *Caenohalictus*

—. Inner hind tibial spur with 5 to 11 teeth; subhorizontal basal zone of propodeum about as long as metanotum 15

15(14). Body uniformly bright metallic green or blue; T2 to T4 with lateral gradular carinae (as in Fig. 66-8a); epistomal suture laterally forming prominent paraocular lobe protruding down into clypeus (Mexico) .... *Paragapostemon*

—. Body not uniformly bright metallic; T2 and T3 and often T4 without lateral gradular carinae (Fig. 66-8b); epistomal suture obtusely angulate laterally (neotropical) .... *Dinagapostemon* (in part)

16(4). Anterior margin of mesoscutum abruptly declivous, well differentiated from dorsal surface; head and thorax coarsely pitted, dorsum of propodeum with coarse irregular rugae; second flagellar segment less than twice as lnog as first; free part or marginal cell less than twice as long as part subtended by submarginal cells (Fig. 66-5b) .... *Sphecodes*

—. Anterior margin of mesoscutum in profile gently convex, not sharply differentiated from dorsal surface; head and thorax finely punctate, dorsum of propodeum with fine radiating ridges; second flagellar segment about twice as long as first; free part of marginal cell about twice as long as part subtended by submarginal cells. (Greater Antilles) .... *Nesosphecodes*

Key to the Genera of the Halictini of the Western Hemisphere (Males)

1. T1 to T4 with apical bands of posteriorly directed, plumose, tomental, pale hairs (Fig. 66-4b), these sometimes limited to extreme sides of terga; terga sometimes also with basal bands; body not or weakly metallic 2

—. T1 to T4 without bands or with basal bands of hairs (Fig. 66-4a), but without apical bands (these sometimes suggested by the density of laterally directed hairs, as in *Mexalictus*); body coloration variable, sometimes bright metallic green or blue 3

2(1). Epistomal suture, at or lateral to tentorial pit, obtusely angulate (North America, rarely south to central Brazil) .... *Halictus*

—. Epistomal suture, at or near tentorial pit, forming acute paraocular lobe protruding down into clypeus (as in Fig. 67-6d) (South America) .... *Pseudagapostemon*

3(1). T2 to T4 with lateral gradular carinae (Fig. 66-8a) (eye with abundant hair as long as ocellar diameter) 4

—. T2 to T4 without lateral gradular carinae (Fig. 66-8b) 5

4(3). Paraocular lobe projecting down into clypeus prominent and narrowly rounded or acutely angulate (as in Fig. 67-6d); propodeal profile with basal subhorizontal portion about as long as metanotum; body bright green or blue (Mexico) .... *Paragapostemon*

—. Paraocular lobe absent or obtuse, because epistomal suture not or obtusely angulate laterally; propodeal profile almost entirely declivous; head and thorax weakly metallic and metasoma not metallic (neotropical) ...... *Rhinetula*

5(3). Gonostylus (not including ventral gonostylus, which is often a large retrorse lobe) usually small and simple, less than half as long as and less than half as wide as gonocoxite (Fig. 66-16a), but if more than half as long as gonocoxite, then slender; distal veins of forewing sometimes weaker than first submarginal crossvein (see couplet 5 in the key for females above; this character is not reliable in males, although all males with the weakened distal veins belong here) .... *Lasioglossum*

—. Gonostylus larger, more elaborate (Fig. 66-10c), often with long processes or folds or bifid, more than half as

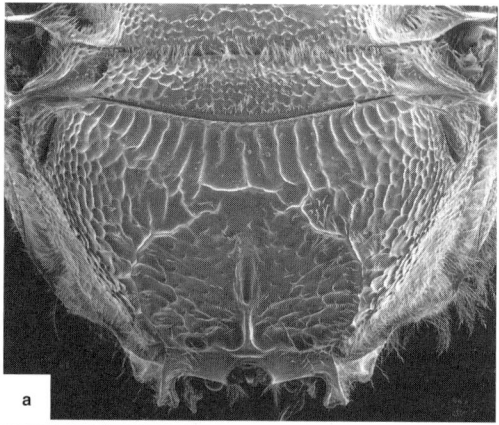

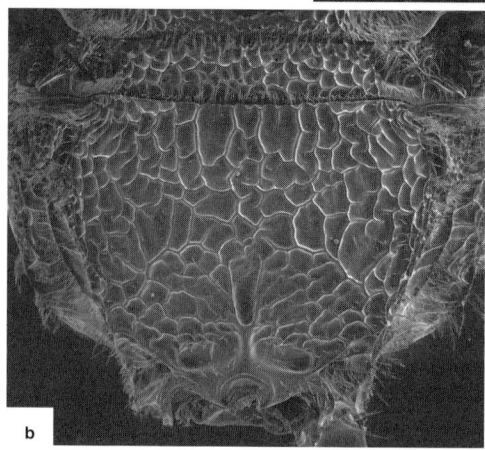
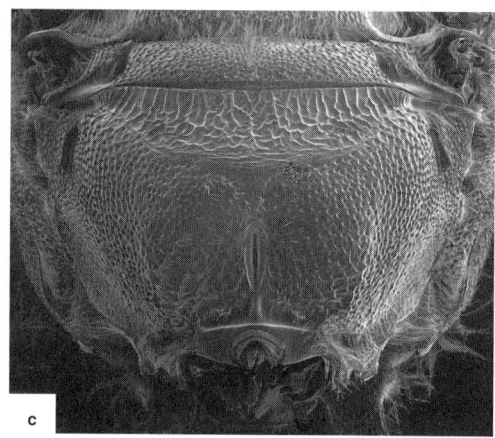

long as or as wide as gonocoxite or both; distal veins of forewing as strong as first submarginal crossvein 6

6(5). Posterior surface of propodeum enclosed by strong carina .... *Agapostemon*

—. Posterior surface of propodeum not or only partially surrounded by strong carina, at least upper transverse part of carina absent 7

7(6). Epistomal suture near tentorial pit forming a paraocular lobe protruding down into clypeus (as in Fig. 67-6d); S7 with four basal apodemes (body weakly metallic or nonmetallic, eyes bare) (Mesoamerica) .... *Mexalictus*

—. Epistomal suture near tentorial pit nearly straight to obtusely angulate (or about right-angular in some *Caenohalictus*, which are bright green with long hairs on eyes); S7 with two basal apodemes, one at each lateral extremity 8

8(7). Metasoma very slender, almost petiolate, T1 over 1.5 times as long as wide; eyes bare or nearly so (neotropical) .... *Habralictus*

—. Metasoma more robust, T1 broader than long or, *if* nearly 1.5 times as long as broad, then eyes usually with abundant long hairs 9

9(8). Body generally coarsely sculptured (Figs. 8-2a, 66-9b) or sometimes with extensive shining impunctate areas; sides of propodeum coarsely reticulate; body nonmetallic, black or partly brown or red; eyes usually hairless or with few short hairs 10

—. Body much more finely sculptured, surface often dull

Figure 66-9. Posterodorsal views of propodeum of male Halictini. **a,** *Lasioglossum zonulum* (Smith); **b,** *Sphecodes monilicornis* (Kirby); **c,** *Halictus rubicundus* (Christ). From Michener, McGinley, and Danforth, 1994.

(Fig. 66-9a, c); sides of propodeum punctate or granular, not coarsely reticulate; head and thorax usually with greenish or bluish reflections or strongly metallic; eyes usually with long hairs 11

10(9). Free part of marginal cell about or more than three times as long as part subtended by submarginal cells (Fig. 66-5a); T1 slightly longer than broad (neotropical) .... *Microsphecodes*

—. Free part of marginal cell about or less than twice as long as part subtended by submarginal cells (Fig. 66-5b); T1 usually broader than long 14

11(9). Hind femur swollen, with angle or tooth on underside; upper sloping part of propodeal profile about as long as metanotum (neotropical) .... *Dinagapostemon*

—. Hind femur slender, edentate; upper subhorizontal part of propodeal profile longer than metanotum 12

12(11). Legs with yellow areas or largely yellow (South America) .... *Ruizantheda*

—. Legs without yellow, or yellow restricted to front tibiae 13

13(12). Body usually metallic blue or green or with metallic areas, although sometimes nonmetallic; scutum and most of body entirely minutely granular and dull because of extremely fine, dense punctation (neotropical) .... *Caenohalictus*

—. Body nonmetallic; scutum and most of body distinctly and in some areas rather coarsely punctate (neotropical) .... *Ptilocleptis*

14(10). Anterior margin of mesoscutum abruptly declivous, well differentiated from dorsal surface; head and thorax coarsely pitted, dorsum of propodeum with coarse irregular rugae; free part of marginal cell less than twice as long as part subtended by submarginal cells (Fig. 66-5b) .... *Sphecodes*

—. Anterior margin of mesoscutum in profile gently convex, not sharply differentiated from dorsal surface; head and thorax finely punctate, dorsum of propodeum with fine radiating ridges; free part of marginal cell about twice as long as part subtended by submarginal cells. (Greater Antilles) .... *Nesosphecodes*

Key to the Genera of Halictini of the Eastern Hemisphere

1. Distal crossveins of forewing as strong as first submarginal crossvein (Fig. 66-6b); gonostylus of male broader or longer than indicated below, sometimes branched or otherwise complex (Fig. 66-11) 2

—. Third and often second submarginal crossvein and second recurrent vein weaker than nearby veins (e.g., first submarginal crossvein) in females (Fig. 66-6a, c) and many males; gonostylus (not including ventral gonostylus or retrorse lobe) of male usually small and simple (as in Fig. 66-16a), less than half as long and less than half as wide as gonocoxite [although long and slender in some *Lasioglossum (Austrevylaeus)* and *Homalictus,* rather large in *Urohalictus,* and quite large and sometimes lobed in some *Lasioglossum (Ctenonomia)* and some Australian subgenera] 10

2(1). Scopa absent (Fig. 8-5a); median finely punctate and finely haired area on T5 of female absent, prepygidial fimbria, if present, thus not divided; S8 of male quadrate, somewhat longer than broad, with basolateral apodeme or arm [this unknown for *Halictus (Paraseladonia)*] [inner hind tibial spur of female ciliate or minutely serrate, or, in *Halictus (Paraseladelina),* coarsely serrate] 3

—. Scopa present (Fig. 8-5b); T5 of female with longitudinal median finely punctate and haired or sometimes smooth area dividing prepygidial fimbria (Fig. 65-1j); S8 of male variable but not as above, disc usually much broader than long 6

3(2). Face fully as long as broad, clypeus produced and protuberant; paraocular lobe extending into clypeus strong, right-angular; labrum of both sexes with apical process (margined by bristles) with longitudinal carina, that of female high and keel-like apically (Madagascar) .... *Parathrincostoma*

—. Face much wider than long, clypeus neither produced nor protuberant; paraocular lobe absent, i.e., lateral segment of epistomal suture nearly straight; labrum of both sexes with apical part truncate or broadly rounded, without longitudinal carina or keel (Fig. 65-1i), or [in *Halictus (Paraseladonia)*] with keel scarcely evident 4

4(3). Inner hind tibial spur of female serrate with about eight teeth; metasomal terga with sparse, pale, apical hair bands, at least in male (body metallic blue-green; basal area of propodeum minutely reticulate) (Africa) .... *Halictus (Paraseladonia)*

—. Inner hind tibial spur of female ciliate or minutely serrate, like other spurs of hind and middle legs; metasomal terga without apical hair bands 5

5(4). First two flagellar segments of male both distinctly broader than long; gonocoxite of male not striate, with dorsolateral depression margined by and usually traversed by flanges or ridges; body with moderate to fine punctation, rarely involving coarse pitting as is usual in *Sphecodes*; mandible of female simple (Africa, oriental) .... *Eupetersia*

—. Second flagellar segment of male usually longer than broad, unlike first; gonocoxite of male usually striate, without depression; body and especially propodeum usually coarsely pitted; mandible of female nearly always with preapical tooth (except in neotropical *Austrosphecodes* group) .... *Sphecodes*

6(2). Female with margin of clypeal truncation, distal to preapical fimbria, extended downward at each side of labrum as a small, rather sharp, impunctate projection (except in some minute Asiatic species of the subgenus *Vestitohalictus* that lack such projections); S4 of male unmodified or at least without coarse apical setae; ventral gonostylus of male absent or directed apically, not retrorse (holarctic, Africa, oriental) .... *Halictus*

—. Female with margin of clypeal truncation, distal to preapical fimbria, extending but little downward at each side of labrum, forming only a low, rounded projection (except in some species of *Thrinchostoma* in which there is a strong projection); S4 of male usually shortened, commonly hidden by S3, frequently with several coarse apical or subapical setae; ventral gonostylus of male usually present, directed ventrally or basally, forming a retrorse lobe 7

7(6). Apical marginal zones of terga with simple, laterally directed hairs usually forming apical bands that are conspicuous only in certain lights; profile of anterior part of scutum gently convex, rising but little above level of pronotum; pronotum with carina or lamella separating dorsal from declivous anterior surface (recurrent veins both entering third submarginal cell, or first recurrent entering extreme apex of second cell) (Africa, oriental) .... *Thrinchostoma*

—. Apical marginal zones of terga variable, but without simple, laterally directed hairs; profile of anterior part of scutum strongly convex, subvertical surface thus rising well above level of pronotum and then curving strongly or angularly onto dorsum of scutum; pronotum medially without carina separating dorsal from anterior surface 8

8(7). Malar area about as long (female) to twice as long (male) as diameter of flagellum; pygidial plate of male not defined (dorsal gonostylus of male not bifurcate) (Asia Minor) .... *Thrincohalictus*

—. Malar area usually linear, rarely about half as long as diameter of flagellum; pygidial plate of male defined at least posteriorly and posterolaterally by a carina 9

9(8). Glossa extremely long and slender, about twice as long as head; dorsal gonostylus of male broad, simple (Africa) .... *Glossodialictus*

—. Glossa shorter than head or rarely about as long as head [somewhat longer in *Patellapis (Zonalictus) concinnula*

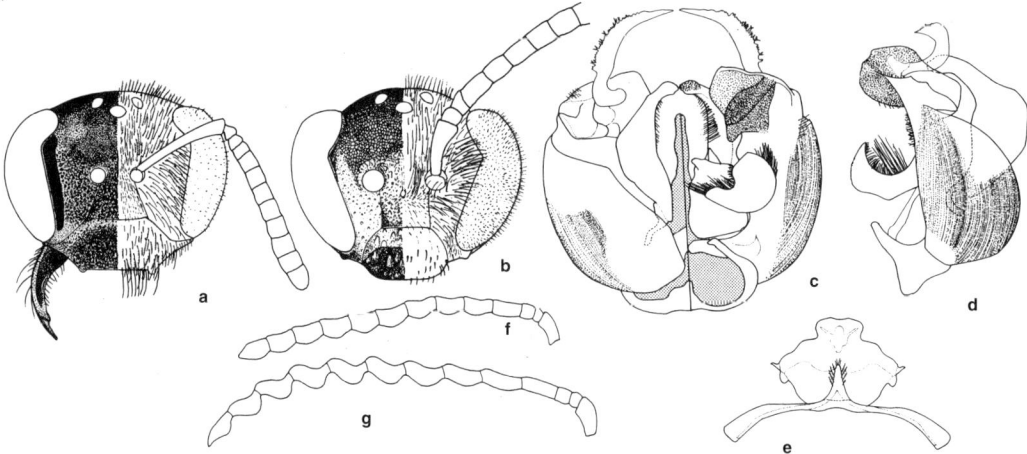

Figure 66-10. Halictini related to *Agapostemon*. **a, b,** Facial views of *A. (Agapostemonoides) hurdi* (Roberts and Brooks), female and male, showing hairy eyes.

c-f, *Dinagapostemon gigas* (Friese), male. **c,** Genitalia, dorsal view on left; **d,** Lateral view of same, showing the hairy lower gonostylus (or retrorse lobe) midventrally; **e,** S7 and S8; **f,** Antenna.

g, Antenna of male of *D. orestes* Roberts and Brooks. Original drawings of a and b by R. W. Brooks; others from Roberts and Brooks, 1987.

(Cockerell)]; dorsal gonostylus of male deeply bifid, one branch sometimes slender and inconspicuous [or, in *P. (Madagalictus) suarezensis* (Benoist), small and simple] (Africa, oriental, Australia) .... *Patellapis*

10(1). S7 and S8 of male, taken together, longer than broad, as long as genital capsule, heavily sclerotized, with large lobes curved up into and around genitalia (Fig. 66-22b-d); pronotum and metanotum with middorsal yellow areas (female unknown) (New Guinea) .... *Urohalictus*

—. S7 and S8 of male transverse, much shorter than genital capsule, weakly sclerotic, without lobes curving upward (as in Figs. 65-1o, p; 66-10e); pronotum and metanotum not yellow middorsally 11

11(10). Femoral scopa reduced to few simple hairs less than two-thirds as long as diameter of femur; dorsolateral angle of pronotum produced to spine; mandible of female simple, acute; inner hind tibial spur of male coarsely pectinate (Samoa) .... *Echthralictus*

—. Femoral scopa normal, of plumose or branched hairs enclosing ventral corbicula; dorsolateral angle or pronotum rounded, angulate (sometimes acute), or rarely lamellate, not produced to spine; mandible of female with preapical tooth; inner hind tibial spur of male finely serrate or ciliate 12

12(11). Metasomal terga of female, especially T1 to T3, usually sharply folded laterally, forming angle at margin between dorsal and ventrolateral parts of terga; scopal hairs plumelike with numerous lateral branches, those on sterna and ventrolateral parts of terga large, those of hind femur arising ventrally and especially at ends of femur forming femoral corbicula; gonobase of male genitalia, as seen from above, usually continuing contours of gonocoxites (Australia, oriental) .... *Homalictus*

—. Metasomal terga rounded from dorsal to ventrolateral parts; scopal hairs with relatively few, longer branches, less developed on metasomal venter, those of hind femur arising subdorsally and curling around femur to form femoral corbicula; gonobase of male genitalia, as seen from above, relatively broad, usually not at all continuing contours of gonocoxites .... *Lasioglossum*

Genus *Agapostemon* Guérin-Méneville

In most species the head and thorax are bright metallic green or blue. In some females the metasoma is the same color, while in others it is black or amber, sometimes with basal tergal bands of white tomentum and very rarely with yellow integumental bands. In males the metasoma is black or weakly metallic, usually with broad yellow or white bands, thus contrasting sharply with the coloration of the head and thorax. Only in a few neotropical species is the body black with strong metallic reflections. The body is robust, 7.0 to 14.5 mm long. The inner hind tibial spur of the female is coarsely pectinate with three to seven large, spatulate teeth, and the hypostomal area is very coarsely striate. A strong carina usually completely surrounds the posterior surface of the propodeum.

Janjic and Packer (2003) made a phylogenetic study of species of *Agapostemon*. They regarded *Agapostemonoides* as a genus distinct from *Agapostemon,* but I think their close relationship is best indicated by the subgeneric level. These authors also recognized two clades in *Agapostemon* as subgenera. No one character distinguishes all species of these clades.

Nests of *Agapostemon* are deep, more or less vertical burrows in flat or sloping soil or sometimes in banks, with lateral burrows extending to the cells. Usually there is one cell per lateral, but sometimes there are two or three in series. In some species, burrows are frequently occupied by communal groups of two to two dozen females, and the main burrow may be branched. Accounts of life history, nests, and behavior are by Eickwort and Eickwort (1969), Roberts (1969), Abrams and Eickwort (1980), and Eickwort (1981).

Key to the Subgenera of *Agapostemon*

1. Eyes with hairs about as long as ocellar radius (Fig. 66-10a, b); epistomal suture very obtusely angulate laterally (neotropical) .... *A. (Agapostemonoides)*
—. Eyes bare or nearly so; epistomal suture, near tentorial pit, forming paraocular lobe protruding down into clypeus, basal margins of this lobe at right or acute angle to each other .... *A. (Agapostemon s. str.)*

Agapostemon / Subgenus *Agapostemon* Guérin-Méneville s. str.

Andrena (Agapostemon) Guérin-Méneville, 1844: 448. Type species: *Apis femoralis* Guérin-Méneville, 1844 = *Apis viridula* Fabricius, 1793, monobasic.
Agapostemon (Notagapostemon) Janjic and Packer, 2003: 109,. Type species: *Agapostemon mourei* Roberts, 1972, by original designation. [**New synonymy.**]

In this subgenus the eyes are bare or with minute scattered hairs, the paraocular lobe is strong and right-angular or acute, and the metasoma of the female lacks yellow markings, although there are sometimes basal tergal bands of pale tomentum. Male genitalia were illustrated by Michener (1954b), Mitchell (1960), and Roberts (1972).

■ *Agapostemon* s. str. is particularly common in North America, ranging from coast to coast and from southern Canada southward to the Bahamas and Antilles and through Mesoamerica. It is less common in South America, but occurs in the Andean countries south to northern Chile, especially the eastern slopes of the Andes, and in Brazil from Bahia, Minas Gerais, and Mato Grosso southward to Paraguay and northeastern Argentina. A total of 43 species are recognized, although some Antillean forms given specific rank may better be considered subspecies. The subgenus was revised by Roberts (1972).

Agapostemon / Subgenus *Agapostemonoides* Roberts and Brooks

Agapostemonoides Roberts and Brooks, 1987: 364. Type species: *Agapostemonoides hurdi* Roberts and Brooks, 1987, by original designation.

This subgenus differs from *Agapostemon* s. str. by the hairy eyes (the hairs being about as long as an ocellar radius), the absence of a paraocular lobe extending into the clypeus, and the presence of basal yellow bands on the metasomal terga of the female as well as the male. Curious features that may be involved in collecting pollen from the flowers of *Piper* (Piperaceae) are the fringes of long curved hairs on the posterior margins of the front and especially the middle trochanters of females. Male genitalia and other structures were illustrated by Roberts and Brooks (1987). Specimens of *Agapostemonoides* are small for *Agapostemon*, 7 to 8 mm long. This subgenus may well be the sister group of *Agapostemon* s. str. rather than a derivative from that subgenus; the paraocular lobes of *Agapostemon* s. str. are likely to be derived relative to *Agapostemonoides*. In either case, *A. hurdi* could be considered as merely an unusual *Agapostemon*, not justifying recognition of the subgenera.

■ *Agapostemonoides* occurs from Costa Rica and Panama south probably along the eastern Andean zone to Bolivia. The only species is *Agapostemon hurdi* (Roberts and Brooks).

Genus *Caenohalictus* Cameron

Caenohalictus Cameron, 1903a: 231. Type species: *Caenohalictus trichiothalmus* Cameron, 1903, monobasic.

The name *Caenohalictus* was misapplied to *Rhinetula* by Michener (1954b). The resultant confusion was explained by Michener (1979b).

Caenohalictus consists of bright-green bees, rarely brassy, red, or nonmetallic black; in the last case the metasoma is usually blackish or partly amber. The body is more slender and usually much smaller than the bright-green members of the *Agapostemon* group of genera (see the account of the tribe Halictini); the length is 5 to 12 mm. The eyes have long hairs, about as long as an ocellar diameter to three times that length, sometimes lacking on the areas near the face; rarely, the eyes are nearly bare. Although there are occasionally areas of coarse punctation or reticulation on the thorax, the head and thorax are otherwise minutely and closely granular. There is much variation among the species; for example, most do not have paraocular lobes, but some of the larger species in which the clypeus is strongly protuberant and produced have strong paraocular lobes, about right-angular. Only the lower margin or lower two-fifths of the clypeus of the male is yellow. The first and second hind tarsal segments appear superficially to be articulated, but a cleared specimen showed the segments to be fused; the joint is rigid and wider than the more distal articulations. Male genitalia and hidden sterna were illustrated by Michener (1979b) and Packer (1993).

■ Although especially abundant and diverse in the Andean countries, *Caenohalictus* is found as far south as Chubut province in Argentina and over much of South America, but seems to be absent from the Amazon Valley and the Guianas; it also occurs uncommonly in Central America and north to the states of San Luis Potosí and Nayarit in Mexico. Rojas and Toro (2000) revised the 15 Chilean species of *Caenohalictus*, seven of which were new. About 55 species of *Caenohalictus* are known.

Because of similar color and hairy eyes, hasty examination is likely to result in confusion of *Caenohalictus* with *Caenaugochlora* s. str. Of course the tribal characters readily differentiate these taxa, as does the dull, granular integument characteristic of *Caenohalictus*.

Nests of some species are similar to those of *Habralictus*, with lateral burrows each leading to a single horizontal cell (Michener and Lange, 1958a; Michener, Breed, and Bell, 1979). Other species, however, construct cells organized into a cluster of more or less horizontal cells in a cavity, suggestive of nests of some Augochlorini. Such nests were illustrated by Claude-Joseph (1926), and copied by Sakagami and Michener (1962, incorrectly associated with the genus *Caenaugochlora*). In view of the morphological and behavioral diversity among the species, it is likely that additional genera (or subgenera) will eventually be recognized.

Genus *Dinagapostemon* Moure and Hurd

Dinagapostemon Moure and Hurd, 1982: 46. Type species: *Halictus sicheli* Vachal, 1901, by original designation.

This genus is a member of the *Agapostemon* group, having the large size (length 11.0-14.5 mm) and robust body form of that group. The color is metallic blue or green to nonmetallic brown or amber. The male antennal flagellum is long and strikingly crenulate (Fig. 66-10f, g), unlike that of all other genera. The hind legs of the male are dark in color and swollen, with femoral, tibial, and basitarsal teeth. I place *Dinagapostemon* near *Paragapostemon* because of their general similarity and especially the slender, tapering process of the male dorsal gonostylus (Fig. 66-10c). The male genitalia of *Dinagapostemon* are among the most complex of halictid genitalia (Fig. 66-10c, d); the lower gonostylus is large and usually projects downward; illustrations were presented by Roberts and Brooks (1987). Because the generic differences are substantial, including the lateral gradular carinae of T2 to T4 in *Paragapostemon* (as in Fig. 66-8a), I hesitantly retain the generic status for *Dinagapostemon*. According to Roberts and Brooks (1987), *Paragapostemon* is the sister group to *Dinagapostemon*.

■ *Dinagapostemon* is found in montane areas from the states of Tamaulipas and Guerrero, Mexico, through Central America to the Andean region of Colombia, Venezuela, and Ecuador. The eight described species were revised by Roberts and Brooks (1987).

Genus *Echthralictus* Perkins and Cheesman

Echthralictus Perkins and Cheesman, 1928: 14. Type species: *Halictus extraordinarius* Kohl, 1908, by original designation.

This is probably a local, parasitic derivative of *Homalictus* s. str. It has, however, lost nearly all of the distinctive generic characters of *Homalictus*. The metasomal scopa is reduced to a few, scattered, very long hairs on the sterna. The femoral scopa is not recognizable except for the row of long, simple hairs on the posterior surface, the longest about two-thirds as long as the maximum femoral diameter. The tibia has a distinct, relatively bare undersurface, but the hairs margining it are not longer than those elsewhere on the tibia and have only short branches. The labral keel is reduced to a strong carina. Perhaps related to the loss of the tergal scopa and the great reduction of the sternal scopa, is the rounded lateral metasomal margin. By contrast, in *Homalictus* the metasoma is depressed so that its lateral margin is a strong bend or crease in the terga where their ventrolateral surfaces join the dorsal surfaces. T5 of the female *Echthralictus* has a bare, broad, shining, apical triangular area instead of the usual narrower area, usually with fine hairs, as is characteristic of nonparasitic Halictini. The body length is 5.5 to 6.0 mm.

■ *Echthralictus* is known only from Samoa, where there are two species; see Perkins and Cheesman (1928).

Since there are no intermediates between *Echthralictus* and *Homalictus*—i.e., there is in fact a major morphological gap—I have chosen to recognize *Echthralictus* as a genus. A cladistic study presumably would show that it is part of *Homalictus* s. str. if the features associated with parasitism in halictids (see Michener, 1978b) were ignored in the analysis. Probably, indeed, *Echthralictus* arose from a *Homalictus* group now found in its island range, as suggested by its metallic green or blue tints, among other characters.

Genus *Eupetersia* Blüthgen

Like *Ptilocleptis,* this presumably parasitic genus may be derived from *Sphecodes,* but the moderate to fine punctation of the body, lacking the coarse pitting and reticulation found in *Sphecodes,* is probably plesiomorphic relative to *Sphecodes*. Exceptions exist, however; *Eupetersia seyrigi* Blüthgen has a coarsely pitted propodeum. *Eupetersia* is tentatively considered a basal branch of the *Sphecodes* clade. The lack of the striae on the gonocoxites of the male is probably also plesiomorphic relative to *Sphecodes*. Each gonocoxite has, however, a dorsolateral depressed area with a lamella or flange on each side and often one or more longitudinal ridges across the depression. As in *Ptilocleptis* the antennae of the male are similar to those of the female, not long and robust as is usual in *Sphecodes*. The mandibles of the female are simple, unlike those of most *Sphecodes*. The coloration, also, is not usually like the black head and thorax and partly or wholly red metasoma of most *Sphecodes;* the body may be black (the thorax sometimes partly red), metallic blue, to wholly reddish yellow. The body length ranges from 5 to 12 mm.

Keys to species were provided by Blüthgen (1928b, 1936) and, to metallic species, by Pauly (1981b). Baker (1974a) gave a list of species with much useful detail. Pauly (1999b) listed the African species of *Eupetersia*, and Brooks and Pauly (in Pauly et al., 2001) revised the eight species from Madagascar.

Callosphecodes, here listed as a synonym of *Sphecodes,* could be a senior synonym of *Eupetersia*. It is known from a single specimen, now lost, believed to be from far outside the known range of *Eupetersia*. The problem of its position is not likely to be solved until more material is collected. See comments under *Sphecodes*.

Key to the Subgenera of *Eupetersia*

1. Scutum rather densely punctate; scutellum bigibbous, the two convexities not or scarcely punctate .... *E. (Eupetersia* s. str.*)*
—. Scutum with widely separated punctures; scutellum gently convex, with punctation similar to that of scutum .... *E. (Nesoeupetersia)*

Eupetersia / Subgenus *Eupetersia* Blüthgen s. str.

Eupetersia Blüthgen, 1928b: 49. Type species: *Eupetersia neavei* Blüthgen, 1928, by original designation. [Also published as new by Blüthgen, 1928c: 165.]
Calleupetersia Cockerell, 1938: 329. Type species: *Halictus lasureus* Friese, 1910, by original designation.

The species placed in *Calleupetersia* seem to differ from the others only in the metallic blue coloration. The male

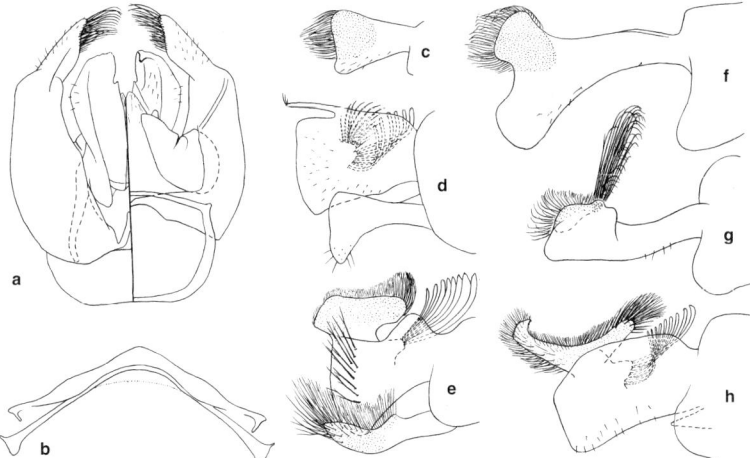

Figure 66-11. Genitalia and hidden sterna of males of *Halictus*. **a**, Genitalia of *H. ligatus* Say (dorsal on left); **b**, S7 and S8 of same. c-h, Posterolateral views of gonostyli: **c**, *H. (Odontalictus) ligatus*; **d**, *H. (Seladonia) confusus* Smith; **e**, *H. (Halictus) quadricinctus* (Fabricius); **f**, *H. (Argalictus) scabiosae* (Rossi); **g**, *H. (Monilapis) patellatus* Morawitz; **h**, *H. (Protohalictus) rubicundus* (Christ). The gonostyli show the diversity within the genus *Halictus*. Only the subgenera *Halictus* s. str. and *Seladonia* have a well-developed lower gonostylus. From Michener, 1978a.

genitalia were illustrated by Michener (1978b) and Pauly (1981b).

■ The range is from Ethiopia and Uganda south to Natal Province in South Africa, and Madagascar. Pauly (1981b) listed 21 species. Blüthgen (1928b, 1936) gave a key to the species then known.

Eupetersia / Subgenus *Nesoeupetersia* Blüthgen

Eupetersia (Nesoeupetersia) Blüthgen, 1936: 182. Type species: *Sphecodes scotti* Cockerell, 1912, by original designation.

■ *Nesoeupetersia* occurs from Zaire and Uganda to Malawi, Madagascar, the Seychelles, and southern India. Pauly (1981b) listed eight species.

Genus *Glossodialictus* Pauly

Glossodialictus Pauly, 1984a: 704. Type species: *Glossodialictus wittei* Pauly, 1984, by original designation.

Because it has strong apical wing venation and occurs in Africa, this genus invites comparison with *Patellapis*. It differs from *Patellapis* in the blue-green metallic tints on the head and thorax; the strong, acute paraocular lobe extending down into the clypeus (but see the comments on *Patellapis braunsella* Michener); the protuberant clypeus; the very long, slender glossa; and the broad, simple, dorsal gonostylus of the male. The body length is 8.5 mm. Other characters include the retrorse ventral gonostylus of the male, the second submarginal crossvein approximately meeting the first recurrent vein, the basal tomentum on T2 and T3, and the lack of apical tergal hair bands although the tergal margins are yellowish brown. In spite of Pauly's (1984a) diagram that shows a substantial malar area, the malar area is said to be linear in the female, a little more developed in the male. Pauly (1984a) sketched the male genitalia and other structures.

■ *Glossodialictus* is known only from Shaba Province in Zaire. The only species is *G. wittei* Pauly.

Genus *Habralictus* Moure

Habralictus consists of small species (length 4.0-6.5 mm) with the head and thorax blackish (usually with metallic tints) to bright green, the surface dull and minutely granular to polished. The metasoma may be brown or black, rarely honey-colored, but in the female usually has yellow basal bands or basal lateral spots on the terga. Males usually lack such markings or have only lateral spots. The metasoma of the male is petiolate, widest at T4 and T5, T1 being much longer than broad. The male clypeus is more than half yellow. The usually bare eyes (although with short hairs in some species), as well as the form of the male metasoma, easily distinguish *Habralictus* from the only other small neotropical Halictini with strong wing venation, i.e., *Caenohalictus*. Male genitalia and hidden sterna were illustrated by Michener (1954b, 1979b).

The nests are burrows, usually in banks, with lateral branches each leading to a single horizontal cell (Michener and Lange, 1958a; Michener, Breed, and Bell, 1979).

Key to the Subgenera of *Habralictus* (Males)

(Two subgenera are tentatively recognized. One of them, *Zikaniella*, is known only in the male; subgeneric characters for females are unknown.)

1. Head not or scarcely wider than thorax; genal area much narrower than eye, as seen from side; clypeal surface not concave or with small median concavity; inner hind tibial spur simple, ciliate .... *H. (Habralictus* s. str.*)*
—. Head much broader than thorax; genal area wider than eye, strongly angulate posteriorly; middle half of clypeal surface broadly concave; inner hind tibial spur pectinate with four teeth .... *H. (Zikaniella)*

Habralictus / Subgenus *Habralictus* Moure s. str.

Habralictus Moure, 1941a: 59. Type species: *Habralictus flavopictus* Moure, 1941, by original designation.

■ *Habralictus* s. str. is seemingly most abundant in Bolivia, but is found from that country and the state of Paraná in Brazil north to the state of Jalisco, Mexico. In the West Indies it is known from St. Vincent. The 21 described species were listed by Moure and Hurd (1987).

Habralictus / Subgenus *Zikaniella* Moure

Zikaniella Moure, 1941a: 57. Type species: *Zikaniella crassiceps* Moure, 1941, by original designation.

Zikaniella is known only from the male, which differs from *Habralictus* s. str. principally in its large head and the associated characters listed in the key to subgenera. Such features are usually not generic characters in bees, sometimes not even specific characters, and if the female of *Zikaniella* is a typical *Habralictus*, then *Zikaniella* should be regarded as a synonym of *Habralictus*, in spite of the page priority of *Zikaniella*; as first reviser, I consider *Habralictus* to have precedence over *Zikaniella*. *Zikaniella* resembles *Habralictus* s. str. not only in the characters indicated in the key to genera but also in the basal yellow spots of T3 and T4, the simple legs and sterna, and the nearly straight lateral segments of the epistomal sutures. Its head and thoracic integument is shining green, largely smooth; this is also true of the basal area of the propodeum, which is minutely roughened in *Habralictus* s. str.

■ This subgenus is known only from the state of Rio de Janeiro, Brazil. The only species is *Habralictus crassiceps* (Moure).

Genus *Halictus* Latreille

Although in this genus the principal metasomal hair bands are on the apices of the terga (Pl. 4), in some species—especially in the subgenus *Seladonia*—there are also basal bands. In other species, especially most species of the subgenus *Vestitohalictus*, white tomentum covers the entire exposed tergal surfaces; the apical bands thus not very distinct or not recognizable. Distinctive attributes of the genus are indicated not only in the key to genera but in the discussion of the tribe Halictini and in Table 66-1. Illustrations of male genitalic and other characters were given by Sandhouse (1941), Mitchell (1960), Ebmer (1969), Pesenko (1985, 1986b), and Sakagami and Ebmer (1987); see also Figure 66-11.

It has been common in the past to recognize three subgenera, *Halictus* s. str., *Seladonia*, and *Vestitohalictus* (see Michener, 1978a), and Pesenko (1984a) recognized them as three genera. *Seladonia* (including *Pachyceble*) was distinguished from *Halictus* s. str. principally by the greenish, bluish, or brassy coloration, at least of the head and thorax; *Halictus* s. str. contained all the nonmetallic species (except for the few nonmetallic *Vestitohalictus*). The elaborate male genitalia of *Seladonia*, however, are similar to those of *Halictus* s. str. in the narrow sense of the present account (compare Fig. 66-11d with e), and there is no reason to make the integumental color decisive in the taxonomic decision. There is such diversity among the black *Halictus* as to overshadow the difference between *Seladonia* and the black forms most similar to it. I have therefore followed Pesenko (1984a) in recognizing several subgenera of black *Halictus* and have retained *Seladonia* and *Pachyceble* as other subgenera among the black ones. The only other greenish *Halictus* are *Paraseladonia* and some species of the subgenus *Vestitohalictus*.

Danforth, Sauquet, and Packer (1999), in a phylogenetic study of 22 species of *Halictus* based on one nuclear gene, found groups mostly supportive of the subgenera recognized by Michener (2000). Recognition of *Vestitohalictus*, however, rendered *Seladonia* paraphyletic. In several analyses *Seladonia* plus *Vestitohalictus* appeared as the sister group of the rest of *Halicus*, but in other cases *Seladonia* plus *Vestitohalictus* arose within the rest of *Halictus*. Pesenko (2004b) made an extensive phylogenetic study of *Halictus* genus based on morphology. He found the genus to be monophyletic, and *Seladonia* s. l. to be the sister group to the rest of *Halictus*; he recognized *Seladonia* and *Halictus* as separate genera. I prefer to recognize *Halictus* as a genus including *Seladonia*; usually the former is black and the latter metallic greenish or bluish. However, a few species of *Seladonia* s. l. are nonmetallic black. The next most consistant difference between *Seladonia* s. l. and *Halictus* is a deep cleft in the upper gonostylus of the male. The lobe mesal to this cleft may be very long and slender, but in a few species it is lost so that there is no obvious cleft. I use the terms *Seladonia* group and *Halictus* group to distinguish the two groups given generic status by Pesenko.

A list of the species of *Halictus* (except *Seladonia* and *Vestitohalictus*), with synonymies, was given by Pesenko (1984b) and a list for the entire genus, 442 names, 182 recognized species, was provided by Pesenko (2004b), updated by Pesenko (2005). Lists of African species of the genus *Halictus* were given by Pauly (1999b). A revision of American species was by Sandhouse (1941).

The key below (modified from Pesenko, 2004b) has one obvious defect; females cannot be run past certain couplets for which only male characters are known. One must try running a female in both alternatives is such cases. Other keys should also be consulted. Michener (2000) gave keys for each sex, but that for females left five subgenera not distinguished. The key by Pesenko (2004b) is a useful alternative.

Key to the Subgenera of *Halictus*
(From Pesenko, 2004b)

1. Integument nonmetallic, usually black; S4 of male often broadly emarginate; dorsal gonostylus of male not cleft [*Halictus* group] 2
—. Integument in at least some areas of head and thorax dull metallic green or bluish (very rarely black); S4 of male with posterior margin straight; dorsal gonostylus of male divided by deep cleft forming a slender inner (median) lobe that may be elongate or rarely lost (so that cleft is not evident) [*Seladonia* group].. 12
2(1). Clypeus of female with rounded tubercle in middle of lower margin; S5 of male with deep, triangular but rounded incision in distal margin; metasoma, at least T1 to T3, red (Indian region) ....*H. (Ramalictus)*

—. Clypeus of female with lower margin straight; S5 of male with margin straight or broadly, weakly emarginate; metasoma usually black 3

3(2). Body length over 16 mm; scutum of female sparsely punctate; antennal flagellum of male flattened with long fringe on lower side, last segment flattened, curved; hind basitarsus of male curved (palearctic) ....
.... *H. (Halictus* s. str.*)*

—. Body length less than 14 mm; scutum of female densely punctate; antennal flagellum of male not flattened, not fringed except rarely with short hairs, last segment unmodified; hind basitarsus of male nearly always straight.
.... 4

4(3). Upper gonostylus of male broad with clump of very coarse bristles on inner surface; lower gonostylus a slender process, much reduced in *H. (Protohalictus) rubicundus* (Christ) and *hedini* Blüthgen 5

—. Upper gonostylus of male elongate, proximal half narrowed, without clump of coarse bristles on inner surface; lower gonostylus absent 7

5(4). Propedeal triangle densely and finely granulate; hind tibia of female black; male antenna reaching scutellum; S4 of male of normal length, posterior margin nearly straight (North America) .... *H. (Nealictus)*

—. Propedeal triangle rugose or rugulose; hind tibia of female usually red or reddish yellow, rarely black; male antenna reaching propodeum; S4 of male short, broadly emarginate 6

6(5). Sides and posterior surface of propedeum nearly smooth, polished; antennal flagellum of male with dense, short pubescence in proximal and distal bands on each segment; head of female as long as wide in frontal view, gena with large tooth (Central Asia) ....
.... *H. (Lampralicutus)*

—. Sides and posterior surface of propodeum granulate or rugose, dull; antennal flagellum with pubescence inconspicuous; head of female shorter than wide in frontal view, genal area without tooth (holarctic) ....
.... *H. (Protohalictus)*

7(4). Hypostomal and lower genal areas of male, at least in ventral view, slightly concave to deeply excavated; first flagellar segment of male broader than long, following segments often convex ventrally and thus moniliform and each with glabrous, shiny area or band mesally; S4 of male with posterior margin broadly and deeply emarginate with extreme lateral angle distinct 8

—. Hypostomal and genal areas of male convex or flat in lateral view; first flagellar segment of male as long as or longer than broad, following segments usually not convex, without glabrous areas; S4 of male with posterior margin straight or shallowly incised medially, without strong angle at side 9

8(7). Propedeal triangle with distinct wrinkles, shiny; malar areas of male more than one third as long as broad; hypostomal and genal areas deeply and sharply excavated (palearctic................................. *H. (Monilapis)*

—. Propedeal triangle obscurely rugulose, silk-shiny; malar area of male less than one fifth as long as broad; hypostomal and genal areas of male only slightly concave (palearctic) .... *H. (Platyhalictus)*

9(7). Male antenna at most reaching scutellum; S7 of male transverse without posterior median process; lower genal area of female with large tooth; inner hind tibial spur of female with one large basal tooth (North America to northern South America) ....
.... *H. (Odontalictus)*

—. Male antenna usually reaching propodeum; S7 of male with median posterior projection; gena of female not toothed; inner hind tibial spur of female serrate with several teeth (palearctic) 10

10(9). Propodeum with carina bordering its posterior surface; last antennal segment of male usually hook-shaped; S4 of male trapezoidal, narrowed, and emarginate posteriorly (palearctic) .... *H. (Hexataenites)*

—. Propodeum without carina; last antennal segment of male of usual form; S4 of male not narrowed posteriorly, margin usually straight 11

11(10). Propodeum and metasomal terga distinctly punctate; metasomal hair bands broadly interrupted medially, consisting of fine hairs; S8 of male with median, apical prominence (palearctic) .... *H. (Tytthalictus)*

—. Propodeum and metasomal terga obscurely punctate; metasomal hair bands unbroken, consisting of coarser hairs; S8 of male without median, apical prominence ....
.... *H. (Argalictus)*

12(1). Scopa almost absent; mandible of female simple; T5 without median specialized areas of minute hairs characteristic of most female Halictini [Cleptoparsitic, male unknown] (Africa) .... *H. (Paraseladonia)*

—. Scopa of female well developed; mandible of female with subapical tooth; T5 with specialized median area of minute hairs as normal for Halictini 13

13(12). Basitibial plate of female slender, pointed, its anterior margin incompletely defined; hairless propodeal triangle small, often not as long as metanotum, area lateral to triangle, and metasomal terga, largely covered with pale tomentum .... *H. (Vestitohalictus)*

—. Basitibial plate of female rounded or pointed, anterior margin defined; propodeal triangle about as long as metanotum, area lateral to triangle usually not densely hairy; terga usually with apical and basal bands of pale plumose hairs 14

14(13). Male antenna usually reaching only to scutellum, second flagellar segment 1.2 to 1.4 times as long as wide; S6 of male flat or with slight longitudinal median depression .... *H. (Seladonia)*

—. Male antenna usually reaching metasoma, second flagellar segment 1.7 to 2.0 times as long as wide; S6 of male with deep triangular depression behind gradulus ....
.... *H. (Pachyceble)*

Halictus / Subgenus *Argalictus* Pesenko

Halictus / Subgenus *Argalictus* Peseko
Halictus (Argalictus) Pesenko, 1984a: 348. Type species: *Hylaeus senilis* Eversmann, 1852, by original designation.

■ The eight species of this subgenus are found in the southern palearctic region, especially the xeric parts of Asia.

Halictus / Subgenus *Halictus* Latreille s. str.

Halictus Latreille, 1804: 182. Type species: *Apis quadricincta* Fabricius, 1776, by designation of Richards, 1935: 170. For other type designations, see Michener, 1997b.

Halictus s. str. includes the largest species of the genus, 16 to 18 mm in body length.

■ This subgenus is widespread in the palearctic region, from Morocco to southern Finland east to northeastern China. Four species were listed by Pesenko (1984a, 2004b).

Nests of this subgenus are unique, so far as is known, in the genus *Halictus* in that the cells are in a dense cluster largely isolated from the surrounding soil matrix by an air space and supported by pillars or rootlets (Fig. 65-4 and Sakagami and Michener, 1962). Thus the architecture suggests that of many Augochlorini and is unusual for Halictini. *H. quadricinctus* (Fabricius) and perhaps the other species are solitary, not eusocial as are many *Halictus* species.

Halictus / Subgenus *Hexataenites* Pesenko

Halictus (Hexataenites) Pesenko, 1984a: 348. Type species: *Apis sexcincta* Fabricius, 1775, by original designation.

This subgenus includes such well known species as *Halictus sexcinctus* (Fabricius) and *H. scabiosae* (Rossi).

■ The eleven species range from westernmost Europe to China.

Halictus / Subgenus *Lampralictus* Pesenko

Halictus (Lampralictus) Pesenko, 1984a: 348. Type species: *Halictus modernus* Morawitz, 1876, by original designation.

Contrary to Michener (2000), who placed this form with *Argalictus*, Pesenko (2004b) showed that this subgenus is related to *Protohalictus*. Like *Odontohalictus*, the female has an angular tooth on the lower genal area.

■ The single species, *Halictus modernus* Morawitz, is found in central Asia.

Halictus / Subgenus *Monilapis* Cockerell

Halictus (Monilapis) Cockerell, 1931f: 529. Type species: *Hylaeus tomentosus* Eversmann, 1852, = *Apis flavipes* Panzer, 1798 (not Fuesslin, 1775; not Fabricius, 1787) = *Andrena compressa* Walckenaer, 1802, by original designation.

Halictus (Acalcaripes) Pesenko, 1984a: 347. Type species: *Halictus patellatus* Morawitz, 1873, by original designation.

Extensive accounts of the identity of the type species of *Monilapis* are by Pesenko (1985, 2004b).

Two species with modified male front tarsi and reduced tibial spurs in the male were separated as a subgenus *Acalcaripes* by Pesenko, but it appears to be a derivative of *Monilapis*, as Pesenko's cladograms suggest. Body length is 9 to 13 mm.

■ This subgenus ranges from western Europe and Morocco to China and Korea. There are about 31 species, revised by Pesenko (1985) and listed by Pesenko (2004b).

Halictus / Subgenus *Nealictus* Pesenko

Halictus (Nealictus) Pesenko, 1984a: 346. Type species: *Halictus parallelus* Say, 1837, by original designation.

Halictus parallelus Say and *farinosus* Smith, the only species of *Nealictus*, may be merely an odd subgroup of *Protohalictus*, which name therefore would become a junior synonym of *Nealictus*. In addition to the differences indicated in the key to subgenera based on males, *Nealictus* differs from *Protohalictus* in having the lower gonostylus of the male reduced to a small, pointed projection. It is much more elongate in at least some of the species of *Protohalictus*. The body length is 10 to 13 mm.

■ *Nealictus* is found from British Columbia and Ontario, Canada, south to California, Texas, and Florida, USA. The two species were included among those treated by Sandhouse (1941); see also Pesenko (2004b).

Halictus / Subgenus *Odontalictus* Robertson

Odontalictus Robertson, 1918: 91. Type species: *Halictus ligatus* Say, 1837, by original designation.

The genal tooth of the female of *Odontalictus* is similar to that of *Halictus (Lampralictus) modernus* Morawitz. The body length is 7 to 14 mm.

■ *Odontalictus* is widespread in North America, from 50° N south through the continent to Colombia, Venezuela, Trinidad, and the Greater Antilles. There are two species, distinguished by molecular rather than morphological characters (Carman and Packer, 1997).

Halictus / Subgenus *Pachyceble* Moure

Pachyceble Moure, 1940: 54. Type species: *Pachyceble lanei* Moure, 1940, by original designation.

This subgenus of small greenish species has usually been included in *Seladonia*; see commentary on that subgenus. It differs from *Seladonia* by the characters indicated in the key to subgenera; see also Pesenko (1904b). Some species have the lower gonostylus of the male reduced or absent.

■ *Pachyceble* is widespread in the palearctic region, from western Europe to China. In the new world it occurs from Canada (British Columbia to New Brunswick) south through Central America to central Brazil. There are 15 Old World species and seven in the New World; one species, *H. (P.) confusus* Smith, is holarctic. The species were listed by Pesenko (2004b). The Asiatic species of *Pachyceble* were revised by Dawut and Tadauchi (2000-2003), the Oriental species by Sakagami and Ebmer (1987), and the European species by Ebmer (1988a). A revision of Chinese species was by Niu, Wu, and Huang (2004). North American species were included in the revision of the genus by Sandhouse (1941).

Halictus / Subgenus *Paraseladonia* Pauly

Paraseladonia Pauly, 1997a: 92. Type species: *Halictus chalybaeus* Friese, 1910, by original designation. [New status.]

This is the only known parasitic *Halictus*, a probable derivative of the subgenus *Seladonia*, as indicated by Pauly (1997a). The strong wing venation, blue-green color, apical tergal hair bands (almost completely absent in females), and the distally directed ventral gonostylus of the male (i.e., no retrorse lobe) are indicative of this relationship. The female clypeus, however, lacks entirely the produced tubercle or angle at the lateral extremity of the clypeal truncation that is characteristic of nearly all *Halictus*. Present, however, is the small translucent clypeal lobe behind the angle, at the side of the labrum, also found in *Halictus*. In females of *Paraseladonia* the scopa and basitibial plate are reduced; the mandibles are simple and pointed; the labral process is broad and rounded, the keel merely indicated; the penicillus of the hind basitarsus is absent; and T5 lacks the specialized area of minute hairs characterisic of most Halictinae. Thus *Paraseladonia* exhibits the usual characteristics of parasitic Halictinae (Michener, 1978b). Pauly (1997a) illustrated various structures.

■ *Paraseladonia* is known across Africa from Togo to Kenya. The only species is *Halictus chalybaeus* Friese. The assumption that it is parasitic is based on its morphology, not on any behavioral observations.

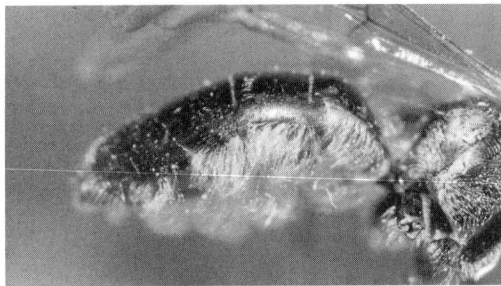

Figure 66-12. Lateral view of metasoma of female of *Homalictus silvestris* Michener, showing the scopa arising from the ventrolateral parts of the terga as well as from the sterna.

Halictus / Subgenus *Platyhalictus* Pesenko

Halictus (Platyhalictus) Pesenko, 1984a: 347. Type species: *Halictus minor* Morawitz, 1876, by original designation.

This subgenus consists of species having body lengths of 8 to 14 mm.

■ *Platyhalictus* is found in the southern mesic zone of Eurasia and in northern India. The 14 known species were revised by Pesenko (1984c) and listed by Pesenko (2004b).

Halictus / Subgenus *Protohalictus* Pesenko

Halictus (Prohalictus) Pesenko, 1984a: 346 (not Armbruster, 1938). Type species: *Apis rubicunda* Christ, 1791, by original designation.

Halictus (Protohalictus) Pesenko, 1986b: 631, replacement for *Prohalictus* Pesenko, 1984. Type species: *Apis rubicunda* Christ, 1791, autobasic.

See comments under *Nealictus,* to which I believe *Protohalictus* is closely related. The body length is 9 to 13 mm.

■ *Protohalictus* occurs throughout the palearctic region, from Ireland to Japan, north to Sweden and eastern maritime Russia, south to Morocco, Egypt, and India. One species, *Halictus rubicundus* (Christ), is holarctic, ranging across the palearctic region and in North America from Alaska, USA, to Nova Scotia, Canada, south to California, Texas, and Florida, USA. Thirteen species were listed by Pesenko (2004b), and the subgenus was revised by Pesenko (1984d).

Halictus / Subgenus *Ramalictus* Pesenko

Halictus (Ramalictus) Pesenko, 1984a: 347. Type species: *Halictus latisignatus* Cameron, 1908, by original designation.

Distinctive features of this subgenus were pointed out, under the heading of *Halictus* Group 4, by Michener (1978a) and were illustrated by Sakagami and Wain (1966). The partly red metasoma results in a bee superficially similar to the very common, sympatric *Apis florea* Fabricius; body length is 7 to 15 mm.

■ *Ramalictus,* found in western peninsular India, contains only one species, *Halictus latisignatus* Cameron.

Halictus / Subgenus *Seladonia* Robertson

Seladonia Robertson, 1918: 91. Type species: *Apis seladonia* Fabricius, 1794, by original designation.

Because of its greenish, bluish, or brassy coloration (the metasoma sometimes nonmetallic) and small to moderate size (length 4.5-10.0 mm), this subgenus and *Pachyceble* have long been recognized, sometimes as a genus *Seladonia* (see Pesenko, 2004b). Although in most species of these subgenera the pale pubescence is not especially dense and the apical, and sometimes also basal, tergal hair bands are distinct, in a few such as *H. (S.) niveocinctulus* Cockerell, pale tomentum tends to spread between basal and apical hair bands, thus suggesting the subgenus *Vestitohalictus.* In most species of *Seladonia* the lower gonostylus of the male is well developed, as in *Halictus* s. str. Another subgeneric character is the clump of anteromedially directed curved setae arising from the inner surface of the principal (upper) gonostylus, as in *Halictus* s. str. and its allies.

■ *Seladonia* occurs throughout the Old World, from Europe and Asia, south into the Oriental region to southern India and Southeast Asia, and south throughout the whole of Africa, including Madagascar and the Mascarene Islands. *Seladonia* does not occur in Australia. Pesenko (2004b) listed 36 species.

Halictus / Subgenus *Tytthalictus* Pesenko

Halictus (Tytthalictus) Pesenko, 1984a: 348. Type species: *Halictus maculatus* Smith, 1848, by original designation.

This subgenus contains small species, 5.0 to 8.5 mm long.

■ *Tytthalictus* occurs in the palearctic region, including northern India; most of the species are in the Mediterranean area. The seven species were revised by Pesenko (1984b); four species were listed by Pesenko (2004b).

Halictus / Subgenus *Vestitohalictus* Blüthgen

Halictus (Vestitohalictus) Blüthgen, 1961: 287. Type species: *Halictus vestitus* Lepeletier, 1841, by original designation.
Placidohalictus Pesenko, 2004b: 102. Type species *Halictus placidus* Blüthgen, 1923, by original designation. **[New synonymy.]**

Mucoreohalictus Pesenko, 2004: 102. Type species: *Hylaeus mucoreus* Eversmann, 1852, by original designation. [**New synonymy.**]

I tentatively recognize *Vestitohalictus* in an older sense, including Pesenko's new subgenera, because the whole group seems distinctive (see the key to the subgenera) and because *Placidohalictus* seems to be paraphyletic in Pesenko's system.

The body in this subgenus is greenish or uncommonly nonmetallic, sometimes with the metasoma red, the metasoma often so densely covered with pale pubescence that the apical bands of the terga are not evident. The body length is 3.5 to 8.0 mm. The upper or principal gonostylus of the male is expanded distally and has a slender, hirsute preapical appendage arising from its upper mesal margin, as in *Halictus (Seladonia) virgatellus* Cockerell (see Sandhouse, 1941, fig. 7). There is also a very small cluster of anteromedially directed setae arising from near the upper margin of the basal part of the gonostylus. The lower gonostylus, when present, is rather slender, parallel-sided; it is sometimes absent.

■ *Vestitohalictus* is widespread in the palearctic region, mostly in xeric regions, from the Canary Islands to China. Pesenko (2004b) listed 35 species, under three subgeneric names.

Genus *Homalictus* Cockerell

In females of this genus the metasoma is flattened, at least T2 and T3 show distinct angles between their dorsal and lateroventral surfaces as in the Nomioidinae, and there are enormous scopal hairs arising on the sterna and ventral parts of the terga (Fig. 66-12), those on the latter directed strongly mesad. The scopal hairs on the hind femora and metasoma are plumose, with numerous branches diverging from a distinct rachis. The hairs forming the hind femoral corbicula arise on the lower surface of the femur, not on the dorsal surface. The large hairs on the basal two-thirds of the lower surface of the hind tibia are pectinate, usually rather erect, and clearly differentiated from the usually short, erect hairs of the lower distal part of the outer surface of the tibia, which is often in a slightly different plane than the rest of the outer tibial surface. (Thus the tibia resembles that of some *Patellapis*, Fig. 66-19a.) All these characters differentiate females from *Lasioglossum*, with which *Homalictus* agrees in the weak third submarginal crossvein and second recurrent vein. Males, however, do not seem to differ from those of *Lasioglossum* in any thoroughly consistent character, although the gonobase character mentioned in the key to the genera usually works. Illustrations of male genitalia and other structures were published by Krombein (1951), Michener (1965b, 1980), and Walker (1986, 1997). For Indonesian and New Guinea species, Pauly (1980b, 1986) published sketches of male genitalia and other structures. In some species the gonostyli are much longer and more slender than those of *Lasioglossum*, as long, for example, as the gonocoxites in *Homalictus eurhodopus* (Cockerell) and *latitarsis* (Friese).

Because of the weakened distal wing venation, prior authors have regarded *Homalictus* as a derivative of *Lasioglossum*; this arrangement would make the latter paraphyletic. In the past I have recognized it nonetheless as a genus because of the huge size of *Lasioglossum* and the distinctiveness of females of *Homalictus*. It now appears, however, that *Lasioglossum* may not be the closest relative of *Homalictus* and that the reduced distal venation of *Homalictus* possibly arose independently from that of *Lasioglossum*. Every one of the characters of *Homalictus* listed above except the weak distal venation is found also among the subgenera of *Patellapis*, which therefore may be the group from which *Homalictus* arose, and thus the group that *Homalictus* may make paraphyletic. In *Patellapis* (*Pachyhalictus* and *Archihalictus*) the metasoma is somewhat flattened, and T2 is somewhat angulate at the sides, especially in *Archihalictus* of the group of *joffrei* (Benoist), suggesting *Homalictus*. The metasomal scopa, in some species of these and other subgenera, is large, in the *joffrei* group fully as well developed as in *Homalictus*. The other characters of the female scopa and hind legs of *Homalictus* can all be duplicated in *Patellapis* subgenera such as *Pachyhalictus*. Although these are a complex set of features, *Homalictus* may none the less be closer to *Lasioglossum* than to *Patellapis*. In this case it could be incorporated into the former, making *Lasioglossum* an even larger and more cumbersome unit.

Unlike *Patellapis*, most *Homalictus* species are metallic blue or green, more rarely bronze or purple, often dull but sometimes as brilliant as in many Augochlorini. The metasoma is sometimes red. S4 of the male is not shortened and is not equipped with long bristles, features that are common in *Patellapis*.

The recognition of the presumably cleptoparasitic *Echthralictus* as a genus makes *Homalictus* paraphyletic. I have decided to recognize *Echthralictus* nevertheless, because there is a major structural gap between it and *Homalictus*. To synonymize it with *Homalictus (Homalictus)*, which is undoubtedly where it goes cladistically, would make irrelevant virtually all of the indications of relations of *Homalictus* to *Patellapis*, *Lasioglossum*, etc., because most of these indications relate to structures that disappear or are modified in parasitic bees.

Key to the Subgenera of *Homalictus* (Females)
(For males, see comments on the subgenera)

1. Hairs of lower distal part of outer surface of hind tibia short, uniform, erect, branched; undersurface of tibia concave .... *H. (Homalictus* s. str.)
—. Hairs of lower distal part of outer surface of hind tibia not differentiated from those on rest of surface, and the hairs not erect; undersurface of tibia not or scarcely concave 2
2(1). T2 to T4 without bands of tomentum; hair on outer surface of hind tibia simple, sparse, almost bristle-like (New Guinea) .... *H. (Papualictus)*
—. T2 to T4 with bands of tomentum, broken medially; hair on outer surface of hind tibia short, minutely plumose, appressed (Australia) .... *H. (Quasilictus)*

Homalictus / Subgenus *Homalictus* Cockerell s. str.

Halictus (Homalictus) Cockerell, 1919b: 13. Type species: *Halictus taclobanensis* Cockerell, 1915, by original designation.

Halictus (Indohalictus) Blüthgen, 1931: 291. Type species: *Halictus buccinus* Vachal, 1894, by original designation.

This is a diverse subgenus of small species, 3.5 to 8.0 mm long. As indicated by Pauly (1980b) and Krombein (1951), *Indohalictus* seems distinct at the subgenus level when Indonesian and Asiatic species are considered, but in the diverse Australian fauna they appear to merge, as noted by Michener (1965b) and Walker (1986).

■ *Homalictus* is most abundant in Australia, where it occurs in all states, including Tasmania; it occurs on South Pacific islands east to Samoa, and in the central Pacific east to the Carolines and Marianas, as well as north through Indonesia to the Philippines, Viet Nam, Thailand, Malaysia, India, and Sri Lanka. There are 44 Australian species, and about 50 other species are found in other areas. Australian species were revised by Walker (1986, 1997), while those of other areas were listed by Pauly (1986), who also gave a key for identification of the species of New Guinea and the Bismarck Archipelago. Pauly (1980b) also gave a key to the species of Indonesia.

Homalictus / Subgenus *Papualictus* Michener

Homalictus (Papualictus) Michener, 1980: 8. Type species: *Homalictus megalochilus* Michener, 1980 = *Halictus lorentzi* Friese, 1911, by original designation.

This subgenus consists of large (body length 8.5-11.0 mm) species with very broad heads and enormous mandibles in the males, as illustrated by Michener (1980). The species are strongly metallic, with largely dark pubescence, as is also true for some *Homalictus* s. str. In males the dorsal surface of the propodeum is elevated on each side to form a shining boss.

■ This subgenus is known from New Guinea and perhaps northern Queensland, Australia. The five New Guinea species were described by Michener (1980); see also Pauly (1986) and Walker (1997).

Either the lone species from Australia should be placed elsewhere, or the description of the subgenus should be changed. It is nonmetallic, only 4.5 mm long, and lacks the propodeal character listed above; see Walker (1997).

Homalictus / Subgenus *Quasilictus* Walker

Homalictus (Quasilictus) Walker, 1986: 166. Type species: *Homalictus brevicornutus* Walker, 1986, by original designation.

This subgenus is so different from other *Homalictus* that subgeneric status may be warranted, even though there is only one species. In addition to the characters indicated in the key to subgenera, which distinguish *Quasilictus* from both other subgenera, it has the following other unique features: The male gonobase is enormous, quadrate, and as wide as the gonocoxites taken together; the volsella is very large and has a ventral projection. The body is metallic blue and green, the metasoma red-brown; the length is 5.4 to 6.2 mm.

■ *Quasilictus* occurs in Northern and Western Australia. The only species is *Homalictus brevicornutus* Walker.

Genus *Lasioglossum* Curtis

Lasioglossum is an enormous genus of morphologically monotonously similar bees. It is the only halictid genus other than *Homalictus* that shows weakened distal wing venation, at least in females. As indicated in the discussion of that genus, *Homalictus* possibly acquired such weakened veins independently from *Lasioglossum*. Older authors and some recent authors (Warncke, 1975b, 1981) included all *Lasioglossum* species in the genus *Halictus*. *Lasioglossum* differs from *Halictus*, however, as indicated in Table 66-1 and in the discussion of the tribe Halictini. Illustrations of male genitalic and other characters can be found in Mitchell (1960), Ebmer (1969-1971), Michener (1979b, 1993a), Do Pham, Plateaux-Quénu, and Plateaux (1984), Sakagami (1989, 1991), Sakagami and Maeta (1990), and others.

The subdivisions and limits of what is here called the genus *Lasioglossum* have been and continue to be subject to differences of opinion. Some authors (e.g., Michener, 1944; Ebmer, 1987a, 1988a, b; Michener, McGinley, and Danforth, 1994) included all or most species in the single genus *Lasioglossum*, while others (Mitchell, 1960; Hurd, 1979; Moure and Hurd, 1987) recognized various genera such as *Evylaeus* and *Dialictus* that are here placed among the subgenera of *Lasioglossum*.

Lasioglossum as here understood can be divided into two series. The *Lasioglossum* series consists of those subgenera in which the second submarginal crossvein is strong, like the first and unlike the third. The *Hemihalictus* series consists of those in which the second submarginal crossvein is weaker than the first, although commonly not as weak as the third (Fig. 66-6c), or rarely the second is completely absent (Fig. 66-1b). Although this character is weak (see below), it is probably of phylogenetic importance, since it separates groups with differing geographical distributions and biologies; e.g., eusociality is frequent in the *Hemihalictus* series but not in the *Lasioglossum* series [although *L. (Lasioglossum) aegyptiellum* (Strand) is probably eusocial; see Packer, 1997]. Additional data on phylogenetic relationships of *Lasioglossum aegyptiellum* were provided by Packer (1998), but no new data on its possibly social behavior are available. However, see comments on other species of *Lasioglossum* s. str. with social tendencies in the introductory account of the subfamily Halictinae. The subgenera belonging to each series are listed in Table 66-2.

It is tempting to separate the two series generically. Unfortunately, I do not know of any character that supports the division except the status of the second submarginal crossvein, and that character often fails completely in males, all the veins being strong. Even in females there are species or individuals that cannot be placed with much

Table 66-2. Subgenera of *Lasioglossum*.

Lasioglossum series	*Hemihalictus* series
Australictus	Acanthalictus
Callalictus	Austrevylaeus
Chilalictus	Dialictus
Ctenonomia	Evylaeus
Glossalictus	Hemihalictus
Lasioglossum s. str.	Paradialictus
Parasphecodes	Sellalictus
Pseudochilalictus	Sphecodogastra
	Sudila

confidence, as noted by Ebmer (1969). McGinley (1986) mentions proportions of antennal segments in males as a supportive character, but it breaks down completely in the eastern hemisphere. Some other characters are helpful in certain areas. Concerning the distinction between males of *Lasioglossum* s. str. and those of *Evylaeus* in the *Hemihalictus* series, McGinley (1986) observed that another useful differentiating feature is clypeal shape. All New World *Evylaeus* males, including *Eickwortia,* and only some *Lasioglossum* [*leucozonium* (Schrank), *zonulum* (Smith)—both holarctic species] have a rounded, protuberant clypeal surface; but most New World *Lasioglossum* s. str. males have a distinctly flattened to slightly depressed clypeal surface. It is likely that the *Lasioglossum* series, defined only by a plesiomorphic wing character state, is paraphyletic.

Weakening of the second submarginal crossvein is presumably an apomorphy of the *Hemihalictus* series, for I do not see evidence of its arising independently from different members of the *Lasioglossum* series. The monophyly of the *Hemihalictus* series and its origin from the *Lasioglossum* series (which in this respect is paraphyletic) is strongly supported by Danforth (1999a). Nonetheless, the weakened crossveins is a frail feature on which to separate major genera. Supporting my hesitation to separate the two series as genera at this time is the name of the *Hemihalictus* series. The oldest applicable genus-group name is *Hemihalictus,* a name that until now has been applied to only one species; it completely lacks the second transverse cubital vein (as also do some other but unrelated species of the *Hemihalictus* series in the subgenus *Dialictus*). To transfer to a genus *Hemihalictus* the many hundreds of species that have never been placed in that genus would be a mistake at this time, because they might soon have to be transferred out of *Hemihalictus.* What is needed is a worldwide phylogenetic study, to be followed by taxonomic decisions likely to have better support and more permanence than can be provided now.

The problems of subgenera within the *Hemihalictus* series are also not subject to satisfying solutions at this time. Some of the smaller subgenera are easily recognizable and quite distinct. Of importance in the present discussion are the two large groups, *Evylaeus* and *Dialictus. Evylaeus* was divided into numerous subgenera in a study of western palearctic species by Warncke (1975b, 1981), but his names also are not relevant to the present discussion. *Dialictus* as usually understood consisted of mostly smaller species with dull greenish or bluish metallic coloration at least on the head and thorax, whereas *Evylaeus* consisted of species of various sizes without metallic coloration. Thus *Evylaeus* and *Dialictus* in this sense differed only in color, and, understandably, some authors synonymized *Evylaeus,* the junior name, under *Dialictus.* The type species of *Evylaeus,* however, belongs to a group that is quite different from typical *Dialictus.* The problem has been discussed in various publications (e.g., Michener, 1979b, 1993). It seems reasonable to restrict the subgenus *Evylaeus* to the "carinate group," i.e., larger species in which carinae usually delimit the posterior surface of the propodeum at least laterally and partway across the angle between the basal area and the declivous posterior surface. Unfortunately, the carinae are missing in some species that seem to belong to this group (Sakagami, Miyanaga, and Maeta, 1994) and, moreover, the carinae are fully developed in a few species of the noncarinate group, including some North American typical greenish *Dialictus.* The carinate group is holarctic and consists of relatively large, nonmetallic species, but to add further to the confusion, L. Packer (in litt., 1997) reports species that are clearly carinate *Evylaeus,* yet possess metallic coloration. See the characterization by Ebmer (1995), and see also the account of the subgenus *Evylaeus* for supportive genitalic characters.

The members of the "noncarinate group" of *Evylaeus* (in which the carinae are commonly—but by no means always—limited to the lower half of the propodeum or at least usually do not curve mesad and separate the basal area from the posterior propodeal surface) are usually smaller, and differ from *Dialictus* as often understood only in lacking the greenish metallic coloration frequently considered characteristic of that group. This character intergrades completely; some species have only a scarcely detectable localized greenish tint (e.g., *Lasioglossum breedi* Michener), and *L. viride* (Brullé) has two forms, one greenish, the other black. Furthermore, metallic coloration is not a major group character in diverse other halictines; there are greenish as well as black species in taxa such as the following: *Halictus (Vestitohalictus), Homalictus* s. str., and *Lasioglossum (Lasioglossum* s. str., *Chilalictus,* and *Ctenonomia).* Within *Dialictus* as here understood there are at least three groups of species containing both greenish and black members or intermediates: *L. breedi* and similar species in the Western Hemisphere as indicated above, *L. viride* on the Canary Islands, and the African group called *Afrodialictus* by Pauly (1984b). The noncarinate group of *Evylaeus* (along with some species that have carinae but are not related to the carinate group) should therefore be transferred to the subgenus *Dialictus,* leaving only the carinate group in the subgenus *Evylaeus.*

Danforth (1999a) provided hypotheses for relationships among subgenera in his phylogenetic treatment based on mitochondrial sequence data from 77 species of *Lasioglossum.* The cleptoparasitic *Paralictus* fell consistently within *Dialictus.* Danforth showed that the acarinate group of *Evylaeus* falls into *Dialictus,* thus supporting the classification used here and in Michener (2000).

Given the morphological similarity of the whole genus and the lack of reliable characters to separate the carinate and noncarinate forms, an alternative procedure would be to synonymize *Evylaeus* under *Dialictus.* This arrangement corresponds to that of Ebmer (1969-71, 1987a, 1988a, b) except that, because the type species of *Dialictus* has only two submarginal cells, he excluded it and placed all the three-celled species (over 99 percent) under the subgeneric name *Evylaeus,* whether carinate or not and whether greenish or black.

Any major transfer of hundreds of species from *Evylaeus* to *Dialictus* would be inappropriate if these groups were considered as genera. Such an arrangement would probably be negated when a world study is made and the species of the noncarinate group reclassified. As indicated above, I use the name *Evylaeus* as the subgenus for the carinate group, even though I cannot at this time place every species as to subgenus. The subgenus *Dialictus* as here defined includes both greenish and nonmetallic species. So segregated, *Evylaeus* includes the species whose eusocial behavior is most complex, including species whose adults come in contact with the larvae and remove feces from larval cells, but it also includes solitary species such as *L. fulvicorne* (Kirby), as noted by L. Packer (in litt., 1997). *Dialictus*, on the other hand, includes many solitary species as well as species that are eusocial; so far as is known, the adults do not contact the larvae. I believe that my use of *Evylaeus* and *Dialictus* combines the recognition of useful groups with avoidance of a great many generic transfers that would probably not persist after future analyses.

The continued use of the generic name *Lasioglossum* makes it possible to conserve a current binomen for every species; thus such behaviorally well-known species as *Lasioglossum zephyrum* (Smith) do not get shifted to genera like *Dialictus, Evylaeus, Hemihalictus*, or *Sudila* (see below), to the certain confusion of nontaxonomists and probably of taxonomists as well. When or if *Lasioglossum* is to be split and species like *L. (Dialictus) zephyrum* are to be transferred to another genus, it should be done once, to some one other genus, something that is not yet possible because of uncertainties.

The keys to the subgenera of *Lasioglossum* below are divided geographically into three regions, thus: the Western Hemisphere; the palearctic, Oriental, and African faunal regions; and the Australian region.

Several particulars concerning the keys bear comment here. For the first key, couplet 2, and second key, couplet 7: see the comments on the subgenera *Evylaeus* and *Dialictus* in the discussion of the genus *Lasioglossum*. The female of *Lasioglossum (Sellalictus) ankaratrense* (Benoist) from Madagascar does not agree with the characters given in the second key, first option of couplet 5, and would run to *Dialictus*, but the male is clearly a *Sellalictus*. Some small species of the subgenus *Parasphecodes* have coarsely serrate or pectinate hind tibial spurs and hence run to the second option of couplet 4 in the third key. They differ in having the metasoma red, and differ from most *Ctenonomia* in lacking basal bands of tomentum on the metasomal terga.

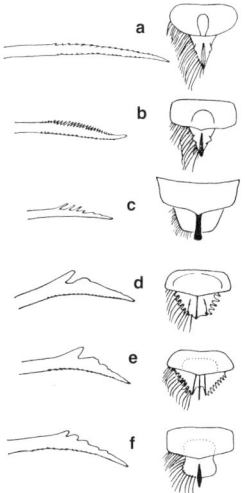

Figure 66-13. Inner hind tibial spurs and labra of females of Australian species of *Lasioglossum*. **a,** *L. (Parasphecodes) musicum* (Cockerell); **b,** *L. (Pseudochilalictus) imitator* Michener; **c,** *L. (Glossalictus) etheridgei* (Cockerell); **d,** *L. (Chilalictus) wahlenbergiae* Michener; **e,** *L. (C.) leai* (Cockerell); **f,** *L. (C.) helichrysi* (Cockerell). From Michener, 1965b.

Key to the Subgenera of *Lasioglossum* of the Western Hemisphere

1. Second submarginal crossvein as strong as first (Figs. 66-1a, 66-6a) [usually rather large and robust, body length 6.8-12.0 mm, usually nonmetallic, although *Lasioglossum pavonotum* (Cockerell) of the Pacific Coast of USA is strongly green] (holarctic) .... *L. (Lasioglossum s. str.)*
—. Second submarginal crossvein, at least in female, weaker than first, although not always as weak as third or as second recurrent vein (Fig. 66-6c) (*Hemihalictus* series) 2
2(1). Posterior surface of propodeum margined at side by vertical carina that extends upward to summit and then mesad at least partway across upper margin, so that there is a sometimes interrupted carina between dorsal and posterior surfaces; body nearly always nonmetallic (holarctic) .... *L. (Evylaeus)*
—. Posterior surface of propodeum in nonmetallic species usually with lateral vertical carina extending only partway to summit, or, *if* attaining summit, then usually not extending mesad across upper margin, so that there is usually no carina between dorsal and posterior surfaces, but in metallic species, with or without propodeal carinae, sometimes even more strongly carinate than in *L. (Evylaeus)*; body black or metallic greenish or bluish, rarely bright metallic green 3
3(2). Scopa sparse, consisting of a row of simple bristles on hind trochanter and lower edge of femur, shorter bristles on upper part of femur, and some long, nearly simple hairs on inner side of tibia (nonmetallic) (nearctic) ......
.... *L. (Sphecodogastra)*
—. Scopa normal, largely plumose (as in Fig. 8-5b) 4
4(3). Nonmetallic and with two submarginal cells; inner hind tibial spur of female serrate (nearctic) ....
.... *L. (Hemihalictus)*
—. Nonmetallic or metallic, with three submarginal cells or, in a few metallic species, with two; inner hind tibial spur of female pectinate with a few large teeth 5
5(4). Mandible of female strongly bidentate at apex; middorsal length of male gonobase less than one-fourth as long as gonocoxite (Mesoamerica) .... *L. (Eickwortia)*
—. Mandible of female with small, dorsal preapical tooth; middorsal length of male gonobase usually over one-third as long as gonocoxite .... *L. (Dialictus)*

Key to the Subgenera of *Lasioglossum* of the Palearctic, Oriental, and African Faunal Regions

1. Second submarginal crossvein as strong as first (to be verified using females) (as in Fig. 66-6a) (*Lasioglossum* series) 2
—. Second submarginal crossvein, at least in female, weaker than first, although not always so weak as third or as second recurrent vein (as in Fig. 66-6c) (*Hemihalictus* series) 3
2(1). Inner hind tibial spur of female serrate or rarely pectinate with five or more teeth (holarctic)..........................
.... *L. (Lasioglossum s. str.)*
—. Inner hind tibial spur of female pectinate with few large teeth, or with one large tooth followed by undulate margin (Africa, Oriental, Australia) .... *L. (Ctenonomia)*
3(1). Mandible of female with two preapical teeth on upper

margin; body length 11 to 13 mm; S6 of male ending in thick, shining, uniformly curved and elevated margin (inner hind tibial spur of female serrate) (Siberia) .... *L. (Acanthalictus)*
—. Mandible of female with one preapical tooth; body smaller; S6 of male not as described above 4
4(3). Basitibial plate of female absent (Fig. 66-17d); scopa of female represented by only a few hairs; propodeum with vertical lunate flange separating lateral from posterior surfaces (male unknown) (Africa) .... *L. (Paradialictus)*
—. Basitibial plate of female distinct; scopa of female well developed, forming corbicula on underside of femur, evident but less well developed in parasitic species of *Dialictus* (Fig. 8-6a); propodeum with carina or rounded angle separating lateral from posterior surfaces 5
5(4). T2 of male with broad, basal band of *erect* white hairs; disc of labrum of female rather uniformly convex; inner hind tibial spur of female serrate or, *if* pectinate, then with five or more teeth, all longer than broad (Africa).... .... *L. (Sellalictus)*
—. T2 of male without basal band of white hairs, or, *if* present, then hairs more or less prostrate as on other terga; disc of labrum of female with elevated area or tubercles, not uniformly convex; inner hind tibial spur of female usually pectinate with less than five long teeth, but sometimes serrate or briefly pectinate with more teeth 6
6(5). Subpleural signum of female usually elevated as distinct tubercle in front of middle coxa; inner hind tibial spur of male briefly pectinate (Oriental) .... *L. (Sudila)*
—. Subpleural signum of female inconspicuous or absent, not elevated as tubercle; inner hind tibial spur of male usually ciliate or minutely dentate 7
7(6). Posterior surface of propodeum margined at side by vertical carina that extends upward to summit and then mesad at least partway across upper margin, so that there is a sometimes interrupted, transverse carina between basal area and posterior surface, i.e., across posterior margin of basal area; rarely transverse carina present but vertical carina reduced or absent; usually rather large, nonmetallic except for some Asian species (holarctic) .... *L. (Evylaeus)*
—. Posterior surface of propodeum usually with lateral vertical carina extending only partway to summit, or, *if* attaining summit, then usually not extending mesad across upper margin, so that there is usually no carina between basal area and posterior surface, i.e., so that basal area is usually not enclosed; body commonly smaller, sometimes metallic .... *L. (Dialictus)*

Key to the Subgenera of *Lasioglossum* of the Australian Region

1. Second submarginal crossvein in both sexes narrower than first (as in Fig. 66-6c); small (less than 7 mm long); inner hind tibial spur of female ciliate, essentially simple, or in some cases bearing a few large oblique teeth (body dull, nonmetallic) (*Hemihalictus* series) .... *L. (Austrevylaeus)*
—. Second submarginal crossvein as strong as first (as in Fig. 66-6a); *if* body less than 7 mm long, then inner hind tibial spur of female pectinate with one to several large teeth, teeth usually not so sloping or oblique as in the above (*Lasioglossum* series) 2

2(1). Females 3
—. Males (unknown for *Pseudochilalictus*) 9
3(2). Apical labral process triangular (Fig. 66-13a, b), widest at base, tapering to pointed apex, apex usually simple except for being keeled (two keels in a few *Parasphecodes*, process broadened preapically in a few *Ctenonomia*) (basitibial plates usually rather elongate and pointed apically)... 4
—. Apical labral process broadened so that it does not taper uniformly to a pointed apex (Fig. 66-13c-f), lateral margins often pectinate, toothed, elevated, or otherwise modified 7
4(3). Inner hind tibial spur finely serrate, ciliate, or essentially simple (Fig. 66-13a, b) 5
—. Inner hind tibial spur pectinate with a few coarse teeth (coarsely serrate in a few species).................................. 6
5(4). Metasomal terga black with broad basal bands of tomentum; teeth of inner margin of inner hind tibial spur about half as long as diameter of spur.......................... .... *L. (Pseudochilalictus)*
—. Metasomal terga red or black, metallic green in one species, without tomentum; teeth of inner hind tibial spur shorter .... *L. (Parasphecodes)*
6(4). Propodeum with triangle smooth, basal area not margined; thorax partially or wholly red or testaceous; metasomal terga without basal bands of tomentum .... *L. (Callalictus)*
—. Propodeum with triangle variously roughened, basal area often margined by a carina; thorax black; metasomal terga usually with tomentum basally or basolaterally .... *L. (Ctenonomia)*
7(3). Inner hind tibial spur of female pectinate; glossa two-thirds as long as face; margin of labral process of female simple (Fig. 66-13c) .... *L. (Glossalictus)*
—. Inner hind tibial spur of female not pectinate; glossa little more than one-half as long as face; margin of labral process of female variously modified 8
8(7). Inner hind tibial spur with a large tooth followed by a wavy margin (Figs. 66-13d, e), tooth in some cases reduced so that margin is almost simple, rarely a second tooth developed sufficiently that spur is almost pectinate; eyes usually bare, but in some cases with rather long hairs .... *L. (Chilalictus)*
—. Inner hind tibial spur very finely serrate or ciliate, thus essentially simple; eyes with a few short hairs .... *L. (Australictus)*
9(2). Basitibial plate present in all but a very few species; mostly small to moderate-sized species (4.5-9.0 mm long, uncommonly to 12 mm); eyes bare...................... 10
—. Basitibial plate absent; large to moderate-sized species (8-11 mm long, uncommonly only 5.5 mm); eyes (except in some *Callalictus*) with scattered, very short hairs 13
10(9). Glossa two-thirds as long as face or longer 11
—. Glossa about one-half as long as face 12
11(10). Glossa about two-thirds as long as face, which is broader than long .... *L. (Glossalictus)*
—. Glossa longer than face, which is longer than broad .... *L. (Ctenonomia)* (in part)
12(10). Gonostylus less than half as long as gonocoxite (males of Australian species not available) .... *L. (Ctenonomia)* (in part)

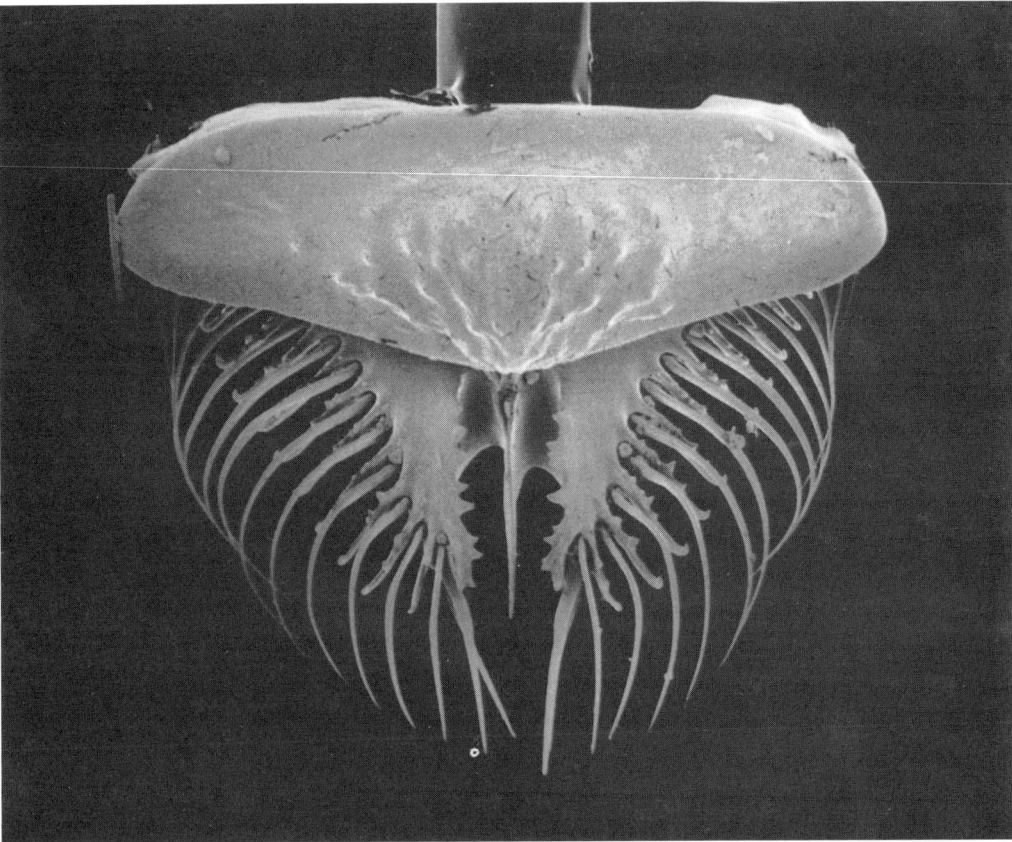

Figure 66-14. The most elaborate labral type among bees: the female of *Lasioglossum (Chilalictus) mirandum* (Cockerell). The function of the pectinations, or indeed of the whole labral process, is unknown. SEM photograph by Kenneth Walker.

—. Gonostylus more than half as long as gonocoxite (except in *L. idoneum* Cockerell), in some cases as long as gonocoxite .... *L. (Chilalictus)*
13(9). Gonostylus almost absent or at most short and triangular .... *L. (Australictus)*
—. Gonostylus distinct and longer than broad (metasomal terga without basal bands of tomentum) 14
14(13). Propodeal triangle smooth, basal area not margined; thorax partly or wholly red or testaceous .... *L. (Callalictus)*
—. Propodeal triangle roughened, basal area usually defined; thorax black .... *L. (Parasphecodes)*

*Lasioglossum / * Subgenus *Acanthalictus* Cockerell

Acanthalictus Cockerell, 1924a: 184. Type species: *Halictus dybowskii* Radoszkowski, 1877, by original designation.

The tridentate female mandibles are unique in the genus. The body is nonmetallic, 9 to 12 mm long. The inner hind tibial spur of the female is finely serrate. The metasoma lacks hair bands, but the bases of T2 and T3 have very scattered bits of white tomentum.

■ This subgenus is known only from Siberia. The two species names are likely to represent queens and workers of a single social species, *Lasioglossum (Acanthalictus) dybowskii* (Radoszkowski).

Acanthalictus is likely to be a derivative of *Evylaeus*; if so, it should be included in that subgenus.

*Lasioglossum / * Subgenus *Australictus* Michener

Lasioglossum (Australictus) Michener, 1965b: 165. Type species: *Halictus peraustralis* Cockerell, 1904, by original designation.

In its size (length 7-10 mm), form, long male antennae, serrate inner hind tibial spur of the female, and genital structure, this subgenus resembles *Parasphecodes*. It differs from that subgenus in the blunt or truncated apex of the apical labral process of the female, the basal tomentum (sometimes weak) on the metasomal terga, and the short, broad, scarcely recognizable male upper gonostylus. The lower gonostylus is a strong retrorse lobe, as in many related bees. The male genitalia, labrum, and other structures were illustrated by Michener (1965b).

■ This subgenus occurs from Tasmania and South Australia to northern Queensland. Eleven specific names are listed by Michener (1965b).

As explained by Michener (1965b), *Australictus* and *Parasphecodes* almost merge and there may be no reason to maintain *Australictus* as a distinct subgenus.

Lasioglossum / Subgenus *Austrevylaeus* Michener

Lasioglossum (Austrevylaeus) Michener, 1965b: 170. Type species: *Halictus sordidus* Smith, 1853, by original designation.

Typical members of this subgenus differ from *Evylaeus* and *Dialictus* in the ciliate hind tibial spurs of the female, the spurs more rarely serrate or weakly pectinate. They are finely punctate, rather dull, nonmetallic species, small in size (3.0-6.5 mm long), and in some of them the metasoma is red.

■ *Austrevylaeus* occurs in New Zealand and Tasmania to southern Queensland and Western Australia. Six described species are placed in this subgenus; 13 others are described in manuscript by K. L. Walker.

This subgenus, with the unplaced relatives mentioned below, is isolated by thousands of kilometers from the Asian and African ranges of other members of the *Hemihalictus* series, such as *Dialictus*. K. L. Walker's recent study reported that the Australian and New Zealand species fall into four groups, one of which, *Austrevylaeus* proper, is found only in New Zealand. The other three are Australian, and among them the hind tibial spurs of females distinguish *Austrevylaeus* from *Evylaeus* or *Dialictus* only with difficulty.

Lasioglossum / Subgenus *Callalictus* Michener

Lasioglossum (Callalictus) Michener, 1965b: 170. Type species: *Parasphecodes tooloomensis* Cockerell, 1929, by original designation.

This subgenus, like *Australictus,* is related to *Parasphecodes,* from which it differs in its shiny integument, including the smooth, undefined propodeal triangle; in the partly red or yellow thorax and sometimes metasoma; in the pectinate inner hind tibial spur of the female (found also in some small *Parasphecodes*); and in the moderately long male gonostylus, longer than in *Parasphecodes.* The body length of *Callalictus* ranges from 7.5 to 9.0 mm. The male genitalia, labrum, and other structures were illustrated by Michener (1965b).

■ *Callalictus* ranges from Victoria to southern Queensland, Australia. The eight specific names were listed by Michener (1965b).

Reddish or yellow coloration has resulted in confusion of some specimens of a new subgenus (to be described by K. W. Walker) with species of *Callalictus.*

Lasioglossum / Subgenus *Chilalictus* Michener

Lasioglossum (Chilalictus) Michener, 1965b: 174. Type species: *Halictus subinclinans* Cockerell, 1915 = *Halictus cognatus* Smith, 1853, by original designation.

The distinguishing feature of *Chilalictus* is the labral process of the female, which is always broad and is usually toothed or pectinate on its lateral margins (Fig. 66-13d-f). Sometimes it is extremely elaborate, as in Figure 66-14. The inner hind tibial spur of the female has a single median tooth followed by an undulate or almost straight margin (Fig. 66-13d, e), rarely, as in *Lasioglossum (Chilalictus) helichrysi* (Cockerell) (Fig. 66-13f), becoming serrate or, as in *L. seductum* (Cockerell), even toothless. Such spurs are also found in Africa in the *Oxyhalictus* group of the subgenus *Ctenonomia.* In appearance, the species of *Chilalictus* vary widely, from small, resembling some *Homalictus,* sometimes bright metallic green and sometimes with the metasoma red, to large, nonmetallic species that superficially resemble the subgenera *Evylaeus* or *Lasioglossum* s. str. The body length varies from 4.5 to 12.0 mm. The metasomal terga usually have basal bands or basolateral areas of pale tomentum. Male genitalia, the labrum of various species, and other structures were illustrated by Michener (1965b) and Walker (1995).

■ This subgenus is found throughout Australia, including Tasmania. One species, *Lasioglossum polygoni* (Cockerell), was collected in New Caledonia in 1927; otherwise, it is found in eastern Australia. This is the largest Australian group of Halictidae; Walker (1995) revised the subgenus and recognized 134 species.

As illustrated in Figure 4-3, at least one species of *Chilalictus, Lasioglossum hemichalceum* (Cockerell) [formerly sometimes identified as *L. erythrurum* (Cockerell)], has dimorphic males. In addition to the ordinary males that can be found on flowers, there are large, flightless males with allometric development of huge heads and jaws; such males do not leave the nests. See Section 4, Kukuk and Schwarz (1988), and Kukuk (1997). Remarkably similar dimorphism of males also occurs in at least two species of *Macrotera* (Panurginae, Perditini, Sec. 58).

Lasioglossum / Subgenus *Ctenonomia* Cameron

Ctenonomia Cameron, 1903b: 178. Type species: *Ctenonomia carinata* Cameron, 1903, monobasic.

Halictus (Nesohalictus) Crawford, 1910: 120. Type species: *Halictus robbii* Crawford, 1910 = *Nomia halictoides* Smith, 1858, by original designation.

Halictus (Oxyhalictus) Cockerell and Ireland, 1935, *in* Cockerell, 1935b: 91. Type species: *Halictus acuiferus* Cockerell and Ireland, 1935, monobasic.

Lasioglossum (Labrohalictus) Pauly, 1981a: 719. Type species: *Lasioglossum saegeri* Pauly, 1981, monobasic.

Lasioglossum (Rubrihalictus) Pauly, 1999b: 158. Type species: *Halictus rubricaudis* Cameron, 1905, by original designation. [**New synonymy.**]

Lasioglossum (Ipomalictus) Pauly, 1999b: 158. Type species: *Halictus nudatus* Benoist, 1962, by original designation. [**New synonymy.**]

As pointed out by Michener (1965b: 338), a group related to *Lasioglossum* s. str., found mostly south of the range of the latter, differs in its frequently small size (length 5-10 mm) and its pectinate inner hind tibial spurs. This is *Ctenonomia,* consisting of black or metallic greenish species with basal bands or basal lateral patches of white tomentum on the metasomal terga. The male upper gonostylus, although usually small, as in most species of the genus *Lasioglossum,* is sometimes more than half as long as the gonocoxite, slender to somewhat broad, or bifid; the lower gonostylus is present to absent, directed apicad to somewhat retrorse; see sketches by Pauly (1980c). Sakagami (1989, 1991) illustrated male genitalia and other structures.

■ *Ctenonomia* is the major group of *Lasioglossum* in Africa and tropical Asia. It occurs throughout sub-Saharan Africa from Senegal to Sudan and Ethiopia, south to

Natal Province in South Africa, east to Madagascar, Yemen, Iraq, Iran, India, Xansi Province of China, northern Honshu in Japan, the Philippines, and Taiwan. From southern Asia it ranges southward through Indonesia to New Guinea, Australia as far south as Victoria, and islets in the Indian Ocean such as the Seychelles. The number of species of *Ctenonomia* is large but unknown, because of the great proliferation of synonymous specific names, especially in Africa. There exist over 100 names for African species with pectinate hind tibial spurs and over 50 for species with more or less unidentate spurs. For Asia and nearby islands there are over 40 names, all for species with pectinate spurs. The named Australian *Ctenonomia* species are six, listed by Michener (1965b: 172, paragraph 3), but K. Walker (in manuscript) reports approximately 20 Australian species. Pauly (1980c) revised the metallic species of Africa and the Arabian peninsula, and Sakagami (1989) gave a key to southern Asiatic subgroups of *Ctenonomia* excluding *Nesohalictus*. Pauly (1984b) revised the species of Madagascar.

The name *Nesohalictus* can be applied to a group of three species from southern Asia and Indonesia that have an elongate face, a glossa about as long as the head, and reduced branching of the scopal hairs of the hind trochanter and femur. They appear to be oligolectic on *Hibiscus* (Malvaceae). See Sakagami (1991) for an account of *Nesohalictus*. He indicates that another character of the group is the comb on the anterior basitarsus of the female that extends to the base of the segment. This comb is also found in some other *Ctenonomia* species, however. The front basitarsus is often produced beyond the base of the second tarsal segment, and the comb continues to the apex of this projection.

The name *Oxyhalictus* can be applied to a large African group that agrees with the characters listed above for *Nesohalictus*, except that the face is not elongate. *Oxyhalictus* differs from both typical *Ctenonomia* and *Nesohalictus* in having a single large tooth, followed by an undulate margin, or by two or three short, rounded teeth, on the inner hind tibial spur of the female, as in the Australian subgenus *Chilalictus*. This character, and especially the tongue length, intergrade with those of more ordinary *Ctenonomia*. For this reason I have included *Oxyhalictus* in *Ctenonomia*, although typical members of the two groups are quite different. Male genitalia seem to be variable in both groups; the dorsal gonostyli can be small and simple to rather large and bifid, and the ventral gonostyli can be large and retrorse to small or absent in both groups. Clearly, the subgenus needs revisional study; changes in classification are likely to result.

The subgeneric names *Rubrihalictus* and *Ipomalictus* were proposed by Pauly (1999b) for groups of species that I have placed within the subgenus *Ctenonomia* of *Lasioglossum*. They represent African groups of considerable size (25 and nearly 40 species, respectively). If recognized as subgenera, *Nesohalictus* and *Oxyhalictus* should also have subgeneric rank, as indicated by Pauly (1999b). Before accepting such an arrangement, the Oriental to Australian species of *Ctenonomia* and *Nesohalictus* should be compared carefully with the large African fauna.

Pauly (1999b) listed the nearly 200 African species of *Ctenonomia* under the subgeneric names *Ctenonomia*, *Oxyhalictus*, *Rubrihalictus*, and *Ipomalictus*. Species from Madagascar allocated to the same groups were revised by Pauly (in Pauly et al., 2001). Pauly (2001) described a species from southern India in the otherwise African group *Ipomalictus*.

The nests of *Ctenonomia* are little known. At least one species, *Lasioglossum albescens* (Smith), is unusual among Halictini in that cells in the branch burrows are arranged in series, end to end (Matsumura and Sakagami, 1971).

Lasioglossum / Subgenus *Dialictus* Robertson

Paralictus Robertson, 1901: 229. Type species: *Halictus cephalicus* Robertson, 1892 (not Morawitz, 1873) = *Halictus cephalotes* Dalla Torre, 1896, by original designation.

Dialictus Robertson, 1902d (Feb. 1): 48. Type species: *Halictus anomalus* Robertson, 1892, by original designation.

Chloralictus Robertson, 1902c (Sept. 10): 245, 248. Type species: *Halictus cressoni* Robertson, 1890, by original designation.

Halictus (Gastrohalictus) Ducke, 1902: 102. Type species: *Halictus osmioides* Ducke, 1902, monobasic.

Halictomorpha Schrottky, 1911: 81. Type species: *Halictomorpha phaedra* Schrottky, 1911, by original designation.

Prosopalictus Strand, 1913b: 26. Type species: *Prosopalictus micans* Strand, 1913 (not *Halictus micans* Strand, 1909, senior homonym in *Lasioglossum*) = *Lasioglossum micante* Michener, 1993, by original designation.

Rhynchalictus Moure, 1947a: 5. Type species: *Rhynchalictus rostratus* Moure, 1947, by original designation.

Halictus (Microhalictus) Warncke, 1975b: 85. Type species: *Melitta minutissima* Kirby, 1802, by original designation.

Halictus (Puncthalictus) Warncke, 1975b: 87. Type species: *Hylaeus punctatissimus* Schenck, 1853, by original designation.

Halictus (Rostrohalictus) Warncke, 1975b: 88. Type species: *Halictus longirostris* Morawitz, 1876, by original designation.

Halictus (Smeathhalictus) Warncke, 1975b: 88. Type species: *Melitta smeathmanella* Kirby, 1802, by original designation.

Halictus (Marghalictus) Warncke, 1975b: 95. Type species: *Hylaeus marginellus* Schenck, 1853, by original designation.

Halictus (Pyghalictus) Warncke, 1975b: 103. Type species: *Andrena pygmaea* Fabricius, 1804, by original designation. [*Pyghalictus* was intended for the group of what is usually called *Lasioglossum politum* (Schenck). The identity of *Andrena pygmaea* Fabricius is in doubt (see Ebmer, 1988b).]

Halictus (Pauphalictus) Warncke, 1981: 87. Type species: *Halictus pauperatus* Brullé, 1832, by original designation.

Habralictellus Moure and Hurd, 1982: 46. Type species: *Halictus auratus* Ashmead, 1900, by original designation.

Lasioglossum (Afrodialictus) Pauly, 1984b: 142. Type species: *Halictus bellulus* Vachal, 1909, by original designation.

Lasioglossum (Mediocralictus) Pauly, 1984b: 143. Type species: *Halictus mediocris* Benoist, 1962, by original designation.

As shown in the above synonymy, the oldest name for this subgenus is *Paralictus*, a name proposed for a parasitic North American species and since applied to four other uncommon parasitic species. In order to avoid changing the names of hundreds of species, including much-studied social species, from *Dialictus* to *Paralictus*, the Commission (Opinion 1882, 1997) has given *Dialictus* (and *Chloralictus*) precedence over *Paralictus* (Michener, 1995c). Thus whenever *Dialictus* (or *Chloralictus*) is used for the same taxon as *Paralictus, Dialictus* (or *Chloralictus*) is to be used instead of *Paralictus*.

Dialictus consists of mostly small (body length 3.5-8.0 mm), black, greenish, or (in the Antilles) bright-green bees; sometimes the metasoma is red. The metasoma is not conspicuously banded (Fig. 66-7), although in a few palearctic and Oriental species females have apical hair bands; basal pale tomentum may be evident, but usually does not form defined bands or basolateral patches and sometimes is spread over whole tergal surfaces. The hind tibial spurs of the female are pectinate with few large teeth. Species with only two submarginal cells occur among the greenish forms (*Dialictus* sensu strictissimo), among the bright-green *Habralictellus* group, as well as among black species (at least *Prosopalictus*), but the vast majority of the species have three cells.

■ *Dialictus* occurs from Alaska and central Canada through the whole Western Hemisphere to Chile (Concepción) and Argentina, including the Antilles and even Bermuda and Juan Fernandez; it also occurs in the palearctic area from the Azores to Japan and north into Finland; southward it ranges to Cape Province, South Africa, and to Madagascar, India, and Taiwan. *Lasioglossum (Dialictus) impavidum* (Sandhouse) from California is now widespread in Hawaii (Snelling, 2003). *Dialictus* is especially common in the holarctic region, but there are many neotropical species; it is relatively scarce in the African and Oriental regions. There are about 340 species in the Western Hemisphere; many of these are probably synonyms but there remain many undescribed species. In the palearctic region, where much of the synonymy is known and many forms are relegated to the subspecies level, there appear to be about 60 species. The number of oriental and African species is less, perhaps 50. Under names here tentatively included in *Dialictus*, Pauly (1984b) listed 15 Malagasy species. These figures are guesses because the line between *Dialictus* and *Evylaeus* remains indefinite. Ebmer (2000) reviewed the group of *Lasioglossum pauperatum* (Brullé) and revised its Asiatic species. Keys to species and revisional works are cited in the account of the tribe Halictini, since most of them include other groups of *Halictus* and *Lasioglossum*. Sandhouse (1923, 1924) gave keys to many North American species of *Dialictus;* for the eastern USA these were replaced by Mitchell (1960), but the species remain difficult to identify.

A major component in the American *Dialictus* is the many greenish species. In the older literature they were mostly placed in *Chloralictus*, which was distinguished from *Dialictus* only by having three instead of two submarginal cells; *Dialictus* has priority over *Chloralictus*, although both were proposed in 1902. Still older names that may ultimately be applicable to this group are *Sudila* (1898) and *Hemihalictus* (1897). I have retained these as separate subgenera for the present in order to avoid premature name changes that might soon be altered again. At least the generic name *Lasioglossum* is not changed; an important reason for retaining *Lasioglossum* for the present as an umbrella genus for its various subgenera is to retain stability of the genus-species combinations until a definitive worldwide analysis and classification can be made. See the discussion above under the genus *Lasioglossum*.

In addition to the large American group of greenish species mentioned in the preceding paragraph and the similar but less numerous palearctic forms, *Dialictus* includes many nonmetallic, mostly small species that have been referred to as noncarinate *Evylaeus*. So far as known group characters go, they differ from greenish *Dialictus* only in color. Moreover, as explained elsewhere, they grade imperceptibly into greenish species, especially in the neotropics and the Eastern Hemisphere. *Prosopalictus* contains such a small, black species, with, however, only two submarginal cells; its characters were reported in detail by Michener (1993). *Habralictellus* includes a few bright-green Antillean species, superficially suggestive of *Augochlorella* in the Augochlorini. Pauly (1999b) placed the African groups which are here retained in *Lasioglossum (Dialictus)* either in the subgenus *Afrodialictus* (about 35 species) or in *Mediocralictus* (2 species). Although these forms are here included in *Dialictus*, recognition of Pauly's groups as subgenera seems desirable. *Mediocralictus* seems to fall among the black *Dialictus*, although it is an African group. *Afrodialictus* has distinctive fine, dull sculpturing on the dorsum of the thorax, at least of females; it is interesting that both black and greenish species fall clearly into this group. *Rostrohalictus* is the name given for a species with an extraordinarily elongate head, the malar area of the male being as long as the eye; Popov (1959b) illustrated the head of both sexes as well as the male genitalia and other structures. A lesser degree of head elongation also occurs in some related species placed by Warncke (1975b) in *Puncthalictus*, and in unrelated species such as that placed in *Rhynchalictus*. *Sudila*, in the female sex, is apparently a large, black *Dialictus;* I have retained *Sudila* as a subgenus, however, partly for nomenclatural reasons (it is senior to *Dialictus*), and because of its extraordinary males.

Lasioglossum / Subgenus *Eickwortia* McGinley

Eickwortia McGinley, 1999: 112. Type species: *Halictus nycteris* Vachal, 1904, by original designation.

Eickwortia can be recognized by the *Evylaeus-Dialictus*-type wing venation combined with the strongly bifid female mandibles and the small male gonobase, as indicated in the key to subgenera. The wings, or at least the anterior marginal areas of the forewings, are heavily infuscated. The propodeum lacks strong carinae such as are found in *Evylaeus*. Body length is 6 to 11 mm. In body size and form, with the anteriorly tapering T1, females of the type species so resemble *Neocorynura* (Augochlorini) that several bee specialists misplaced specimens in that genus on superficial examination.

■ This subgenus occurs on plateaus and mountains from San Luis Potosí and Nayarit, Mexico, to Costa Rica. The two species were distinguished by McGinley (1999).

Lasioglossum / Subgenus *Evylaeus* Robertson

Evylaeus Robertson, 1902c: 247. Type species: *Halictus arcuatus* Robertson, 1893, by original designation.

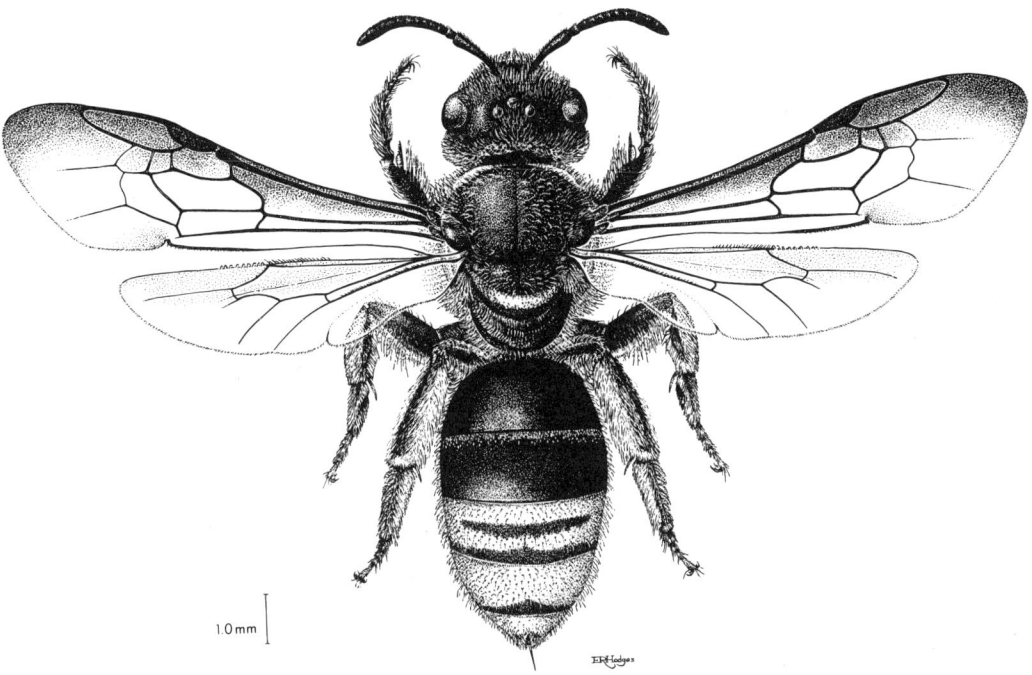

Figure 66-15. Female of *Lasioglossum (Lasioglossum) crocoturum* (Vachal). Drawing by E. R. S. Hodges, from Michener, McGinley, and Danforth, 1994.

Halictus (Calchalictus) Warncke, 1975b: 99. Type species: *Apis calceata* Scopoli, 1763, by original designation.

Halictus (Inhalictus) Warncke, 1975b: 96. Type species: *Hylaeus interruptus* Panzer, 1798, by original designation.

As indicated in the discussion of the genus *Lasioglossum*, a clear distinction between the subgenera *Evylaeus* and *Dialictus* cannot be made, at least for the present. *Evylaeus* as here understood is limited to the "carinate group" of earlier authors. Ebmer (1995), in particular, characterized the group. Additional support for recognition of *Evylaeus* in this sense is provided by the male genitalia, as indicated by L. Packer (1991 and in litt., 1995). The inner dorsal margins of the gonocoxites are parallel in the basal one-third to over one-half of the gonocoxal length. At the end of the parallel region is a strong angle, distal to which the gonocoxal margins diverge strongly and are broadly concave. In other subgenera the angle, if present, is nearer to the base of the gonocoxite and the margin distal to the angle is usually not or gently concave, the main axes of the gonocoxites thus seemingly more divergent than in *Evylaeus*. These characters are shown in the series of plates by Do-Pham, Plateaux-Quénu, and Plateaux (1984).

■ *Evylaeus* ranges throughout the palearctic region from the Azores and Britain to Japan, and from Scandinavia (as far as 68°N) to North Africa and northern India; in Africa to Sudan and Sokotra Island, Yemen. [Pauly (1999b) also listed the group of *L. schubotzi* (Strand) from the Afrotropical region]; in North America from coast to coast, Alaska to Nova Scotia south to Baja California (Mexico) and Florida, USA. Warncke (1981) listed 32 species from the western palearctic, and Ebmer (1995) included over 100 palearctic species. At least about 10 of the 68 North American species listed in *Evylaeus* by Hurd (1979) and of the 80 species listed by Moure and Hurd (1987) belong in this subgenus, but the majority of species listed as *Evylaeus* in these catalogues are nonmetallic species of *Dialictus*.

Packer (1991) made a phylogenetic analysis of eight species of *Evylaeus* based on allozymes, and mapped on this phylogeny what appear to be behavioral changes in the evolution of the group. One species, *Lasioglossum fulvicorne* (Kirby), was the sister group to all the rest and is nonsocial. Since the outgroup, *Dialictus*, contains both solitary and social species, however, *fulvicorne* is not necessarily ancestral in this character, although one can spec-

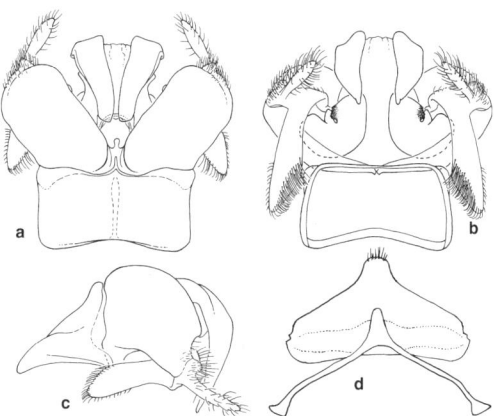

Figure 66-16. Male genitalia and hidden sterna of *Lasioglossum (Lasioglossum) costale* (Vachal). **a-c,** Dorsal, ventral, and lateral views of genitalia, showing the retrorse lobe (reflexed ventral gonostylus); **d,** S7 and S8. From McGinley, 1986.

ulate that social behavior arose within *Evylaeus*. It is a labile character, however, for within species social behavior can be lost, as has occurred in northern high-altitude populations of *L. (Evylaeus) calceatum* (Scopoli). See also Packer (1997), who emphasizes cases of loss of eusocial behavior, and especially Danforth (2002).

Lasioglossum / Subgenus *Glossalictus* Michener

Lasioglossum (Glossalictus) Michener, 1965b: 173. Type species: *Halictus etheridgei* Cockerell, 1916, by original designation.

Glossalictus should perhaps be included within *Ctenonomia*, a subgenus that has long-tongued representatives both in Asia and in Africa. In addition to the characters indicated in the key, *Glossalictus* differs from *Ctenonomia* in the broad labral process of the female (Michener, 1965b: fig. 588), which lacks the tubercle at each side that is frequent in *Ctenonomia* (Fig. 66-13c).

■ *Glossalictus* is found in Western Australia. There is only one species, *Lasioglossum etheridgei* (Cockerell).

Lasioglossum / Subgenus *Hemihalictus* Cockerell

Hemihalictus Cockerell, 1897a: 288 [also proposed as new by Cockerell, 1898c: 216]. Type species: *Panurgus lustrans* Cockerell, 1897, by original designation.

This subgenus contains a single nonmetallic species, with body length 5.5 to 7.5 mm, that has two submarginal cells (Fig. 66-1b). It differs from the subgenus *Dialictus* in having serrate rather than pectinate inner hind tibial spurs in females. The same character distinguishes it from most species of the subgenus *Evylaeus*. The male is robust and has short antennae, thus resembling a female. Placement of the lone species in its own subgenus thus appears to be legitimate.

■ *Hemihalictus* occurs from Texas, New Mexico, and Kansas east to Michigan, Virginia, and Florida, USA. The only species is *Lasioglossum lustrans* (Cockerell).

Lasioglossum lustrans (Cockerell) is oligolectic on flowers of *Pyrrhopappus* (Asteraceae). It is solitary; its nesting biology was described by Daly (1961).

Except for *Parasphecodes*, *Hemihalictus* is the oldest name among the segregates of *Lasioglossum*. It must therefore be considered as a possible generic name when *Lasioglossum* is broken up.

Lasioglossum / Subgenus *Lasioglossum* Curtis s. str.

Lasioglossum Curtis, 1833: pl. 448. Type species: *Lasioglossum tricingulum* Curtis, 1833 = *Melitta xanthopus* Kirby, 1802, by original designation.
Halictus (Lucasius) Dours, 1872: 350 (not Kinahan, 1859). Type species: *Halictus clavipes* Dours, 1872, by designation of Sandhouse, 1943: 566.
Halictus (Lucasiellus) Cockerell, 1905d: 272, replacement for *Lucasius* Dours, 1872. Type species: *Halictus clavipes* Dours, 1872, autobasic.
Halictus (Lucasellus) Schulz, 1911: 202, replacement for *Lucasius* Dours, 1872. Type species: *Halictus clavipes* Dours, 1872, autobasic.
Curtisapis Robertson, 1918: 91. Type species: *Halictus coriaceus* Smith, 1853, by original designation.

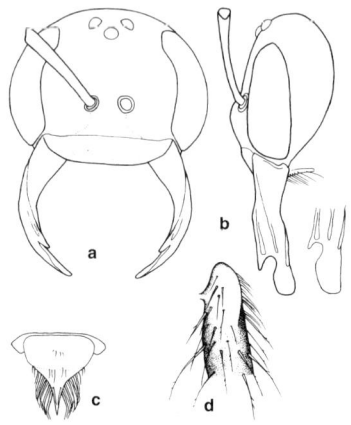

Figure 66-17. Female holotype of *Lasioglossum (Paradialictus) synavei* (Pauly), a presumably cleptoparasitic halictine. **a, b,** Front and lateral views of head; **c,** Labrum; **d,** Base of hind tibia, showing lack of basitibial plate. From Arduser and Michener, 1987.

Halictus (Pallhalictus) Warncke, 1975b: 92. Type species: *Halictus pallens* Brullé, 1832, by original designation.
Halictus (Fahrhalictus) Warncke, 1975b: 95. Type species: *Halictus fahringeri* Friese, 1921, by original designation.
Halictus (Leuchalictus) Warncke, 1975b: 98. Type species: *Apis leucozonia* Schrank, 1781, by original designation.
Lasioglossum (Lophalictus) Pesenko, 1986a: 125. Type species: *Lasioglossum acuticrista* Pesenko, 1986, by original designation.
Lasioglossum (Bluethgenia) Pesenko, 1986a: 136. Type species: *Halictus dynastes* Bingham, 1898, by original designation.
Lasioglossum (Ebmeria) Pesenko, 1986a: 136. Type species: *Halictus costulatus* Kriechbaumer, 1873, by original designation.
Lasioglossum (Sericohalictus) Pesenko, 1986a: 137. Type species: *Halictus subopacus* Smith, 1853, by original designation.

This is a holarctic subgenus of relatively large species (body length 7-12 mm) differing from other Eurasian and African subgenera except the paleotropical *Ctenonomia* in its strong second submarginal crossvein (Fig. 66-6a), a character recognizable at least in females. It differs from *Ctenonomia* in the form of the inner hind tibial spur of the female, which is serrate or rarely pectinate with five or more teeth; in *Ctenonomia* the number of teeth is smaller, sometimes only one. Figure 66-15, illustrates the habitus of the subgenus; Figure 66-16 illustrates the genitalia and hidden sterna, which are representative of much of the genus.

■ *Lasioglossum* s. str. is the major group of the *Lasioglossum* series of subgenera in the holarctic region. It occurs from the Atlantic islands off Europe and North Africa, east to Japan, south to northern India, and north to central Finland. In North America it ranges from coast to coast, and from Northwest Territories, Canada, south to Panama; it is not known in the Antilles. There are about 111 palearctic species and 51 in the New World. Pesenko (1986a) gave a key to females for the entire palearctic re-

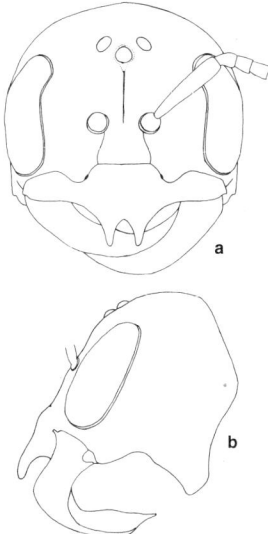

Figure 66-18. Head of a macrocephalic male of *Lasioglossum (Sudila) bidentatum* (Cameron) from Sri Lanka. **a,** Facial view; **b,** Lateral view.

gion. McGinley (1986) revised the species of the nearctic region.

This is a large subgenus as here constituted, and the groups based on females and segregated as subgenera by Pesenko (1986a), or some of them, may advantageously be recognized. I have not attempted to evaluate them because such a classification should consider characters of males and should be applicable to the whole subgenus, not just the palearctic part of it. The American species should be studied with regard to the palearctic classification to see which of the named groups also occur in North America, whether the groups are distinct or merge there, and in general whether division of the subgenus seems justifiable.

Lasioglossum / Subgenus *Paradialictus* Pauly

Paradialictus Pauly, 1984c: 691. Type species: *Paradialictus synavei* Pauly, 1984, by original designation.

This subgenus includes a small (body length 5.5 mm), nonmetallic, presumably cleptoparasitic species, perhaps derived from the nonparasitic group of *Dialictus* called *Afrodialictus* Pauly, but it is much more different from such probable ancestral types than are the parasitic species of *Dialictus*, which are here placed in the same subgenus as nonparasites, although formerly segregated as *Paralictus*. Particularly distinctive features of *Paradialictus* (all based on the female) are the very large, blunt mandible with a large preapical tooth (Fig. 66-17a, b), the flat labrum drawn out into an unkeeled apical spine (Fig. 66-17c), the broad clypeus (about five times as wide as long), the lack of a defined pygidial plate, and other usual features of cleptoparasitic halictines. Details of structures were discussed by Arduser and Michener (1987).

■ *Paradialictus* is known only from Zaire. A single female specimen is known; the species is *Lasioglossum synavei* (Pauly).

Lasioglossum / Subgenus *Parasphecodes* Smith

Parasphecodes Smith, 1853: 39. Type species: *Parasphecodes hilactus* Smith, 1853, by designation of Sandhouse, 1943: 585.

Parasphecodes (Aphalictus) Cockerell, 1930e: 40. Type species: *Parasphecodes bribiensis* Cockerell, 1916, by original designation.

This subgenus consists of mostly large species, although the range in body length, 5.5 to 11.0 mm, includes small species. In many species the metasoma is red, and all of them lack metallic coloration except for the largely bright-green *Lasioglossum permetallicum* Michener from New Guinea. The male of this species is unknown, and it is quite possible that it is not a *Parasphecodes*. The labrum of the female tapers to a point or rarely to a bidentate apex; sometimes there is a tubercle on each side near the base of the process (Fig. 66-13a). The inner hind tibial spur is ciliate to finely serrate (Fig. 66-13a), rarely (in small species) coarsely serrate or pectinate. The metasoma lacks tomentum. The male genitalia, labrum, and other structures were illustrated by Michener (1965b).

■ *Parasphecodes* ranges from Tasmania through eastern Australia to northern Queensland and northwestern Australia; one species, *Lasioglossum permetallicum* Michener (see above), occurs in New Guinea. The 99 specific names were listed by Michener (1965b). Meyer (1919b) gave a key to certain species and German translations of specific descriptions.

Lasioglossum / Subgenus *Pseudochilalictus* Michener

Lasioglossum (Pseudochilalictus) Michener, 1965b: 170. Type species: *Lasioglossum imitator* Michener, 1965, by original designation.

As indicated by the name, bees of this subgenus have exactly the aspect of the larger nonmetallic species of *Chilalictus*, 8 to 9 mm long. Nonetheless, *Pseudochilalictus* agrees with *Parasphecodes* in most of its characters, such as the tapering labral process of the female (Fig. 66-13b), the pointed basitibial plate, and the finely toothed, i.e., ciliate, inner hind tibial spur (Fig. 66-13b). Distinguishing it from *Parasphecodes* are the conspicuous basal bands of tomentum on T2 to T4.

■ This subgenus occurs in the Dividing Range of southern Queensland and northern New South Wales, Australia. There is one species, *Lasioglossum imitator* Michener; its male has not been described.

Lasioglossum / Subgenus *Sellalictus* Pauly

Lasioglossum (Sellalictus) Pauly, 1980a: 120. Type species: *Halictus latesellatus* Cockerell, 1937, by original designation.

This subgenus contains small (length 5-7 mm), nonmetallic species. Females have strong basolateral patches of tergal pubescence, and differ from superficially similar species of *Ctenonomia* in that these patches do not unite to form continuous basal bands. T2 of males has a unique character, a broad basal zone of rather long, erect, white hairs, very different from the tomentum of some taxa, and limited to T2. The apex of the marginal cell is minutely sep-

arated from the wing margin and minutely appendiculate.
- *Sellalictus* occurs from Zaire to South Africa (Natal and Cape Province) and Madagascar. The species were listed by Pauly (1984b), and again, 35 species, by Pauly (1999b).

Lasioglossum / Subgenus *Sphecodogastra* Ashmead

Sphecodogastra Ashmead, 1899a: 92. Type species: *Sphecodes texana* Cresson, 1872, by original designation.

The reduced scopa described in the key is related to the use of the webbed pollen of Onagraceae by this subgenus. The smaller species of this subgenus have ocelli of ordinary size, but the two larger species have enlarged ocelli. Species of *Sphecodogastra* are 7 to 11 mm long, nonmetallic, the metasoma sometimes red.
- *Sphecodogastra* occurs from Ontario (where it is dependent on cultivated plants and may be adventive; Knerer and MacKay, 1969) to Alberta, Canada, and Washington state south to New York state, Indiana, Kansas, Texas, and California, USA, and Coahuila, Veracruz, and Chihuahua, Mexico. The three smaller ones were listed in *Evylaeus* by Moure and Hurd, 1987. McGinley (2003) revised the species of the subgenus *Sphecodogastra* (as a genus), recognizing eight species.

Accounts of the morning and evening foraging, as well as of the nesting behavior of the smaller species, were by Knerer and MacKay (1969) and Bohart and Youssef (1976); an account of one of the larger, more nocturnal species with enlarged ocelli, *Lasioglossum texanum* (Cresson), was by Kerfoot (1967). Kerfoot found this species active at dusk and through moonlit nights. All species collect pollen only or principally from Onagraceae.

Lasioglossum / Subgenus *Sudila* Cameron

Sudila Cameron, 1898: 52. Type species: *Sudila bidentata* Cameron, 1898, by designation of Sandhouse, 1943: 602.
Ceylonicola Friese, 1918: 501. Type species: *Ceylonicola atra* Friese, 1918 = *Sudila bidentata* Cameron, 1898, by designation of Sandhouse, 1943: 536.

Females of *Sudila* are morphologically like large (body length 10-11 mm), black, nonmetallic or slightly greenish *Dialictus* except for the mesepisternal tubercle (the subpleural signum, a minute structure shown in front of the middle coxa in Fig. 10-5a), which is strongest in females but absent in both sexes of *Lasioglossum (Sudila) paralphenum* Sakagami, Ebmer, and Tadauchi. There is almost no pale metasomal pubescence. In at least some species of *Sudila* some males have the head enormous, the mandibles extraordinarily long and pointed, the genal areas triangularly produced, and the clypeal apex strongly bidentate (Fig. 66-18). In other, smaller males of the same species the head is quite ordinary, the clypeus truncate, and the genal area little wider than the eye. Intermediates exist (Sakagami, Ebmer, and Tadauchi, 1996). Thus these male characters, which Cameron and Friese used to justify giving generic names to this group, do not characterize all individuals of any species. The pectinate inner hind tibial spur of both sexes, however, appears to be a useful character of males.
- *Sudila* is known from mountains of Sri Lanka, southern India, and Java. The six species were revised by Sakagami, Ebmer, and Tadauchi (1996).

Sudila appears to consist of extremely modified representatives of *Lasioglossum (Dialictus)*. The older name, *Sudila*, would therefore replace the name *Dialictus* in a potential fusion of these two groups. Because the names of hundreds of species are involved, such decisions should be delayed until phylogenetic hypotheses and a stable classification have been devised. It is likely that the great variation seen in males of some species of *Sudila* is allometric and should not be evaluated as several independent characters.

Genus *Mexalictus* Eickwort

Mexalictus Eickwort, 1978: 567. Type species: *Mexalictus micheneri* Eickwort, 1978, by original designation.
Mexalictus (Georgealictus) Packer, 1993a: 1656. Type species: *Mexalictus polybioides* Packer, 1993, by original designation.

This is a genus of slender, weakly greenish (like many *Dialictus*) or nonmetallic bees, 5 to 11 mm long, that are similar in superficial appearance to some species of *Lasioglossum (Evylaeus* or *Dialictus)* although more slender. It differs from *Lasioglossum* not only in the strong distal wing venation but also in the relatively large, bifid male dorsal gonostylus. The ventral gonostylus is small and not retrorse. A unique feature of the genus is the presence of four long basal projections, presumably apodemal, on S8 of the male. The male genitalia and other structures were illustrated by Eickwort (1978) and Packer (1993).
- This genus occurs in mountains from southern Arizona, USA, to Guatemala; R. Brooks recently collected specimens in Panama. There are five described species and at least three others.

Genus *Microsphecodes* Eickwort and Stage

Sphecodes (Microsphecodes) Eickwort and Stage, 1972: 501. Type species: *Sphecodes kathleenae* Eickwort, 1972, by original designation.

Microsphecodes may be a specialized derivative of the *Austrosphecodes* group of the cleptoparasitic genus *Sphecodes*. Until such a relationship is demonstrated, I have decided to follow Michener (1978b) in giving *Microsphecodes* generic status. It consists of small species, body length 3.5 to 6.0 mm, with more or less extensive yellowish areas on the head and thorax. The venational character (see the key and Fig. 66-5a) is strong. The apical margin of T5 of the female is bare, like the preceding terga, and thus unlike that of nearly all species in the related cleptoparasitic genera. The mandibles of the female are simple, as in *Eupetersia, Ptilocleptis,* and certain *Sphecodes*. But in contrast with these genera, the anterior end of the scutum is gently convex in profile, or strongly so immediately adjacent to the pronotum, and lacks the larger, more or less vertical, less punctate anterior surface found in the other genera.
- *Microsphecodes* occurs from Dominica and St. Kitts in the Lesser Antilles and Costa Rica to Paraná, Brazil. There are eight named species.

Known hosts are *Lasioglossum (Dialictus)* and *Habral-*

ictus (Michener, Breed, and Bell, 1979). Females of *Microsphecodes kathleenae* (Eickwort) appear to enter nests of their host, *Lasioglossum (Dialictus) umbripenne* (Ellis), and remain there, perhaps as social parasites (Eickwort and Eickwort, 1972b), like certain species of *Sphecodes* (Sec. 8).

Genus *Nesosphecodes* Engel

Nesosphecodes Engel. 2006b: 2. Type species: *Nesosphecodes anthracineus* Engel, 2006, by original designation.

Although differentiated at some length from *Sphecodes* in the key to the Genera of Halictini of the Western Hemisphere, *Nesosphecodes* is a close relative of *Sphecodes* and may well make that genus paraphyletic. *Nesospecodes* consists of black species, without the red metasoma common at least in females of *Sphecodes*. The body length is 7 to 9 mm; thus this genus consists of larger forms than *Microsphecodes,* another close relative. In fact its wing characters are intermediate between those of *Sphecodes* and *Microsphecodes* and its convex rather than abruptly declivous anterior margin of the scutum is as in *Microsphecodes*.

■ *Nesosphecodes* is known from Cuba, Hispaniola, and Puerto Rico. The three species are differentiated by Engel (2006b).

Genus *Paragapostemon* Vachal

Paragapostemon Vachal, 1903: 89. Type species: *Halictus podager* Vachal, 1903 = *Nomia coelestina* Westwood, 1875, by designation of Cockerell, 1905a: 354 footnote.

The one species of this genus averages smaller than do species of *Dinagapostemon* (body 9-13 mm long), and the entire body is bright green or blue, unlike that of *Dinagapostemon*. The male antennal flagellum is short and simple; the hind legs are swollen, with a femoral and basitarsal tooth, and the legs are largely yellow. See the comments on similarity to *Dinagapostemon* in the discussion of that genus.

■ *Paragapostemon* inhabits the Mexican plateau from the states of Jalisco, Durango, and San Luis Potosí south to Oaxaca. There is probably only one species, *Paragapostemon coelestinus* (Westwood); see Roberts and Brooks (1987).

Genus *Parathrincostoma* Blüthgen

Parathrincostoma Blüthgen, 1933: 389. Type species: *Parathrincostoma seyrigi* Blüthgen, 1933, by original designation.

To judge by its lack of a scopa, this is a cleptoparasitic genus, probably parasitic on Malagasy species of *Thrinchostoma*, as suggested by its size (length 11-14 mm). Interestingly, because of probably plesiomorphic characters, it is most similar to the subgenus *Eothrincostoma* of *Thrinchostoma*, a subgenus not known from Madagascar. *Parathrincostoma* differs from *Thrinchostoma* in the lack of laterally directed hairs on the depressed tergal marginal zones, the weakly defined basitibial plate of the female, the lack of a median, specialized, largely bare area on T5 of the female, and the unmodified S4 and S5 of the male.

This is the only parasitic halictine with well-developed paraocular lobes extending down into the clypeus. Many structures were described and illustrated by Michener (1978b).

■ *Parathrincostoma* is known only from Madagascar. There are two species.

Genus *Patellapis* Friese

The generic name *Patellapis* is here used in a much broader sense than traditionally. Michener (1978a) expanded the prior application of the name to include a rather diverse group of African halictines and commented that the group then called the genus *Zonalictus* "could easily be incorporated" into *Patellapis*. It now appears that not only is *Zonalictus* morphologically much like *Patellapis,* but also that among African groups intergradations between *Patellapis* s. str. and *Pachyhalictus* exist; it is thus not desirable to separate these groups at the genus level. Michener (1978a) had already commented that the distinctive reticulate sculpturing of *Pachyhalictus* occurs also in a few species of *Patellapis* and *Zonalictus.*

Patellapis consists of nonmetallic species, the metasoma sometimes red, and in the subgenus *Zonalictus* usually with white, yellow, greenish, or bluish apical tergal integumental bands suggestive of *Nomia* (Nomiinae). There is variation among the subgenera in characters often considered of generic importance elsewhere in the Halictinae. For example, some groups have apical tergal hair bands, some have basal tomentum or bands, and some lack tergal hair bands. Ordinarily in *Patellapis* the paraocular lobe into the clypeus is absent or obtuse, the epistomal suture not making a strong bend near the tentorial pit, but in *P. braunsella* Michener the lobe is strong; this characteristic is presumably associated with the long head, elongate glossa, and protuberant clypeus of that species. The dorsal gonostylus of the male is bifid (see illustrations by Ireland, 1935), but in *P. suarezensis* (Benoist), to judge by Pauly's (1984b) illustration, it is small and simple, as is usual in *Lasioglossum*. Thus that species may be out of place in *Patellapis*. One of the problems with the classification of *Patellapis* as here understood is that the males are known for only a small percentage of the species. Discovery of males for more species could revolutionize the classification. *Patellapis* is paraphyletic if *Homalictus* is derived from it (see the discussion of *Homalictus*). It is possible that couplet 1 of the key

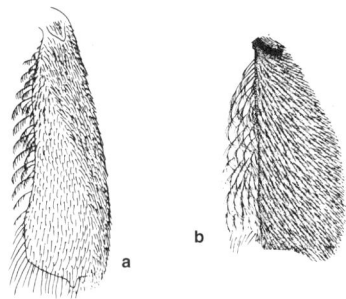

Figure 66-19. Hind tibial vestiture seen laterally in females of two halictine genera. **a**, *Patellapis (Pachyhalictus) merescens* (Cockerell); **b**, *Lasioglossum (Lasioglossum) malachurum* (Kirby).

to subgenera below separates groups that should have generic status, but characters distinguishing the males are unknown.

A feature especially noticeable in the subgenus *Pachyhalictus* is the area of short, nearly erect hairs on the distal half of the hind tibia of the female, this area separated from the rest of the external tibial surface by a slight change in the plane of the surface below a diagonal line across the tibial surface (Fig. 66-19a). Such an area is also evident in *Patellapis (Archihalictus) joffrei* (Benoist), *carinostriata* (Pauly), etc., and in the genus *Homalictus*. *Pachyhalictus* and these species of *Archihalictus* have rather broad hind tibiae, widest apically. The area is recognizable but less distinct, without a noticeable difference in the plane of the surface but with evident differences in hairs, in forms like *P. (Dictyohalictus) plicata* (Pauly). In species with more slender hind tibiae (the distal half parallel-sided, instead of broadest at apex), such as *P. (D.) retigera* (Cockerell), the outer surface of the tibia is uniformly convex, and the area of short hairs is recognizable only because the hairs are nearly erect. The hairs are more sloping in *P. (Archihalictus) laevata* (Benoist) and *P. (Lomatalictus) malachurina* (Cockerell), and the area is completely unrecognizable on the slender tibia of *P. (A.) perineti* (Benoist). The area of short hairs is associated with large hairs on the lower tibial margin that are more or less erect and pectinate or palmate; see couplet 1 in the key to subgenera. In such subgenera as *Zonalictus* these large pectinate hairs are evident even though the area of short, erect hairs is commonly not recognizable. The scopal hairs of the hind femur and, when developed, those of the sterna, tend to be plumose (i.e., with a central rachis and short branches in all directions along the rachis), although in the subgenus *Archihalictus* and in other genera, such as *Lasioglossum* or *Halictus,* they have a few long branches, and in *Lasioglossum* and *Halictus* the central rachis may not be readily recognizable. Plumose scopal hairs, like several other features listed above, are best developed in the subgenera having pectinate or palmate hairs on the lower margin of the hind tibia as well as the area of short hairs on the outer surface of the tibia. The inner hind tibial spur, which in many halictines is a group character, is variable within several of the subgenera of *Patellapis,* as indicated below.

Key to the Subgenera of *Patellapis*

1. Posterior tibia of female with fringe of long, pectinate to palmate hairs along basal two-thirds of lower margin, these hairs not curved strongly posteriorly toward apex of tibia or upward across outer surface of tibia, but *if* pectinate, then branches close together (thus approaching palmate) and tending to be at right angles to rachis; intermediates between large hairs of lower tibial margin and hairs of outer tibial surface very few; metasomal terga frequently with basal areas of pale tomentum, without strong apical hair bands (except in *Lomatalictus*) 2
—. Posterior tibia of female with long hairs of lower margin strongly curved posteriorly toward apex of tibia, or distal parts usually curved upward across outer surface of tibia, with well-separated long branches directed in same general direction as rachis, or usually branched in such a way that the rachis is not identifiable except basally; distal parts of long marginal hairs overlapping hairs of outer tibial surface, and intermediate hair types seemingly present, thus there is no striking superficial contrast between long hairs of lower margin and hairs of outer surface (this is as in *Lasioglossum* and *Halictus*); metasomal terga without basal bands or lateral areas of tomentum, frequently with apical bands of pale hairs 6
2(1). Metasomal terga with apical hair bands, without basal bands or lateral areas of tomentum; claws of female simple or with inner tooth very small, those of male with teeth close together (S4 of male unmodified) (Africa) .... *P. (Lomatalictus)*
—. Metasomal terga without strong apical hair bands, frequently with basal areas of pale tomentum; claws of both sexes toothed, as usual in halictines 3
3(2). Posterior margins of terga, especially T1 to T3, bare, nearly always contrastingly white, yellow, greenish, or bluish (Africa, Arabia) .... *P. (Zonalictus)*
—. Posterior margins of terga brownish or translucent, commonly with weak or very weak apical hair bands 4
4(3). Scutum and scutellum smooth or minutely roughened to coarsely striate, with punctures well separated at least in some areas; metasomal terga without areas of pale tomentum (ignoring pale hairs arising from graduli); S4 of male similar in size and shape to S3 and S5, without very large bristles (Madagascar) .... *P. (Archihalictus)*
—. Scutum and scutellum coarsely and closely punctate or punctures usually so large that surface is coarsely reticulate, rarely, as in *Pachyhalictus (Pachyhalictus) binghami* (Kirby), with partly smooth areas; metasomal terga (at least T2 and T3) with basal areas of pale tomentum, sometimes expanded to cover whole terga; S4 of male short, largely hidden, with a few large apical or preapical bristles 5
5(4). Inner hind tibial spur of female pectinate with three to six long teeth; basitibial plate of male absent or indicated only at apex; S4 of male not produced posteriorly at side but with an enormous and complex lateral bristle, usually hidden by T4 (Fig. 66-20c); first two hind tarsal segments of male fused (Oriental to Australia) .... *P. (Pachyhalictus)*
—. Inner hind tibial spur of female serrate or pectinate with ten or more teeth; basitibial plate of male present; S4 of male produced posteriorly at side with several large bristles usually hidden by T4; first two hind tarsal segments of male apparently articulated, but articulation broader than that between subsequent tarsomeres (Africa) .... *P. (Dictyohalictus)*
6(1). Terga with conspicuous apical hair bands; basitibial plate margined both in front and behind, its apex in female usually rounded (Africa) .... *P. (Patellapis s. str.)*
—. Terga without or with weak apical hair bands; basitibial plate not or incompletely defined on anterior margin, its apex in female angulate or pointed (Africa) .... *P. (Chaetalictus)*

Patellapis / Subgenus *Archihalictus* Pauly

Archihalictus Pauly, 1984b: 132. Type species: *Halictus joffrei* Benoist, 1962, by original designation.
Madagalictus Pauly, 1984b: 125. Type species: *Halictus suarezensis* Benoist, 1962, by original designation.

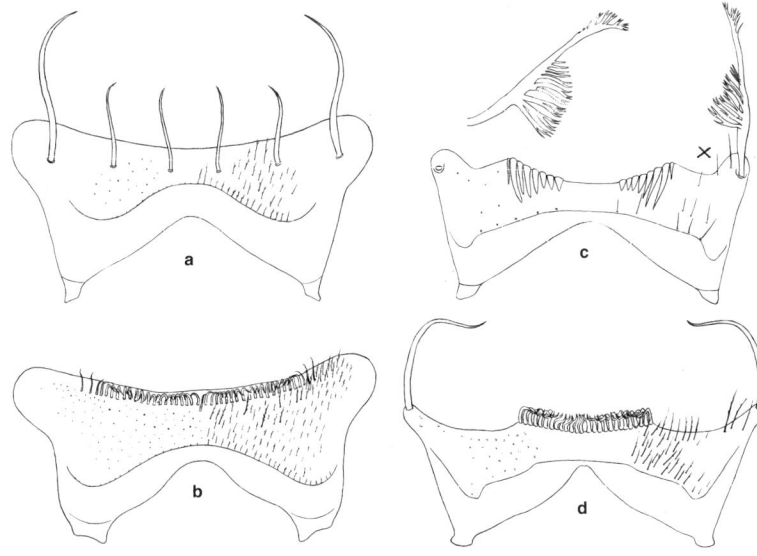

Figure 66-20. Modified fourth sternum (S4) of males of *Patellapis*. S4 of *Patellapis* males is commonly largely hidden under S3 and T3, so that even the large lateral bristles are often invisible without dissection. **a,** *P. (Chaetalictus) pearstonensis* (Cameron); **b,** *P. (Patellapis) schultzei* (Friese); **c,** *P. (Pachyhalictus) merescens* (Cockerell), with one of the giant setae detached (the tergal margin is delicate and X marks a damaged area that was reconstructed in this drawing); **d,** *P. (Zonalictus) albofasciata* (Smith). From Michener, 1978a.

These are very "ordinary looking," unbanded halictines, 4.5 to 7.5 mm long. The tergal margins are concolorous, as is usual also in *Pachyhalictus* and *Dictyohalictus,* not brownish or translucent as in *Chaetalictus, Lomatalictus,* and *Patellapis* s. str. The inner hind tibial spur of the female can be either serrate or pectinate. Male genitalia and other structures were sketched by Pauly (1984b).

■ *Archihalictus* is known only from Madagascar. The sixteen named species were revised by Pauly (in Pauly et al., 2001).

Pauly (1984b) places emphasis on the long hairs of the lower margin of the hind tibia of the female, which in the group called *Madagalictus* are palmate, whereas in *Archihalictus* in his sense they are pectinate, with branches near together and usually projecting almost at right angles to one side. The difference is not sharp, however. Species that should have the hairs pectinate often have a few palmate hairs near the base of the tibia, and *Patellapis perineti* (Benoist), which he placed in *Madagalictus,* has most of its hairs pectinate. Distinctive features of females of *Archihalictus* in the narrow sense are (1) the transverse row or narrow band of very long, loosely branched scopal hairs on S2 to S4, less well developed on S5, and the presence of some similar hairs on S1, and (2) the depressed metasoma, with the ventrolateral extremities of the terga, especially T2, at rather a sharp angle to the dorsal parts of the terga and bearing a few hairs siilar to those of the sterna. These features are similar to those of *Homalictus.* The ventral scopa of long, loosely branched hairs is present also in species placed in *Madagalictus,* although less well developed there, and the metasoma is somewhat depressed in those species [see, for example, *Patellapis (Archihalictus) laevata* (Benoist)]. I have chosen to indicate the close relationship of *Archihalictus* s. str. and *Madagalictus* by uniting them, but the distinctions could also justify separating them.

Pauly (1999b) placed *Archihalictus* s. str. as a subgenus of *Pachyhalicus* but recognized *Madagalictus* as a distinct genus. Pauly's arrangement and mine (as a single subgenus within *Patellapis*) are both rather arbitrary and a detailed phylogenetic study is needed.

Patellapis / Subgenus *Chaetalictus* Michener

Patellapis (Chaetalictus) Michener, 1978a: 509. Type species: *Halictus pearstonensis* Cameron, 1905, by original designation.

In *Chaetalictus* the apical margins of the metasomal terga are brownish or pallid, and apical hair bands are weak or absent; there are no basal bands or lateral basal areas of tomentum. In some species the metasoma is red. The body length is 4 to 8 mm. S4 of the male is commonly short and largely hidden but may be of normal size; the posterior part is armed with several large bristles (Fig. 66-20a) or a dense row of bristles that may be erect or even retrorse. The inner hind tibial spur of the female is finely pectinate, or the teeth are sufficiently short that it could almost be called serrate (Fig. 66-21).

■ This subgenus is common in southern Africa, north at least to Namibia and Zimbabwe, and there are two species (and another undescribed) in Madagascar. The 33 African species names were listed by Michener (1978a).

Patellapis / Subgenus *Dictyohalictus* Michener

Pachyhalictus (Dictyohalictus) Michener, 1978a: 518. Type species: *Halictus retigerus* Cockerell, 1940, by original designation.

Pachyhalictus (Rugalictus) Pauly, 1980a: 121. Type species: *Halictus weenenicus* Cockerell, 1941 = *Halictus retigerus* Cockerell, 1940, by original designation.

Members of this subgenus are less compact and less robust than are species of *Pachyhalictus.* The body length is 6 to 7 mm. *Dictyohalictus* shares with *Pachyhalictus* the reticulate pattern of the scutum and scutellum, but it is less extreme in the former, the cells of the pattern often being smaller than in *Pachyhalictus*. The inner hind tibial spurs of females have ten or more teeth, all usually rather long, so that the spur can be called pectinate, but they are sometimes short enough that it would be called coarsely serrate. Unlike those of *Pachyhalictus,* males of *Dictyohalictus* are relatively well known. All have, at each lateral extremity of S4, a lobe, and, on or near the base of this lobe, three to eight large bristles (hidden by T4). Smaller bristles arise from the sternal margin between the lateral lobes. Male genitalia and sterna were illustrated by Michener (1978a) and Pauly (1989).

■ The subgenus is widespread in tropical Africa, from Ivory Coast and Zaire to Malawi, Zimbabwe, and Natal province, South Africa. The 12 species were revised by Pauly (1989).

Patellapis / Subgenus *Lomatalictus* Michener

Patellapis (Lomatalictus) Michener, 1978a: 509. Type species: *Halictus malachurinus* Cockerell, 1937, by original designation.

This is a subgenus with distinct apical tergal hair bands and broadly pallid, translucent tergal margins. The body length is 5.5 to 7.5 mm. As in *Archihalictus,* S4 of the male is not modified and lacks large bristles. The inner hind tibial spur of the female is coarsely ciliate to pectinate with about nine teeth. Male genitalia and other structures were illustrated by Michener (1978a).

■ *Lomatalictus* is known only from South Africa. There are four specific names, listed by Michener (1978a).

Patellapis / Subgenus *Pachyhalictus* Cockerell

Halictus (Pachyhalictus) Cockerell, 1929d: 589. Type species: *Halictus merescens* Cockerell, 1919, by original designation.

This subgenus consists of robust bees with basal bands, or at least basal lateral areas, of tomentum on the metasomal terga. The body length is 5 to 8 mm. Unlike that in most *Patellapis,* the first recurrent vein is near the distal end of the second submarginal cell or meets the second transverse cubital vein. In most known males, S4 has one enormous bristle at each side (Fig. 66-20c), but the male of *Patellapis intricatus* (Vachal) described by Pesenko and Wu (1997) lacks such a bristle. Male genitalia and other characters were illustrated by Michener (1978a), Walker (1993, 1996), and Pesenko and Wu (1997).

■ *Pachyhalictus* is the only Indo-Australian subgenus of *Patellapis*. It ranges from Sri Lanka and India east through Southeast Asia to Yunnan Province, China, the Philippines, and Taiwan, thence south through Indonesia and New Guinea to Australia, at least as far south as middle Queensland. The 30 specific names were listed by Michener (1978a). Revisional work on the subgenus was included in Blüthgen (1926, 1928a, 1931); the two Australian species were treated by Walker (1996).

Patellapis / Subgenus *Patellapis* Friese s. str.

Halictus (Patellapis) Friese, 1909a: 148. Type species: *Halictus schultzei* Friese, 1909, by designation of Cockerell, 1920b: 311.

The metasomal terga in *Patellapis* s. str. have broad, pallid, translucent margins and apical hair bands; the bands may be present only laterally but are often strong and continuous. The rather elongate body measures 8 to 12 mm. *Patellapis* s. str. differs from *Chaetalictus* in having the basitibial plate of both sexes well defined, including at least part of the anterior margin; in *Chaetalictus* and *Lomatalictus* it is not or weakly defined anteriorly or, in some males, is absent. As in *Chaetalictus,* S4 of the male is of ordinary size to much shortened and hidden under S3; near its posterior margin is a row of more or less close, coarse setae (Fig. 66-20b), the lateral ones enlarged in *Patellapis braunsella* Michener. A distinctive feature of the subgenus is the greatly expanded, helmetlike dorsal crest of the penis valve.

■ This subgenus is known only from South Africa. The five species were listed by Michener (1978a).

Patellapis braunsella Michener is unusual among *Patellapis,* as indicated in the discussion of the genus, in the presence of paraocular lobes that are rounded, ill-defined, but nonetheless somewhat acute, extending down into the clypeus. It is possible that *Mexalictus* is a close relative of *Patellapis;* features of *P. braunsella,* including the paraocular lobe, reduce the differences between *Mexalictus* and *Patellapis.*

Patellapis / Subgenus *Zonalictus* Michener

Zonalictus Michener, 1978a: 513. Type species: *Halictus albofasciatus* Smith, 1879, by original designation.

Though the white, yellow, greenish, or bluish integumental tergal bands are distinct and diagnostic for most species of this subgenus, the bands are sometimes merely yellowish brown and may be quite inconspicuous if the background tergal color is yellow-brown or reddish. The rather elongate body and usually rather long, plumose, yellowish thoracic hairs are characteristic of most species. This subgenus includes the largest *Patellapis,* their lengths ranging from 6.5 to 12.5 mm. In some males the clypeus is partly yellow, a feature not found in other *Patellapis.* The inner hind tibial spur of the female varies from finely to coarsely serrate and even very briefly pectinate, as illustrated by Michener (1978a); see Figure 66-21. S4 of the male is largely retracted under S3, and sometimes has a series of bristles all across the margin, but usually there is a series only medially (usually erect or retrorse) and a giant bristle laterally, hidden under T4 (Fig. 66-20d).

■ *Zonalictus* is found throughout sub-Saharan Africa, and eastward to Socotra, Yemen, and Madagascar. There are 68 named forms, listed by Pauly (1999b); only one of these is from Madagascar (Pauly, 1984b).

Genus *Pseudagapostemon* Schrottky

Almost all species of this neotropical genus are rather small and slender (length 5-11 mm) and have weak greenish, blue, or brassy reflections on the body; a few species, however, are black. The pubescence, mostly pale, forms

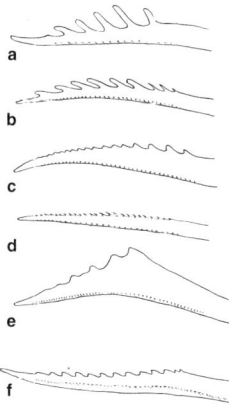

Figure 66-21. Inner hind tibial spurs of females of *Patellapis* and *Thrinchostoma*. **a**, *P. (Chaetalictus) pearstonensis* (Cameron); **b**, *P. (Zonalictus) partita* (Cockerell); **c**, *P. (Z.) albofasciata* (Smith); **d**, *P. (Z.) zacephala* (Cockerell); **e**, *T. (Thrinchostoma) afasciatum* Michener; **f**, *T. (Eothrincostoma) productum* (Smith). From Michener, 1978a.

apical bands, at least laterally, on the metasomal terga. A paraocular lobe is well developed, extending down into the clypeus. The apical tergal bands, the lack of basal tergal tomentum, and the paraocular lobes are among the principal differences from the closely related genus *Ruizantheda*. Male genitalia, sterna, and other structures were illustrated by Moure and Sakagami (1984) and Cure (1989). For most *Pseudagapostemon* the bare or nearly bare eyes further differentiate the genus from *Ruizantheda*, which has long hair on the eyes. The subgenus *Brasilagapostemon*, however, has hairy eyes.

The species *Pseudagapostemon amabilis* and *cyanomelas*, attributed to Moure by Cure, were in fact first described by Cure (1989), who appears to have been responsible for the descriptions and figures.

Pseudagapostemon (Neagapostemon) cyanomelas Cure makes communal nests in earth banks; several females use the same nest entrance, and long laterals lead to cells far from the main burrow (Michener and Lange, 1958d, *P. perzonatus* Cockerell, misidentification). *P. (Pseudagapostemon) brasiliensis* Cure nests in flat ground, four females in the one nest studied; cells are close to the main burrows (Martins, 1993).

Key to the Subgenera of *Pseudagapostemon*

1. Eyes conspicuously hairy; malar area of male about one-third as long as second flagellar segment (Brazil) ....
 .... *P. (Brasilagapostemon)*
—. Eyes nearly bare; malar area of male usually one-fifth as long as second flagellar segment or less 2
2(1). Clypeus usually with longitudinal median elevation or carina, and, in males, fine striae lacking or not converging on midline .... *P. (Neagapostemon)*
—. Clypeus without longitudinal median ridge or carina except sometimes at uppermost end of clypeus, and, in males, with fine oblique striae converging downward to midline of clypeus .... *P. (Pseudagapostemon s. str.)*

Pseudagapostemon / Subgenus *Brasilagapostemon* Moure and Sakagami

Pseudagapostemon (Brasilagapostemon) Moure and Sakagami, 1984: 4. Type species: *Pseudagapostemon fluminensis* Schrottky, 1911, monobasic.

In addition to having hairy eyes, *Brasilagapostemon* differs from other subgenera in having a relatively broad head, the eyes strongly converging below. The male gonostyli are simple and rather broad.

■ This subgenus is known from Minas Gerais to Paraná, Brazil. The three species were revised by Cure (1989).

Pseudagapostemon / Subgenus *Neagapostemon* Cure

Pseudagapostemon (Neagapostemon) Cure, 1989: 295. Type species: *Pseudagapostemon cyanomelas* Cure, 1989, by original designation.

This subgenus includes nonmetallic species as well as species with the greenish or bluish tints usual to the genus. Aside from the median clypeal ridge, there are few common characters for this subgenus, and it may be an artificial assemblage of odd species of *Pseudagapostemon*. S4 of the males varies from unmodified in *P. cyanomelas* Cure to deeply emarginate, and in *P. jenseni* (Friese) it supports large lateral bristles and a series of marginal bristles reminiscent of some *Patellapis* species. The male gonostyli also are very diverse, as illustrated by Cure (1989).

■ *Neagapostemon* occurs from Mendoza and Buenos Aires provinces, Argentina, north to São Paulo, Brazil. The six species were revised by Cure (1989).

Pseudagapostemon / Subgenus *Pseudagapostemon* Schrottky s. str.

Pseudagapostemon Schrottky, 1909a: 145. Type species: *Agapostemon arenarius* Schrottky, 1902, by original designation.

The median clypeal ridge mentioned in the key to subgenera is weak or absent in *Pseudagapostemon* s. str. The convergent striae are part of the fine roughening of the clypeal surface; much coarser punctures, also present on the clypeus, interrupt the microstriae. A digitiform process arising at the base of the male gonostylus is characteristic of this subgenus, according to Cure (1989).

■ *Pseudagapostemon* s. str. ranges from the state of Pará, Brazil, south to the province of Río Negro, Argentina; one species is found in Chile from Aconcagua to Valdivia. The 16 species were revised by Cure (1989).

Genus *Ptilocleptis* Michener

Ptilocleptis Michener, 1978b: 315. Type species: *Ptilocleptis tomentosa* Michener, 1978, by original designation.

This presumably parasitic group may be derived from *Sphecodes*, but it has a number of characters that are probably plesiomorphic relative to *Sphecodes*, so that it may be a basal branch of the *Sphecodes* clade. Such characters include the moderate punctation, without the coarse pitting found in *Sphecodes*; the face not broad as is usual in *Sphecodes*, the eyes being convergent below (Fig. 66-3c) and the clypeus only about twice as broad as long, without a longitudinal median depression; the lack of spinelike setae on the outer surfaces of the hind tibiae; and the lack of striae on the male gonocoxites. Unlike most *Sphecodes,* the mandible of the female is simple. The form and coloration suggest polybiine wasps; in one of the species the pronotum, scutellum, metanotum and much of T1 are yellow, enhancing the wasplike aspect. An unnamed Mexican species, however, has a red metasoma and therefore a *Sphecodes*-like appearance. The body length is 6 to 10 mm. Male genitalia and hidden sterna were illustrated by Michener (1978b).

■ *Ptilocleptis* is known from the state of San Luis Potosí, Mexico, to the eastern foothills of Peru and to the state of Santa Catarina, Brazil. The three named species were revised by Michener (1978b). The species are rarely collected.

The only host record is a specimen of *Ptilocleptis* found by G. Melo inside a nest of *Augochlorodes turrifaciens* Moure in Minas Gerais, Brazil.

Genus *Rhinetula* Friese

Rhinetula Friese, 1922b: 581. Type species: *Rhinetula denticrus* Friese, 1922, by designation of Sandhouse, 1943: 596.

This genus contains a single rather large, robust species, blackish or brown with green or blue tints on the face and thorax and sometimes showing amber areas on the metasoma. The length is 8.5 to 12.0 mm. As in *Paragapostemon,* the metasoma of the male is robust, like that of a female, not slender and parallel-sided as in most other Halictini. The male antennae are short, like those of females. The legs of the male are dark, and there are two teeth on the hind femur and one on the basitarsus. The nearly vertical propodeal profile, which lacks a major subhorizontal basal area, is unique among the Halictini.

■ *Rhinetula* occurs in lowland forests from Honduras to the eastern Peruvian foothills and south to Bolivia. There is only one species, *Rhinetula denticrus* Friese. It was fully described by Roberts and Brooks (1987).

Most records are from Central America and include light-trap catches, indicating nocturnal or matinal flight, in addition to daytime activity. Unlike those of nocturnal bees such as *Megalopta,* the ocelli are not enlarged.

Genus *Ruizantheda* Moure

Ruizantheda Moure, 1964b: 265. Type species: *Halictus proximus* Spinola, 1851, by original designation.
Ruizantheda (Ruizanthedella) Moure, 1964b: 267. Type species: *Halictus mutabilis* Spinola, 1851, by original designation.
Oragapostemon Cure, 1989: 312. Type species: *Halictus divaricatus* Vachal, 1903, by original designation.

Among genera of Halictini with strong wing venation, *Ruizantheda* can be recognized by the strongly emarginate inner ocular orbits (as in Augochlorini), the hairy eyes (although areas near the face are bare), the two or three large, broad teeth of the inner hind tibial spur of the female, and the lateral basal patches of tomentum on T2 and T3 of the female. The body, which is 6 to 12 mm long, is black or weakly metallic blue-green, the metasoma sometimes red; apical tergal pubescent bands, lateral gradular carinae, and paraocular lobes extending down into the clypeus are absent. Male genitalia and other structures were illustrated by Moure (1964b) and by Cure (1989, under the name *Oragapostemon*).

■ This genus occurs in two disjunct areas: (1) Chile (Atacama to Chiloe) and adjacent parts of southern Argentina (provinces of Neuquén to Chubut) and (2) the state of Paraná, Brazil, to Buenos Aires Province, Argentina. It is one of the commonest halictid genera in Chile. The three or perhaps four species were characterized by Moure (1964b) and Cure (1989).

The three taxa synonymized above are rather different but clearly related, as was well shown by Cure (1989). Since there is only one species in each, or in *Ruizanthedella* perhaps two, subgeneric recognition seems unnecessary. The type species, *R. proxima* (Spinola), is unusual in having a deeply emarginate pygidial plate in the male. B. W. Coelho, in studying the genera of the subtribe Caenohalictina, emphasizes the close relationship of *Caenohalictus* and *Ruizantheda.*

Ruizantheda divaricata (Vachal) makes much-branched, usually communal nests inhabited by 1 to 40 females (Fig. 65-2 and Michener and Lange, 1958d as *Pseudagapostemon*); *R. mutabilis* (Spinola) is also sometimes communal (Claude-Joseph, 1926).

Genus *Sphecodes* Latreille

Sphecodes Latreille, 1804: 182. Type species: *Nomada gibba* Fabricius, 1804 = *Sphex gibba* Linnaeus, 1758, monobasic.
Dichroa Illiger, 1806: 46. Type species: *Sphex gibba* Linnaeus, 1758, by designation of Sandhouse, 1943: 545.
Sabulicola Verhoeff, 1890: 328. Type species: *Sabulicola cirsii* Verhoeff, 1890 = *Nomada albilabris* Fabricius, 1793, monobasic.
Thrausmus Buysson, 1900: 177. Type species: *Thrausmus grandidieri* Buysson, 1900, monobasic. [New synonymy.]
Drepanium Robertson, 1903b: 103. Type species: *Sphecodes falcifer* Patton, 1880 = *Sphecodes confertus* Say, 1837, by original designation.
Proteraner Robertson, 1903b: 103. Type species: *Sphecodes ranunculi* Robertson, 1897, monobasic.
Dialonia Robertson, 1903b: 104. Type species: *Sphecodes antennariae* Robertson, 1891, by original designation.
Machaeris Robertson, 1903b: 104. Type species: *Sphecodes stygius* Robertson, 1893, by original designation.
Sphecodium Robertson, 1903b: 104. Type species: *Sphecodium cressonii* Robertson, 1903, by original designation.
Stelidium Robertson, 1903b: 104, *lapsus* for *Sphecodium;* this is not *Stelidium* Robertson, 1902b. Type species: *Sphecodium cressonii* Robertson, 1903, by original designation.
Sphecodes (Callosphecodes) Friese, 1909b: 182. Type species: *Callosphecodes ralunensis* Friese, 1909, monobasic. [Friese described *Callosphecodes* as a subgenus, but treated it as a genus when describing the species.]
Sphegodes Mavromoustakis, 1949: 553, unjustified emendation of *Sphecodes* Latreille, 1804.
Sphecodes (Austrosphecodes) Michener, 1978b: 327. Type species: *Sphecodes chilensis* Spinola, 1851, by original designation.

This is the only common and widespread parasitic taxon among the Halictinae. It is usually black with a partly or wholly red metasoma, but sometimes, especially in males, the metasoma is entirely black; in *Callosphecodes*, which may not be a *Sphecodes* (see below), it is metallic blue. Body length is 4.5 to 15.0 mm. The genus is noteworthy for its coarse sculpturing, the thorax usually being coarsely pitted and the dorsal surface of the propodeum marked by a few, coarse, often irregular longitudinal rugae often delimiting shining spaces (Figs. 8-2; 66-9b). In some species, especially small forms similar to *Microsphecodes,* the punctures are fine but widely separated by smooth, shining interspaces, not like the more densely or finely punctate integument of *Ptilocleptis* and *Eupetersia*. The head is usually much wider than long (Fig. 66-3d) and the clypeus is two or usually three or more times as wide as long, except in the subgenus *Austrosphecodes,* which has a longer clypeus with a longitudinal depression, and is thus biconvex. T5 of the female nearly always has an apical marginal fringe, sometimes interrupted medially, behind the prepygidial fimbria. As in *Lasioglossum,* there are a few species with only two submarginal cells (Robertson, 1903b; Mitchell, 1960). One of the early uses of male genitalia to characterize species of bees was by Hagens (1882), who illustrated the genitalia of many species. Male genitalia were also illustrated by Mitchell (1960) and Tsuneki (1983 and later papers), and were sketched by Warncke (1992b).

■ *Sphecodes* is widespread on all continents except Australia, where it is known only from the northeast. In the palearctic region it occurs from the Canary Islands and Britain to Japan, and as far north as north-central Finland; in North America, it is found from coast to coast and reaches subarctic Alaska and central Canada. To the south the genus extends through the Antilles and continental tropics to Chile and Argentina (at least to the province of Neuquén). In the Eastern Hemisphere, *Sphecodes* occurs south to Cape Province, South Africa, and is present in Madagascar, Southeast Asia, Indonesia, the Philippines, New Guinea, New Britain, and the state of Queensland, Australia. Warncke (1992b) recognized 39 species in the western palearctic region, although he may well have synonymized some valid species. Tsuneki (1983) recognized 35 Japanese species, and published additional species later. There may be 100 named palearctic species. About 120 species are listed for the Western Hemisphere (Moure and Hurd, 1987). Only five species are known from eastern Indonesia to New Britain and to Australia. Sub-Saharan Africa and southern Asia contain an estimated 65 additional species. Pauly (1999b) listed the 49 Afrotropical species of *Sphecodes.* Engel (2006c) recorded only two species in Cuba. The species of the western palearctic region were revised by Warncke (1992b). Other major papers on the taxonomy of Eastern Hemisphere species, including keys to species, were by Blüthgen (1923c, 1924b, 1927, 1928b, c). Lomholdt (1977) revised the Danish species; Sustera (1959), the (then) Czechoslovakian species; and Tsuneki (1983), the Japanese species. Meyer (1919a) provided German translations of specific descriptions of species from all areas.

Blüthgen (1934) and Michener (1978b) gave summaries of information on the hosts of *Sphecodes;* Maeta et al. (1996) listed additional host records. Both solitary and social Halictinae are included among the hosts. Although most *Sphecodes* attack other Halictinae, species exist that parasitize *Nomioides* (Halictidae), *Andrena, Calliopsis, Melitturga,* and *Perdita* (Andrenidae), *Colletes* and *Lonchopria* (Colletidae), and perhaps *Dasypoda* (Melittidae). The behavior of female *Sphecodes* in host nests is considered in Sections 8 and 65, and by Sick et al. (1994).

Several related genera may be derivatives of *Sphecodes.* The neotropical *Ptilocleptis* in particular has characters suggestive of the subgenus *Austrosphecodes*. *Microsphecodes* and *Nesosphecodes* may be derived from the *Austrosphecodes* group of *Sphecodes.* For the present, however, in the absence of a phylogenetic study, I am retaining these genera, as was done by Michener (1978b). Likewise, *Eupetersia* may turn out to be a derivative of *Sphecodes,* as was suggested by earlier authors. I have tentatively retained it also as a genus; possibly it is the sister group to *Sphecodes.*

No detailed study has been published on relationships among species of *Sphecodes*. For the present I have simply listed the potential subgeneric names as synonyms. Some comments on some of the synonyms or subgenera are as follows: *Thrausmus* is based on a rather large species from Madagascar that has a dark metasoma; morphologically, it is a typical *Sphecodes,* according to R. W. Brooks, who has seen the type specimen of the type species. *Callosphecodes* was based on a single specimen from New Britain that seems to have disappeared; the abdomen was metallic blue-black, suggesting *Eupetersia,* and *Callosphecodes* could indeed be a senior synonym of *Eupetersia,* which, however, is known only from Africa and India, the latter about 8,000 km from New Britain. The description of *Callosphecodes* was repeated by Meyer (1919a). *Austrosphecodes* is the principal (or only?) South American group of *Sphecodes,* and was regarded as the most distinct group within the genus by Michener (1978b). Unlike those of most other *Sphecodes,* the mandibles of female *Austrosphecodes* are simple, the legs are slender, the second hind tarsal segment of the male is as narrow at the base as the third, and the male gonocoxites are not or only inconspicuously striate, in contrast to those of the other groups of the genus.

Genus *Thrincohalictus* Blüthgen

Halictus (Thrincohalictus) Blüthgen, 1955: 20. Type species: *Halictus prognathus* Pérez, 1912, by original designation.

This genus includes a nonmetallic black species with apical tergal hair bands and brownish tergal margins. Body length is 9 to 10 mm. The face is elongate, the clypeus being both strongly produced downward and protuberant anteriorly, and the malar area being distinct, half as long as wide in the female and longer than wide in the male. There is a strong paraocular lobe. These features of the head are associated with elongate mouthparts, especially the long glossa but also the galeal blades. The inner hind tibial spur of the female is pectinate with four long teeth. As is common in *Patellapis,* S4 of the male is shorter than the adjacent sterna and its posterior margin supports a dense fringe of long bristles. The upper gonostylus of the male is not bifid as in *Patellapis,* but is equipped with a series of long, curved hairs. The lower gonostylus is large and somewhat retrorse. The genitalia, sterna, and other structures were illustrated by Sakagami, Kato, and Itino (1991).

■ The genus has been found in Israel, islands of the Aegean Sea, Turkey, Iran, and Armenia. There is one species, *Thrincohalictus prognathus* (Pérez).

Genus *Thrinchostoma* Saussure
This is a genus of relatively large (length 8-16 mm), elongate bees. They are nonmetallic, black or with parts of the metasoma and even the whole body yellowish red. The clypeus is strongly produced downward and strongly protuberant forward. The malar area is distinct but variable, from less than one-third as long as wide to four times as long as wide, and as long as or longer than the eye. The paraocular lobe is strongly produced down into the clypeus. Both of the recurrent veins usually enter the third submarginal cell. The hind tibial spur of the female is finely to coarsely serrate or the margin is almost undulate. The hind tibia of the male has a broad yellowish enlargement that carries the tibial spurs. S4 of the male is usually shortened, largely hidden by S3, but the lateral parts extend far posteriorly on either side of a broad concavity. The dorsal gonostylus is large and rather elaborate, the ventral gonostylus erect or retrorse.

Strikingly distinct from most other Halictini, *Thrinchostoma* (with *Parathrincostoma*) has been placed in a separate subtribe, Thrinchostomina; see Figure 61-4. The genus is found in southern Asia and in Africa, with an unusual concentration of species in Madagascar. The Asiatic species of *Thrinchostoma* were reviewed by Blüthgen (1926), the African and Malagasy species by the same author (1930, 1933). Various species have been described since those dates, however, as listed by Michener (1978a). Pauly (in Pauly et al., 2001) revised the twelve Madagascar species, all of which are in the typical subgenus.

Key to the Subgenera of *Thrinchostoma*
(Modified from Michener, 1978a)

1. Forewing without an area of dense hairs along second submarginal crossvein, this vein simple and straight; first submarginal crossvein arising far from base of vein r and margin of stigma; S4 of male with a series of enormous simple bristles arising from margin both medially under S3 and laterally, under the lateral extremity of T4, where the setae are largest (Africa) .... *T. (Eothrincostoma)*
—. Forewing with an area of dense veins near median part of second submarginal crossvein, these hairs forming, in males, a conspicuous dark spot, this crossvein usually angulate or thickened medially, sometimes incomplete (not reaching marginal cell); first submarginal crossvein arising very near margin of stigma; S4 of male without coarse, specialized setae 2
2(1). Head extraordinarily produced below eyes, malar area nearly as long as or longer than eye, several times as long as wide (Fig 66-22b) .... *T. (Diagonozus)*
—. Head only moderately produced below eyes, malar area much shorter than eye, three times as long as wide or less .... *T. (Thrinchostoma s. str.)*

Figure 66-22. Faces of unusual Halictini. **a,** *Urohalictus lieftincki* Michener, holotype male, from Michener, 1980; **b,** *Thrinchostoma (Diagonozus) lettowvorbecki* Blüthgen, from Michener, 1978a.

***Thrinchostoma* / Subgenus *Diagonozus* Enderlein**

Diagonozus Enderlein, 1903: 35. Type species: *Diagonozus bicometes* Enderlein, 1903, monobasic.

The head in this subgenus is so elongate that the entire clypeus is below the lower ocular tangent (Fig. 66-22b). This construction provides space for folding an extremely long proboscis. The inner hind tibial spur of the female is as in *Thrinchostoma* s. str. The pronotum is longer than in the other subgenera, the collar medially being considerably longer than an ocellar diameter.

■ *Diagonozus* is found in tropical West Africa and in Sumatra. The four African species were listed by Michener (1978a).

The Sumatran species was described by Sakagami, Kato, and Itino (1991). Perhaps its development of an extraordinarily long head is independent of the comparable development in Africa.

Thrinchostoma / Subgenus *Eothrincostoma* Blüthgen

Eothrincostoma Blüthgen, 1930: 496, 501. Type species: *Halictus torridus* Smith, 1879, by designation of Sandhouse. 1943: 548.

This subgenus is more like ordinary halictids than are the other subgenera in its wing venation and its lack of a hair spot on the forewing. Moreover, its male antennae and other features do not exhibit the special attributes found in many species of *Thrinchostoma* s. str. The malar area is about as long as wide. The inner hind tibial spur of the female is rather finely and uniformly serrate.

■ *Eothrincostoma* ranges widely over tropical Africa from Uganda to Kenya, south to Natal in South Africa. The six species were listed by Michener (1978a); *Thrinchostoma productum* (Smith) was misplaced in that paper and should be added to the list.

Thrinchostoma / Subgenus *Thrinchostoma* Saussure s. str.

Thrinchostoma Saussure, 1890: 52. Type species: *Thrinchostoma renitantely* Saussure, 1890, monobasic.
Trichostoma Dalla Torre, 1896: 381, also Friese, 1909a: 150, unjustified emendation of *Thrinchostoma* Saussure, 1890.
Thrinchostoma Dalla Torre, 1896: 641, also Blüthgen, 1926, 1928c, 1930, unjustified emendation of *Thrinchostoma* Saussure, 1890.
Trichchostoma Ashmead, 1899a: 91, *lapsus* for *Thrinchostoma* Saussure, 1980.
Rostratilapis Friese, 1914b: 26. Type species: *Halictus macrognathus* Friese, 1914, by designation of Sandhouse, 1943: 597.
Nesothrincostoma Blüthgen, 1933: 364. Type species: *Thrincostoma serricorne* Blüthgen, 1933, monobasic.
Trinchostoma Sandhouse, 1943: 606, incorrectly attributed to Sladen, 1915: 214, unjustified emendation of *Thrinchostoma* Saussure, 1890.

This is the principal subgenus of *Thrinchostoma*. Recognition of *Diagonozus* probably makes *Thrinchostoma* s. str. paraphyletic. A considerable morphological gap exists, however, betweeen the two. The lower margin of the clypeus in the latter is only moderately below the lower ocular tangent. The malar area is one-third to three times as long as wide. The inner hind tibial spur of the female is widened medially by a broad tooth, distal to which the margin is coarsely toothed to almost edentate. S4 of the male often has long setae on posterior lateral projections, but lacks a row of very coarse setae; such a row of setae is often present on the basal thickening of S5.

■ *Thrinchostoma* s. str. is widespread in tropical Africa, from Liberia, Uganda, and Kenya south to Natal, South Africa; species are particularly numerous in Madagascar. Eastward, a few rare species occur from southern India to Assam, Vietnam, Kalimantan, and Java. There are 44 species (Michener, 1978a); 12 of them are from Madagascar. Keys or revisional studies are listed under the genus.

An extraordinary feature of a few species is that the male antennae have 12 instead of 13 segments (see Blüth-

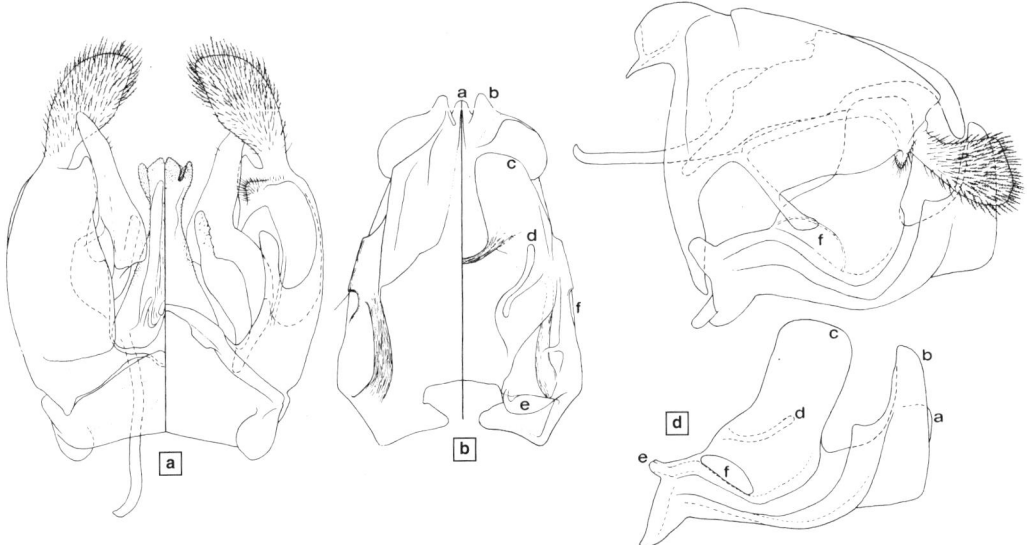

Figure 66-23. Genitalia and hidden sterna of the holotype male of *Urohalictus lieftincki* Michener. **a,** Genitalia, dorsal view on left; **b,** Fused S7 and S8, dorsal view on left; **c,** Lateral view of genitalia and S7 and S8; **d,** Lateral view of S7 and S8. (For the sterna, a given structure seen in different views is marked with the same letter, to facilitate interpretation of the drawings.) From Michener, 1980.

gen, 1930). In other respects such species fall easily into this subgenus.

Genus *Urohalictus* Michener

Urohalictus Michener, 1980: 16. Type species: *Urohalictus lieftincki* Michener, 1980, by original designation.

This genus is known from a single male that differs so greatly from other bees as to justify its generic status and perhaps even tribal or subfamilial status. It is dull greenish blue, 11 mm long, and the clypeus, large supraclypeal area, the undersurface of the scape basally (Fig. 66-22a), and middorsal areas on the pronotum and metanotum are yellow. No other halictine has such yellow areas. The paraocular lobes extend down into the clypeus, the apex of which is strongly emarginate. The terga lack bands of hair. The most remarkable feature, however, is the large size of S7 and S8, totally different from those of all other bees, as indicated in the key and illustrated in Figure 66-23b-d. The upper gonostylus is simple, broad, hairy (Fig. 66-23a), and much larger than usual in *Lasioglossum*. The lower gonostylus is a small, hairy lobe, not retrorse. The penis valves are slender and nearly straight, perhaps replaced functionally by the long, pointed volsellae, which extend beyond the gonocoxites. These structures were illustrated by Michener (1980a).

■ *Urohalictus* is from New Guinea. A single species, *U. lieftincki* Michener, is known from a lone male specimen.

The possibility exists that this genus is parasitic. Until females have been collected, there is no way to know. Because the species is so unusual for a halictid, I have reexamined the type specimen and present here some characters that were not mentioned in the original description. It does indeed appear to be a halictid. The mouthparts, which were not described earlier, appear to be like those of a halictine. Unfortunately, I have not been able to verify that the lacinia is of the characteristic halictid type, but hairs apparently arising on the anterior surface of the labiomaxillary base or tube suggest that the lacinia is in the typical position for halictids. (I do not wish to risk damage by removing and clearing the proboscis.) However, the very elongate prepalpal part of the galea and lack of a galeal comb are as in Halictini. The galeal blade is about twice as long as its width at the base (at the palpus) and has a pencil of bristles at its apex resulting in a stiff, slender, sharply pointed, incurved apical galeal appendage about half as long as the blade. This is an unusual feature, suggesting a possible floral specialization. The glossa is tapering, about four times as long as its basal width and about as long as the clypeus. The distal part of the basal vein is thickened, as in various Halictini (Fig. 66-1b, c).

67. Tribe Augochlorini

This American group was first recognized as the *Halicti hexagoni* by Vachal (1911), whose classification was summarized by Alfken (1926a). Unfortunately, that classification was largely ignored by subsequent workers, probably because it was presented in abbreviated keys, without illustrations, and all within the genus *Halictus*. It was not until Eickwort's (1969a, b) seminal works on the group appeared that the tribe was clearly distinguished from American genera of the tribe Halictini, which, because of bright metallic coloration, were often considered to be among the augochlorine genera.

The major tribal characters are indicated in the key to the tribes of the Halictinae. There are, however, additional features useful in recognition of the tribe. Although some species are not or only weakly metallic, a majority are brilliant metallic green (Pl. 5) or sometimes blue, coppery, or red. In most species the inner orbits of the eyes are more strongly emarginate near the upper third than in most Halictinae (Figs. 67-6, 67-7). This character, although not reliable, is responsible for Vachal's name for the group, *hexagoni,* which refers to the hexagonal appearance of the face. In the male, the genitalia lack the retrorse ventral gonostylus common in the Halictini (although a ventral gonostylus may be present but directed mesad, e.g., Fig. 67-12c, d), the apex of S6 usually has a median notch, and the hind basitarsus is not broadly articulated or fused to the second segment, as is common in male Halictini, but is narrowly articulated, as are the subsequent segments. The distal veins of the forewing are strong (Fig. 67-1), not weaker than other veins as is frequent especially in females of Halictini. Male genitalia, hidden sterna, and other structures of nearly all genera were illustrated by Eickwort (1969b), and of all nearctic genera by Sandhouse (1937) and Mitchell (1960); see also Engel, Brooks, and Yanega (1997).

Augochlorini is principally a tropical American tribe. All genera occur in South America, some only in more or less temperate parts of the continent. Several of the widespread tropical American genera include one or a few species that range into North America, and species of three genera *(Augochlora, Augochlorella,* and *Augochloropsis)* reach southern Canada.

A phylogeny of the genera of Augochlorini was given by Engel (2000b) together with a new generic revision. He divided the tribe into two subtribes, Augochlorina and Corynurina, the latter for the genera *Corynura, Halictillus, Rhectomia,* and *Rhinocorynura* only. The subtribe Corynurina is found only in southern South America and is the only group of the subfamily Halictinae with a well developed galeal comb, a probably plesiomorphic character state, given that identical-looking combs are found in all other short-tongued bee families, but are absent in most Halictinae. The Corynurina also differs from other Augochlorini in lacking the usual apical process of the male labrum.

Engel (2000b) also reviewed the six fossil species of Augochlorini found in the Oligomiocene Dominican amber. They belong to three genera, *Augochlora* (as its sub-

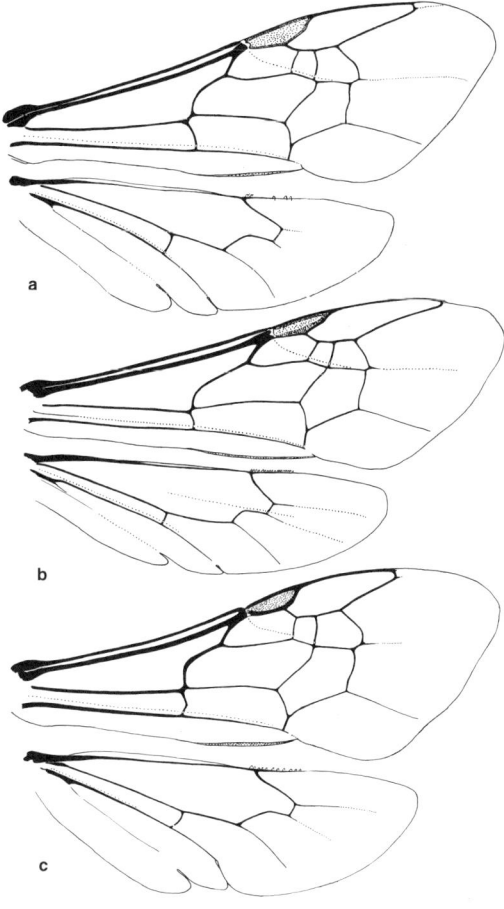

Figure 67-1. Wings of Augochlorini. **a,** *Augochlora pura* (Say); **b,** *Megalopta genalis* Meade-Waldo; **c,** *Augochloropsis metallica* (Fabricius).

genus *Electraugochlora* Engel), *Neocorynura,* and the extinct *Oligochlora* (and its subgenus or synonym *Soliapis* Engel).

Some Augochlorini construct cells isolated in the soil, but many construct groups of closely associated cells, commonly surrounded by an air space or by various burrows, so that the clump of cells can easily be freed from the matrix (Fig. 7-5). The clumps are supported by earthen connections or pillars, or by rootlets. Such clumps are much more common than in the Halictini. Cells are vertical and associated in clumps in *Augochloropsis, Caenaugochlora, Pseudaugochlora* (Fig. 67-2), *Rhinocorynura,* and *Megommation,* more or less horizontal although often variable in orientation in most other genera (Fig. 7-5). In forms that make clumps of cells, the cells may be either excavated into soil from above (Fig. 7-4) or built up in a space excavated in the soil (Fig. 67-3). Because even isolated cells are lined by a layer of earth brought from elsewhere in the nest, the difference be-

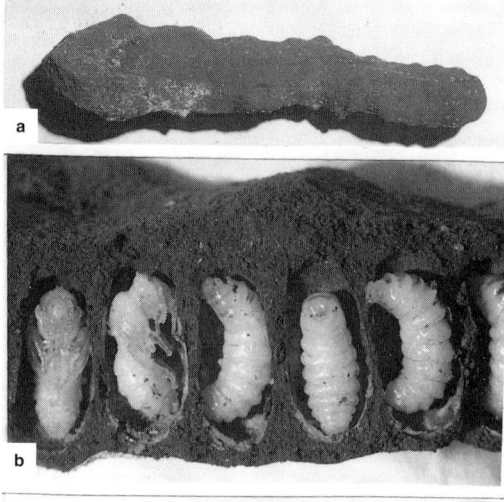

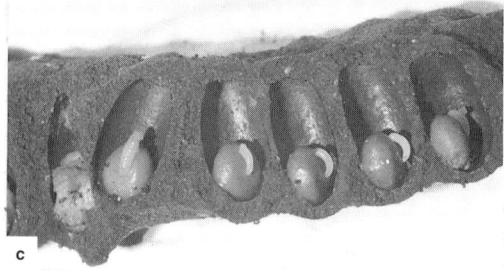

Figure 67-2. Cell cluster of *Pseudaugochlora sordicutis* (Vachal). **a**, Side view; **b**, Portion opened to show prepupae and pupae; **c**, Newer portion opened to show eggs and young larvae on food masses. Photos by C. D. Michener.

tween these two seemingly very distinct modes of construction is minimal. Cells built up in a space simply consist of cell linings. Exceptions to nests in soil occur in *Augochlora* s. str., which makes clumps of cells in rotting wood (Stockhammer, 1966), and *Megalopta*, which makes its cells, irregularly arranged, in rotting wood or vines (Sakagami, 1964). Many features of nest architecture were explained and illustrated by Sakagami and Michener (1962) and Eickwort and Sakagami (1979).

Cleptoparasitic Augochlorini are limited to the genus *Temnosoma* and one subgenus each of *Megalopta* and *Megommation*.

Key to the Genera of Augochlorini (Females)
(Modified from Eickwort, 1969b)

1. Proboscis extraordinarily slender, prementum 10 to over 20 times as long as broad (Fig. 67-4b); proboscidial fossa about as wide as mandibular base 2
—. Proboscis not so slender (Fig. 67-4a, c-e), prementum four to eight times as long as broad; proboscidial fossa much wider than mandibular base 6
2(1). Maxillary palpus much longer than stipes plus galea, reaching metasoma when mouthparts are in repose (basitibial plate moderately developed, much exceeding apex of femur, as in Fig. 67-5c) .... *Ariphanarthra*
—. Maxillary palpus not greatly lengthened 3
3(2). Eye greatly enlarged, projecting above vertex (Fig. 67-7b), separated from lateral ocellus and from vertical part of epistomal suture above tentorial pit by one-fourth of ocellar diameter or less (basitibial plate not short, much surpassing apex of femur; ocelli much enlarged) .... *Megaloptidia*
—. Eyes not projecting above vertex, widely separated from lateral ocellus and from epistomal suture above tentorial pit (Fig. 67-6, 67-13g) 4
4(3). Pronotal dorsal surface swollen, dorsolateral angle and vertical ridge below angle absent; paraocular lobe long, lower end acute (Fig. 67-13a, b); head elongated, owing to greatly lengthened clypeus; inner hind tibial spur serrate .... *Chlerogelloides*
—. Pronotal dorsal surface normal, not swollen; paraocular lobe approximately right-angular (Fig. 67-7a); head not greatly elongated; inner hind tibial spur pectinate or serrate 5
5(4). Basitibial plate very short, scarcely surpassing apex of femur .... *Megommation*
—. Basitibial plate moderately developed (ocelli small; inner hind tibial spur serrate) (placed here on the basis of the description; I have seen no specimens) .... *Micrommation*
6(1). Inner hind tibial spur serrate, teeth shorter than wide, pointed or rounded, or margin of spur undulate or ciliate (Fig. 67-5f, g, k-m) 7
—. Inner hind tibial spur pectinate, teeth longer than wide (Fig. 67-5h-j, n) 12
7(6). Scopa and median specialized area of T5 absent; integument coarsely punctate (Fig. 67-15); labral process truncate, without keel (Fig. 65-1i) .... *Temnosoma*
—. Scopa and median slit or notch and area of fine hairs on T5 present (Fig. 65-1k); integument rarely so coarsely punctate; labral process with strong median keel, usually pointed (Fig. 65-1a, e, g) 8
8(7). Inner hind tibial spur with rounded serrations (Fig. 67-5k-m); preoccipital ridge sharply angled or carinate .. 9
—. Inner hind tibial spur ciliate, with slender sharp teeth, as on outer spur, or with short, narrow teeth (Fig. 67-5f, g); preoccipital ridge sharply angled or rounded (epistomal suture forming obtuse paraocular lobe, Fig. 67-6a) 10
9(8). Epistomal suture forming acute paraocular lobe protruding into clypeus (Fig. 67-6d); clypeus relatively flat, green almost to apex; anterior angle of hypostomal carina usually a sharp right angle or produced into spine; apex of marginal cell narrowly truncate and usually appendiculate (Fig. 67-1a) .... *Augochlora*
—. Epistomal suture forming obtuse or right angle laterally (Fig. 67-6c); clypeus beveled, area below angle not green; anterior angle of hypostomal carina rounded; apex of marginal cell acute or very narrowly truncate .... *Augochlorella*
10(8). Basal area of propodeum completely smooth or granular, without striae; eyes deeply emarginate (Fig. 67-6a) (preoccipital ridge gradually rounded) .... *Corynura (Corynura)*
—. Basal area of propodeum roughened or with weak plicae basally; eyes moderately emarginate 11
11(10). Preoccipital ridge sharply angled; mesoscutum strongly narrowed anteriorly, lip high and sharply angled; labral basal elevation suborbiculate .... *Paroxystoglossa*

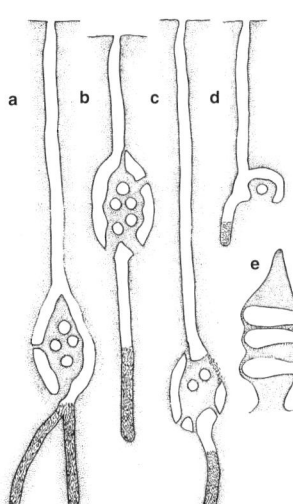

Figure 67-3. Nests of *Paroxystoglossa jocasta* (Schrottky) with cell clusters constructed of soft earth by the bees. **a, b,** Two normal nests; **c, d,** Other nests with abnormal or incomplete clusters. **e,** Sectional view of a cell cluster. From Sakagami and Michener, 1962.

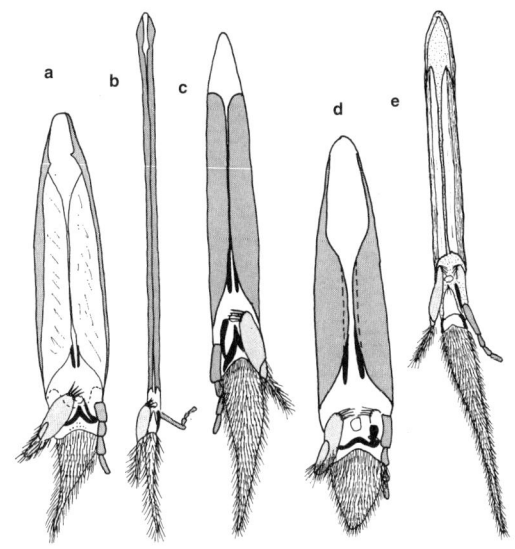

Figure 67-4. Labia (mentum and lorum omitted) of Augochlorini, each with one palpus and one paraglossa omitted. **a,** *Temnosoma smaragdinum* Smith; **b,** *Ariphanarthra palpalis* Moure; **c,** *Augochlora pura* (Say); **d,** *Corynura chilensis* (Spinola); **e,** *Pseudaugochlora graminea* (Fabricius). From Eickwort, 1969b.

—. Preoccipital ridge rounded; mesoscutum very slightly narrowed anteriorly, lip low and rounded; labral basal elevation transverse .... *Megaloptilla*

12(6). Inner hind tibial spur closely pectinate with about ten teeth (Fig. 67-5n) (basitibial plate defined posteriorly but anterior margin evanescent) 13

—. Inner hind tibial spur with not over six teeth, well-separated 14

13(12). Preoccipital carina strong, even almost lamella-like; apex of marginal cell minutely truncate; basal area of propodeum about as long as scutellum .... *Caenaugochlora (Ctenaugochlora)*

—. Preoccipital ridge rounded; apex of marginal cell pointed; basal area of propodeum about as long as scutellum and metanotum taken together (head at least as long as broad) .... *Chlerogella (Ischnomelissa)*

14(12). Malar area as long as wide or longer; head greatly elongate (Fig. 67-7e-g) 15

—. Malar area much shorter than wide; head not greatly elongate (Fig. 67-6) 16

15(14). Pronotum convex dorsally, dorsolateral angle and vertical ridge below angle absent; paraocular lobe acute (Fig. 67-7e); scutellum normal; flagellum ten-segmented .... *Chlerogella (Chlerogella)*

—. Pronotum normal in form; paraocular lobe right-angular (Fig. 67-7f, g); scutellum produced into two convexities; flagellum nine-segmented .... *Chlerogas*

16(14). Epistomal suture acutely angled laterally, forming strong paraocular lobe into clypeus (Fig. 67-7d); apical labral process broad, without coarse marginal bristles and with weak keel (Fig. 67-5a); mandible with preapical teeth on inner surface (Fig. 67-5o) (in addition to preapical tooth on upper margin) except in forms with reduced scopa 17

—. Epistomal suture variously angled, not forming strong lobe; apical labral process with coarse marginal bristles, usually narrow and pointed apically, and with strong keel (Figs. 65-1a, e, g; 67-5b, c); mandible without teeth on inner surface (eyes and ocelli not enlarged) 18

17(16). Eyes and ocelli enlarged, ocellocular distance thus equal to or less than ocellar diameter (Fig. 67-7d, 67-13d); hamuli in a closely packed series (Fig. 67-1b); hind tibia covered with amber setae .... *Megalopta*

—. Eyes and ocelli not enlarged, ocellocular distance twice ocellar diameter or more; hamuli in series broken by gaps (as in most other Augochlorini), some hooks thus isolated; hind tibia and basitarsus largely covered with black setae .... *Xenochlora*

18(16). Tegula with inner posterior angle produced mesally, forming an emargination in the posterior part of the mesal tegular margin (Fig. 67-5t); basitibial plate very short, poorly defined, extending barely past apex of femur (Fig. 67-8a); labrum with distal process expanded and rounded, as broad as body of labrum (Fig. 65-1g), distal keel projecting beyond apex (pronotal dorsal ridge between lateral angle and pronotal lobe lamellate, forming flange from lateral angle to lobe) .... *Augochloropsis*

—. Tegula with posterior end rounded; basitibial plate not short (Fig. 67-5q, r) [except in *Corynura (Callistochlora)* (Fig. 67-5p)]; labrum with distal process usually more slender and pointed (Figs. 65-1a, e; 67-5b, c), but may be expanded distally 19

19(18). Pronotal dorsal ridge lamellate, forming flange from dorsolateral angle to pronotal lobe (scutum strongly produced over pronotum, usually forming carinate or lamellate flange) 20

—. Pronotal dorsal ridge not lamellate 21

20(19). Vertex swollen behind ocelli; clypeus usually armed

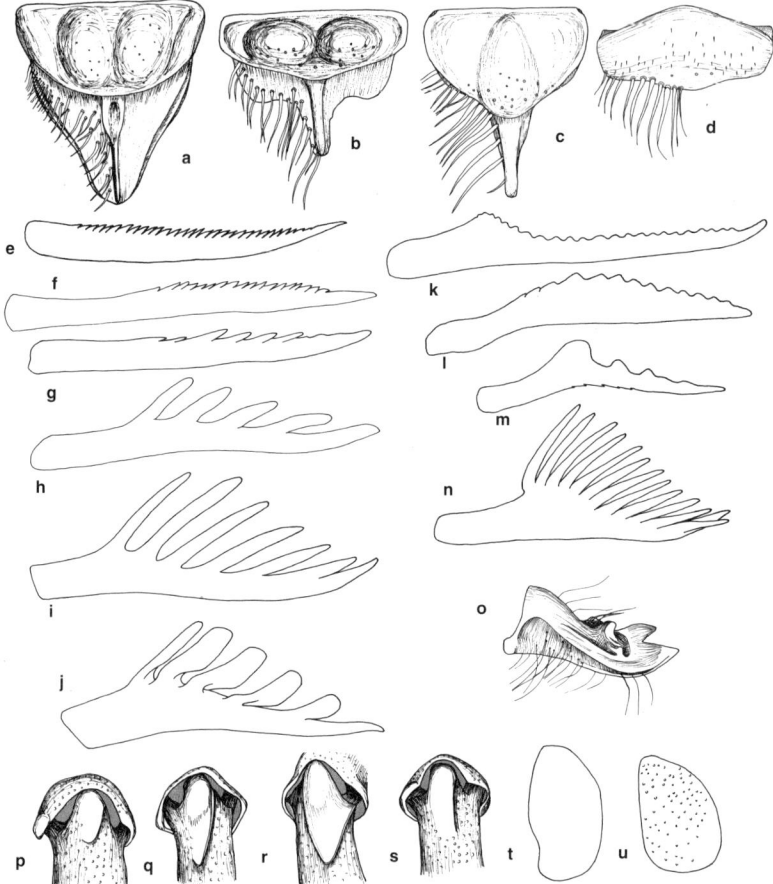

Figure 67-5. Structures of Augochlorini. **a,** Labrum of female, *Megalopta genalis* Meade-Waldo; **b,** Labrum of female, *Rhinocorynura briseis* (Smith); **c, d,** Labra of female and male, *Neocorynura pubescens* (Friese). (For labra of other Augochlorini, see Figure 63-1.)

e-n, Inner hind tibial spurs of females: **e,** *Temnosoma smaragdinum* Smith; **f,** *Corynura (Corynura) chilensis* (Spinola); **g,** *C. (C.) corynogaster* (Spinola); **h,** *C. (Callistochlora) chloris* (Spinola); **i,** *Augochloropsis ignita* (Smith); **j,** *Megalopta genalis* Meade-Waldo; **k,** *Augochlora nigrocyanea* (Cockerell); **l,** *Augochlorella (Ceratalictus) theia* (Schrottky); **m,** *Augochlorella (Pereirapis)* sp.; **n,** *Caenaugochlora (Ctenaugochlora) perpectinata* (Michener).

o, Inner view of female mandible of *Megalopta genalis* Meade-Waldo, showing tooth on inner surface; such modifications are characteristic of wood-nesting Augochlorini.

p-s, Basitibial plates of females: **p,** *Corynura (Callistochlora) chloris* (Spinola); **q,** *Andinaugochlora micheneri* Eickwort; **r,** *Pseudaugochlora graminea* (Fabricius); **s,** *Megalopta genalis* Meade-Waldo.

t, u, Tegular outlines, mesal margins to the left: **t,** *Augochloropsis metallica* (Fabricius); **u,** *Temnosoma smaragdinum* Smith.

From Eickwort, 1969b.

---

with spines or median tubercle (Fig. 67-7h); basal area of propodeum smooth, usually depressed transversely; base of T1 normal .... *Rhinocorynura*
—. Vertex not swollen; clypeus not armed; basal area of propodeum short, striate, completely covered with short, dense pile; T1 with basal enclosure formed by dense plumose hairs near petiole, usually containing large mites .... *Thectochlora*
21(19). Pronotal dorsolateral angle produced, strongly carinate anteriorly and laterally behind angle, dorsal ridge not carinate behind dorsolateral angle, but vertical ridge below angle carinate (mesoscutum slightly produced over pronotum, low and sharply angled; preoccipital ridge rounded) .... *Rhectomia*
—. Pronotal dorsolateral angle not strongly carinate anteriorly, *if* produced, then dorsal ridge carinate 22
22(21). Epistomal suture forming right-angular or slightly acute paraocular lobe laterally (as in Fig. 67-6c), sometimes, as in *Andinaugochlora*, obtuse (about 100°) in laterofrontal view but not or scarcely obtuse in direct frontal view 23
—. Epistomal suture forming distinctly obtuse angle laterally (Fig. 67-6a) 24
23(22). Basitibial plate well-defined posteriorly, obsolescent

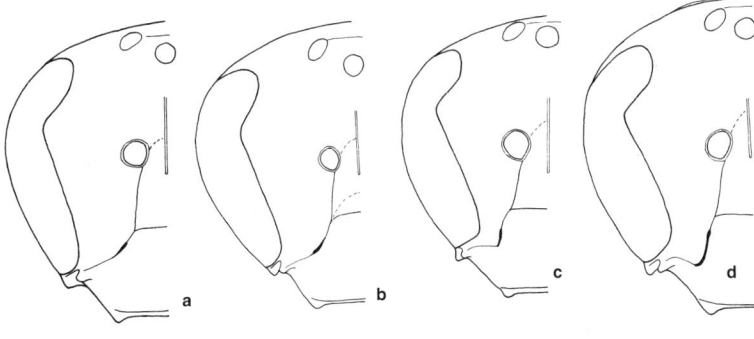

Figure 67-6. Faces of female Augochlorini. **a**, *Corynura chilensis* (Spinola); **b**, *Augochloropsis metallica* (Fabricius); **c**, *Augochlorella striata* (Provancher); **d**, *Augochlora pura* (Say). The paraocular lobe is progressively developed from b to d. From Eickwort, 1969b.

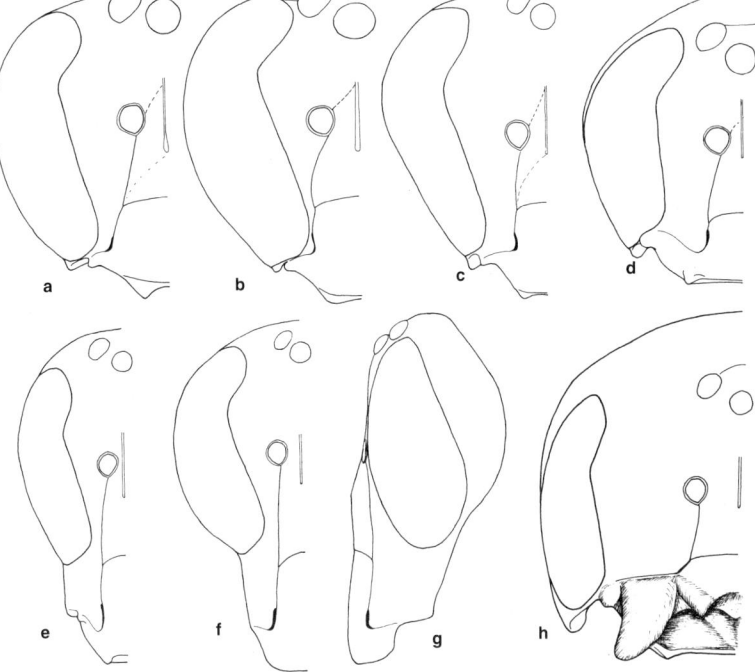

Figure 67-7. Faces of Augochlorini, females except for figures f and g. **a**, *Megommation insigne* (Smith); **b**, *Megaloptidia ?contradicta* (Cockerell); **c**, *Ariphanarthra palpalis* Moure; **d**, *Megalopta genalis* Meade-Waldo; **e**, *Chlerogella elongaticeps* Michener; **f, g**, *Chlerogas hirsutepennis* Cockerell, male; **h**, *Rhinocorynura inflaticeps* (Ducke). Note the well developed malar areas in e to g. From Eickwort, 1969b.

anteriorly (Fig. 67-5q); pronotal dorsolateral angle not produced, obtuse; eye hairs short .... *Andinaugochlora (Andinaugochlora)*
—. Basitibial plate well-defined on all edges (Fig. 67-5r); pronotal dorsolateral angle frequently produced, sometimes obtuse; eye hairs frequently long 29

24(22). Preoccipital ridge carinate or lamellate; mesoscutum *usually* narrowed anteriorly and anterior mesoscutal lip *usually* narrow, high, projecting forward and sharply angled; pronotal dorsolateral angle *usually* strongly produced, lateral ridge carinate or lamellate; propodeum *usually* narrowed posteriorly .... *Neocorynura*
—. Preoccipital ridge rounded; mesoscutum not narrowed anteriorly, the mesoscutal lip *usually* not high or narrow; pronotal dorsolateral angle *usually* not produced, lateral ridge sharply angled or rounded; propodeum not narrowed posteriorly 25

25(24). Head, mesosoma, and metasoma brilliant green or orange or coppery; propodeal basal area weakly striate ..26
—. Head and mesosoma dull metallic green or blue, black, or brownish; metasoma variously colored but not brilliant green; propodeal basal area smooth or roughened 28

26(25). Eyes with long hair; basitibial plate defined only posteriorly, very short (Fig. 67-5p) .... *Corynura (Callistochlora)*
—. Eyes with short hair; basitibial plate well defined on all edges, of normal length 27

27(26). Pronotal dorsal ridge strongly carinate; marginal cell with apex acute; body length approximately 7 mm .... *Augochlorodes*
—. Pronotal dorsal ridge rounded, not carinate; marginal cell with apex minutely rounded and appendiculate; body length 11.5-12.0 mm .... *Andinaugochlora (Neocorynurella)*

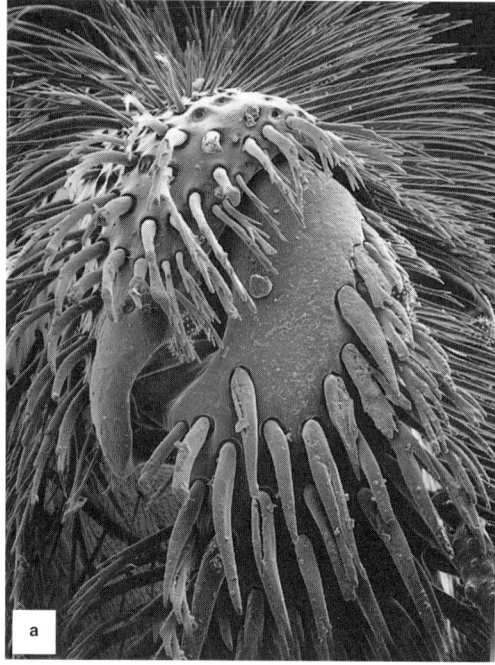
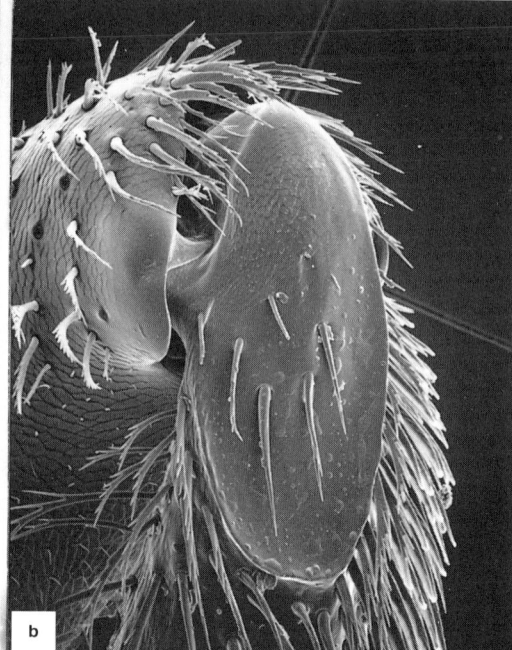

28(25). Basal area of propodeum roughened; length 5-6 mm [body dull metallic, suggesting that of *Lasioglossum (Dialictus)*] .... *Halictillus*
—. Basal area of propodeum smooth or granular; length usually over 6 mm .... *Corynura (Corynura)*
29(23). Vertex produced to rounded ridge behind ocelli (Fig. 67-14a); preoccipital ridge rounded; galea of maxilla with apex pointed, well sclerotized; basal elevation of labrum with apical surface flattened, rimmed (Fig. 65-1b) .... *Pseudaugochlora*
—. Vertex not produced; preoccipital ridge usually sharply angled or carinate; galea normal, with apical lobe; basal elevation of labrum without flattened distal surface .... *Caenaugochlora (Caenaugochlora)*

Key to the Genera of Augochlorini (Males)
(Modified from Eickwort, 1969b; males unknown for *Xenochlora*.)

1. Proboscis extraordinarily slender, prementum 10 to over 20 times as long as broad (Fig. 67-4b); proboscidial fossa about as wide as mandibular base 2
—. Proboscis not so slender, prementum four to eight times as long as broad (Fig. 67-4a, c-e); proboscidial fossa much wider than mandibular base 6
2(1). Maxillary palpus much longer than stipes plus galea, reaching metasoma when mouthparts are in repose .... *Ariphanarthra*
—. Maxillary palpus not greatly lengthened........................ 3
3(2). Eyes greatly enlarged, projecting above vertex, separated from lateral ocellus and from vertical part of epistomal suture above tentorial pit by one-fourth of ocellar diameter or less (Fig. 67-7b); S4 bilobed with deep median notch, the lobes strongly pilose (ocelli much enlarged) .... *Megaloptidia*

Figure 67-8. Apex of hind femur and basitibial plate of females. **a,** *Augochloropsis metallica* (Fabricius); **b,** *Augochlora pura* (Say). From Michener, McGinley, and Danforth, 1994.

—. Eyes not projecting above vertex, widely separated from lateral ocellus and epistomal suture above tentorial pit (Fig. 67-6); S4 not bilobed 4
4(3). Pronotal dorsal surface swollen; paraocular lobe long, acute; head elongate owing to greatly lengthened clypeus; middle femur sometimes greatly swollen and flattened .... *Chlerogelloides*
—. Pronotal dorsal surface normal, not swollen; paraocular lobe approximately right-angular (Fig. 67-7a); head not greatly elongate; middle femur unmodified 5
5 (4). Propodial spiracle surrounded by patch of dense, plumose setae, or ocelli greatly enlarged, or labrial distal process present............ .... *Megommation*
—. Propodeal spiracle not surrounded by dense, plumose setae and ocelli not enlarged, and labrial distal process absent......... .... *Micrommation*
6(1). Malar area as long as wide or longer; head greatly elongate (Fig. 67-7e-g) 7
—. Malar area much shorter than wide; head not greatly elongate 8
7(6). Pronotum convex dorsally, lateral angle a mere convexity and vertical ridge below angle absent; epistomal suture forming acute paraocular lobe laterally; scutellum normal; flagellum 11-segmented .... *Chlerogella* (in part)
—. Pronotum normal; epistomal suture forming right angle laterally (Fig. 67-7f, g); scutellum produced into two large tubercles; flagellum 10-segmented .... *Chlerogas*
8(6). Tegula with inner posterior angle produced mesally (Fig. 67-5t); posterior margins of T1 and T2 frequently

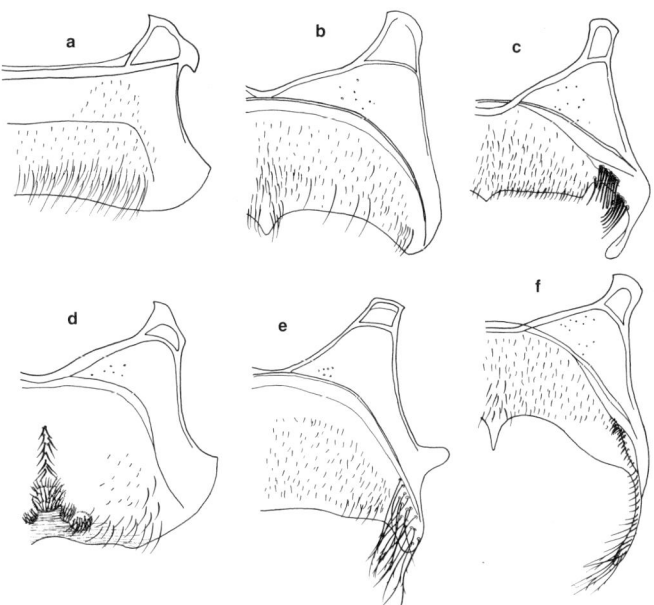

Figure 67-9. S4 of males of Augochlorini. **a,** *Augochlora pura* (Say); **b,** *Augochloropsis (Paraugochloropsis) chloera* (Moure), the species placed in *Glyptobasia* by Moure (1940, 1941a); **c,** *Augochloropsis (Augochloropsis) ignita* (Smith); **d,** *Caenaugochlora (Caenaugochlora) costaricensis* (Friese); **e,** *Augochlorodes turrifaciens* Moure; **f,** *Augochloropsis (Paraugochloropsis) metallica* (Fabricius). From Eickwort, 1969b.

each with a row of strong simple setae (pronotal dorsal ridge lamellate, forming flange from lateral angle to lobe; S4 usually with posterolateral appendage (Fig. 67-9c, f, but see also b), the appendage usually hidden or partly projecting from beneath T4; S4 posteriorly emarginate with median point; marginal cell truncate) .... *Augochloropsis*
—. Tegula with posterior end rounded; posterior margins of T1 and T2 without rows of strong setae 9

9(8). S4 with conspicuous apical or median tufts of specialized setae (Fig. 67-14b) or strong cuticular ridges and depressions 10
—. S4 usually not modified, apical margin at most emarginate or with usually hidden posterior lateral projections 13

10(9). Eyes and ocelli enlarged, ocellocular distance usually less than ocellar diameter (Fig. 67-7d); S3 bilobed; S4 with posterior margin notched laterally, bilobed medially, and with median flap; pronotal dorsal ridge rounded and narrow, anterior edge of dorsolateral angle and vertical ridge below dorsolateral angle both carinate .... *Megalopta*
—. Eyes and ocelli not enlarged (Fig. 67-6); S3 not strongly bilobed, sometimes medially emarginate; S4 not as above; pronotal dorsal ridge carinate or lamellate between dorsolateral angle and lobe, dorsolateral angle not strongly carinate on anterior margin, vertical ridge below dorsolateral angle variable 11

11(10). Pronotal dorsal ridge between dorsolateral angle and lobe, and anterior scutal lip, lamellate, forming flange; antenna long, scape less than 1.5 times length of flagellomere 2; S4 produced into caudally directed lateral processes bearing strong setae (usually hidden), posterior margin depressed and truncate, disc of S4 with median patch of short, erect, stout setae; S5 emarginate with long plumose pubescence .... *Thectochlora* (in part)
—. Pronotum and scutum not lamellate; antenna of variable length, scape more than 1.5 times length of flagellomere 2; S4 and S5 not as above 12

12(11). Epistomal suture forming obtuse paraocular angle laterally (as in Fig. 67-6b); S4 without distinct median or posterior setal patches, posterior margin emarginate and centrally depressed and shiny, or with shiny median apical depression bordered by sharp ridges (S5 with median shiny depression usually bordered by strong ridges) .... *Paroxystoglossa*
—. Epistomal suture forming right-angular paraocular lobe laterally; S4 with distinctive median or apical setal patches 28

13(9). Body very coarsely punctate; T1 to T3 strongly depressed basally (Fig. 67-15); T7 prolonged apically and bilobed .... *Temnosoma*
—. Body not very coarsely punctate; T1 to T3 not strongly depressed basally (Fig. 67-11); T7 not prolonged or bilobed 14

14(13). Pronotal dorsal ridge lamellate from dorsolateral angle to lobe; scutum strongly produced over pronotum, the anterior lip usually carinate or lamellate 15
—. Pronotal dorsal ridge not lamellate; scutum variable but anterior lip not carinate or lamellate 16

15(14). Vertex swollen above ocelli; basal area of propodeum smooth; S4 and S5 unmodified; scape more than twice length of second flagellar segment .... *Rhinocorynura*
—. Vertex not swollen; basal area of propodeum striate; S4 produced into caudally directed lateral process bearing strong setae (usually hidden), posterior margin medially depressed and truncate, disc of S4 with a median patch of short, erect, stout setae; S5 emarginate, with long plumose pubescence; scape less than 1.5 times as long as second flagellar segment .... *Thectochlora* (in part)

16(14). Antenna very long, scape usually shorter than second flagellar segment, never more than 1.25 times length of that segment, which is usually subequal to or longer than preapical segment (preoccipital ridge rounded) 17

—. Antenna of variable length, but scape over 1.25 times length of second flagellar segment, which is usually shorter than preapical segment 19

17(16). Basal area of propodeum about as long as scutellum and metanotum combined; propodeal pit narrow, not enclosed by V-shaped depression; paraocular lobe about right-angular; pronotal dorsal ridge not carinate; gonostylus not divided into dorsal and ventral gonostylar processes .... *Chlerogella (Ischnomelissa)* (in part)

—. Basal area of propodeum about as long as scutellum alone; propodeal pit enclosed by V-shaped depression; paraocular lobe obtuse; pronotal dorsal ridge carinate; gonostylus divided to form dorsal and ventral gonostyli (dorsal gonostylus, however, present as a distinct setose ridge) 18

18(17). Metasoma elongate, not petiolate, dull metallic blue-green; body about 5 to 6 mm long; bridge of penis valves shifted apicad to level of apices of gonocoxites .... *Halictillus*

—. Metasoma petiolate, first two segments long and narrow, *or*, if not, then body usually bright green or orange-green; body length usually over 6 mm; bridge of penis valves in usual position near level of middles of gonocoxites .... *Corynura*

19(16). Dorsolateral angle of pronotum produced and strongly carinate anteriorly and laterally (scutum slightly produced over pronotum; preoccipital ridge rounded) 20

—. Dorsolateral angle of pronotum not strongly carinate anteriorly, if produced, then dorsal ridge between angle and lobe carinate 21

20(19). Mesoscutum weakly narrowed anteriorly; dorsal pronotal ridge between lateral angle and lobe carinate, vertical ridge below angle not carinate; apical margin of S5 unmodified .... *Megaloptilla*

—. Mesoscutum broadly rounded anteriorly; dorsal pronotal ridge between lateral angle and lobe not carinate, vertical ridge below angle carinate; apical margin of S5 with weak median notch .... *Rhectomia*

21(19). Preoccipital ridge rounded 22

—. Preoccipital ridge sharply angled or carinate 23

22(21). Basal area of propodeum smooth; S5 with dense clump of setae medially; S6 shallowly notched along apical margin; dorsal lobes of gonobase weakly defined; venter of penis valve without prong .... *Augochlorodes*

—. Basal area of propodeum striate; S5 with scattered setae; S6 deeply notched along apical margin; dorsal lobes of gonobase strong; penis valve with ventral prong .... *Andinaugochlora (Neocorynurella)*

23(21). Epistomal suture forming distinctly obtuse lateral angle (as in Fig. 67-6b) 24

—. Epistomal suture forming acute or right-angular paraocular lobe laterally (Fig. 67-6c, d), the lobe sometimes, as in *Andinaugochlora*, obtuse (about 100°) in laterofrontal view but nearly right-angular in direct frontal view 26

24(23). Antenna reaching about to scutellum, scape more than 2.5 times length of second flagellar segment, which is subequal to first; scutum not narrowed and not produced over pronotum (integument bright green; tibiae and tarsi orange)...................... *Augochlorella (Ceratalictus)*

—. Antenna long, frequently surpassing propodeum, scape less than 2.5 times length of second flagellar segment, which is longer than first, usually more than 1.5 times length of first; scutum usually narrowed and produced over pronotum 25

25(24). S4 and S5 depressed medially and shiny, posterior margin of S4 broadly emarginate (as in Fig. 67-9a); venter of head and mesosoma and lower surfaces of legs clothed with long plumose pubescence; metasoma not petiolate .... *Paroxystoglossa*

—. S4 and S5 not modified; ventral surfaces without long plumose pubescence; metasoma frequently petiolate, S1 and S2 usually very long and narrow (compare with *Andinaugochlora*) .... *Neocorynura*

26(23). Antenna very long, surpassing propodeum; scape twice length of second flagellar segment or less; flagellum with plate areas containing only sensory plate-organs, these areas without setae; S6 strongly notched with deep concavities on either side of median notch .... *Andinaugochlora (Andinaugochlora)*

—. Antenna of moderate length, usually not surpassing propodeum; scape over three times length of second flagellar segment; flagellum without specialized plate areas (rarely long and with plate areas in *Augochlora*); S6 without deep concavities on either side of median notch 27

27(26). Epistomal suture forming acute lateral angle or paraocular lobe protruding into clypeus (Fig. 67-6d); marginal cell truncate, usually appendiculate (Fig. 67-1a) .... *Augochlora*

—. Epistomal suture forming right angle laterally (Fig. 67-6c); marginal cell acute, rarely very narrowly truncate .... *Augochlorella*

28(12). Apical flagellar segment tapering, hooked (Fig. 67-14c); preoccipital ridge rounded .... *Pseudaugochlora*

—. Apical flagellar segment rounded; preoccipital ridge sharply angled or carinate .... *Caenaugochlora*

Genus *Andinaugochlora* Eickwort

This genus consists of bright metallic green or red, rather hairy bees, 8 to 10 mm long, often with brassy or blue reflections or a metallic red metasoma. They are probably related to *Neocorynura*, but differ in the lack of any pronotal or anterior scutal modifications and in the nonpetiolate metasoma of males. The paraocular lobe is 90° (or 100° in frontolateral view) to 135°, whereas in *Neocorynura* (also in *Paroxystoglossa*, a probable close relative) it is even more obtuse. As in some *Neocorynura* there is sometimes a weak galeal comb.

It is probable that relationships would be best indicated if *Andinaugochlora* and *Paroxystoglossa* were considered as subgenera of *Neocorynura*. A. Smith-Pardo, who has studied *Neocorynura* in detail, considers *Andinaugorchlora* a basal branch (sister-group to the rest) of *Neocorynura* phylogeny.

Key to the Subgenera of *Andinaugochlora*

1. Preoccipital carina present although weak behind vertex; anterior margin of basitibial plate of female absent; paraocular lobe near right-angular (about 100°); penis valve without ventral prong .... *A. (Andinaugochlora s. str.)*
—. Preoccipital carina absent; anterior margin of basitibial plate of female a distinct carina; paraocular lobe strongly obtuse (about 135°); penis valve with ventral prong .... *A. (Neocorynurella)*

Andinaugochlora / Subgenus *Andinaugochlora* Eickwort s. str.

Andinaugochlora Eickwort, 1969b: 407. Type species: *Andinaugochlora micheneri* Eickwort, 1969, by original designation.

The characters of the subgenus are indicated in the key to subgenera.
■ The subgenus is found in the mountains (2,500 m altitude or higher) of Venezuela, Colombia, Ecuador, and Peru. Two species have been described, but Eickwort (1969b) recognized three others. A Costa Rican species recently described (Engel and Smith-Pardo, 2004) is now considered to be a species of *Neocorynura* (A. Smith-Pardo, personal comm.).

Andinaugochlora / Subgenus *Neocorynurella* Engel

Neocorynurella Engel, 1997, *in* Engel and Klein, 1997: 156. Type species: *Neocorynurella seeleyi* Engel and Klein, 1997, by original designation.
Vachalius Moure, 1999b: 74. Type species: *Halictus cosmetor* Vachal, 1911, by original designation.

Although originally described as a genus, *Neocorynurella* seems to differ from *Andinaugochlora* only in the characters indicated in the key to subgenera.
■ *Neocorynurella* is known only from the Venezuelan Andes, at altitudes of 3,900 to 4,300 m for *Andinaugochlora seeleyi* Engel and Klein. Two species were distinguished by Engel and Klein (1997).

Genus *Ariphanarthra* Moure

Ariphanarthra Moure, 1951b: 137. Type species: *Ariphanarthra palpalis* Moure, 1951, by original designation.

This genus includes a bright blue-green species with extraordinarily long maxillary palpi and long and very narrow stipes and prementum (Fig. 67-4b). It is obviously a relative of *Megommation* but does not have the reduced female basitibial plate characteristic of that genus.

■ This genus occurs from Minas Gerais to Paraná in Brazil, and also in Paraguay and eastern Peru. The only species is *Ariphanartha palpalis* Moure.

Genus *Augochlora* Smith

This genus consists of black to bright green, blue, or brassy bees of small to moderate size (length 5-11 mm). From similar forms, such as *Augochlorella*, it differs in its acute (but apically rounded) paraocular lobe protruding into the clypeus (Fig. 67-6d), the minutely truncate apex of the marginal cell (Fig. 67-1a), and the sharp angle or tooth at the bend of the hypostomal carina. Male genitalia were illustrated by Sandhouse (1937), as well as by Eickwort (1969b); see also Figure 67-10c, d.

Species found in the United States were revised by Sandhouse (1937).

The nests include clusters of more or less horizontal cells, or the cells are more scattered and excavated into the substrate. *Augochlora* s. str. builds in rotting wood (Stockhammer, 1966; Eickwort and Eickwort, 1973a), whereas the subgenus *Oxystoglossella* nests in the soil. At least some species of the latter subgenus have primitively social behavior with a worker caste (Michener and Lange, 1958a; Eickwort and Eickwort, 1972a; Sakagami and Moure, 1967). Their nests often have burrows that partly surround a cell cluster; this seems to be the first stage in forming isolated cell clusters. Another work on the nesting behavior of *Augochlora s.str.* is by Wcislo, Gonzalez, and Engel (2003).

Key to the Subgenera of *Augochlora*

1. Female: Mandible normal, preapical tooth not produced, far from apex of mandible; median area of T5 with long setae, not scale-like; S1 usually normal, rarely with slight median ridge or tooth; basitibial plate usually narrowly rounded, occasionally broadly rounded; basal elevation of labrum suborbicular. Male: Ventral gonostylus with long setae, greatly surpassing gonostylus ....
.... *A. (Oxystoglossella)*
—. Female: Mandible bidentate, preapical tooth and occa-

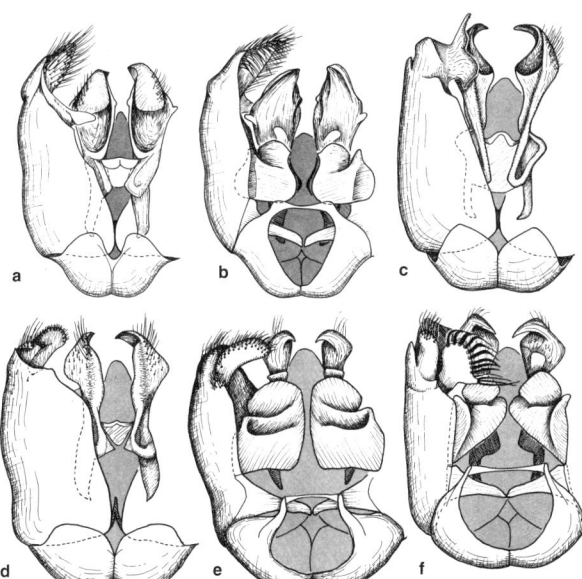

Figure 67-10. Male genitalia of Augochlorini, dorsal and ventral views. **a, b,** *Paroxystoglossa transversa* Moure; **c, f,** *Augochlorella striata* (Provancher); **d, e,** *Augochlora pura* (Say). See Figures 65-1l, m, and 67-12 for genitalia of other augochlorines. From Eickwort, 1969b.

sionally outer ridge produced; median area of T5 with scale-like setae; S1 frequently with tooth or median ridge, sometimes unarmed; basitibial plate broadly rounded (Fig. 67-8b); basal elevation of labrum transverse. Male: Ventral gonostylus with short setae (Fig. 67-10d) .... *A. (Augochlora s. str.)*

Augochlora / Subgenus *Augochlora* Smith s. str.

Augochlora Smith, 1853: 73. Type species: *Halictus purus* Say, 1837, designated by Cockerell, 1923b: 448.
Oxystoglossa Smith, 1853: 83. Type species: *Oxystoglossa decorata* Smith, 1853, monobasic.
Angochlora Schrottky, 1901: 213, incorrect subsequent spelling (repeated) of *Augochlora* Smith, 1853.
Odontochlora Schrottky, 1909a: 141. Type species: *Augochlora mulleri* Cockerell, 1900, by original designation.
Augochlora (Mycterochlora) Eickwort, 1969b: 423. Type species: *Halictus repandirostris* Vachal, 1911, by original designation.

Mycterochlora is based on species of *Augochlora* having unusual derived characters; see Wcislo, Gonzalez and Engel (2003).
■ *Augochlora* s. str. is found from the province of Buenos Aires, Argentina, north through tropical South America (to Fernando Noronha off northeastern Brazil), Mesoamerica, and the Antilles; in North America, from Texas to Florida, north to Minnesota, USA, and Quebec, Canada. The species are particularly numerous in the moist tropics. Moure and Hurd (1987) recognized 86 species.

Augochlora / Subgenus *Oxystoglossella* Eickwort

Augochlora (Oxystoglossella) Eickwort, 1969b: 422. Type species: *Augochlora cordiaefloris* Cockerell, 1907, by original designation.
Augochlora (Aethechlora) Moure and Hurd, 1987: 275. Type species: *Augochlora matucanensis* Cockerell, 1914, by original designation.

This subgenus includes the more "ordinary" species of *Augochlora*. The female mandibles lack the characteristics presumably associated with nesting in rotting wood; nests are in the ground.
■ The range is from Buenos Aires Province, Argentina, north to southernmost Texas and Arizona, USA. Moure and Hurd (1987) listed 27 species.

Genus *Augochlorella* Sandhouse

This genus is understood here in a broader sense than in the past, to include the forms usually placed in *Ceratalictus* and *Pereirapis*. The three taxa are closely related; I believe the classification should serve to indicate relationships of such groups. The labrum of females has a broadly triangular elevated basal area in *Pereirapis* and a rounded, median elevation in the other subgenera, except that a small *Augochlorella* s. str., *A. edentata* Michener, is *Pereirapis*-like in labral structure. Most of the characters cited by Eickwort (1969b) to separate the three subgenera, however, do not intergrade; the three groups are quite distinctive, although sometimes, as a practical matter, difficult to separate without careful study and, for males, dissections.

Augochlorella consists of small (length 4.5 8.0 mm), bright green (Pl. 5), blue, or brassy species. They are common in the eastern and central United States and in Mesoamerica and can be distinguished from the superficially similar *Augochlora* most easily by the pointed rather than minutely truncate marginal cell. In South America they are less common.

Key to the Subgenera of *Augochlorella* (Females)

1. Epistomal suture forming obtuse paraocular lobe (as in Fig. 67-6b); inner hind tibial spur broadest near middle (Fig. 67-5l); basitibial plate poorly defined on anterior edge (preoccipital ridge carinate) .... *A. (Ceratalictus)*
—. Epistomal suture forming right-angular or slightly obtuse paraocular lobe (Fig. 67-6c); inner hind tibial spur broadest at basal one-third to one-fifth; basitibial plate well defined on all edges 2
2(1). Inner hind tibial spur with few short, rounded teeth, basal tooth largest, forming broadest part of spur (Fig. 67-5m); basal area of propodeum strongly granular, striate basally; body length about 5 mm .... *A. (Pereirapis)*
—. Inner hind tibial spur with rounded serrations, basal part broadened and serrate, i.e., not formed from a single tooth; basal area of propodeum not strongly granular; body length usually over 5 mm ...... *A. (Augochlorella s. str.)*

Key to the Subgenera of *Augochlorella* (Males)

1. Epistomal suture forming obtuse paraocular lobe; dorsal gonostylus with long, narrow, hairless basal process extending mesad to midline .... *A. (Ceratalictus)*
—. Epistomal suture forming right-angular paraocular lobe; dorsal gonostylus without such a mesal process 2
2(1). S4 broadly emarginate posteriorly, laterally bearing long, modified setae that are usually hidden; body length about 5 mm; inner lobe of ventral gonostylus without row of coarse setae .... *A. (Pereirapis)*
—. S4 not broadly emarginate, without long lateral setae; body length usually over 5 mm; inner lobe of ventral gonostylus bearing marginal row of coarse, flattened setae (Fig. 67-10f) .... *A. (Augochlorella s. str.)*

Augochlorella / Subgenus *Augochlorella* Sandhouse s. str.

Augochlorella Sandhouse, 1937: 66. Type species: *Augochlora gratiosa* Smith, 1853, by original designation.
Oxystoglossidia Moure, 1943b: 473. Type species: *Oxystoglossidia uraniella* Moure, 1943 = *Oxystoglossa ephyra* Schrottky, 1911, by original designation.

Genitalia of this subgenus were illustrated by Sandhouse (1937) and by Eickwort (1969b); see Figure 67-10e, f.
■ This subgenus ranges from the province of Buenos Aires, Argentina, north through tropical and north-temperate America to Nova Scotia and Alberta, Canada, west to Utah and California, USA. It appears to be absent in the Antilles and in much of the Great Basin, the northwestern United States, and western Canada. Moure and Hurd (1987) listed 16 species. The species of America north of Mexico (with much information on the species in Mexico) were revised by Ordway (1966a). Coelho (2004) revised the subgenus *Augochlorella* s. str. She rec-

ognized 15 species, about equally divided between South American species and North and Central American species.

The biology of two North American species was discussed by Ordway (1965, 1966b) and Mueller (1997); of Brazilian species, by Michener and Lange (1958a) and Sakagami and Moure (1967); and of a Costa Rican species by Eickwort and Eickwort (1973b). Packer (1990) reported on a population of *Augochlorella striata* (Provancher) at the northern limit of the range of the genus in Nova Scotia, Canada, where many nests produced no workers. The average was 0.5 workers per nest, compared to 2+ for the same species in New York state and Kansas. Eusocial colonies are common in the USA, although loss of the foundress and her replacement by a worker, resulting in a semisocial colony, occurs in from 30 to over 60 percent of the nests (Mueller, 1997).

Augochlorella / Subgenus *Ceratalictus* Moure

Ceratalictus Moure, 1943b: 463. Type species: *Oxystoglossa theia* Schrottky, 1911, by original designation.

The bees of this subgenus resemble *Augochlorella* s. str. The principal characters are indicated in the key to subgenera.
■ This group is found in Brazil from São Paulo southward, and in Paraguay, eastern Bolivia, and Peru. Moure and Hurd (1987) listed five species.

Augochlorella / Subgenus *Pereirapis* Moure

Pereirapis Moure, 1943b: 461. Type species: *Pereirapis rhizophila* Moure, 1943 = *Halictus semiauratus* Spinola, 1851, by original designation.

This subgenus consists of small species, usually not much over 5 mm long and thus smaller than most species of *Augochlorella* s. str.
■ The range is from Brazil (state of São Paulo) north to the states of Veracruz and Jalisco, Mexico. Moure and Hurd (1987) listed six species, but M. S. Engel thinks that all the names relate to a single variable species.

The nesting biology of one species was studied in detail by Campos (1980).

Genus *Augochlorodes* Moure

Augochlorodes Moure, 1958a: 53. Type species: *Augochlorodes turrifaciens* Moure, 1958, by original designation.

This is a genus of moderate-sized (length about 7.5 mm), bright-green bees resembling *Augochlorella* and green *Augochlora* in appearance. Although the genitalic and sternal structures of the male (for S4, see Fig. 67-9e) resemble those of *Augochloropsis*, the rounded posterior end of the tegula and pointed marginal cell differ from those of *Augochloropsis*. The noncarinate preocciptal ridge, obtusely angled paraocular lobe (as in *Augochloropsis*), and pectinate inner hind tibial spurs of the female separate it from the genera that it resembles superficially, such as *Augochlorella* and *Augochlora*.
■ This genus is known only from Paraná to Minas Gerais, Brazil. The only species is *Augochlorodes turrifaciens* Moure.

Figure 67-11. *Augochloropsis sumptuosa* (Smith). From Michener, McGinley, and Danforth, 1994.

Nests were described by Michener and Seabra (1959); they are short burrows terminating in clusters of sloping cells surrounded by an air space.

Genus *Augochloropsis* Cockerell

This is a genus of metallic green, blue, golden or red to nonmetallic black bees, 5 to 13 mm in length and more robust than most other Augochlorini (Fig. 67-11); perhaps they should be called euceriform rather than andreniform, although, because they lack long hair, they do not actually resemble other euceriform bees. In no other genus is there the mesal projection of the posterior end of the tegula (also describable as an emargination on the posterior mesal tegular margin) that is characteristic of *Augochloropsis*. The small, short, poorly defined basitibial plate of the female (Fig. 67-8a) suggests *Megommation*, but is actually rather different and presumably independently reduced. Most species have a marginal row of simple bristles on T1 and T2; no other genus has such bristles, but not all *Augochloropsis* have them either. Like *Augochlorodes*, S4 of nearly all males has an apicolateral process bearing long setae (Fig. 67-9c, f, but also b), this usually hidden by T4 unless artificially exposed. The thoracic and labral characters listed in the key to genera are also distinctive. For facial characters, see Figure 67-6b. The subgenera are not easily separated except on the basis of the male sterna and genitalia. Many species have not been placed subgenerically. Male genitalia were illustrated by Sandhouse (1937) as well as by Eickwort (1969b); see also Figure 67-12a, b.
■ The genus ranges from the provinces of Buenos Aires and La Pampa, Argentina, north to Arizona, USA, in western North America, and to Ontario, Canada, in eastern North America. It is not known from the Antilles or from western North America north of Arizona. Moure and Hurd (1987) listed 138 species, most of them from tropical America. The species found in the United States were revised by Sandhouse (1937). Because so many species have not been reliably placed subgenerically, distributional and other information is given here, under the genus.

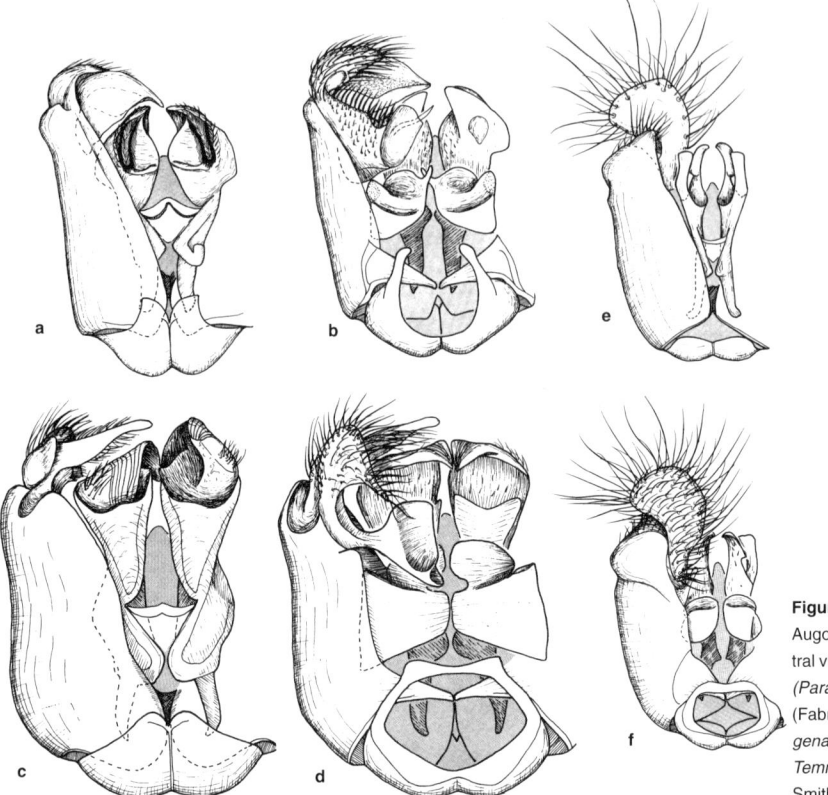

Figure 67-12. Male genitalia of Augochlorini, dorsal and ventral views. **a, b,** *Augochloropsis (Paraugochloropsis) metallica* (Fabricius); **c, d,** *Megalopta genalis* Meade-Waldo; **e, f,** *Temnosoma smaragdinum* Smith. From Eickwort, 1969b.

Nests are in the soil and include clusters of vertical cells opening into a common space usually joined by a horizontal burrow (lateroid) to the main burrow. The pollen masses, rectangular in outline (Fig. 7-4), are attached to the flatter side of the cell. Nests have been described by Michener and Lange (1959), Sakagami and Michener (1962), Sakagami and Moure (1967), and others.

Key to the Subgenera of *Augochloropsis*

1. Basal area of propodeum deeply and regularly pitted or strongly striate basally, smooth apically; S5 of male with apical margin notched, and with long, curved setae centrally; S6 of male with wide apical flange (broken medially) projecting beyond posterior border; distal process of male labrum truncate or notched apically; ventral gonostylus of male bearing row of strong, flattened, modified setae at apex .... *A. (Augochloropsis s. str.)*
—. Basal area of propodeum smooth, irregularly roughened, or with light plicae; S5 of male with apical margin entire, setae not modified; S6 of male with apical flange little developed; distal process of male labrum rounded; ventral gonostylus of male bearing dense unmodified setae at apex (Fig. 67-12b) .... *A. (Paraugochloropsis)*

Augochloropsis / Subgenus *Augochloropsis* Cockerell s. str.

Augochloropsis (Augochloropsis) Cockerell, 1897b: 4. Type species: *Augochlora subignita* Cockerell, 1897 = *Augochlora ignita* Smith, 1861, by original designation.

Angochlora (Angochloropsis) Schrottky, 1901: 213, incorrect subsequent spelling for *Augochloropsis* Cockerell, 1897.

■ *Augochloropsis* s. str. is probably more strictly tropical than is *Paraugochloropsis*; it is found north as far as Veracruz, Mexico (see the account of the genus). I suggest that about one-third of the described species belong here, the rest in *Paraugochloropsis*.

Augochloropsis / Subgenus *Paraugochloropsis* Schrottky

Augochloropsis (Paraugochloropsis) Schrottky, 1906a: 312. Type species: *Augochloropsis lycorias* Schrottky, 1906 = *Augochlora epipyrgitis* Holmberg, 1903, monobasic.
Augochloropsis (Pseudaugochloropsis) Schrottky, 1906a: 313. Type species: *Augochloropsis sthena* Schrottky, 1906, by designation of Sandhouse, 1943: 593. [For an invalid type designation, see Michener, 1997b.]
Augochlora (Tetrachlora) Schrottky, 1909b: 481. Type species: *Halictus multiplex* Vachal, 1903, monobasic.
Paraugochlora Schrottky, 1910: 540. Type species: *Augochlora spinolae* Cockerell, 1900, by original designation.
Rivalisia Strand, 1921: 270. Type species: *Rivalisia metallica* Strand, 1921 = *Augochlora aenigma* Engel, 1996, 2000b, monobasic.
Augochlora (Glyptobasis) Moure, 1940: 48 (not M'Lachlan, 1871). Type species: *Augochlora chloera* Moure, 1940, by original designation.
Augochlora (Glyptobasia) Moure, 1941a: 98, replacement for

Glyptobasis Moure, 1940. Type species: *Augochlora chloera* Moure, 1940, autobasic.

Augochloropsis (*Glyptochlora*) Moure, 1959: 188. Type species: *Megalopta ornata* Smith, 1879, by original designation.

■ This subgenus has the distribution of the genus; see the generic account above.

Two unusual, probably derived species have received the subgeneric names *Glyptobasia* and *Glyptochlora*. Both are unusually coarsely sculptured. Males of *Glyptobasia* lack the lateroapical processes of S4 that are generally characteristic of the genus (Fig. 67-9b), but the genital characters are as in *Paraugochloropsis*. Males of *Glyptochlora* are unknown. In the female the anterior part of the mesoscutum is narrowed, terminating in a forward-directed lamella, unlike that of other *Augochloropsis*. I follow Eickwort (1969b) in including these forms in *Paraugochloropsis*, although Engel (2000b) recognized *Glyptochlora* as a distinct subgenus.

Genus *Caenaugochlora* Michener

Caenaugochlora consists of moderate-sized to large (length 7-14 mm) bees, metallic green or coppery to black, in which the epistomal suture forms a right-angular paraocular lobe laterally and S4 of the male has median or apical setal patches (Fig. 67-9d), sometimes raised on tubercles. The marginal cell is usually narrowly truncate at the apex. *Pseudaugochlora* differs from *Caenaugochlora* principally in certain derived characters; species of *Caenaugochlora* like *C. costaricensis* (Friese) that have large body size and short hairs on the eyes are particularly similar to *Pseudaugochlora*. The male gonostylus is quite different, but the volsellae are remarkably similar. My inclination is to regard *Caenaugochlora* and *Pseudaugochlora* as congeneric, but this view is not supported by M. Engel's phylogenetic study (MS).

Nest cells are mostly vertical, although the orientation may be variable. They are in rather large clumps with thick earthen walls. They have been studied by Michener and Kerfoot (1967), as *Pseudaugochloropsis costaricensis* (Friese). No great differences exist between nests of *Caenaugochlora* (*Caenaugochlora*) *costaricensis* (Friese) and those of species of the genus *Pseudaugochlora*.

Key to the Subgenera of *Caenaugochlora*

1. Basitibial plate of female well defined on all edges; inner hind tibial spur of female pectinate with well separated teeth (as in Fig. 67-5i); eye hair frequently long; S4 of male bearing distinctive setal clumps on raised tubercles on either side of apical median depression (Fig. 67-9d) .... *C.* (*Caenaugochlora* s. str.)
—. Basitibial plate of female obsolescent anteriorly; inner hind tibial spur of female closely pectinate, with over ten teeth (Fig. 67-5n); eye hair short; S4 of male with V-shaped patch of setae bordering slight median depression .... *C.* (*Ctenaugochlora*)

Caenaugochlora / Subgenus *Caenaugochlora* Michener s. str.

Caenaugochlora (*Caenaugochlora*) Michener, 1954b: 76. Type species: *Caenaugochlora macswaini* Michener, 1954, by original designation.

Most species are brilliant metallic green or coppery red, but some rather dull greenish species are included.

■ This subgenus is primarily Mesoamerican, occurring as far north as Sinaloa and San Luis Potosí, Mexico, but is reported to occur also south to Ecuador. There are 13 species (Moure and Hurd, 1987). The species called *Caenaugochlora curticeps* (Vachal) by Michener and Lange (1958a) is in reality a *Caenohalictus* (tribe Halictini) and has nothing to do with *Caenaugochlora*.

Caenaugochlora / Subgenus *Ctenaugochlora* Eickwort

Caenaugochlora (*Ctenaugochlora*) Eickwort, 1969b: 435. Type species: *Neocorynura perpectinata* Michener, 1954, by original designation.

The species of this subgenus are greenish to blue or violet, sometimes nearly black, and superficially resemble some species of *Neocorynura*.

■ The known range is Panama to the state of Veracruz, Mexico. The four species were recognized and reviewed by Engel (1996c).

Genus *Chlerogas* Vachal

Chlerogas Vachal, 1904: 127. Type species: *Halictus chlerogas* Vachal, 1904, monobasic and absolute tautonomy.

In its long head and malar areas, this genus resembles *Chlerogella*, from which it differs in its large size (10.5-14.0 mm long) and bituberculate scutellum. The 12-segmented male and 11-segmented female antennae are unique in Augochlorini; indeed, in no other bees except some species of *Euryglossina* (Colletidae) are the number of antennal segments reduced in both sexes. Coloration varies from bright green to largely blackish.

■ *Chlerogas* is known in the Andes from Venezuela to Bolivia. There are nine described species; see the revision by Brooks and Engel (1999).

Genus *Chlerogella* Michener

This is a genus of rather slender Augochlorini, varying from largely testaceous with greenish tints to dark blue-green, the metasoma sometimes black. Most species are distinguishable from all other Augochlorini except *Chlerogas* and *Chlerogelloides* by their elongate head (Fig. 67-7e). *Chlerogelloides* is very different, the head elongation not involving the malar area (Fig. 67-13a), whereas in *Chlerogella*, if the head is elongate, the malar area is about as long as broad to much longer than broad. Major differences between *Chlerogas* and *Chlerogella* are indicated in the key to genera, in which *Chlerogella* emerges twice because of the variable elongation of the malar areas. In two species, *Chlerogella* (*Ischnomelissa*) *cyanea* Brooks and Engel and *zonata* Engel, the malar areas are linear, i.e., virtually absent.

I here place *Ischnomelissa* as a subgenus of *Chlerogella* because of their numerous common characters, such as the usually elongate malar area, the usually swollen dorsum of the pronotum (with mere convexities for dorsolateral angles), the long propodeum (the length of the basal area being about equal to the combined lengths of the scutellum and the metanotum), and the dull, nonstriate basal area of the propodeum.

Key to the Subgenera of *Chlerogella*

1. Inner hind tibial spur of female pectinate with over ten close teeth (as in Fig. 67-5n); paraocular lobe slightly acute; male gonostylus undivided .... *C. (Ischnomelissa)*
—. Inner hind tibial spur of female with fewer, well-separated teeth (as in Fig. 67-5i); paraocular lobe strongly acute; male gonostylus divided into dorsal and ventral processes .... *C. (Chlerogella s. str.)*

Chlerogella / Subgenus *Chlerogella* Michener s. str.

Chlerogella Michener, 1954b: 75. Type species: *Chlerogella elongaticeps* Michener, 1954, by original designation.

Most species of this subgenus are largely testaceous. The dorsally convex pronotum, with convexities replacing the dorsolateral angles, is usually more prominent than in *Ischnomelissa*. Body length is 6 to 8 mm.

■ This subgenus ranges from Peru to Costa Rica, in the mountains. Three published names are associated with the subgenus (Moure and Hurd, 1987), and about 15 additional species are in a manuscript by M. Engel and R. Brooks.

Chlerogella / Subgenus *Ischnomelissa* Engel

Ischnomelissa Engel, 1997c: 42. Type species: *Ischnomelissa zonata* Engel, 1997, by original designation.

The name *Ischnomelissa* was given to a group of species that resemble *Chlerogella* in most features but differ in the characters given in the key to subgenera, and in their generally dark color, dark blue-green or with the metasoma black. Two species lack malar areas, whereas others have malar areas about as long as broad. The body length is 6.5 to 10.0 mm.

■ This subgenus occurs in the Andes from Colombia to Peru. Seven species have been described, and keys were provided by Brooks and Engel (1998) and Engel and Brooks (2000).

Genus *Chlerogelloides* Engel, Brooks, and Yanega

Chlerogelloides Engel, Brooks, and Yanega, 1997: 3. Type species: *Chlerogelloides femoralis* Engel, Brooks, and Yanega, 1997, by original designation.

In size (5.4-7.0 mm body length), form, coloration, and the elongate form of the head, *Chlerogelloides* resembles species of *Chlerogella* s. str. The cephalic elongation is produced by the greatly lengthened clypeus and supraclypeal region (Fig. 67-13a) in *Chlerogelloides,* whereas in both *Chlerogas* and *Chlerogella* the elongation involves the greatly lengthened malar spaces. Both *Chlerogella* s. str. and *Chlerogelloides* share a peculiar inflation of the pronotal dorsal surface, this surface being glabrous or finely imbricate. *Chlerogelloides* can be seprated from *Chlerogella* by the short malar space, the serrate inner hind tibial spur of the female, the unique sharply pointed and slender paraocular lobes, and the slender proboscis, as in *Megommation*. Although the proboscis is long, the prementum being about as long as the head, the proboscidial fossa is not so narrow as in *Megommation* and the slenderness of the proboscis is probably not homologous to that of *Megommation*. One of the strangest features of *Chlerogelloides* is the long, toothless male mandible, its distal two-fifths strongly bent mesad and extremely attenuate, the apex compressed and blunt.

■ *Chlerogelloides* is found in the state of Amazonas, Brazil, and adjacent parts of Peru, Ecuador, Colombia, and French Guiana. Two species are known.

Genus *Corynura* Spinola

This genus contains diverse-looking elements, from brilliant green to dull metallic or black, the metasoma sometimes red. The body length is 6 to 11 mm. The lack of paraocular lobes is shown in Figure 67-6a, and male genitalia in Figure 65-1l, m. The mouthparts are as described for *Halictillus*. The metasoma of males is petiolate or at least elongate.

The nests of both subgenera of *Corynura* were described and illustrated by Claude-Joseph (1926); see also Sakagami and Michener (1962). The cells, which range from horizontal to almost vertical, are in clusters.

The females of the two subgenera come out at different places in the key to genera because in *Corynura* s. str. the inner hind tibial spur is serrate, whereas in *Callistochlora* it is strongly pectinate.

Key to the Subgenera of *Corynura*

1. Integument bright green or orange-green; inner hind tibial spur of female pectinate (Fig. 67-5h); hairs on eyes long; metasoma of male elongate but not distinctly petiolate .... *C. (Callistochlora)*
—. Integument not bright green, metasoma dull metallic black, brown, or red; inner hind tibial spur of female serrate (Fig. 67-5f, g); hairs on eyes short; metasoma of male usually distinctly petiolate .... *C. (Corynura s. str.)*

Corynura / Subgenus *Callistochlora* Michener

Callochlora Moure, 1964b: 269 (not Packard, 1864). Type species: *Halictus chloris* Spinola, 1851, by original designation.
Callistochlora Michener, 1997b: 12, replacement for *Callochlora* Moure, 1964. Type species: *Halictus chloris* Spinola, 1851, autobasic and original designation.

The male genitalia were illustrated by Moure (1964b).
■ *Callistochlora* contains three or four species (Moure and Hurd, 1987). It is well known in Chile but ranges north to Peru and Ecuador.

Corynura / Subgenus *Corynura* Spinola s. str.

Corynura Spinola, 1851: 296. Type species: *Corynura gayi* Spinola, 1851 = *Halictus rubellus* Haliday, 1836, by designation of Alfken, 1926a: 146.
Corynogaster Sichel, 1867: 146. Type species: *Corynura gayi* Spinola, 1851 = *Halictus rubellus* Haliday, 1836, by designation of Daly, Michener, Moure, and Sakagami, 1987: 104.
Rhopalictus Sichel, 1867: 146. Type species: *Corynura flavofasciata* Spinola, 1851 = *Halictus chilensis* Spinola, 1851, by designation of Sandhouse, 1943: 596.

Sandhouse (1943: 540) designated *gayi* as the type species of both *Corynura* and *Corynogaster*. For *Corynura*,

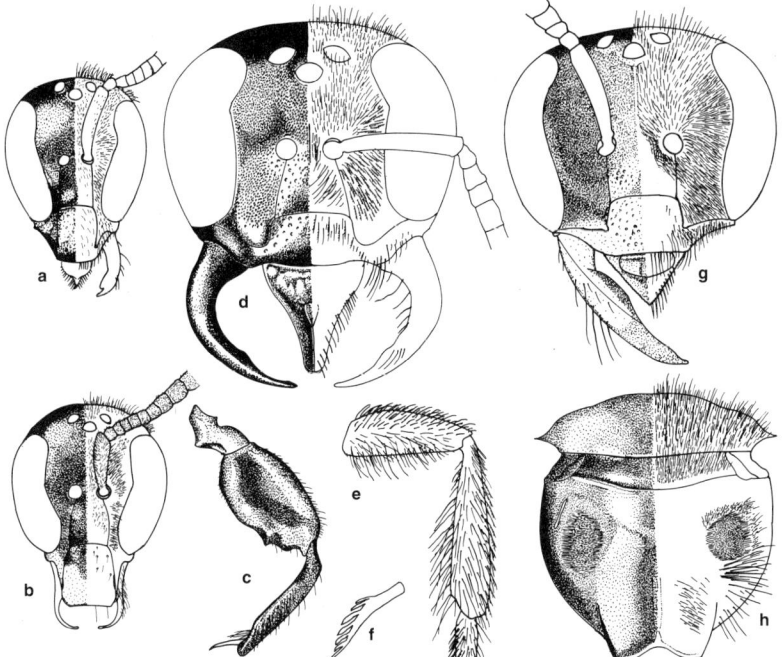

Figure 67-13. Structures of Augochlorini. **a, b,** Faces of *Chlerogelloides femoralis* Engel, Brooks, and Yanega, female and male; **c,** Anterior view of middle leg of same; **d,** Face of *Megalopta (Noctoraptor) noctifurax* Engel, Brooks, and Yanega, female; **e, f,** Hind leg and inner hind tibial spur of female *Megalopta (Noctoraptor) byroni* Engel, Brooks, and Yanega; **g,** Face of *Megommation (Cleptommation) minutum* (Friese), female; **h,** Posterior view of thorax of same. From Engel, Brooks, and Yanega, 1997.

she was anticipated by Alfken (1926a). For *Corynogaster*, her designation is useless because she confused *Halictus gayi* Spinola, 1851, and *Corynura gayi* Spinola, 1851, writing *"Halictus (Corynura) gayi."*

■ This subgenus occurs in Chile and western parts of Patagonia. There are 18 species (Moure and Hurd, 1987). Alfken (1931) gave a key to Chilean species.

Genus *Halictillus* Moure

Halictillus Moure, 1947a: 7. Type species: *Chloralictus loureiroi* Moure, 1941, by original designation.

This genus, like *Rhectomia*, contains species that are the size (5-6 mm long) and color (dull metallic greenish) of common species of *Lasioglossum (Dialictus)* in the Halictini; superficially, these taxa cannot be distinguished. They differ, of course, in the tribal characters explained above. *Halictillus* is most closely related to *Corynura*, with which it agrees in the presence of a strong galeal comb, the short galea (reaching basad only to about the middle of the stipes), and the long male antennae. It could be considered a subgenus of *Corynura*, although the male genitalia are distinctive (Eickwort, 1969b) and the male metasoma is not petiolate as is usual in *Corynura* s. str.

■ *Halictillus* is known from Chile and Río Negro province, Argentina, north to Paraná, Brazil. Two species have been named, but Eickwort (1969b) indicated that there are at least three species.

Genus *Megalopta* Smith

This is a genus of large (10-16 mm long) nocturnal or crepuscular bees, usually largely or partly testaceous with metallic green tints (Pl. 5), but sometimes nonmetallic blackish. The ocelli are enlarged, the lateral one separated from the eye margin by an ocellar diameter or less (Fig. 67-7d, 67-13d). In addition to the characters noted in the key to the genera may be mentioned the long glossa and the narrowly truncate apex of the marginal cell. A unique feature among augochlorines is the closely packed series of hamuli (Fig. 67-1b).

Unlike those of most halictids, nests of *Megalopta*, so far as is known, are in rotting wood (branches, vines, etc.). They have been described by Sakagami (1964) and Sakagami and Moure (1967). Wcislo and Gonzales (2006) gave an account of nests and facultative social behaviors of females of two species. Some nests (25 to 50 percent) are inhabited by more than one female. When this is the case, one is dominant (queen like); she tends to be larger and older than the others and with larger ovaries. Mouth-to-mouth food transfer, commonly from a returning forager to the dominant individual, was repeatedly observed. Such behavior among Halictinae is otherwise known only in *Lasioglossum (Chilalictus) hemichalceum* (Cockerell) in Australia. That species is communal, without the eusocial tendencies seen in the *Megalopta* species.

Three presumably cleptoparasitic species of *Megalopta* have recently been described and placed in a separate subgenus. They lack some derived features of *Megalopta* s. str., such as the supplementary mandibular teeth of females, and may therefore be a sister group to *Megalopta* s. str. rather than a derivative from it.

Key to the Subgenera of *Megalopta*

1. Female with scopa and poorly developed basitibial plate (Fig. 67-5s); mandible of female stout, bidentate, with supplementary teeth on inner surface (Fig. 67-5o); S5 of male narrowly emarginate; clypeus and scape of male partly white .... *M. (Megalopta* s. str.*)*
—. Female without either scopa or basitibial plate (Fig. 67-13e); mandible of female simple, long and slender, without supplementary teeth (Fig. 67-13d); S5 of male broadly notched; clypeus and scape of male black .... *M. (Noctoraptor)*

Megalopta / Subgenus *Megalopta* Smith s. str.

Megalopta Smith, 1853: 83. Type species: *Megalopta idalia* Smith, 1853, designated by the Commission, Opinion 788 (1966). See Michener, 1997b.
Megaloptera Ashmead, 1899a: 92, incorrect spelling.
Megalopta (Megaloptella) Schrottky, 1906a: 312. Type species: *Halictus ochrias* Vachal, 1904, by original designation.
Tmetocoelia Moure, 1943b: 481. Type species: *Megalopta sulciventris* Friese, 1926, by original designation.

■ *Megalopta* s. str. occurs from tropical parts of Mexico (state of Nayarit) to the state of Santa Catarina, Brazil. Of the 28 species listed by Moure and Hurd (1987), two have been transferred to *Xenochlora*. *Megalopta* was revised by Friese (1926), who included *Megaloptidia*. Engel (2006a) reviewed the Central American species and listed the species of the genus.

Megalopta / Subgenus *Noctoraptor* Engel, Brooks, and Yanega

Megalopta (Noctoraptor) Engel, Brooks, and Yanega, 1997: 12. Type species: *Megalopta byroni* Engel, Brooks, and Yanega, 1997, by original designation.

Females of *Noctoraptor* can be distinguished from other *Megalopta* by the common features of cleptoparasites—reduction of the scopa (Fig. 67-13e), large sickle-shaped mandibles (Fig. 67-13d), and the absence of a basitibial plate. Males are more difficult to recognize, but S5 is broadly notched. The ocelli are smaller than in most other *Megalopta*, the lateral one about an ocellar width from the ocular margin (Fig. 67-13d). The ocelli are large in *Megalopta* s. str., except for an undescribed, nonmetallic species from Panama and Costa Rica that has smaller ocelli like those of *Noctoraptor*. In ordinary diurnal augochlorines, including the related genus *Xenochlora*, the ocelli are much smaller, the lateral ocellus separated from the eye by two or three ocellar diameters (Fig. 67-6).
■ *Noctoraptor* is known from Pamana and Guyana to Ecuador. A new record from Guyana was by Hinojosa-Díaz and Engel (2003) who gave a key to the three species of the subgenus.

One specimen was taken in a light trap at Barro Colorado Island in Panama. It is therefore likely that, in spite of its smaller ocelli, *Noctoraptor* is nocturnal or crepuscular, like its possible hosts in the subgenus *Megalopta* s. str. It may be that *Noctoraptor* is a derivative of a *Megalopta* similar to the nonmetallic species from Central America.

Genus *Megaloptidia* Cockerell

Megalopta (Megaloptidia) Cockerell, 1900a: 373, 374. Type species: *Megalopta contradicta* Cockerell, 1900, by original designation.

This is a genus of large, *Megalopta*-like bees which, however, have a slender proboscis with a short glossa, like that of *Megommation*. *Megaloptidia* also differs from *Megalopta* in the approximately right-angular paraocular lobe (Fig. 67-7b), the broad clypeal teeth, the lack of teeth on the inner surface of the mandible of the female, etc.
■ The genus is found from Guyana and Colombia south to the Amazonian parts of Brazil and Peru. The three species were revised by Engel and Brooks (1998).

Genus *Megaloptilla* Hurd and Moure

Megommation (Megaloptilla) Moure and Hurd, 1987: 241. Type species: *Halictus callopis* Vachal, 1911, by original designation.

The name *Megaloptilla* was originally proposed for a species thought to have a slender proboscis, as does *Megommation*. Examination by R. Brooks and M. Engel shows it to have an ordinary proboscis, and to be a genus similar to *Paroxystoglossa*. It contains species 8 to 11 mm in body length, with the inner hind tibial spur of the female ciliate, as in *Paroxystoglossa*. The type species looks superficially like a green *Augochlora*, whereas *M. byronella* Engel and Brooks is larger and nonmetallic, the metasoma and areas on the thorax honey-colored. The combination of produced and strongly carinate dorsolateral pronotal angle, the carina extending to the lobe, with lack of a carina on the ridge below the angle, is distinctive.
■ This genus occurs in Panama and eastern Peru. The known species are quite different in appearance (see above) but similar in structure. Three species were revised by Engel and Brooks (1999).

Genus *Megommation* Moure

Ariphanarthra, *Chlerogelloides*, *Megaloptidia*, *Megommation*, and *Micrommation* constitute a group having a remarkable, slender proboscis (Fig. 67-4b) that fits into an unusually narrow proboscidial fossa. In addition to the large *Megommation* s. str., this genus contains some ordinary-looking species superficially like *Augochlora* or *Augochlorella*. Except for the subgenus *Stilbochlora*, *Megommation* differs from the other genera having a slender proboscis in the greatly shortened basitibial plates of the female (absent in the subgenus *Cleptommation*), obviously a derived feature.

The number of species is small, and a subgeneric classification is probably premature. The morphological diversity is great, however, and I have accepted the subgeneric classification below.

Species of this genus are rarely collected on flowers but appear in flight-intercept traps in forested areas.

Key to the Subgenera of *Megommation*

1. Ocelli much enlarged, ocellocular distance equal to or less than ocellar diameter (Fig. 67-7a); body length 12 mm or more (inner hind tibial spur of female serrate) .... *M. (Megommation s. str.)*
—. Ocelli not enlarged, ocellocular distance more than ocellar diameter (Fig. 67-13g); body length 9 mm or less 2
2(1). Females 3
—. Males 5
3(2) Scopa absent, femoral and tibial hairs simple; mandible long and simple (Fig. 67-13g); inner hind tibial spur serrate .... *M. (Cleptommation)*
—. Scopa present, hairs on outer surfaces of tibia and femur with long apical branches; mandible normal, with subapical tooth; inner hind tibial spur pectinate 4
4(3). Basitibial plate short and rounded, borders not defined; thorax metallic green blue, metasoma dark with green-blue metallic highlights .... *M. (Megaloptina)*
—. Basitibial plate of normal length, narrowly rounded, posterior border well defined; both thorax and metasoma bright metallic green .... *M. (Stilbochlora)*
5(2). Area posterior to propodeal spiracle with evenly spaced sparse setae .... *M. (Stilbochlora)*
—. Area posterior to propodeal spiracle with dense patch of plumose setae (Fig. 67-13h) 6
6(5). Apical margins of S2 and S3 simple; ocellocular distance little more than one ocellar diameter .... *M. (Megaloptina)*
—. Apical margins of S2 and S3 with medial projections; ocellocular distance equal to about 1.5 ocellar diameters .... *M. (Cleptommation)*

Megommation / Subgenus *Cleptommation* Engel, Brooks, and Yanega

Megommation (Cleptommation) Engel, Brooks, and Yanega, 1997: 19. Type species: *Megalopta minuta* Friese, 1926, by original designation.

Females lack the scopa, basitibial plates, and labral keel, and have pointed mandibles; it therefore seems that this subgenus is cleptoparasitic. It is apparently a sister group to *Megaloptina*. The two subgenera share a dense patch of plumose setae behind the propodeal spiracle in the male (Fig. 67-13h) and have similar male genitalia. *Cleptommation* can be separated easily from *Megaloptina* by its color (most of the head and all of the scutum are metallic green, and the rest of the body is yellow-brown to dark brown), the various features associated with parasitism, the serrate inner hind tibial spur of the female, and the projections on the apical margins of S2 to S5 of males.
■ *Cleptommation* occurs from Costa Rica to the state of Amazonas, Brazil, and eastern Peru. The only known species is *Megommation minutum* (Friese).

Megommation / Subgenus *Megaloptina* Eickwort

Megommation (Megaloptina) Eickwort, 1969b: 441. Type species: *Augochlora ogilviei* Cockerell, 1930, by original designation.

This subgenus contains largely black or brownish to strongly metallic green species, 7 to 9 mm long; most commonly, the head and thorax are metallic green and the metasoma dark brown with green highlights.
■ This subgenus ranges from Costa Rica to the Amazon Valley and eastern Bolivia. Moure and Hurd (1987) list two species. Eickwort (1969b) recognized five species, only one of them identified.

Megommation / Subgenus *Megommation* Moure s. str.

Megommation Moure, 1943b: 479. Type species: *Halictus insignis* Smith, 1853, by original designation.

This is a subgenus containing a large (12-15 mm long), dark-brown species. The face is illustrated in Figure 67-7a.
■ The range is from the state of Espírito Santo, Brazil, to Misiones province, Argentina, and Paraguay. The only species is *Megommation insigne* (Smith).

The subterranean nests were described and illustrated by Michener and Lange (1958a) and Sakagami and Moure (1967). The vertical cells form a beautiful cluster in a chamber lateral to the main burrow.

Megommation / Subgenus *Stilbochlora* Engel, Brooks, and Yanega

Megommation (Stilbochlora) Engel, Brooks, and Yanega, 1997: 15. Type species: *Megommation eickworti* Engel, Brooks, and Yanega, 1997, by original designation.

This subgenus can be distinguished by the combination of small ocelli, the bright metallic green thorax and metasoma in both sexes, the absence of a dense tuft of plumose hairs behind the propodeal spiracle in males, and the well-defined posterior margin of the basitibial plate in females.
■ *Stilbochlora* occurs from eastern Bolivia and Mato Grosso, Brazil, to Colombia. The only species is *Megommation eickworti* Engel, Brooks, and Yanega.

Genus *Micrommation* Moure

Micrommation Moure, 1969a: 247. Type species: *Micrommation larocai* Moure, 1969, by original designation.

In its size (body length 10.4 mm) and small ocelli, this genus is similar to *Megommation (Megaloptina)*, but it differs in the female in the more ordinary basitibial plate, better-developed median specialized area of T5, and lack of metallic coloration (thus resembling *Megommation* s. str.). The male of *Micrommation* was unknown until recently, but see Smith-Pardo and Engel (2005).
■ The genus is known only from the state of Paraná, Brazil. The only species is *Micrommation larocai* Moure.

Genus *Neocorynura* Schrottky

Cacosoma Smith, 1879: 39 (not Felder, 1874). Type species: *Cacosoma discolor* Smith, 1879, by designation of Sandhouse, 1943: 532.
Neocorynura Schrottky, 1910: 540, replacement for *Cacosoma* Smith, 1879. Type species: *Cacosoma discolor* Smith, autobasic.

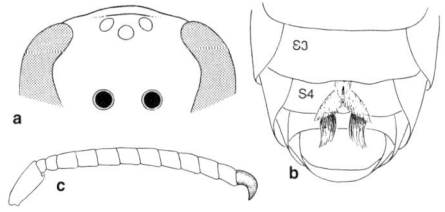

Figure 67-14. *Pseudaugochlora graminea* (Fabricius). **a,** Upper part of face of female, showing ridge behind ocelli; **b,** Underside of metasoma of male; **c,** Antenna of male. From Michener, McGinley, and Danforth, 1994.

Neocorynura (Neocorynuroides) Eickwort, 1969b: 404. Type species: *Halictus rhytis* Vachal, 1904, by original designation.

This is a highly variable genus (see the generic key to females for variations in thoracic structure) of black to metallic green forms, 5.5 to 11.0 mm long, sometimes with yellow markings and darkened costal wing margins suggestive of polybiine wasps, the metasoma rarely red. The metasoma is narrowed basally (T1 narrower than T2) and is petiolate in most males, as in *Corynura*. *Neocorynura* differs from *Corynura* in the carinate or lamellate preoccipital ridge, and in the reduced (six or less teeth) galeal comb, which may be entirely absent.

Two subgenera have been recognized, but *Neocorynuroides* is probably merely a strange and derived species of *Neocorynura*. The carinae of the preoccipital ridge and of the pronotum (lateral angle to the lobe) are expanded into strong lamellae in *Neocorynuroides,* and the male gonostylus is reduced. To recognize *Neocorynuroides* not only would leave *Neocorynura* s. str. probably paraphyletic, but would also invite recognition of the next-oddest species as another subgenus. What is needed is a revision of the whole genus to determine what groups, if any, should receive subgeneric recognition.

■ This genus ranges from as far south as the province of La Rioja, Argentina, north through tropical America to the Lesser Antilles and the state of San Luis Potosí, Mexico. There are 65 species (Moure and Hurd, 1987). It is by far most abundant and diversified in the Andean region, where many undescribed species occur. The Mexican species were revised by Smith-Pardo (2005). A Miocene amber fossil from the Dominican Republic belongs to this genus (Engel, 1996b, d).

Nests ordinarily contain shallow clumps of usually more or less horizontal cells surrounded by air spaces in the soil (Michener and Lange, 1958a; Michener, Kerfoot, and Ramírez, 1966; Michener, 1977a). A few species, such as *N. colombiana* Eickwort, however, make nests in rotting wood (Schremmer, 1979). Interestingly, its mandibles (those of the female) are modified, almost tridentate (Eickwort, 1979); those of the other wood-nesting augochlorines, *Augochlora* s. str., *Megalopta,* and *Xenochlora,* are also modified, although in different ways.

Genus *Paroxystoglossa* Moure

Paroxystoglossa Moure, 1940: 59. Type species: *Oxystoglossa jocasta* Schrottky, 1910, by original designation.

This is a *Neocorynura*-like genus of blackish to metallic green bees, the metasoma sometimes metallic red; the body length is 7 to 12 mm. Unlike those of *Neocorynura,* the inner hind tibial spur of the female is serrate and S4 and S5 of the males are modified, having strong ridges or shiny depressions. Male genitalia were illustrated by Eickwort (1969b); see also Figure 67-10a, b.

■ This group ranges from the state of Minas Gerais, Brazil, to Paraguay and the province of Buenos Aires, Argentina. The nine species were revised by Moure (1960b).

So far as is known, nests are constructed in the ground. The cells are more or less horizontal, closely associated in a cluster largely separated from the surrounding soil (Michener and Lange, 1958a, b; Michener and Seabra, 1959).

Genus *Pseudaugochlora* Michener

Caenaugochlora (Pseudaugochlora) Michener, 1954b: 77. Type species: *Halictus nigromarginatus* Spinola, 1841 = *Megilla graminea* Fabricius, 1804, by original designation.

This taxon was called *Pseudaugochloropsis* by Moure (1940, 1944b), Eickwort (1969b), and Moure and Hurd (1987). This usage was based on the designation of *Halictus nigromarginatus* Spinola as the type species of *Pseudaugochloropsis* by Moure (1940), but *H. nigromarginatus* was not an originally included species, as explained in detail by Michener (1995a). *Pseudaugochloropsis* is actually a synonym of *Augochloropsis (Paraugochloropsis)*.

Pseudaugochlora was originally described as a subgenus of *Caenaugochlora,* to which I still believe it is closely related. *Pseudaugochlora* consists of rather large (8 to 13 mm long), bright-green or occasionally black species, easily recognized by the tapering and hooked last antennal segment of the males (Fig. 67-14c) and the rounded ridge behind the ocelli of females (Fig. 67-14a), the preoccipital ridge being rounded rather than carinate. The apex of the galea is pointed and well sclerotized, a feature not found in other genera, and the glossa is long and attenuate.

■ The genus ranges from northern Argentina to southern Texas, USA. There are seven included species (Moure and Hurd, 1987).

The nests, similar to those of large species of *Caenaugochlora,* have been described by Michener and Lange (1958a), Michener and Kerfoot (1967), and Sakagami and Moure (1967). *Pseudaugochlora sordicutis* (Vachal) is noteworthy for its dichromatism. Some individuals are black, others bright green; and both forms have been found in the same nest (Michener and Kerfoot, 1967).

Genus *Rhectomia* Moure

Rhectomia Moure, 1947a: 9. Type species: *Rhectomia pumilla* Moure, 1947, by original designation.
Corynurella Eickwort, 1969b: 398. Type species: *Corynurella mourei* Eickwort, 1969, by original designation.

Like *Halictillus,* this is a genus of small (length 5-7 mm), dull-green bees sometimes with the body partly testaceous or reddish; some species superficially resemble *Lasioglossum (Dialictus)* in the Halictini, while others

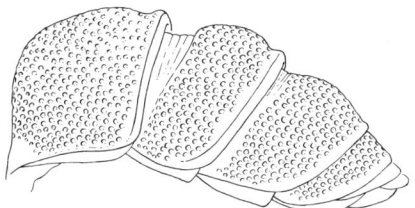

Figure 67-15. Lateral view of metasoma of male, *Temnosoma* sp. From Michener, McGinley, and Danforth, 1994.

have stronger blue-green reflections. The broad base of the labral process of the female suggests *Rhinocorynura*. The epistomal suture forms a slightly acute paraocular lobe protruding into the clypeus. There is a well-developed galeal comb, as in *Corynura*. Male genitalia were illustrated by Engel (1996a).

▪ This genus is known from the province of Misiones, Argentina, and Paraguay north to Minas Gerais, Brazil, and eastern Peru. The four known species were revised by Engel (1996a).

Genus *Rhinocorynura* Schrottky

Corynura (Corynuropsis) Cockerell, 1901a: 220 (not Scott, 1894). Type species: *Corynura darwini* Cockerell, 1901 = *Augochlora briseis* Smith, 1879, by original designation.
Rhinocorynura Schrottky, 1909a: 147. Type species: *Halictus inflaticeps* Ducke, 1906, by original designation.
Ctenocorynura Schrottky, 1914: 628. Type species: *Ctenocorynura vernoniae* Schrottky, 1914 = *Halictus inflaticeps* Ducke, 1906, by original designation.
Corynuroides Sandhouse, 1943: 540, replacement for *Corynuropsis* Cockerell, 1901. Type species: *Corynura darwini* Cockerell, 1901 = *Augochlora briseis* Smith, 1879, autobasic.
Rhynocorynura Sakagami and Moure, 1965: 303, unjustified emendation.
Gnathalictus Moure, 2001c: 493. Type species: *Gnathalictus capitatus* Moure, 2001, by original designation. [**New synonymy.**]

In this genus of mostly black to dark metallic-green species 5 to 14 mm long, the vertex behind the ocelli is swollen in the female, and the clypeus usually has median or lateral tubercles or spines (Fig. 67-7h). The basal part of the labral process of the female is expanded to nearly the full width of the labrum. The galea has a well-developed comb, as in *Corynura* and *Rhectomia*. G. Melo has pointed out to me that *Rhinocorynura* may consist of *Rhectomia*-like bees with exaggerated cephalic structures. *Rhinocorynura* is noteworthy for the variation in female size and the presumably allometric elaboration of cephalic structures in large individuals (Sakagami and Moure, 1965).

▪ This genus is widespread in Brazil and also occurs in tropical parts of eastern Peru, Bolivia, Paraguay, and northern Argentina (Misiones). Five species were listed by Moure and Hurd (1987).

A nest, including a thin-walled cluster of vertical cells, was described by Eickwort and Sakagami (1979).

Genus *Temnosoma* Smith

Temnosoma Smith, 1853: 38. Type species: *Temnosoma metallicum* Smith, 1853, monobasic.
Micraugochlora Schrottky, 1909b: 483 [no included species]; 1909a: 138. Type species: *Micraugochlora sphaerocephala* Schrottky, 1909, monobasic.
Temnosoma (Temnosomula) Ogloblin, 1953: 2. Type species: *Temnosoma platensis* Ogloblin, 1953 = *Micraugochlora sphaerocephala* Schrottky, 1909, by original designation. [Also described as new in Ogloblin, 1954: 5.]

This is a genus of bright-green parasitic bees lacking the scopa and other necessities of nest-making halictids. Some other features of females, probably related to the parasitic way of life, include simple mandibles, the male-like unkeeled labrum (Fig. 65-1i), the lack of basitibial plates and of a specialized median area of T5, and the very coarse punctation (Fig. 67-15). The constricted bases of T2 and T3 are also noteworthy. A peculiar feature of males is the largely membranous S7. The male genitalia, as shown in Figure 67-12e, f, are unusual for Augochlorini, for example in the large, apically rounded gonostylus with long hairs. *Temnosoma sphaerocephala* (Schrottky) has only two submarginal cells, but otherwise, except for its small size, is a typical member of the genus. Specimens of other species of *Temnosoma* also sometimes have only two submarginal cells. This tendency is unique among the Augochlorini, which otherwise consistently have three submarginal cells. The only other known parasitic augochlorines are derivatives of *Megommation* and of *Megalopta*.

▪ *Temnosoma* ranges from Buenos Aires and Mendoza provinces, Argentina, north to Cuba and southern Arizona, USA. Hurd and Moure (1987) list seven species. Friese (1924b) gave a key to species.

What the closest nonparasitic relative of *Temnosoma* may be is not clear. The peculiarly small volsellae of *Temnosoma* are suggestive of those of *Augochloropsis*, which are smaller than those in most other augochlorines. The hosts are probably augochlorines. The largest augochlorine genus, *Augochloropsis*, is absent from the Antilles. *Temnosoma*, therefore, must not require *Augochloropsis* as a host.

Genus *Thectochlora* Moure

Thectochlora Moure, 1940: 51. Type species: *Halictus alaris* Vachal, 1904, by original designation.

This is a genus of moderate-sized (body about 7 mm long) green bees similar to *Augochlorella*. Unusual features are the short, dense hair on the basal area of the propodeum and on the upper surface of the middle tibia of the male. T1 of the female has a basal area surrounded by plumose hairs; this area is usually inhabited by mites. The ridge between the dorsolateral angle of the pronotum and the lobe is strongly lamellate.

▪ This genus is known from Pernambuco to Santa Catarina, Brazil, and Paraguay. The only species is *Thectochlora alaris* (Vachal).

Genus *Xenochlora* Engel, Brooks, and Yanega

Xenochlora Engel, Brooks, and Yanega, 1997: 7. Type species: *Xenochlora ochrosterna* Engel, Brooks, and Yanega, 1997, by original designation.

Xenochlora females suggest nonmetallic macrocephalic *Rhinocorynura* or *Megalopta;* the body is rather robust, 9.5 to 11.5 mm in body length. From the former, *Xenochlora* can be separated by the acutely angled paraocular lobe, the carinate dorsal pronotal ridge, and the lack of a galeal comb. This genus differs from *Megalopta* in the normal-sized ocelli, the dense, stiff black setae on the hind tibia and basitarsus, and the smaller number of hamuli on the hind wing, which exhibit the typical spacing pattern found in other genera of Augochlorini (in *Megalopta* the hamuli are numerous and closely packed, without any uneven spacing among them).

■ This genus occurs from Guyana to Amazonas, Brazil, and Amazonian parts of Peru and Ecuador. The four species were revised by Engel, Brooks, and Yanega (1997).

I would have treated this as a subgenus and probable basal branch of the *Megalopta* clade, but since males of *Xenochlora* are unknown, its position cannot be determined unequivocally; in the absence of evidence from the male, I retain it as a genus.

The extra mandibular teeth indicate that *Xenochlora*, like *Megalopta*, nests in wood and, as noted by Engel, Brooks, and Yanega (1997), *X. ianthina* (Smith) was reported to nest in a tree branch.

68. Family Melittidae

The Melittidae consist of andreniform S-T bees, having the characters of S-T bees listed in Section 19 and in the first couplet of the key to families, Section 33. Nonetheless, other characters are as in L-T bees. No known character is both unique to Melittidae and also a feature of all melittids, but the combination of (1) a (usually) short, pointed glossa (Fig. 73-1) and an unspecialized (i.e., relatively short and not sheathlike) first two segments of the labial palpus, with (2) a slender, V-shaped lorum and a tapering (basally curled) mentum (Fig. 74-1a, b), and (3) an elongate, fully exposed middle coxa (as in Fig. 14-1c), as in L-T bees, separates the Melittidae from other S-T families. The Ctenoplectrini in the Apinae were formerly included in the Melittidae (Michener, 1944) and then were placed as a separate family between Melittidae and the L-T bees (Michener and Greenberg, 1980). The present view, however, is that the Ctenoplectrini are a group of Apinae whose glossa and labial palpi regressed to resemble in some features Melittidae and other S-T bees (Roig-Alsina and Michener, 1993).

The phylogenetic position of the Melittidae was considered in Sections 20 and 21, as well as by Michener and Greenberg (1980), Roig-Alsina and Michener (1993), Alexander and Michener (1995), and Danforth et al. (2006). The more recent studies show that the Melittidae is a paraphyletic or polyphyletic group from which L-T bees arose, or the basal branch or branches from among which all other bees arose. The weakness of characters suggesting monophyly for the Melittidae is explained in Section 21. Alexander and Michener (1995) and Danforth et al. (2006) broke it into three families, here given subfamily status. My reasons for not doing so for now are explained in Section 21.

The melittids are phenetically similar, forming a unit with many common characters (a sample of its wings is shown in Fig. 68-1). I would be satisfied with Melittidae as a paraphyletic unit, given that the evidence that it is holophyletic is weak. There is no evidence known to me that any other family of bees might be derived from any particular group of the Melittidae.

Some principal familial characters are as follows: The labrum is usually much broader than long, the apical margin in both sexes being fringed with bristles. There is one subantennal suture below each antenna. Facial foveae are absent. The lower lateral parts of the clypeus are not much bent posteriorly on either side of the labrum. The glossa is usually shorter than the prementum, its apex being pointed, usually attenuate (Fig. 73-1), and with branched hairs; the flabellum is absent, although there is apical expansion in *Dasypoda* (Fig. 68-2); the posterior surface of

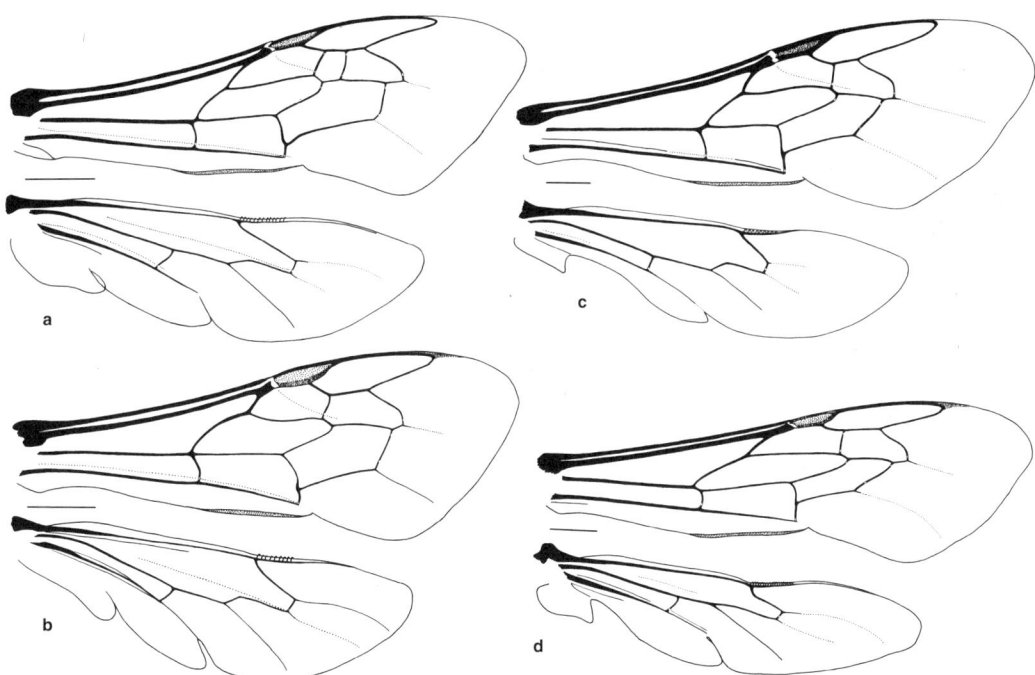

Figure 68-1. Wings of Melittidae (a, b, Melittinae; c, d, Dasypodainae). **a,** *Melitta leporina* (Panzer); **b,** *Macropis europaea* Warncke; **c,** *Hesperapis pellucida* Cockerell; **d,** *Dasypoda panzeri* Spinola. From Michener, 1981a.

 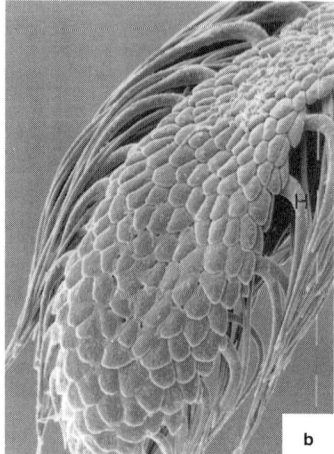 

Figure 68-2. Apices of glossa of Melittidae. **a, b,** Posterolateral views, *Dasypoda panzeri* Spinola; **c,** Posterior view, *Melitta tricincta* Kirby. For melittids whose glossae have been studied, only *Dasypoda* has a glossal structure like that shown in a and b; all others, including *Hesperapis* (also a dasypodine), have a structure similar to that shown in c. SEM photographs by R. W. Brooks, from Michener and Brooks, 1984.

the glossa has a depression or shallow groove within which the surface is disannulate; the glossal rod is absent. The paraglossa is usually small (sometimes absent, Fig. 72-1b) and commonly shorter than its suspensorium, except in the Meganomiinae (Fig. 73-1). As stated above, the first two segments of the labial palpus are more or less cylindrical, not sheathlike. The mentum is elongate, tapering basally, and curled over the lorum to its articulation with the latter (Fig. 74-1a, b) (it is not sclerotized in *Samba*, Fig. 72-1b). The base of the prementum is a detached fragmentum. The lorum is well sclerotized, slender, and strongly v-shaped. The galeal blade is usually equal to or shorter than the stipes; a galeal comb is present [but reduced in *Melitta*, greatly reduced in *Melitta (Dolichochile), Haplomelitta (Atrosamba),* and some *Hesperapis,* and absent or essentially so in *Dasypoda*]. The stipital comb and its concavity are absent, except that, uniquely for an S-T bee, a concavity is distinct in *Eremaphanta* (Fig. 70-2); some hairs on its margin are probably not to be interpreted as a comb. The stigma is usually slender, the jugal lobe of the hind wing shorter than in most S-T bees; there are either two or three submarginal cells (Fig. 68-1). The episternal groove is absent, as is the scrobal groove in front of the pleural scrobe. The middle coxa is fully exposed and nearly reaches the lower metapleural pit. The basitibial plate of the female is usually distinct, that of the male, often distinct. The scopa is largely restricted to the hind tibia and basitarsus (Fig. 68-3). The volsella is present and free (although much reduced or absent in Meganomiinae). The pygidial plate of the female is present; the pygidial and prepygidial fimbriae of the female are also present, except in *Eremaphanta,* but the prepygidial fimbria is reduced to a mere hair band like that of preceding terga in some other Dasypodaini; the pygidial plate of the male is frequently absent. Illustrations of many structures, particularly of the mouthparts and male genitalia and hidden sterna, were provided for all genera by Michener (1981a).

The larvae, similar to those of certain colletids and andrenids, were described by Rozen and McGinley (1974a), Rozen (1977a), and McGinley (1981). No useful larval familial characters are known. The larvae fall into two groups. One has strongly projecting salivary lips and a recognizable although minute galea; the dorsolateral body tubercles, laterally produced from transverse segmental bands, are recognizable. *Meganomia* in the Meganomiinae and *Melitta* and *Macropis* in the Melittinae are in this group; the larvae spin cocoons. The second larval group lacks projecting salivary lips and galea, and the dorsolateral body tubercles are absent or weak. *Dasypoda* and *Hesperapis* in the Dasypodainae are in this group; the larvae do not spin cocoons. Rozen and McGinley (1974a) gave a key to the genera based on mature larvae; but larvae of only four genera were known at that time.

Nest burrows are made in the soil, with laterals leading to cells that are usually isolated but may be in series of two or more. Cells are often horizontal, but the orientation is variable; they are often bilaterally symmetrical (i.e., flatter on the lower surface than elsewhere), not lined with a visible secreted lining, and sometimes rather rough. In *Macropis* they are lined with oil from flowers of *Lysimachia* (Cane et al., 1983). The lack of a visible cell lining like that secreted by most ground-nesting bees may be a plesiomorphic character of the family. Provisions are firm, variable in shape, sometimes subspherical, as in most halictids and andrenids (Malyshev, 1929; Rozen, 1977a; Rozen and Jacobson, 1980).

Members of this rather small and usually uncommon family are found primarily in temperate regions of the Northern Hemisphere and in Africa, the greatest number of genera and species being found in warm xeric areas. Maximal diversity is in southern Africa, where all three subfamilies occur. The palearctic region also supports a moderate number of species and genera (*Dasypoda,*

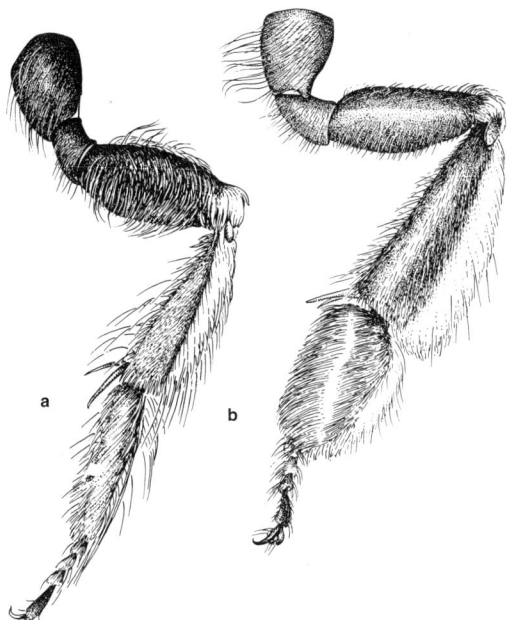

Figure 68-3. Hind legs of female Melittidae, showing the tibial and basitarsal scopae and reduced femoral to coxal scopae. **a,** *Hesperapis arida* Michener (Dasypodainae); **b,** *Macropis patellata* Patton (Melittinae). From Michener, McGinley, and Danforth, 1994.

Eremaphanta, Macropis, Melitta, Promelitta). In the nearctic region the family is rare except for *Hesperapis*, which occurs mostly in the Southwest of the USA; otherwise, the American genera are *Macropis* and *Melitta* (including *Dolichochile*). Melittids are unknown in tropical America and in Australia, as well as in the Indo-Malayan area. Most species are oligolectic or at least have strong floral preferences, and some, *Macropis* and *Rediviva*, collect floral oils.

The family was reviewed to the generic and subgeneric levels by Michener (1981a); this work included a cladistic analysis. The west palearctic species were reviewed by Warncke (1973a), the Chinese species by Wu (1978). Keys to the Iberian species of Melittidae (11 species of *Melitta*, 10 of *Dasypoda*, 2 of *Macropis*) were provided by Ornosa and Ortiz-Sanchez (2003).

The Melittidae are divided into three subfamilies (Michener, 1981a, and below). The relationships among the subfamilies are not clarified by the diverse relationships found among the six melittid genera included in Alexander and Michener's (1995) phylogenetic study of S-T bees. More genera should be included in a new study directed toward learning about melittid relationships. As indicated above, the Melittidae will probably be confirmed as a paraphyletic group from which L-T bees arose. The Meganomiinae are deemed a holophyletic unit on the basis of numerous synapomorphies, and the Dasypodainae are probably holophyletic, for they possess several synapomorphies. The Melittinae, however, may not be holophyletic, and some of Alexander and Michener's (1995) cladograms show *Melitta* and *Macropis* as quite unrelated. One of Michener's (1981a) cladograms shows this subfamily as holophyletic on the basis of one weak synapomorphy. The other cladogram shows the Melittinae as two of the branches of a trifurcation. The numerous characters of the subfamily are mostly plesiomorphies.

Key to the Subfamilies of the Melittidae

1. Yellow or cream markings on head, thorax, and metasoma; body length over 10 mm; mandible of female simple; apex of marginal cell rounded, bent away from wing margin (Africa, Arabia) .... Meganomiinae (Sec. 73)
—. Yellow or cream markings, if present, limited to male face and rarely areas on legs [except for *Eremaphanta* (Central Asia; length 6.5 mm or less) which often has abundant yellow markings, and *Macropis hedini* Popov, which has yellow on supraclypeal area of female]; mandible of female with preapical tooth on upper margin, sometimes two teeth; apex of marginal cell pointed, on or very near wing margin (Fig. 68-1) 2
2(1). Paraglossa largely bare, usually markedly shorter than suspensorium, its hairs largely limited to apex, or paraglossa absent (Fig. 72-1b); forewing with two submarginal cells, second usually shorter than first; first submarginal crossvein (= base of second submarginal cell) usually more or less at right angles to longitudinal veins (Fig. 68-1c, d) and usually close to first recurrent vein; known larvae do not spin cocoons .... Dasypodainae (Sec. 69)
—. Paraglossa densely hairy; forewing with two or three submarginal cells, second (if only two) or second plus third as long as or longer than first; first submarginal crossvein (= base of second submarginal cell) slanting (Fig. 68-1a, b), usually well separated from first recurrent vein; larvae spin cocoons .... Melittinae (Sec. 74)

69. Subfamily Dasypodainae

The body of these bees is minute (4 mm long) to rather large (17 mm long), and lacks yellow markings, except for some *Eremaphanta,* the yellow clypeus of male *Promelitta,* and the partly yellow clypeus of male *Hesperapis rufipes* (Ashmead). The paraglossa is absent (Fig. 72-1b) or much shorter than the suspensorium, except in *Promelitta*. The forewing has two submarginal cells (Fig. 68-1c, d), the second shorter than the first or, in some Sambini, subequal to the first. The apex of the marginal cell is as in the Melittinae or, in *Eremaphanta,* not at all bent away from the costal margin of the wing. S7 of the male is a transverse plate with two apical lobes, not very different from the preceding sterna, or the disc is reduced but carries lateroapical lobes (Fig. 72-1f).

Larvae of *Dasypoda* and various subgenera of *Hesperapis* are known (Rozen and McGinley, 1974a). Although the larva of *Dasypoda* is quite different from that of *Hesperapis,* both exhibit several apomorphies that distinguish them from larvae of the Melittinae and Meganomiinae, and thus support recognition of the subfamily Dasypodainae. Unfortunately, larvae of the dasypodaine tribes Sambini and Promelittini are unknown.

The reduced paraglossae—they are absent in some Sambini—and the presence of only two submarginal cells, the base of the second one transverse, as well as the larval characters listed above, are synapomorphies of the subfamily. Such reduced paraglossae are not known in any other bees. The transverse base of the second submarginal cell (Fig. 63-1c, d), lying more or less at right angles to the longitudinal veins and close to the apex of the second recurrent vein, is an unusual feature. Unfortunately, it is less than satisfying as a key character because the difference between transverse (as in Dasypodainae) and slanting (as in *Macropis* and the genera having three submarginal cells) is small and sometimes bridged by variation within genera or even within species. Nonetheless, the character is a valuable one if used with caution, and does in general distinguish most Dasypodainae from other melittids.

Dasypodainae occur in the palearctic region, in North America (mostly western), and in Africa. They are most abundant in xeric areas but are present in mesic regions such as northern Europe and Japan.

Three tribes are recognized in the Dasypodainae. The Promelittini is perhaps the sister group of the other two tribes. Its paraglossa is intermediate between that of Melittinae and the reduced type found in the other Dasypodainae. (See the discussion of the genus *Promelitta*.) The Dasypodaini and Sambini are sister groups, the most conspicuous synapomorphy of the former being the elevated vertex, whereas the latter has numerous synapomorphies, such as the strong, apically produced, mesal dorsal gonocoxal lobe of the male (Fig. 72-1d), the apically expanded and angulate gonostylus, and the elongate, nonopposable digitus of the volsella (quite different from, and evidently of independent origin from, the elongate, nonopposable digitus of *Melitta*).

The family-group name based on *Dasypoda,* Dasypodidae, is a junior homonym of the mammalian family Dasypodidae, which is based on *Dasypus*. Opinion 1926 (1999) of the Commission establishes the spelling Dasypodaidae, hence Dasypodainae and Dasypodaini for the subfamily and tribal names (Alexander, Michener, and Gardner, 1998).

Key to the Tribes of the Dasypodainae

1. Vertex, as seen from front (except in *Eremaphanta* s. str.), elevated well above summits of eyes, usually convex; gonostylus of male (except in *Dasypoda*) robust, fused to gonocoxite; S7 of male without or (in some *Dasypoda*) with only one pair of straplike lateroapical lobes (Fig. 70-1f); paraglossa small, slender, more or less cylindrical .... Dasypodaini (Sec. 70)
—. Vertex, as seen from front, little elevated above summits of eyes, gently convex to concave; gonostylus of male long, flexibly joined or articulated to gonocoxite; S7 of male with one or two pairs of broad lateroapical lobes (Fig 72-1f, g); paraglossa tapering distally or absent 2
2(1). Clypeus of male largely yellow; metasomal terga with basal zones of pale hair; paraglossa nearly as long as its suspensorium; vertex weakly, uniformly convex, as seen from front (Egypt) .... Promelittini (Sec. 71)
—. Clypeus of male concolorous with rest of head; metasomal terga usually without basal hair bands, but if bands present, then apical; paraglossa much shorter than its suspensorium or usually not recognizable (Fig. 72-1b); vertex straight or concave, as seen from front (Africa) .... Sambini (Sec. 72)

70. Tribe Dasypodaini

The clypeus of these bees is concolorous with the rest of the face, except in *Eremaphanta* from Central Asia; the lower part of the clypeus is yellowish in males of *Hesperapis rufipes* (Ashmead). The head is often narrower than the thorax; the vertex, as seen from in front, is almost always elevated above the summits of the eyes and is usually convex. The metasomal terga usually have apical hair bands. A longitudinal median elevated area of the female pygidial plate is absent except in *Hesperapis (Capicola)*. In the male, lateroapical lobes of S7 are absent or small and slender (Fig. 70-1c, g); the dorsal mesoapical lobe of the gonocoxite is absent and the gonostylus is short, broadly fused to the gonocoxite except in *Dasypoda*, and neither expanded nor angulate apically, and the volsellar digitus is normal (Fig. 70-1a, d).

The genus *Dasypoda*, which has numerous apomorphies, several of which are listed in the generic characterizations below, appears to be the sister group of the other Dasypodaini. Indeed Ascher and Engel (in Engel, 2005) provided the subtribal name Hesperapina to include *Hesperapis* and *Eremaphanta*, in contrast to the Dasypodaina for *Dasypoda*. Apomorphies for the genera of Hesperapina are the reduction of the gonostylus and its fusion to the gonocoxite, the loss of the lateroapical lobes of S7 of the male, and the presence of a broad, transparent apical marginal zone with a deep median cleft on S1. *Eremaphanta*, with many synapomorphies, is probably the sister group to *Hesperapis;* a synapomorphy of the latter appears to be the plumose scopa with its long, emergent, bare hairs. Many illustrations of structures and phylogenetic trees were presented by Michener (1981a).

As shown by Rozen and McGinley (1974a) and Rozen (1978a), the known larvae also support the differentiation of *Dasypoda* from *Hesperapis*. The reduction in cephalic and mouthpart structures that characterizes the subfamily is carried to the extreme in *Hesperapis*, the maxilla and labium being fused and the prementum and postmentum not being separated.

Key to the Genera of the Dasypodaini

1. Yellow integumental markings present, at least on face and legs; stigma large, slightly shorter to slightly longer than costal margin of marginal cell; prepygidial and pygidial fimbriae of female absent (Central Asia) .... *Eremaphanta*
—. Yellow integumental markings absent (except rarely on part of clypeus of male); stigma not enlarged, about half as long as costal margin of marginal cell (Fig. 68-1c, d);

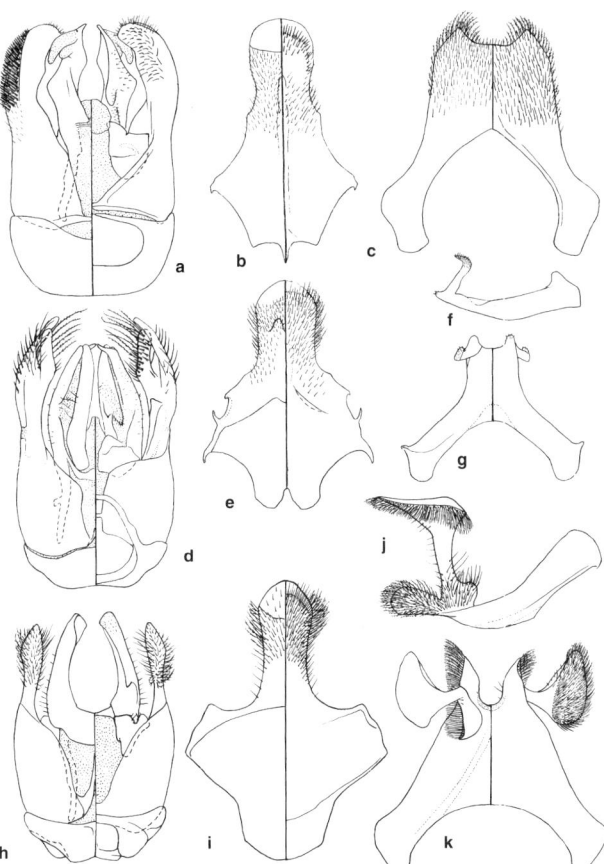

Figure 70-1. Male genitalia, S8, and S7 of Dasypodaini and Promelittini. Lateral views (f and j) are of S7. **a-c,** *Hesperapis (Capicola) braunsiana* (Friese); **d-g,** *Dasypoda hirtipes* (Fabricius); **h-k,** *Promelitta alboclypeata* (Friese). From Michener, 1981a.

prepygidial and pygidial fimbriae of female well developed, or the former represented by a mere hair band similar to that of T4 2

2(1). Basitibial plate absent in both sexes; profile of propodeum all more or less in one plane; galeal comb absent or nearly so; male gonostylus well differentiated, nearly always deeply bifid (Fig. 70-1d) (palearctic) ....
.... *Dasypoda*

—. Basitibial plate of female and nearly all males present, rarely defined only along posterior margin; profile of propodeum more nearly horizontal at base than elsewhere [except in some *Hesperapis (Carinapis)*]; galeal comb present; male gonostylus broad, fully fused to gonocoxite, not bifid (Fig. 70-1a) but often with short inner process or lobe (nearctic, Africa) .... *Hesperapis*

Genus *Dasypoda* Latreille

Dasypoda Latreille, 1802a: 424. Type species: *Andrena hirtipes* Fabricius, 1793 = *Apis altercator* Harris, 1780, by designation of Blanchard, 1840: 414. [*Melitta swammerdamiella* Kirby, 1802, designated as the type species by Curtis (1831: 367), was not originally included in *Dasypoda* but is a synonym of *Andrena hirtipes* Fabricius = *Apis altercator* Harris.]

Podasys Rafinesque, 1815: 123, unnecessary replacement for *Dasypoda* Latreille, 1802. Type species: *Andrena hirtipes* Fabricius, 1793, autobasic.

Microdasypoda Michez, 2004, in Michez, Terzo, and Rasmont, 2004b: 427. Type species: *Dasypoda crassicornis* Friese, 1896, by original designation. **[New synonymy.]**

Megadasypoda Michez, 2004, in Michez, Terzo, and Rasmont, 2004b: 429. Type species: *Dasypoda argentata* Panzer, 1809, by original designation. **[New synonymy.]**

Heterodasypoda Michez, 2004, in Michez, Terzo, and Rasmont, 2004b: 428. Type species *Dasypoda pyrotricha* Förster, 1855, by original designation. **[New synonymy.]**

Dasypoda consists of large bees (length 11 to 17 mm) with abundant long hairs. The scopa consists of long, dense, minutely barbed hairs on both the inner and outer surfaces of the hind tibia and basitarsus; keirotrichia are absent. Male genitalia and hidden sterna were illustrated by Radoszkowski (1887), Warncke (1973a), Wu (1978), and Radchenko and Pesenko (1989); see also Figure 70-1d-g.

■ This genus is widespread in the palearctic region from Portugal and the Canary Islands to Japan and is especially abundant in the Mediterranean basin. About 35 species are recognized. A review of the Spanish species was by Quilis (1928), of west palearctic species by Warncke (1973a), of Chinese species by Wu (1978), and of species of European Russia by Radchenko and Pesenko (1989). An annotated account of species of *Dasypoda* was provided by Baker (2002b), and a revision of the west palearctic species by Michez, Terso, and Rasmont (2004a). In a phylogenetic study, Michez, Terzo, and Rasmont (2004b) recognized four subgenera.

The larva was described by Rozen and McGinley (1974a) and the nesting biology by several authors, including Müller (1884), Malyshev (1927a), Blagoveshchenskaya (1963), Lind (1968), and Radchenko (1987). An interesting feature of the pollen masses is the projections on their undersurfaces that reduce areas of contact with the cell wall and thereby reduce moist contacts and perhaps the likelihood of fungal infestation. Pollen masses of *D. hirtipes* (Fabricius) have three conical projections; those of *D. braccata* Eversmann have two parallel ridges and one conical projection. The cells are unlined, more or less horizontal, and isolated or in series at the ends of lateral burrows.

Genus *Eremaphanta* Popov

Eremaphanta consists of minute bees (4.0-6.5 mm long) whose male genitalia and hidden sterna are similar to those of *Hesperapis*. The body has extensive yellow markings, or at least yellow on the face and legs. The stigma is large and transparent, unlike that of other Melittidae, and more than three times as long as the prestigma. A large concavity in the posterior margin of the maxillary stipes, as in most L-T bees (Fig. 70-2), is a feature unique among S-T bees. Male genitalia, hidden sterna, and other structures were illustrated by Popov (1940, 1957b), Schwammberger (1971a), Michener (1981a), and Michez and Patiny (2006).

The minute size and yellow markings result in a superficial resemblance to *Perdita* (Panurginae), *Nomioides* and *Habralictus* (Halictidae), and certain Euryglossinae. Some common features of small bees, such as the large stigma, increase the resemblance. The genus was revised by Michez and Patiny (2006).

Key to the Subgenera of *Eremaphanta*

1. Terga without apical hair bands; head broader than long; vertex gently convex, scarcely above level of summits of eyes .... *E. (Eremaphanta* s. str.)
—. Terga with apical hair bands; head longer than broad; vertex extending far above summits of eyes...................
.... *E. (Popovapis)*

Eremaphanta / Subgenus *Eremaphanta* Popov s. str.

Eremaphanta Popov, 1940: 53. Type species: *Rhophites vitellinus* Morawitz, 1876, by original designation.

The low vertex is unlike that of other Dasypodaini; presumably, it is a reversion to the condition found in most bees rather than a plesiomorphy.

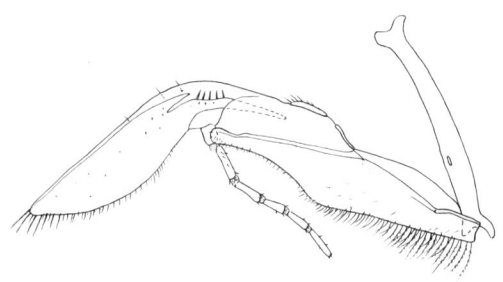

Figure 70-2. Inner view of maxilla of *Eremaphanta dispar* (Morawitz). Note the lower distal concavity of the stipes, otherwise unknown in S-T bees. From Michener, 1981a.

■ The subgenus is known from the Central Asian deserts and Iran south to Oman. It contains six species.

Eremaphanta / Subgenus *Popovapis* Michener

Eremaphanta (Popovapis) Michener, 1981a: 80. Type species: *Rhophites dispar* Morawitz, 1892, by original designation.

The characters indicated above in the key to subgenera are probably plesiomorphic, for they resemble other Dasypodaini rather than the apparently more derived subgenus *Eremaphanta* s. str. The reduced galeal comb (about five bristles) is an apomorphy, but has been observed in only one of the species.
■ *Popovapis,* from Central Asia to Pakistan (Baluchistan), contains two species.

Genus *Hesperapis* Cockerell

Species of this genus look superficially like *Halictus* or *Andrena* but the metasoma is very flat; when pinning specimens, one notes that the integument is soft so that a pin goes through the thorax more easily than with other bees. The male genitalia and hidden sterna were illustrated by Popov (1957b) and Michener (1981a); see also Figure 70-1a-c. As here understood, *Hesperapis* has an extraordinarily disjunct distribution, being found in North America, especially in xeric areas, and in xeric parts of southern Africa. Because of this disjunction, it has been common to recognize two genera, *Capicola* for Africa and *Hesperapis* for North America. There are in fact minor differences between the African and the North American species groups, but the differences are minute and have been hard to find (Cockerell, 1932b). By morphological standards for bee genera, the two groups without doubt belong in the same genus.

The principal differences between African and American *Hesperapis* are indicated in the first couplet of the key to the subgenera. (Michener, 1981a: 80, line 3 of diagnosis, inadvertently says for the African group "with" the inner basal process or lobe of the male gonocoxite; the word should have been "without.") The only other group differences known to me are that the galeal comb is weaker and the propodeal triangle larger in the American than in the African group. Among the few known mature larvae, the American species lack intrasegmental lines on S9 and S10 as well as paired dorsal segmental tubercles, characters found in African species whose larvae are known.

The nests are irregular burrows in the soil with long branches, each branch leading to a single cell or sometimes to a short series of cells (Rozen and McGinley, 1991). The cells are simply spaces, sometimes irregular and neither smoothed nor lined (Rozen, 1987b), but the walls are sometimes smooth and probably slightly firmer than the surrounding soil; they are dull and not waterproofed, and lack evident lining (Rozen and McGinley, 1991). This finding is much as in *Dasypoda*. Unlined cells are unusual because subterranean cells of other bee families are usually beautifully smoothed and lined with "waxlike" or cellophanelike films. The food mass in a cell is spherical, not coated; the egg is laid on top of it.

Xeralictoides has hitherto been considered a distinct genus, but it is similar to *Hesperapis* and its recognition as a genus makes *Hesperapis* unnecessarily paraphyletic. I therefore include *Xeralictoides* as a subgenus within *Hesperapis*. The following treatment of American *Hesperapis* is modified from that of Stage (in Michener, 1981a).

Key to the Subgenera of *Hesperapis*

1. Pygidial plate of female with longitudinal median elevated area, or at least ridges representing edges of such an area; male gonostylus without inner lobe (Fig. 70-1a) (southern Africa) 2
—. Pygidial plate of female without longitudinal median elevated or differentiated area (sometimes with triangular basal area); male gonostylus with inner lobe, usually bearing long, coarse, branched bristles or hairs (North America) 3
2(1). Outer surface of hind tibia of female with large, blunt, spinelike hairs among scopal hairs; S7 of male with disc narrowed, with diverging, densely hairy, lateroapical lobes .... *H. (Capicoloides)*
—. Outer surface of hind tibia of female without spinelike hairs interspersed among scopal hairs; S7 of male with disc broad, without diverging lateroapical lobes (Fig. 70-1c) .... *H. (Capicola)*
3(1). Mesoscutum as long as or longer than minimum intertegular distance; front tibia with series of five or more robust, amber apical and posteroapical spines on outer surface .... *H. (Xeralictoides)*
—. Mesoscutum considerably shorter than minimum intertegular distance; front tibia without such spines 4
4(3). Propodeal triangle contiguously punctured throughout and dull, in contrast to sparsely punctured and shining posterior and lateral propodeal surfaces; pygidial plate of female more or less flat, with mesal area bearing patch of appressed plumose hairs .... *H. (Panurgomia)*
—. Propodeal triangle smooth and shiny, at most anteriorly rugose or irregularly punctured, or, *if* dull and closely punctured, then posterior and lateral propodeal surfaces similar; pygidial plate of female often strongly convex with elevated basal triangle, always without mesal hair patch 5
5(4). Females 6
—. Males 10
6(5). Hairs along upper (i.e., posterior) margin of hind basitarsus parted, margin being narrowly hairless and exposed 7
—. Hairs along upper (i.e., posterior) margin of hind basitarsus not parted, arising on, as well as on both sides of, the margin 8
7(6). Pygidial plate with elevated, triangular basal area margined by carinae that meet posteromedially [except in *H. oliviae* (Cockerell), which lacks these carinae]; body length 7-16 mm .... *H. (Carinapis)*
—. Pygidial plate flat with shallow longitudinal sulcus near apex, without elevated area; body length 4-7 mm .... *H. (Hesperapis s. str.)*
8(6). Head length only slightly less than width (9:10), head thus seemingly elongate; labial palpus with second segment much longer than first .... *H. (Zacesta)*
—. Head length much less than width; labial palpus with second segment little if any longer than first 9
9(8). First flagellar segment at least 1.6 times length of second .... *H. (Disparapis)*

—. First flagellar segment equal to or shorter than second
.... *H. (Amblyapis)*
10(5). Labial palpus with second segment much longer than first; head about as long as broad .... *H. (Zacesta)*
—. Labial palpus with second segment little if any longer than first; head distinctly broader than long (except in some *Hesperapis s. str.*) ....
.... *H. (Amblyapis, Carinapis, Disparapis, Hesperapis s. str.)* (see key below)

Key to the Subgenera of Certain Male *Hesperapis* from North America

Good subgeneric characters are difficult to find among many males. Because the following key, to certain males that run to the second alternative of couplet 10 in the above key, utilizes species-group characters, some subgenera come out in two or more places.

1. Body length 6 mm or less *and* metasomal terga with distinct, white, apical hair bands............ *H. (Hesperapis s. str.)*
—. Body length greater than 6 mm *or* metasomal terga without white apical hair bands, sometimes with pale brown hair bands 2
2(1). Bare, subtriangular pygidial plate present, usually elevated and defined by carinae at least apically ....
.... *H. (Carinapis)* (in part)
—. Pygidial plate absent, T7 covered with pubescence 3
3(2). Posterior lobe of pronotum produced into stout, weakly recurved spine; galea dark, sparsely punctured, glabrous .... *H. (Carinapis)* (in part)
—. Posterior lobe of pronotum normal, not produced into spine; galea either pale, reticulate, heavily punctured, or bearing surface hair or apical fringe 4
4(3). S6 with apical lobes rounded, shiny and relatively free of hair except for apically directed marginal fringe; terga always with distinct pale apical hair bands ....
.... *H. (Disparapis)*
—. S6 with apical lobes acute, dull, and bearing inner, subapical brush of suberect hair; terga with pale apical hair bands occasionally very weakly developed or absent 5
5(4). Terga dull with irregular-sized continuous punctures
.... *H. (Amblyapis)* (in part)
—. Terga shiny with uniform-sized, well-spaced punctures
.... 6
6(5). Large species, more than 7.5 mm in length...............
.... *H. (Carinapis)* (in part)
—. Small species, less than 7.5 mm in length 7
7(6). Apical lobes of S6 with subapical margins nearly straight and inner margin of each lobe bearing elongate, submarginal brush of suberect hairs ....
.... *H. (Carinapis)* (in part)
—. Apical lobes of S6 with apices produced laterally and inner margin of each lobe bearing subovate, submarginal brush of hairs in a whorl .... *H. (Amblyapis)* (in part)

Hesperapis / Subgenus *Amblyapis* Cockerell

Halictoides (Amblyapis) Cockerell, 1910b: 362. Type species: *Halictoides ilicifoliae* Cockerell, 1910b, by original designation.

This subgenus consists of small species (4-9 mm long).
■ *Ambylapis* occurs from west Texas to Nevada and central California, USA, and south to Baja California and Durango, Mexico. It contains six species.

Hesperapis / Subgenus *Capicola* Friese

Capicola Friese, 1911a: 672. Type species: *Capicola braunsiana* Friese, 1911, = *Osmia? capensis* Cameron, 1905 [new synonymy], monobasic.
Rhinochaetula Friese, 1912a: 185. Type species: *Capicola cinctiventris* Friese, 1912, by designation of Cockerell, 1915: 343.

This subgenus has several apomorphies relative to *Capicoloides,* such as an area of dense setae on the outer side of the male gonostylus (Fig. 70-1a) and four apical lobes on S6 of males. S7 of the male has a rather broad disc and a bilobed apex (Fig. 70-1c). The body length is 3.5 to 13.0 mm.
■ This subgenus is found in xeric parts of South Africa and Namibia. It contains about six species; see list by Michener (1981a).

Hesperapis / Subgenus *Capicoloides* Michener

Capicola (Capicoloides) Michener, 1981a: 83. Type species: *Capicola aliciae* Cockerell, 1932, by original designation.

This subgenus is largely characterized by plesiomorphies relative to *Capicola,* but the presence of spinelike hairs in the scopa is a unique apomorphy. The form of S7 of the male may also be an apomorphy, although, if so, it is a reversion toward the structure common in many melittids, colletids, and other bees, but otherwise not seen in the *Eremaphanta-Hesperapis* line. In appearance, the lone species of *Capicoloides* looks like a middle-sized species of the subgenus *Capicola* or of the American subgenus *Panurgomia.* The body length is 7.5 to 9.0 mm.
■ *Capicoloides* occurs in Namibia and Cape Province, South Africa. So far as is known, the subgenus contains a single species, *Hesperapis aliciae* (Cockerell).

Hesperapis / Subgenus *Carinapis* Stage

Hesperapis (Carinapis) Stage, 1981, *in* Michener, 1981a: 98. Type species: *Hesperapis carinata* Stevens, 1919, by original designation.

This subgenus contains moderate-sized to large (7-16 mm long) species.
■ *Carinapis* has the widest range of any subgenus of *Hesperapis,* from Oregon, North Dakota, and Illinois south to northern Florida, USA, and to Baja California and Morelos, Mexico. Of about 15 species, only seven are named.

Hesperapis / Subgenus *Disparapis* Stage

Hesperapis (Disparapis) Stage, 1981, *in* Michener, 1981a: 96. Type species: *Hesperapis arenicola* Crawford, 1917, by original designation.

This subgenus contains species with body lengths of 8 to 14 mm.
■ *Disparapis* occurs in deserts from western Texas, California, and Utah, USA, to Zacatecas and Baja California, Mexico. Only one species has been named, but two others are known.

Hesperapis / Subgenus *Hesperapis* Cockerell s. str.

Hesperapis Cockerell, 1898a: 147. Type species: *Hesperapis elegantula* Cockerell, 1898, monobasic.

This subgenus consists of species 4 to 7 mm long.
- *Hesperapis* s. str. occurs in deserts from western Texas to Idaho, Nevada, and California, USA. Of three species, only one has been described.

Hesperapis / Subgenus *Panurgomia* Viereck

Panurgomia Viereck, 1909: 48. Type species: *Panurgomia fuchsi* Viereck, 1909, monobasic.

This subgenus consists of species 7 to 15 mm long. The propodeal triangle is densely punctured, in contrast to the adjacent areas.
- *Panurgomia* ranges from Sonora and Baja California, Mexico, to California and northern Nevada, USA. There are six species.

Hesperapis / Subgenus *Xeralictoides* Stage

Xeralictoides Stage, 1981, *in* Michener, 1981a: 99. Type species: *Hesperapis laticeps* Crawford, 1917, by original designation.

This subgenus contains species 7 to 10 mm long that superficially resemble, and visit the same flowers (Loasaceae) as, the bees of the rophitine genus *Xeralictus*.
- *Xeralictoides* occurs in deserts of Nevada and California, USA. Of the two species, only one is described.

Hesperapis / Subgenus *Zacesta* Ashmead

Zacesta Ashmead, 1899a: 73. Type species: *Zacesta rufipes* Ashmead, 1899, monobasic.

Zacesta, similar to *Hesperapis* s. str., contains species 4 to 7 mm long.
- The species of *Zacesta* are found in central and southern California, USA. Of the two species, only one is named.

71. Tribe Promelittini

From most other Dasypodainae, this tribe differs in the yellow male clypeus. The vertex, as seen from the front, is only weakly convex, not much elevated above the summits of the eyes. The paraglossa is nearly as long as its suspensorium. The metasomal terga have basal, but not apical, hair bands. A longitudinal median area of the pygidial plate of the female is elevated. The lateroapical lobes of the male S7 are large (Fig. 70-1j, k); the dorsal mesoapical lobe of the gonocoxite is absent, and the gonostylus is long and relatively slender, not fused to the gonocoxite (Fig. 70-1h). Illustrations of these characters are in Michener (1981a).

This tribe occurs in Egypt, Morocco, and also possibly in South Africa.

Promelittini contains a single genus, *Promelitta*, and perhaps also *Afrodasypoda*. Since *Promelitta* does not display the numerous distinctive characters of the Sambini, it is more similar to the Dasypodaini than to the Sambini, but in some features it resembles the Sambini and differs from the Dasypodaini. Some such features are probable plesiomorphies, like the large lateroapical lobes of S7 of the males. The longitudinal median elevation of the pygidial plate of the female, however, is an apomorphy shared by *Promelitta* and the Sambini, but it may have arisen independently, for it is found also in *Hesperapis (Capicola)*.

In various features *Promelitta* resembles the Melittinae at least as much as other Dasypodainae. As in some Sambini, the base of the second submarginal cell slopes more than that in most Dasypodainae but less than that in Melittinae. The tergal graduli, however, are about as in *Melitta*; the free margin of the marginal cell is slightly concave, the distal part of the cell thus slender, as in *Melitta*; and T6 of the male is not bilobed but has a median, apical, hairy area, as in the Melittinae. The yellow clypeus of the male and the long, simple gonostylus are suggestive of the subgenus *Paramacropis* of *Macropis*. Thus on the basis of similarity, *Promelitta* could almost be placed in the Melittinae. Characters indicating its affinities to other Dasypodainae are the lack of a pygidial plate in the male, the submarginal cells and paraglossa, and the strongly elevated lateral apical lobe of S7 of the male (compare Fig. 70-1f with j). None of these is a strong character, but the combination is reasonably convincing. *Promelitta* is probably an archaic type, like several other bees of the palearctic deserts. Perhaps it is a derivative of the group from which Dasypodaini and Sambini arose.

Genus *Promelitta* Warncke

Melitta (Promelitta) Warncke, 1977a: 59. Type species: *Dufourea alboclypeata* Friese, 1900, by original designation.

The type species is a rather ordinary-looking bee 9 mm long.

■ *Promelitta* is known from Egypt and Morocco, and is represented by a single rare species, *Promelitta alboclypeata* (Friese).

Genus *Afrodasypoda* Engel [incertae sedis]

Afrodasypoda Engel. 2005: 16. Type species: *Rhinochaetula plumipes* Friese, 1912, by original designation.

Michener (2000), under the heading of the genus *Promelitta*, wrote, "A South African species, *P. plumipes* (Friese), known from a single female, has been tentatively included [in *Promelitta* and the Promelittini] for lack of a better place to put it (Michener, 1981a) but it may well fall in the Dasypodaini." Although still known from only one female in the Berlin museum that cannot be run with satisfaction through the key to tribes (largely because male characters are unknown), Engel (2005) has placed it in a new genus, *Afrodasypoda*, and tribe, Afrodasypodini, solely on the basis of the published characters. I believe that if adequate material existed, I would agree that a new genus and perhaps tribe were justified, but see Michener (1981a: 59, 60.)

Distinctive characters of the female of *Afrodasypoda* include the following: inner orbits diverging below; hind tibia covered with scopalike hairs, keirotrichia being absent; postpalpal part of maxilla nearly four times as long as broad; basal hair bands of metasoma on T2 and T3 only.

■ The single species, *Afrodasypoda plumipes* (Friese), is from Little Namaland, northwestern Cape Province, South Africa.

72. Tribe Sambini

In both sexes of these bees, the clypeus is concolorous with the rest of the face. The short, broad head and its flat or concave vertex, exaggerated in *Samba,* are characteristic. The paraglossa is absent (Fig. 72-1b) or indistinguishably fused to the suspensorium, or, in *Haplomelitta (Prosamba),* it is distinct but minute, hairless, and tapering. The metasomal terga lack apical hair bands except in *Haplomelitta (Metasamba).* A longitudinal median area of the pygidial plate of the female is elevated. In the male the lateroapical lobes of S7 are large (Fig. 72-1f, g); the dorsal mesoapical lobe of the gonocoxite is strong and produced posteriorly (Fig. 72-1d); the gonostylus is long, not fused to the gonocoxite, and expanded and angulate apically; and the volsellar digitus is produced posteriorly and not opposable to the cuspis. Michener (1981a) illustrated many characters, especially those of the mouthparts, genitalia, and sterna; see also Figure 72-1.

This tribe occurs in eastern and southern Africa. Only the single species of *Samba* occurs in a mesic region; the other species are all found in xeric environments of South Africa and Namibia.

Each known species of Sambini could reasonably be put in a separate genus, for these species differ from one another strikingly in both appearance and structure. I have followed Michener (1981a), however, in recognizing only two genera. One of them, *Haplomelitta,* is probably paraphyletic, because *Samba* was probably derived from it; I know of no synapomorphy for *Haplomelitta* that is not also found in *Samba.* The swollen hind basitarsus of the male suggests that *Samba* is related to the subgenera *Haplomelitta* s. str. and *Metasamba,* as shown in Michener's (1981a) cladograms, but a single character whose manifestation is rather different in the three taxa is not a convincing synapomorphy. A classification based strictly on existing cladograms would either unite *Samba* with *Haplomelitta,* in spite of major morphological di-

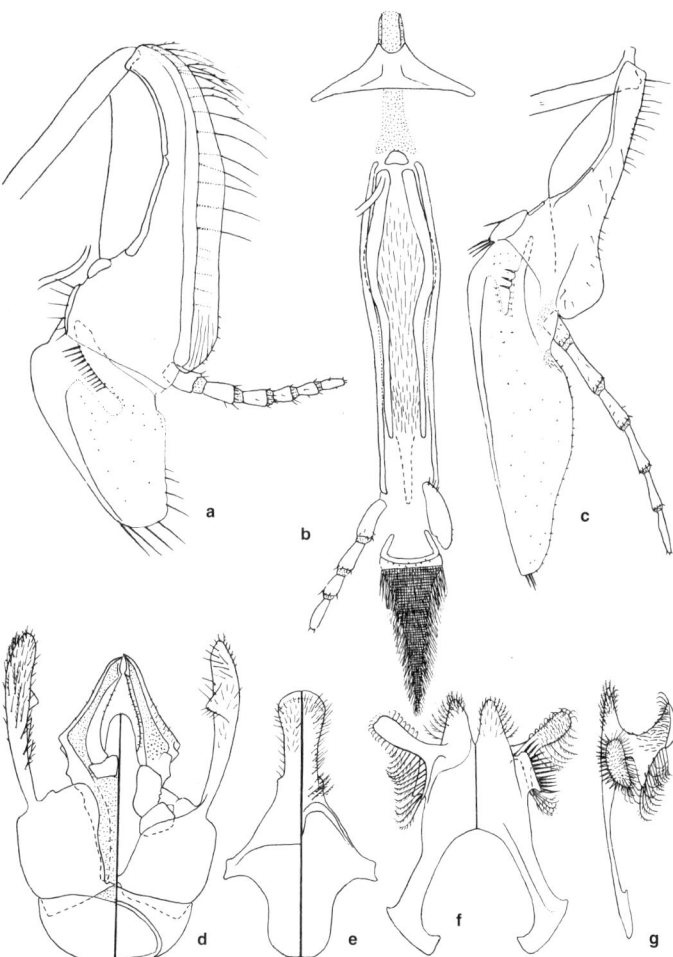

Figure 72-1. Structures of Sambini. **a,** Inner view of maxilla of *Samba calcarata* Friese; **b,** Anterior view of labium of same, one palpus and the opposite paraglossal suspensorium omitted; paraglossae are absent (note the membranous mentum, a structure that is sclerotized in other Sambini); **c,** Inner view of maxilla of *Haplomelitta ogilviei* (Cockerell); **d-g,** Male genitalia, S8, S7, and S7 lateral view of *H. ogilviei* (Cockerell). In the divided drawings, dorsal views are on the left. From Michener, 1981a.

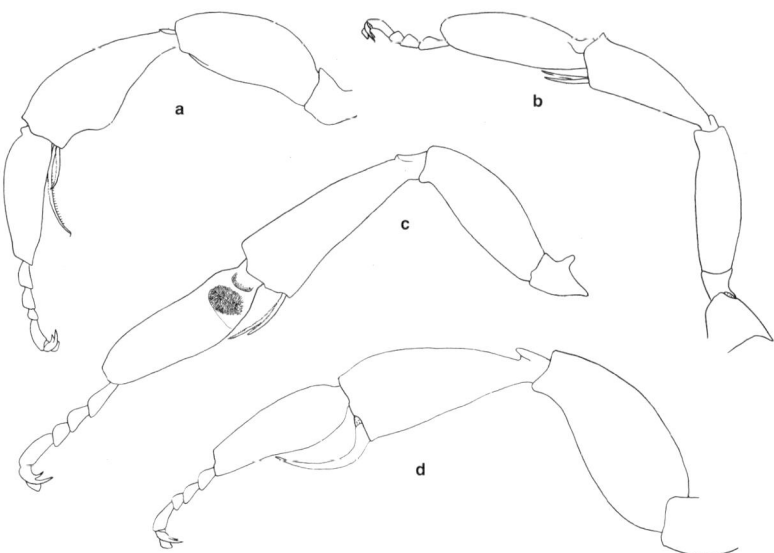

Figure 72-2. Hind legs of Sambini. **a**, *Haplomelitta (Metasamba) fasciata* Michener, male; **b**, *H. (Haplomelitta) ogilviei* (Cockerell), male; **c, d**, *Samba calcarata* Friese, male and female. Note the single tibial spur of the last. From Michener, 1981a.

vergences, or separate the subgenera of *Haplomelitta* at the generic level. I would recommend the latter course, even though the genera would be monotypic, because of the great phenetic differences among them. I have delayed, however, until a phylogenetic analysis directed specifically toward the Sambini is made.

Key to the Genera of the Sambini

1. Female with only one hind tibial spur that is large and curved (Fig. 72-2d); vertex strongly concave, as seen from front; hind basitarsus of male with hairy pocket near base of outer surface (Fig. 72-2c) .... *Samba*
—. Female with two hind tibial spurs; vertex not or scarcely concave; hind basitarsus of male without hairy pocket (Fig. 72-2a, b) .... *Haplomelitta*

Genus *Haplomelitta* Cockerell

This genus is sufficiently characterized in the discussion of the Sambini, above, and in the key to the genera of that tribe. All species of *Haplomelitta* occur in dry parts of western South Africa and Namibia.

Key to the Subgenera of *Haplomelitta*

1. With strong, pale apical tergal hair bands; legs of male with deformed hind and middle basitarsi and inner hind tibial spur (Fig. 72-2a) .... *H. (Metasamba)*
—. Without metasomal hair bands; legs of male not deformed (hind basitarsus sometimes inflated) 2
2(1). Mandible of female tridentate (male unknown) ....
.... *H. (Haplosamba)*
—. Mandible of female bidentate (inner tooth sometimes reduced to a shoulder) 3
3(2). Propodeal triangle strongly differentiated, basal zone (and laterally entire length of triangle) with short longitudinal rugae; hind basitarsus of male strongly inflated (Fig. 72-2b), as broad as tibia .... *H. (Haplomelitta s. str.)*
—. Propodeal triangle differentiated only by weak lines, without rugae on basal zone; hind basitarsus of male slender, much narrower than tibia 4

4(3). Second submarginal cell about as long as first; first recurrent and first submarginal crossvein usually well separated; basal vein meeting cu-v or nearly so; S7 of male with two pairs of pedunculate membranous lobes ....
.... *H. (Atrosamba)*
—. Second submarginal cell distinctly shorter than first; first recurrent and first submarginal crossvein close together, meeting one another or the former on either side of the latter; basal vein much basad of cu-v; S7 of male with one pair of pedunculate membranous lobes and one pair of hairy areas .... *H. (Prosamba)*

Haplomelitta / Subgenus *Atrosamba* Michener

Haplomelitta (Atrosamba) Michener, 1981a: 65. Type species: *Haplomelitta atra* Michener, 1981, by original designation.

This subgenus includes a large (body length 11.0-11.5 mm), robust, dark species without metasomal hair bands. The subapical mandibular tooth (apex of pollex) is weak, often so worn as to be a mere shoulder. The hind basitarsus in both sexes has an apical spine projecting above the base of the second segment.

■ This subgenus is known only from western Cape Province, South Africa. It contains a single described species, *Haplomelitta atra* Michener.

Haplomelitta / Subgenus *Haplomelitta* Cockerell s. str.

Haplomelitta Cockerell, 1934a: 446. Type species: *Rhinochaetula ogilviei* Cockerell, 1932, by original designation.

The body is slender, coarsely punctate, and sparsely hairy, with the metasoma shining, not fasciate, and largely red. The single species is superficially similar to *Haplosamba*. The hind basitarsus of the male is inflated, nearly as long as and about as broad as the tibia. S1 of the female has a broad, deep apical emargination in the middle of which is a slender, posteriorly directed process, which is split and downcurved as two small flaps at its apex.

■ *Haplomelitta* occurs in South Africa. It contains one known species, *Haplomelitta ogilviei* (Cockerell).

This subgenus appears to be an oligolectic visitor to flowers of *Monopsis debilis,* according to S. K. and F. W. Gess.

Haplomelitta / Subgenus *Haplosamba* Michener

Haplomelitta (Haplosamba) Michener, 1981a: 66. Type species: *Haplomelitta tridentata* Michener, 1981, by original designation.

The body is shining, coarsely punctate, rather slender, and 11.5 mm long, and the metasoma is partly red and nonfasciate, like that of *Haplomelitta* s. str. The mandible is tridentate in the female because the apex of the pollex is bidentate. The propodeal triangle is weakly defined.

■ *Haplosamba* occurs in western Cape Province, South Africa. It contains one described species, *Haplomelitta tridentata* Michener, known only from the female.

Haplomelitta / Subgenus *Metasamba* Michener

Haplomelitta (Metasamba) Michener, 1981a: 69. Type species: *Haplomelitta fasciata* Michener, 1981, by original designation.

This subgenus contains a rather small (8.5-9.0 mm long), robust form similar to *Prosamba* but with pale apical hair bands on the metasomal terga, unlike other species of the genus. The hind tibia of the male is thickened beyond the middle. The inner hind tibial spur of the male, about twice as long as the outer spur, is slender and curved away from the basitarsus except at the apex, where it is abruptly bent toward the basitarsus and twisted so that its teeth are directed away from the basitarsus.

■ This subgenus is known only from Namibia. It contains one species, *Haplomelitta fasciata* Michener.

Haplomelitta / Subgenus *Prosamba* Michener

Haplomelitta (Prosamba) Michener, 1981a: 63. Type species: *Haplomelitta griseonigra* Michener, 1981a, by original designation.

Rather small (8 mm long), dark, robust, and lacking metasomal hair bands, the bees of this subgenus are notable for the minute but distinct paraglossa. The basal segments of the maxillary palpus have long hairs. S7 of the male has only one pair of membranous lateroapical processes.

■ This subgenus, known from western South Africa, contains one known species, *Haplomelitta griseonigra* Michener.

Genus *Samba* Friese

Samba Friese, 1908a: 568. Type species: *Samba calcarata* Friese, 1908, monobasic.

The head is extraordinarily short and broad, the vertex concave, as seen from the front, the shortest distance between the eyes being greater than the length of an eye. The mentum is not sclerotized (Fig. 72-1b). The hind basitarsus of the male is swollen, with a large hairy pit on the outer surface near the base. Body length is 9 to 10 mm. This genus has so many unusual features (see the discussion of the tribe, above, and the key to genera; also Michener, 1981a) that recognition at the genus level seems justified, even though so doing may make *Haplomelitta* paraphyletic. The short galeal blade (Fig. 72-1a) is unique in the Sambini.

■ *Samba* occurs in Kenya and Tanzania. The only species is *Samba calcarata* Friese.

73. Subfamily Meganomiinae

This subfamily contains medium-sized to large bees with yellow or cream-colored maculations on the body; the yellow pattern combined with large size differentiate them from all other Melittidae. The mandible of the female is simple. The paraglossa is longer than its suspensorium and densely hairy (Fig. 73-1). The forewing has three submarginal cells, the second and third together being shorter than or as long as the first. The stigma is extremely slender, the sides basal to vein r parallel or even converging apically, and the prestigma is two-thirds as long as the stigma to longer than the stigma. Vein r arises near the apex of the stigma. The apex of the marginal cell is rounded, bent away from the wing margin.

T4 and T5 of males of *Meganomia* and *Ceratomonia* each have two large, oval, finely striate, pregradular areas, probably strigilatory in function (Rozen, 1977a); no other bees have such structures. S7 of the male has a pair of sclerotized apical lobes arising medially and extending laterally or anterolaterally (Fig. 73-2c), often in contact with the disc of the sternum (so that one may not immediately see that the lobes join the sternum only near the midline). The volsella is much reduced, not at all chelate, as shown in the illustrations by Michener (1981a) and by Michener and Brooks (1987), and in Figure 73-2a.

The larva falls in the first of the two groups characterized in describing the larvae of the family, i.e., the larvae spin cocoons. Rozen (1977a: 14) suggested subfamilial status for *Meganomia* on the basis of the larval apomorphies of that genus. Larvae of other meganomiine genera are unknown. The simple, acute mandibular apex, the lack of a dorsal transverse ridge on the perianal area, and the swollen cephalic margin behind the posterior mandibular articulation are among the apomorphies of *Meganomia*. The last may be unique among bees, but the others occur independently in various bees, including the Dasypodainae.

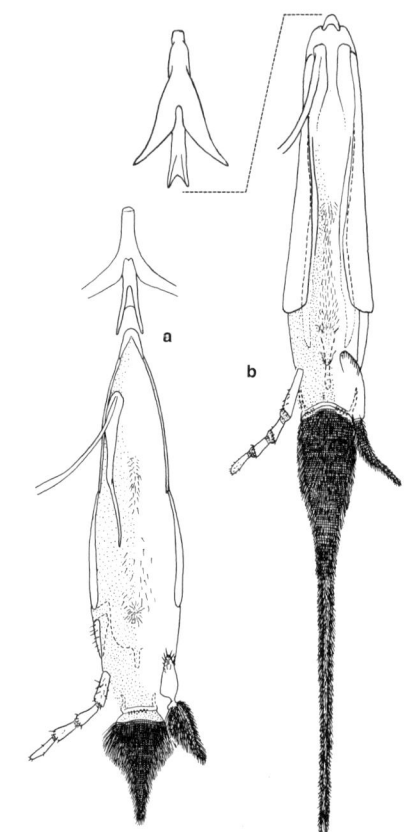

Figure 73-1. Anterior views of labia of Meganomiinae. **a,** *Ceratomonia rozenorum* Michener; **b,** *Pseudophilanthus tsavoensis* (Strand). One labial palpus and one paraglossa are omitted from each. From Michener, 1981a.

The Meganomiinae occur in xeric areas of eastern and southern Africa, but are not known south of the tropic of Capricorn. The subfamily also occurs in Yemen and Madagascar.

This subfamily contains only four small genera. Each genus is about equally different from the other three, owing largely to the derived features of each. Michener (1981a) showed two completely different and almost equally parsimonious cladograms for the genera. It seems best to recognize that we know nothing about their cladistic relationships.

Key to the Genera of the Meganomiinae

1. Arolia absent (although long, hairy, yellow plantae project between the claws); flagellum of male with apical segment or several segments curled and attenuate (Fig. 73-3c); hind basitarsus of male bent or contorted, or at least with curved carina on outer side .... *Meganomia*
—. Arolia present, conspicuously black among associated pale interungual structures; flagellum of male with apex simple or expanded as a plate; hind basitarsus of male simple 2
2(1). Ocelli much in front of posterior edge of vertex, posterior ocellus separated from that edge by more than two ocellar diameters; upper part of head gently convex, as seen from front; front edge of median ocellus little if any nearer to posterior edge of vertex than to antennal bases; glossa as long as prementum or nearly so (Fig. 73-1b) ..
.... *Pseudophilanthus*
—. Posterior ocellus separated from posterior edge of vertex by an ocellar diameter or a little more; upper part of head, as seen from front, feebly convex or flat or with ocellar region slightly elevated; front edge of median ocellus much nearer to posterior edge of vertex than to antennal bases; glossa three-fifths as long as prementum or shorter (Fig. 73-1a) 3
3(2). Flagellum of male not expanded at apex (Fig. 73-3a); T4 and T5 (both sexes) without sublateral stridulating areas; basitibial plate of female not defined (female un-

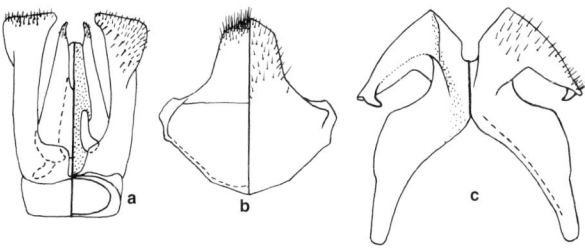

Figure 73-2. Male genitalia, S8, and S7 of *Meganomia binghami* (Cockerell). (Dorsal views are on the left.) From Michener, 1981a.

known in subgenus *Nesomonia*); glossa one-half to three-fifths as long as prementum .... *Uromonia*
—. Flagellum of male with apical segment expanded, plate-like, black (Fig. 73-3d); T4 and T5 (both sexes) with sublateral stridulating areas hidden under preceding tergal margins; glossa less than one-third as long as prementum (Fig. 73-1a) .... *Ceratomonia*

Genus *Ceratomonia* Michener

Ceratomonia Michener, 1981a: 20. Type species: *Ceratomonia rozenorum* Michener, 1981, by original designation.

The body is 13 to 15 mm long and more slender than in most meganomiines. The male flagellum is long, reaching beyond the scutellum, and flat and hairy beneath; the last segment is expanded, discoid (Fig. 73-3d). The labrum is a transverse strip, not produced medially. The basitibial plate of the female is well defined, conspicuous. Structures were illustrated by Michener (1981a).

■ *Ceratomonia* is known only from Namibia. There is one species, *Ceratomonia rozenorum* Michener.

Some aspects of behavior were described by Rozen (1977a) under the heading "*Meganomia* species B."

Genus *Meganomia* Cockerell

Nomia (Meganomia) Cockerell, 1909a: 402. Type species: *Nomia binghami* Cockerell, 1909, monobasic.

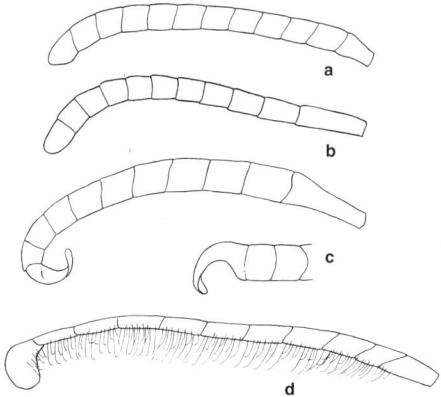

Figure 73-3. Lateral views of antennal flagella of male Meganomiinae. **a,** *Uromonia stagei* Michener (antennae 14-segmented); **b,** *Pseudophilanthus tsavoensis* (Strand); **c,** *Meganomia gigas* Michener, with separated, straightened apex in ventral view; **d,** *Ceratomonia rozenorum* Michener. From Michener, 1981a.

Maxschwarzia Pagliano and Scaramozzino, 1990: 5, unnecessary replacement for *Meganomia* Cockerell, 1909; see Michener, 1992a. Type species: *Nomia binghami* Cockerell, 1909, autobasic.

These are large bees (15-22 mm long). The male flagellum is thickened, without long hairs beneath, and curled apically (Fig. 73-3c), the last one to three segments slender. The labrum is two to three times as wide as long, its apex convex or medially emarginate. The basitibial plate of the female is hidden by hairs, not defined apically. Male genitalia, sterna, and other structures were illustrated by Michener (1981a); see also Figure 73-2.

■ This genus is found from Kenya to Namibia and northern Transvaal, South Africa, and in Yemen. There are four described species, and an additional one in Yemen. The genus was revised by Michener (1981a).

The nests and immature stages of *M. gigas* Michener were described by Rozen (1977a) under the name *M. binghami* (Cockerell). The nests are deep burrows leading to sloping cells that probably have an invisible, somewhat waterproof lining. The provisions are firm and elongate-subrectangular, with the egg on top.

Genus *Pseudophilanthus* Alfken

The body length is 13 to 15 mm. The vertex is convex, as seen from the front, and extends far behind the ocelli, its posterior margin separated from the posterior ocelli by at least two ocellar diameters. The male flagellum is simple. The labrum is less than four times as wide as long.

Key to the Subgenera of *Pseudophilanthus*

1. T7 of male bifid, without pygidial plate; galeal blade broad, shorter than stipes (Madagascar) .... *P. (Dicromonia)*
—. T7 of male simple, with pygidial plate; galeal blade tapering, about as long as stipes (Africa) ....
.... *P. (Pseudophilanthus s. str.)*

Pseudophilanthus / Subgenus *Dicromonia* Michener and Brooks

Agemmonia (Dicromonia) Michener and Brooks, 1987: 100. Type species: *Agemmonia wenzeli* Michener and Brooks, 1987, by original designation.

The second to fourth segments of the male flagellum are broader than long. The labrum is less than three times as wide as long. Numerous other distinctive features are listed in the key to the subgenera, above, and by Michener and Brooks (1987), where illustrations of the male genitalia and hidden sterna were presented.

■ This subgenus is known from southwestern Madagascar. Only a single male specimen of *Pseudophilanthus wenzeli* (Michener and Brooks) has been collected.

Pseudophilanthus / Subgenus *Pseudophilanthus* Alfken s. str.

Pseudophilanthus Alfken, 1939: 121. Type species: *Pseudophilanthus taeniatus* Alfken, 1939, monobasic.
Agemmonia Michener, 1981a: 26. Type species: *Nomia tsavoensis* Strand, 1920, by original designation.

Each of the first four segments of the male flagellum is much longer than broad (Fig. 73-3b). The labrum is over three times as wide as long. The basitibial plate of the female is not defined.

■ This subgenus is from Kenya and the Ethiopia-Kenya border region. The three nominal species, perhaps all variants of one species, were reviewed by Michener (1992a).

Genus *Uromonia* Michener

These are the smallest meganomiines, 10 to 13 mm long. The male flagellum is simple or crenulate. The labrum is over four times as wide as long, the apex convex or broadly truncate. The basitibial plate of the female is not defined (at least in *Uromonia* s. str.).

The two species, quite different from one another, are placed in separate subgenera. The female of *Nesomonia* is unknown.

Key to the Subgenera of *Uromonia*
1. Flagellum of male crenulate, 11-segmented; S8 of male with apical process narrower than disc of sternum (Madagascar) .... *U. (Nesomonia)*
—. Flagellum of male not crenulate, 12-segmented (Fig. 73-3a); S8 of male tapering toward subtruncate apex, without recognizable apical process (Africa) ....
.... *U. (Uromonia s. str.)*

Uromonia / Subgenus *Nesomonia* Michener, Brooks and Pauly

Uromonia (Nesomonia) Michener, Brooks, and Pauly, 1990: 135. Type species: *Nomia flaviventris* Benoist, 1963, by original designation.

In addition to the characters listed in the key to subgenera, *Nesomonia* differs from *Uromonia* s. str. in its simple penis valves and the shape and vestiture of its apical sterna. The male genitalia, hidden sterna, maxilla, and other structures were illustrated by Michener, Brooks, and Pauly (1990).

■ *Nesomonia* is known from southwestern Madagascar. The only known species is *Uromonia flaviventris* (Benoist).

Uromonia / Subgenus *Uromonia* Michener s. str.

Uromonia Michener, 1981a: 23. Type species: *Uromonia stagei* Michener, 1981, by original designation.

This subgenus is unique among all bees in its 14-segmented male antennae. The penis valves are large and complex, with processes projecting both dorsally and ventrally.

■ *Uromonia* s. str. is known only from the Kenya coast. The only species is *Uromonia stagei* Michener.

74. Subfamily Melittinae

These are small to rather large bees (body 7-15 mm long) having the form and often the appearance of an *Andrena*, without yellow maculations, but in *Macropis*, the face of the male, and, in certain Chinese species, the legs are partly yellow; the supraclypeal area of the female of *Macropis hedini* Alfken from China bears a small yellow spot. The mandible of the female (and male) has a subapical inner tooth (apex of the pollex). The paraglossa is about as long as its suspensorium (Fig. 74-1a), not distinctly annulate to rather coarsely annulate, and densely hairy. The forewing has two or three submarginal cells, the second (if there are two) or second plus third being about as long as or slightly longer than the first; the second (if there are two) or third is strongly narrowed toward the costal margin, the second (if there are three) parallel-sided or narrowed toward the costal margin. The base of the second submarginal cell is slanting, not at right angles to the longitudinal veins (Fig. 68-1a, b). The stigma is less slender than that of the Meganomiinae, and the prestigma is over one-third to nearly two-thirds as long as the stigma. Vein r arises near the middle of the stigma. The apex of the marginal cell is narrowly rounded or pointed, bent away from the costal margin only to the extent of one or two vein widths. The metasomal terga usually have apical bands of pale hair arising from depressed apical marginal zones, these bands sometimes represented only at the sides (e.g., in *Melitta haemorrhoidalis* Fabricius), or sometimes expanded to cover much of the terga (e.g., in some *Macropis omeiensis* Wu). Illustrations of many structures are in Michener (1981a) and Snelling and Stage (1995b).

As shown by Rozen and McGinley (1974a) and Rozen and Jacobson (1980), the larvae of *Melitta* and *Macropis* fall in the first of the two groups characterized above in describing the larvae of the family. That is, melittine larvae spin cocoons and have the morphological equipment to do so. They lack various larval synapomorphies of the Meganomiinae. The antenna is a convexity without projecting papilla; it is thus synapomorphic relative to that of the Meganomiinae but plesiomorphic relative to the nearly flat antennal area in Dasypodainae (see also Rozen, 1978a).

The Melittinae were divided in some previous classifications (Michener, 1944) into the Melittinae and Macropidinae, but separation at the subfamily level seems unnecessary; recognition of two tribes, Melittini and Macropidini, would be a possibility, since *Macropis* is rather distinctive. The presence of only two submarginal cells and of extensive yellow areas on the face of the male are apomorphies of *Macropis* ("Macropidini"). The synapomorphies uniting the other Melittinae ("Melittini") (reduced gonostylus, broadly fused to the gonocoxite, reduction of the apicolateral lobes of S7 of the male, and reduction or loss of the pygidial plate of the male) are similarly unimpressive, for all have arisen repeatedly among the bees and could represent convergence. The similarity of *Macropis* and *Melitta* in larval characters is emphasized by Rozen and McGinley

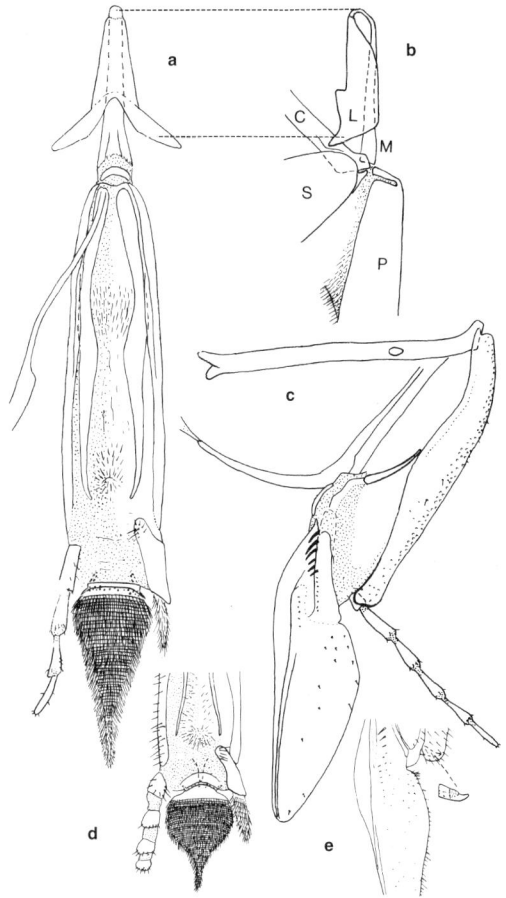

Figure 74-1. Proboscides of *Melitta*. a-c, *Melitta (Melitta) leporina* (Panzer); **a**, Anterior view of labium; **b**, Diagram of lateral view of base of labium and associated parts of maxilla, cardo, and stipes; **c**, Inner view of maxilla. (C, cardo; L, lorum; M, mentum; P, prementum; S, stipes.)

d, e, *M. (Dolichochile) melittoides* (Viereck), apex of labium and base of galeal blade with reduced maxillary palpus. From Michener, 1981a.

(1974a). There is nothing in their study that justifies tribal or subfamilial separation of these genera. Since only one genus would be in the Macropidini, since the monophyly of the Melittini is in some doubt, and since the oil-collecting adaptations of *Macropis* and *Rediviva* (which is close to *Melitta*) are so similar as to suggest common origin (Michener, 1981a), recognition of Macropidini seems undesirable. It should be noted that certain of the synapomorphies of *Macropis* (numbered 55 and 95) shown in the cladograms by Michener (1981a) are invalidated by the subgenus *Paramacropis*, which was not available for study in 1981. Thus *Macropis* is even less distinctive than it appeared to be at that time, when I sunk the Macropidinae into the Melittinae.

The relationships among the genera of Melittinae remain in doubt. Michener (1981a) gave two almost equally parsimonious cladograms, quite different from one another. There is no reliable evidence that one is "better" than the other. Engel (2001b), however, segregated *Rediviva* and *Redivivoides* as a separate tribe, Redivivini.

Key to the Genera of the Melittinae

1. Two submarginal cells (Fig. 68-1b); pygidial plate of male strongly developed; male with yellow face marks; gonostylus of male long, slender at least at base, articulated to gonocoxite; S8 of male not ending in beveled area suggestive of a pygidial plate (holarctic) .... *Macropis*
—. Three submarginal cells (Fig. 68-1a); pygidial plate of male absent or weakly evident; male lacking yellow face marks; gonostylus of male broadly fused to gonocoxite (Fig. 74-2a, d); S8 of male ending in a beveled area suggestive of a pygidial plate 2
2(1). Propodeal triangle dull (finely granular), large (width at upper margin usually at least nearly equal to distance between transmetanotal sutures); second submarginal cell usually wider than long or about as long as wide (Fig. 68-1a); S7 of male a large plate, truncate or emarginate apically, the lobes or processes reduced to small sclerotic structures and sometimes hair patches at posterior lateral angles of sternum (Fig. 74-2f) (holarctic, Africa) .... *Melitta*
—. Propodeal triangle shiny (usually smooth), often ill-defined, small (width at upper margin not over half distance between transmetanotal sutures and often much less); second submarginal cell usually longer than wide; S7 of male with a small disc, apex bifid or with membranous lobes or both (Fig. 74-2c) 3
3(2). Scopa of female consisting of simple bristles; hind tibia and basitarsus of female slender, the latter three or more times as long as wide; anterior tarsus ordinary, with ordinary vestiture; S7 of male with apex weakly emarginate, bearing, at each side, a large, flat, vertically expanded process with long erect hairs on outer surface (Africa).... .... *Redivivoides*
—. Scopa of female consisting of densely plumose understory beneath long simple bristles; hind tibia and basitarsus of female broad, the latter about twice as long as broad; anterior tarsus thickened or elongate, with dense, short vestiture; S7 of male with apex deeply bifid or with two long, slender apical processes, lobes at each side (if present) hairless (although spiculate), not flattened, but coarsely striate or ribbed (Fig. 74-2c) (Africa) .... *Rediviva*

Genus *Macropis* Panzer

As indicated elsewhere, *Macropis* is the most distinctive genus of Melittinae. It differs from other Melittinae in having two submarginal cells, yellow on the face of the male, and a well-developed male pygidial plate elevated above the rest of the surface of T7. The form is rather robust, the body length 7 to 12 mm. Male genitalia and hidden sterna were illustrated by Saunders (1882), Popov (1958a), Mitchell (1960), Michener (1981a), and Snelling and Stage (1995b), and Michez and Patiny (2005). The oil-collecting and -transporting structures of

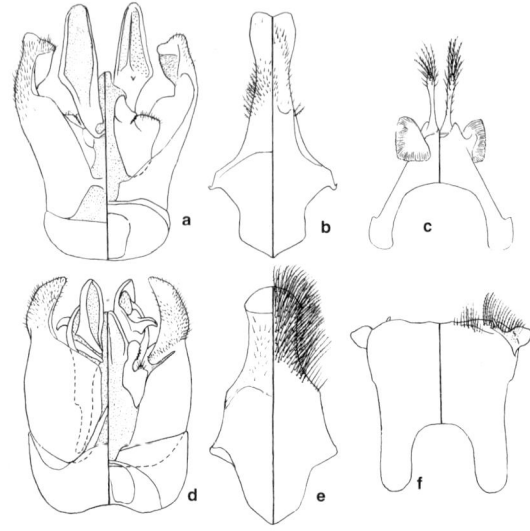

Figure 74-2. Male genitalia, S8, and S7 of Melittinae. **a-c,** *Rediviva peringueyi* (Friese); **d-f,** *Melitta dimidiata* Morawitz. From Michener, 1981a.

females are similar to those of *Rediviva*, although no *Macropis* has elongate front legs like some species of *Rediviva*. These structures of *Macropis* are the short, dense, velvety hairs on the small segments of all the tarsi, for collecting the oil, and the broad hind tibia and basitarsus with fine, dense, plumose scopal hairs (some with emergent simple apices), for oil transport; no function is known for the parted (i.e., diverging) hairs of the truncate apex of the hind basitarsus, leaving a narrow strip of integument visible along the basitarsal truncation.

So far as is known, all species of *Macropis* are dependent for larval food upon oil collected from flowers of *Lysimachia* (Primulaceae) and mixed with pollen from the same plant (Popov, 1958a; Wu, 1965a; Vogel, 1976, 1986; Rozen and Jacobson, 1980; Cane et al., 1983). Cane et al. showed that the oil is also used for lining the brood cells and accounts for the greenish color of the lining. Since *Lysimachia* produces no nectar, adults of *Macropis* visit other flowers for their own energy needs. The nests consist of short burrows in the soil with branches containing short series of subhorizontal cells (Rozen and Jacobson, 1980). The provision mass is firm, and a "foot" or projection holds the front part of the mass away from the floor of the cell, as in *Anthophorula* (Exomalopsini).

Three subgenera of *Macropis* are recognized. *Paramacropis* appears to be the sister group of the other two. *Macropis* s. str. and *Sinomacropis* are united by such synapomorphies as the expanded and bilobed or bifid gonostyli of the males and the hind basitarsal comb of the males. The hairy propodeal triangle of *Sinomacropis* is a striking apomorphy, unique in the Melittidae; the subgenus is clearly monophyletic. It may be, however, that *Macropis* s. str. is made paraphyletic by recognition of *Sinomacropis,* for apomorphies of *Macropis* s. str. are hard to find. However, the bilobed ventroapical process of the male hind basitarsus of *Macropis* s. str. may be an apomorphy of that group; the process is not lobed in *Sinomacropis* and is absent in other bees. Moreover, the long comb on the same basitarsus may be an apomorphy of *Macropis* s. str.; the comb is short in *Sinomacropis* and absent in other bees. The genus was reviewed worldwide by Michez and Patiny (2005).

Key to the Subgenera of *Macropis*

1. Vertex broad behind ocelli, ending in a sharp preoccipital carina; gonostylus of male slender, simple; S6 of male with broadly truncate apical process, S7 with apical process broader than body of sternum, process densely hairy (eastern palearctic) .... *M. (Paramacropis)*
—. Vertex declivous posteriorly, no carina separating it from concave occiput; gonostylus of male apically expanded and bilobed or bifid; S6 of male with attenuate apex, S7 with apical process narrow, pointed, bearing relatively sparse hairs 2
2(1). Propodeal triangle punctate, hairy except for small median area (China) .... *M. (Sinomacropis)*
—. Propodeal triangle impunctate and hairless, smooth or rugose (holarctic) .... *M. (Macropis s. str.)*

Macropis / Subgenus *Macropis* Panzer s. str.

Macropis Panzer, 1809, no. 16. Type species: *Megilla labiata* Fabricius, 1805 = *Megilla fulvipes* Fabricius, 1805, monobasic. [*Macropis* has been attributed to Klug, but as pointed out in Commission Opinion 1383 (1986), Panzer provided the description.]

For a consideration of the name *Megilla* Fabricius (1804) and its relation to the name *Macropis*, see Michener (1983b, 1984) and Commission Opinion 1383 (1986).

■ This is a holarctic subgenus found in mesic (not xeric) regions from western Europe to Japan and in eastern and central North America, from Quebec, Canada, to Georgia and west to Montana and Colorado, with more western records in Idaho and Washington state, USA. This subgenus contains ten species. A key to palearctic species of the genus was given by Wu (1965a), to western palearctic species by Warncke (1973a), and to North American species by Snelling and Stage (1995a). See also Michez and Patiny (2005).

Nesting biology of species of this subgenus has been described by Malyshev (1929), Rozen and Jacobson (1980), and Cane et al. (1983).

Macropis / Subgenus *Paramacropis* Popov and Guiglia

Paramacropis Popov and Guiglia, 1936: 287. Type species: *Ctenoplectra ussuriana* Popov, 1936, monobasic.

The status of this subgenus was reviewed by Wu and Michener (1986), who described the female for the first time and verified the position of *Paramacropis* in the genus *Macropis*. The lone species possesses striking presumably plesiomorphic characters as compared to other species of *Macropis*, such as the simple male gonostylus and the simple hind basitarsus of the male. At the same time it has highly derived features, such as the male S7 and S8 (see the key to the subgenera). The male genitalia were illustrated by Popov (1936b) and Wu and Michener (1986).

■ This subgenus is from northeastern China and the Pacific maritime province of Russia. The only species is *Macropis ussuriana* (Popov).

No observations have been made on the floral behavior of *Paramacropis,* but the female possesses oil-manipulating and -transporting structures like those of other *Macropis.* Presumably, *Paramacropis* uses oil from flowers of *Lysimachia,* as do the other subgenera.

Macropis / Subgenus *Sinomacropis* Michener

Macropis (Sinomacropis) Michener, 1981a: 51. Type species: *Macropis hedini* Alfken, 1936, by original designation.

Sinomacropis is widespread in China and occurs in Laos but is not known from other countries. See a key by Wu (1965a) and a review of the subgenus by Wu and Michener (1986); all four species were reviewed by Michez and Patiny (2005).

The floral biology was considered by Wu (1965a); the relation of *Sinomacropis* to *Lysimachia* appears to be the same as that for *Macropis* s. str.

Genus *Melitta* Kirby

These are melittine bees superficially resembling species of *Andrena;* the body length is 8 to 15 mm. The labrum has a lateral apical lobe, a character not found in other bees. The mouthparts are otherwise ordinary for the group except in the subgenus *Dolichochile.* The scopa on the outer sides of the hind tibia and basitarsus of the female is simple, these segments being slender. The propodeal triangle is large, dull. S7 of the male has a large disc and insignificant lateral apical lobes. The volsella has a long, blunt process extending much posterior to the cuspis and not opposable to it. The male genitalia were sketched by Warncke (1973a) and illustrated along with the hidden sterna by Wu (1978, 1982b), Michener (1981a), and Snelling and Stage (1995b); see also Figure 74-2d-f.

I reluctantly list two subgenera of *Melitta* below, for recognition of *Dolichochile* makes *Melitta* s. str. in its present sense paraphyletic. *Dolichochile* is phenetically remarkable, and I hope that its recognition will encourage investigators to learn the biological significance of its strange mandibular and proboscidial characters. What is needed is a cladistic study of all species of *Melitta* to determine the real relationships of *M. (Dolichochile) melittoides* (Viereck). Then, since it is similar to *Melitta* except for a few remarkable features, the most informative classification might place *M. melittoides* within some species group of *Melitta.*

Key to the Subgenera of *Melitta*

1. Maxillary palpus reduced to two short, fused segments (Fig. 74-1e); mandible of female slightly longer than eye, distal half a long, flat, pointed blade, at base of which on inner side are two small teeth (Fig. 74-3); surface of

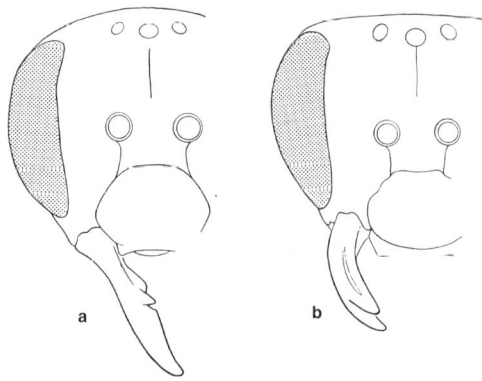

Figure 74-3. Facial views of female *Melitta*, showing mandible. **a**, *M. (Dolichochile) melittoides* (Viereck); **b**, *M. (Melitta) americana* (Smith).

labrum rather uniformly convex, impunctate except near apical margin (nearctic) .... *M. (Dolichochile)*
—. Maxillary palpus six-segmented (Fig. 74-1a); mandible of female shorter than eye, of ordinary form with one subapical inner tooth; labrum with wedge-shaped, slightly depressed median punctate area partially dividing smooth area (holarctic region and Africa) ....
.... *M. (Melitta* s. str.)

Melitta / Subgenus *Dolichochile* Viereck

Dolichochile Viereck, 1909: 49. Type species: *Dolichochile melittoides* Viereck, 1909, by original designation.

Unlike those of all other species of *Melitta*, both palpi of *Dolichochile* are short, the first segment of the maxillary palpus and segments 2 to 4 of the labial palpus being broader than long (Fig. 74-1d).

■ *Dolichochile* occurs in the eastern United States from New Hampshire to Tennessee and Georgia. The only species is *Melitta melittoides* (Viereck).

Melitta / Subgenus *Melitta* Kirby s. str.

Melitta Kirby, 1802: 117. Type species: *Melitta tricincta* Kirby, 1802, by designation of Richards, 1935: 172.
Cilissa Leach, 1812: 155. Type species: *Andrena haemorrhoidalis* Fabricius, 1775, by designation of Westwood, 1840a: 84. [A subsequent designation was listed by Michener, 1997b.]
Kirbya Lepeletier, 1841: 145 (not Robineau-Desvoidy, 1830). Type species: *Melitta tricincta* Kirby, 1802, by designation of Sandhouse, 1943: 561.
Pseudocilissa Radoszkowski, 1891: 241. Type species: *Cilissa robusta* Radoszkowski, 1876 = *Melitta dimidiata* Morawitz, 1876, monobasic.
Brachycephalapis Viereck, 1909: 47. Type species: *Melitta californica* Viereck, 1909, by original designation.

■ This is a holarctic and African subgenus. It is widespread in the Palearctic region (Canary Islands and England to Japan); it also occurs in Cape Province, South Africa, and in Kenya. In North America it occupies two disjunct areas: (1) the Sonoran desert of California and Arizona, USA, and Baja California, Mexico, and (2) the eastern USA from New Hampshire to Florida and Mississippi. There are 20 palearctic species, three nearctic species, and three or four southern African species. Warncke (1973a) gave a key to west palearctic species, Wu (1978) gave a key to Chinese species, and Snelling and Stage (1995b) revised the North American species.

Most species are probably more or less oligolectic. *M. leporina* (Panzer) is a significant pollinator of alfalfa in Europe. Nesting behavior of *Melitta* has been described by Malyshev (1923a) and Tirgari (1968).

Genus *Rediviva* Friese

Andrena (Rediviva) Friese, 1911a: 671. Type species: *Andrena peringueyi* Friese, 1911, by designation of Cockerell, 1931a: 402.
Notomelitta Cockerell, 1933b: 128. Type species: *Rediviva neliana* Cockerell, 1931, by original designation and monobasic.

This genus is similar to *Redivivoides*, but the front tarsus of the female is covered with fine, dense hairs, and the segments are often thickened or elongate. Figure 6-1 illustrates variation in front tarsi related to the flowers from which oil is extracted. The scopa (on the outer sides of the broad hind tibia and basitarsus) is so densely plumose as to hide the surface; it also includes long, simple, emergent hairs (Fig. 68-3b). S7 of the male has a narrow disk that is bifid at the apex and usually has lateroapical lobes (Fig. 74-2c). The body length is 10 to 17 mm. Male genitalia and other structures were illustrated by Michener (1981a) and Whitehead and Steiner (1993).

■ *Rediviva* occurs in South Africa and Lesotho. There are about 23 species. Whitehead and Steiner (2001) revised the 15 species of *Rediviva* found in the winter-rainfall area of South Africa, providing a wealth of distributional, floral, and morphological information.

Although *Andrena*-like in superficial appearance, females of some species have front legs longer than the bee's body (Fig. 6-2). Probably all species collect oil with the front legs from oil-producing flowers, especially *Diascia* (Scrophulariaceae), which has slender floral spurs where oil is secreted, but less commonly from *Hemimeris* and *Bowkeria* (Scrophulariaceae), the latter having a floral pouch where oil is produced, and from flowers of terrestrial Orchidaceae (Vogel, 1984; Vogel and Michener, 1985; Steiner and Whitehead, 1990, 1991; Whitehead and Steiner, 1993). The female *Rediviva* inserts her front legs into the oil-producing spurs or pouches; the oil is collected by the dense tarsal vestiture. The length of the front legs varies among species and populations, parallel with the length of the oil-producing floral spurs of different forms of *Diascia*, suggesting a coevolutionary process. The structures for oil manipulation and transport are similar to those of *Macropis*.

Genus *Redivivoides* Michener

Redivivoides Michener, 1981a: 42. Type species: *Redivivoides simulans* Michener, 1981, by original designation.

These are *Melitta*-like bees the front tarsi of which are ordinary for the group, not densely hairy as in *Rediviva*.

The scopa (on the outer sides of the slender hind tibia and basitarsus) is of short, simple hairs. The propodeal triangle is small and ill-defined. S7 of the male has a broad, subtruncate apex bearing a large, vertical, membranous hairy lobe on each side. The male has a weak pygidial plate. The body length is 10 to 15 mm. This is probably the sister group to *Rediviva*. Some characters, like those of S7 of the male, are unique apomorphies of *Redivivoides*, while the front tarsal and hind-leg characters listed are presumably plesiomorphic compared to *Rediviva*, being widespread among bees.

■ This genus occurs in Cape Province, South Africa. It contains three species, only one of which has been described.

Redivivoides does not collect floral oils and thus lacks the oil-collecting front tarsal and scopal characters of *Rediviva* (Steiner and Whitehead, 1991). It could have lost oil-collecting structures and behavior, as have probably the species of *Centris* (Centridini) that do not collect oil. There is no evidence of evolution from oil-collecting ancestors in the case of *Redivivoides*.

75. Family Megachilidae

In most parts of the world the Megachilidae are among the more easily recognized families of bees; megachilids are L-T bees with a rectangular labrum that is longer than broad and broadly articulated to the clypeus (Fig. 75-1). In recent decades, however, two taxa that do not, or do not always, agree with these megachilid characters have been shown to be relatives of the familiar megachilids. These are the groups often called the Fideliidae and the Pararhophitini. In the past, both have been included in the Anthophoridae or Apidae (Michener, 1944; Popov, 1949a); both are here regarded as tribes of a megachilid subfamily Fideliinae. Rozen (1970a) showed the relationship of Fideliini to megachilids. Rozen, in McGinley and Rozen (1987), also included *Pararhophites* in the Megachilidae. The relationship of fideliines and pararhophitines to other Megachilidae was later confirmed by Roig-Alsina and Michener (1993). Inclusion of these taxa in the Megachilidae makes that family less easy to define than when it included only the taxa here placed in the subfamily Megachilinae. The characteristic thick-headed megachiline appearance is not shared by the Fideliinae, but recognition is a problem only in the limited areas where the Fideliinae occur, i.e., desertic areas in Morocco and Egypt eastward to the Punjab (India) and central Asia; also southern Africa and central Chile. Elsewhere, traditional characters for the Megachilidae will suffice.

The following characters seem to apply to the whole family: These are L-T bees with the principal characters of that group, as indicated in Section 19. The glossal apices, with flabella, are illustrated in Figure 28-2a-f. The dististipital process is present; the stipital comb is absent except in a few Megachilinae; the ligular arms are separated from the prementum; the labrum is longer than broad [broader than long in some *Fidelia (Parafidelia)* and in Pararhophitini], widening at the base to a broad articulation with the clypeus (Figs. 33-2a, 75-1), and there is no apical process (in other bees, including the few Nomadinae in which the labrum is longer than broad, the articulation with the clypeus is narrowed compared to the full labral width). The episternal groove is entirely absent (Fig. 20-5a) except in the tribe Fideliini. The basitibial plate is absent, i.e., not defined (except that in most female Lithurgini it is defined along its posterior margin and at its apex. There are two submarginal cells, except in the Fideliini, which have three; in forms with two submarginal cells, the second is rather long, and it is thus likely that the second submarginal crossvein is the one missing. The metasomal sternal scopa (Fig. 8-7b) is present except in parasitic forms and in Pararhophitini; the scopa on the hind legs is absent except perhaps in *Pararhophites;* see also the discussion of *Aspidosmia* in the Anthidiini. In the Fideliini, long hairs are present on the hind tibia and basitarsus, but they are used to kick soil from the nest entrance, not to carry pollen. In the Pararhophitini, pollen is carried on diverse parts of the body, to judge by pollen on museum specimens, but the hairs of the hind tibia seem to form a sparse scopa.

McGinley and Rozen (1987) and Rozen (1973a) cite

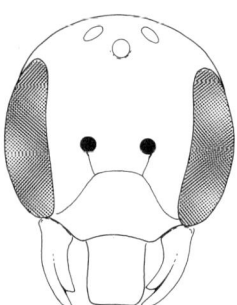

Figure 75-1. Face of male *Chelostoma californicum* Cresson, showing the rectangular labrum broadly articulated to the clypeus. From Michener, McGinley, and Danforth, 1994.

various biological and larval characters that demonstrate the relationship of the major megachilid taxa to one another. A nipple-shaped projection on the cocoon that the mature larva spins is unknown in other bees, but occurs in all megachilid taxa except the Lithurgini. It is associated with an incomplete area in the inner cocoon layers and probably facilitates gas exchange. In the genus *Fidelia* there is a nipple on each end of the cocoon; in others it is at the outer (anterior) end only. Megachilidae, like most Xylocopinae and Melittidae but unlike most other bees, do not line their cells with conspicuous secreted material. Some of them nonetheless have Dufour's glands of substantial size. Williams et al. (1986) found Dufour's gland products, fatty acids and triglycerides, in the cell provisions. Thus the glandular product may provide part of the larval food instead of contributing to the cell lining. A short, broad larval clypeus, about four times as wide as long, and a deeply and broadly emarginate larval labrum are characteristic of Megachilidae, the former otherwise known among bees only in some colletids. Most megachilid larvae are conspicuously setose and spiculate, the body thus seemingly clothed with short hairs. This is not true of the Fideliinae, but even in that subfamily the head has more small setae than does that of most other bees.

In analyses based on adult characters in the phylogenetic study of L-T bees by Roig-Alsina and Michener (1993), the Fideliinae as here constituted appears as the sister group of the Megachilinae (Fig. 75-2). Larval characters do not show such a relationship, and when larval and adult data were combined, *Pararhophites* emerged as the basal branch, thus sister group to all other megachilids, and the next branch was for Fideliini. In other words, the subfamily Fideliinae as here understood was a paraphyletic group from which the Megachilinae arose. But until further data are gathered, I think this result is inconclusive. I include Pararhophitini in the Fideliinae, but could equally easily recognize two separate subfamilies, Pararhophitinae and Fideliinae.

The tribe Lithurgini rather consistently appears as the sister group to all other Megachilinae. Some interesting similarities between Lithurgini and Fideliini may suggest a close relationship between these two groups, in spite of their obvious differences and the more *Megachile*-like aspect of *Lithurgus*. Similarities include the adoral man-

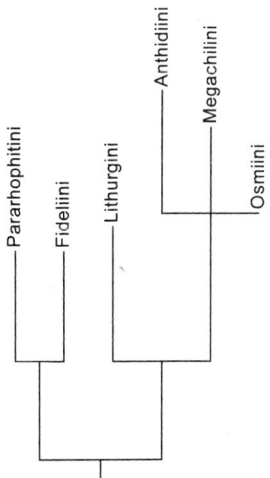

Figure 75-2. Consensus tree for tribes of Megachilidae, based on a phylogenetic study of L-T bees by Roig-Alsina and Michener (1993). The Dioxyini were not included by these authors; they would probably be a part of the multifurcation at the upper right. (Lengths of vertical lines are proportional to the numbers of derived characters.) The data were for particular genera thought to be representative of the tribes. Other analyses showed Pararhophitini as the sister group to all other Megachilidae.

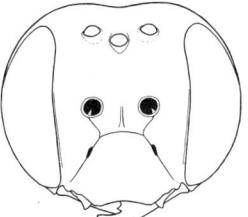

Figure 75-3. Face of male *Anthidium atripes* Cresson, showing subantennal sutures directed toward outer margins of antennal sockets. From Michener, 1944.

dibular tooth and the rather narrow salivary lips of the larva (Rozen, 1973a) and the placement of the egg in the midst of the provisions (Rozen, 1970a, 1973b, 1977c); nearly all other mass-provisioning bees, including *Pararhophites* (McGinley and Rozen, 1987), place the egg on the surface of the provisions.

The Roig-Alsina and Michener (1993) phylogenetic study of L-T bees did not settle relationships among the tribes of Megachilinae, except for the basal position of Lithurgini. Except for *Lithurgus*, only six genera (two Anthidiini, two Megachilini, and two Osmiini) were included in the study, and although in some analyses each of the three tribes was holophyletic, in others the tribes were not supported. I believe that the Anthidiini (also the Dioxyini, which was not included in the study based on adults) and the Megachilini are easily shown to be monolophyletic, but I suspect that the Osmiini are paraphyletic, and that at least the Megachilini arose from them. An attractive solution would be to divide the Osmiini; they are so diverse that such action seems desirable. Just how to do so will not be evident until a detailed analysis is made.

Key to the Subfamilies of the Megachilidae

1. Submarginal cells three (two in *Pararhophites*); mandible of female with preapical tooth (sometimes sufficiently enlarged that mandible is bilobed or bifid) or simple; dorsum of T6 of female hairless at least posteriorly, minutely roughened, often with a lateral ridge or carina suggesting that pygidial plate, if present, occupies most of exposed tergal surface; subantennal suture directed toward middle of lower margin of antennal socket, or usually so short that its direction cannot easily be determined................
....Fideliinae (Sec. 76)
—. Submarginal cells two; mandible of female with three or more teeth (except in Dioxyini and in a few *Megachile* that have a much elongated, bidentate mandible); dorsum of T6 of female usually with hairs, pygidial plate usually absent but, *if* present, then represented only by a narrow midapical process (Fig. 80-2); subantennal suture directed toward outer margin of antennal socket (Fig. 75-3)................................... Megachilinae (Sec. 79)

76. Subfamily Fideliinae

This is an archaic subfamily with a disjunct distribution in xeric areas of Asia, Africa, and South America. It is formed by the union of the Fideliidae of authors and the supposedly anthophorine tribe Pararhophitini (Popov, 1949a). The minutely roughened surface of T6 of the female, commonly surrounded laterally and posteriorly by a ridge or carina in the Fideliini and therefore perhaps constituting the broad pygidial plate, may be a synapomorphy of the group. In the Fideliini, T6 is hairless; in the Pararhophitini, the basal part of T6 supports hairs but the hairless distal part, perhaps only a large marginal zone, projects and suggests the structure of Fideliini. Both inner and outer hind tibial spurs and also the middle tibial spur are coarsely serrate or bear widely separated small teeth, but are sometimes almost toothless, rather than having the usual fine close teeth, i.e., exhibiting the condition called ciliate. In the Megachilinae, the inner hind spur is sometimes coarsely toothed, but the others are ciliate. The premarginal lines of the terga are strongly arcuate forward medially, the hairless marginal zones thus extremely broad middorsally.

The peculiar cocoon of the mature larva, tapering at each end and incorporating sand that had been eaten by the mature larva and voided in strips tending to run on the inside of the cocoon, from one end of the cocoon to the other, is another subfamilial synapomorphy (Rozen, 1970a, 1973b; McGinley and Rozen, 1987). Such a cocoon is unknown in other bees; cocoon structure is therefore the most convincing synapomorphy of the subfamily. The unlined cells with walls no smoother than the burrow walls may also be synapomorphic, since except for Melittidae most other ground-nesting bees have smooth-walled and usually lined cells.

As indicated in Section 75, it may be best to give the Pararhophitini subfamily status; they are clearly quite different from the Fideliini.

Key to the Tribes of the Fideliinae

1. Submarginal cells three; stigma slender, parallel-sided, less than twice as long as prestigma, margin within marginal cell straight or concave (Fig. 76-1b); T7 of male strongly sclerotized with distinct pygidial plate, or with the plate drawn out into a long and sometimes deeply bifid process; episternal groove present above level of scrobal groove and curving posteriorly to form scrobal groove; body densely hairy, without yellow markings except sometimes on T6 of the female and on clypeus of both sexes (Africa, Chile) .... Fideliini (Sec. 78)
—. Submarginal cells two; stigma broad, about twice as long as prestigma, margin within marginal cell convex (Fig. 76-1a); T7 of male weakly sclerotized without distinct pygidial plate; episternal groove absent; body sparsely hairy, yellow or with yellow markings (palearctic) .... Pararhophitini (Sec. 77)

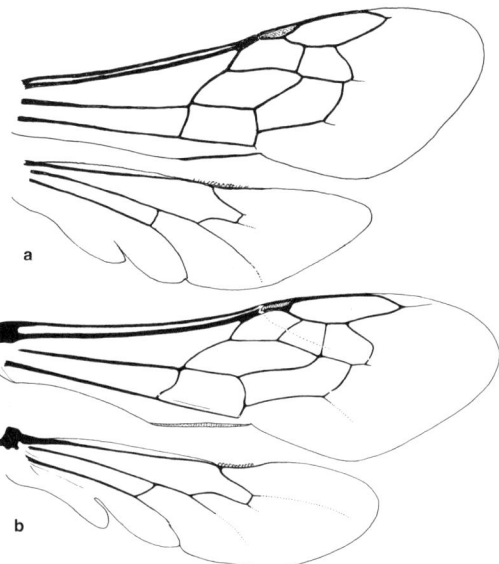

Figure 76-1. Wings of Fideliinae. **a,** *Pararhophites orobinus* (Morawitz); **b,** *Neofidelia profuga* Moure and Michener.

77. Tribe Pararhophitini

This is a tribe of small, euceriform, sparsely haired bees with yellow markings or, in females, with the body largely yellow. The tribe's peculiar features were recognized by Popov (1949a), who compared it with the tribes Ancylini and Exomalopsini in the Apinae, and by McGinley and Rozen (1987), who recognized its megachilid affinities. Unlike that of other Megachilidae, the mandible of the female is simple. There is only one genus.

Genus *Pararhophites* Friese

Rhophites (Pararhophites) Friese, 1898c: 305. Type species: *Pararhophites quadratus* Friese, 1898, monobasic.
Ctenoapis Cameron, 1901a: 116. Type species: *Ctenoapis lutea* Cameron, 1901, by designation of Sandhouse, 1943: 542.

This is a genus of small, sparsely haired bees, 5 to 7 mm long. The females are largely yellow, the males black with yellow markings on the body and legs. This is the most plesiomorphic group of the Megachilidae in the sense that the female has what is probably a tibial scopa and lacks a well-developed sternal scopa; the sterna are, however, more hairy than those in parasitic Megachilinae. The relatively large stigma (Fig. 76-1a) compared to that of other Fideliinae is probably also ancestral. The hind legs, especially the femora, of the male are enlarged. T7 of the male is thin, weakly sclerotized, and produced but rounded posteriorly, without a recognizable pygidial plate. Male genitalia and other structures were illustrated by McGinley and Rozen (1987); see also Figure 77-1. S7 of the male is a little-modified sternum without apical lobes such as are found in some Fideliini. Well-developed volsellae similar to those of S-T bees are a feature of the genitalia not found elsewhere in Megachilidae.

■ This genus is found from Morocco and Egypt to Kazakhstan, and south to Baluchistan (in Pakistan) and northwestern India. There are three species (Popov, 1949a; McGinley and Rozen, 1987); Warncke (1979a) provided a key to the species.

Pararhophites is probably oligolectic on Zygophyllaceae (Popov, 1949a). The nesting biology and larvae were described by McGinley and Rozen (1987). The nests are slanting burrows, each with several laterals leading to isolated, more or less horizontal cells. As in other megachilids there is no evident secreted cell lining, but the distal end of each cell is lined with a "receptacle" of sand grains presumably from elsewhere inside the nest and cemented together, perhaps with nectar. This is the sand that is later ingested by the mature larva and deposited in strips on the inner surface of the cocoon. No other bee constructs such a receptacle; even the Fideliini, whose larvae also ingest sand, appear to take the sand from unmodified cell walls. Pollen is packed into the receptacle and an egg laid, projecting from a small cavity on the surface of the pollen mass.

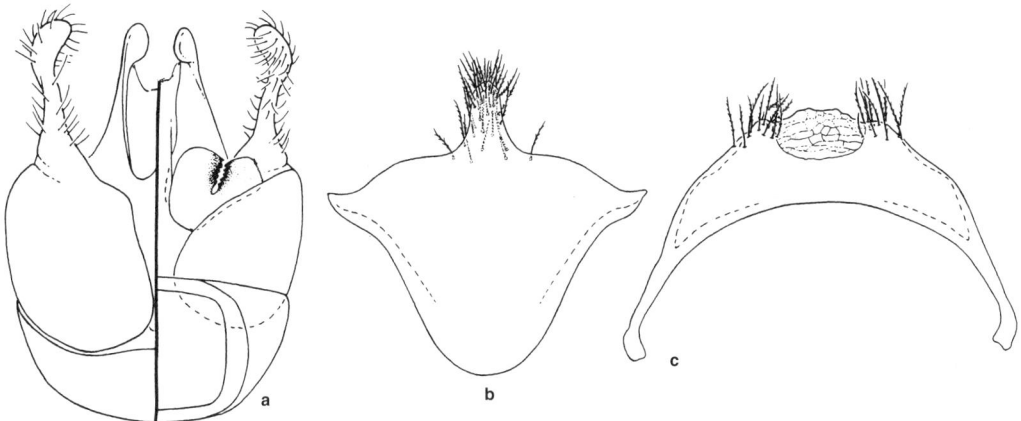

Figure 77-1. Pararhophitini. **a-c,** Male genitalia, S8, and S7 of *Pararhophites orobinus* (Morawitz). (In the divided drawing, the dorsal view is on the left; the sterna are ventral views.) Based on McGinley and Rozen, 1987.

78. Tribe Fideliini

This tribe consists of apiform or euceriform bees with abundant long pale hair, giving them a pale appearance, but the integument is largely black except for the yellow clypeus of some; T6 of the female is sometimes yellow. The peculiar features of the group were described by Popov (1939b), Moure and Michener (1955a), and Rozen (1970a). They include the combination of a ventral metasomal scopa on which pollen is transported, three submarginal cells (Fig. 76-1b), and in some an enlarged and apically lobed S7 of the male, with the disc of the sternum reduced (Fig. 78-1c), suggesting that of many Colletidae. The posterior tibiae and basitarsi of females have long hairs that have sometimes been considered as a scopa. They are not used to carry pollen, however, but appear to function in throwing sand backward (for *Neofidelia;* Rozen, 1973b) or also laterally (for *Fidelia;* Rozen, 1970a) as the bee is digging. The male genitalia and hidden sterna were illustrated by Popov (1939b) and Warncke (1980b); see also Figure 78-1. Perhaps all species are oligolectic, different species on different and frequently unrelated flowers. The floral relations of the species of southern Africa were reported by Whitehead (1984).

Engel (2002b, 2004a) revised the tribe Fideliini and examined the phylogeny and floral relationships of its species. He placed each genus in its own tribe of his subfamily Fideliinae, presumably to emphasize the differences among the genera.

Key to the Genera of the Fideliini

1. Hind basitarsus of female slender, parallel-sided; hind legs of male enlarged, basitarsus forming two enormous curved talons; clypeus separated from antennal bases by much more than diameter of antennal socket (Chile) .... *Neofidelia*
—. Hind basitarsus of female broadest near base, tapering to narrow apex; hind legs of male not enlarged or modified; clypeus separated from antennal bases by less than diameter of socket, subantennal suture thus difficult to observe (Africa) .... *Fidelia*

Genus *Fidelia* Friese

This genus contains robust, hairy, fast-flying bees. Some of the most distinctive generic characters are indicated in the key to the genera. The wings are hairless,

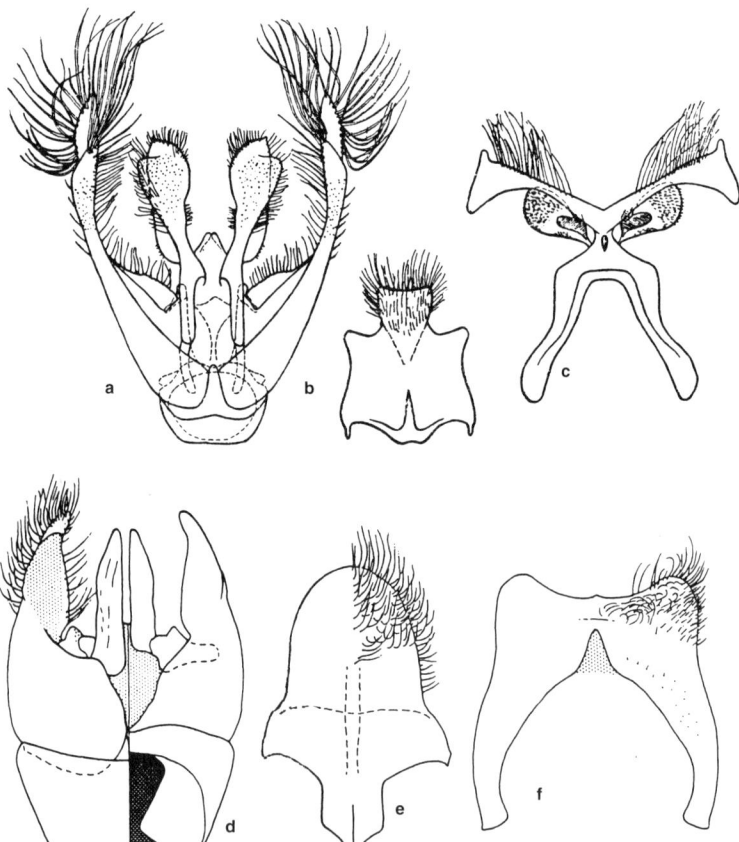

Figure 78-1. Male genitalia, S8, and S7 of Fideliini. **a-c,** *Fidelia villosa* Brauns, modified from Popov, 1939; **d-f,** *Neofidelia longirostris* Rozen, from Rozen, 1970a. (In d, the dorsal view is on the left.)

strongly papillate beyond the venation. The mandible in both sexes is often bifid or deeply bilobed, i.e., the preapical tooth is extended and similar in size and shape to the rutellum. Male genitalia and hidden sterna are illustrated in Figure 78-1a-c.

The nests are burrows in the soil consisting of several branches, each branch ending in an unlined cell. The provisions, which are packed into the distal part of the cell, completely enclose the egg (Rozen, 1970a) or eggs; Rozen (1977c) found that completed food masses contain up to three eggs or larvae, each in a chamber within the food. Thus, as in some *Megachile (Sayapis)*, more than one larva can live and mature within one large cell. The cocoons are distinctive in having a nipple at each end, and as with other Fideliinae contain much sand that is presumably eaten by the mature larva and excreted with feces in cocoon formation.

Whitehead and Eardley (2003) revised the species of *Fidelia*, illustrating genitalia, hidden sterna, and other structures.

Key to the Subgenera of *Fidelia* (modified from Engel, 2002b)

1. Mandible strongly bidentate, teeth long, slender, and diverging from one another; pygidial plate of male present; midapical process of male T7 with apex emarginate .... *F. (Fideliopsis)*
—. Mandible broadly bilobed (teeth broad and flat), or with short, subapical tooth, or simple; male pygidial plate absent; midapical process of male T7 deeply bilobed (essentially forming two processes) or simple 2
2(1). Marginal cell not extending beyond apex of submarginal cells, shorter than combined lengths of second and third submarginal cells; claws of female simple; front basitarsus of male flattened and extended laterally as lamella (only a weak lamella in *F. kobrowi* Brauns) .... *F. (Fidelia* s. str.)
—. Marginal cell extending beyond apex of submarginal cells, as long as or longer than combined lengths of second and third submarginal cells; claws of female bifid or cleft; male front tarsus unmodified, segments cylindrical 3
3(2). Marginal cell about as long as combined lengths of second and third submarginal cells; apical flagellar segment of male antenna hooked, preceding segments ventrally crenulate; female T6 off-white; male T7 deeply emarginate (essentially forming two apical processes)............ .... *F. (Fideliana)*
—. Marginal cell longer than combined lengths of second and third submarginal cells; male antenna simple, unmodified; female T6 black; male T7 with single midapical process, apex undivided .... *F. (Parafidelia)*

Fidelia / Subgenus *Fidelia* Friese s. str.

Fidelia Friese, 1899a: 244. Type species: *Fidelia paradoxa* Friese, 1899, monobasic.

This subgenus has unmodified male antennae, and yellow on the clypeus of both sexes and often on T6 of females. T7 of the male has a median, longitudinally elongate pit and a long, median, deeply bifid process that is somewhat elevated basally and might represent the pygidial plate; in addition, T7 has a large lateroapical tooth, the whole tergum thus quadridentate. The claws of the female are simple. The body length is 12 to 16 mm.

■ This subgenus occurs in western South Africa and Namibia. There are three species. Probably all the species are oligolectic on Mesembryanthemaceae.

Fidelia / Subgenus *Fideliana* Michener

Fidelia (*Fideliana*) Michener, 2000: 422. Type species: *Fidelia braunsiana* Friese, 1905, by original designation.

This subgenus is characterized by modified male antennae [the apical segment is attenuate and hooked, the other flagellar segments slightly crenulate on the lower surface in *F. (F.) braunsiana* Friese or strongly crenulate and greatly thickened in *F. (F.) ulrikei* Warncke]. The clypeus is black, or, in the male of *F. braunsiana* Friese, yellow. The marginal cell is intermediate in length between the very short condition found in the subgenus *Fidelia* s. str. and the elongate condition characteristic of the subgenus *Parafidelia*. T7 of the male is bifid (profoundly so in *F. braunsiana*), lacks evidence of a pygidial plate, and has a lateral tooth that in *F. braunsiana* is largely hidden beneath and partly fused to the broad bifid part of the tergum. The claws of the female (at least in *F. braunsiana*) are bifid, the inner ramus much shorter than the outer, so that it could be described as a strong basal tooth. The mandible of the female is broadly bilobed (in *F. braunsiana*); that of the male, with a preapical tooth. The body length is 13 to 16 mm.

■ The distribution of this subgenus is disjunct, South Africa and Namibia in the south, Morocco in the north. The two species are *Fidelia braunsiana* Friese in southern Africa and *F. ulrikei* Warncke in Morocco.

Fidelia / Subgenus *Fideliopsis* Engel

Fideliopsis Engel, 2002b: 311. Type species: *Fidelia major* Friese, 1911, by original designation.

This subgenus can be recognized by the two long, pointed mandibular teeth of both sexes (in other *Fidelia*, mandible bilobed, or with small subapical tooth, or simple) and the greatly expanded front tarsal segments 1 and 3 (in other subgenera, front tarsi unmodified or only the basitarsus expanded). As in *Parafidelia* the marginal cell is about three fourths as long as the distance from its apex to the wing tip, and the female pygidial plate is black. (It is yellow and covers the entire exposed surface of T6 in the subgenera *Fidelia* s. str. and *Fideliana*.)

■ *Fideliopsis* is found in western South Africa and in Nambia and contains five species (see the key of Whitehead and Eardley, 2003).

There is nothing in the phylogenetic study to indicate whether *Fideliopsis* is best recognized as a genus or as a subgenus of *Fidelia;* according to Engel (2002b) it is the basal branch of *Fidelia* phylogeny, i.e. the sister group to the rest of *Fidelia*. Subjectively it is not more different than the other subgenera of *Fidelia* are from one another, and I have called it a subgenus of *Fidelia* here. Whitehead and Eardley (2003) in their revision of *Fidelia* also considered *Fideliopsis* as a subgenus.

Fidelia / Subgenus *Parafidelia* Brauns

Parafidelia Brauns, 1926: 202. Type species: *Parafidelia friesei* Brauns, 1926, monobasic.

Bees of this subgenus have unmodified male antennae. The clypeus in both sexes varies from black to largely pale yellow. T7 of the male has an elevated probable pygidial plate that is strongly produced posteriorly and simple at the apex; there is also a strong lateral tooth. The claws of the female are bifid, the inner tooth rather small. *Parafidelia* includes the largest fideliines, ranging from 13 to 23 mm in body length.

■ *Parafidelia* occurs principally in Namibia and western South Africa, but also east to Botswana. It contains only two species (Whitehead and Eardley, 2003). The remaining species formerly included in *Parafidelia* have been transferred to *Fideliopsis*.

Genus *Neofidelia* Moure and Michener

Neofidelia Moure and Michener, 1955a: 202. Type species: *Neofidelia profuga* Moure and Michener, 1955, by original designation.

This genus consists of hairy bees 9 to 12 mm long. The wing membrane has scattered hairs, the hairs especially abundant in *Neofidelia longirostris* Rozen, and the distal part of the membrane has papillae, especially in *N. profuga* Moure and Michener. The presumed pygidial plate of the female, like that of *Fidelia*, is large and distinct, and occupies the whole dorsum of the exposed part of T6; that of the male is distinct but narrower than that in the female. T7 of the male lacks spines or processes. The male genitalia, sterna, and other structures were illustrated by Moure and Michener (1955a) and Rozen (1970a); see also Figure 78-1d-f.

■ This genus is found in central Chile. The two species were distinguished by Rozen, 1970a.

The nesting biology of *Neofidelia* was described by Rozen (1973b). The nests are shallow burrows in the soil, each branch ending in an unlined horizontal enlargement or cell. The provisions, packed into the distal part of the cell, have a deep concavity in which the egg is placed. Somewhat similar concavities in the provisions of *Fidelia* are later closed over, so that the eggs are hidden inside the provisions; this is not the case in *Neofidelia*. The cocoon is similar to that of *Fidelia*, containing much sand that is presumably ingested by the mature larva and then defecated during cocoon construction. The cocoon, like that of *Pararhophites*, has a nipple at one end only, unlike that of *Fidelia*, which has a nipple at each end.

79. Subfamily Megachilinae

This subfamily is the equivalent of the Megachilidae of most authors, i.e., those who have not placed the Fideliinae in the Megachilidae. Among the features that help us to recognize the Megachilinae are the hoplitiform to megachiliform (rarely anthophoriform) body, the two submarginal cells of roughly equal length, the strong sternal scopa (Fig. 8-7b) and lack of long hairs on the hind legs of females (except in *Aspidosmia*), the generally quadrate head with subantennal sutures directed toward the outer margins of the antennal sockets (Fig. 75-3), and usually the broad female mandibles with three or more teeth. T6 of the female is largely or entirely hairy, with no indication of a pygidial plate except in the tribe Lithurgini. An important feature of the subfamily is the reduced and often hairless S7 of the male. In Osmiini, Dioxyini, and Megachilini, S7 is a weakly sclerotized transverse band, sometimes divided medially. It is somewhat more developed in many Anthidiini, but in no case are there apical lobes or processes as in many other bees.

This is a large subfamily, common and diversified on every continent. Lithurgini is quite different from the other four tribes, and these four have common synapomorphies. A possible classification, therefore, would be to recognize two tribes, the Lithurgini and the Megachilini, the latter divided into four subtribes. It seems simpler, however, to recognize five tribes, of which the Lithurgini is the sister group to all the rest. The relationships among the remaining four tribes are not known; see the discussion of this topic in Section 75. As indicated there, I suspect that the Osmiini may be a paraphyletic unit from which one or all of the other three tribes (Anthidiini, Dioxyini, Megachilini) arose. Yet the Anthidiini have some features that seem plesiomorphic relative to the features of the other tribes, specifically a more or less recognizable male gonostylus and a better developed male S7. In Osmiini and Megachilini the gonoforceps show no indication of the separation of the gonostylus from the gonocoxites. See the account of *Aspidosmia*, an anthidiine bee, for a form exhibiting some mixture of the usual tribal characters.

The tribes Anthidiini, Dioxyini, Megachilini, and Osmiini contain the great majority of megachilid bees. The characters listed below differentiate these tribes jointly, as a group apart from the Lithurgini. The third and fourth segments of the labial palpus, both present, diverge from the axis of the second, and are not flattened (except that the third is flattened and not diverging in some species of *Chelostoma*). The dorsolateral angle of the pronotum is absent or weakly evident, but produced as a distinct tooth in some *Chelostoma* (Osmiini) and in most Dioxyini. The lower half of the metapleuron is not greatly narrowed (the lower quarter sometimes much narrowed; see the key to the tribes). The jugal lobe of the hind wing is half as long as the vannal lobe or less (Fig. 81-1a, b). The basitibial plate is not defined. The outer surfaces of the tibiae are usually not strongly tuberculate or spiculate, or the spicules terminate in bristles, but in some Anthidiini tubercles are present. The hind basitarsus is often flattened, shorter than the tibia. T1 is well developed, its posterior margin transverse. T6 of the female lacks a posterior lateral spine except in many Anthidiini. The distal sterna of the male are commonly hidden, so that there are only one to five exposed sterna, but S6 is sometimes exposed, e.g., in most Anthidiini. The pygidial plate is absent in both sexes. Distinguishing characters of larvae are described by McGinley (1981), King (1984), and Romasenko (1995), who described and gave a key for mature larvae (prepupae) of 57 species. The characters differentiating these four tribes are summarized in Table 79-1.

The Megachilinae (except Lithurgini) almost always carry foreign materials from outside their nests to form cells, cell walls, partitions, etc. In this respect they differ from nearly all other bees except the corbiculate tribes of the Apinae and some species of *Centris*, also in the Apinae. The Apinae that carry building materials do so using the scopa or corbiculae of the hind tibiae, whereas the Megachilinae carry materials with the mandibles or held by the legs.

Megachiline building materials include leaf or petal pieces, chewed leaf pulp, plant hairs, nectar, resin, pebbles, mud, and various combinations of these; they commonly differ among different genera or species groups. Nectar as a construction material seems improbable, but the sticky substances used for partitions by various Osmiini may be nectar mixed with leaf fragments or other materials. Rozen and Eickwort (1997) report nectar in constructs of *Ashmeadiella* and suggest that it may also be used in construction by other Osmiini. Nests may be in the soil, in holes (made by other insects) in wood, in plant stems, and in diverse cavities, including the shells of dead snails, or may be free-standing constructs situated on rocks, walls, stems, twigs, or even leaves. Some species make their own burrows in soil or in pithy stems but the majority do not do so. In some cases, cells are made merely by partitions of resin, leaf pulp, or mud in a burrow, but in most cases cells are completely lined. Provisions are firm or somewhat sticky masses filling the bottom of each cell; an egg is laid on top of the provisions in each cell, after provisioning is complete. Partitions between cells are rarely omitted. Cocoons almost always have, at the anterior end, a projecting nipple in which the matrix that solidifies to fill the spaces between the silk fibers is incomplete, perhaps permitting gaseous exchange. Accounts of nests and nesting behavior are referenced below, under the tribes and genera.

There are several cleptoparasitic genera, nearly all species of which parasitize other Megachilinae.

Friese (1911b) summarized taxonomic knowledge of the whole subfamily, with keys to species, but species described after the period 1898 to 1901 were not included. The following key is supplemented by Table 79-1.

Key to the Tribes of the Megachilinae

1. Pygidial plate of male present (Fig. 80-4a, b), that of female represented by an apical process or spine (Fig. 80-2); metapleuron with lower half narrow (less than half as wide as upper end and five to six times as long as wide) to

Table 79-1. Characters of Four Tribes of Megachilinae (Tribe Lithurgini omitted).

Character	Megachilini	Osmiini	Anthidiini	Dioxyini
With yellow, white, or red maculations	no	no, except *Ochreriades*, face of male *Aspidosmia*	usually yes	no
Background of body	not metallic	frequently metallic	not metallic except for some *Stelis*	not metallic
Preaxilla[a]	sloping, with long hairs	vertical, nearly hairless	vertical, nearly hairless	vertical, nearly hairless
Stigma and prestigma	long (see key to tribes)	long except in *Aspidosmia* (see key to tribes)	short (see key to tribes)	as in Anthidiini
Arolia	absent (present on front and middle legs of *Megachile* (*Heriadopsis*)	present	present or absent	absent
Claws of female	simple or with basal tooth	simple, except cleft in *Osmia* (*Metallinella*)	cleft or with inner tooth, simple in *Trachusoides*	cleft or with inner tooth
Second recurrent vein	basal to second submarginal crossvein	basal to second submarginal crossvein	usually distal to second submarginal crossvein	basal to second submarginal crossvein
Dorsal lamella of metapleuron[b]	present	absent, except in *Pseudoheriades*	usually absent	absent
T6 of male with preapical transverse flange or carina	present, rarely two spines, absent in some *Megachile* (*Rhodomegachile*)	absent, except in *Hoplosmia*	absent	absent
Vestiture of outer surface of hind tibia	hairs	hairs	usually rather short, robust bristles	hairs
Juxtantennal carina	absent	sometimes present	sometimes present	absent

[a] Often the tegula must be removed to see this character.
[b] This is a narrow lamella or strong carina across the upper end of the metapleuron below the base of the hind wing.

linear; outer surfaces of tibiae, except hind tibiae of some males, with coarse tubercles that do not end in bristles (Fig. 80-3b) .... Lithurgini (Sec. 80)
— Pygidial plate absent; metapleuron with lower half little narrower than upper end (lower one-fourth sometimes considerably narrowed, about twice as long as wide); outer surfaces of tibiae commonly without tubercles, or with tubercles ending in bristles, but sometimes tuberculate as in Lithurgini 2
2(1). Metanotum with median spine or tubercle (except in *Allodioxys* and *Ensliniana*); mandible of female slender apically, bidentate, similar to that of male; pronotum (except in *Prodioxys*) with prominent obtuse or right-angular dorsolateral angle, below which a vertical ridge extends downward; sting and associated structures greatly reduced (scopa absent) .... Dioxyini (Sec. 83)
—. Metanotum without median spine or tubercle; mandible of female usually wider apically, with three or more teeth, except rarely bidentate when mandible is greatly enlarged and porrect and clypeus is also modified; pronotum with dorsolateral angle weak or absent (or produced to a tooth in some *Chelostoma* but without vertical ridge below it); sting and associated structures well developed 3

3(2). Stigma less than twice as long as broad, inner margin basal to vein r usually little if any longer than width, rarely about 1.5 times width; prestigma commonly short, usually less than twice as long as broad; claws of female cleft or with an inner tooth (except in *Trachusoides*); outer surface of hind tibia usually with abundant simple bristles (but with long hairs in *Aspidosmia*); body commonly with yellow or white (sometimes red) integumental marks .... Anthidiini (Sec. 82)
— Stigma over twice as long as broad, inner margin basal to vein r longer than width (Fig. 81-1); prestigma much more than twice as long as broad; claws of female usually simple; outer surface of hind tibia with hairs, these sometimes plumose, but not bristles; body almost always without yellow or white integumental markings 4

4(3). Arolia absent, at least on hind legs, usually on all legs; preaxilla, below posterolateral angle of scutum, sloping and with small patch of hairs, these as long as those of adjacent sclerites; body nonmetallic or nearly so .... Megachilini (Sec. 84)
— Arolia present; preaxilla, below posterolateral angle of scutum, vertical, smooth and shining or with some hairs, these much shorter than those of adjacent sclerites; body sometimes metallic green, blue, or brassy .... Osmiini (Sec. 81)

80. Tribe Lithurgini

This is a small tribe of small to large hoplitiform to megachiliform bees, the females often having the aspect of a *Megachile*. The Lithurgini lack both metallic coloration and yellow maculation. The face of many females and a few males has a median prominence on the upper part of the clypeus or on the supraclypeal area or both. The proboscis is usually long, in repose often reaching the metasoma; the third segment of the labial palpus is flattened, on the same axis as the second, only the fourth segment being directed laterally, or the fourth segment is absent. The lower half of the metapleuron is greatly narrowed (see the key to tribes). The stigma is as in *Megachile* (Fig. 80-1) or sometimes extending farther beyond the base of vein r. The hind basitarsus is slender, almost cylindrical, usually about as long as the tibia. T1 is small and flattened, and its posterior margin is rounded as seen from above, so that the metasoma frequently looks flatter than that of *Megachile*. T6 of the female (Fig. 80-2) is hairy and has a tooth on the posterior lateral margin (hidden in dense hair in *Trichothurgus*) (Fig. 80-2). The male genitalia and hidden sterna are extraordinarily small, the genitalia probably the simplest found among bees (Fig. 80-4c).

Some presumably plesiomorphic characters of the Lithurgini include the following: The jugal lobe of the hind wing is one-half to three-fourths as long as the vannal lobe (Fig. 80-1; generally shorter in other Megachilinae). The basitibial plate of the female is usually defined along the posterior margin and at the apex, although it is not evident in some species of *Trichothurgus*. The metasoma of the male is not curled under apically as in most Megachilinae; T6 lacks a transverse carina or teeth; T7 is fully exposed and directed posteriorly; S1 to S6 are fully exposed. The pygidial plate of the female is represented by a longitudinal median bare zone on T6, extended posteriorly as a projection or spine that is flat or concave on the dorsal surface, sometimes expanded distally (Fig. 80-2). The pygidial plate of the male is a broader plate, margined laterally by carinae, and often pointed at the apex (Fig. 80-4a, b).

On the basis of *Lithurgus* only, the larva resembles that of other Megachilinae in its conspicuous body hairs. Dis-

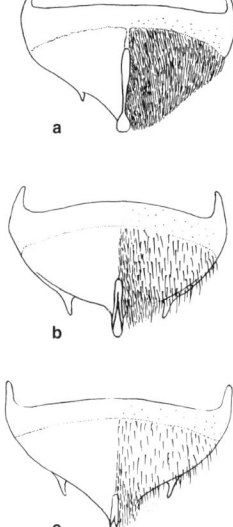

Figure 80-2. T6 of female Lithurgini; the midapical structure is presumably the relictual pygidial plate. **a,** *Trichothurgus wagenknechti* (Moure); **b,** *Microthurge pharcidontus* (Moure); **c,** *Lithurgus apicalis* (Cresson). From Michener, 1983a.

tinguishing characters are described by McGinley (1981) and Roberts (1978).

Engel (2001b) placed *Lithurgus* and its allies in a separate subfamily, the Lithurginae. His phylogenetic pattern does not differ from that shown above in the account of the Family Megachilidae (Fig. 75-2); the classification differs in whether the Lithurgini, the sister group to other Megachilinae, should be included in or excluded from that subfamily.

Nests are burrows excavated in dead, dry, often rotten wood or even in dry cow manure. They often branch. Provisions are placed in the burrows without cell linings, and often without partitions demarcating cells. Eggs are laid before or during provisioning, and are thus often within the firm pollen mass rather than on its surface. Cocoons lack the nipples characteristic of other megachilid cocoons. Nests and nest biology have been described for several species of *Lithurgus* (Malyshev, 1930b; Cros, 1939; Lieftinck, 1939; Brach, 1978; Camillo et al., 1983; Kitamura et al., 2001; Rust et al., 2004; and included references) and one species of *Microthurge* (Garófalo et al., 1981, 1992). Nests of *Microthurge* are sometimes reused by successive generations and are sometimes occupied communally by two or more females, probably provisioning cells in different branch burrows (Garófalo et al., 1992).

The classification of the tribe was considered by Michener (1983a). The following comments are derived from that paper. Most characters of *Trichothurgus* are plesiomorphies or probable plesiomorphies. The large and elongate labrum, however, appears to be a synapomorphy uniting the species of *Trichothurgus*, which can be regarded as a sister group to the other two genera combined. The sparsely hairy wing membrane is probably also a

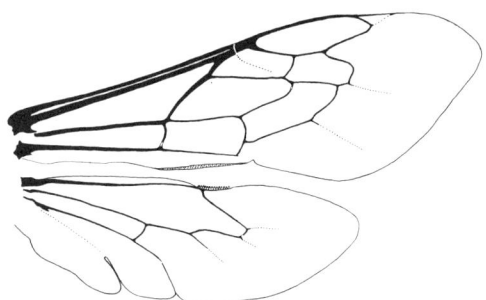

Figure 80-1. Wings of *Trichothurgus dubius* (Sichel).

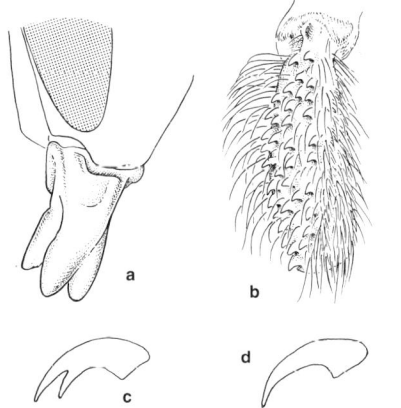

Figure 80-3. Structures of Lithurgini. **a,** Lower part of head and mandible of female *Lithurgus littoralis* Cockerell, showing the tridentate mandible with the middle tooth longest; **b,** Hind tibia of female *L. apicalis* Cresson; **c, d,** Claws of females of *Microthurge pharcidontus* (Moure) and *L. apicalis* Cresson. a and b, from Michener, McGinley, and Danforth, 1994; c and d, from Michener, 1983a.

synapomorphy, although this character arises independently in various other bees.

Lithurgus and *Microthurge* agree in several synapomorphies: the reduced lower mandibular tooth, such that the middle tooth is longest as well as more anterior than the others (Fig. 80-3a), the coarse tibial spicules or tubercles in areas of short and sparse hairs (Fig. 80-3b), and the low profile of the posterior part of the thorax and especially of the propodeum. Compared to *Trichothurgus*, with its many plesiomorphies, *Lithurgus* and *Microthurge* have few, the most noteworthy, perhaps, being the ordinary-sized labrum and the moderate to high density of hairs on the wing membrane.

Microthurge has few apomorphies. These include the short maxillary palpi and perhaps the larger stigma (plesiomorphic for Apoidea as a whole but perhaps derived in connection with small body size in this genus, as well as in some other small megachilids such as *Chelostoma* and *Heriades*). Bifid claws in females (plesiomorphic in bees in general) could be an apomorphy in *Microthurge*. Genes for this character must be retained in all species, because the claws of males are cleft. A regulatory change could therefore cause their reactivation in females, and would be apomorphic. The broad pygidial plate of males of *Microthurge* (Fig. 80-4b) seems to be plesiomorphic, as compared to all other megachilids, but one must question this idea in view of the slender produced plate in both *Trichothurgus* and *Lithurgus* (Fig. 80-4a). If this similarity is due to homology, then the broad plate of male *Microthurge* is a reversion toward the more primitive apoid condition, but for this genus it would be an apomorphy.

For *Lithurgus* the situation is equally confusing. If the stigmal, claw, and pygidial characters listed above are plesiomorphic for *Microthurge,* then the alternative characters are apomorphic for *Lithurgus*.

Key to the Genera of the Lithurgini
(From Michener, 1983a)

1. Labrum longer than clypeus, often much longer; hind tibia of female rather uniformly hairy on outer, anterior, and posterior surfaces, spicules relatively inconspicuous among the hairs; lower mandibular tooth longer than middle tooth, *or,* in some females, lower and middle teeth equal (South America) .... *Trichothurgus*
—. Labrum about as long as clypeus; hind tibia of female with hairs of broad, longitudinal outer zone shorter and sparser than those of anterior and posterior surfaces, spicules large and conspicuous in outer zone (Fig. 80-3b); lower mandibular tooth conspicuously shorter than middle tooth, which is the longest mandibular tooth (Fig. 80-3a) 2
2(1). Body small, slender, hoplitiform; claws of female bifid (Fig. 80-3c); stigma of moderate size, broadest at base of vein r, sides converging toward base; maxillary palpi two-segmented (South America) .... *Microthurge*
—. Body broad, megachiliform; claws of female simple (Fig. 80-3d); stigma small, sides basal to vein r parallel or nearly so [in some small species of *Lithurgus* the stigma is larger, approaching the size of that of *Microthurge.*]; maxillary palpi three- or four-segmented .. .. *Lithurgus*

Genus *Lithurgus* Berthold

Bees of this genus, 8 to 19 mm long, are robust, suggesting a slightly elongated *Megachile,* but the metasoma is somewhat flattened, and commonly has apical tergal and sternal hair bands. The midfacial prominence is almost always present in females, absent in most males. The labial palpus is four-segmented except in certain Old World species that lack the fourth segment. The fore and middle tibiae in the female each have two longitudinal rows of coarse spicules or tubercles extending basad, on the posterior part of the outer surface, from each of the two apical tibial spines, these rows outlining a channel on each tibia that extends basad to the middle of the tibia or beyond, sometimes nearly to the base. There are sometimes some spicules between the rows. In males the spicules are smaller, the rows shorter, sometimes absent. Scattered over the short-haired outer surface of the hind tibia of the female are coarse tubercles (Fig. 80-3b); the tubercles of the males are smaller, absent in many species. The wing membrane is hairy, often densely so. The pygidial plate of the female is a flat or concave dorsal surface of a long apical spine of T6 (Fig. 80-2c); of the male, it is a dorsally concave, broad, blunt projection or robust spine (Fig. 80-4a). The male genitalia and hidden sterna were illustrated by Mitchell (1960), Michener (1983a), and van der Zanden (1986); see also Figure 80-4c-e.

This genus is worldwide in distribution in tropical and warm to moderate temperate zones, except that in the Americas it may be largely absent from the wet tropics. Two subgenera are commonly recognized.

Key to the Subgenera of *Lithurgus*

1. Arolia absent or rudimentary in both sexes (except present in male of *L. rubricatus* Smith from Australia); facial

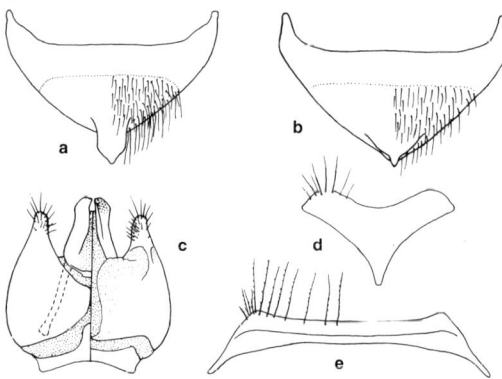

Figure 80-4. Structures of Lithurgini. **a, b,** T7 of males of *Lithurgus apicalis* Cresson and *Microthurge pharcidontus* (Moure), showing pygidial plates; **c–e,** Male genitalia, S8, and S7 of *L. apicalis* Cresson, the dorsal side of the genitalia on the right. The male genitalia of *Lithurgus* are the simplest of any bee, and the smallest relative to the size of the bee. From Michener, 1983a.

prominence of female involving upper part of clypeus and usually also part of supraclypeal area; first flagellar segment about twice as long as broad, more than twice as long as second, which is much broader than long .... *L. (Lithurgus* s. str.)
—. Arolia present in males, absent in females; facial prominence of female entirely supraclypeal (absent in one species); first flagellar segment not or little longer than broad, slightly longer than to shorter than second, which is nearly as long as broad to longer than broad (Western Hemisphere) .... *L. (Lithurgopsis)*

Lithurgus / Subgenus *Lithurgopsis* Fox

Lithurgopsis Fox, 1902: 138. Type species: *Lithurgus apicalis* Cresson, by original designation. The presence of arolia in males is presumably a plesiomorphy showing that this subgenus did not arise from *Lithurgus* s. str.

The presence of arolia in *Trichothurgus* males suggests a relationship to that genus; arolia may have been independently lost in *Microthurge* and within *Lithurgus* s. str.
■ This subgenus is limited to the Western Hemisphere, where it ranges from South Dakota, USA, to Argentina. It is reported from Costa Rica, but is probably absent from broad areas in the tropics. In North America it occurs widely in the west and ranges east to Illinois, North Carolina, and Florida, but is absent from the northeastern part of the continent. These bees collect pollen from Cactaceae; possibly for this reason *Lithurgopsis* is largely restricted to xeric areas. *Lithurgopsis* appears to be absent in Chile as well as Brazil. There are seven North American species and another in the Greater Antilles; about three others that may belong here occur in Argentina. Snelling (1983b) reviewed the North American species; he later presented a new key to species (Snelling, 1986a).

Lithurgus / Subgenus *Lithurgus* Berthold s. str.

Lithurge Latreille, 1825: 463. Type species: *Andrena cornuta* Fabricius, 1787, monobasic. [The name *Lithurge* is French vernacular.]
Lithurgus Berthold, 1827: 467, emendation of *Lithurge* Latreille, 1825. Type species: *Andrena cornuta* Fabricius, 1787, monobasic. [Michener (1997b) listed a subsequent designation.]
Liturgus Ashmead, 1899a: 77, *lapsus* for *Lithurgus* Berthold, 1827.

Since Sandhouse's (1943) list, certain authors have used the French vernacular form *Lithurge*. As noted by Michener (1979b) and others, the correct Latinized form is *Lithurgus*.

The loss of male arolia is an apomorphy for the subgenus, but arolia are found in the males of the Australian *Lithurgus rubricatus* Smith, which at least for the present is in the subgenus *Lithurgus* s. str. *L. rubricatus* does not agree with *Lithurgopsis* in subgeneric characters except in having long arolia in the male. It is difficult to polarize the other subgeneric characters. It is therefore not clear whether *Lithurgus* s. str. and *Lithurgopsis* are sister groups; the alternative is that the former is derived from a paraphyletic *Lithurgopsis*. No species of *Lithurgus* s. str., however, is closely similar to any group or species of *Lithurgopsis;* I therefore hypothesize the sister-group relationship.
■ This subgenus is found in Eurasia, Africa, Australia, and intervening islands and includes all of the Old World species of the genus. In Africa it reaches the Cape, in Australia it reaches New South Wales but is apparently absent from southern parts of the continent, while in Eurasia it extends north to Japan, China, Russia, Germany, Spain, etc. To the east it occurs in the Philippines and as far as Tahiti, perhaps having been carried as nests in timbers of Polynesian boats. One species, *Lithurgus huberi* Ducke, occurs in Brazil; it is probably adventive there and the same as the Old World species *L. atratus* Smith. *Lithurgus (Lithurgus) scabrosus* (Smith) appears to have been introduced in Hawaii (Snelling, 2003). Another species, *L. chrysurus* Fonscolombe, has been recorded by Roberts (1978) as introduced, presumably from the Mediterranean region, into New Jersey. There are no subsequent records of this population, and the species is probably extinct in America. The subgenus is in need of revision; there are probably 15 species. The four species in sub-Saharan Africa were revised by Eardley (1988), and the roughly seven palearctic species by van der Zanden (1986).

Species of the subgenus *Lithurgus* s. str. are floral specialists, some on Malvaceae, others on Asteraceae. The length of the proboscis varies from extremely long (reaching the hind coxae or beyond) for species on Malvaceae and thistles (*Carduus, Cirsium;* see Banaszek and Romasenko, 1998; Pachinger, 2004) to much shorter for species on other Asteraceae (*Centuarea;* see Rust et al., 2004).

Genus *Microthurge* Michener

Microthurge Michener, 1983a: 181. Type species: *Lithurgus pharcidontus* Moure, 1948, by original designation.

This genus consists of small (5-8 mm long) lithurgines having the size and form of a *Heriades* or small *Hoplitis* (Osmiini). The metasomal terga have pale apical hair bands. The supraclypeal facial prominence is present in females, weak or absent in males. The maxillary palpus is extremely short, two-segmented; the labial palpus is three-segmented (the fourth segment absent). The arolia of males are absent and the claws of females bifid (Fig. 80-3c). The stigma is larger than that in other genera, the sides basal to vein r diverging apically. The pygidial plate of the female is represented by the flat upper surface of a short apical spine on T6 (Fig. 80-2b); that of the male, by a broad triangle, nearly pointed apically, not or but little extending beyond the rest of T7 (Fig. 80-4b).

■ *Microthurge* occurs from Buenos Aires province, Argentina, to Bolivia and São Paulo, Brazil. The four species were revised by Griswold (1991).

Genus *Trichothurgus* Moure

Trichothurgus Moure, 1949b: 270. Type species: *Megachile dubia* Sichel, by original designation.

Lithurgomma Moure, 1949b: 277. Type species: *Lithurgomma wagenknechti* Moure, 1949, by original designation.

The two names listed above were published simultaneously. Michener (1983a), as first reviser, selected *Trichothurgus* as the generic name.

These are robust, hairy, megachiliform to euceriform bees, 7 to 21 mm long but usually large, without metasomal hair bands. The facial prominence is commonly absent, but is present in females of a few species. The labrum is longer than the clypeus, sometimes as long as the distance from the clypeal apex to the anterior ocellus. The lower mandibular tooth is longer than the median tooth or, in the females of some species, these two teeth are subequal. The maxillary palpus is three-segmented. The anterior, outer, and posterior surfaces of the tibiae are rather uniformly hairy, the fore and middle tibiae each having a row of small spines or spicules extending basad from each of the two apical tibial spines, these rows limited to the distal part of the tibia and absent in some males. The hind tibia bears similar spicules scattered among hairs of the outer surface in females but not in males. The arolia of males are distinct, long and slender; arolia are absent in females. The claws of females are simple. The pygidial plate of the female is a dorsal strip on T6, usually slightly expanded at the apex, and not projecting as a spine (Fig. 80-2a); that of the male is a dorsally flat or concave, broad, blunt projection or robust spine, similar to that of *Lithurgus* but often hidden by long hair on the adjacent parts of T7. The tooth at each side of T6 of the female is small and hidden in the dense hairs. The male genitalia and hidden sterna were illustrated by Moure (1949b).

■ This genus occurs in generally xeric regions of Argentina (Neuquén to Jujuy), Chile (Santiago to Tarapacá), and Peru. At least some of the species visit Cactaceae for pollen. There are about 13 described species.

81. Tribe Osmiini

The characters of this tribe are indicated in Table 79-1 and the key to tribes (Sec. 79). Osmiini occur on all continents except South America and Australia. Though abundant in north-temperate regions, the tribe is especially diversified in mediterranean and desertic climates, in Eurasia and southern Africa as well as western North America.

Two puzzling groups of fossils from the Eocene Baltic amber have been placed, perhaps incorrectly, as subtribes of the Osmiini. These are the Glyptapina for the genus *Glyptapis* and the Ctenoplectrellina for *Ctenoplectrella* and *Glaesosmia*. Both subtribes differ from most Megachilinae in that the labrum is broader than long. The four species of *Glyptapis* are remarkable for their hairy eyes and foveolate mesepisterna. For the Recent subtribes Heriadina and Osmiina, see below.

Earlier studies (Michener, 1941a) indicated that an elongate thorax (with the scutellum almost flat and the metanotum and base of the propodeum nearly horizontal) was ancestral among the Osmiini, being found also in most S-T bees (the outgroup). Conversely, a more spherical thorax was considered derived. No doubt this is in general true for bees as a whole, but in the Osmiini, evolution can go in both directions, an especially elongate body probably being an adaptation to nesting in narrow burrows in wood or stems. The most decisive evidence is in *Osmia*, subgenus *Pyrosmia*, in which *O. cephalotes* Morawitz has an elongate body and the whole constellation of characters that go with that body form, e.g., the scutellum is flat in profile, the metanotum slopes but little, and the basal zone of the propodeum is much more nearly horizontal than vertical. These features render the thorax elongate. Other species of *Pyrosmia* are robust, as are most *Osmia*. It is almost certain that the slender *Pyrosmia* species reverted toward the body form that outgroup comparison incorrectly shows to be ancestral.

In 1941 I regarded taxa like *Chelostoma* as primitive, not only because of the slender body (Fig. 81-6) but also because of the relatively large stigma (Fig. 81-1c). These forms, however, are mostly small, and it is now known that body size and stigmal size are often inversely related (Danforth, 1989a). *Chelostoma* could well have reverted in both body form and stigmal size as parts of an adjustment to nesting in small holes, instead of being plesiomorphic in these features. The same may be true of *Heriades* and its relatives.

Consideration of the phylogeny of osmiine genera is beset at every stage by the probability that most characters have evolved in at least two directions within the tribe, so that there is no single correct polarity, or that they have evolved convergently two or more times.

Many authors (e.g., Michener, 1941a) have divided the Osmiini into two groups (Table 81-1), often called by the tribal names Heriadini and Osmiini. The first of these has also been called Trypetini, a name that is preoccupied (Engel, 2005). These groups, here called the *Osmia* group and the *Heriades* group, merge; *Protosmia* and *Othinosmia* in particular are intermediate and show relationships

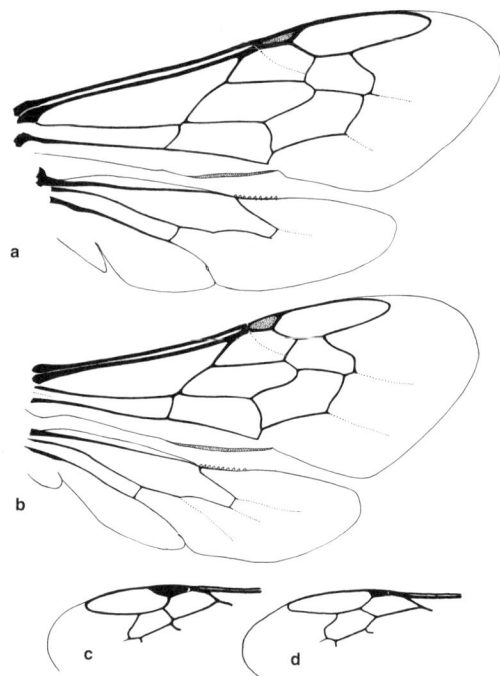

Figure 81-1. Wings of Osmiini. **a,** *Ashmeadiella bucconis* (Say); **b,** *Atoposmia (Hexosmia) copelandica* (Cockerell); **c, d,** Portions of forewings of *Chelostoma (Prochelostoma) philadelphi* (Robertson) and *Osmia (Osmia) lignaria* Say, showing diversity in stigma size.

to both. Nonetheless, since family-group names are in common use, the groups can be called by the subtribal names Heriadina and Osmiina. It is not practical to construct a key that efficiently distinguishes these groups. I have followed Griswold (1985) in excluding *Chelostoma* from the *Heriades* group because it seems closer to *Hoplitis (Alcidamea)* in the *Osmia* group. The following paragraphs are modified from Griswold (1985):

The *Heriades* group, or subtribe Heriadina, can be described as follows: maxillary palpi reduced, two- to four-segmented [four- or five-segmented in the *Osmia* group except two- to five-segmented in *Hoplitis (Proteriades)* and its relatives]; parapsidal lines linear, often longer than in most of the *Osmia* group; scutellum flat or only gently convex; propodeum usually with a basal, horizontal zone (usually declivous in the *Osmia* group); mesopleuron elongate ventrally, as long as its dorsal length (except in *Protosmia* and *Othinosmia*); mesopleural signum linear (except in some *Protosmia*); hind coxa carinate (frequently not carinate in the *Osmia* group); stigma larger than that in most of the *Osmia* group, but similar to that of the genus *Chelostoma* (Fig. 81-1c) in the *Osmia* group; anterior surface of T1 concave (except in *Xeroheriades*). Male: T7 hidden or curled under metasoma (Fig. 81-7); less than six (sometimes only two) sterna visible when metasoma is in repose (except in *Xeroheriades*); S2 and S3

Table 81-1. Grouping of Genera of Osmiini.

Osmia group (Subtribe Osmiina)	Heriades group (Subtribe Heriadina)
Ashmeadiella	Afroheriades
Atoposmia	?Bekilia
Chelostoma	Heriades
Haetosmia	Hofferia
Hoplitis	Noteriades
Hoplosmia	Othinosmia
Ochreriades	Protosmia
Osmia	Pseudoheriades
Stenosmia	Stenoheriades
Wainia	Xeroheriades

without transverse humps; S3 but not S4 with discal hair short, velvety; S3 with lateral hyaline flaps (except in *Pseudoheriades, Noteriades, Afroheriades*); S5 with modified discal or marginal hair and/or its posterior margin emarginate; S6 without basal or lateral hyaline flaps. Female: clypeus not overhanging labrum [except in *Othinosmia (Megaloheriades)* and *Afroheriades primus* Peters]; labrum usually with apical or subapical tuft of hair, usually without marginal fringe of hair; T6 with preapical carina or with wide hyaline apical flange. Alternative features characterize most members of the *Osmia* group, or subtribe Osmiina.

Much of the difficulty in defining the *Heriades* group rests with the intermediate nature of *Protosmia* and *Othinosmia*. The rest of the *Heriades* group can be distinguished from all other Osmiini (except *Chelostoma*) by the ventrally elongate mesopleuron, the narrowly linear mesopleural signum, and the distally diverging stigmal margins. It must be remembered, however, that larger stigmata are commonly associated with relatively smaller body size (Danforth, 1989a); diverging margins characterize a larger stigma and may thus be a reflection of the generally small body size of species of the *Heriades* group, irrespective of relationships. The thoracic structure of *Protosmia* and *Othinosmia* is markedly like that of *Osmia*. All male *Protosmia* and *Othinosmia* have a brush of hair under the margin of S1. Such a brush is also present in some *Heriades*, but not elsewhere in the Megachilidae. A character of most of the *Heriades* group is the emarginate male S5 with its fringe of modified (often capitate) hairs, as found in *Protosmia, Othinosmia, Xeroheriades, Stenoheriades,* and *Heriades*. Such a fifth sternum is not found in other Osmiini. In *Afroheriades* and *Pseudoheriades* the modified hairs arise from the disc of the emarginate sternum, and *Hofferia* seems to lack both modified hairs and the emargination. The combination of a dense brush of hairs covering the disc of S3 with the absence of a similar covering on S4 will also serve to distinguish males of the *Heriades* group from those of the *Osmia* group. The presence of a preapical or apical tuft of hairs on the female labrum is also a character of importance; although not present in all species of the *Heriades* group, this tuft is present in most, and is not found in the *Osmia* group.

Because many authors have placed *Hoplitis* (including *Anthocopa*), *Osmia*, and related taxa together in the genus *Osmia*, various keys and revisional works on *Osmia* of the palearctic region include species in most genera of the *Osmia* group. Rather than mentioning such studies under each genus, I list some of them here: Schmiedeknecht (1885), Ducke (1900), Benoist (1931).

Likewise, various works on nesting biology include material on both *Osmia* and *Hoplitis* under the generic name *Osmia*. Examples are Malyshev (1937) and Westrich (1989).

Key to the Genera of the Osmiini of the Eastern Hemisphere (Males)
(Based on Griswold and Michener, 1998)

1. T7 weakly sclerotized, not exposed, hidden by large T6 2
—. T7 strongly sclerotized, exposed though sometimes small 5
2(1). Scutellum with transverse preapical carina, behind which it is vertical or overhanging metanotum; clypeus with longitudinal carina continued dorsally between antennal sockets (Africa, southern Asia) .... *Noteriades*
—. Scutellum not carinate, without such a posterior marginal zone; clypeus without longitudinal carina, or carina not continued dorsally between antennal sockets 3
3(2). Scutellum curved down posteriorly, posterior part of its surface thus steeply sloping, vertical, or overhanging; metanotum well below level of most of scutellum; posterior lateral angle of scutum frequently acutely produced posteriorly (Africa and Israel to the Philippines) .... *Wainia*
—. Scutellum flat or scarcely convex, not curved down posteriorly and not overhanging metanotum; metanotum medially about on the level of most of scutellum; posterior lateral angle of scutum obtuse 4
4(3). Posterior margins of S2 and S3 notched or stepped laterally; pronotal lobe without carinules in addition to carina (holarctic, Oriental, Africa, Central America) .... *Heriades*
—. Posterior margins of S2 and S3 not notched or stepped laterally; pronotal lobe with several minute carinules behind and parallel to carina, carina rarely absent but carinules present (except carinules absent in subgenus *Dolichosmia* from Burma) (holarctic, Burma) .... *Protosmia*
5(1). Parapsidal line punctiform or rarely short-linear, one-fifth as long as tegula or usually less (Fig. 81-2a); S3 of male commonly emarginate with fringe in emargination, the middle of S3 being largely or wholly covered by an enlarged S2, and S4 neither emarginate nor fringed like S3 (but some have S2 and S3 transverse and similar to one

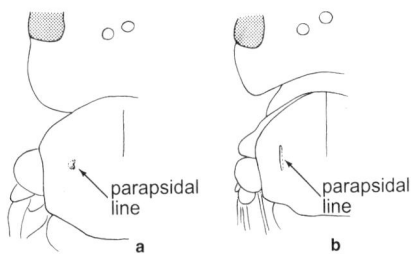

Figure 81-2. Dorsal views of thoraces of Osmiini. **a,** *Osmia* sp.; **b,** *Hoplitis* sp. From Michener, McGinley, and Danforth, 1994.

another); body commonly with some greenish or bluish metallic color (holarctic) .... *Osmia* (in part)
—. Parapsidal line linear (Fig. 81-2b), at least one-fourth as long as tegula; S3 of male neither emarginate nor fringed and not covered medially by S2 [except in *Hoplitis (Jaxartinula* and *Kumobia)* and except sometimes when S4 is similar to S3]; body nonmetallic 6
6(5). T6 with strong transverse preapical carina, the carina sometimes multidentate (Fig. 81-9b, c) 7
—. T6 without preapical carina 10
7(6). Axilla angulate or spined posteriorly (Fig. 81-9a); S3 without dense, velvety discal hair; S5 with hairs unmodified and with reflexed lateral hyaline flaps (palearctic) .... *Hoplosmia*
—. Axilla with posterolateral margin rounded (except pointed in the African *Othinosmia* subgenus *Afrosmia)*; S3 with dense, velvety discal hair; S5 with modified hairs and without lateral hyaline flaps 8
8(7). Preapical carina of T6 smooth, neither crenulate nor spined; basal zone of propodeum not limited posteriorly by carina (Africa) .... *Othinosmia*
—. Preapical carina of T6 crenulate or spined; basal zone of propodeum limited posteriorly by distinct carina, at least laterally 9
9(8). Metanotum vertical in lateral view; apical margin of S1 not produced; T7 emarginate (Mediterranean region to South Africa) .... *Stenoheriades*
—. Metanotum nearly horizontal; apical margin of S1 produced; T7 not emarginate (Mediterranean region) .... *Hofferia*
10(6). T7 more or less quadrate, although sometimes with apical processes, and wrapped around laterally by T6, thus placed in large emargination of T6 11
—. T7 variable in form but not quadrate, not wrapped around laterally by T6 12
11(10). Proboscis short, in repose not or scarcely exceeding fossa; upper end of metapleuron with shiny, shelflike projection; T1 with carina separating anterior and dorsal surfaces; S3 with median apical spike (Africa, India, Mediterranean region) .... *Pseudoheriades*
—. Proboscis long, in repose much exceeding fossa; upper end of metapleuron without shiny projection; T1 without carina separating surfaces; S3 without median apical spike (South Africa) .... *Afroheriades*
12(10). Scutum elongate, its length at least equal to intertegular distance; mesopleural signum narrowly linear; stigma relatively large, its margin within marginal cell convex (Fig. 81-1c) 13
—. Scutum not elongate, its length less than intertegular distance; mesopleural signum not linear; stigma relatively small, its margin within marginal cell usually not convex (Fig. 81-1b, d; but see *Stenosmia*) 14
13(12). Pronotum elevated so that scutum has no anterior surface; body (at least metasoma) with yellow or ivory markings; axilla lobate or angled laterally, extending beyond line from posterolateral angle of scutum to lateral margin of scutellum (Asia Minor, Namibia) .... *Ochreriades*
—. Pronotum medially not attaining level of dorsal scutal surface, so that anterior end of scutum is bent down, forming anterior scutal surface; body without yellow or ivory markings; axilla not produced (holarctic) .... *Chelostoma*
14(12). T6 with lateral tooth [a mere convexity in *Hoplitis (Formicapis)*], sometimes minute but sharp 15
—. T6 without lateral tooth 16
15(14). Pronotal lobe with erect lamella (often hidden by hair); T7 broad, truncate, bulging dorsally; S6 without basal membranous flaps (Mediterranean region, East Africa to central Asia) .... *Haetosmia*
—. Pronotal lobe without lamella but sometimes weakly carinate, usually without carina; T7, if truncate, not bulging dorsally; S6 usually with basal membranous flaps lying appressed against ventral surface of S6 (holarctic, Africa, India) .... *Hoplitis* (in part)
16(14). T7 tridentate; scutellum with profile distinctly convex, sloping so that posterior margin is below anterior margin; S6 with basal membranous flaps (palearctic) .... *Hoplitis (Megahoplitis)*
—. T7 truncate, bilobed, or bidentate; scutellum flat or, if convex, then not sloping; S6 without basal membranous flaps 17
17(16). Stigma about half as long as marginal cell on wing margin; stigmal margin within marginal cell distinctly convex; lower half of omaulus carinate (Egypt to China) .... *Stenosmia*
—. Stigma less than half as long as marginal cell on wing margin (Fig. 81-1d); stigmal margin within marginal cell weakly convex, straight or concave; omaulus sometimes a sharp angle and reinforced by abrupt changes in surface sculpture, but not carinate (holarctic) .... *Osmia* (in part)

Key to the Genera of the Osmiini of the Eastern Hemisphere (Females)
(Based on Griswold and Michener, 1998)

1. Posterolateral corner of scutum with marginal ridge not carinate, a dense patch of long pubescence on lateral surface (T6 with translucent apical lip extending beyond hairs and projecting at strong angle to surface of tergum) (South Africa) .... *Afroheriades*
—. Posterolateral corner of scutum with strong marginal carina, pubescence absent lateral to carina 2
2(1). Apical part of labrum without long, erect hairs; body slender, elongate (Fig. 81-6), scutal length at least equal to intertegular distance 3
—. Apical part of labrum with apical fringe of long, usually erect hairs or preapical tuft of such hairs or with at least some erect hairs on disc; body not elongate and slender (except *Osmia cephalotes* Morawitz, which is metallic), scutal length less than intertegular distance 5
3(2). Pronotum elevated, scutum thus lacking anterior surface; body (at least metasoma) with yellow or ivory markings (Near East, Namibia) .... *Ochreriades*
—. Pronotum medially not attaining level of dorsal scutal surface, anterior end of scutum thus bent down, forming anterior scutal surface; body without yellow or ivory markings 4
4(3). Preoccipital carina complete, reaching hypostomal carina, where it is produced into strong tooth; S6 with acute median apical spine (Mediterranean region) .... *Hofferia*
—. Preoccipital carina incomplete, not reaching hypostomal carina, without strong tooth; S6 without or with small median apical tooth (holarctic) .... *Chelostoma*

5(2). Parapsidal line punctiform (Fig. 81-2a) or rarely short-linear, one-fifth as long as tegula or usually less; body commonly with some greenish or bluish metallic color (holarctic) .... *Osmia* (in part)
—. Parapsidal line linear, at least one-fourth as long as tegula (Fig. 81-2b); body nonmetallic 6
6(5). Mesopleuron elongate ventrally, as long as scutum or nearly so; mesopleural signum narrowly linear; stigma relatively large, margin within marginal cell convex (as in Fig. 81-1c) 7
—. Mesopleuron short ventrally, much shorter than scutum, mesopleural signum short, not slender; stigma smaller, margin within marginal cell usually not convex (as in Fig. 81-1b, d) 10
7(6). Posterior edge of scutellum with premarginal carina, surface behind carina a narrow vertical or overhanging zone; clypeus with strong longitudinal carina (Africa, southern Asia) .... *Noteriades*
—. Posterior margin of scutellum not carinate, usually without vertical or overhanging marginal zone; clypeus without longitudinal carina 8
8(7). Hypostomal area margined laterally by carina and fringe of long curled hair; acetabular interspace of mandible with short, oblique secondary ridge (Africa, India, Mediterranean region) .... *Pseudoheriades*
—. Hypostomal area not carinate or fringed laterally; acetabular interspace of mandible without secondary ridge 9
9(8). Mouthparts long, in repose well exceeding proboscidial fossa [except in palearctic *Stenoheriades asiaticus* (Friese) and *coelostoma* (Benoist), which have medially emarginate clypeus and bidentate mandibles] (Africa, Mediterranean region) .... *Stenoheriades*
—. Mouthparts short, in repose not or scarcely exceeding proboscidial fossa [palearctic species with clypeus not emarginate medially and mandible with at least three teeth] (holarctic, oriental, Africa) .... *Heriades*
10(6). Labrum with apical or preapical tuft of erect hair 11
—. Labrum without apical or preapical tuft of erect hair, though often with marginal fringe 12
11(10). Pronotal lobe carinate (holarctic, Burma) ....
.... *Protosmia*
—. Pronotal lobe usually not carinate (South Africa) ....
.... *Othinosmia* (in part)
12(10). Axilla spined or angulate 13
—. Axilla rounded or slightly lobate 14
13(12). Metapleuron with round dorsal projection; S1 with apical spine (East Africa) .... *Othinosmia (Afrosmia)*
—. Metapleuron planar, without round dorsal projection; S1 without apical spine (palearctic) .... *Hoplosmia*
14(12). Labrum with fringe of hairs near or on margin [weak in *Hoplitis (Bytinskia)* and *(Anthocopa) matheranensis* (Michener)] 15
—. Labrum without fringe of hairs, disc sometimes with erect hairs that may be longest near the margin but are relatively sparse, not forming a distinct fringe 17
15(14). Posterior lateral angle of scutum acute or right-angular, and produced posteriorly or in *Wainia algoensis* (Brauns) upward as strong lamella [except South African *W. elizabethae* (Friese), in which entire propodeum is densely shagreened] (Africa to Israel and east to the Philippines) .... *Wainia* (in part)
—. Posterior lateral angle of scutum not sharp, not produced posteriorly; propodeum at most partly shagreened 16
16(15). Metanotum with medial, dorsal spine, *or* clypeus rugose (palearctic) .... *Osmia (Allosmia,* some *Erythrosmia)*
—. Metanotum not spined medially; clypeus not rugose (holarctic, Africa, India) .... *Hoplitis* (most subgenera)
17(14). Clypeus strongly overhanging labrum 18
—. Clypeus scarcely overhanging labrum 19
18(17). T6 with preapical carina visible through dense pubescence; mouthparts with hooked hairs; clypeus without longitudinal carina (Mediterranean region, East Africa to central Asia) .... *Haetosmia*
—. T6 without preapical carina; mouthparts without hooked hairs; clypeus with longitudinal carina, carina sometimes obscured by dense pubescence (southern Africa) .... *Wainia (Wainiella)*
19(17). Omaulus strongly carinate below nearly to midventral line; clypeus and mandibles obscured by dense plumose hair (Egypt to China) .... *Stenosmia*
—. Omaulus not carinate below; clypeus and mandibles with hair sparse, not obscuring surfaces (Asia) ....
.... *Hoplitis (Kumobia)*

Key to the Genera of the Osmiini of the Western Hemisphere

1. Propodeum with narrow horizontal basal zone, set off from declivous posterior surface by a carina, and divided by carinae into a series of large pits (Fig. 81-3); anterior surface of T1 broadly concave and delimited by strong carina; lines delimiting propodeal triangle absent or feeble (holarctic, Mesoamerica) .... *Heriades*
—. Propodeum without such a horizontal basal zone (but if pits and a zone set off by a carina are evident, then zone is usually sloping, and anterior surface of T1 is neither broadly concave nor delimited by strong carina); lines delimiting propodeal triangle distinct, at least as sharp changes in sculpturing.. 2
2(1). Parapsidal lines punctiform (Fig. 81-2a) or at most three times as long as wide; body usually metallic (holarctic, Mesoamerica) .... *Osmia*
—. Parapsidal lines linear (Fig. 81-2b); body rarely metallic although sometimes strongly so 3
3(2). Thorax elongate (Fig. 81-6), a line tangent to anterior

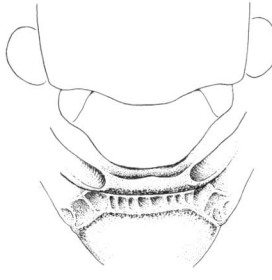

Figure 81-3. Middorsal part of metanotum and propodeum of *Heriades variolosa* (Cresson). Modified from Michener, McGinley, and Danforth, 1994.

ends of tegulae is thus near middle of scutum, and ventral profile of mesepisternum, in front of middle coxa, is as long as dorsoventral thickness of thorax (seen in lateral view); apex of marginal cell curving away from wing margin only at extreme apex (Fig. 81-1c); S6 of male fully exposed; labrum of female with few and inconspicuous erect setae, thus often appearing hairless (holarctic) .... *Chelostoma*
—. Thorax less elongate (Figs. 81-8, 81-10), a line tangent to anterior ends of tegulae is thus considerably anterior to middle of scutum, and ventral profile of mesepisternum, in front of middle coxa, is much shorter than dorsoventral thickness of thorax; apex of marginal cell usually curving away from wing margin in apical one-eighth or more of cell; S6 of male usually concealed or only its apical margin exposed; labrum of female with numerous erect setae 4
4(3). Posterior lobe of pronotum with several small vertical carinulae parallel to posterior margin; clypeus of female with apical, somewhat spatulate, slender horn (Fig. 81-13); T6 of male with median apical rounded flange and broad produced angle laterally (T7 of male not exposed) (palearctic; western North America) .... *Protosmia*
—. Posterior lobe of pronotum usually without carinulae [but with carinulae in *Hoplitis (Alcidamea) pilosifrons* (Cresson) and others]; clypeus of female without apical horn; T6 of male not as described above 5
5(4). S1 of female with strong preapical bilobed process; S6 of female with median and apical spikes; posterior margins of T1 to T3 gently concave middorsally; S5 of male with deep midapical notch, its margin distally bearing a few capitate hairs (S6 of male without flaps such as are found in *Hoplitis*, broadly exposed; S1 convex in profile, its anterior surface not defined) (California) .... *Xeroheriades*
—. S1 of female without strong preapical process; S6 of female without spikes or rarely with midapical spike; posterior margins of T1 to T3 straight middorsally; S5 of male not deeply notched, without capitate hairs on margin 6
6(5). S2 and S3 of male similar to one another; S5 and S6 of male hidden, largely membranous, only transverse apical marginal areas and lateral margins somewhat sclerotized, these sterna with large, basal, minutely hairy areas invaginated anteriorly into metasoma; distance from first submarginal crossvein to first recurrent usually twice distance from second recurrent vein to second submarginal crossvein (Fig. 81-1a); anterior surface of mesepisternum in front of omaulus smooth, shining, lower part usually sharply divided at omaulus from outer surface, lower part of omaulus carinate (except in subgenus *Isosmia*); male T6 with four strong teeth (Fig. 81-4) (except in subgenus *Isosmia*) (nearctic) .... *Ashmeadiella*
—. S2 and S3 of male commonly quite different from one another; S5 or S6 of male, or both, usually exposed, broadly sclerotized, without large, anterior hairy areas; distance from first submarginal crossvein to first recurrent almost always less than twice distance from second recurrent vein to second submarginal crossvein; anterior surface of mesepisternum in front of omaulus usually with at least a few punctures near omaulus, which is not carinate; male T6 not four-toothed 7
7(6). S6 of male with basal, hairless translucent flaps (sometimes joined together) arising at gradulus and extending posteriorly, lying against ventral surface of sternum; body (except for most species of subgenera *Proteriades*, *Penteriades*, and *Acrosmia*) rather elongate (Fig. 81-8), T1 with anterior surface not broadly concave or flat, not margined above by sharp line except sometimes medially, profile of T1 convex, not angulate; parapsidal line often over half as long as tegula (Fig. 81-2b); S2 of male usually not enlarged, not hiding middle of S3 margin, although the fully exposed S2 is sometimes emarginate and fringed, but rarely (subgenera *Penteriades* and *Acrosmia*) S2 and S3 configured as in most species of *Osmia* and *Atoposmia* (holarctic) .... *Hoplitis*
—. S6 of male without basal flaps; body robust, metasoma often almost globose; anterior surface of T1 a broad, concave or flat area margined above by a sharp line usually as long as one-half width of T1 although much less in some [e.g., *A. (Atoposmia) elongata* (Michener)], profile of T1 thus angulate at summit of anterior surface; parapsidal lines usually less than half as long as tegula; S2 of male nearly always much larger than S3, which is broadly emarginate and medially fringed, the fringe or at least sternal margin hidden medially by S2 (western North America, Mexico) .... *Atoposmia*

Genus *Afroheriades* Peters

Pseudoheriades (Afroheriades) Peters, 1970a: 157. Type species: *Pseudoheriades primus* Peters, 1970, by original designation.

Archeriades Peters, 1978a: 337. Type species: *Eriades larvatus* Friese, 1909, by original designation.

This genus contains heriadiform bees 4.0 to 8.5 mm long that differ from *Heriades* in having a relatively long proboscis (extending well beyond the fossa when in repose), in lacking a distinct carina surrounding the broadly concave anterior surface of T1, in lacking also an omaular carina, and in the large, exposed T7 of the male, which lies in a quadrate emargination of T6. As in *Chelostoma* the labrum of the female lacks the preapical tuft of large, erect hairs present in most *Heriades*, although a marginal row of erect hairs may be present. The inner margin of the mandible of the female sometimes bears a fringe of long hairs, as in *Chelostoma*. The preoccipital carina is present dorsally, absent laterally. The pronotal lobes are not carinate. The basal zone of the propodeum is not marked posteriorly by a carina. As in *Pseudoheriades*, the maxillary palpi are reduced to two segments, unlike those of nearly all other Osmiini. Illustrations of male genitalia and sterna are in Peters (1970a, 1978a).

■ *Afroheriades*, known only from South Africa, contains at least eight species, only five of them named (Griswold, 1985).

This genus contains a rather robust species, *Afroheriades primus* (Peters), and several slender species, such as *A. larvatus* (Friese), *dolicocephalus* (Friese), and *geminus* (Peters). The robust species could be called *Afroheriades* s. str., and the slender ones, subgenus *Archeriades*. In *A. primus*, which has a less elongate thorax, the metanotum is sloping and the basal area of the propodeum is well below the level of the scutellum and sloping, whereas, in the slender species, the scutellum, metanotum, and base of the propodeum are almost in the same plane. Somewhat intermediate is *A. capensis* Griswold, MS. Females of *A. primus* are also unique among *Afroheriades* species in having a flat, polished hypostomal area bounded laterally by a carina and a row of curved hairs, thus suggesting females of *Pseudoheriades*, some *Stenoheriades*, *Osmia (Euthosmia)*, *Hoplitis (Anthocopa, Chlidoplitis)*, etc.

Genus *Ashmeadiella* Cockerell

Ashmeadiella consists of small (length 3.5-9.5 mm), robust, nonmetallic species with the metasoma sometimes red and the pubescence entirely pale and usually forming narrow apical fasciae on the terga. The distinctive and unique feature of the genus is the weak sclerotization of the hidden S5 and S6, and the invagination of each to form two minutely hairy flaps, as illustrated by Mitchell (1962), extending forward internally for a distance greater than the length of S4; because S2 and S3 are almost alike, the metasomal venter has four unmodified exposed sterna, S1 to S4. In almost all species the lower part of the omaulus is carinate, separating the smooth, impunctate, hairless preomaular part from the punctate and hairy postomaular part of the mesepisternum. In the subgenus *Isosmia,* although the carina is absent, the sharp change in sculpturing is evident. Some species of *Hoplitis* and *Osmia* have nearly as sharp a change in sculpturing at the omaulus. The venational character mentioned in the key to genera is likewise subject to variation. Except for *Isosmia,* the four-toothed T6 of the male, largely hiding the simple apex of T7 (Fig. 81-4), is also distinctive for the genus. Male genitalia and sterna were illustrated by Mitchell (1962).

The subgenus *Isosmia* has been included among the American group of "Anthocopa," the other members of which have been transferred to *Atoposmia* (Griswold and Michener, 1998). The sternal characters of the male, however, clearly show its relation to *Ashmeadiella*. A reasonable hypothesis is that *Isosmia* is the sister group to the rest of *Ashmeadiella*.

The genus is limited to North America. It was revised by Michener (1939a), and nearly all the species were revised by Hurd and Michener (1955).

Nests are made in holes in wood or stems, in burrows in the ground, or in spaces under rocks. Krombein (1967) reported on nests of several species of the subgenus *Ashmeadiella* s. str. and one of *Arogochila* in trap nests consisting of holes in wood. Cell partitions are made of gummy leaf pulp—probably leaf pulp mixed with nectar, gum, or resin—or less commonly of resin (or dried nectar?) alone, without leaf pulp. Rozen and Eickwort (1997) indicated the use of nectar mixed with leaf pulp as construction material by *Ashmeadiella*. Rozen (1987a) reported a nest of *A. (A.) leucozona* Cockerell in the ground with leaf pulp lining a cell. A nest of *A. (Isosmia) rubrella* (Michener) under a rock consisted of cells made of coarsely chewed flowers of *Dalea* (Yanega, 1994). But Rozen (1987a) reports cells of *A. (Chilosima) holtii* Cockerell in the ground, presumably in burrows made by the bees; the cells were made of soil somewhat harder than the surrounding soil, with no associated plant materials except probably nectar mixed with the soil of the cell walls.

The subgeneric name *Neoashmeadiella* Gupta (1990) is based on a species of *Megachile (Chelostomoda)*.

Key to the Subgenera of *Ashmeadiella*

1. Omaulus not carinate; T6 of male not toothed .... *A. (Isosmia)*
—. Omaulus carinate below; T6 of male with four teeth (Fig. 81-4) 2
2(1). Males 3
—. Females 5
3(2). Mandible tridentate .... *A. (Chilosima)*
—. Mandible bidentate 4
4(3). Lateral margin of T6 distinctly and rather evenly convex, ending at nearly right-angular apex of lateral tooth .... *A. (Cubitognatha, Arogochila)*
—. Lateral margin of T6 straight, feebly convex, or sinuate, ending at acute apex of lateral tooth .... *A. (Ashmeadiella s. str.)*
5(2). Outer side of mandible, as measured along lower carina, at most twice as long as scape [except in the group of *A. (A.) cubiceps* (Cresson)]; clypeus produced over base of labrum, clypeal apex usually truncate; well-developed pair of orange brushes beneath clypeal margin .... *A. (Ashmeadiella s. str.)*
—. Outer side of mandible, as measured along lower carina, more than twice as long as scape; clypeus produced or not over base of labrum, but *if* produced, then not simply truncate; orange brushes beneath clypeal margin reduced or absent 6
6(5). Mandible bidentate, parallel-sided, elbowed; clypeal margin impunctate, upper margin protuberant .... *A. (Cubitognatha)*
—. Mandible three- or four-toothed; clypeus punctured throughout, upper margin not protuberant 7

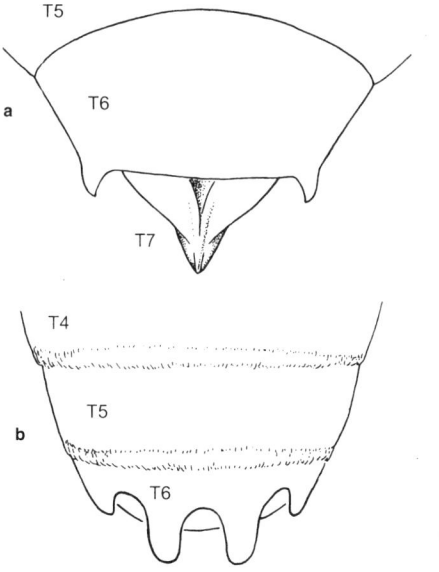

Figure 81-4. Apex of metasoma of male. **a,** *Hoplitis (Alcidamea) producta* (Cresson); **b,** *Ashmeadiella (Ashmeadiella) californica* (Ashmead). From Michener, McGinley, and Danforth, 1994.

7(6). Apex of clypeus irregularly rounded (not truncate or lobed) .... *A. (Chilosima)*
—. Apex of clypeus with apical teeth or lobes, with at least a median broad truncation abruptly produced or with shoulder or small lobe at each side or a produced lateral angle .... *A. (Arogochila)*

Ashmeadiella / Subgenus *Arogochila* Michener

Ashmeadiella (Arogochila) Michener, 1939a: 58. Type species: *Ashmeadiella timberlakei* Michener, 1936, by original designation.

Ashmeadiella (Corythochila) Michener, 1939a: 74. Type species: *Ashmeadiella inyoensis* Michener, 1939, by original designation.

Ashmeadiella (Ramphorhina) Michener, 1939a: 8, *lapsus* for *Arogochila* Michener, 1939.

Ashmeadiella (Rhamphorhina) Michener and Sokal, 1957: 159, unjustified emendation of *Ramphorhina* Michener, 1939.

The relatively slender, three-toothed or occasionally four-toothed female mandible is similar to that of *Chilosima*. The female clypeal apex is often snoutlike or lobed, having at least a small lobe on each side of a truncation, or, in *Ashmeadiella (A.) foxiella* Michener, a truncation set off by an emargination at each side. In the group of *A. (A.) inyoensis* Michener, however, the middle of the clypeus is little produced, but there are strong apicolateral angles. This is the group formerly segregated as *Corythochila*.

■ *Arogochila* is found from western Texas and Idaho west to Baja California and Washington state, USA, and south to Puebla and Durango, Mexico. Most of the 18 species occur in xeric areas.

Ashmeadiella / Subgenus *Ashmeadiella* Cockerell s. str.

Ashmeadiella Cockerell, 1897c: 197. Type species: *Heriades opuntiae* Cockerell, 1897, by original designation.

Titusella Cockerell, 1906b: 445. Type species: *Titusella pronitens* Cockerell, 1906, monobasic.

This is by far the largest and most abundant subgenus of *Ashmeadiella*. Most females are characterized by the punctate, truncately produced clypeus resembling that found in most *Hoplitis* and *Osmia,* and the mandibles, which are somewhat shorter than those in other subgenera and tridentate. In the group of *A. (A.) cubiceps* (Cresson), formerly called subgenus *Titusella,* however, the clypeus is partly or largely impunctate, not produced and apically truncate, and the broad mandibles are longer, four- or five-toothed. This group intergrades completely with and seems to be a specialized derivative of ordinary *Ashmeadiella* s. str.; Figure 81-5 illustrates such an intermediate.

■ *Ashmeadiella* s. str. ranges from British Columbia, Canada, and Baja California, Mexico, east to North Dakota, Wisconsin, Indiana, and Texas, USA, with a disjunct area from North Carolina to Florida, USA; southward the subgenus reaches Yucatan and Morelos, Mexico. It is by far most abundant and speciose in the southwestern USA and northwestern Mexico. There are about 33 species.

Figure 81-5. *Ashmeadiella (Ashmeadiella) occipitalis* Michener. Above, male; Below, female. From Michener, McGinley, and Danforth, 1994.

Ashmeadiella / Subgenus *Chilosima* Michener

Ashmeadiella (Chilosima) Michener, 1939a: 78. Type species: *Ashmeadiella rhodognatha* Cockerell, 1924, by original designation.

Chilosima differs from all other subgenera in having three-toothed male mandibles. The lateral margins of T6 of the male are convex, about as in *Arogochila,* and the female resembles some species of that subgenus in having four-toothed, slender mandibles and lacking the orange brushes beneath the clypeal margin. Thus *Chilosima* shares derived characters with *Arogochila* and could be its sister group, possibly best treated as a group of *Arogochila*.

■ This subgenus ranges from western Texas to eastern California, USA, and south to Baja California, Sonora, and Coahuila, Mexico, in deserts. There are two species.

Ashmeadiella / Subgenus *Cubitognatha* Michener

Ashmeadiella (Cubitognatha) Michener, 1939a: 81. Type species: *Ashmeadiella xenomastax* Michener, 1939, by original designation.

Cubitognatha is easily recognized in the female by the unproduced clypeus and the slender, elbowed, bidentate mandibles. The male resembles that of *Arogochila,* suggesting that *Cubitognatha* may be nothing but a derivative of *Arogochila*. The lateral margins of T6 of the male *Cubitognatha,* however, are less convex than those of *Arogochila,* and *Cubitognatha* males might thus run to *Ashmeadiella* s. str. in the key to subgenera. The male of *Cubitognatha* differs from that of *Ashmeadiella* s. str. in the strongly emarginate apex of its labrum, as in some *Arogochila*.

■ This subgenus occurs in the deserts of eastern California and Nevada, USA, and Baja California and Sonora, Mexico. The only species is *Ashmeadiella xenomastax* Michener.

Ashmeadiella / Subgenus *Isosmia* Michener and Sokal

Anthocopa (Isosmia) Michener and Sokal, 1957: 159. Type species: *Anthocopa rubrella* Michener, 1949, by original designation.

Isosmia lacks the derived features that have long been considered diagnostic for *Ashmeadiella*, i.e., the carina on the omaulus and the four teeth on T6 of the male. Its derived sternal characters clearly show that it is a member of the *Ashmeadiella* clade, however, and it is here included in that genus. It was formerly included in the group of American *"Anthocopa,"* here relegated to the genus *Atoposmia*.

■ This subgenus occurs in deserts from western Texas to Nevada and eastern California, USA, and Coahuila to Sonora, Mexico. Two species are recognized.

Genus *Atoposmia* Cockerell

Atoposmia as here constituted contains most of the American species formerly placed in *Anthocopa*, which is here considered to be a palearctic subgenus of *Hoplitis*. *Atoposmia* differs from all American and most palearctic *Osmia* by the linear parapsidal lines, which are generally longer (one-third to one-half as long as the tegula) than those of the few palearctic species of *Osmia* that have linear parapsidal lines. Except for the subgenus *Hexosmia*, T6 of the male has a lateral tooth; occasional specimens of *Hexosmia* do have a very obtuse, rounded lateral angle. Lateral teeth on T6 are uncommon in *Osmia*. *Atoposmia* was at one time included in the *Anthocopa* group of *Hoplitis* because of the robust body form and linear parapsides, but it lacks the basal flaps on S6 of the male, which are characteristic of nearly all *Hoplitis*. The enlarged S2 and the medially emarginate, fringed, and largely hidden S3 are nearly always configured as in most *Osmia*; these are the principal characters that led me to transfer *Atoposmia* out of *Hoplitis*.

Key to the Subgenera of *Atoposmia*

1. Proboscis in repose extending behind fore coxa; first segment of labial palpus less than one-fourth as long as second .... *A. (Atoposmia* s. str.)
—. Proboscis in repose hardly reaching fore coxa; first segment of labial palpus over one-half as long as second 2
2(1). Body with weak greenish tints; T6 of the male without lateral tooth or angle, or rarely with obtuse angle .... *A. (Hexosmia)*
—. Body without greenish tints; T6 of male with lateral tooth or strong rounded angle .... *A. (Eremosmia)*

Atoposmia / Subgenus *Atoposmia* Cockerell s. str.

Osmia (Atoposmia) Cockerell, 1935a: 50. Type species: *Osmia triodonta* Cockerell, 1935, monobasic.

Atoposmia is black, the pubescence pale to partly black; the body length is 6 to 10 mm. T7 is exposed and basically with a pair of apical lobes between which is a tooth; either the lobes or the tooth, however, may be reduced. The proboscis is unusually long, and in some species the scopa of the anterior and median sterna is unusually long. Mandibles of both sexes are three-toothed; those of the female vary from quite narrow medially to robust. The hind coxae are not or only weakly carinate.

■ This subgenus occurs from Baja California, Mexico, to Washington state, east to Montana, Colorado, and Arizona, USA, mostly in mountains. The 12 species were revised by Michener (1943) and Hurd and Michener (1955).

Atoposmia abjecta (Cresson) constructs clumps of cells of masticated plant material on the undersides of rocks (Parker, 1975b), whereas *A. elongata* Michener constructs clumps of cells of sand and masticated plant material in rock crevices (Parker, 1977a).

Atoposmia / Subgenus *Eremosmia* Michener

Anthocopa (Eremosmia) Michener, 1943: 66. Type species: *Osmia robustula* Cockerell, 1935, by original designation.
Anthocopa (Phaeosmia) Michener, 1943: 77. Type species: *Osmia enceliae* Cockerell, 1935, by original designation.

Eremosmia consists of black, nonmetallic species with pale pubescence. The body length is 5.5 to 9.0 mm. The longitudinal carina on the inner ventral angle of the hind coxa is strong. T7 is rounded or has a median point; it is sometimes retracted and hidden.

■ *Eremosmia* ranges from Oklahoma and western Texas to coastal California, USA, and south to the states of Puebla and Morelos, Mexico. The 14 species were revised by Michener (1943) and Hurd and Michener (1955).

Parker (1975b, 1977a) described nests of *Atoposmia hypostomalis* (Michener) and *enceliae* (Cockerell) in pithy stems. Cell walls as well as partitions were made of a mixture of sand, chewed plant material, and probably secretions that resulted in a hard material. A nest of *A. beameri* (Michener) was in the soil, the cells composed entirely of soil particles possibly hardened with nectar (Yanega, 1994).

The mandibles of males vary from two- to three-toothed; this is one of the characters that led to the segregation of *Phaeosmia*. The reasons for not recognizing two subgenera are indicated by Hurd and Michener (1955).

Atoposmia / Subgenus *Hexosmia* Michener

Anthocopa (Hexosmia) Michener, 1943: 74. Type species: *Osmia copelandica* Cockerell, 1908, by original designation.

Hexosmia is the only group of *Atoposmia* with metallic coloration (weak, greenish). Body length is 5.5 to 7.0 mm. The carina of the hind coxa is weak or nearly absent. T6 of the male is edentate and broadly rounded, rarely with a vague obtuse angle laterally; T7 is hidden, but also broadly rounded.

■ This subgenus occurs from California to British Columbia, east to Wyoming, Colorado, and Arizona, USA. There are two species. A revision, in which probable geographic variants were regarded as species, was by Michener (1943); see also Hurd and Michener (1955).

Nests are in pithy stems, the cell walls and partitions being made of chewed plant material, sometimes mixed with sand (Parker, 1975b).

Genus *Bekilia* Benoist

Bekilia Benoist, 1962: 220. Type species: *Bekilia mimetica* Benoist, 1962, by original designation.

This generic name was proposed for a *Heriades*-like bee presumed to be parasitic because of its lack of a distinct scopa. The type material cannot be found in the Paris Museum and it is not possible to determine its relationships; it is not included in the key to genera. It could be an anthidiine like *Afrostelis* rather than an osmiine.

■ *Bekilia* is from Madagascar. There is a single known species.

Genus *Chelostoma* Latreille

This genus consists of slender (Fig. 81-6), black heriadiform bees, more attenuate and less coarsely sculptured than *Heriades*, mostly 3.4 to 9.0 mm long but some larger, *Chelostoma grande* (Nylander) attaining 14 mm. Although the preoccipital carina is present in the subgenus *Gyrodromella*, *Chelostoma* lacks the various other carinae found in *Heriades*. The parapsidal lines are usually half as long as the tegula to nearly as long as the tegula, but in *Chelostoma* s. str. they are shorter. T1, although somewhat concave on the anterior surface in the subgenus *Gyrodromella*, is usually convex except for the longitudinal depression, as in some *Hoplitis (Alcidamea)*; T1 does not have a carina or even a line defining the anterior surface. The large depression or pit on the dorsal surface of T7 of the male is found also in many *Hoplitis (Alcidamea)*; in some *Chelostoma* such as *C. (Prochelostoma) philadelphi* (Robertson) this depression reaches the tergal margin and is therefore not distinct. Also like some species of *Hoplitis (Alcidamea)*, *Chelostoma* has a median elevation on S2 of the male. Females differ from most other osmiines in their nonfringed labrum, which is often elongate. Unlike *Alcidamea* and most other *Hoplitis*, the base of the labrum is exposed above the closed mandibles, and the mandible of the female is usually fringed with long hairs along the inner margin, and is usually widest at the base and tapering to the rather narrow apex, which is bidentate or tridentate.

An unusual character of most species is that the axis of the third labial palpal segment continues the axis of the second, and only the fourth segment projects laterally. The known exceptions to this character (species with two laterally directed palpal segments) are the North American *Chelostoma philadelphi* (Robertson) (which has been placed in a separate genus or subgenus, *Prochelostoma*, primarily because of this character), the palearctic *C. petersi* (Tkalcŭ) (placed in *Ceraheriades*), and *C. aureocinctum* (Bingham) from Burma and neighboring countries (placed in *Eochelostoma*). These three species show no close relationship to one another, having in common principally the presumably ancestral palpal state.

Chelostoma is widespread in the holarctic region, ranging across North America and western and central Eurasia, but is unknown in China and Japan. It extends south to northern Mexico, northern Africa, and into the montane tropics in northeast India, Burma, and Thailand. In North America only nine native and two introduced species are known, but in Eurasia there are many species. Many and perhaps all species are oligolectic on particular floral taxa.

Nests are made in holes in wood, e.g., old beetle burrows, and in stems (see Westrich, 1989, and references therein; Krombein, 1967; and Parker, 1988). Partitions between cells are made of mud or sand grains, apparently stuck together with salivary secretion or nectar.

Key to the Subgenera of *Chelostoma*
(From Griswold and Michener, 1998)

1. Third segment of labial palpus flattened, its axis a continuation of that of second; T7 of male with dorsal pit 2
—. Third segment of labial palpus not flattened, its axis directed laterally as in most megachilid bees; T7 of male without dorsal pit (male unknown in *Ceraheriades*) 4
2(1). Preoccipital carina present; propodeum with sloping basal zone little more than one-half as long as metanotum; T1 shallowly concave anteriorly (palearctic) .... *C. (Gyrodromella)*
—. Preoccipital carina absent; propodeum with horizontal basal zone at least two-thirds as long as metanotum; T1 with anterior surface convex except for longitudinal groove 3
3(2). Parapsidal line half as long as tegula or more; S2 of male without sloping platform, or, *if* present, then not carinate; labrum of female less than twice as long as broad (longer in *C. isabellinum* Warncke) (holarctic) .... *C. (Foveosmia)*
—. Parapsidal line less than half as long as tegula; S2 of male with median prominence, its posterior surface a flat sloping platform margined by carina; labrum of female elongate, nearly three times as long as broad or more (shorter in *C. nasutum* Pérez) (palearctic, introduced in nearctic) .... *C. (Chelostoma s. str.)*
4(1). Hind coxa not carinate (North America) .... *C. (Prochelostoma)*
—. Hind coxa with longitudinal carina along inner ventral margin 5
5(4). T1 not concave on anterior surface; female mandible

Figure 81-6. Male of *Chelostoma (Foveosmia) californicum* Cresson. From Michener, McGinley, and Danforth, 1994.

elongate, two-toothed; labrum apically thickened (central Asia) .... *C. (Ceraheriades)*

—. T1 shallowly concave on anterior surface; female mandible not elongate, three-toothed; labrum not apically thickened (Himalayas to Thailand) ....
.... *C. (Eochelostoma)*

Chelostoma / Subgenus *Ceraheriades* Tkalcŭ

Archeriades (Ceraheriades) Tkalcŭ, 1984a: 5. Type species: *Archeriades petersi* Tkalcŭ, 1984, by original designation.

This subgenus, known only in the female, resembles *Chelostoma* s. str. in its long mandibles, but differs from that genus in the labial palpi (see the key to subgenera) and in the elongate parapsides (half as long as the tegula or more). The head and other structures were illustrated by Tkalcŭ (1984a).
■ The subgenus is from central Asia. The only species is *Chelostoma petersi* (Tkalcŭ).

Chelostoma / Subgenus *Chelostoma* Latreille s. str.

Chelostoma Latreille, 1809: 161. Type species: *Apis maxillosa* Linnaeus, 1767 = *Apis florisomnis* Linnaeus, 1758, monobasic.

This subgenus differs from all other subgenera in the short parapsidal lines (see the key to subgenera). Most males differ from other subgenera in the carinate projection of S2, which often forms a margined platform on a large prominence, and in the dense covering of plumose or spatulate hairs on S4. The long labrum of the female (see the key to subgenera) (relatively short in *Chelostoma nasutum* Pérez) usually distinguishes this subgenus from others with similar labial palpi, and the elongate mandibles, the upper tooth of which is weak or absent, therefore bidentate, distinguish females from all others except *Ceraheriades*. In females of *C. (Foveosmia) foveolatum* (Morawitz), however, the mandibles are somewhat elongate and bidentate. Length ranges from 6 to 14 mm.
■ This is a palearctic subgenus, one species of which, *C. campanularum* (Kirby), is adventive in New York state, USA. Van der Zanden (in litt., 1993) indicated that there are 27 species.

Some variation among species is of interest. The projection of S2 in males is spatulate in *Chelostoma mocsaryi* Schletterer. In *C. diodon* Schletterer and *nasutum* Pérez, the platform is not margined by a carina, but, rather, weakly carinate, the carina not on the margin. In *C. transversum* (Friese) there is a low transverse carina on S2.

Chelostoma florisomne (Linnaeus) is oligolectic on *Ranunculus* (Ranunculaceae).

Chelostoma / Subgenus *Eochelostoma* Griswold

Chelostoma (Eochelostoma) Griswold, 1998, *in* Griswold and Michener, 1998: 216. Type species: *Heriades aureocincta* Bingham, 1897, by original designation.

This little-known subgenus differs from the others having plesiomorphic labial palpi (i.e., segments three and four directed laterally) in the concave anterior surface of T1 and the apically thickened labrum of the female. It is not possible to compare *Eochelostoma* fully with the subgenus *Ceraheriades*, the other Old World *Chelostoma* with similar labial palpi, because the latter is known only in the female. *Ceraheriades*, however, has elongate, two-toothed mandibles in the female (as does *Chelostoma* s. str.), whereas they are short and three-toothed in *Eochelostoma*. Body length is 6 to 9 mm.
■ *Eochelostoma* is the only *Chelostoma* that occurs in the tropics; it is found in northeast India, Burma, and northern Thailand. The only species is *C. aureocinctum* (Bingham).

Chelostoma / Subgenus *Foveosmia* Warncke

Osmia (Foveosmia) Warncke, 1991c: 267. Type species: *Heriades foveolatus* Morawitz, 1868, by original designation.

Like those of the subgenus *Prochelostoma* but unlike those of *Chelostoma* s. str. and *Gyrodromella*, males of this subgenus have a low, rounded hump on S2. The labrum of the female is usually rather short, not much longer than broad although over 1.5 times as long as broad in *C. foveolatum* (Morawitz); that of the male has a transverse basal hump. The parapsidal lines are long (see the key to the subgenera). The gonostylus is slender apically (not clubbed), the penis valve being enlarged. Most species are small, but body length ranges from 4.5 to 10.0 mm. T7 of males varies from a single quadrate projection to trifid to four-toothed. *C. incisulum* Michener is four-toothed but the two median teeth are partly united, thus approaching the three-toothed condition. Male genitalia of American species were illustrated by Michener (1938b).
■ This subgenus is widespread in the palearctic region, although absent in China and Japan, and occurs also in the western part of North America, from Baja California, Mexico, to Washington and east to Utah, USA. There are 11 species in the palearctic region, according to van der Zanden (in litt., 1993), and eight in North America; all native American *Chelostoma* except *C. (Prochelostoma) philadelphi* (Robertson) fall in this subgenus. American species of the subgenus were revised by Michener (1938b) and Hurd and Michener (1955); Warncke (1991c) gave a key to Turkish species of the subgenus as he defined it.

Warncke (1991c) did not include such species as *Chelostoma campanularum* (Kirby), *distinctum* (Stoeckert), and *ventrale* Schletterer in *Foveosmia*. The pitted basal propodeal zone may distinguish *Foveosmia* from such forms, but since this character varies within *Chelostoma* s. str. and elsewhere, I do not think it alone is an appropriate subgeneric character.

The North American species are probably all oligolectic on *Phacelia* and *Eriodictyon* (Hydrophyllaceae) (personal observation), whereas European species are oligolectic on *Campanula* (Campanulaceae) (see Westrich, 1989).

Chelostoma / Subgenus *Gyrodromella* Michener

Gyrodroma Thomson, 1872: 259 (not Klug, 1807). Type species: *Heriades nigricornis* Nylander, 1848 = *Heriades rapunculi* Lepeletier, 1841, by designation of Cockerell, 1925b: 205.

Chelostoma (Gyrodromella) Michener, 1997: 27, replacement for *Gyrodroma* Thomson, 1872. Type species: *Heriades nigricornis* Nylander, 1848 = *Heriades rapunculi* Lepeletier, 1841, autobasic and by original designation.

This subgenus differs from all the others in the strong preoccipital carina, the short propodeum, and the broadly concave anterior surface of T1. The mandible of the female is relatively short, three-toothed. The body length ranges from 6.5 to 9.0 mm.

■ *Gyrodromella* is found in the palearctic region, and is best known in northern and central Europe. There are about six species. One species, *C. rapunculi* (Lepeletier), is adventive in New York state, USA. The type species has commonly been called *Chelostoma fuliginosum* (Panzer), which, however, is a junior primary homonym in *Apis*.

Like European species of the subgenus *Foveosmia*, *Chelostoma rapunculi* (Lepeletier) is oligolectic on *Campanula* (Campanulaceae) (see Westrich, 1989).

Chelostoma / Subgenus *Prochelostoma* Robertson

Prochelostoma Robertson, 1903a: 167. Type species: *Heriades philadelphi* Robertson, 1891, by original designation.

This subgenus is unusual in lacking a carina on the hind coxa and in lacking the apical fringe on S5 of the male; the latter feature is shared with *Chelostoma (Foveosmia) foveolatum* (Morawitz).

■ This subgenus is found in the eastern half of North America, from New York to Georgia and west to Kansas, USA. The single species is *Chelostoma philadelphi* (Robertson).

Chelostoma philadelphi is oligolectic on *Philadelphus* (Saxifragaceae).

Genus *Haetosmia* Popov

Anthocopa (Haetosmia) Popov, 1952: 104. Type species: *Osmia latipes* Morawitz, 1875 = *Osmia brachyura* Morawitz, 1875, monobasic.

This genus contains small (5-7 mm long), robust, megachiliform bees with abundant pale pubescence. They thus resemble various other groups of desert bees such as *Hoplitis (Pentadentosmia)*. The thorax is rather spherical, as in many *Osmia*-like bees. The parapsidal lines are commonly unusually long, half as long as the tegula or longer. The erect lamella on the pronotal lobe is unique in Osmiinae but unfortunately hidden by hairs in most specimens. The male has a tooth at the side of T6, and T7 is broadly truncate. The female has long, apically curved, capitate hairs arising from the labial palpi, and the front basitarsi are flattened and as broad as the tibia. The male lacks basal flaps on S6 and the inner surface of the gonoforceps has a cluster of long, coarse bristles. The palpal, genitalic, and other characters were illustrated by Peters (1974).

■ These bees range from central Asia to Baluchistan (in Pakistan) and westward to Israel, Egypt, Kenya, and the Canary Islands. Van der Zanden (in litt., 1993) reported three species. The genus was reviewed by Peters (1974).

As for *Hoplosmia*, exclusion of this genus from *Hoplitis*, based on the striking apomorphies of *Haetosmia*, is arbitrary. Although *Hoplitis* in its present status is doubtless paraphyletic, there is no particular group of *Hoplitis* to which *Haetosmia* seems closely related.

Figure 81-7. *Heriades (Neotrypetes) carinata* Cresson. Above, Male; Below, Female. From Michener, McGinley, and Danforth, 1994.

Genus *Heriades* Spinola

This is the largest and commonest genus of small heriadiform megachilids (Fig. 81-7, Pl. 6). Most species are in the range of 4 to 7 mm long, although *Heriades floccifera* Brauns, *mamillifera* Brauns, and *rowlandi* (Cockerell), all from South Africa, attain a length of 10.5 mm. These large forms are in the subgenera *Amboheriades* and *Heriades* s. str.

There is a strong tendency for coarse punctation and strong carinae. The anterior margin of the metanotum is carinate or produced upward as a transverse process, the margin of the carina or process usually attaining or surpassing the level of the scutellum. The anterior surface of T1 is strongly concave, and the omaulus is often strongly carinate. In certain species of three subgenera, *Amboheriades*, *Heriades* s. str., and *Michenerella*, the axillae are produced to angles or spines (as in Fig. 81-9a), but in other species and the other subgenera they are rounded. T7 of the male is entirely hidden by the simple, untoothed T6. The metasoma of the male is curled (Fig. 81-7a) so that at most two sterna are exposed, the apex of T6 being almost in contact with S1 or S2. Females differ from those of *Pseudoheriades* by the lack of an impunctate area on the hypostomal area.

Key to the Subgenera of *Heriades* (Males)
(Modified from Griswold, 1985)

1. S1 produced apically, with a brush of fine hair under margin (except in *Pachyheriades*); S2 depressed basomedially, with a hump or ridge apicolaterally; lateral line of T1 reaching spiracle 2
—. S1 not produced apically, without a brush of hair under margin; S2 not depressed basomedially, without hump or ridge apicolaterally; lateral line of T1 not reaching spiracle 7
2(1). T6 with strong midapical projection below strong rounded carina; S1 produced posteriorly as keeled projection completely covering median part of S2; last antennal segment expanded, wider than preceding segments (central Asia to Israel) .... *H. (Rhopaloheriades)*
—. T6 truncate or emarginate, transverse ridge, if present, apical; S1 produced as unkeeled projection not reaching apex of S2; last antennal segment not wider than other segments 3
3(2). Mandible with two teeth, though upper one may be broad and truncate; gradulus of S4 procurved medially 4
—. Mandible with three teeth; gradulus of S4 straight, not procurved medially 5
4(3). Apex of S6 acutely angled; S2 at most angulate laterally, not notched (North and Central America) ....
.... *H. (Neotrypetes)*
—. Apex of S6 bilobate, truncate, or rounded; S2 distinctly notched laterally (Africa, oriental, palearctic) ....
.... *H. (Michenerella)*
5(3). S2 without distinct fringe, at most with scattered hairs much shorter than hind basitarsal width; apical area of S2 transversely depressed below level of disc; basal zone of propodeum bounded posteriorly by a distinct carina that extends laterally behind spiracle; metanotum shallowly notched medially (Africa) .... *H. (Pachyheriades)*
—. S2 with strong fringe of hair at least as long as width of hind basitarsus; apical area of S2 not depressed below level of disc; basal zone of propodeum not bounded posteriorly by a carina, or carina extremely weak, no carina posterior to spiracle; metanotum not notched medially 6
6(5). T6 with strong longitudinal crest abruptly terminated by a V-shaped, flattened apical area; mandible and hypostoma clothed in dense pubescence obscuring integument (southern Africa) .... *H. (Toxeriades)*
—. T6 without longitudinal crest, contour not abruptly changed or flattened near margin; mandible with sparse hair not concealing surface, but hypostoma with narrow longitudinal strip of short plumose hair obscuring integument, surface otherwise visible (southern Africa) ....
.... *H. (Tyttheriades)*
7(1). Frons and interantennal area with pair of juxtantennal carinae; T6 emarginate medially; T6 with U- or V-shaped raised area medially, without parallel-sided crest; S2 not notched laterally (Africa) .... *H. (Amboheriades)*
—. Frons and interantennal area without juxtantennal carinae; T6 not emarginate medially and/or T6 medially with parallel-sided longitudinal crest (palearctic, Africa, India) .... *H. (Heriades* s. str.)

Key to the Subgenera of *Heriades* (Females)
(Modified from Griswold, 1985)

1. Frons and interantennal area with juxtantennal carinae (as in Fig. 10-3); S6 with short apical spike (Africa) ......
.... *H. (Amboheriades)*
—. Frons and interantennal area without juxtantennal carinae; S6 without apical spike 2
2(1). From North and Central America .... *H. (Neotrypetes)*
—. From the Eastern Hemisphere 3
3(2). Lateral line of T1 short, not reaching spiracle (palearctic, Africa, India) .... *H. (Heriades* s. str.)
—. Lateral line of T1 long, reaching spiracle 4
4(3). Scopa and ventral fringe of hind basitarsus strongly plumose (central Asia to Israel) .... *H. (Rhopaloheriades)*
—. Scopa and ventral fringe of hind basitarsus consisting of simple hairs 5
5(4). Inner surface of fore tibia with a feltlike patch of hair (Africa, oriental, palearctic) ...... *H. (Michenerella)* (in part)
—. Inner surface of fore tibia without a feltlike patch of hair 6
6(5). Basal zone of propodeum horizontal or nearly so, bounded posteriorly by distinct carina, sometimes interrupted medially, that extends laterally behind spiracle 7
—. Basal zone of propodeum sloping, without carina posteriorly, or with indistinct carina not extending laterally behind spiracle 8
7(6). Metanotum with shallow medial notch; scutum without transverse crest of dense, plumose hair anteriorly; acetabular carina prominent on basal half of mandible, abruptly truncated (Africa) .... *H. (Pachyheriades)*
—. Metanotum with strong medial notch; scutum with transverse crest of dense, plumose hair anteriorly; acetabular carina absent or weak, not abruptly truncated (oriental, palearctic) .... *H. (Michenerella)* (in part)
8(6). Mandible with prominent acetabular carina above plane of outer surface of mandible; clypeal margin excavated, exposing base of labrum, thickened; majority of scopal hairs bent posteriorly near apices (southern Africa) .... *H. (Toxeriades)*
—. Mandible without distinct acetabular carina; clypeal margin more or less straight, denticulate, slightly overhanging base of labrum, not thickened; scopal hairs straight, not bent near apices (southern Africa) ....
.... *H. (Tyttheriades)*

Heriades / Subgenus *Amboheriades* Griswold

Heriades (Amboheriades) Griswold, 1998, *in* Griswold and Michener, 1998: 236. Type species: *Heriades canaliculata* Benoist, 1931, by original designation.

This is a subgenus having a particularly elongate, slender body form. The preoccipital carina is present at least laterally. The paired juxtantennal carinae are unique within the genus *Heriades*. The scutum is strongly curved down anteriorly, as in *Heriades* s. str. Illustrations of male genitalia and sterna were given for *H. pogonura* Benoist by Peters (1983).

■ This subgenus is widespread in Africa, from Gabon to Ethiopia, south to Namibia and South Africa. Eleven species were listed by Griswold (1985).

Heriades / Subgenus *Heriades* Spinola s. str.

Heriades Spinola, 1808: 7. Type species: *Apis truncorum* Linnaeus, 1758, designated by Latreille, 1810: 439. [For subsequent designations, see Michener, 1997b.]
Eriades Dalla Torre and Friese, 1895: 69, unjustified emendation of *Heriades* Spinola, 1808.
Trypetes Schenck, 1861: 32, 89. Type species: *Apis truncorum* Linnaeus, 1758, monobasic. [For the date, see Michener, 1986a.]
Heriades (Orientoheriades) Gupta, 1987: 68. Type species: *Heriades orientalis* Gupta, 1987, by original designation.

In this subgenus the preoccipital carina is usually absent and the juxtantennal carinae are absent or indistinct. Illustrations of the male genitalia were provided by Maciel de A. Correia (1980).

■ This subgenus is found throughout Africa, north to northern Europe, east to central Asia, northern India, and the Seychelles. Griswold (1985) listed 42 species, about 35 of them from sub-Saharan Africa. Van der Zanden (in litt., 1993) reported 11 palearctic species. No doubt there is some unrecognized synonymy, but there are many new species, principally in Africa.

In many bee taxa, the shape of the axilla (rounded vs. angulate or produced to a spine) is an important group character. In *Heriades*, however, both spined and obtuse or rounded axillae are found in species of at least three subgenera. Earlier efforts to base subgenera on axillar development did not result in groups that were cohesive on the basis of other characters.

Observations on the biology of *Heriades* s. str. include those by Taylor (1962a), Matthews (1965), Michener (1968b), Macial de A. Correia (1980, 1981), and Westrich (1989) and papers cited therein. The nests, made in preexisting burrows in wood or in stems, consist of series of cells with partitions made of resin, but the partitions are sometimes omitted in *H. (H.) spiniscutis* Cameron (Fig. 7-6; Michener, 1968b). See the summary of nesting biology in Griswold (1985).

Heriades / Subgenus *Michenerella* Krombein

Heriades (Michenerella) Krombein, 1950: 122. Type species: *Heriades paganensis* Yasumatsu, 1942, by original designation.
Heriades (Eutrypetes) Popov, 1955a: 280. Type species: *Heriades turcomanica* Popov, 1955, by original designation.

In this subgenus the preoccipital carina is usually absent and the juxtantennal carinae are absent. The scutal profile is variable, strongly curved down in front or not. A strong fringe on the fore basitarsus of males is common but not universal in *Michenerella*; it is absent in all other Eastern Hemisphere *Heriades*. S6 of the male is bilobed apically, and the gonostylus has a patch of velvetlike hair ventrally. Illustrations of male genitalia and hidden sterna were published by Krombein (1950), Popov (1955a, 1960b, c), and Yasamatsu (1942).

■ This subgenus ranges from Central Asia to Japan, the Mariana Islands, the Philippines, Celebes, and Java westward through India, Turkey, and southeastern Europe [Greece, Yugoslavia, Sicily, etc., *Heriades dalmatica* Maidl] to the whole of Africa (Algeria to Liberia, Gabon, and South Africa). As listed by Griswold (1985), *Michenerella* contains 32 species; 13 were reported as palearctic by van der Zanden (in litt, 1993). Additional species clearly exist.

Heriades (Michenerella) othonis Friese nests in pithy stems (Lieftinck, 1954).

Heriades / Subgenus *Neotrypetes* Robertson

Neotrypetes Robertson, 1918: 91. Type species: *Trypetes productus* Robertson, 1905 = *Megachile variolosa* Cresson, 1872, by original designation.
Heriades (Physostetha) Michener, 1938a: 523. Type species: *Heriades carinatum* Cresson, 1864, by original designation.

Neotrypetes, the only group of *Heriades* in the Americas, like several other subgenera, lacks preoccipital and juxtantennal carinae. The scutum is strongly curved down anteriorly. Illustrations of male genitalia and sterna were presented by Michener (1938a), Timberlake (1947), and Mitchell (1962).

■ This North and Central American subgenus ranges across the continent, and from southern Canada to Panama. There are 13 described species (Griswold, 1985) and about 10 undescribed species, mostly from Mexico, according to T. Griswold. Keys for the identification of North American species were given by Michener (1938a, 1954c), and, for California species, by Hurd and Michener (1955).

The nesting biology of *Heriades carinata* Cresson was investigated by Matthews (1965).

Heriades / Subgenus *Pachyheriades* Griswold

Heriades (Pachyheriades) Griswold, 1998, *in* Griswold and Michener, 1998: 237. Type species: *Eriades langenburgicus* Strand, 1911, by original designation.

As in nearly all *Heriades* s. str., the preoccipital carina and the juxtantennal carinae are absent in this subgenus. The scutum is not strongly bent down anteriorly. The apical depression of the male S2, as though a step down from the rest of the sternum, is unique. Other distinctive features are indicated in the key to subgenera.

■ The subgenus is found in tropical Africa (Equatorial Guinea to Tanzania) south to Natal, in South Africa. Griswold (1985) listed five species and reported five more undescribed species.

Heriades / Subgenus *Rhopaloheriades* Griswold and Michener

Heriades (Rhopaloheriades) Griswold and Michener, 1998: 237. Type species: *Heriades clavicornis* Morawitz, 1875, by original designation.

This subgenus resembles *Pachyheriades*, from which the male differs in having an apical projection on T6, the projection depressed below the rounded carina which at first seems to be the apex of the tergum, and in other characters indicated in the key. Females are unique among *Heriades* in the plumose hair of the scopa and on the hind basitarsus. The distinctive male genitalia and sterna were illustrated by Popov (1956).

■ This subgenus occurs in trans-Caspian Russia and Israel. The only known species is *Heriades clavicornis* Morawitz.

Although it resembles *Pachyheriades* in some features, *Rhopaloheriades* is very different, and moreover comes from a different continent.

Heriades / Subgenus *Toxeriades* Griswold

Heriades (Toxeriades) Griswold, 1998, *in* Griswold and Michener, 1998: 238. Type species: *Heriades apricula* Griswold, 1998, by original designation.

This subgenus includes species that are large and robust by *Heriades* standards; the body length is 7 to 8 mm. As in *Tyttheriades,* the basal zone of the propodeum is poorly defined, there is no carina posterior to the propodeal spiracle, and the fore tibia lacks a velvet patch. The dense plumosity obscuring the clypeus, mandible, and hypostoma in males is unique among *Heriades*. Other characters are indicated in the key to subgenera. Structures were illustrated by Griswold and Michener (1998).

■ *Toxeriades* in known only from South Africa. One species, *Heriades apricula* Griswold, has been named, and three others remain undescribed.

In the type species and two undescribed species, the axilla is spined, the scutellum is acutely angled laterally, and the metanotum is raised medially into an acute point. An undescribed species from the Cape of Good Hope, in South Africa, that has the axilla, scutellum, and metanotum rounded may also belong here.

Heriades / Subgenus *Tyttheriades* Griswold

Heriades (Tyttheriades) Griswold, 1998, *in* Griswold and Michener, 1998: 238. Type species: *Heriades schwarzi* Griswold, 1998, by original designation.

This subgenus includes diminutive forms, 3.5 to 5.0 mm in length, having a poorly defined basal propodeal zone and no carina behind the propodeal spiracle. The fore tibia lacks a velvet patch on the posterior surface. In the males, the mandible is tridentate; the hypostoma has a longitudinal stripe of short, dense, white plumose hairs; and S1 is produced posteriorly and has dense, short hair on the normally hidden dorsal surface. Structures were illustrated by Griswold and Michener (1998).

■ *Tyttheriades* is known from South Africa and Namibia. According to T. Griswold, it contains at least six undescribed species in addition to the type species, *Heriades schwarzi* Griswold.

Genus *Hofferia* Tkalcŭ

Hofferia Tkalcŭ, 1984a: 10. Type species: *Chelostoma schmiedeknechti* Schletterer, 1889, by original designation.

This genus contains elongate (8.5-15.0 mm long) heriadiform bees suggestive of *Chelostoma* in their lack of carinae (except for the complete preoccipital carina), the large exposed T7 of the male, the lack of erect hairs on the labrum, the long proboscis, and other features. T7 of the male lacks the median pit found in *Chelostoma,* but the surface of the tergum has a broad concavity. Distinctive features are the irregular nodulose preapical carina on T6 of the male, the lack of a fringe of long hairs on the inner margin of the mandible of the female (differing in this respect from both *Chelostoma* and some *Afroheriades*), the presence of a median apical spike on S6 of the female (the spike larger than the comparable tooth in some *Stenoheriades* and *Heriades*), and the acute posterior lateral angle of the scutum. The posterior expansion of S1 of the male, with hairs on its upper surface, suggests a relationship to *Protosmia,* the similarity to *Chelostoma* and *Afroheriades* possibly being a result of convergence. Many structures were illustrated by Tkalcŭ (1984a).

Griswold (1985) considered *Hofferia* to be close to *Protosmia*. *Hofferia* differs, however, in the preoccipital carina, the acutely angled posterior lateral scutal angle, etc.

■ This genus occurs in the Mediterranean area, including Morocco, Algeria, Turkey, and southeastern Europe (Bulgaria, Romania, Greece). Four species are known, or, owing to synonymy, only two, according to van der Zanden (in litt., 1993).

As is the case of *Pseudoheriades,* the nests are made in beetle burrows in wood by subdividing the burrow into cells using partitions of resin (Griswold, 1985).

Genus *Hoplitis* Klug

The species of *Hoplitis,* except for a few North American forms, are nonmetallic. Their body form varies from slender and elongate hoplitiform to robust, as in most *Osmia,* i.e., megachiliform; Figure 81-8 shows an intermediate species. A lateral tooth on T6 of the male and a pair of basal flaps on S6 (see below) of the male characterize nearly all species; any bee with such flaps on S6 is a *Hoplitis,* even if it happens to lack the teeth on T6. The scutellum is gently convex or, in forms with a robust thorax, its profile curves down at the posterior margin. The metanotum and the base of the propodeum are sloping or, in forms with a robust thorax, strongly declivous. An account of thoracic and T1 shape was given by Michener (1941a).

Hoplitis is one of the large genera of Osmiini, including among others nearly all of the Old World forms that have been placed in *Anthocopa* and the New World forms placed in *Proteriades*. Michener (1941a) proposed recog-

Figure 81-8. Female of *Hoplitis (Dasyosmia) biscutellae* (Cockerell). From Michener, McGinley, and Danforth, 1994.

nition of two major genera, *Anthocopa* and *Hoplitis*. The former included generally more robust species with the anterior surface of T1 broadly concave and delimited above by a distinct line extending across the upper margin of the concavity. *Hoplitis* in the 1941 sense included more slender species with the anterior surface of T1 convex or flat except for a longitudinal median depression, and the line delimiting the anterior surface weak, short (extending only across the summit of the longitudinal depression), or absent. The form of T1 is associated with the thoracic shape, the robust species having the more declivous metanotum and propodeum and the more spherical thorax, whereas in the more slender species the metanotum and base of the propodeum are more nearly horizontal and the thorax is therefore more cylindrical.

These differences generally segregate two groups. In the Western Hemisphere the distinction between *"Anthocopa"* and *Hoplitis* seemed acceptable in 1941, partly because the American *"Anthocopa"* actually belong nearer *Osmia* and are here placed in a different genus, *Atoposmia*. In the Old World, however, *Anthocopa* and *Hoplitis* seem to merge in such a way as to make recognition of the two as genera arbitrary, although such a classification is favored by specialists on palearctic Osmiini, such as G. van der Zanden. *Anthocopa* proper does not occur in the New World, and is here regarded as a subgenus of *Hoplitis*.

As indicated above, *Hoplitis* as here understood contains all the species in which S6 of the male possesses a pair of basal flaps. These structures are thin, translucent, usually pale, dull or sometimes shiny, and always hairless. Morice (1901) called them "two adjacent flakes of thin white membrane, attached only at their bases." They arise from or near the gradulus, which is near the base of the sternum medially, and lie against the ventral sternal surface, occupying nearly the full width of the sternum. Sometimes, hairs on the sternal surface can be seen through the flaps. They may cover two-thirds or more of the postgradular sternal surface. Such flaps, unknown in other bees, are obviously synapomorphic. The flaps of S6 are present in all subgenera except when stated to be absent.

It might seem reasonable to limit *Hoplitis* to forms possessing these flaps on S6 of males. *Hoplitis* subgenera, however, possess kaleidoscopic variation in their characters. No groups of characters correlated with the flaps on S6 define major sections of the genus. Polarity of characters (except for some autapomorphies) cannot be determined for the whole genus, because even when outgroup comparison identifies a plesiomorphic state, character combinations appear to show that reversion from the derived state to the apparent plesiomorph (really a further apomorphy) has occurred somewhere within the genus. Moreover, most apomorphic states seem to have arisen more than once. When one considers the flaps of S6, the same problems arise. There are species in which the flaps are short; are they ancestrally short, or have they been reduced? Finally, there are a few taxa in which flaps of S6 are absent, yet otherwise the bees' structure is easily within the range of variation of the *Hoplitis* groups that have the flaps. To exclude such taxa from *Hoplitis* on the basis of a single character would be arbitrary, especially since the flaps may well come and go during evolution,

like other characters. Even within the subgenus *Anthocopa*, which ordinarilly has flaps, the flaps were probably lost in *H. (A.) matheranensis* (Michener).

The result is a genus *Hoplitis* that is probably paraphyletic (although in the absence of a phylogeny this is not established). The subgenera *Exanthocopa, Pentadentosmia,* and *Nasutosmia* lack flaps on S6 of males. Indeed, some related taxa called genera could with almost equal justification be included in *Hoplitis*, e.g., *Haetosmia, Hoplosmia,* and *Stenosmia*. These genera lack flaps and have diverse, distinctive characters, some certainly autapomorphies, and show no evidence of close relationship to particular subgenera of *Hoplitis*. Recognizing them as genera rather than as subgenera of *Hoplitis* is thus an arbitrary decision.

Another character of most *Hoplitis* is the lateral tooth of T6 of males, which is absent in most related genera, although present in *Ashmeadiella, Haetosmia,* some *Hoplosmia,* and some *Osmia*. The tooth is sometimes a mere angle in *Hoplitis* subgenera such as *Eurypariella, Formicapis* and *Nasutosmia*. It is completely absent in *Hoplitis (Megahoplitis)*, which, however, is a large form with well-developed flaps on S6.

One section of *Hoplitis* includes the species formerly placed in the genus *Proteriades,* here regarded as the subgenera *Proteriades* and *Penteriades* of *Hoplitis*. They tend to be robust and thus to approach *Anthocopa* in form, although nearly all are small, with red on the metasoma. They have short mouthparts with strong, erect hairs on the galeal blades and first two segments of the labial palpi; and the apices of these hairs are hooked or wavy (though the hooked hairs are nearly absent in males of *Penteriades*). The proboscis, armed with such hairs, serves to pull pollen out of the minute flowers of *Cryptantha* (Boraginaceae). Probably associated with the shortening of the proboscis is the reduction of the maxillary palpi to two to four segments in *Proteriades,* although the palpi retain five segments in *Penteriades*. For many years it seemed that *Proteriades* could be defined as a genus on the strength of its mouthpart characters, even though they are not fully developed in males of *Penteriades*. Similarly robust forms of the subgenus *Acrosmia,* however, as well as more elongate hoplitiform species of the subgenus *Hoplitina,* the latter long regarded as a subgenus of *Hoplitis,* have most of the characters of *Proteriades* except the short mouthparts with the hooked or wavy hairs (Michener and Sokal, 1957). Considering *Acrosmia* and *Hoplitina* along with the five-segmented maxillary palpi and lack of hooked hairs in males of *Penteriades,* one recognizes that there is intergradation between *Proteriades* and *Hoplitis*. The former is therefore included within the latter.

Gogala (1995b) has separated *Hoplitis* s. str. (no doubt along with *Annosmia*) as a genus from *Osmia, Hoplosmia,* and the rest of *Hoplitis,* which he assembled under the genus name *Osmia,* on the basis of the translucent basal lateral structures at the labroclypeal articulation (see couplet 1 of the key to subgenera of the Eastern Hemisphere). In *Osmia* in his sense there are usually two or more pairs of brushes of short, yellowish or reddish, simple hairs under the clypeal margin of females, whereas such brushes are absent in *Hoplitis* in his sense, the marginal hairs being directed downward. Although these are useful observations, the brushes are completely absent in some groups, such as *Hoplitis (Robertsonella),* that lack the translucent labroclypeal projections.

The American species of *Hoplitis* were revised by Michener (1947), by Timberlake and Michener (1950) for *Proteriades* in its traditional sense, and by Hurd and Michener (1955). Wu (1987b) gave a key to Chinese species. Van der Zanden (1988a) listed palearctic species.

Nests of *Hoplitis* are very diverse. Some species of the subgenus *Alcidamea* nest in pithy stems, and make partitions between cells from leaf pulp, sometimes supplemented with pith particles or pebbles (Parker, 1975b). Bees of the subgenus *Formicapis* make similar nests (Clement and Rust, 1975). Other *Hoplitis* nest in holes in wood, sometimes [as in *H. (Monumetha) fulgida* (Cresson)] making partitions from masticated leaf material and pebbles or bits of wood (Clement and Rust, 1976). *Hoplitis (Dasyosmia) biscutellae* (Cockerell) constructs cells of resin and plant parts within abandoned cells of the wasp *Sceliphron* (Rust, 1980a) and in holes made by other insects in earthen banks. Even within the subgenus *Hoplitis* s. str. there is much diversity in nesting behavior, for *H. (H.) adunca* (Panzer) nests in holes in wood or stems, or in the soil, making partitions from sand, pebbles, and clay, while *H. (H.) anthocopoides* (Schenck) makes exposed nests of pebbles and mortar made of dry soil and saliva (Eickwort, 1973, 1975). Nests of at least some species of the subgenus *Anthocopa* are short burrows in the soil, each usually ending in a single cell lined with petals and closed with mud. Species making such nests were reported by Friese (1923), Cros (1937), Michener (1968c), and others; the most famous such species is *H. (Anthocopa) papaveris* (Latreille). Westrich (1989), Eickwort (1975), and Michener (1968a) describe nests and cells of various species, and give references to earlier works. Van der Zanden's (1988a) list will serve to show which of Westrich's "*Osmia*"species belong in *Hoplitis* (and in *Anthocopa,* here considered part of *Hoplitis*). In some cases, more details are given below in the accounts of subgenera.

In couplet 13 of the first key below, the options separate males and females. The female of *Exanthocopa* is not known to me but would run to couplet 30.

Key to the Subgenera of *Hoplitis* of the Eastern Hemisphere
(From Griswold and Michener, 1998)

1. Base of labrum exposed between closed mandibles and lower margin of clypeus; lateral basal extremity of labrum connected to lateral clypeal margin by yellowish membrane extending downward beside labral base (mandible of female rather slender, distinctly narrower at or before middle than at base) 2
—. Base of labrum not or scarcely visible between closed mandibles and lower margin of clypeus [except in females of species like *H. (Anthocopa) cristatula* (Zanden) that have greatly modified clypeus and mandibles; in *H. (Microhoplitis);* and in males of *Anthocopa* in which the basal labral zone is visible]; lateral basal extremity of labrum without yellowish membrane 13
2(1). Males.. 3
—. Females 8
3(2). T7 bidentate, deeply emarginate between teeth 4
—. T7 truncate, pointed, rounded, or tridentate, occasionally very shallowly emarginate medially but without deep medial notch.. 5
4(3). Mandible tridentate; S6 with truncate or bilobed apical extension densely haired throughout (palearctic to Sudan) .... *H. (Annosmia)*
—. Mandible bidentate; S6 with longitudinal row of dense hairs on medial lobate extension (Iran, Baluchistan)...... .... *H. (Coloplitis)*
5(3). T7 tridentate, though lateral teeth may be reduced to right angles so that one sees principally a long median point; S6 with weak longitudinal ridge apically (eastern palearctic) .... *H. (Eurypariella)*
—. T7 rounded or truncate; S6 without longitudinal ridge 6
6(5). T6 truncately produced between lateral teeth, margin of truncation with at least some irregularities, often crenulate; S6 with broad apical hyaline extension (palearctic) .... *H. (Hoplitis s. str.)*
—. T6 roundly produced between lateral teeth, margin not irregular; S6 without differentiated apical extension........ 7
7(6). T7 very shallowly emarginate medially; clypeal margin weakly crenulate without narrow apical band (eastern Mediterranean) .... *H. (Bytinskia)*
—. T7 truncate or rounded, not emarginate; clypeal margin smooth medially with narrow, impunctate apical band (central Asia)..................................... *H. (Kumobia)*
8(2). Mandible four-toothed; *either* clypeus thickened and modified *or* scopa reduced, longest hairs shorter than exposed parts of sterna 9
—. Mandible three-toothed; clypeus not modified and scopa not reduced 10
9(8). Clypeus thickened, modified; scopa not reduced, sternal hairs erect (central Asia) .... *H. (Kumobia)*
—. Clypeus neither thickened nor modified; scopa reduced, sternal hairs appressed, shorter than exposed parts of sterna (eastern Mediterranean; cleptoparasitic).......... .... *H. (Bytinskia)*
10(8). S6 with submarginal carina, sometimes extremely weak and visible only laterally, so that females may not differ appreciably from those of *H. (Annosmia)* (palearctic) .... *H. (Hoplitis s. str.)*
—. S6 without submarginal carina 11
11(10). Labial palpus two-segmented, with abundant long, curled, capitate hair (Iran, Baluchistan) ....*H. (Coloplitis)*
—. Labial palpus four-segmented, without hair or at most with scattered, short, straight hairs 12
12(11). Thorax steeply declivous behind posterior part of scutellum; apical spines of fore and middle tibiae each simple (palearctic to Sudan) .... *H. (Annosmia)*
—. Thorax slanting posteriorly to posterior part of basal area of propodeum, behind which propodeum is steeply declivous; apical spines of fore and middle tibiae robust, each bifid or with preapical shoulder (eastern palearctic) .... *H. (Eurypariella)*
13(1). Males 14
—. Females.. 27
14(13). T6 without lateral tooth; body length 13 mm or more (prestigma longer than stigma; T7 trifid) (Mediterranean region) .... *H. (Megahoplitis)*
—. T6 with lateral tooth, sometimes weak, a mere angle in some small species (e.g., 6-8 mm long) 15

15(14). T7 deeply trifid, margins of middle tooth carinate to base; S6 without basal membranous flaps (T/ without middorsal pit) (palearctic south to Sudan) .... *H. (Pentadentosmia)*
—. T7 pointed, truncate, rounded, bilobed, or four-lobed [except with three strong angles in *H. (Prionohoplitis) curvipes* (Morawitz) and trifid in *H. (Alcidamea) tridentata* (Dufour and Perris), in which middle tooth is not margined]; S6 with basal membranous flaps [except in *Exanthocopa, Nasutosmia*, and *H. (Anthocopa) matheranensis* (Michener)] 16

16(15). T7 medially bilobed, four-lobed, or produced and subtruncate, without middorsal pit 17
—. T7 pointed, trifid, or rounded, commonly with middorsal pit 20

17(16). T1 without basal basin, with longitudinal depression not delimited above by transverse line or carina; T6 with low, obtuse lateral angle; T7 four-lobed, i.e., bilobed medially with large lateral tooth (northern holarctic) .... *H. (Formicapis)*
—. T1 with basal flat or concave area delimited above at least medially by transverse line; T6 with strong, acute lateral tooth; T7 bilobed or produced and subtruncate [almost four-lobed in *H. (A.) matheranensis* (Michener) and *H. (A.) singularis* Morawitz)] 18

18(17). S6 without differentiated triangular area (palearctic, Africa, India) .... *H. (Anthocopa)*
—. S6 with differentiated median triangular area either margined by sharp carinae or projecting apically beyond rest of sternum 19

19(18). Flagellum slender, nearly cylindrical, flagellar segments 2-5 longer than broad, terminal flagellar segment simple; midtarsal segment 2 without elongate apical projection (northern Africa to Turkey) .... *H. (Platosmia)*
—. Flagellum broad, concave ventrally, flagellar segments 2-5 twice as broad as long, terminal flagellar segment with apical buttonlike projection; midtarsal segment 2 with elongate anterior apical projection (Turkey) .... *H. (Chlidoplitis)*

20(16). Exposed part of S2 nearly half as long as metasoma, with strong basal transverse elevation and produced, truncate apex hiding median part of S3, which is strongly emarginate, fringed medially; S8 with two apical horns forming deep median emargination (Central and northeastern Asia) .... *H. (Jaxartinula)*
—. Exposed part of S2 much shorter, without elevation or tubercle at its base, S3 thus exposed medially, neither emarginate nor fringed medially; S8 not emarginate apically 21

21(20). S6 without membranous basal flaps (T7 with median apical point) 22
—. S6 with membranous basal flaps appressed against undersurface of sternum (without transparent lateral inflexed flaps) 23

22(21). T7 with shallow middorsal depression or pit; S6 without lateral flaps; anterior surface of T1 with broad concave basin delimited above by transverse line and containing longitudinal median groove (northern Africa) .. .. *H. (Exanthocopa)*
—. T7 without middorsal depression or pit; S6 with transparent inflexed flaps attached along lateral margins; T1 with anterior surface rather flat except for longitudinal medial groove not delimited above by transverse line (Mediterranean region) .... *H. (Nasutosmia)*

23(21). T7 scarcely exserted beyond T6, broadly rounded, without middorsal pit; S1 to S5 unmodified (Canary Islands) .... *H. (Microhoplitis)*
—. T7 strongly produced beyond T6, variable in shape, often with middorsal pit or depression; sterna variously modified 24

24(23). T1 with distinct angle separating anterior and dorsal surfaces, angle extending across at least median half of width of tergum, anterior surface of T1 a broad basin; preoccipital ridge carinate or sharply angled (mandible bidentate) (palearctic) .... *H. (Prionohoplitis)*
—. T1 without angle separating anterior and dorsal surfaces, or with angle extending less than half of tergal width, anterior surface of T1 without distinct broad basin; preoccipital ridge rounded [except *H. (Alcidamea) tridentata* (Dufour and Perris), in which mandible is tridentate] 25

25(24). S2 bearing strong transverse ridge with rather sharp crest; wings with dark papillae beyond closed cells; body with abundant yellow hair forming broad, distinct metasomal tergal bands (palearctic) .... *H. (Megalosmia)*
—. S2 without strong transverse ridge, unmodified or with large tubercle; wings with weak papillae; body hairs not yellow, not forming broad, dense hair bands 26

26(25). S8 elongate, distal process ligulate, downcurved, undersurface of process with short modified hairs; S6 with midapical tuft of spreading hairs (holarctic) .... *H. (Monumetha)*
—. S8 shorter, without long, downcurved distal process, hairs simple; S6 without apical tuft of spreading hairs (holarctic) .... *H. (Alcidamea)*

27(13). T1 with distinct angle separating anterior and dorsal surfaces, angle extending across at least median half of width of tergum; anterior surface of T1 a broad basin; fore basitarsus often with plumose hair 28
—. T1 without angle separating anterior and dorsal surfaces, or angle extending less than half of tergal width [nearly half in *H. (Pentadentosmia) rufopicta* (Morawitz)]; anterior surface of T1 usually with longitudinal median depression; fore basitarsus without plumose hair 31

28(27). Preoccipital ridge carinate or sharply angled (palearctic) .... *H. (Prionohoplitis)*
—. Preoccipital ridge rounded.. 29

29(28). Hypostoma shiny, with sparse punctures and hairs, but without differentiated area behind mandibular base and without fringe demarcating this area; fore basitarsus without plumose hairs (Canary Islands) .... *H. (Microhoplitis)*
—. Hypostoma with somewhat differentiated, impunctate to sparsely punctate, shiny area near mandibular base, hair in this area sparse or absent [except in *H. (Anthocopa) bisulca* (Gerstaecker)], this area margined laterally by strong fringe of long curled hair [fringe weak in *H. (A.) bisulca* and *singularis* (Morawitz)]; fore basitarsus usually with plumose hair 30

30(29). Clypeal margin with narrow, parallel-sided median notch; hind tibial spurs stout, strongly hooked apically; fore basitarsus with simple hair; outer apex of middle tibia with acute spine (Turkey) .... *H. (Chlidoplitis)*
—. Clypeal margin without notch, rarely with wide, angled

emargination; hind tibial spurs slender, straight; fore basitarsus with plumose hair [but hair absent in *H. (A.) furcula* (Morawitz) and *H. (A.) picicornis* (Morawitz) and sparse and not evident in worn specimens of *H. (A.) matheranensis* (Michener)]; outer apex of middle tibia with narrow to broadly rounded lobe (palearctic, Africa, India) .... *II. (Anthocopa)*

31(27). Clypeus not overhanging labrum, with anteriorly projecting snout 32
—. Clypeus overhanging labrum, without anteriorly projecting snout 33

32(31). Mandible four-toothed; propodeum and anterior surface of T1 shiny (Mediterranean region) ....
.... *H. (Nasutosmia)*
—. Mandible three-toothed (long, straight or undulate edge between middle and upper tooth); propodeum and anterior surface of T1 shagreened (northern holarctic)
.... *H. (Formicapis)*

33(31). Mandible with long, slightly convex margin between second and uppermost teeth, thus possibly four-toothed; median part of metanotum elevated and strongly convex, attaining level of scutellar convexity, so that line tangent to scutal and scutellar convexities is also tangent to metanotum, as seen in profile (Central and northeastern Asia) .... *H. (Jaxartinula)*
—. Mandible without long margin between second and uppermost teeth; median part of metanotum not elevated and thus below line tangent to scutal and scutellar convexities, as seen in profile 34

34(33). Mandible four-toothed 35
—. Mandible three-toothed 36

35(34). Wings with dark papillae beyond closed cells; metasoma with distinct broad yellow hair bands (palearctic)
.... *H. (Megalosmia)* (in part)
—. Wings with weak papillae beyond closed cells, papillae not dark; metasoma with weak white hair bands interrupted medially (holarctic) .... *H. (Monumetha)*

36(34). Mandible greatly broadened apically, twice as wide as at narrowest point or nearly so; abductor swelling of mandible large, light-colored... 37
—. Mandible scarcely broadened apically, little wider than at narrowest point; abductor swelling of mandible small, dark 38

37(36). Wings clear or only lightly stained; pubescence white or off-white; frons below anterior ocellus without impunctate polished area (palearctic south to Sudan) ....
.... *H. (Pentadentosmia)*
—. Wings strongly stained, especially darkened in marginal cell; pubescence yellow; frons below anterior ocellus impunctate and polished (palearctic) ....
.... *H. (Megalosmia)* (in part)

38(36). Apex of front basitarsus and second tarsal segment with strong, upcurved, capitate setae on outer surface; upper mandibular tooth truncate (Mediterranean region) .... *H. (Megahoplitis)*
—. Front tarsus without such modified setae; upper mandicular tooth acute 39

39(38). Proboscis in repose not or scarcely extending beyond fossa; clypeus quite flat, not convex above, somewhat flared outward to broad truncation; suture between clypeus and supraclypeal area weak; body form robust, suggesting subgenus *Anthocopa* (northern Africa to Turkey) .... *H. (Platosmia)*
—. Proboscis in repose extending at least a little beyond fossa; clypeus usually somewhat convex above, usually not flared toward truncation [except in *H. tridentata* (Dufour and Perris), *mitis* (Nylander), etc.]; suture between clypeus and supraclypeal area usually distinct; body form slender (holarctic) .... *H. (Alcidamea)*

Key to the Subgenera of *Hoplitis* of the Western Hemisphere (Males)

1. S6 with longitudinal median hairy ridge, sometimes extending as median process beyond rest of sternum or consisting primarily of a midapical hairy process; T7 bilobed, sometimes only weakly so [except tridentate with lateral spine longer than median spine in *H. (Proteriades) xerophila* (Cockerell)]; metasoma commonly partly or wholly red, never metallic 2
—. S6 without longitudinal median hairy ridge or midapical process (but with a tuft of hairs in *Monumetha*); T7 ending in a truncation, convexity, or point [tridentate with median spine longer than lateral spine in *Cyrtosmia* and *H. (Dasyosmia) biscutellae* (Cockerell)]; metasoma black or brownish black, sometimes metallic 5

2(1). Galeal blade and first two segments of labial palpus with numerous strong hairs that are hooked or wavy apically (western nearctic) .... *H. (Proteriades)*
—. Galeal blade and labial palpus without hooked or wavy hairs (or, in *Penteriades,* with very few)............................ 3

3(2). Basal half of flagellum with some hairs as long as flagellar diameter; apex of last antennal segment produced laterally (western nearctic) .... *H. (Acrosmia)*
—. Flagellum with all hairs short; apex of last antennal segment unmodified or rarely tapering to a curved point....
....4

4(3). Proboscis scarcely extending beyond proboscidial fossa in repose; hind coxa with longitudinal carina on inner ventral angle (western nearctic) .... *H. (Penteriades)*
—. Proboscis extending well beyond proboscidial fossa in repose; hind coxa without longitudinal carina (western nearctic) .... *H. (Hoplitina)*

5(1). Mandible tridentate; T6 with median apical as well as lateral tooth; T7 tapering with three small adjacent teeth (western nearctic) .... *H. (Cyrtosmia)*
—. Mandible bidentate; T6 without median apical tooth except in *Dasyosmia;* T7 pointed, rounded, or truncate or, if tridentate (one species of *Dasyosmia*), then not tapering, teeth widely separated 6

6(5). Metanotum approximately on same level as scutellum, a line tangent to scutum and scutellum in profile thus nearly touching metanotum; stigma distinctly broader than prestigma, as measured to wing margin 7
—. Metanotum depressed below level of scutellum, which curves down posteriorly to meet metanotum, a line tangent to scutum and scutellum in profile thus well above metanotum; stigma smaller, about as broad as prestigma, as measured to wing margin 8

7(6). First recurrent vein nearly meeting first submarginal crossvein; T7 four-lobed (northern holarctic)................
.... *H. (Formicapis)*
—. First recurrent vein considerably distal to first submarginal crossvein; T7 rounded (eastern nearctic) ....
.... *H. (Robertsonella)*

8(6). Hind coxa with large, flattened ventral tooth; T6 with median apical angle or tooth (western nearctic) .... H. (Dasyosmia)
—. Hind coxa without tooth; T6 without median apical angle or tooth 9
9(8). Lateral basal extremity of labrum connected to clypeal margin by yellowish membrane extending downward beside labral base; S1 to S5 not modified (introduced, New York State)... H. (Hoplitis s. str.)
—. Lateral basal angles of labrum without yellowish membrane; one or more of S1 to S5 with projection, spine, or other modification 10
10(9). S8 elongate, distal process downcurved, ligulate, undersurface with short, modified hairs; S6 with midapical tuft of spreading hairs; antennal pedicel exposed [partly hidden in *H. (M.) spoliata* (Provancher)] (holarctic) .... H. (Monumetha)
—. S8 without long, downcurved distal process, its hairs simple; S6 without tuft of spreading hairs; antennal pedicel largely hidden by invagination into apex of scape (holarctic) .... H. (Alcidamea)

Key to the Subgenera of *Hoplitis* of the Western Hemisphere (Females)

Galeal blade and first two segments of labial palpus with numerous strong hairs, these hooked or wavy apically; proboscis short, in repose scarcely extending out of proboscidial fossa (T6 not strongly concave in profile; posterior coxa with longitudinal carina on inner ventral angle; metasoma usually partly or wholly red) 2
—. Galeal blade and labial palpus without hooked hairs; proboscis longer, in repose extending well beyond limit of proboscidial fossa (except in *Acrosmia*)......................... 3
2(1). Maxillary palpus five-segmented; clypeus with upper two-thirds strongly swollen and shining, nearly impunctate medially (western nearctic) .... H. (Penteriades)
—. Maxillary palpus two- to four-segmented; clypeus uniformly convex, punctures moderately dense to crowded (western nearctic) .... H. (Proteriades)
3(1). Metanotum approximately on same level as scutellum, a line tangent to scutum and scutellum in profile thus nearly touching metanotum; stigma distinctly broader than prestigma, as measured to wing margin 4
—. Metanotum depressed below level of scutellum, which curves down posteriorly to meet metanotum, a line tangent to scutum and scutellum in profile thus well above metanotum; stigma smaller, about as broad as prestigma, as measured to wing margin 5
4(3). Mandibular apical width nearly half of mandibular length; mandible tridentate; lower margin of clypeus entire; first recurrent vein considerably distal to first submarginal crossvein (eastern nearctic) .... H. (Robertsonella)
—. Mandibular apical margin more than half of mandibular length; mandible with two lower apical teeth followed by long undulate margin; lower margin of clypeus with median snoutlike projection; first recurrent vein nearly meeting first submarginal crossvein (northern holarctic) .... H. (Formicapis)
5(3). Mandible clearly four-toothed (holarctic) .... H. (Monumetha)
—. Mandible three-toothed, sometimes with a weak convexity between second and third teeth 6

6(5). Median mandibular tooth almost twice as far from upper tooth as from lower tooth, with a weak convexity between median and upper teeth, the convexity sometimes worn so as to yield a long undulating margin above median tooth; S6 with longitudinal median ridge ending in apical tooth (western nearctic) .... H. (Cyrtosmia)
—. Median mandibular tooth more nearly midway between upper and lower teeth; S6 without median longitudinal ridge 7
7(6). Proboscis short, in repose not extending beyond proboscidial fossa (western nearctic) .... H. (Acrosmia)
—. Proboscis longer, in repose extending well beyond proboscidial fossa 8
8(7). T6 nearly straight in profile; metasoma partly red (western nearctic) .... H. (Hoplitina)
—. T6 distinctly concave in profile; metasoma without red 9
9(8). Apex of mandible nearly as broad as eye; distal part of fore wing with minute papillae, median part with very few hairs (western nearctic) .... H. (Dasyosmia)
—. Apex of mandible much narrower than eye; fore wing uniformly finely hairy, without papillate areas 10
10(9). Base of labrum exposed when mandibles are closed; lateral basal extremity of labrum connected to clypeal margin by yellowish membrane extending downward beside labral base (introduced, New York State) .... H. (Hoplitis s. str.)
—. Base of labrum hidden when mandibles are closed; lateral basal angles of labrum without yellowish membrane (holarctic) .... H. (Alcidamea)

Hoplitis / Subgenus *Acrosmia* Michener

Hoplitis (Acrosmia) Michener, 1947: 298. Type species: *Hoplitis plagiostoma* Michener, 1947, by original designation.

Acrosmia, a member of the *Proteriades* group of subgenera, is rather robust (like *Proteriades*), 7.5 to 8.5 mm in body length, usually black but the metasoma sometimes red, and the pubescence pale. The antenna of the male is distinctive, with a few long hairs on at least the basal half of the flagellum and with the last flagellar segment produced to one side. The clypeus of the male is unusually short and broad; that of the female has a large basal convexity suggestive of that of *Penteriades*. Although the proboscis lacks the specialized hairs characteristic of *Proteriades*, it is short as in that subgenus, not extending beyond the proboscidial fossa in repose.

■ This subgenus is known from the mountains of California to Washington state, east to Idaho and Wyoming, USA. The five species were revised by Griswold (1983).

Parker (1978a) described nests of *Hoplitis plagiostoma* Michener in pithy stems; partitions and end plugs were made of masticated plant material containing sand, soil particles, or seeds.

Hoplitis / Subgenus *Alcidamea* Cresson

Alcidamea Cresson, 1864: 385. Type species: *Alcidamea producta* Cresson, 1864, by designation of Michener, 1941a: 158.
Osmia (Acanthosmia) Thomson, 1872: 233. [Species first included by Schmiedeknecht, 1885: 21.] Type species: *Os-*

mia montivaga Morawitz, 1872, by designation of Sandhouse, 1943: 522.

Osmia (Liosmia) Thomson, 1872: 233, no included species; Friese, 1911b: 438, included species. Type species: *Osmia claviventris* Thomson, 1872, by designation of Michener, 1941a: 159.

Osmia (Tridentosmia) Schmiedeknecht, 1885: 21. Type species: *Osmia tridentata* Dufour and Perris, 1840, by designation of Michener, 1941a: 159.

Autochelostoma Sladen, 1916a: 270. Type species: *Autochelostoma canadensis* Sladen, 1916 = *Alcidamea producta* Cresson, 1864, monobasic.

Heriades (Micreriades) Mavromoustakis, 1958: 444. Type species: *Heriades parnesica* Mavromoustakis, 1958, by original designation.

This subgenus consists of small to rather large species (6-14 mm long), the hair pale and usually forming tergal fasciae. Typically, they are slender-bodied, hoplitiform. The mandible of the female is robust, three-toothed, thus unlike that of *Monumetha*, except that there is a fourth or uppermost tooth close to the third in the species that were placed in *Micreriades*. The mandible of the male is short, bidentate or occasionally tridentate. T6 of the male has lateral teeth but is otherwise simple. Male genitalia and sterna were illustrated by Popov (1960d), Mitchell (1962), and Tkalců (1995). T7 of the male is variable, and four groups can be recognized on the basis of T7. They are as follows:

(a) T7 rounded at apex, without a dorsal pit: a small group including *H. simplicicornis* (Morawitz) and *leucomelana* (Kirby).

(b) T7 trifid, without a dorsal pit: *H. tridentata* (Dufour and Perris).

(c) T7 with central projection and lower lateral teeth or shoulders, thus approaching the trifid condition, with a dorsal pit: *H. acuticornis* (Dufour and Perris) and *praestans* (Morawitz).

(d) T7 pointed, with a dorsal pit: the great majority of the species.

■ This subgenus occurs in central Europe but is more abundant and diverse in southern Europe, the Mediterranean basin, and southwestern Asia and ranges eastward to central Asia. In North America, *Alcidamea* ranges from coast to coast and from southern Canada to Florida and Texas, USA, and Sonora and Baja California, Mexico. There are about 37 species in the Old World (Zanden, 1988a), or 63, according to van der Zanden (in litt., 1993), and nine species in North America. American species were revised by Michener (1947), the western American species by Hurd and Michener (1955). Warncke (1991d) gave a key to 24 species from Turkey; some of these species belong elsewhere, e.g., in *Prionohoplitis*.

Both S1 and S2 of the male may be simple, but there is often a median protuberance or spine that can be on either S1 or S2, or on both, as in *Hoplitis pungens* (Benoist). *Hoplitis tridentata* (Dufour and Perris) has a preoccipital carina, not developed in other species. The antennae of males are sometimes unmodified but are sometimes thickened (the scape as well as the flagellum) and the last segment produced to a point, as in *H. producta* (Cresson) and its relatives.

Most species of *Alcidamea* nest in pithy stems and make partitions of chewed leaf material, sometimes supplemented with pith particles or pebbles (Michener, 1955; Parker, 1975b; Westrich, 1989). Large programs of trap nesting using holes in wood (e.g., Krombein, 1967) do not recover *Alcidamea* nests; presumably, stems are preferred sites.

Hoplitis / Subgenus *Annosmia* Warncke

Osmia (Annosmia) Warncke, 1991b: 307. Type species: *Osmia annulata* Latreille, 1811, by original designation.

As in *Bytinskia, Coloplitis,* and *Kumobia,* the mandibular-clypeal-labral organization of this subgenus—including the slender mandibles and the yellowish membrane at the side of the labrum—is similar to that of *Hoplitis* s. str. *Annosmia* differs from *Hoplitis* s. str. in consisting of small (5-11 mm long), generally robust hoplitiform species with all hairs pale, forming strong tergal fasciae. T6 of the male is simple except for the lateral teeth, and is not truncately produced as in *Hoplitis* s. str. S6 of the female lacks the marginal carina present in *Hoplitis* s. str. but the carina is weak to virtually absent in some species such as *H. carinata* (Stanek) and *marchali* (Pérez). Thus females are not always distinguishable from those of *Hoplitis* s. str. T7 of the male is bifid, not truncate or rounded as in *Hoplitis* s. str. The male mandible is long, pale, and three-toothed, not short, dark, and bidentate as in *Hoplitis* s. str. The last flagellar segment of males is slightly hooked apicoventrally, and S6 of most males has a densely hairy, membranous apical extension. The male gonostylus is at least as wide as the penis valves, somewhat flattened apically, and covered with moderately long hairs. The hypostomal area of females, behind the mandible, is sometimes smooth as in the subgenus *Anthocopa;* it is flat and usually surrounded laterally by hairs that curl over it, even if it is punctate. The apex of the clypeus of females is often crenulate.

■ This subgenus occurs from the Mediterranean basin east to central Asia, southeast to Baluchistan (in Pakistan) and south to northern Sudan. Warncke (1991b) reviewed the subgenus, gave a key to Turkish species, and recognized 27 species; van der Zanden (in litt., 1993) reported 31 species.

Hoplitis / Subgenus *Anthocopa* Lepeletier and Serville

Anthocopa Lepeletier and Serville, 1825: 314. Type species: *Apis papaveris* Latreille, 1799, monobasic.

Phyllotoma Duméril, 1860: 842 (not Leach, 1819). Type species: *Apis papaveris* Latreille, 1799, by designation of Michener, 1941a: 160.

Pseudosmia Radoszkowski, 1872a: xviii, no included species; Radoszkowski, 1874b: 152, included species. Type species: *Megachile cristata* Fonscolombe, 1846 (not Dufour, 1841) = *Anthocopa cristatula* Zanden, 1990, by designation of Cockerell, 1922b: 6.

Pseudo-osmia Radoszkowski, 1874a: 137, unjustified emendation of *Pseudosmia* Radoszkowski, 1872.

Pseudoosmia Radoszkowski, 1874b: 152, unjustified emendation of *Pseudosmia* Radoszkowski, 1872.

Osmia (Arctosmia) Schmiedeknecht, 1885: 21. Type species:

Megachile villosa Schenck, 1853 (not Apis villosa Fabricius, 1775) = Osmia platycera Gerstaecker, 1869, monobasic.

Osmia (Furcosmia) Schmiedeknecht, 1885: 22. Type species: Apis papaveris Latreille, 1799, by designation of Cockerell, 1922b: 6.

Pseudocosmia Radoszkowski, 1886: 14, unjustified emendation of Pseudosmia Radoszkowski, 1872.

Osmia (Lithosmia) Alfken, 1935: 188. Type species: Megachile villosa Schenck, 1853 (not Apis villosa Fabricius, 1775) = Osmia platycera Gerstaecker, 1869.

Osmia (Glossosmia) Michener, 1943: 84. Type species: Osmia singularis Morawitz, 1875, by original designation.

These are robust, megachiliform or chalicodomiform, often hairy species, the hairs pale [red in Hoplitis (A.) ursina (Friese)] and often forming metasomal fasciae. The size is variable (7-13 mm long). The anterior surface of T1 is a broad basin demarcated above by a distinct line. The mandible of the female is three-toothed or four-toothed, robust, i.e., not much narrower medially than at the base. In the hypostomal area of the female is a large region behind the mandible that is shiny, largely impunctate, not or sparsely haired, and limited laterally by long hairs that curve over the shiny area. The clypeus of the female is broad, truncate or somewhat rounded, and far overhanging the base of the labrum except in H. (A.) cristatula (Zanden), which has grotesquely modified mandibles and a shortened, protuberant clypeus exposing the base of the labrum. The mandible of the male is three-toothed or, less commonly, two-toothed. Most but not all females have the following additional characteristics: front basitarsus with plumose hairs, hind tibial spurs straight, hind tibia densely haired dorsally, scopal hairs plumose, and S6 flat with dense marginal hairs. In most males S6 is well sclerotized (visible without dissection), S8 is quadrate (except in "Arctosmia"), and the gonostylus is slender, tapering apically, with long hairs on all surfaces (except in "Arctosmia"). Male genitalia and other structures were illustrated by Tkalců (1969a).

■ This subgenus occurs in Central Europe but is especially diverse in the Mediterranean basin and eastward through southwestern Asia; it is much less numerous in central Asia, but one species is described from the region of Lake Baikal. Southward, a few species occur in East Africa and on to South Africa, and an unusual species, Hoplitis matheranensis (Michener), occurs in southern India. Van der Zanden (1988a) listed 57 palearctic species and reported 62 in litt. (1993); more than a dozen others occur in sub-Saharan Africa.

The small groups named Glossosmia, Arctosmia and Pseudosmia are similar to and probably derived from the larger group, Anthocopa in a narrow sense. Hoplitis cristatula (Zanden), as noted above, has greatly modified female clypeus and mandibles and has therefore been placed in a separate subgenus, Pseudosmia. Such modifications occur in some species of diverse other osmiine taxa, e.g., Osmia (Helicosmia and Osmia s. str.) and do not seem to be appropriate bases for dismembering an otherwise rather homogeneous group. Moreover, H. serrilabris (Morawitz) shows much less clypeal modification, but is nonetheless intermediate between H. cristatula and those species that have an unmodified clypeus like that of H. papaveris (Latreille). Hoplitis (Anthocopa) matheranensis (Michener)

from southern India is unusual in almost lacking the basal flaps on S6 of the male, there being only a very narrow thin strip attached to the sternum in place of a flap, and in the simple hairs of the scopa. Its assignment to Anthocopa must be considered tentative.

As noted in the account of the genus above, nests of at least some species of the subgenus Anthocopa are shallow burrows in hard soil or banks, each burrow ending in a single cell. Cells are lined with petals, often brightly colored, as with Hoplitis papaveris (Latreille), which uses red poppy petals. (References in account of genus Hoplitis.)

Hoplitis / Subgenus *Bytinskia* Mavromoustakis

Bytinskia Mavromoustakis, 1954a: 269. Type species: Bytinskia erythrogastra Mavromoustakis, 1954, by original designation.

This appears to be a cleptoparasitic subgenus, to judge from the short scopa. There is no evidence that the name is based on sexually anomalous individuals. The tradition of giving generic rank to parasitic forms (except in Bombus and the Allodapini) is clearly inappropriate when the parasites are similar to and probably derived from existing nonparasitic taxa.

This subgenus consists of small (7-8 mm long), robust species. The hairs, all pale, form metasomal tergal fasciae. The clypeus is not produced over the base of the labrum, its truncation being broad and crenulate as in the subgenus Microplitis. The mandible of the female is pale, long, four-toothed; that of the male is bidentate. The anterior surface of T1 is shallowly concave with a line across the summit of the concavity medially. T6 of the male has a weak lateral angle; this and the basal flaps of S6 indicate placement in the genus Hoplitis. T7 is gently bilobed.

■ This subgenus is known from Turkey, Israel, and Kenya. Four species were reported by van der Zanden (in litt., 1993); the species from Kenya is an undescribed fifth.

According to Mavromoustakis (1954a) the host of Hoplitis erythrogastra (Mavromoustakis) in Israel is H. (Annosmia) sordida (Benoist).

Hoplitis / Subgenus *Chlidoplitis* Griswold

Hoplitis (Chlidoplitis) Griswold, 1998, in Griswold and Michener, 1998: 223. Type species: Hoplitis heinrichi Zanden, 1980, by original designation.

Chlidoplitis consists of robust hoplitiform or almost megachiliform species 7.5 to 9.5 mm in body length. In the past, the species have been included in the subgenus Hoplitis s. str., but although males show similarities to Hoplitis s. str., females more nearly resemble the subgenus Anthocopa. Males resemble those of Hoplitis s. str. in the shape of T6 and T7, the membranous S6, the flattened flagellum and the shortened clypeus that does not overhang the labrum. They differ from Hoplitis s. str. in the lack of yellowish membranous areas at the basal lateral angles of the labrum, the strongly carinate hind coxae, and the strong apical fringe of S4. Females resemble Anthocopa in the distinct anterior basin of T1 with its dorsal carina, the stout mandible, and the presence of a hypostomal basket formed by long hairs curving mesad over the area behind the mandible. They differ from Anthocopa in that this area is closely punctate, not more or less smooth as in *Anthocopa*. They differ further in the simple rather than plumose scopal hairs, the stout and strongly hooked hind tibial spurs, and the pointed rather than rounded outer apical projection of the middle tibia. Unique characters in the male include the buttonlike extension on the last flagellar segment, the arrow-shaped apical extension of S6, and the spinose apical extensions of the first two midtarsal segments. Females are remarkable in that the posterior surface of the propodeum is entirely polished, the posterior lines of the propodeal enclosure are evanescent, and the midclypeal margin is sharply grooved.

■ *Chlidoplitis* is known only from Turkey. The two known species are *Hoplitis heinrichi* Zanden and *illustris* Zanden.

Hoplitis / Subgenus *Coloplitis* Griswold

Hoplitis (Coloplitis) Griswold, 1998, *in* Griswold and Michener, 1998: 223. Type species: *Hoplitis premordica* Griswold, 1998, by original designation.

Coloplitis consists of small, robust hoplitiform species 6 to 7 mm in body length. It belongs to the complex of subgenera within *Hoplitis* that includes *Hoplitis* s. str. and *Annosmia*, the complex characterized by the yellowish membranous material at the basal lateral angles of the labrum (couplet 1 of the key to subgenera of the Eastern Hemisphere). One of the two known species, *Hoplitis persica* Warncke, was formerly included in *Annosmia* (Warncke, 1991b), and males do share the bidentate T7 with *Annosmia*. *Coloplitis*, however, is unique among the Osmiini in the reduction of the labial palpus from four segments to two, the second segment being long, slender, and tapered, and in females provided with hairs that are hooked basad. Such hairs occur also on the distal part of the galea. Similar curled hairs on the labial palpus are found elsewhere in *Haetosmia*, in some *Protosmia (Nanosmia)*, and in a complex of nearctic *Hoplitis* formerly known as *Proteriades*. Males of *Coloplitis* are further distinguishable by the combination of bidentate mandible, bidentate T7, and the nature of S6, with its convex margin and the longitudinal patch of hair on its disc.

■ *Coloplitis* is known from Iran and Baluchistan, Pakistan. The two known species are *Hoplitis persica* (Warncke) and *premordica* Griswold.

Hoplitis / Subgenus *Cyrtosmia* Michener

Hoplitis (Cyrtosmia) Michener, 1947: 292. Type species: *Osmia hypocrita* Cockerell, 1906, by original designation.

This subgenus contains a rather large (10-13 mm long), elongate, hoplitiform species, black with partly black pubescence. Thus it resembles superficially *Hoplitis (Monumetha) albifrons* (Kirby), from which it differs in numerous characters other than those listed in the keys to subgenera. For example, the male antennae are long, reaching the scutellum, the second flagellar segment being twice as long as broad and the apical one being pointed. S5 of the male is short and broad, without specialized hairs. The clypeus of the female is flat and has a longitudinal median impunctate band. S7 of the male tapers to a narrow trifid apex.

■ *Cyrtosmia* is found from British Columbia, Canada, to California, east to Utah, Colorado, and Arizona, USA. The only species is *Hoplitis hypocrita* (Cockerell).

Hoplitis / Subgenus *Dasyosmia* Michener

Hoplitis (Dasyosmia) Michener, 1947: 294. Type species: *Alcidamea biscutellae* Cockerell, 1897, by original designation.

Dasyosmia includes large (9-13 mm long), elongate, hoplitiform, rather hairy species having the form of *Monumetha* and *Cyrtosmia*. Unlike that of those subgenera, the pubescence is all pale. The mandible of the female is tridentate, but sometimes shows a weak convexity between the middle and upper teeth. The distinctly papillate distal parts of the wings and the sparsely hairy or partly bare middle and basal parts are unique for American *Hoplitis*. The flat ventral tooth on the underside of the hind coxa of the male is not the homologue of the hind coxal carina of diverse other Osmiini; it appears to arise from the outer ventral angle of the coxa rather than the inner ventral angle.

■ This subgenus is found in deserts from western Texas to eastern California, north to Utah, USA, and probably south into Mexico. The two species differ strikingly; T7 of the male, for example, is strongly trifid in *Hoplitis biscutellae* (Cockerell) but pointed in *H. paroselae* Michener. Revisions were made by Michener (1947) and Hurd and Michener (1955).

Hoplitis / Subgenus *Eurypariella* Tkalcŭ

Eurypariella Tkalcŭ, 1995: 111. Type species: *Osmia denudata* Morawitz, 1880, by original designation.

This group, recognized as a separate genus by Tkalcŭ (1995), consists of small, slender hoplitiform species, 6 to 7 mm in body length. The stigma is relatively large, suggesting the *Heriades* group, the margins diverging toward the base of vein r. With only the single female at hand it is difficult to be sure of the characters of couplet 1 of the subgeneric key to females, but my judgment appears to be supported by Tkalcŭ's description. The thoracic shape and tibial spur characters (see the key to females) apply to both sexes. The maxillary palpus is five-segmented, with the last segment minute; Tkalcŭ's statement that it is three-segmented is possibly an error. S2-S5 are unmodified, but S6 has rather short, broad basal flaps and a midapical longitudinal ridge. The lateral teeth of T6 are represented by weak convexities, almost absent; T7 has a slender, tapering median apical process and, lateral to it, either a similar process or a right-angular projection.

■ *Euryporiella* is known from China, Mongolia, and Turkestan. The only described species is *Hoplitis denudata* (Morawitz), but two other presumably undescribed species exist.

Hoplitis / Subgenus *Exanthocopa* Tkalcŭ

Anthocopa (Exanthocopa) Tkalcŭ, 1993a: 55. Type species: *Osmia oxypyga* Benoist, 1927, by original designation.

A single megachiliform species that looks like an *Anthocopa* constitutes this subgenus; it is known only in the male, which is 11 to 12 mm long and bears very narrow apical bands of white hair on the metasomal terga. The mandible is less than twice as long as broad, three-toothed. The anterior surface of T1 is strongly concave, the concavity margined above by a line that is fully half as long as the width of the tergum, an *Anthocopa*-like feature. Within the concavity is a longitudinal median depression. The hind coxa is not carinate, S6 completely lacks flaps, and T7 is sharply pointed with a shoulder at each side and with a median depression too shallow to be called a pit; these are features that distinguish it from *Anthocopa*. Van der Zanden (1985) illustrated structures of the type species.

■ *Exanthocopa* is known from Morocco and Algeria. The only species is *Hoplitis oxypyga* (Benoist).

Hoplitis / Subgenus *Formicapis* Sladen

Formicapis Sladen, 1916a: 271. Type species: *Formicapis clypeata* Sladen, 1916 = *Osmia robusta* Nylander, 1848, monobasic.

This subgenus contains a small, holarctic species, 6 to 8 mm long. The pubescence is pale and the tergal bands of pale hair are narrow or absent. The body is hoplitiform, shaped like a small *Alcidamea* except that the head of the female is very large, the clypeus with a median apical projection resembling that of *Nasutosmia*. Unlike that of all other *Hoplitis,* the mandible of the female has an extraordinarily broad apex with two apical teeth, above which is a long, irregular, but roughly straight edge ending at the upper apical angle. The tooth at the side of T6 of the male is a mere angle. T7 of the male has two median apical lobes and a broad apicolateral tooth, the whole structure thus describable as four-lobed. S6 of the male, hidden in repose, possesses basal flaps, as is usual in *Hoplitis*. Numerous structures were illustrated by Popov (1960a) and Mitchell (1962).

■ *Formicapis* occurs in the Alps, Finland, east to Baikal, Yakutia, and the Altai in Russia and the Hantey in Mongolia, and, in North America, Alaska, USA, to Quebec, Canada, south in the mountains to California and Colorado, USA. It contains a single boreal holarctic species, *Hoplitis robusta* (Nylander).

Hoplitis / Subgenus *Hoplitina* Cockerell

Hoplitella Cockerell, 1910d: 169 (not Levinsen, 1909). Type species: *Hoplitella pentamera* Cockerell, 1910 = *Ashmeadiella howardi* Cockerell, 1910, monobasic.
Hoplitina Cockerell, 1913: 34, replacement for *Hoplitella* Cockerell, 1910. Type species: *Hoplitella pentamera* Cockerell, 1910 = *Ashmeadiella howardi* Cockerell, 1910, autobasic and by original designation.

Hoplitina consists of small hoplitiform species (body length 4.5-8.0 mm) with the metasoma partly red and the pubescence pale. As in *Proteriades,* T7 of the male is bilobed (at least feebly so) and S6 has a longitudinal median hairy ridge. The mouthparts, however, are moderately long and lack hooked or wavy hairs. The clypeus of the female is often convex and shiny basally, as in *Penteriades* and *Acrosmia*.

■ This subgenus is known only from California, Arizona, and Nevada, USA. There are six species. Revisions were by Michener (1947) and Hurd and Michener (1955); Parker (1979) gave a key to the species.

A nest of *Hoplitis bunocephala* (Michener) was found by Thorp (1968) in a beetle burrow in an oak twig; the cells formed a series separated by partitions of masticated plant material. Nests of other species were found in pithy stems by Parker (1978a). Construction material for partitions was sand grains and pebbles plus masticated plant parts.

Hoplitis / Subgenus *Hoplitis* Klug s. str.

Hoplitis Klug, 1807b: 225. Type species: *Apis adunca* Panzer, 1798, monobasic.
Osmia (Ctenosmia) Thomson, 1872: 233. Type species: *Apis adunca* Panzer, 1798, by designation of Michener, 1941a: 158.

This subgenus contains small to rather large (7-15 mm long) species with rather slender, parallel-sided bodies (Pl. 6). Nonetheless, the T1 basin is distinct, demarcated above by a distinct line. The mandible of the female is three-toothed, rather slender, distinctly narrower near the middle than at the base. This is the large group of rather ordinary-looking species in which the labral base is exposed when the mandibles are closed in females and in which there is a yellowish membrane at the lateral extremity of the labrum in both sexes. T7 of the male is rounded or subtruncate. Excellent illustrations of this character and of genitalia and sterna were published by Eickwort (1970).

■ This subgenus is found from Central Europe to northern Africa, and from the Canary Islands, Spain, and Britain east to Central Asia. *Hoplitis anthocopoides* (Schenck), a palearctic species, is established in New York State, USA. There are about 43 species (53 were listed by van der Zanden, 1988a, who included species here placed in *Annosmia*). Warncke (1992c) gave a key to about 18 species.

The antennae of males vary considerably. Some are quite ordinary; in others the scape is thickened and the basal part of the flagellum is broad and flattened, then tapering to an ordinary apex [*Hoplitis (H.) adunca* (Panzer)], or the last segment may be tapered and pointed [*H. (H.) pici* (Friese)].

Nests are made in existing hollows, including hollow stems and holes such as those made by other bees in banks [*Hoplitis adunca* (Panzer); see Westrich, 1989] or are constructed in the open on rocks, cliffs, etc. [*H. anthocopoides* (Schenck); see Eickwort, 1973, 1975]. Cells are made of pebbles, sand, and clay, the last serving as mortar. At least the two species mentioned above are oligolectic on *Echium* (Boraginaceae).

Hoplitis / Subgenus *Jaxartinula* Popov

Jaxartinula Popov, 1963: 865. Type species: *Jaxartinula malyshevi* Popov, 1963, by original designation.

Jaxartinula consists of hoplitiform species about 7.5 mm in body length that superficially resemble *Hoplitis (Alcidamea) pilosifrons* (Cresson). The high level of the metanotum (see the key to subgenera) as well as the long margin between the second and the upper mandibular teeth suggest the subgenus *Formicapis*. The metanotum, however, is very strongly convex in profile, unlike that of *Formicapis*. The basal zone of the propodeum is subhorizontal, with short longitudinal striae, but is not defined by a carina or sharp angle. As in *Formicapis* the first recurrent vein meets the first submarginal crossvein. T6 of the male has a large lateral tooth; T7 is produced and rounded, without a dorsal pit. Other features are indicated in the key to subgenera. Illustrations are provided by Popov (1963), Tkalců (1995), and Griswold and Michener (1998).

■ This subgenus is known from Central Asia (Kazakhstan and Uzbekistan) and from Mongolia. The two species were compared by Tkalců (1995).

Hoplitis / Subgenus *Kumobia* Popov

Kumobia Popov, 1962a: 300. Type species: *Osmia tenuicornis* Morawitz, 1875, by original designation.

This subgenus contains small (5.5-7.0 mm long), rather megachiliform species with both unusually large heads and divergent eyes in females. The clypeus of the female is variously modified. The hairs are pale and form apical tergal bands. The mandibles of the female are long and similar to those of *Bytinskia* (see the key to subgenera). The mandibles of males are long, pale, and two-toothed or, in *Hoplitis (Kumobia) tenuicornis* (Morawitz), three-toothed. The labrum of the female lacks the long marginal hairs found in all other *Hoplitis*. T7 of the male is small and either rounded or bearing a thin, translucent, midapical projection; T6 has a lateral tooth. Popov (1962a) and Tkalců (1995) provided illustrations of genitalia and sterna, as well as other structures.

■ *Kumobia* occurs from Mongolia to Central Asia and Turkestan. There are four species.

The large head, modified clypeus, and divergent eyes of the females are not likely to be satisfactory group characters. Elsewhere in the Osmiinae one finds such features in a few but not all species of several subgenera, e.g., *Osmia* s. str. and *Helicosmia* in the genus *Osmia* and *Anthocopa* in the genus *Hoplitis*. Some of the females placed tentatively in *Annosmia* may turn out to be facially unmodified *Kumobia* when males and more material are available. Van der Zanden (1988a) seems to have reached a similar conclusion, for he included a facially unmodified species in *Kumobia*.

Superficially, species of *Kumobia* are extremely similar to the North American genus *Atoposmia*, which they resemble further (at least in the type species) in the *Osmia*-like enlargement of S2 of the male, covering most of S3, which is broadly emarginate and fringed medially. In another specimen, apparently of the type species, S2 appears more transverse and S3 is exposed but emarginate and fringed. *Kumobia* differs from *Atoposmia* in the exposed base of the labrum with the yellowish membrane laterally, as in *Hoplitis* s. str., and in the presence of basal flaps on S6 of the male, as in most of the genus *Hoplitis*.

Hoplitis / Subgenus *Megahoplitis* Tkalců

Hoplitis (Megahoplitis) Tkalců, 1993a: 55. Type species: *Osmia tigrina* Morawitz, 1872, by original designation.

This subgenus includes a large (length 13 mm), robust chalicodomiform, heavily sclerotized, densely punctate (including tegulae) species with all hairs pale and forming distinct tergal fasciae. Its most noteworthy feature is that although the male has the basal flaps of S6, it completely lacks the lateral tooth of T6 characteristic of other *Hoplitis*, T6 being quite simple. T7 is robustly trifid, the median tooth blunt, and there is no dorsal pit. Thus T7 resembles that of *H. (Alcidamea) tridentata* (Dufour and Perris), but the insect is so different in other features that the relation to *tridentata* seems distant. The clypeus of the female is broad and truncate, overhangs the base of the labrum, and has a preapical band of dense orange hair. The mandible of the female is robust, somewhat narrowed medially, and three-toothed; that of the male is bidentate. The anterior surface of T1 is broadly concave, delimited medially by a line above. An unusual feature is that the stigma is small for an osmiine, parallel-sided like that of a megachiline.

■ This subgenus occurs in northern Africa, Greece, and Turkey. The only species is *Hoplitis tigrina* (Morawitz).

Hoplitis / Subgenus *Megalosmia* Schmiedeknecht

Osmia (Megalosmia) Schmiedeknecht, 1885: 23. Type species: *Osmia grandis* Morawitz, 1873, by designation of Sandhouse, 1943: 568.
Hoplitis (Calohoplitis) Tkalců, 1995: 129. Type species: *Osmia laboriosa* Smith, 1878, by original designation.

These are large (10-18mm), robust forms with abundant yellow hairs forming apical tergal fasciae. T1 has a broad basin on the anterior surface, delimited above medially by a line. The clypeus of the female is flat, truncate with rounded angles, the edge thin and somewhat flared outward, the clypeus overhanging the base of the labrum, suggesting that of *Pentadentosmia*, *Platosmia*, and some species of *Anthocopa*. The mandible of the female is narrowed medially, considerably expanded apically, and three-toothed or four-toothed. T6 of the male is simple except for the lateral tooth, and has an apical hair band like that of the preceding terga, an unusual feature. T7 of the male has a slender, sharp midapical point and a dorsal pit (suggesting *Alcidamea*); in some species there are also lateral shoulders, the tergum thus approaching the trifid condition. Figures of the male genitalia, sterna, and other structures were given by Popov (1962b) and Tkalců (1995). Popov emphasized a relationship of *Megalosmia* to *Monumetha*, a relationship supported, for example, by the specialized setae on S8 of males. That sternum in *Megalosmia*, however, lacks the long, curved apical process of *Monumetha*.

■ This subgenus ranges from Eastern Europe to China. There are four species.

Griswold and Michener (1998) justified placement of *Calohoplitis* as a synonym of *Megalosmia*, in spite of differences such as three-toothed rather than four-toothed female mandibles.

Hoplitis / Subgenus *Microhoplitis* Tkalců

Osmia (Microhoplitis) Tkalců, 1993b: 811. Type species: *Osmia hohmanni* Tkalců, 1993 = *Osmia zandeni* Teunissen and van Achterberg, 1992, by original designation.

Microhoplitis consists of a single small species (body length 5.5 mm) with sparse pale pubescence. The head of the female is rather large. The convex, shiny, largely impunctate female clypeus ends in a somewhat thickened, irregular margin bearing some brassy hairs, the margin not covering the labral base, which is slightly exposed above the closed mandibles. The apical part of the female mandible is broad, broader than the base, and much broader than the narrowed premedian region. The four-toothed female mandibles are unusual among small *Hoplitis*. *Microhoplitis* differs from *Nasutosmia* and the *Micreriades* group of *Alcidamea*, which also have four-toothed female mandibles, by the clypeal characters described above, the broadly concave anterior surface of T1, the simple, transverse and unmodified S1 to S5 of the male, the short T7 of the male, and the carinate hind coxa of the male. Tkalcŭ (1993b) and Teunissen and van Achterberg (1992) illustrated the male genitalia and other structures.

■ This subgenus is known only from the Canary Islands, where a single species, *Hoplitis zandeni* Teunissen and van Achterberg, occurs.

Hoplitis / Subgenus *Monumetha* Cresson

Monumetha Cresson, 1864: 387. Type species: *Chelostoma albifrons* Kirby, 1837, by designation of Titus, 1904a: 26.

Andronicus Cresson, 1864: 384. Type species: *Andronicus cylindricus* Cresson, 1864 (not *Osmia cylindrica* Giraud, 1857) = *Osmia spoliata* Provancher, 1888, monobasic.

Chlorosmia Sladen, 1916a: 270. Type species: *Osmia fulgida* Cresson, 1864, monobasic.

This subgenus contains elongate hoplitiform (8-14 mm long) species having the body form of most species of *Alcidamea*, from which *Monumetha* differs in the four-toothed mandible of females and the male sternal characters indicated in the key to subgenera. The body has at least some black or fuscous hairs (at least on the metasoma) in addition to pale hairs. The clypeus of the female is rather convex above, truncate apically, and overlaps the base of the labrum. The female mandible is more robust than in *Hoplitis* s. str., although slightly narrowed medially. The male mandible is bidentate. S2 of the male has a large median prominence in *H. (M.) tuberculata* (Nylander), as in some *Alcidamea*, but such a prominence is lacking in other species. T6 of the male is simple except for its lateral teeth. T7 is rounded or subtruncate with a shallow dorsal pit or, in *H. (M.) albifrons* (Kirby), with a transverse depression. Male genitalia and sterna were illustrated by Mitchell (1962).

■ This subgenus is widespread in North America (north as far as subarctic Alaska) and occurs also in Northern and Central Europe. All but one of the six species are American. American species were revised by Michener (1947); western species were revised by Hurd and Michener (1955).

The single palearctic species, *Hoplitis tuberculata* (Nylander), has a longitudinal median keel on S6 of the female. This character is absent in American species and has led to placement of *H. tuberculata* in the subgenus *Cyrtosmia*, an American subgenus. Its relationship seems to be clearly with *Monumetha*, however, as shown by the four-toothed female mandible, the downcurved, ligulate apical process with modified hairs of S8 of the male, etc.; the carina of S6 of the female must be convergent with that of *Cyrtosmia*. The antennae of males vary from rather elongate and unmodified, as in *H. fulgida* (Cresson), to forms with the scape swollen and the basal flagellar segments expanded, as in *H. spoliata* (Provancher). *H. albifrons* Kirby is intermediate in this respect.

Hoplitis fulgida (Cresson) and its relatives are unique among *Hoplitis* for being brilliant metallic green or blue-green, like the brilliantly metallic species of *Osmia (Melanosmia)*.

Nests are in holes in wood or stems; cell partitions are made of masticated leaf material plus pebbles (Fye, 1965).

Hoplitis / Subgenus *Nasutosmia* Griswold and Michener

Hoplitis (Nasutosmia) Griswold and Michener, 1998: 225. Type species: *Osmia nasuta* Friese, 1899, by original designation.

This subgenus contains small (5-6 mm long), slender hoplitiform species, superficially like small species of the subgenus *Alcidamea*. The pale pubescence sometimes forms narrow apical tergal fasciae. As in many *Alcidamea*, the anterior surface of T1 is smooth, rather flat, and has a longitudinal median groove; this groove is not delimited above by a transverse line. *Nasutosmia* differs from *Alcidamea* and most other *Hoplitis* in the absence of basal flaps on S6 of the male; in this respect it resembles *Pentadentosmia*. T6 of the male is simple, except for the small lateral tooth. T7 is pointed, without the median pit that is common in *Alcidamea*. In the male the more posterior sterna are weakly sclerotized, and the metasoma can be curled so that only S1 and S2 are clearly exposed; this feature suggests *Heriades*, *Protosmia*, and related genera. The mandible of the female is robust and four-toothed (three-toothed in most *Alcidamea*). The clypeus of the female has a strong midapical projection, on the underside of which are the tufts of orange hair (thus suggestive of *Formicapis*).

■ This subgenus occurs in northern Africa and Spain. There are probably two species, *Hoplitis nasuta* (Friese) and *corniculata* Zanden. The species were compared and illustrated by van der Zanden (1989). Some subgeneric characters have not been verified for *H. corniculata*.

Probably because of their small size and slender form, *Nasutosmia* species often have been placed in the subgenus *Micreriades*, here included in *Alcidamea*, but they are quite different from that subgenus.

Hoplitis / Subgenus *Pentadentosmia* Warncke

Osmia (Pentadentosmia) Warncke, 1991a: 14. Type species: *Osmia quinquespinosa* Friese, 1899, by original designation.

This subgenus includes a series of small species (5-11 mm long) that were formerly placed, along with *H. (Alcidamea) tridentata* (Dufour and Perris), in *Tridentosmia* because of the trifid T7 of males. The species of *Pentadentosmia* (along with the very different *Exanthocopa*,

Nasutosmia, and one species of *Anthocopa*), however, unlike other Old World *Hoplitis,* lack the basal flaps of S6 of males that otherwise characterize *Hoplitis.* In *Pentadentosmia* the pubescence is all pale and forms strong tergal bands; the body is broad and *Anthocopa*-like, i.e., megachiliform; the anterior surface of T1 is broadly concave or rather flat and defined above medially by a transverse line. The clypeus of the female is large, flat, often shining; the lower edge is thin, rounded or subtruncate with rounded angles, much overlapping the base of the labrum.

■ This subgenus includes mostly desert forms that range through the Mediterranean basin, west to the Canary Islands, south to the Sudan, eastward to Central Asia and Baluchistan. The 17 west palearctic species were reviewed by Warncke (1991a); van der Zanden (in litt., 1993) indicated that there is a total of 24 species.

In two species, *Hoplitis (P.) karakalensis* Popov and *jejuna* Popov, the integument is red instead of black.

Hoplitis / Subgenus *Penteriades* Michener and Sokal

Proteriades (Penteriades) Michener and Sokal, 1957: 158.
    Type species: *Osmia remotula* Cockerell, by original designation.

This subgenus consists of small (length 5.0-6.5 mm) *Proteriades*-like species with the metasoma usually partly red; hooked hairs are scarcely developed on the proboscis of males, but present on that of females. The upper part of the clypeus of the female is convex and shining, sparsely punctate. The maxillary palpi are five-segmented, as in most members of the *Osmia* group of genera.

■ *Penteriades* occurs from California east to South Dakota, Wyoming, and Arizona, USA. The two species were revised, along with those of *Proteriades,* by Timberlake and Michener (1950) and Hurd and Michener (1955).

A nest of *Hoplitis incanescens* (Cockerell) was found in a pithy stem by Parker (1978a). Partitions were made of sand grains and masticated plant parts.

Hoplitis / Subgenus *Platosmia* Warncke

Osmia (Platosmia) Warncke, 1990: 482. Type species: *Osmia platalea* Warncke, 1990, by original designation.

The species of this subgenus are megachiliform, like the subgenus *Anthocopa,* although small (6-10 mm long); they are pale-haired, and with or without weak metasomal hair bands. As in some *Anthocopa,* the clypeus of the female is large, flat, the lower margin broadly truncate, much overhanging the base of the labrum. *Platosmia* females differ from females of *Anthocopa* in the punctate hypostomal area (except for a small area near the mandibular base), the curved hairs surrounding it laterally neither numerous nor long. The mandibles of the female are narrowed premedially, as in *Hoplitis* s. str., not short and robust as in *Anthocopa.* The mandible of the male is three-toothed. T6 has a feeble, small median emargination, an *Osmia*-like feature not found elsewhere in *Hoplitis.* T7 is sometimes bifid, in a style that is common in *Osmia,* but may also be simple and pointed (see illustrations by Tkalcŭ, 1995). S6 is hidden in repose, and has the basal flaps characteristic of *Hoplitis.*

■ *Platosmia* occurs from Morocco eastward across northern Africa to Turkey, and to Mongolia. Warncke (1990) revised the subgenus; he included eight species.

Hoplitis / Subgenus *Prionohoplitis* Tkalcŭ

Hoplitis (Prionohoplitis) Tkalcŭ, 1993a: 55. Type species: *Osmia curvipes* Morawitz, 1871, by original designation.

In its megachiliform body form this subgenus, particularly in the female, resembles *Anthocopa,* although it is probably more closely related to *Alcidamea,* where it was placed by Warncke (1991d). Apical hair bands on the metasomal terga are conspicuous, and the body length is 10 to 15 mm. This subgenus was proposed for two closely related species, *Hoplitis curvipes* (Morawitz) and *H. insolita* (Benoist), the two characterized by such striking features as an anteriorly greatly broadened proboscidial fossa and the serrate to almost pectinate basal half of the male flagellum. *Prionohoplitis* has been expanded by Griswold and Michener (1998) to include similarly robust, coarsely punctate species such as *H. brachypogon* (Pérez). Males of these species show considerable variation in metasomal structure (comparable diversity is present in *Alcidamea*), but in all the preoccipital ridge is carinate or sharply angled, the anterior and dorsal surfaces of T1 are separated by a strong carina, the scutellum is strongly rounded in profile with the posterior surface vertical, the thorax is broad and short with the mesopleuron ventrally no longer than the length of the middle coxa [a character approached in *H. (Alcidamea) mitis* (Nylander)], and the body is coarsely punctate (this accentuated in females). In females the mandible is tridentate, the teeth equidistant and the lower and upper interspaces similarly notched. In females of *Alcidamea* the upper interspace is broader and more shallowly notched than the lower. In most species the pronotal lobe is carinate. Structures of *H. insolita* (Benoist) were sketched by van der Zanden (1985).

■ *Prionohoplitis* occurs in the Mediterranean basin, from Spain and Algeria eastward, thence to Iran and Kazakhstan. Six species have been placed in this subgenus (Griswold and Michener, 1998).

Hoplitis / Subgenus *Proteriades* Titus

Proteriades Titus, 1904a: 25. Type species: *Proteriades semirubra* Cockerell, 1898, monobasic.
Chelostoma (Cephalapis) Cockerell, 1910e: 23. Type species: *Chelostoma jacintanum* Cockerell, 1910, monobasic.
Anthocopa (Xerosmia) Michener, 1943: 81. Type species: *Osmia xerophila* Cockerell, 1935, by original designation.

This subgenus consists of small (3.0-10.5 mm long), usually robust species with pale pubescence and the metasoma usually with red areas. *Hoplitis (P.) jacintana* (Cockerell), however, has a rather elongate hoplitiform body, partly because of the posterior development of the head. The proboscis is short, and in both sexes the galea and first two labial palpal segments have erect hairs that are hooked or wavy at the tips; the maxillary palpi are two- to four-segmented. Male genitalia, sterna, and other structures were illustrated by Parker (1976, 1977c).

■ *Proteriades* occurs in California, Nevada, and New Mexico, USA, south to Coahuila and Baja California, Mexico. The 22 species were revised by Timberlake and Michener (1950) and by Hurd and Michener (1955).

The names *Cephalapis* and *Xerosmia* were proposed for single species that have numerous special derived features. There seems to be little point in giving such forms subgeneric recognition, although a second *Xerosmia*, *Hoplitis (Proteriades) zuni* (Parker), has been found.

Like species of the subgenus *Penteriades*, those of *Proteriades* visit the minute flowers of *Cryptantha* (Boraginaceae), drawing the pollen out with the hooked hairs on the proboscis. Nests of *Proteriades* have been found in holes in the ground, in old bee cells in a bank, and in holes in galls and stems. Parker (1976, 1977c, 1978a) described numerous nests in pithy stems. Partitions and nest closures are made of sand grains or pebbles and masticated plant parts, or, for a few species, resin and masticated plant material or sand.

Hoplitis Subgenus *Robertsonella* Titus

Robertsonella Titus, 1904a: 22. Type species: *Robertsonella gleasoni* Titus, 1904, by original designation.

Robertsonella contains small (5-8 mm long) black species with pale hair, having about the same slender hoplitiform aspect as small species of the subgenus *Alcidamea*. The thorax, however, has a more elongate form than that of *Alcidamea*, as shown by the elevated metanotum, reaching almost to the level of the scutellum and scutum, as indicated in the keys to subgenera. Among American *Hoplitis*, only *Formicapis* has a similar thorax. In the male, seven terga and six sterna are exposed; T7 is rounded posteriorly. On S6 of the male the basal flaps are very short, although recognizable. An unusual male character is the extremely dense, short white hair of the clypeus, completely concealing the clypeal surface and superficially suggesting the pale clypeus common in males of many other bees.

■ This subgenus is found in eastern North America from Connecticut to Georgia west to central Texas and Kansas. The three species were differentiated by Mitchell (1962).

Genus *Hoplosmia* Thomson

This genus consists of somewhat robust hoplitiform, nonmetallic bees, much like *Hoplitis* but with acutely pointed axillae (Fig. 81-9a). The pale pubescence usually forms tergal fasciae; body length is 6.5 to 10.0 mm. Unlike those of most *Hoplitis*, the maxillary palpi are four-segmented. T6 of the male is sometimes toothed laterally, as in most *Hoplitis*. A distinctive feature is the preapical, commonly nodulose or multidentate carina or thickening on T6 of the male. T7 of the male is variable: pointed, rounded, subtruncate, or bilobed (Fig. 81-9b, c). In females, except *H. fallax* (Pérez), the scopal hairs are crinkled, a feature not found in related taxa. Supporting the exclusion of *Hoplosmia* from *Hoplitis* is the lack of basal flaps on S6 of the male. Tkalců (1974c) illustrated male genitalia and other structures.

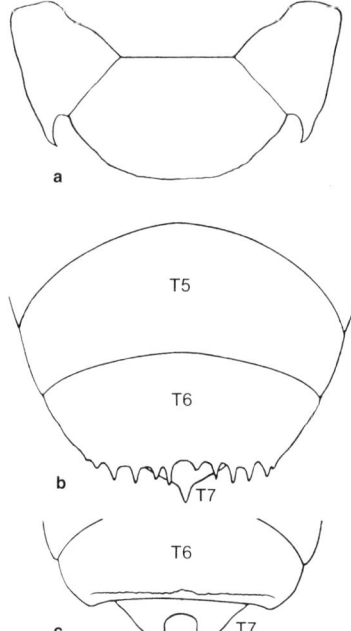

Figure 81-9. Structures of *Hoplosmia*. **a,** Scutellum and axillae of *H. (Hoplosmia) spinulosa* (Kirby); **b,** Apex of male metasoma of *H. (Hoplosmia) spinulosa* (Kirby), the terga numbered; **c,** Same, of *H. (Odontanthocopa) bidentata* (Morawitz). Modified from drawings by B. Tkalců, 1974c.

Recognition of *Hoplosmia* as a genus rather than as a subgenus of *Hoplitis* is arbitrary, given the few apomorphies recognized. *H. scutellaris* Morawitz (and possibly other species) is unusual in that there are carinulae on the pronotal lobe, as in *Protosmia* and *Othinosmia*.

Key to the Subgenera of *Hoplosmia* (Males)

1. Carina of T6 with numerous spines (Fig. 81-9b); T6 with distinct lateral tooth; T7 produced to median spine (Fig. 81-9b) .... *H. (Hoplosmia s. str.)*
—. Carina of T6 simple, irregularly crenulate or nodulose (Fig. 81-9c); lateral tooth of T6 absent or very blunt; T7 not produced to spine 2
2(1). T7 broad, subtruncate, much exserted, exposed part thus as long as T6 .... *H. (Paranthocopa)*
—. T7 bidentate or bilobed with median apical emargination (Fig. 81-9c), or rounded, exposed part much shorter than T6 .... *H. (Odontanthocopa)*

Key to the Subgenera of *Hoplosmia* (Females)

1. Scutellum not overhanging metanotum ....
   .... *H. (Odontanthocopa)* (in part)
—. Scutellum overhanging metanotum 2
2(1). Body clothed with reddish hair that obscures frons and scutum .... *H. (Paranthocopa)*
—. Body clothed with off-white hair that does not obscure frons and scutum 3
3(2). Preoccipital margin carinate dorsally; scopal hairs slender, crinkly, not blunted apically; hind basitarsus at least four times as long as broad .... *H. (Hoplosmia s. str.)*
—. Preoccipital margin not carinate dorsally though rather

abruptly angled; scopal hairs stout, smooth, blunted apically; hind basitarsus three times as long as broad .... *H. (Odontanthocopa)* (in part)

Hoplosmia / Subgenus *Hoplosmia* Thomson s. str.

Osmia (Hoplosmia) Thomson, 1872: 233, no included species; Schmiedeknecht, 1884: 23, included species. Type species: *Apis spinulosa* Kirby, 1802, by designation of Michener, 1941a: 161.

Distinctive characters are indicated in the key to subgenera, and in more detail by Tkalců (1974c).
■ This subgenus is found from Northern Europe to the Mediterranean and Egypt. The three species were included in the revision by Tkalců (1974c).

Hoplosmia / Subgenus *Odontanthocopa* Tkalců

Anthocopa (Odontanthocopa) Tkalců, 1974c: 125. Type species: *Osmia bidentata* Morawitz, 1876, by original designation.
Anthocopa (Odonterythrosmia) Tkalců, 1974c: 131. Type species: *Osmia fallax* Pérez, 1895, by original designation.

The subgeneric characters are indicated in the key to subgenera and in greater detail by Tkalců (1974c).
■ This subgenus is found in the Mediterranean basin and southeastern Europe. There are nine species (Zanden, 1988a).

Tkalců (1974c) separated one species, *Hoplosmia fallax* (Pérez), placing it by itself in a subgenus *Odonterythrosmia*, primarily on the basis of the broader hind basitarsus of the female and the largely red color of T1 to T3. Its relationship seems better indicated by including it in *Odontanthocopa*.

Hoplosmia / Subgenus *Paranthocopa* Tkalců

Anthocopa (Paranthocopa) Tkalců, 1974c: 132. Type species: *Osmia pinquis* Pérez, 1895, by original designation.

The subgeneric characters are indicated in the key to subgenera, and in more detail by Tkalců (1974c).
■ The subgenus is found in northern Africa and Israel. The only species is *Hoplosmia pinquis* (Pérez).

Genus *Noteriades* Cockerell

Heriades (Noteriades) Cockerell, 1931b: 332. Type species: *Megachile tricarinata* Bingham, 1903, by original designation.

This is a genus of short, compact hoplitiform bees, 4.5 to 10.0 mm long, that are characterized by the strong longitudinal carina on the clypeus, the complete preoccipital carina, the vertical propodeal profile (lacking a basal zone), and the posteriorly carinate scutellum. The wide apical hyaline rim of the female T6 is suggestive of *Pseudoheriades* and *Afroheriades*. *Noteriades* is the only osmiine, other than *Osmia* subgenus *Monosmia*, with two tibial spines on the front and middle legs, the anterior spine small and blunt. An unexpected plesiomorphy is the distinct volsella with recognizable digitus and cuspis.
■ *Noteriades* is found in both temperate and tropical parts of sub-Saharan Africa and in southern Asia (Thailand, Burma, India). There are nine recognized species (Griswold, 1985, 1994a) and a few undescribed ones, for a total of possibly 15.

Nests are unknown, but a female was seen at the entrance to a burrow in wood (Griswold, 1985).

Genus *Ochreriades* Mavromoustakis

Ochreriades Mavromoustakis, 1956: 226. Type species: *Eriades fasciatus* Friese, 1899, by original designation.

This genus consists of heriadiform species 7 to 10 mm in body length, even more elongate and slender than *Chelostoma*, to which it is allied. It differs in having yellow or ivory integumental markings and an enlarged pronotum that extends the cylindrical shape of the thorax forward, well in front of the scutum, virtually eliminating the preomaular surface of the mesepisternum as well as the anterior surface of the scutum. These are unique features in the Osmiini. The long mouthparts suggest that these bees may be floral specialists.
■ This rare genus appears to have a disjunct distribution; it is known from the Middle East (Israel, Jordan, Syria) and from Namibia, one species from each area. The two species were reviewed by Griswold (1994b).

Genus *Osmia* Panzer

This large genus includes the common, robust, megachiliform, more or less metallic Osmiinae. In addition, it includes many nonmetallic forms. Some subgenera, such as *Allosmia*, are rather elongate, more or less hoplitiform, but the elongate form in this case is due to the long metasoma, the thorax being rather short, and the posterior edge of the scutellum as well as the metanotum and propodeum being steeply declivous. Thus the thorax of *Osmia* and usually the whole body form is similar to that of *Hoplitis* subgenus *Anthocopa* (Fig. 81-10). An exception is *O. (Pyrosmia) cephalotes* Morawitz, in which the body, including the thorax, is elongate (scutellum nearly flat, metanotum and base of propodeum sloping but nearer horizontal than vertical), although other members of its subgenus are robust, like ordinary species of *Osmia*.

All Osmiini with punctiform or very short parapsidal lines belong to this genus, but the subgenus *Allosmia*, some species of the subgenera *Erythrosmia* and *Tergosmia*,

Figure 81-10. Female of *Osmia (Osmia) lignaria* Say. From Michener, McGinley, and Danforth, 1994.

and *O. (Pyrosmia) cephalotes* Morawitz have linear parapsides sometimes almost as long those in genera such as *Hoplitis*. Distinctive features of most *Osmia* are the enlarged S2 of males, often covering much of S3 and sometimes other sterna, and the fringed emargination in S3, S3 thus being quite different from S4. These features are not found in other genera except *Atoposmia* and, to varying degrees, the *Hoplitis* subgenera *Acrosmia, Jaxartinula, Kumobia, Penteriades,* and a few species of *Proteriades.* Scattered through *Osmia,* however, are a few subgenera or species in which S2 to S4 are similar to one another, transverse, and rather ordinary: *Orientosmia, Ozbekosmia,* and *Tergosmia* in the palearctic region, *Cephalosmia, Euthosmia,* and *Mystacosmia* in North America. Another common feature of *Osmia* is the absence of a lateral tooth on T6 of the male. This character fails in certain subgenera (*Hemiosmia, Diceratosmia,* and *Ozbekosmia,* and some species of *Helicosmia* and *Pyrosmia*) because of the presence of a large tooth at the lateral extremity of the posterior margin of T6 of males. Such lateral teeth seem the same as the lateral teeth of T6 found in nearly all *Hoplitis*. Male genitalia, sterna, and other structures of *Osmia* have been illustrated by Sandhouse (1939), Yasumatsu and Hirashima (1950), Mitchell (1962), Rust (1974), and Rust and Bohart (1986); as in other Osmiini, the gonostylus is unrecognizable except as the distal, hairy part of the gonoforceps, and S7 is transverse, not strongly sclerotized (Fig. 81-11).

The placement of *Allosmia* and *Erythrosmia* in *Osmia* requires comment. These are palearctic taxa that superficially look somewhat like *Hoplitis (Anthocopa)* and similar taxa, and that have linear (although short) parapsidal lines in most species, although the lines are punctiform in some *Erythrosmia*. The lack of basal flaps on S6 of the male and the lack of a lateral tooth on T6, combined with the structure of S2 and S3 of the male of most species, indicate that these taxa do not belong in *Hoplitis,* and that they fall within the genus *Osmia*. If these taxa could be removed from *Osmia,* either together or as separate genera, the definition of that genus would be sharpened, but satisfactory characters for removing them have not been found.

The American taxa *Atoposmia, Eremosmia,* and *Hexosmia,* which formerly were placed in *Anthocopa,* are not now treated in the same way as the above-mentioned palearctic taxa. They also lack flaps on S6 and have more or less *Osmia*-style S2 and S3 in the male, but they have usually more elongate parapsidal lines, as in *Hoplitis,* and they have a lateral tooth on T6 of the male, except for *Hexosmia,* which occasionally has a weak, very obtuse lateral angle on T6. I have arbitrarily placed these three subgenera in a separate genus *Atoposmia,* because assignment to *Osmia* would make that an even more diverse genus than it now is. They would be equally out of place in *Hoplitis*. Clearly, a phylogenetic study would go far toward developing a rational classification of these bees.

Osmia is abundantly represented in Europe, the whole Mediterranean basin, and southwestern Asia, and in western North America. It is less common in eastern North America (27 species east of the Mississippi; Mitchell, 1962) and especially in eastern Asia (7 species in Japan; Hirashima, 1973). *Osmia* does not extend far to the south. One species reaches Costa Rica; none occurs in sub-Saharan Africa, so far as I know, and none in Southeast Asia or Australia. In North America, however, 135 species are listed by Hurd (1979), and for the palearctic region over 160 species are listed by van der Zanden (1988a) in taxa here included in *Osmia.*

North American species were revised by Sandhouse (1939). European species (including those of *Hoplitis, Hoplosmia, Protosmia,* etc.) were revised by Ducke (1900) and listed by van der Zanden (1988a); reviews of various subgenera are indicated in the treatments of those subgenera. The subgeneric classification of American species was treated by Sandhouse (1939), Sinha (1958), and Snelling (1967), and van der Zanden (1988a) summarized that of palearctic species.

Osmia species typically construct nests in cavities that they do not themselves construct—beetle burrows in wood or stems, crevices under or between stones, or in empty snail shells, etc. Sometimes the nests are plastered onto protected surfaces, such as roofs of rock cavities. Some species, such as *O. (Melanosmia) maritima* Friese and *O. (Acanthosmioides) nigrobarbata* Cockerell, nest in burrows in sand or soil that are probably dug by the bees. Cells are made of mud or chewed leaf material, sometimes mixed with mud or resin. Westrich (1989) gives an excellently illustrated review of nesting by Central European species, along with species of *Hoplitis,* which can be identified by reference to Zanden's (1988a) list. Perhaps the salient conclusion is that within various subgenera there is wide diversity in nest sites and in the construction materials used, so that these features do not provide reliable subgeneric characters. Frohlich (1983), however, suggested the contrary, apparently finding group differences among forms here placed in the subgenera *Melanosmia* and *Acanthosmioides*. Nesting in snail shells is a well-known behavior of various European taxa, but is little known in America. The comments on nesting behavior under the subgenera are not exhaustive; references are mostly to recent works that contain references to older publications. Large contributions concerning nesting behavior of Japanese species, including commercial use for orchard pollination, are by Yamada et al. (1971) and Maeta (1978).

There are interesting differences in coloration between

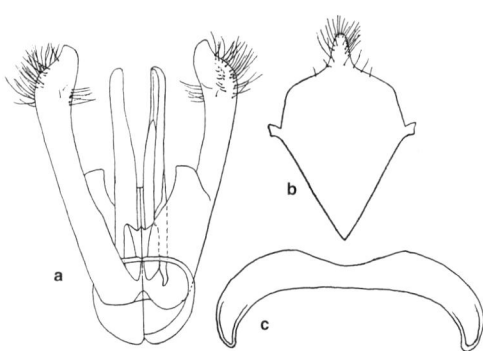

Figure 81-11. *Osmia (Cephalosmia) subaustralis* Cockerell. **a-c,** Male genitalia, S8, and S7. (For the genitalia, the dorsal view is on the left.) a and b, from Sandhouse, 1939.

palearctic and nearctic species of *Osmia*. In Europe, many are nonmetallic, and most of those that are metallic are only weakly so. On the other hand, many of the species in various subgenera are conspicuous because of abundant red or orange hair. American species, except *O. (Diceratosmia) azteca* Cresson, lack abundant red or orange hair; a few species of *Melanosmia* have reddish thoracic hairs. In America there are relatively few nonmetallic species, many are strongly metallic (Pl. 6), and a considerable group (including unrelated sections of the subgenus *Melanosmia* and one species of *Osmia* s. str.) consists of brilliantly metallic green, blue, or even purple species. (The generally nonmetallic genus *Hoplitis* includes three brilliant-green North American species in the subgenus *Monumetha*. Perhaps bees influence the evolution of color in their relatives living in the same area, possibly through mimetic complexes.)

In the two species mentioned in couplet 4 of the first key below, and in some others, the marginal groove of S4 is extremely narrow, the sternal margin is translucent, and a carina is extremely close to the margin. One would never recognize it by the description "two apical carinae," but it is not a simple, thin margin.

Key to the Subgenera of *Osmia* of the Eastern Hemisphere (Males)

1. Malar space with minimum length over half basal width of first flagellar segment; middle femur somewhat broadened basal to middle by rounded angle on undersurface; penis valves attaining apices of gonoforceps, which are slender and parallel-sided medially 2
—. Malar space absent or linear, its minimum length half basal width of first flagellar segment or less; middle femur without rounded premedian angle on undersurface; penis valves not attaining apices of gonoforceps, or, if so, then gonoforceps usually robust and tapering throughout 4
2(1). Proboscis of ordinary length, usually not reaching beyond middle of thorax in repose; anterior surface of T1 shagreened; antenna usually reaching propodeum and middle flagellar segments usually over twice as long as broad .... *O. (Osmia s. str.)*
—. Proboscis extremely long, reaching metasoma in repose; anterior surface of T1 shiny; antenna not reaching propodeum, middle flagellar segments less than twice as long as broad 3
3(2). Front tibia with two apical spines, a shorter one posterior to usual single spine; metasomal terga with diffuse pale hairs but no fasciae; S2 large, nearly completely hiding S3, which is emarginate and fringed; T6 entire, T7 with feeble apical emargination .... *O. (Monosmia)*
—. Front tibia with usual single apical spine; metasomal terga with broad apical bands of dense white plumose hairs; S2 not enlarged, S3 fully exposed, not emarginate, with small, dense midapical fringe; T6 with small apical emargination, T7 entire .... *O. (Orientosmia)*
4(1). S4 with posterior margin not thin, with two apical carinae, at least laterally (the distal one being the sternal margin) and intervening narrow hairless groove along margin [often visible only in posterior view, and absent although margin rather thick in a few species such as *O. (Pyrosmia) versicolor* Latreille and *ferruginea* Latreille] 5
—. S4 with posterior margin thin, without double carinae and groove 7
5(4). T7 ending in long, slender, heavily sclerotized, blunt-tipped process; S2 and S3 similar to one another, both exposed, transverse, each with transverse shining gibbosity, that on S3 broken medially .... *O. (Ozbekosmia)*
—. T7 relatively short, margin medially emarginate, two-, three-, or rarely four-toothed, rarely entire and rounded; S2 large, covering much of S3, which has fringed emargination, these sterna without gibbosities........................ 6
6(5). Margin of T6 medially projecting posteriorly or with median notch or crenulate; S5 simple, without velvety hair on disc or basal or lateral ridges; S6 without longitudinal groove, margin without projecting lobe ....
.... *O. (Helicosmia)*
—. Margin of T6 medially gently convex, not crenulate or notched; S5 with velvety hair on disc or ridges; S6 with longitudinal groove, margin with projecting lobe ....
.... *O. (Pyrosmia)*
7(4). Mandible three-toothed, middle tooth well separated from upper tooth by deep notch, upper tooth sometimes broad, truncate or even concave, so that mandible could be considered four-toothed 8
—. Mandible two-toothed, upper tooth sometimes broad, truncate or even concave, so that mandible could be considered three-toothed, but in the latter case middle angle not separated from upper angle by deep notch 10
8(7). T6 with strong posterolateral tooth; T7 not visible in dorsal view .... *O. (Hemiosmia)*
—. T6 without posterolateral tooth; T7 visible in dorsal view 9
9(8). First flagellar segment at least three times as long as broad; apical margin of S3 with hair-filled notch ....
.... *O. (Neosmia)*
—. First flagellar segment at most two times as long as broad; apical margin of S3, like S2 and S4, transverse, without hair-filled notch .... *O. (Tergosmia)*
10(7). Hind coxa with strong longitudinal carina along inner ventral margin; parapsidal lines usually short-linear (nonmetallic) 11
—. Hind coxa without longitudinal carina along inner ventral margin, or, if with ridge in location of carina, then ridge usually punctate, not raised and shining; parapsidal lines punctiform 12
11(10). T7 apically a simple process, much thickened and heavily sclerotized, flattened on dorsal surface and margined almost like a pygidial plate; S2 with flat-topped, bare longitudinal ridge apically almost to margin; maxillary palpus five-segmented .... *O. (Allosmia)*
—. T7 bifid or bilobed, thin, not particularly heavily sclerotized, not flattened on dorsal surface; S2 apically thin, rounded or feebly emarginate medially, hairy; maxillary palpus four-segmented or with minute apical fifth segment .... *O. (Erythrosmia)*
12(10). Anterior margin of metanotum sharp, elevated approximately to level of scutellar surface and with space between metanotal crest and posterior margin of scutellum; S2 with posterior margin transverse.......... *O. (Metallinella)*
—. Anterior margin of metanotum below level of most of scutellum, which is curved down to metanotal margin; S2 with posterior margin broadly convex, sometimes with small median emargination .... *O. (Melanosmia)*

Key to the Subgenera of *Osmia* of the Eastern Hemisphere (Females)

1. Claws cleft, inner ramus shorter than outer; anterior edge of metanotum sharp, elevated approximately to level of scutellar surface; space present between crest of metanotum and scutellum .... *O. (Metallinella)*
—. Claws simple; anterior edge of metanotum below level of most of scutellum, i.e., scutellar surface curved down to margin of metanotum [except in long-bodied species of *Pyrosmia* (e.g., *O. cephalotes* Morawitz) in which metanotum is horizontal, on level with scutellum]; ordinarily no space between edge of metanotum and scutellum 2
2(1). Head with small, usually bare and shiny depression immediately behind lower end of eye, i.e., below posterior margin of eye (body commonly covered with long, loose hairs rarely forming metasomal fasciae) ....
....*O. (Osmia s. str.)*
—. Head without small depression immediately behind lower end of eye, although depression directly below or in front of lower end of eye, or occupying much of malar area, sometimes present 3
3(2). Depression across mandibular base deep and abruptly set off almost vertically from adjacent flat mandibular surface (Fig. 81-12d); orange hairs beneath clypeal margin arranged in four tufts, except when clypeus is highly modified, in which case there may be two broad tufts, or no orange hairs ....*O. (Helicosmia)*
—. Depression across mandibular base absent or shallower, rarely completely separated from adjacent mandibular surface by vertical wall [some *Melanosmia*, e.g., *O. nigriventris* (Zetterstedt), approach the condition of *Helicosmia*]; orange hairs under clypeal margin in two tufts or brushes, or absent 4
4(3). Malar space, where shortest, as long as width of pedicel; proboscis extremely long, reaching metasoma in repose......5
—. Malar space, where shortest, much shorter than width of pedicel; proboscis not reaching beyond middle of thorax in repose 6
5(4). Front tibia with two apical spines, a short one posterior to usual single spine; metasomal terga with diffuse pale hairs not forming fasciae .... *O. (Monosmia)*
—. Front tibia with single apical spine; metasomal terga with broad apical fasciae of dense, white plumose hairs..
.... *O. (Orientosmia)*
6(4). Hind coxa with strong longitudinal carina along inner margin *and* body nonmetallic, metasoma sometimes red; parapsidal lines usually short-linear, sometimes punctiform; erect tergal hair (middorsally on T2-T3, as seen in profile, ignoring apical tergal margins) sometimes no longer than minimum width of first flagellar segment 7
—. Hind coxa not or weakly carinate, or, if carinate, then body metallic, if with ridge in location of coxal carina (as in some *Hemiosmia, Pyrosmia, Tergosmia,* etc.), then ridge usually punctate, not raised and shining as in above; parapsidal lines punctiform or sometimes short-linear in *Tergosmia;* erect tergal hair much longer than minimum width of first flagellar segment... 8
7(6). Maxillary palpus distinctly five-segmented ....
.... *O. (Allosmia)*
—. Maxillary palpus four-segmented or with minute apical fifth segment .... *O. (Erythrosmia)*
8(6). First flagellar segment nearly three to more than three times as long as wide .... *O. (Neosmia)*
—. First flagellar segment not over 2.5 times as long as wide
.... 9
9(8). Anterior trochanter with longitudinal ventral carina or sharp ridge that is most prominent medially so that lower edge of trochanter, in side view, is convex; mandible in ventrolateral view with strong submedian tooth on inner margin of lower edge (mandible four-toothed or approaching that condition because of distinct cutting edge forming tooth in upper interspace) .... *O. (Hemiosmia)*
—. Anterior trochanter without longitudinal ventral carina or sharp ridge, lower edge of trochanter, in side view, nearly straight; mandible usually without submedian tooth on inner margin of lower edge [but present in *O. (Melanosmia) inermis* (Zetterstedt)] 10
10(9). Hind coxa with longitudinal carina on inner ventral margin; clypeal margin usually with median impression (mandible usually three-toothed without cutting edge in upper interspace) .... *O. (Pyrosmia)*
—. Hind coxa without longitudinal carina on inner ventral margin; clypeal margin without median impression...... 11
11(10). Body frequently slightly to strongly metallic; pubescence often including black or fuscous hairs; mandible commonly four-toothed, usually because of cutting edge in upper interspace .... *O. (Melanosmia)*
—. Body black, nonmetallic; pubescence mostly yellowish white; mandible three-toothed without cutting edge in upper interspace... 12
12(11). Anterior margin of clypeus produced and rounded; impunctate hairless area behind mandibular base overhung by long, curved hairs .... *O. (Ozbekosmia)*
—. Anterior margin of clypeus produced and truncate; impunctate area behind mandibular base limited and not extensively overhung by long hairs .... *O. (Tergosmia)*

Key to the Subgenera of *Osmia* of the Western Hemisphere

1. Hind coxa with strong longitudinal carina along inner ventral angle; parapsidal line somewhat elongate ....
.... *O. (Diceratosmia)*
—. Hind coxa not carinate; parapsidal line punctiform 2
2(1). Malar space as long as width of scape; small shining depression in genal area below and behind lowermost point of eye margin.. *O. (Osmia s. str.)*
—. Malar area shorter than width of scape, except in forms with anterior clypeal margin greatly swollen; no small shiny depression below eye 3
3(2). Males 4
—. Females 10
4(3). S2 transverse, posterior margin feebly concave, straight, to slightly convex; S3 not or weakly emarginate medially, exposed from side to side, posterior margin not or gently concave 5
—. S2 large, posterior margin strongly convex; S3 medially hidden by S2, posterior margin broadly emarginate with fringe of long hairs in emargination 7
5(4). Genal area wider than greatest width of eye, as seen from side; pubescence partly black .... *O. (Cephalosmia)*

—. Genal area narrower than eye, as seen from side; pubescence white 6
6(5). Median flagellar segments 1.5 times as long as broad; T6 prolonged medially over T7 .... *O. (Mystacosmia)*
—. Median flagellar segments 1.9 times as long as broad; T6 with posterior margin evenly convex, not conspicuously prolonged over T7 .... *O. (Euthosmia)*
7(4). S4 with apical margin laterally consisting of two carinae between which is a narrow, hairless, shiny groove .... *O. (Helicosmia)*
—. S4 with apical margin thin, without groove 8
8(7). S2 with protuberance, tooth, or spine on median apical or subapical area (or median band of long hair in *O. integra* Cresson); male gonoforceps with small, slender appendage, usually arising preapically; mandible slender medially, apex (measured from apex of apical tooth to upper apical angle) at least 1.5 times width at constriction and usually wider than base (Fig. 81-12a)..................... .... *O. (Acanthosmioides)*
—. S2 without protuberance, tooth, or spine; male gonoforceps usually tapering from preapical angle to apex, rarely [as in *O. (Melanosmia) tanneri* Sandhouse] with distal preapical process; mandible usually widest at base or subequal at base and apex 9
9(8). Forewing with hairs about half as long as width of stigma; middle flagellar segments twice as long as wide .... *O. (Trichinosmia)*
—. Forewing with hairs much less than half as long as width of stigma; middle flagellar segments less than twice as long as wide (except in *O. bucephala* Cresson) .... *O. (Melanosmia)*
10(3). Clypeal margin with strong lateral tooth and small median tooth; forewing with hairs half as long as width of stigma .... *O. (Trichinosmia)*
—. Clypeal margin not so modified; forewing with hairs much less than half as long as width of stigma 11
11(10). Orange hairs beneath clypeal margin arranged in four tufts; depression across mandibular base deep and set off almost vertically from adjacent mandibular surface and commonly emphasized by mandibular swelling just distal to depression (Fig. 81-12d).............. *O. (Helicosmia)*
—. Orange hairs beneath clypeal margin arranged in two tufts or brushes, or sometimes absent; depression across mandibular base usually less abrupt or absent and not emphasized by swelling distal to it [an exception is *O. (Melanosmia) bucephala* Cresson] 12
12(11). Clypeal punctures large, well separated by smooth ground or largely absent; genal area twice as wide as eye, as seen from side; eyes diverging downward .... *O. (Cephalosmia)*
—. Clypeal punctures smaller, confluent at least on lower half of clypeus; genal area usually less than twice as wide as eye, as seen from side; eyes almost always converging downward or parallel.. 13
13(12). Mandible constricted near base, expanded distally to apex that is nearly twice as wide as constriction (Fig. 81-12a, c) (mandible tridentate or with small intercalary tooth between second and third teeth) ....*O. (Acanthosmioides, Melanosmia)* (in part)
—. Mandible less constricted near base and less expanded apically, apex thus usually less than 1.5 times as wide as constriction... 14
14(13). Clypeus without tufts or brushes of orange hairs beneath margin; scopal hairs restricted on each sternum to a band half as wide as distance from gradulus to sternal margin.. *O. (Euthosmia)*
—. Clypeus with two tufts or brushes of orange hairs beneath margin; scopal hairs occupying larger area on each sternum .... *O. (Melanosmia)* (in part), *O. (Mystacosmia)*

Osmia / Subgenus *Acanthosmioides* Ashmead

Acanthosmioides Ashmead, 1899a: 76. Type species: *Osmia odontogaster* Cockerell, 1897, by original designation.
Acanthosmiades Titus, 1904b: 101, incorrect subsequent spelling of *Acanthosmioides* Ashmead, 1899.

This is a subgenus of metallic green or blue species, almost always with some black hairs; the body length ranges from 6 to 14 mm. *Acanthosmioides* is related to *Melanosmia* and is sometimes difficult to distinguish from that subgenus. The principal distinguishing characters are indicated in the key to subgenera.

■ *Acanthosmioides* is found throughout western North America, from Yukon, Canada, to Baja California, Mexico, east throughout the Rocky Mountain area and to North Dakota. There are 22 species. The subgenus was revised by White (1952), who also included Sinha's (1958) Group II of *Nothosmia*. White had no males of that group; males would have excluded this group from *Acanthosmioides*.

In a group of small desert species [Sinha's (1958) Group II of *Nothosmia*], the males fall in *Melanosmia* but the female mandible is constricted before the middle and broadened into a tridentate apex about twice as wide as the constriction. Thus the female mandible is like that of *Acanthosmioides* (Fig. 81-12d); tentatively I place this group in *Melanosmia*.

Another group that presents a similar problem is the species of *"Centrosmia"* as understood by Sinha and Michener (1958), except for *Osmia bucephala* Cresson, the type species of *Centrosmia*. The females are not distinguishable from those of *Acanthosmioides*, but the males lack the features of S2 and of the gonoforceps that characterize *Acanthosmioides*, although the gonoforceps of some species are rather intermediate.

This subgenus is almost certainly derived from *Melanosmia*, making the latter a paraphyletic group. Since there is no great morphological gap between *Melanosmia* and *Acanthosmioides*, the two could be united. I delay doing so until *Melanosmia* can be more fully studied and the relation of *Acanthosmioides* verified.

Some species of *Acanthosmioides* make shallow nest burrows in the soil (Rozen and Favreau, 1967). Other species excavate burrows in pithy stems or construct cells on sheltered surfaces, e.g., the underside of a rock. The cells are made of mud (Rust, Thorp, and Torchio, 1974) or of chewed plant material (Rozen and Favreau, 1967) or of both (Parker, 1975a). See also Gordon (2003).

Osmia / Subgenus *Allosmia* Tkalců

Osmia (Allosmia) Tkalců, 1974b: 331. Type species: *Osmia rufohirta* Latreille, 1811, by original designation.

In this subgenus the metasoma is elongate and slender, so that in general appearance specimens resemble *Hoplitis* s. str. The lack of metallic coloration and the linear, although short, parapsidal furrows are also suggestive of *Hoplitis*. Features indicating placement in *Osmia* include the large S2 of the male, the medially emarginate and fringed S3, the lack of basal flaps on S6, and the lack of a lateral tooth on T6, as well as the shortness of the parapsidal furrows. The pubescence is pale to red, forming tergal fasciae. The body length is 7.5 to 11.0 mm. The clypeus of the female is truncate apically, covering the base of the labrum. T6 of the male lacks a median emargination, and is strongly curved down laterally around the narrow, produced and pointed T7. Tkalců (1974b) illustrated male genitalia and other structures.

■ This subgenus is found in Central and Southern Europe and northern Africa. I am familiar with only a single species, *Osmia rufohirta* Latreille, but G. van der Zanden (in litt., 1994) listed two others.

An interesting feature of *Osmia rufohirta* is that males vary greatly in size, the larger ones being larger than females. The mating system would be interesting to investigate.

Nests are made in abandoned snail shells, chewed leaf material being used for cell construction (Westrich, 1989). Males of both *Erythrosmia* and *Allosmia* have a pair of broad depressions basally in S6 [not evident in *Osmia (E.) rutila* Erichson]. I am not aware of such structures in other groups of *Osmia*. According to T. Griswold, there is usually a white, crystalline substance in these depressions when dissected. This character supports the view, as do the carinate hind coxae, that these subgenera are close relatives and should probably be merged.

Osmia / Subgenus *Cephalosmia* Sladen

Cephalosmia Sladen, 1916a: 270. Type species: *Osmia armaticeps* Cresson, 1878 = *Osmia montana* Cresson, 1864, monobasic.

This is a subgenus of large (8-17 mm long) black or dark-blue species with some or all of the pubescence black. The mandible of the female is but little narrowed near the base; the apex is broad (1.5-2.0 times the basal width) and four-toothed. There is an angle or tooth on the inner lower margin of the female mandible, where the adductor ridge approaches the mandibular margin; this character is suggestive of *Hemiosmia*. *Cephalosmia* differs from *Hemiosmia*, however, in the dispersed large punctures (or lack of punctures) of the female clypeus, the transverse S2 of the male and scarcely emarginate and fully exposed S3, the lack of lateral teeth on T6, the presence of black hairs, etc. An unusual character of females of *Cephalosmia* is the small T6, without long hairs, in contrast to T5.

■ *Cephalosmia* is found in the western half of North America, rarely eastward to the Great Lakes, and from the Northwest Territories, Canada, to the mountains of Baja California, Mexico. The five species were revised by Rust (1974).

Nests are in holes (usually made by beetles) in wood. Partitions between cells are made of chewed leaf material, mixed with mud in the case of *Osmia californica* Cresson. An unusual feature is that the egg in each cell is placed not on top of the food mass, but in a small pocket in the center of the pollen store (Rust, 1974).

Osmia / Subgenus *Diceratosmia* Robertson

Diceratosmia Robertson, 1903a: 166, 171. Type species: *Osmia 4-dentata* Cresson, 1878 = *Osmia conjuncta* Cresson, 1864, by original designation.

This subgenus contains mostly blue or green metallic species, 4 to 8 mm long, the pubescence mostly pale (red in *Osmia azteca* Cresson). The structure is much like that of *Pyrosmia*, and the two subgenera could well be united. The principal similarities and differences are listed in the discussion of that subgenus. Male genitalia, sterna, and other structures were illustrated by Mitchell (1962).

■ *Diceratosmia* is found principally in eastern and central North America, one species ranging west to eastern California; the subgenus occurs from southeastern Canada to Costa Rica. There are five recognized species. The subgenus was revised by Michener (1949).

Osmia / Subgenus *Erythrosmia* Schmiedeknecht

Osmia (Erythrosmia) Schmiedeknecht, 1885: 20[886]. Type species: *Osmia andrenoides* Spinola, 1808, by designation of Cockerell, 1922b: 6.

Although the type species of this subgenus, *Osmia andrenoides* Spinola, is quite robust, some of the other species, such as *O. sybarita* Smith and *rutila* Erichson, are slender and hoplitiform, suggestive of *Allosmia*. The distinctions from *Allosmia* are indicated in the key to subgenera. The pubescence is pale, sometimes red, forming weak tergal fasciae at least laterally. Body length ranges from 5.5 to 10.0 mm. The body is nonmetallic, the metasoma sometimes largely red. The clypeus of the female is truncate, overlapping the base of the labrum. The parapsidal furrows vary from short-linear, e.g., in *O. andrenoides* Spinola and *rutila* Erichson, to punctiform, e.g., in *O. bischoffi* Atanassov. S2 of the male may be enlarged and S3 emarginate and fringed, as in most species of *Osmia*. In some, however, e.g., *O. andrenoides*, S2 is not enlarged and S3 is not or little emarginate, at most with the margin distinctly fringed medially. T6 of the male is rather long, and entire or with a very feeble median apical emargination. Male genitalia and sterna were illustrated by Popov (1954) for *O. andrenoides*.

■ This subgenus is found in the Mediterranean basin, east to Central Asia. Thirteen species were included by van der Zanden (in litt., 1993).

Except for *Osmia andrenoides* Spinola, species here included in *Erythrosmia* have been placed in *Allosmia* by some authors (Zanden, 1988a), but the arrangement indicated here seems to associate related species.

Several of the species, perhaps all, nest in snail shells and use masticated leaves mixed with bits of snail shells as building material.

Osmia / Subgenus *Euthosmia* Sinha

Osmia (Euthosmia) Sinha, 1958: 235. Type species: *Heriades glaucum* Fowler, 1899, by original designation.

This subgenus contains a small, rather slender species (length 5-7 mm) with blue-green body and white hairs. As in *Mystacosmia,* the male S2 is not enlarged; S2, S3, and S4 are all transverse and similar in shape. The mandible of the female is robust, scarcely narrowed medially, and three-toothed. The clypeal truncation is thin, without tufts of orange hair beneath. In the female the entire hypostomal area is impunctate, shining, and margined laterally by extremely long hairs curled over the hypostomal area; these are supplemented by equally long curved hairs arising near the bases of the maxillary stipites.

■ *Euthosmia* is known only from California. The only species ia *Osmia glauca* (Fowler).

The subgeneric position and characters of this species were clarified by Snelling (1967). In both the palearctic and nearctic regions a minority of *Osmia* taxa lack the usually characteristic (for the genus) form of S2 and S3 of males; instead, these sterna are similar to one another and sometimes similar to S4 also. Considering bees as a whole, such similarity is obviously plesiomorphic, being found in most families as well as in osmiine genera like *Hoplitis.* In *Osmia*, however, such sterna may be reversions in at least some cases; if so, then subgenera like *Euthosmia* and *Mystacosmia* are probably derived from *Melanosmia* and should be united with it.

Nests of *Euthosmia* have been found in hollow stems and in abandoned *Sceliphron* nests. The cells were made of mud (Rust and Clement, 1972).

Osmia / Subgenus *Helicosmia* Thomson

Osmia (Helicosmia) Thomson, 1872: 233, no included species; Schmiedeknecht, 1885: 22, included species. Type species: *Apis aurulenta* Panzer, 1799, by designation of Michener, 1941a: 163.
Osmia (Chalcosmia) Schmiedeknecht, 1885: 20. Type species: *Apis fulviventris* Panzer, 1798 (not Scopoli, 1763) = *Apis niveata* Fabricius, 1804, by designation of Sandhouse. 1939: 13.
Gnathosmia Robertson, 1903a: 165. Type species: *Osmia georgica* Cresson, by original designation.
Osmia (Cryptosmia) Yasumatsu and Hirashima, 1950: 14. Type species: *Osmia satoi* Yasumatsu and Hirashima, 1950, by original designation.

Even though Thomson failed to mention included species, the name *Helicosmia* is valid as of 1872 [Code, article 10(c)(i)] and therefore has priority for this subgenus. *Osmia satoi* Yasumatsu and Hirashima, the type species of *Cryptosmia,* is known from a single female; although characters of its male are therefore unknown, the female seems to fall in *Helicosmia.* It has the depression across the base of the mandible found in other species of *Helicosmia.*

Helicosmia consists of metallic blue or green, or sometimes not or scarcely metallic, moderate-sized to large (7.5-15.0 mm long) species with pale hairs (except sometimes for a black scopa), frequently forming pale apical tergal fasciae. The clypeus of the female is frequently truncate, the lower margin thickened and overhanging the labrum, but it is sometimes much modified (along with the mandibles), exposing the base of the labrum, as in *Osmia latreillei* (Spinola) and *O. satoi* Yasumatsu and Hirashima. There are four tufts of orange hairs beneath the clypeal margin of the female, but they are absent in *O. latreillei* and reduced to two tufts in *O. dimidiata* Morawitz, which has a somewhat modified clypeus; in *O. signata* Erichson the four tufts are almost merged to form two. The mandible of the female is robust, not much narrower medially than at the base (Fig. 81-12d). S4 of the male has two apical carinae, between which is a narrow groove, at least laterally, as in *Diceratosmia.* T7 has a small midapical emargination or is bidentate. T6 may lack a lateral tooth; less commonly it has a small tooth, as in *O. dimidiata* Morawitz, or a large tooth, as in *O. aurulenta* Panzer, *signata* Erichson, and *notata* (Fabricius). Medially, T6 may have a small emargination; it has a broad, median, slightly nodulose truncation, weakly and irregularly emarginate in the middle, in *O. aurulenta,* and a broadly convex, denticulate margin in *O. signata.* In spite of the characters cited above, the three species *O. (H.) aurulenta, notata,* and *signata* seem closely related. The last two, however, have been put in *Chalcosmia,* the first in *Helicosmia. Helicosmia* so limited has a dull propodeum and a less deep basal mandibular groove in females. These characters do not seem to justify recognition of separate subgenera, especially when *Helicosmia* in that limited sense consists of only two species (Zanden, 1988a). Male genitalia and other structures of *Helicosmia* were illustrated by Popov (1935), Mitchell (1962), and Tkalcŭ (1969a, 1975b).

■ *Helicosmia* is found from the Azores, Madeira, Canary Islands, and Britain eastward through Europe and the Mediterranean basin to Korea and China, and from coast to coast in North America south to the state of México. *Osmia coerulescens* (Linnaeus) is holarctic, but in North America it is limited to the eastern half of the USA; probably it was introduced long ago from Europe. About 53 palearctic species were listed by van der Zanden (1988a); he indicated (in litt., 1993) that he knows 76 species. In the nearctic region there are five additional species. Tkalcŭ (1975b) revised the *fulviventris* (= *niveata*) group of European species; Warncke (1988b) revised the species of the western palearctic region; and Rust (1974) revised the American species.

Helicosmia, Diceratosmia, and *Pyrosmia* constitute a group sharing the grooved margin of S4 of the males as well as their general aspect, usually apical tergal fasciae of plumose hairs, etc. Species exist that do not fit well into these subgenera, yet belong to this group of subgenera. An example is *Osmia indigotea* Morawitz, discussed under the subgenus *Pyrosmia.* Perhaps the solution is to unite the three subgenera.

Nests are in holes in wood or stems or in abandoned cells of other bees, either in clay banks or exposed nests. *Osmia (Helicosmia) aurulenta* (Panzer) and *melanogaster* Spinola nest in abandoned snail shells (Pl. 16), and *O. (H.) texana* Cresson appears to specialize in old burrows and cells of *Anthophora occidentalis* Cresson for its nest sites. Cell partitions are made of chewed leaf material (Rust, 1974). The nesting behavior of *O. (H.) coerulescens*

(Linnaeus), a potential pollinator of alfalfa, has been studied in some detail (Tasei, 1972); it nests in holes in wood, and places cells in series, separated by chewed leaf material.

Osmia / Subgenus *Hemiosmia* Tkalců

Osmia (Hemiosmia) Tkalců, 1975a: 34. Type species: *Osmia argyropyga* Pérez, 1879, by original designation.
Osmia (Exosmia) Tkalců, 1979a: 321. Type species: *Osmia difficilis* Morawitz, 1875, by original designation.

This subgenus consists of nonmetallic species with the form of *Helicosmia* or, in the case of *Osmia (H.) balearica* Schmiedeknecht, with longer hairs and a more robust-looking body. The pubescence is pale and often forms tergal fasciae. The body length varies from 7.0 to 13.5 mm. The clypeus of the female is truncate and overhangs the base of the labrum; the mandible is narrower near the middle than at the base, and has a large tooth on the inner lower margin near the middle. The front trochanter of the female has a carina on the lower surface. See the comments below for male characters, many of which, including genitalia, were illustrated by Tkalců (1975a).

■ The subgenus occurs in the Mediterranean area and east to central Asia. Six species were listed by van der Zanden (1988a). The species placed in *Hemiosmia* proper were revised by Tkalců (1975a) and by Haeseler (2005).

Females of the two subgeneric taxa listed in the synonymy above are not distinguishable from one another by group characters. In those placed in *Exosmia*, unlike species placed in *Hemiosmia* in the narrow sense, there are at least some clubbed hairs on the front tarsi. Males differ by characters that elsewhere in the genus are not usually group characters. Clearly, the male of *Hemiosmia* s. str. has some derived characters relative to *Exosmia*, such as the crenulate flagellum; the swollen midtibia, concave and often spiculate on the inner surface; and the small midtibial spur. In *Hemiosmia* s. str., T7 of the male is bifid or bilobed; in *Exosmia*, it is rounded and hidden under T6. In *Hemiosmia* s. str., the apex of the male gonoforceps is bifid; in *Exosmia*, it is spatulate. In spite of these differences, the similarities are impressive, and given the small number of species, subgeneric recognition of the two groups seems unnecessary. The function of subgenera should be to show relationships, not to provide a genus-group name for every unusual species.

Osmia balearica Schmiedeknecht nests in snail shells, using chewed leaf material for building cells (Friese, 1923).

Osmia / Subgenus *Melanosmia* Schmiedeknecht

Osmia (Melanosmia) Schmiedeknecht, 1885: 19[885]. Type species: *Osmia fuciformis* Latreille, 1811 = *Apis xanthomelana* Kirby, 1802, by designation of Sandhouse, 1939: 33.
Nothosmia Ashmead, 1899a: 75. Type species: *Osmia distincta* Cresson, 1864, by original designation.
Centrosmia Robertson, 1903a: 165, 166. Type species: *Osmia bucephala* Cresson, 1864, by original designation.
Leucosmia Robertson, 1903a: 166, 171. Type species: *Osmia albiventris* Cresson, 1864, by original designation.
Xanthosmia Robertson, 1903a: 166, 171. Type species: *Osmia cordata* Robertson, 1902, by original designation.

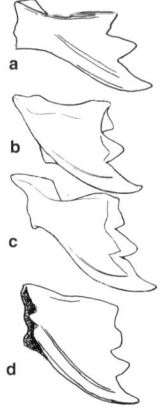

Figure 81-12. Diagrams of mandibles of females of *Osmia*. **a**, *O. (Acanthosmioides) nigrifrons* Cresson; **b**, *O. (Melanosmia) alticola* Benoist; **c**, *O. (Melanosmia) xanthomelana* (Kirby); **d**, *O. (Helicosmia) niveata* (Fabricius), showing depressed (darkened) zone across base of mandible.

Monilosmia Robertson, 1903a: 166, 171. Type species: *Osmia canadensis* Cresson, 1864 = *Osmia simillima* Smith, 1853, by original designation.
Chenosmia Sinha, 1958: 233. Type species: *Osmia penstemonis* Cockerell, 1906, by original designation.

Melanosmia is used here in a broad sense to include robust black to highly metallic species, frequently with black hairs, frequently with impunctate margins on the terga, and usually without apical tergal fasciae of plumose hairs. Length ranges from 6 to 14 mm. The clypeus of the female is truncate and overhangs the base of the labrum. The mandible of the female is ordinarily robust (Fig. 81-12b), but in a few species, such as *Osmia xanthomelana* (Kirby) (Fig. 81-12c) and especially *O. maritima* Friese in the palearctic region, the mandible is much narrowed in the middle. The same disparity is found in two groups of entirely different nearctic forms, the small desert species [Sinha's (1958) Group II of *Nothosmia*] and *Centrosmia* of Sinha and Michener (1958), except for *O. bucephala* Cresson. Although slender mandibles are unusual in *Melanosmia*, they are characteristic of the nearctic subgenus *Acanthosmioides*, which has apically broad and three-toothed (rarely approaching four-toothed) mandibles. It is quite likely that *Acanthosmioides* should also be placed in the synonymy of *Melanosmia* for, at present, clear distinguishing features for these subgenera are unknown. Various *Melanosmia* species have three- or four-toothed mandibles, and every intermediate condition can be found among the numerous species, as the cutting edge between the second and third teeth of a three-toothed mandible appears, becomes a tooth, or recedes. Palearctic species with slender mandibles, like *O. maritima*, have four-toothed mandibles, but Sinha's Group II species have three-toothed mandibles like *Acanthosmioides*. T6 of the male either is entire or, usually, has a small midapical emargination. T7 of the male is bilobed or bidentate. The European species, although all black, are quite diverse and have been placed in different subgenera (*Centrosmia*, *Chenosmia*, and *Melanosmia*) by van der Zanden (1988a). Nonetheless, they form a unified group, as indicated by the descriptions and illustrations of Tkalců (1983). For example, in all species S6 of the male has a median truncate or rounded apical projection not usually seen in other groups of *Osmia*. An exactly similar projection is found, however, in a group of metallic nearctic species placed in *Centrosmia* by Sinha and Michener (1958); such a projection also occurs in *Acanthosmioides*. Male genitalia, sterna, and other structures of many species were illustrated by Sandhouse (1939), Mitchell (1962), and Tkalcŭ (1983).

■ This subgenus occurs in most parts of the holarctic region, north as far as subarctic Alaska and Labrador, south to Florida, USA, and Nuevo Leon and Baja California, Mexico. It is especially well represented in montane regions and is scarce in desertic areas. Seventeen palearctic species are listed by van der Zanden (1988a); nine species were included in Tkalcŭ's (1983) revision of European species. In the nearctic region, this is the major subgenus of *Osmia*, with 91 species listed by Hurd (1979). Sandhouse (1939) revised the nearctic species, mostly under the subgeneric name *Nothosmia*. Sinha and Michener (1958) revised species then placed in *Centrosmia*. Müller (2002) gave a key to the European species of *Melanosmia*.

It now seems likely that *Centrosmia* was an artificial assemblage of species, most of which have enlarged tarsi on the male middle legs. That the character is not a particularly useful group feature is shown by *Osmia (M.) nigriventris* (Zetterstedt), which resembles *O. (M.) bucephala* Cresson, the type species of *Centrosmia*, in many features, but has simple middle tarsi. Van der Zanden (1988a) quite understandably included *nigriventris* in *Centrosmia* in spite of the tarsi. Furthermore, there are a few presumably unrelated species of *Acanthosmioides* that also have such tarsi. Sinha (1958) recognized, for forms here included in *Melanosmia*, not only the subgenus *Centrosmia* but also *Chenosmia*, *Monilosmia*, and *Nothosmia*. I, too, would like to recognize these subgenera, in part because *Melanosmia* as here understood is such a large group in North America, but intergradations appear to exist in all characters. In general, *Nothosmia* has three-toothed female mandibles and no black hairs (though the scopa is sometimes black), whereas the other groups have four-toothed mandibles (but there are intergradations, as indicated above) and some of the hairs are black. The characters separating *Chenosmia* from *Monilosmia* are even less substantial.

Osmia bucephala Cresson is a highly unusual nearctic *Melanosmia*, distinguished by its large size, the strong groove across the base of the female mandible, beyond which is a swelling, the distal edge of the groove thus nearly vertical [as in *Helicosmia* and the palearctic *O. (Melanosmia) nigriventris* (Zetterstedt)], the greatly thickened and protuberant lower part of the female clypeus, etc. As the type species of *Centrosmia*, it could easily constitute a monotypic subgenus like *Trichinosmia*, except that annectant species (related but with much more ordinary features) such as *O. nigriventris* are known.

The nearctic subgenus *Acanthosmioides* mentioned above, and especially the probably monotypic nearctic subgenera *Euthosmia*, *Mystacosmia*, and *Trichinosmia*, appear to be relatives of *Melanosmia* as it is here understood, many of their distinguishing characters probably being derived relative to *Melanosmia*. They may also fall into the synonymy of *Melanosmia*, unless that subgenus is subdivided in some entirely different or more reliable way than has been previously possible.

Even among closely related species of *Melanosmia*, there may be considerable diversity in nesting habits. Thus *Osmia (M.) xanthomelana* (Kirby) makes small exposed cell clusters in grass clumps or exposed on a rock or cliff. The cells are of mud, apparently hardened by the addition of secreted material. The related *O. (M.) maritima* Friese nests in burrows in sand. *O. (M.) inermis* (Zetterstedt) makes large cell clusters constructed by several females on the undersides of stones, the cells being made of chewed leaf material (Westrich, 1989). *Osmia (M.) bucephala* Cresson nests in holes in wood, making partitions of masticated leaf material and wood fibers rasped from the walls of the burrows (Krombein, 1967). Probably most of the American species use chewed leaf material for partitions in holes in wood, as do *O. pumila* Cresson, *gaudiosa* Cockerell (Krombein, 1967), four species studied by Medler (1967), etc. By contrast, *O. (M.) tanneri* Sandhouse makes mud cells on the undersides of stones (Parker, 1975). *O. (M.) sanrafaelae* Parker makes cells of chewed leaf material in diverse sites—cracks in earthen banks, hollow stems, holes in wood; the cells are in series or in clusters depending on the shape of available space (Parker, 1986). *O. (M.) bruneri* Cockerell makes similar nests in holes in stems or other sites (Frohlich, 1983); cell partitions are made of leaf pulp and pith if the latter is present. The inner surfaces of the partitions are coated with a secretion applied by the mouthparts.

Osmia / Subgenus *Metallinella* Tkalcŭ

Metallinella Tkalcŭ, 1966b: 200. Type species: *Osmia atrocaerulea* Schilling, 1849 = *Eucera brevicornis* Fabricius, 1798, by original designation.

Metallinella contains a dull metallic blue or green species; it is megachiliform, superficially resembling the subgenus *Helicosmia*. Length ranges from 6.5 to 12.0 mm. This subgenus has been regarded as a separate genus because of the cleft claws of the female, but this character is certainly a reversion; the male possesses genes for cleft claws that are expressed in all megachilids. A regulatory change could result in the expression of such genes in females. *Metallinella* is, however, a distinctive subgenus on the basis of other characters also, such as the dull, nonshiny propodeum, as in some species of *Helicosmia*. Tkalcŭ (1966b) illustrated the male genitalia.

■ The subgenus is found across Southern Europe (Spain eastward) and thence to Uzbekistan. The only species is *Osmia brevicornis* (Fabricius). Another species, *O. (Helicosmia) mirhyi* Mavromoustakis, was formerly included by error (Zanden, 1992).

The nest is remarkable in the sense that there are no separate cells. A cavity made by a wood-boring insect is filled with pollen, eggs being scattered through the pollen as provisioning progresses. Larvae feed on nearby pollen; cocoons are arranged at random and variously oriented, with larval feces filling the spaces between them (Radchenko, 1978; Bosch, Vicens, and Blas, 1993).

Osmia / Subgenus *Monosmia* Tkalcŭ

Osmia (Monosmia) Tkalcŭ, 1974b: 337. Type species: *Osmia apicata* Smith, by original designation.

This subgenus consists of rather elongate (9-14 mm long), weakly metallic bees with a proboscis as long as the body (reaching basal part of metasoma in repose). This is the only osmiine except *Noteriades* with two spines at the apex of the front tibia; in *Monosmia* the additional spine is behind the usual spine, whereas in *Noteriades* the reverse is true. T6 and T7 of the male are both simple, the latter having a shallow, scarcely noticeable emargination. The clypeus of the female is truncate and overhangs the base of the labrum. Tkalcŭ (1974b) illustrated male genitalia and other structures.

■ This subgenus is known from the eastern Mediterranean area, and eastward in southwestern Asia to western Iran. The only species is *Osmia apicata* Smith.

Osmia / Subgenus *Mystacosmia* Snelling

Osmia (Mystacosmia) Snelling, 1967: 104. Type species: *Osmia nemoris* Sandhouse, 1924, by original designation.

The male of this subgenus is one of those in which S2 is approximately truncate, not hiding the median part of S3, and S3 is likewise transverse, not emarginate. It thus resembles *Euthosmia* and *Cephalosmia* among American subgenera. It differs from both in the prolongation of T6 over T7, the crenulate clypeal truncation, etc. The female does not differ from females of *Melanosmia* in the usual group characters, although Snelling (1967) gives a key couplet that, by "and/or" alternatives, appears to distinguish it from at least part of *Melanosmia*. The presence of long spatulate bristles on the small segments of the anterior tarsi of *Mystacosmia* appears to be distinctive. Except for the black scopa, pubescence is pale; the body is dull metallic blue-green; length is 8 to 12 mm. The mandible of the female is robust, with a distinct transverse basal groove (not as abrupt as in *Helicosmia*), and the apex four-toothed, the third tooth depressed like a "cutting edge."

■ This subgenus occurs in western North America from British Columbia, Canada, to California, east to Montana and Utah, USA. There is one (possibly two) species (Snelling, 1967).

Cells of *Mystacosmia* have been found in hollow stems and in burrows of a bee *(Diadasia)* in the soil. Cells are made of leaf pulp and resin (Rust and Clement, 1972).

Osmia / Subgenus *Neosmia* Tkalcŭ

Osmia (Neosmia) Tkalcŭ, 1974b: 333. Type species: *Osmia gracilicornis* Pérez, by original designation.

This subgenus consists of robust, nonmetallic species, most females with long hairs, suggesting *Osmia* s. str. Hairs are mixed, pale or red and black. The body length ranges from 7.5 to 15.0 mm. The clypeus of the female is truncate, overhanging the base of the labrum, and in the middle of the margin of the truncation is a small depressed area; the disc of the clypeus is rugose. The mandible of the female is narrower medially than at the base. The most distinctive character is the long first flagellar segment of both sexes (see the keys to subgenera). T6 of the male is simple; T7 is weakly or strongly bilobed. Tkalcŭ (1974b, 1977b, 1993b) illustrated male genitalia and other structures.

■ This subgenus occurs from the Canary Islands and Britain eastward through much of Europe and the Mediterranean basin to Israel. An old specimen of *Osmia bicolor* (Schrank) labeled "Angara River" in east central Russia (Tkalcŭ, 1995) may be correctly labeled. Van der Zanden (1988a) listed eight species. The subgenus was revised by Tkalcŭ (1977b).

Osmia (N.) bicolor (Schrank) is perhaps the best known of the snail-shell *Osmia* species. It uses chewed leaf material for the cells, and covers the snail shell with a pile of pine needles or grass stuck together with a secretion (Westrich, 1989). *O. (N.) rufigastra* Lepeletier makes similar nests, armors them with bits of snail shell embedded in leaf material, and moves them under grass or stones.

Osmia / Subgenus *Orientosmia* Peters

Osmia (Orientosmia) Peters, 1978b: 332. Type species: *Osmia maxillaris* Morawitz, 1875, by original designation.

This subgenus contains rather elongate (12-15 mm long), weakly metallic, pale-haired bees with strong white tergal fasciae. As in *Monosmia*, the proboscis is extremely long. The parapsidal line is short-linear, about one-seventh as long as the tegula. T6 and T7 of the male are relatively simple but more heavily sclerotized than those in many other subgenera. S2 to S4 of the male are unmodified, transverse; such sterna are not known in any other palearctic *Osmia* except *Tergosmia* and *Ozbekosmia*.

■ This subgenus is found in Turkey, Iran, and southwestern parts of the former USSR. The only species is *Osmia maxillaris* Morawitz.

Osmia / Subgenus *Osmia* Panzer s. str

Osmia Panzer, 1806: 230. Type species: *Apis bicornis* Linnaeus, 1758 = *Apis rufa* Linnaeus, 1758, by designation of Latreille, 1810: 439. An invalid designation was listed by Michener, 1997b.

Amblys Klug, *in* Illiger, 1807: 198; Klug, 1807b: 226. Type species: *Apis bicornis* Linnaeus, 1758 = *Apis rufa* Linnaeus, 1758, by designation of Latreille, 1811: 577. [Monobasic in Illiger, 1807; Klug, 1807b was published simultaneously and listed two species.]

Osmia (Ceratosmia) Thomson, 1872: 232. Type species: *Apis bicornis* Linnaeus, 1758 = *Apis rufa* Linnaeus, 1758, by designation of Sandhouse, 1939: 9.

Osmia (Aceratosmia) Schmiedeknecht, 1885: 19. Type species: *Osmia emarginata* Lepeletier, 1841, by designation of Sandhouse, 1939: 9.

Osmia (Pachyosmia) Ducke, 1900: 18. Type species: *Apis rufa* Linnaeus, 1758, by designation of Sandhouse, 1939: 9.

Osmia s. str. contains species that are long-haired and robust, 8.5-16.0 mm long, the anterior face of T1 shagreened. Some are completely black but most have weak metallic tints; others are strongly metallic, and the North American species, *O. ribifloris* Cockerell, is brilliant green. In the females of most species the lower lateral parts of the clypeus are produced forward as strong angles or horns, and the base of the labrum is not hidden. In a few species [e.g., *O. ribifloris* Cockerell, *mustelina* Gerstaecker, and *nigrohirta* Friese] the clypeus of the female is quite ordinary, truncate and overhanging the base of the labrum; these are species that would fall in *Acerotosmia* if that subgenus were recognized (see below). Intermediate clypeal structure occurs in *O. (O.) fedtschenkoi* Morawitz and *excavata* Alfken, in which the clypeus overhangs the base of the labrum but is deeply emarginate. T6 and T7 of males are simple, at most with a small median emargination in T7 and rarely in T6; T7 is but little exposed. Male genitalia and other structures were illustrated by Mitchell (1962) and Peters (1978b).

Tkalců (1969a) argues for the recognition of *Aceratosmia* as a separate subgenus because of the four-segmented (rather than five-segmented) maxillary palpi and other characters. It is a distinctive group of four palearctic species and includes also *Osmia ribifloris* Cockerell in North America, but its close relation to *Osmia* s. str. seems clear.

■ This subgenus ranges from westernmost Europe to Japan, south to northern Africa and Baluchistan (Pakistan), and across the whole of North America from Canada to Durango, Mexico. The west palearctic species were revised by Peters (1978b); the American species, by Rust (1974). There are about 20 palearctic species (Zanden, 1988a) and two nearctic species (Sandhouse, 1939; Rust, 1974).

Nests are placed in existing cavities, often in holes in wood or stems, or in abandoned nest burrows of other bees in clay banks. Cells are made of mud, or mud is used for partitions between cells (Pl. 16), if a burrow or hollow stem is divided into a series of cells [*Osmia rufa* (Linnaeus), *cornuta* (Latreille) (Tasei, 1973); *lignaria* Say], or cells are made of masticated leaf material [*O. ribifloris* Cockerell (Rust, 1986)]. *O. emarginata* Lepeletier uses both mud and masticated leaves, according to Malyshev (1937). Some species (*O. mustelina* Gerstaecker) make cells of chewed leaf material placed in crevices in stones.

Osmia / Subgenus *Ozbekosmia* Zanden

Osmia (Heterosmia) Tkalců, 1993a: 56, nomen nudum.

Osmia (Ozbekosmia) Zanden, 1994: 168. Type species: *Osmia avosetta* Warncke, 1988, by original designation.

The lone species of this subgenus was included in the subgenus *Tergosmia* by Warncke (1988a). It does agree with *Tergosmia* in several respects; e.g., it is nonmetallic and pale-haired, the female mandible is tridentate and narrower medially than at the base, S2 and S3 of the male are transverse (with strong, shining transverse gibbosities in *Ozbekosmia*), and S5 of the male is emarginate, with a fringe in the emargination (an unusual character in the Osmiini). It differs from *Tergosmia*, however, in many ways: clypeal margin of the female is produced, not at all truncate; and in the male there is a strong lateral tooth on T6, suggesting *Hoplitis (Anthocopa)*, an enormously produced, heavily sclerotized median apical projection of T7, somewhat suggestive of the subgenus *Allosmia*, and unusually long and slender genitalia, nearly three times as long as the width of the gonobase. The narrow double margin around the apex of S4, except medially, as in *Helicosmia* and *Pyrosmia*, also differs from *Tergosmia*. The punctiform parapsidal lines and lack of basal flaps on S6 of the male exclude *Ozbekosmia* from the genus *Hoplitis*, although it superficially resembles *Hoplitis (Anthocopa)*. Moreover, in the female there are long curved hairs overhanging the impunctate although roughened anterior part of the hypostomal area, as in *H. (Anthocopa)* but unlike *Osmia (Tergosmia)*. The body length is 7 to 13 mm. Various structures were drawn by Warncke (1988a).

■ *Ozbekosmia* is known only from Turkey. The one species is *Osmia avosetta* Warncke.

Osmia / Subgenus *Pyrosmia* Tkalců

Osmia (Pyrosmia) Tkalců, 1975c: 182. Type species: *Osmia ferruginea* Latreille, 1811, by original designation.

Osmia (Viridosmia) Warncke, 1988b: 5. Type species: *Osmia saxicola* Ducke, 1899, by original designation.

Osmia (Caerulosmia) Zanden, 1988a: 123, nomen nudum; Zanden, 1989: 83. Type species: *Osmia gallarum* Spinola, 1808, by original designation.

This subgenus embraces small to moderate-sized (4-8 mm long) metallic blue, green, or sometimes purple or red bees; the pubescence is mostly or entirely pale (red in females of *Osmia ferruginea* Latreille) and often forms apical tergal fasciae. The clypeus of the female is truncate, overlapping the labral base, and the mandibles are robust, but narrower medially than basally. T6 of the male is often somewhat inflated laterally around the rather small T7, which is often deeply trifid but variable, as noted above. The hind coxa has a longitudinal carina on the inner ventral surface, sometimes not developed in males.

■ *Pyrosmia* is found in Central and Southern Europe, the entire Mediterranean basin, and east to India and Mongolia. It was reviewed by van der Zanden (1991a) and a key to the Near Eastern species was given by Warncke (1992d). There are about 33 species.

The name *Pyrosmia* has usually been used for a single colorful species, *Osmia ferruginea* Latreille. *O. mongolica* Morawitz is similarly colorful. Because of their metallic blue, green, red, and purple bodies, the females in particular having red hair, these two species seem out of place when associated with the species recently placed in *Viridosmia*. However, some *Viridosmia* species—e.g., *O. (P.) gemmea* Pérez and *versicolor* Latreille—are sometimes at least partly purplish. A feature of *O. ferruginea* not found in other species of the subgenus as here understood is the entire T7 of the male. In the other species it is usually trifid, but *O. viridana* Morawitz has nearly lost the middle tooth of T7; moreover, it is entirely gone and T7 is bispinose in *O. nana* Morawitz. In the latter, the two remaining spines are laterally thickened in a distinctive way, and in *O. teunisseni* Zanden these lateral expansions are developed into strong teeth, rendering T7 four-toothed. Thus the form of T7 in this group is not a good subgeneric character. Evidence that *O. ferruginea* is related to the species placed in *Viridosmia* includes, for the male, the broadly expanded posterior lateral margin of T6, the longitudinal median groove of S6, and the form of S4, including the weakly indicated double apical carinae at the side; all these features are as in *O. versicolor*, for example. The character of the apical margin of S4 is in fact extremely inconspicuous in some species; see the keys to subgenera and the comment immediately preceding the first key. In the female of *O. ferruginea* the minute median depression in the apical margin of the clypeus is as in many other species of *Pyrosmia*.

Species of this subgenus have also been placed in the American subgenus *Diceratosmia*. Characters common to both *Diceratosmia* and *Pyrosmia* include the doubly carinate posterior margin of the male S4, at least laterally; the carinate hind coxae, at least in females; the slightly elongate rather than strictly punctiform parapsidal lines; and the short, mostly pale pubescence, often forming narrow white apical fasciae on the terga. Although both Warncke and van der Zanden have separated the palearctic species subgenerically from the American species, the differences are not very impressive, and recognition of a single holarctic subgenus is reasonable. The hind coxa in *Pyrosmia* is less strongly carinate or, in males, may lack a carina, whereas in *Diceratosmia* the carina is strong in both sexes and usually continues to the ventral apical angle of the coxa. In *Pyrosmia*, T6 of the male is simple or obtusely angulate laterally and not produced medially, whereas in *Diceratosmia* it is strongly angled laterally and has a broad median apical projection. S6, as noted above, has a longitudinal median groove in *Pyrosmia* that is lacking in *Diceratosmia*.

Osmia (Pyrosmia) cephalotes Morawitz belongs to this subgenus, as shown by the trifid male T7 and other characters, but as noted in the discussions of the tribe and the genus *Osmia*, it has an elongate body, like a *Chelostoma* or slender species of *Hoplitis*.

Nests are made in holes in wood, stems, galls, etc., and *Osmia versicolor* Latreille and *ferruginea* Latreille nest in snail shells. Cell partitions are of leaf pulp.

Osmia / Subgenus *Tergosmia* Warncke

Osmia (Tergosmia) Warncke, 1988a: 390. Type species: *Osmia tergestensis* Ducke, 1897, by original designation.

This subgenus contains rather ordinary-looking, robust, nonmetallic species 6.5 to 13.0 mm long. The clypeus of the female is truncate and overhangs the base of the labrum, as in most *Osmia* species, but lacks the brushes of orange hairs. The mandibles of the female are narrower medially than at the base, and three-toothed. The parapsidal lines are variable, more or less punctiform in *O. (T.) agilis* Morawitz and *lunata* Benoist, but short-linear in *O. (T.) rhodoensis* (Zanden). The most noteworthy character is the simple, transverse S2 to S4 of the male, as in the subgenus *Ozbekosmia* and the very different *Orientosmia*. An unusual character is the strong emargination of S5, which is usually filled with hair. T6 is simple, T7 bilobed to bispinose. Structures were illustrated by Tkalcŭ (1994).

■ This subgenus occurs in the Mediterranean basin and in Uzbekistan and Turkmenistan. The six species were reviewed by Warncke (1988a). One of his species, *Osmia avosetta* Warncke, is out of place here and has been transferred to the subgenus *Ozbekosmia*, but another species was added by Tkalcŭ (1994).

Osmia / Subgenus *Trichinosmia* Sinha

Osmia (Trichinosmia) Sinha, 1958: 244. Type species: *Osmia latisulcata* Michener, 1936, by original designation.

Among the relatives of *Melanosmia*, this subgenus can be recognized by the unusually long and branched hair of the head and thorax as well as the long wing hairs and the female clypeal and male antennal characters indicated in the key to subgenera. The hairs are mixed black and white; the body is metallic green, its length 9 to 10 mm.

■ This subgenus occurs in the southwestern USA, from California and Arizona to Utah. The only species is *Osmia latisulcata* Michener.

Nests have been found in holes in wood blocks; cells are made of sand mixed with leaf pulp and apparently a secretion (Parker, 1984).

Genus *Othinosmia* Michener

This genus is related to *Protosmia* and has been included within that genus (e.g., Griswold, 1985), with which it agrees (except for *Othinosmia* subgenus *Afrosmia*) in the carinulae of the pronotal lobe. It also agrees with *Protosmia (Protosmia* s. str. and *Nanosmia*) in the general shape of the thorax, with its sloping metanotum and the indistinct, sloping basal zone of the propodeum. Distinctive features are the preapical transverse carina on the male T6, the lack of a lateral flap on the same tergum, and the exposed T7 of the male, which may be short and rounded to subtruncate or produced to a rather long point. Except in the subgenus *Afrosmia*, there is a mid-apical notch in S1 of the male; such a notch is absent in *Protosmia*. In both sexes the lateral line on T1 is long, attaining the level of the spiracle, and the summit of the metapleuron lacks a carina. In *Protosmia* the lateral line on T1 is shorter, and an upper metapleural carina is visible from beneath.

Unlike the holarctic genus *Protosmia*, *Othinosmia* is restricted to sub-Saharan Africa.

Key to the Subgenera of *Othinosmia* (Males)
(By Terry L. Griswold)

1. Front tibia with apical spine long, curved posteriorly; S2 not depressed medially, without dense covering of hair; hypostomal area with polished, impunctate area (southern Africa) .... *O. (Othinosmia* s. str.)
—. Front tibia with apical spine short, straight; S2 depressed medially, densely covered with velvety hair; hypostomal area without polished, impunctate area 2
2(1). Metapleuron with large dorsal projection; axilla produced to acute angle (Kenya) .... *O. (Afrosmia)*
—. Metapleuron without dorsal projection; axilla rounded (southern Africa) .... *O. (Megaloheriades)*

Key to the Subgenera of *Othinosmia* (Females)
(By Terry L. Griswold)

1. Metapleuron with dorsal projection; axilla produced to acute angle; body length 9 to 11 mm (pronotal lobe with carina) (Kenya) .... *O. (Afrosmia)*
—. Metapleuron without dorsal projection; axilla rounded; body length 5 to 8 mm 2
2(1). Front tibia with long, curved apical spine; clypeus not overhanging labrum, base of labrum visible when mandibles are closed (pronotal lobe without carina) (southern Africa) .... *O. (Othinosmia* s. str.)
—. Front tibial spine short, nearly straight; clypeus at least slightly overhanging labrum, base of labrum not visible when mandibles are closed (southern Africa) .... *O. (Megaloheriades)*

Othinosmia / Subgenus *Afrosmia* Griswold

Protosmia (Afrosmia) Griswold, 1998, *in* Griswold and Michener, 1998: 239. Type species: *Othinosmia stupenda* Griswold, 1994, by original designation.

The lone species of this subgenus is larger than other *Othinosmia*, and differs further in lacking carinulae behind the carina on the pronotal lobe, and in the angularly produced axilla, and the upper metapleural projection. These features suggest a relationship to *Stenoheriades*, although none is unique to *Stenoheriades* and *Afrosmia*. The preapical carina of T6 of the male, the small (rounded) T7, etc., support placement in *Othinosmia*. Male genitalia and sterna were illustrated by Griswold (1994a).

■ *Afrosmia* is known only from Kenya. The only known species is *Othinosmia stupenda* Griswold.

Othinosmia / Subgenus *Megaloheriades* Peters

Megaloheriades Peters, 1984: 366. Type species: *Osmia schultzei* Friese, 1909, by original designation.

In this subgenus the pronotal lobe is carinate or not (Griswold, 1985). The principal subgeneric characters are indicated in the key to subgenera.

■ This subgenus is known only from Cape Province, South Africa, and Namibia. Griswold (1985) listed seven species.

Nests are made of pebbles and resin, exposed on twigs or rocks, at least in the case of *Othinosmia (M.) globicola* (Stadelmann) (Michener, 1968a).

Othinosmia / Subgenus *Othinosmia* Michener s. str.

Anthocopa (Othinosmia) Michener, 1943: 86. Type species: *Thaumatosoma moniliferum* Cockerell, 1932, by original designation.

The principal subgeneric characters are indicated in the key to subgenera. Others are the convex margin of S4 of the male and the apically widened gonostylus. The receding clypeus of the female, exposing the base of the labrum, is characteristic of the subgenus. Peters (1984) illustrated the genitalia and hidden sterna.

■ This subgenus is known only from Cape Province, South Africa, and Namibia. The five known species were listed by Griswold (1985).

The elongate flagellum of males of some species led to placement of the type species in *Thaumatosoma*, a subgenus of *Megachile*. The only other osmiines with such antennae are the Burmese *Protosmia (Dolichosmia) burmanica* (Bingham) and some of the *Heriades* from southern Asia, such as *H. (Michenerella) testaceicornis* (Cameron).

Nests, so far as is known, are in holes in the ground, the cells lined with resin (Michener, 1968a).

Genus *Protosmia* Ducke

This genus contains more or less heriadiform or nearly megachiliform bees, generally of small size (body length 3.5-9.5 mm). In many ways, *Protosmia* and *Othinosmia* bridge the gap between the *Heriades*-like and *Osmia*-like sections of Osmiini, as indicated in the account of the tribe Osmiini. In some *Protosmia* and some *Othinosmia* the pronotal lobe lacks a carina (as in many other Osmiini) but, except in *Othinosmia stupenda* (Griswold) and *P. burmanica* (Bingham), it has, behind the carina or the location where it would be, several small carinulae. Such carinulae are absent in most other Osmiini. The axillae are rounded. The basal zone of the propodeum is sloping, not delimited by a carina (except it is more horizontal and pitted in the subgenera *Chelostomopsis* and *Dolichosmia*). T1 lacks a carina, the anterior surface being flat or somewhat concave. T6 of the male lacks a preapical carina and has a lateral flap (this not verified for *Dolichosmia*) that is hidden when the metasoma is in repose. As in *Wainia*, T7 of the male is not exposed. Male genitalia and sterna were illustrated by Popov (1961) and Tkalců (1978b).

Four subgenera are recognized, as in the key below. The genus has a disjunct distribution—the Mediterranean region to northern India and Burma, and one species in western North America.

Key to the Subgenera of *Protosmia* (Males)
(By Terry L. Griswold)

1. S1 with large ventral projection; margin of T6 acutely angled laterally; flagellum modified 2
—. S1 without ventral projection; margin of T6 not acutely angled laterally; flagellum not modified 3
2(1). Gena with elongate lateroventral fovea or crease; distal flagellar segments stout (palearctic) .... *P. (Protosmia s. str.)*
—. Gena without lateroventral fovea; distal flagellar segments, except the last, threadlike (Burma) .... *P. (Dolichosmia)*
3(1). T6 in dorsal view with strongly projecting median lobe; pronotal lobe not carinate; S2 with median longitudinal line of short, appressed hairs, these quite different from long, erect hairs of adjacent areas (Mediterranean region, western nearctic) .... *P. (Chelostomopsis)*
—. T6 in dorsal view with weakly projecting median lobe; pronotal lobe carinate; S2 without median longitudinal line of short hairs (Mediterranean region).... *P. (Nanosmia)*

Key to the Subgenera of *Protosmia* (Females)
(By Terry L. Griswold)

1. Mandible elongate, its maximum length approximately equal to maximum clypeal width; gena much wider than eye in lateral view (Mediterranean region, western nearctic) .... *P. (Chelostomopsis)*
—. Mandible not elongate, its maximum length at most two-thirds maximum clypeal width; gena little if any wider than eye in lateral view.. 2
2(1). Body slender, metasoma elongate; pronotal lobe without carinules in addition to carina; posterior surface of front tibia with patch of dense, velvety hair near apex (Burma) .... *P. (Dolichosmia)*
—. Body stout, metasoma not elongate; pronotal lobe with carinules in addition to carina; posterior surface of front tibia sparsely haired 3
3(2). T1 (and usually T2) with narrow, distinct impunctate margins; flagellum clavate (palearctic) ....
.... *P. (Protosmia s. str.)*
—. T1 and T2 without impunctate margins; flagellum nearly cylindrical (Mediterranean region).... *P. (Nanosmia)*

Protosmia / Subgenus *Chelostomopsis* Cockerell

Chelostomopsis Cockerell, 1925b: 205. Type species: *Chelynia rubifloris* Cockerell, 1898, by original designation.
Raphidostoma Cockerell, 1936a: 133. Type species: *Raphidostoma ceanothi* Cockerell, 1936 = *Chelynia rubifloris* Cockerell, 1898, by original designation.

This subgenus differs from the other subgenera in the long mandibles of the female and exposed base of the labrum when the mandibles are closed, suggesting the genus *Hofferia*. *Chelostomopsis* differs from *Hofferia*, of course, in the generic characters of *Protosmia*, e.g., the hidden T7 in the male, the carinulae on the pronotal lobe, a preapical tuft of erect hairs on the labrum of the female, etc. *Chelostomopsis* differs from other subgenera of *Protosmia* in the lack of a carina on the pronotal lobe, the more horizontal, pitted basal zone of the propodeum and the nearly horizontal metanotum that is on a level with the scutellum. The median flagellar segments of the male are longer than wide, not flattened. Illustrations of male genitalia and sterna are by Popov (1961).

■ The distribution is disjunct—the Mediterranean basin (Morocco to Lebanon and Turkey) and western North America from British Columbia, Canada, to Baja California, Mexico, and northern Arizona, USA. Four species were listed by Griswold (1985); only one of them is North American.

The female of the North American species, *Protosmia rubifloris* (Cockerell), is unusual in having a slender, midapical projection on the clypeus (Fig. 81-13). Old World species such as *P. capitata* (Schletterer) lack such a structure but are otherwise remarkably similar to the American species.

Protosmia rubifloris (Cockerell) nests in holes in wood or pinecones. Griswold (1986a) found that overwintering is by young adults.

Protosmia / Subgenus *Dolichosmia* Griswold

Protosmia (Dolichosmia) Griswold, 1998, *in* Griswold and Michener, 1998: 241. Type species: *Thaumatosoma burmanicum* Bingham, 1897, by original designation.

This subgenus differs from other *Protosmia* in the absence of carinules supplementing the carina on the pronotal lobe and in the presence of a dense patch of hair distally on the posterior surface of the fore tibia. The body is elongate and the basal zone of the propodeum is pitted, as in *Chelostomopsis*. Females closely resemble a finely punctate *Heriades*, but the angle between the anterior and dorsal surfaces of T1 is not carinate. The basal zone of the propodeum is less distinctive than in most *Heriades* and does not continue behind the spiracle, and the hind trochanter lacks a ventral carina. In males the distal flagellar segments, except for the greatly expanded last segment, are threadlike. The only other Osmiini with such a condition are a few oriental *Heriades* such as *H. (Michenerella) testaceicornis* (Cameron) and some species of *Othinosmia* s. str. The deep lateral pits on male T6 are also distinctive.

■ This subgenus is known only from a handful of specimens from Burma. The single species is *Protosmia burmanica* (Bingham).

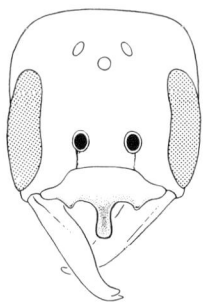

Figure 81-13. Face of *Protosmia (Chelostomopsis) rubifloris* (Cockerell), female. From Michener, McGinley, and Danforth, 1994

Protosmia / Subgenus *Nanosmia* Griswold

Protosmia (Nanosmia) Griswold, 1998, *in* Griswold and Michener, 1998: 241. Type species: *Protosmia asensioi* Griswold and Parker, 1987, by original designation.

This subgenus is similar to *Protosmia* s. str., the females having almost no group distinctions (see the key to subgenera). Males, however, differ strongly in the characters indicated in the key. Moreover, the flagellum is not flattened as in *Protosmia* s. str., its middle segments being longer than wide. Illustrations of male genitalia and sterna are to be found in Griswold and Parker (1987).

■ This subgenus is found in the Mediterranean basin, southwestern Asia, and Kashmir. Griswold (1985) listed six species.

Nests are found in stems and holes in wood (Griswold, 1985).

Protosmia / Subgenus *Protosmia* Ducke s. str.

Osmia (Protosmia) Ducke, 1900: 12. Type species: *Heriades glutinosus* Giraud, 1871, by designation of Michener, 1943: 86 and Sandhouse, 1943: 592.
Anthocopa (Rhodosmia) Michener, 1943: 85. Type species: *Osmia paradoxa* Friese, 1899, by original designation.

Males are separable from those of other subgenera by their flattened flagellum, the middle segments wider than long, and by the genal fovea or crease and the ventral projection of S1. The pronotal lobe is carinate, as in *Nanosmia*. Illustrations of male genitalia and sterna are found in Popov (1961) and Tkalcŭ (1978b).

■ This subgenus occurs in the Mediterranean basin (Spain to Israel), north to Bulgaria, and east through Asia Minor to the Transcaucasus. Griswold (1985) listed 12 species; van der Zanden (in litt., 1993) indicated that he knows 19 species.

In some species the distal margin of the female mandible is made up of multiple small crenulations instead of being three-toothed. The name *Protosmia* was originally proposed for such species, and *Rhodosmia* was intended for those with ordinary tridentate mandibles. No other characters appear to differentiate the two groups, and *Rhodosmia* seems best placed in synonymy with *Protosmia*, following Griswold (1985); Tkalcŭ (1978b) considered them separate, however.

Popov (1961) reviewed the nesting biology. Nests have been reported in various kinds of cavities—snail shells, rock crevices, and abandoned mud nests of other aculeate Hymenoptera (summary by Griswold, 1985).

Genus *Pseudoheriades* Peters

Pseudoheriades Peters, 1970a: 153. Type species: *Eriades moricei* Friese, 1897, by original designation.
Spinasternella Gupta and Sharma, 1993, *in* Gupta, 1993: 104. Type species: *Spinasternella mevatus* Gupta and Sharma, 1993, by original designation.

This is a genus of small (4.0-8.5 mm long) heriadiform bees sharing with *Heriades* coarse punctation and diverse carinae. The preoccipital carina is present or at least represented by a sharp angle laterally, but weak to absent dorsally. The structure of the thorax and T1 is essentially as in *Heriades* except that the summit of the metapleuron has a shiny, projecting shelf and the carina behind the basal zone of the propodeum does not extend laterally behind the propodeal spiracle, as it does in *Heriades*. The maxillary palpi have only two segments. The male differs from that of *Heriades* in the exposed, quadrate T7, sculptured like T6 and quadrately surrounded by T6, and in the medial apical spine on S3. The female differs from that of similar genera (except one species of *Afroheriades*) in the smooth anterior part of the hypostomal area, bounded laterally by a carina and a fringe of long hairs. S1 of the female has a slender, erect spine. The male genitalia and hidden sterna are illustrated by Peters (1970a).

■ *Pseudoheriades* is widespread in sub-Saharan Africa, and ranges north to Algeria, Egypt, Israel, and Iran, and east to the Arabian Peninsula and India. There are seven described species and at least three undescribed species. Gupta (1993) gave a key to four Indian species.

Nests are built in burrows (doubtless made by other insects) in wood, and resin is used for cell partitions (Griswold, 1985).

Genus *Stenoheriades* Tkalcŭ

Pseudoheriades (Stenoheriades) Tkalcŭ, 1984a: 1. Type species: *Pseudoheriades hofferi* Tkalcŭ, 1984, by original designation.

This genus contains small heriadiform bees, 5 to 7 mm in length, that are coarsely punctate and with various carinae, as in *Heriades*, to which genus *Stenoheriades* is probably most closely related. In thoracic and T1 characters, some or all species of *Stenoheriades* agree with the characters of *Heriades*. Males differ from those of *Heriades* and other megachilids in having both of the following features: (1) T6 has a transverse preapical ridge that is strongly dentate or crenulate [*Heriades* has no such ridge, although there is a projection beyond a very different sort of ridge in *H. (Rhopaloheriades)*] and (2) T7 is exposed, notched or bifid apically. *Hofferia* also has a preapical ridge on T6 (but it is nodulose) and an exposed T7 (but it is broad and truncate). The best character distinguishing females of *Stenoheriades* from those of *Heriades* is the longer proboscis of the former, extending well beyond the fossa in repose. Illustrations of male genitalia and sterna can be found in Tkalcŭ (1984a) and Griswold (1994a).

■ This genus ranges from the Mediterranean region (Morocco to southwest Asia) and Bulgaria to South Africa and Madagascar. In addition to ten named species (five of them from sub-Saharan Africa), there are at least six undescribed African species.

Griswold (1985) notes that *Stenoheriades* is divisible into four species groups, possibly as different as subgenera in other genera. Since the number of species is small, as is the number of specimens available (all species are rare), there is no need to invent subgeneric names at this time.

Genus *Stenosmia* Michener

Stenosmia Michener, 1941a: 165. Type species: *Osmia flavicornis* Morawitz, 1878, by original designation.

The relatively large stigma of this genus is suggestive of *Heriades, Chelostoma,* and *Protosmia,* in which the stigmal margin within the marginal cell is usually convex and the length of the stigma is sometimes as much as half the length of the marginal cell on the wing margin (i.e., from the apex of the stigma to the point where the cell bends away from the wing margin) (Fig. 81-1c). *Stenosmia* also resembles those genera in its small size (4.0-6.5 mm long) and slender hoplitiform body. Its relatively fine punctation, lack of carinae except on the lower half of the omaulus (even the pronotal lobe is acarinate or has only a faint carina, without carinulae), relatively broad scutum (minimum intertegular distance much greater than length of scutum), relatively straight (i.e., not strongly curled) male metasoma with exposed (simple or weakly bilobed) T7 and five exposed unmodified sterna (the second not enlarged) are all quite different from the corresponding configurations of the genera listed above, and except for the carinate omaulus are generally suggestive of *Hoplitis*. T6 is not toothed and S6 lacks basal flaps, characters that, along with the large stigma, distinguish *Stenosmia* from *Hoplitis*. The short proboscis, in repose not extending much beyond the fossa, distinguishes *Stenosmia* from several related genera.

■ This genus is found in central Asia, east to the Gobi desert in China; also in Egypt. Van der Zanden (in litt, 1993) indicated that there are 11 species.

Genus *Wainia* Tkalcŭ

This genus contains *Hoplitis*-like to *Heriades*-like bees. Its most distinctive feature is the complete invagination of T7 of the male, forming a largely membranous or weakly sclerotized structure. Unlike that of the other genera having this feature, its scutellum is curved down posteriorly, the metanotum and propodeum being strongly declivous and below the level of the main part of the scutellum; the propodeum lacks a basal pitted zone and carina. A character of all but the subgenus *Caposmia* is the acute or right-angular posterior lateral angle of the scutum. One species of that subgenus, *Wainia algoensis* (Brauns), also has such an angle, but it is a result of a strong marginal lamella and is probably not homologous to the strong angle of the other subgenera. S2 in males is large, covering the following sterna and often meeting T6; S3 and S4 are emarginate, with fringes in the emarginations and S5 and S6 are large, translucent, often coarsely longitudinally wrinkled, and covered with hairs. The male gonoforceps are slender and tapering except in the subgenus *Wainiella*, in which they are broadened apically.

Key to the Subgenera of *Wainia*

1. Metasomal pale hair bands of T2 to T4, except laterally on basal terga, immediately behind graduli, not on posterior marginal zones of terga; mandible of male three-toothed (Africa to southern India and the Philippines) .... *W. (Wainia s. str.)*
—. Metasomal pale hair bands on posterior marginal zones of terga; mandible of male two-toothed 2
2(1). Body heriadiform, 6-8 mm long; omaular carina distinct; female clypeus with longitudinal median carina; tibial spurs strongly hooked; labrum of female without well-formed fringe (Africa) .... *W. (Wainiella)*
—. Body robust hoplitiform, 8.0-10.5 mm long; omaular carina absent; female without clypeal carina; tibial spurs weakly curved at apices; labrum of female with distinct fringe (southern Africa and Israel) .... *Caposmia*

Wainia / Subgenus *Caposmia* Peters

Osmia (Caposmia) Peters, 1984: 378. Type species: *Osmia braunsi* Peters, 1984, by original designation.
Anthocopa (Eremoplosmia) Zanden, 1991b: 164. Type species: *Osmia eremoplana* Mavromoustakis, 1949, by original designation.

This subgenus contains species that look like *Hoplitis (Anthocopa)*, with which group the type species of *Eremoplosmia* was affiliated by van der Zanden (1991b). *Caposmia* differs from *Hoplitis* in the lack of basal flaps on S6 of the male, the lack of lateral teeth on T6 of the male, and the generic characters of *Wainia*. The sterna and genitalia were illustrated by van der Zanden (1991b). Contrary to the original description of *Eremoplosmia*, the claws of the female are simple. The more or less obtuse posterior lateral angle of the scutum differentiates this subgenus from the others, except that in *W. algoensis* (Brauns) the angle is acute because the usual carina is elevated to form a rounded lamella. As in *Wainia* s. str., T6 of the male has a preapical carina, and the margin of the tergum is thus thickened. The body length is 8.0 to 10.5 mm.

■ *Caposmia* is known in Israel, where *Wainia eremoplana* (Mavromoustakis) occurs, and in South Africa, where *W. algoensis* (Brauns), *braunsi* (Peters), and *elizabethae* (Friese) are found. Peters (1984) differentiated these African species.

Wainia / Subgenus *Wainia* Tkalců s. str.

Wainia Tkalců, 1980: 1. Type species: *Wainia lonavlae* Tkalců, 1980, by original designation.
Wainia (Trichotosmia) Tkalců, 1980: 16. Type species: *Wainia consimilis* Tkalců, 1980, by original designation.

This subgenus contains species 4.5 to 6.5 mm long, more robust than those of *Wainiella* and smaller than those of *Caposmia* and lacking clypeal or hind coxal carinae. Either the omaulus is not carinate or the lower portion is weakly carinate. T6 of the male is convex, the apical margin thickened, sometimes carinate, as in *Caposmia*. The male gonoforceps are linear and bear long hairs. Tkalců (1980) illustrated the male genitalia and many other structures.

■ This subgenus occurs from Namibia and South Africa north to Kenya and east to Yemen, Pakistan, India, and, according to T. Griswold (in litt., 1997), the Philippines. Three species have been named.

Wainia / Subgenus *Wainiella* Griswold

Wainia (Wainiella) Griswold, 1998, *in* Griswold and Michener, 1998: 234. Type species: *Heriades sakaniensis* Cockerell, 1936, by original designation.

This subgenus consists of small (6-8 mm long), cylindrical, heriadiform bees that resemble *Noteriades* in the narrow, longitudinal, median clypeal carina of females. The omaulus and hind coxa are carinate. T6 of males is flat, the apical margin simple. The male gonoforceps are enlarged apically, and lack fringes of long hairs.

■ This subgenus is found from South Africa to Kenya. The two known species are *Wainia albobarbata* (Cockerell) and *sakaniensis* (Cockerell).

Genus *Xeroheriades* Griswold

Xeroheriades Griswold, 1986b: 165. Type species: *Xeroheriades micheneri* Griswold, 1986, by original designation.

This is a genus of small, slender bees (length 4.5-7.0 mm) with the metasoma partly red. They lack the carinae (preoccipital, pronotal, omaular, and on T1) common in heriadine bees, but the hind coxa has a longitudinal ventral carina. Carinulae of the pronotal lobe are absent. As in *Chelostoma*, the male metasoma is relatively straight, T7 being exposed. S1 to S6 are all exposed, although S5 and S6 are less sclerotized than the others and only partly exposed. T1 is convex, with no angle between its anterior and dorsal surfaces, and there is a longitudinal median impressed line on the anterior surface. The basal zone of the propodeum is sloping, neither pitted nor defined by a carina.

■ This distinctive genus occurs in the isolated mountain ranges of the Mojave Desert, California. The only species is *Xeroheriades micheneri* Griswold.

82. Tribe Anthidiini

The principal characters of this tribe are indicated in Table 79-1 and in the key to the tribes of Megachilinae. The species are heriadiform, hopliriform, chalicodomiform, megachiliform, to anthophoriform bees, mostly with conspicuous yellow integumental markings on all tagmata (Pl. 8). They are commonly not densely pubescent. The characters of wing venation and stigmal shape are illustrated in Figure 82-1.

There is a tendency in the Anthidiini for the development of numerous, small, morphologically distinct taxa, and in consequence many genera and subgenera have been recognized, on the basis of carination, mandibular dentition, scutellar shape, etc. The diversity in male genitalia is also much greater than that of other tribes of Megachilinae; a sample of this diversity is shown in Figure 82-2. Nonetheless, Warncke (1980a) placed all nonparasitic Anthidiini of the western palearctic region in the genus *Anthidium*. This clearly lumps very dissimilar forms into a paraphyletic group from which parasitic taxa must have evolved.

This tribe is found on all continents, but only one species is known from Australia. Elsewhere, each continent supports many genera and species. A general treatment of the Anthidiini is found in Friese's (1911b) account of the Megachilidae, and treatments for various areas include Warncke (1980a) for the western palearctic region, Friese (1898a) for Europe, Grigarick and Stange (1968) for California, and Pasteels (1972, 1984) for the Indo-Malaysian region and sub-Saharan Africa, respectively. Pasteels (1969a) and Michener and Griswold (1994a) made generic studies for the Old World, and Michener (1948a) considered the generic classification, primarily of New World forms. In numerous publications, Pasteels (1971-1984) gave diagrammatic sketches of structures of Old World Anthidiini. Müller (1996b) developed a phylogeny for 72 western palearctic nonparasitic Anthidiini, for use in a study of their floral relationships. It is the first phylogenetic study of Anthidiini.

The brood cells of the Anthidiini are exposed [on rocks, stems (Fig. 7-1), or leaves] or in preexisting burrows or cavities (including empty snail shells and mud nests of wasps). Females of at least two genera, however, *Trachusa* and *Paranthidium*, may dig their own burrows. Cells are made of materials brought to the nests. Such materials include resin, sometimes mixed with or cementing together pebbles, soil particles, pieces of leaves, chaff, or plant hairs; alternatively, cells may consist of plant hairs or other fibers alone, even though dry plant wool or hairs would seem a difficult material to use in constructing cells. Some Anthidiini have dense, white, plumose hairs on the outer sides of all the female basitarsi. At least in *Anthidium manicatum* (Linnaeus) these hair pads are used to collect glandular products from leaves and stems of plants such as *Pelargonium* (Müller, 1996c). The material is used to moisten and render sticky the plant wool used in cells, thus presumably making it more manageable, perhaps also repellent to water, microorganisms, or

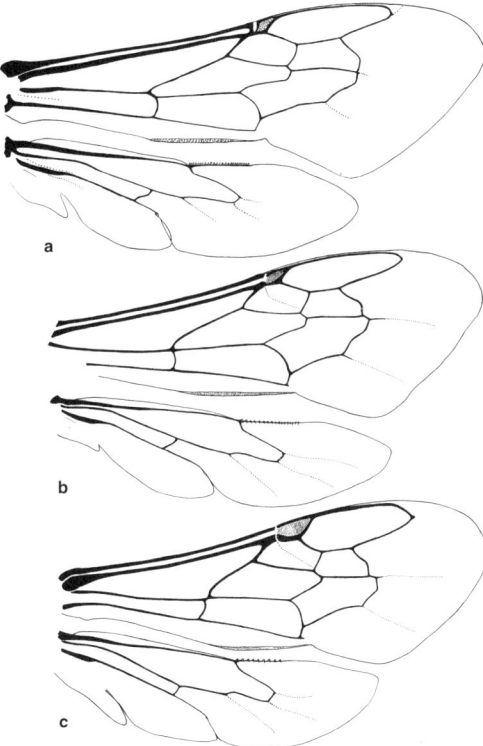

Figure 82-1. Wings of Anthidiini. **a,** *Anthidium manicatum* (Linnaeus); **b,** *Paranthidium jugatorium perpictum* (Cockerell); **c,** *Anthodioctes gualanense* (Cockerell). Note the variation in stigmal size and the position of the second recurrent vein, at or beyond the second submarginal crossvein.

Table 82-1. Genera of Anthidiini in Each Series (see text).

Series A		Series B
Acedanthidium	Hoplostelis	Afranthidium
Afrostelis	Hypanthidiodes	Anthidioma
Anthidiellum	Hypanthidium	Anthidium
Anthodioctes	Icteranthidium	Gnathanthidium
Apianthidium	Larinostelis	Indanthidium
Aspidosmia	Notanthidium	Neanthidium
Aztecanthidium	Pachyanthidium	Pseudoanthidium
Bathanthidium	Paranthidium	Serapista
Benanthis	Plesianthidium	
Cyphanthidium	Rhodanthidium	
Dianthidium	Stelis	
Duckeanthidium	Trachusa	
Eoanthidium	Trachusoides	
Epanthidium	Xenostelis	
Euaspis		

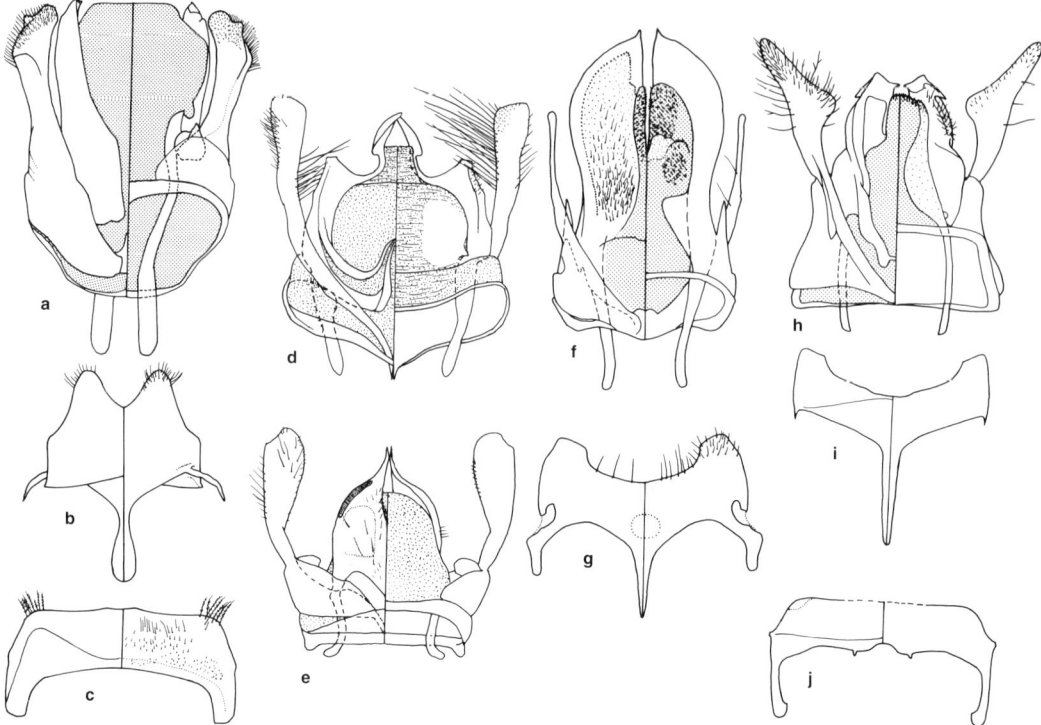

Figure 82-2. Male genitalia and hidden sterna of Anthidiini. **a-c,** Genitalia, S8, and S7 of *Trachusa (Ulanthidium) pueblana* Thorp and Brooks; **d, e,** Genitalia of *Acedanthidium flavoclypeatum* (Gupta) and *Pseudoanthidium (Tuberanthidium) brachiatum* Michener and Griswold; **f, g,** Genitalia and S8 of *Dianthidium (Adanthidium) discophorum* Griswold and Michener; **h-j,** Genitalia, S8, and S7 of *D. (Deranchanthidium) chamela* Griswold and Michener. (In divided illustrations, the dorsal views are on the left.) a-c, from Thorp and Brooks, 1994; d, e, from Michener and Griswold, 1994a; f-j, from Griswold and Michener, 1988.

other insects. Pasteels (1977a) reviewed the nesting habits of Old World species, and the nests of various taxa are described below under the genera or subgenera concerned. Cleptoparasitic Anthidiini mostly parasitize other Megachilidae, although a few deviate and attack other bees; e.g., *Hoplostelis* s. str. parasitizes Euglossini. Larvae of Anthidiini are not known to present conspicuous generic or tribal characters. They have been described and illustrated by Michener (1953a), Grandi (1961), Clement (1976), and others.

Anthidiini can be divided, for convenience, into two series. Series A includes those taxa in which the mandibles of the females have three or four or rarely more teeth joined by shallow or at least rounded concavities, so that, except frequently for the lowermost and uppermost teeth, the teeth are obtuse or rounded and often mere angles on the mandibular margin (Fig. 82-3b). Sometimes the second and third teeth are indistinguishable (Fig. 82-3c) or represented by feeble convexities on a nearly straight margin. Except for the parasitic genera, species in this series make nests with resin, often supplemented by pebbles, earth, leaf fragments, or, in the case of at least one *Pachyanthidium,* plant hairs. Series B includes those in which the mandibles of the female have five or more, commonly sharp teeth, separated by acute, V-shaped notches (Fig. 82-3a). Series B is the group in which nests are made at least largely of fibers such as plant hairs. Series A is the more diverse of the two and is no doubt a paraphyletic group from which Series B arose one or more times. *Pachyanthidium* is placed in Series B but it includes some species that fall in Series A. The first couplet in each of the two keys that follow separates Series A and B, and Table 82-1 lists the genera in each.

Couplet 2 of the first key deals with T5. More anterior terga of both sexes and T6 of males reflect the same features, often less clearly. Taxa that are not clearly separable by this character can be run to either alternative.

Key to the Genera of the Anthidiini of the Eastern Hemisphere

(Modified from Michener and Griswold, 1994a)
(See also the Supplementary Key to Males, below.)

1. Mandible of female with 5 to 18, usually sharp teeth separated by acute notches (Fig. 82-3a) *or* (in one species of *Pachyanthidium*) minutely denticulate; maxillary palpus minute [relatively long in *Afranthidium (Zosteranthidium)*], two-segmented (arolia absent; base of propodeal triangle punctate or finely roughened, nearly always hairy; propodeum without basal series of pits and without fovea behind spiracle; juxtantennal carina absent) (Series B) 2

—. Mandible of female with three or four teeth (Fig. 82-3b), or, *if* with five to ten, then the teeth are rounded and at least some of them separated by rounded emargina-

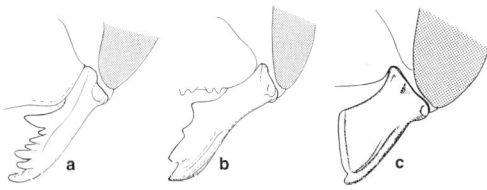

Figure 82-3. Mandible and adjacent structures of female Anthidiini. **a,** *Anthidium maculosum* Cresson; **b,** *Hypanthidium toboganum* (Cockerell); **c,** *Dianthidium (Deranchanthidium) chamela* Griswold and Michener. From Michener, McGinley, and Danforth, 1994.

tions; maxillary palpus commonly three- or four-segmented but sometimes two-segmented (Series A) 11

2(1). T5 with posterior marginal zone depressed (except sometimes medially), more finely punctate than rest of tergum and usually densely so, not over half as wide laterally as medially, this zone ending in very narrow, smooth posterior margin, the anterior margin of depressed zone often obtusely angulate medially, whole zone thus very broadly triangular 3

—. T5 with posterior marginal zone either not depressed or weakly to strongly depressed, but if recognizable, then punctate like rest of tergum to impunctate, often over half as wide laterally as medially, this zone often ending in broad, smooth posterior margin, the anterior margin of depressed zone straight or curved, not angulate, or with small basal median angular projection 4

3(2). T6 of female with margin usually not denticulate, sometimes with lateral tooth, with median apical notch or emargination accommodating sting [margin rarely denticulate, e.g., *A. (Proanthidium) oblongatum* Latreille]; T5 of female with basal edge of depressed marginal zone commonly obtusely angulate medially, zone thus broadly triangular; penis valves of male widely separated basally, united by long, narrow bridge; volsella projecting as lobe at apex of gonocoxite .... *Anthidium*

—. T6 of female with margin denticulate, almost always without lateral tooth, usually without median apical notch; T5 of female with basal edge of depressed marginal zone not angulate medially; penis valves close together or fused basally, bridge therefore short or absent; volsella small or absent (Africa, palearctic) .... *Afranthidium* (in part)

4(2). First recurrent vein joining first submarginal cell; axilla acutely pointed; face with shiny longitudinal median ridge from frons to clypeus (Africa) .... *Serapista*

—. First recurrent vein entering second submarginal cell; axillar margin rounded or straight; face without longitudinal median ridge 5

5(4). Basal area of propodeum hairless except laterally; pale markings absent; scutellum not angulate laterally; terga without impunctate margins (male unknown) (Africa) .... *Anthidioma*

—. Basal area of propodeum with hairs, sometimes very short; pale markings usually present, but *if* not [as in *Gnathanthidium* and *Afranthidium (Immanthidium)*, etc.], then scutellum usually angulate laterally and terga with impunctate marginal zones 6

6(5). Preoccipital ridge dorsally and omaulus produced as translucent lamellae (Africa, Oriental) .... *Pachyanthidium* (in part)

—. Preoccipital ridge and omaulus rounded or at most carinate (except preoccipital ridge lamellate in *Gnathanthidium*) 7

7(6). Subantennal suture straight or weakly arcuate; S4 and S5 of male not strongly concave, rather simple, S3 to S5 without combs or areas of specialized bristles, posterior margins straight or weakly concave (with lateral projections in *Neanthidium* and on S5 in *Gnathanthidium*) 8

—. Subantennal suture distinctly arcuate outward; S3 to S5 of male usually concave, S4 and S5 or at least the latter short and largely hidden (except in *Indanthidium*), at least S5 with posterior lateral projection [except in *Pseudoanthidium (Exanthidium)* and *Indanthidium*], S3 often with comb or area of wavy bristles 10

8(7). T6 and T7 of male each with four large equidistant teeth; S4 and S5 of male with lateral projections; T6 of female with lateral spine and median emargination (length 9-13 mm) (Mediterranean) .... *Neanthidium*

—. T6 of male simple, T7 short, bidentate, bilobed, or tridentate; S4 and S5 without lateral projections or (in *Gnathanthidium*) S5 with such projections; T6 of female without or with very weak lateral spine and median emargination 9

9(8). Mandible of female with 13 or 14 teeth, apex broad, lower two teeth and upper one large, others small and subequal; scutellum transverse, truncate, carinate; tibiae coarsely tuberculate (Africa) .... *Gnathanthidium*

—. Mandible of female with eight teeth or less; scutellum rounded as seen from above, sometimes weakly emarginate medially, not or incompletely carinate; tibiae tuberculate or not (Africa, palearctic) .... *Afranthidium* (in part)

10(7). T7 of male nearly as wide as T6, multidentate; subantennal suture arising at upper end of tentorial pit (S3 of male without wavy bristles; S5 with margin strongly concave but no lateral projections) (India) .... *Indanthidium*

—. T7 of male markedly narrower than T6, two- or three-toothed or lobed; subantennal suture arising from epistomal suture well above tentorial pit [except in *P. (Royanthidium) reticulatum* Mocsáry] .... *Pseudoanthidium*

11(1). Hind tibia of female with scopa-like hairs, some of them at least as long as tibial width; basal vein of forewing curved (as in *Halictus* but less strongly) and meeting vein Cu at a right angle; jugal lobe of hind wing about half as long as vannal lobe (yellow markings absent except on clypeus of male) (Africa)................................. *Aspidosmia*

—. Hind tibia of female with relatively short hairs or bristles; basal vein of forewing nearly straight and meeting vein Cu at acute angle; jugal lobe of hind wing less than half as long as vannal lobe 12

12(11). Omaulus lamellate, continued onto venter of thorax and there separated from middle coxa by less than width of middle trochanter 13

—. Omaulus lamellate or not, if lamellate then often not continued onto venter of thorax, but *if* so, then mesepisternum between middle coxa and omaulus (however recognized) as wide as or wider than width of middle trochanter.. 14

13(12). Propodeum with fovea defined by carina behind spiracle (as in Fig. 82-8); preoccipital ridge dorsally rounded or with low carina .... *Anthidiellum* (in part)
—. Propodeum without fovea behind spiracle; preoccipital ridge behind vertex lamellate (Africa, oriental) .... *Pachyanthidium* (in part)

14(12). Lower part of preoccipital carina sloping forward and continuing directly to lower mandibular articulation; axilla frequently pointed posteriorly (anterior coxa with lamella in most species; hind trochanter of male with preapical ridge, carina, lamella or tooth on inner surface; arolia absent) (palearctic, Sahel) .... *Icteranthidium*
—. Lower part of preoccipital carina absent, or, *if* present and extending to lower part of head, then ending below and mesal to lower mandibular articulation, or, *if* reaching mandibular articulation [in *Anthidiellum (Chloranthidiellum)*], directed below it and then curving up to articulation; axilla not pointed posteriorly (except in some parasitic genera that lack a scopa) 15

15(14). Face with three longitudinal ridges or carinae, two of them juxtantennal carinae and the third—a median longitudinal one on frons and supraclypeal area—often only a shiny ridge (body without yellow markings)......... .... *Euaspis*
—. Face without a longitudinal median ridge or carina and usually without juxtantennal carinae 16

16(15). Vein cu-v of hind wing usually half as long as second abscissa of M+Cu or longer, oblique; middle tibia as broad as hind tibia or nearly so (T7 of male simple or bilobed) 17
—. Vein cu-v of hind wing less than half as long as second abscissa of M+Cu (Fig. 82-1b, c), oblique or transverse; middle tibia usually narrower than hind tibia................ 19

17(16). Claws of female simple (male unknown) (India) .... *Trachusoides*
—. Claws of female cleft or with inner median or preapical tooth... 18

18(17). T7 of male curled under, dorsal surface thus facing downward; mandible of female dull, minutely roughened and bearing very short hairs, carinae absent on basal half of mandible; middle tibia with anterior margin strongly convex, at lowermost extremity usually at right angle to line across distal end of tibia (Fig. 82-4c) .... *Trachusa*
—. T7 of male directed posteriorly although small, short, and transverse; mandible of female slightly shining, carinae strongly shining; middle tibia with anterior margin less strongly convex, at acute angle to line across distal end of tibia (Borneo) .... *Apianthidium*

19(16). Anterior part of axilla produced to a point or lobe directed laterally, behind which margin is concave; margin of T7 of male with median point and two lobes on each side, thus with five apical projections (India) .... *Acedanthidium*
—. Axilla rounded, or sometimes pointed posteriorly, or, *if* with basal lateral projection, then curved posteriorly; margin of T7 of male with less than five apical projections 20

20(19). Axilla positioned and produced laterally, thus almost abutting posterior end of tegula; arolia absent; scopa absent (Africa) .... *Larinostelis*
—. Axilla not abutting tegula; arolia present [except in *Eoanthidium (Salemanthidium)*]; scopa present (except in *Stelis* and *Afrostelis*) 21

21(20). Scopa present; front and middle tibiae each with one apical spine or angle (except *Cyphanthidium* and some *Eoanthidium*, which have two spines on middle tibia) 22
—. Scopa absent; front and middle tibiae each with two spines, one midapical and one posterior apical 28

22(21). Juxtantennal carinae present although sometimes weak; interantennal distance usually less than, rarely equal to, antennocular distance; S6 of female (except in subgenus *Clistanthidium*) with spine or premarginal ridge, sometimes weak and lateral only, sternal margin thus appearing thick, sometimes elevated to lateral tooth (T7 of male over half as wide as T6) (palearctic, oriental) .... *Eoanthidium*
—. Juxtantennal carinae completely absent; interantennal distance usually greater than antennocular distance; S6 of female unmodified, margin thin 23

23(22). Scutoscutellar suture superficially similar to scutoaxillar suture, usually closed, but *if* smooth shining floor of groove visible, then usually not divided; subantennal suture approximately straight or only slightly arcuate; fovea behind propodeal spiracle absent; body usually over 10 mm long (although in *Cyphanthidium* as little as 6.5 mm long) 24
—. Scutoscutellar suture open to shiny bottom or fovea (as in Fig. 82-7a, b), thus very different from scutoaxillar suture, shiny area divided into two parts medially, or, *if* suture closed (as in *Anthidiellum* s. str.), then subantennal suture strongly arcuate outward (Fig. 82-5c); fovea behind propodeal spiracle present, defined posteriorly by carina, but fovea sometimes not larger than spiracle; body usually 8 mm long or less... 27

24(23). T6 of male with median apical tooth or small projection; body length 8.5 mm or less and metasoma with continuous yellow bands [form and coloration as in *Afranthidium (Oranthidium)*] (Africa) ...... *Cyphanthidium*
—. T6 of male without median apical tooth; body length usually 8.5 mm or more, but if less [as in *Benanthis* and some *Plesianthidium (Spinanthidium)*], then metasoma without yellow, or yellow bands broken 25

25(24). Yellow or cream markings absent or limited to face of male; T3 and other terga with depressed premarginal zone sublaterally nearly one-half length of exposed part of tergum; T6 of male with median lobe (often subtruncate and elevated) and lateral tooth, thus trifid or (in subgenus *Spinanthidiellum*) truncate with a longitudinal median ridge at apex (Africa) .... *Plesianthidium*
—. Body with yellow or reddish-yellow markings; T3 and other terga with depressed premarginal zone sublaterally one-third length of exposed part of tergum or less; T6 of male simple, bilobed, or with short, broad, rounded median lobe, sometimes (in *Rhodanthidium* s. str.) also with lateral tooth, thus trifid... 26

26(25). Apex of T7 of male strongly bilobed; S4 and S5 of male simple, without combs or lateral teeth, margin of S6 with lateral lobe resembling a tooth in side view; ocellocipital distance about equal to ocellar diameter (Madagascar) .... *Benanthis*
—. Apex of T7 of male with a median projection, thus with three projections (or five in subgenus *Meganthidium*); S4 or S5 of male frequently with marginal comb, S5 often

with lateral tooth, S6 without lateral projection; ocellocipital distance two or more ocellar diameters (palearctic) .... *Rhodanthidium*

27(23). Omaular carina absent or extending down only to middle of mesepisternum; T7 of male, if trilobed, then with median lobe much longer than lateral lobe or spine; subantennal suture straight (eastern Asia).......... *Bathanthidium*

—. Omaular carina strong, sometimes lamellate, and extending onto ventral surface of thorax, sometimes across venter (except in subgenus *Clypanthidium*, in which omaular carina does not reach lower part of mesepisternum); T7 of male, if trilobed, then with median lobe small, either not separated from lateral lobe by emargination or not longer than lateral lobe; subantennal suture usually arcuate outward (Fig. 82-5c) .... *Anthidiellum* (in part)

28(21). Tegula of ordinary size and shape, widest medially and longer than wide (scutum wider than long, only moderately so in subgenus *Stelidomorpha*) .... *Stelis* (in part)

—. Tegula enlarged, especially posteriorly, width posteriorly almost as great as length 29

29(28). Scutum longer than wide; body without yellow markings (Africa) .... *Afrostelis*

—. Scutum slightly broader than long; metasoma with yellow markings (Yeman) .... *Xenostelis*

Supplementary Key to Males of the Anthidiine Genera of the Eastern Hemisphere
(From Griswold and Michener, 1994a)

The preceding key will be frustrating for various reasons, chief among which will likely be that couplet 1 is based largely on a character of females; supplementary characters will help, but as indicated within the couplet, they are not always decisive. The following supplementary key for males leads either to certain genera or to numbered couplets in the main key, thus bypassing couplet 1. In reality, its main function is to help identify males of taxa that should run to 11 in the main key and that lack arolia, as do all taxa that run to 2.

A. Arolia absent.. B
—. Arolia present .... go to 11
B(A). Paleotropical species C
—. Palearctic species .... go to 2
C(B). Vein cu-v of hind wing more than half as long as second abscissa of M+Cu, oblique (as in Fig. 82-1a); middle tibia as broad as hind tibia or nearly so .... D
—. Vein cu-v of hind wing less than half as long as second abscissa of M+Cu (as in Fig. 82-1b, c), oblique or transverse; middle tibia narrower than hind tibia..................... E
D(C). T7 curled under, dorsal surface thus facing downward; middle tibia with anterior margin strongly curved, at lowermost extremity thus usually at right angle to line across distal end of tibia (Fig. 82-4c) .... *Trachusa* (in part)
—. T7 directed posteriorly although small, short, and transverse; middle tibia with anterior margin less strongly convex, thus at acute angle to line across distal end of tibia (as in Fig. 82-4d, e) .... *Apianthidium*
E(C). Axilla almost entirely lateral to lateral margin of scutum; outer, apical margins of fore and middle tibiae each with two minute spines (placed on basis of female characters; male unknown) .... *Larinostelis*

—. Axilla at most extending slightly lateral to lateral margin of scutum; outer, apical margins of fore and middle tibiae each with at most one spine .... F
F(E). Omaulus carinate for at least three-fourths of distance from upper end to midventral line .... G
—. Omaulus not carinate, or carinate for no more than one-half of distance from upper end to midventral line .... go to 2
G(F). Preoccipital carina present dorsally, behind vertex .... H
—. Preoccipital carina absent dorsally, behind vertex I
H(G). Hind tibia tuberculate on outer surface; scutellum very short, width greater than four times length, only slightly overhanging metanotum (for one-third its length) .... *Gnathanthidium*
—. Hind tibia not tuberculate; scutellum moderately long, width equal to or less than three times length, greatly overhanging metanotum (for one-half its length) .... *Pachyanthidium* (in part)
I(G). Juxtantennal carinae present (as in Fig. 82-5a); T7 broadly truncate with small median projection.............. .... *Eoanthidium* (in part)
—. Juxtantennal carinae absent (as in Fig. 82-5c); T7 with three apical spines .... *Serapista*

Key to the Genera of the Anthidiini of the Western Hemisphere

1. Mandible of female with five or more teeth separated by acute notches (Fig. 82-3a); with the following *combination* of other characters: arolia absent; basal vein of forewing several vein widths basal to cu-v; base of propodeal triangle minutely roughened, punctate, hairy, without series of pits; postspiracular fovea of propodeum absent; juxtantennal carina absent .... *Anthidium*
—. Mandible of female with three or four (rarely five) teeth, at least some of them separated by obtuse or rounded emarginations (Fig. 82-3b), rarely distal margin edentate except for small tooth near lower margin (Fig. 82-3c); without the combination of other characters listed above 2

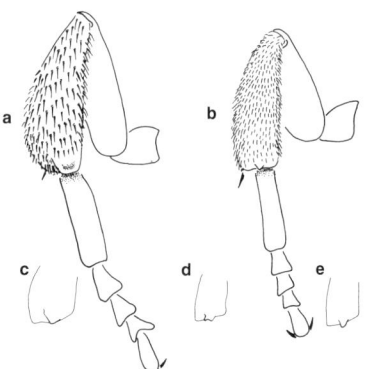

Figure 82-4. Middle legs, showing apices of tibiae, of anthidiine bees. **a**, *Paranthidium jugatorium perpictum* (Cockerell); **b**, *Dianthidium (Dianthidium) curvatum* (Smith); **c**, *Trachusa (Trachusomimus) gummifera* Thorp; **d**, *D. (Adanthidium) arizonicum* (Rohwer); **e**, *Hypanthidiodes (Anthidulum) currani* (Schwarz). a, b, from Michener, McGinley, and Danforth, 1994; c-e, from Griswold and Michener, 1988.

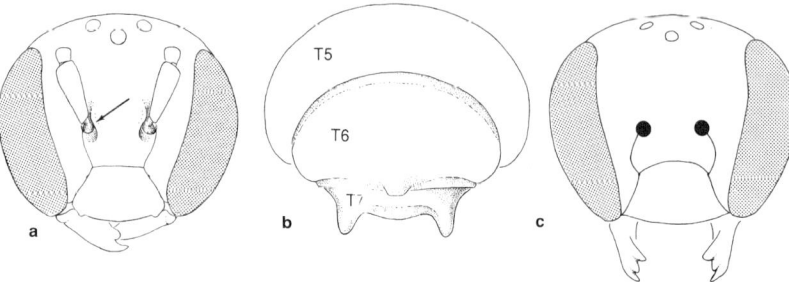

Figure 82-5. Anthidiini. **a,** Face of *Hypanthidiodes (Anthidulum) currani* (Schwarz), the juxtantennal carina marked by arrow; **b,** Apex of metasoma of *H. (Anthidulum)* sp.?; **c,** Face of *Anthidiellum (Anthidiellum) notatum robertsoni* (Cockerell), showing the arcuate subantennal sutures. From Michener, McGinley, and Danforth, 1994.

2(1). Middle tibia with two apical spines (Fig. 82-6a) (for small specimens with abundant hairs, examine in distal view); scopa absent .... *Stelis*
—. Middle tibia with one apical spine, or spine reduced to a sharp angle or rounded margin (Figs. 82-4, 82-6b); scopa present (except in *Hoplostelis*) 3

3(2). Propodeum with fovea delimited posteriorly by a carina behind spiracle (Fig. 82-8); omaular carina present; base of propodeum frequently with row of pits across upper margin connecting postspiracular foveae (Fig. 82-8), but sometimes row present only laterally 4
—. Propodeum without fovea behind spiracle (or, in *Notanthidium*, fovea sometimes present); omaular carina absent or weak [or, in *Hoplostelis (Rhynostelis)*, lamellate]; row of pits across upper margin of propodeum absent, weak, or present only laterally 13

4(3). Apex of middle tibia without tibial spine or angular vestige of spine, but with convex carina (with dense short hairs beneath) across tibial apex, curving basad anteriorly, forming edge of broad, scoop-shaped concavity on apical tibial surface (Fig. 82-4a); apical margin of mandible of female strongly oblique, about half as long as mandible (North and Central America) .... *Paranthidium* (in part)
—. Apex of middle tibia with angle or short spine on outer side, without preapical concavity (Figs. 82-4b-e, 82-6b); apical margin of mandible of female usually less oblique and often less than half as long as mandible 5

5(4). Subantennal suture long and distinctly arcuate outward (Fig. 82-5c); preoccipital ridge behind vertex strongly carinate and produced posteriad, covering pronotum; scutellum ending as sharply margined truncation (with small median emargination) overhanging metanotum and propodeum (Fig. 82-7c, e) (North and Central America) .... *Anthidiellum*
—. Subantennal suture more or less straight (Fig. 82-5a) (or, in some *Hypanthidioides*, arcuate although rather short); preoccipital ridge behind vertex not or moderately carinate, not produced posteriad; scutellum not truncate (Fig. 82-7d), or, if so, then sharply margined only laterally 6

6(5). Juxtantennal carina present, sometimes a low and elongate, curved ridge rather than a sharp carina, always extending downward as well as upward from level of middle of antennal socket (Fig. 82-5a) [a welt immediately mesal to antennal base, as in *Dianthidium (D.) marshi* Grigarick and Stange and some species of *D. (Adanthidium)*, is not considered a juxtantennal carina] 7
—. Juxtantennal carina absent or, if present (as in *Epanthidium*), then arising abruptly at innermost margin of antennal socket and extending only upward 11

7(6). Preoccipital carina strong laterally, behind eye, and extending down to join posterior end of hypostomal carina; scutoscutellar suture open to form two strong foveae with smooth bottoms, each five times as long as broad or less (Fig. 82-7a) (arolia present in both sexes) (neotropical) .... *Anthodioctes*
—. Preoccipital carina absent, or, if present, then not approaching hypostomal carina; scutoscutellar suture usually narrower, but *if* with two foveae, then each usually more than five times as long as broad (Fig. 82-7b) 8

8(7). Stigma small, its width about equal to its length as measured on costal margin; S5 and S6 of male both without lateral tooth or lobe; T6 of male produced posteriorly to large rounded lobe at each side, with broad emargination and sometimes median spine between lobes; preoccipital carina sometimes present behind upper two-thirds of eye (South America) .... *Duckeanthidium*
—. Stigma larger, its width less than its length on costal margin; S5 and S6 of male each with lateral tooth or small lobe; T6 of male without widely separated lateral lobes; preoccipital carina absent 9

9(8). Scopa absent; T7 of male small, not or scarcely bilobed (arolia present in both sexes) (neotropical) 10
—. Scopa present; T7 of male distinctly bilobed or with two apicolateral projections separated by straight or concave margin (neotropical) .... *Hypanthidioides*

10(9). Base of female mandible without protuberance; mandible of female unmodified, 4-toothed like that of *Hypanthidioides*; body somewhat elongate, as in *Hypanthidioides* .... *Austrostelis*
—. Female mandible with strong protuberance near anterior articulation; mandible of female with fourth (uppermost) tooth shifted basad, nearer to base of mandible than to apex, or absent; body robust, metasoma sometimes almost globose .... *Hoplostelis*

11(6). Anterior margin of scutum abruptly declivous, steeply sloping or vertical, in contrast to dorsal surface (Fig. 82-7f); posterior margin of metanotum, lateral to metanotal pit, with area or strip of short, white hairs (rarely absent); mandible of female three-toothed, sometimes modified by fusion of upper two teeth to form long, nearly straight, untoothed margin (Fig. 82-3c) (nearctic, Mexico) .... *Dianthidium*
—. Anterior margin of scutum with surface a continuation

of curvature of dorsal surface or at least not deviating from that curvature by more than 45°; metanotum without area or strip of distinctive short hairs; mandible of female four-toothed (teeth sometimes low or badly worn; mandibles often must be open to evaluate this character) 12

12(11). Arolia absent; pronotal lobe with strong carina or rarely with short lamella; juxtantennal carina absent (neotropical).. *Hypanthidium*

—. Arolia present in males and most females; pronotal lobe with high lamella (as in Fig. 82-7d); juxtantennal carina present, arising on inner margin of antennal socket and extending upward, not at all downward, or, if virtually absent, then often showing origin on inner margin of antennal socket (neotropical) .... *Epanthidium*

13(3). Omaulus lamellate; juxtantennal carina present (as in Fig. 82-5a); base of mandible of female with protuberance near anterior articulation; clypeus of female with basal median projection; scopa absent (South America) .... *Hoplostelis (Rhynostelis)*

—. Omaulus rounded or weakly carinate; base of mandible and of clypeus of female unmodified; scopa present 14

14(13). Apex of middle tibia without angular tibial spine, but with strongly convex carina across posterior part of tibial apex, the carina (beneath which are dense, short hairs) curving basad anteriorly, forming apex of gentle concavity in tibial surface, at least in females (Fig. 82-6a); apical margin of female mandible strongly oblique, about half as long as mandible (North and Central America) .... *Paranthidium* (in part)

—. Apex of middle tibia with angle or short spine on outer side, without preapical concavity; apical margin of female mandible less oblique, usually less than half as long as mandible... 15

15(14). Vein cu-v of hind wing half as long as second abscissa of M+Cu or more, oblique (as in Fig. 82-1a); T7 of male small, exposed part often half as long as exposed part of T6, often complexly angled but not bilobed; spine of front and middle tibiae a blunt projection extending little if any beyond apical tibial margin but extending anterobasally as carina on tibial surface (Fig. 82-4c) (nearctic, Mexico) .... *Trachusa*

—. Vein cu-v of hind wing less than half as long as second abscissa of M+Cu, usually more or less transverse (as in Fig. 82-1b, c); T7 of male larger, relatively flat with two or four strong apical lobes; spine of front and middle tibiae an acute projection (as in Figs. 82-4b, e; 82-6b) 16

16(15). T6 of female and T5 and T6 of male each with strong lateral tooth (Fig. 82-10); preoccipital carina strong; interocellar distance about half of ocelloccipital distance or less (Mexico) .... *Aztecanthidium*

—. T5 and T6 both without lateral teeth; preoccipital carina absent; interocellar distance equal to or greater than ocelloccipital distance (South America) ...... *Notanthidium*

Genus *Acedanthidium* Michener

Acanthidium Michener and Griswold, 1994a: 305 (not Lowe, 1839). Type species: *Acanthidium batrae* Michener and Griswold, 1994 = *Dianthidium flavoclypeatum* Gupta, 1993, by original designation.

Acedanthidium Michener, 2000: 480. Replacement for *Acanthidium* Michener and Griswold, 1994. Type species *Acanthidium batrae* Michener and Griswold, 1994 = *Di-*

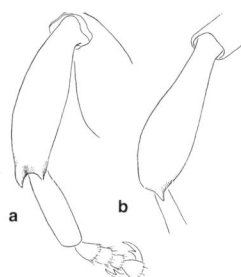

Figure 82-6. Middle tibiae of Anthidiini, showing apical spines. **a**, *Stelis rubi* Cockerell, male; **b**, *Anthidium maculosum* Cresson, male. In small species the spines may be minute and, because of nearby hairs, visible only in apical view. From Michener, McGinley, and Danforth, 1994.

anthidium flavoclypeatum Gupta, 1993, autobasic and by original designation.

This is a genus of bees having the body form and size (7.5-8 mm long) of *Eoanthidium* or *Pseudoanthidium*, i.e., somewhat slender megachiliform, with conspicuous yellow markings. The basal part of the axilla is produced laterally and pointed, suggesting the shape of the axilla of some *Rhodanthidium (Asianthidium)*, which led to the naming of *Axillanthidium* Pasteels, and likewise suggesting that shape in *Stelis (Malanthidium) malaccensis* (Friese). Thus this striking character has evidently arisen at least three times; if one ignores it, *Acedanthidium* still does not fit into any other genus. The foveate scutoscutellar suture suggests *Eoanthidium* and *Bathanthidium*, but from the first *Acedanthidium* differs in its lack of juxtantennal carinae and the simple S6 of the female, among other features; from the latter it differs in its lack of the postspiracular fovea of the propodeum and the absence of the row of pits across the base of the propodeum. From all other genera *Acedanthidium* differs in the short, broad T7 of the male, with its five apical projections, the lateral ones broadly rounded, the sublateral ones low, and the median one slender but blunt. Carinae are lacking on the head and thorax except for one on the pronotal lobe; the upper half of the omaulus is sharply angular but not truly carinate. The scutellum is rounded posteriorly, not emarginate, rounded in profile, and scarcely overhanging the metanotum. The sterna of the male lack combs. The male genitalia were illustrated by Gupta (1993) and Michener and Griswold (1994a); see also Figure 82-2d.

■ This genus occurs in India. The single known species is *Acedanthidium flavoclypeatum* (Gupta), new combination.

Genus *Afranthidium* Michener

This genus contains a large group of relatives of *Anthidium*, generally small and megachiliform, with multidentate mandibles in the female, as in other members of Series B. Some species agree with *Anthidium* also in the depressed, medially widened premarginal zone of T5 (see the key to genera for more details); the anterior margin of this zone, however, is not angulate as in *Anthidium*. In other species there is no such zone, and the impunctate marginal zone is sometimes broad, as in the subgenus *Immanthidium*. These and the other external characters indicated in the key to genera are generally distinctive but have exceptions. For example, in *Anthidium (Proanthidium) oblongatum* Latreille, T6 of the female is denticulate, without a lateral tooth, and has a small midapical notch, thus combining the usual features of *Anthidium* and *Afranthidium*. The wide separation of the penis valves of *Anthidium* and the long bridge between their bases is the only known character consistently distinguishing *Anthidium* from *Afranthidium*, and as more species are dissected and examined, it too may fail.

Afranthidium differs from *Pseudoanthidium* as indicated in the key to genera, and in the simple male metasomal sterna. In *Afranthidium* the sterna are not strongly concave, do not bear lateral processes, and lack combs or specialized bristles. Usually, S1 to S6 are exposed, although S5 and rarely S4 may be largely hidden.

It is quite possible that *Anthidium* is derived from a paraphyletic taxon, *Afranthidium*, but there is no particular group of *Afranthidium* to which *Anthidium* appears most closely related. The characters of *Afranthidium* that distinguish it from *Anthidium* are possibly all plesiomorphies. It is a phenetically distinct group and recognition of it at this time seems reasonable. Recognition of the *Afranthidium* subgenera as genera seems undesirable because of their similarity and consequent difficulty of identification; moreover, there are a few, often unidentified and probably undescribed species that combine characters of different subgenera. Perhaps worse, placement of the *Afranthidium* subgenera as subgenera of *Anthidium* would make the latter an enormously diverse genus.

All *Afranthidium* are sub-Saharan except for the subgenus *Mesanthidium* and at least two species of *Capanthidium*, all of which are palearctic.

Key to the Subgenera of *Afranthidium*
(From Michener and Griswold, 1994a)

1. Marginal zones of T2 to T5 broadly impunctate, median lengths of impunctate zones usually one-sixth length of exposed parts of terga or more, these zones transparent, pale brown or cream-colored; male gonostylus enormous, broad, flat, and almost membranous; pronotal lobe not carinate, sometimes with small punctate and hairy ridge in position of carina (Africa) .... *A. (Immanthidium)*
—. Marginal zones of T2 to T5 punctate or narrowly impunctate, median lengths of impunctate zones about one-seventh lengths of exposed parts of terga or less, these zones dark or translucent brownish; male gonostylus neither broad nor flat nor almost membranous although sometimes broadly paddle-shaped; pronotal lobe usually with transverse carina or low lamella that is hairless even when it extends only partway across lobe 2
2(1). Scutellum and axilla laterally not much overhanging as seen obliquely so as to show profiles; scutellum rounded in profile, not or scarcely overhanging metanotum; outer surfaces of tibiae not more coarsely punctate than mesepisternum, not tuberculate............................... 3
—. Scutellum laterally and usually axilla overhanging large fossa and usually acute or narrowly rounded, as seen obliquely so as to show profile of lateral part of scutellum and of axilla; scutellum acute, right-angular, or sometimes rounded in profile, medially often strongly overhanging metanotum; outer surfaces of tibiae, especially of female, coarsely, irregularly punctate, sometimes with extensive smooth areas between punctures, the punctures commonly coarser than those of mesepisternum, surfaces usually tuberculate (but not or weakly so in *Capanthidium* and *Mesanthidium*).................................... 7

3(2). T2 to T5 with apical bands, broken medially, of white plumose hair; propodeal triangle with punctures (and hairs) widely separated from one another, surface between punctures strongly shagreened, dull; body and legs black, without pale markings (Africa)............................
.... *A. (Zosteranthidium)*
—. T2 to T5 without apical hair bands; propodeal triangle rather densely punctate and hairy, surface between punctures shining; body and legs usually with yellow or white markings, but *if* not, then at least anterior surfaces of front and middle tibiae brownish yellow 4
4(3). Hind basitarsus of female with apical projection over base of second tarsal segment; hind trochanter of male with apicoventral denticle; sterna of male with distinct basal hair fasciae arising at graduli; body black, with yellow or cream color usually limited to minute streak along inner margin of eye of both sexes (Africa).....................
.... *A. (Nigranthidium)*
—. Hind basitarsus of female with apex truncate, without apical projection; hind trochanter of male without apicoventral denticle; sterna of male without basal hair fasciae; body brown or black with yellow or cream-colored areas at least on face of male and metasomal terga of both sexes 5
5(4). T5 of female without lateral spine; T1 of female with carina separating anterior from dorsal surfaces abruptly strengthened laterally and thence extended lateroposteriorly; apex of T7 of male with two lobes, each lobe two or three times as broad as long, emargination between them with small median spine; S6 of male with small, pointed midapical process at base of which are two spines directed forward (Africa).................... *A. (Domanthidium)*
—. T5 of female with lateral spine; T1 carina of female unmodified; apex of T7 of male two-lobed, usually with median angle or tooth, or three-lobed, each lateral lobe about as long as broad or longer; S6 of male without small apical process and spines 6
6(5). Gonostylus of male tapering, not or scarcely longer than gonocoxite, not attaining middle of penis valve; ventral surface of mesepisternum of female covered with strong, backward-directed, golden to black bristles that appear flattened, minutely barbed, and blunt or abruptly tapered at apices (S6 of male elongate, produced to narrowly rounded or bidentate apex) (Africa) ....
.... *A. (Oranthidium)*
—. Gonostylus of male expanded apically, about twice as long as gonocoxite, attaining apex of penis valve or nearly so; ventral surface of mesepisternum of female with pale, gradually tapering hairs similar to hairs of adjacent areas (Africa) .... *A. (Afranthidium s. str.)*
7(2). Posterior margins of metasomal terga not curved upward, lying near surfaces of following terga; axilla extending laterally beyond scutal margin; male gonostylus greatly reduced, attaining about middle of penis valves, which are completely fused to one another (palearctic) ....
.... *A. (Mesanthidium)*
—. Posterior margins of at least some metasomal terga, as seen in profile, curved upward away from following terga; axilla not extending laterally beyond scutum (except in some *Branthidium*); male gonostylus reaching to or beyond level of apical one-fourth of penis valves 8
8(7). T6 of female with preapical denticulate ridge parallel to denticulate apical margin (scutellum distinctly cari-

nate except for small midapical emargination) (male unknown) (Africa) .... *A. (Xenanthidium)*
—. T6 of female without preapical denticulate ridge 9
9(8). T6 of male with preapical, usually denticulate transverse ridge at least laterally; tibiae coarsely punctate but not or weakly tuberculate on outer surfaces (Africa, palearctic) .... *A. (Capanthidium)*
—. T6 of male without preapical ridge; tibiae strongly tuberculate on outer surfaces 10
10(9). Preoccipital carina present laterally; male S3 with trapezoidal apical projection; T5 and T6 of male with lobate lateral carinae, T5 and T6 of female with lateral longitudinal carinae (Africa).................. *A. (Mesanthidiellum)*
—. Preoccipital carina absent; male S3 margin not produced; T5 (usually) and T6 without lateral carinae in either sex (Africa) .... *A. (Branthidium)*

Afranthidium / Subgenus *Afranthidium* Michener s. str.

Anthidium (Afranthidium) Michener, 1948a: 24. Type species: *Hypanthidium halophilum* Cockerell, 1936, by original designation.

Afranthidium s. str. consists of small to medium-sized (6-10 mm long) robust species, black with the metasoma often brown, with limited cream or yellow markings, those of the metasoma forming narrow, unbroken tergal bands as in *Oranthidium* and *Domanthidium*. Unlike the male gonostyli of those subgenera, the gonostyli of *Afranthidium* s. str. are slender, slightly expanded apically, and about twice as long as the gonocoxites, and they attain the apices of the penis valves or nearly so. The penis valves are broad, adjacent, and heavily sclerotized distally with retrorse preapical points, much as in *Oranthidium*. The lateral metasomal spines are strong, as in *Oranthidium*.
■ This subgenus is known from Cape Province, South Africa, and from Namibia. Pasteels (1984) included nine species, but some of them may belong elsewhere.

Afranthidium murinum (Pasteels) does not completely fit the characterization of any recognized subgenus. Because the species is known only in the female, a firm decision on its placement is premature. The pronotal lobe lacks a carina, but in other respects *A. murinum* runs to couplet 3 in the key to subgenera, or to *Afranthidium* s. str. if its tergal hair bands are ignored. It differs from *Afranthidium* s. str. in its absence of lateral metasomal spines, although T3 to T5 have small lateral lobes, and from *Zosteranthidium* in the presence of yellow maculations and other characters.

Afranthidium / Subgenus *Branthidium* Pasteels

Branthidium Pasteels, 1969a: 88. Type species: *Anthidium braunsi* Friese, 1904, by original designation.
Honanthidium Pasteels, 1969a: 88. Type species: *Anthidium honestum* Cockerell, 1936, by original designation.

Branthidium contains small (5-7 mm long), stout species, black with narrow pale metasomal bands, the bands often not interrupted, or with the metasoma largely yellow. *Branthidium* could reasonably be included in the subgenus *Capanthidium* but is retained for reasons given in the account of the latter. In most males the fore femur is flattened basoventrally and the mesosternum is covered with long, silky, posteriorly directed, decumbent hair. Of all the subgenera of *Afranthidium*, *Branthidium* and the closely related *Mesanthidiellum* are the ones that most resemble the genus *Pseudoanthidium*.
■ This subgenus is found from Lesotho to western Cape Province, South Africa, north to Shaba Province in Zaire and to Kenya. Pasteels (1984) recognized ten species and revised the group.

Reasons for including *Honanthidium* in *Branthidium* are given by Michener and Griswold (1994a). As noted by those authors, *Afranthidium (Branthidium) guillarmodi* (Mavromoustakis) resembles *Mesanthidium* in its laterally projecting axillae and *Mesanthidiellum* in its laterally carinate female T5.

Afranthidium / Subgenus *Capanthidium* Pasteels

Capanthidium Pasteels, 1969a: 85. Type species: *Anthidium "capicole* Friese," *lapsus* for *capicola* Brauns, 1905, by original designation.

Capanthidium consists of small (5-8 mm long) species, black with limited light markings on the head and thorax; the metasoma exhibits narrow unbroken light tergal bands or is almost entirely yellow. This subgenus is closely related to *Branthidium* and *Mesanthidiellum*, and the three could be regarded as a single subgenus. Characters previously used to distinguish among them do not hold true. For example, in his key to genera and subgenera, Pasteels (1984) indicates that the second recurrent vein almost meets the second submarginal crossvein in *Capanthidium* but is three to four vein widths beyond it in *Branthidium*, but the full range of this variation occurs even within the type species of *Capanthidium*. Further the lateral part of the scutellar margin and the margin of the axilla are sharper and irregularly crenulate in species placed in *Branthidium* and *Mesanthidiellum*, unlike those of species placed in *Capanthidium*, but this character, too, varies, and the difference is not consistent. In *Mesanthidiellum* and some *Branthidium* the marginal expansion of the axilla is sufficient that the axilla extends laterad behind the tegula. The male gonostyli are long and parallel-sided, although flat, in *Afranthidium (Capanthidium) capicola* (Brauns) and *naefi* (Benoist), and long, apically expanded, and paddle-shaped, although sometimes bearing a marginal tooth, in species placed in *Branthidium*. In some species that have been included in *Capanthidium*, however, e.g., *A. (C.) rubellulum* (Cockerell), the gonostyli are as broadly expanded as those of *Branthidium*.

A number of characters, however, differentiate *Capanthidium* from *Branthidium* and *Mesanthidiellum*. The more flattened body (*Anthidium*-like rather than *Pseudoanthidium*-like) is distinctive. Other characters of *Capanthidium* not mentioned in the key to subgenera include the depressed marginal zones of terga that are finely and densely punctate and lack distinct impunctate margins except in *Afranthidium capicola* (Brauns) and *naefi* (Benoist); the long, apical fringes of S7 and S8 of the male; the penis valves, which are flat dorsoventrally; and the semicircular disc of S6, usually with a lateral tooth near the margin. In addition, most females of *Capanthidium* have a median impunctate line on the clypeus (as in *Afranthidium* s. str. and *Oranthidium*) and a blunt lateral tooth on T5, conditions never found in *Branthidium* and *Mesanthidiellum*. The presence or absence of lateral spines on the metasomal terga may provide the best separation for the groups; T6 of the male of *Capanthidium* has such a spine, but it is absent in *Branthidium* and represented by a carina or lobe in *Mesanthidiellum*. And in *Mesanthidiellum* the preoccipital carina is present laterally, behind the eye, a feature not found in *Branthidium* or *Capanthidium*.

■ *Capanthidium* has a disjunct distribution: (1) Cape Province of South Africa and Namibia and (2) Morocco and Spain. *A. (C.) schulthessii* (Friese) and *naefi* (Benoist) from Morocco and Spain were placed in *Mesanthidium* by Pasteels (1969a) and Warncke (1980a), but are closely related to *A. (C.) capicola* (Brauns). Ten species from southern Africa were recognized by Pasteels (1984).

Afranthidium (?*Capanthidium*) *poecilodontum* (Mavromoustakis) is an anomalous species. Both recurrent veins are received by the second submarginal cell, as in *A. (Nigranthidium) concolor* (Friese). The male is further distinguished by the poorly sclerotized S6, the absence of long hairs on the male gonostyli, and greatly enlarged penis valves.

Afranthidium / Subgenus *Domanthidium* Pasteels

Domanthidium Pasteels, 1969a: 95. Type species: *Anthidium abdominale* Friese, 1904, by original designation.

In appearance, *Domanthidium* resembles *Oranthidium* and *Afranthidium* s. str., as well as some species of *Capanthidium*, being rather robust, 5.5 to 7.5 mm long, with a black or brown metasoma. The pale yellow markings are limited, the tergal bands being narrow and unbroken, and the marginal zones of the terga are translucent brownish.

Domanthidium was synonymized with *Oranthidium* by Pasteels (1984). Although the gonostyli are greatly reduced, slender and tapering in both subgenera, they extend farther toward the apices of the penis valves in *Domanthidium* than in *Oranthidium* because of the larger gonocoxites. The penis valves are not upturned at the apices as in *Oranthidium* and *Afranthidium* s. str. *Domanthidium* further differs from *Oranthidium* in its lack of a keel or lamella on the lateral ventral margin of T6 of the female, next to the lateral margin of S6, and in the ordinary hairs on the ventral surface of the mesothorax of the female. T7 of the male has a single small, acute median point and a broad lobe, the lobe more than twice as wide as long, on each side; this is unlike the trifid T7 of males of *Oranthidium*. The lateral spines of the metasoma are small in males, absent in females; they are strong in *Oranthidium*. *Domanthidium* differs from *Capanthidium* in the minute male gonostyli and in the finely punctate, not tuberculate outer surfaces of the tibiae. T7 of the male, as described above, differs from that of most *Capanthidium*; in *Afranthidium* (*Branthidium*) *guillarmodi* (Mavromoustakis), however, T7 is similar to that of *Domanthidium* (see fig. 89, Pasteels, 1984).

■ This subgenus occurs in Namibia and in Cape Province to Transvaal, probably also in Natal, South Africa. There is a single species, *Afranthidium abdominale* (Friese), of which *A. nigritarse* (Friese) is a probable synonym.

Afranthidium / Subgenus *Immanthidium* Pasteels

Immanthidium Pasteels, 1969a: 89. Type species: *Anthidium immaculatum* Smith, 1854, by original designation.

This subgenus consists of small to medium-sized (5-10 mm long), robust (anthophoriform) species, largely lacking yellow markings in females, but in most males the clypeus, sometimes markings on the thorax, and broken or continuous bands on the terga are yellow or white. Thus males are very different from females in coloration. In many individuals of some species, the expanded, impunctate posterior marginal zones of the terga are ivory-colored; there are thus apical tergal bands, not homologous to the bands of most anthidiines, which are in the middles of the terga. Marginal bands occur in both sexes but not in all individuals of any species. *Immanthidium* lacks carinae, although there is sometimes a weak ridge on the pronotal lobe. The decisive characters of the subgenus are those of the male terminalia, which are only feebly sclerotized and flat, unlike those of other Anthidiini. The gonostylus is broad and flat, almost membranous, rounded apically, and nearly twice as long as the gonocoxite. The penis valves are adjacent and perhaps broadly fused to one another. S8 is broad and thin, like the gono-stylus, and rounded apically except for a small median emargination. The apex of the male metasoma is not strongly curled. T7 of the male is not strongly exserted; its posterior margin has a median emargination between submedian teeth, sometimes with a median tooth producing a tridentate condition; lateral teeth are sometimes present, producing, in the absence of a median tooth, a quadridentate metasomal apex. T6 lacks lateral teeth.

There are two species groups of *Immanthidium*. In one group [*Afranthidium junodi* (Friese), *repetitum* (Schulz), *sjoestedti* (Friese)] males have a unique patch of dense, velvety hair on the middle femur posterobasally. The other group [*A. immaculatum* (Smith), *mlanjense* (Mavromoustakis)] lacks this feature. Most females of this subgenus could be mistaken for osmiines because of the absence of yellow markings and the long, dense reddish hair covering the body.

■ *Immanthidium* is widespread in eastern Africa from Sudan to South Africa (Natal west to Cape Province) and Namibia. Unlike other African subgenera, it is uncommon in western Cape Province and other xeric areas. Pasteels (1984) recognized five species and revised the group.

Michener (1968a) described the nests of two species under the generic name *Anthidium*. *Afranthidium* (*Immanthidium*) *junodi* (Friese) nested in burrows probably made by other insects in a clay bank. The other species, *A. (I.) repetitum* (Schulz), evidently communal, constructed a huge mass of cells, an estimated 1,750, in a box. The cells were made of plant hairs, as in related genera with multidentate mandibles.

Afranthidium / Subgenus *Mesanthidiellum* Pasteels

Mesanthidiellum Pasteels, 1969a: 83. Type species: *Mesanthidiellum amoenum* Pasteels, 1969, by original designation.

This subgenus consists of small (6-8 mm long), stout bees similar to *Branthidium* in form; in fact, it should probably be considered a synonym of *Branthidium*. In addition to the characters given in the key to subgenera are two others: the propodeum is shagreened in *Mesanthidiellum*, shiny in all *Branthidium* except *Afranthidium (B.) guillarmodi* (Mavromoustakis); and the trapezoidal extension of male S3 is unique.

■ *Mesanthidiellum* is widespread in sub-Saharan Africa from Senegal and Sudan to Namibia and Natal Province, South Africa. Pasteels (1984) listed three species.

Afranthidium / Subgenus *Mesanthidium* Popov

Mesanthidium Popov, 1950a: 316. Type species: *Anthidium pentagonum* Gussakovskij, 1930, by original designation.

Species of this subgenus are somewhat more elongate than those of *Afranthidium* s. str., 7 to 9 mm long, and although sometimes largely black, usually have considerable yellow ornamentation on the body. The yellow or cream-colored metasomal bands are usually unbroken, and the tergal margins translucent brownish. The reduced gonostyli of the male, though suggestive of those of *Oranthidium*, are quite different, laterally compressed and fused to the gonocoxites; perhaps the reduction in *Mesanthidium* was independent from that of *Oranthidium*. The penis valves are fused to one another for most of their lengths, as in *Immanthidium*. The male genitalia and sterna were illustrated by Popov (1950a).

■ This palearctic subgenus is found from Morocco eastward through the Mediterranean basin to Egypt, Turkey, Bulgaria, and on to central Asia. Pasteels (1969a) listed ten species, but at least two of the ten, *Afranthidium schulthessii* (Friese) and *naefi* (Benoist), are palearctic representatives of *Capanthidium*.

Male genitalia presumably provide the most reliable feature for distinguishing *Mesanthidium* from *Capanthidium*. In the former the gonostyli are reduced, in the latter they are long and either parallel-sided or paddle-shaped.

Afranthidium / Subgenus *Nigranthidium* Pasteels

Melanthidium Pasteels, 1969a: 90 (not Cockerell, 1947).
  Type species: *Anthidium concolor* Friese, 1913, by original designation.
Nigranthidium Pasteels, 1984: 57, replacement for *Melanthidium* Pasteels, 1969. Type species: *Anthidium concolor* Friese, 1913, by original designation and autobasic.
Warnckeia Pagliano and Scaramozzino, 1990: 6, replacement for *Melanthidium* Pasteels, 1969. Type species: *Anthidium concolor* Friese, 1913, autobasic.

This is a subgenus of almost entirely black bees, even the clypeus of the male being black. Light marks, the same in both sexes, are restricted to an inconspicuous yellow streak along the upper inner orbit and others on the anterior surfaces of the fore and mid tibiae. Body length is 7 to 9 mm; the body form is similar to that of *Oranthidium*. The face is unusually short and broad. The second submarginal cell receives both recurrent veins in *Afranthidium (Nigranthidium) concolor* (Friese), a character that is rare in nonparasitic Anthidiini. In *A. (N.) willowmorense* (Brauns), the second recurrent vein is distal to, but within a vein width of, the second submarginal crossvein. Further, the stigma is sometimes more elongate than in most Anthidiini, the length from the base to vein r in *A. (N.) concolor* being about 1.5 times as long as the width of the stigma; in *A. willowmorense*, however, the length so measured is about equal to the width, but the stigma tapers to a point distal to vein r. Males have a small apical tubercle on the hind trochanter. The margins of the sterna of males are unmodified, but there are strong fasciae of hairs along the graduli. The apical sterna (S6 to S8) and the genitalia are as described for *Capanthidium*. The gonostylus and penis valves are sculptured with fine, dense, longitudinal grooves.

■ *Nigranthidium* is found in Namibia and in Cape Province, South Africa. There are two named species (Michener and Griswold, 1994a) and a third, undescribed species.

Afranthidium / Subgenus *Oranthidium* Pasteels

Oranthidium Pasteels, 1969a: 95. Type species: *Anthidium oraniense* Brauns, 1905 = *Anthidium folliculosum* Buysson, 1897, by original designation.

This is a subgenus of small to medium-sized (5.5 to 10.0 mm long), robust species similar in appearance—including unbroken and usually narrow tergal bands and limited pale markings on the head and thorax—to *Afranthidium* s. str. and *Domanthidium*. The greatly reduced male gonostylus and unusual vestiture of the thoracic venter in the female are diagnostic. This is the only group in which, in most females, the ventral lamella of T6, adjacent to the lateral margin of S6, is produced posteriorly as a spine, one on each side of the narrowly rounded or pointed apex of S6. The true margin of S6 of the female is narrowly, angularly produced. As in *Afranthidium* s. str., T3 to T6 each have a large lateral spine in males; T3 or T4 to T5 have much smaller spines in females. The scutum is clothed, at least in part, with decumbent light hair. The clypeus has a distinct impunctate line medially. S6 of the male is longer than in *Afranthidium* s. str. and has a strong pair of longitudinal carinae. The penis valve has an oblique carina basally on the dorsal surface.

■ *Oranthidium* occurs in Namibia and in Cape Province, South Africa, east to the Transvaal. There are five or six species, of which three are named and were revised by Pasteels (1984).

Afranthidium / Subgenus *Xenanthidium* Pasteels

Xenanthidium Pasteels, 1984: 33. Type species: *Xenanthidium biserratum* Pasteels (described as *biserrata*), 1984, by original designation.

Xenanthidium should probably be considered a synonym of *Capanthidium*, but since it is known only from the female and has rather distinctive characters, it seems premature to synonymize it. In appearance, including its robust form, unbroken metasomal bands, and small size (length 6 mm), it suggests a species of the subgenus *Branthidium*. The most distinctive feature is the presence of a preapical denticulate carina on T6; and because the apical margin of T6 is also denticulate there are two transverse series of teeth across the apex of the tergum. The scutellum, except for the small midapical emargination, is carinate all the way across. The outer surfaces of the tibiae are coarsely punctate, tuberculate on the hind legs. Pasteels' emphasis on the proximity of the second recurrent vein and the second submarginal crossvein ("subinterstitielle") is misplaced; they are separated by about two vein widths, and this character varies in other subgenera. Except for the rather narrow impunctate margin of T5, the depressed marginal zone is well defined and strongly punctate, but distinctly more finely so than the rest of the tergum. T5 has a lateral tooth, as it does in *Capanthidium, Oranthidium,* etc.; *Branthidium* and *Mesanthidiellum* have no such tooth.

■ *Xenanthidium* is from northern Cameroon. The only species is *Afranthidium biserratum* (Pasteels), known from one specimen.

Afranthidium / Subgenus *Zosteranthidium* Michener and Griswold

Afranthidium (Zosteranthidium) Michener and Griswold, 1994a: 310. Type species: *Nigranthidium tergofasciatum* Pasteels, 1984, by original designation.

Zosteranthidium consists of an entirely black species, lacking pale areas even on the legs. The body is chalicodomiform or hoplitiform, more elongate than that of other *Afranthidium,* and the metasoma has hair bands; the general appearance of this bee is exactly that of some *Hoplitis (Alcidamea)* (Osmiini). The body length is 9 to 10 mm. Distinctive features are the broken hair bands on the metasomal terga; the relatively long, although only two-segmented, maxillary palpus, as long as the width of the maxilla at the point of palpal attachment; and the strongly shagreened although sparsely punctate propodeum. Unlike that of the superficially similar *Nigranthidium,* the hind basitarsus of the female ends in a truncation rather than in a process extending beyond the base of the second tarsal segment.

■ This subgenus is found in Cape Province, South Africa. There is only one species, *Afranthidium tergofasciatum* (Pasteels).

Genus *Afrostelis* Cockerell

Afrostelis Cockerell, 1931b: 340. Type species: *Afrostelis tegularis* Cockerell, by original designation.

Afrostelis contains small (4-6 mm long), black, presumably cleptoparasitic, hoplitiform or heriadiform species with large, punctate tegulae and with the scutum extraordinarily elongated, slightly longer than broad, narrowed and rounded posteriorly (see the key to genera for details). Otherwise it is *Stelis*-like, including the narrow pitted zone across the base of the propodeum and the exposed male sterna. There is a preoccipital carina midlaterally, the pronotal lobe is weakly carinate, and the omaulus is almost carinate above. Aside from the tegular and scutal characters, the most distinctive features are in the male genitalia. The gonostyli are slender, straight, minutely capitate at the apices. The volsellar lobes are extraordinarily elongate, almost reaching the apices of the gonostyli, and appear to be articulated at the bases, forming distinct volsellae.

■ *Afrostelis* occurs from Tanzania and Congo (Kinshasa) to Cape Province, South Africa, and Namibia. There are five specific names (Pasteels, 1984; Baker, 1996a). The hosts are unknown, but the elongate body (especially the elongation of the thorax, as illustrated by the scutum) is likely to be an adaptation for entering nests of *Heriades* in burrows in wood or stems.

The two spines on the front and middle tibiae and other characters indicate that this genus is closely related to *Stelis;* it might be merely a specialized derivative of *Stelis* not warranting generic rank. The male gonostylar form, however, is probably ancestral to that of any *Stelis.* That fact and the striking thoracic characters lead me to recognize *Afrostelis* as a genus, as did Cockerell.

Genus *Anthidiellum* Cockerell

Anthidiellum consists of rather small, robust, megachiliform to bombiform bees (Pl. 8) with a strong omaular carina, often more or less lamellate, extending well below the middle of the mesepisternum (sometimes to the discrimen) or only to the middle of the lateral surface in the subgenus *Clypanthidium.* The carina of the pronotal lobe is also frequently expanded into a lamella. The scutellum (Fig. 82-7d, f) is carinate or lamellate, truncate or medially emarginate as seen from above, acute as seen in profile, but rounded in the subgenus *Clypanthidium.* There is a fovea, margined by a carina, behind the propodeal spiracle; sometimes it is not much larger than the spiracle. The row of pits across the upper margin of the propodeum is evident only laterally, except that in the subgenus *Chloranthidiellum* it is strongly developed, horizontal, margined by a carina laterally and absent or

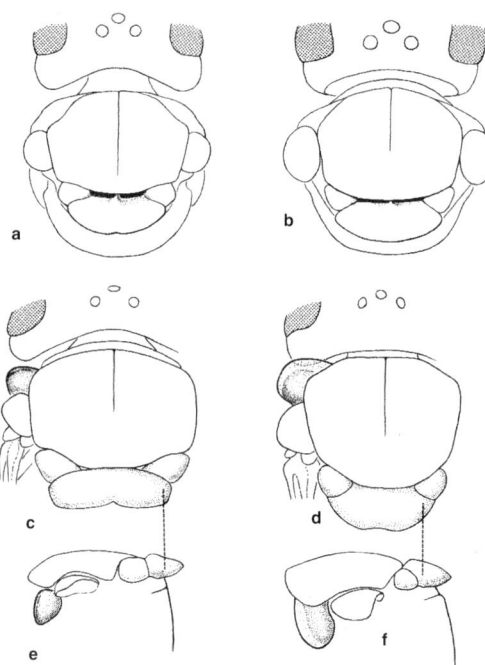

Figure 82-7. Diagrams of thoraces of anthidiine bees. **a, b,** *Anthodioctes* sp. and *Hypanthidiodes currani* (Schwarz), showing scutoscutellar foveae; **c, d,** *Anthidiellum (Anthidiellum) notatum* (Latreille) and *Dianthidium (Dianthidium) curvatum sayi* Cockerell, showing the truncate scutellum of *Anthidiellum* and the lamellate pronotal lobes of *Dianthidium* s. str.; **e, f,** Side views of c and d. From Michener, McGinley, and Danforth, 1994.

nearly so only in the median third. The inner surface of the hind tibia is rather flat; the upper edge, at least on the apical half of the tibia, is sharply marked by the limit of the keirotrichiate area, this limit sometimes carinate. This character is least evident in American species of *Anthidiellum* s. str., but even in these there is not a gradation from one hair type to the other as there is in the genus *Bathanthidium*. The subantennal sutures are usually strongly arcuate (Fig. 82-5c). The male gonostyli are expanded apically, commonly truncate or bilobed, or sometimes [*A. (Anthidiellum) strigatum* (Panzer)] extended to one side as though one lobe of a bilobed structure had been lost. The genitalia and hidden sterna were illustrated by Popov (1935) and Mitchell (1962).

Nests of *Anthidiellum* s. str. are constructed in the open, on stems, leaves, or rocks, and consist of isolated single cells made of resin (Pl. 8; Schwarz, 1928; Grigarick and Stange, 1968) or groups of cells (Friese, 1923). Bees of the subgenus *Ranthidiellum*, however, make nests in the soil, using resin for cells and entrance tubes (Pasteels, 1972).

Key to the Subgenera of *Anthidiellum*

1. Postgradular parts of T2 to T5 of females and to T6 of males swollen laterally, so that, from above, sides of metasoma seem lobed; S5 of male without comb; mandible of female with preapical shoulder on lower margin, below lower tooth 6
—. Terga not swollen laterally; S5 of male with margin broadly concave and armed with comb of black teeth, at least laterally (male unknown in *Ananthidiellum*); mandible of female without preapical shoulder on lower margin (Eastern Hemisphere) 2
2(1). Mandible of female minutely sculptured, dull, almost without carinae, its apex expanded, 1.5 times as wide as basal width; subantennal suture straight; T1 to T4 without pale markings (oriental) .... *A. (Ananthidiellum)*
—. Mandible of female somewhat shining, often coarsely punctate (somewhat dull distally in *Ranthidiellum* and *Clypanthidium*), with carinae, its apex but little wider than base (except in *Ranthidiellum*); subantennal suture arcuate [not clearly recognizable and perhaps not arcuate in *A. (Clypanthidium) bimaculatum* (Friese)]; T1 to T4 with yellow, cream, or reddish bands (except in subgenus *Clypanthidium* and some *Ranthidiellum*) 3
3(2). Preoccipital carina present at sides, behind eyes, but absent behind vertex; lamella on pronotal lobe tapering but extending nearly as far mesad from lateral margin of scutum as laterad; axilla extending laterally beyond margin of scutum (basal zone of propodeum laterally horizontal with well-developed pits) (Africa) .... *A. (Chloranthidiellum)*
—. Preoccipital carina complete, or present only behind vertex, or absent; lamella or carina on pronotal lobe usually extending little mesad from lateral margin of scutum; axilla not extending laterally beyond lateral margin of scutum 4
4(3). Tegula narrowly rounded posteriorly; omaular carina reaching about halfway down omaulus (oriental) .... *A. (Clypanthidium)*
—. Tegula broadly rounded or almost transverse posteriorly; omaular carina extending to venter (in some *Pycnanthidium* very weak or perhaps absent on lower half of mesepisternum) 5
5(4). T1 with carina separating anterior and dorsal surfaces; apex of mandible but little wider than base; body black with yellow or cream markings on all tagmata (Africa, oriental) .... *A. (Pycnanthidium)*
—. T1 without carina between anterior and dorsal surfaces; apex of mandible of female about 1.5 times as wide as base; body brown with red markings, or yellowish and black but without yellow on T1 to T4 (oriental) .... *A. (Ranthidiellum)*
6(1). Preoccipital ridge strongly carinate behind vertex; keirotrichiate area along upper margin of hind tibia not sharply defined (Western Hemisphere) .... *A. (Loyolanthidium)*
—. Preoccipital ridge not or scarcely carinate; keirotrichiate area along upper margin of hind tibia sharply defined (palearctic region, India) .... *A. (Anthidiellum* s. str.)

Anthidiellum / Subgenus *Ananthidiellum* Pasteels

Anthidiellum (Ananthidiellum) Pasteels, 1969a: 49. Type species: *Anthidium anale* Friese, 1914, by original designation.

Like *Ranthidiellum*, this subgenus consists of species that do not resemble most anthidiines but are superficially suggestive of some species of *Trigona*. Yellow markings are greatly reduced, the apical terga are white, and the wings are infuscated apically and clear basally [in *Anthidiellum anale* (Friese)] or the reverse (in an undescribed species). In addition to the characters given in the key to subgenera, a distinctive feature is the impunctate vertical anterior margin of the scutum, sharply separated by a right angle from the punctate dorsal surface; no other subgenus has such a sharply differentiated and vertical anterior scutal margin. The length is 6.5 to 7.0 mm.

■ This subgenus is known from Sikkim and Malaysia. The only described species is *Anthidiellum anale* (Friese), known from only one specimen. An undescribed species is also known from one specimen.

Anthidiellum / Subgenus *Anthidiellum* Cockerell s. str.

Anthidium (Anthidiellum) Cockerell, 1904c: 3. Type species: *Trachusa strigata* Panzer, 1805, by original designation.
Anthidium (Cerianthidium) Friese, 1923: 304. Type species: *Trachusa strigata* Panzer, 1805, by designation of Cockerell, 1925a: 361.

The lobate aspect (as seen from above) of the sides of the metasoma is not found in other Old World subgenera. This is the only Old World subgenus that lacks the combs of small black teeth on the metasomal sterna of males. The species are mostly larger (5-7 mm long) than those of the other speciose subgenus, *Pycnanthidium*.

■ This subgenus occurs throughout Europe, north to Finland (63°N), but is better represented in the Mediterranean basin (Morocco and Portugal to the Balkan peninsula, Turkey, and Israel), south to Eritrea, and east to Tadjikistan and India. There are about seven species.

Anthidiellum / Subgenus *Chloranthidiellum* Mavromoustakis

Anthidiellum (Chloranthidiellum) Mavromoustakis, 1963b: 491. Type species: *Anthidium flavescens* Friese, 1925, by original designation.

Anthidiellum (Chloranthidium) Pasteels, 1969a: 48, *lapsus* for *Chloranthidiellum* Mavromoustakis, 1963.

This subgenus contains a small (6 mm long) species with black-and-yellow coloration. It differs from all other subgenera in having the preoccipital ridge carinate laterally, behind the eyes, but not dorsally, behind the vertex. The posterior lobe of the pronotum is strongly lamellate (see the key to subgenera). S5 of the male is truncate with a marginal comb on the middle third.

■ This subgenus occurs from Kenya to Zimbabwe. The only species is *Anthidiellum flavescens* (Friese).

Anthidiellum / Subgenus *Clypanthidium* Pasteels

Bathanthidium (Clypanthidium) Pasteels, 1968a: 1060. Type species: *Anthidium ruficeps* Friese, 1914, by original designation. [Also described as new by Pasteels, 1969a: 53.]

Clypanthidium includes somewhat robust bees about 7 to 8 mm long. The body may be black with yellow marks only on T5 (female), as in *Anthidiellum (C.) bimaculatum* (Friese), or the head may be largely red or reddish yellow, the thorax with red or yellow marks, and the metasoma black with large, light-red or yellow areas at least on T5 and T6 (female). The scutellum is less sharply margined than that in other subgenera, but is large, distinctly emarginate medially, as seen from above, and strongly overhanging the metanotum. The wings of *A. (C.) ruficeps* (Friese) are dusky basally, whitish apically, as in some *Trigona* and one species of *A. (Ananthidiellum)* occurring in the same area.

■ This subgenus occurs in Malaysia. Three species have been described.

Discovery of males of this subgenus might show that it is out of place in *Anthidiellum*. The *Anthidiellum*-like features of *Clypanthidium,* by which it differs from *Bathanthidium,* where it was placed by Pasteels (1968a, 1969a, 1972), include (1) the presence of a carina on the upper half or more of the omaulus (it extends farther down in other *Anthidiellum*), (2) the enlarged scutellum, which strongly overhangs the metanotum (but is not sharply angled, carinate, or lamellate apically, as in other *Anthidiellum*), (3) the rather abrupt line between the area bearing keirotrichia and the area with other hairs along the upper margin of the hind tibia (but the line is not so abrupt as in other *Anthidiellum*), (4) the robust hind basitarsus (it is about three times as long as wide, about as in other *Anthidiellum,* but is over four times as long as wide in *Bathanthidium*), and (5) the robust body form.

Anthidiellum / Subgenus *Loyolanthidium* Urban

Anthidiellum (Loyolanthidium) Urban, 2001: 64. Type species: *Anthidium apicale* Cresson, 1878, by original designation.

Although I earlier decided that the differences between American and Palearctic species then placed in *Anthidiellum* s. str. did not justify subgeneric distinction, the name *Loyolanthidium* is now available and appropriate for the American species. Distinguishing characters are indicated in the keys. Some species of *Loyolanthidium* are larger than those of *Anthidiellum* s. str., up to 10 mm in body length.

■ *Loyolanthidium* is primarily North American, ranging from British Columbia to Quebec, Canada, south to Florida and Baja California and through Mesoamerica, rarely as far south as Pará, Brazil, and Beni, Bolivia [*A. (L.) bolivianum* (Urban)]. About eight species are known.

Anthidiellum / Subgenus *Pycnanthidium* Krombein

Pycnanthidium Krombein, 1951: 292. Type species: *Pycnanthidium solomonis* Krombein, 1951, by original designation.

Pygnanthidium Mavromoustakis, 1963b: 491, unjustified emendation of *Pycnanthidium* Krombein, 1951.

Pygnanthidium (Pygnanthidiellum) Mavromoustakis, 1963b: 492. Type species: *Anthidium zebra* Friese, 1904, by original designation.

Pycnanthidium includes some of the smallest nonparasitic anthidiines, 4.5 to 9.0 mm long. They are largely black, but with limited yellow markings on all tagmata. The broadly and strongly concave margin of S5 of the male, armed with a row of minute black teeth, differentiates this from other subgenera.

■ This subgenus occurs from Natal, South Africa, and Namibia north to Mali, the Central African Republic, and Kenya, as well as in Madagascar, Sri Lanka, India, Burma, Thailand, and on to the Philippines, New Guinea, the Solomon Islands, and Queensland, Australia. There are about 13 African species and nine Indo-Australian spe-cies.

The name *Pygnanthidiellum* was proposed for the African species and contrasted to the Indo-Australian species. In the latter group the hind tibia and basitarsus are finely punctate, the omaular carina is weak or absent below, and the pronotal lobe is carinate or narrowly lamellate. In the African group the hind tibia and basitarsus are coarsely punctate, the omaular carina is complete, and the pronotal lobe is sometimes strongly lamellate. Further, in the African group, the hind basitarsus of the female is enlarged, nearly as wide as the tibia. It now appears that both groups occur in Sri Lanka, India, and Burma (Pasteels, 1972), and I do not regard the differences between the groups as justification for subgeneric distinction.

Anthidiellum / Subgenus *Ranthidiellum* Pasteels

Anthidiellum (Ranthidiellum) Pasteels, 1969a: 48. Type species: *Protoanthidium rufomaculatum* Cameron, 1902, by original designation. [For aspects of this type designation, see Michener and Griswold, 1994a.]

Anthidiellum (Rhanthidiellum) Pasteels, 1972: 102, unjustified emendation of *Ranthidiellum* Pasteels, 1969.

Ranthidiellum resembles *Ananthidiellum* in size and appearance, the metasoma either having red bands or being largely black, the apical terga sometimes yellow or covered with white hair. Moreover, the apically broad female mandible, about 1.5 times as wide as the base, is as in *Ananthidiellum* and unlike that of other subgenera. *Ranthidiellum* differs from *Ananthidiellum*, however, in lacking the preoccipital carina, lacking a carina separating the anterior and dorsal surfaces of T1, and having arcuate subantennal sutures.

■ This subgenus occurs in Malaysia, Borneo, and Sumatra. There are three species; two are described in some detail by Pasteels (1969a, 1972).

Genus *Anthidioma* Pasteels

Anthidioma Pasteels, 1984: 34. Type species: *Anthidioma chalicodomoides* Pasteels, 1984, by original designation.

Anthidioma (males are not known) consists of nonmaculate, black, chalicodomiform species 8 to 13 mm long, with rather abundant, long, gray and white hairs. They have none of the usual carinae, but do have the usual features of Series B (see the discussion of the tribe). The posterior surface of the propodeum, including the whole triangle, is hairless, the triangle being dull and minutely roughened; there is no fovea behind the spiracle, and the pits across the base of the propodeum are completely absent; the propodeal triangle is almost vertical. The posterior part of the head (both the vertex and the genal area) is well developed. The scutellum is short, rounded posteriorly as seen from above, and the axilla does not extend at all laterally, behind the tegula. In profile the scutellum is rounded, right-angular laterally, and only slightly overhangs the metanotum. The hind basitarsus is long and slender. The terga lack lateral spines and apical depressed zones, and T6 lacks a ventral marginal lamella such as is found in *Afranthidium* (*Afranthidium* s. str. and *Oranthidium*).

■ This genus occurs in Namibia and western Cape Province, South Africa. The only described species is *Anthidioma chalicodomoides* Pasteels, but there is an additional species, undescribed.

In the absence of males, the true position of *Anthidioma* is not clear; for the present its generic rank is maintained.

Genus *Anthidium* Fabricius

Species of this genus are mostly robust, moderate-sized, and characteristic in shape, with a rather parallel-sided but broad, somewhat flattened metasoma, i.e., they are megachiliform. The body is sometimes black but usually exhibits a conspicuous yellow pattern. In Asia, however, there are species as small as 6 mm long, in some of which the body is largely yellow or red, and in Mexico two species attain 19 mm in length. The metasoma in the African subgenus *Severanthidium* tapers posteriorly, T4 being only four-fifths as wide as T1 or T2. The subgenus *Proanthidium* and some others, in which the metasoma tapers less obviously, have the body form of *Dianthidium*, i.e., they are more nearly chalicodomiform. The distinctions between *Anthidium* and the remaining genera of Series B, as a group (see the discussion of the tribe), are rather subtle and tend to break down among probably derived subgenera that seem to have lost one or another of the characters of their genus. The combination of characters remains distinctive, however. In *Anthidium* the subantennal suture is usually straight. T6 of the female has an apical depressed rim, usually smooth and shining, often hidden by hairs, and sometimes unrecognizable. This rim and usually the tergum as a whole have a median apical notch or emargination, sometimes small or largely hidden by hairs, but sometimes large and conspicuous, especially in the subgenus *Callanthidium*. Laterally, T6 of the female nearly always has a tooth, angle, or shoulder, mesal to which there is an emargination, sometimes very weak. In the subgenus *Proanthidium* the lateral emargination and tooth are absent but the impressed margin and notch are present medially. T1 to T5 of females and T1 to T6 of males have narrow, smooth, apical margins of uniform width, usually flat or nearly so. Anterior to each margin but behind the elevated midtergal or disc zone (often distinct only laterally) is the depressed marginal zone, differentiated (commonly by finer and closer punctation) from the rest of the tergum. Problems with this character are found, among others, in *A.* (*Nivanthidium*) *niveocinctum* Gerstaecker from Africa, in which the depressed zone of T5 of the female is sparsely punctate medially, but laterally is as described. In *A.* (*Severanthidium*) *severini* Vachal, also from Africa, the punctures of the depressed zone are sparse and shallow and this zone merges into the smooth margin; the shape of the depressed zone, however, is as in other *Anthidium*. The depressed marginal zones are wider (at least on T5) medially than laterally; this distinction is usually evident even when the zone is not well differentiated medially. The anterior margin of the marginal zone is usually angled medially, the whole zone thus forming a very broad triangle. These characters of the tergal margins are best examined on T5 of both sexes, but are often evident on more anterior terga. In the other genera having similar female mandibles, the smooth apical tergal margins are usually convex and the marginal zones are not recognizable, except laterally, or are scarcely wider medially than laterally. In some species of the subgenus *Proanthidium* the marginal zones are also convex. As noted above, *Proanthidium* differs from other subgenera of *Anthidium* in ways suggestive of other genera. The male genitalia and sterna, however, are as in *Anthidium*. S1 to S5 are unmodified, commonly fringed, S5 sometimes being quite short and broadly emarginate. The body of S8 is moderately elongate, commonly longer than broad. The male gonostyli are finger- or club-shaped, hairy; basal "volsellar lobes" are present, and, most indicative of *Anthidium*, the penis valves are widely separated but connected by a long, slender bridge. The male genitalia and hidden sterna were illustrated by Mitchell (1960), these and many other structures by Moure and Urban (1964) and Toro and Rojas (1970b).

This genus is found on all continents except Australia. It is rather poorly represented in sub-Saharan Africa, where species of other genera are numerous, and is absent from Madagascar and the tropics of southern Asia and Indonesia.

Revisions of the species of *Anthidium* are found in various works on the Anthidiini, e.g., Warncke (1980a), mostly in his subgenus *Anthidium* s. str., for western palearctic species; Grigarick and Stange (1968) for California species; Pasteels (1984) for sub-Saharan Africa; Moure and Urban (1964) for Brazil; etc. Toro and Rodriguez (1998) revised the Chilean species of Anthidiini, including 15 species of *Anthidium (Anthidium)*. North American species were reviewed by Schwarz (1927).

Anthidium makes nests of plant hairs in preexisting cavities in the soil, walls, wood, stems, etc., or in cavities excavated in loose soil. Masuda (1933) gave an extensive account of the nesting biology of *A. (Anthidium) japonicum* Smith and summarized many of the earlier accounts. The fragmentary information on nests of North American species is summarized by Grigarick and Stange (1968). Parker (1987) described nests of the subgenus *Callanthidium,* which are not strikingly different from those of *Anthidium* s. str. Westrich (1989) gave an excellent account of the nesting behavior of German species, with appropriate references to earlier work. [He includes other genera in *Anthidium;* his "Harzbienen" and *Anthidium lituratum* (Panzer) belong in other genera, according to the present classification.] Westrich (1989) also summarized accounts of territoriality and mating behavior in *Anthidium*. Unlike those of most bees, males are often larger than females, and large individuals hold territories that include food resources (Alcock, Eickwort, and Eickwort, 1977; Severinghaus, Kurtak, and Eickwort, 1981; and, for the subgenus *Callanthidium,* Alcock, 1977).

Key to the Subgenera of *Anthidium*
(From Michener and Griswold, 1994a)

1. T6 of female with deep median emargination about one-fifth as wide as tergum, and strong lateral tooth about midway between base and apex of exposed part of tergum, T6 thus appearing strongly four-toothed; S6 of male with two large, erect, heavily sclerotized, bare apical lobes between which lie the greatly elongated, retrorsely bent apices of the penis valves (nearctic) .... *A. (Callanthidium)*
—. T6 of female with much smaller midapical emargination, lateral to which margin is entire or uncommonly denticulate, often with a relatively small lateral tooth; S6 of male variable but not as described above; penis valves not greatly elongate, not bent retrorsely 2
2(1). Scutellum rounded in profile, not carinate or lamellate, not greatly overhanging metanotum and propodeum; pronotal lobe with or without carina .... *A. (Anthidium s. str.)*
—. Scutellum angulate in profile (at least as seen obliquely to show profile of lateral part of scutellum), strongly carinate or lamellate at least laterally, greatly overhanging metanotum and propodeum; pronotal lobe carinate or lamellate 3
3(2). Pronotal lobe carinate; axillar suture weak; scutoscutellar suture not in deep depression, scutellum thus nearly continuing profile of scutum (scutellum strongly produced posteriorly as rather flat structure ending in lamella) (Africa) .... *A. (Nivanthidium)*
—. Pronotal lobe with more or less anteriorly directed or erect, translucent lamella; axillar suture strong; scutoscutellar suture in depression, scutellum thus independently convex in profile 4
4(3). Scutellum ending in lamella or large carina almost all the way across; hind basitarsus with longitudinal carina on outer surface; omaulus sharply angulate or weakly carinate 5
—. Scutellar margin with broad median part neither carinate nor lamellate; hind basitarsus not carinate; omaulus rounded or forming rounded angle 6
5(4). Posterior scutellar margin subtruncate as seen from above, its lateral part curved forward and becoming more or less longitudinal; antennae arising below level of middles of eyes, which converge strongly below, rendering clypeus unusually small (Africa) .... *A. (Severanthidium)*
—. Posterior scutellar margin broadly rounded as seen from above, with small median emargination, laterally oblique, only at extreme end next to axilla sometimes becoming longitudinal; antennae arising near level of middles of eyes, which converge slightly to moderately (palearctic) .... *A. (Gulanthidium)*
6(4). Scutellum as seen from above with margin gradually curved forward at each side to axillar margin; hind basitarsus of female less than four times as long as broad (palearctic) .... *A. (Turkanthidium)*
—. Scutellum as seen from above with margin more or less transverse, curved forward rather abruptly or angled forward at each side to axillar margin, angle often protruding posteriorly; hind basitarsus of female four or more times as long as broad (palearctic) .... *A. (Proanthidium)*

Anthidium / Subgenus *Anthidium* Fabricius s. str.

Anthidium Fabricius, 1804: 364. Type species: *Apis manicata* Linnaeus, 1758, by designation of Latreille, 1810: 439.
Stenanthidium Moure, 1947a: 16. Type species: *Anthidium espinosai* Ruiz, 1938, by original designation.
Tetranthidium Moure, 1947a: 15. Type species: *Anthidium latum* Schrottky, 1902, by original designation.
Melanthidium Cockerell, 1947: 106, not *Melanthidium* Pasteels, 1969. Type species: *Melanthidium carri* Cockerell, 1947, by original designation.
Anthidium (Melanoanthidium) Tkalců, 1967: 91. Type species: *Anthidium montanum* Morawitz, 1864, by original designation.
Anthidium (Echinanthidium) Pasteels, 1969a: 101. Type species: *Anthidium echinatum* Klug, 1832, by original designation.
Anthidium (Pontanthidium) Pasteels, 1969a: 105. Type species: *Anthidium pontis* Cockerell, 1933, by original designation.
Anthidium (Ardenthidium) Pasteels, 1969a: 103. Type species: *Anthidium ardens* Smith, 1879, by original designation.
Anthidium (Morphanthidium) Pasteels, 1969b: 423. Invalid because no type species designated; three included species. (Synonym of *Ardenthidium,* to judge by two of the included species, and was clearly intended by Pasteels to be the same as *Ardenthidium.*)

The metasoma in this subgenus is usually rather parallel-sided, T4 being about as wide as T1, and the whole perhaps a little flatter than in some related taxa, giving a characteristic form whereby the subgenus can often be tentatively recognized. The body length is variable, 8 to 19 mm. Females of most species differ from those of other subgenera in the strong outer ridge of the mandible, which extends from the mandibular acetabulum (upper articulation) almost to the middle of the tooth row (fig. 28, Michener and Fraser, 1978; Fig. 82-3a). Although a few *Anthidium* s. str. are all black, most are more or less extensively patterned with yellow to cream-colored markings. The metasomal bands are often broken medially, and the posterior or anterior margin at each side is often emarginate; if the emargination cuts through the band, the band becomes four spots, a pattern that is rather common. In males the apex of the metasoma is usually curved downward, and T7 is strongly developed and usually trifid, with a median apical spine and a slender to broad lobe on either side. In some species, such as *A. punctatum* Latreille, the median spine is reduced to an angle, and in others, such as *A. tesselatum* Klug and the species placed in *Ardenthidium* by Pasteels, the median spine is absent and the lateral projections are acute lobes or spines, T7 thus bifid.

■ Species of the subgenus *Anthidium* s. str. are found throughout the range of the genus, i.e., on all continents except Australia, and are absent also in the Indo-Malayan tropics. In the palearctic region they occur from the Canary Islands to Japan, and in the nearctic they occur from coast to coast and north to subarctic Alaska but are absent except for the introduced *A. manicatum* (Linnaeus) in the east north of Virginia. *Anthidium* s. str. ranges south through the tropics (although probably absent in Amazonian regions) to Paraguay, Bolivia, and central Chile. The introduced (from Europe) *A. manicatum* (Linnaeus) occurs in southern Brazil and south to Buenos Aires Province, Argentina. *Anthidium* s. str. includes 18 species in the western palearctic region, according to the conservative account of Warncke (1980a); 25 in America north of Mexico; 21 in the neotropical region; and a few in Africa and eastern Asia. The total must be about 90. Urban (2002) revised the South American species of *Anthidium,* recognizing 39 species.

This is a large and rather diverse subgenus. Unusual species or small groups have been given subgeneric or generic names, but such names seem unnecessary, being apparently based on one or very few species derived from among the "ordinary" species of *Anthidium* s. str. Most Old World species and most South American species have a longitudinal carina on the lower margin of the outer surface of the hind tibia. However, most North American species, as well as *Anthidium montanum* Morawitz *(Melanoanthidium)* and *A. echinatum* Klug *(Echinanthidium)* in the palearctic region, lack this carina; other characters correlated with the presence or absence of this carina are not known. The name *Ardenthidium* has been applied to a group of unrelated species all having about ten mandibular teeth in the female, compared to five to seven in most other species. In other respects the type species, *A. ardens* Smith, and the quite dissimilar *A. undulatiforme* Friese appear to be *Anthidium* s. str. *A. echinatum* Klug, the type species of *Echinanthidium,* is perhaps better differentiated from most other *Anthidium* s. str. It is one of the few palearctic forms without a hind tibial carina, and the clypeal margin of the female is not thickened as it is in most *Anthidium* s. str. In the latter respect it resembles *Turkanthidium* and most *Proanthidium*. Like many desert bees (Morocco to Pakistan), the species placed in *Echinanthidium* have a pallid aspect due to the largely yellow metasoma with its preapical tergal fringes of dense white hairs. *A. pontis* Cockerell *(Pontanthidium)* differs from other *Anthidium* s. str. in having a protuberant clypeus. Other variants of *Anthidium* s. str. include the Brazilian *A. latum* Schrottky *(Tetranthidium),* in which the median tooth of T7 of the male is double, the tergum thus quadridentate, and the Chilean *A. espinosai* Ruiz *(Stenanthidium),* in which T7 of the male is narrowed apically, the two lateral lobes being close together beneath the median spine. Each of the two proposed South American subgenera consists of only a single species. The two Mesoamerican species placed in *Melanthidium* are very large, and the metasoma is almost entirely black; the strongly reflexed, trilobed S6 of the male is unusual; and a large lateral spine on T6 of the male may be present or absent. In *A. (A.) tesselatum* Klug the second recurrent vein is basad to the second submarginal crossvein, a character found elsewhere in nonparasitic Anthidiini in one species of *Afranthidium (Nigranthidium)* and in two subgenera of *Trachusa.*

Anthidium / Subgenus *Callanthidium* Cockerell

Callanthidium Cockerell, 1925a: 365. Type species: *Anthidium illustre* Cresson, 1879, by original designation.

This subgenus of large (11-18 mm long), robust bees with yellow markings can be recognized by the deep median emargination of T6 of the female and the strong lateral tooth that renders T6 quadridentate. T7 of the male has a midapical spine and lateral lobes. S5 is shorter medially and thus more emarginate than that of *Anthidium* s. str. S6 has lateral lobes and in one species a median spine, thus mirroring the form of T7. The extraordinarily long, bladelike, reflexed apices of the penis valves are distinctive, but suggest those of *A. (A.) edwini* Ruiz from Chile.

■ This North American subgenus ranges from Baja California, Mexico, to New Mexico and north to Oregon and Montana, USA. The two species were revised by Grigarick and Stange (1968).

Anthidium / Subgenus *Gulanthidium* Pasteels

Anthidium (Gulanthidium) Pasteels, 1969a: 101. Type species: *Anthidium anguliventre* Morawitz, 1888, by original designation.

Gulanthidium consists of robust or somewhat elongate species, 8.5 to 12.0 mm long, black with extensive yellow markings. This is one of the most carinate (or lamellate) groups of *Anthidium* (omaulus, scutellum, hind tibia, hind basitarsus), being similar within the genus *Anthidium* only to *Severanthidium* in these features. T7 of the male, little exserted, is rounded or has a small median emargination.

■ *Gulanthidium* occurs from Israel to Turkey, Iran, Oman, and the Baluchistan area of Pakistan. Pasteels (1969a) mentioned three species, only one of which, *Anthidium anguliventre* Morawitz, is named.

Anthidium / Subgenus *Nivanthidium* Pasteels

Anthidium (Nivanthidium) Pasteels, 1969a: 106. Type species: *Anthidium niveocinctum* Gerstaecker, 1857, by original designation.

Nivanthidium contains a large (9-14 mm long), robust, shiny black species with a large, nearly flat scutellum and very restricted cream-colored markings. Both sexes have areas of dense, white hairs along the lateral margins of the metasoma. T7 of the male is bispinose, with long hairs occupying the space between the spines; there are strong lateral spines on T4 to T6.

■ This subgenus is known from eastern Africa (Mozambique, Malawi). The only species is *Anthidium niveocinctum* Gerstaecker.

Anthidium / Subgenus *Proanthidium* Friese

Anthidium (Proanthidium) Friese, 1898a: 101. Type species: *Anthidium oblongatum* Latreille, 1809 = *Anthophora oblongata* Illiger, 1806, by designation of Cockerell, 1909b: 269. [Pasteels (1969a) incorrectly listed *A. undulatum* Dours as the type species of *Proanthidium*.]

This subgenus consists of species that are more convex than those of *Anthidium* s. str., having a form more like that of *Dianthidium*. The body is black with yellow markings, the metasomal bands often broken but without emarginations. The length is 7 to 12 mm. Unlike that of *Anthidium* s. str., the carina of the pronotal lobe is elevated to form a lamella and the scutellum is carinate laterally, commonly produced to a tooth or angle close to the axilla, as in many *Pseudoanthidium*. Not all *Proanthidium* species, however, have a distinct lateral scutellar angle; for example, *A. (P.) amabile* Alfken, which was placed by Pasteels in the subgenus *Echinanthidium*, seems to be a species of *Proanthidium* without such an angle. Another feature not shared with most *Anthidium* but found in the subgenus *Severanthidium* and the genus *Pseudoanthidium* is the form of T6 of the female, which in *Proanthidium* is concave in profile. T7 of the male is little exserted, the deep median emargination forming two rounded lobes.

■ *Proanthidium* ranges from Portugal through the Mediterranean basin north to 52°N in Europe and east through Asia Minor to Central Asia. Pasteels (1969a) listed eight species. *Anthidium (Proanthidium) oblongatum* (Illiger) is adventive in the eastern United States, known from Maryland to New York State (Hoebeke and Wheeler, 1999).

Anthidium / Subgenus *Severanthidium* Pasteels

Anthidium (Severanthidium) Pasteels, 1969a: 106. Type species: *Anthidium severini* Vachal, 1903, by original designation.

Severanthidium consists of moderate-sized to large (7-13 mm long), very robust species with a wide, flat head and the tip of the metasoma tapering more than that of most *Anthidium* s. str., T4 being narrower than T1 to T3. The coloration is black with yellow markings, these sometimes quite restricted on the head and thorax; on the metasoma the bands are broken, without emarginations, the ground color being black, yellow-brown, to red. In body form and coloration, *Severanthidium* resembles the genus *Pachyanthidium*. A lamella on the pronotal lobe is stronger than that in other subgenera. This is the only group of *Anthidium* with an omaular carina on the upper half of the mesepisternum, except for *A. gratum* Morawitz, a species placed by Warncke (1980a) in *Turkanthidium* but whose proper position remains in doubt. The apex of the male metasoma in *Severanthidium* is not much curled. T7 of the male, like that of *Proanthidium*, is medially emarginate, forming two rounded lobes.

■ This subgenus is widespread in sub-Saharan Africa, from eastern Cape Province in South Africa north to Senegal and Sudan; it is also known from Oman on the Arabian Peninsula. The ten recognized species were revised by Pasteels (1984).

Anthidium / Subgenus *Turkanthidium* Pasteels

Anthidium (Turkanthidium) Pasteels, 1969a: 103. Type species: *Anthidium unicum* Morawitz, 1875, by original designation.

This is a subgenus of small (5-7 mm long), robust bees with a globose metasoma. The body is black with extensive yellow areas or largely reddish brown. The broad hind basitarsus of the female is unusual in the genus; in some species it is as wide as the tibia. The carina of the pronotal lobe is elevated to form a narrow lamella, and the scutellum is extended posteriorly, forming a carina at the sides and posteriorly except in the median emargination. T7 of the male is deeply bilobed, sometimes with a small median spine.

■ *Turkanthidium* is known from Central Asia and Afghanistan. There are about five species.

Genus *Anthodioctes* Holmberg

This neotropical genus consists of rather small (4.5-10.5 mm long) hoplitiform to slender megachiliform, coarsely punctate, strongly carinate and foveate (Fig. 82-8), dark species, usually with rather limited yellow markings, these sometimes lacking on the thorax. A few species, however, have extensive yellow markings. A noteworthy feature is the strong preoccipital carina at the side of the

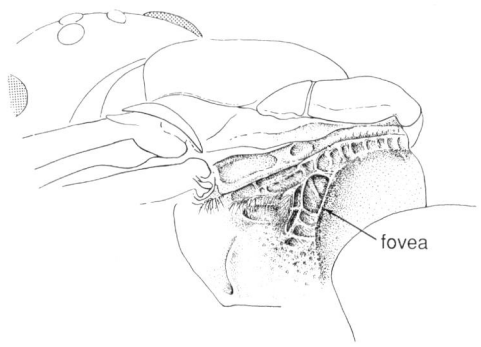

Figure 82-8. *Anthodioctes* sp., posterodorsal view of propodeum, showing the postspiracular fovea and the series of pits across the basal zone. From Michener, McGinley, and Danforth, 1994.

head behind the eye, the partitions that form punctures in front of the carina extending up onto the anterior surface of the carina, which is thereby subdivided into numerous small segments. All species have arolia in both sexes, unlike most of the superficially similar species of *Hypanthidium* and *Hypanthidioides,* but like the equally similar *Hoplostelis (Austrostelis).* T7 of the male is small, rounded to bilobed; the terga lack spines.

Key to the Subgenera of *Anthodioctes*

1. Pronotal lobe shorter anteroposteriorly than tegula; stigma about twice as long as prestigma .... *A. (Anthodioctes* s. str.*)*
—. Pronotal lobe as long anteroposteriorly as tegula; stigma about as long as prestigma .... *A. (Bothranthidium)*

*Anthodioctes / * Subgenus *Anthodioctes* Holmberg s. str.

Anthodioctes Holmberg, 1887b: 36, nomen nudum.
Anthodioctes Holmberg, 1903: 435. Type species: *Anthodioctes megachiloides* Holmberg, 1903, by designation of Cockerell, 1927b: 2. [*A. dasygastrinus* Holmberg, 1903, designated by Isensee (1927: 375), was not an originally included species name and was not a published name, so far as is known.]
Nananthidium Moure, 1947a: 26. Type species: *Nananthidium bettyae* Moure, 1947, by original designation.

Like the related genus *Hypanthidioides,* this subgenus has an unusually large stigma for an anthidiine. T7 of the male is entire or weakly bilobed. The slender, hoplitiform or heriadiform body of the species placed in *Nananthidium* seems to intergrade with the more robust form of most species of *Anthodioctes* s. str. Since no other characters distinguish *Nananthidium,* it is best regarded as a synonym of *Anthodioctes.*

■ *Anthodioctes* s. str. ranges through the tropics from as far north as Tamaulipas, Chihuahua, and Sinaloa, Mexico, south to Bolivia and Buenos Aires Province, Argentina. Urban (1998c) revised the nine species of the slender *(Nananthidium)* group of *Anthodioctes* and gave a key to the species. Urban (1999b) also gave a key to species of *Anthodioctes* s. str., exclusive of the the slender species formerly placed in *Nananthidium.* She recognized 27 species, including 19 new ones.

The ecology and nesting biology of *Anthodioctes* in the Amazonian forest were described by Morato (2001). The nests, in trap nests (holes in wood blocks) mostly placed 15 m above the ground, consisted of series of cells made of resin mixed with particles of wood. Nests of another species in similar trap nests, much nearer to ground level, were described by Alves-dos-Santos (2004) along with observations on the nest-making behavior of females.

*Anthodioctes / * Subgenus *Bothranthidium* Moure

Bothranthidium Moure, 1947a: 23. Type species: *Bothranthidium lauroi* Moure, 1947, by original designation.

The hoplitiform body of species of this subgenus suggests the slender species of *Anthodioctes* s. str. formerly placed in *Nananthidium.* *Bothranthidium* differs from such species in having an enormous combined prothoracic lobe and dorsolateral angle of the pronotum and a much smaller stigma (like that of most Anthidiini), as indicated in the key to subgenera. Additional distinctive features of *Bothranthidium* are the very broad scutoscutellar foveae and other differentiating features cited by Moure (1947a: 27, 28). T7 of the male is strongly bilobed, the emargination between the lobes being deeper than a semicircle.

■ This subgenus occurs in Brazil (Esperito Santo to Santa Catarina), Paraguay, Bolivia, and Peru. The only species in *Anthodioctes lauroi* (Moure).

Genus *Apianthidium* Pasteels

Apianthidium Pasteels, 1969a: 41. Type species: *Anthidium apiforme* Meade-Waldo, 1914, by original designation.

Apianthidium appears to be a relative of *Trachusa,* as suggested by its megachiliform body and size (length 12-13 mm), the oblique, long cu-v (fully half the length of the second abscissa of M+Cu) of the hind wing, and the small T7 of the male. In the absence of arolia and the absence of all the usual anthidiine carinae, it agrees with some subgenera of *Trachusa.* In addition to the characters given in the key to distinguish it from *Trachusa,* it differs in the strongly hooked apices of the hind tibial spurs, the relatively slender hind basitarsus of the female (over three times as long as broad), and the yellow or reddish-yellow body with the posterior half of each tergum black.

■ *Apianthidium* occurs in Borneo. The only species is *A. apiforme* (Meade-Waldo).

Genus *Aspidosmia* Brauns

Osmia (Aspidosmia) Brauns, 1926: 208. Type species: *Osmia arnoldi* Brauns, 1926, monobasic.

This genus consists of dark, nonmetallic, robust bees, 8 to 10 mm long, having about the form of species of *Osmia* s. str. They are long-haired, without tergal fasciae. *Aspidosmia* was long included among the little-known genera related to *Osmia,* partly because the second recurrent vein enters the second submarginal cell, but Peters (1972a) demonstrated its relationship to the Anthidiini. The clypeus of the male is partly yellow, but there are no other yellow markings. The maximum width of the stigma is almost as great as the inner margin of the stigma basal to vein r. Peters (1972a) cited another character that *Aspidosmia* shares with Anthidiini, namely, a shortened thorax with a large propodeal triangle, but variations in both Anthidiini and Osmiini render this character weak. The prestigma is longer than the stigma. This is usually an osmiine feature but occurs also in a few anthidiines; thus the stigmal and prestigmal areas of *Aspidosmia* are similar to those of *Trachusa (Heteranthidium) larreae* (Cockerell). The cleft claws of the female, a presumed plesiomorphy, also resemble those of nearly all Anthidiini, but cleft claws also appear in females of *Osmia,* subgenus *Metallinella,* and simple claws appear in *Trachusoides,* an anthidiine. Because males all have cleft claws, the genes for such claws are present in every female, and a regulatory change could cause such claws to appear in females. On the preaxilla, below the posterior lateral angle of the scutum, are some long hairs similar to those of adjacent areas, as is the case in Megachilini. The jugal lobe is about half as long as the vannal lobe of the hind wing, as is the case in the Lithurgini. Recognizable gonostyli of males, though partly fused to the gonocoxites, are similar to those found in Anthidiini. S7 and S8 of males are also similar to those of Anthidiini, and better developed than those found in Osmiini (see the illustrations by Peters, 1972a). Peters (1972a) gave a detailed description and illustrations of other features of the genus.

The hind tibial scopa (which, like that of the sterna, bears abundant pollen in museum specimens of both species) was illustrated by Peters (1972a). If it is truly a scopa, i.e., if it functions for carrying pollen to the nest, it is unique in the Megachilidae, which otherwise do not have a tibial scopa, except possibly for *Pararhophites*. The scopa-like tibial hairs of the Fideliini are not used for carrying pollen.

■ *Aspidosmia* occurs in Namibia and Cape Province, South Africa. It contains two species, one of which, *A. arnoldi* (Brauns), has grotesquely modified female clypeus, mandibles, and hypostomal areas. Peters (1972a) described each species in detail.

If the supposed scopal hairs of the hind tibiae are ancestral rather than a derived feature, *Aspidosmia* may be an ancestral type, surviving from near the inception of the megachiline tribes Anthidiini, Osmiini, and Megachilini. Peters (1972a) tends to support this view. Several of the characters listed above are possible or probable plesiomorphies relative to other members of these tribes. In various features, including the lack of yellow markings except on the clypeus of the male, *Aspidosmia* resembles *Plesianthidium,* especially the subgenus *Carinanthidium,* but the resemblence is probably superficial and convergent.

Brauns (1926) illustrated a nest, consisting of several cells, made of pebbles in a matrix of plant material, probably made from a resinous plant. It was on the underside of a stone.

Although Peters (1972a) made clear the anthidiine characters of *Aspidosmia,* Griswold (1985) and Griswold and Michener (1997) retained it in the Osmiini. The cladogram on which our decision was based utilized too few characters to be reliable, and I now consider *Aspidosmia* to be clearly an anthidiine.

Genus *Austrostelis* Michener and Griswold

Hoplostelis (Austrostelis) Michener and Griswold, 1994b: 676. Type species: *Stelis aliena* Cockerell, 1919, by original designation.

Like *Hoplostelis, Austrostelis* contains species that lack the scopa and were therefore formerly placed in *Stelis.* Species range from 5.5 to 8.5 mm in body length and have the shape and coloration of many *Hypanthidioides (Dichanthidium).* The flat projection at the apex of the hind tibia of the female is found only in *Austrostelis.* For additional comparisons see the account of *Hoplostelis.*

■ *Austrostelis* ranges from Buenos Aires and Salta provinces, Argentina, north through Brazil to western Colombia (Valle Province) and to the state of San Luis Postosí, Mexico. Eight species were listed by Griswold and Michener (1988).

Figure 82-9. *Aztecanthidium tenochtitlanicum* Snelling. Above, Male; Below, Female. From Michener, McGinley, and Danforth, 1994.

Hosts are unknown but cannot all be euglossine bees, the hosts of *Hoplostelis,* since *Austrostelis* occurs outside the range of Euglossini. It is very probable that *Austrostelis* parasitizes related Anthidiini, such as *Hypanthidioides.*

Genus *Aztecanthidium* Michener and Ordway

Aztecanthidium Michener and Ordway, 1964: 70. Type species: *Aztecanthidium xochipillium* Michener and Ordway, 1964, by original designation.

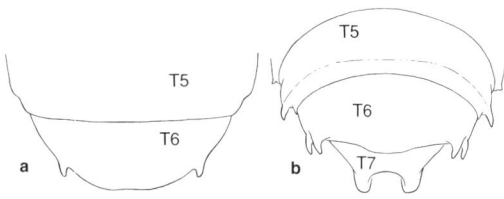

Figure 82-10. Apices of metasomas of *Aztecanthidium tenochtitlanicum* Snelling. **a,** Female; **b,** Male. From Michener, McGinley, and Danforth, 1994.

Aztecanthidium contains large (11-21 mm long), chalicodomiform species (Fig. 82-9), either black with yellow markings or red-brown with or without reddish-yellow markings. The clypeus of the female is more than twice as broad as long, sometimes protuberant. The preoccipital carina is complete, laterally joining the posterior end of the hypostomal carina. T5 of the male (and sometimes T4) has a lateral spine; T6 has two strong lateral spines, and T7 is deeply bilobed (Fig. 82-10b). The female is unusual in having a sublateral spine on T6 (Fig. 82-10a). The male genitalia, hidden sterna, and heads were illustrated by Snelling (1987).

■ This genus is known only from Mexico—Oaxaca to Puebla and Nayarit. The three species were revised by Snelling (1987).

Genus *Bathanthidium* Mavromoustakis

Bathanthidium is a small, little-known, Asiatic genus, its species all black to black with yellow on all tagmata, the yellow bands being broken on the anterior terga. These forms are united generically by the medially divided, smooth-floored scutoscutellar sulcus (although the sulcus is narrow in the subgenus *Manthidium*), the absence of an omaular carina, the presence of at least a weak carina on the pronotal lobe, and the lack of other carinae. There is a fovea (small in *Manthidium*) behind the propodeal spiracle. The preoccipital ridge is not carinate. The scutellum is rounded or medially emarginate, in profile rounded or at least not sharp and not overhanging the metanotum. The inner surface of the hind tibia is rather convex, curving onto the upper margin, there being no sharp edge or carina marking the upper edge of the keirotrichiate area. Arolia are present. T6 of the male is simple; T7 ranges from simple (in *Bathanthidium* s. str.) to trilobed, with the median lobe longest.

This genus contains three taxa that have genus-group names, but because only four species are involved, it may seem unreasonable to recognize subgenera. I have done so partly because there is no assurance that the genus is monophyletic. The subgenera *Bathanthidium* s. str. and *Stenanthidiellum*, slender and hoplitiform, are quite clearly close relatives; e.g., both have a small median comb on S4 of the male. The subgenus *Manthidium* is more robust, megachiliform, resembling *Anthidiellum* subgenus *Ranthidiellum*, and may be misplaced in *Bathanthidium*. All, however, have a comb on S5 of the male occupying almost the entire width of the segment; this is an unusual feature suggesting a mutual relationship.

Key to the Subgenera of *Bathanthidium*

1. Fovea behind propodeal spiracle rounded, delimited by strong carina; T7 of male trilobed, median lobe longest (upper margin of propodeum without row of pits except laterally) (oriental) .... B. (*Bathanthidium* s. str.)
—. Fovea behind propodeal spiracle elongate, weakly delimited; T7 of male simple or with slightly produced median lobe.. 2
2(1). Basal zone of propodeum not distinct, with irregular, sculptured, sloping pits laterally; T6 of male with elevated median section and concave lateral sections (oriental) .... B. (*Manthidium*)
—. Basal zone of propodeum distinct laterally, with regular, shiny, nearly horizontal pits; T6 of male with surface convex (eastern palearctic) .... B. (*Stenanthidiellum*)

Bathanthidium / Subgenus *Bathanthidium* Mavromoustakis s. str.

Dianthidium (*Bathanthidium*) Mavromoustakis, 1953a: 837. Type species: *Dianthidium bifoveolatum* Alfken, 1937, by original designation.

This subgenus consists of a single small (8-9 mm long), slender hoplitiform species with yellow marks on all tagmata. It differs from the others in the dull mandibular surface of the female.

■ This subgenus occurs in southeastern China and Taiwan. The only known species is *Bathanthidium bifoveolatum* (Alfken).

Bathanthidium / Subgenus *Manthidium* Pasteels

Manthidium Pasteels, 1969a: 43. Type species: *Anthidium binghami* Friese, 1901 (first described under the homonymous name *Anthidium fraternum* Bingham, 1897), by original designation.

Manthidium consists of a megachiliform species, 8 mm long, without yellow marks on the head and thorax except on the face of the male, but with dark yellow bands on the terga, broken on T1 and T2. Contrary to Pasteels' (1969a) observations, the scutoscutellar suture has a deeply invaginated part with a shiny bottom, divided medially to form two foveae. T6 of the male differs from both other subgenera as indicated in the key to subgenera; T7 has a weak midapical convexity, on each side of which is a shoulder.

■ This subgenus is known from northeastern India and Malaysia. The only species is *Bathanthidium binghami* (Friese).

Bathanthidium / Subgenus *Stenanthidiellum* Pasteels

Bathanthidium (*Stenanthidiellum*) Pasteels, 1968a; 1059. Type species: *Anthidium sibiricum* Eversmann, 1852, by original designation. [Also described as new by Pasteels, 1969a: 53.]
?*Lasanthidium* Romankova, 1988: 26. Type species: *Stelis malaisei* Popov, 1941, by original designation.

Like *Bathanthidium* s. str., this is a subgenus containing small (5.5-8.0 mm long), slender, hoplitiform species with broken yellow tergal bands. It differs from the other subgenera in the evenly convex, simple T6 and T7 of the male and in the complete series of pits across the upper margin of the propodeum.

■ This subgenus occurs in eastern Siberia and Korea. There are apparently two species. Romankova (1988) considered the type species of *Stenanthidiellum* and *Lasanthidium* to be in different genera. The identity of *Anthidium sibiricum* Eversmann may be uncertain (see synonymy above). The genitalia of *Stelis malaisei*, the type species of *Lasanthidium*, are similar to those of *Bathanthidium* (*B.*) *bifoveolatum* Alfken (see Popov, 1941a), and it may be that *Bathanthidium* and *Stenanthidiellum* should be united.

The slender body form of *Stenanthidiellum* and *Bathanthidium* s. str. suggests other megachilids that live in narrow burrows in wood. Romankova's (1988) account of nests consisting of series of resin cells in preexisting burrows in wood or hollow plant stems supports this idea.

Genus *Benanthis* Pasteels

Benanthis Pasteels, 1969a: 61. Type species: *Anthidium madagascariensis* Benoist, 1963, by original designation.

Benanthis consists of a rather small species (body length 9 mm) of rather slender form with yellow markings, including medially broken metasomal tergal bands. The usual carinae are absent except for a carina on the pronotal lobe. The subantennal sutures are nearly straight. The scutellum is rounded in profile, scarcely overhanging the metanotum. The fore and middle tibiae each have one apical spine. Arolia are present. T6 of the male is unmodified, T7 is deeply bilobed without lateral teeth, the exposed part about as long as T6. The long last segment of the maxillary palpus illustrated by Pasteels (1969a) probably consists of three segments, and the palpus is thus probably four-segmented. The small projection on the posterior side of the apex of the hind trochanter mentioned by Pasteels seems almost insignificant.

■ The genus is known only from Madagascar. Two species are reported (Pauly, in Pauly et al., 2001).

Genus *Cyphanthidium* Pasteels

Cyphanthidium Pasteels, 1969a: 57. Type species: *Cyphanthidium intermedium* Pasteels, 1969, by original designation.

Trianthidiellum Pasteels, 1969a: 58. Type species: *Hypanthidium sheppardi* Mavromoustakis, 1937, by original designation.

This genus consists of species having the size (7.5-8.5 mm long), megachiliform bodies, and general coloration of *Afranthidium (Oranthidium) folliculosum* (Buysson) and many other species of *Afranthidium*. The unbroken yellow tergal bands, narrow at least on the more anterior terga, are of the style common in *Afranthidium* s. str. and the subgenera *Oranthidium*, *Domanthidium*, and *Capanthidium*. As in some of these taxa, the gonostyli are rather slender and paddle-shaped and the penis valves are simple, not curved downward. *Cyphanthidium* is readily separated from *Afranthidium*, however, by the four-toothed female mandibles and the presence of arolia; thus *Cyphanthidium* is a member of Series A while *Afranthidium* is in Series B (see the discussion for the tribe). The only carinae are one on the pronotal lobe (where the carina is enlarged into a lamella in *C. intermedium* Pasteels), a weak carina (sometimes absent) on the upper part of the omaulus, and the sharp, thin margin of the scutellum in the same species. There is no fovea behind the propodeal spiracle, and a row of pits across the upper edge of the propodeum is absent except at the extreme sides. T7 of the male is trilobed. Sternal combs are absent in males, but S5 is broadly emarginate with a large lateral lobe, the lobe of *C. sheppardi* (Mavromoustakis) having a small

Figure 82-11. Female Anthidiini, showing the similarity between nonparasitic and cleptoparasitic forms. Above, *Dianthidium (Dianthidium) ulkei* (Cresson); Below, *Stelis (Dolichostelis) laticincta* Cresson. From Michener, McGinley, and Danforth, 1994.

tooth. The same species has a small lateral tooth on T6 of the male, this tooth absent in *C. intermedium*.

■ *Cyphanthidium* is known from Cape Province, South Africa, and from Namibia and Zimbabwe. There are two described species and another, undescribed form.

Genus *Dianthidium* Cockerell

The chalicodomiform species (Fig. 82-11a) constituting this genus are black or sometimes reddish with limited or extensive yellow to cream markings; in the subgenus *Mecanthidium* the body is almost wholly red. The following are some probable synapomorphies of the genus as here understood, following Griswold and Michener (1988): (a) there is a patch of short plumose hairs on the posterior margin of the metanotum lateral to the metanotal pit; (b) the scutum has a transverse anterior crest or angle separating its vertical anterior margin from the horizontal dorsal surface (Figs. 82-7f and 82-11a show the substantial vertical anterior part of the scutum); (c) the pronotal lobe has an anteriorly sloping translucent lamella (Fig. 82-7f) and just mesal to the inner end of the lamella, the dorsal surface is sloping and not horizontal and on the level of the scutal surface; (d) the tegula is as wide as long, the widest point behind the middle; and (e) impunctate tergal margins (rather broad) end abruptly at the side of the metasoma. In some other genera, such as *Hypanthidioides* and *Hypanthidium*, the anterior margin of the scutum bends down at an angle of about 45° to the adjacent scutal surface. This bend is not so abrupt, nor is the margin so nearly vertical, as in *Dianthidium*. Other characters of *Dianthidium* include the carinate omaulus, at least dorsally, the tridentate female mandible (in some species the preapical tooth is reduced to an angle at the end of a long cutting edge), the preapical crest or ridge of T6 of the female, and the apical fringe or tuft of hair (sometimes paired) on S3 of the male. The subgenus *Mecanthidium* was described in the genus *Paranthidium* because of the long, oblique apex of the female mandible, but it possesses the characters listed above, although the lateral feltlike patches of the metanotum and the transverse anterior crest of the scutum are less well developed than in other subgenera. Its relationship to *Dianthidium* was first noted by Moure (1965). The extreme male genitalic diversity among the subgenera (as in *Afranthidium*) is illustrated by Figures 82-2f and h.

Nests of *Dianthidium* usually consist of several cells and are made of pebbles stuck together by a matrix of resin and attached to rock surfaces or twigs of bushes or trees. Some species, however, make resin cells using pebbles and chaff in holes in banks or in holes in wood or stems. Exposed nests have been described for the subgenera *Mecanthidium* (Parker, 1977b), *Adanthidium* (Melander, 1902; Middleton, 1916), and *Dianthidium* s. str. (Fig. 7-1; also Grigarick and Stange, 1968; Clement, 1976). Subterranean nests have been described for *Dianthidium* s. str. by Custer and Hicks (1927), Fischer (1951), and Michener (1975a). [Through an error, the last were reported as nests of *Paranthidium jugatorium* (Say), but the bees were in fact *Dianthidium curvatum* (Smith).] Nests in holes in wood were described for *Dianthidium* s. str. (Krombein, 1967); one species, *D. (D.) ulkei* (Cresson) nests in holes in the ground, among rocks, and in pithy stems (Frohlich and Parker, 1985). In all cases cells are made of resin, usually intermixed with pebbles, chaff, seeds, and the like.

Key to the Subgenera of *Dianthidium*
(Modified from Griswold and Michener, 1988)

1. Hind coxa with elongate ventral, apical spine in male, short tooth (best seen in lateral view) in female; hypostomal area dulled by fine, dense punctation (nearctic) .... *D. (Dianthidium s. str.)*
—. Hind coxa without tooth or spine; hypostomal area shiny between punctures 2
2(1). Arolia absent; mouthparts in repose scarcely exceeding proboscidial fossa (male S5 without sclerotized apical comb) (Mexico) .... *D. (Deranchanthidium)*
—. Arolia present; mouthparts in repose considerably exceeding proboscidial fossa 3
3(2). Male S5 with sclerotized apical comb; female mandible short, its length considerably less than maximum clypeal width (North America) .... *D. (Adanthidium)*
—. Male S5 without sclerotized apical comb; female mandible long, its length equal to maximum clypeal width (Mexico) .... *D. (Mecanthidium)*

Dianthidium / Subgenus *Adanthidium* Moure

Adanthidium Moure, 1965: 29. Type species: *Anthidium texanum* Cresson, 1878, by original designation.

This subgenus contains species 8 to 10 mm long that superficially resemble those of *Dianthidium* s. str. The comb on S5 of the male of *Adanthidium* is not found in other subgenera. The male genitalia, hidden sterna, and other structures were illustrated by Griswold and Michener (1988); see also Figure 82-2f, g.
■ The subgenus is found from Arizona to Texas (with a record for Kansas), USA, south to Oaxaca, in Mexico. There are four species.

Dianthidium / Subgenus *Deranchanthidium* Griswold and Michener

Dianthidium (Deranchanthidium) Griswold and Michener, 1988: 34. Type species: *Dianthidium chamela* Griswold and Michener, 1988, by original designation.

Species of *Deranchanthidium*, by their somewhat elongate form and restricted thoracic markings, superficially resemble *Anthodioctes, Hypanthidium,* or *Hypanthidioides* more than *Dianthidium* s. str. Body length is 7.5 to 9.0 mm. The male genitalia, hidden sterna, and other structures were illustrated by Griswold and Michener (1988); see also Figure 82-2h-j.
■ This subgenus is found in western Mexico from Jalisco to Oaxaca. There are two species.

Dianthidium / Subgenus *Dianthidium* Cockerell s. str.

Anthidium (Dianthidium) Cockerell, 1900b: 412. Type species: *Anthidium curvatum* Smith, 1854, by original designation.

This is the principal group of *Dianthidium*, consisting of small to moderate-sized (5.5-13.0 mm long) species, mostly with abundant yellow markings. Male genitalia and hidden sterna were illustrated by Mitchell (1962) and Grigarick and Stange (1968).
■ *Dianthidium* s. str. is found from southern British Columbia and Ontario, Canada, and Maine south to Florida, USA, and the states of Nuevo Leon, Durango, and Baja California, Mexico. The 20 species were reviewed in papers by Timberlake (1943a) and Grigarick and Stange (1968).

Dianthidium / Subgenus *Mecanthidium* Michener

Paranthidium (Mecanthidium) Michener, 1942a: 278. Type species: *Paranthidium sonorum* Michener, 1942, by original designation.

Mecanthidium contains large species (body length 11-23 mm). The elongate body is largely red-brown, sometimes with yellowish-red markings. Because of size and color *Mecanthidium* superficially resembles *Aztecanthidium* more than it does other subgenera of *Dianthidium*.
■ This subgenus ranges through western Mexico from

Oaxaca and Morelos north to Sonora and to southern Arizona, USA. There are two species.

Genus *Duckeanthidium* Moure and Hurd

Duckeanthidium Moure and Hurd, 1960: 2. Type species: *Anthidium megachiliforme* Ducke, 1907, by original designation.

Grafanthidium Urban, 1995b: 435. Type species: *Grafanthidium amazonense* Urban, 1995, by original designation.

Atropium Pasteels, 1984: 132. Type species: *Megachile atropos* Smith, 1853, by original designation.

Ketianthidium Urban, 1999a: 160. Type species: *Ketianthidium zanolae* Urban, 1999, by original designation. (doubtful synonymy).

Urban (1999a) described *Ketianthidium* on the basis of one male from Argentina. She suggested that it is close to *Duckeanthidium*, differing by the very short omaular carina, the weakly emarginate apex of T7, etc., features that can be duplicated among species of *Duckeanthidium* of the the group called *Grafanthidium*. Unfortunately the stigma was not described, so that its agreement with the small stigma of *Duckeanthidium* cannot be determined. It probably runs to 8 in the key to genera of the Western Hemisphere. Michener (2002c) showed that *Atropium* Pasteels is a junior synonym, having been previously incorrectly included in the African fauna.

Duckeanthidium consists of robust, chalicodomiform bees 8.5 to 16.0 mm in length. They vary from wholly black, as in the female of *D. megachiliforme* (Ducke), with limited yellow markings in males, to forms with extensive yellow markings and the ground color of the metasoma brown. The juxtantennal carinae are distinct, unlike those of other large and robust American anthidiines except *Hoplostelis*. Arolia are well developed in males, absent or nearly so in females. There is a fovea margined by a carina behind the propodeal spiracle. T7 of males is produced to lateral lobes, often strongly developed, and sometimes a median spine, all projecting posteriorly. S1 to S6 of males are exposed and lack combs.

■ This genus is known from Costa Rica to Amazonas and Rondônia, Brazil, to Peru and probably Argentina. Five species were recognized by Urban (1995b) in *Duckeanthidium* and *Grafanthidium*.

Duckeanthidium megachiliforme (Ducke), the type species of the genus, was rather fully described by Moure and Hurd (1960), see also Michener (2002c). *D. cibele* Urban is clearly congeneric with the species placed in *Grafanthidium* by Urban (1995b). *D. megachiliforme* and *thielei* Michener could be generically or subgenerically different from the other four species because they are large (13-16 mm long) and the male has a strong median carinate spine between the lobes of T7. The only equivalent of this spine in other species is a small median angle in *D. rondonicola* (Urban). The species placed in *Grafanthidium* by Urban have a weak lateral angle on the margin of the scutellum, but only a suggestion of such an angle can be seen in *D. cibele*. Thus *Grafanthidium* and *Duckeanthidium* show a tendency to grade one into the other.

Thiele (2002) found the nests of *Duckeanthidium thielei* Michener, a close relative of *D. megachiliforme* (Ducke), in holes in wood blocks placed high in the forest canopy in Costa Rica; the pollen collected was almost exlusively that of *Bauhinia* (Fabaceae, Caesalpinioideae).

Genus *Eoanthidium* Popov

This is a genus of yellow-and-black bees (the metasoma yellow and brown in the subgenus *Clistanthidium*), mostly more elongate and slender than those of *Anthidiellum*. The most distinctive feature is the juxtantennal carinae, which in some species, however, are small, short ridges mesal to and slightly above the antennal bases. As indicated in the key to genera, there are certain other features that seem associated with the juxtantennal carinae. Nonetheless, the subgenera are quite different from one another (see the key to subgenera) and could have evolved the juxtantennal carinae independently. Generic characters other than those indicated above include the following: the preoccipital ridge is not carinate; the pronotal lobe has a strong carina or low lamella; the omaular carina is present at least on the upper half of the mesepisternum; the scutellum is rounded posteriorly with a small midapical emargination as seen from above, and is produced well over the metanotum, being narrowly rounded or acute as seen in profile; the fovea behind the propodeal spiracle is small (larger in the subgenus *Clistanthidium*), only partly enclosed, or absent; the row of pits across the upper margin of the propodeum is absent except as indicated at the extreme sides; S6 of the female is modified, with either a spine or a premarginal carina; the male sterna lack combs; T6 of the male is simple except for a tooth at each side; the male gonostylus is rather simple, slightly flattened distally, and often has a median bend or angle. The gonostylus is more broadened apically in the subgenera *Clistanthidium* and *Hemidiellum* than in others, and both have an apical brush of dense gonostylar hairs that is absent in the others. The penis valves are large (especially in some *Clistanthidium*) and pointed apically.

Key to the Subgenera of *Eoanthidium*
(Modified from Michener and Griswold, 1994a)

1. Subantennal suture straight; inner surface of hind tibia curving onto upper margin without sharp line between keirotrichiate area and region of longer hairs (arolia present; scutoscutellar suture closed, similar to scutoaxillar suture; T4 to T6 of female and T5 and T6 of male with small lateral spines) (India) .... *E. (Hemidiellum)*
—. Subantennal suture strongly arcuate outward; inner surface of hind tibia flat, keirotrichiate area ending abruptly at sharp, often carinate line along upper margin of tibia 2
2(1). Arolia absent; profile of T6 of female convex; T4 to T6 with lateral spines (scutoscutellar suture closed, similar to scutoaxillar suture) (India) .... *E. (Salemanthidium)*
—. Arolia present; profile of T6 of female concave (convex distally in subgenus *Clistanthidium*); T4 and T5 without lateral spines 3
3(2). Front coxa with strong carina or lamella transverse to axis of body or bending posteriorly and becoming longitudinal mesally; S6 of female not thickened apically, without lateral carina or spine; T6 of male with lateral tooth; scutoscutellar suture sometimes not very different from scutoaxillar suture (palearctic, Africa) ....
.... *E. (Clistanthidium)*
—. Front coxa without transverse carina or lamella; S6 of

female thickened apically, with lateral carina or spine; T6 of male without lateral tooth; scutoscutellar suture open, with shiny impunctate floor divided medially to form two foveae (palearctic, Africa) .... *E. (Eoanthidium s. str.)*

Eoanthidium / Subgenus *Clistanthidium* Michener and Griswold

Eoanthidium (Clistanthidium) Michener and Griswold, 1994a: 315. Type species: *Dianthidium turnericum* Mavromoustakis, 1934, by original designation.

This subgenus differs from its closest relative, *Eoanthidium* s. str., in the often closed scutoscutellar suture (see the key to subgenera); the strong juxtantennal carina; the strong transverse carina or lamella, produced to a median angle in most males, on the front coxa, or, in *Eoanthidium nasicum* (Friese), the mesal part of the lamella extending distad toward the coxal spine and thus longitudinal; and the form of S6 of the female, which lacks marginal thickening but has a small preapical median spine. The body length is 8 to 9 mm.
■ *Clistanthidium* occurs in Turkey, Israel, Iran, and Pakistan, thence south in eastern Africa to Natal in South Africa, west to Shaba Province in Zaire, and to Namibia. There are probably six species; five named species were listed by Michener and Griswold (1994a).

Eoanthidium / Subgenus *Eoanthidium* Popov s. str.

Dianthidium (Eoanthidium) Popov, 1950a: 316. Type species: *Anthidium insulare* Morawitz, 1873, by original designation.
Eoanthidium (Eoanthidiellum) Pasteels, 1969a: 51. Type species: *Anthidium elongatum* Friese, 1897 = *Anthidium clypeare* Morawitz, 1873, by original designation.

Eoanthidium s. str. consists of strongly marked, yellow-and-black species 8 to 10 mm long. Pasteels (1969a) divided this group into two, as shown in the synonymy above. The differences, however, are weak or incorrect. Contrary to Pasteels' account, *Eoanthidium clypeare* (Morawitz), the type species of *Eoanthidiellum*, has juxtantennal carinae, although they are weak. In *E. insulare* (Morawitz), the type species of *Eoanthidium*, the omaular carina continues onto the thoracic venter, a feature that might serve to distinguish it from *Eoanthidiellum*.
■ This subgenus occurs in the eastern Mediterranean basin, including Greece and the Balkans, southwestern Asia, thence east to Iran and southern Russia and south to Kenya. There are about five species.

Eoanthidium / Subgenus *Hemidiellum* Pasteels

Eoanthidium (Hemidiellum) Pasteels, 1972: 112. Type species: *Eoanthidium semicarinatum* Pasteels, 1972, monobasic.

Hemidiellum consists of a small species (6.0-6.5 mm long) sufficiently different from other *Eoanthidium* that it may deserve generic rank. As in some *Eoanthidium* s. str., the omaular carina is present only on the upper half of the mesepisternum.
■ This subgenus is known only from southern India. The lone species is *Eoanthidium semicarinatum* Pasteels.

Eoanthidium / Subgenus *Salemanthidium* Pasteels

Eoanthidium (Salemanthidium) Pasteels, 1969a: 51. Type species: *Hypanthidium salemense* Cockerell, 1919, by original designation.

Salemanthidium consists of species with more limited and paler yellow marks than is usual in *Eoanthidium* s. str.; body length is 8 to 9 mm. The substantial morphological differences from *Eoanthidium* s. str. are indicated in the key to subgenera. Lack of arolia, the most unusual feature, is duplicated among those Old World forms that do not have multidentate female mandibles only in *Icteranthidium, Apianthidium,* and some subgenera of *Trachusa*. The upper and lower parts of the base of the mandible are elevated, resulting in a broad concavity between them that is more conspicuous than in other subgenera. T7 of the male, rather large and long, has a longitudinal median carina; the apex is subtruncate except for a small median projection.
■ *Salemanthidium* is known only from southern India. There are two species (Pasteels, 1972).

Genus *Epanthidium* Moure

This genus consists of bees with a body form similar to that of *Dianthidium* s. str. or more elongate, such that the only Mexican species resembles an *Aztecanthidium*. Some species are black with yellow or cream maculations but in most some of the pale areas or the background of parts of the metasoma are red. The pronotal lobe has an anteriorly sloping lamella as in *Dianthidium*; otherwise, the lettered (a to e) characters in the discussion of *Dianthidium* do not apply to *Epanthidium*. As noted in the key to genera, most *Epanthidium* have at least a minute indication of a juxtantennal carina, arising abruptly at the middle of the mesal margin of the antennal socket and extending upward. This carina thus differs from the generally larger juxtantennal carina of genera such as *Anthodioctes* and *Hypanthidioides,* which extends both up and down from the middle of the inner margin of the antennal socket and does not necessarily involve the margin of the socket proper. Although arolia are distinct in most species, they are extremely small in females (not males) of *E. (Epanthidium) tigrinum* (Schrottky) and could sometimes be considered absent. An interesting feature of all species except those in the subgenus *Ananthidium* is the strong submarginal carina of S6 of the female, which has one to four teeth. Male genitalia and other structures were illustrated by Stange (1983, 1995).

Epanthidium, except for *Ananthidium,* was revised by Stange (1983); *Ananthidium* was revised by Urban (1991). A nest, according to Stange (1983), consisted of cells made of clay, not resin as in *Dianthidium* and many other Anthidiini. Stange (personal communication, 1993) questioned this and reported exposed nests of *Epanthidium* each consisting of a single cell made of resin with embedded plant fibers. G. Melo (in litt., 1996) reported multicellular nests of *E. tigrinum* (Schrottky) made of resin and constructed in cavities.

Key to the Subgenera of *Epanthidium* (Males)

1. T7 three-lobed or with large median spine between two lobes (lamella of pronotal lobe translucent) (South America) .... *E. (Epanthidium s. str.)*

—. T7 two-lobed 2
2(1). Tegula widest in front of middle; body length 9 to 12 mm; lamella of pronotal lobe opaque, black or dusky (South America, Mexico) .... *E. (Carloticola)*
—. Tegula widest near middle; body length 6.5-8.0 mm; lamella of pronotal lobe translucent (South America) ....
.... *E. (Ananthidium)*

Key to the Subgenera of *Epanthidium* (Females)

1. T6 without longitudinal carina; lamella of pronotal lobe opaque, blackish (S6 with premarginal carina enlarged to form tooth on each side of apex of sternum) ....
.... *E. (Carloticola)*
—. T6 with longitudinal median carina; lamella of pronotal lobe translucent 2
2(1). Clypeal apex truncate, nodulose, only slightly overhanging base of labrum, labrum therefore exposed when mandibles are closed; juxtantennal carina of *Epanthidium* type present although sometimes minute; S6 with premarginal carina enlarged to form one to four premarginal teeth .... *E. (Epanthidium* s. str.)
—. Clypeal apex broadly rounded, rather thin, scarcely nodulose, overhanging base of labrum, labrum therefore hidden when mandibles are closed; juxtantennal carina absent or virtually so; S6 without premarginal carina or teeth, margin thin .... *E. (Ananthidium)*

Epanthidium / Subgenus *Ananthidium* Urban

Ananthidium Urban, 1991: 73. Type species: *Anthidium inerme* Friese, 1908, by original designation.

This is the most distinctive of the subgenera of *Epanthidium*; both Urban (1991) and Stange (1995) gave it generic status. The principal characters are indicated in the key to subgenera and in Stange's (1995) key to genera. It consists of small species, 6.5 to 7.5 mm long. Male genitalia and hidden sterna were illustrated by Urban (1991) and Stange (1995).

■ *Ananthidium* is known from Argentina (provinces of Misiones and Salta to Chubut), Paraguay, and the state of Minas Gerais, Brazil. There are two species, distinguished by Urban (1991) and Stange (1995).

Epanthidium / Subgenus *Carloticola* Moure and Urban

Carloticola Moure and Urban, 1990: 90. Type species: *Dianthidium paraguayense* Schrottky, 1908, by original designation.

This subgenus contains relatively large bees, 9 to 12 mm long, having a somewhat more elongate form than most species of *Epanthidium* s. str. Genitalia and other structures were illustrated by Moure and Urban (1990).

■ *Carloticola* occurs in two disjunct areas: (1) the states of São Paulo and Mato Grosso, Brazil, to Paraguay and the provinces of Catamarca and Misiones, Argentina; and (2) Mexico from Chiapas to Tamaulipas and Jalisco. There are three species, two in South America, one in Mexico.

Epanthidium / Subgenus *Epanthidium* Moure s. str.

Epanthidium Moure, 1947a: 33. Type species: *Hypanthidium tigrinum* Schrottky, 1905, by original designation.

This subgenus includes species 6 to 10 mm long.

■ *Epanthidium* s. str. is found from the states of Pará and Paraíba, Brazil, south to Bolivia, Paraguay, and the provinces of Buenos Aires and Mendoza, Argentina. There are about 18 species. In the southern parts of its range, species of *Epanthidium* s. str. are the most common Anthidiini, particularly in xeric areas. Stange (1983) and Urban (1995c) gave keys to the species.

Genus *Euaspis* Gerstaecker

Euaspis Gerstaecker, 1857: 460. Type species: *Thynnus abdominalis* Fabricius, 1793, by original designation.
Dilobopeltis Fairmaire, 1858: 266. Type species: *Dilobopeltis fuscipennis* Fairmaire, 1858 = *Thynnus abdominalis* Fabricius, 1793, monobasic.
Parevaspis Ritsema, 1874: lxxi. Type species: *Parevaspis basalis* Ritsema, 1874, by designation of Sandhouse, 1943: 585.

This is a genus of moderate-sized to large (6-17 mm long) cleptoparasites. They are black, sometimes faintly bluish, and lack yellow markings, but the metasoma is commonly red. The interantennal area is elevated, with a strong, curved juxtantennal carina mesal to each antennal base and also with a smooth, longitudinal, median ridge or carina. The preoccipital carina is present laterally, behind the genal area, but not behind the vertex. The pronotal lobe has a carina, as does the omaulus. The hind coxa has a strong longitudinal carina on the inner margin; the outer surface is strongly expanded laterad to form a projecting lobe. The scutellum is strongly produced posteriorly, overhanging the propodeum, and forms a thin lamella in African species, but is thicker, punctate and hairy, in Asiatic species. The scopa is absent. T7 of the male is trilobed, small, extending little beyond T6. The male gonostylus is slender at the base, enlarged and flattened beyond the base, as illustrated by Viereck (1924b) and Baker (1995c). The volsellar lobe of the gonocoxite is large. A fuller generic description was given by Pasteels (1968a).

■ *Euaspis* is widespread in Africa, from Nigeria to Kenya and south to South Africa, and in southern and eastern Asia, from Nepal, India, and Sri Lanka east through Indonesia to the Moluccas and Kai (Key) and north through the Philippines, Taiwan, Japan, Korea, and China at least as far as Beijing. Of 12 species, two are African and ten Asian. Pasteels (1980) revised the genus and Baker (1995c) reviewed the Asiatic species.

Because of the scutellar character cited above, *Parevaspis* could be recognized as a subgenus for the Asiatic species, but no other distinguishing characters are known for the Asiatic group and recognition of subgenera seems unnecessary.

Euaspis species with red metasomas superficially resemble the similarly colored species of *Pachyanthidium*, with which they also agree in the produced, flat scutellum and the distally enlarged but rounded male gonostyli, the trilobed male S7, the presence of two apical spines on the fore and middle tibiae (as in *Stelis*), and so forth. In fact, *Pachyanthidium* is the only nonparasitic anthidiine that usually has the last feature. *Euaspis*, however, differs from *Pachyanthidium* in *lacking* a scopa, a carina or lamella on the preoccipital ridge behind the vertex, and a broad toothed apex of the mandible of the female, and in *having* the juxtantennal carinae and a vertical carina on the mesepisternum in front of the middle coxa. It is not clear whether *Euaspis* was derived from a *Pachyanthidium*-like ancestor. *Euaspis* also resembles *Stelis,* but in view of the differences listed in the key to genera as well as in the male genitalia, it may be that the similarities are convergent, resulting from the common parasitic habit.

Euaspis parasitizes other Megachilidae (*Lithurgus* and chalicodomiform *Megachile*), but unlike most cleptoparasitic bees, the female parasite enters the host nest by making a hole in the closure and burrowing through the nest (making holes in cell partitions), then throws out the host larvae and probably eats host eggs. Then, working toward the entrance, she reworks the pollen masses and lays an egg on each, often skipping cells near the entrance, and finally she closes the entrance with resin or pollen (Iwata, 1976: 420). *Stelis (Dolichostelis)* appears to have similar behavior (Parker et al., 1987).

Genus *Gnathanthidium* Pasteels

Gnathanthidium Pasteels, 1969a: 92. Type species: *Pachyanthidium prionognathum* Mavromoustakis, 1935, by original designation. [This is not *Gnathanthidium* Urban, 1994 = Hypanthidioides (Michanthidium).]

Gnathanthidium consists of a largely black species with only parts of the face of the male and lateral tergal spots of the female yellow. The metasoma tapers posteriorly so that T5 is much narrower than T1 or T2. The body is 9.5 to 10.0 mm long, coarsely punctate, and the tibiae are strongly tuberculate on the outer surfaces; there is a carina on the upper margin of the hind tibia. *Gnathanthidium* is a member of Series B related to *Pseudoanthidium* and should perhaps be regarded as a subgenus of that genus, related most closely to the subgenus *Micranthidium*, as indicated by the strongly carinate, almost lamellate preoccipital ridge, pronotal lobe, omaulus, and scutellar truncation; the lack of lateral teeth on T6 and T7 of the male; and the bilobed T7 of the male. It differs, however, in the nearly straight subantennal suture and the not particularly concave S3 to S5 of the male; the only unusual setae are on S3 and S4, which have median (not marginal) patches of dense white hairs, and the posterior margins of which are convex, translucent, and hairless. S5 of the male is reduced to a narrow band medially but laterally is produced to a short, hairless apical projection. An unusual feature is the broad, flat hind basitarsus of the female, conspicuously broader than the tibia and less than twice as long as broad.

■ This genus is found in eastern Africa from northern Natal Province, South Africa, to Kenya. The only species is *Gnathanthidium prionognathum* (Mavromoustakis).

Genus *Hoplostelis* Dominique

Austrostelis and *Hoplostelis* comprise the neotropical parasitic Anthidiini, i.e., the neotropical Anthidiini that lack scopae. *Hoplostelis* has been recognized as a distinctive form, often at the genus level, for many years, formerly under the name *Odontostelis*. Various other neotropical bees have been described in the genus *Stelis* because they are anthidiines less distinctive than *Hoplostelis* and lack a scopa in the female and are therefore presumably cleptoparasites. However, they have only one apical spine on each front and middle tibia, and in other respects also do not agree with *Stelis*. For example, the male gonostyli do not have the characteristic angulate form found in *Stelis*. In fact, these neotropical *Stelis*-like bees are related to *Hoplostelis* and are now placed in the genus *Austrostelis*. They are also related to *Hypanthidioides. Austrostelis, Hoplostelis* and *Hypanthidioides* have lateral teeth on S5 and S6 of the male. The presence of arolia in both sexes of the parasitic genera suggests the subgenera *Dichanthidium* and *Ctenanthidium* of *Hypanthidioides*. The small and simple or weakly bilobed T7 of males of the parasitic genera is more similar to the bilobed T7 of *Ctenanthidium, Dichanthidium*, and *Saranthidium* than to other subgenera of *Hypanthidioides*. The male parasites do not have a comb on S5 as in *Saranthidium* or on S3 as in *Ctenanthidium*. Thus the parasites cannot be placed close to any one of the subgenera of *Hypanthidioides*.

Rozen (1966c), in describing the larva of *Hoplostelis* (as *Odontostelis*), regarded it as perhaps generically distinct from *Stelis*, a view that I strongly support. Young larvae of at least some species of *Stelis* have simple, pointed mandibles with which they kill eggs or larvae of the host. Larvae of *Hoplostelis*, however, have bidentate mandibles, as in most Anthidiini, and the adult, not the larva, destroys the host egg or larva.

Key to the Subgenera of *Hoplostelis*

1. Scutellum bigibbous; scutum bigibbous posteriorly; propodeum without foveae or pits; omaulus lamellate almost to lower end; T1 without transverse carina at summit of anterior surface; T4 to T6 (at least in female) each with longitudinal median carina (Brazil) .... *H. (Rhynostelis)*
—. Neither scutellum nor scutum bigibbous; propodeum with well-defined fovea, divided by transverse carinae, behind spiracle, continued as basal propodeal zone divided into many pits or small foveae; omaulus carinate in upper one-half or one-third, rounded below; T1 with transverse carina at summit of anterior surface; terga without longitudinal median carinae .... *H. (Hoplostelis* s. str.)

Hoplostelis / Subgenus *Hoplostelis* Dominique s. str.

Hoplostelis Dominique, 1898: 60. No valid included species. Type species: *Stelis abnormis* Friese, 1925 = *Anthidium bivittatum* Cresson, 1878, by inclusion and designation of Griswold and Michener, 1988: 36.
Odontostelis Cockerell, 1931d: 542. Type species: *Stelis abnormis* Friese, 1925 = *Anthidium bivittatum* Cresson, 1878, by original designation.

This subgenus includes relatively large (8-11 mm long), robust, chalicodomiform species having the special features indicated in the key as well as metasomal punctation that is fine compared to that of the genus *Aus-*

trostelis. The clypeus of the female has an apical tubercle or two apical processes. Unlike those of *Austrostelis*, T1 and T2 are commonly black; the yellow of the remaining, smaller terga therefore contrasts sharply with the black metasomal base.

■ *Hoplostelis* s. str. occurs from Bolivia and Santa Catarina, Brazil, north through the tropics as far as Jalisco, Mexico. Moure and Urban (1994) gave a key to the three species of this subgenus. One of them, *H. cornuta* (Bingham), was described from Burma, no doubt in error, for it is known from Trinidad.

So far as is known, *Hoplostelis* s. str. parasitizes euglossine bees of the genus *Euglossa* (Bennett, 1966). The female parasite opens a host cell, kills the host egg or larva, lays her egg on the food mass, and reseals the cell. She then drives the adult host from its nest. Moreover, over a few days the parasite comes and goes from the nest, closing the nest entrance at each departure, all the while parasitizing the younger cells and killing larger larvae, pupae, and even unemerged adults in the older cells.

Hoplostelis s. str. has the body form of *Duckeanthidium cibele* Urban and its congeners. It is tempting to believe that *Hoplostelis* s. str. is derived from *Duckeanthidium* and that the genus *Austrostelis* is derived from *Hypanthidiodes,* but other characters do not support a derivation from *Duckeanthidium.*

Hoplostelis / Subgenus *Rhynostelis* Moure and Urban

Rhynostelis Moure and Urban, 1994: 297. Type species: *Anthidium multiplicatum* Smith, 1879, by original designation.

This subgenus consists of a large (14 mm long), robust species similar to *Hoplostelis* s. str. Distinctive features are indicated in the key to subgenera; further, the clypeus of the female has a basomedian tubercle and the apex of the clypeus is simple. D. Urban (personal communication, 1994) writes, of a character not mentioned in the published description, that the juxtantennal carina is present but shorter than that in *Hoplostelis* s. str., arising at the level of the middle of the antennal alveolus and extending upward.

■ *Rhynostelis* occurs in Amazonas, Brazil. The only known species is *Hoplostelis multiplicata* (Smith), new combination.

This subgenus is known from only two female specimens; the male is unknown. One of the specimens was taken at the entrance to a nest site that had been occupied by *Eufriesea pulchra* (Smith); it is therefore likely that *Rhynostelis,* like *Hoplostelis* s. str., parasitizes Euglossini.

Genus *Hypanthidiodes* Moure

This is a neotropical genus of chalicodomiform or hoplitiform bees superficially similar to *Hypanthidium, Anthodioctes,* and *Austrostelis.* Most species are dark, i.e., with limited yellow markings and dusky wings, but a few have extensive yellow areas. Many are smaller than most *Hypanthidium;* body length is from 5 to 9 mm. The presence of juxtantennal carinae distinguishes *Hypanthidiodes* from *Hypanthidium;* the lack of lateral preoccipital carinae distinguishes it from *Anthodioctes.* The genus that seems closest to *Hypanthidioides* is *Austrostelis,* which lacks the scopa and is presumably cleptoparasitic. *Austrostelis* is essentially a *Hypanthidioides* in the sense in which that generic name is used here. There is a strong tendency for loss of arolia in *Hypanthidioides;* only the subgenera *Ctenanthidium* and *Dich-anthidium* have arolia in both sexes.

Key to the Subgenera of *Hypanthidioides*

1. Mandible of female with long, smooth apical margin between two small, lower apical teeth and upper apical angle or tooth; surface of distal half of mandible of female dull and minutely granular, without longitudinal carinae or with slender carinae that do not extend onto basal half of mandible; mandible of male with upper tooth separated from middle tooth by broad concave margin, upper interspace nearly twice as long as lower interspace; first segment of labial palpus of female with coarse, erect, curved or hooked hairs on undersurface.......................... 2
—. Mandible of female with three or four more or less evenly spaced teeth; surface of female mandible not or less dull, with strong longitudinal carinae extending onto basal half; mandible of male with teeth more evenly spaced; first segment of labial palpus of female without erect, curved or hooked hairs.. 3
2(1). Mandible of female without distinct upper apical tooth, surface without carinae; hind coxa of male with small apical spine directed mesad; distal margin of T6 of male with lateral tooth but otherwise unmodified .... *H. (Michanthidium)*
—. Mandible of female with strong upper apical tooth separated from long smooth margin by deep emargination, surface with usual carinae but carinae weak; hind coxa of male unarmed; distal margin of T6 of male with obtuse lateral tooth, margin medially expanded, elevated, sometimes bilobed .... *H. (Larocanthidium)*
3(1). Tegula large, with depressed translucent marginal area expanded anterolaterally, much wider there than midlaterally; body hoplitiform (arolia present in both sexes; T7 of male bilobed) .... *H. (Dichanthidium)*
—. Tegula normal, its depressed translucent margin not much broader anterolaterally than midlaterally; body more robust, hoplitiform to chalicodomiform 4
4(3). Mandible of female three-toothed; S4 of male with small but strong basal median comb............................... .... *H. (Mielkeanthidium)*
—. Mandible of female four-toothed; S4 of male without comb 5
5(4). Arolia well developed in both sexes; S3 of male with median marginal comb of coarse black setae that are shorter than hairs of fringe on same margin laterally...... .... *H. (Ctenanthidium)*
—. Arolia absent or extremely minute in females and some males; S3 of male without apical comb 6
6(5). Arolia of male absent; T7 of male bilobed, emargination between lobes about as wide as a lobe; S5 of male with apical comb; tegula widest in front of middle .... *H. (Saranthidium)*
—. Arolia of male present; T7 of male with narrow apicolateral lobes, space between them much wider than a lobe or, rarely [as in *H. (Moureanthidium) capixaba* (Urban)], lobes triangular and separated by broad emargination; S5

of male without apical comb (but with rows of coarse setae in *Dicranthidium*); tegula widest near middle 7
7(6). T6 and T7 of male curled anteriorly, their surfaces thus directed ventrad and processes of T7 overlapping S2; S3 or S4 to S6 of male hidden; hind coxa of male with one or more coarse peglike setae on mesal surface; omaular carina extending onto venter of thorax (scutellum produced to translucent apical carina except medially)........ .... *H. (Dicranthidium)*
—. T6 and T7 of male not so strongly curled (except in *Moureanthidium*), T6 not clearly directed ventrad; S3 to S5 or S6 ordinarily visible; hind coxa of male without peglike setae; omaular carina on upper half of mesepisternum, disappearing or becoming irregular because of punctation below 8
8(7). Scutellum swollen, biconvex; omaular carina extending onto lower half of mesepisternum as irregular line (hind tarsus of male elongate, first two segments taken together as long as hind tibia) .. *H. (Hypanthidioides s. str.)*
—. Scutellum not biconvex; omaular carina not extending onto lower one-half or one-third of mesepisternum 9
9(8). Hind tarsus of male elongate, first two segments taken together as long as hind tibia; S3 of male with small median bilobed projection; T7 of male strongly curled, surface thus directed ventrad; juxtantennal carina longer than diameter of antennal socket .... *H. (Moureanthidium)*
—. Hind tarsus of male shorter, first two segments taken together shorter than hind tibia; S3 of male without median, bilobed projection; T7 of male not so strongly curled; juxtantennal carina about as long as diameter of antennal socket .... *H. (Anthidulum)*

Hypanthidioides / Subgenus *Anthidulum* Michener

Dianthidium (Anthidulum) Michener, 1948a: 19. Type species: *Dianthidium currani* Schwarz, 1933, by original designation.

Anthidulum contains small (4.5-6.5 mm long), moderately robust species, richly marked with yellow. S2 of the male is somewhat enlarged, suggesting *Dicranthidium*, but the metasoma is not tightly curled as in that subgenus, S3 to S6 being exposed. The male antennae are not elongate as in *Dicranthidium*, the middle flagellar segments being little longer than broad.
■ This subgenus is found from Costa Rica to the state of Paraná, Brazil, and the provinces of Misiones to Tucumán, Argentina. The four species were revised by Urban (1993b).

Hypanthidioides / Subgenus *Ctenanthidium* Urban

Ctenanthidium Urban, 1993a: 85. Type species: *Ctenanthidium gracile* Urban, 1991, by original designation.

Ctenanthidium includes species related to *Dicranthidium* and especially to *Anthidulum*, and similar in appearance to those subgenera, although the body is rather slender; the length is 5.5 to 7.0 mm. It differs from both of those subgenera in the form of the male T7, which is bilobed and lacks apicolateral projections, in addition to other characters indicated in the key to subgenera and in Urban (1993a).
■ The range is from Uruguay, the province of Córdoba, Argentina, and Bolivia north to the state of Paraná, Brazil. The four species were revised by Urban (1993a).

Hypanthidioides / Subgenus *Dichanthidium* Moure

Dichanthidium Moure, 1947a: 30 (January); also described as new by Moure, 1947b: 235 (June). Type species: *Dichanthidium exile* Moure, 1947, by original designation.

Dichanthidium includes species having the size (body length 7 mm) and slender form of some *Anthodioctes* (*Bothranthidium* and the species formerly placed in *Nananthidium*) but with all the principal features of *Hypanthidioides*, except that it has arolia in both sexes. In this respect *Dichanthidium* resembles *Hypanthidioides* (*Ctenanthidium*) and *Austrostelis*.
■ This subgenus is known from the provinces of Salta, Argentina, and Santa Cruz, Bolivia. The only described species is *Hypanthidioides exile* (Moure); the Bolivian species appears to be different.

Hypanthidioides / Subgenus *Dicranthidium* Moure and Urban

Dicranthidium Moure and Urban, 1975: 837. Type species: *Anthidium arenarium* Ducke, 1907, by original designation.

Dicranthidium, which was synonymized with the subgenus *Anthidulum* by Griswold and Michener (1988), appears to be a clearly distinct subgenus, though perhaps the sister group to *Anthidulum*, sharing with it small body size and the enlarged S2. The body is robust, the most robust of the subgenera of *Hypanthidioides*; length is 5 to 7 mm. The yellow markings are pale and limited in extent. As in *Moureanthidium*, the scutellum (except medially) and the axilla are produced to a translucent apical carina or lamella. Numerous structures, including male genitalia and sterna, were illustrated by Moure and Urban (1975).
■ This subgenus ranges from the states of Paraná and Paraíba, Brazil, to Trinidad (West Indies) and the province of Valle, Colombia. The six species were reviewed by Urban (1993b).

Some species differ from one another in certain striking features. For example, the male of *Hypanthidioides arenaria* (Ducke) lacks the hind tibial spurs that are present in other species.

Laroca and Rosado Neto (1975) described nests of *Hypanthidioides arenaria* (Ducke) consisting of one to several resin cells constructed in small cavities of various sorts.

Hypanthidioides / Subgenus *Hypanthidioides* Moure s. str.

Hypanthidiodes Moure, 1947a: 35. Type species: *Anthidium flavofasciatum* Schrottky, 1902, by original designation.
Hypanthidioides Michener, 1948a: 22; Moure and Urban, 1975: 837; Urban, 1993b: 28; error for *Hypanthidiodes* Moure, 1947.
Hypanthidiodes Moure, 1947a: 18. Type species: *Anthidium flavofasciatum* Schrottky, 1902, by original designation.

The name for this taxon was spelled in two different ways in the original publication, *Hypanthidioides* on p. 18 and *Hypanthidiodes* on p. 35. The first has been used in general, the second was used by Michener (2000); I was then unaware of the two spellings in the original publication, although I had earlier used *-ioides* (Michener (1948a). The Code, article 24.2.4, in such cases, allows the original author (Moure) to function as First Reviser and select the correct spelling simply by using one of the spellings. Moure (in Moure and Urban, 1975: 837) used *Hypanthidioides,* thus fixing the proper spelling.

This subgenus includes a lone, rather elongate chalicodomiform species (body length 6-10 mm) of *Hypanthidioides* that resembles superficially the species of the genus *Hypanthidium.* In the type species, *Hypanthidioides flavofasciata* (Schrottky), T7 of the male is unusually short and transverse, its posterolateral projections small, and has large posterior lateral projections on S6. Urban (1993b) has further characterized the subgenus (as a genus).

■ *Hypanthidioides* s. str. ranges through eastern Brazil from the state of Paraíba to Rio Grande do Sul. Only *Hypanthidioides flavofasciata* (Schrottky) has been assigned to the subgenus.

According to Schrottky (1902a) the resin nest, consisting of several cells, is exposed, attached to a stem.

Hypanthidioides / Subgenus Larocanthidium Urban

Larocanthidium Urban, 1997a: 299. Type species: *Larocanthidium emarginatum* Urban, 1997, by original designation.

This subgenus resembles *Michanthidium* in the dull mandibles of the female, the long, smooth mandibular margin above the two small lower apical teeth, and the presence of arolia in males but not in females. Some of the differences from *Michanthidium* are indicated in the key to subgenera. The lamella of the pronotal lobe is straight; in *Michanthidium* it is arcuate. The body, 5 to 8 mm in length, is more robust than in *Michanthidium,* suggestive of *Anthidulum* or *Saranthidium* in form. The sterna and curvature of the metasoma are as in *Michanthidium.*

■ *Larocanthidium* is found from Paraná to Pará, Brazil, west to Bolivia. The ten species were revised by Urban (1997a).

Hypanthidioides / Subgenus Michanthidium Urban

Gnathanthidium Urban, 1994b: 337 (not Pasteels, 1969). Type species: *Gnathanthidium sakagamii* Urban, 1994, by original designation.
Michanthidium Urban, 1994a: 281, replacement for *Gnathanthidium* Urban, 1994.

Like various other subgenera, *Michanthidium* lacks arolia in females but they are well developed in males. The appearance is similar to that of *Saranthidium, Anthidulum,* and similar genera; body length is 7.0 to 8.5 mm. The mandibular characters of both sexes are distinctive. Interestingly, the apex of the mandible of the female is similar to that of the Mexican *Dianthidium (Deranchanthidium);* it also resembles that of the subgenus *Larocanthidium.* The sterna of the males lack both combs and fringes of long hair. T7, which curls under but not strongly so as in *Dicranthidium,* is bilobed, the lobes moderately close together and sometimes with a small protruding median angle between them.

■ *Michanthidium* is found in southern Brazil (states of Santa Catarina and Rio Grande do Sul) and northern Argentina (provinces of Misiones and Tucumán). The two species were described by Urban (1994b).

The curved or hooked hairs on the underside of the first segment of the labial palpus of females of this subgenus and *Larocanthidium* suggest a special floral relationahip. In *Michanthidium,* strongly hooked hairs occur throughout the length of the galeal blade of both sexes; such hairs are absent in *Larocanthidium.*

Hypanthidioides / Subgenus Mielkeanthidium Urban

Mielkeanthidium Urban, 1996: 121. Type species: *Mielkeanthidium nigripes* Urban, 1996, by original designation.

Mielkeanthidium contains hoplitiform species 6 to nearly 8 mm in body length and similar in appearance to species of *Michanthidium, Moureanthidium,* and *Saranthidium.* Of these, it resembles *Michanthidium* in that T7 is not curled forward but is directed more or less downward. It is bilobed, the lobes as wide as or wider than the emargination between them. The female mandible is unique in the genus in being three-toothed, with the upper interspace about twice as long as the lower interspace. The arolia are small in the male, absent in the female. The omaular carina is very strong and continues ventrad almost to the discrimen; it is even stronger than in *Dicranthidium.* As in *Moureanthidium* the scutellum is rounded posteriorly as seen from above, but as seen laterally the margin is thin and lamellate, overhanging the metano-tum.

■ This subgenus occurs in Santa Catarina and Rio Grande do Sul, Brazil. The two species were treated by Urban (1996).

Hypanthidioides / Subgenus Moureanthidium Urban

Moureanthidium Urban, 1995a: 37. Type species: *Dianthidium subarenarium* Schwarz, 1933, by original designation.

Moureanthidium consists of rather elongate, hoplitiform species 6 to 8 mm long, superficially resembling species of the subgenera *Michanthidium* and *Saranthidium.* As in several other subgenera, males have arolia, whereas females lack them. The metasoma of the male is curled but not so strongly as in the subgenus *Dicranthidium;* T6 is not facing downward like T7, but the form of T7 suggests *Dicranthidium.* In this and other features, *Moureanthidium* is intermediate between *Dicranthidium* and other subgenera. The inner surface of the hind coxa of the male sometimes has a spine or lobe; this is the same area that in *Dicranthidium* is modified and with peglike setae. The elongate hind tarsi of the male are as in the subgenus *Hypanthidioides* s. str.

■ This subgenus occurs in Brazil from Santa Catarina to Bahía. Urban (1995a) revised the five species.

Hypanthidioides / Subgenus *Saranthidium* Moure and Hurd

Hypanthidium (Saranthidium) Moure and Hurd, 1960: 6. Type species: *Anthidium flavopictum* Smith, 1854, by original designation; incorrectly stated to be *Anthidium furcatum* Ducke, 1908, by Griswold and Michener, 1988: 31.

This subgenus consists of forms with limited yellow markings, an exception being *Hypanthidioides (S.) panamense* (Cockerell), which has much yellow, suggesting *H. (Anthidulum) currani* (Schwarz) from the same region. The body form is as in *Hypanthidioides* s. str.; the length is 6 to 7 mm. This is the only subgenus in which males as well as females lack arolia. It resembles *Dichanthidium* in the two moderately broad lobes of T7 of the male.

■ *Saranthidium* ranges from Paraguay and Santa Catarina, Brazil, north through the tropics to Oaxaca and Veracruz, Mexico. There are seven named species.

Genus *Hypanthidium* Cockerell

Although this neotropical genus comes out in the key next to *Epanthidium*, it does not resemble that genus superficially because of its chalicodomiform body and usually dark coloration; its species are thus similar to most *Anthodioctes*. Rarely, the metasoma is entirely red. The body length is 7.0 to 10.5 mm. The lack of juxtantennal carinae, of a strong carina around the basin of T1, and of arolia readily distinguishes *Hypanthidium* from *Anthodioctes*. T7 of the male is rather large, broadly rounded, subtruncate or weakly emarginate midapically, the exposed part about as long as that of T6, the distal margin often translucent.

Key to the Subgenera of *Hypanthidium*

1. Preoccipital carina present as a sharp although punctate ridge laterally, behind eye; mandible of male bidentate, with strong, broad emargination between teeth; S2 of male with strong projection on each side .... *H. (Tylanthidium)*
—. Preoccipital carina absent; mandible of male with indications of three teeth; S2 of male with at most weak convexity on each side...................... *H. (Hypanthidium* s. str.)

Hypanthidium / Subgenus *Hypanthidium* Cockerell s. str.

Hypanthidium Cockerell, 1904b: 292. Type species: *Anthidium flavomarginatum* Smith, 1879, = *Anthidium divaricatum* Smith, 1954, by original designation.

■ *Hypanthidium* s. str. occurs through the tropics from the states of Sonora, San Luis Potosí, and Yucatan, in Mexico, to Misiones, in Argentina, and to Paraguay and Bolivia. Urban (1998b) gave a key to the 16 species.

Hypanthidium / Subgenus *Tylanthidium* Urban

Tylanthidium Urban, 1994a: 277. Type species: *Tylanthidium tuberigaster* Urban, 1994, by original designation. [New status.]

In the male of *Tylanthidium* the lateral margins of T7 diverge apically to a broad apex that has a deeper midapical emargination than that in most other *Hypanthidium*, and the undersurface of T7 is densely hairy. In general, however, the distinctive form of this tergum of *Tylanthidium* is as in *Hypanthidium* s. str. The hind tarsus of the male is more thickened than is usual in *Hypanthidium*, the second segment being about as long as broad and about as broad as the basitarsus, the fourth segment much broader than long because of its dorsal lobe. Other characters, including the carinate pronotal lobe, are as in *Hypanthidium* s. str., and *Tylanthidium* should probably be regarded as a synonym of *Hypanthidium*.

■ This subgenus occurs in the state of Amazonas, Brazil. It is based on a single species, *Hypanthidium tuberigaster* (Urban).

Genus *Icteranthidium* Michener

Icteranthidium Michener, 1948a: 25. Type species: *Anthidium limbiferum* Morawitz, 1875, by original designation.

Icteranthidium contains species with rich yellow markings or with the body largely yellow; the body form is similar to that of *Dianthidium*, and the size is variable (7.5-15.0 mm long). Arolia are absent. The preoccipital carina is absent behind the vertex but is strong laterally, and unlike that of other bees, its lower end extends nearly straight to the posterior mandibular articulation instead of extending more posteriad and approaching or joining the hypostomal carina or, rarely, curving forward to the mandibular articulation. The mandible of the female has four teeth separated by shallow concavities. The pronotal lobe has a strong lamella and the omaulus is carinate. The scutellum is produced to a lamella except medially, and the axilla is frequently produced to a posterior angle or point projecting beyond the contour of the scutellum. A strong carina or lamella between the ventral and anterior surfaces of the front coxa characterizes most species. T6 of the female is concave in profile, the margin convex with a median notch. The truncate T7 of the male has a midapical projection, sometimes also a lateral projection or tooth, and is thus trifid. Male genitalia and other structures were illustrated by Popov (1967).

■ This genus ranges from Morocco and Portugal to Mongolia. It is particularly well represented in the xeric areas of Asia, but occurs in southern Europe (north to Hungary), northern Africa south in the Sahel to Senegal, Mali, Chad, and northern Kenya, and southeast to Baluchistan in Pakistan. Warncke (1980a) reported 15 west palearctic species but did not include Central Asia, Mongolia, the Sahel, Pakistan, etc., in his study. There are probably about 25 species.

Genus *Indanthidium* Michener and Griswold

Indanthidium Michener and Griswold, 1994a: 315. Type species: *Indanthidium crenulaticauda* Michener and Griswold, 1994, by original designation.

A member of Series B (see the discussion of the tribe), *Indanthidium* resembles a small (length 6.0-7.5 mm) *Anthidium*, largely because the yellow metasomal bands are divided into four spots each, as they are in some species of *Anthidium*. But in contrast to *Anthidium*, the depressed marginal zone of T5 is punctured like the rest of the tergum, or a little more densely so in the male, the zone not differentiated and its margin not angulate medially; and the posterior margin of T5 and other terga is rather broadly impunctate. *Indanthidium* is similar to *Pseudoanthidium,* with which it agrees in its distinctly arcuate subantennal sutures. It differs from that genus in the unmodified S1 to S6 of the male, which lack combs or processes, and in S4 and S5, which are not especially concave. An unusual feature is T7 of the male, which is nearly as broad as T6 and has strong median and lateral spines (the latter the longest) and irregular smaller teeth between the median and lateral ones. Another distinctive feature is the maxillary palpus, which appears to consist of only one segment. The male genitalia were illustrated by Michener and Griswold (1994a).

■ This genus occurs in southern India. The only known species is *Indanthidium crenulaticauda* Michener and Griswold.

Genus *Larinostelis* Michener and Griswold

Larinostelis Michener and Griswold, 1994a: 317. Type species: *Larinostelis scapulata* Michener and Griswold, 1994, by original designation.

This genus, known only from the female, consists of a small (6.5 mm long), robust, black bee with extensive yellow markings. Like that of *Stelis,* the female lacks a scopa. *Larinostelis* is the most carinate of all *Stelis*-like bees, with juxtantennal carinae between the antennal sockets, and with carinae also on the pronotal lobe, the omaulus, the axilla, the scutellum (except for the median notch), the basal zone of the propodeum (except medially) and continuing behind the spiracle, the dorsal margin of the metapleuron, longitudinally on the hind basitarsus, transversely on S1, and across the base of S6. Arolia are absent. The axilla is greatly produced laterally, thus behind the tegula. The two apical spines of the fore and middle tibiae are minute.

■ *Larinostelis* is known only from Kenya. The single species is *L. scapulata* Michener and Griswold, known from a single female specimen.

It is not clear whether *Larinostelis* is a derivative of *Stelis* or an independently cleptoparasitic form. In the former case, it would be best regarded as a subgenus of *Stelis.* When discovered, the male should help in resolving this question.

Genus *Neanthidium* Pasteels

Neanthidium Pasteels, 1969a: 93. Type species: *Anthidium octodentatum* Pérez, 1895, by original designation.

This genus consists of a rather elongate species (9-13 mm long) with abundant yellow markings; the broad metasomal bands are broken medially, the aspect being that of some *Rhodanthidium. Neanthidium* differs from that genus, however, in the characters of Series B (see the discussion of the tribe), e.g., the five-toothed female mandible and lack of arolia. Unique features of the male are the strongly four-toothed T6 and T7 and the strong, sharp lateral teeth of both S4 and S5; S1 to S6 are otherwise unmodified and exposed. The male gonostylus is slender, slightly broadened with numerous long hairs medially, and much longer than the short, wide gonocoxite; the penis valves are in contact through most of their lengths.

■ *Neanthidium* occurs in North Africa (Algeria, Morocco) and is represented by a single species, *N. octodentatum* (Pérez).

Some *Pseudoanthidium* species have lateral apical processes on S4 and S5 of the male, but the processes are blunt and often have modified setae, rather than being smooth and spinelike as in *Neanthidium.* T6, T7, and other features of *Neanthidium* are very different from those of *Pseudoanthidium.*

Genus *Notanthidium* Isensee

Notanthidium is used here in a broader than usual sense, to include *Allanthidium. Notanthidium* s. str. differs from *Allanthidium* principally in its slender, hoplitiform body. Equal variation in body form also occurs in other genera, e.g., *Anthodioctes* s. str., where heriadiform to chalicodomiform bodies are found among species of the same subgenus. The modified clypeus and mandibles of the female of *Notanthidium* s. str. do not seem to justify generic distinction for the single species of that subgenus. In both *Notanthidium* s. str. and *Allanthidium* there are individuals in which the postspiracular foveae of the propodeum are recognizable. In other individuals, they are completely absent or the area concerned is coarsely sculptured with ridges or carinae suggesting the carinae that delimit or subdivide the fovea when it is well defined. Likewise, the row of pits across the base of the propodeum is variable, recognizable if at all only laterally. Thus two characters other than body form used to differentiate *Notanthidium* and *Allanthidium* by Michener (1948a) do not work.

Notanthidium has yellow or cream markings at least on the metasoma, often also on the head and thorax. The pubescence is often but not always partly or wholly black. The body length is 7.5 to 11.5 mm. The pronotal lobe is strongly carinate, but the omaulus is not carinate or rarely feebly so. The scutellum is rounded posteriorly, not at all carinate, and not or weakly overhanging the metanotum. T7 of the male is moderately to strongly curled forward, and bilobed. S1 to S6 are all exposed, S4 to S6 or at least one of them with a lateral tooth. Male genitalia, sterna, and other characters were illustrated by Toro and Rojas (1970b).

Key to the Subgenera of *Notanthidium*

1. Body hoplitiform; clypeus of female largely impunctate, upper part tumescent, lower margin with sublateral projection; clypeus in both sexes not at all overhanging base of labrum; mandible of female porrect, with deep emargination in upper margin just beyond middle, width of its apex less than basal width, three-toothed .... *N. (Notanthidium* s. str.)
—. Body chalicodomiform; clypeus of female punctate, not tumescent, without apical projections; apical margin of clypeus in both sexes overhanging base of labrum; mandible of female of the usual shape, width of its apex at least as great as basal width, vaguely four-toothed or with long straight margin above two lowermost teeth, sometimes undulate, thus producing a vaguely five-toothed margin 2

2(1). Omaulus carinate; mandible of female obscurely five-toothed .... *N. (Chrisanthidium)*

—. Omaulus rounded; mandible of female obscurely four-toothed .... *N. (Allanthidium)*

Notanthidium / Subgenus *Allanthidium* Moure

Allanthidium Moure, 1947a: 21. Type species: *Anthidium rodolfi* Ruiz, 1938, by original designation.

Trichanthidium Moure, 1947a: 20 (not Cockerell, 1930). Type species: *Anthidium subpetiolatum* Schrottky, 1910, by original designation.

Allanthidium (Anthidianum) Michener, 1948a: 13, replacement for *Trichanthidium* Moure, 1947. Type species: *Anthidium subpetiolatum* Schrottky, 1910, autobasic.

Moure (1947a) and Michener (1948a) recognized two taxa, here united under the name *Allanthidium*. The principal difference is that those then placed in *Allanthidium* are generally smaller, have pale pubescence, and have yellow areas on the head and thorax as well as the metasoma. Species formerly placed in *Anthidianum* are generally larger, have abundant black hair, and lack yellow marks on the head and thorax except on the face of males. T7 of the male is two-lobed in the first group, two-lobed or four-lobed in the second; contrary to Michener (1948), then, this character does not distinguish the groups.

■ *Allanthidium* occurs from the provinces of Puno in Peru and Jujuy in Argentina south to south-central Chile and the province of Neuquén in Argentina. There are six described species and at least one undescribed.

The type species, *Notanthidium rodolfi* (Ruiz), differs from others in the short jugal lobe of the hind wing, little more than one-fourth as long as the vannal lobe. If it is considered subgenerically distinct, the name *Anthidianum* is available for the remaining species.

Notanthidium / Subgenus *Chrisanthidium* Urban

Chrisanthidium Urban, 1997b: 181. Type species: *Anthidium bidentatum* Friese, 1908, by original designation. [New status.]

This subgenus resembles in appearance the well-marked, pale-haired species of *Allanthidium* such as *Notanthidium rodolfi* (Ruiz). It differs in the characters indicated in the key to subgenera.

■ *Chrisanthidium* occurs from Valparaíso to Atacama, Chile, and from Mendoza to Jujuy, Argentina. Urban (1997b) distinguished three species.

Notanthidium / Subgenus *Notanthidium* Isensee s. str.

Dianthidium (Notanthidium) Isensee, 1927: 373. Type species: *Anthidium steloides* Spinola, 1851, by original designation.

This slender-bodied subgenus, although characterized by partly black pubescence and a lack of pale marks on the thorax, does not present a black aspect, as do the dark species of *Allanthidium*.

■ This subgenus occurs in central and southern Chile and in the provinces of Neuquén and Chubut, Argentina. The one species is *Notanthidium steloides* (Spinola).

Genus *Pachyanthidium* Friese

These are robust, compact, megachiliform bees, 5 to 12 mm long. The metasoma tapers such that T5 is little more than half as wide as T1 and T2. The mandibles of the female are short and variable in dentition, from four-toothed (with at least some of the spaces between the teeth gently concave) to five- to ten-toothed (the teeth short and blunt, and the notches between them sometimes acute), to minutely denticulate with multiple, minute teeth. Thus some species fall in Series A, most in Series B (see the discussion of the tribe). For this reason the genus comes out twice in the key to genera. The color pattern is also variable, from black-and-yellow. as in most anthidiines, to black with a red metasoma, to wholly black. The arolia are absent except in males of the subgenus *Trichanthidium* and in both sexes of *Ausanthidium*. Pasteels' (1969a, 1984) statements that arolia are present but minute in *Pachyanthidium* seem to be based on the projecting, minutely setose plantae. *Pachyanthidium* is among the most carinate of the Anthidiini; the following are not merely carinate but are strongly lamellate: the preoccipital ridge behind the vertex, the pronotal lobe, the omaulus, the lateral margin of the axilla, and the posterior margin of the scutellum. The posterior margin of the scutellum is nearly straight (broadly rounded laterally in the subgenus *Ausanthidium*) and greatly overhangs the propodeum, which lacks basal pits and foveae (but see *Ausanthidium*). T7 of the male is trifid. The apices of the front and middle tibiae have two spines, as in *Stelis*, the two often reduced to one in the subgenus *Trichanthidium*.

Key to the Subgenera of *Pachyanthidium*

1. Eyes hairless; preoccipital carina absent laterally, behind eyes 2
—. Eyes with abundant, short hairs; preoccipital carina present laterally, behind eyes (mandible of female four-toothed) 3
2(1). Arolia present; mandible of female four-toothed (Namibia) .... *P. (Ausanthidium)*
—. Arolia absent; mandible of female usually five- to ten-toothed or denticulate (four-toothed only in *Pachyanthidium micheneri* Pasteels) (Africa) .... *P. (Pachyanthidium s. str.)*
3(1). T3 to T5 each with slender lateral spine; ocelli small, diameter about equal to width of base of first flagellar segment; hind coxa not carinate; arolia present in male, absent in female (Africa, oriental) .... *P. (Trichanthidium)*
—. T3 to T5 without lateral spines; ocelli of ordinary size, diameter greater than width of base of first flagellar segment; hind coxa carinate; arolia absent (possibly with very minute arolia in both sexes) (Africa, Arabia) .... *P. (Trichanthidioides)*

Pachyanthidium / Subgenus *Ausanthidium* Pasteels

Ausanthidium Pasteels, 1969a: 60. Type species: *Anthidiellum ausense* Mavromoustakis, 1934, by original designation.

In having arolia in both sexes, this subgenus differs from other *Pachyanthidium*. The body has abundant yellow markings, and the background of the metasoma is partly red-brown. The length is 7.0 to 7.5 mm. The terga lack the strong lateral spines found in *Trichanthidium*, although T6 of the male has a moderate-sized sublateral spine. T6 of the female has a large apical emargination; females of the other subgenera do not.

■ This subgenus is known only from Namibia. The single species is *Pachyanthidium ausense* (Mavromoustakis).

This subgenus was placed in *Anthidiellum* by Pasteels (1969a), but nearly all its characters support affiliation with *Pachyanthidium*; see Michener and Griswold (1994a).

Pachyanthidium / Subgenus *Pachyanthidium* Friese s. str.

Anthidium (Pachyanthidium) Friese, 1905a: 66. Also described as new by Friese, 1910b: 158. Type species: *Anthidium bicolor* Lepeletier, 1841, by designation of Cockerell, 1920b: 298.

Pachyanthidium consists of species 7.5 to 12.0 mm long, with yellow markings or a red metasoma, and without lateral spines on T2 to T5. The hypostomal carina is lamellate. The apices of the front and middle tibiae have two spines (as in *Stelis*). S5 of the male has a broad apical comb. A remarkable feature of the subgenus is the variability in dentition of the female mandibles, from four-toothed to multi-toothed and to minutely serrate. Males also are variable in mandibular dentition, from four-toothed to eight-toothed. In spite of this variation, the species are similar in most of their characters and constitute a morphologically rather homogeneous taxon.

■ The species of this subgenus are widespread in Africa, from Senegal to Ethiopia and south to Cape Province and Natal in South Africa. The 11 species were revised by Pasteels (1984).

The nest of *P. bicolor* (Lepeletier) consists of cells constructed adjacent to one another on leaves. The cells consist of resin or gum with intermixed plant hairs (Michener, 1968a). Females of another species were seen gathering solidifying latex from injured tips of a cactuslike *Euphorbia*.

Pachyanthidium / Subgenus *Trichanthidioides* Michener and Griswold

Pachyanthidium (Trichanthidioides) Michener and Griswold, 1994a: 319. Type species: *Pachyanthidium semiluteum* Pasteels, 1981, by original designation.

Trichanthidioides resembles the subgenus *Trichanthidium* in its small size (body length 5 mm), hairy eyes, continuation of the omaular carina across the thoracic venter, and lack or near lack of arolia in females (also in males). It differs from *Trichanthidium* in its extensive yellow areas, full-sized ocelli, lack of lateral tergal spines, and lack of a comb on S4 of the male, but presence of a comb on S5.

■ This subgenus is known from Kenya and Saudi Arabia. The only species is *Pachyanthidium semiluteum* Pasteels.

Pachyanthidium / Subgenus *Trichanthidium* Cockerell

Pachyanthidium (Trichanthidium) Cockerell, 1930b: 52. Type species: *Pachyanthidium occipitale* Cockerell, 1930 = *Anthidium benguelense* Vachal, 1903, monobasic.

Trichanthidium consists of small bees, 5 to 8 mm long, that are entirely black or have very restricted yellow areas, occurring only as sublateral spots on metasomal terga. The transverse carina of T1 is doubled laterally. The sides of T3 to T5 and sometimes T2 have long spines. The apices of the front and middle tibiae sometimes have two spines but usually only one is recognizable. The wings are strongly darkened as in *Pachyanthidium* s. str.

■ This subgenus ranges in Africa from the Ivory Coast to southern Egypt, south to Angola and Natal Province in South Africa, and in Asia from India to Yunnan Province, China. There are at least three species.

Genus *Paranthidium* Cockerell and Cockerell

Paranthidium (not to be confused with *Paraanthidium*, a subgenus of *Trachusa*) agrees with *Dianthidium* s. str. in general form and coloration; the body length is 7.5 to 11.0 mm. Most species have distinct postspiracular foveae; at the same time, in all species, the propodeal triangle is punctate and completely lacks a series of pits across its base. *Paranthidium* differs from *Dianthidium* not only in the characters indicated in the key but also in the lack of omaular or preoccipital carinae and the short lamella of the pronotal lobe, which scarcely extends mesad from the lateral scutal margin. T7 of the male is three-lobed, the median lobe always larger and longer than the small lateral lobes. Arolia are present. Male genitalia and hidden sterna were illustrated by Mitchell (1960).

In no other anthidiines are the spines of the front and middle tibiae modified as they are in *Paranthidium*. Each of these tibiae terminates on the outer surface in a convex carina, subtended by short hairs, that forms the apex of a shallow apical concavity on the tibia's outer surface (Fig. 82-4a). In the subgenus *Rapanthidium* the carina is strongly convex, and in females is produced as a projection with a rounded apex. In males of *Rapanthidium* the projection (carina) is limited to the posterior half of each tibia, and the concavity basad to the carina is weak and broken up by hairs.

Evans (1993) described nests of *Paranthidium (P.) jugatorium perpictum* (Cockerell). They are burrows made by the bees in sandy soil, with cells, each made of a thin layer of resin or gum, in series in the burrows.

Key to the Subgenera of *Paranthidium*

1. Propodeum with postspiracular fovea; T6 of male without lateral lobe; S4 of male with median comb, S5 with comb laterally (nearctic, Mesoamerica) .... *P. (Paranthidium* s. str.*)*
—. Propodeum without postspiracular fovea or with fovea only about the size of the spiracle; T6 of male with lateral lobe; sterna of male without combs (Mesoamerica) .... *P. (Rapanthidium)*

Paranthidium / Subgenus *Paranthidium* Cockerell and Cockerell s. str.

Paranthidium Cockerell and Cockerell, 1901: 50. Type species: *Anthidium perpictum* Cockerell, 1898, by original designation.

■ This subgenus occurs from New York to Utah, south to Georgia, USA, and through Mexico to Panama. There are about four species. Forms found in the USA were treated by Schwarz (1926), along with *Dianthidium (Adanthidium)*.

Paranthidium / Subgenus *Rapanthidium* Michener

Paranthidium (Rapanthidium) Michener, 1948a: 11. Type species: *Anthidium vespoides* Friese, 1925, by original designation.

The lone described species, *Paranthidium vespoides* (Friese), and one undescribed species have limited white or yellow markings but another undescribed species has abundant yellow markings like those of the species of *Paranthidium* s. str.

■ *Rapanthidium* occurs from the state of Sinaloa, Mexico, south to Costa Rica. There are at least three species, two of them undescribed.

Genus *Plesianthidium* Cameron

Plesianthidium consists of four subgenera that agree in their robust megachiliform or (in the subgenus *Carinanthidium*) chalicodomiform bodies, lack of pale markings except on the face of the male, and in other characters indicated below and in the key to genera. The subgenera differ enough from one another that they have been given generic status. Common features, however, other than those indicated in the key to genera, are (1) a lack of carinae except sometimes on the pronotal lobe and sometimes on the preoccipital ridge behind the vertex, (2) the straight subantennal sutures arising at or near the tentorial pits, (3) the rounded scutellum, not or little overhanging the metanotum, in profile rounded or (in the subgenus *Spinanthidiellum*) angled, (4) the presence of arolia, (5) the tendency of T6 of the male to be trifid or trilobed (not in the subgenus *Spinanthidiellum*), (6) the small and little-exserted male T7, which is three-toothed or (in the subgenera *Plesianthidium* s. str. and *Carinanthidium*) has the middle tooth reduced to a minor convexity between two long, widely separated teeth, (7) the minimal modifications of the male sterna (S5 has an apical comb, absent in the subgenus *Spinanthidium*, and S6 is characteristically lobed or shows various convexities), and (8) the simple male gonostyli, which are slightly expanded and hairy apically, and in *Plesianthidium* s. str. have two small teeth at the apex. As in many other genera, the mandible of the female is four-toothed, that of the male three-toothed. The apical spines of the front and middle tibiae, particularly in females, are broad, the apices convex in *Carinanthidium* but truncate or emarginate (so that there are two spines) in *Spinanthidium* and *Plesianthidium* s. str.

Key to the Subgenera of *Plesianthidium*

1. Preoccipital ridge with carina behind vertex; T6 of male truncate, without lateral tooth, with longitudinal ridge distally, highest at posterior margin of tergum; mandible less than twice as long as broad .... *P. (Spinanthidiellum)*
—. Preoccipital ridge not carinate; T6 of male with median truncate or rounded to pointed projection, rarely weakly produced, and a strong lateral tooth; mandible over twice as long as broad 2
2(1). S6 of female with strong longitudinal median carina; hind trochanter of male with mesal subapical spine; pubescence all black except for white on face of male .... *P. (Carinanthidium)*
—. S6 of female not carinate; hind trochanter of male not spined; pubescence brown to gray, or with whitish on face and venter 3
3(2). T7 of male strongly trifid, median tooth or lobe exceeding lateral ones; S5 of male without apical comb; maxillary palpus two-segmented .... *P. (Spinanthidium)*
—. T7 of male with median tooth reduced to low prominence, tergum thus essentially bifid; S5 of male with apical comb; maxillary palpus three-segmented .... *P. (Plesianthidium s. str.)*

Plesianthidium / Subgenus *Carinanthidium* Pasteels

Carinanthidium Pasteels, 1969a: 42. Type species: *Megachile cariniventris* Friese, 1904, by original designation.

This subgenus consists of a wholly black, chalicodomiform species 12 to 14 mm long. Except for the white hairs on the lower face of the male, the hairs are all black. The *Megachile*-like appearance of this bee makes it the most distinctive *Plesianthidium*, although morphologically it shares various features with *Spinanthidium* and *Plesianthidium* s. str.

■ This subgenus is found in western Cape Province, South Africa, but the type specimen was reported to be from northern Transvaal, South Africa. The only species is *Plesianthidium cariniventre* (Friese).

Plesianthidium / Subgenus *Plesianthidium* Cameron s. str.

Plesianthidium Cameron, 1905: 256. Type species: *Plesianthidium fulvopilosum* Cameron, 1905, by original designation.

This subgenus contains the largest species of the genus; it has brown or gray hair, and the face of the male is partly cream-colored. The body is robust (length 15-16 mm) but more elongate than that of *Spinanthidium*. Structurally, this subgenus is similar to *Spinanthidium*.

■ *Plesianthidium* s. str. occurs in northwestern Cape Province, South Africa. The only species is *Plesianthidium fulvopilosum* Cameron.

Plesianthidium / Subgenus *Spinanthidiellum* Pasteels

Spinanthidium (Spinanthidiellum) Pasteels, 1969a: 59. Type species: *Anthidium volkmanni* Friese, 1909, by original designation.

This is in some ways the morphologically most distinctive subgenus of *Plesianthidium*, although in general aspect the specimens look like small (7.0-8.5 mm long) *Spinanthidium*. The body, short and robust, has an almost globular metasoma, and the pubescence, brown or gray, is abundant. The face of the male has cream-colored areas. Because the three teeth of T7 of the male are short, the tergum is only weakly trifid.

■ This subgenus is found in western Cape Province, South Africa. The two species are *Plesianthidium volkmanni* (Friese) and *rufocaudatum* (Friese), the latter incorrectly synonymized by Pasteels (1984) (T. Griswold, in litt., 1995).

Plesianthidium / Subgenus *Spinanthidium* Mavromoustakis

Dianthidium (Spinanthidium) Mavromoustakis, 1951: 977. Type species: *Anthidium trachusiforme* Friese, 1913, by original designation.

This subgenus contains gray- or brown-haired species, larger than *Spinanthidiellum* but smaller than *Plesianthidium* s. str. (8-12 mm long) and almost as robust as the former, with cream-colored areas on the face of the male. T7 of the male is deeply trifid, the median projection either truncate or pointed. Unlike that of the other subgenera, S5 of the male lacks a comb, but S4 has a comb in one species, *Plesianthidium calescens* (Cockerell) (T. Griswold, in litt., 1995).

■ This subgenus is found in Cape Province, South Africa. The five species were revised by Pasteels (1984).

This subgenus could reasonably be synonymized with *Plesianthidium* s. str.

Genus *Pseudoanthidium* Friese

Pseudoanthidium is here used in a broad sense to include numerous species having compact megachiliform bodies, commonly having the form of *Anthidium* s. str. Except in the subgenus *Micranthidium* the head is thick and the tibiae are tuberculate on the outer surfaces; but even in *Micranthidium* the tibiae are coarsely punctate. The integument is black with yellow markings, the metasomal bands broken or reduced to lateral spots. Commonly, the subantennal suture is arcuate outward (although not always visible); the pronotal lobe is lamellate; T6 of the male lacks lateral teeth except in the subgenus *Exanthidium*; and T7, not much exserted except in *Exanthidium*, is usually bilobed, the emargination between the lobes sometimes feeble or absent, the margin then being convex and sometimes undulate. The tergal margins, except in the subgenus *Royanthidium*, are rather broadly impunctate, slightly to strongly flared as seen in lateral view, and often translucent. S3 to S5 of the male are concave, the posterior margin of S3 with an area of wavy bristles except in *Royanthidium*, these bristles weak in *Pseudoanthidium* (*Tuberanthidium*) *brachiatum* Michener and Griswold; the margin of at least S5 is concave and has an apical lateral process (merely a lobe in *Exanthidium*), the process frequently armed at the tip with a small comb.

Key to the Subgenera of *Pseudoanthidium*
(From Michener and Griswold, 1994a)

1. Gena margined posteriorly by distinct preoccipital carina (gena narrower than eye as seen from side) 2
—. Gena without preoccipital carina (scutellum rounded or medially emarginate as seen from above; tibiae tuberculate on outer surfaces at least in female; male without lamellate lateral lobe on S3; clypeus of female with discal hairs straight) 3
2(1). Preoccipital carina behind vertex strongly produced back over front of thorax as a lamella; posterior basitarsus of male more than twice as long as broad (scutellum broadly truncate in dorsal view, posterior margin angulate laterally near axilla; tibiae not tuberculate but coarsely punctate; lateral lobe of S3 of male lamellate; clypeus of female with discal hairs bent down apically) (Africa, Arabia) .... *P. (Micranthidium)*
—. Preoccipital carina behind vertex rather weakly carinate; posterior basitarsus of male less than twice as long as broad (Africa) .... *P. (Semicarinella)*
3(1). Propodeum shagreened; T6 of male with strong apical, medially emarginate flange; fore and middle basitarsi of female with long plumose hair (Africa)..................... *P. (Tuberanthidium)*
—. Propodeum shiny where not punctate; T6 of male without emarginate flange; fore and middle basitarsi of female with long but not plumose hair 4
4(3). T6 of female with surface broadly and conspicuously excavated; T5 of female with small, midapical, marginal projecting lobe; exposed part of T7 of male longer than exposed part of T6, very deeply bilobed; S5 of male without strong lateral tooth or lobe (palearctic, Sudan) .... *P. (Exanthidium)*
—. T6 of female with surface largely convex; T5 of female without midapical lobe; exposed part of T7 of male shorter than exposed part of T6, weakly to strongly bilobed; S5 of male with strong lateral process 5
5(4). Clypeal apex of female fully exposed, coarsely denticulate, protruding; S3 and S4 of male broadly emarginate, or S3 with median V-shaped emargination, and S5 with apicolateral projection, all without combs or specialized hairs (palearctic) .... *P. (Royanthidium)*
—. Clypeal apex of female largely hidden by hair, without shining premarginal ridge; S3 of male with comb of long, wavy bristles and S5 with comb at apex of apicolateral projection (palearctic, oriental) .... *P. (Pseudoanthidium s. str.)*

Pseudoanthidium / Subgenus *Exanthidium* Pasteels

Exanthidium Pasteels, 1969a: 82. Type species: *Anthidium eximium* Giraud, 1863, by original designation.

This is a subgenus of small anthidiines, 7-9 mm long, having yellow areas on all tagmata. T6 of the female, in addition to exhibiting the excavated surface indicated in the key to subgenera, has a broadly rounded or truncate, somewhat denticulate apex with a pair of small, shiny, depressed median lobes on either side of a notch. In *Pseudoanthidium eximium* (Giraud), the notch contains a pair of slender black projections. *Exanthidium* differs from *Anthidium* in characters of the male genitalia, S8, and the marginal zone of S5, which in the female is narrow but expanded posteriorly in the middle on a small midapical lobe but in the male is broader and somewhat expanded anteriorly in the middle, suggesting *Anthidium*. T7 of the male is strongly exserted, unlike that of other *Pseudoanthidium*, and is very deeply bilobed; S3 is large and ends in a median lobe or broad truncation and an area of dense, long, wavy hairs; S4 and S5 are short, largely hidden by the lobe and hairs of S3. In addition to all these distinctive characters, males of *Exanthidium* differ from those of other *Pseudoanthidium* in the lack of, or the very blunt, lateral tooth of S5. In the male there is a tubercle near the base of the hind trochanter and a posterior basal tooth on the hind femur.

■ *Exanthidium* occurs in the Mediterranean basin, thence east to Tadzhikistan, and south to Sudan. There are four species.

In *Pseudoanthidium (Exanthidium) eximium* the scutellum is truncate and angulate laterally, as in the subgenus *Micranthidium*, but in *P. (E.) wahrmanicum* (Mavromoustakis) the posterior margin of the scutellum is rounded laterally as seen from above.

Pseudoanthidium / Subgenus *Micranthidium* Cockerell

Pachyanthidium (Micranthidium) Cockerell, 1930b: 45. Type species: *Anthidium truncatum* Smith, 1854, by original designation.

This is a subgenus of robust species, 6 to 8 mm long, the metasoma of which tapers posteriorly, leaving T5 little over half as wide as T1. The limited yellow markings are sometimes absent on the thorax; on the metasomal terga they are lateral spots or broadly broken bands. The metasomal form and the long, dark wings suggest *Pachyanthidium (Trichanthidium)*, from which *Micranthidium* is immediately distinguishable by its bare eyes. The preoccipital carina is complete, behind the genal area as well as behind the vertex, but is not lamellate. The pronotal lobe is lamellate and the omaulus is strongly carinate. The transverse, truncate, sharply margined scutellum suggests that of *Gnathanthidium* and *Anthidium (Proanthidium)*, but differs from all other *Pseudoanthidium* except certain species of *Exanthidium*. The presence on the upper part of the metapleuron of a shiny impunctate area demarcated anteriorly by a hairy zone is unique among anthidiines. Also unique among males are an oval shining impunctate area on the undersurface of the fore tibia and a deep transverse pocket across the base of S4.

■ *Micranthidium* is widespread in tropical Africa, from Ivory Coast to Angola east to Sudan and Natal, South Africa. One species is reported from the Arabian peninsula. The three species were revised by Pasteels (1984).

Pseudoanthidium / Subgenus *Pseudoanthidium* Friese s. str.

Anthidium (Pseudoanthidium) Friese, 1898a: 101. Type species: *Anthidium alpinum* Morawitz, 1873, designated by Sandhouse, 1943: 593.
Paranthidiellum Michener, 1948a: 25. Type species: *Anthidium cribratum* Morawitz, 1875, by original designation.
Pseudoanthidium (Paraanthidiellum) Pasteels, 1969a: 79, unnecessary emendation of *Paranthidiellum* Michener.
Pseudoanthidium (Carinellum) Pasteels, 1969a: 80. Type species: *Anthidium ochrognathum* Alfken, 1932, by original designation.
Trachusa (Orientotrachusa) Gupta, 1993: 50. Type species: *Anthidium orientale* Bingham, 1897, by original designation.

This subgenus consists of small (6-8 mm long), black species commonly having well-developed yellow markings, the tergal bands being broadly broken. But in the species placed in *Carinellum* by Pasteels (1969a), e.g., *Pseudoanthidium ochrognathum* (Alfken), the body of the female is entirely black except for lateral yellow spots on the terga. The posterior lobe of the pronotum has a low lamella or (in *P. ochrognathus*) a strong lamella. The scutellum is sometimes medially emarginate; laterally, the margin is thin and carinate. In *P. ochrognathum* the omaulus is lamellate. Features of *Pseudoanthidium* s. str. not found in males of other *Pseudoanthidium* include a comb on each side of the midline of S5, sometimes on a tubercle, in addition to the comb on the lateral arm. *Carinellum* differs from other *Pseudoanthidium* s. str. only in the characters listed above, and is therefore placed in synonymy. It contains only two species; the males are unknown to me.

■ The range of *Pseudoanthidium* s. str. is the Canary Islands, Europe (north to Germany), and the Mediterranean basin, including northern Africa, east to Central Asia, India, and Southeast Asia. Pasteels (1969a) lists 18 species in the groups here included under *Pseudoanthidium* s. str.

The application of the name *Pseudoanthidium* s. str. is probably correct, but a possible doubt is justified by the insufficient knowledge of the type species. The matter was discussed by Michener and Griswold (1994a).

Pseudoanthidium / Subgenus *Royanthidium* Pasteels

Royanthidium Pasteels, 1969a: 86. Type species: *Anthidium melanurum* Klug, 1832, by original designation.
Reanthidium Pasteels, 1969a: 87. Type species: *Anthidium reticulatum* Mocsáry, 1884, by original designation.

This subgenus contains robust species, 8 to 9 mm long, having the form and coloration of the subgenus *Tuberanthidium*, with additional yellow spots on the vertex. The carinae are also as in *Tuberanthidium*, except that the pronotal lobes are strongly lamellate. The scutellum is distinctly emarginate medially. Unlike those of the other subgenera, the margins of T1 to T4 are smooth, narrow, black, and scarcely flared. Sternal ornamentation is limited to a small medial comb or series of bristles and combless lateral arms on S5.

■ *Royanthidium* is found in Central Europe and the

Mediterranean basin east to Central Asia. Pasteels (1969a) lists six species.

The species formerly placed in *Reanthidium* differ from other *Royanthidium* in having the preoccipital ridge carinate behind the vertex.

Pseudoanthidium / Subgenus *Semicarinella* Pasteels

Pseudoanthidium (Semicarinella) Pasteels, 1984: 32. Type species: *Pseudoanthidium latitarse* Pasteels, 1984, by original designation.

This subgenus may be based on a small species (length 6.5 mm) of *Micranthidium*, from which it appears to differ in the "subcarinate" preoccipital carina behind the vertex (it is lamellate in *Micranthidium*), the carina (rather than a lamella) on the pronotal lobe, the laterally rounded (rather than angulate) posterior margin of the scutellum, and the broad (less than twice as long as broad) hind basitarsus of the male. It should be noted that in *Pseudoanthidium (Micranthidium) truncatum* (Smith) the hind basitarsus is somewhat broadened and that the scutellar character listed above is doubtful because Pasteels' description, unlike his diagram, indicates that the scutellum is as in *Micranthidium*.

■ *Semicarinella* occurs in Senegal. The name is based on a single male specimen that I have not seen.

Pseudoanthidium / Subgenus *Tuberanthidium* Pasteels

Tuberanthidium Pasteels, 1969a: 87. Type species: *Anthidium tuberculiferum* Brauns, 1905, by original designation.

This subgenus consists of rather robust bees (showing the form of *Royanthidium*) 7 to 11 mm long. They are black with yellow markings on all tagmata, including lateral spots or broken bands on the metasoma. The dorsal bulge on the female mandible is unique in the genus, but absent in *Pseudoanthidium (Tuberanthidium) brachiatum* Michener and Griswold. The pronotal lobes are carinate; otherwise, the usual carinae are absent. The scutellum is rounded or feebly emarginate in the middle, as seen from above; in profile it is rounded or somewhat acute, but laterally it forms a sharp overhanging margin. S5 of the male is largely hidden medially beneath S4; its lateroapical projection is bare or in *P. brachiatum* with a comb. As in *Royanthidium* there is a medially emarginate flange on the margin of the male T6 and a semicircular emargination on T7. Whereas in *Tuberanthidium* as it was understood by Pasteels, S3 of the male is medially produced and carries long, wavy bristles, in *P. (T.) brachiatum* it is not produced and has much shorter, relatively inconspicuous wavy bristles. Male genitalia of that species were illustrated by Michener and Griswold (1994a).

■ *Tuberanthidium* is found in Namibia, South Africa (Cape Province to Natal), Lesotho, Botswana, and Tanzania. Pasteels (1984) treats three species of *Tuberanthidium*; a fourth is the more *Pseudoanthidium*-like species *P. (T.) brachiatum* Michener and Griswold.

Genus *Rhodanthidium* Isensee

This is a palearctic genus of moderate-sized to very large, chalicodomiform or euceriform bees with abundant yellow or reddish-yellow markings (Pl. 8), although some specimens of *Rhodanthidium (Asianthidium) glasunovii* (Morawitz) lack such markings on the thorax and metasoma. The metasomal color bands, when present, are broken medially at least on the anterior terga. Common features are (1) a lack of carinae, except commonly a carina or even a lamella on the pronotal lobe and sometimes a weak upper omaular carina, (2) the straight subantennal sutures, (3) the rounded or laterally sharp scutellum not or little overhanging the metanotum, (4) the presence of arolia, (5) the strongly trifid male T7 [lateral teeth rather small in *R. (Asianthidium) aculeatum* (Klug)], (6) the male gonostyli, which are about as described for *Plesianthidium*, their apices simple or having two small teeth, and (7) the male sterna, which are not greatly modified, although S4 and S5 sometimes carry median apical combs and S5 has a lateral tooth in *Rhodanthidium* s. str.

The number of segments in the maxillary palpus varies from two to three and perhaps four. It is difficult to determine the number; segments sometimess seem to be only partly separated, so that one must make arbitrary decisions.

Key to the Subgenera of *Rhodanthidium*

1. T6 of male with median, produced truncation or rounded process and lateral tooth or strong shoulder; S5 of male with strong lateral tooth and median marginal comb; omaulus weakly carinate above; margin of scutellum with sharp edge laterally; female with apical projection or spine of fore and mid tibiae narrowly bidentate .... *R. (Rhodanthidium* s. str.)
—. T6 of male simple or with scarcely produced broad truncation, with or without strong shoulder or weak lateral tooth; S5 of male without lateral tooth, without comb; omaulus not carinate [or weakly so above in *R. (Asianthidium) caturigense* (Giraud) and its relatives]; margin of scutellum rounded, not forming sharp edge [except in *R. (Asianthidium) caturigense*]; female with apical projection of fore and mid tibiae either not notched or widely and shallowly emarginate 2
2(1). Trifid apex (median tooth sometimes broad and rounded or truncate) of T7 of male occupying full width of tergum; pronotal lobe with lamella or carina [weak in *R. (Asianthidium) glasunovi* (Morawitz)]; head with background color black .... *R. (Asianthidium)*
—. Trifid apex of T7 of male much narrower than tergum, which has lateral angles rendering tergum five-toothed; pronotal lobe without or with very weak carina; head largely yellow .... *R. (Meganthidium)*

Rhodanthidium / Subgenus *Asianthidium* Popov

Meganthidium (Asianthidium) Popov, 1950a: 315. Type species: *Anthidium glasunovii* Morawitz, 1894, by original designation.
Rhodanthidium (Trianthidium) Mavromoustakis, 1958: 435. Type species: *Anthidium caturigense* Giraud, 1863, by original designation.
Meganthidium (Oxyanthidium) Mavromoustakis, 1963a:

653. Type species: *Anthidium aculeatum* Klug, 1832, by original designation.

Axillanthidium Pasteels, 1969a: 39. Type species: *Axillanthidium axillare* Pasteels, 1969 = *Anthidium ducale* Morawitz, 1876, by original designation.

Bees of this subgenus have unusually heavily sclerotized penis valves, each with an apical or preapical spine, lobe, or process. Body length is 11 to 18 mm.

■ The range is from Spain to southwestern Asia and to Central Asia. There are seven or more species.

The synonymy indicated above is based on mutual similarity and the small number of species involved. *Asianthidium* (in its original narrow sense), *Axillanthidium,* and *Oxyanthidium* each contain only a single species. *Asianthidium* in the sense of *Rhodanthidium (A.) glasunovi* (Morawitz) seems the most deserving of separate subgeneric recognition if the subgenus as here understood were to be divided. Its weak rather than strong carina on the pronotal lobe and the elevated median process of T7 of the male (suggesting *Rhodanthidium* s. str.) differentiate it from other species. *R. (A.) aculeatum* (Klug), the species placed in *Oxyanthidium,* differs from other species in the strongly denticulate and laterally angulate margin of T6 of the female and the translucent preapical lamella above the apex of S6 of the female, the sternum thus having a double margin.

Rhodanthidium / Subgenus *Meganthidium* Popov

Meganthidium Popov, 1950a: 315. Type species: *Anthidium christophi* Morawitz, 1884 = *Anthidium superbum* Radoszkowski, 1876, by original designation.

This subgenus is easily recognized by the characters given in the key to subgenera. Aside from its great size (19-22 mm long), characters include a comb on S4 but not on S5 of the male, the simple T6 of the male, and the three teeth arising from a small, elevated area on T7. The apex of the male middle tibia appears to have a pair of apical angles as in females, but the angles are preapical and connected by a carina; there is no corresponding pair of angles on the fore tibia, which has a transverse preapical carina. Popov (1950a) illustrated the male genitalia and sterna.

■ *Meganthidium* ranges from Turkey to Central Asia. There is probably only one species, though there may be two similar species.

Rhodanthidium / Subgenus *Rhodanthidium* Isensee s. str.

Anthidium (Rhodanthidium) Isensee, 1927: 374. Type species: *Anthidium siculum* Spinola, 1838, by original designation.

Bellanthidium Pasteels, 1969a: 38. Type species: *Anthidium infuscatum* Erichson, 1835, by original designation.

As is evident from the key to subgenera, the subgeneric characters of females do not always hold. An additional character for males is the apically notched, thus bidentate, gonostylus; in this respect *Rhodanthidium* s. str. differs from the other subgenera. Length ranges from 8.5 to 19.0 mm. Although for some species, like *Rhodanthidium siculum* (Spinola) and *sticticum* (Fabricius), the pale areas are yellowish red, for others, such as *R. septemdentatum* (Latreille), they are bright yellow.

■ *Rhodanthidium* s. str. occurs in the Mediterranean area from Morocco and Portugal to Central Europe, eastward to Turkey. There are about five species.

Most of the species form a unified group, but *Rhodanthidium infuscatum* (Erichson) is quite distinctive and has been given the genus name *Bellanthidium.* Since there is only one such species and its relationship to the others is clear, a subgenus for it seems unnecessary. Among its distinctive characters are strong shoulders rather than spines at the sides of T6 of the male, a marginal comb on S5 but not on S4 of the male (in other species there is a comb on S4 and sometimes also S5), and especially an enormous emargination in T6 of the female. Thus those who wish to recognize *Bellanthidium* as a monotypic subgenus will have no difficulty in doing so.

Genus *Serapista* Cockerell

Serapis Smith, 1854: 218 (not Link, 1830). Type species: *Serapis denticulatus* Smith, 1854, monobasic.

Serapista Cockerell, 1904a: 357, replacement for *Serapis* Smith, 1854. Type species: *Serapis denticulatus* Smith, 1854, autobasic.

These are large, robust, megachiliform bees (9-15 mm long) superficially similar to *Anthidium (Nivanthidium) niveocinctum* Gerstaecker. The body is black with a pattern of white spots caused by dense white hairs; yellow integument is limited to the clypeus and mandibles of males. In addition to the characters indicated in the key to genera, *Serapista* has the following features: the preoccipital carina is absent; the fore tibia of the female has a longitudinal row of coarse, spiculate setae; the pronotal lobe is provided with a large lamella; the tegula is acutely angled anteriorly; the omaular carina is present at least on the upper half; and the scutellum is strongly produced as a lamella overhanging the propodeum. The apex of the male metasoma is about as in ordinary species of *Anthidium* s. str., but the preapical zones of the terga are of uniform width, not widened medially as in *Anthidium.* All terga except the first are spined laterally, and T6 of females and T7 of males have a longitudinal median carina.

■ *Serapista* ranges widely in Africa, from Cameroon to Tanzania south to Cape Province, South Africa. The four species were revised by Pasteels (1984). Nests, masses of plant down often intermixed with animal hairs or even feathers and placed on plant stems, have been described by several authors; a summary and full description of one such nest is by Michener (1968a).

Genus *Stelis* Panzer

This is a genus of small to rather large (4-14 mm long) cleptoparasitic bees varying from all black to richly marked with cream, yellow, or orange, and in North America to strongly metallic blue or green, with or without pale markings. Unlike that of *Euaspis,* the interantennal area lacks juxtantennal carinae. The preoccipital carina is absent, as are thoracic carinae, or there may be weak carinae on the lateral part of the preoccipital ridge

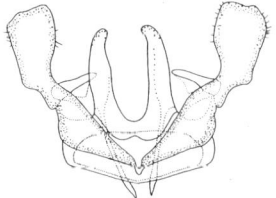

Figure 82-12. Male genitalia of *Stelis (Dolichostelis) louisae* Cockerell. Modified from Parker and Bohart, 1979.

and on the pronotal lobe, or in some species of the subgenus *Protostelis* the pronotal and omaular carinae are strong. The scutellum, little produced, is rounded, except in some oriental species. Behind the propodeal spiracle is sometimes a fovea, which forms the lateral extremity of a series of pits (often broken medially) commonly forming a narrow subhorizontal or slanting zone across the base of the propodeum (Fig. 82-14); this basal zone is often merely roughened, not pitted, or is pitted only laterally. The scopa is absent. T7 of the male is small and only a little exposed; its apex is rounded, weakly bilobed, or weakly trilobed, or it has a median angle exceeding a bilobed base. The male gonostylus has a slender base and an expanded, angulate distal part often directed mesad at an angle to the basal part (Fig. 82-12); there is a strong projecting angle, marked by an arrow in the figure, on the outer margin at the base of the expanded portion. The volsellar lobe is present. Popov (1938) and Mitchell (1962) illustrated the genitalia of diverse American species; see also Figure 82-12. Warncke (1992a) revised the west palearctic species. The genus *Afrostelis* agrees with the above comments except for the male genitalic characters.

The genus *Stelis* is primarily holarctic in distribution but extends south to Costa Rica (subgenus *Dolichostelis*), Malaysia (*Malanthidium*), and Kenya (*Stelidomopha*). Griswold and Parker (2003) described a species of unknown subgeneric relationships on the basis of one female from western Cape Province, South Africa.

Pasteels (1968a, 1969a) believed that the closest nonparasitic relative of *Stelis* is *Bathanthidium*, partly because of its divided scutoscutellar fovea. The suggestion has been made that the subgenus *Protostelis* might be derived from *Trachusa*, and that *Stelis* therefore might be polyphyletic. *Protostelis*, however, does not have the reflexed male T7 characteristic of *Trachusa* and does have *Stelis*-like features such as the two spines on the apices of the front and middle tibiae (Fig. 82-6a). I therefore regard *Protostelis* as a subgenus of *Stelis*, not related to *Trachusa*.

The hosts of *Stelis* are other Megachilinae. In most cases the parasite, after locating a host nest, returns to it repeatedly to place an egg in each of several host cells before they are closed, as noted by Michener (1955) and others. In the subgenus *Dolichostelis*, however, the female opens the resin seal of the finished nest of the host and reseals it with the same resin on departure (Parker et al., 1987). Probably the behavior is similar to that of another parasitic genus, *Euaspis*. Larvae of *Stelis* are active and have sharp mandibles with which they destroy eggs or larvae of the host. A possible exception is in the subgenus *Dolichostelis*, whose adults may destroy the host egg or larva.

Key to the Subgenera of *Stelis* of the Eastern Hemisphere

1. Clypeus produced well over mandibles, its apex bilobed, strongly so in females; anterior spine of front and middle tibiae conspicuous, enlarged, curved posteriorly ....
   .... *S. (Stelidomorpha)*
— Clypeus not greatly produced over mandibles, truncate or subtruncate; anterior spine of front and middle tibiae less than twice the size of posterior spine 2
2(1). Axilla projecting laterally beyond lateral margin of scutum; scutellum strongly projecting over metanotum and propodeum (oriental) .... *S. (Malanthidium)*
—. Axilla not projecting laterally beyond margin of scutum; scutellum at most weakly projecting over metanotum and propodeum 3
3(2). Scutellum not carinate laterally; head and thorax without light markings; hind tibial apex with two spines or angles (Fig. 82-13b), one near outer middle of apical tibial margin (if with only one spine, as in *S. simillima* Morawitz, then not on posterior apical angle) or the spines united to form truncate margin, the area sparsely hairy and the structure thus easily seen; omaulus not carinate .... *S. (Stelis s. str.)*
—. Scutellum carinate laterally; head and thorax with light markings; hind tibial apex with a single spine (sometimes a mere angle) largely hidden in hairs (Fig. 82-13a) near the posterior apical angle of the tibia, in front of which the apex of the tibia presents a convex margin; omaulus usually carinate 4
4(3). Hind basitarsus with carina along inner dorsal angle (Fig. 82-13a); middle tibia flattened, apically enlarged, twice as wide apically as basally; S1 with transverse carina overhanging apical margin .... *S. (Heterostelis)*
—. Hind basitarsus without carina; middle tibia not flat-

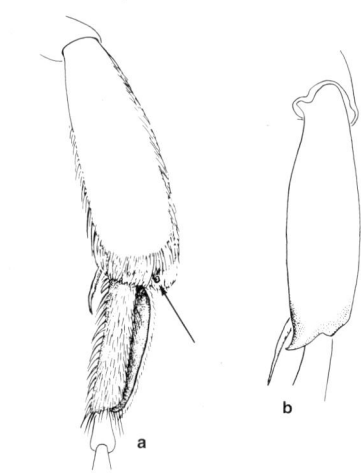

Figure 82-13. Hind tibiae of *Stelis*. **a**, *S. (Heterostelis) australis* Cresson. (The arrow indicates the spine largely and sometimes completely hidden by dense hairs); **b**, *S. (Stelis) rubi* Cockerell. Modified from Michener, McGinley, and Danforth, 1994.

tened, little enlarged apically, at most 1.5 times as wide apically as basally; S1 without transverse carina 5
5(4). Pronotum without notch between lobe and rest of pronotum; omaulus not carinate; carinate pitted basal zone of propodeum extending behind spiracle ....
.... *S. (Protostelis)*
—. Pronotum with notch between lobe and rest of pronotum; omaulus usually carinate; carinate pitted basal zone of propodeum not extending behind spiracle ....
.... *S. (Pseudostelis)*

Key to the Subgenera of *Stelis* of the Western Hemisphere

1. Hind tibia with a single prominent tooth or tibial spine (usually hidden among hairs) on margin near apex of posterior margin of tibia; hind basitarsus with strong lamella-like carina along posterior margin, separated by longitudinal depression from longitudinal thickening of outer surface of basitarsus (Fig. 82-13a) .... *S. (Heterostelis)*
—. Hind tibia with an apical median tooth or tibial spine and a less prominent, rounded projection near apex of posterior margin of tibia; hind basitarsus unmodified 2
2(1). Base of propodeum with zone set off by carina and divided into a series of pits (Fig. 82-14), this zone projecting subhorizontally behind vertical metanotum; anterior surface of mesepisternum impunctate at least below and set off from lateral surface by sharp angle or weak omaular carina .... *S. (Dolichostelis)*
—. Base of propodeum vertical or sloping, rarely subhorizontal, without series of pits or such pits usually present only laterally, this zone with about same slope as metanotum; anterior surface of mesepisternum punctate, omaular carina absent .... *S. (Stelis s. str.)*

Stelis / Subgenus *Dolichostelis* Parker and Bohart

Dolichostelis Parker and Bohart, 1979: 138. Type species: *Stelis laticincta* Cresson, 1878, by original designation.

Dolichostelis, the species of which were formerly incorrectly placed in *Protostelis* (e.g., Hurd, 1979), contains rather elongate, chalicodomiform, black species, 6 to 10 mm long, richly marked with yellow, orange, or cream. As shown by Figure 82-11b, the aspect is similar to that of *Dianthidium*, but the costal part of the forewing is dark. Parker and Bohart (1979) regarded *Dolichostelis* as a genus allied to neotropical forms unrelated to *Stelis* (presumably *Hoplostelis* species), but this appears to be an error; the clubbed and angulate gonostylus (Fig. 82-12) is as in *Stelis*, as are the two apical spines on the front and middle tibiae (Fig. 82-6a). The characters of the subgenus all appear to be derivable from those of *Stelis* s. str. Distinctive features of *Dolichostelis* are (1) the angulate or weakly carinate omaulus, the surface in front of it more shiny and less closely punctate, at least below, than the surface behind it, (2) the subhorizontal, pitted to foveate base of the propodeum, extending back from the vertical metanotum (Fig. 82-14), (3) the rather sharp posterior marginal carina of the scutellum, and (4) the pair of broad translucent lobes forming much of the posterior margin of S3 of the male. The genitalic and sternal characters were illustrated by Parker and Bohart (1979); see also Figure 82-12.

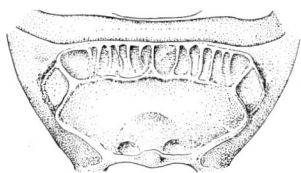

Figure 82-14. Posterodorsal view of propodeum of *Stelis (Dolichostelis) rudbeckiarum* Cockerell. From Michener, McGinley, and Danforth, 1994.

■ This subgenus occurs across North America from Maine to California, USA, and south to Costa Rica. The six species were revised by Parker and Bohart (1979).

So far as is known, this subgenus is cleptoparasitic in nests of *Megachile (Chelostomoides)* (Parker et al., 1987). For further details, see the account of the genus.

Stelis / Subgenus *Heterostelis* Timberlake

Stelis (Heterostelis) Timberlake, 1941: 125. Type species: *Stelis anthidioides* Timberlake, 1941, by orginal designation.
Doxanthidium Pasteels, 1969a: 28. Type species: *Anthidium paradoxum* Mocsáry, 1884, by original designation.

The subgeneric name *Heterostelis* replaces *Protostelis* of Michener (2000: 514) because the type species of *Protostelis* was incorrect in that work (see Sandhouse 1943).

This subgenus contains robust species 10 to 12 mm long with strong yellow markings on all tagmata. Unlike most *Stelis*, which have two-segmented maxillary palpi, those of *Heterostelis* appear to be three-segmented. The pronotal lobe is strongly carinate. The two surfaces of the mesepisternum are at right angles to one another and the omaulus is carinate, except in *S. hurdi* Thorp. The axillae are usually not angulate as in many palearctic *Stelis* s. str., but the margin often projects roundly as seen from above, not being a continuation of the curve of the scutellar margin; in *S. annulata* (Lepeletier) the axillae are somewhat angulate. The combined shape of the scutellum and axillae varies from semicircular to quadrate. The series of pits along the basal margin of the propodeum is broadly interrupted, and the median part of the propodeal surface is wholly vertical. The propodeum sometimes lacks a distinct pit or fovea behind the spiracle. Male genitalia, sterna, and other structures were illustrated by Thorp (1966).

■ *Heterostelis* is found throughout the southern USA from California to Florida, north to New Jersey, and south to Morelos, Mexico. In the palearctic region it is found in southern Europe north to Germany, in the Mediterranean basin, and east to Iran. There are about three palearctic species and six in North America. The latter were revised by Thorp (1966).

Heterostelis species may all be parasites of *Trachusa*. One species has been definitely associated with *Trachusa (Trachusomimus)* (see Thorp, 1966). A new species was observed entering nests of *Trachusa (Heteranthidium) catinula* Brooks and Griswold (T. Griswold, personal communication, 1995).

Stelis / Subgenus *Malanthidium* Pasteels

Malanthidium Pasteels, 1969a: 26. Type species: *Anthidium malaccense* Friese, 1914, by original designation.

This subgenus includes rather elongate (8-11 mm long), parallel-sided species with the wings dark, at least in part, the body black with yellow markings, or the metasoma red. As in *Larinostelis* the axillae are produced laterally beyond the lateral margin of the scutum, but they do not approach the tegulae as they do in that genus. In *Stelis malaccensis* (Friese) the anterolateral part of the axilla is produced to a posteriorly hooked, pointed projection. Such a process is not developed in other species, although the lateral margin of the axilla is a rounded but approximately right-angular projection. The scutellum is elongate, clearly overhanging the propodeum, and acute although rounded in profile. There is a strong preapical carina on S1 as in *Heterostelis*, but it does not overhang the margin.

■ This subgenus is known only from the Malay Peninsula. The only described species is *Stelis malaccensis* (Friese), but two undescribed species also exist.

Stelis / Subgenus *Protostelis* Friese

Stelis (Protostelis) Friese, 1895: 25. Type species: *Anthidium signatum* Latreille, 1809, by designation of Popov, 1933: 389.

In Michener (2000) the type species of *Protostelis* was incorrectly given as *Stelis freygessneri* Friese, 1885 = *Anthidium annulatum* Lepeletier, 1841 (a species here placed in the subgenus *Heterostelis*). The wording of Popov's (1938: 41) designation, which was recognized by Sandhouse (1943), is dubious and in any event, as pointed out by Baker (1999: 238) and indicated above, there is an earlier valid type designation (Popov, 1933b).

The type species of *Protostelis*, *Stelis signata* (Latreille), was tentatively placed in the subgenus *Pseudostelis* by Michener (2000). It now seems best, at least until a detailed study is made, to remove *S. signata* from *Pseudostelis* and place it alone (among the species that I have examined) in the subgenus *Protostelis*.

■ *Protostelis* is uncommon but ranges widely in Europe (Finland to Sicily and east into Russia). The only included species is *Stelis (Protostelis) signata* (Latreille) which is a cleptoparasite of *Anthidiellum* (see Westrich, 1989).

The omaulus is not carinate, the distinctive notch between the pronotal lobe and the rest of the pronotum is absent, and there is a carinate pitted basal propodeal zone that continues behind the spiracle. More important, the triangular gonostylus is unlike that of any other *Stelis*.

Stelis / Subgenus *Pseudostelis* Popov

Pseudostelis Popov, 1956: 167. Type species: *Stelis strandi* Popov, 1935, by original designation.

This subgenus includes the rather small (5-8 mm long) species formerly included in *Protostelis* (now *Heterostelis*). They are robust, and display abundant light markings on the head and thorax as well as the metasoma. As in *Heterostelis* the omaulus is often carinate and the scutellum is carinate laterally. The hind tibia has two apical angles, or, if one, it is on the outer face and thus not homologous with the dorsal angle of *Heterostelis*. The similarities to and differences from *Heterostelis* are indicated in the discussion of that subgenus. The male genitalia and hidden sterna were illustrated by Popov (1956).

■ *Pseudostelis* is found from the Mediterranean basin to Tadzhikistan. There are three species.

Stelis / Subgenus *Stelidomorpha* Morawitz

Stelidomorpha Morawitz, 1875: 131. Type species: *Anthidium nasutum* Latreille, 1809, monobasic.

This subgenus consists of rather elongate (7.0-10.5 mm long), parallel-sided, chalicodomiform species with yellow markings on all tagmata. The clypeus is produced over the mandibles in both sexes, but more prominently in females; the deep clypeal emargination is less evident or even almost absent in males. The scutellum is half as long as broad, the posterior margin seen from above uniformly curved and continued uninterruptedly by the margin of the axilla. Except at the extreme sides, close to the postspiracular foveae, the narrow basal zone of the propodeum is finely roughened, not divided into shining pits, and nearly vertical. On the hind tibia, the usual two spines found in many *Stelis* are united by a flared margin, forming a subtruncate marginal projection rather than two separate spines. Subgeneric characters were treated in detail by Noskiewicz (1961).

■ *Stelidomorpha* ranges widely in southern and central Europe (north to central Germany), North Africa (Morocco to Egypt), south to Kenya, and east to southwest Asia and Uzbekistan. There are three or four species.

As summarized by Westrich (1989), this subgenus parasitizes species of *Megachile,* including the *Chalicodoma* group.

Stelis / Subgenus *Stelis* Panzer s. str.

Trachusa Jurine, 1801: 164 (not Panzer, 1804). Type species: *Apis aterrima* Panzer, 1798, by designation of Morice and Durrant, 1915: 426. Suppressed by Commission Opinion 135, 1939 (Direction 4).
Stelis Panzer, 1806: 246. Type species: *Apis aterrima* Panzer, 1798 (not Christ, 1791) = *Apis punctulatissima* Kirby, 1802, monobasic.
Gyrodroma Klug in Illiger, 1807: 198; Klug, 1807b: 225. Type species: *Apis aterrima* Panzer, 1798 (not Christ, 1791) = *Apis punctulatissima* Kirby, 1802, designated by Sandhouse, 1943: 555. [Sandhouse incorrectly considered *Gyrodroma* to be monobasic; two species were listed by Klug in Illiger, 1807, which has page priority over Klug, 1807b.]
Gymnus Spinola, 1808: 9. Type species: *Apis aterrima* Panzer, 1798 (not Christ, 1791) = *Apis punctulatissima* Kirby, 1802, monobasic.
Ceraplastes Gistel, 1848: x, unjustified replacement for *Stelis* Panzer, 1806. Type species: *Apis aterrima* Panzer, 1798 (not Christ, 1791) = *Apis punctulatissima* Kirby, 1802, autobasic.
Chelynia Provancher, 1888: 322. Type species: *Chelynia labiata* Provancher, 1888, monobasic.
Melanostelis Ashmead, 1898: 283. Type species: *Melanostelis betheli* Ashmead, 1898 = *Stelis rubi* Cockerell, 1898, by original designation.

Stelidium Robertson, 1902b: 323. Type species: *Stelidium trypetinum* Robertson, 1902, monobasic. [See Michener, 1997b.]

Microstelis Robertson, 1903a: 170, 175. Type species: *Stelis lateralis* Cresson, 1864, by original designation.

Stelis (Pavostelis) Sladen, 1916b: 313. Type species: *Stelis montana* Cresson, 1864, monobasic.

Stelis (Stelidina) Timberlake, 1941a: 131. Type species: *Stelis hemirhoda* Linsley, 1939, by original designation.

Stelis (Stelidiella) Timberlake, 1941a: 133. *Lapsus* for *Stelidina* Timberlake, 1941.

Stelis (Leucostelis) Noskiewicz, 1961: 126, 132. Type species: *Gyrodroma ornatula* Klug, 1807, by original designation (p. 132).

This is the principal subgenus of *Stelis*. There is usually a carina on the pronotal lobe, but elsewhere no carinae on the body. The axillae are sometimes angulate posteriorly. The body length is 4 to 11 mm. The coloration in the palearctic region is black, sometimes with cream-colored or yellow spots or tergal margins or bands on the metasoma. In the nearctic area the diversity among species is greater than elsewhere; some are wholly black or with cream or yellow markings, the metasoma sometimes red, while other species are strongly metallic blue or green, with or without cream or yellow markings. In North America, in certain species usually placed in the subgenus *Chelynia*, the second recurrent vein meets the second submarginal cell instead of lying beyond the apex of that cell as in most Anthidiini. In *S. (S.) labiata* (Provancher), the type species of *Chelynia*, however, the venation is intermediate, the second recurrent vein being at about the apex of the second submarginal cell. Likewise, the presence and development of postspiracular foveae varies. The various quite different groups of North American *Stelis* appear to merge, it is therefore not practical to recognize subgenera (T. Griswold, personal communication, 1992). Sterna of various species were illustrated by Mitchell (1962); together with the genitalia they show considerable homogeneity among the species of *Stelis* s. str.

■ Species of this subgenus are found almost throughout Europe and northern Africa, including the Canary Islands, and east at least to Mongolia. In North America *Stelis* s. str. ranges across the continent and from Canada south to the northern border states of Mexico. There are some 20 Eurasian species and 55 North American species. Warncke (1992a) gave a key to the west palearctic species, including species of other subgenera.

Hosts are other megachilids, species of *Anthidiellum, Anthidium, Ashmeadiella, Chelostoma, Heriades, Hoplitis, Megachile,* and *Osmia*. So far as is known, a *Stelis* egg is introduced while a cell is being provisioned by the host, or after it is provisioned but not yet closed. After hatching, the resultant *Stelis* larva kills the host egg or larva (Michener, 1955; Rust and Thorp, 1973). As with some other cleptoparasitic bees such as *Nomada*, two or more eggs are often found in a single host cell. Although in many other cleptoparasitic bees such as the Nomadinae, young larvae kill the host egg or larva, it may be that in *Stelis* it is older larvae that are most likely to do so (Rozen, 1987a).

Genus *Trachusa* Panzer

This genus consists of moderate-sized to large megachiliform anthidiines (Pl. 8; Fig. 82-15) without the usual carinae or with carinae or even lamellae on the pronotal lobe and the omaulus. The posterior part of the head is well-developed, the lateral ocellus thus nearer to the eye than to the preoccipital margin, except in the subgenus *Metatrachusa*, in which these distances are about equal. The middle tibia is broad, usually nearly as broad as the hind tibia, and both anterior and posterior margins are convex; sometimes, however, in spite of these convexities, the middle tibia is distinctly narrower than the hind tibia. Vein cu-v of the hind wing is oblique and usually nearly half as long as the second abscissa of M+Cu or longer. In some *Metatrachusa*, however, it is distinctly less than half as long as the abscissa of M+Cu, although oblique. T7 of the male is small, not strongly exserted, and flexed forward such that its morphologically dorsal surface is directed ventrad; sometimes (e.g., in subgenera *Archianthidium* and *Metatrachusa*) T7 has a basal or median projection a small area of which is directed apicad or dorsad. Several characters that are often stable within a genus vary in *Trachusa*. Arolia are commonly present but sometimes greatly reduced or absent. Combs on S4 and S5 of the male are sometimes present but also often absent, and may vary even within a species (of the subgenus *Heteranthidium*; see Brooks and Griswold, 1988). The number of mandibular teeth of the female is commonly four but may be three, as in *Trachusa* s. str., or five, six, or seven, as in the subgenera *Massanthidium* and *Congotrachusa*. The number of segments in the maxillary palpus may be four or reduced to three.

So far as is known, species of *Trachusa* nest in the ground; unlike most anthidiines they make their own burrows. Cells are made from resin and pieces of green leaves by bees of the subgenera *Trachusa* s. str. and *Trachusomimus* (Hachfeld, 1926; Michener, 1941b; Westrich, 1989). In the subgenus *Heteranthidium*, cells are made of resin (Cane, 1996).

Trachusa is found in the holarctic, oriental, and African regions. Keys to the subgenera are separated for the Old World and New World forms; no subgenus occurs in both hemispheres.

Key to the Subgenera of *Trachusa* of the Eastern Hemisphere

1. Mandible of female with three more or less equidistant teeth; maxillary palpus as long as maximum width of galea, four-segmented (yellow markings absent except for face of male) (palearctic) .... *T. (Trachusa* s. str.*)*
—. Mandible of female four- to seven-toothed; maxillary palpus shorter than width of galea, three- or four-segmented 2
2(1). Second recurrent vein entering second submarginal cell basal to second submarginal crossvein; T7 of male with median basal projection (palearctic) .... *T. (Archianthidium)*
—. Second recurrent vein meeting or distal to second submarginal crossvein; T7 of male without basal projection 3
3(2). Subantennal suture distinctly arcuate outward; gonoforceps of male deeply bifid, Y-shaped (male unknown in *Orthanthidium*) 4

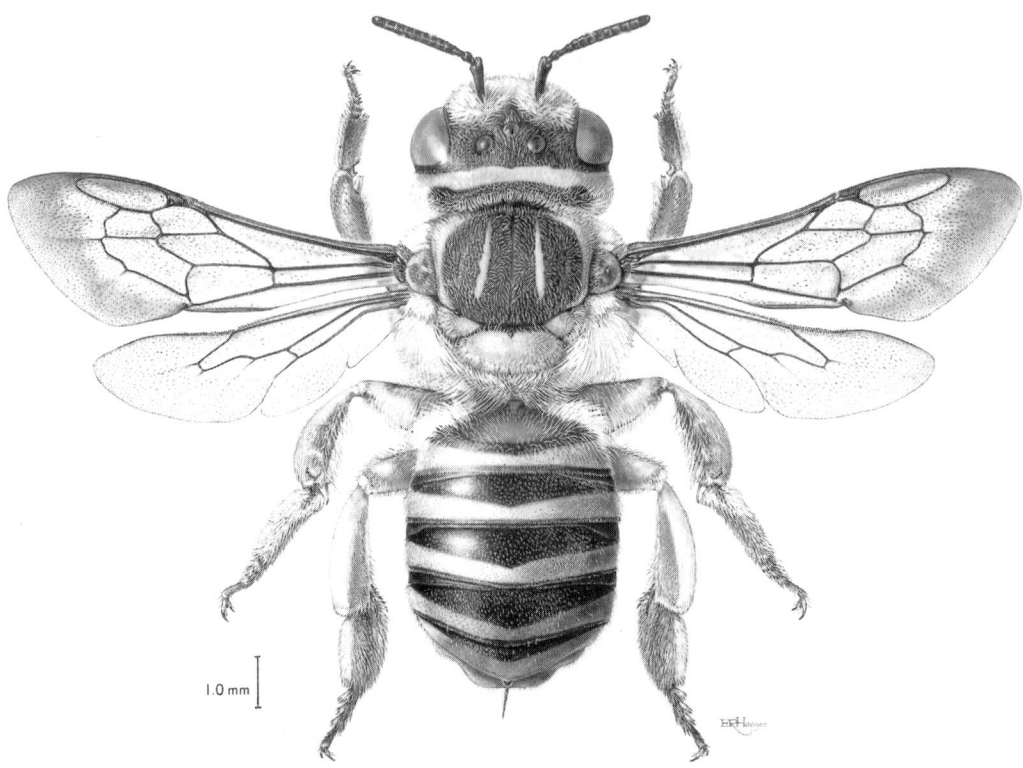

—. Subantennal suture nearly straight; gonoforceps of male not Y-shaped 5

4(3). Scutellum truncate posteriorly as seen from above, margin curved sharply forward laterally to meet longitudinal margin of axilla; eyes of female conspicuously diverging below; interocellar distance of female much less than half the ocellocciptal distance (male unknown) (oriental) .... *T. (Orthanthidium)*

—. Scutellum broadly rounded but medially emarginate; lateral margin of axilla convex but in general slanting; eyes of female subparallel; interocellar distance about half of ocellocciptal distance or somewhat greater (palearctic, oriental) .... *T. (Paraanthidium)*

5(3). Omaulus distinctly carinate, carina extending onto ventral surface of thorax close to middle coxa (Africa) .... *T. (Massanthidium)*

—. Omaulus not carinate 6

6(5). Arolia present in female, absent in male; S4 and S5 of male without combs (oriental) .... *T. (Metatrachusa)*

—. Arolia minute if present; S4 and S5 of male with combs as in *Paraanthidium* (Africa) .... *T. (Congotrachusa)*

Key to the Subgenera of *Trachusa* of North America

1. All tagmata with abundant yellow or cream-colored markings, including median tergal bands on metasoma .... *T. (Heteranthidium)*

—. Integument without yellow or cream-colored markings except usually on face of male and, in *Legnanthidium*, on apical metasomal tergal margins 2

2(1). Arolia absent .... *T. (Ulanthidium)*

—. Arolia present 3

3(2). Second recurrent vein joining second submarginal cell basal to second submarginal crossvein; mandible of female four-toothed .... *T. (Legnanthidium)*

—. Second recurrent vein distal to second submarginal crossvein; mandible of female three-toothed .... *T. (Trachusomimus)*

Figure 82-15. *Trachusa (Heteranthidium) bequaerti* (Schwarz), female. Drawing by E. R. S. Hodges, from Michener, McGinley, and Danforth, 1994.

Trachusa / Subgenus *Archianthidium* Mavromoustakis

Archianthidium Mavromoustakis, 1939: 91. Type species: *Anthidium laticeps* Morawitz, 1873, by original designation.

Although this subgenus was not included in *Trachusa* by Pasteels (1969a), it falls easily within that genus. Superficially, because of its size (12.5-18.0 mm long), form, and abundant yellow markings, it resembles *Heteranthidium* and well-marked species of *Paraanthidium*, although, unlike those on the former, the yellow metasomal bands at least on T1 are broken medially. *Archianthidium* is indeed similar to and probably closely related to *Heteranthidium*, as indicated especially by the large, volsella-like process from the apex of the male gonocoxite. All of its characters fall within the range of variation for *Heteranthidium* except the broken metasomal bands, the position of the second recurrent vein (see the key to subgenera), and the midbasal, retrorse, blunt or truncate process of T7 of the male. Arolia are present in both sexes, and the female mandible often does not show four recognizable teeth, the upper two being united in a sometimes undulate margin. Male genitalia were diagrammed by Mavromoustakis (1939).

■ *Archianthidium* occurs in the Mediterranean region from Spain and Morocco north to Hungary and east through southwestern Asia to Baluchistan (in Pakistan) and Central Asia. Pasteels (1969a) recognized six to probably eight species.

Trachusa / Subgenus Congotrachusa Pasteels

Trachusa (Congotrachusa) Pasteels, 1969a: 24. Type species: *Anthidium schoutedeni* Vachal, 1910, by original designation.

Congotrachusa is based on a single remarkable species. It has abundant yellow markings, including an unbroken tergal band on T1 and broken ones on T2 to T4. Body length is 13.5 mm. The mandible of the female is broad, with six teeth. Unique features are (1) the hind basitarsus of the female, which is broader than the tibia; (2) the strongly protuberant clypeus of the male, the protuberant part being broad above, almost pointed below, and the surface lateral to this triangle being excavated; and (3) a long, slender spine arising from the anterior part of the hypostomal carina of the male.

■ This subgenus is known from Congo (Brazzaville), Congo (Kinshasa), and "Portuguese Congo" = ?Cabinda. The single species is *Trachusa schoutedeni* (Vachal).

Trachusa / Subgenus Heteranthidium Cockerell

Anthidium (Heteranthidium) Cockerell, 1904b: 292. Type species: *Anthidium dorsale* Lepeletier, 1841, by original designation.

This is a rather variable subgenus with abundant yellow or cream-colored maculations (Pl. 8), the tergal bands being continuous but gradually narrowed medially. Length ranges from 10 to 15 mm. Arolia are well developed in some, but in the group of *Trachusa occidentalis* (Cresson) they are often reduced and are absent in most females. The comb on S4 of males can be present or absent, even within the same species, *T. occidentalis*. Male genitalia and sterna were illustrated by Mitchell (1962), Snelling (1966d), and Brooks and Griswold (1988).

■ *Heteranthidium* ranges from New Jersey, South Dakota and Oregon southward to Florida and California, USA, and through Mexico to the state of Chiapas. There are 13 species. Keys to the species were given by Snelling (1966d) and Brooks and Griswold (1988).

Trachusa / Subgenus Legnanthidium Griswold and Michener

Trachusa (Legnanthidium) Griswold and Michener, 1988: 27. Type species: *Anthidium ridingsii* Cresson, 1878, by original designation.

This subgenus is noteworthy for lacking yellow or cream markings of the usual sort except on the face of the male, but having, instead, cream-colored or sometimes merely translucent bands on the posterior marginal zones of the metasomal terga, reminding one of *Afranthidium (Immanthidium)*. The body length is 12 to 13 mm. Disregarding the character of forewing venation used in the key to subgenera, *Legnanthidium* runs to the vicinity of *Ulanthidium*, from which it differs in the presence of arolia, and of *Heteranthidium*, from which it differs in the lack of yellow markings on the thorax and metasoma. As shown by Mitchell (1962), the male genitalia are quite different from those of *Heteranthidium*, the basal half of the gonostylus being extremely slender and the volsella or equivalent lobe, which is very large in *Heteranthidium*, being absent.

■ *Legnanthidium* is known from the southern USA, from North Carolina to Florida and Texas. The only species is *Trachusa ridingsii* (Cresson).

Trachusa / Subgenus Massanthidium Pasteels

Trachusa (Massanthidium) Pasteels, 1969a: 24. Type species: *Trachusa flavorufula* Pasteels, 1969, by original designation.

Massanthidium, known only from females, consists of largely dark species (in one the metasoma is yellow-orange), the cream-colored markings restricted and on the metasoma limited to the sides of the terga. The body length is 11 to 14 mm. The female mandible is broad, and has five to seven teeth. *Massanthidium* resembles *Paraanthidium* in its lamellate pronotal lobe and the carinate omaulus (not always in *Paraanthidium*, in which the carina does not extend ventrally as in *Massanthidium*); it differs from *Paraanthidium* in the minute arolia (doubtfully present) and the straight subantennal sutures. An unusual feature is the location of the omaular carina on the thoracic venter, extremely close to the middle coxa. Such a position for it is otherwise known among unrelated forms with an omaular lamella—*Pachyanthidium* and *Anthidiellum*. The relationships of *Massanthidium* will remain in doubt until males are found.

■ *Massanthidium* is known from Kenya and Eritrea. Pasteels (1984) described three species. An additional species is represented by a specimen from Namibia.

Trachusa / Subgenus Metatrachusa Pasteels

Trachusa (Metatrachusa) Pasteels, 1969a: 22. Type species: *Anthidium pendleburyi* Cockerell, 1927, by original designation.

Metatrachusa includes some of the smallest species of *Trachusa*, 8.5 to 11.5 mm long. Their pubescence is largely black and the integument is black except for yellow areas on the face of males and some females and narrow yellow bands on at least the more apical metasomal terga of females. (I have seen only the male of *Trachusa orientalis* Pasteels and the female of *T. pendleburyi* Cockerell.) Arolia are present in the female, absent in the male. The omaulus lacks a carina; the pronotal lobe is carinate in the male but not in the female (probably this should read "carinate in *T. orientalis* but not in *pendleburyi*"). The mandible of the female is five-toothed. T7 of the male has a median (not marginal) discal tooth directed somewhat basad, not, as in *Archianthidium*, arising from the extreme base and directed strongly basad.

■ This subgenus is known from Malaysia. The two species were distinguished and described by Pasteels (1972).

Trachusa / Subgenus *Orthanthidium* Mavromoustakis

Paraanthidium (Orthanthidium) Mavromoustakis, 1953a: 837. Type species: *Anthidium formosanum* Friese, 1917, by original designation.

This subgenus contains a very large (18-20 mm) species, known only in the female, that may be a *Paraanthidium*. The thorax and metasoma have buff-colored maculations, those of the metasomal terga being unbroken and limited to or strongest on T1 and T2. The omaulus is carinate and the pronotal lobe lamellate, as in some *Paraanthidium*. There are five mandibular teeth, as in most *Paraanthidium*. Otherwise, the characters are as in that subgenus, except as noted in the key to subgenera. The male will be decisive in the placement of *Orthanthidium*.

■ This subgenus occurs in southeastern China and in Taiwan. The only species is *Trachusa formosana* (Friese).

Trachusa / Subgenus *Paraanthidium* Friese

Anthidium (Paraanthidium) Friese, 1898a: 101. Type species: *Apis interrupta* Fabricius, 1781, by designation of Cockerell, 1909b: 269.
Protanthidium Cockerell and Cockerell, 1901: 49. Type species: *Megachile steloides* Bingham, 1896 (preoccupied in *Anthidium*; see Michener and Griswold, 1994a) = *Anthidium longicorne* Friese, 1902, by original designation.
Protoanthidium Cameron, 1902b: 125. Type species: *Protoanthidium rufobalteatum* Cameron, 1902, by designation of Sandhouse, 1943: 591.
Trachusa (Philotrachusa) Pasteels, 1969a: 22. Type species: *Anthidium aquiphilum* Strand, 1912 (= *Trachusa aquifilum*, Pasteels, 1969, misspelling), by original designation.

Paraanthidium (not to be confused with the North American genus *Paranthidium*) contains species 11 to 14 mm long, of diverse coloration. Some, including the type species, *Trachusa interrupta* (Fabricius), as well as *T. aquiphilum* (Strand), have extensive yellow markings and look like bees of the subgenus *Heteranthidium* of the nearctic region, except that the more anterior yellow metasomal bands are often broken. *Trachusa ovata* (Cameron) from Borneo, however, lacks pale markings and is covered with black hair, suggesting the subgenus *Metatrachusa*. *Trachusa longicornis* (Friese), the type species of *Protanthidium*, is intermediate, having yellow bands only on the more posterior terga (T5 and T6 of the female). *Paraanthidium* has more conspicuously arcuate subantennal sutures than any other subgenus of *Trachusa*, though it is approached in this character by *Orthanthidium*. The arolia are present. The omaular carina is also present, the upper part lamellate in *T. longicornis* (Friese), but in some males [e.g., *T. interrupta* (Fabricius)], the carina is absent. The carina of the pronotal lobe is present, sometimes lamellate. The female mandible is four- or five-toothed. T4 and T5 of the males have black combs. A very distinctive feature is the bifid, Y-shaped gonoforceps of the male, as diagrammed by Mavromoustakis (1939) and Warncke (1980a).

■ *Paraanthidium* occurs in the Mediterranean region from Morocco and Spain eastward to Macedonia and Turkey, and onward to Sikkim, Borneo, and southeastern China. One species occurs in Namibia and South Africa. Four species are known from the oriental region (Pasteels, 1972); two other species occur in Europe, North Africa, and southwestern Asia (Warncke, 1980a).

The subgenus *Orthanthidium*, known only from females, could be considered a synonym of *Paraanthidium*. If the male of *Orthanthidium* proves to have the characteristic features of *Paraanthidium*, the two subgenera should be united.

Trachusa / Subgenus *Trachusa* Panzer s. str.

Trachusa Panzer, 1804b: expl. pl. 14-15. Type species: *Trachusa serratulae* Panzer, 1804 = *Apis byssina* Panzer, 1798, by designation of Sandhouse, 1943: 605. [For *Trachusa* Jurine, 1801, see *Stelis*.]
Diphysis Lepeletier, 1841: 307. Type species: *Diphysis pyrenaica* Lepeletier, 1841 = *Apis byssina* Panzer, 1798, monobasic.
Megachileoides Radoszkowski, 1874b: 132. Type species: *Trachusa serratulae* Panzer, 1804 = *Apis byssina* Panzer, 1798, by designation of Michener, 1995a: 375.
Megachiloides Saussure, 1890: 35, incorrect subsequent spelling of *Megachileoides* Radoszkowski, 1874; see Michener, 1995a.

This subgenus consists of a species 10 to 12 mm long with pale markings only on the face of the male, and with abundant, rather long hair. It thus superficially resembles a small *Anthophora* or a small, pale-haired species of *Megachile (Chalicodoma)*. Arolia are well developed, the omaulus and pronotal lobe are not carinate, and males have no sternal combs. Wing venation in the lone species, *Trachusa byssina* (Panzer), is variable; the second recurrent vein is basal to the second submarginal crossvein in some specimens, but is distal, as in most anthidiines, in others. The genitalia and hidden sterna were illustrated by Popov (1964).

■ This subgenus is found in the mountains and the north of Europe, north to 64°N, eastward to the Urals and the Caucasus and beyond to Siberia. The only species is *Trachusa byssina* (Panzer).

Trachusa / Subgenus *Trachusomimus* Popov

Trachusomimus Popov, 1964: 406. Type species: *Trachusa perdita* Cockerell, 1904, by original designation.

Trachusomimus is a nearctic relative of *Trachusa* s. str. It differs from *Trachusa* s. str. in its slightly larger size (13-14 mm long) and in the characters indicated in the key to subgenera. Popov (1964) pointed out other differences and concluded that the male genitalia of *Trachusa* and *Trachusomimus* are so different that these bees are not closely related. The similarities in other characters, e.g., the relatively long maxillary palpi, indicate close relationship, but these two genera probably do not constitute a monophyletic group, since the common characters are probable plesiomorphies. Male genitalia, sterna, and other structures were illustrated by Popov (1964) and Thorp and Brooks (1994).

■ This subgenus is found in coastal California, USA, and Baja California, Mexico. The two species were revised by Thorp (1963), Grigarick and Stange (1968), and Thorp and Brooks (1994).

Trachusa / Subgenus *Ulanthidium* Michener

Ulanthidium Michener, 1948a: 13. Type species: *Ulanthidium mitchelli* Michener, 1948, by original designation.

Ulanthidium (Olmecanthidium) Peters, 1972b: 377. Type species: *Ulanthidium interdisciplinaris* Peters, 1972, by original designation.

Like two other North American subgenera, *Trachusomimus* and *Legnanthidium*, this subgenus has yellow integumental markings only on the face of the male, and in *Trachusa interdisciplinaris* (Peters) even the face of the male is black. The size (11-14 mm long) is also similar to that of those two subgenera. A lack of arolia distinguishes *Ulanthidium* from the other, largely unmarked subgenera. The male gonostyli are much less elaborate than those in *Heteranthidium* and *Legnanthidium;* see Peters (1972b), Michener and Ordway (1964), Thorp and Brooks (1994), and Figure 82-2a-c for illustrations of the genitalia and other structures. In some species the second recurrent vein meets the second submarginal crossvein, a feature unusual among anthidiines.

■ *Ulanthidium* ranges from southern Arizona and New Mexico, USA, south to the states of Jalisco, Oaxaca, and Puebla, Mexico. The six species were revised by Thorp and Brooks (1994).

Genus *Trachusoides* Michener and Griswold

Trachusoides Michener and Griswold, 1994a: 324. Type species: *Trachusoides simplex* Michener and Griswold, 1994, by original designation.

This genus contains a single, large, megachiliform species (length 12 mm) similar in form to *Apianthidium* and *Trachusa*. The head of the female (the male is unknown) is black, the thorax has limited pale-yellow markings, and the metasoma is yellowish red. A unique character is the simple female claws, breaking down an otherwise constant tribal character of cleft or toothed claws in female Anthidiini. Except for a strong carina on the pronotal lobe, there are no carinae on the head and thorax, and there are no basal or postspiracular pits or foveae on the propodeum. The middle tibia (female) is narrower than the hind tibia, and its anterior and posterior margins are equally and symmetrically convex as seen laterally, the apex of the tibia as narrow as its base; these tibial characters differentiate it from both *Trachusa* and *Apianthidium*.

■ *Trachusoides* is found in southern India and known from a single female of *T. simplex* Michener and Griswold.

Genus *Xenostelis* Baker

Xenostelis Baker, 1999: 232. Type species: *Xenostelis polychroma* Baker, 1999, by original designation.

Xenostelis was based on a female specimen that cannot now be placed in any other anthidiine genus. At couplet 21 in the key to Old World genera of Anthidiini, there may be a problem because the two apical spines of the front and middle tibiae may not be homologous to those of *Stelis*. However, there are two spines and one can go to couplet 28. The large tegulae are similar in *Afrostelis* and *Xenostelis*. Aside from the tegulae, one of the main features that led to acceptance of *Afrostelis* as a genus is its male genitalia, quite unlike those of *Stelis*. The genitalic characters of *Xenostelis* are unknown because the male is unkown. Thus the status of *Xenostelis* remains in doubt; if its genitalia are similar to those of *Afrostelis*, *Xenostelis* would be best placed as a synonym or subgenus of *Afrostelis*. However, its less elongate thorax and presence of yellow markings (sides of T2 and T3, discs of T5 and T6) are *Stelis*-like features.

■ This genus is known only from the island of Socotra, Yemen. The only species is *Xenostelis polychroma* Baker.

83. Tribe Dioxyini

Thanks to the careful work of Popov (1936a, 1947, 1953), much is known about the morphological peculiarities and diversity of this tribe. Dioxyines are cleptoparasitic bees without a scopa, and with the metasoma of the female somewhat tapering posteriorly. Although in some genera the apex of the female metasoma is slender and pointed, the tapering of the whole metasoma (Fig. 83-4) is frequently less conspicuous than that in *Coelioxys,* a parasitic megachiline genus. The extremely long (much longer than the mandible), parallel-sided, truncate labrum, lacking long hairs, and the strong preoccipital carina are other tribal characters. A unique feature of the tribe among megachilids is the median tubercle or spine on the metanotum (Fig. 83-1); two genera, nonetheless, lack this structure. (A similar structure exists in the nomiine genus *Mellitidia.*) A dioxyine feature that is very rare in megachilids, although common in many S-T bees, is the distinct dorsolateral angle of the pronotum, from which a vertical ridge descends; this structure, however, is entirely absent in *Prodioxys*. The rather small, slender bidentate mandible of the female, similar to that of the male, is not found in other Megachilinae. It could be an ancestral feature or an apomorphy acquired by the female following the male's developmental pattern. The greatly reduced sting and associated structures are unique among bees. The reduction is far more extreme than that in Meliponini (Apidae), the sting itself being essentially absent. The structures involved were illustrated for various dioxyine genera by Popov (1953). Such reduction is remarkable because in most parasitic bees the sting is as large as or commonly larger than that of their nonparasitic relatives.

The Dioxyini have sometimes (Michener, 1944) been included in the Anthidiini, with which they agree in their stigmal and prestigmal characters (Fig. 83-2) and the cleft claws of the female. Dioxyini could be derived from Anthidiini, which would make the latter a paraphyletic group, but evidence for such a relationship is lacking; Anthidiini and Dioxyini are more likely to be sister groups. One derived anthidiine character not shared by the Dioxyini is the position of the second recurrent vein beyond the second submarginal cell. There are exceptions in the Anthidiini, as subgeneric or specific characters in unrelated genera, in *Afranthidium (Nigranthidium), Aspidosmia, Trachusa (Archianthidium* and *Legnanthidium),* and some species of *Anthidium* and *Stelis.* No doubt these exceptions are reversals to the usual megachilid character state found in Lithurgini, Megachilini, and Osmiini, as well as Dioxyini. Gogala (1995a) made a phylogenetic study of major groups of Megachilinae and concluded that *Dioxys* was the basal branch, i.e., sister group to all other Megachilinae. Some of the characters used in the study are highly variable and some were incorrectly polarized; moreover, no outgroup was included. The distinctness of the Dioxyini from the Anthidiini and other tribes, however, is supported.

Relationships among the dioxyine genera are not clear, but were discussed by Michener (1996b). *Allodioxys* may be the first branch, i.e., sister group to all other Dioxyini; *Aglaoapis* or *Metadioxys* is the next. The genera *Dioxys, Paradioxys, Prodioxys, Eudioxys,* and *Ensliniana* share elongate derived male genitalia and probably constitute a derived monophyletic group.

Armature of the thorax, although common in parasitic bees and perhaps related to defense of the neck and petiolar regions against attacks by the hosts, is remarkable in Dioxyini because spines or carinae have arisen on (and been lost from?) diverse sclerites. In different genera, spines occur on the posterior lateral angles of the scutum, the axillae, the sides of the scutellum, and the middle of the metanotum. No one of these sites is spined in every dioxyine genus. Further, the pronotal lobes may be simple or carinate, even lamellate in one species of *Aglaoapis.* Table 83-1 shows the distribution of these structures by genus.

Summaries of our knowledge of the hosts of Dioxyini were published by Popov (1936a, 1953) and Hurd (1958a). The hosts are other megachilids of the tribes Megachilini, Osmiini, and Anthidiini. The dioxyine parasites are not very host-specific; various species have been reared from cells of two of the tribes listed above, and *Dioxys pomonae* Cockerell has been reared from cells of all three. None of the Old World species appears to have been reared from Anthidiini, however. Rozen (1967b) noted that larvae of *Dioxys* in the first three stages have simple, sickle-shaped mandibles and presumably may kill host larvae in any of these stages (in the Nomadinae, only the first-stage larvae are equipped for killing). Immature stages of the palearctic *Dioxys cincta* (Jurine) were de-

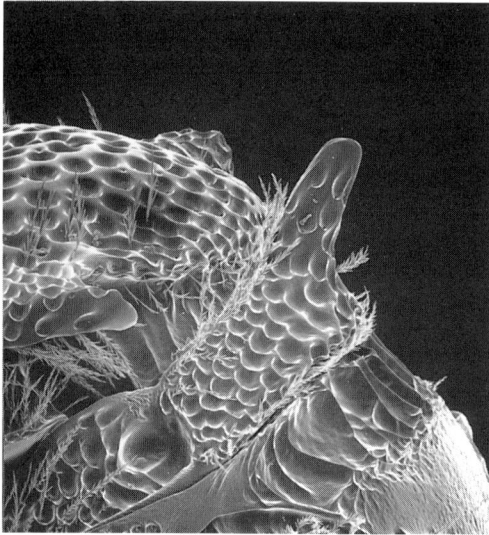

Figure 83-1. Lateral view of metanotum, showing median metanotal spine, of *Dioxys productus cismontanicus* Hurd. The coarsely punctured scutellum is at the left, the propodeum at the right. From Michener, McGinley, and Danforth, 1994.

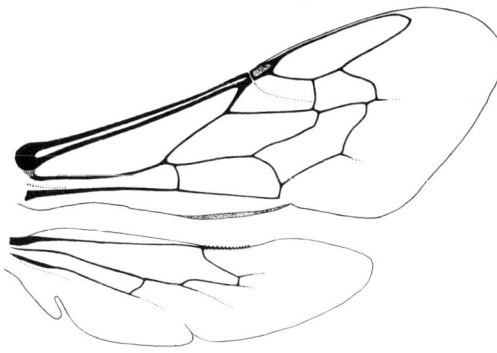

Figure 83-2. Wings of *Dioxys productus subruber* (Cockerell).

scribed by Rozen and Ozbek (2004), who noted various differences from the nearctic *D. pomonae* as well as pharate first instar larvae, frequently unrecognized in earlier accounts. Eggs of *Dioxys pomonae* are inserted through the cell wall of the host after the host female has closed the cell (Rozen and Favreau, 1967), but eggs of *D. cincta* (Jurine) appear to be deposited in open host cells (Rozen and Osbek, 2005a).

Keys to the species of Dioxyini of the western palearctic region were provided by Warncke (1977b); for the nearctic region by Hurd (1958a).

Key to the Genera of the Dioxyini
(From Michener, 1996b)

1. Scutellum with posterior margin rounded 2
—. Scutellum produced laterally to form posteriorly directed tooth or spine (Fig. 83-3a) 3
2(1). Metanotum with median tooth; posterior lateral angle of scutum produced to a strong spine (as in Fig. 83-1); S5 of female without teeth (palearctic) .... *Allodioxys*
—. Metanotum without or with scarcely developed median tubercle or tooth; posterior lateral angle of scutum not produced; S5 of female with tooth on each side of base of S6 (palearctic) .... *Ensliniana*
3(1). Axilla produced to a strong, curved spine; metanotum without median tubercle or tooth (palearctic) ...... *Eudioxys*
—. Axilla not spined or produced to a small, straight spine that lies close to and usually slightly below lateral margin of scutellum; metanotum with median tubercle or tooth, rarely weakly developed 4
4(3). Dorsolateral angle of pronotum unrecognizable; first recurrent vein entering first submarginal cell or nearly meeting first transverse cubital vein; jugal lobe of hind wing about one-seventh as long as vannal lobe, as measured from wing base; body covered with long red hair, branches short and inconspicuous (palearctic) .... *Prodioxys*
—. Dorsolateral angle of pronotum distinct; first recurrent vein entering second submarginal cell; jugal lobe of hind wing about one-third as long as vannal lobe, as measured from wing base; body with shorter, usually pale hairs, some of them densely plumose and superficially almost scale-like 5
5(4). Scutellum with strong carina, arcuate posteriorly, between lateral teeth; exposed part of T5 of female 2.5 to 4.0 times as long as exposed part of T6, tapering posteriorly to subtruncate apex only about one-seventh as wide as base of tergum; T6 of female with exposed part two or more times as long as broad, smooth and shining, hairless; S6 of female needle-like, extending well beyond apex of T6 (palearctic) .... *Paradioxys*
—. Scutellum without carina between lateral teeth; exposed part of T5 of female shorter than exposed part of T6, similar in form to preceding terga, its apex over two-thirds as wide as base of exposed part; T6 of female with exposed part about as broad as long or broader than long, punctured and setose like other terga; S6 of female broad, not or slightly exceeding T6, thus no needle-like metasomal apex (Fig. 83-4) 6
6(5). Labrum without transverse basal carina; male genitalia as a whole more than twice as long as width of gonobase; front coxa with anterior surface rounded (holarctic) .... *Dioxys*
—. Labrum with transverse carina close to base (visible above closed mandibles); male genitalia as a whole much less than twice as long as width of gonobase (Fig. 83-3b); front coxa with tubercle or strong carina at summit of anterior surface 7
7(6). Front coxa with strong carina at summit of smooth anterior surface; metanotum with small median tubercle; third valvula (gonostylus) of female with row of coarse bristles (palearctic) .... *Metadioxys*
—. Front coxa with tubercle at summit of punctate anterior surface; metanotum with strong median spine (as in Fig.

Table 83-1. Positions of Spines and Carinae on Thoracic Sclerites of Dioxyini.
sp, spine; sm sp, small spine; c, carina; and l, lamella; —indicates no spine or carina present.

Taxon	Position				
	pronotum	scutum	axilla	scutellum	metanotum
Ensliniana	c	—	—	—	—
Eudioxys	c	—	sp	sp	—
Prodioxys	—	—	—	sp	sp
Paradioxys	c	—	—	sp	sp
Dioxys	c	—	— or sm sp	sp	sp
Allodioxys	c	sp	—	—	sp
Metadioxys	c	?	—	sp	sm sp
Aglaoapis	c	—	sm sp	sp	sp

83-1); third valvula (gonostylus) of female with only minute hairs (palearctic, India, Africa).................. *Aglaoapis*

Genus *Aglaoapis* Cameron

Aglaoapis Cameron, 1901b: 262. Type species: *Aglaoapis brevipennis* Cameron, 1901, monobasic.
Dioxoides Popov, 1947: 89. Type species: *Coelioxys tridentata* Nylander, 1848, by original designation.

In external features *Aglaoapis* is similar to *Dioxys*, although the labral and other characters indicated in the key to genera (couplet 6) are distinctive. The male genitalia and female sting apparatus, however, are very different from those of *Dioxys*. The male genitalia are broader than long (Fig. 83-3b). In dorsal view the rather robust, curved gonostylus is easily distinguished from the gonocoxite and is hairy throughout its length. The male genital and hidden sterna were illustrated by Popov (1936a) and Michener (1996b); see also Figure 83-3b-d. In the female, the large, discoid T7 hemitergite is quite different from that of all other Dioxyini [the female of *A. alata* (Michener) is unknown, and that of *A. brevipennis* Cameron has never been dissected]. Unlike those of *Dioxys*, the female gonostyli are covered with minute hairs (see Popov, 1953). The body length is 10 to 12 mm.

■ The distribution of *Aglaoapis* is disjunct, Europe and western Asia, western India, and South Africa. In the palearctic region it occurs north to 62°N in Finland, south to Spain, Italy, Greece, and Cyprus, and east to the Caucasus, the Urals, and Kazakhstan. In India it is known from Bombay. In Africa it is known only from the western Cape Province, South Africa; it is the only dioxyine from sub-Saharan Africa. Three species are recognized.

Genus *Allodioxys* Popov

Allodioxys Popov, 1947: 87. Type species: *Dioxys schulthessi* Popov, 1936, by original designation.

The most noteworthy character of *Allodioxys* is the posterior lateral angle of the scutum, which is produced as a strong curved spine, whereas the axilla is small, its lateral margin convex, not angulate. The scutal spines and the median metanotal spine are the only thoracic spines. The propodeum has a sloping basal zone with irregular, anastomosing rugae. The last three metasomal segments of the female taper strongly, the exposed part of T6 at its base being about or more than half as wide as the base of T5, S6 having only the slender, pointed apex exposed and exceeding T6. The male genitalia and S7 and S8 were illustrated by Popov (1936a: fig. 16); the short and robust probable gonostylus is much shorter than the gonocoxite, unlike that of all other Dioxyini. The body length is 6 to 11 mm.

■ This genus is found in Israel and North Africa (Algeria and Libya). Warncke (1977b) listed four species.

Genus *Dioxys* Lepeletier and Serville

Dioxys Lepeletier and Serville, 1825: 109. Type species: *Trachusa cincta* Jurine, 1807, monobasic.
Hoplopasites Ashmead, 1898: 284. Type species: *Phileremus productus* Cresson, 1879, by original designation.
Chrysopheon Titus, 1901: 256. Type species: *Chrysopheon aurifuscus* Titus, 1901, monobasic.

In this genus the arcuate ridge from the pronotal angle to the pronotal lobe and the omaulus are both strongly carinate. The axilla is rounded or slightly angulate, its posterior lateral angle acute and sometimes produced as a small spine close to or in contact with the lateral margin of the scutellum. The scutellum has a strong lateral tooth or spine. The metanotum has a median spine, laterally compressed. The basal zone of the propodeum, which slopes steeply, is demarcated posteriorly by a carina and divided by numerous rugae into a series of shiny pits (Fig. 83-1). T6 of the female is broad, sculptured like T5, and its apex is broadly rounded or slightly angulate to weakly emarginate, the exposed part usually longer than the exposed part of T5. The exposed part of S6, not or slightly exceeding T6, is similar in shape to that of T6. Thus although the metasoma of the female tapers, it lacks the needle-like apex of some other genera (Fig. 83-4). Popov (1936a) illustrated the male genitalia and hidden sterna of several species, and Popov (1953) illustrated the structures associated with the sting; the female gonostyli or third valvulae are much reduced and hairless, and the T7 hemitergites are triangular or quadrate. The body length is 6 to 12 mm.

■ *Dioxys* ranges from the Canary Islands, Morocco, and Spain eastward through northern Africa and Europe (as far north as 49°N), and eastward into Central Asia. It is not known in China and Japan, but occurs in western North America from Oregon eastward to Wyoming, Colorado, and western Texas, USA, south to Baja California and Durango, Mexico. There are about ten palearctic species and five nearctic species. Warncke (1977b) gave a key to west palearctic species and Hurd (1958a) revised the nearctic species.

Genus *Ensliniana* Alfken

Ensliniana Alfken, 1938: 431. Type species: *Ensliniana cuspidata* Alfken, 1938 = *Stelis bidentata* Friese, 1899, by original designation.

This genus is similar to *Allodioxys* but the posterior lateral angle of the scutum is not produced and the meta-

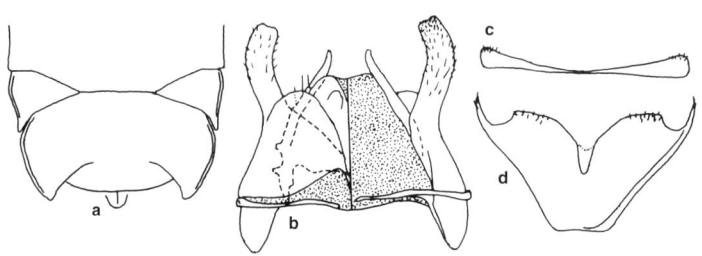

Figure 83-3. *Aglaoapis alata* (Michener). **a**, Dorsal view of axillae and scutellum; **b-d**, Male genitalia, S7, and S8, dorsal view of genitalia at left. From Michener, 1996b.

Figure 83-4. *Dioxys productus cismontanicus* Hurd. Above, Male; Below, Female. From Michener, McGinley, and Danforth, 1994.

notum has no median tubercle or spine. Thus the thorax is the least spinose of any dioxyine. T6 and S6 of the female, more elongate and slender than those of *Allodioxys*, are smooth and nearly hairless and curved downward; S5 has a preapical tooth at each side of the base of S6. S6 of the female is much longer than that of *Allodioxys*, the exposed part about as long as S5, sharply pointed, and much exceeding T6. The terminal structures of the female were illustrated by Popov (1936a: figs. 10, 11). The structures associated with the sting are much elongated; the third valvula (with minute hairs) is fused by way of a long neck to what is probably the second valvifer (see Popov, 1953). In the male genitalia, which are only moderately elongate, the gonocoxite and gonostylus are indistinguishably fused, as in most dioxyines. The body length is 8.0 to 8.5 mm.

■ This genus ranges from northern Africa (Algeria, Tunisia), Israel, and Syria to Turkmenistan. There are probably three species.

Genus *Eudioxys* Mavromoustakis

Eudioxys Mavromoustakis, 1963c: 443. Type species: *Dioxys quadrispinosa* Friese, 1899, by original designation.

This is the only dioxyine genus in which the axilla has a large, curved lateral spine, similar in form to and only slightly smaller than the lateral spine of the scutellum. The pronotum and omaulus are carinate as in *Dioxys*. The metanotum lacks a median spine; this is the only genus with scutellar spines but without a metanotal tubercle or spine. The basal zone of the propodeum is subhorizontal, delimited posteriorly by a carina, and divided by a few rugae into large, shiny pits. The exposed parts of T5 and T6 of the female are broader than long, and T6 has a blunt midapical projection; S6 is broadly rounded and extends slightly beyond the apex of T6. The male genitalia are shaped as in *Dioxys*, but the apices of the penis valves are simple, pointed. The body length is 6.5 to 7.0 mm. Male genitalia and various thoracic and metasomal structures were illustrated by Mavromoustakis (1968a).

■ This genus is known from Egypt and Iran. The two species were compared by Mavromoustakis (1968a).

Genus *Metadioxys* Popov

Metadioxys Popov, 1947: 88. Type species: *Dioxys formosa* Morawitz, 1875, by original designation.

Like *Aglaoapis*, *Metadioxys* is similar externally to *Dioxys*. The male genitalia and female sting apparatus, however, are very different from those of both *Dioxys* and *Aglaoapis*. The male genitalia are slightly longer than broad. The gonostylus is indistinguishably fused to the gonocoxite, and the slender, curved gonostylar part, which is pointed apically and bears short hairs, is only about as long as the broad gonocoxal part. Popov (1936a: figs. 12, 13) illustrated the genitalia and hidden sterna of two species. The female sting-associated sclerites include the rather large, tapering gonostylus or third valvula, with a row of coarse setae, and the large T7 hemitergite, which has a lateral process (Popov, 1953: fig. 4). The body length is from 10 to 11 mm.

■ The range of this genus is from Morocco and Greece east through southwestern Asia to Uzbekistan. There are three species. A key was provided by Warncke (1977b).

Genus *Paradioxys* Mocsáry

Paradioxys Mocsáry, 1894: 35. Type species: *Dioxys pannonica* Mocsáry, 1877, monobasic.

This genus is similar to *Dioxys* but is separated primarily because of the very different apex of the female metasoma. *Paradioxys* differs from *Dioxys* (see the comments on that genus) as follows: The pronotal and omaular carination is weak, but the posterior scutellar margin between the teeth is more carinate than is usual in *Dioxys*; the basal zone of the propodeum has more numerous rugae than are usual in *Dioxys*. And unlike those of *Dioxys*, the apical metasomal terga of the female taper strongly; T5 has a smooth, shiny subtruncate apex; T6 is smooth, shiny, hairless, the exposed part two or three times as long as broad, less than half as long as the exposed part of T5, narrowly rounded at the apex; the exposed part of S6, extending beyond the apex of T6, is needle-like; and S6 and T6 are both straight, not curved downward as in the similarly tapering apical segments of *Ensliniana*. The sting-associated structures of the female are elongate and hairless; the third valvulae are very slender and attached to the end of what is probably a long, nearly straight, longitudinal second valvifer. The male genitalia are similar to those of *Dioxys* but more elongate and slender (see Popov, 1936a: fig. 9). The body length is about 7.5 to 9.0 mm.

■ *Paradioxys* is known from Austria, Hungary, Israel, and Iran. There are two species.

Genus *Prodioxys* Friese

Prodioxys Friese, 1914a: 221. Type species: *Prodioxys cinnabarina* Friese, 1914 = *Dioxys longiventris* Pérez, 1895, monobasic.

This genus contains the only conspicuously hairy dioxyines; on the head, thorax, and T1 are long, erect red hairs that have short, sparse branches; the remaining metasomal hairs are also relatively long and red. There are none of the short, densely plumose, almost scale-like hairs found at least locally on most other dioxyines. The lateral pronotal angle, present in all other dioxyines, is absent. There is no sharp curved ridge from the angle to the lobe of the pronotum, and the lobe as well as the omaulus are rounded and completely lack carinae. The lateral scutellar tooth or spine is of only moderate size, but there is a strong, sharp median metanotal spine. The propodeum has a steeply sloping basal zone with irregular anastomosing rugae. The front wing is unique for the tribe in that the first recurrent vein enters the first submarginal cell, sometimes very near its apex; the hind wing is also unique for the tribe in the very small size of its jugal lobe (see the key to genera). T5 of the female tapers strongly; T6 is small, the exposed part shorter than T5, the apex rounded; the exposed part of S6, narrowly rounded apically, is similar in size to T6 but somewhat narrower, and not or scarcely exceeding T6. The male genitalia are elongate, similar to those of *Dioxys*. The body length is 9 to 11 mm.

■ *Prodioxys* occurs from Algeria to Egypt and Israel. Warncke (1977b) recognized three species.

84. Tribe Megachilini

This vast tribe is distinguished by the characters indicated in the key to tribes and in Table 79-1. The more elongate stigma that differentiates the Megachilini from the Anthidiini is shown in Figure 84-1. In numerous illustrations of structures of Megachilini, Mitchell (1934-1980) and Pasteels (1965-1982) gave an idea of the diversity within the tribe. The aspect of some Megachilini is indicated by Figures 8-3, 8-7, 84-2, 84-3, 84-8, and 84-9, and by Plate 7.

The Megachilini are found in large numbers on all continents. And although the number of genera recognized is small, compared to those recognized in the Osmiini and Anthidiini, the morphological diversity within the major genus *Megachile* is great, perhaps as great as that found in all the Anthidiini. Nonetheless, the tribe Megachilini (thus *Megachile*) is not as readily divisible into distinctive types as are the other two tribes. In Anthidiini in particular, the presence or absence of various carinae and lamellae provide convenient characters that led to the tradition of recognizing numerous genera. In Megachilini, the tradition has been to recognize only one nonparasitic genus *(Megachile)* or three *(Chalicodoma, Creightonella,* and *Megachile;* Michener, 1962a), though Mitchell (1980) added several other genera to the list. Although my original manuscript for this group used several nonparasitic genera, I reluctantly retreated to the single genus *Megachile* for reasons that will be explained below.

Brood cells of the Megachilini made of leaf pieces (usually not masticated leaves as is common in Osmiini) are common in burrows in the ground or in cavities in wood, stems, or manmade objects. Other species make cells of mud, probably combining salivary material, or of resin, either in cavities or exposed on the surfaces of stones or walls or on branches of bushes or trees.

Cleptoparasitic species, mostly *Coelioxys,* are numerous; most of them attack bees of the related genus *Megachile,* but a few parasitize distantly related bees in the family Apidae.

Key to the Genera of the Megachilini (Females)

1. Scopa present on S2 to S5 or S6 (Fig. 8-7); metasoma not tapering throughout its length (Fig. 8-3) .... *Megachile*
—. Scopa absent; metasoma tapering from near base to narrow, often acutely pointed, apex (Figs. 8-2, 84-2, 84-3) 2
2(1). Omaular carina present; scutellum with dorsal and posterior surfaces usually separated by distinct, sometimes carinate, angle; axilla almost always produced posteriorly to angle or spine (Fig. 84-6a) .... *Coelioxys* (in part)
—. Omaular carina absent; scutellum with no distinct angle between dorsal and posterior surfaces; axilla rounded (Fig. 8-2) 3
3(2). Preoccipital carina absent; axilla and scutellum forming a single rounded posterior margin, not separated by incision; base of propodeum with row of strong pits (Africa) .... *Coelioxys: C. gracillima* Pasteels
—. Preoccipital carina present at sides of head; axilla separated by incision from scutellar margin; base of propodeum finely wrinkled (palearctic) .... *Radoszkowskiana*

Key to the Genera of the Megachilini (Males)

1. T6 appearing multispinose because of two pairs of long, preapical spines, each spine of upper pair sometimes divided into two, or crenulate, rounded, or fused to other spine of pair (Figs. 84-6f, g; 84-7) .... *Coelioxys*
—. T6 with large preapical carina (apparent apex of metasoma), often crenulate, often emarginate medially, sometimes reduced to two spines or rarely absent (Figs. 84-14, 84-18) 2
2(1). Posterior lobe of pronotum with strong transverse lamella extending posterolaterally as flat spine (palearctic) .... *Radoszkowskiana*
—. Posterior lobe of pronotum usually with weak transverse ridge, sometimes with carina or low lamella, but without spine .... *Megachile*

Genus *Coelioxys* Latreille

This genus is easily recognizable by the characters indicated in the key to genera. Most species also differ noticeably from all other Megachilini by having abundant hairs on the eyes. Figures 84-2 and 84-3 illustrate the distinctive aspect of species of this genus. Species of *Coelioxys* are black, though the legs and metasoma are sometimes partly or wholly red, and, rarely, the metasoma is faintly bluish. Length varies from 5 to 22 mm, some of the largest species being in the subgenera *Torridapis* and *Liothyrapis*. Illustrations of male genitalia and, in some cases, metasomal apices of females were published by Erlandsson (1955), Mitchell (1962), and Tkalců (1974b); see also Figure 84-4. Mitchell (1973) and Baker (1975) illustrated many structures of species from the Western Hemisphere.

Coelioxys consists of cleptoparasites related to *Megachile*. A series of apomorphies indicates that *Coelioxys* is a monophyletic unit; the transverse postgradular grooves of T2 and T3 are suggestive of hoplitiform taxa of *Megachile,* such as *Chelostomoides.* These hoplitiform taxa are probably ancestral forms of *Megachile,* and suggest either that *Coelioxys* is a sister group of *Megachile* or—if the grooves are homologous, i.e., constituting a single

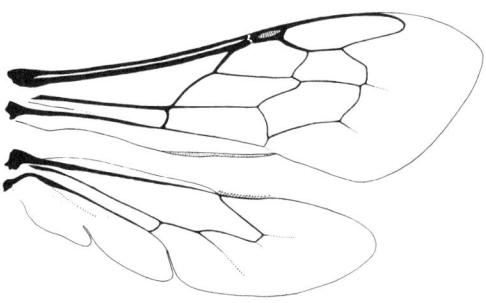

Figure 84-1. Wings of *Megachile chrysopyga* Smith.

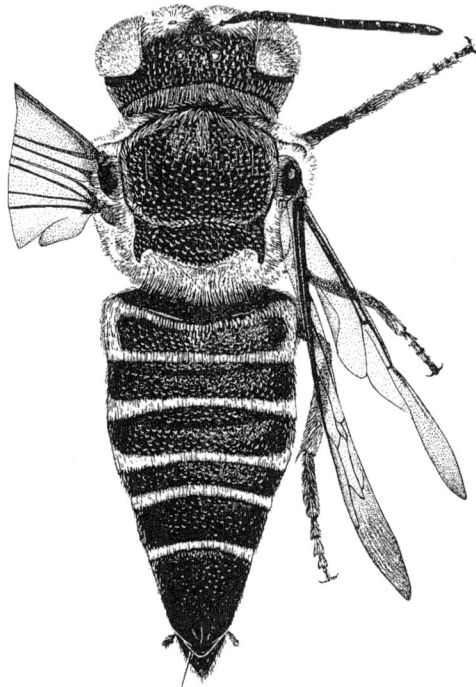

Figure 84-2. *Coelioxys mesae* Cockerell, female. Drawing by J. R. Baker, from Baker, 1975.

synapomorphy—that *Coelioxys* arose from this group of *Megachile,* rendering that genus paraphyletic.

Coelioxys is represented by many species on all continents except Australia, for which only five specific names have been proposed (Michener, 1965b). By contrast, nearly 200 names have been proposed for South American species (Mitchell, 1973); in the absence of a revisional study, the number of South American species is doubtful, but it is likely that the number of new species there will equal the number that become junior synonyms. In North America north of Mexico there are about 44 species (Hurd, 1979). In the palearctic region there is probably a similar number; Warncke (1986) listed 16 Central European species and Friese (1895) listed 28 European species. In Africa there are over 80 species (Pasteels, 1968b). It seems clear that South America has the world's richest *Coelioxys* fauna, with, again, nearly 200 named species. Although many neotropical species were examined by Mitchell (1973), the number that were identified was relatively small. Therefore, the numbers of named neotropical species reported below for the various subgenera are small, since many species remained unidentified, though probably new, in the material studied by Mitchell. Revisional studies include those for African species (Pasteels, 1968b, 1977b), North American species (Baker, 1975), Swedish species (Erlandsson, 1955), west palearctic species (Warncke, 1992e), and Argentine species (Holmberg, 1916).

Coelioxys has been divided into numerous subgenera in the New World (Mitchell, 1973), but a comparable subgeneric classification has not been made for the Old World. The treatment here is therefore different for the two areas, although I have synonymized certain North American subgeneric names that seemed unnecessary. The accounts of New World subgenera below are modified from Mitchell's treatment. The 139 infrageneric names used by Holmberg (1917, 1918) for groups of South American *Coelioxys* were suppressed by Opinion number 2069 of the International Commission on Zoological Nomenclature published in the Bulletin of Zoological Nomenclature 61: 67-69 (2004). See Michener, 2002d.

I believe that most of Mitchell's New World subgenera are natural groups, although some seemingly distinctive probable apomorphies, such as the pair of foveae on T2 and sometimes T3 of males, the lateral notches on S6 of females, and the excavated hypostomal areas of males, appear in various subgenera, presumably having arisen independently. Old World species include some that clearly fall in Mitchell's subgenera, as indicated in the subgeneric accounts below, but there are various groups that do not agree with his subgeneric characterizations. Some species could well be placed in his subgenera if they were recharacterized somewhat more broadly, but then distinctions between some subgenera become dubious. Clearly, these bees need study on a worldwide basis; they are a fascinating group with striking characters and deserve large-scale phylogenetic study.

There are two Old World subgenera of *Coelioxys* that differ strikingly from the rest of the genus in having hairless eyes. Further, T7 of the males is more exposed than that in the hairy-eyed taxa. Presumably, both the eye and T7 characters are plesiomorphies, but each of the two groups has abundant apomorphies. These groups taken together are likely to be the sister group of hairy-eyed *Coelioxys*. Popov (1955b) and Pasteels (1982) regarded them jointly as a separate genus, *Liothyrapis*. For the present I regard them as two subgenera of *Coelioxys,* because I do not know of strong characters associated with lack of ocular hairs, but generic status for each of the two subgenera with hairless eyes is a possibility. They seem more different from each other and from hairy-eyed *Coelioxys* than are the many subgenera in the Western Hemisphere.

Species of *Coelioxys* are among the commonest cleptoparasitic bees; and their parasitic behavior has been studied by various authors. Most species are cleptoparasites of

Figure 84-3. *Coelioxys alternata* Say, female. From Michener, McGinley, and Danforth, 1994.

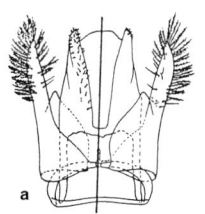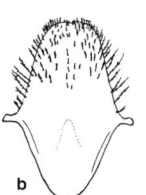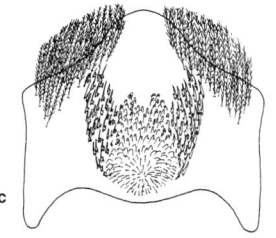

Figure 84-4. *Coelioxys (Xerocoelioxys) mesae* Cockerell. **a-c,** Male genitalia, S8, and S6. S7 is scarcely sclerotized, often only two minute remnants, in *Coelioxys*. From Baker, 1975.

Megachile, but there are also records of *Trachusa* s. str. (Anthidiini) and *Hoplitis (Anthocopa)* (Osmiini) serving as hosts of species that also attack *Megachile*. Baker (1975) summarized host records for North American species, and Westrich (1989) summarized European host records. Two European species known to attack *Megachile* are also reported as parasites of the apid, *Anthophora*, usually *A. (Clisodon) furcata* (Panzer), which nests in wood. Some such reports appear to be correct, but one has to wonder about the possibility that *Megachile* species nested in old *Anthophora* burrows, and were the true hosts. Some *Coelioxys*, however, do attack other apids, such as *Eucera*, *Tetraloniella*, *Centris*, and *Xylocopa*. Iwata (1976) reviewed host records worldwide up to that time. The most recent addition to the list of *Coelioxys* host genera is *Euglossa*; Ramírez-Arriaga, Cuadriello-Aguilar, and Martínez (1996) recorded *Coelioxys costaricensis* Cockerell laying eggs in provisioned but still open cells, as well as emerging as adults from cells of *Euglossa atroveneta* Dressler.

The long, tapering metasomal apex of females is usually used to insert the eggs through the food mass in an open cell into the cell wall, between or through leaf pieces in the case of leafcutter bees. In most cases, oviposition occurs while the host is foraging, and is into cells that have not yet been closed; see illustrations by Iwata (1939) and Baker (1971). In some species, however, eggs are laid after host cell closure, between the leaf pieces that form the closure. Little is known about oviposition sites for parasites of non-leafcutting hosts. Perhaps by analogy with the Nomadinae, in which first-stage larvae have a sclerotized head and large mandibles, authors have assumed a similar arrangement for *Coelioxys*. Iwata (1939) and Baker (1971), however, both show that the brief first stage of *Coelioxys* is unsclerotized and equipped with rather short mandibles (Fig. 84-5c). It is the second or second and third stages that have the enormous jaws on a sclerotized head (Fig. 84-5a, d) with which they churn the food mass and destroy eggs or larvae of the host and of any conspecific competitors that may be in the cell. Older *Coelioxys* larvae, by contrast, have an ordinary head and are very *Megachile*-like (Fig. 84-5b).

Key to the Subgenera of *Coelioxys* of the Western Hemisphere
(Modified from Mitchell, 1973)

1. Postgradular grooves of T2 and T3 complete, clearly evident at midline (Fig. 84-2) 2
—. Postgradular grooves of T2 and T3 interrupted medially, clearly evident only laterally 4
2(1). S6 of female with lateral subapical notch; T2 of male with sublateral pit or fovea just posterior to postgradular groove (Fig. 84-7) (holarctic) ....
.... *C. (Boreocoelioxys)* (in part)
—. S6 of female without lateral notches; T2 of male without foveae 3
3(2). Carina of pronotal tubercle conspicuously elevated, forming lamella anterior to tegula; pubescence of thorax shorter, scutum nearly bare but with marginal lines or

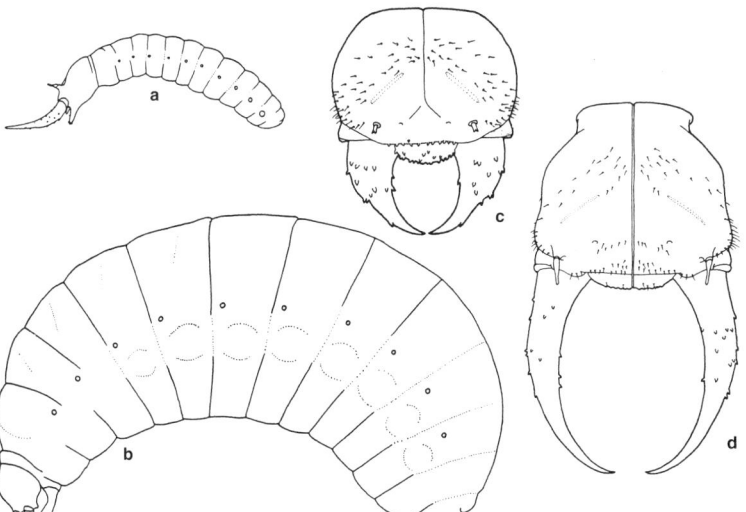

Figure 84-5. Larvae of *Coelioxys octodentata* Say. **a, b,** Side views of probable third-stage and mature larvae, showing the sclerotized head and falcate mandibles of the young larvae (compare with Fig. 4-1b, d); **c, d,** Dorsal views of heads of first- or second- and third-stage larvae. a and b, from Michener, 1953b; c and d, from Baker, 1971.

spots of pale tomentum; gradulus of T2 approaching apical margin of tergum medially (nearctic) .... *C. (Xerocoelioxys)*
—. Carina of pronotal tubercle not elevated, often obscured by long, dense, thoracic pubescence; scutum and scutellum more conspicuously pubescent, but without pale tomentose lines or spots; gradulus of T2 more nearly straight, not closely approaching apical tergal margin (holarctic) .... *C. (Coelioxys s. str.)*
4(1). Median ocellus partially surrounded by somewhat swollen and usually smooth portion of subocellar frontal area 5
—. Subocellar area of frons usually closely punctate and quite flat, but *if* with smooth swollen area, then usually with a few punctures or divided medially by punctate zone 6
5(4). S6 of female rounded or broadly angulate apically, with marginal fringe of short hairs, without subapical notch; metasoma of male with four exposed sterna, dorsal processes of T6 forming a transverse, irregularly crenulate plate (North and Central America) .... *C. (Synocoelioxys)*
—. S6 of female narrowed apically, without marginal fringe, with lateral subapical notch; metasoma of male with five exposed sterna, dorsal processes of T6 normal .... *C. (Neocoelioxys)*
6(4). Scutellum smooth, punctures well separated or sparse, hind margin carinate and usually angulate, projecting over metanotum and propodeum .... *C. (Acrocoelioxys)*
—. Scutellum usually well punctured, hind margin not carinate and not overlying metanotum except sometimes for a median angle or tubercle 7
7(6). Concavity of T1 lacking marginal carina; T1 without dorsal basal fascia of pale hairs (S6 of female usually notched laterally) 8
—. Concavity of T1 with dorsal marginal carina; T1 usually with more or less conspicuous pale basal fascia of hairs (i.e., on dorsum immediately behind basal concavity) 9
8(7). Axilla short, not reaching posterior transverse tangent of scutellum; scutellum broadly rounded or subtriangular; T2 of male with sublateral, transverse, elongate fovea on surface behind gradulus (Fig. 84-7) (holarctic) .... *C. (Boreocoelioxys)* (in part)
—. Axilla usually prominent; scutellum usually angulate or with median ridge or apical tubercle; T2 of male not foveate .... *C. (Glyptocoelioxys)*
9(7). S6 of female with small, lateral subapical notch; gena of male much narrowed below, hypostomal area lacking usual concavity (clypeus of female flattened, its apical margin straight and simple) .... *C. (Haplocoelioxys)*
—. S6 of female without lateral subapical notch; hypostomal concavity present in male, though sometimes obscured by pubescence 10
10(9). Scutellum subtriangular, or, if broadly rounded posteriorly, then dorsal and posterior surfaces usually only indefinitely separated (S6 of female usually fringed with setae) .... *C. (Cyrtocoelioxys)*
—. Hind margin of scutellum nearly or quite straight as seen from above, or, if broadly rounded, then dorsal and posterior surfaces separated by distinct or subcarinate edge (punctures of dorsal scutellar surface usually distinct, coarse and to some degree separated; S6 acute, without lateral notches or fringes) 11
11(10). S5 of female greatly expanded, hiding all but tip of S6; postgradular grooves of T3 to T5 in male not or only slightly fasciate; body length 6.5-7.2 mm (neotropical) .... *C. (Platycoelioxys)*
—. S5 of female normal, S6 very narrow and elongate; postgradular grooves of T3 to T5 in male usually conspicuously fasciate; body usually much larger (clypeus in female variously modified) (neotropical) .... *C. (Rhinocoelioxys)*

Key to the Subgenera of *Coelioxys* of the Eastern Hemisphere

1. Eyes bare 2
—. Eyes hairy 3
2(1). S5 of female greatly elongated, forming part of apical metasomal elongation, exposed part far longer than exposed parts of S4 and S6 (Fig. 84-6c); S6 of female narrow, without long hairs, not fringed; T6 of female with-

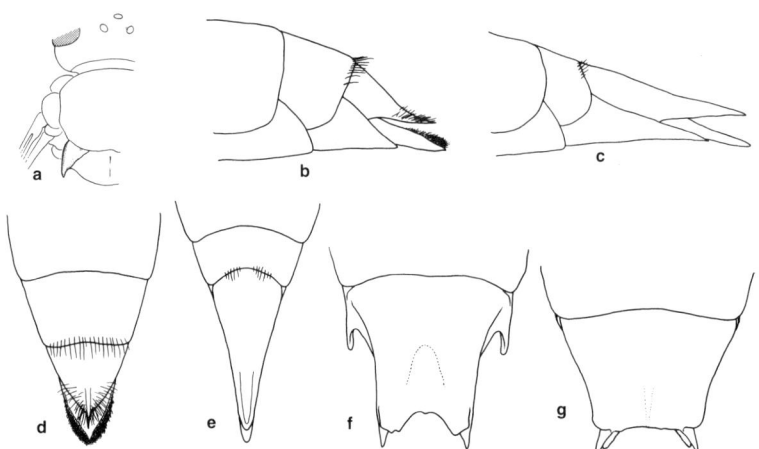

Figure 84-6. *Coelioxys.* **a,** Thoracic dorsum of *C.* sp., showing pointed axilla; **b, c,** Lateral views of apices of metasomas of females of *C. (Liothyrapis) scioensis* Gribodo and *C. (Torridapis) analis* Friese; **d, e,** Dorsal views of same; **f, g,** Dorsal views of T6 of males of *C. (Liothyrapis) decipiens* Spinola and *C. (Torridapis) analis* Friese. Drawings by D. Yanega.

out keel, without long erect hairs; T6 of male without lateral tooth (Fig. 84-6g) (Africa, oriental, Australia) .... *C. (Torridapis)*
—. S5 of female not attenuate, exposed part not much if any longer than exposed parts of S4 and S6; S6 of female broad, its apex bluntly pointed or rounded, fringed; T6 of female with keel ending in bare apical spine, with long erect hairs lateral to keel (Fig. 84-6b); T6 of male with lateral basal tooth (Fig. 84-6f) .... *C. (Liothyrapis)*
3(1). Transverse subocular carina (extending posteriorly from posterior mandibular articulation) joining preoccipital carina; subgenal fossa absent; T6 of male with eight teeth, two lateral, six apical .... *C. (Allocoelioxys)*
—. Transverse subocular carina (extending posteriorly from posterior mandibular articulation) absent or ending free, without approaching preoccipital carina, except in some males in which subocular carina is extended as margin of subgenal fossa, the margin continuing upward as preoccipital carina; T6 of male with six teeth or projections, two lateral, four apical 4
4(3). Front and middle femora each with two strong but irregular longitudinal carinae on undersurfaces, one of the carinae of front femur elevated and keel-like, spaces between carinae densely covered with white, scale-like hairs; white areas or fasciae of body formed by white scales; T6 and S6 of female produced and slender but apices truncate .... *C. (Mesocoelioxys)*
—. Front and middle femora unmodified; white areas of body formed by patches or fasciae of dense hairs or scale-like hairs; T6 and S6 of female usually pointed 5
5(4). T2 of male with sublateral pit or fovea on each side behind postgradular groove (Fig. 84-7); S6 of female with lateral subapical notch .... *C.(Boreocoelioxys)*
—. T2 of male without foveae behind postgradular groove; S6 of female without lateral notch .... *C. (Coelioxys s. str.)*

Coelioxys / Subgenus *Acrocoelioxys* Mitchell

Coelioxys (Acrocoelioxys) Mitchell, 1973: 71. Type species: *Coelioxys otomita* Cresson, 1878, by original designation.
Coelioxys (Melanocoelioxys) Mitchell, 1973: 78. Type species: *Coelioxys tolteca* Cresson, 1878, by original designation. [New synonymy.]

The medially interrupted postgradular grooves of T2 and T3, the flattened and punctate subocellar area of the face, and the smooth, shining, and largely impunctate scutellum, with its conspicuously carinate and produced posterior margin, characterize this subgenus.
■ This primarily neotropical subgenus ranges from Argentina north to Sonora, Mexico, and North Carolina, USA. There are 25 or more species, of which Mitchell (1973) lists 18; only *Coelioxys dolichos* Fox occurs in the USA.

Coelioxys / Subgenus *Allocoelioxys* Tkalců

Allocoelioxys Tkalců, 1974b: 340. Type species: *Coelioxys afra* Lepeletier, 1841, by original designation.
Coelioxita Pasteels, 1977b: 180. Type species: *Coelioxys afra* Lepeletier, 1841, by original designation.
Coelioxula Pasteels, 1982: 110. Type species: *Coelioxys ruficaudata* Smith, 1854 = *Coelioxys ruficauda* Lepeletier, 1841, by original designation.
Coelioxys (Intercoelioxys) Ruszkowski, in Ruszkowski, Biliński, and Gosek, 1986: 117. Type species: *Coelioxys ruficaudata* (sic) Smith, 1854 = *C. echinata* Förster, 1853, monobasic.
Coelioxys (Lepidocoelioxys) Ruszkowski, in Ruszkowski, Biliński, and Gosek, 1986: 117. Invalid because no type species was designated; according to Přidal and Tkalců (2001) the included species all would fall in *Allocoelioxys*.
Coelioxys (Tropicocoelioxys) Gupta, 1991: 425. Type species: *Coelioxys genoconcavitus* Gupta, 1991, by original designation. [New synonymy.]
Coelioxys (Orientocoelioxys) Gupta, 1992: 73. Type species: *Coelioxys quadrifasciatus* Gupta, 1992, by original designation. [New synonymy.]
Coelioxys (Nigrocoelioxys) Gupta, 1993: 235. Type species: *Coelioxys fuscipennis* Smith, 1854, by original designation. [New synonymy.]

Přidal and Tkalců (2001) gave a useful translation from Polish to English of part of Ruszkowski's work. I see no justification for their belief that *Coelioxula* Pasteels is a *nomen nudum,* since a character and type species were specified by Pasteels (1982: 110).

The late Dr. Donald B. Baker (in litt., Nov. 2002) indicated that the three subgeneric names proposed by Gupta and synonymized under *Coelioxys* s. str. by Michener (2000) should all be placed instead as synonyms of the subgenus *Allocoelioxys*.
■ This subgenus is widespread in the Old World, including the palearctic region from Europe to China and Taiwan; to the south it occurs in all of Africa and southern Asia, at least to Java. Pasteels (1977b) reports 21 species in sub-Saharan Africa, and Warncke (1992e) reports 15 species in the western palearctic region. There are at least nine European species and a few additional species in southern Asia. The total may be about 40 species.

Coelioxys / Subgenus *Boreocoelioxys* Mitchell

Coelioxys (Boreocoelioxys) Mitchell, 1973: 37. Type species: *Coelioxys rufitarsus* Smith, 1854, by original designation.
Coelioxys (Schizocoelioxys) Mitchell, 1973: 50. Type species: *Coelioxys funeraria* Smith, 1854, by original designation. [The spelling *Schizococoelioxys* in the heading of Mitchell's description (p. 50) is corrected on the same page and elsewhere in the work.]

In *Boreocoelioxys* the postgradular grooves of T2 and T3 in both sexes vary from complete and often deep to widely interrupted. The subgenus therefore appears twice in the key to subgenera. The carina of the pronotal tubercle is elevated and conspicuous. S6 of the female has a distinct, though sometimes small, pair of lateral, subapical notches, and in the males T2 has a pair of usually small pits or foveae just posterior to the gradular groove and about midway between the median line and lateral margins (Fig. 84-7). In general, the included species are coarsely sculptured bees with abundant and often rather long pubescence on the head and thorax.
■ This subgenus is mainly holarctic. In America it occurs from northern Canada to Costa Rica, and from the Atlantic to the Pacific coast; in Eurasia it is known from Europe to Japan. The 11 North American species were re-

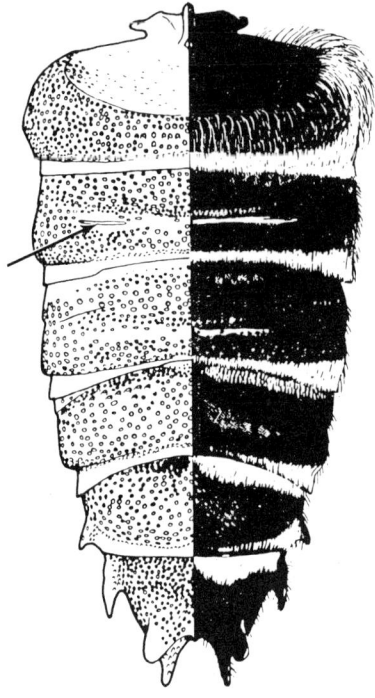

Figure 84-7. Dorsum of metasoma of male of *Coelioxys (Boreocoelioxys) rufitarsis* Smith, showing punctation at left and vestiture at right. Note the fovea (arrow) on T2. From Baker, 1975.

vised by Baker (1975). At least six palearctic species are probable members of this subgenus.

T. Griswold suggested the synonymy indicated above. Some of the species placed in *Xerocoelioxys* also may fall in this subgenus.

Coelioxys / Subgenus *Coelioxys* Latreille s. str.

Coelioxys Latreille, 1809: 166. Type species: *Apis conica* Linnaeus, 1758 = *Apis quadridentata* Linnaeus, 1758, by designation of Curtis, 1831: pl. 349. [See note by Michener (1997b).]

Coelioxys (Paracoelioxys) Gribodo, 1884: 274. Type species: *Coelioxys montandoni* Gribodo, 1884 = *Coelioxys alata* Förster, 1853, monobasic.

The postgradular grooves of T2 and T3 are complete; the carina of the pronotal lobe is low and inconspicuous; and the pubescence is usually copious and elongate. S6 of the female is not notched laterally, but the apex is usually constricted. In the male T2 and T3 are not foveate.

■ This subgenus, as constituted by Mitchell (1973), occurs in most parts of the USA north to northern Canada and Alaska, and south to Coahuila and Sonora, Mexico, as well as in Europe. The five nearctic species were revised by Baker (1975), and at least one palearctic species (Mitchell, 1973) belongs in this subgenus in Mitchell's narrow sense. In the Old World numerous other species are placed in this subgenus, which has not been so finely subdivided as in the New World. For example, in some of the Old World species the postgradular grooves of T2 and T3 are broadly interrupted medially (Pasteels, 1977b), unlike those of New World species placed in this subgenus. In its broad sense the subgenus occurs throughout the holarctic region, throughout Africa (31 species; Pasteels, 1977b) as well as southern Asia, Indonesia, and Australia, with an estimated total of 50 species in the Eastern Hemisphere.

Coelioxys / Subgenus *Cyrtocoelioxys* Mitchell

Coelioxys (Cyrtocoelioxys) Mitchell, 1973: 106. Type species: *Coelioxys costaricensis* Cockerell, 1914, by original designation.

In both sexes of this subgenus, the subocellar area of the face is flat or nearly so, its surface dull and rather closely punctate; the scutum and scutellum are usually quite coarsely or closely punctate; the hind margin of the scutellum is broadly rounded or to some degree obtusely angulate; and the postgradular grooves of T2 and T3 are interrupted medially. In females, S6 is more or less conspicuously fringed with hairs or setae, and the apex is acute or spinose. Many of the species are small, down to 5 mm long, but in the neotropics some are nearly 20 mm long.

■ This subgenus ranges from Argentina to North America, as far north as Quebec, Canada, and Utah and California, USA. The majority of the species are neotropical. Mitchell (1973) and Toro and Fritz (1993) list 39 species, and there are undoubtedly many more. Eight occur in the USA (Hurd, 1979). Argentine species were revised by Toro and Fritz (1993).

The type species, *Coelioxys costaricensis* Cockerell, is a cleptoparasite of *Euglossa* (Ramírez-Arriaga, Cuadriello-Aguilar, and Martínez, 1996), but other species live far outside the range of the Euglossini and are probably cleptoparasites of *Megachile*.

Coelioxys / Subgenus *Glyptocoelioxys* Mitchell

Coelioxys (Glyptocoelioxys) Mitchell, 1973: 92. Type species: *Coelioxys vidua* Smith 1854, by original designation.

Coelioxys (Dasycoelioxys) Mitchell, 1973: 99. Type species: *Coelioxys pergandei* Schletterer, 1890, by original designation.

The subocellar area of the face is closely sculptured, and either flat or slightly convex; the scutellum is usually well punctured, but in several of the included species there is a median triangular shining and impunctate area that protrudes slightly beyond the apical margin, which is not otherwise carinate; the margin of the basal concavity of T1 is neither carinate nor fasciate; the postgradular grooves of T2 and T3 are interrupted medially; S6 of the female usually has a pair of distinct lateral subapical notches; S5 of the male is normally retracted and covered by S4.

■ This subgenus occurs from Chile and Argentina (as far south as Tierra del Fuego) to the eastern USA, north as far as Illinois and New Jersey. There are no doubt over 50 species, of which Mitchell (1973) lists 23 in *Glyptocoelioxys* and 17 in *Dasycoelioxys;* only *Coelioxys germana* Cresson occurs in the USA.

In the Andean region and southern South America

some species are rather large and densely hairy, often lacking the usual areas of dense, scale-like hairs. It is these species that were segregated by Mitchell as the subgenus *Dasycoelioxys*, here considered a synonym of *Glyptocoelioxys*. Toro and Fritz (1991) gave a key to such species.

Coelioxys / Subgenus *Haplocoelioxys* Mitchell

Coelioxys (Haplocoelioxys) Mitchell, 1973: 85. Type species: *Coelioxys mexicana* Cresson, 1878, by original designation.

In this group the postgradular grooves of T2 and T3 are at least slightly interrupted medially; the subocellar area is flat or slightly convex, usually well punctured; the concavity of T1 is distinctly carinate dorsally, but only obscurely fasciate, if at all; and S6 in the female is laterally notched toward the apex.

■ This subgenus ranges from Argentina north to Sonora, Mexico, and North Carolina and Texas, USA. Only five species are known, one of which, *Coelioxys mexicana* Cresson, reaches the USA.

Coelioxys / Subgenus *Liothyrapis* Cockerell

Coelioxys (Liothyrapis) Cockerell, 1911b: 246. Type species: *Coelioxys apicata* Smith, 1854 = *Coelioxys decipiens* Spinola, 1838, by original designation.
Coelioxys (Liothygrapis) Cockerell, 1932c: 26, incorrect subsequent spelling of *Liothyrapis* Cockerell, 1911.
Liothgraphis Sandhouse, 1943: 564, incorrect subsequent spelling of *Liothgrapis* Cockerell, 1932.
Coelioxys (Hemicoelioxys) Pasteels, 1968b: 133. Type species: *Coelioxys gracilis* Pasteels, 1968 (not Schenck, 1868) = *Hemicoelioxys gracillima* Pasteels, 1977, by original designation.

The following, in addition to pertinent elements in the key, is provided for contrast with *Torridapis*, the other *Coelioxys* subgenus with hairless eyes: The first flagellar segment of the female is more than half as long as the second. The axillae are angular but short, often only attaining a transverse line through the middle of the scutellum, or, in *Coelioxys gracillima* Pasteels, rounded. The apex of the female metasoma is relatively short; the upper surface of S6 has a carina extending from the middle of the apex to the basolateral part of the exposed surface of the sternum, the area lateral to this carina being densely hairy, the hairs extending laterally and appearing from beneath as a fringe (Fig. 84-6d); T6 of the male has a lateral tooth formed from the posterior lateral extremity of the gradulus (Fig. 84-6f).

■ This subgenus occurs throughout Africa, from Cape Province, South Africa, to Egypt, Algeria, and Senegal, northward to the eastern Mediterranean basin (e.g., Crete, Israel), thence eastward to the Trans-Caspian region and through India (northern and southern) to Southeast Asia and Indonesia (Java). There are about 26 species in sub-Saharan Africa (Pasteels, 1968b) and several more in Asia; an estimate of the total number of species is 35.

Pasteels (1968b) recognized this group as the section *decipiens*. As noted above, Pasteels (1977b) regarded this subgenus together with *Torridapis* as generically distinct from *Coelioxys*. The hairless eyes are strikingly different from those of most subgenera of *Coelioxys*, but in other features there are no major differences. Probably hairless eyes are plesiomorphic, since other Megachilini lack hairs on the eyes.

As noted by Pasteels (1968b), the head and thorax of *Coelioxys gracillima* Pasteels are similar to those of *Megachile* (except that the female mandibles are not *Megachile*-like), while the metasoma is *Coelioxys*-like. This may indicate that the special metasomal features of *Coelioxys*—in females related to egg deposition in the host nest, and in males to problems of mating with such females—evolved first, and were followed by the special cephalic and thoracic features of *Coelioxys*, at least some of which are presumably defenses against host stings and jaws.

Hemicoelioxys, based on *Coelioxys gracillima* Pasteels, was said by Pasteels (1968b) to differ from other *Coelioxys* in the absence of a preoccipital carina, absence of an omaular carina, lack of a carina or lamella on the pronotal lobe, the flat scutellum without strong demarcation of dorsal and posterior surfaces, the rounded axillae whose margins continue the curvature of the scutellar margin as seen from above, and the row of pits across the base of the propodeum. Maximillian Schwarz (personal communication, 1992) indicates that these characters are weak. The carinae and axillar angles or spines of *Coelioxys* are poorly developed in small species. He considers *Hemicoelioxys* to be a synonym of *Liothyrapis*, *C. gracillima* being the smallest species of that subgenus. Even in large species of *Liothyrapis* the axillar spines are short and the angle between dorsal and posterior surfaces of the scutellum is rounded.

Coelioxys / Subgenus *Mesocoelioxys* Ruszkowski

Coelioxys (Mesocoelioxys) Ruszkowski, in Ruszkowski, Biliński, and Gosek, 1986: 117. Type species: *Coelioxys argentea* Lepeletier, 1841, monobasic.
Coelioxys (Argocoelioxys) Warncke, 1992e: 39. Tye species: *Coelioxys argentea* Lepeletier, 1841, by original designation. **[New synonymy.]**

This name was proposed for an unusual species that could be retained in *Coelioxys* s. str., but at least until groups or subgenera of the genus are studied, I retain the name. Its scale-like vestiture and short eye hairs are suggestive of *Allocoelioxys*. The apex of S5 of the female is truncate and greatly thickened, unlike that of other groups.

■ *Argocoelioxys* is found in the Mediterranean region, east to Turkey. The only species is *Coelioxys argentea* Lepeletier.

Coelioxys / Subgenus *Neocoelioxys* Mitchell

Coelioxys (Neocoelioxys) Mitchell, 1973: 64. Type species: *Coelioxys assumptionis* Schrottky, 1911, by original designation.

The postgradular grooves of T2 and T3 are interrupted medially; the subocellar area of the face is swollen, only sparsely punctate at most, and partially surrounds the median ocellus. In the female, S6 has a distinct lateral subapical notch on each side. In the male, S5 is not retracted.

■ This subgenus is widespread in the neotropical region, ranging from Argentina and Tarapacá, Chile, to Cuba and the USA as far north as North Carolina, Texas, and Arizona. Numerous species presumably belong here, but Mitchell (1973) listed only seven, two of which occur in the nearctic region.

Coelioxys / Subgenus *Platycoelioxys* Mitchell

Coelioxys (Platycoelioxys) Mitchell, 1973: 120. Type species: *Coelioxys spatuliventer* Cockerell, 1927, by original designation.

The female of this monotypic subgenus is easily recognizable by the peculiar form of S5. It is both broad and elongate, the lateral margins nearly parallel, and its median length is about one-third of the total length of the metasoma. Only the tip of S6 projects slightly beyond the apex of S5; S6 is thus nearly hidden between T6 and S5. Males, though not so easily recognized, agree with females in their small size and in the coarseness of the sculpture. The hind margin of the scutellum is nearly straight, abruptly bent downward but not carinate. The postgradular grooves of T2 and T3 are interrupted medially, and those of T3 to T5 are not fasciate. The postgradular areas of T2 and T3 are very narrow, hardly broader than the apical rims. The subocellar area of the frons is flat, dull, and densely sculptured, and the hypostomal area is bare and rather broadly flattened but hardly concave. The concavity of T1 is carinate dorsally, though not conspicuously fasciate.

■ This subgenus ranges from Bolivia and the state of São Paulo in Brazil to San Luis Potosí, Mexico. A single species, *Coelioxys spatuliventer* Cockerell, is known. Mitchell (1973) speculated that it parasitizes *Megachile (Ptilosarus)*.

Coelioxys / Subgenus *Rhinocoelioxys* Mitchell

Coelioxys (Rhinocoelioxys) Mitchell, 1973: 113. Type species: *Coelioxys zapoteca* Cresson, 1878, by original designation.

In both sexes the postgradular grooves of T2 and T3 are interrupted medially, the subocellar area of the frons is flat, dull, and closely sculptured. The scutellum is coarsely sculptured, its midline short, and its posterior margin usually nearly straight, quite abruptly bent down but usually not at all carinate. The dorsal margin of the basal concavity of T1 is carinate and usually pale fasciate. In females the clypeus is rather curiously modified and S6 is narrow and elongate, its lateral margin neither notched nor fringed with setae. S5, however, which is usually elongate and tapers apically, often bears a conspicuous fringe along the lateral margin. In males the postgradular grooves of T3 to T5 usually are conspicuously pale fasciate, and the apical fasciae of T2 to T5 are broadly interrupted medially.

■ This is a strictly neotropical subgenus, known from Veracruz and Oaxaca, Mexico, to Brazil. Mitchell (1973) listed five species and indicated that he had about an equal number of undescribed species.

Coelioxys / Subgenus *Synocoelioxys* Mitchell

Coelioxys (Synocoelioxys) Mitchell, 1973: 57. Type species: *Coelioxys texana* Cresson, 1872, by original designation.

In *Synocoelioxys* the postgradular grooves of T2 and T3 are widely interrupted medially and usually densely fasciate and the subocellar area of the frons, partially surrounding the median ocellus, is more or less swollen and impunctate. S6 of the female is rounded or broadly angulate, with a dense marginal fringe of short hairs. In the male the usual four sterna are exposed, S5 is normally retracted, and the dorsal processes of T6 are more or less united, forming a transverse, irregularly crenulate plate.

■ This subgenus is widespread from Canada to Costa Rica. Baker (1975) recognized five species.

The known hosts are *Megachile* of the subgenus *Sayapis*.

Coelioxys / Subgenus *Torridapis* Pasteels

Liothyrapis (Torridapis) Pasteels, 1977b: 195. Type species: *Coelioxys torrida* Smith, 1854, by original designation.

This group was first recognized and briefly characterized as the section *torrida* by Pasteels (1968b), who later provided a subgeneric name. The first flagellar segment of the female, on the shortest side, is about half as long as the second or less. The axillae are rather long, extending beyond a transverse line through the middle of the scutellum and sometimes nearly attaining the transverse tangent of the posterior scutellar margin. The apex of the metasoma of the female is extremely attenuate (Fig. 84-6c, e), T6 being drawn out to a narrowly rounded or vaguely and minutely tridentate apex and S6 being narrow and pointed, not extending much beyond T6. T6 of the male lacks a lateral tooth, although the posterolateral extremity of the gradulus of T6 is sometimes somewhat elevated [as expected, the tooth is absent in *Coelioxys (Torridapis) analis* Friese, although Pasteels, 1968b, illustrated such a tooth for this species]; the upper median spines of T6 of the male are fused to the lower median spines and to each other, sometimes scarcely projecting, and in that case not recognizable as separate spines (Fig. 84-6g).

■ This subgenus is widespread in Africa (Ethiopia and Ivory Coast to Cape Province, South Africa), and ranges eastward to Aden, presumably southern Asia, and on to southern China, Indonesia, New Guinea, and Queensland, Australia. There are seven African species and one from Mauritius (Pasteels, 1968b, 1977b) and a similar number of Asiatic and Indo-Australian species. There is probably only one species in New Guinea and Australia, although there are two names, *Coelioxys weinlandi* Schulz and *albiceps* Friese. Specimens from Sulawesi (Indonesia) identified as the Indian *C. ducalis* Smith belong to this subgenus.

Coelioxys / Subgenus *Xerocoelioxys* Mitchell

Coelioxys (Xerocoelioxys) Mitchell, 1973: 44. Type species: *Coelioxys edita* Cresson, 1872, by original designation.

In both sexes the postgradular grooves of metasomal T2 and T3 are complete and often deep, and the carina of the pronotal tubercle is elevated and conspicuous. S6 in the females is not notched laterally but may be narrowed preapically. In the males neither T2 nor T3 is foveate.

■ This subgenus is widespread in the nearctic region, from coast to coast and from southern Alberta, Canada, to Jalisco and Zacatecas, Mexico. Baker (1975) recognized ten species.

Genus *Megachile* Latreille

The bees here included in *Megachile* constitute an enormous and morphologically as well as behaviorally diverse group. While recognizing various subgenera, most authors placed them in a single genus until Michener (1962a, 1965b), because of the great differences between major groups of subgenera, divided the genus into three. He recognized two large genera, *Megachile* and *Chalicodoma,* and a third, much smaller genus, *Creightonella.* A similar division into the two large groups was recognized at least as early as Radoszkowski (1874a, b).

With fuller knowledge of the bees of the Eastern Hemisphere, I no longer consider the recognition of three genera as appropriate, primarily because many intermediates exist, as was indicated already by Michener, McGinley, and Danforth (1994). Clearly among forms here included in *Megachile* there are species fully as different as are genera among other bees. No doubt several genera will ultimately be recognized. To facilitate discussion I recognize three groups, each equivalent to one of the three genera previously recognized. Group 1 (*Megachile* of Michener, 1962a) and Group 3 *(Creightonella)* each have synapomorphic characters and are presumably monophyletic, although no cladistic analysis has been made. Group 2, *Chalicodoma* (in the sense of Michener, 1962a), however, appears to have no unique synapomorphies, is highly diverse, and should eventually be divided into several genera. Some of the problems one faces in doing so are explained below. The three groups of *Megachile* in its present sense are considered in the following paragraphs:

Figure 84-8. *Megachile (Xanthosarus) latimanus* Say; Above, Male; Below, Female. From Michener, McGinley, and Danforth, 1994.

Group 1. *Megachile* **sensu Michener (1962a)** consists largely of megachiliform species (Figs. 8-3b, 84-8; Pl. 7), i.e., those with a somewhat flattened metasoma, in females usually widest at segment 2 or 3 and tapering posteriorly. Some exceptions, i.e., Group 1 taxa that are chalicodomiform or almost hoplitiform, roughly parallel-sided, are the subgenera *Eumegachile, Sayapis, Schrottkyapis,* and *Stelodides.* Females of Group 1 have partial or complete cutting edges in mandibular tooth interspaces 2 or 3 (or both) (counting from the apex of the mandible upward; Figs. 84-11, 84-12); the neotropical subgenera *Chrysosarus, Schrottkyapis,* and *Stelodides,* however, lack such cutting edges. Moreover, in the subgenus *Megachile* s. str., *M. montivaga* Cresson lacks cutting edges, though they are present but sometimes small in other species of the same subgenus. Species of Group 1 lack hairs on the lateral margins of S8 of males, although the disc of the sternum often has short hairs that occasionally project beyond the sternal margins. An exception is *Eumegachile,* which has one marginal hair arising at the base of the lateral margin of S8. Species of Group 1 have on or near the posterior margin of S6 of the female a fringe of short, dense, plumose hairs, different from the unbranched scopal hairs. Such a fringe of hairs is ordinarily absent in Group 2. A similar fringe, however, is well developed in *M. (Callomegachile) sculpturalis* Smith, and some plumose hairs are present near the apex of S6 in *M. (Pseudomegachile)* species such as *lanata* (Fabricius) and *ericetorum* (Lepeletier), all members of Group 2. There are other characters associated with those enumerated above, but that are even less constant. A series of features of the posterior end of the thorax was described by Michener (1962a, 1965b), but although they are tendencies, they are not constant, as was noted in the earlier papers cited. Also burdened with exceptions is the shape of S6 of males; in Group 1 it is short and wide, its width being six or more times its minimum length, measured in the sublateral constriction. This character fails, however, in various species of the subgenera *Leptorachis, Pseudocentron, Megachiloides,* and *Sayapis,* in which the sternum is about five times as wide as its minimum length, as it is in many members of Group 2. In nearly all species of Group 1 the volsella is represented by a lobe fused to the gonoforceps, but in the subgenus *Tylomegachile* it is a separate sclerite, as it is in many members of Group 2. There is thus a complex of characters that ordinarily makes *Megachile* sensu Michener (1962a), i.e., Group 1, recognizable. They are nearly all apomorphies, as judged by comparison with other Megachilinae (especially Osmiini).

Species of Group 1 normally use leaf pieces for constructing cells, another derived feature. In most cases the leaf pieces—oblong for the cell bases and walls, circular for the cell closures—are uniformly cut with smooth edges. Most species make use of preexisting burrows or cavities for their nests. Thus their cells, made of leaf pieces, can be found in old beetle burrows in wood, in hollow stems, in diverse small cavities (e.g., keyholes) in manmade structures, or in banks of soil or between stones in walls. The nesting biology has been described in detail for certain such species (see Michener, 1953b; Klostermeyer and Gerber, 1969; and Trostle and Torchio, 1994). Some species, however, dig their own burrows for cells in the ground (Eickwort, Matthews, and Carpenter, 1981; Neff and Simpson, 1991). Williams et al. (1986) studied a species, *Megachile (Megachiloides) integra* Cresson, that burrows in sand and constructs only one cell per nest, at least under conditions of disturbance by ants.

Group 1 subgenera that have only a small cutting edge in the second interspace of the female mandibles (Fig. 84-12c) have, so far as is known, reduced leafcutting behavior. Thus species of the subgenus *Sayapis* use very few leaf pieces, along with chewed leaf material and soil or small pebbles, to make partitions between cells in unlined burrows (Medler, 1964a; Krombein, 1967; Frohlich and Parker, 1983). (Compare these findings with the behavior of the Group 2 subgenus *Chelostomoda*, below.) The parallel-sided body form (chalicodomiform) of bees with such mandibles is presumably related to the use of burrows so narrow that leaf or resin cell walls are unnecessary.

Some species of Group 1 that altogether lack cutting edges on the female mandibles use flower petals or other delicate materials instead of leaves. *M. montivaga* Cresson, a species of *Megachile* s. str. lacking the usual mandibular cutting edges, makes cells of petals and uses pith for the nest closure (Hicks, 1926). *Stelodides*, a Chilean subgenus that lacks cutting edges, also uses petals for its cells (Claude-Joseph, 1926). *Schrottkyapis*, another South American subgenus that lacks cutting edges, lines its cells with chewed leaf material but fills in burrow space with leaf pieces and mud (Martins and Almeida, 1994). The nest habits of the large neotropical subgenus *Chrysosarus*, still another that lacks cutting edges, differ from those of most *Megachile*. I have seen a species in Mexico collecting mud. This, of course, does not mean that it does not also use petals or leaves. Friese (1924a) described a nest of "*Megachile azteca* Cr." from Costa Rica that used a layer of clay between layers of petals to make its cells. This is likely to have been a *Chrysosarus*, although the specific identity is unreliable; *M. azteca* Cresson actually belongs to the subgenus *Pseudocentron*. *Megachile (Chrysosarus) tapytensis* Mitchell, in Brazil, was reported by Laroca (1971a) to make cells with a thick (1.5-2.0 mm) layer of mud between outer and inner layers of leaves. The partitions between cells in a series included a similar mud layer between the closing (probably round) leaf pieces of one cell and the pieces forming the inner lining of the next cell in the series. Another species of the subgenus *Chrysosarus*, *Megachile pseudoanthidioides* Moure, constructs cells and closures with an outer layer of leaves, followed by a layer of mud, and an inner layer of petals (Zillikens and Steiner, 2004). Some of the nests were in boxes where the cell series were constructed even in the absence of confining walls.

Group 2. *Chalicodoma* sensu Michener (1962a) was the other frequently recognized large genus. In general it can be distinguished by the obverse of the characters listed for Group 1. Thus the metasoma is strongly convex and rather parallel-sided, although sometimes broader and tapering, for example in the females of the subgenera *Chalicodoma*, *Thaumatosoma*, and some species of *Gronoceras*. An exception is *Matangapis* which is megachiliform. Otherwise, the body form is chalicodomiform, or, for those that are most slender, hoplitiform (Fig. 84-9; Pl. 7) or even heriadiform. The mandibles of females lack cutting edges between the teeth (Fig. 84-12e, f) except for a partial cutting edge in the second interspace in certain subgenera (as in Fig. 84-12c and as explained below). All species of Group 2 have marginal hairs on S8 of males (Fig. 84-10j, n), only one on each side in some individuals of *Rhodomegachile*. (*Eumegachile*, tentatively placed in Group 1, also has one such hair at the base of the disc of S8; but other members of Group 1 entirely lack marginal hairs on S8.) In Group 2 the width of S6 of males is ordinarily five or fewer times its length, as measured at the sublateral constriction, but as shown in Figure 84-10 this is variable. In some species the volsella is separable from the gonoforceps; and in some species of the subgenus *Gronoceras* the digitus and cuspis are distinct, presumably a plesiomorphy.

Nests of Group 2 *Megachile* are ordinarily made with resin or mud and do not include leaf or petal pieces such as are characteristic of Group 1. This character is correlated with the lack of cutting edges on the mandibles of females; compare Figures 84-11 and 84-12. In the subgenus *Chelostomoda*, however, cell walls are omitted (nests are in burrows of other insects in wood) but cell closures are made of very irregularly cut leaf pieces plus leaf pulp, and the nest may be closed with mud (Yamamoto, 1944; Iwata, 1976). As noted above, unlike most other Group 2 taxa, *Chelostomoda* has a small cutting edge in the second interspace, and its cells in some ways resemble those of a few Group 1 taxa such as *Sayapis* that have similar mandibles.

Megachile (Callomegachile) monticola Smith, a member of Group 2 with no cutting edges in the female mandibles, makes cells of resin in burrows of other insects, such as *Xylocopa*, in wood; to close a cell series, however, it uses irregular leaf pieces, followed by clay; sometimes a leaf piece is also found between cells (Piel, 1930). It seems that in Group 2 the potential exists for specialization in the use of leaf pieces, as is found in Group 1.

The morphological and behavioral characters of Group 2 listed above are all probable plesiomorphies relative to Group 1. It is therefore likely, as suggested above, that Group 2 is paraphyletic, but it is not at all clear from what part of Group 2 the Group 1 subgenera may have arisen. It is likely that some chalicodomiform Group 1 taxa such as *Stelodides* reverted to this body form, perhaps in the course of accommodating to narrow nest burrows that require only partitions to make cells. Likewise for other characters, reversals are probable.

Group 3. *Creightonella*. This is the third genus recognized by Michener (1962a, 1965b). It consists of species with a convex and parallel-sided metasoma, as in taxa such as *Callomegachile* in Group 2. Unlike the Group 2 mandibles, however, those of Group 3 females typically have five or six subequal teeth and incomplete cutting edges in interspaces 2, 3, and sometimes 4 (Fig. 84-17d). There are five or six exposed sterna in males, but even where largely hidden, S5 and S6 are broad sternal plates, the margin of S6 convex posteriorly and the sternum thus

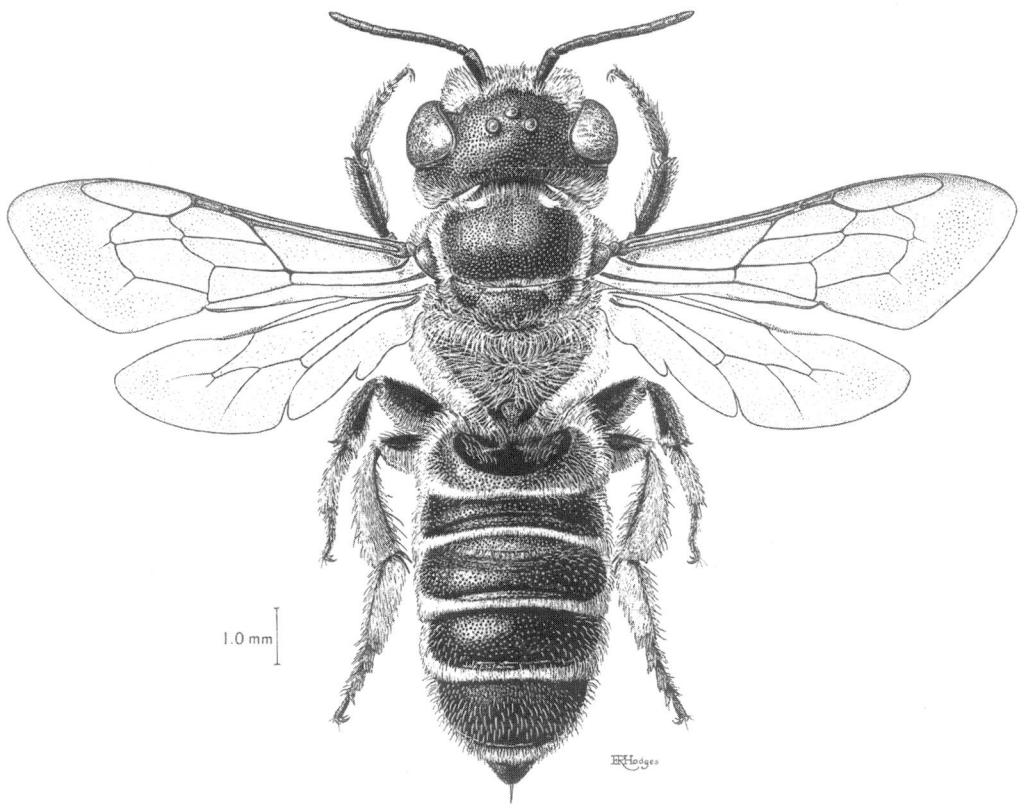

Figure 84-9. *Megachile (Chelostomoides) occidentalis* Fox, female. Drawing by E. R. S. Hodges, from Michener, McGinley, and Danforth, 1994.

with no sublateral constriction (Fig. 84-17c). These sterna lack the capitate, spatulate, or thickened hairs that are common in the other groups, as shown in Figure 84-10. Such hairs, as well as sublateral constrictions in S6, are also lacking, however, in *Megachile (Callomegachile) mystaceana* (Michener), among others that are members of Group 2. As in Group 2, S8 of the male in Group 3 has long marginal hairs.

Creightonella cells are constructed with leaf fragments, as could be predicted from the presence of cutting edges on the female mandibles. The leaf pieces, however, are irregularly cut, often with serrate margins, unlike the regularly shaped, smooth-margined pieces cut by most members of Group 1. Moreover, extensive use is made of leaf pulp or mud and possibly resin or a secretion to make a firm cell wall (Ferton, 1901, as *Megachile sericans* Fonscolombe = *albisecta* Klug; Michener and Szent-Ivany, 1960; Michener, 1968a).

Discussion. Table 84-1 summarizes some characters of groups here placed in *Megachile*, with separate lines for some "problem" or intermediate subgenera, i.e., those that confuse the characters of the three groups. The following comments on the subgenera concerned relate to my suggestion that Group 2 should be subdivided into several genera if a rational classification of nonparasitic Megachilini is to be achieved.

Chalicodoma s. str. might be segregated as a genus recognized by the convex, unthickened, denticulate clypeal margin, the long, oblique, edentate, female mandibular margin above the first interspace, and the multidentate carina of T6 of the male. *Parachalicodoma*, however, although resembling *Chalicodoma*, has shorter five-toothed female mandibles (as in Fig. 84-12e, f) not at all like those of *Chalicodoma*; *Parachalicodoma* thus connects *Chalicodoma* to more ordinary members of Group 2 such as *Callomegachile*.

The subgenera *Chelostomoides*, *Hackeriapis*, *Heriadopsis* (in spite of arolia), and *Chelostomoda* (in spite of a partial cutting edge in the second interspace) resemble one another in having postgradular, often fasciate depressions (suggesting *Coelioxys*) and other characters. In Australia, however, they seem to intergrade with *Callomegachile* and its relatives.

Matangapis is remarkable for having arolia on all the legs (the only other subgenus with arolia is *Heriadopsis*). I consider it a member of group 2 despite its megachiliform body.

Mitchellapis, like *Sayapis* and *Eumegachile*, has a partial cutting edge in the second interspace and could be a member of Group 1 in spite of an elongate body. But because it has marginal hairs on S8 of the male, I place it in Group 2.

Megella has broad female mandibles that suggest Group 1, with a cutting edge behind the third interspace

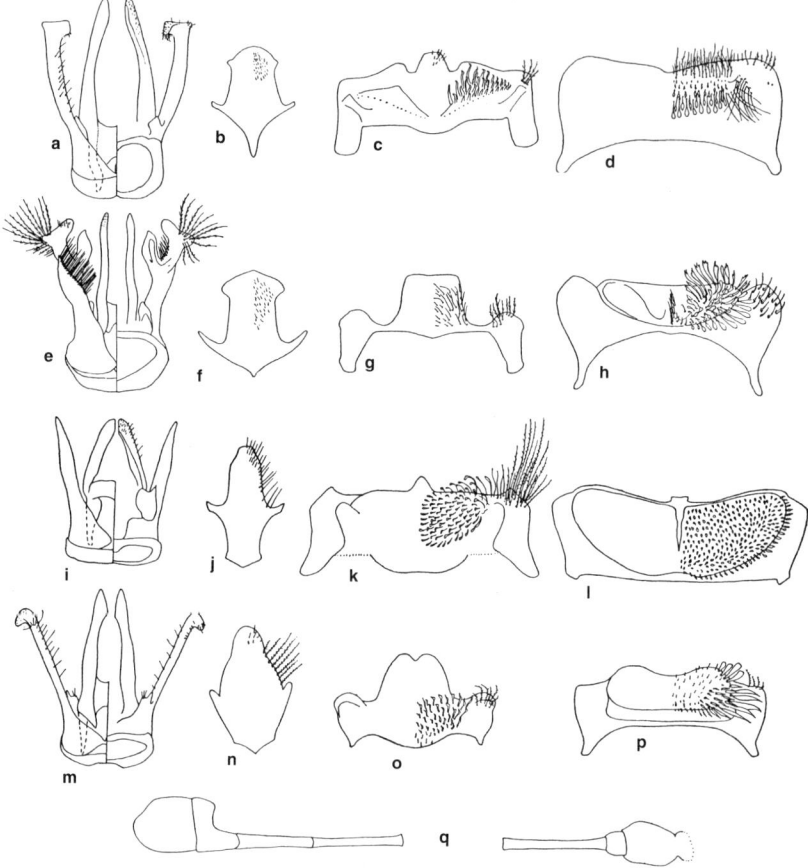

Figure 84-10. Structures of male *Megachile*. a-p, Genitalia (dorsal views are on left), S8, S6, and S5. **a-d,** *M. (Amegachile) mystacea* (Fabricius); **e-h,** *M. (Eutricharaea) chrysopyga* Smith; **i-l,** *M. (Austrochile) resinifera* Meade-Waldo; **m-p,** *M. (Chelostomoda) spissula* Cockerell. **q,** First three (at right) and last four antennal segments of male *M. (Thaumatosoma) duboulaii* Smith. By no means all species here placed in the subgenus *Eutricharaea* have the elaborate gonoforceps shown in e. In *Megachile*, S7 is a scarcely sclerotic transverse band and lacks interesting characters. From Michener, 1965b.

(Fig. 84-11h), but the elongate body and especially the hairs on the margins of S8 of the male are as in Group 2. This is one of the principal problem subgenera that bridges Groups 1 and 2; I have arbitrarily included it in Group 2.

Eumegachile is much like *Sayapis* and I place it in Group 1 even though there is one lateral hair on the basal lateral margin of S8, thus suggesting Group 2.

Members of Group 1 that lack (and perhaps have lost) cutting edges on the mandibles and thus resemble Group 2 have been considered above. The principal example is *Chrysosarus*, which has a body form indistinguishable from that of ordinary members of Group 1.

Not only do Groups 1 and 2 run together, but Group 3 also presents problems. What appeared to me in 1962 to be quite a distinct genus, *Creightonella*, on the basis of the female mandibles and the male sterna, seems not clearly diagnosed when the diverse African fauna is considered. The male sternal characters of *Creightonella* are plesiomorphies (to judge by sterna in the Osmiini), and given the considerable retraction of S5 and S6 in some species, approaching the condition in some *Callomegachile*, this character is not strong. Furthermore, the female mandibular characters fail. Thus in *Megachile (Creightonella) cornigera* Friese, the only well-formed partial cutting edge is in the second interspace, even though the mandibles are six-toothed and clearly of the *Creightonella* type. More important, in *M. adeloptera* Schletterer and presumably in its close relatives (the *adeloptera* group of Pasteels, 1965) the female mandible has only four teeth, the cutting edge in the second interspace is much larger than that in other *Creightonella* although not complete, and the other cutting edge is free-standing in a broad interspace between the third and fourth teeth. If this cutting edge represents a modified tooth, then the mandible is still of the basic *Creightonella* type. In any event, the distinction between *Creightonella* (Group 3) and other Megachilini is about as weak as that between Groups 1 and 2.

The proposal of Mitchell (1980) to divide *Megachile* s. l. into eight genera, two of them subdivisions of *Chalicodoma* (in the sense of Michener, 1962a) and five of them

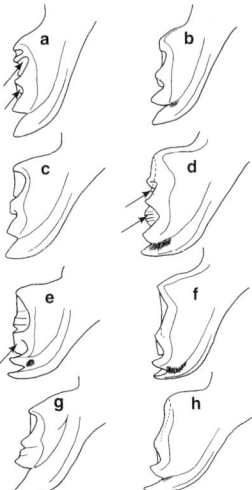

Figure 84-11. Diversity of female mandibles of leafcutting *Megachile* (h is probably not a leafcutter). **a,** *M. (Xanthosarus) perihirta* Cockerell; **b,** *M. (X.) melanophaea* Smith; **c,** *M. (X.) addenda* Cresson; **d,** *M. (Paracella) semivenusta* Cockerell; **e,** *M. (Amegachile) bituberculata* Ritsema; **f,** *M. (Neocressoniella) anthracina* Smith; **g,** *M. (Eutricharaea) eurymera* Smith, a member of the group called *Eurymella;* **h,** *M. (Megella) malimbana* Strand. (In d, f, and h, the cutting edges in the area of the upper interspace are partly or entirely hidden behind the main tooth row and are shown by broken lines; some of the cutting edges are marked by arrows.) Drawings by D. Yanega.

subdivisions of *Megachile*, the other being *Creightonella*, does not now seem to me practical. Intermediates resulting from convergence or plesiomorphies destroy nearly all the "generic" lines, so that the "genera" are not clearly recognizable, even though some of them represent natural groups. (Some do not; I think *Phaenosarus* and *Xan-*

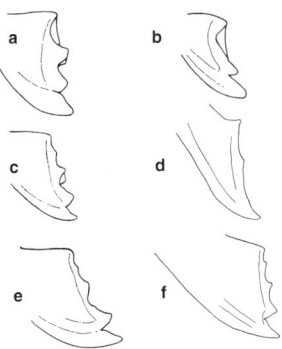

Figure 84-12. Mandibular apices of females of *Megachile*. **a,** *M. (Megachiloides) integra* Cresson; **b,** *M. (Megachiloides) oenotherae* (Mitchell); **c,** *M. (Sayapis) pugnata* Say; **d,** *M. (Chalicodomoides) aethiops* (Smith); **e,** *M. (Pseudomegachile) lanata* (Fabricius); **f,** *M. (Callomegachile) mystaceana* (Michener). Mandibles d to f lack cutting edges. a-c and e, from Mitchell, 1962; d and f, from Michener, 1965b.

thosarus are close relatives, but Mitchell put them in separate genera.)

Given the problems cited above, I am placing all of the nonparasitic Megachilini in the genus *Megachile*. I would have liked to present a classification that breaks up forms that fall in Group 2, somewhat as follows:

Matangapis
Mitchellapis
Megella
Chelostomoides (including *Callomegachile* and perhaps *Gronoceras; Thaumatosoma* may be a member of this group and that name has priority)
Chalicodoma (? including *Parachalicodoma*)

To formalize such a classification would result in many new combinations and, given its lack of objectivity, would be an inappropriate move at this time. What is needed is a proper character analysis and phylogenetic analysis before major classificatory changes are made. The grouping of the subgenera here recognized is indicated in Table 84-2.

It is important to note a series of characters that are conspicuous in males, might seem important in indicating relationships, and have often been used in the definition of subgenera. These characters, which probably relate to mating behavior, are as follows: (1) Enlarged, often elaborately fringed front tarsi, pale and sometimes translucent but often having dark spots on the concave inner surfaces; these may possibly fit over the eyes of the female during mating. (2) A large, erect spine on the front coxa. (3) Minute reddish bristles on the anterior surface of the front coxa, which is usually largely bare when the bristles are present; characters (2) and (3) may relate to positioning and holding the female. (4) A dentate or emarginate carina on T6. All of these are apomorphies relative to other tribes of the Megachilinae, and the list could be extended by other characters of T6, T7, and the sterna. But they all appear independently in various subgenera throughout the genus *Megachile*. Presumably, the genus has a tendency to produce these characters; perhaps they evolve in connection with certain developments (as yet unstudied) in mating behavior. In any event, if our understanding of the subgenera of *Megachile* is at all reliable, these features are not appropriate as the principal justification for recognizing groups, even though some groups do exhibit consistency in such characters. I am therefore doubtful about the recognition of certain subgenera whose principal characters are these male features. Such subgenera include *Trichurochile* and *Tylomegachile*, whose females seem to me doubtfully distinguishable from related South American subgenera, and *Phaenosarus*, whose females seem to be *Xanthosarus*.

Although the lack of arolia is a usual tribal character of the Megachilini, there are two taxa that possess arolia. *Megachile (Heriadopsis)* from Africa, a relative of Group 2 subgenera such as *Chelostomoda* and *Chelostomoides*, has arolia on the front and middle legs. Second, *Megachile (Matangapis)* from Borneo, unrelated to *Heriadopsis*, is megachiliform and has arolia on all legs; it, too, is a member of group 2.

Male genitalia and other structures of *Megachile* were illustrated by Mitchell (1934-1937, 1962, 1980), and

Table 84-1. Comparison of Some Characters of Certain Subgenera of *Megachile*.

Unlisted subgenera agree with Group 1 or Group 2; the "problem taxa" fall between "Most of Group 1" and "Most of Group 2" in the list below.

Subgenus	Character			
	Female		Male	
	Shape of metasoma	Mandibular apical margin	Number of exposed sterna	Long marginal hairs on S8
Most of Group 1, such as *Megachile* s. str.	megachiliform	3-5 teeth, with cutting edges in second or third interspace or both (Figs. 84-11; 84-12a, b)	4	Absent[a]
Chrysosarus	megachiliform	4-5 teeth, no cutting edges	4	Absent
Stelodides	chalicodomiform	5 teeth, no cutting edge	4	Absent
Schrottkyapis	chalicodomiform	4 teeth, no cutting edge	4	Absent
Grosapis	short chalicodomiform	4 teeth, cutting edge probably in second interspace only	3	Absent
Sayapis	chalicodomiform	4-5 teeth, cutting edge in second interspace only (Fig. 84-12c)	4	Absent
Eumegachile	chalicodomiform	4 teeth, cutting edge in second interspace only	4	Absent or one on each side
Mitchellapis	chalicodomiform	5 teeth, cutting edge in second interspace only	4	Present
Megella	chalicodomiform	4 teeth, cutting edge hidden behind third interspace (Fig. 84-11h)	4	Present
Chelostomoda	heriadiform or hoplitiform	5 teeth, cutting edge in second interspace only	3	Present
Matangapis	megachiliform	4-teeth, no cutting edges	3?	Present (weak)
Heriadopsis	heriadiform or hoplitiform	4-teeth, no cutting edges	3	Present

(*continues*)

Table 84-1. (*continued*)

Subgenus	Character			
	Female		Male	
	Shape of metasoma	Mandibular apical margin	Number of exposed sterna	Long marginal hairs on S8
Most of Group 2, such as *Chalicodoma* and *Chelostomoides*	chalicodomiform to heriadiform	2-5 teeth, no cutting edges (Fig. 84-12d-f)	3-4	Present
Group 3 *Creightonella*	chalicodomiform	4-6 nearly equally spaced teeth; incomplete cutting edges usually in interspaces 2-4 (Fig. 84-17d)	5-6	Present

[a] Mitchell's (1980) depiction (fig. 33) of marginal hairs in *Megachile oenotherae* Mitchell is an error, as shown by his accompanying description.

Michener (1965b) and in numerous works by Tkalcŭ and others; see also Figure 84-10. S7 is extremely reduced, as in *Coelioxys*, and is not illustrated.

Raw (2002) gave subgeneric positions for many of the neotropical species of *Megachile* that previously had not been assigned to subgenera. Thus the numbers of species in certain subgenera were augmented.

Some keys and revisional works on *Megachile* s. l. are as follows: Mitchell (1934-1937) for America north of Mexico, Schrottky (1913b) for Brazil, Mitchell (1930) for the neotropics, Perkins (1925) for Britain, Benoist (1940) for France, Hirashima and Maeta (1974) for northern Japan, and Yasumatsu and Hirashima (1965) for Taiwan.

Floral relationships of *Megachile* are diverse. Many species are moderately polylectic. For example, *M. brevis* Cresson visits whitish, blue, purple, or pink flowers of many families for pollen, although it rarely collects pollen from yellow Asteraceae such as *Helianthus* (Michener, 1953b). Other species, however, are strongly oligolectic. Examples are *Megachile fortis* Cresson, which collects pollen from sunflowers (*Helianthus*) (Neff and Simpson,

Table 84-2. Subgenera of *Megachile,* Placed in the Three Groups Discussed in the Text.

Group 1		Group 2	Group 3
Acentron	Ptilosarus	Austrochile	Creightonella
Amegachile	Rhyssomegachile	Callomegachile	
Argyropile	Sayapis	Cestella	
Austromegachile	Schrottkyapis	Chalicodoma	
Chrysosarus	Stelodides	Chalicodomoides	
Cressoniella	Trichurochile	Chelostomoda	
Dasymegachile	Tylomegachile	Chelostomoides	
Eumegachile	Xanthosarus	Cuspidella	
Eutricharaea	Zonomegachile	Gronoceras	
Grosapis		Hackeriapis	
Leptorachis		Heriadopsis	
Litomegachile		Largella	
		Matangapis	
Megachile s. str.		Maximegachile	
Megachiloides		Megella	
Melanosarus		Mitchellapis	
Moureapis		Parachalicodoma	
Neochelynia		Pseudomegachile	
Paracella		Rhodomegachile	
Platysta		Schizomegachile	
Pseudocentron		Stenomegachile	
Ptilosaroides		Thaumatosoma	

1991), and *M. campanulae* (Robertson), which collects pollen from *Campanula*.

The keys are divided geographically as follows: Western Hemisphere, palearctic and oriental, sub-Saharan Africa, and Australian and Papuan. For *Neochalicodoma* and *Stellenigris*, see the end of the section on the genus *Megachile*.

Zonomegachile is omitted from the following key to females of the Western Hemisphere because of lack of material. Females thought to be associated with males of *Zonomegachile* would run to couplet 23 *(Dasymegachile, Litomegachile, Megachiloides)* in the key, and differ from the last two in being South American rather than nearctic. There is a large but incomplete cutting edge in the second interspace (Mitchell, 1980, fig. 51); this character excludes *Zonomegachile* from both alternatives of couplet 23.

In keys to males, the number of exposed metasomal sterna is often important. In some cases, however, owing to artificial straightening of the metasoma, sterna that are ordinarily not or but little exposed become broadly exposed. Exposed sterna are usually punctate and hairy in a manner rather similar to one another, whereas hidden sterna, including those that become artificially exposed by straightening the metasoma, are more delicate, less punctate, with shorter and often flattened or otherwise modified hairs.

Key to the Subgenera of *Megachile* of the Western Hemisphere (Females)
(Partly from Mitchell, 1943, 1980)

1. S6 with at least posterior half bare or nearly so, except for subapical row of short hairs, behind which is bare, smooth rim directed posteriorly (body megachiliform) .. 2
—. S6 with well-dispersed scopal hairs, *or, if* partly bare, then without bare apical rim behind transverse fringe of short hairs, *or* (in *Argyropile*) rim directed upward, *or* rim narrow and barely recognizable 6
2(1). Mandible five-toothed, a long cutting edge in second interspace, none elsewhere .... *M. (Melanosarus)*
—. Mandible four-toothed, a well-formed cutting edge in third interspace 3
3(2). Second interspace distinct, with cutting edge usually present 4
—. Second interspace lacking or small, without cutting edge 5
4(3). Inner angle of mandible truncate, *or* apical margin of clypeus impressed medially (S6 rather narrowly truncate) (neotropical) .... *M. (Moureapis)*
—. Inner angle of mandible acute or rounded; clypeal margin straight and entire .... *M. (Pseudocentron)*
5(3). Mandible more robust, apical tooth more protuberant, much broader than other teeth; gena usually broader than eye in lateral view .... *M. (Acentron)*
—. Mandible less robust, apical tooth not much broader than second or third; gena usually narrow .... *M. (Leptorachis)*
6(1). Mandible without cutting edges between teeth (Fig. 84-12e, f), or with incomplete cutting edge in second interspace only (Fig. 84-12c); mandible with less than five teeth, *or, if* five-toothed, then upper two teeth (4 and 5) usually closer together than teeth 3 and 4 7
—. Mandible with cutting edges between teeth, if in second interspace only, then edge complete (in three-toothed mandible; Fig. 84-12b), *or* mandible clearly five-toothed, with teeth 4 and 5 about as far apart as 3 and 4 (body megachiliform) 16
7(6). Mandible with incomplete cutting edge in second interspace, and no cutting edges elsewhere (Fig. 84-12c) 8
—. Mandible without cutting edges (Fig. 84-12e, f) 9
8(7). Body very large and robust (20 X 10 mm); vestiture entirely rufous; profile of T6 straight (Mexico) .... *M. (Grosapis)*
—. Body not so large and robust; vestiture not entirely rufous, that of metasomal terga dark with pale apical bands; profile of T6 strongly concave distally because of projecting apical margin .... *M. (Sayapis)*
9(7). Megachiliform, the metasoma less than twice as long as wide unless unusually extended 10
—. Heriadiform, hoplitiform, or chalicodomiform, the metasoma twice as long as wide or more 11
10(9). T6 distinctly concave in profile, without conspicuous erect pubescence except near base (holarctic) .... *M. (Megachile s. str.)* (in part)
—. T6 nearly straight in profile, with abundant erect pubescence (neotropical) .... *M. (Chrysosarus)*
11(9). Pronotal lobe with transverse hairless lamella hidden among hairs; mandible with third interspace narrowly U-shaped and much deeper than others (South America) .... *M. (Schrottkyapis)*
—. Pronotal lobe with transverse, unusually hairy ridge, sometimes with shiny but low carina; mandible with third interspace not narrower and deeper than others.... 12
12(11). Pubescence largely white, not fulvous, forming narrow white apical bands on metasomal terga and sometimes narrow bands on postgradular grooves .... *M. (Chelostomoides)*
—. Pubescence with large areas of black or fulvous, forming striking color pattern (gray in American *Gronoceras*) 13
13(12). Mandible five-toothed; pubescence of thorax and metasoma black except for broad white band on T3 (Chile) .... *M. (Stelodides)*
—. Mandible four-toothed; pubescence otherwise (introduced, Caribbean area) 14
14(13). Apex of front tibia with three distinct sharp spines or teeth on outer surface; clypeus with longitudinal elevation, highest at lower clypeal margin (?Jamaica) .... *M. (Gronoceras)*
—. Apex of front tibia with two teeth or spines on outer surface; clypeus unmodified or not modified as above 15
15(14). Mandibular carinae minutely roughened, sometimes dull .... *M. (Callomegachile)*
—. Mandibular carinae shining and smooth (at 40 X) .... *M. (Pseudomegachile)*
16(6). S6 with apical rim directed upward beyond fringe of hairs, this rim conspicuous if tergum and sternum are spread apart (nearctic) .... *M. (Argyropile)*
—. S6 without apical rim, *or,* if rim present, then directed posteriorly and usually inconspicuous 17
17(16). Mandible three-toothed (Fig. 84-12b) or median tooth weakly divided and mandible thus obscurely four-toothed, with cutting edge limited to upper interspace (second if mandible tridentate, third if mandible quadridentate) (nearctic) .... *M. (Megachiloides)* (in part)

—. Mandible four- or five-toothed, with cutting edges in third and usually second interspaces, or rarely in second only (Figs. 84-11a-c, 84-12a) 18
18(17). Thoracic venter, including leg bases and S2, with dense covering of fine, plumose hairs, sharply differentiated from other scopal hairs (neotropical) .... *M. (Ptilosarus)*
—. Thoracic venter and leg bases with ordinary hairs, and scopal hairs all unbranched 19
19(18). Metasomal sterna with entire and conspicuous white apical hair fasciae beneath scopa 20
—. Metasomal sterna with white hair fasciae absent or broadly interrupted medially 21
20(19). Mandible four-toothed, no cutting edge in second interspace (introduced, North America, Antilles, Chile, Argentina) .... *M. (Eutricharaea)*
—. Mandible with fourth tooth emarginate, thus five-toothed, second interspace with conspicuous but incomplete cutting edge (South America) .... *M. (Trichurochile)*
21(19). Mandible four-toothed, upper tooth acute or right-angular (Fig. 84-12a) 22
—. Mandible four- or five-toothed, but *if* four-toothed, then upper tooth rounded, truncate, or incised (sometimes only minutely) and thus approaching the five-toothed condition (Fig. 84-11a-c) 25
22(21). Metasoma broadly conical, T3 narrower than T1 or T2 (neotropical) .... *M. (Tylomegachile)* (in part)
—. Metasoma more ovoid, T3 as broad as T1 23
23(22). T6 straight in profile; mandible with second tooth often rounded or obtuse (Fig. 84-12a), usually no cutting edge in second interspace (nearctic) .... *M. (Megachiloides)* (in part)
—. T6 usually concave in profile; mandible with second tooth acute, a small beveled cutting edge in second interspace 24
24(23). Scopa black; body usually covered with long, dense hairs (South America) .... *M. (Dasymegachile)*
—. Scopa white except on S6; body not densely covered with long hairs (nearctic) [The female of *Megachile (Xanthosarus) addenda* Cresson (couplet 24 of the first key) runs to *Litomegachile,* from which, usually, it is immediately distinguishable by its large size, 12-17 mm in body length] .... *M. (Litomegachile)*
25(21). Mandible clearly five-toothed, distance between upper two teeth not or only slightly less than distance between other pairs of teeth (holarctic) .... *M. (Megachile s. str.)* (in part)
—. Mandible four-toothed but upper tooth rounded, truncate, or itself bidentate (sometimes minutely), mandible thus five-toothed but distance between upper two teeth short compared to distances between other pairs of teeth (Fig. 84-11a-c) 26
26(25). Metasoma distinctly conical, T1 and T2 broader than T3 27
—. Metasoma more ovoid, T3 as broad as or broader than T1 28
27(26). Metasomal sterna with widely interrupted apical white hair fasciae; posterior apical angle of hind basitarsus slightly produced, that of segment 2 more conspicuously so (neotropical) .... *M. (Austromegachile)* (in part)
—. Metasomal sterna not at all fasciate; segments 1 and 2 of hind tarsus not or little produced apically (neotropical) .... *M. (Tylomegachile)* (in part)
28(26). Median area of clypeus somewhat elevated and strongly flattened, sloping away on each side (apical margin of clypeus medially emarginate) (neotropical) .... *M. (Austromegachile)* (in part)
—. Clypeus broadly convex or nearly flat, neither elevated nor flat medially 29
29(28). T6 with much conspicuous, erect pubescence visible in profile 30
—. Pubescence of T6 largely decumbent, with few or no erect hairs visible in profile 31
30(29). Lateral ocellus considerably nearer to posterior margin of vertex than to eye .... *M. (Cressoniella)*
—. Lateral ocellus usually as near as or nearer to eye than to posterior margin of vertex (holarctic)...... *M. (Xanthosarus)*
31(29). Preoccipital carina distinct behind gena; cutting edges of mandible obsolescent (lateral ocellus widely removed from occipital margin) (South America) .... *M. (Rhyssomegachile)*
—. Preoccipital margin of gena usually not carinate, but *if* so, then cutting edges of mandible well formed 32
32(31). Thorax and metasoma above densely and minutely punctate throughout, largely covered with appressed or suberect tomentum (neotropical) .... *M. (Ptilosaroides)*
—. Punctures of thorax and metasoma distinctly separated, surface not tomentose to any considerable degree .... *M. (Neochelynia)*

Key to the Subgenera of *Megachile* of the Western Hemisphere (Males)

1. Middle tibial spur absent or much shorter than apical width of tibia, sometimes immovably fused to tibia, and middle basitarsus not or little modified (body megachiliform) 2
—. Middle tibial spur present, articulated to tibia, about as long as apical tibial width, or, *if* absent (as in some species of *Xanthosarus*), then middle basitarsus modified and swollen 6
2(1). Middle tibial spur present, articulated, but small .... *M. (Leptorachis)*
—. Middle tibial spur absent or represented by prong immovably fused to tibia 3
3(2). Middle tibia with a spurlike apical prong (spur presumably fused to tibia), prong sometimes reduced to large, acute tooth .... *M. (Pseudocentron)*
—. Middle tibia without such a process 4
4(3). Front and middle tibia and tarsus simple and unmodified; front tarsus slender, usually black (neotropical) .... *M. (Moureapis)*
—. Front and middle tibia and tarsus modified, middle tibia broadened apically or angulate on lower margin; basitarsus usually excavated along anterior margin; front tarsus dilated and brightly colored............................ 5
5(4). Scutum finely and densely rugoso-punctate, punctures not individually distinguishable .... *M. (Acentron)*
—. Punctures of scutum usually well separated, but, if close, then individually distinguishable .... *M. (Melanosarus)*
6(1). S4 not exposed or only its posterior margin exposed (Fig. 84-13a); punctation and vestiture of S4 (except sometimes for posterior margin) reduced and different from those of S3 7

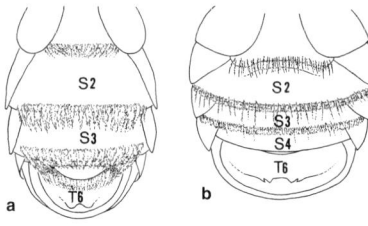

Figure 84-13. Undersurfaces of male metasomas of *Megachile*. **a,** *M. (Chelostomoides) campanulae* (Robertson); **b,** *M. (Litomegachile) mendica* Cresson.

—. S4 exposed, thus four exposed sterna (Fig. 84-13b), punctation and vestiture of S4 similar to those of S3 8

7(6). Mandible three-toothed, toothed margin much shorter than distance from upper tooth to base of mandible; body not covered with rufous hairs .... *M. (Chelostomoides)*

—. Mandible four-toothed, elongate, toothed margin as long as distance from upper tooth to base of mandible; body covered with long, rufous hairs (Mexico) .... *M. (Grosapis)*

8(6). Pronotal lobe with erect, hairless transverse lamella (clypeus protuberant medially; mandible as described for *Grosapis* in couplet 7) (South America) .... *M. (Schrottkyapis)*

—. Pronotal lobe rounded or with transverse, usually hairy ridge, sometimes with shiny, bare, but low carina 9

9(8). S8 with hairs on lateral margin (as in Fig. 84-10j, n); body chalicodomiform, with large areas of black or fulvous hairs forming a striking color pattern (except in our species of *Gronoceras*) (introduced into Caribbean region) 10

—. S8 without marginal hairs (as in Fig. 84-10b, f) but discal hairs sometimes extending laterally beyond margin; body usually megachiliform, usually without a striking color pattern (but see *Stelodides*, couplet 25) 12

10(9). T6 with preapical brush of long hairs and two long, slender spines representing preapical carina (Fig. 84-18) .... *M. (Gronoceras)*

—. T6 without brush of long hairs and without long spines..11

11(10). T6 with carina short, low, not or shallowly emarginate, not denticulate (Fig. 84-14c) ...... *M. (Callomegachile)*

—. T6 with carina strong, strongly dentate or denticulate (Fig. 84-14a), or sometimes scarcely undulate (Fig. 84-14b) .... *M. (Pseudomegachile)*

12(9). Mandible lacking inferior projection or tooth 13

—. Mandible with definite projection, tooth, or angle on lower margin 24

13(12). Front coxal spine present 14

—. Front coxal spine absent 19

14(13). Front coxal spine short, inconspicuous; basal segment of flagellum shorter than pedicel; T6 projecting posteriorly, thus nearly horizontal above carina (small, slender species) .... *M. (Neochelynia)* (in part)

—. Front coxal spine longer, conspicuous; basal segment of flagellum usually longer than pedicel (but much shorter in *Rhyssomegachile*); T6 more nearly vertical, usually not visible from above 15

15(14). Carina of T6 without emargination but with small median apical point (neotropical) .... *M. (Tylomegachile)*

—. Carina of T6 with a deep, rounded emargination (as in Fig. 84-14a) 16

16(15). Preoccipital carina strong behind genal area (front tarsus slender, dark) (South America) .... *M. (Rhyssomegachile)*

—. Preoccipital carina absent 17

17(16). Anterior end of hypostomal area, immediately behind mandible, with strong, angular projection (South America) .... *M. (Zonomegachile)*

—. Anterior end of hypostomal area unmodified 18

18(17). Front tarsus slender, black; carina of T6 represented principally by two spines, one on each side of emargination; mandible four-toothed (neotropical) .... *M. (Ptilosaroides)*

—. Front tarsus enlarged, pale; carina of T6 better developed, not represented merely by two spines; mandible three-toothed (neotropical) ...... *M. (Chrysosarus)* (in part)

19(13). First and second segments of flagellum subequal (mandible three-toothed, middle tooth sometimes notched, suggesting a four-toothed condition) (South America) .... *M. (Dasymegachile)*

—. First segment of flagellum shorter than second 20

20(19). Carina of T6 with pair of acute spines or teeth 21

—. Carina of T6 not bispinose, lateral portions on each side of emargination obtuse, if present 23

21(20). Large emargination between teeth of carina of T6 filled by dense fringes of long, plumose hairs largely arising from the teeth (mandible three-toothed) (South America) .... *M. (Trichurochile)*

—. Emargination between teeth of carina of T6 not filled by fringe 22

22(21). Mandible four-toothed; body length often 12 mm or more; pubescence erect and rather long...... *M. (Cressoniella)*

—. Mandible three-toothed; body smaller, length about 7 mm; pubescence short, appressed (neotropical) .... *M. (Ptilosarus)*

23(20). T6 more nearly horizontal, carina either deeply emarginate, with dorsal surface markedly concave, or surface convex, carina low, with only a small median notch .... *M. (Neochelynia)* (in part)

—. T6 vertical, completely hidden in dorsal view of metasoma, carina low and entire or with small median emargination (neotropical) .... *M. (Austromegachile)*

24(12). Metasoma about twice as long as wide (carina of T6 usually emarginate medially; front tarsus usually enlarged and pale; front coxa with spine and usually with red bristles) 25

—. Metasoma less than twice as long as wide 26

25(24). Pubescence of thorax and metasoma black except for broad white band on T3; mandible with small preapical inferior angle (Chile) .... *M. (Stelodides)*

—. Pubescence not forming above color pattern; mandible with large basal inferior projection .... *M. (Sayapis)*

26(24). Carina of T6 entire or crenulate, median part the most produced, with no trace of a median emargination 27

—. Carina of T6 commonly crenulate, median part emarginate (as in Fig. 84-14a) or sometimes irregular but not produced 29

27(26). Front tarsus slender and simple, black or fuscous;

front coxa pubescent anteriorly, without red bristles, spine short and slender; lower tooth of mandible slender and acute; apical segment of antenna not at all dilated, fully three times as long as broad (nearctic) .... *M. (Argyropile)* (in part)

—. Front tarsus usually dilated, ferruginous or yellowish; front coxa usually bare anteriorly, with broad, flat conspicuous spine and patch of red bristles; lower tooth of mandible usually robust; apical segment of antenna usually dilated, about twice as long as broad 28

28(27). S4 with small but distinct median tubercle on apical margin (large, robust species) (holarctic) .... *M. (Xanthosarus)* (in part)

—. S4 without median apical tubercle, apical margin usually broadly membranous (nearctic) .... *M. (Megachiloides)*

29(26). Mandible four-toothed 30

—. Mandible three-toothed 31

30(29). Front coxa usually largely bare anteriorly, often with patch of short red bristles in front of spine; front tarsus frequently modified, pallid (holarctic) .... *M. (Xanthosarus)* (in part)

—. Front coxa hairy, without red bristles; front tarsus simple, dark-colored (nearctic) .... *M. (Argyropile)* (in part)

31(29). Mandible with low median or preapical inferior angle in place of usual tooth (neotropical) .... *M. (Chrysosarus)* (in part)

—. Mandible with strong inferior basal tooth 32

32(31). Front tarsus broadly dilated, pale (holarctic) .... *M. (Xanthosarus)* (in part)

—. Front tarsus simple, black or nearly so 33

33(32). Front coxal spine reduced to inconspicuous tubercle or absent (holarctic) .... *M. (Megachile s. str.)*

—. Front coxal spine well developed 34

34(33). Morphological apical margin (not carina) of T6 without evident teeth (introduced) .... *M. (Eutricharaea)*

—. Morphological apical margin of T6 with four small but distinct teeth (nearctic) .... *M. (Litomegachile)*

Key to the Palearctic and Oriental Subgenera of *Megachile* (Females)

1. Mandible five- or six-toothed, teeth (except the first) similar and with similarly shaped incomplete cutting edges in second and third (and sometimes fourth) interspaces (Fig. 84-17d); apices of mandibular teeth roughly equidistant from nearest neighbors; preapical transverse mandibular groove distinct and filled with short, fine, pale hairs (these not found in other subgenera with cutting edges)... *M. (Creightonella)*

—. Mandible three- to five-toothed, without cutting edges, or, *if* with cutting edges, then teeth above first of different shapes and cutting edges often of different shapes, or only one cutting edge present; apices of mandibular teeth commonly separated from nearest neighbors by different distances; preapical transverse mandibular groove, in forms with cutting edges, absent, or, *if* present, then not filled with short, pale hairs 2

2(1). Mandible without cutting edges (Fig. 84-12d-f) (body chalicodomiform)... 3

—. Mandible with cutting edge in at least one interspace, sometimes hidden behind margin of interspace (Fig. 84-11) 8

3(2). Mandible with five fully distinct teeth, about equidistant (northern Africa) .... *M. (Parachalicodoma)*

—. Mandible with three or four teeth 4

4(3). Anterior margin of clypeus rounded (rarely weakly emarginate medially), strongly crenulate, produced well over base of labrum, not thickened; head little developed posteriorly, ocelloccipital distance thus not greater than interocellar distance (palearctic) ....*M. (Chalicodoma)*

—. Anterior margin of clypeus truncate or highly modified, usually not crenulate, often not much produced over base of labrum, but *if* rounded and somewhat crenulate (as in some *Pseudomegachile*), then margin thickened and impunctate; head usually much developed posteriorly, ocelloccipital distance thus greater than interocellar distance 5

5(4). Apex of front tibia with three spines, posterior one less acute and hidden by dense, short hairs; mandible strongly expanded apically, outer margin thus concave in basal half, narrowest part little more than half as wide as apical margin (oriental)................... *M. (Largella)*

—. Apex of front tibia with one or usually two spines, posterior one absent; mandible not so expanded apically, outer margin not or weakly concave, narrowest part more than half as wide as apical margin 6

6(5). Mandible elongate, more or less parallel-sided or narrowest preapically, with teeth across apex *and* mandibular ridges smooth and shining (at 40X) (southwestern Asia) .... *M. (Maximegachile)*

—. Mandible of ordinary shape or if elongated as above (as in some *Callomegachile*), then ridges minutely roughened and dull 7

7(6). Mandible with ridges largely shiny, not dulled by microsculpturing (at 40X) .... *M. (Pseudomegachile)*

—. Mandible with ridges dulled by microsculpturing........ .... *M. (Callomegachile)*

8(2). Mandible with cutting edge only in second interspace (as in Fig. 84-12c); body heriadiform, hoplitiform (as in Fig. 84-9), or chalicodomiform 9

—. Mandible with cutting edge in third interspace and frequently also in second; body megachiliform (Fig. 84-8) except metasoma over twice as long as broad in *Megella*) 10

9(8). Metasomal sterna at least laterally with apical fasciae of white hair under scopa; mandible of ordinary shape, five-toothed .... *M. (Chelostomoda)*

—. Metasomal sterna without apical fasciae; mandible over twice as long as basal width, four-toothed (palearctic) .... *M. (Eumegachile)*

10(8). Mandible with five teeth, distance between fourth and fifth little less than distance between other teeth (palearctic)... *M. (Megachile s. str.)*

—. Mandible with four teeth, or *if* with five, then distance between fourth and fifth less than half distance between third and fourth (Fig. 84-11a) 11

11(10). Metasoma over twice as long as broad; body over 20 mm long (cutting edge in third interspace, behind interspace margin and not or only slightly exposed in frontal view; Fig. 84-11h) (oriental) .... *M. (Megella)*

—. Metasoma less than twice as long as broad; body 16 mm long or less 12

12(11). Cutting edges large, completely filling second and third interspaces, teeth thus not extending beyond cutting edges and, together with adjacent teeth, forming a thin, generally straight although irregular mandibular margin (Fig. 84-11e); cutting edge in second interspace at least in part indistinguishably fused to third tooth (oriental) .... *M. (Amegachile)*
—. Cutting edges variable but not completely filling interspaces, teeth extending beyond cutting edges at least in some parts of mandibular margin (Fig. 84-11a-d, f, g); cutting edges usually nowhere indistinguishably fused to teeth 13
13(12). Upper mandibular tooth broad (convex or truncate) or emarginate so that the mandible is five-toothed (Fig. 84-11a); sterna without apical hair bands beneath scopa; third interspace of mandible frequently deeper than others (palearctic) .... *M. (Xanthosarus)*
—. Upper mandibular tooth acute, sharply right-angular, or narrowly rounded; sterna frequently with apical white hair bands beneath scopa; third interspace of mandible not or little deeper than others 14
14(13). Mandible without or with only hidden or very small cutting edge in second interspace but with cutting edge in third interspace (rarely hidden behind interspace margin) (Fig. 84-11g) .... *M. (Eutricharaea)*
—. Mandible with well-developed cutting edges in second and third interspaces (Fig. 84-11d) 15
15(14). Mandible with upper interspace as long as or longer than rest of apical margin, cutting edge in upper interspace complete, occupying full length of interspace and over twice as long as incomplete cutting edge in second interspace (oriental) .... *M. (Neocressoniella)*
—. Mandible with upper interspace shorter than rest of apical margin, cutting edge in upper interspace complete or incomplete, sometimes hidden, less than twice as long as incomplete cutting edge in second interspace (oriental) .... *M. (Paracella)*

Key to the Palearctic and Oriental Subgenera of *Megachile* (Males)

(*Megella* is omitted because the male of the oriental species is unknown to me.)

1. S5 (sometimes also S6) exposed and generally similar to preceding sterna (sometimes S5 largely hidden but S6 exposed); lateral extremity of carina of T6 directed basad, away from apical margin of tergum ...... *M. (Creightonella)*
—. S5 and S6 retracted, variously modified, less sclerotized, less punctate, and less hairy than S1 to S4; lateral extremity of carina of T6 absent or directed toward lateral extremity of apical margin of tergum 2
2(1). S8 with lateral marginal hairs (Fig. 84-10j, n) (only one in *Eumegachile*); metasoma commonly strongly convex and twice as long as wide or more 3
—. S8 without marginal hairs (Fig. 84-10b, f) but discal hairs sometimes extending beyond margin laterally; metasoma usually less strongly convex and usually less than twice as long as wide 12
3(2). S4 largely or wholly retracted (as in Fig. 84-13a), less sclerotized, less punctate, and less hairy than S2 and S3 (body heriadiform or hoplitiform) ...... *M. (Chelostomoda)*
—. S4 exposed, similar in punctation and pubescence to preceding sterna (as in Fig. 84-13b) 4
4(3). T6 with carina dentate or denticulate, sometimes with median emargination but denticulate lateral to it (Fig. 84-14a), sometimes bilobed but margin with at least a few faint irregularities (Fig. 84-14b); surface of T6 above carina usually without median depression 5
—. T6 with carina bilobed or sometimes simple, not at all toothed (Fig. 84-14c); surface of T6 above carina usually with median depression 8
5(4). T7 with narrow, median, apically truncate projection extending well beyond teeth of T6 carina (northern Africa) .... *M. (Parachalicodoma)*
—. T7 a low sclerite largely hidden behind T6, sometimes produced to small median spine 6
6(5). Front coxa with erect spine ....
.... *M. (Pseudomegachile)* (in part)
—. Front coxa without spine or with tubercle or short spine 7
7(6). Toothed margin of mandible (three- to four-toothed) strongly oblique, nearly as long as distance from upper tooth to mandibular base .... *M. (Chalicodoma)*
—. Toothed margin of mandible (three-toothed) less oblique, much shorter than distance from upper tooth to mandibular base .... *M. (Pseudomegachile)* (in part)
8(4). Hairs of middle of T6 above carina extremely long, extending well beyond carina; carina of T6 high, strongly bilobed (T6 conspicuously acute in lateral view) 9
—. Hairs of middle of T6 immediately above carina not very long, not extending far beyond carina, although dorsum of T6 may have very long hairs laterally; carina of T6 usually lower, not or rather weakly bilobed 11
9(8). Posterior margin of T6 with strong lateral tooth (partly obscured by hair); apex of clypeus with small median nodule continued up clypeus for a distance as a carina (Asia Minor) .... *M. (Maximegachile)*
—. Posterior margin of T6 without lateral tooth; clypeus without median nodule and carina 10
10(9). Front tibia with apical posterior angle, which is obtuse but distinct and covered with short hair (oriental) .... *M. (Largella)*
—. Front tibia without apical posterior angle, or, *if* weakly evident, then angle not covered with short hair .... *M. (Pseudomegachile)* (in part)
11(8). Mandible with strong premedian projection from lower margin at right angles to axis of mandible; T7 with large, exposed, biconvex, punctate surface; mandibular carinae shining and smooth (at 40×) (palearctic) .... *M. (Eumegachile)* (in part)
—. Mandible without inferior projection or with hairy convexity or basal lobe; T7 hidden or exposed as narrow rim; mandibular carinae usually dulled by minute sculpturing .... *M. (Callomegachile)*
12(2). T7 with large, exposed, biconvex, punctate surface; body chalcidomiform (palearctic) .... *M. (Eumegachile)* (in part)
—. T7 hidden or exposed as narrow rim or crescentic sclerite, sometimes prolonged to median spine; body megachiliform .... *M. (Amegachile, Eutricharaea, Megachile* s. str., *Neocressoniella, Paracella, Xanthosarus)*

Key to the Sub-Saharan Subgenera of *Megachile* (Females)

(See the long note following the key to males, below.)

1. Mandible four- to six-toothed, all except lowermost teeth similar in shape with incomplete, similarly shaped cut-

ting edges in second and third (sometimes also fourth) interspaces (Fig. 84-17d), rarely [*M. (C.) cornigera* (Friese)] in second interspace only, and *if* mandible four-toothed, then cutting edges sometimes dissimilar; metasoma strongly convex, often twice as long as broad, parallel sided, body chalicodomiform; preapical transverse mandibular groove distinct and filled with short, fine, pale hairs (this is diagnostic in combination with presence of cutting edges) .... *M. (Creightonella)*

—. Mandible three- to five toothed, rarely seven-toothed, without cutting edges, or, *if* with cutting edges, then second and higher teeth of different shapes and cutting edges of different shapes, or only one cutting edge present (Fig. 84-11); metasomal shape variable; preapical transverse mandibular groove, in forms with cutting edges, absent, or, *if* present, then not filled with short, pale hairs.. 2

2(1). Mandible without cutting edges between teeth (Fig. 84-12e, f); S6 usually with apical hairs like scopal hairs of nearby surface of sternum, sometimes with bare rim; body chalicodomiform, metasoma strongly convex, more or less parallel-sided, and commonly two or more times as wide as long (as in Fig. 84-9) 3

—. Mandible usually with cutting edge in interspace between at least one pair of teeth, cutting edge rarely hidden behind interspace margin (Fig. 84-11); S6 with apical (or preapical if there is bare rim) fringe of dense, short, often plumose hairs different from scopal hairs; body megachiliform (except in *Megella*), thus metasoma more or less flattened, cordate, tapering to apical point, usually less than twice as long as wide 12

3(2). Apex of front tibia on outer surface with three distinct, sharp, fully exposed teeth or spines, spaces between them shining .... *M. (Gronoceras)*

—. Apex of front tibia with one or two spines, or, *if* (rarely) with three spines, then posterior spine a mere tubercle or covered by a patch of short hairs, or *if* distinct and pointed, then spaces between spines with punctures and hairs, not noticeably shining 4

4(3). Arolia present on front and middle legs ....
.... *M. (Heriadopsis)*

—. Arolia absent 5

5(4). Anterior margin of clypeus rounded (rarely weakly emarginate medially), strongly crenulate, produced well over base of labrum, not thickened; head little developed posteriorly, ocelloccipital distance thus not greater than interocellar distance .... *M. (Chalicodoma)*

—. Anterior margin of clypeus truncate or highly modified, usually not crenulate, often not much produced over base of labrum, but *if* rounded and somewhat crenulate (as in some *Pseudomegachile*), then margin usually thickened and impunctate; head much developed posteriorly, ocelloccipital distance thus usually greater than interocellar distance 6

6(5). Apex of front tibia with three spines, posterior spine less acute than others and hidden by dense, short hairs; mandible strongly expanded apically, outer margin thus concave in basal half and narrowest part little over half as wide as apical margin (Zanzibar) .... *M. (Largella)*

—. Apex of front tibia with one or usually two spines, posterior spine as found in *Largella* being absent; mandible not so expanded apically, outer margin not or weakly concave, narrowest part more than half as wide as apical margin 7

7(6). Mandible elongate, approximately parallel-sided or narrowest preapically, *and* mandibular ridges smooth and shining (at 40×); posterior hypostomal area with strong tooth 8

—. Mandible of ordinary shape, or, *if* elongate as above (as in some *Callomegachile*), then ridges minutely roughened and dull; posterior hypostomal area without a tooth, sometimes with obtuse angle 9

8(7). Mandible three-toothed .... *M. (Maximegachile)*

—. Mandible four-toothed .... *M. (Stenomegachile)*

9(7). Mandible with ridges largely shiny, not dulled by microsculpturing (at 40×) 10

—. Mandible with ridges dulled by microsculpturing 11

10(9). Apical margin of clypeus broadly and deeply emarginate, lower part of clypeus strongly depressed, separated from upper part by curved ridge ...... *M. (Cuspidella)*

—. Clypeus not or little modified, truncate, margin sometimes crenulate .... *M. (Pseudomegachile)*

11(9). Mandible five-toothed, with deep, rounded emargination between third and fourth teeth, fourth and fifth teeth (interpretable jointly as angularly truncate fourth tooth) connected by straight margin basal to level of first three teeth; clypeus with large, deep, triangular, shining, hairless area in middle of which, arising from impunctate surface, is a large tubercle (Madagascar) .... *M. (Cestella)*

—. Mandible three- to seven-toothed, without especially deep emargination and with no teeth displaced basad; clypeus truncate to highly modified, but not as above ..
.... *M. (Callomegachile)*

12(2). (See note at end of key to African male *Megachile*, below.) Mandible without or with only hidden cutting edge in second interspace but with cutting edge in third interspace, this edge sometimes hidden behind interspace margin (in a few species in the *Eurymella* group of *Eutricharaea*, such as *Megachile michaelis* Cockerell and *dolichognatha* Cockerell, second and third teeth fused, the long margin below upper tooth then with cutting edge on upper part only) 13

—. Mandible with distinct cutting edges in second and usually third interspaces 15

13(12). Mandible five-toothed, upper two teeth close together (interpretable jointly as emarginate fourth tooth) but with small cutting edge between them, i.e., in fourth interspace, in addition to large but rather incomplete cutting edge in third interspace .... *M. (Platysta)*

—. Mandible four-toothed or rarely three-toothed, with cutting edge only in uppermost interspace or sometimes also with faint indication of cutting edge (or cutting edge hidden behind interspace margin) in second interspace, attached to lower edge of third tooth 14

14(13). Mandible much broadened apically, outer margin thus strongly concave; body length 18 mm or more; metasoma more than twice as long as wide .... *M. (Megella)*

—. Mandible of ordinary shape, outer margin not strongly concave; body length much less than 18 mm; metasoma less than twice as long as wide .... *M. (Eutricharaea)*

15(12). Mandible with third tooth truncate because fused to cutting edge of second interspace, fourth tooth acute; cutting edges usually crenulate (Fig. 84-11e) ....
.... *M. (Amegachile)*

—. Mandible with third tooth pointed or rounded, fourth tooth usually rather broad, rounded or truncate, rarely acute, rarely emarginate so that mandible five-toothed; cutting edges not crenulate .... *M. (Paracella)*

Key to the Sub-Saharan Subgenera of *Megachile* (Males)
(See the long note following this key.)

1. S5 (sometimes also S6) exposed and generally similar to preceding sterna (sometimes S5 largely hidden but S6 exposed); lateral extremity of carina of T6 directed basad, away from apical margin of tergum .... *M. (Creightonella)*
—. S5 and S6 retracted, variously modified (less sclerotized, less punctate, and less hairy than S2 to S4); lateral extremity of carina of T6 absent or directed toward lateral extremity of posterior margin of tergum 2
2(1). S8 with lateral marginal hairs (Fig. 84-10j, n); metasoma commonly strongly convex and twice as long as wide or more, chalicodomiform 3
—. S8 without marginal hairs (Fig. 84-10b, f) but discal hairs sometimes extending beyond margin laterally; metasoma usually less strongly convex and usually less than twice as long as wide, megachiliform 15
3(2). Front tibia on outer surface ending distally in three bare spines or teeth, the posteriormost extended as carina along much of outer posterior margin of tibia; T6 with carina represented by long spines or long lobes, partly hidden by extremely long hairs arising before carina (Fig. 84-18) .... *M. (Gronoceras)*
—. Front tibia on outer surface ending in two spines or teeth (posterior tooth of *Gronoceras* sometimes represented by angle, which if strong is covered by short hair, see *Largella*); tibia usually without carina along outer posterior margin; T6 with carina variable, not represented by long spines, but if bilobed, then lobes much broader than long, surface of T6 usually without extremely long hairs, but *if* present, then such hairs sparse enough that carina almost always easily seen 4
4(3). Carina of T6 strongly produced medially to truncate, untoothed process with longitudinal median carina (T6 with strong lateral tooth) .... *M. (Cuspidella)*
—. Carina of T6 dentate or medially emarginate, not produced medially, without longitudinal median carina but sometimes with broad ridge 5
5(4). T6 with carina dentate or denticulate, sometimes with median emargination but denticulate lateral to it, sometimes bilobed but margin with at least a few irregularities (Fig. 84-14a, b), rarely with median emargination forming two large teeth and an additional large lateral tooth (T6 thus four-toothed); surface of T6 above carina commonly without median depression 6
—. T6 with carina bilobed (Fig. 84-14c) or sometimes simple, not at all toothed, or sometimes with small lateral tooth; surface of T6 above carina usually with median depression 9
6(5). Front coxa with erect spine ....
.... *M. (Pseudomegachile)* (in part)
—. Front coxa without spine or with tubercle or short spine 7
7(6). Toothed margin of mandible (three- to four-toothed) strongly oblique, nearly as long as distance from upper tooth to mandibular base .... *M. (Chalicodoma)*
—. Toothed margin of mandible (three-toothed) less oblique, much shorter than distance from upper tooth to mandibular base 8
8(7). Eyes unusually large, ocellocular distance thus much less than interocellar distance; T6 without lateral spine (Madagascar) .... *M. (Cestella)*
—. Eyes of ordinary size, ocellocular distance about equal to interocellar distance; T6 with strong lateral spine ......
.... *M. (Pseudomegachile)* (in part)
9(5). Arolia present on front and middle legs ....
.... *M. (Heriadopsis)*
—. Arolia absent 10
10(9). Hairs of T6 above middle of carina extremely long, extending well beyond carina; carina of T6 high, strongly bilobed, T6 conspicuously acute in lateral view 11
—. Hairs of T6 above middle of carina not very long, not extending far beyond carina, although laterally dorsum of T6 may have very long hairs; carina of T6 commonly lower, not or weakly bilobed [in *M. (Megella?) exsecta* Pasteels with deep median emargination leaving two teeth, not lobes] 14
11(10). Front tibia with longitudinal carina along outer posterior angle; hypostoma with large tooth close behind mandibular base; hairs of T6 so long and dense as to almost hide carina .... *M. (Stenomegachile)*
—. Front tibia without longitudinal carina; no large tooth behind mandibular base; hairs of T6 not at all obscuring carina 12
12(11). Posterior margin of T6 with strong lateral tooth (partly obscured by hair); apex of clypeus with small median nodule continued up clypeus as carina ....
.... *M. (Maximegachile)*
—. Posterior margin of T6 without lateral tooth; clypeus without median nodule and carina 13
13(12). Apex of front tibia with posterior angle (indicating third apical spine) obtuse but distinct and covered with short hair .... *M. (Largella)*
—. Apex of front tibia without posterior angle, or, if weakly evident, then not covered with short hair ....
.... *M. (Pseudomegachile)* (in part)
14(10). Posterior margin of T6 simple .... *M. (Callomegachile)*
—. Posterior margin of T6 with slender, mesally directed spine at each extreme side and a mediolateral convexity representing a tooth .... *M. (Megella)*

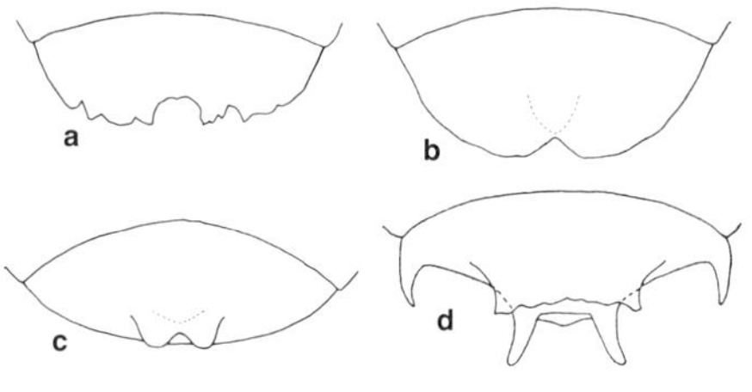

Figure 84-14. T6 of males of *Megachile*. **a**, *M. (Pseudomegachile) ericetorum* Lepeletier; **b**, *M. (P.) lanata* (Fabricius); **c**, *M. (Callomegachile) torrida* Smith; **d**, *M. (Austrochile) resinifera* Meade-Waldo. In a-c the actual posterior margin of T6, commonly with two or four small teeth, is out of sight behind the transverse carina or convexity. In d, however, the carina is the median truncate-undulate structure and the four marginal teeth are large. a-c, drawings by D. Yanega; d, from Michener, 1965b.

15(2). Carina of T6 short, occupying little over one-fifth width of tergum, high, rounded-truncate, slightly crenulate, not emarginate medially .... *M. (Platysta)*
—. Carina of T6 much longer, usually occupying most of width of tergum, commonly crenulate, commonly emarginate medially .... *M. (Amegachile, Eutricharaea, Paracella)*

NOTE: The African subgenera of megachiliform, leafcutting species are much confused. Pasteels (1965) provided a subgeneric classification, but the diversity within some of his subgenera is so great that distinctions between subgenera almost vanish. I have eliminated some of his subgenera by synonymy, but I believe that a detailed study would result in recognition of additional groups that would be called subgenera by the standards used elsewhere. Moreover, various species are not assigned to subgenus at all, and one of the great, overriding problems is proper association of the sexes. The keys given above will certainly be found quite unsatisfactory. I hope that they suggest some useful major groupings. Further details are provided in accounts of the subgenera.

Key to the Australian and Papuan Subgenera of *Megachile* (Females)

1. Mandible with six teeth, incomplete cutting edges of similar shape (Fig. 84-17d) in second, third, and fourth interspaces, the last small and often inconspicuous (metasoma parallel-sided) .... *M. (Creightonella)*
—. Mandible with two to five teeth, cutting edges absent or in second and third interspaces only, usually of quite different shapes in the two interspaces or absent in one or the other 2
2(1). Mandible without cutting edges between teeth (Fig. 84-12d-f); metasoma parallel-sided, body thus heriadiform or chalicodomiform (as in Fig. 84-9) 3
—. Mandible with at least one cutting edge between teeth; metasoma usually broad, less than twice as long as broad, body thus megachiliform (but see *Mitchellapis* and *Chelostomoda*) 9
3(2). S1 with large midapical spine .... *M. (Austrochile)*
—. S1 without apical spine 4
4(3). Claws each with two teeth on underside; proboscidial fossa closed posteriorly by process from lower side of each genal area .... *M. (Schizomegachile)*
—. Claws without or with but one ventral tooth each; proboscidial fossa open posteriorly 5
5(4). Metasomal integument red, and T2 and T3 without deep transverse postgradular grooves; glossa broad, ligulate .... *M. (Rhodomegachile)*
—. Metasomal integument black, or, *if* red, then with deep postgradular grooves; glossa linear or only slightly broadened 6
6(5). Mandible bidentate; T6, as seen from above, with strong, rounded, basolateral shoulders, lateral margins thus strongly concave immediately posterior to shoulders .... *M. (Thaumatosoma)*
—. Mandible with three or more teeth; T6 without shoulders, lateral margins gently convex to gently and rather uniformly concave 7
7(6). T2 and T3 usually with deep, transverse postgradular grooves, these absent from some middle-sized and large species in which claws have strong basal tooth (except in *M. semiluctuosa* Smith); pubescence usually giving a gray aspect, often forming apical white tergal fasciae; fulvous pubescence often present but confined to apical part of metasoma, rarely (*M. ustulata* Smith) metasoma with extensive fulvous pubescence .... *M. (Hackeriapis)*
—. Metasomal terga without deep, transverse postgradular grooves [except in *C. (Callomegachile) mcnamerae* Cockerell and others from New Guinea northwestward]; claws without basal teeth; pubescence black, fulvous, or with white patches, not grayish in aspect and not forming tergal fasciae; fulvous pubescence, *if* present, not confined to apical part of metasoma 8
8(7). Mandible shining (although reticulate), apical margin very oblique, as long as distance from basal tooth to base of mandible; sharp hypostomal tooth behind base of mandible .... *M. (Chalicodomoides)*
—. Mandible dull with minute roughening, apical margin shorter; no tooth behind base of mandible .... *M. (Callomegachile)*
9(2). Mandible with incomplete cutting edge in second interspace, and without other cutting edges (as in Fig. 84-12c); body heriadiform or chalicodomiform (Fig. 84-9) 10
—. Mandible with cutting edge in third interspace and often also in second (Fig. 84-11); body megachiliform 11
10(9). S6 with large smooth hairless area before apical fringe; T6 with many long hairs visible in profile, its apex produced and shallowly emarginate; pronotal lobe with rounded transverse ridge .... *M. (Mitchellapis)*
—. S6 uniformly punctate and hairy; T6 with only very short hairs visible in profile, its apex rounded; pronotal lobe with strong transverse carina .... *M. (Chelostomoda)*
11(9). Species with fulvous and black coloration suggestive of *Callomegachile*; third mandibular tooth broad and irregularly truncate because of fusion with cutting edge of second interspace (Fig. 84-11e), second (in Australian area) truncate (sometimes obliquely) or sinuate at apex .... *M. (Amegachile)*
—. Species usually dull-colored, gray, often with pale metasomal bands of hair; third mandibular tooth as well as second angulate, cutting edge in second interspace usually present although incomplete and not indistinguishably fused to third tooth .... *M. (Eutricharaea)*

Key to the Australian and Papuan Subgenera of *Megachile* (Males)

1. S5 and usually S6 exposed (metasoma parallel-sided) .... *M. (Creightonella)*
—. S5 and S6, and sometimes S4, hidden under preceding sterna 2
2(1). S8 with lateral hairs (Fig. 84-10j, n); body heriadiform or chalicodomiform, metasoma ordinarily twice as long as wide (Fig. 84-9) 3
—. S8 without lateral hairs but sometimes discal hairs may extend slightly beyond lateral margins (Fig. 84-10b, f); body megachiliform, metasoma ordinarily less than twice as long as wide 12
3(2). S1 with large midapical spine; carina of T6 very high and weakly crenulate to serrate medially, usually not emarginate, abruptly disappearing laterally so that general aspect is a large, truncate projection (Fig. 84-14d) .... *M. (Austrochile)*
—. S1 at most apically tuberculate; carina of T6 low (rarely

even absent), not crenulate or serrate, medially slightly emarginate (except in *Mitchellapis, Rhodomegachile,* and a few others) 4

4(3). Hind tibial spurs absent; S6 divided into two lateral sclerites by broad, median, membranous region .... *M. (Schizomegachile)*
—. Two hind tibial spurs present (only one in *Thaumatosoma*); S6 continuous, not divided into lateral hemisternites 5

5(4). Carina of T6, near the untoothed tergal margin, broadly rounded except for median tooth at apex of low, longitudinal median ridge .... *M. (Mitchellapis)*
—. Carina of T6 variable, without median tooth 6

6(5). S4 retracted, or rear margin in some cases exposed (Fig. 84-13a) 7
—. S4 more or less fully exposed (Fig. 84-13b) 10

7(6). Carina of T6 reduced to short, non-emarginate ridge or absent; metasomal integument red, without strong postgradular grooves on T2 and T3; glossa broad, ligulate .... *M. (Rhodomegachile)*
—. Carina of T6 evident, in almost every case medially emarginate [merely an emarginate swelling, not carinate, in *M. (Hackeriapis) cliffordi* Rayment; a scarcely emarginate ridge in *M. (H.) apposita* Rayment]; metasomal integument usually black, but *if* red, then with strong transverse postgradular grooves on T2 and T3; glossa linear, rarely somewhat broadened 8

8(7). Distance between apices of first and third mandibular teeth nearly equal to distance from third tooth to base of mandible; apex of S1 produced as a broad, nearly hairless, median, suberect flap; large, robust species; metasoma without indications of pale tergal bands .... *M. (Chalicodomoides)*
—. Distance between first and third (or second if mandible only bidentate) mandibular teeth much less than distance from uppermost tooth to base of mandible; apex of S1 not as above; usually smaller and more slender species; metasoma usually with indications of pale tergal bands 9

9(8). Region of carina of T6 swollen except at median emargination (northern Australia and northward) .... *M. (Chelostomoda)*
—. Region of carina of T6 not swollen (common throughout Australia and Tasmania, rare in New Guinea, not known elsewhere) .... *M. (Hackeriapis)* (in part)

10(6). Only one hind tibial spur; clypeus with group of coarse, quill-like bristles arising near middle; flagellum exceedingly attenuate, first segment longer than others, last two segments expanded (Fig. 84-10q) .... *M. (Thaumatosoma)*
—. Two hind tibial spurs; clypeus without such bristles; flagellum rarely attenuate, first segment shorter than second, last segment rarely expanded 11

11(10). T2 and T3 usually without deep transverse postgradular grooves; pubescence all black or with pale or ocher areas not forming metasomal bands; posterior margin of T6 without teeth .... *M. (Callomegachile)*
—. Terga usually with deep postgradular grooves; pubescence usually giving a gray appearance, in some cases fulvous on metasoma, usually forming metasomal bands; posterior margin of T6 usually with four teeth, median teeth absent in some cases .... *M. (Hackeriapis)* (in part)

12(2). Body with fulvous-and-black pubescence suggestive of *Callomegachile;* apical margin of T6 with four widely separated small teeth .... *M. (Amegachile)*
—. Body usually appearing dull-colored, gray, often with pale metasomal tergal hair bands; apical margin of T6 with small lateral tooth only, or sometimes with weak indication of two submedian teeth .... *M. (Eutricharaea)*

Megachile / Subgenus *Acentron* Mitchell

Megachile (Acentron) Mitchell, 1934: 307. Type species: *Megachile albitarsis* Cresson, 1872, by original designation.

This is one of the group of megachiliform subgenera of Group 1 related to *Pseudocentron*. The female differs from the others in the broad base of the apical mandibular tooth, which is twice as broad as the second tooth. The male lacks the middle tibial spur, and the middle tibia has an apical angle on the undersurface. The subgenus was fully characterized and male genitalia and sterna illustrated by Mitchell (1937b).

■ This subgenus ranges from the southern USA (North Carolina to Arizona) south to Argentina. Only one species, *Megachile albitarsis* Cresson, occurs in the USA but there are at least ten additional species in the neotropical region.

Megachile / Subgenus *Amegachile* Friese

Megachile (Amegachile) Friese, 1909a: 326. Type species: *Megachile sjoestedti* Friese, 1901 = *Megachile bituberculata* Ritsema, 1880, by designation of Cockerell, 1931c: 167. [Mitchell's (1934: 298) designation of *Megachile nasicornis* Friese, 1903, as the type species is subsequent and therefore invalid.]
Megachile (Callochile) Michener, 1962a: 27. Type species: *Megachile ustulatiformis* Cockerell, 1910 = *Apis mystacea* Fabricius, 1775, by original designation.
Megachile (Platychile) Michener, 1965b: 205, nomen nudum. Type species: *Megachile foliata* Smith, 1861, monobasic. [*Platychile* was introduced by error; *Callochile* was intended.]

This subgenus in Group 1 consists of large (12-20 mm long), broad-bodied megachiliform bees without pale metasomal bands, sometimes with areas of orange hair. The mandible of the female (Fig. 84-11e) is unusually broad, four-toothed, the second and third teeth broad, thin, and more or less truncate, often irregular or oblique. The third interspace has a large, complete cutting edge, the second interspace often appearing to be without a cutting edge, but in reality a large cutting edge seems to be fused with the third tooth to make a broad, thin, more or less truncate margin; an indication of the line between the third tooth and this edge can be seen in unworn mandibles of some species, but Pasteel's (1965) drawings are unrealistic in showing a clear separation of the tooth and cutting edge. Michener's (1965b) contention that the broad, thin margin arose from a cutting edge in *Amegachile* but from the tooth itself in *Callochile* seems to be an error; the two subgenera are synonymous. The cutting edges are often crenulate. The clypeal margin of the female is shallowly emarginate between two raised marginal areas; if this emargination represents the truncation, then it is much shorter than in most *Megachile*. The metasomal sterna lack fringes beneath the scopa. The male has a three- to four-toothed mandible with an inferior basal projection. T6 has a large, crenulate carina, usually emarginate medially, in some species strongly bilobed; the posterior margin has small lateral teeth and sometimes lacks submedian teeth. Fuller descriptions were given by Michener (1965b) for Australian species and Pasteels (1965) for African species. Male genitalia and sterna were illustrated by Michener (1965b); see also Figure 84-10a-d.

■ *Amegachile* is widespread in Africa (Senegal to Ethiopia south to Namibia and Natal, South Africa), including Madagascar, and ranges eastward through India, Southeast Asia, and the Philippines, south through Indonesia and New Guinea to the Solomon Islands and northern Australia (as far as southern Queensland). Pasteels (1965) recognized nine African species, including one from Madagascar. Michener (1965b) listed 13 specific names applied to Australian and South Pacific species. There are probably eight other names for species from Asia, western Indonesia, the Philippines, etc., making a total of about 30.

Megachile / Subgenus *Argyropile* Mitchell

Megachile (Argyropile) Mitchell, 1934: 308. Type species: *Megachile parallela* Smith, 1853, by original designation.

This subgenus consists of megachiliform Group 1 bees 9 to 16 mm long. S6 of the female is bare or sparsely haired apically, and there is a bare rim behind the preapical fringe of hairs. In this respect it resembles *Pseudocentron* and related subgenera, but the rim is directed upward, unlike that of *Pseudocentron* and its relatives. Otherwise, *Argyropile* does not show a close relationship to *Pseudocentron;* for example, the middle tibial spur of the male is fully developed. Mitchell (1980) indicated a relation to *Megachiloides,* but the reason for this view is not obvious. The subgenus was fully characterized and genitalia and sterna illustrated by Mitchell (1937b).

■ *Argyropile* occurs in North America, from southern Canada (British Columbia to Saskatchewan) to Chiapas, Mexico. The species are uncommon east of the Great Plains, but there are North Carolina and Florida records. There are seven species; a revision was by Mitchell (1937b).

Megachile / Subgenus *Austrochile* Michener

Chalicodoma (Austrochile) Michener, 1965b: 202. Type species: *Megachile resinifera* Meade-Waldo, 1915, by original designation.

This is a subgenus of chalicodomiform Group 2 species, more robust than most *Hackeriapis,* 9 to 11 mm in length, with distinct but nonfasciate postgradular grooves on T2 and T3. S1 of both sexes possesses a large subapical process. In the male, T6 has its carina absent laterally but truncately produced medially; the carina is crenulate, rarely emarginate, and usually exceeded by the four strong spines of the tergal margin (Fig. 84-14d). S4 of the male is exposed but thin and less punctate than S3; this subgenus could be considered, on this basis, as related to *Hackeriapis* and *Chelostomoides*. This subgenus was more fully characterized by Michener (1965b), who illustrated the male genitalia and sterna; see also Figure 84-10i-l.

■ *Austrochile* is found from one side of Australia to the other, but is probably absent in tropical parts of that continent. Ten specific names have been proposed, but the number of species is possibly only half that.

Megachile / Subgenus *Austromegachile* Mitchell

Megachile (Austromegachile) Mitchell, 1943: 666. Type species: *Megachile montezuma* Cresson, 1878, by original designation.

Megachile (Holcomegachile) Moure, 1953c: 119. Type species: *Megachile giraffa* Schrottky, 1913, by original designation.

The most distinctive feature of most species of this subgenus of moderate-sized (7.5-16.0 mm long) megachiliform Group 1 bees is the somewhat elevated but flat median section of the female clypeus, extending apically to an emarginate clypeal margin. The elevated area, which is about half as wide as the clypeus, is not sharply defined, but the clypeal surfaces slope laterally away from the elevated area to the sides of the clypeus. The lower part of the supraclypeal area is also flat. The clypeal character fails in *Megachile giraffa* Schrottky, the only species of *Holcomegachile,* but the shape is nonetheless vaguely indicated. Other characters show the close relation of that species to the rest of *Austromegachile*. The presence of a preoccipital carina behind the gena in both sexes is rather unusual in megachiliform *Megachile;* it varies in distinctness, but there is always at least a sharp margin. *Rhyssomegachile* also has such a carina and may be a close relative of *Austromegachile.* The mandible of the female is four-toothed, the upper tooth narrowly truncate; there is a complete cutting edge in the third interspace and a long but incomplete one in the second interspace, this one sometimes narrow and thus hidden in direct facial view, as illustrated by Mitchell (1980) (but see Mitchell, 1943). The third tooth is often acute, not always obtuse as described and illustrated by Mitchell (1943). The apical fasciae of white hair under the scopa of the female are broadly broken, present only laterally in *M. giraffa.* They are an unusual feature that emphasizes the relationship of *Holcomegachile* to *Austromegachile.*

■ *Austromegachile* is principally South American, occurring from as far south as Buenos Aires Province, Argentina, north through Brazil and on to Central America and Mexico as far north as Hidalgo and Sinaloa. Mitchell (1943) listed 15 species and undoubtedly there are several more, plus the single species placed in *Holcomegachile;* there are perhaps 25 species in all.

Megachile / Subgenus *Callomegachile* Michener

Chalicodoma (Callomegachile) Michener, 1962a: 21. Type species: *Chalicodoma mystaceana* Michener, 1962, by original designation.

Chalicodoma (Eumegachilana) Michener, 1965b: 191. Type species: *Megachile clotho* Smith, 1861, by original designation.

Chalicodoma (Carinella) Pasteels, 1965: 447 (not Johnston,

1833). Type species: *Megachile torrida* Smith, 1853, by original designation.

Chalicodoma (Morphella) Pasteels, 1965: 537. Type species: *Megachile biseta* Vachal, 1903, by original designation.

Cressoniella (Orientocressoniella) Gupta, 1993: 165. Type species: *Megachile relata* Smith, 1879, by original designation.

Megachile (Carinula) Michener, McGinley, and Danforth, 1994: 174, replacement for *Carinella* Pasteels, 1965. Type species: *Megachile torrida* Smith, 1853, autobasic.

It is not clear what species Gupta (1993) intended by the name *Megachile relata,* the type species *Orientocressoniella;* it appears to have been quite unlike the true *M. relata* Smith which, according to the late Dr. D. B. Baker (in litt., 2002, and Baker and Engel, 2006), is a *Callomegachile* from China, not western India. I agree with those authors that it is best to consider the nominal species as the type species of *Orientocressoniella* (ICZN, Edition 4, Article 70.3.1). As a result, Gupta's name is a synonym of *Callomegachile.*

This is a subgenus of elongate species of Group 2, often with brightly colored (red, yellow, or black) hair patterns, perhaps properly called heriadiform or hoplitiform although their large size (8-39 mm long) leads them to be more often considered chalicodomiform. The ridges of the female mandibles are minutely roughened, rather shiny in the former subgenus *Carinula* but otherwise dull. The female mandible is three- to seven-toothed. The female clypeus varies from quite ordinary to greatly modified; in the latter case the mandible may be long and parallel-sided (Fig. 84-15) as in *Maximegachile* and *Stenomegachile.* There is a preoccipital carina at the sides of the head. On T6 of the male the carina is weakly bilobed or lacks a median emargination, and is usually short (Fig. 84-14c); just basal to the carina is a strong median depression, and the posterior margin of T6 is simple, without teeth. Male genitalia and sterna were illustrated by Michener (1965b).

■ This largely tropical subgenus is widespread in sub-Saharan Africa, eastward through southern Asia to China and Japan, where it occurs north of the tropics; southward, it reaches South Africa and northern Queensland, Australia. It is perhaps native on islands such as the New Hebrides, New Caledonia, and Mauritius, but no doubt is adventive in Hawaii and the Greater Antilles. In the Antilles, species of both the *Carinula* group (see below) and ordinary *Callomegachile* can be found. The wide ranges outside of Madagascar of the two species of *Callomegachile* known from that island (Pauly in Pauly et al., 2001) suggest that they may not be natives of Madagascar. In that case *Callomegachile* is unknown as a native of Madagascar. Pasteels (1965) listed 34 African species; Michener (1965b) listed 37 other names for species of New Guinea and nearby islands, the South Pacific, and Australia. At least 20 more species from Asia (including China and Japan) and Indonesia belong to this subgenus. Except in eastern Asia, where *Megachile sculpturalis* Smith occurs as far north as southern Hokkaido (Japan), and in the eastern USA (Georgia to New York), where the same species is adventive (Mangum and Brooks, 1997; Hinojosa-Diaz et al., 2005), *Callomegachile* is a tropical subgenus; a single species, *Megachile breviceps* Friese 1898, was described from one specimen from Spain (Pasteels, 1966), possibly mislabeled.

The group (five species) called *Carinella* by Pasteels, now *Carinula,* is distinctive, having (1) a complete longitudinal median clypeal carina in the female, (2) less dull female mandibles than those of other *Callomegachile,* (3) no front coxal spine in the male (other *Callomegachile* have a small tubercle to a large spine), (4) the male gonoforceps broadened in the distal half (slender in most other *Callomegachile*), (5) the volsella absent, as is the concavity for reception of the volsella in the basal part of the

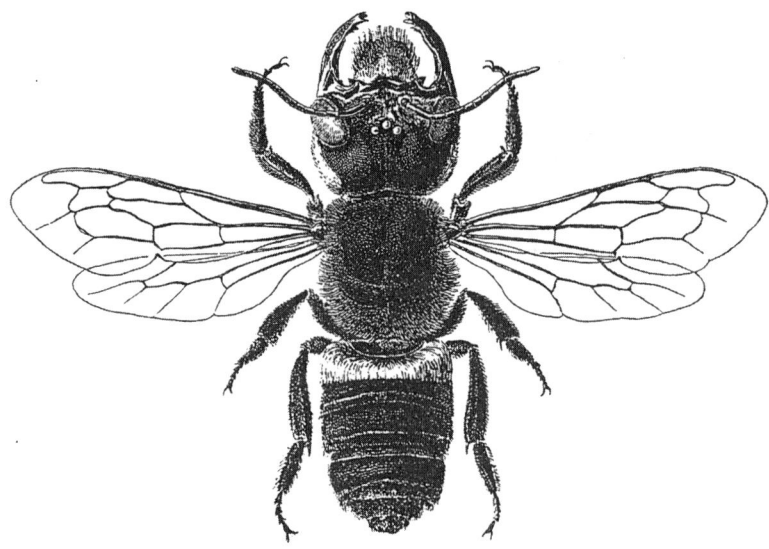

Figure 84-15. The longest known bee (39 mm), *Megachile (Callomegachile) pluto* Smith, female. From Friese, 1911b.

gonoforceps [these structures are usually present in other *Callomegachile*, but absent in *M. (C.) disjuncta* (Fabricius)], and (6) the penis valve slender apically (apex bent mesad, somewhat expanded and translucent in other *Callomegachile*). Thus *Carinula* could be recognized as a subgenus. Because of its similarity to other *Callomegachile* and variability among its species, I have chosen not to do so.

Morphella was proposed by Pasteels for two species that agree with "other *Callomegachile*" in the characters discussed above. The type species, *Megachile biseta* Vachal, but not *M. (C.) ambigua* (Pasteels), is unusual in having T7 of the male in the form of a large flat plate, resembling S4 in punctation and pubescence and superficially resembling an exposed fifth sternum. The front tibia of the female of *M. biseta* has a small but sharp posterior apical tooth, so that there are three tibial spines.

Another unusual species is *Megachile devexa* Vachal, which has a strong and complete omaular carina. Some other species have an omaular carina only on the lower part of the thorax, where it is often weak; most species lack this carina entirely.

The group that was previously called *Eumegachilana* consists of very large bees (18-39 mm long, the females usually over 23 mm) remarkable for the enormous heads and long jaws of females (Fig. 84-15). *Megachile pluto* Smith from the north Moluccas, Indonesia, is the longest bee in the world, females attaining 39 mm. In the South Pacific area (Michener, 1965b) north to the Himalayas, China, and Japan, females of the *Eumegachilana* group are mostly very distinct from other oriental *Callomegachile* because of the long, parallel-sided mandible with a small apical margin. In Africa, however, as Pasteels (1965) pointed out, there are intergradations in these structures, and it seems best to unite *Eumegachilana* with *Callomegachile*. Moreover, the eastern palearctic *M. (Callomegachile) sculpturalis* Smith also has rather intermediate female mandibles. Michener (1965b) discussed variability among species of the oriental and Australian regions; Pasteels (1965) gave similar information for Africa.

Megachile pluto Smith nests communally in arboreal termite nests, and uses its enormous labrum and mandibles to collect and carry the resin used to make termite-proof nest walls, as well as, when the resin is mixed with wood fibers, to make brood cells (Messer, 1984). Another *Callomegachile*, the oriental *M. monticola* Smith of China and Japan, was reported by Piel (1930) to nest in burrows made by *Xylocopa*. Its cells are made of resin, but between cells or principally as a closure after construction of a short cell series, it uses irregular leaf pieces and also clay. Tsuneki (1970) records a nest of *M. (C.) sculpturalis* Smith made of a mass of solid resin enclosing cells, and another nest in a hollow cane consisted of a series of cells, the partitions being made of both resin and mud.

Megachile / Subgenus *Cestella* Pasteels

Chalicodoma (Cestella) Pasteels, 1965: 547. Type species: *Megachile cestifera* Benoist, 1954, by original designation.

The dull mandible of the female of this subgenus suggests another subgenus of Group 2, *Callomegachile*, although the curious dentition and the clypeal structure are distinctive (see the key to subgenera). The denticulate carina of T6 of the male is quite unlike that of any *Callomegachile* but resembles instead some species of *Pseudomegachile*. The combination of characters supports recognition of *Cestella* as a subgenus. The body length is 14 to 17 mm.

■ This subgenus is known only from Madagascar. Of the two known species, one is undescribed.

Megachile / Subgenus *Chalicodoma* Lepeletier

Chalicodoma Lepeletier, 1841: 309. Type species: *Apis muraria* Olivier, 1789 (as *Xylocopa muraria* Fabricius, 1804) = *Apis parietina* Geoffroy, 1785, by designation of Girard, 1879: 778.

Chalicodoma (Euchalicodoma) Tkalcŭ, 1969b: 358. Type species: *Megachile asiatica* Morawitz, 1875, by original designation.

Chalicodoma (Allochalicodoma) Tkalcŭ, 1969b: 359. Type species: *Chalicodoma lefebvrei* Lepeletier, 1841, by original designation.

Chalicodoma (Parachalicodoma) Tkalcŭ, 1969b: 363 (not Pasteels, 1966). Type species: *Chalicodoma rufitarsis* Lepeletier, 1841, by original designation.

Megachile (Heteromegachile) Rebmann, 1970: 41. Type species: *Chalicodoma lefebvrei* Lepeletier, 1841, by original designation.

Megachile (Allomegachile) Rebmann, 1970: 42. Type species: *Megachile asiatica* Morawitz, 1875, by original designation.

Megachile (Katamegachile) Rebmann, 1970: 43. Type species: *Megachile manicata* Giraud, 1861, by original designation.

Chalicodoma (Xenochalicodoma) Tkalcŭ, 1971: 34, replacement for *Parachalicodoma* Tkalcŭ, 1969. Type species: *Chalicodoma rufitarsis* Lepeletier, 1841, autobasic.

This is a subgenus of usually large (11-20 mm long), hairy, chalicodomiform bees, commonly without apical hair bands on the metasomal terga. They are members of *Megachile* Group 2. Like *Chalicodomoides*, *Largella*, and *Parachalicodoma*, these bees are more robust than *Callomegachile* and *Pseudomegachile*. In both sexes the mandible is rather slender, the apical margin strongly oblique and about as long as the distance from the upper tooth to the base; above the first and second teeth in the female is a long margin extending to the upper distal angle of the mandible, often with no indication of teeth but sometimes with weak, very obtuse third and fourth teeth. The clypeal margin is convex, usually denticulate, not thickened, more finely punctate than the disc except for the impunctate, usually crenulate, margin proper. The male mandible is at least weakly three-toothed. T6 of the male has a large carina, usually strongly spinose, sometimes denticulate, the median concavity sometimes present, the spines rarely limited to a tooth at the side of the concavity and one laterally, the carina thus four-toothed. Tkalcŭ (1969b), Rebmann (1970), and Mitchell (1980) describe and illustrate many details of structure, including male genitalia and hidden sterna.

■ The subgenus occurs in central Europe (Germany, Poland) but is abundant in the Mediterranean region from the Canary Islands to southwestern Asia, and ranges east to Mongolia and China and also southward to Baluchistan (in Pakistan) and throughout Africa. There are at least 11 palearctic species, and at least 20 in sub-Saharan Africa.

As is evident from the above synonymy, Tkalců (1969b, 1971) and Rebmann (1970) independently divided the subgenus in the same way: *Allochalicodoma* Tkalců = *Heteromegachile* Rebmann, *Euchalicodoma* Tkalců = *Allomegachile* Rebmann, and *Katamegachile* Rebmann = *Parachalicodoma* Tkalců = *Xenochalicodoma* Tkalců. Although, as shown by both authors, the groups are quite different from one another, it seems to me useful to show their relationships; moreover, if they were recognized, the African species would have to be divided into several subgenera.

The problem of the name for the type species of *Chalicodoma* was discussed by Alfken (1941). A species included in *Chalicodoma* by Lepeletier (1841) and later designated as the type species by Girard (1879) was *Xylocopa muraria* as that name was used by Fabricius in 1804. But *Apis muraria* was used by earlier authors (see Dalla Torre, 1896), as Fabricius indicated. The name *muraria* Retzius, 1783, had long been in use, until Alfken showed that it was based on a species of *Osmia*. Alfken therefore resurrected the specific name *parietina* Geoffroy, 1785, for the species generally known as *muraria* Retzius, 1783, and recent authors have accepted this action. Retzius' name, moreover, must be ignored because he did not use binominal nomenclature.

The hard clay nests, clumps of cells adhering to stones and buildings as well as branches and twigs of trees and bushes, are well known for indestructability and for covering and hiding Egyptian hieroglyphs. They must include some secreted materials that harden and perhaps waterproof the nest material.

Megachile / Subgenus *Chalicodomoides* Michener

Chalicodoma (Chalicodomoides) Michener, 1962a: 24. Type species: *Megachile aethiops* Smith, 1853, by original designation.

This subgenus contains robust chalicodomiform species of Group 2, 13 to 18 mm long, that are superficially similar to *Chalicodoma*. They also agree with *Chalicodoma* in the slender mandible, the apical margin of which in the female is very oblique and as long as the distance from the upper tooth to the mandibular base. The female (also male) mandible is three-toothed (Fig. 84-12d), unlike that of *Chalicodoma*. Moreover, the clypeus of the female is produced far over the labral base to a rather narrow and somewhat emarginate truncation, quite unlike the rounded, denticulate clypeus of *Chalicodoma*. The female has a sharp tooth below the eye, immediately behind the lower mandibular condyle. S4 of the male is retracted, suggesting that in spite of its appearance, this subgenus may be related to *Hackeriapis*. The carina of the male T6 is bilobed, without teeth or denticles; the tergum has no deep depression above the carina, and has a small lateral tooth. A fuller description, including illustrations of male genitalia and sterna, is by Michener (1965b).

■ This subgenus occurs in northern Australia from Queensland to Western Australia. The two species were revised by King and Exley (1985c).

Megachile / Subgenus *Chelostomoda* Michener

Chalicodoma (Chelostomoda) Michener, 1962a: 24. Type species: *Megachile spissula parvula* Strand, 1913 = *M. spissula* Cockerell, 1911, by original designation.
Ashmeadiella (Neoashmeadiella) Gupta, 1990: 56. Type species: *Ashmeadiella indica* Gupta, 1990, by original designation. [For characters of *M. indica* (Gupta) see Michener, 2000.]

This subgenus has exactly the appearance of ordinary species of *Hackeriapis* or *Chelostomoides* (e.g., Fig. 84-9), being small (6.5-10.0 mm long) and heriadiform with strong postgradular grooves in T2 and T3; as in most *Hackeriapis*, these grooves are not fasciate. Unlike the female of other subgenera of Group 2, that of *Chelostomoda* has a large but incomplete cutting edge in the second mandibular interspace, thus resembling the unrelated subgenera *Eumegachile*, *Mitchellapis*, and *Sayapis*. The female mandible of *Chelostomoda* is five-toothed. Unlike those of nearly all *Hackeriapis*, females of *Chelostomoda* have apical hair bands beneath the scopa on S2 to S4, at least laterally. Michener (1965b) gave a fuller description and illustrations, including figures of male genitalia and sterna.

■ *Chelostomoda* ranges from China and Japan south to the Solomon Islands and northern Queensland (Australia), westward throughout Indonesia and Southeast Asia to India. The number of species is probably small, but at least 14 specific names have been proposed.

Megachile / Subgenus *Chelostomoides* Robertson

Chelostomoides Robertson, 1901: 231. Type species: *Megachile rufimanus* Robertson, 1891 = *Chelostoma rugifrons* Smith, 1854, monobasic.
Oligotropus Robertson, 1903a: 168. Type species: *Oligotropus campanulae* Robertson, 1903, monobasic.
Gnathodon Robertson, 1903a: 168 (not Oken, 1816, etc.). Type species: *Megachile georgica* Cresson, 1878, monobasic.
Sarogaster Robertson, 1918: 92, replacement for *Gnathodon* Robertson, 1903. Type species: *Megachile georgica* Cresson, 1878, autobasic.
Chalicodoma (Chelostomoidella) Snelling, 1990: 36. Type species: *Megachile spinotulata* Mitchell, 1934, by original designation.

Partly because of great variation in female clypeal and mandibular structure, sometimes involving grotesque modifications (Fig. 84-16), several genus-group names have been applied to species of this subgenus, all of them monobasic.

Chelostomoides contains the only native Western Hemisphere members of Group 2 and is the only Western Hemisphere subgenus of heriadiform or hoplitiform *Megachile* (Fig. 84-9), although *Sayapis*, *Schrottkyapis*, and *Stelodides* are also rather slender. It is also the only Western Hemisphere subgenus with three rather than four exposed metasomal sterna in the male. The mandibles of females lack cutting edges. The deep transverse basal grooves on the metasomal terga are commonly fasciate because of

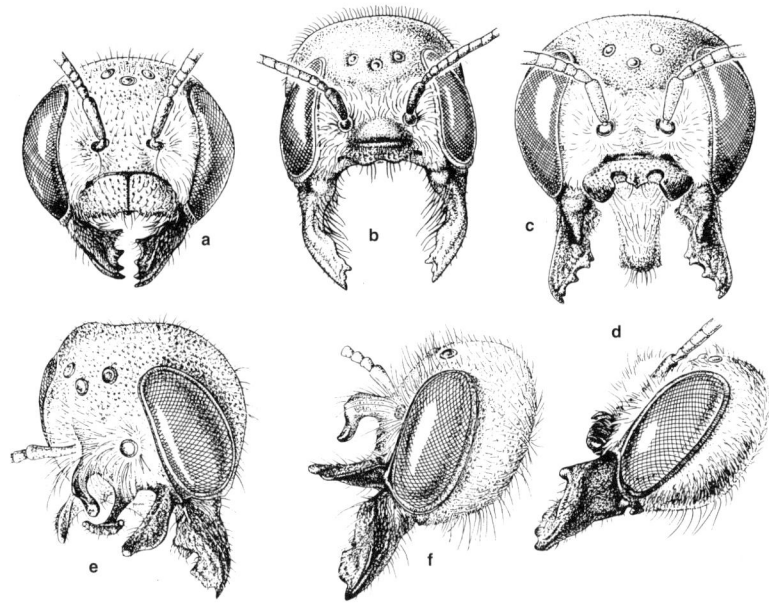

Figure 84-16. Cephalic diversity in females of *Megachile* subgenus *Chelostomoides*. **a, b,** Frontal views of *M. (C.) exilis* Cresson and *chilopsidis* Cockerell; **c, d,** Frontal and lateral views of *M. (C.) occidentalis* Fox; **e, f,** Oblique and lateral views of *M. (C.) armaticeps* Cresson. From Snelling, 1990.

pale hairs. The body length ranges from 7 to 17 mm. Mitchell (1937d) characterized the subgenus more fully. Male genitalia and other structures were illustrated by Mitchell (1937d, 1962) and Snelling (1990).

■ *Chelostomoides* is widespread in North America, from British Columbia, Canada, and New Jersey, USA, southward. It is abundant in the arid Southwest and extends to Cuba, Jamaica, and Costa Rica; it also occurs in Colombia and Peru. Mitchell (1937d) revised the species of the USA, and Snelling (1990) revised the subgenus for North and Central America, including Cuba, but for some reason not Jamaica; he recognized 26 species. Two others occur in Jamaica and three in South America, making a total of 31 species.

So far as is known, nests are in holes probably usually made by beetles in wood or stems, partitioned into cells with resin (Medler, 1966; Krombein, 1967). Nests of certain species of the subgenus *Chelostomoides* were described by Armbrust (2004). As in other species they consist of series of cells in burrows in wood or twigs. The entrance closures are highly variable, within as well as among species; those of *M. prosopidis* Cockerell are particularly variable, consisting of layers of sand (or sand mixed with resin), pebbles (often mixed with bits of wood, stones, etc.), chewed leaves, etc. *M. discorrhina* Cockerell differs from the other species in using a white liquid, applied by the proboscis, to stick pebbles and sand grains together.

Megachile / Subgenus *Chrysosarus* Mitchell

Megachile (Chrysosarus) Mitchell, 1943: 664. Type species: *Megachile guaranitica* Schrottky, 1908, by original designation.

Megachile (Dactylomegachile) Mitchell, 1943: 670. Type species: *Megachile parsonsiae* Schrottky, 1914, by original designation.

This is the only large group of megachiliform Group 1 species that completely lacks cutting edges between the mandibular teeth of the female. The mandible is five-toothed, the upper two teeth close together and sometimes scarcely distinguishable, but sometimes, in the smaller species *(Dactylomegachile)*, separated by a distance almost equal to the distance between the third and fourth teeth. In the male, the carina of T6 has a median emargination. The front coxa of the male has a rather short spine or only a tubercle and sometimes has a few short, reddish bristles. *Dactylomegachile* differs from *Chrysosarus* principally in color: *Dactylomegachile* is like most *Megachile* in having dark tegulae, clear wings, and dark wing veins, whereas *Chrysosarus* has testaceous or reddish tegulae, yellowish wings, and testaceous veins. *Dactylomegachile* is usually smaller, and males have a slight angle on the lower margin of the mandible that is lacking in *Chrysosarus* s. str. Body length for the subgenus is 8 to 17 mm.

■ *Chrysosarus* ranges from central Mexico (Jalisco) to Argentina (Buenos Aires province) and central Chile. Mitchell (1943) listed 15 species under the two subgeneric names that he used, and I suppose that there must be 25 or more species in all.

As indicated above, brood cells of *Chrysosarus* differ from those of most *Megachile* of group 1 in having a mud layer between two layers of petals or leaves. The synonymy of *Dactylomegachile* is supported by the observation that a species of the *Dactylomegachile* group makes such cells, in this case with mud between layers of petals (Laroca et al., 1992).

Megachile / Subgenus *Creightonella* Cockerell

Megachile (Creightonella) Cockerell, 1908b: 146. Type species: *Megachile mitimia* Cockerell, 1908 = *Megachile cognata* Smith, 1853, by original designation.

Creightoniella Pasteels, 1965: 10, unjustified emendation of *Creightonella* Cockerell, 1908.

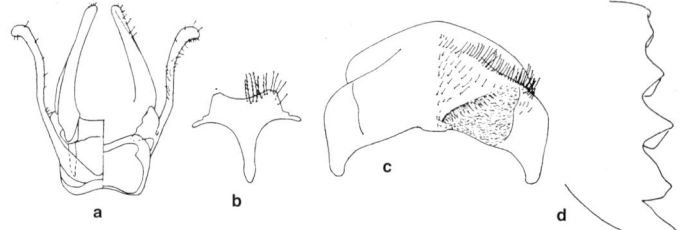

Figure 84-17. Structures of *Megachile (Creightonella) frontalis* (Fabricius). **a-c,** Male genitalia (the dorsal view is on the left), S8, and S6; **d,** Tooth row on apex of mandible of female. From Michener, 1965b.

Megachile (Metamegachile) Tkalců, 1967: 102. Type species: *Anthophora albisecta* Klug, 1817, by original designation.

This subgenus constitutes Group 3 of the genus *Megachile;* see Table 84-2. It consists of large species (11 to 23 mm long) with a strongly convex, parallel-sided metasoma. Although some have pale pubescence and white metasomal bands, most are black-haired or brightly colored, superficially resembling species of the subgenus *Callomegachile.* Females differ from those of other chalicodomiform subgenera in the presence of incomplete cutting edges, usually in mandibular interspaces 2 to 4. Most species have five or six mandibular teeth (Fig. 84-17d) (exceptions are discussed under the genus). Males are unique among subgenera of *Megachile* in having five or six exposed sterna that are rather ordinary sternal plates. This appears to be a plesiomorphy relative to all other subgenera. The carina of T6 is strongly produced, denticulate, sometimes emarginate medially, and bent forward laterally so that lateral parts are almost parallel to one another and separate a broad central part of the tergum (often with a longitudinal median ridge or carina) from vertical lateral parts. Tkalců (1967) provided good illustrations of some structures of *M. albisecta* (Klug), and male genitalia and sterna of a very different species, *M. frontalis* (Fabricius), were illustrated by Michener (1965b); see also Figure 84-17a-c.

■ The subgenus occurs from southern Europe to South Africa, east across Asia to China, Taiwan, the Philippines, Indonesia, and the Solomon Islands. There is one record for northeastern Australia. Pasteels (1965, 1970) reports 40 species from sub-Saharan Africa, and it is in Africa that great diversity occurs. In tropical Asia there are few species, perhaps only one, *Megachile frontalis* (Fabricius), in all of eastern Indonesia, New Guinea, the Solomons, the Philippines, etc. From Spain through the Mediterranean basin to central Asia there may be only one species, *M. albisecta* (Klug), with others farther south, e.g., in the Arabian peninsula. The total number of species is probably about 50.

Noteworthy groups within *Creightonella* include numerous African species the females of which have the vestiture of the mesothoracic venter and leg bases consisting of stiff, clubbed hairs. Another group of both African and Asian species is unusual in having a short, apically truncate or concave male S8, as illustrated for *Megachile frontalis* (Fabricius) in Figure 84-17b. For additional discussion of the diversity within the subgenus, see the account under the genus, as well as Tkalců (1969c).

Megachile / Subgenus *Cressoniella* Mitchell

Megachile (Cressoniella) Mitchell, 1934: 307. Type species: *Megachile* "*zapoteka* Cresson" = *M. zapoteca* Cresson, 1878, by original designation.

These are robust megachiliform Group 1 bees 9 to 15 mm long, sometimes lacking the usual pale metasomal tergal hair bands. *Cressoniella* belongs to a group of subgenera in which the female mandible is four-toothed (the upper tooth usually notched, thus approaching the five-toothed condition), with cutting edges in interspaces 2 and 3, and the male fore tarsi are usually unmodified. Male genitalia and hidden sterna were illustrated by Mitchell (1935a).

■ *Cressoniella* ranges from southern Arizona, USA, and San Luis Potosí, Mexico, to Tarapacá, Chile, and Buenos Aires province, Argentina, apparently primarily in xeric areas, although present also in mesic sites, e.g., in Costa Rica. *Cressoniella* contains probably a dozen species (six are listed by Mitchell, 1943); only *Megachile zapoteca* Cresson reaches the USA.

Megachile / Subgenus *Cuspidella* Pasteels

Chalicodoma (Cuspidella) Pasteels, 1965: 544. Type species: *Chalicodoma quadraticauda* Pasteels, 1965, by original designation.

This subgenus could be based on a modified *Pseudomegachile;* it agrees with that subgenus of Group 2 in most characters. The female differs in having a broad clypeal emargination (see the key to subgenera) and broad (almost lobelike) mandibular teeth. It is the remarkable T6 of the male, however, that would make assignment to *Pseudomegachile* arbitrary. The carina is neither denticulate nor bilobed, but is greatly produced in the middle to a narrowly truncate process. Laterally, T6 has a tooth, as in the *Neglectella* group of *Pseudomegachile.* T7 is not exposed. The body length is 16 mm.

■ *Cuspidella* is from Katanga in Congo (Kinshasa). *Megachile quadraticauda* (Pasteels) is the only known species.

Megachile / Subgenus *Dasymegachile* Mitchell

Megachile (Dasymegachile) Mitchell, 1943: 669. Type species: *Megachile saulcyi* Guérin, 1845, by original designation.
Cressoniella (Chaetochile) Mitchell, 1980: 63. Type species: *Cressoniella golbachi* Schwimmer, 1980, by original designation.

This subgenus contains robust, megachiliform Group 1 species, 10 to 15 mm long, commonly with unusually long and dense pubescence, sometimes obscuring the integument and not forming pale apical tergal fasciae, producing a *Bombus*-like aspect. Some species, however, have pale apical tergal fasciae. The mandible of the female is four-toothed and has an acute upper angle, a long cutting edge in the third interspace, and an incomplete cutting edge in the second.

■ *Dasymegachile* occurs primarily in the Andean region of Argentina, Peru, and Chile; most species are Chilean. It also ranges eastward to Paraguay and southern Brazil. Mitchell (1943) included five species, but with the addition of *Chaetochile* and unnamed species, there must be at least 20 in all, 12 of them from Chile.

Two species, *Megachile* (*D.*) *golbachi* (Schwimmer) and an unnamed species, are unusual in the male because the second mandibular tooth (sometimes notched) is much nearer the first than the third, and in the female because the clypeus and supraclypeal area are rather flat and dull, and have abundant erect, rather short and partly hooked hairs. Because of these characters, Mitchell (1980) erected the subgenus *Chaetochile* for these species. But such a clypeus, flat with erect hooked or wavy hairs, has arisen independently in various bee taxa (*Anthophora, Caupolicana, Tetraloniella, Trachusa*, etc.) probably in adaptation to the use of *Salvia*-like flowers, and there is no reason to consider *M. golbachi* as other than a derived species of *Dasymegachile*. The *Chaetochile* species have pale metasomal fasciae and do not look *Bombus*-like, but this is also true of some (unidentified) species that lack the special features of *Chaetochile*.

Megachile / Subgenus *Eumegachile* Friese

Eumegachile Friese, 1898b: 198, no included species; Friese, 1899c: 36, included species. Type species: *Megachile bombycina* Radoszkowski, 1874, by designation of Cockerell, 1930c: 209.

The lone species of this subgenus is chalicodomiform but is included in Group 1; it looks much like *Sayapis*, with pale apical tergal fasciae. Because the female mandibles are long and parallel-sided, a number of unrelated species in *Callomegachile* and *Hackeriapis* have been incorrectly referred to *Eumegachile* in the past. There is a large but incomplete cutting edge in the second mandibular interspace of the female, separating *Eumegachile* proper from such forms of Group 2. Because of this cutting edge, the dense fringe of plumose hairs at the apex of T6 of the female, and the general similarity to *Sayapis*, I regard *Eumegachile* as a close relative of that subgenus. S8 of the male lacks hairs, but in one specimen there is a marginal hair on one side only; the subgenus appears twice in the key to males because of the weakness of this character. The carina of T6 of the male is rather small and bilobed, and there is a depression above it, as there is in most *Callomegachile*; there are no teeth on the apical margin. T7 is unusually large, suggesting *Megachile* (*Callomegachile*) *biseta* Vachal but not other *Megachile*. Pasteels (1966) and Mitchell (1980) described and illustrated this subgenus and its species in detail.

■ *Eumegachile* occurs from Hungary and Finland to Siberia. The only species is *Megachile bombycina* Radoszkowski.

Megachile / Subgenus *Eutricharaea* Thomson

Megachile (*Eutricharaea*) Thomson, 1872: 228. Type species: *Apis argentata* Fabricius, 1793, monobasic.
Megachile (*Paramegachile*) Friese, 1898b: 198. Type species: *Apis argentata* Fabricius, 1793, by designation of Mitchell, 1934: 298.
Megachile (*Paramegalochila*) Schulz, 1906: 71, unjustified emendation of *Paramegachile* Friese, 1898.
Androgynella Cockerell, 1911c: 313. Type species: *Megachile detersa* Cockerell, 1910, by original designation.
Perezia Ferton, 1914: 233, not Léger and Dubosc, 1909. Type species: *Perezia maura* Ferton, 1914 = *Megachile leachella* Curtis, 1828, monobasic.
Fertonella Cockerell, 1920a: 257, replacement for *Perezia* Ferton, 1914. Type species: *Perezia maura* Ferton, 1914 = *Megachile leachella* Curtis, 1828, autobasic and by original designation.
Megachile (*Eurymella*) Pasteels, 1965: 64. Type species: *Megachile eurimera* Smith, 1854, by original designation.
Megachile (*Digitella*) Pasteels, 1965: 191. Type species: *Megachile digiticauda* Cockerell, 1937, by original designation.
Megachile (*Neoeutricharaea*) Rebmann, 1967a: 36. Type species: *Apis rotundata* Fabricius, 1787, by original designation.
Megachile (*Melaneutricharaea*) Tkalcŭ, 1993b: 803. Type species: *Megachile hohmanni* Tkalcŭ, 1993, by original designation.
Megachile (*Anodonteutricharaea*) Tkalcŭ, 1993b: 807. Type species: *Megachile larochei* Tkalcŭ, 1993 = *Megachile lanigera* Alfken, 1933, by original designation.

For *Eutricharaea, Paramegachile*, and *Neoeutricharaea* there are problems of misidentified type species (Hurd, 1967, Rebmann, 1967b). In no case, however, is there a problem in application of the subgeneric name, thanks (for *rotundata*) to Opinion 1093 of the Commission (1977). *Androgynella* appears to be based on sexually anomalous individuals of *Eutricharaea* (Michener, 1965b). The names *Perezia* and *Fertonella* were based on a sexually anomalous specimen of *Megachile* (*Megachile*) *centuncularis* (Linnaeus), according to Pasteels (1969c). Van der Zanden (1988b), however, showed that they were based on a specimen of *M.* (*Eutricharaea*) *leachella* Curtis.

This is a subgenus of small to moderate-sized (5-16 mm long) megachiliform bees, members of Group 1 related to *Megachile* s. str.; in many areas the smallest megachiliform *Megachile* species belong here. For practical purposes (see the key), there is ordinarily only one cutting edge on the female mandible; it is in the uppermost interspace, i.e., the third interspace in the ordinary four-toothed mandible (Fig. 84-11g); occasionally, it is hidden behind the margin of the interspace, and the mandible, as seen from in front, thus seems to lack cutting edges. Rarely, among species of Australia, Pacific islands, and southern Asia, there are species with a distinct although incomplete cutting edge in the second interspace, or there is a small cutting edge hidden behind the main mandibular margin of the second interspace. Rarely, as in *M. semierma* Vachal and *zambesica* Pasteels, the cutting edge is crenulate, suggesting some species of *Amegachile*. Also rarely, as in the African *M. dolichognatha* Cockerell and *michaelis* Cockerell, the third tooth is absent and a long, nearly straight edge extends from the second to the fourth tooth, with the cutting edge limited to the upper part of this margin, between the "third" and fourth teeth. This structure suggests that of some species of the North American subgenus *Megachiloides,* in which, however, the cutting edge is much longer (Fig. 84-12b). Most species have white apical sternal fasciae under the scopa, but as described by Michener (1965b) there are intermediates between these species that have fully developed continuous bands and the minority that lack them completely. Males are quite diverse in structure, the mandible being three- or four-toothed with or without a ventral projection; the front coxa have a spine or do not, and the front tarsus is pale and expanded or not. The carina of T6 is strong, usually toothed or denticulate, and usually has a median emargination that is sometimes scarcely discernible in irregular and coarse marginal dentition. Male genitalia and sterna were illustrated by Michener (1965b), Rebmann (1967a, 1968), and Tkalců (1993b); see also Figure 84-10e-h).

■ *Eutricharaea* ranges throughout the palearctic, African, oriental, and Australian areas, from cool-temperate regions to deserts and to the moist tropics. In the Pacific it occurs as far east as Tahiti. It has been introduced into North America and the Antilles, as well as Argentina, Chile, and New Zealand (one species each in most of the countries listed, but three in the USA). In numbers of species, the subgenus is enormous. In his revision of the African species, Pasteels (1965) recognized nearly 120 species. In the western palearctic there are about 25 species (Rebmann, 1967a, 1968). For species in Australia and the South Pacific islands, 66 names have been proposed (Michener, 1965b); there is no revision. The oriental region contains at least 25 additional species. Keys for the identification of three species introduced into North America were given by Parker (1978b) and Mitchell (1980). Four species of *Eutricharaea* are adventive in Hawaii (Snelling, 2003).

There is great diversity among species here placed in *Eutricharaea*, and various authors have subdivided the group, as shown by the synonymy above. Eventually, such subdivision should be helpful, but for the present I have synonymized several taxa. Pasteels (1965) put emphasis on tiny brushes of short, dense, coppery hairs in the apices of the grooves of the acetabular interspace of the female mandible. In *Eutricharaea* and *Paracella*, as these subgenera are here understood, such brushes are commonly present in (1) the outer groove of the acetabular interspace that ends at the notch between the first and second mandibular teeth and (2) the lower groove that ends about in the middle of the base of the first tooth. In each of these subgenera, however, there are species that lack the brush in the lower groove. Within the subgenus *Eutricharaea* as here understood, the groups called *Eurymella* and *Digitella* by Pasteels (as well as in the subgenera *Platysta* and *Neocressoniella*) are among those that lack a brush in the lower groove. In fact, this character is almost the only one that separates *Eutricharaea* in Pasteels' sense from his *Eurymella*, and I tentatively conclude that it is insufficient to justify separation of the subgenera. For practical purposes, one may also note that these brushes are hard to see, often being covered with detritus or worn off. Another character that can be gleaned from Pasteels' work is that in *Eutricharaea* (in Pasteels' sense) the first two segments of the labial palpus are equal in length, whereas in *Eurymella* the second segment is longer than the first. This character is variable and does not clearly separate the two groups. The males of both *Eurymella* and *Eutricharaea* are variable, and no subgeneric differences are known. The lone species placed in *Digitella* by Pasteels has a swollen front tibia and femur in the male, also a character of the group called *Melaneutricharaea*. The similarly swollen front tibia and femur of *Megachile (Eutricharaea) aliceae* Cockerell diminishes the usefulness of this character.

In both of Pasteels' large groups, *Eurymella* and *Eutricharaea*, as well as in *Amegachile*, the apex of the gonoforceps in some of the species is strongly bifid but is not in others. It is possible that with fuller analysis this will prove to be a group character, as it has been considered in Europe with the separation of *Neoeutricharaea* from *Eutricharaea* (Rebmann, 1967a). These groups *(Eutricharaea* and *Neoeutricharaea)* appear to be useful in the palearctic region, and the former was revised by Rebmann (1968), but until the vast African and Indo-Australian faunas are carefully studied, it is difficult to justify their recognition in a world view.

The classification used here, involving the recognition of *Eutricharaea* and *Paracella* for large and diverse subgenera, and the synonymy indicated above under *Eutricharaea,* is arbitrary and simply recognizes ignorance of useful groupings that probably exist. Association of sexes and study of species throughout the Eastern Hemisphere are badly needed.

Megachile edwardsi Friese was placed in *Eurymella* by Pasteels (1965), but its female has a large cutting edge in the second interspace as well as the third. It thus resembles the subgenus *Paracella*, from which it differs in its acute fourth mandibular tooth and lack of a brush of orange hairs in the apex of the lower acetabular groove. Its proper position remains doubtful.

A species that could easily be missed by future students of this group is *Megachile auriculata* (Gupta, 1989) from India, described in the genus *Anthocopa* [*Anthocopa* is here included in *Hoplitis* (Osmiini)]. This species is an obvious *Megachile;* Gupta's statement that arolia are present is an error. The holotype, apparently the only specimen in the Indian Agricultural Research Institute, New Delhi, is a male. T1 to T3 of the holotype are in situ, and the genitalia and the sterna illustrated by Gupta are on a slide. The rest of the metasoma is missing. The specimen is thus difficult to place but is probably a *Eutricharaea*. The front tarsus is simple and dark, the front coxa has a strong apical spine, and the surface in front of the spine is only sparsely hairy.

For pollination of alfalfa, *Megachile rotundata* (Fabricius) has been purposefully introduced into Argentina, Australia, Canada, Chile, and New Zealand, as well as into other areas. Although originally a palearctic species, it is used commercially in western North America, having been introduced accidentally into the USA. For solitary bees, some species of this subgenus are extraordinarily prolific. For example, Maeta and Kitamura (2005) found that marked females of *M. rotundata* (Fabricius) in cages with plenty of flowers made and oviposited in 51 to 77 cells during lives of 45 to 57 days.

Megachile / Subgenus *Gronoceras* Cockerell

Gronoceras Cockerell, 1907b: 65. Type species: *Gronoceras wellmani* Cockerell, 1907 = *Megachile bombiformis* Gerstaecker, 1857, by original designation.
Megachile (Berna) Friese, 1911c: 668. Type species: *Berna africana* Friese, 1911 (combination as though *Berna* were a genus) = *Megachile africanibia* Strand, 1912, monobasic.
Gronoceras (Digronoceras) Cockerell, 1931c: 134. Type species: *Megachile combusta* Smith, 1853 = *Apis cincta* Fabricius, 1781, by original designation.

This is a subgenus of large, elongate (17-29 mm long), chalicodomiform species of Group 2, sometimes gray but often with areas of bright-red pubescence. In the female the mandibles are dull, as in *Callomegachile*. They are four-toothed, the two lower teeth being close together. The three anterior and middle tibial spines are similar to but better developed than those of other subgenera having three such spines, i.e., the subgenus *Largella* and females of *Megachile (Callomegachile) biseta* Vachal. The spines are all fully exposed and the spaces between them are shiny, with few hairs or punctures. In the male the long dense hairs of T6 (Fig. 84-18) suggest those of the subgenus *Stenomegachile;* no other subgenera have such hairs.

■ *Gronoceras* is widespread in Africa, from Senegal to Eritrea south to Namibia and Natal, South Africa. An old and doubtful record from Jamaica suggests an introduction that might have occurred at about the same time that *Pseudomegachile* and *Callomegachile* were introduced; there is no evidence of a surviving population in Jamaica. Pasteels (1965) included ten species in *Gronoceras*.

There is much variability within the subgenus. The lower margin of the male mandible may be simple or may have a projection. The hypostomal area may be variously modified. The antennal flagellum of males of some species is enlarged and on one side deeply concave. The front coxa of the male may have a spine, and the front tarsus may be enlarged and yellowish. The carina of T6 of the male may be represented by two very long spines or by two long lobes. These characters are not well correlated, and splitting of the subgenus does not seem to be needed in spite of the major differences among species.

Megachile / Subgenus *Grosapis* Mitchell

Eumegachile (Grosapis) Mitchell, 1980: 46. Type species: *Megachile cockerelli* Rohwer, 1923, by original designation.

Although having only three exposed sterna in the male, this subgenus is not closely related to *Chelostomoides*. It is probably a member of Group 1 (see Table 84-1); it is regarded as chalicodomiform, but the metasoma is much less than twice as long as wide. Presumably because the female mandible has a distinct (although incomplete) cutting edge in the second interspace, and no other distinct cutting edge, Mitchell (1980) placed *Grosapis* near *Sayapis*. But unlike that of *Sayapis,* the mandibular tooth row of the female is much shorter than the distance from the upper tooth to the base of the mandible. Moreover, there is a feeble suggestion of a cutting edge in the third interspace, on the side of the fourth tooth, and the latter is slightly emarginate, the mandible thus approaching the five-toothed condition. The mandible of the male is very long, attenuate, and four-toothed, and lacks a projection on the lower margin. The carina of T6 of the male consists of two broad, shiny lobes; T7 has a strong median apical projection, as in *Megachile (Sayapis) frugalis* Cresson and *M. (Schrottkyapis) assumptionis* (Schrottky), and the male mandibles are similar to those of the latter. It is not clear whether these characters indicate a relationship to *Schrottkyapis* or possibly *Sayapis;* in its large, robust body (length 20.0 mm; head width 6.5 mm), covered with long red hairs, *Grosapis* does not resemble these other Western Hemisphere subgenera. Some of its structures were illustrated by Mitchell (1930).

■ *Grosapis* is known only from the Cordillera Occidental of northern Mexico. The single species, *Megachile cockerelli* Rohwer, was described in 1923 from a series of 25 specimens from Meadow Valley, Mexico, which according to Labougle (1990) is Río Piedras Verdes, 6 km south of Colonía Garcia, in Chihuahua. It is surprising that such a large and conspicuous bee has not been taken subsequently.

Megachile (Grosapis) cockerelli is superficially similar to certain large, red-haired African species such as *M. (Gronoceras) praetexta* Vachal. *Grosapis* differs from superficially similar African bees in many structures, including the lack of marginal hairs on S8 of the male. This feature shows its relationship to the leafcutting subgenera (Group 1) rather than to Group 2.

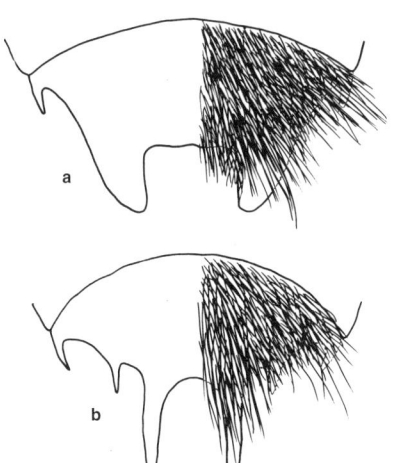

Figure 84-18. T6 of males of *Megachile (Gronoceras).* **a,** *M. (G.) cincta* (Fabricius); **b,** *M. (G.) felina* Gerstaecker. Drawings by D. Yanega.

Megachile / Subgenus *Hackeriapis* Cockerell

Megachile (Hackeriapis) Cockerell, 1922c: 267. Type species: *Megachile rhodura* Cockerell, 1906, by original designation.

Like the New World subgenus *Chelostomoides,* the primarily oriental *Chelostomoda,* and the African *Heriadopsis, Hackeriapis* is a subgenus of Group 2 that contains usually small (5-18 mm long) heriadiform or hoplitiform bees, usually with postgradular grooves on T2 and T3 and with only three exposed sterna, S4 being retracted or sometimes its posterior margin being exposed. S4 (except its posterior margin when exposed) is less sclerotized, less hairy, and less punctate than the exposed sterna. The American *Chelostomoides* and Australian *Hackeriapis* are very similar, both quite diversified; in Australia, *Hackeriapis* or its antecedents probably gave rise to the subgenera *Rhodomegachile* and *Schizomegachile,* and possibly others. In the Australian *Hackeriapis* the postgradular grooves of T2 and T3, which are usually strong, are not fasciate, whereas in the American *Chelostomoides* they contain at least some pale hairs, so that there are basal as well as (usually) apical pale hair bands on these terga. Michener (1965a) and King (1994) listed other differences, none of which is entirely constant. The penis valves exceed the gonoforceps in males of *Hackeriapis* and are exceeded by the gonoforceps in nearly all *Chelostomoides,* and the metanotal-propodeal suture is more conspicuous in *Hackeriapis* than in *Chelostomoides.* In *Hackeriapis* the mandible of the female usually has shining ridges and three to five teeth. There are no sternal hair bands beneath the scopa except laterally in *Megachile (H.) atrella* Cockerell.

Some of the larger species lack postgradular tergal grooves or have them only weakly developed, and separation from *Callomegachile* therefore becomes a problem. Some small *Hackeriapis* have a small tooth near the base of each claw, and the large species [except *Megachile (H.) semiluctuosa* Smith] have a large or very large tooth in this position. In *Callomegachile* such teeth are absent. *Hackeriapis* species [except for *M. (H.) ustulata* Smith] are generally grayish because of pale or mixed pubescence over a black integument, and, if they have fulvous pubescence on the metasoma, it is confined to the apical terga (again except in *M. ustulata*). In *Callomegachile* the color is usually black, fulvous, or black-and-white, or all three colors, owing to pubescence that rarely gives a grayish color to the head and thorax as it does in *Hackeriapis.* In *Callomegachile* the posterior margin of T6 of the male is not toothed; it has lateral and usually median teeth (often weak) in *Hackeriapis.* The mandibles of the females of *Callomegachile* are minutely roughened and dull; they are shining although irregularly reticulate in *Hackeriapis* and all other Australian subgenera. *M. (H.) ustulata* is intermediate in this respect. Michener (1965b) gave a fuller description of the subgenus. Male genitalia and sterna were illustrated by Michener (1965b) and King (1994).

■ *Hackeriapis* is widespread in Australia and occurs also in Tasmania and in savanna regions of New Guinea. About 90 specific names have been proposed. The 17 smaller species (28 names) were retained in *Hackeriapis* and revised by King (1994). She would presumably remove the other species to undescribed subgenera.

Michener (1965b) gave a brief account of variability within the subgenus. The following is extracted from that statement. Some of the larger species, excluded from the subgenus by King (1994) but not placed elsewhere, differ so greatly from the usual sort of *Hackeriapis* that they might be placed in separate subgenera were there not intermediates or confusing combinations of characters among the various species. The heads of the largest species are especially large and much developed posteriorly. *Megachile fumipennis* Smith and *macleayi* Cockerell are coarsely punctate. In *M. mackayensis* Cockerell there is a row of pits across the base of the propodeum (the same is true of *M. nigrovittata* Cockerell, a smaller species); in *M. macleayi* there is a smooth transverse channel in this location. In the latter species the front and middle tibiae are coarsely roughened and tuberculate on their outer surfaces, suggesting *Lithurgus.* The front tarsi of the male may be expanded, although the commonly associated modifications of the coxae and other structures are lacking, and in other species, e.g., *M. ferox* Smith, the front tarsi are enlarged, the front coxae smooth anteriorly and spined apically, and the mandible has a median inferior prominence. The front coxae are also spined (but not smooth) in males of some forms that have unmodified tarsi. *M. ustulata* Smith is worthy of special note because its coloration (black, the metasoma covered with orange hair) is like that of *M. (Callomegachile) mystaceana* Michener and *M. (Amegachile) mystacea* (Fabricius). These three superficially similar species can be collected together in Queensland.

Megachile subserricauda Rayment, known only in the male, differs from *Hackeriapis* in having several long teeth on the carina of T6 of the male. S4 is rather broadly exposed and seems to have punctation and pubescence similar to that on the preceding sterna. The mandibles are largely pale yellow. The deep grooves across the bases of the metasomal terga are occupied by distinct bands of pale hairs, as in the American subgenus *Chelostomoides.* Although it resembles *Hackeriapis* in appearance, it probably belongs in a different subgenus.

Megachile / Subgenus *Heriadopsis* Cockerell

Heriadopsis Cockerell, 1931b: 338. Type species: *Heriadopsis striatulus* Cockerell, 1931, by original designation.

This subgenus contains a heriadiform species of Group 2, 8 to 9 mm long, with metasomal fasciae of white hairs in basal grooves of T2 and T3, as well as laterally on the apices of the terga. S4 of the male is nearly impunctate, having only minute hairs, so that there are probably only three regularly exposed sterna, i.e., S1 to S3. Thus in general appearance and in structure, as demonstrated by Peters (1970b), *Heriadopsis* resembles *Chelostomoda* from Asia, *Chelostomoides* from North America, and *Hackeriapis* from Australia. The presence of arolia on the front and middle legs, however, is unique in the Megachilini except for an undescribed megachiliform taxon from Borneo (D. Baker, personal communication, 1995). Otherwise, the tribe lacks arolia. The four-toothed female mandible of *Heriadopsis* lacks cutting edges. Features supporting placement in *Megachile* are the erect spine on the male front coxa, the expanded and pale front tarsus of the male, and the preapical carina on T6 of the male, this carina broadly bilobed, with a broad emargination between the lobes. The rather simple clypeus somewhat overhangs the base of the labrum. The male genitalia, sterna, and other structures were illustrated by Peters (1970b).

■ *Heriadopsis* occurs in Congo (Kinshasa), Zimbabwe, and Malawi. There is a single species, *Megachile striatula* (Cockerell).

Megachile / Subgenus *Largella* Pasteels

Chalicodoma (Largella) Pasteels, 1965: 534. Type species: *Chalicodoma semivestita* Smith, 1853, by original designation.

This subgenus and both *Chalicodoma* and *Chalicodomoides* look superficially alike, being robust and truly chalicodomiform members of Group 2, substantially more robust than such subgenera as *Pseudomegachile* and *Callomegachile*. *Largella*, 14 to 16 mm long, differs from *Chalicodoma* in having, in the female, a truncate, rather than rounded-and-denticulate clypeal margin and, in the male, a bilobed carina on T6, which lacks lateral teeth. The female mandible is expanded distally, very different from the slender mandible of *Chalicodoma*, with its strongly oblique margin.

■ *Largella* is found in southern Asia and Indonesia. Tkalcŭ (1969b) recorded the subgenus from as far north and west as Afghanistan. One species, *Megachile semivestita* (Smith), has been introduced into Zanzibar (Pasteels, 1965). At least two Asiatic species appear to exist.

Megachile / Subgenus *Leptorachis* Mitchell

Megachile (Leptorachis) Mitchell, 1934: 301, 308. Type species: *Megachile petulans* Cresson, 1878, by original designation.

Pseudocentron (Leptorachina) Mitchell, 1980: 56. Type species: *Megachile laeta* Smith, 1853, by original designation.

Pseudocentron (Grafella) Mitchell, 1980: 56. Type species: *Pseudocentron crotalariae* Schwimmer, 1980, by original designation.

This megachiliform subgenus of Group 1 bees, with a body length of 8 to 16 mm, agrees with *Pseudocentron* in having, in the female, a four-toothed mandible with a complete cutting edge in the third interspace; it disagrees with *Pseudocentron* in having no cutting edge, or almost none, in the second interspace. The male differs from that of *Pseudocentron* in having a small but articulated middle tibial spur. The second interspace of the female mandible is variably reduced, weakest in *Megachile laeta* Smith, which Mitchell (1980) segregated as the monotypic subgenus *Leptorachina*. Intermediates occur, and even in *M. laeta* the interspace is often better developed than in Mitchell's illustration (he may have drawn a worn specimen). At the other extreme, the single species that Mitchell placed in *Grafella* has a distinct second interspace, sometimes with a minute indication of a cutting edge, but when that edge is absent, the mandible is not different from that of typical *Leptorachis*. (The captions for the drawings of female mandibles of *Pseudocentron* and *Grafella* in Mitchell, 1980, seem to be reversed.) Male characters also do not demonstrate the need for the two monotypic subgenera. *M. laeta* Smith has an expanded, pale front tarsus; others in general do not, but some *Leptorachis* have slightly expanded front tarsi. In *M. crotalariae* (Schwimmer) the median emargination of the carina of T6 is broad and deep, the resulting projections thus narrow, almost spinelike. They are broader in other species, but this need not be a subgeneric character. Male genitalia and sterna were illustrated by Mitchell (1937b).

■ This subgenus is principally neotropical, ranging from Argentina northward to the USA. One species, *Megachile petulans* Cresson, occurs in the southeastern USA, thence north to New Jersey and west to Arizona. There are about 30 named species, and many more probably exist in the American tropics.

Megachile / Subgenus *Litomegachile* Mitchell

Megachile (Litomegachile) Mitchell, 1934: 301, 308. Type species: *Megachile brevis* Say, 1837, by original designation.

This subgenus consists of moderate-sized (8-17 mm long) species of Group 1 having the general appearance (broad abdomen, pale hair, white metasomal bands) of species of the subgenera *Megachile* s. str. and *Eutricharaea*. The erect spine on the front coxa of the male and the conspicuously concave profile of T6 of the female are among the characters distinguishing it from *Megachile* s. str. From the originally Old World *Eutricharaea* it differs in lacking bands of white hair beneath the scopa of the female. I do not know of consistent differences between *Litomegachile* and *Eutricharaea* in males. The subgenus was described in detail and genitalia and hidden sterna were illustrated by Mitchell (1935a).

■ *Litomegachile* includes some of the commonest *Megachile* species in most parts of North America, ranging north into southern Canada (British Columbia to Quebec) and south as far as the state of Oaxaca, Mexico. The subgenus is rare or absent in the neotropical region; Mitchell (1943) reported a specimen supposedly from Peru. *Megachile (Litomegachile) gentilis* Cresson of the western United States is now widespread in Hawaii (Snelling, 2003). The seven species of *Litomegachile* were revised by Mitchell (1935a).

Megachile / Subgenus *Matangapis* Baker and Engel

Megachile (Matangapis) Baker and Engel, 2006: 2. Type species: *Megachile alticola* Cameron, 1902, by original designation.

Although superficially rather ordinary looking, megachiliform, 7.5 to 8 mm in body length, this subgenus is extraordinary in the presence of arolia on all the legs and the 4-toothed female mandibles, the teeth separated only by shallow concavities and without traces of cutting edges.

■ Known only from Borneo, this subgenus contains a single species, *Megachile alticola* Cameron.

Megachile / Subgenus *Maximegachile* Guiglia and Pasteels

Megachile (Maximegachile) Guiglia and Pasteels, 1961: 27. Type species: *Megachile maxillosa* Guérin, 1845, by original designation.

This is a subgenus of large (14-25 mm long), robust, black-and-white-haired, chalicodomiform members of Group 2. In the female, the mandible is long, shiny, parallel-sided, and three-toothed. Also in the female there is a large, transverse, truncate tooth at the posterior end of each hypostomal area and a small tooth on the preoccipital carina at about the level of the lower end of the eye. The labrum of the female tapers to an apex only about one-third as wide as the base. The mandible of the male is four-toothed. Guiglia and Pasteels (1961) published diagrams of various structures.

■ This subgenus ranges through sub-Saharan Africa from Senegal to Sudan and south to Namibia and Natal (South Africa); it enters the palearctic region in Egypt and Israel. Pasteels (1965) included two species; G. van der Zanden (in litt., 1993) said that there are three.

Megachile / Subgenus *Megachile* Latreille s. str.

Megachile Latreille, 1802a: 434. Type species: *Apis centuncularis* Linnaeus, 1758, by designation of Curtis, 1828, pl. 218. [A subsequent designation, *Xylocopa muraria* Fabricius, 1804 = *Apis parietina* Fourcroy, 1785, was by Blanchard, 1840: 408.]

Megalochila Schulz, 1906: 263, replacement for *Megachile* Latreille, 1802. Type species: *Apis centuncularis* Linnaeus, 1758, autobasic.

Anthemois Robertson, 1903a: 168, 172. Type species: *Megachile infragilis* Cresson, 1878 = *Apis centuncularis* Linnaeus, 1758, by original designation.

Cyphopyga Robertson, 1903a: 169, 172. Type species: *Megachile montivaga* Cresson, 1878, by original designation.

This subgenus contains ordinary-looking species of Group 1 having the approximate size (7-20 mm long) and appearance of *Litomegachile*. *Megachile* s. str. differs from *Litomegachile* in that the front coxa of the male is hairy and has no apical spine, or the spine is represented by a small tooth; there is no group of rufescent bristles. T6 of the female is usually straight in profile. The subgenus was described in detail and male genitalia and sterna were illustrated by Mitchell (1935b).

■ *Megachile* s. str. is holarctic, mostly in cool climates, north to Newfoundland and subarctic Alaska, but also occurs south to Florida and California, USA, and Coahuila and Jalisco, Mexico. There are five North American species, and five are listed for the palearctic region by Mitchell (1935b); one species, *M. centuncularis* (Linnaeus), is common to the two regions. North American species were revised by Mitchell (1935b).

Megachile (*Megachile*) *montivaga* Cresson is unusual for members of Group 1 in lacking cutting edges on the mandibles and using petals rather than leaves for cell construction. It was given the generic name *Cyphopyga* by Robertson. Some other species, however, such as *M. relativa* Cresson, nearly lack cutting edges, and I see no virtue in recognizing a supraspecific taxon for *M. montivaga* alone. As illustrated by Mitchell (1935b), there are five mandibular teeth in all species, not four in *M. montivaga* as specified by Mitchell (1980).

Megachile / Subgenus *Megachiloides* Mitchell

Megachiloides Mitchell, 1924: 154. Type species: *Megachiloides oenotherae* Mitchell, 1924, by original designation.

Megachile (*Xeromegachile*) Mitchell, 1934: 302, 309. Type species: *Megachile integra* Cresson, by original designation.

Megachile (*Derotropis*) Mitchell, 1936: 156. Type species: *Megachile pascoensis* Mitchell, 1934, by original designation.

Megachiloides Saussure (1890) is an incorrect subsequent spelling of *Megachileoides* Radoszkowski and does not preoccupy *Megachiloides* Mitchell (see *Trachusa* s. str. in the Anthidiini).

This subgenus of Group 1 consists of megachiliform bees 9 to 17 mm long. In males the carina of T6 is entire, i.e., without a median emargination. The major group of species was placed in *Xeromegachile* by Mitchell (1937a). Females of that group have four-toothed mandibles with a long cutting edge in the upper (third) interspace. The group called *Derotropis* has three-toothed mandibles with a very long cutting edge in the upper (second) interspace. *Megachiloides* Mitchell (1924, 1936) is similar except for long mouthparts; its mandibles are four-toothed, but the two median teeth are scarcely separable, and in this respect it is intermediate between the *Xeromegachile* and *Derotropis* groups. I believe that these three groups should be united under the name *Megachiloides* in spite of the differences in female mandibles. Male genitalia and sterna were illustrated by Mitchell (1936, 1937a).

■ *Megachiloides* ranges from southern Alberta, Canada, south throughout the USA and to Oaxaca, Mexico, and is particularly frequent in xeric areas. This is the largest subgenus of *Megachile* in North America, with about 60 species, 44 of them in the group called *Xeromegachile* by Mitchell. The species were revised by Mitchell (1936, 1937a); several of them appear to be oligolectic on flowers of various families of plants.

Megachile / Subgenus *Megella* Pasteels

Megachile (*Megella*) Pasteels, 1965: 167. Type species: *Megachile malimbana* Strand, 1911, by original designation.

This subgenus contains large (15-22 mm long), elongate species with a body form suggestive of *Megachile* (*Callomegachile*) *devexa* Vachal or *M.* (*Pseudomegachile*) *armatipes* Friese. Moreover, as shown in Table 84-1, S8 of the male has long marginal hairs as in the Group 2 subgenera. Nonetheless, S6 of the female has an apical fringe of plumose hair, and the four-toothed mandible of the female has a cutting edge hidden behind the margin of the third interspace (Fig. 84-11h), or, in the Asiatic species, barely visible in facial view. The presence of four teeth (the lateral ones mesally directed spines, the median ones low convexities) on the posterior margin of T6 of the male is a common character in Group 1 but rare in *Chalicodoma*-like subgenera, i.e., Group 2. *Megella* is one of the taxa that makes separation of *Megachile* and *Chalicodoma* at the genus level difficult. In *Megachile* (*Megella*) *malimbana* Strand the carina of T6 of the male is broadly rounded, and there is a broad median depres-

sion basal to the carina. Pasteels (1965) described a second species (male only), *M. (M.) exsecta* Pasteels, that has a semicircular emargination in the carina. The Asiatic species, *M. (M.) pseudomonticola* Hedicke, is known to me only in the female, which is morphologically similar to *M. (M.) malimbana,* although the mandibular surface is duller and the cutting edge is slightly exposed in frontal view.
■ This subgenus occurs in West Africa from Liberia to the Congo basin and Katanga, and in Southeast Asia, including southern China. Pasteels (1965) reviewed two species from Africa; a third is Asiatic.

Megachile / Subgenus *Melanosarus* Mitchell

Megachile (Melanosarus) Mitchell, 1934: 303, 307. Type species: *Megachile xylocopoides* Smith, 1853, by original designation.

This is a megachiliform subgenus of the *Pseudocentron* subgroup of Group 1, as shown by the rim of S6 of the female (see the key to the subgenera) and the lack of the middle tibial spur in the male. It is the only member of the *Pseudocentron* group of subgenera that has five-toothed mandibles in the female. It consists of large (10-16 mm long) megachiliform bees, some species having the pubescence entirely or largely black. The subgenus was fully characterized and male genitalia and sterna illustrated by Mitchell (1937b).
■ *Melanosarus* ranges from the coastal southeastern USA (Maryland to Texas) southwest to Jalisco, Mexico, and southward through the tropics, including the Antilles, to the province of Misiones, Argentina. There are about eight species; the two that occur in the USA were revised by Mitchell (1937b).

Megachile / Subgenus *Mitchellapis* Michener

Megachile (Mitchellapis) Michener, 1965b: 211. Type species: *Megachile fabricator* Smith, 1868, by original designation.

This is a subgenus of chalicodomiform bees of rather large size (12-16 mm long). As in *Sayapis* and *Eumegachile,* there is only one cutting edge in the female mandible, located in the second interspace. Contrary to these two subgenera, there are lateral hairs on S8 of the male, as in Group 2 (Table 84-1). *Mitchellapis* has apical bands of sternal hairs underneath the scopa, as in *Eutricharaea* and *Chelostomoda,* and a median spine on the carina of T6 of the male. The subgenus was more fully described by Michener (1965b), who also illustrated the male genitalia and sterna.
■ *Mitchellapis* is found from Queensland to Western Australia. Six specific names have been proposed.

Megachile / Subgenus *Moureapis* Raw

Pseudocentron (Moureana) Mitchell, 1980: 56; not *Moureana* Zajciw, 1967, a cerambycid beetle. Type species: *Megachile anthidioides* Radoszkowski, 1874, by original designation.
Megachile (Willinkella) Laroca, Cure, and Bortoli, 1982: 97, *nomen nudum.*
Megachile (Acentrina) Schlindwein, 1995: 97, *nomen nudum.*
Megachile (Moureapis) Raw, 2002: 23, replacement for *Moureana* Mitchell, 1980. Type species: *Megachile anthidioides* Radoszkowski, 1874, autobasic and by original designation.

This subgenus of ordinary-looking, megachiliform bees of Group 1, 7 to 12 mm long, is a member of the *Pseudocentron* group of subgenera. The male lacks the middle tibial spur, as do *Acentron* and *Melanosarus,* but differs from them in having simple front and middle legs. The female mandible is much like that of *Pseudocentron,* but the upper mandibular tooth is often truncate rather than acute.
■ This subgenus is abundant in southern Brazil and Argentina but ranges from Buenos Aires province, Argentina, through the tropics to the states of Yucatan and Hidalgo, Mexico. There are at least eight described species.

Specimens of this subgenus are found in many collections under a patronymic given by J. S. Moure honoring A. Willink, a well-known Argentine hymenopterist. See the synonymy above.

Megachile / Subgenus *Neochelynia* Schrottky

Neochelynia Schrottky, 1920: 187. Type species: *Neochelynia paulista* Schrottky, 1920, monobasic.
Megachile (Neomegachile) Mitchell, 1934: 302, 306. Type species: *Megachile chichimeca* Cresson, 1878, by original designation.

The small species (length 6.5-10.0 mm) constituting this subgenus are slender for a Group 1 megachiliform bee, and the costal part of the forewing is often conspicuously darkened. The subgenus is near *Rhyssomegachile* but differs in the bidentate median mandibular tooth of the male, such that the mandible could well be considered four-toothed, as well as in other characters. From *Ptilosaroides* it differs in appearance (*Ptilosaroides* is more robust and has abundant golden tomentum like that of *Ptilosarus*) as well as other features. Males of *Neochelynia* are unusual in that the metasoma is elongate and tapering, the carina of T6 is directed posteriorly instead of ventrally, and the surface basal to that carina is more or less dorsal and horizontal, rather than posterior and vertical or even posteroventral. These features are suggestive of *Coelioxys,* and no doubt explain Schrottky's failure to recognize the correct genus of *Megachile paulista* (Schrottky), new combination, which was based on a male. Male genitalia and hidden sterna were illustrated by Mitchell (1935a).
■ *Neochelynia* ranges from southern Texas, USA, and Sonora and Yucatan, Mexico, at least to the states of São Paulo and Mato Grosso, Brazil. There are at least five species. *Megachile chichimeca* Cresson is common in Mexico, but elsewhere the subgenus is known from rather few specimens.

The identity of *Neochelynia* as a *Megachile* was pointed out to me by A. Roig-Alsina. *Megachile uniformis* Mitchell is a junior synonym of *M. paulista* (Schrottky).

Megachile / Subgenus *Neocressoniella* Gupta

Cressoniella (Neocressoniella) Gupta, 1993: 172. Type species: *Megachile carbonaria* Smith, 1853, = *Anthophora barbata* Fabricius, 1804, by original designation.

This subgenus consists of large (12-21 mm in body length), megachiliform but relatively elongate species of Group 1, the two best known species being entirely black in females and largely so in males. Baker and Engel (2006) regarded *Neocressoniella* as a synonym of *Xanthosarus*. Proper study of the group may well show that they are correct. For the present, because of the characters of females indicated in the key to palearctic and oriental subgenera above, I retain *Neocressoniella*. The structure is similar to that of *Paracella*, differing in the female as indicated in the key to palearctic and oriental subgenera. As in *Paracella*, there is a strong preoccipital carina behind the gena. The upper tooth of the four-toothed female mandible is sharply right-angular to acute; the mandibular teeth were illustrated by Baker (1993: 141). As in many *Eutricharaea* and *Paracella*, there is a tuft of short orange hairs near the apex of the outer acetabular groove (ending between the first and second mandibular teeth) of the female mandible. There are no apical sternal hair fasciae beneath the scopa. The mandible of the male is also four-toothed, and has no inferior projection. The front coxae and tarsi of the male are not or little modified. The carina of T6 is large, not appreciably crenulate, and has a small median emargination; the posterior margin of T6 has a slender, sharp lateral tooth and a broad submedian tooth.

■ *Neocressoniella* occurs in India. The two black species are *Megachile carbonaria* Smith and *M. anthracina* Smith, the latter being the most common. Two other species probably belong here.

Megachile / Subgenus *Paracella* Michener

Megachile (Paracella) Pasteels, 1965: 277, no type species designated.
Megachile (Paracella) Michener, 1997b: 44. Type species: *Megachile semivenusta* Cockerell, 1931, by original designation.

Because Pasteels failed to designate a type species for his subgenus *Paracella*, his subgeneric name is invalid. In order to reduce confusion, I used the same name to reestablish the subgenus.

This subgenus consists of moderate-sized (9-13 mm long) megachiliform members of Group 1. The mandible of the female is four-toothed, the upper tooth generally broad, sometimes truncate, sometimes emarginate (so that there are five teeth, the upper two close together and only shallowly separated). A cutting edge, large but sometimes incomplete, occupies the second interspace, and another is in the third interspace, the latter sometimes hidden behind the margin of the interspace. As in many *Eutricharaea*, there is a brush of short orange hairs in the apex of the lower acetabular groove of the female mandible, in addition to a similar brush in the apex of the outer acetabular groove; the brush in the lower groove is often absent, however. Also as in *Eutricharaea*, most species have apical sternal fasciae beneath the scopa. The mandible of the male is four-toothed, has no inferior projection, but is sometimes swollen with a dense fringe below. The front tarsus of the male is not or slightly broadened although often pale. The carina of T6 of the male is large, crenulate or not; its median emargination is sometimes lost when the margin is coarsely spinose-crenulate. The posterior margin of T6 varies from simple to one having four small teeth. The subgenus was more fully characterized by Pasteels (1965).

■ *Paracella* is found throughout sub-Saharan Africa, from Liberia to Kenya, south to South Africa. In his revision of the group, Pasteels (1965) recognized 39 species. Unidentified specimens that appear to belong to this subgenus are from India and Indonesia. Unfortunately, many *Megachile* species from tropical Asia have not been studied in detail or placed in subgenera; see the comments under *Eutricharaea*.

Megachile / Subgenus *Parachalicodoma* Pasteels

Chalicodoma (Parachalicodoma) Pasteels, 1966: 13 (not Tkalců, 1969). Type species: *Megachile incana* Friese, 1898, by original designation.

This subgenus, a member of Group 2, superficially resembles *Chalicodoma*, and consists of robust, hairy, chalicodomiform bees without tergal fasciae. It also resembles *Chalicodoma* in the large, multi-spined carina of the male T6 and the irregularly rounded, unthickened clypeal apex. It differs from *Chalicodoma*, however, in the female mandible, which is broad, shiny, and coarsely reticulate, and lacks carinae except apically, the apical margin not as oblique as in *Chalicodoma* and distinctly five-toothed, the teeth about equidistant. An unusual feature is that T7 of the male has a long, truncate median process that extends well beyond the teeth of T6. S6 of the female is apically emarginate.

■ This subgenus is found in northern Africa (Egypt, Algeria). The single species is *Megachile incana* Friese.

Megachile / Subgenus *Platysta* Pasteels

Megachile (Platysta) Pasteels, 1965: 171. Type species: *Megachile platystoma* Pasteels, 1965, by original designation.

Like *Eutricharaea*, *Platysta* consists of species of Group 1. The female of this subgenus has five-toothed mandibles, the fourth and fifth teeth being close together and easily interpretable as an emarginate fourth tooth, such as is found in a few species of *Eutricharaea*. But in *Platysta* there is a small cutting edge in the fourth interspace as well as a larger one in the third; there is none in the second (I have not seen females and base these comments on Pasteels, 1965.) In its large size (18-22 mm long), *Platysta* differs from most megachiliform species. The male is distinctive: T6 has a large median projection, about one-fifth as wide as the tergum, representing the usual carina; it is rounded but denticulate apically, and flattened on the upper (= posterior) surface, and the tergum in front of the carina is convex, not depressed. The posterior margin of T6 has a small lateral tooth hidden in hair and a strong lateromedian tooth. The posterior margins of the front and middle tibiae are carinate, as in *Gronoceras*.

■ This African subgenus occurs from Senegal to Botswana. There are two species in Pasteels' (1965) revision.

In some species here included in *Eutricharaea*, T6 of the male resembles that of *Platysta* except that the carina extends laterally from the median projection. *Megachile konowiana* Friese, in particular, seems to link *Platysta* to

Eutricharaea. It seems quite possible that, with further study, such species, perhaps all of Pasteels' (1965) *patellimana* group of *Eutricharaea* (actually placed in his subgenus *Eurymella*), would be removed to *Platysta*.

Megachile / Subgenus *Pseudocentron* Mitchell

Pseudocentron Mitchell, 1934: 303, 307. Type species: *Megachile pruina* Smith, 1853, by original designation.

This is a subgenus of megachiliform Group 1 species of moderate to large size (8-16 mm long). The female mandible is four-toothed with a complete and long cutting edge in the third interspace and a rather large but incomplete edge in the second interspace. The latter feature differentiates the mandible from that of *Leptorachis*. The male is unique in having, in place of the middle tibial spur, a slender, unarticulated prong. The subgenus was fully characterized and male genitalia and sterna illustrated by Mitchell (1937b).

■ *Pseudocentron* is primarily neotropical, ranging from Malleco, Chile, and Neuquén, Argentina, north through the tropics to Cuba and the southern USA (southern California to North Carolina). This is the major South American subgenus of *Megachile*. Mitchell (1943) listed 41 neotropical species and there are subsequent additions (about 55 named species), as well as undescribed species; the total is probably near 80. Probably only three species occur in the USA; see the revision by Mitchell (1937b).

Megachile / Subgenus *Pseudomegachile* Friese

Megachile (Pseudomegachile) Friese, 1898b: 198; species first included by Friese, 1899c: 36. Type species: *Megachile ericetorum* Lepeletier, 1841, designated by Alfken, 1933a: 56.
Megachile (Pseudomegalochila) Schulz, 1906: 71, unjustified emendation of *Pseudomegachile* Friese, 1898.
Megachile (Archimegachile) Alfken, 1933: 56. Type species: *Megachile flavipes* Spinola, 1838, by original designation.
Chalicodoma (Neglectella) Pasteels, 1965: 431. Type species: *Megachile armatipes* Friese, 1909, by original designation.
Chalicodoma (Dinavis) Pasteels, 1965: 549. Type species: *Megachile muansae* Friese, 1911, by original designation.
Megachile (Xenomegachile) Rebmann, 1970: 44. Type species: *Megachile albocincta* Radoszkowski, 1874, by original designation.

Species of this subgenus of Group 2 have in general the appearance of *Callomegachile* or are slightly more robust; length ranges from 10 to 22 mm. Some are gray but most are highly colored, like most *Callomegachile*. The females differ from those of *Callomegachile* in their smooth, shiny mandibular ridges. The clypeus is not as grotesquely modified, and the mandibles are not as long and parallel-sided, as in some species of *Callomegachile*. The mandibles of females are three- to five-toothed. Males usually differ from *Callomegachile* and its relatives in the high, usually multidentate or denticulate carina of T6, sometimes emarginate medially (Fig. 84-14a). In such species as *Megachile lanata* (Fabricius), however, the carina is often merely bilobed, although in some specimens of that species there are small irregularities in the margin suggesting the denticulate condition (Fig. 84-14b). Males of such species can be distinguished from *Callomegachile* by the higher carina and lack of a deep median depression above the carina, and usually by the more shiny mandibular carinae, though this character is not as useful as in females.

■ *Pseudomegachile* occurs from Europe (as far west as Spain, as far north as middle Finland), especially the Mediterranean region, east to China, south through Africa to Cape Province, South Africa; also south in India and Southeast Asia, and to islands such as Madagascar and Reunion. It has been introduced in the Antilles, and one species, *Megachile lanata* (Fabricius), has been taken in southern Florida, USA. Pasteels (1965) includes about 50 species from sub-Saharan Africa. The number of additional species in the palearctic and oriental regions is not clear but is probably 30 or more.

Pasteels (1965) regarded 11 species as constituting a subgenus *Neglectella*, which he perhaps separated because of their five-toothed female mandibles (but they are also five-toothed in *Megachile louisae* Brauns), the lateral tooth on T6 of the male, and the presence of volsellae. Volsellae can be present or absent in other subgenera, and I think need not be regarded as a subgeneric character in this case. Males of *Pseudomegachile* are so variable that in the key to subgenera they come out in several places, and species probably exist that will not run properly in the key. Even within *Pseudomegachile* as limited by Pasteels (1965), although front coxal spines are usually large, they are small in *M. lanata* (Fabricius); the mandible can be three- or four-toothed, with or without an inferior projection; the front tarsi can be modified or not; and T6 varies as indicated above. Within the *Neglectella* group, the front coxae vary even more, having a very large spine in some species, none in others. Although the mandibles are always three-toothed, they may or may not have a ventral convexity; the front tarsi and T6 vary at least as much, and in the same ways, as those in *Pseudomegachile* sensu Pasteels.

Dinavis was also recognized as a distinct subgenus by Pasteels; from his descriptions and figures I see no compelling reason to separate it from *Pseudomegachile*, but it is known only from males and I have not seen specimens. My placement of it here is therefore tentative.

Alfken (1933) separated two groups of palearctic species under the names *Pseudomegachile* and *Archimegachile*. I follow Pasteels in not recognizing them at the subgeneric level. Rebmann (1970) recognized another palearctic form, *Megachile albocincta* Radoszkowski, as the subgenus *Xenomegachile*. It differs from most *Pseudomegachile* in having the head less extended behind the eyes, so that the ocellocipital distance is less than the interocellar distance, but this seems to me not to warrant its recognition as a subgenus.

Megachile / Subgenus *Ptilosaroides* Mitchell

Cressoniella (Ptilosaroides) Mitchell, 1980: 63. Type species: *Megachile neoxanthoptera* Cockerell, 1933, by original designation.

This Group 1 subgenus has the distinctive appearance of *Ptilosarus* because of the abundant yellow, often appressed pubescence, yellowish wing bases and dusky costal margins, producing a color pattern similar also to that of the vespid genus *Brachygastra*. The body is megachiliform, 8 to 9 mm long. *Ptilosaroides* differs from *Ptilosarus* in lacking the peculiar scopal features of the latter and in lacking a strong preoccipital carina. The carina of S6 of the male is represented by two spines, often more slender than the two teeth of *Ptilosarus*, and the front coxa has a spine, which is absent in *Ptilosarus*.

■ This is a neotropical subgenus ranging at least from southern Brazil to Panama. Only *Megachile neoxanthoptera* Cockerell has been formally included in the subgenus, but specimens of at least two other species are available.

Megachile / Subgenus *Ptilosarus* Mitchell

Megachile (Ptilosarus) Mitchell, 1943: 667. Type species: *Megachile bertonii* Schrottky, 1908, by original designation.

This genus consists of small (7-10 mm long) megachiliform bees of Group 1 with areas of short, appressed golden pubescence or tomentum. The costal margins of the forewings are ordinarily dark, and this, together with the areas of yellow pubescence, suggests the vespid genus *Brachygastra*. An uncommon feature in *Megachile* is the strong preoccipital carina, not only laterally but extending across behind the vertex. The females are unique in the long, dense, plumose pubescence on the thoracic venter and replacing ordinary unbranched scopal hairs on S2. This feature may be associated with their pollen-collecting behavior on the erect inflorescences of the genus *Piper*. The bees crawl forward both up and down the inflorescences, moving the metasoma from side to side (personal observations, Panama and French Guiana). In the male the carina of T6 consists of two small, sharp teeth; otherwise it is scarcely evident.

■ This subgenus ranges from Veracruz, Mexico, to Misiones, Argentina, mostly in moist tropics. Mitchell (1943) listed nine species; several more are tentatively recognized.

Megachile / Subgenus *Rhodomegachile* Michener

Chalicodoma (Rhodomegachile) Michener, 1965b: 201. Type species: *Megachile abdominalis* Smith, 1853, by original designation.

This is a subgenus of small (5-10 mm long), heriadiform or chalicodomiform species of Group 2 with the metasoma entirely red. T2 and T3 lack both postgradular grooves and apical fasciae. An unusual feature is the broad, ligulate glossa. T6 of the male either lacks a carina (thus differing from other Megachilini) or has only a short, median, transverse, non-emarginate ridge. S4 of the male is retracted, and there are thus only three exposed sterna. Male genitalia and sterna were illustrated by Michener (1965b) and King and Exley (1985a).

■ This subgenus occurs throughout the more or less temperate parts of Australia, although it is relatively abundant only in southern Queensland. Two of the three species were reviewed by King and Exley (1985a).

Rhodomegachile was more fully characterized by Michener (1965b) and King and Exley (1985a). The latter authors excluded one species, *Megachile deanii* Rayment, in part because it does not have attenuate front tarsi in the male, as do the other two species. But many subgenera of *Megachile* include species both with and without modified front tarsi (and intergrading). Thus this character is often not of subgeneric importance, and for the present I retain *M. deanii* in *Rhodomegachile*.

Megachile / Subgenus *Rhyssomegachile* Mitchell

Cressoniella (Rhyssomegachile) Mitchell, 1980: 63. Type species: *Megachile simillima* Smith, 1853, by original designation.

This subgenus consists of megachiliform Group 1 species 8 to 11 mm long, quite ordinary in general appearance. Because the upper mandibular tooth is strongly incised, the mandible can be called five-toothed, but the upper two teeth are close together. The second and third interspaces both have incomplete and scarcely evident cutting edges arising from the lower margins of the third and fourth teeth. [The larger cutting edge in the second interspace shown by Mitchell (1980: fig. 44) appears to be an error, as indicated by his text where he says the cutting edges are obsolescent.] An unusual feature for megachiliform subgenera is the strong preoccipital carina of both sexes, suggestive of *Ptilosarus* and *Austromegachile*, from which females of *Rhyssomegachile* differ by their complete lack of white fasciae beneath the scopa and the very incomplete cutting edge in the third mandibular interspace. Males that I have seen have a front coxal spine, but Mitchell's key (1980) indicates that this character is variable. The carina of T6 of the male has only a shallow emargination, and the posterior margin of T6 has only a small lateral tooth.

■ *Rhyssomegachile* occurs in Brazil from Pará to Paraná. I know of only one species, *Megachile simillima* Smith.

Megachile / Subgenus *Sayapis* Titus

Gnathocera Provancher, 1882: 232 (not Kirby, 1825). Type species: *Gnathocera cephalica* Provancher, 1882 = *Megachile pugnata* Say, 1837, monobasic.
Ceratias Robertson, 1903a: 172 (not Kroyer, 1845). Type species: *Megachile pugnata* Say, 1837, by original designation.
Sayapis Titus, 1906: 154, replacement for *Gnathocera* Provancher, 1882, and *Ceratias* Robertson, 1903. Type species: *Megachile pugnata* Say, 1837, autobasic.

This is one of the two major Western Hemisphere subgenera of more or less parallel-sided *Megachile*, the other being *Chelostomoides*. Its members are in general chalicodomiform, less slender than *Chelostomoides* but often more slender than the Old World subgenus *Chalicodoma*; see Plate 7. It differs from *Chelostomoides*, as well as from the small Western Hemisphere subgenera with parallel-sided bodies, *Stelodides* and *Schrottkyapis*, in having an incomplete cutting edge in the second interspace of the female mandible (but no other cutting edges). Unlike *Chelostomoides*, it has four exposed metasomal sterna in the male. As indicated in Table 84-1, *Sayapis* is a member of Group 1; it is probably related to the palearctic subgenus *Eumegachile*, and I seriously considered synonymizing these two subgenera, but decided to delay such action until a phylogenetic study is made. The short female mandible, with its dentate margin at least half as long as the mandible, distinguishes *Sayapis* from *Eumegachile*. [The character states for female mandibles are reversed in Mitchell's (1980) key to the subgenera (his couplet 4, p. 46) that he placed under the generic name *Eumegachile*.] The male genitalia and sterna were illustrated by Mitchell (1937c); the same work illustrates the variation among species in the front tarsi of males, from unmodified to much expanded.

■ *Sayapis* ranges across North America, and from Canada to Argentina. Nine species are known in the USA, and ten additional species occur in tropical America. Because one species, *Megachile dentipes* Vachal, ranges from Texas, USA, to Argentina, the total is 18. The species found in the USA were revised by Mitchell (1937c); those found in Argentina, by Durante and Diaz (1996) under the generic name *Eumegachile*.

Megachile / Subgenus *Schizomegachile* Michener

Chalicodoma (Schizomegachile) Michener, 1965b: 199. Type species: *Megachile monstrosa* Smith, 1868, by original designation.

This subgenus contains a large (17-22 mm long), chalicodomiform species of Group 2 with numerous extraordinary features, as detailed by Michener (1965b). In the female each claw has two teeth on the undersurface, T6 has a shining flat-topped longitudinal ridge, and S6 has a median ridge ending in an acute sternal apex. In the male, the mandible is bidentate and has an inferior, transparent, basal lamella; the hind tibial spurs are represented by a single large spine; S4 is exposed but less hairy and punctate than S3, thus possibly derived from a retracted S4; and S5 and S6 are unique, illustrated with genitalia by Michener, 1965b.

■ *Schizomegachile* occurs in temperate parts of both eastern and western Australia. Although there are two specific names, this subgenus probably contains only a single species, *Megachile monstrosa* Smith.

Megachile / Subgenus *Schrottkyapis* Mitchell

Eumegachile (Schrottkyapis) Mitchell, 1980: 46. Type species: *Megachile assumptionis* Schrottky, 1908, by original designation.

This subgenus looks like a large (14-15 mm long) *Chelostomoides*, with which it agrees in the lack of cutting edges between the four mandibular teeth of the female (the fourth tooth is slightly incised at the apex). The lack of marginal hairs on S8, however, suggests that *Schrottkyapis* is not related to *Chelostomoides*, but is a member of Group 1. It seems to be a *Sayapis* that has lost the cutting edge of the female mandible and the inferior projection on the male mandible. Peculiar features, other than those indicated in the key to subgenera, include the strong, bifid median process of the female clypeus, extending down over the base of the labrum, and the thickened, protuberant male clypeus, which is short and does not overhang the base of the labrum.

■ *Schrottkyapis* is found from Paraguay north to Minas Gerais, Brazil. The only species is *Megachile assumptionis* Schrottky.

Megachile assumptionis Schrottky may be a specialist in nest-site selection, using abandoned burrows of *Ptilothrix plumatus* Smith (Emphorini). Each *Ptilothrix* nest is a shallow burrow ending in one cell. The *Megachile* likewise constructs only one cell per burrow, lining the *Ptilothrix* cell with masticated leaf material, which is also used for the cell closure, then filling the burrow above the cell with leaf pieces, another closure of masticated leaf material, and finally mud (Martins and Almeida, 1994).

Megachile / Subgenus *Stelodides* Moure

Stelodides Moure, 1953c: 123. Type species: *Megachile euzona* Pérez, 1899, by original designation.

This subgenus resembles *Chrysosarus* and is thus a member of Group 1 in spite of its chalicodomiform body. The color pattern (described in the key) is distinctive. It seems likely that this is a specialized derivative of *Chrysosarus*; if so, it presumably does not deserve subgeneric rank, but a decision about such synonymy should await a fuller study.

■ *Stelodides* occurs from Santiago, Chile, south to Neuquén, Argentina. The only species is *Megachile euzona* Pérez.

Megachile / Subgenus *Stenomegachile* Pasteels

Chalicodoma (Stenomegachile) Pasteels, 1965: 507. Type species: *Megachile chelostomoides* Gribodo, 1894, by original designation.

Although smaller (11-17 mm) and much more slender, this subgenus resembles *Maximegachile,* also a member of Group 2, in the long, shining mandibles and toothed posterior hypostomal areas of the female. The bilobed carina of T6 of the male and the lateral tooth of T6 also are as in *Maximegachile.* The female differs from that of *Maximegachile* in the four-toothed mandible and the very long but more nearly parallel-sided labrum; the male differs in the dense, long hairs of T6 and its four-toothed mandible.

■ This subgenus occurs from Eritrea to Zambia and in Madagascar. Pasteels (1965) included four species.

Megachile dawensis (Pasteels) is so different (the mandible of the female is shorter and three-toothed) from the type species that its position in *Stenomegachile* is doubtful. The Malagasy species, *M. dolichosoma* Benoist, is so slender as to suggest a large species of *Chelostoma* (Osmiini), and is unusual in that the median part of the scutum is drawn out anteriorly, then abruptly bent down to form an anterior-facing area. The male lacks lateral teeth on T6. The resemblance to *Stenomegachile* thus may be superficial. These slender species resemble one another in the sloping profile of the propodeum (about 45% to the axis of the body); in most other *Megachile* the profile is more nearly vertical, although often sloping to some degree.

Megachile / Subgenus *Thaumatosoma* Smith

Thaumatosoma Smith, 1865: 394. Type species: *Thaumatosoma duboulaii* Smith, 1865, monobasic.

Thaumatosoma was long regarded as a genus because of the strange characters of the male: swollen scape, attenuate flagellum (most segments about ten times as long as broad), broadly flattened flagellomeres 10 and 11 (Fig. 84-10q), a single hind tibial spur, and central clypeal tuft of long, coarse bristles (see illustrations by Michener, 1965b, and King and Exley, 1985b). An unusual feature of the female is the long, bidentate mandible. Unusual as these features are, they seem to indicate derivation from other Group 2 *Megachile,* not a sister-group status for *Megachile* and *Thaumatosoma.* I therefore follow my earlier decision (1965b) to regard *Thaumatosoma* as a subgenus. The body is chalicodomiform, 9 to 12 mm long, rather robust in the female. T2 and T3 have strong postgradular grooves as in *Hackeriapis,* but S4 of the male is exposed. Since the last of these characters is presumably a plesiomorphic condition compared to retraction of S4, *Thaumatosoma* may have a sister-group relation to *Hackeriapis.* Male genitalia and sterna were illustrated by Michener (1965b) and King and Exley (1985b).

■ *Thaumatosoma* is widespread across the southern third of Australia, in desertic areas and along the western coast. The two species were revised by King and Exley (1985b).

Megachile / Subgenus *Trichurochile* Mitchell

Cressoniella (Trichurochile) Mitchell, 1980: 63. Type species: *Megachile thygaterella* Schrottky, 1913, by original designation.

This subgenus consists of megachiliform members of Group 1, 10 to 12 mm long. The fringe of long plumose hairs arising from the inner margins of the teeth of the carina of T6 of the male (and largely filling the emargination between the teeth) is distinctive. In the female, the white apical fasciae beneath the sternal scopa, as in *Eutricharaea,* are distinctive; other *Megachile* native to the Western Hemisphere lack such fasciae or, less commonly (in *Austromegachile*), have them broadly broken.

■ *Trichurochile* is known from Brazil (Santa Catarina, Minas Gerais), Paraguay, and Bolivia; it doubtless also occurs in northern Argentina. The only known species is *Megachile thygaterella* Schrottky.

Megachile / Subgenus *Tylomegachile* Moure

Megachile (Tylomegachile) Moure, 1953c: 120. Type species: *Megachile orba* Schrottky, 1913, by original designation.

This subgenus consists of robust, megachiliform species of Group 1, 9 to 13 mm long. Although larger and more robust than that of *Neochelynia,* the metasoma of the female is commonly tapering, segment 3 being narrower than the more anterior segments. Unfortunately, this character is influenced by the position in which the bee dies. Specimens that do not show this character will run to *Cressoniella* in the key; they differ from that subgenus in the very obtuse angle of the third mandibular tooth and the lack or near lack of a cutting edge in the second interspace. The male is unusual in that the carina of T6 lacks any indication of an emargination and has, instead, a small median tooth or angle. The apical margin of T6 has, however, two rather large submedian teeth, much larger than the lateral teeth. The female mandible is strikingly similar to that of *Austromegachile,* and the relationship may be close. The lack of a preoccipital carina, lack of white fasciae under the scopa, and the rather flat female clypeus in *Tylomegachile* support the distinction, but further study may well indicate that relationships (rather than differences) are best indicated by uniting these subgenera.

■ This subgenus ranges from Misiones, Argentina, to Sonora and Tamaulipas, Mexico. Only *Megachile orba* Schrottky has been formally included, but *M. simplicipes* Friese, rather common in Mexico and Central America, is also a member of this subgenus. I have seen four other South and Central American species, all probably unnamed.

Megachile / Subgenus *Xanthosarus* Robertson

Xanthosarus Robertson, 1903a: 168, 169, 172. Type species: *Megachile latimanus* Say, 1823, by original designation.
Megachile (Delomegachile) Viereck, 1916: 745. Type species: *Megachile vidua* Smith, 1853 = *M. latimanus* Say, 1823, monobasic.
Megachile (Phaenosarus) Mitchell, 1934: 303, 309. Type species: *Megachile fortis* Cresson, 1872, by original designation.
Megachile (Macromegachile) Noskiewicz, 1948: 48. Type species: *Apis lagopoda* Linnaeus, 1761, by original designation. [Mitchell (1980:23) incorrectly lists *Apis maritima* Kirby, 1802, as the type species.]
Megachile (Addendella) Mitchell, 1980: 24. Type species: *Megachile addenda* Cresson, 1878, by original designation.

Xanthosarus consists of large (8-18 mm long), robust, megachiliform species of Group 1. The front tarsi of males are frequently pale, fringed, and greatly expanded (Fig. 84-19). The subgenus was characterized (along with *Phaenosarus* and *Delomegachile*) by Mitchell (1936). Females are similar to and sometimes not readily separable from those of *Cressoniella* (see the key to subgenera), but males differ from *Cressoniella* in having a large basal projection on the lower margin of the mandible and, on the front coxa, a bare anterior surface (except sometimes for a clump of red bristles) and a large, erect, apical spine. Male genitalia and sterna were illustrated by Mitchell (1935b, 1936) and Marikovskaya (1984).

■ *Xanthosarus* is holarctic, ranging across North America from Canada and Alaska to Florida and California, USA, south to Oaxaca, Mexico, and from Spain and Britain to Japan. There are 13 North American species and perhaps a similar number in Eurasia. American species were revised by Mitchell (1935b, 1936).

The type species of *Delomegachile* is a synonym of the type species of *Xanthosarus* (Mitchell, 1962: 157). The species formerly placed in *Delomegachile,* its genotype having been misidentified, could now be placed in *Addendella.* In North America, males placed in *Addendella* (= *Delomegachile* auctorum) and *Xanthosarus* are separable; for example, the former has four-toothed mandibles and some species have red bristles on the front coxae, whereas the latter has three-toothed mandibles and lacks red bristles. But in the palearctic region, species like *Megachile lagopoda* (Linnaeus) and *maritima* (Kirby) have a different combination of characters: three-toothed mandibles and red bristles. Other palearctic species, such as *M. circumcincta* (Kirby) and *willughbiella* (Kirby),

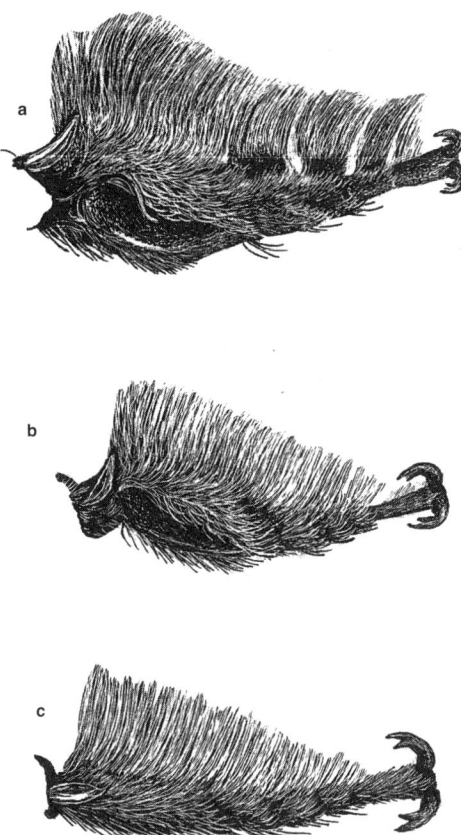

Figure 84-19. Inner surfaces of front tarsi of males of *Megachile* (*Xanthosarus*) spp. **a,** Much broadened, *M. latimanus* Say; **b,** Intermediate, *M. gemula* Cresson; **c,** Unmodified, *M. addenda* Cresson. Drawings by S. Taliaferro.

agree with *Addendella* in both the four-toothed male mandible and the red bristles on the fore coxae. The male gonoforceps are complex (enlarged and bifid distally) in American *Xanthosarus* s. str., much more simple in the *Addendella* group. They are complex in the palearctic *M. lagopoda* and *maritima* but simple in *M. circumcincta*. Moreover, as noted below, they are enlarged but not bifid distally in species formerly placed in *Phaenosarus*. The mandible of the female causes *Xanthosarus* s. str. and *Addendella* to come out far apart in the key to subgenera (also in Mitchell, 1934, 1962); in *Xanthosarus* s. str. the mandible is five-toothed (Fig. 84-11a) whereas in the *Addendella* group it is four-toothed (Fig. 84-11b, c). Every intergrade occurs, however; as noted in the keys, the upper tooth of the four-toothed mandible may be incised or bidentate, sometimes only to the extent of a mere dimple that becomes invisible with a little wear, sometimes as a strong emargination. Sometimes the upper tooth is rather broad, as in *M. frigida* Smith (see Mitchell, 1962, fig. 43), but in species such as *M. circumcincta* it is rather narrow but bidentate, not appreciably different from the condition in American *Xanthosarus* s. str. and the palearctic *M. lagopoda*, except that a single cutting edge extends continuously from the third tooth behind the fourth to the fifth. As shown by Mitchell (1936, 1962), the third interspace in the female mandible of American *Xanthosarus* s. str. is very deep (although filled with a cutting edge) (Fig. 84-11a). In *Addendella* it is often no deeper than the second interspace, but in *M. addenda* Cresson the third is much deeper than the second (Fig. 84-11c). Considering the complex of characters listed above, it seems best to unite *Addendella* (= *Delomegachile* auctorum) with *Xanthosarus*.

Mitchell (1980) separated *Megachile addenda* Cresson subgenerically as *Addendella* from the rest of the group then called *Delomegachile*. It is true that it is the most distinctive species of the group, as shown by the simple front tarsi of the male (Fig. 84-19c) and the near absence of the cutting edge in the second interspace of the female mandible, but some other species, such as *M. melanophaea* Smith and *mucida* Cresson, have an almost equally reduced cutting edge. Intermediacy also occurs in the male front tarsi; they are only moderately enlarged in *M. gemula* Cresson (Fig. 84-19b) and *ingenua* Cresson. I see no need to separate *M. addenda* from the rest of *Xanthosarus* by placing it in its own monotypic subgenus. The female of *M. addenda*, however, differs from that of *Litomegachile* principally in its large size; it does not run to *Xanthosarus* because the upper mandibular tooth is acute (Fig. 84-11c). The male differs from that of *Litomegachile* principally in its four-toothed mandibles. Thus the type species of *Addendella* may be as near to *Litomegachile* as to *Xanthosarus*.

Mitchell (1936, 1980) noted the close similarity of females that he placed in *Phaenosarus* to those of *Xanthosarus*, but regarded the males of *Phaenosarus* as quite different, and in view of male characters considered that taxon to be related to *Xeromegachile*. This view was probably based largely on the lack of an emargination in the crenulate carina of T6 in males of *Phaenosarus*. However, the emargination is sometimes weakly indicated in *Phaenosarus*, and is sometimes also weak in *Xanthosarus*, leaving occasional specimens essentially alike, a weak emargination being largely obscured by irregular crenulations. To me, it seems clear that female mandibular structure is decisive, and that the three species of *Phaenosarus* are best regarded as a subgroup of *Xanthosarus*.

Megachile / Subgenus *Zonomegachile* Mitchell

Chrysosarus (*Zonomegachile*) Mitchell, 1980: 72. Type species: *Megachile mariannae* Dalla Torre, 1896, by original designation.

This subgenus contains megachiliform Group 1 species, 9 to 14 mm long, that resemble species of *Chrysosarus*. The male agrees with that of *Chrysosarus* in the three-toothed mandible without a ventral projection, the enlarged front tarsi, the presence of a front coxal spine, and the strong emargination in the carina of T6. The distinctive feature of the male *Zonomegachile* is the strong hypostomal projection behind the mandibular base. The male could easily be considered an unusual *Chrysosarus*. The female, however, if the association of sexes is correct, differs from that of *Chrysosarus* in the four-toothed mandible with incomplete cutting edges in the second and third interspaces (Mitchell, 1980, fig. 51); see the note at the beginning of the key to subgenera.

■ This subgenus is known from Amazonas, Brazil, to northern Argentina (Santiago del Estero). Mitchell (1980) indicates that it is a small subgenus; most species are undescribed; I am familiar with two species.

Megachile mariannae Dalla Torre, the only species formally included in the subgenus, is relatively large, with rufotestaceous wings, tegulae, and wing veins, thus resembling *Chrysosarus* s. str., whereas an undescribed species from Santiago del Estero, Argentina, is smaller with clear wings and dark tegulae and wing veins, like the group of *Chrysosarus* called *Dactylomegachile*.

Two Unplaced Subgenera of *Megachile*, Probably Group 2

Megachile / Subgenus *Neochalicodoma* Pasteels

Chalicodoma (Neochalicodoma) Pasteels, 1970: 231. Type species: *Chalicodoma pseudolaminata* Pasteels, 1965, by original designation.

The two species of this subgenus are known from Liberia and Tanzania. Only males are known.

Megachile / Subgenus *Stellenigris* Meunier [incertae sedis]

Stellenigris Meunier, 1888: 152. Type species: *Stellenigris vandeveldii* Meunier, 1888, monobasic.

This generic name has never been properly placed. It was probably based on one of the larger species of Group 2 of *Megachile*, such as are found in the subgenera *Creightonella*, *Callomegachile*, *Gronoceras*, etc., although Meunier did not give the size for his specimen. He did provide a considerable number of details of structure and coloration; I have not found a species that agrees with his description.

Genus *Radoszkowskiana* Popov

Paracoelioxys Radoszkowski, 1893a: 53 (not Gribodo, 1884). Type species: *Paracoelioxys barrei* Radoszkowski, 1893, by designation of Sandhouse, 1943: 583. [For an invalid designation, see Michener, 1997b.]

Radoszkowskiana Popov, 1955b: 547, replacement for *Paracoelioxys* Radoszkowski, 1893. Type species: *Paracoelioxys barrei* Radoszkowski, 1893, autobasic and by original designation.

Radoszkowskiana consists of cleptoparasitic species, 10 to 13 mm in body length, having the metasoma of the female red, that of the male blackish. In the female the metasoma is widest at the second segment but tapers posteriorly, giving it a *Coelioxys*-like aspect. In the male, the metasoma is blunt, suggesting *Megachile (Chelostomoides)*. Other *Coelioxys*-like features are the lack of a scopa, the presence of a preoccipital carina at the side of the head (its lower extremity fuses with the posterior end of the hypostomal carina, as in most *Coelioxys*), and the lamellate pronotal lobe. Other *Megachile*-like features include the rounded axilla (though there is a notch between the posterior axillar and scutellar margins, as seen from above), the strong, crenulate, and medially emarginate transverse preapical carina on T6 of the male, and only four exposed metasomal sterna in the male. T7 of the male, exposed from side to side, is flat, with no teeth or projections. The apex of the metasoma of the female is flattened, the exposed part of T6 longer than broad, the apex broadly rounded with a small median emargination. S6 is also rounded and slightly surpasses T6.

■ The range is from Algeria to the Trans-Caspian region. The genus *Radoszkowskiana* was revised by Schwarz (2001) who recognized four species and showed that *R. barrei* (Radoszkowski), the type species, is not the same as *R. rufiventris* (Spinola), as had been alleged.

85. Family Apidae

Apidae is one of the most diverse families of bees, containing more tribes than any other family. The unfortunate diversity of meanings that can justifiably be attributed to the word "Apidae" is explained in Section 29. As indicated in Sections 19-21, Apidae as here understood and as understood by Michener (1944) is one of the two families of L-T bees, the other being the Megachilidae. One can say, then, that Apidae includes all those L-T bees that are not Megachilidae, a statement suggesting that Apidae is a polyphyletic or paraphyletic unit. Roig-Alsina and Michener (1993), however, in their phylogenetic study of L-T bees, considered it holophyletic. Although there are exceptions to all of the external characters of the Apidae, as listed in Section 21, many and perhaps all such exceptions are derived. An example of a derived character that makes definition of the family difficult is reduction of the labial palpi, such that the first two segments are as short as the remaining segments. Thus the palpi are like those of S-T bees; and a major division of the bees seems to have been broken down. This condition is found in some parasitic forms of Allodapini; similar and also probably reduced palpi occur in the Ctenoplectrini and in the genus *Ancyla* (Ancylini). Larval characters, also, do not clearly define the Apidae.

One internal derived character is unique to all the Apidae, i.e., the number of ovarioles per ovary in females and of tubules per testis in males. In all other families, this number is three. In the Apidae, there are four or more; the basic number is four (Fig. 85-1), although larger numbers (up to 13) are sometimes found in parasitic bees. Some highly social bees are also exceptions. In most meliponine queens (family Apidae) the number is four, but in certain species of *Trigona, Lestrimelitta,* and *Plebeia* the queens have 13, 12, and 8 to 10 ovarioles, respectively (J. M. F. de Camargo, in litt., 1996). The number sometimes attains 15 (Cruz-Landim, Reginato, and Imperatriz-Fonesca, 1998.) In other species of *Trigona* and in *Plebeia* s. str. the number is four. Finally, in queens of *Apis* the number may exceed 150, and in workers ranges from 2 to 12. Thus the exceptions in no way suggest a link to those families in which the number is three.

The biology of the Apidae is highly variable: from solitary to highly social behavior with dissimilar castes, from nest provisioners to social parasites and cleptoparasites, from mass provisioners to progressive provisioners of brood cells to forms that completely lack brood cells, from nest excavators in soil or in wood to forms that occupy preexisting cavities or construct nests in the open, and so forth. Such behavioral attributes will be described in the discussions of the subfamilies and tribes.

The Apidae are divided into three subfamilies (see Roig-Alsina and Michener, 1993). The large and diverse subfamily Apinae appears to be the sister group of either the Xylocopinae or the Nomadinae. Xylocopinae is undoubtedly a holophyletic group in spite of the widely different aspects of small, slender species of tribes like the Ceratinini and large, robust members of the tribe Xylocopini. In certain of the analyses published by Roig-Alsina and Michener (1993), the Xylocopinae appear among the tribes of Apinae rather than as a basal group. For this reason their status as a subfamily is not entirely clear, but in other analyses they appear as the most basal branch of the Apidae, i.e., sister to Apinae + Nomadinae; see Figure 102-1. This is the arrangement that seems preferable.

The Nomadinae consist entirely of cleptoparasites. Since there is no tribe of nonparasitic bees to which they seem most closely related, their immediate nonparasitic ancestors are unknown, probably extinct. Three tribes often included in the Nomadinae are here excluded from it; they are the Isepeolini, Osirini, and Protepeolini. For lack of a better place to put them, these tribes are included in Apinae, but proper evidence for such a placement is weak. In the females of most Apinae, S6 has a vestiture of slender hairs, especially dense near the convex apical margin of the sternum. In the Nomadinae the vestiture of slender hairs is reduced or absent from the truncate, bilobed or bifid, or sometimes pointed apical margin, on which in most tribes a few hairs are modified into coarse, blunt, often curved bristles, as illustrated by Roig-Alsina (1991); see Figure 91-1. Presumably, this form of S6 is related to the insertion of eggs into the walls of open, unprovisioned, or recently provisioned host cells; similar behavior of Protepeolini (Rozen, Eickwort, and Eickwort, 1978) is probably facilitated by similar but nonhomologous tergal rather than sternal bristles. The characteristic S6 of female Nomadinae is not found in the Isepeolini, Osirini, or Protepeolini, and its absence is one of the reasons for excluding these tribes from the Nomadinae. These tribes also differ from the Nomadinae proper in the absence of a glandular pouch on each side of the common oviduct (Alexander, 1996); such a pouch is known only in the Nomadinae, and characterizes all members of that

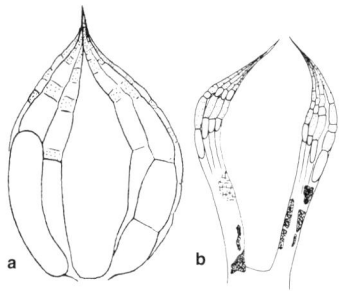

Figure 85-1. Sketches of ovaries. **a,** *Lasioglossum (Dialictus) rhytidophorum* (Moure) (Halictidae); **b,** *Allodapula melanopus* (Cameron) (Apidae). Both sketches were selected to show the full number of ovarioles, three in each ovary for all families except Apidae; four per ovary is the basic number for Apidae. Commonly, not all ovarioles are visible in any one view. Dotted areas represent groups of nurse cells. The *Lasioglossum* was in laying condition; the *Allodapula* was in a reproductively resting stage, the dark areas representing resorbed oocytes or groups of nurse cells.

subfamily. Of these three excluded tribes, larvae are known for the Isepeolini and Protepeolini. These larvae are quite different from those of Nomadinae, at least partly because the mature larvae spin cocoons, unlike those of Nomadinae. Although recent authors such as Roig-Alsina and Michener (1993) and Roig-Alsina and Rozen (1994) have excluded these tribes from the Nomadinae, I believe there is still the possibility that they are basal groups of that subfamily, as was argued by Rozen, Eickwort, and Eickwort (1978).

The great diversity of the subfamily Apinae is an invitation to those who would like to subdivide it into several subfamilies, but it is not clear how such subdivision should be done. Although the tribes are mostly stable groups whose members remain together in various analyses, the arrangement of the tribes is different in different analyses (Roig-Alsina and Michener, 1993; Silveira, 1993b). Three tribes that might be transferred to the Nomadinae are discussed above; possible recognition of the corbiculate tribes as a separate subfamily is discussed under the subfamily Apinae. See also Section 21.

Because of frequent exceptions to group characters, it is difficult to construct a key to the subfamilies of Apidae. I have therefore prepared instead a key to the tribes of the family, with the subfamily subsuming each tribe indicated; a separate key to the tribes of Xylocopinae will be found under that subfamily, although the same tribes appear in the key below.

Considering that most of the tribes in the key are recognizable to an experienced melittologist without microscopic examination, the difficulties of the key are surprising. A particular problem concerns the venation of the hind wing, which presents a continuum, as shown by Figures 106-2, 108-2, 108-3, 111-3, 112-4, 114-1, 116-4, and 117-3. Couplets in the key subdivide this continuum, but some species are intermediate and such couplets must be used with care.

Key to the Subfamilies and Tribes of the Apidae

1. Submarginal crossveins and second recurrent vein weak compared with other veins, commonly absent (Fig. 120-1); marginal cell open at apex or closed by weakened vein (hind tibial spurs absent).... Apinae, Meliponini (Sec. 120)
—. All veins well developed, conspicuous; marginal cell closed by strong vein 2
2(1). Scopa of female, when present, forming a corbicula on posterior tibia (Figs. 10-11, 102-2a, 118-3, 121-1a); inner apical margin of posterior tibia of nonparasitic females (except in queens of Apini) with row of stiff bristles (the rastellum, Fig. 102-2); pygidial and basitibial plates absent; eyes hairy *or* jugal lobe of hind wing absent 3
—. Scopa of female not forming a tibial corbicula [except in *Canephorula* and *Hamatothrix* (Eucerini) from Argentina], and scopa sometimes absent; inner apical margins of posterior tibiae bare or hairy, without comb; pygidial and basitibial plates frequently present; eyes very rarely hairy *and* jugal lobe of hind wing almost always present 5
3(2). Eye hairy; jugal lobe of hind wing present although notch delimiting it shallow; hind tibial spurs absent; arolia present .... Apinae, Apini (Sec. 121)
—. Eye bare; jugal lobe of hind wing absent; hind tibial spurs present; arolia absent 4
4(3). Proboscis in repose reaching beyond base of metasoma; body usually at least partly metallic; hind tibia of male with deep, hairy groove on upper surface (Fig. 118-4a-c); comb or group of bristles in position of jugal lobe of hind wing (Fig. 118-5) (neotropical) .... Apinae, Euglossini (Sec. 118)
—. Proboscis in repose not reaching base of metasoma; body not metallic; hind tibia of male not grooved; no comb of bristles on base of hind wing .... Apinae, Bombini (Sec. 119)
5(2). Stigma absent (Fig. 88-2a); middle and hind basitarsi usually longer than tibiae; clypeus nearly flat, lower lateral areas not curved backward (Figs. 21-3c, g, 88-5b) .... Xylocopinae, Xylocopini (Sec. 88)
—. Stigma present, although sometimes (e.g., in *Anthophora, Centris;* Figs. 113-1, 114-1) very small; middle and hind basitarsi usually shorter than tibiae; clypeus variable but usually more convex with lower lateral parts bent backward (Fig. 21-3d, h) 6
6(5). Pygidial plate absent, sometimes represented by spine in pygidial fimbria of female; scopa present (but reduced in parasitic Allodapini); epistomal suture between lateral extremity and subantennal suture usually bent mesad such that upper part of clypeus is almost parallel-sided; clypeus not or weakly protuberant 7
—. Pygidial plate present in females and most males, or if absent in female, then scopa also absent; epistomal suture not or rarely bent mesad such that upper part of clypeus is nearly parallel-sided; clypeus commonly strongly protuberant, lower lateral parts bent strongly posteriorly (Fig. 21-3d, h) 9
7(6). Submarginal cells two (Fig. 90-11); clypeus typically slightly constricted at level of tentorial pits (Fig. 90-2), this level near middle of clypeus (Eastern Hemisphere) .... Xylocopinae, Allodapini (Sec. 90)
—. Submarginal cells three (Fig. 89-4); clypeus not constricted at level of tentorial pits, this level above middle of clypeus 8
8(7). Jugal lobe of hind wing less than one-fifth as long as vannal lobe; female with spinelike pygidial plate arising from pygidial fimbria (Chile, Argentina) ....Xylocopinae, Manueliini (Sec. 87)
—. Jugal lobe over one-third as long as vannal lobe; pygidial plate and pygidial fimbria absent .... Xylocopinae, Ceratinini (Sec. 89)
9(6). Labrum longer than broad (Fig. 96-3a) or rarely about as long as broad; scopa absent; body rather short-haired, or, if with long hairs, then metasoma usually with pattern of white, appressed hair-spots or bands 10
—. Labrum broader than median length (Fig. 96-3b) [except rarely (e.g., *Epicharis*) longer than broad, in which case scopa present, body with long hairs, and metasoma without spots or bands of appressed white hair] 15
10(9). Distal parts of wings, beyond venation, hairless and coarsely papillate (Fig. 85-2a) .... Apinae, Melectini (in part) (Sec. 117)
—. Distal parts of wings, beyond venation, with hairs (as in Fig. 85-2b), not or weakly papillate (most strongly so in *Ammobatoides*) 11
11(10). Arolia absent; preoccipital carina strong behind eye,

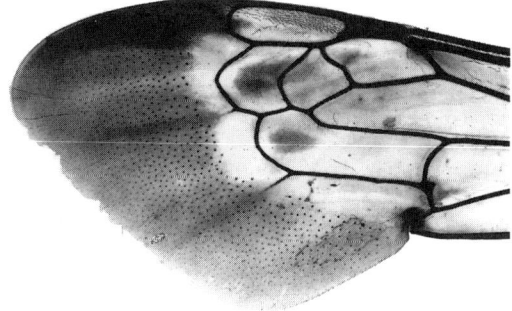 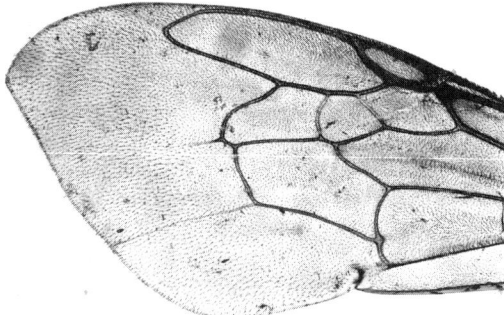

Figure 85-2. Papillate, hairless wing vs. nearly uniform hairy wing. Left, *Xeromelecta californica* (Cresson); Right, *Paratrapedia lugubris* (Cresson). From Michener, McGinley, and Danforth, 1994.

upper end curved forward to summit of eye; mandible with large tooth near middle of inner margin, projecting almost at right angle to mandibular axis (neotropical) .... Apinae, Tetrapediini, *Coelioxoides* (Sec. 109)
—. Arolia present; preoccipital carina absent, or, *if* present, then not curved to summit of eye; mandible simple or with subapical tooth 12
12(11). Middle coxa much shorter than distance from its summit to posterior wing base; body coarsely punctured with patches of squamiform pubescence; each submarginal cell usually receiving a recurrent vein 13
—. Middle coxa at least nearly as long as distance from its summit to posterior wing base; body finely punctate without patches of squamiform hairs; second submarginal cell receiving both recurrent veins 14
13(12). Episternal groove extending well below scrobal groove; scutellum and axilla lamellate or produced laterally; T5 of female without longitudinal ridge; preoccipital carina present (South America) .... Nomadinae, Caenoprosopidini (Sec. 101)
—. Episternal groove not extending below scrobal groove; scutellum and axilla neither lamellate nor produced; T5 of female with median, rounded longitudinal ridge; preoccipital carina absent (holarctic).................................. .... Nomadinae, Ammobatoidini (in part) (Sec. 96)
14(12). Pygidial plate of female absent or feebly defined; S6 of female produced to simple or bifid apex, without spinelike bristles (Fig. 91-2i); eyes of male not convergent above .... Nomadinae, Ammobatini (Sec. 100)
—. Pygidial plate of female well defined; S6 of female broadly emarginate apically, lateral lobes with many blunt, spinelike bristles (as in Fig. 91-2a); eyes of male converging above .... Nomadinae, Ammobatoidini, *Ammobatoides* (Sec. 96)
15(9). Marginal cell at most but little larger than stigma (Fig. 99-1), truncate apically; one or two submarginal cells, the second, when present, much wider than long (nearctic) .... Nomadinae, Neolarrini (Sec. 99)
—. Marginal cell far larger than stigma; two or three submarginal cells, if two, then second usually at least as long as broad 16
16(15). Second abscissa of vein M+Cu of hind wing shorter than cu-v (Figs. 116-4, 117-3), sometimes virtually absent, or as long as cu-v in some Rhathymini; scopa absent (posterior angle of mandible beneath middle of lower end of eye; claws of at least hind leg usually with inner ramus a broad lobe, Fig. 117-6b) 17
—. Second abscissa of vein M+Cu of hind wing as long as to much longer than cu-v, but *if* about equal, then hairy bees with scopae 19
17(16). Wings hairy throughout, not papillate (as in Fig. 85-2b) neotropical [Some specimens of *Tetralonioidella* in Melectini have hairy, nonpapillate wings. They are Oriental whereas Rhathymini are Neotropical.] ....Apinae, Rhathymini (part) (Sec. 115)
—. Wings bare beyond veins, coarsely papillate (Fig. 85-2a) 18
18(17). Middle tibial spur large, bifid or multidentate apically (Fig. 116-3); male gonostyli short, broad, often with slender projections (Fig. 116-5a, d, g); T6 of female with pygidial plate defined only at apex; marginal cell considerably exceeding last submarginal cell (Western Hemisphere) .... Apinae, Ericrocidini (Sec. 116)
—. Middle tibial spur unmodified; male gonostyli elongate; T6 of female with well-defined pygidial plate; marginal cell in common genera not or scarcely exceeding last submarginal cell, but in certain rare Eastern Hemisphere genera much exceeding last submarginal cell .... Apinae, Melectini (in part) (Sec. 117)
19(16). Jugal lobe of hind wing small, less than one-fourth as long as vannal lobe (Figs. 108-2a, 109-1); scopa absent [except in *Caenonomada* (Tapinotaspidini) and *Tetrapedia* (Tetrapediini)] 20
—. Jugal lobe of hind wing at least one-fourth as long as vannal lobe; scopa usually present 29
20(19). Hind tibial spur curved, coarsely pectinate (Fig. 109-2d), only one spur on hind tibia (scopa present; arolia absent) (neotropical) .... Apinae, Tetrapediini, *Tetrapedia* (Sec. 109)
—. Tibial spurs unmodified, relatively straight, ciliate or serrate, two on hind tibia 21
21(20). Maxillary palpus absent and axilla small, not at all produced; epistomal suture absent below anterior tentorial pit so that clypeus and lower paraocular areas are fused .... Apinae, Rhathymini (part) (Sec. 115)
—. Maxillary palpus present, if reduced to minute subspherical body as in some *Odyneropsis* (Epeolini), then axilla produced to point; epistomal suture usually complete 22
22(21). Arolia extremely small; scopa present; third submarginal cell larger than others (Fig. 108-2a) (neotropical) ...... Apinae, Tapinotaspidini, *Caenonomada* (Sec. 108)

—. Arolia of ordinary size; scopa absent; when three submarginal cells present, first largest 23

23(22). Scape, excluding basal bulb, less than twice as long as broad; T6 of female with apical margin broadly concave, without pygidial plate (except plate present in *Rhopalolemma*); sting sheath (third valvula) thickened (usually two submarginal cells) (holarctic) .... Nomadinae, Biastini (Sec. 97)

—. Scape, excluding basal bulb, more than twice as long as broad; T6 of female not broadly concave apically, usually with pygidial plate; sting sheath slender, more or less parallel-sided 24

24(23). Submarginal cells two, second much shorter than first and receiving only second recurrent vein (Fig. 97-3a), or sometimes also first recurrent very near its base; S6 of female broadly emarginate, lobes at either side of emargination armed with a series of large, spinelike bristles (Fig. 91-2c) (nearctic) .... Nomadinae, Townsendiellini (Sec. 98)

—. Submarginal cells three, or, *if* two, then second at least nearly as long as first and receiving both recurrent veins (except very rarely in genus *Nomada,* in which marginal cell is pointed on wing margin, unlike that of Townsendiellini); S6 of female not broadly emarginate, although sometimes with spinelike bristles 25

25(24). Front coxae quadrate, trochanters arising from outer distal angles of coxae, trochanters thus far apart; axilla nearly always produced to acute angle or spine (Fig. 93-1c); S6 of female largely invaginated, disc reduced and lateral distal portions produced to form a pair of long, dentate or spinose processes (Fig. 95-3c, d) .... Nomadinae, Epeolini (Sec. 95)

—. Front coxae somewhat triangular, trochanters (except in *Paranomada,* Brachynomadini) arising close to one another from apices of coxae (except lateral to apical coxal spine if spine present); axilla usually rounded; S6 of female not much invaginated, disc not greatly reduced, spinose processes, if present, short or mere lobes (Fig. 91-2e, f) .. 26

26(25). Mesal margin of front coxa with carina that usually bends laterad at base and extends across base of coxa; S6 of female curved to form tubular guide for sting, without blunt, spinelike bristles .... Apinae, Osirini (in part) (Sec. 104)

—. Mesal and basal margins of front coxa not carinate; S6 of female bifid to subtruncate with blunt, spinelike bristles 27

27(26). S6 of female with apex elongate, bifurcate (Fig. 91-2f); maxillary palpus usually less than half as long as maxillary blade (body without yellow markings) (Western Hemisphere) .... Nomadinae, Brachynomadini (Sec. 93)

—. S6 of female truncate to bilobed (Fig. 91-2d, e); maxillary palpus over half as long as maxillary blade 28

28(27). Apex of marginal cell sharply pointed (Fig. 94-2); metasomal terga without distinct apical hair bands (Fig. 94-1); S6 of female subtruncate or feebly emarginate (Fig. 91-2e); body frequently with white or yellow markings .... Nomadinae, Nomadini (Sec. 94)

—. Apex of marginal cell rounded; metasomal terga with strong apical hair bands (Fig. 92-1); S6 of female emarginate (Fig. 91-2d); body without pale markings .... Nomadinae, Hexepeolini (Sec. 92)

29(19). First two segments of labial palpus somewhat flattened but neither long nor shearhlike (Fig. 110-1c, d), thus as in S-T bees; S2 to S5 of female with oblique areas of long, sigmoid, oil-collecting hairs (Fig. 110-3) (rudiments on S3 to S5 even in parasitic species); inner hind tibial spur of female a fine comb of long teeth, in nonparasitic species with base greatly broadened across tibia (Fig. 110-2a, c, d) (Africa, oriental, eastern palearctic) .... Apinae, Ctenoplectrini (Sec. 110)

—. First two segments of labial palpus long, flattened, sheathlike as in L-T bees [except in *Ancyla* (Ancylini)]; sterna of female without oil- collecting hairs; inner hind tibial spur variable but not greatly broadened at base 30

30(29). Scopa present; basitibial plate present except in some males; S6 of female not curled upward laterally; body euceriform to anthophoriform, with abundant plumose hairs 31

—. Scopa absent; basitibial plate absent; S6 of female curled to form guide for sting; body epeoliform, not noticeably hairy 38

31(30). Paraglossa as long as first two segments of labial palpus taken together (except in *Eucerinoda* from Chile); jugal lobe of hind wing about half as long as vannal lobe or sometimes more; antenna of males commonly elongate .... Apinae, Eucerini (Sec. 112)

—. Paraglossa much shorter than first segment of labial palpus; jugal lobe of hind wing usually less than half as long as vannal lobe (more than half as long in *Ancyloscelis*); antenna of males not unusually elongate 32

32(31). Stigma small, usually shorter than prestigma, parallel-sided, vein r arising near its apex (Figs. 113-1, 114-1); distal parts of wings usually strongly papillate, basal parts usually with large bare areas (as in Fig. 85-2a) 33

—. Stigma larger, longer than prestigma, tapering beyond vein r, which arises near middle of stigma (Fig. 111-3); wings usually pubescent throughout and only weakly papillate apically (as in Fig. 85-2b, but see Isepeolini) 34

33(32). First submarginal cell larger than second, posterior margin longer than that of second (Fig. 113-1); scopal hairs of female largely simple, sometimes plumose along upper margin of tibia .... Apinae, Anthophorini (Sec. 113)

—. First submarginal cell usually smaller than second, posterior margin shorter than that of second (Fig. 114-1); scopal hairs of female mostly plumose, often intermixed with simple hairs or with projecting simple apices (Western Hemisphere) .... Apinae, Centridini (Sec. 114)

34(32). Vertex seen in facial view uniformly convex (Fig. 111-2a); second abscissa of vein M+Cu of hind wing not much longer than vein cu-v and half as long as vein M (Fig. 111-3b, c); first submarginal cell on posterior margin commonly longer than either of the others (Fig. 111-3), second shortest, third intermediate or sometimes equal to first (mandible simple) (Western Hemisphere) .... Apinae, Emphorini (in part) (Sec. 111)

—. Vertex seen in facial view depressed between eye and ocelli, the ocellar area a little elevated (as in Fig. 111-2b); second abscissa of vein M+Cu of hind wing much longer than vein cu-v and usually more than half as long as vein M (Figs. 106-2a, 107-1, 108-2b, 108-3); first submarginal cell on posterior margin often not conspicuously longest when there are three submarginal cells, usually shorter than or subequal to third (Figs. 108-2, 108-3) 35

35(34). Clypeus strongly protuberant, in side view protruding in front of eye by about width of eye; vertex gently convex as seen from front; hind leg of male (femur to basitarsus) enlarged (Fig. 111-2c) (Western Hemisphere) .... Apinae, Emphorini *(Ancyloscelis)* (Sec. 111)
—. Clypeus rather flat, in side view protruding in front of eye by much less than width of eye; vertex not or scarcely convex as seen from front; hind leg of male almost never enlarged 36
36(35). Second abscissa of vein M+Cu of hind wing half as long as vein M or less (Fig. 107-1) (palearctic) .... Apinae, Ancylini (Sec. 107)
—. Second abscissa of vein M+Cu of hind wing over half as long as vein M (Figs. 106-2a, 108-2, 108-3) (Western Hemisphere) 37
37(36). Second abscissa of vein M of posterior wing less than twice as long as vein cu-v (Figs. 108-2b, 108-3); mandible with preapical tooth, rarely with two such teeth; scopa with simple hairs extending beyond plumosity .... Apinae, Tapinotaspidini (in part) (Sec. 108)
—. Second abscissa of vein M of posterior wing more than twice as long as vein cu-v (Fig. 106-2a); mandible nearly always simple; scopal hairs plumose to apices .... Apinae, Exomalopsini (Sec. 106)
38(30). Arolia much shorter than claws, scarcely capitate; pygidial plate present in both sexes .... Apinae, Osirini, *Epeoloides* (Sec. 104)
—. Arolia unusually large, capitate, nearly as long as claws or longer; pygidial plate absent in both sexes (but in female of Protepeolini represented by narrow, flat, process) 39
39(38). Wings with hairs sparse or largely absent, papillate beyond veins (as in Fig. 85-2a) (South America)............ .... Apinae, Isepeolini (Sec. 103)
—. Wings rather uniformly hairy, not or weakly papillate distally (as in Fig. 85-2b) (Western Hemisphere) .... Apinae, Protepeolini (Sec. 105)

86. Subfamily Xylocopinae

An assemblage of bees, very diverse in size and appearance but united by a series of common characters, constitutes the Xylocopinae. As noted in Section 85 on the family Apidae, in some of Roig-Alsina and Michener's (1993) cladograms based on L-T bees, the tribes here placed in the subfamily Xylocopinae appeared together but among tribes of Apinae. In other analyses (when cleptoparasites were omitted), however, the Xylocopinae appeared as the sister group to the Apinae or to Apinae + Nomadinae; we therefore decided to recognize the Xylocopinae as a subfamily, and I have followed that decision here.

Xylocopinae includes two superficially very different sorts of bees; species of the tribe Xylocopini are large to very large, robust euceriform to anthophoriform bees, whereas the other three tribes consist of small, slender, andreniform to almost hylaeiform bees. Except in parasitic species that lack it, the scopa is reduced and slender compared to that of most pollen-collecting noncorbiculate Apinae. Presumably, some of the pollen carried to the nest is in the crop rather than on the scopa. The front coxae are considerably wider than long, with a tendency to be quadrate. This character, although variable, is to some degree evident in all Xylocopinae, but not in other bees except in some Nomadinae. The basitibial plates are often much modified, scarcely recognizable (Fig. 88-3b, c). The prepygidial fimbria is absent or evident only at the sides of T5 of the female. The pygidial plate of the female is usually absent (Fig. 89-2) or reduced to a spine (Fig. 88-3e, f), although in *Xylocopa* subgenus *Xylocopoides* the possible homologue of a broad plate is indicated by series of small teeth on the surface of T6 (Fig. 88-3d), and in the subgenus *Proxylocopa* there is a possible homologue of a broad plate. S7 of the male is reduced to a transverse bar, without or with a minute to small median disc area (largest in Manueliini) but no apical lobes. S8 is similarly reduced, without an apical process but with a spiculum, except in the Manueliini, in which the disc is large and an apical process is present; see Fig. 88-6c, d. The long, slender, scarcely flattened hind basitarsus, lacking both an apical process and a penicillus, is characteristic of the subfamily. So also is the relatively flat face, with the clypeus not or little protuberant (Fig. 88-5b), its lower lateral parts not curved back or only small areas curved back on either side of the labrum. As in most L-T bees the flabellum at the apex of the glossa is well developed (Fig. 86-1). The Xylocopini are unusual in that the flabellum is at the apex of a bare shank (Fig. 86-1a-c). In McGinley's (1981) study of larvae, the Xylocopinae (he did not have Manueliini) came out together in a nearest-neighbor analysis. The circular or oval salivary opening, without lips, is unique among larvae of nonparasitic bees. Mature larvae do not spin cocoons.

Unifying features of the nesting biology include the following: Except for *Xylocopa (Proxylocopa)*, all species nest in dead plant material—hollow stems or galls or burrows made in pithy stems or galls or in rotten or solid wood. The cells (the Allodapini make no cells) are unlined or the secreted lining is weak and scarcely detectable, except in the ground-nesting *X. (Proxylocopa)*. Provisions (except for the progressive feeders, the Allodapini) are loaf-shaped, firm, rather dry pollen masses, often partly separated from the cell walls, being supported in part by the egg (Figs. 89-3; 90-5a). There is a tendency toward social behavior, two or more adults being frequently found in nests (Michener, 1990b). Although one to several young adults commonly are present in a nest with the old mother, only a minority of nests in most species contain two or more older bees, with division of labor and eusocial or semisocial relationships. Some storage of food for adult consumption, a practice found in no other bees except the corbiculate Apinae, is common in the Xylocopinae. The view that Xylocopinae is the sister group to the corbiculate Apinae, as indicated by Sakagami and Michener (1987), was strengthened by this attribute (food storage for adult consumption), but is no longer a tenable idea, as shown by Roig-Alsina and Michener (1993).

The relationships among the four tribes of Xylocopinae were dealt with by Sakagami and Michener (1987), Roig-Alsina and Michener (1993), and Engel (2001b). Roig-Alsina and Michener, in a separate analysis of Xylocopinae, derived two minimum-length cladograms differing in the positions of *Manuelia* and *Xylocopa*. In agreement with Sakagami and Michener, one of the cladograms showed *Manuelia* as the first branch, *Xylocopa* next. The other reversed these positions. It seems likely that *Manuelia* is the sister group to all the others because of its less reduced and presumably plesiomorphic S7 and S8 of the male (Fig. 87-1b, c). If the ground-nesting behavior and associated structures of *Xylocopa (Proxylocopa)* were plesiomorphic, as advocated by Malyshev (1913) and Hurd (1958b), the basal position of *Xylocopa* rather than *Manuelia* would be probable, but it seems likely that in *Proxylocopa* these are not ancestral features. This matter is discussed in more detail in Section 88, on the Xylocopini.

Engel (2001b) proposed the tribe Boreallodapini for the three Eocene Baltic amber species of *Boreallodape*. This appears to be the sister group of the Recent Allodapini, from which it differs most conspicuously by having an ordinary shaped clypeus with the lateral margins concave so that it is much narrower above the tentorial pits than below.

A key to the Recent tribes of Xylocopinae is incorporated into the key to tribes of the Apidae in Section 85; a separate key to the xylocopine tribes nevertheless may be useful.

Key to the Tribes of the Xylocopinae
(Modified from Sakagami and Michener, 1987)

1. Stigma virtually absent (Fig. 88-2a); wings distally strongly papillate; flagellar segment 1 as long as or longer than 2 and 3 taken together; arolia absent; robust forms usually more than 13 mm long .... Xylocopini (Sec. 88)

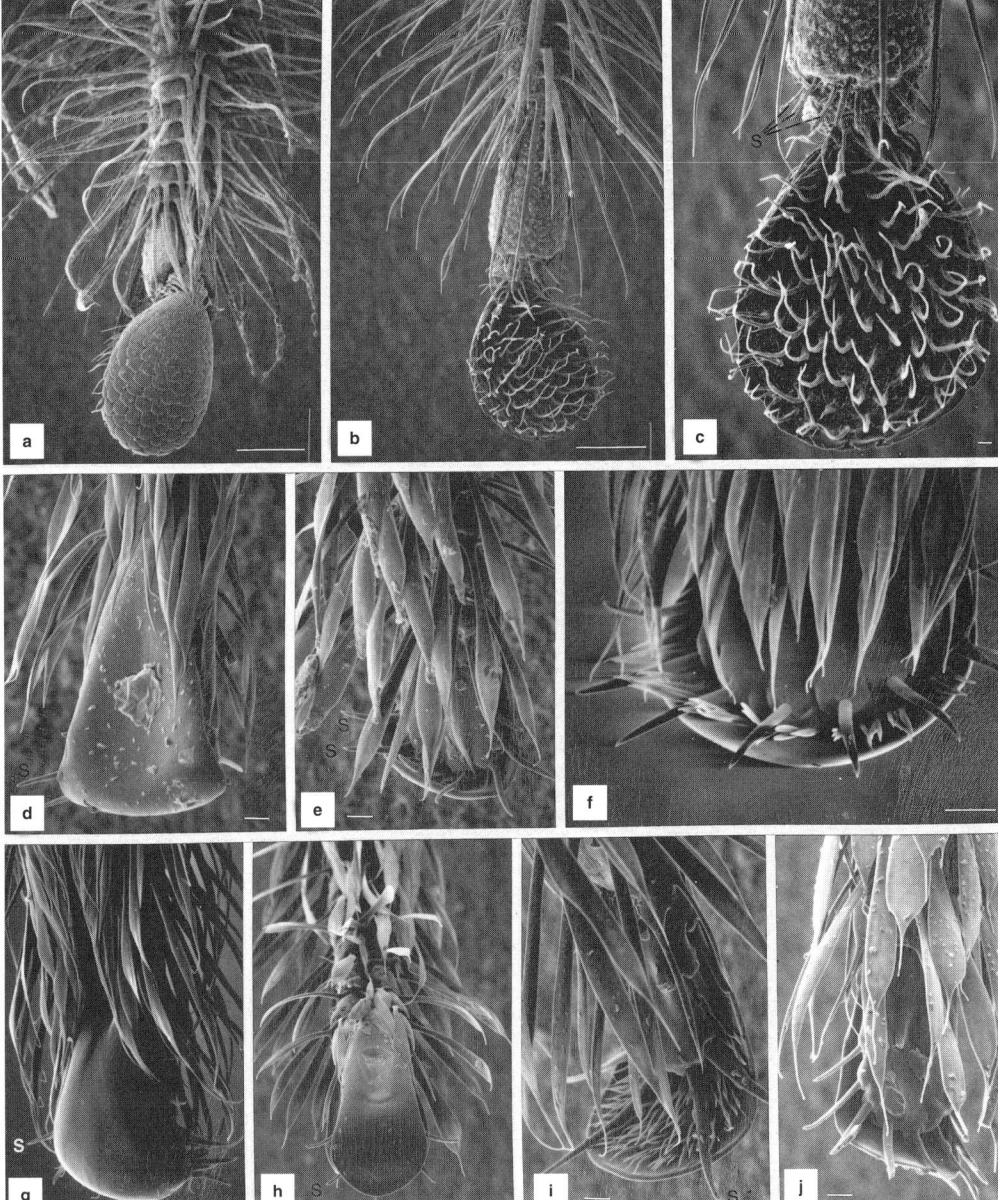

Figure 86-1. SEM photos of apices of glossae of Xylocopinae. **a-c,** *Xylocopa augusti* Lepeletier, posterolateral view and two anterior views; **d, e,** *Ceratina neomexicana* Cockerell, posterior and anterior views; **f, g,** *Manuelia gayi* (Spinola), anterior and posterior views; **h, i,** *Allodape ceratinoides* Gribodo, posterior and anterior views; **j,** *Macrogalea candida* (Smith), anterior view. (For a and b, Scale lines = 0.1 mm; for c-j, 0.01 mm; S indicates seta.) SEM photos by R. W. Brooks, from Michener and Brooks, 1984.

—. Stigma large (Figs. 28-4, 89-4, 90-11a-c, 90-13) or at least recognizable (smallest in *Megaceratina* and *Macrogalea*, Fig. 90-12); wings distally pubescent but not papillate; flagellar segment 1 shorter than 2 and 3 taken together; arolia present; usually slender forms, usually less than 12 mm long 2

2(1). Submarginal cells two (Figs. 28-4, 89-4, 90-11a-c, 90-13) (or only one in *Eucondylops reducta* Michener); clypeus above tentorial pits not greatly narrower than below (Fig. 90-2); apical metasomal terga of female depressed (Eastern Hemisphere) .... Allodapini (Sec. 90)

—. Submarginal cells three (Fig. 89-4); clypeus above tentorial pits much narrower than below; apical metasomal terga strongly convex, not depressed 3

3(2). S8 of male simple, without apical process (Fig. 89-5c); male gonostylus [except in *Ceratina (Euceratina)*] short,

more or less fused with gonocoxite (Fig. 89-5a, b); female labrum without basal elevated area; pygidial plate absent (Fig. 89-2) .... Ceratinini (Sec. 89)
—. S8 of male robust, with distinct apical process (Fig. 87-1b); male gonostylus several times as long as broad, not fused with gonocoxite (Fig. 87-1a); female labrum with basal smooth elevated area; T6 of female with apical spine, flat or concave on dorsal surface, that may represent pygidial plate (South America) .... Manueliini (Sec. 87)

87. Tribe Manueliini

This probably relictual tribe contains small bees (5-9 mm long) similar in appearance to dark-colored Ceratinini, one species being bluish black, the others nonmetallic. Useful tribal characters are the rather flat clypeus, its upper part contrasting with the raised supraclypeal area, its lower lateral part only obliquely bent back at the side of the labrum, its lateral margin not strongly concave such that the clypeus is neither in the form of an inverted T (as in Ceratinini) nor hourglass-shaped (as in Allodapini). The mandible is bidentate. As in the Ceratinini, the wings are hairy, not papillate, and the stigma is broad. T6 of the female has an apical spine, perhaps representing the pygidial plate, with dense plumose hairs of the pygidial fimbria at each side, as in many *Xylocopa*. S8 of the male, which has a substantial, heavily sclerotized disc, is presumably plesiomorphic relative to the reduced S8 of other Xylocopinae. The jugal lobe of the hind wing is small, only one-fifth as long as the vannal lobe, unlike that of all other Xylocopinae. The male genitalia and other structures were illustrated by Daly et al. (1987); see also Figure 87-1.

The nests are branched burrows, and the cells, at least in *Manuelia gayi* (Spinola), are narrowed at the ends. These are *Xylocopa*-like features not found in the Ceratinini. Knowledge of nesting biology is summarized by Daly et al. (1987).

Manueliini contains only a single genus, *Manuelia*.

Genus *Manuelia* Vachal

Manuelia Vachal, 1905b: 25. Type species: *Halictus gayi* Spinola, 1851, by designation of Daly, Michener, Moure, and Sakagami, 1987: 104. [See Michener, 1997b.]

Presbia Sandhouse, 1943: 590. Nomen nudum. [See Daly et al., 1987.]

This genus contains three species, so different that they could well be placed in different genera or at least subgenera. There seems to be no need, however, to recognize three monotypic genera or subgenera.

■ *Manuelia* occurs in Chile from Coquimbo to Llanquihue and in adjacent parts of Río Negro and Chubut, Argentina. The three species were revised by Daly et al., 1987.

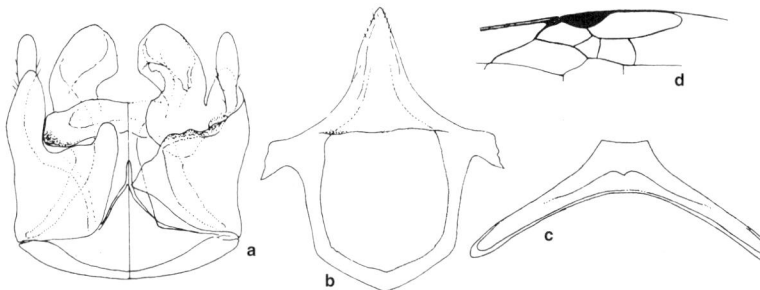

Figure 87-1. Structures of *Manuelia gayi* (Spinola). **a,** Male genitalia, dorsal view on the left; **b, c,** S8 and S7 of male; **d,** Part of forewing. Drawings by S. Sakagami, from Sakagami and Michener, 1987.

88. Tribe Xylocopini

The Xylocopini are large, robust bees (Pl. 10; Fig. 88-1), 13 to 30 mm long, that are altogether unlike other Xylocopinae in form and appearance. They are often confused with bumblebees (*Bombus,* in the Apinae) by the uninitiated. Some of the principal characters of the Xylocopini are probably associated with large size. Among these are the loss of the stigma (Fig. 88-2a), the very long prestigma and marginal cell (see Danforth, 1989a), and the strongly papillate distal parts of the wings. Another distinctive feature is the long first flagellar segment, longer than the second and third taken together; this character occurs in various unrelated large bees and may be somehow related to large size or fast flight. The rather short proboscis is distinctive, the parts being strongly sclerotized, the postpalpal part of the galea bladelike (Fig. 88-2b) and presumably used to cut into the corollas of tubular flowers to rob the nectar. The Xylocopini typically have three submarginal cells, but the first and second are sometimes partly or wholly fused owing to the disappearance of the posterior part or the whole of the first submarginal crossvein. Unlike other Xylocopinae, the bees of the tribe have no arolia, although the densely hairy plata projects somewhat between the claws (Fig. 10-10c). Numerous structures were illustrated by Maa (1938-1970), Yu (1954), Lieftinck (1955, 1956b, 1957a, b), Hurd and Moure (1963), Wu (1983b, c), and Eardley (1983).

Although it has been usual in the past to recognize *Lestis* and *Proxylocopa* as genera, *Xylocopa* is the only genus of Xylocopini here recognized. *Lestis* has been differentiated principally by having (1) four-segmented maxillary palpi (in other *Xylocopa* they are six-segmented, or apparently five-segmented because of the minute sixth segment) and (2) the first and second submarginal crossveins approximately parallel, the second submarginal cell thus approximately a parallelogram. In other *Xylocopa* this cell is over twice as wide on the posterior side as on the anterior side, because of the strongly slanting first submarginal crossvein (Fig. 88-2a), which, however, is sometimes partly or wholly absent. In *X. (Nodula)* species such as *X. amethystina* (Fabricius) the first submarginal crossvein is less slanting than in other *Xylocopa*; thus its venation is slightly more *Lestis*-like. Minckley (1998), in a cladistic analysis of the Xylocopini using many characters, found *Lestis* to be consistently near the base of the cladogram but among the groups of *Xylocopa*, not the sister group to *Xylocopa*. In an unpublished later study, when the more "ordinary" second submarginal cell of *Lestis* was considered a plesiomorphy with the apomorphic alternative characterizing all other Xylocopini,

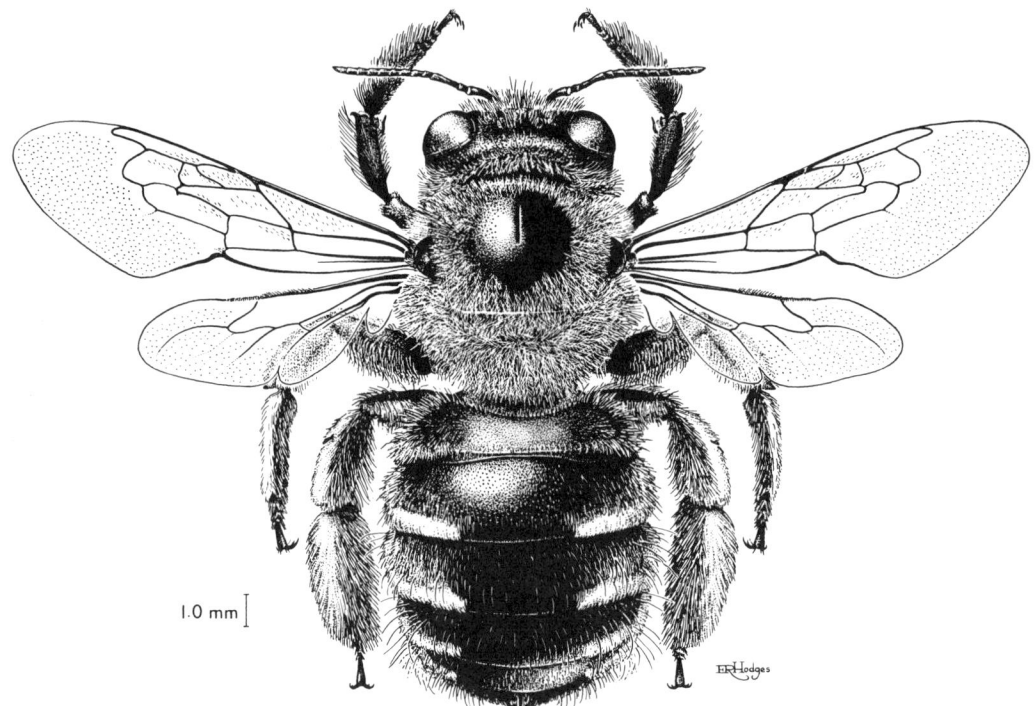

Figure 88-1. *Xylocopa (Notoxylocopa) tabaniformis* Smith, female. Drawing by E. R. S. Hodges, from Michener, McGinley, and Danforth, 1994.

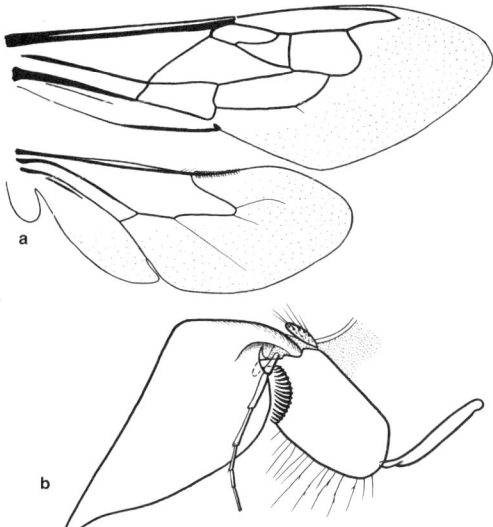

Figure 88-2. Structures of *Xylocopa tabaniformis orpifex* Smith. **a,** Wings; **b,** Outer view of maxilla, showing the broad galeal blade. From Michener, 1944.

Minckley still found *Lestis* in about the same position as in the earlier analyses (R. Minckley, in litt., 1996). In view of the great similarity of *Lestis* to other Xylocopini, as well as Minckley's results, I follow him in relegating *Lestis* to subgeneric status.

Leys, Cooper, and Schwarz (2000) made a study using mitochondrial DNA to investigate the phylogeny of the subgenera of *Xylocopa* (mostly based on one species per subgenus). They verified that *Lestis* and *Proxylocopa* are not basal groups in the Xylocopini; their recognition as genera would make *Xylocopa* paraphyletic. The mitochondrial evidence supports, in a general way, most of the phylogenetic conclusions of Minckley's (1998) morphological study.

Proxylocopa has probably received generic status in the past largely because it is the only group of the subfamily Xylocopinae known to nest in the ground. Probably associated with this behavior is the fact that females have basitibial plates (Fig. 88-3a) and a raised pygidial plate. These character states suggest those of ordinary ground-nesting anthophorine bees. The basitibial plate is a callosity, not margined by carinae, and may not be homologous to basitibial plates in other ground-nesting bees. The pygidial plate is relatively narrow but extends basad across the disc of T6. As noted by Daly et al. (1987), Malyshev (1913) and Hurd (1958b) attached much importance to the ground-nesting habits and the basitibial and pygidial plates of *Proxylocopa*; see also the summary by Hurd and Moure (1963: 55-56). Because *Proxylocopa* is a member of the tribe Xylocopini, Malyshev regarded that tribe as more closely related to the ground-nesting pre-xylocopine ancestor than is Ceratinini. This view is supported by the clustered cells and constructed cell walls of *Proxylocopa*, described and illustrated by Gutbier (1916), and by brood-cell linings secreted by Dufour's gland also found in *Proxylocopa* but not visible in other Xylocopinae (Kronenberg and Hefetz, 1984a). Such cell clusters and cells resemble those of various nonxylocopine ground-nesting bees.

Because *Proxylocopa* shares with other Xylocopini remarkable wing and mouthpart structure, the mesosomal gland of males mentioned below, as well as details of male genitalia, sterna, etc., it is clearly a close relative of other *Xylocopa*. Minckley's (1998) cladistic analysis shows it to be reasonably close to *Xylocopa* s. str. and not the sister group to all other *Xylocopa*. It therefore seems probable that *Proxylocopa* reverted to soil nesting in Asiatic deserts where there is little wood in which to nest. *Manuelia* is much richer in features ancestral to the Xylocopini than is *X. (Proxylocopa)*.

An unusual feature of most male Xylocopini, not known in any other bees, is a large gland opening on the metanotal-propodeal line (Fig. 88-4e-h). Its products seem to play a role in courtship. Its presence results in unusual sexual differences in the form and structure of the posterior part of the thorax, which becomes elongated when the gland is large (Minckley, 1994 and Fig. 88-4h). This mesosomal gland is present in such divergent groups as the subgenera *Lestis* (Fig. 88-4f) and *Proxylocopa*, reinforcing the view that they belong in *Xylocopa*. As illustrated especially well by Eardley (1983), the profile of the posterior part of the thorax provides many useful characters, only some of which are related to the presence and size of this gland (Fig. 88-4a-d).

Females of Old World *Xylocopa* of the subgenera *Kop-*

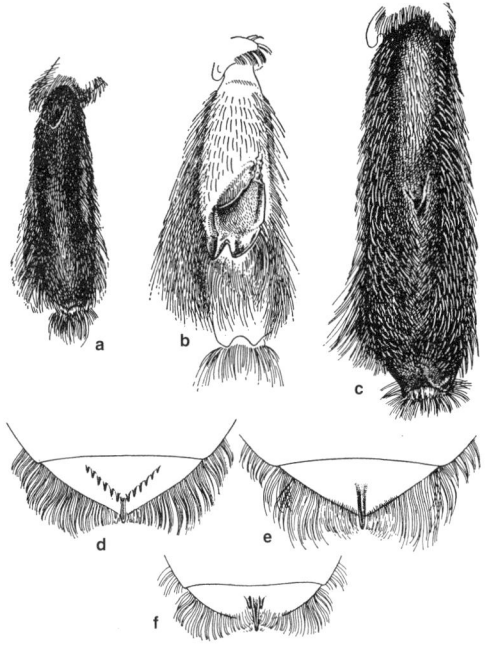

Figure 88-3. Structures of female *Xylocopa*. **a-c,** Hind tibiae, showing basitibial plates, of *X. (Proxylocopa) nitidiventris* Smith, *X. (Xylocopoides) californica* Cresson, and *X. (Hoplitocopa) assimilis* Ritsema; **d-f,** Dorsum of T6 of *X. (Xylocopoides) californica* Cresson, *X. (Mesotrichia) flavorufa* (DeGeer), and *X. (Notoxylocopa) tabaniformis* (Smith). From Hurd and Moure, 1963.

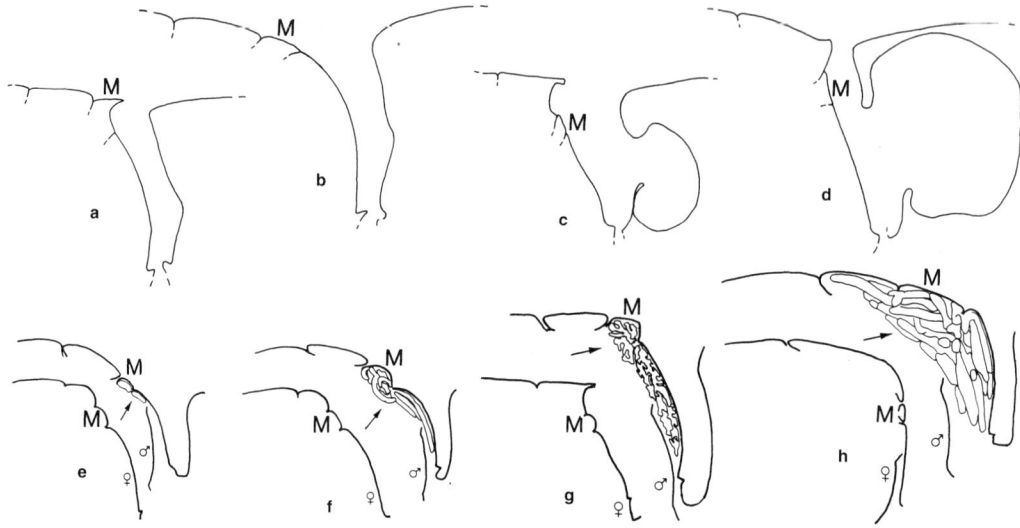

Figure 88-4. Sagittal sections of *Xylocopa*. (For orientation, the metanotum is indicated by an M.) a-d, Posterior part of thorax and T1, of: **a**, *X. (Koptortosoma) caffra* (Linnaeus), male; **b**, *X. (K.) nigrita* (Fabricius), male; **c**, *X. (K.) caffra* (Linnaeus), female; **d**, *X. (K.) nigrita* (Fabricius), female. In c and d, the mite pouch occupies much of the interior of T1.

e-h, Posterior part of thorax, showing males above, females below, and, in the males, the glands (marked by arrows) that open on the metanotal-propodeal line. **e**, *X. (Schonnherria) micans* Lepeletier; **f**, *X. (Lestis) bombylans* (Fabricius); **g**, *X. (Koptortosoma) watmoughi* Eardley; **h**, *X. (Neoxylocopa) varipuncta* Patton. Note that in g and h the second phragma of males (represented by a vertical line) is separated from the posterior wall of the propodeum by gland enlargement. a-d, from Eardley, 1983; e-h, from Minckley, 1994.

tortosoma and *Mesotrichia* are remarkable for the often enormous invagination in the anterior surface of T1, forming a chamber, or **acarinarium**, inhabited by mites of the genus *Dinogamasus*. Eardley (1983) illustrated these chambers diagrammatically for certain species; see Figure 88-4c, d. Hurd and Moure (1963) list the groups (in their classification, falling into eight subgenera, here reduced to two) possessing such mite pouches. Certain species in the same subgenera have thoracic acarinaria (O'Connor, 1993) in addition to that in T1.

A feature of some Xylocopini is the pale, usually yellow or testaceous, coloration of males, in striking contrast to dark-colored females. In other species the sexes are similarly colored, usually largely dark. In most cases the pale coloration of males is a result of yellow or testaceous hairs, the integument being black or brownish black. In two groups, however, the integument of males is reddish brown (some species of the paleotropical subgenus *Koptortosoma*) or even testaceous and partly yellowish (the largely neotropical subgenus *Neoxylocopa*), and the pubescence is often even paler. It seems that such males attract females to their flight and perching places, which are independent of nest or foraging locations (Anzenberger, 1977; Alcock and Smith, 1987).

There are many accounts of nesting behavior by species of *Xylocopa*. Except for the subterranean nests of the subgenus *Proxylocopa*, and except when in stems or similarly restricted substrates, the burrows are in wood, often solid wood, and usually branch into parallel passages running with the grain of the wood, with barrel-shaped unlined cells in short series in the branches. Some of the smaller species of *Xylocopa*, such as *X. (Copoxyla) cyanescens* Brullé (= *iris* auct.) and *X. (Nanoxylocopa) ciliata* Burmeister, typically nest in dead stalks of large herbaceous plants, in which there is no space for branch burrows. Some species of the subgenera *Biluna* and *Stenoxylocopa*, and *X. (Schonnherria) bambusae* Schrottky, nest in bamboo stalks. The female must first cut a hole through the hard wall of the bamboo, but then makes cells in the nearly empty cavity; thus the nest-making behavior is rather different from that of species that nest in wood or stems. As reviewed by Michener (1990d), young adults are regularly fed by the mother, and daughters commonly remain in their mothers' nests with some social interactions. Watmough (1974) gave an account of the ecology and behavior of various species. He indicated that each female lays about ten eggs; this low reproductive potential is, of course, associated with excellent maternal care, the mothers commonly being still active at the time of maturation of their offspring. An interpretation of social interactions in *X. pubescens* Spinola was presented by Dunn and Richards (2003). The following are some of the major accounts of nesting biology: Malyshev, 1931, 1947; Hurd, 1958b; Sakagami and Laroca, 1971; Anzenberger, 1977; Ben Mordechai et al., 1978; Gerling, Hurd, and Hefetz, 1983; and Houston, 1992a; a review is by Gerling, Velthuis, and Hefetz, 1989.

Genus *Xylocopa* Latreille

Because it is the only genus of Xylocopini, *Xylocopa*'s principal characters are indicated above in the account of the tribe. Typical features are illustrated in Figures 88-2, 88-5, and 88-6. Keys to the subgenera are presented separately for the Eastern and Eestern hemispheres; no subgenus occurs in both hemispheres except for *Neoxylocopa*, which has been introduced through commerce from the

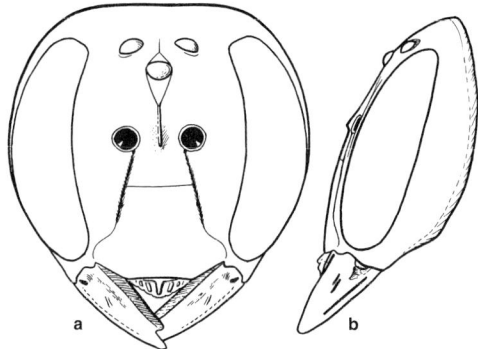

Figure 88-5. Head of *Xylocopa tabaniformis orpifex* Smith, female. **a,** Frontal view; **b,** Lateral view, showing the flat clypeus. See also Figure 21-3. From Michener, 1944.

Western Hemisphere to certain Pacific islands that lie in the Eastern Hemisphere.

Much of this account of *Xylocopa* is based on Minckley (1998) and Hurd and Moure (1963). Minckley's phylogenetic analysis has been useful in indicating that various small or monotypic subgenera of Hurd and Moure are based on unusual species that, on the basis of other characters, fit into larger subgenera. One result is a key (modified from Hurd and Moure, 1963) in which several subgenera come out in two or more places. Revisional studies of *Xylocopa* include the following: Hurd and Moure, 1963, for the world; Eardley, 1983, for southern Africa; Leys, 2000, for Australia; Ma, 1938, for India; Maa, 1954, for Afghanistan; Lieftinck, 1955, for the Lesser Sunda Islands and Tanimbar; Lieftinck, 1956b, for the Moluccas; Vecht, 1953, for Sulawesi (Celebes); Brèthes, 1916, for Argentina; Ackerman, 1916, for the USA; and Hurd, 1955, for California. Wu (1982c) gave a key to Chinese species. Pérez (1901) and Maidl (1912) made contributions toward a monograph of *Xylocopa*. Hurd (1978a) provided a detailed catalogue of subgenera and species of the Western Hemisphere, Ospina (2000) presented a list of the neotropical species, and Eardley (1987) published a catalogue for sub-Saharan Africa.

Key to the Subgenera of *Xylocopa* of the Western Hemisphere (Males)
(Modified from Hurd and Moure, 1963)
(Males are unknown for the subgenera *Monoxylocopa* and *Diaxylocopa*.)

1. Clypeus black, or, if with pale area, then (1) vertex in frontal view at most only scarcely raised above level of eye summits, usually at or below this level; (2) apex of scape not or scarcely surpassing level of eye summits unless T1 to T4 integument maculated with yellowish white; (3) occipitocular distance nonexistent or much shorter than first flagellar segment; and (4) eyes frequently enlarged and often markedly convergent above 2
—. Clypeus invariably maculated with pale coloration and with (1) vertex in frontal view strongly elevated above level of eye summits; (2) apex of scape much surpassing level of eye summits; (3) occipitocular distance longer than first flagellar segment; and (4) eyes always small, never strongly convergent above 10

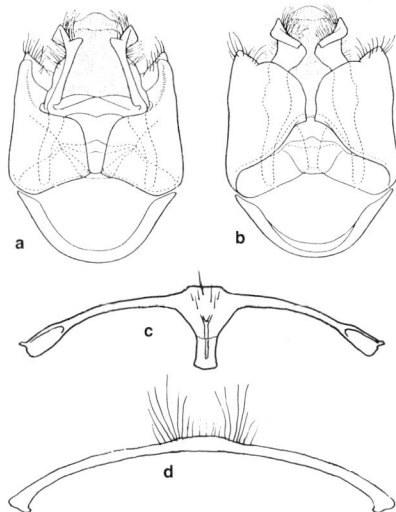

Figure 88-6. Structures of male *Xylocopa virginica* (Linnaeus). **a, b,** Dorsal and ventral views of genitalia; **c, d,** S8 and S7. a and b, from Hurd and Moure, 1963.

2(1). Tegula punctate throughout or thickly pubescent circummarginally 3
—. Tegula impunctate, glabrous and shining on at least posterolateral third 4
3(2). T1 to T4 integument broadly banded with yellowish white; interantennal distance shorter than interocellar distance; subantennal suture longer than interantennal distance; scape surpassing upper eye margins; posterolateral margin of tegula thickly pubescent; apical margin of S1 entire .... *X. (Schonnherria)* (in part)
—. Metasomal tergal integument entirely black; interantennal distance longer than interocellar distance; subantennal suture much shorter than interantennal distance; scape not surpassing upper eye margins; posterolateral margin of tegula not pubescent; apical margin of S1 deeply emarginate medially (South America) .... *X. (Nanoxylocopa)*
4(2). Scutellum flat, not convex in profile, exposed on same subhorizontal plane as median portion of metanotum and anterodorsal margin of propodeum; propodeum more or less horizontal anterodorsally before declivous posterior surface 5
—. Scutellum convex in profile, well above metanotal and propodeal levels, and median portion of metanotum steeply inclined; propodeum declivous, not horizontal along anterodorsal margin 6
5(4). Hind tibia with high, thin lamella beneath, beginning near base of inner hind tibial spur and extending obliquely forward across ventral surface of tibia; inner distal end of hind tibia not modified (South America) .... *X. (Xylocopoda)*
—. Hind tibia without lamella beneath and without inner hind tibial spur; inner distal end of hind tibia bearing an enlarged, flattened, highly polished tubercle (South America) .... *X. (Cirroxylocopa)*
6(4). Outer distal end of hind tibia with two denticles or spines 7

—. Outer distal end of hind tibia with one denticle or spine 8

7(6). T7 armed with two well-separated, posterolateral, dentiform projections; antennal sockets closer to upper clypeal margin than to anterior margin of median ocellus .... *X. (Notoxylocopa)*

—. T7 without posterolateral, dentiform projections; antennal sockets nearly midway between upper clypeal margin and anterior margin of median ocellus (North and Central America) .... *X. (Xylocopoides)* (in part)

8(6). Triangular area of propodeum well developed, broadly transverse, its surface nonmetallic; anterior half of T3 broadly depressed on each side of basally carinate median line, anterior margin of each depression bearing a transverse row of very long, posteriorly directed plumose hairs that overlie the depression (North and Central America) .... (*Calloxylocopa* group) *X. (Xylocopoides)* (in part)

—. Triangular area of propodeum not recognizable; T3 normal, not modified as above 9

9(8). T3 to T5 with graduli deeply incised, medially interrupted, and near anterior tergal borders; integument weakly metallic (South America) .... *X. (Xylocopsis)*

—. Metasomal terga without graduli; integument frequently brilliantly metallic ...... *X. (Schonnherria)* (in part)

10(1). Mandible, as seen from side, expanded at apex, its lower apical tooth markedly projected or expanded below lower mandibular margin 11

—. Mandible, as seen from side, not expanded at apex, its lower apical tooth not projected or expanded below lower mandibular margin 12

11(10). Body integument extensively ferruginous or yellowish, richly clothed with ferruginous or fulvous pubescence; tegula ferruginous; scape entirely ferruginous or with longitudinal yellow frontal stripe .... *X. (Neoxylocopa)*

—. Body integument chiefly black at least on dorsal surface, with predominantly blackish, whitish, or griseous pubescence; tegula black; scape largely or entirely black, without longitudinal frontal stripe .... *X. (Stenoxylocopa)* (in part)

12(10). Upper lateral margin of clypeus abruptly and steeply elevated above adjacent paraocular area; tegula transversely expanded, posterolateral margin greatly thickened in contrast with anterolateral margin (*Xylocopina* group) .... *X. (Stenoxylocopa)* (in part)

—. Upper lateral margin of clypeus scarcely raised, rounded, not abruptly and steeply elevated above adjacent paraocular area; tegula normal, not transversely expanded or modified as above 13

13(12). Propodeum steeply inclined, almost without a basal horizontal surface; hind basitarsus longer than tibia (South America) .... *X. (Dasyxylocopa)*

—. Propodeum broadly arched and greatly exposed anterodorsally, thus subhorizontal basally; hind basitarsus shorter than tibia .... *X. (Schonnherria)* (in part)

Key to the Subgenera of *Xylocopa* of the Western Hemisphere (Females)
(Modified from Hurd and Moure, 1963)

1. Pygidial spine armed on each side with subapical spine preceded anteriorly by strongly divergent row of spinelike teeth (Fig. 88-3d), the two rows outlining broad pygidial area on disc of T6; basitibial plates usually markedly excavated subapically (Fig. 88-3b) (North and Central America) .... *X. (Xylocopoides)*

—. Pygidial spine simple or armed on each side with a single subapical spine (Fig. 88-3e, f), without rows of teeth or other indication on disc of T6; basitibial plates at most only weakly excavated subapically 2

2(1). Head with greatly elevated ridges or prominences adjacent to ocelli; posterior margin of posterior ocellus above level of eye summits; clypeus separated from eye by more than minimum diameter of first flagellar segment; anterolateral corners of clypeus strongly protuberant (*Megaxylocopa* group) .... *X. (Neoxylocopa)* (in part)

—. Head without greatly elevated ridges or prominences adjacent to ocelli; posterior margin of posterior ocellus below level of eye summits; clypeus separated from eye by much less than minimum diameter of first flagellar segment; anterolateral corners of clypeus flat, not protuberant 3

3(2). Lower tooth of mandible much narrower than upper tooth or teeth, as measured across mandible at level of apex of notch(es) between teeth .... *X. (Stenoxylocopa)*

—. Lower tooth of mandible at least equal to, usually much wider than, upper tooth or teeth, as measured across mandible at level of apex of notch(es) between teeth 4

4(3). Longitudinal midventral carina of metasoma strongly developed, raised above integumental surface as high, sharply carinate ridge; dorsolateral margin of clypeus strongly raised above adjacent paraocular area, separated from clypeal punctation by prominent, subcarinate polished ridge that extends upward on each side of clypeus, and curves mesad to join its fellow beneath transverse section of epistomal suture; apical margin of S1 entire (*Neoxylocopa* group) .... *X. (Neoxylocopa)* (in part)

—. Longitudinal midventral carina of metasoma absent, or, if present, then not raised above integumental surface as high, sharply carinate ridge and not evident on all sterna; dorsolateral margin of clypeus on same level as adjacent paraocular area, or, if elevated, then rounded and not modified as above, ridge following upper clypeal margin discontinuous or absent; apical margin of S1 usually indented or incurved medially 5

5(4). Eyes more convergent above than below; body integument often brilliantly metallic 6

—. Eyes equally convergent above and below or, usually, more convergent below than above; body integument at most feebly metallic as seen in bright light 9

6(5). Propodeal triangle absent; antennocular distance less than interantennal distance .... *X. (Notoxylocopa)*

—. Propodeal triangle present, well differentiated, and margined; antennocular distance equal to or more than interantennal distance 7

7(6). Flagellum at most approximately as long as maximum length of eye; T2 to T6 without graduli; body integument brightly metallic.............. *X. (Schonnherria)* (in part)

—. Flagellum much exceeding maximum length of eye; T3 and T4 each with gradulus; body integument feebly metallic as seen in bright light 8

8(7). Labrum with five tubercles or multituberculate; clypeus much less than half as long as broad; dorsum of metasoma clothed with very short hairs, not shaggy in appearance (South America) .... *X. (Xylocopsis)*

—. Labrum with three tubercles; clypeus about half as long as broad; dorsum of metasoma clothed with very long hairs, decidedly shaggy in appearance (South America) .... *X. (Dasyxylocopa)*

9(5). Clypeus less than half as long as broad; median clypeal length shorter than clypeocellar distance; gradulus of T1 not continued posteriorly near lateral margin of tergum 10

—. Clypeus at least half as long as broad; median clypeal length longer than clypeocellar distance; gradulus of T1 continued posteriorly near lateral margin of tergum 13

10(9). Parapsidal line less than half as long as first flagellar segment; subantennal sutures parallel or diverging above, meeting antennal sockets at their lower margins 11

—. Parapsidal line more than half as long as first flagellar segment; subantennal sutures converging above, meeting antennal sockets near their inner margins 12

11(10). Apex of basitibial plate strongly asymmetrical, dentiform anteriorly, lobate posteriorly; subantennal sutures parallel, nearly as long as interantennal distance; upper portion of gena broadly glabrous and pruinose, very sparsely punctate; scutum broadly impunctate medially, punctation adjacent to parapsidal line sparse .... *X. (Schonnherria)* (in part)

—. Apex of basitibial plate nearly symmetrical, produced into two dentiform processes; subantennal sutures divergent above, much shorter than interantennal distance; upper portion of gena narrowly impunctate, hirsute, without a pruinose bloom; scutum narrowly impunctate medially, punctation adjacent to parapsidal line dense, nearly confluent (South America) .... *X. (Xylocopoda)*

12(10). Mandible tridentate apically, with prominent preapical internal tooth on upper margin; clypeus conspicuously depressed transversely near apical margin, depression emphasized by bulging upper portion of clypeus (Brazil) .... *X. (Diaxylocopa)*

—. Mandible bidentate apically, without preapical internal tooth on upper margin; clypeus flat, neither transversely depressed subapically nor bulging above (Brazil) .... *X. (Monoxylocopa)*

13(9). Frontal carina short, transverse, tuberculiform, not extending between antennal sockets, lower end considerably closer to anterior ocellus than to upper margin of clypeus; metasomal tergal discs sparsely pubescent; outer apex of hind tibia with two dentiform processes, i.e., tibial spines (South America) .... *X. (Nanoxylocopa)*

—. Frontal carina elongate, longitudinal, cariniform, extending between antennal sockets, lower end considerably closer to upper margin of clypeus than to anterior ocellus; metasomal tergal discs densely hirsute; outer apex of hind tibia with single dentiform process, i.e., tibial spine (South America) .... *X. (Cirroxylocopa)*

Key to the Subgenera of *Xylocopa* of the Eastern Hemisphere (Males)
(Modified from Hurd and Moure, 1963)
(The subgenus *Neoxylocopa* has been introduced into some Pacific islands, such as the Marianas, that lie in the Eastern Hemisphere; it is not included in this key.)

1. T1 with subhorizontal dorsal surface sloping or rounding into declivous anterior surface; gradulus of T1 curved posteriorly at side, becoming more or less longitudinal at lateral extremity 2

—. T1 with subhorizontal dorsal surface abruptly and angulately separated from declivous anterior surface or at least subangulately rounded *(Alloxylocopa)* into that surface; gradulus of T1 transverse, lateral extremities not or scarcely directed posteriorly.. 18

2(1). Parapsidal line short, at most three times as long as broad, situated anterior to posterior end of median scutal line (palearctic) .... *X. (Proxylocopa)*

—. Parapsidal line usually elongate-linear, but *if* short, then situated well behind posterior end of median scutal line .. 3

3(2). First and second submarginal crossveins nearly parallel; undersurface of flagellum crenulate; maxillary palpus four-segmented (Australia) .... *X. (Lestis)*

—. First submarginal crossvein strongly slanting, second submarginal cell thus more than twice as long on posterior margin as on anterior margin; undersurface of flagellum not crenulate; maxillary palpus six-segmented or apparently five-segmented 4

4(3). Antennocular distance approximately equal to, or usually much less than, length of subantennal suture; ocellocular distance less than length of subantennal suture 5

—. Antennocular distance nearly two or more times length of subantennal suture; ocellocular distance much greater than length of subantennal suture 13

5(4). Apex of basitibial plate strongly bifid; first flagellar segment at most scarcely longer than combined lengths of succeeding two segments.. 6

—. Apex of basitibial plate entire; first flagellar segment considerably longer than combined lengths of succeeding two segments 7

6(5). Posterior lobe of pronotum prolonged posteriorly well onto mesepisternum; spiracle of T3 with an elevated scale-like process near its mesal border; ocelli normal, not globose; ocelloccipital distance greater than occipitocular distance (Africa, Asia)...................... *X. (Ctenoxylocopa)*

—. Posterior lobe of pronotum not prolonged posteriorly onto mesepisternum; spiracle of T3 without elevated, scale-like process; ocelli greatly enlarged, globose; ocelloccipital distance equal to or shorter than occipitocular distance (oriental) .... *X. (Nyctomelitta)*

7(5). Eyes enlarged, maximum width of eye in frontal view equal to or usually greater than minimum upper interocular distance; face narrow, minimum upper interocular distance shorter than length of antennal scape 8

—. Eyes small, maximum width of eye in frontal view considerably less than minimum upper interocular distance; face broad, minimum upper interocular distance equal to or greater than length of antennal scape 12

8(7). Outer apex of hind tibia with two teeth (tibial spines); subantennal suture longer than interantennal distance; apical margin of labrum entire 9

—. Outer apex of hind tibia with one tooth; subantennal suture shorter than interantennal distance; apical margin of labrum emarginate 10

9(8). Face narrowest at middle; inner orbit convex, bowed toward antennal socket; frontal line sulciform, not elevated at lower end; apex of scape much surpassing upper eye margin; scutellum modified, dorsal surface rounding into vertical posterior surface (Madagascar) .... *X. (Prosopoxylocopa)*

—. Face narrowest near eye summits; inner orbit concave, curved toward central eye axis; frontal line cariniform, elevated at lower end above supraclypeal integument; apex of scape not surpassing upper eye margin; scutellum unmodified, convex in profile (eastern Asia) .... *X. (Bomboixylocopa)* (in part)

10(8). Disc of scutum punctate throughout; antennal sockets about midway between upper margin of clypeus and anterior margin of median ocellus; tegula punctate nearly throughout, marginally thickly pubescent; ocelli without adjacent swellings or ridges (oriental) ...... *X. (Zonohirsuta)*

—. Disc of scutum narrowly impunctate medially; antennal sockets very close to upper margin of clypeus; tegula with posterolateral third impunctate, glabrous and shining; lateral ocellus usually bounded below by conspicuous transverse ridge (oriental, palearctic) 11

11(10). Thoracic declivity beginning as abrupt although narrowly rounded angle in scutellar profile .... *X. (Maaiana)*

—. Beginning of thoracic declivity not defined, scutellum evenly rounded in profile .... *X. (Nodula)*

12(7). Clypeus black; propodeal spiracle asymmetrical, complex, lower extremity broadened, anterior margin limited by deeply impressed line forming elongate triangular flap with lower anterior margin projecting strongly forward; outer apex of hind tibia with acute tooth (palearctic) .... *X. (Rhysoxylocopa)*

—. Clypeus blackish, stained with ferruginous; propodeal spiracle simple, not modified as above; outer apex of hind tibia unarmed (Africa) .... *X. (Xylomelissa)* (in part)

13(4). Outer apex of hind tibia with at least one tooth; punctation of supraclypeal area and upper margin of clypeus so strong as to obscure or virtually obliterate transverse section of epistomal suture 14

—. Outer apex of hind tibia unarmed; punctation of supraclypeal area and upper margin of clypeus at most only moderate, not obscuring transverse section of epistomal suture 17

14(13). Outer apex of hind tibia with two teeth; posterior margin of metanotum straight, not angulately lengthened medially; median length of metanotum less than combined lengths of second and third flagellar segments; apex of gonocoxite simple 15

—. Outer apex of hind tibia with one tooth; posterior margin of metanotum angulately lengthened medially; median length of metanotum greater than combined lengths of second and third flagellar segments; apex of gonocoxite complex 16

15(14). Clypeus black; scutum narrowly impunctate medially; subantennal suture much shorter than interantennal distance; malar area short, linear; propodeum entirely declivous; interantennal distance less than interocellar distance (palearctic)............................ *X. (Xylocopa s. str.)*

—. Clypeus blackish, maculated with yellow; scutum punctate throughout; subantennal suture longer than interantennal distance; malar area long, minimum length equal to length of second flagellar segment; propodeum subhorizontal along upper margin; interantennal distance greater than interocellar distance (eastern Asia) .... *X. (Bomboixylocopa)* (in part)

16(14). Apex of basitibial plate bifid; tegula punctate nearly throughout and marginally thickly pubescent; clypeus long, median length greater than clypeocellar distance;

apex of scape much surpassing summit of eye; disc of scutum punctate throughout (Africa)...... *X. (Gnathoxylocopa)*

—. Apex of basitibial plate simple; tegula with posterolateral third impunctate, glabrous and shining; clypeus short, median length less than clypeocellar distance; apex of scape not or very slightly surpassing eye summit; disc of scutum narrowly impunctate medially (palearctic) .... *X. (Copoxyla)*

17(13). Basitibial plate present; anterior ocellus not bordered laterally by impunctate, crescent-shaped swellings; outer apex of middle tibia with acute tooth (Africa) .... *X. (Xylomelissa)* (in part)

—. Basitibial plate absent; anterior ocellus bounded on each side by impunctate, crescent-shaped swelling; outer apex of middle tibia rounded, without tooth (oriental) .... *X. (Biluna)*

18(1). Eyes enlarged, maximum width of eye in frontal view equal to or usually greater than minimum upper interocular distance; face narrow, minimum upper interocular distance less than length of scape; apex of scape not surpassing eye summit; occipitocular distance short or nonexistent, at most scarcely longer than second flagellar segment 19

—. Eyes small, maximum width of eye in frontal view much less than minimum upper interocular distance; face broad, minimum upper interocular distance greater than length of scape; apex of scape much surpassing eye summit; occipitocular distance about as great as or greater than length of first flagellar segment 23

19(18). Ocelli high on face, ocelloccipital distance equal to or less than ocellar diameter .... *X. (Koptortosoma)* (in part)

—. Ocelli low on face, ocelloccipital distance considerably greater than ocellar diameter 20

20(19). Posterior trochanter with inner apex produced into long spine; scutellum on approximately same subhorizontal plane as metanotum; scutellar profile not angled, thus not divided into horizontal and vertical surfaces 21

—. Posterior trochanter not produced into a long spine; scutellum conspicuously above metanotum, its profile angled posteriorly and thus divided into a long horizontal surface and a short vertical surface; surface of metanotum vertical 22

21(20). Metasomal terga brilliantly metallic, without gradulus on T3; clypeus black, densely punctate; punctation on supraclypeal area and upper part of clypeus so strong as to obscure or virtually obliterate transverse section of epistomal suture; middle trochanter with inner apex produced as acute spine .... *X. (Koptortosoma)* (in part)

—. Metasomal terga dull, scarcely metallic, with widely interrupted gradulus on T3; clypeus blackish, maculated with yellow, sparsely to moderately punctate; punctation of supraclypeal area and upper part of clypeus sparse, not obscuring transverse section of epistomal suture; middle trochanter unarmed (Africa, oriental) .... *X. (Mesotrichia)* (in part)

22(20). Tegula elongate, posterolateral third impunctate, glabrous, and shining; tarsi of middle and hind legs greatly flattened, expanded, and maculated (Africa, oriental) .... *X. (Mesotrichia)* (in part)

—. Tegula normal, rounded behind, punctate throughout,

thickly pubescent marginally; tarsi of all legs normal, not modified as above (oriental) .... *X. (Alloxylocopa)*

23(18). Face less broad, maximum interocular distance less than eye length; paraocular area not modified as below .... X. *(Koptortosoma)* (in part)

—. Face very broad, maximum interocular distance greater than eye length; paraocular area strongly elevated along inner eye margin, forming groove adjacent to inner orbit that continues for a short distance around summit of eye 24

24(23). Face black; mandible edentate at apex; labrum trapezoidal in outline, much longer than clypeus; metanotum obliquely inclined, shortened medially; tegula with posterolateral third impunctate, glabrous, and shining; interantennal distance greater than interocellar distance (Africa, oriental) .... *X. (Mesotrichia)* (in part)

—. Face blackish, maculated with yellow; mandible bidentate at apex; labrum transverse, much shorter than clypeus; metanotum subhorizontal, lengthened medially; tegula punctate nearly throughout, thickly pubescent marginally; interantennal distance less than interocellar distance (Africa) .... *X. (Xenoxylocopa)*

Key to the Subgenera of *Xylocopa* of the Eastern Hemisphere (Females)
(Modified from Hurd and Moure, 1963)

1. Pygidial plate or spine armed on each side with subapical spine 2
—. Pygidial plate or spine unarmed 17

2(1). Outer apex of hind tibia rounded or armed with single tooth or spiniform process 3
—. Outer apex of hind tibia armed with two teeth or two spiniform processes 12

3(2). Ocelli with adjacent cariniform ridges or anterior ocellus bounded on either side by an impunctate, crescent-shaped swelling, but *if* occasionally cariniform ridges vestigial or absent (some species of subgenus *Nodula*), then ventral surface of hind femur impunctate, glabrous, and highly polished 4
—. Ocelli without adjacent cariniform ridges; anterior ocellus without lateral impunctate crescent-shaped swellings; ventral surface of hind femur not as above 11

4(3). Mandible tridentate at apex; rear basal angle of mandible behind posterior margin of eye 5
—. Mandible bidentate at apex; rear basal angle of mandible before posterior margin of eye 6

5(4). Malar area short, minimum length much less than length of pedicel; clypeus less than half as long as broad, median length less than clypeocellar distance (upper clypeal margin to median ocellus); clypeocular distance less than minimum diameter of first flagellar segment; first flagellar segment at most about as long as next two segments taken together; triangular area of propodeum present (Africa, Asia) .... *X. (Ctenoxylocopa)*
—. Malar area long, minimum length greater than length of pedicel; clypeus longer, more than half as long as broad, median length longer than clypeocellar distance; clypeocular distance greater than minimum diameter of first flagellar segment; first flagellar segment about as long as or longer than next three segments taken together; triangular area of propodeum absent (Africa) .... *X. (Xylomelissa)* (in part)

6(4). Apex of basitibial plate entire, much before middle of tibia; outer apex of middle tibia unarmed, with cushion of short, densely compacted, bristle-like hairs; anterior ocellus bounded on each side by impunctate, crescent-shaped swelling (oriental) .... *X. (Biluna)*
—. Apex of basitibial plate bifid, well beyond middle of tibia; outer apex of middle tibia with tooth or spiniform process; anterior ocellus not bounded laterally by impunctate, crescent-shaped swellings 7

7(6). Scutellum convex in profile, not divided into dorsal and posterior surfaces 8
—. Scutellum angulate or subangulate in profile, divided into a subhorizontal dorsal surface and a vertical posterior surface (oriental, palearctic) 10

8(7). Punctation adjacent to mesal border of spiracle of T4 very fine and dense, markedly differentiated from punctation elsewhere on that tergum (Africa) .... *X. (Xylomelissa)* (in part)
—. Punctation adjacent to mesal border of spiracle of T4 not markedly differentiated from punctation elsewhere on that tergum 9

9(8). Discal hairs of T1 simple or sparsely plumose; first flagellar segment shorter than combined lengths of next three segments; median longitudinal groove of propodeum complete, clearly impressed (Africa) .... *X. (Xylomelissa)* (in part)
—. Discal hairs of T1 thickly plumose; first flagellar segment equal to or longer than combined lengths of next three segments; median longitudinal groove of propodeum indistinct, virtually obliterated (palearctic) .... *X. (Rhysoxylocopa)*

10(7). Scutellum, in profile forming rounded projection, overhanging entire metanotum .... *X. (Nodula)*
—. Scutellum not bulging over metanotum .... *X. (Maaiana)*

11(3). Mandible bidentate at apex, lower inferior surface not modified as below; ocelli greatly enlarged, globose; eyes more convergent above than below; maximum interocular distance less than eye length; antennal sockets approximately midway between upper margin of clypeus and anterior tangent of median ocellus; disc of scutum narrowly impunctate medially; body integument chiefly reddish brown (oriental) .... *X. (Nyctomelitta)*
—. Mandible tridentate at apex, lower surface bearing a long, stout, curved hook; ocelli normal, not globose; eyes more convergent below than above; maximum interocular distance greater than eye length; antennal sockets below midpoint between upper margin of clypeus and anterior tangent of median ocellus; disc of scutum punctate throughout; body integument chiefly black (Africa) .... *X. (Gnathoxylocopa)*

12(2). Scutellum convex in profile, not divided into subhorizontal dorsal surface and subvertical posterior surface; apical margin of S1 emarginate medially; dorsolateral margin of clypeus and side of supraclypeal area not or at most scarcely raised above adjacent paraocular area 13
—. Scutellum subangulate or angulate in profile, divided into subhorizontal dorsal surface and subvertical posterior surface; apical margin of S1 entire, not emarginate medially; dorsolateral margin of clypeus and lateral side of supraclypeal area conspicuously raised above adjacent paraocular area 15

13(12). Malar area long, minimum length half as long as first flagellar segment; clypeocular distance greater than minimum diameter of first flagellar segment; median length of clypeus shorter than clypeocellar distance; triangular area of propodeum absent; vertical fold of T1 narrowly and deeply sulcate (eastern Asia) .... *X. (Bomboixylocopa)* (in part)
—. Malar area short, minimum length much less than half as long as first flagellar segment; clypeocular distance less than minimum diameter of first flagellar segment; median length of clypeus longer than clypeocellar distance; triangular area of propodeum present; vertical fold of T1 linear, shallowly impressed 14

14(13). Mandible bidentate at apex; apex of basitibial plate before middle of tibia; maximum interocular distance less than length of eye (Africa) .... *X. (Xylomelissa)* (in part)
—. Mandible tridentate at apex; apex of basitibial plate well beyond middle of tibia; maximum interocular distance greater than length of eye (palearctic).... *X. (Xylocopa s. str.)*

15(12). Rear basal angle of mandible slightly behind posterior margin of eye; apex of basitibial plate considerably before middle of tibia; vertical fold of T1 broadly and deeply sulcate; gradulus of T1 virtually coincident with anteroventral margin of metasoma, not continued posteriorly adjacent to lateral margin of T1 (Africa)............ .... *X. (Xenoxylocopa)*
—. Rear basal angle of mandible before posterior margin of eye; apex of basitibial plate beyond middle of tibia; vertical fold of T1 linear, shallowly impressed; gradulus of T1 obliquely curved away from anteroventral margin of metasoma and continued posteriorly adjacent to lateral margin of T1.. 16

16(15). Subantennal suture longer than interantennal distance; disc of scutum narrowly impunctate medially; apex of basitibial plate simple; supraclypeal area strongly swollen, bigibbose; antennal sockets slightly above midpoint between upper margin of clypeus and anterior tangent of median ocellus (Madagascar) ....*X. (Prosopoxylocopa)*
—. Subantennal suture conspicuously shorter than interantennal distance; disc of scutum punctate virtually throughout; apex of basitibial plate feebly bifid; supraclypeal area only moderately elevated, not swollen or bigibbose; antennal sockets below midpoint between upper margin of clypeus and anterior tangent of median ocellus (oriental) .... *X. (Zonohirsuta)*

17(1). Scutellum convex in profile, not divided into a subhorizontal dorsal surface and a subvertical posterior surface; vertical fold of T1 sometimes inconspicuous, without a foveate depression or an invaginated orifice 18
—. Scutellum with subhorizontal dorsal surface rounding into or abruptly and angulately separated from subvertical posterior surface; vertical fold of T1 with a foveate depression or an invaginated orifice at or near its summit 21

18(17). Pygidial plate raised above surface of T6 and extending basad across that surface, its margins entire; basitibial plate with apex at or before basal fourth of tibia, its apex not a raised scale; parapsidal line at most three times as long as broad, anterior to end of median scutal line (palearctic)......................... *X. (Proxylocopa)*
—. Pygidial plate a dorsally flat or concave apical spine, sometimes hidden among hairs; basitibial plate with apex beyond basal one-fourth of tibia, commonly elevated as a scale; parapsidal line usually elongate-linear, but *if* short, then behind posterior end of medial scutal line.... 19

19(18). Paraocular area below antenna with strong protuberance; first and second submarginal crossveins almost parallel; maxillary palpus four-segmented (Australia) .... *X. (Lestis)*
—. Paraocular area without protuberance; first submarginal crossvein strongly slanting, second submarginal cell thus more than twice as long on posterior margin as on anterior margin; maxillary palpus six-segmented or apparently five-segmented 20

20(19). Mandible bidentate at apex; metasomal terga at most only faintly metallic in bright light; malar area with minimum length half as long as first flagellar segment or more; outer apex of hind tibia with two teeth; apex of scape much surpassing summit of eye (eastern Asia) .... *X. (Bomboixylocopa)* (in part)
—. Mandible tridentate at apex; metasomal terga brilliantly metallic; malar area with minimum length considerably less than half length of first flagellar segment; outer apex of hind tibia with one tooth; apex of scape not surpassing summit of eye (palearctic) .... *X. (Copoxyla)*

21(17). Posterodorsal margin of scutellum projecting beyond posterior margin of metanotum.... *X. (Koptortosoma)*
—. Posterodorsal margin of scutellum not surpassing posterior margin of metanotum 22

22(21). Posterodorsal margin of scutellum not acutely angled in profile and not projecting posteriorly beyond posterior surface of scutellum as a thin-edged flange; T1 without an invaginated chamber and hence without an opening on anterior surface (oriental)....... *X. (Alloxylocopa)*
—. Posterodorsal margin of scutellum in profile sharply and acutely angled, projecting posteriorly beyond posterior surface of scutellum as a thin-edged flange; T1 with an invaginated chamber opening at or near summit of vertical fold on anterior surface (Africa, oriental) .... *X. (Mesotrichia)*

Xylocopa / Subgenus *Alloxylocopa* Hurd and Moure

Xylocopa (Alloxylocopa) Maa, 1939: 155. Nomen nudum because no characters given, although a type species was designated; see ICZN, 3rd ed., art. 13(a)(i).

Xylocopa (Alloxylocopa) Hurd and Moure, 1963: 239. Type species: *Xylocopa appendiculata* Smith, 1852, by original designation.

Alloxylocopa resembles *Xenoxylocopa,* but the female lacks the lateral spine at the base of the pygidial spine and the male is dark-colored with a fovea on the anterior surface of T1.

■ This subgenus is found from Japan and China south to Sumatra. Hurd and Moure (1963) listed six species.

Xylocopa / Subgenus *Biluna* Ma

Xylocopa (Biluna) Ma, 1938: 276. Type species: *Xylocopa nasalis* Westwood, 1842, by original designation.

This is a group of rather large species with elongate bodies that nest, so far as is known, only in stems of bamboo. According to Minckley (1998) it is related to *Xy-*

lomelissa, Nodula, and *Rhysoxylocopa*. Males can be distinguished from all other subgenera by the combined lack of basitibial plates and of spines on the outer apex of the hind tibia. The anterior ocellus of the male is bounded on either side by a crescentic, impunctate swelling. Females have a dense mat of short setae on the middle tibia, and lack an apical middle tibial spine.

■ *Biluna* ranges from India and Sri Lanka to the Lesser Sunda Islands, the Philippines, and Taiwan north to Kashmir and central China and Japan. There are about five species. Hurd and Moure (1963) gave a key to the species, based on Maa's (1946) revision.

Xylocopa / Subgenus *Bomboixylocopa* Maa

Xylocopa (Bomboixylocopa) Maa, 1939: 155. Type species: *Xylocopa bomboides* Smith, 1879, by original designation.
Xylocopa (Mimoxylocopa) Hurd and Moure, 1963: 203. Type species: *Xylocopa rufipes* Smith, 1852, by original designation.

The combination of a malar area longer than the flagellar width and two spines on the outer apex of the hind tibia distinguishes females of *Bomboixylocopa* from that of all other *Xylocopa*. Males have appressed setae on pregradular sternal areas and near the metasomal spiracles and, as in the female, two hind tibial spines.

■ *Bomboixylocopa* is found in China and Taiwan. Five species were listed by Hurd and Moure (1963), and a revision of the species of *Bomboixylocopa* s. str. was by Maa (1939).

Xylocopa / Subgenus *Cirroxylocopa* Hurd and Moure

Xylocopa (Cirroxylocopa) Hurd and Moure, 1963: 102. Type species: *Xylocopa vestita* Hurd and Moure, 1963, by original designation.

This subgenus is probably a basal branch of the *Schonnherria* clade or of the *Neoxylocopa* clade (Minckley, 1998). It consists of a small, densely pubescent, weakly metallic species.

■ *Cirroxylocopa* is found in Minas Gerais, Goiás, and Mato Grosso, Brazil, and in Paraguay. The only species is *Xylocopa vestita* Hurd and Moure.

Xylocopa / Subgenus *Copoxyla* Maa

Xylocopa (Copoxyla) Maa, 1954: 211. Type species: *Apis bomb. iris* Christ, 1791, = *Xylocopa cyanescens* Brullé, 1832, by original designation. [Christ's name has sometimes been regarded as an invalid trinominal.]

This is a subgenus of small, bright-metallic species morphologically similar to *Ctenoxylocopa*. Males differ from those of that subgenus by the presence of a propodeal triangle and small body size; the face is entirely dark and there is a single spine at the outer apex of the hind tibia. Females can be distinguished from those of *Ctenoxylocopa* by the combination of smoothly rounded thorax, three-toothed mandibles, and lack of a small tooth beside the base of the pygidial spine.

■ *Copoxyla* occurs from the Mediterranean basin north to Slovakia, east to Russia. There are three to possibly six species. Popov and Ponomareva (1961) gave a key to Russian species.

Xylocopa / Subgenus *Ctenoxylocopa* Michener

Xylocopa (Ctenopoda) Ma, 1938: 285 (not McAtee and Malloch, 1933). Type species: *Apis fenestrata* Fabricius, 1798, by original designation.
Xylocopa (Ctenoxylocopa) Michener, 1942a: 282, replacement for *Ctenopoda* Ma, 1938. Type species: *Apis fenestrata* Fabricius, 1798, autobasic.
Baana Sandhouse, 1943: 530, replacement for *Ctenopoda* Ma, 1938. Type species: *Apis fenestrata* Fabricius, 1798, autobasic and by original designation.

Minckley (1998) found this subgenus to be closest to some species of *Xylocopa* s. str. The prolonged posterior pronotal lobes and an elevated process on the spiracles of T3 are unique characters of males. Females can be recognized by the combination of a row of tubercles along each margin of the basitibial plate, three-toothed mandibles, and a single spine on the outer apex of the hind tibia.

■ This subgenus occurs from Natal, South Africa, north to Gambia and Ethiopia, northeast to the Arabian Peninsula, Israel, Iraq, trans-Caspian Russia, and southeast to Madagascar, Mauritius, Sri Lanka, Pakistan, India, and Burma. The six or more species were listed by Hurd and Moure (1963) and revised by Maa (1970).

Xylocopa / Subgenus *Dasyxylocopa* Hurd and Moure

Xylocopa (Dasyxylocopa) Hurd and Moure, 1963: 113. Type species: *Xylocopa bimaculata* Friese, 1903, by original designation.

This subgenus is similar to *Schonnherria*. Males have graduli on T1 to T5, females on T1 to T4; in *Schonnherria* there is a gradulus only on T1. This may be a basal branch of or sister group to *Schonnherria,* and could well be synonymized.

■ *Dasyxylocopa* occurs from Goiás to Santa Catarina in Brazil, thence to Paraguay and Misiones, Argentina. The one species is *Xylocopa bimaculata* Friese.

Xylocopa / Subgenus *Diaxylocopa* Hurd and Moure

Xylocopa (Diaxylocopa) Hurd and Moure, 1963: 129. Type species: *Xylocopa truxali* Hurd and Moure, 1963, by original designation.

This taxon, known only in the female, is a close relative of *Monoxylocopa* and of *Schonnherria*. In the absence of more information, I maintain it as a subgenus.

■ *Diaxylocopa* is known only from the central Brazilian state of Goiás. The only species is *Xylocopa truxali* Hurd and Moure.

Xylocopa / Subgenus *Gnathoxylocopa* Hurd and Moure

Xylocopa (Gnathoxylocopa) Hurd and Moure, 1963: 182. Type species: *Xylocopa sicheli* Vachal, 1898, by original designation.

Females of this subgenus differ from those of all other *Xylocopa* in the long, curved process arising from the lower margin of the mandible. Males have an elongate thorax (the posterior declivity beginning in the metanotum) and bifid apices of the basitibial plates.

■ *Gnathoxylocopa* occurs in southern Africa, north as far as Zimbabwe and Namibia. The three species names listed by Hurd and Moure (1963) relate to a single species (Eardley, 1983), *Xylocopa sicheli* Vachal.

Xylocopa / Subgenus *Koptortosoma* Gribodo

Koptortosoma Gribodo, 1894: 271. Type species: *Koptortosoma gabonica* Gribodo, 1894, by designation of Sandhouse, 1943: 561. [See Michener, 1997b.]
Koptorthosoma Dalla Torre, 1896: 202, unjustified emendation of *Koptortosoma* Gribodo, 1894.
Cyaneoderes Ashmead, 1899a: 70. Type species: *Cyaneoderes fairchildi* Ashmead, 1899 = *Bombus coeruleus* Fabricius, 1804, by original designation.
Coptorthosoma Pérez, 1901: 3, unjustified emendation of *Koptortosoma* Gribodo, 1894.
Xylocopa (Orbitella) Ma, 1938: 305 (not Douvillé, 1915). Type species: *Xylocopa confusa* Pérez, 1901, by original designation.
Xylocopa (Maiella) Michener, 1942a: 282, replacement for *Orbitella* Ma, 1938. Type species: *Xylocopa confusa* Pérez, 1901, autobasic.
Euryapis Sandhouse, 1943: 551, replacement for *Orbitella* Ma, 1938. Type species: *Xylocopa confusa* Pérez, 1901, autobasic.
Xylocopa (Eoxylocopa) Sakagami and Yoshikawa, 1961: 413, nomen nudum.
Xylocopa (Cyphoxylocopa) Hurd and Moure, 1963: 283. Type species: *Xylocopa ocularis* Pérez, 1901, by original designation.
Xylocopa (Afroxylocopa) Hurd and Moure, 1963: 264. Type species: *Apis nigrita* Fabricius, 1775, by original designation.
Xylocopa (Oxyxylocopa) Hurd and Moure, 1963: 275. Type species: *Xylocopa varipes* Smith, 1854, by original designation.
Xylocopa (Lieftinckella) Hurd and Moure, 1963: 286. Type species: *Xylocopa smithii* Ritsema, 1876, by original designation.

This subgenus resembles *Mesotrichia*, and, according to Minckley (1998), is the sister group of that subgenus. As in *Mesotrichia*, the female scutellum has a sharp transverse truncation overhanging the metanotum. Males have unmodified tegulae, unlike the elongate tegulae of *Mesotrichia*. *Xylocopa smithii* Ritsema is remarkable for its brilliant metallic coloration, suggesting some species of the subgenus *Schonnherria*.

■ This subgenus ranges throughout sub-Saharan Africa including Madagascar as well as the Mediterranean countries of Africa (Morocco to Egypt), Dalmatia, southwestern Asia, and southern Asia east to the Philippines, Taiwan, and Japan, and south through Indonesia, New Guinea, and the Bismarck Archipelago to southernmost Australia. This largest subgenus of *Xylocopa* contains about 196 species.

The species of the Papuan region were revised by Lieftinck (1957a), those of Africa by Eardley (1983).

Xylocopa / Subgenus *Lestis* Lepeletier and Serville

Lestis Lepeletier and Serville, 1828: 799. Type species: *Apis bombylans* Fabricius, 1775 (misidentified as *Apis muscaria* Fabricius, 1775; see Hurd and Michener, 1961), monobasic, fixed by Commission Opinion 657 (1963).

Lestis consists of rather small, bright-green or blue-green species. Salient features are the four-segmented maxillary palpi, the nearly parallel first and second submarginal crossveins, and a large tubercle on the paraocular area of the female below the antenna.

■ This subgenus is found in eastern Australia from Victoria to northern Queensland and west into South Australia. Houston (1992a) recognized only two species.

Houston (1992a) gives an account of the nesting biology. *Lestis* is probably an old, long-isolated Australian element, unlike the only other Australian Xylocopini (subgenus *Koptortosoma*), which are close relatives of species from southern Asia and the intervening islands.

Xylocopa / Subgenus *Maaiana* Minckley

Xylocopa (Maaiana) Minckley, 1998: 32. Type species: *Xylocopa bentoni* Cockerell, 1919, by original designation.

The subgenus *Nodula* has been described as consisting of two groups. Minckley's (1999) phylogenetic analysis failed to show common derived characters for these groups, and he therefore described *Maaiana* as a new subgenus for the *bentoni* group of the old *Nodula*. Principal characters that distinguish *Maaiana* from *Nodula* are in thoracic shape (see the keys to subgenera), the lack of a propodeal triangle in males, and the presence of a mesosomal gland reservoir in males (lacking in *Nodula*).

■ This subgenus occurs in Sri Lanka and India. The species are the first six in the key to males of *Nodula* prepared by Hurd and Moure (1963).

Xylocopa / Subgenus *Mesotrichia* Westwood

Mesotrichia Westwood, 1838: 112. Type species: *Mesotrichia torrida* Westwood, 1838, monobasic.
Xylocopa (Platynopoda) Westwood, 1840b: 271. Type species: *Apis latipes* Drury, 1773, by designation of Ashmead, 1899a: 71.
Xylocopa (Audinetia) Lepeletier, 1841: 203. Type species: *Apis latipes* Drury, 1773, by designation of Sandhouse, 1943: 529.
Platinopoda Dalla Torre, 1896: 202, *lapsus* for *Platynopoda* Westwood, 1840.
Andineta Ashmead, 1899a: 71, incorrect subsequent spelling for *Audinetia* Lepeletier, 1841.
Audineta Ashmead, 1899a: 97, incorrect subsequent spelling for *Audinetia* Lepeletier, 1841.
Xylocopa (Hoplitocopa) Lieftinck, 1955: 27. Type species: *Xylocopa assimilis* Ritsema, 1880, by original designation.
Xylocopa (Hoploxylocopa) Hurd and Moure, 1963: 260. Type species: *Xylocopa acutipennis* Smith, 1854, by original designation.

This subgenus, like *Koptortosoma*, can be distinguished from other subgenera by the right-angular or acute scutellar beginning of the female thoracic declivity. Females have craterlike supraocellar pits. Males have posteriorly elongate tegulae. Most males have greatly widened front or middle tarsi, but *Xylocopa assimilis* Ritsema and *acutipennis* Smith have unmodified tarsi and were therefore excluded from *Mesotrichia* by some authors. *Mesotrichia* in a narrow sense contains species with the anterior coxal spine of the male short. In the group commonly called *Platynopoda,* however, males have a long anterior coxal spine.

■ *Mesotrichia* in a narrow sense is found throughout sub-Saharan Africa and northeast as far as Iran. The *Platynopoda* group is chiefly oriental, occurring in Sri Lanka north to Kashmir, eastward through the Lesser Sunda Islands, Sumatra, Borneo, and Southeast Asia to the Philippines. Hurd and Moure (1963) listed 23 species and gave a key to those of the *Platynopoda* group, which was revised by Maa (1940c).

Xylocopa / Subgenus *Monoxylocopa* Hurd and Moure

Xylocopa (Monoxylocopa) Hurd and Moure, 1963: 127. Type species: *Xylocopa abbreviata* Hurd and Moure, 1963, by original designation.

This taxon, known only in the female, is a close relative of *Diaxylocopa* and *Schonnherria.* In the absence of more information, I maintain it as a subgenus.

■ *Monoxylocopa* is known from the Brazilian states of Goiás and Mato Grosso. The only species is *Xylocopa abbreviata* Hurd and Moure.

Xylocopa / Subgenus *Nanoxylocopa* Hurd and Moure

Xylocopa (Nanoxylocopa) Hurd and Moure, 1963: 99. Type species: *Xylocopa ciliata* Burmeister, 1876, by original designation.

This subgenus is close to *Schonnherria* and may be united to that subgenus as one of the basal phylogenetic branches. As in *Xylocopsis,* graduli are present only on T1 and T2 of the female but they are on T1 to T5 of the male.

■ *Nanoxylocopa* ranges from Paraná and Rio Grande do Sul in Brazil to the provinces of Chaco to Buenos Aires, Argentina; also Uruguay, Paraguay, and Bolivia. The only species is *Xylocopa ciliata* Burmeister.

Xylocopa / Subgenus *Neoxylocopa* Michener

Apis (Ancylosoma) Dalla Torre, 1896: 206. [This name appeared in the synonymy of *Xylocopa brasilianorum* (Linnaeus) as a result of several errors; see Michener (1997b).]
Xylocopa (Neoxylocopa) Michener, 1954b: 157. Type species: *Apis brasilianorum* Linnaeus, 1767, by original designation.
Xylocopa (Megaxylocopa) Hurd and Moure, 1963: 151. Type species: *Apis frontalis* Olivier, 1789, by original designation.

This is a subgenus of large bees with black females and yellow or testaceous males. In the female the apex of the mandible is bidentate and a raised impunctate strip accompanies the whole epistomal suture. In males the posterior thoracic declivity begins on the base of the propodeum and the mesosomal gland is extremely large. R. Minckley (in litt., 1996) notes that an excellent character for male *Neoxylocopa* is the internal shape of the scutellum: the posterior scutellar inflection is reduced, thus allowing the mesosomal gland to extend forward to occupy the scutellar space.

■ This subgenus ranges from the provinces of Buenos Aires, La Pampa, and Mendosa, Argentina, and Tarapacá, Chile, north throughout tropical America, reaching Texas to California in the USA. It occurs also in the Antilles, Bahamas, and Bermuda. In the Pacific, it is presumably native to the Galapagos and Revillagigedo islands and introduced by human activity to Hawaii, the Marianas, etc. The 49 species were listed by Hurd (1978a).

Megaxylocopa is differentiated only by very large size and a strong carina below each lateral ocellus of the female.

Xylocopa / Subgenus *Nodula* Ma

Xylocopa (Nodula) Ma, 1938: 290. Type species: *Apis amethystina* Fabricius, 1793, by original designation.

Nodula consists of small to moderate-sized *Xylocopa* that are related to *Maaiana, Xylomelissa,* and *Rhysoxylocopa.* Males differ from those of *Xylomelissa* in the presence of a spine on the apex of the hind tibia and from those of *Rhysoxylocopa* by the unmodified propodeal spiracles. For differentiating characters of *Maaiana,* see the key to subgenera and the account of that subgenus.

■ *Nodula* occurs in southern Asia from India east to the Lesser Sunda Islands, the Philippines, and Taiwan, and north to central China and the Turkmenian part of the palearctic region. The seven species were listed by Hurd and Moure (1963), who gave a key for their identification.

Xylocopa / Subgenus *Notoxylocopa* Hurd

Xylocopa (Notoxylocopa) Hurd, 1956: 2. Type species: *Xylocopa tabaniformis* Smith, 1854, by original designation.

This is a distinctive subgenus of rather small, nonmetallic species. A unique feature of the males is an elongation of the spatha. Females differ from those of other Western Hemisphere subgenera by the lack of a propodeal triangle.

■ This subgenus occurs from Oregon, Utah, New Mexico, and southern Texas, USA, south through Mexico and Central America to Colombia and Ecuador. Hurd (1978a) included the various named forms in two species. O'Brien and Hurd (1965) revised the subgenus.

Xylocopa / Subgenus *Nyctomelitta* Cockerell

Xylocopa (Nyctomelitta) Cockerell, 1929e: 303. Type species: *Bombus tranquebaricus* Fabricius, 1804, by original designation.

This subgenus consists of large species rather thickly covered with reddish pubescence. Noteworthy characters are the large eyes, convergent above in both sexes, and the large ocelli, considerably larger than the antennal sockets.

■ *Nyctomelitta* occurs from Sri Lanka and India to Sumatra and Borneo (perhaps Java) and to Thailand and Laos. Hurd and Moure (1963) provided a key to the three species; they were revised by Maa (1940b).

Species of this subgenus are nocturnal. *Nyctomelitta* is not clearly related to any other subgenus of *Xylocopa.*

Xylocopa / Subgenus *Prosopoxylocopa* Hurd and Moure

Xylocopa (Prosopoxylocopa) Hurd and Moure, 1963: 215. Type species: *Xylocopa mirabilis* Hurd and Moure, 1963, by original designation.

The lateral projection from the ventral distal part of the male gonocoxite is a unique feature of *Prosopoxylocopa;* the narrow face of the male, with its continuous yellow from the labrum to the supraclypeal area, is unusual. The female can be recognized by the basitibial plate, which is represented by a single, very strongly projecting carina.

■ *Prosopoxylocopa* contains a single species, *Xylocopa mirabilis* Hurd and Moure, from Madagascar. Two African species were tentatively and probably incorrectly included by Hurd and Moure (1963).

Minckley (1994a) found *Prosopoxylocopa* to be the probable sister group of all the rest of a clade culminating in *Mesotrichia* and *Koptortosoma*.

Xylocopa / Subgenus *Proxylocopa* Hedicke

Xylocopa (Proxylocopa) Hedicke, 1938: 192. Type species: *Xylocopa olivieri* Lepeletier, 1841, by original designation.
Proxylocopa (Ancylocopa) Maa, 1954: 198. Type species: *Xylocopa nitidiventris* Smith, 1878, by original designation.

This subgenus contains rather small, dull-colored species, notable for their long faces and, in the female, the basitibial plates which lie at the bases of the tibiae as in ground-nesting bees, and the elevated pygidial plate. See the discussion of the tribe Xylocopini for further information on these characters and the possible generic status of *Proxylocopa*. Males can be recognized by the combination of short, ovate parapsidal lines, not over three times as long as wide, and two spines on the outer apex of the hind tibia.

■ *Proxylocopa* ranges from Albania, Greece, and Israel east to western China, in desert areas. About 16 species are recognized. Hurd and Moure (1963) and Wu (1983b) gave keys to the species of the *Ancylocopa* group, and Wu (1983c) gave a key to the species of the *Proxylocopa* s. str. group.

This subgenus contains ground-nesting species at least some of which are crepuscular. The name *Ancylocopa* was provided by Maa (1954) for species with ordinary ocelli; others, i.e., *Proxylocopa* s. str., have enlarged ocelli as do other crepuscular or nocturnal bees. Minckley (1998) found intergradation in this and other characters and concluded that recognition of *Ancylocopa* was unnecessary.

Xylocopa / Subgenus *Rhysoxylocopa* Hurd and Moure

Xylocopa (Rhysoxylocopa) Hurd and Moure, 1963: 178. Type species: *Xylocopa cantabrita* Lepeletier, 1841, by original designation.

This subgenus is related to *Xylomelissa, Nodula,* and *Biluna*. Males differ from those of all other *Xylocopa* in their small, asymmetrical, propodeal spiracles. Females differ from their relatives by the lack of a median groove on the propodeal triangle.

■ *Rhysoxylocopa* is found in the Mediterranean area from Spain and north Africa east to Central Asia and possibly south into the Sahel of Africa. Hurd and Moure (1963) listed eight species.

Xylocopa / Subgenus *Schonnherria* Lepeletier

Xylocopa (Schonnherria) Lepeletier, 1841: 207. Type species: *Xylocopa micans* Lepeletier, 1841, by designation of Sandhouse, 1943: 598.
Xylocopa (Schönherria) Dalla Torre, 1896, 202, unjustified emendation of *Schonnherria* Lepeletier, 1841 (not Burmeister, 1855).
Shornherria Ashmead, 1899a: 71, error for *Schonnherria* Lepeletier, 1841.
Xylocopa (Schoenherria) Hurd and Moure, 1963: 118, unjustified emendation of *Schonnherria* Lepeletier, 1841.
Xylocopa (Ioxylocopa) Hurd and Moure, 1963: 116. Type species: *Xylocopa chrysopoda* Schrottky, 1902, by original designation.
Xylocopa (Xylocospila) Hurd and Moure, 1963: 109. Type species: *Xylocopa bambusae* Schrottky, 1902, by original designation.

As shown above, the subgeneric name has been spelled in various ways. There is nothing in the original publication to indicate any error in spelling and Smith (1854) used the original spelling. Dalla Torre (1896), however, indicated that the name was a patronymic for C. A. Schönherr and accordingly emended the spelling to *Schönherria*, which was emended by Hurd and Moure (1963) to *Schoenherria*.

Most species of this subgenus are metallic green and bluish, some of them brightly metallic. *Schonnherria* differs from other New World groups in that graduli are absent on all terga except T1 of both males and females (present on T2 of the male in one species), and the apex of the male gonostylus is bifid; there is a large spine on the ventral margin of the male gonocoxite.

■ This subgenus occurs from Virginia to Florida, west to Texas and Arizona, USA, south through tropical America to the province of Chubut, Argentina, and Tarapacá, Chile. The 29 species were listed by Hurd (1978a).

Xylocospila and *Ioxylocopa* appear to be derived monotypic branches from *Schonnherria*. They were synonymized by Minckley (1998). The first is for a single remarkable species with broad yellowish integumental tergal bands in the male; the second is for a species that is separated from the rest of *Schonnherria* by the characters of couplet 1 in the key to subgenera based on males. Hurd and Moure (1963) were aware of *Ioxylocopa*'s close relationship to *Schonnherria*.

Xylocopa / Subgenus *Stenoxylocopa* Hurd and Moure

Xylocopa (Stenoxylocopa) Hurd and Moure, 1960: 809. Type species: *Xylocopa artifex* Smith, 1874, by original designation.
Xylocopa (Xylocopina) Hurd and Moure, 1963: 160. Type species: *Xylocopa ruficollis* Hurd and Moure, 1963, by original designation.

This subgenus contains rather small, elongate, nonmetallic or very feebly metallic species. Females differ from those of all other *Xylocopa* by the upper mandibular tooth, which is as wide as or wider than the lower tooth. Males can be recognized by the combination of two characters, the impunctate posterolateral one-third of the tegula and the two spines on the outer apex of the hind tibia. Male genitalia and other structures were illustrated by Hurd and Moure (1960). Both Hurd and Moure (1963) and Minckley (1998) regarded *Stenoxylocopa* as a near relative of *Neoxylocopa*.

■ *Stenoxylocopa* ranges from Buenos Aires and La Pampa provinces, Argentina, north through South and Central America to Tamaulipas and Chihuahua, Mexico, and southern Arizona, USA. The five species were revised by Hurd (1978b). An additional species, *Xylocopa ruficollis* Hurd and Moure, the only species placed in *Xylocopina*, is transferred to *Stenoxylocopa* on the authority of Minckley (1998).

The specificity of various *Xylocopa* species for particular nesting substrates has been greatly exaggerated. Thus *Stenoxylocopa* has been regarded as especially adapted for nesting in bamboo. However, its range extends well beyond that of bamboolike plants in Arizona, and in Brazil also other substrates are used (Silveira, 2002).

Xylocopa / Subgenus *Xenoxylocopa* Hurd and Moure

Xylocopa (Xenoxylocopa) Hurd and Moure, 1963: 243. Type species: *Mesotrichia chiyakensis* Cockerell, 1908, by original designation.

Xenoxylocopa is intermediate between (1) the group of subgenera *(Koptortosoma* and *Mesotrichia)* in which the posterior thoracic declivity of the female (beginning on the scutellum with a sharp truncation) and the anterior metasomal declivity of the female are sharply defined and (2) the group in which these declivities are rounded. In *Xenoxylocopa,* but not in *Koptortosoma* or *Mesotrichia,* there is no median fovea in the anterior surface of T1 and the posterior truncation of the scutellum is present but not strong. *Xenoxylocopa* is unique, among the subgenera having sharp female declivities, in having a small tooth on each side of the base of the median pygidial spine; such small spines are common features in the subgenera with rounded declivities.

■ This subgenus occurs from Senegal to the Transvaal and Cape Province, South Africa, perhaps mostly in xeric habitats. Hurd and Moure (1963) listed three species.

Xylocopa / Subgenus *Xylocopa* Latreille s. str.

Xilocopa Latreille, 1802a: 432. Suppressed by Commission Opinion 743 (1965).
Xylocopa Latreille, 1802b: 379. Type species: *Apis violacea* Linnaeus, 1758, by designation of Westwood, 1840a: 86.

Within this small subgenus there is considerable diversity, and Minckley (1998) suggested relations of different species to both *Xylocopoides* and *Ctenoxylocopa*. Males differ from those of other Eastern Hemisphere groups by the combination of lack of yellow on the face, rounded profile of the posterior part of the thorax, and two spines on the outer apex of the posterior tibia. Females show the same hind tibial character and three-toothed mandibles.

■ *Xylocopa* s. str. is found in Europe (north as far as Germany) and east as far as Russia, Afghanistan, and the western Himalayas. The eight species were listed by Hurd and Moure (1963), who gave a key to most of them.

Xylocopa / Subgenus *Xylocopoda* Hurd and Moure

Xylocopa (Xylocopoda) Hurd and Moure, 1963: 105. Type species: *Xylocopa elegans* Hurd and Moure, 1963, by original designation.

A unique feature of this subgenus is the lamella on the undersurface of the hind tibia of the male (see the key to subgenera). Females can be recognized by the paraocular areas, which are abruptly elevated near the lower margin of the eye.

■ This subgenus is known from Minas Gerais and Paraná, Brazil, to the province of Misiones, Argentina. There are two species.

Xylocopa / Subgenus *Xylocopoides* Michener

Xylocopa (Xylocopoides) Michener, 1954b: 155. Type species: *Apis virginica* Linnaeus, 1771, by original designation.
Xylocopa (Calloxylocopa) Hurd and Moure, 1963: 142. Type species: *Xylocopa tenuata* Smith, 1874, by original designation.

Females of this subgenus of dark-blue or greenish species differ from others in having a broad pygidial plate indicated by rows of teeth diverging from the base of the apical spine. Males differ from those of all other subgenera except some *Xylocopa* s. str. by the two spines on the outer apex of the hind tibia. The lack of recognizable male gonostyli is also distinctive. Male genitalia and other structures were illustrated by Hurd (1961).

■ *Xylocopoides* is widespread in North America from Maine to Florida west (including southern Ontario, Canada) to Nebraska and Oregon, USA, thence south through Mexico and on to Costa Rica. The five species (excluding *Calloxylocopa*) were revised by Hurd (1961).

Calloxylocopa includes a single, highly derived species that appears to have the basic features of *Xylocopoides*.

Xylocopa / Subgenus *Xylocopsis* Hurd and Moure

Xylocopa (Xylocopsis) Hurd and Moure, 1963: 124. Type species: *Xylocopa funesta* Maidl, 1912, by original designation.

Like *Nanoxylocopa* and *Dasyxylocopa,* this may be a basal branch of the *Schonnherria* clade. *Xylocopsis* differs from its relatives in having graduli on T1 and T2 of the female.

■ This subgenus ranges from the state of Paraná to Rio Grande do Sul in Brazil and to Paraguay and the province of Misiones, Argentina. The only species is *Xylocopa funesta* Maidl.

Xylocopa / Subgenus *Xylomelissa* Hurd and Moure

Xylocopa (Xylomelissa) Hurd and Moure, 1963: 219. Type species: *Xylocopa carinata* Smith, 1874 = *Xylocopa hottentotta* Smith, 1854 (synonymy according to Eardley, 1983), by original designation.
Xylocopa (Epixylocopa) Hurd and Moure, 1963: 223. Type species: *Xylocopa rufitarsis* Lepeletier, 1841, by original designation.

Xylocopa (Apoxylocopa) Hurd and Moure, 1963: 226. Type species: *Xylocopa lugubris* Gerstaecker, 1857, by original designation.

Xylocopa (Dinoxylocopa) Hurd and Moure, 1963: 230. Type species: *Xylocopa absurdipes* Enderlein, 1903 = *Xylocopa io* Vachal, 1898, by original designation.

Xylocopa (Perixylocopa) Hurd and Moure, 1963: 232. Type species: *Xylocopa erythrina* Gribodo, 1894, by original designation.

Xylocopa (Euxylocopa) Hurd and Moure, 1963: 234. Type species: *Xylocopa fraudulenta* Gribodo, 1894 = *Xylocopa erythrina* Gribodo, 1894, by original designation.

Xylocopa (Acroxylocopa) Hurd and Moure, 1963: 236. Type species: *Xylocopa capitata* Smith, 1854, by original designation.

Males of this subgenus differ from those of other *Xylocopa* by their combination of lack of spines on the outer side of the hind tibial apex with presence of a basitibial plate. Females differ from those of the most similar subgenus, *Rhysoxylocopa,* by the presence of a median groove on the propodeum.

■ *Xylomelissa* is found throughout sub-Saharan Africa. There are about 65 species.

Hurd and Moure recognized the relationship of their six subgenera here synonymized to *Xylomelissa;* most of these subgenera are based on one or a very few species with characters that seem to be derived relative to the large group of species included by those authors in *Xylomelissa.* Eardley (1983), in revising the species of southern Africa, has dealt with several of these matters, and Minckley (1998) in a cladistic study, showed the proximity of the taxa listed in the above synonymy, the subgenus *Nodula* also being sometimes associated with *Xylomelissa.*

Xylocopa / Subgenus *Zonohirsuta* Ma

Xylocopa (Zonohirsuta) Ma, 1938: 300. Type species: *Xylocopa collaris* Lepeletier, 1841 (not *Apis collaris* Olivier, 1789) = *Xylocopa dejeanii* Lepeletier, 1841, by original designation.

Zonohirsuta consists of rather small species, similar to *Prosopoxylocopa* but lacking some of the unusual features of that subgenus. Both lateral margins of the basitibial plate of the female are strongly carinate and project perpendicularly from the leg.

■ This subgenus is found in Sri Lanka, India, and Tibet, thence eastward through Burma to Southeast Asia and the Philippines, and through Indonesia to Sulawesi. Hurd and Moure (1963) gave a key to four species but list 14 valid names, most of which are geographical variants; Maa (1940a) revised the subgenus.

89. Tribe Ceratinini

The Ceratinini consists of rather slender, usually small andreniform species with generally shining, superficially nearly hairless bodies (Fig. 89-1) that vary from black to brilliantly metallic green, rarely with the metasoma red or metallic red. Most species have yellow markings at least on the face, and some (especially in eastern and southern Asia) have extensive yellow maculation. In females of many species the body is almost wholly dark except for a robust, vertically elongate yellow bar in the middle of the clypeus. This facial pattern is very rare in other bees. Some tribal characters other than those indicated in the key to tribes include the strongly concave lateral margins of the clypeus, the clypeus thus shaped like a thick, inverted T; the mandibular shape, which is broad at the base and abruptly narrowed medially; the complete lack of a female pygidial fimbria (it is only weakly developed in the Allodapini), and the lack of the spine that possibly represents the pygidial plate in females of the other tribes (Fig. 89-2). The basal part of S2 and often of S3 of females often has a semilunar area called a "wax plate."

The Ceratinini nest in pithy dead stems or twigs that they enter at broken ends. The cells are unlined, usually cylindrical (Figs. 89-3, 90-5a), and formed merely by partitions in the unbranched burrow; they are not barrel-shaped as in Xylocopini. The partitions are made of pith particles, loosely held together without any obvious ad-

Figure 89-1. *Ceratina (Zadontomerus) timberlakei* Daly; **a,** Male; **b,** Female. Drawings by B. B. Daly, from Daly, 1973.

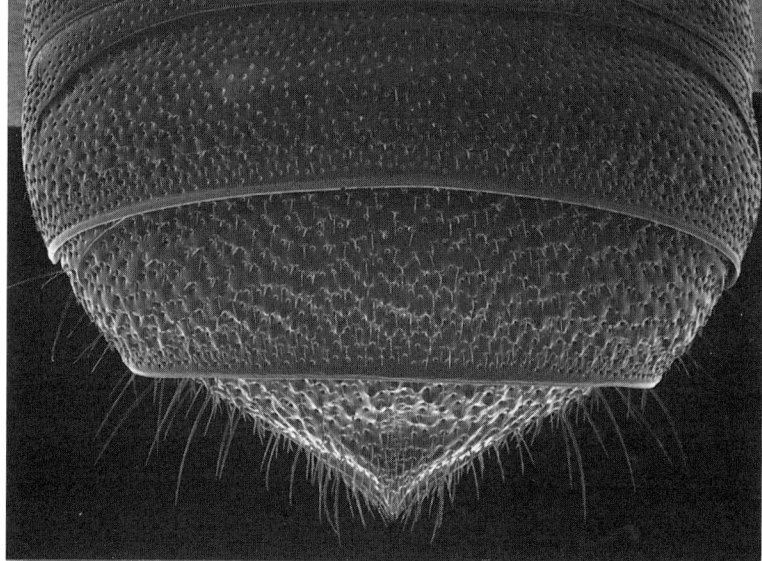

Figure 89-2. Apex of metasoma (T4 to T6, dorsal view) of female of *Ceratina dupla* Say or *calcarata* Robertson, showing lack of a pygidial plate. From Michener, McGinley, and Danforth, 1994.

hesive. Accounts of nesting biology are by Malyshev (1913), Daly (1966), Kislow (1976), Maeta and Katayama (1978), Sakagami and Maeta (1985), Okazaki (1987, 1992), Westrich (1989), and others. Although the Ceratinini are traditionally regarded as solitary bees whose females never meet their progeny, S. F. Sakagami, Y. Maeta, and others have shown that in many species females work their way through the completed cell series, penetrating the partitions between cells and incorporating feces or dead larvae into the partitions that they reconstruct. Moreover, in various species, two or rarely more adult females sometimes work in the same nest, with division of labor. Michener (1990d) summarized these matters and gave references to the numerous relevant papers; Maeta and Sakagami (1995) included references to more recent papers. Perhaps in all species the majority of nests are constructed and cared for by a single female, but for some species, if a nest is reused it is likely to be inhabited by a mother-daughter or daughter-daughter combination, sometimes with division of labor between the two or more adult inhabitants, one being rather queenlike, the other(s) workerlike.

Parthenogenesis is known in some species; apparently *Ceratina dalletorreana* Friese is almost wholly parthenogenetic (Daly, 1966, 1973, 1983a), and *C. acantha* Provancher has parthenogenetic populations in Southern California (Daly, 1973).

The tribe includes numerous groups, most of which usually have been and are here regarded as subgenera of *Ceratina*. In addition, it contains morphologically distinctive taxa, *Ctenoceratina*, *Megaceratina*, and *Pithitis*, that have been regarded as genera. Terzo (2000) showed in a phylogenetic analysis that these three distinctive taxa are derived from more ordinary *Ceratina;* their recognition as genera would make *Ceratina* paraphyletic. Furthermore, the apical setae of terga and sterna, the principal characteristic of *Ctenoceratina*, are sometimes largely absent [e.g., in *C. (Ctenoceratina) ericia* Vachal]. Moreover, *Protopithitis* is in various ways intermediate between

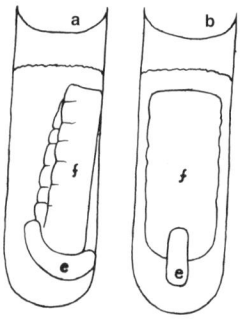

Figure 89-3. The lowest cell in a nest of *Ceratina australensis* Perkins in a pithy stem, closed by a partition of compacted bits of pith. **a,** Side view showing the egg (e) helping to support the food mass (f); **b,** Frontal view of the same. From Michener, 1962c.

Pithitis and other *Ceratina,* since it lacks the pointed axillae, the carina delimiting the basal area of the propodeum, and other features that led to recognition of *Pithitis* as a genus, while having the gradular and other characters found in *Pithitis. Ctenoceratina* also is *Pithitis*-like in the coarse punctation, long notauli, and pointed axillae of some species, and in the lack of graduli on T4 and behind. *Megaceratina,* also, is morphologically very distinctive but phylogenetically part of *Ceratina*.

Genus *Ceratina* Latreille

This is the only genus of the tribe Ceratinini. Figure 89-1 shows the characteristic form. Body length ranges from 2.2 to 12.5 mm. Male genitalia and other structures were illustrated by Vecht (1952), Mitchell (1962), Shiokawa (1963), Daly (1973, 1983a), Shiokawa and Hirashima (1982), and Terzo (2000); see also Figure 89-5, but note that, unlike other subgenera, the subgenus *Euceratina* has well-developed gonostyli.

Ceratina is found on all continents, although it is rare and limited in distribution in Australia. Since no subgenus occurs naturally in both Eastern and Western hemispheres, separate keys to subgenera are provided for these regions.

There remain species, at least in the neotropical region,

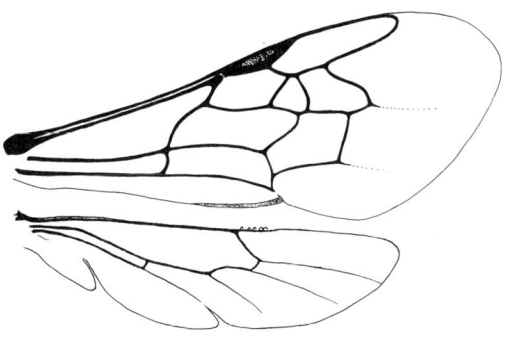

Figure 89-4. Wings of *Ceratina dupla* Say.

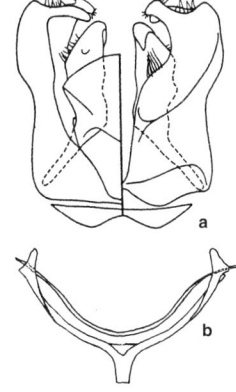

Figure 89-5. Structures of male *Ceratina (Zadontomerus) calcarata* Robertson. **a**, Genitalia, dorsal view at the left; **b**, S8 (with truncate median spiculum) and S7. These two sterna are in contact, not separated as here shown. a, from Daly, 1973.

that do not fall in any recognized subgenus. An example is *Ceratina laticeps* Friese, which was incorrectly included in *Zadontomerus* by Michener (1954b). In addition, many of the 120 or more names proposed for neotropical species have never been placed to subgenus.

Regional studies of *Ceratina* are as follows: Vecht (1952), oriental species; Yasumatsu and Hirashima (1969), Japan; Hirashima (1971a), southern and eastern Asia and the Pacific; Daly (1983a), North Africa and the Iberian peninsula; Daly (1973), North America north of Mexico; Friese (1896a), the palearctic region; Terzo and Rasmont (2004), Turkestan.

Key to the Subgenera of *Ceratina* of the Eastern Hemisphere
(In part modified from Hirashima, 1971a)

1. Prestigma about as long as distance from base of stigma to vein r; stigma not wider than prestigma, as measured to wing margin; margins of stigma basal to vein r parallel, margin in marginal cell not convex; distal halves of middle and hind femora with sharp, bladelike edges ventrally (Africa) .... *C. (Megaceratina)*
—. Prestigma much shorter than distance from base of stigma to vein r (Fig. 89-4); stigma wider than prestigma, as measured to wing margin (except in large species like *Ceratina chalcites* Germar); margins of stigma basal to vein r convergent basally, margin in marginal cell convex except in largest species; middle and hind femora without sharp ventral edges 2
2(1). T2 to T6 without graduli (punctation very strong; punctures of lower part of paraocular area usually large, as close as they can be, flat-bottomed so that the pattern is a network of ridges; supraclypeal area with transverse carina below antennae; basal area of propodeum not longer than metanotum; notaulus extending nearly to posterior margin of scutum; hind tibia of female with no evidence of basitibial plate) 3
—. At least T2 and T3, and sometimes subsequent terga, with graduli 4
3(2). Axilla produced to posterior angle separated from scutellum; basal area of propodeum separated from posterior surface by carina; T7 of male rounded or pointed (Africa, oriental) .... *C. (Pithitis)*
—. Axilla not angulate, abutting scutellum; basal area and posterior surface of propodeum not separated by carina; T7 of male bidentate (Africa) .... *C. (Protopithitis)*
4(2). Paraocular area above antennal socket impunctate; body usually with yellow, or yellow and ferruginous, or rarely red, at least paraocular area with yellow marking *in both sexes;* pregradular areas of metasomal terga often yellow or ferruginous (usually medium-sized to large species) 5
—. Paraocular area above antennal socket punctate, at least with a few distinct punctures on space between eye and ocellus *or* with row of fine punctures along inner margin of eye, *or*, if this area impunctate, then body small, without pale markings except sometimes on head, pronotal lobe, and legs; paraocular areas without yellow marking *in female* (except for *Ceratinidia*); pregradular areas of metasomal terga neither yellow nor ferruginous 8
5(4). Basal area of propodeum horizontal, sharply separated from posterior surface, which is steeply declivous; antenna of male long, third segment of flagellum longer than broad (apex of T7 of male bluntly tridentate, the median tooth large, triangular; S6 of male with large concavity in middle of subapical portion; head and thorax largely impunctate; dark parts of body slightly to distinctly metallic blue or green (Philippines) ....
.... *C. (Chloroceratina)*
—. Basal area of propodeum usually strongly slanting, not abruptly separated from posterior surface; antenna of male short, third segment of flagellum broader than long
.... 6
6(5). Preoccipital carina present, especially strong in male, forming angle between vertex and occiput (vertex impunctate or nearly so; large, robust, black species with more or less rich yellow markings on head, thorax, and metasoma) (oriental) .... *C. (Catoceratina)*
—. Preoccipital carina absent, vertex rounded onto occiput
.... 7
7(6). Basitibial plate of female rudimentary; at least median part of clypeus finely coriaceous in female; basal area of propodeum also finely coriaceous, not coarsely sculptured; genitalia of male without bundles of hairs (oriental) .... *C. (Lioceratina)*
—. Basitibial plate of female distinct, although small; clypeus polished, not coriaceous, or rarely finely coriaceous in female as in *Lioceratina;* basal area of propodeum often slightly more coarsely sculptured than in *Lioceratina;* genitalia of male with four bundles of hairs (oriental) .... *C. (Xanthoceratina)*
8(4). Posterior margins of T2 to T5 and S2 to S5 each with

row of coarse, posteriorly directed setae that are usually thickened, sometimes scalelike, the rows on terga sometimes interrupted middorsally; graduli limited to T2, T3, S2, and S3 and sometimes S6 of male 9

—. Terga and sterna without apical rows of specialized setae; graduli present on T2, T3, and usually T4 of female and on T2 to T4 of male, also usually present behind S3 but sometimes weak on S4, etc. 10

9(8). Scutellum strongly convex in profile, its posterior part nearly vertical; profile of metanotum and propodeum strongly declivous (Africa) .... *C. (Simioceratina)*

—. Scutellum gently convex, its posterior part and metanotum and propodeum forming a single slope as seen in profile (Africa) .... *C. (Ctenoceratina)*

10(8). T5 with distinct gradulus in both sexes 11

—. T5 without gradulus (or with gradulus in male of *Copoceratina* which can be recognized by leg characters listed in couplet 16) 14

11(10). Black with yellow markings on head, thorax, metasoma, and legs (frons and vertex densely and rather coarsely punctate; mesopleura densely punctate; medium-sized to large, robust species) (oriental, east palearctic) .. ..*C. (Ceratinidia)*

—. Black or metallic, with only a few pale markings, if any, on head, pronotal lobe, and legs; metasoma without yellow markings 12

12(11). Genal and frontal areas smooth, largely impunctate except on upper part of genal area, which is punctate, sometimes scattered punctures along frontal margin of eye [But see note on *C. parvula* Smith under the subgenus *Ceratina* s. str.] .... *C. (Ceratinula)*

—. Genal and frontal areas punctate, at least a row of dense punctures along frontal margin of eye and scattered punctures on most of genal area 13

13(12). Maxillary palpus five- or six-segmented; S5 of female with gradulus; T6 of male without gradulus; T7 of male truncate or rounded or pointed posteriorly; S2 of male without tubercle; gonostylus of male without down-curved projection (palearctic, Africa, oriental) .... *C. (Ceratina s. str.)*

—. Maxillary palpus five-segmented; S5 of female without gradulus; T6 of male usually with gradulus; T7 of male usually extending posteriorly as long projection that is simple or bidentate at apex; S2 of male usually with tubercle in middle; gonostylus of male with down-curved projection (palearctic, oriental, Australia) .... *C. (Neoceratina)*

14(10). Palearctic. Male gonostylus several times as long as broad, simple, hairy, well separated from gonocoxite; T6 with median longitudinal keel; T7 of male strongly extending posteriorly, simple or bidentate at apex ....*C. (Euceratina)*

—. Africa or Madagascar. Male gonostylus short, less than twice as long as broad, with one or two pointed processes or hooks and without hairs or with hairs on restricted areas; T6 not keeled; T7 variable, not so extended 15

15(14). Body dark metallic blue; posterior femur of male without tooth or comb of hairs (Madagascar)................ ....*C. (Malgatina)*

—. Body nonmetallic black or with weak bronze reflections; posterior femur of male with ventrobasal tooth or with ventral comb of hairs.................................... 16

16(15). Posterior femur of male with ventrobasal tooth, without median ventral comb of hairs; T6 of male without median convexity; terga without apical plumose hairs (Africa, Madagascar).............................. *C. (Copoceratina)*

—. Posterior femur of male without tooth, with median ventral (sometimes interrupted) comb of hairs; T6 of male with large median convexity with punctuation much finer than on nearby areas; T1 to T4 (or at least T1 and T2) with apical bands of white plumose hairs laterally (Africa, Madagascar) .... *C. (Hirashima)*

Key to the Subgenera of *Ceratina* of the Western Hemisphere

1. Pronotum with distinct transverse carina sometimes interrupted medially, in front of posterior margin of pronotum, curving strongly downward at each side to front coxa 2

—. Pronotum without transverse carina (or such carina almost coincident with posterior margin of pronotum), and carina usually not extending downward to front coxa 3

2(1). Female without yellow or white on paraocular area; posterior tibia of female and some males on outer side basally usually with oblique carina representing posterior margin of basitibial plate; body usually brilliantly metallic, at least in some areas; maxillary palpus five- or six-segmented (neotropical) .... *C. (Calloceratina)*

—. Female with yellow or white on paraocular area; posterior tibia of both sexes without oblique carina, usually with protuberant angle or tooth representing apex of basitibial plate; body dark metallic; maxillary palpus six-segmented (South and Central America) ...... *C. (Crewella)*

3(1). Extensive impunctate regions on genal and frontal areas; maxillary palpus five- or six-segmented (preoccipital carina present or absent; paraocular area with pale marks in males and some females) .... *C. (Ceratinula)*

—. Genal and frontal areas with punctures; maxillary palpus six-segmented 4

4(3). Preoccipital carina absent or weak; male gonostylus several times as long as broad, hairy (California; introduced) .... *C. (Euceratina)*

—. Preoccipital carina strong; male gonostylus little longer than broad 5

5(4). Paraocular area with pale marking in both sexes; T5 without gradulus; frons punctate throughout; T6 of female with strong carina extending from apex anterolaterally, separating dorsal from lateroventral surface of tergum (South America) .... *C. (Rhysoceratina)*

—. Paraocular area dark in females and nearly all males; T5 with gradulus; frons with smooth convexity on each side of midline; T6 of female without carina (North and Central America) .... *C. (Zadontomerus)*

Ceratina / Subgenus *Calloceratina* Cockerell

Ceratina (Calloceratina) Cockerell, 1924d: 77. Type species: *Ceratina amabilis* Cockerell, 1897 = *C. exima* Smith, 1862, by original designation.

The species of *Calloceratina* are usually brilliantly metallic green or blue, or the metasoma sometimes red, pale areas limited to the face, pronotal lobe, and legs. Even when the body appears only weakly metallic, like *Zadontomerus*, as it does in *Ceratina (Calloceratina) capitosa*

Smith, there are usually small areas of bright color. The pronotal structure is as in the related subgenus *Crewella*. Females and some males differ from those of all other *Ceratina* in that the basitibial plate is represented by an oblique carina representing the posterior margin of the plate, not by a protuberance or spine representing its apex. The body length is 6.5 to 13.5 mm.

■ The subgenus is found in the New World tropics from Tamaulipas and Nayarit, Mexico, to Bolivia and Rio Grande do Sul, Brazil; a recent record is from central Texas (J. L. Neff). There are perhaps ten species; this is nothing but a guess, in view of the lack of studies.

Ceratina / Subgenus *Catoceratina* Vecht

Ceratina (Catoceratina) Vecht, 1952: 30. Type species: *Ceratina perforatrix* Smith, 1879, by original designation.

This subgenus consists of species with yellow markings on all tagmata, although they are more restricted than those in most *Ceratinidia*. As in *Lioceratina* and *Chloroceratina*, the sculpturing is delicate; unlike those subgenera and like *Ceratinidia*, the preoccipital carina is present. There is a scopa on the metasomal venter of the female, as in *Megaceratina*. T7 of the male is bluntly bituberculate. The body length is 8.0 to 10.5 mm.

■ *Catoceratina* occurs from Burma, Thailand, and the Philippine Islands south to Sumatra, Java, and Borneo. The one species is *Ceratina perforatrix* Smith; see Vecht (1952) and Hirashima (1971a).

Ceratina / Subgenus *Ceratina* Latreille s. str.

Clavicera Latreille, 1802a: 432. Type species: *Hylaeus albilabris* Fabricius, 1793 = *Apis cucurbitina* Rossi, 1792, monobasic. Suppressed by Commission Opinion 1011 (1973).
Ceratina Latreille, 1802b: 380. Type species: *Hylaeus albilabris* Fabricius, 1793 = *Apis cucurbitina* Rossi, 1792, monobasic. Placed on Official List of Generic Names in Zoology by Commission Opinion 1011 (1973). [Later type designations were listed by Michener (1997b).]

Ceratina s. str. consists of nonmetallic species with pale coloration limited to the head, paranotal lobe, and legs. Body length is 5 to 9 mm. I suspect that this subgenus grades into *Neoceratina*; the distinguishing features, such as they are, appear in the key to subgenera.

■ This subgenus occurs from France, Switzerland, Bulgaria, and Turkey south through Africa to Cape Province, South Africa, and east through Eurasia to Japan, Taiwan, and Thailand. Of the more than 20 species, some are listed by Hirashima (1971a).

The smallest species of the genus *Ceratina* (forewing length about 2.2 mm), *C. parvula* Smith, found in the Mediterranean basin and west as far as the Canary Islands, has the head largely smooth and impunctate, as in species of the subgenus *Ceratinula*. For this reason it was placed in *Ceratinula* by Michener (2000) with the suggestion that it might have been introduced long ago from South America. This idea is no longer tenable. *C. parvula* has not been found in the Americas, and except for its small size and smooth integument, it has no clear resemblance to *Ceratinula*. For example, the inner orbits of the female are subparallel (convergent below in *Ceratinula*) and T7 of the male is broadly truncate with two slender, submedian apical spines (unlike that of any *Ceratinula*). For the present I place *C. paravula* in *Ceratina* s. str. although Terzo (2000) in an as-yet-unpublished work places it with certain larger African species in a new subgenus.

Ceratina / Subgenus *Ceratinidia* Cockerell and Porter

Ceratina (Ceratinidia) Cockerell and Porter, 1899: 406. Type species: *Ceratina hieroglyphica* Smith, 1854, by original designation.

This is the commonest of the subgenera of *Ceratina* with abundant yellow markings on all tagmata. It differs from the other such colored subgenera in the relatively strong punctation (see couplet 3 of the key to subgenera), although there are conspicuous impunctate areas. The body length is 5.0 to 11.5 mm.

■ *Ceratinidia* is an oriental subgenus found from Sri Lanka and India throughout southeastern Asia, north through China to the maritime province of Siberia, also including all of Japan, Taiwan, the Philippines, and Indonesia east to the western tip of New Guinea. Twenty-one species were recognized and revised by Vecht (1952); about five additional species are recognized. Shiokawa and Hirashima (1982) revised the *flavipes* group. Baker (2002a) gave a key to Asian mainland species of the subgenus.

This is the subgenus to which belong *Ceratina flavipes* Smith, *japonica* Cockerell, and *okinawana* Matsumura and Uchida, whose social and other behavior has been elucidated in a series of papers by S. F. Sakagami and Y. Maeta; a reasonably inclusive paper in this series is Sakagami and Maeta (1989).

Ceratina / Subgenus *Ceratinula* Moure

Ceratinula Moure, 1941a: 78. Type species: *Ceratina lucidula* Smith, 1854, by original designation.

This subgenus consists of minute (3-6 mm long) species, metallic or not (rarely with the metasoma red). Hirashima (1971a) and Michener (1965b) noted their similarity to *Ceratina* s. str. and included them in that subgenus. They differ from typical species of *Ceratina* s. str. in having very extensive smooth areas, for example lacking punctures on the paraocular area above the antenna and on the gena, if not on the whole head, and in the shape of the second submarginal cell, which is narrowed, sometimes to a point, anteriorly. This venational character may be merely a reflection of small size, for it is duplicated in some other minute species of *Ceratina*. The primarily neotropical distribution of *Ceratinula*, in contrast to the palearctic and African distribution of *Ceratina* s. str., supports recognition of the separate subgenus *Ceratinula*, but see below.

■ *Ceratinula* occurs from North Carolina, Texas, and California, USA, south through the tropics to Argentina, and one species has been inadvertently introduced to Hawaii. One can only guess at the total number of species—perhaps 30. Two occur in the USA, and nine others, with a key, were reported for Panama (Michener, 1954b).

Ceratina / Subgenus *Chloroceratina* Cockerell

Ceratina (Chloroceratina) Cockerell, 1918a: 143. Type species: *Ceratina cyanura* Cockerell, 1918, by original designation.

Chloroceratina has yellow markings on all tagmata, and some of the blackish areas are metallic blue or green. It is the only subgenus with a long third flagellar segment (see the key to subgenera). The body length is 7 to 8 mm.
■ This subgenus is known only from northern Luzon, Philippine Islands. The two species are perhaps not specifically distinct (see Vecht, 1952, and Hirashima, 1971a).

Ceratina / Subgenus *Copoceratina* Terzo and Pauly

Ceratina (Copoceratina) Terzo and Pauly, in Pauly et al., 2001: 292. Type species: *Ceratina madecassa* Friese, 1900, by original designation.

This subgenus contains two somewhat finely punctured species, with the basal area of the propodeum minutely granular and longer than the metanotum. The supraclypeal area has an irregular transverse ridge or carina somewhat nearer to the upper end of the clypeus than to the antennal bases.
■ One species of the subgenus, *C. minuta* Friese, is found from Kenya to the Transvaal, South Africa. The other, *C. madecassa* Friese, occurs in Madagascar and the Seychelles.

Ceratina / Subgenus *Crewella* Cockerell

Ceratina (Crewella) Cockerell, 1903b: 202. Type species: *Ceratina titusi* Cockerell, 1903, by original designation.

This subgenus consists of strongly punctured, weakly metallic species with pale coloration limited, as in all Western Hemisphere *Ceratina,* to the face, pronotal lobes, and legs. The paraocular areas ordinarily have pale markings. This subgenus and *Calloceratina* differ from all others in having a strong carina across the dorsum of the pronotum, well in front of the posterior margin of the pronotum, the carina curved sharply down at the side and extending to the front coxa. Certain species of *Crewella* have an unusually long proboscis, but others have one of ordinary length. The body length is 6 to 13 mm.
■ *Crewella* occurs from Costa Rica to Bolivia, Paraguay, and Buenos Aires Province, Argentina; a doubtful record, without details, is from Mexico. About 12 species are in this subgenus, although in the complete absence of revisional papers, this number is only an estimate.

Ceratina / Subgenus *Ctenoceratina* Daly and Moure

Ctenoceratina Daly and Moure, 1988, *in* Daly, 1988: 12. Type species: *Ceratina armata* Smith, 1854, by original designation. [New status.]

This subgenus consists of strongly punctate black species, sometimes with metallic reflections, having the pale areas limited to the head, pronotal lobes, and legs. As in *Simioceratina,* there is an apical row of coarse, sometimes scalelike setae on T2 to T5 and S2 to S5. Unlike that of *Simioceratina,* the scutellum is wholly dorsal and slopes slightly posteriorly to join the sloping metanotum and propodeum. Body length is 5 to 9 mm.
■ *Ctenoceratina* is widespread in Africa, from Senegal to Ethiopia, south to Cape Province, South Africa. The ten species were revised by Daly (1988).

Ceratina / Subgenus *Euceratina* Hirashima, Moure, and Daly

Ceratina (Euceratina) Hirashima, Moure, and Daly, 1971, *in* Hirashima, 1971a: 369. Type species: *Apis callosa* Fabricius, 1794, by original designation.

Euceratina consists of rather weakly metallic, strongly punctate species. The pale markings are limited to at most the face, pronotal lobe, and legs, as in *Ceratina* s. str., *Zadontomerus,* etc. The most distinctive feature is the male gonostylus, which is articulated to the gonocoxite, and is several times as long as wide, and hairy; it was illustrated for various species by Daly (1983a). Presumably, this is an ancestral characteristic, lost in all other subgenera; as shown in Hirashima's (1971a) illustration, the genitalia as a whole are very different from those of other *Ceratina.* The body length is 6 to 12 mm.
■ This subgenus occurs from Britain, Spain, and Morocco east through Europe and the Mediterranean basin to southern Russia, Pakistan, and Somalia. One species, the parthenogenetic *Ceratina dallatorreana* Friese, is established in California, where it was no doubt introduced by commerce. Hirashima (1971a) listed 16 species, and at least four more exist. Daly (1983a) revised the 11 species of North Africa and the Iberian peninsula.

Ceratina / Subgenus *Hirashima* Terzo and Pauly

Ctenoceratina (Hirashima) Terzo and Pauly, in Pauly et al., 2001: 298. Type species: *Ceratina nyassensis* Strand, 1911, by original designation.

This subgenus of strongly punctured species resembles *Ctenoceratina* in some respects, as emphasized by Terzo and Pauly (in Pauly et al., 2001). *Hirashima* lacks the spinelike tergal and sternal setae of *Ctenoceratina;* on T1 to T4 (or perhaps worn and thus on T1 and T2 only) it has lateroapical bands of white plumose hairs, an unusual feature in *Ceratina.* Contrary to the statements of Terzo and Pauly, the axillae are not spinelike; they are shaped exactly as in *Copoceratina, Malgatina,* etc. Some of the leg and metasomal characters of males are much as in *Ctenoceratina; Hirashima* seems to bridge the differences between *Ctenoceratina* and *Ceratina,* s. str.
■ *Hirashima* ranges from Tanzania and Nigeria south to South Africa, and to Madagascar and Aldabra. Of the four species, one, *C. nyassensis* Strand, occurs in Africa (Malawi) as well as Madagascar and Aldabra.

Ceratina / Subgenus *Lioceratina* Vecht

Ceratina (Lioceratina) Vecht, 1952: 32. Type species: *Ceratina flavopicta* Smith, 1858, by original designation.

Lioceratina consists of delicately sculptured species with abundant yellow markings, usually with much of the face finely coriaceous. T6 of the male has two or four long spines and T7 has median and lateral angles or is emarginate. The male gonostylus lacks tufts of long hairs. The body length is 6.5 to 12.0 mm.

■ This subgenus ranges from India through Southeast Asia, Indonesia as far east as Bali and Sulawesi, and to the Philippines. Hirashima (1971a) and Vecht (1952) indicated that there are about seven species.

Ceratina / Subgenus *Malgatina* Terzo and Pauly

Ceratina (Malgatina) Terzo and Pauly, in Pauly et al., 2001: 288. Type species: *Ceratina azurea* Benoist, 1955, by original designation.

Malgatina contains a rather dark metallic blue, strongly punctured species. As indicated by Terzo and Pauly (in Pauly et al., 2001), it resembles *Euceratina* in some features, such as the distribution of tergal graduli, but it differs strongly in the male genitalia and other features (see key above).

■ This subgenus contains a single species, *Ceratina azurea* Benoist, known only from Madagascar.

Ceratina / Subgenus *Megaceratina* Hirashima

Megaceratina Hirashima, 1971b: 251. Type species: *Ceratina bouyssoui* Vachal, 1903 = *Ceratina sculpturata* Smith, 1854, by original designation.

In addition to the key characters (see above), *Megaceratina* has other interesting features. The apical one-fourth or more of the marginal cell is bent away from the wing margin. Larger females have a strong projection from the posterior end of the hypostomal area and nearby gena. S2 and S3 of the female, and to a lesser extent the following sterna, have a scopa of suberect hairs. The metasomal terga are black to red, sometimes with orange or yellow spots. The body length is 10 to 12 mm. Male genitalia and other structures were illustrated by Daly (1985).

■ *Megaceratina* occurs in tropical Africa from Senegal to Zaire and east to Uganda. There is one species, *M. sculpturata* (Smith), if Daly (1985) is correct that the extensive color variation is intraspecific.

The wing venation of *Megaceratina* may have diverged from that of other *Ceratina* because of the large body size, but the almost equally large species such as *C. (Euceratina) chalcites* Germar have only moderately reduced stigmal size and a short prestigma, thus not closely resembling *Megaceratina* in this respect.

Ceratina / Subgenus *Neoceratina* Perkins

Neoceratina Perkins, 1912: 117. Type species: *Neoceratina australensis* Perkins, 1912, monobasic.

This is a group of black or weakly metallic, strongly punctate species, the pale markings limited to the face, pronotal lobe, and legs, as in *Ceratina* s. str. and other groups. The size is small, 4 to 6 mm long. It is with some hesitation that I maintain the distinction between *Neoceratina* and *Ceratina* s. str.; the weakness of the differentiation is suggested in the key to subgenera.

■ *Neoceratina* is known from Turkey and Cyprus east through southwest and southern Asia and Indonesia to southern China, the Ryukyu Islands, Micronesia, the Philippines, and south to the Bismarck Archipelago, the Solomon Islands, and eastern Australia as far as New South Wales. An unidentified species of the subgenus *Neoceratina* is now widespread in Hawaii (Snelling, 2003). Hirashima (1971a) listed eight species and gave a key for their separation.

Ceratina / Subgenus *Pithitis* Klug

Pithitus Klug, *in* Illiger, 1807: 198; Klug, 1807b: 225. Type species: *Apis smaragdula* Fabricius, 1787, monobasic. [The papers by Illiger and Klug were published simultaneously; Illiger credited *Pithitis* to Klug and it is appropriate to do so.]

This subgenus consists of metallic and often brilliantly metallic species that are cylindrical in form and coarsely and densely punctate. The preoccipital carina is present, the axillae are each produced to an angle or spine, and the short basal area of the propodeum is defined by a carina posteriorly. The body length is 5 to 10 mm. Interesting features of the males of some species are the paired, broad, depressed dull-black areas on T4 to T6.

■ *Pithitis* is found from Senegal to Egypt south throughout Africa to Cape Province, north perhaps to Crete, eastward in Saudi Arabia, Yemen, Pakistan, India, and Sri Lanka, throughout southeast Asia to the Philippines, Taiwan, the Ryukyu Islands, and southeast China, and through Indonesia east as far as Ambon. About nine species were revised by Hirashima (1969), the Oriental ones by Vecht (1952) and Shiokawa and Sakagami (1969), and those of northeastern Africa and the Arabian peninsula by Daly (1983b). The subgenus *Pithitis*, specifically *Ceratina (P.) smaragdula* (Fabricius), is now apparently established on Oahu, Hawaii (Arakaki et al., 2002; Snelling, 2003).

Ceratina smaragdula (Fabricius) has been reared in greenhouses in Utah, California, and Florida, USA, as a possible pollinator, but when liberated outdoors it did not become established (S. Batra, personal communication).

Ceratina / Subgenus *Protopithitis* Hirashima

Pithitis (Protopithitis) Hirashima, 1969: 651. Type species: *Ceratina aereola* Vachal, 1903, by original designation.

Protopithitis consists of a dark blue-green, coarsely punctate, *Pithitis*-like species that lacks the spiniform axillae and the carina delimiting the basal propodeal area that are characteristic of *Pithitis*. Body length is about 8 mm.

■ The subgenus occurs in Zaire. The single species, *Ceratina aereola* Vachal, was redescribed by Hirashima (1969).

Ceratina / *Rhysoceratina* Michener

Ceratina (Rhysoceratina) Michener, 2000: 599. Type species: *Ceratina montana* Holmberg, 1886, by original designation.

This subgenus consists of weakly metallic species having the body form of *Zadontomerus*. The following are its principal characters:

Body strongly punctate, without impunctate areas on face, without the two smooth convexities on frons found in *Zadontomerus,* with or without smooth area on gena, with smooth area on scutum. Paraocular area and clypeus with yellow maculation in both sexes. Preoccipital carina strong. Pronotum without a dorsal, premarginal carina; scutellum dorsal; metanotum and base of propodeum somewhat slanting; basal area of propodeum much longer than metanotum, slightly depressed, strigose. Basitibial plate of female indicated by strong tooth, that of male absent. Graduli present on T2 to T4 and S2 and S3 of both sexes; T6 of female with strong carina extending anterolaterally from apex and separating dorsal from ventrolateral surfaces of tergum; T6 of male not toothed, T7 simple, subtruncate except for small midapical marginal projection. Body length 5.0 to 7.5 mm.

The strong, sometimes almost lamellate carinae delimiting the dorsal surface of T6 of the female are not found in any other subgenus.

■ *Rhysoceratina* occurs from Santa Catarina, Brazil, to the provinces of Catamarca, Córdoba, and Buenos Aires, Argentina. Of at least five species, two that have names are *Ceratina montana* Holmberg (= *volitans* Schrottky) and *C. stilbonota* Moure.

Ceratina / Subgenus *Simioceratina* Daly and Moure

Ctenoceratina (Simioceratina) Daly and Moure, 1988, *in* Daly, 1988: 42. Type species: *Ceratina moerenhouti* Vachal, 1903, by original designation.

This subgenus agrees in general appearance, and in the apical rows of coarse setae on terga and sterna, with *Ctenoceratina.* It differs from other *Ceratina* in the strongly convex and elevated scutellum, the posterior part of which is nearly vertical, and the steeply declivous, nearly vertical, metanotum and propodeum as seen in profile. The body length is 5 to 8 mm.

■ *Simioceratina* ranges from Liberia to Kenya south to Namibia and Natal Province, South Africa. The three species were revised by Daly (1988).

Ceratina / Subgenus *Xanthoceratina* Vecht

Ceratina (Xanthoceratina) Vecht, 1952: 39. Type species: *Ceratina cladura* Cockerell, 1919, by original designation.

This subgenus, closely related to and perhaps a synonym of *Lioceratina,* consists of delicately sculptured species usually with yellow markings, the body sometimes largely red. T6 of the male ends in two rounded teeth and T7 lacks angles or teeth. From the male gonostylus arise two tufts of long hairs. The body length is 5.0 to 10.5 mm.

■ *Xanthoceratina* is found in Sri Lanka, Burma, Southeast Asia including Indonesia as far east as Java, the Philippines, and southern China. Nine species were tentatively recognized by Hirashima (1971a); see also Vecht (1952).

Ceratina / Subgenus *Zadontomerus* Ashmead

Zadontomerus Ashmead, 1899a: 69. Type species: *Ceratina tejonensis* Cresson, 1864, by original designation.
Zaodontomerus Cockerell and Porter, 1899: 406, unjustified emendation of *Zadontomerus* Ashmead, 1899.

This subgenus consists of weakly metallic species with pale areas limited to the head, pronotal lobe, and legs. The paraocular areas and genae, which are punctate, and the preoccipital carina, which is present, are features that distinguish the few small species from *Ceratinula.* There is a pair of impunctate swellings on the frons, a feature not found in other subgenera. As in *Calloceratina,* graduli are present on T2 to T5 and S2 and S3 in both sexes. The body length is 4 to 12 mm, usually 5 to 7 mm.

■ *Zadontomerus* occurs from Quebec to British Columbia, Canada, south throughout North and Central America to northern Colombia and Venezuela (J. Ascher, personal comm.). Of about 25 species, 18 occur north of Mexico. Species of America north of Mexico were revised by Daly (1973).

90. Tribe Allodapini

Most Allodapini are rather slender andreniform to hylaeiform bees superficially resembling *Ceratina* except that the cuticle is soft and delicately sculptured (Fig. 90-1). The pubescence is usually short and inconspicuous. Only *Macrogalea* is rather robust andreniform and hairy. The most obvious tribal character is the presence of only two submarginal cells (rarely one), the second usually not much over half as long as the first (Figs. 28-4, 90-1, 90-7, 90-13). The clypeus is rather flat, its lower lateral angle bent back as a large tooth on each side of the labrum; the upper half of the clypeus, above the level of the tentorial pits, is not greatly narrower than the lower half (Fig. 90-2), and the clypeus thus cannot be described as an inverted T as it is in Ceratinini. Moreover, the clypeus extends high up to about the middle of the face, so that the subantennal sutures are rarely much longer than the diameter of an antennal socket; *Eucondylops* is an exception. The last three metasomal terga of the female are somewhat flattened; these flattened terga block the nest entrance when there is a disturbance but are either not flattened or less flattened in the social parasites such as *Eucondylops*. Basitibial plates are absent or indicated only by a carina demarcating the posterior side of a small basal plate, the distal end of the carina slightly elevated and marking the presumptive distal end of the plate, which is not over one-sixth as long as the tibia. Australian forms may have a more complete plate of the same small size. The dorsal surface of the propodeum is dull, minutely and uniformly roughened, and much longer than the metanotum. The marginal cell is pointed or nar-

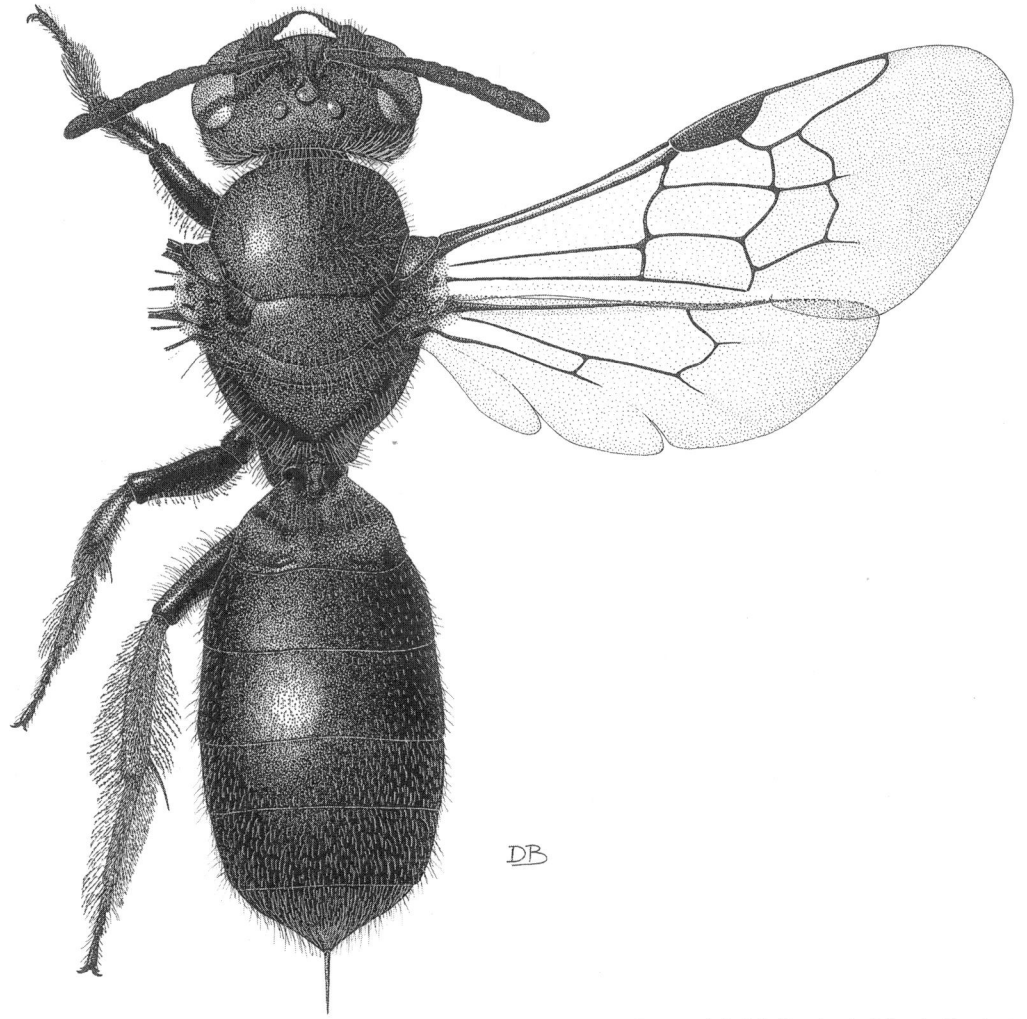

Figure 90-1. *Braunsapis facialis* (Gerstaecker), female. Drawings by D. J. Brothers, from Michener, 1975b.

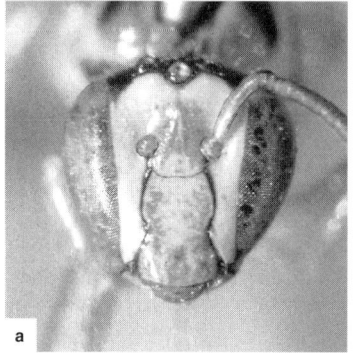
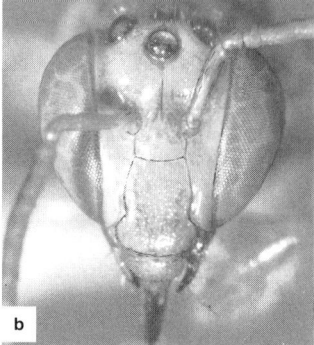
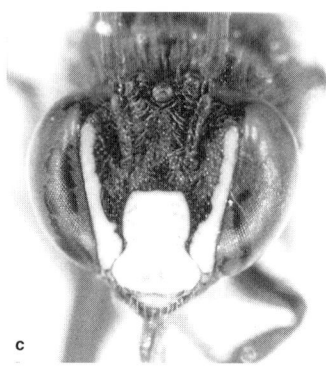

rowly rounded, the apex on or close to the wing margin (Figs. 28-4, 90-1, 90-7, 90-13). The stigma, except in *Macrogalea* (Fig. 90-12a), is broad; and even in *Macrogalea*, where it is only as wide as the prestigma (measured to the wing margin), vein r arises midway between the base and apex of the stigma. T6 of the female sometimes has a minute apical spine that may represent the pygidial plate; the pygidial fimbria is small and sparse, often scarcely recognizable. Among the distinctive features of the male are the short, broad, roundly truncate to feebly emarginate apex of T7 with a subapical fringe of long hairs that are longest and curved medially at the sides; and the large membranous roof of the genital chamber (Fig. 90-3). This roof part of T8 is generally biconvex as seen from above when the metasoma is dissected, but of course is biconcave when seen from the ventral, i.e., genital chamber, viewpoint. The concavity, which contains the genitalia, is provided with spicules that are numerous, long, and hairlike in most species. Because the membrane is usually wrinkled, the convexities appear brainlike.

A possibly useful key to African species of Allodapini, prepared at a time when they were all placed in the genus *Allodape*, was by Strand (1914b). A more recent treatment for the same continent was by Michener (1975b).

The nests of Allodapini are in pithy or hollow stems or in other cavities such as plant galls. The Allodapini are almost unique among bees in having nests without cells (Fig. 90-4). In *Compsomelissa (Halterapis) nigrinervis* (Cameron), food masses are located along the burrow, each with an egg or a larva, in such a way that the nest resembles a series of cells with the partitions omitted (Fig. 90-5b). In most allodapines, however, the larvae live together and are fed progressively, so that when a nest is examined, it frequently contains no stored food. *Compsomelissa (Compsomelissa)* is not well known, but small larvae are fed progressively; large ones are seemingly mass-provisioned.

In *Allodapula* several small eggs of about the same age are attached, in a pattern distinctive for each species (longitudinal row, horizontal ring), to the inner wall of the nest burrow. When the larvae hatch they maintain their positions in the nest burrow by continued attachment to the egg chorion, or later, when they are larger, by forming a clump that fills the burrow (Fig. 90-5d); they are fed progressively from a common food mass. Small egg size

Figure 90-2. Faces of Allodapini, showing characteristic clypeal shape. **a, b,** *Compsomelissa zaxantha* (Cockerell) and *C. ocellata* (Michener), females; **c,** *Allodape quadrilineata* (Cameron), male. From Michener, 1966b and 1975b.

and the laying of several eggs at about the same time must be apomorphic features, for large eggs are characteristic of related groups (Fig. 90-5a-c), not only in Allodapini but also in Ceratinini and Xylocopini. In *Allodape* and *Braunsapis* the eggs are large and produced at longer intervals, so that young of various ages (egg to pupa) are present at the same time (Figs. 90-4, 90-5c). Each larva is provided with its own food. Instead of the young in a burrow being arranged from oldest at the bottom to youngest near the nest entrance, as in mass provisioners like *Ceratina* and *Compsomelissa (Halterapis) nigrinervis*, the order is reversed (Figs. 90-4, 90-5c). Therefore, all food for young larvae has to be carried past pupae and older larvae. Accounts of nests and nest contents of many species are contained in Michener (1971a) and papers cited therein. More recent accounts of the Australian *Exoneura*

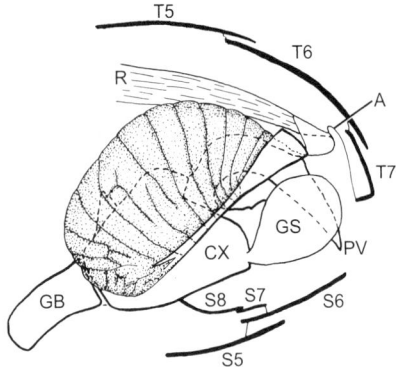

Figure 90-3. Diagram of a longitudinal dissection of the apex of the metasoma of a male *Allodape stellarum* Cockerell. Terga and sterna are shown in sagittal section. The stippled area represents the elaborated and wrinkled membranous roof of the genital chamber, characteristic of the Allodapini, probably derived from T8. It is margined posteriorly on each side by sclerotized lateral rods of T8, whose upper ends are connected by membrane above the anus. (A, anus; CX, gonocoxite; GB, gonobase; GS, gonostylus; PV, penis valve; R, rectum.) From Michener, 1975b.

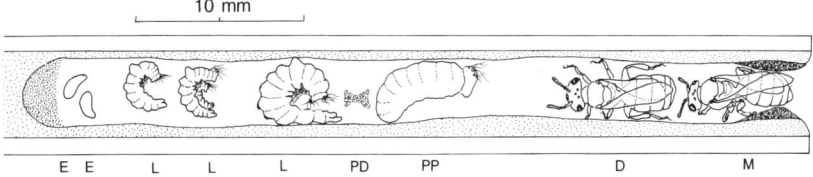

are by M. P. Schwarz and associates, e.g., Schwarz (1986), Schwarz and O'Keefe (1991), Silberbauer and Schwarz (1995), and Silberbauer (1997).

Probably as an evolutionary result of their not being enclosed individually in cells, but instead having to support themselves in an often vertical burrow and interact with one another and with adults, the larvae of Allodapini are more diverse in gross morphology than those of all other bee taxa taken together (Fig. 90-6; the letters in this paragraph refer to larvae illustrated in this figure). Presumably they evolved from larvae like those of *Ceratina* (k). Different genera have different structures for supporting themselves in the frequently vertical burrows and for handling pollen or sensing activity in the nest. Such structures include elongated antennae (u), a cephalic prominence (s), enormous diverging curved hairs on the head or on the prothorax (a-g), eversible tubercles in many positions on the body, including armlike branched ones on the thorax (p-r), and hairs in various locations on the body. These larval features were described in a series of papers, the last of which (Michener, 1976) included a key to African genera based on larvae. Although not well shown in (j), many of the minute hairs of *Macrogalea* larvae are hooked; they may support the larvae by hooking into the pith walls of the nests. Pupae also have special features, mostly hairs that probably help to support them. A key to genera based on pupae was included in Michener and Scheiring (1976).

Figure 90-4. Colony of *Braunsapis* in nest in pithy stem. The entrance, at the right, is narrowed, and the oldest female is closing it with her metasoma. (E, egg; L, larvae, the oldest toward the nest entrance; PD, pollen deposit; PP, prepupa; D, daughter; M, mother.) From Maeta, Sakagami, and Michener, 1992.

Although most nests of nearly all allodapine species contain only a single adult female and her immature offspring, in many species from 4 percent to nearly 50 percent or more of the nests, or in some populations of *Exoneura bicolor* Smith in Australia, nearly all nests, contain two or more cooperating, competing adult females. Such individuals characteristically show division of labor, including queenlike and workerlike aspects, even though often only a minority of the population lives in such colonies (for summary, see Michener, 1990d; see also Maeta, Sakagami, and Michener, 1992). An excellent review of aspects of social behavior in Allodapini is by Schwarz, Bull, and Hogendoorn (1998). See also Table 8-1 and Schwarz et al. (2005). Melna and Schwarz (1994) show that even in autumnal pre-reproductive colonies of

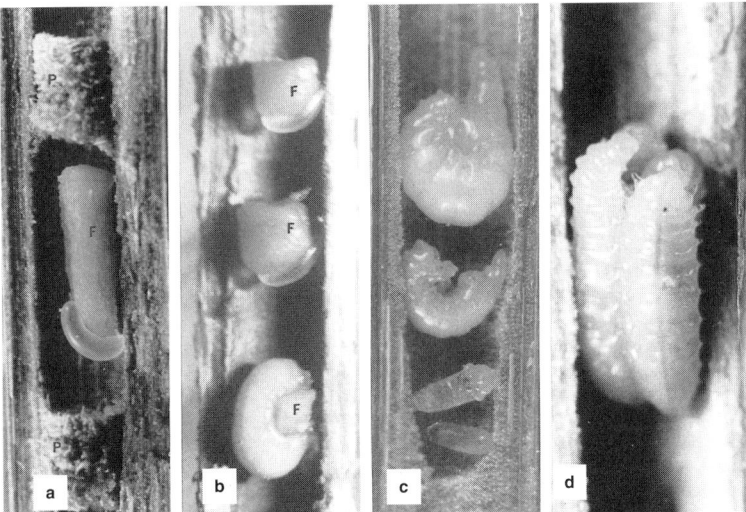

Figure 90-5. Nests of Ceratinini and Allodapini in burrows in pithy stems. **a,** A single cell, out of a series, of *Ceratina* sp.?, showing food mass (F) between two partitions (P) made of compacted bits of pith; **b,** Section of a nest of *Compsomelissa (Halterapis) nigrinervis* (Cameron), showing food masses (F) not separated by partitions, supported by eggs or an egg chorion, and the oldest immature near the bottom of the nest; **c,** Nest of *Allodape ceratinoides* Gribodo, showing an egg at the bottom and progressively fed larvae, the older ones toward the nest entrance; **d,** Section of a nest of *Allodapula (Dalloapula) acutigera* Cockerell, showing a clump of larvae feeding from a common food mass that is not visible. Photos by C. Michener, from Michener, 1977c.

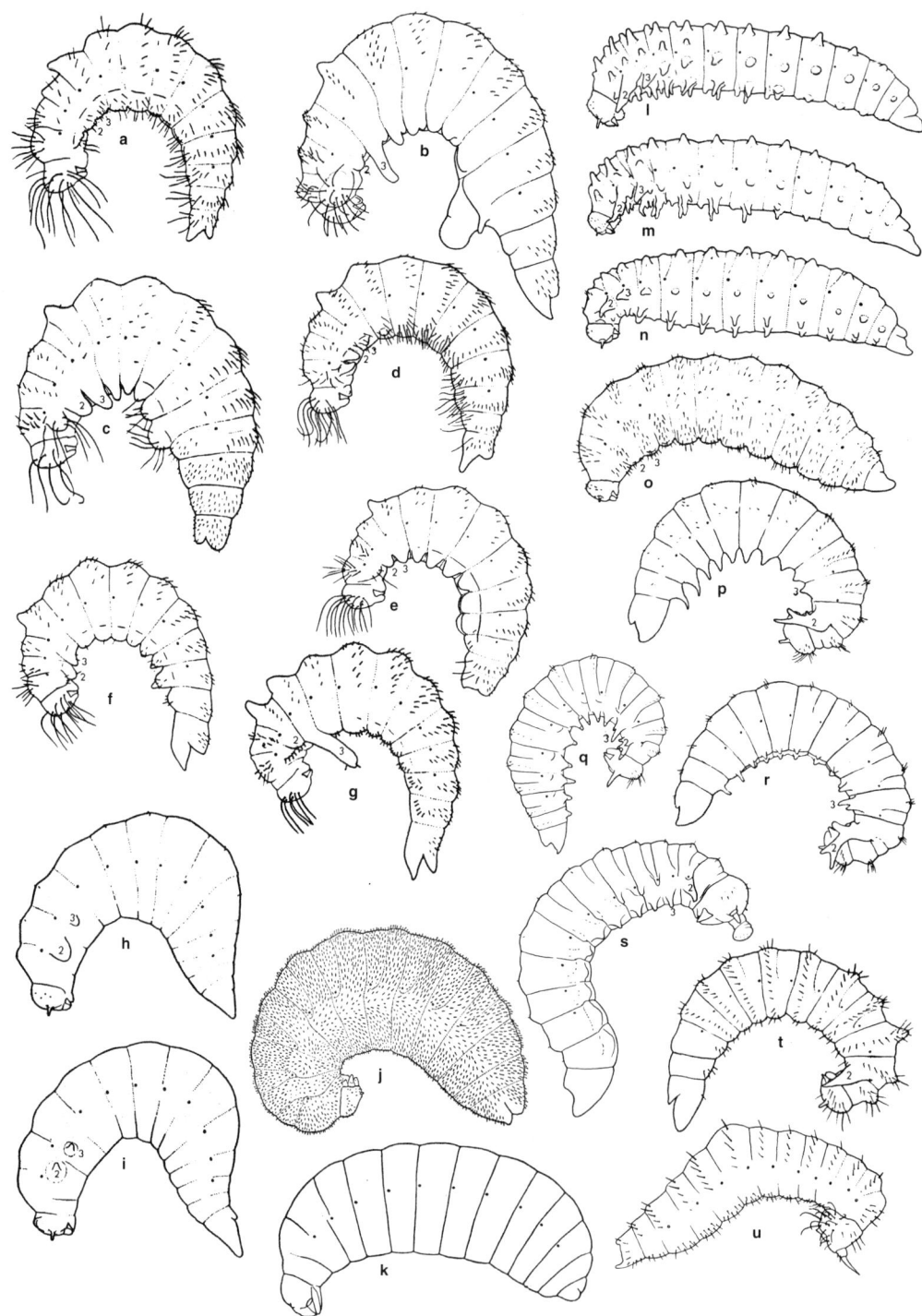

Figure 90-6. Mature larvae of Allodapini and Ceratinini. **a**, *Nasutapis straussorum* Michener; **b**, *Allodape mucronata* Smith; **c**, *Braunsapis trochanterata* (Gerstaecker); **d**, *B. natalica* Michener; **e**, *B. simplicipes* Michener; **f**, *B. foveata* (Smith); **g**, *B. leptozonia* (Vachal); **h**, *Compsomelissa (Compsomelissa) stigmoides* (Michener); **i**, *C. (Halterapis) nigrinervis* (Cameron); **j**, *Macrogalea candida* (Smith); **k**, *Ceratina dupla* Say; **l**, *Eucondylops konowi* Brauns; **m**, *Allodapula (Allodapula) melanopus* (Cameron), a close relative of the host of *E. konowi*; **n**, *A. (Dalloapula) acutigera* Cockerell; **o**, *A. (Allodapulodes) hessei* Michener; **p**, *Exoneura (Inquilina) excavata* Cockerell; **q**, *E. (Exoneura) obscuripes* Michener; **r**, *E. (E.) variabilis* Rayment; **s**, *E. (E.) subbaculifera* Rayment; **t**, *E. (Brevinura) concinnula* Cockerell; **u**, *Exoneurella lawsoni* (Rayment). All are Allodapini except k, which is Ceratinini. Allodapini facing to the left are African; those facing to the right are members of the *Exoneura* group from Australia. (For recognition of the segmental homologies among tubercles and pseudopods of the ventrolateral series, the second and third members of this series, when present, are numbered 2 and 3.) From Michener, 1977c.

E. bicolor there is division of labor into four classes of individuals, the males participating in two of these classes.

Although there are no known parasitic forms of Xylocopini or Ceratinini, several socially parasitic species are known or surmised in the Allodapini (Michener, 1961b, 1970, 1977b, 1983c; Reyes and Michener, 1990; Batra, Sakagami, and Maeta, 1993). The parasites have reduced scopa, shortened mouthparts, and frequently projections on the head as though for pushing hosts. They do not feed from flowers, so far as is known; they live more or less as colony members in nests of hosts. Batra et al. found that they not only eat from food stores placed in the nest by hosts, but solicit food from hosts and participate in many host activities such as grooming larvae, feeding larvae, and guarding at the nest entrance. In all known cases, the parasites and hosts are related, i.e., in the same genus or the same clade.

The parasites differ from their hosts to various degrees. Three parasitic species, *Macrogalea mombasae* Cockerell, *Braunsapis breviceps* (Cockerell), and *B. kaliago* Reyes and Sakagami, and four probable but little-known parasites, *Allodapula guillarmodi* Michener, *B. natalica* Michener and *pallida* Michener, and *Allodape greatheadi* Michener, are sufficiently similar to their host or probable host genera that they have not received generic names of their own. It has been the custom among melittologists to give parasites such generic names, however, in spite of similarity to host genera, as for example *Psithyrus* (here placed as a subgenus within *Bombus*) and *Paralictus* (here placed within *Dialictus*, a subgenus of *Lasioglossum*). In most cases, such parasitic taxa probably make the host genera paraphyletic. I continue to recognize parasitic and host genera as distinct when they are markedly different and lack intermediates, but when similar and differing principally in loss or reduction of the scopa and similar features, I relegate the parasitic taxon to synonymy or subgeneric status. Thus, as noted above, *Paralictus* is a synonym of *Dialictus* and *Psithyrus* is a subgenus of *Bombus*. For allodapines, rather arbitrarily I consider *Inquilina* a subgenus of *Exoneura*, but I recognize *Effractapis*, *Nasutapis*, and *Eucondylops* as genera. *Braunsapis breviceps* (Cockerell) and *kaliago* Reyes and Sakagami could also be placed in a separate genus or subgenus, for in numerous characters they differ from their nonparasitic relatives in the genus *Braunsapis*. The same can be said for *Braunapis bislensis* Michener and Borges. Contrarywise, persons so inclined could reasonably synonymize *Effractapis* and *Nasutapis* under *Braunsapis*, and *Allodapula* under *Eucondylops*, to eliminate probable paraphyly of the nonparasitic genera.

An old and informal phylogenetic study of the genera of Allodapini was by Michener (1977c). A recent phylogenetic study by Reyes (1998) was based on nine larval characters, ten characters of adult morphology, and three characters of adult behavior or life history. Both of these authors considered *Compsomelissa* in the present broad sense (i.e., as those species that use partial or complete mass provisioning of larvae) as the sister group to all other Allodapini. The Allodapini other than *Compsomelissa* were then shown by Reyes to consist of two clades. One was the Australian genera plus the Near Eastern *Exoneuridia* and the African *Macrogalea*; the other consisted of the African *Allodapula* and its relatives that lay small eggs, plus the large-egged *Allodape* (Africa) and *Braunsapis* (widespread). Similarity to the cell series of mass provisioned larvae found in many bees (such as the Ceratinini) seemed to support the idea (Michener, 1977c; Reyes, 1998) that *Compsomelissa* is the sister group to all other Allodapini, conserving as a plesiomorphy the behavior that results in series of young with the oldest at the bottom of the nest and youngest near the nest entrance. See *C. (Halterapis) nigrinervis* (Cameron) (Michener, 1971a). More recent phylogenetic studies (Schwarz, Bull, and Cooper, 2005) show, however, that *Macrogalea* is the sister to all the rest, and Engel (2005) has even provided a subtribal name, Macrogaleina, for *Macrogalea*. It seems that the sequence of stages in nests of *C. nigrinervis*, with mass provisioned larvae, is an apomorphy derived from the behavior of progressive feeders. Viewed in this light, laying of eggs before provisioning and firmly attaching the eggs to the nest wall, as does *C. nigrinervis*, are features not found in ordinary bees like *Ceratina* and are supportive of a relationship between *Allodapula* and *Composmelissa*.

Reyes (1998) was not always certain whether a social parasite is derived from within a nonparasitic genus, or is a sister group to such a genus. Regardless of phylogenetic details, it is clear that larvae provide important information about diversity and phylogeny among genera with similar adults, because they have changed but little even when adults diverge strikingly as parasites.

Some of the nonparasitic genera of Allodapini are so similar in adult characters that they might not have been recognized as genera in the absence of larval characters. Particularly *Compsomelissa (Halterapis) nigrinervis* (Cameron) and its relatives would have been included in *Braunapis* if nests and male genitalia were unknown. I have therefore given a key to genera based on larvae, in addition to that based on adults.

Key to the Genera of the Allodapini Based on Adults

1. Body rather robust, head and thorax hairy, female with transverse fasciae of short, appressed plumose hairs on T2 to T5; stigma slender (Fig. 90-12b), basal part parallel-sided; prestigma about as long as distance from base of stigma to base of vein r; jugal lobe of hind wing enormous (Fig. 90-12b), almost as long as vannal lobe (Africa) .... *Macrogalea*
—. Body slender, inconspicuously hairy and lacking metasomal fasciae of plumose hairs; stigma broad, not parallel-sided (Figs. 28-4, 90-11a-c, 90-13); prestigma usually half as long as distance from base of stigma to base of vein r; jugal lobe of hind wing of ordinary size, extending little if any beyond vein cu-v [except in *Braunsapis trochanterata* (Gerstaecker)] 2
2(1). Labial palpus with apical segments not diverging from axis of first two segments; maxillary palpus three- to five-segmented (proboscis short; no functional scopa, no midfemoral brush in female) 3
—. Labial palpus with apical segments small and in ordinary position sharply divergent from axis of first two segments; maxillary palpus five- to six-segmented 5
3(2). Labial palpus three-segmented; frons and apex of clypeus without projections (Madagascar) .... *Effractapis*

—. Labial palpus four-segmented; frons or apex of clypeus with projection 4
4(3). Second recurrent vein absent (Figs. 90-10b, 90-13d); facial projection on frons, bilobed or paired (Africa)...... .... *Eucondylops*
—. Second recurrent vein present (Fig. 90-13c); facial projection on apex of clypeus, snoutlike, not bilobed or paired (Africa) .... *Nasutapis*
5(2). Second recurrent vein absent 6
—. Second recurrent vein present 9
6(5). Malar space nearly as long as broad; T5 and T6 of female each with sharp, lateral lamella or strong carina (southwestern Asia) .... *Exoneuridia*
—. Malar space much wider than long; T5 and T6 without lateral lamellate carinae, at most T5 with lateral part of gradulus as on preceding terga 7
7(6). Maxillary palpus six-segmented; thorax and metasoma with yellow or white markings (Africa, Arabia).............. .... *Compsomelissa* (in part)
—. Maxillary palpus five-segmented; thorax and metasoma without yellow or white markings except sometimes on pronotum, tegula, and axillary sclerites, and rarely bands on metasomal terga 8
8(7). Costal margin of second submarginal cell short, often less than half as long as first submarginal crossvein [except almost as long as that crossvein in *E. tridentata* (Houston)]; basitibial plate not indicated; T6 of female with apex strongly produced, upturned, bidentate or tridentate, not at all hidden by pubescence (Australia) .... *Exoneurella*
—. Costal margin of second submarginal cell usually almost as long as first submarginal crossvein; basitibial plate of female usually indicated by carina strongly raised across apex of plate; T6 of female with apex little produced, not or weakly upturned, not or minutely bidentate, not tridentate, and usually hidden by pubescence denser at apex of tergum than elsewhere (Australia) .... *Exoneura*
9(5). T6 of female with lateral parts bent under rather abruptly, usually a rather abrupt change in vestiture along this bend, dorsal part of tergum rather flat or even concave [*Compsomelissa* (*Halterapis*) *nigrinervis* (Cameron) shows these features only weakly]; apex of T6 of female without minute, median, apical pygidial plate, or plate completely covered by hairs; gonostylus of male much reduced or slender, if thin and flattened, then more than three times as long as broad 10
—. T6 of female with lateral parts bent under in a broadly rounded way (i.e., bend broad in transverse section), change in vestiture at bend gradual, dorsal part of tergum convex [exceptions are in those species in which T6 is laterally projecting or angulate scoop-shaped: *Allodape mucronata* Smith, *Braunsapis angolensis* (Cockerell), *associata* (Michener), and *paradoxa* (Strand), etc.]; apex of T6 of female usually with minute, median, apical, elevated spine or pygidial plate (absent in the species with modified T6 listed above); gonostylus of male broad, flattened, concave on inner surfaces, flaplike, less than twice as long as broad 11
10(9). Ventroapical plate across apex of male gonocoxite with long, apically directed projection arising mesally; upper margin of clypeus often strongly concave between subantennal sutures (Africa, Arabia).............................. .... *Compsomelissa* (in part)

—. Ventroapical plate across apex of male gonocoxite without projection from mesal portion; upper margin of clypeus not or less strongly concave (Africa) .... *Allodapula*
11(9). Outer surface of tibia of female, on distal half, with brush of dense, usually coarse hairs, commonly coarser or of different form and color than hairs of adjacent areas, and indications of brush present in most males; distance from first submarginal crossvein to first recurrent vein much greater than from second recurrent to second submarginal crossvein in most species; yellow along inner and outer orbits of most females (Africa)......... *Allodape*
—. Outer surface of tibia of both sexes without dense brush [except in *Braunsapis trochanterata* (Gerstaecker)]; distance from first submarginal crossvein to first recurrent vein subequal to that from second recurrent to second submarginal crossvein in most species; yellow along inner and outer orbits absent (except in female *B. aureoscopa* Michener) .... *Braunsapis*

Key to Genera of Allodapini Based on Mature Larvae
(Illustrated in Figure 90-6; larvae of *Effractapis* and *Exoneuridia* are unknown)

1. Antenna tapering to attenuate, acute point; ventrolateral area of head swollen, tending to hang down on either side of mouthparts, bearing large curved hairs; body sharply bent in middle rather than uniformly curved (Australia) .... *Exoneurella*
—. Antenna blunt or almost absent; lateral area of head, *if* enlarged and with large hairs, then dorsolateral, not lobed down on either side of mouthparts; body uniformly curved or rather straight 2
2(1). Head or lateral protuberances of first body segment with enormous curved hairs; antenna represented by feeble convexity... 3
—. Anterior end of body without large hairs; head hairs, if present, usually not much if any longer than those found elsewhere on body; antenna strongly projecting.............. 4
3(2). Large, curved hairs on upper and lateral parts of head 10 to 18 or more; first ventrolateral projection of body without such hairs [except that *Braunsapis trochanterata* (Gerstaecker) has three to five]; ventral or ventrolateral parts of body with hairs (limited to first and last two segments in *B. simplicipes* Michener) .... *Braunsapis, Nasutapis*
—. Large, curved hairs on head eight or less, on vertex; first ventrolateral projection of body with seven to about 20 much larger hairs; ventral and ventrolateral parts of body hairless except for first ventrolateral projections (Africa) .... *Allodape*
4(2). Body curled, covered all over with short hairs, some of those of dorsum hooked (Africa) .... *Macrogalea*
—. Body naked or nearly so, or, *if* hairy, then nearly straight and without hooked hairs 5
5(4). Body curled, hairless or nearly so, with lateral projections corresponding to second and third body segments but otherwise without tubercles; maxilla and labium without projecting sensory areas (Africa, Arabia) .... *Compsomelissa*
—. Body rather straight, hairy or multituberculate; maxilla and labium with projecting sensory areas representing palpi 6
6(5). Australian .... *Exoneura*
—. African .... *Allodapula, Eucondylops*

Genus *Allodape* Lepeletier and Serville

Allodape Lepeletier and Serville, 1825: 18. Type species: *Allodape rufogastra* Lepeletier and Serville, 1825, monobasic.
Allodapa Schulz, 1906: 244, unjustified emendation of *Allodape* Lepeletier and Serville, 1825.

Allodape differs from its close relative, *Braunsapis*, in its commonly larger size (6-13 mm long); the presence in both sexes of most species of paraocular and genal yellow or white marks and of the tibial brushes mentioned in the key to genera; and in the male, in having a strongly sclerotized hairless process of the ventroapical plate (?lower gonostylus) mesad of the gonostylus, in lacking the peglike setae on the ventroapical plate (Fig. 90-9c), and in the presence of a thickening and angle at the base of the hind femur, next to the apex of the trochanter. The male genitalia and other structures were illustrated by Michener (1975b); see also Figure 90-9c-e.

■ This genus occurs from southernmost Africa north to Tanzania, Uganda, Central African Republic, and west to Guinea (Conakry). The approximately 30 species were revised by Michener (1975b).

One species, *Allodape greatheadi* Michener (1970), known from one specimen, has a reduced scopa and is presumably parasitic. The male of *Allodape mirabilis* Schulz has strange modifications of ordinary apid structures, some of which are illustrated in Figure 90-7. A nest and immature stages are shown in Figure 90-5c.

Genus *Allodapula* Cockerell

Black-bodied species of this genus, such as *A. acutigera* Cockerell exactly resemble species of *Braunsapis* in appearance, having similar ivory- colored facial marks and pronotal lobe. Most species of *Allodapula*, however, are slightly more slender and have a red or partly red metasoma, an unusual feature in *Braunsapis* and one not found in *Braunsapis* species occurring within the range of *Allodapula*. As shown by the male genitalia, *Allodapula* is not particularly close to *Braunsapis*; for example, *Allodapula* lacks the broad, thin, often scoop-shaped male gonostylus and the peglike setae on the ventroapical plate found in *Braunsapis*. The male genitalia and other structures were illustrated by Michener (1975b); see also Figure 90-8a-f.

The larvae of *Allodapula* are relatively straight (Fig. 90-6l-o) and lack the wide head and large head hairs of *Braunsapis*, and batches of larvae of about the same age (Fig. 90-5d) are fed from common food masses. In *Braunsapis* there are no groups of larvae of similar age, since the large eggs are laid at long intervals and larvae are

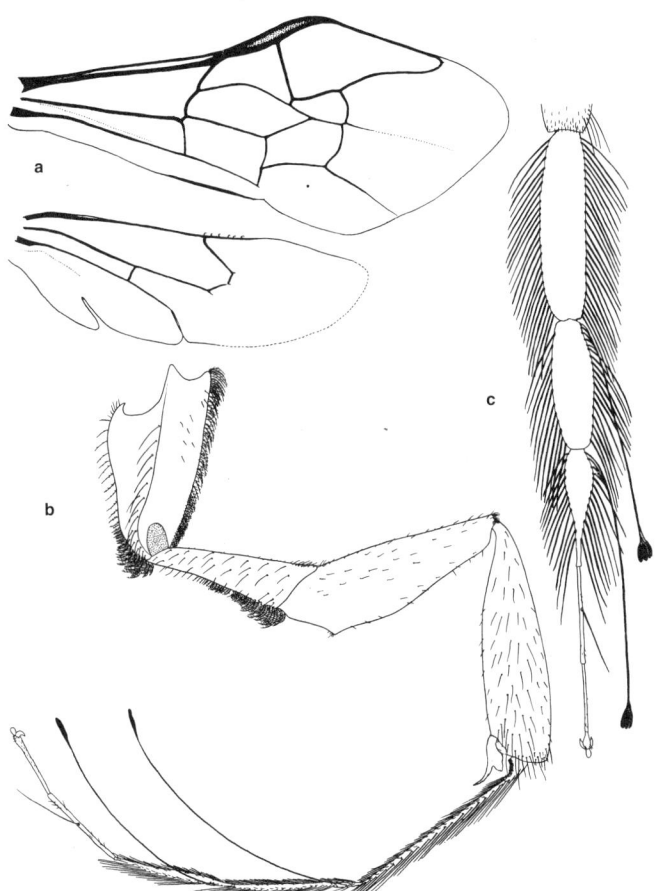

Figure 90-7. Extraordinary male structures of *Allodape mirabilis* Schulz. **a,** Wings; **b,** Foreleg; **c,** Front tarsus from above. Females do not share these strange features. From Michener, 1977d.

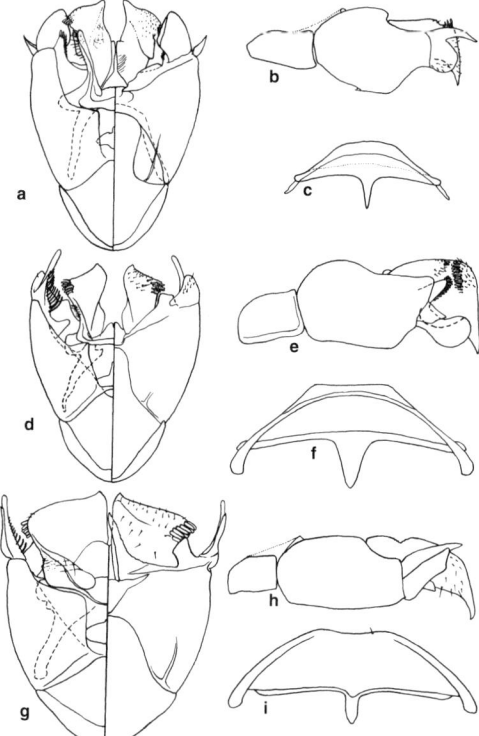

Figure 90-8. Apical metasomal structures of male Allodapini. **a-c,** Genitalia, in dorsal (at left), ventral, and lateral views, and S7 and S8 taken together, of *Allodapula (Dalloapula) dichroa* (Strand); **d-f,** Same structures, *A. (Allodapulodes) xerica* Michener; **g-i,** Same structures, *Compsomelissa (Halterapis) nigrinervis* (Cameron). From Michener, 1975b.

fed individually. Larvae of *Allodapula* were illustrated by Michener (1975c).

Allodapula was revised by Michener (1975b). Since that revision did not take into account the subgenera, the following listing may be useful: *Allodapulodes: hessei* Michener, *maculithorax* Michener, *palliceps* (Friese), *xerica* Michener. *Dalloapula: acutigera* Cockerell, *dichroa* (Strand). *Allodapula* s. str.: all other species.

Key to the Subgenera of *Allodapula* Based on Adult Males

1. Gonostylus of male minute, stylus-shaped (Fig. 90-8a), with short, slender hairs; penis valve and ventroapical plate much exceeding gonostylus; roof of genital chamber weakly convex, not or scarcely folded, each half over twice as long as broad .... *A. (Dalloapula)*
—. Gonostylus variously shaped, not styluslike, with hairs as long as width of gonostylus and often robust (Fig. 90-8d); penis valve and projections from gonocoxite not exceeding gonostylus; convexities of roof of genital chamber strong, about as in *Braunsapis,* each half broad, folded or not 2
2(1). Gonobase reduced, consisting of narrow sclerotized bands; penis valves slender, only weakly downcurved, without coarse setae .... *A. (Allodapula* s. str.*)*
—. Gonobase of ordinary form; penis valves robust, strongly downcurved, with cluster of coarse setae on outer angle (Fig. 90-8d) .... *A. (Allodapulodes)*

Key to the Subgenera of *Allodapula* Based on Mature Larvae

1. Body without conspicuous tubercles, with scattered hairs (Fig. 90-6o) .... *A. (Allodapulodes)*
—. Body with several longitudinal rows of conspicuous tubercles or projections (Fig. 90-6l-n); hairs few and minute or absent 2
2(1). Labium with median apical projection; maxilla not bulbous; projections of body without basal shoulders or branches (Fig. 90-6n) .... *A. (Dalloapula)*
—. Labium with median apical emargination; maxilla bulbous; several anterior projections of lateral and ventrolateral series with basal shoulders or branches (Fig. 90-6l, m) .... *A. (Allodapula* s. str.*)* and genus *Eucondylops*

Allodapula / Subgenus *Allodapula* Cockerell s. str.

Allodape (Allodapula) Cockerell, 1934c: 220. Type species: *Allodape variegata* Smith, 1854, by original designation.

This subgenus consists of relatively slender species, 4.5 to 7.5 mm long, with a red or rarely reddish-black metasoma and nearly always a yellow band or spot on the scutellum.

■ *Allodapula* s. str. is widespread in South Africa, and occurs as far north as Zimbabwe, with a doubtful record for Tanzania. The nine species were included in the revision by Michener (1975b).

Allodapula guillarmodi Michener was based on a female with reduced scopa, presumably a social parasite in nests of another species.

Allodapula / Subgenus *Allodapulodes* Michener

Allodapula (Allodapulodes) Michener, 1969b: 291. Type species: *Allodape palliceps* Friese, 1924, by original designation.

Like *Dalloapula,* this subgenus includes rather robust species, 5 to 7 mm long. The metasoma is black to red, and the scutellum sometimes has a cream-colored or yellow spot. In males of some species the legs are wholly yellow.

■ This subgenus is known only from Cape Province, South Africa. The four or five species were included in the revision by Michener (1975b).

Allodapula / Subgenus *Dalloapula* Michener

Allodapula (Dalloapula) Michener, 1975c: 246. Type species: *Allodape dichroa* Strand, 1915, by original designation.

This subgenus consists of somewhat robust species 4.5 to 9.0 mm long, with the metasoma black or partly red, and without yellow on the scutellum.

■ This subgenus occurs in Cape Province and Natal, South Africa. The two species were included in the revision by Michener (1975b).

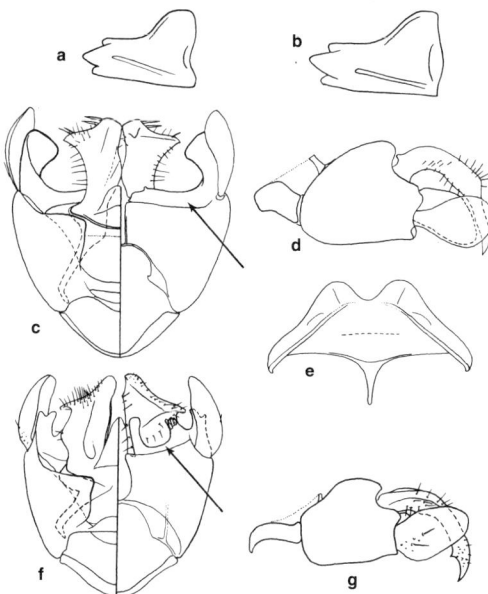

Figure 90-9. Structures of Allodapini. **a, b,** Mandibles of male and female of *Allodape obscuripennis* Strand. **c, d,** Male genitalia of *Allodape obscuripennis* Strand, in dorsal (at left), ventral, and lateral views; **e,** S7 and S8 of male of *A. mucronata* Smith; **f, g,** Male genitalia of *Braunsapis leptozonia* (Vachal), in dorsal (at left), ventral, and lateral views. The arrows indicate the ventroapical plate, which frequently supports peglike setae, as in f. The ventroapical plate appears to be a mesad extension of the lower gonostylus. From Michener, 1975b.

Genus *Braunsapis* Michener

Allodape (Braunsapis) Michener, 1969b: 290. Type species: *Allodape facialis* Gerstaecker, 1857, by original designation.

Most species of *Braunsapis* are black with ivory or yellow marks on the face and often on the pronotal lobes and tarsi. Females lack such marks on the paraocular and genal areas except for strong marks in *B. aureoscopa* Michener, and occasionally very small paraocular marks, e.g., in *B. maculata* Reyes. Rarely the metasoma is red, and in *B. pallida* Michener the body is largely testaceous. The size is commonly smaller than that of *Allodape*, body length being 3 to 9 mm. The gonostylus of the male is flat, usually broad (Fig. 90-9f, g), and there is no projecting lower gonostylus or process mesal to it, as in *Compsomelissa* and *Allodape*. Except in *B. aureoscopa* there are one to several peglike setae on the ventroapical plate (?lower gonostylus) (Fig. 90-9f). The hind femur of the male is rather slender, not angulate at its extreme base, but sometimes with a more distal angle. The male genitalia and other structures were illustrated by Michener (1975b, 1977b) and Reyes (1991a, c, 1993, 1998); see also Figure 90-9f, g.

■ *Braunsapis* ranges from southernmost Africa north to Senegal, Ethiopia, and Madagascar and eastward from Pakistan, India, and Sri Lanka to southernmost China, Taiwan, the Philippines, the Solomon Islands, and Australia south as far as New South Wales and South Australia. This is by far the largest genus of the Allodapini. There are well over 45 species in Africa and Madagascar, 19 in the oriental region, and 23 in the Australian region. The African figure is clearly low, for unplaceable specimens are numerous, and many names thought to be synonymous probably represent recognizable species. Revisions were by Michener (1975b) for Africa, and by Reyes (1991a, b, 1993, 1998) for Madagascar, the oriental region, and the Australian region. Australian species were listed by Cardale (1993).

Studies of nesting behavior (Figs. 90-4) and social behavior were presented by Michener (1962b, 1971a) on Austrralian and African species and in a series of papers on a Taiwan species, of which one is Maeta, Sakagami, and Michener (1992). Social behavior with two or occasionally more females working in a single nest is common. As indicated in the discussion of the tribe, several species of *Braunsapis* are known or probable social parasites of other species of *Braunsapis*; behavioral information on *B. breviceps* (Cockerell) and *kaliago* Reyes and Sakagami was provided by Reyes and Michener (1990) and Batra, Sakagami, and Maeta (1993). In addition social parasites probably derived from *Braunsapis* constitute the genera *Effractapis* and *Nasutapis*. Limited biological information on the latter was presented by Michener (1971a).

Genus *Compsomelissa* Alfken

This genus is here used in a broad sense to include species formerly placed in *Halterapis*. The mature larvae, robust and nearly hairless, have weak lateral tubercles on two thoracic segments, but otherwise lack tubercles (Fig. 90-6h, i). The male gonostylus consists of two elements, the upper and lower gonostyli, appressed together (Fig. 90-8g, h). They are rather slender, or, if broad, are broadly attached to the gonocoxite. The structure that I call the ventral gonostylus with some confidence is continuous with the ventroapical plate, which is probably part of the ventral gonostylus. It is produced as a long mesal process parallel to and beneath the penis valve. The male genitalia and other structures were illustrated by Michener (1975b) and Reyes and Michener (1992); see also Figure 90-8g-i).

The two subgenera are weakly separated on adult and larval characters, although the manner of feeding larvae differs, being by mass provisioning in the subgenus *Halterapis* and by progressive feeding of young larvae and mass provisioning of older larvae in the subgenus *Compsomelissa* s. str., so far as is known.

Key to the Subgenera of *Compsomelissa*

1. All tagmata with some yellow or white integument, or integument largely pale, but metasoma sometimes red .... *C. (Compsomelissa* s. str.)
—. Pale areas limited to head, pronotal lobe and legs, but metasoma sometimes red .... *C. (Halterapis)*

Compsomelissa / Subgenus *Compsomelissa* Alfken s. str.

Compsomelissa Alfken, 1924a: 251. Type species: *Compsomelissa borneri* Alkfen, 1924, monobasic.
Exoneurula Michener, 1966b: 573. Type species: *Exoneurula*

zavattarii Michener, 1966 = *Compsomelissa borneri* Alfken, 1924, by original designation.

This subgenus consists of small (length 4-6 mm) species with yellow or whitish markings on all tagmata, sometimes with nearly the whole body yellow or with the metasoma largely red. In most species the second recurrent vein is absent (Fig. 90-13a), as in *Exoneura* and *Exoneuridia,* but in one largely yellow species, *Compsomelissa zaxantha* (Cockerell), this vein is present.

■ This subgenus occurs from Cape Province and Natal, South Africa, north to Ethiopia and west to Nigeria, and east to Yemen. There are six species. The five African species were revised by Michener (1975b).

Although larval structure is essentially the same as that in the subgenus *Halterapis,* at least the younger larvae of *C. (C.) borneri* Alfken are fed progressively and several eggs may be found in a nest, instead of only one or two with their large masses of food as in *Halterapis.*

Compsomelissa / Subgenus *Halterapis* Michener

Halterapis Michener, 1969b: 289. Type species: *Allodape nigrinervis* Cameron, 1905, by original designation. [New status.]

This subgenus consists of species superficially similar to common species of *Braunsapis,* the body being black (to red brown in Madagascar), the face and sometimes genal area, pronotal lobe, and parts of legs yellow or whitish; sometimes the metasoma is red. The second recurrent vein, present in all species (Fig. 90-13b), is a character that distinguishes this subgenus from all but one species of *Compsomelissa* s. str. The African species have body lengths of 3.5 to 5.0 mm; Malagasy species are 5 to 8 mm long.

■ This subgenus occurs in South Africa from Cape Province to Natal, north to Zimbabwe, and in Madagascar. The three or four African species and seven Malagasy species were revised by Michener (1975b) for Africa and by Michener (1977b, supplemented by 1992b) for Madagascar; Brooks and Pauly (in Pauly et al., 2001) placed a total of 18 species from Madagascar in *Halterapis.* Larvae remain unknown for nearly all Madagascar species, and males are unknown for most, so that the generic status of most of these species remains in doubt. To judge by the illustrations of male genitalia, some of the species have the mesal projection of the ventroapical plate of the gonocoxite as in *Compsomelissa.* Other species lack such a projection and therefore fall in another genus or indicate that the understanding of *Compsomelissa (Halterapis)* must be modified. Perhaps they should be placed in *Allodapula.* The discovery of mature larvae, as well as additional males, together with more adequate descriptions, should make clarification of the problem possible.

Several of the Malagasy species were placed in *Allodapula* by Michener (1977b) on the basis of adult females only. A single male indicated that its species belongs in *Halterapis* (Reyes and Michener, 1992).

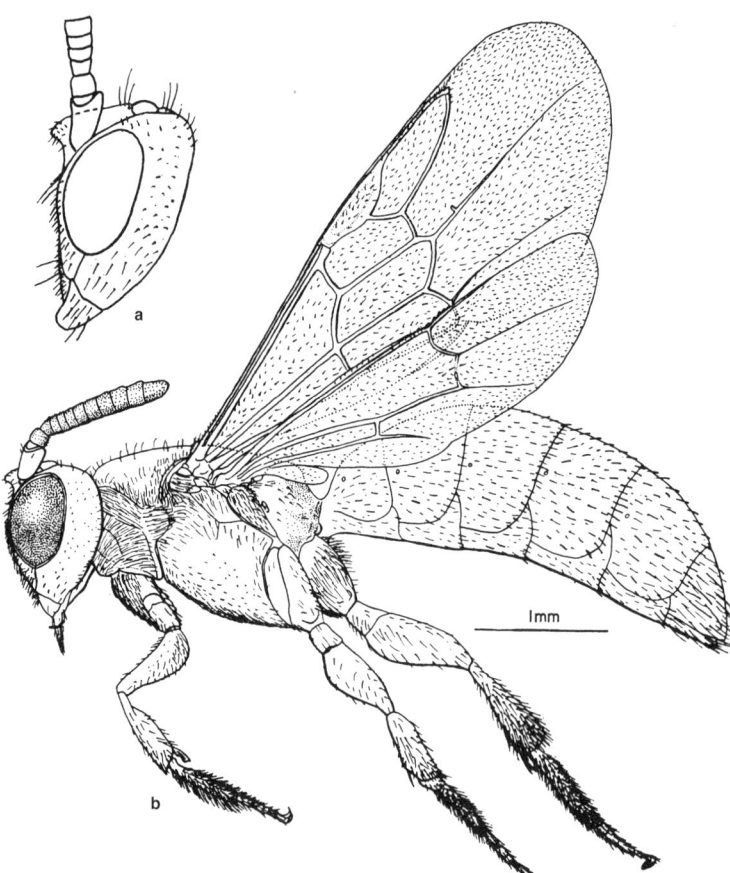

Figure 90-10. *Eucondylops reducta* Michener, female. **a,** Head; **b,** Lateral view of entire insect. From Michener, 1970.

At least in *Compsomelissa (Halterapis) nigrinervis* (Cameron) and probably in *C. (H.) angustula* (Cockerell), larvae are fully and individually mass provisioned, as described by Michener (1971a). This practice differs from that of all other allodapine bees. One collection of nests of *C.(H.) minuta* (Brooks and Pauly) in Madagascar revealed an extraordinary, and so far as known, unique mode of feeding larvae. Three to eleven eggs are glued to the wall of the nest burrow, near the bottom of the nest. Then a firm pollen-honey mixture is provided, covering the eggs and extending up the burrow wall toward the nest entrance. The larvae move slowly upward, consuming the common food mass as they go (Schwarz et al., 2005). It is not certain whether food is added to this mass by the one to several adult females in the nest, as the larvae feed.

Genus *Effractapis* Michener

Effractapis Michener, 1977b: 6. Type species: *Effractapis furax* Michener, 1977, by original designation.

Effractapis is the generic name used for a small species (length 4 mm) no doubt derived from *Braunsapis* and presumably parasitic on species of that genus. Interesting characters in addition to those indicated in the key to genera are the entirely black face of both sexes, the gently convex whole face, and the presence of two setae, longer than the scape, extending forward from near the upper clypeal margin, and of a few long hairs on the front tibiae and basitarsi. Male genitalia and other structures were illustrated by Michener (1977b).

■ This genus occurs in Madagascar. The only known species is *Effractapis furax* Michener.

Genus *Eucondylops* Brauns

Eucondylops Brauns, 1902: 377. Type species: *Eucondylops konowi* Brauns, 1902, monobasic.

Eucondylops consists of elongate (5.5-7.0 mm), dark brown or blackish parasitic species that differ from nearly all other Allodapini in lacking pale markings on the face in both sexes, a character also found in another social parasite, *Effractapis furax* Michener. The head is unusually small, the eyes too are small, and the malar area is therefore large (Fig. 90-10); a bilobed projection or two projections arise from the upper part of the frons. The reduction of the proboscis is extreme, the mentum being a small, weakly sclerotized structure, wider than long, and the lorum being a transverse bar, not V- or Y-shaped as in all other L-T bees. One species, *Eucondylops reducta* Michener, has only one submarginal cell (Fig. 90-10b), an unusual feature among bees, but the type species has two submarginal cells (Fig. 90-13c). Male gonostyli are absent; penis valves are slender and nearly straight, not curved strongly downward. Male genitalia were illustrated by Michener (1975b).

■ This genus occurs in South Africa in nests of *Allodapula*. The two species were revised by Michener (1970). Although larvae (Fig. 90-6l) look like those of *Allodapula* s. str., suggesting that *Eucondylops* is derived from that subgenus, the adult structure is so extraordinary that I do not hesitate to regard *Eucondylops* as a genus, even though the result may be a paraphyletic *Allodapula*.

Genus *Exoneura* Smith

Exoneura differs from most other Allodapini in lacking the second recurrent vein (Fig. 90-11a-b). In most species the metasoma is red, but some are black; because most have limited yellowish marks on the face, sometimes also the pronotal lobe, those with a black metasoma superficially resemble the common species of *Braunsapis*. The gonostyli of males are absent or at least unrecognizable (as in Fig. 90-11d). The larvae (Fig. 90-6q-t) tend to have a pair of large and often branched thoracic appendages.

Exoneura is the common allodapine genus in the temperate parts of Australia and does not occur on other continents. Its nesting biology has been discussed by various authors, e.g., Michener (1965a) and a series of papers by M. P. Schwarz and coauthors, of which Schwarz and O'Keefe (1991) and Silberbauer and Schwarz (1995) are examples. The subgenus *Inquilina* consists of social parasites in nests of *Exoneura* s. str.

Key to the Subgenera of *Exoneura*

1. Female with scopa reduced to mass of short, dense hairs; mandible of female with ventral tooth reduced to mere

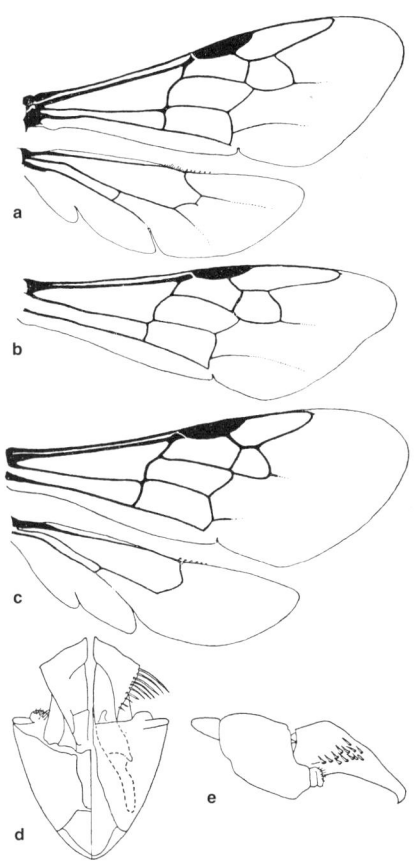

Figure 90-11. Australian Allodapini. **a,** Wings of female *Exoneura (Exoneura) bicolor* Smith; **b,** Forewing of male of same species; **c,** Wings of *Exoneurella lawsoni* (Rayment); **d, e,** Male genitalia of *Exoneurella lawsoni* (Rayment), in dorsal (at left), ventral, and lateral views, showing the greatly reduced gonostyli. From Michener, 1965b.

convexity so that mandible is bidentate; face concave; middle and hind tibial spurs robust, in female strongly curved apically, less than one-half as long as basitarsi; anterior margin of pronotum in both sexes with a sharp angle or small spine at level of base of fore coxa .... *E. (Inquilina)*
—. Female with ordinary tibial scopa; mandible of female three-toothed (as in Fig. 90-9b); face not noticeably concave; middle and hind tibial spurs slender, in female gently curved, one-half as long as basitarsi or longer; anterior margin of pronotum with an inconspicuous rounded angle at level of base of fore coxa 2
2(1). Vein Cu_1 of forewing usually long and gradually tapering (Fig. 90-11a, b); clypeus usually widely separated from antennal sockets, subantennal suture thus about as long as diameter of socket; fore tarsus of male attenuate, first two segments taken together longer than tibia; hind basitarsus of male dilated (scarcely so in species such as *E. asimillima* Rayment), nearly as wide as tibia (except in *E. asimillima*), not parallel-sided .... *E. (Exoneura s. str.)*
—. Vein Cu_1 of forewing a short stub, ending abruptly (as in Fig. 90-11c), although in some cases continued as a pigmented line; clypeus usually closely approaching antennal sockets, subantennal suture thus much shorter than diameter of socket; fore tarsus of male of ordinary form, first two segments taken together shorter than or as long as tibia; hind basitarsus of male much more slender than tibia, parallel-sided .... *E. (Brevineura)*

Exoneura / Subgenus *Brevineura* Michener

Exoneura (Brevineura) Michener, 1965b: 224. Type species: *Exoneura concinnula* Cockerell, 1913, by original designation.

Brevineura includes mostly smaller species (body length 3.5-7.0 mm), none of which have the enlarged eyes and associated features found in most males of *Exoneura* s. str.

■ The distribution is Australia, north to the latitude of southern Queensland. The 26 species names were listed by Michener (1965b) and Cardale (1993).

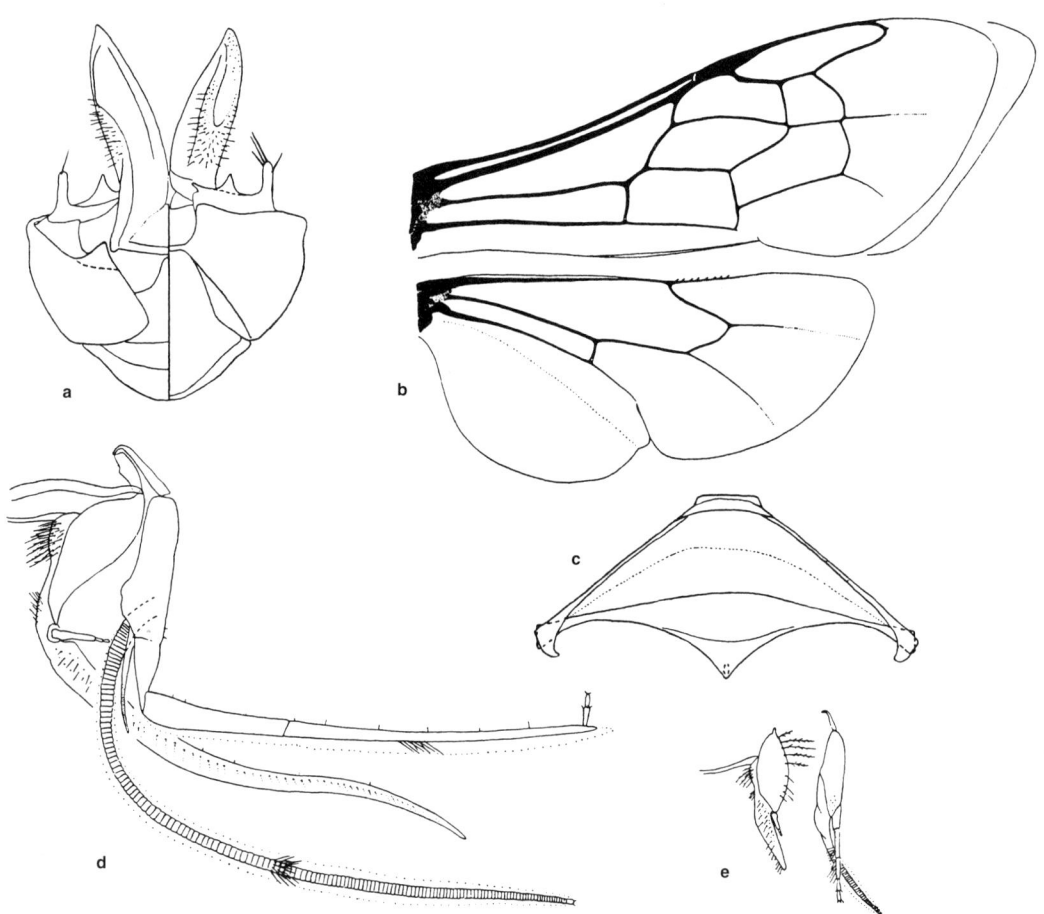

Figure 90-12. Structures of *Macrogalea candida* Smith. **a,** Dorsal (at left) and ventral views of male genitalia; **b,** Wings of female and (at right) the outline of forewing of male; **c,** S7 and S8 of male; **d,** Lateral view of proboscis of female; **e,** Proboscis of male (cardo and lorum omitted) to same scale. From Michener, 1971b.

Exoneura / Subgenus *Exoneura* Smith s. str.

Exoneura Smith, 1854: 232. Type species: *Exoneura bicolor* Smith, 1854, monobasic.

This subgenus contains the larger species (body length 5-9 mm) of the genus, many of which have distinctive males that form swarms in the air and probably mate in or near such swarms. Such males have enlarged eyes, long dense clypeal hairs, short antennae, and relatively pointed wings. Sexual variation in wing shape, rare in bees, is shown in Figure 90-11a, b. The features of these males are also found in males of *Macrogalea*. In other species, however, these features are less developed or are absent, thus reducing the differences from *Brevineura* (see Michener, 1965b, for further discussion of these characters).
■ This subgenus occurs throughout southern Australia, north as far as southern Queensland, and south to Tasmania. Forty names have been proposed for species of this subgenus; they are listed by Michener (1965b) and Cardale (1993).

Exoneura / Subgenus *Inquilina* Michener

Inquilina Michener, 1961b: 179. Type species: *Exoneura excavata* Cockerell, 1922, by original designation.

Inquilina has hitherto been regarded as a genus, but since it is very similar to and an obvious derivative of *Exoneura* s. str., I prefer to regard it as at most a subgenus. Yellow facial marks are absent in both sexes. The body length is 6 to 8 mm.
■ This parasitic subgenus is found in eastern Australia from Victoria to southern Queensland, in nests of *Exoneura* s. str. There are two species, distinguished by Michener (1983c) and listed by Cardale (1993).

Lowe and Crozier (1997) sequenced the mitochondrial cytochrome b gene of the two species of *Inquilina* and of their two hosts, *Exoneura* s. str. The results strongly indicated the monophyly of *Inquilina*.

Genus *Exoneurella* Michener

Exoneurella Michener, 1963: 257. Type species: *Exoneura lawsoni* Rayment, 1946, by original designation.

Although larvae of this genus are very different from those of *Exoneura*, adults of black species are separated with difficulty from those of *Exoneura (Brevineura)*. In the absence of larvae, the species of *Exoneurella* would have been included in *Brevineura*. Certain species have ivory metasomal tergal bands, a feature not found in *Exoneura*. Numerous structures of all stages were illustrated by Michener (1965b) and by Houston (1976); see also Figure 90-11c-e. The body length ranges from 3.5 to 10.0 mm, but only the giant, probably nonflying morph females of one species, *Exoneurella tridentata* (Houston), exceed 5.5 mm.
■ This genus occurs in central and eastern Australia: Northern Territory to South Australia and Queensland to Victoria. There are four species, revised by Houston (1976).

Although I have sometimes considered *Exoneurella* to be a subgenus of *Exoneura* (Michener, 1965b), the totally different larval structure of all species (Michener, 1964a, Houston, 1976) leads me to reassert its generic distinctness (see Fig. 90-6u), a step also taken by Reyes (1998).

Three species are probably usually solitary, i.e., their nests usually contain a female and her progressively fed larvae (Michener, 1964a; Houston, 1977). The remaining species, *Exoneurella tridentata* (Houston), commonly

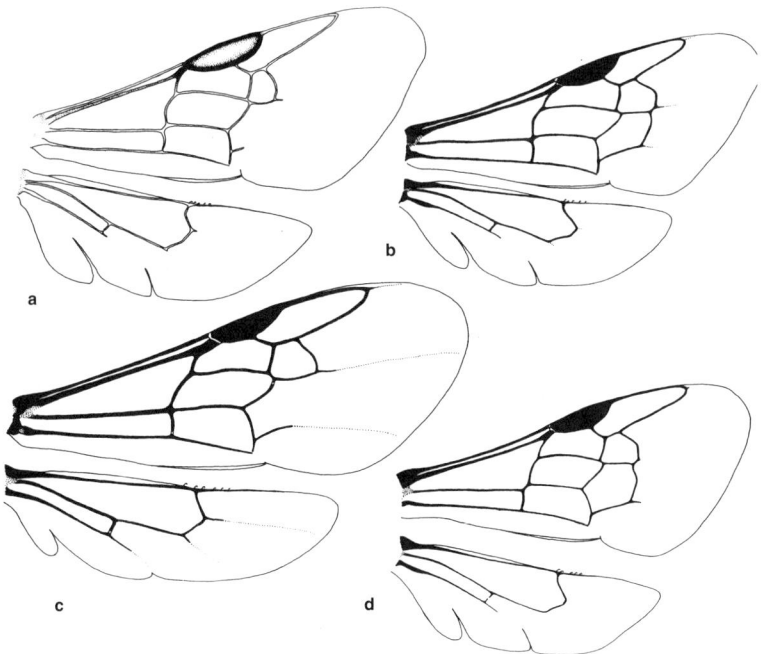

Figure 90-13. Wings of Allodapini. **a**, *Compsomelissa (Compsomelissa) stigmoides* (Michener); **b**, *C. (Halterapis) nigrinervis* (Cameron); **c**, *Eucondylops konowi* Brauns; **d**, *Nasutapis straussorum* Michener. From Michener, 1971b.

lives in small colonies and has large and small female morphs (Houston, 1977). Reyes and Schwarz (1997) gave an account of the phylogeny and behavior of the four species of *Exoneurella*, which differ greatly in their social levels. Reyes, Cooper, and Schwarz (2000) revisited the phylogeny and the relations of *Exoneurella* to similar genera.

Genus *Exoneuridia* Cockerell

Exoneura (Exoneuridia) Cockerell, 1911d: 232. Type species: *Exoneura libanensis* Friese, 1899, by original designation.

Like *Exoneura, Exoneurella,* and most species of *Compsomelissa* s. str., *Exoneuridia* lacks the second recurrent vein (as in Fig. 90-11a-c). The long malar space and the strong lateral carina on T5 and T6 of the female distinguish *Exoneuridia* from its relatives. The body length is 6 to 9 mm. Unfortunately, its larval characters remain unknown. It is the only palearctic genus of Allodapini.

Terzo (1999) has placed the three species of *Exoneuridia* in two subgenera. Although this seems like excessive splitting, the subgenera are quite different and I tentatively recognize them:

Key to the Subgenera of *Exoneuridia*

1. All tagmata richly marked with yellow; mandible of female with median tooth longer than others; maxillary palpus 5-segmented; gonobase of male nearly as long as midventral length of gonocoxite..............*E. (Alboneuridia)*
—. Body dark except for clypeus and limited pale marks elsewhere (thus coloration as in most *Braunsapis*); mandible of female with lower tooth longer than others; maxillary palpus 6-segmented; gonobase much shorter than midventral length of gonocoxite............................
.... *E. (Exoneuridia* s str.)

Exoneuridia / Subgenus *Alboneuridia* Terzo

Exoneuridia (Alboneuridia) Terzo, 1999: 147. Type species: *Allodape oriola* Warncke, 1979, by original designation.

The principal subgeneric characters are indicated in the key to the subgenera above.
■ This subgenus is known only from southwestern Iran. The only species is *Exoneuridia oriola* (Warncke); see Terzo (1999).

Exoneuridia / Subgenus *Exoneuridia* Cockerell s. str.

Exoneura (Exoneuridia) Cockerell, 1911: 232. Type species: *Exoneura libanesis* Friese, 1899, by original designation.

The principal subgeneric characters are indicated in the key above to subgenera.
■ *Exoneuridia* s. str is found in Turkey, Lebanon, Israel, and Syria. The two species were included in the revision by Terzo (1999).

Genus *Macrogalea* Cockerell

Macrogalea Cockerell, 1930f: 291. Type species: *Allodape candida* Smith, 1879, by original designation.

This is the only hairy, robust allodapine. Body length is 7.5 to 10.5 mm. The integument is blackish with a pale clypeal mark in most females. The female metasoma is flattened. The males, which are extremely rare in collections, have enlarged eyes, very short antennae, and extremely long, dense, erect hair on the clypeus and hypostomal areas; the male mandible tapers to the apex and has a small tooth on the upper margin; the male proboscis is one-half to one-third (depending on what structures are measured) the size of the female's (Fig. 90-12d, e) and does not project out of the proboscidial fossa. The maxillary palpus is three-segmented, the first segment being elongate. The male genitalia and numerous other structures were illustrated by Michener (1971b); see also Figure 90-12. As in *Exoneura*, the male upper gonostyli appear to be absent; the gonostylus-like projection is fused with the ventroapical plate and is presumably part of the lower gonostylus.
■ This genus occurs in Madagascar and in Africa from Ethiopia to Tanzania and Namibia. Like *Liotrigona* and certain other taxa, the genus *Macrogalea* is widespread in Africa but represented there by only one nonparasitic species (plus one known only from Zanzibar and one parasitic species), but it has speciated in Madagascar, where nine species are known (Brooks and Pauly, in Pauly et al., 2001). The genus was reviewed by Michener (1971b), with supplementary information by Michener (1977b).

Tierney et al. (2002) provided an account of nests of *Macrogalea zanzibarica* Michener and its frequently social organization. When other genera are also considered, implications exist for the monophyletic origin of social behavior in the ancestral Allodapini.

As shown by Michener (1971b), one species of *Macrogalea, M. mombasae* Cockerell, is a social parasite in nests of *M. candida* (Smith). The parasite shares features of various parasitic bees, such as a reduced scopa; see Michener (1970). Larvae of *Macrogalea*, unlike those of other Allodapini, have abundant short hairs, many of them hooked, all over the body (Fig. 90-6j) and lack the tubercles and large hairs found in other allodapines. Presumably, the short hairs serve to hold the larvae, which are typically curled, in their positions in vertical burrows in pithy stems.

Genus *Nasutapis* Michener

Nasutapis Michener, 1970: 208. Type species: *Nasutapis straussorum* Michener, 1970, by original designation.

Like *Effractapis, Nasutapis* is a probable parasitic derivative of *Braunsapis* and could well be included in that genus as was done by Reyes (1998); its larvae do not differ from those of *Braunsapis* by any known group characters, and its wing venation (Fig. 90-13d) is not different from that of small species of *Braunsapis*. Unusual characters of adults include a broad head, a snoutlike lower median clypeal projection, and the yellow clypeus and paraocular areas of the female, which are similar to those of the male. Body length is 4.5 to 5.5 mm.
■ *Nasutapis* is known from Natal Province, South Africa, in nests of *Braunsapis facialis* (Gerstaecker). The only species is *N. straussorum* Michener.

91. Subfamily Nomadinae

This subfamily consists of nomadiform and epeoliform cleptoparasitic bees, often resembling wasps rather than bees in habitus. As indicated in the account of the family Apidae, certain groups frequently included in the Nomadinae have been excluded here. So restricted, the monophyly of the subfamily is supported by the apical structure of S6 of the female, which is commonly armed with strong, blunt, spinelike bristles (Fig. 91-1), the sternum sometimes modified to a slender, bifid or even simple, pointed apex without strong bristles, or, in the Cacnoprosopidini, longitudinally split into two separate rods or plates. S6 of various taxa was illustrated by Grütte (1935), Linsley and Michener (1939), and Roig-Alsina (1991b); see Figures 91-2 and 95-3c, d. Additional adult characters considered by Roig-Alsina (1991b) as synapomorphies of the subfamily, although lost in some species, are (1) the lateral clypeal carina of the male along the lower epistomal suture and the smooth depression above it, and (2) a basiventral projection of the penis. The presence of a glandular pouch on each side of the common oviduct appears to be a unique synapomorphy of the subfamily (Alexander, 1996). Probably plesiomorphic characters of the subfamily include the distinct separation of the male gonostylus from the gonocoxite (except in the Caenoprosopidini, Fig. 101-2) and the strong apical process of S8. S7 of the male lacks apical lobes. Figures 91-3 and 91-4 illustrate some of the diversity in these structures.

A series of larval characters also separates the Nomadinae s. str. from other bees (Rozen, Eickwort, and Eickwort, 1978; Rozen, 1996a). A majority of these characters, however, are reductions probably associated with the fact that the larvae do not spin cocoons, whereas those of related groups such as the Protepeolini and nearly all other Apinae do spin cocoons. Since similar characters differ in cocoon-spinning and nonspinning members of various taxa such as Rophitinae, it is evident that loss of spinning and of associated larval structures is rather com-

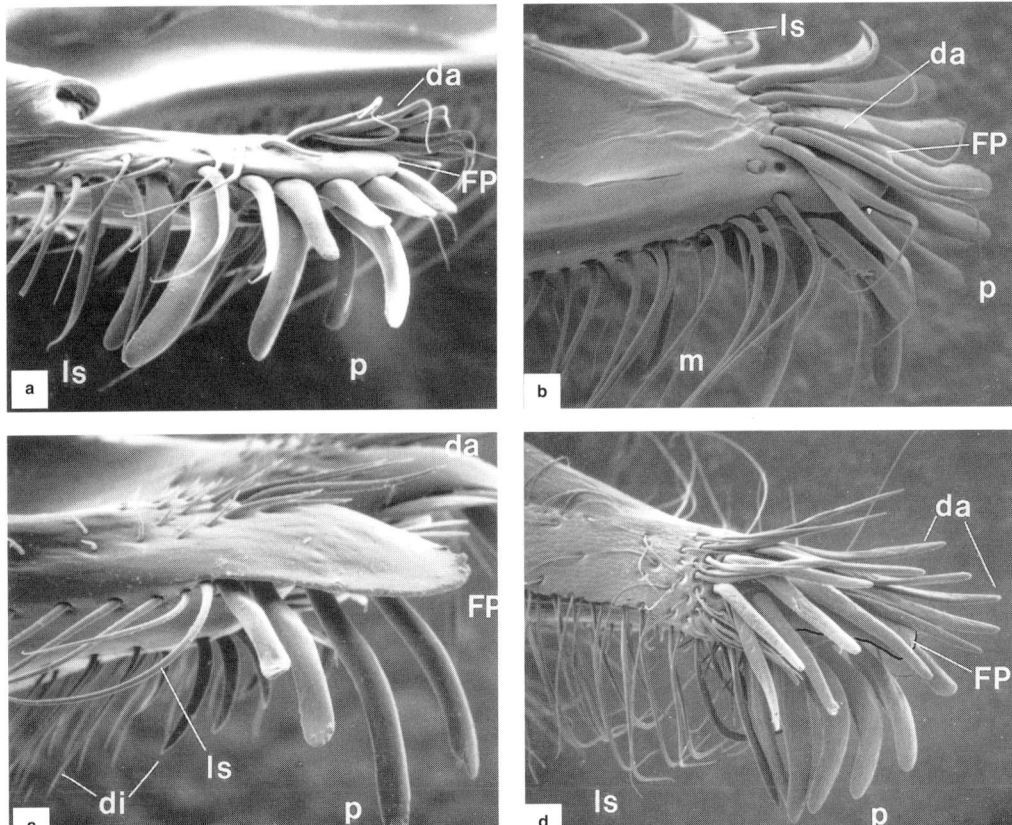

Figure 91-1. Apices of lobes of S6 of female Nomadinae, showing the coarse blunt setae. **a,** *Holcopasites calliopsidis* (Linsley), left lobe, lateral view; **b,** Same species, right lobe, dorsal view; **c,** *Kelita* sp., left lobe, lateral view; **d,** *Odyneropsis veseyi* (Cockerell), left lobe, lateral view. (da, dorsoapical setae; FP, flattened apical projection of S6; P, principal spinelike setae; ls, lateral series of setae; m, marginal setae; di, discal setae.) From Roig-Alsina, 1991b.

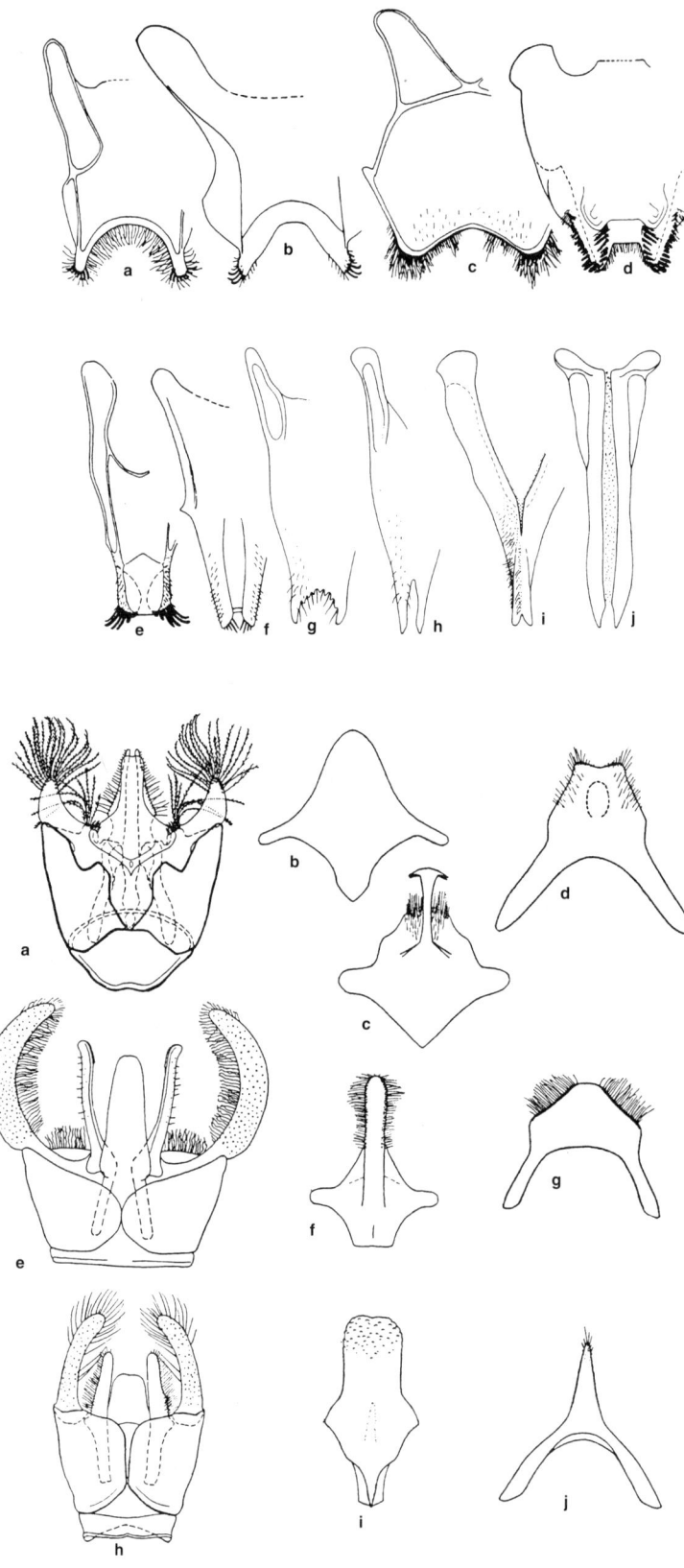

Figure 91-2. Ventral views of S6 of female Nomadinae. **a,** *Holcopasites illinoiensis* (Robertson); **b,** *Neopasites fulviventris* (Cresson); **c,** *Townsendiella californica* Michener; **d,** *Hexepeolus rhodogyne* Linsley and Michener; **e,** *Nomada (Holonomada)* sp., **f,** *Brachynomada (Melanomada) melanantha* (Linsley); **g,** *Neolarra vigilans* (Cockerell); **h,** *Neolarra verbesinae* (Cockerell); **i,** *Oreopasites vanduzeei* Cockerell; **j,** *Caenoprosopis crabronina* Holmberg. a-f and i, from Linsley and Michener, 1939; g and h, from Roig-Alsina, 1991b; j, from Roig-Alsina, 1987.

Figure 91-3. Male genitalia (dorsal views), S8 (center column), and S7 (righthand column) of Nomadinae. **a,** *Nomada (Holonomada* group) sp.?; **b,** *N. crotchii* Cresson, S8 only; **c, d,** S8 and S7 of *N. (Holonomada* group) sp.?; **e-g,** *Neopasites fulviventris* (Cresson); **h-j,** *Oreopasites* sp.? From Linsley and Michener, 1939.

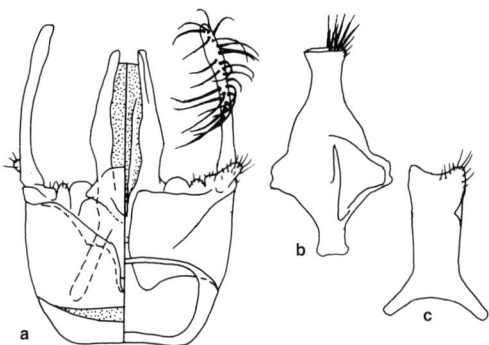

Figure 91-4. *Trichonomada roigella* Michener. **a-c,** Male genitalia (dorsal on left), S8, and S7. From Michener, 1996a.

mon among bees, and the many structures involved should not be totaled as a number of independent characters in a cladistic analysis.

Nomadinae is the largest and most diverse taxon of cleptoparasitic bees (see also Sec. 8). It does not exhibit a relationship to any particular group of nonparasitic bees; we therefore know nothing of its origin except that it appears to be holophyletic, probably demonstrating a single origin of parasitic behavior (Roig-Alsina and Michener, 1993). Its great diversity both in morphology and in hosts suggests great antiquity as a parasitic group. It seems likely, however, that the parasitic habit is related to unusual, perhaps unusually rapid, morphological divergence as compared to other bees (see Sec. 8). Structures involved include the apical metasomal features of females probably having to do with insertion of eggs into host cell walls, male genitalic features probably related to mating with apically modified females, and strong cuticle with various features probably serving for defense against irate hosts. It is probably because of such characters that 11 tribes are recognized; there are more distinctive structures than in many other taxa.

An unusual character of some tribes of Nomadinae is the elongate labrum, usually much longer than broad. It suggests the labrum of the Megachilinae, but is narrowed at the extreme base (Fig. 96-3a). Such a labrum is found in the tribes Ammobatini, Ammobatoidini, and Caenoprosopini, which, as shown in Figure 91-5, may not all be close relatives.

The relationships of the tribes of Nomadinae to one another have been studied by several authors, using larval characters (Rozen, 1966a; Rozen, Eickwort, and Eick-

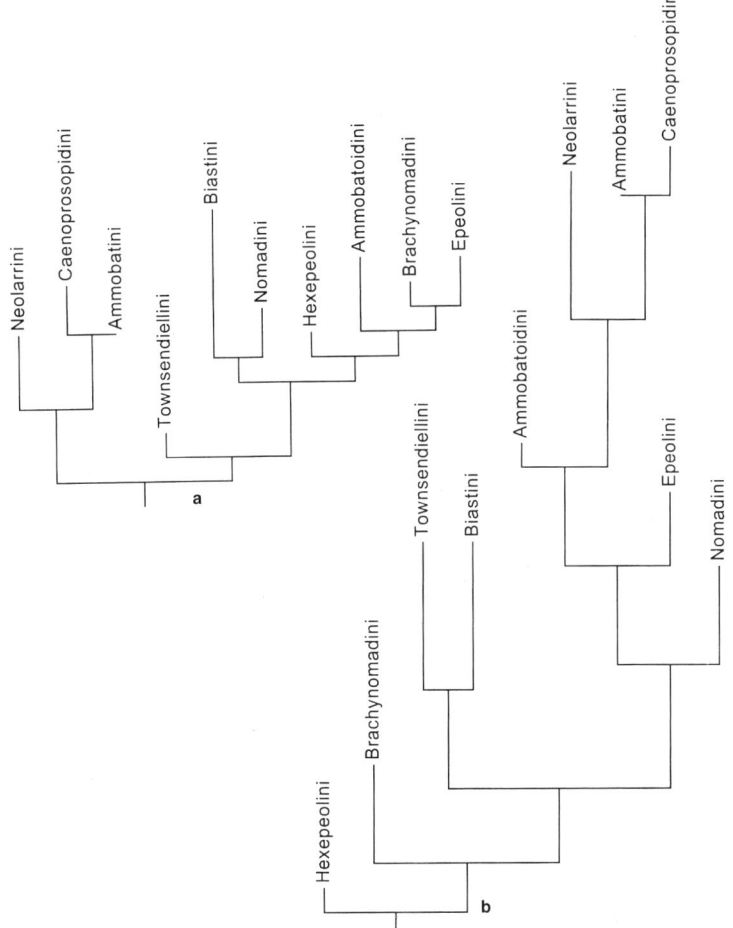

Figure 91-5. Cladograms of tribal relationships of nomadine tribes. Vertical line lengths are proportional to the number of derived characters. **a,** Based on a study of adult characters by Roig-Alsina, 1991b; **b,** Based on adult and larval characters by Rozen, Roig-Alsina, and Alexander, 1997. These studies were made with particular genera; the use of tribal names here summarizes the results.

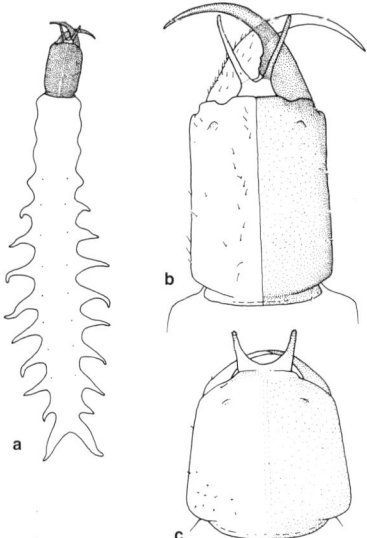

Figure 91-6. First-stage larvae of Nomadinae. **a**, *Triepeolus grandis* (Friese); **b**, Dorsal view of head of same; **c**, Dorsal view of head of *Townsendiella pulchra* Crawford. The long paired processes above the mandibles are the greatly prolonged labral tubercles. From Rozen, 1991a.

wort, 1978) and mostly adult characters (Alexander, 1990, 1996; Roig-Alsina, 1991b; Roig-Alsina and Michener, 1993). A reanalysis included in the last work resulted in a single cladogram similar to that of Roig-Alsina (1991b), differing in topology only in that the positions of Nomadini and Hexepeolini were reversed; Figure 91-5a summarizes his cladogram. There does not seem to be strong intuitive or biogeographical support for this cladogram, although the close relationship of Caenoprosopidini and Ammobatini is very probable. It is likely that these two tribes should be united; this cladogram shows the latter as paraphyletic. Finally, Rozen (1996a) and Rozen, Roig-Alsina, and Alexander (1997) have made extensive phylogenetic studies based especially on larvae but also on previously used adult characters. Rozen offered numerous alternative cladograms based on different sets of characters and different analyses; Figure 91-5b is a summary of one, based on both larval and adult characters. Torchio (1986) proposed a division of the Nomadinae based on the egg structure—the tribes Epeolini, Ammobatini, and Biastini (also Protepeolini in the Apinae) having an operculum surrounded by a flange at one end of the egg, the tribes Nomadini and Ammobatoidini lacking such a flange. Adult characters do not support such a dichotomy.

Rozen (1991a) gave an excellent account of the mode of parasitism in the Nomadinae and other parasitic Apidae, together with the morphology of first-stage larvae. In all cases the parasitic female enters the open cell of the host while its owner is away from the nest. There the parasite inserts its egg, which is unusually small for the size of the bee, into the wall or lining of the cell. The female parasite then departs, probably to return later to parasitize subsequent host cells as they are built and provisioned. The first-stage parasite larva emerges after the host has completed provisioning and closed the cell. Its large mandibles (Fig. 91-6), elongate and sclerotized head capsule, and the very long, divergent, probably sensory labral tubercles together with mobility facilitated by modified apical body segments, enable it to find and kill the host egg or young larva. As explained in Section 8, different genera of Nomadinae deposit their eggs in different positions and with different degrees of exposure. One would suppose that the partly exposed egg position of *Nomada* would be plesiomorphic relative to the various types of completely hidden eggs of other Nomadinae. Such eggs sometimes have an operculum at one end flush with the cell surface, as in *Doeringiella* in the Epeolini (Torchio, 1986), or the chorion of one side is thickened and roughened and lies flush with the cell surface, as in Biastini and Hexepeolini (Secs. 92, 97).

The key to the tribes of the Nomadinae is incorporated into that for Apidae as a whole, in Section 85.

92. Tribe Hexepeolini

This tribe consists of epeoliform bees 7 to 8 mm in length with bands of appressed white hairs on the metasomal terga, which are sometimes partly red; these bands are in the apical depressed marginal zones of the terga (Fig. 92-1). Bands in the Epeolini are on the discs of the terga, although sometimes extending onto the marginal zones. Other differences from the Epeolini include the long six-segmented maxillary palpi; straight-margined, nonprojecting axillae; the triangular rather than quadrate front coxae; the basal horizontal zone of the propodeal triangle; and the exposed and only moderately emarginate S6 of the female (Fig. 91-2d), with series of spinelike setae on the apex partly surrounding a lateral depressed area. The male genitalia and apical sterna of both sexes were illustrated by Linsley and Michener (1939).

The tribe Hexepeolini contains a single genus, *Hexepeolus*. It was considered to be closest to *Nomada* by Linsley and Michener (1939) and Michener (1944), largely because of the long maxillary palpi and other plesiomorphies, but it differs from *Nomada* in the rather hairy body, rounded apex of the marginal cell, emarginate S6 of the female, and many other characters. Roig-Alsina (1991b) showed that *Hexepeolus* is only distantly related to other Nomadinae, and Roig-Alsina and Michener (1993) placed it in its own tribe after a phylogenetic analysis. Rozen (1996a) considers it likely to be the basal branch of the Nomadinae, i.e., sister group to all other Nomadinae.

Genus *Hexepeolus* Linsley and Michener

Hexepeolus Linsley and Michener, 1937: 77, 81. Type species: *Hexepeolus mojavensis* Linsley and Michener, 1937 = *Hexepeolus rhodogyne* Linsley and Michener, 1937, by original designation.

■ *Hexepeolus* occurs in the xeric parts of California and Arizona, USA, and no doubt also in northwestern Mexico. The only species is *Hexepeolus rhodogyne* Linsley and

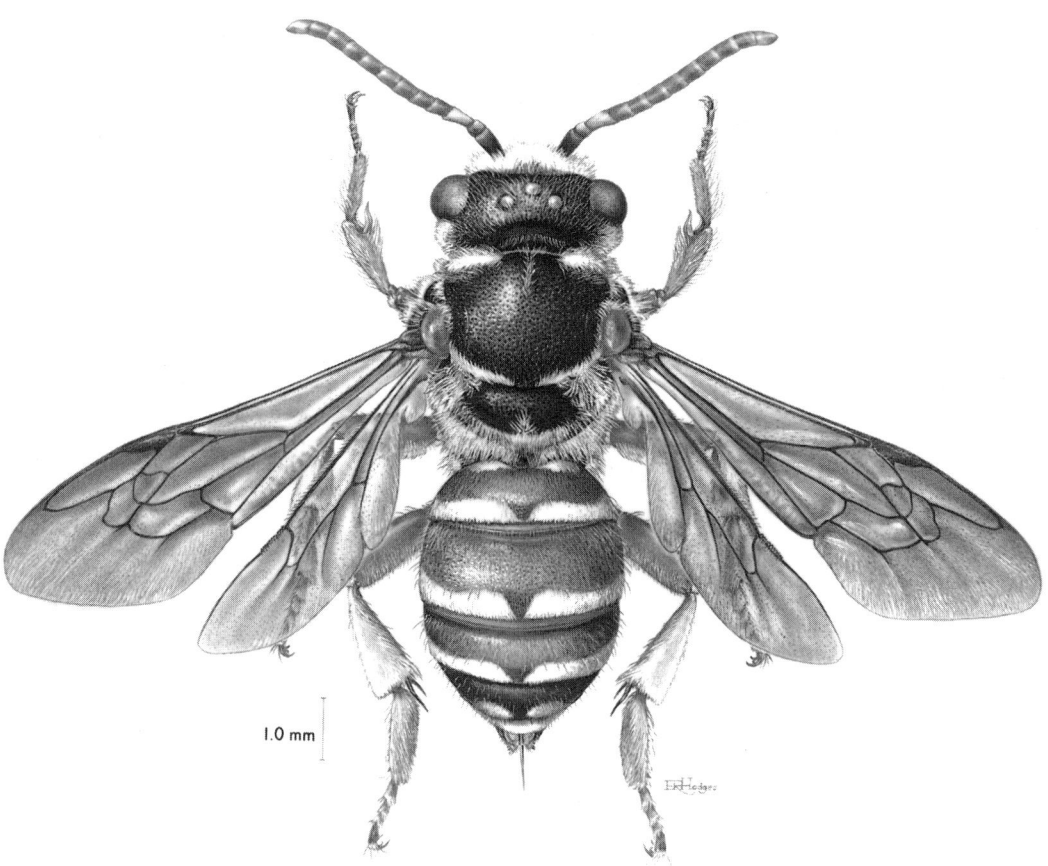

Figure 92-1. *Hexepeolus rhodogyne* Linsley and Michener, female. Drawing by E. R. S. Hodges, from Michener, McGinley, and Danforth, 1994.

Michener, the status of which was reviewed by Shanks Gingras (1983).

As long suspected, and verified by Rozen (1992a), *Hexepeolus* is a cleptoparasite in nests of *Ancylandrena* (Andreninae); its larva was described by Rozen (1994a, 1996b). As in Biastini, the egg is placed in a groove in the host cell surface. One side is thus exposed but is papillate and made inconspicuous by a flange all around the egg that obscures any fissure between the egg surface and the cell lining.

93. Tribe Brachynomadini

This American group was recognized as the melanomadine complex by Rozen (1977b), Ehrenfeld and Rozen (1977), Snelling and Rozen (1987), and Alexander (1990), and given the present tribal status after phylogenetic analysis by Roig-Alsina and Michener (1993).

The bees of this tribe have the general aspect of small Nomadini or of Ammobatini and have usually been placed in the Nomadini. Brachynomadini are nomadiform or small epeoliform bees, black or with the metasoma or much of the body red. The yellow or white integumental markings found in most *Nomada* species are absent. The two basal mandibular articulations are about equidistant from the eye (Fig. 93-1b), except that in some *Brachynomada* and especially in *Trichonomada* the anterior articulation is more distant; the anterior angle is farther from the eye in Nomadini (Fig. 93-1a). The maxillary palpus is three- to six-segmented and usually less than half as long as the blade of the galea; it is six-segmented and longer in Nomadini. Except in *Kelita* the pseudopygidial area of females is clothed with minute posterolaterally directed hairs, parted on the midline, such that in most positions half of the area appears silvery, the other half dark. Except in *Trichonomada*, S5 of the female is produced to a median hairy process or lobe, not found in Nomadini. S6 of the female has a deep midapical cleft between the two pointed or narrowly rounded lobes, and the lobes are close together and armed mesally with a few coarse setae (Fig. 91-2f) or, in *Kelita*, armed apically with a few short, blunt setae; this configuration is altogether unlike the subtruncate apex with coarse blunt setae of Nomadini.

Bees of this tribe are found only in the Western Hemisphere. Larvae were described by Rozen (1977b, 1994b, 1997a), and Ehrenfeld and Rozen (1977). Hosts include Colletinae, Exomalopsini, and Panurginae. Eggs are thought to be inserted into the cell wall as in Nomadini.

Kelita appears to be the most distinctive brachynomadine taxon. The genera *Brachynomada*, *Paranomada*, and *Triopasites* are similar to one another; unique specimens in various collections indicate that there are undescribed species of this group in Mexico and South America. As such additional species become known in both sexes, it is likely that these three genera will merge and be synonymized or recognized as subgenera.

Key to the Genera of the Brachynomadini
(From Rozen, 1997a)

1. Hind coxa normal, not expanded, with carina along mesal dorsal edge but without carina along outer dorsal edge; antennal scape, exclusive of basal bulb, shorter than twice maximum diameter; graduli of terga and sterna simple (South America) .... *Kelita*
—. Hind coxa broadly expanded laterally, with pronounced longitudinal carina along mesal dorsal edge and strong carina along outer dorsal edge, the two separated by strongly concave, mostly glabrous surface; antennal scape, exclusive of basal bulb, distinctly longer than twice maximum diameter; graduli of T2 to T4 and S2 to S4 and others produced as distinct, clear, thin, posteriorly directed lamellae (not just carinae) appressed against postgradular areas 2

2(1). Anterior mandibular articulation separated from eye by distance almost as great as ocellar diameter 3
—. Anterior mandibular articulation almost as close to eye as posterior articulation, separated from eye by distance no greater than half ocellar diameter (Fig. 93-1b) 4

3(2). Thorax not strongly flattened, width about equal to depth; eyes with sparse setae about half as long as ocellar diameter; frons in front of ocelli punctate, not particularly shiny; scutal disc punctate; maxillary palpus five-segmented (counting minute segment l); S5 of female broadly rounded apically; clypeus of male with low lateral carina in addition to deep suture immediately above it, extending from anterior tentorial pit to anterior mandibular articulation (South America) .... *Trichonomada*
—. Thorax strongly flattened, width distinctly greater than depth; eyes with sparse minute setae, scarcely noticeable; frons in front of ocelli nearly glabarous, polished; scutal disc virtually glabrous, highly polished; maxillary palpus usually four-segmented (counting minute segment 1); S5 of female with apical margin produced as round, median, hairy projection; clypeus of male without lateral carina but with very deep suture from anterior tentorial pit to anterior mandibular articulation (North America) .... *Paranomada*

4(2). Median apical triangle of S1 projecting downward; male with dorsal gonostylus short, laterally flattened, bladelike near base, tapering evenly toward apex as seen in lateral view (North America) .... *Triopasites*

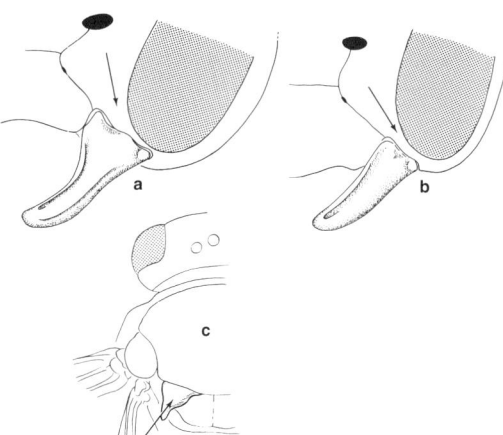

Figure 93-1. Cephalic and thoracic structures of Nomadinae. **a, b,** Mandibular articulations of *Nomada imbricata* Smith and *Brachynomada (Melanomada) grindeliae* (Cockerell), with arrows indicating distinguishing feature; **c,** Dorsum of thorax showing produced axilla (arrow) of *Doeringiella (Triepeolus)* sp. From Michener, McGinley, and Danforth, 1994.

—. Median apical triangle of S1 scarcely produced; male with dorsal gonostylus short or attenuate, but slender, sometimes widening near apex, neither laterally flattened nor bladelike.. *Brachynomada*

Genus *Brachynomada* Holmberg

The number of submarginal cells is two or three, differing among species of this genus, and the number of segments in the maxillary palpus is five or six. Nonetheless, the genus is morphologically rather homogeneous, with punctures on all parts of the body although widely separated by shiny ground in some parts of some species. The body length is 3.5 to 9.0 mm. The scutellum is biconvex. T5 of the female has a large median pseudopygidial area about twice as wide as long, covered with extremely fine, dense, simple hairs. The male genitalia and apical sterna of both sexes were illustrated by Linsley and Michener (1939) and Rozen (1994b, 1997a).

Key to the Subgenera of *Brachynomada*

1. Hind femur of male without basal thornlike projection or other modification; S8 of male elongate, strongly produced distally, its apex broadening before ending abruptly (South America)............ *B. (Brachynomada s. str.)*
—. Hind femur of male with thornlike projection ventrally and, in some species, with other modifications on ventral surface; S8 of male less elongate, not strongly produced, tapering to apex (North America) .... *B. (Melanomada)*

Brachynomada / Subgenus *Brachynomada* Holmberg s. str.

Brachynomada Holmberg, 1886b: 233, 239, 272. Type species: *Brachynomada argentina* Holmberg, 1886, designated by Sandhouse, 1943: 531.

■ This subgenus occurs from Peru and the state of São Paulo, Brazil, to the province of Buenos Aires, Argentina. Probable members of this subgenus are reported from areas as distant as Ceará, Brazil, and Mendoza, Argentina, but I have not seen specimens from such distant areas; they might belong to similar-looking bees such as *Trichonomada*. Eight species have been described. Several species of this subgenus were misplaced in *Doeringiella* (Epeolini) by Friese (1908b), who gave a key to certain species.

Species of *Brachynomada* s. str. are cleptoparasites of *Exomalopsis* (Exomalopsini), as shown by Rozen (1997a), and *Psaenythia* (Panurginae), as shown by Rozen (1994b). The use of such dissimilar hosts, belonging to different families of bees, is of interest. The larva of *B. scotti* Rozen was described by Rozen (1997a).

Brachynomada / Subgenus *Melanomada* Cockerell

Nomada (Melanomada) Cockerell, 1903d: 587. Type species: *Nomada grindeliae* Cockerell, 1903, by original designation.
Hesperonomada Linsley, 1939b: 5. Type species: *Hesperonomada melanantha* Linsley, 1939, by original designation.

■ *Brachynomada (Melanomada)* occurs from California and Nebraska south to Texas, USA, and Durango and Jalisco, Mexico. The seven species were revised by Snelling and Rozen (1987), supplemented by Rozen (1994b).

Species of *Melanomada* are cleptoparasites in nests of *Anthophorula (Anthophorisca)* (Exomalopsini) (Rozen, 1977b, 1997a; Rozen and Snelling, 1986). Host records are summarized by Snelling and Rozen (1987), supplemented by Rozen (1994b, 1997a).

Genus *Kelita* Sandhouse

This is a distinctive genus of Brachynomadini, as indicated by the first couplet of the key to genera. Moreover, the antennal scape is not flattened as it is in the other genera, and the species are often smaller, the body length being from 2.8 to 5.8 mm. The pseudopygidial area is partly covered by broad, scalelike, posteriorly directed, appressed hairs; in all other Brachynomadini the hairs are slender and directed posterolaterally.

Key to the Subgenera of *Kelita*
(From Rozen, 1997a)

1. Submarginal cells two; maxillary palpus six-segmented; hind tibia of male normal in shape; hind tibial spurs of male normal in size, subequal in length; dorsal and median processes of gonostylus shorter than gonocoxite (Chile) .... *K. (Kelita s. str.)*
—. Submarginal cells three; maxillary palpus five-segmented; hind tibia of male broadly expanded apically with anterior edge forming apical curved spine; hind tibial spurs of male very different, inner one normal-sized, outer one extremely short or absent; dorsal and median processes of gonostylus elongate, distinctly longer than gonocoxite (Argentina) .... *K. (Spinokelita)*

Kelita / Subgenus *Kelita* Sandhouse, s. str.

Herbstiella Friese, 1916: 168 (not Stimpson, 1871). Type species: *Herbstiella chilensis* Friese, 1916, monobasic.
Kelita Sandhouse, 1943: 561, replacement for *Herbstiella* Friese, 1916. Type species: *Herbstiella chilensis* Friese, 1916, autobasic.

Numerous structures were excellently illustrated by Ehrenfeld and Rozen (1977); see also Rozen (1997a).
■ *Kelita* s. str. occurs from Atacama province to Nuble, Chile. The four species were revised by Ehrenfeld and Rozen (1977).

Two species of *Kelita* s. str. are known to be cleptoparasites in nests of *Liphanthus* and *Protandrena (Parasarus)* (Panurginae), as reported by Ehrenfeld and Rozen (1977); one of the *Liphanthus* host species was placed incorrectly in *Psaenythia* by these authors. A third species parasitizes *Leioproctus (Perditomorpha)* (Colletinae) (Rozen, 1994b).

Kelita / Subgenus *Spinokelita* Rozen

Kelita (Spinokelita) Rozen, 1997a: 5. Type species: *Kelita argentina* Rozen, 1997, by original designation.

The male genitalia and other structures were illustrated by Rozen (1997a).
■ *Spinokelita* is known from Santa Cruz to San Juan provinces, Argentina. The only described species is *Kelita argentina* Rozen.

Figure 93-2. *Paranomada velutina* Linsley, female. Drawing by F. Abernathy, from Linsley, 1943a.

Genus *Paranomada* Linsley and Michener

Paranomada Linsley and Michener, 1937: 82. Type species: *Paranomada nitida* Linsley and Michener, 1937, by original designation.

Paranomada appears to be closely related to *Brachynomada* but differs in its curiously flattened and elongated body form (Fig. 93-2), shiny and nearly impunctate integument, and dense apical hair bands on the metasomal terga. The scutellum is flat, but curved down near the posterior margin, not biconvex. The body length is 5 to 11 mm. The wings are similar to those of other Brachynomadini, the marginal cell pointed almost on the wing margin (Fig. 93-3). The male genitalia and apical sterna were illustrated by Linsley and Michener (1939).

▪ This genus is known from Arizona and Southern California, USA, and Baja California, Chihuahua, and Zacatecas, Mexico. The three species were reviewed by Linsley (1945; see also Linsley, 1943a). A South American species was incorrectly placed in *Paranomada* by Michener (1996a). It appears to be a relatively smooth species of *Brachynomada* s. str. with a somewhat flattened thorax; the anterior mandibular articulation is only a little farther from the eye than the posterior articulation, unlike that of *Paranomada*. It is probably the form described as *Doeringiella thoracica* Friese.

The host of *Paranomada* is *Exomalopsis*, as discovered by Rozen (1977b); he suggested that the flat body may permit the parasite to appress itself against the burrow wall and thus avoid detection by passing host bees. Because the *Exomalopsis* are communal, the avoidance of hosts may be more difficult than with solitary hosts.

Genus *Trichonomada* Michener

Trichonomada Michener, 1996a: 90. Type species: *Trichonomada roigella* Michener, 1996, by original designation

This genus consists of a species 5.8 to 6.5 mm long, resembling in form and red coloration species of *Pasites* (Ammobatini) such as *P. maculatus* Jurine, as well as certain species of *Brachynomada*. *Trichonomada* is of special interest because it destroys some of what had seemed to be strong tribal characters differentiating Brachynomadini from Nomadini, e.g., it lacks the hairy median lobe or process on S6 of the female, and the anterior mandibular articulation is far from the eye. *Trichonomada* is not, however, similar to Nomadini; rather, it lacks some features otherwise typical of Brachynomadini. Structures including male genitalia were illustrated by Michener (1996a); see also Figure 91-4.

▪ *Trichonomada* is known from Minas Gerais, Brazil, to Entre Ríos, Argentina. The single species is *T. roigella* Michener.

Genus *Triopasites* Linsley

Triopasites Linsley, 1939b:8. Type species: *Triopasites timberlakei* Linsley, 1939, by original designation.

Triopasites consists of species similar to small *Brachynomada* that differ in the characters listed in the key to genera. The habitus is shown in Figure 93-4. There are three

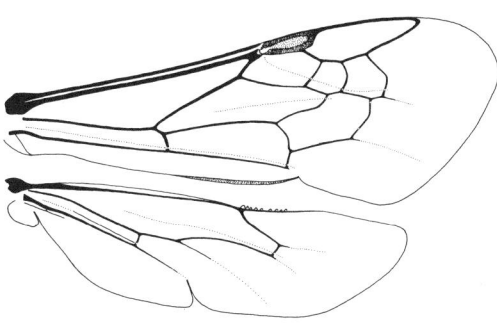

Figure 93-3. Wings of *Paranomada velutina* Linsley.

Figure 93-4. *Triopasites penniger* (Cockerell), female. Drawing by F. Abernathy, from Linsley, 1943a.

or sometimes two submarginal cells; body length is 4.0 to 6.5 mm.

■ This genus occurs from western Texas to Southern California, USA, and in Tamaulipas, Morelos, and Guerrero to Baja California, Mexico. Several species (see Linsley, 1943a) were synonymized, and only two were recognized by Rozen (1997a).

Hosts of this cleptoparasitic bee are species of *Anthophorula (Anthophorula)* (Exomalopsini) (Rozen, 1977b). The larva was described in the same work.

94. Tribe Nomadini

Nomada, the only genus of Nomadini, consists of slender, sparsely haired, wasplike bees (Pl. 2; Fig. 94-1); although some are black or largely so, the majority have yellow or white integumental markings, often on all tagmata, or the whole body or at least the metasoma is red; sometimes yellow or white integumental markings are on a red background. Pubescent tergal fasciae and other conspicuous patterns of pubescence, as are prevalent in the Epeolini and Hexepeolini, are absent in the Nomadini. The body length is from little over 3 mm to 16 mm. As shown in Figure 93-1a, the anterior mandibular articulation is well separated from the eye. The only Nomadinae having the apex of the marginal cell sharply pointed and on the wing margin are the tribe Nomadini (Fig. 94-2) and some Brachynomadini (Fig. 93-3). In most species there are three submarginal cells, the first nearly as long as the second and third taken together. In some species, however, the second submarginal crossvein is absent, leaving two subequal submarginal cells. In rare individuals or possibly species, the first submarginal crossvein is lost, resulting in a long first submarginal cell and a small second, as in bees such as most *Hylaeus*. S5 of the female is emarginate with a tuft of bristles on each resultant lobe. T6 of the female has a lateroapical tuft of bristles on each side of the pygidial plate. The apex of S6 of the female is subtruncate, with apicolateral groups of blunt spinelike setae (Fig. 91-2e), not emarginate or bifid as in most other Nomadinae. The male genitalia and apical sterna of males or of both sexes were illustrated by Linsley and Michener (1939), Rodeck (1947, 1949), Mitchell (1962), Schwarz (1967 and other papers), Snelling (1986b), Eardley and Schwarz (1991), Alexander (1994), and Celary (1995); see also Figure 91-3a-d.

Genus *Nomada* Scopoli

Nomada Scopoli, 1770: 44. Type species: *Apis ruficornis* Linnaeus, 1758, by designation of Curtis, 1832: pl. 419. [Invalid designations are listed by Michener, 1997b.]

Hypochrotaenia Holmberg, 1886b: 234, 273. Type species: *Hypochrotaenia parvula* Holmberg, 1886, monobasic.

Nomadita Mocsáry, 1894: 37. Type species: *Nomadita montana* Mocsáry, 1894, monobasic.

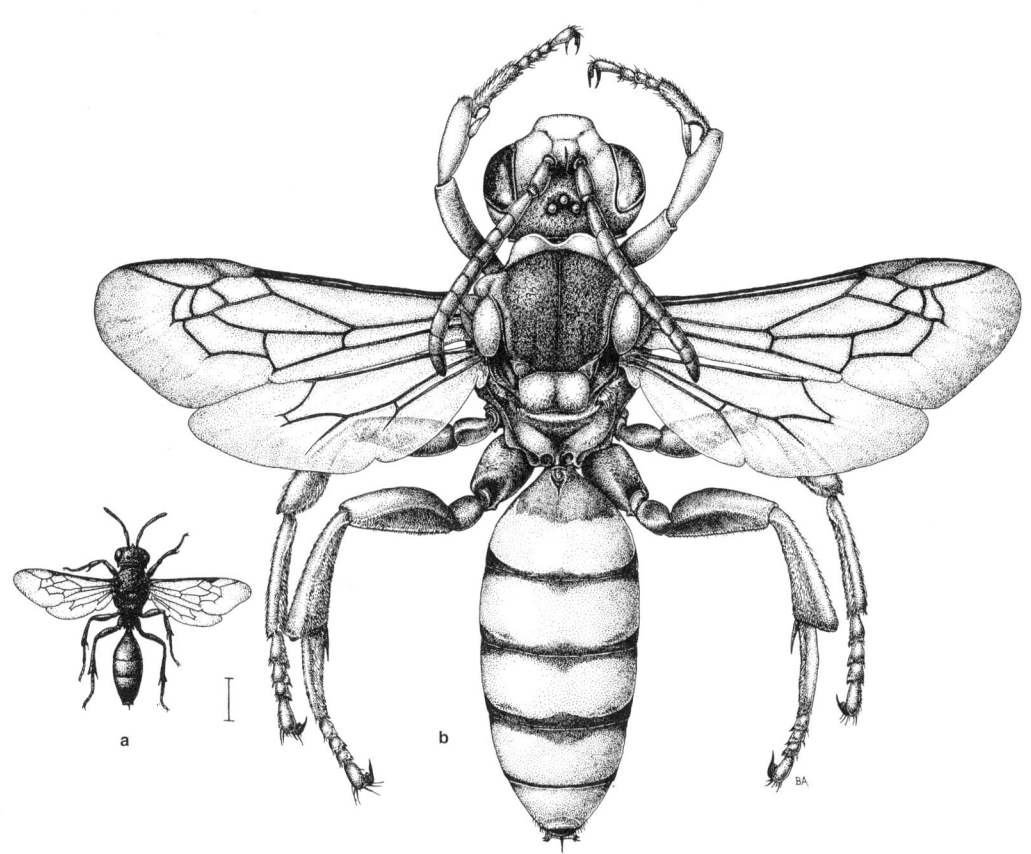

Figure 94-1. Small and large *Nomada* species, females. **a**, *N. penangensis* Cockerell; **b**, *N. coxalis* Morawitz. Drawing by B. Alexander. Scale line = 1mm.

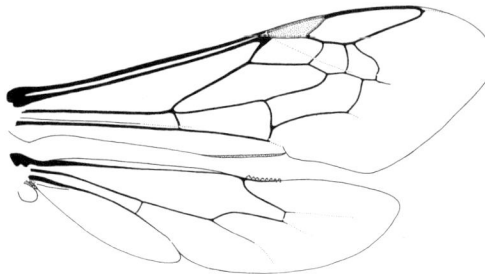

Figure 94-2. Wings of *Nomada annulata* Smith.

Lamproapis Cameron, 1902a: 419. Type species: *Lamproapis maculipennis* Cameron, 1902, monobasic.
Nomada (Heminomada) Cockerell, 1902, *in* Cockerell and Atkins, 1902: 42, footnote. Type species: *Nomada obliterata* Cresson, 1863, by original designation.
Nomada (Micronomada) Cockerell and Atkins, 1902: 44. Type species: *Nomada modesta* Cresson, 1863, by original designation.
Centrias Robertson, 1903d: 174, 176. Type species: *Nomada erigeronis* Robertson, 1897, by original designation.
Cephen Robertson, 1903d: 174, 176. Type species: *Nomada texana* Cresson, 1872, by original designation.
Gnathias Robertson, 1903d: 173, 174, 175. Type species: *Nomada bella* Cresson, 1863, by original designation.
Holonomada Robertson, 1903d: 174, 175, 176. Type species: *Nomada superba* Cresson, 1863, by original designation.
Xanthidium Robertson, 1903d: 174, 175, 177 (not Ehrenberg, 1833). Type species: *Nomada luteola* Olivier, 1811, by original designation. [This name is preoccupied but since it is a synonym, has not been replaced.]
Phor Robertson, 1903d: 174, 175, 176. Type species: *Nomada integra* Robertson, 1893 (not Brullé, 1832) = *Nomada integerrima* Dalla Torre, 1896, by original designation.
Nomada (Nomadula) Cockerell, 1903d: 611. Type species: *Nomada americana* Kirby of Robertson, 1903d = *Nomada articulata* Smith, 1854, by original designation. [See Code, ed. 3, art. 70(c).]
Nomadosoma Rohwer, 1911: 24. Type species: *Pasites pilipes* Cresson, 1865, by original designation.
Polybiapis Cockerell, 1916d: 208. Type species: *Polybiapis mimus* Cockerell, 1916, by original designation.
Nomada (Callinomada) Rodeck, 1945: 181. Type species: *Nomada antonita* Cockerell, 1909, by original designation.
Nomada (Pachynomada) Rodeck, 1945: 180. Type species: *Nomada vincta* Say, 1837, by original designation.
Nomada (Laminomada) Rodeck, 1947: 266. Type species: *Nomada hesperia* Cockerell, 1903, by original designation.
Acanthonomada Schwarz, 1966: 383. Type species: *Nomada odontophora* Kohl, 1905, by original designation.
Nomada (Phelonomada) Snelling, 1986b: 24. Type species: *Nomada belfragei* Cresson, 1878, by original designation.
Hypochrotaenia (Alphelonomada) Snelling, 1986b: 9. Type species: *Nomada cruralis* Moure, 1960, by original designation.
Nomada (Asteronomada) Broemeling, 1988: 336. Type species: *Nomada adducta* Cresson, 1878, by original designation.
Nomada (Adamon) Hirashima and Tadauchi, 2002: 47. Type species: *Nomada koikensis* Tsuneki, 1973. [new synonymy]

This is the only genus of its tribe; the tribal and generic characters are indicated in the key to tribes and in the account of the Nomadini.

It is the custom to recognize subgenera in large and morphologically diverse bee genera like *Nomada*. Snelling (1986b), following this custom and using genus-group names proposed from 1886 to 1947, broke *Nomada* of the Western Hemisphere into three genera, *Centrias, Hypochrotaenia*, and *Nomada*, the last two of which, taken together, were subdivided into nine subgenera. Because these taxa tend to grade into one another, it is difficult to construct keys that decisively separate them; Snelling's keys employ considerable wording in order to accommodate exceptions to various stated characters. In fact, exceptions exist to every character in Snelling's key separating *Nomada* from *Hypochrotaenia*, and only one invariant character separates *Centrias*. Alexander's (1994) table 3 illustrates the same thing: there are no characters that reliably separate the three genera. Nevertheless, Snelling's three genera seem reasonably distinct, and I might have accepted them except for the recent worldwide phylogenetic study of groups of *Nomada* by Alexander (1994). He showed that, even though constructing a workable key is a challenge, two of Snelling's genera, *Hypochrotaenia* and *Centrias*, are holophyletic groups; the third, *Nomada*, is a paraphyletic group from which the others arose. Sometimes, recognition of a paraphyletic group is important, even if subjective, in maintaining a classification that is useful for information storage and retrieval, but this is so only if there is a large and easily recognizable morphological gap differentiating the paraphyletic taxon. This is not true in the case of *Nomada*, and I have reluctantly followed Alexander's lead in recognizing no subgeneric or generic subgroups within *Nomada* s. l.

Alexander (1994), however, did recognize a series of species groups to which a less cautious author might have given subgeneric names. The following list relates his species groups to previously used genus-group names:

gigas group: southern Africa
integra group: palearctic
adducta group: *Asteronomada*—nearctic
vincta group: *Pachynomada*—nearctic
odontophora group: *Acanthonomada*—palearctic
vegana group: *Cephen, Aphelonomada, Hypochrotaenia, Micronomada, Nomadosoma, Polybiapis*—neotropical, nearctic
roberjeotiana group: *Nomadita, Callinomada*—holarctic, sub-Saharan (paraphyletic according to Alexander, 1994)
erigeronis group: *Centrias, Nomadula*—nearctic
ruficornis group: *Gnathias, Heminomada, Lamproapis, Nomada* s. str., *Phor, Xanthidium*:—holarctic (?paraphyletic; see Alexander, 1994).
armata group: palearctic
belfragei group: *Phelonomada*—nearctic
superba group: *Holonomada*—holarctic
basalis group: *Laminomada*—holarctic
bifasciata group: palearctic
trispinosa group: palearctic
furva group: palearctic, oriental, to Australia

■ *Nomada* is abundant throughout the holarctic region, north to Alaska and Finland. It is less abundant in the neotropical region, the *vegana* group occurring from the Antilles and Mexico south to Buenos Aires Province, Argentina. There are only ten species in sub-Saharan Africa, all in the *gigas* and *roberjeotiana* groups, ranging as far south as Cape Province, South Africa. In the oriental region only the *furva* group occurs in tropical India, Southeast Asia, and on as far as the Philippines, the Solomon Islands, and Queensland, Australia. Alexander (1994) studied over 800 described and undescribed species of *Nomada;* the list of species names (Alexander and Schwarz, 1994) totals 795. Revisional or review papers include Swenk (1912) for species of Nebraska, Evans (1972) for the North American *superba* group *(Holonomada),* Broemeling (1988) for the North American *roberjeotiana* group *(Nomadita),* Broemeling and Moalif (1988) for the *vincta* group *(Pachynomada),* Schmiedeknecht (1882) for palearctic species, Perkins (1919) for British species, Smit (2004) for the species of the Netherlands, Stöckhert (1941, 1943) for certain European groups, Pittioni (1953) for palearctic species, Schwarz (1967) for the palearctic *cinctiventris* (= *integra*) group, Celary (1995) for Polish species, Eardley and Schwarz (1991) for sub-Saharan species, Tsuneki (1973) for Japanese species, and Schwarz (1990) for the tropical Asian species. Alexander and Schwarz (1994) listed the species in each group; Snelling (1986b) listed the species in the American groups. A new subgeneric name, *Adamon,* which for the present I list as a synonym of *Nomada,* has recently been proposed. Probably subgenera will eventually be recognized in *Nomada;* this one is remarkable for having 12-segmented antennae in the male. This group includes two species, both from Japan. One has only two submarginal cells.

Although most species of *Nomada* are cleptoparasites in nests of *Andrena,* some are parasites of *Agapostemon, Halictus, Lasioglossum,* and *Lipotriches* in the Halictidae, *Panurgus* in the Andrenidae, *Melitta* in the Melittidae, probably *Colletes* in the Colletidae, and *Exomalopsis* and *Eucera* in the Apidae (Apinae). Using such hosts, *Nomada* has extended its geographical range far beyond that of its principal host, *Andrena.* Snelling (1986b) summarized what is known about hosts of *Nomada* in the Western Hemisphere; Westrich (1989) reviewed host information especially for European species, and Alexander (1991b) provided a world list of host associations of *Nomada.* Linsley and MacSwain (1955) gave an account of the habits of certain species of *Nomada* that parasitize *Andrena* species; two or sometimes more (up to four) *Nomada* eggs and an *Andrena* egg are found in most parasitized cells. The first *Nomada* to hatch destroys the other *Nomada* egg or eggs, then the *Andrena* egg, before feeding on the provisions in the cell. Radchenko (1981) shows how the egg of the palearctic *Nomada ruficornis* (Linnaeus) projects from the wall of an *Andrena* cell, being only half inserted into the wall. This behavior seems to characterize American species also, and is quite unlike that of many other Nomadinae, whose eggs are completely hidden in the cell wall and even provided with a lamella forming a smooth surface where the egg is hidden. One species, *Nomada japonica* Smith, is obligately parthenogenetic (thelytokous) in Japan (Maeta, Kubota, and Sakagami, 1987), although males are known from China; otherwise among bees, obligate thelytoky is known only in *Ceratina.*

I have suggested in the past that *Nomada,* like so many other bee taxa, may originally have been neotropical, and that when it spread into North America it encountered and became adapted to *Andrena* as a host. It was then able to spread across the holarctic region as a parasite of *Andrena* and other S-T bees. This scenario would suggest that the *vegana* group, i.e., *Hypochrotaenia,* parasitizing Exomalopsini and perhaps other Apinae in tropical America, is the basal group of *Nomada.* Alexander's (1994) phylogenetic study did not support this idea, but placed the South African *gigas* group at the base of *Nomada* phylogeny. I am not satisfied with this conclusion, although parasitization of a more closely related host is the principal support for the suggestion of neotropical origin.

95. Tribe Epeolini

The Epeolini consist of relatively robust (i.e., epeoliform) parasitic bees. They lack the integumental yellow or white areas of most Nomadini, but usually have areas of dense, short, appressed, white or pale, plumose hairs that produce a conspicuous pattern (Pl. 2), at least on the metasoma and usually also on the other tagmata (Figs. 8-4a, 95-1). In *Thalestria* these white areas are much reduced, but much of the body is covered with appressed, plumose or scalelike, metallic blue or blue-green hairs. The background integument is black, sometimes partly or largely red-brown. Unlike the long, six-segmented maxillary palpi of the Nomadini, those of Epeolini are short, two- or three-segmented. The axilla is produced to a posteriorly directed point (Fig. 93-1c) except in a few small species of *Epeolus* in which it is rounded, but protrudes as a lobe, not continuing the contour of the scutellum. The marginal cell is rounded at the apex (Fig. 95-2), or, if pointed, the apex is well away from the wing margin. S6 of the female is retracted and largely or wholly invisible without dissection, profoundly emarginate medially because of a long, posteriorly directed process on each side (Figs. 8-10f; 95-3c, d); frequently the disc of S6 is reduced to a transverse bar.

Rightmyer (2004) made a phylogenetic study and reclassification of Epeolini. The classification is summarized as follows:

Subtribe Odyneropsina, *Odyneropsis*
Subtribe Rhogepeolina, *Rhogepeolus*

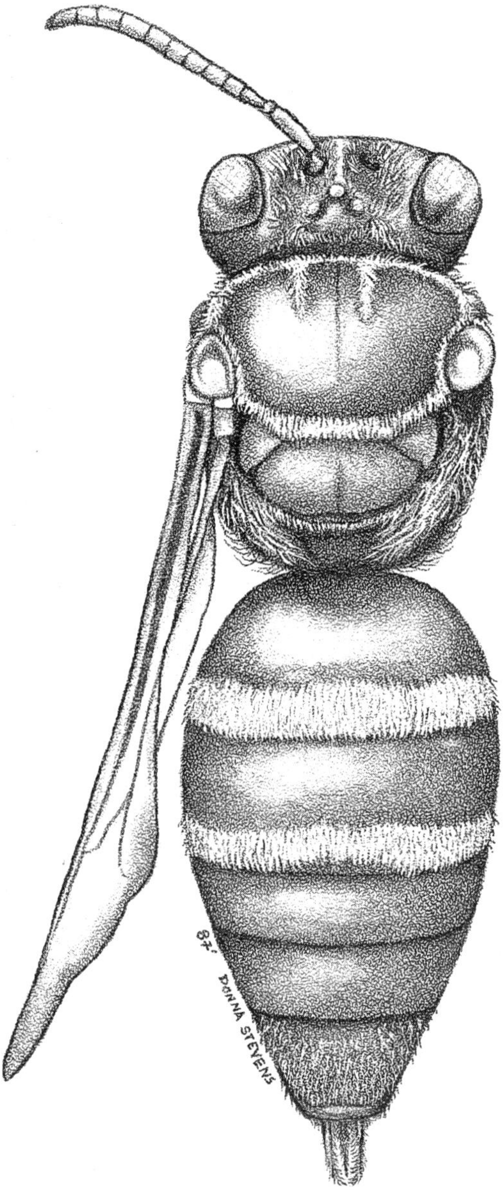

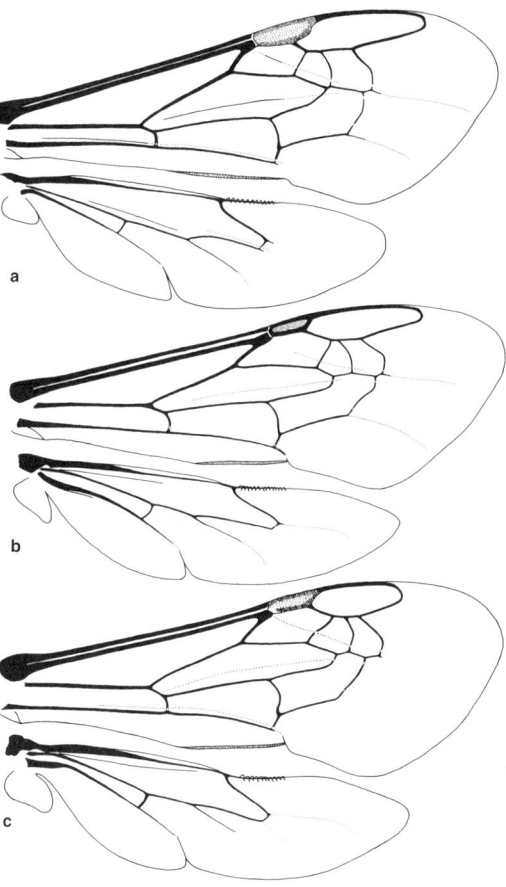

Figure 95-1. *Doeringiella bizonata* Holmberg, female. Drawing by D. Stevens, from Roig-Alsina, 1989b.

Figure 95-2. Wings of Epeolini. **a,** *Odyneropsis* sp.?; **b,** *Triepeolus verbesinae* (Cockerell); **c,** *Epeolus cruciger* Panzer.

Subtribe Epeolina, *Epeolus*
Subtribe Thalestriina, *Doeringiella*, Pseudepeolus*, Rhinepeolus, Thalestria, Triepeolus*

*Considered subgenera of *Doeringiella* by Michener (2000).

The monogeneric subtribes are characterized by the generic characters indicated below. The subtribe Thalestriina is characterized especially by the form of S6 of the female (Figs. 8-10f and 95-3c; Rightmyer, Figs. 7 and 10).

As in other Nomadinae, first-stage larvae have large, heavily sclerotized heads and jaws (Fig. 91-6a, b); Rozen (1989b) illustrated the larvae and distinguished those of *Epeolus* from *Triepeolus*.

The Epeolini are found on all continents except Australia and are cleptoparasites in the nests of various groups of bees.

Key to the Genera of the Epeolini
(part of this key is based on that of Rightmyer, 2004)

1. Body largely covered by minute, appressed, plumose, scalelike, metallic blue or green hairs, and similar white hairs on metasoma limited to small lateral patches; eyes closest at upper ends, strongly converging above in male; T5 of female with apical transverse area densely covered with long, stiff, black hairs but without distinct pseudopygidial area; preoccipital carina behind vertex strong but on posterior surface of head much below level of ocelli (neotropical) .... *Thalestria*
—. Body nonmetallic, usually with extensive areas of minute, appressed, plumose, white to pale-tan hairs on metasoma; eyes usually converging below (but in *Odyneropsis* slightly closer above than below in both sexes); T5 of female with apical area covered with short brown to white hairs (but long black hairs in *Rhogepeolus*), often well enough defined to be called a pseudopygidial area; preoccipital carina behind vertex (upper sector, see Fig. 95-4) absent or at summit of posterior surface of head, little below level of ocelli 2
2(1). Inner margins of eyes subparallel, usually slightly closer above than below; metasoma without conspicuous areas of pale pubescence except sometimes on T1; stigma rather large, vein r arising near middle (Fig. 95-2a); marginal cell as long as three submarginal cells combined; T5 of female with small middorsal depressed area often surrounded by ridges (often hidden by T4) (neotropical) .... *Odyneropsis*
—. Inner margins of eyes distinctly converging below; metasoma usually with conspicuous areas of pale pubescence; stigma smaller, with vein r arising well beyond middle (Fig. 95-2b, c) [except in *Epeolus* (*Trophocleptria* group)]; marginal cell shorter than submarginal cells combined [except in *E.* (*Trophocleptria* group)]; T5 of female without middorsal depressed area 3
3(2). Scutellum with median longitudinal strip of appressed, pale setae between convexities; pseudopygidial area of female with apical margin strongly concave, with medioapical slit, apical margin fringed with relatively long, curved, simple setae; first flagellar segment longer than second .... *Rhogepeolus*
—. Scutellum rarely with distinct median longitudinal strip of appressed setae; pseudopygidial area of female variable but apex rarely strongly concave, without median apical slit (except in *T. roni* Genaro); first flagellar segment not longer than second 4
4(3). Males 5
—. Females 9
5(4). Scape dramatically swollen (Fig. 95-4b) or, if not, forming subbasal angle on lateral surface (seen with antenna directed upward) so that only basal third or fourth tapers toward basal bulb; metafemur with setae on undersurface (rarely lacking); basitibial plate completely bordered by carina; S3 with elongate, curled setae on apical margin .... *Doeringiella*
—. Scape not swollen, not forming subbasal angle on lateral surface, basal half or more of scape tapering toward basal bulb; metafemur very rarely with conspicuous setae on undersurface; basitibial plate absent to completely bordered by carina; S3 with apical setae variable, rarely curled 6
6(5). Supraclypeal area produced into bulbous protrusion with weak median carina; preoccipital carina continuous; scutum anteriorly with median longitudinal band of appressed setae (sometimes faint); scutellum relatively flat but bearing two mammiform tubercles; second abscissa of hindwing vein M+Cu over twice as long as cu-v .... *Rhinepeolus*

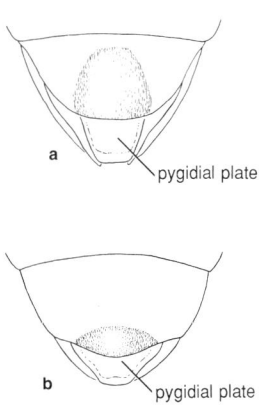 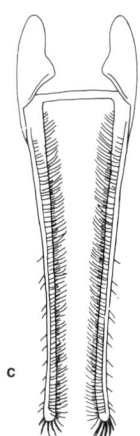 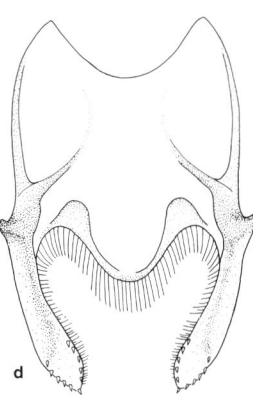

Figure 95-3. Apical metasomal structures of female Epeolini. **a,** Dorsal view of T5 and T6, indicating (by shading) the pseudopygidial area on T5, of *Triepeolus texanus* (Cresson) (see also Fig. 93-5); **b,** Same, of *Epeolus compactus* Cresson; **c,** S6 of *Triepeolus concavus* (Cresson); **d,** S6 of *Epeolus* (*Trophocleptria* group) sp.? From Michener, McGinley, and Danforth, 1994.

—. Supraclypeal area and median carina variable, rarely produced into bulbous protrusion; preoccipital carina absent at least on upper corners of head; scutum usually lacking median longitudinal band of appressed setae (sublateral bands can be present); scutellum variable but not bearing mammiform tubercles; second abscissa of hindwing vein M+Cu usually less than twice as long as cu-v 7

7(6). Scape relatively short (length approximately 1.5 times width), flattened on condylar surface; basitibial plate incompletely or completely bordered by carinae; S3 with mostly straight, elongate setae on apical margin (curved in *Pseudepeolus willinki* Roig-Alsina) ....
.... *Pseudepeolus*

—. Scape length variable, not flattened on condylar surface; basitibial plate absent or rarely incompletely bordered by carinae; S3 with setae usually undifferentiated on apical margin, rarely elongate or curled 8

8(7). Mandible lacking distinct preapical tooth; pygidial plate usually with median constriction, often apically differentiated to form distinct, down-turned, posterior surface; S7 usually with median emargination on distal margin, apicolateral lobes not attaining apical margin between lobes, and apical setae mostly ventral, forming distinct pocket near apicolateral lobe; gonostylus lacking basal lobe; penis usually lacking lateral projections or sometimes with subapical, lamellate projection; dorsobasal lobe of penis valve covering basolateral margin of penis; antennal pedicel usually set into scape ....
.... *Triepeolus*

—. Mandible usually with preapical tooth; pygidial plate almost always all in one plane, broadly rounded posteriorly; S7 usually lacking median emargination on distal margin, apicolateral lobes exceeding apical margin between lobes, and with apical setae mostly dorsal, on surface leading to lateral lobe; gonostylus with basal angle or lobe; penis with widely divergent, fleshy lateral lobe, lacking in *Trophocleptria* group; dorsobasal lobe of penis valve not covering basolateral margin of penis; antennal pedicel usually largely exposed .... *Epeolus*

9(4). Processes of S6 spatulate, with apical principal setae indicated by small denticles (Fig. 95-3d); pseudopygidial area forming wide apical lunule of silvery setae (Fig. 95-3b); apical ventral surface of pygidial plate with two medial, flattened, rounded processes, sometimes very reduced .... *Epeolus*

—. Processes of S6 rodlike, with apical principal setae elongate and hooked (Fig. 95-3c); pseudopygidial area variable, very rarely forming wide lunule of silvery setae on apical margin; apical ventral surface of pygidial plate with lateral, curled, scroll-like processes 10

10(9). Supraclypeal area produced into bulbous protrusion with weak median carina; scutellum relatively flat but bearing two mammiform tubercles; pseudopygidial area with median longitudinal row of dark, stout setae, and with apical margin convex; preoccipital carina continuously curved, not angulate at upper corners of head ......
.... *Rhinepeolus*

—. Supraclypeal area not bulbous, with strong or weak protrusion and carina; scutellum variable but not bearing mammiform tubercles; pseudopygidial area variable but lacking median, longitudinal row of dark, stout setae; preoccipital carina forming angles or broken at upper corners of head (Fig. 95-4a) 11

11(10). Scape length about twice width, forming subbasal angle on lateral surface (seen with antenna directed upward) so that only basal third or less tapers toward basal bulb; preoccipital carina complete or absent at upper corners of head; basitibial plate completely bordered by carina; first and second flagellar segments of about same length .... *Doeringiella*

—. Scape length usually only 1.5 times its width, rarely twice, not forming subbasal angle on lateral surface, basal half or more tapering toward basal bulb; preoccipital carina absent at upper corners of head or along entire upper border of head; basitibial plate absent to completely bordered by carina; first flagellar segment usually shorter than second, rarely the same length 12

12(11). Scutum almost always with elongate longitudinal bands of appressed setae reaching middle; preoccipital carina absent on upper corners of head or entire sector absent .... *Triepeolus*

—. Scutum with longitudinal bands of appressed setae often reduced, usually restricted to anterior fourth; preoccipital carina absent at upper corners of head only ....
.... *Pseudepeolus*

Genus *Doeringiella* Holmberg

Doeringiella Holmberg, 1886a: 151; 1886b: 233. Type species: *Doeringiella bizonata* Holmberg, 1886, monobasic.

Doeringiella (Orfilana) Moure, 1954b: 266. Type species: *Doeringiella variegata* Holmberg, 1886 = *Epeolus holmbergi* Schrottky, 1913, by original designation.

This is the *Doeringiella* s. str. of Michener (2000). Superficially, the species of *Doeringiella* are similar in appearance to those of *Triepeolus*. The distinguishing characters are indicated in the key to genera.

In some males of this genus the scape is greatly enlarged, with a cavity opening on the lower surface (Fig. 95-4b). Moure (1945b) proposed the name *Orfilana* for the species of *Doeringiella* that do not have the antennal scape of the male thus enlarged. Illustrations of various structures were provided by Roig-Alsina (1989b) and Rightmyer (2004).

■ *Doeringiella* occurs from Cautín, Chile, north to Lima, Peru, and from Chubut province north to Formosa province, Argentina, to Bolivia and the state of Pará, Brazil. The 31 species then known were revised by Roig-Alsina (1989b). Compagnucci and Roig-Alsina (2003) reviewed *Doeringiella*, recognizing 35 species, and prepared a new phylogenetic analysis of the species. This analysis indicated that, contrary to earlier conclusions, swollen male antennal scapes arose only once.

So far as known, species of this genus are cleptoparasites in nests of Apinae, mostly Eucerini (Roig-Alsina, 1989b). The only firm host record is *Svastrides* (Eucerini); probable hosts, also Eucerini, are in the genera *Svastra* and *Melissoptila*. Other possible hosts are in the genera *Diadasia* (Emphorini) and *Caupolicana* (Colletidae).

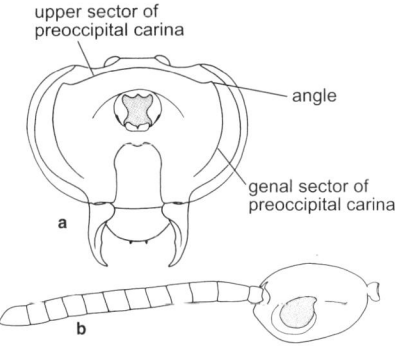

Figure 95-4. Structures of *Doeringiella*. **a,** *D. singularis* (Friese), posterior view of head to illustrate the preoccipital carina; **b,** Male antenna of *D. arechavaletai* (Brèthes), showing the swollen scape and cavity on underside (shaded). From Roig-Alsina, 1989b.

Genus *Epeolus* Latreille

Epeolus Latreille, 1802a: 427. Type species: *Apis variegata* Linnaeus, 1758, monobasic.

Trophocleptria Holmberg, 1886b: 233, 275. Type species: *Trophocleptria variolosa* Holmberg, 1886, monobasic.

Epeolus (Diepeolus) Gribodo, 1894: 79. Type species: *Epeolus giannellii* Gribodo, 1894, monobasic.

Epeolus (Monoepeolus) Gribodo, 1894:80. Type species: *Apis variegata* Linnaeus, monobasic.

Pyrrhomelecta Ashmead, 1899a: 66. Type species: *Epeolus glabratus* Cresson, 1878, by original designation.

Argyroselenis Robertson, 1903c: 284. Type species: *Triepeolus minimus* Robertson, 1902, by original designation.

Oxybiastes Mavromoustakis, 1954: 260. Type species: *Oxybiastes bischoffi* Mavromoustakis, 1954, by original destination.

The minute conical setae on the distal margins of the processes of S6 of the female, making these margins seem denticulate (Fig. 95-3d), are diagnostic for *Epeolus*. Various authors have sought characters to distinguish males of *Epeolus* from those of *Triepeolus*. *Epeolus* species are usually smaller than *Triepeolus* and usually have two-segmented rather than three-segmented maxillary palpi, but these characters do not hold. A few species of *Epeolus* have two submarginal cells instead of three; the same is true of *Triepeolus*.

Oxybiastes was based on an unusually hairy species that lacks the usual pattern of short, pale hairs. *Argyroselenis* was based on a species with three-segmented maxillary palpi. Richards (1937), Linsley and Michener (1939), Pittioni (1945), Lith (1956), Iuga (1958), Mitchell (1962), Brumley (1965), and Rightmyer (2004) illustrated male genitalia, sterna, and other structures of both sexes.

In Michener (2000) *Trophocleptria* was treated as a subgenus of *Epeolus* with the remark that it is so distinctive that it might well be considered a genus. Rightmyer (2004), however, found that *Trophocleptria* renders the rest of *Epeolus* paraphyletic, so that to retain the subgeneric status of *Trophocleptria*, one should divide the remaining *Epeolus* into subgenera in ways that are not clear. To retain the information on *Trophocleptria*, I keep it distinct here but under the heading "Trophocleptria group," with "Epeolus group" for the rest of the genus.

Unlike other Epeolini, *Epeolus* species, so far as is known, are all cleptoparasites of species of *Colletes*. Possibly the processes of S6 of the female differ from those of other genera because they have to cut through the cellophane-like cell lining of *Colletes* to lay eggs between the outer and inner linings, rather than inserting them into the cell wall as is done by most Nomadinae. Some details

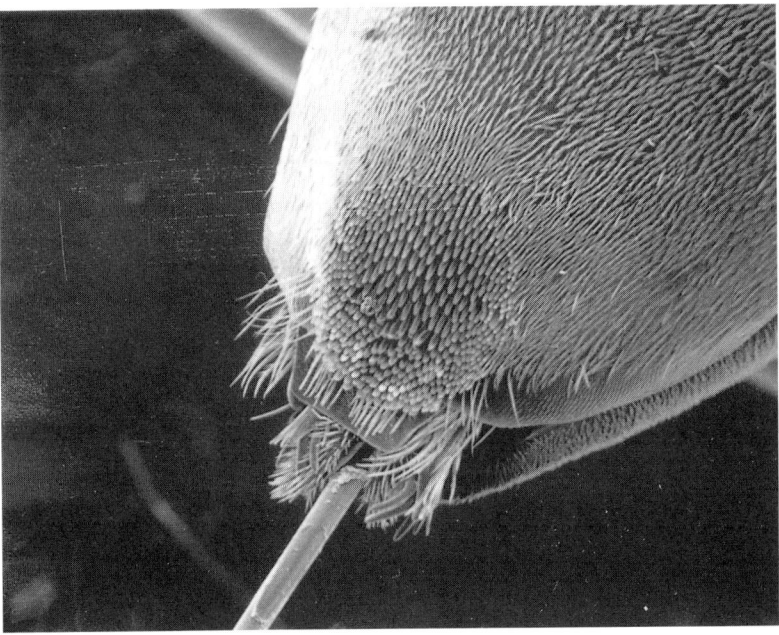

Figure 95-5. *Doeringiella guttata* Roig-Alsina, apex of female metasoma, showing the pseudopygidial area on T5; T6 is largely hidden under T5. From Roig-Alsina, 1989b.

of the egg and other aspects of cleptoparasitic behavior of *Epeolus* have been published by Rozen and Favreau (1968) and Torchio and Burdick (1988).

Key to the Groups of *Epeolus*

1. Scape, excluding basal bulb, 1.5 times as long as broad or little more; venter of mesothorax with anterior vertical part set off from horizontal part by a transverse carina, the lower part of omaular carinae continued across the venter, often irregular or broken medially and rarely weak; lateral angle of pronotum rounded but well developed, its anteroposterior length about equal to diameter of pronotal lobe (but *if* only moderately developed, then transverse ventral mesothoracic carina strong); scutellum coarsely punctate or reticulate, often produced to two posteriorly directed teeth .... *Trophocleptria* group
—. Scape, excluding basal bulb, twice as long as broad or nearly so; venter of mesothorax with anterior vertical part rounding onto horizontal part with no intervening transverse carina; lateral angle of pronotum weakly developed, its anteroposterior length much less than diameter of pronotal lobe; scutellum punctured, biconvex .... *Epeolus* group

The *Trophocleptria* group differs from the *Epeolus* group not only in the characters listed in the key above, but also in the larger stigma and the longer marginal cell, typically longer than the distance from its apex to the wing tip. These wing characters suggest the very different genus *Odyneropsis*. *Epeolus bifasciatus* Cresson, the only *Trophocleptria* in temperate North America, is intermediate in wing characters. It also has areas of white pubescence on the metasoma that are poorly developed in other species. It was regarded as intermediate between *Epeolus* s. str. and *Trophocleptria* by Michener (1954b). Male genitalia and hidden sterna of the *Trophocleptria* group were illustrated by Michener (1954b) and Rightmyer (2004).

■ The *Epeolus* group occurs across the entire holarctic region, north to the latitudes of subarctic Alaska and Finland, from coast to coast in North America, and in the Eastern Hemisphere from the Canary Islands, Portugal, and Britain to Japan. Southward, the *Epeolus* group is not known from tropical India or southeast Asia but occurs through Africa to the Cape and through the Americas to Cuba, Central America, and Colombia. About 55 species are known from North and Central America, about 35 from the palearctic region, and 11 from sub-Saharan Africa. The species of western North America were revised by Brumley (1965), those of sub-Saharan Africa were revised by Eardley (1991b). Yasumatsu (1933) and Hirashima (1955) gave keys to Japanese species; Bischoff (1930) and Lith (1956) gave accounts of palearctic species. Richards (1937) revised British species and gave an extensive account of the relations of those species and their morphs to their *Colletes* hosts.

■ The *Trophocleptria* group occurs from Connecticut, Minnesota, and Colorado, USA, south through the tropics to the province of Buenos Aires, Argentina. There are about eight species.

Genus *Odyneropsis* Schrottky

Odyneropsis consists of moderate-sized to rather large (body length, 9 to 17 mm), slender species without areas of white, appressed pubescence, except sometimes on the pronotum, propodeum, and anterior surface of T1. The entire body is sericeous because of short, appressed, brown hairs. The body shape and long wings cause large species to closely resemble *Polistes* wasps, and some species have yellowish areas reinforcing the resemblance. An unusual feature is the near absence of the maxillary palpus in most species of *Odyneropsis* s. str.; it is probably represented by a minute subspherical body in approximately the position of the palpus. By contrast, Moure (1955) observed one specimen with three segments on one side, only the basal one on the other, and some species of the subgenus *Parammobates* have three segments. The pseudopygidial area of T5 of the female is rather large, covered with dark hairs, and the posterior margin has a broad notch, from the apex of which a narrow, smooth, shining, longitudinal line extends forward, bisecting the area. This line is posterior to the median depression of T5, which is surrounded by a ridge (absent anteriorly in the subgenus *Parammobates*). The structure of T5 suggests that of *Rhogepeolus*, which, however, is a much more ordinary epeoline. Unlike that of *Rhogepeolus*, the frontal carina is strong along the summit of the interantennal prominence. S6 of the female has a relatively broad disc, as in *Epeolus* and *Rhogepeolus*; the apicolateral process is hairy and bears at its apex a row of curved, thickened setae, enlarged and broadened at their blunt apices. Illustrations of numerous structures were provided by Rightmyer (2004).

Key to the Subgenera of *Odyneropsis*

1. T5 of female with small oval middorsal depression surrounded by an apparent carina formed by short, blunt, multiridged setae; body length 14 mm or more .... *O. (Odyneropsis)* s. str.
—. T5 of female with such middorsal depression not well defined, entirely open anteriorly; body length 11 mm or less .... *O. (Parammobates)*

Odyneropsis / Subgenus *Odyneropsis* Schrottky, s. str.

Odyneropsis Schrottky, 1902a: 432. Type species: *Odyneropsis holosericea* Schrottky, 1902 = *Rhathymus armatus* Friese, 1900, by original designation.

This subgenus contains the large species of the genus that often resemble vespid wasps of the genus *Polistes*.

■ This subgenus occurs from southern Arizona, USA, and the states of Puebla and Jalisco, Mexico, south through the tropics to Bolivia, Tucumán province, Argentina, and the state of Santa Catarina, Brazil. The ten species were listed and annotated by Moure (1955).

Known hosts are species of the colletid genus *Ptiloglossa* (Rozen, 1966a, 1994b).

Odyneropsis / Subgenus *Parammobates* Friese

Parammobates Friese, 1906e: 118. Type species: *Parammobates brasiliensis* Friese, 1906, monobasic.

Parammobates consists of the smaller species of *Odyneropsis* referred to by Michener (2000).

■ The four species placed in *Parammobates* are known from Ecuador, Colombia, Trinidad, and Brazil. The diversity among species is described by Rightmyer (2004).

Genus *Pseudepeolus* Holmberg

Pseudepeolus Holmberg, 1886b: 234, 284. Type species: *Pseudepeolus fasciatus* Holmberg, 1886, monobasic.
Pseudopeolus Ashmead, 1899a: 80, unjustified emendation of *Pseudepeolus* Holmberg, 1886.
Doeringiella (Stenothisa) Moure, 1954b: 277. Type species: *Doeringiella angustata* Moure, 1954, by original designation.

Pseudepeolus consists of species that differ from *Doeringiella* in that the preoccipital carina at its upper corners is absent, rounded, not angulate. See also the key to genera; the short scape is usually distinctive.

Pseudepeolus fasciata (Holmberg) has only two submarginal cells, whereas *P. angustata* Moure has three, the second unusually small and almost pointed on the marginal cell. There is considerable variation in this character, even from right wing to left wing of a single individual. In the male the fringes of long hairs on S3 to S5 that are prevalent in the Epeolini are rudimentary in most species. Roig-Alsina (1989b) indicated the synonymy of *Stenothisa* and *Pseudepeolus;* I have not seen *P. fasciata,* the type species of the latter name.

■ *Pseudepeolus* is known from the province of Formosa, Argentina, north to Peru and to Pará, Brazil. Roig-Alsina (2003) revised the genus, recognizing five species.

Genus *Rhinepeolus* Moure

Rhinepeolus Moure, 1955: 115. Type species: *Epeolus rufiventris* Friese, 1908, by original designation.

Like that of other members of the subtribe Thalestriina, S6 of the female *Rhinepeolus* consists largely of the two longitudinal processes with their blunt, curved, thickened setae apically, in addition to many slender hairs. The most easily seen distinctive feature of *Rhinepeolus* is the bituberculate scutellum, which is one of the characters that led A. Roig-Alsina (in litt., 1995) to consider *Rhinepeolus* as more closely related to *Thalestria* than to *Doeringiella*, a view not supported by Rightmyer's (2004) phylogenetic analysis. Illustrations of male genitalia, hidden sterna, and other structures were given by Rightmyer (2004). T1 and T2 are red; the body length is 8.0 to 9.5 mm.

■ *Rhinepeolus* occurs from Río Negro and Buenos Aires provinces, Argentina, north to Tucumán province and Paraguay. The only species is *R. rufiventris* (Friese).

Genus *Rhogepeolus* Moure

Rhogepeolus Moure, 1955: 117. Type species: *Rhogepeolus bigibbosus* Moure, 1955, by original designation.
Coptepeolus Moure, 1955: 120. Type species: *Coptepeolus emarginatus* Moure, 1955, by original designation.

Roig-Alsina (1996) established the synonymy of *Coptepeolus* with *Rhogepeolus* and selected the latter as the name for the genus, the two having been published simultaneously. This genus consists of unusually robust Epeolini, 7.5 to 9.0 mm in body length. S6 of the female has a substantial disc, as in *Epeolus*. The lateroapical processes, however, are not at all spatulate but taper toward blunt apices, which are armed with several large, straight, pointed, spinelike setae; the processes also have numerous long hairs. The scutellum is more strongly biconvex than in related genera. Roig-Alsina (1996) and Rightmyer (2004) illustrated the male genitalia, sterna, and other structures.

■ *Rhogepeolus* is known from the province of Río Negro, Argentina, north to the state of Ceará, and west to Pará, Brazil. The four species then known were revised, including a key to the species, by Roig-Alsina (1996); Rightmyer (2003) updated the key with a fifth species. The hosts are unknown.

Genus *Thalestria* Smith

Thalestria Smith, 1854: 283. Type species: *Thalestria smaragdina* Smith, 1854 = *Euglossa spinosa* Fabricius, 1804, monobasic.

This genus consists of a large (12-19 mm long) metallic blue or green species that superficially resembles a rather slender *Mesoplia* (Ericrocidini). S6 of the female has a reduced disc, as in *Doeringiella,* and extremely long, slender apicolateral processes, about as long as the hind tarsi, armed at the tips with tapering, pointed, spinelike setae. As is *Rhinepeolus,* the scutellum is strongly bituberculate.

■ *Thalestria* occurs from Misiones province, Argentina, and Bolivia north to Costa Rica. It has not been found in Panama and is primarily South American. There is probably only one species, *T. spinosa* (Fabricius).

Thalestria is a cleptoparasite in nests of *Oxaea* (Oxaeinae) (Bertoni, 1911).

Genus *Triepeolus* Robertson

Triepeolus Robertson, 1901: 231. Type species: *Epeolus concavus* Cresson, 1878, by original designation.
Triepeolus (Synepeolus) Cockerell, 1921b: 6. Type species: *Triepeolus insolitus* Cockerell, 1921, monobasic.

This is the *Doeringiella,* subgenus *Triepeolus,* of Michener (2000). It is by far the largest genus of the subtribe Thalestriina and the only one represented in the Eastern Hemisphere. Rightmyer's phylogenetic analysis indicates that if *Triepeolus* is included within an expanded genus *Doeringiella,* then *Thalestria* must also be included. *Thalestria* is the oldest name, so species of *Doeringiella, Triepeolus,* as well as the small genera *Rhinepeolus* and *Pseudepeolus,* would all have to be transferred to *Thalestria. Thalestria* is so distinctive that this seems undesirable; moreover the phylogeny should be verified before changing the scientific names of most of the species of Epeolini. I therefore follow Rightmyer in recognizing *Triepeolus* as a genus.

Triepeolus consists of strikingly marked (by areas of pale plumose hairs) species closely resembling species of *Doeringiella* and the like. As indicated in the key to genera, small species of *Triepeolus* closely resemble *Epeolus* s. str.; males cannot always be distinguished by any generic characters except those of the genitalia (see the key to genera). Females, however, differ strikingly in the reduced disc and very long processes of S6 (Fig. 95-3c), armed at and near their apices with several course, curved, blunt bristles in addition to hairs, much as in other genera of the subtribe Thalestriina. Also as in the others, T5 of the female nearly always differs from that of the genus *Epeolus* in the relatively large, apically dark (not silvery) pseudopygidial area (Fig. 95-3a, 95-5) (but see comment below on *T. epeolurus* Rightmyer). Three-segmented maxillary palpi usually distinguish most *Triepeolus* from most *Epeolus*, which usually has two-segmented palpi, but this character is variable in both *Epeolus* and *Triepeolus*. Male genitalia and other structures of both sexes were illustrated by Linsley and Michener (1939), Michener (1954b), Iuga (1958), Mitchell (1962), Rozen (1989a), and Rightmyer (2004).

■ The genus *Triepeolus* ranges from coast to coast in North America, and from southern Canada south through the Antilles, Mesoamerica, and South America to western Peru (not Chile), Paraguay, and the province of Río Negro, Argentina. In the palearctic region, *Triepeolus* occurs in central and southern Europe east through Russia to Japan. About 120 species are found in North America, including nearctic Mexico; but in the neotropics only about 19 species are known, including six from the Antilles. From Eurasia only two species are known. Moure (1955) listed and described the seven South American species. In the palearctic region, *T. tristis* (Smith) and *ventralis* (Meade-Waldo) were differentiated by Bischoff (1930). They do not always cluster with New World *Triepeolus* in Rightmyer's (2004) phylogenetic analyses. An unusual Mesoamerican species, *T. epeolurus* Rightmyer, has a subapical transverse band of silvery hairs on T5 of females, suggesting *Epeolus*.

Most species of *Triepeolus* are cleptoparasites in nests of Eucerini (*Melissodes, Peponapis, Svastra, Synhalonia, Tetraloniella, Xenoglossa*), and the abundance and diversity of these bees in North America must be related to the great diversity of species of *Triepeolus* on that continent. Some species of *Triepeolus*, however, parasitize other hosts, including *Anthophora, Centris,* and *Melitoma* in the Apinae, *Ptiloglossa* in the Colletidae, *Protoxaea* in the Oxaeinae, and *Dieunomia* in the Halictidae (Hurd, 1979; Rozen 1989a). The egg of *Triepeolus*, as illustrated by Torchio (1986) and Rozen (1989a), is inserted directly into the wall of the host cell, only the subtruncate end and a surrounding flange being exposed. The hatching of the larva, with its large head and jaws, was described by Torchio.

96. Tribe Ammobatoidini

This tribe was called the Neopasitini by Linsley and Michener (1939) and Michener (1944), and the Holcopasitini by Rozen (1966a). Recognition of Neopasitini (*sensu* Michener, 1944) or Holcopasitini suggests the placement of *Ammobatoides* in a different tribe from *Holcopasites* and its relatives, but the differences lie mostly in size and size-related characters such as the stigmal size; *Ammobatoides* is large with a relatively small stigma. The name Ammobatoidini has priority over Holcopasitini (Michener, 1986a), and the name Neopasitini is not applicable to members of this tribe, since *Neopasites* is a genus of the Biastini. See the account of the subgenus *Neopasites* (Sec. 97) for details.

These are pasitiform bees, often coarsely punctate, commonly with a red metasoma, and with short, sparse pubescence that commonly includes patches of pale, often white, scalelike appressed hairs, the metasoma thus often having a spotted rather than banded appearance (Fig. 96-1a). In banded species, however, the metasomal bands are usually at the bases of terga rather than at the apices as they are in most bees, and are usually broken medially or, in *Aethammobates,* there are both basal and apical bands; in *Ammobatoides,* however, the bands are apical, although in females largely missing dorsally except on T4. The labrum appears longer than broad (Fig. 96-3), although sometimes only as long as broad, particularly in small species. In both sexes, there is a pygidial plate, that of the male well defined on a strongly projecting process, that of the female almost always hidden by T5, which has a longitudinal median rounded ridge and no recognizable pseudopygidial area. The ridge and sometimes adjacent parts of T5 are minutely punctate and may represent the pseudopygidial area. The marginal cell is rounded at the apex, which is away from the wing margin (Fig. 96-2); the second submarginal cell is much smaller than the first, and the first recurrent vein joins either cell. S5 of the female is apically emarginate. S6 of the female (Fig. 91-2a) is similar to that of Biastini. (The female of *Aethammobates* is unknown.)

Except for a South African *Ammobatoides* species, this is a holarctic tribe of four genera, one or two of which are so similar to *Schmiedeknechtia* that they could reasonably be considered as subgenera.

Key to the Genera of the Ammobatoidini

1. Body length 10 mm or more; eyes of male strongly converging above; inner orbits of female gently concave (palearctic) .... *Ammobatoides*
—. Body length 7.5 mm or less; eyes of male diverging above; inner orbits of female straight or slightly convex 2
2(1). Pronotum with well-developed, horizontal dorsal collar surface extending across between dorsolateral angles; scutum without vertical anterior surface; T1 with anterior surface deeply concave, separated from dorsal surface by strongly arcuate carina that forms strong anterolateral tergal angles (Egypt) .... *Aethammobates*
—. Pronotum with dorsal surface of collar broadly interrupted medially, where scutum curves down to form vertical anterior surface; T1 with anterior surface shallowly to deeply concave, without or with only suggestion of carina between anterior and dorsal surfaces, anterolateral part of T1 rounded 3
3(2). Male antenna 12-segmented; middle flagellar segments broader than long; metasomal terga with graduli strong, shelflike, a deep transverse furrow immediately behind each gradulus on at least T2 to T4; clypeus and nearby structures dark (nearctic) .... *Holcopasites*
—. Male antenna 13-segmented; middle flagellar segments as long as or longer than broad; metasomal terga with weaker graduli, not emphasized by furrow behind each gradulus; clypeus (at least lower margin), bases of mandibles, labrum, and antennal scape yellow (palearctic) .... *Schmiedeknechtia*

Genus *Aethammobates* Baker

Aethammobates Baker, 1994: 155. Type species: *Aethammobates prionogaster* Baker, 1994, by original designation.

This genus, known from a single male specimen, differs from the other genera in the striking characters indicated in the key. Some of the other characters suggest that it may be more like a large species (body length 7.5 mm) of the otherwise small *Schmiedeknechtia* than was recognized when *Aethammobates* was described. The middle femora are much expanded below; an approach to this feature occurs in large species of *Holcopasites*. The posterior margins of the discs of the terga are elevated and strongly denticulate; this feature is also approached in large *Holcopasites*. Thus some of the apparent generic characters may be merely features of large species of the *Holcopasites-Schmiedeknechtia* group. In the 13-segmented, rather long antennae and probably in the lack of a groove behind each tergal gradulus, *Aethammobates* resembles *Schmiedeknechtia*.

■ *Aethammobates* is known only from Egypt. The one species is *A. prionogaster* Baker.

Genus *Ammobatoides* Radoszkowski

Ammobatoides Radoszkowski, 1867: 82 (not Schenck, 1869). Type species: *Philerus abdominalis* Eversmann, 1852, by designation of Sandhouse, 1943: 525.
Phiarus Gerstaecker, 1869: 147. Type species: *Philerus abdominalis* Eversmann, 1852, monobasic.
Euglages Gerstaecker, 1869: 149. Type species: *Euglages scripta* Gerstaecker, 1869, monobasic.
Paidia Radoszkowski, 1872b: 10 (not Herrich-Schaffer, 1847), unnecessary replacement for *Ammobatoides* Radoszkowski, 1868. Type species: *Philerus abdominalis* Eversmann, 1852, autobasic.
Paedia Dalla Torre, 1891: 147, unjustified emendation of *Paidia* Radoszkowski, 1872.

This genus contains the largest pasitiform bees except for those in the Malagasy genus *Melanempis* and some large species of *Pasites* in the Ammobatini; the body

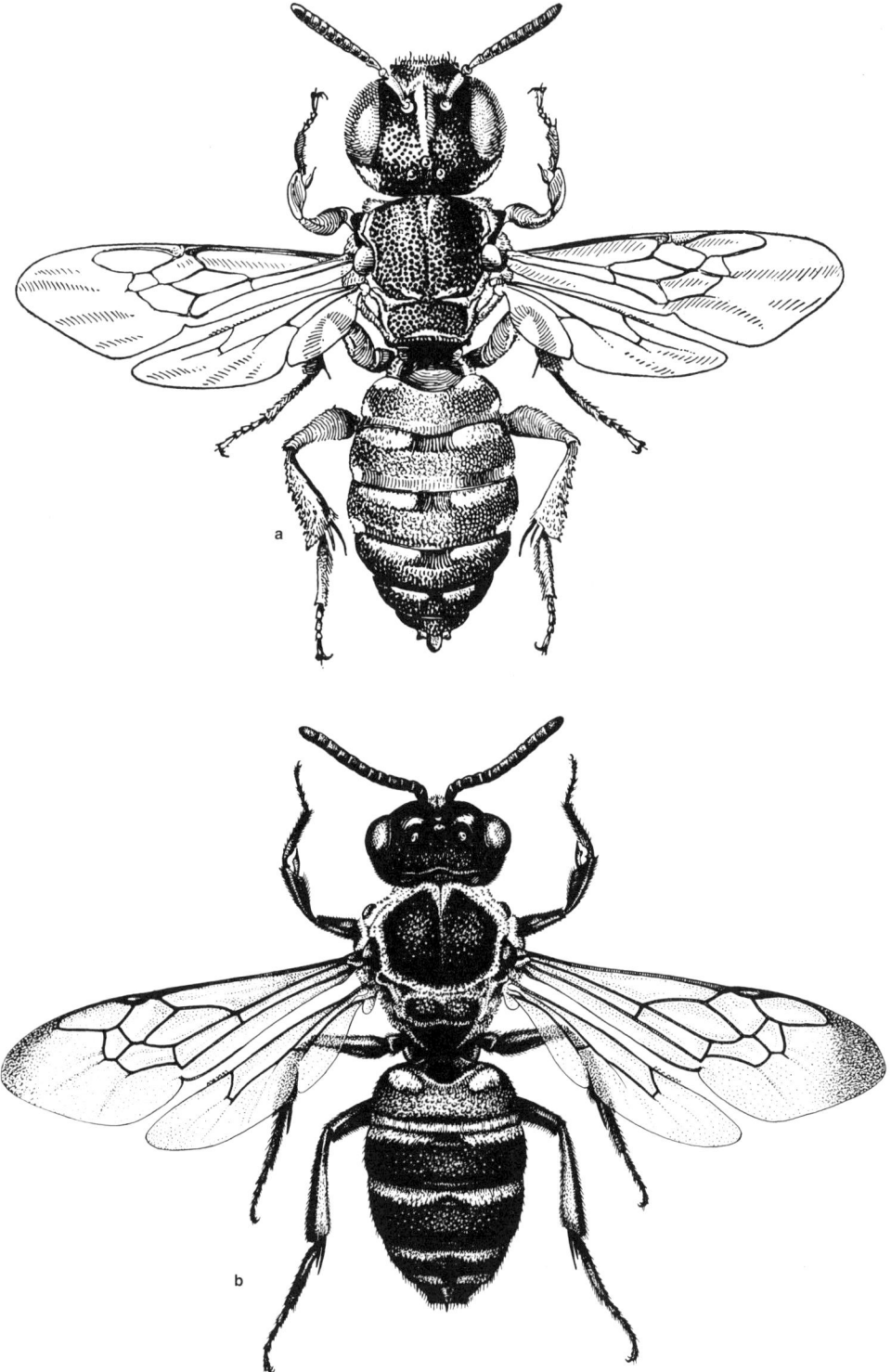

Figure 96-1. *Holcopasites.* **a,** Male of *H. arizonicus* (Linsley); **b,** Female of *H. bigibbosus* Hurd and Linsley. a, drawing by F. Abernathy, from Linsley, 1943b; b, drawing by C. Green, from Hurd and Linsley, 1972.

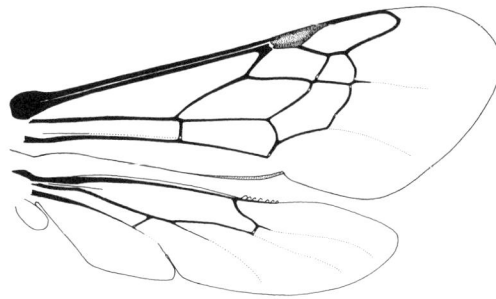

Figure 96-2. Wings of *Holcopasites heliopsis* (Robertson).

length is 10 to 14 mm. Unlike other Ammobatoidini, there are no median bands of pale pubescence on the metasomal terga, but only apical bands, frequently reduced and evident only laterally. The eyes of the male, converging upward and at most two ocellar diameters from the lateral ocelli, are unique in the Nomadinae. The male gentialia and hidden sterna were illustrated by Radoszkowski (1885), Popov (1933), and Schwarz (1993a). The apical tergum and sternum of the female were illustrated by Iuga (1958).

■ *Ammobatoides* occurs from Morocco and Spain eastward through Turkey and Russia to China with a disjunct species in Cape Province, South Africa (Bischoff, 1923). There are six species. The palearctic species were revised by Popov (1933), those of the western palearctic region by Warncke (1982).

The hosts of *Ammobatoides* are *Melitturga* and *Meliturgula* (Panurginae), as summarized by Popov (1933).

Genus *Holcopasites* Ashmead

Holcopasites Ashmead, 1899a: 82, no species. Type species: *Phileremus illinoiensis* Robertson, 1891, by designation of Crawford, 1915: 123.

Neopasites (Trichopasites) Linsley, 1942: 127. Type species: *Neopasites insoletus* Linsley, 1942, by original designation. [New synonymy.]

Neopasites (Odontopasites) Linsley, 1942: 128. Type species: *Neopasites arizonicus* Linsley, 1942, by original designation. [New synonymy.]

For comments on the confusion with *Neopasites,* see the text for the subgenus *Neopasites* in Section 97 and Table 97-1.

Odontopasites and *Trichopasites* seem to represent small derived groups of *Holcopasites* remarkable principally for their hairy eyes; the eyes are bare in other Ammobatoidini. *Holcopasites* and *Schmiedeknechtia* are so similar that they could well be given subgeneric rather than generic status. In addition to the generic characters indicated in the key to genera, most *Holcopasites* differ from *Schmiedeknechtia* in having the clypeus protuberant, in side view extending in front of the eye by about half an eye width. In the minute species of *Holcopasites,* however, this character fails, the head shape being more like that of *Schmiedeknechtia,* which consists of minute species. In the minute species, also, the usually elongate labrum (Fig. 96-3a) is sometimes only about as long as broad. Male genitalia, hidden sterna, and S6 of the female were illustrated by Linsley and Michener (1939); see also Popov (1933), Mitchell (1962), and Figures 91-1a and 91-2a. The body length varies from 2.5 to 8.0.

■ *Holcopasites* occurs from Alberta, Canada, and Maine to Georgia, west to Idaho and California, USA, and south to Guerrero and Veracruz, Mexico. The 16 species were revised by Linsley (1943b), as *Neopasites,* and by Hurd and Linsley (1972).

Hosts, listed by Hurd and Linsley (1972), include *Calliopsis (Calliopsis* s. str., *Calliopsima, Hypomacrotera), Pseudopanurgus,* and *Protandrena (Heterosarus, Metapsaenythia, Pterosarus)* (all Panurginae). Observations on eggs and immature stages of *Holcopasites* were by Rozen (1965b, 1989c), and the larva was described by Rozen (1966a).

Genus *Schmiedeknechtia* Friese

Schmiedeknechtia Friese, 1896b: 277. Type species: *Schmiedeknechtia oraniensis* Friese, 1896, monobasic.

Schmiedeknechtia (Cyrtopasites) Mavromoustakis, 1963d: 753. Type species: *Schmiedeknechtia verhoeffi* Mavromoustakis, 1959, by original designation.

As indicated in the discussion of *Holcopasites,* this palearctic genus is similar to the nearctic *Holcopasites.* The account of that genus and the key to genera indicate the principal differences. All known *Schmiedeknechtia* species are small, 4 to 5 mm in body length; *Aethammobates* may be merely a large *Schmiedeknechtia* with structures similar to those of large *Holcopasites.* Male genitalia and hidden sterna were illustrated by Popov (1933) and Schwarz (1993a).

■ *Schmiedeknechtia* is found in Algeria, Tunisia, and possibly Spain east to Turkey and Uzbekistan. The five species were revised by Schwarz (1993a).

The hosts of *Schmiedeknechtia* are said to be *Camptopoeum* (Panurginae), in some cases of the subgenus *Epimethea* (Schwarz, 1993a).

I am indebted to K. W. Cooper of Riverside, California, and the late D. B. Baker of Oxford University for clarifying and emphasizing characters of *Holcopasites* and *Schmiedeknechtia.* Schwarz (1993a) makes a strong case for the separation of *S. verhoeffi* Mavromoustakis in a separate subgenus, *Cyrtopasites,* since it is quite different, especially in the form of the scutellum, from the other species.

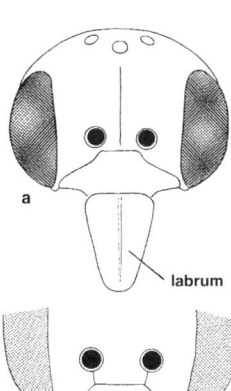

Figure 96-3. Shape of labrum in "pasitine" bees. **a,** *Holcopasites arizonicus* (Linsley), female; **b,** *Neopasites* sp.? From Michener, McGinley, and Danforth, 1994.

97. Tribe Biastini

This is a holarctic tribe of small, black epeoliform bees, often with the metasoma red; thus the appearance is as in various groups of pasitine bees (Fig. 97-1). The pubescence is not largely scalelike, as it is in the Townsendiellini and Neolarrini, although the hairs may be short and plumose. The labrum is much broader than long (Fig. 96-3b). The antennae are short, the scape less than twice as long as broad. T5 of the female has a pseudopygidial area or a median marginal lunule with slightly distinctive short pubescence. T6 usually lacks a pygidial plate in the female, but in *Rhopalolemma* the plate is recognizable. The marginal cell is pointed or narrowly rounded, the apex bent away from the wing margin (Figs. 97-2, 97-3b), and there are two submarginal cells, the second at least two-thirds as long as the first and receiving both recurrent veins, but in some specimens of *Rhopalolemma* there are three submarginal cells. The apex of S6 of the female has two lobes, each bearing a few coarse, blunt bristles, these lobes separated by a broad concavity, as illustrated by Linsley and Michener (1939) and Roig-Alsina (1991b); see also Figure 91-2b. The sting is reduced, sometimes bifurcate at the tip, and the gonostylus of the female is enlarged medially (in *Rhopalolemma* apically), unlike that of other bees.

The recent discovery of the genus *Rhopalolemma* demolished some of the long-recognized tribal characters. For example, the Biastini were unique among small pasitiform bees because of their preapical mandibular tooth, but *Rhopalolemma* has simple mandibles.

So far as is known, all Biastini are cleptoparasites in nests of Rophitinae; known hosts for *Biastes* are *Dufourea, Rophites,* and *Systropha;* for *Neopasites, Dufourea;* and for *Rhopalolemma, Protodufourea.*

Key to the Genera of the Biastini

1. Mandible simple; T6 of female produced medially, with pygidial plate; gonostylus of female clubbed; scape 1.75 times as long as its maximum diameter (basal bulb excluded) (nearctic) .... *Rhopalolemma*
—. Mandible with preapical tooth; T6 of female without indication of pygidial plate, produced to sharp point at each side, the margin between lateral projections strongly, broadly concave, hairy; gonostylus of female thickened medially; scape about 1.5 times as long as its maximum diameter (basal bulb excluded) 2
2(1). T5 of female with small, raised, semicircular, sloping pseudopygidial area covered with appressed pale hairs; metanotum with median elevation well above level of adjacent part of scutellum; second submarginal cell longer than first (palearctic) .... *Biastes*
—. T5 with ill-defined lunule of pale or black hairs, not raised, not sloping, and usually broader than semicircular; metanotum without median elevation, not or little above level of adjacent part of scutellum; second submarginal cell as long as or shorter than first (nearctic) .... *Neopasites*

Genus *Biastes* Panzer

Biastes Panzer, 1806: 239. Type species: *Tiphia brevicornis* Panzer, 1798, monobasic.

Rhineta Illiger, 1807: 198. Type species: *Nomada schottii* Fabricius, 1804 = *Tiphia brevicornis* Panzer, 1798, monobasic.

Melittoxena Morawitz, 1873: 154. Type species: *Nomada truncata* Nylander, 1848, monobasic.

Biastoides Schenck, 1874: 252. Type species: *Pasites punctatus* Schenck, 1870 = *Phileremus emarginatus* Schenck, 1853, monobasic.

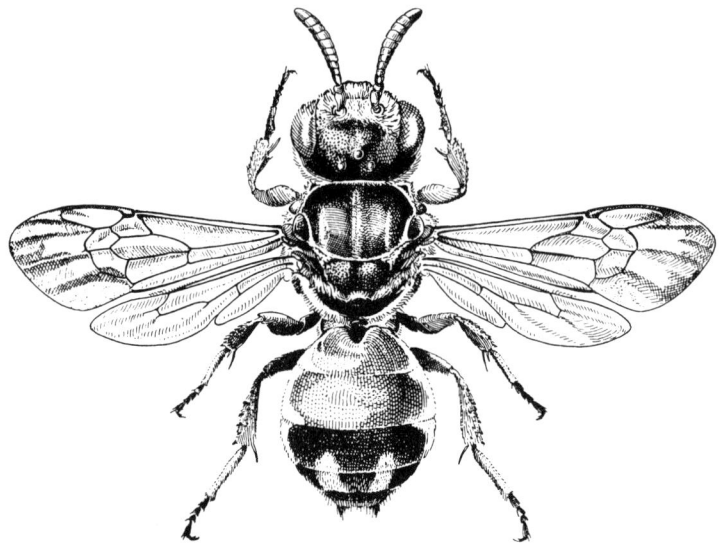

Figure 97-1. *Neopasites fulviventris* (Cresson), female. Length 8 mm. Drawing by F. Abernathy, from Linsley, 1943c.

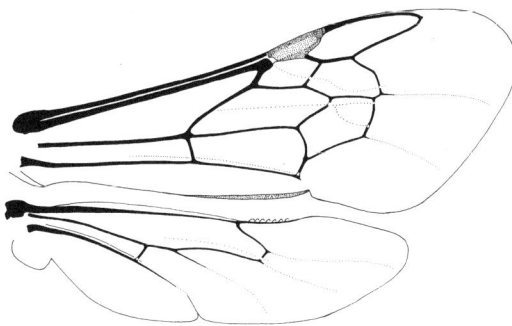

Figure 97-2. Wings of *Biastes brevicornis* (Panzer).

Biastes is similar to *Neopasites;* the two could be considered as only subgenerically distinct. The principal differences are indicated in the key to genera. In some species the male antennae have only 12 segments. The body length is 5 to 9 mm. In species that I have studied, the maxillary palpus is less reduced than that in *Neopasites,* four- or five-segmented. Male genitalia and hidden sterna were illustrated by Popov (1933a), who recognized *Biastoides* and *Melittoxena* as subgenera. These structures were also illustrated by Radoszkowski (1885) and Iuga (1958). For comments on S6 of the female, see the account of *Neopasites.*

■ This genus occurs in Europe, north to 55°N in Finland, and east to the Urals and Caucasus. The four species were revised by Warncke (1982).

Species of this genus are believed to parasitize *Dufourea, Rophites,* and *Systropha* (Rophitinae). The larva was briefly described by Rozen (1993a). The egg is laid lying in a groove in the host cell wall, parallel to the surface. Thus one whole side of the egg, rather than just one end, is exposed.

Genus *Neopasites* Ashmead

This is the North American counterpart of *Biastes,* from which it differs in the characters indicated in the key to genera. In *Neopasites* the punctation of the body is finer than that in some *Biastes,* but the general appearance is similar. Michener (1944) indicated that S6 of the female in *Biastes* might differ from that of *Neopasites,* because Grütte (1935) illustrated the sternal disc in *Biastes* as reduced to a transverse bar while Linsley and Michener (1939) illustrated a moderate-sized disc for *Neopasites.* Actually, in both genera there is a well sclerotized, curved bar along the emargination between the two apical lobes, and the rest of the disc, which is rather membranous, was omitted by Grütte. The structure is alike in the two genera; see Figure 91-2b. The male genitalia and apical sterna of both sexes were illustrated by Linsley and Michener (1939); see Figure 91-3e-g.

Neopasites was revised by Linsley (1943c) under the name *Gnathopasites.*

Key to the Subgenera of *Neopasites*

1. Maxillary palpus shorter than or as long as greatest galeal width, one- or two-segmented; second submarginal cell about as long as first; body length 3.5 to 6.0 mm .... *N. (Micropasites)*
—. Maxillary palpus much longer than greatest galeal width, four-segmented; second submarginal cell much shorter than first; body length 6 to 8 mm. .... *N. (Neopasites* s. str.)

Neopasites / Subgenus *Micropasites* Linsley

Gnathopasites (Micropasites) Linsley, 1942: 130. Type species: *Neopasites cressoni* Crawford, 1916, by original designation.

■ *Micropasites* is found from California to New Mexico, USA. The three species were revised by Linsley (1943c); at least one undescribed species is known.

The hosts of *Micropasites* are in the genus *Dufourea* (Rophitinae); Torchio et al. (1967) gave an account of the biology of two species, and Rozen (1966a) described the larva.

Neopasites / Subgenus *Neopasites* Ashmead s. str.

Neopasites Ashmead, 1898: 284. Type species: *Phileremus fulviventris* Cresson, 1878, by original designation.
Gnathopasites Linsley and Michener, 1939: 272. Type species: *Phileremus fulviventris* Cresson, 1878, by original designation.

■ *Neopasites* s. str. is known only from California. The two species were revised by Linsley (1943c).

Neopasites s. str. is a parasite of *Dufourea* (Rophitinae), according to Torchio et al. (1967).

Ashmead's (1898) description of *Neopasites,* as well as the specimen on which he based it, show that he actually had before him a specimen of what is now called *Holcopasites,* which he misidentified as *Phileremus fulviventris* Cresson. Linsley and Michener (1939) therefore transferred the name *Neopasites* to *Holcopasites* and proposed a new name, *Gnathopasites,* for what is now called *Neopasites,* as shown in Table 97-1. The Code, 3rd ed., art. 70, however, specifies that the *name* used in designating a type species is to be followed, unless the Commission rules otherwise. There is no need to request a Commission opinion for uncommon and little-known insects like these, especially since for many years the Linsley and Michener proposal has been rejected by melittologists and the names have been applied as they are here.

Genus *Rhopalolemma* Roig-Alsina

Rhopalolemma Roig-Alsina, 1991b: 33. Type species: *Rhopalolemma robertsi* Roig-Alsina, 1991, by original designation.

Table 97-1. Equivalents of Certain Generic Names in the Nomadinae.

Tribe	Linsley & Michener (1939)	Other authors and current usage
Biastini	*Gnathopasites*	*Neopasites*
Ammobatoidini	*Neopasites*	*Holcopasites*

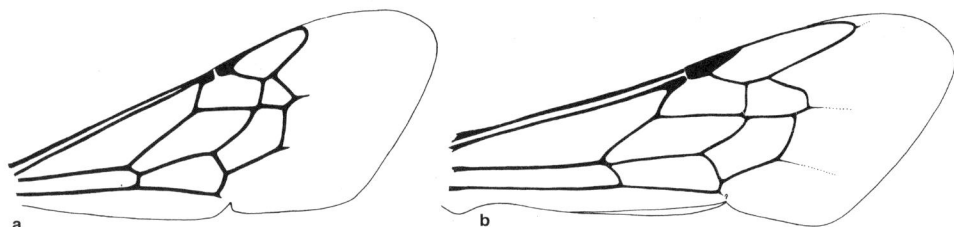

Figure 97-3. Forewings of "pasitine" bees. **a,** *Townsendiella californica* Michener; **b,** *Rhopalolemma robertsi* Roig-Alsina. From Michener, McGinley, and Danforth, 1994.

Rhopalolemma differs from other Biastini in several characters that are plesiomorphic relative to the other genera. Examples are the longer scape, larger sting valves, the presence of a pygidial plate on a more or less ordinarily shaped T6 in the female and occupying most of the dorsum of T7 in the male, and the presence of three submarginal cells in some individuals. The wing of a two-celled individual is shown in Figure 97-3b. Roig-Alsina (1991b) and Rozen, Roig-Alsina, and Alexander (1997) illustrated diverse structures. The body length is 4.8 to 8.0 mm.

■ This genus occurs in desertic areas of southern California and Arizona, USA. The two species were distinguished by Rozen, Roig-Alsina, and Alexander (1997).

The host is *Protodufourea* (Rophitinae) (Rozen, Roig-Alsina, and Alexander, 1997). As in *Biastes* and *Hexepeolus* (Hexepeolini), the relatively straight egg is laid in a groove in the host cell wall, and one side, the exposed surface, is corrugated.

98. Tribe Townsendiellini

This is a North American tribe of pasitiform bees (Fig. 98-1) having short, scalelike, appressed pubescence suggesting that of Neolarrini. Unlike that of the Neolarrini, however, the marginal cell is far larger, longer than the stigma (Fig. 97-3a). There are two submarginal cells, the second much shorter than the first but not extremely shortened as it is in Neolarrini. Both sexes have a well-developed pygidial plate, and there is no indication of a pseudopygidial area on T5 of the female. S6 of the female is broadly emarginate at the apex, the lobes on either side of the emargination bearing spinelike setae (Fig. 91-2c).

The mature larva was described by Rozen and McGinley (1991), and the first-stage larva was illustrated by Rozen (1991a). As in other Nomadinae, the head of the first-stage larva is sclerotized; the mandibles are large and acute (Fig. 91-6c), although not so excessively developed as they are in Epeolini.

The only genus is *Townsendiella*.

Genus *Townsendiella* Crawford

Townsendiella Crawford, 1916: 136, 138. Type species: *Townsendiella pulchra* Crawford, 1916, monobasic.

Townsendiella (Xeropasites) Linsley, 1942: 130. Type species: *Townsendiella rufiventris* Linsley, 1942, by original designation.

Townsendiella (Eremopasites) Linsley, 1942: 131. Type species: *Townsendiella californica* Michener, 1936, by original designation.

Townsendiella exhibits the tribal characters cited above. It consists of small species, 4 to 6 mm in body length. Although there are only three species, each is quite different from the others, and each has been placed in a subgenus of its own. I hesitantly reject these subgenera as unnecessary. *T. rufiventris* Linsley has four-segmented maxillary palpi, unlike the six- segmented palpi of the other species. In *T. californica* Michener the marginal cell is much shorter than the distance from its apex to the wing tip (Fig. 97-3a), whereas it is as long as or longer than that distance in the other species. The male genitalia and apical sterna of both sexes were illustrated by Linsley and Michener (1939).

■ *Townsendiella* occurs from central California to New Mexico, USA, and Baja California to Coahuila, Mexico, in arid areas. The three species were revised by Linsley (1943a).

Townsendiella californica Michener was collected in a nest aggregation of *Hesperapis (Zacesta) rufipes* (Ashmead) (Dasypodainae), and *T. pulchra* Crawford is a cleptoparasite of *Hesperapis (Panurgomia) larreae* Cockerell (Dasypodainae).

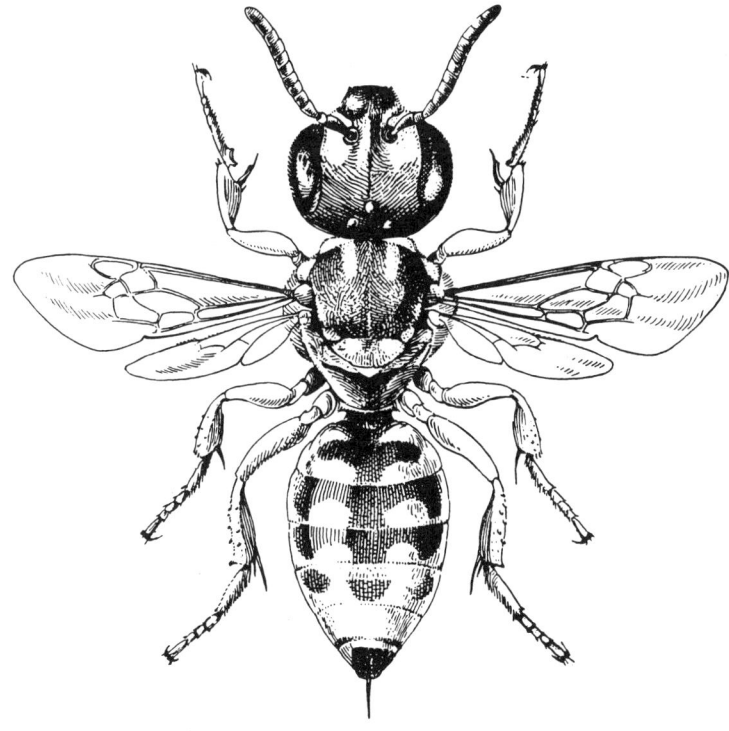

Figure 98-1. *Townsendiella californica* Michener, female. Drawing by F. Abernathy, from Linsley, 1943a.

99. Tribe Neolarrini

This nearctic tribe, consisting of the single genus *Neolarra*, contains minute to small, slender, pasitiform bees, black or with the metasoma red, the body pruinose because of being more or less covered with short, appressed, scalelike, pale pubescence. The wing venation is distinctive, often not extending beyond the middle of the forewing (Fig. 99-1); the broadly truncate marginal cell is little if any longer than the stigma, and the second submarginal cell is either absent or very short, receiving only the second recurrent vein, which may be absent. Veins C and R of the forewing are very close together. In small species, e.g., of the subgenus *Phileremulus,* these veins are in contact, leaving no costal cell between them; therefore they do not run properly in most keys to superfamilies of Hymenoptera. T5 of the female is sometimes notched or cleft midapically, but it has nothing resembling a pseudopygidial area. S6 of the female has a narrow apical emargination, and the lobe on each side lacks coarse setae (Fig. 91-2g, h). The male lacks a pygidial plate but T7 is produced to a short, bare process. The male genitalia and apical sterna of both sexes were illustrated by Linsley and Michener (1939).

The tribe Neolarrini is found only in North America; its species are cleptoparasites in the nests of *Perdita* and perhaps *Calliopsis (Micronomadopsis)* (Panurginae) (Shanks, 1978). The larva was described by Rozen (1966a).

Neolarrini is not particularly similar to any other tribe; its closest relative is best considered undecided. Michener (1944) thought it was closest to Townsendiellini but gave no analysis of characters. Rozen (1966a), on the basis of larval characters, placed it as the sister group of *Neopasites* (Biastini). Alexander (1990), using adult characters, reached the same conclusion, but other phylogenetic studies (Roig-Alsina, 1991b; Rozen, Roig-Alsina, and Alexander, 1997; see Fig. 87-5) placed it far from the Biastini, e.g., as the sister group of Ammobatini + Caenoprosopidini.

Genus *Neolarra* Ashmead

The principal characters of the genus are those of the tribe. The two subgenera are separable by the following key. Revisions were by Michener (1939b) and Shanks (1978).

Key to the Subgenera of *Neolarra*

1. Scutellum widest at or in front of middle; outer margin of axilla rounded .... *N. (Neolarra s. str.)*
—. Scutellum widest behind middle; axilla produced to acute tooth .... *N. (Phileremulus)*

Neolarra / Subgenus *Neolarra* Ashmead s. str.

Neolarra Ashmead, 1890: 8. Type species: *Neolarra pruinosa* Ashmead, 1890, monobasic.

Neolarra s. str. contains the larger species of *Neolarra,* 3 to 7 mm in body length; most species have two submarginal cells (Fig. 99-1a), although some have only one.
■ The range of *Neolarra* s. str. is from California north and east to Idaho, Montana, North Dakota, and Texas, USA, and south to Jalisco and Durango, Mexico. The 11 species were revised by Shanks (1978).

Neolarra / Subgenus *Phileremulus* Cockerell

Phileremulus Cockerell, 1895: 9. Type species: *Phileremulus vigilans* Cockerell, 1895, by original designation.

This subgenus contains minute species, 2.0 to 3.7 mm in body length, having only one submarginal cell, and lacking also the second recurrent vein (Fig. 99-1b).
■ *Phileremulus* occurs from southern Alberta and Saskatchewan, Canada, to Tennessee, Georgia, Texas, and California, USA, and Hidalgo to Baja California, Mexico. The three species were revised by Shanks (1978).

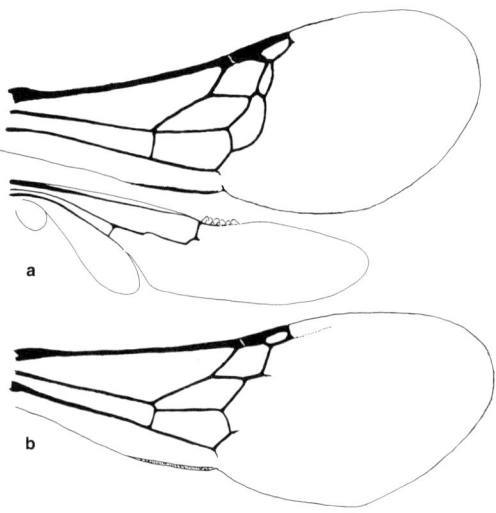

Figure 99-1. Wings of *Neolarra.* **a,** *N. (Neolarra) verbesinae* Cockerell; **b,** Forewing of *N. (Phileremulus) vigilans* (Cockerell).

100. Tribe Ammobatini

In this tribe of cleptoparasitic pasitine bees the body is typically black, the metasoma usually wholly or partly red; rarely, the head and thorax also are reddish. The pubescence is not scalelike, usually sparse, although apical patches or broken bands of white, appressed hair are common on the metasomal terga (Fig. 100-1). The long labrum (longer than broad, as in Figure 96-3, or about as long as broad) is unusual in Apidae but is shared with Ammobatoidini and Caenoprosopidini. The antennae are not so short as in the Biastini, the scape (excluding the basal bulb) being 2.5 or more times as long as broad. T5 of the female completely lacks a pseudopygidial area, and T6 usually lacks a recognizable pygidial plate, although there is a distinct area usually having erect hairs or bristles; sometimes, as in *Melanempis,* there is an area, defined by a carina, that is probably not homologous to a pygidial plate. In males there may be a well-defined projecting pygidial plate or a flat, rounded or truncate but otherwise undefined projection of T7. S5 of the female is usually strongly curved up laterally, around the apical process of S6 and the sting, and covering the lateral parts of T6. S6 of the female has a slender apical process, sometimes completely retracted; sometimes it is a simple spine but usually it is notched (Fig. 91-2i) or bifurcate; it lacks the coarse setae found in most Nomadinae.

Interesting features found in the tribe include 12-segmented male antennae in *Pasites, Melanempis,* and *Parammobatodes,* and extreme variability in the maxillary palpi. For example, in the genus *Pasites,* these palpi vary from absent to five-segmented; in *Ammobates* from one- to six-segmented (one- to five-segmented in the subgenera *Euphileremus* and *Parammobatodes*); and in *Oreopasites,* from four- to six-segmented. Both Popov (1951c) and Warncke (1983) tabulated the number of segments in palpi of various species. In *O. barbarae* Rozen the labial palpi are three-segmented instead of four-segmented. A distinctive feature of males of the tribe, not found in other bees and also absent in *Melanempis,* is a tuft or fringe of long, pale hairs arising from the lower lateral part of the clypeus and curved back adjacent to the side of the labrum.

Popov (1951c) recognized two tribes, Ammobatini and Pasitini, for the bees here included in the Ammobatini. Rozen (1992b) discussed the tribal characters in detail without deciding whether to recognize the two tribes; Rozen and McGinley (1974b) had shown that larval characters do not seem to support recognition of two tribes. In a phylogenetic study of genera and a revision of afrotropical species, Eardley and Brothers (1997) recognized only one tribe. I believe that recognition of two tribes would obscure the close relationships among all genera here included in the Ammobatini. Warncke (1983) placed the entire tribe in one genus, *Pasites.*

Key to the Genera of the Ammobatini

1. Mandibles in repose directed posteromesally, so that, except when badly worn, they cross one another well before their apices, their anterior margins forming angle of 90° to 145° 2
—. Mandibles in repose directed mesally, so that they overlap one another and cross, if at all, at extremely obtuse angle 4
2(1). Vein Rs of hind wing transverse or directed basad from costal margin, discal cell (R) thus not extending distad from base of vein Rs; first recurrent vein joining distal half of second submarginal cell, second medial cell thus greatly narrowed toward costa (palearctic, oriental) ......
.... *Parammobatodes*
—. Vein Rs of hind wing directed strongly distad, discal cell (R) thus extending beyond base of vein Rs; first recurrent vein joining basal half of second submarginal cell, second medial cell thus not greatly narrowed toward costa 3
3(2). Labrum pointed apically, entirely exposed in front of crossed mandibles in repose; male without defined pygidial plate; S5 of female without preapical collar of long hairs (Africa) .... *Sphecodopsis*
—. Labrum truncate or broadly rounded apically, extend-

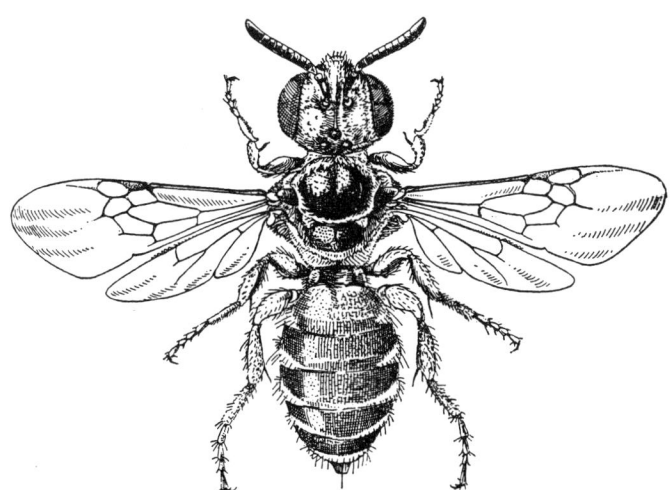

Figure 100-1. *Oreopasites euphorbiae* Cockerell, female. Drawing by F. Abernathy, from Linsley, 1941.

ing under and beyond crossed mandibles; male with pygidial plate; S5 of female with preapical V- or U-shaped collar of long hairs (nearctic) .... *Oreopasites*

4(1). S5 of female with posterior margin (curved up and sometimes around process of S6 and sting) unmodified, covered with fine hairs; labrum not extending under or posterior to mandibles in repose (male antenna 12-segmented) (palearctic, oriental, Africa) .... *Pasites*

—. S5 of female with posterior margin bare at least medially, usually with median protuberance or troughlike projection that may look superficially like another sternum; labrum commonly extending under and sometimes beyond mandibles in repose 5

5(4). Female with nearly circular, hairless pygidial plate margined by circular carina, on posterior subvertical surface of T6; jugal lobe of hind wing essentially absent; maxillary palpus absent; body length 15 to 18 mm; male antennae 12-segmented (Madagascar) .... *Melanempis*

—. Female without circular pygidial plate; jugal lobe of hind wing present, but small as in most Nomadinae; maxillary palpus one- to six-segmented; body length 4 to 13 mm; male antenna 13-segmented (male unknown in *Spinopasites*) 6

6(5). S6 of female ending in a single spine; S5 of female with posterior part bent upward and not visible from below, preapical concavity V-shaped and not defined by hairy ridge (northwen Africa) .... *Spinopasites*

—. S6 of female bifurcate; S5 of female with posterior part not so strongly bent upward, visible from below, preapical concavity more U-shaped and defined by hairy ridge (palearctic, oriental, southern Africa) .... *Ammobates*

Genus *Ammobates* Latreille

This is the largest genus of the Ammobatini. Its species have the usual pasitiform aspect; usually the metasoma is partly or wholly red, the whole body rarely red-brown. Unlike those of *Pasites,* the male antennae are 13- segmented; S5 of the female exhibits generic characters as indicated in the key to genera.

Revisional studies were by Popov (1951c) and Warncke (1983); the former includes excellent illustrations of male genitalia and hidden sterna as well as habitus drawings; the latter includes sketches of diverse structures. Good illustrations were also published by Radoszkowski (1885), Iuga (1958), and Mavromoustakis (1968b).

There are three groups here considered as subgenera; certain of them have been considered as genera by other authors in the past.

Key to the Subgenera of *Ammobates*

1. Second submarginal crossvein meeting second recurrent vein (S5 of female with deep longitudinal channel, the apex deeply emarginate; maxillary palpus six-segmented; S8 of male with disc and spiculum forming triangle; apical process of S8 broad, membranous, sometimes spiculate, not cleft) .... *A. (Xerammobates)*

—. Second submarginal crossvein distal to second recurrent vein 2

2(1). T6 of female more or less rooflike with longitudinal median ridge, sides sloping, densely hairy; labrum relatively short, 1.5 times as long as its basal width ....
.... *A. (Euphileremus)*

—. T6 of female flat or convex, not longitudinally ridged, sparsely or at most only partially hairy, often with specialized median area, or posterior part bent down and vertical, or both; labrum more than twice as long as its basal width .... *A. (Ammobates s. str.)*

Ammobates / Subgenus *Ammobates* Latreille s. str.

Ammobates Latreille, 1809: 169. Type species: *Ammobates rufiventris* Latreille, 1809, by designation of Latreille, 1810: 439. [A subsequent and invalid designation was listed by Michener, 1997b.]

Philéremus Latreille, 1809: 169. Type species: *Epeolus punctatus* Fabricius, 1804, by designation of Latreille, 1810: 439. [A subsequent and invalid designation was listed by Michener, 1997b.]

Ammobatoides Schenck, 1869: 349 (not Radoszkowski, 1867). Type species: *Ammobates bicolor* Lepeletier, 1825 = *Epeolus punctatus* Fabricius, 1804, monobasic.

Ammobates (Caesarea) Friese, 1911d: 142. Type species: *Ammobates depressa* Friese, 1911, by designation of Sandhouse, 1943: 532.

In species of this subgenus the labrum is over twice as long as its basal width, and the face, especially the clypeus, is strongly protuberant. The body length is 4 to 11 mm.

■ *Ammobates* s. str. occurs from Portugal and Morocco north in Europe to 53° latitude, east to Uzbekistan, and south to southern India, with one disjunct species in Namibia and South Africa (Eardley and Brothers, 1997). Of a total of about 30 species, those of the palearctic region were revised by Warncke (1983).

Probable hosts include *Ancyla* (Ancylini), *Tetraloniella* (Eucerini), and *Anthophora* (Anthophorini) (Warncke, 1983).

Ammobates / Subgenus *Euphileremus* Popov

Ammobates (Euphileremus) Popov, 1951c: 906. Type species: *Philéremus oraniensis* Lepeletier, 1841, by original designation.

In *Euphileremus,* as in *Xerammobates* and the genus *Parammobatodes,* the labrum is about 1.5 times as long as the basal width and the face is almost flat. T6 of the female is more or less horizontal, commonly with a median ridge and thus roof-shaped. The body length is 5.0 to 8.5 mm.

■ *Euphileremus* occurs from the Canary Islands, Morocco, and Spain east through southwestern Asia and southern Russia to Uzbekistan. About seven species were revised by Warncke (1983).

So far as is known, species of this subgenus parasitize *Tetraloniella* (Eucerini) (Warncke, 1983).

Ammobates / Subgenus *Xerammobates* Popov

Ammobates (Xerammobates) Popov, 1951c: 904. Type species: *Ammobates oxianus* Popov, 1951, by designation of the Commission, Opinion 1853 (1996). [See Michener, 1997b. Popov (1951c) misidentified the original type species of *Xerammobates* as *Ammobates biastoides* Friese, 1895, which is actually a species of *Ammobates* s. str. (Baker, 1995b).]

Pasites (Micropasites) Warncke, 1983: 283 (not Linsley, 1942). Type species: *Ammobates minutissimus* Mavromoustakis, 1959, renamed *tunensis* Warncke, 1983, because of homonymy in *Pasites*.

Pasites (Ebmeriana) Pagliano and Scaramozzino, 1990: 127, replacement for *Micropasites* Warncke, 1983. Type species: *Ammobates minutissimus* Mavromoustakis, autobasic.

The labrum is short for the genus, little over 1.5 times as long as its basal width. The clypeus is nearly flat in lateral view. The body length is 4.0 to 7.5 mm.

■ *Xerammobates* is found disjunctly in Tunisia and Turkmenistan. Warncke (1983) treated the three species.

Genus *Melanempis* Saussure

Phileremus (Melanempis) Saussure, 1890: 84. Type species: *Phileremus ater* Saussure, 1890, monobasic.

This is the most distinctive of the ammobatine genera. The species are larger and more elongate than other Ammobatini, almost nomadiform or actually suggesting *Polistes*, black or dark reddish, sometimes with green tints; the body length is 15 to 22 mm. The first recurrent vein meets the first submarginal crossvein; the hind wing lacks a jugal lobe; the scutellum is strongly bituberculate; and the maxillary palpus is absent (as it is in two species of *Pasites*). The relatively straight vertex, as seen from the front, and the slanting or appressed simple hairs on various parts of the body suggest *Sphecodopsis* and led Eardley and Brothers (1997), in their phylogenetic analysis, to place *Melanempis* as the sister group to *Sphecodopsis*. Several of the characters supporting this position for *Melanempis*, however, are characters of the vestiture that could be fused into one character; I am therefore not certain that the relation to *Sphecodopsis* is firm.

■ *Melanempis* occurs in Madagascar. The five species of *Melanempis* were revised by Brooks and Pauly (in Pauly et al., 2001).

The lack of a jugal lobe on the hind wing is unusual, especially in a large bee like *Melanempis*.

Genus *Oreopasites* Cockerell

Oreopasites Cockerell, 1906b: 442. Type species: *Oreopasites scituli* Cockerell, 1906, monobasic.

Oreopasites (Perditopasites) Rozen, 1992b: 49. Type species: *Oreopasites linsleyi* Rozen, 1992, by original designation.

Oreopasites is the only American genus of Ammobatini. The species have the typical ammobatine aspect (Fig. 100-1), usually with the metasoma red, and vary from 2.2 to 7.5 mm in length. In addition to the characters indicated in the key to genera, the male antennae are 13-segmented and S8 of the male tapers to a rounded or truncate apex that is not or scarcely spiculate. S5 of the female does not have an apical projection but in some species has a longitudinal median ridge. S6 of the female usually is notched at the apex of the apical process, but in *O. arizonica* Linsley, the process is simple, as in *Sphecodopsis (Pseudodichroa)* and *Ammobates (Spinopasites)*. The male genitalia and apical sterna of both sexes were illustrated by Linsley and Michener (1939).

■ This genus occurs from Washington state, Idaho, and western Nebraska south to western Texas, USA, and Durango and Baja California, Mexico. The 11 species were revised by Rozen (1992b); an earlier study was by Linsley (1941).

Oreopasites species are cleptoparasites in the nests of *Calliopsis* (*Hypomacroptera, Macronomadopsis, Micronomadopsis,* and *Nomadopsis*) and *Perdita* (*Epimacrotera, Hexaperdita,* and *Perdita* s. str.) (all Panurginae). Rozen (1992b) listed known hosts.

I have not recognized *Perditopasites*, described by Rozen as a subgenus for two minute species. The reduced venation of the hind wing may be a correlate of minute body size; the smallest species, *O. barbarae* Rozen, also almost completely lacks the jugal lobe of the hind wing. I suspect that *Perditopasites* may be a derived group rather than the sister group of the other *Oreopasites* as Rozen believed. The genus has too few known specific characters for reliable phylogenetic analysis.

Genus *Parammobatodes* Popov

Parammobatodes Popov, 1932: 462. Type species: *Phiarus minutus* Mocsáry, 1878, by original designation.

This genus seems closest to *Ammobates*, which it resembles in appearance. The body length is 2.5 to 8.0 mm; two of the minute species are entirely red-brown. As in the subgenera *Xerammobates* and *Euphileremus* of *Ammobates*, the labrum is short, about as long as broad to 1.5 times as long as broad, and does not extend beneath the closed mandibles. The maxillary palpus has one to five segments. S5 of the female is convex, the apex straight or concave. S8 of the male has a slender, tapering basal spiculum and two membranous, spiculate apical lobes separated by a deep cleft. The male genitalia, hidden sterna, and other structures were illustrated by Popov (1932) and Iuga (1958).

■ *Parammobatodes* occurs from the Canary Islands and Morocco east to Egypt, Afghanistan, and India, and from Hungary and Greece east to the Caucasus and Ukraine. The six palearctic species were reviewed by Warncke (1983); at least one additional Indian species, *P. indicus* (Cockerell), belongs to this genus.

The presumed hosts of *Parammobatodes* are in the genus *Camptopoeum* (Panurginae) (Warncke, 1983).

Minute species of a new genus near *Parammobatodes* known from southwestern Asia to Sri Lanka are cleptoparasites of *Nomioides* (Halictidae) (D. Baker and J. Rozen, personal communications, 1993, 1998). The new genus, like *Melanempus,* is remarkable for lacking the jugal lobe of the hind wing.

Genus *Pasites* Jurine

Pasites Jurine, 1807: 224. Type species: *Pasites maculata* Jurine, 1807, by original designation. [A subsequent designation is listed by Michener, 1997b.]

Morgania Smith, 1854: 253. Type species: *Pasites dichroa* Smith, 1854, monobasic.

Omachthes Gerstaecker, 1869: 154. Type species: *Omachthes carnifex* Gerstaecker, 1869, designated by Sandhouse, 1943: 580.

Homachthes Dalla Torre, 1896: 499, unjustified emendation of *Omachthes* Gerstaecker, 1869.

Omachtes Friese, 1909a: 436, unjustified emendation of *Omachthes* Gerstaecker, 1869.

Pasitomachthes Bischoff, 1923: 596. Type species: *Pasitomachthes nigerrimus* Bischoff, 1923, by original designation. [See note by Michener, 1997b.]

Pasitomachtes Sandhouse, 1943: 586, unjustified emendation of *Pasitomachthes* Bischoff, 1923.

The species of *Pasites* resemble superficially those of *Ammobates,* usually having a red metasoma and varying in body length from 2.3 to 12.5 mm. The labrum is variable, from about as long as broad to much longer than broad; its apex is pointed or truncate and visible in front of the closed mandibles which usually cross, forming a very obtuse angle. The pygidial plate is recognizable or not in each sex. The male genitalia and sterna and female sting were illustrated by Radoszkowski (1885), Popov (1931b), and Iuga (1958).

■ *Pasites* ranges from Portugal and Morocco to Mongolia and Japan, south to India and through the whole of Africa. Of about 21 species, two are palearctic, 15 are from sub-Saharan Africa, three from Madagascar, and one is from India. African species were revised by Bischoff (1923) and Eardley and Brothers (1997); Malagasy species, by Eardley and Pauly (in Pauly et al., 2001).

Pasites maculatus Jurine parasitizes *Pseudapis* (Nomiinae) (Rozen, 1986); and Nomiinae are probable hosts of other species. The egg of *Pasites* is folded into a U and inserted in the cell wall of the host, the truncate anterior end flush with the cell wall, all this much as in *Oreopasites* (Rozen, 1986).

The palearctic species, *Pasites maculatus* Jurine, differs enough from some of the sub-Saharan and Indian species that the latter have usually been regarded as generically distinct, being placed in *Morgania* or *Omachthes.* Eardley and Brothers (1997) united them, and their phylogenetic analysis showed *P. maculatus* Jurine falling among African species. The swollen upper paraocular area and the depression above the antennal sockets to accommodate the scapes are features of *P. maculatus* not well developed in other species; further, in some African species the antennal pedicel is firmly set in the apex of the scape, whereas it seems to be more freely articulated in others, including *P. maculatus*. Moreover, *P. maculatus* lacks maxillary palpi, whereas in other species (except *P. gnoma* Eardley) they are present, although highly variable. Baker (1974b) emphasized a distinction between *Pasites* and *Morgania* (including *Omachthes*) based on T6 and S5 of the female. In *Morgania* the pygidial plate, or at least its posterior margin, is clearly defined by a carina behind which the tergum is vertical, covered with short, dense, dark-colored, specialized hairs, and the apical lobes of S5 are bent upward, forming with the posterior surface of T6 a truncate apex of the metasoma. *Pasites maculatus* has almost these same features but they are less developed. T6 has the pygidial plate weakly defined, not margined by a carina, and the surface behind it is relatively short, less vertical, and the specialized dark hairs are limited to a small space, while the apical lobes of S5 are less strongly bent upward. Thus the apex of the metasoma is less truncate than that of *Morgania.* Comparable structures are found in each group, and intergradation in various features exists; I agree with Eardley and Brothers (1997) in not recognizing *Morgania* as a genus.

Genus *Sphecodopsis* Bischoff

This southern African genus consists of rather small, black forms 4 to 9 mm long, the metasoma usually partly or wholly red. The male antennae are 13-segmented. The scutellum is gently and evenly curved as seen in profile. S5 is either strongly or shallowly concave apically when viewed from behind and has a midapical prominence, sometimes weak or even included in an apical emargination.

Eardley (1994a) was entirely correct in emphasizing the close relationship of *Pseudodichroa* to *Sphecodopsis.* He did not recognize subgenera, but I retain *Pseudodichroa* for the present as a subgenus, since it is quite different and he did not show how it is related to *Sphecodopsis* s. str.; it could be the sister group to that subgenus.

Key to the Subgenera of *Sphecodopsis*

1. S6 of female produced to a long, slender, median spine; S5 of female with midapical process that is truncate and longer than broad .... S. (*Pseudodichroa*)
—. S6 of female produced to a process with bifurcate apex; S5 with midapical projection rounded or truncate, much broader than long, sometimes scarcely recognizable, sometimes in emargination that is broader than projection .... S. (*Sphecodopsis* s. str.)

Sphecodopsis / Subgenus *Pseudodichroa* Bischoff

Pseudodichroa Bischoff, 1923: 595. Type species: *Omachtes capensis* Friese, 1915, by designation of Sandhouse, 1943: 593.

Males of species probably belonging to this subgenus do not exhibit distinctive subgeneric characters. The body length is 6 to 11 mm. Diverse structures were well illustrated by Rozen (1968a).

■ *Pseudodichroa* is known only in western Cape Province, South Africa. The two species were revised by Rozen (1968a).

The hosts are species of *Scrapter* (Colletinae); the behavior and mode of parasitism were described by Rozen and Michener (1968). The egg is inserted into the cell wall of the host; a flange around the outer end fits against the cell wall so that the egg would probably be difficult to detect.

Sphecodopsis / Subgenus *Sphecodopsis* Bischoff s. str.

Sphecodopsis Bischoff, 1923: 593. Type species: *Omachthes capicola* Strand, 1911, by original designation.

Sphecodopsis (*Pseudopasites*) Bischoff, 1923: 593. Type species: *Pasites pygmaeus* Friese, 1922, by designation of Sandhouse, 1943: 594.

The body length is 4 to 9 mm.

■ *Sphecodopsis* s. str. is found in southern Africa from Namibia and Zimbabwe south through South Africa; most species are in western Cape Province and Namibia. The eight species were revised by Bischoff (1923) and Eardley and Brothers (1997).

Genus *Spinopasites* Warncke

Pasites (Spinopasites) Warncke, 1983: 281. Type species: *Pasites spinotus* Warncke, 1983, by original designation.

In this genus the second submarginal crossvein is distal to the second recurrent vein, and the first recurrent vein meets the middle of the second submarginal cell. Thus in venation it is much like *Ammobates (Euphileremus)*. Its distinctive feature is the simple rather than bifurcate or notched apical process of S6 of the female; it projects as a curved spine from the apex of the metasoma. The maxillary palpus is five-segmented. The male is unknown. Body length is 3.5 to 4.0 mm.

■ *Spinopasites* is known only from Tunisia. The only species is *Spinopasites spinotus* (Warncke).

Eardley and Brothers (1997) place *Spinopasites* in their cladogram at the base of the *Ammobates-Sphecodopsis* branch. I am indebted to C. D. Eardley for certain particulars about the genus.

101. Tribe Caenoprosopidini

The Caenoprosopidini consist of strongly punctate bees with a red metasoma and short, appressed, plumose hairs forming narrow, white, apical tergal bands. T5 of the female lacks a pseudopygidial area. The antennae arise far below the middle of the face. The preoccipital carina is strong. Unlike that of other L-T bees, the episternal groove is very long, extending to the thoracic venter; it is perhaps not homologous to the comparable groove frequently found in S-T bees. The omaulus is angular, the upper part almost carinate but often hidden by hair. An unusual feature among the small groups of Nomadinae is the presence of small basitibial plates in both sexes. The sting is reduced. S6 of the female is elongate, consisting of two longitudinal rods pointed distally and connected by a narrow membrane (Fig. 91-2j). The male genitalia are small and simple, the gonostyli, gonocoxites, and gonobase being fused into a single unit (Roig-Alsina, 1987; Fig. 101-2).

All studies of relationships of the tribes of Nomadinae have shown this tribe to be affiliated with the Ammobatini. Rozen and Roig-Alsina (1991) found larvae to be almost identical. Phylogenetic studies (Roig-Alsina, 1987, 1991b; Alexander, 1990; Eardley and Brothers, 1997) show *Caenoprosopis* or the Caenoprosopidini as the sister group to the Ammobatini. The major differences, such as fusion of male genitalic parts and division of S6 of the female into two parts, lead me to recognize the two tribes.

Bees of this tribe occur in Argentina, Paraguay, and southern Brazil. They have no relatives in South America; the Ammobatini are found in the palearctic region, Africa, and western North America.

Key to the Genera of Caenoprosopidini

1. Total length of marginal cell much greater than that of stigma, about as long as distance from its apex to wing tip (Fig. 101-1); second submarginal cell about two-thirds as long as first; axillae and scutellum not spined, but with projecting convex lateral edges; middle flagellar segments about as long as broad .... *Caenoprosopis*
—. Total length of marginal cell as long as stigma, much less than distance from its apex to wing tip; second submarginal cell less than half as long as first; axillae and posterior lateral angles of scutellum produced to posteriorly directed spines; middle flagellar segments much broader than long .... *Caenoprosopina*

Genus *Caenoprosopina* Roig-Alsina

Caenoprosopina Roig-Alsina, 1987: 312. Type species: *Caenoprosopina holmbergi* Roig-Alsina, 1987, by original designation.

This genus consists of a species that superficially resembles a small *Holcopasites*. The body length is 4 to nearly 5 mm. Many structures were illustrated by Roig-Alsina (1987).

■ *Caenoprosopina* is known from Catamarca to Salta, Argentina. The only species is *C. holmbergi* Roig-Alsina.

Rozen and Roig-Alsina (1991) found *Caenoprosopina* to be a parasite of *Callonychium* (Panurginae).

Genus *Caenoprosopis* Holmberg

Caenoprosopis Holmberg, 1886b: 235, no included species; 1887a: 22, one included species. Type species: *Caenoprosopis crabronina* Holmberg, 1887, first included species.
Austrodioxys Cockerell, 1916a: 432. Type species: *Austrodioxys thomasi* Cockerell, 1916 = *Caenoprosopis crabronina* Holmberg, 1887, monobasic.

This genus consists of a species that looks superficially like a *Holcopasites*. The body length is 6.0 to 8.5 mm. Many structures were illustrated by Roig-Alsina (1987). The fusion of the genitalic parts (Fig. 101-2), shared also with *Caenoprosopina,* is one of the unusual features that emphasize the distinctness of the tribe from the Ammobatini.

■ *Caenoprosopis* occurs from the province of Neuquén, Argentina, to the state of São Paulo, Brazil. There is a single species, *C. crabronina* Holmberg.

Rozen and Roig-Alsina (1991) found *Caenoprosopis* to be a cleptoparasite in nests of *Arhysosage* (Panurginae).

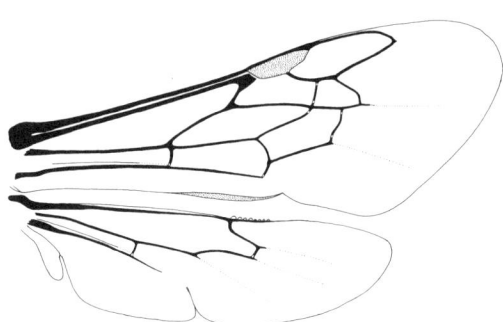

Figure 101-1. Wings of *Caenoprosopis crabronina* Holmberg.

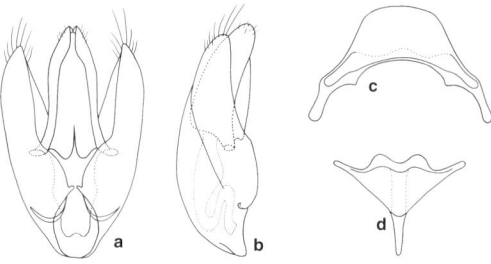

Figure 101-2. Structures of male of *Caenoprosopis crabronina* Holmberg. **a, b,** Genitalia, in dorsal and lateral views; **c, d,** S7 and S8. From Roig-Alsina, 1987.

102. Subfamily Apinae

This subfamily consists of the corbiculate Apidae (i.e., tribes Apini, Bombini, Euglossini, and Meliponini) plus most of the taxa sometimes included in the Anthophoridae or Anthophorinae. Thus it includes not only the Apinae of authors such as Michener (1944) but also his Anthophorinae minus the cleptoparasitic forms here placed in the subfamily Nomadinae. Apinae in the present sense was recognized in the phylogenetic (parsimony-based) study by Roig-Alsina and Michener (1993).

Distinctive characters for the Apinae that are not lost in some of its tribes or genera do not exist, so far as is known. That is, there are exceptions to all recognized subfamilial characters. The protuberant clypeus, with the lower lateral parts bent back on either side of the labrum (Fig. 21-3d, h), as in Nomadinae, breaks down as a character in *Apis, Centris, Exomalopsis,* etc. The presence of the stipital comb and the concavity in which it lies is a reasonably good synapomorphy of Apinae but is lost in various parasitic taxa, in Meliponini, and in others. Conversely, the presence of these structures in the Xylocopinae (Fig. 88-2b) and weakly in a few megachilids also detracts from their strength as a synapomorphy of Apinae. The presence in females of a pygidial plate, of well-developed pygidial and prepygidial fimbriae, and of basitibial plates breaks down in some cleptoparasitic tribes as well as in the corbiculate tribes, i.e., in those that have lost the plesiomorphic soil-nesting and cell-making behavior. The scopa is restricted to the hind tibia and basitarsus (Fig. 6-4), as elsewhere in the family, and is lost in the cleptoparasitic tribes. The anterior coxae are usually little if any broader than long, but similar coxae can be found in some Xylocopinae, a group whose front coxae are usually broader. Nonparasitic Apinae are usually robust and hairy, euceriform (Figs. 111-1, 112-1, 112-2), anthophoriform, or bombiform (Fig. 88-1), but andreniform or trigoniform Tetrapediini and some Tapinotaspidini (Fig. 108-1) are exceptions. The mature larvae differ from those of the other subfamilies of Apidae in retaining cocoon-spinning behavior and the associated complex of structures, such as the lips of the salivary openings. The tribe Anthophorini, however, is an exception; its larvae do not spin cocoons. Larvae of the Apinae did not form a single cluster in McGinley's (1981) nearest-neighbor analysis. The characters mentioned above are all probable plesiomorphies at the subfamily level, with the possible exception of the stipital comb and concavity, which could be apomorphic if similar structures in Xylocopinae are not homologous to those of the Apinae.

What are additional apomorphies that support the weak stipital comb and concavity in suggesting holophyly for the subfamily? All such characters seem full of exceptions and problems about polarity. One example will suffice—the absence of a pygidial plate in males, one of the characters of the subfamily as shown in some of the cladograms resulting from the phylogenetic analysis by Roig-Alsina and Michener (1993). The plate reappears (or at least its posterior extremity is indicated by a carina) in males of Isepeolini, Exomalopsini, etc. Presence of a pygidial plate on T6 of females is no doubt plesiomorphic among bees; it is present in most S-T bees and many Apidae. It is used in the construction of the plesiomorphic types of nest and cells. The plate on T7 of males is probably homologous to that on T6 of females in the sense that it is controlled by the same genes, for its presence in males is associated with that in females, although with exceptions, as in many Apinae. So far as is known, the plate of the male has no function, and it seems both to disappear and to reappear. It might reappear as a regulatory change in any clade whose females have a pygidial plate, for genes that specify it must be present in males as well as females. An interesting example that must represent reappearance can be seen in *Anthophora (Heliophila) squammulosa* Dours, whose male has what appears to be a pygidial plate, although males of most other species of its subgenus, as well as of most related subgenera and genera, lack the plate. Thus absence of the pygidial plate in males can be a subfamily synapomorphy subject to reversals. It is clear, however, that the only honest conclusion one can reach is that we do not know of good synapomorphies for Apinae, and we therefore do not know whether it is a monophyletic group, although it was recognized as such by Roig-Alsina and Michener (1993).

The cladograms of Roig-Alsina and Michener (1993) and of Silveira (1993b) mostly divide the Apinae into two major groups (Fig. 102-1) called by Silveira the eucerine line (E) and the apine line (A). They are not well differentiated; a few tribes appear in one or the other in different cladograms, and the eucerine line sometimes appears as a paraphyletic group from which the apine line arose. Nonetheless, in a general way two groups were made evident by these studies. In the following paragraphs the annotations (E) and (A) indicate the positions of tribes in one or the other of these two groups, as indicated in Analysis C (which omits five characters associated with parasitism) by Roig-Alsina and Michener (1993); see Figure 102-1. The groupings indicated below are not all based on phylogenetic or any other analysis, but are merely useful groups for certain commentaries.

The Isepeolini (E?), Osirini (E?), and Protepeolini (E?) were formerly included among the tribes of Nomadinae. As indicated in the discussions of that subfamily and of the family Apidae (Secs. 85 and 91), they do not belong in the Nomadinae in the strict sense, and fall in the Apinae in the phylogenetic analysis by Roig-Alsina and Michener (1993). I believe, however, that the possibility still exists that one or more of them is a basal nomadine. Additional characters or taxa could easily influence the analyses, for the basal part of the apid phylogeny, as presented by Roig-Alsina and Michener, is based on few characters.

The Ctenoplectrini (E) were formerly included among the S-T bees because, like *Ancyla,* they do not have and presumably lost the elongate, sheathlike condition of the first two segments of the labial palpi.

The Exomalopsini (E), Ancylini (E), Tapinotaspidini (E), and Tetrapediini (A?) include rather small species that constitute most of the "primitive nonparasitic anthophorine bees" of Michener and Moure (1957). One parasitic genus, *Coelioxoides,* is now recognized as a tetrapediine.

The Emphorini (E), Eucerini (E), Anthophorini (A), and Centridini (A) are, in antithesis to the above, mostly larger and constitute the derived nonparasitic anthophorine bees. At this level the eucerine and apine lines seem quite distinct.

Three tribes of parasitic bees, the Rhathymini (A), Ericrocidini (A), and Melectini (A), are associated—both as cleptoparasites and phylogenetically—with the Centridini and Anthophorini. The phylogenetic relationships of these three parasitic tribes were not clearly established by Roig-Alsina and Michener's (1993) analyses. Convergence among parasites independently derived from nonparasitic ancestors probably involves many characters, and one therefore doubts the reliability of phylogenetic analyses based on parsimony. Both geographical and morphological considerations suggest that the Melectini are related to the Anthophorini, and the other two parasitic tribes to the Centridini. Host relationships support the same view, since cleptoparasitic bees usually attack their relatives. It is not clear whether the nonparasitic tribes are paraphyletic and ancestral to the parasites, or are sister groups to the parasites. The latter conclusion is supported by some characters, such as the larger stigma, that are believed to be more plesiomorphic in the parasitic tribes than in the others. The prevalence, in various other cleptoparasitic taxa, of characters that are plesiomorphic relative to the most closely related nonparasites suggests, however, that parasites tend to regress in some features while they are evolving the specialist features of parasites. If this is so, these "plesiomorphies" are in fact apomorphies, and the parasites could be derived from paraphyletic host taxa.

The remaining tribes, the Euglossini (A), Bombini (A), Meliponini (A), and Apini (A), constitute the apine clade of Silveira (1993b) and are here called the corbiculate Apidae. If one ignores the cleptoparasites, the closest relatives of the corbiculate Apidae are the Anthophorini and the Centridini (Fig. 102-1). Unlike the groups discussed above, the four corbiculate tribes clearly constitute a holophyletic unit, one that has often been called the family Apidae (e.g., Michener, 1990a). Michener (1944) and Roig-Alsina and Michener (1993) considered that this group does not deserve separate familial status and have united it with the Anthophoridae under the family name Apidae. The latter authors showed that the corbiculate bees are associated with the whole apine line, and more specifically with the Centridini and Anthophorini. If the corbiculate tribes were given family rank, then the rest of

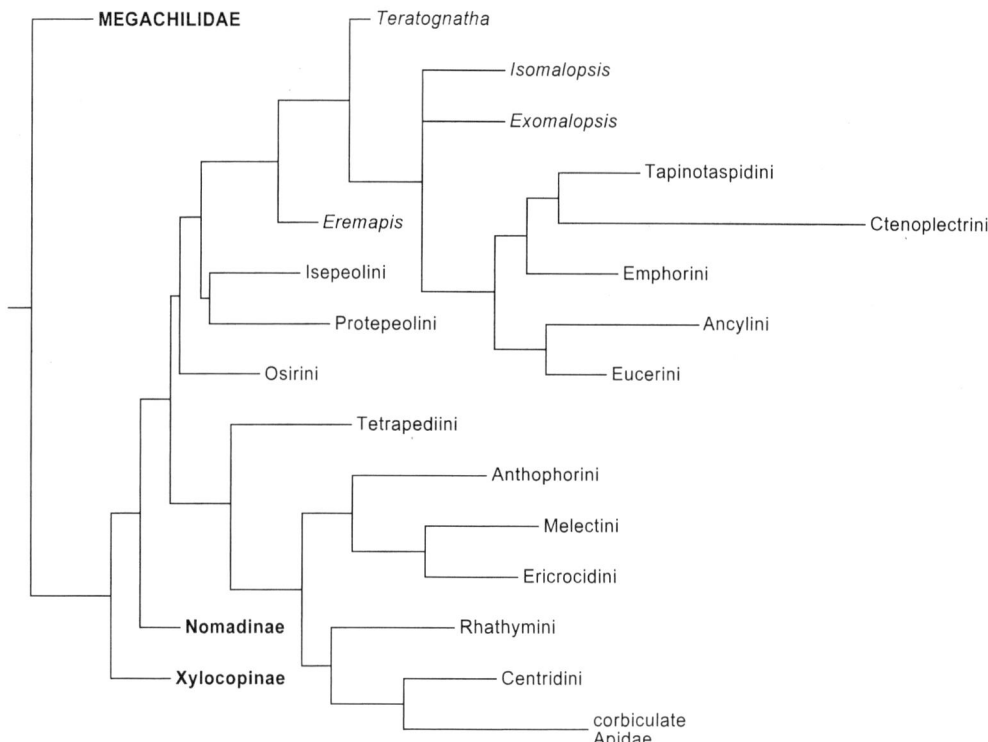

Figure 102-1. Cladogram of major groups of Apinae derived from parsimony-based phylogenetic analysis C of Roig-Alsina and Michener (1993). Characters used were those of adults, and five characters associated with parasitism were omitted in the hope of associating cleptoparasitic taxa with their nonparasitic relatives instead of with one another. The analysis was made using particular genera and species, and is summarized here in terms of tribes, except that, above, four genera usually placed in Exomalopsini are shown separately because they did not constitute a monophyletic group. (Lengths of the horizontal lines are proportional to the number of characters.)

the Apinae would have to be divided into numerous families. Nonetheless, it is convenient to discuss the four tribes under the heading "corbiculate Apidae."

Several special terms are associated with the pollen-carrying and pollen-manipulating structures of the corbiculate Apidae; all are unique to females of this clade or to certain of its branches, and are synapomorphies of the clade or its branches. These structures are absent in most parasitic or robber taxa, and in the queens of highly social taxa; they are illustrated in Figure 102-2. The basitarsi lack scopal hairs and pollen-carrying function and are articulated near the anterior end of the apex of the tibia, leaving room for the rastellum and auricle behind the articulation (Fig. 102-2). The tibial **corbicula** is the smooth, concave or sometimes flat area, surrounded by long hairs, on the outer surface of the hind tibia. The **rastellum** is the comb of strong, usually blunt-tipped bristles across the inner surface of the apex of the hind tibia. It is reduced to ordinary hairs in some Meliponini; I have described it in such cases as a rastellum made up of hairs, or of tapering (pointed) hairs. I could have said "rastellum absent," but since there are all degrees of reduction it is hard to say at what point such a comment becomes appropriate. The **auricle** is the posterior expansion of the base of the hind basitarsus, used for pushing pollen up into the corbicula. It is absent in the Meliponini. The **penicillum** is a compact tuft of strong, usually curved bristles arising near the front (or lower extremity) of the apical margin of the hind tibia, usually directed posteriorly, sometimes almost parallel to the apical tibial margin. It is found only in the Meliponini. The **anterior** and **posterior parapenicilla** are groups of bristles arising at the anterior and posterior (lower and upper) apical angles, respectively, of the hind tibia of some Meliponini. There are always hairs in these positions; how coarse they must be to be called parapenicilla is a matter of judgment. The posterior parapenicillum is developed only in the genus *Meliponula*. The anterior parapenicillum is more widespread but is not present in Figure 102-2. Since the parapenicilla are difficult to define, I have used the terms only sparingly. Behaviors associated with the pollen-carrying and -manipulating structures of corbiculate Apidae were discussed by Michener, Winston, and Jander (1978). Other synapomorphies of the apine clade are the complete absence of basitibial and pygidial plates and the reduction of the maxillary palpus to one or two segments.

Indicated below are some internal and larval characters showing that the four tribes of the corbiculate Apidae are related to one another, i.e., that they have not convergently evolved the external features of the group. As evidence, the internal characters suffer from not having been examined in many species; yet they are known for various species of each tribe and appear to be characteristics of all corbiculate Apidae. The cephalic salivary glands are

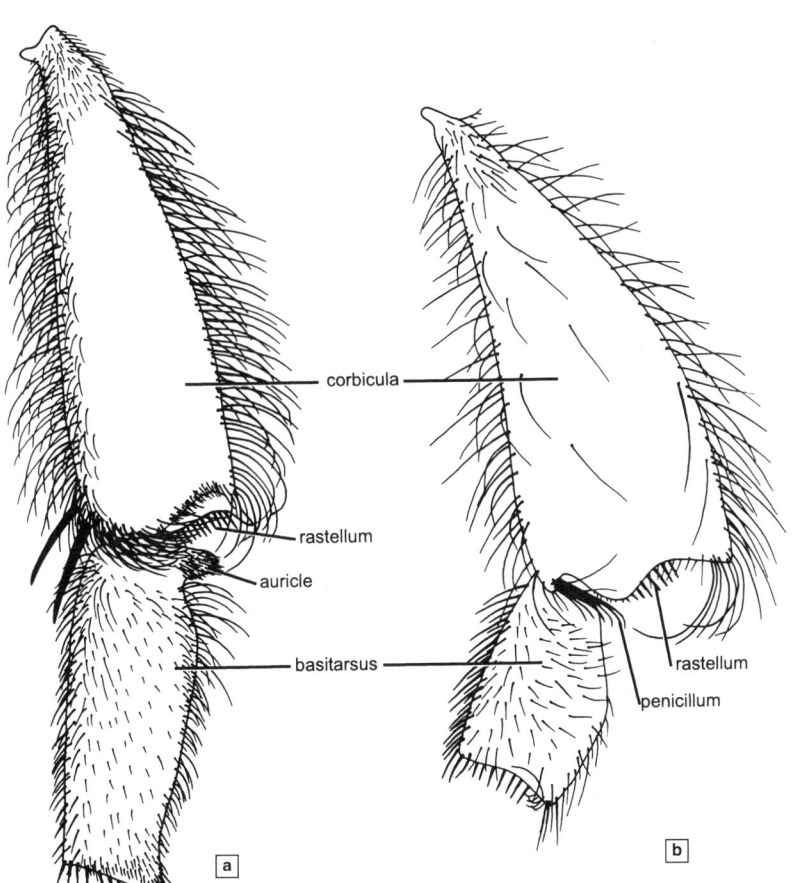

Figure 102-2. Outer surfaces of hind tibiae and basitarsi of workers of corbiculate Apidae. **a**, *Bombus pennsylvanicus* (Degeer); **b**, *Plebeia (Plebeia) frontalis* (Friese).

present and formed of many alveoli. In other bees these glands are tubular (sometimes locally expanded) or absent (Cruz-Landim, 1967). The hypopharyngeal glands, attenuate and usually pedunculate, discharge through a common duct on each side. In other bees these glands are usually shorter, sessile, and crowded against the hypopharyngeal plate or rods, the single duct at each side absent or possibly hidden among the acini (Cruz-Landim, 1967). The thoracic salivary glands of corbiculate Apidae are made up of short, simple tubes, quite distinct from the collecting ducts; the cells are cubical, and the secretory tubes are formed by cuboid cells. In other bees the tubes are of diverse types, but not as in the corbiculate Apidae (Cruz-Landim, 1967, 1973). The metasomal wax glands, tergal and sternal in Bombini, tergal in Meliponini, sternal in Apini, and perhaps at the base of T6 in Euglossini, are also common features of corbiculate Apidae (Cruz-Landim, 1967), except perhaps the Euglossini. Except in *Apis* the larvae of the corbiculate Apidae have small, conical, dorsolateral tubercles on at least the thoracic segments (Michener, 1953a). They differ in this respect from other bee larvae. A phylogenetic analysis of larval characters by Roig-Alsina and Michener (1993), as well as a nearest-neighbor analysis of larvae by McGinley (1981), both associated the larvae of the corbiculate apid tribes.

The nests of corbiculate Apidae are usually constructed in large or irregular cavities, the cells being built up rather than excavated in the substrate. In most other bee groups, although not in many Megachilidae, cells and burrows are excavated into a substrate. Sakagami (1966) pointed out that the ability to excavate in substrates, so obvious in most Apidae as well as in S-T bees, is virtually absent in the corbiculate Apidae. This is presumably a synapomorphy, although the Bombini and Meliponini often enlarge subterranean cavities, and young queens of the former excavate hibernacula in the soil.

Relationships among the four tribes—the Apini, Bombini, Euglossini, and Meliponini—have been considered for many years and are discussed in several recent papers (Winston and Michener, 1977; Kimsey, 1984b; Plant and Paulus, 1987; Michener, 1990a; Cameron, 1991, 1993; Sheppard and McPheron, 1991; Prentice, 1991; Chavarría and Carpenter, 1994). Molecular approaches do not agree with morphological approaches, especially because the former suggest a sister-group relationship between Bombini and Meliponini (Fig. 102-3c). No known morphological character supports this conclusion. Molecular studies were based on both mitochondrial and ribosomal genes, and if they reflect the true phylogeny, the discussion below must be taken as an indication of how incorrect phylogenies based on morphology can be. The "total evidence" analysis by Chavarría and Carpenter (1994) agrees in topology (except that Euglossini and Bombini were sometimes regarded as sister groups) with the views expressed by Michener (1944), Prentice (1991), and Roig-Alsina and Michener (1993); see Figure 102-3a. Recent papers by Schultz, Engel, and Prentice (1999) and Schultz, Engel, and Ascher (2001) contrasted divergent phylogenies, morphological vs. molecular, proposed for the corbiculate Apidae, and supported the former (Fig. 102-3a), as did a phylogenetic study based on behaviorial characters by Noll (2002).

Cameron and Mardulyn (2001, 2003), however, further supported the supposed sister group relationship of Bombini and Meliponini on the basis of molecular analysis. But employing largely data from fossils, Engel (2001a, b) supported the phylogeny shown in Figure 102-3a, and demonstrated fossil taxa, especially in the extinct tribe Melikertini, intermediate morphologically between the Apini and Meliponini. Thus the Bombini-Meliponini relationship was not supported.

Engel (2001b) characterized three fossil (Baltic amber) tribes of corbiculate bees, Electobombini for *Electrobombus*, Electrapini for *Electrapis, Protobombus* and *Thaumastobombus,* and Melikertini for *Succinapis, Melikertes, Melissites,* and *Roussyana.* Thus in the late Eocene of Europe there was an unexpected diversity of corbiculate Apinae (four tribes, 10 genera, including two genera of Meliponini).

In various characters, however, the Meliponini rather than Euglossini have apparent plesiomorphies relative to the other tribes. The slender base of the hind basitarsus of workers of Meliponini, i.e., the lack of an auricle, in contrast to the condition in the other three corbiculate tribes (Fig. 102-2), seems at first to be a plesiomorphy, because noncorbiculate bees also lack an auricle and have a relatively slender basitarsal base. But in all pollen-collecting corbiculate Apidae, the basitarsus arises from near the anterior (or lower) distal angle of the tibia, i.e., from near the anterior (or lower) end of the apical tibial margin (Fig. 102-2). In other bees the basitarsus arises nearer the median axis of the tibia. I believe it likely that this characteristic of the corbiculate Apidae, i.e., the anterior position of the hind tibiotarsal articulation, permitted development of the auricle by providing space for it. The auricle was later lost in ancestors of the Meliponini when a different pollen-manipulating device (involving the penicillum) evolved, although the position of the tibiotarsal articulation was retained. Loss of the auricle is therefore a possible autapomorphy of the Meliponini. Of course if this interpretation is not correct, the meliponine condition probably would be plesiomorphic relative to

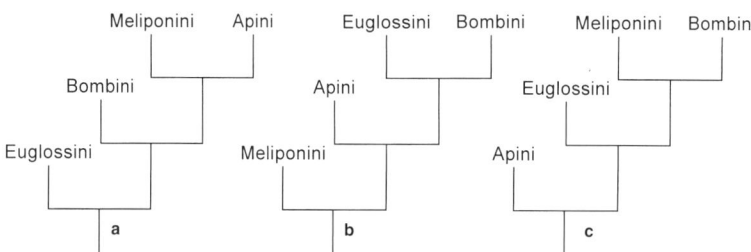

Figure 102-3. Alternate cladograms for the tribes of corbiculate Apidae. **a,** In agreement with the preferred cladograms of Roig-Alsina and Michener (1993) and Chavarría and Carpenter (1994) and others; **b,** In agreement with Kimsey (1984b); **c,** In agreement with Cameron (1993).

other tribes of the clade, and could suggest that the Meliponini are the sister group of the other three tribes (Fig. 102-3b).

Other characters that seem to be plesiomorphic in Meliponini relative to the other corbiculate tribes include the simple main axis of the malus of the strigilis of Meliponini, contrasted with the expanded lamella or prong (anterior velum of Schönitzer, 1986) found on the velum of the other tribes (Fig. 119-5i, j). An approach to such a structure occurs in some other Apinae, such as *Epicharis* (Centridini), although in most bees it is simple. Likewise, the large stigma of the Meliponini (Fig. 120-1) seems to be plesiomorphic relative to the reduced stigma of other corbiculate tribes, since large stigmata generally characterize the more primitive bees. As is well known, however, relative stigmal size is commonly negatively correlated with body size in aculeate Hymenoptera (Danforth, 1989a). The reduced wing venation of the Meliponini suggests that this group originated as minute bees, which may therefore have had an enlarged stigma as a reversion that has been retained along with reduced venation even in larger meliponines.

The larval mandibles of Apini and Meliponini are weakly sclerotized, the apices simple or with fine denticles. The larval mandibles of Bombini and Euglossini are robust, with considerable sclerotization; they lack denticles but have a strong tooth on the upper margins (Michener, 1953a). Neither type of mandible occurs in the Xylocopinae, considered as the appropriate outgroup by Sakagami and Michener (1987), Michener (1990a), and Cameron (1991, 1993), but mandibles similar in general form to those of Bombini and Euglossini occur in various noncorbiculate Apidae such as the Anthophorini and Centridini, now considered as appropriate outgroups (Roig-Alsina and Michener, 1993) for the corbiculate Apidae; see Figure 102-1. The weak larval mandibles of Apini and Meliponini are quite different from one another, and their weakness could be either convergent or homologous.

In spite of the characters discussed above, some of which suggest that the tribe Meliponini is the basal branch of the corbiculate Apidae, other and more impressive morphological evidence seems to indicate that Apini and Meliponini are sister groups (Michener, 1990a; Prentice, 1991) (Fig. 102-3a). Some common character states for Apini and Meliponini, i.e., states that support the sister-group relationship of these tribes, are losses. However striking, they might be independently evolved. Examples are reduction of mandibular grooves (Fig. 119-5a-d), loss of hind tibial spurs, and reduction of hidden sterna and most male external genitalic parts. Some synapomorphies in the prosternum of Meliponini and Apini taken together, however, such as its slender disc, large anterior region and anterolateral processes, and elongate groove (instead of round apophyseal pit), are not loss features and seem unlikely to have evolved convergently. Another such synapomorphy of Apini and Meliponini is the broad pharyngeal plate with its transverse (rather than longitudinal) fields of sensilla.

The relationships of the Euglossini and Bombini remain in doubt. Rather than as they are shown in Figure 102-3a, they could be sister groups. The loss of the jugal lobe of the hind wing in both of these tribes, very rare among bees, supports a sister-group relationship for these tribes, although the actual condition is rather different in the two. Likewise, the great reduction in Bombini and loss in Euglossini of the arolium and associated structures support the sister-group relationship of these taxa. These losses could of course have evolved independently.

Considering fossils and biogeography, a cladogram with Meliponini as the lowest branch (sister group to all other tribes), like that in Figure 102-3b, seems reasonable, because we know of fossil Meliponini older than any other bees (Cretaceous; Michener and Grimaldi, 1988a, b; see Sec. 22). There are also late Eocene meliponines from Baltic amber; Apini, Bombini, and Euglossini are not represented until the Miocene. Meliponini is now the most widely distributed tribe of Apidae; in appropriate climatic zones it occurs worldwide. The other three tribes of the apine clade are all geographically limited—Apini to Eurasia and Africa before being dispersed by humans, Bombini to the holarctic, oriental (montane), and neotropical regions, and Euglossini to the neotropics. Such distributions might indicate groups less old than the Meliponini.

The fossils of corbiculate Apidae have been reviewed by Zeuner and Manning (1976) and in part by Wille (1977). Engel (2001b) examined the little-known fossil corbiculate Apinae from the late Eocene, mostly from Baltic amber. Aside from two genera of minute Meliponini, these fossils fall in three tribes, as follows:

Electrobombini for *Electrobombus samlandensis* Engel (2001b). Large, robust (18 mm long), differing from Bombini in presense of small jugal lobe on hind wing; hind tibial spurs present, as in Bombini.

Electrapini for *Electrapis, Protobombus,* and *Thaumastobombus* with 5, 5, and 1 species respectively. Closest to Bombini, differing by presence of jugal lobe on hind wing, presence of only one, small hind tibial spur; and stigma much larger than prestigma, as in Electrobombini.

Melikertini for *Succinapis, Melikertes, Melissites,* and *Roussyana* with 3, 3, 1, and 1 species, respectively. Similar in appearance to *Trigona* in Meliponini but wing venation complete, hind tibia with one apical spur (none in Meliponini). *Succinapis* differs from *Melikertes* by the remarkable basal clypeal projection; otherwise it is not different from *Melikertes*

Bibliographical references for the above genera are in Section 22 on Fossil Bees.

The key to the tribes of Apinae is incorporated into the account of the family Apidae (Sec. 85). The 19 tribes of Apinae, in the sequence employed below, are as follows: Isepeolini, Osirini, Protepeolini, Exomalopsini, Ancylini, Tapinotaspidini, Tetrapediini, Ctenoplectrini, Emphorini, Eucerini, Anthophorini, Centridini, Rhathymini, Ericrocidini, Melectini, Euglossini, Bombini, Meliponini, and Apini.

103. Tribe Isepeolini

The Isepeolini have frequently been placed among the Nomadinae. The tribe consists of South American cleptoparasitic epeoliform bees, usually with conspicuous patches of white or rarely blue pubescence on the metasoma (Fig. 103-2g). Like the Protepeolini, it could be a basal branch of the Nomadinae, but in the absence of synapomorphies showing this relationship, it is here regarded as a tribe of the subfamily Apinae. This position is supported by Rozen's (1991a) suggestion, which is based on larval characters, of a relationship between Isepeolini and Ericrocidini, another tribe of Apinae.

Some characters of Isepeolini include the apically fimbriate and emarginate T6 of the female, without spinelike setae, and the apically pointed S6 of the female, sometimes with a series of spinelike or peglike preapical setae laterally (Fig. 103-2a-c). The axillae are rounded, not pointed as in Epeolini. As in Protepeolini the three submarginal cells are subequal in the lengths of their posterior margins (Fig. 103-1). The wings have bare areas basally; distally they are strongly papillate. The pygidial plate is entirely absent in both sexes. S7 of the male has a hairless median apical projection but no lobes (Fig. 103-2e); S8 sometimes has a median apical projection, sometimes does not. The male genitalia usually have both dorsal and ventral gonostyli (Fig. 103-2d); sometimes both are much reduced and fused to the gonocoxite.

This tribe was revised by Roig-Alsina (1991a), who provided numerous illustrations of structures, including the male genitalia and hidden sterna. Male genitalia were also illustrated by Toro and Rojas (1968), who provided a key to Chilean species.

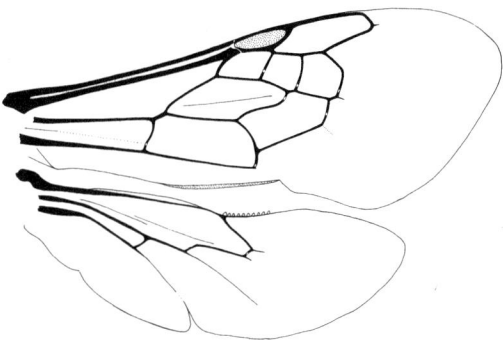

Figure 103-1. Wings of *Isepeolus viperinus* (Holmberg).

Key to the Genera of the Isepeolini
(Modified from Roig-Alsina, 1991a)

1. Vertex of head between lateral ocellus and eye slightly rounded to flat as seen in frontal view, at most with shallow depression; first flagellar segment of male conspicuously widened apically, as in female; sclerotized apex of female S6 bordered by long spinelike setae (Fig. 103-2c), and lateral apical margin folded down, partially covering such setae; apex of female T5 prolonged by membranous rim bordered by upcurved hairs; male S6 without preapical tubercle .... *Melectoides*
—. Vertex of head with conspicuous depression between eye and ocelli as seen in frontal view; first flagellar segment of male slightly widened apically, contrasting with conspicuous widening in female; sclerotized apex of female S6

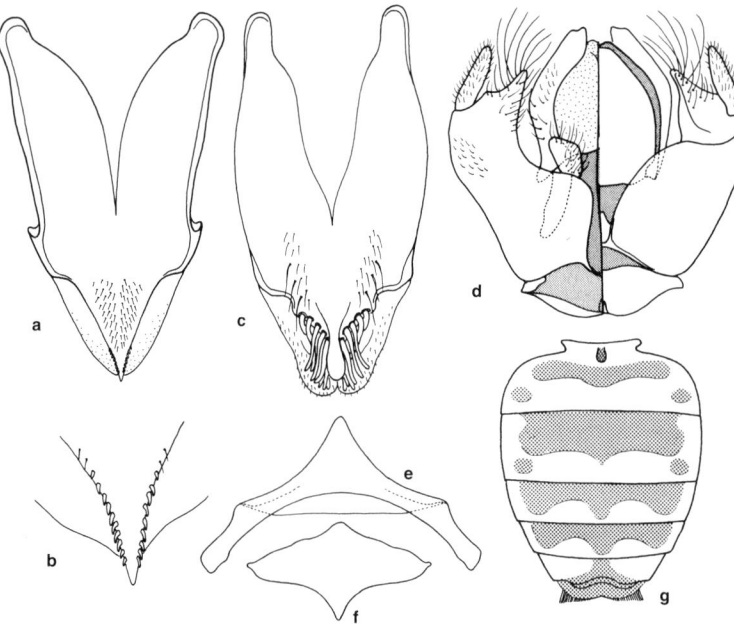

Figure 103-2. Structures of Isepeolini. **a,** S6 of female of *Isepeolus viperinus* (Holmberg); **b,** Apex of same, enlarged; **c,** S6 of female of *Melectoides kiefferi* (Jšrgensen); **d,** Male genitalia of *Melectoides triseriatus* (Friese), dorsal view on the right; **e, f,** S7 and S8 of the same; **g,** Dorsal metasomal pattern of *Melectoides politus* Roig-Alsina. From Roig-Alsina, 1991a).

with spinelike setae absent or short (Fig. 103-2a), barely visible at low magnification, lateral apical margin not expanded, not folded down; apex of female T5 without membranous rim; male S6 with large preapical tubercle
.... *Isepeolus*

Genus *Isepeolus* Cockerell

Isepeolus Cockerell, 1907b: 64. Type species: *Isepeolus albopictus* Cockerell, 1907 = *Epeolus viperinus* Holmberg, 1886, monobasic.
Palinepeolus Holmberg, 1909b: 77. Type species: *Epeolus viperinus* Holmberg, 1886, by original designation.
Calospiloma Brèthes, 1909b: 68 footnote. Type species: *Epeolus viperinus* Holmberg, 1886, by original designation.

Isepeolus consists of species with the metasoma white spotted, 6 to 11 mm in body length.
■ This genus occurs from Santa Cruz province, Argentina, and Magallanes, Chile, north to the state of Ceará, Brazil; there is a possibly erroneous record from Cali, Colombia. Eleven species were recognized in the revision by Roig-Alsina (1991a).

Isepeolus has been reared from nests of *Colletes* (Michener, 1957; Lucas de Oliveira, 1966); the first-stage larva, with its sclerotized head and large, curved mandibles, was described in those papers and by Rozen (1991a); mature larvae were described by Lucas de Oliveira and by Rozen (1966a).

Genus *Melectoides* Taschenberg

Melectoides Taschenberg, 1883: 75. Type species: *Melectoides senex* Taschenberg, 1883, by original designation.

Most species in this genus are nonmetallic, white-spotted or banded, and rather small in size (length 7.5-11.0), thus resembling species of *Isepeolus*. Two species, however, *Melectoides senex* Taschenberg and *tucumanus* (Friese), have metallic blue or green metasomal pubescence, in the latter species lacking white patches; these two species, both large (length 12.5-16.0 mm), resemble *Mesoplia* or *Mesonychium* (Ericrocidini) in superficial appearance.
■ *Melectoides* occurs in Chile from Atacama to Cautín and in Argentina from Salta province to Chubut province. The ten species were revised by Roig-Alsina (1991a).

The large species of *Melectoides* are too large to parasitize *Colletes*; perhaps they attack diphaglossine Colletidae. However, *Melectoides bellus* (Jörgensen) is a cleptoparasite of *Canephorula apiformis* (Friese) in the Eucerini, as shown by Michelette, Camargo, and Rozen (2000).

104. Tribe Osirini

The tribe Osirini is composed of cleptoparasitic genera that were long included in the Nomadinae. Roig-Alsina (1989a), however, found that these genera do not possess some of the principal apomorphies of the Nomadinae, such as (1) the largely retracted, truncate, emarginate to bifid or pointed S6 of the female, usually with spinelike apical or preapical setae, (2) the specialized median area usually present on T5 of the female, sometimes clearly defined and called the pseudopygidial area, and (3) the simple stipital margin presumably resulting from loss of the stipital comb. The Osirini possess various features also found in nonparasitic Apinae but absent in the Nomadinae, such as a little-modified (although attenuate and tubular in *Osiris*) S6 of the female; a ridge on the outer surface of the stipes; and a translucent, impunctate lamella at the posterior end of the metasternum. It therefore seems likely that the Osirini evolved parasitic habits independently from the Nomadinae, a conclusion supported by the phylogenetic analysis of Roig-Alsina and Michener (1993); see Figure 102-1. Other distinctive characters of Osirini include a small, round, central sclerite in the cervical membrane (unique among bees); a ventral carina along the inner and basal margins of the fore coxa; a large stigma, nearly three to about eight times as long as the prestigma; and a marginal cell that is narrowly separated from the wing margin throughout its length (Fig. 104-1), so that there is a membranous rim in front of the marginal cell. There are three submarginal cells. The body is nomadiform (Fig. 104-2a) or epeoliform. The male genitalia, diverse in structure, usually have both dorsal and ventral gonostyli, but only one is present in *Epeoloides* and *Osiris*.

It is likely that Osirini (except *Epeoloides*) are all cleptoparasites of Tapinotaspidini. Rozen (1984a) found a

Figure 104-2. Cleptoparasitic Apidae. Above, *Osiris* sp., female; Below, *Rhathymus* sp., male. From Michener, McGinley, and Danforth, 1994.

Parepeolus (Parepeolus) niger Roig-Alsina in a nest of *Tapinotaspoides tucumana* (Vachal), and Roig-Alsina (1989a) published records of occurrence of *Parepeolus (Ecclitodes) stuardi* (Ruiz) around nest sites of *Chalepogenus caeruleus* (Friese). Rozen et al. (2006) studied the biology of *Parepeolus minutus* Roig-Alsina in nests of *Chalepogenus (Lanthanomelissa) betinae* (Urban) and *Protosiris gigos* Melo in nests of *Monoeca haemorrhoidalis* (Smith). *Osiris* has been reared from nests of *Paratetrapedia* by Camillo, Garófalo, and Serrano (1993). In contrast, the holarctic genus *Epeoloides* is a parasite of *Macropis* in the Melittidae.

Key to the Genera of the Osirini
(Modified from Roig-Alsina, 1989a)

1. Jugal lobe of hind wing rounded; basal vein of fore wing basal to cu-v or sometimes meeting it; lateral margin of clypeus not continued by paraocular carina above level of anterior mandibular articulation; labrum with pair of preapical tubercles or two to four teeth near middle of disc; ventral gonostylus of male bifid, dorsal gonostylus large, flattened (South America) .... *Parepeolus*
—. Jugal lobe of hind wing elongate (Fig. 104-1); basal vein of fore wing apical to cu-v or sometimes meeting it; lateral margin of clypeus continued by paraocular carina above level of anterior mandibular articulation; labrum with single preapical tubercle or several small preapical denticles; ventral gonostylus of male simple, dorsal gonostylus small, cylindrical or absent 2

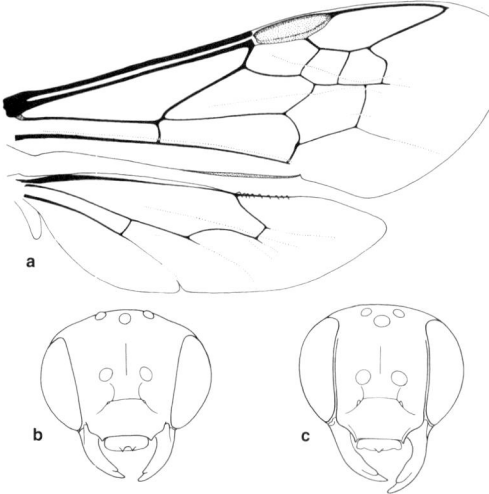

Figure 104-1. Osirini. **a**, Wings of *Osiris* sp.?; **b, c,** Faces of females of *Parepeolus minutus* Roig-Alsina and *Epeoloides pilosula* (Cresson). b, c, from Roig-Alsina, 1989a.

2(1). Apex of marginal cell curved away from wing margin (Fig. 104-3a); paraocular carina almost reaching summit of eye (Fig. 104-1c); eyes of male strongly converging above, those of female slightly so; labrum with strong preapical tubercle; stigma 2.8-3.2 times as long as prestigma; rami of male claws similar and adjacent (holarctic) .... *Epeoloides*
—. Apex of marginal cell terminating close to wing margin (Fig. 104-1a); paraocular carina fading near lower margin of eye; eyes of both sexes parallel or diverging above; labrum with several small denticles; stigma more than 5.5 times as long as prestigma; inner ramus of male claws shorter than outer, flattened 3
3(2). Collar of pronotum dorsally bulging, with a distinct subhorizontal portion connecting swollen, rounded dorsolateral angles; inner margin of eye with shallow emargination near upper third; mandible with two subapical teeth; S6 of female usually exceeding T6 (Fig. 104-4c); apex of pygidial plate of female not reaching margin of tergum, a wide rim present apical to plate; pygidial plate of male reduced to sclerotization on tergal margin; male gonostylus simple (neotropical) .... *Osiris*
—. Transverse middorsal part of pronotal collar short or absent, dorsolateral lobes or angles thus not connected by elevated collar; inner margin of eye almost straight; mandible with one subapical tooth or simple (an inner angle sometimes present); S6 of female not exceeding T6; apex of pygidial plate of female constituting apex of tergum; pygidial plate of male present; male gonostylus double, i.e., upper and lower gonostyli well separated 4
4(3). Inner margins of eyes almost parallel; mandible long, strongly curved, outer basal width 0.33 to 0.35 times length of mandible; mesepisternum with hairs simple or at most with one or two basal barbs; legs slender; strigilar concavity 0.26 to 0.27 times length of basitarsus; scutum as long as intertegular distance (neotropical) .... *Protosiris*
—. Inner margins of eyes convergent below; mandible short, outer basal width 0.40 to 0.45 times length of mandible; at least anterior part of mesepisternum with plumose hairs; legs short, stout; strigilar concavity shallow, 0.38 to 0.41 times length of basitarsus; scutum shorter than intertegular distance (South America)........ .... *Osirinus*

Genus *Epeoloides* Giraud

Epeoloides Giraud, 1863: 44. Type species: *Epeoloides ambiguus* Giraud, 1863 = *Apis coecutiens* Fabricius, 1775, monobasic.
Viereckella Swenk, 1907: 298. Type species: *Viereckella obscura* Swenk, 1907 = *Nomada pilosula* Cresson, 1878, by original designation.

In this genus the vestiture of the head and thorax is relatively long and plumose (although not hiding the surface), unlike that of other Osirini. The body is epeoliform, black or with the metasoma red; length is 7 to 10 mm. A striking character of the male is the small, elongate, strongly elevated pygidial plate margined by strong carinae that nearly enclose the plate anteriorly as well as posteriorly.

Because it is parasitic on the melittid genus *Macropis*

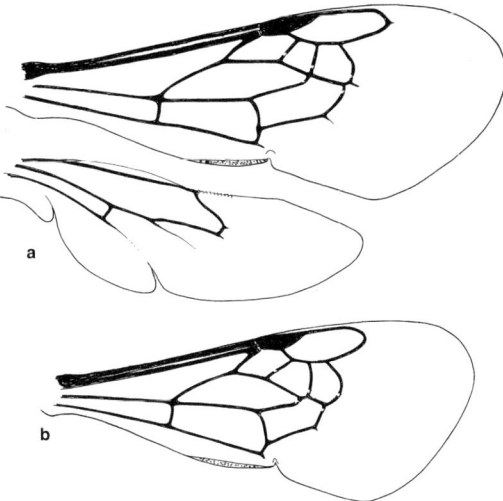

Figure 104-3. Wings of Osirini. **a,** *Epeoloides coecutiens* (Fabricius); **b,** Forewing of *Parepeolus niger* Roig-Alsina. From Roig-Alsina, 1989a.

rather than on Tapinotaspidini, and because it is holarctic (the other Osirini are strictly neotropical), *Epeoloides* might be convergent with the other Osirini rather than phylogenetically closely related. Roig-Alsina (1989a) was the first to note the similarity; *Epeoloides* was previously placed by itself in a tribe Epeoloidini. At present there is no good reason to disagree with Roig-Alsina's decision to include *Epeoloides* in the Osirini; convergence in characters like the separation of the marginal cell from the wing margin (Fig. 104-3) seems unlikely.

■ The range of *Epeoloides* includes Northern and Central Europe from the Netherlands and France to western Russia, and North America from Quebec, Canada, to Georgia west to North Dakota and Nebraska, USA. There are probably only two species, one in Europe, the other in America. Linsley and Michener (1939) gave a description of the genus; that work, Iuga (1958), Popov (1958a), Mitchell (1962), and Roig-Alsina (1989a) all illustrated the male genitalia and other structures.

The American species, *Epeoloides pilosula* (Cresson), seemed not to have been collected in the USA since 1942 and in Canada since 1960. However, it was rediscovered in 2002 in Nova Scotia, Canada (Sheffield et al., 2004). The biology of the European *E. coerutiens* (Fabricius) was dealt with by Bogush (2005). Its presumed host, *Macropis*, remains widespread but localized to patches of its required flower, *Lysimachia* (Primulaceae).

Genus *Osirinus* Roig-Alsina

Osirinus Roig-Alsina, 1989a: 17. Type species: *Osirinus lemniscatus* Roig- Alsina, 1989, by original designation.
Compsoclepta Moure, 1995b: 143. Type species: *Compsoclepta fasciata* Moure, 1995 = *Osirinus lemniscatus* Roig-Alsina, 1989, by original designation.

Osirinus and *Protosiris* resemble *Osiris*, with which they have usually been confused, in their nomadiform body, shiny reddish, yellow, or testaceous integument (dark in some *Osirinus*), and pointed marginal cell. They differ from *Osiris* in the shorter head, absence of a malar space, more abundant plumose hairs, and other characters described in the key to genera. S6 of the female is relatively short and blunt (Fig. 104-4a), not attenuate and pointed as in *Osiris*. The body length is 5.5 to 10.0 mm.

■ This genus occurs from the provinces of Córdoba and Entre Ríos, Argentina, north to northeastern Brazil. The seven species were revised by Melo and Zanella (2003).

Genus *Osiris* Smith

Osiris Smith, 1854: 288. Type species: *Osiris pallidus* Smith, 1854, by designation of Sandhouse, 1943: 580.
Euthyglossa Radoszkowski, 1884a: 21. Type species: *Euthyglossa fasciata* Radoszkowski, 1884 = *Eucera euthyglossa* Dalla Torre, 1896, monobasic.

Osiris consists of slender, nomadiform, smooth, shiny, almost completely impunctate species (Fig. 104-2a) 6.2 to 18.0 mm long, the whole body often yellow or testaceous but sometimes blackish. The head is somewhat elongate compared to other Osirini, the length of the malar area being one-fifth to one-half as long as the basal width of the mandible. Behind the ocelli is a preoccipital carina. S6 of the female is elongate, usually exceeding T6, the lateral margins upturned to form a guide for the enormous sting (Fig. 104-4b, c), which at least in death is often exserted and reflexed above the metasoma. Male genitalia and hidden sterna were illustrated by Popov (1939c), Michener (1954b), and Shanks (1986).

■ *Osiris* occurs from the states of Nayarit and San Luis Potosí, Mexico, south through the tropics to Bolivia, Misiones Province, Argentina, and the state of Santa Catarina, Brazil. Shanks (1986, 1987) revised the 21 species of the genus.

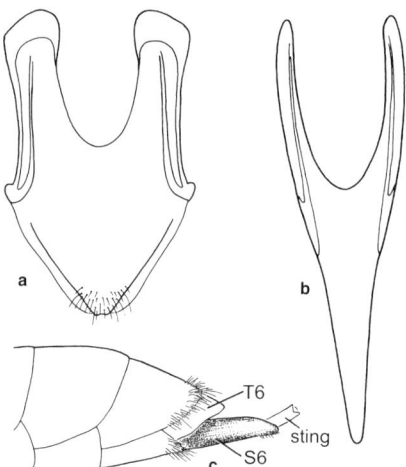

Figure 104-4. Apical structures of female Osirini. **a**, S6 of *Osirinus rutilans* (Friese); **b**, S6 of *Osiris variegatus* Smith; **c**, Side view of apex of metasoma of *Osiris* sp., the distal part of the sting omitted. a, b, from Roig-Alsina, 1989a; c, from Michener, McGinley, and Danforth, 1994.

Genus *Parepeolus* Ducke

This South American genus includes epeoliform species 6 to 13 mm in length. The two subgenera have hitherto been regarded as genera, but are so similar and contain so few species that subgeneric rank, which emphasizes their close relationship, seems preferable.

Key to the Subgenera of *Parepeolus*

1. Propodeal triangle microstriate; posterior margin of first submarginal cell about twice as long as apical margin; maxillary palpus six-segmented; metasomal terga with apical bands of pale hairs; suprategular carina curved cephalad of tegula, following tegular margin, reaching mesoscutum-pronotal lobe boundary close to tegula .... *P. (Ecclitodes)*
—. Propodeal triangle polished, smooth or with few scattered punctures; posterior margin of first submarginal cell less than 1.5 times as long as apical margin (Fig. 104-3b); maxillary palpus four- to five-segmented; metasomal terga black or variously patterned with pale hairs, an apical band, when present, appearing only on T1; suprategular carina slightly curved, directed anteriorly, diverging from tegular margin and reaching mesoscutum-pronotal lobe boundary near middle of that margin .... *P. (Parepeolus s. str.)*

Parepeolus / Subgenus *Ecclitodes* Roig-Alsina

Ecclitodes Roig-Alsina, 1989a: 9. Type species: *Epeolus stuardi* Ruiz, 1935, by original designation.

Ecclitodes includes bees with apical zones of white, appressed hairs on T1 to T5. They therefore superficially resemble *Epeolus* in the Nomadinae.

■ This subgenus occurs from Nuble to Valdivia, Chile, and in the province of Río Negro, Argentina. The only described species is *Parepeolus (Ecclitodes) stuardi* (Ruiz); Roig-Alsina (1989a) gave a full account of the subgenus (as a genus) and indicated that there may be two species.

Parepeolus / Subgenus *Parepeolus* Ducke s. str.

Parepeolus Ducke, 1912: 71, 102. Type species: *Leiopodus lecointei* Ducke, 1907 = *Epeolus aterrima* Friese, 1906, by designation of Sandhouse, 1943: 585.

Parepeolus s. str. consists of largely black species, sometimes having spots of pale pubescence on the metasomal terga, the pubescence sometimes forming an apical pale band on T1 only. To the distinguishing characters listed in the key to subgenera can be added the usually rounded rather than somewhat truncate apex of the marginal cell (Fig. 104-3b) and the occurrence of the subgenus in warmer environments.

■ *Parepeolus* s. str. occurs from the provinces of Buenos Aires and La Pampa, Argentina, north to the state of Pará, Brazil. The four species were revised by Roig-Alsina (1989a).

Genus *Protosiris* Roig-Alsina

Protosiris Roig-Alsina, 1989a: 20. Type species: *Osiris obtusus* Michener, 1954, by original designation.

The characters that differentiate this genus from *Osiris* are listed not only in the key to genera but also in the account of *Osirinus*. *Protosiris* differs from *Osirinus* in its generally larger size (9.5-17.0 mm long), in the more slender legs (like those of *Osiris*), in the minute, scattered punctures in the integument, and in other characters indicated in the key to genera. The similarities to *Osirinus* in male genitalia and other characters are well shown by Roig-Alsina (1989a). These similarities, however, are largely plesiomorphies relative to *Osiris*, and Roig-Alsina shows *Protosiris* as the sister group to *Osiris*, and *Osirinus* as sister group to the combined *Osiris* + *Protosiris*.

■ *Protosiris* is known from Bolivia and the state of São Paulo, Brazil, north through the tropics to the state of Puebla, Mexico. There are four described species (the Mexican species is undescribed); they are those that run to couplet 7 in Shanks' (1986) key to species of *Osiris*. They were described in that work, and structures of *P. obtusus* (Michener) were illustrated by Roig-Alsina (1989a).

105. Tribe Protepeolini

Protepeolini consists of Western Hemisphere epeoliform bees 6 to 12 mm in length, with areas of pale appressed pubescence on the metasoma (Fig. 105-1), suggesting *Epeolus*. Like Osirini and Isepeolini, Protepeolini has traditionally been included in the Nomadinae. Some authors (Alexander, 1990; Rozen, Eickwort, and Eickwort, 1978) regarded the Isepeolini and Protepeolini as basal groups of Nomadinae, whereas Roig-Alsina and Michener (1993) separated these two tribes from the Nomadinae and considered them as parasitic tribes of Apinae (see Fig. 102-1). The Protepeolini differ from the Nomadinae and especially from its tribe Epeolini in having (1) S6 of the female tapering, its lateral margins curved upward to form a guide for the sting (not as long as in *Osiris*), neither emarginate nor with two processes apically and without spinelike setae; (2) T6 of the female lacking a typical pygidial plate, although it has a slender, median, parallel-sided or spatulate flat projection, on each side of which is a series of long spines, (3) T7 of the male lacking a pygidial plate, and (4) the three submarginal cells subequal in the lengths of their posterior margins (Fig. 105-2). The axillae are rounded, not produced and angular as in Epeolini. The arolia are extraordinarily large, often longer than the claws, and swollen apically. Illustrations of male genitalia and other structures were provided by Linsley and Michener (1939) and Eickwort and Linsley (1978); immature stages were described by Rozen, Eickwort, and Eickwort (1978) and Roig-Alsina and Rozen (1994).

As in the Nomadinae, the egg is hidden in the wall of an open host cell. The head of the first-stage larva is only slightly prognathous, the labiomaxillary region is membranous, and the mandible is extremely slender and pointed but not enormous (Fig. 105-3) (Rozen, 1991a; Roig-Alsina and Rozen, 1994).

There is only one genus of Protepeolini.

Figure 105-1. *Leiopodus singularis* (Linsley and Michener), female. Drawing by C. Green, from Michener, McGinley, and Danforth, 1994.

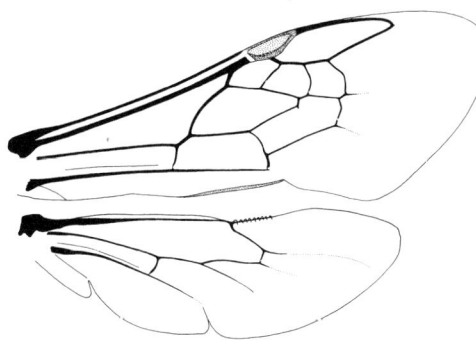

Figure 105-2. Wings of *Leiopodus lacertinus* Smith.

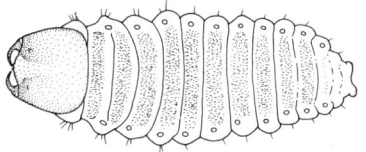

Figure 105-3. First-stage larva of *Leiopodus singularis* (Linsley and Michener). From Rozen, 1991a.

Genus *Leiopodus* Smith

Leiopodus Smith, 1854: 252. Type species: *Leiopodus lacertinus* Smith, 1854, monobasic.

Liopodus Schulz, 1906: 258, unjustified emendation of *Leiopodus* Smith, 1854.

Protepeolus Linsley and Michener, 1937: 75. Type species: *Protepeolus singularis* Linsley and Michener, 1937, monobasic and by original designation.

The characters are those of the tribe.

■ *Leiopodus* occurs from Southern California to western and southern Texas, USA, south through the tropics to the province of Río Negro, Argentina. The North American *L. singularis* (Linsley and Michener) and the Argentine *L. abnormis* (Jörgensen) are almost identical. The five species, only one of which is known from North America, were revised by Roig-Alsina and Rozen (1994).

Leiopodus is a cleptoparasite in nests of the emphorine genera *Diadasia, Melitoma,* and *Ptilothrix* (Rozen, Eickwort, and Eickwort, 1978; Roig-Alsina and Rozen, 1994).

106. Tribe Exomalopsini

The name Exomalopsini was long used, for example by Michener (1944) and Michener and Moure (1957), in a broad sense to include the genera now placed in the Tapinotaspidini, as well as *Ancyloscelis,* now in the Emphorini. These groups were removed from Exomalopsini by Roig-Alsina and Michener (1993). As shown in Figure 102-1, the removal of these groups did not result in a monophyletic Exomalopsini, and Silveira (1995a) removed *Teratognatha* and *Chilimalopsis* from the Exomalopsini to form a new tribe, the Teratognathini. The two last-named genera are clearly closely related to one another, but their separation from the Exomalopsini s. str. seems to me premature. Although Silveira's phylogenetic analysis justifies the tribe Teratognathini, tribal characters to distinguish it from Exomalopsini s. str. are neither convincing nor particularly practical. The Teratognathini have the plesiomorphic preapical mandibular tooth, sometimes scarcely evident, whereas Exomalopsini s. str. lack such a tooth (Fig. 108-4a), but there is an exception: *Eremapis* in the Exomalopsini retains a tooth. The erect hairs along the inner orbits of Exomalopsini are distributed among taxa in exactly the same way, being absent in both *Eremapis* and the Teratognathini. The unique synapomorphies indicated by Silveira for the Teratognathini are (1) the sparse and not or sparsely branched scopa, a character that frequently varies according to the type of pollen collected (see *Tetralonia,* in the Eucerini); (2) the more or less pentagonal disc of S7 of the male (Fig. 106-3l), a character that seems less significant when one notes its great variability within the Exomalopsini s. str. (Fig. 106-4), and (3) the rather slender and apparently basally elbowed male gonostyli, probably really straight but projecting posteriorly from an apicolateral process of the gonocoxite (Fig. 106-3j). With the objective of having readily recognizable tribes, I have decided to include *Teratognatha* and *Chilimalopsis* in the Exomalopsini for the present. Their distinctiveness can be indicated by placing them in a subtribe Teratognathina.

The Exomalopsini are minute to moderate-sized, rather hairy, anthophoriform bees, commonly having pale metasomal hair bands (Fig. 106-1) and lacking yellow or white integumental markings, except sometimes for the clypeus, labrum, and mandibles. Sometimes the metasomal integument is red. The maxillary palpi are six-segmented, but five-segmented in *Eremapis.* Except in *Chilimalopsis, Eremapis,* and *Teratognatha,* there is a row of long, erect, well-separated hairs along each inner ocular orbit; this feature is unique to this tribe. The marginal cell is obliquely truncate (Fig. 106-2), but in some specimens of *Eremapis* the truncation is not evident, the apex of the cell being bent away from the costa and pointed; the total length of the cell is usually greater than the distance from its apex to the wing tip, but the length of the cell on the costa is less than the distance to the wing tip. There are usually three submarginal cells, but there are species with only two in *Anthophorula, Chilimalopsis,* and *Exomalopsis.* Silveira (1995a) has provided phylogenetic hypotheses at the generic and subgeneric levels.

Since the genus *Anthophorula* has usually been included in *Exomalopsis,* published keys and revisions pertain to the two together; such works are Timberlake

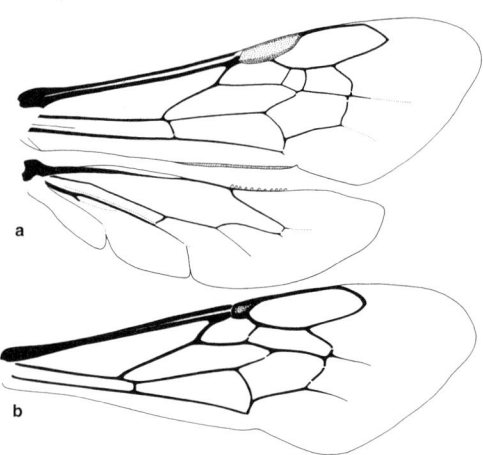

Figure 106-2. Wings of Exomalopsini. **a,** *Exomalopsis (Exomalopsis) zexmeniae* Cockerell; **b,** *Anthophorula (Anthophorula) compactula* Cockerell.

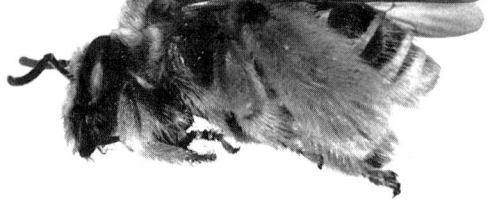

Figure 106-1. Lateral views of *Exomalopsis*. Above, *E. similis* Cresson, male; Below, *E. solani* Cockerell, female. From Michener, McGinley, and Danforth, 1994.

(1947, 1980b). Silveira (1995b) has catalogued the species of Exomalopsini.

This tribe is found only in the New World and is primarily neotropical, although one species of *Anthophorula* extends north to the central USA.

Key to the Genera of the Exomalopsini

1. Hind tibial and basitarsal scopa consisting of sparse, simple or sparsely branched hairs; mandible with preapical tooth that is sometimes minute; S7 of male with disc pentagonal (Fig. 106-3l), median apex separating two apical margins, apicolateral angle with short, robust process; male gonocoxite with apicolateral process from apex of which arises rather long gonostylus, the process and gonostylus sometimes appearing together as basally elbowed gonostylus (Fig. 106-3j) (subtribe Teratognathina) 2
—. Scopa consisting of strongly plumose hairs; mandible simple (but with preapical tooth in *Eremapis*); S7 of male variable, not as above; male gonocoxite without apicolateral process from which gonostylus arises (Figs. 106-3, 106-4) (subtribe Exomalopsina) 3
2(1). Stigma broad, three times as long as prestigma; pronotal lobe produced to horizontal lamella in female, strong ridge in male, rounded as seen from above (Argentina) .... *Teratognatha*
—. Stigma small, less than twice as long as prestigma; pronotal lobe rounded (South America) .. *Chilimalopsis*
3(1). Second medial cell of forewing about as long as second cubital; labrum and clypeus of female yellow; margin of T7 of male produced into a broad, sharp lamella, rounded as seen from above; S7 of male broadly trilobed; hind basitarsus of female without apical process and penicillus (Argentina) .... *Eremapis*
—. Second medial cell of forewing much shorter than second cubital (Fig. 106-2); labrum and clypeus of female entirely dark; margin of T7 of male not produced into a carina or lamella, or, *if* carina present, then its posterior margin elevated as margin of pygidial plate; S7 of male not trilobed; hind basitarsus of female with apical process and penicillus 4
4(3). Basitibial plate of female small, with surface planar, or, *if* margins of plate raised and central area with velvety pilosity, then transverse carina of T1 of female absent; labrum and clypeus of male yellow or white [except dark in *A. (Anthophorisca) levigata* Timberlake and *linsleyi* Timberlake and in *A. (Isomalopsis) niveata* (Friese)]; outer side of penis valve without or with small lateral process (Fig. 106-3a, d, g); dorsal flange of male gonocoxite absent .... *Anthophorula*
—. Basitibial plate of female large, central area with velvety pilosity separated from raised margin by groove, and transverse carina of T1 of female present; labrum and clypeus of male entirely dark; outer side of penis valve with strong lateral process; dorsal flange of male gonocoxite present .... *Exomalopsis*

Genus *Anthophorula* Cockerell

This genus has usually been included in *Exomalopsis* but was differentiated by Silveira (1995a, b). In addition to the characters indicated in the key to genera, it differs from *Exomalopsis* in having the areas between the ocelli and the eyes convex. The stigma is commonly smaller than in *Exomalopsis,* less than half as long as the length of the marginal cell on the wing margin, but this character is consistent only for the subgenus *Anthophorula* s. str. In *Anthophorisca* the stigma is often as large as in those *Exomalopsis* having smaller stigmas, although never as large as those in some *Exomalopsis* s. str. The body length ranges from 2.5 to 8.0 mm; the smallest species are in the subgenus *Anthophorula* s. str., but one species of *Anthophorisca* is only 3.5 mm long. Male genitalia and hidden sterna were illustrated by Michener and Moure (1957), Mitchell (1962), Timberlake (1980b), and Silveira (1995a, b).

As in *Exomalopsis* but not *Eremapis*, nests of *Anthophorula* are usually communally occupied burrows, each being inhabited by several females (Rozen, 1984c). Cells are oval, with a thin waterproof lining. As in the genus *Exomalopsis,* the provision mass is partly lifted off the cell surface by a projection, called a "foot" by Rozen (1977b, 1984a); it presumably reduces the area of moist contact between the provision mass and the cell surface, and thus may reduce the danger of mold. In *A. (Anthophorisca) sidae* (Cockerell) and *nitens* (Cockerell), larvae of the autumn generation spin cocoons, but those of the summer generation do not (Rozen, 1984a; Rozen and Snelling, 1986), although they retain the cephalic structures characteristic of cocoon-spinning larvae.

Key to the Subgenera of *Anthophorula*

1. Transverse carina of T1 at summit of anterior surface weak or absent in female, that of male absent; peglike setae present on mesal side of male gonostylus (Fig. 106-3d) [not visible in *A. linsleyi* (Timberlake)] (North America) .... *A. (Anthophorisca)*
—. Transverse carina of T1 of female strong, that of male present; peglike setae absent from male gonostylus 2
2(1). Stigma large, four times as long as prestigma, its inner breadth (i.e., not counting marginal veins) much greater than breadth of its marginal veins; yellow marks on clypeus of male absent or restricted to apical transverse line; S6 of male with two apicolateral flanges that bear a series of spicules on their inner margins (Argentina) .... *A. (Isomalopsis)*
—. Stigma small, no more than three times as long as prestigma, its inner breadth (i.e., not counting marginal veins) as great as or less than breadth of its marginal veins; clypeus of male entirely yellow; S6 of male simple (North America) .... *A. (Anthophorula* s. str.)

Anthophorula / Subgenus *Anthophorisca* Michener and Moure

Exomalopsis (Anthophorisca) Michener and Moure, 1957: 433. Type species: *Melissodes pygmaea* Cresson, 1872, by original designation.
Exomalopsis (Panomalopsis) Timberlake, 1980b: 82. Type species: *Exomalopsis linsleyi* Timberlake, 1980, by original designation.

I follow Silveira (1995a, b) in regarding *Panomalopsis* as a derived *Anthophorisca;* it has unusual male genitalia but does not seem to require subgeneric status.

■ This subgenus occurs across the southern USA from

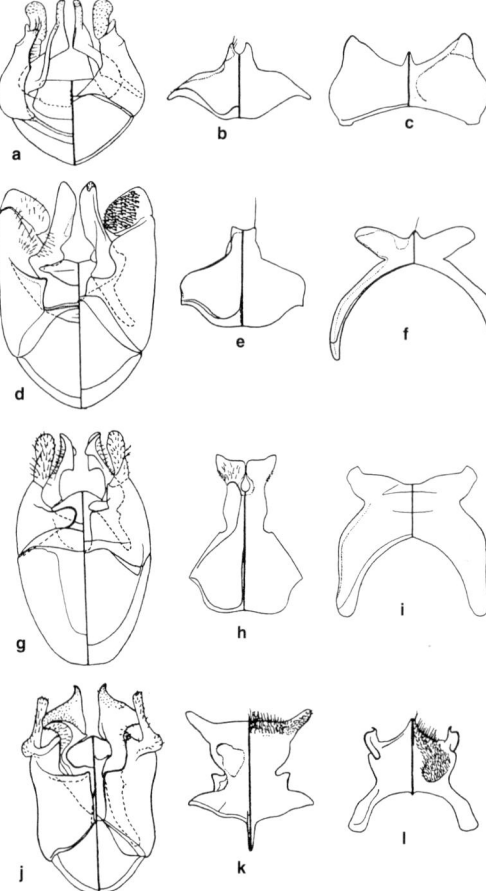

Figure 106-3. Male genitalia, S8, and S7 of Exomalopsini. **a-c**, *Anthophorula (Isomalopsis) niveata* (Friese); **d-f**, *A. (Anthophorisca) pygmaea* (Cresson); **g-i**, *A. (Anthophorula) compactula* Cockerell; **j-l**, *Teratognatha modesta* Ogloblin. (Dorsal views are on the left.) From Michener and Moure, 1957.

California to Mississippi, north as far as Utah, Nebraska, and Indiana and south to Oaxaca, Mexico. The 29 species were revised by Timberlake (1980b) within *Exomalopsis*, and were listed by Silveira (1995b).

Anthophorula / Subgenus *Anthophorula* Cockerell s. str.

Athrophorula Cockerell, 1897d: 44. Type species: *Anthophorula compactula* Cockerell, 1897, monobasic.
Diadasiella Ashmead, 1899a: 64. Type species: *Diadasiella coquilletti* Ashmead, 1899 = *Synhalonia albicans* Provancher, 1896, by original designation.
Exomalopsis (Pachycerapis) Cockerell, 1922e: 4. Type species: *Exomalopsis cornigera* Cockerell, 1922, by original designation.

■ This subgenus occurs from Oregon, Utah, and Texas, USA, south to Baja California, Jalisco, and Veracruz, Mexico. The 29 species were revised by Timberlake (1980b) within *Exomalopsis* and were listed by Silveira (1995b).

Anthophorula / Subgenus *Isomalopsis* Michener and Moure

Isomalopsis Michener and Moure, 1957: 434. Type species: *Tetralonia niveata* Friese, 1908, by original designation.

A distinctive character of *Isomalopsis* is the form of S7 of the male (Fig. 106-3c), which is similar to that of more anterior sterna, with a broad disc and very short apodemes, as illustrated by Michener and Moure (1957). In the North American subgenera, S7 is short and broad, and its basolateral apodemes are much longer than the length of the disc (Fig. 106-3f, i).

■ *Isomalopsis* is the only South American *Anthophorula*. The two species occur in central and northern Argentina.

Genus *Chilimalopsis* Toro

Chilimalopsis Toro, 1976: 73. Type species: *Chilimalopsis parvula* Toro, 1976, by original designation.

In *Chilimalopsis*, which consists of small (3.5-4.0 mm long) blackish bees, the labrum and sometimes the lower margin of the clypeus are whitish in the male, and there are sparse pale hairs and apical white hair bands on the metasomal terga. As in *Teratognatha*, the second medial cell is shorter than that of *Eremapis* but longer than that of *Exomalopsis*. The male genitalia and other structures were illustrated by Toro (1976) and Roig-Alsina (1992).

■ This genus is known from Coquimbo, Chile, and from Buenos Aires province to Jujuy province, Argentina. There appear to be three species, one of them undescribed.

Genus *Eremapis* Ogloblin

Eremapis Ogloblin, 1956: 149. Type species: *Eremapis parvula* Ogloblin, 1956, by original designation.

This is a genus of small bees, 4.0 to 4.5 mm in body length. Additional characters are the large interocellar distance, nearly twice the ocellocular distance; the punctured, not concave, upper paraocular areas; the convex vertex as seen in frontal view, almost as in Emphorini; and, in the female, the lack of an apical process and brush of the hind basitarsus and presence of a sternal scopa of coarse, strongly hooked hairs. The long second medial cell (see the key to genera) is unique in the Exomalopsini. Genitalia and hidden sterna were illustrated by Ogloblin (1956) and Michener and Moure (1957).

■ *Eremapis* occurs in Argentina from the provinces of Catamarca and Santiago del Estero to Río Negro. There is a single species, *E. parvula* Ogloblin.

Nests are burrows in sandy soil, each built by a single female (Neff, 1984). The horizontal, oval cells lack obvious secreted linings, unlike most related bees.

Genus *Exomalopsis* Spinola

Because of the removal of *Anthophorula*, *Exomalopsis* is here recognized in a narrower than usual sense. The differentiating characters are indicated in the key to genera and in the account of the genus *Anthophorula*. The body length is 4 to 12 mm. Genitalia and hidden sterna were illustrated by Michener (1954b), Michener and Moure

(1957), Mitchell (1962), Timberlake (1980b), and Silveira (1995a, b).

As in *Anthophorula*, nests, so far as is known, are communal. Zucchi (1973) showed that nests of *Exomalopsis (Exomalopsis) auropilosa* Spinola may contain hundreds of females (up to 884) and attain a depth of over 5 meters. Probably females of generation after generation return to the nest and extend it. The cells, which as in most ground-nesting bees are lined with "waxlike" material, are usually vertical, isolated at the ends of lateral, downcurved burrows, but in other species the cells are horizontal or variable in orientation, sometimes seemingly in series (Norden, Krombein, and Batra, 1994). The food mass often lacks the "foot," well known in the genus *Anthophorula*. Larvae may or may not spin cocoons, and those species that do not, such as *E. bruesi* Cockerell, have recessed labiomaxillary regions and reduced salivary lips so that they cannot do so (Rozen, 1997a). There is evidence, summarized by Rozen (1984a), of cooperative provisioning of cells by females; if that is verified, such colonies are probably semisocial or quasisocial. The evidence for such behavior is that females with undeveloped ovaries sometimes collect pollen, and that there are sometimes more pollen gatherers inhabiting a nest than cells being provisioned. The nesting biology of two additional species of *Exomalopsis* with 2 to 12 and 2 to 19 bees per nest was described by Raw (1976, 1977).

Key to the Subgenera of *Exomalopsis*

1. Vertex in frontal view convex; area between lateral ocellus and eye not excavated (except in *E. arcuata* Timberlake) and lateral ocelli below level of summit of head; marginal zones of T1 and T2 of female smooth and glabrous; T2 to T4 of female with white apical fringes sometimes interrupted medially; S7 of male with apical sclerotization forming short transverse sclerite across apex of sternum and fused to arms of disc; S8 of male with apical process a single bare lobe (North and Central America) .... *E. (Stilbomalopsis)*
—. Vertex in frontal view straight; area between lateral ocellus and eye excavated and/or lateral ocelli above level of summit of head; marginal zones of T1 and/or T2 of female punctate and pilose; T2 to T4 of female with apical fringes absent; S7 of male with apical process absent or complex and with two free basolateral lobes under ventral surface; S8 of male with apical process bearing two apical arms (short or long), or, *if* consisting of a single broad lobe, then lobe hairy (Fig. 106-4) 2
2(1). Submarginal cells two (vertex of female, between ocellus and eye, not excavated; T1 of female with premarginal line not depressed; S6 of male entirely flat; S7 and S8 of male with peglike setae) .... *E. (Diomalopsis)*
—. Submarginal cells three 3
3(2). T1 of female with premarginal line depressed, forming transverse sulcus, and marginal zone between dorsolateral convexities smooth and shining, comprising no more than two-thirds of dorsal surface of tergum; S6 of male entirely planar .... *E. (Phanomalopsis)*
—. T1 of female with premarginal line not depressed, or, *if* depressed, then marginal zone between dorsolateral convexities punctate and/or comprising much more than two-thirds of dorsal surface of tergum; S6 of male with median elevated area that broadens toward apex of sternum, forming carina or spine at each side (S7 and S8 of male without peglike setae; vertex of female, between ocellus and eye, excavated) .... *E. (Exomalopsis s. str.)*

Almeida and Silveira (1999) presented an alternative key to the subgenera of *Exomalopsis*.

Exomalopsis / Subgenus *Diomalopsis* Michener and Moure

Exomalopsis (Diomalopsis) Michener and Moure, 1957: 431. Type species: *Exomalopsis bicellularis* Michener and Moure, 1957, by original designation.

In the genus *Anthophorula*, species with two submarginal cells are close relatives of those with three submarginal cells. In *Exomalopsis*, however, the only two-celled species is distinctive in other respects and appears to warrant its own subgenus. Distinctive characters are the large stigma, longer than the length of the marginal cell on the wing margin, and the pedunculate, apically broad and truncate process of S8 of the male (Fig. 106-4b). The genitalia and hidden sterna were illustrated by Michener and Moure (1957) and Silveira (1995a, b); see also Figure 106-4a-c.

■ This subgenus is known from Paraná and São Paulo, Brazil, and Paraguay. The two species were distinguished by Almeida and Silveira (1999).

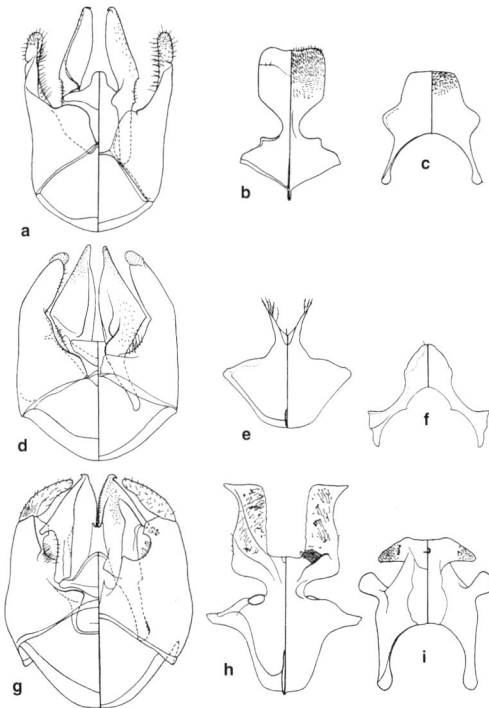

Figure 106-4. Male genitalia, S8, and S7 of *Exomalopsis*. **a-c**, *E. (Diomalopsis) bicellularis* Michener and Moure; **d-f**, *E. (Exomalopsis) aureopilosa* Spinola; **g-i**, *E. (Phanomalopsis) aureosericea* Friese. From Michener and Moure, 1957.

Exomalopsis / Subgenus *Exomalopsis* Spinola s. str.

Exomalopsis Spinola, 1853: 89. Type species: *Exomalopsis fulvo-pilosa* Spinola, 1853 = *Exomalopsis auropilosa* Spinola, 1853, by designation of Taschenberg, 1883: 82.
Epimonispractor Holmberg, 1887b: 225, nomen nudum.
Epimonispractor Holmberg, 1903: 426. Type species: *Epimonispractor gratiosus* Holmberg, 1903 = ?*Exomalopsis analis* Spinola, 1853, by designation of Sandhouse, 1943: 548. [See note by Michener, 1997b.]
Megomalopsis Michener and Moure, 1957: 430. Type species: *Exomalopsis diversipes* Cockerell, 1949 = *Exomalopsis mellipes* Cresson, 1878, by original designation.

I follow Silveira (1995a, b) in relegating *Megomalopsis* to synonymy under *Exomalopsis* s. str. In *Exomalopsis* s. str. the stigma, although variable, is usually large, about as long as the length of the marginal cell on the wing margin (Fig. 106-2a). The vertex in facial view is nearly straight, but the area between the ocelli and the summit of the eye is concave. The preoccipital ridge is sharp behind the vertex, at least lateral to the ocelli. S7 and S8 of the male have small discs and the apical process is short or unrecognizable, that of S8 notched or bifid at the apex (Fig. 106-4e, f). The genitalia and hidden sterna were illustrated by Michener and Moure (1957), Timberlake (1980b), and Silveira (1995a, b); see also Figure 106-4d-f.
■ This subgenus is found from Baja California and Sonora, Mexico, southern Texas and southern Florida, USA, and the Bahamas south through Mesoamerica, the Antilles, and South America to Tarapacá, Chile, Misiones province, Argentina, and Paraguay. There are about 55 species in this subgenus (Silveira, 1995b). Timberlake (1980b) revised 41 of these from North and Central America and the Antilles.

Exomalopsis / Subgenus *Phanomalopsis* Michener and Moure

Exomalopsis (Phanomalopsis) Michener and Moure, 1957: 430. Type species: *Exomalopsis jenseni* Friese, 1908, by original designation.

Silveira (1995a, b) removed most of the North American species previously included in *Phanomalopsis* to the subgenus *Stilbomalopsis*. The vertex of those remaining in *Phanomalopsis*, unlike that of *Stilbomalopsis*, is straight as seen in frontal view. The apical processes of S7 and S8 of the male are relatively large and complex (that of S8 deeply bilobed), unlike those of *Stilbomalopsis*. Illustrations of male genitalia and hidden sterna were provided by Michener and Moure (1957) and Silveira (1995a, b); see also Figure 106-4g-i.
■ *Phanomalopsis* occurs from Neuquén and Río Negro, Argentina, north to Sonora, Mexico, and southern Texas, USA. The 15 species were listed, and those of the *jenseni* group revised, by Silveira (1995b); species from Mesoamerica were included in a revision of the genus by Timberlake (1980b).

Exomalopsis / Subgenus *Stilbomalopsis* Silveira

Exomalopsis (Stilbomalopsis) Silveira, 1995a: 450. Type species: *Exomalopsis solani* Cockerell, 1896, by original designation.

Species of this subgenus have usually been mixed with those of *Phanomalopsis*. They differ by the convexity of the vertex (as seen from the front), the presence of white apical fringes on T2 to T4 of the female, and the simpler apical ornamentation of S7 and S8 of the male. Male genitalia and hidden sterna were illustrated by Timberlake (1980b) and Silveira (1995a, b).
■ *Stilbomalopsis* occurs from Southern California to Texas, USA, south to Nicaragua. At least 11 and probably 13 of the species included in *Phanomalopsis* by Timberlake (1980b) belong to *Stilbomalopsis*. The group was revised by Timberlake (1980b), and the species were listed by Silveira (1995a, b).

Genus *Teratognatha* Ogloblin

Teratognatha Ogloblin, 1956: 154. Type species: *Teratognatha modesta* Ogloblin, 1956, by original designation.

Teratognatha contains small bees, 3.5 to 4.5 mm long, blackish with the labrum of the male whitish, and with sparse pale hairs and apical bands of white hairs on the terga. As indicated by the characters listed in the first couplet of the key to genera, this genus is related to *Chilimalopsis*. Interesting additional characters are in the mouthparts. Because the last segment of both maxillary and labial palpi is reflexed, the palpi are hooked, and the hairs on the preceding segments are broad and blunt, or, on long segments of the labial palpi, hooked. Structures were illustrated by Ogloblin (1956) and Michener and Moure (1957); see also Figure 106-3j-l.
■ This genus is known from xeric areas of Tucumán and Salta provinces, Argentina. The only species is *Teratognatha modesta* Ogloblin.

107. Tribe Ancylini

The Ancylini are rather small to moderate-sized (5-13 mm long, mostly under 10 mm), euceriform to almost anthophoriform palearctic Apinae, similar in appearance to and perhaps closely related to the Eucerini or the American tribe Exomalopsini. These relationships, as well as those between the two included genera (*Ancyla* and *Tarsalia*, regarded as congeneric by Warncke, 1979c), were investigated by Silveira (1993a, b), who reached no decisive conclusions. Silveira (1995a, b), however, in another analysis, considered the Ancylini to be probably holophyletic. Baker (1998) restudied the two genera with much more abundant material and considered them not closely related; he removed *Tarsalia* from the Ancylini and placed it as a basal member of the Eucerini, leaving *Ancyla* as the only genus of Ancylini. But since he used only two eucerine genera in his study, and since parsimony analyses are often sensitive to the taxa included, and especially since the characters that he used are not known to me, I have not accepted his conclusions, although they may be correct. Some of the distinctive features of *Ancyla* involve its reduced glossa; some other characters of *Ancyla* may also be loss characters.

In males and some females the clypeus and lower paraocular areas are partly yellow. The Ancylini differ from the Exomalopsini (other than *Anthophorula* s. str.) by the small stigma, little broader than the prestigma as measured to the wing margin, the stigmal margin in the marginal cell being not or scarcely convex. The base of the propodeum has a subhorizontal zone, best developed in *Tarsalia*, and better developed than in most Exomalopsini. The wings have three submarginal cells, and the marginal cell bends gradually away from the wing margin distally (Fig. 107-1), not being obliquely truncate as in most Exomalopsini. The scopa on the hind tibia is rather large and consists of plumose hairs. T7 of the male is medially produced and weakly bidentate.

The male genitalia and hidden sterna are similar in *Ancyla* and *Tarsalia* and fundamentally similar to those of the Exomalopsini and Eucerini that have long gonostyli, a somewhat elaborate distal part of S7, and a bilobed apex of S8; these structures are asymmetrical in *Tarsalia*. Silveira (1993b) called attention to a character that appeared to distinguish *Ancyla* and *Tarsalia* from Exomalopsini, namely, a sulcus dividing the male gonocoxite (see Roig-Alsina and Michener, 1993). This sulcus, however, also occurs in some *Exomalopsis*; for example, in *E. mellipes* (Cresson) it is very strong, whereas it is very weak in *Ancyla* and absent in many Exomalopsini.

Key to the Genera of the Ancylini

1. Glossa and labial palpus much shorter than prementum (Fig. 107-2); labial palpus with first and second segments not sheathlike, second segment not much longer than broad; clypeus nearly flat, scarcely bent back at side of labrum; T7 of male densely hairy, pygidial plate narrow, triangular, pointed posteriorly, strongly elevated, nearly bare; clypeus of female black; flagellum of male crenulate on undersurface .... *Ancyla*
—. Glossa and labial palpus longer than prementum; labial palpus with first two segments long, sheathlike; clypeus more protuberant, strongly bent back at side of labrum; T7 of male sparsely hairy throughout, pygidial plate, if correctly interpreted, occupying most of dorsum of tergum, defined laterally by strong carinae, and ending in truncate or bidentate apex; clypeus of female often yellow; flagellum of male not crenulate .... *Tarsalia*

Genus *Ancyla* Lepeletier

Ancyla Lepeletier, 1841: 294. Type species: *Ancyla oraniensis* Lepeletier, 1841, monobasic.
Plistotrichia Morawitz, 1874: 134. Type species: *Nomia flavilabris* Lucas, 1846 = *Ancyla oraniensis* Lepeletier, 1841, monobasic.

Ancyla is extraordinary among L-T bees in that the glossa and labial palpi are short (Fig. 107-2), the mouthparts thus suggesting those of the Melittidae. The labial palpi are robust, little longer than the glossa, and the second segment is but little longer than the third (Silveira, 1993a), an unusual character for an L-T bee. Nonetheless, *Ancyla* has basic features of L-T bees, such as the presence of a stipital comb and the lack of a galeal comb; presumably, the short glossa and labial palpi result from reduction, possibly in response to regular use of shallow flowers such as those of Apiaceae. Silveira (1993a) described these and other characters of the mouthparts. The body length is 5 to 10 mm. The male genitalia were illustrated by Silveira (1995a, b).

■ *Ancyla* is found in the Mediterranean area from Spain (one specimen) and Morocco east through northern

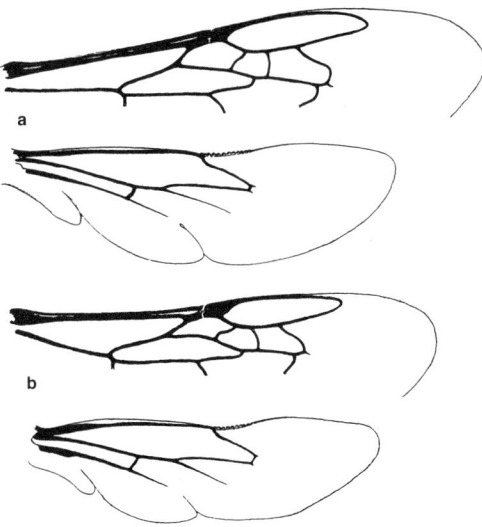

Figure 107-1. Wings (posterior parts of forewings omitted) of Ancylini. **a,** *Tarsalia ancyliformis mediterranea* Pittioni; **b,** *Ancyla holtzi* Friese. Based on drawings by D. B. Baker.

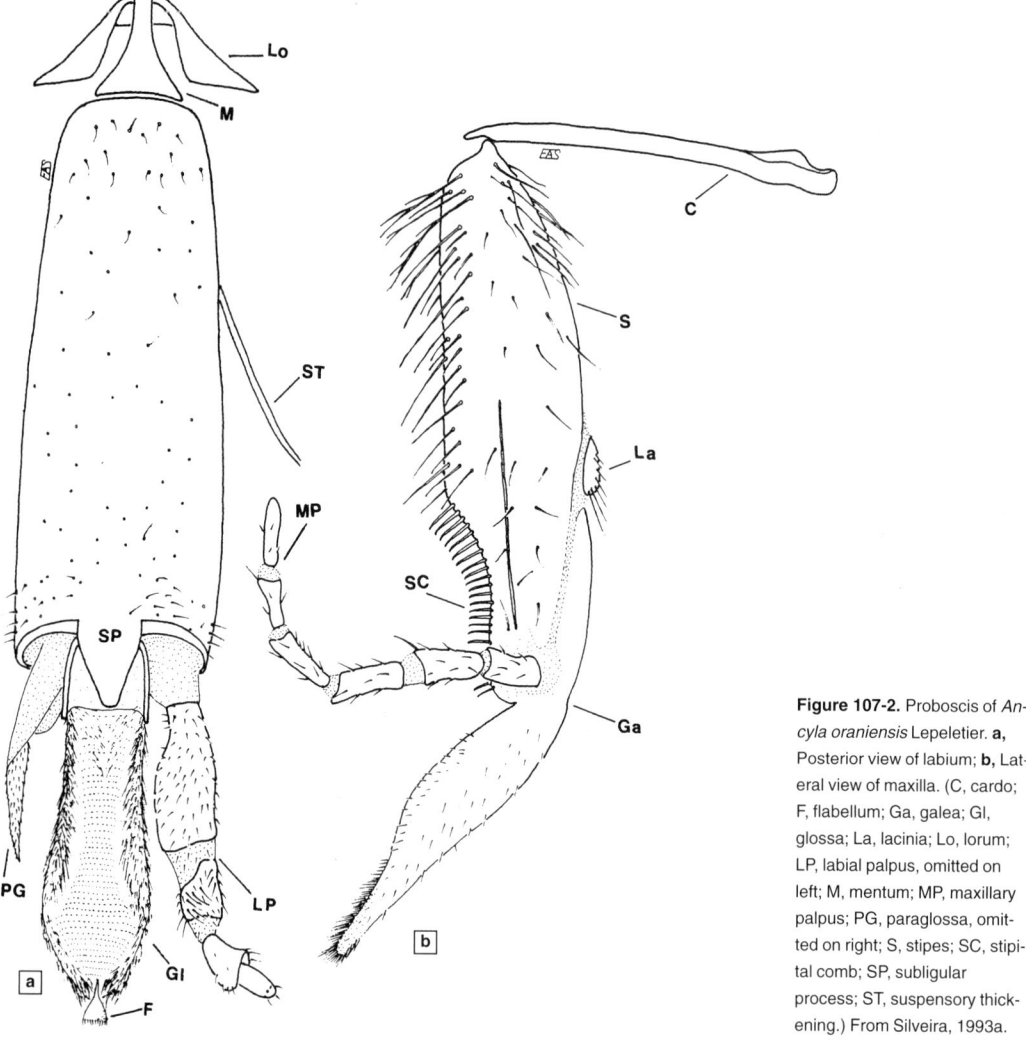

Figure 107-2. Proboscis of *Ancyla oraniensis* Lepeletier. **a,** Posterior view of labium; **b,** Lateral view of maxilla. (C, cardo; F, flabellum; Ga, galea; Gl, glossa; La, lacinia; Lo, lorum; LP, labial palpus, omitted on left; M, mentum; MP, maxillary palpus; PG, paraglossa, omitted on right; S, stipes; SC, stipital comb; SP, subligular process; ST, suspensory thickening.) From Silveira, 1993a.

Africa, southeastern Europe (Greece, Bulgaria) and southwestern Asia to Iran and the Caucasus, and south to Sudan. Warncke (1979c) gave a key to ten species; Baker (1998) recognized eight species.

Genus *Tarsalia* Morawitz

Tarsalia Morawitz, 1895: 9. Type species: *Tarsalia hirtipes* Morawitz, 1895, monobasic.

Tarsalia is adequately characterized in the key to the genera of the Ancylini. Unlike *Ancyla*, it has normal mouthparts for L-T bees. As indicated by Silveira (1993b), species that he studied differ in a number of features from *Ancyla*, and Warncke's (1979c) placement of *Tarsalia* as a subgenus of *Ancyla* is unjustified. Females have dense hairs on the metasomal sterna, suggesting a scopa, but specimens with a large pollen load on the hind legs have almost no pollen on the sterna. The body length varies from 7 to 13 mm. Male genitalia and hidden sterna were illustrated by Popov (1935) and Baker (1972b, 1998). Unlike *Ancyla* and other bees, these sterna and the male genitalia are strongly asymmetrical.

■ *Tarsalia* is found principally in Central Asia (Turkmenistan, Uzbekistan, Tadzhikistan) but occurs westward to Iran, Turkey, and the islands of Cyprus and Sardinia, and southward to Sudan and a possibly disjunct area in southern India. The species were reviewed by Warncke (1979c) as a subgenus of *Ancyla;* Baker (1998) recognized seven species.

108. Tribe Tapinotaspidini

Species of this tribe were included in the Exomalopsini until Roig-Alsina and Michener (1993) and Moure (1994a, b) segregated them. Tapinotaspidini consists of euceriform or trigoniform bees (Fig. 108-1) that are less conspicuously hairy than the Exomalopsini, and usually lack metasomal hair bands. The integument is black, rarely bluish, to red or yellow, sometimes with extensive yellow markings on a dark background. Many of the species are superficially very *Trigona*-like in appearance. The wings are commonly rather large; the stigma is well developed and the marginal cell elongate (Figs. 108-2, 108-3), unlike the usually broad and obliquely truncate marginal cell of most Exomalopsini. The mandible has a preapical tooth (Fig. 108-4b; rarely in some species of *Paratetrapedia,* two such teeth), whereas in Exomalopsini the mandible is nearly always simple (Fig. 108-4a). The scopa on the hind tibia and basitarsus includes long, simple hairs extending beyond the plumose hairs; in Exomalopsini the hairs are plumose to the apices or nearly so. The distal transverse row of ventral bristles on the middle and hind basitarsi is made up of very large bristles; these basitarsi, or at least the hind pair, are broad and flat, and the row of large bristles is therefore long. In the Exomalopsini these bristles are of ordinary size and the basitarsi are not broad and flat. Except in the genera *Monoeca* and *Caenonomada* (Figs. 108-5e, 108-6b), S8 of the male in the Tapinotaspidini is deeply bifid with a pair of large, commonly hairy apical lobes (Figs. 108-5, 108-6, 108-7); S7 has a pair of large, complex, and often lobate apical lobes. These features are suggestive of Emphorini. Further suggesting a relationship to Emphorini is a unique feature of the spatha, which has a ventral thickening that locks onto the basal margins of the penis valves (character 128-2, Roig-Alsina and Michener, 1993).

Figure 108-1. *Paratetrapedia lugubris* (Cresson), male. From Michener, McGinley, and Danforth, 1994.

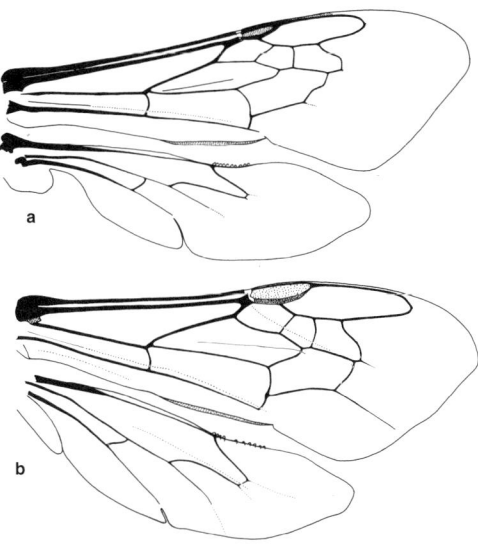

Figure 108-2. Wings of Tapinotaspidini. **a,** *Caenonomada bruneri* Ashmead; **b,** *Paratetrapedia calcarata* (Cresson).

Nests of *Chalepogenus (Lanthanomelissa), Monoeca, Tapinotaspis, Trigonopedia,* and *Paratetrapedia (Xanthopedia)* are in soil, either in flat ground or vertical banks (Rozen, 1984a; Rozen and Michener, 1988). At least two species of *Paratetrapedia* s. str. nest in rotting wood or beetle holes in wood, and I suspect that this is the usual substrate for this and related subgenera of *Paratetrapedia.* In *Tapinotaspis* and some *Paratetrapedia* the cells are subhorizontal, but they are vertical in *Chalepogenus (Lanthanomelissa), Monoeca,* and *Trigonopedia* (Michener and Lange, 1958c; Rozen, 1984a; Sakagami and Laroca, 1988). Cells are lined with a thin, secreted layer more or less impervious to water. Provisions can be a flattened sphere (*Paratetrapedia;* Raw, 1984) or variously shaped in *Tapinotaspis* and *Monoeca,* as shown by Rozen (1984a). Accounts of the nesting biology of tapinotaspidine bees were given by Aguiar et al. (2004) and Rozen et al. (2006).

Like most Centridini, most or probably all Tapinotaspidini collect floral oils, presumably to provision larval cells. This seems to be a characteristic of the tribe, but the principal collecting structure seems to have shifted. The collecting organs of females consist of combs or pads of setae. Much additional information on oil-collecting structures and behavior of Tapinotaspidini was provided, with excellent illustrations, by Cocucci, Sérsic, and Roig-Alsina (2000). They noted such structures on the forelegs (most genera), on the mid legs (*Tapinotaspis*), on both fore and mid legs (*Monoeca*), or on the metasomal sterna (*Tapinotaspoides*). In *Monoeca* the combs, as in the Centridini, are on the inner margins of the basitarsi, next to the strigilis (Fig. 108-4c). In other genera they are on the outer margins, no doubt independently evolved (Fig.

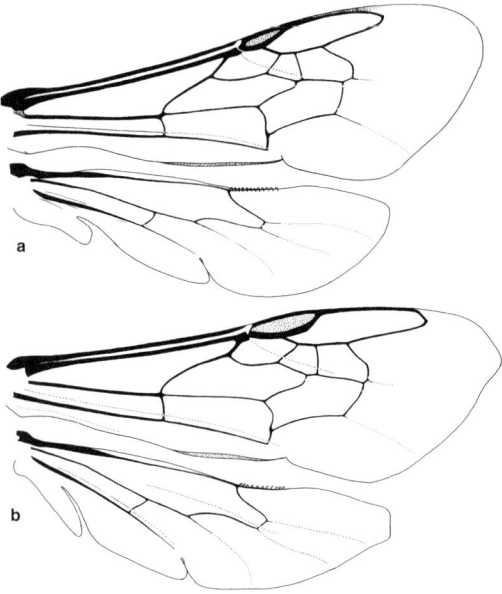

Figure 108-3. Wings of Tapinotaspidini. **a**, *Chalepogenus caeruleus* (Friese); **b**, *Trigonopedia* sp.

108-4d). Males also have combs, in *Paratetrapedia* as well developed as those of females, but presumably functionless.

The Tapinotaspidini were revised to the subgeneric level as part of the Exomalopsini by Michener and Moure (1957); Roig-Alsina (1997) reevaluated the classification as part of a phylogenetic study.

All the genera are South American, but *Paratetrapedia* and *Monoeca* also occur widely in Mesoamerica.

Key to the Genera of the Tapinotaspidini
(by Arturo Roig-Alsina)

1. Middle tibial spur with angle close to apex, apex thus notched; fore basitarsus with comb of strong setae along margin following strigilar concavity (Fig. 108-4c); middle basitarsus with similar comb on apical half; venter of female thorax, coxae, and trochanters with hooked bristles .... *Monoeca*
—. Middle tibial spur tapering apically; fore basitarsus without comb, or with comb along margin opposite to strigilar concavity (Fig. 108-4d); middle basitarsus without comb; venter of female thorax and leg bases with hairs branched or simple, not hooked 2
2(1). Shaft of inner hind tibial spur strongly curved basally, sinuous; spur coarsely pectinate; second to fourth tarsal segments of middle leg hairier than those of other legs, with dense brushes of hairs of uniform length; antenna of male elongate, with first flagellar segment shorter than second, and second 1.5 or more times longer than wide .... *Tapinotaspis*
—. Shaft of inner hind tibial spur not curved basally, sometimes sinuous apically; spur either pectinate or serrate; second to fourth tarsal segments of middle leg not hairier than those of other legs [in a few cases with brushes *(Tapinotaspoides)*, but then brushes asymmetrical, with hairs longer posteriorly]; antenna of male with first flagellar segment as long as or longer than second, and second as long as or shorter than its apical width 3
3(2). First flagellar segment of female twice as long as its apical width and over half of length of scape; face of male in frontal view with area between lateral ocellus and eye, and also vertex, conspicuously depressed; jugal lobe of hind wing short, 0.3 times as long as vannal lobe or less 4
—. First flagellar segment of female at most 1.5 times as long as its apical width, less than half of length of scape; face of male in frontal view with area between lateral ocellus and eye slightly convex to flat, vertex not depressed; jugal lobe of hind wing over 0.4 times as long as vannal lobe 5
4(3). Integument of legs and metasoma extensively marked with yellow; S2 to S4 of female with hairs of apical fringes branched, those of S2 shorter than hairs of S3 or S4; scape of male swollen; hind leg of male with one tibial spur .... *Caenonomada*

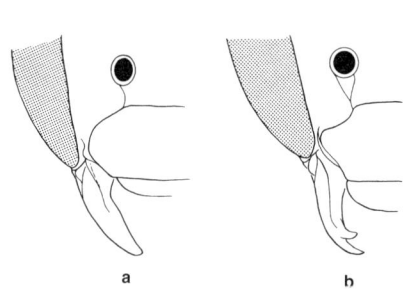
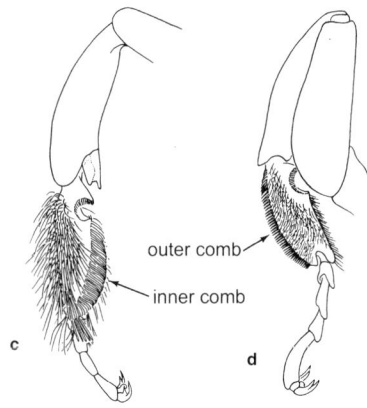

Figure 108-4. Structures of Exomalopsini and Tapinotaspidini. **a, b,** Faces of females of *Exomalopsis solani* Cockerell and *Paratetrapedia* sp.; **c, d,** Fore basitarsi of females of *Monoeca lanei* (Moure) and *P. lugubris* (Cresson). From Michener, McGinley, and Danforth, 1994.

—. Integument of legs and metasoma black; S2 to S4 of female with hairs of fringes ribbonlike, wavy, and apical fringes of S2 to S4 of similar length; scape of male slender; hind leg of male with two tibial spurs .... *Tapinotaspoides*

5(3). Vertex of head behind ocelli carinate; hind basitarsus of female scarcely projecting above articulation of second segment, truncate at apex 6

—. Vertex of head behind ocelli rounded or sloping, not carinate; hind basitarsus of female distinctly projecting at apex above articulation of second tarsal segment, projection rounded or ending obliquely 7

6(5). Face with fine punctures, scarcely wider than hairs arising from punctures; paraocular areas flat next to orbits .... *Trigonopedia*

—. Face with at least some strong punctures on supraclypeal area and clypeus, such punctures several times wider than hairs arising from punctures or frequently giving rise to no hairs at all; paraocular areas convex next to orbits .... *Paratetrapedia*

7(5). Inner surface of fore basitarsus flattened, with polished longitudinal area bearing minute setae, the polished area margined basally and dorsally with rows of short flattened setae; forewing with two submarginal cells; pygidial plate of male distinct, glabrous .... *Chalepogenus (Lanthanomelissa)*

—. Inner surface of fore basitarsus convex, evenly covered with long setae; forewing usually with three submarginal cells (but with two in a few *Chalepogenus*); pygidial plate of male either absent or upper surface covered by dense hairs 8

8(7). Scutum evenly covered by extremely short, dense setae (0.1-0.2 times flagellar diameter); marginal cell 1.25 times as long as distance from apex of cell to wing apex, or longer; scutellum strongly convex, with distinct dorsal and posterior surfaces; pygidial plate of male absent .... *Arhysoceble*

—. Scutum usually with long hairs (as long as diameter of flagellum or longer); in a few species most hairs extremely short, but some scattered hairs long and marginal cell length subequal to distance from its apex to apex of wing (Fig. 108-3a); scutellum evenly rounded, not forming distinct posterior surface; pygidial plate of male at least indicated apically by sclerotized, rounded margin .... *Chalepogenus* s. str.

Genus *Arhysoceble* Moure

Arhysoceble Moure, 1948: 335. Type species: *Arhysoceble xanthopoda* Moure, 1948, by original designation.

Roig-Alsina (1997) has called attention to the very distinctive features of *Arhysoceble;* in his phylogenetic analyses it appeared in diverse positions but not near *Paratetrapedia,* where it was placed by Michener and Moure (1957). *Arhysoceble* has yellow markings on the slender, almost nomadiform, black body, the markings sometimes lacking on the metasoma. The body length is 6 to 8 mm. The pronotum lacks a transverse carina. The front basitarsus has a strong comb on the outer margin; the outer surface, moreover, is covered by a dense pad of fine hairs as in *Chalepogenus* s. str., the pad more extensive than that in *Paratetrapedia.* The inner hind tibial spur of the female is more broadly pectinate than the outer spur. The pygidial plate of the female is concave laterally, and the apical part is thus parallel-sided or slightly expanded. Male genitalia and hidden sterna were illustrated by Michener and Moure (1957); see Figure 108-5a-c.

■ This genus occurs from Tucumán province, Argentina, and Rio Grande do Sul, Brazil, north to the state of Ceará, Brazil. The five known species were listed by Michener and Moure (1957).

Genus *Caenonomada* Ashmead

Caenonomada Ashmead, 1899a: 68. Type species: *Caenonomada bruneri* Ashmead, 1899, by original designation.
Chacoana Holmberg, 1887b: 225, nomen nudum.
Chacoana Holmberg, 1903: 432. Type species: *Chacoana melanoxantha* Holmberg, 1903 = *Caenonomada bruneri* Ashmead, 1899, monobasic.

Caenonomada consists of yellow-and-black to largely yellow euceriform or apiform bees with a rather broad and flat metasoma; the posterior margin of T1 is convex, and the exposed part of T2 is thus shorter medially than laterally. The body length is 9 to 14 mm. Although mor-

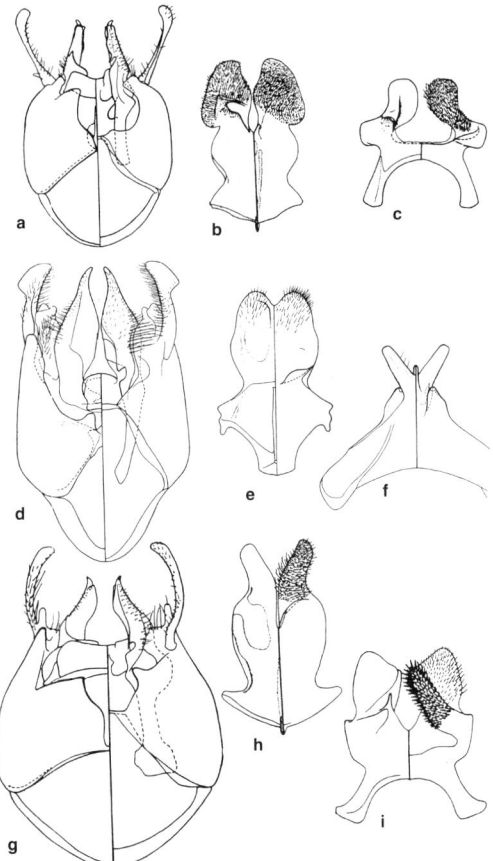

Figure 108-5. Male genitalia, S8, and S7 of Tapinotaspidini. **a-c,** *Arhysoceble melampoda* Moure; **d-f,** *Caenonomada bruneri* Ashmead; **g-i,** *Chalepogenus goeldiana* (Friese). (Dorsal views are on the left.) From Michener and Moure, 1957.

phologically very distinctive, this genus is the sister group to *Tapinotaspoides,* according to a cladistic analysis by Roig-Alsina (1997). Additional features include the sharply pointed and somewhat hooked last flagellar segment of the male, the the third submarginal cell which is larger than either of the others, the lack of basitarsal combs (although some modifications suggest that this, like other Tapinotaspidini, is an oil-collecting bee), and the broadly pectinate inner hind tibial spur of the female. The stigma is usually not wider than the prestigma (measured to the wing margin) but extends well beyond vein r, the margin of the stigma in the marginal cell being convex (Fig. 108-2a). Male genitalia and hidden sterna were illustrated by Michener and Moure (1957); see Figure 108-5d-f.

■ This genus ranges from Buenos Aires and Tucumán provinces, Argentina, to the state of Ceará, Brazil. In his revision of the genus, Zanella (2002a) recognized three species.

Genus *Chalepogenus* Holmberg

This genus is used here in a broader sense than in the past, to include *Lanthanomelissa* and the species formerly placed in the subgenus *Tapinorhina* of *Tapinotaspis. Chalepogenus* consists of small, euceriform species some of which have yellow tergal bands on the metasoma. As in *Tapinotaspis* the comb on the outer margin of the front basitarsus is weakly developed or absent, unlike that of related genera, in which it is conspicuous. The outer surface of the front basitarsus is covered by a broad, dense pad of fine branched setae, as in *Arhysoceble.* The species of the subgenus *Lanthanomelissa* differ from those placed in *Chalepogenus* s. str. in the shape and vestiture of the front basitarsus of both sexes, as indicated in the key below. There are almost no other differences between *C. discrepans* (Holmberg), the type species of *Lanthanomelissa,* and some species of *Chalepogenus* s. str., such that if the front tarsi are broken off, it is difficult to determine the subgenus. A. Roig-Alsina (in litt., 1996) found that Michener and Moure (1957) based their account of *Lanthanomelissa* on species of *Chalepogenus* s. str. In Roig-Alsina'a (1997) phylogenetic analysis, *Lanthanomelissa* and *Chalepogenus* come out in such a way that if united, the resultant genus is paraphyletic. I nonetheless unite them, (1) because of their great similarity and (2) because unlike most of the rest of Roig-Alsina's cladograms, this part is not strongly supported, each clade having only one or two, not necessarily unique, synapomorphies. Additional characters or taxa might easily modify this portion of the cladogram.

A key to the subgenera is provided for convenience and to emphasize certain characters, even though the subgenera come out separately in the key to genera.

Key to the Subgenera of *Chalepogenus*

1. Inner surface of anterior basitarsus in both sexes flat or gently concave, shiny, with a few short hairs; hairy upper margin of front basitarsus with rows of small, flat hairs above the nearly bare, smooth area; fore wing with two submarginal cells .... *C. (Lanthanomelissa)*
—. Inner surface of anterior basitarsus convex and hairy, much as on adjacent parts of basitarsus; small, flat hairs of front basitarsus absent; fore wing in most species with three submarginal cells .... *C. (Chalepogenus s. str.)*

Chalepogenus / Subgenus *Chalepogenus* Holmberg s. str.

Chalepogenus Holmberg, 1903: 416. Type species: *Chalepogenus incertus* Holmberg, 1903 = *Tetrapedia muelleri* Friese, 1899, by designation of Sandhouse, 1943: 537. [See note by Michener, 1997b.]

Schrottkya Friese, 1908c: 170, no included species. Friese, 1908b: 58, included a species. Type species: *Tetrapedia goeldiana* Friese, 1899, first included species.

Desmotetrapedia Schrottky, 1909d: 223. Type species: *Tetrapedia muelleri* Friese, 1899, by original designation.

Lanthanomelissa (Lanthanella) Michener and Moure, 1957: 417. Type species: *Lanthanomelissa completa* Michener and Moure, 1957 = *Tetrapedia goeldiana* Friese, 1899, by original designation.

Tapinotaspis (Tapinorhina) Michener and Moure, 1957: 421. Type species: *Exomalopsis caerulea* Friese, 1906, by original designation.

Tapinorrhina Moure, 1994a: 274, unjustified emendation of *Tapinorhina* Michener and Moure, 1957.

Like *Lanthanomelissa,* some *Chalepogenus* s. str. have basal yellow fasciae on the metasomal terga. Others lack such yellow markings; *C. herbsti* (Friese) has strong white tergal hair bands, and in *C. caeruleus* (Friese) the metasomal terga are dark metallic blue. Male genitalia and hidden sterna were illustrated by Michener and Moure (1957) under the generic names listed in the synonymy above as well as *Lanthanomelissa;* see also Figure 108-5g, h. The body length is 4 to 9 mm.

■ This subgenus is known from Valdivia, Chile, and Chubut, Argentina, north to Paraíba, Brazil and in the Andes to Ecuador. Roig-Alsina (1999) revised the 21 species included in the subgenus.

Chalepogenus / Subgenus *Lanthanomelissa* Holmberg

Lanthanomelissa Holmberg, 1903: 418. Type species: *Lanthanomelissa discrepans* Holmberg, 1903, monobasic.

Lanthanomelissa consists of small species, 4 to 6 mm in body length, exactly resembling small species of *Chalepogenus* s. str. that have yellow tergal bands.

■ The known range is Buenos Aires province to Tucumán, Argentina, north to Paraná, Brazil. The five species were reviewed by Urban (1995e). Rozen et al. (2006) gave an account of nesting behavior of *Lanthanomelissa,* together with descriptions of larvae.

Genus *Monoeca* Lepeletier and Serville

Monoeca Lepeletier and Serville, 1828: 528. Type species: *Monoeca brasiliensis* Lepeletier and Serville, 1828, monobasic.

Epeicharis Radoszkowski, 1884a: 18. Type species: *Epeicharis mexicanus* Radoszkowski, 1884, monobasic.

Fiorentinia Dalla Torre, 1896: 334, replacement for *Epeicharis* Radoszkowski, 1884 (not *Epicharis* Klug). Type species: *Epeicharis mexicanus* Radoszkowski, 1884, autobasic.

Florentina Ashmead, 1899a: 67, error for *Fiorentinia* Dalla Torre, 1896.

Pachycentris Friese, 1902: 186. Type species: *Pachycentris schrottkyi* Friese, 1902, monobasic.

Chaetostetha Michener, 1942a: 281. Type species: *Exomalopsis pyropyga* Friese, 1925, by original designation.

Monoeca consists of robust black bees, sometimes with the metasoma red or with yellow integumental bands on terga. The body length varies from 8 to 12 mm. This genus is not close to any other. As in *Epicharis* and most *Centris*, the front basitarsus has a strong comb on the inner margin (next to the strigilis) (Fig. 108-4c), and the middle basitarsus has a comb in the equivalent position. The thoracic venter, leg bases, and S2 to S4 of the female are covered with strong, hooked bristles, a feature unique to this genus. The inner hind tibial spur is broadly and coarsely pectinate, as in *Caenonomada*. Male genitalia and hidden sterna were illustrated by Michener and Moure (1957); see Figure 108-6a-c.

■ This genus occurs from San Luis Potosí and Jalisco, Mexico, south to southernmost Brazil. Six named species were listed by Michener and Moure (1957). There are at least two or three additional species.

Cunha (2002) described and illustrated the nests, which are branching burrows in the soil. More extensive information is in Rozen et al. (2006), along with descriptions of larvae.

Genus *Paratetrapedia* Moure

Paratetrapedia is by far the largest genus of the Tapinotaspidini, and the only one except the relatively rare *Monoeca* that is found throughout the moist tropics of the Americas. All the species, whether all black, all red, black with a red metasoma, or with extensive yellow or reddish-yellow markings, resemble species of *Trigona* to such an extent that one must often catch specimens to determine whether they are *Paratetrapedia* or *Trigona*; females of the former sting, those of *Trigona* do not! The lack of long hairs and the smooth and largely hairless margins of the metasomal terga enhance the resemblance to *Trigona*. The body length is 6 to 12 mm. The comb on the outer margin of the front basitarsus is strongly developed, and the outer surface of the basitarsus has a rather narrow zone of fine, dense hairs, unlike the broad pad of such hairs found in *Arhysoceble* and *Chalepogenus*. Male genitalia and hidden sterna were illustrated by Michener (1954b) and Michener and Moure (1957); see also Figure 108-6d-f.

Many of the species described in *Tetrapedia* by earlier authors belong in *Paratetrapedia*, but some of the types have not been reexamined to determine their proper placements. Thus more named species fall in this genus than are indicated in the following accounts of subgenera.

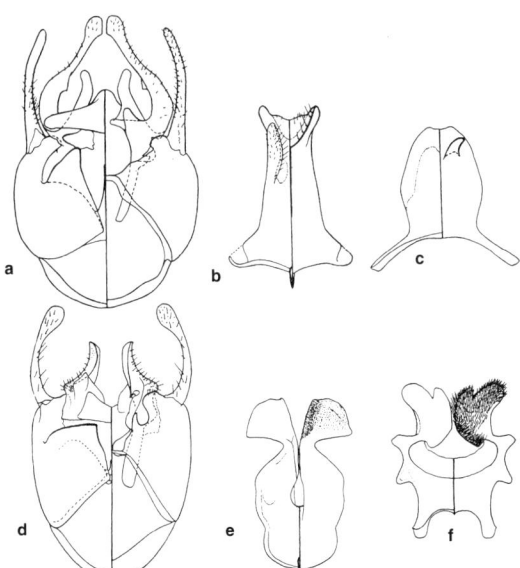

Figure 108-6. Male genitalia, S8, and S7 of Tapinotaspidini. **a-c,** *Monoeca lanei* (Moure); **d-f,** *Paratetrapedia (Lophopedia) pygmaea* (Schrottky). (Dorsal views are on the left.) From Michener and Moure, 1957.

Key to the Subgenera of *Paratetrapedia*

1. Inner hind tibial spur much more broadly (but finely) pectinate or ciliate basally than outer; basitibial plate of female large (or small in *Amphipedia*), dull, without shining excavated zone; basitibial plate of male distinct, with lateral margins clearly defined; pygidial plate of female variable, only rarely with depressed spatulate apical portion and, *if* so, then with marginal carinae of anterior portion meeting in obtuse angle at base of spatulate portion.. 2
—. Inner hind tibial spur finely ciliate and not or little broader than outer; basitibial plate of female very small, with shining excavated submarginal zone; basitibial plate of male obsolete, with lateral margins not clearly defined; pygidial plate of female with depressed spatulate apical portion, marginal carinae of anterior portion meeting in acute angle at base of spatulate portion 4
2(1). Pronotum with narrowly rounded transverse subapical ridge but without transverse carina; S4 and S5 of male with strong, continuous, apical fringes; lower margin of supraclypeal area unmodified .... *P. (Xanthopedia)*
—. Pronotum with transverse carina; S4 and S5 of male almost unfringed or with remnants of fringe sublaterally; lower margin of supraclypeal area with transverse carina, at least in male 3
3(2). Pygidial plate of female broadly triangular; basitibial plate of female large; preoccipital carina not extending behind eyes; comb of anterior basitarsus of male a mere fringe, inconspicuous because of nearby hairs ....
.... *P. (Tropidopedia)*
—. Pygidial plate of female with apical spatulate portion (much as in *Paratetrapedia* s. str.); basitibial plate of female small; preoccipital carina extending behind eyes; comb of anterior basitarsus of male distinct ....
.... *P. (Amphipedia)*
4(1). Preoccipital carina separated from eye by antennal diameter or less; pronotum with high, translucent, transverse lamella, surface immediately anterior to lamella thus concave in profile; second anterior tarsal segment of female without hooked bristle .... *P. (Lophopedia)*
—. Preoccipital carina separated from eye by more than antennal diameter; pronotum with strong transverse carina, surface immediately anterior to carina convex; second anterior tarsal segment of female with greatly thickened hooked bristle on outer margin .... *P. (Paratetrapedia s. str.)*

Paratetrapedia / Subgenus *Amphipedia* Michener and Moure

Paratetrapedia (Amphipedia) Michener and Moure, 1957: 413. Type species *Tetrapedia haeckeli* Friese, 1910, by original designation.

This subgenus contains black species with yellow markings on all tagmata. It most closely resembles *Tropidopedia* but differs in the characters indicated in the key to subgenera.
■ *Amphipedia* occurs in Brazil; details are not known. The only known species is *Paratetrapedia haeckeli* (Friese).

Paratetrapedia / Subgenus *Lophopedia* Michener and Moure

Paratetrapedia (Lophopedia) Michener and Moure, 1957: 413. Type species: *Tetrapedia pygmaea* Schrottky, 1902, by original designation.

This subgenus resembles *Paratetrapedia* s. str. in most features, including the variable coloration, which runs from black to red, sometimes with yellow markings on all tagmata. Distinctive features other than those indicated in the key to subgenera include the strongly convex scutellum with its steeply declivous posterior portion.
■ *Lophopedia* occurs from Bolivia and Santa Catarina, Brazil, north through the tropics to San Luis Potosí and Nayarit, Mexico. Seven species were listed by Michener and Moure (1957), but there are many others.

Paratetrapedia / Subgenus *Paratetrapedia* Moure s. str.

Paratetrapedia Moure, 1941b: 517. Type species: *Ancyloscelis lineata* Spinola, 1851, by original designation.
Chalepogenoides Michener, 1942a: 279. Type species: *Chalepogenus leucostoma* Cockerell, 1923, by original designation.

The body is usually black, rarely ferruginous or with yellow markings on all tagmata. The subgenus is most similar to *Lophopedia*, but differs as indicated in the key to subgenera and in having an only moderately convex scutellum, the posterior part of which is slanting downward but not so steeply as in *Lophopedia*.
■ *Paratetrapedia* s. str. occurs from the states of Tamaulipas and Nayarit, Mexico, south through the tropics to Bolivia, Tucumán Province, Argentina, and Santa Catarina, Brazil. Michener and Moure (1957) listed 14 species; many other species presumably belong here.

Paratetrapedia / Subgenus *Tropidopedia* Michener and Moure

Paratetrapedia (Tropidopedia) Michener and Moure, 1957: 411. Type species: *Paratetrapedia seabrai* Michener and Moure, 1957, by original designation.

In *Tropidopedia* the body is yellow or red with black dorsal areas, dusky metasomal areas, and yellow markings on the head and thorax.
■ This subgenus occurs from the state of Rio de Janeiro to the state of Pará, Brazil. Two named species and at least one other are included.

Paratetrapedia / Subgenus *Xanthopedia* Michener and Moure

Paratetrapedia (Xanthopedia) Michener and Moure, 1957: 411. Type species: *Paratetrapedia tricolor* Michener and Moure, 1957 = *Tetrapedia iheringii* Friese, 1899, by original designation.
Lissopedia Moure, 1994b: 306. Type species: *Tetrapedia globulosa* Friese, 1899, by original designation.

The two genus-group names above represent forms so similar that I see no need for two subgenera; it is more useful to indicate their close relationship.

In *Xanthopedia* the body is black to red, with yellow markings on the head and thorax and sometimes on the metasoma. Unlike other *Paratetrapedia*, the male has a moderately distinct pygidial area.
■ This subgenus occurs from Bolivia and Santa Catarina, Brazil, north to Jamaica and Yucatán, Mexico. Five named species and at least two others are included.

The nesting biology and larvae of the Jamaican species, *Paratetrapedia swainsonae* (Cockerell), were described by Rozen and Michener (1988). The nests are small burrows in earthen banks and form dense aggregations.

Genus *Tapinotaspis* Holmberg

Tapinotapsis [sic] Holmberg, 1887b: 225, nomen nudum.
Tapinotaspis Holmberg, 1903: 413. Type species: *Tapinotaspis chacabucensis* Holmberg, 1903 = *Exomalopsis chalybaea* Friese, 1899, by designation of Sandhouse, 1943: 603.

The body in this genus lacks yellow markings except for the yellow clypeus of most males; the metasoma may have bluish reflections. The frons and vertex of the male, especially the ocellocular area, are depressed but not distinctly concave as in *Tapinotaspoides*. The male flagellum, which reaches the base of the metasoma, looks annulate because of a basal pubescent zone on each segment; segments 2 to 10 are almost twice as long as broad. The front basitarsus is elongate, not broad and flat, and is covered on all sides with hairs; it lacks both combs and pads of fine hairs. Genitalia and hidden sterna of males were illustrated by Michener and Moure (1957). *Tapinotaspis* contains dark-haired and a pale-haired species, superficially quite different. The body length is 7.5 to 10.0 mm.
■ This genus occurs from Río Negro north to Tucumán, Argentina. A record from Pará, Brazil, is presumably an error, possibly for the state of Paraná. Three species were listed by Roig-Alsina (1997).

Genus *Tapinotaspoides* Moure

Tapinotaspoides Moure, 1944c: 10. Type species: *Tetrapedia serraticornis* Friese, 1899, by original designation.

This genus consists of largely black species with the pubescence mostly dark. The long fringes of the metasomal sterna of the female consist of ribbonlike, flat hairs that are so minutely wavy that they glitter. Unlike in other

Tapinotaspidini, the sternal hairs of females are used to mop up oily material from extrafloral sources on a variety of plants (Melo and Gaglianone, 2005). As in *Tapinotaspis,* the front basitarsus is not broad and flat, and is covered with hairs; it lacks both combs and pads of fine hairs. From *Tapinotaspis* this genus differs in having shorter male antennae, flagellar segments 2 to 10 being less than 1.5 times as long as broad, and in the presence of a preoccipital carina behind the ocelli of the female. The clypeus is ordinarily black, unlike that of male *Tapinotaspis.* The male genitalia and hidden sterna were illustrated by Michener and Moure (1957); see Figure 108-7a-c. The body length is 8 to 11 mm.

■ *Tapinotaspoides* occurs from central Argentina to Paraguay and north through Brazil at least to the state of Paraíba. The four species were listed by Michener and Moure (1957).

Genus *Trigonopedia* Moure

Trigonopedia Moure, 1941b: 518. Type species: *Trigonopedia oligotricha* Moure, 1941, by original designation.

This genus was placed as a subgenus of *Paratetrapedia* by Michener and Moure (1957), but it is quite distinct morphologically, and Roig-Alsina's (1997) phylogenetic analysis showed it to be well separated from *Paratetrapedia* but the probable sister group of *Monoeca*. *Trigonopedia* consists of black to partly reddish-yellow trigoniform bees that usually lack a transverse pronotal carina. The inner hind tibial spur of the female is broadly pectinate, and the margins of the pygidial plate are nearly straight, converging posteriorly. As in *Paratetrapedia,* the outer margin of the front basitarsus has a strong comb. Male genitalia and hidden sterna were illustrated by Michener and Moure (1957); see Figure 108-7d-f. The body length is 8 to 10 mm.

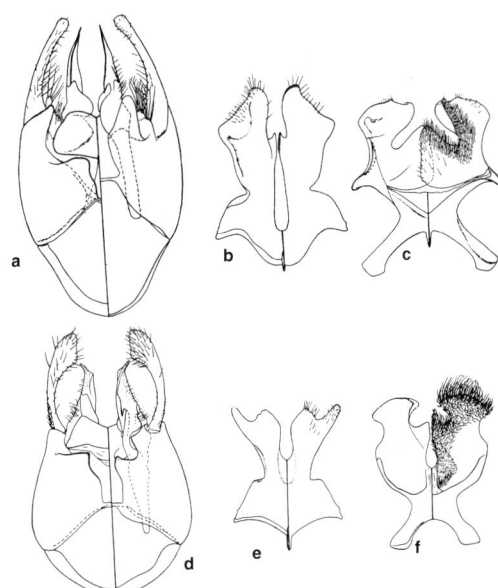

Figure 108-7. Male genitalia, S8, and S7 of Tapinotaspidini. **a-c,** *Tapinotaspoides tucumana* (Vachal); **d-f,** *Trigonopedia oligotricha* Moure. (Dorsal views are on the left.) From Michener and Moure, 1957.

■ *Trigonopedia* is found in eastern Brazil from the state of Espírito Santo to São Paulo. The four named species were listed by Michener and Moure (1957).

The nests are in earthen banks and form dense aggregations (Michener and Lange, 1958c); see Figure 5-1.

109. Tribe Tetrapediini

Tetrapediini includes two very different genera, one *(Coelioxoides)* being cleptoparasitic, the other *(Tetrapedia)* consisting of nest-making species. Roig-Alsina (1990), who first recognized that *Coelioxoides* is related to *Tetrapedia,* lists features that indicate this relationship and constitute tribal characters. They include bending of the marginal cell away from the costa for much of its length, the small jugal lobe of the hind wing (Fig. 109-1), the long antennal pedicel, 1.3 to 1.7 times as long as wide, combined with a short scape, little more than 2.5 times as long as the pedicel (in most other Apinae the pedicel is at most as long as wide, but if elongate, the scape is also long); the slanting propodeal profile, with no separation into horizontal and vertical surfaces; the hairy propodeal triangle (bare medially in *Coelioxoides*); the short, broad middle tibial spur with elongate points forming pectination on both margins, the teeth largest near the apex of the spur (Fig. 109-2b, c); the lack of arolia (although in *Coelioxoides* the compressed inner rami of the claws are often approximate, superficially resembling an arolium); and the simple S7 of the male, lacking apical lobes.

The nesting biology and immature stages of both genera of Tetrapediini were treated by Alves-dos-Santos, Melo, and Rozen. (2002). *Coelioxoides* was established as a cleptoparasite of *Tetrapedia; Coelioxoides* places its eggs into cells shortly after they are closed by the host.

Key to the Genera of the Tetrapediini

1. Scopa absent; metasoma tapering posteriorly, in the female to a point (Fig. 109-3), suggesting *Coelioxys;* metepisternum almost linear except at upper end (Fig. 109-2a), pits united to form one large pit; hind tibia with two spurs in female, none in male .... *Coelioxoides*
—. Scopa present on hind tibia and basitarsus of female; metasoma not tapering; metepisternum narrow but not linear, pits separate although close together; hind tibia in both sexes with one short, strongly pectinate spur (Fig. 109-2d) .... *Tetrapedia*

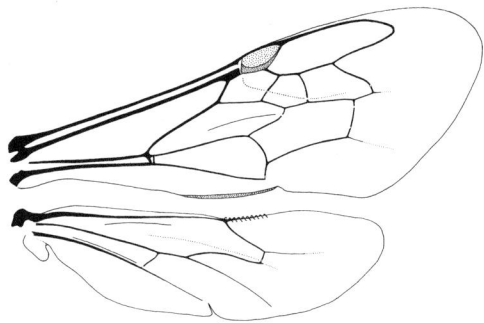

Figure 109-1. Wings of *Tetrapedia* sp.

Genus *Coelioxoides* Cresson

Coelioxoides Cresson, 1878: 94. Type species: *Coelioxoides punctipennis* Cresson, 1878, monobasic.

Among apine bees, *Coelioxoides* can be recognized immediately by the *Coelioxys*-like aspect, particularly of females. The wings are dark with a hyaline spot or transverse band distal to the closed cells (Fig. 109-4). The body length varies from 7.5 to 13.0 mm. Some derived characters include the one-segmented maxillary palpus, the enormous sting, and the produced, sharply pointed apex of the metasoma of the female, accentuated by S6, which extends beyond T6 (Fig. 109-3) and forms a nearly closed tube for the sting. Characters that are probably ancestral relative to *Tetrapedia* include the double (upper and lower) male gonostyli (Fig. 109-5), the complete gonobase, and the presence of a volsella. These features suggest that *Coelioxoides* and *Tetrapedia* are sister groups, not that *Coelioxoides* is derived from *Tetrapedia.*

■ *Coelioxoides* occurs from the states of Jalisco and San Luis Potosí in Mexico to Bolivia and the province of Córdoba, Argentina. The three species were revised by Roig-Alsina (1990).

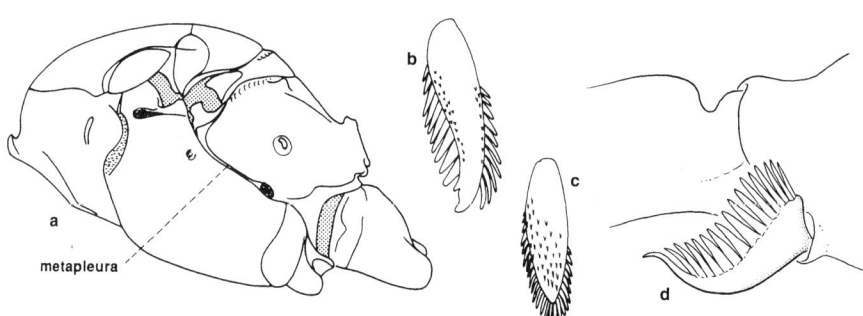

Figure 109-2. Structures of Tetrapediini. **a,** Lateral view of thorax of *Coelioxoides punctipennis* (Cresson), showing the reduced metapleuron; **b, c,** Middle tibial spurs of females of *Tetrapedia clypeata* Friese and *C. waltheriae* Ducke; **d,** Hind tibial spur of *T. peckoltii* Friese, female. b, c, from Roig-Alsina, 1990; d, from Michener, McGinley, and Danforth, 1994.

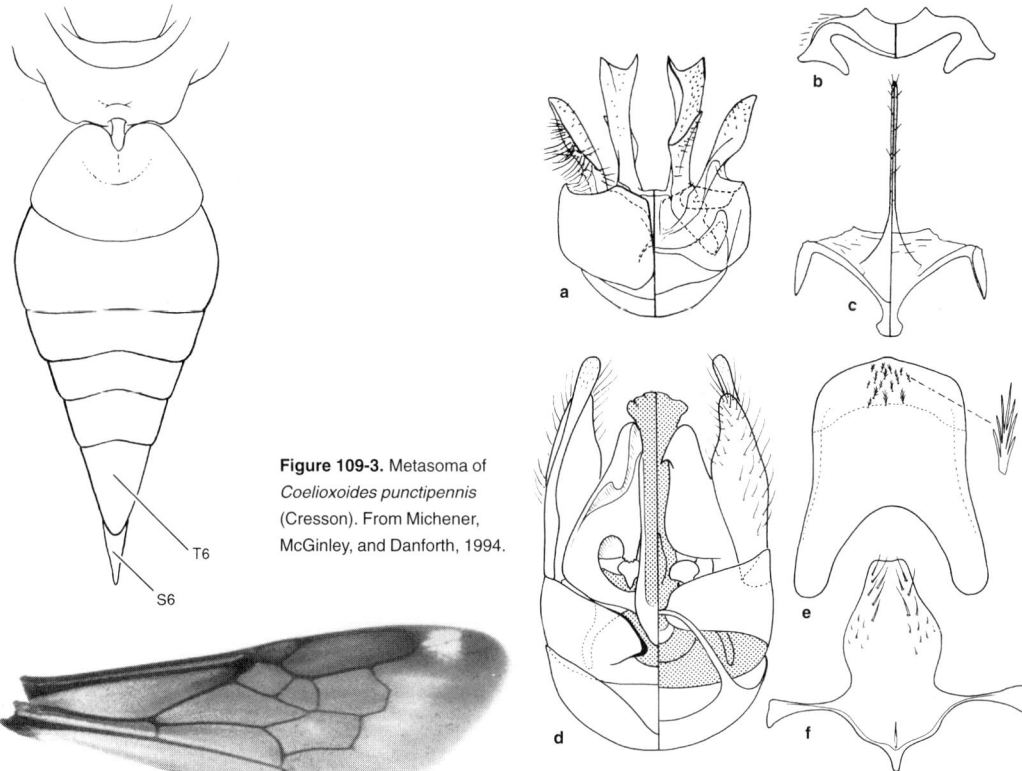

Figure 109-3. Metasoma of *Coelioxoides punctipennis* (Cresson). From Michener, McGinley, and Danforth, 1994.

Figure 109-4. Forewing of *Coelioxoides waltheriae* Ducke, showing the color pattern and the unusually large third submarginal cell. From Roig-Alsina, 1990.

Figure 109-5. Male genitalia and hidden sterna of Tetrapediini. a-c, *Tetrapedia diversipes* Klug; d-f, *Coelioxoides exulans* (Holmberg). (In the divided drawings, dorsal views are on the left.) a-c, from Michener and Moure, 1957; d-f, from Roig-Alsina, 1990.

Genus *Tetrapedia* Klug

Tetrapedia Klug, 1810: 33. Type species: *Tetrapedia diversipes* Klug, 1810, monobasic.
Tetrapedium Berthold, 1827: 468, unjustified emendation of *Tetrapedia* Klug, 1810.
Lagobata Smith, 1861: 151. Type species: *Lagobata diligens* Smith, 1861 = *Ancyloscelis ornata* Spinola, 1853, monobasic.
Tetrapaedia Dalla Torre, 1896: 299, unjustified emendation of *Tetrapedia* Klug, 1810.

The use of the name *Tetrapedia* (and secondarily also *Exomalopsis*) was endangered by the finding that Klug's illustration and description of *Tetrapedia diversipes* were based on a species of *Exomalopsis,* verified by Klug's type specimen in Berlin. Not only the generic names, but also the two tribal names based on the generic names, were endangered. Michener and Moure (2002) therefore requested the International Commission on Zoological Nomenclature to designate a replacement neotype for *T. diversipes* Klug to conserve these names in their traditional senses. Opinion 2070 of the Commission, published in the Bulletin of Zoological Nomenclature 60: 70-71 (2004), designated this replacement neotype.

This genus includes rather slender species, usually black; sometimes the clypeus is yellow or the body has extensive reddish-yellow markings. Superficially, they closely resemble species of *Paratetrapedia* and of *Trigona*. The body length is 8 to 13 mm. The females collect floral oils (Neff and Simpson, 1981); oil-manipulating and -carrying structures include a strong, dense comb on the outer margin (away from the strigilis) of the front basitarsus as in *Paratetrapedia,* a densely pectinate hind tibial spur (Fig. 109-2d), and a scopa on the hind tibia and basitarsus consisting of dense, strongly plumose hairs on the outer surfaces of the segments and longer, simple hairs emerging above the plumose hairs, suggesting the scopa of some other oil-carrying bees such as *Rediviva* and *Macropis* in the Melittidae. Males, which presumably do not collect oil, nonetheless have the same structures but with very few of the long, simple emergent hairs on the tibia and basitarsus. Other distinctive structures include a slender, well-defined pygidial plate in the female; a weakly emarginate T7 without a pygidial plate in the male; very dense, erect, plumose hairs in the male on the distal parts of T5 and T6, with emergent simple hairs; and long hind legs, with the basitarsi often modified, in the male. Male genitalia and hidden sterna were illustrated by Michener (1954b) and Michener and Moure (1957); see also Figure 109-5a-c.

■ *Tetrapedia* occurs from the states of San Luis Potosí and Jalisco in Mexico to Salta, Tucumán, and Misiones in Argentina and Santa Catarina in Brazil. About 13 species were listed by Michener and Moure (1957).

Although Michener and Moure (1957) regarded *Lagobata* as a recognizable subgenus, it now seems that it includes one distinctive species, *Tetrapedia ornata* (Spinola), and another intermediate species, *T. clypeata* Friese. Maintenance of a subgeneric name seems unnecessary under these circumstances.

Tetrapedia nests are found in old burrows in wood. Cells are in series, constructed of material (gum, resin?) evidently carried into the nest on the scopa (Michener and Lange, 1958c). G. Melo has observed female *Tetrapedia* landing on sandy areas in forest roads, perhaps to collect sand particles to carry to nest sites on their sticky (oily?) scopae.

110. Tribe Ctenoplectrini

Ctenoplectra has usually been placed among the S-T bees, as a subfamily of the Melittidae (Michener, 1944) or as a distinct family, the Ctenoplectridae (Michener and Greenberg, 1980). The latter authors showed that, in spite of the relatively short glossa and labial palpi (Fig. 110-1c), similar in that respect to those of Melittidae and other S-T bees, *Ctenoplectra* has other features characteristic of L-T bees. These include the presence of a stipital concavity and comb (Fig. 110-1a, b), the absence of a galeal comb, and a glossa with an invaginated disannulate surface (Fig. 110-1c), its seriate hairs minute and converging. Michener and Greenberg (1980) therefore considered *Ctenoplectra* to be the sister group to all L-T bees. Roig-Alsina and Michener (1993) and Silveira (1993b, 1995a, b), however, in phylogenetic studies, found that *Ctenoplectra* fell within the Apinae (Fig. 102-1), in the group that Silveira called the eucerine line or near the base of that line, and thus in the midst of L-T bees. If this position is correct, *Ctenoplectra* is an L-T bee that has a relatively short glossa and has lost the elongate and flattened aspect of the first two segments of the labial palpi. It is not unique in having lost these hallmarks of L-T bees; see *Ancyla* (Ancylini) (Fig. 107-2) and the parasitic taxa of Allodapini. A more recent phylogenetic analysis by Alexander and Michener (1995) resulted in *Ctenoplectra* appearing again as the sister group to L-T bees, but this was an analysis of the S-T bees; *Ctenoplectra* was merely one of the outgroups, and because its presumed relatives, such as Eucerini, among the Apinae were not represented in the analysis, its grouping with them could not be shown. The position of *Ctenoplectra* in the Apidae is supported by Rozen's (2003b) finding that the number of ovarioles is four per ovary. In other families of bees the basic number is three.

Ctenoplectrini is a paleotropical and East Asian group. The V-shaped lorum and tapering mentum (Fig. 110-1c) are as in Melittidae and L-T bees. Glossal and palpal structures are indicated above. The marginal cell curves gradually away from the costa toward its pointed or narrowly rounded apex, and there are two submarginal cells of roughly equal length (Fig. 110-4a). Except as weakly developed in parasitic species, small but distinct, elevated basitibial plates characterize both sexes (Fig. 110-2b); this is unusual for bees that do not nest, so far as is known, in the ground. The inner hind tibial spur of females, except in parasitic species, is enormously broadened at the base,

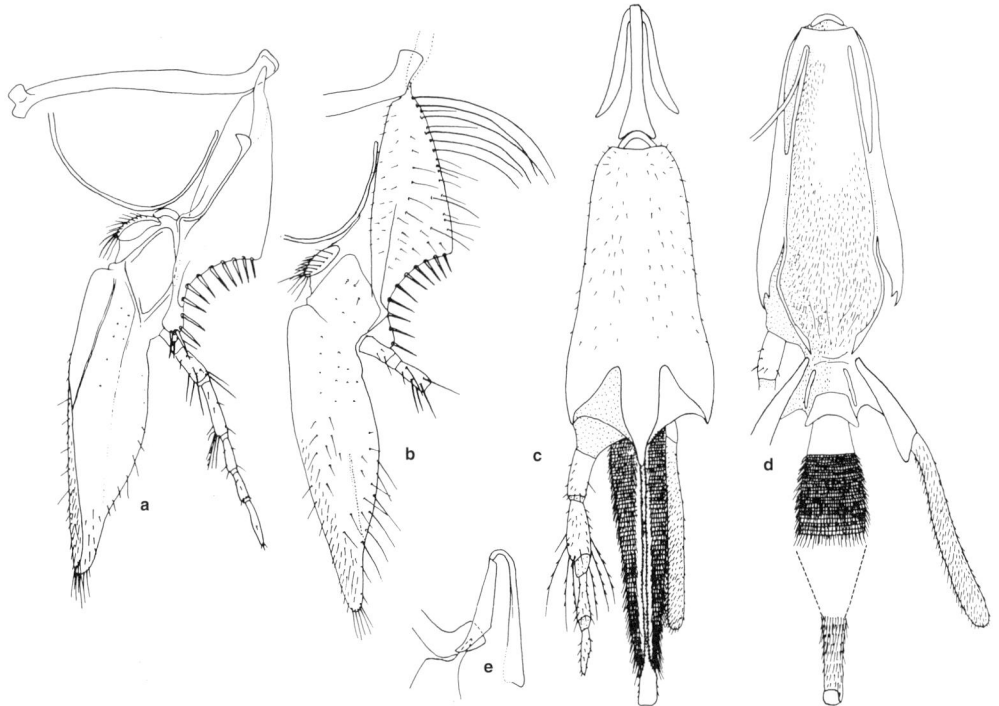

Figure 110-1. Proboscis of *Ctenoplectra fuscipes* Friese. **a, b,** Inner and outer views of maxilla; **c, d,** Posterior and anterior views of labium, in d the mentum and lorum omitted and the glossa and paraglossae pulled out to show the basal structures; **e,** Lateral view of mentum (on right), lorum, and associated maxillary sclerites. From Michener and Greenberg, 1980.

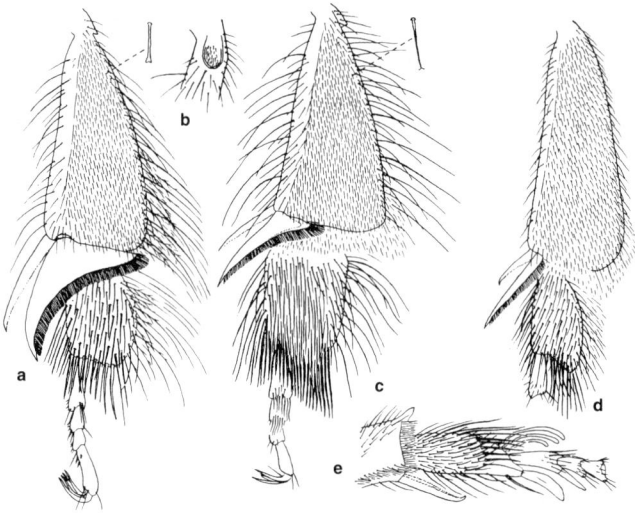

Figure 110-2. Leg structures of female Ctenoplectrini. **a,** Inner surface of posterior tibia and tarsus of *Ctenoplectra albolimbata* Magretti; **b, c,** Basitibial plate and inner surface of posterior tibia and tarsus of *C. bequaerti* Cockerell (note the enlargements of the keirotrichia); **d,** Inner surface of posterior tibia and tarsus of *Ctenoplectrina politula* (Cockerell); **e,** Inner view of middle tarsus of *Ctenoplectra bequaerti* Cockerell, showing modified setae. From Michener and Greenberg, 1980.

thus crescentic (Fig. 110-2a, c); the margin is finely comblike. Even in parasitic species (Fig. 110-2d) the spur is broadly and finely comblike. The term "pectinate" often used for these spurs is confusing for one familiar with that term as it is used elsewhere among bees, especially in the Colletinae and Halictidae, where it means having a very few long, coarse teeth. In the Ctenoplectrini the spur has many long fine teeth, presumably prolongations of the processes of the ciliate spurs found in many bees. To avoid changing established, even if none too appropriate, terminology, I am simply calling ctenoplectrine inner hind tibial spurs comblike. Arolia are absent. The distal part of the pygidial plate of the female is slender, parallel-sided, or, in parasitic species, scarcely evident. S2 to S5 of females have oblique bands, broken medially, of long, coarse, distally curved and serrate or squamose hairs (Fig. 110-3) used in wiping floral oils from the cucurbitaceous genera *Thladiantha* and *Momordica* (Vogel, 1981, 1990). These hairs are present but reduced in the parasitic *Ctenoplectrina* (Fig. 110-3b). Many structures were illustrated by Michener and Greenberg (1980); genitalia and hidden sterna were illustrated by Popov and Guiglia (1936) and Wu (1978). As shown in Figure 110-4b, the spatha is well developed and there are two gonostyli; the dorsal gonostylus is greatly reduced in some species, however.

Ctenoplectra is known to nest in beetle holes in old wood (Williams, 1928) and in abandoned mud-and-resin megachilid nests (Rozen, 1978); probably they utilize small holes of various kinds. Floral oil may be added to earth that is carried to the nest on the scopa, as observed by Williams. Rozen described the larva and the cocoon spun by the mature larva.

Key to the Genera of the Ctenoplectrini
(Females Only)

1. Hind tibial and basitibial scopa consisting largely of long simple hairs, many of them nearly as long as tibial width; inner hind tibial spur expanded at base across apex of tibia, sometimes to full width of tibia, so that spur is crescentic, its margin finely comblike (Fig. 110-2a, c) .... *Ctenoplectra*
—. Hind tibia and basitarsus without scopa, hairs much shorter than width of tibia; inner hind tibial spur not expanded at base, relatively slender, not crescentic, but finely comblike (Fig. 110-2d) .... *Ctenoplectrina*

Genus *Ctenoplectra* Kirby

Ctenoplectra Kirby *in* Kirby and Spence, 1826: 681, no species. Type species: *Ctenoplectra chalybea* Smith, 1857, by inclusion and designation of Sandhouse, 1943: 542.

Ctenoplectra Smith, 1857: 44. Type species: *Ctenoplectra chalybea* Smith, 1857, monobasic.

This genus includes the nonparasitic species of the tribe. In addition to the characters indicated in the key to genera, females of *Ctenoplectra* have long dense sternal fringes (Fig. 110-3a), as described in the account of the tribe, the apical margins of S4 and S5 are deeply emarginate, and the setae on the midtarsal segments 1 and 2 (Fig. 110-2e) are curiously modified. The basitibial and pygidial plates of females are sharply defined, and the prepygidial fimbria on T5 is strong. The hind tibia of the female is triangular, broadest apically, and tapering to-

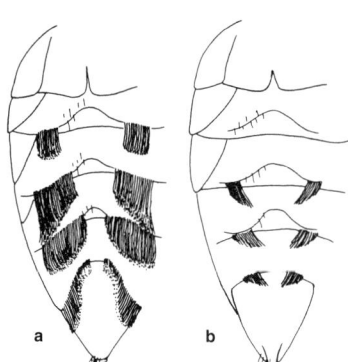

Figure 110-3. Undersurfaces of metasomas of female Ctenoplectrini. **a,** *Ctenoplectra bequaerti* Cockerell, showing oil-collecting hairs; **b,** *Ctenoplectrina politula* Cockerell, showing the reduction in oil-collecting hairs. From Michener and Greenberg, 1980.

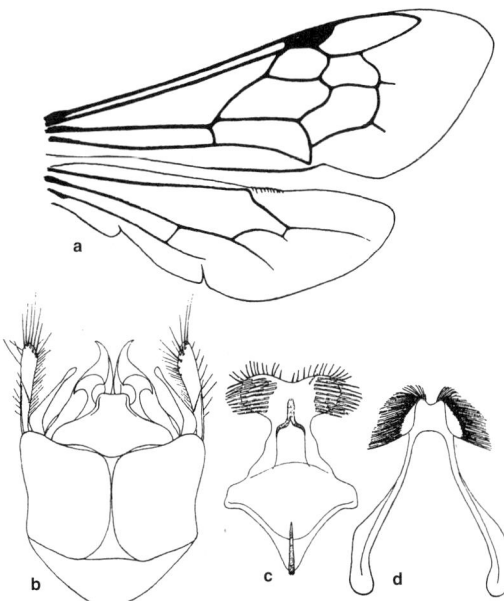

Figure 110-4. Structures of *Ctenoplectra*. **a,** Wings of *Ctenoplectra* sp.; **b-d,** Male genitalia (dorsal view), S8, and S7 of *C. armata* Magretti. b- d, from Popov and Guiglia, 1936.

ward the base; a scopa of long, superficially simple hairs occupies nearly the whole outer surface. The hind basitarsus is 1 to 1.5 times as long as wide. (For illustrations of the above characters, see Michener and Greenberg, 1980, as well as Fig. 110-2a-d.) There is much variation among the species of *Ctenoplectra*. Length ranges from 4.5 to 15.0 mm. Larger species tend to have the propodeal profile subvertical and the metanotum nearly vertical, whereas small African species have a slanting basal propodeal zone and a subhorizontal metanotum. These characters, and all those used by Popov and Guiglia (1936) to distinguish two species groups, intergrade, so that recognition of taxonomic units seems impractical. Although most species are black, the metasomal terga of the larger species are often bright blue, green, or purple; in small species there terga are sometimes red or with apical integumental white fasciae. Male genitalia and hidden sterna were illustrated by Popov and Guiglia (1936); see also Figure 110-4b-d.

■ *Ctenoplectra* occurs in Africa from Liberia, Uganda, and Ethiopia south to Namibia and Natal, South Africa; and from the Russian Far East and north China south to Taiwan, the Philippines, New Guinea, and north Queensland in Australia, and west to Sumatra and West Malaysia. The genus is not known from India or Sri Lanka. About 24 species names fall in this genus. The species were listed by Vogel (1990). Friese (1910b) gave a key to the species; in earlier works (e.g., Friese, 1909a) he used the generic name *Scrapter* for this genus. African species were revised by Eardley (2003).

Genus *Ctenoplectrina* Cockerell

Ctenoplectra (Ctenoplectrina) Cockerell, 1930a: 360. Type species: *Ctenoplectra politula* Cockerell, 1930, monobasic.

This genus consists of two presumably cleptoparasitic species 5 to 7 mm in length. Probably its recognition makes *Ctenoplectra* paraphyletic; it appears to be most closely related to the group of *Ctenoplectra* containing the small African species. It differs from *Ctenoplectra* in many more characters than differentiate some other cleptoparasites, such as *Bombus (Psithyrus),* from their close relatives and hosts. I therefore retain *Cleptoplectrina* as a genus for the present. In addition to the characters indicated in the key to genera, females differ from *Clenoplectra* in the greatly reduced fringes (lacking on S3) on the transverse rather than concave margins of the metasomal sterna (Fig. 110-3b) and the lack of modified setae on the middle tarsi. The basitibial and pygidial plates are shaped as in *Ctenoplectra* but only slightly elevated and not sharply defined, the basitibial plate lacking the marginal carina as well as the vestiture of short hairs found in *Ctenoplectra*. The posterior margin of T5 is smooth like those of preceding terga, there being no prepygidial fimbria. The hind tibia is broader than in most males of *Ctenoplectra* and broadest apically, but is much narrower than in females of *Ctenoplectra*. The hind basitarsus is much narrower than the apex of the tibia, over twice as long as wide, and obliquely truncate at the apex.

■ *Ctenoplectrina* is known from Somalia to Natal, South Africa, and west to Cameroon. The two described species were reviewed by Eardley (2003).

111. Tribe Emphorini

This is the tribe that has often been called the Melitomini. The decision to use the name Emphorini was explained by Michener (1997b).

The Emphorini are mostly euceriform bees restricted to the Western Hemisphere. Nearly all species are hairy, the metasomal terga often having pale bands or covered uniformly with pale hairs, so that they superficially resemble many Eucerini (Fig. 111-1). The male antennae are short, like those of females, in contrast to those of nearly all Eucerini. The convex vertex, as seen in frontal view (Fig. 111-2a), distinguishes the Emphorini from similar bees, but is not well developed in *Ancyloscelis*. In typical members of this tribe, vein cu-v of the hind wing is usually oblique and is only a little if any shorter than the second abscissa of vein M+Cu, which is much shorter than vein M (Fig. 111-3b, c). In *Ancyloscelis,* however, vein cu-v is usually transverse and is less than half as long as the second abscissa of M+Cu, which is three-fourths as long as M to as long as M (Fig. 111-3a). In this respect *Ancyloscelis* resembles the Exomalopsini. *Meli-philopsis* is intermediate in hind-wing venation, vein cu-v being transverse and half as long as the second abscissa of M+Cu, which is about half as long as M. As in Eucerini, the marginal cell is long and pointed, its apex bent gradually away from the costal margin of the wing (Fig. 111-3); this configuration is unlike Exomalopsini, in nearly all of which the cell bends abruptly away from the wing margin and is thus obliquely truncate. S8 of males possesses two large apical lobes (Fig. 111-4b, e), possibly an indication of relationship to the Tapinotaspidini. Except in *Ancyloscelis,* S7 also ends in two such lobes, sometimes complex and divided into sublobes (Fig. 111-4f). The result is that beneath the male genitalia there are two layers of large lobes.

As indicated in the key to the subfamilies and tribes of Apidae (Sec. 85) and in the discussion above, *Ancyloscelis* is quite different from other Emphorini and is placed in a subtribe, Ancyloscelina, separate from the subtribe Emphorina (sometimes called Melitomina). *Ancyloscelis* was formerly placed in the Exomalopsini (Michener, 1944, 1954b; Michener and Moure, 1957), but Roig-Alsina and Michener (1993) showed it to be the sister group to the Emphorina and erected the subtribe Ancyloscelina for it.

Nests of Melitomini are shallow, often branching burrows in banks or in flat ground, often in aggregations, and frequently with entrance turrets. At least some species of *Ptilothrix* and *Melitoma* nest in hard soil and carry water

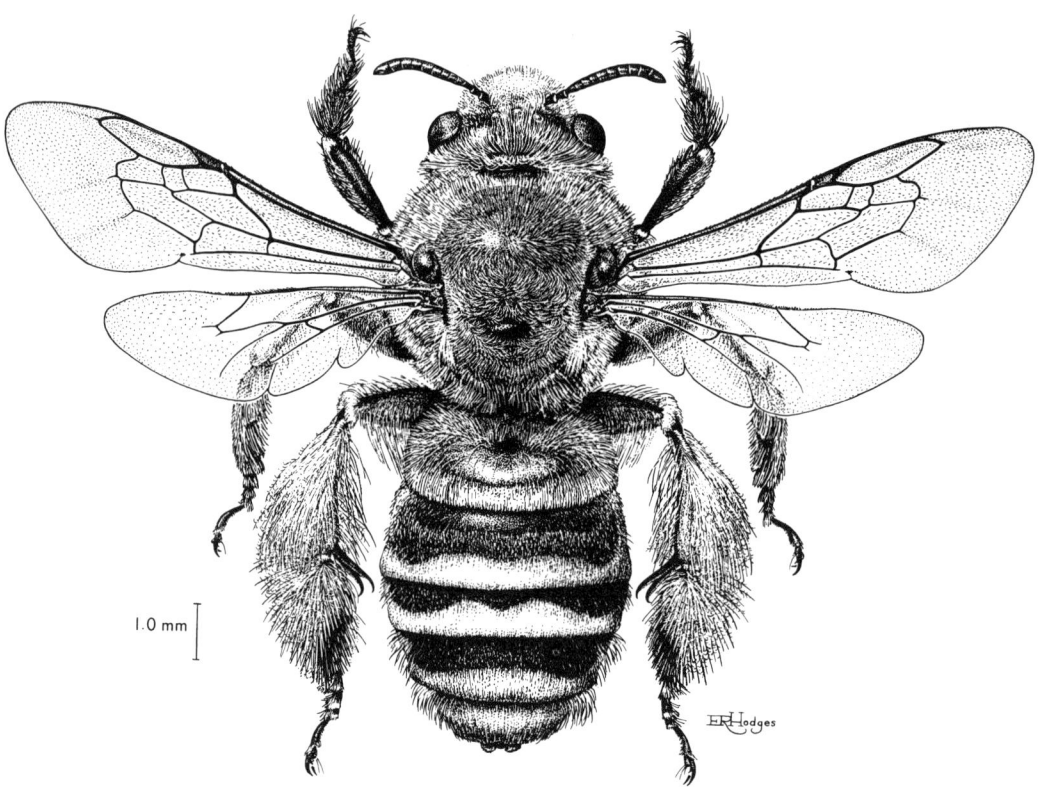

Figure 111-1. Female of *Diadasia rinconis* Cockerell. Drawing by E. R. S. Hodges, from Michener, McGinley, and Danforth, 1994.

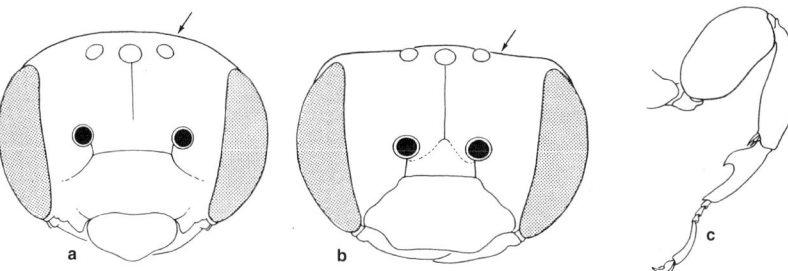

Figure 111-2. Structures of Apinae. **a,** Face of female of *Diadasia sphaeralcearum* Cockerell (Emphorini); **b,** Face of female of *Melissodes lupina* Cresson (Eucerini); **c,** Hind leg of male of *Ancyloscelis toluca* (Cresson) (Emphorini). Arrows indicate the vertex, the shape of which is characteristic of Emphorini and Eucerini. From Michener, McGinley, and Danforth, 1994.

from ponds and puddles to soften the soil for excavation and cell construction. The cells are urn-shaped, with smooth earthen walls such that in some forms, such as *Melitoma,* they can easily be separated intact from the earthen matrix in which they are constructed. They are lined with a very thin secreted membrane, or such a lining may appear to be absent. The cells are either isolated, one in each branch of a nest, or alone in an unbranched nest, or more commonly they are in short series; in most species their orientation is variable, horizontal to vertical. The provisions are firm masses occupying the distal ends of the cells. After the provisions for a cell are collected and shaped, an egg is laid, underneath the provisions in the subtribe Emphorina; this is a unique feature of the subtribe. It is likely that the observation by D. Dias (in Michener and Lange, 1958c) of eggs on top of food masses of *Ptilothrix plumatus* Smith (see also Michener, 1974c) is an error. Studies of several *Ptilothrix* species by Hazeldine (1997a) show eggs beneath food masses as in other Emphorina. The females show no evident apical structures for inserting eggs under the pollen masses. In the subtribe Ancyloscelina, the egg is laid on top of the food mass (Rozen, 1984a). The emphorine larva is unusually elongate, almost wormlike, and curls around the food mass, eating its way around it. After feeding is completed, the larva deposits a layer of fecal material covering the whole interior of the cell except sometimes for the closure, and then covers it with a thin cocoon. Such a layer of feces, which appears as a layer of pollen exines without recognizable fecal pellets, is also a unique emphorine characteristic. Various distinctive features of the nest, as described above, support the placement of *Ancyloscelis* in the Emphorini.

Perhaps all species of Emphorini are oligolectic, but as indicated in the accounts of the genera, the tribe as a whole and even different species of the genus *Diadasia* visit a wide variety of flowers for pollen.

Key to the Genera of the Emphorini
(Modified from a key by A. Roig-Alsina, 1998a)

1. Paraocular carina present along most of inner eye margin; vein cu-v of hind wing less than half as long as second abscissa of vein M+Cu (Fig. 111-3a); second abscissa of vein M+Cu of hind wing at least three-fourths as long as vein M; maxillary palpus with sparse, short hairs; T7 of male rounded apically; hind leg of male modified, femur dilated, at least twice as thick as middle femur (Fig. 111-5); S7 of male with broad disc and two or four small apical lobes, shorter than disc (Fig. 111-4c) (subtribe Ancyloscelina) .... *Ancyloscelis*
—. Paraocular carina absent except sometimes along lower or upper extremity of eye; vein cu-v of hind wing about half as long as (in *Meliphilopsis*), or nearly as long as, second abscissa of vein M+Cu (Fig. 111-3b, c); second abscissa of vein M+Cu of hind wing half as long as vein M or less; maxillary palpus, at least in female, with brush of hairs on one side of segment 3 and usually also 2 and 4; T7 of male with two apical points or angles; hind leg of male not or moderately modified, femur less than twice as thick as middle femur (except about twice as broad but flattened in some *Melitoma*); S7 of male with short and often narrow disc and large apical lobes, much longer than disc (Fig. 111-4f) (subtribe Emphorina) 2

2(1). Second segment of labial palpus 1.1-3.0 times as long as first, and usually longer than eye (except in Melitomella, in which it is 0.95 times eye length); proboscis in repose reaching well beyond front coxae and usually surpassing middle coxae 3
—. Second segment of labial palpus shorter, 0.5-0.8 times as long as first, and usually shorter (0.3-0.6 times) than eye; proboscis in repose reaching at most front coxae 6

3(2). T2 to T4 with broad, bare, median apical areas occupying much of terga in female; anterior surface of T1 largely bare; T7 of male apically truncate, with apical points far apart, separated by long straight margin; center of T7 bare, polished (South America) ...... *Meliphilopsis*
—. T3 and T4, usually also T2, with pubescence reaching apices, frequently forming conspicuous apical hair bands; anterior surface of T1 hairy; T7 of male apically with two close points or lobes; center of T7 hairy 4

4(3). Clypeus with yellow area; middle tibia with strong outer apical spine easily seen because of sparse vestiture of tibia; first flagellar segment as long as apical width (South America) .... *Toromelissa*
—. Integument of clypeus dark; middle tibia with small apical point hidden by dense, appressed vestiture of tibia; first flagellar segment over 1.5 times as long as apical width 5

5(4). Second labial palpal segment two to three times as long as first; labrum of female parallel-sided, more or less quadrate, apical margin with a median tubercle and one to three smaller tubercles on each side; hind femur of male with lower margin strongly carinate .... *Melitoma*

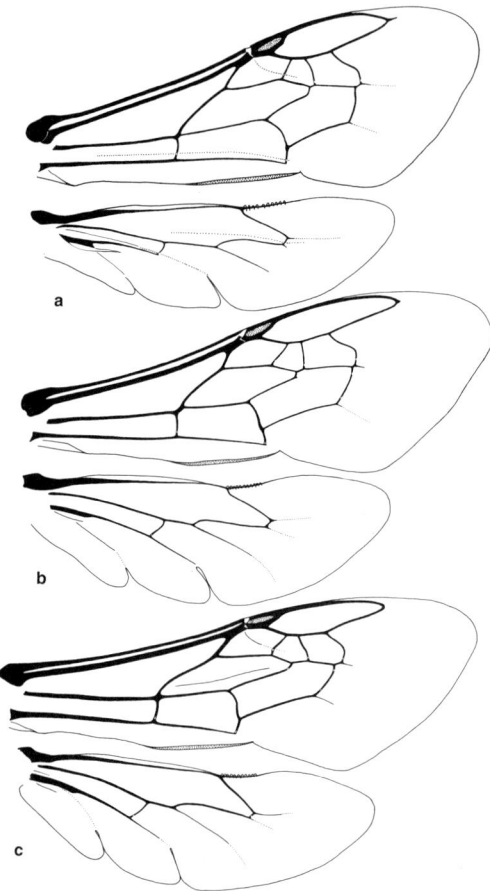

Figure 111-3. Wings of Emphorini. **a,** *Ancyloscelis panamensis* Michener; **b,** *Diadasia afflicta* (Cresson); **c,** *Ptilothrix fructifer* (Holmberg).

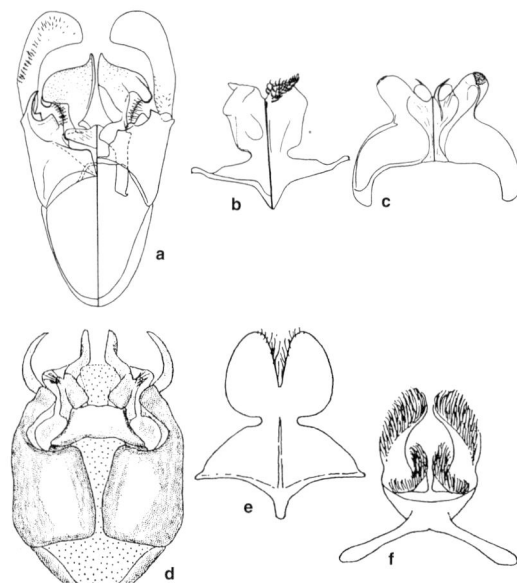

Figure 111-4. Male genitalia, S8 and S7 of Emphorini. **a-c,** *Ancyloscelis ursinus* Haliday, dorsal view on the left; **d-f,** *Melitoma taurea* (Say), dorsal view in d, ventral views in e and f. a-c, from Michener and Moure, 1957; d-f, from Mitchell, 1962.

—. Second labial palpal segment 1.5 times as long as first; labrum of female with sides diverging apically, pentagonal, apical margin without tubercles; hind femur of male with lower margin narrowly rounded .... *Melitomella*
6(2). Arolia absent; first flagellar segment usually two or more times as long as apical width; branches of tibial scopal hairs mostly directed distad, most of them not curved basad .... *Ptilothrix*
—. Arolia present; first flagellar segment less than twice as long as apical width; branches of tibial scopal hairs mostly curved basad at their apices [except in *Alepidosceles* and *Diadasina (Leptometriella)*] 7
7(6). Gradulus of S2 medially angulate posteriorly, graduli of S3 and S4 often weakly so; basal half of front femur of female commonly with dense brush of long hairs; propodeal triangle of male bare or, rarely, with some hairs close to posterior margin of triangle, but then posterior claws with rounded apices .... *Diadasia*
—. Graduli of metasomal sterna with median parts gently recurved; front femur of female with long hairs not forming dense basal brush; propodeal triangle of male hairy [or in *Diadasina (Leptometriella)* sometimes largely bare]; claws pointed 8

8(7). Basitibial plate of female absent; mandible of male broadened preapically; apex of labrum of male medially with bilobate tubercle .... *Alepidosceles*
—. Basitibial plate of female distinct; mandible of male tapering apically (or, in subgenus *Leptometriella*, slightly broadened preapically); apex of labrum of male with or without simple median tubercle .... *Diadasina*

Genus *Alepidosceles* Moure

Alepidosceles Moure, 1947b: 244. Type species: *Ancyloscelis imitatrix* Schrottky, 1909, by original designation.

Alepidosceles is the only emphorine genus that lacks a basitibial plate in the female. The metasoma is covered with yellowish hairs, sometimes denser on the apical parts of terga than elsewhere, forming weak bands. The body length is 7 to 11 mm. The branches of the hairs of the tibial scopa are directed distad. In both sexes the propodeal triangle is hairy. The male upper and lower gonostyli are both rather long and slender, the lower exceeding the upper. Male genitalia and hidden sterna were illustrated by Moure (1947b).

■ This genus occurs from Mendoza province, Argentina, north to Minas Gerais, Brazil. Six species have been described; those of Argentina were revised by Roig-Alsina (1998b).

Genus *Ancyloscelis* Latreille

Ancyloscelis Berthold, 1827: 466, nomen nudum. Latinization of Ancyloscele Latreille, 1825: 463.
Ancyloscelus Berthold, 1827: 565, nomen nudum.
Ancyloscelis Latreille, 1829: 355, no included species; Hali-

day, 1836: 320, included one species. Type species: *Ancylosceles* [sic] *ursinus* Haliday, 1836, first included species. [For a subsequent type designation, see Michener, 1997b.]

Ancylosceles Haliday, 1836: 320, unjustified emendation of *Ancyloscelis* Latreille, 1829.

Leptergates Holmberg, 1887b: 224, nomen nudum.

Leptergatis Holmberg, 1903: 422. Type species: *Leptergatis halictoides* Holmberg, 1903, by designation of Lutz and Cockerell, 1920: 592.

Dipedia Friese, 1906d: 92. Type species: *Ancyloscelis armatus* Smith, 1854, by designation of Lutz and Cockerell, 1920: 592.

As in similar cases, there is no certainty that the *nomina nuda* listed above actually were based on this genus, but the probabilities are good that their listing here is correct. Of course they have no nomenclatural status. They are listed only to provide a historical account of the generic name.

Ancyloscelis contains species 6 to 10 mm in length that differ from other genera of Emphorini not only in the characters indicated in the key to genera, but also in the strongly protuberant clypeus, which in side view extends forward from the eye margin for a distance about equal to the eye width. The granular propodeal triangle is bare medially but usually carries hairs laterally. The swollen hind leg of the male is distinctive (Figs. 111-2c, 111-5). The dense scopa on the hind tibia and basitarsus of the female consists of long, coarsely plumose hairs, their abundant branches directed distad. The male gonostylus is large, curved, and as long as the gonocoxite or longer (Fig. 111-4a). Male genitalia and hidden sterna were illustrated by Michener (1954b) and Michener and Moure (1957); see also Figure 111-4a-c.

■ This genus ranges from Texas, Colorado, and Arizona, USA, through tropical America to Buenos Aires province, Argentina. There are about 25 species. The species found in the USA were revised by Michener (1942b).

Nests of at least some species are in aggregations in banks or flat ground. *Ancyloscelis apiformis* (Fabricius) commonly nests in the same banks as *Melitoma euglossoides* Lepeletier and Serville and its close relatives. The cells are alone or in short series at the ends of branch burrows, the inner surfaces beautifully smooth but without visible linings. Cocoons are thin and delicate, perhaps sometimes absent. Accounts of nesting biology are by Torchio (1974) and Michener (1974b). Probably all species are oligolectic; the group of *A. apiformis,* including all the North and Central American species as well as some of the South American species, seems largely restricted to *Ipomoea* (Convolvulaceae) for pollen collecting. Some Brazilian species are oligolectic on certain Pontederiaceae and have very long proboscides with many hooked hairs for pulling pollen out of the flowers (Alves-dos-Santos, 1999a,b).

Genus *Diadasia* Patton

Diadasia Patton, 1879a: 475. Type species: *Melissodes enavata* Cresson, 1872, by original designation.

Dasiapis Cockerell, 1903c: 450. Type species: *Dasiapis ochracea* Cockerell, 1903, monobasic.

Leptometria Holmberg, 1903: 409. Type species: *Leptometria pereyrae* Holmberg, 1903, by designation of Brèthes, 1910: 213.

Coquillettapis Viereck, 1909: 47. Type species: *Coquillettapis melittoides* Viereck, 1909 = *Melissodes nigrifrons* Cresson, 1878, by original designation.

Diadasia is the largest genus of the Emphorini, being rather speciose in both North and South America. It differs from all others in the course of the gradulus of S2, the median part of which is produced posteriorly, angulate or narrowly rounded. In the other genera the gradulus of S2 is transverse or gently convex posteriorly. Males differ from other genera except some *Diadasina (Leptometriella)* in having a bare or largely bare propodeal triangle. In males of most species the claws of the posterior legs are rather broad and rounded at their apices, a character not found in other genera. The species of *Diadasia* vary in appearance, some having the metasoma uniformly hairy, others having distinct apical tergal bands (Fig. 111-1). Body length ranges from 5 to 20 mm. The apices of the

Figure 111-5. Male 11of *Ancyloscelis* sp. From Michener, McGinley, and Danforth, 1994.

branches of the hairs of the tibial scopa commonly curve basad. Males mostly have only one long gonostylus; a few species, e.g., *D. enavata* (Cresson), have a second (dorsal) gonostylus about half as long as the first. Male genitalia and hidden sterna were illustrated by Adlakha (1969), Toro and Ruz (1969), and Snelling (1994).

■ *Diadasia* has an amphitropical distribution, from Washington, Utah, Nebraska, and California to Texas, USA, south through Mesoamerica to Costa Rica and from Colombia and Rio Branco, Brazil south to Bío-Bío, Chile, and Río Negro, Argentina. Only the *Dasiapis* group (see below) is known from moist tropical locations. About 30 species occur in North America; about 15 additional species are known from South America. It is difficult to know from descriptions which named South American species actually belong in *Diadasina* instead of *Diadasia*; revisional studies involving examination of types will be needed to clarify such problems. Timberlake (1941b) gave a key to species found in the United States and Adlakha (1969) revised the same species. Toro and Ruz (1969) revised the species (including the genus *Toromelissa*) of Chile. Snelling (1994) revised the North American species of the *Dasiapis* group.

The species are rather diverse in appearance. Those that are oligolectic on *Helianthus* (Asteraceae) and on Cactaceae are large and distinctly banded. Those oligolectic on Malvaceae and Onagraceae are mostly small and tend to have a uniform covering of pale hairs on the metasoma. The one species on Convolvulaceae, *Diadasia bituberculata* (Cresson), is intermediate in size and has more dark hair. The diversity among species led to the recognition of four subgenera, characters of which were tabulated, along with those of the genus *Diadasina,* by Michener (1954b). Unfortunately, certain Holmberg genus-group names were misinterpreted at that time by both Moure and Michener, but the group characters remain of interest. Subgenera may not be desirable for *Diadasia,* and need not be recognized until the genus is more fully studied; the names are shown as synonyms above. The only subgenus usually recognized is that having yellow on the clypeus or at least on the bases of the mandibles, i.e., *Dasiapis* = *Leptometria*; I term this the *Dasiapis* group.

Nests of *Diadasia* are rather shallow burrows, sometimes in large aggregations, commonly with earthen turrets at the entrances, leading to one to several cells at the ends of branch burrows. Unlike *Melitoma* and *Ptilothrix,* the bees do not alight on water and do not carry water to nest sites; they may soften hard earth with nectar. Cells are made of hardened earth but do not come free from the substrate, like those of *Melitoma.* Accounts of nesting biology are by Linsley, MacSwain, and Smith (1952), Linsley and MacSwain (1957), Adlakha (1969), Snyder, Barrows, and Chabot (1976), Eickwort, Eickwort, and Linsley (1977), Neff, Simpson, and Dorr (1982), Ordway (1984), and Hazeldine (1997b).

Genus *Diadasina* Moure

Diadasina is here used in a broad sense to include *Diadasia*-like species in which the gradulus of S2 is straight or gently convex medially, not angulate as in *Diadasia;* and the claws of the male are pointed, not broadened and rounded on at least the posterior legs as in many *Diadasia*. S6 of the male has a broad, rounded, longitudinal ridge densely covered with long hair in the subgenus *Diadasina* s. str.; such a ridge is also found in *Melitomella grisescens* (Ducke). *Diadasina* is probably paraphyletic, to judge from the preliminary results of a phylogenetic study by A. Roig-Alsina (manuscript). His cladograms show the two subgenera (see the key below) as separate, usually basal branches of the clade that includes *Alepidosceles* and *Ptilothrix.* Since the phylogeny remains speculative, I prefer to indicate the similarity and relationship of the two subgenera by including them in a single genus.

All species of *Diadasina* are South American.

Key to the Subgenera of *Diadasina*

1. Propodeal triangle with hairs covering entire surface, except sometimes a narrow longitudinal median bare band; apex of labrum of female with margin rounded, lacking denticle; S6 of male with hairs longer along median longitudinal convexity, frequently forming dense tuft ....
.... *D. (Diadasina s. str.)*
—. Propodeal triangle with upper bare area bordering metanotum, or triangle largely bare; labrum of female with distal margin elevated, smooth, frequently with median denticle; S6 of male evenly covered with short hairs
.... *D. (Leptometriella)*

Diadasina / Subgenus *Diadasina* Moure s. str.

Diadasina Moure, 1950d: 392. Type species: *Melitoma paraensis* Ducke, 1912, by original designation.
Diadasiana Michener, 1954b: 130. Incorrect subsequent spelling.

Diadasina consists of small species 6 to 8 mm in length.
■ This subgenus is found from Pará, Brazil, to Buenos Aires province, Argentina. At least four species are included; probably other little-known species currently included in *Diadasia* will be found to be species of *Diadasina* s. str. or *Leptometriella*.

The nesting biology of one species, *Diadasina distincta* (Holmberg), has been described by Martins and Antonini (1994). The nests were in aggregations in more or less flat ground. Each burrow ended in a single cell; each female must make several nests.

Diadasina / Subgenus *Leptometriella* Roig-Alsina

Leptometriella Roig-Alsina, 1998a: 23. Type species: *Leptometria tucumana* Brèthes, 1910, by original designation.

Like *Diadasina* s. str., subgenus *Leptometriella* consists of small species 6 to 8 mm in length. The branches of the scopal hairs are mostly directed distad rather than retrorsely curved as they are in many other Emphorina.
■ This subgenus is known from Salta to Mendoza, Argentina. Three named species are included.

Genus *Meliphilopsis* Roig-Alsina

Meliphilopsis Roig-Alsina, 1994: 183. Type species: *Meliphilopsis melanandra* Roig-Alsina, 1994, by original designation.

This genus is one of the most distinctive of the Emphorini. As indicated in the account of the tribe, its hind-wing venation is intermediate between that of the subtribes Ancyloscelina and Emphorina. Its reduced hairiness (see the key to genera) is unique in the Emphorini. The propodeal triangle is bare in both sexes, unlike that of other Emphorina. The hind basitarsus of the male is densely hairy and straight; it is less hairy and arcuate in other Emphorina. The body length is 5.8 to 11.0 mm. The small, slender, almost andreniform *Meliphilopsis ochrandra* Roig-Alsina is unusual in the Emphorini, but the clearly congeneric *M. melanandra* Roig-Alsina is similar in form to other Emphorini. Male genitalia and hidden sterna were illustrated by Roig-Alsina (1994).

■ *Meliphilopsis* is found from La Rioja to Salta province, Argentina. The two species were distinguished by Roig-Alsina (1994).

Nests of *Meliphilopsis melanandra* Roig-Alsina were found by Hazeldine (1996) in a vertical bank. The branched burrows contained series of up to five adjacent horizontal cells, provisioned with pollen of Convolvulaceae.

Genus *Melitoma* Lepeletier and Serville

Melitome Latreille, 1825: 464, nomen nudum.
Melitoma Berthold, 1827: 468, nomen nudum for *Melitome* Latreille.
Melitoma Lepeletier and Serville, 1828: 529. Type species: *Melitoma euglossoides* Lepeletier and Serville, 1828, monobasic.
Entechnia Patton, 1879a: 476. Type species: *Anthophora taurea* Say, 1837, by original designation.
Meliphila Schrottky, 1902b: 310. Type species: *Meliphila ipomoeae* Schrottky, 1902 = *Melitoma euglossoides* Lepeletier and Serville, 1828, by original designation.
Eutechnia Holmberg, 1903: 400. *Lapsus* for *Entechnia* Patton, 1879.

Of the emphorine genera, *Melitoma* is the one with the most elongate proboscis. As indicated in the key to genera, it also has other distinctive characters; a less decisive one is the areas of dark and pale hair on the mesoscutum of most species, resulting in a characteristic pattern of pale and dark areas. *Melitoma* is distinguished from *Diadasia* by the straight rather than medially angulate gradulus of S2 and the hairy propodeal triangle of the male. The posterior claws are acutely pointed, unlike those of most male *Diadasia*, Body length is from 8 to 15 mm. Male genitalia and hidden sterna were illustrated by Michener (1954b) and Mitchell (1962); see also Figure 111-4d-f.

■ *Melitoma* ranges from South Dakota to New Jersey, south to Florida and New Mexico, USA, southwest to Sonora, Mexico, and south through both arid and humid tropics to western Peru and Buenos Aires province, Argentina. *Melitoma* is not found in the Antilles. There are about ten species.

Most or all species appear to be oligolectic on *Ipomoea* or related Convolvulaceae, and possibly on some Malvaceae. Nests of several species are typically in aggregations in hard clay banks or adobe walls. The bees carry water to the nest sites to soften the clay. The cells are readily separable from the matrix. Papers containing information on nesting biology, as well as references to earlier works on this topic, include Michener and Lange (1958c) and Linsley, MacSwain, and Michener (1980).

Genus *Melitomella* Roig-Alsina

Melitomella Roig-Alsina, 1998a: 19. Type species: *Podalirius grisescens* Ducke, 1907, by original designation.

This genus suggests a relatively large (7-11 mm body length), long-tongued *Diadasina*. S6 of the male has long hair on the median longitudinal ridge, as in *Diadasina* s. str. In various other features, *Melitomella* resembles *Melitoma* rather than *Diadasina*. Not only is the proboscis long (see the key to genera), but the first flagellar segment is more than twice as long as its apical width and the pygidial plate of the female has a rather narrow apical region because the lateral margins are concave, rather than nearly straight as in *Diadasina*.

■ *Melitomella* is known from Bahia to Pará, Brazil, and Panama. The named species are *grisescens* (Ducke) and *murihirta* (Cockerell) from Brazil and *schwarzi* (Michener) from Panama.

Genus *Ptilothrix* Smith

Ptilothrix Smith 1853: 131. Type species: *Ptilothrix plumatus* Smith, 1853, monobasic.
Ptilothryx Marschall, 1873: 269, unjustified emendation of *Ptilothrix* Smith.
Emphor Patton, 1879a: 476. Type species: *Melissodes bombiformis* Cresson, 1878, by original designation.
Teleutemnesta Holmberg, 1887b: 10, no included species; Holmberg, 1903: 400, five included species. Type species: *Teleutemnesta fructifera* Holmberg, 1903, by designation of Cockerell, 1918b: 36.
Energoponus Holmberg, 1903: 406. Type species: *Energoponus strenuus* Holmberg, 1903 = *Ptilothrix plumatus* Smith, 1853, by designation of Sandhouse, 1943: 547.

Holmberg's description of his genus *Energoponus* includes information on the labial palpi showing that he was describing a species now placed in *Melitoma*; it was *M. ameghinoi* (Holmberg). This was no doubt the principal species on which the generic description was based, for Holmberg records parts on a slide (prep. micr. n. 30). However, the type species of *Energoponus, E. strenuus* Holmberg, was described without comment on the mouthparts and Holmberg may or may not have examined them; they are much shorter than those of *M. ameghinoi*. The type of *E. strenuus* is lost (A. Roig-Alsina, in litt., 1997), but the assumption that it was a *Ptilothrix* (*relatus* Holmberg or *plumatus* Smith) is reasonable and agrees with past judgments on the identity of *E. strenuus*.

Ptilothrix is unique among the Emphorini in lacking arolia. The propodeal triangle is hairy in both sexes. In most species the metasomal terga bear strong apical bands of pale hair, but in some the hair is dark and bands are not in evidence. Most species are rather large, but body length ranges from 7 to 15 mm. The body is typically more slender than usual in *Diadasia*, more nearly apiform than euceriform. The gradulus of S2 is transverse or gently curved in the middle, not angulate as in *Diadasia*. Male genitalia and hidden sterna were illustrated by Mitchell (1962).

■ *Ptilothrix* is an amphitropical genus. In North America it occurs from New Jersey to Kansas, south to Florida, Texas, and Arizona, USA, and to Oaxaca, Mexico. J. Ascher (personal comm.) reports a single specimen from Nicaragua. In South America it is found from Pará, Brazil, south to Bolivia, Paraguay, and Córdoba and Entre Ríos provinces, Argentina. There are three North American species, one of them unnamed, and about ten South American species.

The lone species of eastern North America, *Ptilothrix bombiformis* (Cresson), is an oligolectic visitor of *Hibiscus* (Malvaceae), but the floral associations of other species are not well known. According to J. Rozen (in litt.) the unnamed species from Arizona collects pollen from cotton.

Ptilothrix are noteworthy for their ability, in spite of the large size of these bees, to alight on water surfaces, supported only by their tarsi, when they take up water to soften the hard soil in which they nest. There are no obvious morphological features of their tarsi that permit such behavior. The nests are in banks or often in flat ground, and are shallow burrows each leading to one or a series of very few cells. Nest entrances are commonly provided with short turrets. The nesting biology, including comparisons with other genera and references to earlier works, was described by Linsley, MacSwain, and Smith (1956). More recent accounts are those of Michener and Lange (1958c) and Martins, Guimarães, and Dias (1996) on South American species and Butler (1967) and Rust (1980b) on North American species.

Genus *Toromelissa* Roig-Alsina

Toromelissa Roig-Alsina, 1998a: 22. Type species: *Diadasia nemaglossa* Toro and Ruz, 1969, by original designation.

This genus consists of a species that is superficially *Diadasia*-like but differs from that genus not only in its long proboscis (see the key to genera) but in the simple transverse gradulus of S2 and the distally directed rather than retrorse branches of the tibial scopal hairs. The propodeal triangle of the male is largely bare, as in *Diadasia*. The posterior claws are acutely pointed, unlike those of most *Diadasia*. The body length is 7 to 8 mm. The yellow area on the clypeus distinguishes *Toromelissa* from all *Diadasia* except those of the *Dasiapis* group. Male genitalia and hidden sterna were illustrated by Toro and Ruz (1969).

■ *Toromelissa* is found in the Atacama region, Chile. The only species is *T. nemaglossa* (Toro and Ruz).

112. Tribe Eucerini

The Eucerini consist of euceriform, rather hairy bees, the metasoma often having pale hair bands (Fig. 112-1). Males of many species are easily recognized because of their long antennae, often reaching well beyond the base of the metasoma (Pl. 11; Fig. 112-2a), sometimes to the apex of the metasoma, although species with short antennae, similar to those of females, occur in various groups. Except for *Eucerinoda,* the Eucerini differ from all other bees in having long paraglossae, often reaching the distal end of the second segment of the labial palpus. Also except for *Eucerinoda,* the Eucerini have paraocular carinae. Except for *Canephorula,* vein cu-v of the hind wing is transverse to moderately oblique and at most a little over half as long as the second abscissa of M+Cu (Fig. 112-4). Roig-Alsina and Michener (1993) recognized three subtribes, Eucerinodina and Canephorulina, with one species each, and Eucerina, with all the rest. Characters applicable to the whole tribe are the dense short hairs on at least the distal part of the labrum; the presence of a branch of the anterior tentorial arm fused to the upper wall of the antennal socket; the long marginal cell, about as long as the distance from its apex to the wing tip, its apex bent gradually away from the wing margin (Fig. 112-4); the short stigma, extending little beyond the base of vein r; the two or four apical lobes on S7 of the male; the presence in most cases of a clump of usually blunt setae at the ventral base of the male gonostylus; the presence of a spatha; and the strong cobblestone pattern of the flabellum (Fig. 112-3). (I am not now able to verify one of the characters above—the tentorial arm fused to the antennal socket—for *Eucerinoda,* but Roig-Alsina and Michener (1993, table 2) reported this feature for *Eucerinoda* as well as other Eucerini.)

Because of the homogeneity of the subtribe Eucerina, which includes all but two species of Eucerini, some authors have placed all of them in the genus *Eucera* (Dalla Torre, 1896). By tradition, however, the Eucerini, especially in the Western Hemisphere (Robertson, 1905; Moure and Michener, 1955b; LaBerge, 1957), seem to have been divided more finely at the genus level than are most tribes of bees. I have sought to show relationships by uniting certain genera, recognizing various taxa as sub-

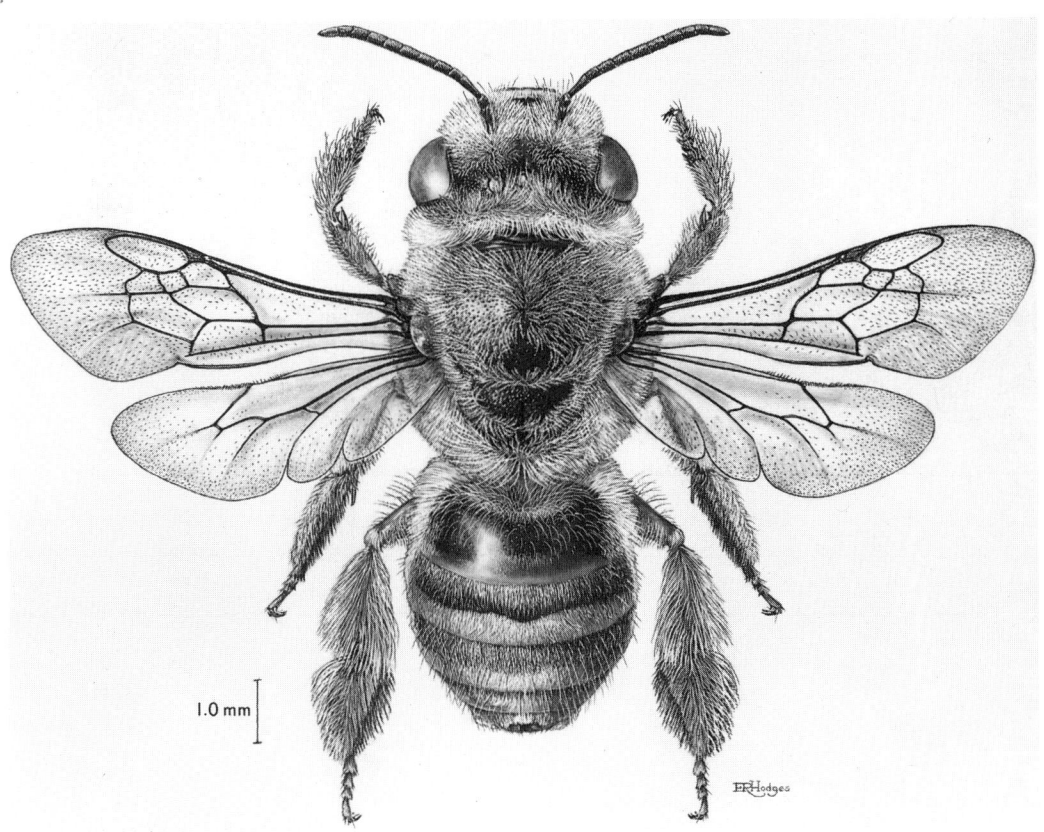

Figure 112-1. *Gaesischia (Gaesischiana) exul* Michener, LaBerge, and Moure, female. Drawing by E. R. S. Hodges, from Michener, McGinley, and Danforth, 1994.

Figure 112-2. Eucerini. Above and Center, *Melissodes bimaculata* (Lepeletier), male and female; Below, *Simanthedon linsleyi* Zavortink, female. From Michener, McGinley, and Danforth, 1994.

genera rather than as genera. For the most part, however, this process has not been successful, for relationships among the current genera are often not clear. More detailed studies are needed to distinguish resemblance due to convergence from that due to homology. I strongly suspect that, in spite of differences in some characters that are often considered important and possibly phylogenetically revealing, certain genera should be united to better clarify relationships. Some examples: *Peponapis* should be a subgenus of *Xenoglossa, Simanthedon* should be a subgenus of *Martinapis, Platysvastra* should be a subgenus of

Gaesochira. The currently known characters do not conclusively support these changes in the classification and I therefore have not made them.

Especially confusing genera are those commonly called *Synhalonia, Tetralonia, Tetraloniella, Xenoglossodes,* and *Eucera*. The last as usually understood consisted of bees with only two submarginal cells (Fig. 112-4a), thereby clearly different from almost all other Eucerini. *Synhalonia*, however, is distinguishable from two-celled *Eucera* almost only by having three submarginal cells and is therefore here regarded as a subgenus of *Eucera*. Moreover, the large taxa *Tetraloniella* (= *Xenoglossodes*) and *Eucera (Synhalonia)* are difficult to separate and have often been considered together under the name *Tetralonia*. Some species seem to belong to one genus on the basis of certain characters, yet in other characters they agree with another taxon. Largely for historical reasons, and to avoid changes that may have to be reversed, I have retained *Tetraloniella* as a distinct genus. It would have been appropriate to include it among the subgenera of *Eucera*. *Tetralonia* proper is here considered to be a different small genus, a subgenus of which is *Eucara*. It should be clear that my arrangement is arbitrary and will probably be changed with further study, which should include phylogenetic analysis of representatives of the world fauna.

The temperate South American genera *Eucerinoda* and *Canephorula,* here each placed in its own subtribe as noted above, have sometimes been placed in their own tribes separate from the Eucerini. *Eucerinoda* differs from other Eucerini mostly by ancestral states of various characters. It appears to be a sister group to all other Eucerini, according to Roig-Alsina and Michener's (1993) analysis. *Canephorula*, however, differs from typical Eucerini mostly in unique, derived features. But it has at least one striking character that I believe is ancestral, the simple recurved gradulus of S2 of the female (character 109 of Roig-Alsina and Michener, 1993), like that of *Eucerinoda* and unlike the bilobed gradulus of the Eucerina (Fig. 112-5a, b). I therefore suspect that *Canephorula* is the sister group to all Eucerina, even though it appears among genera of Eucerina in some of Roig-Alsina and Michener's analyses; it is shown as that sister in their figure 3b.

No comprehensive phylogenetic study of the Eucerini has been made, but some suggestive observations are possible. The tribe is far more diverse in the Western Hemisphere than in the Eastern Hemisphere; this diversity is especially notable in South America, where the two monotypic, perhaps relictual subtribes, Eucerinodina and Canephorulina, are found. If large and elaborate lobes on S7, such as are especially common in the Colletidae, are ancestral, the genera *Alloscirtetica* (Fig. 112-7c) and *Eucerinoda* (Fig. 112-17c) are suggested as basal branches of eucerine phylogeny. In contrast, the simple form of S7 with reduced apical lobes characterizes many genera, including all those of the Eastern Hemisphere. Western Hemisphere genera such as *Florilegus* and *Gaesischia* are intermediate in the development of the lobes of S7 (Fig. 112-7f, i).

Nests and larvae of *Eucerinoda* are unknown. Known nests of Eucerina and Canephorulina are burrows in the ground, and are sometimes, especially for *Thygater*, horizontal burrows in banks, but usually more or less vertical

Figure 112-3. Apices of glossae of Eucerini, showing cobblestoned pattern of flabella. **a, b,** *Peponapis pruinosa* (Say), posterior and anterior views; **c,** *Svastra obliqua* (Say), anterior view of the extreme apex, not showing anterior surface of flabellum; **d,** *Melissoptila (Ptilomelissa)* sp., posterior view; **e,** *Thygater analis* (Lepeletier), posterior view; **f,** *Svastra obliqua* (Say), posterior view. (S, seta.) SEM photos by R. W. Brooks, from Michener and Brooks, 1984.

burrows in flat ground. Each cell is at the end of a rather long lateral burrow, and the cells are vertical, elongate, and lined with "waxlike" secreted material. Contrary to this description, Thorp and Chemsak (1964) reported that in *Melissodes pallidisignata* Cockerell each burrow ends in a single cell; thus each cell is in a separate nest. Since eucerines usually fill each lateral with earth excavated from the succeeding lateral, so that at any one time the nest burrow is open to only one cell, this observation should be verified; see the comments by Cameron et al. (1997). Unlike many cells of Anthophorini and Emphorini, the constructed earthern cell wall is not readily separable from the matrix; the cells therefore cannot be separated intact from the surrounding earth. The cell cap is not lined. The egg is placed on top of the soft provisions that fill the bottom of the cell. The mature larva places its feces against the cell cap, and then spins a thin cocoon. Some works on the nesting biology of Eucerini are as follows: Malyshev (1924a, 1930a), Linsley, MacSwain, and Smith (1955), Michener and Lange (1958c), Rozen (1964b, 1969b, 1974, 1991b), Mathewson (1968), Wafa and Mohamed (1970), Clement (1973), Miliczky (1985), Triplett and Gittins (1988), Popova (1990), and Cameron et al. (1997). From these works and the papers cited therein, it is evident that something is known of nesting biology of *Eucera* (including *Synhalonia*), *Florilegus*, *Melissodes*, *Peponapis*, *Svastra* (including *Idiomelissodes*), *Thygater*, and *Xenoglossa*.

An additional work on eucerine biology (Michelette, Camargo, and Rozen, 2000) concerns the subtribe Canephorulina [*Canephorula apiformis* (Friese)]. The nest cells seem somewhat separable from the matrix; this may not differ from the condition in some other Eucerini. Although *Canephorula* has been placed in the separate subtribe Canephorulina, its nesting and other behavior do not differ greatly from those of the familiar subtribe Eucerina.

Larvae of the Eucerini are rather homogeneous and similar to those of related tribes, although they do have some distinctive tribal characters, as listed by Rozen (1991b). They have been described in detail by Rozen (1965a, 1991b) and in publications cited therein.

Keys to the genera are provided separately for South America (both sexes jointly), North and Central America (females and males separately), and the Eastern Hemisphere (both sexes jointly).

Concerning the key to South American Eucerini, the genus *Svastrina* is unknown in the female but will prob-

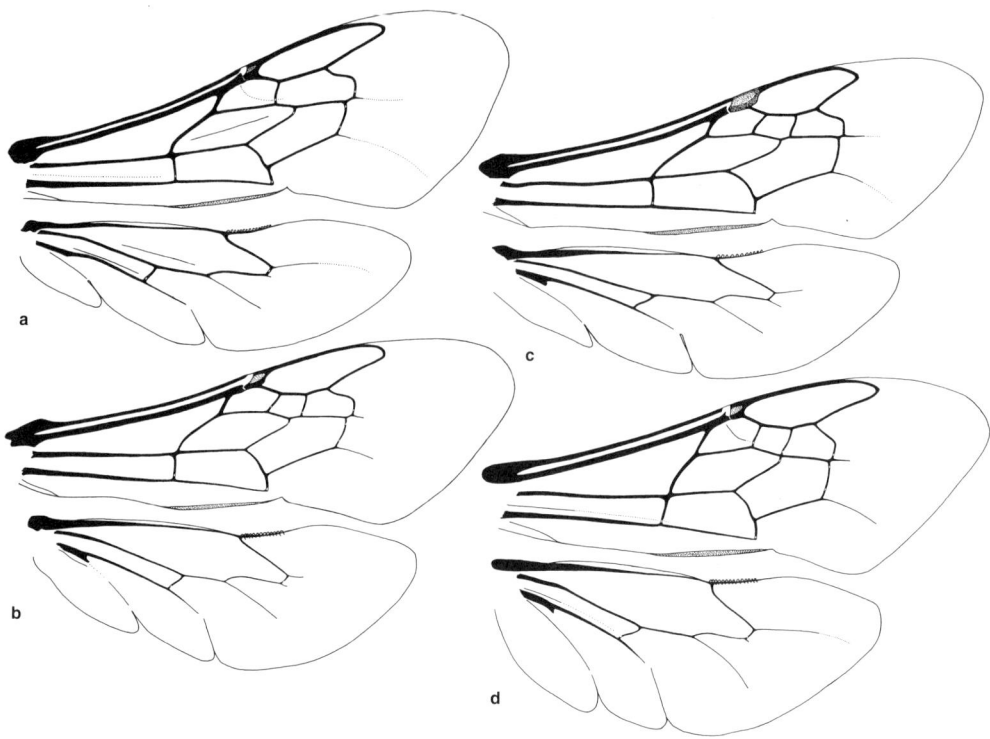

Figure 112-4. Wings of Eucerini. **a**, *Eucera chrysopyga* Pérez; **b**, *Melissodes agilis* Cresson; **c**, *Melissoptila (Ptilomelissa)* sp.; **d**, *Thygater analis* (Lepeletier).

ably be easily distinguished from other Eucerini, except *Pachysvastra*, by the lack of arolia. *Platysvastra* is known only in the female; the scutellar characters described in the key to females probably distinguish males also.

Users will find difficulties with portions of these keys, particularly in distinguishing *Tetraloniella* from *Eucera (Synhalonia)*. These genera should be restudied, on a worldwide basis, before further reclassification; perhaps the subgenera of *Tetraloniella* should be included in *Eucera*.

Key to the South American Genera of Eucerini

1. Paraglossa much shorter than first segment of labial palpus; paraocular area of male with yellow (gradulus of S2 simple, recurved) (subtribe Eucerinodina) ...... *Eucerinoda*
—. Paraglossa attaining apex of second segment of labial palpus; paraocular area of male not yellow 2
2(1). Outer surface of hind tibia of female with corbicula on distal half, corbicula margined posteriorly by comb of long, stiff, closely spaced bristles; wings nearly bare, strongly papillate beyond veins; vein cu-v of hind wing well over half as long as second abscissa of M+Cu (Fig. 112-16); gradulus of S2 simple, recurved (subtribe Canephorulina)... *Canephorula*
—. Scopa formed of more or less uniformly spaced hairs, not forming a corbicula except for bare area on distal fourth of tibia, and without row or comb of long, stiff bristles; wings hairy more or less throughout, not strongly papillate; vein cu-v of hind wing commonly more or less oblique, but little if any more than half as long as second abscissa of M+Cu; gradulus of S2 birecurved, i.e., with two convexities to the rear, one on either side of midline (Fig. 112-5a, b) (subtribe Eucerina) 3
3(2). Males 4
—. Females 23
4(3). Clypeocular distance at least as great as minimum diameter of first flagellar segment (Fig. 112-6e); lower part of paraocular carina completely absent (indicated by ridge in a few *Thygater*); lateral parts of clypeus, parts of paraocular area above latter, and part of paraocular area adjacent to eye all nearly in one plane; labrum at least three-fourths as long as broad; clypeus very strongly protuberant, i.e., lower end, as seen in side view, in front of lower anterior eye margin by about 1.5 times maximum eye width in same view (Fig. 112-6e) [except in *Thygater (Nectarodiaeta)*] 5
—. Clypeocular distance usually less than minimum diameter of first flagellar segment (Fig. 112-6a-d); lower part of paraocular carina present, or, *if* absent, then indicated by a ridge such that a narrow zone adjacent to eye is in a very different plane from adjacent regions (this may not be evident in forms in which clypeus, and hence paraocular carina, is very close to eye); labrum less than three-fourths as long as broad (In *Mirnapis*, labrum nearly three fourths as long as broad, apex deeply emarginate.); clypeus weakly to strongly protuberant, usually not very strongly so, i.e., lower end, as seen in side view, in front

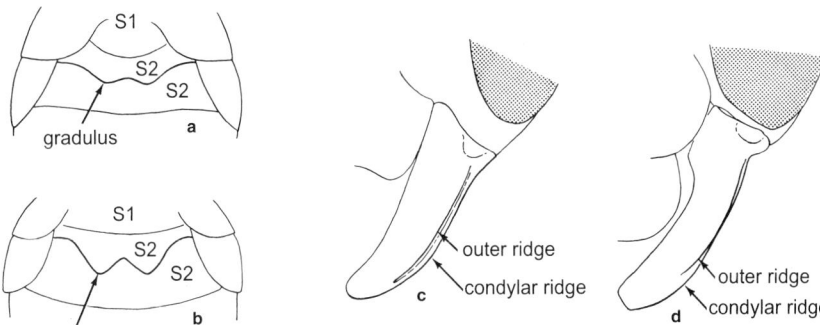

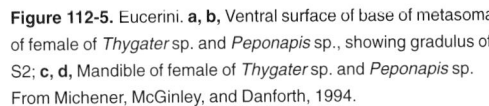

of lower anterior eye margin by about maximum width of eye or less (except for *Santiago* and some *Peponapis* that have very protuberant clypeus) 6

5(4). T7 ending in two lobes or blunt teeth, merely weakly emarginate in one species; pygidial plate unrecognizable or indicated only by weak, incomplete, lateral carinae; segments of antennal flagellum of approximately uniform diameter, without long hairs .... *Thygater*

—. T7 not bilobed; pygidial plate clearly defined; eighth and following flagellar segments more slender than preceding ones, with row of hairs on one side .... *Trichocerapis*

6(4). Flagellum tapering, its apical segment thus less than half as broad as its second segment; vertex with strong preoccipital carina .... *Lophothygater*

—. Flagellum not or less strongly tapering; vertex without preoccipital carina 7

7(6). Pygidial plate absent, reduced and ending in an acute apex, or with lateral carinae ending preapically, the apical portion of the plate truncate or rounded and without raised margins; S7 usually with median apical projection between elaborate, hairy apical lobes (Fig. 112-7c) .... *Alloscirtetica*

—. Pygidial plate with distinct raised margin or carina laterally and posteriorly (broken posteriorly only in a few forms with the apex of the plate bilobed or bidentate); S7 without median apical projection 8

Figure 112-5. Eucerini. **a, b,** Ventral surface of base of metasoma of female of *Thygater* sp. and *Peponapis* sp., showing gradulus of S2; **c, d,** Mandible of female of *Thygater* sp. and *Peponapis* sp. From Michener, McGinley, and Danforth, 1994.

8(7). T7 with gradular tooth or strong angle on each side of pygidial plate, sometimes hidden in dense hair or by T6 9

—. T7 without lateral teeth 13

9(8). Arolia absent; pygidial plate deeply notched medially; antenna scarcely reaching beyond scutellum, median flagellar segments less than twice as long as broad and about 1.5 times broader than ocellar diameter.......... *Pachysvastra*

—. Arolia present; pygidial plate entire; antenna usually much longer, median flagellar segments more than twice as long as broad and little broader than ocellar diameter 10

10(9). First segment of flagellum at least one-fourth as long as second; maxillary palpus two- to three-segmented; stigma large for the tribe, longer than prestigma; second submarginal cell large for the tribe, posterior margin often almost as long as that of first; inner and outer apical lobes of S7 almost indistinguishably fused and usually hairless (Fig. 112-9i); spatha over four times as wide as long .... *Melissoptila*

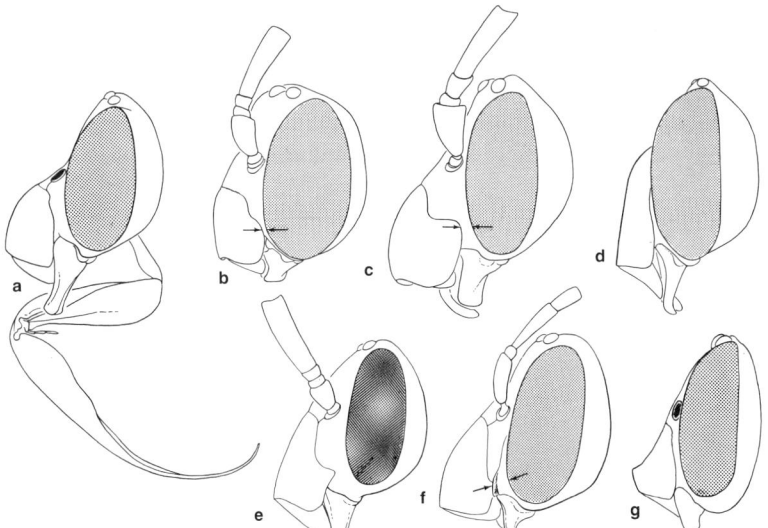

Figure 112-6. Heads of Eucerini. **a,** *Tetraloniella (Loxoptilus) longifellator* (LaBerge), male; **b,** *T. (Tetraloniella) albata* (Cresson), male; **c,** *Eucera (Synhalonia) atriventris* (Smith), male; **d,** *Tetraloniella (Pectinapis)* sp., male; **e,** *Thygater analis* (Lepeletier), male; **f,** *Tetraloniella (Loxoptilus) longifellator* (LaBerge), female; **g,** *Simanthedon linsleyi* Zavortink, male. (Arrows indicate clypeocular distance.) From Michener, McGinley, and Danforth, 1994.

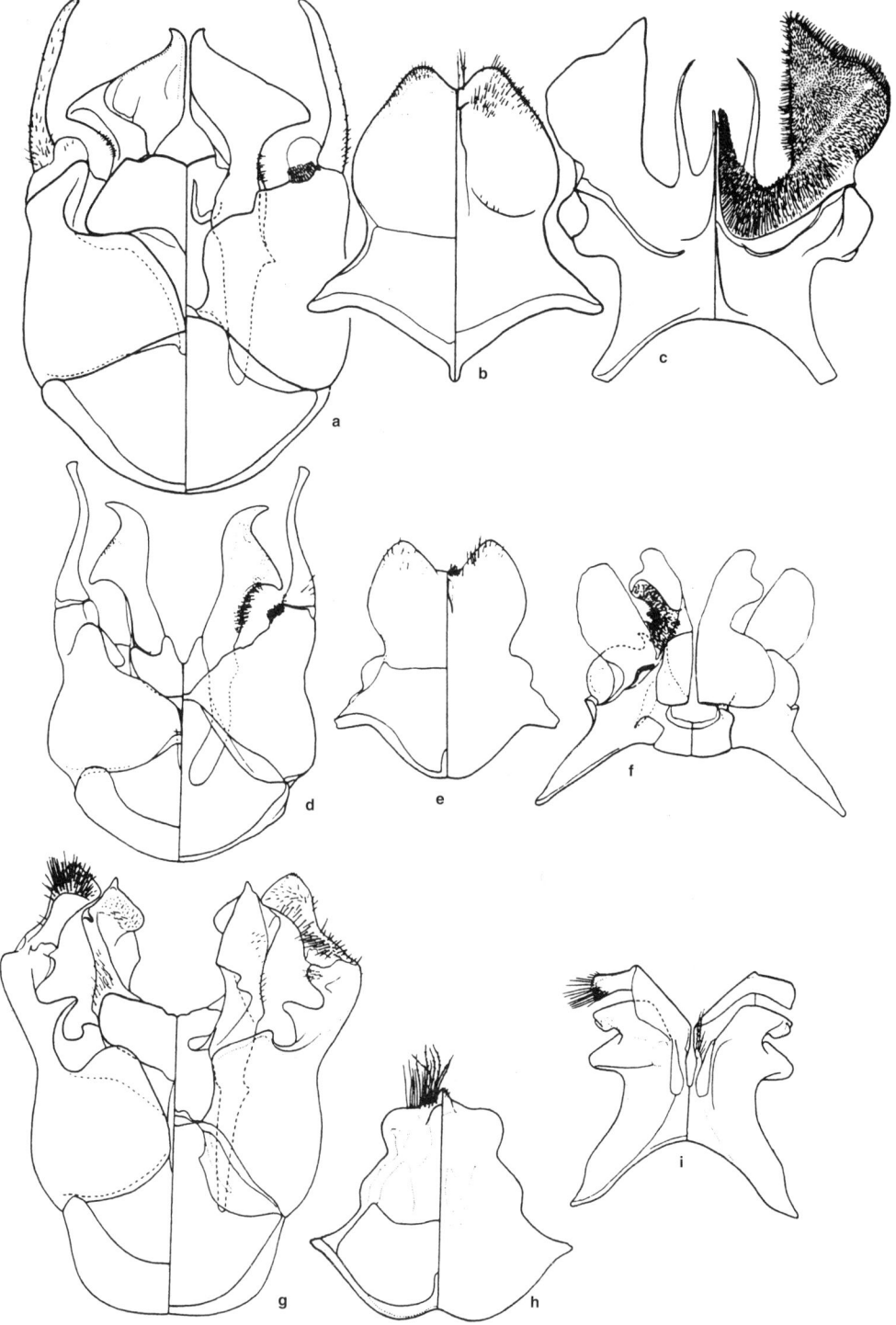

Figure 112-7. Male genitalia, S8, and S7 of Eucerini. **a-c,** *Alloscirtetica (Megascirtetica) mephistophelica* (Schrottky); **d-f,** *Trichocerapis (Trichocerapis) mirabilis* Smith; **g-i,** *Florilegus (Euflorilegus) barbiellinii* Moure and Michener. From Moure and Michener, 1955b.

—. First segment of flagellum not over one-fifth as long as second (Fig. 112-8e); maxillary palpus four- to five-segmented, very rarely three-segmented; stigma and second submarginal cell usually smaller; inner and outer apical lobes of S7 separated by transparent zone, the inner lobe either curled or reflexed or large with scattered hairs (Figs. 112-7i, 112-9c, f); spatha less than four times as wide as long 11

11(10). Pygidial plate with strong, transverse, preapical carina separating depressed apical part of plate from main part of plate; scutellum flat in profile ....*Gaesochira*

—. Pygidial plate without transverse preapical carina; scutellum convex in profile [except in *Florilegus (Floriraptor)*] 12

12(11). Anterior femur slender, distinctly more than three times as long as broad, and broadest near base; maxillary palpus four-segmented, rarely three-segmented; anterolateral margin of tegula (commonly hidden by hair) gently concave (Fig. 112-8a) .... *Melissodes*

—. Anterior femur somewhat robust, nearly three times as long as broad, and broadest near or beyond middle; maxillary palpus five-segmented; lateral margin of tegula continuously convex (as in Fig. 112-8b) .... *Florilegus*

13(8). Anterior mandibular articulation nearly twice as far from eye margin as posterior one; clypeocular distance about half of minimum width of first flagellar segment 14

—. Anterior mandibular articulation little farther from eye margin than posterior one; clypeocular distance less than half (more than half in *Mirnapis*) of minimum width of first flagellar segment (inner apical lobes of S7 except in *Svastra*, large, much exceed-ing outer lobes) 15

14(13). Gonostylus bent downward near middle; clypeus protuberant by eye width in front of eye margin as seen in lateral view; S6 with pair of strong converging carinae on lateral margins (Fig. 112-10d); S7 with inner apical lobes small, not much exceeding outer lobes ...... *Peponapis*

—. Gonostylus not bent near middle; clypeus protuberant by about 1.4 times eye width; S6 with converging carinae weak and confused with margin; S7 with inner apical lobes much exceeding outer lobes .... *Santiago*

15(13). Arolia absent; posterior trochanters with strong ventral prominence .... *Svastrina*

—. Arolia present; posterior trochanters usually without such a prominence 16

16(15). Pygidial plate ending in two upturned points; middle and hind legs thickened and somewhat contorted, tibial spurs robust, inner hind spur only about four times as long as its greatest breadth .... *Micronychapis*

—. Pygidial plate rounded or truncate; legs not so modified, tibial spurs of the usual slender form 17

17(16). Distal process of gonocoxite as long as rest of gonocoxite and extending much beyond distal margin of spatha (S6 without converging carinae, with median basal elevated area; labral emargination scarcely evident; middle femur hairy, without tuft of short dense hair; flagellum scarcely wider than diameter of middle ocellus) .... *Hamatothrix*

—. Distal process of gonocoxite shorter than rest of gonocoxite and extending little if any beyond margin of spatha 18

18(17). Labrum nearly three fourths as long as broad, apex deeply emarginate, sides of emargination at about right angle to one another and lobe on each side of emargination narrowly rounded (S6 with pair of strong, converging carinae near tergal margins; first flagellar segment on shortest side slightly longer than broad) .... *Mirnapis*

—. Labrum much less than three fourths as long as broad, broadly emarginate, lobes gently convex 19

19(18). S6 without a pair of converging carinae but with median, basal, somewhat elevated area or longitudinal median lamella; labral emargination deep and broad, oc-

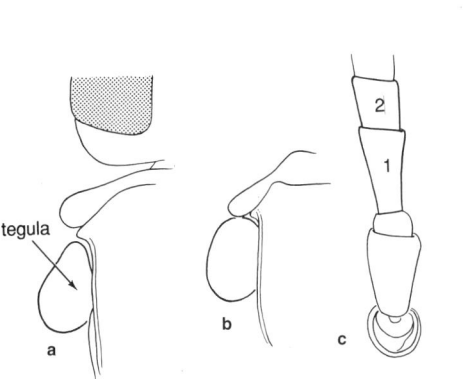
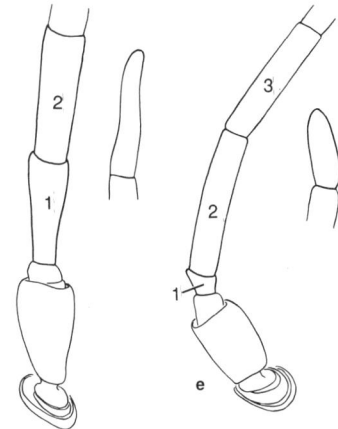

Figure 112-8. Tegula and antennal structures of Eucerini. **a, b,** Tegulae of *Melissodes* sp. and *Eucera (Synhalonia) atriventris* (Smith); **c,** Antennal base of male of *Xenoglossa kansensis* Cockerell; **d, e,** Antennal bases and apices of males of *Martinapis luteicornis* (Cockerell) and *Gaesischia exul* Michener, LaBerge, and Moure. (The flagellar segments are numbered.) From Michener, McGinley, and Danforth, 1994.

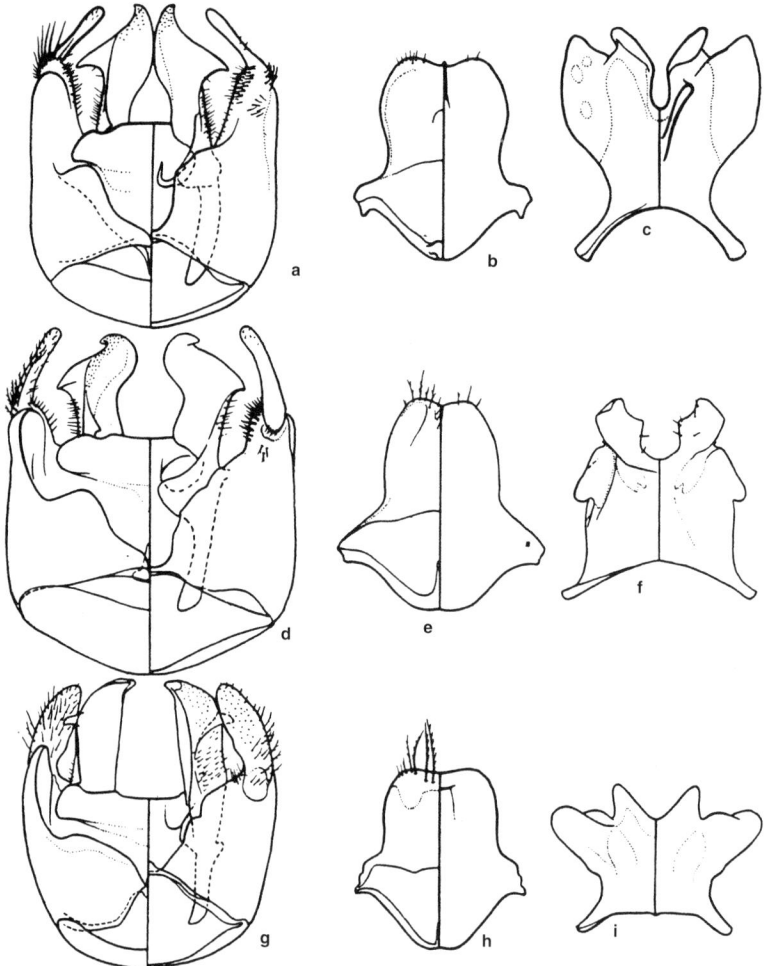

Figure 112-9. Male genitalia, S8, and S7 of Eucerini. **a-c**, *Melissodes (Melissodes) tepaneca* Cresson; **d-f**, *Melissodes (Ecplectica) nigroaenea* (Smith); **g-i**, *Melissoptila (Ptilomelissa) bonaerensis* Holmberg. (Dorsal views are on the left.) From Moure and Michener, 1955b.

cupying well over one-third of apical margin; middle femur usually nearly bare beneath except for a mass or tuft of short, dense hairs; flagellum much thicker than diameter of ocellus .... *Gaesischia* (in part)
—. S6 with pair of strong converging carinae but without basal elevated area; labral emargination shallow and occupying less than one-third of margin; middle femur without hair mass or tuft; flagellum little thicker than ocellar diameter 20
20(19). First flagellar segment longer than broad, more than one-fifth as long as second; inner apical lobes of S7 not reflexed basally, not diverging apically; vertex usually not elevated behind ocelli, which are on dorsal surface of head; clypeus almost reaching eye and almost without lateroclypeal carina 21
—. First flagellar segment much broader than minimum length, less than one-sixth as long as second; inner apical lobes of S7 strap-shaped or expanded apically, reflexed near bases and diverging apically; vertex elevated to form a weak ridge behind ocelli, which are thus to some extent on anterior slope of head; clypeus distinctly separated from eye and with distinct lateroclypeal carina 22
21(20). First flagellar segment more than half as long as second (Fig. 112-8d); inner apical lobes of S7 large, straplike, expanded and converging apically, much exceeding lateral lobes; metasoma without spatuloplumose hairs; lateral carina of pygidial plate with tooth subapically .... *Martinapis (Svastropsis)*
—. First flagellar segment less than half as long as second; inner apical lobes of S7 very small, much exceeded by lateral lobes; basal hair bands of T2 and sometimes T3 with spatuloplumose hairs (Fig. 112-10b); lateral carina of pygidial plate not toothed subapically........................ *Svastra*
22(20). T2 and T3 without appressed pale pubescence, sometimes with basal pale bands of semierect pubescence, without median or apical pale bands; length over 10 mm; median apical lobe of S7 usually straplike and not expanded apically; gonostylus with coarse plumose hairs ...... *Svastrides*
—. T2 and T3 with appressed pale pubescence, usually

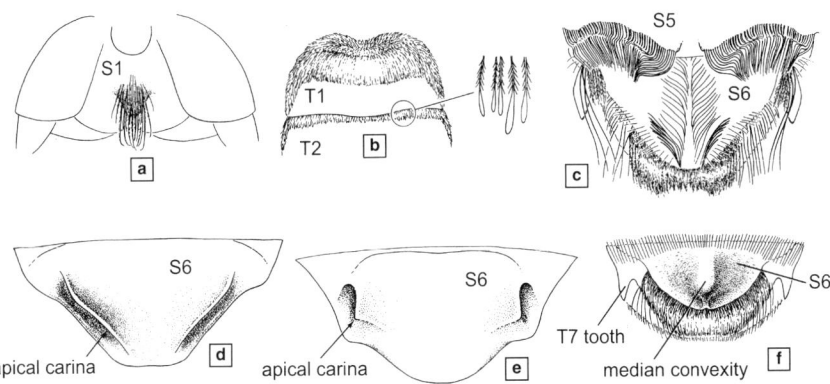

Figure 112-10. Structures of Eucerini. **a,** Ventral surface of base of metasoma of *Syntrichalonia exquisita* (Cresson), male, showing the hairy tubercle of S1; **b,** Dorsal surface of base of metasoma of *Svastra (Epimelissodes) obliqua* (Say), with enlargment of spatuloplumose hairs; **c,** S5 and S6 of male of *Svastra (Idiomelissodes) duplocincta* (Cockerell); **d, e,** S6 of males of *Peponapis pruinosa* (Say) and *Eucera (Synhalonia) atriventris* (Smith); **f,** S6 and adjacent sclerites of male of *Florilegus condignus* (Cresson). From Michener, McGinley, and Danforth, 1994.

forming apical as well as basal bands; length under 10 mm; median lobe of S7 expanded apically; gonostylus without coarse plumose hairs .... *Gaesischia* (in part)
23(3). Gradulus of S2 weakly biconvex (Fig. 112-5a); blade of galea at least 1.4 times as long as eye; mandible with condylar ridge expanded forward and thus at least as salient as and usually more salient than outer ridge; gradulus of T6 without lateral parts; labrum usually two-thirds as long as broad 24
—. Gradulus of S2 strongly biconvex, forming angle of 140° or less between two convexities (Fig. 112-5b); blade of galea as long as eye or slightly longer (nearly 1.5 times as long as eye in *Santiago*); mandible normal, with condylar ridge less salient than outer ridge; gradulus of T6 often with lateral parts; labrum usually less than two-thirds as long as broad 26
24(23). Vertex with strong preoccipital carina; scape almost as long as interantennal distance; first flagellar segment slightly shorter than scape .... *Lophothygater*
—. Vertex without preoccipital carina; scape much shorter than interantennal distance; first flagellar segment as long as or longer than scape 25
25(24). Scape little if any more than twice as long as broad; clypeus black, without pair of ridges diverging below; pygidial plate narrower, margins at angle of about 65°; metasomal terga without bands of appressed pubescence [or, in group of *Thygater analis* (Lepeletier), with basal bands completely hidden by preceding terga] ...... *Thygater*
—. Scape more than twice as long as broad; clypeus usually with a pair of yellowish spots and with a pair of weak ridges or carinae diverging below; pygidial plate very broad, margins at angle of 80° or more; T2 to T4 with basal bands of sparse, appressed, pale, plumose pubescence .... *Trichocerapis*
26(23). Apical bare area on outer surface of hind tibia nearly one-fourth as long as tibia, delimited posteriorly near base by clump of strong, amber-colored bristles; hairs of metasomal sterna strongly hooked (clypeus almost entirely yellow) .... *Hamatothrix*
—. Apical bare area on outer surface of hind tibia absent or not over one-fifth as long as tibia, delimited posteriorly by ordinary hairs; hairs of metasomal sterna not hooked 27
27(26). Posterior basitarsus with hairs of inner surface sparse except for a narrow band of dense hairs near lower margin (Fig. 112-11a); clypeus strongly protuberant; labrum about two-thirds as long as broad .... *Peponapis*
—. Posterior basitarsus densely hairy on inner surface (Fig. 112-11b), a band of especially dense hairs near lower margin; clypeus variable but usually moderately protuberant (strongly so in *Santiago*); labrum usually nearly one-half as long as broad or broader 28
28(27). Basitibial plate with margin entirely exposed (Fig. 112-12a), surface often bare; T6 with lateral arm of gradulus lamelliform and ending in strong tooth .... *Florilegus*
—. Basitibial plate with margin exposed, if at all, only posteriorly, surface covered with hair (Fig. 112-12b); T6 with gradulus variable, usually not ending in tooth 29
29(28). Scutellum half as long as scutum, flattened, projecting well beyond metanotum over whole propodeum, profile of which is vertical .... *Platysvastra*

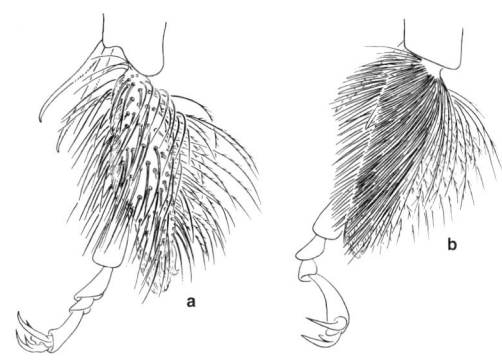

Figure 112-11. Inner views of hind tarsi of females. **a,** *Peponapis pruinosa* (Say); **b,** *Melissodes desponsa* Smith. From Michener, McGinley, and Danforth, 1994.

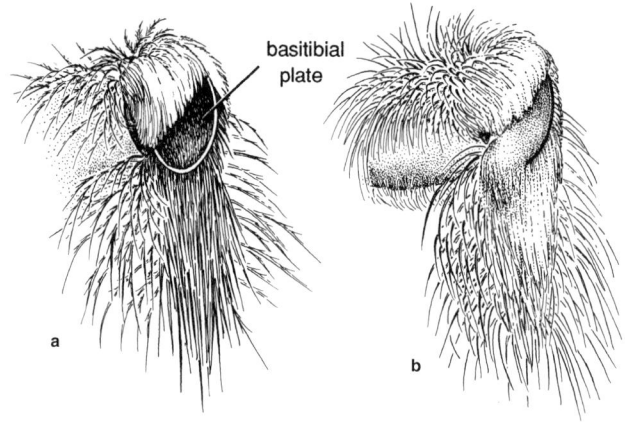

Figure 112-12. Basitibial plates of female Eucerini. **a,** *Florilegus condignus* (Cresson); **b,** *Tetraloniella albata* (Cresson). From Michener, McGinley, and Danforth, 1994.

—. Scutellum nearly always less than half as long as scutum, rarely flattened, not projecting beyond metanotum and over propodeum, profile of which has subhorizontal to steeply sloping basal zone 30

30(29). Arolia absent; flagellum thicker than ocellar diameter, first segment about one and one-half times as long as thick (gradulus of T6 with lateral part lamelliform, ending in strong tooth; maxillary palpus five-segmented) 31

—. Arolia present; flagellum little if any thicker than ocellar diameter, first segment over twice as long as thick (this antennal character not verified for *Platysvastra*) 32

31(30). Hairs of prementum sparse, simple; first flagellar segment slender at base, twice as long as greatest width .... *Svastrina*

—. Hairs of prementum dense, undulate, or bent at apices; first flagellar segment robust from base, about 1.5 times as long as greatest width .... *Pachysvastra*

32(30). Maxillary palpus two- or three-segmented; gradulus of T6 with lateral part elevated and terminating in strong tooth; stigma usually longer than prestigma (Fig. 112-4c); second submarginal cell distinctly longer than broad .... *Melissoptila*

—. Maxillary palpus four- to six-segmented, rarely three-segmented; gradulus of T6 with lateral part weak, cariniform, or, *if* lamelliform, then not ending in tooth or with tooth low and usually rounded; stigma usually shorter than prestigma (Fig. 112-4b); second submarginal cell variable, usually shorter than or as long as broad............33

33(32). Clypeus strongly protuberant, extending forward for distance greater than width of eye as seen in lateral view; blade of galea about 1.5 times as long as eye .... *Santiago*

—. Clypeus protuberant for less than width of eye as seen in lateral view; blade of galea not much if any longer than eye 34

34(33). Maxillary palpus four-segmented, rarely three-segmented; anterior coxa without spine; paraocular carina, when lower part is visible, independent of lateroclypeal carina; anterolateral margin of tegula straight or gently concave (Fig. 112-8a).. *Melissodes*

—. Maxillary palpus usually five- or six-segmented, rarely four-segmented; anterior coxa with large apical spine (Fig. 112-13c); paraocular carina, when lower part is visible, connected to lateroclypeal carina (except in *Svastra*); lateral margin of tegula continuously convex (Fig. 112-8b) 35

35(34). Basal parts of T2 and T3 with dense pilose bands, in strong contrast to remainder of these terga; T6 with lateral parts of gradulus lamellate or strongly carinate; maxillary palpus five-segmented 36

—. Basal parts of T2 and T3 without dense basal bands of pubescence or with median or apical bands as well (very broad basal bands only, in some *Gaesischia*); T6 with lateral parts of gradulus absent, cariniform, or occasionally lamellate; maxillary palpus four-, five-, or six-segmented 37

36(35). Pubescent bands of T2 and T3 with spatuloplumose hairs (Fig. 112-10b); eyes converging below; clypeus closely approaching eye; lateroclypeal carina not connected to paraocular carina .... *Svastra*

—. Pubescent bands of T2 or T3 or both with plumose hairs; eyes subparallel; clypeocular distance about half flagellar width or more; lateroclypeal carina connected to paraocular carina when latter is traceable............ *Svastrides*

37(35). Claws very small, outer rami little exceeding inner teeth; S6 deeply and narrowly notched at apex; middle basitarsus less than two-thirds as long as tibia .... *Micronychapis*

—. Claws normal, outer rami much exceeding inner teeth (latter rarely absent); S6 only slightly notched or emarginate; middle basitarsus about as long as tibia 38

38(37). Scutellum more than one-third as long as scutum, flattened and nearly on same plane as posterior part of scutum; clypeus extraordinarily flat; middle and posterior tibial spurs robust and almost as long as basitarsi .... *Gaesochira*

—. Scutellum less than one-third as long as scutum, convex; clypeus moderately to strongly protuberant; tibial spurs normal, two-thirds as long as basitarsi or less 39

39(38). Vertex elevated behind ocelli, which are therefore on anterior surface; paraocular carina strong; anterior coxa usually with strong apical spine (Fig. 112-13c); maxillary palpus four- or five-segmented, rarely six-segmented .... *Gaesischia* (in part)

—. Vertex not elevated behind ocelli, which are dorsal; paraocular carina variable, weak if vertex slightly elevated

behind ocelli; anterior coxa without strong apical spine; maxillary palpus six-segmented, occasionally five-segmented 40
40(39). Gradulus of T6 with lateral parts long and strongly carinate or lamellate .... *Gaesischia* (in part)
—. Gradulus of T6 with lateral parts usually absent or very short (indicated as a long but weak carina only in some *Alloscirtetica*) 41
41(40). Mandible with apex strongly bidentate (perhaps sometimes simple due to wear); first flagellar segment nearly as long as scape (Fig. 112-8d); maxillary palpus five-segmented .... *Martinapis*
—. Mandible with apex simple; first flagellar segment clearly shorter than scape; maxillary palpus six- or rarely five-segmented .... *Alloscirtetica*

Key to the North and Central American Genera of the Eucerini (Females)

(Often, hairs must be removed to see the first character in couplet 5(4). In *Melissodes stearnsi* Cockerell, although the tegula is shaped much as in other *Melissodes*, the relevant tegular margin is feebly convex; this species runs to couplet 19 and fails to agree with either alternative.)

1. Gradulus of S2 weakly biconvex (Fig. 112-5a); mandible with condylar ridge expanded forward, at least as salient as and usually more salient than outer ridge (Fig. 112-5c); gradulus of T6 without lateral parts; labrum two-thirds as long as broad or longer (tropical) .... *Thygater*
—. Gradulus of S2 strongly biconvex (Fig. 112-5b), forming angle of 140° or less between two convexities; mandible normal, with condylar ridge less salient than outer ridge (Fig. 112-5d); gradulus of T6 usually with lateral parts; labrum usually less than two-thirds as long as broad 2
2(1). Apical clypeal margin trilobed, with median lobe short, broad, and often slightly emarginate (Fig. 112-13d) (eastern and central USA) .... *Cemolobus*
—. Apical clypeal margin truncate 3
3(2). Inner margin of mandible with tooth near base (Fig. 112-13e) .... *Xenoglossa*
—. Inner margin of mandible without basal tooth (Fig. 112-13f) 4
4(3). Posterior basitarsus with inner surface sparsely hairy except for narrow band of dense hairs near lower margin ((Fig. 112-11a) .... *Peponapis*
—. Posterior basitarsus with inner surface uniformly densely hairy or more densely hairy near lower margin (Fig. 112-11a) 5
5(4). Tegula narrowed anteriorly (Fig. 112-8a), lateral margin slightly concave or straight in anterior half or less; maxillary palpus usually four-segmented, rarely three- or five-segmented 6
—. Tegula not narrowed anteriorly, lateral margin convex (Fig. 112-8b); maxillary palpus three- to six-segmented 7
6(5). Mandible simple or scarcely notched at apex, widest preapical part less than three-fourths as wide as base (Fig. 112-14a); last antennal segment much less than twice as long as wide .... *Melissodes*
—. Mandible strongly notched and therefore bilobed at apex (but often worn, so that this structure is lost), ex-

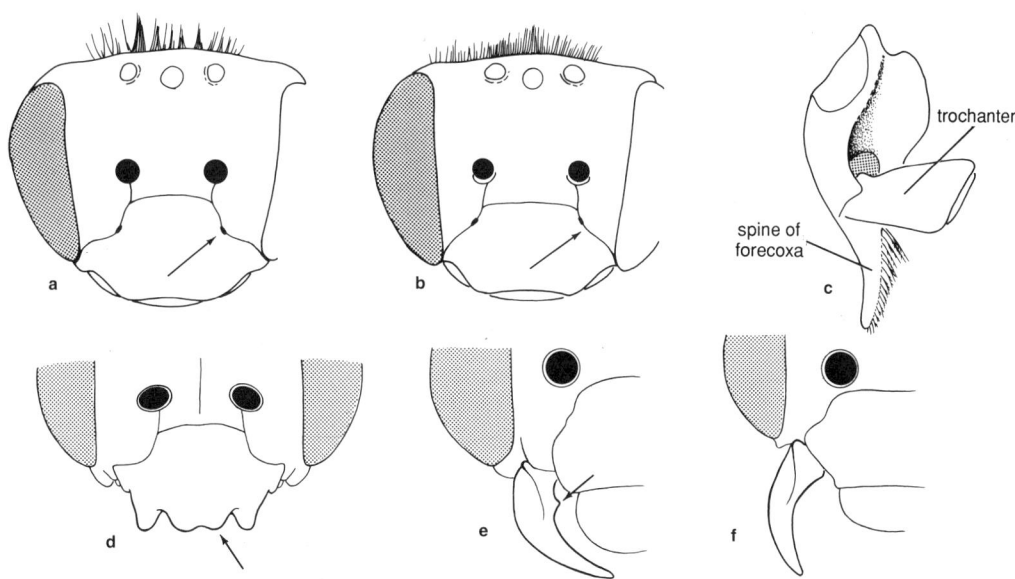

Figure 112-13. Eucerini. **a, b,** Faces of females of *Gaesischia (Gaesischiana) exul* Michener, LaBerge, and Moure and *Tetraloniella (Tetraloniella) albata* (Cresson), showing the angle in the epistomal suture at the tentorial pit (arrows) in the former; **c,** Fore coxa and trochanter of female of *Gaesischia (Gaesischiopsis) flavoclypeata* Michener, LaBerge, and Moure, showing the coxal spine; **d,** Lower face of male of *Cemolobus ipomoeae* (Robertson), showing the lobate clypeus; **e, f,** Mandible of female of *Xenoglossa strenua* (Cresson) and *Peponapis pruinosa* (Say), showing basal tooth (arrow) of the former. From Michener, McGinley, and Danforth, 1994.

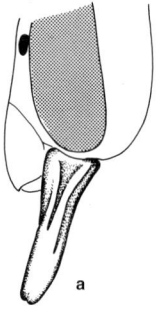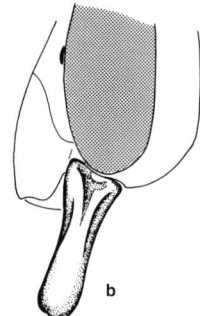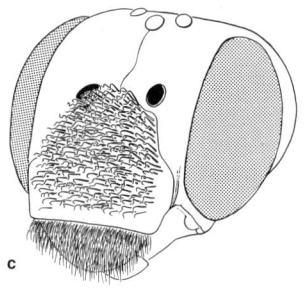

Figure 112-14. Structures of female Eucerini. **a, b,** Mandibles of *Melissodes agilis* Cresson and *Martinapis luteicornis* (Cockerell); **c,** Face of *Tetraloniella (Pectinapis)* sp., showing hooked clypeal hairs. From Michener, McGinley, and Danforth, 1994.

panded apically and preapical part thus nearly as wide as base (Fig. 112-14b); last antennal segment about twice as long as broad (deserts of southwestern USA and Mexico) .... *Martinapis*
7(5). Scopal hairs simple or with minute barbs 8
—. Scopal hairs with branches 14
8(7). Clypeal hairs short, erect, robust especially basally, bristle-like, and apically hooked or wavy (Fig. 112-14c) .... *Tetraloniella (Pectinapis)*
—. Clypeal hairs slender, not hooked, not bristle-like 9
9(8). Pale pubescent bands of metasomal terga with abundant, basally plumose, apically spatulate hairs; maxillary palpus four-segmented .... *Svastra (Anthedonia)*
—. Pale pubescent bands of metasomal terga without spatuloplumose hairs; maxillary palpus five- or six-segmented, rarely four-segmented 10
10(9). Minimum oculoclypeal distance much greater than minimum width of first flagellar segment (Fig. 112-6f); clypeus strongly protuberant; stipes with long, dense, coarse, apically hooked or wavy hairs (Mexico) .... *Tetraloniella (Loxoptilus)*
—. Minimum oculoclypeal distance not greater than minimum width of first flagellar segment; clypeus variable; stipes without area of hooked hairs 11
11(10). Blade of galea longer than eye; clypeus protuberant 12
—. Blade of galea not longer than eye; clypeus flat to slightly protuberant (slightly less so than in male) 13
12(11). Middle ocellus not as broad as flagellum [or, in *Eucera venusta* (Cresson), as wide as flagellum]; maxillary palpus six-segmented; pygidial plate rather broad, rounded apically, apicolateral margin convex (except in *E. venusta* and others) .... *Eucera (Synhalonia)*
—. Middle ocellus broader than flagellum; maxillary palpus five-segmented; pygidial plate tapering and bluntly pointed apically, apicolateral margin concave (deserts of southwestern USA and Mexico) .... *Simanthedon*
13(11). Scopal hairs with minute barbs; clypeus with margin indented at anterior tentorial pit to form almost right-angular notch (Fig. 112-13a) (tropical to Arizona) .... *Gaesischia* (in part)
—. Scopal hairs simple; clypeus with margin at level of anterior tentorial pits straight or slightly concave (Fig. 112-13b) .... *Tetraloniella* (in part)
14(7). Fore coxa with inner apical hairy spine (Fig. 112-13c) (tropical to Arizona) .... *Gaesischia* (in part)
—. Fore coxa without spine... 15
15(14). Vertex elevated, median ocellus thus below summit in facial view; hairs of upper and outer parts of scopa with abundant, uniform, short branches, mostly with ten or more branches on each side of rachis and often with as many as 15; apical part of rachis extending beyond last branch usually shorter than average length of branches (Mexico, southwestern USA) .... *Syntrichalonia*
—. Vertex weakly elevated if at all, median ocellus thus near or on summit in facial view; scopal hairs mostly with six to eight branches on each side of rachis, rarely with as many as ten; apical part of rachis long, extending beyond last branch by at least average length of branches 16
16(15). Tibial spurs weak, on middle leg less than half as long as tibia, as measured from base of spur to anterior tibiofemoral articulation; lateral arm of hypostomal carina prominent, sublamelliform; T2 and T3 with short, dense, white pubescence in broad basal bands, with short, relatively simple, dark, appressed hairs from basal bands almost to apices of terga (Baja California, California) .... *Agapanthinus*

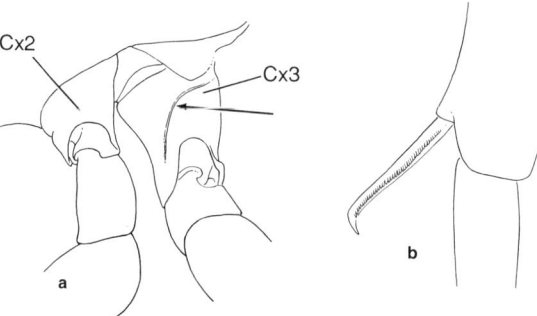

Figure 112-15. Structures of females of Eucerini. **a,** Middle and hind coxae of *Melissoptila (Ptilomelissa)* sp., the lateral hind coxal carina marked by an arrow; **b,** Middle tibial spur of *Svastra (Idiomelissodes) duplocincta* (Cockerell). From Michener, McGinley, and Danforth, 1994.

—. Tibial spurs strong, on middle leg more than half as long as tibia; lateral arm of hypostomal carina weak, cariniform; T3 and usually T2 without basal pale pubescent bands, or with distal pale band in addition, or entirely covered by pale pubescence 17

17(16). Prestigma shorter than stigma (as in Fig. 112-4c); lateral hind coxal carina sharp, bent strongly posteriad basally to form a rounded angle of almost 90° (Fig. 112-15a); maxillary palpus two- or three-segmented (tropical to Texas) .... *Melissoptila*

—. Prestigma as long as or longer than stigma (as in Fig. 112-4b); lateral hind coxal carina absent or reduced to short apical portion, straight or only slightly curved toward rear; maxillary palpus four- to six-segmented 18

18(17). Maxillary palpus usually four-segmented, but *if* five-segmented, then basal pubescent band of T2 with at least a few spatuloplumose hairs (Fig. 112-10b); lateral arms of gradulus of T6 lamelliform, often with a small tooth .... *Svastra*

—. Maxillary palpus five- or six-segmented; basal pubescent band of T2 without spatuloplumose hairs; lateral arms of gradulus of T6 cariniform to lamelliform 19

19(18). Basitibial plate with margin entirely exposed, surface often bare; T6 with lateral parts of gradulus lamelliform and ending in strong tooth .... *Florilegus*

—. Basitibial plate with anterior and apical part of margin hidden, surface usually hairy; T6 with lateral parts of gradulus cariniform, never toothed if sublamelliform .... *Tetraloniella* (in part)

Key to the North and Central American Genera of the Eucerini (Males)

(*Peponapis timberlakei* Hurd and Linsley runs *either* to *Svastra* in couplet 14, except that the last antennal segment is not tapering and acuminate, *or* to couplet 15, except that the last flagellar segment is more than twice as long as broad.)

1. Minimum length of malar area greater than minimum width of first flagellar segment (Fig. 112-6e); pygidial plate unrecognizable or indicated by weak lateral carinae, mostly covered by long, appressed hairs, T7 bidentate apically; labrum at least three-fourths as long as broad (tropical) .... *Thygater*

—. Minimum length of malar area equal to or less than minimum width of first flagellar segment; pygidial plate prominent, exposed, with short hairs or bare, T7 truncate or rounded apically; labrum variable but usually less than three-fourths as long as broad 2

2(1). Clypeal margin trilobed, median lobe broad and often shallowly emarginate medially (Fig. 112-13d); first flagellar segment as long as second segment; S6 with large, laterally directed, lateral tooth (eastern and central USA) .... *Cemolobus*

—. Clypeal margin truncate; first flagellar segment usually shorter than second segment; S6 usually without lateral teeth 3

3(2). T7 with lateral gradular tooth or strong angle on each side of pygidial plate, sometimes hidden in dense hair or by T6 4

—. T7 without lateral teeth (occasionally S6 with lateral teeth that can be seen from above and may be confused with tergal teeth) 7

4(3). Stigma slightly longer than prestigma (Fig. 112-4c); maxillary palpus two- or three-segmented; lateral hind coxal carina prominent, curved (Fig. 112-15a) (tropical to Texas) .... *Melissoptila*

—. Stigma small, usually as short as or shorter than prestigma (Fig. 112-4b); maxillary palpus usually four- or five-segmented, rarely three-segmented; lateral hind coxal carina reduced or absent 5

5(4). S6 with prominent median convexity; fore femur broadest about one-fourth or one-third of its length from apex; maxillary palpus five-segmented .... *Florilegus*

—. S6 flat or with exceedingly shallow, longitudinal median depression; fore femur broadest basal to middle; maxillary palpus usually three- or four-segmented, rarely five-segmented 6

6(5). Tegula narrowed anteriorly, lateral margin slightly concave or straight in anterior one-half or one-third (Fig. 112-8a) (often hidden by hairs); clypeus little or moderately protruding, extending in front of eye by eye width or less as seen in lateral view .... *Melissodes*

—. Tegula not narrowed anteriorly, lateral margin continuously convex (Fig. 112-8b); clypeus strongly protuberant, extending in front of eye by more than eye width as seen in lateral view .... *Peponapis* (in part)

7(3). First flagellar segment longer than second; inner margin of mandible with tooth near base (Fig. 112-13e) .... *Xenoglossa*

—. First flagellar segment no longer than second segment and often much shorter (longer in *Peponapis timberlakei*, Hurd and Lindsey); inner margin of mandible without tooth near base 8

8(7). Clypeus strongly protuberant, abruptly beveled and snoutlike apically (Fig. 112-6g), its profile forming distinct preapical angle and concave above angle (rare, southwestern deserts) .... *Simanthedon*

—. Clypeus uniformly convex or straight in profile 9

9(8). Tibial spurs weak, middle tibial spur about half or less than half as long as tibia, as measured from base of spur to anterior tibiofemoral articulation 10

—. Tibial spurs strong, middle tibial spur more than half as long as tibia 13

10(9). First flagellar segment only slightly shorter than second (Fig. 112-8d); last flagellar segment tapering to apex; flagellum bright yellow (deserts of southwestern USA and Mexico) .... *Martinapis*

—. First flagellar segment half as long as second or less (Fig. 112-8e); last flagellar segment not tapering; flagellum tan to black 11

11(10). Last flagellar segment with short, pointed, hooked apex twisted slightly laterad (Baja California, California) .... *Agapanthinus*

—. Last flagellar segment with rounded apex 12

12(11). Hind basitarsus flattened, shining, largely hairless on outer surface; distal two flagellar segments often slightly compressed (tropical to Arizona) .... *Gaesischia*

—. Hind basitarsus normal, hairy; distal two flagellar segments not compressed (central and western USA and Mexico) .... *Tetraloniella* (in part)

13(9). Maximum length of first flagellar segment as great as or slightly greater than minimum length of second segment; last flagellar segment at least twice as long as broad 14

—. Maximum length of first flagellar segment usually much

less than length of second segment or, *if* about the same, then last flagellar segment less than twice as long as broad and rounded apically 15
14(13). S1 with hairy median convexity (Fig. 112-10a), profile thus convex; last flagellar segment rounded apically (Mexico, southwestern USA) .... *Syntrichalonia*
—. S1 relatively flat, profile straight and lacking prominent median eminence; last flagellar segment tapering and acuminate apically .... *Svastra (Anthedonia)*
15(13). Maxillary palpus usually four-segmented, but *if* five-segmented, then T2 with basal pubescent band at least a few hairs of which are basally plumose and apically spatulate (Fig. 112-10b) .... *Svastra*
—. Maxillary palpus five- or six-segmented, and T2 without spatuloplumose hairs 16
16(15). Blade of galea twice as long as eye or longer (Fig. 112-6a); clypeus strongly protuberant; lower part of paraocular carina prominent; antenna long, reaching stigma or beyond in repose (Mexico) .... *Tetraloniella (Loxoptilus)*
—. Blade of galea 1.5 times as long as eye or shorter; clypeus variable, often flat; paraocular carina variable, lower part often obsolete; antenna variable in length 17
17(16). Oculoclypeal distance extremely short (Fig. 112-6b), not more than about one-fourth of minimum width of first flagellar segment .... *Tetraloniella* (in part)
—. Oculoclypeal distance short to long, equal to one-third of minimum width of first flagellar segment or more 18
18(17). S6 with oblique lateral apical carina straight, sternum not toothed or angled laterally (Fig. 112-10d); antenna of moderate length, not reaching stigma in repose .... *Peponapis* (in part)
—. S6 almost always with oblique lateral apical carina curved outward and thickened basally, ending in a lateral blunt tooth or obtuse angle of sternum (Fig. 112-10e); antenna usually long, reaching stigma or beyond in repose 19
19(18). Profile of clypeus distinctly convex .... *Eucera (Synhalonia)*
—. Profile of clypeus nearly straight, in spite of rather strong protuberance (Mexico) .... *Tetraloniella (Pectinapis)*

Key to the Genera of the Eucerini of the Eastern Hemisphere

1. Forewing with two submarginal cells (Fig. 112-4a) 2
—. Forewing with three submarginal cells 3
2(1). Maxillary palpus six-segmented, very rarely five-segmented; first submarginal cell of front wing smaller than second (Fig. 112-4a), and distance from first submarginal crossvein to first recurrent vein greater than distance from second recurrent to second submarginal crossvein; first segment of flagellum of male short, 0.9 to 2.5 times as long as broad, second segment 1.7 to 5.0 times as long as first; gradulus of S6 of male arcuate; medial area of S7 of male more or less flat (palearctic) .... *Eucera* (in part)
—. Maxillary palpus three- or four-segmented; first submarginal cell of front wing equal to second, and distance from first submarginal crossvein to first recurrent vein about equal to that from second recurrent to second submarginal crossvein; first segment of flagellum of male long, 3.3 to 4.0 times as long as broad, second segment 0.7 to 1.2 times as long as first; gradulus of S6 of male sharply angulate to rear near lateral extremity (sometimes weakly so in subgenus *Cubitalia* s. str.); medial area of S7 of male produced downward posteriorly (palearctic) .... *Cubitalia*
3(1). Maxillary palpus three- to four-segmented or with minute fifth segment, often not longer than maximum width of galeal blade; keirotrichiate area of female apically at least half as wide as hind tibia and extending basad beyond middle of tibia; scopa consisting of sparse, strongly plumose hairs on outer surface of tibia, each arising from tubercle separated from its neighbors by full width of tubercle .... *Tetralonia*
—. Maxillary palpus five- or six-segmented, often longer than maximum width of galeal blade; keirotrichiate area and associated finely and closely punctate surface of female usually less than half as wide as hind tibia and extending basad only to middle of tibia, rarely absent; scopa consisting of denser hairs, often simple, *if* plumose, then arising from tubercles less widely separated from one another 4
4(3). Blade of galea more than twice as long as eye; first flagellar segment of male much longer than scape and as long as second flagellar segment, that of female about as long as scape and distinctly longer than second and third flagellar segments taken together (Turkmenistan).......... .... *Notolonia*
—. Blade of galea shorter; first flagellar segment of male shorter than scape and much shorter than second flagellar segment, that of female shorter than scape and not or scarcely longer than second and third flagellar segments taken together 5
5(4). S6 of male with the two converging carinae each usually angulate near anterior end, forming lateral angle usually reflecting marginal sternal angle or lobe (Fig. 112-10e), which is often strengthened by lateral branch carina arising at angle of converging carina, sometimes carinae absent except for branch and mesally directed anterior end of converging carina, together forming transverse carina directed anteromesally from marginal sternal angle or lobe; clypeus usually protuberant for width of eye or more as seen in lateral view; blade of galea 1.2 or more times length of eye; middle femur of male on undersurface commonly with area of dense, appressed, red-brown hair hiding surface .... *Eucera (Synhalonia)*
—. S6 of male with the two converging carinae each usually simple and more or less straight; clypeus usually protuberant for less than width of eye as seen in lateral view; blade of galea usually not much longer than length of eye; middle femur of male sometimes with area of sparse, appressed whitish hair not hiding surface, more commonly with unspecialized sloping or erect hair .... *Tetraloniella*

The genus *Ulugombakia* Baker (known only in the female) runs to 3 in the Key to the Genera of Eucerini of the Eastern Hemisphere. If one ignores characters of males and the number of maxillary palpal segments (justified by the palpal length even though there are only four segments), it runs to *Tetraloniella,* from which it can be distinguished by characters cited in the account of the genus.

Genus *Agapanthinus* LaBerge

Agapanthinus LaBerge, 1957: 35. Type species: *Melissodes callophila* Cockerell, 1923, by original designation.

Although *Melissodes*-like in general appearance, 9 to 11 mm in length, *Agapanthinus* is not closely related to *Melissodes*. The small midtibial spur and modified apex of the male antenna (see the key to genera) suggest a relation to *Gaesischia*. Other characters include the large and complex apical lobes of S7 of the male, somewhat as in *Gaesischia*. The male gonostylus is unique among Eucerini, being shorter than the apical process of the gonocoxite, almost as broad as long, tapering, and bearing long hairs. Male genitalia and hidden sterna were illustrated by LaBerge (1957).

■ *Agapanthinus* occurs in the desert of Baja California and California. The single species is *A. callophilus* (Cockerell).

Genus *Alloscirtetica* Holmberg

In this South American genus the pygidial plate of the male is reduced, at least the lateral carinae ending before the apex of the pygidial area and the plate often unrecognizable or ending in a point. Elsewhere in the Eucerini such reduction of the male pygidial plate occurs in *Thygater*, a very different genus. S7 of the male nearly always has an unpaired median projection between the apical lobes, which are rather large, delicate, hairy, and elaborate (Fig. 112-7c). S6 of the male lacks the carinae common thereon in many Eucerini. Except for the subgenus *Megascirtetica*, which looks like a large black *Thygater*, the species of *Alloscirtetica* look rather like *Melissodes*. The six-segmented maxillary palpi of most species resulted in many of them being described in the genus *Tetralonia*.

Alloscirtetica is largely restricted to temperate and Andean South America, with a few species occurring in tropical Brazil. Urban (1971, 1982) revised the genus and provided keys to its species, except for *Megascirtetica*. Michener, LaBerge, and Moure (1955a), Moure and Michener (1955b), and Urban (1971, 1977) recognized up to four subgenera in *Alloscirtetica* (not including *Megascirtetica*). In those works *Alloscirtetica* s. str. consisted simply of those species that did not fall into any other subgenus. As more species became known, it became increasingly difficult to recognize the subgenera, and Urban (1982), after studying all known species, decided that the subgenera could not be maintained. I agree with her decision.

Megascirtetica has hitherto been regarded as a genus related to *Alloscirtetica*. Its distinctive features, however, are all probably derived; it is thus a specialized derivative of *Alloscirtetica*, and its relationships are best indicated by its inclusion in that genus.

Key to the Subgenera of *Alloscirtetica*

1. Claws of female simple; arolia unusually small; mandible with large, right-angular preapical tooth separated from apex by deep notch; head strongly elevated behind ocelli; body length over 15 mm .... *A. (Megascirtetica)*
—. Claws of female with inner tooth; arolia of ordinary size; mandible simple, preapical tooth vestigial or absent; head not or weakly elevated behind ocelli; body length 12 mm or less .... *A. (Alloscirtetica s. str.)*

Alloscirtetica / Subgenus *Alloscirtetica* Holmberg s. str.

Scirtetica Holmberg, 1903: 389 (not Saussure, 1884). Type species: *Scirtetica antarctica* Holmberg, 1903, monobasic.

Alloscirtetica Holmberg, 1909a: 77, replacement for *Scirtetica* Holmberg, 1903. Type species: *Scirtetica antarctica* Holmberg, 1903, autobasic.

Neoscirtetica Schrottky, 1913a: 256, replacement for *Scirtetica* Holmberg, 1903. Type species: *Scirtetica antarctica* Holmberg, 1903, autobasic.

Holmbergiapis Cockerell, 1918b: 36, replacement for *Scirtetica* Holmberg, 1903. Type species: *Scirtetica antarctica* Holmberg, 1903, autobasic.

Alloscirtetica (Dasyscirtetica) Michener, LaBerge, and Moure, 1955a: 218. Type species: *Tetralonia gilva* Holmberg, 1884, by original designation.

Alloscirtetica (Ascirtetica) Moure and Michener, 1955b: 260. Type species: *Eucera herbsti* Friese, 1906, by original designation.

Alloscirtetica (Scirteticops) Moure and Michener, 1955b: 261. Type species: *Tetralonia gayi* Spinola, 1851, by original designation.

This is a highly variable taxon having a generally *Melissodes*-like form, although its members are highly variable in hair color. The body length is 8 to 12 mm. Some of the characters of males that vary, and that led to recognition of four subgenera in what is here called *Alloscirtetica* s. str., are enumerated below. Females are less variable and the subgenera recognized on the basis of males were not easily (or at all) separable among females. T7 of males may be rounded, the dorsal surface almost without evidence of a pygidial plate, but usually there is at least a well-defined bare area, often margined by carinae, these margins converging posteriorly to a point on or in front of the tergal margin. In other species the carinae do not meet, and sometimes the apical part of a subtruncate pygidial plate is recognizable, but the carinae do not extend to its apex. Urban (1982) diagrammed the variation in T7 of males. The length of the first flagellar segment of males varies from very short to almost two-thirds as long as the second segment. Urban (1982) illustrated this variation also. The posterior claws are sometimes asymmetrical. In the form formerly placed in *Scirteticops,* the profile of the propodeum is wholly declivous, whereas in others there is a sloping zone behind the metanotum. The jugal lobe of the hind wing is only about half as long as cell Cu in the former *Scirteticops,* whereas it is variably longer in others. In spite of such variation, *Alloscirtetica* s. str. is a distinctive, unified taxon. Male genitalia and hidden sterna as well as other characters were illustrated by Michener, LaBerge, and Moure (1955a), Moure and Michener (1955b), and Urban (1971, 1977, 1982). The hairs of the male S7 are particularly elaborate, and Urban illustrated many of them.

■ This subgenus ranges from Magallanes, Chile, and the provine of Chubut, Argentina, north to Pará, Brazil, and in the Andean uplift to Ecuador. Most species occur in Chile and the xeric parts of Argentina. The 36 species were revised by Urban (1971, 1982); the species from Chile by Vivallo (2003).

Alloscirtetica / Subgenus *Megascirtetica* Moure and Michener

Megascirtetica Moure and Michener, 1955b: 264. Type species: *Macrocera mephistophelica* Schrottky, 1902, by original designation. [New status.]

This subgenus consists of a large (17 mm long), black-haired species superficially resembling a large black *Thygater*. Characters (other than those in the key to subgenera) that distinguish *Megascirtetica* from *Alloscirtetica* s. str. are the closely spaced ocelli, the interocellar distance being about half of the ocellocular distance, and the clear inner apical lobe of the male gonocoxite. Male genitalia and hidden sterna were illustrated by Moure and Michener (1955b); see also Figure 112-7a-c.

■ *Megascirtetica* is known only from the state of São Paulo, Brazil. The only species is *Alloscirtetica mephistophelica* (Schrottky).

Genus *Canephorula* Jörgensen

Corbicula Friese, 1908b: 59, also 1908c: 170 (not Megerle, 1811). Type species: *Corbicula apiformis* Friese, 1908, monobasic.

Canephora Friese, 1908b: 94 (not Hübner, 1822), replacement for *Corbicula* Friese, 1908. Type species: *Corbicula apiformis* Friese, 1908, autobasic.

Canephorula Jörgensen, 1909: 212, replacement for *Canephora* Friese, 1908. Type species: *Corbicula apiformis* Friese, 1908, autobasic.

Canephorula is the only genus of the subtribe Canephorulina. It is pallid- or tan-haired, the metasoma largely covered with appressed hair, the body length 10.0 to 13.5 mm. Its most noteworthy character is the bare area or corbicula on the outer surface of the distal half of the hind tibia of the female. (The genus *Hamatothrix* has a similar but much smaller corbicula.) These are the only bees other than the corbiculate tribes of Apinae that have such a structure. In *Canephorula* the posterior margin of the corbicula is formed by a row of long, close-set setae forming a strong comb; otherwise, the margin is formed by long scopal hairs, some of which curve over and partly obscure the corbicula. Some characters are indicated in the account of Eucerini and in the key to South American genera. Other characters are the black face of the male; male antennae reaching the scutellum, the first flagellar segment shorter than the second; the broad midapical notch in the clypeus of both sexes, exposing the labral base; the absence of arolia (arolia are present in all other Eucerini except *Pachysvastra* and *Svastrina*); the enlarged and blunt inner claw of the hind leg; and the presence of basitibial and pygidial plates in the male. T7 lacks the distal lobes found in other Eucerini. The gonostylus consists of three long lobes, quite unlike that in all other Eucerini. As in some Eucerina, the stigma is very short, broader than long (Fig. 112-16). Vein cu-v of the hind wing is longer than that in other Eucerini. Details of the corbicula, genitalia, and other structures were illustrated by Friese (1920) and Michener, LaBerge, and Moure (1955b).

■ This genus is found in xeric areas of Argentina from Salta to Mendoza province. The only species is *Canephorula apiformis* (Friese).

The nesting and floral biology of *Canephorula apiformis* (Friese) was described by Michelette, Camargo, and Rozen (2000). The mature larva resembles that of other Eucerini, thus supporting the placement of *Canephorula* in the tribe Eucerini.

Genus *Cemolobus* Robertson

Cemolobus Robertson, 1902b: 324. Type species: *Xenoglossa ipomoeae* Robertson, 1891, by original designation.

The lobate clypeal apex (Fig. 112-13d) and lateral tooth of the male S6, as described in the key to genera, are unique features of this genus. In appearance, it suggests a *Peponapis* or *Xenoglossa (Eoxenoglossa)*, 13 to 16 mm long. The hairs of the inner surface of the hind basitarsus of the female are not as sparse as those in most *Peponapis* and in *Xenoglossa*, but suggest *P. fervens* (Smith) of South America. The long first flagellar segment of the male suggests *Xenoglossa*. The male genitalia and hidden sterna were illustrated by LaBerge (1957) and Mitchell (1962).

■ *Cemolobus* occurs from Pennsylvania to Georgia, west to Missouri. The single species is *C. ipomoeae* (Robertson), an oligolege on flowers of *Ipomoea* (Convolvulaceae).

Genus *Cubitalia* Friese

Cubitalia and *Eucera* are the two Old World eucerine genera that have only two submarginal cells. It is not clear whether the loss of a submarginal crossvein, resulting in only two cells, is homologous or not in the two genera; I am inclined to believe that they are not closely related, although Pesenko and Sitdikov (1988) show *Cubitalia* arising from a paraphyletic *Eucera*. The three subgenera of *Cubitalia* are quite different from one another, but their relationships are quite clear, as shown by Pesenko and Sitdikov; I have chosen to emphasize this relationship by including all of them in the genus *Cubitalia*. Except for size, there are no known subgeneric characters for females.

Risch (1999) placed *Cubitalia* as a subgenus of *Eucera* and recognized as other subgenera several of the taxa syn-

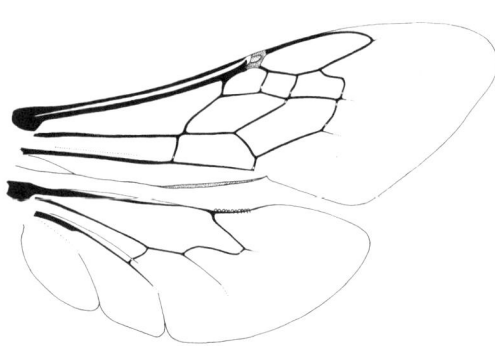

Figure 112-16. Wings of *Canephorula apiformis* (Friese).

onymized by Michener (2000) under *Eucera (Hetereucera)*. (Note that *"Donatica,"* which appears in Risch's abstract as though it were a genus, is in fact a specific name, the type species of *Opacula*.)

Key to the Subgenera of *Cubitalia* (Males Only)
1. Postgradular area of S5 with two tufts of bristly hairs directed laterally; front basitarsus slightly curved, of normal length, about 1.8 times as long as three following segments taken together; gonostylus sharply curved medially, with pointed ventral lamellar projection; clypeus gently convex in profile, lower part extended down and facing forward; middle tibia slightly curved ....*C. (Opacula)*
—. Postgradular area of S5 more or less uniformly pubescent; front basitarsus straight, shortened, 1.2 to 1.5 times length of three following segments taken together; gonostylus broadened at apex, not bent medially, without ventral projection; clypeus obtusely angulate in profile, lower part curved back and facing somewhat downward; middle tibia not curved 2
2(1). Mandible longer than eye, bidentate, acetabular carina not forming preapical tooth; inner orbits diverging downward; antenna reaching propodeum; front margins of middle and hind basitarsi with broad excisions or emarginations; S7, as seen from beneath, with large convexity on each side of median carina ....*C. (Cubitalia s. str.)*
—. Mandible shorter than eye, acetabular carina ending in preapical tooth, mandibular apex thus three-toothed; inner orbits parallel; antennae long, reaching metasoma; front margins of middle and hind basitarsi unmodified; S7 without large ventral convexities .... *C. (Pseudeucera)*

Cubitalia / Subgenus *Cubitalia* Friese s. str.

Friese, 1911d: 136. Type species: *Eucera breviceps* Friese, 1911, monobasic.

Cubitalia s. str. consists of large (16-18 mm long), robust bees. The head of the male, and especially the mandibles, is large. Hidden sterna and other structures were illustrated by Tkalců (1984b) and Pesenko and Sitdikov (1988).
■ This subgenus occurs in the Balkans and Greece east to the Caucasus. The four species were revised by Tkalců (1984b).

Cubitalia / Subgenus *Opacula* Pesenko and Sitdikov

Opacula Pesenko and Sitdikov, 1988: 849. Type species: *Opacula donatica* Sitdikov, 1988, by original designation. [New status.]

This subgenus, like *Pseudeucera*, consists of species smaller than those of *Cubitalia* s. str., body length 13 mm. As in *Pseudeucera*, the male mandible is three-toothed, because the acetabular carina ends in a preapical tooth; and the middle and hind basitarsi do not have emarginations in their anterior margins. Illustrations of male genitalia, sterna, legs, etc., were provided by Pesenko and Sitdikov (1988).
■ *Opacula* is known only from Kirghizia, in Central Asia. The only species is *Cubitalia donatica* (Sitdikov).

Cubitalia / Subgenus *Pseudeucera* Tkalců

Pseudeucera Tkalců, 1978a: 157, 158. Type species: *Eucera parvicornis* Mocsáry, 1878, by original designation. [New status.]

In addition to the characters indicated in the key, *Pseudeucera* differs from *Opacula* in having relatively short, straight male gonostylus; the distal one-third is broadened, quite unlike the fanlike enlargement of the extreme apex in *Cubitalia* s. str. and the incurved, slender apex in *Opacula*. The gonostylus and various other structures were illustrated by Pesenko and Sitdikov (1988). The body length is 12 to 14 mm.
■ *Pseudeucera* occurs from Hungary, Italy, and Greece east to the Ukraine. The only species is *Cubitalia parvicornis* (Mocsáry), an oligolege on Boraginaceae (Müller, 1995).

Genus *Eucera* Scopoli

Eucera includes most of the palearctic Eucerini that have two submarginal cells, *Cubitalia* being the only other such group, as shown in the key to genera. *Eucera* also includes, however, many species with three submarginal cells, commonly placed in *Tetralonia* or *Synhalonia*, but here considered to constitute the subgenus *Synhalonia* of *Eucera*. The loss of the second submarginal crossvein, resulting in two instead of three submarginal cells, is widespread among bees and frequently is not a generic character. Given the variability in both two-celled *Eucera* and in three-celled *Synhalonia*, there is no character other than the number of submarginal cells that separates the two-celled from the three-celled groups. *Eucera* as usually recognized may be polyphyletic, its species derived from different three-celled ancestors. This possibility is supported by the similar variability of other characters within two-celled and three-celled groups. For example, in both groups the male gonostylus is sometimes elbowed and sometimes arcuate or nearly straight, suggesting that two-celled *Eucera* with the elbowed gonostylus may have arisen from *Synhalonia* with the elbowed gonostylus, and two-celled *Eucera* with the straight gonostylus, from comparable *Synhalonia*. The structure of the male gonostylus, however, is particularly useless for recognition of groups in this section of Eucerini, and appears to have led various authors astray. In the related genus *Tetraloniella*, within its *ruficornis* species group, there are species with gently curved gonostyli and others with weakly and strongly elbowed gonostyli (Tkalců, 1979b). On the basis of other characters, at least certain groups of *Eucera* do not appear to have immediate ancestors among *Synhalonia* species. For example, in the large subgenus *Eucera* s. str., the keirotrichiate area of the hind tibia of females is large, occupying most of the inner surface of the hind tibia, whereas in related Eucerini it is small, limited to the distal half of the tibia and occupying less than one-third of the tibial width; the large area is presumably plesiomorphic.

Evidence for the close relationship of two-celled *Eucera* to *Synhalonia* is found in the strongly protuberant clypeus, the long proboscis, such that the blade of the galea is conspicuously longer than the eye, and the lack of paraocular carinae or, if present, the carinae not reaching the carinate lateral margins of the clypeus. Further, S6 of the male usually has a lateral marginal projection commonly strengthened by the anterior end of one of the converging carinae of S6, or by a lateral branch from that carina, the main axis of which often bends mesad near its anterior end, producing an angle that is often produced laterally. Sometimes, in both two-celled and three-celled species, the convergent carina is largely absent, but its anterior end remains, combined with its lateral branch, as a transverse carina on each side of the sternum strengthening the lateral marginal projection of the sternum. With minor adjustments, these are all characters of *Eucera* that help to distinguish it from *Tetraloniella;* as noted elsewhere, I recognize *Tetraloniella* as a genus largely for historical reasons, awaiting proper studies of the species worldwide, for all of the characters that separate it from *Eucera (Synhalonia)* break down. *Synhalonia* and most two-celled *Eucera* resemble one another superficially; they are all large, and the males often lack pale metasomal hair bands.

The two large two-celled subgenera, *Eucera* s. str. and *Hetereucera*, which contain the great majority of two-celled species of Eucerini, are clearly congeneric with *Synhalonia*. The subgenera *Oligeucera*, which I have not seen, and *Pteneucera*, however, the two together comprising only five species, are quite different from the large subgenera and may deserve generic rank. The latter has paraocular carinae near the eye margins, reaching the lateral clypeal carinae or nearly so. I include them in *Eucera* only because it is not clear where else they might be placed. Neither seems to be a *Tetraloniella*.

Key to the Subgenera of *Eucera*

1. Forewing with three submarginal cells (holarctic) .... *E. (Synhalonia)*
—. Forewing with two submarginal cells (palearctic) 2
2(1). Males 3
—. Females 6
3(2). Gonostylus, as seen in lateral view, straight or gently arcuate .... *E. (Hetereucera)*
—. Gonostylus as seen in lateral view angled, i.e., base directed dorsoapically, then bent ventroapically 4
4(3). Hind femur with angular protuberance in middle of lower margin; body length 7.5 to 8.0 mm; first flagellar segment broader than long .... *E. (Oligeucera)*
—. Hind femur with middle of lower margin more or less straight; body length 10.0 to 16.5 mm; first flagellar segment longer than broad 5
5(4). Gonostylus with retrorse ventral spine; S4 and S5 with lateral areas of strong yellowish bristles; body length 10.0 to 12.5 mm .... *E. (Pteneucera)*
—. Gonostylus without ventral spine; S4 and S5 without lateral tufts of bristles (except in *Eucera interrupta* Bär); body length 12.0 to 16.5 mm .... *E. (Eucera s. str.)*
6(2). Hind tibia with keirotrichiate area occupying most of inner surface; scopal hairs limited to outer surface and margins of inner surface of tibia (scopal hairs simple; body length 12-17 mm) .... *E. (Eucera s. str.)*
—. Keirotrichiate area a longitudinal band or triangle less than half as wide as tibia and limited to distal half of tibia; scopal hairs occupying not only outer surface of tibia but also inner surface, except for keirotrichiate area 7
7(6). Lower half of clypeus largely yellow; S2 to S5 with twisted hairs; body length 7 mm (scopal hairs simple) .... *E. (Oligeucera)*
—. Clypeus entirely dark; S2 to S5 with straight hairs; body length 8 to 17 mm 8
8(7). Scopal hairs on outer surface of hind tibia coarsely branched .... *E. (Pteneucera)*
—. Scopal hairs simple (or, in *E. furfurea* Vachal, with very short branches) .... *E. (Hetereucera)*

Eucera / Subgenus *Eucera* Scopoli s. str.

Eucera Scopoli, 1770: 8. Type species: *Apis longicornis* Linnaeus, 1758, by designation of Latreille, 1810: 439.

Eucera s. str. is a distinctive subgenus of large bees (see the key to subgenera) with long and usually dark-colored male antennae, like those of *Synhalonia*. The large keiro-trichiate area of the hind tibia of the female, in particular, distinguishes *Eucera* s. str. from related Eucerini. In view of the distribution of keirotrichiae in other bees, the large area here may be a plesiomorphy, not an apomorphy as interpreted by Sitdikov and Pesenko (1988). Male genitalia, hidden sterna, and other structures were illustrated by Iuga (1958), Tkalců (1978a, 1984c), and Sitdikov and Pesenko (1988).

■ *Eucera* s. str. occurs from the Canary Islands, Spain, and Britain to Japan. Of over 130 named species of two-celled *Eucera*, only about 20, representing about 10 species, have been shown to belong to *Eucera* s. str., but about 50 of the proposed species names of *Eucera* probably relate to this subgenus.

Eucera / Subgenus *Hetereucera* Tkalců

Eucera (Hetereucera) Tkalců, 1978a: 167. Type species: *Eucera hispana* Lepeletier, 1841, by original designation.
Eucera (Pareucera) Tkalců, 1978a: 164. Type species: *Eucera caspica* Morawitz, 1873, by original designation.
Eucera (Stilbeucera) Tkalců, 1978a: 162. Type species: *Eucera clypeata* Erichson, 1835, by original designation.
Eucera (Atopeucera) Tkalců, 1984c: 71. Type species: *Eucera seminuda* Brullé, 1832, by original designation.
Eucera (Agatheucera) Sitdikov and Pesenko, 1988: 87. Type species: *Eucera bidentata* Pérez, 1887, by original designation.
Eucera (Hemieucera) Sitdikov and Pesenko, 1988: 88. Type species: *Eucera paraclypeata* Sitdikov, 1988, by original designation.
Eucera (Pileteucera) Sitdikov and Pesenko, 1988: 87. Type species: *Eucera cineraria* Eversmann, 1852 = ?*E. cinerea* Lepeletier, 1841, by original designation.
Eucera (Rhyteucera) Sitdikov and Pesenko, 1988: 87. Type species: *Eucera parvula* Friese, 1895, by original designation.

As here understood, *Hetereucera* is a second major subgenus of two-celled *Eucera*, *Eucera* s. str. being the first. Sitdikov and Pesenko broke it up into eight subgenera, seven of them arising in their phylogeny from a paraphyletic *Hetereucera* in a narrow sense. *Hetereucera* in that sense was highly variable, and although the seven other taxa were quite distinct and certainly represent recognizable groups, I have chosen to join them with *Hetereucera* as species groups. The distinguishing characters for these groups are mostly in the male genitalia and hidden sterna, although other characters exist and Sitdikov and Pesenko (1988) were able to give keys to females as well as males. Most *Hetereucera* species look superficially much like *Eucera* s. str. The distinguishing character for males indicated in the key to subgenera, i.e., the more or less straight gonostylus, is suspect because it varies among related species in other eucerine groups. Male genitalia, sterna, and other structures were illustrated by Tkalců (1978a, 1984c, 1993b), Sitdikov and Pesenko (1988) and Risch (2003).

■ *Hetereucera* occurs from the Canary Islands, Morocco, and Spain north as far as Hungary and east to Central Asia. Many specific names in the two-celled groups of *Eucera* have not been placed to subgenus; perhaps 60 names will fall in *Hetereucera*. About 35 have been properly placed in the subgenus as here understood. Risch (2001) revised the eight species of the group of *Pareucera*, which he recognized as a subgenus of *Eucera* rather than as a group within the subgenus *Hetereucera*. Risch (2003) revised the groups of *Stilbeucera*, *Atopeucera*, and *Hemieucera* (22 species), all recognized as subgenera of *Eucera*.

Eucera / Subgenus *Oligeucera* Sitdikov and Pesenko

Eucera (*Oligeucera*) Sitdikov and Pesenko, 1988: 83. Type species: *Eucera popovi* Sitdikov, 1988, by original designation.

Oligeucera contains the smallest *Eucera* (see the key to subgenera). Its genitalia and hidden sterna were illustrated by Sitdikov and Pesenko (1988).

■ This subgenus is from Tadjikistan; only one species, *Eucera popovi* Sitdikov, is known.

Eucera / Subgenus *Pteneucera* Tkalců

Pteneucera Tkalců, 1984c: 72. Type species: *Eucera eucnemidea* Dours, 1873, by original designation.

Illustrations of male sternal and genitalic structures were presented by Iuga (1958), Tkalců (1984c), and Sitdikov and Pesenko (1988).

■ *Pteneucera* occurs from Spain to Iran and Central Asia. The species were revised by Sitdikov (1988) and by Risch (1997) who recognized eight species.

Eucera / Subgenus *Synhalonia* Patton

Synhalonia Patton, 1879a: 473. Type species: *Melissodes fulvitarsis* Cresson, 1878, by original designation.
Eusynhalonia Ashmead, 1899a: 63. Type species: *Melissodes edwardsii* Cresson, 1878, by original designation.
Synalonia Robertson, 1905: 365, unjustified emendation of *Synhalonia* Patton, 1879.

This is the only subgenus of *Eucera* having three submarginal cells. Like *Eucera* s. str., it consists of rather large species, body length 9 to 19 mm. Most palearctic species have on the underside of the middle femur of the male a well-defined elongate area of dense (so dense that the femoral surface is completely hidden), appressed, red-brown hair directed posteriorly at a right angle to the main axis of the femur. In nearctic species, and in a few palearctic species such as *E. (Synhalonia) floralia* (Smith), the hairs are similarly appressed but white and sparse enough that the cuticle is visible among them. The same is true in species of the two-celled subgenera and in a few species of the genus *Tetraloniella*, although in most species of that genus the hairs are sloping or erect, not appressed. Male genitalia and hidden sterna of *Synhalonia* were illustrated by LaBerge (1957) and Mitchell (1962), as *Tetralonia*; sterna were illustrated by Timberlake (1969b).

■ In the nearctic area *Synhalonia* occurs from British Columbia, Canada, to Massachusetts, USA, south to California, Texas, and Georgia, USA, and Oaxaca, Mexico. In the palearctic region it occurs from the Canary Islands

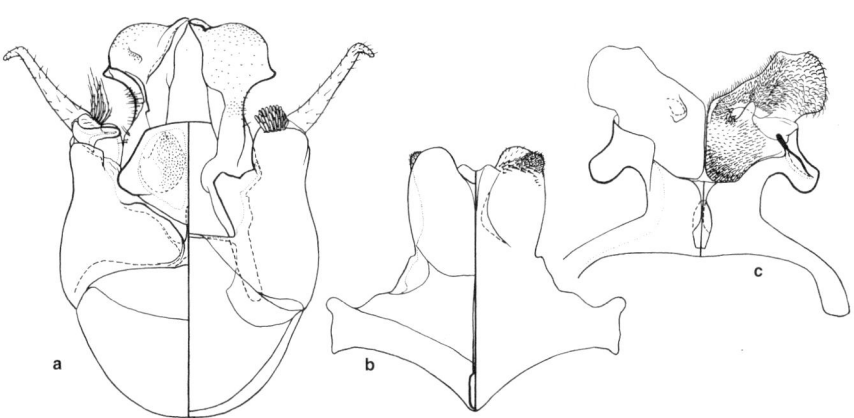

Figure 112-17. a-c, Male genitalia, S8, and S7 of *Eucerinoda gayi* (Spinola). From Michener and Moure, 1957.

and Spain eastward throughout the Mediterranean countries, north to Slovakia, south to India (Punjab), and through Asia to Japan. There are 54 North American species and perhaps as many palearctic species. Many species from the latter region have not been reliably placed to genus, because *Tetralonia, Tetraloniella,* and *Eucera (Synhalonia)* have been commonly regarded as a single unit usually called *Tetralonia*. The nearctic species of *Synhalonia* were revised by Timberlake (1969b).

The nesting biology of *Eucera hamata* (Bradley) was described by Miliczky (1985).

Genus *Eucerinoda* Michener and Moure

Eucerinoda Michener and Moure, 1957: 445. Type species: *Anthophora gayi* Spinola, 1851, by original designation.

This genus constitutes the subtribe Eucerinodina. Its distinctive features are indicated not only in the key to South American genera, but also in the account of the tribe Eucerini. It consists of a rather robust species with a body length of 10.5 to 11.0 mm. The pubescence is pale, and forms broad bands on the metasomal terga, especially apically. It differs from most but not all other Eucerini in having a rather flat clypeus, the lateral parts of which are not much bent posteriad, and in its stigma, which, although small, is longer than the prestigma with vein r arising near its middle, about as in *Melissoptila*. In the male the clypeus and paraocular areas below the antennae are yellow; the antennae reach the base of the scutellum, the first flagellar segment being almost as long as the second and third segments taken together; and the hind leg is enlarged, the basitibial plate and tibial spurs absent, and the basitarsus highly modified. A full description and figures of genitalia and sterna were provided by Michener and Moure (1957); see also Figure 112-17, presented here to demonstrate the similarity of *Eucerinoda* to other Eucerini (compare Figs. 112-7, 112-9).
■ *Eucerinoda* occurs in central Chile. There is only one species, *E. gayi* (Spinola).

Genus *Florilegus* Robertson

Florilegus consists of *Melissodes*-like bees of North and South America having weak iridescent reflections on the metasomal terga and five-segmented maxillary palpi. Females differ from those of all similar genera in the exposed and often bare margin of the basitibial plate (Fig. 112-12a), and from those of most others in the lamellate lateral arm of the gradulus of T6, ending in a strong tooth. The male differs from that of *Melissodes* and other similar bees by the relatively large and complex apical lobes of S7 (Fig. 112-7i), suggesting those of some *Gaesischia*. Unlike the male gonostyli of that genus, those of *Florilegus* are relatively short, robust, and expanded apically. The male genitalia and hidden sterna were illustrated by Michener (1954b), Moure and Michener (1955b), LaBerge (1957), Mitchell (1962), and Urban (1970); see also Figure 112-7g-i.

This genus was revised by Urban (1970).

Key to the Subgenera of *Florilegus* (Males)
(Modified from Moure and Michener, 1955b)

1. Hind tibia at least slightly contorted with band of dense hairs on otherwise largely bare inner surface; hind femur concave beneath; inner apical lobe of S7 much longer than broad, terminating in relatively short, sparse hairs (South America) .... *F. (Euflorilegus)*
—. Hind tibia unmodified; hind femur convex beneath; inner apical lobe of S7 broader than long, terminating in fringe of long hairs 2
2(1). Large black forms without pale fasciae on first three metasomal terga; scutellum relatively flat, its profile nearly a continuation of that of scutum (neotropical) .... *F. (Floriraptor)*
—. Smaller forms with pale fasciae (sometimes broken) on T2 and T3; scutellum more convex ...... *F. (Florilegus s. str.)*

Key to the Subgenera of *Florilegus* (Females)
(Modified from Moure and Michener, 1955b)

1. Hairs of maxilla and mentum erect, hooked; marginal cell slightly shorter than distance from apex to wing tip (South America) .... *F. (Euflorilegus)*
—. Hairs of maxilla and mentum simple, inconspicuous; marginal cell longer than distance from apex to wing tip 2
2(1). Scutellum more than one-third as long as scutum, flattened and nearly on same plane as posterior part of scutum; first flagellar segment as long as distance between posterior ocelli; first three terga largely glabrous (neotropical).. *F. (Floriraptor)*
—. Scutellum less than one-third as long as scutum, convex; first flagellar segment distinctly shorter than distance between posterior ocelli; at least T3 with band (sometimes broken) of pale pubescence .... *F. (Florilegus s. str.)*

Florilegus / Subgenus *Euflorilegus* Ogloblin

Florilegus (Euflorilegus) Ogloblin, 1955: 231. Type species: *Florilegus riparius* Ogloblin, 1955, by original designation.

Members of this subgenus look much like those of *Florilegus* s. str. in size and coloration.
■ *Euflorilegus* occurs from Buenos Aires Province, Argentina, north to Pará and Amazonas, Brazil. The five species were revised by Urban (1970).

Florilegus / Subgenus *Florilegus* Robertson s. str.

Florilegus Robertson, 1900: 53. Type species: *Melissodes condigna* Cresson, 1879, monobasic.

Like *Euflorilegus*, species of this subgenus are relatively small, 9 to 11 mm long, with much yellowish hair.
■ This subgenus occurs from New Jersey to Colorado, USA, southward through the Antilles (at least Cuba), Mexico, and the American tropics to Buenos Aires and La Rioja provinces, Argentina. The five species were revised by Urban (1970).

LaBerge and Ribble (1966) gave an account of nesting biology, larval structure, and the like, for *Florilegus condignus* (Cresson).

Florilegus / Subgenus *Floriraptor* Moure and Michener

Florilegus (Floriraptor) Moure and Michener, 1955b: 268. Type species: *Melissodes atropos* Smith, 1879 = *Tetralonia melectoides* Smith, 1879, by original designation.

This subgenus consists of large species, 12 to 14 mm long, with much black pubescence, therefore rather different in appearance from members of the other subgenera.

■ *Floriraptor* occurs from Panama south to Santiago del Estero, Argentina, and Paraná, Brazil. The single species, *Florilegus melectoides* (Smith), was treated by Urban (1970).

Genus *Gaesischia* Michener, LaBerge, and Moure

A substantial group of neotropical Eucerini is here placed in the genus *Gaesischia*. They are superficially *Melissodes*-like bees (Fig. 112-1), 9 to 16 mm long. They were separated into two genera, *Gaesischia* and *Dasyhalonia*, by Moure and Michener (1955b) and Urban (1967c, 1968a) and come out in separate couplets of the key to South American genera of Eucerini, above. However, as shown by LaBerge (1958), they are much alike and merge, the Mexican subgenus *Prodasyhalonia* being intermediate in various features. Common characters that differentiate them from genera such as *Melissodes, Melissoptila,* and *Svastra* are (1) the broad, complex apical lobes of S7 of the male, the basolateral part of the lateral lobe usually elaborated, and without the midapical projection found in most *Alloscirtetica,* and (2) the long and usually weakly sigmoid male gonostyli. *Svastrides, Santiago,* and *Hamatothrix* have the same features and might also be included in *Gaesischia*.

Species formerly placed in *Gaesischia* (i.e., in the three subgenera *Gaesischia* s. str., *Gaesischiopsis*, and *Gaesischiana*) have the head elevated behind the ocelli, which face somewhat forward. Among Eucerini, similar construction is found in some *Alloscirtetica,* especially in the subgenus *Megascirtetica,* and to a lesser degree in the genus *Svastrides*. The three subgenera of *Gaesischia* listed above also have a pair of converging carinae on S6 of the male, as in many Eucerini. Species formerly placed in *Dasyhalonia* (i.e., in the subgenera *Dasyhalonia* and *Pachyhalonia*), on the contrary, have the head horizontal behind the ocelli, which face upward. The males lack converging carinae on S6. The subgenus *Prodasyhalonia* is intermediate in head formation and has its own distinctive type of male S6, with a longitudinal lamella ending in a posteriorly directed spine. In the subgenera *Dasyhalonia* and *Pachyhalonia,* as well as *Prodasyhalonia,* S4 and S5 of the male, sometimes also S3, have long, coarse, curled subapical hairs, not found in other subgenera. Long, straight hairs are present in the same position on S5 of *Gaesischiopsis*. The male flagellum of *Gaesischiana* and *Prodasyhalonia* is tapering, the last segment strongly compressed and sometimes broadened. In the extreme case, *G. (Gaesischiana) patellicornis* (Ducke), the antenna is suggestive of that of *Trichocerapis* but less threadlike preapically and is certainly an independent development. LaBerge (1958) gave a useful tabulation of subgeneric characters. Male genitalia were illustrated by Moure and Michener (1955b) and Urban (1967c, 1968a, 1989b).

Key to the Subgenera of *Gaesischia* (Males)

1. S6 with pair of carinae converging posteriorly [except in *Gaesischia patellicornis* (Ducke)]; labral emargination shallow or occupying less than one-third of distal labral margin; middle femur without mass or tuft of dense hair; flagellum little thicker than diameter of ocellus 2
—. S6 without such carinae, but with elevated basal area or median lamella; labral emargination deep and occupying over one-third of distal labral margin; middle femur nearly bare beneath except sometimes for mass or tuft of short, dense hairs; flagellum much thicker than diameter of ocellus 4
2(1). Maxillary palpus four-segmented; S5 with subapical row of long, bristle-like hairs (South America) ....
 .... *G. (Gaesischiopsis)*
—. Maxillary palpus five- or six-segmented; S5 without long bristle-like hairs 3
3(2). Posterior leg modified, femur enlarged with area of short, dense hairs near posterior margin of bare undersurface, basitarsus with row of long hairs on anterodistal margin; flagellum tapering, last segment compressed and sometimes expanded (neotropical, Arizona) ....
 .... *G. (Gaesischiana)*
—. Posterior leg unmodified; flagellum unmodified (South America) .... *G. (Gaesischia s. str.)*
4(1). S6 with longitudinal median lamella, ending posteriorly in a tooth; penultimate flagellar segment more than twice as long as broad (Mexico) .... *G. (Prodasyhalonia)*
—. S6 with elevated basal area sometimes ending in a tooth, but without lamella; penultimate flagellar segment less than twice as long as broad 5
5(4). Antennal flagellum greatly thickened, thickest from sixth to eighth segments, whence it tapers both basally and apically; lateral carinae of pygidial plate dentate subapically; underside of middle femur with small dense hair tuft (South America) .... *G. (Dasyhalonia)*
—. Antennal flagellum of approximately uniform thickness or second and third segments thickest; lateral carinae of pygidial plate not dentate; underside of middle femur with large oblique band of dense hairs (South America)
 .... *G. (Pachyhalonia)*

Key to the Subgenera of *Gaesischia* (Females)

1. Anterior coxa with apical spine reaching to middle of trochanter or beyond (Fig. 112-13c) 2
—. Anterior coxa without apical spine or with short, triangular projection 3
2(1). Anterior coxal spine reaching apex of trochanter (Fig. 112-13c); maxillary palpus four-segmented; lateral part of gradulus of T6 absent or nearly so (South America)
 .... *G. (Gaesischiopsis)*
—. Anterior coxal spine reaching about or somewhat beyond middle of trochanter; maxillary palpus five- or six-segmented; lateral part of gradulus of T6 evident as carina (South America, Mexico) ....
 .... *G. (Gaesischia s. str., Prodasyhalonia)*
3(1). Head elevated behind ocelli such that ocelli face strongly forward (clypeus closely approaching eye margin) (neotropical, Arizona) .... *G. (Gaesischiana)*
—. Head not elevated behind ocelli, which face upward 4
4(3). Clypeus separated from eye margin by about one-third of minimum flagellar diameter; margins of basitibial plate entirely hidden except posteriorly (South America) .... *G. (Pachyhalonia)*
—. Clypeus very closely approaching eye margin; margins of basitibial plate evident although covered with hairs except posteriorly (South America) .... *G. (Dasyhalonia)*

Gaesischia / Subgenus *Dasyhalonia* Michener, LaBerge, and Moure

Dasyhalonia Michener, LaBerge, and Moure, 1955a: 226. Type species: *Tetralonia spiniventris* Friese, 1910 = *Tetralonia mimetica* Brèthes, 1910, by original designation.

The dense, yellowish, appressed metasomal pubescence gives the included species the appearance of certain species of the genus *Alloscirtetica* and of some species of *Melissoptila (Ptilomelissa)*, as well as of the emphorine taxon *Diadasia (Dasiapis* group). The lateral carina of the pygidial plate of the male forms a subapical tooth on each side of the plate. The genitalia and sterna were illustrated by Michener, LaBerge, and Moure (1955a) and Urban (1967c).

■ This subgenus is found in Argentina from the provinces of Buenos Aires and Mendoza to Catamarca. It contains two species and was revised by Urban (1967c).

Gaesischia / Subgenus *Gaesischia* Michener, LaBerge, and Moure s. str.

Gaesischia Michener, LaBerge, and Moure, 1955a: 220. Type species: *Svastra fulgurans* Holmberg, 1903, by original designation.

Genitalia and hidden sterna of the male were illustrated by Michener, LaBerge, and Moure (1955a) and Urban (1968a, 1989b).

■ This subgenus is found from Misiones and perhaps Córdoba, Argentina, Paraguay, and Rio Grande do Sul, Brazil, to Ceará and Pará, Brazil. The 19 species were revised by Urban (1968a), and a new key was published by Urban (1989b).

Gaesischia / Subgenus *Gaesischiana* Michener, LaBerge, and Moure

Gaesischia (Gaesischiana) Michener, LaBerge, and Moure, 1955a: 224. Type species: *Gaesischia exul* Michener, LaBerge, and Moure, 1955, by original designation.
Gaesischia (Agaesischia) Moure and Michener, 1955b: 273. Type species: *Eucera patellicornis* Ducke, 1910, by original designation.

This subgenus contains the only *Gaesischia* in which the clypeus is well separated, by one-third of the flagellar diameter, from the eye margin. The scopal hairs are plumose (in the former *Agaesischia*) or simple (in the former *Gaesischiana*). The body length is 8 to 12 mm. The male genitalia, hidden sterna, and other structures were illustrated by Michener, LaBerge, and Moure (1955a) and by Urban (1968a).

■ *Gaesischiana* is known from two areas and its range may be disjunct: first, southern Arizona, USA, to Costa Rica with a report from Colombia, and second, central Brazil from Mato Grosso and Bahia north to Pará. Two species were revised by Urban (1968a); a third species and a key to the species were published by Urban (1989b).

Gaesischia / Subgenus *Gaesischiopsis* Michener, LaBerge, and Moure

Gaesischia (Gaesischiopsis) Michener, LaBerge, and Moure, 1955a: 221. Type species: *Gaesischia flavoclypeata* Michener, LaBerge, and Moure, 1955, by original designation.

This is the only subgenus of *Gaesischia* with four-segmented maxillary palpi; other subgenera have five or six segments. The male genitalia and hidden sterna of this subgenus were illustrated by Michener, LaBerge, and Moure (1955a) and by Urban (1968a, 1989b). The body length is 7 to 10 mm.

■ This subgenus is found from the states of Pará and Ceará, Brazil, to Rio Grande do Sul, Brazil, and Paraguay. It contains seven species and was revised by Urban (1968a, 1989b).

Gaesischia / Subgenus *Pachyhalonia* Moure and Michener

Dasyhalonia (Pachyhalonia) Moure and Michener, 1955b: 281. Type species: *Dasyhalonia justi* Moure and Michener, 1955 = *Svastra sapucacensis* Cockerell, 1918, by original designation.
Dasyhalonia (Zonalonia) Moure and Michener, 1955b: 250, 282. Type species: *Eucera cearensis* Ducke, 1910, by original designation.
Dasyhalonia (Zonolonia) Moure and Michener, 1955b: 280. Printer's error for *Zonalonia* Moure and Michener, 1955; see Michener, 1997b.
Dasyhalonia (Zonohalonia) Moure and Michener, 1955b: 255, *lapsus* for *Zonalonia* Moure and Michener, 1955.

The subgenus includes both rather small species *(Zonalonia)* and large, *Svastra*-like forms; body length ranges from 9.5 to 16.0 mm. I follow Urban (1967c) in the synonymy; LaBerge (1958) tabulated some differences, such as plumose scopal hairs in *Zonalonia* and serrate but unbranched hairs in *Pachyhalonia* proper. Male genitalia and hidden sterna were illustrated by Moure and Michener (1955b) and Urban (1967c).

■ *Pachyhalonia* ranges from the state of Ceará, Brazil, south to Paraguay and the province of San Luis, Argentina. The three species were revised by Urban (1967c).

Gaesischia / Subgenus *Prodasyhalonia* LaBerge

Gaesischia (Prodasyhalonia) LaBerge, 1958: 199. Type species: *Gaesischia mexicana* LaBerge, 1958, by original designation.

Prodasyhalonia was included in *Gaesischia* s. str. by Urban (1968a). It differs from *Gaesischia* s. str., however, in the less strongly elevated posterior part of the vertex, the presence of long curled hairs on S4 and S5 of the male (as in *Dasyhalonia* and *Pachyhalonia*), and the lack of convergent carinae and presence of a longitudinal median lamella on S6 of the male, among other characters. I have chosen to retain *Prodasyhalonia*. The male genitalia and hidden sterna were illustrated by LaBerge (1958).

■ This subgenus is known from the states of Chiapas, Oaxaca, and Veracruz, Mexico. The only known species is *Gaesischia mexicana* LaBerge.

Genus *Gaesochira* Moure and Michener

Gaesochira Moure and Michener, 1955b: 283. Type species: *Gaesochira complanata* Moure and Michener, 1955 = *Eucera obscurior* Dalla Torre, 1896, by original designation.

This genus has a broad, flat scutellum suggestive of that of *Florilegus (Floriraptor)*. The female differs from that of *Florilegus* in the hidden basitibial plates and the lack of gradular teeth on T6. The tibial spurs, curved apically, are very large, almost reaching the apices of the basitarsi, and the clypeus is very flat, longitudinally strigose; these two features are unique among the Eucerini. The body length is 10 to 12 mm. Male genitalia and hidden sterna were illustrated by Michener and Moure (1956) and Urban (1974b).

■ *Gaesochira* is known from Colombia and from Amazonas, Pará, and Rondônia, Brazil. The single species is *G. obscurior* (Dalla Torre). In view of the Code, 3rd ed., art. 59(b), Dalla Torre's replacement name is to be used rather than *obscura* Smith; this disposition is contrary to the decision of Urban (1974b).

Genus *Hamatothrix* Urban

Hamatothrix Urban, 1989c: 121. Type species: *Hamatothrix silvai* Urban, 1989, by original designation.

This genus is based on a rather small species (8-9 mm long) from Argentina. As in some species of both *Alloscirtetica* and *Gaesischia*, S6 of the male lacks the usual convergent carinae. The well-developed pygidial plate of the male, as well as the lack of a median apical process on S7 of the male, show that *Hamatothrix* is not *Alloscirtetica* but is close to *Gaesischia*; on the basis of the male alone I would have included it in that genus, although its combination of characters differs from that of any *Gaesischia* subgenus. Thus the lack of carinae and presence of a basal median elevated area on S6 and the more or less uniform width of the flagellum suggest the subgenus *Pachyhalonia*, but the labral emargination is so small and weak as to be nearly absent, and the middle femur lacks an area of dense hairs. It is the remarkable features of the female that may justify recognition of *Hamatothrix* as a genus (see the key to genera). The clump of strong bristles behind the small tibial corbicula (or apical bare space, as in other Eucerini) suggests the single row of stronger bristles behind the larger corbicula in *Canephorula*. Other characters of *Canephorula*, however, such as the simple curvature of the gradulus of S2 of the female, the long, oblique vein cu-v of the hind wing, and the papillate wings, suggest that *Canephorula* and *Hamatothrix* are not closely related and evolved strong bristles behind the bare areas of the hind tibiae independently. The completely exposed marginal carina of the basitibial plate of the female suggests *Florilegus*, but in most respects, *Hamatothrix* females, like males, resemble *Gaesischia*. In the key to the subgenera of *Gaesischia*, the female would run to *Dasyhalonia* except for the exposed margin of the basitibial plate. Male genitalia and other structures were illustrated by Urban (1989c).

■ *Hamatothrix* occurs in the provinces of Santiago del Estero and La Rioja, Argentina. The only species is *H. silvai* Urban.

Genus *Lophothygater* Moure and Michener

Lophothygater Moure and Michener, 1955b: 313. Type species: *Tetralonia decorata* Smith, 1879, by original designation.

Lophothygater is a member of the *Thygater-Trichocerapis* group, as shown by the gradulus of S2 of the female and other characters. It differs in having a strong preoccipital carina across the posterior margin of the vertex, and in the long scape of the female, which is almost as long as the interantennal distance. The male differs from other members of the *Thygater* group in the reduced clypeocular distance, etc., so that it runs to 6 at couplet 4 of the key to genera. The male genitalia, hidden sterna, and other structures were illustrated by Urban (1967b). Body length is from 11.0 to 13.5 mm.

■ This genus is known only from the Amazon valley. There is a single species, *Lophothygater decorata* (Smith). Its characters and distribution were reported by Urban (1967b).

Genus *Martinapis* Cockerell

This is a genus of bees that are similar in form to species of *Svastra* or large *Melissodes;* the body length is 12 to 16 mm. *Martinapis* differs in the long first flagellar segment of the male, more than half as long as the second (Fig. 112-8d), and in the elongate median apical lobes of S7 of the male (these lobes are small and short in *Svastra* and in most *Melissodes*). The outer margin of the tegula, anteriorly, is rather broadly expanded, translucent, and impunctate, even more so than in *Simanthedon*. As in *Melissodes* but unlike *Simanthedon* and other Eucerini, this margin is slightly concave, at least in North American species.

■ The distribution of *Martinapis* is disjunct, with one species in Argentina and two in the southwestern United States and northern Mexico. The Argentine species is so different from the others that a different subgeneric name has been proposed for it.

Key to the Subgenera of *Martinapis*

1. Antenna of male rather short, reaching propodeum, entirely yellow, first flagellar segment over 0.7 times as long as second; last flagellar segment of male tapering to blunt point (Fig. 112-8d) (nearctic) .... *M. (Martinapis s. str.)*
—. Antenna of male reaching beyond propodeum, black, flagellum red-yellow beneath, first flagellar segment nearly 0.6 times as long as second; last flagellar segment of male with ordinary rounded apex (South America) .... *M. (Svastropsis)*

Martinapis / Subgenus *Martinapis* Cockerell s. str.

Melissodes (Martinella) Cockerell, 1903c: 450 (not Jousseaume, 1887). Type species: *Melissodes luteicornis* Cockerell, 1896, monobasic.

Martinapis Cockerell, 1929f: 19, replacement for *Martinella* Cockerell, 1903. Type species: *Melissodes luteicornis* Cockerell, 1896, autobasic.

The male genitalia and hidden sterna were illustrated by LaBerge (1957) and Zavortink and LaBerge (1976).

■ *Martinapis* s. str. occurs in deserts from western Texas to California, USA, and south to Puebla and Morelos,

Mexico. The two species were revised by Zavortink and LaBerge (1976).

Martinapis / Subgenus *Svastropsis* Moure and Michener

Martinapis (Svastropsis) Moure and Michener, 1955b: 291. Type species: *Tetralonia bipunctata* Friese, 1908, by original designation.

The male genitalia and hidden sterna were illustrated by Moure and Michener (1955b).
■ This subgenus occurs in Mendoza, Argentina. The only known species is *Martinapis bipunctata* (Friese).

Genus *Melissodes* Latreille

This is the major North American genus of Eucerini, although there are also a few South American species. Males differ from those of *Svastra* in the strong lateral arm of the gradulus of T7, ending in a tooth at the side of the pygidial plate. Both sexes differ from those of nearly all *Svastra* in lacking the spatuloplumose hairs. Both sexes differ from other Eucerini except *Martinapis* in tegular shape, although the character is subtle, and removal of hairs is often necessary in order to see it. Although the anterior part of the lateral tegular margin is usually gently concave, in some species it is straight or even feebly convex. It is not, however, simply a convex continuation of the rest of the tegular margin, as in nearly all other Eucerini. The males are also similar to those of *Florilegus*. The maxillary palpi have four or rarely three segments whereas those of *Florilegus* have five. The characters of the male genitalia and hidden sterna also differ from those of *Florilegus;* in *Melissodes* S7 has small apical lobes and the gonostyli are small and simple (Fig. 112-9a-f). In the female *Melissodes*, the basitibial plate is largely hidden by short hair and T6 lacks gradular teeth. Male genitalia and hidden sterna were illustrated by Michener (1954b), LaBerge (1956a, b, 1961), and Mitchell (1962); see also Figure 112-9a-f.

Key to the Subgenera of *Melissodes* (Males) (Modified from LaBerge, 1961)

1. Clypeus protuberant in front of eye for one-half to three-fourths of eye width, as seen in lateral view and antenna long, minimum length of first flagellar segment less than one-third maximum length of second segment; T2 to T5 fringed with narrow apical bands of appressed white pubescence, bands much narrower than basal areas (nearctic) .... *M. (Apomelissodes)*
—. Clypeus usually not protruding in front of eye by as much as half of eye width, as seen in side view, or, *if* protruding by half of eye width or more, then minimum length of first flagellar segment one-third or more of maximum length of second segment; terga often not fringed by apical pubescent bands, bands when present interrupted medially and/or preapical 2
2(1). Posterior margin of S4 (and usually S3) broadly convex, or produced into a broad, thin, hyaline, colorless flap (nearctic) .... *M. (Callimelissodes)*
—. Posterior margins of S3 and S4 straight to slightly concave, never produced into flaps 3
3(2). Clypeus protuberant in front of eye by at least three-fourths of width of eye, as seen in side view; maximum length of first flagellar segment 0.4 or more of maximum length of second segment (nearctic) .. *M. (Heliomelissodes)*
—. Clypeus usually protruding one-half or less of width of eye, as seen in side view, or, *if* protruding more, then maximum length of first flagellar segment less then 0.4 of maximum length of second segment 4
4(3). Maximum length of first flagellar segment as great as or almost as great as maximum length of second segment and longer than third segment (antenna female-like); clypeus wholly black (nearctic)............ *M. (Psilomelissodes)*
—. Maximum length of first flagellar segment less than maximum length of second segment and distinctly less than that of third; clypeus usually pale, occasionally partly or wholly black 5
5(4). Minimum length of first flagellar segment distinctly more than half maximum length of second segment; T2 to T5 with pubescent bands apical, of nearly uniform width across each tergum, and subequal in width to each other (nearctic) .... *M. (Tachymelissodes)*
—. Minimum length of first flagellar segment half of maximum length of second segment or less; T2 to T5 with pale pubescent bands usually not all apical or not subequal in width, often interrupted medially and usually subapical 6
6(5). Median apical lobes of S7 without hairs on ventral surfaces, usually small, curled ventrally along an oblique axis to form half or more of an oblique cylinder or scroll (Fig. 112-9c), but often secondarily flattened and expanded, or secondarily reduced in size .... *M. (Melissodes s. str.)*
—. Median apical lobes of S7 thin, hyaline, with short to moderately long hairs on ventral surfaces, not curled ventrally, relatively large 7
7(6). Gonostylus often less than half as long as gonocoxite, in lateral view at least twice as broad near base as near apex, narrowing abruptly near middle, not capitate; median lobes of S7 relatively small, with several short hairs ventrally (Fig. 112-9f) (neotropical) .... *M. (Ecplectica)*
—. Gonostylus at least half as long as gonocoxite, in lateral view not twice as broad near base as near apex, often somewhat capitate; median lobes of S7 large, with abundant short to moderately long hairs ventrally (North and Central America) .... *M. (Eumelissodes)*

Key to the Subgenera of *Melissodes* (Females) (Modified from LaBerge, 1961)

1. Scopal hairs simple or, *if* weakly branched, then clypeus in profile protruding in front of eye by at least two-thirds width of eye, as seen in lateral view (pygidial plate not narrow) (nearctic) .... *M. (Apomelissodes)*
—. Scopal hairs branched, usually abundantly so, but if weakly branched, then clypeus in profile not protruding in front of eye by as much as two-thirds width of eye, as seen in lateral view 2
2(1). Clypeus protruding anteriorly in front of eye by one-half to two-thirds width of eye, as seen in lateral view; inner orbits of eyes often parallel; inner surface of hind basitarsus with hairs dark brown to black (scopal hairs highly plumose, often yellowish) (nearctic)............*M. (Heliomelissodes)*
—. Clypeus protruding in front of eye by less than half width of eye, as seen in lateral view, or, *if* protruding by as much as half width of eye, then inner orbits distinctly

converging below *and/or* inner surface of hind basitarsus with hairs bright red to yellow (scopal hairs occasionally only weakly branched) 3

3(2). Scopal hairs weak, with few branches, not hiding outer surfaces of hind basitarsus and tibia; metasomal terga very sparsely and weakly punctate, dulled by dense, fine shagreening and weakly banded with sparse pubescence; pygidial plate V-shaped with broadly rounded apex (nearctic) .... *M. (Psilomelissodes)*

—. Scopal hairs strongly branched and hiding outer surfaces of hind basitarsus and tibia; or, *if* weak and with few branches, then terga coarsely punctate at least basally, or moderately shiny to shiny and strongly banded with abundant pubescence, *or* pygidial plate narrowly U- shaped 4

4(3). T2 to T4 with distal pale pubescent bands reaching apical margins of terga, of about same width across each tergum and subequal in width to one another, as narrow as or narrower than basal areas of dark pubescence (nearctic) .... *M. (Tachymelissodes)*

—. T2 to T4 with distal pale pubescent bands (at least on T2) not reaching apices of terga, or, *if* reaching apices of terga, then diffuse over entire tergum or much wider than basal area of dark pubescence or of varying width across each tergum or between terga ....*M. (Callimelissodes, Ecplectica, Eumelissodes, Melissodes s. str.)*

Melissodes / Subgenus *Apomelissodes* LaBerge

Melissodes (Apomelissodes) LaBerge, 1956a: 1175. Type species: *Melissodes fimbriata* Cresson, 1878, by original designation.

In *Apomelissodes* the clypeus is protuberant, extending in front of the eye by much more than half the eye width in lateral view, and often by the full eye width. The blade of the galea is over twice to nearly three times as long as the clypeus. The scopal hairs are simple to weakly plumose, with two to four branches on each side. The body length is 9 to 14 mm. Male genitalia and hidden sterna were illustrated by LaBerge (1956a, b).
■ This subgenus occurs primarily in eastern North America from Maine to Florida, USA, but ranges westward as far as Kansas. The four species were revised by LaBerge (1956b).

Melissodes / Subgenus *Callimelissodes* LaBerge

Melissodes (Callimelissodes) LaBerge, 1961: 294. Type species: *Melissodes lupina* Cresson, 1878, by original designation.

This subgenus is similar to *Eumelissodes,* from which it differs in the broad, hyaline apical flap on S4 and usually S3 of the male. On T6 of the female, the lateral arm of the gradulus is lamelliform and often ends in a small tooth, whereas in *Eumelissodes* it is short and carinate. Body length is 7.5 to 16.0 mm. Male genitalia and hidden sterna were illustrated by LaBerge (1961).
■ *Callimelissodes* is found from Alberta, Canada, Washington state, Wisconsin, and Indiana south to Texas and North Carolina, USA, and Baja California, Mexico. The 14 species were revised by LaBerge (1961).

Melissodes / Subgenus *Ecplectica* Holmberg

Ecplectica Holmberg, 1884: 123. Type species: *Ecplectica tintinnans* Holmberg, 1884, monobasic.

Ecplectia is the only entirely neotropical subgenus of *Melissodes*. It is similar to *Melissodes* s. str. but differs in the faint violaceous reflections nearly always evident on the metasomal terga, the short male gonostylus, often not over one-third as long as the gonocoxite and twice as broad basally as in the narrowed apical section, and the characters indicated in the key. The body length is 7.5 to 11.0 mm. Male genitalia and hidden sterna were illustrated by LaBerge (1956a) and Urban (1973); see also Figure 112-9d-f.
■ This subgenus occurs from Veracruz, Mexico, and Puerto Rico in the Antilles south to Tarapacá, Chile, Buenos Aires Province, Argentina, and Uruguay. There are about eight species; three found in Mesoamerica and the Antilles were revised by LaBerge (1956a) and five found in South America were revised by Urban (1973).

Melissodes / Subgenus *Eumelissodes* LaBerge

Melissodes (Eumelissodes) LaBerge, 1956a: 1177. Type species: *Melissodes agilis* Cresson, 1878, by original designation.

This is the largest and most abundant subgenus of *Melissodes*. The clypeus is not or little protuberant, usually yellow or white in the male. The labrum of the male is variable in color. The first flagellar segment of the male is less than half, usually less than one-third, as long as the second segment. The scopa of the female is plumose, usually strongly so. The body length is 8 to 16 mm. The male genitalia and hidden sterna were illustrated by LaBerge (1956a, 1961).
■ *Eumelissodes* is found from British Columbia, Canada, to Maine, USA, and south throughout North and Central America to Panama and to Cuba. The 72 species were revised by LaBerge (1961).

Melissodes / Subgenus *Heliomelissodes* LaBerge

Melissodes (Heliomelissodes) LaBerge, 1956a: 1172. Type species: *Melissodes desponsa* Smith, 1854, by original designation.

In this small subgenus the clypeus is protuberant in front of the eye by at least half the width of the eye as seen in lateral view, and the blade of the galea is more than twice as long as the clypeus. The labrum of the male is black. The first flagellar segment of the male is one-third the length of the second segment or slightly longer. Body length is 9 to 17 mm. Male genitalia and hidden sterna were illustrated by LaBerge (1956a, b).
■ *Heliomelissodes* occurs from Nova Scotia to British Columbia, Canada, south to North Carolina, Alabama, Texas, and Arizona, USA. The two species were revised by LaBerge (1956b).

The species of *Heliomelissodes* are oligolectic on flowers of thistles of the genus *Cirsium* (Asteraceae).

Melissodes / Subgenus *Melissodes* Latreille s. str.

Melissode Latreille, 1825: 464, nomen nudum.
Melissodes Berthold, 1827: 468, nomen nudum, emendation of *Melissode* Latreille. [There is no certainty that the two names above refer to *Melissodes* Latreille, 1829. They are listed here to avoid possible confusion.]
Melissodes Latreille, 1829: 354, no included species; Romand, 1841: 6, first included species *Melissodes fonscolombei* Romand, 1841, suppressed by Commission Opinion 750 (1965). Type species: *Melissodes leprieuri* Blanchard, 1846, first valid included species and designated as type species by Opinion 750. [A designation of *Macrocera rustica* Say, 1837 = *Apis druriella* Kirby, 1802, by Taschenberg (1883: 78) was thereby invalidated.]

In this subgenus the labrum of the male is usually pale. The first flagellar segment of the male is one-tenth to one-third the length of the second; the flagellum reaches at least to T1 and usually to its apex or beyond. The body length is 7.5 to 16.0 mm. Male genitalia and hidden sterna were illustrated by LaBerge (1956a); see also Figure 112-9a-c.

■ *Melissodes* s. str. occurs from Maine to Florida, USA, west to British Columbia, Canada, to Baja California, Mexico, and south throughout Mesoamerica to Panama and through the Bahamas and the Greater and Lesser Antilles to Trinidad, the Guianas and Roraima in Brazil. There are 23 species. The subgenus was revised by LaBerge (1956a); the single South American species, by Urban (1973).

Melissodes / Subgenus *Psilomelissodes* LaBerge

Melissodes (Psilomelissodes) LaBerge, 1956a: 1173. Type species: *Melissodes intorta* Cresson, 1872, by original designation.

In this subgenus the male antennae are similar to those of the female, not reaching beyond the propodeum, and the first flagellar segment is about equal to the length of the second or slightly more. The labrum and clypeus of the male are black, also as in the female. The scopal hairs are weakly plumose, with only one to three branches on each side. The body length is 11 to 13 mm. Male genitalia and hidden sterna were illustrated by LaBerge (1956a).

■ *Psilomelissodes* occurs in the southern Great Plains from Texas to Kansas, USA. The only species, *Melissodes intorta* Cresson, is oligolectic on the malvaceous genus *Callirhoe*.

Melissodes / Subgenus *Tachymelissodes* LaBerge

Melissodes (Tachymelissodes) LaBerge, 1956a: 1170. Type species: *Melissodes dagosa* Cockerell, 1909, by original designation.

This small subgenus from the xeric parts of North America is distinctive because of the strong apical white pubescent bands on T2 to T4, the basal bands being weak or absent. The male antennae are short for *Melissodes*, not or scarcely reaching the base of T1, the first flagellar segment being more than half as long as the second. The body length is 8 to 12 mm. Male genitalia and sterna were illustrated by LaBerge (1956a, b).

■ *Tachymelissodes* occurs in western North America, from Washington state and Idaho to California and eastern Texas, USA, and south to the state of México, Mexico. The three species were revised by LaBerge (1956b).

Genus *Melissoptila* Holmberg

Melissoptila Holmberg, 1884: 119. Type species: *Melissoptila tandilensis* Holmberg, 1884, monobasic.
Thyreotremata Holmberg, 1887b: 225, nomen nudum.
Thyreothremma Holmberg, 1903: 391. Type species: *Thyreothremma rhopalocera* Holmberg, 1903 = *Melissoptila tandilensis* Holmberg, 1884, by designation of Sandhouse, 1943: 604.
Thyreotremata Sandhouse, 1943: 604. Type species: *Thyreothremma rhopalocera* Holmberg, 1903 = *Melissoptila tandilensis* Holmberg, 1884, by designation of Moure and Michener, 1955b: 305. Published as a junior synonym and therefore not available according to the Code, 3rd ed., art. 11(e). [See Moure and Michener, 1955b: 305 footnote.]
Ptilomelissa Moure, 1943b: 482. Type species: *Melissoptila bonaerensis* Holmberg, 1903, by original designation.
Melissoptila (Comeptila) Moure and Michener, 1955b: 304. Tye species: *Thyreothremma paraguayensis* Brèthes, 1909, by original designation.

This is a genus of *Melissodes*-like bees; most species are smaller than common species of *Melissodes*. It differs superficially from most *Melissodes* in its appressed tan metasomal pubescence and absence of basal hair bands on the metasomal terga. The maxillary palpi are short, two- or three-segmented. The stigma is larger than that of most Eucerini (see the key to genera). In the female the gradulus of T6 is lamellate laterally and ends in a strong tooth completely hidden among dense hairs. In the male the first flagellar segment varies from 0.3 to 0.9 times the length of the second. The male genitalia and hidden sterna were illustrated by Michener (1954b), Moure and Michener (1955b), and Urban (1968b); see also Figure 112-9g-i.

With justification, Urban (1998a) did not use the three commonly recognized subgenera of *Melissoptila*, largely because two of them (*Comeptila* and *Melissoptila* s. str.) contain only one or two species each, whereas the other (*Ptilomelissa*) contains all the rest. The two small subgenera merely segregate two or three species with odd males from a large and variable group. Both of the small segregates have the middle leg broadened in the male, the basitarsis little if any more than twice as long as broad. *Melisopitila* s. str. lacks the middle tibial spur in the male, and the last antennal segment is broadened, while *Comeptila* males have rather short antennae for a eucerine, reaching little beyond the wing bases, and S6 has the convergent carinae long and lamelliform. These segregates contain moderate-sized species, 10 to 13.5 mm in body length, whereas more ordinary forms (formerly *Ptilomelissa*) are among the smallest Eucerini, 6.5 to 10 mm long.

■ *Melissoptila* occurs from central Chile, and Chubut Province, Argentina, north through the tropics to the Greater Antilles and southern Texas, USA. Urban (1968b) revised the genus and Urban (1998a) revised the South American forms; she recognized 52 species.

Genus *Micronychapis* Moure and Michener

Micronychapis Moure and Michener, 1955b: 286. Type species: *Tetralonia duckei* Friese, 1908, by original designation.

The lone species of this genus is robust, anthophoriform, and resembles a large *Peponapis* or *Xenoglossa*, or a *Centris* with a red abdomen. The body length is about 13.5 mm. Distinctive characters are the short middle basitarsus, less than two-thirds the tibial length; the robust tibial spurs; the minute arolium and claws, the rami of the claws almost equal in length; the lamelliform lateral arm of the gradulus of S6 of the female, forming a rounded tooth; and the deeply notched apex of S6 of the male. The male genitalia and hidden sterna were illustrated by Moure and Michener (1955b).

■ This genus is found in the states of Pará and Rio de Janeiro, Brazil. The only species is *Micronychapis duckei* (Friese).

Genus *Mirnapis* Urban

Mirnapis Urban, 1998d: 565. Type species: *Mirnapis inca* Urban, 1998, by original designation.

This genus, known only in the male, seems related to *Gaesischia, Hamatothrix,* and *Santiago,* and like those genera, has the general aspect of a *Melissodes.* The body length is 10 to 11 mm. The first flagellar segment is much shorter than the second but longer than broad. The ocelli are nearly in a straight line, the anterior margin of the median one being only slightly in front of the anterior tangent of the lateral ocelli. S6 has submarginal convergent carinae that are well separated at their apices; otherwise the surface of S6 is rather flat. T6 has a distinct lateral tooth while T7 lacks teeth.

■ The only known species, *Mirnapis inca* Urban, is from the Andes of Peru.

Genus *Notolonia* Popov

Notolonia Popov, 1962a: 294. Type species: *Notolonia astragali* Popov, 1962, by original designation.

In this genus the clypeus is produced in front of the eye by a distance greater than the width of the eye as seen in lateral view, and the proboscis is extremely long (see the key to genera). The male antennae scarcely reach the base of the metasoma; the long first flagellar segment is noted in the key to genera. T7 of the male lacks lateral teeth. The body length is 15 to 16 mm. Male genitalia and other structures were illustrated by Popov (1962a).

■ This genus is known only from Turkmenistan. The only species is *Notolonia astragali* Popov.

Genus *Pachysvastra* Moure and Michener

Pachysvastra Moure and Michener, 1955b: 284. Type species: *Tetralonia leucocephala* Bertoni and Schrottky, 1910, by original designation.

Like *Svastrina, Pachysvastra* lacks arolia. Another unusual character is the vestiture of the stipes and galea of the female, which consists of abundant hairs hooked and undulate at their apices. The body length is 9.0 to 11.5 mm. The male genitalia and hidden sterna were illustrated by Moure and Michener (1955b) and Urban (1974c).

■ This genus is known from São Paulo to Mato Grosso and Minas Gerais, Brazil. The only species is *Pachysvastra leucocephala* (Bertoni and Schrottky).

Genus *Peponapis* Robertson

Peponapis Robertson, 1902b: 324. Type species: *Macrocera pruinosa* Say, 1837, by original designation.
Peponapis (*Colocynthophila*) Moure, 1948: 342. Type species: *Tetralonia fervens* Smith, 1879, by original designation.
Peponapis (*Eopeponapis*) Hurd and Linsley, 1970: 20. Type species: *Xenoglossa utahensis* Cockerell, 1905, by original designation.
Peponapis (*Austropeponapis*) Hurd and Linsley, 1970: 21. Type species: *Tetralonia melonis* Friese, 1925, by original designation.
Peponapis (*Xeropeponapis*) Hurd and Linsley, 1970: 28. Type species: *Peponapis timberlakei* Hurd and Linsley, 1964, by original designation.
Peponapis (*Xenopeponapis*) Hurd and Linsley, 1970: 29. Type species: *Melissodes crassidentata* Cockerell, 1949, by original designation.

Peponapis consists of robust species, 11 to 16 mm long, nearly anthophoriform rather than euceriform. The clypeus is strongly protuberant; the anterior articulation of the mandible is twice as far from the eye margin as the posterior articulation. The scopa is sparse, with very few branches to coarsely plumose; the inner surface of the hind basitarsus of the female except in *P. fervens* (Smith) from South America, has especially sparsely placed large hairs but, as in other genera, along the lower margin of that surface is a band of dense, shorter hairs. Unlike many male Eucerini, in which frequently the whole clypeus is yellow, in male *Peponapis* there is usually a midapical yellowish area, commonly small. There is considerable morphological diversity among the species of *Peponapis;* for example, the maxillary palpi range from four- to six-segmented (four-segmented only in southern South America). One species, *P. crassidentata* (Cockerell), has distinct lateral gradular teeth on T7 of the male, unlike all others of the genus. In another, *P. timberlakei* Hurd and Linsley, the first flagellar segment of the male is about as long as the second, as in *Xenoglossa.* Nonetheless, I have chosen not to recognize the subgenera (Hurd and Linsley, 1970); six subgenera for 13 species appears to be unnecessary splitting. Male genitalia and hidden sterna were illustrated by Michener (1954b), Moure and Michener (1955b), and Mitchell (1962).

■ *Peponapis* occurs from Ontario, Canada, and Maine to California and Georgia, USA, and throughout Mesoamerica to Trinidad, Venezuela, the Andean countries (except Chile), and in eastern South America from Bahia, Brazil, to Río Negro and Jujuy provinces, Argentina. The 13 species, mostly from Mexico and the southwestern United States, were revised by Hurd and Linsley (1964, North America; 1966, Mexico; 1967b, South America; 1970, subgenera).

Peponapis is dependent on the pollen of *Cucurbita* (Cucurbitaceae).

Genus *Platysvastra* Moure

Platysvastra Moure, 1967b: 148. Type species: *Platysvastra macraspis* Moure, 1967, by original designation.

Platysvastra can be recognized in the female and probably also in the unknown male by the robust body, 13.5 mm long, with the flat scutellum that projects over the metanotum and propodeum. In this character it suggests, but is more extreme than, *Gaesochira*. The female differs further from *Gaesochira* in its shorter middle tibial spur, only half as long as the basitarsus, and in the strong gradular tooth on T6 (it is lamellate but not toothed in *Gaesochira*). The male will be needed to learn whether *Platysvastra* is actually closely related to *Gaesochira* or convergent in scutellar form. Moure (1967b) illustrated various characters, including the thoracic profile.

■ This genus is known from Guyana. There is one species, *Platysvastra macraspis* Moure.

Genus *Santiago* Urban

Santiago Urban, 1989c: 117. Type species: *Santiago mourei* Urban, 1989, by original designation.

Santiago is probably close to *Gaesischia*. It differs from that genus in having a strongly protuberant clypeus and long proboscis, as indicated in the key to genera. The somewhat tapering male flagellum resembles that of many *Gaesischia*, as does the large mesal apical lobe of S7 of the male; it bears an apical straplike process, expanded apically. *Santiago* resembles *Svastrides* in its large size (body length 10.0-14.5 mm) and most of the generic characters indicated, but it lacks the fuscous or black hairs found in all known *Svastrides*. In the female the basitibial plate is completely hidden by short, dense, appressed hair; the plate is largely but incompletely hidden in *Svastrides*. The male gonostylus lacks the strongly plumose hairs found in *Svastrides*. The male genitalia and other structures of *Santiago* were illustrated by Urban (1989c).

■ *Santiago* was described from Minas Gerais, Brazil, on the basis of *S. mourei* Urban. A second species, from Peru, was described by Urban (2003).

Genus *Simanthedon* Zavortink

Simanthedon Zavortink, 1975: 232. Type species: *Simanthedon linsleyi* Zavortink, 1975, by original designation.

This genus contains a species 12 to 15 mm long that is similar in appearance and many other features to *Martinapis* s. str. For example, the male antenna tapers to a rounded point, the scape of the male is yellow, and the outer margin of the tegula is broadly impunctate and translucent. It differs from *Martinapis* s. str., however, in the longer male antennae, convex outer tegular margin, the narrow V-shaped pygidial plate of the female, lack of a gradular tooth on T6 of the male, the strongly sclerotized lateral apical lobe of S7 of the male, and the tapering male gonostylus. Male genitalia and other structures were illustrated by Zavortink (1975).

■ *Simanthedon* occurs in New Mexico and Arizona, USA, south to Durango, Mexico. The only species is *S. linsleyi* Zavortink.

Genus *Svastra* Holmberg

This genus contains species that are usually larger (length 8.5-20.0 mm) than most *Melissodes*. Spatuloplumose hairs (Fig. 112-10b) are usually present among other hairs of the basal band (sometimes also in the apical band) of T2 and sometimes T3 and T4. The tegula is of the usual shape, the outer margin convex throughout, not as configured in *Melissodes*. The lateral arm of the gradulus of T6 of the female is lamellate, ending in a weak tooth. T7 of the male lacks gradular teeth. The male gonostylus is long compared to that of most *Melissodes*.

Most species of this genus are North and Central American, but a small group (*Svastra* s. str.) is found on the other side of the tropics, in temperate South America.

Key to the Subgenera of *Svastra*

1. Scopal hairs simple; first flagellar segment of male as long as second, last segment tapering, attenuate (nearctic) .... *S*. (*Anthedonia*)
—. Scopal hairs plumose; first flagellar segment of male much shorter than second, last segment cylindrical with rounded apex 2
2(1). Middle tibial spur hooked near apex (Fig. 112-15b); front tibial spur of male as long as basitarsus or slightly longer; posterior margin of male S5 with shallow lateral emargination bordered by long, posteriorly directed, hooked hairs overlying shallow, bare depression of S6 (Fig. 112-10c) (nearctic) .... *S*. (*Idiomelissodes*)
—. Middle tibial spur not hooked; front tibial spur of male shorter than basitarsus; S5 of male without lateral emarginations, without hooked hairs 3
3(2). Hairs of basal band of T2 and elsewhere plumose, spatuloplumose hairs absent; antenna of male scarcely reaching scutellum, undersurface of scape bright yellow .... *Brachymelissodes*
—. Spatuloplumose hairs present in basal band of T2 and elsewhere; antenna of male reaching beyond scutellum, scape without yellow 4
4(3). Maxillary palpus five-segmented; scutoscutellar suture without, and basal band of T3 with, spatuloplumose hairs like those of T2; S6 of male with pair of converging carinae (South America) .... *S*. (*Svastra* s. str.)
—. Maxillary palpus usually four-segmented; scutoscutellar suture with, and T3 without, spatuloplumose hairs like those of T2; S6 of male without pair of converging carinae, sometimes with ridges instead (North and Middle America) .... *S*. (*Epimelissodes*)

Svastra / Subgenus *Anthedonia* Michener

Anthedon Robertson, 1900: 53 (not Agassiz, 1847). Type species: *Melissodes compta* Cresson, 1878, monobasic.
Anthedonia Michener, 1942a: 282, replacement for *Anthedon* Robertson, 1900. Type species: *Melissodes compta* Cresson, 1878, autobasic.
Abda Sandhouse, 1943: 521, replacement for *Anthedon* Robertson, 1900. Type species: *Melissodes compta* Cresson, 1878, autobasic.

Anthedonia consists of large (13-18 mm long), robust species similar to other *Svastra*, although usually given separate generic status. The female differs from other *Svastra* in having the scopal hairs simple. Males differ from other *Svastra* in several respects: the first flagellar segment is as long as the second; the last flagellar segment is tapering, attenuate; the gonostylus has a bend at or before the middle in one species and weakly in the other; and on S7 the inner apical lobe is broadly expanded across and beyond the outer apical lobe. Like most *Svastra*, *Anthedonia* has areas of spatuloplumose hairs; their locations are along the scutoscutellar suture, on the metanotum, and in areas of pale hair on T1 to T4. Genitalia and hidden sterna were illustrated by LaBerge (1957) and Mitchell (1962).

■ *Anthedonia* occurs from New Jersey to Georgia, thence west to California, USA, and south to Durango, Mexico. The two species were revised by LaBerge (1955). Species of this subgenus appear to be oligolectic on flowers of Onagraceae.

Svastra / Subgenus *Brachymelissodes* LaBerge

Melissodes (Brachymelissodes) LaBerge, 1956a: 926. Type species: *Eucera cressonii* Dalla Torre, 1896, by original designation.

In this subgenus the male antennae are short, not or scarcely reaching the scutellum, and the first flagellar segment is more than half as long as the second. The eyes of the male are bulging. The hairs that are spatuloplumose in most *Svastra* are plumose, not spatulate. The body length is 8.5 to 13.0 mm. Male genitalia and sterna were illustrated by LaBerge (1956a).

■ *Brachymelissodes* occurs from Nebraska to Texas, USA, and Durango, Chihuahua, and Sonora, Mexico. The two species were revised by LaBerge (1956a).

Svastra / Subgenus *Epimelissodes* Ashmead

Epimelissodes Ashmead, 1899a: 63. Type species: *Melissodes atripes* Cresson, 1872, by original designation.

The scutoscutellar suture in this subgenus is marked by a fascia that includes spatuloplumose hairs, like those of T2, but T3 lacks such hairs. The maxillary palpus is usually four-segmented. The male antennae are of moderate length, scarcely reaching T1, and the first flagellar segment is more than one-third, often more than one-half, as long as the second. The metanotum has a coarsely punctured median prominence covered with long hairs. The body length is 9.5 to 20.0 mm. Illustrations of male genitalia are in LaBerge (1956a) and Mitchell (1962).

■ *Epimelissodes* occurs from Quebec, Canada, to Florida, west to Washington state, USA, and Baja California and south to Veracruz and Oaxaca, Mexico; two species are also reported from Costa Rica. The 13 species were revised by LaBerge (1956a).

The name *Epimelissodes* was long misused for certain South American species of *Melissoptila*.

Svastra / Subgenus *Idiomelissodes* LaBerge

Melissodes (Idiomelissodes) LaBerge, 1956a: 1027. Type species: *Melissodes duplocincta* Cockerell, 1905, by original designation.

Idiomelissodes has been regarded as a distinct genus, although its relation to *Epimelissodes* is clear. It has the aspect of a *Melissodes* 10 to 12 mm long. Spatuloplumose hairs are abundant on the scutoscutellar suture, on the anterior surface of T1, and in the basal hair band of T2. The extremely slender male gonostyli are unique among American Eucerini. The male genitalia and hidden sterna were illustrated by LaBerge (1956a).

■ This subgenus occurs in the deserts from Baja California Sur to Durango and Coahuila, Mexico, north to California to New Mexico, USA. The single species is *Svastra duplocincta* (Cockerell), which appears to be an oligolectic visitor to flowers of cactus, especially *Ferocactus*.

Svastra / Subgenus *Svastra* Holmberg s. str.

Svastra Holmberg, 1884: 127. Type species: *Svastra bombilans* Holmberg, 1884, by designation of Sandhouse, 1943: 602.

In this subgenus, the scutoscutellar suture lacks spatuloplumose hairs but the basal hair band of T3 has such hairs, as does the band of T2. The maxillary palpus is five-segmented. The body length is 10 to 15 mm. The genitalia and hidden sterna of the male were illustrated by Michener, LaBerge, and Moure (1955a).

■ *Svastra* s. str. occurs in central Chile and from Río Negro to Misiones, Argentina. The three species were listed by Michener, LaBerge, and Moure (1955a).

Genus *Svastrides* Michener, LaBerge, and Moure

Svastrides Michener, LaBerge, and Moure, 1955a: 220. Type species: *Tetralonia melanura* Spinola, 1851, by original designation.

This genus contains rather large, elongate species (10-15 mm long) with abundant fuscous or black pubescence. The lack of apical hair bands on T1 to T3, the presence of basal zones, often weak, of pale hair on T2 and T3, and the large second submarginal cell suggest *Svastra*, a very different genus having much smaller distal lobes of S7 of the male, spatuloplumose hairs, and a longer first flagellar segment in the male but shorter flagellum as a whole. The flagellum of male *Svastrides* is very long, most of the segments flattened, the distal ones slightly crenulate, tapering. The male gonostyli have coarse plumose hairs; the genitalia and hidden sterna were illustrated by Michener, LaBerge, and Moure (1955a) and Urban (1972, 1975). These structures are similar to those of *Gaesischia*, and it is possible that *Svastrides* should become a subgenus of *Gaesischia*.

■ *Svastrides* occurs from Aisen, Chile, and Neuquén, Argentina, north to Santiago del Estero, Argentina, and in the Andean uplift to Peru. The four species were revised by Urban (1972); a new key was published by Urban (1975).

Genus *Svastrina* Moure and Michener

Svastrina Moure and Michener, 1955b: 276. Type species: *Svastrina anaroliata* Moure and Michener, 1955 = *Svastra subapicalis* Brèthes, 1910, by original designation.

In form and coloration, *Svastrina* suggests *Svastrides*, i.e., as in that genus, most of the pale hairs on the metasoma of the male are near its apex (on T4 to T6). The body length is 11 to 12 mm. Like *Pachysvastra*, *Svastrina* lacks arolia. Other characters are the deeply notched apex of the labrum and the rather long vein cu-v of the hind wing, two-thirds as long as the second abscissa of M+Cu. The male differs from that of *Pachysvastra* in the lack of lateral (gradular) teeth on T7, the long antennae, and the unnotched pygidial plate. The male genitalia and hidden sterna were illustrated by Moure and Michener (1955b) and by Urban (1974a). The female differs from that of *Pachysvastra* in the lack of hooked and undulate hairs on the prementum and the slender base of the first flagellar segment (see key). The female was first described by Urban (2003); I have not seen specimens.

■ The genus is known from the provinces of Sante Fe and Formosa, Argentina, and the state of Mato Grosso, Brazil. The only species is *Svastrina subapicalis* (Brèthes).

Genus *Syntrichalonia* LaBerge

Syntrichalonia LaBerge, 1957: 10. Type species: *Melissodes exquisita* Cresson, 1878, by original designation.

This genus includes large (body length 14-17 mm), robust species well covered with erect ochraceous to fulvous pubescence, the weak metasomal bands being formed by longer and less erect apical hairs of the same color. The male antennae are relatively short, not reaching the posterior end of the thorax, and the first flagellar segment is subequal to the second, characters that distinguish *Syntrichalonia* from most of its relatives. The clypeus is little protuberant, the blade of the galea is no longer than the eye, and on S6 of the male the convergent carinae are simple; these are characters of *Tetraloniella*, to which *Syntrichalonia* may be related. Male genitalia and hidden sterna were illustrated by LaBerge (1957).

■ *Syntrichalonia* occurs from southern Arizona to Texas, USA, south to Oaxaca, Mexico. The two species were differentiated by LaBerge (1994).

Genus *Tetralonia* Spinola

For many years, especially early in this century, the name *Tetralonia* was used for almost all Eucerini with three submarginal cells. More recently it has been restricted, for example, to the forms here placed in *Tetralonia, Tetraloniella*, and *Eucera (Synhalonia)*, or only to the last, and it is now further restricted to a small group of species including *T. malvae* (Rossi) plus the species of the related subgenera *Eucara* and *Thygatina*, both of which have usually been given generic status.

The sparse tibial scopa, consisting of branched hairs, often with most of the branches on one side, is distinctive, as are the other characters indicated in the key to genera. The male antennae are relatively short but may reach the base of the metasoma. The males usually have unusually large claws, particularly on the middle and hind legs. In side view the clypeus protrudes forward usually less than the width of the eye, less in *Tetralonia* s. str. than in the other subgenera. The body length is 10 to 16 mm, the larger species being in the subgenus *Eucara*.

The American genera *Peponapis* and *Xenoglossa* share various features—especially the sparse scopa, which consists of coarsely branched hairs, and the relatively short antennae of males—with *Tetralonia*. If these features are

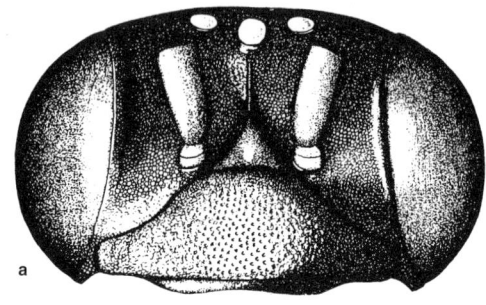

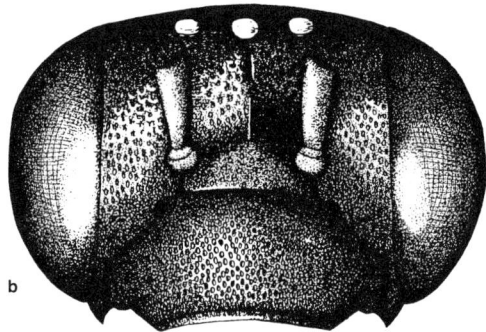

Figure 112-18. Faces of female Eucerini. **a,** *Tetralonia (Eucara) macrognatha* (Gerstaecker); **b,** *Tetraloniella (Tetraloniella) junodi* (Friese). From Eardley, 1989.

determined to be homologous rather than convergent, the relations could be indicated by placing those genera as subgenera of *Tetralonia*.

At least some species of this genus are specialist collectors of pollen of Malvaceae, perhaps also of *Ipomoea* (Convolvulaceae), and the distinctive scopa may be an adaptation to the coarse pollen of these plants.

Key to the Subgenera of *Tetralonia*

1. T7 of male with lateral tooth; male gonostylus gently arcuate in side view; maxillary palpus five-segmented, fifth segment minute and indistinct; antenna of male reaching base of metasoma (palearctic) ....
   .... *T. (Tetralonia* s. str.*)*
—. T7 of male without lateral tooth; male gonostylus elbowed as seen in lateral view; maxillary palpus three- to four-segmented; antenna of male reaching scutellum 2
2(1). Lower clypeal margin mesal from mandibular articulation nearly transverse and rounding onto transverse or weakly concave median part of clypeal margin; lower lateral part of clypeus, as seen from beneath, curved back on either side of labrum in such a way that margins are at angle of about 45° to long axis of body (Africa).... *T. (Eucara)*
—. Lower clypeal margin mesal from mandibular articulation diagonal, separated by distinct angle from transverse median part of clypeal margin; lower lateral part of clypeus, as seen from beneath, curved back on either side of labrum in such a way that margin at each side is nearly parallel to long axis of body (Africa, oriental) ....
   .... *T. (Thygatina)*

Tetralonia / Subgenus *Eucara* Friese

Anthophora (Eucara) Friese, 1905b: 241. Type species: *Anthophora laticeps* Friese, 1905, by designation of Cockerell, 1933a: 465.

The extraordinarily short, broad face (Fig. 112-18a) and strongly diverging eyes of *Eucara* are distinctive, although closely approached in some Indian species of the subgenus *Thygatina*. The less-than-usually-curved-backward lower lateral parts of the clypeus are probably a result of the broad lower face, and are unique in the Eucerini. Most species have, in males, somewhat modified hind legs, unlike those of males of *Eucera* and *Tetraloniella*, some of which have modified middle legs. Various structures were illustrated by Eardley (1989).

■ *Eucara* occurs in Africa from eastern Cape Province, South Africa, north to Ethiopia and Burkina Faso. The seven known species were revised by Eardley (1989).

Tetralonia / Subgenus *Tetralonia* Spinola s. str.

Macrocera Latreille, 1810: 339, 439 (not Meigen, 1803). Type species: *Eucera antennata* Fabricius, 1793 = *Apis malvae* Rossi, 1790, by original designation.

Tetralonia Spinola, 1839: 538, replacement for *Macrocera* Latreille, 1810. Type species: *Eucera antennata* Fabricius, 1793 = *Apis malvae* Rossi, 1790, autobasic. [The identity of the type species was clarified by Michener (1997a) and Baker (1996b).]

Tetralonia s. str. is the most *Tetraloniella*-like subgenus of *Tetralonia*, in that the head is of an ordinary shape with the eyes commonly converging below; the pubescence of the legs is commonly pale; and the maxillary palpi and male antennae are long compared to those of other subgenera of *Tetralonia*. This subgenus is unlike most related Eucerini but like the subgenus *Thygatina* in having the margin of the basitibial plate of the female exposed.

■ This subgenus is found in the Mediterranean basin, north to Germany and Russia. I know only one species, *Tetralonia malvae* (Rossi), although D. B. Baker (personal communication, 1995) indicated that there are two or three other species in the group.

Tetralonia / Subgenus *Thygatina* Cockerell

Thygatina Cockerell, 1911f: 237. Type species: *Thygatina fumida* Cockerell, 1911, monobasic.

This subgenus contains species most of which have dusky or black hairs on the legs and sometimes other parts of the body. The group is intermediate between *Eucara* and *Tetralonia* s. str. Thus the head is wider than in the latter, the interocular distance being about equal to or more than the eye length, but is usually not so extraordinarily broad as in *Eucara*; and the inner orbits are parallel or diverging below. The lower clypeal curvature is as in *Tetralonia* s. str.; *Eucara*-like characters are listed in the first couplet of the key to subgenera. The hind leg of the male lacks the modifications found in most species of *Eucara*. An unnamed species from southern India has a relatively broad head and diverging eyes, and is thus quite *Eucara*-like. Various structures were illustrated by Eardley (1989), who placed the African species in *Tetralonia*, as distinct from *Eucara*.

■ *Thygatina* occurs in southern India and Sri Lanka, and in Africa from eastern Cape Province north to Kenya and Ivory Coast. There are nine named species, six of them African. The African species were revised by Eardley (1989) under the name *Tetralonia*; the Indian and Sri Lankan species were reviewed by Engel and Baker (2006).

Genus *Tetraloniella* Ashmead

This genus consists of species most of which are smaller than *Eucera (Synhalonia)* and have pale metasomal hair bands; most males of *Eucera (Synhalonia)* do not. Thus its species can often be distinguished on sight from *Eucera (Synhalonia)*. Nonetheless, intermediates seem to occur in small numbers, and the distinguishing features are not satisfying. In *Tetraloniella* the clypeus is less protuberant than in *Eucera*, extending in front of the eye by less than the width of the eye as seen in lateral view, and the blade of the galea is little longer than the eye. Exceptions to these characters occur, however, in such species as *T. mastrucata* (Morawitz) and *intermedia* (Morawitz), in males of which the clypeus is protuberant by more than the eye width as seen in lateral view, and in which the blade of the galea is 1.3 to 1.5 times as long as the eye. The size and metasomal vestiture of those species are as in *Tetraloniella*, however, and the convergent carinae of S6 of the male are small and simple, as in *Tetraloniella*. In the subgenus *Loxoptilus*, also, the blade of the galea is very long and the clypeus is strongly protuberant. The paraocular carinae, at least in most females of *Tetraloniella*, extend with little or no interruption to the lateroclypeal carinae, a feature not found in *Eucera (Synhalonia)*. This character is quite useful. The lower part of the paraocular carina is sometimes irregular as a result of punctation but at least in females a small ridge extends almost to the upper end of the lateroclypeal carina. In *Eucera (Synhalonia)* the scopal hairs are simple; in most *Tetraloniella* they are either plumose or mostly simple but with plumose hairs along the upper margin of the tibial scopa. In some *Tetraloniella*, however, including species of the subgenus *Pectinapis*, the scopal hairs are all simple. Similar problems exist with all characters that might distinguish *Tetraloniella* from three-celled *Eucera*; yet the genera here recognized seem to be useful units, at least until an adequate study of the group is made. Male genitalia and hidden sterna were illustrated by LaBerge (1957), Iuga (1958), Mitchell (1962), Popov (1962a), Eardley (1989), and Tkalců (1993b), some of them as *Xenoglossodes*.

Tetraloniella may well be a paraphyletic group from which *Eucera* arose. The subgenera recognized below are simply groups that are distinctive and that have been named; *Tetraloniella* s. str. consists of all of the other species. This classification is anything but satisfying but represents the current stage of knowledge of the group.

Baker (1996b) gave an account of the problem concerning the identity of *Tetraloniella graja* Eversmann, the type species of the genus. It is likely that whatever the solution to this problem, *T. graja* will be a species of *Tetraloniella* s. str. as that taxon is recognized below, and thus a correct identification of *T. graja* would not modify the nomenclature here presented.

Key to the Subgenera of *Tetraloniella*

1. Gonostylus of male about three times as long as basal width, as seen in side view; basitibial plate of female almost bare, margin fully exposed and not at all raised but curving down to level of rest of tibia; claws of female with inner tooth minute, its axis parallel to that of distal part of claw (palearctic Asia) .... *T. (Glazunovia)*
—. Gonostylus of male five or more times as long as basal width; basitibial plate of female with small, more or less appressed hairs that usually hide part of margin, which is often minutely raised; claws of female with inner tooth larger, diverging from axis of distal part of claw 2
2(1). Clypeus of female broadly flattened, covered with short, erect, robust hairs with hooked or wavy apices (Fig. 112-14c) (oculoclypeal distance of male 0.5 to 1.0 times minimum width of first flagellar segment) (Mexico, Texas)) .... *T. (Pectinapis)*
—. Clypeus of female convex, with ordinary vestiture..........3
3(2). Maxillary stipes of female with longitudinal band of coarse bristles having hooked or wavy apices; oculoclypeal distance greater than minimum width of first flagellar segment (Fig. 112-6a, f) (Mexico) .... *T. (Loxoptilus)*
—. Maxillary stipes without such bristles; oculoclypeal distance variable but often minute, less than one-fourth of minimum width of first flagellar segment (holarctic and Africa) .... T. (*Tetraloniella s. str.*)

Tetraloniella / Subgenus *Glazunovia* Baker

Glazunovia Baker, 1998: 846. Type species: *Tetralonia nigriceps* Morawitz, 1895, by original designation.

Glazunovia consists of a black species with abundant white hair forming basal white bands on T2 to T4 and apical white bands on T3 and T4 of the female, T2 to T6 of the male. In both sexes, the face lacks yellow or white areas. The body length is 9 to 12 mm. The following are some characters other than those indicated above and in the key to subgenera: the paraocular carina is broken by punctation before reaching the lateroclypeal carina; the maxillary palpus is six-segmented, about as long as the basal width of the galeal blade, the last segment minute; the marginal cell is unusually short and broad; the scopa is plumose, the keirotrichiate area absent (in this respect, *Glazunovia* is unlike other *Tetraloniella,* which have such an area on the distal half of the tibia); T6 and T7 of the male each have a rather small, right-triangular, lateral tooth; and S6 of the male lacks a lateral tooth but has two rather short carinae converging posteriorly, as is usual in *Tetraloniella.*

■ This subgenus occurs in Central Asia. The only known species is *Tetraloniella nigriceps* (Morawitz).

Popov (1962a) recognized the distinctiveness of the species now placed in *Glazunovia* and illustrated the male genitalia, hidden sterna, wing, and faces of both sexes. He placed it in the genus *Melissina,* which he had not seen, and which is a synonym of *Tetraloniella* s. str., as shown by its slender, elbowed male gonostyli, the presence of a keirotrichiate area on the female hind tibia, and other characters indicated in the key to subgenera.

Tetraloniella / Subgenus *Loxoptilus* LaBerge

Loxoptilus LaBerge, 1957: 28. Type species: *Loxoptilus longifellator* LaBerge, 1957, by original designation. [New status.]

This is a probable derivative of *Tetraloniella* s. str. The strongly protuberant clypeus, in males produced by a distance nearly equal to the eye width as seen in lateral view, the long blade of the galea, much exceeding the eye length, and the great oculoclypeal distance, greater than the minimum width of the first flagellar segment (Fig. 112-6a, f), are not typical of *Tetraloniella,* although each of these characters can be found in one or another species of *Tetraloniella* s. str. The strong paraocular carinae, joining the lateroclypeal carinae, and the simple converging carinae on S6 of the male, are features of *Tetraloniella.* The only unique feature of *Loxoptilus* is the strong, hooked or wavy bristles on the stipes of the female. These bristles and the long proboscis suggest a specialist on some deep-flowered plant, as yet unknown. The body length is 9 to 15 mm. The male genitalia and hidden sterna were illustrated by LaBerge (1957).

■ *Loxoptilus* is found from Sonora to Morelos, Puebla, and Yucatan, Mexico. There are two species, differentiated by LaBerge (1957).

Tetraloniella / Subgenus *Pectinapis* LaBerge

Pectinapis LaBerge, 1970: 322. Type species: *Pectinapis fasciata* LaBerge, 1970, by original designation. [New status.]

Pectinapis has hitherto been given generic rank. It is readily recognized by the clypeal characters of the female given in the key to subgenera. Females of *Tetraloniella fasciata* (LaBerge) but not other species are further remarkable for the comb of erect bristles across the supraclypeal area below the antennae, and females of that species and of *T. auricauda* (LaBerge) are noteworthy for the expansion of the clypeus upward almost to the antennal bases. In various other groups of bees, specialist collectors of *Salvia* pollen have a flattened clypeus with erect hooked or wavy bristles, as in *Pectinapis.* These features by themselves are not elsewhere regarded as generic characters, and I see no reason why they should be generic characters in the Eucerini. The greater oculoclypeal distance as compared to most *Tetraloniella,* is duplicated in some palearctic species. The body length of *Pectinapis* varies from 10 to 15 mm. Male genitalia and hidden sterna were illustrated by LaBerge (1970).

■ *Pectinapis* occurs from Texas, USA, and Jalisco, Mexico, south to Oaxaca, Mexico. The four species were revised by LaBerge (1989b).

Tetraloniella / Subgenus *Tetraloniella* Ashmead s. str.

Tetraloniella Ashmead, 1899a: 61. Type species: *Macrocera graja* Eversmann, 1852, by original designation.
Xenoglossodes Ashmead, 1899a: 63. Type species: *Melissodes albata* Cresson, 1872, by original designation.
Melissina Cockerell, 1911e: 670. Type species: *Melissina viator* Cockerell, 1911, monobasic.

Tetraloniella s. str. contains all of the species of the genus that have not been placed in other subgenera. The body length ranges from 7 to 14 mm. Among its species, there exist many combinations of the characters that vary. As in *Eucera (Synhalonia)*, the male gonostylus varies from elbowed near the base to gently curved. No other character seems to be correlated with this gonostylar character. Variations in other characters discussed under the generic heading all occur within the subgenus *Tetraloniella* s. str. *Tetraloniella dentata* (Klug) is unusually large for the subgenus (body length 11-14 mm), and T7 of the male, unlike that of other species, has a blunt tooth on each side of the pygidial plate. It is further remarkable for the broadly truncate distal part of the female mandible, which is about three-fourths as broad as the mandibular base. Male genitalia and hidden sterna of the subgenus were illustrated by Mitchell (1962, as *Xenoglossodes*) and Tkalců (1979b).

■ This subgenus occurs in North America from South Dakota and Illinois to Texas, west to California, USA, and south to Panama. In the palearctic region it occurs from Spain to India and Central Asia, ranging north as far as northern Germany. In sub-Saharan Africa it occurs from northern Nigeria and Ethiopia south through east Africa to South Africa and Namibia, and also occurs in Madagascar. There are about 35 North American species (mostly Mexican), revised by LaBerge (2001); about 30 sub-Saharan species, revised by Eardley (1989); four species from Madagascar were revised by Pauly (in Pauly, et al., 2001); and an unknown number [because of confusion with *Eucera (Synhalonia)*], perhaps 50, of palearctic species.

Melissina, from Pakistan, was characterized by Cockerell (1911e), who noted the robust five-segmented maxillary palpi with the long fourth segment. His count, based on more than one specimen, was probably accurate, but on examining the type (London) I could not be certain of the segmentation beyond the two basal segments. The clypeus of the female, largely white, is protuberant for a distance little over one-third of the eye width as seen in lateral view. The tibial and basitarsal scopa is dense, hiding the surface or nearly so, the hairs simple except plumose along the upper margins of both segments. The margin of the basitibial plate of the female is entirely exposed, unlike that of other *Tetraloniella* s. str.; this is the only character known to me that might justify recognition of *Melissina* as a taxon distinct from the variable *Tetraloniella*. When *Tetraloniella* is well studied, however, *Melissina* may well be the appropriate name for one of its segregates.

Genus *Thygater* Holmberg

This is one of the more distinctive genera of Eucerini, as indicated in the key to genera. Useful characters not mentioned there are the very long male antennae, the black male clypeus often contrasted with a white or yellowish labrum, the strongly protuberant clypeus, and the long first flagellar segment of the female, longer than the scape. Genitalia and hidden sterna of males were illustrated by Michener (1954b) and, for both subgenera, by Moure and Michener (1955b) and Urban (1961, 1962, 1967a). The body length is 10 to 16 mm.

The genus was revised by Urban (1967a).

Key to the Subgenera of *Thygater* (Males)
(Modified from Moure and Michener, 1955b)

1. First flagellar segment one-sixth as long as second or less; flagellum reaching far beyond stigma, greatest breadth nearly equal to basal width of mandible; malar space one-third as long as wide or longer (neotropical) .... *T. (Thygater* s. str.)
—. First flagellar segment one-fourth to one-third as long as second; flagellum reaching stigma, greatest breadth about two-thirds basal width of mandible; malar space less than one-fourth as long as wide (South America) .... *T. (Nectarodiaeta)*

Key to the Subgenera of *Thygater* (Females)
(Modified from Moure and Michener, 1955b)

1. Eyes strongly diverging below; antenna less than twice as long as eye; middle flagellar segments usually less than twice as long as broad (neotropical) ...... *T. (Thygater* s. str.)
—. Eyes not or scarcely diverging below; antenna nearly twice as long as eye or longer; middle flagellar segments almost twice as long as broad or longer (South America) .... *T. (Nectarodiaeta)*

Thygater / Subgenus Nectarodiaeta Holmberg

Nectarodiaeta Holmberg, 1887b: 225, nomen nudum.
Nectarodiaeta Holmberg, 1903: 420. Type species: *Nectarodiaeta oliveirae* Holmberg, 1903, monobasic.

In this subgenus the clypeus is less protuberant than in *Thygater* s. str., the inner orbits of the female are subparallel, and the male flagellum is somewhat shorter and less flattened than in *Thygater* s. str. Thus it is less extreme in its differences from typical Eucerini than is *Thygater* s. str.

■ *Nectarodiaeta* occurs in Buenos Aires Province, Argentina, and the state of Paraná, Brazil. There are two species, revised by Urban (1961).

Urban (1962, 1967a) placed under the subgeneric name *Nectarodiaeta* all those species of *Thygater* having four rather than three segments in the maxillary palpi. No other characters are perfectly correlated with the palpal segmentation. This characterization seems arbitrary, since the fourth palpal segment is sometimes minute, as was noted by Urban, and since other characters, for example the lobes of S7 of the male, suggest that some species with four-segmented maxillary palpi are close relatives of species with three-segmented maxillary palpi. On the whole, I think that the characterization of this subgenus by Moure and Michener (1955b) results in natural groups probably worth recognizing. If characterized as by Urban, *Nectarodiaeta* should be regarded as a synonym of *Thygater*.

Thygater / Subgenus Thygater Holmberg s. str.

Thygater Holmberg, 1884: 133. Type species: *Tetralonia terminata* Smith, 1854 = *Macrocera analis* Lepeletier, 1841, monobasic.
Macroglossa Radoszkowski, 1884a: 17 (not Ochsenheimer, 1816). Type species: *Macroglossa oribazi* Radoszkowski, 1884 = *Macrocera analis* Lepeletier, 1841, monobasic.
Macroglossapis Cockerell, 1899c: 14, replacement for

Macroglossa Radoszkowski, 1884. Type species: *Macroglossa oribazi* Radoszkowski, 1884 = *Macrocera analis* Lepeletier, 1841, autobasic.

■ This subgenus occurs from Córdoba Province, Argentina, and the state of Rio Grande do Sul, Brazil, north through the tropics to the states of San Luis Potosí, Chihuahua, and Sonora, Mexico. It contains 23 species, which were revised, partly under the subgeneric name *Nectarodiaeta,* by Urban (1967a).

Genus *Trichocerapis* Cockerell

This genus is used here in a broader sense than in prior studies, in order to demonstrate the close relationship of *Dithygater* and *Trichocerapis* as these taxa were understood by Moure and Michener (1955b). The genus differs in many characters from *Thygater.* Such characters are the yellow clypeus of the male; the short apical spine on the front coxa of the female; and the slender segments 8 to 11 of the male flagellum, each with a fringe of hairs on one side, in contrast to the more robust segments 1 to 7. The genitalia and hidden sterna of both subgenera were illustrated by Moure and Michener (1955b); see also Figure 112-7d-f.

Key to the Subgenera of *Trichocerapis* (Males)
(Modified from Moure and Michener, 1955b)

1. Pygidial plate rounded posteriorly, marginal carina uniform; last flagellar segment longer than any of the others, not broadened; submarginal cells two .... *T. (Dithygater)*
—. Pygidial plate narrowed subapically and ending in small upturned process, marginal carina elevated subapically; last flagellar segment shorter than others except first, broad and flat; submarginal cells three .... *T. (Trichocerapis s. str.)*

Key to the Subgenera of *Trichocerapis* (Females)
(Modified from Moure and Michener, 1955b)

1. Clypeus without carinae but strongly declivous in upper lateral areas forming a pair of rounded ridges diverging below; malar area one-fifth as long as wide; submarginal cells two .... *T. (Dithygater)*
—. Clypeus with two distinct carinae diverging below; malar area about one-third as long as wide; submarginal cells three .... *T. (Trichocerapis s. str.)*

Trichocerapis / Subgenus *Dithygater* Moure and Michener

Dithygater Moure and Michener, 1955b: 309. Type species: *Dithygater seabrai* Moure and Michener, 1955, by original designation. [New status.]

This is the only New World eucerine taxon with two submarginal cells. Although it is related to *Thygater* and *Trichocerapis* s. str., it is by no means merely a *Thygater* or *Trichocerapis* having two submarginal cells. It differs from both in the five-segmented (rather than three- or four-segmented) maxillary palpus and the broad, weak ridges diverging from the upper median part of the clypeus toward the lower lateral angles (such ridges are present but carinate in *Trichocerapis* s. str.). *Dithygater* looks like a small (11.5- 15.0 mm long) *Trichocerapis* s. str.

■ This subgenus is known from the states São Paulo to Minas Gerais, Brazil. There is only one species, *Trichocerapis seabrai* (Moure and Michener) [new combination].

Trichocerapis / Subgenus *Trichocerapis* Cockerell s. str.

Trichocerapis Cockerell, 1904b: 292. Type species: *Tetralonia mirabilis* Smith, 1865, monobasic and by original designation.

The threadlike eighth to tenth flagellar segments of the male and the broad, flat, black eleventh segment are unique and unmistakable. S6 of the male has a strong sublateral tubercle and a pair of carinae converging posteriorly (S6 is simple in *Dithygater*). The labrum is as long as wide or, in the male, nearly so; in *Dithygater* it is much wider than long. The body length is 11 to 15 mm.

■ *Trichocerapis* occurs from Pernambuco, Brazil, to Paraguay and northwest to Peru. A key to the five species was published by Urban (1989a).

Genus *Ulugombakia* Baker

Ulugombakia Baker, 2003: 124. Type species: *Ulugombakia platytarsus* Baker, 2003, by original designation.

Because the male is unknown, this genus is not included in the Key to the Genera of Eucerini of the Eastern Hemisphere. See the note following the key.

Ulugombakia, based on a single female specimen, is most similar to *Tetraloniella* s. str. It differs in the 4-segmented maxillary palpus which, however, as in *Tetraloniella,* is longer than the maximum width of the galeal blade, and in the relatively large stigma, longer than the prestigma and (from Baker's photograph) extending well beyond the base of vein r. Such a stigma is unusual in Eucerini although approached in the unrelated American genus *Melissoptila.* The paraocular carinae become lamelliform and convergent anteriorly," unlike the simple and anteriorly reduced carinae of *Tetraloniella.* Baker (2003) enumerated other distinctive features.

■ *Ulugombakia* is known from West Malaysia. The single species is listed above.

Genus *Xenoglossa* Smith

Like *Peponapis, Xenoglossa* consists of robust, nearly anthophoriform rather than euceriform species (Pl. 11). The male antennae are shorter than those of *Peponapis,* not surpassing the tegulae. The first flagellar segment of the male is longer than the second (Fig. 112-8c); this condition is rare in *Peponapis.* The mandible of both sexes has a tooth on its inner margin near the base (Fig. 112-13e). The clypeus of the female is frequently maculated with yellow; it is black in *Peponapis.* In the female the inner surface of the hind basitarsus is tan with black spots representing the bases of the widely dispersed large hairs, except for a densely hairy band along the lower margin. This presumably derived condition is as in *Peponapis* and supports the view that *Peponapis* should be regarded as a subgenus of *Xenoglossa.* The possibility exists, however, that these characters are adaptive, related to the extremely coarse pollen of *Cucurbita,* and that the two genera originated from different ancestors. The arrangement used by

LaBerge (1957) suggests that this was his view. Male genitalia and hidden sterna were illustrated by LaBerge (1957) and Mitchell (1962).

Like species of *Peponapis*, those of *Xenoglossa* are oligolectic collectors of pollen of *Cucurbita*. *Xenoglossa* has been divided into two subgenera; they seem more distinct than the subgenera that have been proposed for *Peponapis*.

Key to the Subgenera of *Xenoglossa*

1. Eyes weakly diverging below, antennal socket closer to eye margin than to lateral ocellus; ocelli scarcely larger than antennal sockets .... *X. (Eoxenoglossa)*
—. Eyes markedly diverging below, antennal socket farther from eye margin than from lateral ocellus; ocelli enlarged, much larger than antennal sockets .... *X. (Xenoglossa s. str.)*

Xenoglossa / Subgenus *Eoxenoglossa* Hurd and Linsley

Xenoglossa (Eoxenoglossa) Hurd and Linsley, 1970: 34. Type species: *Melissodes strenua* Cresson, 1878, by original designation.

Species of this subgenus look much like *Peponapis*; their body length is 11 to 18 mm.

■ *Eoxenoglossa* occurs from Maryland to Florida west to Wisconsin and California, USA, and south to San Luis Potosí, Durango, and Baja California, Mexico. The two species were revised by Hurd and Linsley (1964, 1967a).

Xenoglossa / Subgenus *Xenoglossa* Smith s. str.

Xenoglossa Smith, 1854: 315. Type species: *Xenoglossa fulva* Smith, 1854, monobasic.

This subgenus includes large Eucerini, 16 to 24 mm long. The enlarged ocelli, large body size, and, for some species, the largely fulvous body (Pl. 11) suggest earlier matinal activity than for either *Eoxenoglossa* or the genus *Peponapis*.

■ *Xenoglossa* s. str. occurs from California to Texas, USA, and south through Mesoamerica to Nicaragua and possibly Panama and Venezuela. The five species were revised by Hurd and Linsley (1964, 1967a).

113. Tribe Anthophorini

This is a tribe of robust, fast-flying, anthophoriform (Pls. 12, 13) or rarely euceriform or apiform, pollen-collecting bees, apparently most closely related to the Centridini. The wings are largely bare, the distal parts beyond the veins being strongly papillate (as in Fig. 85-2a). The stigma is small, usually ending at the base of vein r. The marginal cell is slightly shorter than or about as long as the distance from its apex to the wing tip, and rounded and often appendiculate at the apex (Fig. 113-1). The first submarginal cell is short, much shorter than the combined lengths of the second and third submarginal cells, all measured along the posterior margins. The jugal lobe of the hind wing is less than one-half as long as the vannal lobe, usually about one-third as long.

Some authors divide this tribe to form the separate tribes Habropodini and the Anthophorini s. str., distinguishing them in part by the characters indicated in the first couplet of the key to genera. To me these groups of genera appear very similar and I do not recognize a tribe Habropodini. It seems important that the classification show the relationship of *Habropoda* and its allies to *Anthophora*.

Nests of Anthophorini are burrows in the soil, either in banks or in flat ground, with the exception of *Anthophora* subgenus *Clisodon*, which nests in rotten wood or pithy stems. The best-known species nest in aggregations in vertical clay banks, but probably the majority of species make nonaggregated nests in flat ground. The bees may facilitate excavation by moistening clay with regurgitated water or nectar. The barrel-shaped cells are either isolated or in series in the burrows, and are made of earth or clay in such a way that they can be separated from the surrounding matrix. Separation is possible because the cell wall is constructed in a larger cavity from clay particles moistened by Dufour's gland secretion and tamped into place with the pygidial plate. In the case of *Clisodon*, similar cells are made from chewed pith or wood particles. The cells are lined with more or less liquid triglyceride-rich material from Dufour's gland, which solidifies into a solid diglyceride-rich membranous "waxlike" lining on contact with salivary secretion in the food mass. This lining waterproofs the cell but is later eaten by the large larva. The provisions are a viscous or quite liquid mixture of pollen and nectar and Dufour's gland and salivary gland products that has a distinctive yeasty smell. This odor has been attributed to the action of yeast, but Batra and Norden (1996) noted it (they say the odor is that of butyric acid) from the moment that the triglyceride lining begins to be converted to solid diglycerides; the odor must therefore be a product of that reaction. After a cell is provisioned and an egg laid on the provisions, the cell is closed with a spirally constructed mud or earthen closure, in the center of which a small hole remains. The bee may insert the glossa through this hole to apply the lining material to the inner surface of the plug, then finally close the central hole with earth. Thus the entire inner surface of the cell—except for a small plug, less then 1 mm in diameter, in the center of the cell closure—is lined with the "waxlike" diglyceride material. Gess and Gess (1996) do not accept this explanation, at least for certain *Amegilla* species. Since there appears to be wax on the inner surface of the central closure, they believe that wax is added with the mud of the closure. Some of the works on anthophorine nesting behavior in earth are Malyshev (1928), Stephen (1961), Torchio and Youssif (1968), Rozen (1969c), Thorp (1969c), Torchio (1971), Brooks (1983), Norden (1984), Torchio and Trostle (1986), Houston (1991b), Batra and Norden (1996), and Alcock (1996b, 1997a,b, 1999). For nesting in rotten wood or pith (*Anthophora* subgenus *Clisodon*), see Medler (1964b).

Amegilla dawsoni (Rayment) in Australia has males of two different sizes, as do some species of *Centris* in the Western Hemisphere. Tomkins, Simmons, and Alcock (2001) relate this phenomenon to the provisioning strategy of females.

Unlike those of most Apinae, the mature larvae of Anthophorini do not spin cocoons and have lost much of the labial structure of cocoon-spinning larvae. Presumably, overwintering is usually as prepupae, but some species of *Anthophora (Lophanthophora)* (Brooks, 1988) and *Habropoda* (Stephen, 1961) pass the winter as adults in natal cells.

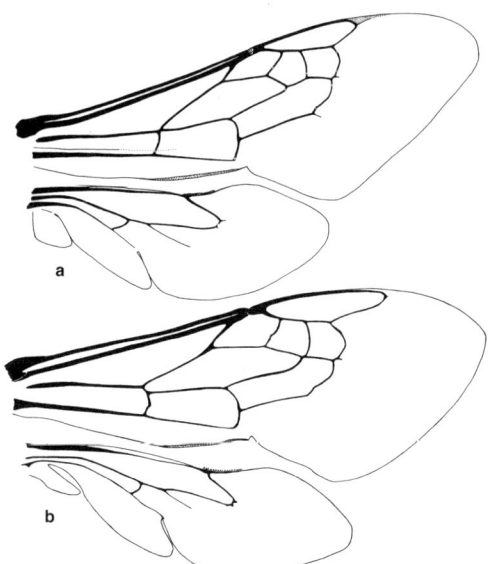

Figure 113-1. Wings of Anthophorini. **a,** *Anthophora occidentalis* Cresson; **b,** *Deltoptila montezumia* (Smith).

Key to the Genera of the Anthophorini

1. First recurrent vein joining second submarginal cell near middle; third submarginal cell subquadrate, with front and rear margins of about equal length, and basal and distal margins of about equal length (Fig. 113-1a); male gonostylus usually less than one-third as long as gono-

coxite (Fig. 113-2a, d), often not double, sometimes reduced to almost nothing 2
—. First recurrent vein terminating at or near apex of second submarginal cell (Fig. 113-1b); third submarginal cell (except in *Habrophorula*) with front margin much shorter than rear margin because of curvature of third submarginal crossvein, such that distal margin of the cell is longer than basal margin; male gonostylus double, dorsal and ventral gonostyli usually both one-third as long as gonocoxite or more 3
2(1). Arolia present (worldwide except Australian region) .... *Anthophora*
—. Arolia absent (Eastern Hemisphere) .... *Amegilla*
3(1). Stigma almost absent, represented by minute area broader than long; T2 to T5 or at least T3 with lateral, longitudinal parts of graduli strong, sometimes reaching posterior marginal zones of terga; pygidial plate of male present as a well-defined apical process of T7, margined by carinae across apex and at sides, at least apically (Africa, Madagascar) .... *Pachymelus*
—. Stigma at least as long as broad, usually much longer than broad (Fig. 113-1b); T2 to T5 commonly without lateral longitudinal parts of graduli; pygidial plate of male absent or indicated by smooth bare area margined by carina only across apex 4
4(3). Malar space at least twice as long as antennal pedicel; vein cu-v of hind wing nearly transverse, at angle of 50° or more to first abscissa of M+Cu (Mesoamerica) .... *Deltoptila*
—. Malar space linear to about as long as antennal pedicel; vein cu-v of hind wing usually conspicuously oblique, at angle of 45° or less to first abscissa of M+Cu (except nearly transverse in *Habrophorula*) (holarctic and oriental) 5
5(4). Mandible with two preapical teeth, thus tridentate; T7 and S6 of male somewhat attenuate apically, apices rounded or pointed (S7 of male weakly sclerotized, without transverse ridge at base of apical process; hind leg of male enlarged, trochanter with broad rounded projection) (oriental) .... *Elaphropoda*
—. Mandible with one preapical tooth, thus bidentate; T7 and S6 of male not attenuate, apex of T6 nearly always bidentate or with emarginate apical truncation 6
6(5). Third submarginal cell about as wide on anterior margin as on posterior margin; male with first flagellar segment about as broad as long and shorter than second; S7 of male weakly sclerotized, transverse, disc broader than long, without apical process (oriental) .... *Habrophorula*
—. Third submarginal cell with anterior margin much shorter than posterior margin; male with first flagellar segment much longer than broad and longer than second; S7 of male strongly sclerotized, disc giving rise to large apical process, base of which often bears transverse ridge (holarctic, oriental) .... *Habropoda*

Genus *Amegilla* Friese

Podalirius (Amegilla) Friese, 1897a: 18, 24. Type species: *Apis quadrifasciata* Villers, 1789, by designation of Cockerell, 1931e: 277.
Alfkenella Börner, 1919: 168. Type species: *Apis quadrifasciata* Villers, 1789, by original designation.
Asaropoda Cockerell, 1926b: 216. Type species: *Saropoda bombiformis* Smith, 1854, by original designation.
Amegilla (Aframegilla) Popov, 1950b: 260. Type species: *Anthophora nubica* Lepeletier, 1841, by original designation.
Amegilla (Zonamegilla) Popov, 1950b: 260. Type species: *Apis zonata* Linnaeus, 1758, by original designation.
Amegilla (Zebramegilla) Brooks, 1988: 502. Type species: *Anthophora albigena* Lepeletier, 1841, by original designation.
Amegilla (Dizonamegilla) Brooks, 1988: 505. Type species: *Megilla sesquicincta* Erichson and Klug, 1842, by original designation.
Amegilla (Megamegilla) Brooks, 1988: 505. Type species: *Apis acraensis* Fabricius, 1793, by original designation.
Amegilla (Ackmonopsis) Brooks, 1988: 508. Type species: *Anthophora mimadvena* Cockerell, 1916, by original designation.
Amegilla (Micramegilla) Brooks, 1988: 508. Type species: *Anthophora niveata* Friese, 1905, by original designation.
Amegilla (Notomegilla) Brooks, 1988: 511. Type species: *Anthophora aeruginosa* Smith, 1854, by original designation.
Amegilla (Glossamegilla) Brooks, 1988, 512. Type species: *Anthophora mesopyrrha* Cockerell, 1930, by original designation.

These bees have the form of *Anthophora*. Some common groups have metallic blue or green pubescence, especially on the metasoma (Pl. 13). Such coloration does not occur in *Anthophora*, and species of *Amegilla* lacking such coloration are most easily separated from *Anthophora* by the lack of arolia. The face of both sexes usually shows yellow or white markings. The tibial scopa includes a band of plumose hairs near the upper margin of the tibia, as in most *Anthophora*. The hind leg of the male is ordinarily unmodified, without a basitarsal tooth, and the middle tarsus lacks brushes. The pygidial plate of the male is absent. The mentum has a submedian tooth on its anterior surface except in *Amegilla nonconforma* Brooks. Illustrations of male genitalia and other structures for all species groups were presented by Brooks (1988); Eardley (1994b) illustrated the species of southern Africa; additional such illustrations were published by Lieftinck (1944, under the genus *Anthophora*; 1956a, 1975), Popov (1950b), Iuga (1958), Michener (1965b), and Wu (1983d); see Figure 113-2d-f. The body length is 8 to 24 mm.

I have arbitrarily decided not to recognize the numerous subgenera of *Amegilla* defined by Brooks (1988). They are largely indistinguishable in females, and in males differ from one another considerably less than do most subgenera of *Anthophora*. They probably do constitute natural groups and are available for use by anyone wishing to use them; the names and type species are listed in the synonymy above. The great number of species of *Amegilla* supports recognition of groups or subgenera.

■ This genus is found throughout Africa (including Madagascar) and the Mediterranean basin and from the Canary Islands east across southern Europe to Japan, Korea and northeast China, south to Yemen, Sri Lanka, Indonesia, New Guinea, and the whole of Australia (including Tasmania), and east to the Solomon Islands.

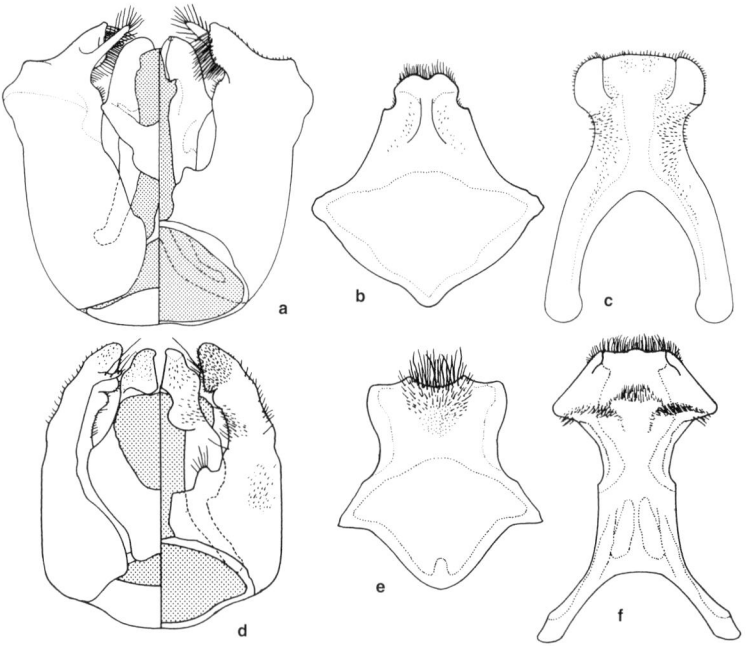

Figure 113-2. Male genitalia, S8, and S7 of Anthophorini. **a-c,** *Anthophora (Anthophoroides) phaceliae* Brooks; **d-f,** *Amegilla elsei* Brooks. (In the divided drawings, dorsal views are on the left.) From Brooks, 1988.

Brooks (1988) lists 253 species. Keys and revisional works on *Amegilla* include the following: Lieftinck (1956a) for the oriental region, Rayment (1944, 1947, 1951) for Australia, and Eardley (1994) for southern Africa. Priesner (1957) included *Amegilla* in his revision of Egyptian *Anthophora* species. Brooks (1988) listed the species and placed them into groups, as did Cardale (1993) for Australian species.

Genus *Anthophora* Latreille

These are robust (Pl. 13), fast-flying bees with the same form as *Amegilla*. None has the metallic blue or green pubescence that occurs in some *Amegilla*. The hind leg of the male is often modified, often with a basitarsal tooth, and the middle tarsus frequently has brushes of hair, some of which are illustrated in Figure 113-3. The base of the outer hind tibial spur is not or only partly isolated by a sclerotized process from the base of the inner spur. The mentum usually lacks a submedian tooth on its anterior surface. Illustrations of male genitalia, sterna, and other structures were provided for all subgenera by Brooks (1988); see also Iuga (1958), Mitchell (1962), Lieftinck (1966), Marikovskaya (1979), Wu (1982b, 1986), Eardley and Brooks (1989), and Tkalců (1993b); see Figure 113-2a-c.

Anthophora is abundant in the holarctic and African regions, scarce in the neotropics and Southeast Asia, and absent in the Indo-Australian area and Madagascar. The species of *Anthophora* found in southern Africa were revised by Eardley and Brooks (1989). Most earlier keys and revisions included *Amegilla* within *Anthophora*. Aside from keys in general regional accounts of bees, keys and revisions can be found in Priesner (1957) for Egypt, Ortiz-Sánchez and Jiménez-Rodriguez (1991) for Spain,

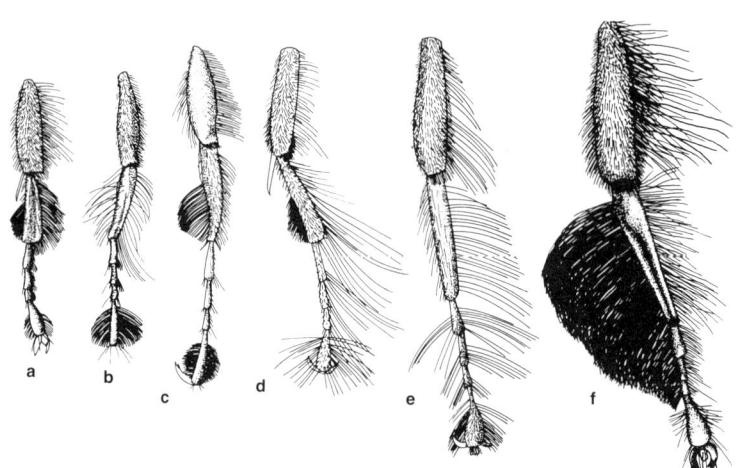

Figure 113-3. Middle tibiae and tarsi of males of *Anthophora*, showing variation in ornamentation. **a,** *A. (Pyganthophora) retusa* (Linnaeus); **b,** *A. (P.) crotchii* Cresson; **c,** *A. (Caranthophora) dufourii* Lepeletier; **d,** *A. (Anthophora) plumipes* (Pallas); **e,** *A. (Lophanthophora) porterae* Cockerell; **f,** *A. (L.) hispanica* (Fabricius). From Brooks, 1988.

and Cockerell (1906e) for North America. Brooks (1988) listed the species for the world, placing them into subgenera and in many cases species groups. Marikovskaya (1976, 1979, 1980) proposed a very different classification, using the name *Clisodon* for a major segment of what is here called the genus *Anthophora*.

Key to the Subgenera of *Anthophora* (Males) (Modified from Brooks, 1988)

1. Middle distitarsus with brush of dense hairs, middle basitarsus often with brush; apex of S7 with lateral margin forming two low lobes, one sometimes partly above the other 2
—. Middle tarsus simple, without any brushes; apex of S7 laterally simple or with two lobes 5
2(1). Basitibial plate absent; pygidial plate absent 3
—. Basitibial plate present; pygidial plate present 4
3(2). Flabellum apically entire; S7 apically narrowed, elongate, apodemes at least 1.3 times as long as length of disc; apex of gonocoxite deeply bilobed (palearctic) ....
.... *A. (Anthophora s. str.)*
—. Flabellum apically with fingerlike projections (as in Fig. 113-4a-c); S7 apically broad, short, apodemes no more than 0.4 times as long as length of disc; apex of gonocoxite weakly trilobed (palearctic) ...... *A. (Caranthophora)*
4(2). Distal half of S6 with a pair of apically diverging oblique ridges; apex of gonocoxite simple, flattened, hooked ventrad (holarctic) .... *A. (Lophanthophora)*
—. Distal half of S6 simple, without ridges; apex of gonocoxite bilobed or triangular, not flattened and hooked ventrad (holarctic, Africa) ...... *A. (Pyganthophora)* (in part)
5(1). Mandible tridentate because of preapical convexity or tooth on lower margin in addition to usual preapical tooth on upper margin; gonostylus concolorous with apex of gonocoxite, perhaps absent; flabellum apically simple (holarctic) .... *A. (Clisodon)*
—. Mandible bidentate or sometimes simple; gonostylus lightly sclerotized, thus clearly demarcated from gono-

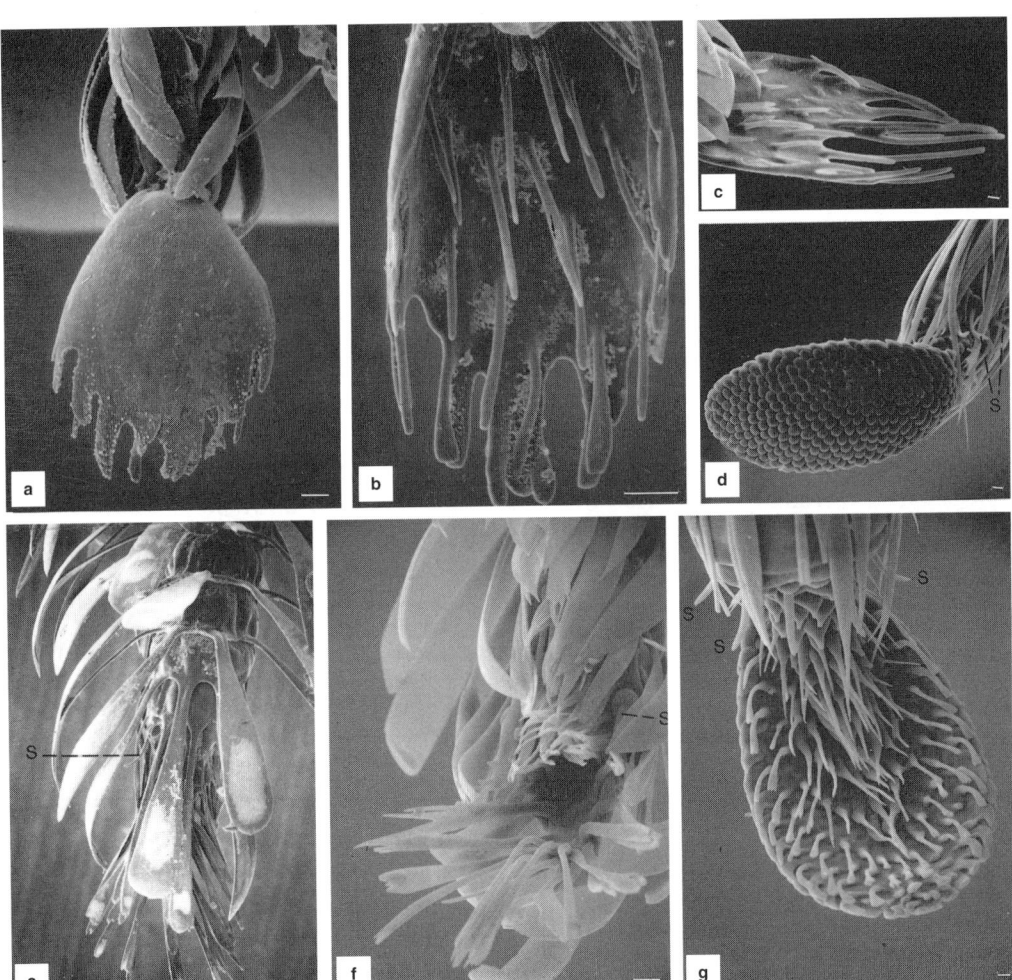

Figure 113-4. Apices of glossae of Anthophorini. **a, b,** *Anthophora (Heliophila) cockerelli* Timberlake, posterior and anterior views; **c,** *A. (Mystacanthophora) walchii* Cresson, anterior view;; **d,** *Deltoptila* n. sp., posterolateral view; **e,** *Habropoda miserabilis* (Cresson), anterior view;; **f,** *H. pallida* (Timberlake), posteroapical view; **g,** *Deltoptila* n. sp., anterior view.

Habropoda is extraordinary for an L-T bee in that it lacks a flabellum. From Michener and Brooks, 1984.

coxite; flabellum apically with fingerlike lobes (Fig. 113-4a-c) 6

6(5). Mentum with submedian tooth on anterior surface; apex of S7 with one lateral lobe (Fig. 113-2c) (hind basitarsus with anterior tooth or strong ridge) 7

—. Mentum without tooth near middle of anterior surface, or, *if* tooth present, then hind basitarsus simple, almost parallel-sided, and without anterior tooth or ridge; apex of S7 with two lateral lobes on each side, sometimes only weakly developed, or, *if* apex simple laterally or with one lobe, then body small (7.5-13.5 mm long) and/or S6 with thick pad of hair 8

7(6). Body length at most 13 mm; hind tibial spur arising near apex of an apical tibial spine; apex of gonocoxite heavily sclerotized with a distinct melanized demarcation (seen only in cleared preparations of genital capsule) between apex and base; face with white integumental markings (yellow in *Anthophora vallorum* Cockerell and *californica* Cresson) (North and Central America) .... *A. (Anthophoroides)*

—. Body length greater than 13 mm; hind tibia simple, without apical spine; apex of gonocoxite not more heavily sclerotized than base, no demarcation between the two; face with yellow integumental markings (holarctic) .... *A. (Melea)*

8(6). Malar space more than one-third as long as wide; flabellum elongate, apically entire or with numerous small apical lobes 9

—. Malar space almost always short and linear, at most one-third as long as wide; flabellum short and wide, apically with three to ten fingerlike projections (Fig. 113-4a) or if entire, then apical two-thirds not narrowed and flabellum lacking numerous short apical lobes (S6 almost always medioapically emarginate) 10

9(8). Flabellum apically entire, elongate, with apical two-thirds narrowed; T7 of male with longitudinal median ridge not reaching apex, with two small, weakly developed, submedian projections; penis valve bridge short, basally emarginate (palearctic) .... *A. (Anthomegilla)*

—. Flabellum apically with numerous small lobed divisions, broom-shaped, lateral margins almost parallel; T7 of male apicomedially, dorsoventrally flattened with longitudinal median ridge reaching apex; penis valve bridge long, basally pointed (Asia) .... *A. (Rhinomegilla)*

10(8). Body length usually 7.5-13.5 mm *and/or* S6 with thick pad of hair (often hidden until sternum is relaxed and pulled apically); gradular process of T7 often prolonged into tooth when body size is small; apex of disc of T7 laterally simple, not with two lobes, lateral margins of disc usually parallel or apically converging, rarely narrowed medially (holarctic, Africa) .... *A. (Heliophila)*

—. Body length usually 9-24 mm; S6 simple, without pad of hair; gradular process of T7 at most obtusely projecting, usually not developed; apex and base of disc of S7 wider than middle, apex laterally angulate or with two small lobes on each side 11

11(10). Pygidial plate present, lateral tooth below pygidial plate sometimes also present; gradular process of T7 sometimes present 12

—. Pygidial plate and lateral tooth both absent, apex of T7 with two submedian teeth; gradular process of T7 absent 13

12(11). Basitibial plate present, and hind basitarsus simple, nearly parallel-sided; pygidial plate with margin carinate and not concave; gradular process of T7 sometimes present (holarctic, Africa) .... *A. (Pyganthophora)* (in part)

—. Basitibial plate absent, or, *if* present, then hind basitarsus not parallel-sided, with anterior process dilated; pygidial plate with margin rounded, almost never carinate, but *if* carinate, then lateral margin concave; gradular process of T7 absent (holarctic, Africa) .... *A. (Paramegilla)* (in part)

13(11). S7 with disc broad, its minimum width over half of distance between tips of apodemes; S8 with broad, deep, apical emargination (weak in *A. calcarata* Lepeletier), lateral margin on apical half concave 14

—. S7 with disc narrow, its minimum width less than half of distance between tips of apodemes; S8 with narrow shallow apical emargination, lateral margin on apical half angulate or concave 15

14(13). S8 with apicolateral margin angulate, spiculum projecting somewhat basally; S7 with apex shallowly emarginate, with two weakly developed apicolateral lobes, basal apodeme with lateral angle or process; apex of gonocoxite somewhat bilobed in profile, lower lobe weakly developed, not carinate, sometimes with a tooth on lower inner margin of gonocoxite; face almost always with yellow markings (palearctic) .... *A. (Petalosternon)*

—. S8 with apicolateral margin prolonged and rounded, spiculum not easily visible from ventral aspect but projecting dorsally at a 90° angle; S7 with apex distinctly emarginate, with two strong apicolateral lobes, basal apodeme without lateral process or angle; apex of gonocoxite complex in profile with dorsoventral and lateral carinae; face with white markings (Western Hemisphere, palearctic) .... *A. (Mystacanthophora)*

15(13). T7 with two short submedian teeth, their bases separated by convex lamellate or angular apical margin, and median longitudinal carina often present; apical zone of T7, as seen from ventral view, narrow where bent ventroanteriad; hind basitarsus anteriorly with tooth, blunt process, or longitudinal ridge; apodemes of S7 almost always short, 0.34-0.78 as long as length of disc, lateral process absent; apex of gonocoxite in profile simple to weakly bilobed (holarctic, Africa) .... *A. (Paramegilla)* (in part)

—. T7 with two long submedian teeth, their bases separated by rounded apical margin; apical zone of T7, as seen from ventral view, wide where bent ventroanteriad; hind basitarsus almost parallel-sided, anterior margin simple; apodemes of S7 long, 1.00-1.75 times as long as length of disc, lateral process present; apex of gonocoxite in profile deeply bilobed (palearctic) .... *A. (Dasymegilla)*

Key to the Subgenera of *Anthophora* (Females) (Modified from Brooks, 1988)

1. Apex of mandible with lower subapical tooth and thus tridentate; basitibial plate about twice as long as broad, tapering to pointed apex (flabellum entire) (holarctic) .... *A. (Clisodon)*

—. Apex of mandible without lower subapical tooth, bidentate; basitibial plate less slender, apex rounded 2

2(1). Flabellum narrowly elongate, apically entire or with numerous short, small lobes; first flagellar segment equal in length to next 3.2-4.0 segments taken together 3

—. Flabellum wide, with few long fingerlike apical lobes

(Fig. 113-4a-c); first flagellar segment equal in length to next 2.2-5.5 segments taken together 5

3(2). Malar area almost always short, rarely well developed, at most one-third as long as wide; flabellum more or less parallel-sided or lanceolate, not attenuate in apical two-thirds, apex entire; upper margin of hind tibia with scopal hairs plumose (palearctic) .... *A. (Anthophora s. str.)*

—. Malar area more than one-third as long as wide; flabellum elongate with apical two-thirds narrowed or flabellum with numerous small apical lobes; upper margin of hind tibia with scopal hairs simple 4

4(3). Flabellum apically entire, elongate, with apical two-thirds narrowed (palearctic) .... *A. (Anthomegilla)*

—. Flabellum apically with numerous small lobes (Asia) .... *A. (Rhinomegilla)*

5(2). Mentum with a well-developed submedian tooth on anterior surface; first flagellar segment equal in length to next 2.4-3.3 segments taken together 6

—. Mentum without tooth on anterior surface [rarely with tooth in old world species having upper zone of scopa plumose, e.g., *Anthophora (Dasymegilla) excisa* Morawitz]; first flagellar segment equal in length to next 2.2-5.5 segments taken together 7

6(5). Hind basitarsus with upper distal process acute; S6 with subapical tooth; upper margin of hind tibia with simple hair; body 13 mm long or longer (holarctic) .... *A. (Melea)*

—. Hind basitarsus with upper distal process blunt; S6 without subapical tooth; upper margin of hind tibia with plumose hair; body no longer than 13 mm (North and Central America) .... *A. (Anthophoroides)*

7(5). Upper margin of hind tibia with simple hair; length 6-15 mm 8

—. Upper margin of hind tibia with plumose hair; length 10-24 mm 9

8(7). Face black; body length 12-14 mm (palearctic) .... *A. (Paramegilla)* (in part)

—. Face almost always with yellow to white integumental markings and body length 6-15 mm, but *if* face black, then length 11-13 mm and southern African in distribution (holarctic, Africa) .... *A. (Heliophila)*

9(7). Clypeus with pale integumental marking in shape of inverted T, and first flagellar segment equal in length to next 2.2-3.0 segments taken together (palearctic) .... *A. (Caranthophora)* (in part)

—. Clypeus almost always black, rarely with pale integumental marking, or, *if* marking present, then not in shape of inverted T, or, *if* in shape of inverted T, then first flagellar segment longer than combined lengths of next three segments 10

10(9). Metasoma with apical bands of hair interrupted medially or surface covered evenly with appressed hair (holarctic, Africa) .... *A. (Paramegilla)* (in part)

—. Metasoma with apical bands of hair complete and/or surface covered with semierect to erect hair 11

11(10). Clypeus normal, without apically hooked or wavy hairs or basal pecten; first flagellar segment equal in length to next 2.2-5.5 segments taken together .... *A. (Caranthophora* [in part], *Dasymegilla, Lophanthophora, Mystacanthophora* [in part], *Paramegilla* [in part], *Petalosternon, Pyganthophora*)

—. Clypeus flattened, with hairs apically hooked or wavy, sometimes with a basal pecten; first flagellar segment equal in length to next 2.2-3.4 segments taken together (Western Hemisphere, palearctic) .... *A. (Mystacanthophora)* (in part)

Anthophora / Subgenus *Anthomegilla* Marikovskaya

Anthomegilla Marikovskaya, 1976: 688. Type species: *Anthophora arctica* Morawitz, 1883, by original designation.

Anthomegilla has a well-developed malar space, two or three times as wide as long. The flabellum is widest near the base, distally attenuate, four or more times as long as wide, rounded or subtruncate at the apex, and not divided into distal processes; it is thus quite different from that of the other subgenera that lack an apically divided flabellum, *Anthophora* s. str. and *Clisodon*. As in *Melea, Rhinomegilla,* and *Heliophila,* the scopa of the female hind tibia consists entirely of simple hairs. The male lacks basitibial and pygidial plates. The body length is 9 to 14 mm. Wu (1986) and Brooks (1988) illustrated male genitalia and other structures.

■ This subgenus occurs in the Siberian subarctic south to the mountains of Central Asia and China to Tibet. There are eight species (Brooks, 1988). Wu (1986) gave a key to species.

Anthophora / Subgenus *Anthophora* Latreille s. str.

Podalirius Latreille, 1802a: 430. Type species: *Apis pilipes* Fabricius, 1775 = *Apis plumipes* Pallas, 1772. *Polalirius* was suppressed by Commission Opinion 151 (1944) (Direction 4).

Anthophora Latreille, 1803: 167, replacement for *Podalirius* Latreille, 1802. Type species: *Apis pilipes* Fabricius, 1775 = *Apis plumipes* Pallas, 1772, designated by Commission Opinion 151 (1944). [See Michener, 1997b.]

Lasius Panzer, 1804b: tab. 16. Type species: *Lasius salviae* Panzer, 1804 = *Anthophora crinipes* Smith, 1854, monobasic. Suppressed by Commission Opinion 151 (1944) (Direction 4). [According to Opinion 151 (1944), the correct date for *Lasius* Panzer is [1801-1802]; see Literature Cited.]

Megilla Fabricius, 1804: 328. Type species: *Apis pilipes* Fabricius, 1775 = *Apis plumipes* Pallas, 1772, designated by Commission Opinion 1383 (1986). [For previous type designations, see Sandhouse (1943) and Michener (1997b).]

Anthophora s. str. differs from other subgenera in having an elongate, simple flabellum, without the deep sinuses forming apical lobes found in most other subgenera. The basitibial plate of the male is weakly delimited or absent, and the pygidial plate is absent in the male. The apex of the male gonocoxite is strongly bifid. The body length is 10 to 19 mm.

■ This subgenus occurs from Spain and Britain to Korea and eastern China. Brooks (1988) listed 11 species.

Most of the studies on nesting biology of this subgenus concerned *Anthophora acervorum* of authors, which is now called *A. plumipes* Pallas because *Apis acervorum* Linnaeus is a species of *Bombus* (see Brooks, 1988).

Anthophora / Subgenus *Anthophoroides* Cockerell and Cockerell

Anthophoroides Cockerell and Cockerell, 1901: 48. Type species: *Podalirius vallorum* Cockerell, 1896, by original designation.

As indicated under *Melea,* that subgenus and *Anthophoroides* are closely related, one of their common apomorphies being the submedian tooth on the anterior surface of the mentum. It would be reasonable to synonymize *Melea* under *Anthophoroides;* differences are indicated in the key to subgenera and in the discussion of *Melea.* Species of *Anthophoroides* are usually smaller than those of *Melea,* the body length being 10 to 13 mm.

■ *Anthophoroides* ranges from southern Oregon to Colorado, USA, and thence southward through Mexico to Honduras. There are six described species, according to Brooks (1988), in addition to a similar number of undescribed species.

Anthophora / Subgenus *Caranthophora* Brooks

Anthophora (Caranthophora) Brooks, 1988: 470. Type species: *Anthophora dufourii* Lepeletier, 1841, by original designation.

In this subgenus the basitibial and pygidial plates of the male are absent. S7 and S8 of the male are remarkable for their large discs, that of S7 being about as broad as long. An unusual feature is the presence in the female of an inverted, T-shaped pale (usually white) mark on the clypeus (absent in *Anthophora hedini* Alfken). The modification of the hind legs of the male is extreme, the femur and tibia being dilated and the basitarsus flattened and with both a basal and an apical tooth. Body length is 8.5 to 18.0 mm.

■ *Caranthophora* is found from the Mediterranean basin and eastern Europe to northern India and eastern China. There are six species (Brooks, 1988).

Anthophora / Subgenus *Clisodon* Patton

Clisodon Patton, 1879a: 479. Type species: *Anthophora terminalis* Cresson, 1869, by original designation.

The noteworthy character of this subgenus is the tridentate mandible of the female, a feature presumably associated with nesting in rotting wood or in pithy stems rather than in the ground. In other characters also, *Clisodon* is distinctive. The flabellum is relatively broad, i.e., not slender as in *Anthophora* s. str. and *Anthomegilla,* but is entire as in those subgenera. It thus differs from all other subgenera. The basitibial plate of the female is slender, tapering toward the pointed apex, in contrast to that of other genera. The basitibial and pygidial plates of the male are absent. The body length is 9 to 13 mm.

■ This is a boreal holarctic subgenus that ranges far into the subarctic woodlands of both Eurasia and North America (including subarctic Alaska), and south in coniferous forests to northern California and into the mountains to Southern California, Arizona, and New Mexico in North America and to the Pyrenees, Alps, and Caucasus in Eurasia. Brooks (1988) listed four species but Davydova and Pesenko (2002) recognized only two, the holarctic *Anthophora terminalis* Cresson and the palearctic *A. furcata* (Panzer). The variation and distribution of the subgenus was discussed by Popov (1951b).

Clisodon has often been given generic rank, but it fits easily into *Anthophora.*

Although *Clisodon* is usually reported to nest in decomposing logs and stumps, *Anthophora terminalis* Cresson also readily nests in pithy stems; Medler (1964b) reported 60 nests in pieces of sumac *(Rhus)* stems.

Anthophora / Subgenus *Dasymegilla* Brooks

Lasius Jurine, 1801: 164. Type species: *Apis quadrimaculata* Panzer, 1798, monobasic. Suppressed by Commission Opinion 135 (1939).
Anthophora (Dasymegilla) Brooks, 1988: 486. Type species: *Apis quadrimaculata* Panzer, 1798, by original designation.

This subgenus lacks the pygidial plate in males as well as (except in *Anthophora excisa* Morawitz) the basitibial plate. S7 of the male has a much smaller and more slender disc and more elongate and slender apodemes than those in *Petalosternon.* The apex of the male gonocoxite is bilobed as seen in side view, as in that subgenus, but the slender gonostylus arises near the base of the upper lobe instead of at its apex. S6 of the male has a pad of dense hair occupying the basal half of the sternum. The body length is 10.0 to 12.5 mm.

■ *Dasymegilla* occurs from Britain and the Mediterranean basin east to China. Brooks (1988) lists six species.

Anthophora / Subgenus *Heliophila* Klug

Heliophila Klug, 1807, *in* Illiger, 1807: 197; Klug, 1807b: 227. Type species: *Apis bimaculata* Panzer, 1798, monobasic. [For comments, see Michener, 1997b.]
Saropoda Latreille, 1809: 177, unnecessary replacement for *Heliophila* Klug, 1807. Type species: *Apis bimaculata* Panzer, 1798, autobasic. [For comments, see Michener, 1997b.]
Micranthophora Cockerell, 1906e: 66. Type species: *Anthophora curta* Provancher, 1895, by original designation.

Heliophila consists of generally small species (6-10 mm long, rarely in southern Africa 15 mm), nearly always with white or yellow markings on the face of the female. Although the basitibial plate of the male is usually present, the pygidial plate is absent in most species although well developed in some, including *Anthophora curta* Provancher, the type species of *Micranthophora.* The tibial scopa of the female is of simple hairs, without plumose hairs along the upper margin; in this respect, *Heliophila* agrees with the subgenera *Clisodon, Melea, Rhinomegilla,* and *Anthomegilla* and differs from all other *Anthophora* and from *Amegilla.* The vestiture of the metasomal terga tends to be appressed, giving a pale color to the whole metasoma or to tergal bands.

■ This subgenus occurs in the Mediterranean basin (including the Canary Islands), south through East Africa to South Africa, and east to Central Asia, Tibet, and India, and in North America from Wyoming, Kansas, and Texas west to the Pacific Coast and south through Mexico to Honduras. A single specimen supposedly from Ecuador (Brooks, 1988) is probably mislabeled. Brooks (1988) lists 65 species from the Old World; 26 other species occur in North America.

One species, *Anthophora (H.) peritomae* Cockerell, frequently nests communally, with up to six females using a single entrance, but each with her own branch burrow usually diverging from a subsurface communal chamber (Torchio, 1971).

Anthophora / Subgenus *Lophanthophora* Brooks

Anthophora (Lophanthophora) Brooks, 1988: 464. Type species: *Anthophora porterae* Cockerell, 1900, by original designation.

Like those of *Pyganthophora*, males of *Lophanthophora* have pygidial and basitibial plates and a flabellum with apical fingerlike lobes. The apex of the gonocoxite of the male is simple or rarely weakly bilobed. S6 of the male is diagnostic, differing from that of all other subgenera in having a pair of narrow ridges that diverge apically from near the middle of the sternum and that nearly always are continued basad. The body length is 11 to 20 mm.

■ This subgenus is widespread in North America (Massachusetts to Georgia, USA, west to British Columbia, Canada, south to Coahuila to Baja California, Mexico) and in Europe, northern Africa, and east at least to Central Asia. The 33 species were listed by Brooks (1988).

Anthophora pacifica Cresson overwinters as unemerged adults in their cells, instead of as prepupae (Brooks, 1988).

Anthophora / Subgenus *Melea* Sandhouse

Anthemoessa Robertson, 1905: 372 (not Agassiz, 1847). Type species: *Anthophora abrupta* Say, 1837, by original designation.
Melea Sandhouse, 1943: 569, replacement for *Anthemoessa* Robertson, 1905. Type species: *Anthophora abrupta* Say, autobasic.

In *Melea* the basitibial and pygidial plates of males are entirely absent. The hairs of the tibial scopa are all simple, an unusual feature shared with the very different subgenera *Clisodon, Rhinomegilla, Anthomegilla,* and *Heliophila*. This subgenus is related to *Anthophoroides*, with which it agrees (and differs from other *Anthophora*) in the presence of a tooth on the anterior surface of the mentum and in the relatively short first flagellar segment. *Anthophoroides* differs from *Melea*, however, in having a strip of plumose scopal hairs near the posterior margin of the hind tibia of the female, as in most subgenera. The body length is 13 to 17 mm.

■ *Melea* ranges across North America from Ontario and Northwest Territories, Canada, and subarctic Alaska, south to Florida and California, USA, and in the Old World from Siberia and China to Spain and Algeria. Brooks (1988) lists nine species. The North American species were revised by Brooks (1983).

Species of this subgenus appear to be *Bombus* mimics and vary geographically in hair coloration, participating in mimetic complexes of *Bombus*. *Melea* is the subgenus of *Anthophora* whose nest burrows, commonly aggregated, have downcurved entrance chimneys of hardened mud projecting from hard-clay nesting banks.

Anthophora / Subgenus *Mystacanthophora* Brooks

Anthophora (Mystacanthophora) Brooks, 1988: 466. Type species: *Anthophora montana* Cresson, 1869, by original designation.

In this subgenus the pygidial plate of the male is not defined, but there is an apical truncate or emarginate process, presumably representing the plate, on T7; the basitibial plates are absent or poorly defined, although well defined in two species. S5 and S6 of the male are almost always modified, S5 with tufts or a band of black hair, S6 often with an apical emargination; in *Anthophora urbana* Cresson, however, these sterna are unmodified. The body length is 9 to 16 mm. In many species the clypeus of the female is flat or weakly convex, little protuberant, and has erect, robust, commonly hooked hairs, sometimes also a transverse row of blunt, spinelike projections across the base. These features are suggestive of *Pectinapis* in the Eucerini, some *Trachusa (Ulanthidium)* in the Anthidiini, and other bees that visit flowers of *Salvia* (Lamiaceae).

■ *Mystacanthophora* is found from British Columbia to Manitoba, Canada, and east to Massachusetts, USA, south through North America (but absent in the southeastern USA) to Oaxaca, Mexico; also the Bahamas and the Greater and Lesser Antilles, and in South America from Venezuela to central Chile, thence east across Argentina to southern Brazil as far north as the state of Minas Gerais. Two Old World species range from Central Europe to Iran and eastward into Siberia. The distribution appears to be doubly disjunct because of the absence of the subgenus, so far as is known, from southern Mexico and Central America as well as from Alaska and eastern Siberia. Brooks (1988) listed 19 species. Brooks (1999) revised the seven Antillean (including Bahaman) species of *Anthophora*, all of which are in the subgenus *Mystacanthophora*.

Anthophora / Subgenus *Paramegilla* Friese

Podalirius (Paramegilla) Friese, 1897a: 18, 24. Type species: *Apis ireos* Pallas, 1773, designated by Sandhouse, 1943: 584.
Solamegilla Marikovskaya, 1980: 650. Type species: *Anthophora prshewalskyi* Morawitz, 1880, by original designation.

The pygidial plate of the male varies from almost unrecognizable to distinct; on each side, lateral to its apex, is almost always a distinct tooth. The basitibial plate also varies from absent to distinctly defined. The upper margin of the tibial scopa usually has a band of plumose hairs, but in certain species such as *Anthophora larvata* Giraud the scopal hairs are all simple. The hind legs of males are almost always modified, the basitarsus usually with an anterior projection or spine. On S6 of the male the distal margin is flexed ventrally, usually concave. A distinctive feature of some species is the large median clump of dense hairs on S4 of the male; another is the strong but medially interrupted pale tergal hair bands. The body length is 9.5 to 24.0 mm.

■ *Paramegilla* is widespread and diverse from Europe to China and southward to southern South Africa. In addition are two species from California and Nevada to New Mexico, USA. Brooks (1988) listed 66 species.

Anthophora / Subgenus *Petalosternon* Brooks

Anthophora (Petalosternon) Brooks, 1988: 484. Type species: *Anthophora rivolleti* Pérez, 1895, by original designation.

In *Petalosternon* the basitibial and pygidial plates are absent. S7 of the male is short, the apodemes broad basally and abruptly narrowed apically. The gonocoxite is bilobed at the apex as seen in lateral view, the slender gonostylus arising from the apex of the upper lobe. Body length is 10.0 to 12.5 mm.

■ This subgenus is found largely in the Mediterranean basin but extends thence eastward to Mongolia. There are 21 species (Brooks, 1988).

Anthophora / Subgenus *Pyganthophora* Brooks

Anthophora (Pyganthophora) Brooks, 1988: 460. Type species: *Apis retusa* Linnaeus, 1758, by original designation.

In this subgenus the male has well-developed pygidial and basitibial plates. The flabellum has several long, fingerlike apical lobes. The gonocoxite of the male is weakly to strongly bifid. Body length is 12 to 16 mm.

■ This subgenus is widespread in the palearctic region, from Western Europe to China, with numerous species in the Mediterranean basin; there is a possibly disjunct group of seven species in southern Africa and six species (representatives of two species groups) on the Pacific Coast of North America from British Columbia, Canada, to Baja California, Mexico, and extending east to Utah and New Mexico, USA. The 66 species were listed by Brooks (1988).

Anthophora / Subgenus *Rhinomegilla* Brooks

Anthophora (Rhinomegilla) Brooks, 1988: 482. Type species: *Anthophora megarrhina* Cockerell, 1910, by original designation.

This subgenus is related to *Anthomegilla*; it has an even more elongate malar area (slightly longer than wide) and proboscis, and more protuberant clypeus. The flabellum is almost four times as long as wide, nearly parallel-sided, with many small filaments at the apex, thus somewhat different from the distally attenuate flabellum with a simple apex found in *Anthomegilla*. The gonostylus of the male is slender and distinct; it is absent or nearly so in *Anthomegilla*. The body length is 8 to 20 mm.

■ *Rhinomegilla* is known from Sikkim, Xigang (Tibet), and Sichuan Province, China. The four species were listed by Brooks (1988).

Genus *Deltoptila* LaBerge and Michener

Deltoptila LaBerge and Michener, 1963: 211. Type species: *Habropoda montezumia* Smith, 1879, by original designation.

This genus consists of anthophoriform bees 10 to 15 mm long, with abundant long hair often forming color patterns like those of local *Bombus* species, and with a strongly protuberant clypeus and very long proboscis. The flabellum has a cobblestone pattern on the posterior surface and short capitate hairs on the anterior surface (Fig. 113-4d, g). As in *Habropoda*, the mandible has one preapical tooth, and the anterior margin of the third submarginal cell is considerably shorter than the posterior margin (Fig. 113-1b). S7 of the male is not heavily sclerotized; it has a flat, distally broadened, apical process. The genitalia and hidden sterna were illustrated by LaBerge and Michener (1963).

■ *Deltoptila* occurs in the mountains from the states of Veracruz and México, Mexico, south to Panama. About ten species were revised by LaBerge and Michener (1963).

Genus *Elaphropoda* Lieftinck

Elaphropoda Lieftinck, 1966: 148. Type species: *Habropoda impatiens* Lieftinck, 1944, by original designation.

Elaphropoda consists of apiform species, frequently with the basal terga or sometimes all terga red-brown, enhancing a similarity to *Apis*. The metasomal pubescence is short, forming weak, pale, apical hair bands. The body length is 12 to 19 mm. As in *Deltoptila* the clypeus is strongly protuberant and the proboscis extremely long. The clypeus has a longitudinal median carina. Unlike the similar-looking *Habrophorula*, the third submarginal cell is shorter on the anterior than on the posterior margin, and vein cu-v of the hind wing is strongly oblique. The distinction from *Habropoda* is enhanced by the presence of a flabellum. Male genitalia and other structures were illustrated by Lieftinck (1944, as *Habropoda*; 1966) and Wu (1979, 1991).

■ *Elaphropoda* occurs in mountains from northern India eastward to southeast China and Taiwan, and south to Java. The six species were revised by Lieftinck (1966); Wu (1979) included the Chinese species in keys to species of *Habropoda*; another key is by Wu (1991).

Genus *Habrophorula* Lieftinck

Habrophorula Lieftinck, 1974: 217. Type species: *Habropoda nubilipennis* Cockerell, 1930, by original designation.

This genus resembles non-reddish individuals of *Elaphropoda* in its apiform body and the short metasomal pubescence forming weak apical tergal bands. The body length is 10 to 12 mm. *Habrophorula* differs from *Elaphropoda* in its only moderately convex clypeus and moderately long mouthparts, its slender and unmodified male legs, and the mandibular, sternal, and flagellar characters indicated in the key to genera. Male genitalia and other structures were illustrated by Lieftinck (1974) and Wu (1991). *Habrophorula* is the only anthophorine genus except *Amegilla* and *Anthophora* in which the anterior and posterior sides of the third submarginal cell are of about equal length; the cell is considerably longer than broad, however, unlike that of those genera.

■ *Habrophorula* is known only from China. The three species were included in a key by Wu (1991).

Genus *Habropoda* Smith

Habrophora Smith, 1854: 318 (not Erichson, 1846). Type species: *Habrophora ezonata* Smith, 1854 = *Tetralonia tarsata* Spinola, 1838, inversed autobasic, because of type designation for replacement name, *Habropoda*, by Patton, 1879a: 477.

Habropoda Smith, 1854: 320, replacement for *Habrophora* Smith, 1854. Type species: *Habrophora ezonata* Smith, 1854 = *Tetralonia tarsata* Spinola, 1838, by designation of Patton, 1879a: 477.

Emphoropsis Ashmead, 1899a: 60, no included species; Cockerell and Cockerell, 1901: 48, included species. Type species: *Anthophora floridana* Smith, 1854 = *Bombus laboriosus* Fabricius, 1804, by designation of Cockerell and Cockerell, 1901: 48.

Meliturgopsis Ashmead, 1899a: 62, no included species; Cockerell, 1909c: 414, included a species while synonymizing *Meliturgopsis* under *Emphoropsis.* Type species: *Emphoropsis murihirta murina* Cockerell, 1909, first included species, monobasic. [For a subsequent designation, see Michener 1997b.]

Psithyrus (Laboriopsithyrus) Frison, 1927: 69. Type species: *Bombus laboriosus* Fabricius, 1804, by original designation. [The specific name was misapplied by Frison and others, who considered it to be a species of *Psithyrus,* a subgenus of *Bombus.*]

Habropoda consists of anthophoriform bees; indeed, various species are so *Anthophora*-like in form and coloration, matching color patterns of diverse species of *Anthophora,* that the generic characters of wing venation mentioned in couplet 1 of the key to genera of Anthophorini must be examined in order to place specimens to genus correctly. The body length is 10 to 18 mm. *Habropoda* appears to differ from all other genera of Anthophorini, and indeed from most other L-T bees, in the absence of the flabellum (Michener and Brooks, 1984; Fig. 113-4e, f); although many species of *Habropoda* have been examined for this character, others, particularly additional palearctic species, should be examined for verification. The hind leg of the male varies from simple to enlarged and variously modified. Likewise, the front coxa of the male sometimes has a long apical spur. Such modifications are found especially in palearctic species and are not or rarely developed in oriental and nearctic species. Male genitalia, hidden sterna, and other structures were illustrated by Iuga (1958), Mitchell (1962), Lieftinck (1974), and Wu (1979, 1983a).

■ This genus is found in the Eastern Hemisphere from southern France and Algeria eastward through Europe as far north as Hungary, and through Asia Minor, Central Asia, to northeastern China, and southward to northern India, Thailand, Viet Nam, and Taiwan. In the Western Hemisphere it occurs from British Columbia, Canada, to Baja California, Mexico, east across the continent to Connecticut to Florida, USA. Most of the American species are Californian; only one occurs east of Texas and the Great Plains. There are approximately 50 species, equally divided between the Eastern and Western hemispheres. Most of the Asiatic species were revised by Lieftinck (1974); Wu (1979, 1991) gave a key to Chinese species.

According to Stephen (1961), *Habropoda miserabilis* (Cresson) nests in sand and makes only one cell at the end of each burrow. The young metamorphose in autumn, pass the winter as quiescent adults in their natal cells, and emerge in the spring.

Genus *Pachymelus* Smith

This genus of large to very large anthophoriform bees from Africa and Madagascar differs from all other Anthophorini in having a greatly reduced stigma. The marginal cell is longer and more slender than that of most Anthophorini, longer than the distance from its apex to the wing tip. Numerous structures, including male genitalia, were illustrated by Eardley (1993).

The two subgenera of *Pachymelus* have not been distinguished in the past. Morphological diversity among the species is sufficiently great that multiple genera or subgenera may eventually be proposed.

Key to the Subgenera of *Pachymelus*

1. Arolia absent; first flagellar segment shorter than scape; labrum with transverse preapical ridge, often broken; T2 with lateral gradulus reaching about middle of exposed part of tergum .... *P. (Pachymelopsis)*
—. Arolia present; first flagellar segment as long as or usually longer than scape; labrum without transverse preapical ridge; T2 with lateral gradulus usually strong and reaching posterior marginal zone of tergum .... *P. (Pachymelus s. str.)*

Pachymelus / Subgenus *Pachymelopsis* Cockerell

Pachymelus (Pachymelopsis) Cockerell, 1905a: 331. Type species: *Pachymelus conspicuus* Smith, 1879, by original designation.

Pachymelopsis consists of species 14 to 19 mm long with the metasoma pale-fasciate or covered with tan or orange appressed hair. In both sexes the clypeus and usually the labrum and mandibles are marked with yellow. The clypeus is moderately convex, as seen in profile.

■ This subgenus occurs from Namibia and South Africa north through East Africa to Ethiopia. The five species were revised [with *Pachymelus (Pachymelus) peringueyi* (Friese) included] by Eardley (1993).

The African *"Habropoda"* mentioned by Lieftinck (1966) was a *Pachymelopsis.*

Pachymelus / Subgenus *Pachymelus* Smith s. str.

Pachymelus Smith, 1879: 116. Type species: *Pachymelus micrelephas* Smith, 1879, by designation of Sandhouse, 1943: 581. [For comments, see Michener, 1997b.]

Pachymelus s. str. includes species 15 to 30 mm in length, the latter the largest of the Anthophorini. The appressed metasomal pubescence varies from pale apical tergal fasciae to all-black, largely red, or various shades of brown.

■ This subgenus is found only in Madagascar, with the exception of *Pachymelus peringueyi* (Friese), which occurs in western Cape Province, South Africa, and in Namibia. There are about 15 species. Saussure (1890) gave a key to the species then known. Brooks and Pauly (in Pauly et al., 2001) revised the Malagasy species.

The type species is one of the enormous species with the clypeus of the female angularly produced, as seen in profile, and without yellow facial marks. Among the very large species with an angulate clypeal profile, females of some, *Pachymelus micrelephas* Smith and *ocularis* Saussure, have unusually long, essentially edentate mandibles, and *P. hova* Saussure has mandibles with a preapical tooth. Some other species are mostly smaller (although *P. heydenii* Saussure is 22 to 27 mm long) and like the subgenus *Pachymelopsis* have a merely convex clypeus with yellow areas. As Saussure (1890) reported, mandibular dentition also varies among females of the smaller species. *Pachymelus (Pachymelus) unicolor* Saussure and *bicolor* Saussure are remarkable in that the jugal lobe of the hind wing is minute, almost absent. There are very few other bees with this characteristic.

Pachymelus (Pachymelus) peringueyi (Friese) is the only African species of its subgenus and its position is somewhat doubtful, for in some ways it is intermediate between *Pachymelus* s. str. and *Pachymelopsis*. Its existence is a major reason for not regarding *Pachymelus* and *Pachymelopsis* as generically distinct. It looks much like *P. (Pachymelopsis) reichardti* Stadelmann. Its first flagellar segment is about as long as the scape; in *Pachymelopsis* it is shorter, whereas in other *Pachymelus* s. str., it is longer. Moreover, the lateral gradulus of T2 of *P. peringueyi* is as in *Pachymelopsis*. The distribution of that species, also, is consistent with *Pachymelopsis*, not *Pachymelus* s. str. Nonetheless, its arolium and labrum support placement in *Pachymelus* s. str., and its rounded, densely hairy male volsella is similar to that of at least some *Pachymelus* s. str.

114. Tribe Centridini

This is a largely tropical Western Hemisphere tribe of medium-sized to very large, robust, anthophoriform to eucериform, hairy, fast-flying bees. Arolia are absent, a feature that is uncommon in American anthophoriform Apidae. The first flagellar segment is often longer than the scape. The wings are conspicuously papillate apically, usually partly bare basally; the stigma is very short and no wider than the prestigma, measured to the wing margin (Fig. 114-1). The scopa, on the hind tibia and basitarsus, is extraordinarily large and dense. As shown in Figure 116-2h-j, the flabellum is elongate and has a cobblestone pattern on the posterior surface.

Most Centridini collect floral oil from Malpighiaceae, Krameriaceae, or *Calceolaria* (Scrophulariaceae), etc., presumably for use in provisioning cells. Oil collection is facilitated by beautiful combs of flattened blunt bristles on the margins (next to the tibial spurs) of both front and middle basitarsi (Fig. 6-3) (Neff and Simpson, 1981). Such combs are also present, although weakly developed, in males, presumably unused. Some groups of *Centris*, e.g., the subgenus *Xerocentris*, appear to have lost such combs and are not known to collect oil; they occupy desertic areas in both North and South America where oil-producing flowers must be at least locally absent; the loss of combs may have been independent in the two continents. The subgenus *Exallocentris* also appears to have lost the combs, but they are replaced by pads of fine setae that may absorb oils.

Nests of Centridini are frequently burrows in the soil, either in banks or in flat ground, but as noted below, some species of *Centris* nest in holes in old logs or branches.

Coville, Frankie, and Vinson (1983) and Frankie et al. (1993) reviewed nesting habits for the genus *Centris*. The latter authors gave an account of populations and nest-site preferences of 17 species of Costa Rican *Centris*. *Centris (Ptilotopus)* and at least one species of *C. (Trachina)* make nests in termitaria of *Nasutitermes,* often in trees (see also Gaglianone, 2001). Species of *C. (Heterocentris)* and *(Xanthemisia)* nest in preexisting burrows in old wood, the soil, or other substrates (Michener and Lange, 1958c; Coville, Frankie, and Vinson, 1983; Snelling, 1984). Because they carry mud on the scopae to aboveground sites, the cells in holes in old wood are often nonetheless in or constructed of soil. Some *Epicharis,* such as *E. rustica* (Olivier) and *elegans* Smith, nest in aggregations in deeply shaded banks, as in cave or mine entrances (Camargo, Zucchi, and Sakagami, 1975); other *Epicharis* nest in flat ground. In most Centridini the cells are in irregular clumps or isolated, not in distinct series. In some species each cell is at the end of a different burrow from the surface, i.e., there is a separate nest for each cell (Vinson and Frankie, 1988); nests of *Centris pallida* Fox frequently contain only one cell (Alcock, Jones, and Buchmann, 1976).

The cells in soil are usually more or less vertical. The cell walls are often strengthened and rendered waterproof by the admixture of resin or other materials, including floral oils of Malpighiaceae (Hiller and Wittmann, 1994). Cells of *Epicharis zonata* (Smith) in French Guiana were so well waterproofed that immature stages survived in sand that was well below the water table for months during the wet season (Roubik and Michener, 1980); the adults were active only during the dry season. Other species, such as *Centris pallida* Fox in southwestern deserts of the USA, make cells of compacted soil that are not waterproofed with resin or oil, but as in all species, are lined with a thin, secreted, "waxlike" lining (Rozen and Buchmann, 1990). *Centris mixta* Friese of the Chilean deserts makes similar cells of compacted soil particles, presumably held together by a secretion of the bee (Chiappa and Toro, 1994). The nesting biologies of *Centris (Centris) flavofasciata* (Friese) and *C. (C.) aenea* Lepeletier were described by Vinson and Frankie (2000a) and Aguir and Gaglianone (2003). They use pollen and floral oil, perhaps no nectar, to provision cells, and some individuals complete a nest with one cell, or two cells in a single nest, in one day. Mature larvae of *Centris* spin cocoons with a terminal nipple as in anthidiine Megachilidae. Citations to earlier works on *Centris* nesting behavior are in Vinson, Frankie, and Coville (1987) and Rozen and Buchmann (1990); see also Aguiar and Gaglianone (2003).

The larvae of Centridini are rather similar to those of *Bombus* or *Anthophora,* the robust mandible having a strong scoop-shaped inner apical concavity. Details of larval structure are given by Rozen (1965a) and Rozen and Buchmann (1990). The larvae of the two genera of the tribe, *Centris* and *Epicharis,* differ markedly, for *Centris*

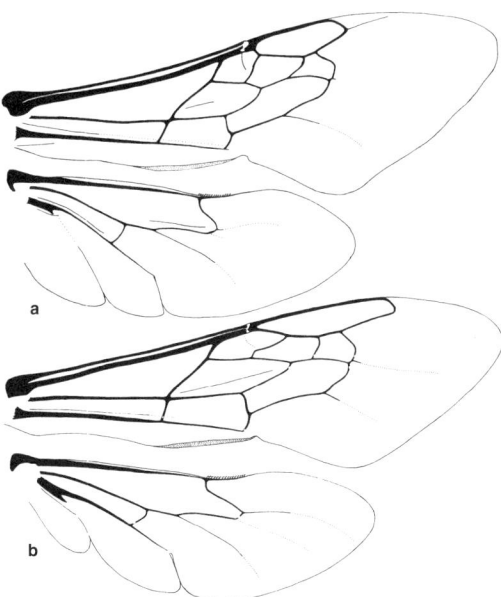

Figure 114-1. Wings of Centridini. **a,** *Centris (Centris) poecila* Lepeletier; **b,** *Epicharis (Epicharana) elegans* Smith.

larvae spin cocoons but *Epicharis* larvae do not, and have the reduced labiomaxillary region and antennae characteristic of larvae of all taxa that have lost cocoon-spinning behavior.

Key to the Genera of the Centridini

1. Marginal cell longer than distance from its apex to wing tip (Fig. 114-1b); a few long, whiplike setae arising from preoccipital ridge behind summit of eye and usually reaching anterior margin of tegula (but scarcely reaching anterior margin of scutum in subgenus *Triepicharis*) .... *Epicharis*
—. Marginal cell shorter than distance from its apex to wing tip (Fig. 114-1a); whiplike setae arising behind summit of eye absent, or, *if* present (subgenus *Ptilotopus*), then reaching only to anterior margin of scutum .... *Centris*

Genus *Centris* Fabricius

In *Centris* not only is the marginal cell short, as indicated in the key to genera, but its apical part is usually abruptly bent away from the costal margin of the wing, and is thus obliquely truncate (Fig. 114-1a), a shape not found in *Epicharis*. Species of most subgenera are very compact, strikingly robust, anthophoriform, and capable of hovering as well as extremely rapid flight. The largest species, however, those of the subgenera *Melacentris* and *Ptilotopus*, are more elongate and less adept at both hovering and rapid flight. The genus contains some of the largest neotropical bees, often with colorful pubescence: black, yellow and black, fulvous, white, etc. The integument is black or black and yellow, sometimes with the metasoma red or metallic bluish or greenish, often with yellow markings. Male genitalia and hidden sterna were illustrated by Michener (1954b), Moure (1969c), and Snelling (1984); see also Figures 114-2 and 114-3a-f.

At one stage (e.g., Michener, 1944) this genus was called *Hemisia,* and the tribe, Hemisiini, because the type species of *Centris* was considered to be a species now placed in *Eulaema* in the Euglossini. *Centris,* however, was conserved in its traditional sense by Commission Opinion 567 (1957). *Ptilotopus,* given generic rank by Snelling (1984), is here regarded as a subgenus of *Centris,* since it is no more distinctive than some of the other subgenera.

A cladistic study by Ayala (1998) indicates that *Centris* is divisible into three groups, as follows: (1) *Acritocentris, Centris* s. str., *Exallocentris, Paracentris, Xanthemisia, Xerocentris* (*Exallocentris,* lacking giant setae on the male genitalia in a group otherwise characterized by such setae, may be out of place in group 1); (2) *Heterocentris, Trachina;* (3) *Melacentris, Ptilocentris, Ptilotopus, Wagenknechtia.* These three groups could be regarded as genera, although such a division poses some problems, as indicated above for *Exallocentris.* Ayala's work resulted in two new subgenera (Ayala, 2002); to incorporate them I present a new key to the subgenera kindly made available by Dr. Ayala. An additional reason for the new key is that the old key (in Michener, 2000) contains errors.

Papers reviewing the subgenera of *Centris* are by Michener (1951b), Snelling (1974, 1984), and Ayala (1998). Revisions at the species level are found in the above papers by Snelling for North and Central America, and in

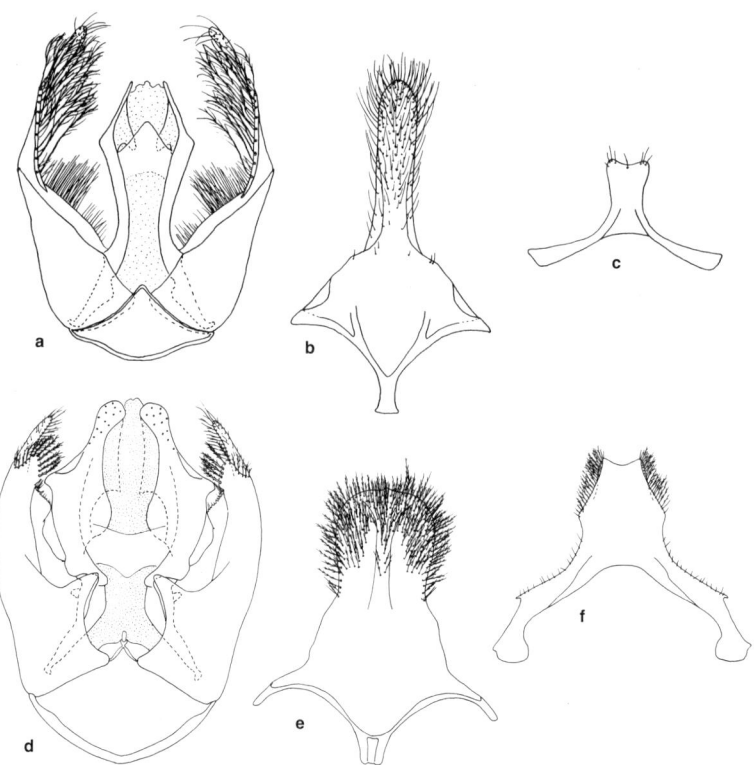

Figure 114-2. Dorsal views of genitalia and ventral views of S8 and S7, of male *Centris*. **a-c**, *C. (Centris) aethiocesta* Snelling; **d-f**, *C. (Ptilotopus) zonata* Mocsáry. From Snelling, 1984.

Friese (1900). Roig-Alsina (2000) revised the 31 Argentine species of *Centris*.

In various species of *Centris*, in diverse subgenera, some males are much larger than the usual form and commonly have more extensive yellow markings, more enlarged hind legs, and other distinctive features. Such "metanders" or "Beta males" have often been described as distinct species. Moure (1969c) illustrated the nearly identical male genitalia of "normal" males and metanders of *C. (Centris) aenea* Lepeletier and *flavifrons* (Fabricius). The behavioral correlates involved in the production of such individuals, as well as their mating and other activities, are in need of study.

Mating systems have been investigated in several species of *Centris* and vary greatly among the species. In some species, like *C. pallida* Fox and *caesalpiniae* Cockerell, males dig into the ground where females are emerging (Alcock, 1989; Rozen and Buchmann, 1990). The largest males have an advantage in mating with the virgin females, for they can push aside smaller, weaker males, which must, however, be at an advantage somewhere, perhaps where nests are widely scattered, or they would probably not be produced generation after generation. Males of other species establish territories, probably for mating, marked by pheromones secreted either by the mandibular glands (*Centris* s. str.) or by glands in the enlarged hind legs of some species (subgenera *Heterocentris, Trachina,* and *Xanthemisia*) (Coville et al., 1986; Frankie, Vinson, and Williams, 1989). Coville et al. show such a gland as occupying part of the interior of both the femur and the tibia and opening through a pore at the apex of the tibia.

Key to the Subgenera of *Centris*
(By Ricardo Ayala)

1. Scutellum with two lobes, tubercles, or convexities (except in males of *Ptilocentis* in which scutellum is convex with lobes not well defined) 2
—. Scutellum rounded, without lobes, tubercles or separate convexities 6
2(1). Hypoepimeral area with strong tubercle on anterior lower part (except in *C. americana* Klug); scutellum, and sometimes scutum with glabrous areas; female with basitibial plate simple, clearly without secondary plate (body length 30-32 mm) .... *C. (Ptilotopus)*
—. Hypoepimeral area without tubercle; scutellum, and scutum hairy, without glabrous areas except sometimes apices of scutellar tubercles; female with secondary basitibial plate on disc of primary plate 3
3(2). Mandible of female 5-toothed, of male 4-toothed, the two basal teeth flat and blunt; pygidial plate of female with apex notched .... *C. (Schisthemisia)*
—. Mandible of female 4-toothed, of male 3-toothed, the basal teeth acute (sometimes robust) in female (in some *Melacentris* basal tooth obtuse, somewhat divided in two); pygidial plate of female with apex acute or obtuse, not notched 4
4(3). Secondary basitibial plate of female with apex broadly rounded, behind central axis of primary plate; wings dark, almost black, with blue, violet, or green iridescence (metasomal terga with abundant pubescence and usually without metallic tints) .... *C. (Melacentris)*
—. Secondary basitibial plate of female with apex acute, near central axis of primary plate; wings dark but not black, somewhat translucent, with magenta and green iridescence 5
5(4). Clypeus in both sexes yellow with median black area narrowed toward lower margin; ocellocular distance equal to median ocellar diameter or less; metasomal terga without long plumose hairs (terga sometimes strongly metallic) .... *C. (Aphemisia)*
—. Clypeus black in female, in male with median yellow area; ocellocular distance greater than median ocellar diameter; T2 and T3 with abundant plumose hairs .... *C. (Ptilocentris)*
6(1). Maxillary palpus 3- or 4-segmented (counting robust basal segment, i.e., with two or three free segments); hind femur and tibia of male usually swollen; hind basitarsus of male usually with spine or elevated carina on posterior margin 7
—. Maxillary palpus 5- or 6-segmented (4 or 5 free segments); hind leg of male usually not swollen; hind basitarsus of male without spine or elevated carina 8
7(6). Male labrum broader than long with sparse or abundant hairs on whole surface; female basitibial plate simple, without secondary plate or with tendency to form secondary plate covered with short, dense hairs .... *C. (Heterocentris)*
—. Male labrum longer than broad with abundant hairs on distal margin only, sparse hairs elsewhere; female basitibial plate with elevated secondary plate not hidden by hairs, apex of secondary plate near posterior margin of primary plate .... *C. (Trachina)*
8(6). Females 9
—. Males 15
9(8). Mandible 5-toothed, one of these teeth on inner surface at base of distal tooth; pygidial plate with longitudinal groove from apex of secondary plate toward spatulate apex of pygidial plate .... *C. (Xanthemisia)*
—. Mandible 3- or 4-toothed, without tooth on inner surface at base of distal tooth; pygidial plate without longitudinal groove and spatulate apex 10
10(9). Front and mid tibiae without combs for collecting floral oil, or in South America sometimes with vestigal combs 11
—. Front and mid tibiae with well developed combs for collecting oil 12
11(10). Pubescence completely black; mandible 4-toothed; mid basitarsus with area of dense, spatulate hairs; T2 and T3 with sparse, usually short pubescence; basitibial plate traversed by two more or less transverse lines .... *C. (Exallocentris)*
—. Pubescence largely pale; mandible 3-toothed; mid basitarsus with hairs simple, not spatulate (South American species with vestigal comb); T2 and T3 with abundant short pale pubescence; basitibial plate variable but without transverse lines .... *C. (Xerocentris)*
12(10). Front basitarsus with comb for collecting floral oil only on distal half; tergal integument black with intense metallic blue reflections .... *C. (Wagenknechtia)*
—. Front basitarsus with comb on full length of anterior margin; terga with or without metallic reflections, sometimes with transverse yellow bands 13
13(12). Basitibial plate with defined secondary plate that lacks sharp projecting margin, margin of secondary plate extending basad near anterior margin of basitibial plate

and distad near posterior margin of plate ....
.... *C. (Paracentris)*
—. Basitibial plate with defined secondary plate with sharp projecting margins; margins of secondary plate extending basad both anteriorly and posteriorly 14
14(13). Clypeus without yellow or white markings............
....*C. (Acritocentris)*
—. Clypeus with yellow or white inverted T or Y ....
.... *C. (Centris* s. str.)
15(8). Middle basitarsus with comb presumably for collecting oil (as also in some *Paracentris*); pygidial plate well defined with exposed, elevated margins ....
.... *C. (Acritocentris)*
—. Middle basitarsus without comb for collecting oil (except in some *Paracentris,* which lack pygidial plate); pygidial plate absent or not well defined, if present lateral margins covered by pubescence and not elevated 16
16(15). Mandible with acetabular goove ending between apical and subapical teeth (acetabular carina ending between the two most proximal teeth); posterior femur swollen .... *C. (Xanthemisia)*
—. Mandible with acetabular groove ending between proximal and subapical (median) teeth; posterior femur not swollen 17
17(16). Basitibial plate well defined, distal margin elevated; pygidial plate well defined, lateral margins not elevated, covered by pubescence medially (integument black, pubescence black and whitish).................... *C. (Exallocentris)*
—. Basitibial plate absent; pygidial plate absent, apical margin of T7 bilobed 18
18(17). Clypeus yellow with upper lateral margins black ..
.... *C. (Centris* s. str.)
—. Clypeus black or reddish brown or if yellow, then its upper lateral margins yellow, not black 19
19(18). Metasomal terga black with strong, usually dark blue, reflections (clypeus black) .... *C. (Wagenknechtia)*
—. Metasomal terga black or brown without metallic reflections 20
20(19). Lateral ocellus separated from eye by less than ocellar diameter; T2 to T4 with abundant whitish pubescence .... *C.(Xerocentris)*
—. Lateral ocellus separated from eyes by at least ocellar diameter; T2 to T4 usually with dark pubescence, rarely whitish .... *C. (Paracentris)*

Centris / Subgenus *Acritocentris* Snelling

Centris (Acritocentris) Snelling, 1974: 36. Type species: *Centris ruthannae* Snelling, 1966, by original designation.

This subgenus consists of species 15 to 19 mm in body length that resemble large *Paracentris,* to which males run in the key. Females, however, differ from *Paracentris* in their three-toothed mandibles and the overhanging margin of the secondary basitibial plate, although the overhang may be worn off in old specimens. It should be noted that in the related subgenus *Xerocentris* some species have three-toothed female mandibles whereas others have four-toothed mandibles. Further, *Centris (Paracentris) autrani* Vachal from Chile has a partially overhanging margin of the secondary basitibial plate, thus resembling *Acritocentris.*

■ *Acritocentris* occurs from southern Arizona, USA, to Tamaulipas and southward to Chiapas, Mexico. The four species were revised by Snelling (1984).

Centris / Subgenus *Aphemisia* Ayala

Centris (Aphemisia) Ayala, 2002: 1. Type species: *Centris plumipes* Smith, 1854, by original designation.

The species of this subgenus resemble *Melacentris* but differ in the more translucent wings. In the female the acute apex of the secondary basitibial plate, more or less on the longitudinal axis of the primary plate, distinguishes it from *Melacentris*. The apical process of S8 of the male is widest medially, not basally as in related groups; the articulation of the male gonostylus, slightly preapically on the gonocoxite, also differs from related groups.
■ *Aphemisisa* occurs from Panama and French Guiana south to Bolivia. The three species were reviewed by Moure (2002). (The replacement name given by Moure to the type species is unnecessary since *Centris plumipes* Smith is, in fact, not a junior homonym of *Apis plumipes* Fabricius, which, although listed in *Centris* by Fabricius in 1804, was transferred to *Anthophora* in 1854.)

Centris / Subgenus *Centris* Fabricius s. str.

Centris Fabricius, 1804: 354. Type species: *Apis haemorrhoidalis* Fabricius, 1775, designation by Commission Opinion 567 (1959). [See Michener, 1997b.]
Hemisia Klug, 1807, in Illiger, 1807: 198, nomen nudum.
Hemisia Klug, 1807b: 213, 227. Type species: *Apis haemorrhoidalis* Fabricius, 1775, by designation of Cockerell, 1906e: 105.
Centris (Cyanocentris) Friese, 1900: 244, 251. Type species: *Apis versicolor* Fabricius, 1775, by designation of Sandhouse, 1943: 543.
Centris (Poecilocentris) Friese, 1900: 244, 252. Type species: *Centris fasciatella* Friese, 1900, by designation of Sandhouse, 1943: 589.

Centris s. str. consists of species usually having metallic metasomal background color. Yellow metasomal markings occur in many species and in some, e.g., *C. eisenii* Fox, the yellow bands are so extensive that the metallic background is essentially eliminated. Rarely, the metasoma is red and nonmetallic. The body length is 12 to 24 mm. A unique character of the subgenus is the long, slender, apical projection of the male gonocoxite, extending parallel to the gonostylus; this gonocoxal projection bears giant branched setae (Fig. 114-2a).
■ This subgenus occurs from Baja California, Mexico, and southern Arizona to southern Florida, USA, and the Bahamas, south through the Antilles and the continental tropics to Santa Catarina, Brazil. Of perhaps 35 species in all, about 20 occur in North and Central America and the Antilles. The species of Mexico and Central America were reviewed by Snelling (1984). Moure (1945c) listed 11 mostly South American species.

Centris / Subgenus *Exallocentris* Snelling

Centris (Exallocentris) Snelling, 1974: 35. Type species: *Centris anomala* Snelling, 1966, by original designation.

This is a subgenus of medium-sized (length 14-18 mm) bees suggestive of *Melacentris* but probably nearer to *Paracentris*. They lack the combs of setae on the fore and middle basitarsi; instead, there are pads of dense, branched hairs that may serve, like the combs of most other subgenera, for oil collecting. The scutellum is simply convex rather than biconvex, and the metasoma is metallic bluish, unlike that of most *Paracentris* and *Melacentris*.

■ *Exallocentris* is found from Sonora and Hidalgo south to Oaxaca, Mexico. The only species is *Centris anomala* Snelling.

Centris / Subgenus *Heterocentris* Cockerell

Gundlachia Cresson, 1865: 195 (not Pfeiffer, 1850). Type species: *Centris ? cornuta* Cresson, 1865, monobasic.

Heterocentris Cockerell, 1899a: 14, replacement for *Gundlachia* Cresson, 1865. Type species: *Centris ? cornuta* Cresson, 1865, autobasic.

Centris (Rhodocentris) Friese, 1900: 244, 250. Type species: *Centris difformis* Smith, 1854, by designation of Sandhouse, 1943: 596. [*Centris difformis* and *C. cornuta* have been considered synonymous, making *Rhodocentris* and *Heterocentris* isotypic, but Snelling (1984) argues that they are distinct.]

Hemisiella Moure, 1945c: 407. Type species: *Apis lanipes* Fabricius, 1775, by original designation.

This is a subgenus of small to rather large (9-19 mm long) *Centris*, the metasoma often red. It is most closely related to the subgenus *Trachina*, but differs as indicated in the key to subgenera. Unique features of *Heterocentris* are (1) some scopal hairs, especially near the base of the tibia, that are simple (they are plumose in other subgenera); and (2) a groove on the inner ventral edge of the front coxa. *Heterocentris* has been considered to differ from *Hemisiella* because of the modified labrum, mandibles, and clypeus of the female of the former. These are derived features that possibly arose more than once; *Hemisiella* would be almost certainly paraphyletic relative to *Heterocentris* in the narrow sense.

■ *Heterocentris* occurs from southern Arizona, USA, Tamaulipas, Mexico, and the Antilles southward through the tropics to eastern Bolivia and the provinces of La Rioja and Misiones in Argentina. Of about 17 species, nine are found in North and Central America and were revised by Snelling (1984) under the subgeneric names *Heterocentris* and *Hemisiella*. Under the same subgeneric names, Moure (1945c) listed ten mostly South American species. The Central American species of *Heterocentris* in the narrow sense, excluding *Hemesiella*, were reviewed by Thiele (2003), who also described male dimorphism in *Centris labrosa* Friese.

Vinson and Frankie (2000b) discuss nest structure and usurpation, possible intraspecific cleptoparasitism, as well as nest defense, of *Centris (Heterocentris) bicornuta* Mocsáry. Like other *Heterocentris*, this species nests in holes in wood. On completion, such a nest is closed with a plug of wood chips, followed by a custardlike material. Usurpation was not observed after the latter material was in place.

Centris / Subgenus *Melacentris* Moure

Centris (Melacentris) Moure, 1995: 947. Type species: *Centris dorsata* Lepeletier, 1841, by original designation.

Since 1900, this subgenus has been known as *Melanocentris* Friese. Moure (1995c) corrected the identification of *Centris atra* Friese, the type species of *Melanocentris*, showing it to be a species of *Ptilotopus*. He therefore renamed *Melanocentris* of authors as *Melacentris* and placed *Melanocentris* Friese in the synonymy of *Ptilotopus*.

Melacentris consists of medium-sized to large *Centris* (body length 18-28 mm), nonmetallic black or the metasoma dark reddish, rarely with metallic reflections. The mandible of the female, four-toothed, is broad and bent preapically. The scutellum is usually gently biconvex. Male genitalia and hidden sterna are illustrated in Figure 114-3a-c.

■ This subgenus occurs from the Mexican states of Nayarit and San Luis Potosí south through the moist tropics to Paraguay and the state of Santa Catarina, Brazil. There are about 18 species (see Michener, 1951b); nine occur in North and Central America and were revised by Snelling (1984).

Centris / Subgenus *Paracentris* Cameron

Paracentris Cameron, 1903a: 235. Type species: *Paracentris fulvohirta* Cameron, 1903, monobasic.

Centris (Penthemisia) Moure, 1950d: 390. Type species: *Hemisia chilensis* Spinola, 1851, by original designation.

Centris (Trichocentris) Snelling, 1956: 4. Type species: *Centris rhodoleuca* Cockerell, 1923 = *Centris caesalpiniae* var. *rhodopus* Cockerell, 1897, by original designation.

This is a subgenus that resembles *Wagenknechtia*, *Xerocentris*, and *Xanthemisia* in appearance and in having giant setae on the male genitalia limited to a region close to the base of the gonostylus. *Paracentris* consists of relatively small species, mostly 9 to 15 mm long, although occasional individuals attain 20 mm. Its principal characters are indicated in the key to subgenera. Along with *Xerocentris*, which may be a mere offshoot of *Paracentris*, this is the principal group of *Centris* found in the northern and southern parts of the range of the genus.

■ *Paracentris* occurs from California to Kansas and Florida, USA, south to Chile and Argentina. It is found principally in xeric areas and is rare in the moist tropics. There are 15 North and Central American species and perhaps ten additional South American species. Revisions of North American species are by Snelling (1974, 1984). Zanella (2002b) revised the South American species of the subgenus *Paracentris* (including *Penthemisia*). He regarded *Acritocentris*, *Exallocentris*, and *Xerocentris* as synonyms of *Paracentris*, and recognized 15 South American species of *Paracentris* and four of *Penthemisisa*. The Chilean species of *Paracentris* were revised by Vivallo, Zanella, and Toro (2003). These authors also regarded *Penthemisia* as a subgenus distinct from *Paracentris*.

Centris / Subgenus *Ptilocentris* Snelling

Centris (Ptilocentris) Snelling, 1984: 22. Type species: *Centris festiva* Smith, 1854, by original designation.

This subgenus is similar to *Centris* s. str., from which it differs especially in the lack of giant setae on the male genitalia and the short apical process of the gonocoxite. The body length is about 19 mm. T1 to T4 have long, dense, erect plumose hair. The male gonostylus is attached preapically to the gonocoxite and is little more than twice as long as wide. Snelling (1984) illustrated the male genitalia and hidden sterna.

■ *Ptilocentris* occurs from the Mexican Distrito Federal to Peru. The only species is *Centris festiva* Smith.

Centris / Subgenus Ptilotopus Klug

Ptilotopus Klug, 1810: 32. Type species: *Ptilotopus americanus* Klug, 1810, monobasic.
Centris (Melanocentris) Friese, 1900: 244. Type species: *Centris atra* Friese, 1899, by designation of Sandhouse, 1943: 569.

See the treatment of subgenus *Melacentris* for an explanation of the synonymy of *Melanocentris*.

Ptilotopus contains the largest and most spectacular species of *Centris*, with body lengths of 20 to 32 mm. Well-defined bare areas on the scutellum and sometimes on the scutum characterize the subgenus; similar but less well-defined areas are found in some *Melacentris*. Snelling (1984) enumerated the distinctive features of *Ptilotopus*, arguing that it should have generic rank. Contrary to Snelling, the male genitalia have giant branched setae, as in various other subgenera of *Centris*, but they are relatively small giants, as illustrated for *C. zonata* Mocsáry by Snelling (1984); see also Figure 114-2d.

■ This subgenus occurs from the state of São Paulo, Brazil, north to Costa Rica. Of at least 12 species, nine were listed by Michener (1951b).

Centris / Subgenus Schisthemisia Ayala

Centris (Schisthemisia) Ayala, 2002: 5. Type species: *Centris flavilabris* Mocsáry, 1899, by original designation.

Like *Aphemisia*, this subgenus contains *Melacentris*-like species. It differs from similar subgenera in the 5-toothed mandibles of the female, 4-toothed in the male, and the notched apex of the well-formed pygidial plate of both sexes.

■ *Schisthemisia* is known from the state of Amazonas, Brazil, to Bolivia. There are two species.

Centris / Subgenus Trachina Klug

Trachina Klug, 1807b: 226. Type species: *Centris longimana* Fabricius, 1804, by designation of Cockerell, 1906e: 105.
Paremisia Moure, 1945c: 406. Type species: *Centris lineolata* Lepeletier, 1841 = *Bombus similis* Fabricius, 1804, by original designation.

Like *Heterocentris*, *Trachina* has only three segments in the maxillary palpus, and the metasoma is usually red. Most species are larger than most *Heterocentris*; the body length is 13 to 20 mm. A longitudinal median yellow line on the clypeus, sometimes lost in a largely yellow clypeus, distinguishes both sexes of *Trachina* from *Heterocentris*, which has apical spots or an apical band, although in some males the clypeus is yellow except at the base. Male genitalia and hidden sterna are illustrated in Figure 114-3d-f; see also Snelling, 1984.

■ *Trachina* ranges from Sinaloa and Veracruz, Mexico, south through the tropics to eastern Bolivia and São Paulo, Brazil. Of about 15 species, nine occur in Mexico and Central America and were revised by Snelling (1984). Four South American species were listed by Moure (1945c) under the name *Paremisia*.

Centris / Subgenus Wagenknechtia Moure

Centris (Wagenknechtia) Moure, 1950d: 389. Type species: *Centris cineraria* Smith, 1854, by original designation.

Species of this subgenus have the size (11-16 mm long) and appearance of *Paracentris*, and *Wagenknechtia* may be a close relative of that subgenus. It differs from *Paracentris* in lacking the following: (1) giant setae on the male genitalia, (2) a raised, median, secondary pygidial plate in the female, and (3) a raised secondary basitibial plate. These may be ancestral characters, and as suggested by Michener (1951b), *Wagenknechtia* may be a basal branch of *Centris* phylogeny.

■ *Wagenknechtia* occurs from Antofagasta to Asién, Chile, and in the provinces of Neuquén and Río Negro, Argentina. The species of the subgenus *Wagenknechtia* were revised by Vivallo, Zanella, and Toro (2002). These authors recognized five species.

Centris / Subgenus Xanthemisia Moure

Xanthemisia Moure, 1945c: 401. Type species: *Centris bicolor* Lepeletier, 1841, by original designation.

This subgenus is related to *Paracentris* and perhaps should be considered a group derived from that subgenus. It has the size (14-21 mm long) and appearance of that subgenus; the principal differentiating characters are indicated in the key to subgenera.

■ *Xanthemisia* occurs from the states of Guerrero and Oaxaca, Mexico, south through the tropics to the state of Santa Catarina, Brazil. Of at least four species, three found in North and Central America were revised by Snelling (1984).

Centris / Subgenus Xerocentris Snelling

Centris (Xerocentris) Snelling, 1974: 3. Type species: *Centris californica* Timberlake, 1940, by original designation.

This subgenus, which occurs principally in xeric areas where plants producing floral oils are scarce, probably lost the oil-collecting combs of the fore and middle basitarsi. In other characters *Xerocentris* is similar to *Paracentris*, and may consist of specialized species of that subgenus. The body length is 9 to 17 mm.

■ *Xerocentris* as here understood has a disjunct amphitropical distribution, from California to New Mexico, USA, south to Baja California, Sonora, and Guerrero, Mexico, and in Antofagasta and Tarapacá, Chile. The seven North American species were revised by Snelling (1974). According to Neff and Simpson (1981) the one Chilean species, *Centris mixta* Friese, is a derivative of

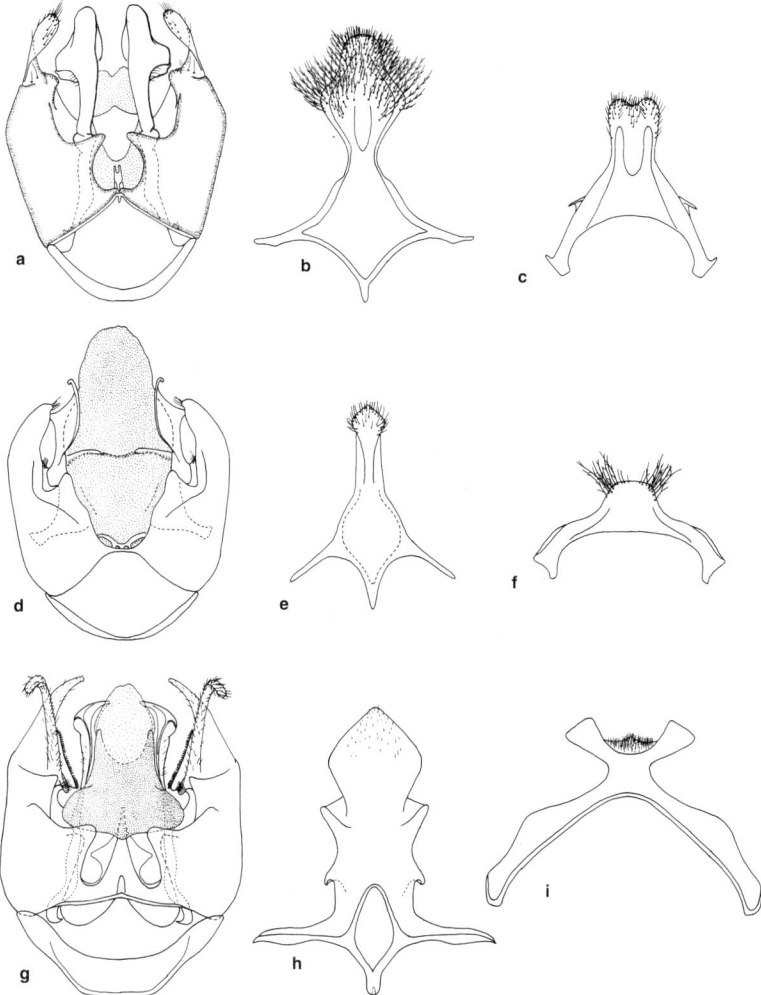

Figure 114-3. Dorsal views of genitalia and ventral views of S8 and S7, of male Centridini. **a-c,** *Centris (Melacentris) agiloides* Snelling; **d-f,** *C. (Trachina) eurypatana* Snelling; **g-i,** *Epicharis (Epicharoides) albofasciata* Smith. From Snelling, 1984.

South American *Paracentris* and not closely related to the North American species. In that case *Xerocentris* is exclusively North American.

The most common and best-known species is *Centris pallida* Fox, whose nesting behavior and mating system have been much studied (Alcock, Jones, and Buchmann, 1976; Alcock, 1989; Rozen and Buchmann, 1990). The nests are in dense aggregations but not in the same sites year after year. Males dig out emerging females for mating, large males having an advantage because they are able to push away smaller competitors (see the account for the genus *Centris*).

Genus *Epicharis* Klug

Epicharis contains large, *Centris*-like bees that have a long marginal cell, bent away from the wing margin at the apex (Fig. 114-1b) and usually rounded, not obliquely truncate as in *Centris*. The body is nonmetallic and the metasoma is sometimes red, frequently with yellow metasomal bands, the band of T2 or its lateral remnants usually being stronger than the others and often the only one developed. The body is more elongate than usual in *Centris*, like that of *C. (Melacentris)* and *(Ptilotopus)* or even more so, and thus euceriform. Most *Epicharis* are 15 to 25 mm long, but the subgenera *Epicharitides* and *Epicharoides* are generally smaller, 10 to 16 mm long. Male genitalia and hidden sterna of various species were illustrated by Michener (1954b) and Snelling (1984); see Figure 114-3g-i. The first couplet of the key to subgenera separates two groups. In addition to the characters listed in the key, in at least some species in each of the subgenera of the first group (*Epicharis* s. str., etc.) the claws of the hind legs of the female are simple, whereas, so far as I know, all species in the second group have a tooth on each claw.

Epicharis is the sister group of *Centris* (Ayala, 1998). The long marginal cell and perhaps the presence of two male gonostyli (Fig. 114-3g) in *Epicharis* are plesiomorphic relative to *Centris*. At the same time, the short, two- or three-segmented maxillary palpi and the reduced larval mouthparts and loss of cocoon-spinning activity are derived relative to *Centris*. Although most *Centris* have four- or five-segmented palpi, two of its subgenera have three-segmented maxillary palpi, but these subgenera are in other ways derived and their palpal reduction is presumably independent from that of *Epicharis*.

The subgenera of *Epicharis* were reviewed by Moure (1945b), Snelling (1984), and Ayala (1998). The three subgenera *Epicharis* s. str., *Epicharana,* and *Hoplepicharis* taken together were the sister group to all others in Ayala's cladistic analyses.

Key to the Subgenera of *Epicharis*
(Modified from Snelling, 1984)

1. Female with secondary basitibial plate; male with obsolete pygidial plate, its margins more or less coextensive with margins of T7, its apex thin and bilobate or bidentate 2
—. Female without secondary basitibial plate; male with distinct, sharply margined pygidial plate, no more than one-half as wide as T7, its apex narrowly truncate or rounded 4
2(1). Lateral margins of female pygidial plate nearly straight, apex narrowly truncate; female hind tibia longer than basitarsus; hind leg of male with long, scopa-like hairs and basitarsus with prominent tooth at basal one-third .... *E. (Hoplepicharis)*
—. Lateral margins of female pygidial plate distinctly concave as seen in dorsal view, apex broadly truncate; female hind tibia no longer than basitarsus; male without scopa-like hairs on hind leg and hind basitarsus without tooth at basal one-third 3
3(2). Female: Frontal carina ending more than diameter of anterior ocellus below that ocellus; apical pygidial truncation narrower than diameter of anterior ocellus; disc of T5 with short hairs mostly simple or barbed. Male: Mesosternal tubercles absent; hind basitarsus without carinate ridge on outer margin; hind trochanter and femur without ventral patch of short, dark, plumose setae (South America) .... *E. (Epicharis* s. str.)
—. Female: Frontal carina to anterior ocellus ending less than diameter of anterior ocellus below that ocellus; apical pygidial truncation broader than diameter of anterior ocellus; disc of T5 with short hairs distinctly plumose. Male: Two mesosternal tubercles or teeth projecting posteriorly between middle coxae; hind basitarsus with carinate ridge on outer margin, terminating in toothlike process; hind trochanter and femur with ventral patch of short, dark, plumose setae .... *E. (Epicharana)*
4(1). First flagellar segment of female equal to length of second to fourth segments combined, of male longer than scape and longer than following two segments combined; posterior margin of dorsal surface of scutellum deeply impressed in middle; maxillary palpus two-segmented .... *E. (Parepicharis)*
—. First flagellar segment short, in female usually no longer than second and third segments combined and always shorter than second to fourth combined, and in male shorter than scape; posterior margin of dorsal surface of scutellum not or weakly impressed in middle, or, *if* deeply impressed, then maxillary palpus three-segmented 5
5(4). Lateral ridges of clypeal disc absent *or,* on basal one-half or less of clypeus, weakly evident; male hind basitarsus with carinate ridge on inner margin, terminating in large tooth at about midlength 6
—. Lateral ridges of clypeal disc strong and sharply defined to near apical margin of clypeus; male hind basitarsus without carinate ridge or tooth on inner margin 7
6(5). Maxillary palpus two-segmented; flagelliform preoccipital setae distinct and extending beyond anterior margin of mesoscutum; ocellocular distance in male less than diameter of lateral ocellus (South America) .... *E. (Anepicharis)*
—. Maxillary palpus three-segmented; flagelliform preoccipital setae short, not projecting beyond occipital hairs and scarcely reaching anterior margin of mesoscutum; ocellocular distance in male equal to diameter of lateral ocellus (South America) .... *E. (Triepicharis)*
7(5). Dorsal surface of scutellum bigibbous; prepygidial fimbria of female not preceded by shiny, glabrous area (Brazil) .... *E. (Cyphepicharis)*
—. Dorsal surface of scutellum flat; prepygidial fimbria of female preceded by a shiny, nearly glabrous area.............. 8
8(7). Jugal lobe of hind wing about one-half as long as vannal lobe, nearly attaining apex of cubital cell; female hind basitarsus about twice as long as broad, its posterior margin nearly straight; male pygidial plate broad, covering most of T7.. *E. (Epicharoides)*
—. Jugal lobe of hind wing about one-third as long as vannal lobe, its apex well short of that of cubital cell; female hind basitarsus about one-third longer than broad, its posterior margin strongly curved; male pygidial plate narrow, covering less than half of T7 ...... *E. (Epicharitides)*

Epicharis / Subgenus *Anepicharis* Moure

Anepicharis Moure, 1945b: 302. Type species: *Epicharis dejeanii* Lepeletier, 1841, by original designation.

■ *Anepicharis* is found in Brazil from the state of Rio Grande do Sul to Goiás. Three species were recognized by Moure (1945b).

Epicharis dejeani Lepeletier nests in aggregations in flat ground. Floral oils from Malpighiaceae are used to waterproof the cells and not as a larval food. Overwintering in the temperate climate of southern Brazil is as prepupae (Hiller and Wittmann, 1994).

Epicharis / Subgenus *Cyphepicharis* Moure

Cyphepicharis Moure, 1945b: 306. Type species: *Cyphepicharis borgmeieri* Moure, 1945, by original designation.

■ *Cyphepicharis* is known from the states of São Paulo and Minas Gerais, Brazil. The only species is *Epicharis borgmeieri* (Moure).

Epicharis / Subgenus *Epicharana* Michener

Epicharis (*Epicharana*) Michener, 1954b: 144. Type species: *Apis rustica* Olivier, 1789, by original designation.

This is the group that Moure (1945b) considered to be *Epicharis* s. str.; for an explanation, see Michener

(1997b). The distinction between *Epicharis* s. str. and *Epicharana* is easy in males, but females are very similar. In addition to the characters indicated in the key, in *Epicharis* s. str. the scutellum is more or less depressed in the middle, with vestigial carinae laterally, features not found in *Epicharana*. Unique features of males of *Epicharana* include a strong apical mesal process on the fore coxa and a mesosternal tooth or process directed posteriorly between the middle coxae.

■ *Epicharana* occurs from the states of Jalisco and Tamaulipas, Mexico, south to Bolivia and to the states of Mato Grosso and São Paulo, Brazil. About six species are known (Moure, 1945b; Snelling, 1984); Snelling gave a key to the North and Central American species.

At least *Epicharis rustica* (Olivier) and *elegans* Smith nest in aggregations in vertical earthen banks that are heavily shaded or almost in the dark in mouths of caverns or mines or other such situations; see the account of the tribe Centridini.

Epicharis / Subgenus *Epicharis* Klug s. str.

Epicharis Klug, 1807, *in* Illiger, 1807: 197, nomen nudum.
Epicharis Klug, 1807b: 226. Type species: *Centris umbraculata* Fabricius, 1804, by designation of Lutz and Cockerell, 1920: 562. [See Michener, 1997b, for an account of other type designations.]
Eucharis Dalla Torre, 1896: 300. Listed as a junior synonym of *Epicharis* and *Eucharis hirtipes* listed in synonymy on p. 301, and both attributed to Lepeletier [and Serville], 1825 (correctly 1828): 794 and 795. [See Michener, 1997b.]
Xanthepicharis Moure, 1945b: 297. Type species: *Epicharis bicolor* Smith, 1854, by original designation.

■ This subgenus occurs from southern Brazil to the Guianas. Three species were included by Moure (1945b).

Epicharis / Subgenus *Epicharitides* Moure

Epicharitides Moure, 1945b: 311. Type species: *Epicharis cockerelli* Friese, 1900, by original designation.

Like *Epicharoides*, this subgenus includes relatively small species, usually with bold yellow or whitish metasomal markings.

■ *Epicharitides* occurs from Panama south to Bolivia and the state of São Paulo, Brazil. There are seven named species. Moure (1954b) revised the subgenus.

Epicharis / Subgenus *Epicharoides* Radoszkowski

Epicharoides Radoszkowski, 1884a: 20. Type species: *Epicharoides bipunctatus* Radoszkowski, 1884 = *Epicharis maculatus* Smith, 1874, monobasic.

■ This subgenus occurs from the states of Sinaloa and Veracruz, Mexico, south to Bolivia, the province of Misiones, Argentina, and Uruguay. It contains four species. The two that occur in Mexico and Central America were revised by Snelling (1984).

Epicharis / Subgenus *Hoplepicharis* Moure

Hoplepicharis Moure, 1945b: 300. Type species: *Epicharis fasciata* Lepeletier and Serville, 1825, by original designation.

■ *Hoplepicharis* occurs from the states of Nayarit and Veracruz, Mexico, south to Santa Catarina, Brazil. There are four species. The species of North and Central America were revised by Snelling (1984).

Epicharis / Subgenus *Parepicharis* Moure

Parepicharis Moure, 1945b: 307. Type species: *Epicharis zonata* Smith, 1854, by original designation.

■ *Parepicharis* occurs from Bolivia and the state of Paraná, Brazil, north to Costa Rica. The two species were revised by Snelling (1984).

Epicharis / Subgenus *Triepicharis* Moure

Triepicharis Moure, 1945b: 304. Type species: *Epicharis analis* Lepeletier, 1841, by original designation.

■ *Triepicharis* occurs in Brazil from the states of Mato Grosso and São Paulo to Pará. There are two species, as indicated by Moure (1954b).

115. Tribe Rhathymini

This tribe of tropical American cleptoparasitic bees consists of large (13-28 mm long), elongate, nomadiform species (Fig. 104-2b). Various studies, e.g., Michener (1944) and Roig-Alsina and Michener (1993), place the Rhathymini among the apine bees, not in the Nomadinae, in spite of superficial resemblance to the large species of the epeoline genus *Odyneropsis*. The tribe is easily distinguished from *Odyneropsis* by tribal characters, such as the oblique vein cu-v of the hind wing, longer than to slightly shorter than vein M+Cu; the short first submarginal cell, little if any longer than the second on its posterior margin; the tapering rather than quadrate anterior coxa with a slender, hairy, apical spine mesal to the articulation of the trochanter; and the tapering, somewhat tubular (because of upturned lateral margins) S6 of the female, which is not produced into processes armed with specialized setae as in Epeolini. From the Ericrocidini, the Rhathymini differs in its slender body; the hairy, not papillate wings; the presence of arolia; and the unmodified middle tibial spur. Other characters are the simple mandibles, the absence of maxillary palpi, the often biconvex scutellum, its posterior part declivous, and the slender, parallel-sided middle and hind tarsi, the former and sometimes the latter as long as the corresponding tibia. S4 and S5 of the male are strongly fringed; in most species the margin of S5 is broadly concave and produced posterolaterally under T5 as a hairy projection, resulting in preapical metasomal tufts of long hair that are visible from above, as shown in Figure 104-2b. A fuller characterization of the tribe was by Engel, Michener, and Rightmyer (2004a).

Larvae of Rhathymini were described by Rozen (1969a) and Camargo, Zucchi, and Sakagami (1975). The absence of galeae and the elongate labiomaxillary region are derived characters of mature larvae in common with those of Ericrocidini and Melectini. The first-stage larva has a less sclerotized and more globose and hypognathous head than that in other cleptoparasitic Apinae, and the antennae are scarcely noticeable convexities, unlike those of Melectini and Ericrocidini (Rozen, 1991a). The egg is evidently laid through an aperture opened in the closure of the host cell, and the young larva presumably kills the egg or young larva of the host.

Key to the Genera of Rhathymini

1. Mesepisternum with large submedian tubercle; vein cu-v of hind wing strongly oblique and distinctly longer than second abscissa of M+Cu; supraclypeal area strongly elevated, crested medially, not continuing convexity of clypeus .... *Rhathymus*
—. Mesepisternum without tubercle; vein cu-v of hind wing less strongly oblique and shorter than or subequal to second abscissa of M+Cu; supraclypeal area with surface, in general, a continuation of convexity of clypeus although with small frontal tubercle at lower end of frontal line .... *Nanorhathymus*

Genus *Nanorhathymus* Engel, Michener, and Rightmyer

Rhathymodes Engel, Michener, and Rightmyer, 2004a: 6 (not Turner, 1911). Type species: *Rhathymus acutiventris* Friese, 1906, by original designation.
Nanorhathymus Engel, Michener, and Rightmyer, 2004b: 316, replacement for *Rathymodes* Engel, Michener, and Rightmyer. Type species: *Rathymus acutiventris* Friese, 1906, autobasic.

This genus includes species smaller than most *Rhathymus* (13-18 mm in body length; 16-28 mm for *Rhathymus*). The most prominent morphological differentiating characters are indicated in the key above. Others include the nearly flat (rather than strongly depressed) area between the lateral ocellus and the eye and the position of the second recurrent vein near the middle or apical third (rather than near the apex) of the second submarginal cell.

■ *Nanorhathymus* is known from central Mexico to Paraguay. The two species were differentiated by Engel, Michener, and Rightmyer (2004a).

Nanorhathymus superficially resembles large *Osirinus* (tribe Osirini) but differs in the much smaller stigma, absence of maxillary palpi, lack of separation of the anterior wing margin from vein R1, absence of a mesal carina on the anterior coxa, etc.

Genus *Rhathymus* Lepeletier and Serville

Colax Lepeletier and Serville, 1825: 4, 213, nomen nudum.
Colax Lepeletier and Serville, 1828: 448 (not Hübner, 1819). Type species: *Rhathymus bicolor* Lepeletier and Serville, 1828, monobasic. [See Michener, 1997b.]
Rhathymus Lepeletier and Serville, 1828: 448. Type species: *Rhathymus bicolor* Lepeletier and Serville, 1828, monobasic.
Liogastra Perty, 1833: 146. Type species: *Liogastra bicolor* Perty, 1833 = *Rhathymus bicolor* Lepeletier and Serville, 1825, monobasic.
Rathymus Smith, 1854: 278 (not Dejean, 1831), unjustified emendation of *Rhathymus* Lepeletier and Serville, 1828.
Bureauella Dominique, 1898: 61. Type species: *Bureauella insignis* Dominique, 1898, monobasic.

The characters of *Rhathymus* are indicated in the discussion of the tribe and in the key to genera. Species vary from black, the metasoma frequently red, to largely yellow.

■ The range is from the state of San Luis Potosí, Mexico, south through the tropics to eastern Bolivia and the province of Misiones, Argentina. There are about eight species. Friese (1912b) gave a key to the species, along with *Odyneropsis* in the Epeolini.

Rhathymus species are cleptoparasites of *Epicharis* in the Centridini (Rozen, 1969a; Camargo, Zucchi, and Sakagami, 1975; Hiller and Wittmann, 1994).

116. Tribe Ericrocidini

This is a Western Hemisphere tribe formerly sometimes called Ctenioschelini. It consists of anthophoriform to euceriform cleptoparasitic bees having the size and form of species of the largely Eastern Hemisphere tribe Melectini (Fig. 116-1). In the phylogenetic study of Roig-Alsina and Michener (1993) these tribes appear as sister groups, but this apparent relationship could be a result of convergence as parasites rather than genuine phylogenetic relationship; that work emphasized the major influence of adaptations to cleptoparasitism on adult morphology and apparent phylogeny. It is a reasonable assumption that the Ericrocidini is actually related to its principal host group, the Centridini, and Melectini to *its* host group, the Anthophorini. A unique feature of the Ericrocidini is the bipartite flabellum, consisting of an elongate preflabellum with a rounded apex curved anteriorly and a smaller postflabellum arising preapically from the anterior surface of the preflabellum and curved posteriorly (Fig. 116-2e-g). The convex side of each has a cobblestone pattern. Another unique feature of the tribe is the large, bifurcate or multidentate apex of the middle tibial spur (Fig. 116-3). This character is approached in some Centridini (species of *Epicharis*) by the preapical shoulder of the same spur, a fact supporting the suggested relationship of Ericrocidini to the Centridini. In the ericrocidines *Epiclopus gayi* Spinola and *wagenknechti* (Ruiz), there is merely a preapical group of larger teeth on the midtibial spur rather than the usual ericrocidine structure. Ericrocidini differs further from the Melectini in its long marginal cell, extending well beyond the third submarginal cell (Fig. 116-4); the short, broad male gonostylus, sometimes having a slender dorsal process that in some *Hopliphora* arises at the base of the gonostylus like a separate dorsal gonostylus (Fig. 116-5); and the flattened middle basitarsus, usually with a carina (Fig. 116-3c). The middle and hind tibiae have scattered spinelike setae on the outer surfaces. Arolia are absent. Parts of the body commonly appear metallic blue or green; this is usually because of appressed metallic hairs or scales, but in *Epiclopus lendliana* (Friese) the integument itself is metallic. Conversely, the genus *Ericrocis* and some species of *Hopliphora* and *Epiclopus* lack metallic coloration entirely.

Snelling and Brooks (1985) presented a generic revision of the tribe, including illustrations of male genitalia, hidden sterna, and other structures (see also Fig. 116-5). Larvae have been described by Rozen (1969a) and Rozen and Buchmann (1990). First-stage larvae described by Rozen (1991a) suggested a possible close relationship between Ericrocidini and Isepeolini, but are not similar to those of Melectini or Rhathymini.

So far as is known, Ericrocidini, like Melectini and Rhathymini, introduce their eggs into closed cells of the host by breaking a hole in the cell closure. The egg is then left attached to the closure or nearby cell wall, or in *Ericrocis* it may drop onto the food mass, where such eggs, already killed by a young larva from another egg, were found by Rozen (1991a).

Key to the Genera of the Ericrocidini
(Modified from Snelling and Brooks, 1985)

1. Third submarginal cell of forewing large, receiving both recurrent veins; hind tarsus very long, in male with dense brush of long, dark, plumose hairs; maxillary palpus absent (South America) .... *Acanthopus*
—. Second and third submarginal cells each receiving a recurrent vein, or first recurrent vein and second submarginal crossvein meeting; hind tarsus without dense brush of long, dark, plumose hairs; maxillary palpus present, although sometimes very short 2
2(1). Juncture of anterior and dorsal surfaces of T1 slightly humped and subangulate in middle (Fig. 116-6b); middle basitarsus without distal, flattened process; mandible simple 3
—. Juncture of anterior and dorsal surfaces of T1 evenly rounded, not subangulate (Fig. 116-6a); middle basitarsus almost always with distal, flattened process on posterior margin, often continued basad as a raised, cariniform ridge (Fig. 116-3c); mandible usually with preapical tooth 5
3(2). Scutellum with two flat and platelike processes directed caudad (Fig. 116-7b); forewing with an apical cloud in marginal cell in addition to that at wing apex .... *Mesocheira*

Figure 116-1. Female Ericrocidini. Above, *Ericrocis lata* (Cresson); Below, *Mesoplia azurea* (Lepeletier and Serville). From Michener, McGinley, and Danforth, 1994.

—. Scutellum bituberculate, the projections stout, subconical, and suberect; forewing with apex dark but without dark area in marginal cell 4
4(3). Pronotum carinate between collar and lobe; omaulus lamellate; male antenna normal, not extending much beyond tegula .... *Aglaomelissa*
—. Pronotum not carinate between collar and lobe, lateral end of collar clearly defined; omaulus abruptly rounded; male flagellar segments greatly elongate, flagellum extending well beyond apex of metasoma (Fig. 116-7a) .... *Ctenioschelus*
5(2). Middle and hind distitarsi with a group of appressed, black setae on each side (inner eye margins divergent above; male gonostylus without dorsal lobe or process) 6
—. Middle and hind distitarsi with normal setae 7
6(5). Metasomal scales contrasting black and white (may be somewhat tawny); labrum with erect preapical median tubercle; scutellum without tubercles (nearctic, Mexico) .... *Ericrocis*
—. Metasomal scales or hairs iridescent blue or green, with or without small, contrasting, whitish hair patches; labrum with transverse preapical ridge; scutellum with pair of mammiform tubercles (South America) .... *Mesonychium*
7(5). Eyes diverging above; scutellum not bituberculate (in *E. gayi* Spinola with median depression, thus weakly biconvex) (South America) .... *Epiclopus*
—. Eyes not diverging above, inner orbits usually more or less parallel; scutellum with two large tubercles 8
8(7). Metasoma covered with minute blue-green scales, often with small patches of white or yellow pubescence .... *Mesoplia*
—. Metasoma black or faintly metallic, without patches of pale pubescence (South America) .... *Hopliphora*

Genus *Acanthopus* Klug

Acanthopus Klug, 1807, *in* Illiger, 1807: 199; Klug, 1807b: 226. Type species: *Apis splendida* Fabricius, 1793 = *Apis palmata* Olivier, 1789, monobasic. [See the annotation by Michener, 1997b.]

Acanthopus consists of large (body length 20-25 mm), dark blue species, the color resulting from appressed blue pubescence. These bees have extremely long hind legs, the hind tarsus of the male being densely hairy and thus appearing broader than the tibia. Unlike that of other Ericrocidini, the inner ramus of the claw of *Acanthopus* is a strong tooth, not a compressed lobe as in many other parasitic bees. Unlike the mentum of almost all long-tongued bees, as indicated by Snelling and Brooks (1985), that of *Acanthopus* is a slender transverse bar, and the lorum is a stronger bar extending between the apices of the cardines, but not fused to them.

■ This genus occurs from the state of Goiás, Brazil, north to the Guianas and Trinidad, with a doubtful report from Panama. There are at least two species, although five names have been proposed (Snelling and Brooks, 1985). *Acanthopus* parasitizes *Centris* (*Ptilotopus*) (Rozen, 1969a), which nests in arboreal termite nests.

Genus *Aglaomelissa* Snelling and Brooks

Aglaomelissa Snelling and Brooks, 1985: 25. Type species: *Melissa duckei* Friese, 1906, by original designation.

Aglaomelissa resembles a rather small *Mesoplia* (length 9-10 mm), from which it differs in having a carinate omaulus and a subangulate profile of T1, the angle separating the anterior from the dorsal surface of this tergum (as in Fig. 116-6b). The metasoma is green because of its metallic scales.

■ This genus occurs from Costa Rica to Colombia, Venezuela, Trinidad, and the state of Pará, Brazil. The only species is *Aglaomelissa duckei* (Friese).

Genus *Ctenioschelus* Romand

Ctenioschelus Romand, 1840: 336. Type species: *Acanthopus goryi* Romand, 1840, monobasic. Taschenberg, 1883: 78, designated *Melissodes latreillei* Lepeletier, 1841 = *Acanthopus goryi* Romand, 1840.
Ischnocera Shuckard, 1840: 166, no included species; Smith, 1854: 284, included *Ischnocera* as a synonym of *Ctenioschelus*, with the single species *latreillii* Lepeletier. Type species: *Melissoda latreillii* Lepeletier, 1841 = *Acanthopus goryi* Romand, 1840.
Melissoda Lepeletier, 1841: 508. Type species: *Melissoda latreillii* Lepeletier, 1841 = *Acanthopus goryi* Romand, 1840, monobasic.

Various authors, from Smith (1854) to Snelling and Brooks (1985), have recognized the existence of the name *Ischnocera*, while calling this genus *Ctenioschelus*. Shuckard's (1840) brief comments could apply to no other bee. Smith recognized the priority of *Ctenioschelus*, on the basis of Romand's (1841) publication. Romand (1840), however, proposed *Ctenioschelus* earlier, probably before the date of Shuckard's work. Romand's paper appeared in November, 1840. Shuckard's work is in a volume whose preface includes a note dated "Nov. 1840." It is therefore unlikely to have been published before December, 1840, and a copy in the Natural History Museum, London, bears the handwritten notation, "Dec. 1840" (this information thanks to J. Harvey, Librarian). Romand's name is here considered to have priority, as well as universal usage; *Ischnocera* has not appeared in print except as a synonym since 1840, so far as I know. Article 69(a)(i)(1) and (vii) of the Code, ed. 3, appears to require the type designation indicated above for *Ischnocera*.

The male of *Ctenioschelus* is unmistakable because of its antennae, which look like those of a cerambicid beetle and extend well beyond the apex of the metasoma (Fig. 116-7a). The female looks superficially like a *Mesoplia* about 15 mm long with the metasoma more green or gold than in the usually blue *Mesoplia*. It differs from *Mesoplia* in its carinate omaulus, backward-directed scutellar tubercles extending over the propodeum (but not flat as in *Mesocheira*), the angle between the anterior and dorsal surfaces of T1, and the lack of slender apical points on the male gonostyli.

■ *Ctenioschelus* ranges from the state of Jalisco, Mexico, south through tropical America to Uruguay and Paraguay. There are two species.

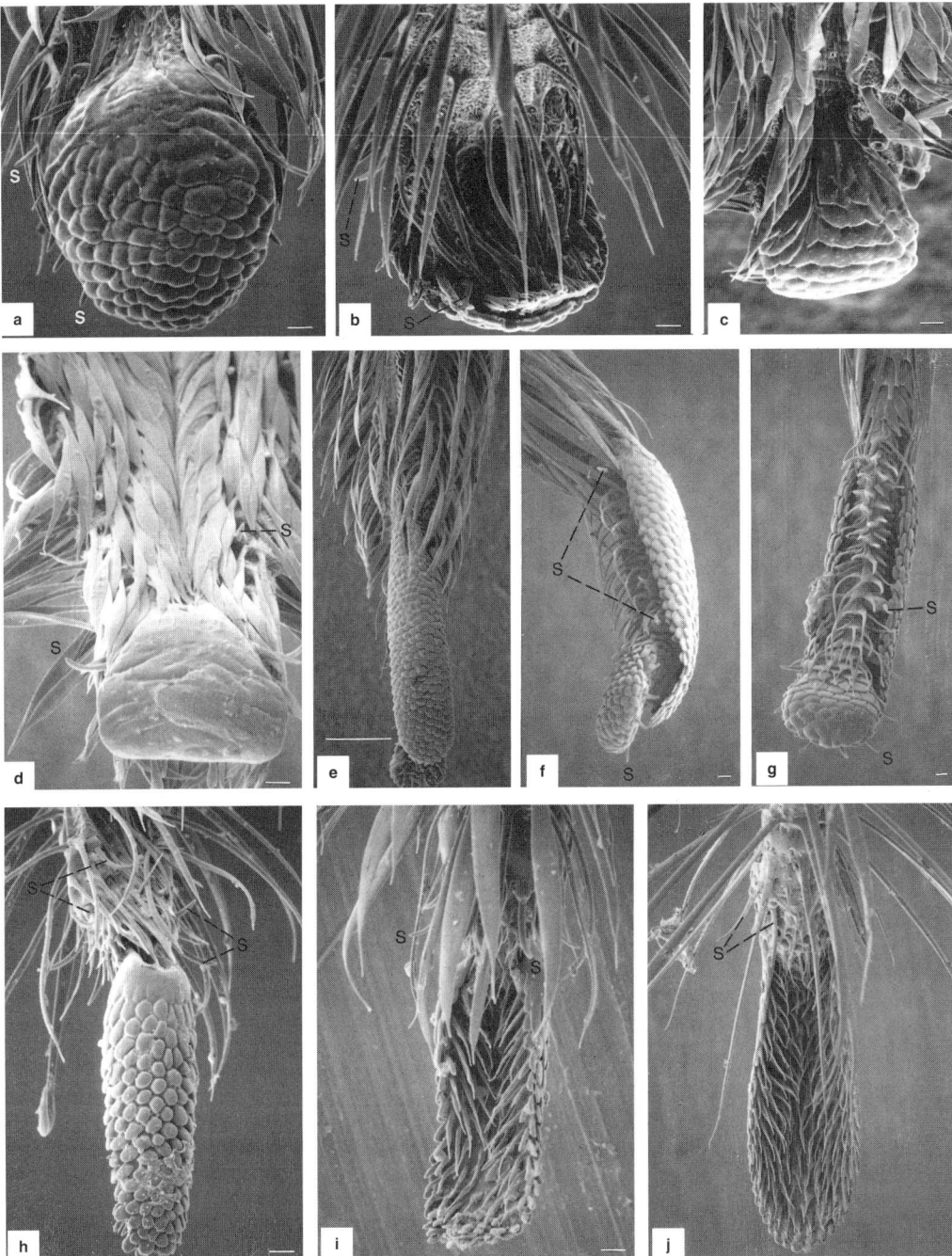

Figure 116-2. Flabella of Melectini (a-d), Ericrocidini (e-g), and Centridini (h-j). **a, b,** *Melecta albifrons* (Forster), posterior and anterior views; **c,** *Thyreus ramosus* (Lepeletier), posterior view; **d,** *Xeromelecta (Melectomorpha) californica* (Cresson), posterior view; **e,** *Hopliphora funerea* (Smith), posterior view; **f,** *Mesocheira bicolor* (Fabricius), lateral view; **g,** *Mesoplia azurea* (Lepeletier and Serville), anterior view, some hairs of the basal half broken off; **h, i,** *Centris (Paracentris) atripes* Mocsáry, posterior and anterior views; **j,** *Epicharis (Epicharana) rustica* (Olivier), anterior view. From Michener and Brooks, 1984.

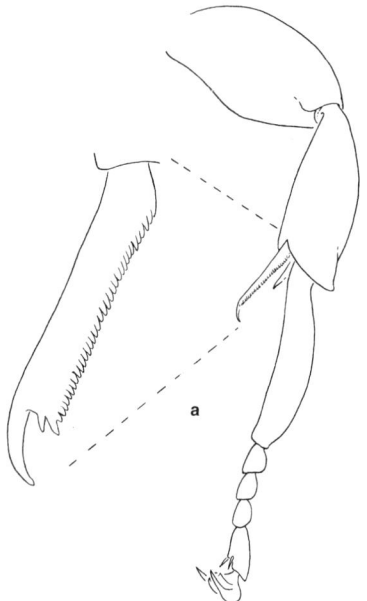

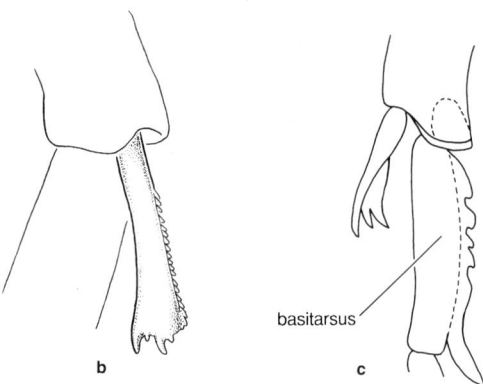

Figure 116-3. Leg structures of Ericrocidini. **a,** Middle leg of *Ericrocis pintada* Snelling and Zavortink, female, with an enlargement of the tibial spur; **b,** Middle tibial spur of female of *Mesocheira bicolor* (Fabricius); **c,** Middle basitarsus and tibial spur of female of *Mesoplia azurea* (Lepeletier and Serville). From Michener, McGinley, and Danforth, 1994.

Genus *Epiclopus* Spinola

Epiclopus Spinola, 1851: 183. Type species: *Epiclopus gayi* Spinola, 1851, monobasic.
Epicolpus Ashmead, 1899a: 61, *lapsus* for *Epiclopus* Spinola, 1851.
Abromelissa Snelling and Brooks, 1985: 23. Type species: *Melissa lendliana* Friese, 1910, by original designation.

This genus contains a group of unusually robust species, 10 to 15 mm long, with long white hair on the thorax, on T1, and sometimes on T2. The metasoma otherwise is black in *Epiclopus wagenknechti* (Ruiz), with appressed dark-blue hairs in *E. gayi* Spinola. The integument is weakly blue in *E. lendlianus* (Friese), a character unique in the Ericrocidini. The first two species listed above were included in *Mesonychium* by Snelling and Brooks; they differ from that genus in their much longer maxillary palpi and the lack of a group of short setae on each side of the middle and hind distitarsi. *E. lendlianus*, which was put in its own genus *Abromelissa* by Snelling and Brooks, is distinctive in the presence of a dorsal process on the male gonostylus and the blue metasomal cuticle, but its relationships seem best shown by inclusion in *Epiclopus*.

■ *Epiclopus* occurs from Coquimbo to Aisén, Chile, and in Neuquén, Argentina. The three species are indicated by name above.

Wagenknecht Huss (1969) found circumstantial evidence that *Epiclopus* species parasitize *Centris*.

Genus *Ericrocis* Cresson

Ericrocis Cresson, 1887: 131, 134. Type species: *Crocisa? lata* Cresson, 1878, monobasic.

The species of this genus are nonmetallic black with a conspicuous pattern of white to tawny pubescence; body length is 9 to 15 mm. All other Ericrocidini either have metallic hairs or cuticle on the metasoma, or lack a pattern of appressed pale hairs on the metasoma, or both. The male genitalia and hidden sterna were illustrated by Mitchell (1962) and Snelling and Brooks (1985).

■ *Ericrocis* occurs from Southern California and Nevada east to Kansas and northern Florida (rare east of the Great Plains), USA, south to the state of Oaxaca, Mexico. The two species were revised by Snelling and Zavortink (1984).

Throughout its range, *Ericrocis* is a suspected parasite of *Centris*. It has been reared from *Centris* cells, and the larva was described by Rozen and Buchmann (1990). The report of *Anthophora* as a host (Linsley and MacSwain *in* Michener, 1944) needs verification.

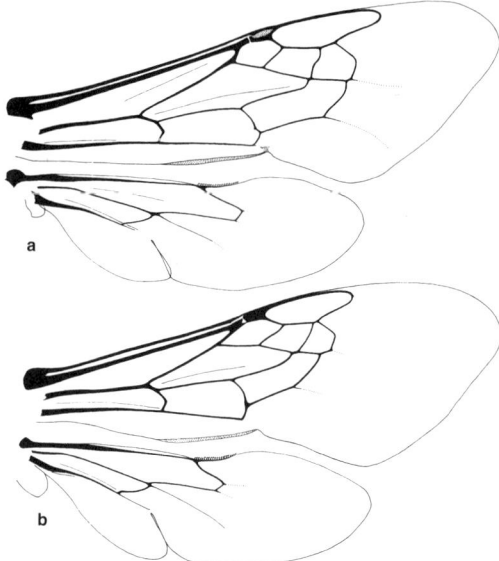

Figure 116-4. Wings of Ericrocidini. **a,** *Mesocheira bicolor* (Fabricius); **b,** *Mesonychium garleppi* (Schrottky).

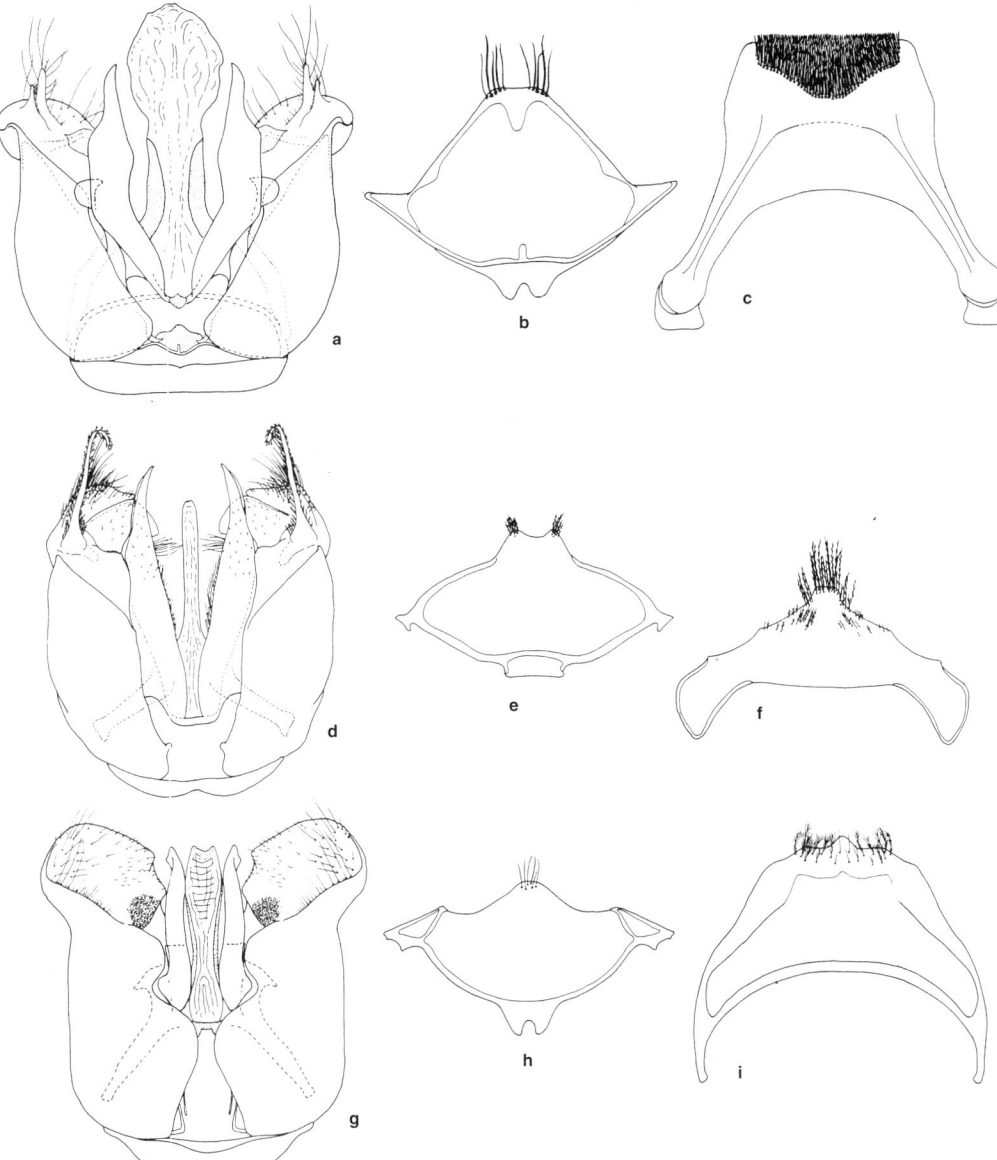

Figure 116-5. Dorsal views of genitalia and ventral views of S8, and S7 of male Ericrocidini. **a-c,** *Mesoplia azurea* (Lepeletier and Serville); **d-f,** *Hopliphora velutina* (Lepeletier and Serville); **g-i,** *Mesocheira bicolor* (Fabricius). From Snelling and Brooks (1985).

Genus *Hopliphora* Lepeletier

Hopliphora Lepeletier, 1841: 458. Type species: *Mesocheira velutina* Lepeletier and Serville, 1825, monobasic.

Eurytis Smith, 1854: 279. Type species: *Eurytis funereus* Smith, 1854, monobasic.

Oxynedys Schrottky, 1902a: 491. Type species: *Oxynedys beroni* Schrottky, 1902 = *Mesocheira velutina* Lepeletier and Serville, 1825, by original designation.

Cyphomelissa Schrottky, 1902a: 493. Type species: *Cyphomelissa pernigra* Schrottky, 1902 = *Melissa diabolica* Friese, 1900, by original designation.

Oxynedis Moure, 1946d: 18, 27, 31, pl. 3, unjustified emendation of *Oxynedys* Schrottky, 1902.

Hopliphora consists of large (16-23 mm long), not or faintly metallic species without patches of appressed pale metasomal pubescence, although yellow apical tergal hair bands characterize one form. See Figure 116-5d-f.

■ This genus ranges from the Brazilian state of São Paulo to the Guianas. There are perhaps as many as seven species; specific names were listed by Snelling and Brooks (1985).

Genus *Mesocheira* Lepeletier and Serville

Mesocheira Lepeletier and Serville, 1825: 106. Type species: *Melecta bicolor* Fabricius, 1804, by designation of Taschenberg, 1883: 72.

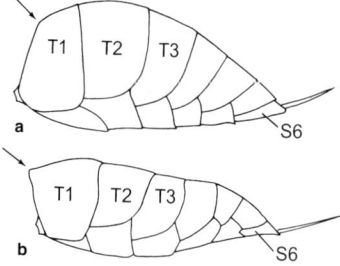

Figure 116-6. Lateral views of metasomas of Ericrocidini. **a,** *Mesoplia imperatrix* (Friese); **b,** *Mesocheira bicolor* (Fabricius). (The angle of T1 is marked by an arrow.) From Michener, McGinley, and Danforth, 1994.

M[eso]chira Agassiz, 1846: 20; Schulz, 1906: 257, unjustified emendation of *Mesocheira* Lepeletier and Serville, 1825.

A unique character of this genus is the broad, flat, shelflike scutellar lobes that extend horizontally back over the base of T1 (Fig. 116-7b). The wings are more strongly maculate than in other genera, there being separate, well-defined separate dark areas in the apex of the marginal cell and at the wing tip. The metasoma is covered with metallic blue scales. This genus includes some of the smallest Ericrocidini (body length 9-14 mm). Male genitalia and hidden sterna were illustrated by Michener (1954b) and Snelling and Brooks (1985); see Figure 116-5g-i.

■ *Mesocheira* occurs from the Mexican states of Sonora and Veracruz and the Greater Antilles south through the tropics to Paraguay. There is probably only one species, *M. bicolor* (Fabricius), although six names have been proposed, as listed by Snelling and Brooks (1985).

Genus *Mesonychium* Lepeletier and Serville

Mesonychium Lepeletier and Serville, 1825: 107. Type species: *Mesonychium coerulescens* Lepeletier and Serville, 1825, monobasic.

In size (10-16 mm long) and coloration, including the metallic scale-like hairs on the metasoma, this genus is similar to *Mesoplia,* and the two genera have usually been confused. A distinctive character of the genus is the greatly reduced maxillary palpus consisting of an obscure (doubtful) first segment and an ovoid second segment not over twice as long as broad; related genera have a much longer palpus and the second segment is parallel-sided and several times as long as broad.

■ *Mesonychium* occurs from central Chile and Argentina north to Peru, and through Brazil to French Guiana. The 12 species names in this genus, as here understood, were listed by Snelling and Brooks (1985).

Two species included in *Mesonychium* by Snelling and Brooks (1985) are here removed to *Epiclopus.* These species are responsible for the morphological diversity indicated for *Mesonychium* in their key and description.

Circumstantial evidence indicates that *Mesonychium* species parasitize *Centris* (Wagenknecht Huss, 1969).

Genus *Mesoplia* Lepeletier

In *Mesoplia* the metasoma is bright metallic blue or green-

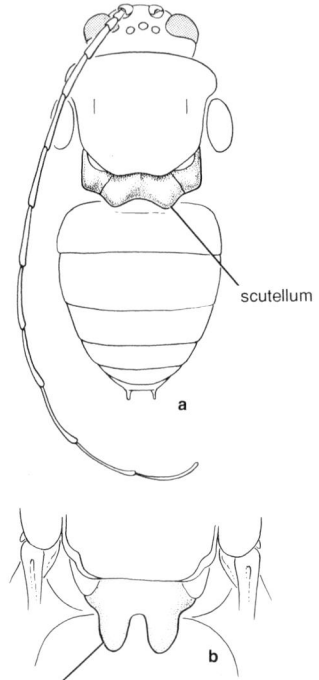

Figure 116-7. Ericrocidini. **a,** *Ctenioschelus goryi* (Romand), male; **b,** Dorsal view of scutellum of *Mesocheira bicolor* (Fabricius). From Michener, McGinley, and Danforth, 1994.

ish blue because of colored scales, each about the size of a puncture; sometimes there are also small patches of pale hairs on the terga. The maxillary palpus is elongate, three- or four-segmented. S5 of the male is broadly emarginate and largely hidden under S4. The body length is 11.5 to 20.0 mm. Male genitalia and hidden sterna were illustrated by Michener (1954b) as well as Snelling and Brooks (1985); see also Figure 116-5a-c.

This genus is known to parasitize both *Centris* and *Epicharis* (Rozen, 1969a). As with Melectini and Rhathymini, the parasite commonly attains access to closed cells by opening a hole in the cell closure and inserting an egg, which is attached to the upper cell walls (Vinson, Frankie, and Coville, 1987). In some cases an egg was found in a cell that showed no signs of having been opened; presumably in these cases the egg was laid before the cell was closed.

Key to the Subgenera of *Mesoplia*

1. Hind tibial spurs of male not reaching beyond basal third of basitarsus, sometimes only one spur present; hind tibia of male with inner distal patch of dense black hairs; pygidial plate of female broad, occupying exposed dorsal surface of T6, covered with hairs or scales .... *M. (Mesoplia* s. str.*)*
—. Hind tibial spurs of male both present, reaching to or beyond middle of basitarsus; hind tibia of male without inner distal hair patch; pygidial plate of female narrow, bare .... *M. (Eumelissa)*

Mesoplia / Subgenus *Eumelissa* Snelling and Brooks

Mesoplia (Eumelissa) Snelling and Brooks, 1985: 21. Type species: *Melissa decorata* Smith, 1854, by original designation.

■ *Eumelissa* occurs from Costa Rica southward through Brazil to eastern Bolivia. Snelling and Brooks (1985) listed the five specific names.

Mesoplia / Subgenus *Mesoplia* Lepeletier s. str.

Mesoplia Lepeletier, 1841: 457. Type species: *Mesocheira azurea* Lepeletier and Serville, 1825, monobasic.

Melissa Smith, 1854: 279. Type species: *Mesocheira azurea* Lepeletier and Serville, 1825, by designation of Sandhouse, 1943: 570.

■ *Mesoplia* s. str. occurs from southern Arizona, USA, the state of Tamaulipas, Mexico, and the Greater Antilles south through tropical America to the state of Paraná, Brazil. Snelling and Brooks (1985) listed 18 species names.

117. Tribe Melectini

This tribe consists of apiform to anthophoriform cleptoparasitic bees (Pl. 12; Figs. 117-1, 117-2). They replace the largely neotropical Ericrocidini in temperate North America and in the Eastern Hemisphere. The largely bare and apically papillate wings characterize a small group of tribes—Anthophorini, Ericrocidini, and Melectini (Fig. 85-2), as well as the Isepeolini. The Anthophorini are nest-making bees with a tibial scopa in females. The tribes Ericrocidini and Melectini lack a scopa, and the second abscissa of vein M+Cu of the hind wing is shorter than the oblique cu-v (Figs. 116-4, 117-3) and sometimes almost absent. The marginal cell of the Melectini is rounded apically, and in the commonest genera scarcely if at all exceeds the last submarginal cell (Fig. 117-3); the middle tibial spur is unmodified; and T7 of the male has a pygidial plate in *Zacosmia* as well as in some *Melecta*. S6 of the female is tapering, and the lateral margins are upturned to form a guide for the sting. Unlike the Ericrocidini, the flabellum is simple and rounded, with a strong to weak cobblestone pattern (Fig. 116-2a-d).

The Melectini are most diverse in the palearctic region but occur also in the nearctic, and range south into Africa, the oriental and Australian areas, and the northern parts of the neotropics in Mexico and the Antilles. General accounts of the tribe include Linsley (1939a) for North America, Hurd and Linsley (1951) for California, and Lieftinck (1972) for genera of the Eastern Hemisphere. The last gives an account of known host relationships. Melectines are parasites of species of Anthophorini. *Thyreus* is primarily a parasite of *Amegilla*, with some records of *Anthophora* (e.g., Rozen, 1969c) and of *Eucera (Synhalonia)* (Eucerini) (Wafa and Mohamed, 1970) as hosts. *Melecta* is usually a parasite of *Anthophora*, although the subgenus *Eupavlovskia* may be associated with *Habropoda*. *Tetralonioidella* is a probable parasite of *Habropoda* and *Elaphropoda*, *Xeromelecta (Melectomorpha)* is a parasite of *Anthophora*, and *Zacosmia* is a parasite of *Anthophora (Heliophila)*. Larvae of various melectine genera were described by Rozen (1969a), and the egg of *Thyreus*, by Rozen and Özbek (2005b).

Unlike those of Nomadinae, females of Melectini break into closed cells of their hosts, oviposit (on the cell cap or upper cell wall), and reclose the cells with earth moistened with a secretion. Accounts of their behavior are by Torchio and Youssef (1968), Thorp (1969b), and Torchio and Trostle (1986). *Xeromelecta* eggs were once thought to be

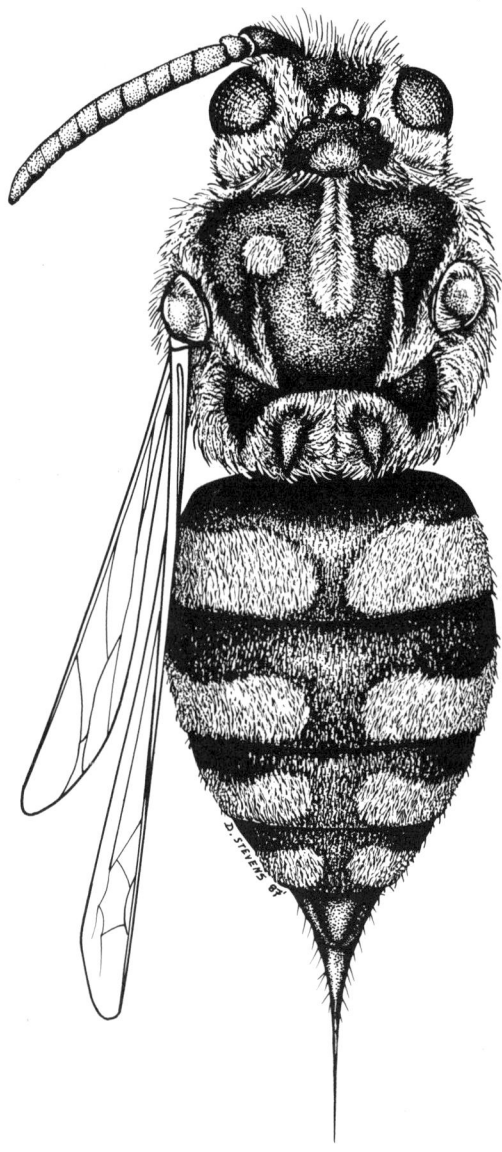

Figure 117-1. *Xeromelecta (Melectomorpha) californica* (Cresson), female. From Michener, McGinley, and Danforth, 1994.

Figure 117-2. *Xeromelecta (Nesomelecta) alayoi* Michener, female, from Cuba. Drawing by D. Stevens, from Michener, 1988.

laid before the host cell was closed, but Torchio and Trostle showed that this was an erroneous interpretation.

Popov (1955c) and Lieftinck (1968) suggest that the Melectini are polyphyletic, *Thyreus* being derived from *Amegilla*-like ancestors and *Melecta* and its allies from *Anthophora*-like ancestors. This may be correct, but the evidence is weak because loss of arolia and palpal reduction, cited as common characteristics of *Thyreus* and *Amegilla*, are both also found in some *Melecta*-like taxa.

Separate keys are provided for genera of the Western and of the Eastern hemispheres.

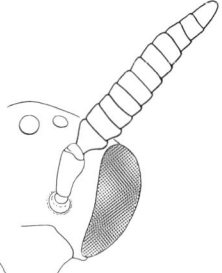

Figure 117-4. Antenna of *Zacosmia maculata* (Cresson), male. From Michener, McGinley, and Danforth, 1994.

Key to the Genera of the Melectini of the Western Hemisphere

1. Wings with two submarginal cells; metasoma densely and uniformly covered with appressed, cinereous pubescence .... *Brachymelecta*
—. Wings with three submarginal cells; metasoma all black with erect hair or with patches of appressed hair producing a pattern of pale spots or bands 2
2(1). Marginal cell scarcely longer than stigma (Fig. 117-3b); maxillary palpus one- to three-segmented; flagellum of male greatly thickened, median segments thus more than twice as wide as long (Fig. 117-4); pygidial plate of male present; arolia absent .... *Zacosmia*
—. Marginal cell much longer than stigma (Fig. 117-3a); maxillary palpus four- to six-segmented; flagellum of male not thickened, median segments at most twice as broad as long; pygidial plate of male usually absent; arolia present 3
3(2). Inner ramus of claws of middle and posterior legs broad, vertically expanded, lobelike although subtruncate or pointed (Fig. 117-6d); T1 without or almost without long hair similar to that of thorax .... *Xeromelecta*
—. Inner ramus of claws of middle and posterior legs pointed, shaped somewhat like outer ramus (Fig. 117-6b) (does not hold for species of the Eastern Hemisphere); T1 with long hair like that of thorax (except in subgenus *Melectomimus*) .... *Melecta*

Key to the Genera of the Melectini of the Eastern Hemisphere

1. Scutellum flat or nearly so, produced posteriorly over metanotum, propodeum, and, in some positions, base of T1, as sharply margined plate, bidentate with broad V- or U-shaped emargination between teeth, and posterior part of scutellar surface on underside of plate, facing downward; body with areas of appressed plumose hairs forming white, blue, or green spots or broken bands (arolia absent; maxillary palpus absent or minute, zero- to four-segmented .... *Thyreus*
—. Scutellum convex, biconvex, bituberculate, or bispinose, surface curving downward posteriorly, and posterior part of scutellar surface thus declivous, facing posteriorly; body with or without areas of appressed plumose hairs, but *if* present, then these hairs white, not at all bluish or greenish 2
2(1). Marginal cell longer than distance from its apex to wing tip; body without spots of white pubescence, metasomal pubescence brown, orange, or yellowish (oriental) .... *Tetralonioidella*
—. Marginal cell equal to or shorter than distance from its apex to wing tip; body commonly with spots of white, appressed pubescence, metasomal pubescence entirely black or with such white spots, or pale and forming apical fasciae on terga 3
3(2). Submarginal cells two; body without patches of appressed white hair; metasomal terga with apical white fasciae; first flagellar segment over 1.5 times as long as broad and about twice as long as second segment (China) .... *Sinomelecta*
—. Submarginal cells three; body usually with patches of appressed white hair and without pale apical tergal fasciae (but *Melecta oreina* Baker agrees with *Sinomelecta* in these features of vestiture); first flagellar segment less than 1.5 times as long as broad and less than twice as long as second segment 4
4(3). T1 with dorsal surface longer than to scarcely shorter than T2 in normal position; arolia absent or nearly so 5
—. T1 with dorsal surface distinctly shorter than T2 in normal position; arolia present (palearctic) .... *Melecta*
5(4). Marginal cell exceeding third submarginal cell and only slightly shorter than distance from its apex to wing tip; apex of labrum with median emargination (Africa) .... *Afromelecta*
—. Marginal cell not or scarcely extending beyond third submarginal cell and distinctly shorter than distance

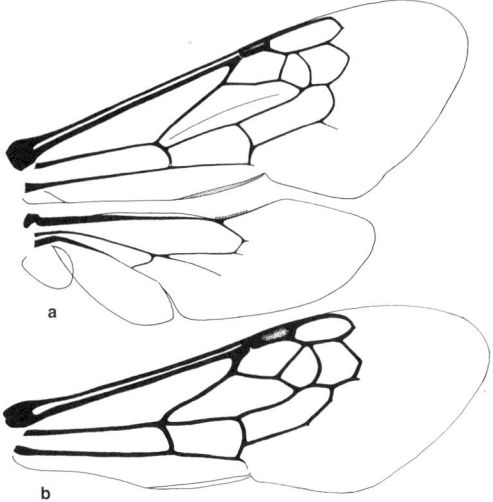

Figure 117-3. Wings of Melectini. **a**, *Xeromelecta (Melectomorpha) californica* (Cresson), see also Figure 83-1a; **b**, Forewing of *Zacosmia maculata* (Cresson).

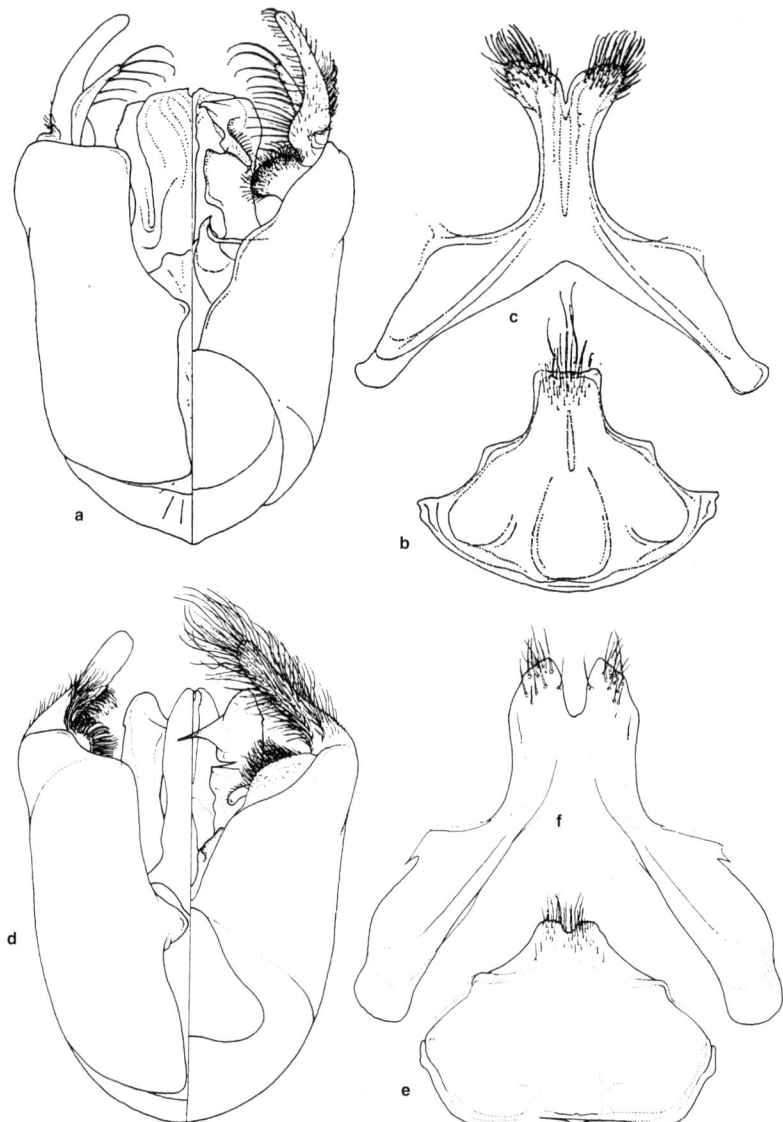

Figure 117-5. Male genitalia (dorsal views on the left side), S8, and S7 of Melectini. **a-c,** *Afromelecta (Acanthomelecta) bicuspis* (Stadelmann); **d- f,** *Melecta (Melecta) fumipennis* Lieftinck. From Lieftinck, 1972, 1980.

from its apex to wing tip; apical margin of labrum gently convex (Palearctic Asia and North Africa) ....
.... *Thyreomelecta*

Genus *Afromelecta* Lieftinck

This is a *Melecta*-like genus from sub-Saharan Africa. The scutellar spines are strongly lateral in position. The marginal cell exceeds the third submarginal cell and is only slightly shorter than the distance from its apex to the wing tip. The general appearance and areas of white appressed pubescence on the metasoma are as in *M. (Paracrocisa)* or even *Xeromelecta (Melectomorpha).* Body length is 13 to 16 mm. Male genitalia and other structures were illustrated by Lieftinck (1972); see also Figure 117-5a.

Key to the Subgenera of *Afromelecta*

1. Maxillary palpus six-segmented; scutellar spines as long as middorsal length of scutellum .... *A. (Acanthomelecta)*

—. Maxillary palpus minute, one- to two-segmented; scutellar spines at most half as long as middorsal length of scutellum .... *A. (Afromelecta s. str.)*

Afromelecta / Subgenus *Acanthomelecta* Lieftinck

Acanthomelecta Lieftinck, 1972: 316. Type species: *Crocisa bicuspis* Stadelmann, 1898, by original designation. [New status.]

The six-segmented maxillary palpi are strikingly different from the short palpi of *Afromelecta* s. str., but variation within taxa of Melectini is often considerable. The dorsal branch of the male gonostylus in *Acanthomelecta* is slender, pointed, and sparsely haired (Fig. 117-5a), in contrast to the robust, densely hairy structure in *Afromelecta* s. str.; see Lieftinck (1972).

■ *Acanthomelecta* occurs in Tanzania. The only species is *Afromelecta bicuspis* (Stadelmann), which was fully described and its genitalia and hidden sterna illustrated by Lieftinck (1972); see also Figure 117-5a-c.

Afromelecta / Subgenus *Afromelecta* Lieftinck s. str.

Afromelecta Lieftinck, 1972: 309. Type species: *Crocisa fulvohirta* Cameron, 1905, by original designation.

The contrast with *Acanthomelecta* is indicated in the text for that subgenus. The maxillary palpus of *Afromelecta* s. str. is no longer than the two apical segments of the labial palpus taken together. Unlike better known taxa, the marginal cell extends well beyond the third submarginal cell.

■ *Afromelecta* s. str. occurs from Cape Province and Natal, South Africa, north to Kenya and Ethiopia. Lieftinck (1972) revised the subgenus, and Eardley (1991c) gave a key to the two named species.

Genus *Brachymelecta* Linsley

Brachymelecta Linsley, 1939a: 458. Type species: *Melecta mucida* Cresson, 1879, by original designation.

This genus contains a small anthophoriform species (9 mm in body length) lacking all indication of the spots or broken bands of appressed pale metasomal pubescence found in most Melectini. Instead the metasoma is uniformly covered with short, apppressed, pale hair. Unlike that of the similarly small *Zacosmia,* the male flagellum is slender, the first segment cylindrical and scarcely longer than the second, and the pygidial plate of the male is absent. The marginal cell, rounded at the apex, is short, about attaining the apex of the last (second) submarginal cell. The male genitalia and hidden sterna have never been dissected and are therefore unknown.

■ *Brachymelecta* is known from a single male specimen collected before 1878 in "Nevada."

Genus *Melecta* Latreille

This large holarctic genus includes some rather diverse forms, some of which have been given generic status by Popov (1955c) and Lieftinck (1972). These are small, divergent groups that seem likely to be derived from among the large mass of *Melecta,* making the latter paraphyletic. In the absence of a phylogenetic analysis, however, this interpretation is not well founded, for they are clearly so similar to *Melecta* that differentiating them at the generic level is not easily justified, given the nature of most apoid genera. The problem is illustrated by Lieftinck's (1972) "descriptive key," which separates *Eupavlovskia, Paracrocisa,* and *Pseudomelecta* from *Melecta* at the generic level. For very few characters is the state of one taxon matched by an alternative in others; rather, the combination of characters of one taxon differs from combinations or from variability found in others. I have elected to recognize the taxa listed below as subgenera.

Melecta consists of robust, anthophoriform species. These bees are black, frequently with patches of appressed white hairs on the metasomal terga, and usually with long hair like that of the thorax on T1, at least laterally.

Key to the Subgenera of *Melecta*

1. S3 to S5 (sometimes also S2 and S6) of male with subapical fringes of strong, backward-directed, black bristles often forming dense brushes; T1 without long, erect pale hair either basally or laterally (middle and hind tarsal claws with inner rami shorter and broader than outer, often subtruncate).. 2
—. S2 to S6 of male without subapical brushes of black, posteriorly directed bristles, although often with abundant subapical hairs; T1 usually with long, erect pale hair, at least basolaterally 3
2(1). Marginal cell short, 2.2 to 3.0 times as long as broad; hind tibia of male unmodified; inner hind tibial spur of male straight or gently curved, not undulate, little if any longer than outer spur (palearctic) ...... *M. (Pseudomelecta)*
—. Marginal cell 3.3 to 4.0 times as long as broad; hind tibia of male expanded toward apex with spiculate ventroapical process extending laterally beyond bases of spurs; inner hind tibial spur of male much longer than outer, gently curved in different directions, thus weakly undulate (palearctic) .... *M. (Paracrocisa)*
3(1). Hind tibia of male expanded toward apex with strong ventroapical process extending laterally beyond bases of spurs; inner hind tibial spur of male much longer than outer, gently curved in different directions so as to be weakly undulate; hind basitarsus of male broadened distally, widest about two-thirds of distance from base to apex; S7 of male with disc slender, over twice as long as broad (palearctic).................................. *M. (Eupavlovskia)*
—. Hind tibia of male not or weakly expanded apically; inner hind tibial spur of male usually not modified [but in *M. albifrons* (Forster) and its relatives modified somewhat as in *Eupavlovskia*]; hind basitarsus less broadened [although somewhat so in group of *M. albifrons*(Forster)]; S7 of male with disc ordinarily less than twice as long as broad, often little longer than broad 4
4(3). Male flagellar segments about 1.5 times as long as broad; pygidial plate of female broadly triangular, occupying much of distal part of T6; metasoma without patches of white appressed hairs; T1 without long hairs like those of thorax (nearctic) .... *M, (Melectomimus)*
—. Male flagellar segments mostly little longer than broad, often broader than long; pygidial plate of female narrowed apically, nearly parallel-sided to spatulate (but in *M. megaera* Lieftinck and *aegyptiaca* Radoszkowski about as in *Melectomimus*); metasoma usually with patches of appressed white pubescence; T1 usually with long pale hairs like those of thorax, often especially well developed laterally (holarctic)................................ *M. (Melecta* s. str.)

Melecta / Subgenus *Eupavlovskia* Popov

Eupavlovskia Popov, 1955c: 330. Type species: *Melecta funeraria* Smith, 1854, by original designation.

This subgenus consists of robust species 12.5 to 16.0 mm long with a densely hairy thorax. The metasomal terga usually have lateral patches of white appressed hair. The scutellum overhangs the metanotum and bears two straight, divergent spines about two-thirds as long as the scutellum and completely hidden by hair. The forewing is broad distally, the apical margin strongly convex. In its modified hind legs of the male, *Eupavlovskia* resembles *Paracrocisa*, although the ventroapical process on the tibia is not spiculate. The slender S7 of the male also resembles that of *Paracrocisa*. In the subgenus *Melecta*, the group of *M. albifrons* (Forster) approaches *Eupavlovskia* in the hind leg characters of the male. Popov (1955c) and Lieftinck (1969) illustrated the male genitalia, hidden sterna, wings, and male hind tibiae.

■ *Eupavlovskia* occurs from Spain eastward on the north side of the Mediterranean to Turkey and the Caucasus, north in Europe to Austria. The two species were revised by Lieftinck (1969).

Melecta / Subgenus *Melecta* Latreille s. str.

Melecta Latreille, 1802a: 427. Type species: *Apis punctata* Fabricius, 1775 = *Apis albifrons* Forster, 1771, by designation of Latreille, 1810: 439.
Symmorpha Illiger, 1807: 198, nomen nudum.
Symmorpha Klug, 1807b: 227. Type species: *Apis punctata* Fabricius, 1775 = *Apis albifrons* Forster, 1771, monobasic.
Bombomelecta Patton, 1879b: 370. Type species: *Melecta thoracica* Cresson, 1876, monobasic.

Melecta s. str. is a large and diverse group consisting of all the congeners that have not been segregated into different subgenera. It is no doubt paraphyletic. I do not favor segregation into further subgenera until a study of the phylogeny of the whole genus is made.

Species of *Melecta* s. str. are robust, usually with lateral patches of white appressed hair on the metasomal terga and with long hair like that of the thorax at the sides of T1, sometimes, particularly in American species, extending across T1 from side to side. The antennal scape is usually provided with long hairs, unlike that of other subgenera. The body length is 6 to 20 mm, usually in the range 10 to 18 mm. Many illustrations of male genitalia and other structures are to be found in Radoszkowski (1893b), Iuga (1958), Mitchell (1962), and Lieftinck (1980); see also Figure 117-5d-f.

■ This subgenus in the Eastern Hemisphere ranges from the Canary Islands and Britain east to eastern China. Although especially abundant in the Mediterranean basin and southwestern Asia, *Melecta* s. str. occurs north to about 60°N in Scandanavia; to the south it does not occur beyond the coastal countries of North Africa, but in Asia it reaches Baluchistan in Pakistan. In the Western Hemisphere *Melecta* s. str. occurs from British Columbia, Canada, to Southern California, east to North Dakota and Texas, very rarely to New Jersey and Georgia, USA. There are about 44 palearctic species and four additional nearctic species. Revisions are by Lieftinck (1980) for palearctic species and Linsley (1939a) and Hurd and Linsley (1951) for nearctic species.

Melecta / Subgenus *Melectomimus* Linsley

Melecta (Melectomimus) Linsley, 1939a: 448. Type species: *Melecta edwardsii* Cresson, 1878, by original designation.

This subgenus consists of a species that is so different from *Melecta* s. str. that it may warrant subgeneric rank, at least until the phylogeny of *Melecta* is studied. It is robust, like *Melecta* s. str., and 11 to 14 mm long. The scutellum has two very long spines, almost as long as the scutellar hairs.

■ *Melectomimus* is found in Nevada and California, USA, and Baja California, Mexico. The one species is *Melecta edwardsii* Cresson.

Melecta / Subgenus *Paracrocisa* Alfken

Paracrocisa Alfken, 1937: 173. Type species: *Paracrocisa sinaitica* Alfken, 1937, monobasic.

Paracrocisa consists of somewhat elongate *Melecta* species, 12 to 16 mm long, with dorsolateral areas of white pubescence on the terga, including T1. The scutellum is posteriorly declivous between the spines, which are at most one-third the length of the scutellum. The hind leg of the male is modified somewhat as in *Eupavlovskia*, but the ventroapical process of the tibia is strongly spiculate and the basitarsus is less broadened and is widest medially. Popov (1955c) and Lieftinck (1972, 1977) illustrated various characters, including the genitalia and hidden sterna.

■ This subgenus occurs in Morocco, Egypt, and southwestern Asia, including the Arabian Peninsula and Turkey, east to Afghanistan and Uzbekistan. The three species were revised by Lieftinck (1972).

Melecta / Subgenus *Pseudomelecta* Radoszkowski

Pseudomelecta Radoszkowski, 1865: 55. Type species: *Melecta diacantha* Eversmann, 1852, by designation of Sandhouse, 1943: 594.

Pseudomelecta includes moderately robust bees, 8.0 to 12.5 mm long, having dorsolateral areas of white, appressed pubescence on the terga of most species. The hind legs of the male are slightly modified, the tibia and basitarsus slightly broader than in the female and the basitarsus flattened; this pattern is suggestive of the hind-leg modification in *Melecta (Melecta) albifrons* (Forster). An unusual feature is the small subquadrangular labrum at most half as broad as the lower interorbital distance. S7 of the male is rounded or tuberculate at the apex, not emarginate, and the apodemal arm is broad with a strong basal angle. The male gonostylus is broad, hairy throughout, and has a broad, rounded, hairy, ventral lobe or ventral gonostylus, as illustrated by Popov (1955c) and Lieftinck (1972).

■ This subgenus is found from eastern Turkey through Central Asia to Mongolia. The five species (one, *Melecta chalybeia* Lieftinck, unusual in having a blue to purplish and greenish luster on the metasoma, and unknown in the male) were revised by Lieftinck (1972).

Genus *Sinomelecta* Baker

Sinomelecta Baker, 1997: 245. Type species: *Sinomelecta oreina* Baker, 1997, by original designation.

Rather slender for a melectine, 11.0 to 11.5 mm long and without patches of white, appressed hair, *Sinomelecta* suggests *Tetralonioidella*, but does not have the brown colors and relatively long marginal cell of that genus. In the key to subgenera of *Melecta*, *Sinomelecta* runs to 4 and to *Melecta* s. str. except that the pygidial plate of the female is narrowly triangular. The inner ramus of the claws is pointed, not broad as in many species of *Melecta*. As in *Melecta,* the horizontal part of T1 is shorter than the exposed part of T2. Male geneitalia and other structures were illustrated by Baker (1997).

■ *Sinomelecta* occurs in southwestern China. The only known species is *S. oreina* Baker.

Melecta (Melecta) emodi Baker from Tibet is similar to *Sinomelecta* in appearance, fasciate metasoma, complete lack of patches of appressed white hairs, and acute inner rami of the claws. If closely related to *Sinomelecta,* it may show that the latter should be recognized only as a subgenus of *Melecta.* However, *M. emodi* has the robust body, three submarginal cells, shorter first flagellar segment, and parallel-sided distal part of the female pygidial plate, as in other species of *Melecta.*

Genus *Tetralonioidella* Strand

Tetralonioidella Strand, 1914a(Apr.-May): 140. Type species: *Tetralonia (?) hoozana* Strand, 1914, monobasic.
Protomelissa Friese, 1914c(June): 322. Type species: *Protomelissa iridescens* Friese, 1914 = *Tetralonia (?) hoozana* Strand, 1914, by designation of Sandhouse, 1943: 592. [See also Michener, 1997b.]
Callomelecta Cockerell, 1926c: 621. Type species: *Callomelecta pendleburyi* Cockerell, 1926, by original designation.

This is a genus of rather slender melectines without patches of appressed white pubescence, but sometimes with tergal bands of yellowish to brownish hair; thoracic hairs are red, yellowish, brown, or blackish. The body length is 8.5 to 13.5 mm. The marginal cell extends well beyond the third submarginal cell, unlike that of the better known taxa of Melectini. Illustrations of male genitalia and other structures were published by Lieftinck (1944, 1972, 1983).

Certain specimens lack the characteristic strong papillae on the hairless distal parts of the wings found on other Melectini and have instead hairy, nonpapillate wings. Other specimens, seemingly of the same species, have the large papillae and bare wings characteristic of the tribe.

■ *Tetralonioidella* occurs from northern India eastward through southern Asian mountains to Thailand, southeast China, and Taiwan, south as far as Java. The ten species were revised by Lieftinck (1972, 1983).

Genus *Thyreomelecta* Rightmyer and Engel

Thyreomelecta Rightmyer and Engel, 2003: 3. Type species: *Thyreomelecta kirghisia* Rightmyer and Engel, 2003, by original designation.

This genus resembles *Thyreus* in the lack of arolia and lack of pupal cocoons. (So far as known, other Melectini pupate in cocoons). Unlike *Thyreus,* the scutellum is not flattened as a plate projecting posteriorly, but is more like that of *Afromelecta,* i.e., with strong, laterally placed, posteriorly directed spines.

■ *Thyreomelecta* is known from North Korea west through southern Siberia and central Asia to Armenia, thence to northern Egypt and Libya. Rightmyer and Engel (2003) provided a key to the seven species, as well as illustrations of the male genitalia and other structures.

A host of *Thyreomelecta kirghisia* Rightmyer and Engel is *Anthophora albifascies* Alfken. The pupa was described by Rozen (2000) and eggs by Rozen and Özbek (2003).

Genus *Thyreus* Panzer

Crocisa Jurine, 1801: 164. Type species: *Nomada scutellata* Jurine, 1801 = *Melecta histrionica* Illiger, 1806, by designation of Morice and Durrant, 1915: 423. Suppressed by Commission Opinion 135 (1939).
Thyreus Panzer, 1806: 263. Type species: *Nomada scutellaris* Fabricius, 1781, monobasic.
Crocissa Panzer, 1806: 263. Type species: *Nomada scutellaris* Fabricius, 1781, by designation of Sandhouse, 1943: 541. [See notation by Michener, 1997b.]
Crocisa Jurine, 1807: 239. Type species: *Nomada histrio* Fabricius, 1775, by designation of Latreille, 1810: 439. [Two subsequent designations were listed by Michener, 1997b. Furthermore, *Melecta histrionica* Illiger, 1806, was designated as the type species by Morice and Durrant (1915: 423) but was not an originally included species. Morice and Durrant's comments concern the suppressed *Crocisa* Jurine, 1801, not 1807.]

This Old World genus is very different from other Melectini, as shown by the characters given in the key to genera. The body is less robust than that of most *Melecta*, probably best called apiform or euceriform, with a striking pattern often including pale to bright-blue or greenish areas of appressed hairs (Pl. 12). The body length is 8 to 14 mm. Male genitalia, hidden sterna, and other characters were illustrated by Radoszkowski (1893b), Iuga (1958), Lieftinck (1958, 1959, 1962, 1968), and Schwarz (1993b).

■ *Thyreus* occurs from the Canary Islands and Portugal to China and Japan, north as far as the Netherlands, Germany, Mongolia, and Manchuria, thence south throughout Africa, including Madagascar and through southern

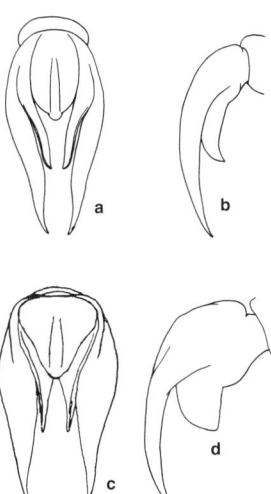

Figure 117-6. Claws of hind legs of females of Melectini, distal and lateral views with setae omitted. **a, b,** *Melecta (Melecta) pacifica* Cresson; **c, d,** *Xeromelecta (Melectomorpha) californica* (Cresson).

Asia, including the islands of Taiwan, the Philippines, and Indonesia, eastward to the Solomon Islands and southward in Australia as far as the state of Victoria and central Western Australia. In the palearctic and Indo-Australian areas a total of 83 species are recognized, after revision and elimination of many synonyms and subspecies. For sub-Saharan Africa nearly 40 species have been recognized, after revision. Revisions are by Lieftinck (1958, 1959, 1962 for Indo-Australian species, 1968 for palearctic species), Vergés (1967, for Spanish species), and Eardley (1991c, for sub-Saharan species). Meyer (1921) listed described species and reproduced descriptions.

Genus *Xeromelecta* Linsley

Xeromelecta is an American genus closely related to *Melecta,* although the body is more slender than that of most species of *Melecta.* The body length is 8 to 16 mm. As in *M. (Pseudomelecta* and *Paracrocisa),* the integument of much of the scutum and scutellum is clearly exposed, with only short hairs or rather sparse hairs. This sparsity of pubescence is in contrast to that of most *Melecta* species, as well as *Tetralonioidella,* in which the thoracic hair is long and rather dense, more or less completely hiding the scutellar spines even when they are long. *Xeromelecta* also differs from most *Melecta,* although not from the subgenera *Pseudomelecta, Paracrocisa,* and *Melectomimus,* in its lack of long hair like that of the thorax on T1 (Fig. 117-2). Such hair is especially well developed at the sides of T1 in most Old World *Melecta (Melecta)*; in most American species it extends across T1 from side to side. As the genera are here understood, *Xeromelecta* differs from *Melecta* in the absence of a dorsobasal lobe or process of the male gonostylus [not verified for *Xeromelecta (Nesomelecta)*]. Except for *X. (Xeromelecta) larreae* (Cockerell), *Xeromelecta* differs from *Melecta* in its shorter maxillary palpi, four- or five-segmented in *X. (Melectomorpha)* and much shortened, three-segmented or possibly sometimes two-segmented in *X. (Nesomelecta).* The male genitalia and hidden sterna were illustrated by Mitchell (1962). The broad inner ramus of the claws of the middle and hind legs (see the key to North American genera above and Figs. 28-5, 117-6c, d) is a common feature in this genus, found for some reason in many other groups of cleptoparasitic bees.

Although *Xeromelecta* resembles *Melecta (Pseudomelecta)* in some characters, it differs in having a longer marginal cell, and unmodified hind legs in the male, and in lacking brushes of black bristles on the male metasomal sterna; see the key to the subgenera of *Melecta.* Except for the marginal cell, the same characters differentiate it from the subgenus *Paracrocisa.*

Xeromelecta was revised by Linsley (1939a) and Hurd and Linsley (1951).

The subgeneric names *Xeromelecta* and *Melectomorpha* were published simultaneously. The precedence of *Xeromelecta* was established by Hurd and Linsley (1951), and the genus name is *Xeromelecta.*

Key to the Subgenera of *Xeromelecta*

1. Scutellum biconvex, with only the faintest suggestion of a posterior tooth on each convexity; maxillary palpus more than half as long as first segment of labial palpus, six-segmented; body without patches of appressed pale pubescence; wings blackish .... *X. (Xeromelecta s. str.)*
—. Scutellum convex, with two distinct, posteriorly directed teeth or spines; maxillary palpus less than half as long as first segment of labial palpus, three- to five-segmented (possibly two-segmented); body with patches of appressed, pale pubescence; wings rather clear with apical dusky areas 2
2(1). Scutellar lobes each ending in acute spine or tooth; maxillary palpus three-segmented (possibly sometimes two-segmented), less than one-fourth as long as first segment of labial palpus (Antilles) .... *X. (Nesomelecta)*
—. Scutellar lobes each ending in approximately right-angular tooth; maxillary palpus four- to five-segmented, little less than one-half as long as first segment of labial palpus .... *X. (Melectomorpha)*

Xeromelecta / Subgenus Melectomorpha Linsley

Melecta (Melectomorpha) Linsley, 1939a: 451. Type species: *Melecta californica* Cresson, 1878, by original designation.

This subgenus includes the commonest American melectines. They have patches of appressed white or rarely pale brownish pubescence, forming medially broken bands on the metasomal terga.

■ *Melectomorpha* occurs from British Columbia, Canada, to California, east to Minnesota, Wisconsin, Illinois, and Texas, USA, and south to Baja California, Sonora, Zacatecas, and Puebla, Mexico. There are two species.

Xeromelecta / Subgenus Nesomelecta Michener

Melecta (Nesomelecta) Michener, 1948b: 15. Type species: *Melecta haitensis* Michener, 1948, by original designation.

Nesomelecta resembles *Melectomorpha* in its areas of white appressed pubescence, although the pattern is more elaborate on the head and thorax (Michener, 1988; Fig. 117-2) and in one species the pale tergal bands are not broken medially.

■ This subgenus is found on Puerto Rico, Hispaniola, and Cuba. The three species were differentiated by Michener (1988).

This is the only tropical American group of Melectini. The Puerto Rican species has been known as *Xeromelecta pantalon* (Dewitz), but according to D. B. Baker (in litt., 1992) a senior synonym is *Nomada tibialis* Fabricius, 1793, hence *X. tibialis* (Fabricius).

Xeromelecta / Subgenus Xeromelecta Linsley s. str.

Melecta (Xeromelecta) Linsley, 1939a: 450. Type species: *Bombomelecta larreae* Cockerell, 1900, by original designation.

This subgenus contains a slender species, black-haired except for fulvous hairs on the thoracic dorsum.

■ *Xeromelecta* occurs in xeric areas from California to New Mexico, USA. The single species is *Xeromelecta larreae* (Cockerell).

Genus *Zacosmia* Ashmead

Zacosmia Ashmead, 1898: 282. Type species: *Melecta maculata* Cresson, 1879, by original designation.

Micromelecta Baker, 1906: 143. Type species: *Melecta maculata* Cresson, 1879, by original designation.

The fusiform male flagellum, minute one- to three-segmented maxillary palpus, and marginal cell only about half as long as the distance from its apex to the wing tip (Fig. 117-3b) are characteristic of this genus. *Zacosmia* contains the smallest Melectini; body length is 5 to 9 mm, anthophoriform. The metasomal terga are covered with short, appressed hair, white with patches or large areas of varying shades of brown on both the bases and the apices of terga, the white thus reduced to an undulating median band middorsally across most terga.

■ *Zacosmia* occurs from Washington state, USA, and southern Alberta, Canada, south to the states of Baja California, Chihuahua, and Durango, Mexico. The one species was reviewed by Linsley (1939a) and Hurd and Linsley (1951).

The hosts of *Zacosmia* are *Anthophora (Heliophila)*, which are parasitized in the same manner as larger *Anthophora* species are parasitized by *Melecta* (Torchio and Youssif, 1968); see the account of the tribe.

118. Tribe Euglossini

These are the orchid bees of the American tropics, so called because the males are pollinators of the larger orchids of that region. The name is also appropriate because of the large size and gaudy coloration of many of the bees themselves (Pl. 15).

A long proboscis, characteristic of the tribe, reaches its extreme in *Euglossa* species of the *Glossura* group, in which even when folded in repose, it extends well beyond the apex of the metasoma (Fig. 118-1a). As illustrated by Zucchi, Lucas de Oliviera, and Camargo (1969), the proboscis of the *Glossura* pupa is coiled in order to fit within the cocoon. The following characters elaborate on those listed in the key to tribes (Sec. 85): The body is of moderate size (8.5 mm long) to very large (29 mm long), moderately to densely hairy, usually anthophoriform (Figs. 118-1, 118-2a), euceriform (Fig. 118-2b), or, in *Aglae*, even less robust. The claws of the female have a basal tooth, and those of the male are similar or cleft; arolia are absent. The hind tibial spurs are present; the strigilis has a prong on the anterior side. The hind basitarsus of nonparasitic females is broadest at the base, with a posterior basal angle or auricle. The hind tibia of nonparasitic females is greatly expanded with an immense corbicula (Fig. 118-3), but lacks a penicillum. The hind tibia of males is greatly swollen in nonparasitic genera; in all genera it has a large hairy slit on the upper margin distally (Fig. 118-4a-c). The forewing has complete, strong venation, the marginal cell being less than twice as long as the distance from its apex to the wing tip. The stigma is minute, vein r arising near the middle, the margin within the marginal cell being straight or weakly concave (Fig. 118-5). The hind wing lacks a jugal lobe but has a comb of bristles in its place. The clypeus is strongly protuberant; the labrum is much less than twice as wide as long, thus longer than in other corbiculate Apidae. Unlike those of the Bombini, the mandibles of females have strong apical teeth (Fig. 119-5g, h). The flabellum is elongate, smooth to imbricate on the posterior surface (Fig. 118-6). The male S8 is large, strongly sclerotic, and longer than broad, with a strong, usually pointed apical process (Fig. 118-7b). The male genitalia are strongly sclerotized with a distinct gonobase, a small to moderate-sized, sometimes bifid, hairy, upper gonostylus, and a minute to large, hairy, lower gonostylus (Fig. 118-7a, 119-4a); the volsella is rather small and minutely hairy, as illustrated by Michener (1990a and Fig. 119-4a).

The larva has small, pointed dorsolateral tubercles on the thoracic segments and at least sometimes on the first abdominal segment and a pair of similar tubercles on the vertex. The mandible is heavily sclerotized and blunt, and

Figure 118-1. *Euglossa*. Above, *E. gorgonensis* Cheesman, male; Below, *E. dodsoni* Moure, female. From Michener, McGinley, and Danforth, 1994.

Figure 118-2. Euglossini. Above, *Eulaema polychroma* (Mocsáry), female; Below, *Exaerete smaragdina* (Guérin-Méneville), male. From Michener, McGinley, and Danforth, 1994.

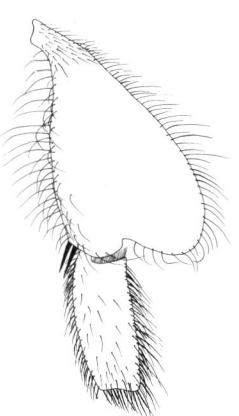

Figure 118-3. Outer surfaces of hind tibia and basitarsus of *Euglossa imperialis* Cockerell, female, showing the enormous tibial corbicula.

there is a large apical concavity on its inner surface. Larvae of *Eufriesea* were described by Michener (1953), those of *Euglossa* by Roberts and Dodson (1967) and by Zucchi, Lucas de Oliveira, and Camargo (1969).

Except for *Aglae*, which is known from South America and perhaps eastern Panama, each genus ranges from Mexico to Argentina, mostly in the moist forests. The eastern Panama report of *Aglae* is doubtful but the genus certainly occurs in Colombia. The number of species of these magnificent bees in tropical forests is sometimes large; 57 species are known from the forests of central Panama alone (Roubik and Ackerman, 1987).

In spite of the long proboscis characteristic of the tribe, female euglossines visit various flowers for pollen and for nectar, including some that contain nectar in a short corolla where the long proboscis would appear to be a hindrance. Male euglossines are attracted to flowers of orchids, certain Araceae and Gesneriaceae, and Solanaceae. Plate 15 shows pollinia of orchids on male Euglossini. They are also attracted to rotting logs, certain fungi, and occasionally other objects in tropical forests, and some species appear not to be attracted by the fragrances of floral origin. Indeed Whitten, Young, and Stern (1993) suggest that floral sources may be secondary in importance and subsequent in use, evolutionarily speaking, to other sources. Artificially, the male bees can be attracted to various aromatic compounds (e.g., 1,8-cineole, eugenol, eucalyptol, methyl salicylate, and skatol) that are similar to those in the natural sources listed above (Dodson et al., 1969; Dressler, 1982a). These fragrances are picked up along with salivary secretions of the bees (Whitten, Young, and Williams, 1989), using moplike hair masses on the front tarsi, and are transferred to the hairy groove or slit on each hind tibia. Sakagami (1965), Vogel (1966), and Kimsey (1984a) illustrated the structures and the movements involved in collecting these compounds. Males of one species, *Eufriesea purpurata* (Mocsáry), even collect DDT from buildings sprayed for malaria control, apparently with no ill effect (Roberts et al., 1982) in spite of the presence of DDT in various parts of the bees' bodies. *E. purpurata* is not known to be attracted to other compounds, although presumably there is a compound similar to DDT somewhere in the Amazonian forests where it lives. Peruquetti (2000) reported interesting observations and summarized the work of others on possible functions of the aromatic substances collected by male euglossine bees. The internal structure of the male hind tibia involves an extremely elaborate invagination (Cruz-Landim et al., 1965; Vogel, 1966; Cruz-Landim and Franco, 2001) that presumably serves to absorb the aromatic compounds introduced through the tibial slit. The compounds in the labial salivary glands, secreted onto the source by the bees, are similar to those in the hind tibial organs and probably serve as solvents for the fragrances (Whitten, Young, and Williams, 1989). The fragrances may be precursors of sex pheromones although as collected they do not attract females.

Bembe (2004) postulated that material from the hind tibial slits (and associated internal tibial gland or gland-like tissues) of male Euglossini is somehow transferred to

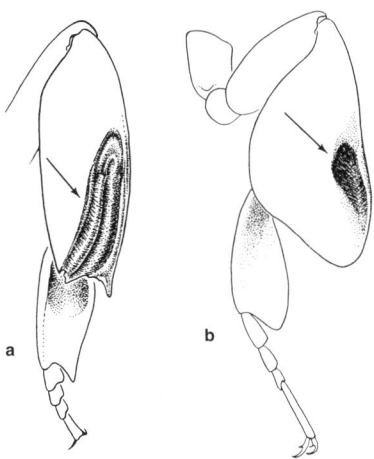
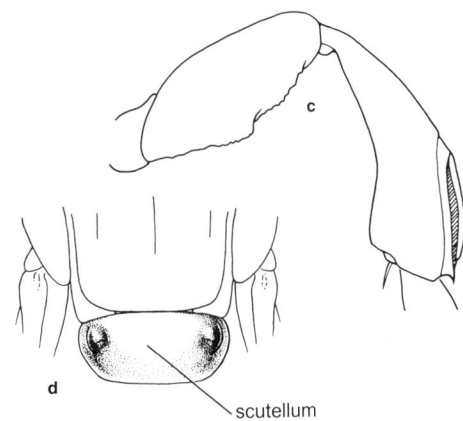

Figure 118-4. Structures of Euglossini. **a-c,** Hind legs of males, showing tibial slit or opening (see arrows) of *Eulaema cingulata* (Fabricius), *Euglossa imperialis* Cockerell, and *Exaerete smaragdina* (Guérin-Méneville) and femoral teeth of the last; **d,** Scutellum of *Exaerete frontalis* (Guérin-Méneville). From Michener, McGinley, and Danforth, 1994.

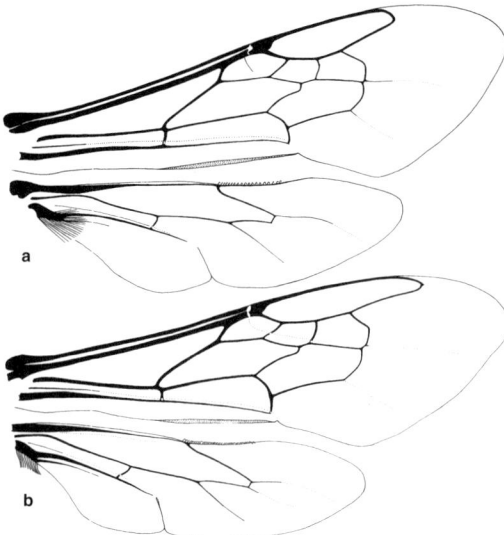

Figure 118-5. Wings of Euglossini. **a**, *Euglossa cordata* (Linnaeus); **b**, *Exaerete smaragdina* (Guérin-Méneville).

the feltlike patches of short plumose hairs on the outer surfaces of the middle tibiae, specifically to those on the bases of the middle tibiae (Fig. 118-9). He noted that the combs of bristles that replace the jugal lobes of the hind wings in Euglossini are much more prominent in males than in females. He demonstrated that these combs, in flight, with the mid legs held in appropriate positions, can brush a spray of minute particles of liquid into the air. Such a hypothetical spray could consist of the aromatic compounds collected by males, modified or mixed with other substances in the hind tibial glands, transferred to the felt patches of the middle tibiae, and might be involved in courtship.

Nests are in some cases exposed (some species of *Euglossa*, Eberhard, 1989) but are usually in cavities in banks, tree trunks, logs, old buildings, etc. Aside from the outside covering of exposed nests, the construction consists principally of brood cells, which are mass-provisioned. There are no storage pots or storage cells, as there are in other tribes of corbiculate Apidae. Construction materials are resin *(Euglossa)*, resin often mixed with bark fragments *(Eufriesea,* Fig. 118-10), and mud or feces perhaps mixed with resin *(Eulaema,* Fig. 118-8).

The Euglossini are the only corbiculate Apidae whose species are neither clearly eusocial nor social parasites. Some species are solitary *(Euglossa, Eufriesea)*. Some species of *Eufriesea* often produce aggregations of cells in protected places, constructed by several to many females. Some species of *Euglossa* and perhaps all *Eulaema* frequently have several females per nest. The nature of the interactions among them is diverse and little known, but the colonies are mostly not eusocial; see Zucchi, Sakagami, and Camargo (1969). The colonies are started by lone females, at least in *Eulaema nigrita* Lepeletier (Santos and Garófalo, 1994; Garófalo, Camillo, and Serrano, 1996) and *Euglossa atroveneta* Dressler. In the latter, of 51 nests started in boxes by single females who constructed up to 11 cells each, nine of the nests were reactivated after emergence of female offspring, forming colonies. Sometimes the foundress and daughters, and sometimes only daughters, a maximum of ten females, worked in a nest, and a maximum of 25 cells were constructed by such colonies. One bee, the mother or oldest daughter, was dominant and rarely left the nest (Ramírez-Arriaga, Cuadriello-Aguilar, and Martínez, 1996). Thus while most nests in this sample contained only one adult female, others contained small colonies with some kind of parasocial or eusocial organization.

Garófalo et al. (1998) found both solitary and communal nests of *Euglossa annectans* Dressler. One nest was started by two females; others inhabited by two individ-

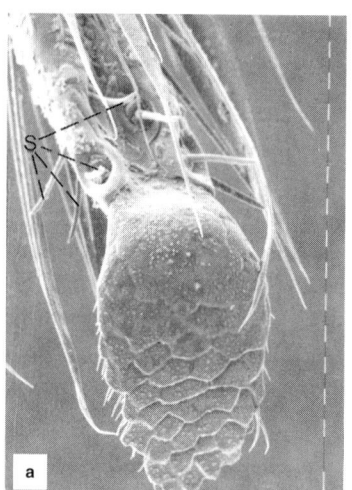

Figure 118-6. Flabella of *Eufriesea*. **a**, *E. surinamensis* (Linnaeus), posterior view; **b**, *E.* sp., posterior view; **c**, *E. surinamensis* (Linnaeus), anterior view. (s, setae.) From Michener and Brooks, 1984.

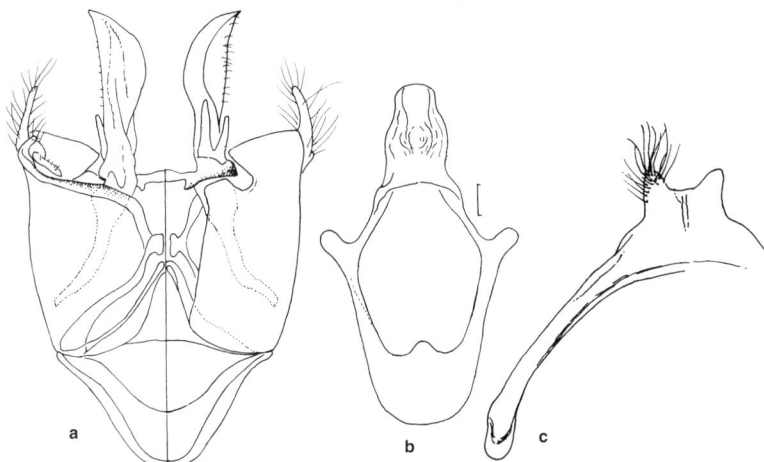

Figure 118-7. Male genitalia (dorsal view on the right) and hidden sterna of *Eulaema*. **a, b,** Genitalia and S8 of *E. nigrita* Lepeletier; **c,** S7 of *E. seabrai bennetti* Moure. From Sakagami and Michener, 1987.

uals were being reused by bees reared in those nests. Of thirty-five occupied nests of *E. hyacinthina* Dressler, eleven contained two or more females apparently living communially (Soucy, Giray, and Roubik, 2003). Ramírez, Dressler, and Ospina (2002) gave a comprehensive catalogue of euglossine species, reporting for each species attractants, plants used for food and attractant sources, relevant literature references, and taxonomic position.

The genus *Aglae* is cleptoparasitic in the nests of *Eulaema*; *Exaerete* is cleptoparasitic in nests of *Eulaema* and *Eufriesea*.

Aspects of euglossine biology were summarized by Dodson (1966), Dressler (1982a), Roubik (1989), and Rebêlo (2001), and lists of species were published by Moure (1967a) and Kimsey and Dressler (1986). The species found at Tambopata, Peru, were reviewed with keys by Dressler (1985); those in Colombia were reviewed, with keys and illustrations of structures, by Bonilla-Gómez and Nates-Parra (1992). Roubik and Hanson (2004) gave a general account in Spanish and English of the Euglossini, with emphasis on those of Central America and northern South America. This book contains keys to species and many colored illustrations.

The genera of this tribe, although rather diverse in appearance, are remarkably uniform in many features, and all the nonparasitic species were included in *Euglossa* by most authors early in the last century. Females of the parasitic forms lack corbiculae and other pollen-carrying and -manipulating structures. The hind tibiae are therefore slender, not excessively broadened as in females of nest-making genera. Interestingly, males of the parasitic genera also have slender hind tibiae, not swollen like those of the nonparasites. Since the hind tibiae of males receive the aromatic compounds collected by males, and since this function is probably related somehow to mating behavior, one wonders why males of parasitic genera would not have hind tibiae as enlarged as those of other genera. Perhaps the slender hind tibiae of males are plesiomorphic features preserved in parasitic forms, as occurs in various other groups of parasitic bees.

Cladistic relations among euglossine genera were discussed by Kimsey (1987) and Michener (1990a) and thers. Otero (1996) speculated on the phylogeny of the use of different materials in nest construction by euglossine bees.

Engel (1999f) made a new cladistic analysis of the Euglossini which resulted in the following arrangement: (*Aglae, Eulaema*) (*Eufriesea* (*Euglossa, Exaerete*)). He also described a new species of Oligomiocene *Euglossa* from Dominican amber. *Paleoeuglossa* Poinar (1999) from the same deposits may be a *Eufriesea*, as suggested by Engel (1999f). Neither *Euglossa* nor *Eufriesea* now occurs in Hispaniola. Cameron (2004) gave a review of phylogenetic hypotheses and behavioral aspects.

A study of the phylogeny using both morphology and DNA evidence (4 genes) was made by Michel-Salzat, Cameron, and Olivera (2004). Curiously they believe that the parasitic genera are not derived from among the nonparasitic genera but are sisters of the latter group, i.e., basal branches of the tree. Unless they arose as parasites of now extinct host genera, this seems unlikely.

The five genera of Euglossini can be distinguished by the key given below. For the most part they are easily sep-

Figure 118-8. Cells constructed of mud or mammalian feces in a preexisting cavity by *Eulaema meriana terminata* (Smith). Photo by F. D. Bennett.

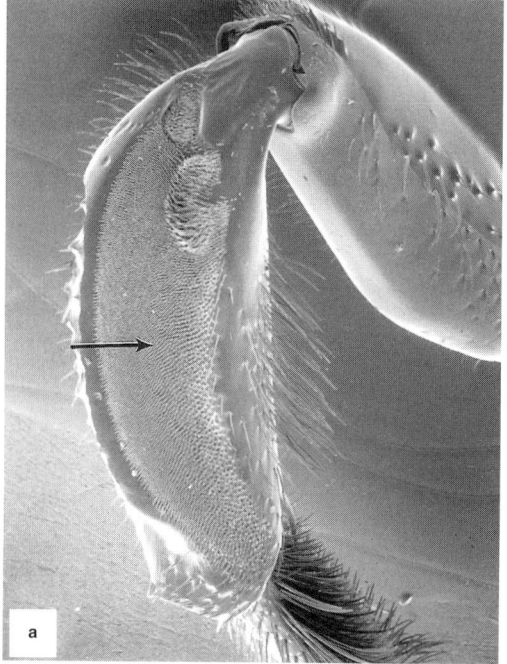 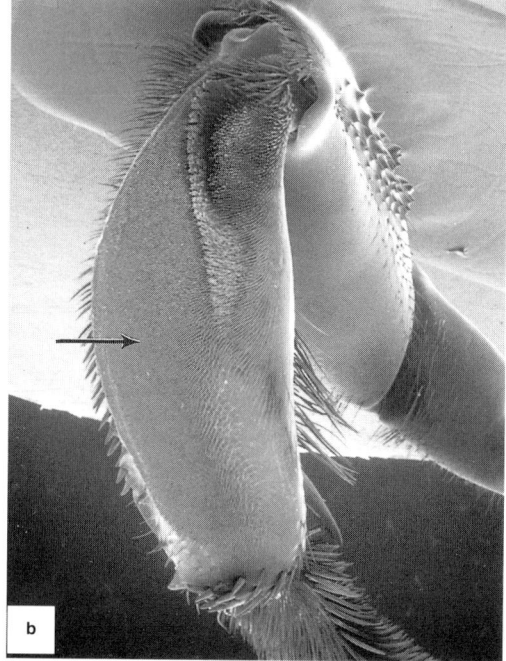

Figure 118-9. Middle tibiae of males of Euglossini, with small felt patches at bases of large felt areas marked by arrows. **a**, *Euglossa cordata* (Linnaeus); **b**, *Eufriesea concava* (Friese). From Michener, McGinley, and Danforth, 1994.

arable in general appearance also: *Exaerete* species are large, all green (rarely all purple), not conspicuously hairy; *Aglae* is slender, blue; and *Euglossa* is small to moderate-sized and usually brilliantly metallic (but there are a few dull-colored species). *Eufriesea* and *Eulaema*, similar to one another superficially, are large, robust, and conspicuously hairy, except that some species of *Eufriesea* are less hairy and are brilliantly metallic, resembling *Euglossa*. There are cases of probable Müllerian mimicry involving *Eufriesea* and *Eulaema* (Dressler, 1979), and close examination may be needed to recognize the usually rarer *Eufriesea*.

Key to the Genera of the Euglossini
(Modified from Kimsey, 1987)

1. Hind tibia three or more times as long as broad in both sexes (Fig. 118-4c); female hind tibia somewhat inflated, without corbicula; cleptoparasitic species 2
—. Hind tibia twice as long as broad or less in females (Fig. 118-3) and most males (Fig. 118-4a, b); female hind tibia flat and shieldlike with enormous corbicula (Fig. 118-3); nonparasitic species 3
2(1). Hind femur swollen and usually denticulate ventrally (Fig. 118-4c); hind tibia curved and expanded apically; scutellum dorsally convex, with sublateral tubercle or welt (Fig. 118-4d) .... *Exaerete*
—. Hind femur slender and unmodified; hind tibia straight and apically narrowed; scutellum flat........................ *Aglae*
3(1). Labrum whitish with two large, dark oval spots; male hind tibial slit short, not reaching apical margin of tibia, and basally curved (Fig. 118-4b); male middle tibia with two, or less commonly one or three, small felty patches in basal end of large patch (Fig. 118-9a); female usually with median, black scutellar tuft .... *Euglossa*
—. Labrum dark in color; male hind tibial slit long, reaching apical margin, broad and not curved basally (Fig. 118-4a); male middle tibia with one relatively large basal felty patch adjacent to large patch (Fig. 118-9b); female with *(Eulaema)* or without *(Eufriesea)* scutellar tuft 4
4(3). Labial palpus four-segmented; face metallic without white markings; clypeal ridging various, usually without single medial ridge... *Eufriesea*
—. Labial palpus two-segmented; face black or brown, often with white markings; clypeus with single strong medial ridge .... *Eulaema*

Genus *Aglae* Lepeletier and Serville

Aglae Lepeletier and Serville, 1825: 105. Type species: *Aglae coerulea* Lepeletier and Serville, 1825, monobasic.
Aglaa Schulz, 1906: 258, unjustified emendation of *Aglae* Lepeletier and Serville, 1825.

This is a monotypic genus of relatively slender, steel-blue bees 20 to 28 mm in length, reported to be cleptoparasites of *Eulaema*. It is the most distinctive euglossine genus, as indicated by Kimsey (1982), who considered it the sister group of all the other genera, and by Kimsey (1987), who documents its many autapomorphies. In the latter work she regards it as the sister group of *Eulaema*, but see my comments, in the account of the Tribe Euglossini, on relationships among the genera.

■ *Aglae* occurs in the moist forests from Bolivia to Colombia with a doubtful report for eastern Panama. There is only one species, *A. coerulea* Lepeletier and Serville.

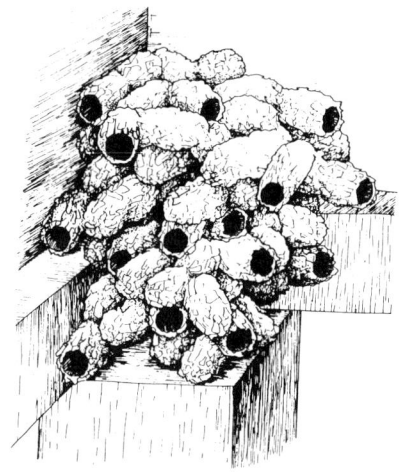

Figure 118-10. Cells of *Eufriesea surinamensis* (Linnaeus) in a corner beneath the floor of a building. From Kimsey, 1982.

Genus *Eufriesea* Cockerell

Plusia Hoffmannsegg 1817: 52 (not Hübner, 1806). Type species: *Plusia superba* Hoffmannsegg, 1817, monobasic.

Eumorpha Friese, 1899b: 126 (not Hübner, 1807). Type species: *Euglossa pulchra* Smith, 1854, by designation of Cockerell, 1908a: 41 footnote.

Eufriesea Cockerell, 1908a: 41, replacement for *Eumorpha* Friese, 1899. Type species: *Euglossa pulchra* Smith, 1854, autobasic.

Eufriesia Lutz and Cockerell, 1920: 544, unjustified emendation of *Eufriesea* Cockerell, 1908.

Euplusia Moure, 1943a: 189, replacement for *Plusia* Hoffmannsegg, 1817. Type species: *Plusia superba* Hoffmannsegg, 1817, autobasic.

Another probable synonym of *Eufriesea* is *Paleoeuglossa* Poinar (1999) based on an Oligomiocene amber fossil from the Dominican Republic that was evidently broken and misinterpreted. For reference, see Table 22-1.

Kimsey (1979b) showed that the broad, flat scutellum formerly thought to separate *Eufriesea* from *Euplusia* cuts across natural groups and is not a useful generic or subgeneric character state. Many species are brilliantly metallic but others have colorful yellow and black hairs and superficially resemble *Eulaema* except for weak metallic reflections at least on the face. The body length is 13 to 27 mm. Male genitalia and hidden sterna were illustrated by Sakagami and Michener (1965) and Sakagami and Sturm (1965); see also Fig. 119-6.

■ Although found primarily in tropical forests, *Eufriesea* occurs south to the province of Córdoba, Argentina, with a record for Chubut, and north to the states of Sinaloa, Chihuahua, and San Luis Potosí, Mexico. There are about 52 species, most of which were placed in *Euplusia* from 1943 to about 1980. The genus was revised by Kimsey (1982).

Nests consist of linear or branched series of cylindrical cells of resin often mixed or covered with bark fragments (Fig. 118-10). Numerous small pieces of bark are sometimes carried to a nest stuck in resin in the bee's corbiculae. Nests are located in protected crevices or cavities, under overhangs of rocky banks, under buildings, in preexisting burrows or rot cavities in wood, in termite nests, etc. Although such nests are sometimes aggregated, there is no evidence of social organization. Kimsey (1982) summarizes what is known about the nests; see also Sakagami and Michener (1965), and Sakagami and Sturm (1965).

Genus *Euglossa* Latreille

Euglossa Latreille, 1802a: 436. Type species: *Apis cordata* Linnaeus, 1758, by designation of Taschenberg, 1883: 85.

Cnemidium Perty 1833: 148 (not Goldfuss, 1826). Type species: *Cnemidium viride* Perty, 1833, monobasic.

Euglossa (Glossura) Cockerell 1917b: 144. Type species: *Euglossa piliventris* Guérin-Méneville, 1845, by original designation.

Euglossa (Euglossella) Moure 1967: 401, replacement for *Cnemidium* Perty, 1833. Type species: *Cnemidium viride* Perty, 1833, autobasic.

Euglossa (Dasystilbe) Dressler, 1978: 193. Type species: *Euglossa villosa* Moure, 1968, by original designation.

Euglossa (Glossurella) Dressler, 1982b: 131. Type species: *Euglossa bursigera* Moure, 1970, by original designation.

Euglossa (Glossuropoda) Moure, 1989c: 387. Type species: *Euglossa intersecta* Latreille, 1837, by original designation.

This genus consists of moderate-sized (the smallest of the Euglossini) to rather large (body length 9-19 mm), usually brilliantly metallic species, usually green but in some species blue, purple, coppery, or partly red. Dressler (1978) provided a classification of the species, placing them in 12 species groups organized into four subgenera (*Dasystilbe* Dressler, *Glossura* Cockerell, *Euglossa* Latreille s. str., and *Euglossella* Moure). The subgenera are seemingly natural groups, but there is some intergradation among them. Later Dressler (1982b) and Moure (1989c) elevated two other groups to subgeneric status. I have found it difficult to make a key to the named subgenera, and I list them above as synonyms without prejudice to the idea that they may well be as recognizable and useful as subgenera elsewhere.

■ *Euglossa* occurs from Paraguay north through the tropics to the states of Sonora and Tamaulipas, Mexico; also Jamaica. With 103 species described, it is the largest genus of the tribe Euglossini. Revisions of the subgenera *Glossura* and *Euglossella* in Central America are by Moure (1969b, 1970). Note also the keys and lists of Euglossini referred to in the account of the tribe.

Nests of some species are constructed of resin and located on stems or twigs in the open. The cells are packed into the interior of a more or less spherical resinous envelope. Other species construct cells, isolated or in small clumps, in small cavities in tree branches or trunks, earthen banks, or in buildings. Some nests are built and occupied by lone females, but others contain several females seemingly living more or less cooperatively (Roberts and Dodson, 1967). As an example, Otero (1996) described nests of *Euglossa nigropilosa* Moure constructed in spaces within walls of buildings. The nest cavities were sometimes partly lined or restricted by an involucrum of resin and the small round entrance was commonly of the same material and frequently closed at night. Some nests, probably in recently occupied sites, contained less resin construction than others. Nests varied greatly in size, from six resin cells and two adult females to about 130 cells and 22 adults. The cells were mostly in contact with one another, forming clusters, and tended to have vertical axes. See also the account of the tribe Euglossini.

Genus *Eulaema* Lepeletier

Eulaema Lepeletier, 1841: 11. Type species: *Apis dimidiata* Fabricius, 1793 = *Apis meriana* Olivier, 1789, by designation of Smith, 1874:22.

Eulaenia Spinola, 1851: 167. *Lapsus* for *Eulaema* Lepeletier, 1841.

Eulema Smith, 1854: 380, unjustified emendation of *Eulaema* Lepeletier, 1841.

Eulaema (Apeulaema) Moure, 1950b: 184. Type species: *Eulaema fasciata* Lepeletier, 1841 = *Centris cingulata* Fabricius, 1804, by original designation.

This genus was at one time called by the name *Centris* Fabricius (1804) (Sandhouse, 1943; Michener, 1944) because of an early and commonly ignored type designation for *Centris*. The usual meanings of the names *Centris* and *Eulaema* have been preserved, however, thanks to Commission Opinion 567 (1959).

These large bees (length 18-31 mm) are all black or have conspicuous patterns of yellow or orange hair, sometimes with limited metallic tints, usually on the metasoma.

■ *Eulaema* occurs from Santa Catarina, Brazil; Misiones, Argentina; and Paraguay, north through the tropics to Sonora and Tamaulipas, Mexico, with an old record for southernmost Texas, USA. A presumably vagrant specimen of *E. polychroma* (Mocsáry) was recently taken by R. Minckley in southern Arizona. There are 25 species. The genus was revised by Moure (1950b, 2003) and Dressler (1979); the Central American species, by Moure (1963).

Nests consist of clusters of oval cells made of mud or feces, intermixed with glandular secretions or resin, located in cavities in soil, banks, tree trunks, etc. Often two or more females of the same generation work simultaneously in a single nest (Michener, 1974a), each making, provisioning, and ovipositing in her own cells (Santos and Garófalo, 1994). A very small nest was illustrated by Sakagami and Michener (1965); larger nests are illustrated in the papers cited above. Cameron and Ramírez (2002) gave much new information on nest sites and within-nest behavior of *Eulaema meriana* (Olivier). It seems to prefer above-ground sites in hollow trunks or man-made buildings, whereas *E. nigrita* Lepeletier frequently nests in cavities in the ground.

Moure (1950b, 1963) gives the distinctions in detail between the subgenera *Eulaema* s. str. and *Apeulaema*. R. L. Dressler (in litt., 1995) indicated that there are four species groups; it is not clear whether any or all of them justify subgeneric status.

Genus *Exaerete* Hoffmannsegg

Exaerete Hoffmannsegg, 1817: 53. Type species: *Apis dentata* Linnaeus, 1758, monobasic.

Chrysantheda Perty, 1833: 147. Type species: *Chrysantheda nitida* Perty, 1833 = *Apis dentata* Linnaeus, 1758, monobasic.

Caliendra Gistel 1848: viii, unjustified replacement for *Chrysantheda* Perty, 1833. Type species: *Chrysantheda nitida* Perty, 1833 = *Apis dentata* Linnaeus, 1758, autobasic.

This is a genus of large, brilliant-green or rarely purple bees, 18 to 28 mm in body length.

■ *Exaerete* occurs from Tucumán province, Argentina, north through the tropics to the states of Nuevo León, Chihuahua, and Nayarit, Mexico. Six species were revised by Kimsey (1979a); seven are now known and a key was provided by Anjos-Silva and Rebêlo (2006).

The species of *Exaerete* are cleptoparasites of *Eufriesea* and *Eulaema*. Bennett (1972) observed that a female of *Exaerete* opened a recently closed host cell, destroyed the host egg, replaced it with her own, and then resealed the cell with wood particles and resin. Thus the method of parasitization employed resembles that of *Sphecodes* (Halictinae) and *Stelis (Hoplostelis)* (Anthidiini) and differs from that of most cleptoparasitic bees, whose young larvae destroy the host eggs or young larvae. Garófalo and Rozen (2001) gave an extensive account of the cleptoparasitic behavior of *Exaerete smaragdina* (Guérin-Méneville) in nests of its host, *Eulaema nigrita* (Lepeletier); they included descriptions of the five larval stages.

119. Tribe Bombini

This tribe consists of the bumble bees and, as here classified, consists of a single genus, *Bombus*. Of course they are bombiform, but worn and partly hairless individuals sometimes superficially resemble the Anthophorini or Eucerini. The head shape, with long malar spaces (Fig. 119-1b), and the corbiculate tibiae of females immediately distinguish Bombini. Except for the parasitic species, and except for spring nests in which queens care for their immature offspring, all are primitively eusocial. The approximately 250 species (or 239, according to the rather conservative list of Williams, 1998) are morphologically monotonous compared to the Euglossini and especially to the Meliponini. Nonetheless, there is interesting diversity in a few structures, especially the male genitalia, the female stings, and the mandibles. Most of the classification of the group is based on the male genitalia.

The Bombini are middle-sized (9 mm long) to very large (22 mm long), bombiform, hairy bees (Pl. 14). The claws of the female are cleft; the arolia are small but present and the hind tibial spurs are present. Except for social parasites, the hind tibia and basitarsus of both female castes (Fig. 102-2a) are similar to those of workers of Apini; likewise, the flabellum (Fig. 119-2) resembles that of *Apis*. The wings have complete strong venation (Fig. 119-1a). The marginal cell is somewhat longer than the distance from its apex to the wing tip. The stigma is small, little if any longer than the prestigma, vein r arising near or beyond the middle, the margin within the marginal cell being straight or usually concave. The hind wing lacks a jugal lobe. S8 of the male is well developed and sclerotized, and has a median apical hairy process (Fig. 119-3b); the disc of the sternum is not thickened and excavated as in most Euglossini. The male genitalia are well sclerotized with the gonobase distinct. The gonocoxite is large and supports the short and broad, often variously angulate gonostylus ("squama" in much *Bombus* literature) and the broad, hairy volsella ("gonostylus" of Ito, 1985; "lacinia" of Richards, 1968) which attains the apex of the gonostylus or exceeds it and also extends far toward the base of the gonocoxite on the lower surface (Fig. 119-4b). A re-

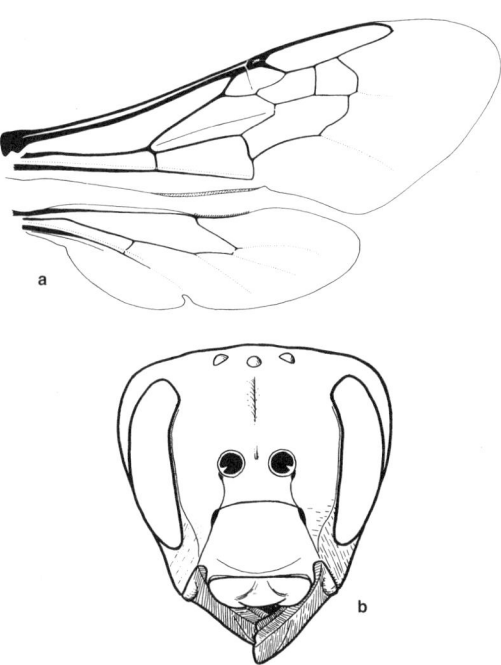

Figure 119-1. Structures of *Bombus*. **a,** Wings of *B. pennsylvanicus* (Degeer); **b,** Head of *B. vosnesenskii* Radoszkowski. b, from Michener, 1944.

view of homologies and terminology of the male genitalic structures of Bombini was given by Michener (1990a); see Table 119-1, which is derived from that work, and see also Kopelke (1981) and Williams (1985, 1991). Male genitalia and hidden sterna have been illustrated for many species in the regional studies listed under the genus *Bombus* below; see also Michener, 1990a. Figure 119-8 shows some of the diversity. Radoszkowski, as early as 1884b, illustrated male genitalia and hidden sterna of many species.

Additional special terminology that has been widely

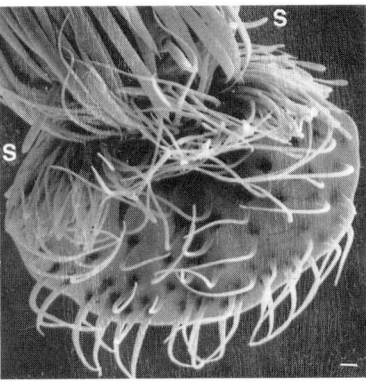

Figure 119-2. Flabellum of worker of *Bombus morrisoni* Cresson. Left, Posterior view; Right, Anterior view (s, setae). From Michener and Brooks, 1984.

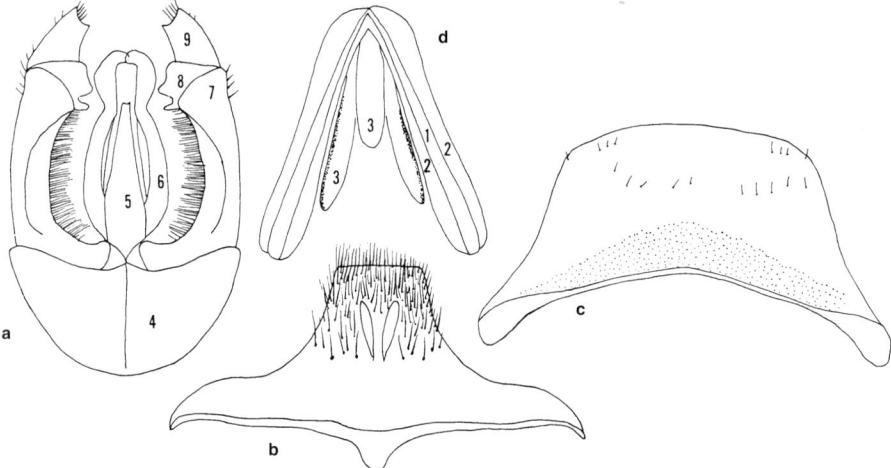

used in studies of *Bombus* concerns the mandibles of females. They are apically subtruncate or broadly rounded (Fig. 119-5e, f). The lowermost part of the apical margin commonly forms the apical angle or tooth. This tooth is sometimes made more evident and toothlike by a small notch or incision in the margin above it; this is often called the "incisura," although I have not used this term below, and it is not present in species shown in Figure 119-5. The ridges or keels on the outer surface of the mandible provide some characters. Most used is the "sulcus obliquus," illustrated in various works, e.g., Williams (1991); in terms of structures of other bees, it is the lower

Figure 119-3. Structures of *Bombus (Alpigenobombus) sikkimi* Friese. **a-c,** Male genitalia (dorsal view) and S8 and S7 (ventral views); **d,** Dorsobasal view of female sting apparatus. (1, ramus of first valvula; 2, outer and inner sides of ramus of second valvula; 3, folds of membrane; 4, gonobase; 5, spatha; 6, penis valve; 7, gonocoxite; 8, gonostylus; 9, volsella.) Modified from Richards, 1968.

ramus of the outer groove (illustrated for *Bombus* by Michener and Fraser, 1978, and in Figure 119-5e).

Bombus is one of the few groups of bees in which characters of the female sting apparatus have been widely used

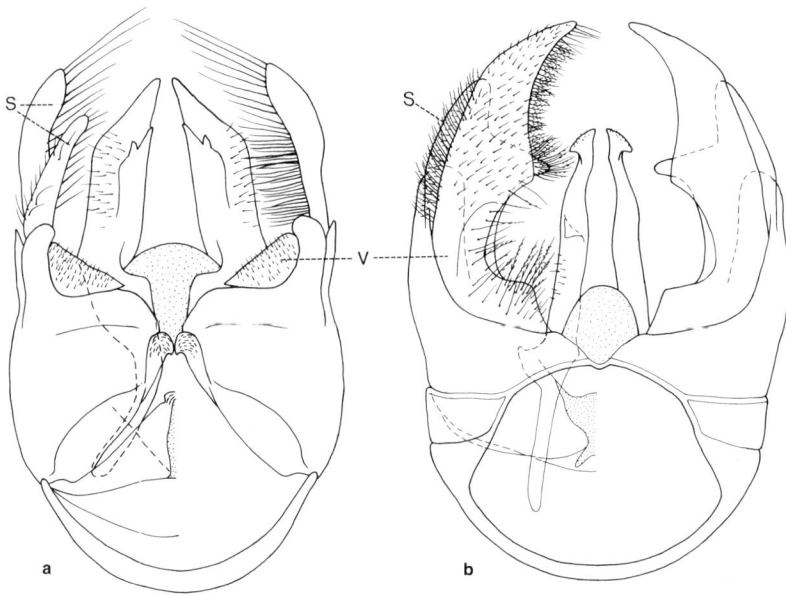

Figure 119-4. Male genitalia of Euglossini and Bombini, ventral views. **a,** *Eufriesea pulchra* (Smith); **b,** *Bombus (Psithyrus) variabilis* Cresson. If the homologies are properly understood, the volsella in Bombini is larger and extends farther apically than in any other bee. (S, gonostylus, upper and lower in *Eufriesea*; V, volsella.) From Michener, 1990a.

Table 119-1. Terminology of Male Genitalia of Bombini, as Employed in Four Major Works.

Michener (1944)	Richards (1968)	Ito (1985)	Williams (1985) and present work
gonobase	cardo	gonobase	gonobase
gonocoxite	stipes	gonocoxite	gonocoxite
gonostylus	squama	squama	gonostylus
gonostylus	lacinia	gonostylus	volsella
spatha	spatha	spatha	spatha
penis valve	sagitta	penis valve	penis valve

in taxonomic studies. The structures described are commonly called the inner and outer "thickenings of the sting sheath." This terminology is quite inappropriate. The structures do not sheath the sting; moreover, the female gonostyli are often called sting sheaths in *Apis* and other Hymenoptera. The structures concerned are the rami of the first and second valvulae. They are described and illustrated for each subgenus in the keys and descriptions by Richards (1927, 1968) and were illustrated for most American species by Hazeltine and Chandler (1964), and for Argentine species by Abrahamovich, Diaz, and Lucia (2005); see also Kopelke (1981). On each side there are three longitudinal elements of the rami. The middle one (1 of Fig. 119-3d) is the ramus of the first valvula, which is slender and varies little among species of *Bombus*. It lies firmly in a groove in the broad ramus of the second valvula, which extends laterally on both sides of the first, thus producing inner and outer elements (2 and 2' of Fig. 119-3d) that vary among subgenera. The inner margin provides most of the characters of interest.

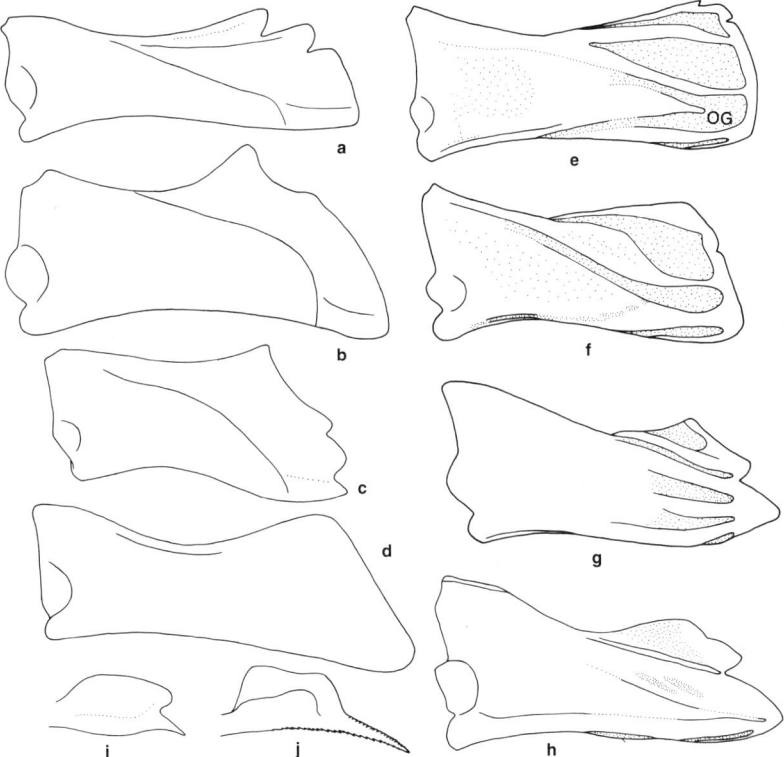

Figure 119-5. Structures of corbiculate Apidae. a-h, Mandibles of females (workers in social species). **a,** *Meliponula bocandei* (Spinola) (the two upper denticles are smaller in many Meliponini, and there is sometimes only one); **b,** *Cephalotrigona capitata* (Smith); **c,** *Trigona (Trigona) cilipes* (Fabricius); **d,** *Apis mellifera* Linnaeus; **e,** *Bombus pennsylvanicus* (Degeer), OG, the outer groove, the lower ramus of which is the *sulcus obliquus* of *Bombus* literature; **f,** *B. (Psithyrus) variabilis* (Cresson); **g,** *Eufriesea violacea* (Blanchard); **h,** *Exaerete smaragdina* (Guérin-Méneville). **i, j,** Anterior tibial spur (strigilis) of workers, inner surface, of *Melipona rufiventris* Lepeletier and *Bombus pennsylvanicus* (Degeer), showing the inner lamella (anterior velum of Schönitzer, 1986) of *Bombus*. From Michener, 1990a.

The larva has small pointed dorsolateral tubercles on the thoracic segments. The mandible is heavily sclerotized, its apex bluntly rounded or minutely denticulate, acute in the subgenus *Psithyrus,* and has a small preapical tooth on the upper margin and a large apical concavity on the inner surface. Larvae have been described and illustrated in detail by Ritcher (1933), Cumber (1949), Michener (1953a), Stephen and Koontz (1973), and others.

Bumble bees for the most part occur in cool climates and are most abundant in the holarctic region, with many more species and subgenera in Eurasia than in North America. Williams (1985) indicated that for nonparasitic *Bombus* there are 199 species in Asia, 58 in Europe, 41 in North America, and 43 in Mexico, Central America, and South America. Northward, they range in small numbers as far as there is land. Southward, they occur in North Africa and west as far as the Canary Islands but not in sub-Saharan Africa. [There is one specimen of a *Fervidobombus,* an American subgenus and probably a South American species, reported from Cameroon (*B. abditus* Tkalcŭ, 1966); it was no doubt mislabeled or introduced.] To the east, *Bombus* species are numerous in the Himalayas but absent below 1000 meters altitude in India (Williams, 1985); to the southeast a few species occur in the mountains of Southeast Asia, as far as Java, Taiwan, and the Philippines, but they are absent from the lowlands. *Bombus* is apparently absent from the mountains of Borneo, the single specimen supposedly from there being a mislabeled specimen of the Mesoamerican *B. ephippiatus* Say. With the intervening land mass, Borneo, lacking *Bombus,* it is not surprising that the *Bombus* of the Philippines are entirely different species from those of Sumatra and Java (Starr, 1989). In the western hemisphere there is a rather small *Bombus* fauna all the way to Tierra del Fuego, mostly in montane areas or south-temperate latitudes, but whereas African and Asian species do not, a few New World species do occur in the moist lowland tropics, such as the Amazon valley.

A summary of nest architecture was given by Michener (1974a). Bumble bee nests are commonly in rodent nests, bird nests, cavities under bunch grass or other vegetation, etc. Queens searching for nest sites can commonly be recognized because they alternate short, slow flights with periods of crawling into vegetation or cavities. Some species nest primarily on or above the surface of the ground, others in holes such as those made by rodents in the ground. A requirement is fine material such as dead grass, hair, small leaves, or moss that the bee can move about to form a nest cavity. There may be a thin wax (and pollen) covering over the nest. As the nest grows, additional material is often added by workers to maintain the roof. The cells, the first batch constructed by the queen, the later ones by workers, are totally different from the cells of any other bees, for they are closed (sometimes incompletely) but grow with the growing larvae, commonly contain several eggs or larvae, develop a separate bulge for each larva, and become divided into several cells each as the larvae mature (Fig. 119-6). Larvae are fed progressively, either by food introduced through the tops of the cells or by food pressed in through pockets at the bases of the cells and forming the cell floors. Both methods may occur in the same species, usually at different seasons or for different castes. When mature, the larvae spin cocoons and the

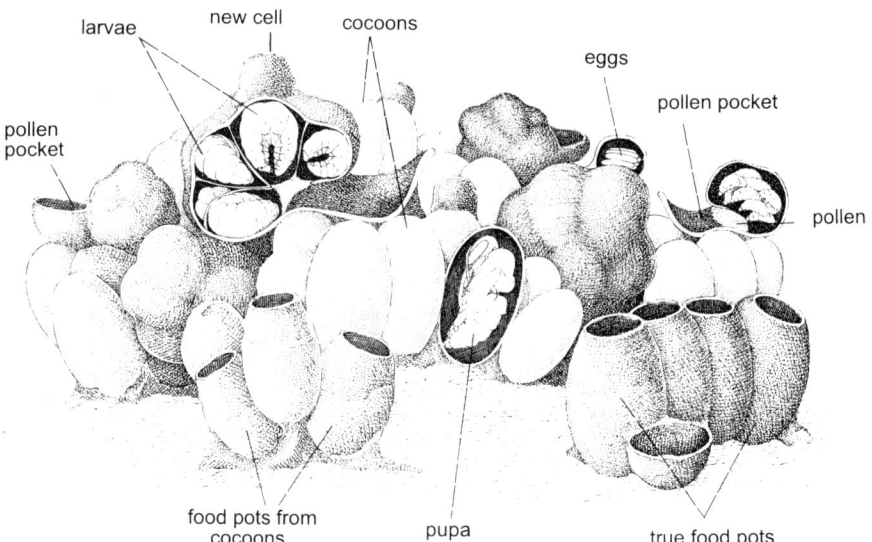

Figure 119-6. Nest structure of *Bombus (Fervidobombus) niger* Franklin. This is a species that feeds each cluster of larvae by pushing pollen into a pollen pocket where it forms a base on which the larvae feed; at the upper right is a group of small larvae, and at the upper left of center are large larvae that have a small amount of pollen still accessible to most of them. Original drawing by J. M. F. de Camargo.

wax of the cell walls is removed for reuse, leaving the cocoons in clumps, each of which represents the young that originated in a single cell. The cells and cocoons do not form neat combs as in Apini and most Meliponini. Nonetheless they tend to form irregular horizontal layers, for, as the nest grows, new cells are usually established on the tops of old cocoons. In small colonies such layers may never grow over 8 to 9 cm in diameter, but species that develop large colonies may make layers that attain over 20 cm in diameter. Both honey and pollen are stored in pots separate from brood cells; the pots are often made of old cocoons but may be wholly constructed of wax by the bees (the true food pots of Fig. 119-6). Indeed, the first action of a queen establishing a new nest is to make a small wax pot and supply it with honey.

Except for the social parasites, the Bombini all produce eusocial colonies. These colonies are usually annual. Each is started as a subsocial unit by a single gyne (potential or actual queen); in temperate and arctic climates this occurs in the spring. Her early progeny are all workers, usually much smaller than the gyne but morphologically similar. Males and young gynes are produced later; they mate, the colony fades away, and the young mated gynes hibernate until the following spring. In the tropics this sort of cycle is sometimes modified, but it never attains the major features of highly social bees whose queens never live alone and are morphologically very different from workers. One study of tropical *Bombus* described year-round activity of *B. atratus* Franklin in the Andes, over 2000 m altitude. The same species also occurs in the lowlands (Gonzalez, Mejia, and Rasmussen, 2004). Additional works on the biology of tropical *Bombus* are by Dias (1958) and Zucchi (1973). Other features of *Bombus* behavior were described by Sladen (1912), Free and Butler (1959), Michener (1974a), Alford (1975), Morse (1982), and others.

Some species of Bombini are workerless social parasites in nests of other species. Most of the parasites are in what is here called the subgenus *Psithyrus;* females of this subgenus lack the pollen-gathering and pollen-manipulating structures characteristic of other *Bombus,* and the apex of the metasoma, pointed and curled downward, houses a powerful sting. The female social parasite dominates or kills the host queen and may become, in essence, the queen of a colony consisting of herself and workers of the host species. Fisher (1987) made a study of one species of *Psithyrus;* his work contains references to older studies.

In addition, there are a few parasitic species in other subgenera. Thus *Bombus (Alpinobombus) arcticus* (Quenzel) (usually known as *B. hyperboreus* Schönherr) is a parasite of *B. (A.) polaris* Curtis (K. Richards, 1973), and *B. (Thoracobombus) inexspectatus* (Tkalců) is a parasite of other species of *Thoracobombus* (Yarrow, 1970; Müller, 2006). In such parasites the pollen-manipulating and -carrying structures may be slightly reduced, although not nearly to the degree found in the subgenus *Psithyrus.* The most notable difference from nonparasitic relatives is the lack of a worker caste. Any *Bombus* species known principally from large queenlike females (and, of course, males) is a possible parasitic species.

It has been traditional to recognize *Psithyrus* as a genus consisting of most of the workerless social parasites that inhabit nests of other species of *Bombus.* Many authors have speculated on the relation of *Psithyrus* to nonparasitic *Bombus.* Recent studies by Ito (1985), Ito and Sakagami (1985), and Williams (1985, 1994), placing emphasis on male genitalia and hidden sterna, which are unlikely to evolve convergently, show decisively that *Psithyrus* is a monophyletic unit related to certain groups of *Bombus.* Chen and Wang (1997) agreed that *Psithyrus* is monophyletic but found it to be the sister group of all the other *Bombus.* They weighted characters (from 1 to 3) without adequate explanation, and included characters associated with parasitism; parasitism itself was involved in two different characters (39, 40), both weighted as 2. Thus I believe that parasitism was overemphasized and resulted in a phylogenetically incorrect position for *Psithyrus.* Electrophoretic studies of genetic relationships support the cohesiveness among *Psithyrus* species (Pamilo, Pekkarinen, and Varvio, 1987; Obrecht and Scholl, 1981). More recently, Kawakita et al. (2003) have further investigated the phylogeny of *Bombus* species on the basis of three nuclear genes and 66 species. For the most part subgenera based on morphological features were supported; *Psithyrus* was placed among *Bombus* taxa. Similar results based on a mitochondrial and a nuclear gene and 26 European species had been obtained by Pedersen (2002).

Ito (1985) and Williams (1985) provided a wealth of information on group characters of Bombini as well as interpretations of relationships. Ito gives excellent illustrations of male sterna and genitalia of nearly all groups. His analyses (based on genital character states of males) are phenetic; they show *Psithyrus* in the midst of *Bombus* groups, and closest to the subgenera *Mucidobombus, Eversmannibombus,* and *Orientalibombus.* A cladistic study of the problem by Ito and Sakagami (1985) indicated a relationship of *Psithyrus* to *Orientalibombus* and some species of *Fervidobombus,* especially the Chilean *Bombus dahlbomii* Guérin-Méneville.

Williams (1985) developed a cladogram for representative species of groups of Bombini, based on male genitalic character states. *Mendacibombus* appears in his study as the sister group of all other Bombini; *Psithyrus* is the next branch, the sister group of all but *Mendacibombus.* On this basis Williams recognized three genera, *Mendacibombus, Psithyrus,* and *Bombus.* Delmas (1976) had recognized the same three genera. Ito (1985) also recognized *Mendacibombus* as a distinctive group, but in all but one of his phenograms it was closely associated with other groups of *Bombus.* Richards (1968) did not indicate that he considered *Mendacibombus* unusual.

With additional species Williams (1991, 1994) found *Mendacibombus* to represent more than one clade. In 1994 he presented a fuller cladistic analysis, using not only male genitalic characters but also those from all parts of the body and both sexes. *Psithyrus* comes out not so close to *Mendacibombus* as in his prior study, but as the sister group to *Bombus persicus* Radoszkowski *(Eversmannibombus);* see Figure 119-7. I follow Williams' recommendation that *Psithyrus* be considered a subgenus of *Bombus,* which seems especially appropriate in view of the workerless parasitic species that do not belong to *Psithyrus.* The numerous characters on the *Psithyrus* stem,

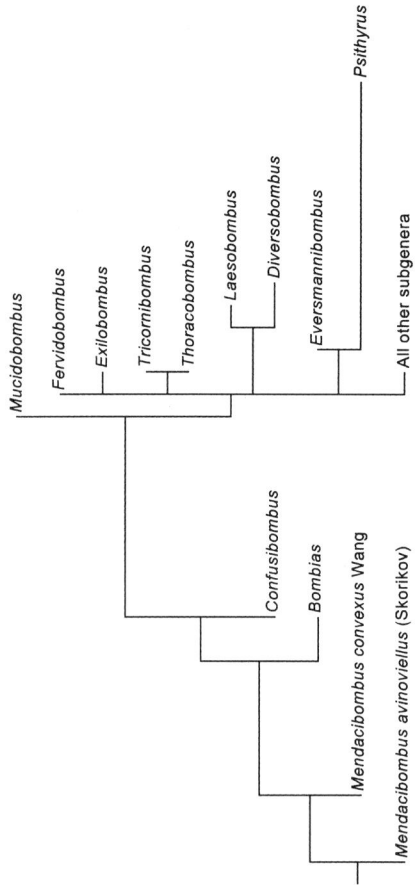

Figure 119-7. Cladogram of basal subgenera of *Bombus*, showing relationships of *Mendacibombus* and *Psithyrus*; from Williams' (1994) maximally resolved tree based on one or rarely two species of each subgenus. *Mendacibombus* is obviously paraphyletic if this phylogeny is correct; other subgenera may present similar problems when more species are included in the analysis. (Lengths of the lines are proportional to the number of characters.)

as indicated by its length in Figure 119-7, could be used to justify a genus *Psithyrus* and a paraphyletic genus *Bombus*.

Probably none of the fossils that have been attributed to the Bombini can be reliably placed there except the Miocene species placed in *Bombus* (Zeuner and Manning, 1976; Rasnitsyn and Michener, 1991). The Eocene *Probombus* Piton (1940), listed in the Appendix of Zeuner and Manning's work, was presumably not a corbiculate apid. Piton's illustration shows a female without a corbicula; the wing venation appears to be crudely represented and was misinterpreted, but does not seem *Bombus*-like. The Oligocene *Calyptapis* Cockerell (1906a), synonymized by Zeuner and Manning with *Bombus*, seems to me probably something else. Miocene true *Bombus* fossils are known from China, the Russian Far East, and Washington state, USA (Rasnitsyn and Michener, 1991).

Genus *Bombus* Latreille

The generic characters are those of the tribe. About 35 subgenera of *Bombus* are recognized, and some of them, such as *Fervidobombus*, are quite diversified and may not represent unified groups. Ito (1983, 1985) gave a detailed and useful historical account of subgenera, and a briefer one was provided by Richards (1968). Given the morphological homogeneity of *Bombus*, most authors have continued to use that generic name for all the nonparasitic species. Efforts to find a few recognizable natural units have failed because of intergradation and discordance in character-state distributions. See the discussion under the tribe Bombini.

The named subgenera seem less different from one another than are subgenera in most groups of bees. In fact, the homogeneity of the species in the genus is outstanding. Careful searches for group character states have been made by diverse specialists (among more recent authors, Richards, 1968; Sakagami and Ito, 1981; Ito, 1983, 1985; Williams, 1985, 1994). The only key to subgenera (except regional ones) is that of Richards (1968). The key to subgenera below is that of Richards, modified by Michener (2000); a regional key that illustrates the subgeneric characters photographically is by Intoppa et al. (2003). One might reasonably hope to find a few large units into which the genus could be divided, for the sake of convenient treatment of the species. Milliron (1961) divided nonparasitic *Bombus* into three genera and Tkalcŭ (1972) recognized eight genera. These genera, however, are hard to distinguish, they intergrade, and they have little usefulness. The smaller supraspecific units (subgenera or species groups), however, based primarily on male genitalic and sternal character states, are more stable, and for a *Bombus* specialist are useful. Figure 119-8 gives some idea of the variation in these features. The discussions above under the tribe Bombini offer some account of the phylogeny of subgenera, and the most recent proposed phylogeny (Williams, 1994) of species in the "basal" subgenera (showing the position of *Psithyrus*) is summarized in Figure 119-7.

Further studies of behavioral characters like those of Hobbs (1964), Sakagami (1976), and Katayama (1989) may help in the delineation of more useful units. Katayama (1989) recorded in admirable detail the often subgenus-specific behaviors in cell construction and egg laying, but the sampling of species was necessarily limited. Particularly, data are needed for more subgenera on the earliest stages of colony development, when some of the most important characters are manifest.

As molecular methods of determining genetic relatedness have evolved, various authors have used such methods to indicate relationships among species of *Bombus*. Such studies usually reinforce the groups or subgenera based on morphological differences, showing that at least some of these groups are useful monophyletic or paraphyletic units (Pamilo, Pekkarinen, and Varvio, 1987).

Some important regional treatments, mostly with keys, illustrations of genitalia as well as color and other characters, and mostly including *Psithyrus* as a genus separate from *Bombus*, are as follows:

WESTERN HEMISPHERE: Franklin, 1913a, b; Chandler, 1950 (Indiana, USA); Stephen, 1957 (western

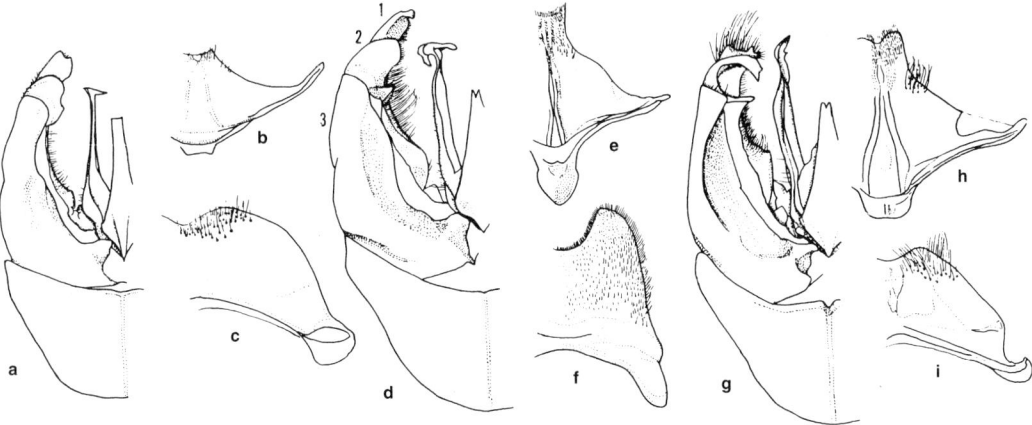

Figure 119-8. A sample of genitalic and sternal variation in male *Bombus*; dorsal views of genitalia and ventral views of S8 and S7. **a-c**, *B. (Melanobombus) lapidarius* (Linnaeus); **d-f**, *B. (Fraternobombus) fraternum* (Smith); **g-i**, *B. (Rubicundobombus) rubicundus* Smith. See also Figures 117-4b and 117-9c, d. (1, volsella; 2, gonostylus; 3, gonocoxite.) From Ito, 1985.

North America); LaBerge and Webb, 1962 (Nebraska, USA); Mitchell, 1962 (eastern North America); Moure and Sakagami, 1962 (Brazil); Medlar and Carney, 1963 (Wisconsin, USA); Milliron, 1970-1973; Thorp, Horning, and Dunning, 1983 (California, USA); Labougle, 1990 (Mesoamerica); Laverty and Harder, 1998 (eastern Canada); Rasmussen, 2003 (Peru); Abrahamovich and Diaz, 2002 (neotropics); Abrahamovich, Diaz, and Lucia, 2005 (Argentina).

EUROPE: Friese and Wagner, 1910 (Germany); Krüger, 1917, 1920 (Central Europe); Richards, 1927 (Britain); Pittioni, 1937 (east Tirol), 1938, 1939a (Balkans); Kruseman, 1947 (Netherlands); Knechtel, 1955 (Romania); May, 1959 (Czechoslovakia); Yarrow *in* Free and Butler, 1959 (Britain); Elfving, 1960 (Finland); Hammer and Holm, 1970 (Denmark); Løken, 1973, 1984, 1985 (Scandanavia); Alford, 1975 (Britain); Delmas, 1976 (France); Rasmont, 1983 (west palearctic region); D'Albore, 1986 (Italy); Prŷs-Jones and Corbet, 1987 (Britain); Hagen, 1990 (Central Europe); Rasmont and Adamski, 1995 (Corsica).

ASIA: Frison, 1928 (Philippines), 1930 (Indonesia); Maa, 1948 (*Psithyrus* of eastern Asia); Özbek, 1983 (eastern Turkey); Williams, 1991 (Kashmir); Starr, 1992 (Taiwan); Pawlikowski, 1996, 1999 (Poland); Özbek, a series on the bumble bee fauna of Turkey, 1997, 1998, 2000, 2002.

Of course, keys to species of Bombini are also found in the various regional works on bees or on Hymenoptera, such as are listed in Section 30.

A few "couplets" in the keys that follow offer three alternatives instead of two. These are "couplet" 25 in the key to males and "couplets" 21 and 22 in the key to females; each of these is marked "triplet" to assure proper use. Given the intricacy of Richards' (1968) key on which the key below is based, I have not risked introducing errors by converting these items to a dichotomous format. *Exilobombus* is omitted from the key to males.

Key to the Subgenera of *Bombus* (Males)
(Modified from Richards, 1968, with additions based on Williams, 1991)

1. Volsella weakly or strongly sclerotized, inner corner near midpoint of its length, without any inwardly directed hooks, distal half of volsella thus usually nearly triangular (Fig. 119-9d); gonostylus without an inner basal process, or, *if* process present, then associated with many long, branched hairs; apex of penis valve nearly straight in dorsal view 2
—. Volsella always strongly sclerotized, inner corner usually much closer to apex than to base, forming an inner apical corner often bearing two inwardly directed hooks or a single inwardly directed process; gonostylus usually with distinct inner basal process or shelf, not associated with long hairs; apex of penis valve much modified, often curved strongly toward midline as sickle-shaped hook, otherwise curved outward in some individuals, but *if* nearly straight in dorsal aspect, then volsella with pronounced inner hooks 3

2(1). Volsella weakly sclerotized, yellowish in color; gonostylus with pronounced inner basal process associated with many long, branched hairs (Fig. 119-9d); apex of penis valve, as defined by an outer lateral ridge, accounting for less than one-fourth of total length of penis valve, not strongly curved but shaped like slender arrowhead in lateral aspect (holarctic) .... *B. (Psithyrus)*
—. Volsella strongly sclerotized, dark brown in color; gonostylus without inner basal process or hairs; apex of penis valve, as defined by an outer lateral toothed ridge, accounting for nearly one-half of total length, ventrally curved and sabre-shaped in lateral aspect (Eurasian mountains) .... *B. (Mendacibombus)*

3(1). Apex of mandible with two preapical teeth (palearctic) .... *B. (Alpigenobombus)*
—. Apex of mandible with one preapical tooth 4

4(3). Penis valve narrow and pointed; ocellocular distance equal to 3.0 to 3.5 ocellar diameters; hind tibia bare on disc; second flagellar segment 1.1 to 1.4 times as long as first (Himalayas to Southeast Asia) .... *B. (Orientalibombus)*
—. If penis valve narrow and pointed, then lateral ocellus

much closer to eye, *or* hind tibia hairy, *or,* if with a small, bare disc, then second flagellar segment about half as long as first 5

5(4). Penis valve curved outward into a rounded, recurved hook; second and third flagellar segments 1.2 and 1.5 times as long as first segment, and middle and hind basitarsi without long fringes (palearctic) .... *B. (Kallobombus)*

—. Penis valve not curved outward into a recurved hook; second and third flagellar segments relatively short (second shorter than first) except in a few species with short basitarsal fringes 6

6(5). Second and third flagellar segments 0.36 to 0.56 and 0.48 to 0.78 times as long as first segment; penis valve narrow and pointed, or finger-shaped, only in one group curved inward with flange on outside of curve; ocellocular distance one ocellar diameter or less 7

—. Second and third flagellar segments relatively longer and penis valve usually of a different shape, particularly if antenna approaches above description; ocellocular distance usually more than one ocellar diameter (although sometimes less, e.g., in *Brachycephalibombus*) 9

7(6). Penis valve curved inward with small flange on outside of curve; hind basitarsus with short dorsal fringe (North America) .... *B. (Fraternobombus)*

—. Penis valve pointed or finger-shaped; hind basitarsus with long dorsal fringe 8

8(7). Gonostylus V-shaped in section, formed of two plates set at an angle and joined at outer edges; S7 crescentic but with rounded-triangular median process; S8 crescentic, little produced, apex subtruncate; hind tibia with fringe of long hairs (Europe) .... *B. (Confusibombus)*

—. Gonostylus not V-shaped in section, all angles rounded; S7 broadly triangular, apex feebly emarginate; S8 broadly triangular, apex deeply emarginate; hind tibia with fringe of very short hairs (North America) .... *B. (Bombias)*

9(6). Penis valve in form of wide sinuate vertical plate (holarctic).. *B. (Bombus s. str.)*

—. Penis valve of quite different form 10

10(9). Penis valve ending in strong, inwardly directed, recurved hook; middle basitarsus usually with short fringe 11

—. Penis valve rarely ending in strong, inwardly directed, recurved hook, but *if* so, then middle basitarsus with long fringe 17

11(10). Ocellocular distance less than one ocellar diameter; malar area shorter than second flagellar segment; sides of spatha strongly convergent posteriorly.......................... 12

—. Ocellocular distance greater than one ocellar diameter; malar area longer than second flagellar segment, as long as or longer than first flagellar segment; sides of spatha subparallel, little convergent posteriorly......................... 14

12(11). S7 trapezoidal, apex broadly emarginate; S8 with apical process broadly emarginate; penis valve ending in broadened apex with acute point directed mesad (Mesoamerica) .... *B. (Brachycephalibombus)*

—. S7 trapezoidal or crescentic, apex not or scarcely emarginate; S8 with apical process narrower, apex convex; penis valve ending in broadened apex with blunt or slender point 13

13(12). Penis valve ending in rounded hook, tip not acute; impression of gonocoxite wide but ill-defined; third flagellar segment shorter than first; S7 trapeziform, apex feebly emarginate; S8 subtriangular (North America).... .... *B. (Separatobombus)*

—. Penis valve ending in somewhat smaller hook, tip acute; impression of gonocoxite wide and well-defined; third flagellar segment longer than first; S7 subcrescentic; S8 with parallel-sided, tongue-like projection (California, Mexico) .... *B. (Crotchiibombus)*

14(11). Gonostylus considerably longer than broad, inner side emarginate; impression of gonocoxite strong, sharp-edged; third flagellar segment clearly shorter than first (palearctic) .... *B. (Sibiricobombus)*

—. Gonostylus not clearly longer than broad, inner side not emarginate; impression of gonocoxite feebly defined; third flagellar segment longer than first 15

15(14). Volsella scarcely projecting beyond gonostylus (more so in *Festivobombus*) (gonostylus small and rounded-triangular); ocellocular distance at least three ocellar diameters; S8 with parallel-sided, tongue-like projection .... *B. (Pyrobombus, Festivobombus)*

—. Volsella very long, projecting well beyond gonostylus; ocellocular distance less than three ocellar diameters; S8 broadly subtriangular 16

16(15). Gonostylus transverse with inner side emarginate; hind basitarsus with short or long fringe; malar area not longer than first flagellar segment; ocellocular distance one and three-fourths to two and one-fourth ocellar diameters (holarctic) .... *B. (Cullumanobombus)*

—. Gonostylus S-shaped, inner end pointed, almost the whole of it lying inside inner margin of gonocoxite; hind basitarsus with short fringe; malar area longer than first flagellar segment; ocellocular distance two and one-half ocellar diameters (Southeast Asia to Java and Taiwan).... .... *B. (Rufipedibombus)*

17(10). Gonostylus almost as long as gonocoxite, attaining midline of genitalia, directed mesad at right angle to gonocoxite, without lobe or process; volsella produced as slender, hairy fingerlike apical projection almost as long as gonostylus (Mesoamerica) .... *B. (Dasybombus)*

—. Gonostylus less than half as long as gonocoxite, not approaching midline of genitalia, and *if* directed mesad, then commonly with lobe or process; volsella without long, slender apical projection, but with tooth or lobe or apically pointed, not fingerlike 18

18(17). Middle and hind basitarsi with long fringes 19

—. Middle and hind basitarsi with short fringes 25

19(18). Penis valve not curved inward or outward at apex, but with two external teeth, one of which is sometimes small; second flagellar segment not much shorter than first, third segment 1.25 times longer than first; malar area in most species distinctly longer than first flagellar segment (arctic and alpine, holarctic) .... *B. (Alpinobombus)*

—. Penis valve sometimes curved inward at apex, never with more than one outer tooth; second flagellar segment not more than three-fourths as long as first, third segment not more than 1.1 times longer than first; malar area rarely longer than first flagellar segment 20

20(19). Volsella projecting far beyond gonostylus; malar area as long as first flagellar segment or a little longer 21

—. Volsella projecting at most a moderate distance beyond gonostylus (gonostylus with rounded angles and an inner emargination); malar area usually shorter than first flagellar segment 22

21(20). Gonostylus completely fused with gonocoxite, produced into long acute process directed downward and backward; penis valve at apex with small pointed flange on inside and large pointed one on outside; S7 broadly trapezoidal; S8 with parallel-sided projection; gonocoxite with broad, sharp-edged impression; volsella twisted, its plane thus oblique distally, with dense short hairs at apex and on inner side; ocellocular distance equal to three ocellar diameters (Himalayas) .... *B. (Pressibombus)*

—. Gonostylus separate from gonocoxite, rounded-transverse with small, proximal inner process; penis valve with acute apex and acute inner flange (half-arrowhead-shaped); S7 crescentic; S8 broadly triangular; gonocoxite narrow, without inner impression; volsella not twisted, with few hairs; ocellocular distance variable, equal to one to three ocellar diameters (palearctic) .... *B. (Melanobombus)*

22(20). Third flagellar segment a little longer than first; impression of gonocoxite wide and deep 23

—. Third flagellar segment distinctly shorter than first; impression of gonocoxite not more than half as wide as long 24

23(22). Penis valve curved inward at apex with serrate flange on outside; gonostylus markedly transverse, widening inward, inner edge emarginate; volsella with rounded-angular projection on inner edge, inner corner of apex with long curved hook; S7 crescentic; S8 with tonguelike projection the sides of which converge posteriorly; malar area hardly more than half as long as first flagellar segment; ocellocular distance equal to less than one ocellar diameter (Central and South America) .... *B. (Robustobombus)*

—. Penis valve slightly curved inward at apex and with small inner tooth; gonostylus rounded-quadrangular with large subcircular inner emargination; volsella wide, posterior end with a short hook, apex with dense tuft of relatively long bristles; S7 transverse with central one-fourth produced; S8 subtriangular; malar area a little shorter than first flagellar segment; ocellocular distance equal to 1.5 ocellar diameters (western South America) .... *B. (Rubicundobombus)*

24(22). Gonocoxite with strong impressions; volsella short, apex with small posteriorly directed hook (outer and ventral sides with dense short hairs); S7 transverse, apex biemarginate; S8 with long, parallel-sided midapical process; malar area slightly longer than first flagellar segment; ocellocular distance equal to two ocellar diameters (South America) .... *B. (Coccineobombus)*

—. Gonocoxite with small, ill-defined depressions; volsella moderately long, apex with small hooklike projection at inner end; S7 crescentic, weakly emarginate; S8 subtriangular; malar area three-fourths as long as first flagellar segment; ocellocular distance less than two ocellar diameters (South America) .... *B. (Funebribombus)*

25(18) [triplet]. Volsella long and narrow, produced inward at apex as process shaped like toe and heel; penis valve narrow, outer side distally strongly serrate [a character otherwise seen only in *B. (Thoracobombus) pascuorum* (Scopoli)], no tooth beneath; gonostylus forming a large, elongate, vertical, curved plate on inner side two-thirds enclosing an oval space, anterior lower corner produced upward as sharp spike; malar area as long as combined lengths of pedicel and first and second flagellar segments (palearctic) .... *B. (Megabombus)*

—. Volsella long and narrow, produced inward at apex as spurlike process with several points; penis valve narrow, sinuate, with apical outer flange forming a small tooth at its proximal end, emarginate beneath but not forming a real tooth; gonostylus narrow and transverse, posterior inner corner produced into thumb-shaped process, anterior inner corner with long curved hook extending as far back as posterior margin; malar area as long as pedicel and first flagellar segment combined (Asia) .... *B. (Diversobombus)*

—. Differing in character combinations from both *Megabombus* and *Diversobombus;* volsella at apex produced at most into a small hook or serrate, rounded lobe (in *Senexibombus* somewhat similar to *Megabombus* but penis valve serrate only on recurved distal quarter) 26

26(25). Length of third flagellar segment 1.5 to 1.9 times length of first segment 27

—. Length of third flagellar segment 1.0 to 1.3 times length of first segment 29

27(26). Second flagellar segment less than 0.7 times as long as first; spatha very narrow (Malaysia to Philippines) .... *B. (Senexibombus)*

—. Second flagellar segment as long as or longer than first segment; spatha very broad with convergent sides 28

28(27). Volsella very wide, not extending far beyond outer part of gonostylus, on inner side proximally with very long acute hook, inner edge behind hook straight-truncate, somewhat serrate, with long bristles; gonostylus sclerotized, with elongate outer lobe set in oblique plane, on inner side at a lower level produced into two very long acute processes directed obliquely forward and backward respectively; penis valve with feebly serrate outer flange apically, with a tooth beneath; malar area about as long as first and second flagellar segments combined (metasoma very closely punctured) (Asia) .... *B. (Tricornibombus)*

—. Volsella very long, broadly digitiform, inner side near center produced into strong, parallel-sided process with expanded end with sharp angles; gonostylus with large, outer part pale and submembranous, generally transverse but posteriorly produced on inside into rounded lobe, before this deeply emarginate and then produced into a large subcircular lobe (mainly in a vertical plane) with its dorsal edge serrate and the whole attached to gonostylus by narrow stalk; penis valve narrow, simply pointed at apex, not toothed beneath; malar area as long as second flagellar segment (palearctic) .... *B. (Laesobombus)*

29(26). Penis valve at end somewhat hooked inward, with two large teeth on outer side, midpoint beneath with bifid or trifid tooth (holarctic) .... *B. (Subterraneobombus)*

—. Penis valve at end curved or hooked outward, or pointed, not toothed, or in one species serrate 30

30(29). Malar area normally a little longer than combined lengths of first and second flagellar segments; penis valve curved outward at end [except in *B. (Fervidobombus) brevivillus* Franklin], this end-piece serrate or truncate, with central tooth beneath; gonocoxite with rather well-defined inner impressions 31

—. Malar area a little shorter than combined lengths of first

and second flagellar segments; penis valve with small, pointed external lobe, or pointed, or serrate; gonocoxite without distinct inner impressions 32
31(30). Mandible without beard; gonostylus on inside produced obliquely downward and backward as twisted anterior plate; volsella with apex produced on inside to small lobe with small tooth; volsellar region with no conspicuous bristles (palearctic) .... *B. (Rhodobombus)*
—. Mandible with beard; gonostylus on inside usually produced into a vertical platelike anterior lamella and a wider, more dorsal, rounded process (details vary considerably among species); volsella with apex produced on inside to small hook or angular process; volsellar region with dense bristles (Western Hemisphere) ....
.... *B. (Fervidobombus)*
32(30). Volsella wide but not very long, apex more or less pointed, center of inner edge produced into spike or at least narrow, truncate process, pubescence widespread but not dense; penis valve variable but with central tooth beneath (palearctic) .... *B. (Thoracobombus)*
—. Volsella long and broad, finger-shaped, center of inner edge produced as wide lobe defined at each end by small tooth or else simple, inner and often ventral surface with dense, quite long pubescence; penis valve simple or emarginate beneath, with no distinct tooth 33
33(32). Volsella with center of inner edge produced into wide lobe defined at each end by small tooth; penis valve simply pointed; fringe of hind tibia long (palearctic) ....
.... *B. (Mucidobombus)*
—. Volsella with inner edge straight or feebly concave, with no lobe or process; penis valve at end hardly acute, with slight outer, feebly serrate flange; fringe of hind tibia short (palearctic) .... *B. (Eversmannibombus)*

Key to the Subgenera of *Bombus* (Females)
(Modified from Richards, 1968, with additions based on Williams, 1991. This key is based primarily on gynes and may not work for some workers.)

1. Hind tibia convex and hairy on outer surface, corbicula thus absent (Fig. 119-9b), and rastellum absent; hind basitarsus without auricle; apex of metasoma curved downward; S6 with lateral carina; worker caste absent (holarctic) .... *B. (Psithyrus)*
—. Hind tibia with corbicula (Fig. 119-9a) and rastellum; hind basitarsus with strong auricle; apex of metasoma not curved downward; S6 without carinae; worker caste usually present 2
2(1). Apex of mandible with six teeth; hind basitarsus proximally near ventral margin often with a number of bristles almost as long as corbicular bristles (palearctic) ......
.... *B. (Alpigenobombus)*
—. Apex of mandible with one or two small dorsal teeth and sometimes a small preapical notch setting off an apical tooth; hind basitarsus with no bristles as long as corbicular bristles (except in *Mendacibombus* and *Pressibombus*)
.... 3
3(2). Middle basitarsus with posterior apical angle obtuse, more or less rounded [except for some *Sibiricobombus*, most of which have ocellocular distance equal to about two ocellar diameters and differ further from *Subterraneobombus*, which they often most resemble, in having whole surface of auricle of hind basitarsus densely hairy; *B. (Melanobombus) tanguticus* Morawitz has apical angle of middle basitarsus produced but has a strong, bare subcircular boss on T6 such as is not found in alternative] 4
—. Midsle basitarsus with posterior apical angle right-angular or acute and usually coarsely bristled [see exceptions in first alternative of this couplet] 22
4(3). Outer surface of hind tibia densely reticulate, dull, some long bristles arising from upper half of corbicular disc; malar area very elongate; first flagellar segment fully three times as long as median width (Eurasian mountains) .... *B. (Mendacibombus)*
—. Outer surface of hind tibia less coarsely reticulate, rarely if ever so dull, long bristles confined to corbicular margins except very near base, or, *if* bristles more widespread in corbicula, then malar area transverse; first flagellar segment nearly always shorter 5
5(4). Malar area longer than combined lengths of pedicel and first flagellar segment; first flagellar segment about four times as long as broad; middle basitarsus with posterior apical angle somewhat produced (some species difficult to separate from some *Subterraneobombus;* see couplet 3) (palearctic) .... *B. (Sibiricobombus)* (in part)
—. Malar area clearly shorter, or, *if not,* then first flagellar segment about 2.5 times as long as broad; middle basitarsus with posterior apical angle quite obtuse 6
6(5). Corbicular hairs unusually dense, tibial surface between them covered throughout with rather sparse but quite distinct, very short feathered hairs (southeast Asia to Java and Taiwan) .... *B. (Rufipedibombus)*
—. Corbicular hairs less dense, tibial surface between them without short feathered hairs [except in a few *Melanobombus* (couplet 8)] 7
7(6). Frons with very large unpunctured areas, most of area for some distance in front of ocelli being unpunctured and only narrow band of punctures between ocelli and eyes; area immediately behind ocelli also unpunctured; ocellocular distance equal to three ocellar diameters (Himalayas to Southeast Asia) .... *B. (Orientalibombus)*
—. Frons without large unpunctured area, especially in front of ocelli, or else densely punctured right up to ocelli posteriorly; ocelli closer to eyes 8
8(7). T6 with a bare, convex, more or less rounded boss not divided by furrow; hind tibia with dorsal apical inner corner not or rarely somewhat produced; hind basitarsus as a rule with unusually dense, short, feathery hairs (palearctic) .... *B. (Melanobombus)*
—. T6 without a bare, convex, rounded boss or with boss divided by furrow; hind tibia with dorsal apical inner corner more or less strongly, angulately produced [except in *Kallobombus* and a few *Pyrobombus*—e.g., *Bombus pratorum* (Linnaeus) and *atrocinctus* Smith]; hind basitarsus rarely so densely haired 9
9(8). Whole corbicular surface of hind tibia with scattered but quite numerous, short, unbranched hairs; frons crossed by wide transverse band of microscopic punctures across unpunctured area; T6 with raised boss (almost as in *Melanobombus*) divided by deep, well-defined furrow (South America) .... *B. (Coccineobombus)*
—. Hind tibia with at least a considerable distal corbicular

area bare; frons without band of microscopic punctures across unpunctured area; T6 without deep, well-defined furrow 10

10(9). Malar area distinctly transverse; either a definite band of close punctures along inner margin of eye, *or* ocellocular distance equal to about two ocellar diameters, *or* both 11

—. Malar area variable, i.e., elongate, quadrate, or transverse (if distinctly transverse as in *Pyrobombus,* then without fine punctures along inner eye margin *and* ocellocular distance equal to three ocellar diameters) 16

11(10). Ocellocular distance equal to three ocellar diameters, *and* corbicular surface of hind tibia entirely bare and shining; mandible with strong notch above apical tooth; clypeus strongly punctured on almost whole surface (holarctic) .... *B. (Bombus s. str.)*

—. Ocellocular distance not more than two ocellar diameters, *or* proximal one-half or one-third of corbicular surface bristly and whole surface more or less strongly reticulate, *or* both; mandible with notch above apical tooth weak or absent; clypeus sometimes with some sparse, large punctures, but *if* punctures are close, then they are small 12

12(11). Hind tibia with corbicular surface bare (North and Central America) 13

—. Hind tibia with proximal one-half or one-third of corbicular surface bristly (Central and South America) 15

13(12). Frons with rather strong, close punctures along inner margin of eye; clypeus elongate, impressions feeble, finely and closely punctured; labral furrow deep and wide, nearly as wide as length of first flagellar segment (North America) .... *B. (Separatobombus)*

—. Frons rather sparsely punctured all around ocelli except for a densely punctured area immediately behind them, unpunctured areas large and ill-defined, without band of fine punctures near eye (although in *Crotchiibombus* fine punctures may be seen, but not close to eye margin); clypeus short, impressions with coarse punctures; labral furrow deep and narrower 14

14(13). Malar area clearly shorter than first flagellar segment, which is less than twice length of second; clypeus more coarsely though shallowly punctured, impressions weaker; labral tubercle more convex but less angular, furrow deep and wider than length of first flagellar segment; corbicular hairs shorter than half tibial width and dense (North America) .... *B. (Fraternobombus)*

—. Malar area clearly longer than first flagellar segment, which is nearly twice length of second; clypeus more finely punctured, impressions stronger; labral tubercle less raised but more angular at inner end, furrow deeper but much narrower than length of first flagellar segment; corbicular hairs mostly longer than half tibial width and less dense (California, Mexico) .... *B. (Crotchiibombus)*

15(12). Frons rather closely punctured, with large but well-defined unpunctured areas and a band of fine punctures along inner margin of eye; ocellocular distance equal to three ocellar diameters; ocelli a little in front of postocular tangent; clypeus swollen, with lower third flattened, with numerous, mostly rather large punctures and apical impressions; labral furrow narrow; hind basitarsus not unusually bristly (western South America) .... *B. (Rubicundobombus)*

—. Frons with large unpunctured or very sparsely punctured area in front of and around ocellus, without especially defined unpunctured areas and without band of punctures along inner margin of eye; ocellocular distance equal to two ocellar diameters; ocelli well in front of postocular tangent; clypeus strongly swollen, sometimes somewhat flattened ventrally, little or moderately punctured, apical impressions feeble; labral furrow wide or very wide; hind basitarsus with bristles on outer surface longer and more numerous than usual (Central and South America) .... *B. (Robustobombus)*

16(10). Malar area distinctly longer than broad, at least as long as combined lengths of pedicel and first flagellar segment, often as long as first and second flagellar segments taken together (arctic and alpine, holarctic) .... *B. (Alpinobombus)*

—. Malar area transverse, quadrate or scarcely longer than broad, at most a little longer than first flagellar segment 17

17(16). Mandible with no notch above apical tooth but with ventral apical angle produced as small process (South America) .... *B. (Funebribombus)*

—. Mandible sometimes with notch above apical tooth, ventral apical angle not produced as short process 18

18(17). Labral tubercles little raised and much rounded, furrow shallow and ill-defined; ocellocular distance about 2.5 times ocellar diameter or less (ocelli well in front of postocular tangent); first flagellar segment as long as or slightly longer than combined lengths of second and third segments 19

—. Labral tubercles more or less raised and flattened, inner ends more or less angled, furrow deeper and well-defined; lateral ocellus more widely separated from eye (except in some *Cullumanobombus*); first flagellar segment clearly shorter than combined lengths of second and third segments 20

19(18). Malar area a little longer than broad and a little longer than first flagellar segment; clypeus with wide flattened disc; frons mostly closely and rather finely punctured, unpunctured areas well-defined, a wide band of close fine punctures along inner margin of eye (Europe) .... *B. (Confusibombus)*

—. Malar area roughly quadrate, about as long as first flagellar segment; clypeus long, strongly swollen; frons moderately strongly and closely punctured, unpunctured areas large and ill-defined, a narrow band of rather fine sculpture along inner margin of eye (North America) .... *B. (Bombias)*

20(18). Ocellocular distance somewhat or distinctly less than three ocellar diameters; ocelli well in front of postocular tangent (holarctic) .... *B. (Cullumanobombus)*

—. Ocellocular distance equal to fully three ocellar diameters; ocelli scarcely in front of postocular tangent 21

21(20) [triplet]. Clypeus rather strongly and evenly punctured, impressions strong but not more closely punctured; frons rather closely punctured, unpunctured areas small and well-defined, a wide band of fine sculpture along inner margin of eye; mandible without preapical notch; hind basitarsus with sparse pubescence and no long bristles (palearctic) .... *B. (Kallobombus)*

—. Clypeus, except impressions, largely unpunctured [more strongly and closely punctured in *B. lapponicus* (Fabri-

cius)]; frons mostly rather sparsely punctured, unpunctured areas ill-defined, fine punctures along inner margin of eye absent, margin largely shining; mandible with well-marked notch above apical tooth; hind basitarsus not usually very densely haired, without long bristles ....B. (*Pyrobombus, Festivobombus*)
—. Clypeus with fairly numerous scattered punctures, mostly small but some large; frons neither closely nor coarsely punctured, unpunctured areas ill-defined, not large; fine punctures along inner margin of eye absent, margin largely shining; mandible with feeble notch above apical tooth; hind basitarsus densely pubescent along lower edge for its whole length and its disc in part with long bristles (Himalayas) .... B. (*Pressibombus*)
22(3) [triplet]. Malar area twice as long as first flagellar segment 23
—. Malar area less than 1.3 times as long as first flagellar segment 25
—. Malar area 1.4 to 1.8 times as long as first flagellar segment 29
23(22). First flagellar segment a little shorter than combined lengths of second and third segments (10.5:11.0); furrow between labral tubercles narrower; middle basitarsus acutely spinosely produced at apex (palearctic) .... B. (*Megabombus*)
—. First flagellar segment clearly shorter than combined lengths of second and third segments (at most ll.5:13.0); furrow between labral tubercles wider; middle basitarsus with apical projection wider, scarcely spinose 24
24(23). Inner dorsal angle of hind tibia not or scarcely produced apically; much of clypeus rather strongly and closely punctured, midline on upper third with several rows of fine punctures (Western Hemisphere) .... B. (*Fervidobombus*) (in part)
—. Inner dorsal angle of hind tibia pointed apically though point short and broad; clypeus finely or little punctured, midline dorsally without rows of punctures (holarctic) .... B. (*Subterraneobombus*) (in part)
25(22). Ocellocular distance about 1.6 to 1.8 ocellar diameters; ocelli well in front of postocular tangent 26
—. Ocellocular distance about three ocellar diameters; ocelli just in front of postocular tangent 27
26(25). Impunctate area of frons small, ill-defined, punctured area reaching anterior margins of ocelli, a few coarse punctures between lateral ocellus and eye (Mesoamerica) ....B. (*Dasybombus*)
—. Impunctate area of frons larger, well-defined, large area in front of ocelli impunctate, only few minute punctures between lateral ocellus and eye (Mesoamerica) .... B. (*Brachycephalibombus*)
27(25). Frons with large unpunctured area separated from eye by narrow band of close, fine punctures (palearctic) .... B. (*Laesobombus*)
—. Frons with small unpunctured area separated from eye by wide band of fine punctures 28
28(27). Middle basitarsus broadly produced at apex; band of fine punctures along inner edge of eye not spreading over unpunctured area of frons; labral lamella not prominent; malar area with many fine punctures (palearctic) .... B. (*Eversmannibombus*)
—. Middle basitarsus spinosely produced at apex; band of fine punctures along inner edge of eye spreading halfway across unpunctured area of frons; labral lamella considerably thickened; malar area unpunctured (northeastern Asia) .... B. (*Exilobombus*)
29(22). Metasomal terga with very close, coarse punctures, except T6, which has coarse granules (Asia) .... B. (*Tricornibombus*)
—. Metasomal terga without close, coarse punctures 30
30(29). Frons with no band of close, fine punctures along inner margin of eye (Malaysia to Philippines)................ .... B. (*Senexibombus*)
—. Frons with band of close, fine punctures along inner margin of eye 31
31(30). Malar area very long; first flagellar segment about four times as long as broad; auricle of hind basitarsus with dense brown pile even on surface not apposed to hind tibia (palearctic) .... B. (*Sibiricobombus*) (in part)
—. Malar area shorter; first flagellar segment shorter; auricle with dense pile, if present, limited to surface apposed to end of hind tibia (and present only in a few species of *Subterraneobombus* that are similar to *Sibiricobombus*) 32
32(31). Middle basitarsus acutely spinosely produced 33
—. Middle basitarsus acutely but not spinosely produced 37
33(32). Malar area about as long as combined lengths of pedicel and first two flagellar segments 34
—. Malar area not longer than combined lengths of first and second flagellar segments... 35
34(33). Third flagellar segment clearly longer than second, which is transverse rather than quadrate; hind basitarsus little produced apically (Asia)............. B. (*Diversobombus*)
—. Third flagellar segment very little longer than second, which is at least quadrate; hind basitarsus distinctly produced apically (palearctic) .... B. (*Thoracobombus*)
35(33). Large area in front of median ocellus unpunctured; lateral unpunctured areas large but well-defined, band of fine sculpture along inner margin of eye narrow; first flagellar segment slightly longer than combined length of second and third segments; inner dorsal angle at apex of hind tibia not or scarcely produced (palearctic) .... B. (*Rhodobombus*) (in part)
—. Area in front of median ocellus coarsely but not closely punctured, lateral unpunctured areas of moderate size and fairly well-defined, band of fine sculpture along inner margin of eye wide or rather wide; first flagellar segment shorter than or as long as combined lengths of second and third segments; inner dorsal angle at apex of hind tibia more acutely produced 36
36(35). Malar area as long as combined lengths of first and second flagellar segments; apical impressions of clypeus rather strong, with close, moderately coarse punctures; labral tubercles flattened, furrow shallow (palearctic) .... B. (*Thoracobombus*) (in part)
—. Malar area not quite as long as combined lengths of pedicel and first flagellar segment; apical impressions of clypeus weak, with narrow deeper strip of close, moderately coarse punctures; labral tubercles somewhat raised and angular at inner ends, furrow moderately deep (palearctic) .... B. (*Mucidobombus*)
37(32). Clypeus swollen with no furrow or lines of punc-

tures on upper third of midline; hind tibia with inner dorsal apical angle sharp, although projection rather broad (holarctic) .... *B. (Subterraneobombus)* (in part)
—. Clypeus with slight furrow or distinct lines of punctures on upper third of midline; hind tibia with inner dorsal apical angle not or scarcely produced 38

38(37). Clypeus with dispersed, sparse, fine punctures; middle basitarsus more spinosely produced; frons less punctured, with a larger unpunctured area in front of and at sides of ocelli (palearctic) .... *B. (Rhodobombus)* (in part)
—. Clypeus generally with closer and coarser punctures; middle basitarsus often not very distinctly produced; frons more punctured, with unpunctured areas smaller (Western Hemisphere) .... *B. (Fervidobombus)* (in part)

Bombus / Subgenus *Alpigenobombus* Skorikov

Alpigenobombus Skorikov, 1914a: 128. Type species: *Alpigenobombus pulcherrimus* Skorikov, 1914 = *Bombus kashmirensis* Friese, 1909, by designation of Williams, 1991. [See Williams (1991) for explanation of other supposed type species.]
Alpigenibombus Skorikov, 1938a: 145, unjustified emendation of *Alpigenobombus* Skorikov, 1914.
Bombus (Mastrucatobombus) Krüger, 1917: 66. Type species: *Bombus mastrucatus* Gerstaecker, 1869 = *Bombus wurflenii* Radoszkowski, 1859, monobasic.
Nobilibombus Skorikov, 1933a: 62. Invalid because no type species was designated. [For subsequent designations in synonymy, see Michener (1997b) and Williams (1991).]
Nobilibombus Richards, 1968: 216, 222. Type species: *Bombus nobilis* Friese, 1904, by original designation.

■ *Alpigenobombus* occurs from northern Spain and Norway east to the Himalayas and China. Six species were listed by Williams (1998).

Bombus / Subgenus *Alpinobombus* Skorikov

Bombus (Alpinobombus) Skorikov, 1914a: 122. Type species: *Apis alpinus* Linnaeus, 1758, by designation of Frison, 1927: 66.
Alpinibombus Skorikov, 1937: 53, unjustified emendation of *Alpinobombus* Skorikov, 1914.

■ This subgenus occurs in the high arctic almost as far north as there is land, and southward in arctic-alpine habitats as far as California, New Mexico, and the European Alps. Five species were listed by Williams (1994, 1998).
Bombus (Alpinobombus) arcticus (Quenzel) (= *B. hyperboreus* Schönherr) is a workerless social parasite in nests of *B. (A.) polaris* Curtis, at least in some areas (K. Richards, 1973).

Bombus / Subgenus *Bombias* Robertson

Bombias Robertson, 1903a: 176. Type species: *Bombias auricomus* Robertson, 1903, by original designation.
Nevadensibombus Skorikov, 1922b: 149. Type species: *Bombus nevadensis* Cresson, 1874, by designation of Frison, 1927: 64.
Bombus (Boopobombus) Frison, 1927: 62. Type species: *Bombias auricomus* Robertson, 1903, by designation of Williams, 1994: 339. [A sectional name that, according to Article 10(e) of the Code (3rd ed.) must be treated as a subgenus.]

■ This subgenus is widespread across North America from southern Canada south to Texas and California, USA, and the state of Hidalgo, Mexico. The two closely related forms of *Bombias* are *Bombus nevadensis* Cresson and its subspecies or close relative, *B. auricomus* (Robertson).
Bombias is unusual among bumble bees in that each egg is laid in a separate small cell; thus eggs are not laid in clumps (Michener, 1974a) as is usual for *Bombus* species.

Bombus / Subgenus *Bombus* Latreille s. str.

Bremus Jurine, 1801: 164. Type species: *Apis terrestris* Linnaeus, 1758, by designation of Morice and Durrant, 1915: 428. Invalidated by Commission Opinion 135 (1939).
Bombus Latreille, 1802a: 437. Type species: *Apis terrestris* Linnaeus, 1758, monobasic. [Westwood, 1840a: 86, designated *Apis muscorum* Linnaeus, 1758, as type species; it was not an originally included species.]
Bremus Panzer, 1804a: 19. Type species: *Apis terrestris* Linnaeus, 1758, by designation of Benson, Ferrière, and Richards, 1937: 93. [For other designations and a comment on the date, see Michener, 1997b.]
Bombus (Leucobombus) Dalla Torre, 1880: 40. Type species: *Apis terrestris* Linnaeus, 1758, by designation of Sandhouse, 1943: 564.
Bombus (Terrestribombus) Vogt, 1911: 55. Type species: *Apis terrestris* Linnaeus, 1758, by designation of Frison, 1927: 67.

■ *Bombus* s. str. ranges from the Canary Islands, Madeira, Spain, and Britain to Japan, south to the Himalayas, Southeast Asia, and Taiwan, north to Norway, Finland, and northeastern Siberia; in North America, Alaska to Nova Scotia, south to California to Florida. Ten species were listed by Williams (1998).

Bombus / Subgenus *Brachycephalibombus* Williams

Bombus (Brachycephalibombus) Williams, 1985: 247. Type species: *Bombus brachycephalus* Handlirsch, 1888, by original designation.

■ This subgenus occurs from Nayarit to San Luis Potosí, Mexico, south to Honduras. Two species were included by Williams (1998).
Of the two species of this subgenus, the rare *Bombus haueri* Handlirsch has been placed in *Crotchiibombus* by Laboulge (1990), who did not know the male, and by G. Chavarría (in litt., 1995). Williams' (1994) analysis, however, shows its close relationship to *B. brachycephalus* Handlirsch.

Bombus / Subgenus *Coccineobombus* Skorikov

Alpigenobombus (Coccineobombus) Skorikov, 1922b: 157. Type species: *Bombus coccineus* Friese, 1903, by designation of Sandhouse, 1943: 539.

■ This subgenus occurs from western Argentina and Bolivia north to Ecuador. Two species were listed by Williams (1994, 1998).

Bombus / Subgenus *Confusibombus* Ball

Bombus (Confusibombus) Ball, 1914: 78. Type species: *Bombus confusus* Schenck, 1859, monobasic.

Bombus (Sulcobombus) Krüger, 1917: 65. Type species: *Bombus confusus* Schenck, 1859, designated by Sandhouse, 1943: 502. [This was a sectional name, but is treated as a subgeneric name in view of the Code, 3rd ed., art 10(e).]

Confusobombus Skorikov, 1922b: 156. Type species: *Bombus confusus* Schenck, 1859, by designation of Richards, 1968: 214.

■ This subgenus contains a single species, *Bombus confusus* Schenck. It is widespread in Europe with a disjunct population in Central Asia.

Bombus / Subgenus *Crotchiibombus* Franklin

Bombus (Crotchiibombus) Franklin, 1954: 51. Type species: *Bombus crotchii* Cresson, 1878, by original designation.

■ This subgenus occurs in California, USA, and Baja California, Mexico. The only species is *Bombus crotchii* Cresson.

Bombus / Subgenus *Cullumanobombus* Vogt

Bombus (Cullumanobombus) Vogt, 1911: 57. Type species: *Apis cullumana* Kirby, 1802, by designation of Frison, 1927: 66.

Cullumanibombus Skorikov, 1938a: 145, unjustified emendation of *Cullumanobombus* Vogt, 1911.

Bremus (Rufocinctobombus) Frison, 1927: 78. Type species: *Bombus rufocinctus* Cresson, 1863, monobasic.

■ This subgenus ranges from Britain and Spain eastward across northern Eurasia and in North America from British Columbia to Nova Scotia, Canada, south to Illinois, Kansas, New Mexico, and California, USA, with a probably isolated population in Hidalgo, the state of México, and the Distrito Federal, Mexico. Four species were listed by Williams (1998).

Bombus / Subgenus *Dasybombus* Labougle and Ayala

Bombus (Dasybombus) Labougle and Ayala, 1985: 49. Type species: *Bombus macgregori* Labougle and Ayala, 1985, by original designation.

■ This subgenus occurs in Guerrero and Jalisco, Mexico, and in Guatemala. If Williams (1994, 1998) is correct in placing *Bombus handlirschi* Friese here, *Dasybombus* also occurs in Peru. G. Chavarría (in litt., 1995), however, considers *B. handlirschi* to be a *Rubicundobombus*. There are two species if *B. handlirschi* is included.

Bombus / Subgenus *Diversobombus* Skorikov

Bombus (Diversobombus) Skorikov, 1914b: 406. Type species: *Bombus diversus* Smith, 1869, by designation of Sandhouse, 1943: 546.

Diversibombus Skorikov, 1938b: 1, unjustified emendation of *Diversobombus* Skorikov, 1914.

■ This subgenus is found from Pakistan, India, and Malaysia to the Russian Pacific maritime provinces and Japan. Four species were listed by Williams (1998).

Bombus / Subgenus *Eversmannibombus* Skorikov

Agribombus (Eversmannibombus) Skorikov, 1938a: 145. Type species: *Mucidobombus eversmanniellus* Skorikov, 1922b = *Bombus eversmanni* Friese, 1911 (not *B. modestus eversmanni* Skorikov, 1910) = *Bombus persicus* Radoszkowski, 1881, by designation of Richards, 1968: 214.

■ This subgenus occurs in Eastern Europe and western Asia and contains a single species, *Bombus persicus* Radoszkowski.

Bombus / Subgenus *Exilobombus* Skorikov

Mucidobombus (Exilobombus) Skorikov, 1922b: 150. Type species: *Mucidobombus exil* [misprinted *exiln*] Skorikov, 1922, monobasic.

Megabombus (Exilnobombus) Milliron, 1973: 81, emendation of *Exilobombus*.

The male genitalia and other structures were illustrated by Tkalců (1974d), who rendered the specific name *exul*.

■ This subgenus occurs in eastern Siberia, Mongolia, and Pacific maritime Russian provinces. The single species is *Bombus exil* (Skorikov).

Bombus / Subgenus *Fervidobombus* Skorikov

Fervidobombus Skorikov, 1922b: 123, 153. Type species: *Apis fervida* Fabricius, 1798, by designation of Frison, 1927: 69.

Bombus (Digressobombus) Laverty, Plowright, and Williams, 1984: 1051. Type species: *Megabombus digressus* Milliron, 1962, by original designation.

■ This rather variable subgenus ranges from southern Canada south through the Americas to Tierra del Fuego, sparsely inhabiting the moist tropics as well as montane and temperate areas. Williams (1998) listed 20 species. *Bombus niger* Franklin (= *B. atratus* Franklin, 1913, not Friese, 1911) and perhaps other tropical species form colonies that survive for several seasons, unlike most *Bombus* (Zucchi, 1973; reviewed by Michener, 1974a).

Bombus / Subgenus *Festivobombus* Tkalců

Atrocinctob[ombus] Skorikov, 1933b: 244, nomen nudum.

Pyrobombus (Festivobombus) Tkalců, 1972: 27. Type species: *Bombus festivus* Smith, 1861, by original designation.

■ This subgenus occurs in northern India and Sikkim to China. There is one species, *Bombus festivus* Smith.

Bombus festivus Smith was included in *Pyrobombus* by Richards (1968) who, however, pointed out its differences from *Pyrobombus* using the synonyous specific name *atrocinctus* Smith.

Bombus / Subgenus *Fraternobombus* Skorikov

Alpigenobombus (Fraternobombus) Skorikov, 1922b: 156. Type species: *Apathus fraternus* Smith, 1854, by designation of Frison, 1927: 63.

■ This North American subgenus occurs from New Jersey to Florida, west to North Dakota to New Mexico, USA, and Chihuahua, Mexico. The single species is *Bombus fraternus* (Smith).

Bombus / Subgenus *Funebribombus* Skorikov

Alpigenobombus (Funebribombus) Skorikov, 1922b: 157. Type species: *Bombus funebris* Smith, 1854, monobasic.

■ This subgenus occurs in the Andean countries of South America from Colombia to Chile. Williams (1998) listed two species.

Bombus / Subgenus *Kallobombus* Dalla Torre

Bombus (Kallobombus) Dalla Torre, 1880: 40. Type species: *Apis soroeensis* Fabricius, 1776, by designation of Sandhouse, 1943: 561.
Bombus (Callobombus) Dalla Torre, 1896: 503, unjustified emendation of *Kallobombus* Dalla Torre, 1880.
Bombus (Soroeensibombus) Vogt, 1911: 63. Type species: *Apis soroeensis* Fabricius, 1776, monobasic.

■ This subgenus occurs in Europe and western Asia. The only species is *Bombus soroeensis* (Fabricius).

Bombus / Subgenus *Laesobombus* Krüger

Agrobombus (Laesobombus) Krüger, 1920: 350. Type species: *Bombus laesus* Morawitz, 1875, monobasic.
Agrobombus (Laesobombus) Skorikov, 1922b: 150. Type species: *Bombus laesus* Morawitz, 1875, monobasic. [Probably proposed without knowledge of Krüger's earlier use of the same name.]
Laesibombus Skorikov, 1938a: 145, unjustified emendation of *Laesobombus* Krüger, 1920.

This subgenus is sometimes included in *Thoracobombus* but may constitute a separate clade, as indicated by Williams' (1994) phylogenetic study.

■ *Laesobombus* occurs from Spain and Morocco east to China. Williams (1994) listed four species, but synonymy reduced the number to one (Williams, 1998).

Bombus / Subgenus *Megabombus* Dalla Torre

Bombus (Megabombus) Dalla Torre, 1880: 40. Type species: *Bombus ligusticus* Spinola, 1805 = *Apis argillacea* Scopoli, 1763, monobasic.
Bombus (Megalobombus) Schulz, 1906: 267, unjustified emendation of *Megabombus* Dalla Torre, 1880.
Bombus (Hortobombus) Vogt, 1911: 56. Type species: *Apis hortorum* Linnaeus, 1761, by designation of Sandhouse, 1943: 559.
Hortibombus Skorikov, 1938a: 146, unjustified emendation of *Hortobombus* Vogt, 1911.
Bombus (Odontobombus) Krüger, 1917: 61. Type species: *Apis argillacea* Scopoli, 1763, by designation of Williams, 1994: 339. [A sectional name, to be treated as a subgenus according to the Code, 3rd ed., art. 10(e).]

■ *Megabombus* is found from the Azores, Europe, and northern Africa to Japan and eastern Siberia. *Bombus ruderatus* (Fabricius) has been introduced for clover pollination into New Zealand and Chile. From Chile it has spread to Río Negro province, Argentina. (*B. ruderatus* was recently redescribed as a new species from Chile!) *Megabombus* is not known in Southeast Asia, but there is a record from Sumatra. Williams (1998) listed 14 species.

Williams (1994) found that species of *Megabombus, Senexibombus,* and *Diversobombus* fall in a single clade. Perhaps these subgenera should be united, but his cladograms show only a single synapomorphy uniting the three taxa.

Bombus / Subgenus *Melanobombus* Dalla Torre

Bombus (Melanobombus) Dalla Torre, 1880: 40. Type species: *Apis lapidaria* Linnaeus, 1758, by designation of Sandhouse, 1943: 569.
Bombus (Lapidariobombus) Vogt, 1911: 58. Type species: *Apis lapidaria* Linnaeus, 1758, by designation of Sandhouse, 1943: 562.
Kozlovibombus Skorikov, 1922b: 152. Type species: *Bombus kozlovi* Skorikov, 1909 = *Bombus keriensis* Morawitz, 1886, by designation of Sandhouse, 1943: 561.
Bombus (Kozlowibombus) Bischoff, 1936: 10, unjustified emendation of *Kozlovibombus* Skorikov, 1922.
Lapidariibombus Skorikov, 1938a: 145, unjustified emendation of *Lapidariobombus* Vogt, 1911.
Bombus (Tanguticobombus) Pittioni, 1939b: 201. Type species: *Bombus tanguticus* Morawitz, 1886, by original designation.

■ This subgenus is widespread from Britain and Morocco to the Himalayas, China, Taiwan, and eastern Siberia. The 14 species were listed by Williams (1998).

Bombus / Subgenus *Mendacibombus* Skorikov

Bombus (Mendacibombus) Skorikov, 1914a: 125. Type species: *Bombus mendax* Gerstaecker, 1869, by designation of Sandhouse, 1943: 572.

■ This subgenus is found in the high mountains of Europe and Asia. Williams (1998) listed 12 species. Williams (1994) recognized two clades in this paraphyletic subgenus.

Nests, so far as is known, differ from those of other *Bombus* in that cocoons are not long-lasting, but are destroyed and not used for food storage. Storage is in cells that tend to be clustered, hexagonal, and away from the immediate vicinity of brood (Haas, 1976).

Bombus / Subgenus *Mucidobombus* Krüger

Mucidobombus Krüger, 1920: 350. Type species: *Bombus mucidus* Gerstaecker, 1869, monobasic. Krüger gave no subgeneric characters, but according to the Code, 3rd ed., art. 12(b)(5), the name is nonetheless valid.
Mucidobombus Skorikov, 1922b: 149. Type species: *Bombus mucidus* Gerstaecker, 1869, by designation of Sandhouse, 1943: 574. [Homonym and junior synonym of *Mucidobombus* Krüger.]

■ This subgenus occurs in Europe and western Asia. The only species is *Bombus mucidus* Gerstaecker.

Bombus / Subgenus *Orientalibombus* Richards

Bombus (Orientalibombus) Richards, 1929: 378. Type species: *Bombus orientalis* Smith, 1854 = *B. haemorrhoidalis* Smith, 1852, by original designation.
Bombus (Orientalobombus) Kruseman, 1952: 102, unjustified emendation of *Orientalibombus* Richards, 1929.

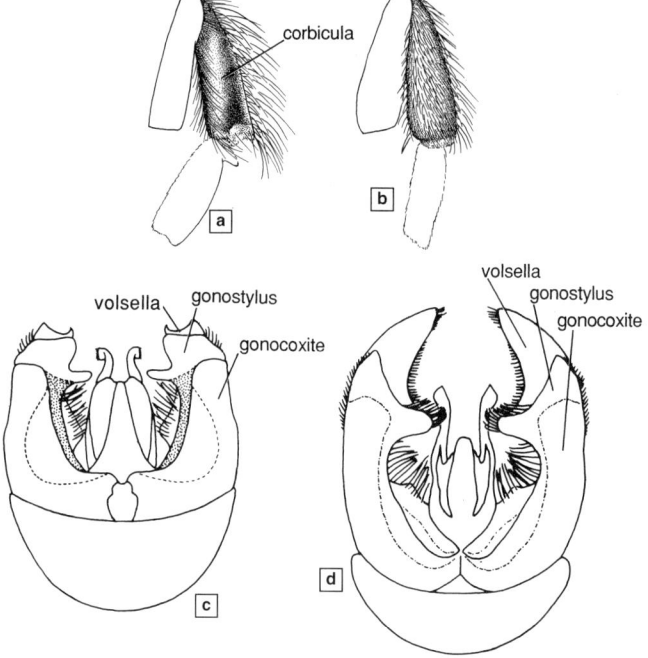

Figure 119-9. Distinguishing features of the subgenus *Psithyrus*. **a**, Hind tibia of worker of *Bombus (Pyrobombus) impatiens* Cresson; **b**, Same, of female of *B. (Psithyrus) fernaldae* Franklin; **c**, Male genitalia of *B. (Fervidobombus) fervidus* (Fabricius), dorsal view; **d**, Same, of *B. (Psithyrus) variabilis* (Cresson). From Michener, McGinley, and Danforth, 1994.

■ This subgenus occurs from the Himalayas to Vietnam and southern China. Three species were listed by Williams (1998).

Bombus / Subgenus *Pressibombus* Frison

Bremus (Pressibombus) Frison, 1935: 342. Type species: *Bremus pressus* Frison, 1935, by original designation.

■ This species is from the Himalayas. The only species is *Bombus pressus* (Frison).

Williams (1994) united this subgenus with *Bombus* s. str.; *B. pressus* (Frison) may be a derived form of that subgenus. Williams (1998), however, retained *Pressibombus* as a subgenus.

Bombus / Subgenus *Psithyrus* Lepeletier

Psithyrus Lepeletier, 1833: 373. Type species: *Apis rupestris* Fabricius, 1793, by designation of Curtis, 1833: pl. 468.
Apathus Newman, 1834: 404 footnote, unjustified replacement for *Psithyrus* Lepeletier, 1833. Type species: *Apis rupestris* Fabricius, 1793, autobasic.
Bremus Kirby, 1837: 272 footnote (not Panzer, 1804). Type species: *Apis rupestris* Fabricius, 1793, by designation of Milliron, 1961: 59.
Psithyrus (Ashtonipsithyrus) Frison, 1927: 69. Type species: *Apathus ashtoni* Cresson, 1864, by original designation.
Psithyrus (Fernaldaepsithyrus) Frison, 1927: 70. Type species: *Psithyrus fernaldae* Franklin, 1911, monobasic.
Psithyrus (Allopsithyrus) Popov, 1931a: 136. Type species: *Apis barbutella* Kirby, 1802, by original designation.
Psithyrus (Eopsithyrus) Popov, 1931a: 134. Type species: *Apathus tibetanus* Morawitz, 1886, by original designation.
Psithyrus (Metapsithyrus) Popov, 1931a: 135. Type species: *Apis campestris* Panzer, 1801, by original designation.
Psithyrus (Ceratopsithyrus) Pittioni, 1949: 270. Type species: *Psithyrus klapperichi* Pittioni, 1949 = *Psithyrus cornutus* Frison, 1933, monobasic.
Citrinopsithyrus Thorp, 1983, *in* Thorp, Horning, and Dunning, 1983: 50. Type species: *Apathus citrinus* Smith, 1854, by original designation. [Substitute for *Laboriopsithyrus* Frison, 1927, the type species of which turns out to be an anthophorine bee of the genus *Habropoda*].

Psithyrus, as explained elsewhere, is usually regarded as a genus of social parasites. Lack of a worker caste is its most notable feature, although duplicated in one or two unrelated species of *Bombus*. As indicated by Milliron (1961) and shown by both Ito (1983, 1985) and Williams (1985), the subgenera sometimes recognized within *Psithyrus* are even more similar to one another than are those of nonparasitic *Bombus*; I consider them synonyms of *Psithyrus*. They were treated in detail by Popov (1931a), who illustrated male genitalia, hidden sterna, and other structures.

■ *Psithyrus* is a holarctic subgenus found in the areas where nonparasitic *Bombus* are most abundant, but does not range into the high arctic; in the New World, it is not known south of Guatemala and Honduras. Williams (1994, 1998) lists 29 species. Species of eastern Asia were revised by Maa (1948).

Bombus / Subgenus *Pyrobombus* Dalla Torre

Bombus (Pyrobombus) Dalla Torre, 1880: 40. Type species: *Apis hypnorum* Linnaeus, 1758, monobasic.
Bombus (Pyrrhobombus) Dalla Torre, 1882: 28, unjustified emendation of *Pyrobombus* Dalla Torre, 1880.

Bombus (Pocilobombus) Dalla Torre, 1882: 23. Type species: *Bombus sitkensis* Nylander, 1848, by designation of Sandhouse, 1943: 589.

Bombus (Pratobombus) Vogt, 1911: 49. Type species: *Apis pratorum* Linnaeus, 1761, by designation of Frison, 1927: 67.

Pratibombus Skorikov, 1937: 59, unjustified emendation of *Pratobombus* Vogt, 1911.

Bombus (Anodontobombus) Krüger, 1917: 61. Type species: *Apis hypnorum* Linnaeus, 1758, by designation of Williams, 1991: 69. [A sectional name treated as a genus-group name in view of the Code, 3rd ed., art. 10(e).]

Bombus (Uncobombus) Krüger, 1917: 65. Type species: *Apis hypnorum* Linnaeus, 1758, by designation of Williams, 1991: 69. [A sectional name, treated as a genus-group name in view of the Code, 3rd ed., art. 10(e); for a note on the authorship, see Michener, 1997b.]

Bombus (Hypnorobombus) Quilis, 1927: 97. Type species: *Apis hypnorum* Linnaeus, 1758, monobasic.

Bombus (Lapponicobombus) Quilis, 1927: 19, 22, 63. Type species: *Apis lapponica* Fabricius, 1793, by designation of Milliron, 1961: 58.

■ This is the largest subgenus of *Bombus*. It ranges widely in Europe and Asia, from Iceland and Norway to Spain to eastern Siberia, south to Malaysia, Taiwan, and the Philippines; in North America, from Baffin Island and Newfoundland, Canada, to Alaska, and south to Florida and California, USA, and through Mexico and Central America to the mountains of western Panama. Forty-three species were listed by Williams (1998), about equally divided between the Eastern and Wstern hemispheres.

Bombus / Subgenus *Rhodobombus* Dalla Torre

Bombus (Rhodobombus) Dalla Torre, 1880: 40. Type species: *Bremus pomorum* Panzer, 1804, by designation of Sandhouse, 1943: 596.

Bombus (Pomobombus) Krüger, 1917: 65. Type species: *Bremus pomorum* Panzer, 1804, by designation of Sandhouse, 1943: 589.

Pomibombus Skorikov, 1938a: 145, unjustified emendation of *Pomobombus* Krüger, 1917.

■ This subgenus occurs from Europe to central Asia, south to Iran. Three species were listed by Williams (1994, 1998).

Bombus / Subgenus *Robustobombus* Skorikov

Alpigenobombus (Robustobombus) Skorikov, 1922b: 157. Type species: *Bombus robustus* Smith, 1854, by designation of Sandhouse, 1943: 597.

Voucellobombus Skorikov, 1922b: 123, 149. Type species: *Bombus voucelloides* Gribodo, 1891, monobasic.

■ This subgenus occurs from Costa Rica to Venezuela and south in the Andes to Argentina. Williams (1998) listed five species.

Bombus / Subgenus *Rubicundobombus* Skorikov

Fervidobombus (Rubicundobombus) Skorikov, 1922b: 154. Type species: *Bombus rubicundus* Smith, 1854, by designation of Sandhouse, 1943: 597.

■ This subgenus occurs in the Andean countries of South America. One species is *Bombus rubicundus* Smith (Williams, 1998); perhaps a second species is *B. handlirschi* Friese (see subgenus *Dasybombus*).

Bombus / Subgenus *Rufipedibombus* Skorikov

Rufipedibombus Skorikov, 1922b: 156. Type species: *Bombus rufipes* Lepeletier, 1836, monobasic.

Bombus (Rufipedobombus) Kruseman, 1952: 102, unjustified emendation of *Rufipedibombus*.

■ This subgenus occurs in Southeast Asia, southern China, south to Java, as well as in Taiwan. Two species were listed by Williams (1994).

Bombus / Subgenus *Senexibombus* Frison

Bremus (Senexibombus) Frison, 1930: 3. Type species: *Bombus senex* Vollenhoven, 1873, by original designation.

■ This subgenus is found from Sumatra to the Philippines and Taiwan. The four species were listed by Williams (1998).

Bombus / Subgenus *Separatobombus* Frison

Bremus (Separatobombus) Frison, 1927: 64. Type species: *Bombus separatus* Cresson, 1863 = *Apis griseocollis* DeGeer, 1773, by original designation.

■ This North American subgenus occurs from Quebec to British Columbia, Canada, south to Florida and northern California, USA. The two species (Williams, 1994, 1998) may not be parts of one clade.

Bombus / Subgenus *Sibiricobombus* Vogt

Bombus (Sibiricobombus) Vogt, 1911: 60. Type species: *Apis sibirica* Fabricius, 1781, by designation of Sandhouse, 1943: 599.

Sibiricibombus Skorikov, 1938a: 145, unjustified emendation of *Sibiricobombus* Vogt, 1911.

Bombus (Obertobombus) Reinig, 1930: 107. Type species: *Bombus oberti* Morawitz, 1883, monobasic.

Bombus (Obertibombus) Reinig, 1934: 167, unjustified emendation of *Obertobombus* Reinig, 1930.

■ This subgenus is found from Eastern Europe through northern Asia and south to the Himalayas. Williams (1998), listed seven species.

Williams (1994) found that *Obertobombus*, or at least one of its species, *Bombus morawitzi* Radoszkowski, is more closely related to *Melanobombus* than to *Sibiricobombus* proper. Williams (1998) retained *Obertobombus* as a subgenus.

Bombus / Subgenus *Subterraneobombus* Vogt

Bombus (Subterraneobombus) Vogt, 1911: 62. Type species: *Apis subterranea* Linnaeus, 1758, by designation of Frison, 1927: 68.

Subterraneibombus Skorikov, 1938a: 145, unjustified emendation of *Subterraneobombus* Vogt, 1911.

■ This subgenus is found from Europe and southwestern Asia to the Himalayas and the Pacific maritime provinces of Russia, and in North America across the continent in southern Canada, south to New Jersey, South Dakota, and the mountains of New Mexico and California, USA. Williams (1998) listed nine species. One species, *Bombus subterraneus* (Linnaeus), has been introduced into New Zealand for red clover seed production.

Bombus / Subgenus *Thoracobombus* Dalla Torre

Bombus (Thoracobombus) Dalla Torre, 1880: 40. Type species: *Apis sylvarum* Linnaeus, 1761, by designation of Sandhouse, 1943: 604.

Bombus (Chromobombus) Dalla Torre, 1880: 40. Type species: *Apis muscorum* Linnaeus, 1758, by designation of Sandhouse, 1943: 538.

Bombus (Agrobombus) Vogt, 1911: 52. Type species: *Apis agrorum* Fabricius, 1787 (not Schrank, 1781) = *Apis pascuorum* (Scopoli), 1763, by designation of Sandhouse, 1943: 523.

Agribombus Skorikov, 1938a: 145, unjustified emendation of *Agrobombus* Vogt, 1911.

Bombus (Ruderariobombus) Krüger, 1920: 350. Type species: *Apis ruderaria* Müller, 1776, by designation of Yarrow, 1971: 27.

Agrobombus (Adventoribombus) Skorikov, 1922a: 25. Type species: *Apis sylvarum* Linnaeus, 1761, by designation of Yarrow, 1971: 28. [See Yarrow, 1971, and Michener, 1997b, for comments on a subsequent designation.]

■ This subgenus occurs from Europe and northern Africa to Japan and eastern Siberia, probably not entering tropical areas. Williams (1998) listed 19 species.

Bombus (Thoracobombus) inexpectatus Tkalců from the European Alps is presumed to be a *Psithyrus*-like social parasite (Yarrow, 1970; Müller, 2006).

Bombus / Subgenus *Tricornibombus* Skorikov

Agrobombus (Tricornibombus) Skorikov, 1922b: 151. Type species: *Bombus tricornis* Radoszkowsky, 1888, monobasic.

■ *Tricornibombus* occurs in Central and northeastern Asia. Three species were listed by Williams (1998). Tkalců (1968) revised the subgenus.

Williams (1994) found species of this subgenus to fall in the same clade as *Thoracobombus*, on the basis of a single synapomorphy.

120. Tribe Meliponini

These are the stingless honey bees found in tropical and southern subtropical areas throughout the world. They are found in colonies ranging from a few dozen to 100,000 or more workers and are the only highly social bees other than the true honey bees, tribe Apini. Like the Apini and unlike all other bees, they live in "permanent" colonies and have morphologically as well as behaviorally very different female castes, queen and worker. There are several hundred species, an approximation to the real number being impossible because of the abundance of cryptic species, differing from their relatives only on the bases of seemingly trivial characters. Most genera in most areas have not been adequately analyzed for recognition of such forms; good starts have been made by Camargo (1980, for part of *Partamona*), Camargo and Moure (1994, 1996, for *Paratrigona* and *Geotrigona*), Pedro and Camargo (2003, for *Partamona* s. str.), and Sakagami (1978, for the *Tetragonula* group of *Trigona* subgenus *Heterotrigona*).

In many areas in tropical America, Meliponini are the most common bees and therefore presumably play a major role as pollinators of native vegetation. The many lists of bee species found in limited areas in Brazil and elsewhere in South America show Meliponini as prominent elements and the abundance of individuals often overshadows that of other bees. Because of their behavioral complexity and diversity, as well as their abundance and probable importance as pollinators, Meliponini are much studied in tropical America. Evidence is contained in the annals of a major meeting on bees in Brazil, in which nearly 40% of the 265 contributions were on Meliponini (Garófalo and Fritas, 2002).

The following characters elaborate on those listed in the key to tribes. The body is trigoniform to apiform (Pls. 9, 10), sparsely hairy or short-haired to moderately hairy like true honey bees, and 1.8 to 13.5 mm in length. The claws of females are simple, arolia are present, and hind tibial spurs are absent. The strigilis lacks a prong on the anterior side (Fig. 119-5i), the hind basitarsus is rather slender at the base, without an auricle (Fig. 120-11), and the hind tibia of the worker has a penicillum (Fig. 120-11); in all these features the Meliponini differ from other corbiculate Apidae. The wing venation is reduced, the marginal cell often being open apically, or at least the distal parts of its veins are much narrower than their basal parts, near the stigma (Fig. 120-1). The stigma is large to moderate-sized, vein r arising near the middle of the stigma. The prestigma is short, often almost absent. The second recurrent vein is absent; the first, when present, is short and often strongly angled near its anterior end (Fig. 120-1a). The first and second submarginal cells are often unrecognizable and at most weakly defined; the third is not defined. The hind wing has a well-developed jugal lobe one-third to two-thirds as long as the vannal lobe. The clypeus is flat, not protuberant. The maxillary palpus is minute, one-segmented. The male gonobase is absent or represented by a narrow ribbon or weak lateral sclerites. The gonostylus is long, usually slender and simple, and the volsella is absent; the penis valve is heavily sclerotized, tapering, curved, clawlike. S7 is reduced to a small plate, sometimes without the expected basolateral apodemal arms, and is best developed, so far as is known, in *Lestrimelitta* (Schwarz, 1948). S8 is a mere elongate longitudinal sclerite or apparently absent (Figs. 120-7 to 120-9). Reduction of these structures parallels that of the Apini but seems quite different and likely independent; for example, S8 is a longitudinal rudiment in Meliponini, a transverse rudiment in Apini. The sting and associated structures of the female are greatly reduced (Fig. 120-6). Michener (1990a) provided illustrations of male genitalia and female sting rudiments for all genera and subgenera; see also Figures 120-6 to 120-9. Male genitalia and other structures were also illustrated by Schwarz (1939a, 1948), Camargo (1980), Sakagami and Inoue (1985, 1987, 1989), and Camargo and Moure (1994, 1996). Camargo, Kerr, and Lopes (1967) gave an excellent, well-illustrated morphological account of *Melipona marginata* Lepeletier.

The larva lacks strong tubercles but has a small, conical, dark dorsolateral tubercle on each side of the first three to ten body segments; such tubercles are characteristic of most corbiculate Apidae. The mandible is attenuate, its apex blunt, the inner surface concave, its margins and often the apex denticulate but lacking large teeth. Larvae were described and illustrated by Michener (1953a) and Lucas de Oliveira (1965).

The Meliponini are found in the tropics of the world (although not east of the Solomon Islands in the Pacific). To the south they extend into temperate regions (about 35°S in Australia and South America, 28°S in Africa). To the north they extend little beyond the Tropic of Cancer (23.5°N).

Nests. The nests of most species occupy cavities that the bees find, and may limit by walling off unused areas; such walls are part of the batumen that typically surrounds the nest, and are called *batumen plates*. Some excavating is probably done by certain of the species that nest in the ground, and must be done by some species that regularly establish their nests within nests of *Nasutitermes* or ants. The cavities used vary from small, e.g., an abandoned cerambycid beetle burrow, to large hollows in a tree trunk or cavities in the soil (Figs. 7-7, 120-2). Some species, however, do not occupy cavities but make exposed nests on tree branches or on walls or cliff faces.

In contrast to the Apini, new nests are begun by workers going back and forth from an existing colony, carrying building materials and food to the new site. Ultimately, a young queen goes to the new site, workers stay there, and independence from the old colony is gradually attained over a period of weeks or months. Long-distance dispersal by individual reproductives or by swarms is therefore impossible.

Nests are made of wax secreted from the metasomal terga mixed with resins and gums collected by the bees. Some species add mud, feces, or other materials to certain parts of the construct. In all species the composition and

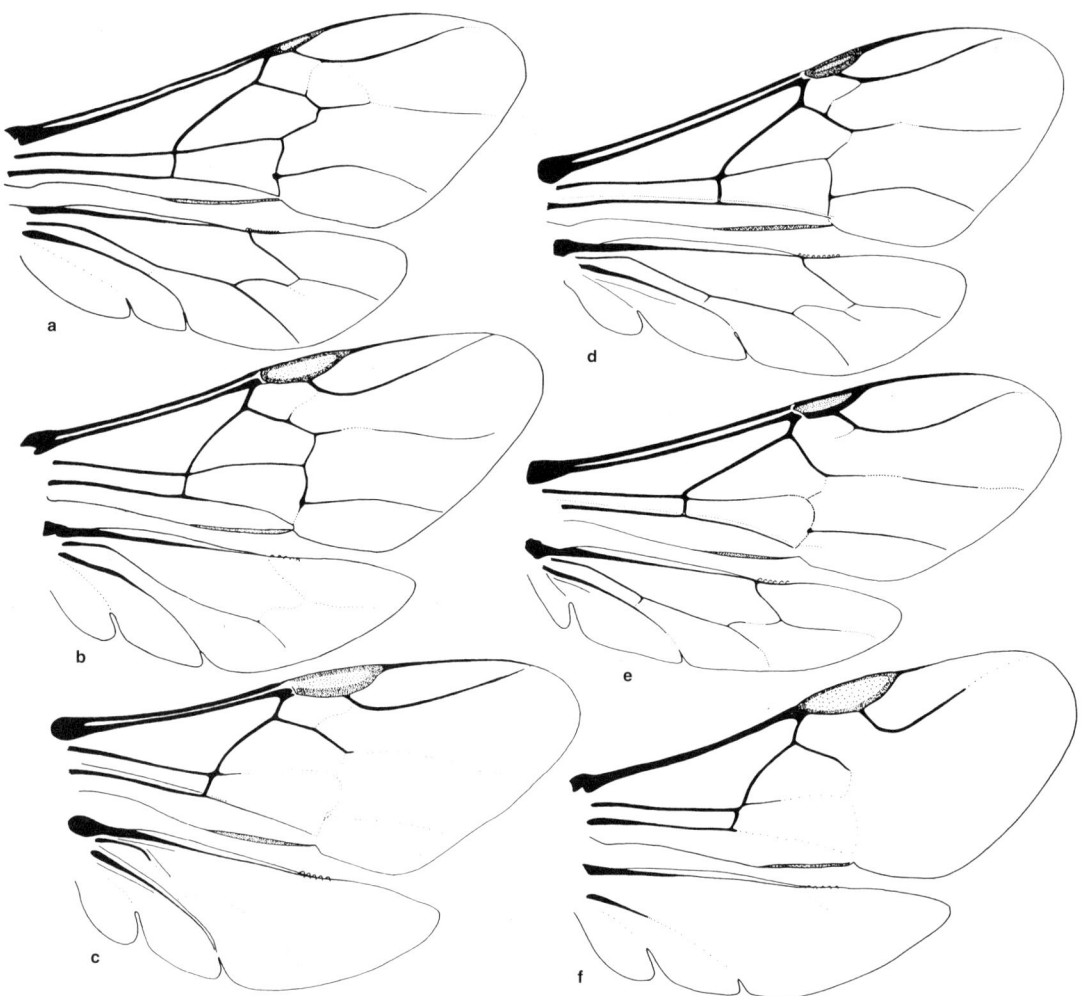

Figure 120-1. Wings of Meliponini (workers). **a,** *Melipona fasciata* Latreille; **b,** *Plebeia (Plebeia) frontalis* (Friese); **c,** *P. (P.) schrottkyi* (Friese); **d,** *Meliponula (Meliponula) bocandei* (Spinola); **e,** *Dactylurina schmidti* (Stadelmann); **f,** *Trigonisca buyssoni* (Friese). From Michener, 1990a.

texture differ in different parts of the nest. Brood cells and often the sheets of involucrum around the brood chamber and the storage pots are of soft material, presumably largely wax. Some of the supports and batumen plates are of tough, rather hard material, and the external sheet of batumen around some exposed nests is hard and brittle. For the terminology of nest parts, see Section 7 and Figure 7-8.

The brood cells are mass provisioned and either clustered (Figs. 120-3, 120-4b) or arranged in combs (Figs. 7-7, 7-8, 120-2, 120-4a) that are usually horizontal. The cells open upward (rarely horizontally) and are closed after an egg is laid. The egg is positioned standing up on the semiliquid provisions, which consist of pollen, hypopharyngeal-gland secretion, and nectar or honey. The cells are destroyed after use and cannot be reused as they are in *Apis*. Food is stored in pots that are quite different from and larger than brood cells (Figs. 7-7, 120-2, 120-3). Nogueira-Neto (1970, 1997) gave much practical information about the culture of meliponines, including, of course, biological information about them; he reviewed much of their biology. Details of nesting behavior and nest architecture are dealt with by Schwarz (1948), who made a massive review of all earlier literature; Michener (1961a, 1974a); Kerr et al. (1967); Camargo (1970); Wille and Michener (1973); Roubik (1979, 1983, 1989, 1992); Fletcher and Crewe (1981); Sakagami, Yamane, and Hambali (1983); Sakagami et al. (1983); and works cited therein. Reviews are by Wille and Michener (1973); Michener (1974a); Sakagami (1982); and Wille (1983).

It is not unreasonable to suppose that brood cells in a disorganized cluster are plesiomorphic relative to cells arranged in combs. Most Meliponini arrange cells in horizontal (sometimes spiral) combs. Cells in clusters, however, characterize some species of the genus *Trigona*, i.e., all species of the subgenus *Frieseomelitta*, *T. (Heterotrigona) canifrons* Smith, and most but not all species of the

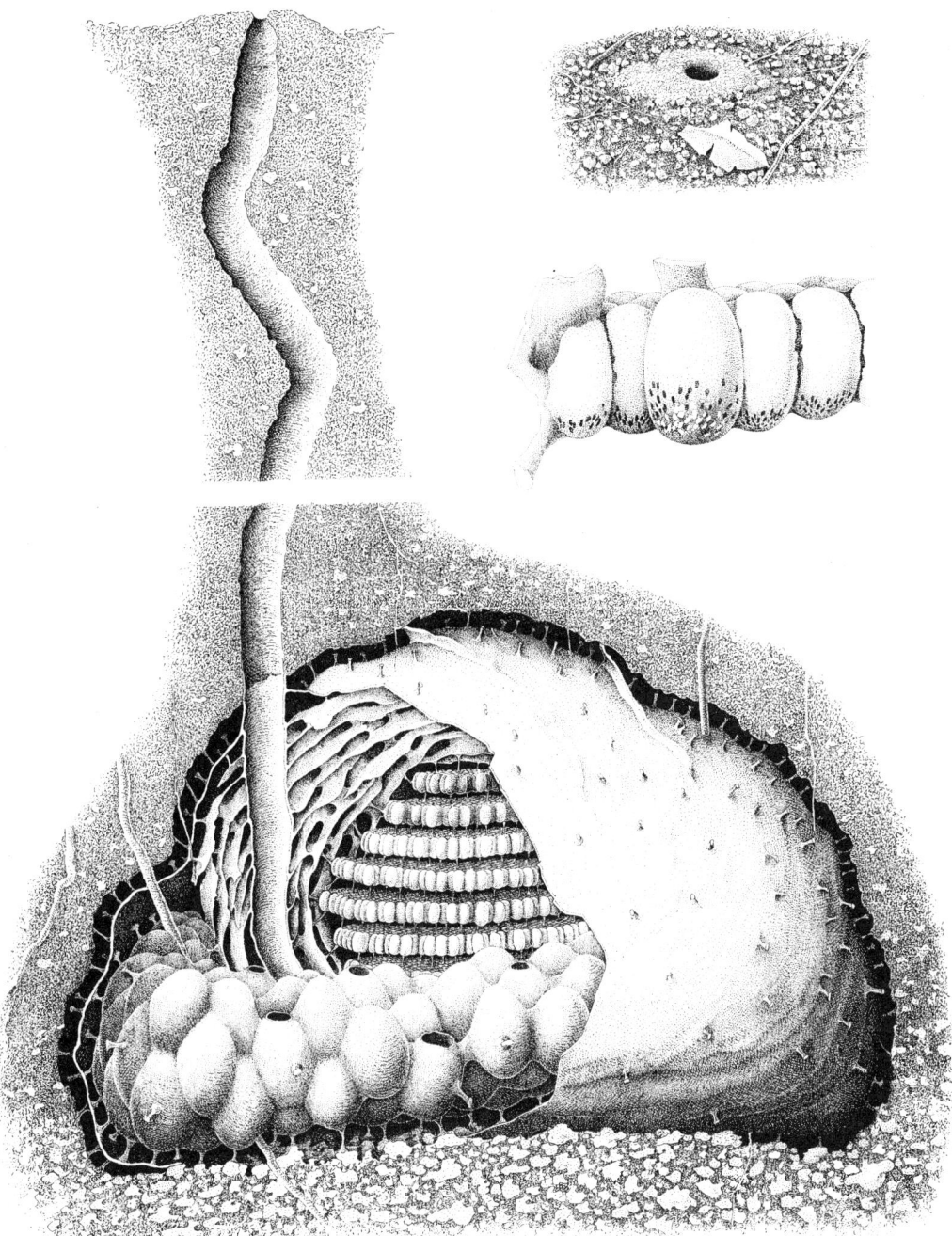

Figure 120-2. Nest of *Plebeia (Plebeia) caerulea* (Friese) in a soil cavity, the entrance and a portion of a comb of cocoons at the upper right. The brood cells are in horizontal combs, the brood chamber surrounded by a multilayered involucrum; the whole nest, including the food pots, is surrounded by a single thin layer of batumen (intact on the right side of the figure) supported by short pillars extending to the cavity walls. See also Figures 7-7 and 7-8. Original drawing by C. M. F. de Camargo.

Tetragonula group of *T. (Heterotrigona)*. Moreover, cells are also placed in clusters by all species of the genera *Austroplebeia* [although layered, approaching combs, in *A. cincta* (Mocsáry); Michener, 1961a], *Cleptotrigona, Hypotrigona, Liotrigona, Trichotrigona,* and *Trigonisca,* and by some but not all species of both *Plebeia (Plebeia)* and *P. (Scaura)*.

As suggested by Michener (1961a), clustering may be the ancestral cell arrangement for the Meliponini, perhaps retained by *Austroplebeia*. But clustering is probably

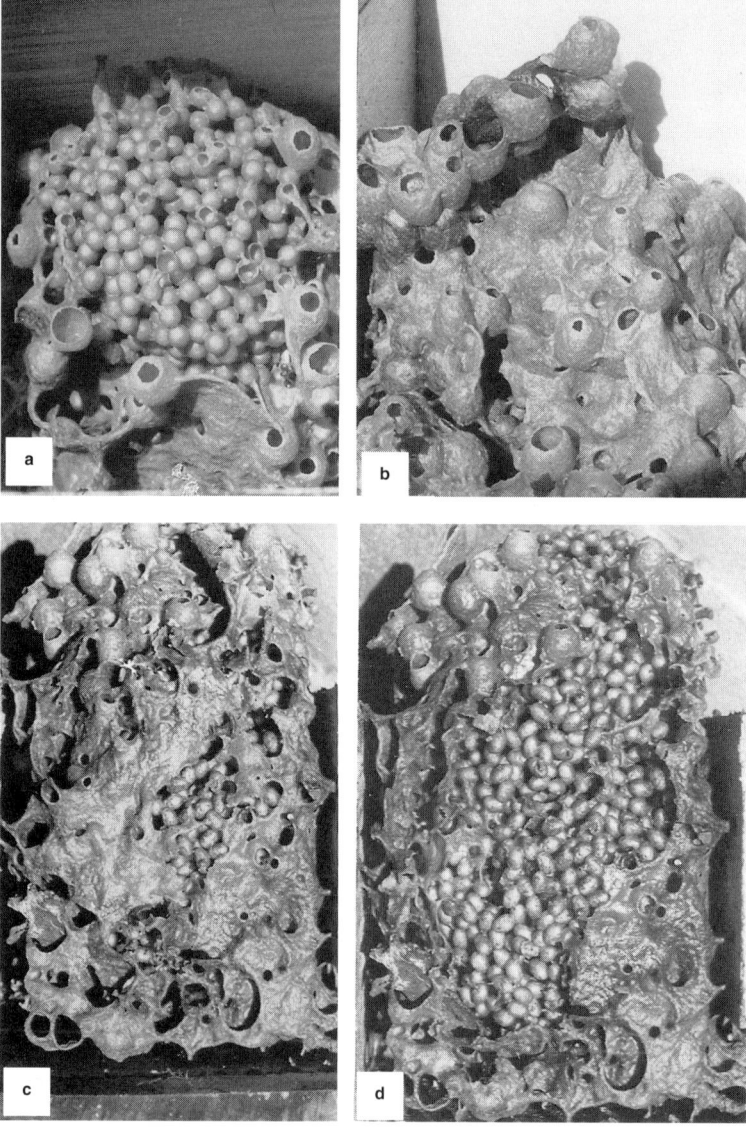

Figure 120-3. Nest structures of *Austroplebeia australis* (Friese). **a,** Advancing front of cells in a cluster (no combs); **b,** Food pots; **c,** Cluster of cocoons except for cells at top, largely hidden by involucrum; **d,** Same, the involucrum largely removed to show the cluster of cocoons. From Michener, 1961a.

derived for species like *Trigona (Heterotrigona) fuscobalteata* Cameron that nest in small, irregular cavities where combs would be impractical. Indeed, most cluster-makers are small to minute bees. Michener (1961a) contended that nearly spherical cells in clusters (as in *Austroplebeia*) are probably ancestral, that elongate cells resulted from packing cells into combs, and that therefore species that make elongate cells in clusters are derived from ancestors that made combs. Unfortunately for this theory, spherical cells are almost unknown in other groups of bees, so there is no good evidence for the polarity of this character.

Dactylurina is unique among Meliponini in that its combs, instead of being horizontal or nearly so, with the cells opening upward, are vertical, with cells on both sides opening laterally, as in the combs of *Apis*. *Dactylurina* is not *Apis*-like in other features, and this architecture must have evolved independently.

Interesting aspects of social behavior of the Meliponini are the oviposition rituals and associated activities, much studied and described in a series of papers by Sakagami and others (reviewed by Sakagami, 1982; see also Sakagami, Yamane, and Inoue, 1983, Sakagami and Yamane, 1987, Zucchi, 1993; Benthem, Imperatriz-Fonseca, and Velthuis, 1995, and other papers cited in Yamane, Heard, and Sakagami, 1995). These rituals are often group-specific and often accompanied by the laying of trophic eggs (usually queen food) by workers. The behavior of queens and workers during laying might provide characters of phylogenetic significance. Polarization of most of these behaviors is doubtful, however, since there are no counterparts in the outgroups. Zucchi (1993), however, has

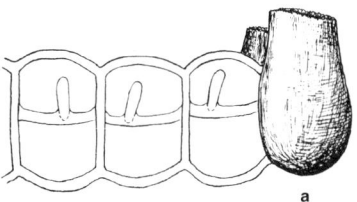 

Figure 120-4. Brood cells of Meliponini, some viewed in section to show the provisions and an erect egg in each. **a,** Portion of a comb of *Trigona (Heterotrigona) carbonaria* Smith, the cell at the right ready for provisioning and closure, those in sectional view each with an egg; **b,** Portions of a cell cluster of *Austroplebeia australis* (Friese), the cell at the left provisioned and closed, next to it a cell ready to receive provisions, next an artificially opened cell and a cell in sectional view, each with an egg. From Michener, 1961a.

not only summarized the oviposition behaviors but interpreted most of them as ritualized dominance behaviors. If this is correct, then aggressive, less ritualized activity such as characterizes *Plebeia* must be ancestral relative to less aggressive, more ritualized interactions. Material on the oviposition rituals not included in the above-listed reviews was given by Silva-Matos, Zucchi, and Yamane (1997). Considerable variation is now known in the oviposition rituals among species here included in *Plebeia* (Drumond, Zucchi, and Oldroyd, 2000). Bego et al. (1999) provided data on ovipostion rituals of *Trigona (Tetragonisca)*.

Caste systems. Female caste differentiation is quite different from that familiar in *Apis*, which results from differences in food quality. Three different systems are known in the Meliponini. In the first system, found in most Meliponini, queens are produced in small numbers in large cells, often at the margins of combs, and the large quantity of food placed in such cells seems to be the factor responsible for the development of queens in them rather than workers. The fact that workers are reared in adjacent cells in combs is not known to lead to any breakdown of cell walls, robbing of food, and consequent production of queens in fused workers' cells. A modification of the first system, suggestive of the third, is found in *Plebeia (Schwarziana)*, in which queen cells are only slightly larger than worker cells and are scattered rather than marginal. The second system is found in *Trigona (Frieseomelitta) varia* Lepeletier and perhaps other species that rear brood in clusters of usually well-separated cells rather than in combs; *T. varia* is not known to construct large queen cells. If two cells happen to be contiguous, an older larva may break into the other's cell and eat the food stored there, and become a queen. Or in the absence of a queen, the workers construct new cells in contact with certain cells containing large larvae. Such an auxillary cell is provisioned, but closed without an egg. The cells then become united and the larva consumes the provisions in its own as well as the auxillary cell, producing a queen (Faustino et al, 2002). Finally, in the third system, found in the genus *Melipona*, no special queen cells are present. The queens are worker-sized on maturation and are produced in relatively large numbers, up to one-quarter of the maturing females at certain seasons, nearly all of which are soon killed or ejected by the workers. The mechanisms for caste determination in this case are briefly explained under the genus *Melipona*. An excellent review of caste determination in Meliponini is by Velthuis and Sommeijer (1991); see also Buschini and Campos (1995).

Phylogeny and classification. The classification of stingless honey bees has been presented very differently by different authors (see review by Sakagami, 1982). Michener (1944) and Schwarz (1948) recognized only two principal genera, *Melipona* and *Trigona*. Moure, however, in 1961 recognized 23 genera (no subgenera) from the Old World and 10 from the New World. In the New World he later recognized 27 supraspecific taxa (genera and subgenera) and in 1971 he elevated some of the subgenera (and by inference the others) to the genus level. Camargo and Pedro (1992b) and especially Camargo (1989) presented detailed summaries of these classifications and that of Wille (1979b). Sakagami (1975, 1982) presented an intermediate system, in some ways similar to that used in the present work, which in general follows Michener (1990a).

Michener (1990a) made a cladistic analysis of meliponine genera. It was based, however, on only 17 characters (all 2-state except four 3-state characters) and doubtless will be modified when more synapomorphies are found. Camargo and Pedro (1992a) repeated the analysis, omitting characters of males because they are unknown for some taxa, repolarizing certain characters, adding two characters, and finishing with only 12 characters. Figure 120-5a summarizes the preferred cladogram by Michener, and Figure 120-5b, that by Camargo and Pedro. In Michener's cladogram, *Melipona* (neotropical) appears to be the sister group to all the other Meliponini. *Trigona* and a group of related genera are restricted to tropical America, except that *Trigona* also occurs from southern Asia to Australia. Apparently derived from the neotropical group is a group of Old World genera that includes one New World genus, *Trigonisca*. The African genera except for *Hypotrigona* constitute a group having several common characters in spite of superficial resemblance to different neotropical genera; the African genera appear to have been ultimately derived from among New World genera. In contrast, the Camargo and Pedro cladogram (Fig. 120-5b) indicates a major division of the Meliponini into African (left side of the figure) and non-African (right side of the figure) genera and places *Melipona* in a large group of mostly neotropical genera. Another "ex-

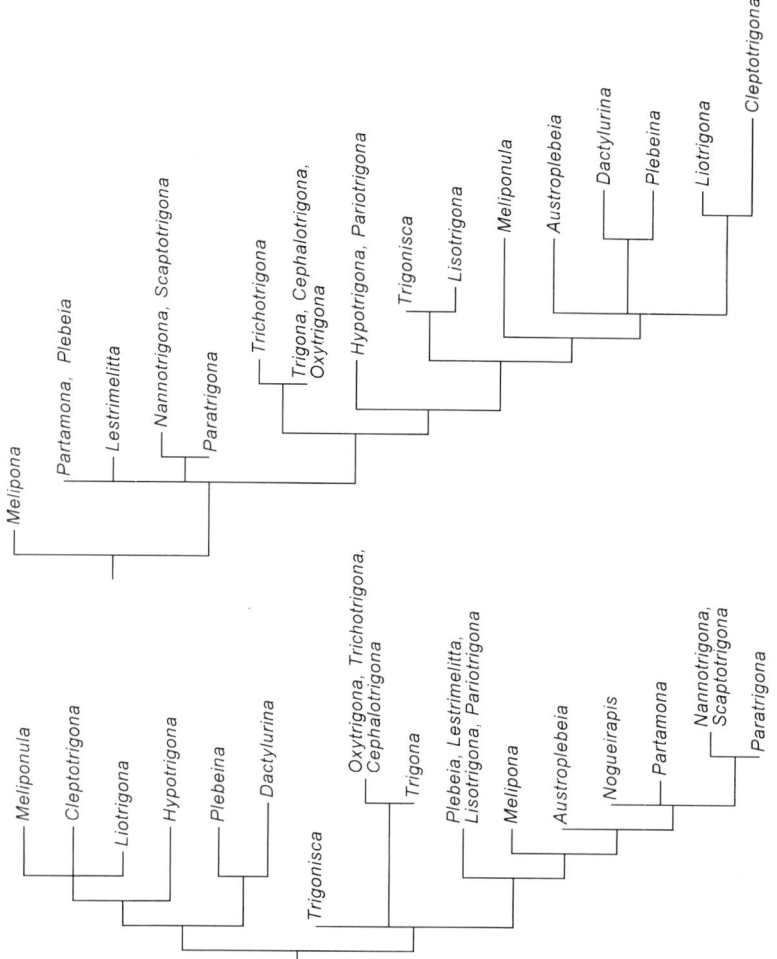

Figure 120-5. Alternative cladograms for the genera of Meliponini based on morphology. **a,** Cladogram of genera of Meliponini based on Michener (1990a). *Nogueirapis* was not included separately in the study, but would presumably be placed with *Partamona* and *Plebeia*; **b,** Cladogram of genera of Meliponini based on Camargo and Pedro (1992a), using mostly reworked characters from Michener (1990a) and using generic concepts here accepted. As indicated in the text, a study based on mitochondrial DNA supported the position of *Melipona* in the middle of the tree as in b, not as sister to all other Meliponini. (Lengths of vertical lines are proportional to numbers of synapomorphies, in both cladograms.)

ploratory phylogeny," including the fossil genus *Cretotrigona*, was prepared, using mostly the same characters, by Engel (2000a). An independent phylogenetic study by Costa et al. (2003) was based on mitochondrial DNA sequences. These studies support the position of *Melipona* among American groups, not near the base of the tree.

The small numbers of characters used in the phylogenetic analyses based on morphology indicate that more characters are needed to strengthen the conclusions. Cytological and molecular characters may help. Rocha et al. (2003) provided data on the number (n = 9 to n = 18) and morphology of chromosomes of 74 species falling in 31 genera and subgenera of Meliponini, but the possible contribution of these data to a phylogeny based on data from all sources (morphology, behavior, and molecules) has not been tested.

The common characters of the African Meliponini were first recognized by Wille (1979b); he regarded the African groups as ancestral and placed them in five genera while including all other Meliponini except *Melipona* and *Lestrimelitta* in the genus *Trigona*. A curious observation, first made by Wille (1979b) but see also Michener (1990a), is that in all African meliponine taxa except *Hypotrigona* the worker gonostyli arise close together; they are flattened, and they have minute hairs and few or localized larger setae (Fig. 120-6e-g). Moreover, in most cases the remnant of the sting stylet is acute, as in *Melipona* but unlike that of other taxa of Meliponini. It seems unlikely that these essentially internal structures would be convergent among African taxa; the larger sting remnant is a plesiomorphy. These characters are partly responsible for the grouping of African taxa in Figure 120-5.

Nonetheless, African taxa show remarkable external similarities to different taxa found elsewhere, mostly in the Americas. Thus the African *Dactylurina* resembles *Trigona* in diverse characters, the African *Plebeina* and some subgenera of *Meliponula* (according to the present classification) resemble *Plebeia*, the African *Liotrigona* resembles *Trigonisca*, and the African *Meliponula* s. str. resembles *Melipona*. Camargo and Pedro (1992b), in spite of their cladogram (1992a), describe for all Meliponini (1) the *Tetragonisca-Tetragona* line, including *Trigona* and *Dactylurina*, (2) the *Hypotrigona* line (including *Liotrig-*

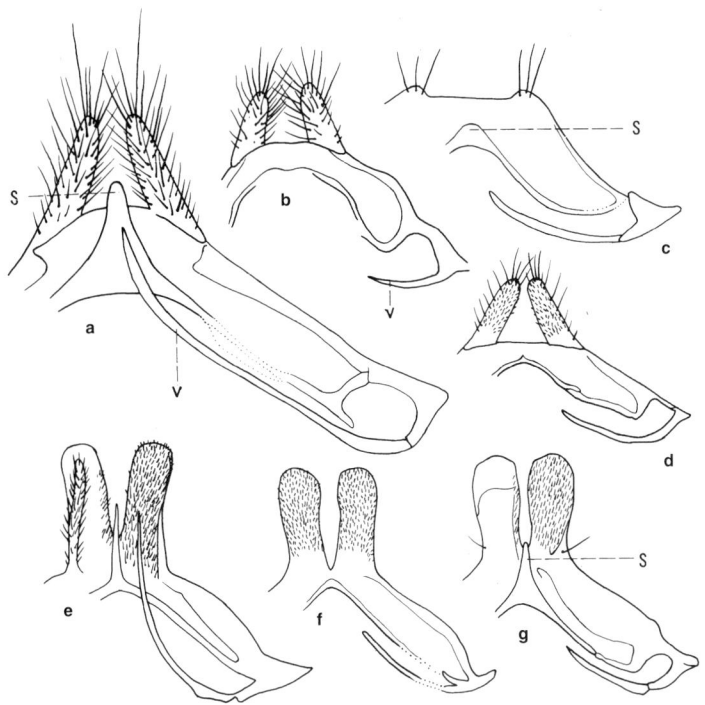

Figure 120-6. Sting rudiments of workers of Meliponini (a-d, American genera; e-g, African genera). **a,** *Melipona rufiventris* Lepeletier; **b,** *Scaptotrigona mexicana* Guérin-Méneville; **c,** *Partamona bilineata* (Say); **d,** *Trigonisca longicornis* (Friese); **e,** *Meliponula (Meliponula) bocandei* (Spinola); **f,** *Dactylurina schmidti* (Stadelmann); **g,** *Meliponula (Meliplebeia) beccarii* (Gribodo). (s, sting stylet; v, first valvula.) From Michener, 1990a.

ona and *Trigonisca*), and (3) the *Plebeia* line (including *Plebeina, Meliponula,* and *Melipona*). Recognition of these three "lines" suggests that the authors consider the common internal characters of African worker gonostyli and stings (except for *Hypotrigona*) as in fact convergent, for the grouping into three lines is based on the external characters. In a special section below I explain my view that it is the *external* characters that are convergent. Details of these matters are presented by Michener (1990a).

No genus of the Meliponini occurs in both Africa and South America. Therefore the meliponine faunas of these continents probably date from after the origin of the South Atlantic Ocean in the late Cretaceous. Figure 120-5a suggests that the Meliponini arose in tropical America (which at that time extended far into North America). We know nothing of when the group moved between the American continents, but, as noted below, there is a late Cretaceous *Trigona* or *Trigona*-like form from New Jersey (Michener and Grimaldi, 1988a, b). The dissimilarity of the neotropical and African faunas could suggest that meliponines reached South America from North America later, after considerable separation of South America from Africa.

Following the idea of Kerr and Maule (1964), the meliponines (including *Trigona*) may have spread through what is now the holarctic region when it was warmer. The Eocene *Kelneriapis* from Baltic amber (see below) and the Cretaceous *Trigona* from New Jersey are evidence of meliponines in the holarctic region. With climatic deterioration during the Tertiary, *Trigona* came to be limited to southern Asia (south to Australia) and the neotropical region. The African fauna must have evolved when Africa was substantially isolated from American and Eurasian invasions (Michener, 1990a). This outline seems to be in general agreement with the views expressed by Camargo and Pedro (1992b). The problem of the Australian *Austroplebeia* will be discussed under that genus.

Male genitalia. The gonocoxites of the male genitalia of Meliponini sometimes open basad, in a more or less straight line across the base of the genital capsule, as in most other bees. In some genera and in some preparations of others, however, the capsule is split longitudinally from the base by a V-shaped incision, and the gonocoxites open mesad along the arms of the V. To avoid repeated wordy descriptions, I call the former *rectigonal,* the latter, *schizogonal,* and forms believed able to exhibit both conformations are called *amphigonal.* These terms represent different positions of main genitalic structures. Rectigonal, with the lumina of the two gonocoxites opening basally, is illustrated by Figure 120-7b. Schizogonal, with the lumina opening into a crevice between the two gonocoxites, is illustrated for similar genitalia in Figure 120-7a. Amphigonal species, such as those illustrated in Figure 120-7a, b, can have either arrangement. These three terms do not necessarily represent phylogenetically significant conditions, although rectigonal is like other bees and therefore plesiomorphic, and schizogonal, unique to the Meliponini, is derived. At least in certain genera, however, specimens of the same species can be either rectigonal or schizogonal. The genital capsule in such cases is so loosely put together that the gonocoxites are hinged on the median points where they meet and can fold basad to take the schizogonal position; sometimes, then, the originally basal margins of the gonocoxites almost meet one another in preparations. The heavily sclerotized prong-like penis valves, commonly associated with this move-

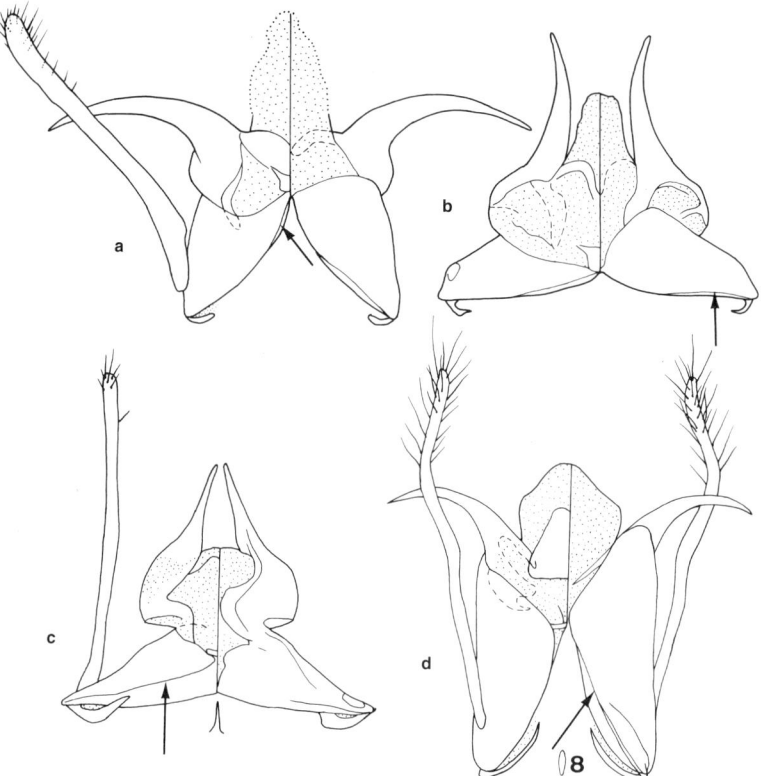

Figure 120-7. Male genitalia of American Meliponini, dorsal views at the left. **a**, *Scaptotrigona barrocoloradensis* (Schwarz), schizogonal conformation, the gonostylus omitted in the ventral view; **b**, *S. mexicana* (Guérin-Méneville), rectogonal conformation, the gonostyli omitted but the articulation shown at left; note the similar shape of the gonocoxites in the two conformations shown in a and b; **c**, *Trigonisca buyssoni* (Friese), the gonostylus omitted in the ventral view; **d**, *Melipona fulva* Lepeletier. (8, a probable remnant of S8; arrows indicate the openings of the gonocoxites into the metasomal cavity. The gonobase is probably represented by the small basolateral structures.) From Michener, 1990a.

ment, rotate and at the same time flex laterad; compare Figure 120-7a with b.

Some of the conspicuous differences among genitalic preparations result from mobility of parts. Apparently the mobility varies; it is great in most meliponines, those that are amphigonal (Fig. 120-7a-c). Such genitalia in otherwise very different genera are illustrated in Figure 120-7a *(Scaptotrigona)* and 120-7c *(Trigonisca)*. Such mobility can apparently be negligible in the permanently rectogonal forms like *Cleptotrigona* and *Lestrimelitta* (Fig. 120-8a-c, j-l), forms that I hesitate to classify as schizogonal like *Liotrigona* (Fig. 120-8g-i), and the probably permanently schizogonal forms like *Melipona* (Fig. 120-7d).

The genera *Liotrigona* and *Cleptotrigona* require special comment because the penis valves apparently extend laterally (probably to hold the female) by crossing one another (Fig. 120-8g, j) instead of by divergently spreading. *Lestrimelitta* and *Hypotrigona* (Fig. 120-8a-f) have dissimilar but distinctive genitalia, very different from those of most Meliponini. The other African genera also have distinctive genitalia, as shown in Figure 120-9.

Convergence. As noted above, unless there has been convergence in such structural characters as worker gonostyli, which differentiate major groups of genera, there has been remarkable convergence in the external features of workers of various meliponine bees. Wille (1979b) deals with this matter in some detail. The following paragraphs summarize the main points:

Melipona (neotropical) and *Meliponula* s. str. (Africa). Robust, thorax and head densely hairy, integument dull, basal propodeal area hairy, dorsal vessel arched between longitudinal indirect muscles of flight. The arch of the dorsal vessel, characteristic of many large, fast-flying bees, may be related to the robust body and fast flight of *Melipona* and *Meliponula* s. str.; the form of the dorsal vessel was documented for various taxa by Wille (1958, 1963, 1979b). In spite of these similarities, the hidden characters of the worker stings place these genera far apart (Fig. 120-6a, e).

Hypotrigona (Africa), *Liotrigona* (Africa), *Lisotrigona* (Asia), *Pariotrigona* (Asia), and *Trigonisca* (neotropical). Minute, sparsely haired, the pterostigma relatively large, wing venational characters as listed in the first alternative of couplet 1 of each key to genera, below. All except possibly the two little-known Asiatic taxa are attracted to perspiration. At least *Hypotrigona*, *Liotrigona*, and *Trigonisca* are quite unrelated to one another to judge by the sting and male genitalic characters, although superficially almost indistinguishable. *Cleptotrigona* (Africa) also falls in this group, being a relative of *Liotrigona*, but has the special features of robbers.

Dactylurina (Africa) and *Trigona* (neotropical; Asia to Australia). Typically rather elongate and long-legged, although in some American forms (like the subgenus *Geotrigona* and some species of *Trigona* s. str.) the metasoma is short and broad. The inner surface of the hind tibia of workers has a longitudinal band of keirotrichia on an elevated ridge, usually little if any wider than the depressed, shining upper zone of the tibia. The upper fringe of the hind tibia of workers includes plumose hairs except in some small subgenera of *Trigona* (Fig. 120-10a). In spite of the similarities of the two genera listed above, the

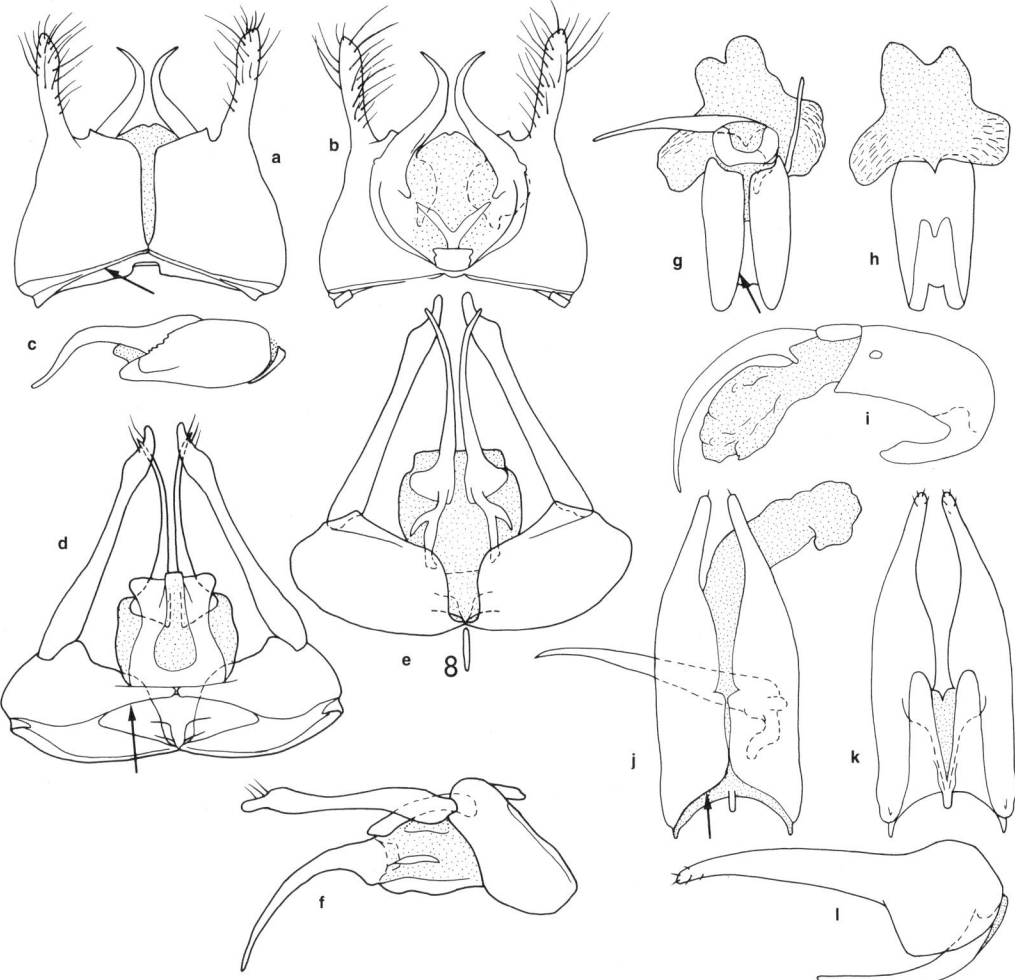

Figure 120-8. Dorsal, ventral, and lateral views of male genitalia of Meliponini. In the genera here illustrated, the gonocoxites are probably unable to assume different positions; they are rectogonal except for *Liotrigona*. **a-c,** *Lestrimelitta limao* (Smith); **d-f,** *Hypotrigona braunsi* (Kohl); **g-i,** *Liotrigona mahafalya* Brooks and Michener; **j-l,** *Cleptotrigona cubiceps* (Friese). In g, the left penis valve and left gonostylus are omitted; in h, both penis valves and gonostyli are omitted; in i, the gonostylus is omitted; in j, the left penis valve is omitted; in k and l, both penis valves are omitted. (8, a probable remnant of S8; arrows indicate the margins of the openings of the gonocoxites into the metasomal cavity.) From Michener, 1990a.

worker sting rudiments differ greatly, those of *Dactylurina* resembling those of various other African genera.

Austroplebeia (Australia), *Meliponula* (except s. str.) (Africa), *Nannotrigona, Paratrigona,* and *Plebeia* (neotropical), and *Plebeina* (Africa). Mostly small, robust bees of superficially similar aspect, often with restricted dull yellowish (or bright yellow in *Paratrigona*) markings on the head and thorax. The upper margin of the inner surface of the hind tibia is commonly shiny, often depressed (Fig. 120-10b), but keirotrichia sometimes reach the margin.

Cleptotrigona (Africa), *Lestrimelitta* (neotropical). Robber bees with shiny, sparsely haired bodies. The following features are presumably somehow related to robbing behavior: vertex and genal areas broad; proboscidial fossa greatly narrowed posteriorly; eyes small; clypeus small; labrum concave between lateral prominences; corbicula absent, penicillum and rastellum reduced to tapering hairs. Wille (1979b) correctly showed that in spite of their similarities, these genera are not closely related, as shown, for example, by the male genitalia (Fig. 120-8a-c, j-l).

Partamona, Scaptotrigona, and *Trigona* s. str. (*spinipes* group), and *T. (Geotrigona)* (all neotropical). Robust, often black bees with short metasomas, superficially similar in form and color, but well differentiated by the generic characters.

Classificatory questions. Regardless of one's methods, decisions on classificatory levels are subjective, and there will be some reasonable disagreements with my decisions. Some would regard all genus-group names as appropriate for genera. I believe that this obscures relationships that are useful to show in the classification, but even if one accepts in a general way the classification presented below,

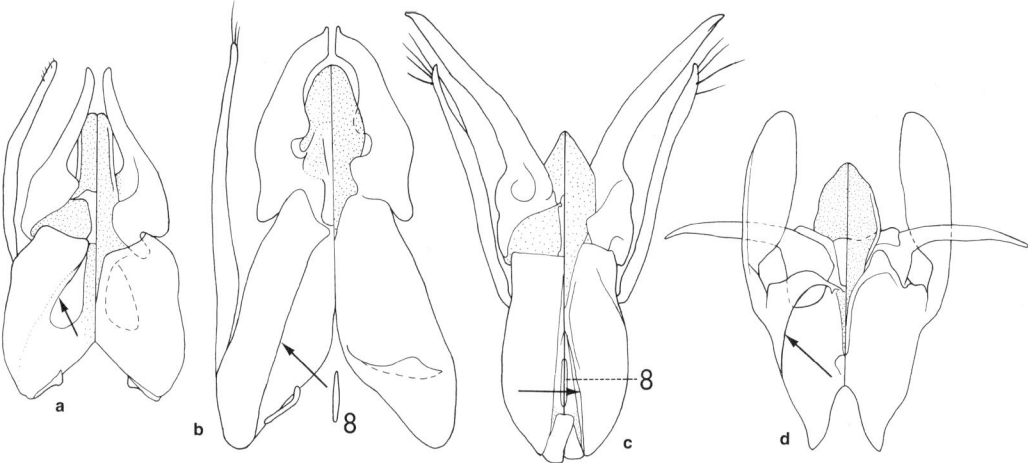

Figure 120-9. Male genitalia of African Meliponini, dorsal views at the left. **a**, *Meliponula (Meliplebeia) beccarii* (Gribodo); **b**, *M. (Meliponula) bocandei* (Spinola); **c**, *Plebeina denoiti* (Vachal); **d**, *Dactylurina schmidti* (Stadelmann). (8, a probable remnant of S8; arrows indicate the openings of gonocoxites into the metasomal cavity.) From Michener, 1990a.

there are decisions that I had to make arbitrarily and that could equally well have been different. Chief among these are the following:

Scaura could have been given generic status with *Schwarzula* as a monotypic subgenus if desired, instead of placement of *Scaura* as a subgenus of *Plebeia*.

Nogueirapis could have been placed as a subgenus of *Partamona* or of *Plebeia;* it has been placed in each of these positions in the past. I regard it as a separate genus.

Scaptotrigona could have been placed as a subgenus of *Nannotrigona*. They are sister groups whose relationship might well be indicated by the classification, but they are different in appearance and easily distinguished, and I have hesitantly treated both of them as genera.

Cephalotrigona, Oxytrigona, and *Trichotrigona* could have been considered subgenera of *Trigona*. It is not clear whether these taxa make *Trigona* paraphyletic. A phylogenetic study of *Trigona* and its close relatives would help to decide on a useful classification.

Ptilotrigona could have been given subgeneric status in *Trigona*, instead of being synonymized with the subgenus *Tetragona*.

Five groups of the subgenus *Heterotrigona* of *Trigona* could have been given subgeneric status, as suggested in the discussion of *Heterotrigona*. Subjectively, these groups seem less distinct than the subgenera here recognized.

The four named groups of *Trigonisca* could have been given subgeneric status.

The three named groups of *Meliponula* subgenus *Meliplebeia* could have been given subgeneric status.

Fossil Meliponini. There are five genus-group names based on fossils. They are *Cretotrigona* Engel, *Kelneriapis* Sakagami, *Liotrigonopsis* Engel, *Meliponorytes* Tosi, and *Proplebeia* Michener; for references see Table 22-1.

The true position of *Cretotrigona prisca* (Michener and Grimaldi) must be viewed as doubtful, even though it was originally described as a species of *Trigona* s. str., with which it agrees in toothed mandibles and other external characteristics except its probably simple hairs of the upper margin of the hind tibia. In the latter feature it resembles the subgenus *Lepidotrigona* and perhaps *Papuatrigona*. The upper hind tibial fringe may have included a very few branched hairs, as in some *T. (Geotrigona)*.

Engel (2000a) re-examined the type specimen and recognized certain characters not clarified in the original description. His phylogenetic hypothesis, placing *Cretotrigona* and *Dactylurina* at the summit as the most derived Meliponini, is difficult to accept. Thus there is little in its external characters to exclude it from the genus *Trigona* in spite of its great antiquity. The problem is that the same can be said of the African *Dactylurina*, yet the worker gonostyli and male genitalia show that *Dactylurina* is only distantly related to *Trigona*. The problem is accentuated, moreover, by the realization that in the Cretaceous, when *C. prisca* was living, its location (New Jersey) was not far from Africa, where *Dactylurina* now lives.

A new account of the genus *Proplebeia* and descriptions of three species, all from Dominican amber, were provided by Camargo, Grimaldi, and Pedro (2000).

Recent Genera of the Meliponini. The supraspecific taxa of the Meliponini have been described in detail and included in keys by Moure (1951a, 1961) as well as by Wille (1959a) and Michener (1990a). Useful regional keys to supraspecific taxa and to species, often with valuable illustrations, are included in the following works: Ducke (1916, 1924), Brazil; Schwarz (1934), Panama; Schwarz (1937), Borneo; Schwarz (1938), Guyana; Schwarz (1939a), Indo-Malayan region; Schwarz (1948), neotropical region; Schwarz (1949), Mexico; Sakagami, Inoue, and Salmah (1985, 1990), Sumatra; Ayala (1999), Mexico; Roubik (1992), Panama; Nates-Parra (1996), Colombia; Eardley (2004), Africa. Nearly all these works treat the whole tribe; references to them are not repeated under the genera below.

The keys below are based primarily on workers. Male characters have been added in various couplets. When "workers" are not specified, the character states given apply to males as well, but are often less well developed in males, and the identification of a male, not accompanied by workers, will often be difficult. Fortunately, males are almost always found with workers. Queens have been available for only a few taxa, and their characters have not been incorporated into the keys. Keys are presented separately for the neotropics, Africa, Asia (including the Sunda Islands), and Australia (including New Guinea).

Key to the Neotropical Genera of the Meliponini

1. Base of marginal cell broad, basal angle (between stigmal margin and vein r, within marginal cell) slightly acute (not under 68°) to right-angular (Fig. 120-1f); marginal cell, at apex of stigma, broader than submarginal cell area; forewing less, usually much less, than 4 mm long .... *Trigonisca*
—. Base of marginal cell of usual shape, basal angle strongly acute (not over 50°) (Fig. 120-1a-e) (except about 80% in *Nogueirapis*); marginal cell, at apex of stigma, little if any broader than submarginal cell area; forewing usually over 4 mm long.. 2
2(1). Inner surface of hind tibia with strongly depressed, shining, upper marginal zone, which at least apically is usually about as broad as longitudinal median keirotrichiate ridge, and midway of tibial length is at least half as wide as keirotrichiate ridge (Fig. 120-10a)............ 3
—. Inner surface of hind tibia with depressed upper marginal zone more narrow (much less than half as wide as area with keirotrichia) or absent, keirotrichia extending to or close to margin (Fig. 120-10b) 6
3(2). Eyes hairy; rastellum reduced to tapering hairs (Brazil) .... *Trichotrigona*
—. Eyes bare; rastellum strongly developed 4
4(3). Face short and broad, minimum distance between eyes much greater than length of eye; clypeus less than twice as broad as long; malar space almost twice as long as flagellar diameter; keirotrichiate zone on inner surface of worker hind tibia nearly twice as wide as depressed upper marginal zone at midlength of tibia .... *Oxytrigona*
—. Face of ordinary shape, minimum distance between eyes little more than to less than length of eye; clypeus usually more than twice as broad as long; malar space little over 1.5 times as long as flagellar diameter or usually much less; keirotrichiate zone on inner surface of worker hind tibia usually narrower (Fig. 120-10a), rarely over 1.5 times as wide as depressed upper marginal zone at mid length of tibia 5
5(4). Preoccipital carina strong and shining across full width behind vertex; lower face and genal area shining and coarsely punctate in contrast to dull, densely, minutely punctate upper face, genal area and scutum....*Cephalotrigona*
—. Preoccipital carina absent; lower face and genal area finely sculptured like upper part of head and scutum .... *Trigona*
6(2). First flagellar segment of worker nearly as long as second plus third taken together, of male nearly as long as second; outer surface of hind tibia convex, without corbicula, lower margin convex like upper margin; penicillum absent; rastellum consisting of tapering hairs .... *Lestrimelitta*
—. First flagellar segment of worker shorter than second plus third taken together, of male much shorter than second; outer surface of hind tibia of worker (and some males) flat or concave at least distally, forming corbicula, lower margin gently convex to concave, unlike largely or

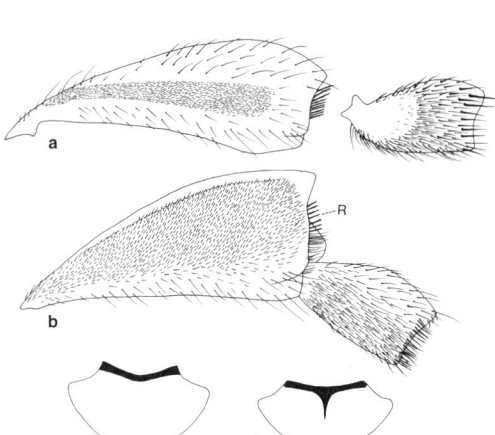

Figure 120-10. Structures of workers of Meliponini. **a,** Inner surfaces of hind tibia and basitarsus of *Trigona (Trigona) amalthea* (Olivier), showing the keirotrichia on the tibial ridge and the sericeous area on the base of the basitarsus (for the outer surfaces, see Fig. 120-11); **b,** Inner surfaces of hind tibia and basitarsus of *Plebeia (Plebeia) frontalis* (Friese), showing the keirotrichiate area extending nearly to the upper tibial margin (for the outer surfaces, see Fig. 10-11); **c, d,** Scutellum of *Partamona bilineata* (Say) and *Scaptotrigona mexicana* (Guérin-Méneville) (R, rastellum). From Michener, 1990a.

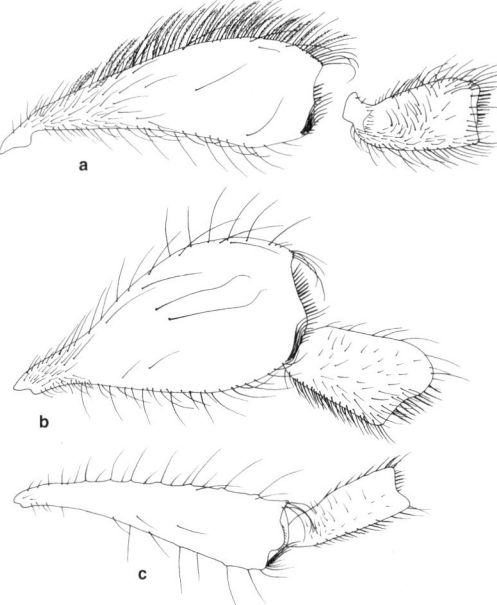

Figure 120-11. Outer surfaces of hind tibiae and basitarsi of workers of Meliponini. **a,** *Trigona (Trigona) amalthea* (Olivier) (for the inner surfaces, see Fig. 120-10); **b,** *Partamona bilineata* (Say); **c,** *Trigonisca longitarsis* (Ducke). From Michener, 1990a.

wholly convex upper margin (Fig. 120-11); penicillum present (Fig. 120-11); rastellum variable 7
7(6). Hamuli 9-14 (rarely 8); wings extending little if any beyond apex of metasoma; stigma with margin within marginal cell straight or weakly concave (Fig. 120-1a) (body apiform; basal propodeal area dull, hairy) ....
.... *Melipona*
—. Hamuli 5-7, rarely up to 9 or even 10; wings long, extending well beyond apex of metasoma; stigma with margin within marginal cell slightly convex (Fig. 120-1b-f)
.... 8
8(7). Anterior part of scutellum with shining, longitudinal V- or U-shaped median depression opening anteriorly into scutoscutellar fossa (Fig. 120-10d); preoccipital carina present, extending far down on each side of head 9
—. Anterior part of scutellum without such a shining, median depression (Fig. 120-10c); preoccipital carina absent or with transverse part only, behind vertex and weakly indicated, except in *Paratrigonoides* 10
9(8). Head and thorax, or at least scutellum, with extremely coarse, cribriform punctation; posterior margin of scutellum notched or emarginate medially as seen from above; anterior margin of pronotal lobe with strong, transverse carina... *Nannotrigona*
—. Head and thorax with fine punctation; posterior margin of scutellum entire; anterior margin of pronotal lobe rounded .... *Scaptotrigona*
10(8). Mandible of worker with four apical teeth (lower two sometimes united by translucent septum but teeth still recognizable); scutellum, as seen in lateral view projecting posteriorly as thin shelf over median part of metanotum .... *Paratrigona*
—. Mandible of worker with (rarely without) one or two denticles at upper end of apical margin, otherwise without teeth; scutellum, as seen in lateral view, rather thick and rounded, not projecting as thin shelf over metanotum 11
11(10). Hind tibia of worker greatly broadened, spoonshaped (Fig. 120-11b), about three times as wide as femur, outer surface largely occupied by corbicula, lower margin of tibia with distal one-half convex; basal area of propodeum densely hairy .... *Partamona*
—. Hind tibia of worker not greatly broadened, less than three times as wide as femur (Fig. 10-11), corbicula extending but little if at all basad of middle of tibia, lower margin of tibia convex only in distal one-fourth or less; basal area of propodeum usually hairless 12
12(11). Malar space about one-fifth as long as eye; upper margin of hind basitarsus strongly convex medially; yellow markings absent .... *Meliwillea*
—. Malar space much less than one-fifth as long as eye; upper margin of hind basitarsus gently convex (Fig. 120-10b); yellow markings almost always present, at least on face 13
13(12). Upper margin of inner surface of hind tibia strongly depressed, shiny, in sharp contrast to keirotrichiate area [except not depressed in apical third of tibia of *Plebeia (Scaura) timida* (Silvestri)]; concave surface of corbicula usually not occupying whole distal half of hind tibia ....
.... *Plebeia*
—. Upper margin of inner surface of hind tibia not depressed, although shiny and in contrast to keirotrichiate area; concave surface of corbicula occupying full width of distal half of hind tibia 14
14(13). Integument of head and thorax dull, microreticulate; preoccipital carina lamellate across upper part of head, with row of coarse setae, branched apically; supraclypeal area expanded laterally, forming flange partly covering antennal socket .... *Paratrigonoides*
—. Integument largely shiny; preoccipital carina absent, without row of course setae; supraclypeal area not expanded laterally .... *Nogueirapis*

Key to the African Genera of the Meliponini

1. Forewing length less than 3.5 mm; hind wing without closed cells (as in Fig. 120-1f), veins closing cells R and Cu, if visible at all, clear and unpigmented; forewing with submarginal crossveins almost always completely absent, thus indications of submarginal cells absent; at least distal part of cell second Cu of forewing undefined or defined by completely unpigmented vein traces; vein M of forewing terminating without bend at about position of anterior end of first recurrent vein (as in Fig. 120-1f), which, however, is absent 2
—. Forewing length about 4 mm or more; hind wing commonly with cells R and Cu closed by at least weakly brownish veins; forewing with one or two submarginal crossveins usually weakly indicated, first submarginal cell thus usually recognizable; cell second Cu of forewing completely indicated, at least by faint veins; vein M of forewing extending at least slightly beyond position of anterior end of first recurrent vein and angulate at end of that vein (Fig. 120-1d, e; as in 120-1c), which is usually at least faintly visible 4
2(1). Outer surface of hind tibia of worker convex, without corbicula; penicillum absent; clypeus much more than twice as wide as long .... *Cleptotrigona*
—. Distal part of outer surface of hind tibia of worker flat or concave, margined by long hair, forming corbicula; penicillum present; clypeus twice as wide as long or less 3
3(2). Upper apical part of hind tibia of worker forming distinct angle; gonostyli of worker much longer than broad, flat, adjacent or separated by less than one gonostylar diameter, without setae but with minute hairs (as in Fig. 120-6e-g) .... *Liotrigona*
—. Upper apical part of hind tibia of worker rounded; gonostyli of worker minute, tuberculiform, separated by several gonostylar diameters, with setae but without minute hairs (as in Fig. 120-6c) .... *Hypotrigona*
4(1). Inner surface of hind tibia with strongly depressed, shining, upper marginal area nearly as broad apically as longitudinal median keirotrichiate ridge, and about half as wide as keirotrichiate ridge midway of tibial length; first metasomal segment longer than broad .... *Dactylurina*
—. Inner surface of hind tibia with depressed upper marginal area narrow (much less than half as wide as broad area with keirotrichia) or absent, keirotrichia extending to or close to margin; first metasomal segment broader than long 5
5(4). Hind tibia of worker rather spoon-shaped, upper apical angle rounded but with coarse, amber-colored to blackish bristles (posterior parapenicillum); sting stylet of worker distinct, acute (Fig. 120-6e, g) .... *Meliponula*
—. Hind tibia of worker slender, triangular with distinct up-

per apical angle supporting long, pale hairs (not especially coarse); sting stylet of worker a mere rounded con-vexity .... *Plebeina*

Key to the Genera of the Meliponini of Asia and the Sunda Islands

1. Forewing length commonly (but not always) over 4 mm, wing venation usually not greatly reduced (for a meliponine), but if minute and with some of the wing reduction characters listed below, then upper margin of hind tibia with plumose hairs intermixed with simple ones (as in Fig. 120-11a); hind wing commonly with cells R and Cu closed by at least weakly brownish veins; forewing with one or two submarginal crossveins usually weakly indicated, first submarginal cell usually recognizable; cell second Cu of forewing completely indicated at least by faint veins; vein M of forewing usually extending at least slightly beyond position of anterior end of first recurrent vein and angulate at end of that vein (as in Fig. 120-1c), which is usually at least faintly visible .... *Trigona*
—. Forewing length less than 3 mm, wing venation greatly reduced; upper margin of hind tibia without plumose hairs; hind wing without closed cells, veins closing cells R and Cu, if visible at all, clear and unpigmented; forewing with submarginal crossveins almost always completely absent, thus without indications of submarginal cells; at least distal part of cell second Cu of forewing undefined or defined by completely unpigmented vein traces; vein M of forewing terminating without bend at about position of anterior end of first recurrent vein which, however, is absent (as in Fig. 120-1f) 2
2(1). Malar space almost one-fifth as long as eye, much longer than flagellar diameter; gonostylus of worker with setae but without minute hairs .... *Pariotrigona*
—. Malar space shorter than flagellar diameter; gonostylus of worker with many minute hairs, in addition to setae along outer and distal margins .... *Lisotrigona*

Key to the Genera of the Meliponini of Australia and New Guinea

1. Scutellum and usually face and scutum with well-developed yellow markings; inner surface of hind tibia with keirotrichiate area broad, nearly reaching upper margin of tibia (as in Fig. 120-10b) .... *Austroplebeia*
—. Head and thorax without distinct yellow markings; inner surface of hind tibia with strong longitudinal keirotrichiate ridge above which is a broad depressed, shining marginal area (Fig. 120-10a) .... *Trigona*

Genus *Austroplebeia* Moure

Austroplebeia Moure, 1961: 195. Type species: *Trigona cassiae* Cockerell, 1910, by original designation.

This genus includes bees that closely resemble species of the neotropical group *Plebeia (Plebeia)* as well as the African *Plebeina*. *Austroplebeia* consists of rather robust species, 3 to 4 mm long, with distinct yellow areas on the scutellum and axillae, usually also on the lateral margins of the scutum and on the face. It differs from *Plebeia* and resembles African genera in the abundant, minute hairs and few small setae on the worker gonostyli and the slender and pointed bristles of the rastellum. The sting stylet, however, is a rounded projection, not pointed as in most African genera. On the inner side of the hind tibia, the keirotrichiate area closely approaches the upper margin of the tibia, leaving a narrow shiny margin as in *Plebeia* s. str. but less depressed or not depressed.

■ *Austroplebeia* is found in the northern half of Australia and in New Guinea. There are several species; nine names were listed by Michener (1965b) and Cardale (1993).

Nests are found in cavities in rather small tree trunks. An outstanding feature is the cluster of spherical brood cells (Michener, 1961a).

No bees similar to *Austroplebeia* are found in Asia or Indonesia north and west of New Guinea; the relationship of this genus to African or possibly American genera gives rise to a biogeographical puzzle—how and when did such bees reach Australia? Almost certainly they have been in Australia longer than the Australian *Trigona* species, which are close relatives of Indo-Malaysian forms.

Genus *Cephalotrigona* Schwarz

Trigona (Cephalotrigona) Schwarz, 1940: 10. Type species: *Trigona capitata* Smith, 1854, by original designation.

This genus is closely related to *Trigona* and could be considered a subgenus of that genus. The body seems more strongly sclerotized than in *Trigona* and is larger than most species of that genus (body length 6-10 mm). Unique features include (1) the strongly, coarsely punctate and shining clypeus, lower supraclypeal area, lower paraocular area, and lower genal area contrasting with the dull, reticulate frons, the dull, closely punctate vertex and scutum, and the extremely dull and minutely punctate upper genal area; and (2) the distal margin of the mandible of the worker, which has a single large tooth at the upper extremity, separated by a broad, shallow concavity from the rest of the margin, which is edentate (Fig. 119-5b). The propodeal triangle has conspicuous hair bases and, like the rest of the propodeum, abundant plumose hairs. The characters of the genus are well illustrated by Schwarz (1940, 1948) and described comparatively by Moure (1951a).

■ *Cephalotrigona* is found from Jalisco and Tamaulipas, Mexico, to Santa Catarina, Brazil, and Misiones, Argentina. It contains at least three species (listed as varieties by Schwarz, 1948).

Nests are typically in cavities in large tree trunks, although also recorded from the ground. The large honey pots contain excellent honey.

Genus *Cleptotrigona* Moure

Lestrimelitta (Cleptotrigona) Moure, 1961: 219. Type species: *Lestrimelitta cubiceps* Friese, 1912, by original designation.

This African robber genus is strongly convergent with the American robber, *Lestrimelitta,* and Moure (1961) regarded them as congeneric; common characters are listed above in the subsection on convergence in the Meliponini. Wille (1979b), however, emphasized their distinctness and recognized their resemblances as convergent. This genus contains small (3.5-4.0 mm long) black bees. In workers the gonostyli are flat and divergent, the bases separated by about half a gonostylar width; they bear many minute hairs in addition to several setae along the edges of a dorsal thickening. The male gonocoxites are longer than broad, broadly fused apically to gonostyli that are broad and flattened at the bases and tapering apically, or if the gonostyli are absent, then the gonocoxites are produced apically (Fig. 120-8j-l). Such gonostyli are unique in the Meliponini and strikingly different from the extremely slender, easily deciduous gonostyli of *Liotrigona* (Fig. 120-8g- i). The base of the genital capsule is curled under the rest of the genitalia, extending apicad as two slender lobes as far as the bases of the gonostyli. This suggests the structure of *Liotrigona*, in which, however, the lobes are fused to one another except apically. In both genera the penis valves, when extended laterad, cross at their bases, so that the lefthand penis valve is directed to the right, and the righthand one to the left. This configuration is unlike that of other Meliponini and is a probable synapomorphy supporting the relationship of *Liotrigona* and *Cleptotrigona*. The rather flat scutellum, not at all overhanging the metanotum and elevated but little above the level of the dorsal surface of the propodeum, is unique among the corbiculate Apidae. Among bees as a whole this feature is a plesiomorphy, but within this clade it must be an apomorphy, i.e., a reversion to the condition found in many other bees.

■ *Cleptotrigona* is known from Liberia to Tanzania, south to Angola and South Africa (northern Transvaal). The two species were listed by Moure (1961).

Cleptotrigona forages in nests of *Hypotrigona* and probably *Liotrigona*. It is not known to visit flowers. The robbing behavior of *C. cubiceps* (Friese) was described by Portugal-Araújo (1958). Its nests are in tree cavities, with brood cells in clusters.

Genus *Dactylurina* Cockerell

Dactylurina Cockerell, 1934b: 47. Type species: *Trigona staudingeri* Gribodo, 1893, by original designation.

This is the only African group that has the long legs, slender body (length 5-7 mm), plumose hairs on the upper margin of the hind tibia, and narrow keirotrichiate ridge on the inner surface of the hind tibia, such as are found in *Trigona*. As in most other African Meliponini, however, the worker gonostyli are covered with minute hairs. The male genitalia as well as the worker gonostyli of *Dactylurina* show that it is only distantly related to *Trigona*, in spite of its resemblance to species of that genus. Unusual features include the slender, fingerlike metasoma and the presence of only one denticle on the upper part of the apical mandibular margin.

■ The genus is widespread in tropical Africa, from Kenya and Tanzania to Zaire and Liberia. The two species were differentiated by Michener (1990a).

The vertical double brood combs suggesting those of *Apis*, although surrounded by a batumen envelope, are unique in the Meliponini; see the illustration in Michener (1974a). The nests are exposed, on tree branches.

Genus *Hypotrigona* Cockerell

Trigona (Hypotrigona) Cockerell, 1934b: 47. Type species: *Trigona gribodoi* Magretti, 1884, by original designation.

This is one of the genera of minute stingless bees (body length 1.9-4.0 mm) that exhibit (convergently) the character states listed in couplet 1 of the above keys to the genera of the African and oriental regions. It is, however, a distinct and isolated genus (with the possible exception that *Pariotrigona*, unknown in the male, may be related). Character states unique within the Meliponini include the apical process of S5 of the male, which is not especially heavily sclerotized and lies horizontally in the concavity of S6; the U-shaped S6 with strong basolateral apodemes like those of more anterior sterna; and the form of S7, which is a transverse bar with a small, median, basal angle. The sternal characteristics were illustrated by Brooks and Michener (1988), who interpreted what I believe is the apical process of S5 as S6 fused to S5. The male genital capsule is illustrated by the same authors, and by Michener (1990a); see Figure 120-8d-f. It is rectional but unique among Meliponini in the completely dorsal basal opening of its gonocoxites and the largely membranous basal bulb of the penis valve. The male gonostyli are freely articulated but do not break off easily. The gonostyli of workers are minute to papilliform, not flattened, separated by several times their lengths, with several setae but without minute hairs; those of all other African Meliponini have minute hairs. The worker sting stylus is a blunt convexity, not acute as in most African genera of Meliponini. The mesoscutum is typically dull; in workers the upper apical angle of the hind tibia is absent, i.e., broadly rounded; and the scutellum is wholly dark. These are character states that usually distinguish *Hypotrigona* from other minute Meliponini.

■ *Hypotrigona* is widespread and abundant in tropical Africa, from Ghana to Kenya and south to Angola and Natal, South Africa. Unlike the similarly minute African genus *Liotrigona*, *Hypotrigona* does not occur in Madagascar. It is represented by five species, listed by Moure (1961), and one additional species was shown to belong here by Michener (1990a).

Nests are to be found in almost any small cavities, in tree bark or tree cavities, or in small spaces in manmade structures. In complex spaces, cell clusters are often divided, connected only by passageways.

Genus *Lestrimelitta* Friese

Trigona (Lestrimelitta) Friese, 1903a: 361. Type species: *Trigona limao* Smith, 1863, monobasic.

Like *Cleptotrigona*, this is a robber genus. *Lestrimelitta* consists of shiny black species 4 to 7 mm long, somewhat more slender than most black Meliponini. Unlike that of nearly all other Meliponini, S6 of the male lacks a median apical process and has instead a small notch. S7 has a broad, squarish disc. Both S6 and S7 have long basolateral apodemes (see the illustrations by Schwarz, 1948; Camargo and Moure, 1989). These features seem ancestral (i.e., more like more basal sterna) relative to the specialized or reduced aspects of these sterna in other Meliponini. It seems unlikely, however, that a robber genus would preserve archaic structure, for its habits and the related features such as loss of the corbicula and rastellum are obviously derived. Another explanation is that since genes for ordinary sterna are obviously present, a developmental change could lead to their control of the more apical sterna as well as the preceding ones.

■ *Lestrimelitta* is widespread in the neotropical region,

from Nayarit and San Luis Potosí, Mexico, to Rio Grande do Sul, Brazil, and Misiones, Argentina. Camargo and Moure (1989) gave a key to five species, one of which appeared to be a complex of similar forms; Oliveira and Marchi (2005) gave a key to the eight known species.

The workers do not forage except in nests of other bees, especially species of the genera *Plebeia* and *Nannotrigona*, more rarely *Melipona, Scaptotrigona,* and *Trigona;* they even attack weak colonies of *Apis*. As in other forms that do not forage from flowers, there are only short, straight setae on the labial palpus. The robbing behavior of *Lestrimelitta* has been described by various authors, most fully by Sakagami, Roubik, and Zucchi (1993); see also Wittmann et al. (1990). *Lestrimelitta* colonies, which can be large and long-lived, are usually found in tree cavities, with unusually large entrance tubes; the cells are in comb formation.

Genus *Liotrigona* Moure

Liotrigona Moure, 1961: 223. Type species: *Trigona bottegoi* Magretti, 1895 = ? *Trigona madecassa* Saussure, 1891, by original designation.

This is one of the genera of minute bees (body length 1.8-3.0 mm) segregated in couplet 1 of the above key to African genera. Unlike the superficially similar *Hypotrigona*, *Liotrigona* is a member of the African group having flattened worker gonostyli that bear numerous minute hairs. The male genital capsule is elongate, permanently schizogonal, the gonocoxites are much longer than broad and broadly fused ventrally, and the bases (plus possibly the gonobase) are curled under and directed apicad, fused to one another except distally (Fig. 120-8g-i). The gonostyli, arising near the apices of the gonocoxites, are slender, easily detached from the gonocoxites during dissection. As in *Cleptotrigona*, the flexion of the penis valves is contralateral, so that they cross one another when flexed. The male sterna differ greatly from those of *Hypotrigona;* the male genitalia and sterna were illustrated by Brooks and Michener (1988). Workers of *Liotrigona* can usually be distinguished without dissection from those of *Hypotrigona* by the shiny mesoscutum, with only minute well-separated punctures, the distinct upper apical angle of the hind tibia, and the presence of pale or yellowish streaks on the preaxilla and the posterior margin of the scutellum; these character states occasionally fail, however.

■ *Liotrigona* is widespread but not very common in Africa, ranging from Ethiopia to Natal (South Africa), and from Ghana to Angola, but is common in Madagascar. *Liotrigona* is the only genus of Meliponini found in Madagascar. Pauly (in Pauly et al., 2001) reviewed the seven species from that island. Thus there are at least eight species, only one of them, *L. bottegoi* (Magretti), African; it may be the same as one of the Malagasy species (Brooks and Michener, 1988; Michener, 1990a).

Nests are similar to those of *Hypotrigona*, with cells in irregular and often divided clusters; see Brooks and Michener (1988).

Genus *Lisotrigona* Moure

Lisotrigona Moure, 1961: 194. Type species: *Melipona cacciae* Nurse, 1907, by original designation.

This genus of minute bees (body length 2.5-3.0 mm) shares the character states listed in couplet 1 of the above keys to the genera from Asia. It is known only from workers; hence its position relative to other genera is in doubt. The gonostyli of workers are flat, separated by a median concavity about as wide as a gonostylus, and are covered with abundant minute hairs; on the outer and distal margin of each there are several long, delicate setae. Thus the gonostylar vestiture is similar to that of *Trigonisca*. The hind tibia has a much rounded upper apical angle, the hairs on the inner side of the hind basitarsus are not in noticeable rows, and the base of the marginal cell is acute; in these features *Lisotrigona* differs from *Trigonisca*.

■ *Lisotrigona* is known from Sri Lanka and Madhya Pradesh in India to Vietnam, Borneo, and Sumatra. Engel (2000c) recognized three species and revised the genus.

Genus *Melipona* Illiger

Melipona Illiger, 1806: 157. Type species: *Apis favosa* Fabricius, 1798, by designation of Latreille, 1810: 439.
Melipona (Micheneria) Kerr, Pisani, and Aily, 1967: 139 (not Orfila and Rossi, 1956). Type species: *Melipona scutellaris* Latreille, 1811, by original designation.
Melipona (Michmelia) Moure, 1975: 621, replacement for *Micheneria* Kerr, Pisani, and Aily, 1967. Type species: *Melipona scutellaris* Latreille, 1811, autobasic.
Melipona (Melikerria) Moure, 1992: 34. Type species: *Apis compressipes* Fabricius, 1804, by original designation.
Melipona (Eomelipona) Moure, 1992: 35. Type species: *Melipona marginata* Lepeletier, 1836, by original designation.

This is the most distinctive meliponine genus. It consists of rather large (8-15 mm long) apiform species, mostly somewhat more robust than the workers of *Apis*, with wings that extend little if any beyond the apex of the metasoma, a slender stigma that is not convex within the marginal cell (Fig. 120-1a), and 9 to 14 hamuli. The sting stylet of the worker is right-angular or acute (Fig. 120-6a), and the male genital capsule is schizogonal (Fig. 120-7d). In the sting stylet and male genital characters, *Melipona* resembles the African group of genera. Unlike the rastellum of that group, however, that of *Melipona* is strongly developed, many of the bristles being blunt or abruptly narrowed at the apices, and the worker gonostyli are widely separated, setose, and lacking minute hairs (Fig. 120-6a). This is the only Recent genus with such gonostyli, and at the same time with a commonly acute sting and long male gonocoxites. All three of these character states are probable plesiomorphies. The male genitalia, sterna, and other structures were illustrated by Moure and Kerr (1950), Camargo, Moure, and Roubik (1988), Michener (1990a), Rego (1990), and Moure and Camargo (1994).

■ *Melipona* ranges from Sinaloa and Tamaulipas, Mexico, to Tucumán and Misiones, Argentina. It contains about 40 species. The species were revised by Schwarz (1932), and important reviews of the classification were by Moure and Kerr (1950) and Moure (1992).

Nests of most *Melipona* species are typically in tree cavities, although *M. quadrifasciata* Lepeletier nests in the ground. In nesting biology, *Melipona* is unique among the Meliponini in rearing numerous small queens in comb cells that are among and identical to worker cells. In all other genera only a few large queens are produced, usually in special large cells. Caste determination of *Melipona* is presumably partly genetic, in the sense that a combination of trophic and genetic factors determines caste. The long-established idea was that some females lack the potential to become queens, while others have such a potential but do so only under favorable trophic conditions. Velthuis and Sommeijer (1991) reviewed the matter and presented a more complex and probably more realistic theory of the interactions of genetic and trophic factors.

In *Melipona*, the production of numerous gynes, nearly all of which were believed to be killed by workers, has long been puzzling. It seemed to indicate wastage of space, provisions, and energy for producing nearly 25% of the female progeny in some cases (see review in Michener, 1974, p. 105-107). Recent studies of *Melipona favosa* (Fabricius) indicate that many virgin gynes leave the nest alive, that they stay alive for some time and visit flowers outside the nest, that they search for and find nests of their own species, and that they can sometimes pass the guards and enter such nests. Possible acceptance and reproductive behavior remain unknown but the existence of searching and entering behavior suggests that the young gynes sometimes survive and reproduce (Sommeijer and de Bruijn, 2003; Sommeijer, de Bruijn, Meeuwsen, and Slaa, 2003).

The method of recruitment to resources by *Melipona* workers is also different from that of other Meliponini; see review by Michener (1974a).

Although four subgenera have been recognized and constitute recognizable groups, the species of *Melipona* are morphologically similar and I have not recognized the subgenera. Male genitalia of 25 species studied by Rego (1990) are monotonously similar, as shown by her illustrations; her preferred cladogram (Rego, 1992), based on male genitalia, suggests that *Melikerria* and *Eomelipona* might constitute clades different from that of *Melipona* s. str. Fernandez-Salomão et al. (2005) prepared a phylogeny for eight species, two of each subgeneric group of *Melipona*, based on nuclear DNA in an internal transcribal spacer. For whatever it means with only two species per group, all four groups were supported, and *Melikerria* was basal, i.e., sister group to all the others.

Genus *Meliponula* Cockerell

This genus includes not only the type species, *Meliponula bocandei* (Spinola), but also a series of smaller African forms placed in other genera (see the subgenera, below) by Wille (1979b) and Moure (1961); see Michener (1990a). *Meliponula* shares with most other African Meliponinae the flattened worker gonostyli with their many minute hairs (Fig. 120-6e-g) (sometimes also with a few setae), the presence of a conspicuous remnant of the gonobase attached to the male gonocoxite lateroventrally (Fig. 120-9), the schizogonal (presumably permanently) male genital capsule with the gonocoxites longer than broad, at least in ventral view, and the strongly reflexed median apical process of S6 of the male. *Meliponula* differs from *Plebeina* in the reduction of the rastellum to slender hairs, no coarser than those of similar length on adjacent parts of the tibial apex; the rounded or very obtuse upper apical angle of the worker hind tibia, the tibia thus being rather spoon-shaped; the presence of coarse, amber or blackish bristles arising from or near this angle, forming what Wille (1979b) called the posterior parapenicillum in *M. bocandei* (Spinola); the presence of hairs, at least laterally, on the basal area of the propodeum; and the acute worker sting stylet (Fig. 120-6e, g). Like *Plebeina* and unlike other African genera, *Meliponula* has a broad area of keirotrichia on the inner surface of the worker hind tibia; the upper margin may be depressed and shining, almost as in *Plebeia* s. str., only slightly depressed (e.g., in *Meliponula* s. str.), or poorly defined and not at all depressed (subgenus *Axestotrigona*). The smaller species of *Meliponula* resemble the larger species of *Plebeia* s. str. superficially, but differ from most *Plebeia* not only in the character states indicated above but in the very fine, dense punctation at least of the mesoscutum.

Key to the Subgenera of *Meliponula*

1. Propodeal profile largely vertical; corbicula occupying less than distal half of hind tibia; apical reflexed process of S6 of male short and rounded; metasomal terga dull, minutely sculptured .... *M. (Meliponula* s. str.)
—. Propodeal profile with slanting dorsal portion rounding onto vertical portion; corbicula occupying more than distal half of hind tibia; apical reflexed process of S6 of male longer than body of sternum; metasomal terga at least partly shining 2
2(1). Head and thorax without yellow markings; inner surface of worker hind tibia without well-defined, shining, depressed upper margin, although keirotrichiate area does not reach margin at least distally.... *M. (Axestotrigona)*
—. Head and thorax with yellow markings; inner surface of worker hind tibia with shining upper margin, at least slightly depressed .... *M. (Meliplebeia)*

Meliponula / Subgenus *Axestotrigona* Moure

Axestotrigona Moure, 1961: 237. Type species: *Melipona ferruginea* Lepeletier, 1836, by original designation.

This subgenus consists of moderate-sized (length 5.5-7.0 mm) robust species. There are delicate plumed hairs among the marginal bristles at the upper apical angle of the worker hind tibia and across the apex of the tibia.

■ The subgenus ranges from Gambia to Kenya, south to Angola and the Transvaal, South Africa. The 12 species names, some probably representing only color forms of others, were listed by Moure (1961).

Nest sites include both tree cavities and terrestrial termite nests; brood cells form combs.

Meliponula / Subgenus *Meliplebeia* Moure

Meliplebeia Moure, 1961: 229. Type species: *Trigona beccarii* Gribodo, 1879, by original designation.
Plebeiella Moure, 1961: 226. Type species: *Trigona lendliana* Friese, 1900, by original designation.
Apotrigona Moure, 1961: 233. Type species: *Trigona nebulata* Smith, 1854, by original designation.

This subgenus contains moderate-sized [e.g., *Meliponula beccarii* (Gribodo)] to small [*M. lendliana* (Friese)] species, the body length varying from 4.5 to 7.0 mm. In *M. beccarii* there are delicate plumed hairs among the bristles near the upper apical angle of the worker hind tibia, as in the subgenus *Axestotrigona*. Such hairs are absent in the other groups of *Meliplebeia*.

■ *Meliplebeia* ranges from Senegal to Ethiopia, south to Namibia and Natal, South Africa. The 12 species names were listed by Moure (1961).

It is tempting to recognize two subgenera, *Meliplebeia* and *Plebeiella*, with *Apotrigona* placed as a synonym of *Plebeiella*. Unfortunately, the male of *Apotrigona* is unknown to me. Until its character states are known, I have decided that the best classification is one that shows the close relationship of the three groups given generic names by Moure, especially in view of the small number of species involved.

Nests are in the ground or in terrestrial termite nests; brood cells form combs.

Meliponula / Subgenus *Meliponula* Cockerell s. str.

Trigona (Meliponula) Cockerell, 1934b: 47. Type species: *Melipona bocandei* Spinola, 1851, by original designation.

This subgenus contains the largest (6.5-8.0 mm long) and most *Melipona*-like species of the genus. These bees are robust, compact in form, and lack yellow markings, although they have yellowish-brown areas on, for example, the scutellum and axillae. On the upper and distal margins of the hind tibia there are bristles, but no plumose hairs. The inner surface of the worker hind tibia has a well-defined but rather dull, slightly depressed upper margin. An interesting feature emphasized by Wille (1963) is the arch of the aorta between the longitudinal muscles of the thorax. This feature is as in *Melipona*, but unlike other stingless bees [including *Meliponula (Meliplebeia) beccarii* (Gribodo), see Wille, 1958]. It is, however, a feature of many moderate-sized and large, fast-flying bees and is doubtless a convergence in *Melipona* and *Meliponula* s. str., not an indication of close relationship.

■ The subgenus ranges from Liberia, the Central African Republic, and Uganda south to Angola. The single species is *Meliponula bocandei* (Spinola).

Nests are found in tree cavities. The brood cells are arranged in irregular combs.

Genus *Meliwillea* Roubik, Segura, and Camargo

Meliwillea Roubik, Segura, and Camargo, 1997: 67. Type species: *Meliwillea bivea* Roubik, Segura, and Camargo, 1997, by original designation.

This genus contains a small (4.5-5.5 mm body length), shiny, entirely black species with the robust body form of *Scaptotrigona*, to which genus it is presumably related. It lacks some of the distinctive derived characters of *Scaptotrigona* and *Nannotrigona*, such as the anterior median scutellar depression and the preoccipital carina. The scutellum is not produced over the metanotum, as it is in those genera. *Meliwillea* differs from *Plebeia* in its black face and the upper margin of the hind tibia, which on the inner surface is narrowly bare as in *Plebeia* but not depressed as in that genus. It resembles *Scaptotrigona* in the long sinuous hairs of the hind coxa. The original description is beautifully illustrated with Camargo's drawings of characters of males as well as workers.

■ *Meliwillea* is found in the mountains of Costa Rica and western Panama. The only species is *M. bivea* Roubik, Segura, and Camargo.

Genus *Nannotrigona* Cockerell

Nannotrigona Cockerell, 1922a: 9. Type species: *Melipona testaceicornis* Lepeletier, 1836, by original designation.

Nannotrigona, being 3 to 5 mm long, looks like a deeply punctate, pitted *Plebeia* s. str. Thus it is quite different in appearance from the larger, robust *Scaptotrigona*. As emphasized by Schwarz (1938: 483) and Wille (1959a, 1979b), however, *Nannotrigona* and *Scaptotrigona* share numerous character states, and I do not doubt their close relationship. The scutellum is produced posteriorly as a thin shelf (as seen in side view) hiding the median part of the metanotum (in dorsal view), the apex notched or emarginate (dorsal view). As in *Scaptotrigona* the anterior margin of the scutellum has a V- or U-shaped median depression projecting posteriorly into the disc of the scutellum (Fig. 120-10d). The broad keirotrichiate area extends to or nearly to the upper margin of the hind tibia, which is not abruptly depressed as in *Plebia*.

■ *Nannotrigona* ranges from Sonora, Chihuahua, and San Luis Potosí, Mexico, to Santa Catarina, Brazil, and Paraguay. There are about nine species.

Nests are in tree cavities or in artificial sites, and the brood cells are in combs. The nest entrances are rather large for such small bees and, when not disturbed, are lined by workers looking outward. These openings are in contrast to the entrances of many other small Meliponini, which are so small that one or a few workers can block them.

Genus *Nogueirapis* Moure

Partamona (Nogueirapis) Moure, 1953b: 247. Type species: *Trigona butteli* Friese, 1900, by original designation.

Nogueirapis has sometimes been placed as a subgenus of *Partamona* (Moure, 1953b, Moure and Camargo, 1982), but Michener (1990a) placed it as a subgenus of *Plebeia*. Probably both associations are errors. *Nogueirapis* differs from *Partamona* in the only slightly spoon-shaped, not greatly enlarged hind tibia of the worker as well as small size (body length 3.5-5.5 mm), abundant yellow markings, the few and mostly curved (none sinuous) large setae of the labial palpi (as in various species of *Plebeia* s. str.), and the shining and hairless basal propodeal area, as was indicated by Wille (1964). The inner surface of the hind tibia of *Nogueirapis* has a narrow, bare, shiny, but not depressed upper margin, whereas it is depressed in *Plebeia*. Other differences from *Plebeia* include (1) the larger corbicula (but not as large as in *Partamona*) with one or two very large bristles arising from its surface (as in *Partamona*) and (2) the near absence of longish hairs arising from the frons and scutum.

■ This genus ranges from Costa Rica to Bolivia. In addition, it includes the Miocene fossil species *Plebeia (Nogueirapis) silacea* (Wille) from Chiapas, Mexico. There are three Recent species.

So far as is known, *Nogueirapis* species nest in the ground and the brood cells form combs.

Genus *Oxytrigona* Cockerell

Trigona (Oxytrigona) Cockerell, 1917a: 124. Type species: *Trigona flaveola mediorufa* Cockerell, 1913, by original designation.

This genus is closely related to *Trigona* and could be considered a subgenus of that genus. *Oxytrigona* has several striking character states of its own, however, and *Trigona* has at least one probable synapomorphy (the narrower, better-defined keirotrichiate band) that usually distinguishes it from *Oxytrigona* and suggests that *Oxytrigona* may be the sister genus of *Trigona* plus *Cephalotrigona*. Noteworthy features of *Oxytrigona* include the short, wide face (see the key to genera); the small clypeus, widely separated from the eyes; the shining but distinctly punctate clypeus and lower paraocular areas; the long malar space of the worker (about one-third as long as the eye); the convex genal area of the worker, providing space for the posterior ramus of the deeply bifid mandibular gland (Michener, 1974a); and the abundant, large, straight setae on the labial palpus of the worker (Fig. 120-12b), a feature unique in the Meliponini. The body length is 4.0 to 5.5 mm.

■ *Oxytrigona* ranges from Chiapas, Mexico, to Bolivia and Santa Catarina, Brazil. There are about eight species. Schwarz (1948) listed the included species (as varieties) and provided detailed descriptions and illustrations of *Oxytrigona;* Camargo (1984) added species and raised some varieties to the species level.

Oxytrigona nests are in tree cavities, often in large tree trunks. The brood cells form combs. An attribute of interest is the secretion of the mandibular glands, which contain formic acid (Roubik, Smith and Carlson, 1987). Workers, in nest defense, bite this liquid into the skin of an intruder, causing, in human skin, painful and long-lasting lesions (Michener, 1974a). The name "fire bee" is therefore in wide use for this insect.

Genus *Paratrigona* Schwarz

Trigona (Paratrigona) Schwarz, 1938: 487. Type species: *Melipona prosopiformis* Gribodo, 1893, by original designation.
Paratrigona (Aparatrigona) Moure, 1951a: 60. Type species: *Melipona impunctata* Ducke, 1916, by original designation.

This genus contains small species (body length 4-5 mm), the head and thorax (often also the metasoma) of which are dull with extremely minute punctation; they have conspicuous yellow to white markings on the thorax and usually on the face. As in *Nannotrigona* and *Scaptotrigona,* the scutellum is produced posteriorly as a thin shelf hiding the median part of the metanotum, as seen from above. The scutellum, however, lacks the anteromedian depression characteristic of those genera. The inner surface of the hind tibia has a broad keirotrichial area extending nearly to the upper margin, which is not depressed. The four-toothed mandibles of workers distinguish this genus from all Meliponini except *Trigona (Trigona);* the two lower teeth may be united by a thin

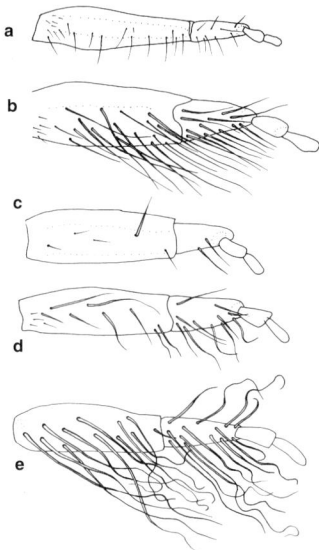

Figure 120-12. Posterior surfaces of the labial palpi of Meliponini; marginal hairs and small hairs are omitted, the mesal margins uppermost. **a,** *Cephalotrigona capitata* (Smith); **b,** *Oxytrigona obscura* (Friese); **c,** *Trigona (Tetragona) clavipes* (Fabricius); **d,** *T. (Duckeola) ghilianii* Spinola; **e,** *T. (Frieseomelitta) savannensis* Roubik. From Michener, 1990a.

septum, however, and the mandible can thus look like that of an ordinary meliponine with two unusually large denticles on the upper part of the apical margin. Many structures were illustrated by Schwarz (1948) and Camargo and Moure (1994).

■ *Paratrigona* ranges from Veracruz and Chiapas, Mexico, to Uruguay; Salta province, Argentina; and Bolivia. There are about 28 species; see the revisions by Schwarz (1948) and Camargo and Moure (1994).

Nest sites of *Paratrigona* are unusually diverse. Some species regularly nest in the ground, probably in abandoned mammal or ant burrows. Others nest in termite or ant nests, or make exposed nests on tree branches. Probably the plesiomorphic site is tree cavities. Wille and Michener (1973) indicated the full diversity of sites. Brood cells form combs.

Moure (1951a) separated two unusually robust species as a subgenus *Aparatrigona*. There is no doubt that they constitute a monophyletic unit that appears to be the sister group to the rest of the genus; *Aparatrigona,* however, may not be sufficiently different from *Paratrigona* s. str. to necessitate recognition of the subgenus for just these two species.

In *Paratrigona opaca* (Cockerell), S6 of the male is a plate entirely lacking the apical process usual in meliponines and found in other species of *Paratrigona*. A somewhat intermediate condition exists in *P. prosopiformis* (Gribodo); here S6 is broader than usual and the apical process is triangular (see the illustrations by Schwarz, 1948).

Genus Paratrigonoides Camargo and Roubik

Paratigonoides Camargo and Roubik, 1905; 34. Type species: *Paratigonoides mayri* Camargo and Roubik, 2005, by original designation.

As the generic name suggests, *Paratrigonoides* resembles species of *Paratrigona*, for example in the dull integument and yellow markings on the face and thorax. Indeed Camargo and Roubik (2005) in a phylogenectic analysis found it to be the probable sister group to *Paratrigona* (both *Aparatrigona* and *Paratrigona* proper, which might well be recognized as subgenera). However, it differs from *Paratrigona* in the two denticles (instead of four teeth) on the mandible, the relatively short scutellum, the lamellate upper part of the preoccipital carina, the laterally expanded supraclypeal area, etc. It differs from *Nogueirapis* in the features indicated in the key to genera.
■ *Paratrigonoides* is known only from the Cauca Valley region of Colombia. The only species is *P. mayri* Camargo and Roubik.

Genus *Pariotrigona* Moure

Pariotrigona Moure, 1961: 192. Type species: *Trigona pendleburyi* Schwarz, 1939, by original designation.

This is another genus of minute Meliponini (body length 2.5-3.0 mm), sharing the character states listed in couplet 1 of the above keys to the genera of Asia and Africa. Because it is known only from workers, its generic status and its position relative to other genera remain in doubt. The gonostyli of workers bear setae and lack minute hairs. The only other minute genus with such gonostyli is *Hypotrigona*. Those of *Pariotrigona*, however, are quite different, being broadened at the bases and separated by little more than a stylar width. The hind tibia has a distinct although rounded upper apical angle in *Pariotrigona* that is lacking in *Hypotrigona*. The long malar space of *Pariotrigona* suggests some species of the neotropical genus *Trigonisca*, a similarity strengthened by the nearly right-angular basal angle of the marginal cell and the transverse rows of hairs on the inner surface of the hind basitarsus. *Trigonisca*, however, has quite different worker gonostyli.
■ *Pariotrigona* is found from the Malay Peninsula to Vietnam, Borneo, and Sumatra. The male of *Pariotrigona* was described, and the two specific names synonymized, by Michener (2002a); thus the genus now contains only one species.

Genus *Partamona* Schwarz

This genus is composed of relatively robust forms, 4.5 to 7.0 mm long, that are superficially suggestive of *Scaptotrigona*, *Trigona (Geotrigona)*, and some species of *Trigona* s. str. *Partamona* differs from such forms in having yellowish face marks (often restricted), in the rather dense covering of erect hair on the basal propodeal area, and in the hind tibial structure of workers. This tibia is spoon-shaped and greatly broadened, about three times as wide as the femur, the outer surface mostly occupied by the enormous corbicula (Fig. 120-11b). On the inner surface of the tibia the broad keirotrichiate area extends nearly to the upper margin, which is shining but not depressed.

Key to the Subgenera of *Partamona*

1. Cuticle of thorax shining with minute, widely separated punctures; yellow of face pale and inconspicuous; metasomal terga without yellow maculations; worker gonostylus a rounded tubercle with few setae (Fig. 120-6c) .... *P. (Partamona* s. str.*)*
—. Cuticle of thorax dull and minutely roughened; paraocular areas largely bright yellow; metasomal terga usually with yellow bands or lateral spots; worker gonostylus about 1.5 times as long as broad, and setose .... *P. (Parapartamona)*

Partamona / Subgenus *Parapartamona* Schwarz

Trigona (Parapartamona) Schwarz, 1948: 428. Type species: *Trigona zonata* Smith, 1854, by original designation. [New status.]

Parapartamona differs from *Partamona* not only in the characters indicated in the key to subgenera, but also in having conspicuously long wings and an elongate propodeum, its dorsal surface being about twice as long as the scutellum.
■ This subgenus occurs in the mountains from Colombia to Peru. The group was dealt with by Schwarz (1948), and the seven species were revised by Moure (1995a); Bravo (1992), however, recognized only two species.

Partamona / Subgenus *Partamona* Schwarz s. str.

Trigona (Patera) Schwarz, 1938: 475 (not Lesson, 1837). Type species: *Melipona testacea* Klug, 1807, by original designation.
Trigona (Partamona) Schwarz, 1939b: 23, replacement for *Patera* Schwarz, 1938. Type species: *Melipona testacea* Klug, 1807, autobasic and original designation.

This subgenus lacks the special features listed for *Parapartamona*. In the species examined, the worker gonostylus is more reduced than that in other Meliponini.
■ *Partamona* ranges from Sonora, Chihuahua, and San Luis Potosí, Mexico, to Santa Catarina, Brazil, and eastern Peru, but appears to be absent from southernmost Brazil and adjacent countries. The species with testaceous bodies were revised by Camargo (1980), and the entire subgenus was revised (as a genus) by Pedro and Camargo (2003). They recognized 34 species. An account by Camargo and Pedro (2003a) provided a further review, phylogenetic and biogeographic patterns for the species, accounts of nests, and magnificent illustrations of the nests and nest entrances.

Although some *Partamona* species frequently occupy tree cavities, the commonly found nests are in partly or fully exposed sites, such as cavities in bank or cliff surfaces, or on walls of buildings, frequently below protecting eaves. The bees are usually highly aggressive when disturbed. The brood cells are arranged in combs.

Partamona grandipennis (Schwarz) and *xanthogastra* Pedro and Camargo from Costa Rica and Panama have unusually long wings, suggesting those of *Parapartamona*, but are otherwise like *Partamona* s. str. Moure (1995a) excluded *P. grandipennis* from *Parapartamona*, but by error placed it in *Parapartamona* in the same paper.

Genus *Plebeia* Schwarz

Bees of this genus have shiny cephalic and thoracic integument with minute [somewhat larger in *Plebeia (Plebeia) caerulea* (Friese)], well-separated punctures, varying to dull, densely and minutely punctate in the subgenus *Schwarziana* and in *P. (P.) schrottkyi* (Friese). The scutellum is rounded in lateral view, not shelflike, and often but not always overhangs and hides the median part of the metanotum, as seen from above. The scutellum lacks a median depression in the anterior margin like that of *Nannotrigona* and *Scaptotrigona,* but in some species there is a weak indication of such a depression. The broad area with keirotrichia on the inner side of the hind tibia extends nearly to the upper margin of the tibia (Fig. 120-10b) but the margin is shiny, largely bare, and abruptly depressed, except on the distal third of the tibia of *P. (Scaura) timida* (Silvestri), where it is not depressed.

The forms listed above with dull cephalic and thoracic integument are not closely related to one another as judged by tibial and other characters. In *Trigona,* also, there is a subgenus with fine, dull thoracic integument *(Lepidotrigona)* and a subgenus with a somewhat more shining and less closely punctate thoracic integument *(Papuatrigona),* in addition to the majority, which are shiny with well-separated, minute punctures. In *Partamona,* also, there are dull as well as shiny species. The idea that integumental dullness is necessarily a generic or subgeneric character should be abandoned.

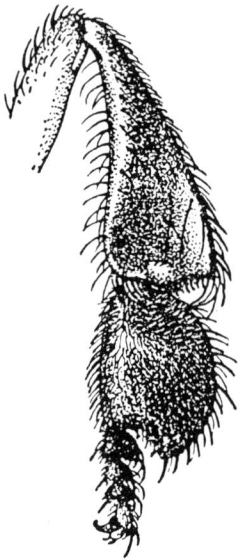

Figure 120-13. Hind tibia and tarsus of *Plebeia (Scaura) latitarsis* (Friese), worker. From Schwarz, 1948.

Key to the Subgenera of *Plebeia*

1. Hind basitarsus thickened, nearly as broad as or broader than tibia (Fig. 120-13) (face without yellow markings) .... *P. (Scaura)*
—. Hind basitarsus flat, much narrower than tibia (Fig. 120-10b) 2
2(1). Body (including metasomal terga) dull, minutely and closely punctured; forewing length about 6 mm; S3 of male with enormous procurved band of erect hooked hairs, behind which is a concave membranous area with erect hairs (South America) .... *P. (Schwarziana)*
—. Body (or at least metasoma) shining; forewing length less than 5 mm, usually 4 mm or less; S3 of male unmodified .... *P. (Plebeia s. str.)*

Plebeia / Subgenus Plebeia Schwarz s. str.

Trigona (Plebeia) Schwarz, 1938: 480. Type species: *Trigona mosquito* Smith, 1863, by original designation.
Mourella Schwarz, 1946, *in* Moure, 1946a: 442. Type species: *Melipona caerulea* Friese, 1900, by original designation.
Friesella Moure, 1946a: 441; 1946b: 611. Type species: *Melipona schrottkyi* Friese, 1900, by original designation.

This is a subgenus of small bees (body length 3-6 mm), mostly with whitish or yellow markings on the face and thorax. Unlike *Scaura,* which commonly has a more slender metasoma, that of *Plebeia* s. str. is as broad as the thorax. Special features of two species resulted in each receiving a genus-group name, as shown in the above synonymy. I doubt if subgeneric names are needed for them. *Mourella* was described and illustrated by Schwarz (1948).

■ *Plebeia* s. str. ranges from Sinaloa and Tamaulipas, Mexico, to San Luis province, Argentina. There are about 30 species; several occur in most areas. Two unusual species, *P. (Plebeia) schrottkyi* (Friese) and *caerulea* (Friese) [*Friesella* and *Mourella*], occur in southern Brazil and adjacent countries; *P. (P.) intermedia* (Wille), an unusually large species, is from Bolivia. Thus most of the diversity in the subgenus is found in southern South America. The possibility exists that *P. schrottkyi* and *caerulea* are distinct taxa convergent with *Plebeia* s. str.

Species of this subgenus nest in tree cavities, artificial containers, or in the ground. Brood cells of most species form combs, but in a small species, *Plebeia schrottkyi* (Friese), the combs are irregular, and in *P. tica* (Wille) and related minute species the cells are in clusters. The former is an unusual species constituting *Friesella* but the latter is morphologically a rather ordinary although small *Plebeia* s. str.

Plebeia / Subgenus Scaura Schwarz

Trigona (Scaura) Schwarz, 1938: 479. Type species: *Trigona latitarsis* Friese, 1900, by original designation.
Schwarzula Moure, 1946a: 439. Type species: *Trigona timida* Silvestri, 1902, by original designation.

This subgenus differs from other Meliponini in the form of the hind basitarsi, which are nearly as broad as to broader than the tibia (Fig. 120-13) and convex on the outer surfaces, at least apically, as illustrated along with other character states (including those of male genitalia and sterna) of the *latitarsis* group by Schwarz (1948). These are small bees (body length 4.0-5.5 mm). *Plebeia timida* (Silvestri) has the form of a *Plebeia* s. str.; the others have variably more slender metasomas. An interesting feature, best developed in *P. latitarsis* (Friese), is the series of flat, curved bristles on the posterior margins of S4 and S5 of the worker. The head and thorax of *Scaura* are black, without the yellow markings characteristic of nearly all species of *Plebeia* s. str. The body surface is shining with small, scattered punctures like those of most *Plebeia* s. str.

■ *Scaura* ranges from Veracruz, Mexico, to Paraná, Brazil, and Bolivia; there are five species.

At least in *Plebeia latitarsis* (Friese) and presumably to varying degrees in other species of *Scaura,* the enlarged hind basitarsi are used to rub floral and leaf surfaces to glean pollen that falls from anthers during the activities of other bees (Laroca and Lauer, 1973). Nests are in tree cavities or, for *P. latitarsis,* in the arboreal nests of *Nasutitermes.* The brood cells are arranged in combs except for those of *P. timida* (Silvestri), which are in clusters.

The subgenus consists of two units that differ considerably. *Plebeia timida* (Silvestri), sometimes placed in a monotypic genus or subgenus *Schwarzula,* has more plesiomorphies, e.g., two denticles on the upper part of the apical mandibular margin and less expanded hind basitarsi, which are only weakly convex on the outer surfaces. Unlike that of other species of the genus *Plebeia,* the upper margin of the inner surface of the hind tibia is depressed only basally and is not depressed in the apical third or fourth of the tibia. The other species of *Scaura* have untoothed mandibles (or the two denticles near the upper end of the apical margin are barely perceptible), broader and more convex hind basitarsi, and a fully depressed inner upper margin of the hind tibia as in *Plebeia* s. str.

Plebeia / Subgenus *Schwarziana* Moure

Trigona (Schwarziana) Moure, 1943a: 147. Type species: *Melipona quadripunctata* Lepeletier, 1836, by original designation.

I have retained this name for a large species (body length 6.0-7.5 mm) that suggests in body form and pale markings a large *Plebeia* s. str. with a dull, minutely punctate body (including the metasomal terga). The dorsal propodeal area has a few hairs. The most remarkable features are those of the male sterna, which are highly modified, as illustrated by Schwarz (1948); see also the above key to the subgenera. The possibility exists that *Schwarz-iana* is not phylogenetically closely related to *Plebeia* s. str.

■ *Schwarziana* is found from Goiás and Minas Gerais, Brazil, southward to Paraguay, and Misiones, Argentina. Two species are known and Melo (2003) indicated that there appear to be two others.

The biology was described in detail by Camargo (1974); nests are in the ground, and brood cells are arranged in spiral combs. Queens are unusually small, little larger than workers when they emerge from queen cells, which are scattered among the worker cells.

Genus *Plebeina* Moure

Plebeina Moure, 1961: 228. Type species: *Melipona denoiti* Vachal, 1903, by original designation.

This African genus is similar to the neotropical *Plebeia (Plebeia)* in body form, size (body length 4-5 mm), presence of limited yellowish marks at least on the face, and the narrow, depressed, shining upper margin of the inner surface of the worker hind tibia. It differs in the form of the worker gonostyli, which are flattened, diverge apically, lack setae, and have numerous minute hairs; in the male gonocoxites, which are schizogonal (probably permanently) and much longer than wide (Fig. 120-9c); and in the gonobase remnants, which are rather large. These features are as in most African Meliponini. In spite of its appearance, this genus is evidently not closely related to *Plebeia,* for it belongs to the group of African genera. *Plebeina* differs from *Meliponula* in the roughly right-angular to acute upper apical angle of the worker hind tibia, this angle bearing long, slender whitish or yellowish hairs; in the strong although pointed bristles of the rastellum; in the hairless basal area of the propodeum; and especially in the sting stylet of the worker, which is merely a rounded, membranous prominence.

■ *Plebeina* ranges from Kenya and Uganda to eastern Zaire, Botswana, and northern Transvaal and Natal, South Africa. It may be absent from West Africa. This genus contains one variable species, *P. denoiti* (Vachal), which may be divisible into several closely related species.

Nests are found in cavities in terrestrial termite nests; the brood cells form combs.

Genus *Scaptotrigona* Moure

Trigona (Scaptotrigona) Moure, 1942: 315. Type species: *Trigona postica* Latreille, 1807, by original designation.
Sakagamilla Moure, 1989b: 681. Type species: *Sakagamilla affabra* Moure, 1989, by original designation. [New synonymy.]

Scaptotrigona is among the most robust of the Meliponini; the body length is 5 to 7 mm. The head and thorax are rather strongly punctate, more coarsely so than in *Plebeia* and its relatives [although approached in coarseness by *P. caerulea* (Friese)], and there is shining ground between the punctures. As in *Nannotrigona* the scutellum is produced as a thin shelf (as seen in side view) hiding the median part of the metanotum from above. Also as described for *Nannotrigona* there is a shining V- or U-shaped median depression on the anterior margin of the scutellum (Fig. 120-10d); and the preoccipital carina and hind tibia are as described for *Nannotrigona.* *Scaptotrigona* differs from *Nannotrigona* not only in form and punctation but also in the rounded apex of the scutellum (as seen from above) and the rounded anterior margin of the pronotal lobe.

■ *Scaptotrigona* ranges from Sinaloa, Durango, and Tamaulipas, Mexico, through the tropics to Salta, Argentina, and contains about 24 species.

I have not studied specimens of *Sakagamilla,* but J. M. F. de Camargo has indicated characters that show it as a derived species of *Scaptotrigona.*

Nests are found in tree cavities, often in trunks of large trees. The brood cells form combs.

Genus *Trichotrigona* Camargo and Moure

Trichotrigona Camargo and Moure, 1983: 421. Type species: *Trichotrigona extranea* Camargo and Moure, 1983, by original designation.

This genus has many features of *Trigona (Frieseomelitta),* including the cluster rather than comb arrangement of its brood cells. The keirotrichiate ridge on the inner surface of the hind tibia is as in *Frieseomelitta* and most other subgenera of *Trigona,* with the shiny concave channel that marks its upper margin extending onto the basal fourth of the tibia. The body length is 5 mm. Extraordinary features of *Trichotrigona* are many: (1) the hairy eyes and unusually hairy body and wings, most of the hairs of the body being coarse, almost bristle-like, not plumose; (2) the short, broad second segment of the labial palpus, which is about as broad as long and only somewhat over one-fourth as long as the first segment [large setae on these segments are few (about five) and straight]; (3) the rudimentary penicillum and the replacement of the rastellar bristles with slender, tapering hairs; (4) the slender, parallel-sided, hind basitarsus of the worker, nearly four times as long as broad, with all its hairs directed apicad, i.e., without the posteriorly directed hairs and associated ridges near the base that contribute to the pollen-press function in most Meliponini (Wille, 1979a); (5) the lack of plumose hairs on the upper margin of the hind tibia except at its apex, where it rounds onto the convex apical margin of the tibia, which has numerous plumose hairs; (6) the presence of numerous, scattered, rather short hairs on the surface of the corbicula, in addition to a few longer hairs; (7) the robust front tibia covered on the outer surface with coarse, weakly spatulate hairs; and (8) the flattened, pointed, bare projection on T6, suggestive of a pygidial plate. Males of *Trichotrigona* are unknown.

■ *Trichotrigona* is known only from Amazonas, Brazil. The only species is *T. extranea* Camargo and Moure.

The reduced rastellum and penicillum and the lack of pollen-manipulating structures of the hind basitarsus, and the presence of rather numerous hairs on the surface of the corbicula and perhaps also the structure of the labial palpus, as described above, suggest that *Trichotrigona* is a robber bee like *Lestrimelitta* and *Cleptotrigona*, or possibly the only known meliponine social parasite. The first colony found was discovered while a nest of *Trigona (Frieseomelitta) paranigra* Schwarz was being opened; the relation, if any, to the *paranigra* colony is unknown; no connection was observed.

Additional nests of *Trichotrigona extranea* Camargo and Moure were reported by Camargo and Pedro (2003b). No association with colonies of *Trigona (Frieseomelitta)* or other Meliponini was found. Males were obtained and a phylogenetic analysis placed *Trichotrigona* as the sister group to *Duckeola* + *Frieseomelitta*. Since the last two taxa are here considered to be subgenera of *Trigona*, *Trichotrigona* should be placed as a subgenus of *Trigona* if further study verifies the phylogenetic conclusions.

Genus *Trigona* Jurine

This is the largest and most widely distributed genus of Meliponini. It is distinguished from all other genera except *Oxytrigona*, *Cephalotrigona*, *Trichotrigona*, and *Dactylurina* by the inner surface of the hind tibia of the worker, which has a longitudinal elevation covered with keirotrichia, above which is a depressed, shining marginal zone without keirotrichia, usually about as wide as the elevated zone, at least toward the apex of the tibia (Fig. 120-10a). In *Trigona* the slope separating the keirotrichiate ridge from the smooth zone above it is abrupt (except in the subgenera *Lepidotrigona* and *Papuatrigona*) and extends as a shining channel nearly to the base of the tibia.

Trigona is found in the neotropics from Mexico to Argentina, and in the Indo-Australian region from India and Sri Lanka to Taiwan, east to the Caroline Islands (introduced?), the Solomon Islands, and south throughout Indonesia and New Guinea to about latitude 34°S in Australia.

Because of biogeographical considerations and the startling cases of convergence in the Meliponini, the Asiatic to Australian groups of *Trigona* are probably only distantly related to the American groups. No character is known to support such an idea, but the probability should be further investigated.

In several subgenera of *Trigona* there is a well-defined sericeous area of short, dense, easily lost hairs on the base of the inner surface of the hind basitarsus (Fig. 120-10a, shown as a bare area). Such an area is present in workers and, in some subgenera, in males also. It is not a generic characteristic, since in some subgenera it is altogether absent, the surface being uniformly setose. Any bee with the sericeous area, however, belongs to the genus *Trigona* or, in Africa, to one species of *Dactylurina*.

In the great majority of species of *Trigona* and in all those in the Americas, some of the hairs along the upper margin of the hind tibia are plumose (Fig. 120-10a). No other Meliponini except for the African genus *Dactylurina* have such hairs. (There are a few on the distal part of the margin in some species of *Meliponula* and *Plebeia*.) This characteristic is therefore useful in generic recognition of most species of the genus.

Workers of the genus *Oxytrigona* and of *Trigona (Lepidotrigona)* have unusually broad hind basitarsi, convex on the outer surfaces, thus suggesting *Plebeia (Scaura)*. In *Oxytrigona* and in species of *Lepidotrigona* without greatly broadened tibiae, such as *T. (Lepidotrigona) terminata* Smith, the basitarsal width is about equal to the width of the tibia at midlength. In view of the dissimilarity of *Oxytrigona*, *Lepidotrigona*, and *Scaura*, I regard their large hind basitarsi as convergent rather than as a synapomorphy.

In most species of *Trigona* the scutum is shining with minute, widely separated punctures. In the subgenus *Lepidotrigona* it is dull, with minute, dense punctures, as in the genus *Cephalotrigona*. *Papuatrigona* is intermediate in this regard, with minute punctures separated by about a puncture's width of shiny ground.

Trigona (Tetragona) lurida Smith in the neotropics and *T. (Heterotrigona) planifrons* Smith, *flaviventris* Friese, *keyensis* Friese, and *canifrons* Smith in southeast Asia are unusual in the hairy basal area of the propodeum, a feature found also in the related genus *Cephalotrigona*, and one that crops up several times in unrelated Meliponini and does not even characterize all species of the *T. planifrons* group. The American and Asiatic species listed above show no other significant similarities and probably evolved the hairs independently.

An architectural character that has received attention in the classification of the genus *Trigona* is the arrangement of the brood cells. Most species arrange the cells in combs. In the *iridipennis* group *(Tetragonula)* of the subgenus *Heterotrigona*, however, most species arrange cells in clusters, although some (e.g., *T. carbonaria* Smith in Australia) make combs and others (e.g., *T. hockingsi*

Cockerell) are intermediate (see Michener, 1961a). *Trigona (H.) canifrons* Smith and the whole subgenus *Frieseomelitta* also arrange their cells in clusters. Probably the cluster arrangement within the genus *Trigona* is derived, presumably independently in the Indo-Australian and American taxa, the ancestral pattern for the genus being horizontal combs.

Separate keys to the subgenera are provided for the Western Hemisphere and for the Indo-Australian region. Oliveira (2002) gave a new key to the subgenera of *Trigona*, using the presence or absence of the middle tibial spur as an important character.

Key to the Neotropical Subgenera of *Trigona*

1. Mandible of worker with four or five teeth along distal margin (Fig. 119-5c); inner surface of hind basitarsus of males and workers with basal sericeous area (Fig. 120-10a) .... *T. (Trigona s. str.)*
—. Mandible of worker with lower half or two-thirds of distal margin edentate, upper part of margin with one or usually two teeth; inner surface of hind basitarsus of males without basal sericeous area, that of workers, variable 2
2(1). Metasoma short, about as wide as thorax, dorsoventrally flattened; upper margin of hind tibia of worker usually with few plumose hairs, most of them with only two to six scattered branches not concentrated toward apices; yellow markings absent; vein M of forewing dark almost to wing margin .... *T. (Geotrigona)*
—. Metasoma usually narrower than thorax, often noticeably elongate; upper margin of hind tibia of worker with numerous strongly plumose hairs (Fig. 120-11a), usually with abundant branches toward apices; yellowish or reddish markings present on face of some species; vein M of forewing usually fading away near widest part of wing 3
3(2). Inner surface of hind basitarsus of worker with basal sericeous area covered with minute setae or sometimes lacking setae (Fig. 120-10a) .... *T. (Tetragonisca)*
—. Inner surface of hind basitarsus of worker without basal sericeous area, rather uniformly setose 4
4(3). Posterior margin of vertex elevated as strong, hairy ridge between summits of eyes; upper distal angle of hind tibia of worker acute (Brazil) .... *T. (Duckeola)*
—. Posterior margin of vertex not elevated; upper distal angle of hind tibia of worker broadly rounded 5
5(4). Labial palpi with large, sinuous setae on first two segments (Fig. 120-12e) .... *T. (Frieseomelitta)*
—. Labial palpi with setae no longer than palpal width and straight or nearly so (Fig. 120-12c) .... *T. (Tetragona)*

Key to the Indo-Australian Subgenera of *Trigona*

1. Hairs along upper margin of hind tibia of workers and males all simple, or some plumose only on apical fifth or sixth of margin; elevated, keirotrichiate median zone of inner surface of hind tibia separated from shining upper marginal zone by gentle slope 2
—. Hairs along upper margin of hind tibia of workers (Fig. 120-11a) and some males partly plumose; elevated, keirotrichiate median zone of inner surface of hind tibia separated from shining upper marginal zone by abrupt slope 3
2(1). Head and thorax dull, with minute close punctures; propodeal dorsum finely reticulate; upper margin of hind tibia of worker without plumose hairs; scutum margined with whitish, densely plumose ("scale-like") hairs (Southeast Asia) .... *T. (Lepidotrigona)*
—. Head and thorax shining, although with minute, rather close punctures; propodeal dorsum smooth, shining; upper margin of hind tibia of worker with plumose hairs among bristles on apical one-fifth or one-sixth of margin; scutum without conspicuous plumose hairs (New Guinea) .... *T. (Papuatrigona)*
3(1). Inner surface of hind basitarsus of worker with basal sericeous area covered with minute setae or sometimes lacking setae (Fig. 120-10a) .... *T. (Heterotrigona)*
—. Inner surface of hind basitarsus of worker without basal sericeous area, rather uniformly setose (Southeast Asia) .... *T. (Homotrigona)*

Trigona / Subgenus *Duckeola* Moure

Duckeola Moure, 1944a: 71. Type species: *Trigona huberi* Friese, 1901 = *Trigona ghilianii* Spinola, 1853, by original designation.

This subgenus consists of large (8-9 mm long), rather robust species so different from other *Tetragona*-like bees that they must be placed in their own subgenus. It resembles *Tetragona* in lacking a sericeous area on the inner surface of the hind basitarsus of the worker. It differs from *Tetragona* and all other subgenera in the strong, hairy ridge on the posterior margin of the vertex between the summits of the eyes (this differs from the shiny, hairless carina in a similar position in the genus *Cephalotrigona*); in the rather slender hind tibia of the worker, its upper apical angle strongly produced apicad and acute; and in the roughly 20 large setae of the labial palpi, which are at most only about 1.5 times as long as the palpal width, yet are mostly curved or slightly sinuous (Fig. 120-12d). Males are unknown.

■ *Duckeola* is found in Brazil and Colombia. There are two species.

Trigona / Subgenus *Frieseomelitta* Ihering

Frieseomelitta Ihering, 1912: 5. Type species: *Trigona silvestrii* Friese, 1902, monobasic.

This subgenus consists of slender, delicate-looking species 4.0 to 6.5 mm long that agree with *Tetragona* in subgeneric attributes except that the labial palpus of the worker possesses many (19-23 in species examined) large, sinuous setae (Fig. 120-12e). Such setae are absent in *Tetragona*, although present in *Tetragonisca*. *Frieseomelitta* is further distinguished from all other American groups of the genus *Trigona* in the arrangement of the brood cells—in clusters rather than in combs. This architectural character does not separate subgenera among the Indo-Australian *Trigona* and in the genus *Plebeia*, but appears to do so in American *Trigona*. Another architectural character state of *Frieseomelitta* is the elongate storage pots, which are also found in the genus *Trigonisca* and in some species of *Trigona (Heterotrigona)*.

■ *Frieseomelitta* ranges from Sinaloa and Veracruz, Mexico, to Minas Gerais, Brazil. There are about ten species.

Nests are commonly in tree cavities, but some species favor cavities in large dead roots.

Trigona / Subgenus *Geotrigona* Moure

Trigona (Geotrigona) Moure, 1943a: 146. Type species: *Trigona mombuca* Smith, 1863, by original designation.

Geotrigona consists of robust black species 5.0 to 6.5 mm long, which, because of their short, broad metasoma, superficially resemble some of the black species of *Partamona, Scaptotrigona,* and *Trigona (Trigona)*. The *Geotrigona* group was included in *Tetragona* by Wille (1979b) and others, and it is closely related to that subgenus. In both, the hairs of the labial palpi are short and straight. *Geotrigona* differs, however, not only in the body form but in the relatively short legs (the hind tibia is much shorter than cell R of the forewing) and the sparseness of branched hairs, and of the branches themselves, on the upper margin of the hind tibia. Male genitalia and other structures were illustrated by Camargo (1996b).

■ This subgenus ranges from Michoacán, Mexico, to Santiago del Estero, Argentina. The 16 species were revised by Camargo (1996b).

Nests are typically in cavities in the ground.

Trigona / Subgenus *Heterotrigona* Schwarz

Trigona (Heterotrigona) Schwarz, 1939a: 96. Type species: *Trigona itama* Cockerell, 1918, by original designation.
Platytrigona Moure, 1961: 203. Type species: *Trigona planifrons* Smith, 1864, by original designation.
Lophotrigona Moure, 1961: 205. Type species: *Trigona canifrons* Smith, 1857, by original designation.
Tetragonula Moure, 1961: 206. Type species: *Trigona iridipennis* Smith, 1854, by original designation.
Tetragonilla Moure, 1961: 210. Type species: *Trigona atripes* Smith, 1857, by original designation.
Geniotrigona Moure, 1961: 212. Type species: *Trigona thoracica* Smith, 1857, by original designation.
Odontotrigona Moure, 1961: 213. Type species: *Trigona haematoptera* Cockerell, 1919, by original designation.
Tetrigona Moure, 1961: 215. Type species: *Trigona apicalis* Smith, 1857, by original designation.
Trigonella Sakagami and Moure, 1975, *in* Sakagami, 1975: 57 (not da Costa, 1778). Type species: *Trigona moorei* Schwarz, 1937, monobasic.
Trigona (Sundatrigona) Inoue and Sakagami, 1993: 769, replacement for *Trigonella* Sakagami and Moure, 1975. Type species: *Trigona moorei* Schwarz, 1937, by original designation and autobasic.

Heterotrigona contains minute to moderate-sized (3.0-7.5 mm long) Indo-Australian bees with a sericeous area on the base of the inner side of the hind basitarsus of workers (as in Fig. 120-10a) but not of males. The unity of *Heterotrigona* and *Homotrigona* is indicated by the frequently concave surface of the posterobasal part of the hind basitarsus, this area being delimited anteriorly by a low ridge bearing a row of hairs (the "additional" row of hairs of the pollen press, Wille, 1979a). This structure is weakly developed or unrecognizable in small species. Another character that is useful although not decisive in separating the Indo-Australian species from *Tetragona* is the setae of the labial palpus. In *Tetragona* there are no large palpal setae. In *Heterotrigona* and *Homotrigona*, except in one species, there are large setae, and at least one or two are curved; frequently, most are curved or sinuous.

The type species of *Heterotrigona, Trigona (H.) itama* Cockerell, is aberrant relative to nearly all of the rest of the species. Workers have only one denticle instead of two on the upper part of the apical mandibular margin. Males have a greatly enlarged and apically pointed hind tibia; much shortened small segments of the hind tarsus; long, thickened, and only briefly cleft hind claws; and long, fingerlike lateroapical processes on S5 (Schwarz, 1939a). In males of the group of *T. (H.) moorei* Schwarz (*Trigonella*), however, the hind legs and other features are somewhat modified in the direction of *T. (H.) itama* (Sakagami and Inoue, 1989). The same is true for *T. thoracica* Smith (*Geniotrigona*), at least insofar as the hind tarsi are concerned (Sakagami and Inoue, 1989). These findings support the placement of other groups in the same subgenus as *T. itama*.

■ *Heterotrigona* is abundant in Southeast Asia, including Borneo and Sumatra. The number of species diminishes westward to only three in India and one in Sri Lanka; few species are found east and south to the Philippines, Solomon Islands, and Australia (south to about 34°S). The subgenus also occurs in the Caroline Islands, where it is likely to have been introduced. In total there are about 36 species of *Heterotrigona*. The six Australian species were revised by Dollin, Dollin, and Sakagami (1997). The Asiatic species of the *iridipennis* group (*Tetragonula*) were revised by Sakagami (1978), modified by Sakagami and Inoue (1985).

Nests are found in tree cavities, in the ground, or, for some of those species that make clusters rather than combs of brood cells, in small and often flat spaces in or under tree bark or in manmade constructs, e.g., the space between two boards.

Of the names in synonymy with *Heterotrigona*, several are monotypic or probably so; only the *iridipennis* group (*Tetragonula*) contains more than two or three species. The named taxa in the synonymy appear to represent natural groups or single distinctive species, but do not seem different enough to recognize at the subgenus level. See Michener (1990a) for further discussion of this matter. S. F. Sakagami (in litt., 1989) suggested the recognition of the following five subgenera: *Heterotrigona* (including *Sundatrigona* and *Geniotrigona*); *Odontotrigona* (including *Tetrigona*); *Lophotrigona*; *Tetragonula* (including *Tetragonilla*); and *Platytrigona*.

Trigona / Subgenus *Homotrigona* Moure

Homotrigona Moure, 1961: 200. Type species: *Trigona fimbriata* Smith, 1857, by original designation.

Like *Duckeola, Frieseomelitta, Geotrigona, Heterotrigona,* and *Tetragonisca,* this subgenus is separated from *Tetragona* with some hesitation. The male shows plesiomorphic features (especially in S6) that suggest *Heterotrigona* as a possible outgroup for the rest of *Trigona*. It therefore seems reasonable to give it subgeneric status. Other phenetically distinctive groups, such as the group of *T. (Heterotrigona) itama* Cockerell, are distinguished by clearly derived, autapomorphic features and are not recognized here at the subgenus level. *Homotrigona* consists of rather large (7.5-8.0 mm long), robust species with the metasoma short. The two mandibular teeth are large, occupying the upper half of the distal mandibular margin, and are along the mandibular axis so that a line between their apices is almost parallel to the long axis of the mandible. *Homotrigona* is the only Indo-Australian subgenus of *Trigona* except *Lepidotrigona* and *Papuatrigona* that lacks the sericeous area on the base of the inner side of the hind basitarsus; in this respect it resembles the American subgenera *Duckeola*, *Frieseomelitta*, and *Tetragona*. The most unusual feature is S6 of the male, which is a rather ordinary-looking sternum, without traces of the heavily sclerotized median apical process usual in the Meliponini.

■ *Homotrigona* occurs from western Malaysia to Vietnam and south to Sumatra and Borneo. The single species, *Trigona fimbriata* Smith, has variants that are probably subspecies.

Nests are found in cavities in large trees.

Trigona / Subgenus *Lepidotrigona* Schwarz

Trigona (Lepidotrigona) Schwarz, 1939a: 132. Type species: *Trigona nitidiventris* Smith, 1857, by original designation.

These are delicate bees, 4.0 to 5.5 mm long, that are similar to many species of the subgenus *Heterotrigona* but differ not only as indicated in the key but also in the mostly dull, minutely roughened integument and, in males, the pair of long spines on S5 (but see *Trigona (Heterotrigona) itama* Cockerell). As in *Homotrigona*, S6 is more like an ordinary sternum than is that in most other males of *Trigona*, having a moderately large disc with the median process short and triangular. The hind tibia of the worker, especially of *T. nitidiventris* Smith and *trochanterica* Cockerell, is slender, expanded apically, and thus "racket-shaped," a convergence with species of the New World subgenus *Frieseomelitta*.

■ This subgenus occurs from India to the Philippines and Taiwan, south to Sumatra, Borneo, and Java. There are four species, some of which have named color variants that may represent subspecies or additional species.

Trigona / Subgenus *Papuatrigona* Michener and Sakagami

Trigona (Papuatrigona) Michener and Sakagami, 1990, *in* Michener, 1990a: 153. Type species: *Trigona genalis* Friese, 1908, by original designation.

Papuatrigona consists of a testaceous species 4.5 to 5.0 mm long. At first it seems to lack plumose hairs on the upper margin of the hind tibia, like *Lepidotrigona*, but such hairs are present among simple hairs on the distal one-fifth or one-sixth of the tibial margin. The keirotrichiate ridge on the inner surface of the hind tibia is nearly twice as wide as the depressed marginal zone, and the slope from the ridge to the marginal zone is gentle, the ridge not being high; distally there is almost no slope, the ridge being undefined and scarcely higher than the zone above it; proximally the slope is more distinct and extends well into the basal one-fourth of the tibia. The outer surface of the hind basitarsus has a posterior basal concavity behind a longitudinal, curved, hairy ridge, as in *Lepidotrigona* and many species of *Heterotrigona*, suggesting a relationship with those subgenera, and also with *Homotrigona*, which also has such a concavity. Such a basitarsal concavity also occurs in the American genus *Oxytrigona*, which *Papuatrigona* resembles in certain features. The broad face, long malar space, and broad clypeocular space, so unusual and distinctive of *Oxytrigona*, are all approached in *Papuatrigona*. I presume this resemblance is a result of convergence; all these characteristics are apparently results of a single tendency in facial development.

■ This subgenus is known only from New Guinea and nearby islands. The only species is *Trigona genalis* Friese.

The only nest recorded was in a cavity in a rather large living tree.

Trigona / Subgenus *Tetragona* Lepeletier and Serville

Trigona (Tetragona) Lepeletier and Serville, 1828: 710. Type species: *Trigona elongata* Lepeletier and Serville, 1828 = *Centris clavipes* Fabricius, 1804, by original designation.
Trigona (Ptilotrigona) Moure, 1951a: 47. Type species: *Trigona heideri* Friese, 1900 = *Trigona lurida* Smith, 1854, by original designation.
Camargoia Moure, 1989a: 72. Type species: *Camargoia camargoi* Moure, 1989, by original designation.

This subgenus as here limited consists of long-legged bees similar in form to *Tetragonisca* and *Frieseomelitta* but somewhat less delicate. Body length varies from 5 to 8 mm. The hind tibia is nearly as long as cell R of the forewing. *Ptilotrigona* is based on a distinctive species, differing from the rest of the subgenus in its large size, large mandibular teeth (two, on the upper part of the apical margin), and hairy propodeal triangle. The difference is probably not sufficient to justify subgeneric rank for a single divergent species, but J. M. F. de Camargo (in litt., 1993) tells me of two undescribed species of *Ptilotrigona*. One could justify recognizing it as a subgenus on the basis of its differences from other *Tetragona*. Similar comments apply to *Camargoia*. Camargo's (1996a) phylogenetic treatment shows *Tetragona*, *Ptilotrigona*, and *Camargoia* constituting a single clade. It is therefore a matter of judgment whether to unite them, as I have done, or to separate them (at the genus level), as Camargo did. *Tetragona* differs from the Indo-Australian *Heterotrigona* in the uniformly setose inner surface of the hind basitarsus of the worker and in the lack of large setae on the labial palpi (palpal setae are short and straight, Fig. 120-12c). See also the discussion of *Heterotrigona*. Male genitalia and other structures were illustrated by Moure (1989a) and Camargo (1996a).

■ *Tetragona* occurs from Tabasco, Mexico, to Minas Gerais, Brazil. About 16 species are recognized. The three species of the *Camargoia* group were revised by Camargo (1996a).

Camargo and Pedro (2004) revised the species of the *Ptilotrigona* group, here included in *Trigona (Tetragona)*, but characterized as a genus by these authors. They recognized three species; the range of the group is from Peru and central Brazil to southernmost Costa Rica. Nests as well as characters of the adults were illustrated and described. Species of this group are remarkable in that they store pollen in great quantities (3 kg in one nest) in assosciation with yeasts, and store little honey.

Nests are made in tree cavities or among roots at the bases of trees.

Trigona / Subgenus *Tetragonisca* Moure

Tetragonisca Moure, 1946a: 438. Type species: *Trigona jaty* Smith, 1863 = *Trigona angustula* Latreille, 1811, by original designation.

This is one of a group of subgenera separated from *Tetragona* with some hesitation. It consists of small (body length 4-5 mm), slender species, and is the only neotropical group other than *Trigona* s. str. with a sericeous area on the base of the inner side of the hind basitarsus of the worker (as in Fig. 120-10a), but not of the male. The labial palpus differs from that of the subgenus *Tetragona* in having 12 to 15 large, sinuous setae; there are no such large setae in *Tetragona*. In this respect *Tetragonisca* resembles *Frieseomelitta*. A characteristic of workers of *Tetragonisca* is the extremely small corbicula, the concavity being limited to the apical one-fifth of the tibia and not occupying the full tibial width. The same feature characterizes the subgenus *Frieseomelitta*, which, however, lacks the sericeous area on the inner surface of the hind basitarsus. The African genus *Dactylurina* has equally reduced corbiculae. This must be an independently derived feature in *Trigona* and *Dactylurina*, probably also independent in the *Trigona* subgenera *Tetragonisca* and *Frieseomelitta*. Intermediates between such small corbiculae and large ones occur in various groups of the subgenera *Tetragona* and *Heterotrigona*.

■ *Tetragonisca* ranges from Veracruz, Mexico, to Misiones, Argentina. It contains four species.

Nests are found in tree cavities, in the ground, or, for *Trigona angustula* Latreille, in almost any small artificial cavity.

Trigona / Subgenus *Trigona* Jurine s. str.

Trigona Jurine, 1807: 245. Type species: *Apis amalthea* Olivier, 1789, by designation of Latreille, 1810: 439.
Amalthea Rafinesque, 1815: 123, unjustified replacement for *Trigona* Jurine. Type species: *Apis amalthea* Olivier, 1789, autobasic.
Aphaneura Gray, 1832: 575. Type species: *Aphaneura rufescens* Gray, 1832, = *Apis pallens* Fabricius, 1798, monotypic. [See Michener, 1995a.]
Alphaneura Gray, 1832: pl. 111, based on *Aphaneura*. Type species: *Aphaneura rufescens* Gray, autobasic.

D. Baker stated (in litt., 1995) that he had seen the type of *Aphaneura rufescens* Gray in the Oxford University collection. It could not be found earlier (see Michener, 1995a) and has not reappeared. Even if it exists, it should not render invalid the neotype on which the current understanding of *Aphaneura* depends. Such a change could result in *Aphaneura* or *Alphaneura* being the name for some other meliponine taxon, although neither has been used for any insect since 1832.

In only two taxa of Recent Meliponini—*Trigona* s. str. and *Paratrigona*—is the distal margin of the mandible of the worker toothed in its lower part and often for its full length. The teeth are best developed in the former, and are either four (Fig. 119-5c) or five in number. Sometimes in *T. (T.) cilipes* (Fabricius) the distal teeth are united by a thin septum, but the dentate margin is still distinct from the convex margin, with at most two denticles near its upper end, as in most Meliponini. The labrum of *Trigona* s. str. differs from that of other Meliponini in that the apex is produced to a distinct angle. Both males and workers have a basal sericeous area on the inner side of the hind basitarsus (Fig. 120-10a). Such an area occurs elsewhere only in workers of the subgenera *Tetragonisca* and *Heterotrigona*. The body length is 5.5 to 11.0 mm.

■ *Trigona* s. str. is found from Nayarit to Veracruz, Mexico, south to Santa Catarina, Brazil, and Misiones, Argentina. It contains about 30 species, or more as sibling species are recognized. The subgenus was revised by Schwarz (1948); the necrophagous species (group of *T. hypogea* Silvestri) were revised by Camargo and Roubik (1991).

Nests of different species are found in the ground, among roots at the foot of a tree, in tree cavities, or exposed on the branches of trees. Some species, such as *Trigona fulviventris* Guérin-Méneville, are unaggressive, but those that make exposed nests, like *T. corvina* Cockerell, are extremely aggressive and unpleasant to work with. Various species of this subgenus are attracted to carrion (Baumgartner and Roubik, 1989) and probably use it as a supplementary protein source as well as perhaps for nest construction.

The three species of the group of *Trigona hypogea* Silvestri, however, do not collect pollen and they use carrion as their only known protein source. The storage pots and brood food of this group of species contain no pollen or only scattered, no doubt accidentally introduced, pollen grains. Noll et al. (1997) have shown that *T. hypogea* does not even visit flowers for nectar, but collects fruit juices and secretions of extrafloral nectaries. Storage pots contain either honeylike material or a paste made of honeylike material, masticated carrion, and bacteria. In *T. hypogea* the amount of protein in such pots is small but indicates the addition of some carrion (Serrão, Cruz-Landim, and Silva-de-Moraes, 1997), although the content is still honeylike. The food in larval cells contains more protein, suggesting that glandular secretions of workers contribute to the larval food. In the related *T. necrophaga* Camargo and Roubik, some storage pots, instead of containing honeylike material, contain a paste consisting largely of masticated carrion with some honeylike material. In the environment of sugars and appropriate bacteria, the proteins break down and the carrion-based mixture becomes a yellowish viscous fluid. The lack of large palpal setae and the reduced corbiculae and relatively narrow hind tibiae in all three obligately necrophagous species appear to be associated with loss of pollen-collecting and -carrying needs (Schwarz, 1948; Roubik, 1982; Camargo and Roubik, 1991).

Mateus and Noll (2004) reported that *Trigona (Trigona) hypogea* Silvestri not only feeds on carrion, but also can be predaceous on living wasp larvae in nests abandoned by the adult wasps, and on the eggs of toads (*Bufo*) stranded by low water. Possibly other soft proteinaceous sources are also attacked.

Genus *Trigonisca* Moure

Hypotrigona (Trigonisca) Moure, 1950a: 249. Type species: *Trigona duckei* Friese, 1900, by original designation.

Hypotrigona (Leurotrigona) Moure, 1950a: 244. Type species: *Trigona muelleri* Friese, 1900, by original designation.

Hypotrigona (Celetrigona) Moure, 1950a: 246. Type species: *Trigona longicornis* Friese, 1903, by original designation.

Hypotrigona (Dolichotrigona) Moure, 1950a: 248. Type species: *Trigona longitarsis* Ducke, 1916, by original designation.

Trigonisca is one of the minute (body length 2-4 mm), *Hypotrigona*-like genera. It is the principal American group having the reduced wing venation (Fig. 120-1f) characteristic of other minute Meliponini, as summarized in couplet 1 of the keys to genera. A few species of *Plebeia,* especially *P. schrottkyi* (Friese) (Fig. 120-1c), also have such reduced venation. The right-angular or weakly obtuse basal angle of the marginal cell, associated with the broad base of this cell, as indicated in couplet 1 of the key to genera, is variable and requires some explanation. In *T. muelleri* (Friese) the marginal cell is much more as in other Meliponini than is that of other *Trigonisca*. Nonetheless, in *T. muelleri* the basal angle of the cell (between the stigmal margin and vein r) is about 68°, and the width of the marginal cell at the apex of the stigma is greater than the distance across the submarginal cell area from vein Rs to vein M. In other genera (but not in *Nogueirapis*) the basal angle is usually less than 50° and the width of the marginal cell at the apex of the stigma is not greater than the distance across the submarginal cell area. These characters, among others, show the relationship of the four groups listed in the generic synonymy. *Trigonisca* differs from all other American Meliponini in the minute hairs that are widespread on the worker gonostyli (Fig. 120-6d). In addition there are setae, most or all of them along the outer margin of each gonostylus. The gonostyli are adjacent or separated by somewhat over one gonostylar width, and converge so that, when adjacent at the bases, they overlap distally. Some of the large hairs on the upper margin of the hind tibia arise from tubercles that are particularly conspicuous in the species having the slender tibiae (Fig. 120-11c). Thus the upper edge of the tibia appears weakly nodulose to strongly tuberculate, instead of smooth as in other Meliponini. Various structures including male genitalia were illustrated by Moure, Camargo, and Garcia (1988); see also Figure 120-7c.

■ *Trigonisca* ranges from Jalisco and Veracruz, Mexico, to Paraguay. There are about 30 species. The 19 species of *Trigonisca* in the narrow sense were revised by Albuquerque (1990) and the ten species included in *Dolichotrigona* were revised by Camargo and Pedro (2005).

Like the minute species of Meliponini in Africa, species of *Trigonisca* are attracted to perspiration and are sometimes pests.

Species of *Trigonisca* typically nest in small cavities such as a hollow stem or a cerambycid beetle burrow in wood or in a stem. Of course there is no space for combs of brood cells, and the cells are always in clusters.

Unlike most species, which have two small denticles at the upper end of the apical mandibular margin, *Trigonisca longitarsis* (Ducke) and *schulthessi* (Friese) have only one. If more species of these groups are found, it will be reasonable to recognize the names in the above synonymy as subgenera. It is important, however, to indicate their close relationship (shown by the worker gonostyli, the broad base of the marginal cell, the tuberculate hind tibiae, etc.) to the rest of *Trigonisca*.

121. Tribe Apini

These are the true honey bees. Probably because of the familiarity and importance of *Apis mellifera* Linnaeus, there has been considerable proliferation of names, considering that *Apis* is a small and morphologically and behaviorally unified group. Maa (1953) made a careful revision of the subfamily, recognizing four genus-group names and many species. Although he clearly split more than necessary at both the genus-group and species levels, his work is the basic one for a study of apine systematics. More recent papers were reviewed by Ruttner (1987), whose morphometric studies are important at the infraspecific level. Phylogenetic treatments were provided by Alexander (1991a, c), Engel and Schultz (1997), and others cited in the latter paper.

The Apini consists of small (7 mm long) to large (19 mm long), moderately hairy, rather elongate, apiform bees (Pls. 9, 10). The eyes are hairy. The mandibles of workers lack teeth and carinae (Fig. 119-5d). The claws of females are cleft and arolia are present. The hind tibial spurs are absent (Fig. 121-1); otherwise the hind tibia of workers is similar to that of *Bombus* and, likewise, the flabellum is similar to that of *Bombus* (see Fig. 119-2). The wings have complete, strong venation (Fig. 121-2). The marginal cell is nearly four times as long as the distance from its apex to the wing tip; the stigma is small and slender, scarcely recognizable in large species, vein r arising near the middle, the margin within the marginal cell being straight to concave. The prestigma is almost as long as or longer than the stigma. The second and third submarginal crossveins are directed posteriodistad and form acute angles with vein M. Both jugal and vannal incisions of the hind wing are shallow (Fig. 121-2). The male genitalia, compared to those found in other Hymenoptera, are greatly reduced (Fig. 121-3), largely replaced by the huge and elaborate endophallus, as illustrated by Snodgrass (1956) and many others.

The larva lacks the small conical tubercles found in other corbiculate Apidae but has transverse dorsolateral elevations on segments 1 to 4, strongest on 1 and progressively weaker to the rear. Larvae were described and illustrated by Michener (1953a), Torchio and Torchio (1975), and others.

Nests are exposed or in cavities such as hives or hollow trees, sometimes in cavities in the ground, and consist primarily of combs of cells made of wax secreted by the sternal wax glands of workers. The cells, like the brood cells of *Dactylurina* (Meliponini), are subhorizontal, forming vertical combs of two layers of cells opening in opposite directions, their bases constituting a median vertical wax sheet. In contrast to cell provisioning in the Meliponini, food for larvae is provided progressively; cells are not closed until the larva has finished feeding. A nest may consist of a single exposed comb or of multiple combs, usually in a cavity. Unlike those of other social Apidae, brood cells for workers and storage cells for honey or pollen are alike, hexagonal; brood cells for males are similar but larger. Queen-producing cells are not in combs; they are irregular, not hexagonal, and tend to hang individually from brood combs of worker cells.

The tribe Apini, like the Meliponini, has "permanent" colonies and morphologically very different female castes. New colonies are formed by fission, the old queen and a swarm of workers leaving to find a new site. Colony sizes range from a few thousand to 60,000 or more workers. Accounts of behavior can be found in Michener (1974a) and Ruttner (1987), and in innumerable books on the honey bee (*Apis mellifera* Linnaeus); an excellent recent one with numerous references to others is by Win-

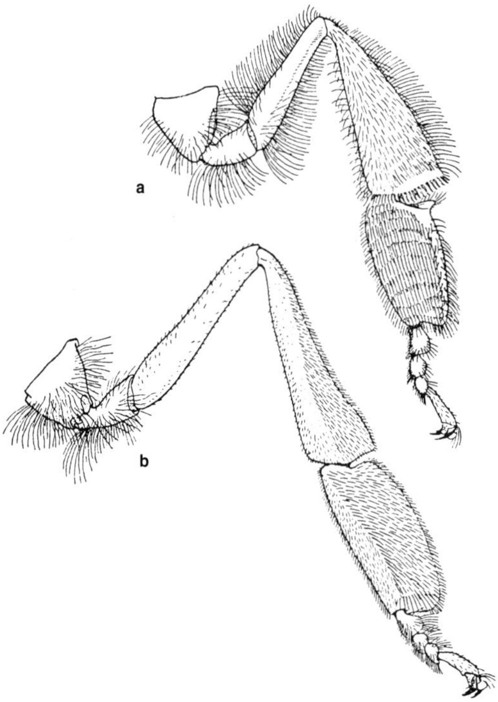

Figure 121-1. Inner surfaces of hind legs of *Apis mellifera* Linnaeus. **a,** Worker; **b,** Male. From Snodgrass, 1942.

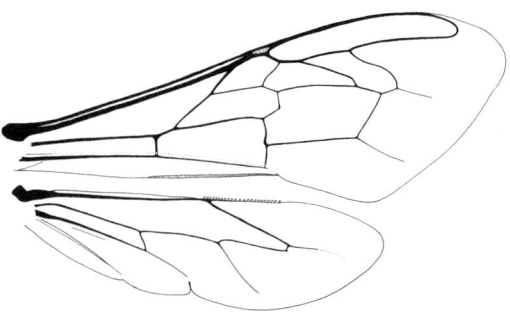

Figure 121-2. Wings of *Apis mellifera* Linnaeus, worker. From Michener, 1990a.

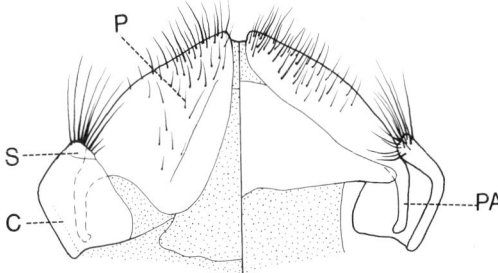

Figure 121-3. Male genitalia of *Apis mellifera* Linnaeus, dorsal view on the left, the endophallus omitted. (c, gonocoxite; p, penis valve; pa, apodeme of penis valve; s, gonostylus.) This figure was prepared to be comparable to genitalic illustrations of other bees; the endophallus of *Apis* (not illustrated) is so enormous and other parts so reduced that *Apis* genitalia on first examination appear to be entirely different from those of other Hymenoptera. From Michener, 1990a.

ston (1987). There are no parasitic or obligate robber species in the tribe Apini. The only genus is *Apis*.

Genus *Apis* Linnaeus

Apis Linnaeus, 1758: 343, 574. Type species: *Apis mellifica* Linnaeus, 1761 = *A. mellifera* Linnaeus, 1758, by designation of Latreille, 1810: 439.

Apicula Rafinesque, 1814: 27, unjustified replacement for *Apis* Linnaeus, 1758. Type species: *Apis mellifera* Linnaeus, 1758, autobasic.

Apiarus Rafinesque, 1815: 123, unjustified replacement for *Apis* Linnaeus, 1758. Type species: *Apis mellifera* Linnaeus, 1758, autobasic.

Megapis Ashmead, 1904a: 120. Type species: *Apis dorsata* Fabricius, 1793, by original designation.

Micrapis Ashmead, 1904a: 122. Type species: *Apis florea* Fabricius, 1787, by original designation.

Apis (Synapis) Cockerell, 1907a: 229. Type species: *Apis henshawi* Cockerell, 1907 (fossil), monobasic.

Hauffapis Armbruster, 1938: 37. Type species: *Hauffapis scheuthlei* Armbruster, 1938 = *Apis armbrusteri* Zeuner, 1931 (fossil), by designation of Zeuner and Manning, 1976: 243. [*Hauffapis* is not a valid name; see Michener 1997b.]

Apis (Sigmatapis) Maa, 1953: 556. Type species: *Apis cerana* Fabricius, 1793, by original designation.

Apis (Cascapis) Engel, 1999b: 187. Type species: *Apis armbrusteri* Zeuner, 1931, by original designation. [**New synonymy.**]

Apis (Priorapis) Engel, 1999b: 188. Type species: *Apis retusa* Engel, 1998, by original designation. [**New synonymy.**]

■ The genus *Apis* is found in the palearctic region north to southern Norway and the Pacific maritime provinces of Russia and in the entire African and Oriental regions. *Apis mellifera* Linnaeus has been distributed worldwide by human activity. The very many specific and subspecific names have been reduced to about 11 species. Maa (1953) recognized numerous species; more recent treatments are by Alexander (1991a, c) and Otis (1997). A taxonomic review of the genus *Apis* was by Engel (1999b). He recognized *Megapis*, *Micrapis*, *Synapis*, *Cascapis* (formerly *Hauffapis*), and *Priorapis* as subgenera; the last three are based on fossils.

Apis is primarily tropical, and was restricted to the Old World until *Apis mellifera* Linnaeus was introduced worldwide. Unlike the Meliponini, the Apini spread primarily northward from the tropics, *A. mellifera* Linnaeus probably being native as far north as southern Norway and *A. cerana* Fabricius, as far as northern China and the Pacific maritime provinces of Russia. Only in Africa does the original range of *Apis* extend into the south-temperate zone, to the southernmost part of the continent. In the tropical Asiatic islands, *Apis* ranges south to Java and east to Timor and the Philippines, but did not reach the area of New Guinea and Australia before *A. mellifera* was introduced.

Phylogenetic analysis by Alexander (1991a) indicated that (1) the small species (group of *Apis florea* Fabricius) constitute the basal branch, i.e., sister group to the rest, and that (2) the giant species (group of *A. dorsata* Fabricius) are the next branch, i.e., sister to (3) the medium-sized species (group of *A. mellifera* Linnaeus). Engel and Schulz (1997), using an architectural character and numerous larval characters in addition to adult characters and DNA sequence data, reached the same conclusion when total evidence was considered, as did Engel (1998b, 1999b). The members of group (3) construct combs in cavities; hence construction of combs in the open, as in the other two groups must be ancestral. The species in the three groups are as follows:

1. Small species with single exposed combs; dances on expanded horizontal base of comb: *Apis florea* Fabricius, *andreniformis* Smith.
2. Large species with single exposed combs; dances on vertical curtains of bees or on comb: *Apis dorsata* Fabricius, *laboriosa* Smith, *binghami* Cockerell, *breviligula* Maa. The last two are probably allopatric segregates of *A. dorsata* and may not represent distinct species.
3. Middle-sized species with multiple combs in cavities; dances on vertical surfaces of combs in the dark: *Apis mellifera* Linnaeus, *cerana* Fabricius, *koschevnikovi* Buttel-Reepen, *nigrocincta* Smith, *nuluensis* Tinget, Koeniger, and Koeniger. The last two are only recently recognized as specifically distinct from *cerana;* see Otis (1997) and Tinget, Koeniger, and Koeniger (1996).

Literature Cited

Included here are citations for relevant publications through 2005, with additions as they became available to 2006. Literature references given by author and date in the text, tables, synonymies, and figure legends are included. A date immediately following the name of an author of a specific name (e.g., "*Apis mellifera* Linnaeus, 1758") serves only to specify the species by indicating the year of its original description. This does not constitute a literature reference. But an author-date sequence not immediately following a specific name is a literature reference. In the references below, plates included in the pagination are not indicated but plates that are not numbered pages are cited.

Abrahamovich, A. H., and N. B. Díaz. 2002. Bumblebees of the neotropical region. *Biota Colombiana* 3(2): 199-214.

Abrahamovich, A. H., N. B. Diaz, and M. Lucia. 2005. Las especies del género *Bombus* Latreille en Argentina (Hymenoptera: Apidae). Estudio taxonómico y claves para su identificación. *Neotropical Entomology* 34: 235-250.

Abrams, J., and G. C. Eickwort. 1980. Biology of the communal sweat bee *Agapostemon virescens* (Hymenoptera: Halictidae) in New York State. *Search: Agriculture* [Cornell University Agricultural Experiment Station] 1980(1): 1-19.

Ackerman, A. J. 1916. The carpenter-bees of the United States of the genus *Xylocopa*. *Journal of the New York Entomological Society* 24: 196-232, pl. 10.

Adlakha, R. L. 1969. *A systematic revision of the bee genus Diadasia Patton in America north of Mexico.* 187 pp. Davis: Ph.D. thesis, University of California.

Agassiz, L. [1847]. *Nomenclatoris Zoologici Index Universalis . . . .* viii + 393 pp. Solduri: Jent and Gassmann. [This work is usually dated 1846, but the wrapper was dated 1847, according to D. B. Baker.]

Aguiar, A. C., G. A. R. Melo, J. G. Rozen, and I. Alves-dos-Santos. 2004. Synopsis of the nesting biology of Tapinotaspidini (sic) bees, *Proceedings of the 8th IBRA International Congress and VI Encontro sobre Albelhas* [Ribeirão Preto], pp. 80-85.

Aguiar, C. M. L., and M. C. Gaglianone. 2003. Nesting biology of *Centris (Centris) aenea* Lepeletier. *Revista Brasileira de Zoologia* 20: 601-606.

Albans, K. R., R. T. Aplin, J. Brehcist, J. F. Moore, and C. O'Toole. 1980. Dufour's gland and its role in secretion of nest cell lining in bees of the genus *Colletes*. *Journal of Chemical Ecology* 6: 549-564.

Albuquerque, P. M. C. de. 1990. *Revisão do gênero Trigonisca Moure, 1950.* [4] + 175 pp. Ribeirão Preto, Brazil: Thesis, Faculdade de Filosofia, Ciências e Letras.

Alcock, J. 1977. Patrolling and mating by males of *Callanthidium illustre*. *Southwestern Naturalist* 22: 554-557.

———. 1989. Size variation in the anthophorid bee *Centris pallida:* New evidence on its long term maintenance. *Journal of the Kansas Entomological Society* 62: 484-489.

———. 1996a. Site fidelity and homing ability of males of Dawson's burrowing bee (*Amegilla dawsoni*). *Journal of the Kansas Entomological Society* 69: 182-190.

———. 1996b. The relation between male body size, fighting, and mating success in Dawson's burrowing bee, *Amegilla dawsoni*. *Journal of Zoology* [London] 239: 663-674.

———. 1997a. Competition from large males and the alternative mating tactics of small males of Dawson's burrowing bee (*Amegilla dawsoni*). *Journal of Insect Behavior* 10: 99-113.

———. 1997b. Small males emerge earlier than large males in Dawson's burrowing bee (*Amegilla dawsoni*), *Journal of Zoology* [London] 242: 453-462.

———. 1999. The nesting behavior of Dawson's Burrowing bee, *Amegilla dawsoni* (Hymenoptera, Anthophorini), and the production of offspring of different sizes, *Journal of Insect Behavior* 12: 363-384.

Alcock, J., and J. P. Alcock. 1983. Male behavior in two bumblebees, *Bombus nevadensis auricomus* and *B. griseicollis*. *Journal of Zoology* 200: 561-570.

Alcock, J., E. M. Barrows, G. Gordh, L. J. Hubbard, L. Kirkendall, D. W. Pyle, T. L. Ponder, and F. G. Zalom. 1978. The ecology and evolution of male reproductive behaviour in the bees and wasps. *Zoological Journal of the Linnean Society* [London] 64: 293-325.

Alcock, J., G. C. Eickwort, and K. R. Eickwort. 1977. The reproductive behavior of *Anthidium maculosum* (Hymenoptera: Megachilidae) and the evolutionary significance of multiple copulations by females. *Behavioral Ecology and Sociobiology* 2: 385-396.

Alcock, J., and T. F. Houston. 1996. Mating systems and male size in Australian hylaeine bees. *Ethology* 102: 591-610.

Alcock, J., C. E. Jones, and S. L. Buchmann. 1976. The nesting behavior of three species of *Centris* bees. *Journal of the Kansas Entomological Society* 49: 469-474.

Alcock, J., and A. P. Smith. 1987. Hilltopping, leks and female choice in the carpenter bee *Xylocopa (Neoxylocopa) varipuncta*. *Journal of Zoology* 211: 1-10.

Alexander, B. A. 1990. A cladistic analysis of the nomadine bees. *Systematic Entomology* 15: 121-152.

———. 1991a. Phylogenetic analysis of the genus *Apis*. *Annals of the Entomological Society of America* 84: 137-149.

———. 1991b. *Nomada* phylogeny reconsidered. *Journal of Natural History* 25: 315-330.

———. 1991c. A cladistic analysis of the genus *Apis*, pp. 1-28 *in* D. R. Smith, ed., *Diversity in the Genus Apis*. Boulder, Colo.: Westview Press.

———. 1992. An exploratory analysis of cladistic relationships within the superfamily Apoidea, with special reference to sphecid wasps. *Journal of Hymenoptera Research* 1: 25-61.

———. 1994. Species-groups and cladistic analysis of the cleptoparasitic bee genus *Nomada*. *University of Kansas Science Bulletin* 55: 175-238.

———. 1996. Comparative morphology of the female reproductive system of nomadine bees. *Memoirs of the Entomological Society of Washington* no. 17: 14-35.

Alexander, B. A., and C. D. Michener. 1995. Phylogenetic studies of the families of short-tongued bees. *University of Kansas Science Bulletin* 55: 377-424.

Alexander, B. A., C. D. Michener, and A. L. Gardner. 1998. Dasypodidae Börner, 1919 (Insecta, Hymenoptera): Proposed emendation of spelling to Dasypodaidae, so removing the homonymy with Dasypodidae Gray, 1821 (Mammalia, Xenarthra). *Bulletin of Zoological Nomenclature* 55: 24-28.

Alexander, B. A., and J. G. Rozen, Jr. 1987. Ovaries, ovarioles, and oocytes in parasitic bees. *Pan-Pacific Entomologist* 63: 155-164.

Alexander, B. A., and M. Schwarz. 1994. A catalog of the species of *Nomada* (Hymenoptera: Apoidea) of the world. *University of Kansas Science Bulletin* 55: 239-270.

Alfken, J. D. 1912. Die Bienenfauna von Westpreussen. *Bericht des Westpreussischen Botanisch-Zoologischen Vereins* [Danzig] 34: 1-94, pls. I-II.

―――. 1924. Wissenschaftliche Ergebnisse der mit unterstutzung der Akademie der Wissenschaften in Wien aus der Erbschaft Treitl von F. Werner unternommenen zoologischen Expedition nach dem Anglo-Ägyptischen Sudan (Kordofan) 1914, XVI. Hymenoptera F., Apidae. *Denkschriften der Akademie der Wissenschaften in Wien, Mathematische-Naturwissenschaftliche Klasse* 99: 247-257.

―――. 1926a. Die mir bekannten chilenischen Arten der Bienengattung *Corynura* M. Spinola. *Deutsche Entomologische Zeitschrift* 1926: 145-163.

―――. 1926b. Fauna Buruana, Hymenoptera, Fam. Apidae. *Treubia* 7: 259-275.

―――. 1931. Ein weiterer Beitrag zur Kenntnis der chilenischen Arten der Bienengattung *Corynura* M. Spin. *Stettiner Entomologische Zeitung* 92: 211-218.

―――. 1932. Die chilenischen Arten der Gattung *Caenohalictus* Cam. *Archiv für Naturgeschichte* (n.f.) 1: 654-659.

―――. 1933. Beitrag zur Kenntnis der Untergattung *Pseudomegachile* Friese. *Konowia* 12: 55-59.

―――. 1935. Beitrag zur Kenntnis der Bienenfauna von Palästina. *Veröffentlichungen des Deutschen Kolonial- und Uebersee-Museums Bremen* 2: 169-192.

―――. 1937. Zur Unterscheidung der Bienengattungen *Crocisa* Jur. und *Melecta* Latr. *Konowia* 16: 172-175.

―――. 1938. Ein weiterer Beitrag zur Kenntnis der Bienenfauna von Palästina mit Einschluss des Sinai-Gebirges. *Deutsche Entomologische Zeitschrift* 1938: 418-433.

―――. 1939. Hymenoptera, Apidae, pp. 111-122 *in Missione Biologica nel Paese dei Borana*, Vol. 3, Raccolte Zoologiche, Parte 2. Roma: Reale Accademia d'Italia.

―――. 1941. Welchen wissenschaftlichen Namen hat die schwarze Mortelbiene zu führen? *Mitteilungen der Münchner Entomologischen Gesellschaft* 31: 89-92.

Alford, D. V. 1975. *Bumblebees*. xxi + 352 pp. London: Davis-Poynter.

Almeida, E. A. B., and F. A. Silveira. 1999a. Revision of the species of the subgenera of *Exomalopsis* Spinola, 1853, occurring in South America, I. *Diomalopsis* Michener and Moure, 1957. *University of Kansas Natural History Museum Special Publication* 24: 167-170.

Almeida, E. A. B., and F. A. Silveira. 1999b. Revision of the species of the subgenera of *Exomalopsis* Spinola, 1853, occurring in South America. I: *Diomalopsis* Michener and Moure, 1957 (Hymenoptera: Apidae), and a revised key to the subgenera, pp. 167-170 *in* G. W. Byers, R. H. Hagen, and R. W. Brooks, eds., *Entomological Contributions in Memory of Byron A. Alexander, University of Kansas Natural History Museum Special Publication* no. 24.

Alves-dos-Santos, I. 1999a. Aspectos morphologicos e comportamentais dos machos de *Ancyloscelis* Latreille. *Revista Brasileira de Zoologia* 16(suppl. 2): 37-43.

―――. 1999b. The proboscis of the long-tongued *Ancyloscelis* bees (Anthophoridae/ Apoidea), with remarks on flower visits and pollen collecting with the mouthparts. *Journal of the Kansas Entomological Society* 72: 277-288.

―――. 2004. Biologia de nidificação de *Anthodioctes megachiloides* Holmberg. *Revista Brasileira de Zoologia* 21: 739-744.

Alves-dos-Santos, I., G. A. R. Melo, and J. G. Rozen, Jr. 2002. Biology and immature stages of the bee tribe Tetrapediini. *American Museum Novitates* no. 3377: 1-45.

Amiet, F. 1996. *Insecta Helvetica 12, Apidae, 1. Teil*, 98 pp. Neuchâtel: Schweizerische Entomologische Gesellschaft.

Amiet, F. (with A. Müller and R. Neumeyer). 1999. *Fauna Helvetica 4, Apidae 2*. 219 pp. Neuchâtel: Schweizerische Entomologische Gesellschaft.

Amiet, F., M. Herrmann, A. Müller, and R. Neumeyer. 2001 and 2004. *Fauna Helvetica 6, Apidae 3*, 208 pp. and *Fauna Helvitica 9, Apidae 4*, 273 pp. Neuchâtel: Schweizerische Entomologische Gesellschaft.

Andersson, M. 1984. The evolution of eusociality. *Annual Review of Ecology and Systematics* 15: 165-189.

Anjos-Silva, E. J. dos, and J. M. M. Rebélo. 2006. A new species of *Exaerete* Hoffmannsegg (Hymenoptera: Apidae: Euglossini) from Brazil. *Zootaxa* no. 1105: 27-35.

Anonymous. 2006. *Bibliografia Brasileira de Polinização e Polinizadores*, 243 pp., Brasília: Ministério de Meio Ambientes.

Anzenberger, G. 1977. Ethological study of African carpenter bees of the genus *Xylocopa*. *Zeitschrift für Tierpsychologie* 44: 337-374.

Arakaki, K. T., W. A. Perreira, D. J. Preston, and J. W. Beardsley. 2002. *Pithitis smaragdula* (Fabricius), an Asiatic bee (Hymenoptera: Apidae): now apparently established in Oahu. *Proceedings of the Hawaiian Entomological Society* 35: 151.

Arduser, M. S., and C. D. Michener. 1987. An African genus of cleptoparasitic halictid bees. *Journal of the Kansas Entomological Society* 60: 324-329.

Armbrust, E. A. 2004. Resource use and nesting behavior of *Megachile prosopidis* and *M. chilopsidis* with notes on *M. discorhina*. *Journal of the Kansas Entomological Society* 77: 89-98.

Armbruster, L. 1938. Versteinerte Honigbienen aus dem obermiocänen Randecker Maar. *Archiv für Bienenkunde* 19: 1-48, 97-133.

Ascher, J. S. 2001. *Hylaeus hyalinatus* Smith, a European bee new to North America, with notes on other adventive bees. *Proceedings of the Entomological Society of Washington* 103: 184-190.

―――. 2003. Evidence for the phylogenetic position of *Nolanomelissa* from nuclear EF-1α sequence data, pp. 107-108 in Rozen, 2003c.

Ascher, J. S., and S. Patiny. 2002 A new name for the bee subgenus *Stenostylus*. *Entomological News* 113: 140.

Ashmead, W. H. 1890. On the Hymenoptera of Colorado; Descriptions of new species, notes, and a list of the species found in the state. *Bulletin of the Colorado Biological Association* no. 1: 1-47.

———. 1894. The habits of aculeate Hymenoptera, II. *Psyche* 7: 40-46.

———. 1898. Some new genera of bees. *Psyche* 8: 282-285.

———. 1899a. Classification of the bees, or the superfamily Apoidea. *Transactions of the American Entomological Society* 26: 49-100.

———. 1899b. A generic table of the family Panurgidae: A reply to Mr. Cockerell's critique on the segregation of *Perdita* Cockerell [sic]. *Psyche* 9: 372-376.

———. 1904a. Remarks on honey bees. *Proceedings of the Entomological Society of Washington* 6: 120-122.

———. 1904b. A list of the Hymenoptera of the Philippine Islands, with descriptions of new species. *Journal of the New York Entomological Society* 12: 1-22.

Ayala B., R. 1990. Abejas silvestres (Hymenoptera: Apoidea) de Chamela, Jalisco, México. *Folia Entomológica Mexicana* no. 77(1988): 395-493.

———. 1998. *Sistematica de los taxa supraespecificos de las abejas de la tribu Centridini*. iv + 180 pp. México, D.F.: Doctoral thesis, Universidad Nacional Autónoma de México.

Ayala B., R., T. L. Griswold, and D. Yanega. 1996. Apoidea, pp. 423-464 *in* J. Llorente B., A. N. García A., and E. González S., eds., *Biodiversidad Taxonomía y Biogeografía de Artrópodos Méxicanos*. México: Universidad Nacional Autónomo de México.

———. 1999. Revision de las abejas sin aguijon de México, *Folia Entomologica Méxicano* 106: 1-23.

———. 2002. Two new subgenera of bees in the genus *Centris*. *Scientific Papers, Natural History Museum, The University of Kansas* no. 25: 1-8.

Baker, C. F. 1906. American bees related to *Melecta*. *Invertebrata Pacifica* 1: 142-145.

Baker, D. B. 1972a. A revision of the genus *Plesiopanurgus* Cameron, with notes on some Arabian and African Panurginae. *Journal of Entomology* (B)41: 35-43.

———. 1972b. A new *Tarsalia* (Hym., Apoidea) from southern India. *Entomologist's Monthly Magazine* 107: 246-248.

———. 1974a. *Eupetersia*, a genus of parasitic halictine bees (Hym., Apoidea) new to the oriental region. *Entomologist's Monthly Magazine* 110: 59-63, pl. II.

———. 1974b. Two genera of nomadine bees new to India. *Entomologist's Monthly Magazine* 110: 237-240.

———. 1993. *The type material of the nominal species of exotic bees described by Frederick Smith*. vi + 312 pp. Oxford: Thesis, Oxford University.

———. 1994. A new genus of nomadine bees from North Africa. *Tijdschrift voor Entomologie* 137: 155-159.

———. 1995a. A new Malayan *Andrena*. *Deutsche Entomologische Zeitschrift* (n.f.)42: 67-69.

———. 1995b. *Xerammobates* Popov, 1951 (Insecta, Hymenoptera): Proposed designation of *Ammobates (Xerammobates) oxianus* Popov, 1951 as the type species. *Bulletin of Zoological Nomenclature* 52: 157-158.

———. 1995c. A review of Asian species of the genus *Euaspis* Gerstäcker. *Zoologische Mededelingen* [Leiden] 69: 281-302.

———. 1996a. An annotated list of the nominal species assigned to the genus *Afrostelis* Cockerell. *Deutsche Entomologische Zeitschrift* 1996: 155-157.

———. 1996b. Hymenoptera collections of Boyer de Fonscolombe: Apoidea in the University Museum, Oxford. *Journal of Natural History* 30: 537-550.

———. 1996c. Notes on some palearctic and oriental *Systropha*, with descriptions of new species and a key to the species. *Journal of Natural History* 30: 1527-1547.

———. 1997. New Melectini from western China. *Entomologist's Gazette* 48: 245-256.

———. 1998. Taxonomic and phylogenetic problems in Old World eucerine bees, with special reference to the genus *Tarsalia* Morawitz, 1895. *Journal of Natural History* 32: 823-860.

———. 1999. On new stelidine bees from S. W. Asia and N. W. Africa, with a list of the Old World taxa assigned to the genus *Stelis* Panzer, 1806. *Deutsche Entomologische Zeitschrift* 46: 231-242.

———. 2002a. On the identity of *Ceratina hieroglyphica* Smith. *Reichenbachia* 34: 357-373.

———. 2002b. A provisional, annotated, list of the nominal taxa assigned to the genus *Dasypoda* Latreille, 1802, with the description of an additional species. *Deutsche Entomologische Zeitschrift* 49: 89-103.

———. 2002c. On palaearctic and oriental species of the genera *Pseudapis* W. F. Kirby, 1900, and *Nomiapis* Cockerell, 1919. *Beitrag zur Entomologie* 52: 1-83.

———. 2003. *Ulugombakia*, a new eucerine bee from Malaya. *Beitrag zur Entomologie* 53: 123-129.

Baker, D. B., and M. S. Engel. 2006. A new subgenus of *Megachile* from Borneo with arolia. *American Museum Novitates* no. 3505:1-12.

Baker, H. G., and P. D. Hurd. 1968. Intrafloral ecology. *Annual Review of Entomology* 13: 385-414.

Baker, J. R. 1971. Development and sexual dimorphism of larvae of the bee genus *Coelioxys*. *Journal of the Kansas Entomological Society* 44: 225-235.

———. 1975. Taxonomy of five nearctic subgenera of *Coelioxys*. *University of Kansas Science Bulletin* 50: 649-730.

Ball, F. J. 1914. Les bourdons de la Belgique. *Annales de la Société Entomologique de Belgique* 58: 77-108, 1 pl.

Banaszak, J., ed. 1995. *Changes in Fauna of Wild Bees in Europe*. 220 pp. Bydgoszcz, Poland: Pedagogical University.

Banaszak, J. 2000. A checklist of the bee species (Hymenoptera, Apoidea) of Poland, with remarks on their taxonomy and zoogeography: revised edition. *Fragmenta Faunistica* 43: 135-193.

Banaszak, J., and L. Romasenko. 1998. *Megachilid Bees of Europe*. 237 pp. Bydgoszcz, Poland: Pedagogical University.

Barrows, E. M. 1975. Occupancy by *Hylaeus* of subterranean halictid nests. *Psyche* 82: 74-77.

Barth, F. G. 1991. *Insects and Flowers, the Biology of a Partnership*. x + 408 pp. Princeton: Princeton University Press.

Barthell, J. F., and H. V. Daly. 1995. Male size variation and mating site fidelity in a population of *Habropoda depressa* Fowler. *Pan-Pacific Entomologist* 71: 149-156.

Barthell, J. F., and R. W. Thorp. 1995. Nest usurpation among females of an introduced leaf-cutter bee, *Megachile apicalis*. *Southwestern Entomologist* 20: 117-184.

Batra, S. W. T. 1966a. Nests and social behavior of halictine bees of India. *Indian Journal of Entomology* 28: 377-393.

———. 1966b. Social behavior and nests of some nomiine bees in India. *Insectes Sociaux* 13: 145-154.

———. 1980. Ecology, behavior, pheromones, parasites and management of the sympatric vernal bees *Colletes inaequalis, C. thoracicus* and *C. validus. Journal of the Kansas Entomological Society* 53: 509-538.

Batra, S. W. T., and B. B. Norden. 1996. Fatty food for their brood: How *Anthophora* bees make and provision their cells. *Memoirs of the Entomological Society of Washington* 17: 36-44.

Batra, S. W. T., S. F. Sakagami, and Y. Maeta. 1993. Behavior of the Indian allodapine bee *Braunsapis kaliago*, a social parasite in nests of *B. mixta. Journal of the Kansas Entomological Society* 66: 345-360.

Baumgartner, D. L., and D. W. Roubik. 1989. Ecology of necrophilous and filth-gathering stingless bees (Apidae: Meliponinae) of Peru. *Journal of the Kansas Entomological Society* 62: 11-22.

Bawa, K. S. 1990. Plant-pollinator interactions in tropical rain forests. *Annual Review of Ecology and Systematics* 21: 399-422.

Bego, L. R., A. F. Grosso, R. Zucchi, and S. F. Sakagami. 1999. Oviposition behavior of the stingless bees XXIV. Ethological relationships of *Tetragonisca angustula angustula* to other Meliponinae taxa, *Entomological Science* 2:473-482.

Bembe, B. 2004. Functional morphology in male euglossine bees and their ability to spray fragrances. *Apidologie* 55:283-291.

Benedek, P. 1973. An undescribed dufoureine bee from the Carpathian basin. *Acta Zoologica Academiae Scientiarum Hungaricae* 19: 271-276.

Ben Mordechai, Y., R. Cohen, D. Gerling, and E. Moscovitz. 1978. The biology of *Xylocopa pubescens* Spinola (Hymenoptera: Anthophoridae) in Israel. *Israel Journal of Entomology* 12: 107-121.

Bennett, B., and M. D. Breed. 1985. The nesting biology, mating behavior, and foraging ecology of *Perdita opuntiae. Journal of the Kansas Entomological Society* 58: 185-194.

Bennett, F. D. 1966. Notes on the biology of *Stelis (Odontostelis) bilineolata* (Spinola), a parasite of *Euglossa cordata* (Linnaeus). *Journal of the New York Entomological Society* 74: 72-79.

———. 1972. Observations on *Exaerete* spp. and their hosts *Eulaema terminata* and *Euplusia surinamensis* (Hymen., Apidae, Euglossinae) in Trinidad. *Journal of the New York Entomological Society* 80: 118-124.

Benoist, R. 1931. Les osmies de la faune française. *Annales de la Société Entomologique de France* 100: 23-60.

———. 1940. Remarques sur quelques espèces de mégachiles principalement de la faune française. *Annales de la Société Entomologique de France* 109: 41-88.

———. 1942. Les Hyménoptères qui habitent les tiges de ronce aux environs de Quito (Equateur). *Annales de la Société Entomologique de France* 111: 75-90.

———. 1959. Les *Prosopis* de France. *Cahiers des Naturalistes, Bulletin des Naturalistes Parisiens* 15: 75-87.

———. 1962. Nouvelles espèces d'apides malgaches. *Bulletin de la Société Entomologique de France* 67: 214-223.

Benson, R. B., C. Ferrière, and O. W. Richards. 1937. The Generic Names of British Insects, Part 5. Hymenoptera Aculeata, pp. 81-149. London: Royal Entomological Society.

Benthem, F. D. J. van, V. L. Imperatriz-Fonseca, and H. H. W. Velthuis. 1995. Biology of the stingless bee *Plebeia remota* (Holmberg): Observations and evolutionary implications. *Insectes Sociaux* 42: 71-87.

Bernhardt, P., and K. Walker. 1996. Observations on the foraging preferences of *Leioproctus (Filiglossa)* Rayment (Hymenoptera: Colletidae) in eastern Australia. *Pan-Pacific Entomologist* 72: 130-137.

Berthold, A. A. 1827. *Latreille's . . . Naturliche Familien des Thierreichs aus dem Französischen mit Anmerkungen und Zusätzen*. x + 604 pp. Weimar: Landes-Industr.-Compt.

Bertoni, A. de W. 1911. Contribución á la biología de las avispas y abejas del Paraguay. *Anales del Museo Nacional de Buenos Aires* 15: 97-146.

Bienvenu, R. J., F. W. Atchison, and E. A. Cross. 1968. Microbial inhibition by prepupae of the alkali bee, *Nomia melanderi. Journal of Invertebrate Pathology* 12: 278-282.

Biesmeijer, J. C., S. P. M. Roberts, M. Reemer, R. Ohiemüller, M. Edwards, T. Peeters, A. P. Schaffers, S. G. Potts, R. Kleukers, C. D. Thomas, J. Settele, and W. E. Kumin. 2006. Parallel declines in pollinators and insect pollinated plants in Britain and the Netherlands. *Science* 313: 351-354.

Bingham, C. T. 1897. *The Fauna of British India Including Ceylon and Burma*, Hymenoptera, Vol. I. Wasps and Bees. xxix + 577 pp., 4 pls. London: Taylor and Francis. [Reprinted, 1975, New Delhi: Today and Tomorrow's.]

Bischoff, H. 1923. Zur Kenntnis afrikanischer Schmarotzerbienen. *Deutsche Entomologische Zeitschrift* 1923: 585-603.

———. 1927. Biologie der Hymenopteren. vii + 598 pp. Berlin: Springer.

———. 1930. Beitrag zur Kenntnis paläarktischer Arten der Gattung *Epeolus. Deutsche Entomologische Zeitschrift* 1930: 1-15.

———. 1934. Gedanken zu einem natürlichen System der Bienen. *Deutsche Entomologische Zeitschrift* 1934: 324-331.

———. 1936. Schwedisch-chinesische wissenschaftliche Expedition nach den nordwestlichen Provinzen Chinas, unter Leitung von Dr. Sven Hedin und Prof. Sü Pingchang. Insekten gesammelt vom schwedischen Arzt der Expedition Dr. David Hummel 1927-1930, 56. Hymenoptera, 10. Bombinae. *Arkiv för Zoologi* 27: 1-27.

Blagoveshchenskaya, N. N. 1963. Giant colony of the solitary bee *Dasypoda plumipes* Panz. *Entomologicheskoe Obozrenie* 42: 115-117. [In Russian.]

Blair, B. H. 1935. The bees of the group *Dieunomia. Journal of the New York Entomological Society* 43: 201-214, pl. XVI.

Blanchard, E. 1840. Hyménoptères, pp. 219-415, pls. 1-7, *in* F. L. N. Laporte de Castelnau, *Histoire Naturelle des Insectes . . .*, Vol. 3. Paris: Duméril. [Castelnau is commonly cited as the name of the editor of this work, but for consistency with Lepeletier de Saint-Fargeau, Laporte seems more appropriate, as indicated by D. B. Baker, in litt. (1996).]

———. 1845-1849. Hyménoptères, pp. 113-227, pls. 107-129, *in* G. L. C. F. D. Cuvier, *Le Règne Animal . . .*, ed. 3, Vol. 2. Paris: Fortin, Masson. [Usually dated 1849; but see F. C. Cowan, 1976, *Journal of the Society for the Bibliogra-*

phy of Natural History 8: 32-64. The text on bees (Les Mellifères, pp. 201-227) appeared in 1846. Plates and their years of publication are as follows: 125, 1847; 126, 1846; 127, 1846; 128, 1847; 128 bis, 1846; 129, 1845.]

Blüthgen, P. 1920, 1921. Die deutschen Arten der Bienengattung *Halictus* Latr. *Deutsche Entomologische Zeitschrift* 1920: 81-132; 1921: 267-302.

———. 1923a. Beiträge zur Kenntnis der Bienengattung *Halictus* Latr. *Archiv für Naturgeschichte,* Abt. A 89(5): 232-332.

———. 1923b. Beiträge zur Systematik der Bienengattung *Halictus* Latr. *Konowia* 2: 65-142.

———. 1923c. Beiträge zur Systematik der Bienengattung *Sphecodes* Latr. *Deutsche Entomologische Zeitschrift* 1923: 441-514.

———. 1924a. Contribución al conocimiento de las especies españolas de *"Halictus."* *Memorias de la Real Sociedad Española de Historia Natural* 11: 331-544.

———. 1924b. Beiträge zur Systematik der Bienengattung *Sphecodes* Latr. *Deutsche Entomologische Zeitschrift* 1924: 457-516.

———. 1924c. Beiträge zur Systematik der Bienengattung *Halictus* Latr. *Konowia* 3: 53-64, 76-95, 253-284.

———. 1925. Die Bienengattung *Nomioides* Schenck. *Stettiner Entomologische Zeitung* 86: 1-100.

———. 1926. Beiträge zur Kenntnis der indo-malayischen *Halictus-* und *Thrincostoma-*Arten. *Zoologische Jahrbücher, Abteilung für Systematik, Geographie und Biologie der Tiere* 51: 375-698, pls. 4-5.

———. 1927. Beiträge zur Systematik der Bienengattung *Sphecodes* Latr., III. *Zoologische Jahrbücher, Abteilung für Systematik, Geographie und Biologie der Tiere* 53: 23-112.

———. 1928a. Beiträge zur Kenntnis der indo-malayischen *Halictus-* und *Thrincostoma-*Arten, 1. Nachtrag. *Zoologische Jahrbücher, Abteilung für Systematik, Geographie und Biologie der Tiere* 54: 343-406.

———. 1928b. 2. Beitrag zur Kenntnis der äthiopischen Halictinae. *Deutsche Entomologische Zeitschrift* 1928: 49-72.

———. 1928c. Beiträge zur Kenntnis der afrikanischen Halictinae. *Zoologische Jahrbücher, Abteilung für Systematik, Geographie und Biologie der Tiere* 55: 163-252.

———. 1930. Beitrag zur Kenntnis der äthiopischen Halictinae. *Mitteilungen aus dem Zoologischen Museum in Berlin* 15: 495-542.

———. 1931. Beiträge zur Kenntnis der indomalayischen *Halictus-* und *Thrincostoma-*Arten. *Zoologische Jahrbücher, Abteilung für Systematik, Geographie und Biologie der Tiere* 61: 285-346.

———. 1933. Beitrag zur Kenntnis der äthiopischen Halictinae. *Mitteilungen aus dem Zoologischen Museum in Berlin* 18: 363-394.

———. 1934. Die Wirte der paläarktischen *Sphecodes*-Arten. *Zeitschrift für Wissenschaftliche Insektenbiologie* 27: 205-214.

———. 1936. Beitrag zur Kenntnis der äthiopischen Halictinae. *Deutsche Entomologische Zeitschrift* 1935: 177-190.

———. 1937. Halictinae (Hymenoptera; Apidae) von den Kanarischen Inseln. *Commentationes Biologicae [Societas Scientiarum Fennica]* 6(11): 1-11.

———. 1949. Neues oder Wissenswertes uber mitteleuropaische Aculeaten und Goldwespen, I. Apidae. *Beiträge zur Taxonomischen Zoologie* 1: 77-100.

———. 1955. The Halictinae (Hymen., Apoidea) of Israel, I. Genus *Halictus* (subgenera *Halictus* s. str. and *Thrincohalictus*). *Bulletin of the Research Council of Israel* (B) 5: 5-23.

———. 1961. Ergebnisse der Deutschen Afghanistan-Expedition 1956 der Landessammlungen für Naturkunde Karlsruhe. *Beiträge zur Naturkundlichen Forschung in Südwestdeutschland* 19: 277-287.

Bogusch, P. 2005. Biology of the parasitic bee *Epeoloides coecutiens*. *Journal of the Kansas Entomological Society* 78: 1-12.

Bohart, G. E. 1965. A new genus of dufoureine bee from Texas (Hymenoptera: Halictidae). *Annals of the Entomological Society of America* 58: 319-321.

———. 1970. *The evolution of parasitism among bees.* ii + 30 pp. Logan: 41st Honor Lecture, Utah State University.

Bohart, G. E., and T. L. Griswold. 1987. A revision of the dufoureine genus *Micralictoides* Timberlake. *Pan-Pacific Entomologist* 63: 178-193.

———. 1997. A revision of the rophitine genus *Protodufourea*. *Journal of the Kansas Entomological Society* 69(1996, suppl.): 177-184.

Bohart, G. E., and N. N. Youssef. 1976. The biology and behavior of *Evylaeus galpinsiae* Cockerell. *Wasmann Journal of Biology* 34: 185-234.

Bohart, R. M., and A. S. Menke. 1976. *Sphecid Wasps of the World.* 695 pp. Berkeley: University of California Press.

Bonilla-Gómez, M. A., and G. Nates-Parra. 1992. Abejas euglosinas de Colombia (Hymenoptera: Apidae), 1. Claves ilustradas. *Caldasia* 17: 149-172.

Börner, C. 1919. Stammesgeschichte der Hautflügler. *Biologisches Zentralblatt* 39: 145-185.

Bosch, J., N. Vicens, and M. Blas. 1993. Análisis de los nidos de algunos Megachilidae nidificantes en cavidades preestablecidas. *Orsis* 8: 53-63.

Bourmeister-Radoschkowsky. See Radoszkowski.

Bouseman, J. K., and W. E. LaBerge. 1979. A revision of the bees of the genus *Andrena* of the Western Hemisphere, Part IX. Subgenus *Melandrena*. *Transactions of the American Entomological Society* 104: 275-389.

Brach, V. 1978. Notes on the biology of *Lithurgus gibbosus* Smith in Florida. *Bulletin of the Southern California Academy of Sciences* 77: 144-147.

Brady, S. G., and B. N. Danforth. 2004. Recent intron gain in elongation factor-1α of colletid bees. *Molecular Biology and Evolution* 21: 691-696.

Braue, A. 1913. Die Pollensammelapparate der biensammelnden Bienen. *Jenaischen Zeitschrift der Naturwissenschaft* 50: 1-96.

Brauns, H. 1902. *Eucondylops* n. g. Apidarum. *Zeitschrift für Systematische Hymenopterologie und Dipterologie* 2: 377-380.

———. 1926. V. Nachtrag zu "Friese, Bienen Afrikas." *Zoologische Jahrbücher, Abteilung für Systematik, Geographie und Biologie der Tiere* 52: 187-230, pl. 5.

———. 1929. Neue und auffallende Apiden aus Sud-Afrika. *Zeitschrift für Wissenschaftliche Insektenbiologie* 24: 130-143.

Bravo, F. 1992. Sistematica e distribução de *Parapartamona* Schwarz. *Revista Brasileira de Entomologia* 36: 863-878.

Breed, M. D., and B. Bennett. 1987. Kin recognition in highly eusocial insects, pp. 243-285 *in* D. J. C. Fletcher and C. D. Michener, eds., *Kin Recognition in Animals.* Chichester, U.K.: Wiley.

Brèthes, J. 1909a. Hymenoptera Paraguayensis. *Anales del Museo Nacional de Buenos Aires* 19: 225-256.

———. 1909b. Himenópteros nuevos de las Repúblicas del Plata y del Brasil. *Anales del Museo Nacional de Buenos Aires* 19: 49-69.

———. 1909c. Himenópteros de Mendoza y de San Luis. *Anales del Museo Nacional de Buenos Aires* 17: 455-463.

———. 1910. Sur les *Ancylocelis* et genres voisins [Hym.]. *Bulletin de la Société Entomologique de France* 1910: 211-213.

———. 1916. Le genre *"Xylocopa"* Latreille dans la Republique Argentine. *Physis* [Buenos Aires] 2: 407-421.

———. 1922. Himenópteros y Dípteros de varias procedencias. *Anales de la Sociedad Científica Argentina* 93: 119-146.

Bridwell, J. C. 1919. Miscellaneous notes on Hymenoptera. *Proceedings of the Hawaiian Entomological Society* 4: 109-165.

Broemeling, D. K. 1988. A revision of the *Nomada* subgenus *Nomadita* of North America. *Pan-Pacific Entomologist* 64: 321-344.

Broemeling, D. K., and A. S. Moalif. 1988. A revision of the *Nomada* subgenus *Pachynomada. Pan-Pacific Entomologist* 64: 201-227.

Brooks, R. W. 1983. Systematics and bionomics of *Anthophora*: The *bomboides* group and species groups of the new world. *University of California Publications in Entomology* 98: i-x + 1-86.

———. 1988. Systematics and phylogeny of the anthophorine bees. *University of Kansas Science Bulletin* 53: 436-575.

———. 1999. Bees of the genus *Anthophora* Latreille 1803 (Hymenoptera, Apidae, Anthophorini) of the West Indies. *Tropical Zoology* 12: 105-124.

Brooks, R. W., and M. S. Engel. 1998. New bees of the genus *Ischnomelissa. Deutsche Entomologische Zeitschrift* 45: 181-189.

———. 1999. A revision of the augochlorine bee genus *Chlerogas* Vachal. *Zoological Journal of the Linnean Society* [London] 125: 463-486.

Brooks, R. W., and T. L. Griswold. 1988. A key to the species of *Trachusa* subgenus *Heteranthidium* with descriptions of new species from Mexico. *Journal of the Kansas Entomological Society* 61: 332-346.

Brooks, R. W., and C. D. Michener. 1988. The Apidae of Madagascar and nests of *Liotrigona. Sociobiology* 14: 299-323.

Brothers, D. J. 1975. Phylogeny and classification of the aculeate Hymenoptera, with special reference to the Mutillidae. *University of Kansas Science Bulletin* 50: 483-648.

———. 1976. Modifications of the metapostnotum and origin of the 'propodeal triangle' in Hymenoptera Aculeata. *Systematic Entomology* 1: 177-182.

———. 1999. Phylogeny and evolution of wasps, ants and bees, pp. 233-249 *in* F. Ronquist, Phylogeny of the Hymenoptera (Insecta): The state of the art. *Zoologica Scripta* 28.

Brothers, D. J., and J. M. Carpenter. 1993. Phylogeny of Aculeata: Chrysidoidea and Vespoidea. *Journal of Hymenoptera Research* 2: 227-304.

Brumley, R. L. 1965. A revision of the bee genus *Epeolus* of western America north of Mexico. 92 pp. Logan: MS. thesis, Utah State University.

Buchmann, S. L. 1987. The ecology of oil flowers and their bees. *Annual Review of Ecology and Systematics* 18: 343-369.

Budrys, E. 2001. On the origin of nest building behavior in digger wasps. *Norwegian Journal of Entomology* 48: 45-49.

Bugnion, E. 1928. *Les Glandes Salivaires de l'Abeille et des Apiaires en Général.* 64 pp. Montfavet (Vaucluse): Librairie de Vulgarisation Apicole.

Buschini, M. L. T., and L. A. de O. Campos. 1995. Caste differentiation in *Trigona spinipes* (Hymenoptera; Apidae): Influence of the available food and the juvenile hormone. *Revista Brasileira de Biologia* 55(suppl.): 121-129.

Butler, G. D. 1967. Biological observations on *Ptilothrix sumichrasti* (Cresson) in southern Arizona. *Pan-Pacific Entomologist* 43: 8-14.

Buysson, R. du. 1900. Sur quelques hyménoptères de Madagascar. *Annales de la Société Entomologique de France* 69: 177-180.

Camargo, J. M. F. 1970. Ninhos e biologia de algumas espécies de melipoideos (Hymenoptera: Apidae) da região de Pôrto Velho, Território de Rondônia, Brasil. *Revista de Biologia Tropical* 16: 207-239.

———. 1974. Notas sobre a morfologia e biologia de *Plebeia (Schwarziana) quadripunctata quadripunctata. Studia Entomologica* 17: 433-470.

———. 1980. O grupo *Partamona (Partamona) testacea* (Klug): Suas espécies, distribuição e diferenciação geográfica. *Acta Amazonica* (suplemento) 10(4): 1-175.

———. 1984. Notas sobre o gênero *Oxytrigona* (Meliponinae, Apidae, Hymenoptera). *Boletim do Museu Paraense Emílio Goeldi.* (Série Zoologia) 1: 115-124.

———. 1989. Comentários sobre a sistemática de Meliponinae. *Anais do XIV Simpósio Anual da Academia de Ciências do Estado de São Paulo* [São Carlos], suplemento, [21 pp.].

———. 1996a. Meliponini neotropicais: O gênero *Camargoia* Moure, 1989. *Arquivos de Zoologia* [São Paulo] 33: 71-92.

———. 1996b. Meliponini neotropicais: O gênero *Geotrigona* Moure, 1943 (Apinae, Apidae, Hymenoptera) con especial referência à filogenia e biogeografia. *Arquivos de Zoologia* [São Paulo] 33: 93-161.

Camargo, J. M. F., D. A. Grimaldi, and S. R. M. Pedro. 2000. The extinct fauna of stingless bees (Hymenoptera: Apidae: Meliponini) in Dominican amber: Two new species and redescription of the male of *Proplebeia dominicana* (Wille and Chandler). *American Museum Novitates* no. 3293: 1-24.

Camargo, J. M. F., W. E. Kerr, and C. R. Lopes. 1967. Morfologia externa de *Melipona (Melipona) marginata* Lepeletier. *Papéis Avulsos de Zoologia, São Paulo* 20: 229-258, pls. A-P + unlettered colored plate.

Camargo, J. M. F., and J. S. Moure. 1983. *Trichotrigona*, um novo gênero de Meliponinae (Hymenoptera, Apidae), do Rio Negro, Amazonas, Brasil. *Acta Amazônica* 13: 421-429.

———. 1989. Duas espécies novas de *Lestrimelitta* Friese (Meliponinae, Apidae, Hymenoptera) da região Amazônica. *Boletim do Museu Paraense Emílio Goeldi.* (Série Zoologia) 5: 195-212.

———. 1994. Meliponinae neotropicais: Os gêneros *Paratrigona* Schwarz, 1938 e *Aparatrigona* Moure, 1951. *Arquivos de Zoologia, Museu de Zoologia da Universidade de São Paulo* 32: 33-109.

———. 1996. Meliponini neotropicais: O gênero *Geotrigona* Moure, 1943 (Apinae, Apidae, Hymenoptera), com especial referência à filogenia e biogeografia. *Arquivos de Zoologia, Museu de Zoologia da Universidade de São Paulo* 33: 95-161.

Camargo, J. M. F., J. S. Moure, and D. W. Roubik. 1988. *Melipona yucatanica* new species (Hymenoptera: Apidae: Meliponinae): Stingless bee dispersal across the Caribbean arc and post-Eocene vicariance. *Pan-Pacific Entomologist* 64: 147-157.

Camargo, J. M. F., and S. R. de M. Pedro. 1992a. Sistemática de Meliponinae (Hymenoptera, Apidae): Sobre a polaridade e significado de alguns caracteres morfológicos, pp. 45-49 *in* Anais do Encontro Brasileiro sobre Biologia de Abelhas e Outros Insetos Sociais. *Naturalia,* edição especial.

———. 1992b. Systematics, phylogeny and biogeography of the Meliponinae (Hymenoptera: Apidae): A mini-review. *Apidologie* 23: 509-522.

———. 2003a. Meliponini neotropicais: O genero *Partamona* Schwarz, 1939 (Hymenoptera, Apidae, Apinae) bionomia e biogeografia. *Revista Brasileira de Entomologia* 47: 311-372.

———. 2003b. Sobre as relações filogenéticas de *Trichotrigona* Camargo and Moure, pp. 109-122 *in* Melo and Alves-dos-Santos (2003).

———. 2004. Meliponini Neotropicais: o genero *Ptilotrigona* Moure. *Revista Brasileira de Entomologia* 48: 353-377.

———. 2005. Meliponini neotropicais: o genero *Dolichotrigona* Moure. *Revista Brasileira de Entomologia* 49: 69-92.

Camargo, J. M. F., and D. W. Roubik. 1991. Systematics and bionomics of the apoid obligate necrophages: The *Trigona hypogea* group. *Biological Journal of the Linnean Society* 44: 13-39.

———. 2005. Neotropical Meliponini: *Paratrigona mayri* new genus and species from western Colombia (Hymenoptera, Apidae) and phylogeny of related genera. *Zootaxa* no. 1081: 33-45.

Camargo, J. M. F., R. Zucchi, and S. F. Sakagami. 1975. Observations on the bionomics of *Epicharis (Epicharana) rustica flava* (Olivier) including notes on its parasite, *Rhathymus* sp. *Studia Entomologica* 18: 313-340.

Cameron, P. 1898. Hymenoptera Orientalia, or contributions to a knowledge of the Hymenoptera of the Oriental zoological region, Part VII. *Memoirs, Manchester Literary and Philosophical Society* 42(11): 1-84, pl. 4.

———. 1901a. Descriptions of three new genera and seven new species of Hymenoptera from eastern Asia and Australia. *Annals and Magazine of Natural History* (7)8: 116-123.

———. 1901b. Description of a new genus of bees from India. *Entomologist* 34: 262-263.

———. 1902a. Descriptions of new genera and species of Hymenoptera collected by Major C. G. Nurse at Deesa, Simla and Ferozepore, Part II. *Journal of the Bombay Natural History Society* 14: 419-449, 1 pl.

———. 1902b. On the Hymenoptera collected by Mr. Robert Shelford at Sarawak, and on the Hymenoptera of the Sarawak Museum. *Journal of the Straits Branch of the Royal Asiatic Society* 37: 29-140.

———. 1903a. Descriptions of new species of Hymenoptera taken by Mr. Edward Whymper on the "Higher Andes of the Equator." *Transactions of the American Entomological Society* 29: 225-238.

———. 1903b. Descriptions of new genera and species of Hymenoptera taken by Mr. Robert Shelford at Sarawak, Borneo. *Journal of the Straits Branch of the Royal Asiatic Society* 39: 89-181.

———. 1905. On the Hymenoptera of the Albany Museum, Grahamstown, South Africa. *Records of the Albany Museum* 1: 185-265.

———. 1907. On a new genus and some new species of aculeate Hymenoptera collected by Lieut.-Col. C. G. Nurse in Baluchistan. *Journal of the Bombay Natural History Society* 18: 130-136.

Cameron, S. A. 1991. A new tribal phylogeny of the Apidae inferred from mitochondrial DNA sequences, pp. 71-87 *in* D. R. Smith, ed., *Diversity in the Genus* Apis. Boulder, Colo.: Westview Press.

———. 1993. Multiple origins of advanced eusociality in bees inferred from mitochondrial DNA sequences. *Proceedings of the National Academy of Sciences USA* 90: 8687-8691.

———. 2004. Phylogeny and biology of neotropical orchid bees. *Annual Review of Entomology* 49: 377-404.

Cameron, S. A., and P. Mardulyn. 2001. Multiple molecular data sets suggest independent origins of highly eusocial behavior in bees. *Systematic Biology* 50: 194-214.

———. 2003. The major opsin gene is useful for inferring higher level phylogenetic relationships of the corbiculate bees. *Molecular Phylogenetics and Evolution* 28: 610-613.

Cameron, S. A., and S. Ramírez. 2002. Nest architecture and nesting ecology of the orchid bee *Eulaema meriana. Journal of the Kansas Entomological Society* 74(2001): 142-165.

Cameron, S. A., J. B. Whitfield, C. L. Hulslander, W. A. Cresko, S. P. Isenberg, and R. W. King. 1997. Nesting biology and foraging patterns of the solitary bee *Melissodes rustica* (Hymenoptera: Apidae) in northwest Arkansas. *Journal of the Kansas Entomological Society* 69(1996 suppl.): 260-273.

Camillo, E., C. A. Garófalo, M. J. de O. Campos, and J. C. Serrano. 1983. Preliminary notes on the biology of *Lithurgus huberi. Revista Brasileira de Biologia* 43: 151-156.

Camillo, E., C. A. Garófalo, and J. C. Serrano. 1993. Habitos de nidificação de *Melitona segmentaria, Centris collaris, Centris fuscata,* e *Paratetrapedia gigantea. Revista Brasileira de Entomologia* 37: 145-156.

Campos, M. J. de O. 1980. *Aspectos da sociologia e fenologia de* Pereirapis semiauratus. [10] + 188 pp. São Carlos, Brasil: Dissertação (Mestre) da Universidade Federal de São Carlos.

Cane, J. H. 1979. The hind tibiotarsal and tibial spur articulations in bees. *Journal of the Kansas Entomological Society* 52: 123-137.

———. 1981. Dufour's gland secretions in the cell linings of bees. *Journal of Chemical Ecology* 7: 403-410.

———. 1983a. Preliminary chemosystematics of the Andrenidae and exocrine lipid evolution of the short-tongued bees. *Systematic Zoology* 32: 417-430.

———. 1983b. Chemical evolution and chemosystematics of the Dufour's gland secretions of the lactone-producing bees. *Evolution* 37: 657-674.

———. 1996. Nesting resins obtained from *Larrea* pollen host by an oligolectic bee, *Trachusa larreae* (Cockerell). *Journal of the Kansas Entomological Society* 69: 99-102.

Cane, J. H., G. C. Eickwort, F. R. Wesley, and J. Spielholz. 1983. Foraging, grooming and mate-seeking behaviors of *Macropis nuda* (Hymenoptera, Melittidae) and use of *Lysimachia ciliata* (Primulaceae) oils in larval provisions and cell linings. *American Midland Naturalist* 110: 257-264.

Cane, J. H., S. Gerdin, and G. Wife. 1983. Mandibular gland secretions of solitary bees (Hymenoptera: Apoidea): Potential for nest cell disinfection. *Journal of the Kansas Entomological Society* 56: 199-204.

Cane, J. H., D. Schiffhauer, and L. J. Kervin. 1996. Pollination, foraging, and nesting ecology of the leaf-cutter bee *Megachile (Delomegachile) addenda* (Hymenoptera: Megachilidae) on cranberry beds. *Annals of the Entomological Society of America* 89: 361-367.

Cane, J. H., and J. O. Tengö. 1981. Pheromonal cues of male *Colletes cunicularius*. *Journal of Chemical Ecology* 7: 427-436.

Cardale, J. C. 1993. Hymenoptera: Apoidea, *in* W. W. K. Houston and G. V. Maynard, eds., *Zoological Catalogue of Australia*, Vol. 10: ix + 405 pp. Canberra: Government Publishing Service.

Carman, G. M., and L. Packer. 1997. A cryptic species allied to *Halictus ligatus* Say (Hymenoptera: Halictidae) detected by allozyme electrophoresis. *Journal of the Kansas Entomological Society* 69(1996, suppl.): 168-176.

Cebellos, G. 1956. *Catalogo do los Himenópteros de España*. 544 pp., 1 map. Madrid: Instituto Español de Entomologia.

Celary, W. 1995. Nomadini (Hymenoptera, Apoidea, Anthophoridae) of Poland. *Monografie Fauny Polski* [Kraków] No. 20: 1-281.

Chandler, L. 1950. The bumblebees of Indiana. *Proceedings of the Indiana Academy of Sciences* 60: 167-177.

Chavarría, G., and J. M. Carpenter. 1994. "Total evidence" and the evolution of highly social bees. *Cladistics* 10: 229-258.

Chen, X., and S. Wang. 1997. A study on phylogenetic relationships among the subgenera of bumblebees. *Entomologica Sinica* 4: 324-336.

Chiappa, E., M. Rojas G.-L., and H. Toro. 1990. Clave para los géneros de abejas de Chile. *Revista Chilena de Entomologia* 18: 67-81.

Chiappa, E., and H. Toro. 1994. Comportamiento reproductivo de *Centris mixta tamarugalis* (Hymenoptera: Anthophoridae). II Parte: Nidificacion y estados inmaduros. *Revista Chilena de Entomologia* 21: 99-115.

Claude-Joseph, F. 1926. Recherches biologiques sur les Hyménoptères du Chile (Melliferes). *Annales des Sciences Naturelles, Zoologie* (10)9: 114-268. [Translated into Spanish by M. Etcheverry and A. Valenzuela, 1960, Investigaciones biológicas sobre himenópteros de Chile (Melíferos) de Claude-Joseph, *Publicaciones del Centro de Estudios Entomologicos, Universidad de Chile,* no. 1: 1-60, pls. 1-18.]

Clement, S. L. 1973. The nesting biology of *Melissodes (Eumelissodes) rustica* (Say), with a description of the larva. *Journal of the Kansas Entomological Society* 46: 516-525.

———. 1976. The biology of *Dianthidium heterulkei heterulkei* Schwarz, with a description of the larva. *Wasmann Journal of Biology* 34: 9-22.

Clement, S. L., and R. W. Rust. 1975. The biology of *Hoplitis robusta*. *Entomological News* 86: 115-120.

———. 1976. The nesting biology of three species of *Hoplitis* Klug. *Pan-Pacific Entomologist* 52: 110-119.

Cockerell, T. D. A. 1895. New species of bees. *Psyche* 7, supplement I: 9-11.

———. 1896. The bees of the genus *Andrena* found in New Mexico. *Annals and Magazine of Natural History* (6)18: 78-92.

———. 1897a. On the generic position of some bees hitherto referred to *Panurgus* and *Calliopsis*. *Canadian Entomologist* 29: 287-290.

———. 1897b. On the Mexican bees of the genus *Augochlora*. *Canadian Entomologist* 29: 4-6.

———. 1897c. A new genus of bees. *Entomological News* 8: 197.

———. 1897d. Life-zones in New Mexico. *Agricultural Experiment Station, New Mexico College of Agriculture, Bulletin* No. 24: 1-44.

———. 1898a. New bees from New Mexico. *Canadian Entomologist* 30: 146-148.

———. 1898b. On some panurgine and other bees. *Transactions of the American Entomological Society* 25: 185-198.

———. 1898c. Another yellow *Perdita*. *Entomological News* 9: 215-216.

———. 1899a. Notes on nomenclature of some Hymenoptera. *Entomologist* 32: 14.

———. 1899b. Some synonymy. *Psyche* 8: 311-312.

———. 1899c. *Catálogo de las Abejas de México*. 20 pp. Ciudad de México: Secretaria de Fomento.

———. 1900a. Descriptions of new bees collected by Mr. H. H. Smith in Brazil, I. *Proceedings of the Academy of Natural Sciences of Philadelphia* 52: 356-377.

———. 1900b. Observations on bees collected in Las Vegas, New Mexico, and in the adjacent mountains. *Annals and Magazine of Natural History* (7)5: 401-416.

———. 1901a. Descriptions of new bees collected by Mr. H. H. Smith in Brazil, II. *Proceedings of the Academy of Natural Sciences of Philadelphia* 53: 216-222.

———. 1901b. A peculiar new type of halictine bees. *Entomological News* 12: 208-209.

———. 1903a. A catalogue of the bees of California. *Psyche* 10: 74-78.

———. 1903b. New American Hymenoptera, mostly of the genus *Nomada*. *Annals and Magazine of Natural History* (7)12: 200-214.

———. 1903c. North American bees, and a new homopteron. *Annals and Magazine of Natural History* (7)12: 442-455.

———. 1903d. Bees of the genus *Nomada* from California. *Proceedings of the Academy of Natural Sciences of Philadelphia* 55: 559-614.

———. 1904a. The genus *Apista*, etc. *Canadian Entomologist* 36: 357.

———. 1904b. New genera of bees. *Entomological News* 15: 292.

———. 1904c. The bees of southern California—1. *Bulletin of the Southern California Academy of Sciences* 3: 3-6.

———. 1904d. New records of bees. *Entomologist* 37: 231-236.

———. 1905a. Notes on some bees in the British Museum. *Transactions of the American Entomological Society* 31: 309-364.

———. 1905b. Descriptions and records of bees—III. *Annals and Magazine of Natural History* (7)16: 301-308.

———. 1905c. Some American Hymenoptera. *Entomological News* 16: 9-11.

———. 1905d. New Australian bees in the collection of the British Museum. *Entomologist* 38: 270-273, 302-304.

———. 1906a. Fossil Hymenoptera from Florissant, Colorado. *Bulletin of the Museum of Comparative Zoology, Harvard University* 50: 33-58.

———. 1906b. The bees of Florissant, Colorado. *Bulletin of the American Museum of Natural History* 22: 419-455.

———. 1906c. Four interesting Australian bees, in the collection of the British Museum. *Entomologist* 39: 15-18.

———. 1906d. North American bees of the genera *Andrena* and *Melitta* in the British Museum. *Psyche* 13: 33-37.

———. 1906e. The North American bees of the family Anthophoridae. *Transactions of the American Entomological Society* 32: 63-116.

———. 1907a. A fossil honey-bee. *Entomologist* 40: 227-229.

———. 1907b. Descriptions and records of bees—XV. *Annals and Magazine of Natural History* 7(20): 59-68.

———. 1907c. On a collection of Australian and Asiatic bees. *Bulletin of the American Museum of Natural History* 23: 221-236.

———. 1908a. Notes on the bee-genus *Exaerete*. *Psyche* 15: 41-42.

———. 1908b. A new subgenus of African bees. *Entomologist* 41: 146-147.

———. 1908c. Descriptions and records of bees—XIX. *Annals and Magazine of Natural History* (8)1: 337-344.

———. 1908d. Descriptions and records of bees—XX. *Annals and Magazine of Natural History* (8)2: 323-334.

———. 1909a. Descriptions and records of bees—XXIII. *Annals and Magazine of Natural History* (8)4: 393-404.

———. 1909b. Some European bees. *Entomologist's Record* 21: 269-270.

———. 1909c. Descriptions of some bees in the U.S. National Museum. *Proceedings of the United States National Museum* 36: 411-420.

———. 1909d. Some European fossil bees. *Entomologist* 42: 313-317.

———. 1909e. Descriptions of Hymenoptera from Baltic amber. *Schriften der Physikalisch-ökonomischen Gesellschaft zu Königsberg* 50: 1-20.

———. 1909f. Two fossil bees. *Entomological News* 20: 159-161.

———. 1909g. Some additional bees from Prussian amber. *Schriften der Physikalisch-ökonomischen Gesellschaft zu Königsberg* 50: 21-25.

———. 1910a. New and little-known bees. *Transactions of the American Entomological Society* 36: 199-249.

———. 1910b. Descriptions of records of bees—XXXIII. *Annals and Magazine of Natural History* (8)6: 356-366.

———. 1910c. Some very small Australian bees. *Entomologist* 43: 262-264.

———. 1910d. Some new American bees. *Canadian Entomologist* 42: 169-171.

———. 1910e. Descriptions and records of bees—XXIV. *Annals and Magazine of Natural History* (8)5: 22-30.

———. 1910f. The North American bees of the genus *Nomia*. *Proceedings of the United States National Museum* 38: 289-298.

———. 1911a. A new genus of Australian bees. *Entomologist* 44: 140-142.

———. 1911b. Bees in the collection of the United States National Museum, 2. *Proceedings of the United States National Museum* 40: 241-264.

———. 1911c. Descriptions and records of bees—XXXV. *Annals and Magazine of Natural History* (8)7: 310-319.

———. 1911d. Descriptions and records of bees—XXXIV. *Annals and Magazine of Natural History* (8)7: 225-237.

———. 1911e. Descriptions and records of bees—XXXIX. *Annals and Magazine of Natural History* (8)8: 660-673.

———. 1911f. New and little known bees. *Transactions of the American Entomological Society* 37: 217-241.

———. 1912. New and little-known bees. *Entomologist* 45: 175-178.

———. 1913. The bee genus *Hoplitella*. *Canadian Entomologist* 45: 34.

———. 1914. Bees from Ecuador and Peru. *Journal of the New York Entomological Society* 22: 306-328.

———. 1915. Descriptions and records of bees—LXVI. *Annals and Magazine of Natural History* (8)15: 341-350.

———. 1916a. Descriptions and records of bees—LXXII. *Annals and Magazine of Natural History* (8)17: 428-435.

———. 1916b. Descriptions and records of bees—LXXIII. *Annals and Magazine of Natural History* (8)18: 44-53.

———. 1916c. New and little known bees from California. *Pomona Journal of Entomology and Zoology* 8: 43-64.

———. 1916d. Some neotropical parasitic bees (Hym.). *Entomological News* 27: 208-210.

———. 1917a. New social bees. *Psyche* 24: 119-128.

———. 1917b. Some euglossine bees. *Canadian Entomologist* 49: 144-146.

———. 1917c. New records of bees from Natal. *Annals of the Durban Museum* 1: 460-468.

———. 1918a. The megachilid bees of the Philippine Islands. *Philippine Journal of Science* (D)13: 127-144.

———. 1918b. Neotropical bees, principally collected by Professor Bruner in Argentina. *Transactions of the American Entomological Society* 44: 25-38.

———. 1919a. Bees in the collection of the United States National Museum—3. *Proceedings of the United States National Museum* 55: 167-221.

———. 1919b. The metallic-colored halictine bees of the Philippine Islands. *Philippine Journal of Science* 15: 9-13.

———. 1920a. On South African bees, chiefly collected in Natal. *Annals of the Durban Museum* 2: 247-262.

———. 1920b. On South African bees, chiefly collected in Natal. *Annals of the Durban Museum* 2: 286-318.

———. 1921a. Descriptions and records of bees—XC. *Annals and Magazine of Natural History* (9)7: 201-212.

———. 1921b. The epeoline bees of the American Museum Rocky Mountain expeditions. *American Museum Novitates* no. 23: 1-16.

———. 1922a. Bees in the collection of the United States

———. National Museum—4. *Proceedings of the United States National Museum* 60: 1-20.

———. 1922b. Notes on some western bees. *American Museum Novitates* no. 40: 1-7.

———. 1922c. Descriptions and records of bees—XCV. *Annals and Magazine of Natural History* (9)10: 265-269.

———. 1922d. Bees of the genus *Panurginus* obtained by the American Museum Rocky Mountain expeditions. *American Museum Novitates* no. 36: 1-10.

———. 1922e. Two new subgenera of North American bees. *American Museum Novitates* no. 47: 1-5.

———. 1923a. Two nocturnal bees and a minute *Perdita*. *American Museum Novitates* no. 66: 1-4.

———. 1923b. Some bees from British Guiana. *Annals and Magazine of Natural History* (9)11: 442-459.

———. 1924a. Descriptions and records of bees—CI. *Annals and Magazine of Natural History* (9)14: 179-185.

———. 1924b. A new genus of bees from California. *Entomological News* 35: 169-171.

———. 1924c. A new bee from Oregon. *Psyche* 31: 243-244.

———. 1924d. Notes on the structure of bees. *Proceedings of the Entomological Society of Washington* 26: 77-85.

———. 1925a. Anthidiine bees in the collection of the California Academy of Sciences. *Proceedings of the California Academy of Sciences* (4)14: 345-367.

———. 1925b. Bees in the collection of the California Academy of Sciences. *Proceedings of the California Academy of Sciences* (4)14: 185-215.

———. 1926a. Descriptions and records of bees—CVIII. *Annals and Magazine of Natural History* (9)17: 214-226.

———. 1926b. Descriptions and records of bees—CXII. *Annals and Magazine of Natural History* (9)18: 216-227.

———. 1926c. Descriptions and records of bees—CXIII. *Annals and Magazine of Natural History* (9)18: 621-627.

———. 1927a. Two new types of desert bees. *Pan-Pacific Entomologist* 4: 41-44.

———. 1927b. Megachilid bees from Bolivia collected by the Mulford biological expedition, 1921-22. *Proceedings of the United States National Museum* 71(12): 1-22.

———. 1929a. Bees in the Australian Museum collection. *Records of the Australian Museum* 17: 199-243.

———. 1929b. Bees in the Queensland Museum. *Memoirs of the Queensland Museum* 9: 298-323.

———. 1929c. Descriptions and records of bees—CXV. *Annals and Magazine of Natural History* (10)3: 354-360.

———. 1929d. Descriptions and records of bees—CXX. *Annals and Magazine of Natural History* (10)4: 584-594.

———. 1929e. Descriptions and records of bees—CXIX. *Annals and Magazine of Natural History* (10)4: 296-304.

———. 1929f. New name for a genus of bees. *Entomologist* 62: 19.

———. 1930a. African bees of the family Ctenoplectridae from the Belgian Congo and Liberia. *Revue de Zoologie et de Botanique Africaines* 18: 359-363.

———. 1930b. African bees of the family Megachilidae, (Anthidiinae, *Serapista* and *Lithurgus*) mainly from Liberia and the Belgian Congo. *Revue de Zoologie et de Botanique Africaines* 19: 43-55.

———. 1930c. The bees of Australia. *Australian Zoologist* 6: 205-236.

———. 1930d. A new subgenus of andrenine bees. *Pan-Pacific Entomologist* 7: 5-8.

———. 1930e. New Australian bees. *Memoirs of the Queensland Museum* 10: 37-50.

———. 1930f. A new African genus of Ceratinidae (Hymenoptera, Apoidea). *Revue de Zoologie et de Botanique Africaines* 18: 291-293.

———. 1930g. Descriptions and records of bees—CXXIV. *Annals and Magazine of Natural History* (10)6: 48-57.

———. 1931a. Some African bees. *Annals and Magazine of Natural History* (10)8: 400-405.

———. 1931b. Heriadine and related bees from Liberia and the Belgian Congo. *Revue de Zoologie et de Botanique Africaines* 20: 331-341.

———. 1931c. African bees of the family Megachilidae from Liberia and the Belgian Congo. *Revue de Zoologie et de Botanique Africaines* 20: 132-179.

———. 1931d. Descriptions and records of bees—CXXX. *Annals and Magazine of Natural History* (10)8: 537-553.

———. 1931e. Descriptions and records of bees—CXXVI. *Annals and Magazine of Natural History* (10)7: 273-281.

———. 1931f. Descriptions and records of bees—CXXVIII. *Annals and Magazine of Natural History* (10)7: 529-536.

———. 1932a. *Scrapter*, a misunderstood genus of bees. *Entomologist* 65: 10-12.

———. 1932b. Descriptions and records of bees—CXXXV. *Annals and Magazine of Natural History* 10(10): 166-176.

———. 1932c. Bees from the Belgian Congo. *Revue de Zoologie et de Botanique Africaines* 23: 18-27.

———. 1933a. Descriptions and records of bees—CXLIII. *Annals and Magazine of Natural History* (10)11: 456-468.

———. 1933b. Descriptions and records of bees—CXLV. *Annals and Magazine of Natural History* (10)12: 126-136.

———. 1933c. Bees collected at Borego, California, by Wilmatte Cockerell and Milene Porter. *Pan-Pacific Entomologist* 9: 25-28.

———. 1934a. Descriptions and records of bees—CXLVIII. *Annals and Magazine of Natural History* (10)13: 444-456.

———. 1934b. Some African meliponine bees. *Revue de Zoologie et de Botanique Africaines* 26: 46-62.

———. 1934c. Some new or little-known South African bees of the genus *Allodape* in the British Museum. *Annals and Magazine of Natural History* (10)14: 220-242.

———. 1935a. Some California bees. *Pan-Pacific Entomologist* 11: 49-54.

———. 1935b. Scientific results of the Vernay-Lang Kalahari expedition, March to September, 1930, Hymenoptera (Apoidea). *Annals of the Transvaal Museum* 17: 63-94.

———. 1936a. Bees from northern California. *Pan-Pacific Entomologist* 12: 133-164.

———. 1936b. Descriptions and records of bees—CLIV. *Annals and Magazine of Natural History* (10)17: 477-483.

———. 1937. *African Bees of the Genera* Ceratina, Halictus *and* Megachile. xvi + 254 pp. London: British Museum.

———. 1938. Bees of the genus *Sphecodes* from the Belgian Congo. *Revue de Zoologie et de Botanique Africaines* 30: 327-329.

———. 1947. A new genus of bees from Honduras. *Proceedings of the Entomological Society of Washington* 49: 106.

Cockerell, T. D. A., and E. Atkins. 1902. Contributions from the New Mexico Biological Station—XIII. On the bees of the family Nomadidae of Ashmead. *Annals and Magazine of Natural History* (7)10: 40-46.

Cockerell, T. D. A., and B. H. Blair. 1930. Rocky Mountain bees, I. *American Museum Novitates* no. 433: 1-19.

Cockerell, T. D. A., and W. P. Cockerell. 1901. Contributions from the New Mexico Biological Station—IX. On certain genera of bees. *Annals and Magazine of Natural History* (7)7: 46-50.

Cockerell, T. D. A., and W. Porter. 1899. Contributions from the New Mexico Biological Station—VII. Observations on bees, with descriptions of new genera and species. *Annals and Magazine of Natural History* (7)4: 403-421.

Cocucci, A. A., A. Sérsic, and A. Roig-Alsina. 2000. Oil-collecting structures in the Tapinotaspidini: their diversity, function, and probable origin. *Mitteilungen der Münchner Entomologischen Gesellschaft* 90: 51-74.

Coelho, B. W. T. 2002. The biology of the primitively eusocial *Augochloropsis iris* (Schrottky, 1902). *Insectes Sociaux* 49: 181-190.

———. 2004. A review of the bee genus *Augochlorella*. *Systematic Entomolgy* 29: 282-323.

Compagnucci, L. A. 2004. El genero *Acamptopoeum* Cockerell: tres nuevas especies y clave para la indentificacion de las especies argentinas. *Revista del Museo Argentino de Ciencias Naturales* (n. s.) 6: 321-332.

Compagnucci, L. A., and A. Roig-Alsina. 2003. Cuatro nuevas especies y análisis filogenético de *Doeringiella* Holmberg *sensu stricto*, pp. 123-133 *in* Melo and Alves-dos-Santos (2003).

Comstock, J. H. 1924. *An Introduction to Entomology.* xix + 1044 pp. Ithaca, N.Y.: Comstock Publishing Company.

Constantinescu, A. 1973, 1974b. Contributions à l'étude de l'armature génitale du mâle chez les espèces de *Prosopis* F. *Traveaux du Museum d'Histoire Naturelle "Grigore Antipa"* 13: 253-263; 14: 195-208.

———. 1974a. Contributions à la connaissance de la morphologie des espèces *Systropha curvicornis* Scop. et *S. planidens* Giraud: Précisions concernant la position systématique du genre *Systropha* Latr. *Travaux du Museum d'Histoire Naturelle "Grigore Antipa"* 15: 283-292.

Costa, J. T., and T. D. Fitzgerald. 1996. Developments in social terminology: Semantic battles in a conceptual war. *Trends in Ecology and Evolution* 11: 285-289.

Costa, M. A., M. A. Del Lama, G. A. R. Melo, and W. S. Sheppard. 2003. Molecular phylogeny of stingless bees (Apidae, Apinae, Meliponini) inferred from mitochondrial 16S rDNA sequences. *Apidologie* 34: 73-84.

Coville, R. E., G. W. Frankie, S. L. Buchmann, S. B. Vinson, and H. J. Williams. 1986. Nesting and male bevavior of *Centris heithausi* (Hymenoptera: Anthophoridae) in Costa Rica with chemical analysis of the hindleg glands of males. *Journal of the Kansas Entomological Society* 59: 325-336.

Coville, R. E., G. W. Frankie, and S. B. Vinson. 1983. Nests of *Centris segregata* (Hymenoptera: Anthophoridae) with a review of the nesting habits of the genus. *Journal of the Kansas Entomological Society* 56: 109-122.

Crane, P. R., and P. S. Herendeen. 1996. Cretaceous floras containing angiosperm flowers and fruits from eastern North America. *Review of Palaeobotany and Palynology* 90: 319-337.

Crawford, J. C. 1903. A new genus of bees. *Canadian Entomologist* 35: 337-338.

———. 1907. New North American Hymenoptera. *Journal of the New York Entomological Society* 15: 177-192.

———. 1910. New Hymenoptera from the Philippine Islands. *Proceedings of the United States National Museum* 38: 119-133.

———. 1915. The bee genus *Holcopasites* Ashmead. *Insecutor Inscitiae Menstruus* 3: 123-126.

———. 1916. Some new American Hymenoptera. *Insecutor Inscitiae Menstruus* 4: 135-144.

———. 1926. North American bees of the genus *Panurginus*. *Proceedings of the Entomological Society of Washington* 28: 207-214.

Crepet, W. L. 1996. Timing of the evolution of derived floral characters: Upper Cretaceous (Turonian) taxa with tricolpate and tricolpate-derived pollen. *Review of Paleobotany and Palynology* 90: 339-357.

Crespi, B. J., and D. Yanega. 1995. The definition of eusociality. *Behavioral Ecology* 6: 109-115.

Cresson, E. T. 1864. On the North American species of several genera of Apidae. *Proceedings of the Entomological Society of Philadelphia* 2: 373-411.

———. 1865. On the Hymenoptera of Cuba. *Proceedings of the Entomological Society of Philadelphia* 4: 1-200.

———. 1875. Report upon the collections of Hymenoptera made in portions of Nevada, Utah, Colorado, New Mexico, and Arizona. *Report of the U.S. Geographical Survey West of the 100th Meridian* 5: 705-728.

———. 1878. Descriptions of new North American Hymenoptera in the collection of the American Entomological Society. *Transactions of the American Entomological Society* 7: 61-136.

———. 1887. Synopsis of the families and genera of the Hymenoptera of America, north of Mexico. *Transactions of the American Entomological Society,* suppl. Vol., 1887: 1-350 + errata page.

Cros, A. 1937. *Osmia longispina* Pérez, étude biologique. *Bulletin et Annales de la Société Entomologique de Belgique* 77: 175-185.

———. 1939. Considérations générales sur le genre *Lithurgus* Latreille et biologie du *Lithurgus tibialis* Mor. *Bulletin de la Société Fouad 1er d' Entomologie* 23: 37-59.

Cross, E. A. 1958. A revision of the bees of the subgenus *Epinomia* in the new world. *University of Kansas Science Bulletin* 38: 1261-1301.

Cross, E. A., and G. E. Bohart. 1960. The biology of *Nomia (Epinomia) triangulifera* with comparative notes on other species of *Nomia*. *University of Kansas Science Bulletin* 41: 761-792.

Cruz-Landim, C. da. 1967. Estudo comparativo de algumas glândulas das abelhas (Hymenoptera, Apoidea) e respectivas implicações evolutivas. *Arquivos de Zoologia* [São Paulo] 15: 177-290.

———. 1973. Tipos de glândulas salivares do tórax presentes em abelhas. *Studia Entomologica* 16: 209-213.

Cruz-Landim, C. da, and A. C. Franco. 2001. Light and electron microscopic aspects of glands and pseudoglandular structures in the legs of bees (Hymenoptera, Apinae, Euglossini). *Brazilian Journal of Morphological Sciences* 18: 81-90.

Cruz-Landim, C. da, R. D. Reginato, and V. L. Imperatriz-Fonseca. 1998. Variation on ovariole number in Meliponinae (Hymenoptera, Apidae) queen's ovaries, with comments on ovary development and caste differentiation. *Papéis Avulsos de Zoologia* 40(18): 289-296.

Cruz-Landim, C. da, and J. E. do Serrao. 1994. The evolutive significance of pollen use as protein resource by Trigonini bees. *Journal of Advanced Zoology* 15(1): 1-5.

Cruz-Landim, C. da, A. C. Stort, M. A. da Costa-Cruz, and E. W. Kitajima. 1965. Órgão tibial dos machos de Euglossini, estudo ao microscópio óptico e eletrônico. *Revista Brasileira de Biologia* 25: 323-342.

Cumber, R. A. 1949. Larval specific characters and instars of English Bombidae. *Proceedings of the Royal Entomological Society of London* (A)24: 14-19.

Cunha, R. da. 2002. Estrutura de ninhos de *Monoeca xanthopyga* (Hymenoptera, Apoidea, Tapinotaspidini) no planalto das Araucárias, Rio Grande do Sul, Brasil. *Biociências* [Porto Alegre] 10: 25-33.

Cure, J. R. 1989. Revisão de *Pseudagapostemon* Schrottky e descrição de *Oragapostemon* gen. n. *Revista Brasileira de Entomologia* 33: 229-335.

Cure, J. R., and D. Wittmann. 1990. *Callonychium petuniae*, a new panurgine bee species (Apoidea, Andrenidae), oligolectic on *Petunia*. *Studies on Neotropical Fauna and Environment* 25: 153-156.

Curtis, J. 1824-1839. *British Entomology; ...* , Vols. 1-16. London: privately published. [Parts cited are 1826, Vol. 3, pls. 99-146; 1828, Vol. 5, pls. 195-241; 1831, Vol. 8, pls. 338-383; 1832, Vol. 9, pls. 384-433; and 1833, Vol. 10, pls. 434-481.]

Custer, C. P. 1928. The bee that works in stone, *Perdita opuntiae* Cockerell. *Psyche* 35: 67-83.

Custer, C. P., and C. H. Hicks. 1927. Nesting habits of some anthidiine bees. *Biological Bulletin* 52: 258-277.

D'Albore, G. R. 1986. *Bombus* Latr. e *Psithyrus* Lep. in Umbria. *Redia* 69: 171-256, pls. 1-8.

Dalla Torre, K. W. von. 1880. Unsere Hummel-*(Bombus)* Arten. *Die Naturhistoriker* 2: 30, 40-41. [N.B.: K. W. von Dalla Torre was the same person as C. G. de Dalla Torre.]

———. 1882. Bemerkungen zur Gattung *Bombus* Latr., II. *Bericht des Naturwissenschaftlich-Medezinischen Vereins in Innsbruck* 12: 14-31.

———. 1891. Die Gattungen und Arten der Phileremiden. *Bericht des Naturwissenschaftlich-Medezinischen Vereins in Innsbruck* 19: 137-159.

Dalla Torre, C. G. de. 1896. *Catalogus Hymenopterorum,* Vol. 10, Apidae (Anthophila). viii + 643 pp. Leipzig: Engelmann.

Dalla Torre, C. G. de, and H. Friese. 1895. Synonymischer Katalog der europäischen Sammelbienen. *Entomologische Nachrichten* 21: 21-26, 37-50, 53-62, 69-80.

Daly, H. V. 1961. Biological observations on *Hemihalictus lustrans,* with a description of the larva. *Journal of the Kansas Entomological Society* 34: 134-141.

———. 1966. Biological studies on *Ceratina dalletorreana,* an alien bee in California which reproduces by parthenogenesis. *Annals of the Entomological Society of America* 59: 1138-1154

———. 1973. Bees of the genus *Ceratina* in America north of Mexico. *University of California Publications in Entomology* 74: 1-114.

———. 1983a. Taxonomy and ecology of Ceratinini of North Africa and the Iberian Peninsula. *Systematic Entomology* 8: 29-62.

———. 1983b. Bees of the genus *Pithitis* from northeast Africa and the Arabian Peninsula. *Journal of the Kansas Entomological Society* 56: 499-505.

———. 1985. Bees of the genus *Megaceratina* in equatorial Africa. *Pan- Pacific Entomologist* 61: 339-344.

———. 1988. Bees of the new genus *Ctenoceratina* in Africa south of the Sahara. *University of California Publications in Entomology* 108: i-ix + 1-69.

Daly, H. V., and K. N. Magnacca. 2003. Hawaiian *Hylaeus (Nesoprosopis)* bees. *Insects of Hawaii* 17: [4] + 1-234.

Daly, H. V., C. D. Michener, J. S. Moure, and S. F. Sakagami. 1987. The relictual bee genus *Manuelia* and its relation to other Xylocopinae. *Pan-Pacific Entomologist* 63: 102-124.

Danforth, B. N. 1989a. The evolution of hymenopteran wings: The importance of size. *Journal of Zoology* [London] 218: 247-276.

———. 1989b. Nesting behavior of four species of *Perdita*. *Journal of the Kansas Entomological Society* 62: 59-79.

———. 1990. Provisioning behavior and estimation of investment ratios in a solitary bee, *Calliopsis (Hypomacrotera) persimilis* (Cockerell). *Behavioral Ecology and Sociobiology* 27: 159-168.

———. 1991a. Phylogeny of the bee genus *Perdita*. viii + 202 pp. Lawrence: Ph.D. thesis, University of Kansas.

———. 1991b. The morphology and behavior of dimorphic males of *Perdita portalis*. *Behavioral Ecology and Sociobiology* 29: 235-247.

———. 1991c. Female foraging and intranest behavior of a communal bee, *Perdita portalis*. *Annals of the Entomological Society of America* 84: 537-548.

———. 1994. Taxonomic review of *Calliopsis* subgenus *Hypomacrotera* (Hymenoptera: Andrenidae), with special emphasis on the distributions and host plant associations. *Pan-Pacific Entomologist* 70: 283-300.

———. 1996. Phylogenetic analysis and taxonomic revision of the *Perdita* subgenera *Macrotera, Macroteropsis, Macroterella* and *Cockerellula*. *University of Kansas Science Bulletin* 55: 635-692.

———. 1999a. Phylogeny of the bee genus *Lasioglossum* (Hymenoptera: Halictidae) based on mitochondrial COI sequence data. *Systematic Entomology* 24: 377-393.

———. 1999b. Emergence dynamics and bet hedging in a desert bee, *Perdita portalis*. *Proceedings of the Royal Society of London* (B) 266: 1985-1994.

———. 2002. Evolution of sociality in a primitively eusocial lineage of bees. *Proceedings of the National Academy of Sciences USA* 99: 286-290.

Danforth, B. N., S. G. Brady, S. D. Sipes, and A. Pearson. 2004. Single copy nuclear genes recover Cretaceous-age divergences in bees. *Systematic Biology* 53: 309-326.

Danforth, B. N., L. Conway, and S. Ji. 2003. Phylogeny of eusocial *Lasioglossum* reveals multiple losses of eusociality within a primitively eusocial clade of bees. *Systematic Biology* 52: 23-36.

Danforth, B. N., and C. D. Michener. 1988. Wing folding in Hymenoptera. *Annals of the Entomological Society of America* 81: 342-349.

Danforth, B. N., and J. L. Neff. 1992. Male polymorphism and polyethism in *Perdita texana*. *Annals of the Entomological Society of America* 85: 616- 626.

Danforth, B. N., J. L. Neff, and P. Burretto-Ko. 1996. Nestmate relatedness in a communal bee, *Perdita texana* (Hymenoptera: Andrenidae), based on DNA fingerprinting. *Evolution* 50: 276-284.

Danforth, B. N., H. Sauquet, and L. Packer. 1999. Phylogeny of the bee genus *Halictus* (Hymenoptera: Halictidae) based on parsimony and likelihood analyses of nuclear EF-α sequence data, *Molecular Phylogenetics and Evolution* 13: 605-618.

Danforth, B. N., S. Sipes, J. Fang, and S. G. Brady. 2006. The history of early bee diversification based on five genes plus morphology. *Proceedings of the National Acadamy of Science USA* 103: 15118–15123.

Danks, H. V. 1971. Biology of some stem-nesting aculeate Hymenoptera. *Transactions of the Royal Entomological Society of London* 122: 323-390.

Dathe, H. H. 1979. Der Gattungsname der Maskenbienen: *Hylaeus* versus *Prosopis*. *Linzer Biologische Beiträge* 11: 147-154.

———. 1980a. Die Arten de Gattung *Hylaeus* F. in Europa. *Mitteilungen aus dem Zoologischen Museum in Berlin* 56: 207-294.

———. 1980b. Die *Hylaeus*-Arten einer apidologischen Sammelreise in den Iran. *Entomologische Abhandlungen* [Dresden] 43: 77-97.

———. 1986. Die Bienengattung *Hylaeus* Fabricius in der Mongolei. *Annales Historico-Naturales Musei Nationalis Hungarici* 78: 265-300.

———. 1993. Taxonomie und Verbreitung der Gattung *Hylaeus* F. auf den Kanarischen Inseln, pp. 743-760 *in* H. Hohmann, F. LaRoche, G. Ortega, and J. Barquin, eds., Bienen, Vespen und Ameisen der Kanarischen Inseln. *Veröffentlichungen aus dem Übersee-Museum Bremen, Naturwissenschaften* 12.

Davis, L. R., Jr., and W. E. LaBerge. 1975. The nest biology of the bee *Andrena (Ptilandrena) erigeniae* Robertson. *Illinois Natural History Survey Biological Notes* no. 75: 1-16.

Davydova, N. G., and Y. A. Pesenko. 2002. Bees of the subgenus *Clisodon* (*Anthophora*, Hymenoptera, Apidae) in the fauna of Russia. *Zoologicheskii Zhurnal* 81: 1350-1353 [in Russian], English version *in Entomological Review* 82: 621-624.

Dawut, A., and O. Tadauchi. 2000-2003. A systematic study of the subgenus *Seladonia* of the genus *Halictus* in Asia. *Esakia* no. 40: 63-79; no. 41: 106-128; no. 42: 121-150; no. 43: 97-131.

Delmas, R. 1976. Contribution à l'étude de la faune française des Bombinae. *Annales de la Société Entomologique de France* (n.s.)12: 247-290.

Dias, D. 1958. Contribuição para o conhecimento da bionomia de *Bombus incarum* Franklin da Amazônia. *Revista Brasileira de Entomologia* 8: 1- 20, pls. 1-10.

Diniz, M. de A. 1962. Claves para la identificación de los géneros de Apidos de la Península Ibérica. *Graellsia* 19: 113-135, pls. 3-14.

———. 1963. A nervaçao alar dos Himenópteros. *Memórias e Estudos do Museu Zoológica da Universidade de Coimbra* No. 282: 1-26.

Dodson, C. H. 1966. Ethology of some bees of the tribe Euglossini. *Journal of the Kansas Entomological Society* 39: 607-629.

Dodson, C. H., R. L. Dressler, H. C. Hills, R. M. Adams, and N. H. Williams. 1969. Biologically active compounds in orchid fragrances. *Science* 164: 1243-1249.

Dollin, A. E., L. J. Dollin, and S. F. Sakagami. 1997. Australian stingless bees of the genus *Trigona*. *Invertebrate Taxonomy* 11: 861-896.

Dominique, J. 1898. Coup d'oeil sur les mellifères sud-américains du muséum de Nantes. *Bulletin de la Société des Sciences Naturelles de l'Ouest de la France* [Nantes] 8(1): 57-63.

Donovan, B. J. 1977. A revision of North American bees of the subgenus *Cnemidandrena*. *University of California Publications in Entomology* 81: i-ii + 1-107.

———. 1983a. Comparative biology of native Apoidea of New Zealand and New Caledonia. *GeoJournal* 7.6: 511-516.

———. 1983b. The occurrence of the Australian bee *Hyleoides concinna* (Hymenoptera: Colletidae) in New Zealand. *New Zealand Journal of Zoology* 10: 345-348.

Do-Pham, T. T., C. Plateaux-Quénu, and L. Plateaux. 1984. Étude comparative des genitalia males de quelques Halictinae (Hymenoptera) répercussion éventuelle sur la classification. *Annales de la Société Entomologique de France* (n.s.)20: 3-46.

Dours, L. 1872. Hyménoptères nouveaux du bassin méditerranéen. *Revue et Magasin de Zoologie* (2)23: 293-311, 349-359, 396-399, 419-434, pl. 28.

———. 1873. Hyménoptères nouveaux du bassin méditerranéen. *Andrena* (suite). *Revue et Magasin de Zoologie* (3)1: 274-325, pl. 14.

Dressler, R. L. 1968. Pollination by euglossine bees. *Evolution* 22: 202- 210.

———. 1978. An infrageneric classification of *Euglossa*, with notes on some features of special taxonomic importance. *Revista de Biología Tropical* 26: 187-198.

———. 1979. *Eulaema bombiformis, E. meriana*, and Mullerian mimicry in related species. *Biotropica* 22: 144-151.

———. 1982a. Biology of the orchid bees (Euglossini). *Annual Review of Ecology and Systematics* 13: 373-394.

———. 1982b. New species of *Euglossa*. III. The *bursigera* species group. *Revista de Biología Tropical* 30: 131-140.

———. 1985. Euglossine bees (Hymenoptera: Apidae) of the Tambopata reserved zone, Madre de Dios, Peru. *Revista Peruana de Entomología* 27: 75-79.

Drumond, P. M., R. Zucchi, and B. P. Oldroyd. 2000. Description of the cell provisioning and oviposition process of seven species of *Plebeia* Schwarz (Apidae, Meliponini), with notes on their phylogeny and taxonomy. *Insectes Sociaux* 47: 99-112.

Ducke, A. 1900. Die Bienengattung *Osmia* Panz. *Bericht des Naturwissenschaftlich-Medezinischen Vereins in Innsbruck* 25: 1-323.

———. 1902. Ein neues subgenus von *Halictus* Latr. *Zeitschrift für Systematische Hymenopterologie und Dipterologie* 2: 102-103.

———. 1906. Neue Beobachtungen über die Bienen der Amazonasländer. *Zeitschrift für Wissenschaftliche Insektenbiologie* 2: 51-60.

———. 1907. Beitrag zur Kenntnis der Solitärbienen Brasiliens. *Zeitschrift für Systematische Hymenopterologie und Dipterologie* 7: 361-368.

———. 1912. Die natürlichen Bienengenera Südamerikas. *Zoologische Jahrbücher, Abteilung für Systematik, Geographie und Biologie der Tiere* 34: 51-116.

———. 1916. Enumeração dos hymenopteros colligidos pela Commissão e revisão das espécies de abelhas do Brasil. *Commissão de Linhas Telegraphicas Estrategicas de Matto*

Grosso ao Amazonas, Publicacão No. 35, Annexo No. 5, Historia Natural, Zoologia, pp. 1-175 + [3] + 7 double pls., 5 pp. errata.

———. 1924. Die stachellosen Bienen *(Melipona)* Brasiliens. *Zoologische Jahrbücher, Abteilung für Systematik, Geographie und Biologie der Tiere* 49: 335-448, double pls. 3, 4.

Duméril, A. M. C. 1860. Entomologie analytique. *Mémoires de l'Académie des Sciences de l'Institut de France* 31: xxii + 1339. Paris.

Dunn, T., and M. H. Richards. 2003. When to bee social: interactions among environmental constraints, incentives, guarding, and relatedness in a facultatively social carpenter bee. *Behavioral Ecology* 14: 417-424.

Durante, S. P., and N. B. Diaz. 1996. Estudio taxonómico de las espécies argentinas del genero *Eumegachile* Friese. *Revista Brasileira de Entomologia* 40: 327-336.

Dylewska, M. 1974. *Klucze do Oznaczania Owadów Polski* [Keys for the Identification of Polish Insects]. Część 24, Zeszyt 68d, Apidae, Andreninae, pp. 1-79. Warsaw: Polish Entomological Society.

———. 1987. Die Gattung *Andrena* Fabricius (Andrenidae, Apoidea) in Nord- und Mitteleuropa. *Acta Zoologica Cracoviensia* 30: 359-708.

Eardley, C. D. 1983. A taxonomic revision of the genus *Xylocopa* Latreille (Hymenoptera: Anthophoridae) in southern Africa. *Entomological Memoir, Department of Agriculture, Republic of South Africa* No. 58: 1-67.

———. 1987. Catalogue of Apoidea (Hymenoptera) in Africa south of the Sahara, Part I, The genus *Xylocopa* Latreille. *Entomological Memoir, Department of Agriculture and Water Supply, Republic of South Africa* no. 70: 1-20.

———. 1988. A revision of the genus *Lithurge* Latreille (Hymenoptera: Megachilidae) in sub-Saharan Africa. *Journal of the Entomological Society of Southern Africa* 51: 251-263.

———. 1989. The afrotropical species of *Eucara* Friese, *Tetralonia* Spinola, and *Tetraloniella* Ashmead. *Entomological Memoir, Department of Agriculture and Water Supply, Republic of South Africa* no. 75: 1-61.

———. 1991a. The southern African Panurginae (Andrenidae: Hymenoptera). *Phytophylactica* 23: 115-136.

———. 1991b. The genus *Epeolus* from subsaharan Africa. *Journal of Natural History* 25: 711-731.

———. 1991c. The Melectini in subsaharan Africa. *Entomological Memoir, Department of Agriculture and Water Supply, Republic of South Africa* no. 82: 1-49.

———. 1993. The African species of *Pachymelus* Smith. *Phytophylactica* 25: 217-229.

———. 1994. The genus *Amegilla* Friese (Hymenoptera: Anthophoridae) in southern Africa. *Entomological Memoir, Department of Agriculture and Water Supply, Republic of South Africa* no. 91: 1-67.

———. 1996. The genus *Scrapter* Lepeletier & Serville. *African Entomology* 4: 37-92.

———. 2003. Revision of the afrotropical Ctenoplectrini. *African Plant Protection* 9(1): 5-18.

———. 2004. Taxonomic revision of the African stingless bees. *African Plant Protection* 10(2): 63-96.

Eardley, C. D., and R. W. Brooks. 1989. The genus *Anthophora* Latreille in southern Africa. *Entomology Memoir, Department of Agriculture and Water Supply, Republic of South Africa* no. 76: 1-55.

Eardley, C. D., and D. J. Brothers. 1997. Phylogeny of the Ammobatini and revision of the Afrotropical genera. *Journal of Hymenoptera Research* 6: 353-418.

Eardley, C. D., and M. Schwarz. 1991. The afrotropical species of *Nomada* Scopoli. *Phytophylactica* 23: 17-27.

Eberhard, W. G. 1989. Group nesting in two species of *Euglossa* bees. *Journal of the Kansas Entomological Society* 61: 406-411.

Ebmer, A. W. 1969, 1970, 1971, 1974. Die Bienen des Genus *Halictus* Latr. s. l. im Grossraum von Linz. *Naturkundliches Jahrbuch der Stadt Linz* 1969: 133-183; 1970: 1-82; 1971: 63-156; 1974: 123-163.

———. 1984. Die westpaläarktischen Arten der Gattung *Dufourea* Lepeletier 1841 mit illustrierten Bestimmungstabellen. *Senckenbergiana Biologica* 64: 313-379.

———. 1987a. Die europäischen Arten der Gattungen *Halictus* Latreille 1804 und *Lasioglossum* Curtis 1833 mit illustrierten Bestimmungstabellen, 1. *Senckenbergiana Biologica* 68: 59-148.

———. 1987b. Die westpaläarktischen Arten der Gattung *Dufourea* Lepeletier 1841 mit illustrierten Bestimmungstabellen. *Linzer Biologische Beiträge* 19: 43-56.

———. 1988a. Die europäischen Arten der Gattungen *Halictes* Latreille 1804 und *Lasioglossum* Curtis 1833 mit illustrierten Bestimmungstabellen, 2. *Senckenbergiana Biologica* 68: 323-375.

———. 1988b. Kritische Liste der nicht-parasitischen Halictidae Osterreichs mit Berücksichtigung aller mitteleuropäischen Arten. *Linzer Biologische Beiträge* 20: 527-711.

———. 1989. Die westpaläarktischen Arten der Gattung *Dufourea* Lepeletier 1841 mit illustrierten Bestimmungstabellen. *Linzer Biologische Beiträge* 21: 193-210.

———. 1993a. Die Bienengattung *Rophites* Spinola 1808— Erster Nachtrag. *Linzer Biologische Beiträge* 25: 3-14.

———. 1993b. Die westpaläarktischen Arten der Gattung *Dufourea* Lepeletier 1841 mit illustrierten Bestimmungstabellen, Dritte Nachtrag. *Linzer Biologische Beiträge* 25: 15-42.

———. 1994. *Systropha difformis* Smith, 1879 und *Systropha inexpectata* n. sp., die beiden östlichen Vertreter der altweltlichen Gattung *Systropha* Illiger, 1806. *Linzer Biologische Beiträge* 26: 807-821.

———. 1995. Asiatische Halictidae, 3. Die Artengruppe de *Lasioglossum* carinate-*Evylaeus*. *Linzer Biologische Beiträge* 27: 525-652.

———. 2000. Asiatische Halictidae-9. Die Artengruppe des *Lasioglossum pauperatum*. *Linzer Biologische Beiträge* 32: 399-453.

Ebmer, A. W., and K. Schwammberger. 1986. Die Gattung *Rophites* Spinola, 1808. *Senckenbergiana Biologica* 66: 271-304.

Ehrenfeld, J., and J. G. Rozen, Jr. 1977. The cuckoo bee genus *Kelita,* its systematics, biology and larvae. *American Museum Novitates* no. 2631: 1- 24.

Eickwort, G. C. 1967. Aspects of the biology of *Chilicola ashmeadi* in Costa Rica. *Journal of the Kansas Entomological Society* 40: 42-73.

———. 1969a. Tribal positions of western hemisphere green sweat bees, with comments on their nest architecture. *Annals of the Entomological Society of America* 62: 652-660.

———. 1969b. A comparative morphological study and generic revision of the augochlorine bees. *University of Kansas Science Bulletin* 48: 325-524.

———. 1970. *Hoplitis anthocopoides*, a European mason bee established in New York state. *Psyche* 77: 190-201.

———. 1973. Biology of the European mason bee, *Hoplitis anthocopoides* (Hymenoptera: Megachilidae), in New York state. *Search: Agriculture* (Cornell University Agricultural Experiment Station) 3(2): 1-31.

———. 1975. Nest-building behavior of the mason bee *Hoplitis anthocopoides*. *Zeitschrift für Tierpsychologie* 37: 237-254.

———. 1978. *Mexalictus,* a new genus of sweat bees from North America (Hymenoptera, Halictidae, Halictinae). *Journal of the Kansas Entomological Society* 51: 567-580.

———. 1979. A new species of wood-dwelling sweat bee in the genus *Neocorynura*, with description of its larva and pupa. *Entomologia Generalis* 5: 143-148.

———. 1981. Aspects of the nesting biology of five nearctic species of *Agapostemon*. *Journal of the Kansas Entomological Society* 54: 337-351.

Eickwort, G. C., J. M. Eickwort, J. Gordon, and M. A. Eickwort. 1996. Solitary behavior in a high-altitude population of the social sweat bee *Halictus rubicundus*. *Behavioral Ecology and Sociobiology* 38: 227-233.

Eickwort, G. C., and K. R. Eickwort. 1969. Aspects of the biology of Costa Rican halictine bees, I. *Agapostemon nasutus*. *Journal of the Kansas Entomological Society* 42: 421-452.

———. 1972a. Aspects of the biology of Costa Rican halictine bees, IV. *Augochlora (Oxystoglosella)*. *Journal of the Kansas Entomological Society* 45: 18-45.

———. 1972b. Aspects of the biology of Costa Rican halictine bees, III. *Sphecodes kathleenae*, a social cleptoparasite of *Dialictus umbripennis*. *Journal of the Kansas Entomological Society* 45: 529-541.

———. 1973a. Notes on the nests of three wood-dwelling species of *Augochlora* from Costa Rica. *Journal of the Kansas Entomological Society* 46: 17-22.

———. 1973b. Aspects of the biology of Costa Rican halictine bees, V. *Augochlorella edentata*. *Journal of the Kansas Entomological Society* 46: 3-16.

Eickwort, G. C., K. R. Eickwort, and E. G. Linsley. 1977. Observations on nest aggregations of the bees *Diadasia olivacea* and *D. diminuta*. *Journal of the Kansas Entomological Society* 50: 1-17.

Eickwort, G. C., and H. S. Ginsberg. 1980. Foraging and mating behavior in Apoidea. *Annual Review of Entomology* 25: 421-446.

Eickwort, G. C., P. F. Kukuk, and F. R. Wesley. 1986. The nesting biology of *Dufourea novaeangliae* (Hymenoptera: Halictidae) and the systematic position of the Dufoureinae based on behavior and development. *Journal of the Kansas Entomological Society* 59: 103-120.

Eickwort, G. C., and E. G. Linsley. 1978. The species of the parasitic bee genus *Protepeolus*. *Journal of the Kansas Entomological Society* 51: 14-21.

Eickwort, G. C., R. W. Matthews, and J. Carpenter. 1981. Observations on the nesting behavior of *Megachile rubi* and *M. texana* with a discussion of the significance of soil

nesting in the evolution of megachilid bees. *Journal of the Kansas Entomological Society* 54: 557-570.

Eickwort, G. C., and S. F. Sakagami. 1979. A classification of nest architecture of bees in the tribe Augochlorini (Hymenoptera: Halictidae; Halictinae) with description of a Brazilian nest of *Rhinocorynura inflaticeps*. *Biotropica* 11: 28-37.

Eickwort, G. C., and G. I. Stage. 1972. A new subgenus of neotropical *Sphecodes* cleptoparasitic upon *Dialictus* (Hymenoptera: Halictidae, Halictinae). *Journal of the Kansas Entomological Society* 45: 500-515.

Elfving, R. 1951. Die Gattung *Prosopis* Fabr. (Hym., Apidae) in Finnland. *Notulae Entomologicae* 31: 67-92.

———. 1960. Die Hummeln und Schmarotzerhummeln Finnlands. *Fauna Fennica* No. 10: 1-43.

Enderlein, G. 1903. Drei neue Bienen mit rüsselartiger Verlängerung des Kopfes. *Berliner Entomologische Zeitschrift* 48: 35-40.

Engel, M. S. 1996a. The bee genus *Rhectomia* (Hymenoptera: Halictidae): Discovery of the male and two new species. *Journal of the New York Entomological Society* 103(1995): 302-310.

———. 1996b. *Neocorynura electra*, a new fossil bee species from Dominican amber. *Journal of the New York Entomological Society* 103(1995): 317-323.

———. 1996c. Three new species of *Caenaugochlora (Ctenaugochlora)*. *Journal of the New York Entomological Society* 103(1995): 281-286.

———. 1997a. New augochlorine bees (Hymenoptera: Halictidae) in Dominican amber, with a brief review of fossil Halictidae. *Journal of the Kansas Entomological Society* 69(1996, suppl.): 334-345.

———. 1997b. A new fossil bee from the Oligo-Miocene Dominican amber. *Apidologie* 28: 97-102.

———. 1997c. *Ischnomelissa*, a new genus of augochlorine bees (Halictidae) from Colombia. *Studies on Neotropical Fauna and Environment* 32: 41-46.

———. 1998a. A new species of the Baltic amber bee genus *Electrapis*. *Journal of Hymenoptera Research* 7: 94-101.

———. 1999a. *Megachile glaesaria*, the first megachilid bee fossil from amber. *American Museum Novitates* no. 3276: 1-13.

———. 1999b. The taxonomy of recent and fossil honey bees. *Journal of Hymenoptera Research* 8: 165-196.

———. 1999c. Augochlorini Beebe, 1925 (Insecta Hymenoptera): corrected authorship and date (not Moure, 1943). *Bulletin of Zoological Nomenclature* 56(3): 1.

———. 1999d. A new species of the bee genus *Neocorynura* from the Andes of Ecuador. *Spixiana* 22: 173-178.

———. 1999e. A new xeromelissine bee in Tertiary amber of the Dominican Republic. *Entomologica Scandinavia* 30: 453-458.

———. 1999f. The first fossil *Euglossa* and phylogeny of the orchid bees. *American Museum Novitates* no. 3272: 1-14.

———. 2000a. A new interpretation of the oldest fossil bee. *American Museum Novitates* no 3296: 1-11.

———. 2000b. Classification of the bee tribe Augochlorini. *Bulletin of the American Museum of Natural History* no. 250: 1-89.

———. 2000c. A review of the Indo-malayan meliponine genus *Lisotrigona*, with two new species. *Oriental Insects* 34: 229-237.

———. 2000d. A revision of the panurgine bee genus *Arhysosage*. *Journal of Hymenoptera Research* 9: 182-208.

———. 2001a. Monophyly and extensive extinction of advanced eusocial bees: Insights from an unexpected Eocene diversity. *Proceedings of the National Academy of Sciences USA* 98: 1661-1664.

———. 2001b. A monograph of the Baltic amber bees and evolution of the Apoidea. *Bulletin of the American Museum of Natural History* no. 259: 1-192.

———. 2002a. Halictine bees from the Eocene-Oligocene boundry of Florissant, Colorado. *Neues Jahrbuch für Geologie und Paläontologie, Abhandlungen* 225: 251-273.

———. 2002b. Phylogeny of the bee tribe Fideliini (Hymenoptera: Megachilidae), with the description of a new genus from southern Africa. *African Entomology* 10: 305-313.

———. 2004a. Fideliine phylogeny and classification revisited. *Journal of the Kansas Entomological Society* 77: 821-836.

———. 2004b. Geological history of the bees. *Tecnologia e Ambiente* [Criciuma, Brazil] 10: 9-33.

———. 2005. Family-group names for bees. *American Museum Novitates* no. 3476: 1-33.

———. 2006a. A new nocturnal bee of the genus *Megalopta*, with notes on other Central American species, *Mitteilungen des Internationalen Entomologischen Vereins* 31: 37-49.

———. 2006b. A new genus of cleptoparasitic bees from the West Indies. *Acta Zoologica Cracoviensia* 49B: 1-8.

———. 2006c. The *Sphecodes* of Cuba. *Acta Zoologica Cracoviensia* 49B: 73-78.

Engel, M. S., and D. B. Baker. 2006. A new species of *Thygatina* from India, with notes on the Oriental fauna. *American Museum Novitates*, in press.

Engel, M. S., and R. W. Brooks. 1998. The nocturnal bee genus *Megaloptidia*. *Journal of Hymenoptera Research* 7: 1-14.

———. 1999. The augochlorine bee genus *Megaloptilla*, pp. 9-15 *in* G. W. Byers, R. H. Hagen, and R. W. Brooks, eds., *Entomological Contributions in Memory of Byron A. Alexander, University of Kansas Natural History Museum Special Publication* no. 24.

———. 2000. A new *Chlerogelloides* from French Guiana, with comments on the genus, *Journal of the Kansas Entomological Society* 72(1999): 160-166.

———. 2002. A new bee of the genus *Ischnomelissa* with a key to the known species. *Entomological News* 113: 1-5.

Engel, M. S., R. W. Brooks, and D. Yanega. 1997. New genera and subgenera of augochlorine bees. *Scientific Papers, Natural History Museum, University of Kansas* no. 5: 1-21.

Engel, M. S., and B. A. Klein. 1997. *Neocorynurella*, a new genus of augochlorine bees from South America. *Deutsche Entomologische Zeitschrift* 44: 155-163.

Engel, M. S., C. D. Michener, and M. G. Rightmyer. 2004a. The cleptoparasitic bee tribe Rhathymini (Hymenoptera: Apidae): a new genus and tribal review. *Journal of Hymenoptera Research* 13: 1-12.

———. 2004b. A replacement name for the cleptoparasitic bee genus *Rhathymodes*. *Journal of Hymenoptera Research* 13: 316.

Engel, M. S., and T. R. Schultz. 1997. Phylogeny and behavior in honeybees. *Annals of the Entomological Society of America* 90: 43-53.

Engel, M. S., and A. H. Smith-Pardo. 2004b. The bee genus *Andinaugochlora* in Central America. *Journal of the Kansas Entomological Society* 77: 116-120.

Enslin, E. 1921. Beiträge zur Kenntnis der Hymenopteren. *Deutsche Entomologische Zeitschrift* 1921: 59-64, pl. 1.

Erlandsson, S. 1955. Die schwedischen Arten der Gattung *Coelioxys* Latr. *Opuscula Entomologica* 20: 174-191.

Espelie, K. E., J. H. Cane, and D. S. Himmelsbach. 1992. Nest cell lining of the solitary bee *Hylaeus bisinuatus*. *Experientia* 48: 414-416.

Evans, D. L. 1972. A revision of the subgenus *Holonomada* of the genus *Nomada*. *Wasmann Journal of Biology* 30: 1-34.

Evans, H. E. 1958. Studies on the larvae of digger wasps (Hymenoptera, Sphecidae), Part IV: Astatinae, Larrinae, and Pemphredoninae. *Transactions of the American Entomological Society* 84: 109-139.

———. 1993. Observations on the nests of *Paranthidium jugatorium perpictum* (Cockerell). *Pan-Pacific Entomologist* 69: 319-322.

Exley, E. M. 1968a. Revision of the genus *Euryglossula* Michener. *Australian Journal of Zoology* 16: 203-217.

———. 1968b. Revision of the genus *Euryglossella* Cockerell. *Australian Journal of Zoology* 16: 219-226.

———. 1968c. *Quasihesma*—a new genus of Australian bees. *Australian Journal of Zoology* 16: 227-235.

———. 1968d. Revision of the genus *Euryglossina* Cockerell. *Australian Journal of Zoology* 16: 915-1020.

———. 1968e. Revision of the genus *Brachyhesma* Michener. *Australian Journal of Zoology* 16: 167-201.

———. 1968f. Three new species of *Brachyhesma*. *Journal of the Australian Entomological Society* 7: 135-141.

———. 1969a. A new species of *Euryglossula*. *Journal of the Australian Entomological Society* 8: 137-138.

———. 1969b. Revision of the subgenus *Xenohesma* Michener. *Australian Journal of Zoology* 17: 535-551.

———. 1969c. *Argohesma*—A new genus of Australian bees. *Australian Journal of Zoology* 17: 527-534.

———. 1969d. Revision of the genus *Xanthesma* Michener. *Australian Journal of Zoology* 17: 515-526.

———. 1969e. A supplement to the revision of *Euryglossina*. *Journal of the Australian Entomological Society* 8: 139-144.

———. 1972. Revision of the genus *Pachyprosopis* Perkins. *Australian Journal of Zoology*, suppl. series no. 10: 1-43.

———. 1974a. A contribution to our knowledge of Australia's smallest bees with descriptions of new species. *Journal of the Australian Entomological Society* 13: 1-9.

———. 1974b. Revision of the subgenus *Callohesma* Michener. *Australian Journal of Zoology*, suppl. series no. 26: 1-58.

———. 1974c. A contribution to our knowledge of the bee fauna (Colletidae: Euryglossinae) of remote areas of Australia with descriptions of new species. *Proceedings of the Royal Society of Queensland* 85: 95-110.

———. 1975a. New species and records of *Brachyhesma* Michener. *Journal of the Australian Entomological Society* 14: 139-144.

———. 1975b. Revision of the genus *Hyphesma* Michener. *Australian Journal of Zoology* 23: 277-291.

———. 1976a. New species and records of *Pachyprosopis*

Perkins. *Journal of the Australian Entomological Society* 14: 399-407.

———. 1976b. Revision of the subgenus *Euryglossa* Smith. *Australian Journal of Zoology,* suppl. series no. 41: 1-72.

———. 1976c. Notes on flying characteristics of *Euryglossa (Xenohesma)* bees and how a gynandromorph resolves a taxonomic problem. *Journal of the Australian Entomological Society* 15: 469-470.

———. 1976d. New species and records of *Euryglossina* Cockerell. *Journal of the Australian Entomological Society* 15: 273-279.

———. 1977. The Australian genus *Brachyhesma* (Apoidea: Colletidae) revised and reviewed. *Australian Journal of Zoology,* suppl. series no. 53: 1-54.

———. 1978a. *Chaetohesma*—A new genus of Australian bees. *Australian Journal of Zoology* 26: 373-397.

———. 1978b. A new species of *Xanthesma* from Australia. *Journal of the Kansas Entomological Society* 51: 781-786.

———. 1980. New species and records of *Quasihesma* Exley. *Journal of the Australian Entomological Society* 19: 161-170.

———. 1982. The genus *Euryglossella* Cockerell. *International Journal of Entomology* 1: 21-29.

———. 1983. The genus *Heterohesma* Michener. *Journal of the Australian Entomological Society* 22: 219-221.

———. 1996. *Tumidihesma*, a new genus of Australian bees. *Australian Journal of Entomology* 35: 253-255.

———. 2001. The walkeraina species-group of *Euhesma* Michener. *Australian Journal of Entomology* 40: 102-112.

———. 2002. Bees of the *Euhesma crabronica* species-group. *Records of the Western Australian Museum* 21: 203-211.

———. 2004. Revision of the genus *Dasyhesma* Michener. *Records of the Western Australian Museum* 22: 129-146.

Fabricius, J. C. 1775. *Systema Entomologiae, Sistens Insectorum Classes, Ordines, Genera, Species, Adiectis Synonymis, Locis, Descriptionibus, Observationibus.* xxviii + 832 pp. Flensburgi et Lipsiae: Korte.

———. 1781. *Species Insectorum,* vol. 1, 8 + 552 pp. Hamburg und Kilon: Bohn.

———. 1793. *Entomologia Systematica Emendata et Aucta . . . .,* Vol. 2, viii + 519 pp. Hafniae: Proft.

———. 1804. *Systema Piezatorum. . . .* xiv + [15]-[440] + [1]-30 pp. Brunsvigae: Reichard. [The index (pp. [1]-30) appeared in 1805, according to Hedicke, 1941, *Mitteilungen der Deutschen Entomologischen Gesellschaft* 10: 82-83.]

Fairmaire, L. 1858. Ordre Hyménoptères, *in* J. Thomson, Voyage au Gabon, Histoire Naturelle des Insectes et des Arachnides. *Archives Entomologiques* 2: 263-267, figs. on pl. 10.

Faustino, C. D., E. V. Silva-Matos, S. Mateus, and R. Zucchi. 2002. First record of emergency queen rearing in stingless bees. *Insectes Sociaux* 49: 181-190.

Fernandez-Salomão, T. M., R. B. Rocha, L. A. O. Campos, and E. F. Araújo. 2005. The first internal transcribed spacer (ITS-1) of *Melipona* species (Hymenoptera, Apidae, Meliponini): characterization and phylogenetic analysis. *Insectes Sociaux* 52: 11-18.

Ferton, C. 1901. Notes détachées sur l'instinct des Hyménoptères mellifères et ravisseurs avec la description de quelques espèces. *Annales de la Société Entomologique de France* 70: 83-148, pls. 1-3.

———. 1914. *Perezia maura*—nouveau genre d'apiaires parasites d'Algérie et observations de ce genre. *Annales de la Société Entomologique de France* 83: 233-237.

Field, J. 1992. Intraspecific parasitism as an alternative reproductive tactic in nest-building wasps and bees. *Biological Reviews* 67: 79-126.

———. 1996. Patterns of provisioning and iteroparity in a solitary halictine bee, *Lasioglossum (Evylaeus) fratellum* (Pérez), with notes on *L. (E.) calceatum* (Scop.) and *L. (E.) villosulum* (K.). *Insectes Sociaux* 43: 167-182.

Fischer, R. L. 1951. Observations on the nesting habits of megachilid bees. *Journal of the Kansas Entomological Society* 24: 46-50.

Fisher, R. M. 1987. Queen-worker conflict and social parasitism in bumble bees. *Animal Behaviour* 35: 1026-1036.

Fletcher, D. J. C., and R. M. Crewe. 1981. Nest structure and thermoregulation in the stingless bee *Trigona (Plebeina) denoiti* Vachal (Hymenoptera: Apidae). *Journal of the Entomological Society of Southern Africa* 44: 183-196.

Forbes, S. H., R. M. M. Adams, C. Bitney, and P. F. Kukuk. 2002. Extended parental care in communal social groups. *Journal of Insect Science* 2(22): 1-18.

Fox, W. J. 1902. *Lithurgopsis,* a new genus of bees. *Entomological News* 13: 137-140.

Frankie, G. W., L. Newstrom, S. B. Vinson, and J. F. Barthell. 1993. Nesting habitat preferences of selected *Centris* bee species in Costa Rican dry forest. *Biotropica* 25: 322-333.

Frankie, G. W., S. B. Vinson, L. E. Newstrom, J. F. Barthell, W. A. Haber, and J. K. Frankie. 1990. Plant phenology, pollination ecology, pollinator behavior and conservation of pollinators in neotropical dry forest, pp. 37-47 *in* K. S. Bawa and M. Hadley, eds., *Reproductive Ecology of Tropical Forest Plants, Man in the Biosphere Series,* Vol. 7. Paris: UNESCO.

Frankie, G. W., S. B. Vinson, and H. Williams. 1989. Ecological and evolutionary sorting of 12 sympatric species of *Centris* bees in Costa Rican dry forest, pp. 535-549 *in* J. H. Bock and Y. B. Linhart, eds., *The Evolutionary Ecology of Plants.* Boulder, Colo.: Westview Press.

Franklin, H. J. 1913a, b. The Bombinae of the New World. *Transactions of the American Entomological Society* 38(1912): 177-486, 39(1913): 73-200, pl. 1-22.

———. 1954. The evolution and distribution of American bumble-bee kinds. *Transactions of the American Entomological Society* 80: 43-51.

Free, J. B., and C. G. Butler. 1959. *Bumblebees.* xiv + 208 pp. London: Collins. [Appendices by I. H. H. Yarrow.]

Freitas, B. M., and J. O. P. Pereira, eds. 2004. *Solitary Bees: Conservation, Rearing and Management for Pollination,* 285 pp. Fortaleza, Brasil: Univ. Federal do Ceará.

Frey-Gessner, E. 1899-1912. Hymenoptera, Apidae, *in Fauna Insectorum Helvetiae,* Vol. I, viii + 392 pp.; Vol. II, vi + 319 pp. Bern: published as supplements to *Mitteilungen der Schweizerischen Entomologischen Gesellschaft.*

Friese, H. 1895. *Die Bienen Europa's,* Vol. 3, Theil I. 128 pp. Berlin: Friedländer.

———. 1896a. Monographie der Bienengattung *Ceratina* (Latr.) (Palaearktischen Formen). *Természetrajzt Füzetek* 19: 34-65.

———. 1896b. Neue oder wenig bekannte südeuropäische Apiden. *Természetrajzt Füzetek* 19: 277-284.

———. 1896c. *Die Bienen Europa's*, Vol. 3, Theil II. 216 pp. Berlin: Friedländer.

———. 1897a. *Die Bienen Europa's*, Vol. 3, Theil III. vi + 1-316 pp. Berlin: Friedländer.

———. 1897b. Monographie der Bienengattung *Nomia* (Latr.) (Palaearctische Formen), pp. 45-84 *in Festschrift zur Feier des fünfzigjährigen Bestehens des Vereins für schlesische Insektenkunde in Breslau.* Breslau: Verein für Schlesische Insektenkunde.

———. 1898a. *Die Bienen Europa's*, Vol. 3, Theil IV. 303 pp., 1 pl. Innsbruck: Lampe.

———. 1898b. Species aliquot novae vel minus cognitae generis *Megachile* Latr. (et *Chalicodoma* Lep.). *Természetrajzt Füzetek* 21: 198-202.

———. 1898c. Beiträge zur Bienenfauna von Aegypten. *Természetrajzt Füzetek* 21: 303-313.

———. 1899a. Monographie der Bienengattungen *Megacilissa, Caupolicana* und *Oxaea*. *Annalen des K. K. Naturhistorischen Hofmuseums* [Wien] 14: 239-246.

———. 1899b. Monographie der Bienengattung *Euglossa*. *Természetrajzt Füzetek* 22: 117-170.

———. 1899c. *Die Bienen Europa's*, Vol. 3, Theil V. 228 pp. Innsbruck: Lampe.

———. 1900. Monographie der Bienengattung *Centris* (s. lat.). *Annalen des K. K. Naturhistorischen Hofmuseums* [Wien] 15: 237-350.

———. 1901. *Die Bienen Europa's*, Vol. 3, Theil VI. 284 pp. Innsbruck: Lampe.

———. 1902. Zwei neue Bienengattungen. *Zeitschrift für Systematische Hymenopterologie und Dipterologie* 2: 186-187.

———. 1903a. Neue Meliponiden II. *Zeitschrift für Systematische Hymenopterologie und Dipterologie* 3: 359-361.

———. 1903b. *Meliturgula*, eine neue Bienengattung aus Süd-Afrika. *Zeitschrift für Systematische Hymenopterologie und Dipterologie* 3: 33-34.

———. 1905a. Die Wollbienen Afrikas, Genus *Anthidium*. *Zeitschrift für Systematische Hymenopterologie und Dipterologie* 5: 65-75.

———. 1905b. Neue *Anthophora*-Arten aus Afrika (Hym.). *Zeitschrift für Systematische Hymenopterologie und Dipterologie* 5: 233-242.

———. 1906a. Die Bienengattung *Oediscelis* Philipp. [sic] und *Pseudiscelis* Friese. *Zeitschrift für Systematische Hymenopterologie und Dipterologie* 6: 225-228.

———. 1906b. Neue Bienenarten aus Chile und Argentina. *Zeitschrift für Systematische Hymenopterologie und Dipterologie* 6: 169-176.

———. 1906c. Eine neue Bienengattung aus Chile und Argentinien. *Zeitschrift für Systematische Hymenopterologie und Dipterologie* 6: 374-380.

———. 1906d. Resultate einer Reise des Herrn A. C. Jensen-Haarup in die Gegend von Mendoza (Argentina). *Flora og Fauna* 8: 89-102.

———. 1906e. Neue Schmarotzerbienen aus der neotropischen Region. *Zeitschrift für Systematische Hymenopterologie und Dipterologie* 6: 118-121.

———. 1908a. Neue Bienenarten aus Ostafrika. *Deutsche Entomologische Zeitschrift* 1908: 567-572.

———. 1908b. Die Apidae (Blumenwespen) von Argentina nach den Reisenergebnissen der Herren A. C. Jensen-Haarup und P. Jörgensen in den Jahren 1904-1907. *Flora og Fauna* 10: 331-425; reprint 1-94.

———. 1908c. Eine neue Bienengattung *Corbicula* aus Argentina. *Zeitschrift für Systematische Hymenopterologie und Dipterologie* 8: 170-174.

———. 1909a. Die Bienen Afrikas nach dem Stande unserer heutigen Kenntnisse, pp. 83-476, pls. ix-x, *in* L. Schultz, *Zoologische und Anthropologische Ergebnisse einer Forschungsreise im westlichen und zentralen Südafrika ausgefuhrt in den Jahren 1903-1905,* Band 2, Lieferung 1, X Insecta (ser. 3) [Jenaische Denkschriften Vol. 14]. Jena: Fischer.

———. 1909b. Die Bienenfauna von Neu-Guinea. *Annales Musei Nationalis Hungarici* 7: 179-288.

———. 1910a. Zur Bienenfauna des südlichen Argentinien (Hym.). *Zoologische Jahrbücher, Abteilung für Systematik, Geographie und Biologie der Tiere* 29: 641-660.

———. 1910b. Apidae, pp. 119-168 *in* Y. Sjöstedt, *Wissenschaftliche Ergebnisse der schwedischen zoologischen Expedition nach dem Kilimanjaro, dem Meru und den umgebenden Massaisteppen Deutsch-Ostafrikas 1905-1906 ....,* Band 2, Abt. 8 (Hymenoptera). Stockholm: Palmquist.

———. 1911a. Neue Bienen der äthiopischen Region. *Zoologische Jahrbücher, Abteilung für Systematik, Geographie und Biologie der Tiere* 30: 671-678.

———. 1911b. Apidae I, Megachilinae. *Das Tierreich,* Lieferung 28: xxvi + 440 pp. Berlin: Friedländer.

———. 1911c. Nachtrag zu "Bienen Afrikas." *Zoologische Jahrbücher, Abteilung für Systematik, Geographie und Biologie der Tiere* 30: 651-670.

———. 1911d. Neue Bienen-Arten der palaearktischen Region (Hym.). *Archiv für Naturgeschichte* 77(1): 135-143.

———. 1912a. Neue und wenig bekannte Bienen Süd-Afrikas. *Archiv für Naturgeschichte,* Abt. A 78(5): 181-189.

———. 1912b. Neue und wenig bekannte Bienenarten der neotropischen Region. *Archiv für Naturgeschichte,* Abt. A 78(6): 198-226.

———. 1913a. Neue Bienenarten aus Afrika. *Deutsche Entomologische Zeitschrift* 1913: 573-578.

———. 1913b. II. Nachtrag zu "Bienen Afrikas." *Zoologische Jahrbücher, Abteilung für Systematik, Geographie und Biologie der Tiere* 35: 581-598.

———. 1914a. Neue Apiden der palaearktischen Region. *Stettiner Entomologische Zeitung* 75: 218-233.

———. 1914b. Die Bienenfauna von Java. *Tijdschrift voor Entomologie* 57: 1-61, 2 pls.

———. 1914c. Neue Bienenarten der orientalischen Region. *Deutsche Entomologische Zeitschrift* 1914: 320-324.

———. 1916. Neue Bienen-Arten aus Chile und Südamerika .... *Stettiner Entomologische Zeitung* 77: 163-174.

———. 1918. Bienen aus Sumatra, Java, Malakka und Ceylon gesammelt von Herrn. Prof. Dr. v. Buttel-Reepen .... *Zoologische Jahrbücher, Abteilung für Systematik, Geographie und Biologie der Tiere* 41: 489-520.

———. 1920. *Canephorula apiformis* Friese (Hym.), eine Bienen-Art mit dem Beginne der Körbchen-bildung aus Argentinien. *Zoologische Jahrbücher, Abteilung für Systematik, Geographie und Biologie der Tiere* 43: 461-470.

———. 1921a. Eine neue Bienengattung *Melittoides* n.

gen., pp. 177-180 *in* J. Fahringer and H. Friese, Eine Hymenopteren-Ausbeute aus dem Amanusgebirge (Kleinasien und Nord-Syrien, südl. Armenien). *Archiv für Naturgeschichte,* Abt. A 87(3): 150-180.

———. 1921b. Nachtrag zur Bienenfauna von Costa Rica. *Stettiner Entomologische Zeitung* 82: 74-98.

———. 1922a. Eine neue Gattung der Urbienen: *Brachyglossa. Zoologische Jahrbücher, Abteilung für Systematik, Geographie und Biologie der Tiere* 45: 577-580.

———. 1922b. Eine neue Bienengattung aus Südamerika: *Rhinetula. Zoologische Jahrbücher, Abteilung für Systematik, Geographie und Biologie der Tiere* 45: 581-586.

———. 1923. *Die Europäischen Bienen.* xvii + 456 pp., 33 pls. Berlin and Leipzig: Gruyter.

———. 1924a. Über die Nestzellen der Blattschneiderbiene *Megachile azteca* Cr. (bei San José de Costa Rica). *Zeitschrift für Wissenschaftliche Insektenbiologie* 19: 193-194.

———. 1924b. Über die Urbienengattung *Temnosoma* Sm. *Zoologische Jahrbücher, Abteilung für Systematik, Geographie, und Biologie der Tiere* 49: 534-536.

———. 1924c. Die Urbienengattung *Polyglossa* Fr. in Südafrika. *Zoologische Jahrbücher, Abteilung für Systematik, Geographie, und Biologie der Tiere* 49: 513-519.

———. 1925a. Neue neotropische Bienenarten, zugleich II. Nachtrag zur Bienenfauna von Costa Rica. *Stettiner Entomologische Zeitung* 86: 1-41.

———. 1925b. IV. Nachtrag zu "Bienen Afrikas." *Zoologische Jahrbücher, Abteilung für Systematik, Geographie und Biologie der Tiere* 49: 501-512.

———. 1926. Die nachtbienen-Gattung *Megalopta* Sm. *Stettiner Entomologische Zeitung* 87: 111-135.

———. 1930. Corrigenda zu "Neue neotropische Bienenarten" in Stettin. Entomol. Zeitg. vol. 86, 1925. *Stettiner Entomologische Zeitung* 91: 127.

Friese, H., and F. v. Wagner. 1910. Zoologische Studien an Hummeln, I. Die Hummeln der deutschen Fauna. *Zoologische Jahrbücher, Abteilung für Systematik, Geographie und Biologie der Tiere* 29: 2-104, 7 pls.

Frisch, K. von. 1967. *The Dance Language and Orientation of Bees.* xiv + 566 pp. Cambridge, Mass.: Harvard University Press.

Frison, T. H. 1927. A contribution to our knowledge of the relationships of the Bremidae of America north of Mexico (Hymenoptera). *Transactions of the American Entomological Society* 53: 51-78, pls. XVI, XVII.

———. 1928. The bumblebees of the Philippine Islands. *Philippine Journal of Science* 37: 273-281.

———. 1930. The bumblebees of Java, Sumatra and Borneo. *Treubia* 12: 1-22.

———. 1935. Records and descriptions of *Bremus* from Asia. *Records of the Indian Museum* 37: 339-363.

Frohlich, D. R. 1983. On the nesting biology of *Osmia (Chenosmia) bruneri. Journal of the Kansas Entomological Society* 56: 123-130.

Frohlich, D. R., and F. D. Parker. 1983. Nest building behavior and development of the sunflower leafcutter bee: *Eumegachile (Sayapis) pugnata* (Say). *Psyche* 90: 193-209.

———. 1985. Observations on the nest building and reproductive behavior of a resin-gathering bee: *Dianthidium ulkei.* Annals of the Entomological Society of America 78: 804-810.

Fye, R. E. 1965. Biology of Apoidea taken in trap nests in northwestern Ontario (Hymenoptera). *Canadian Entomologist* 97: 863-877.

Gadagkar, R. 1995. Why the definition of eusociality is not helpful to understand its evolution and what we should do about it. *Oikos* 70: 485- 488.

Gaglianone, M. C. 2001. Nidificação e forrageamento de *Centris (Ptilotopus) scopipes* Friese. *Revista Brasileira de Zoologia* 18(suppl. 1): 107-117.

———. 2003. Abelhas de tribo Centridini na Estação Ecológica de Jataí (Luiz Antônio, SP): composição de espécies e interações com flores de Malpighiaceae, pp. 279-284 *in* G. A. R. Melo and I. Alves-dos-Santos, eds., *Apoidea Neotropica: Homenagem aos 90 Anos de Jesus Santiago Moure.* Criciúma: Editoria UNESC.

Garófalo, C. A., E. Camillo, S. C. Augusto, B. M. Vieira de Jesus, and J. C. Serrano. 1998. Nest structure and communal nesting in *Euglossa (Glossura) annectans* Dressler. *Revista Brasileira de Zoologia* 15: 589-596.

Garófalo, C. A., E. Camillo, M. J. O. Campos, and J. C. Serrano. 1992. Nest re-use and communal nesting in *Microthurge corumbae* (Hymenoptera, Megachilidae), with special reference to nest defense. *Insectes Sociaux* 39: 301-311.

Garófalo, C. A., E. Camillo, M. J. O. Campos, R. Zucchi, and J. C. Serrano. 1981. Bionomical aspects of *Lithurgus corumbae* (Hymenoptera, Megachilidae), including evolutionary considerations on the nesting behavior of the genus. *Revista Brasileira de Genetica* 4: 165-182.

Garófalo, C. A., E. Camillo, and J. C. Serrano. 1996. Fundações de ninhos por *Eulaema nigrita,* p. 331 *in* C. A. Garófalo et al., eds., *Anais do II Encontro sobre Abelhas.* Ribeirão Preto, Brazil: Faculdade de Filosofia, Ciências e Letras.

Garófalo, C. A., and G. Freita, eds. 2002. *Anais do V Encontro sobre Abelhas,* xxv + 355 pp. Ribeirão Preto: Departmento de Biologia, Universidade de São Paulo–Ribeirão Preto.

Garófalo, C. A., and J. G. Rozen, Jr. 2001. Parasitic behavior of *Exaerete smaragdina* with descriptions of its mature cocyte and larval instars. *American Museum Novitates* no. 3349: 1-26.

Gauld, I., and B. Bolton. 1988. *The Hymenoptera.* xi + 332 pp. Oxford: Oxford University Press.

Gebhardt, M., and G. Röhr. 1987. Zur bionomie der Sandbienen *Andrena clarkella* (Kirby), *A. cineraria* (L.), *A. fuscipes* (Kirby) und ihrer Kuckucksbienen. *Drosera* 87: 89-114.

Genaro, J. A., and L. Packer. 2005. A revision of the genus *Xenochilicola* (Hymentoptera: Apoidea, Colletidae), with the description of a new species. *Zootaxa* no. 1054: 1-14.

Genise, J. F., and T. M. Bown. 1996. *Uruguay* Roselli 1938 and *Rosellichnus,* n. ichnogenus: Two ichnogenera for clusters of fossil bee cells. *Ichnos* 4: 199-217.

Gennerich, J. 1922. Morphologische und biologische Untersuchungen der Putzapparate der Hymenopteren. *Archiv für Naturgeschichte,* Abt. A 88(12): 1-63.

Gerber, H. S., and R. D. Akre. 1969. The external morphology of *Megachile rotundata* (Fabricius). *Melanderia* 1: 1-36.

Gerling, D., P. D. Hurd, and A. Hefetz. 1983. Comparative

behavioral biology of two middle east species of carpenter bees (*Xylocopa* Latreille). *Smithsonian Contributions to Zoology*, no. 369: frontispiece + 1-31.

Gerling, D., H. H. W. Velthuis, and A. Hefetz. 1989. Bionomics of the large carpenter bees of the genus *Xylocopa*. *Annual Review of Entomology* 34: 163-190.

Germar, E. F. 1839. Die versteinerten Insecten Solenhofens. *Verhandlungen der Kaiserlichen Leopoldinisch-Carolinische Deutsche Akademie der Naturforscher* 19: 187-222, pls. 21-23.

———. 1849. Ueber einige Insekten aus Tertiärbildungen. *Zeitschrift der Deutschen Geologischen Gesellschaft* 1: 52-66, pl. II.

Gerstäcker. See Gerstaecker.

Gerstaecker, A. 1858. [Bees and wasps collected in Mozambique.] *Monatsberichte, Akademie der Wissenschaften, Berlin,* 29 October 1857, pp. 460-464.

———. 1868. *Psaenythia*, eine neue Bienengattung mit gezahnten Schienensporen. *Archiv für Naturgeschichte* 34: 111-137.

———. 1869. Beiträge zur näheren Kenntniss einiger Bienen-Gattungen. *Entomologische Zeitung* 30: 139-184, 315-367.

Gess, F. W., and S. K. Gess. 1996. Nesting and flower visiting of some southern African Anthophorini. *Annals of the Cape Provincial Museums, Natural History* 19: 347-373.

Girard, M. J. A. 1879. *Les Insectes.* Traité Élémentaire d'Entomologie, Vol. 2. [i] + 1028 pp. Paris: Baillière.

Giraud, J. 1863. Hyménoptères recueillis aux environs de Suse, en Piémont, et dans le département des Hautes-Alpes, en France; et description de quinze espèces nouvelles. *Verhandlungen der Zoologisch-Botanischen Gesellschaft in Wien* 13: 11-46.

Gistel [Gistl], J. N. F. X. 1848. *Naturgeschichte des Thierreichs, für Höhere Schulen.* xvi + 216 pp., 32 pls. Stuttgart: Hoffmann.

———. 1850. Erläuternde Anmerkungen zu dieser Dichtung. *Isis, Encyclopädische Zeitschrift für Naturgeschichte, Physiologie, etc.* no. 6: 82-83.

Gogala, A. 1995a. Phylogenetic relationships of the bee genus *Dioxys* Lepeletier and Serville, 1825. *Acta Entomologica Slovenica* 3: 43-48.

———. 1995b. Partial generic revision of the bee tribe Osmiini. *Acta Entomologica Slovenica* 3: 37-41.

Gogala, A. 1999. Bee fauna of Slovenia: checklist of species. *Scopolia* no. 42: 1-79.

Golick, D., and M. Ellis. 2002. *A Guide to Identifying Nebraska Bumble Bee Species B Bumble Boosters,* 91 pp. Lincoln, Neb.: Cooperative Extension Institute, Univ. of Nebraska. [A 2000 edition lacks photographic illustrations.]

Gonzalez, V. H., A. Mejia, and C. Rasmussen. 2004. Ecology and nesting behavior of *Bombus atratus* Franklin in Andean highlands. *Journal of Hymenoptera Research* 13: 234-242.

Gonzalez, V. H., and G. Nates-Parra. 2004. *Trigona* subgenus *Duckeola* in Colombia. *Journal of the Kansas Entomological Society* 77: 292.

Gordon, D. M. 2003. Life history & nest biology of the mason bee *Osmia (Acanthosmioides) integra* Cresson in coastal dunes. *Pan-Pacific Entomologist* 79: 45-53.

Goulet, H., and J. Huber. 1993. *Hymenoptera of the World, an Identification Guide to Families.* vii + 668 pp. Ottawa: Agriculture Canada.

Graeffe, E. 1902. Die Apiden-Fauna des osterreichischen Küstenlandes. *Verhandlungen der Zoologisch-Botanischen Gesellschaft in Wien* 52: 113-135.

Graf, V. 1972. Contribuição ao estudo da anatomia da cabeça dos Apoidea II—A musculatura do complexo labio-maxilar. *Boletim da Universidade Federal do Paraná, Zoologia* 5: 139-173.

Grandi, G. 1961. Studi di un entomologo sugli imenotteri superiori. *Bollettino dell'Istituto di Entomologia dell'Università di Bologna* 25: i-xv + 1-659.

Gray, G. 1832. [New genera and species], *in* E. Griffith and E. Pidgeon, *Class Insecta, Arranged by the Baron Cuvier, with Supplementary Additions to Each Order, and Notices of New Genera and Species by George Gray, Esq.,* Vol. 2. London. [This is Volume 15 of the series: Griffith, E., 1832, *The Animal Kingdom Arranged in Conformity with Its Organization by the Baron Cuvier, Member of the Institute of France, etc., with Supplementary Additions to Each Order.*]

Gribodo, G. 1884. Sopra alcune specie nuove o poco conosciute di Imenotteri Antofili. *Bollettino della Società Entomologica Italiana* [Firenze] 16: 269-274.

———. 1893, 1894. Note Imenotterologiche, Nota II. Nuovi generi e nuove specie di Imenotteri antofili ed osservazioni sobra alcune specie gia conosciute. *Bollettino della Società Entomologica Italiana* [Firenze] 25 (1893): 248-287, 389-428; 26 (1894): 76-135, 262-314.

Grigarick, A. A., and L. A. Stange. 1968. The pollen-collecting bees of the Anthidiini of California. *Bulletin of the California Insect Survey* no. 9: 1-113.

Grimaldi, D. 1999. The co-radiation of pollinating insects and anigiosperms in the Cretaceous. *Annals of the Missouri Botanical Garden* 86: 373-406.

Grimaldi, D. and M. S. Engel. 2005. *Evolution of the insects,* xiii + 755 pp. New York: Cambridge University Press.

Grimaldi, D., C. W. Beck, and J. J. Boon. 1989. Occurrence, chemical characteristics, and paleontology of the fossil resins from New Jersey. *American Museum Novitates* no. 2948: 1-27.

Griswold, T. L. 1983. Revision of *Proteriades* Subgenus *Acrosmia* Michener. *Annals of the Entomological Society of America* 76: 707-714.

———. A generic and subgeneric revision of the *Heriades* genus-group. xiii + 165 pp. Logan: Ph.D. thesis, Utah State University.

———. 1986a. Notes on the nesting biology of *Protosmia (Chelostomopsis) rubifloris* (Cockerell). *Pan-Pacific Entomologist* 62: 84-87.

———. 1986b. A new heriadine bee from the Mojave Desert. *Southwestern Entomologist* 11: 165-169.

———. 1991. A review of the genus *Microthurge*. *Pan-Pacific Entomologist* 67: 115-118.

———. 1994a. Taxonomic notes on some heriadines, with descriptions of three new species. *Journal of the Kansas Entomological Society* 67: 17-28.

———. 1994b. A review of *Ochreriades*. *Pan-Pacific Entomologist* 70: 318-321.

Griswold, T. L., and C. D. Michener. 1988. Taxonomic observations on Anthidiini of the western hemisphere. *Journal of the Kansas Entomological Society* 61: 22-45.

———. 1998. The classification of the Osmiini of the eastern hemisphere. *Journal of the Kansas Entomological Society* 70(1997): 207-253.

Griswold, T. L., and F. D. Parker. 1987. A new species of *Protosmia* Ducke from Spain with notes on related species. *Psyche* 94: 51-56.

———. 2003. *Stelis rozeni*, new species, the first record of the parasitic bee genus *Stelis* from southern Africa. *Journal of the Kansas Entomological Society* 76: 282-285.

Grozdanić, S., and V. Vasić. 1968. Biološka posmatranja na *Systropha planidens* Gir. *Glasnik Prirodnjackog Muzeja u Beogradu (Bulletin du Muséum d'Histoire Naturelle Belgrade)* (B)23: 169-178. [In Serbo-Croatian, German summary.]

Grünwaldt, W. 1976. *Andrena grossella* n. sp., eine Insekten-Art mit 9 gliedrigen Maxillar- und Labialpalpen. *Nachrichtenblatt der Bayerischen Entomologen* 25(4): 65-70.

Grütte, E. 1935. Zur Abstammung der Kuckucksbienen. *Archiv für Naturgeschichte* (n.f.)4: 449-534.

Guérin-Méneville, F. E. 1830, 1831. Crustacés, arachnides et insectes, Zool. 2, div. 1 *in* L. I. Duperrey, *Voyage Autour du Monde, Exécuté par Ordre du Roi, sur la Corvette . . . La Coquille.* 471 pp., 22 pls. in Atlas. Paris: Bertrand. [The introduction is dated 1838, but the text was published in 1830 (pp. 1-24) and 1831 (pp. 25-471).]

———. 1844. *Iconographie du Règne Animal de G. Cuvier*, Vol. 7, Insects. 576 pp. Paris: Baillière.

Guiglia, D., and J. J. Pasteels. 1961. Aggiunte ed osservazioni all'elenco delle specie di imenotteri descritte da Guérin-Méneville che si trovano nelle collezioni del museo di Genova. *Annali del Museo Civico di Storia Naturale di Genova* 72: 17-30.

Gupta, R. K. 1987. On a new subgenus *Orientoheriades* and two new species of genus *Heriades* Spinola from India. *Reichenbachia* 25(15): 67-71.

———. 1990. On a new subgenus *Neoashmeadiella* and two new species of genus *Ashmeadiella* Cock. from India. *Reichenbachia* 29(10): 55-59.

———. 1991. A new subgenus of *Coelioxys* Latreille (Hymenoptera: Apoidea: Megachilidae) from India. *Journal of the Bombay Natural History Society* 88: 425-428.

———. 1992. On a new subgenus *Orientocoelioxys* and a new species of genus *Coelioxys* Latr. from India. *Reichenbachia* 29: 73-76.

———. 1993. *Taxonomic Studies on the Megachilidae of North-Western India.* [4] + 288 pp. New Delhi: Indian Council of Agricultural Research. Also published 1998, Jodhpur: Scientific Publishers (India).

Gusenleitner, F., and M. Schwarz. 2002. Weltweite Checkliste der Bienengattung *Andrena* mit Bemerkungen und Ergänzungen zu Paläarktischen Arten. *Entomofauna* Suppl. 12: 1-1280.

Gusenleitner, F., M. Schwarz, J. Ascher, and E. Scheuchl. 2005. Korrekturen und Nachträge zu Gusenleiter & Schwarz (2002): Weltweite Checkliste der Bienengattung *Andrena* mit Bemerkungen und Ergänzungen zu paläarktischen Arten. *Entomofauna* 26: 437-472.

Gutbier, A. 1916. Essai sur la classification et sur le développement des nids des guêpes et des abeilles. *Horae Societatis Entomologicae Rossicae* 41(7): 1-57, pls. 1, 2. [In Russian.]

Haas, A. 1949. Arttypische Flugbahnen von Hummelmännchen. *Zeitschrift für Vergleichende Physiologie* 31: 281-307.

———. 1976. Paarungsverhalten und Nestbau der alpinen Hummelart *Bombus mendax. Entomologica Germanica* 3: 248-259.

Hachfeld, G. 1926. Zur Biologie der *Trachusa byssina* Pz. *Zeitschrift für Wissenschaftliche Insektenbiologie* 21: 63-84.

Haeseler, V. 2005. *Osmia chrysolepta* sp. nov., a new bee of the subgenus *Hemiosmia* Tkalců, 1975 with notes on the identification and distribution of species closely related to *O. (Hemiosmia) balearica* Schmiedeknecht, 1885. *Entomofauna* 26: 473-492.

Hagen, E. von. 1990. *Hummeln Bestimmen, Ansiedeln, Vermehren, Schützen.* 256 pp. Augsburg: Natur Verlag.

Hagens, D. von. 1882. Ueber die männlichen Genitalien der Bienen-Gattung *Sphecodes. Deutsche Entomologische Zeitschrift* 26: 209-228, pls. VI, VII.

Haliday, A. H. 1836. Descriptions, etc. of the Hymenoptera, pp. 316-331 *in* J. Curtis, A. H. Haliday, and F. Walker, eds., Descriptions, etc. of the insects collected by Captain P. P. King, R. N., F. R. S., in the survey of the Straits of Magellan. *Transactions of the Linnean Society of London* 17: 315-359.

Hammer, K., and S. N. Holm. 1970. Danske humlebier og snyltehumler. *Natur og Museum* [Århus] 14(2-3): 1-22, 2 color pls.

Handlirsch, A. 1906-1908. *Die Fossilen Insekten und die Phylogenie der Rezenten Formen.* ix + 1430 + li pp., 51 pls. Leipzig: Engelmann.

Harder, L. D. 1983. Functional differences of the proboscides of short- and long-tongued bees. *Canadian Journal of Zoology* 61: 1580-1586.

Hazeldine, P. L. 1996. Comportamiento de nidificación de *Meliphilopsis melandra* Roig Alsina. *Extra*, nueva serie no. 138: 1-4. [Buenos Aires: Museo Argentino de Ciencias Naturales "Bernadino Rivadavia."]

———. 1997a. *Comportamiento de nidificación de cuatro especies de* Ptilothrix *Smith. Physis* [Buenos Aires] (C)54(1996): 27-41.

———. 1997b. Comportamiento de nidificación de *Diadasia* Patton. *Physis* [Buenos Aires] (C)54(1996): 43-48.

Hazeltine, W. E., and L. Chandler. 1964. A preliminary atlas for the identification of female bumble bees. *Journal of the Kansas Entomological Society* 37: 77-87.

Hedicke, H. 1922. Nomina nova. *Deutsche Entomologische Zeitschrift* 1922: 427-428.

———. 1926. Beiträge zur Apidenfauna der Philippinen. *Deutsche Entomologische Zeitschrift* 1926: 413-423.

———. 1930. 16. Ordnung: Hautflügler, Hymenoptera, *in* P. Brohmer, P. Ehrmann, and G. Ulmer, eds., *Die Tierwelt Mitteleuropas, Insekten,* Vol. 5, 2. Teil, pp. 1-246. Leipzig: Quelle and Meyer.

———. 1933. Beiträge zur Systematik der Gattung *Andrena* F *Mitteilungen aus dem Zoologischen Museum in Berlin* 19: 199-220.

———. 1938. Über einige Apiden vom Hindukusch. *Deutsche Entomologische Zeitschrift* 1938: 186-196.

Heer, O. 1849. Die Insektenfauna der Tertiärgebilde von

Oenigen und von Radoboj in Croatien. *Neue Denkschriften der Allgemeinen Schweizerischen Gesellschaft für die gesammten Naturwissenschaften* [Zurich] 11(1850): v + 264, pls. I-XVII.

Hefetz, A., H. M. Fales, and S. W. T. Batra. 1979. Natural polyesters: Dufour's gland macrocyclic lactones form brood cell laminesters in *Colletes* bees. *Science* 204: 415-417.

Heider, K. 1926. *Nomenclator Animalium Generum et Subgenerum*, Vol. 1, Lief. 2, pp. 161-320. Berlin: Preussischen Akademie der Wissenschaften.

———. 1935. *Nomenclator Animalium Generum et Subgenerum*, Vol. 4, Lief. 16, pp. 2185-2248. Berlin: Preussischen Akademie der Wissenschaften.

Hennig, W. 1966. *Phylogenetic Systematics*. viii + 264 pp. Urbana: University of Illinois Press.

Hensen, R. V. 1987. *Hylaeus (Metylaeus) mahafaly* sp. n., a new Malagasy bee. *Entomologische Berichten* 47: 152-154.

Herbst, P. 1922. Zur Biologie der Gattung *Chilicola* Spin. *Entomologische Mitteilungen* 11: 63-68.

Hicks, C. H. 1926. Nesting habits and parasites of certain bees of Boulder County, Colorado. *University of Colorado Studies* 26(8): 217-252.

Hiller, B., and D. Wittmann. 1994. Seasonality, nesting biology and mating behavior of the oil-collecting bee *Epicharis dejeanii*. *Biociências* [Porto Alegre, Brazil] 2: 107-124.

Hinojosa-Díaz, I. A., and M. S. Engel. 2003. *Megalopta (Noctoraptor) furunculosa* sp. n., a new nocturnal, cleptoparasitic bee from Guyana. *Folia Heyrovskyana* 11: 137-141.

Hinojosa-Diaz, I. A., O. Yañez-Ordóñez, G. Chen, A. T. Peterson, and M. S. Engel. 2005. The North American invasion of the giant resin bee. *Journal of Hymenoptera Research* 14: 69-77.

Hirashima, Y. 1952. Description of *Andrena yasumatsui* n. sp., with a provisional key to the subgenera of palearctic *Andrena*. *Mushi* 24: 59-65, pl. 4.

———. 1955. A new species of the genus *Epeolus* from Japan. *Insecta Matsumurana* 19: 40-43.

———. 1956. Some bees of the genus *Nomia* Latreille from Formosa. *Insecta Matsumurana* 20: 29-33.

———. 1957. Descriptions and records of bees of the genus *Andrena* from eastern Asia, IV. *Mushi* 30: 59-66.

———. 1961. Monographic study of the subfamily Nomiinae of Japan. *Acta Hymenopterologica* 1: 241-303.

———. 1962. Systematic and biological studies of the family Andrenidae of Japan (Hymenoptera, Apoidea) Part 1, Biology. *Journal of the Faculty of Agriculture, Kyushu University* 12: 1-20.

———. 1963. Systematic and biological studies of the family Andrenidae of Japan (Hymenoptera, Apoidea) Part 2, Systematics, 2. *Journal of the Faculty of Agriculture, Kyushu University* 12: 241-263.

———. 1964a. Systematic and biological studies of the family Andrenidae of Japan (Hymenoptera, Apoidea) Part 2, Systematics, 3. *Journal of the Faculty of Agriculture, Kyushu University* 13: 41-69.

———. 1964b. Systematic and biological studies of the family Andrenidae of Japan (Hymenoptera, Apoidea) Part 2, Systematics, 4. *Journal of the Faculty of Agriculture, Kyushu University* 13: 71-97.

———. 1965a. Systematic and biological studies of the family Andrenidae of Japan (Hymenoptera, Apoidea) Part 2, Systematics, 5. *Journal of the Faculty of Agriculture, Kyushu University* 13: 461-491.

———. 1965b. Systematic and biological studies of the family Andrenidae of Japan (Hymenoptera, Apoidea) Part 2, Systematics, 6. *Journal of the Faculty of Agriculture, Kyushu University* 13: 493-517.

———. 1966a. Systematic and biological studies of the family Andrenidae of Japan (Hymenoptera, Apoidea) Part 2, Systematics, 7. *Journal of the Faculty of Agriculture, Kyushu University* 14: 89-131.

———. 1966b. Comments on *Nomia (Ptilonomia)* with descriptions of two new species from New Guinea. *Kontyû* 34: 293-297.

———. 1967a. A new subgenus and species of *Hylaeus* from New Guinea. *Kontyû* 35: 134-138.

———. 1967b. Discovery of *Gephyrohylaeus* from the Philippines. *Kontyû* 35: 62-64.

———. 1967c. Metallic forms of *Nomia (Mellitidia)* of New Guinea in the collection of Bishop Museum. *Journal of the Faculty of Agriculture, Kyushu University* 14: 311-331.

———. 1969. Synopsis of the genus *Pithitis* Klug of the world (Hymenoptera: Anthophoridae). *Pacific Insects* 11: 649-669.

———. 1971a. Subgeneric classification of the genus *Ceratina* Latreille of Asia and West Pacific, with comments on the remaining subgenera of the world (Hymenoptera, Apoidea). *Journal of the Faculty of Agriculture, Kyushu University* 16: 349-375.

———. 1971b. *Megaceratina*, a new genus of bees of Africa (Hymenoptera, Anthophoridae). *Journal of Natural History* 5: 251-256.

———. 1973. Two new species of the genus *Osmia* from Japan and N. China. *Journal of the Faculty of Agriculture, Kyushu University* 18: 63-68.

———. 1975a. Revision of the bee genus *Xenorhiza* Michener of New Guinea, with descriptions of two new species. *Esakia* no. 9: 1-14.

———. 1975b. New species of bees of the genus *Palaeorhiza* Perkins from New Guinea mainly in the collection of the Rijksmuseum van Natuurlijke Historie, Leiden. *Journal of the Faculty of Agriculture, Kyushu University* 20: 27-46.

———. 1977. Revision of the Japanese species of *Nesoprosopis*, with descriptions of two new species. *Esakia* no. 10: 21-43.

———. 1978a. A synopsis of the bee genus *Palaeorhiza* Perkins (Hymenoptera, Colletidae) of New Guinea, Part II. Subgenera *Gressittapis, Noonadania, Sphecogaster, Anchirhiza, Ceratorhiza*. *Esakia* no. 12: 63-87.

———. 1978b. A synopsis of the genus *Palaeorhiza* Perkins (Hymenoptera, Colletidae) of New Guinea, Part I. Subgenus *Palaeorhiza* s. str. *Esakia* no. 11: 89-119.

———. 1978c. Some Asian species of *Austronomia*, a subgenus of *Nomia*, with descriptions of three new species from Sri Lanka. *Esakia* no. 12: 89-101.

———. 1980. A synopsis of the bee genus *Palaeorhiza* Perkins (Hymenoptera, Colletidae) of New Guinea, Part III. Subgenera *Trachyrhiza, Paraheterorhiza, Hadrorhiza*. *Journal of the Faculty of Agriculture, Kyushu University* 25: 99-117.

———. 1981a. A synopsis of the bee genus *Palaeorhiza* Perkins (Hymenoptera, Colletidae) of New Guinea, Part V. Subgenus *Cnemidorhiza*. *Esakia* no. 17: 1-48.

———. 1981b. A synopsis of the bee genus *Palaeorhiza* Perkins (Hymenoptera, Colletidae) of New Guinea, Part IV. Subgenus *Cheesmania*. *Kontyû* 49: 27-36.

———. 1982a. A synopsis of the bee genus *Palaeorhiza* Perkins (Hymenoptera, Colletidae) of New Guinea, Part VI. Subgenus *Zarhiopalea*. *Kontyû* 50: 57-66.

———. 1982b. A synopsis of the bee genus *Palaeorhiza* Perkins (Hymenoptera, Colletidae) of New Guinea, Part VII. Subgenus *Cercorhiza*. *Journal of the Faculty of Agriculture, Kyushu University* 26: 87-109.

———. 1988. Bees of the genus *Palaeorhiza* Perkins (Hymenoptera, Colletidae) of Papua New Guinea collected by the Kyushu University expedition, I. *Esakia* no. 26: 5-19.

———. 1989. A synopsis of the bee genus *Palaeorhiza* Perkins (Hymenoptera, Colletidae) of New Guinea, Part VIII. Subgenus *Callorhiza*. *Esakia* no. 28: 1-9, pl. 1.

———. 1992. A new name of *Palaeorhiza*. *Japanese Journal of Entomology* 60: 395.

———. 1996. Subgeneric classification of the genus *Xenorhiza* Michener of Papua New Guinea. *Memoirs of the Entomological Society of Washington* no. 17: 78-81.

———. 1997. Review of My Study on the New Guinea Bees for the Years 1966-1996. 88 pp. Fukuoka: Published by the author. [In Japanese.]

Hirashima, Y., and Y. Haneda. 1973. New or little known species of the genus *Andrena* from Japan. *Mushi* 47: 67-73.

Hirashima, Y., and M. A. Lieftinck. 1982. Systematic studies on the genus *Palaeorhiza* of New Guinea collected by the third Archbold expedition (I). *Esakia* no. 19: 1-50.

———. 1983. Systematic studies on the genus *Palaeorhiza* of New Guinea collected by the third Archbold expedition (II). *Esakia* no. 20: 93-129.

Hirashima, Y., and Y. Maeta. 1974. Bees of the genus *Megachile sensu lato* (Hymenoptera, Megachilidae) of Hokkaido and Tohoku District of Japan. *Kontyû* 42(2): 157-173.

Hirashima, Y., and H. Roberts. 1986. Discovery of the bee genus *Pharohylaeus* Michener from Papua New Guinea, with description of a new species. *Esakia* no. 24: 63-66.

Hirashima, Y., and O. Tadauchi. 1975. A new subgenus of the genus *Andrena* (Hymenoptera, Andrenidae) from Japan and allied areas. *Journal of the Faculty of Agriculture, Kyushu University* 19: 175-186.

———. 1984. Two new species of *Prosopisteroides* from Papua New Guinea. *Esakia* no. 21: 59-65.

———. 2002. *Adamon*, a new subgenus of the genus *Nomada* Scopoli from Japan. *Esakia* no. 42: 47-54.

Hobbs, G. A. 1964. Phylogeny of bumble bees based on brood-rearing behaviour. *Canadian Entomologist* 96: 115-116.

Hoebeke, E. R., and A. G. Wheeler. 1999. *Anthidium oblongatum* (Illiger): an Old World bee (Hymenoptera: Megachilidae) new to North America, and new North American records for another adventive species, *A. manicatum*, pp. 21-24 *in* G. W. Byers, R. H. Hagen, and R. W. Brooks, eds., *Entomological Contributions in Memory of Byron A. Alexander, University of Kansas Natural History Museum Special Publication* no. 24.

Hoffmannsegg, G. J. C. von. 1817. Entomologische Bemerkungen bei Gelegenheit der Abhandlungen über amerikanische Insekten . . . . *Zoologisches Magazin* [Kiel] 1: 8-56.

Holmberg, E. L. 1884, 1886a. Viajes al Tandil y a La Tinta, 2nd Parte, Zoologia, Insectos, I. Himenópteros-Hymenoptera. *Actas de la Academia Nacional de Ciencias de la República Argentina en Córdoba* 5: 117-136 (1884); 137-184, 2 pls. (1886).

———. 1886b, 1887a. Sobre ápidos Nómadas de la República Argentina. *Anales de la Sociedad Científica Argentina* 22(1886): 231-240, 272-286; 23(1887): 17-33, 67-82.

———. 1887b. Viaje á Misiones. *Boletín de la Academia Nacional de Ciencias en Córdoba* 10: 5-391.

———. 1903. Delectus Hymenopterologicus Argentinus . . . . *Anales del Museo Nacional de Buenos Aires* (3)2: 377-512.

———. 1909a. *Alloscirtetica*, Holmb. n. nom. *Apuntes de Historia Natural* [Buenos Aires] 1: 77.

———. 1909b. *Palinepeolus*, Holmb. nov. nom. *Apuntes de Historia Natural* [Buenos Aires] 1: 77-78.

———. 1916. Las especies argentinas de *Coelioxys*. *Anales del Museo Nacional de Historia Natural de Buenos Aires* 28: 541-591.

———. 1918. Suplemento I á las especies argentinas de *Coelioxys*. *Physis* [Buenos Aires] 4: 1-13, 145-166.

———. 1921. Apidae argentinae, Generis *Psaenythia* Gerstaecker. *Anales del Museo Nacional de Historia Natural de Buenos Aires* 31: 249-354.

Hong, Y-c. 1984. New fossil insects of the Laiyang Group from the Laiyang Basin, Shandong Province. *Professional Papers on Stratigraphy and Palaeontology* [Beijing] 11: 31-41. [In Chinese; English summary.]

Houston, T. F. 1969. Observations on the nests and behaviour of some euryglossine bees. *Journal of the Australian Entomological Society* 8: 1-10.

———. 1970. Discovery of an apparent male soldier caste in a nest of a halictine bee (Hymenoptera: Halictidae), with notes on the nest. *Australian Journal of Zoology* 18: 345-351.

———. 1975a. A revision of the Australian hylaeine bees. *Australian Journal of Zoology*, suppl. series no. 36: 1-135.

———. 1975b. Nests, behavior and larvae of the bee *Stenotritis pubescens* (Smith) and behavior of some related species. *Journal of the Australian Entomological Society* 14: 145-154.

———. 1976. New Australian allodapine bees (subgenus *Exoneurella* Michener) and their immatures. *Transactions of the Royal Society of South Australia* 100: 15-28.

———. 1977. Nesting biology of three allodapine bees in the subgenus *Exoneurella* Michener. *Transactions of the Royal Society of South Australia* 101: 99-113.

———. 1981a. A revision of the Australian hylaeine bees, II. *Australian Journal of Zoology*, suppl. series no. 80: 1-128.

———. 1981b. Alimentary transport of pollen in a paracolletine bee (Hymenoptera: Colletidae). *Australian Entomological Magazine* 8: 57-59.

———. 1983a. A revision of the bee genus *Ctenocolletes*

(Hymenoptera: Stenotritidae). *Records of the Western Australian Museum* 10: 269-306.

———. 1983b. A new species of *Ctenocolletes* (Hymenoptera: Stenotritidae). *Records of the Western Australian Museum* 10: 307-313.

———. 1983c. An extraordinary new bee and adaptation of palpi for nectar-feeding in some Australian Colletidae and Pergidae. *Journal of the Australian Entomological Society* 22: 263-270.

———. 1984. Biological observations of bees in the genus *Ctenocolletes*. *Records of the Western Australian Museum* 11: 153-172.

———. 1985. Supplement to a revision of the bee genus *Ctenocolletes* (Hymenoptera: Stenotritidae). *Records of the Western Australian Museum* 12: 293-305.

———. 1987. A second contribution to the biology of *Ctenocolletes* bees. *Records of the Western Australian Museum* 13: 189-201.

———. 1989. *Leioproctus* bees associated with Western Australian smoke bushes (*Conospermum* spp.) and their adaptations for foraging and concealment. *Records of the Western Australian Museum* 14: 275-292.

———. 1990. Descriptions of new paracolletine bees associated with flowers of *Eremophila*. *Records of the Western Australian Museum* 14: 583-621.

———. 1991a. Two new and unusual species of the bee genus *Leioproctus* Smith (Hymenoptera: Colletidae), with notes on their behavior. *Records of the Western Australian Museum* 15: 83-96.

———. 1991b. Ecology and behavior of the bee *Amegilla (Asaropoda) dawsoni* (Rayment) with notes on a related species. *Records of the Western Australian Museum* 15: 591-609.

———. 1992a. Biological observations of the Australian green carpenter bees, genus *Lestis*. *Records of the Western Australian Museum* 15: 785-798.

———. 1992b. Three new, monolectic species of *Euryglossa (Euhesma)* from Western Australia. *Records of the Western Australian Museum* 15: 719-728.

Houston, T. F., and R. W. Thorp. 1984. Bionomics of the bee *Stenotritus greavesi* and ethological characteristics of Stenotritidae (Hymenoptera). *Records of the Western Australian Museum* 11: 375-385.

Hurd, P. D., Jr. 1955. The carpenter bees of California. *Bulletin of the California Insect Survey* 4: 35-72.

———. 1956. Notes on the subgenera of the new world carpenter bees of the genus *Xylocopa*. *American Museum Novitates* no. 1776: 1-7.

———. 1958a. American bees of the genus *Dioxys* Lepeletier and Serville. *University of California Publications in Entomology* 14: 275-302.

———. 1958b. Observations on the nesting habits of some new world carpenter bees with remarks on their importance in the problem of species formation. *Annals of the Entomological Society of America* 51: 365-375.

———. 1961. A synopsis of the carpenter bees belonging to the subgenus *Xylocopoides* Michener. *Transactions of the American Entomological Society* 87: 247-257, pls. VII, VIII.

———. 1967. The identity of *Megachile rotundata* (Fabricius) and *M. argentata* (Fabricius). *Entomologiske Meddelelser* 35: 3-10.

———. 1978a. *An Annotated Catalog of the Carpenter Bees (Genus Xylocopa Latreille) of the Western Hemisphere.* [5] + 106 pp. Washington: Smithsonian Institution Press.

———. 1978b. Bamboo-nesting carpenter bees (genus *Xylocopa* Latreille) of the subgenus *Stenoxylocopa* Hurd and Moure. *Journal of the Kansas Entomological Society* 51: 746-764.

———. 1979. Superfamily Apoidea, pp. 1741-2209 *in* K. V. Krombein, P. D. Hurd, Jr., D. R. Smith, and B. D. Burks, eds., *Catalog of Hymenoptera in America North of Mexico,* Vol. 2. Washington: Smithsonian Institution Press.

Hurd, P. D., Jr., W. E. LaBerge, and E. G. Linsley. 1980. Principal sunflower bees of North America with emphasis on the southwestern United States. *Smithsonian Contributions to Zoology* no. 310: iv + 158 pp.

Hurd, P. D., Jr., and E. G. Linsley. 1951. The melectine bees of California. *Bulletin of the California Insect Survey* 1: 119-[140].

———. 1964. The squash and gourd bees—genera *Peponapis* Robertson and *Xenoglossa* Smith—inhabiting America north of Mexico. *Hilgardia* 35: 373-477.

———. 1966. The Mexican squash and gourd bees of the genus *Peponapis*. *Annals of the Entomological Society of America* 59: 835-851.

———. 1967a. Squash and gourd bees of the genus *Xenoglossa*. *Annals of the Entomological Society of America* 60: 988-1007.

———. 1967b. South American squash and gourd bees of the genus *Peponapis*. *Annals of the Entomological Society of America* 60: 647-661.

———. 1970. A classification of the squash and gourd bees *Peponapis* and *Xenoglossa*. *University of California Publications in Entomology* 62: i- iv + 1-39.

———. 1972. Parasitic bees of the genus *Holcopasites* Ashmead. *Smithsonian Contributions to Zoology* no. 114: 1-41.

———. 1975. The principal *Larrea* bees of the southwestern United States. *Smithsonian Contributions to Zoology* no. 193: 1-74.

———. 1976. The bee family Oxaeidae with a revision of the North American species. *Smithsonian Contributions to Zoology* no. 220: 1-75.

Hurd, P. D., Jr., and C. D. Michener. 1955. The megachiline bees of California. *Bulletin of the California Insect Survey* 3: 1-248.

———. 1961. *Lestis* Lepeletier and Serville, 1828 (Insecta: Hymenoptera): Proposed designation of a type species under the plenary powers. *Bulletin of Zoological Nomenclature* 18: 201-202.

Hurd, P. D., Jr., and J. S. Moure. 1960. A new-world subgenus of bamboo-nesting carpenter bees belonging to the genus *Xylocopa* Latreille. *Annals of the Entomological Society of America* 53: 809-821.

———. 1963. A classification of the large carpenter bees (Xylocopini). *University of California Publications in Entomology* 29: i-vi + 1-365.

Ihering, H. von. 1912. Zur Biologie der brasilianischen Meliponiden. *Zeitschrift für Wissenschaftliche Insektenbiologie* 8(1): 1-5, (2): 43-46.

Ikudome, S. 1989. A revision of the family Colletidae of Japan. *Bulletin of the Institute of Minami-kyûshû Regional Science, Kagoshima Women's Junior College* no. 5: 43-314.

———. 1991. Two new subgenera of *Hylaeus*. *Japanese Journal of Entomology* 59: 789-792.

———. 1994. A list of the bee taxa of Japan and their Japanese names. *Bulletin of Kagoshima Women's Junior College* no. 29: 1-23.

Illiger, K. 1801. Namen der Insekten-Gattungen, ihr Genitiv, ihr grammatisches Geschlecht, ihr Silbenmass, ihre Herleitung: Zugleich mit den Deutschen Benennungen. *Magazin für Insektenkunde* 1: 125-155.

———. 1806. William Kirby's Familien der Bienenartigen Insekten mit Zusätzen, Nachweisungen und Bemerkungen. *Magazin für Insektenkunde* 5: 28-175.

———. 1807. Vergleichung der Gattungen der Hautflügler, Piezata Fabr. Hymenoptera Linn. Jur. *Magazin für Insektenkunde* 6: 189-199.

Imperatriz-Fonseca, V. L., A. M. Saraiva, and D. DeJong, eds. 2006. *Bees as pollinators in Brazil*, 96 pp + [14 pls.]. Ribeirão Preto: Conservation International, Holos Editora.

Inoue, T., and S. F. Sakagami. 1993. A new name of *Trigona*. *Japanese Journal of Entomology* 61: 769.

Intoppa, F., M. G. Piazza, and G. Bolchi Serini. 2003. Repertorio dei caratteri morfologici per una chiave dicotomica dei sottogeneri di Bombinae presenti in Italia. *Redia* 86(Appendice): 1-23.

Ireland, L. M. 1935. Scientific results of the Vernay-Lang Kalahari expedition, March to September, 1930—Morphology of male Halictidae. *Annals of the Transvaal Museum* 17: 95-107.

Isensee, R. 1927. A study of the male genitalia of certain anthidiine bees. *Annals of the Carnegie Museum* 17: 371-382, pl. xxxii.

Ito, M. 1985. Supraspecific classification of bumblebees based on the characters of the male genitalia. *Contributions from the Institute of Low Temperature Science, Hokkaido University* (ser. B)20: 1-143.

Ito, M., and S. F. Sakagami. 1985. Possible synapomorphies of the parasitic bumblebees (*Psithyrus*) with some nonparasitic bumblebees (*Bombus*). *Sociobiology* 10: 105-119.

Iuga, V. G. 1958. Hymenoptera Apoidea, Fam. Apidae, Subfam. Anthophorinae, in *Fauna Republicii Populare Romîne, Insecta*, Vol. IX, fasc. 3: 1-270.

Iwata, K. 1939. Biology of *Coelioxys elongata* Lepeletier. *Mushi* 12: 34-40.

———. 1976. *Evolution of Instinct: Comparative Ethology of Hymenoptera*. xii + 536 pp. New Delhi: Amerind Publishing Co.

Iwata, K., and S. F. Sakagami. 1966. Gigantism and dwarfism in bee eggs in relation to the modes of life, with notes on the number of ovarioles. *Japanese Journal of Ecology* 16: 4-16.

Jander, R. 1976. Grooming and pollen manipulation in bees (Apoidea): The nature and evolution of movements involving the foreleg. *Physiological Entomology* 1: 179-194.

Jang, Y., C. T. Wuellner, and C. S. Scott. 1996. Floating and fidelity in nest visitation by *Crawfordapis luctuosa*. *Journal of Insect Behavior* 9: 493-504.

Janjic, J., and L. Packer. 2003. Phylogeny of the bee genus *Agapostemon*. *Systematic Entomology* 28: 101-123.

Jones, C. E., and R. J. Little, eds. 1983. *Handbook of Experimental Pollination Biology*. xviii + 558 pp. New York: Van Nostrand Reinhold.

Jörgensen, P. 1909. Beobachtungen über blumenbesuch, biologie, verbreitung usw. der Bienen von Mendoza, Teil I, Teil II. *Deutsche Entomologische Zeitschrift* 1909: 53-65, 211-228.

———. 1912. Revision der Apiden der Provinz Mendoza, Republica Argentina. *Zoologische Jahrbücher, Abteilung für Systematik, Geographie und Biologie der Tiere* 32: 89-162.

Jurine, L. 1801. In [G. W. F. Panzer], Nachricht von einem neuen entomologischen Werke, des Hrn. Prof. Jurine in Geneve. *Intelligenzblatt der Litteratur-Zeitung* [Erlangen] 1: 160-165. [This work was suppressed for nomenclatural purposes by ICZN Opinion 135 (Direction 4) (1939).]

———. 1807. *Nouvelle Méthode de Classer les Hyménoptères et les Diptères*, Vol. 1, Hyménoptères. iv + 320 + 4 pp, 14 pls. Geneva: Paschoud.

Katayama, E. 1989. Comparative studies on the egg-laying habits of some Japanese species of bumblebees. *Entomological Society of Japan Occasional Publication* no. 2: 1-161.

Kawakita, A., T. Sota, J. S. Ascher, M. Ito, H. Tanaka, and M. Kato. 2003. Evolution and phylogenetic utility of alignment gaps within intron sequences of three nuclear genes in bumble bees. *Molecular Biology and Evolution* 20: 87-92.

Kelner-Pillault, S. 1970. Une melipone (s. l.) de l'ambre Balte. *Annales de la Société Entomologique de France* (n.s.)6: 437-441.

Kerfoot, W. B. 1967. The lunar periodicity of *Sphecodogastra texana*, a nocturnal bee. *Animal Behaviour* 15: 479-486.

Kerr, W. E., and V. Maule. 1964. Geographic distribution of stingless bees and its implications. *Journal of the New York Entomological Society* 72: 2-17.

Kerr, W. E., J. F. Pisani, and D. Aily. 1967. Aplicação de princípios modernos a sistematica do genero *Melipona* Illiger, com a divisão em dois subgeneros. *Papéis Avulsos de Zoologia, São Paulo* 20: 135-145.

Kerr, W. E., S. F. Sakagami, R. Zucchi, V. de Portugal Araújo, and J. M. F. de Camargo. 1967. Observações sôbre a arquitetura dos ninhos e comportamento de algumas espécies de abelhas sem ferrão das vizinhanças de Manaus, Amazonas. *Atas do Simposio sôbre a Biota Amazônica (Zoologia)* 5: 255-309.

Kevan, P. G., and V. L. Imperatriz-Fonseca. 2002. *Pollinating bees, the Conservation Link between Agriculture and Nature*, xviii + 313 pp. Brasilia: Ministry of Environment.

Kimsey, L. S. 1979a. An illustrated key to the genus *Exaerete* with descriptions of male genitalia and biology. *Journal of the Kansas Entomological Society* 52: 735-746.

———. 1979b. Synonymy of the genus *Euplusia* Moure under *Eufriesia* Cockerell. *Pan-Pacific Entomologist* 55: 126.

———. 1982. Systematics of bees of the genus *Eufriesea*. *University of California Publications in Entomology* 95: i-ix + 1-125.

———. 1984a. The behavioural and structural aspects of grooming and related activities in euglossine bees. *Journal of Zoology* [London] 204: 541-550.

———. 1984b. A re-evaluation of the phylogenetic relationships in the Apidae. *Systematic Entomology* 9: 435-441.

———. 1987. Generic relationships within the Euglossini. *Systematic Entomology* 12: 63-72.
Kimsey, L. S., and R. L. Dressler. 1986. Synonymic species list of Euglossini. *Pan-Pacific Entomologist* 62: 229-236.
King, J. 1984. Immature stages of some Megachilidae. *Journal of the Australian Entomological Society* 23: 51-57.
———. 1994. The bee family Megachilidae (Hymenoptera: Apoidea) in Australia, I. Morphology of the genus *Chalicodoma* Lepeletier, and a revision of the subgenus *Hackeriapis* Cockerell. *Invertebrate Taxonomy* 8: 1373-1419.
King, J., and E. M. Exley. 1985a. A revision of *Chalicodoma (Rhodomegachile)* Michener. *Journal of the Australian Entomological Society* 24: 199-204.
———. 1985b. A reinstatement and revision of the genus *Thaumatosoma* Smith. *Journal of the Australian Entomological Society* 24: 87-92.
———. 1985c. A revision of *Chalicodoma (Chalicodomoides)* Michener. *Journal of the Australian Entomological Society* 24: 187-191.
Kirby, W. 1802. *Monographia Apum Angliae;* . . . Vol. 1: xxii + 258 pp.; Vol. 2: 388 pp., 18 pls. Ipswich, U.K.: privately published.
———. 1837. Part IV, Insects. *In* J. Richardson, *Fauna Boreali-Americana; or the Zoology of the Northern Parts of British America.* xxxix + 325 pp., pls. I-VIII. London: Longman.
Kirby, W., and W. Spence. 1826. *An Introduction to Entomology,* Vol. 3. 732 pp. London: Longman.
Kirby, W. F. 1900. Descriptions of the new species of Hymenoptera, *in* The Expedition to Sokotra. *Bulletin of the Liverpool Museums* 3: 13-24.
Kislow, C. J. 1976. *The comparative biology of two species of small carpenter bees,* Ceratina strenua *F. Smith and* C. calcarata *Robertson.* iii + 221 pp. Athens: Ph.D. thesis, University of Georgia.
Kitamura, K., Y. Maeta, K. Takahashi, and R. Miyanuga. 2001. Nest architecture of a Japanese lithurgine bee, *Lithurgus (Lithurgus) collaris* Smith (Hymenoptera, Megachilidae), with notes on the relationship between burrow diameter and thickness of the thorax in various taxa of bees. *Japanese Journal of Entomology* NS 4: 49-61. [in Japanese, English abstract.]
Klostermeyer, E. C., and H. S. Gerber. 1969. Nesting behavior of *Megachile rotundata* (Hymenoptera: Megachilidae) monitored with an event recorder. *Annals of the Entomological Society of America* 62: 1321-1325.
Klug, F. 1807a. *Oxaea,* eine neue Gattung aus der Ordnung der Piezaten. *Gesellschaft Naturforschender Freunde zu Berlin, Magazin für Neuesten Entdeckungen in der Gesammten Naturkunde* 1: 261-263.
———. 1807b. Kritische Revision der Bienengattungen in Fabricius neuem Piezatensysteme . . . . *Magazin für Insektenkunde* 6: 200-228.
———. 1810. Einige neue Piezatengattungen. *Gesellschaft Naturforschender Freunde zu Berlin, Magazin für Neuesten Entdeckungen in der Gesammten Naturkunde* 4: 31-45, pl. 1.
Knechtel, W. K. 1955. Hymenoptera, subfamilia Apinae, *in Fauna Republicii Populare Romîne,* Insecta, Vol. IX, fasc. 1: 1-111.
Knerer, G. 1980. Biologie und Sozialverhalten von Bienenarten der Gattung *Halictus* Latreille. *Zoologische Jahrbücher, Abteilung für Systematik, Ökologie und Geographie der Tiere* 107: 511-536.
Knerer, G., and P. MacKay. 1969. Bionomic notes on the solitary *Evylaeus oenotherae* (Stevens) (Hymenoptera, Halictinae), a matinal summer bee visiting cultivated Onagraceae. *Canadian Journal of Zoology* 47: 289-294.
Kocourek, M. 1966. *Andrena,* in Prodromus der Hymenopteren der Tschechoslowakei, pars 9, Apoidea, 1: 1-122, 1 map. Praha: Národní Muzeum.
Kopelke, J.-P. 1981. Funktion der Genitalstrukturen bei *Bombus-*Arten am Beispiel von *B. lapidarius* (Linnaeus 1758) und deren Bedeutung für die Systematik. *Senckenbergiana Biologica* 62: 267-286.
Koster, A. 1986. Het genus *Hylaeus* in Nederland. *Zoologische Bijdragen* [Leiden], no. 36: 1-120.
Krombein, K. V. 1950. The aculeate Hymenoptera of Micronesia, II. Colletidae, Halictidae, Megachilidae, and Apidae. *Proceedings of the Hawaiian Entomological Society* 14: 101-142.
———. 1951. Additional notes on the bees of the Solomon Islands. *Proceedings of the Hawaiian Entomological Society* 14: 277-295.
———. 1967. *Trap-Nesting Wasps and Bees: Life Histories, Nests, and Associates.* vi + 570 pp. Washington: Smithsonian Press.
———. 1979. Superfamily Sphecoidea, pp. 1573-1740 *in* K. V. Krombein, P. D. Hurd, Jr., D. R. Smith, and B. D. Burks, eds., *Catalog of Hymenoptera in America North of Mexico,* Vol. 2. Washington, D.C.: Smithsonian Institution Press.
Krombein, K. V., and B. B. Norden. 1997a. Nesting behavior of *Krombeinictus nordenae* Leclercq, a sphecid wasp with vegetarian larvae. *Proceedings of the Washington Entomological Society* 99: 42-49.
———. 1997b. Bizarre nesting behavior of *Krombeinictus nordenae* Leclercq. *Journal of South Asian Natural History* 2: 145-154.
Kronenberg, S., and A. Hefetz. 1984a. Comparative analysis of Dufour's gland secretions of two carpenter bees (Xylocopinae: Anthophoridae) with different nesting habits. *Comparative Biochemistry and Physiology* B 79: 421-425.
———. 1984b. Role of labial glands in nesting behavior of *Chalicodoma sicula. Physiological Entomology* 9: 175-179.
Krüger, E. 1917. Zur Systematik der mitteleuropäischen Hummeln. *Entomologische Mitteilungen* 6: 55-66.
———. 1920. Beiträge zur Systematik und Morphologie der mittel-europäischen Hummeln. *Zoologische Jahrbücher, Abteilung für Systematik, Geographie und Biologie der Tiere* 42: 289-464.
Kruseman, G. 1947. Tabellen tot het bepalen van de Nederlandsche soorten der Genera *Bombus* Latr. en *Psithyrus* Lep. *Tijdschrift voor Entomologie* 88 1945 [1947]: 173-188.
———. 1952. Subgeneric division of the genus *Bombus* Latr. *Transactions of the Ninth International Congress of Entomology, Amsterdam* 1: 101-103.
Kukuk, P. F. 1997. Male dimorphism in *Lasioglossum (Chilalictus) hemichalceum:* The role of larval nutrition. *Journal of the Kansas Entomological Society* 69(1996, suppl.): 147-157.
Kukuk, P. F., and M. P. Schwarz. 1988. Macrocephalic male

bees as functional reproductives and probable guards. *Pan-Pacific Entomologist* 64: 131-137.

LaBerge, W. E. 1955. Bees of the genus *Anthedonia* Michener in North America. *Journal of the Kansas Entomological Society* 28: 132-135.

———. 1956a, b. A revision of the bees of the genus *Melissodes* in North and Central America, Parts I and II. *University of Kansas Science Bulletin* 37: 911-1194; 38: 533-578.

———. 1957. The genera of bees of the tribe Eucerini in North and Central America. *American Museum Novitates* no. 1837: 1-44.

———. 1958. Notes on the genus *Gaesischia* Michener, LaBerge and Moure, with descriptions of a new species and subgenus from Mexico. *Pan-Pacific Entomologist* 34: 195-201.

———. 1961. A revision of the bees of the genus *Melissodes* in North and Central America, Part III. *University of Kansas Science Bulletin* 42: 283-663.

———. 1964. Prodromus of American bees of the genus *Andrena*. *Bulletin of the University of Nebraska State Museum* 4: 279-316.

———. 1967. A revision of the bees of the genus *Andrena* of the western hemisphere, Part I. *Callandrena*. *Bulletin of the University of Nebraska State Museum* 7: 1-316.

———. 1969. A revision of the bees of the genus *Andrena* of the western hemisphere, Part II. *Plastandrena, Aporandrena, Charitandrena*. *Transactions of the American Entomological Society* 95: 1-47.

———. 1970. A new genus with three new species of eucerine bees from Mexico. *Journal of the Kansas Entomological Society* 43: 321-328.

———. 1971a. A new subgenus of *Andrena* found in California and Oregon. *Pan-Pacific Entomologist* 47: 47-57.

———. 1971b. A revision of the bees of the genus *Andrena* of the western hemisphere, Part IV. *Scrapteropsis, Xiphandrena* and *Rhaphandrena*. *Transactions of the American Entomological Society* 97: 441-520.

———. 1973. A revision of the bees of the genus *Andrena* of the western hemisphere, Part VI. Subgenus *Trachandrena*. *Transactions of the American Entomological Society* 99: 235-371.

———. 1977. A revision of the bees of the genus *Andrena* of the western hemisphere, Part VIII. Subgenus *Thysandrena, Dasyandrena, Psammandrena, Rhacandrena, Euandrena*, and *Oxyandrena*. *Transactions of the American Entomological Society* 103: 1-143.

———. 1980. A revision of the bees of the genus *Andrena* of the western hemisphere, Part X. Subgenus *Andrena*. *Transactions of the American Entomological Society* 106: 395-525.

———. 1986. A revision of the bees of the genus *Andrena* of the western hemisphere, Part XI. Minor subgenera and subgeneric key. *Transactions of the American Entomological Society* 111: 441-567.

———. 1987. A revision of the bees of the genus *Andrena* of the western hemisphere, Part XII. Subgenera *Leucandrena, Ptilandrena, Scoliandrena*, and *Melandrena*. *Transactions of the American Entomological Society* 112: 191-248.

———. 1989a. A revision of the bees of the genus *Andrena* of the western hemisphere, Part XIII. Subgenera *Simandrena* and *Taeniandrena*. *Transactions of the American Entomological Society* 115: 1-56.

———. 1989b. A review of the bees of the genus *Pectinapis*. *Journal of the Kansas Entomological Society* 62: 524-527.

———. A new species of *Syntrichalonia* from Mexico. *Journal of the Kansas Entomological Society* 67: 283-287.

———. 2001. Revision of the bees of the genus *Tetraloniella* in the New World. *Illinois Natural History Survey Bulletin* 36: 63-162.

LaBerge, W. E., and J. K. Bouseman. 1970. A revision of the bees of the genus *Andrena* of the western hemisphere, Part III. *Tylandrena*. *Transactions of the American Entomological Society* 96: 543-605.

LaBerge, W. E., and P. D. Hurd, Jr. 1965. A new subgenus and species of matinal *Andrena* from the flowers of *Sicyos* (Cucurbitaceae) in Mexico. *Pan-Pacific Entomologist* 41: 186-193.

LaBerge, W. E., and C. D. Michener. 1963. *Deltoptila*, a Middle American genus of anthophorine bees. *Bulletin of the University of Nebraska State Museum* 4: 211-226.

LaBerge, W. E., and D. W. Ribble. 1966. Biology of *Florilegus condignus* (Hymenoptera, Anthophoridae), with a description of its larva, and remarks on its importance in alfalfa pollination. *Annals of the Entomological Society of America* 59: 944-950.

———. 1972. A revision of the bees of the genus *Andrena* of the western hemisphere, Part V. *Gonandrena, Geissandrena, Parandrena, Pelicandrena*. *Transactions of the American Entomological Society* 98: 271-358.

———. 1975. A revision of the bees of the genus *Andrena* of the western hemisphere, Part VII. Subgenus *Euandrena*. *Transactions of the American Entomological Society* 101: 371-446.

LaBerge, W. E., and R. W. Thorp. 2005. A revision of the bees of the genus *Andeena* of the western hemisphere. Part XIV. Subgenus *Onagrandrena*. *Illinois Natural History Survey Bulletin* 37: i–iv, 1–63.

LaBerge, W. E., and M. C. Webb. 1962. The bumblebees of Nebraska. *University of Nebraska College of Agriculture Research Bulletin* 205: 1-38.

Labougle, J. M. 1990. *Bombus* of Mexico and Central America (Hymenoptera, Apidae). *University of Kansas Science Bulletin* 54: 35-73.

Labougle, J. M., and R. Ayala. 1985. A new subgenus and species of *Bombus* (Hymenoptera: Apidae) from Guerrero, Mexico. *Folia Entomológica Mexicana* no. 66: 47-55.

Lanham, U. N. 1949. A subgeneric classification of the new world bees of the genus *Andrena*. *University of California Publications in Entomology* 8: 183-238.

———. 1950. A preoccupied subgeneric name in *Andrena*. *Entomological News* 61: 140.

———. 1960. A neglected diagnostic character of the Apoidea. *Entomological News* 71: 85-86.

———. 1981. Evolutionary origin of bees. *Journal of the New York Entomological Society* 88: 199-209.

Laroca, S. 1971a. Notas sôbre a nidificação de *Chrysosarus tapytensis* Mitchell. *Boletim da Universidade Federal do Paraná, Zoologia* 4(8): 39-44.

———. 1971b. Notas sobre a biologia de *Hylaeus cecidonastes* Moure. *Revista Brasileira de Biologia* 32: 285-290.

———. 1983. Bioceoenotics of wild bees (Hymenoptera,

Apoidea) at three nearctic sites, with comparative notes on some neotropical assemblages, pp. 8-194. Lawrence: Ph.D. thesis, University of Kansas.

Laroca, S., E. Corbella, and G. Varela. 1992. Biologia de *Dactylomegachile affabilis* (Hymenoptera, Apoidea): 1. Descrição do ninho. *Acta Biologia Paranaense* [Curitiba] 21: 23-29.

Laroca, S., J. R. Cure, and C. de Bortoli. 1982. A associação de abelhas silvestres (Hymenoptera, Apoidea) de uma área restrita no interior da cidade de Curitiba (Brasil): uma abordagem biocenótica. *Dusenia* 13: 93-117.

Laroca, S., and M. C. de Almeida. 1985. Adaptação dos palpos labiais de *Niltonia virgilii* (Hymenoptera, Apoidea, Colletidae) para coleta de néctar em *Jacaranda puberula* (Bignoniaceae), com descrição do macho. *Revista Brasileira de Entomologia* 29: 289-297.

Laroca, S., and S. Lauer. 1973. Adaptação comportamental de *Scaura latitarsis* para coleta de pólen. *Acta Biológica Paranaense* 2: 147-152.

Laroca, S., C. D. Michener, and R. M. Hofmeister. 1989. Long mouthparts among "short-tongued" bees and the fine structure of the labium in *Niltonia*. *Journal of the Kansas Entomological Society* 62: 400-410.

Laroca, S., and G. H. Rosado Neto. 1975. Notas bionômicas: *Hypanthidoides* [sic] *arenaria*. *Revista Brasileira de Biologia* 35: 847-853.

Latreille, P. A. 1802a. *Histoire Naturelle des Fourmis*.... xvi + 445 pp., 12 pls. Paris: Crapelet.

———. 1802b. *Histoire Naturelle Générale et Particulière des Crustacés et des Insectes,* Vol. 3, xii + 467 pp. Paris.

———. 1803. [Several articles] *in Nouveau Dictionnaire d'Histoire Naturelle*..., Vol. 18. Paris: Déterville.

———. 1804. Tableau méthodique des insectes, pp. 129-200 *in Nouveau Dictionnaire d'Histoire Naturelle,* Vol. 24. Paris: Déterville.

———. 1809. *Genera Crustaceorum et Insectorum*...., Vol. 4, 399 pp. Argentorati: Koenig.

———. 1810. *Considérations Générales... des Crustacés, des Arachnides, et des Insectes.* 444 pp. Paris: Schoell.

———. 1811. [Articles] *in* M. Olivier, ed., *Encyclopedie Méthodique, Histoire Naturelle, Insectes,* Vol. 8. Paris: Agasse.

———. 1825. *Familles Naturelles du Règne Animal*.... v + 570 pp. Paris: Baillière.

———. 1829. Les Insectes, *in* G. C. L. D. Cuvier, *Le Règne Animal,* 2nd ed., Vol. 5. xxiv + 556 pp. Paris: Déterville.

Laverty, T. M., and L. D. Harder. 1998. The bumblebees of eastern Canada. *Canadian Entomologist* 120: 965-987.

Laverty, T. M., R. C. Plowright, and P. H. Williams. 1984. *Digressobombus,* a new subgenus with description of the male of *Bombus digressus* (Milliron). *Canadian Entomologist* 116: 1051-1056.

Leach, W. E. 1812. Entomology, pp. 57-172 *in Brewster's Edinburgh Encyclopaedia,* Vol. 9. Edinburgh [date according to Ornosa and Ortiz-Sánchez, 1998].

Lello, E. de. 1976. Adnexal glands of the sting apparatus of bees: Anatomy and histology, V. *Journal of the Kansas Entomological Society* 49: 85-99.

Lepeletier de Saint-Fargeau, A. L. M. 1833. Observations sur l'ouvrage intitulé: Bombi Scandinaviae monographice tractato, etc., à Gustav Dahlbom.... *Annales de la Société Entomologique de France* 1(1832): 366-382.

———. 1835, 1841. *Histoire Naturelle des Insectes— Hyménoptères.* 1835, Vol. 1, 1-547; 1841, Vol. 2, 1-680. Paris: Roret. [See also the Atlas (bound as Vol. 5). Vol. 1 is labeled 1836, but according to D. Baker was published in December, 1835.]

Lepeletier de Saint-Fargeau, A. L. M., and A. Serville. 1825, 1828. [Articles] *in* M. Diderot et al., *Encyclopédie Méthodique, Histoire Naturelle.* Insectes, Vol. 10, P. A. Latreille, ed. Paris. [Pp. 1-344, 1825; pp. 345-832, 1828, according to C. D. Sherborn and B. B. Woodward, 1906, *Annals and Magazine of Natural History* (7)17: 578.]

Ler, P. A., ed. 1993. *Identification of Insects of Oriental Russia,* Vol. 4, Neuroptera, Mecoptera, and Hymenoptera, part 1, 602 pp. St. Petersburg: Russian Academy of Sciences. [Bees on pp. 480-580, sections by different authors: A. Z. Osychnyuk, T. G. Romankova, and A. N. Kupyanskaya. In Russian.]

Leys, R. 2000. A revision of the Australian carpenter bees, genus *Xylocopa* Latreille, subgenera *Koptortosoma* Gribodo and *Lestis* Lepeletier and Serville. *Invertibrate Taxonomy* 14: 115-136.

Leys, R., S. J. B. Cooper, and M. P. Schwarz. 2000. Molecular phylogeny of the large carpenter bees, genus *Xylocopa* (Hymenoptera: Apidae), based on mitochondrial DNA sequences. *Molecular Phylogenetics and Evolution* 17: 407-418.

Lieftinck, M. A. 1939. Uit het leven van *Lithurgus atratus,* een indisch houtbijtje. *Tropische Natuur* 28: 193-201.

———. 1944. Some Malaysian bees of the family Anthophoridae. *Treubia* (Dobutu Gaku-Iho), hors serie: pp. 57-138, pl. 42.

———. 1954. Bij het nest van een Javaans harsbijtje. *Idea* 10: 20-25.

———. 1955. The carpenter-bees of the Lesser Sunda Islands and Tanimbar (Hymenoptera, Apoidea). *Verhandlungen der Naturforschenden Gesellschaft in Basel* 66: 5-32.

———. 1956a. Revision of some Oriental anthophorine bees of the genus *Amegilla* Friese. *Zoologische Verhandelingen* no. 30: 1-41.

———. 1956b. Revision of the carpenter-bees (*Xylocopa* Latreille) of the Moluccan Islands, with notes on other Indo-Australian species. *Tidjschrift voor Entomologie* 99: 55-73.

———. 1957a. Revision of the carpenter-bees (*Xylocopa* Latr., subgenus *Maiella* Michener) of the Papuan region. *Nova Guinea* (n.s.)8: 325-376.

———. 1957b. The identity of some Fabrician types of bees (Hymenoptera, Apoidea), 1A and 1B. *Koninklijke Nederlandse Akademie van Wetenschappen—Amsterdam* (C)60: 432-450.

———. 1958, 1959. Revision of the Indo-Australian species of the genus *Thyreus* Panzer (= *Crocisa* Jurine) (Hym., Apoidea, Anthophoridae): Part 1, Introduction and list of species; Part 2, *Thyreus nitidulus* Fabricius. *Nova Guinea* (n.s.)9: 21-30 (1958); 10: 99-130, 2 pls. (1959).

———. 1962. Revision of the Indo-Australian species of the genus *Thyreus* Panzer (= *Crocisa* Jurine) (Hym., Apoidea, Anthophoridae): Part 3, Oriental and Australian species. *Zoologische Verhandelingen* no. 53: 1- 212, 3 pls.

———. 1966. Notes on some anthophorine bees, mainly from the Old World. *Tijdschrift voor Entomologie* 109: 125-161.

———. 1968. A review of old world species of *Thyreus*

Panzer (= *Crocisa* Jurine) (Hym., Apoidea, Anthophoridae): Part 4, palearctic species. *Zoologische Verhandelingen* no. 98: 1-139, pls. 1-4.

———. 1969. The melectine genus *Eupavlovskia* Popov, 1955, with notes on its distribution and host relations. *Tijdschrift voor Entomologie* 112: 101-122.

———. 1972. Further studies on old world melectine bees, with stray notes on their distribution and host relationships. *Tijdschrift voor Entomologie* 115: 253-324, 2 pls.

———. 1974. A review of the central and east Asiatic *Habropoda* F. Smith, with *Habrophorula,* a new genus from China. *Tijdschrift voor Entomologie* 117: 157-224.

———. 1975. Bees of the genus *Amegilla* Friese from Korea with a new species. *Annales Historico-Naturales Musei Nationalis Hungarici* 67: 279-292.

———. 1977. Notes on the melectine genus *Paracrocisa* Alfken, with a new record of *P. sinaitica* Alfken. *Entomologische Berichten* 37: 125-127.

———. 1980. Prodrome to a monograph of the palaearctic species of the genus *Melecta* Latreille 1802. *Tijdschrift voor Entomologie* 123: 129- 349.

———. 1983. Notes on the nomenclature and synonymy of old world melectine and anthophorine bees. *Tijdschrift voor Entomologie* 126: 269-284.

Lin, N., and C. D. Michener. 1972. Evolution of sociality in insects. *Quarterly Review of Biology* 47: 131-159.

Lind, H. 1968. Nest provisioning cycle and daily routine of behavior in *Dasypoda plumipes. Entomologiske Meddelelser* 36: 343-372.

Linnaeus, C. 1758. *Systema Naturae,* Vol. 1, ed. 10. 824 pp. Holmiae: Salvii.

Linsley, E. G. 1939a. A revision of the nearctic Melectinae. *Annals of the Entomological Society of America* 32: 429-468.

———. 1939b. Some new genera and species of epeoline and nomadine bees. *Pan-Pacific Entomologist* 15: 1-11.

———. 1941. A revision of the genus *Oreopasites. Transactions of the American Entomological Society* 66: 307-318.

———. 1942. Notes and descriptions of some North American parasitic bees. *Pan-Pacific Entomologist* 18: 127-132.

———. 1943a. Revisions of the genera *Townsendiella, Triopasites* and *Paranomada. Transactions of the American Entomological Society* 69: 93- 106.

———. 1943b. A revision of the genus *Neopasites. Transactions of the American Entomological Society* 69: 119-140.

———. 1943c. A revision of the genus *Gnathopasites. Transactions of the American Entomological Society* 69: 141-149.

———. 1945. A new species of *Paranomada* with notes on *Melecta thoracica* Cresson. *Entomological News* 56: 150-153.

Linsley, E. G., and J. W. MacSwain. 1955. The habits of *Nomada opacella* Timberlake with notes on other species. *Wasmann Journal of Biology* 13: 253-276.

———. 1956. Further notes on the taxonomy and biology of the andrenine bees associated with *Oenothera. Pan-Pacific Entomologist* 32: 111-121.

———. 1957. The nesting habits, flower relationships, and parasites of some North American species of *Diadasia. Wasmann Journal of Biology* 15: 199-235.

Linsley, E. G., J. W. MacSwain, and C. D. Michener. 1980. Nesting biology and associates of *Melitoma. University of California Publications in Entomology* 90: i-viii + 1-45.

Linsley, E. G., J. W. MacSwain, P. H. Raven, and R. W. Thorp. 1973. Comparative behavior of bees and Onagraceae, V. *University of California Publications in Entomology* 71: 1-68, pls. 1-6.

Linsley, E. G., J. W. MacSwain, and R. F. Smith. 1952. The bionomics of *Diadasia consociata* Timberlake and some biological relationships of emphorine and anthophorine bees. *University of California Publications in Entomology* 9: 267-290, pls. 1-6.

———. 1955. Biological observations on *Xenoglossa fulva* Smith with some generalizations on biological characters of other eucerine bees. *Bulletin of the Southern California Academy of Sciences* 54: 128-141.

———. 1956. Biological observations on *Ptilothrix sumichrasti* (Cresson) and some related groups of emphorine bees. *Bulletin of the Southern California Academy of Sciences* 55: 83-101.

Linsley, E. G., and C. D. Michener. 1937. Some new genera and species of North American parasitic bees. *Pan-Pacific Entomologist* 13: 75-84.

———. 1939. A generic revision of the North American Nomadidae. *Transactions of the American Entomological Society* 65: 265-305, pls. xv- xviii.

Lith, J. P. van. 1956. Notes on *Epeolus. Tijdschrift voor Entomologie* 99: 31-45.

Løken, A. 1973. Studies on Scandinavian bumble bees. *Norsk Entomologisk Tidsskrift* 20: 1-218.

———. 1984. Scandinavian species of the genus *Psithyrus* Lepeletier. *Entomologica Scandinavica* suppl. no. 23: 1-45.

———. 1985. Humler, tabell til norske arter. *Norske Insekttabeller* no. 9: 1-39.

Lomholdt, O. 1977. De danske blodbier *Sphecodes. Entomologische Meddelelser* 45: 99-108.

———. 1982. On the origin of the bees. *Entomologica Scandinavica* 13: 185-190.

Louis, J. 1973. La nomenclature de l'aile des hyménoptères, essai de normalisation. *Beiträge zur Entomologie* 23: 275-289.

Lowe, R. M., and R. H. Crozier. 1997. The phylogeny of bees of the socially parasitic Australian genus *Inquilina* and their *Exoneura* hosts. *Insectes Sociaux* 44: 409-414.

Lucas de Oliveira, B. 1960. Mudas ontogenéticas em larvas el *Melipona nigra schencki* Gribodo. *Boletim da Universidade do Paraná, Zoologia* no. 2: 1-16.

———. 1965. Observações em larvas e pupas de *Plebeia* (*Plebeia*) *droryana* (Friese, 1900). *Papéis Avulsos de Zoologia, São Paulo* 18: 29-38.

———. 1966. Descrição de estádios imaturos de *Isepeolus viperinus* (Holmberg) e confrontações com outras larvas de Anthophoridae parasitas conhecidas. *Boletim da Universidade Federal do Paraná, Zoologia* II, no. 11: 163-176.

Lucas de Oliveira, B., and J. S. Moure. 1963. Nova posição sistemática para *Rhophitulus testaceus* Ducke, 1907. *Anais da Academia Brasileira de Ciências* 35: 575-584.

Lutz, F. E., and T. D. A. Cockerell. 1920. Notes on the distribution and bibliography of North American bees of the families Apidae, Meliponidae, Bombidae, Euglossidae, and Anthophoridae. *Bulletin of the American Museum of Natural History* 42: 491-641.

Lutz, H. 1993. *Eckfeldapis electrapoides* nov. gen. n. sp., eine "Honigbiene" aus dem Mittel-Eozän des "Eckfelder Maares" bei Manderscheid/Eifel, Deutschland. *Meinzer Naturwissenschaftliches Archiv* 31: 177-199.

Ma, T.-c. 1938. The Indian species of the genus *Xylocopa* Latr. (Hymenoptera). *Records of the Indian Museum* 40: 265-329.

Maa, T.-c. 1939. *Xylocopa* orientalia critica (Hymen.), I. Subgenus *Bomboixylocopa* novum. *Lingnan Science Journal* 18: 155-160.

———. 1940a. *Xylocopa* orientalia critica (Hymen.), II. Subgenus *Zonohirsuta* Maa. *Lingnan Science Journal* 19: 383-402.

———. 1940b. *Xylocopa* orientalia critica (Hymen.), IV. Subgenus *Nyctomelitta* Ckll. *Lingnan Science Journal* 19: 577-582.

———. 1940c. *Xylocopa* orientalia critica (Hymen.), III. Subgenus *Platynopoda* Westw. *Lingnan Science Journal* 19: 565-575.

———. 1946. *Xylocopa* orientalia critica (Hymen.), V. Subgenus *Biluna* Maa. *Biological Bulletin, Fukien Christian University* [Foochow] 5: 67-92.

———. 1948. On some eastern Asiatic species of the genus *Psithyrus* Lepel. *Notes d'Entomologie Chinoise, Musée Heude* 12(3): 19-37.

———. 1953. An inquiry into the systematics of the tribus Apidini or honeybees. *Treubia* 21: 525-640.

———. 1954. The xylocopine bees (Insecta) of Afghanistan. *Videnskabelige Meddelelser fra Dansk Naturhistorisk Forening i Kjøbenhavn* 116: 189-231.

———. 1970. A revision of the subgenus *Ctenoxylocopa*. *Pacific Insects* 12: 723-752.

Maciel de Almeida Correia, M. de L. 1980. Contribution à l'étude de la biologie d'*Heriades truncorum* L., I—Aspect biologique et morphologique. *Apidologie* 11: 309-339.

———. 1981. Contribution à l'étude de la biologie d'*Heriades truncorum* L., III—Aspect ethologique. *Apidologie* 12: 221-256.

Mader, D. 1992. *Beiträge zu Paläoökologie und Paläoenvironment des Buntsandsteins sowie ausgewählte Bibliographie von Buntsandstein und Keuper in Thüringen, Franken und Umgelbung*. xxi + 628 pp. Stuttgart: Fischer.

———. 1999. *Geologische und Biologische Entomoökologie der Rezenten Seidenbiene Colletes*, Band I, xliii + 807pp. Köln: Logabook.

Maeta, Y. 1978. Comparative studies on the biology of the bees of the genus *Osmia* of Japan, with special reference to their managements for pollinations of crops. *Bulletin of the Tohoku National Agricultural Experiment Station* no. 57: 1-221 + errata 1-7. [In Japanese, English summary.]

Maeta, Y., M. Fujiwara, and K. Kitamura. 2004. Notes on the bionomics of *Andrena (Plastandreana) japonica* (Smith). *Japanese Journal of Entomology* (NS) 7: 155-171.

Maeta, Y., K. Gôukon, N. Sugiura, and R. Miyanaga. 1996. Host records of cleptoparasitic bees in Japan. *Japanese Journal of Entomology* 64: 830-842.

Maeta, Y., and E. Katayama. 1978. Life history and nesting habits of a small carpenter bee, *Ceratina megastigmata*. *Bulletin of the Tohoku National Agricultural Experiment Station* no. 58: 231-260.

Maeta, Y., and K. Kitamura. 2005. On the number of eggs laid by one individual of females in the alfalfa leaf-cutting bee *Megachile (Eutricharaea) rotundata* (Fabricius). *Chugoku Kontya* no. 19: 39-43.

Maeta, Y., N. Kubota, and S. F. Sakagami. 1987. *Nomada japonica* as a thelytokous cleptoparasitic bee, with notes on egg size and egg complement in some cleptoparasitic bees. *Kontyû* 55: 21-31.

Maeta, Y., and S. F. Sakagami. 1995. Oophagy and egg replacement in artificially induced colonies of a basically solitary bee, *Ceratina (Ceratinidia) okinawana* (Hymenoptera, Anthophoridae, Xylocopinae), with a comparison of social behavior among *Ceratina, Xylocopa* and the halictine bees. *Japanese Journal of Entomology* 63: 347-375.

Maeta, Y., S. F. Sakagami, and C. D. Michener. 1992. Laboratory studies on the behavior and colony structure of *Braunsapis hewitti*, a xylocopine bee from Taiwan. *University of Kansas Science Bulletin* 54: 298-333.

Maeta, Y., A. Yoshida, and R. Miyanaga. 2001. Some notes on the bionomics and social structure of *Lasioglossum (Lasioglossum) scitulum* Smith, *Chugoku Kontyu* no. 14: 11-19. [in Japanese, English summary.]

Maidl, F. 1912. Die Xylocopen (Holzbienen) des Wiener Hofsmuseums, ein Beitrag zu einer Monographie dieser Gattung. *Annalen des K.K. Naturhistorischen Hofsmuseums* [Wien] 26: 249-330.

Malyshev, S. I. 1913. Life and instincts of some *Ceratina*-bees. *Horae Societatis Entomologicae Rossicae* 40(8): 1-58, pl. 3. [In Russian, English summary.]

———. 1923a. The nesting habits of *Melitta leporina* Pz. *Izvestiya Leningradskovo Nauchnovo Instituta imeni P. F. Leshafti [Bulletin de l'Institut Leshafti]* 6: 1-7. [In Russian, English summary.]

———. 1923b. The nesting of *Colletes* Latr. *Russkoe Entomologicheskoe Obozrenie* 18: 103-124. [In Russian.]

———. 1924a. The nesting habits of long-horned bees of the subgenus *Macrocera* Latr. (*Tetralonia* Spin. . . . .). *Izvestiya Leningradskovo Nauchnovo Instituta imeni P. F. Leshafti* 8: 251-266. [In Russian, English summary.]

———. 1924b. The nesting of *Panurginus* Nyl. *Izvestiya Leningradskovo Nauchnovo Instituta imeni P. F. Leshafti* 9: 196-200. [In Russian, English summary.]

———. 1925a. The nesting habits of *Rhophites* Spin. *Russkoe Entomologicheskoe Obozrenie* 19: 105-110. [In Russian, English summary.]

———. 1925b. The nesting habits of *Meliturga* Latr. *Izvestiya Leningradskovo Nauchnovo Instituta imeni P. F. Leshafti* 11: 67-73. [In Russian, English summary.]

———. 1925c. The nesting of bees of the genus *Systropha* Latr. *Russkoe Entomologicheskoe Obozrenie* 19: 21-26. [In Russian, English summary.]

———. 1926. Nesting of the mining bees, *Andrena* F. *Trudy Leningradskoe Obshchestvo Estestvoispytatelei* 56(2): 25-78. [In Russian, English summary.]

———. 1927a. The nesting habits of *Dasypoda*. *Trudy Leningradskoe Obshchestvo Estestvoispytatelei* 57: 123-146. [In Russian.]

———. 1927b. Lebensgeschichte des *Colletes cunicularius* L. *Zeitschrift für Morphologie und Ökologie der Tiere* 9: 390-401.

———. 1928. Lebensgeschichte der *Anthophora acervorum* L. *Zeitschrift für Morphologie und Ökologie der Tiere* 11: 763-781.

———. 1929. The nesting habits of *Macropis* Pz. *Eos* 5: 97-109.

———. 1930a. Lebensgeschichte der *Tetralonia malvae* Rossi. *Zeitschrift für Morphologie und Ökologie der Tiere* 16: 541-558.

———. 1930b. Nistgewohnheiten der Steinbienen *Lithurgus* Latr. *Zeitschrift für Morphologie und Ökologie der Tiere* 19: 116-134.

———. 1931. Lebensgeschichte der Holzbienen *Xylocopa* Latr. *Zeitschrift für Morphologie und Ökologie der Tiere* 23: 754-809.

———. 1935. The nesting habits of solitary bees. *Eos* 11: 201-309, pls. III-XV.

———. 1937. Lebensgeschichte der Osmien (*Osmia* Latr.). *Zoologische Jahrbücher, Abteilung für Systematik, Geographie und Biologie der Tiere* 69: 107-176, 2 double pls.

———. 1947. The life and instincts of the dwarf carpenter bee *Xylocopa iris* Christ. *Isvestia Akademii Nauk CCCP, Seriia Biologicheskaia* 1: 53-77. [In Russian, English summary.]

———. 1968. *Genesis of the Hymenoptera and the Phases of Their Evolution.* viii + 319 pp. London: Methuen and Co.

Mangum, W. A., and R. W. Brooks. 1997. First records of *Megachile (Callomegachile) sculpturalis* Smith (Hymenoptera, Megachilidae) in the continental United States. *Journal of the Kansas Entomological Society* 70: 140-142.

Manning, F. 1961. A new fossil bee from Baltic amber. *Verhandlungen XI. Internationaler Kongress für Entomologie* [Wien, 1960] 1: 306-308, pl. 5, fig. 1.

Marikovskaya, T. P. 1976. On the systematics of the tribe Anthophorini. *Entomologicheskoe Obozrenie* 55: 684-690. [In Russian; translation into English, Entomological Review 55: 126-130.]

———. 1979. On the structure and zoogeography of the genus *Clisodon* Patton. *Isvestiya Serii Biologicheskaya, Akademia Nauk Kazakhskoi SSR* [Alma-Ata] 17(2): 40-48.

———. 1980. A new genus of bees of the family Anthophoridae. *Entomologicheskoe Obozrenie* 59: 650-653. [In Russian; translation into English, Entomological Review 59: 126-130.]

———. 1984. Contributions to morphology and ecology of the *Megachile*-bees of the subgenus *Xanthosarus* Robertson, pp. 64-73 in Y. A. Pesenko, ed., *Systematics and Ecology of Bees. Proceedings of the Zoological Institute, USSR Academy of Sciences, Leningrad,* Vol. 128. [In Russian.]

Marschall, A. F. von. 1873. *Nomenclator Zoologicus Continens Nomina Systematica Generum Animalium tam Viventium quam Fossilium, secundum ordinem alphabeticum disposita....* [v] + 482 pp. Vindobonae: Ueberreuter (M. Salzer). [Hymenoptera, pp. 254-273.]

Martins, R. P. 1993. The nesting behavior of a quasisocial or semisocial bee *Pseudagapostemon (Neagapostemon) brasiliensis* Cure. *Ciencia e Cultura* 45: 133-134.

Martins, R. P., and D. A. O. de Almeida. 1994. Is the bee, *Megachile assumptionis* (Hymenoptera: Megachilidae), a cavity-nesting specialist? *Journal of Insect Behavior* 7: 759-765.

Martins, R. P., and Y. Antonini. 1994. The biology of *Diadasina distincta* (Holmberg, 1903). *Proceedings of the Entomological Society of Washington* 96: 553-560.

Martins, R. P., F. G. Guimarães, and C. M. Dias. 1996. Nesting biology of *Ptilothrix plumata* Smith, with a comparison to other species in the genus. *Journal of the Kansas Entomological Society* 69: 9-16.

Masuda, H. 1933. Bionomic and ecological notes on *Anthidium japonicum* Smith (tomon hanabachi) belonging to the Hymenoptera and the family Megachilidae. *Mushi* 11: 133-156. [In Japanese, U.S. Dept. of Agriculture translation.]

———. 1943. Bionomics of *Nomia punctata* Westwood. *Mushi* 15: 16-28. [In Japanese; U.S. Dept. of Agriculture translation.]

Mateus, S., and F. B. Noll. 2004. Predatory behavior in a necrophagous bee *Trigona hypogea. Naturwissenschaften* 91: 94-96.

Matheson, A., S. L. Buchmann, C. O'Toole, P. Westrich, and I. H. Williams, eds. 1996. *The Conservation Biology of Bees.* x + 252 pp. London: Academic Press.

Mathewson, J. A. 1968. Nest construction and life history of the eastern cucurbit bee, *Peponapis pruinosa. Journal of the Kansas Entomological Society* 41: 255-261.

Matsumura, T. 1970. Nesting habits of three species of *Andrena* in Hokkaido. *Journal of the Faculty of Science, Hokkaido University* (ser. VI, Zoology)17: 520-538.

Matsumura, T., and S. F. Sakagami. 1971. Bees from Nepal, II. *Lasioglossum albescens,* with some bionomic notes. *Annotationes Zoologicae Japonenses* 44: 47-55.

Matthews, R. W. 1965. The biology of *Heriades carinata* Cresson. *Contributions of the American Entomological Institute* 1(3): 1-33.

Mavromoustakis, G. A. 1939. New and little-known bees of the subfamily Anthidiinae (Apoidea)—Part I. *Annals and Magazine of Natural History* (11)3: 88-97.

———. 1949, 1950, 1952, 1953b, 1954b, 1955, 1957a, 1957b. On the bees (Hymenoptera, Apoidea) of Cyprus, Parts I-VIII. *Annals and Magazine of Natural History* (12)1(1948): 541-587; (12)4: 334-354; (12)5: 814-843; (12)6: 769-781; (12)7: 578-588; (12)8: 97-105; (12)10: 321-337 and 843-850.

———. 1951. Further contributions to our knowledge of the Ethiopian Anthidiinae (Hymenoptera: Apoidea) and their classification. *Annals and Magazine of Natural History* (12)4: 962-981.

———. 1953a. New and little-known bees of the subfamily Anthidiinae (Apoidea)—Part VI. *Annals and Magazine of Natural History* (12)6: 834-840.

———. 1954a. New and interesting bees from Israel. *Bulletin of the Research Council of Israel* 4: 256-275.

———. 1956. On the bees (Hymenoptera, Apoidea) of Siria, Part 1. *Eos* 32: 215-229.

———. 1958. Bees (Hymenoptera, Apoidea) of Attica (Greece), Part I. *Annals and Magazine of Natural History* (13)1: 433-447.

———. 1963a. On the bees (Hymenoptera, Apoidea) of Lebanon, Part III. *Annals and Magazine of Natural History* (13)5: 647-655.

———. 1963b. Further contributions to our knowledge of the Ethiopian Anthidiinae (Hymenoptera: Apoidea) and their classification—Part 2. *Annals and Magazine of Natural History* (13)6: 481-499.

———. 1963c. A new genus of parasitic bees. *Annals and Magazine of Natural History* (13)5[1962]: 443-445.

———. 1963d. On some parasitic bees. *Annals and Magazine of Natural History* (13)5[1962]: 751-754.

———. 1968a. New and little known bees of the family Megachilidae. *Bollettino del Museo Civico di Storia Naturale di Venezia* 18(1965): 125-149.

———. 1968b. New and little known bees of the genus *Ammobates. Polskie Pismo Entomologiczne* 38: 141-157.

May, D. G. K. 1972. Water uptake during larval development of a sweat bee, *Augochlora pura. Journal of the Kansas Entomological Society* 45: 439-449.

May, J. 1959. *Čmeláci v ČSR [The Bumblebees of Czechoslovakia].* 170 pp. Praha: Ceskoslovenská Akademie Zemědělských Věd. [In Czech, Russian and German summaries.]

Maynard, G. V. 1991. Revision of *Leioproctus (Protomorpha)* Rayment (Hymenoptera: Colletidae) with description of two new species. *Journal of the Australian Entomological Society* 30: 67-75.

———. 1992. Revision of *Leioproctus (Cladocerapis)* Cockerell. *Journal of the Australian Entomological Society* 31: 1-11.

———. 1993. Revision of *Leioproctus (Ceratocolletes)* Michener. *Journal of the Australian Entomological Society* 32: 313-316.

———. 1994. Revision of *Leioproctus (Filiglossa)* Rayment. *Journal of the Australian Entomological Society* 33: 299-304.

———. 1996. Revision of *Chrysocolletes* Michener stat. nov. (Hymenoptera: Colletidae), and descriptions of four new species. *Journal of the Australian Entomological Society* 35: 1-8.

———. 1997. Revision of *Leioproctus (Anacolletes)* Michener, and description of a new subgenus *Leioproctus (Odontocolletes). Australian Journal of Entomology* 36: 137-148.

McCorquodale, D. B., and I. D. Naumann. 1988. A new Australian species of communal ground nesting wasp, in the genus *Spilomena* Shuckard. *Journal of the Australian Entomological Society* 27: 221-231.

McGinley, R. J. 1980. Glossal morphology of the Colletidae and recognition of the Stenotritidae at the family level. *Journal of the Kansas Entomological Society* 53: 539-552.

———. 1981. Systematics of the Colletidae based on mature larvae with phenetic analysis of apoid larvae. *University of California Publications in Entomology* 91: i-xvi, 1-307.

———. 1986. Studies of Halictinae (Apoidea: Halictidae), I: Revision of new world *Lasioglossum* Curtis. *Smithsonian Contributions to Zoology* no. 429: i-vi + 1-294.

———. 1987. [Apoidea], pp. 689-704 *in* F. W. Stehr, ed., *Immature Insects*. Dubuque, Iowa: Kendell/Hunt.

———. 1989. A catalog and review of immature Apoidea. *Smithsonian Contributions to Zoology* no. 494: 1-24.

———. 1999. *Eickwortia* (Apoidea: Halictidae), a new genus of bees from Mesoamerica. *University of Kansas Natural History Museum Special Publication* 24: 111-120.

———. 2003. Studies of Halictinae (Apoidea: Halictidae), II: Revision of *Sphecodogastra* Ashmead, floral specialists of Onagraceae. *Smithsonian Contributions to Zoology* no. 610: i-iii + 1-55.

McGinley, R. J., and J. G. Rozen, Jr. 1987. Nesting biology, immature stages, and phylogenetic placement of the palearctic bee *Pararhophites. American Museum Novitates* no. 2903: 1-21.

Meade-Waldo, G. 1914a. Notes on the Hymenoptera in the collection of the British Museum, with descriptions of new species. *Annals and Magazine of Natural History* (8)14: 450-464.

———. 1914b. Notes on the Apidae (Hymenoptera) in the collection of the British Museum, with descriptions of new species. *Annals and Magazine of Natural History* (8)13: 399-405.

———. 1923. Hymenoptera, fam. Apidae, subfam. Prosopidinae, fasc. 181, pp. 1-45 *in* P. Wytsman, ed., *Genera Insectorum*. Brussels.

Medler, J. T. 1964a. A note on *Megachile (Sayapis) pugnata pugnata* Say in trap-nests in Wisconsin. *Canadian Entomologist* 96: 918-921.

———. 1964b. *Anthophora (Clisodon) terminalis* Cresson in trap-nests in Wisconsin. *Canadian Entomologist* 96: 1332-1336.

———. 1966. A resin bee using trap nests in Wisconsin, and a note on other resin bees. *Entomological News* 77: 228-230.

———. 1967. Biology of *Osmia* in trap nests in Wisconsin. *Annals of the Entomological Society of America* 60: 338-344.

Medler, J. T., and D. W. Carney. 1963. Bumblebees of Wisconsin. *University of Wisconsin Research Bulletin* no. 240: 1-47.

Méhelÿ, L. 1935. *Naturgeschichte der Urbienen.* 214 pp., 60 pls. Budapest: Stephaneum Nyomda és Könyvkiadó R. T.

Melander, A. L. 1902. The nesting habits of *Anthidium. Biological Bulletin* 3: 27-32.

Mello, M. L. S., and C. A. Garófalo. 1986. Structural dimorphism in cocoons of a solitary bee, *Lithurgus corumbae* (Hymenoptera, Megachilidae) and its adaptive significance. *Zoologische Anzeiger* 217: 195-206.

Melna, P. A., and M. P. Schwarz. 1994. Behavioural specialization in pre-reproductive colonies of the allodapine bee *Exoneura bicolor. Insectes Sociaux* 41: 1-18.

Melo, G. A. R. 1996. Two new Brazilian bees of the genus *Protodiscelis. Revista Brasileira de Entomologia* 40: 97-100.

———. 1999. Phylogenetic relationships and classification of the major lineages of Apoidea (Hymenoptera), with emphasis on the crabronid wasps. *Scientific Papers, Natural History Museum, the University of Kansas* no. 14: 1-55.

———. 2003. Notas sobre meliponineos neotropicais, com a descrição de tres novas especies, pp. 85-91, in Melo and Alves-dos-Santos (2003).

Melo, G. A. R., and I. Alves-dos-Santos, eds. 2003. *Apoidea Neotropica, homenagem aos 90 anos de Jesus Santiago Moure,* xiv + 320 pp, Criciúma (Brazil): Univ. do Extremo Sul Catarinense.

Melo, G. A. R., and M. C. Gaglianone. 2005. Females of *Tapinotaspoides,* a genus in the oil-collecting bee tribe Tapinotaspidini, collect secretions from non-floral trichomes. *Revista Brasileira de Entomologia* 49: 167-168.

Melo, G. A. R., and R. B. Gonçalves. 2005. Higher level bee classifications. *Revista Brasileira de Zoologia* 22:153-159.

Melo, G. A. R., and F. C. V. Zanella. 2003. The species of the parasitic bee genus *Osirinus. Journal of Natural History* 37: 2911-2929.

Menke, A. 1993. Crabroninae vs. Larrinae (Sphecidae). *Sphecos* 26: 6-7.

Menke, A. S., and C. D. Michener. 1973. *Sericogaster* Westwood, a senior synonym of *Holohesma* Michener. *Journal of the Australian Entomological Society* 12: 173-174.

Messer, A. C. 1984. *Chalicodoma pluto:* The world's largest bee rediscovered living communally in termite nests. *Journal of the Kansas Entomological Society* 57: 165-168.

Metz, C. W. 1910. Bees of the genus *Colletes* from Mexico. *Pomona Journal of Entomology* 2: 191-208.

———. 1911. A revision of the genus *Prosopis* in North America. *Transactions of the American Entomological Society* 37: 85-156, pls. II-IX.

Meunier, F. 1888. Megachillidae [sic]. *Naturalista Siciliano* 7: 152.

Meyer, R. 1919a. Apidae—Sphecodinae. *Archiv für Naturgeschichte,* Abt. A 1919(1): 79-160; (2): 161-242.

———. 1919b. Apidae—Halictinae I, Gatt. *Parasphecodes* Sm. *Archiv für Naturgeschichte,* Abt. A 1919(11): 112-137.

———. 1921. Apidae—Nomadinae I, Gattung *Crocisa* Jur. *Archiv für Naturgeschichte,* Abt. A 1921(1): 67-178.

Meyer-Holzapfel, M. 1984. Zur Ethologie des Männchens der Trugbiene (*Panurgus banksianus* Kirby). *Zeitschrift für Tierpsychologie* 64: 221- 252.

Michelette, E. R. F., J. M. F. Camargo, and J. G. Rozen, Jr. 2000. Biology of the bee *Canephorula apiformis* and its cleptoparasite *Melectoides bellus:* nesting habits, floral preferences, and mature larvae. *American Museum Novitates* no. 3308: 1-23.

Michel-Salyat, A., S. A. Cameron, and M. L. Oliveira. 2004. Phylogeny of the orchid bees (Hymenoptera: Apinae: Euglossini): DNA and morphology yield equivalent patterns. *Molecular Phylogenetics and Evolution* 32: 309-323.

Michener, C. D. 1936. Some Pacific coast *Panurginus. Canadian Entomologist* 67: 275-278.

———. 1938a. American bees of the genus *Heriades. Annals of the Entomological Society of America* 31: 514-531.

———. 1938b. American bees of the genus *Chelostoma. Pan-Pacific Entomologist* 14: 36-45.

———. 1939a. A revision of the genus *Ashmeadiella. American Midland Naturalist* 22: 1-84.

———. 1939b. A revision of the genus *Neolarra. Transactions of the American Entomological Society* 65: 347-362.

———. 1941a. A generic revision of the American Osmiinae with descriptive notes on Old World genera. *American Midland Naturalist* 26(1): 147-166.

———. 1941b. A synopsis of the genus *Trachusa* with notes on the nesting habits of *T. perdita. Pan-Pacific Entomologist* 17: 119-125.

———. 1942a. Taxonomic observations on bees with descriptions of new genera and species (Hymenoptera: Apoidea). *Journal of the New York Entomological Society* 50: 273-282.

———. 1942b. North American bees of the genus *Ancyloscelis. Pan-Pacific Entomologist* 18: 108-113.

———. 1943. The American bees of the genus *Anthocopa* with notes on Old World subgenera. *Annals of the Entomological Society of America* 36: 49-86.

———. 1944. Comparative external morphology, phylogeny, and a classification of the bees. *Bulletin of the American Museum of Natural History* 82: 151-326.

———. 1947. A revision of the American species of *Hoplitis. Bulletin of the American Museum of Natural History* 89: 257-318.

———. 1948a. The generic classification of the anthidiine bees. *American Museum Novitates* no. 1381: 1-29.

———. 1948b. Notes on the Amcrican bees of the genus *Melecta. Proceedings of the Entomological Society of Washington* 50: 15-18.

———. 1949. A revision of the American species of *Diceratosmia. Annals of the Entomological Society of America* 42: 258-264.

———. 1951a. Superfamily Apoidea, pp. 1043-1255 *in* C. F. W. Muesebeck and K. V. Krombein, eds., *Hymenoptera of America North of Mexico—Synoptic Catalog.* Washington: U.S. Dept. of Agriculture Monograph no. 2. [Supplements 1, 1958, and 2, 1967, by K. V. Krombein.]

———. 1951b. Subgeneric groups of *Hemisia. Journal of the Kansas Entomological Society* 24: 1-11.

———. 1952. A new genus of panurgine bee from Arizona. *Journal of the Kansas Entomological Society* 25: 24-28.

———. 1953a. Comparative morphological and systematic studies of bee larvae with a key to the families of hymenopterous larvae. *University of Kansas Science Bulletin* 35: 987-1102.

———. 1953b. The biology of a leafcutter bee *(Megachile brevis)* and its associates. *University of Kansas Science Bulletin* 35: 1659-1748.

———. 1954a. Observations on the pupae of bees. *Pan-Pacific Entomologist* 30: 63-70.

———. 1954b. Bees of Panamá. *Bulletin of the American Museum of Natural History* 104: 1-176.

———. 1954c. Records and descriptions of North American megachilid bees. *Journal of the Kansas Entomological Society* 27: 65-78.

———. 1955. Some biological observations on *Hoplitis pilosifrons* and *Stelis lateralis. Journal of the Kansas Entomological Society* 28: 81-87.

———. 1957. Notes on the biology of a parasitic bee, *Isepeolus viperinus. Entomological News* 68: 141-146.

———. 1960. Notes on the behavior of Australian colletid bees. *Journal of the Kansas Entomological Society* 33: 22-31.

———. 1961a. Observations on the nests and behavior of *Trigona* in Australia and New Guinea. *American Museum Novitates* no. 2026: 1-46.

———. 1961b. A new parasitic genus of Ceratini from Australia. *Journal of the Kansas Entomological Society* 34: 178-180.

———. 1962a. Observations on the classification of the bees commonly placed in the genus *Megachile. Journal of the New York Entomological Society* 70: 17-29.

———. 1962b. Biological observations on the primitively social bees of the genus "Allodapula" in the Australian region. *Insectes Sociaux* 9: 355- 373.

———. 1962c. The genus *Ceratina* in Australia, with notes on its nests. *Journal of the Kansas Entomological Society* 35: 414-421.

———. 1963a. The bee genus *Eulonchopria. Annals of the Entomological Society of America* 56: 844-849.

———. 1963b. New Ceratinini from Australia. *University of Kansas Science Bulletin* 44: 257-261.

———. 1964a. The bionomics of *Exoneurella,* a solitary relative of *Exoneura. Pacific Insects* 6: 411-426.

———. 1964b. Evolution of the nests of bees. *American Zoologist* 4: 227- 239.

———. 1965a. The life cycle and social organization of bees

of the genus *Exoneura* and its parasite, *Inquilina*. *University of Kansas Science Bulletin* 46: 335-376.

———. 1965b. A classification of the bees of the Australian and South Pacific regions. *Bulletin of the American Museum of Natural History* 130: 1-362, pls. 1-15.

———. 1965c. A generic review of the Dufoureinae of the western hemisphere. *Annals of the Entomological Society of America* 58: 321-326.

———. 1966a. The classification of the Diphaglossinae and North American species of the genus *Caupolicana*. *University of Kansas Science Bulletin* 46: 717-751.

———. 1966b. A new genus of ceratinine bees from the Red Sea area. *Journal of the Kansas Entomological Society* 39: 572-576.

———. 1968a. Nests of some African megachilid bees, with description of a new *Hoplitis*. *Journal of the Entomological Society of Southern Africa* 31: 337-359.

———. 1968b. *Heriades spiniscutis,* a bee that facultatively omits partitions between rearing cells. *Journal of the Kansas Entomological Society* 41: 484-493.

———. 1969a. Notes on the nests and life histories of some African halictid bees with description of a new species. *Transactions of the American Entomological Society* 94: 473-497.

———. 1969b. African genera of allodapine bees. *Journal of the Kansas Entomological Society* 42: 289-293.

———. 1970. Social parasites among African allodapine bees. *Zoological Journal of the Linnean Society* [London] 49: 199-215, pls. 1, 2.

———. 1971a. Biologies of African allodapine bees. *Bulletin of the American Museum of Natural History* 145: 219-302.

———. 1971b. The bee genus *Macrogalea*, pp. 61-71 *in* Entomological Essays to Commemorate the Retirement of Professor K. Yasumatsu. Tokyo: Hokuryukan.

———. 1974a. *The Social Behavior of the Bees.* xii + 404 pp. Cambridge: Harvard University Press.

———. 1974b. Further notes on nests of *Ancyloscelis*. *Journal of the Kansas Entomological Society* 47: 19-22.

———. 1975a. Nests of *Paranthidium jugatorium* in association with *Melitoma taurea*. *Journal of the Kansas Entomological Society* 48: 194-200.

———. 1975b. A taxonomic study of African allodapine bees. *Bulletin of the American Museum of Natural History* 155: 67-240.

———. 1975c. Larvae of African allodapine bees—3. The genera *Allodapula* and *Eucondylops*. *Journal of the Entomological Society of Southern Africa* 38: 243-250.

———. 1976. Larvae of African allodapine bees—4. *Halterapis, Compsomelissa, Macrogalea*, and a key to African genera. *Journal of the Entomological Society of Southern Africa* 39: 33-37.

———. 1977a. Nests and seasonal cycle of *Neocorynura pubescens* in Colombia. *Revista de Biologia Tropical* 23: 39-41.

———. 1977b. Allodapine bees of Madagascar. *American Museum Novitates* no. 2622: 1-18.

———. 1977c. Discordant evolution and the classification of allodapine bees. *Systematic Zoology* 26: 32-56.

———. 1977d. Supplementary taxonomic observations on African allodapine bees. *Journal of the Kansas Entomological Society* 50: 422-430.

———. 1978a. The classification of halictine bees: Tribes and old world nonparasitic genera with strong venation. *University of Kansas Science Bulletin* 51: 501-538.

———. 1978b. The parasitic groups of Halictidae. *University of Kansas Science Bulletin* 51: 291-339.

———. 1979a. Biogeography of the bees. *Annals of the Missouri Botanical Garden* 66: 277-347.

———. 1979b. New and little-known halictine bees from Colombia. *Journal of the Kansas Entomological Society* 52: 180-208.

———. 1980. The large species of *Homalictus* and related Halictinae from the New Guinea area. *American Museum Novitates* no. 2693: 1-21.

———. 1981a. Classification of the bee family Melittidae with a review of species of Meganomiinae. *Contributions of the American Entomological Institute* 18(3): i-iii + 1-135.

———. 1981b. Comparative morphology of the middle coxae of Apoidea. *Journal of the Kansas Entomological Society* 54: 319-326.

———. 1982. A new interpretation of fossil social bees from the Dominican Republic. *Sociobiology* 7: 37-45.

———. 1983a. The classification of the Lithurginae. *Pan-Pacific Entomologist* 59: 176-187.

———. 1983b. Proposal to suppress the first designation of a type species for the generic name *Megilla* . . . . *Bulletin of Zoological Nomenclature* 40: 207-208.

———. 1983c. The parasitic Australian allodapine genus *Inquilina*. *Journal of the Kansas Entomological Society* 56: 555-559.

———. 1984. Proposal to emend ZN(S) 2401 by designating *Apis pilipes* as type species of *Megilla* Fabricius. *Bulletin of Zoological Nomenclature* 41: 138-139.

———. 1985a. A comparative study of the mentum and lorum of bees. *Journal of the Kansas Entomological Society* 57: 705-714.

———. 1985b. From solitary to eusocial: Need there be a series of intervening species?, pp. 293-305 *in* B. Hölldobler and M. Lindauer, eds., Experimental Behavioral Ecology and Sociobiology. *Fortschritte der Zoologie* 31.

———. 1986a. Family-group names among bees. *Journal of the Kansas Entomological Society* 59: 219-234.

———. 1986b. A review of the tribes Diphaglossini and Dissoglottini. *University of Kansas Science Bulletin* 53: 183-214.

———. 1986c. New Peruvian genus and a generic review of Andreninae. *Annals of the Entomological Society of America* 79: 62-72.

———. 1988. The parasitic anthophorid genus *Xeromelecta* in Cuba. *Annals of the Entomological Society of America* 81: 377-379.

———. 1989. Classification of the American Colletinae. *University of Kansas Science Bulletin* 53: 622-703.

———. 1990a. Classification of the Apidae. *University of Kansas Science Bulletin* 54: 75-153.

———. 1990b. *Liotrigona* from Malagasy rain forests. *Journal of the Kansas Entomological Society* 63: 444-446.

———. 1990c. Reproduction and castes in social halictine bees, pp. 77-121 *in* W. Engels, ed., *Social Insects, an Evolutionary Approach to Castes and Reproduction*. Berlin: Springer Verlag.

———. 1990d. Castes in xylocopine bees, pp. 123-146 *in* W. Engels, ed., *Social Insects, an Evolutionary Approach to Castes and Reproduction.* Berlin: Springer Verlag.

———. 1992a. Nomenclatural problems in the Meganomiinae and a review of the genus *Pseudophilanthus. Journal of the Kansas Entomological Society* 65: 146-150.

———. 1992b. The genus *Halterapis* Michener 1967 in Madagascar. *Tropical Zoology* 5: 249-253.

———. 1992c. Sexual dimorphism in the glossa of Colletidae. *Journal of the Kansas Entomological Society* 65: 1-9.

———. 1993. The status of *Prosopalictus,* a halictine bee from Taiwan. *Japanese Journal of Entomology* 61: 67-72.

———. 1994. Mexican and Central American species of *Chilicola. Folia Entomológica Mexicana,* no. 85 (1992): 77-93.

———. 1995a. Some genus-group names of bees. *Journal of the Kansas Entomological Society* 67(1994): 373-377.

———. 1995b. A classification of the bees of the subfamily Xeromelissinae. *Journal of the Kansas Entomological Society* 68: 332-345.

———. 1995c. *Dialictus* Robertson, 1902 and *Chloralictus* Robertson, 1902 (Insecta, Hymenoptera): Proposed precedence over *Paralictus* Robertson, 1901. *Bulletin of Zoological Nomenclature* 52: 316-318.

———. 1996a. A review of the genera of Brachynomadini and a new South American genus. *Journal of the Kansas Entomological Society* 69: 87-96.

———. 1996b. The first South African dioxyine bee and a generic review of the tribe Dioxyini, pp. 142-152 *in* B. B. Norden and A. S. Menke, eds., Contributions on Hymenoptera and Associated Insects Dedicated to Karl V. Krombein. *Memoirs of the Entomological Society of Washington* no. 17.

———. 1997a. The identity of the type species of the bee genus *Tetralonia. Journal of the Kansas Entomological Society* 70: 18-20.

———. 1997b. Genus-group names of bees and supplemental family group names. *Scientific Papers, Natural History Museum, University of Kansas* no. 1: 1-81.

———. 2000. *The Bees of the World,* xiv + [1] + 913 pp. Baltimore: Johns Hopkins University Press.

———. 2001. A high Andean subgenus and species of *Hylaeus. Journal of the Kansas Entomological Society* 73(2000): 1-5.

———. 2002a. Comments on minute Meliponini and the male of the genus *Pariotrigona. Journal of the Kansas Entomological Society* 74: 231-236.

———. 2002b. The bee genus *Chilicola* in the tropical Andes, with observations on nesting biology and phylogenetic analysis of the subgenera. *Scientific Papers, Natural History Museum, the University of Kansas* no. 26: 1-47.

———. 2002c. *Duckeanthidium,* a genus new to Central America, with generic synonymy and a new species. *Journal of the Kansas Entomological Society* 75: 233-240.

———. 2002d. E. L. Holmberg (1917, 1918), 'Las especies argentinas de *Coelioxys'* (Insecta, Hymenoptera): proposed suppression of 139 names applied to groups of species. *Bulletin of Zoological Nomenclature* 59: 121-124.

Michener, C. D., and F. D. Bennett. 1977. Geographical variation in nesting biology and social organization of *Halictus ligatus. University of Kansas Science Bulletin* 51: 233-260.

Michener, C. D., M. D. Breed, and W. J. Bell. 1979. Seasonal cycles, nests, and social behavior of some Colombian halictine bees. *Revista de Biología Tropical* 27: 13-34.

Michener, C. D., and R. W. Brooks. 1984. Comparative study of the glossae of bees. *Contributions of the American Entomological Institute* 22(1): i-vi + 1-[74].

———. 1987. The family Melittidae in Madagascar. *Annales de la Société Entomologique de France* (n.s.)23: 99-103.

———. 2003. Dispersal of brood cells in a Mesoamerican hylaeine bee: a possible risk-spreading behavior, pp. 151-152 *in* Melo and Alves-dos-Santos (2003).

Michener, C. D., R. W. Brooks, and A. Pauly. 1990. Little-known meganomiine bees with a key to the genera. *Journal of African Zoology* 104: 135-140.

Michener, C. D., M. S. Engel, and R. Ayala. 2003. The bee genus *Caupolicana* in Central America. *Journal of the Kansas Entomological Society* 76: 160-171.

Michener, C. D., and A. Fraser. 1978. A comparative anatomical study of mandibular structure in bees. *University of Kansas Science Bulletin* 51: 463-482.

Michener, C. D., and L. Greenberg. 1980. Ctenoplectridae and the origin of long-tongued bees. *Zoological Journal of the Linnean Society* [London] 69: 183-203.

———. 1985. The fate of the lacinia in the Halictidae and Oxaeidae. *Journal of the Kansas Entomological Society* 58: 137-141.

Michener, C. D., and D. A. Grimaldi. 1988a. A *Trigona* from Late Cretaceous amber of New Jersey. *American Museum Novitates* no. 2917: 1-10.

———. 1988b. The oldest fossil bee: Apoid history, evolutionary stasis, and antiquity of social behavior. *Proceedings of the National Academy of Sciences USA* 85: 6424-6426.

Michener, C. D., and T. L. Griswold. 1994a. The classification of old world Anthidiini. *University of Kansas Science Bulletin* 55: 299-327.

———. 1994b. The neotropical *Stelis*-like cleptoparasitic bees. *Proceedings of the Entomological Society of Washington* 96: 674-678.

Michener, C. D., and W. B. Kerfoot. 1967. Nests and social behavior of three species of *Pseudaugochloropsis. Journal of the Kansas Entomological Society* 40: 214-232.

Michener, C. D., W. B. Kerfoot, and W. Ramírez B. 1966. Nests of *Neocorynura* in Costa Rica. *Journal of the Kansas Entomological Society* 39: 245-258.

Michener, C. D., W. E. LaBerge, and J. S. Moure. 1955a. Some American Eucerini bees. *Dusenia* 6: 213-230.

———. 1955b. Canephorulini, a tribe of South American bees. *Dusenia* 6: 207-212.

Michener, C. D., and R. B. Lange. 1957. Observations on the ethology of some Brazilian colletid bees. *Journal of the Kansas Entomological Society* 30: 71-80.

———. 1958a. Observations on the behavior of Brasilian halictid bees, III. *University of Kansas Science Bulletin* 39: 473-505.

———. 1958b. Observations on the behavior of Brazilian halictid bees, II: *Paroxystoglossa jocasta. Journal of the Kansas Entomological Society* 31: 129-138.

———. 1958c. Observations on the ethology of neotropical anthophorine bees. *University of Kansas Science Bulletin* 39: 69-96.

———. 1958d. Observations on the behavior of Brazilian halictid bees (Hymenoptera, Apoidea), I. *Pseudagapostemon. *Annals of the Entomological Society of America* 51: 155-164.

———. 1959. Observations on the behavior of Brazilian halictid bees (Hymenoptera, Apoidea), IV. *Augochloropsis*, with notes on extralimital forms. *American Museum Novitates* no. 1924: 1-41.

Michener, C. D., R. J. McGinley, and B. N. Danforth. 1994. *The Bee Genera of North and Central America*. viii + 209 pp. Washington: Smithsonian Institution Press.

Michener, C. D., and J. S. Moure. 1956. The generic positions of certain South American eucerine bees. *Dusenia* 7: 277-290.

———. 1957. A study of the classification of the more primitive non- parasitic anthophorine bees. *Bulletin of the American Museum of Natural History* 112: 395-452.

———. 2002. *Tetrapedia* Klug, 1810, *T. diversipes* Klug, 1819 (sic) and *Exomalopsis* Spinola, 1853 (Insecta, Hymenoptera): proposed conservation of usuage of the names by the designation of a neotype for *T. diversipes*. *Bulletin of Zoological Nomenclature* 59: 34-37.

Michener, C. D., and E. Ordway. 1963. The life history of *Perdita maculigera maculipennis*. *Journal of the Kansas Entomological Society* 36: 34-45.

———. 1964. Some anthidiine bees from Mexico. *Journal of the New York Entomological Society* 72: 70-78.

Michener, C. D., and G. Poinar. 1997. The bee fauna of the Dominican amber. *Journal of the Kansas Entomological Society* 69 (1996, suppl.): 353-361.

Michener, C. D., and C. W. Rettenmeyer. 1956. The ethology of *Andrena erythronii* with comparative data on other species. *University of Kansas Science Bulletin* 37: 645-684.

Michener, C. D., and J. G. Rozen, Jr. 1999. A new ground-nesting genus of xeromelissine bees from Argentina and the tribal classification of the subfamily. *American Museum Novitates* no. 3281: 1-10.

Michener, C. D., and J. F. Scheiring. 1976. Pupae of allodapine bees. *Journal of the Australian Entomological Society* 15: 63-70.

Michener, C. D., and C. A. C. Seabra. 1959. Observations on the behavior of Brasilian halictid bees, VI. Tropical species. *Journal of the Kansas Entomological Society* 32: 19-28.

Michener, C. D., and B. H. Smith. 1987. Kin recognition in primitively eusocial insects, pp. 209-242 *in* D. J. C. Fletcher and C. D. Michener, eds., *Kin Recognition in Animals*. Chichester, U.K.: Wiley.

Michener, C. D., and R. R. Sokal. 1957. A quantitative approach to a problem in classification. *Evolution* 11: 130-162.

Michener, C. D., and J. J. H. Szent-Ivany. 1960. Observations on the biology of a leaf-cutter bee *"Megachile frontalis,"* in New Guinea. *Papua and New Guinea Agricultural Journal* 13: 22-35.

Michener, C. D., M. L. Winston, and R. Jander. 1978. Pollen manipulation and related activities and structures in bees of the family Apidae. *University of Kansas Science Bulletin* 51: 575-601.

Michener, G. R., and C. D. Michener. 1999. Mating behavior of *Dianthidium curvatum* (Hymenoptera, Megachilidae) at a nest aggregation. *University of Kansas Natural History Museum Special Publication* 24: 37-43.

Michez, D., and S. Patiny. 2005. World revision of the oil-collecting bee genus *Macropis* Panzer 1809 (Hymenoptera: Apoidea: Melittidae) with a description of a new species from Laos. *Annales de la Société Entomologique de France* (n.s.) 41: 15-28.

———. 2006. Review of the bee genus *Eremaphanta* Popov 1940 (Hymenoptera, Melittidae), with the description of a new species. *Zootaxa* 1148: 47-68.

Michez, D., M. Terzo, and P. Rasmont. 2004a. Révision des espèces-paléarctiques du genre *Dasypoda*. *Linzer Biologische Beitrage* 36: 847-900.

———. 2004b. Phylogénie, biogéographic et choix floraux des abeilles oligolectique du genre *Dasypoda* Latreille 1802, *Annales de la Société Entomologique de France*, (n.s.) 40: 421-435.

Middleton, W. 1916. Notes on *Dianthidium arizonicum* Rohwer. *Proceedings of the Entomological Society of Washington* 18: 193-195.

Miliczky, E. R. 1985. Observations on the nesting biology of *Tetralonia hamata* Bradley with a description of its mature larva. *Journal of the Kansas Entomological Society* 58: 686-700.

Milliron, H. E. 1961. Revised classification of the bumblebees—A synopsis. *Journal of the Kansas Entomological Society* 34: 49-61.

———. 1970-1973. A monograph of the western hemisphere bumblebees. *Memoirs of the Entomological Society of Canada* no. 65: i-lii (1970); no. 82: 1-80 (1971); no. 89: 81-237 (1973); no. 91: 239-333 (1973).

Minckley, R. L. 1994. Comparative morphology of the mesosomal 'gland' in male large carpenter bees. *Biological Journal of the Linnean Society* [London] 53: 291-308.

———. 1998. A cladistic analysis and classification of the subgenera and genera of large carpenter bees, Tribe Xylocopini. *Scientific Papers, Natural History Museum, University of Kansas* no. 9: 1-47.

Minckley, R. L., J. H. Cane, and L. Kervin. 2000. Origin and ecological consequences of pollen specialization among desert bees. *Proceedings of the Royal Society* [London] (B) 267: 265-271.

Minckley, R. L., W. T. Wcislo, D. Yanega, and S. L. Buchmann. 1994. Behavior and phenology of a specialist bee *(Dieunomia)* and sunflower *(Helianthus)* pollen availability. *Ecology* 75: 1406-1419.

Mitchell, T. B. 1924. New megachilid bees. *Journal of the Elisha Mitchell Scientific Society* 40: 154-165.

———. 1930. A contribution to the knowledge of neotropical *Megachile* with descriptions of new species. *Transactions of the American Entomological Society* 56: 155-305, pls. X-XIV.

———. 1934-1937. A revision of the genus *Megachile* in the nearctic region. *Transactions of the American Entomological Society* 59: 295-361, pls. XX-XXI (part I, 1934); 61: 1-44, pl. I (part II, 1935a); 61: 155-205, pls. VIII-IX (part III, 1935b); 62: 117-166, pls. VIII-XI (part IV, 1936); 62: 323-382, pls. XXII-XXVI (part V, 1937a); 63: 45-83, pls. V-VI (part VI, 1937b); 63: 175-206, pls. XII-XIII (part VII, 1937c); 63: 381-426, pls. XXVI-XXIX (part VIII, 1937d).

———. 1943. On the classification of neotropical *Megachile*. *Annals of the Entomological Society of America* 36: 656-671.

———. 1944. New species and records in *Megachile. Pan-Pacific Entomologist* 20: 132-143. [Key to species of subgenus *Derotropis*.]

———. 1960, 1962. Bees of the eastern United States, 1: 1-538 (1960); 2: 1-557 (1962). *North Carolina Agricultural Experiment Station Technical Bulletin* nos. 141, 152.

———. 1973. *A Subgeneric Revision of the Bees of the Genus* Coelioxys *of the Western Hemisphere*. iii + 129 pp. Raleigh: Department of Entomology, North Carolina State University.

———. 1980. *A Generic Revision of the Megachiline Bees of the Western Hemisphere.* [ii] + 95 pp. Raleigh: Department of Entomology, North Carolina State University.

Miyanaga, R., Y. Maeta, and K. Hoshikawa. 2000. Nesting biology and occurance of social nests in a bivoltine basically solitary halictine bee, *Lasioglossum (Lasioglossum) scitulum* Smith. *Entomological Science* 3: 291-302.

Miyanaga, R., Y. Maeta, and G. Mizuta. 1998. Discovery of social nests in a bivoltine, basically solitary halictine bee, *Lasioglossum (Lasioglossum) mutilum* (Vachal). *Entomological Science* 1: 165-169.

Mjelde, A. 1983. The foraging strategy of *Bombus consobrinus. Acta Entomologica Fennica* 42: 51-56.

Mocsáry, A. 1894. E fauna apidarum hungariae. *Természetrajzt Füzetek* 17: 34-37.

Móczár, M. 1957-1967. Hymenoptera III, *in Magyarország Állatvilága [Fauna Hungariae]* 19: 1-76 (1957); 35: 1-78 (1958); 51: 1-64 (1960); 85: 1-116 (1967). [In Magyar.]

Moldenke, A. R. 1976a. California pollination ecology and vegetation types. *Phytologia* 34: 305-361.

———. 1976b. Evolutionary history and diversity of the bee faunas of Chile and Pacific North America. *Wasmann Journal of Biology* 34: 147-178.

Moldenke, A. R., and J. L. Neff. 1974. The bees of California, a catalogue with special reference to pollination and ecological research. v + 245 pp. *Origin and Structure of Ecosystems, Technical Reports* 74-1 to 74-6. Santa Cruz: University of California.

Morato, E. F. 2001. Biologia e ecologia de *Anthodioctes moratoi* Urban (Hymenoptera, Megachilidae, Anthidiini) em matas continuas e fragmentos na Amazônia Central, Brasil. *Revista Brasileira de Zoologia* 18: 729-736.

Morawitz, F. 1873. Nachtrag zur Bienenfauna des Gouvernements von St. Petersburg. *Horae Societatis Entomologicae Rossicae* 9: 151-159.

———. 1874. Die Bienen Daghestans. *Horae Societatis Entomologicae Rossicae* 10(1873): 129-189.

———. 1875. Pchely (Mellifera Latr.), Tetrad 1, Apidae Genuinae, pp. 1-160 *in* A. P. Fedtschenko, *Puteshestvie v Turkestan (Reisen in Turkestan), Zoogeografischeskaya Izsledovaniya*, part V, sec. 7. [Also as *Izvestiya Imperatorskago Obshchestva Lyubiteley Yestestvoznaniya Antropologii i Etnografii* 19(2): 1-160.]

———. 1876. Zur Bienenfauna der Caucasusländer. *Horae Societatis Entomologicae Rossicae* 12: 3-69.

———. 1895. Beitrag zur Bienenfauna Turkmeniens. *Horae Societatis Entomologicae Rossicae* 29: 1-76.

Morice, F. D. 1901. Illustrations of the 6th male ventral segment in 17 *Osmia*-species of the *adunca*-group, with a note on the synonymy of four species, and descriptions of four which seem new. *Transactions of the Entomological Society of London* 1901: 161-178, pls. VII, VIII.

———. 1904. Illustrations of the male terminal segments and armatures in thirty-five species of the hymenopterous genus *Colletes. Transactions of the Entomological Society of London* 1904: 25-63, pls. VI-IX.

Morice, F. D., and J. H. Durrant. 1915. The authorship and first publication of the "Jurinean" genera of Hymenoptera .... *Transactions of the Entomological Society of London* 1914: 339-436.

Morse, D. H. 1982. Behavior and ecology of bumblebees, pp. 245-322 *in* H. R. Hermann, ed., *Social Insects,* Vol. III. New York: Academic Press.

Motschulsky, V. 1856. Lettres... à M. Ménétries, 3. *Études Entomologiques* [Helsinki] 5: 3-38, 1 pl. [Fossil bee, p. 28.]

Moure, J. S. 1940. Apoidea neotropica. *Arquivos de Zoologia do Estado de São Paulo* 2: 39-64, pls. I-III. [Also *Revista do Museu Paulista* 25: 39-64.]

———. 1941a. Apoidea neotropica, III. *Arquivos do Museu Paranaense* 1: 41-99, 1 pl.

———. 1941b. Notas sobre abelhas do grupo *Tetrapedia* Klug. *Revista de Entomologia* [Rio de Janeiro] 12: 515-521.

———. 1942. Abelhas de Salobra. *Papéis Avulsos do Departamento de Zoologia, São Paulo* 2: 291-321.

———. 1943a. Abelhas de Batatais. *Arquivos do Museu Paranaense* 3: 145-203, pl. XI.

———. 1943b. Notas sôbre abelhas da coleção Zikán. *Revista de Entomologia* [Rio de Janeiro] 14: 447-484.

———. 1944a. Abejas del Perú. *Boletin del Museo de Historia Natural "Javier Prado"* 8(28-29): 67-75.

———. 1944b. Notas sôbre abelhas da coleção Zikán, II. *Revista de Entomologia* [Rio de Janeiro] 15: 273-291.

———. 1944c. Apoidea da coleção do Conde Amadeu A. Barbiellini. *Revista de Entomologia* [Rio de Janeiro] 15: 1-18.

———. 1945a. Contribução para o conhecimento dos Diphaglossinae, particularmente *Ptiloglossa. Arquivos do Museu Paranaense* 4: 137-178, pl. XV.

———. 1945b. Notas sobre os Epicharitina. *Revista de Entomologia* [Rio de Janeiro] 16: 293-314.

———. 1945c. Apoidea da coleção do Conde Amadeu A. Barbiellini, II. *Revista de Entomologia* [Rio de Janeiro] 16: 394-414.

———. 1946a. Contribução para o conhecimento dos Meliponinae. *Revista de Entomologia* [Rio de Janeiro] 17: 437-443.

———. 1946b. Meliponas do Brasil. *Chacaras e Quintais* 74: 609-612.

———. 1946c. Notas sobre abelhas da sub familia Chilicolinae, pp. 243-250, pl. XII, *in Livro de Homenagem a Romualdo Ferreira d'Almeida*. São Paulo: Sociedade Brasileira de Entomologia.

———. 1946d. Notes sôbre as mamangabas. *Boletim Agrícola* 4(12-13): 1-32, pls. 1-3.

———. 1947a. Novos agrupamentos genéricos e algumas espécies novas de abelhas sulamericanas. *Museu Paranaense Publicações Avulsas* no. 3: 1-37.

———. 1947b. Notas sobre algunas abejas de la provincia de Salta. *Revista de la Sociedad Entomológica Argentina* 13: 218-253.

———. 1948. Notas sobre algumas abelhas de Tacanas, Tucumán, Argentina. *Revista de Entomologia* [Rio de Janeiro] 19: 313-346.

———. 1949a. Notas sobre algunas abejas de Tacanas, Tucumán, Argentina, II. *Revista de Entomologia* [Rio de Janeiro] 20: 437-460, pls. X, XI.

———. 1949b. Las espécies chilenas de la sub-familia Lithurginae. *Arquivos do Museu Paranaense* 7: 265-286, pls. II-V.

———. 1950a. Contribuição para o conhecimento das espécies Brasileiras de *Hypotrigona* Cockerell. *Dusenia* 1: 241-260.

———. 1950b. Contribuição para o conhecimento do gênero *Eulaema* Lepeletier. *Dusenia* 1: 181-200.

———. 1950c. Euherbstiinae nova sub-familia de Andrenidae. *Dusenia* 1: 303-306.

———. 1950d. Alguns agrupamentos novos de abelhas neotropicais. *Dusenia* 1: 385-396.

———. 1951a. Notas sobre Meliponinae. *Dusenia* 2: 25-70.

———. 1951b. *Ariphanarthra,* um novo gênero de Halictidae. *Dusenia* 2: 137-140.

———. 1951c. Notas sôbre abelhas do antigo gênero *Pasiphae. Dusenia* 2: 189-198.

———. 1953a. Notas sôbre Colletidae sul-americanos. *Dusenia* 4: 61-78.

———. 1953b. *Nogueirapis,* no grupo de Trigonini da região neotropical. *Ciencia e Cultura* 5: 247-249.

———. 1953c. Notas sobre Megachilidae de Bolivia, Perú y Chile. *Dusenia* 4: 113-124.

———. 1954a. Novas notas sôbre abelhas do antigo gênero *Pasiphae. Dusenia* 5: 165-190.

———. 1954b. Notas sôbre Epeolini sul-americanos. *Dusenia* 5: 259-286.

———. 1955. Notas sôbre Epeolini sulamericanos. *Dusenia* 6: 115-138.

———. 1956. Tres espécies novas de Paracolletini do sul do Brasil. *Dusenia* 7: 305-315.

———. 1958a. *Augochlorodes,* a new genus of Halictinae from Brasil. *Journal of the Kansas Entomological Society* 31: 53-55.

———. 1958b. On the genus *Arhysosage* Brèthes from Argentina. *Entomological News* 69: 43-48.

———. 1959. On the species of *Megalopta* described by F. Smith (Hymenoptera, Apoidea). *Journal of the New York Entomological Society* 66(1958): 179-190.

———. 1960a. Os tipos das espécies neotropicais de *Hylaeus* da coleção Vachal no Museu de História Natural de Paris. *Boletim do Museu Paraense Emílio Goeldi* (Nova Série), Biologia no. 27: 1-104.

———. 1960b. A review of the genus *Paroxystoglossa. University of Kansas Science Bulletin* 40: 121-133.

———. 1960c. Notes on the types of the neotropical bees described by Fabricius. *Studia Entomologica* 3: 97-160.

———. 1961. A preliminary supra-specific classification of the Old World meliponine bees. *Studia Entomologica* 4: 181-242.

———. 1963. Una nueva especie de *Eulaema* de Costa Rica. *Revista de Biología Tropical* 11: 211-216.

———. 1964a. Os espécies de *Zikanapis,* com a descrição de dois novos subgêneros e duas espécies novas. *Studia Entomologica* 7: 417-458.

———. 1964b. Two new genera of halictine bees from the Araucanian subregion of South America. *Journal of the Kansas Entomological Society* 37: 265-275.

———. 1964c. Um novo Paracolletini do Litoral Paulista. *Boletim da Universidade do Paraná (Zoologia)* 2: 51-56.

———. 1965. New placements for some species of *Paranthidium. Proceedings of the Entomological Society of Washington* 67: 29-31.

———. 1967a. A check-list of the known euglossine bees, pp. 395-415 *in Atas do Simpósio sôbre a Biota Amazônica,* Vol. 5 (Zoologia).

———. 1967b. A new genus and two new species of eucerine bees from South America (Hymenoptera, Apoidea). *Dusenia* 8: 147-152.

———. 1969a. *Micrommation,* nôvo genero de Halictidae do Paraná. *Atas da Sociedade de Biología do Rio de Janeiro* 12: 247-249.

———. 1969b. The Central American species of *Euglossa* subgenus *Glossura* Cockerell, 1917. *Revista de Biología Tropical* 15: 227-247.

———. 1969c. Notas sôbre algumas espécies de *Centris* da Guiana. *Anais da Academia Brasileira de Ciências* 41: 113-123.

———. 1970. The species of euglossine bees of Central America belonging to the subgenus *Euglossella. Anais da Academia Brasileira de Ciências* 42: 147-157.

———. 1971. Descrição de uma nova espécie de *Tetragona* do Brasil Central. *Boletim da Universidade Federal do Paraná, Zoologia* 4:47-50.

———. 1972. Descrição de duas espécies novas de *Hylaeus* do sul do Brasil. *Revista Brasileira de Biologia* 32: 275-284.

———. 1975. Notas sobre as espécies de *Melipona* descritas por Lepeletier em 1836. *Revista Brasileira de Biologia* 35: 615-623.

———. 1989a. *Camargoia,* um novo gênero neotropical de Meliponinae. *Boletim do Museu Paraense Emílio Goeldi,* Série Zoologia 5: 71-78.

———. 1989b. *Sakagamilla affabra* gen. n. e sp. n. de Meliponinae (Hymenoptera, Apidae) de Rondônia. *Revista Brasileira de Zoologia* 6: 681-684.

———. 1989c. *Glossuropoda,* novo subgênero de *Euglossa,* e duas espécies novas da Amazônia, do mesmo subgênero. *Memorias do Instituto Oswaldo Cruz* 84: 387-389.

———. 1992. *Melikerria* e *Eomelipona,* dois subgêneros novos em *Melipona* Illiger, 1806, pp. 32-38 *in* Anais do Encontro Brasileiro sobre Biologia de Abelhas e Outros Insectos Sociais, *Naturalia,* edição especial.

———. 1994a. Sobre a posição sistemática de *Exomalopsis latitarsis* Friese. *Revista Brasileira de Zoologia* 9(1992): 273-276.

———. 1994b. *Lissopedia,* gen. n. de Paratetrapediini para a região neotropical, com as descrições de três espécies novas. *Revista Brasileira de Zoologia* 9[1992]: 305-317.

———. 1995a. Notas sobre o gênero *Parapartamona* Schwarz. *Acta Biologica Paranaense* 21(1992): 121-134.

———. 1995b. *Compsoclepta fasciata,* antoforídeo parasita do noroeste da Argentina. *Acta Biologica Paranaense* 21(1992): 143-148.

———. 1995c. Reestudo de alguns tipos de abelhas neotropicais descritos por Friese e conservados no museu de Berlin. *Revista Brasileira de Zoologia* 12: 939-951.

———. 1999a. Espécies de *Parapsaenythia* de Costa Rica descritas por Friese. *Revista Brasileira de Zoologia* 16(suppl. 1): 105-111.

———. 1999b. Tipos de Halictidae de Vachal no

Naturkunde Museum, Berlin. *Revista Brasileira de Zoologie* 16(suppl. 1): 73-89.

———. 2001. Uma pequena abelha com cabeça e mandibulas excepcionas. *Revista Brasileira de Zoologia* 18: 493-497.

———. 2002. O subgênero *Centris (Aphemisia)* Ayala: notas complementares e descrição de uma nova espécie. *Revista Brasileira de Entomologia* 46: 257-261.

———. 2003. As espécies do gênero *Eulaema* Lepeletier, 1841. *Acta Biologíca Paranaense* 29 (2000): 1-70.

Moure, J. S., and J. M. F. Camargo. 1978. A fossil stingless bee from Copal. *Journal of the Kansas Entomological Society* 51: 560-566.

———. 1982. *Partamona (Noqueirapis) minor*, nova espécie de Meliponinae (Hymenoptera, Apidae) do Amazonas e notas sobre *Plebeia variicolor* (Ducke). *Boletim do Museu Paraense Emílio Goeldi.* (Nova Série), Zoologia no. 120: 1-10.

———. 1994. *Melipona (Michmelia) capinaba*, uma nova espécie de Meliponinae (Hymenoptera, Apidae) de sudeste do Brasil. *Revista Brasileira de Zoologia* 11: 289-296.

Moure, J. S., J. M. F. Camargo, and M. V. B. Garcia. 1988. Uma nova espécie de *Leurotrigona*. *Boletim do Museu Paraense Emílio Goeldi.* (Série Zoologia) 4: 145-154.

Moure, J. S., V. Graf, and D. Urban. 1999. Catálogo de Apoidea da região neotropical (Hymenoptera, Colletidae). 1. Paracolletini. *Revisita Brasileira de Zoologia* 16(Suppl. 1): 1-46.

Moure, J. S., and P. D. Hurd, Jr. 1960. On the systematic position of three anthidiine bees described by Adolfo Ducke. *Boletim do Museu Paraense Emílio Goeldi.* (Nova Série), Zoologia no. 28: 1-13.

———. 1982. On two new groups of neotropical halictine bees. *Dusenia* 13: 46.

———. 1987. *An Annotated Catalog of the Halictid Bees of the Western Hemisphere.* vii + 405 pp. Washington, D.C.: Smithsonian Institution Press.

Moure, J. S., and W. E. Kerr. 1950. Sugestões para a modificação da sistemática do gênero *Melipona*. *Dusenia* 1: 105-129, pls. II-III.

Moure, J. S., and B. Lucas de Oliveira. 1962. Novo gênero de Panurginae para a América do Sul. *Boletim da Universidade do Paraná* (Zoologia) no. 15: 1-14.

Moure, J. S., and C. D. Michener. 1955a. The bee family Fideliidae in South America. *Dusenia* 6: 199-206.

———. 1955b. A contribution toward the classification of neotropical Eucerini. *Dusenia* 6: 239-331.

Moure, J. S., and S. F. Sakagami. 1962. As mamangabas sociais do Brasil (*Bombus* Latr.). *Studia Entomologica* 5: 65-194.

———. 1984. Notes on *Pseudagapostemon,* a neotropical halictine bee genus, with descriptions of a new subgenus and two new species from southern Brazil. *Dusenia* 14: 1-9.

Moure, J. S., and C. A. C. Seabra. 1962a. Una nueva especie de *"Zikanapis"* de Argentina y Paraguay. *Neotropica* 8: 31-36.

———. 1962b. A new species of the genus *Oxaea* from Brazil. *Journal of the New York Entomological Society* 70: 235-238.

Moure, J. S., and D. Urban. 1963. Uma nova espécie de *"Oxaea"* de Mato Grosso, Brasil. *Revista Brasileira de Biologia* 23: 361-364.

———. 1964. Revisão das espécies Brasileiras do género *Anthidium* Fabricius, 1804. *Anais do II Congresso Latino-Americano de Zoologia* (S. Paulo, 1962), Vol. 1: 93-114.

———. 1975. *Dicranthidium* novo subgenero de *Hypanthidioides* [sic] Moure, 1947. *Revista Brasileira de Biologia* 35: 837-842.

———. 1990. *Carloticola* gen. n. e *Carloticola trichura* sp. n. da Argentina. *Acta Biologica Paranaense* 19: 89-99.

———. 1994. *Rhynostelis,* gen. n. e notas sobre *Hoplostelis* Dominique. *Revista Brasileira de Zoologia* 11: 297-302.

———. 1995. *Sarocolletes* Michener, 1989, status nov. e espécies novas. *Acta Biologica Paranaense* 21(1992): 113-120.

———. 2002a. Catálogo de Apoidea da região neotropical (Hymenoptera, Colletidae). III Colletini, *Revista Brazileira de Zoologia* 19: 1-30.

———. 2002b. Catálogo de Apoidea da Região Neotropical (Hymenoptera, Colletidae. V. Xeromelissinae. *Revista Brasileira de Zoologia* 19(suppl. 1): 1-25.

Moure, J. S., D. Urban, and V. Graf. 1999. *Baptonedys,* novo gênero para *Lonchopria bicellularis* Ducke. *Acta Biologica Paranaense* 28: 11-17.

Mueller, U. G. 1997. Life history and social evolution of the primitively eusocial bee *Augochlorella striata*. *Journal of the Kansas Entomological Society* 69(1996, suppl.): 116-138.

Müller, A. 1995. Morphological specializations in central European bees for the uptake of pollen from flowers with anthers hidden in narrow corolla tubes. *Entomologia Generalis* 20: 43-57.

———. 1996a. Convergent evolution of morphological specializations in central European bee and honey wasp species as an adaptation to the uptake of pollen from nototribic flowers. *Biological Journal of the Linnean Society* 57: 235-252.

———. 1996b. Host plant specialization in western palearctic anthidiine bees. *Ecological Monographs* 66: 235-257.

———. 1996c. Collection of extrafloral trichome secretions for nest wool impregnation in the solitary bee *Anthidium manicatum*. *Naturwissenschaften* 83: 230-232.

———. 2002. *Osmia (Melanosmia) steinmanni* sp. n., a new bee from the Swiss Alps. *Revue Suisse de Zoologie* 109: 803-812.

———. 2006. A scientific note on *Bombus inexspectatus* (Tkalcŭ; 1963): evidence for a social parasitic mode of life. *Apidologie* 37: 1-2.

Müller, A., A. Krebs, and F. Amiet. 1997. *Bienen-Mitteleuropäishe Gattungen, Lebenwiese, Beobachtung,* 384 pp. München: Naturbuch-Verlag.

Müller, H. 1884. Ein Beitrag zur Lebensgeschichte der *Dasypoda hirtipes. Verhandlungen des Naturhistorischen Vereins der Preussischen Rheinlande* [Bonn] 41: 1-52, pls. 1, 2.

Muller, J. 1970. Palynological evidence on early differentiation of angiosperms. *Biological Reviews* [Cambridge, U.K.] 45: 417-450.

Münster-Swendsen, M. 1968. On the biology of the solitary bee *Panurgus banksianus* Kirby (Hymenoptera, Apidae), including some ecological aspects. *Royal Veterinary and Agricultural College* [Copenhagen] *Yearbook* 1968: 215-241.

———. 1970. Nesting behavior of the bee *Panurgus banksianus* Kirby. *Entomologica Scandanavica* 1: 93-101.

Nates-Parra, G. 1996. Abejas sin aguijón (Hymenoptera: Meliponini) de Colombia, pp 181-268 *in* G. Amat, G. Andrade and F. Fernandez, eds., *Insectos de Colombia-Estudios Escogidos*. Santa Fé de Bogotá: Academia Colombiana de Ciencias Exactas, Fisicas y Naturales, Centro Editorial Javeriano, Pontificia Universidad Javeriana.

Neff, J. L. 1984. Observations on the biology of *Eremapis parvula* Ogloblin, an anthophorid bee with a metasomal scopa. *Pan-Pacific Entomologist* 60: 155-162.

———. 2003. Nest provisioning biology of the bee *Panurginus polytrichus* Cockerell (Hymenoptera: Andrenidae), with a description of a new *Holcopasites* species (Hymenoptera, Apidae), its probable nest parasite. *Journal of the Kansas Entomological Society* 76: 203-216.

Neff, J. L., and B. N. Danforth. 1992. The nesting and foraging behavior of *Perdita texana* (Cresson). *Journal of the Kansas Entomological Society* 64: 394-405.

Neff, J. L., and J. G. Rozen, Jr. 1995. Foraging and nesting biology of the bee *Anthemurgus passiflorae* (Hymenoptera: Apoidea), descripti ons of its immature stages, and observations on its floral host (Passifloraceae). *American Museum Novitates* no. 3138: 1-19.

Neff, J. L., and B. B. Simpson. 1981. Oil-collecting structures in the Anthophoridae (Hymenoptera): Morphology, function, and use in systematics. *Journal of the Kansas Entomological Society* 54: 95-123.

———. 1991. Nest biology and mating behavior of *Megachile fortis* in central Texas. *Journal of the Kansas Entomological Society* 64: 324-336.

———. 1997. Nesting and foraging behavior of *Andrena rudbeckiae* Robertson (Hymenoptera: Apoidea: Andrenidae) in Texas. *Journal of the Kansas Entomological Society* 70: 100-113.

Neff, J. L., B. B. Simpson, and L. J. Dorr. 1982. The nesting biology of *Diadasia afflicta* Cress. *Journal of the Kansas Entomological Society* 55: 499-518.

Newman, E. 1834. Attempted division of British insects into natural orders. *Entomological Magazine* 2: 379-431.

Niemelä, P. 1949. Mitteilungen über die Apiden (Hym.) Finnlands. *Annales Entomologici Fennici* 15: 101-120.

Nilsson, L. A. 2003. *Prerevisional Checklist and Synonoymy of the Bees of Sweden (Hymenoptera: Apoidea)*, 111pp. Uppsala: Lantbruksuniversitet ArtDatabanken.

Niu, Z.-Q., Y.-R. Wu, and D.-W. Huang. 2004. A taxonomic study on the subgenus *Seladonia* (Hymenoptera: Halictidae: *Halictus*) in China with a description of a new species. *Zoological Studies* 43: 647-670.

Nogueira-Neto, P. 1970. *A Criação de Abelhas Indígenas sem Ferrão*, 2nd ed. 365 pp. São Paulo: Edição Technapis, Chácaras e Quintais.

———. 1997. *Vida e Criação de Abelhas Indígenas sem Ferrão*. 446 pp. São Paulo: Editora Nogueirapis.

Noll, F. B. 2002. Behavioral phylogeny of corbiculate Apidae (Hymenoptera; Apinae), with special reference to social behavior. *Cladistics* 18: 137-153.

Noll, F. B., R. Zucchi, J. A. Jorge, and S. Mateus. 1997. Food collection and maturation in the necrophagous stingless bee, *Trigona hypogea*. *Journal of the Kansas Entomological Society* 69(1996, suppl.): 287-293.

Norden, B. B. 1984. Nesting biology of *Anthophora abrupta*. *Journal of the Kansas Entomological Society* 57: 243-262.

Norden, B. B., K. V. Krombein, and S. W. T. Batra. 1994. Nests and enemies of *Exomalopsis (Phanamalopsis) solani* Cockerell. *Proceedings of the Entomological Society of Washington* 96: 350-356.

Norden, B. B., K. V. Krombein, and B. N. Danforth. 1992. Taxonomic and bionomical observations on a Floridian panurgine bee, *Perdita (Hexaperdita) graenicheri* Timberlake. *Journal of Hymenoptera Research* 1: 107-118.

Norden, B.B., K.V. Krombein, M. A. Deyrup, and J. P. Edirisinghe. 2003. Biology and behavior of a seasonally aquatic bee, *Perdita (Alloperdita) floridensis* Timberlake. *Journal of the Kansas Entomological Society* 76: 236-249.

Noskiewicz, J. 1936. *Die Palearktischen* Colletes-*Arten*. v + 532 pp., 28 pls. Lwowie: Prace Naukowe.

———. 1948. Remarques sur quelques espèces du genre *Megachile* de la Silésie et des pays limitrophes. *Polskie Pismo Entomologiczne* 18: 31-59. [In Polish, summary in French.]

———. 1961. Beiträge zur Kenntnis der paläarktischen Arten der Gattung *Stelis*. *Polskie Pismo Entomologiczne* 31: 113-133.

Nylander, W. 1848. Adnotationes in expositionem monographicam apum borealium. *Notiser ur Sällskapets pro Fauna et Flora Fennica Förhandlingar* 1: 165-282, pl. III.

Obrecht, E., and A. Scholl. 1981. Enzymelektrophoretische Untersuchungen zur Analyse der Verwandtschaftsgrade zwischen Hummel- und Schmarotzerhummelarten. *Apidologie* 12: 257-268.

O'Brien, L. B., and P. D. Hurd, Jr. 1965. Carpenter bees of the subgenus *Notoxylocopa*. *Annals of the Entomological Society of America* 58: 175- 196.

O'Connor, B. M. 1993. The mite community associated with *Xylocopa latipes* (Hymenoptera: Anthophoridae: Xylocopinae) with description of a new type of acarinarium. *International Journal of Acarology* 19: 159-166.

Ogloblin, A. A. 1948. Descripción de dos géneros nuevos de Paracolletini argentinos. *Notas del Museo de la Plata* 13: 165-177, pls. I-IV.

———. 1953. Un nuevo Halictidae (Hym. Apodidae [sic]): Un nuevo subgénero de *Temnosoma*. *Boletín Sociedad Entomológica Argentina* no. 2: [2-3].

———. 1954. Un nuevo subgenero de *"Temnosoma"* F. Smith. *Neotropica* 1: 5- 8.

———. 1955. Descripción de un nuevo subgénero de Anthophorinae. *Dusenia* 6: 231-237.

———. 1956. Dos géneros nuevos de la tribu Exomalopsini (Apidae, Hym.) de la República Argentina. *Dusenia* 7: 149-158.

Okazaki, K. 1987. Life cycle of a subtropical xylocopine bee, *Ceratina okinawana*, with some related problems. *Kontyû* 55: 1-8.

———. 1992. Nesting habits of the small carpenter bee, *Ceratina dentipes*, in Hengchun peninsula, southern Taiwan. *Journal of the Kansas Entomological Society* 65: 190-195.

Oliveira, F. F. de. 2002. The mesotibial spur in stingless bees: a new character for the systematics of Meliponini. *Journal of the Kansas Entomological Society* 75: 194-202.

Oliveira, F. F. de, and P. Marchi. 2005. Três espécies novas de *Lestrimelitta* Friese (Hymenoptera, Apidae) da Costa Rica, Panamá e Guiana Francesa. *Revista Brasileira de Entomologia* 49: 1-6.

Ordway, E. 1965. Caste differentiation in *Augochlorella*. *Insectes Sociaux* 12: 291-308.

———. 1966a. Systematics of the genus *Augochlorella* (Hymenoptera, Halictidae) north of Mexico. *University of Kansas Science Bulletin* 46: 509-624.

———. 1966b. The bionomics of *Augochlorella striata* and *A. persimilis* in eastern Kansas. *Journal of the Kansas Entomological Society* 39: 270-313.

———. 1984. Aspects of the nesting behavior and nest structure of *Diadasia opuntiae* Ckll. *Journal of the Kansas Entomological Society* 57: 216-230.

Ornduff, R. 1991. Size classes, reproductive behavior, and insect associates of *Cycas media* (Cycadaceae) in Australia. *Botanical Gazette* 152: 203-207.

Ornosa, C., and F. J. Ortiz-Sánchez. 1998. Contribución al conocimiento de los melítidos ibéricos. *Boletín de la Asociación Española de Entomología* 22: 181-202.

———. 2003. Claves de identificación para las especies ibéricas de Melittidae. *Linzer Biologische Beiträge* 35: 555-579.

———. 2004. *Hymenoptera, Apoidea I; Fauna Iberica,* vol. 23, 553 pp. Madrid: Museo Nacional de Ciencias Naturales.

Ortiz-Sánchez, F. J., and A. J. Jiménez-Rodriquez. 1991. Actualización del catálogo de las especies españolas de Anthophorini. *Boletin de la Asociatión Española de Entomología* 15: 297-315.

Osgood, E. A. 1989. Biology of *Andrena crataegi* Robertson (Hymenoptera: Andrenidae), a communally nesting bee. *Journal of the New York Entomological Society* 97: 56-64.

Ospina, M. 2000. Abejas carpinteras (Hymenoptera: Apidae: Xylocopinae: Xylocopini) de la región neotropical. *Biota Colombiana* 1: 239-252.

Osychnyuk [= Osytshnjuk], A. Z. 1970, 1977. Apoidea—Colletidae (1970) and Andrenidae (1977), *in Fauna of the Ukraine* 12(4): 1-158; 12(5): 1-328. Kiev: Zoological Institute, Academy of Sciences of Ukraine. [In Ukrainian.]

———. 1983. A new subgenus and new species of bees of the genus *Andrena* (Hymenoptera, Andrenidae) from the asiatic part of the USSR. *Entomologicheskoe Obozrenie* 62: 794-799. [In Russian; translation into English in *Entomological Review* 62: 124-129.]

———. 1984a. A new palearctic subgenus and a new species of genus *Andrena*. *Vestnik Zoologii* 1984(2): 23-30. [In Russian.]

———. 1984b. A new east-Mediterranean subgenus and a new species of *Andrena* bees. *Vestnik Zoologii* 1984(1): 19-24. [In Russian.]

———. 1984c. New and little known bees of the subgenus *Carandrena* War. (Hymenoptera, Andrenidae, *Andrena* F.) in the fauna of the USSR, pp. 3-15 *in* Y. A. Pesenko, ed., Systematics and Ecology of Bees. *Proceedings of the Zoological Institute, USSR Academy of Sciences, Leningrad,* Vol. 128. [In Russian.]

———. 1993a, b. New subgenera and new species of palearctic *Andrena* bees (Hymenoptera, Andrenidae), Communication 1 and Communication 2. *Vestnik Zoologii* 1993a: 17-23; 1993b: 60-66. [In Russian, English summaries.]

———. 1994. New subgenera and new species of palearctic *Andrena* bees. *Vestnik Zoologii* 1994: 17-23 and 34.

Osychnyuk, A. Z., D. V. Panfilov, and A. A. Ponomareva. 1978. Apoidea, pp. 279-519 *in* V. I. Tobias, ed., *Species of Insects of the European Region of the USSR,* Vol. 3, Hymenoptera. Leningrad: Zoological Institute, Academy of Sciences of the USSR. [In Russian.]

Osycknyuk, A. Z., L. Romasenko, J. Banaszak, and T. Cierzniak. 2005. Andreninae of the central and eastern Palearctic, Part 1, pp. 1-235. Polish Entomoligical Society, Poznań, Bydgoszez, Poland.

Otero, J. T. 1996. Biologia de *Euglossa nigropilosa* Moure (Apidae: Euglossinae) I: Caracteristicas de nidificacion en la Reserva Natural La Planada. *Boletin del Museo de Entomología, Universidad del Valle* [Cali, Colombia] 4: 1-19.

Otis, G. W. 1997. Distributions of recently recognized species of honey bees (Hymenoptera: Apidae: *Apis*) in Asia. *Journal of the Kansas Entomological Society* 68(1996, suppl.): 311-333.

O'Toole, C. 1993. Diversity of native bees and agroecosystems, pp. 169-196 *in* J. LaSalle and D. Gauld, eds., *Hymenoptera and Biodiversity.* Wallingford, U.K.: CAB International.

O'Toole, C., and A. Raw. 1991. *Bees of the World.* 192 pp. London: Blandford.

Özbek, H. 1983. *Doğu Anadolu'nun Bazi Yörelerindeki Bombinae (Hymenoptera: Apoidea, Bombidae) Türleri Üzerinde Taksonomik ve Bazi Biyolojik Çalişmalar [Taxonomic and some biological studies on Bombinae (Hymenoptera: Apoidea, Bombidae) from some parts of eastern Anatolia].* [iv] + 70 pp. Erzurum: Ataturk Üniversitesi, Ziraat Fakültesi Yayinlari No. 287. [In Turkish.]

———. 1997. Bumblebees fauna of Turkey with distribution maps (Hymenoptera: Apidae: Bombinae) Part 1: *Alpigenobombus* Skorikov, *Bombias* Roberstson and *Bombus* Latreille. *Turkish Journal of Entomology* 21: 37-56.

———. 1998. On the bumblebee fauna of Turkey: II. The genus *Pyrobombus, Zoology in the Middle East* 16: 89-106.

———. 2000. On the bumblebee fauna of Turkey: III. The subgenus *Thoracobombus* D. T. *Journal of the Entomological Research Society* 2(2): 43-61.

———. 2002. On the bumblebee fauna of Turkey: IV. The subgenera *Megabombus, Eversmannibombus, Laesobombus, Rhodobombus,* and *Subterraneobombus. Zoology in the Middle East* 25: 79-98.

Pachinger, B. 2004. Über das Vorkommen der Steinbienen *Lithurgas* Latr. (Hymenoptera: Apoidea, Megachilidae) in Österreich B Ökologie, Verbreitung und Gefährdung. *Linzer Biologische Beiträge* 36: 559-566.

Packer, L. 1990. Solitary and eusocial nests in a population of *Augochlorella striata* (Provancher) (Hymenoptera; Halictidae) at the northern edge of its range. *Behavioral Ecology and Sociobiology* 27: 339-344.

———. 1991. The evolution of social behavior and nest architecture in sweat bees of the subgenus *Evylaeus* (Hymenoptera: Halictidae): A phylogenetic approach. *Behavioral Ecology and Sociobiology* 29: 153-160.

———. 1992. The social organisation of *Lasioglossum (Dialictus) laevissimum* (Smith) in southern Alberta. *Canadian Journal of Zoology* 70: 1767-1774.

———. 1993a. Two distinctive new species of halictine bees from high altitude in the new world tropics. *Canadian Journal of Zoology* 71: 1653-1662.

———. 1993b. Multiple-foundress associations in sweat

bees, pp. 215-233 *in* L. Keller, ed., *Queen Number and Sociality in Insects*. Oxford: Oxford University Press.

———. 1997. The relevance of phylogenetic systematics to biology: Examples from medicine and behavioral biology. *Mémoires du Muséum National d'Histoire Naturelle* [Paris] 173: 11-29.

———. 1998. A phylogenetic analysis of western European species of the *Lasioglossum leucozonium* species-group (Hymenoptera: Halictidae): sociobiological and taxonomic implications. *Canadian Journal of Zoology* 76: 1611-1621.

———. 2003. Comparative morphology of the skeletal parts of the sting apparatus of bees. *Zoological Journal of the Linnean Society* 138: 1-38.

———. 2004a. Morphological variation in the gastral sterna of female Apoidea. *Canadian Journal of Zoology* 82: 130-152.

———. 2004b. Taxonomic and behavioral notes on Patagonian Xeromelissinae with the description of a new species. *Journal of the Kansas Entomological Society* 77: 805-820.

Packer, L., A. Dzinas, K. Strickler, and V. Scott. 1995. Genetic differentiation between two host "races" and two species of cleptoparasitic bees and between their two hosts. *Biochemical Genetics* 33: 97-109.

Pagliano, G., and P. Scaramozzino. 1990. Elenco dei generi di Hymenoptera del mondo. *Memoria della Società Entomologica Italiana* 68(1989): 1-210.

Pallas, P. S. 1772. *Spicilegia Zoologica* vol. 1, fasc 9: 1-36.

Pamilo, P., A. Pekkarinen, and S.-L. Varvio. 1987. Clustering of bumblebee subgenera based on interspecific genetic relationships. *Annales Zoologici Fennici* 24: 19-27.

Panzer, G. W. F. 1804a. *Faunae Insectorum Germanicae*, heft 85. Nürnberg: Felssecker. [The actual date of publication may have been as early as 1801; see C. D. Sherborn, 1923, *Annals and Magazine of Natural History* (9)11: 567 and Sandhouse, 1943: 532.]

———. 1804b. *Faunae Insectorum Germanicae*, heft 86. Nürnberg: Felssecker. [According to Commission Opinion 151 (1944), the correct date for this heft is [1801-1802], but Sherborn (ref. *supra*) gives the date as 1804.]

———. 1806. *Kritische Revision der Insektenfauna Deutschlands*, Vol. 2. [14] + 271 pp., 2 pls. Nürnberg: Felssecker.

———. 1809. *Faunae Insectorum Germanicae*, heft 107. Nürnberg: Felssecker. [The actual date of publication may have been as early as 1806; see Panzer, 1804a.]

Parker, F. D. 1975a. Nests of the mason bees *Osmia tanneri* Sandhouse and *O. longula* Cresson with a description of the female of *O. tanneri*. *Pan-Pacific Entomologist* 51: 179-183.

———. 1975b. Nest descriptions and associates of three American bees of the genus *"Anthocopa"* Lepeletier. *Pan-Pacific Entomologist* 51: 113-122.

———. 1976. A new *Proteriades* reared from trap stems, its biology and nest associates. *Pan-Pacific Entomologist* 52: 73-80.

———. 1977a. Nests of *Anthocopa enceliae* (Cockerell) and *A. elongata* (Michener). *Pan-Pacific Entomologist* 53: 47-52.

———. 1977b. Biological notes on some Mexican bees. *Pan-Pacific Entomologist* 53: 189-192.

———. 1977c. A new *Proteriades (Xerosmia)* from New Mexico with biological notes. *Journal of the Kansas Entomological Society* 50: 317-323.

———. 1978a. Biology of the bee genus *Proteriades* Titus. *Journal of the Kansas Entomological Society* 51: 145-173.

———. 1978b. An illustrated key to alfalfa leafcutter bees *Eutricharaea*. *Pan-Pacific Entomologist* 54: 61-64.

———. 1979. A new *Proteriades* with distributional notes and a key to its subgenus *(Hoplitina)*. *Pan-Pacific Entomologist* 55: 51-56.

———. 1984. The nesting biology of *Osmis (Trichinosmia) latisulcata* Michener. *Journal of the Kansas Entomological Society* 57: 430-436.

———. 1986. Nesting, associates, and mortality of *Osmia sanrafaelae* Parker. *Journal of the Kansas Entomological Society* 59: 367-377.

———. 1987. Nests of *Callanthidium* from block traps. *Pan-Pacific Entomologist* 63: 125-129.

———. 1988. Nesting biology of two North American species of *Chelostoma*. *Pan-Pacific Entomologist* 64: 1-7.

Parker, F. D., S. W. T. Batra, and V. J. Tepedino. 1987. New pollinators for our crops. *Agricultural Zoology Reviews* 2: 279-307.

Parker, F. D., and G. E. Bohart. 1979. *Dolichostelis*, a new genus of parasitic bees. *Journal of the Kansas Entomological Society* 52: 138-153.

Parker, F. D., J. H..Cane, G. W. Frankie, and S. B. Vinson. 1987. Host records and nest entry by *Dolichostelis*, a kleptoparasitic anthidiine bee. *Pan-Pacific Entomologist* 63: 172-177.

Pasteels, J. J. 1965. Révision des Megachilidae (Hymenoptera Apoidea) de l'Afrique Noire, 1. Les Genres *Creightoniella* [sic], *Chalicodoma* et *Megachile* (s. str.). *Annales Musée Royal de l'Afrique Central* [Tervuren], *Sciences Zoologiques* (IN-8%)137: ix + 579 pp.

———. 1966. Megachilidae (genres: *Creightoniella* [sic], *Megachile* et *Chalicodoma*) peu connues ou nouvelles des régions paléarctique et africaine. *Bulletin et Annales de la Société Royale d'Entomologie de Belgique* 102: 1-19.

———. 1968a. Statut, affinités et origines des Anthidiinae parasites. *Naturaliste Canadien* 95: 1055-1063.

———. 1968b. Révision des Megachilidae (Hymenoptera Apoidea) de l'Afrique Noire, II. Le genre *Coelioxys*. *Annales Musée Royal de l'Afrique Central* [Tervuren], *Sciences Zoologiques* (IN-8%)167: [iii] + 139 pp.

———. 1969a. La systématique générique et subgénérique des Anthidiinae (Hymenoptera, Apoidea, Megachilidae) de l'ancien monde. *Mémoires de la Société Royale d'Entomologie de Belgique* 31: 1-148.

———. 1969b. New Anthidiinae (Hymenoptera, Apoidea, Megachilidae) from the Mediterranean area and from the Near East. *Israel Journal of Entomology* 4: 409-434.

———. 1969c. *Perezia maura* Ferton: Une forme intersexuée de *Megachile centuncularis* L. *Bulletin de la Société Entomologique de France* 74: 248.

———. 1970. Espèces nouvelles de Megachilidae (Hymenoptera, Apoidea) d'Afrique Noire, I. Genres *Creightonella* Cockerell et *Chalicodoma* Lepeletier. *Revue de Zoologie et de Botanique Africaines* 82: 205-232.

———. 1972. Révision des Anthidiinae (Hymenoptera Apoidea) de la région indo-malaise. *Bulletin et Annales de la Société Royale Belge d'Entomologie* 108: 72-128.

———. 1977a. Une revue comparative de l'éthologie des Anthidiinae nidificateurs de l'ancien monde. *Annales de la Société Entomologique de France* (n.s.)13: 651-667.

———. 1977b. Les Megachilini parasites (*Coelioxys* s. l.) d'Afrique noire. *Revue de Zoologie Africaine* 91: 161-197.

———. 1980. Révision du genre *Euaspis* Gerstaecker. *Bulletin et Annales de la Société Royale Belge d'Entomologie* 116: 73-89.

———. 1982. Les *Coelioxys* Latreille (Hymenoptera Apoidea) du groupe *afra* Lepeletier d'Europe et du proche-orient. *Bulletin et Annales de la Société Royale Belge d'Entomologie* 118: 109-120.

———. 1984. Révision des Anthidiinae (Hymenoptera, Apoidea, Megachilidae) de l'Afrique subsaharienne. *Mémoires de la Classe des Sciences, Académie Royale de Belgique* IN-4-2 19(1): 1-165.

Pasteels, J. M., J. J. Pasteels, and L. de Vos. 1983. Étude au microscope électronique à balayage des scopas collectrices de pollen chez les Panurginae. *Archives de Biologie* [Bruxelles] 94: 53-73.

Patiny, S. 1997. Notes préliminaires à la révision des sous-genre *Melanapis* Cameron, 1902 et du statut d'*Andrena fuscosa* Erichson, 1835. *Entomofauna* 18: 529-548.

———. 1998. Description d'un sous-genre nouveau de *Meliturga* Latreille, 1809. *Bembix* 10: 29-33.

———. 1999a. Systématique générique et subgénérique des *Melitturga* Latreille-*Meliturgula* Friese-*Flavomeliturgula* Warncke (sic). *Bulletin de la Société Entomologique de France* 104: 241-256.

———. 1999b (July 30). Etude phylogénétique des Panurginae de l'Ancien Monde. *Linzer Biologische Beiträge* 31: 249-275.

———. 1999c. Révision des Panurginae ouest-paléarctiques n'appartenant pas à la tribu des Melitturgini Michener, 1944, Partie I: *Panurgus* Panzer, 1806 et *Camptopoeum* Spinola, 1843. *Entomofauna* 20: 309-328.

———. 1999d. Descriptions of two new panurgine species from the Palearctic. *Bembix* 12: 28-31.

———. 2000. Description d'un genre nouveau de Panurginae: *Borgatomelissa* g. nov. *Notes Fauniques de Gembloux* no. 41: 101-104.

———. 2001. A new panurgine genus from Iran: *Gasparinahla* g. nov. described on base (sic) of a new species: *Gasparinahla megapalpae* sp. nov. *Linzer Biologische Beiträg* 33: 309-313.

———. 2002a. Nouvellas espècies de Panurginae (Hymenoptera, Apoidea, Andrenidae) du sud de l'ouest-paléarctique. *Notes Fauniques de Gembloux* no. 47: 41-46.

———. 2002b. A new *Clavipanurgus* (Andrenidae, Panurginae) species from Georgia, *Clavipanurgus eurystylus* sp. nov. *Linzer Biologische Beiträge* 34: 1263-1266.

———. 2003a. Revision of the subgenus *Dufourea* (*Flavodufourea*) Ebmer, 1984 (Hymenoptera, Halictidae, Rophitinae) and description of a new species *D*. (*Flavodufourea*) *ulkenkalkana* sp. nov. from Kazakhstan. *Zootaxa* no. 255: 1-8.

———. 2003b. Contemporary distributions of *Panurginus* species and subspecies in Europe. *Proceedings of the 13th International Colloquium, European Invertebrate Survey* [Leiden] 2001: 115-121.

———. 2003c. Phylogénie des espèces de *Clavipanurgus* Warncke, 1972. *Annales de la Société Entomolgique de France* (n.s.) 39: 229-234.

———. 2004a. Two new panurgine bee (Hymenoptera: Andrenidae) species from the Near-and Middle East. *Zootaxa* no. 715: 1-7.

———. 2004b. Descriptions of males of two recently described South African Panurginae (Hymenoptera: Andrenidae), with updated keys to the African species of *Melitturga* and *Meliturgula*. *Zootaxa* no. 669: 1-12.

Patiny, S., and D. Michez. 2006. Phylogenetic analysis of the *Systropha* Illiger 1806 (Hymenoptera: Apoidea: Halictidae) and description of a new subgenus. *Annales de la Société Entomologique de France* ns 42: 27-44.

Patiny, S., and P. Rasmont. 1999. Description d'un nouveau sous-genre de *Plesiopanurgus* Cameron, 1907. *Notes Fauniques de Gembloux* no. 37: 77-80.

Patton, W. H. 1879a. Generic arrangement of the bees allied to *Melissodes* and *Anthophora*. *Bulletin of the United States Geological and Geographical Survey of the Territories* 5: 471-479.

———. 1879b. List of a collection of aculeate Hymenoptera made by Mr. S. W. Williston in northwestern Kansas. *Bulletin of the United States Geological and Geographical Survey of the Territories* 5: 349-370.

Pauly, A. 1980a. Descriptions préliminaires de quelque sous-genres afrotropicaux nouveaux dans la famille des Halictidae. *Revue de Zoologie Africaine* 94: 119-125.

———. 1980b. Les espèces Indonesiennes du genre *Homalictus* Cockerell. *Zoologische Mededelingen* 55(2): 13-28.

———. 1980c. Les espèces métalliques afrotropicales du sous-genre *Ctenonomia* du genre *Lasioglossum*. *Revue de Zoologie Africaine* 94: 1-10.

———. 1981a. *Lasioglossum* (*Labrohalictus*) *saegeri*, nouveau sous-genre et nouvelle espèce de Halictidae du Parc National de la Garamba (Zaïre). *Revue de Zoologie Africaine* 95: 717-720.

———. 1981b. Note sur les abeilles cleptoparasites du genre *Eupetersia* avec une révision du sous-genre *Calleupetersia*. *Bulletin et Annales de la Société Royale Belge d'Entomologie* 117: 263-274.

———. 1984a. *Glossodialictus wittei*, nouveau genre et nouvelle espèce de Halictides récolté au Parc National de l'Upemba (Zaïre). *Revue de Zoologie Africaine* 98: 703-708.

———. 1984b. Classification des Halictidae de Madagascar et des îles voisines, I. Halictinae. *Verhandlungen der Naturforschenden Gesellschaft in Basel* 94: 121-156.

———. 1984c. *Paradialictus*, un nouveau genre cleptoparasite récolté au Parc National des Virungas (Zaïre). *Revue de Zoologie Africaine* 98: 689-692.

———. 1984d. Contribution à l'étude des genres afrotropicaux de Nomiinae. *Revue de Zoologie Africaine* 98: 693-702.

———. 1986. Les abeilles de la sous-famille des Halictinae en Nouvelle-Guinée et dans l'archipel Bismarck. *Zoologische Verhandelingen* no. 227: 1-58.

———. 1989. Les espèces afrotropicales de *Pachyhalictus* Cockerell du sous-genre *Dictyohalictus* Michener. *Revue de Zoologie Africaine* 103: 41-49.

———. 1990. Classification des Nomiinae Africains. *Annales Musée Royal de l'Afrique Central* [Tervuren], *Sciences Zoologiques* 261: 1-206.

———. 1991. Classification des Halictidae de Madagascar,

II. Nomiinae. *Annales de la Société Entomologique de France* (n.s.)27: 287-321.

———. 1997a. *Paraseladonia,* noveau genre cleptoparasite afrotropical. *Bulletin et Annales de la Société Royale Belge d'Entomologie* 133: 91-99.

———. 1997b. *Pronomia,* un nouveau genre de Nomiinae de Madagascar. *Bulletin et Annales de la Société Royale Belge d'Entomologie* 133: 101-112.

———. 1999a. Classification des Nomiinae africains, Le genre *Trinomia* Pauly. *Belgian Journal of Entomology* 1: 101-136.

———. 1999b. Classification des Halictini de la région Afrotropicale. *Bulletin de l'Institute Royal des Sciences Naturelles de Belgique, Entomologie* 69: 137-196.

———. 2000. Classification des Nomiinae africains: Le genre *Leuconomia* Pauly, 1980. *Bulletin de l'Institut Royal des Sciences Naturelles de Belgique, Entomologie* 70: 165-188.

———. 2001. *Ipomalictus* Pauly, 1999, sous-génre de *Lasioglossum,* nouveau pour la Région Orientale, avec des notes sur quelques espèces afrotropicales. *Bulletin de l'Institute Royale des Sciences Naturelles de Beligique, Entomologie* 71: 145-154.

———. 2003. Classification des Nomiinae africains: le genre *Nubenomia* Pauly. *Annales du Musée Royal de l'Afrique Central (Sciences Zoologiques)* 291: 99-116.

Pauly, A., R. W. Brooks, L. A. Nilsson, Y. A. Pesenko, C. D. Eardley, M. Terzo, T. Griswold, M. Schwarz, S. Patiny, J. Munzinger, and Y. Barbier. 2001. Hymenoptera Apoidea de Madagascar et los iles voisines. *Annales Sciences Zoologiques, Musée Royal d l'Afrique Centrale* [Tervuren], 286: 1-390, pls. 1-16.

Pawlikowski, T. 1996. Apinae *in* Klucze do Oznaczania Owadów Polski. *Polskie Towarzystwo Entomologiczne, Toruń* no. 148: 1-56.

———. 1999. *A field guide to the identification of bumblebees in Poland,* pp. 1-30. Toruń: Wydawnictwo Universytetu Mikolaja Kopernika.

Paxton, R. J., and T. Tengö. 1996. Intranidal mating, emergence, and sex ratio in a communal bee *Andrena jacobi* Perkins 1921. *Journal of Insect Behavior* 9: 421-440.

Paxton, R. J., P. A. Thorén, N. Gyllenstrand, and J. Tengö. 2000. Microsatellite DNA analysis reveals low diploid male productionn in communal bee with inbreeding. *Biological Journal of the Linnean Society* [London] 69: 483-502.

Pedersen, B. V. 2002. European bumblebees (Hymenoptera: Bombini)—phylogenetic relationships inferred from DNA sequences. *Insect Systematics and Evolution* 33: 361-386.

Pedro, S. R. de M. 1996. Lista preliminar das espécies de abelhas (Hymenoptera, Apoidea) que ocorrem na região de Ribeirão Preto e Cajuru, S.P., pp. 248-258 *in* C. A. Garófalo et al., eds., *Anais do II Encontro sobre Abelhas.* Ribeirão Preto, Brazil: Faculdade de Filosofia, Ciências e Letras.

Pedro, S. R. de M., and J. M. F. Camargo. 1999. Apoidea Apiformes, pp. 195-211 *in* C. R. F. Brandão and E. M. Cancello, eds, Vol. 5, Invertebrados Terrestres, *in* C. A. Joly and C. E. de M. Bicudo, eds., *Biodiversidade do Estado de São Paulo, Brasil.* São Paulo: FAPESP.

———. 2003. Meliponini neotropicais: o gênero *Partamona* Schwarz, 1939. *Revista Brasileira de Entomologia* 47(suppl): 1-117.

Pereira, M., and C. A. Garófalo. 1996. Aprovisionamento de céllulas por *Centris (Hemisiella) vittata* Lepeletier, p. 329 *in* C. A. Garófalo et al., eds., *Anais do II Encontro sobre Abelhas.* Ribeirão Preto, Brazil: Faculdade de Filosofia, Ciências e Letras.

Pérez, J. 1890. Catalogue des mellifères du sud-ouest. *Actes de la Société Linnéenne de Bordeaux* 44: 133-192.

———. 1897a. Quelques espèces de mégachiles nouvelles ou mal connues. *In Comptes Rendus des Séances, Actes de la Société Linnéenne de Bordeaux* 52: lviii-lxvii.

———. 1897b. Sur une forme nouvelle de l'appareil buccal des hyménoptères. *Comptes Rendus de l'Académie des Sciences* [Paris] 125: 259-260.

———. 1901. Contribution à l'étude des xylocopes. *Actes de la Société Linnéenne de Bordeaux* (6)6(= Vol. 56): 1-128.

Perkins, R. C. L. 1899. Hymenoptera Aculeata [except ants], pp. 1-115, pls. I, II, *in* R. C. L. Perkins and A. Forel, Hymenoptera Aculeata, *in* D. Sharp, ed., *Fauna Hawaiiensis,* Vol. 1, pt. 1. London: Cambridge University Press.

———. 1908. Some remarkable Australian Hymenoptera. *Proceedings of the Hawaiian Entomological Society* 2: 27-35.

———. 1910. Hymenoptera (Supplement), pp. 600-686 *in* D. Sharp, ed., *Fauna Hawaiiensis,* Vol. 2, pt. 12. London: Cambridge University Press.

———. 1912. Notes, with descriptions of new species, on aculeate Hymenoptera of the Australian Region. *Annals and Magazine of Natural History* (8)9: 96-121.

———. 1913. Introduction, pp. xv-lxxviii, pls. i-xvi, *in* D. Sharp, ed., *Fauna Hawaiiensis,* Vol. 1, pt. vi. London: Cambridge University Press.

———. 1919. The British species of *Andrena* and *Nomada. Transactions of the Entomological Society of London* 1919: 218-319, pls. XI-XV.

———. 1922. The British species of *Halictus* and *Sphecodes. Entomologist's Monthly Magazine* 58: 46-52.

———. 1925. The British species of *Megachile* . . . . *Entomologist's Monthly Magazine* 61: 95-101.

Perkins, R. C. L., and L. E. Cheesman. 1928. *Insects of Samoa and Other Samoan Terrestrial Arthropoda,* Part V. Hymenoptera, fasc. 1, Apoidea, Sphecoidea, and Vespoidea, pp. 1-32. London: British Museum (Natural History).

Perty, J. A. M. 1833. *Delectus Animalium Articulatorum . . . Brasiliam,* fasc. 3, pp. 1-224, pls. 1-40. Monachii. [Hymenoptera on pp. 129-151, pls. 26-28.]

Peruquetti, R. C. 2000. Function of frangrances collected by Euglossini males. *Entomologia Generalis* 25: 33-37.

Pesenko, Y. A. 1983. Tribe Nomioidini (in the palearctic fauna), pp. 1-198 *in Fauna of the USSR, Insecta-Hymenoptera, Halictid Bees (Halictidae), subfamily Halictinae,* 17(1). Leningrad: Nauka. [In Russian.]

———. 1984a. A subgeneric classification of bees of the genus *Halictus* Latreille sensu stricto. *Entomologicheskoe Obozrenie* 63: 340-357. [In Russian, English summary; English translation in *Entomological Review* 63(3): 1-20.]

———. 1984b. Synonymic annotated catalog of species names of the bees of the genus *Halictus* Latreille sensu stricto (Hymenoptera, Halictidae) in the world fauna, pp. 16-32 *in* Y. A. Pesenko, ed., *Systematics and Ecology of*

Bees, *Trudy Zoologicheskova Instituta, Akademii Nauk SSSR* 128. [In Russian].

———. 1984c. Systematics of the bees of the genus *Halictus* Latreille (Hymenoptera, Halictidae) with a description of the 7th and 8th metasomal sterna of males: Subgenus *Platyhalictus,* pp. 33-48 *in* Y. A. Pesenko, ed., *Systematics and Ecology of Bees, Trudy Zoologicheskova Instituta, Akademii Nauk SSSR* 128. [In Russian].

———. 1984d. The bees of the genus *Halictus* Latreille sensu stricto (Hymenoptera, Halictidae) of Mongolia and north-western China, with a review of publications on Halictini of this region and a revision of the subgenus *Prohalictus* of the world fauna, pp. 446-481 *in* B. A. Korotyaev, ed., *Insects of Mongolia,* no. 9. Leningrad: Nauka. [In Russian.]

———. 1985. Systematics of the bees of the genus *Halictus* Latreille (Hymenoptera, Halictidae) with a description of the 7th and 8th metasomal sterna of males: Subgenus *Monilapis* Cockerell, pp. 77-105 *in* Y. A. Pesenko, ed., *News of Insect Systematics: Hymenoptera, Trudy Zoologicheskova Instituta, Akademii Nauk* SSSR 132. [In Russian.]

———. 1986a. An annotated key to females of the palaearctic species of the genus *Lasioglossum* sensu stricto (Hymenoptera, Halictidae), with descriptions of new subgenera and species, pp. 113-151 *in* Y. A. Pesenko, ed., *Systematics of Hymenopterous Insects, Trudy Zoologicheskova Instituta, Akademii Nauk* SSSR 159. [In Russian.]

———. 1986b. Systematics of the bees of the genus *Halictus* Latreille (Hymenoptera, Halictidae) with description of the 7th and 8th metasomal sterna of males: Subgenus *Tytthalictus* Pesenko. *Entomologicheskoe Obozrenie* 65: 618-632. [In Russian.]

———. 1993. A new halictine bee of the tribe Nomioidini from Madagascar. *Journal of the Kansas Entomological Society* 66: 1-5.

———. 1996. Madagascan bees of the tribe Nomioidini. *Entomofauna* 17: 493-516.

———. 1998. New and little known bees of the genus *Dufourea* Lepeletier (Hymenoptera, Halictidae) from the Palaearctic. *Entomologicheskoe Obozrenie* 77: 670-686. [in Russian; English translation (2000) in *Entomological Review* 78: 598-612.]

———. 2000a. Phylogeny and classification of the family Halictidae revised. *Journal of the Kansas Entomological Society* 72(1999): 104-123.

———. 2000b. Phylogeny and classification of bees of the tribe Nomioidini. *Entomologicheskoe Obozrenie* 79: 210-226 [in Russian; English translation (2000) in *Entomological Review* 80: 171-184.]

———. 2004a. New data on the taxonomy and distribution of bees of the subfamily Nomioidinae (Hymenoptera: Halictidae) of Middle Asia and Kazakhstan. *Proceedings of the Russian Entomological Society* 75: 283-295.

———. 2004b. The phylogeny and classification of the tribe *Halictini,* with special reference to the *Halictus* genus-group. *Zoosystematica Rossica* 13: 83-113.

———. 2005. New data on the taxonomy and distribution of the palaearctic halicids: genus *Halictus. Entomofauna* 26: 313-348.

Pesenko, Y. A., and Y. V. Astafurova. 2003. *Annotated Bibliography of Russian and Soviet Publications on the Bees (Hymenoptera: Apoidea; excluding Apis mellifera):* 1771-2002, 616 pp. Linz, Austria: Denisia 11.

Pesenko, Y. A., J. Banaszak, V. G. Radchenko, and C. Ciesznniak. 2000. *Bees of the family Halictidae (excluding Sphecodes) of Poland; taxonomy, ecology, bionomics,* ix + 348 pp. Bydgoszcz: Pedagogical University.

Pesenko, Y. A., and I. M. Kerzhner. 1981. *Nomioides* Schenck, 1866: Proposed designation of type species. *Bulletin of Zoological Nomenclature* 38: 225-227.

Pesenko, Y. A., and A. Pauly. 2005. Monograph of the bees of the subfamily Nomioidinae (Hymenoptera: Halictidae) of Africa (excluding Madagascar). *Annales de la Société Entomologique de France* 41 (2): 129-236.

Pesenko, Y. A., and A. A. Sitdikov. 1988. Classification and phylogenetic relations of the genera of the tribe Eucerini (Hymenoptera, Anthophoridae) with two submarginal cells. *Entomologicheskoe Obozrenie* 67: 846-860. [In Russian; English translation in *Entomological Review* 69: 88-104 (1990).]

Pesenko, Y. A., and Wu Yan-ru. 1997. Chinese bees of the genus *Pachyhalictus. Zoosystematica Rossica* 6: 287-296.

Pesotskaya, E. A. 1929. The role of the glandular apparatus in the instinctive activity of bees. *Trudy Leningradskogo Obshchestva Estestvoispytatelei* 59(2): 21-44. [In Russian.]

Petanidou, T., W. N. Ellis, and A. C. Ellis-Adam. 1995. Ecogeographical patterns in the incidence of brood parasitism in bees. *Biological Journal of the Linnean Society* [London] 55: 261-272.

Peters, D. S. 1969. Phänokopierende Missbildungen des Flügelgeäders bei akuleaten Hymenopteren. *Deutsche Entomologische Zeitschrift* (n.f.) 16: 367-374.

———. 1970a. *Pseudoheriades* n. gen., *Afroheriades* n. subgen., *Pseudoheriades primus* n. sp., neue Formen aus der Familie Megachilidae. *Entomologische Zeitschrift* 80(16): 153-160.

———. 1970b. Die Stellung von *Heriadopsis* Cockerell 1931. *Senckenbergiana Biologica* 51: 199-203.

———. 1972a. Über die Stellung von *Aspidosmia* Brauns 1926 nebst allgemeinen Erörterungen der phylogenetischen Systematik der Megachilidae. *Apidologie* 3: 167-186.

———. 1972b. Drei neue Arten der Megachilidae aus Mexico. *Senckenbergiana Biologica* 53: 373-382.

———. 1974. Über die Untergattung *Haetosmia* Popov 1952. *Senckenbergiana Biologica* 55: 293-309.

———. 1978a. *Archeriades* gen n., eine verhältnismässig ursprüngliche Gattung der Megachilidae. *Entomologica Germanica* 4: 337-343.

———. 1978b. Systematik und Zoogeographie der westpaläarktischen Arten von *Osmia* Panzer 1806 s. str., *Monosmia* Tkalců 1974 und *Orientosmia* n. subgen. *Senckenbergiana Biologica* 58: 287-346.

———. 1983. Einige von H. Priesner am Gebel Elba gesammelte Megachilidae sowie eine neue Art der Gattung *Heriades* aus Sudafrika. *Linzer Biologische Beiträge* 14: 95-109.

———. 1984. Revision der von H. Brauns beschriebenen oder behandelten afrikanischen *Heriades-* und *Osmia-*Arten. *Entomofauna* 5: 359-387.

Philippi, R. A. 1866. Einige Insekten von Chile. *Stettiner Entomologische Zeitung* 27: 109-117, Taf. II.

Piel, O. 1930. Nidification de *Megachile monticola* Smith, chez *Xylocopa rufipes* Sm. [♀], *pictifrons* Sm. [♂]. *Notes d'Entomologie Chinoise, Musée Heude,* fasc. 5: 1-8, 2 pls.

Piton, L.-E. 1940. *Paléontologie du Gisement Éocène de Menat (Puy-de-Dôme) (Flore et Faune).* vii + 303 pp. Clermont-Ferrand: Vallier; Paris: Lechevalier.

Pittioni, B. 1937. Die Hummelfauna des Kalsbachtales in Ost-Tirol, pp. 64-127 *in Festschrift für Prof. Dr. Embrik Strand,* Vol. 3. Riga, Latvia.

———. 1938, 1939a. Die Hummeln und Schmarotzerhummeln der Balkan-Halbinsel. *Mitteilungen aus den Koniglichen Naturwissenschaftlichen Instituten in Sofia* 11: 12-69, 12: 49-115, pls. I-V.

———. 1939b. *Tanguticobombus* subg. nov. (Hymenopt., Apidae). *Zoologischer Anzeiger* 126: 201-205.

———. 1945. Beiträge zur Kenntnis paläarktischer Apiden, I. Die Gruppe des *Epeolus tarsalis* Mor. *Zeitschrift der Wiener Entomologischen Gesellschaft* 30: 128-147, pl. 6.

———. 1948a. Beiträge zur Kenntnis Italienischer Bienen, I. Über einige Arten der Gattung *Andrena* Fabr. *Bollettino dell'Istituto de Entomologia della Università di Bologna* 17: 46-61.

———. 1948b. *Andrena (Andrenella) enslinella* Stckht. und ihre Verwandten. *Annalen des Naturhistorischen Museums in Wien* 56: 130-149.

———. 1949. Beiträge zur Kenntnis der Bienenfauna SO-Chinas . . . . *Eos* 25: 241-284.

———. 1950. Die westpaläarktischen Arten der Gattung *Spatulariella* Pop. *Bollettino della Società Veneziana di Storia Naturale* 5: 76-113.

———. 1953. Die *Nomada*-Arten der Alten Welt: Bestimmungstabelle der Männchen. *Annalen des Naturhistorischen Museums in Wien* 59: 223-291.

Pittioni, B., and R. Schmidt. 1942. Die Bienen des südöftlichen Niederdonau, I. Apidae, Podaliriidae, Xylocopidae und Ceratinidae, Niederdonau. *Natur und Kultur,* heft 19: 1-69. pls. I-VII.

Plant, J. D., and H. F. Paulus. 1987. Comparative morphology of the postmentum of bees (Hymenoptera: Apoidea) with special remarks on the evolution of the lorum. *Zeitschrift für Zoologische Systematik und Evolutionsforschung* 25: 81-103.

Poinar, G., Jr. 1999. *Paleoeuglossa melissiflora* gen. n., sp. n. (Euglossinae, Apidae), fossil orchid bees in Dominican amber. *Journal of the Kansas Entomological Society* 71(1998): 29-34.

Ponomareva, A. A. 1967. Notes sur les espèces paléarctiques du genre *Systropha* Ill. *Polskie Pismo Entomologiczne* 37: 677-698. [In Russian, French summary.]

Popov, V. B. 1931a. Zur Kenntnis der paläarktischen Schmarotzerhummeln (*Psithyrus* Lep.). *Eos* 7: 131-209.

———. 1931b. To the knowledge of the genera *Pasites* Jurine and *Parammobatodes* gen. nov. *Ezhegodnik Zoologicheskovo Muzeya Akademii Nauk SSSR [Annuaire du Musée Zoologique de l'Académie des Sciences de l'URSS]* 32: 453-467. [In Russian, English summary; 1931 yearbook perhaps published in 1932.]

———. 1933a. Notes on the parasitic bees allied to the genus *Biastes* Panz. *Trudy Instituta Zoologii, Akademii Nauk SSSR [Traveaux de l'Institut Zoologique de l'Académie des Sciences de l'URSS]* 2: 51-75. [In Russian, English summary.]

———. 1933b. On the palearctic forms of the tribe Stelidini Roberts. *Trudy Zoologicheskogo Instituta, Akademiya Nauk SSSR Leningrad* 1: 375-414. [in Russian, English summary.]

———. 1935. Beiträge zur Bienenfauna von Tadjikstan. *Trudy Akademiya Nauk SSSR, Tadzhikskoi Bazi* no. 5: 351-408. [In Russian, German summary.]

———. 1936a. Toward knowledge of the genus *Dioxys* Lep. *Trudy Zoologicheskova Instituta, Akademii Nauk SSSR* 3: 3-32. [In Russian.]

———. 1936b. A new bee of the genus *Ctenoplectra* Sm. *Proceedings of the Royal Entomological Society of London* (B)5: 78-80.

———. 1938. Notiz über die Gattung *Chelynia* Prov. und einige Untergattungsgruppierungen der Gattung *Stelis* Panz. *Konowia* 17: 36-41.

———. 1939a. Subgeneric groupings of the genus *Prosopis* F. *Comptes Rendus (Doklady) de l'Académie des Sciences de l'URSS* (n.s.)25: 167-170.

———. 1939b. Family Fideliidae and morphological convergence among bees. *Comptes Rendus (Doklady) de l'Académie des Sciences de l'URSS* (n.s.)22: 640-643.

———. 1939c. Relationships of the genus *Osiris* F. Smith and its position in the system of bees. *Comptes Rendus (Doklady) de l'Académie des Sciences de l'URSS* (n.s.)25: 163-166.

———. 1940. A new genus of bees from Turkestan. *Trudy Zoologicheskova Instituta, Akademii Nauk SSSR* 6: 53-60. [In Russian.]

———. 1941a. Notes on *Dianthidium sibericum* (Eversm.) and a new species of *Stelis* Panz. *Entomologisk Tidskrift* 62: 222-224.

———. 1941b. Family Oxaeidae and processes of morphological reduction in bees. *Comptes Rendus (Doklady) de l'Académie des Sciences de l'URSS* (n.s.)30: 82-85.

———. 1946. Notes on the nomenclature of the bees. *Proceedings of the Royal Entomological Society of London* (B)15: 106-109.

———. 1947. Generic groupings of the subfamily Dioxynae [sic]. *Entomologicheskoe Obozrenie* 29: 84-92. [In Russian, English summary.]

———. 1949a. The tribe Pararhophitini (Hymenoptera, Anthophorinae), an early Tertiary element of the contemporary fauna of the desert of central Asia and Egypt. *Doklady Akademii Nauk SSSR* 66: 507-510. [In Russian.]

———. 1949b. Notes on the bee fauna of Tadjikstan. *Trudy Zoologicheskova Instituta, Akademii Nauk SSSR* 8: 688-699. [In Russian.]

———. 1949c. The subgenus *Plastandrena* Hedicke and its new species. *Entomologicheskoe Obozrenie* 30: 389-404. [In Russian.]

———. 1950a. Generic groupings of the mid-Asian bees of the subfamily Anthidiinae. *Doklady Akademii Nauk SSSR* 70: 315-318. [In Russian.]

———. 1950b. Concerning the genus *Amegilla* Friese. *Entomologicheskoe Obozrenie* 31: 257-261. [In Russian.]

———. 1951a. On species of *Poecilomelitta* Friese and *Melitturga* Friese (Hymenoptera, Apoidea) from Arabia. *Entomologicheskoe Obozrenie* 31: 577-586. [In Russian.]

———. 1951b. Geographic distribution and evolution of the bee genus *Clisodon* Patton. *Zoologicheskii Zhurnal* 30: 243-252. [In Russian.]

———. 1951c. The parasitic bee genus *Ammobates* Latr., I.

Tribes Ammobatini and Pasitini, their size and taxonomic position. *Trudy Instituta Zoologii, Akademii Nauk SSSR* 9: 895-949. [In Russian.]

———. 1952. Bee fauna (Hymenoptera, Apoidea) of southwestern Turkmenii and its landscape distribution. *Trudy Zoologicheskova Instituta, Akademii Nauk SSSR* 10: 61-117. [In Russian.]

———. 1953. The reduction of the sting apparatus of Dioxynae [sic], a parasitic subfamily of bees. *Trudy Zoologicheskova Instituta, Akademii Nauk SSSR* 13: 337-351. [In Russian.]

———. 1954. About the geographic diffusion and evolution of the bee subgenus *Erythrosmia*. *Zoologicheskii Zhurnal* 33: 443-451. [In Russian.]

———. 1955a. A new subgenus of bees of the genus *Heriades*. *Entomologicheskoe Obozrenie* 34: 280-286. [In Russian.]

———. 1955b. On the parasitic genus *Radoszkowskiana* (Hymenoptera, Megachilidae) and its origins. *Zoologicheskii Zhurnal* 34: 547-556. [In Russian.]

———. 1955c. Generic groupings of the palearctic Melectinae. *Trudy Zoologicheskova Instituta, Akademii Nauk SSSR* 21: 321-334. [In Russian.]

———. 1956. New and little-known bees from central Asia. *Entomologicheskoe Obozrenie* 35: 159-171. [In Russian.]

———. 1957a. On the genera *Morawitzella*, gen. nov., and *Trilia* Vach. *Entomologicheskoe Obozrenie* 36: 916-924. [In Russian.]

———. 1957b. New species and peculiarities of geographical distribution of the bees of the genus *Eremaphanta* Popov. *Zoologicheskii Zhurnal* 36: 1704-1715. [In Russian, English summary.]

———. 1958a. Peculiar features of correlated evolution of two genera of bees—*Macropis* and *Epeoloides* (Hymenoptera, Apoidea) and a plant genus *Lysimachia* (Primulaceae). *Entomologicheskoe Obozrenie* 37: 499-519. [In Russian.]

———. 1958b. On three subgenera of the genus *Andrena*. *Trudy Vsesoyuznovo Entomologicheskovo Obshchestva (Horae Societatis Entomologicae Unionis Soveticae)* 46: 109-161. [In Russian.]

———. 1958c. Zoogeographical peculiarities of central Asiatic species of the genus *Halictoides*. *Doklady Akademii Nauk Tadzhikskoe SSR* 1: 47-51. [In Russian.]

———. 1959a. New species of the genera *Dufourea* and *Halictoides* from eastern Asia. *Entomologicheskoe Obozrenie* 38: 225-237. [In Russian.]

———. 1959b. *Halictus longirostris* F. Mor. (Hymenoptera, Halictidae) as an oligolectic bee. *Zoologicheskii Zhurnal* 38: 426-431. [In Russian, English summary.]

———. 1960a. *Formicapis* Sladen—holarctic genus of bees. *Zoologicheskii Zhurnal* 39: 1044-1049. [In Russian.]

———. 1960b. New and little-known megachilid bees (Hymenoptera) from middle Asia. *Entomologicheskoe Obozrenie* 39: 424-437. [In Russian.]

———. 1960c. New and little-known species of bees (Hymenoptera, Apoidea) from Turkmenia. *Trudy Zoologicheskova Instituta, Akademii Nauk SSSR* 27: 247-263. [In Russian.]

———. 1960d. The new mountain forms of the subgenus *Liosmia* (Hymenoptera) from middle Asia. *Izvestiya Otdeleniya Sel'skochozyaistvennich i Biologicheskich Nauk* 2: 73-78. [In Russian.]

———. 1961. On the evolution of the bee-genera *Protosmia* Ducke and *Chelostomopsis* Cockerell. *Zoologicheskii Zhurnal* 40: 359-371. [In Russian, English summary.]

———. 1962a. New genera of bees from middle Asia. *Trudy Instituta Zoologii, Akademii Nauk SSSR* 30: 291-309. [In Russian.]

———. 1962b. Bees of the subgenus *Megalosmia*. *Trudy Instituta Zoologii, Akademii Nauk Kazakh SSR* 18: 155-162. [In Russian.]

———. 1963. A new genus of Megachilidae (Hymenoptera) from central Asia. *Zoologicheskii Zhurnal* 42: 865-869. [In Russian, English summary.]

———. 1964. On the bee genera *Trachusa* Panzer and *Trachusomimus*, gen. n. *Entomologicheskoe Obozrenie* 43: 405-417. [In Russian; English translation in *Entomological Review* 43: 207-214.]

———. 1967. The bees (Hymenoptera, Apoidea) of Iran. *Trudy Zoologicheskova Instituta, Akademii Nauk SSSR* 43: 184-216.

Popov, V. B., and D. Guiglia. 1936. Note sopra i gen. *Ctenoplectra* Sm. e *Macropis* Panz. *Annali del Museo Civico di Storia Naturale di Genova* 59: 275-288.

Popov, V. B., and A. A. Ponomareva. 1961. Notes on the carpenter bee fauna of the Soviet Union. *Entomologicheskoe Obozrenie* 40: 393-404.

Popova, L. M. 1990. Nesting habits of some species of anthophorid bees (Hymenoptera, Anthophoridae) in the middle Volga region. *Entomologicheskoe Obozrenie* 69: 23-35. [In Russian, English translation in *Entomological Review* 69: 81-93.]

Portugal-Araújo, V. 1958. A contribution to the bionomics of *Lestrimelitta cubiceps*. *Journal of the Kansas Entomological Society* 31: 203-211.

Prentice, M. 1991. Morphological analysis of the tribes of Apidae, pp. 51-69 in D. R. Smith, ed., *Diversity in the Genus* Apis. Boulder, Colo.: Westview Press.

Přidal, A. 2004. Checklist of the bees in the Czech Republic and Slovakia with comments on their distribution and taxonomy. *Acta Universitatis Agriculturae et Silviculturae Mendelianae Brunensis* 52: 29-65.

Přidal, A., and B. Tkalců. 2001. Commentary on the subgenera of the genus *Coelioxys* described by Roszkowski 1986. *Entomofauna* 22: 357-363.

Priesner, H. 1957. A review of the *Anthophora* species of Egypt. *Bulletin de la Société Entomologique d'Égypte* 41: 1-115.

Provancher, L. 1882. Faune Canadienne. Les Insectes Hyménoptères. *Naturaliste Canadien* 13: 225-242.

———. 1888, in 1885-1889. Additions et Corrections au Volume II de la Faune Entomologique du Canada Traitant des Hyménoptères, pp. 1-475. Quebec: Darveau. [The parts on bees were published in 1888.]

Prŷs-Jones, O. E., and S. A. Corbet. 1987. *Bumblebees*. [v] + 86 pp., 4 pls. Cambridge: Cambridge University Press.

Quilis P., M. 1927. Los ápidos de España. Genero *Bombus* Latr. *Trabajos del Laboratorio de Historia Natural de Valencia* 16: 1-119, 10 pls.

———. 1928. Estudio monografico de las *Dasypoda* Latr. *Eos* 4: 173-241, pls. III-V.

Radchenko, V. G. 1978. A new type of nest without cells in *Metallinella atrocaerulea*. *Entomologicheskoe Obozrenie* 57: 515-517. [In Russian; English translation in *Entomological Review* 57: 353-355.]

———. 1979. Nesting of *Nomioides minutissimus* (Rossi). *Entomologicheskoe Obozrenie* 58: 762-765. [In Russian; English translation in *Entomological Review* 58(4): 71-74.]

———. 1981. Nesting of four species of bees of the genus *Andrena*. *Entomologicheskoe Obozrenie* 60: 766-774. [In Russian.]

———. 1987. Nesting of *Dasypoda braccata* Eversm. (Hymenoptera, Melittidae) in the southwestern Ukraine. *Entomologicheskoe Obozrenie* 67: 299-301. [In Russian; English translation in *Entomological Review* 67: 57-60.]

———. 1993. *The Eusocial Colony of the Halictine Bees*. 60 pp. Kiev: Academiya Nauk Ukrania, Institut Zoologii. [In Ukrainian.]

———. 1995. Evolution of nest building in bees. *Entomologicheskoe Obozrenie* 74: 342-354. [In Russian; English translation in *Entomological Review* 75: 20-32, 1996.]

Radchenko, V. G., and Y. A. Pesenko. 1989. A key to the bees of the genus *Dasypoda* Latreille (Hymenoptera, Melittidae) of the European part of the USSR. *Trudy Zoologicheskova Instituta, Akademii Nauk SSSR* 188: 114-121. [In Russian.]

———. 1994a. *Biology of Bees*. 350 pp. St. Petersburg: Russian Academy of Sciences. [In Russian; English summary, pp. 314-331.]

———. 1994b. Protobee and its nests: A new hypothesis concerning the early evolution of Apoidea. *Entomologischeskoe Obozrenie* 73: 913-933. [In Russian; English translation in *Entomological Review* 75: 140-162, 1996.]

Radoschkowsky, see Radoszkowski. Radoskovskij, see Radoszkowski.

Radoszkowski, O. 1865. [Tribu des Mélictides.] *Horae Societatis Entomologicae Rossicae* 3: 53-60.

———. 1867. Matériaux pour servir à l'étude des insectes de la Russie, IV. Notes sur quelques Hyménoptères de la tribu des Apides. *Horae Societatis Entomologicae Rossicae* 5: 73-90, pl. III.

———. 1872a. [Résultats de ses études hyménoptèrologique, Megachilidae.] *In* Séance du 3 (15) Mai. *Horae Societatis Entomologicae Rossicae* 8: Bulletin pp. xvi-xviii.

———. 1872b. Supplément indispensable à l'article publié par M. Gerstaecker, en 1869, sur quelques genres d'hyménoptères. *Bulletin de la Société Imperiale des Naturalistes de Moscou* 45: 1-40, pl. 1.

———. 1874a. Supplément indispensable à l'article publié par M. Gerstaecker en 1869, sur quelques genres d'hyménoptères. *Bulletin de la Société Imperiale des Naturalistes de Moscou* 46(3)(1873): 133-151, 1 pl. [This issue was misnumbered vol. 47, which otherwise does not exist, according to D. B. Baker, in litt., 1992; Baker also determined that the actual year of publication was 1874.]

———. 1874b. Supplément indispensable à l'article publié par M. Gerstaecker en 1869, sur quelques genres d'hyménoptères. *Bulletin de la Société Imperiale des Naturalistes de Moscou* 48(1)(1873): 132-164.

———. 1884a. Quelques nouveaux hyménoptères d'Amérique. *Horae Societatis Entomologicae Rossicae* 18: 17-22, 1 pl.

———. 1884b. Révision des armures copulatrices des mâles du genre *Bombus*. *Bulletin de la Société Imperiale des Naturalistes de Moscou* 59: 51-92, pls. I-IV.

———. 1885. Révision des armures copulatrices des mâles de la tribu Philérémides. *Bulletin de la Société Imperiale des Naturalistes de Moscou* 61: 359-370, pls. II, III.

———. 1886. Faune Hyménoptérologique Transcaspienne. *Horae Societatis Entomologicae Rossicae* 20: 3-56, pls. 1-11.

———. 1887. Révision du genre *Dasypoda* Latr. *Horae Societatis Entomologicae Rossicae* 20: 179-194, pls. XIII-XV.

———. 1891. Révision des armures copulatrices des mâles des genres *Cilissa* et *Pseudocilissa*. *Horae Societatis Entomologicae Rossicae* 25: 236-243, pl. 2 (part).

———. 1893a. Faune Hyménoptérologique Transcaspienne. *Horae Societatis Entomologicae Rossicae* 27: 38-81.

———. 1893b. Revue des armures copulatrices des mâles des genre: *Crocisa* Jur., *Melecta* Lat., . . . , *Eulema* Lep., *Acanthopus* Klug. *Bulletin de la Société Imperiale des Naturalistes de Moscou* 68: 163-188, pls. IV-VII.

Rafinesque-Schmaltz, C. S. 1814. *Principes Fondamentaux de Somiologie* . . . . 52 pp. Palermo: Privately printed.

———. 1815. *Analyse de la Nature*. 224 pp. Palermo: Privately printed.

Ramírez, S., R. L. Dressler, and M. Ospina. 2002. Abejas euglosinas (Hymenoptera: Apidae) de la region neotropical: Listado de especies con notas sobre su biología. *Biota Colombiana* 3(1): 7-118.

Ramírez-Arriaga, E., J. I. Cuadriello-Aguilar, and E. Martínez Hernández. 1996. Nest structure and parasite of *Euglossa atroveneta* Dressler (Apidae: Bombinae: Euglossini) at Unión Juárez, Chiapas, México. *Journal of the Kansas Entomological Society* 69: 144-152.

Ramos, K. dos Santos, C. G. dos Santos, C. Schlindwein, and B. Blochtein. 2004. Tegumentar glands associated to foveae in the second metasomal terguna of *Panurgillus* Moure. *Revista Brasileira de Entomologia* 48: 163-167.

Rasmont, P. 1983. Catalogue commenté des bourdons de la région ouest-paléarctique. *Notes Fauniques de Gembloux* no. 7: 1-72.

Rasmont, P., and A. Adamski. 1995. Les bourdons de la Corse. *Notes Fauniques de Gembloux* no. 31: 1-87.

Rasmont, P., A. Ebmer, J. Banaszak, and G. van der Zanden. 1995. Hymenoptera Apoidea Gallica. *Bulletin de la Société Entomologique de France* 100(hors série): 1-98.

Rasmussen, C. 2003. Clave de identificación par las especies peruanus de *Bombus* Latreille, 1809 (Hymenoptera, Apidae), con notas sobre su biología y distribución. *Revista Peruana de Entomologia* 43: 31-45.

Rasnitsyn, A. P. 1980. Origin and evolution of the hymenopterous insects. *Trudy Paleontologicheskii Institut, Akademii Nauk SSSR* 174: 1-192. [In Russian.]

Rasnitsyn, A. P., and C. D. Michener. 1991. Miocene fossil bumblebee from the Soviet far east with comments on the chronology and distribution of fossil bees. *Annals of the Entomological Society of America* 84: 583-589.

Raven, P. H. 1963. Amphitropical relations in the flora of North and South America. *Quarterly Review of Biology* 29: 151-177.

Raven, P. H., and D. I. Axelrod. 1974. Angiosperm biogeography and past continental movements. *Annals of the Missouri Botanical Garden* 61: 539-673.

Raw, A. 1976. Seasonal changes in numbers and foraging activities of two Jamaican *Exomalopsis* species. *Biotropica* 8: 270-277.

———. 1977. The biology of two *Exomalopsis* species (Hymenoptera: Anthophoridae) with remarks on sociality in bees. *Revista de Biología Tropical* 25: 1-11.

———. 1984. The nesting biology of nine species of Jamaican bees. *Revista Brasileira de Entomologia* 28: 497-506.

———. 2002. New combinations and synonymies of leafcutter and mason bees of the Americas. *Zootaxa* 71: 1-43.

Rebêlo, J. M. M. 2001. *História Natural das Euglossíneas, as Abelhas das Orquídeas,* 152 pp. São Luis, Maranhão, Brazil: Litograf Editora.

Rayment, T. 1930a. *Microglossa* and *Melitribus,* new genera of Australian bees. *Proceedings of the Royal Society of Victoria* (n.s.)42: 211-220, pl. 21. [*Microglossa* is a pemphredonine wasp *(Spilomena),* not a bee.]

———. 1930b. New and remarkable bees. *Proceedings of the Royal Society of Victoria* (n.s.)43: 42-61.

———. 1931. Bees in the collections of the Western Australian Museum and the Agricultural Department, Perth. *Journal of the Royal Society of Western Australia* 17: 157-190.

———. 1935. *A Cluster of Bees.* 752 pp. Sydney: Endeavour.

———. 1944. A critical revision of species in the *zonata* group of *Anthophora* by new characters (Part I). *Treubia* (Japanese hors serie), pp. 1-30. [Although dated 1942, this paper was issued May 29, 1944, according to M. A. Lieftinck.]

———. 1947. A critical revision in [sic] the *zonata* group of *Anthophora* by new characters. *Treubia* 19: 46-73.

———. 1950. New bees and wasps—Part XIII. *Victorian Naturalist* 67: 20-25.

———. 1951. A critical revision of species in the genus *Asaropoda* by new characters. *Memoirs of the National Museum of Victoria* no. 17: 65-80, pls. I-V.

———. 1954. Remarkable bees from the rain forest. *Australian Zoologist* 12: 46-56.

———. 1956. The *Nomia australica* Sm. complex—Its taxonomy, morphology and biology with description of a new mutillid wasp. *Australian Zoologist* 12: 176-200, pls. 23-25.

———. 1959a. A new genus of bees in the family Colletidae. *Australian Zoologist* 12: 324-329, pl. 38.

———. 1959b. A new and remarkable colletid bee. *Australian Zoologist* 12: 334-336, pl. 11.

Rebmann, O. 1967a. Beitrag zur Kenntnis der Gattung *Megachile* Latr. (Hym., Apidae): Subgenus *Eutricharaea* Thoms. und *Neoeutricharaea* nov. subg. *Entomologische Zeitschrift* 77: 33-38.

———. 1967b. 2. Beitrag zur Kenntnis der Gattung *Megachile* Latr. (Hym., Apidae): Was ist "*Megachile argentata* auct." und "*Megachile rotundata* auct."? *Entomologische Zeitschrift* 77: 169-171.

———. 1968. 3. Beitrag zur Kenntnis der Gattung *Megachile* Latr. (Hym. Apidae): Subgenus *Eutricharaea* und seine bisher bekanntgewordenen Arten. *Deutsche Entomologische Zeitschrift* (n.f.)15: 21-48.

———. 1970. 6. Beitrag zur Kenntnis der Gattung *Megachile* Latr. *Nachrichtenblatt der Bayerischen Entomologen* 19: 37-47.

Reed, E. C. 1892. Revision de las abejas chilenas descritas en la obra de Gay. *Actes de la Société Scientifique du Chili* [Santiago, Chile] 2: 223-239.

———. 1894. Entomolojía [sic] Chilena. *Anales de la Universidad* [Santiago, Chile] 85: 599-653.

Rego, M. M. C. 1990. *Revisão do Gênero* Melipona *Illiger, 1806: Genitalia e Esternos Pré-genitais de Machos.* 76 pp. Ribeirão Preto, Brazil: Dissertação de Mestre, Faculdade de Filosofia, Ciências e Letras.

———. 1992. Morfologia das estruturas genitais de machos na sistemática de *Melipona,* pp. 73-78 *in* Anais do Encontro Brasileiro sobre Biologia de Abelhas e Outros Insetos Sociais, *Naturalia,* edição especial.

Reinig, W. F. 1930. Untersuchungen zur Kenntnis der Hummelfauna des Pamir-Hochlandes. *Zeitschrift für Morphologie und Ökologie der Tiere* 17: 68-123.

———. 1934. Entomologische Ergebnisse der deutsch-russischen Alti-Pamir-Expedition, 1928 (III), Hymenoptera VIII (Gen. *Bombus* Fabr.), Nachtrag. *Deutsche Entomologische Zeitschrift* 1933: 163-174.

Reyes, S. G. 1991a. A review of the bee genus *Braunsapis* in Madagascar. *Journal of African Zoology* 105: 125-130.

———. 1991b. Revision of the bee genus *Braunsapis* in the oriental region. *University of Kansas Science Bulletin* 54: 179-207.

———. 1993. Revision of the bee genus *Braunsapis* in the Australian region. *University of Kansas Science Bulletin* 55: 97-122.

———. 1998. A cladistic analysis of the bee tribe Allodapini. *Philippine Entomologist* 12: 55-83.

Reyes, S. G., S. J. B. Cooper, and M. P. Schwarz. 2000. Species phylogeny of the bee genus *Exoneurella* Michener (Hymenoptera: Apidae: Allodapini): Evidence from molecular and morphological data sets. *Annals of the Entomological Society of America* 92: 20-29.

Reyes, S. G., and C. D. Michener. 1990. Observations on a parasitic allodapine bee and its hosts in Java and Malaysia. *Tropical Zoology* 3: 139-149.

———. 1992. The genus *Halterapis* Michener 1969 in Madagascar. *Tropical Zoology* 5: 249-253.

Reyes, S. G., and M. P. Schwarz. 1997. Social evolution in the bee genus *Exoneurella* and related taxa (Hymenoptera: Apidae: Allodapini): a phylogenetic perspective. *Proceedings of the International Collaquia on Social Insects, Russian Language Section of the International Union for the Study of Social Insects* (St. Petersburg), V. E. Kipyatkov, ed., 3-4: 267-276.

Ribble, D. W. 1965. A revision of the banded subgenera of *Nomia* in America. *University of Kansas Science Bulletin* 45: 277-357.

———. 1967. The monotypic North American subgenus *Larandrena* of *Andrena. Bulletin of the Nebraska State Museum* 6: 27-42.

———. 1968a. Revisions of two subgenera of *Andrena*: *Micrandrena* Ashmead and *Derandrena,* new subgenus. *Bulletin of the University of Nebraska State Museum* 8: 237-394.

———. 1968b. A new subgenus, *Belandrena*, of the genus *Andrena*. *Journal of the Kansas Entomological Society* 41: 220-236.

———. 1974. A revision of the bees of the genus *Andrena* of the western hemisphere, subgenus *Scaphandrena*. *Transactions of the American Entomological Society* 100: 101-189.

Richards, K. W. 1973. Biology of *Bombus polaris* Curtis and *B. hyperboreus* Schönherr at Lake Hazen, Northwest Territories (Hymenoptera: Bombini). *Quaestiones Entomologicae* 9: 115-157.

———. 1993. Non-*Apis* bees as crop pollinators. *Revue Suisse de Zoologie* 100: 807-822.

Richards, M. H. 1994. Social evolution in the genus *Halictus*: A phylogenetic approach. *Insectes Sociaux* 41: 315-325.

———. 2000. Evidence for geographic variation in colony social organization in an obligately social sweat bee, *Lasioglossum malachurum* Kirby. *Canadian Journal of Zoology* 78: 1259-1266.

———. 2001. Nesting biology and social organization of *Halictus sexcinctus* (Fabricius) in southern Greece. *Canadian Journal of Zoology* 79: 2210-2220.

Richards, M. H., E. J. von Wettberg, and A. C. Rutgers. 2003. A novel social polymorphism in a primitively eusocial bee. *Proceedings of the National Academy of Sciences USA* 100: 7175-7180.

Richards, O. W. 1927. The specific characters of the British bumblebees. *Transactions of the Entomological Society of London* 1927: 233-268, pls. 24, 25.

———. 1929. A revision of the humble-bees allied to *Bombus orientalis* Smith, with the description of a new subgenus. *Annals and Magazine of Natural History* (10)3: 378-386.

———. 1932. On species of *Panurginus*, Nyl. *Annals and Magazine of Natural History* (10)9: 84-91.

———. 1935. Notes on the nomenclature of the aculeate Hymenoptera, with special reference to the British genera and species. *Transactions of the Royal Entomological Society of London* 83: 143-176.

———. 1937. A study of the British species of *Epeolus* Latr. and their races, with a key to the species of *Colletes*. *Transactions of the Society for British Entomology* 4: 89-130.

———. 1968. The subgeneric divisions of the genus *Bombus* Latreille. *Bulletin of the British Museum (Natural History), Entomology* 22: 211-276.

———. 1978. Part 4, Hymenoptera, ix + 150 pp. *in* G. S. Kloet and W. D. Hincks, *A Checklist of British Insects*, 2nd ed. London: Royal Entomological Society of London. [Bees on pp. 134-140.]

Ricketts, T. H., G. C. Daily, P.R. Ehrlich, and C. D. Michener. 2004. Economic value of tropical forest to coffee production. *Proceedings of the National Academy of Sciences USA* 101: 12579-12582.

Rightmyer, M. G. 2003. A new species of the bee genus *Rhogepeolus* Moure from Peru. *Journal of the Kansas Entomological Society* 76: 290-294.

———. 2004. Phylogeny and classification of the parasitic bee tribe Epeolini. *Scientific papers, Natural History Museum, the University of Kansas* no. 33: 1-51.

Rightmyer, M. G., and M.S. Engel. 2003. A new palearctic genus of melectine bees. *American Museum Novitates* no. 3392: 1-22.

Risch, S. 1997. Die Arten der Gattung *Eucera* Scopoli 1770 (Hymenoptera, Apidae), Die Untergattung *Pteneucera* Tkalců 1984. *Linzer Biologische Beiträge* 29: 555-580.

Risch, S. 1999. Neue und wenig bekannte Arten der Gattung *Eucera* Scopoli 1770. *Linzer Biologische Beiträg* 31: 115-145.

Risch, S. 2001. Die Arten des Genus *Eucera* Scopoli 1770 (Hymenoptera, Apidae), Untergattung *Pareucera* Tkalců 1979. *Entomofauna* 22: 365-376.

Risch, S. 2003. Die Arten der Gattung *Eucera* Scopoli 1770 (Hymenoptera, Apidae). Die Untergattungen *Stilbeucera* Tkalců 1979, *Atopeucera* Tkalců, 1984 und *Hemieucera* Sitdikov and Pesenko 1988. *Linzer Biologische* Beiträge 35: 1241-1292.

Ritcher, P. O. 1933. The external morphology of larval Bremidae and key to certain species. *Annals of the Entomological Society of America* 26: 53-63.

Ritsema, C. 1873. Beschrijving van een nieuw Hymenopteren-genus uit de onder-familie der Andrenidae Acutilingues. *Tijdschrift voor Entomologie* 16: 224-228, pl. 10 (part).

———. 1874. [*Zodion*, Hymenoptera, ...]. *Tijdschrift voor Entomologie* 17: lxviii-lxxv.

Roberts, D. P., W. D. Alecrim, J. M. Heller, S. R. Ehrhardt, and J. B. Lima. 1982. Male *Eufriesea purpurata*, a DDT-collecting bee in Brazil. *Nature* 297: 62-63.

Roberts, R. B. 1969. Biology of the bee genus *Agapostemon*. *University of Kansas Science Bulletin* 48: 689-719.

———. 1971. Biology of the crepuscular bee *Ptiloglossa guinnae* n. sp. with notes on associated bees, mites and yeasts. *Journal of the Kansas Entomological Society* 44: 283-294.

———. 1972. Revision of the bee genus *Agapostemon*. *University of Kansas Science Bulletin* 49: 437-590.

———. 1973. Nest architecture and immature stages of the bee *Oxaea flavescens* and the status of Oxaeidae. *Journal of the Kansas Entomological Society* 46: 437-446.

———. 1978. The nesting biology, behavior and immature stages of *Lithurge chrysurus*, an adventitious wood-boring bee in New Jersey. *Journal of the Kansas Entomological Society* 51: 735-745.

Roberts, R. B., and R. W. Brooks. 1987. Agapostemonine bees of Mesoamerica (Hymenoptera: Halictidae). *University of Kansas Science Bulletin* 53: 357-392.

Roberts, R. B., and C. H. Dodson. 1967. Nesting biology of two communal bees, *Euglossa imperialis* and *Euglossa ignita* including description of larvae. *Annals of the Entomological Society of America* 60: 1007-1014.

Roberts, R. B., and S. R. Vallespir. 1978. Specialization of hairs bearing pollen and oil on the legs of bees. *Annals of the Entomological Society of America* 71: 619-627.

Robertson, C. 1897. North American bees—Descriptions and synonyms. *Transactions of the Academy of Science of St. Louis* 7: 315-356.

———. 1900. Some Illinois bees. *Transactions of the Academy of Science of St. Louis* 10: 47-55.

———. 1901. Some new or little-known bees. *Canadian Entomologist* 33: 229-231.

———. 1902a. Synopsis of Andreninae. *Transactions of the American Entomological Society* 28: 187-194.

———. 1902b. Some new or little-known bees—IV. *Canadian Entomologist* 34: 321-331.

———. 1902c. Synopsis of Halictinae. *Canadian Entomologist* 34: 243-250.

———. 1902d. Some new or little-known bees—II. *Canadian Entomologist* 34: 48-49.

———. 1903a. Synopsis of Megachilidae and Bombinae. *Transactions of the American Entomological Society* 29: 163-178.

———. 1903b. Synopsis of Sphecodinae. *Entomological News* 14: 103-107.

———. 1903c. Synopsis of Epeolinae. *Canadian Entomologist* 35: 284-288.

———. 1903d. Synopsis of Nomadinae. *Canadian Entomologist* 35: 172-179.

———. 1904. Synopsis of Anthophila. *Canadian Entomologist* 36: 37-43.

———. 1905. Synopsis of Euceridae, Emphoridae and Anthophoridae. *Transactions of the American Entomological Society* 31: 365-372.

———. 1918. Some genera of bees. *Entomological News* 29: 91-92.

———. 1928. *Flowers and Insects*. 221 pp. Carlinville, Illinois: Published by the author.

Rocha, M. P., S. das G. Pompolo, and L. A. de O. Campos. 2003. Citogenética da tribo Meliponini, pp. 311-320 *in* Melo and Alves-dos-Santos (2003).

Rodeck, H. G. 1945. Two new subgenera of *Nomada* Scopoli. *Entomological News* 56: 179-181.

———. 1947. *Laminomada*, a new subgenus of *Nomada*. *Annals of the Entomological Society of America* 40: 266-270.

———. 1949. North American bees of the genus *Nomada*, subgenus *Callinomada*. *Annals of the Entomological Society of America* 42: 174-186.

Rohwer, S. A. 1911. A new genus of nomadine bees. *Entomological News* 22: 24-27.

Roig-Alsina, A. 1987. The classification of the Caenoprosopidini. *Journal of the Kansas Entomological Society* 60: 305-315.

———. 1989a. The tribe Osirini, its scope, classification, and revisions of the genera *Parepeolus* and *Osirinus*. *University of Kansas Science Bulletin* 54: 1-23.

———. 1989b. A revision of the bee genus *Doeringiella*. *University of Kansas Science Bulletin* 53: 576-621.

———. 1990. *Coelioxoides* Cresson, a parasitic genus of Tetrapediini. *Journal of the Kansas Entomological Society* 63: 279-287.

———. 1991a. Revision of the cleptoparasitic bee tribe Isepeolini. *University of Kansas Science Bulletin* 54: 257-288.

———. 1991b. Cladistic analysis of the Nomadinae s. str. with description of a new genus. *Journal of the Kansas Entomological Society* 64: 23-37.

———. 1992. La presencia de *Chilimalopsis* Toro en la Argentina y descripcion de una nueva especie. *Neotropica* 38: 149-153.

———. 1993. The evolution of the apoid endophallus, its phylogenetic implications, and functional significance of the genital capsule. *Bollettino di Zoologia* 60: 169-183.

———. 1994. *Meliphilopsis*, a new genus of emphorine bees, and notes on the relationships among the genera of Emphorina. *Reichenbachia* 30: 181-188.

———. 1996. Las especies del genero *Rhogepeolus* Moure. *Neotropica* 42: 55-59.

———. 1997. A generic study of the bees of the tribe Tapinotaspidini, with notes on the evolution of their oil-collecting structures. *Mitteilungen der Münchner Entomologischen Gesellschaft* 87: 3-21.

———. 1998a. Sinopsis genérica de la tribu Emphorini, con la descripción de tres nuevos géneros. *Physis (Buenos Aires)* (C)56:17-25.

———. 1998b. Las especies del género *Alepidosceles* Moure en Argentina. *Neotrópica* 44: 69-74.

———. 1999a. Revisión de las abejas colecteoras de aceites del género *Chalepogenus* Holmberg. *Revista del Museo Argentino de Ciencias Naturales* (n.s.) 1: 67-101.

———. 1999b. Revisión de las abejas colectoras de aceites del género *Chalepogenus* Holmberg. *Revista del Museo Argentino de Ciencias Naturales* (n.s.) 1: 67-101.

———. 2000. Claves para las especies argentinas de *Centris* (Hymenoptera, Apidae), con descripción de nuevas especies y notas sobre distribución. *Revista del Museo Argentino de Ciencias Naturales* (n.s.)2: 171-193.

———. 2003. The bee genus *Doeringiella* Holmberg (Hymenoptera: Apidae): A revision of the subgenus *Pseudepeolus* Holmberg. *Journal of Hymenoptera Research* 12: 136-147.

Roig-Alsina, A., and L. A. Compagnucci. 2003. Description, phylogenetic relationships, and biology of *Litocalliopsis adesmiae*, a new genus and species of South American calliopsine bees. *Revista del Museo Argentino de Ciencias Naturalles* (n.s.)5: 99-112.

Roig-Alsina, A., and C. D. Michener. 1993. Studies of the phylogeny and classification of long-tongued bees. *University of Kansas Science Bulletin* 55: 124-162.

Roig-Alsina, A., and J. G. Rozen, Jr. 1994. Revision of the cleptoparasitic bee tribe Protepeolini, including biologies and immature stages. *American Museum Novitates* no. 3099: 1-27.

Rojas-A., F. and H. Toro-G. 2000. Revisión de las especies de *Caenohalictus* (Halictidae-Apoidea) presentes en Chile, *Boletin del Museo Nacional de Historia Natural* [Chile] 49: 163-214.

Romand, B. de. 1840. Sur l'Hyménoptère nommé *Acanthopus Goryi* (voy. cette Revue, 1840, p. 248). *Revue Zoologique par la Société Cuvierienne* 1840: 335-336.

———. 1841. Notice sur divers insectes Hyménoptères de la famille des mellifères. *Magazin de Zoologie* (2)3: 1-8, pls. 68-70.

Romankova, T. G. 1988. A new far-eastern bee of the tribe Anthidiini. *Vestnik Zoologii* 1988(4): 25-30. [In Russian.]

Romasenko, L. P. 1995. A comparative morphological description of the prepupae of the family Megachilidae. *Entomologicheskoe Obozrenie* 74: 186-208, 267-268. [In Russian; English translation in *Entomological Review* 75: 1-23 (1996).]

Roubik, D. W. 1979. Nest and colony characteristics of stingless bees from French Guiana. *Journal of the Kansas Entomological Society* 52: 443-470.

———. 1982. Obligate necrophagy in a social bee. *Science* 217: 1059-1060.

———. 1983. Nest and colony characteristics of stingless bees from Panama. *Journal of the Kansas Entomological Society* 56: 327-355.

———. 1989. *Ecology and Natural History of Tropical Bees.* x + 514 pp. Cambridge: Cambridge University Press.

———. 1992. Stingless bees: A guide to Panamanian and Mesoamerican species and their nests, pp. 495-524 *in* D. Quintero and A. Aiello, eds., *Insects of Panama and Mesoamerica.* New York: Oxford University Press.

Roubik, D. W., and J. D. Ackerman. 1987. Long-term ecology of euglossine orchid-bees (Apidae: Euglossini) in Panama. *Oecologia* 73: 321-333.

Roubik, D. W., and P. E. Hanson. 2004. *Orchid bees of tropical America biology and field guide,* 370 pp. Santo Domingo de Heredia, Costa Rica: Instituto Nacional de Biodiversidad (INBio). [in Spanish and English.]

Roubik, D. W., and C. D. Michener. 1980. The seasonal cycle and nests of *Epicharis zonata,* a bee whose cells are below the wet-season water table. *Biotropica* 12: 56-60.

———. 1985. Nesting biology of *Crawfordapis* in Panamá. *Journal of the Kansas Entomological Society* 57: 662-671.

Roubik, D. W., B. H. Smith, and R. G. Carlson. 1987. Formic acid in caustic cephalic secretions of a stingless bee, *Oxytrigona. Journal of Chemical Ecology* 13: 1079-1086.

Rozen, J. G., Jr. 1958. Monographic study of the genus *Nomadopsis* Ashmead. *University of California Publications in Entomology* 15: i-iv + 1-202.

———. 1964a. Phylogenetic-taxonomic significance of last instar of *Protoxaea gloriosa* Fox, with descriptions of first and last instars. *Journal of the New York Entomological Society* 72: 223-230.

———. 1964b. The biology of *Svastra obliqua obliqua* (Say), with a taxonomic description of its larva. *American Museum Novitates* no. 2170: 1-13.

———. 1965a. The larvae of Anthophoridae (Hymenoptera, Apoidea), Part 1. Introduction, Eucerini, and Centridini. *American Museum Novitates* no. 2233: 1-27.

———. 1965b. Biological notes on the cuckoo bee genera *Holcopasites* and *Neolarra. Journal of the New York Entomological Society* 73: 87-91.

———. 1965c. The biology and immature stages of *Melitturga clavicornis* (Latreille) and of *Sphecodes albilabris* (Kirby) and the recognition of the Oxaeidae at the family level. *American Museum Novitates* no. 2224: 1-18.

———. 1966a. The larvae of Anthophoridae (Hymenoptera, Apoidea), Part 2. The Nomadinae. *American Museum Novitates* no. 2244: 1-38.

———. 1966b. Systematics of the larvae of North American panurgine bees. *American Museum Novitates* no. 2259: 1-22.

———. 1966c. Taxonomic descriptions of the immature stages of the parasitic bee, *Stelis (Odontostelis) bilineolata* (Spinola). *Journal of the New York Entomological Society* 74: 84-91.

———. 1967a. Review of the biology of panurgine bees, with observations on North American forms. *American Museum Novitates* no. 2297: 1-44.

———. 1967b. The immature instars of the cleptoparasitic genus *Dioxys. Journal of the New York Entomological Society* 75: 236-248.

———. 1968a. Review of the South African cuckoo-bee genus *Pseudodichroa. American Museum Novitates* no. 2347: 1-10.

———. 1968b. Biology and immature stages of the aberrant bee genus *Meliturgula. American Museum Novitates* no. 2331: 1-18.

———. 1969a. The larvae of Anthophoridae (Hymenoptera, Apoidea), Part 3. The Melectini, Ericrocini, and Rhathymini. *American Museum Novitates* no. 2382: 1-24.

———. 1969b. Biological notes on the bee *Tetralonia minuta* and its cleptoparasite *Morgania histrio transvaalensis. Proceedings of the Entomological Society of Washington* 71: 102-107.

———. 1969c. The biology and description of a new species of African *Thyreus,* with life history notes on two species of *Anthophora. Journal of the New York Entomological Society* 77: 51-60.

———. 1970a. Biology, immature stages, and phylogenetic relationships of fideliine bees, with the description of a new species of *Neofidelia. American Museum Novitates* no. 2427: 1-25.

———. 1970b. Biology and immature stages of the panurgine bee genera *Hypomacrotera* and *Psaenythia. American Museum Novitates* no. 2416: 1-16.

———. 1970c. Department of Entomology Report 102, *Annual Report of the American Museum of Natural History* no. 102: 4-5.

———. 1971a. Systematics of the South American bee genus *Orphana. American Museum Novitates* no. 2462: 1-15.

———. 1971b. Biology and immature stages of Moroccan panurgine bees. *American Museum Novitates* no. 2457: 1-37.

———. 1973a. Immature stages of lithurgine bees with descriptions of the Megachilidae and Fideliidae based on mature larvae. *American Museum Novitates* no. 2527: 1-14.

———. 1973b. Life history and immature stages of the bee *Neofidelia. American Museum Novitates* no. 2519: 1-14.

———. 1974. Nest biology of the eucerine bee *Thygater analis. Journal of the New York Entomological Society* 82: 230-234.

———. 1977a. Biology and immature stages of the bee genus *Meganomia. American Museum Novitates* no. 2630: 1-14.

———. 1977b. Immature stages and ethological observations on the cleptoparasitic bee tribe Nomadini. *American Museum Novitates* no. 2638: 1-16.

———. 1977c. The ethology and systematic relationships of fideliine bees, including a description of the mature larva of *Parafidelia. American Museum Novitates* no. 2637: 1-15.

———. 1978. The relationships of the bee subfamily Ctenoplectrinae as revealed by its biology and mature larva. *Journal of the Kansas Entomological Society* 51: 637-652.

———. 1984a. Comparative nesting biology of the bee tribe Exomalopsini. *American Museum Novitates* no. 2798: 1-37.

———. 1984b. Nesting biology of diphaglossine bees. *American Museum Novitates* no. 2786: 1-33.

———. 1986. The natural history of the old world nomadine parasitic bee *Pasites maculatus* (Anthophoridae: Nomadinae) and its host *Pseudapis diversipes* (Halictidae, Nomiinae). *American Museum Novitates* no. 2861: 1-8.

———. 1987a. Nesting biology of the bee *Ashmeadiella holtii* and its cleptoparasite, a new species of *Stelis*. *American Museum Novitates* no. 2900: 1-10.

———. 1987b. Nesting biology and immature stages of a new species in the bee genus *Hesperapis*. *American Museum Novitates* no. 2887: 1-13.

———. 1988. Ecology, behavior, and mature larva of a new species of the old world bee genus *Camptopoeum*. *American Museum Novitates* no. 2925: 1-10.

———. 1989a. Two new species and the redescription of another species of the cleptoparasitic bee genus *Triepeolus* with notes on their immature stages. *American Museum Novitates* no. 2956: 1-18.

———. 1989b. Morphology and systematic significance of first instars of the cleptoparasitic bee tribe Epeolini. *American Museum Novitates* no. 2957: 1-19.

———. 1989c. Life history studies of the "primitive" panurgine bees. *American Museum Novitates* no. 2962: 1-27.

———. 1991a. Evolution of cleptoparasitism in anthophorid bees as revealed by their mode of parasitism and first instars. *American Museum Novitates* no. 3029: 1-36.

———. 1991b. Nesting biology and mature larva of the bee *Idiomelissodes duplocincta*. *American Museum Novitates* no. 3012: 1-11.

———. 1992a. Biology of the bee *Ancylandrena larreae* (Andrenidae: Andreninae) and its cleptoparasite *Hexepeolus rhodogyne* (Anthophoridae: Nomadinae) with a review of egg deposition in the Nomadinae. *American Museum Novitates* no. 3038: 1-15.

———. 1992b. Systematics and host relationships of the cuckoo bee genus *Oreopasites*. *American Museum Novitates* no. 3046: 1-56.

———. 1993a. Nesting biologies and immature stages of the rophitine bees (Halictidae) with notes on the cleptoparasite *Biastes* (Anthophoridae). *American Museum Novitates* no. 3066: 1-28.

———. 1993b. Phylogenetic relationships of *Euherbstia* with other short-tongued bees. *American Museum Novitates* no. 3060: 1-17.

———. 1994a. Biologies of the bee genera *Ancylandrena* (Andrenidae: Andreninae) and *Hexepeolus* (Apidae: Nomadinae) and phylogenetic relationships of *Ancylandrena* based on its mature larva. *American Museum Novitates* no. 3108: 1-19.

———. 1994b. Biology and immature stages of some cuckoo bees belonging to Brachynomadini, with descriptions of two new species. *American Museum Novitates* no. 3089: 1-23.

———. 1996a. Phylogenetic analysis of the cleptoparasitic bees belonging to the Nomadinae based on mature larvae. *American Museum Novitates* no. 3180: 1-39.

———. 1996b. First and last larval instars of the cleptoparasitic bee *Hexepeolus rhodogyne*. *Memoirs of the Entomological Society of Washington* no. 17: 188-193.

———. 1997a. New taxa of brachynomadine bees. *American Museum Novitates* no. 3200: 1-26.

———. 1997b. South American rophitine bees. *American Museum Novitates* no. 3206: 1-27.

———. 2000. Pupal descriptions of some cleptoparasitic bees (Apidae), with a preliminary generic key to pupae of parasitic bees. *American Museum Novitates* no. 3289: 1-19.

———. 2001. A taxonomic key to mature larvae of cleptoparasitic bees. *American Museum Novitates* no. 3309: 1-27.

———. 2003a. Eggs, ovariole numbers, and modes of parasitism of cleptoparasitic bees, with emphasis on neotropical species. *American Museum Novitates* no. 3413: 1-36.

———. 2003b. Ovarian formula, mature oocyte. and egg index of the bee *Ctenoplectra*. *Journal of the Kansas Entomological Society* 76: 640-642.

———. 2003c. A new tribe, genus, and species of South American panurgine bee (Andrenidae, Panurginae) oligolectic on *Nolana* (Nolanaceae), pp. 93-108 *in* Melo and Alves-dos-Santos (2003).

Rozen, J. G., Jr., and S. L. Buchmann. 1990. Nesting biology and immature stages of the bees *Centris caesalpiniae, C. pallida,* and the cleptoparasite *Ericrocis lata*. *American Museum Novitates* no. 2985: 1-30.

Rozen, J. G., Jr., and G. C. Eickwort. 1997. The entomological evidence. *Journal of Forensic Sciences* 42: 394-397.

Rozen, J. G., Jr., K. R. Eickwort, and G. C. Eickwort. 1978. The bionomics and immature stages of the cleptoparasitic bee genus *Protepeolus*. *American Museum Novitates* no. 2640: 1-24.

Rozen, J. G., Jr., and M. S. Favreau. 1967. Biological notes on *Dioxys pomonae pomonae* and its host, *Osmia nigrobarbata*. *Journal of the New York Entomological Society* 75: 197-203.

———. 1968. Biological notes on *Colletes compactus compactus* and its cuckoo bee, *Epeolus pusillus*. *Journal of the New York Entomological Society* 76: 106-111.

Rozen, J. G., Jr., and N. R. Jacobson. 1980. Biology and immature stages of *Macropis nuda*, including comparisons to related bees. *American Museum Novitates* no. 2702: 1-11.

Rozen, J. G., Jr., and R. J. McGinley. 1974a. Phylogeny and systematics of Melittidae based on mature larvae. *American Museum Novitates* no. 2545: 1-31.

———. 1974b. Systematics of ammobatine bees based on their mature larvae and pupae. *American Museum Novitates* no. 2551: 1-16.

———. 1976. Biology of the bee genus *Conanthalictus*. *American Museum Novitates* no. 2602: 1-6.

———. 1991. Biology and larvae of the cleptoparasitic bee *Townsendiella pulchra* and nesting biology of its host *Hesperapis larreae*. *American Museum Novitates* no. 3005: 1-11.

Rozen, J. G., Jr., G. A. R. Melo, A. J. C. Aguiar, and I. Alves-dos-Santos. 2006. Nesting biologies and immature stages of the Tapinotaspine bee genera *Monecca* and *Lanthanomelissa* and of their osirine cleptoparasites *Protosiris* and *Parapeolus*, *American Museum Novitates* no. 3501: 1-60.

Rozen, J. G., Jr., and C. D. Michener. 1968. The biology of *Scrapter* and its cuckoo bee, *Pseudodichroa*. *American Museum Novitates* no. 2335: 1-13.

———. 1988. Nests and immature stages of the bee *Paratetrapedia swainsonae*. *American Museum Novitates* no. 2909: 1-13.

Rozen, J. G., Jr., and H. Özbek. 2003. Oocytes, eggs, and ovarioles of some long-tongued bees. *American Museum Novitates* no. 3393: 1-35.

———. 2004. Immature stages of the cleptoparasitic bee *Dioxys cincta*. *American Museum Novitates* no. 3443: 1-12.

———. 2005a. Egg deposition of the cleptoparasitic bee *Dioxys cincta*. *Journal of the Kansas Entomological Society* 78: 221-226.

———. 2005b. Notes on the egg and egg deposition of the cleptoparasitic *Thyreus ramosus*. *Journal of the Kansas Entomological Society* 78: 34-40.

Rozen, J. G., Jr., and A. Roig-Alsina. 1991. Biology, larvae and oocytes of the parasitic bee tribe Caenoprosopidini. *American Museum Novitates* no. 3004: 1-10.

Rozen, J. G., Jr., A. Roig-Alsina, and B. A. Alexander. 1997. The cleptoparasitic bee genus *Rhopalolemma*, with reference to other Nomadinae (Apidae) and the biology of its host, *Protodufourea*. *American Museum Novitates* no. 3194: 1-28.

Rozen, J. G., Jr., and L. Ruz. 1995. South American panurgine bees (Andrenidae: Panurginae), Part II. Adults, immature stages, and biology of *Neffapis longilingua*, a new genus and species with an elongate glossa. *American Museum Novitates* no. 3136: 1-15.

Rozen, J. G., and R. R. Snelling. 1986. Ethology of the bee *Exomalopsis nitens* and its cleptoparasite. *Journal of the New York Entomological Society* 94: 480-488.

Rozen, J. G., Jr., and D. Yanega. 1999. Nesting biology and immature stages of the South American bee genus *Acamptopoeum*, pp. 59-67 *in* G. W. Byers, R. H. Hagen, and R. W. Brooks, eds., *Entomological Contributions in Memory of Byron A. Alexander*, University of Kansas Natural History Museum Special Publication no. 24.

Rust, R. W. 1974. The systematics and biology of the genus *Osmia*, subgenera *Osmia*, *Chalcosmia*, and *Cephalosmia*. *Wasmann Journal of Biology* 32: 1-93.

———. 1976. Notes on the biology of North American species of *Panurginus*. *Pan-Pacific Entomologist* 52: 159-166.

———. 1980a. Nesting biology of *Hoplitis biscutellae* (Cockerell). *Entomological News* 91: 105-109.

———. 1980b. The biology of *Ptilothrix bombiformis*. *Journal of the Kansas Entomological Society* 53: 427-436.

———. 1986. Biology of *Osmia (Osmia) ribifloris* Cockerell. *Journal of the Kansas Entomological Society* 59: 89-94.

———. 1988. Biology of *Nomadopsis larreae* (Hymenoptera, Andrenidae), with an analysis of yearly appearance. *Annals of the Entomological Society of America* 81: 99-104.

Rust, R. W., G. Cambon, J.-P. Torre Grossa, and B. E. Vaissière. 2004. Nesting biology and foraging ecology of the wood-boring bee *Lithurgus chrysarus*. *Journal of the Kansas Entomological Society* 77: 269-279.

Rust, R. W., and G. E. Bohart. 1986. New species of *Osmia* (Hymenoptera, Megachilidae) from the southwestern United States. *Entomological News* 97: 147-155.

Rust, R. W., and S. L. Clement. 1972. The biology of *Osmia glauca* and *Osmia nemoris*. *Journal of the Kansas Entomological Society* 45: 523-528.

Rust, R. W., and R. W. Thorp. 1973. The biology of *Stelis chlorocyanea*, a parasite of *Osmia nigrifrons*. *Journal of the Kansas Entomological Society* 46: 548-562.

Rust, R. W., R. W. Thorp, and P. F. Torchio. 1974. The ecology of *Osmia nigrifrons* with a comparison to other species of *Acanthosmioides*. *Journal of Natural History* 8: 29-47.

Ruszkowski, A., M. Bilin'ski, and J. Gosek. 1986. Rośliny pokarmowe i gospodarze pasozytniczych pszczól miesiarkowatych (*Coelioxys* Latr., *Stelis* Pz., *Dioxys* Lep. et Serv., *Dioxoides* Pop. oraz *Paradioxys* Mocs. *Pszczelnicze Zeszyty Naukowe* 30: 111-131.

Ruttner, F. 1987. *Biogeography and Taxonomy of Honeybees.* xii + 284 pp. Berlin: Springer.

Ruz, L. 1980. *Pseudosarus*, nuevo genero de Panurginae Chileno *Revista Chilena de Entomología* 10: 25-28.

———. 1986. *Classification and phylogenetic relationships of the panurgine bees.* iii + 312 pp., 67 figs. Lawrence: Ph.D. thesis, University of Kansas.

———. 1990. Redefinición del género *Xenopanurgus* (Hymenoptera, Andrenidae) y descripción de una nueva especie de México. *Folia Entomológica Mexicana* no. 79: 151-161.

———. 1991. Classification and phylogenetic relationships of the panurgine bees: The Calliopsini and allies. *University of Kansas Science Bulletin* 54: 209-256.

Ruz, L., and E. Chiappa. 2004. *Protandrena evansi*, a new panurgine bee from Chile. *Journal of the Kansas Entomological Society* 77: 788-795.

Ruz, L., and G. A. R. Melo. 1999. Reassessment of the bee genus *Chaeturginus* (Apoidea; Andrenidae, Panurginae), with the description of a new species from southeastern Brazil, pp. 231-236 *in* G. W. Byers, R. H. Hagen, and R. W. Brooks, eds., *Entomological Contributions in Memory of Byron A. Alexander*, University of Kansas Natural History Museum Special Publication no. 24.

Ruz, L., and J. G. Rozen, Jr. 1993. South American panurgine bees (Apoidea: Andrenidae: Panurginae), Part I. Biology, mature larva, and description of a new genus and species. *American Museum Novitates* no. 3057: 1-12.

Ruz, L., and H. Toro. 1983. Revision of the bee genus *Liphanthus*. *University of Kansas Science Bulletin* 52: 235-299.

Sakagami, S. F. 1961. *Nomia umesaoi* sp. nov., an aberrant bee from Thailand. *Insecta Matsumurana* 24: 43-51.

———. 1964. Wiederentdeckung des Nestes einer Nachtfurchenbiene, *Megalopta* sp., am Amazonas. *Kontyû* 32: 457-463.

———. 1965. Über dem Bau der männlichen Hinterschiene von *Eulaema nigrita* Lepeletier. *Zoologischer Anzeiger* 175: 347-354.

———. 1966. Comparative ethology of Apidae. *Japanese Society for Systematic Zoology*, circ. 35: 1-6. [In Japanese.]

———. 1975. Stingless bees (excl. *Tetragonula*) from the continental southeast Asia in the collection of the Bernice P. Bishop Museum, Honolulu. *Journal of the Faculty of Science, Hokkaido University*, ser. VI, Zoology 20: 49-76.

———. 1976. Specific differences in the bionomic characters of bumblebees. A comparative review. *Journal of the Faculty of Science, Hokkaido University*, ser. VI, Zoology 20: 390-447.

———. 1978. *Tetragonula* stingless bees of the continental Asia and Sri Lanka. *Journal of the Faculty of Science, Hokkaido University*, ser. VI, Zoology 21: 165-247.

———. 1982. Stingless bees, pp. 361-423 *in* H. R. Hermann, ed., *Social Insects*, Vol. III. New York: Academic Press.

———. 1989. Taxonomic notes on a Malesian bee *Lasioglossum carinatum*, the type species of the subgenus *Ctenonomia*, and its allies. *Journal of the Kansas Entomological Society* 62: 496-510.

———. 1991. The halictine bees of Sri Lanka and the vicinity, II. *Nesohalictus*. *Zoological Science* 8: 169-178.

Sakagami, S. F., and A. W. Ebmer. 1987. Taxonomic notes on oriental halictine bees of the genus *Halictus* (subgen. *Seladonia*). *Linzer Biologische Beiträge* 19: 301-357.

Sakagami, S. F., A. W. Ebmer, and O. Tadauchi. 1996. The halictine bees of Sri Lanka and the vicinity, III. *Sudila* (Hymenoptera, Halictidae) Part 1. *Esakia* no. 36: 143-189.

Sakagami, S. F., and T. Inoue. 1985. Taxonomic notes on three bicolorous *Tetragonula* stingless bees in southeast Asia. *Kontyû* 53: 174-189.

———. 1987. Stingless bees of the genus *Trigona* (subgenus *Trigonella*) with notes on the reduction of spatha in male genitalia of the subgenus *Tetragonula*. *Kontyû* 55: 610-627.

———. 1989. Stingless bees of the genus *Trigona* (subgen. *Geniotrigona*) (Hymenoptera, Apidae), with description of *T. (G.) incisa* sp. nov. from Sulawesi. *Japanese Journal of Entomology* 57: 605-620.

Sakagami, S. F., T. Inoue, and S. Salmah. 1985. Key to the stingless bee species found or expected from Sumatra, pp. 37-43 *in* R.-i. Ohgushi, ed., *Evolutionary Ecology of Insects in Humid Tropics, Especially in Central Sumatra*. Kanazawa University, Japan: Sumatra Nature Study (Entomology).

———. 1990. Stingless bees of central Sumatra, pp. 125-137 *in* S. F. Sakagami, R.-i. Ohgushi, and D. W. Roubik, eds., *Natural History of Social Wasps and Bees in Equatorial Sumatra*. Sapporo: Hokkaido University Press.

Sakagami, S. F., T. Inoue, S. Yamane, and S. Salmah. 1983. Nesting habits of Sumatran stingless bees, pp. 38-45 *in* R.-i. Ohgushi, ed., *Ecological Study on Social Insects in Central Sumatra with Special Reference to Wasps and Bees*. Kanazawa University, Japan: Sumatra Nature Study (Entomology).

Sakagami, S. F., and M. Ito. 1981. Specific and subgeneric variations in tibial corbiculation of male bumblebees (Hymenoptera: Apidae), an apparently functionless character. *Entomologica Scandinavica*, suppl. 15: 365-376.

Sakagami, S. F., M. Kato, and T. Itino. 1991. *Thrinchostoma (Diagonozus) asianum* sp. nov.: Discovery of an African subgenus of long-malared halictine bees from Sumatra, with some observations on its oligotrophy to *Impatiens*. *Tropics* 1: 49-58.

Sakagami, S. F., and S. Laroca. 1971. Observations on the bionomics of some neotropical xylocopine bees, with comparative and biofaunistic notes. *Journal of the Faculty of Science, Hokkaido University* (VI) 18: 57-127.

———. 1988. Nests of an exomalopsine bee *Lanthanomelissa goeldiana*. *Journal of the Kansas Entomological Society* 61: 347-349.

Sakagami, S. F., and Y. Maeta. 1985. Multifemale nests and rudimentary castes in the normally solitary bee *Ceratina japonica*. *Journal of the Kansas Entomological Society* 57(1984): 639-656.

———. 1989. Compatibility and incompatibility of solitary life with eusociality in two normally solitary bees *Ceratina japonica* and *Ceratina okinawana* (Hymenoptera, Apoidea) with notes on the incipient phase of eusociality. *Japanese Journal of Entomology* 57: 417-439.

———. 1990. *Lasioglossum (Lasioglossum) primavera* sp. nov., a Japanese halictine bee which overwinters in both female and male adults. *Bulletin of the Faculty of Agriculture, Shimane University*, no. 24: 52-59.

Sakagami, S. F., T. Matsumura, and Y. Maeta. 1985. Bionomics of the halictine bees in northern Japan, III. *Lasioglossum (Evylaeus) allodalum*, with remarks on the serially arranged cells in the halictine nests. *Kontyû* 53: 409-419.

Sakagami, S. F., and C. D. Michener. 1962. The Nest Architecture of the Sweat Bees (Halictinae), a Comparative Study. [v] + 135 pp. Lawrence: University of Kansas Press.

———. 1965. Notes on the nests of two euglossine bees, *Euplusia violacea* and *Eulaema cingulata*. *Annotationes Zoologicae Japonensis* 38: 216-222.

———. 1987. Tribes of Xylocopinae and origin of the Apidae. *Annals of the Entomological Society of America* 80: 439-450.

Sakagami, S. F., R. Miyanaga, and Y. Maeta. 1994. Discovery of a eusocial halictine bee, *Lasioglossum (Evylaeus) subtropicum* sp. nov. from Iriomote Is., southernmost Japan, with a morphometric comparison of castes in some social halictines. *Bulletin of the Faculty of Agriculture, Shimane University*, no. 28: 5-21.

Sakagami, S. F., and J. S. Moure. 1965. Cephalic polymorphism in some neotropical halictine bees. *Anais de Academia Brasileira de Ciências* 37: 303-313.

———. 1967. Additional observations on the nesting habits of some Brazilian halictine bees. *Mushi* 40: 119-138.

Sakagami, S. F., D. W. Roubik, and R. Zucchi. 1993. Ethology of the robber stingless bee, *Lestrimelitta limao*. *Sociobiology* 21: 237-277.

Sakagami, S. F., and H. Sturm. 1965. *Euplusia longipennis* (Friese) und ihre merkwürdigen Brutzellen aus Kolumbien. *Insecta Matsumurana* 28: 83-92, pls. XI-XVI.

Sakagami, S. F., and F. L. Wain. 1966. *Halictus latisignatus* Cameron: A polymorphic Indian halictine bee with caste differentiation. *Journal of the Bombay Natural History Society* 63: 57-73.

Sakagami, S. F., and S. Yamane. 1987. Oviposition behavior and related notes of the Taiwanese stingless bee *Trigona (Lepidotrigona) ventralis hoozana*. *Journal of Ethology* 5: 17-27.

Sakagami, S. F., S. Yamane, and G. G. Hambali. 1983. Nests of some Southeast Asian stingless bees. *Bulletin of the Faculty of Education, Ibaraki University (Natural Sciences)*, no. 32: 1-21.

Sakagami, S. F., S. Yamane, and T. Inoue. 1983. Intranidal behaviors of *Trigona (Tetragonula) laeviceps* and *T. (Trigonella) moorei*, pp. 46-51 *in* R.-i. Ohgushi, ed., *Ecological Study on Social Insects in Central Sumatra with Special Reference to Wasps and Bees*. Kanazawa University, Japan: Sumatra Nature Study (Entomology).

Sakagami, S. F., and K. Yoshikawa. 1961. Bees of Xylocopinae and Apinae collected by the Osaka City University biological expedition to southeast Asia 1957-58, with some biological notes. *Nature and Life in Southeast Asia* [Tokyo] 1: 409-444.

Sakagami, S. F., and R. Zucchi. 1978. Nests of *Hylaeus (Hylaeopsis) tricolor*: The first record of non-solitary life in colletid bees, with notes on communal and quasisocial colonies. *Journal of the Kansas Entomological Society* 51: 597-614.

Sandhouse, G. A. 1923. The bee genus *Dialictus*. *Canadian Entomologist* 55: 193-195.

———. 1924. New North American species of bees belonging to the genus *Halictus (Chloralictus)*. *Proceedings of the U.S. National Museum* 65(19): 1-43.

———. 1937. The bees of the genera *Augochlora, Augochloropsis* and *Augochlorella* (Hymenoptera; Apoidea) occurring in the United States. *Journal of the Washington Academy of Sciences* 27: 65-79.

———. 1939. The North American bees of the genus *Osmia*. *Memoirs of the Entomological Society of Washington* 1: [ii] + 1-167.

———. 1941. The American bees of the subgenus *Halictus*. *Entomologica Americana* (n.s.)21: 23-38, pl. II.

———. 1943. The type species of the genera and subgenera of bees. *Proceedings of the United States National Museum* 92: 519-619.

Santos, M. L., and C. A. Garófalo. 1994. Nesting biology and nest re-use of *Eulaema nigrita*. *Insectes Sociaux* 41: 99-110.

Saunders, E. 1882, 1884. Synopsis of British Hymenoptera, Diploptera and Anthophila, Part I to end of Andrenidae; Part II, Apidae. *Transactions of the Entomological Society of London* 1882: 165-290, pls. VII-XI; 1884: 159-246, pls. V-XII.

———. 1891. On the tongues of the British Hymenoptera. *Journal of the Linnean Society* [London] (Zoology) 23: 410-432, pls. 3-10.

Saussure, H. de. 1890. Histoire Naturelle des Hyménoptères, vol. 20, xxi + 590 pp., 27 pls., *in* A. Grandidier, *Histoire Physique, Naturelle et Politique de Madagascar.* Paris: Imprimerie Nationale. [Actually published in 1891?]

Scheloske, H.-W. 1974. Untersuchungen über das Vorkommen, die Biologie und den Nestbau der Seidenbiene *Colletes daviesanus* Sm. *Zoologische Jahrbücher, Abteilung für Systematik, Ökologie und Geographie der Tiere* 101: 153-172.

Schenck, A. 1861. Die Nassauischen Bienen. *Jahrbücher des Vereins für Naturkunde im Herzogthum Nassau* 14: 1-414. [For date, see Michener, 1986a.]

———. 1867. Verzeichniss der naussauischen Hymenoptera aculeata mit Hinzufügung der übrigen deutschen Arten. *Berliner Entomologische Zeitschrift* 10(1866): 317-369.

———. 1869. Beschreibung der nassauischen Bienen, Zweiter Nachtrag. *Jahrbücher des Nassauischen Vereins für Naturkunde* 21-22: 1[269]-114[382]. [Preprint, dated 1868, published 1869, pp. 1-114; Jahrbücher for 1867-68, published in 1870, pp. 269-382. For dates, see Michener, 1986a.]

———. 1874. Ueber einige streitige und zweifelhafte Bienen-Arten. *Berliner Entomologische Zeitschrift* 17(1873): 243-259.

Scheuchl, E. 1996. *Illustrierte Bestimmungstabellen der Wildbeinen Deutschlands und Osterreichs, Band II, Megachilidae-Melittidae,* 116 pp. Velden: published by the author.

———. 1997. *Illustrierte Bestimmungstabellen der Wildbeinen Deutschlands und Osterreichs, Band III, Andrenae,* 180 pp. Velden: published by the author.

———. 2000. *Illustrierte Bestimmungstabellen der Wildbeinen Deutschlands und Osterreichs, Band I, Anthophoridae,* 158 pp. Velden: published by the author.

Schlindwein, C. 1995. *Wildbienen und ihre Trachtpflanzen in einer südbrasilianischen Buschlandschaft: Fallstudie Guaritas. Bestäubung bei Kakteen und Loasaceen.* 148 pp. Stuttgart: Verlag Ulrich E. Grauer.

Schlindwein, C., and J. S. Moure. 1998. *Panurgillus* gênero novo de Panurginae, com a descrição de quatorze espécies do sul do Brasil. *Revista Brasileira de Zoologia* 15: 397-439.

———. 1999. Espécies de *Panurgillus* Schlindwein & Moure (Hymenoptera, Andrenidae) depositados no Naturkunde Museum, Berlin. *Revista Brasileira de Zoologia* 16: 113-133.

Schmiedeknecht, H. L. O. 1882-1884. *Apidae Europaeae [Die Bienen Europa's],* Vol. 1, [xiv] + 866 pp., pls. 1-15. Berlin: Gumperdae and Berolini. [Pp. 1-314 published in 1882; 315-550, 1883; and 551-866, 1884.]

———. 1885-1886. *Apidae Europaeae [Die Bienen Europa's],* Vol. 2, [iv] + 1-205 [867-1071] + [2, index], pls. 16, 17. Berlin: Friedländer. [Pp. 1-110 published in 1885, 111-205 in 1886.]

———. 1930. *Die Hymenopteren Nord- und Mitteleuropas.* 2nd ed., x + 1062 pp. Jena: Fischer.

Schönitzer, K. 1986. Comparative morphology of the antenna cleaner in bees. *Zeitschrift für Zoologische Systematik und Evolutionsforschung* 24: 35-51.

Schönitzer, K., W. Grünwaldt, F. Gusenleitner, A. Z. Osytshnjuk, and J. Schuberth. 1995. Klärung von *Andrena forsterella,* mit Hinweisen zu den anderen Arten der *Andrena labialis*-Gruppe. *Linzer Biologische Beiträge* 27: 823-850.

Schönitzer, K., and C. Klinksik. 1990. The ethology of the solitary bee *Andrena nycthemera* Imhoff, 1866. *Entomofauna* 11: 377-427.

Schönitzer, K., and M. Renner. 1980. Morphologie der antennenputzapparate bei Apoidea. *Apidologie* 11: 113-130.

Schrader, M. N., and W. E. LaBerge. 1978. The nest biology of the bees *Andrena (Melandrena) regularis* Malloch and *Andrena (Melandrena) carlini* Cockerell. *Biological Notes, Illinois Natural History Survey,* no. 108: 1-24.

Schremmer, F. 1979. Zum Nest-Aufbau der neuen neotropischen Furchenbienen-Art *Neocorynura colombiana*. *Entomologia Generalis* 5: 149-154.

Schrottky, C. 1901. Biologische Notizen solitärer Bienen von S. Paulo (Brasilien). *Allgemeine Zeitschrift für Entomologie* 6: 209-216.

———. 1902a. Ensaio sobre as abelhas solitarias do Brazil. *Revista do Museo Paulista* 5: 330-613, pls. XII-XIV.

———. 1902b. Hyménoptères nouveau de l'Amérique Méridionale. *Anales del Museo Nacional de Buenos Aires* 7: 309-315.

———. 1905. Al conocimento de los Himenópteros del Paraguay. *Anales Científicos Paraguayos* (ser. 1)no. 4: 1-14.

———. 1906a. Neue und wenig bekannte südamerikanische Bienen. *Zeitschrift für Systematische Hymenopterologie und Dipterologie* 6: 305-316.

———. 1906b. Zur Synonymie der Apiden. *Zeitschrift für Systematische Hymenopterologie und Dipterologie* 6: 115-118.

———. 1909a. Nuevos himenópteros sudamericanos. *Revista del Museo de La Plata* 16: 137-149.

———. 1909b. Synonymische Bemerkungen über einige

südamerikanische Halictinae. *Deutsche Entomologische Zeitschrift* 1909: 479-485.
———. 1909c. Himenópteros de Catamarca. *Anales de la Sociedad Científica Argentina* 68: 233-272.
———. 1909d. Hymenoptera nova. *Anales de la Sociedad Científica Argentina* 67: 209-228.
———. 1910. Berichtigung (Hym.). *Deutsche Entomologische Zeitschrift* 1910: 540.
———. 1911. Descripção de abelhas novas do Brazil e de regiões visinhas. *Revista do Museo Paulista* 8: 71-88.
———. 1913a. La distribución geográfica de los Himenópteros argentinos. *Anales de la Sociedad Científica Argentina* 75: 115-144, 180-286.
———. 1913b. As especies brazileiras do genero *Megachile. Revista do Museu Paulista* 9: 134-223.
———. 1914. Einige neue Bienen aus Süd-Amerika. *Deutsche Entomologische Zeitschrift* 1914: 625-630.
———. 1920. Himenópteros nuevos o poco conocidos sudamericanos. *Revista do Museo Paulista* 12: 179-227.
Schuberth, J., and K. Schönitzer. 1993. Vergleichende Morphologie der Fovea facialis und der Stirnseitendrüse bei Apoidea und Sphecidae. *Linzer Biologische Beiträge* 25: 205-277.
Schultz, T. R., M. S. Engel, and J. S. Ascher. 2001. Evidence for the origin of eusociality in the corbiculate bees. *Journal of the Kansas Entomological Society* 74: 10-16.
Schultz, T. R., M. S. Engel, and M. Prentice. 1999. Resolving conflict between morphological and molecular evidence for the origins of eusociality in the "corbiculate" bees (Hymenoptera: Apidae): A hypothesis-testing approach, pp. 125-138 *in* G. W. Byers, R. H. Hagen, and R. W. Brooks, eds., *Entomoligical Contributions in Memory of Byron A. Alexander, University of Kansas Natural History Museum Special Publications* no. 24.
Schulz, W. A. 1906. *Spolia Hymenopterologica*. 356 pp. Paderborn: Pape.
———. 1911. Zweihundert alte Hymenopteren. *Zoologische Annalen Würzburg* 4: 1-220.
Schwammberger, K.-H. 1971a. Zwei neue Bienen-Arten aus Iran. *Stuttgarter Beiträge zur Naturkunde* no. 255: 1-4.
———. 1971b. Beitrag zur Kenntnis der Bienengattung *Rhophites* Spinola. *Bulletin des Recherches Agronomiques de Gembloux* (n.s.)6: 578-584.
———. 1975a. Die bisher bekanntgewordenen Arten der Bienengattung *Rhophitoides* Schenck. *Senckenbergiana Biologica* 56: 57-63.
———. 1975b. Zur Kenntnis der Bienengattung *Morawitzia* Friese. *Senckenbergiana Biologica* 56: 65-68.
Schwarz, H. F. 1926. North American *Dianthidium, Anthidiellum*, and *Paranthidium. American Museum Novitates* no. 226: 1-25.
———. 1927. Additional North American bees of the genus *Anthidium. American Museum Novitates* no. 253: 1-17.
———. 1928. Bees of the subfamily Anthidiinae, including some new species and varieties, and some new locality records. *Journal of the New York Entomological Society* 36: 369-419.
———. 1932. The genus *Melipona:* The type genus of the Meliponidae or stingless bees. *Bulletin of the American Museum of Natural History* 63: 231-460, pls. I-X.
———. 1934. The social bees (Meliponidae) of Barro Colorado Island, Canal Zone. *American Museum Novitates* no. 731: 1-23.
———. 1937. Results of the Oxford University Sarawak (Borneo) expedition: Bornean stingless bees of the genus *Trigona. Bulletin of the American Museum of Natural History* 73: 281-328, pls. II-VII.
———. 1938. The stingless bees (Meliponidae) of British Guiana and some related forms. *Bulletin of the American Museum of Natural History* 74: 437-508, pls. LII-LXII.
———. 1939a. The Indo-Malayan species of *Trigona. Bulletin of the American Museum of Natural History* 76: 83-141.
———. 1939b. A substitute name for *Patera* Schwarz. *Entomological News* 50: 23.
———. 1940. Additional species and records of stingless bees (Meliponidae) from British Guiana based on specimens collected by the Terry-Holden Expedition. *American Museum Novitates* no. 1078: 1-12.
———. 1948. Stingless bees (Meliponidae) of the Western Hemisphere. *Bulletin of the American Museum of Natural History* 90: i-xvii + 1-546.
———. 1949. The stingless bees (Meliponidae) of Mexico. *Anales del Instituto de Biología, México* 20: 357-370.
Schwarz, M. 1966. Beitrag zur Subfamilie Nomadinae. *Polskie Pismo Entomologiczne* 36: 383-394.
———. 1967. Die Gruppe der *Nomada cinctiventris* Fr. (= *stigma* auct. nec F.). *Polskie Pismo Entomologiczne* 37: 263-339.
———. 1990. Beitrag zur Kenntnis orientalischer *Nomada*-Arten. *Entomofauna* suppl. 5: 1-56.
———. 1993a. Revision der Gattung *Schmiedeknechtia* Friese, 1896, stat. rev. *Entomofauna* 14: 429-463.
———. 1993b. Eine neue *Thyreus*-Art von den Kanarischen Inseln, pp. 869- 873, pl. 8, *in* H. Hohmann, F. LaRoche, G. Ortega, and J. Barquin, Bienen, Wespen und Ameisen der Kanarischen Inseln. *Veröffentlichungen aus dem Übersee-Museum Bremen Naturwissenschaften* 12.
———. 2001. Revision der Gattang *Radoszkowskiana* Popov 1955 und ein Beitrag zur Kenntnis der Gattung *Coelioxys* Latreille 1809. *Linzer Biologische Beitrag* 33: 1267-1286.
Schwarz, M., F. Gussenleitner, and T. Kopf. 2005. Weitere Angaben zur Bienenfauna Österreichs somie Beschreibung einer neuen *Osmia*-Art Vorstudie zu einer Gesantbearbeitung der Bienen Österreichs VIII. *Entomofauna* 26: 117-164.
Schwarz, M., F. Gusenleitner, P. Westrich, and H. H. Dathe. 1996. Katalog der Bienen Österreichs, Deutschlands und der Schweiz. *Entomofauna* suppl. 8: 1-398.
Schwarz, M. P. 1986. Persistent multi-female nests in an Australian allodapine bee, *Exoneura bicolor. Insectes Sociaux* 33: 258-277.
Schwarz, M. P., N. J. Bull and S. J. B. Cooper. 2003. Molecular phylogenetics of allodapine bees, with implications for the evolution of sociality and progressive rearing, *Systematic Biology* 52: 1-14.
Schwarz, M. P., N. J. Bull, and K. Hogendoorn. 1998. Evolution of sociality in the alloapine bees: a review of sex allocation, ecology and evolution. *Insectes Sociaux* 45: 349-368.
Schwarz, M. P., and K. J. O'Keefe. 1991. Order of eclosion and reproductive differentiation in a social allodapine bee. *Ethology, Ecology and Evolution* 3: 233-245.

Schwarz, M. P., S. M. Tierney, J. Zammit, P. M. Schwarz, and S. Fuller. 2005. Brood provisioning and colony composition of a Malagasy species of *Halterapis:* Implications for social evolution in the allodapine bees. *Annals of the Entomological Society of America* 98: 126-133.

Schwenninger, H. R. 1999. *Die Wildbienen Stuttgarts,* [8] + 151 + [4] + 2 maps, Stuttgart: Schriftenreihe des Amtes für Umweltschutz-Heft 5/1999.

Scopoli, J. A. 1770. Dissertatio de Apibus, pp. 7-47 *in Annus IV Historico-Naturalis.* 152 pp. Lipsiae: Christ. Gottlob. Hilscher.

Scott, V. 1997. Pollen selection by three species of *Hylaeus* in Michigan. *Journal of the Kansas Entomological Society* 69 (1996, suppl.): 195-200.

Scudder, G. G. E. 1971. Comparative morphology of insect genitalia. *Annual Review of Entomology* 26: 379-406.

Serrão, J. E., C. da Cruz-Landim, and R. L. M. Silva-de-Moraes. 1997. Morphological and biochemical analysis of the stored and larval food of an obligate necrophagous bee, *Trigona hypogea. Insectes Sociaux* 44: 337-344.

Severinghaus, L. L., B. H. Kurtak, and G. C. Eickwort. 1981. The reproductive behavior of *Anthidium manicatum* (Hymenoptera, Megachilidae) and the significance of size for territorial males. *Behavioral Ecology and Sociobiology* 9: 51-58.

Shanks, S. S. 1978. A revision of the cleptoparasitic bee genus *Neolarra. Wasmann Journal of Biology* 35: 212-246.

———. 1986. A revision of the neotropical bee genus *Osiris. Wasmann Journal of Biology* 44: 1-56.

———. 1987. Two new species of *Osiris*, with a key to the species from Mexico. *Wasmann Journal of Biology* 45: 1-5.

Shanks Gingras, S. 1983. Taxonomic notes on the bee genus *Hexepeolus. Wasmann Journal of Biology* 41: 50-52.

Sheffield, C. S., S. M. Rigby, R. F. Smith, and P. G. Kevan. 2004. The rare cleptoparasitic bee *Epeoloides pilosula* (Hymenoptera: Apoidea: Apidae) discovered in Nova Scotia, Canada, with distributional notes. *Journal of the Kansas Entomological Society* 77: 161-164.

Sheppard, W. S., and B. A. McPheron. 1991. Ribosomal DNA diversity in Apidae, pp. 87-102 *in* D. R. Smith, ed., *Diversity in the Genus* Apis. Boulder, Colo.: Westview Press.

Sherman, P. W., E. A. Lacey, H. K. Reeve, and L. Keller. 1995. The eusociality continuum. *Behavioral Ecology* 6: 102-108.

Shinn, A. F. 1964. The bee genus *Xenopanurgus. Entomological News* 75: 73-78.

———. 1965. The bee genus *Acamptopoeum:* Diagnosis, key, and a new species. *Journal of the Kansas Entomological Society* 38: 278-284.

———. 1967. A revision of the bee genus *Calliopsis* and the biology and ecology of *C. andreniformis. University of Kansas Science Bulletin* 46: 753-936.

Shiokawa, M. 1963. Redescriptions of *Ceratina flavipes* Smith and *C. japonica* Cockerell. *Kontyû* 31: 276-280.

Shiokawa, M., and Y. Hirashima. 1982. Synopsis of the *flavipes*-group of the bee genus *Ceratina* of eastern Asia. *Esakia* no. 19: 177-184.

Shiokawa, M., and S. F. Sakagami. 1969. Additional notes on the genus *Pithitis* or green metallic small carpenter bees in the oriental region, with descriptions of two species from India. *Nature and Life in Southeast Asia* [Tokyo] 6: 139-149.

Shuckard, W. E. 1840. [Bees], pp. 158-171 *in* W. Swainson and W. E. Shuckard, *On the History and Natural Arrangement of Insects.* ii + 406 pp. London: Longman.

Sichel, J. 1867. Hymenoptera mellifera, pp. 143-156 *in* H. Saussure, ed., Hymenoptera . . . , *in Reise der Österreichischen Fregatte Novara um die Erde in den Jahren 1857, 1858, 1859 . . . ,* Zoologischer Theil 2. Wien: Gerold.

Sick, M., M. Ayasse, J. Tengö, W. Engels, G. Lübke, and W. Francke. 1994. Host-parasite relationships in six species of *Sphecodes* bees and their halictid hosts: Nest intrusion, intranidal behavior, and Dufour's gland volatiles. *Journal of Insect Behavior* 7: 101-117.

Silberbauer, L. X. 1997. The effect of non-synchronous dispersal on brood production in an allodapine bee, *Exoneura bicolor* Smith. *Insectes Sociaux* 44: 95-107.

Silberbauer, L. X., and M. P. Schwarz. 1995. Life cycle and social behavior in a heathland population of the allodapine bee, *Exoneura bicolor. Insectes Sociaux* 42: 201-218.

Silva-Matos, E. V. da, R. Zucchi, and S. Yamane. 1997. Oviposition behavior of stingless bees, XXII. *Oxytrigona tataira* and its successive oviposition process in the presence of generalized agitation. *Natural History Bulletin of Ibaraki University* 1: 121-134.

Silveira, F. A. 1993a. The mouthparts of *Ancyla* and the reduction of the labiomaxillary complex among long-tongued bees. *Entomologica Scandinavica* 24: 293-300.

———. 1993b. Phylogenetic relationships of the Exomalopsini and Ancylini. *University of Kansas Science Bulletin* 55: 163-173.

———. 1995a. Phylogenetic relationships and classification of Exomalopsini with a new tribe Teratognathini. *University of Kansas Science Bulletin* 55: 425-454.

———. 1995b. *Phylogenetic relationships and classification of Exomalopsini (Insecta: Apidae), with a revision of the* Exomalopsis (Phanomalopsis) jenseni *species-group and a catalog of the species of Exomalopsini.* vii + 168 pp. Lawrence: Ph.D. thesis, University of Kansas.

———. 2002. The bamboo-nesting carpenter bee, *Xylocopa (Stenoxylocopa) artifex* Smith (Hymenoptera: Apidae), also nests in fibrous branches of *Vellozia* (Velloziaceae). *Lundiana* 3: 57-60.

Silveira, F. A., G. A. R. Melo, and E. A. B. Almeida. 2002. *Albelhas Brasileiras, Sistemática e Identificação,* 251 pp. Belo Horizonte: Silveira.

Sinha, R. N. 1958. A subgeneric revision of the genus *Osmia* in the Western Hemisphere. *University of Kansas Science Bulletin* 39: 211-261.

Sinha, R. N., and C. D. Michener. 1958. A revision of the genus *Osmia*, subgenus *Centrosmia. University of Kansas Science Bulletin* 39: 275-303.

Sitdikov, A. A. 1988. Systematics of the bee genus *Eucera* Scopoli (Hymenoptera, Anthophoridae) of the USSR and neighboring countries: Subgenus *Pteneucera* Tkalcû. *Trudy Zoologicheskii Institut, Akademiia Nauk SSSR* 175: 102-111. [In Russian.]

Sitdikov, A. A., and Y. A. Pesenko. 1988. A subgeneric classification of bees of the genus *Eucera* Scopoli (Hymenoptera, Anthophoridae) with a scheme of the phylogenetic relationships between the subgenera. *Trudy Zoologicheskii Institut, Akademiia Nauk SSSR* 175: 75-101. [In Russian.]

Skorikov, A. S. 1914a. Les formes nouvelles des bourdons. *Russkoe Entomologicheskoe Obozrenie* 14: 119-129. [In Russian.]

———. 1914b. Contribution à la faune des bourdons de la partie meridionale de la province maritime. *Russkoe Entomologicheskoe Obozrenie* 14: 398-407. [In Russian.]

———. 1922a. The bumblebees of the Petrograd district. *Petrogradskii Agronomicheskii Institut, Entomologicheskaya Stantsiya. Fauna Petrogradskoi Gubernii* 2(c) no. 11: 1-51. [In Russian.]

Skorikov, A. S. 1922b. Les bourdons de la faune paléarctique, Partie 1, Biologie générale. *Izvestiya Severnoi Oblastnoi Stantsii Zashchity Rastnii ot Vreditelei [Bulletin de la Station Régionale Protectrice des Plantes à Petrograd]* 4: 1-160, 15 maps. [In Russian.]

———. 1933a. Zur Hummelfauna Japans und seiner Nachbarlandes. *Mushi* 6: 53-65, 2 figs.

———. 1933b. Zur Fauna und Zoogeographie der Hummeln des Himalaya. *Doklady Akademii Nauk SSSR* (n.s.)1(5): 243-248. [In Russian, German summary.]

———. 1937. Die gronlandischen Hummeln im Aspekte der Zirkumpolarfauna. *Entomologiske Meddelelser* 20: 37-64.

———. 1938a. Zoogeographic uniformity of the bumblebee fauna of the Causasus, Iran and Anatolia. *Entomologicheskoe Obozrenie* 27: 145-151. [In Russian, German summary.]

———. 1938b. Vorläufige Mitteilung über die Hummelfauna Burmas. *Arkiv för Zoologi* 30B: 1-3.

Sladen, F. W. L. 1912. *The Humble-Bee.* xiii + 283 pp. London: Macmillan.

———. 1915. The bee genus *Thrinchostoma* in India. *Canadian Entomologist* 47: 213-215.

———. 1916a. Bees of Canada—Fam. Megachilidae. *Canadian Entomologist* 48: 269-272.

———. 1916b. Canadian species of the bee genus *Stelis* Panz. *Canadian Entomologist* 48: 312-314.

Smit, J. 2004. De Wesbijen (*Nomada*) van Nederland. *Nederlandse Faunistische Mededelingen* 20: 33-125.

Smith, A. G., and J. C. Briden. 1977. *Mesozoic and Cenozoic Paleocontinental Maps.* Cambridge: Cambridge University Press.

Smith, A. G., D. G. Smith, and B. M. Funnell. 1994. *Atlas of Mesozoic and Cenozoic Coastlines.* ix + 99 pp. Cambridge: Cambridge University Press.

Smith, F. 1853, 1854. *Catalogue of Hymenopterous Insects in the Collection of the British Museum,* Part 1: [i] + 1-198, pls. i-vi, 1853; Part 2: 199-465, pls. vii-xii, 1854. London: British Museum.

———. 1857, 1858. Catalogue of the hymenopterous insects collected at Sarawak, Borneo; Mount Ophir, Malacca; and at Singapore by A. R. Wallace. *Journal of the Proceedings of the Linnean Society of London, Zoology* 2: 42-88 and pls. 1, 2, 1857; 89-130, 1858.

———. 1861. Descriptions of new genera and species of exotic Hymenoptera. *Journal of Entomology* [London] 1: 146-155.

———. 1865. Descriptions of some new species of hymenopterous insects belonging to the families Thynnidae, Masaridae and Apidae. *Transactions of the Entomological Society of London* (3)2: 389-399, pl. 21.

———. 1868a. Descriptions of aculeate Hymenoptera from Australia. *Transactions of the Entomological Society of London* 1868: 231-258.

———. 1868b. [A new name for *Oestropsis.*] *Proceedings of the Entomological Society of London,* p. xxxix.

———. 1874. Revision of the genera *Epicharis, Centris, Eulema* (sic) and *Euglossa. Annals and Magazine of Natural History* (4)13: 317-321, 357-372, 440-446.

———. 1875. Descriptions of new species of Indian aculeate Hymenoptera, collected by Mr. G. R. James Rothney, member of the Entomological Society. *Transactions of the Entomological Society of London* 1875: 33-50.

———. 1879. *Descriptions of New Species of Hymenoptera in the Collection of the British Museum.* xxi + 240 pp. London: British Museum.

Smith-Pardo, A. H. 2003. A preliminary account of the bees of Colombia (Hymenoptera: Apoidea): present knowledge and future directions. *Journal of the Kansas Entomological Society* 76: 335-344.

———. 2005. The bees of the genus *Neocorynura* of Mexico. *Folia Entomológica Mexicana* 44: 165-193.

Smith-Pardo, A. H., and M. S. Engel. 2005. The bee genus *Micrommation* (Hymenoptera: Halictidae), a new diagnosis and description of the male. *Folia Heyroveskyana* 12: 179-189.

Snelling, R. R. 1956. Bees of the genus *Centris* in California. *Pan-Pacific Entomologist* 32: 1-8.

———. 1966a. Studies on North American bees of the genus *Hylaeus* 3. The nearctic subgenera. *Bulletin of the Southern California Academy of Sciences* 65: 164-175.

———. 1966b. Studies on North American bees of the genus *Hylaeus,* 1. Distribution of the western species of the subgenus *Prosopis* with description of new forms. *Contributions in Science, Los Angeles County Museum of Natural History* no. 98: 1-18.

———. 1966c. Studies on North American bees of the genus *Hylaeus,* 2. Description of a new subgenus and species. *Proceedings of the Biological Society of Washington* 79: 139-144.

———. 1966d. A new species of *Heteranthidium* from California. *Contributions in Science, Los Angeles County Museum of Natural History* no. 97: 1-8.

———. 1967. Description of a new subgenus of *Osmia. Bulletin of the Southern California Academy of Sciences* 66: 103-108.

———. 1968. Studies on North American bees of the genus *Hylaeus,* 4. The subgenera *Cephalylaeus, Metziella* and *Hylaeana. Contributions in Science, Los Angeles County Museum of Natural History* no. 144: 1-6.

———. 1969. The Philippine subgenus *Hoploprosopis* of *Hylaeus. Contributions in Science, Los Angeles County Museum of Natural History* no. 171: 1-5.

———. 1970. Studies on North American bees of the genus *Hylaeus,* 5. The subgenera *Hylaeus* s. str. and *Paraprosopis. Contributions in Science, Los Angeles County Museum of Natural History* no. 180: 1-59.

———. 1974. Notes on the distribution and taxonomy of some North American *Centris. Contributions in Science, Natural History Museum of Los Angeles County* no. 259: 1-41.

———. 1975. Taxonomic notes on some colletid bees of western North America with descriptions of new species. *Contributions in Science, Natural History Museum of Los Angeles County* no. 267: 1-9.

———. 1980. New bees of the genus *Hylaeus* from Sri

Lanka and India. *Contributions in Science, Natural History Museum of Los Angeles County* no. 328: 1-18.

———. 1982. The taxonomy of some neotropical *Hylaeus* and descriptions of new taxa. *Bulletin of the Southern California Academy of Sciences* 81: 1-25.

———. 1983a. Studies on North American bees of the genus *Hylaeus*, 6. On adventive Palaearctic species in southern California. *Bulletin of the Southern California Academy of Sciences* 82: 12-16.

———. 1983b. North American species of the bee genus *Lithurge*. *Contributions in Science, Natural History Museum of Los Angeles County* no. 343: 1-11.

———. 1984. Studies on the taxonomy and distribution of American centridine bees. *Contributions in Science, Natural History Museum of Los Angeles County* no. 347: 1-69.

———. 1985. The systematics of the hylaeine bees (Hymenoptera: Colletidae) of the Ethiopian zoological region: The genera and subgenera with revisions of the smaller groups. *Contributions in Science, Natural History Museum of Los Angeles County* no. 361: 1-33.

———. 1986a. The taxonomic status of two North American *Lithurge*. *Bulletin of the Southern California Academy of Sciences* 85: 29-34.

———. 1986b. Contributions toward a revision of the new world nomadine bees: A partitioning of the genus *Nomada*. *Contributions in Science, Natural History Museum of Los Angeles County* no. 376: 1-32.

———. 1987. A revision of the bee genus *Aztecanthidium*. *Pan-Pacific Entomologist* 63: 165-171.

———. 1990. A review of the native North American bees of the genus *Chalicodoma*. *Contributions in Science, Natural History Museum of Los Angeles County* no. 421: 1-39.

———. 1994. *Diadasia*, subgenus *Dasiapis*, in North America. *Contributions in Science, Natural History Museum of Los Angeles County* no. 448: 1-8.

———. 2003. Bees of the Hawaiian Islands, exclusive of *Hylaeus (Nesoprosopis)*. *Journal of the Kansas Entomological Society* 76: 343-356.

Snelling, R. R., and R. W. Brooks. 1985. A review of the genera of cleptoparasitic bees of the tribe Ericrocini. *Contributions in Science, Natural History Museum of Los Angeles County* no. 369: 1-34.

Snelling, R. R., and B. N. Danforth. 1992. A review of *Perdita*, subgenus *Macrotera*. *Contributions in Science, Natural History Museum of Los Angeles County* no. 436: 1-12.

Snelling, R. R., and J. G. Rozen, Jr. 1987. Contributions toward a revision of the new world nomadine bees, 2. The genus *Melanomada*. *Contributions in Science, Natural History Museum of Los Angeles County* no. 384: 1-12.

Snelling, R. R., and G. I. Stage. 1995a. Systematics and biology of the bee genus *Xeralictus*. *Contributions in Science, Natural History Museum of Los Angeles County* no. 451: 1-17.

———. 1995b. A revision of the nearctic Melittidae: The subfamily Melittinae. *Contributions in Science, Natural History Museum of Los Angeles County* no. 451: 19-31.

Snelling, R. R., and T. J. Zavortink. 1984. A revision of the cleptoparasitic bee genus *Ericrocis*. *Wasmann Journal of Biology* 42: 1-26.

Snodgrass, R. E. 1941. The male genitalia of Hymenoptera. *Smithsonian Miscellaneous Collections* 99(14): 1-86, 33 pls.

———. 1942. The skeleto-muscular mechanisms of the honey bee. *Smithsonian Miscellaneous Collections* 103(2): 1-120.

———. 1956. *Anatomy of the Honey Bee*. xiv + 334 pp. Ithaca, New York: Cornell University Press.

Snyder, T. P. 1977. A new electrophoretic approach to biochemical systematics of bees. *Biochemical Systematics and Ecology* 5: 133-150.

Snyder, T. P., E. M. Barrows, and M. R. Chabot. 1976. Nests of *Diadasia afflicta* Cresson. *Journal of the Kansas Entomological Society* 49: 200-203.

Sommeijer, M. J., and L. L. M. de Bruijn. 2003. Why do workers of *Melipona favosa* chase their sister-gynes out of the nest? *Proceedings, Experimental and Applied Entomology, Nederlandse Entomologische Vereniging* 14: 45-48.

Sommeijer, M. J., L. L. M. de Bruijn, and F. J. A. J. Meeuwsen. 2003. Reproductive behavior of stingless bees: solitary gynes of *Melipona favosa* (Hymenoptera: Apidae, Meliponini) can penetrate existing nests. *Entomologische Berichten* 63(2): 31-35.

Sommeijer, M. J., L. L. M. de Bruijn, F. J. A. J. Meeuwsen, and E. J. Slaa. 2003. Reproductive behavior of stingless bees: nest departures of non-accepted gynes and nuptial flights in *Melipona favosa*. *Entomologische Berichten* 63(1): 7-13.

Soucy, S. L. 2002. Nesting biology and socially polymorphic behavior of the sweat bee *Halictus rubicundus*. *Annals of the Entomological Society of America* 95: 57-65.

Soucy, S. L., and B. N. Danforth. 2002. Phylogeography of the socially polymorphic sweat bee *Halictus rubicundus*. *Evolution* 56: 330-341.

Soucy, S. L., T. Giray, and D. W. Roubik. 2003. Solitary and group nesting in the orchid bee *Euglossa hyacinthina*. *Insectes Sociaux* 50: 248-255.

Spessa, A., M. P. Schwarz, and M. Adams. 2000. Sociality in *Amphylaeus morosus*. *Annals of the Entomological Society of America* 93: 684-692.

Spinola, M. 1808. *Insectorum Liguriae Species Novae aut Rariores...*, Vol. 2, fasc. 2-4, 262 pp. Genuae: Printed for the author. [Fasc. 2, pp. 1-81, published in 1807, according to D. Baker.]

———. 1839. Compte-rendu des Hyménoptères recueillis par M. Fischer pendant son voyage en Egypte.... *Annales de la Société Entomologique de France* 7(1838): 437-546.

———. 1843. Sur quelques Hyménoptères peu connus, recueillis en Espagne, pendant l'annee 1842, par M. Victor Ghiliani, voyageur-naturaliste. *Annales de la Société Entomologique de France* (2)1: 111-144.

———. 1851. Hymenópteros, pp. 153-569 *in* C. Gay, *Historia Fisica y Politica de Chile...*, Zoologia, Vol. 6. Paris: Casa del autor.

———. 1853. Compte rendu des hyménoptères inédits provenants du voyage entomologique de M. Ghiliani dans le Para en 1846. *Memorie della Reale Accademia delle Scienze di Torino* (2)13: 19-94.

Stange, L. A. 1983. A synopsis of the genus *Epanthidium* Moure with the description of a new species from northeastern Mexico. *Pan-Pacific Entomologist* 59: 281-297.

———. 1995. Further description of *Ananthidium* Urban, with keys to Argentine Anthidiini. *Insecta Mundi* 9: 11-16.

Starr, C. K. 1979. Origin and evolution of insect sociality: A review of modern theory, pp. 35-79 *in* H. R. Hermann, ed., *Social Insects,* Vol. 1. New York: Academic Press.

———. 1989. *Bombus folsomi* and the origin of Philippine bumble bees. *Systematic Entomology* 14: 411-415.

———. 1992. The bumble bees (Hymenoptera: Apidae) of Taiwan. *Bulletin of the National Museum of Natural Science* [Taiwan] no. 3: 139-157.

Steiner, K. E., and V. B. Whitehead. 1990. Pollinator adaptation to oil-secreting flowers—*Rediviva* and *Diascia. Evolution* 44: 1701-1707.

———. 1991. Oil flowers and oil bees: Further evidence for pollinator adaptation. *Evolution* 45: 1493-1501.

Stephen, W. P. 1954. A revision of the bee genus *Colletes* in America north of Mexico. *University of Kansas Science Bulletin* 36: 149-527.

———. 1957. Bumble bees of western America. *Oregon Agricultural Experiment Station Technical Bulletin* no. 40: 1-163.

———. 1961. Biological observations on *Emphoropsis miserabilis* (Cresson), with comparative notes on other anthophorids. *Annals of the Entomological Society of America* 54: 687-692.

———. 1966. *Andrena (Cryptandrena) viburnella,* I. Bionomics. *Journal of the Kansas Entomological Society* 39: 42-51.

Stephen, W. P., G. E. Bohart, and P. F. Torchio. 1969. *The Biology and External Morphology of Bees.* ii + 140 pp. Corvallis: Agricultural Experiment Station, Oregon State University.

Stephen, W. P., and T. Koontz. 1973. The larvae of the Bombini, I and II. *Melanderia* 13: 1-12, 13-29.

Stockhammer, K. A. 1966. Nesting habits and life cycle of a sweat bee, *Augochlora pura. Journal of the Kansas Entomological Society* 39: 157-192.

———. 1967. Some notes on the biology of the blue sweat bee, *Lasioglossum coeruleum. Journal of the Kansas Entomological Society* 40: 177-189.

Stöckhert, E. 1941. Über die Gruppe der *Nomada zonata* Panz *Mitteilungen der Münchner Entomologischen Gesellschaft* 31: 1072-1122.

———. 1943. Über die Gruppe der *Nomada furva* Panz. *Deutsche Entomologische Zeitschrift* 1943: 89-126.

Stoeckhert, F. K. 1933. Die Bienen Frankens. *Beihefte, Deutsche Entomologische Zeitschrift* 1932: i-vii + 1-294.

———. 1954. Fauna Apoideorum Germaniae. *Abhandlungen der Bayerischen Akademie der Wissenschaften* (n.f.)65: 1-87. [Anhang by H. Bischoff, pp. 70-73.]

Strand, E. 1910. Apidologisches aus dem Naturhistorischen Museum zu Wiesbaden. *Jahrbücher des Nassauischen Vereins für Naturkunde* 63: 37-45.

———. 1913a. Apidae von Ceylon, gesammelt 1899 von Herrn. Dr. W. Horn. *Archiv für Naturgeschichte,* Abt. A 79(2): 135-150.

———. 1913b. H. Sauter's Formosa-Ausbeute, Apidae I. *Supplementa Entomologica* no. 2: 23-67.

———. 1913c. Bestimmungstabelle nebst weiteren Beiträgen zur Kenntnis afrikanischer *Nomia*-Arten. *Archiv für Naturgeschichte,* Abt. A 79(10): 121-144.

———. 1914a. H. Sauter's Formosa-Ausbeute, Apidae III. *Archiv für Naturgeschichte,* Abt. A 80(1): 136-144.

———. 1914b. Zur Kenntnis afrikanischer Arten der Bienengattung *Allodape* Lep. *Archiv für Naturgeschichte,* Abt. A 80(12): 34-60.

———. 1921. Apidologisches, insbesondere über paläarktische *Andrena*-Arten, auf Grund von Material des Deutschen Entomologischen Museums. *Archiv für Naturgeschichte,* Abt. A 87(3): 266-304.

———. 1926. Miscellanea nomenclatorica zoologica et palaeontologica. *Archiv für Naturgeschichte,* Abt. A 92(8): 30-75.

———. 1932. Miscellanea nomenclatorica zoologica et palaeontologica. *Folia Zoologica et Hydrobiologica* 4: 193-196.

Strohl, J. 1908. *Die Copulationsanhange der solitären Apiden und de Artentstehung durch "Physiologische Iso-lierung."* 52 pp., 3 double pls. Freiburg: Inaugural-Dissertation, Albert-Ludwigs-Universität.

Sustera, O. 1958. Übersicht des Systems der paläarktischen und mitteleuropäischen Gattungen der Superfamilie Apoidea. *Acta Entomologica Musei Nationalis Prague* 32: 443-463.

———. 1959. Bestimmungstabelle der Tschechoslowakischen Arten der Bienengattung *Sphecodes* Latr. *Časopis Československé Společnosti Entomologické* 56: 169-180.

Svensson, B. G., and J. Tengö. 1976. *Andrena* (Hym., Apoidea) on the island of Öland, Sweden. *Entomologisk Tidskrift* 97: 78-89.

Swenk, M. H. 1907. The bees of Nebraska, III. *Entomological News* 18: 293- 300.

———. 1908. Specific characters in the bee genus *Colletes. University of Nebraska Contributions from the Department of Entomology* no. 1: 43-102, 3 pls.

———. 1912. Studies of North American bees, I. Family Nomadidae. *University Studies* [University of Nebraska, Lincoln] 12: 1-113.

Tadauchi, O. 1985. Synopsis of *Andrena (Micrandrena)* of Japan. *Journal of the Faculty of Agriculture, Kyushu University* 30: 59-76, 77-94.

Tadauchi, O., and Y. Hirashima. 1983. New or little known bees of Japan (Hymenoptera, Apoidea), IV. Supplement to *Andrena (Simandrena). Esakia* no. 20: 81-92.

———. 1984a. Synopsis of *Andrena (Euandrena)* of Japan. *Esakia* no. 22: 107-113.

———. 1984b. New or little known bees of Japan (Hymenoptera, Apoidea), V. Supplement to *Andrena (Hoplandrena). Kontyû* 52: 278-285.

———. 1988. Synopsis of *Andrena (Stenomelissa)* with a new species from Japan. *Journal of the Faculty of Agriculture, Kyushu University* 33: 67-76.

Tadauchi, O., Y. Hirashima, and T. Matsumura. 1987. Synopsis of *Andrena (Andrena)* of Japan. *Journal of the Faculty of Agriculture, Kyushu University* 31: 11-35, 37-54.

Tadauchi, O., and Xu H.-l. 1995. A revision of the subgenus *Simandrena* of the genus *Andrena* of eastern Asia with a key to palearctic species. *Esakia* no. 35: 201-222.

———. 1998. A revision of the subgenus *Holandrena* of the genus *Andrena* of eastern Asia. *Entomological Science* 1: 137-143.

———. 2000. A revision of the subgenus *Poecilandrena* of the genus *Andrena* of the eastern Asia. *Insecta Koreana* 17: 79-90.

———. 2002. A revision of the subgenus *Cnemidandrena* of the genus *Andrena* of Eastern Asia. *Esakia* no. 42: 75-119.

———. 2003. A revision of the subgenus *Taeniandrena* of the genus *Andrena* of eastern Asia. *Esakia* no. 43: 65-95.

———. 2004. The subgenus *Cordandrena* of the genus *Andrena* newly recorded from eastern Asia, with a new species. *Esakia* no. 44: 81-90.

Taschenberg, E. 1883. Die Gattungen der Bienen (Anthophila). *Berliner Entomologische Zeitschrift* 27: 37-100.

Tasei, J.-N. 1972. Observations préliminaires sur la biologie d'*Osmia (Chalcosmia) coerulescens* L. (Hymenoptera Megachilidae), pollinisatrice de la luzerne (*Medicago sativa* L.). *Apidologie* 3: 149-165.

———. 1973. Le comportement de nidification chez *Osmia (Osmia) cornuta* Latr. et *Osmia (Osmia) rufa* Latr. *Apidologie* 4: 195-225.

Taylor, J. S. 1962a. Notes on *Heriades freygessneri* Schletterer. *Journal of the Entomological Society of Southern Africa* 25: 133-139.

———. 1962b. A note on *Nothylaeus heraldicus* (Smith), the membrane bee. *Pan-Pacific Entomologist* 38: 244-248.

Taylor, T. N., and E. L. Taylor. 1993. *The Biology and Evolution of Fossil Plants.* xxii + 982 pp. Englewood Cliffs, N.J.: Prentice Hall.

Tengö, J., and G. Bergström. 1976. Odor correspondence between *Melitta* females and males of their nest parasite *Nomada flavopicta* K. *Journal of Chemical Ecology* 2: 57-65.

———. 1977. Cleptoparasitism and odor mimetism in bees: Do *Nomada* males imitate the odor of *Andrena* females? *Science* 196: 1117-1119.

Terzo, M. 1999. Révision du genre *Exoneuridia* Cockerell, 1911. *Belgian Journal of Entomology* 1: 137-152.

———. 2000. *Classification phylogénétique des cératines du monde et monographie des espèces de la région ouest-paléarctique et de l'Asie Centrale,* pp. 4 + 263 + I-XXII, Thesis, Universite de Mons-Hainaut, Belgium.

Terzo, M., and P. Rasmont. 2004. Biogéographie et systématique des abeilles rubicoles du genre *Ceratina* Latreille au Turkestan. *Annals de la Société Entomologique de France* (n.s.) 40: 109-130.

Teunissen, H. G. M., and C. van Achterberg. 1992. *Osmia zandeni,* a new species from Fuerteventura, Canary Islands. *Zoologische Mededelingen* 66: 313-315.

Thiele, R. 2002. Nesting biology and seasonality of *Duckeanthidium thielei* Michener (Hymenoptera: Megachilidae), an oligolectic rainforest bee. *Journal of the Kansas Entomological Society* 75: 274-282.

———. 2003. A review of Central American *Centris (Heterocentris)* and evidence for male dimorphism in *C. labrosa. Mitteilungen aus dem Museum for Naturkunde in Berlin, Deutsche Entomologische Zeitschrift* 50: 237-242.

Thomson, C. G. 1872. *Skandinaviens Hymenoptera,* Vol. 2: pp. 1-286. Lund: Berling.

Thorp, R. W. 1963. A new species of the genus *Trachusa* from California with a key to the known species. *Pan-Pacific Entomologist* 39: 56-58.

———. 1966. A synopsis of the genus *Heterostelis* Timberlake. *Journal of the Kansas Entomological Society* 39: 131-146.

———. 1968. Ecology of a *Proteriades* and its *Chrysura* parasite, with larval descriptions. *Journal of the Kansas Entomological Society* 41: 324-331.

———. 1969a. Systematics and ecology of bees of the subgenus *Diandrena. University of California Publications in Entomology* 52: [iii] + 1-146.

———. 1969b. Ecology and behavior of *Melecta separata callura. American Midland Naturalist* 82: 338-345.

———. 1969c. Ecology and behavior of *Anthophora edwardsii. American Midland Naturalist* 82: 321-337.

———. 1979. Structural, behavioral, and physiological adaptations of bees (Apoidea) for collecting pollen. *Annals of the Missouri Botanical Garden* 66: 788-812.

Thorp, R. W., and R. W. Brooks. 1994. A revision of new world *Trachusa,* subgenera *Ulanthidium* and *Trachusomimus. University of Kansas Science Bulletin* 55: 271-297.

Thorp, R. W., and J. A. Chemsak. 1964. Biological observations on *Melissodes (Eumelissodes) pallidisignata. Pan-Pacific Entomologist* 40: 75-83.

Thorp, R. W., D. S. Horning, Jr., and L. L. Dunning. 1983. Bumble bees and cuckoo bumble bees of California. *Bulletin of the California Insect Survey* 23: viii + 79 pp.

Thorp, R. W., and W. E. La Berge. 2005. A revision of the bees of the genus *Andrena* of the western hemisphere. Part XV. Subgenus *Hesperandrena. Illinois Natural History Survey Bulletin* 37: i-ii + 65-93.

Tierney, S. M., M. P. Schwarz, T. Neville, and P. M. Schwarz. 2002. Sociality in the phylogenetically basal allodapine bee genus *Macrogalea* (Apidae: Xylocopinae): implications for social evolution in the tribe Allodapini. *Biological Journal of the Linnean Society* 76: 211-224.

Timberlake, P. H. 1939. New species of bees of the genus *Dufourea* from California (Hymenoptera, Apoidea). *Annals of the Entomological Society of America* 32: 395-414.

———. 1941a. Ten new species of *Stelis* from California. *Journal of the New York Entomological Society* 49: 123-137.

———. 1941b. Synoptic table of North American species of *Diadasia. Bulletin of the Brooklyn Entomological Society* 36: 1-11.

———. 1943a. Racial differentiation in nearctic species of *Dianthidium. Journal of the New York Entomological Society* 51: 71-109.

———. 1943b. Bees of the genus *Colletes* chiefly from Colorado. *Bulletin of the American Museum of Natural History* 81: 385-410.

———. 1947. A revision of the species of *Exomalopsis* inhabiting the United States. *Journal of the New York Entomological Society* 55: 85-106.

———. 1952a. Descriptions of new species of *Nomadopsis* from California and Texas, and of a new allied genus from South America. *Annals of the Entomological Society of America* 45: 104-118.

———. 1952b. A new name for the bee genus *Ruiziella. Annals of the Entomological Society of America* 45: 528.

———. 1953a. Erratum. *Annals of the Entomological Society of America* 46: 598.

———. 1953b. Bees of the genus *Perdita* in the collection of the University of Kansas. *University of Kansas Science Bulletin* 35: 961-985.

———. 1954. A revisional study of the bees of the genus

Perdita F. Smith, with special reference to the fauna of the Pacific coast, Part I. *University of California Publications in Entomology* 9: 345-432.

———. 1955a. A new genus for two new species of dufoureine bees from California. *Pan-Pacific Entomologist* 31: 105-108.

———. 1955b. Notes on the species of *Psaenythia* of North America. *Bollettino del Laboratorio di Zoologia Generale e Agraria "Filippo Silvestri"* [Portici] 33: 398-409.

———. 1956. A revisional study of the bees of the genus *Perdita* F. Smith, with special reference to the fauna of the Pacific coast, Part II. *University of California Publications in Entomology* 11: 247-350.

———. 1958. A revisional study of the bees of the genus *Perdita* F. Smith, with special reference to the fauna of the Pacific coast, Part III. *University of California Publications in Entomology* 14: 303-410.

———. 1960. A revisional study of the bees of the genus *Perdita* F. Smith, with special reference to the fauna of the Pacific coast, Part IV. *University of California Publications in Entomology* 17: 1-156.

———. 1961. A review of the genus *Conanthalictus*. *Pan-Pacific Entomologist* 37: 145-160.

———. 1962. A revisional study of the bees of the genus *Perdita* F. Smith, with special reference to the fauna of the Pacific coast, Part V. *University of California Publications in Entomology* 28: 1-124.

———. 1964a. A revisional study of the bees of the genus *Perdita* F. Smith, with special reference to the fauna of the Pacific coast, Part VI. *University of California Publications in Entomology* 28: 125-388.

———. 1964b. Some new species of *Pseudopanurgus* of the subgenus *Heterosarus* Robertson. *American Museum Novitates* no. 2185: 1-26.

———. 1967. New species of *Pseudopanurgus* from Arizona. *American Museum Novitates* no. 2298: 1-23.

———. 1968. A revisional study of the bees of the genus *Perdita* F. Smith, with special reference to the fauna of the Pacific coast, Part VII (including index to parts I to VII). *University of California Publications in Entomology* 49: 1-196.

———. 1969a. *Metapsaenythia*, a new panurgine bee genus. *Entomological News* 80: 89-92.

———. 1969b. A contribution to the systematics of North American species of *Synhalonia*. *University of California Publications in Entomology* 57: i-vi + 1-76.

———. 1971. Supplementary studies on the systematics of the genus *Perdita*. *University of California Publications in Entomology* 66: i-vi + 1-63.

———. 1973. Revision of the genus *Pseudopanurgus* of North America. *University of California Publications in Entomology* 72: i-vi + 1-58.

———. 1975. The North American species of *Heterosarus* Robertson. *University of California Publications in Entomology* 77: i-vi + 1-56, pls. 1-8.

———. 1976. Revision of the North American bees of the genus *Protandrena* Cockerell. *Transactions of the American Entomological Society* 102: 133-227.

———. 1980a. Supplementary studies on the systematics of the genus *Perdita*, Part II. *University of California Publications in Entomology* 85: i-vii + 1-65.

———. 1980b. Review of North American *Exomalopsis*, Parts I-IV. *University of California Publications in Entomology* 86: i-vi + 1-158.

Timberlake, P. H., and C. D. Michener. 1950. The bees of the genus *Proteriades*. *University of Kansas Science Bulletin* 33: 387-440.

Tingek, S., G. Koeniger, and N. Koeniger. 1996. Description of a new cavity nesting species of *Apis* (*Apis nuluensis* n. sp.) from Sabah, Borneo, with notes on its occurrence and reproductive biology. *Senkenbergiana Biologica* 76: 115-119.

Tirgari, S. 1968. La choix du site de nidification par *Melitta leporina* (Panz.) et *Melitturga clavicornis* (Latr.). *Annales de l'Abeille* 11: 79-103.

Titus, E. S. G. 1901. A new genus in the Coelixinae [sic]. *Canadian Entomologist* 33: 256.

———. 1904a. Notes on Osmiinae with descriptions of new genera and species. *Journal of the New York Entomological Society* 12: 22-27.

———. 1904b. Some new Osmiinae in the United States National Museum. *Proceedings of the Entomological Society of Washington* 6: 98-102.

———. 1906. Some notes on the Provancher Megachilidae. *Proceedings of the Entomological Society of Washington* 7: 149-165.

Tkalců, B. 1966a. *Megabombus* (*Fervidobombus*) *abditus* sp. n. aus äquatorial-Afrika. *Časopis Moravského Musea* 51: 271-274.

———. 1966b. *Metallinella* gen. n. der Familie Megachilidae. *Acta Entomologica Bohemoslovaca* 63: 200-202.

———. 1967. Bemerkungen zur Taxonomie einiger paläarktischer Arten der Familie Megachilidae. *Acta Entomologica Bohemoslovaca* 64: 91-104.

———. 1968. Revision der Arten der Untergattung *Tricornibombus* Skorikov. *Zborník Slovenského Národného Muzea, Prírodné Vedy [Acta Rerum Naturalium Musei Nationalis Slovaci, Bratislava]* 14: 79-94.

———. 1969a. Beiträge zur Kenntnis der Fauna Afghanistans—Osmiini, Megachilidae, Apoidea, Hym. *Acta Musei Moraviae* 54(suppl.): 327-346.

———. 1969b. Beiträge zur Kenntnis der Fauna Afghanistans—*Chalicodoma* Lep., Megachilidae, Apoidea, Hym. *Acta Musei Moraviae* 54(suppl.): 347-384, 6 unnumbered plates.

———. 1969c. Sur la position systématique du sous-genre *Metamegachile*. *Bulletin de la Société Entomologique de Mulhouse,* July-August 1969: 65-67.

———. 1971. *Xenochalicodoma*, nomen novum. *Bulletin de la Société Entomologique de Mulhouse* 1971: 34.

———. 1972, 1974a. Arguments contre l'interprétation traditionnelle de la phylogénie des abeilles. *Bulletin de la Société Entomologique de Mulhouse,* April-June 1972: 17-28; April-June 1974: 17-40.

———. 1974b. Ergebnisse der Albanien-Expedition 1961 des "Deutschen Entomologischen Institutes," Hymenoptera: Apoidea V (Megachilidae). *Beiträge zur Entomologie* [Berlin] 24: 323-348.

———. 1974c. Revision und Klassifikation der bisher zur Untergattung *Hoplosmia* Thomson gestellten *Anthocopa*-Arten. *Acta Entomologica Bohemoslovaca* 71: 114-135.

———. 1974d. Ergebnisse der 1. und 2. mongolisch-tschechoslowakischen entomologisch-botanischen Expedition in der Mongolei, Nr. 29: Hymenoptera, Apoidea, Bombinae. *Acta Faunistica Entomologica Musei Nationalis Pragae* 15: 25-58.

———. 1974e. Ergebnisse der 1. und 2. mongolisch-tschechoslowakischen entomologisch-botanischen Expedition in der Mongolei, Nr. 28: Hymenoptera, Apoidea, *Melitturga* Latr. *Acta Faunistica Entomologica Musei Nationalis Pragae* 15: 21-24.

———. 1975a. Die *Osmia*-Arten der Untergattung *Hemiosmia* subgen. n. *Acta Entomologica Bohemoslovaca* 72: 34-49.

———. 1975b. Revision der europäischen *Osmia (Chalcosmia)*-Arten der *fulviventris*-gruppe. *Věstník Československé Společnosti Zoologické* 39: 297-317.

———. 1975c. Sammelergebnisse der von RNDr. A. Hoffer geleiteten Algerien-Expeditionen in den Jahren 1971 und 1972 (Hymenoptera: Apoidea), 1. Teil: Megachilidae. *Zborník Slovenského Národného Muzea, Prírodné Vedy [Acta Rerum Naturalium Musei Nationalis Slovaci, Bratislava]* 21: 165-190.

———. 1977a. Taxonomisches zu einigen paläarktischen Bienenarten. *Věstník Československé Společnosti Zoologické* 41: 223-239.

———. 1977b. Die *Osmia*-Arten der Untergattung *Neosmia* Tkalců. *Acta Entomologica Bohemoslovaca* 74: 85-102.

———. 1978a. Beiträge zur Kenntnis der Fauna Afghanistans: *Melitturga* Latr., *Eucera* Scop., Apidae; *Lithurge* Latr., *Stelis* Pz., *Creightonella* Cockll., Megachilidae, Apoidea, Hym. *Časopis Moravského Musea* 63: 153-181.

———. 1978b. Fünf neue paläarktische Arten der Familie Megachilidae. *Časopis Slezského Muzea vedy Prírodné* (A)27: 153-169.

———. 1979a. Neue paläarktische Taxa der Familie Megachilidae. *Acta Entomologica Bohemoslovaca* 76: 318-329.

———. 1979b. Revision der europäischen Vertreter der Artengruppe von *Tetralonia ruficornis* (Fabricius). *Časopis Moravského Musea* 64: 127-152, 2 pls.

———. 1980. Zwei neue Arten der Gattung *Wainia* gen. n. aus Vorderindian. *Annotationes Zoologicae et Botanicae* [Bratislava], no. 135: 1-20.

———. 1983. Die europäischen *Osmia*-Arten der Untergattung *Melanosmia*. *Věstník Československé Společnosti Zoologické* 47: 140-159.

———. 1984a. Neue paläarktische Arten der Gattungen *Pseudoheriades* und *Archeriades* mit Beschreibung von *Hofferia* gen. n. *Annotationes Zoologicae et Botanicae* [Bratislava], no. 158: 1-22.

———. 1984b. Revision der gattung *Cubitalia* Friese, 1911. *Annotationes Zoologicae et Botanicae* [Bratislava], no. 161: 1-15.

———. 1984c. Systematisches Verzeichnis der westpaläarktischen *Tetralonia*- und *Eucera*-Arten, deren Männchen als Blütenbesucher verschiedener *Ophrys*-Arten festgestellt wurden; mit Beschreibung neuer Taxa. *Nova Acta Regiae Societatis Scientiarum Upsaliensis* (Ser. V:C)3: 57-77.

———. 1993a. Quatre nouveaux sous-genres paléarctiques de la tribu des Osmiini. *Bulletin de la Société Entomologique de Mulhouse,* July- September: 55-56.

———. 1993b. Neue Taxa der Bienen von den Kanarischen Inseln, pp. 791-858 *in* H. Hohmann, F. LaRoche, G. Ortega, and J. Barquin, Bienen, Wespen und Ameisen der Kanarischen Inseln. *Veröffentlichungen aus dem Übersee-Museum Bremen Naturwissenschaften* 12.

———. 1994. Comment on the supraspecific taxon *Tergosmia* and redescription of *Osmia agilis*. *Acta Societatis Zoologicae Bohemicae* 58: 217-220.

———. 1995. Die Bienen der Tribus Osmiini der Mongolei. *Entomologische Abhandlungen* 57: 109-147.

Tompkins, J. L., L. W. Simmons, and J. Alcock. 2001. Brood-provisioning strategies in Dawson's burrowing bee, *Amegilla dawsoni*. *Behavioral Ecology and Sociobiology* 50: 81-89.

Torchio, P. F. 1965. Observations on the biology of *Colletes ciliatoides*. *Journal of the Kansas Entomological Society* 38: 182-187.

———. 1971. The biology of *Anthophora (Micranthophora) peritomae* Cockerell. *Contributions in Science, Los Angeles County Museum of Natural History* no. 206: 1-14.

———. 1974. Notes on the biology of *Ancyloscelis armata* Smith and comparisons with other anthophorine bees. *Journal of the Kansas Entomological Society* 47: 54-63.

———. 1975. The biology of *Perdita nuda* and descriptions of its immature forms and those of its *Sphecodes* parasite. *Journal of the Kansas Entomological Society* 48: 257-279.

———. 1984. The nesting biology of *Hylaeus bisinuatus* Forster and development of its immature forms. *Journal of the Kansas Entomological Society* 57: 276-297.

———. 1986. Late embryogenesis and egg eclosion in *Triepeolus* and *Anthophora* with a prospectus of nomadine classification. *Annals of the Entomological Society of America* 79: 588-596.

———. 1991. Bees as crop pollinators and the role of solitary species in changing environments. *Acta Horticulturae* 288: 49-61.

Torchio, P. F., and D. J. Burdick. 1988. Comparative notes on the biology and development of *Epeolus compactus* Cresson, a cleptoparasite of *Colletes kincaidii* Cockerell. *Annals of the Entomological Society of America* 81: 626-636.

Torchio, P. F., and B. Burwell. 1987. Notes on the biology of *Cadeguala occidentalis* (Hymenoptera: Colletidae) and a review of colletid pupae. *Annals of the Entomological Society of America* 80: 781-789.

Torchio, P. F., J. G. Rozen, Jr., G. E. Bohart, and M. S. Favreau. 1967. Biology of *Dufourea* and its cleptoparasite, *Neopasites*. *Journal of the New York Entomological Society* 75: 132-146.

Torchio, P. F., and D. M. Torchio. 1975. Larvae of the Apidae (Hymenoptera, Apoidea), Part I. Apini, *Apis*. *Agricultural Experiment Station, Utah State University, Research Report* 20: 1-36.

Torchio, P. F., and G. E. Trostle. 1986. Biological notes on *Anthophora urbana urbana* and its parasite, *Xeromelecta californica* (Hymenoptera: Anthophoridae), including descriptions of late embryogenesis and hatching. *Annals of the Entomological Society of America* 79: 434-447.

Torchio, P. F., G. E. Trostle, and D. J. Burdick. 1988. The nesting biology of *Colletes kincaidii* Cockerell (Hymenoptera: Colletidae) and development of its immature forms. *Annals of the Entomological Society of America* 81: 605-625.

Torchio, P. F., and N. N. Youssef. 1968. The biology of *Anthophora (Micranthophora) flexipes* and its cleptoparasite,

Zacosmia maculata, including a description of the immature stages of the parasite. *Journal of the Kansas Entomological Society* 41: 289-302.

Toro, H. 1973a. Contribucion al estudio de las especies chilenas del genero *Leioproctus*. *Revista Chilena de Entomología* 7: 145-172.

———. 1973b. Tres neuvas especies chilenas del genero *Leioproctus*. *Anales del Museo de Historia Natural de Valparaíso* 6: 205-212.

———. 1976. *Chilimalopsis*, nuevo genero chileno de Exomalopsini. *Anales del Museo de Historia Natural de Valparaíso* 9: 73-76.

———. 1980. *Austropanurgus* nuevo genero de Panurginae chileno. *Anales del Museo de Historia Natural de Valparaíso* 13: 209-212.

———. 1981. Contribucion al conocimiento de los Xeromelissinae chilenos (Hymenoptera, Apoidea). *Anales del Museo de Historia Natural de Valparaíso* 14: 217-224.

———. 1985. Ajuste mecanico para la copula de *Callonychium chilense*. *Revista Chilena de Entomología* 12: 153-158.

———. 1989. Contribucion al conocimiento de los Panurginae chilenos. *Acta Entomologica Chilena* 15: 229-232.

———. 1997. Nuevas especies chilenas de Xeromelissinae. *Acta Entomologica Chilena* 21: 7-12.

———. 2000. Una neuva especie chilena de *Leioproctus (Spinolapis)*. *Acta Entomologica Chilena* 24: 65-68.

Toro, H., and V. Cabezas. 1977. Nuevos generos y especies de Colletini sudamericanos, primera parte. *Anales del Museo de Historia Natural de Valparaíso* 10: 45-64.

Toro, H., and V. Cabezas. 1978. Nuevos generos y especies de Colletini sudamericanos, segunda parte. *Anales del Museo de Historia Natural de Valparaíso* 11: 131-148.

Toro, H., Y. Frederick, and A. Henry. 1989. Hylaeinae (Hymenoptera, Colletidae), nueva subfamilia para la fauna Chilena. *Acta Entomologica Chilena* 15: 201-204.

Toro, H., and M. Fritz. 1991. Contribucion al conocimiento de *Dasycoelioxys* Mitchell. *Acta Entomologica Chilena* 16: 69-80.

———. 1993. Las especies argentinas del genero *Coelioxys (Cyrtocoelioxys)*. *Acta Entomologica Chilena* 18: 147-161.

Toro, H., and M. Herrera. 1980. Las especies chilenas del genero *Callonychium* (Andrenidae–Apoidea) y descripcion de un nuevo genero. *Anales del Museo de Historia Natural de Valparaíso* 13: 213-225.

Toro, H., and E. de la Hoz. 1976. Factores mecanicos en la aislacion reproductiva de Apoidea. *Revista de la Sociedad Entomologica Argentina* 35: 193-202.

Toro, H., and J. C. Magunacelaya. 1987. Estructura muscular femoral de Xeromelissinae. *Acta Entomologica Chilena* 14: 13-24.

Toro, H., and C. D. Michener. 1975. The subfamily Xeromelissinae and its occurrence in Mexico. *Journal of the Kansas Entomological Society* 48: 351-357.

Toro, H., and A. Moldenke. 1979. Revision de los Xeromelissinae chilenos. *Anales del Museo de Historia Natural de Valparaíso* 12: 95-182.

Toro, H., and S. Rodriguez. 1998. Los anthidiini de Chile: clave para especies. *Acta Entomologica Chilena* 22: 63-78.

Toro, H., and F. Rojas A. 1968. Dos nuevas especies de *Isepeolus* con clave de las especies chilenas. *Revista Chilena de Entomología* 6: 55-60.

———. 1970a. Contribucion al estudio de las especies del genero *Leioproctus (Bicolletes)* en Chile. *Anales del Museo de Historia Natural de Valparaíso* 3: 85-109.

———. 1970b. Los Anthidiinae (Hymenoptera–Apoidea) de la provincia de Valparaíso. *Boletin del Museo Nacional de Historia Natural* [Chile] 31: 125-184.

Toro, H., and L. Ruz. 1969. Contribucion al conocimiento del genero *Diadasia*. *Anales del Museo de Historia Natural de Valparaíso* 2: 117-134, 2 pls.

———. 1972. Revision del genero *Spinoliella*. *Anales del Museo de Historia Natural de Valparaíso* 5: 137-171, 293-295.

Tosi, A. 1896. Di nuovo genere di Apiaria fossile nell'ambra di Sicilia *Meliponorytes succini–M. sicula)*. *Rivista Italiana di Paleontologia* 2: 352-356.

Triplett, D. C., and A. R. Gittins. 1988. Nesting, mating and foraging habits of *Melissodes (Melissodes) tepida tepida* Cresson in Idaho. *Proceedings of the Entomological Society of Washington* 90: 462-470.

Trostle, G., and P. F. Torchio. 1994. Comparative nesting behavior and immature development of *Megachile rotundata* (Fabricius) and *Megachile apicalis* Spinola. *Journal of the Kansas Entomological Society* 67: 53-72.

Tsuneki, K. 1970. Bionomics of some species of *Megachile*, *Dasypoda*, *Colletes* and *Bombus*. *Etizenia* no. 48: 1-20.

———. 1973. Studies on *Nomada* of Japan. *Etizenia* no. 66 (I and II): 1-141.

———. 1983. A contribution to the knowledge of *Sphecodes* Latreille of Japan. *Special Publications of the Japan Hymenopterists Association* no. 26: 1-72.

Urban, D. 1961. As espécies de *Thygater (Nectarodiaeta)* Holmberg, 1903. *Boletim da Universidade do Paraná, Zoologia* no. 8: 1-14.

———. 1962. Novas notas sobre *Thygater (Nectarodiaeta)* Holmberg, 1903. *Boletim da Universidade do Paraná, Zoologia* no. 17: 1-13.

———. 1967a. As espécies do gênero *Thygater* Holmberg, 1884. *Boletim da Universidade Federal do Paraná, Zoologia* II no. 12: 177-309.

———. 1967b. O gênero *Lophothygater* Moure and Michener, 1955. *Dusenia* 8: 135-145.

———. 1967c. As espécies do gênero *"Dasyhalonia"* Michener, LaBerge e Moure, 1955. *Revista Brasileira de Biologia* 27: 247-266.

———. 1968a. As espécies de *Gaesischia* Michener, LaBerge e Moure, 1955. *Boletim da Universidade Federal do Paraná, Zoologia* III, no. 4: 79-129.

———. 1968b. As espécies do gênero *Melissoptila* Holmberg, 1884. *Revista Brasileira de Entomologia* 13: 1-94.

———. 1970. As espécies do gênero *Florilegus* Robertson, 1900. *Boletim da Universidade Federal do Paraná, Zoologia* III, no. 12: 245-280.

———. 1971. As espécies de *Alloscirtetica* Holmberg, 1909. *Boletim da Universidade Federal do Paraná, Zoologia* III, no 16: 307-369.

———. 1972. As espécies de *Svastrides* Michener, LaBerge e Moure. *Revista Brasileira de Biologia* 32: 485-498.

———. 1973. As espécies sulamericanas do gênero *Melis-*

sodes (Latreille, 1829). *Revista Brasileira de Biologia* 33: 201-220.

———. 1974a. O gênero *Svastrina* Moure & Michener, 1955. *Revista Brasileira de Biologia* 34: 309-314.

———. 1974b. O gênero *Gaesochira* Moure & Michener, 1955. *Revista Brasileira de Biologia* 34: 315-321.

———. 1974c. O gênero *Pachysvastra* Moure & Michener, 1955. *Revista Brasileira de Biologia* 34: 323-330.

———. 1975. Uma espécie nova de *Svastrides* da Argentina. *Revista Brasileira de Biologia* 35: 113-116.

———. 1977. Espécies novas de *Alloscirtetica* Holmberg, 1909. *Dusenia* 10: 1-14.

———. 1982. Sobre o gênero *Alloscirtetica* Holmberg, 1909. *Dusenia* 13: 65-80.

———. 1989a. Duas espécies novas do gênero *Trichocerapis* Cockerell, 1904. *Revista Brasileira de Zoologia* 6: 457-462.

———. 1989b. Espécies novas e notas sobre o gênero *Gaesischia* Michener, LaBerge & Moure, 1955. *Revista Brasileira de Entomologia* 33: 75-102.

———. 1989c. Dois gêneros novos de Eucerinae neotropicais. *Revista Brasileira de Zoologia* 6: 117-124.

———. 1991. *Ananthidium*, um gênero novo di Dianthidiini neotropical. *Revista Brasileira de Zoologia* 7: 73-78.

———. 1993a. *Ctenanthidium*, gen n. de Dianthidiini com quatro espécies novas da América do Sul. *Revista Brasileira de Zoologia* 8(1991): 85-93.

———. 1993b. Considerações sobre *Anthidulum* Michener, stat. n. e *Dicranthidium* Moure & Urban, stat. n. e descrições de espécies novas. *Revista Brasileira de Zoologia* 9(1992): 11-28.

———. 1994a. *Tylanthidium*, gen. n. de Anthidiinae da América do sul (Hymenoptera, Megachilidae) e nota taxonômica. *Revista Brasileira de Zoologia* 11: 277-281.

———. 1994b. *Gnathanthidium*, gen. n. de Anthidiinae da América do Sul. *Revista Brasileira de Zoologia* 9(1992): 337-343.

———. 1995a. *Moureanthidium*, gen. n. de Dianthidiini do Brasil. *Revista Brasileira de Zoologia* 12: 37-45.

———. 1995b. *Grafanthidium*, gen. n. de Dianthidiini do Brasil e uma espécie nova de *Duckeanthidium* Moure & Hurd. *Revista Brasileira de Zoologia* 12: 435-443.

———. 1995c. Espécies novas de *Epanthidium* Moure. *Acta Biologica Paranaense* 21: 1-21.

———. 1995d. Espécies novas de Paracolletini e Panurgini do sul do Brasil e Argentina. *Revista Brasileira de Zoologia* 12: 397-405.

———. 1995e. Espécies novas de *Lanthanomelissa* Holmberg e *Lanthanella* Michener and Moure. *Revista Brasileira de Zoologia* 12: 767-777.

———. 1996. *Mielkeanthidium*, gen. n. de Dianthidiini da América do Sul. *Revista Brasileira de Zoologia* 13: 121-125.

———. 1997a. *Larocanthidium* gen. n. de Anthidiinae do Brasil. *Revista Brasileira de Zoologia* 14: 299-317.

———. 1997b. *Chrisanthidium*, um novo gênero sulamericano de Dianthidiini. *Revista Brasileira de Zoologia* 14: 181-185.

———. 1998a. Espécies novas de *Melissoptila* Holmberg da America do Sul e notas taxonómicas. *Revista Brasileira de Zoologia* 15: 1-46.

———. 1998b. Notas taxônomicas e espécies novas de *Hypanthidium* Cockerell. *Acta Biologica Paranaense* 26(1997): 95-123.

———. 1998c. Notas taxônomicas e espécies novas de *Nananthidium* Moure e descrição do macho de *Bothranthidium* Moure. *Revista Brasileira de Zoologia* 15: 621-632.

———. 1998d. *Mirnapis inca*, gênero e espécie novas de Eucerinae da América do Sul. *Revista Brasileira de Zoologia* 14(1997): 565-569.

———. 1999a. Gênero novo e espécies novas de Anthidiinae e nota taxonômica. *Acta Biologica Paranaense* 28: 159-169.

———. 1999b. Espécies novas e notas sobre *Anthodioctes* Holmberg. *Revista Brasileira de Zoologia* 16(suppl. 1): 135-169.

———. 1999c. Sobre a gênero *Austrostelis* Michener and Griswold stat. n. (Hymenoptera, Megachilidae), com algumas modificações nomenclaturais. *Revista Brasileira de Zoologia* 16 (Suppl 1): 181-187.

———. 2001. *Loyolanthidium* gen. n. e tres espécies novas neotropicais. *Revista Brasileira de Zoologia* 18: 63-70.

———. 2002. O gênero *Anthidium* Fabricius na América do Sul: chave para as espécies, notas descritivas e de distribuição geográfica. *Revista Brasileira de Entomologia* 46: 495-513.

———. 2003. *Santiago wittmanni* sp. nov. do Peru e notas sobre Eucerini. *Revista Brasileira de Zoologia* 20: 201-205.

———. 2005. Espécies novas de *Anthrenoides* Ducke do Brasil. *Revista Brasileira de Entomologia* 49: 36-62.

Urban, D., and V. Graf. 2000. *Albinapis gracilis* gen. n. e sp. n. e *Hexantheda enneomera* sp. n. do Sul do Brasil. *Revista Brasileira de Zoologia* 17: 595-601.

Urban, D., and J. S. Moure. 1993. *Ceblurgus longipalpis* gen. e sp. n., primeiro representante de Dufoureinae do Brasil. *Anais da Academia Brasileira de Ciências* 65: 101-106.

———. 2001. Católogo de Apoidea da região neotropical, II Diphaglossinae. *Revista Brasileira de Zoologia* 18: 1-34.

———. 2002. Católogo de Apoidea da região neotropical (Hymenoptera, Colletidae). IV Hylaeinae. *Revista Brasileira de Zoologia* 19: 31-56.

Vachal, J. 1897. Éclaircissements sur de genre *Scrapter* et description d'une espèce nouvelle de *Dufourea*. *Bulletin de la Société Entomologique de France* 1897: 61-64.

———. 1900. Contributions Hyménoptiques. *Annales de la Société Entomologique de France* 68(1899): 534-539.

———. 1903-1904. Étude sur les *Halictus* d'Amérique. *Miscellanea Entomologica* [Narbonne] 11: 89-104, 121-136 (1903); 12: 9-24, 113-128, 137-144 (1904).

———. 1905a. *Lonchopria*, un nouveau genre d'Hyménoptères, de la famille Apidae. *Bulletin de la Société Entomologique de France* 1905: 204.

———. 1905b. *Manuelia*, un nouveau genre d'Hyménoptères mellifères. *Bulletin de la Société Entomologique de France* 1905: 25-26.

———. 1908, 1909a, 1910. Espèces nouvelles ou litigieuses d'Apidae du haut bassin du Parana et des régions contiguës et délimitation d'une nouvelle sous-famille Diphaglossinae. *Revue d'Entomologie* [Caen] 27: 221-244 (1908); 28: 5-64 (1909); 65-70 (1910).

———. 1909b. Sur le genre *Melitoma* S. F. et Serv. et sur les genres voisins de la sous-famille Anthophorinae. *Annales de la Société Entomologique de France* 78: 5-14.

———. 1911. Étude sur les *Halictus* d'Amérique. *Miscellanea Entomologica* [Narbonne] 19: 9-24, 41-56, 107-116.

Vecht, J. van der. 1928. Hymenoptera Anthophila (Q XI-IIm) A. *Andrena*, pp. 1- 144 *in* H. Boschma et al., eds., *Fauna van Nederland*, Aflevering IV. Leiden: Sijthoff.

———. 1952. A preliminary revision of the oriental species of the genus *Ceratina*. *Zoologische Verhandelingen* no. 16: ii + 1-85.

———. 1953. The carpenter bees (*Xylocopa* Latr.) of Celebes, with notes on some other Indonesian *Xylocopa* species. *Idea* 9: 57-69.

Velthuis, H. H. W., and M. J. Sommeijer. 1991. Roles of morphogenetic hormones in caste polymorphism in stingless bees, pp. 347-383 *in* A. P. Gupta, ed., *Morphogenetic Hormones of Arthropoda*. New Brunswick, N.J.: Rutgers University Press.

Vergara, C. H., and C. D. Michener. 2004. A new species of *Caupolicana* s. str. from the Tehuacán-Culicatlán valley, Mexico, and a key to the North American species of the subgenus. *Journal of the Kansas Entomological Society* 77: 783-787.

Vergés, F. 1967. Estudo monográfico de los *Thyreus* Panzer (*Crocisa* Jurine) de España. *Miscelanea Zoologica, Museo de Zoologia de Barcelona* 11: 101-110.

Verhoeff, C. 1890. Ein Beitrag zur deutschen Hymenopteren-Fauna. *Entomologische Nachrichten* 16: 321-335.

Viereck, H. L. 1904a. A bee visitor of *Pontederia* (pickerelweed). *Entomological News* 15: 244-246.

———. 1904b. American genera of the bee family Dufoureidae. *Entomological News* 15: 261-262.

———. 1909. Descriptions of new Hymenoptera. *Proceedings of the Entomological Society of Washington* 11: 42-51.

———. 1912. Contributions to our knowledge of bees and ichneumon-flies, including the descriptions of twenty-one new genera and fifth-seven new species of ichneumon-flies. *Proceedings of the United States National Museum* 42: 613-648.

———. 1916. The Hymenoptera, or wasp-like insects of Connecticut. *Connecticut State Geological and Natural History Survey Bulletin* 22: 1-824, pls. I-X.

———. 1917a. New species of North American bees of the genus *Andrena* contained in the collections of the Academy of Natural Sciences of Philadelphia. *Transactions of the American Entomological Society* 43: 365-407.

———. 1917b. Contributions to our knowledge of the bee genus *Perdita* Smith. *Bulletin of the American Museum of Natural History* 37: 241-242.

———. 1922. New bees of the genus *Andrena*. *Occasional Papers of the Boston Society of Natural History* 5: 35-45, pl. 4.

———. 1924a. Prodromus of *Andrena*, a genus of bees. *Canadian Entomologist* 56: 19-24.

———. 1924b. The Philippine species of *Parevaspis*, a genus of bees. *Philippine Journal of Science* 24: 745-747.

Vinson, S. B., and G. W. Frankie. 1988. A comparative study of the ground nests of *Centris flavifrons* and *Centris aethiocesta*. *Entomologia Experimentalis et Applicata* 49: 181-187.

———. 2000a. Nesting behavior of *Centris flavofasciata* (Hymenoptera: Apidae) with respect to the source of the cell wall, *Journal of the Kansas Entomological Society* 72(1999): 46-59.

———. 2000b. Nest selection, usurpation, and a function for the nest entrance plug of *Centris bicornuta*. *Annals of the Entomological Society of America* 93: 254-260.

Vinson, S. B., G. W. Frankie, and R. E. Coville. 1987. Nesting habits of *Centris flavofasciata* Friese (Hymenoptera: Apoidea: Anthophoridae) in Costa Rica. *Journal of the Kansas Entomological Society* 60: 249-263.

Visscher, P. K., and B. N. Danforth. 1993. Biology of *Calliopsis pugionis* (Hymenoptera: Andrenidae): Nesting, foraging, and investment sex ratio. *Annals of the Entomological Society of America* 86: 822-832.

Vitzthum, H. G. 1930. Acarologische Beobachtungen (14 Reihe). Zoologische *Jahrbücher, Abteilung für Systematik, Ökologie und Geographie der Tiere* 59: 281-350.

Vivallo, F. 2003. Las especies chilenas de *Alloscirtetica* Holmberg, pp. 67-76 *in* Melo and Alves-dos-Santos (2003).

Vivallo, F., F. Zanella, and H. Toro. 2002. Las especies chilenas de *Centris (Wagenknechtia)* Moure, 1950. *Acta Entomológica Chilena* 26: 59-80.

Vivallo, F., F. C. Zanella, and H. Toro. 2003. Las especies chilenas de *Centris (Paracentris)* Cameron y *Centris (Penthemisia)* Moure, pp. 77-83 *in* Melo and Alves-dos-Santos (2003).

Vogel, M. E., and P. F. Kukuk. 1994. Individual foraging effort in the facultatively social halictid bee, *Nomia (Austronomia) australica* (Smith). *Journal of the Kansas Entomological Society* 67: 225-235.

Vogel, S. 1966. Parfümsammelnde Bienen als Bestäuber von Orchidaceen und *Gloxinia*. *Österreichischen Botanischen Zeitschrift* 113: 302-361.

———. 1974. Ölblumen und ölsammelnde Bienen. *Tropische und Subtropische Pflanzenwelt* [Mainz], no. 7: 1-267.

———. 1976. *Lysimachia*: Ölblumen der Holarktis. *Naturwissenschaften* 63: 44.

———. 1981. Abdominal oil-mopping—a new type of foraging by bees. *Naturwissenschaften* 67: 627.

———. 1984. The *Diascia* flower and its bee—an oil-based symbiosis in southern Africa. *Acta Botanica Neerlandica* 33: 509-518.

———. 1986. Ölblumen und ölsammelnde Bienen, zweit Folge. *Tropische und Subtropische Pflanzenwelt* [Mainz], no. 54: 1-168.

———. 1988. Die ölblumensymbiosen–parallelismus und andere Aspekte ihrer Entwicklung in Raum und Zeit. *Zeitschrift für Zoologische Systematik und Evolutionsforschung* 26: 341-362.

———. 1990. Ölblumen und ölsammelnde Bienen—Dritte Folge, *Mormodica, Thladiantha* und die Ctenoplectridae. *Tropische und Subtropische Pflanzenwelt* [Mainz] no. 73: 1-186.

Vogel, S., and C. D. Michener. 1985. Long bee legs and oil-producing floral spurs, and a new *Rediviva*. *Journal of the Kansas Entomological Society* 58: 359-364.

Vogt, O. 1911. Studien über das Artproblem. Mitteilung 2, Teil 2. *Sitzungsberichte der Gesellschaft Naturforschender Freunde zu Berlin* 1911: 31-74.

Wafa, A. K., and M. I. Mohamed. 1970. The life-cycle of *Tetralonia lanuginosa* [sic] Klug. *Bulletin de la Société Entomologique d'Égypte* 54: 259-267.

Wagenknecht Huss, R. 1969. Contribución a la biologia de los Apoidea chilenos. *Anales del Museo de Historia Natural, Valparaíso* 2: 171-176.

Walker, K. L. 1986. Revision of the Australian species of the genus *Homalictus* Cockerell (Hymenoptera: Halictidae). *Memoirs of the Museum of Victoria* 47: 105-200.

———. 1993. *Pachyhalictus stirlingi* (Cockerell) (Hymenoptera: Halictidae)—A unique Australian bee. *Australian Entomologist* 20: 59-65.

———. 1995. Revision of the Australian native bee subgenus *Lasioglossum (Chilalictus)*. *Memoirs of the Museum of Victoria* 55(1): 1-423.

———. 1996. A new species of Australian *Pachyhalictus* Cockerell. *Australian Entomologist* 23: 125-131.

———. 1997. Supplement to a revision of Australian members of the bee genus *Homalictus* (Cockerell). *Memoirs of the Museum of Victoria* 56: 69-82.

Wappler, T., and M. S. Engel. 2003. The middle Eocene bee faunas of Eckfeld and Messel, Germany *Journal of Paleontology* 77: 908-921.

Warncke, K. 1965. Beitrag zur Kenntnis der Bienengattung *Andrena* Fabricius in Griechenland. *Beiträge zur Entomologie* 15: 27-76.

———. 1967. Beitrag zur Klärung paläarktischer *Andrena*-Arten. *Eos* 43: 171-318.

———. 1968. Die untergattungen der westpaläarktischen Bienengattung *Andrena* F. *Memórias e Estudos do Museu Zoológico da Universidade de Coimbra* no. 307: 1-111.

———. 1972. Westpaläarktische Bienen der Unterfamilie Panurginae. *Polskie Pismo Entomologiczne* 42: 53-108.

———. 1973a. Die westpaläarktischen Arten der Bienenfamilie Melittidae. *Polskie Pismo Entomologiczne* 43: 97-126.

———. 1973b. Zur Systematik und Synonymie der mitteleuropäischen Furchenbienen *Halictus* Latreille. *Bulletin de la Société Royale des Sciences de Liège* 42: 277-295.

———. 1974. Die Sandbienen der Türkei, Teil I. *Mitteilungen der Münchner Entomologischen Gesellschaft* 64: 81-116.

———. 1975a. Die Sandbienen der Türkei, Teil B. *Mitteilungen der Münchner Entomologischen Gesellschaft* 65: 29-102.

———. 1975b. Beitrag zur Systematik und Verbreitung der Furchenbienen in der Türkei (Hymenoptera, Apoidea, *Halictus*). *Polskie Pismo Entomologiczne* 45: 81-123.

———. 1976a. Zur Systematik und Verbreitung der Bienengattung *Nomia* Latr. in der Westpaläarktis und dem turkestanischen Becken. *Reichenbachia* 16: 93-120.

———. 1976b. Beitrag zur Bienenfauna des Iran, 2. Die Gattung *Systropha*. *Bollettino del Museo Civico di Storia Naturale de Venezia* 28: 93-97.

———. 1977a. Ideen zum natürlichen System der Bienen. *Mitteilungen der Münchner Entomologischen Gesellschaft* 67: 39-63.

———. 1977b. Beitrag zur Systematik der westpaläarktischen Bienengattung *Dioxys* Lep. & Serv. *Reichenbachia* 16: 265-282.

———. 1978. Über die westpaläarktischen Arten der Bienengattung *Colletes* Latr. *Polskie Pismo Entomologiczne* 48: 329-370.

———. 1979a. Beiträge zur Bienenfauna des Iran: II. Die Gattung *Pararhophites* Fr. *Bollettino del Museo Civico di Storia Naturale di Venezia* 30: 197-198.

———. 1979b. Beiträge zur Bienenfauna des Iran: 3. Die Gattung *Rophites* Spin., mit einer revision des Westpaläarktischen arten der Bienengattung *Rophites* Spin. *Bollettino del Museo Civico di Storia Naturale di Venezia* 30: 111-155.

———. 1979c. Beiträge zur Bienenfauna des Iran: 10. Die Gattung *Ancyla* Lep., mit einer Revision der Bienengattung *Ancyla* Lep. *Bollettino del Museo Civico di Storia Naturale di Venezia* 30: 183-195.

———. 1980a. Die Bienengattung *Anthidium* Fabricius, 1804 in der Westpaläarktis und im turkestanischen Becken. *Entomofauna* 1: 119-209.

———. 1980b. *Fidelia*, eine für die Westpaläarktis neue Bienengattung. *Mitteilungen der Münchner Entomologischen Gesellschaft* 70: 89-94.

———. 1980c. Die bienengattungen *Nomia* und *Systropha* im Iran mit erganzungen zu den *Nomia*-Arten der Westpaläarktis. *Linzer Biologische Beiträge* 12: 363-384.

———. 1981. Beitrag zur Bienenfauna des Iran 14. Die gattung *Halictus* Latr., mit Bemerkungen über bekannte und neue *Halictus*-Arten in der Westpaläarktis und Zentralasian. *Bollettino del Museo Civico di Storia Naturale di Venezia* 32: 67-166.

———. 1982. Zur Systematik der Bienen—Die Unterfamilie Nomadinae. *Entomofauna* 3: 97-126.

———. 1983. Zur Kenntnis der Bienengattung *Pasites* Jurine, 1807, in der Westpaläarktis. *Entomofauna* 4: 261-347.

———. 1985. Beiträge zur Bienenfauna des Iran 19-20: Die Gattungen *Panurgus* Pz. und *Melitturgula* Fr. *Bollettino del Museo Civico di Storia Naturale di Venezia* 34(1983): 221-235.

———. 1986. Die Wildbienen Mitteleuropas ihre gültigen Namen und ihre Verbreitung. *Entomofauna*, suppl. 3: 1-128.

———. 1987. Ergänzende Untersuchungen an Bienen der Gattungen *Panurgus* und *Melitturgal* Andreninae, Apidae, vor allem aus dem turkischen Raum. *Bollettino del Museo Civico di Storia Naturale di Venezia* 36(1985): 75-107.

———. 1988a. Die Bienengattung *Osmia* Panzer, 1806, ihre Systematik in der Westpaläarktis und ihre Verbreitung in der Türkei 2 und 3: Die Untergattungen *Tergosmia* und *Exosmia*. *Entomofauna* 9: 389-403.

———. 1988b. Die Bienengattung *Osmia* Panzer, 1806, ihre Systematik in der Westpaläarktis und ihre Verbreitung in der Türkei 1: Untergattung *Helicosmia* Thomson, 1872. *Entomofauna* 9: 1-48.

———. 1990. Die Bienengattung *Osmia* Panzer, 1806, ihre Systematik in der Westpaläarktis und ihre Verbreitung in der Türkei 4. Die Untergattung *Platosmia* subgen. nov. *Entomofauna* 11: 481-495.

———. 1991a. Die Bienengattung *Osmia* Panzer, 1906, ihre Systematik in der Westpaläarktis und ihre Verbreitung in der Türkei 5. Die Untergattung *Pentadentosmia* subg. nov. *Entomofauna* 12: 13-32.

———. 1991b. Die Bienengattung *Osmia* Panzer, 1906, ihre Systematik in der Westpaläarktis und ihre Verbreitung in der Türkei 9. Die Untergattung *Annosmia* subg. n. *Linzer Biologische Beiträge* 23: 307-336.

———. 1991c. Die Bienengattung *Osmia* Panzer, 1906, ihre Systematik in der Westpaläarktis und ihre Verbreitung in der Türkei 7. Die Untergattung *Foveosmia* nov. *Linzer Biologische Beiträge* 23: 267-281.

———. 1991d. Die Bienengattung *Osmia* Panzer 1806, ihre Systematik in der Westpaläarktis und ihre Verbreitung in der Türkei 10. Die Untergattung *Alcidamea* Cress. *Linzer Biologische Beiträge* 23: 701-751.

———. 1992a. Die westpaläarktischen Arten der Bienengattung *Stelis* Panzer, 1806. *Entomofauna* 13: 341-376.

———. 1992b. Die westpaläarktischen Arten der Bienengattung *Sphecodes* Latr. *Bericht der Naturforschende Gesellschaft Augsburg* no. 52: 9-64.

———. 1992c. Die westmediterranen Arten der Bienen *Osmia* Subg. *Hoplitis. Linzer Biologische Beiträge* 24: 103-121.

———. 1992d. Die Bienen *Osmia* Panzer 1806, ihre Systematik in der Westpaläarktis und ihre Verbreitung in der Türkei 11. Die Untergattung *Pyrosmia* Tkalců 1975. *Linzer Biologische Beiträge* 24: 893-921.

———. 1992e. Die westpaläarktischen Arten der Bienengattung *Coelioxys* Latr. *Bericht der Naturforschenden Gesellschaft Augsburg* no. 53: 31-77.

Watmough, R. H. 1974. Biology and behavior of carpenter bees in southern Africa. *Journal of the Entomological Society of Southern Africa* 37: 261-281.

Wcislo, W. T. 1987. The roles of seasonality, host synchrony, and behavior in the evolution and distributions of nest parasites in Hymenoptera (Insecta), with special reference to bees. *Biological Reviews* 62: 515-543.

———. 1993. Communal nesting in a North American pearly-banded bee, *Nomia tetrazonata*, with notes on nesting behavior of *Dieunomia heteropoda. Annals of the Entomological Society of America* 86: 813-821.

———. 1997a. Are behavioral classifications blinders to studying natural variation?, pp. 8-13 *in* J. C. Choe and B. J. Crespi, eds., *The Evolution of Social Behavior in Insects and Arachnids.* Cambridge: Cambridge University Press.

———. 1997b. Invasion of nests of *Lasioglossum imitatum* by a social parasite, *Paralictus asteris. Ethology* 103: 1-11.

———. 1999a. Male territoriality and nesting behavior of *Calliopsis hondurasicus* Cockerell. *Journal of the Kansas Entomological Society* 72: 91-98.

———. 1999b. Transvestism hypothesis: a cross-sex source of morphological variation for the evolution of parasitism among sweat bees. *Annals of the Entomological Society of American* 92: 239-242.

Wcislo, W. T., and V. H. Gonzalez. 2006. Social and ecological contexts of trophallaxis in facultatively social sweat bees, *Megalopta genalis* and *M. ecudoria. Insectes Sociaux* 53: 220-225.

Wcislo, W. T., V. H. Gonzalez, and M. S. Engel. 2003. Nesting and social behavior of a wood-dwelling neotropical bee, *Augochlora isthmii* (Schwarz), and notes on a new species, *A. alexanderi* Engel. *Journal of the Kansas Entomological Society* 26: 588-602.

Wcislo, W. T., and S. L. Buchmann. 1995. Mating behaviour in the bees, *Dieunomia heteropoda* and *Nomia tetrazonata*, with a review of courtship in Nomiinae. *Journal of Natural History* 29: 1015-1027.

Wcislo, W. T., and J. H. Cane. 1996. Floral resource utilization by solitary bees (Hymenoptera: Apoidea) and exploitation of their stored foods by natural enemies. *Annual Review of Entomology* 41: 257-280.

Wcislo, W. T., and B. N. Danforth. 1997. Secondarily solitary: The evolutionary loss of social behavior. *Trends in Ecology and Evolution* 12: 468-474.

Wcislo, W. T., and M. E. Engel. 1997. Social behavior and nest architecture of nomiine bees. *Journal of the Kansas Entomological Society* 69(1996, suppl.): 158-167.

Wcislo, W. T., A. Wille, and E. Orozco. 1993. Nesting biology of tropical solitary and social sweat bees, *Lasioglossum (Dialictus) figueresi* Wcislo and *L. (D.) aeneiventre* (Friese) (Hymenoptera: Halictidae). *Insectes Sociaux* 40: 21-40.

Westerkamp, C. 1996. Pollen in bee-flower relations. *Botanica Acta* 109: 325-332.

Westrich, P. 1984. Kritisches Verzeichnis der Bienen der Bundesrepublik Deutschland. *Courier Forschungsinstitut Senckenberg* 66: 1-86.

———. 1989. *Die Wildbienen Baden-Württembergs:* Allgemeiner Teil, pp. 1-431; Spezieller Teil, pp. 437-972. Stuttgart: Eugene Ulmer.

Westwood, J. O. 1835. [Characters of new genera and species of hymenopterous insects.] *Proceedings of the Zoological Society of London* 3: 51-54, 68-72.

———. 1838. Description of a new genus of exotic bees. *Transactions of the Entomological Society of London* 2: 112-113, pl. XI, fig. 7.

———. 1840a. Synopsis of the genera of British insects, pp. 1-154 (1838-1840), *in An Introduction to the Modern Classification of Insects . . .* , Vols. 1 and 2, 1838-1840. London: Longman. [Commission Direction 63 (1957) gives dates for the parts of this work.]

———. 1840b. *In* J. Duncan, *The Natural History of the Bees, in* W. Jardine, ed., The Naturalists' Library vol. 26 Entomology Vol. 6. viii + 17-301, 30 pls. Edinburgh: Lizars.

———. 1875. Descriptions of some new species of short-tongued bees belonging to the genus *Nomia. Transactions of the Entomological Society of London* 1875: 207-222, pls. IV, V.

White, J. R. 1952. A revision of the genus *Osmia*, subgenus *Acanthosmioides. University of Kansas Science Bulletin* 35: 219-307.

Whitehead, V. B. 1984. Distribution, biology and flower relationships of fideliid bees of southern Africa. *South African Journal of Zoology* 19: 87-90.

Whitehead, V. B., and C. D. Eardley. 2003. African Fidelini: Genus *Fidelia* Friese. *Journal of the Kansas Entomological Society* 76: 250-276.

Whitehead, V. B., and K. E. Steiner. 1993. A new *Rediviva* bee (Hymenoptera: Apoidea: Melittidae) that collects oil from orchids. *African Entomology* 1: 159-166.

———. 2001. Oil-collecting bees of the winter rainfall area of South Africa. *Annals of the South African Museum* 108(2): 143-277 + 2 pls.

Whitfield, G. H., K. W. Richards, and T. M. Kveder. 1987. Number of instars of larvae of the alfalfa leafcutter bee, *Megachile rotundata* (F.). *Canadian Entomologist* 119: 859-865.

Whitten, W. M., A. M. Long, and D. L. Stern. 1993. Nonfloral sources of chemicals that attract male euglossine bees. *Journal of Chemical Ecology* 19: 3017-3027.

Whitten, W. M., A. M. Young, and N. H. Williams. 1989. Function of glandular secretions in fragrance collection by male euglossine bees. *Journal of Chemical Ecology* 15: 1285-1295.

Wilkaniec, Z., F. Wójtowski, and B. Szymaś. 1985. Some investigations on solitary bee *Rhophitoides canus* Ev. (Apoidea, Halictidae) nesting in alfalfa seed plantations. *Zoologica Poloniae* 32: 139-151.

Wille, A. 1958. A comparative study of the dorsal vessels of bees. *Annals of the Entomological Society of America* 51: 538-546.

———. 1959a. Comparative morphology and classification of the stingless bees. 217 pp. Lawrence: Ph.D. Thesis, University of Kansas.

———. 1959b. A new fossil stingless bee (Meliponini) from the amber of Chiapas, Mexico. *Journal of Paleontology* 33: 849-852, pl. 119.

———. 1963. Phylogenetic significance of an unusual African stingless bee, *Meliponula bocandei* (Spinola). *Revista de Biología Tropical* 11: 25-45.

———. 1964. Notes on a primitive stingless bee *Trigona (Nogueirapis) mirandula*. *Revista de Biología Tropical* 12: 117-151.

———. 1977. A general review of fossil stingless bees. *Revista de Biología Tropical* 25: 43-46.

———. 1979a. A comparative study of the pollen press and nearby structures in the bees of the family Apidae. *Revista de Biología Tropical* 27: 217-221.

———. 1979b. Phylogeny and relationships among the genera and subgenera of the stingless bees (Meliponinae) of the world. *Revista de Biología Tropical* 27: 241-277.

———. 1983. Biology of the stingless bees. *Annual Review of Entomology* 28: 41-64.

Wille, A., and L. Chandler. 1964. A new stingless bee from the tertiary amber of the Dominican Republic. *Revista de Biología Tropical* 12: 187-195.

Wille, A., and C. D. Michener. 1971. Observations on the nests of Costa Rican *Halictus* with taxonomic notes on neotropical species. *Revista de Biología Tropical* 18: 17-31.

———. 1973. The nest architecture of stingless bees with special reference to those of Costa Rica. *Revista de Biología Tropical,* suppl. 1, 21: 1-278.

Williams, F. X. 1928. The natural history of a Philippine nipa house with descriptions of new wasps. *Philippine Journal of Science* 35: 58-118, pls. 1-8.

Williams, H. J., M. R. Strand, G. W. Elzen, S. B. Vinson, and S. J. Merritt. 1986. Nesting behavior, nest architecture, and use of Dufour's gland lipids in nest provisioning by *Megachile integra* and *M. mendica mendica*. *Journal of the Kansas Entomological Society* 59: 588-597.

Williams, P. H. 1985. A preliminary cladistic investigation of relationships among the bumble bees. *Systematic Entomology* 10: 239-255.

———. 1991. The bumblebees of the Kashmir Himalaya. *Bulletin of the British Museum (Natural History), Entomology* 60: 1-204.

———. 1994. Phylogenetic relationships among bumblebees (*Bombus* Latr.): A reappraisal of morphological evidence. *Systematic Entomology* 19: 327-344.

———. 1998. An annotated checklist of bumblebees with an analysis of patterns of description. *Bulletin of the Natural Hisstory Museum* [London] (Entomology) 67: 79-152.

Wilson, E. O. 1971. *The Insect Societies.* x + 548 pp. Cambridge, Mass.: Harvard University Press.

Winston, M. L. 1979. The proboscis of the long-tongued bees: A comparative study. *University of Kansas Science Bulletin* 51: 631-667.

———. 1987. *The Biology of the Honeybee.* x + 281 pp. Cambridge, Mass.: Harvard University Press.

Winston, M. L., and C. D. Michener. 1977. Dual origin of highly social behavior among bees. *Proceedings of the National Academy of Sciences USA* 74: 1135-1137.

Wittmann, D., and B. Blochtein. 1995. Why males of leafcutter bees hold the females' antennae with their front legs during mating. *Apidologie* 26: 181-195.

Wittmann, D., R. Radtke, J. Zeil, G. Lübke, and W. Franke. 1990. Robber bees *(Lestrimelitta limao)* and their host-chemical and visual cues in nest defense by *Trigona (Tetragonisca) angustula*. *Journal of Chemical Ecology* 16: 631-641.

Wu, Y.-r. 1965a. A study of Chinese *Macropis* with descriptions of two new species. *Acta Entomologica Sinica* 14: 591-599. [In Chinese, English summary.]

———. 1965b. *Hymenoptera Apoidea, Chinese Economic Insect Fauna,* Vol. 9, i-ix + 1-83, pls. i-vii. Beijing: Science Press. [In Chinese.]

———. 1978. A study of Chinese Melittidae with descriptions of new species. *Acta Entomologica Sinica* 21: 419-428. [In Chinese, English summary.]

———. 1979. A study on the Chinese *Habropoda* and *Elaphropoda* with descriptions of new species. *Acta Entomologica Sinica* 22: 343-348. [In Chinese, English summary.]

———. 1982a. Description of a new subgenus of *Nomia*. *Zoological Research* 3: 275-280. [In Chinese, English summary.]

———. 1982b. Hymenoptera: Apoidea, pp. 379-426 *in Insects of Xizang,* Vol. 2. Beijing: Science Press. [In Chinese, English summary.]

———. 1982c. A study on Chinese *Xylocopa* with description of a new species. *Zoological Research* 3: 193-200. [In Chinese, English summary.]

———. 1982d. Studies on Chinese *Andrena (Chrysandrena)* with descriptions of a new species and a new subspecies. *Sinozoologia* 2: 63-66. [In Chinese, English summary.]

———. 1983a. Three new species of *Habropoda* from China. *Acta Zootaxonomica Sinica* 8: 91-94. [In Chinese, English summary.]

———. 1983b. A study of Chinese *Proxylocopa* with descriptions of a new species. *Entomotaxonomia* 5: 1-6. [In Chinese, English summary.]

———. 1983c. A study of Chinese *Proxylocopa* with descriptions of two new species. *Entomotaxonomia* 5: 129-132. [In Chinese, English summary.]

———. 1983d. Two new species of *Amegilla* from China. *Acta Entomologica Sinica* 26: 222-225. [In Chinese, English summary.]

———. 1983e. Four new species of the genus *Nomia* from China. *Acta Zootaxonomica Sinica* 8: 274-279. [In Chinese, English summary.]

———. 1983f. Two new species of *Halictoides* from Yunnan, China. *Acta Entomologica Sinica* 26: 344-347. [In Chinese, English summary.]

———. 1985. A study on the genus *Rhopalomelissa* of China with descriptions of new subgenus and new species. *Zoological Research* 6: 57-68. [In Chinese, English summary.]

———. 1986. A study on *Anthomegilla* from China with descriptions of two new species. *Sinozoologia* 4: 209-212. [In Chinese, English summary.]

———. 1987a. A study on Chinese *Halictoides* with descriptions of three new species. *Sinozoologia* 5: 187-201. [In Chinese, English summary.]

———. 1987b. A study on Chinese *Hoplitis* with descriptions of new species. *Acta Entomologica Sinica* 30: 441-449. [In Chinese, English summary.]

———. 1991. Studies on Chinese Habropodini with descriptions of new species, pp. 215-233 *in* Zhang G.-X., ed., *Treatise on Systematic and Evolutionary Zoology*. Beijing: Chinese Science and Technology Press. [In Chinese, English summary.]

———. 2000. Melittidae-Apidae in *Fauna Sinica, Insecta*, vol. 20, xiv + 442pp., ix pls. Beijing: Science Press.

Wu, Y.-r., and C. D. Michener. 1986. Observations on Chinese *Macropis*. *Journal of the Kansas Entomological Society* 59: 42-48.

Wyman, L. M., and M. H. Richards. 2003. Colony social organization of *Lasioglossum malachurum* Kirby (Hymenoptera, Halictidae) in southern Greece. *Insectes Sociaux* 50: 201-211.

Xu H.-l and O. Tadauchi. 1995. A revision of the subgenus *Calomelissa* of the genus *Andrena* (Hymenoptera, Andrenidae) of eastern Asia. *Japanese Journal of Entomology* 63: 621-631.

———. 1999. A revision of the subgenus *Tarsandrena* of the genus *Andrena* of eastern Asia. *Esakia* no. 39: 31-46.

———. 2002. A revision of the sugenus *Chlorandrena* of the genus *Andrena* of Eastern Asia. *Esakia* no. 42: 55-73.

———. 2005. A revision of the subgenus *Hoplandrena* of the genus *Andrena* of eastern Asia, *Esakia* no. 45: 19-40.

Xu, H.-l., O. Tadauchi, and Y.-r. Wu. 2000. A revision of the subgenus *Oreomelissa* of the genus *Andrena* of eastern Asia. *Esakia* no. 40: 41-61.

Yamada, M., N. Oyama, N. Sekita, S. Shirasaki, and C. Tsugawa. 1971. The ecology of megachilid bee, *Osmia cornifrons* (Radoszkowski) (Hym.: Apidae) and its utilization for apple pollination. *Bulletin of the Aomori Apple Experiment Station* no. 15: 1-80. [In Japanese, English summary.]

Yamamoto, D. 1944. The habits of *Megachile spissula* Ckll. *Hati* 2(6): 29-34 [from Iwata, 1976, q.v.]. [In Japanese.]

Yamane, S., T. A. Heard, and S. F. Sakagami. 1995. Oviposition behavior of the stingless bees (Apidae, Meliponinae) XVI. *Trigona (Tetragonula) carbonaria* endemic to Australia, with a highly integrated oviposition process. *Japanese Journal of Entomology* 63: 275-296.

Yanega, D. 1988. Social plasticity and early-diapausing females in a primitively social bee. *Proceedings of the National Academy of Science USA* 85: 4374-4377.

———. 1989. Caste determination and differential diapause in the first brood of *Halictus rubicundus* in New York. *Behavioral Ecology and Sociobiology* 24: 97-107.

———. 1993. Environmental influences on male production and social structure in *Halictus rubicundus* (Hymenoptera: Halictidae). *Insectes Sociaux* 40: 169-180.

———. 1994. Nests and hosts of three species of megachilid bees (Hymenoptera: Apoidea: Megachilidae) from Coahuila, Mexico. *Journal of the Kansas Entomological Society* 67: 415-417.

Yarrow, I. H. H. 1970. Is *Bombus inexpectatus* (Tkalcu) a workerless obligate parasite? *Insectes Sociaux* 17: 95-111.

———. 1971. The author and date of certain subgeneric names in *Bombus*. *Journal of Entomology* (B) 40: 27-29.

Yasumatsu, K. 1933. Die Schmuckbienen *(Epeolus)* Japans. *Transactions of the Kansai Entomological Society* no. 4: 1-6, 3 pls.

———. 1942. Apoidea of Micronesia, III. Records of the genera *Megachile, Heriades, Ceratina* and *Prosopis*. *Tenthredo* 3: 335-348, pl. vii.

Yasumatsu, K., and Y. Hirashima. 1950. Revision of the genus *Osmia* of Japan and Korea. *Mushi* 21: 1-18, pls. 1-3.

———. 1965. Two new species of *Megachile* from Taiwan. *Kontyû* 33: 373-384.

———. 1969. Synopsis of the small carpenter bee genus *Ceratina* of Japan. *Kontyû* 37: 61-70.

Youssef, N. N., and G. E. Bohart. 1968. The nesting habits and immature stages of *Andrena (Thysandrena) candida* Smith. *Journal of the Kansas Entomological Society* 41: 442-455.

Yu, F.-L. 1954. The carpenter or xylocopine bees of Formosa. *Memoirs of the College of Agriculture, National Taiwan University* 3(3): 1-12. [In Chinese, English summary.]

Zanden, G. van der. 1985. Ergebnisse der Untersuchungen der von R. Benoist beschriebenen *Osmia*-Arten, mit Liste seiner Schriften. *Reichenbachia* 23: 47-72.

———. 1986. Die paläarktischen Arten der Gattung *Lithurgus* Latreille, 1825. *Mitteilungen aus dem Zoologischen Museum in Berlin* 62: 53-59.

———. 1988a. Beitrag zur Systematik und Nomenklature der palaarktischen Osmiini, mit Angaben über ihre Verbreitung. *Zoologische Mededelingen* 62: 113-133.

———. 1988b. Nomenklatorische und taxonomische Bemerkungen zu einigen paläarktischen Arten der Familie Megachilidae. *Reichenbachia* 26: 55-64.

———. 1989. Neue oder wenig bekannte Arten und Unterarten der palaearktischen Megachiliden. *Entomologische Abhandlungen* [Dresden] 53: 71-86.

———. 1991a. Systematik und Verbreitung der paläarktischen Arten der Untergattung *Caerulosmia* van der Zanden 1989. *Linzer Biologische Beiträge* 23: 37-78.

———. 1991b. Neue oder wenig bekannte Arten der Osmiini aus dem paläarktischen Gebiet. *Reichenbachia* 28: 163-171.

———. 1992. Neue oder unvollständig bekannte Arten paläarktischer Bauchsammler. *Linzer Biologische Beiträge* 24: 65-74.

———. 1994. Neue Arten und Unterarten, eine neue Untergattung und einige neue Fälle von Synonymie der paläarktischen Bauchsammler. *Reichenbachia* 30: 167-172.

Zanella, F. C. V. 2002a. Systematics and biogeography of the bee genus *Caenonomada*. *Studies in Neotropical Fauna and Environment* 37: 249-261.

———. 2002b. Sistemática, filogenia e distribuição geografica das espécies sul-americanas de *Centris (Paracentris)* Cameron, 1903 e de *Centris (Penthemisia)* Moure, 1950, incluindo uma análise filogenética do "grupo *Centris*" sensu Ayala, 1998. *Revista Brasileira de Entomologia* 46: 435-488.

Zavortink, T. J. 1972. A new subgenus and species of *Megandrena* from Nevada, with notes on its foraging and mating behavior. *Proceedings of the Entomological Society of Washington* 74: 61-75.

———. 1974. A revision of the genus *Ancylandrena*. *Occasional Papers of the California Academy of Sciences* no. 109: 1-36.

———. 1975. A new genus and species of eucerine bee from North America. *Proceedings of the California Academy of Sciences* (4)40: 231-242.

Zavortink, T. J., and W. E. LaBerge. 1976. Bees of the genus *Martinapis* Cockerell in North America. *Wasmann Journal of Biology* 34: 119-145.

Zeuner, F. E., and F. J. Manning. 1976. A monograph on fossil bees. *Bulletin of the British Museum (Natural History), Geology* 27: 149-268, pls. 1-4.

Zillikens, A., and J. Steiner. 2004. Nest architecture, life cycle and cleptoparasite of the neotropical leaf-cutting bee *Megachile (Chrysosarus) pseudoanthidioides* Moure. *Journal of the Kansas Entomological Society* 77: 193-202.

Zucchi, R. 1973. *Aspectos bionômicos de* Exomalopsis aureopilosa *[sic] e* Bombus atratus *incluindo considerações sobre a evolução do comportamento social.* viii + 172 pp. Ribeirão Preto, Brazil: thesis, Faculdade de Filosofia, Ciências e Letras.

———. 1993. Ritualized dominance, evolution of queen-worker interactions and related aspects in stingless bees, pp. 207-249 *in* T. Inoue and S. Yamane, eds., *Evolution of Insect Societies.* Tokyo: Hakuhin-sha.

Zucchi, R., B. Lucas de Oliveira, and J. M. F. Camargo. 1969. Notas bionômicas sôbre *Euglossa (Glossura) intersecta* Latreille 1838 e descrição de suas larvas e pupa. *Boletim da Universidade Federal do Paraná, Zoologia* 3: 203-224.

Zucchi, R., S. F. Sakagami, and J. M. F. Camargo. 1969. Biological observations on a neotropical parasocial bee, *Eulaema nigrita,* with a review on the biology of Euglossinae. *Journal of the Faculty of Science, Hokkaido University* (VI, Zoology) 17: 271-380.

Addenda

The following bibliographic references and notations concern recent works that would have been referenced in the text, had they been available earlier. Items, when appropriate, are provided with full references in small type for citations of new genera and subgenera. Sections to which each item is referable are indicated in boldface. The sequence corresponds to that of the text.

- Poinar, G. O., Jr., and B. N. Danforth. 2006. A fossil bee from early Cretaceous Burmese amber, *Science* 314: 614 + supporting online material: pp. 1–3 and Fig. S1.

Mellitosphex Poinar and Danforth, 2006: 614. Type species: *Mellitosphex burmensis,* 2006, by original designation.

This early Cretaceous amber fossil is by far the oldest bee or beelike insect known, *Critotrigona* from late Cretaceous being the only other Mesozoic bee. (Cretaceous was a very long period compared with the Tertiary periods.) The specimen of *M. burmensis* is minute (less than 3 mm long); it is a male, so lacks a scopa and other features that might be associated with manipulating pollen. The forewing and face seem very beelike. Abundant short hair is said to be plumose; the plumosity is not clear in photographs but the authors are convinced and presumably correct. The hind leg lacks a strigil such as is found in crabronid wasps, and the claws are cleft as in many bees. Features that are not beelike and that are consistent with some or all crabronid wasps are slender hind basitarsi, scarcely wider than subsequent tarsal segments, and the two middle tibial spurs.

In the absence of mouthpart characters, one can only say that *Mellitosphex* seems not to fit into any extant family of bees. Recognition of a family Mellitosphecidae is appropriate. In view of the mixture of beelike and crabronid characters, Poinar and Danforth place Mellitosphecidae in the same clade with Crabronidae and the bees, and as the sister group to all other bees. **Mellitosphecidae (Secs. 12–15, 20–23)**

- Almeida, E. A. B. 2007. *Systematics and Biogeography of Colletidae.* xi + 225 pp. Ithaca, N.Y.: PhD thesis, Cornell University.

This as yet unpublished work provides, among other things, phylogenetic analyses for 144 species representing most colletid genera, based largely on sequence data from four nuclear loci. The resultant reclassification (with keys to genera) involves recognition as genera of many taxa hitherto called subgenera, and especially the predicted breakup of *Leioproctus* into numerous genera. A minor technicality: Almeida uses the spelling Scrapterinae, as do Melo and Gonçalves (2005). Believing that the stem is *Scraptr-*, I follow Engel (2005) in writing Scraptrinae (or Scraptrini). **Colletidae (Secs. 37–48, esp. 39)**

- Packer, L. 2006. A new *Leioproctus* with unique wing venation in males

(Hymenoptera: Colletidae: Paracolletinae) with comments on unusual wing modifications in bees, *Zootaxa* no. 1104: 47–57.

Leioproctus idiotropoptera Packer from Australia does not fall in any of the recognized subgenera. The extraordinary forewing venation of the male (one specimen), not shared by females, suggests an abnormality because the two recurrent veins are close together rather than widely spread across the wing membrane. However, wing structure is presumably the same in the left and right wings. **Paracolletini (Sec. 39)**

- Kuhlmann, M. 2006. Scopa reduction and pollen collecting of bees of the *Colletes fasciatus*-group in the winter rainfall area of South Africa, *Journal of the Kansas Entomological Society* 79: 165–175.

Eleven species of *Colletes* in Southern Africa have a reduced scopa and do not transport pollen externally. Kuhlmann assumes that, like the Hylaeinae, they transport it to their nests in the crop. A less likely possibility is that they are cleptoparasitic, presumably on scopate forms of the same species group. **Colletini (Sec. 40)**

- Packer, L. E. 2007. Phylogeny and classification of the Xeromelissinae (Hymenoptera: Apoidea, Colletidae) with special emphasis upon the genus *Chilicola, Systematic Entomology,* in press.

Using 248 morphological characters (many duplicated because of being recorded separately for males and females), Packer prepared a phylogenetic reconstruction and classification. An interesting finding is that *Xeromelissa* renders *Chilimelissa* paraphyletic; they are synonymous. *Chilicola* consists of 15 subgenera, four of which are new. **Xeromelissinae (Sec. 46)**

- Magnacca, K. N., and B. N. Danforth. 2006. Evolution and biogeography of native Hawaiian *Hylaeus* bees, *Cladistics* 22: 393–411. **Hylaeinae (Sec. 47)**

- A new species of *Oxaea* from the northern Andean region has short maxillary palpi (T. Griswold, personal comm.). This is of interest because other species of the genus lack maxillary palpi, while other genera of Oxaeinae have long 6-segmented maxillary palpi. **Oxaeinae (Sec. 60)**

- Smith-Pardo, A. H. 2005. The bees of the genus *Neocorynura* of Mexico, *Folia Entomologica Mexicana* 44: 165–193. **Augochlorini (Sec. 67)**

- Michez, D., and S. Patiny. 2006. Review of the bee genus *Eremaphanta* Popov 1940 (Hymenoptera: Melittidae) with a description of a new species, *Zootaxa* no. 1148: 47–68. **Dasypodaini (Sec. 70)**

■ Wu, Y.-r. 2006. Hymenoptera, Megachilidae, *in* *Fauna Sinica, Insecta,* vol. 44. Beijing: Science Press [24] + 1–474 pp. + pls. 1–4.

This is an account of all species of Megachilidae known from China. Although in Chinese, all keys and descriptions of new species are also provided in English (pp. 384–443). **Megachilidae (Secs. 75–84).**

■ Engel, M. S., and D. B. Baker. 2006. A remarkable new leaf-cutter bee from Thailand, *Beitrag zur Entomologie* 56: 69–74.

Megachile (Aethomegachile) Engel and Baker, 2006: 70. Type species: *Megachile trichorhytisma* Engel, 2006, by original designation.

Although females of this subgenus are unknown, it appears to be a member of Group 1 (Michener, 2000) like *Megachile* s. str. **Megachilini (Sec. 84)**

■ John S. Asher (personal comm.) reports *Megachile (Pseudomegachile) lanata* (Fabricius) adventive in Trinidad and French Guiana, as well as in all the Greater Antilles, and in St. Vincent and St. Lucia in the Lesser Antilles, thus extending the known range of the subgenus. **Megachilini (Sec. 84)**

■ Rozen, J. G., Jr., and S. M. Kamel. 2006. Interspecific variation in larvae of the cleptoparasitic bee genus *Coelioxys, Journal of the Kansas Entomological Society* 79: 348–358.

Substantial interspecific or intersubgeneric differences in structure of third stage larvae were observed and illustrated. **Megachilini (Sec. 84)**

■ Smith, J. A., and M. P. Schwarz. 2006. Sociality in a Malagasy allodapine bee, *Macrogalea antanosy,* and the impacts of the facultative social parasite, *Macrogalea maizina, Insectes Sociaux* 53: 101–107. **Allodapini (Sec. 90)**

■ Engel, M. S. 2006. A new genus of minute ammobatine bees, *Acta Entomologica Slovenica* 14: 113–121.

Chiasmognathus Engel, 2006: 114. Type species: *Parammobatodes gussakovskii* Popov, 1951, by original designation.

This paper provides a generic name for the minute species of a new genus mentioned in the discussion of *Parammobatodes* by Michener (2000: 643). There are three named species and additional unnamed species, probably all cleptoparasites of Nomioidinae. **Ammobatini (Sec. 100)**

■ Nates-Parra, G. 2006. *Abejas corbiculadas de Colombia.* Bogotá: Univ. Nacional de Colombia, 156 pp.

A well-illustrated account of the corbiculate Apinae, considering biology and nests as well as morphology; 240 species. **Apinae (Secs. 102, 118–121)**

■ Rasmussen, C., and S. A. Cameron. 2007. A molecular phylogeny of the Old World stingless bees (Hymenoptera: Apidae: Meliponini) and the non-monophyly of the large genus *Trigona, Systematic Entomology* 32: 26–39.

Based on DNA sequences from four genes (one mitochondrial, three nuclear), this work verifies the prediction that species from Asia to Australia placed in *Trigona* are not closely related to the American *Trigona*. The analysis recognizes three major clades of Meliponini as follows: (1) all American genera, (2) all African genera plus the minute Asiatic and Australian forms, and (3) non-minute Asiatic and Australian genera. A possibility exists that the African *Hypotrigona* constitutes a separate clade. **Meliponini (Sec. 120)**

Index of Terms

Note: Page numbers followed by *t* and *f* indicate tables and figures, respectively. **Boldface** page numbers indicate explanations or definitions.

abundance of bees, **102-104**. *See also* biogeography
acarinarium, **598**
acetabulum, of mandible, 45*f*
acrosternite, 52
acrotergite, 52
adults, 6*f*, 8; delayed emergence, 7; key to families based on, **122-125**; notes on certain couplets in, **126**
Africa: bee fauna of, 103; distribution of bees in, 103-104, 106, 108-109; East, bee fauna of, 104; North, bee fauna of, 104; northern. *See* Palearctic region; southern, bee fauna of, 102; sub-Saharan, distribution of bees in, 106, 107*t*-108*t*; tropical, bee fauna of, 103
aggregations, of nests, 13, 13*f*-14*f*
agriculture, bee species important in, 4
alar fenestrae, **51**, 51*f*
alkaline gland. *See* Dufour's gland
alveoli (antennal), **45**
alveolocellar distance, 46*f*
alveolocular distance, 46*f*
amber, bees in, 98, 100-101, 357, 671, 812
amphigonal genitalia, **56**, 809, 811*f*
andreniform body, 42
angiosperms, 106; evolution of, 100
annular areas, of colletid glossa, 132, 132*f*
annular hairs, of colletid glossa, 89*f*, 112, 113*f*, 132, 132*f*
annulate surface, of glossa, 84*f*-85*f*
annulets, larval: caudal, 57*f*, 58; cephalic, 57*f*, 58
annuli, of colletid glossa, 89*f*, 132, 132*f*
Antarctica, and bee dispersal, 108-109
antecostal suture, **52**, 54*f*
antenna, 87
antenna (pl., antennae), 45, 46*f*, 95; flagellum of, 46*f*; larval, 57*f*; length of, 112-114; pedicel of, 46*f*; scape of, 46*f*
antennal sockets, 44*f*, **45**
antennocellar distance, 46*f*
antennocular distance, 46*f*
antennomeres, number of, sex differences in, 43
anterior velum, **51**
anthophoriform body, 42
Antilles: bee fauna of, 105; distribution of bees in, 106, 107*t*-108*t*
anus, 55*f*
apical annular area, of colletid glossa, 132, 132*f*
apical process, of labrum, **45**
apiculture, 1
apiform body, 42
apodeme(s), 52, 54*f*, **55**; of penis valve, 56*f*
apomorphy, **59**, 87
Araucanian region, distribution of bees in, 106, 107*t*-108*t*
Arctic, bee fauna of, 103
Argentina: bee fauna of, 102, 106. *See also* Araucanian region
arolium (pl., arolia), **52**, 53*f*, 116*f*; loss or reduction of, 111
Asia: bee fauna of, 103; Central, bee fauna of, 102; distribution of bees in, 106, 108-109; tropical. *See* Orient
auricle, **669**
Australia: bee fauna of, 102, 104-105; distribution of bees in, 106, 107*t*-108*t*, 108-109
Austria, bee fauna of, 103
axillae, **48**, 49*f*; angularly produced, 116
axillar suture, **48**

bacular plate, 84*f*
Baja California, bee fauna of, 102
basal annular area, of colletid glossa, 132, 132*f*
basal area: of labrum, **45**, 243; of propodeum, **49**, 49*f*
basal vein, 50*f*, **51**, 51*t*
basal zone, of propodeum, **49**, 49*f*
basicoxite, 63, 63*f*
basiglossal sclerite, 84*f*-85*f*, 132*f*
basiglossal sensilla, 84*f*-85*f*, 132*f*
basistipital process, 86
basitarsal brushes, reduction or loss of, 110
basitarsi, of male bees, 9
basitarsus (pl., basitarsi), 18, 18*f*-19*f*, 38*f*, **52**, 52*f*-53*f*, 62; hind, 60-61; of parasitic bees, 36
basitibial plate(s), 37, 40, 40*f*, **52**, 52*f*, 61, 96; reduction or loss of, 110; secondary, 52
basolateral angles, 122
batumen, 27*f*, **28**, 28*f*
Bering Straits, and bee dispersal, 109
biogeography of bees, 90, 102-104, **105**, **106-109**
biology of bees, 1-58
Bismarck Archipelago. *See* Australia
body form, 42
body parts. *See* tagmata
bombiform body, 42
Brazil, bee fauna of, 103-104, 106
bristles, tibial, 61
brood cells, 23-24, 23*f*, 25*f*, 27*f*; fossil, 101; of sphecoid wasps, 23-24
brood combs, **26**, 27*f*-28*f*
brush(es), 52; basitarsal, reduction or loss of, 110; glossal, 85*f*, 90, 132, 132*f*; mid-femoral, **52**, 61, 86; midtibial, **52**, 61, 86
burrows, excavation of, 23-24, 23*f*-24*f*
buzz-pollination, 20

California, bee fauna of, 102-103
cardines (sing., cardo), **46**, 47*f*, 83*f*-84*f*, 85, 94*f*, 95, 96*f*, 124*f*
carina (pl., carinae), 40, 114-116; gradular, lateral, **53**; hypostomal, 44*f*, **46**; interalveolar. *See* carina, juxtantennal; interantennal. *See* carina, juxtantennal; juxtantennal, **45**, 46*f*, 114; omaular, 114-116; paraocular, 44*f*; preoccipital, **46**, 114; of pronotal lobe and dorsolateral angle of pronotum, 114; transverse: on scutellum, 116; of T1, 116
carrion-eating bees, 21-22, 29, 60, 828-829
castes, 1, 807
cell(s), **23**, 23*f*; brood. *See* brood cells; functions of, 28; heteromorphic, **25**; homomorphic, **25**
cell linings, 24, 63, 88, 90, 103
cerumen, **26-28**, 27*f*-28*f*
chalicodomiform body, 42

Chihuahuan desert, bee fauna of, 102-103, 106
Chile: bee fauna of, 102-103, 108. *See also* Araucanian region
China. *See* Orient; Palearctic region
chorion, 6
clade, **59**; bees and sphecoid wasps as, 59; bee-sphecoid, classification of, 65, 65*t*
cladistic analysis, **59**, 76
cladograms, **59**, 76, 88*f*, 89, 89*f*, 635-636, 635*f*, 667, 668*f*, 670, 670*f*, 671, 789-790, 790*f*, 807-809, 808*f*
classification of bees, 1, 66, 67*t*-75*t*, 88; based on Ashmead, 79, 79*t*; based on Bischoff, 80; based on Börner, 79-80, 80*t*; based on Latrielle, 77, 77*t*; based on Lepeletier, 77, 78*t*; based on Michener, 80-81, 80*t*; based on Robertson, 79, 79*t*; based on Schenck, 77, 79*t*; based on Schmiedeknecht, 78, 78*t*; based on Thomson, 77-78, 78*t*; based on Warncke, 81, 81*t*; family-level, 93-97; history of, **77-82**; meliponine bees in, 807-809, 808*f*, 811-812; methods of, **76**; molecular, 120; morphological, 120
classification of bee-sphecoid clade, 65, 65*t*
claws: cleft, 61-62; form of, 116, 116*f*; tarsal, **52**, 52*f*, 116, 116*f*
cleaning behavior, 59, 61
cleptoparasites, 6, **31**, 31-41, 32*t*, 60, 127, 674, 674*f*; tarsal claws of, **52**, 52*f*, 116, 116*f*
clypeoantennal distance, 46*f*
clypeocular distance, 46*f*
clypeus, 44*f*, **45-46**, 45*f*, 96*f*, 97, 122-123; larval, 57*f*; length of, 46*f*
cocoons, 7-8, 95, 633-635
colonies, **12**, 12-13; communal, 12, 15; dispersal of, 105; eusocial, **12**, 13-15; parasocial, **13**; quasisocial, **12-13**; semisocial, **12**; subsocial, 12, 12
comb, 52; galeal, 83, 85, 94*f*, 95, 122; for oil collecting, 18, 18*f*, 95; stipital, 83, 83*f*, 95. *See also* midfemoral brush (comb); midtibial brush (comb)
combs, brood, **26**
communal bees: male morphology in, 9; mating behavior, 9
communal behavior, 12, 14-15
communal colonies, 12, 15
communication, 1
compound eye, 45, 45*f*
condylar groove, of mandible, 45*f*
condylar ridge, of mandible, 44, 45*f*
condyle, mandibular, 44, 45*f*
conjunctival thickenings, **48**, 83*f*-84*f*, 95
conservation, of pollinating insects, 5
corbicula (pl., corbiculae), 18-19, 30, 51, **52**, 92, **669**; femoral, 1916*f*; of parasitic bees, 36; tibial, 18-19, 53*f*
corbiculate Apinae, 97
corium, mandibular, larval, 57*f*
coxa (pl., coxae), **48**, 49*f*, 52*f*, **91**, 91*f*; front, 49*f*; hemicryptic, 91, 91*f*; hind, 49*f*, 63, 63*f*; middle, 49*f*, 63, 63*f*, 91, 91*f*
crops, pollination of, 4
cubital cell, 50*f*

907

cuckoo bees. *See* cleptoparasites; social parasites
cuspis, of volsella, **56**, 90
cuticle, 40
Czech Republic, bee fauna of, 102

declivous surface, of propodeum, **49**
defecated larva, 7
defecation, larval, 7, 8*f*
dendrograms, 59
deserts, bee fauna of, 4, 102-103, 106
development of bees, 6-11
diapause, 7
digitus, of volsella, **56**, 90
diphyletic taxa, **59**, 62
diploidy, in male bees, 6
disannulate surface, of glossa, 84*f*-85*f*, 86, 90
disc, **53**, 54*f*, 91
dispersal of bees, **105**; long-distance, 105-109. *See also* biogeography
disticoxite, 63, 63*f*
distitarsus, 52*f*-53*f*
distribution of bees, 102-104, **105**; amphitropical, 106-108; disjunct, 106-109
diversity of bees, 1, **102-104**; local, human activities as threats to, 4-5. *See also* biogeography
DNA analysis, 92, 101; of Colletidae, 93
dorsolateral angle, of pronotum, **48**, 49*f*
dorsolateral convexities, of terga, **53**, 54*f*
dorsolateral tubercle, larval, 57*f*
Dufour's gland, **28**, 35, 63, 95

eggs, of bees, 57; of cleptoparasites, 31-32; fertilization of, 6; food with, 6; hatching of, 6-7; numbers laid, 6; parasitism, 60; size of, 6, 57
enclosure, of propodeum, **49**, 49*f*
endophallus, **56**, 90*f*, 91
epeoliform body, **42**
epidermal glands, of male bees, 9
episternal groove, **48**, 49*f*, 89-90, 91*f*, 95, 97, 123
epistomal suture, 44*f*, **45**, 45*f*; larval, 57*f*
euceriform body, **42**
Europe: bee fauna of, 103; distribution of bees in, 102-103
eusocial behavior, **12**, 13-15, 17
evolution of bees, 100; convergences in, 110, 112, 810-811; loss of structures during, 110-111; new and modified structures in, 112-116; resurrection of structures in, 111; and social behavior, 14
eyes of bees: hairs on, 112; and male mating strategies, 9; size of, 114

face(s), 116, 122, 122*f*, 126; of male Hymenoptera, 10-11; measurements, 46, 46*f*
facial fovea (pl., foveae), 44*f*, **45**, 94, 123, 126
facial quadrangle, 243
families of bees: classification at level of, **93-97**; key to, based on adults, **122-125**; notes on certain couplets in, 126
family-group names, 117
family-group taxa, key to, based on females, **127-128**
female bees, 43; antenna, 46*f*; division of labor among, 12; glossa, 90-91; mating behavior, 9; parasitic, body form of, 35*f*-41*f*, 36-41; as pollinators, 16; social behavior of, 12
femur (pl., femora), 38*f*, **52**, 52*f*, 61
Fiji, bee fauna of, 105

flabellum, 83*f*-84*f*, 85-86, 112, 113*f*, 123
flagellum, 112; antennal, 46*f*
flexion lines (in wings), **51**, 62
floccus (flocculus), **243**; complete (perfect), **243**; incomplete (imperfect), **243**
flooding, 103
floral constancy, **17**
floral relationships of bees, 1, 4, 16-18, 100, 103
food, 18, 28-29; storage: for adult consumption, 29; for larval consumption, 28-29
foraging behavior, 1
foramen magnum, 44*f*
forewing base, 49*f*
forewing cells, 50-51, 50*f*, 51*t*
forewing measurements, 51, 51*f*
forewing veins, 50-51, 50*f*, 51*t*
fossil record, 91, 98-99, 98*t*-99*t*, 100-101, 105, 353, 357, 393, 448, 670-671, 781, 783, 790, 809, 812, 819, 831, 895, Plate 17-Plate 20
fovea (pl., foveae), facial, 44*f*, **45**, 94, 123
France, bee fauna of, 102
Frisch, Karl von, 1
frons, 44*f*, **45**
frontal line (of head), 44*f*
fruit trees, pollination of, 4
fungi, 103-104

Galápagos Islands, bee fauna of, 105
galeal blades, 46, 47, 47*f*, 93, 122
galeal comb, 83, 85, 93, 94*f*, 95, 122
galeal rib, 47
galeal setae, 85, 86*f*
galeal velum, 47, 93-94, 94*f*
galea (pl., galeae), **46**, 47*f*, 61, 63, 83, 83*f*-84*f*, 93, 94*f*, 95
gaster, 42-43
genal area, 44*f*-45*f*, 46
genera, bee: and subgenera, total number of, 75*t*; total number of, 75*t*
genitalia, male, 55-56, 56*f*; amphigonal, **56**; rectigonal, **56**; schizogonal, **56**
genus-group names, **117**, 118
geological history of bees, 100-101
geological time scale, 100*t*
Germany, bee fauna of, 102-103
glossal brush, 85*f*, 90, 93, 132, 132*f*
glossal canal, 83, 84*f*
glossal groove, 83, 84*f*
glossal hairs, 83, 89*f*, 132, 132*f*
glossal lobe, 85*f*, 90, 93, 132, 132*f*
glossal rod, 83, 84*f*
glossa (pl., glossae), **46**, 47*f*, 83-86, 83*f*-86*f*, 88-91, 89*f*, 93, 123, 126; colletid, 132, 132*f*; convergent characteristics in, 112, 113*f*-114*f*; evolution of, 91-92
gonobase, **55**, 56*f*; male, 87
gonocoxites: female, 54. *See also* valvifers; male, **55**, 56*f*
gonoforceps, **55**
gonostylus (pl., gonostyli): female, **54**, 55*f*; male, **55**, 55-56, 56*f*; lower (ventral), **55**; upper (dorsal), **55**
gradular carinae, lateral, **53**
gradular lamellae, lateral, **53**
gradular spines, lateral, **53**
gradulus (pl., graduli), **52-53**, 54*f*; lateral parts (arms), **53**, 54*f*
Great Plains, bee fauna of, 104
ground-nesting bees, 23-25, 28, 61, 110; biogeography of, 105
Guatemala, 106
gymnosperms, 100
gyne, **12**, 13

hair(s), 126; annular, of colletid glossa, 89*f*, 112, 113*f*, 132, 132*f*; branched, 60; on eyes, 112; facial, for pollen collection, 20, 116; glossal, 83, 89*f*, 132, 132*f*; marginal (of glossa), 83, 84*f*-85*f*; oil-collecting, on metasomal sterna, 127; plumose, 60, 60*f*, 62; for pollen collection, 20; scopal, 60, 60*f*, 92; seriate, 83, 84*f*-85*f*, 86, 89*f*, 90, 95, 112, 113*f*; of sternal scopa, 60, 60*f*; structure of, 17-18; tibial, 96; of tibial scopa, 60, 60*f*
hamuli, 50*f*
haplodiploid insects, 6, 9, 14
haploidy, in male bees, 6, 9
Hawaii, bee fauna of, 105
head, **42**, 44-48, 44*f*-46*f*; measurements, 46, 46*f*
hemitergites, 54, **54**, 61
heriadiform body, 42
heteromorphic cells, **25**
hibernaculae, 8
hidden sterna, 55-56
highly eusocial bees, **12**; biogeography of, 103, 105
hind wing base, 49*f*
holophyletic group(s), **3**, 59, 85; bees as, 60-62
homomorphic cells, **25**
honey, storage, for adult consumption, 29
honey bees: feral, 4; larval instars, 7; materials collected by, 22; parasites of, 4; as pollinators, 4-5; social behavior of, 12
hoplitiform body, **42**
humid areas, bee fauna of, 103-104, 106, 108
hylaeiform body, **42**
Hymenoptera: evolution of, 6; males, face coloration of, 10-11
hypoepimeral area, **48**
hypopharynx, larval, 57*f*
hypostoma, 44*f*, 95, 96*f*
hypostomal area, **46**
hypostomal carina, 44*f*, **46**

identification of bees, **121**, 121*t*
Iles du Salut, bee fauna of, 103
Illinois (Carlinville), bee fauna of, 103
inbreeding, 6, 9
incisura, 786
inclusive fitness, 13-14
India: northern, bee fauna of, 104. *See also* Orient; Palearctic region
individual recognition, 30
Indonesia: bee fauna of, 105. *See also* Orient
inner hind tibial spur, **51**, 116
inner orbit, **46**
instars, 7
interalveolar carinae. *See* juxtantennal carinae
interalveolar distance, 46*f*
interantennal carinae. *See* juxtantennal carinae
interantennal distance, 46*f*
intermediate forms, 120
interocellar distance, 46*f*
interocular distance, 46*f*
intersegmental line, larval, 58
intersegmental suture, 59
involucrum, 27*f*, **28**, 28*f*
islands: bee fauna of, 105; dispersal of bees to, 103

Japan. *See* Palearctic region
Java, bee fauna of, 103

Index of Terms

jugal lobe (of hind wing), 50*f*, **51**; reduction or loss of, 110
juxtantennal carinae, **45**, 46*f*, 114

keirotrichia, **52**, 53*f*

labial palpus (pl., palpi), **46**, 47*f*, 83-84, 83*f*, 84-85, 84*f*, 85-86, 86*f*, 122, 126; larval, 57*f*; number of segments of, 110-111
labial sclerites, 47; basal, 124*f*-125*f*
labiomaxillary tube, **48**, 94*f*, 95, 123, 124*f*
labium (pl., labia), 47*f*, 83*f*, 86*f*, 122*f*; larval, 57*f*, 58
labral tubercle, larval, 57*f*
labrum (pl., labra), 37, 40*f*, 86, 95-97, 122, 122*f*, 123, 126; adult, 37, 44*f*, 45, 45*f*; apical process of, 45, **45**; basal area of, 45, **243**; larval, 57*f*; process of, 243
lacinia, 47*f*, 83*f*, 86*f*, 94*f*, 95, 123
lactones, 95
lamellae, 40, 114-116
lancets (first valvulae), **54**, 55*f*
lant hosts, bee coevolution with, 16
larva (pl., larvae), 6, 6*f*, 7, 8*f*, **57**, 57*f*, 87; andrenine, 94; atypical, 7; cephalic structures of, 57, 57*f*; of cleptoparasites, 32-33; defecation by, 7, 8*f*, 28; feeding by, 7, 28-29, 60, 62; food mass for, 6-7, 28-29; halictid, 95; identification of, 121; molting by, 7; nutrition for, 17; oxaeine, 94-95; structure of, 57-58, 57*f*; terminology for, 57-58, 57*f*
lateral line, of T1 (first metasomal tergum), **54**, 54*f*
legs, 51-52, 52*f*
life cycle of bees, 6, 6*f*
life span of bees, 8
long-tongued bees, 47, 47*f*, **48**, 77, 80-81, **83-87**, 83*f*-84*f*, 88-89, 89*f*, 91, 95-97, 110-112, 122-123, 124*f*, 126, 317, 321*f*, 418, 434-435, 435*f*, 587, 592, 666, 685-686, 686*f*, 697; fossil, 101
loral apron, **48**, 123, 124*f*
lorum, 47-48, 47*f*, **48**, 83*f*-84*f*, 86*f*, 95, 123, 124*f*-125*f*

Madagascar: bee fauna of, 105; distribution of bees in, 106, 107*t*-108*t*
malar area, 44*f*, **45**, 45*f*; length of, 46*f*
male bees: antennal segments of, 43; body size of, and mating, 9-10; diploid, 6; eye size in, 114; face coloration of, 10-11; flightless, 9, 10*f*, 297, 299*f*; genitalia of, 9, 55-56, 56*f*; glossa, 90-91; gonobase, 87; gonocoxites of, **55**, 56*f*; gonostylus of, **55**, 55-56, 56*f*; haploid, 6, 9; hind legs of, 9, 116; large-eyed, 9; large-headed, 9, 10*f*; mandible of, 9; mating strategies of, 9-10; mating territory, 9-10; morphology of, 9, 10*f*; parasitic, cephalic secretions of, 35-36; pollination by, 21; as pollinators, 16; size, 9-10; territoriality in, 9-10; triploid, 6; ventral armature, 10; wings of, 9, 10*f*
male-female interactions, 9-11
male-male interactions, 9-11
malus, **51**
mandible, 40; acetabular groove of, 45*f*; acetabulum of, 45*f*; adult, 44-45, 44*f*-45*f*; condylar groove of, 45*f*; condylar ridge of, **44**, 45*f*; condyle of, 44, 45*f*; larval, 57*f*, 61; of male bees, 9; outer groove of, 45*f*; outer ridge of, **45**, 45*f*; rutellum of, **44**
mandibular sockets, 96*f*

mandibular tooth, 44
margin, **53**
marginal cell, 50*f*, **51**, 51*f*; costal edge (margin) of, **51**; length of, 51*f*; length of, 51*f*
marginal hairs, 84*f*-85*f*
marginal line, 84*f*-85*f*
marginal zone, **53**, 54*f*
mass provisioning, **28**
mating, 9-11
maxilla (pl., maxillae), 47*f*, 63, 83*f*-84*f*, 86*f*, 94, 94*f*, 122*f*, 123; larval, 57*f*, 58, 61
maxillary galeae, 83, 83*f*
maxillary palpus, 47*f*, 83*f*-84*f*, 85, 94*f*; larval, 57*f*, 61; number of segments of, 110
maxillary stipes, 83*f*-84*f*
meat use by bees, 21-22, 29, 60
medial cells, of wings, 50*f*
mediotarsus, **51**
Mediterranean basin, bee fauna of, 102, 108
megachiliform body, **42**
meliponine bees (Meliponini), 3-5, 7; flower damage by, 16; flower preferences, 17; robbing among, 30
melittology, **1-2**
mentum, **48**, 83*f*-84*f*, 86*f*, 87, 95, 123, 124*f*-125*f*; adult, **47**, 47-48, 47*f*
mesepisternum, **48**, 49*f*
mesic temperate areas: bee fauna of, 108; distribution of bees in, 102-103
mesopleural suture, 59
mesopleuron, 48
mesoscutum, 59. *See also* scutum
mesosoma, 42
mesothorax, 48
metamorphosis of bees, 6-8, 6*f*
metanotum, **48**, 49*f*, 112
metapleuron, 48
metapostnotum, **49**
metasoma, 37, 39*f*, **42**, 52-56, 54*f*-55*f*
metasomal bands (fasciae), **54**
metasomal scopa, 96
metasomal sterna, 52-56, 54*f*, 55, 55*f*
metasomal terga, 52-56, 54*f*-55*f*
metathorax, **48**, 49*f*
metepisternum, **48**, 49*f*
Mexico: bee fauna of, 102, 106; tropical. *See* neotropical region
Micronesia, bee fauna of, 105
Middle East: bee fauna of, 103. *See also* Palearctic region
midfemoral brush (comb), **52**, 61, 86
midtibial brush (comb), **52**, 61, 86
mites (Acarina), 4
molecular genetics, 101, 120; of bees, 92; of Colletidae, 93
monolectic bees, **19-20**
monophyletic group(s), **3**, 59; bees as, 60-62
Morocco, bee fauna of, 108
mouthparts, 83, 83*f*, 95, 127

Nearctic region, distribution of bees in, 106, 107*t*-108*t*, 108
nearest-neighbor analysis. *See* phenetic classification
necrophagous species, 21-22, 29
nectar: ingestion of, 17; larval consumption of, 60; storage of: for adult consumption, 29; for larval consumption, 28; utilization of, 16-17
neotropical region, distribution of bees in, 106, 107*t*-108*t*
nest: architecture, 1, 23-26, 26*f*-28*f*; excavation, 61; fossil, 101; guarding, 30;

mating in, 9; odors, 30, 35; recognition, 30; robbing, 30; usurpation, 30, **30**
New Guinea. *See* Australia
New Zealand: bee fauna of, 105; distribution of bees in, 106, 107*t*-108*t*
nomadiform body, **42**
nomenclature, 118-119
noncorbiculate Apinae, 97
North America, bee fauna of, 102, 104, 106-109
notaulus, 49*f*
nurse cells, 587*f*

occipital sulcus, 44*f*
occiput, 44*f*
ocelloccipital distance, **46**
ocellocular distance, 46*f*
ocellus, 44*f*-45*f*
ocular tangent: lower, **46**; upper, **46**
oil, floral: manipulation of, 116; utilization of, 17-18, 60
oligolectic bees, **19**, 21; broadly, **19**; narrowly, **19**
oligolecty, **19**, 19-21
omaular carina (pl., carinae), 114-116
omaulus, **48**, 91*f*
oocytes, 40-41
operculum, 6
orbit, **46**
orchids, pollination of, 21
Orient, distribution of bees in, 106, 107*t*-108*t*
ostia, 97; numbers of, 90
outbreeding, 9
outer groove, of mandible, 45*f*
outer orbit, **46**
outer ridge, of mandible, **45**, 45*f*
ovaries, 587, 587*f*
ovarioles, 41, 587, 587*f*
overwintering, 8

Pakistan: below Himalayas. *See* Orient; Palearctic region, distribution of bees in
Palearctic region, distribution of bees in, 106, 107*t*-108*t*
palpus (pl., palpi): labial, **46**, 47*f*, 83-84, 83*f*, 84-85, 84*f*, 85-86, 86*f*, 122, 126; larval, 57*f*; maxillary, 47*f*, 83*f*-84*f*, 85, 94*f*; number of segments of, 110-111
Panama, bee fauna of, 103-104
papilla (pl., papillae), 112; maxillary, larval, 61
paraglossa (pl., paraglossae), **46**, 47*f*, 83*f*-84*f*, 86*f*, 95
paramandibular process, **46**
paraocular areas, 44*f*, **45**
paraocular carina, 44*f*
paraocular lobe, **46**, 46*f*
parapenicillum (pl., parapenicilla), 669
paraphyletic groups, **3**, 59, 63-65, 83, 85, 89, 95
paraphyletic taxa, 76
parapsidal line, 49*f*
parasites: of honey bees, 4; size classes of, 34-35; social, 30-31, 31*t*, 127
parasitic bees, 1, 84, 587; classification of, 77-81; loss or reduction of structures in, 110; number of, geographical distribution and, 41; as pollinators, 16; spurs of, 116; sting apparatus of, 111
parasocial colony, **13**
parietal band, larval, 57*f*
parsimony analysis, 88
parthenogenesis, 645
pasitine bees, classification of, 77-80
pectinate spurs, 116
pedicel, antennal, 46*f*

pencil, 85, 86f
penicillum, 52, **669**
penicillus, 36, 38f; **52**, 52f
penis, 56f
penis valve, **56**, 56f; apodeme of, 56f
Perkins-McGinley hypothesis of proto-bee, 24, 90-93
phenetic classification, 59, 93-94
phenograms, 59, 94
pheromones, 9
Philippines: bee fauna of, 105. *See also* Orient
phylogenetic analysis, 76, 92-93, 110, 112
phylogenetic trees, **89**, 92
phylogeny of bees, 1, 66, 87-89; meliponine bees in, 807-809
phytophagous species, 60
plant hosts, bee coevolution with, 100
plesiomorphy, **59**, 85, 89
Poland, bee fauna of, 102
pollen: collection by bees, 3, 18, 19f, 100; larval consumption of, 60; manipulation of, 16-17, 60-62, 116; on scopae, 16, 18, 60-61, 90; sources of, evolution of, 100; storage of: for adult consumption, 29; for larval consumption, 28; transport of, 16-18, 60-62, 90; by proto-bee, 92
pollen baskets, 18, 19f
pollen thievery, 16
pollex, 44
pollination, 100; of orchids, 21
pollinators, bees as, 1, 4, 16; factors affecting, 16-17. *See also* floral relationships of bees
polylectic bees, **19**, 21
polylecty, **19**, 20
polyphyletic taxa, **59**
postmentum, 48; adult, 47; larval, 57f
postoccipital pouch, 48
postoccipital suture, 44f
postocciput, 44f
postocellar ridge, 46
postocular tangent, 46
prairie, bee species in, 4
preannular area, 84f-85f; of colletid glossa, 132f
preapical fringe, 85f, 90; of colletid glossa, 132, 132f
preapical teeth of mandible, 44, 44-45, 45f
pre-episternal groove. *See* episternal groove
pregradular area, **53**, 54f
pregradular disc, **53**
premarginal line, **53**, 54f
prementum, 48, **48**, 83f-84f, 85, 94-95, 124f-125f; adult, 46, 47, 47f; basal apodeme of, 124f; basal fragmentum of, 124f; larval, 57f
preoccipital carina, 46, 114
preoccipital ridge, 45f, **46**
preomaular area, **48**
prepectus, 48
prepupa (pl., prepupae), 7, 8f, 57-58. *See also* larva
prepygidial fimbria, **54**, 54f, 96; reduction or loss of, 110
prestigma, 50f-51f; length of, 51, 51f; width of, 51, 51f
primitively eusocial bees, 12, **12**, 15; biogeography of, 105
proboscidial fossa, 44f, **46**, 96f
proboscidial lobe, **48**, 94f, 123
proboscis (pl., proboscides), **46**, 47f, 83, 83f-84f, 85, 86f, 93-95, 97, 122, 122f; length, and pollination, 4; and nectar ingestion, 17; of parasitic bees, 36; structure of, 34f, 46-47, 47f

progressive feeding, **28**
pronotal collar, **48**
pronotal lobe, **48**, 49f; carina of, 114
pronotum, **48**, 49f, 59; dorsolateral angle of, **48**, 49f; carina of, 114
propleura, 48
propodeal pit, **49**, 49f
propodeal spiracle, 49f
propodeal triangle, **49**, 49f, 59
propodeum, 42, **48**, 49, 49f, 112; basal area (zone) of, 49, 49f; enclosure of, 49, 49f; posterior surface of, 49, 49f
prosoma, 42
prosternum, 48
prothoracic lobe, 59
prothorax, 48
proto-bee, 24, 88-92
pseudopygidial area, **54**
pterostigma. *See* stigma
pupa (pl., pupae), 6, 6f, **8**, 58
pygidial fimbria, 40, **54**, 54f, 96; reduction or loss of, 110
pygidial plate, 37, 40, 40f, **54**, 54f, 62, 96; reduction or loss of, 110

quasisocial behavior, 12-13, **12-13**
queen, **12**, 13, 30

radial cell, 50f
radiomedial veins, 51
rastellum, **669**
rectigonal genitalia, **56**, 809, 811f
reproduction, of bees, 6-8
retrorse lobe, **56**
risk-spreading strategy, in reproduction, 7-9
robber bees, loss or reduction of structures in, 110
robbing, 30, **30**
rod (of glossa), 83, 84f
rutellum, 44

S1 (first metasomal sternum), **42**, 54f
S6 (sixth metasomal sternum), 54f-55f, 61
S7 (seventh metasomal sternum), 61, 61f, 96; of males, 55, 90-91, 93
S8 (eighth metasomal sternum), 54f, 55, 61
salivary gland secretions, 28-39, 63
salivary lips, larval, 57f
salts, collection by bees, 21
Samoa, bee fauna of, 105
scape, 123; antennal, 46f
schizogonal genitalia, **56**, 809, 811f
sclerites, 95
scopa (pl., scopae), 16, 18, 30, 38f-39f, 40, 52, 53f, 60-61, 90, 96-97, 122-123, 1916f; femoral, 92, 123; metasomal, 96, 126; of parasitic bees, 36; pollen transport on, 9, 16, 18, 60-61; reduction or loss of, 110; sternal, 60, 60f, 92; tibial, 60, 60f, 92, 93, 123; trochanteral, 92
scopal hairs, 60, 60f, 92
scrobal groove, **48**, 49f, 90, 95, 97, 123
scrobe, **48**, 49f
scutellum, **48**, 49f, 112; transverse carina on, 116
scutum, **48**, 49f
semisocial behavior, **12**, 13
seriate hairs, 83, 84f, 85f, 86, 89f, 90, 95, 112, 113f
seriate line (ridge), 84f-85f
setae (sing., seta), 84f-85f, 87, 114f; combs of, for oil collecting, 18, 18f; galeal, 85, 86f
sex determination, in bees, 6, 9
sexual dimorphism, 43

short-tongued bees, 47, 47f, 52, 64, 77, 79, **83-87**, 84f, 86f, 88, 88f, 89-94, 94f, 95-97, 110, 122, 129, 136, 319, 321f, 418, 587, 666-667, 697; fossil, 101
sister group, 59
Slovakia, bee fauna of, 102
Slovenia, bee fauna of, 102
social bees, 77-78, 78t
social behavior of bees, 1, 12-15; evolution of, 14-15; origin of, 14-15; and reproductive potential, 6
social parasites, 30-31, 31t, 127
solitary bees, **12**, 12-15, 41, 77-78, 78t; biogeography of, 105; as floral specialists, 17; reproductive potential of, 6
Solomon Islands, bee fauna of, 105
sonication, by bees, 20
Sonoran Desert, bee fauna of, 102, 106
South America: bee fauna of, 106, 108-109. *See also* neotropical region
Spain, bee fauna of, 102
spatha, 56
species-group names, **117**
species of bees, number of, 1, 67, 75t, 118
spermatheca, 6
spermatocytes, 61
sperm cells, 61
sperm preference, 9
sphecoid wasps, 3, 60-61, 65, 85, 88f, 89-91, 106, 110, 664; brood cells, secreted lining of, 24-25; nests of, 23-24
spicules, 87; tibial, 37, 38f
spiculum, of S8, **55**, 56f
spines, 40, 116; coxal, 8; gradular, lateral, **53**; pupal, 8, 58; tibial, **51**, 52f
spiracles, 54f; larval, 57f, 58
spurs: pectinate, 116; tibial, **51**, 52f, 62, 116
squama, 785. *See also* gonostylus
Sri Lanka. *See* Orient
stem-nesting bees, 26, 63, 90
sternum, 37, 41f. *See also* hidden sterna
stigma, **50**, 50f, 51, 51f, 112, 115f, 123; length of, 51f; width of, 51f
sting apparatus, 40, 43, 54-55; loss or reduction of, 111, 111f; terminology for, 55, 55t
sting sheaths, 54, 787
stipital comb, 83, 83f, 95; reduction or loss of, 110
stipital concavity, 122
stipites (sing., stipes), 46, 47f, 63, 83f-84f, 85, 94f, 95, 122
storage pots, 27f-28f, 29
strigil, posterior, 61-62
strigilis, **51**
stylus (stylet), of females, **54**, 55f
subantennal area, **45**, 123, 126
subantennal suture, 44f, **45**, 94-95, 123, 126; length of, 46f
subgalea, **47**
subgenal coronet, **46**, **243**
subgenera of bees, 66, 67t-75t
submarginal cells, 50, 50f, 111, 122-123
submarginal crossveins (first, second, and third), **50**, 50f, 111
submentum, 48
subsocial behavior, 12, **12**
sulcus obliquus, 786, 787f
Sumatra, bee fauna of, 105
supra-antennal area (frons), 44f, **45**
supraclypeal area, 44f, **45**
suspensorium, 95; paraglossal, 132f; of prementum, 48
swarms, dispersal by, 105
sweat collection by bees, 21
Sweden, bee fauna of, 102

symplesiomorphy, 59
synapomorphy(ies), 59-64, 85-86, 91; based on loss of structures, 110-111; based on novel structures, 110, **112-116**; familial, 93; and reappearance of structures, 111
synonymy, 118-119

T1 (first metasomal tergum), **42**; lateral line of, **54**, 54*f*; transverse carina on, 116
T2 (second metasomal tergum), fovea of, **54**, 54*f*
T4 (fourth metasomal tergum), **54**, 54*f*
T5 (fifth metasomal tergum), **54**, 54*f*
T6 (sixth metasomal tergum), **54**, 54*f*-55*f*
T7 (seventh metasomal tergum), **54**, 54*f*-55*f*, 61
T7 hemitergites, **54**
T8 (eighth metasomal tergum), **54**, 55*f*
T8 hemitergites, **54**
tagmata (sing., tagma), **42**, 42-43
Taiwan. *See* Orient
tarsal claws, **52**, 52*f*, 116, 116*f*
tarsi, of male bees, 9
tarsi (sing., tarsus), 53*f*
Tasmania. *See* Australia
taxon (pl., taxa), **59**; of bees, 66-67, 67*t*-75*t*; problematic, **120**
taxonomy, 118-119
tegula, 49*f*, 59
temperate forests, pollination of, 4
temperate xeric areas. *See* xeric regions

tentorial pit: anterior: adult, 44*f*; larval, 57*f*; posterior: adult, 44*f*; larval, 57*f*
tentorium, 95, 96*f*
testis, 587
thelytoky, obligate, 645
thorax, **42**, 48-49, 49*f*, 91*f*; shape of, 112, 115*f*
tibia (pl., tibiae), 38*f*-39*f*, 51, 52*f*-53*f*
tibial spines, **51**, 52*f*
tibial spurs, **51**, 52*f*, 62, 116
transmetanotal suture, **48**
transverse cubital veins. *See* submarginal crossveins
triangular plate. *See* valvifers, first
trigoniform body, 42
Trinidad. *See* neotropical region
triploidy, in male bees, 6
trochanter, 52, 52*f*, 61
tropical areas: American, bee fauna of, 104; bee fauna of, 4, 102-104; pollination of, 4
tubercles: dorsolateral, larval, 57*f*; larval, 58; ventrolateral, larval, 57*f*
Turkey. *See* Palearctic region

United States, bee fauna of, 102-103
usurpation, **30**

valve, of lancet, **54**, 55*f*
valvifers: first, **55**, 55*f*; second, **54**, 55*f*
valvulae, 786*f*, 787; first, **54**; rami of, 55*f*; second, **54**; third, **54**
vannal lobe, 50*f*, **51**

velum: of anterior tibial spur, **51**; galeal, 47, 93-94, 94*f*
Venezuela, bee fauna of, 104
ventrolateral tubercle, larval, 57*f*
vertex, of adult bee head, 44*f*, **45**, 45*f*
vibrating behavior, 20
Vietnam. *See* Orient
vision, and mating, 9
volsella (pl., volsellae), **56**, 56*f*, 90, 96, 122; cuspis of, 122; digitus of, 122

wasps: and bees, comparison of, 3; origin of bees from, 63-64; pemphredonine, 100. *See also* sphecoid wasps
water: collection by bees, 21; and larval survival, 103; relations, in brood cells, 28
water barriers, dispersal across, 105-106
wild bees: human activities as threats to, 4-5; as pollinators, 4
wind dispersal of bees, 105
wing bases, 48, 49*f*
wings, 49-51; anterior, **50**; cell terminology for, 50, 50*f*; convergent characteristics in, 112, 115*f*; distal, **50**; of male bees, 9, 10*f*; vein terminology for, 49-50, 50*f*; morphologically noncommital, 50, 50*f*, 51*t*
wood-nesting bees, biogeography of, 105
workers, **12**; fossil, 100-101; inclusive fitness of, 13-14; morphology of, 12

xeric regions: distribution of bees in, 102-104, 106-109; pollinators in, 4

Index of Taxa

Note: Page numbers followed by *f*, *k*, and *t* indicate figures, keys, and tables, respectively. Page numbers in **boldface** indicate principal accounts.

Abda, 734
Abromelissa, 766
Abrupta, 67t, 198t, 200k, 201f, 201k, **203**, 209
Acalcaripes, 368
Acamptopoeum, 69t, 307k, **308-309**, 309, 314
Acamptopoeum submetallicum, 308f
Acanthalictus, 70t, 371f, 374k, **375**
Acanthidium, 497
Acanthidium batrae, 497
Acanthomelecta, 75t, **772-773**, 772f, 772k
Acanthonomada, 644
Acanthopus, 74t, 763k, **764**
Acanthopus goryi, 764
Acanthosmia, 466
Acanthosmiades, 479
Acanthosmioides, 71t, 476, **479**, 479k, 482-483, 482f, 483
Acedanthidium, 71t, 491t, 494k, **497**
Acedanthidium flavoclypeatum, 492f, 497
Acentrina, 578
Acentron, 72t, 557t, 558k-559k, **566**, 579
Aceratosmia, 484-485
Aciandrena, 68t, 252k, 254k, **259**
Ackmonopsis, 743
Acritocentris, 74t, 754, **756**, 756k, 757
Acrocoelioxys, 72t, 546k, **547**
Acrosmia, 71t, 452, 462, 465k, **466**, 466k, 470, 476
Acroxylocopa, 610
Actenosigynes, 67t, 147, 147t, **150**
Aculeata (Section), 3, 43, 101
Acunomia, 69t, 332, 332f, 333, 333f, 339, **339-340**, 339k, 341, 343
Adamon, 644-645
Adanthidium, 71t, 492f, 495f, 496, **513**, 513k, 525
Addendella, 584-585
Adventoribombus, 802
Aeganopria, 162
Aenandrena, 68t, 252k, 255k, **259**
Aethammobates, 73t, 653, **653**, 653k, 655
Aethammobates prionogaster, 653
Aethechlora, 402
Aethomegachile, 906
Aframegilla, 743
Afranthidium, 491t, 493k, **497-502**, 498k-499k, 512-513
Afranthidium (Afranthidium s. str.), 71t, 498k, 499, **499**, 500-501, 505, 512
Afranthidium (Branthidium), 71t, 498, **499**, 499k, 500-502

Afranthidium (Capanthidium), 71t, 498, 498k, **499-500**, 499k, 501-502, 512
Afranthidium (Domanthidium), 71t, 498k, 499, **500**, 501, 512
Afranthidium (Immanthidium), 71t, 493, 497, 498k, **500**, 501, 535
Afranthidium (Mesanthidiellum), 71t, 499, 499k, 500, **500-501**, 502
Afranthidium (Mesanthidium), 71t, 498, 498k, 499-500, **501**
Afranthidium (Nigranthidium), 71t, 498k, 500, **501**, 502, 507, 538
Afranthidium (Oranthidium), 71t, 494, 498k, 499-501, **501**, 502, 505, 512
Afranthidium (Xenanthidium), 71t, 499k, **501-502**
Afranthidium (Zosteranthidium), 71t, 492, 498k, 499, **502**
Afranthidium abdominale, 500
Afranthidium biserratum, 502
Afranthidium capicola, 499-500
Afranthidium concolor, 500-501
Afranthidium folliculosum, 512
Afranthidium guillarmodi, 499-501
Afranthidium immaculatum, 500
Afranthidium junodi, 500
Afranthidium mlanjense, 500
Afranthidium murinum, 499
Afranthidium naefi, 499-501
Afranthidium nigritarse, 500
Afranthidium poecilodontum, 500
Afranthidium repetitum, 500
Afranthidium rubellulum, 499
Afranthidium s. str., 71t, 498k, 499, **499**, 500-501, 505, 512
Afranthidium schulthessii, 500-501
Afranthidium sjoestedti, 500
Afranthidium tergofasciatum, 502
Afranthidium willowmorense, 501
Afrodasypoda, 70t, **422**
Afrodasypoda plumipes, 422
Afrodialictus, 372, 377-378, 381
Afrodufourea, 326
Afroheriades, 71t, 449, 449t, 450k, **452-453**, 461, 475, 489
Afroheriades capensis, 453
Afroheriades dolicocephalus, 452
Afroheriades geminus, 452
Afroheriades larvatus, 452
Afroheriades primus, 449, 452-453

Afromelecta, 771k, **772**, 772k, 775
Afromelecta (Acanthomelecta), 75t, **772-773**, 772f, 772k
Afromelecta (Afromelecta s. str.), 75t, 772k, **773**
Afromelecta bicuspis, 772f, 773
Afromelecta s. str., 75t, 772k, **773**
Afronomia, 69t, 335, **336**, 336k, 338
Afrosmia, 71t, 450, 451k, 486, 486k, **487**
Afrostelis, 32t, 34, 71t, 456, 491t, 494, 495k, **502**, 530, 537
Afrostelis tegularis, 502
Afroxylocopa, 606
Agaesischia, 728
Agandrena, 68t, 252k, 254k, **259**
Agapanthinus, 74t, 718k-719k, **721**
Agapanthinus callophilus, 721
Agapostemon, 80t, 105, 319-320, 348-350, 354-357, 358k, 360k, **362-363**, 363k, 364, 645
Agapostemon (Agapostemon s. str.), 70t, 363k
Agapostemon (Agapostemonoides), 70t, 112, 362f, **363**, 363k
Agapostemon (Notagapostemon), 363
Agapostemon arenarius, 387
Agapostemon hurdi, 362f, 363
Agapostemon mourei, 363
Agapostemon s. str., 70t, **363**, 363k
Agapostemon texanus, 355f
Agapostemonoides, 70t, 112, 362, 362f, **363**, 363k
Agapostemonoides hurdi, 363
Agatheucera, 724
Agemmonia, 428
Agemmonia (Dicromonia), 427
Agemmonia wenzeli, 427
Aglaa, 78t, 782
Aglae, 33t, 75t, 79t-80t, 778-781, **781**, 782, 782k
Aglae coerulea, 782-783
Aglaoapis, 72t, 108, 538, 539k-540k, **540**, 541
    positions of spines and carinae on thoracic sclerites of, 539t
Aglaoapis alata, 540, 540f
Aglaoapis brevipennis, 540
Aglaomelissa, 74t, **764**, 764k
Aglaomelissa duckei, 764
Agogenohylaeus, 67t, 187f, 188, 190f, **191**, 191k, 192f, 207
Agribombus, 802
Agribombus (Eversmannibombus), 798
Agrobombus, 802
Agrobombus (Adventoribombus), 802

Agrobombus (Laesobombus), 799
Agrobombus (Tricornibombus), 802
Alastor (wasp), 203
Alayoapis, 67t, **174**, 174k
Albinapis, 67t, 143k, 147t, **150**
Albinapis gracilis, 150
Albocolletes, 167
Alboneuridia, 73t, **632**, 632k
Alcidamea, 42, 71t, 448, 452, 453f, 456, 462, 464, 464k-465k, 466, **466-467**, 466k, 470-474, 502
Alcidamea biscutellae, 469
Alcidamea producta, 466-467
Alepidoscles, 74t, 702, **702**, 702k, 704
Alfkenella, 743
Alfkenomia, 337
Alfkenylaeus, 67t, 198t, 202k, **203**, 204, 210
Allanthidium, 72t, 522, **523**, 523f
Allanthidium (Anthidianum), 523
Allochalicodoma, 569-570
Allocoelioxys, 72t, 547, 547k, 549
Allodapa, 625
Allodape, 30, 73t, 77, 79, 79t-81t, 112, 115f, 620, 623, 624k, **625**, 627
Allodape (Allodapula), 626
Allodape (Braunsapis), 627
Allodape candida, 632
Allodape ceratinoides, 593f, 621f
Allodape dichroa, 626
Allodape exoloma, 115f
Allodape facialis, 627
Allodape greatheadi, 31t, 623, 625
Allodape interrupta, 115f
Allodape mirabilis, 625, 625f
Allodape mucronata, 115f, 622f, 624, 627f
Allodape nigrinervis, 628
Allodape obscuripennis, 627f
Allodape oriola, 632
Allodape palliceps, 626
Allodape quadrilineata, 620f
Allodape rufogastra, 625
Allodape stellarum, 620f
Allodape variegata, 626
allodapine bees. *See* Allodapini (Tribe)
Allodapini (Tribe), 6-7, 13, 15, 26, 28-30, 31t, 34f, 36-37, 39f, 57-58, 73t, 83-84, 110, 112, 114, 121, 126, 128k, 468, 587-588, 588k, 592, 593k, 595, 611, **619-632**, 620f-622f, 626f-627f, 629f, 631f, 697, 906
    based on adults, 623k-624k
    based on mature larvae, 624k
    biogeography of, 107t, 109

913

Allodapula, 6, 30, 31*t,* 620, 623, 624*k,* **625-626,** 626*k,* 628-629
Allodapula (Allodapula s. str.), 73*t,* 622*f,* 626, **626,** 626*k*
Allodapula (Allodapulodes), 73*t,* 622*f,* 626, **626,** 626*f,* 626*k*
Allodapula (Dalloapula), 73*t,* 621*f-*622*f,* 626, **626,** 626*f,* 626*k*
Allodapula acutigera, 621*f-*622*f,* 625-626
Allodapula dichroa, 626, 626*f*
Allodapula guillarmodi, 31*t,* 623, 626
Allodapula hessei, 622*f,* 626
Allodapula maculithorax, 626
Allodapula melanopus, 34*f,* 587*f,* 622*f*
Allodapula palliceps, 626
Allodapula s. str., 73*t,* 622*f,* 626, **626,** 626*k*
Allodapula xerica, 626, 626*f*
Allodapulodes, 73*t,* 622*f,* 626, **626,** 626*f,* 626*k*
Allodioxys, 72*t,* 442, 538, 539*k,* **540**
positions of spines and carinae on thoracic sclerites of, 539*t*
Allomacrotera, 69*t,* 300, 301*k,* **302**
Allomegachile, 569-570
Alloperdita, 69*t,* 299, 301*k,* **302,** 305
Allopsithyrus, 800
Alloscirtetica, 708, 711*f,* 717, 717*k,* **721-722,** 721*k,* 727-729
Alloscirtetica (Alloscirtetica s. str.), 74*t,* 721, **721-722,** 721*k*
Alloscirtetica (Ascirtetica), 721
Alloscirtetica (Dasyscirtetica), 721
Alloscirtetica (Megascirtetica), 74*t,* 712*f,* 721, 721*k,* **722,** 727
Alloscirtetica (Scirteticops), 721
Alloscirtetica mephistophelica, 712*f,* 722
Alloscirtetica s. str., 74*t,* 721, **721-722,** 721*k*
Allosmia, 71*t,* 451, 475-476, 477*k-*478*k,* **480,** 485
Alloxylocopa, 73*t,* 601, 602*k-*603*k,* **604,** 604*k*
Alocandrena, 68*t,* 94, 129, 236, **238**
Alocandrena porteri, 238, 238*f*
Alocandreninae (Subfamily), 68*t,* 91, 94, 127*k,* 235-237, 237*k,* **238**
biogeography of, 107*t*
Alphaneura, 828
Alphelonomada, 644
Alpigenibombus, 797
Alpigenobombus, 75*t,* 786*f,* 791*k,* 794*k,* **797**
Alpigenobombus (Coccineobombus), 797
Alpigenobombus (Fraternobombus), 798
Alpigenobombus (Funebribombus), 799
Alpigenobombus (Robustobombus), 801

Alpigenobombus pulcherrimus, 797
Alpinobombus, 797
Alpinobombus, 30, 31*t,* 75*t,* 789, 792*k,* 795*k,* **797**
Alpinodufourea, 326
Amalthea, 828
Amblyapis, 70*t,* **420,** 420*k*
Amblys, 484
Amblyspatulariella, 213
Amboheriades, 71*t,* 458, **459,** 459*k*
Amegachile, 72*t,* 554*f-*555*f,* 557*f,* 562*k-*563*k,* 565*k,* 566-567, 566*k,* 574, 576
Amegilla, 20, 111, 742, **743-744,** 743*k,* 744, 748, 750, 770-771, Plate 12
Amegilla (Ackmonopsis), 743
Amegilla (Aframegilla), 743
Amegilla (Dizonamegilla), 743
Amegilla (Glossamegilla), 743
Amegilla (Megamegilla), 743
Amegilla (Micramegilla), 743
Amegilla (Notomegilla), 743
Amegilla (Zebramegilla), 743
Amegilla (Zonamegilla), 743, Plate 13
Amegilla dawsoni, 7, 9, 742
Amegilla elsei, 744*f*
Amegilla nonconforma, 743
Amegilla quadrifasciata, Plate 12
Ammobates, 661, **662-663,** 662*k,* 663-664
Ammobates (Ammobates s. str.), 73*t,* **662,** 662*k*
Ammobates (Caesarea), 662
Ammobates (Euphileremus), 73*t,* 661, **662,** 662*k,* 663, 665
Ammobates (Spinopasites), 663
Ammobates (Xerammobates), 73*t,* **662-663,** 662*k,* 663
Ammobates biastoides, 662
Ammobates bicolor, 662
Ammobates depressa, 662
Ammobates minutissimus, 663
Ammobates oxianus, 662
Ammobates rufiventris, 662
Ammobates s. str., 73*t,* **662,** 662*k*
Ammobates tunensis, 663
Ammobatini (Tribe), 43, 73*t,* 81, 97, 589*k,* 635-636, 639, 641, 653, 660, **661-665,** 661*k-*662*k,* 666, 906
biogeography of, 108*t*
Ammobatoides, 73*t,* 588, 589*k,* 653, **653-655,** 653*k,* 662
Ammobatoidini (Tribe), 43, 58, 73*t,* 77, 112, 589*k,* 635, 640, **653-655,** 653*k,* 657*t,* 661
biogeography of, 108*t*
Amphipedia, 74*t,* 688, 691, 691*k,* **692**
Amphylaeus, 91, 93, 132, 188-189, 189*k-*190*k,* **191,** 191*k,* 214
Amphylaeus (Agogenohylaeus), 67*t,* 187*f,* 188, 190*f,* **191,** 191*k,* 192*f,* 207
Amphylaeus (Amphylaeus s. str.), 67*t,* 187*f,* 188, 188*f,* 189, 190*f,* **191,** 191*k,* 192*f,* 196*f*
Amphylaeus morosus, 125*f,* 187*f-*

188*f,* 190*f,* 191, 192*f,* 196*f*
Amphylaeus nubilosellus, 190*f,* 192*f*
Amphylaeus obscuriceps, 187*f*
Amphylaeus s. str., 67*t,* 187*f,* 188, 188*f,* 189, 190*f,* **191,** 191*k,* 192*f,* 196*f*
Ampulicidae (Family) (wasps), 64, 65*t,* 100
Ampulicinae (Subfamily) (wasps), 62-63
Anacolletes, 155-157
Analastoroides, 67*t,* 188, 190*f,* 193, 195*k,* 198*t,* **203,** 214
Analastoroides foveata, 203
Ananthidiellum, 71*t,* **503,** 503*k,* 504
Ananthidioma, 71*t*
Ananthidium, 72*t,* 515, **516,** 516*k*
Anchandrena, 68*t,* 245*k-*246*k,* 249*k,* **259**
Anchirhiza, 68*t,* 190*f,* **216,** 216*t*
Ancyla, 73*t,* 79*t-*80*t,* 81, 81*t,* 84, 127, 587, 590, 662, 667, **685-686,** 685*k,* 686, 697
Ancyla holtzi, 685*f*
Ancyla oraniensis, 685, 686*f*
Ancylandrena, 68*t,* 236, 239, 239*k,* **240,** 240*k,* 267, 269, 638
Ancylandrena atoposoma, 240, 241*f*
Ancylandrena larreae, 239*f*
Ancylini (Tribe), 73*t,* 80*t,* 81, 84, 127, 437, 587, 590, 591*k,* 662, 667, 668*f,* 671, **685-686,** 685*f,* 685*k,* 697
biogeography of, 108*t*
Ancylocopa, 608
Ancylosceles, 703
Ancylosceles ursinus, 703
Ancylloscelina (Subtribe), 700-701, 705
Ancyloscelis, 74*t,* 81*t,* 590-591, 680, 700-701, 701*k,* **702-703,** 703*f*
Ancyloscelis apiformis, 703
Ancyloscelis armatus, 703
Ancyloscelis imitatrix, 702
Ancyloscelis lineata, 692
Ancyloscelis ornata, 695
Ancyloscelis panamensis, 702*f*
Ancyloscelis toluca, 701*f*
Ancyloscelis ursinus, 702*f,* 703
Ancyloscelus, 702
Ancylosoma, 607
Andinaugochlora, 396, **400-401,** 400*k*
Andinaugochlora (Andinaugochlora s. str.), 70*t,* 397*k,* 400*k,* **401**
Andinaugochlora (Neocorynurella), 70*t,* 397*k,* 400*k,* **401**
Andinaugochlora micheneri, 396*f,* 401
Andinaugochlora s. str., 70*t,* 397*k,* 400*k,* **401**
Andinaugochlora seeleyi, 401
Andineta, 606
Andrena, 8, 17, 20-21, 35-36,

41-42, 45-48, 52, 56, 66, 77*t,* 78, 78*t-*81*t,* 91-92, 108, 120, 126, 132, 136, 138, 144, 155, 166-167, 188-189, 214, 235*f,* 236-239, 239*k,* **240-267,** 240*k,* 242*t,* 267, 278, 306, 316, 323, 353, 389, 419, 429, 431, 645, Plate 3
of Japan, 256*k-*258*k*
of North and Central America, 243*k-*251*k*
of Western Palearctic region, 251*k-*256*k*
Andrena (Aciandrena), 68*t,* 252*k,* 254*k,* **259**
Andrena (Aenandrena), 68*t,* 252*k,* 255*k,* **259**
Andrena (Agandrena), 68*t,* 252*k,* 254*k,* **259**
Andrena (Agapostemon), 363
Andrena (Anchandrena), 68*t,* 245*k,* 249*k,* **259**
Andrena (Ancylandrena), 240
Andrena (Andrena s. str.), 68*t,* 245*k-*246*k,* 249*k-*250*k,* 254*k,* 256*k-*258*k,* **259**
Andrena (Andrenella), 263
Andrena (Aporandrena), 68*t,* 243*k,* 248*k,* **259**
Andrena (Archiandrena), 68*t,* 244*k-*245*k,* 249*k,* **259**
Andrena (Augandrena), 68*t,* 246*k,* 250*k,* **259**
Andrena (Avandrena), 68*t,* 251*k-*252*k,* 255*k,* **259**
Andrena (Belandrena), 68*t,* 244*k,* 246*k-*249*k,* **259**
Andrena (Biareolina), 68*t,* 252*k,* 254*k,* **259,** 266
Andrena (Brachyandrena), 68*t,* 252*k,* 254*k,* **259**
Andrena (Bythandrena), 263
Andrena (Calcarina), 264
Andrena (Callandrena), 68*t,* 83, 126, 240, 240*f-*241*f,* 243, 244*k,* 246*k-*248*k,* 250*k,* **259**
Andrena (Calomelissa), 68*t,* 257*k-*258*k,* **260**
Andrena (Campylogaster), 68*t,* 254*k,* 256*k,* **260**
Andrena (Carandrena), 68*t,* 253*k,* 256*k,* **260**
Andrena (Carinandrena), 68*t,* 251, **260**
Andrena (Celetandrena), 68*t,* 244*k,* 248*k,* **260**
Andrena (Charitandrena), 68*t,* 243*k,* 248*k,* 252*k,* 254*k,* **260**
Andrena (Chaulandrena), 261
Andrena (Chlorandrena), 68*t,* 251*k,* 254*k,* 256*k,* 258*k,* **260**
Andrena (Chrysandrena), 68*t,* 252*k,* 255*k,* **260**
Andrena (Cnemidandrena), 68*t,* 245*k,* 250*k,* 254*k,* 256*k,* 258*k,* **260**
Andrena (Conandrena), 68*t,* 245*k,* 248*k,* **260,** 261
Andrena (Cordandrena), 68*t,* 253*k,* 255*k,* **260**
Andrena (Cremnandrena), 68*t,* 246*k,* 248*k,* **260**
Andrena (Cryptandrena), 68*t,*

Index of Taxa

251*k*, 253*k*-254*k*, **260**, 263
Andrena (Cubiandrena), 68*t*, 252*k*, 254*k*, **261**
Andrena (Dactylandrena), 68*t*, 245*k*, 247*k*, **261**
Andrena (Dasyandrena), 68*t*, 246*k*, 251*k*, **261**
Andrena (Derandrena), 68*t*, 244*k*, 246*k*, 249*k*, **261**
Andrena (Diandrena), 68*t*, 244*k*, 246*k*-247*k*, **261**
Andrena (Didonia), 68*t*, 252*k*-254*k*, 256*k*, 260, **261**
Andrena (Distandrena), 68*t*, 252*k*, 254*k*, **261**
Andrena (Elandrena), 265
Andrena (Erandrena), 68*t*, 244*k*, 248*k*, **261**
Andrena (Eremandrena), 265
Andrena (Euandrena), 68*t*, 243*k*, 246*k*-248*k*, 251*k*-252*k*, 255*k*, 257*k*-258*k*, **261**
Andrena (Fumandrena), 68*t*, 252*k*, 254*k*, **261**
Andrena (Fuscandrena), 68*t*, 251, **261**
Andrena (Geandrena), **261**
Andrena (Geissandrena), 68*t*, 245*k*, 249*k*, **261**
Andrena (Genyandrena), 68*t*, 246*k*, 249*k*, **262**
Andrena (Glyphandrena), 265
Andrena (Gonandrena), 68*t*, 245*k*, 249*k*, **262**, 264
Andrena (Graecandrena), 68*t*, 252*k*, 254*k*, **262**
Andrena (Gymnandrena), 263
Andrena (Habromelissa), 68*t*, 256*k*, 258*k*, **262**
Andrena (Hesperandrena), 68*t*, 244*k*, 246*k*, 248*k*, **262**
Andrena (Holandrena), 68*t*, 246*k*, 249*k*, 251, 253*k*-254*k*, 256*k*-258*k*, **262**
Andrena (Hoplandrena), 68*t*, 253*k*, 255*k*-258*k*, **262**
Andrena (Hyperandrena), 68*t*, 252*k*, 255*k*, **262**
Andrena (Iomelissa), 68*t*, 243, 244*k*, 248*k*, 262, **262**
Andrena (Larandrena), 68*t*, 245*k*, 249*k*, 254*k*, 256*k*, **262**, 264
Andrena (Leimelissa), 68*t*, 251, **262**
Andrena (Lepidandrena), 68*t*, 251*k*, 255*k*, **262-263**
Andrena (Leucandrena), 68*t*, 243*k*, 245*k*-246*k*, 248*k*-249*k*, 251*k*, 253*k*, 255*k*, **263**
Andrena (Longandrena), 68*t*, 251, **263**
Andrena (Malayapis), 68*t*, 251, 253*k*, **263**
Andrena (Margandrena), 68*t*, 253*k*, 256*k*, **263**
Andrena (Megandrena), 268
Andrena (Melanelissa), 68*t*, 252*k*, 254*k*, **263**, 895
Andrena (Melandrena), 68*t*, 247*k*, 250*k*-251*k*, 253*k*, 255*k*, 257*k*-258*k*, **263**
Andrena (Melittoides), 68*t*, 253*k*, 255*k*, **263**, 269*f*

Andrena (Micrandrena), 68*t*, 246*k*, 250*k*, 252*k*, 253, 253*k*-256*k*, 258*k*, 261, **263**
Andrena (Mimandrena), 266
Andrena (Mitsukuriapis), 265
Andrena (Mitsukuriella), 265
Andrena (Nemandrena), 68*t*, 245*k*-246*k*, 248*k*, **263**
Andrena (Nobandrena), 68*t*, 253*k*, 255*k*, **263-264**
Andrena (Notandrena), 68*t*, 245*k*, 249*k*, 253*k*, 256*k*-258*k*, 262, **264**
Andrena (Oligandrena), 68*t*, 243*k*, 249*k*, **264**
Andrena (Onagrandrena), 68*t*, 243, 245*k*, 247*k*, 250*k*, **264**
Andrena (Orandrena), 68*t*, 251*k*, 255*k*, **264**
Andrena (Oreomelissa), 68*t*, 251, 254*k*, 258*k*, **264**
Andrena (Osychnyukandrena), 68*t*, 251, **264**, xiv
Andrena (Oxyandrena), 68*t*, 245*k*, 251*k*, **264**
Andrena (Pallandrena), 68*t*, 252*k*, 255*k*, **264**
Andrena (Parandrena), 68*t*, 244*k*, 247*k*, 256*k*, 262, **264**
Andrena (Parandrenella), 68*t*, 252*k*, 255*k*, **264**
Andrena (Pelicandrena), 68*t*, 246*k*-247*k*, **264**
Andrena (Planiandrena), 68*t*, 251, **265**
Andrena (Plastandrena), 68*t*, 243*k*, 248*k*, 250*k*, 252*k*, 254*k*, 256*k*-258*k*, **265**
Andrena (Platandrena), 266
Andrena (Poecilandrena), 68*t*, 253*k*, 255*k*, 257*k*-258*k*, **265**
Andrena (Poliandrena), 68*t*, 253*k*, 255*k*-256*k*, **265**, 266
Andrena (Psammandrena), 68*t*, 247*k*, 251*k*, **265**
Andrena (Ptilandrena), 68*t*, 244*k*, 249*k*-250*k*, 253*k*, 256*k*, **265**
Andrena (Rediviva), 432
Andrena (Rhacandrena), 68*t*, 246*k*-247*k*, 251*k*, **265**
Andrena (Rhaphandrena), 68*t*, 245*k*, 249*k*, **265**
Andrena (Rufandrena), 68*t*, 251*k*, 254*k*, **265**
Andrena (Scaphandrena), 69*t*, 246*k*-247*k*, 250*k*-251*k*, 253*k*, 255*k*, **265**
Andrena (Schizandrena), 265
Andrena (Scitandrena), 69*t*, 252*k*, 254*k*, **265**
Andrena (Scoliandrena), 69*t*, 244*k*, 250*k*, **265-266**
Andrena (Scrapteropsis), 69*t*, 244*k*, 250*k*, 259, **266**
Andrena (Simandrena), 69*t*, 246*k*-247*k*, 250*k*-252*k*, 255*k*, 257*k*-258*k*, **266**
Andrena (Stenandrena), 266
Andrena (Stenomelissa), 69*t*, 251, 256*k*, 258*k*, 264, **266**
Andrena (Suandrena), 69*t*, 252*k*, 255*k*, **266**

Andrena (Taeniandrena), 69*t*, 247*k*, 250*k*, 253*k*-254*k*, 257*k*-258*k*, **266**
Andrena (Tarsandrena), 69*t*, 253*k*, 255*k*-256*k*, **266**
Andrena (Thysandrena), 69*t*, 247*k*, 251*k*, 253*k*, 255*k*, **266**
Andrena (Trachandrena), 69*t*, 244*k*, 250*k*, 252*k*, 254*k*, 256*k*, 258*k*, 259, 266, **266**
Andrena (Troandrena), 69*t*, 253*k*-254*k*, **266**
Andrena (Tropandrena), 262
Andrena (Truncandrena), 265
Andrena (Tylandrena), 69*t*, 244*k*, 249*k*, **266**
Andrena (Ulandrena), 69*t*, 252*k*, 255*k*, **266**
Andrena (Xanthandrena), 261
Andrena (Xiphandrena), 69*t*, 245*k*, 249*k*, **266-267**
Andrena (Zonandrena), 69*t*, 253*k*, 255*k*, **267**
Andrena accepta, 240*f*, 259
Andrena aciculata, 259
Andrena aeniventris, 259
Andrena aerinifrons, 260
Andrena amplificata, 265
Andrena angustella, 259
Andrena anisochlora, 260
Andrena atypica, 264
Andrena aulica, 265
Andrena auricoma, 261
Andrena australis, 338
Andrena avara, 259
Andrena bairucumensis, 262
Andrena banksi, 259
Andrena barbilabris, 263
Andrena bellidoides, 260
Andrena bicolor, 261
Andrena bradleyi, 260
Andrena brevipalpis, 265
Andrena brevipennis, 291
Andrena caerulea, 261
Andrena caliginosa, 261
Andrena candida, 266
Andrena carinifrons, 260
Andrena carlini, 263
Andrena cercocarpi, 265
Andrena chalybeata, 155
Andrena chrysochersonesus, 263
Andrena clarkella, 259
Andrena coactipostica, 259
Andrena cochlearicalcar, 264
Andrena coitana, 251, 264
Andrena colletiformis, 259
Andrena compressa, 368
Andrena cordialis, 260
Andrena cornuta, 446
Andrena cratagei, 243
Andrena cressonii, 262
Andrena cubiceps, 261
Andrena curiosa, 263
Andrena curvipes, 340
Andrena curvungula, 262
Andrena dentiventris, 264
Andrena enceliae, 268
Andrena erberi, 260
Andrena erigeniae, 265
Andrena erythrogaster, 94*f*, 124*f*
Andrena escondida, 262
Andrena fenningeri, 266
Andrena flavipes, 267
Andrena fragilis, 262
Andrena fumida, 261
Andrena fuscicollis, 261

Andrena fuscosa, 263
Andrena graecella, 262
Andrena grossella, 110-111
Andrena haemorrhoa, 266
Andrena haemorrhoidalis, 432
Andrena halictoides, 266
Andrena hattorfiana, 260
Andrena heterodoxa, 240
Andrena hirtipes, 418
Andrena humeralis, 241*f*
Andrena humilis, 260
Andrena illinoiensis, 240*f*
Andrena imitatrix, 266
Andrena jacobi, 9, 243
Andrena labialis, 243
Andrena labiata, 265
Andrena lagopus, 259
Andrena lobata, 289
Andrena longibarbis, 261
Andrena longiceps, 263
Andrena longifovea, 264
Andrena mackiae, 262
Andrena macrocephala, 264
Andrena marginata, 263
Andrena mariae, 239*f*
Andrena maura, 261
Andrena maurula, 281
Andrena melanochroa, 263
Andrena melittoides, 263, 269*f*
Andrena mendica, 266-267
Andrena micheneriana, 83, 126
Andrena mimetica, 53*f*, 96*f*, 115*f*, 122*f*, 241*f*
Andrena miserabilis, 262
Andrena mitakensis, 264
Andrena modesta, 281
Andrena montrosensis, 265
Andrena mucida, 261
Andrena nasonii, 266
Andrena nasuta, 20
Andrena neglecta, 259
Andrena nemophilae, 259
Andrena nitidiuscula, 264
Andrena nobilis, 263
Andrena obscuripostica, 261
Andrena oenotherae, 264
Andrena omogensis, 262
Andrena oralis, 264
Andrena osmioides, 20, 265
Andrena pallidicincta, 265
Andrena peringueyi, 432
Andrena persimulata, 262
Andrena planirostris, 265
Andrena plumiscopa, 259
Andrena polita, 265
Andrena porterae, 261
Andrena prima, 265
Andrena principalis, 261
Andrena propinqua, 266
Andrena prostomias, 260
Andrena pulchella, 259
Andrena pygmaea, 377
Andrena rudbeckiae, 243
Andrena rufiventris, 265
Andrena rugosa, 266
Andrena s. str., 68*t*, 245*k*-246*k*, 249*k*-250*k*, 254*k*, 256*k*-258*k*, **259**
Andrena schulzi, 266
Andrena scita, 265
Andrena scotia, 9
Andrena solenopalpa, 261
Andrena spiralis, 330
Andrena suerinensis, 266
Andrena surda, 235*f*
Andrena tarsata, 266
Andrena toluca, 260

Andrena torulosa, 263
Andrena trevoris, 261
Andrena troodica, 266
Andrena truncatilabris, 265
Andrena tscheki, Plate 3
Andrena vandykei, 261
Andrena variegata, 346
Andrena ventricosa, 260
Andrena viburnella, 243
Andrena vidalesi panamensis, 259
Andrena vinnula, 260
Andrena violae, 243, 262
Andrena wilkella, 235*f,* 266
Andrena wrisleyi, 101
Andrenella, 263
Andrenetae (Family), 77*t*
Andrenetes (Family), 77*t*
Andrenidae (Family), 20, 25, 28, 34-35, 43, 45, 61, 65*t,* 68*t*-69*t,* 78*t*-79*t,* 80, 80*t,* 81, 81*t,* 91-95, 101, 103, 111-112, 116, 122*f,* 123*k,* 124*f,* 126, 127*k*-128*k,* 129, 131, 133, 169, 188, **235-237,** 237*k,* 270, 316-317, 319, 323, 345, 389, 645, Plate 3, Plate 5
 biogeography of, 107*t*
Andrenidae (Subfamily), 78*t*
andrenids, 54
Andrenina (Tribe), 78*t*
Andreninae (Subfamily), 9, 20, 45, 52, 68*t*-69*t,* 79*t*-80*t,* 81, 81*t,* 83, 87, 91, 93-95, 123*k,* 126, 127*k,* 131, 156, 235-237, 237*k,* **239-269,** 239*f,* 239*k*-240*k,* 270, 316-317, 323, 638, 895
 biogeography of, 107*t*
 fossil, 99*t*
Andrenini (Tribe), 239
Andrenites (Tribe), 78*t*
Andrenoidea, 79*t*
Andrenopsis, 67*t,* 144, 145*f,* 147*t,* 149*k,* **150**
Andrenopsis flavorufus, 150
Andrenopsis velutinus, 152
Androgynella, 573
Andronicus, 472
Andronicus cylindricus, 472
Anepicharis, 74*t,* **760,** 760*k*
Angochlora, 402
Angochloropsis, 404
Anhylaeus, 210
Annosmia, 71*t,* 462, 463*k,* **467,** 468-469, 471
Anodonteutricharaea, 573
Anodontobombus, 801
Anoediscelis, 67*t,* 181, 181*f,* **182,** 182*f,* 182*k,* 183*f,* 184
Anomalohesma, 68*t,* **215,** 224, 224*k*
Anthedon, 734
Anthedonia, 20, 74*t,* 718*k,* 720*k,* **734-735,** 734*k*
Anthemoessa, 749
Anthemois, 578
Anthemurgini (Tribe), 273
Anthemurgus, 69*t,* 236, 274, **274-275,** 274*k*
Anthemurgus passiflorae, 20, 270*f,* 274-275, 279*f*
Anthidianum, 523
Anthidiellum, 491*t,* 494*k*-496*k,* 502-504, 502-505, 503*k,* 514, 524, 532-533, 535
Anthidiellum (*Ananthidiellum*), 71*t,* **503,** 503*k,* 504
Anthidiellum (*Anthidiellum* s. str.), 25, 71*t,* 494, 496*f,* 502*f,* **503,** 503*k,* 504
Anthidiellum (*Chloranthidiellum*), 71*t,* 494, 502, 503*k,* **504**
Anthidiellum (*Chloranthidium*), 504
Anthidiellum (*Clypanthidium*), 71*t,* 495, 502-503, 503*k,* **504**
Anthidiellum (*Loyolanthidium*), 71*t,* **504**
Anthidiellum (*Pycnanthidium*), 71*t,* 503, 503*k,* **504**
Anthidiellum (*Ranthidiellum*), 71*t,* 503, 503*k,* **504-505,** 511
Anthidiellum (*Rhanthidiellum*), 504
Anthidiellum anale, 503
Anthidiellum ausense, 523
Anthidiellum bimaculatum, 503-504
Anthidiellum bolivianum, 504
Anthidiellum flavescens, 504
Anthidiellum notata robertsoni, Plate 8
Anthidiellum notatum, 502*f*
Anthidiellum notatum robertsoni, 496*f*
Anthidiellum perplexum, 114*f*
Anthidiellum ruficeps, 504
Anthidiellum s. str., 25, 71*t,* 494, 496*f,* 502*f,* **503,** 503*k,* 504
Anthidiellum strigatum, 503
Anthidiinae (Subfamily), 79*t*
Anthidiini (Tribe), 21, 32*t,* 66, 71*t*-72*t,* 79*t*-80*t,* 111-112, 117, 126, 434-435, 435*f,* 441, 442*t,* 443*k,* **491-537,** 491*f,* 491*t,* 492*f*-493*f,* 495*f*-497*f,* 512*f,* 538, 543, 545, 578, 749, 784, 895, Plate 8
 biogeography of, 107*t*
 of Eastern Hemisphere, 492*k*-495*k*
 fossil, 99*t*
 Series A, 492
 Series B, 492
 of Western Hemisphere, 495*k*-497*k*
anthidine bees, 23*f*
Anthidioma, 71*t,* 491*t,* 493*k,* **505**
Anthidioma chalicodomoides, 505
Anthidium, 20, 42, 66, 78*t*-81*t,* 111, 491, 491*t,* 493*k,* 495*k,* 497-498, **505-508,** 506*k,* 521, 527, 529, 533, 538, 895
Anthidium (*Afranthidium*), 499
Anthidium (*Anthidiellum*), 503
Anthidium (*Anthidium* s. str.), 71*t,* 505-506, **506-507,** 506*k,* 508, 526
Anthidium (*Ardenthidium*), 506-507
Anthidium (*Callanthidium*), 71*t,* 505-506, 506*k,* **507**
Anthidium (*Cerianthidium*), 503
Anthidium (*Dianthidium*), 513
Anthidium (*Echinanthidium*), 506-508
Anthidium (*Gulanthidium*), 71*t,* 506*k,* **507**
Anthidium (*Heteranthidium*), 535
Anthidium (*Melanoanthidium*), 506-507
Anthidium (*Morphanthidium*), 506
Anthidium (*Nivanthidium*), 71*t,* 505, 506*k,* **508,** 529
Anthidium (*Pachyanthidium*), 524
Anthidium (*Paraanthidium*), 536
Anthidium (*Pontanthidium*), 506-507
Anthidium (*Proanthidium*), 71*t,* 493, 497, 505, 506*k,* 507, **508,** 527
Anthidium (*Pseudoanthidium*), 527
Anthidium (*Rhodanthidium*), 529
Anthidium (*Severanthidium*), 71*t,* 505, 506*k,* 507-508, **508**
Anthidium (*Turkanthidium*), 71*t,* 506*k,* 507, **508**
Anthidium abdominale, 500
Anthidium aculeatum, 529
Anthidium alpinum, 527
Anthidium amabile, 508
Anthidium anale, 503
Anthidium anguliventre, 507
Anthidium annulatum, 532
Anthidium apicale, 504
Anthidium apiforme, 509
Anthidium aquifilum, 536
Anthidium ardens, 506-507
Anthidium arenarium, 519
Anthidium atripes, 53*f,* 91*f,* 122*f,* 435*f*
Anthidium benguelense, 524
Anthidium bicolor, 524
Anthidium bidentatum, 523
Anthidium binghami, 511
Anthidium bivittatum, 517
Anthidium braunsi, 499
Anthidium capicole, 499
Anthidium caturigense, 528
Anthidium christophi, 529
Anthidium clypeare, 515
Anthidium concolor, 501
Anthidium cribratum, 527
Anthidium curvatum, 513
Anthidium divaricatum, 521, 895
Anthidium dorsale, 535
Anthidium ducale, 529
Anthidium echinatum, 506-507
Anthidium edwini, 507
Anthidium elongatum, 515
Anthidium espinosai, 506-507
Anthidium eximium, 526
Anthidium flavescens, 504
Anthidium flavofasciatum, 519
Anthidium flavomarginatum, 521, 895
Anthidium flavopictum, 521
Anthidium folliculosum, 501
Anthidium formosanum, 536
Anthidium fraternum, 511
Anthidium furcatum, 521
Anthidium glasunovii, 528
Anthidium gratum, 508
Anthidium honestum, 499
Anthidium illustre, 63*f,* 507
Anthidium immaculatum, 500
Anthidium inerme, 516
Anthidium infuscatum, 529
Anthidium insulare, 515
Anthidium japonicum, 506
Anthidium laticeps, 534
Anthidium latum, 506-507
Anthidium limbiferum, 521
Anthidium lituratum, 506
Anthidium longicorne, 536
Anthidium maculosum, 493*f,* 497*f*
Anthidium madagascariensis, 512
Anthidium malaccense, 532
Anthidium manicatum, 10, 491, 491*f,* 507
Anthidium megachiliforme, 514
Anthidium melanurum, 527
Anthidium montanum, 506-507
Anthidium multiplicatum, 518
Anthidium nasutum, 532
Anthidium niveocinctum, 505, 508, 529
Anthidium oblongatum, 493, 497, 508
Anthidium ochrognathum, 527
Anthidium octodentatum, 522
Anthidium oraniense, 501
Anthidium orientale, 527
Anthidium paradoxum, 531
Anthidium pendleburyi, 535
Anthidium pentagonum, 501
Anthidium perpictum, 525
Anthidium pontis, 506-507
Anthidium punctatum, 507
Anthidium reticulatum, 527
Anthidium ridingsii, 535
Anthidium rodolfi, 523
Anthidium ruficeps, 504
Anthidium s. str., 71*t,* 505-506, **506-507,** 506*k,* 508, 526
Anthidium schoutedeni, 535
Anthidium severini, 505, 508
Anthidium sibiricum, 511
Anthidium siculum, 529
Anthidium steloides, 523
Anthidium subpetiolatum, 523
Anthidium superbum, 529
Anthidium tesselatum, 507
Anthidium texanum, 513
Anthidium trachusiforme, 526
Anthidium truncatum, 527
Anthidium tuberculiferum, 528
Anthidium undulatiforme, 507
Anthidium undulatum, 508
Anthidium unicum, 508
Anthidium vespoides, 525
Anthidium volkmanni, 526
Anthidium zebra, 504
Anthidulum, 72*t,* 495*f*-496*f,* **519,** 519*k,* 520-521
Anthocharessa, 259
Anthocopa, 453, 455
Anthocopa, 71*t,* 449, 451, 453, 455, 461-464, 464*k,* 465, **467-468,** 468-471, 473, 475-476, 485, 490, 545, 574
Anthocopa (*Eremoplosmia*), 490
Anthocopa (*Eremosmia*), 455, 476

Anthocopa (Exanthocopa), 469
Anthocopa (Haetosmia), 458
Anthocopa (Hexosmia), 455
Anthocopa (Isosmia), 455
Anthocopa (Odontanthocopa), 475
Anthocopa (Odonterythrosmia), 475
Anthocopa (Othinosmia), 487
Anthocopa (Paranthocopa), 475
Anthocopa (Phaeosmia), 455
Anthocopa (Rhodosmia), 488
Anthocopa (Xerosmia), 473
Anthocopa cristatula, 467
Anthocopa rubrella, 455
Anthodioctes, 491*t*, 496*k*, 502*f*, **508-509**, 508*f*, 509*k*, 513, 515, 518, 521
Anthodioctes (Anthodioctes s. str.), 71*t*, **509**, 509*k*, 522
Anthodioctes (Bothranthidium), 71*t*, **509**, 509*k*, 519
Anthodioctes (Nananthidium), 509
Anthodioctes dasygastrinus, 509
Anthodioctes gualanense, 491*f*
Anthodioctes lauroi, 509
Anthodioctes megachiloides, 509
Anthodioctes s. str., 71*t*, **509**, 509*k*, 522
Anthoglossa, 67*t*, 165, **165**, 165*k*
Anthoglossa plumata, 165
Anthomegilla, 74*t*, 746*k*, 747, 747*k*, 748-750
Anthophila, 3
Anthophora, 20, 28, 34, 44, 77*t*-81*t*, 105, 536, 545, 573, 588, 652, 662, 742-743, 743*k*, **744-750**, 745*k*-747*k*, 750-751, 753, 756, 766, 770-771, 777, 780, 895
Anthophora (Anthomegilla), 74*t*, 746*k*, 747, 747*k*, 748-750
Anthophora (Anthophora s. str.), 74*t*, 744*f*, 745*k*, 747, 747*k*, 748
Anthophora (Anthophoroides), 74*t*, 744*f*, 746*k*-747*k*, **748**, 749
Anthophora (Caranthophora), 74*t*, 744*f*, 745*k*, 747*k*, **748**
Anthophora (Clisodon), 74*t*, 545, 742, 745*k*-746*k*, 747, **748**, 749
Anthophora (Dasymegilla), 74*t*, 746*k*, 747, 747*k*, **748**
Anthophora (Eucara), 737
Anthophora (Heliophila), 74*t*, 109, 667, 745*f*, 746*k*, 747, **748-749**, 749, 770, 777
Anthophora (Lophanthophora), 74*t*, 742, 744*f*, 745*k*, 747*k*, **749**
Anthophora (Melea), 74*t*, 746*k*, 747, 747*k*, 748, **749**
Anthophora (Mystacanthophora), 74*t*, 745*f*, 746*k*-747*k*, **749**
Anthophora (Paramegilla), 74*t*, 746*k*-747*k*, **749**
Anthophora (Petalosternon), 74*t*, 746*k*-747*k*, 748, **750**
Anthophora (Pyganthophora), 74*t*, 744*f*, 745*k*-747*k*, 749, **750**

Anthophora (Rhinomegilla), 74*t*, 746*k*, 747, 747*k*, 748-749, **750**
Anthophora abrupta, 749
Anthophora acervorum, 747
Anthophora aeruginosa, 743
Anthophora albifascies, 775
Anthophora albigena, 743
Anthophora albisecta, 572
Anthophora arctica, 747
Anthophora barbata, 579
Anthophora bomboides stanfordiana, Plate 13
Anthophora calcarata, 746
Anthophora californica, 746
Anthophora cockerelli, 745*f*
Anthophora crinipes, 747
Anthophora crotchii, 744*f*
Anthophora curta, 748
Anthophora dufourii, 744*f*, 748
Anthophora edwardsii, 53*f*, 83*f*, 96*f*, 122*f*
Anthophora excisa, 747-748
Anthophora floridana, 751
Anthophora fulvitarsis, Plate 13
Anthophora furcata, 545, 748
Anthophora gayi, 726
Anthophora hedini, 748
Anthophora hispanica, 744*f*
Anthophora larvata, 749
Anthophora laticeps, 737
Anthophora megarrhina, 750
Anthophora mesopyrrha, 743
Anthophora mimadvena, 743
Anthophora montana, 749
Anthophora niveata, 743
Anthophora nubica, 743
Anthophora oblongata, 508
Anthophora occidentalis, 124*f*, 481, 742*f*
Anthophora pacifica, 749
Anthophora peritomae, 749
Anthophora phaceliae, 744*f*
Anthophora plumipes, 744*f*, 747
Anthophora porterae, 744*f*, 749
Anthophora prshewalskyi, 749
Anthophora retusa, 744*f*
Anthophora rivolleti, 750
Anthophora s. str., 74*t*, 744*f*, 745*k*, 747, 747*k*, 748
Anthophora squammulosa, 667
Anthophora taurea, 705
Anthophora terminalis, 748
Anthophora urbana, 749
Anthophora vallorum, 746
Anthophora walchii, 745*f*
Anthophoridae (Family), 77, 79, 79*t*, 81, 97, 434, 667-668
Anthophoridae (Subfamily), 78*t*
Anthophorinae (Subfamily), 80, 80*t*, 81, 81*t*, 97, 144, 667
Anthophorini (Tribe), 9, 33*t*, 35, 74*t*, 80*t*, 117, 590*k*, 662, 667-668, 668*f*, 671, 709, **742-752**, 742*f*, 742*k*-743*k*, 744*f*-745*f*, 763, 770, 785
biogeography of, 108*t*
fossil, 98*t*
Anthophorisca, 73*t*, 640, 681, **681-682**, 681*k*, 682*f*
Anthophorites, fossil, 98*t*
Anthophorites (Tribe), 78*t*

Anthophorites mellona, fossil, 98*t*
Anthophoroidea, 79*t*
Anthophoroides, 74*t*, 744*f*, 746*k*-747*k*, **748**, 749
Anthophorula, 106, 430, 680-681, **681-682**, 681*k*, 682-683
Anthophorula (Anthophorisca), 73*t*, 640, 681, **681-682**, 681*k*, 682*f*
Anthophorula (Anthophorula s. str.), 73*t*, 642, 680*f*, 681, 681*k*, **682**, 682*f*, 685
Anthophorula (Isomalopsis), 73*t*, 681, 681*k*, **682**, 682*f*
Anthophorula compactula, 680*f*, 682, 682*f*
Anthophorula levigata, 681
Anthophorula linsleyi, 681
Anthophorula nitens, 681
Anthophorula niveata, 681, 682*f*
Anthophorula pygmaea, 682*f*
Anthophorula s. str., 73*t*, 642, 680*f*, 681, 681*k*, **682**, 682*f*, 685
Anthophorula sidae, 681
Anthrena, 259
Anthrenoides, 69*t*, 272*k*-274*k*, **275**, 279, 283
Anthrenoides alfkeni, 275
Anthrenoides meridionalis, 274*f*
Anthrophorula, 682
Anylaeus, 210
Aparatrigona, 820-821
Apathus, 800
Apathus ashtoni, 800
Apathus citrinus, 800
Apathus fraternus, 798
Apathus tibetanus, 800
Apeulaema, 784
Aphalictus, 381
Aphaneura, 828
Aphaneura rufescens, 828
Aphemisia, 74*t*, 755*k*, **756**, 758
Apianthidium, 71*t*, 491*t*, 494*k*-495*k*, **509**, 515, 537
Apianthidium apiforme, 509
Apiares (Family), 77*t*
Apiaria, fossil, 98*t*
Apiaria dubia, fossil, 98*t*
Apiariae (Family), 77*t*
Apiarides (Family), 78*t*
Apiarites (Tribe), 78*t*
Apiarus, 831
Apicula, 831
Apidae (Family), 18, 20, 28, 31*t*, 32, 32*t*, 33-35, 33*t*, 37, 37*f*, 39*f*, 41, 51, 56, 59, 65, 65*t*, 72*t*-75*t*, 78, 78*t*-79*t*, 80, 80*t*, 81, 81*t*, 82, 92, 96-97, 110-112, 116-117, 122*f*, 123*k*, 124*f*, 126-127, 127*k*-128*k*, 271, 434, 538, **587-591**, 587*f*, 588*k*-591*k*, 633, 636, 645, 661, 667-668, 671, 700, Plate 1-Plate 2, Plate 9-Plate 14
biogeography of, 107*t*-108*t*
corbiculate, 81-82, 110, 117, 667-669, 668*f*-669*f*, 670-671, 778, 780, 787*f*, 801, 816, 906

fossil, 671, Plate 19-Plate 20
fossil, 101, 671, Plate 19-Plate 20
Apidae (Subfamily), 78*t*
Apiformes, 3, 65*t*, 81
Apina (Tribe), 78*t*
Apinae (Subfamily), 18, 21, 28, 31*t*, 33, 33*t*, 34, 37*f*, 40, 48, 51-52, 56, 73*t*-75*t*, 77, 79*t*-80*t*, 81, 81*t*, 82, 93, 97, 110, 116-117, 127, 128*k*, 171, 413, 437, 441, 587-588, 588*k*-591*k*, 592, 633, 645, 648, 652, **667-671**, 668*f*, 669, 672, 674, 678, 685, 694, 697, 701*f*, 742, 906, Plate 10-Plate 12
biogeography of, 108*t*
corbiculate, **97**
fossil, Plate 19-Plate 20
noncorbiculate, **97**
apine bees, 61, 82, 667-668, 668*f*
Apini (Tribe), 9, 12, 18-19, 26, 30, 52, 75*t*, 80*t*, 81-82, 96-97, 102, 104-105, 112, 117, 236, 588, 588*k*, 667-668, 670-671, 785, 803, **830-831**
biogeography of, 108*t*
fossil, 98*t*-99*t*, 670, 670*f*, 671, Plate 20
Apis, 5-7, 12-13, 17, 21, 28-29, 56, 65, 75*t*, 77, 77*t*-81*t*, 87, 103, 105, 112, 114, 117, 587, 667, 670, 750, 785, 787, 804, 806-807, 816-817, 830-831, **831**
Apis (Ancylosoma), 607
Apis (Cascapis), 831
fossil, 98*t*
Apis (Prioriapis), 831
fossil, 99*t*
Apis (Sigmatapis), 831
Apis (Synapis), 831
fossil, 99*t*
Apis acervorum, 747
Apis acraensis, 743
Apis adunca, 470
Apis agilissima, 259
Apis agrorum, 802
Apis albifrons, 774
Apis alpinus, 797
Apis altercator, 418
Apis amalthea, 828
Apis amethystina, 607
Apis andreniformis, 831
Apis annulata, 207
Apis argentata, 573
Apis argillacea, 799
Apis armbrusteri, 831
fossil, 98*t*-99*t*, 831
Apis aterrima, 532
Apis aurulenta, 481
Apis banksiana, 289
Apis barbutella, 800
Apis bicolorata, 262
Apis bicornis, 484
Apis bimaculata, 748
Apis binghami, 831
Apis bomb. iris, 605
Apis bombylans, 606
Apis brasilianorum, 607
Apis breviligula, 831
Apis byssina, 536
Apis calcarata, 289

Apis calceata, 379
Apis calendarum, 167
Apis callosa, 616
Apis campestris, 800
Apis carbonaria, 265
Apis centuncularis, 578
Apis cerana, 1, 831, Plate 10
Apis cincta, 575
Apis coecutiens, 675
Apis collaris, 610
Apis compressipes, 817
Apis conica, 548
Apis cordata, 783
Apis cucurbitina, 615
Apis cullumana, 798
Apis cunicularia, 167
Apis dentata, 784
Apis dimidiata, 784
Apis dorsata, 26, 831
Apis druriella, 732
Apis favosa, 817
Apis femoralis, 363
Apis fenestrata, 605
Apis fervida, 798
Apis flavipes, 368
Apis florea, 26, 369, 831, Plate 9
Apis florisomnis, 457
Apis frontalis, 607
Apis fulvago, 260
Apis fulviventris, 481
Apis griseocollis, 801
Apis haemorrhoidalis, 756
Apis helvola, 259
Apis henshawi, fossil, 99*t*, 831, Plate 20
Apis hortorum, 799
Apis hypnorum, 800-801
Apis interrupta, 536
Apis ireos, 749
Apis koschevnikovi, 831
Apis laboriosa, 831
Apis lagopoda, 584
Apis lanipes, 757
Apis lapidaria, 799
Apis lapponica, 801
Apis latipes, 606
Apis leucozonia, 380
Apis longicornis, 724
Apis malvae, 737
Apis manicata, 506
Apis maritima, 584
Apis maxillosa, 457
Apis meliponoides, fossil, 98*t*
Apis mellifera, 1, 9, 19-20, 42, 44, 787*f*, 830-831, 830*f*-831*f*, 831, 831*f*. See also honey bees
Apis mellifica, 831
Apis meriana, 784
Apis minutissima, 347
Apis muraria, 569-570
Apis muscaria, 606
Apis muscorum, 797, 802
Apis mystacea, 566
Apis nigrita, 606
Apis nigrocincta, 831
Apis niveata, 481
Apis nuluensis, 831
Apis pallens, 828
Apis palmata, 764
Apis palmnickenensis, fossil, 99*t*
Apis papaveris, 467-468
Apis parietina, 569, 578
Apis pascuorum, 802
Apis pilipes, 747
Apis plumipes, 747, 756

Apis pratorum, 801
Apis punctata, 774
Apis punctulatissima, 532
Apis quadricincta, 367
Apis quadridentata, 548
Apis quadrifasciata, 743
Apis quadrimaculata, 748
Apis retusa, 750, 831
Apis rotundata, 573
Apis rubicunda, 369
Apis ruderaria, 802
Apis rufa, 484
Apis ruficornis, 643
Apis rupestris, 800
Apis rustica, 760
Apis seladonia, 369
Apis sericea, 263
Apis sexcincta, 368
Apis sibirica, 801
Apis smaragdula, 617
Apis soroeensis, 799
Apis spinulosa, 475
Apis splendida, 764
Apis subterranea, 801
Apis succincta, 167
Apis sylvarum, 802
Apis terrestris, 797
Apis thoracica, 263
Apis truncorum, 460
Apis variegata, 649
Apis versicolor, 756
Apis vestuta, fossil, 99*t*
Apis villosa, 468
Apis violacea, 609
Apis virginica, 609
Apis viridula, 363
Apis xanthomelana, 482
Apis zonata, 743
Apista, 179
Apista gaullei, 142
Apista opalina, 179
Apoidea (superfamily), 3, 59, 62, 65, 65*t*, 79*t*, 101, 445
Apomelissodes, 74*t*, 730*k*, **731**
Aporandrena, 68*t*, 243*k*, 248*k*, **259**, 262-263
Apotrigona, 818-819
Apoxylocopa, 610
Apygidialia, 62, 79, 79*t*
Archeriades, 452
Archeriades (*Ceraheriades*), 457
Archeriades petersi, 457
Archiandrena, 68*t*, 244*k*-245*k*, 249*k*, **259**
Archianthidium, 72*t*, 533, 533*k*, **534-535**, 535, 538
Archihalictus, 70*t*, 370, 384, **384-385**, 384*k*, 386
Archimegachile, 581
Arctosmia, 467-468
Ardenthidium, 506-507
Argalictus, 70*t*, 365*f*, **367**, 367*k*, 368
Argocoelioxys, 549
Argohesma, 68*t*, **233**, 233*k*
Argohesma eremica, 233
Argyropile, 72*t*, 557*t*, 558*k*, 561*k*, **567**
Argyroselenis, 649
Arhysoceble, 73*t*, **689**, 689*k*, 690-691
Arhysoceble melampoda, 689*f*
Arhysoceble xanthopoda, 689
Arhysosage, 69*t*, 306, 307*k*, **309**, 315, 666
Arhysosage johnsoni, 309

Ariphanarthra, 70*t*, 394*k*, 398*k*, **401**, 408
Ariphanarthra palpalis, 395*f*, 397*f*, 401
Arogochila, 71*t*, 453, 453*k*-454*k*, 454, **454**
Asaropoda, 743
Ascirtetica, 721
Ashmeadiella, 108, 441, 449*t*, 452*k*, **453-455**, 453*k*-454*k*, 462, 533
Ashmeadiella (*Arogochila*), 71*t*, 453, 453*k*-454*k*, 454, **454**
Ashmeadiella (*Ashmeadiella* s. str.), 71*t*, 453, 453*f*, 453*k*, **454**, 454*f*
Ashmeadiella (*Chilosima*), 71*t*, 453, 453*k*, **454**
Ashmeadiella (*Corythochila*), 454
Ashmeadiella (*Cubitognatha*), 71*t*, 453*k*, **454-455**
Ashmeadiella (*Isosmia*), 71*t*, 452-453, 453*k*, **455**
Ashmeadiella (*Neoashmeadiella*), 453, 570
Ashmeadiella (*Ramphorhina*), 454
Ashmeadiella (*Rhamphorhina*), 454
Ashmeadiella (*Titusella*), 454
Ashmeadiella bigeloviae, 114*f*
Ashmeadiella bucconis, 448*f*
Ashmeadiella californica, 453*f*
Ashmeadiella cubiceps, 453-454
Ashmeadiella femorata, 116
Ashmeadiella foxiella, 454
Ashmeadiella holtii, 453
Ashmeadiella howardi, 470
Ashmeadiella indica, 570
Ashmeadiella inyoensis, 454
Ashmeadiella leucozona, 453
Ashmeadiella occipitalis, 454*f*
Ashmeadiella rhodognatha, 454
Ashmeadiella rubrella, 453
Ashmeadiella s. str., 71*t*, 453, 453*f*, 453*k*, **454**, 454*f*
Ashmeadiella timberlakei, 454
Ashmeadiella xenomastax, 454-455
Ashtonipsithyrus, 800
Asianthidium, 72*t*, 497, 528, **528-529**, 528*k*
Aspidosmia, 71*t*, 126, 434, 441, 442*t*, 443, 491*t*, 493*k*, **509-510**, 538
Aspidosmia arnoldi, 510
Astatinae (Subfamily) (wasps), 63
Asteronomada, 644
Atopeucera, 724-725
Atoposmia, 120, 449*t*, 452*k*, 453, 455, **455-456**, 455*k*, 462, 471, 476
Atoposmia (*Atoposmia* s. str.), 71*t*, 452, **455**, 455*k*
Atoposmia (*Eremosmia*), 71*t*, **455**, 455*k*
Atoposmia (*Hexosmia*), 71*t*, 448*f*, **455-456**, 455*k*, 476
Atoposmia abjecta, 455
Atoposmia beameri, 455
Atoposmia copelandica, 448*f*
Atoposmia elongata, 452, 455
Atoposmia enceliae, 455
Atoposmia hypostomalis, 455

Atoposmia s. str., 71*t*, 452, **455**, 455*k*
Atrocinctob[ombus], 798
Atrodufourea, 326
Atronomioides, 70*t*, **346**, 346*k*
Atropium, 514
Atrosamba, 70*t*, 414, **424**, 424*k*
Audineta, 606
Audinetia, 606
Augandrena, 68*t*, 246*k*, 250*k*, **259**
Augochlora, 45, 79*t*-80*t*, 327, 393, 394*k*, 400, 400*k*, **401-402**, 401*k*-402*k*, 403, 408
Augochlora (*Aethechlora*), 402
Augochlora (*Angochloropsis*), 404
Augochlora (*Augocha* s. str.), 348
Augochlora (*Augochlora* s. str.), 70*t*, 394, 401, **402**, 402*k*, 410
Augochlora (*Augochloropsis*), 404
Augochlora (*Electroaugochlora*), fossil, 98*t*, 393
Augochlora (*Glyptobasia*), 404-405
Augochlora (*Glyptobasis*), 404-405
Augochlora (*Mycterochlora*), 402
Augochlora (*Oxystoglossella*), 70*t*, 401, 401*k*, **402**
Augochlora (*Tetrachlora*), 404
Augochlora aenigma, 404
Augochlora briseis, 411
Augochlora chloera, 404-405
Augochlora cordiaefloris, 402
Augochlora epipyrgitis, 404
Augochlora gratiosa, 402
Augochlora ignita, 404
Augochlora leptoloba, fossil, 98*t*
Augochlora matucanensis, 402
Augochlora mulleri, 402
Augochlora nigrocyanea, 396*f*
Augochlora ogilviei, 409
Augochlora pura, 393, 395*f*, 397*f*-399*f*, 401*f*
Augochlora s. str., 70*t*, 348, 394, 401, **402**, 402*k*, 410
Augochlora spinolae, 404
Augochlora subignita, 404
Augochlorella, 15, 378, 393, 394*k*, 400*k*, 401, **402-403**, 402*k*, 408
Augochlorella (*Augochlorella* s. str.), 70*t*, **402-403**, 402*k*
Augochlorella (*Ceratalictus*), 70*t*, 396*f*, 400*k*, 402, 402*k*, **403**
Augochlorella (*Pereirapis*), 70*t*, 396*f*, 402, 402*k*, **403**
Augochlorella edentata, 402
Augochlorella pomoniella, Plate 5
Augochlorella s. str., 70*t*, **402-403**, 402*k*
Augochlorella striata, 25*f*, 352*f*, 397*f*, 401*f*, 403
Augochlorella theia, 396*f*
Augochlorina (Subtribe), 393
augochlorine bees. See Augochlorini (Tribe)
Augochlorini (Tribe), 15, 32*t*, 43, 70*t*, 81, 101, 104, 112, 116, 345, 348, 350, 352, 352*f*, 353, 353*k*, 354-355, 370, 378, 388, **393-412**, 393*f*, 394*k*-

400*k*, 395*f*-397*f*, 399*f*,
401*f*, 404*f*, 407*f*, 895,
905, Plate 5
biogeography of, 107*t*, 109
fossil, 98*t*-99*t*, Plate 17
Augochlorodes, 70*t*, 397*k*, 400*k*,
403
Augochlorodes turrifaciens, 388,
399*f*, 403
Augochloropsis, 393, 395*k*,
399*k*, **403**-404, 404*k*, 411
Augochloropsis (*Augochloropsis* s.
str.), 70*t*, 399*f*, **404**, 404*k*
Augochloropsis (*Glyptochlora*),
405
Augochloropsis (*Paraugochlorop-
sis*), 70*t*, 399*f*, **404**-405,
404*f*, 404*k*, 410
Augochloropsis (*Pseudaugo-
chloropsis*), 404
Augochloropsis chloera, 399*f*
Augochloropsis ignita, 396*f*, 399*f*
Augochloropsis iris, 350
Augochloropsis lycorias, 404
Augochloropsis metallica, 320*f*,
349*f*, 393*f*, 396*f*-399*f*,
404*f*
Augochloropsis s. str., 70*t*, 399*f*,
404, 404*k*
Augochloropsis sparsilis, 25*f*, 351
Augochloropsis sthena, 404
Augochloropsis sumptuosa, 403*f*
Auricularia, 212
Ausanthidium, 72*t*, 523, **523-
524**, 523*k*
Australictus, 70*t*, 371*t*, 374*k*,
375, 375*k*
Australomelitturga, 292
Austrandrena, 281
Austrevylaeus, 70*t*, 361, 371*t*,
374*k*, **376**
Austrochile, 72*t*, 554*f*, 557*t*,
564*f*, 565*k*, **567**
Austrodioxys, 666
Austrodioxys thomasi, 666
Austromegachile, 72*t*, 557*t*,
559*k*-560*k*, **567**, 582, 584
Austronomia, 69*t*, 332-333,
332*f*, 333, 333*f*, 335, **336-
337**, 336*k*, 338, 342*f*
Austropanurgini (Tribe), 273
Austropanurgus, 69*t*, 273,
278-279, 279*k*, **280**, 280*k*
Austropanurgus punctatus, 280
Austropeponapis, 733
Austroplebeia, 75*t*, 805-806,
809, 811, **815**, 815*k*
Austroplebeia australis,
806*f*-807*f*
Austroplebeia cincta, 805
Austrosphecodes, 361, 382, 388-
389
Austrostelis, 71*t*, 496*k*, 509,
510, 517-519
Autochelostoma, 467
Autochelostoma canadensis, 467
Avandrena, 68*t*, 251*k*-252*k*,
255*k*, **259**
Avpanurgus, 69*t*, 285, 285*k*, **286**
Avpanurgus flavofasciatus, 286,
286*f*
Axestotrigona, 75*t*, **818**, 818-
819, 818*k*
Axillanthidium, 497, 529
Axillanthidium axillare, 529
Aztecanthidium, 71*t*, 491*t*,
497*k*, **510-511**, 513, 515

*Aztecanthidium tenochtit-
lanicum*, 510*f*
Aztecanthidium xochipillium,
510
Baana, 605
Baeocolletes, 67*t*, 145*f*, 147*t*,
148*k*, **150-151**
Baptonedys, 156-157
Barbata, 209
Bathanthidium, 491*t*, 495*k*,
497, 503-504, **511**, 511*k*,
530
Bathanthidium (*Bathanthidium*
s. str.), 71*t*, **511**, 511-512,
511*k*
Bathanthidium (*Clypanthid-
ium*), 504
Bathanthidium (*Manthidium*),
71*t*, 511, **511**, 511*k*
Bathanthidium (*Stenanthidiel-
lum*), 71*t*, 511, **511-512**,
511*k*
Bathanthidium bifoveolatum,
511
Bathanthidium binghami, 511
Bathanthidium s. str., 71*t*, **511**,
511-512, 511*k*
Bekilia, 32*t*, 71*t*, 449*t*, **456**
Bekilia mimetica, 456
Belandrena, 68*t*, 244*k*, 246*k*-
249*k*, **259**
Bellanthidium, 529
Belopria, 151, 157
Belopria zonata, 157
Benanthis, 71*t*, 491*t*, 494, **512**
Benanthis madagascariensis, 512
Berna, 575
Berna africana, 575
Betheliella, 326
Betheliella calocharti, 326
Biareolina, 68*t*, 252*k*, 254*k*,
259, 266
Biastes, 43, 73*t*, 81*t*, 656, **656-
657**, 656*k*, 657-658
Biastes brevicornis, 657*f*
Biastinae (Subfamily), 81
Biastini (Tribe), 43, 73*t*, 590*k*,
636, 638, 653, **656-658**,
656*k*, 657*k*, 660-661
biogeography of, 108*t*
Biastoides, 656
Bicolletes, 157-158
Bicolletes neotropica, 157
Bicornelia, 179
Bicornelia andina, 177
Bicornelia serrata, 179
Biglossa, 67*t*, 151, 158*f*, 161,
161*f*, **162-163**, 162*k*,
163-164
Biglossa chalybaea, 162
Biglossa laticeps, 151
Biglossa thoracica, 162
Biglossidia, 162-163
Biluna, 73*t*, 598, 602*k*-603*k*,
604-605, 608
Binghamiella, 140
Binghamiella (*Pachyodonta*),
140
Binghamiella fulvicornis, 140
Birkmania, 287
Birkmania andrenoides, 287
Bluethgenia, 380
Bombias, 75*t*, 792*k*, 795*k*, 797
Bombias auricomus, 797
Bombidae (Family), 78, 78*t*-
79*t*, 81, 97

Bombides (Family), 78*t*
Bombina (Tribe), 78*t*
Bombinae (Subfamily), 81
Bombini (Tribe), 18-19, 26,
29-30, 31*t*, 52, 56, 75*t*,
76, 80*t*, 81-82, 97, 110,
117, 127, 588*k*, 667-668,
670-671, 778, **785-802**,
786*f*
biogeography of, 108*t*
fossil, 98*t*-99*t*, 670, 670*f*, 671
male genitalia of, terminology of, 787*t*
Bomboixylocopa, 73*t*,
601*k*-602*k*, 604*k*, **605**
Bombomelecta, 774
Bombomelecta larreae, 776
Bombus, 6, 9, 16-17, 23, 28-30,
31*t*, 36, 41-42, 76-77, 77*t*,
78, 78*t*-81*t*, 91*f*, 468, 573,
596, 623, 747, 749-750,
753, **785-790**, 785*f*, **790-
802**, 830, Plate 14
of Asia, 791
of Europe, 791
females, 794*k*-797*k*
fossil, 790
males, 791*f*, 791*k*-794*k*
of Western Hemisphere,
790-791
Bombus (*Agrobombus*), 802
Bombus (*Alpigenobombus*), 75*t*,
786*f*, 791*k*, 794*k*, **797**
Bombus (*Alpinobombus*), 30,
31*t*, 75*t*, 789, 792*k*, 795*k*,
797
Bombus (*Anodontobombus*), 801
Bombus (*Bombias*), 75*t*, 792*k*,
795*k*, **797**
Bombus (*Bombus* s. str.), 75*t*,
792*k*, 795*k*, **797**
Bombus (*Boopobombus*), 797
Bombus (*Brachycephalibombus*),
75*t*, 792*k*, 796*k*, **797**
Bombus (*Callobombus*), 799
Bombus (*Chromobombus*), 802
Bombus (*Coccineobombus*), 75*t*,
793*k*-794*k*, **797**
Bombus (*Confusibombus*), 75*t*,
792*k*, 795*k*, **798**
Bombus (*Crotchiibombus*), 75*t*,
792*k*, 795*k*, 797, **798**
Bombus (*Cullumanobombus*),
75*t*, 792*k*, 795, 795*k*, **798**
Bombus (*Dasybombus*), 75*t*,
792*k*, 796*k*, **798**, 801
Bombus (*Digressobombus*), 798
Bombus (*Diversobombus*), 75*t*,
793*k*, 796*k*, **798**, 799
Bombus (*Eversmannibombus*),
75*t*, 789, 794*k*, 796*k*, **798**
Bombus (*Exilobombus*), 75*t*,
796*k*, **798**
Bombus (*Fervidobombus*), 75*t*,
788, 788*f*, 789-790, 793,
793*k*, 796*k*-797*k*, **798**,
800*f*
Bombus (*Festivobombus*), 75*t*,
792*k*, 796*k*, **798**
Bombus (*Fraternobombus*), 75*t*,
791*f*, 792*k*, 795*k*, **798**
Bombus (*Funebribombus*), 75*t*,
793*k*, 795*k*, **799**
Bombus (*Hortobombus*), 799
Bombus (*Hypnorobombus*), 801
Bombus (*Kallobombus*), 75*t*,
792*k*, 794, 795*k*, **799**

Bombus (*Kozlowibombus*), 799
Bombus (*Laesobombus*), 75*t*,
793*k*, 796*k*, **799**
Bombus (*Lapidariobombus*), 799
Bombus (*Lapponicobombus*),
801
Bombus (*Leucobombus*), 797
Bombus (*Mastrucatobombus*),
797
Bombus (*Megabombus*), 75*t*,
793*k*, 796*k*, **799**
Bombus (*Megalobombus*), 799
Bombus (*Melanobombus*), 75*t*,
791*f*, 793*k*, 794, 794*k*,
799, 801
Bombus (*Mendacibombus*), 75*t*,
789, 791*k*, 794, 794*k*, **799**
Bombus (*Mucidobombus*), 75*t*,
789, 793*k*, 796*k*, **799**
Bombus (*Obertibombus*), 801
Bombus (*Obertobombus*), 801
Bombus (*Odontobombus*), 799
Bombus (*Orientalibombus*), 75*t*,
789, 791*k*, 794*k*, **799-800**
Bombus (*Orientalobombus*), 799
Bombus (*Poecilobombus*), 801
Bombus (*Pomobombus*), 801
Bombus (*Pratobombus*), 801
Bombus (*Pressibombus*), 75*t*,
793*k*, 794, 794*k*, 796*k*,
800
Bombus (*Psithyrus*), 30, 31*t*, 37,
41, 75*t*, 699, 786*f*-787*f*,
788-791, 791*k*, 794*k*,
800, 800*f*
Bombus (*Pyrobombus*), 75*t*,
792*k*, 794-795, 796*k*,
800-801, 800*f*
Bombus (*Pyrrhobombus*), 800
Bombus (*Rhodobombus*), 75*t*,
793*k*, 796*k*-797*k*, **801**
Bombus (*Robustobombus*), 75*t*,
793*k*, 795*k*, **801**
Bombus (*Rubicundobombus*),
75*t*, 791*f*, 793*k*, 795*k*,
798, **801**
Bombus (*Ruderariobombus*), 802
Bombus (*Rufipedibombus*), 75*t*,
792*k*, 794*k*, **801**
Bombus (*Rufipedobombus*), 801
Bombus (*Senexibombus*), 75*t*,
793, 793*k*, 796*k*, 799, **801**
Bombus (*Separatobombus*), 75*t*,
792*k*, 795*k*, **801**
Bombus (*Sibericobombus*), 75*t*
Bombus (*Sibiricobombus*), 792*k*,
794, 794*k*, 796, 796*k*,
801
Bombus (*Soroeensibombus*), 799
Bombus (*Subterraneobombus*),
75*t*, 793*k*, 794, 796,
796*k*-797*k*, **801-802**
Bombus (*Sulcobombus*), 798
Bombus (*Tanguticobombus*), 799
Bombus (*Terrestribombus*), 797
Bombus (*Thoracobombus*), 30,
31*t*, 75*t*, 789, 793, 793*k*,
796*k*, 799, 802, **802**
Bombus (*Tricornibombus*), 75*t*,
793*k*, 796*k*, **802**
Bombus (*Uncobombus*), 801
Bombus abditus, 788
Bombus arcticus, 31*t*, 789, 797
Bombus atratus, 789, 798
Bombus atrocinctus, 794, 798
Bombus brachycephalus, 797
Bombus brevivillus, 793

Bombus coccineus, 797
Bombus coeruleus, 606
Bombus confusus, 798
Bombus consobrinus, 17
Bombus crotchii, 798
Bombus dahlbomii, 178, 789
Bombus diversus, 798
Bombus ephippiatus, 788
Bombus eversmanni, 798
Bombus exil, 798
Bombus fernaldae, 800*f*
Bombus fervidus, 800*f*
Bombus festivus, 798
Bombus fraternum, 791*f*
Bombus fraternus, 798
Bombus funebris, 799
Bombus haemorrhoidalis, 799
Bombus handlirschi, 798, 801
Bombus haueri, 797
Bombus hyperboreus, 789, 797
Bombus hypnorum, Plate 14
Bombus impatiens, 800*f*
Bombus inexspectatus, 31*t,* 789, 802
Bombus kashmirensis, 797
Bombus keriensis, 799
Bombus kozlovi, 799
Bombus laboriosus, 751
Bombus laesus, 799
Bombus lapidarius, 791*f*
Bombus lapponicus, 795
Bombus ligusticus, 799
Bombus macgregori, 798
Bombus mastrucatus, 797
Bombus mendax, 799
Bombus modestus eversmanni, 798
Bombus morawitzi, 801
Bombus morrisoni, 785*f*
Bombus mucidus, 799
Bombus nevadensis, 797
Bombus niger, 788*f,* 798
Bombus nobilis, 797
Bombus oberti, 801
Bombus orientalis, 799
Bombus pascuorum, 793
Bombus pennsylvanicus, 669*f,* 785*f,* 787*f*
Bombus persicus, 789, 798
Bombus polaris, 789, 797
Bombus pratorum, 794
Bombus pressus, 800
Bombus robustus, 801
Bombus rubicundus, 791*f,* 801
Bombus ruderatus, 799
Bombus rufipes, 801
Bombus rufocinctus, 798
Bombus s. str., 75*t,* 792*k,* 795*k,* **797**
Bombus senex, 801
Bombus separatus, 801
Bombus sikkimi, 786*f*
Bombus similis, 758
Bombus sitkensis, 801
Bombus soroeensis, 799
Bombus subterraneus, 802
Bombus sylvarum, Plate 14
Bombus tanguticus, 794, 799
Bombus terrestris, 4
Bombus tranquebaricus, 607
Bombus tricornis, 802
Bombus variabilis, 786*f*-787*f,* 800*f*
Bombus volucelloides, 801
Bombus vosnesenskii, 785*f*
Bombus wurflenii, 797
Bombusoides, fossil, 98*t*
Bombusoides mengei, fossil, 98*t*

Boopobombus, 797
Boreallodape, fossil, 98*t,* 591, Plate 17-Plate 18
Boreallodape baltica, fossil, 98*t*
Boreallodape mollyae, fossil, Plate 17
Boreallodape striebichi, fossil, Plate 18
Boreallodapini (Tribe), fossil, 98*t,* 591
Boreocoelioxys, 72*t,* 545*k*-546*k,* **547-548,** 547*k,* 548*f*
Boreopsis, 208-209
Borgatomelissa, 69*t,* **291,** 291*k*
Bothranthidium, 71*t,* **509,** 509*k,* 519, 895
Bothranthidium lauroi, 509
Brachyandrena, 68*t,* 252*k,* 254*k,* **259**
Brachycephalapis, 432
Brachycephalibombus, 75*t,* 792*k,* 796*k,* **797**
Brachygastra [vespid], 581
Brachyglossa, 139
Brachyglossa bouvieri, 140*f*
Brachyglossa rufocaerulea, 139
Brachyglossula, 67*t,* 138*k,* **139-140,** 145, 151, 165
Brachyglossula bouvieri, 147*f*
Brachyhesma, 133, 220-222, 222*k,* **224-225,** 224*k,* 230
Brachyhesma (Anomalohesma), 68*t,* **215,** 224, 224*k*
Brachyhesma (Brachyhesma s. str.), 68*t,* **215,** 223*f,* 224, 224*k*
Brachyhesma (Henicohesma), 68*t,* **215,** 224, 224*k*
Brachyhesma (Microhesma), 68*t,* **215,** 222*f*-223*f,* 224, 224*k*
Brachyhesma incompleta, 215, 222*f*-223*f*
Brachyhesma macdonaldensis, 215
Brachyhesma s. str., 68*t,* **215,** 223*f,* 224, 224*k*
Brachyhesma scapata, 215
Brachyhesma sulphurella, 223*f*
Brachymelecta, 75*t,* 771*k,* **773**
Brachymelissodes, 74*t,* 734*k,* **735**
Brachynomada, 639, **640,** 640*k,* 641
Brachynomada (Brachynomada s. str.), 73*t,* **640,** 640*k*
Brachynomada (Melanomada), 73*t,* 634*f,* 639*f,* **640,** 640*k*
Brachynomada argentina, 640
Brachynomada grindeliae, 639*f*
Brachynomada melanantha, 634*f*
Brachynomada s. str., 73*t,* **640,** 640*k*
Brachynomada scotti, 640
Brachynomadini (Tribe), 73*t,* 112, 590, 590*k,* **639-642,** 639*k*-640*k,* 643
biogeography of, 108*t*
Brachyspatulariella, 213
Branthidium, 71*t,* 498, **499,** 499*k,* 500-502
Brasilagapostemon, 70*t,* 387, **387,** 387*k*
Braunsapis, 6, 30, 31*t,* 73*t,* 106, 620, 621*f,* 623, 624*k,* 625-626, **627,** 628-629, 632

Braunsapis angolensis, 624
Braunsapis associata, 624
Braunsapis aureoscopa, 624, 627
Braunsapis bislensis, 31*t,* 623
Braunsapis breviceps, 30, 31*t,* 39*f,* 623, 627
Braunsapis facialis, 619*f,* 632
Braunsapis foveata, 622*f*
Braunsapis kaliago, 30, 31*t,* 623, 627
Braunsapis leptozonia, 622*f,* 627
Braunsapis maculata, 627
Braunsapis natalica, 31*t,* 622*f,* 623
Braunsapis pallida, 31*t,* 623, 627
Braunsapis paradoxa, 624
Braunsapis simillima, 39*f*
Braunsapis simplicipes, 622*f,* 624
Braunsapis trochanterata, 622*f,* 623-624
Bremus, 797, 800
Bremus (Pressibombus), 800
Bremus (Rufocinctobombus), 798
Bremus (Senexibombus), 801
Bremus (Separatobombus), 801
Bremus pomorum, 801
Bremus pressus, 800
Brevineura, 73*t,* 622*f,* **630,** 630*k,* 631
bumble bees (Bombini), 12. See also *Bombini (Tribe); Bombus*
Bureauella, 762
Bureauella insignis, 762
Bythandrena, 263
Bytinskia, 32*t,* 71*t,* 451, 463*k,* 467, **468**
Bytinskia erythrogastra, 468

Cacosoma, 409
Cacosoma discolor, 409
Cadeguala, 67*t,* **177,** 177*k*
Cadeguala albopilosa, 177*f*
Cadeguala occidentalis, 177, 177*f*
Cadegualina, 67*t,* **177-178,** 177*f*
Cadegualina andina, 177*f*
Caenaugochlora, 112, 393, 400*k,* **405,** 405*k*
Caenaugochlora (Caenaugochlora s. str.), 70*t,* 363-364, 398*k,* 399*f,* **405,** 405*k*
Caenaugochlora (Ctenaugochlora), 70*t,* 395*k,* 396*f,* **405,** 405*k*
Caenaugochlora (Pseudaugochlora), 410
Caenaugochlora costaricensis, 399*f,* 405
Caenaugochlora curticeps, 405
Caenaugochlora macswaini, 405
Caenaugochlora perpectinata, 396*f*
Caenaugochlora s. str., 70*t,* 363-364, 398*k,* 399*f,* **405,** 405*k*
Caenohalictina (Subtribe), 320, 354, 357, 388
Caenohalictus, 70*t,* 112, 320, 348, 354, 356-357, 359*k,* 360, 361*k,* **363-364,** 365, 388, 405
Caenohalictus serripes, 278

Caenohalictus trichiothalmus, 363
Caenonomada, 73*t,* 81*t,* 589, 687, 688*k,* **689-690,** 691
Caenonomada bruneri, 687*f,* 689, 689*f*
Caenoprosopidini (Tribe), 73*t,* 77, 128*k,* 589*k,* 633, 636, 660-661, **666,** 666*k*
biogeography of, 108*t*
Caenoprosopina, 73*t,* **666,** 666*k*
Caenoprosopina holmbergi, 666
Caenoprosopini, 635
Caenoprosopis, 73*t,* **666,** 666*f,* 666*k*
Caenoprosopis crabronina, 634*f,* 666, 666*f*
Caerulosmia, 485
Caesarea, 662
Calcarina, 264
Calchalictus, 379
Caliendra, 784
Callalictus, 70*t,* 371*t,* 374*k*-375*k,* **376**
Callandrena, 68*t,* 83, 126, 240, 240*f*-241*f,* 243, 244*k,* 246*k*-248*k,* 250*k,* **259**
Callanthidium, 71*t,* 505-506, 506*k,* **507**
Calleupetersia, 364
Callimelissodes, 74*t,* 730*k*-731*k,* **731**
Callinomada, 644
Calliopsima, 69*t,* 309, **310,** 310*k,* 312-314, 655
Calliopsini (Tribe), 69*t,* 112, 270, 271*k,* 287, **306-315,** 307*f,* 307*k,* 308*f,* 314*f*
biogeography of, 107*t*
Calliopsis, 87, 112, 235, 271, 279, 306, 306*f,* 307*k,* **309-314,** 309*k*-310*k,* 314, 353, 389, 655, 663
Calliopsis (Calliopsima), 69*t,* 309, 309*k,* **310,** 310*k,* 312-314
Calliopsis (Calliopsis s. str.), 69*t,* 306*f*-307*f,* 308, 308*f,* 309, 309*k*-310*k,* **311,** 311*f,* 312, 314, 655
Calliopsis (Ceroliopoeum), 69*t,* 306*f,* 310*k,* **311**
Calliopsis (Hypomacrotera), 69*t,* 309, 310*k,* 312, **312,** 312*f,* 663
Calliopsis (Leiopoeodes), 306*f*
Calliopsis (Liopoeodes), 69*t,* 310*k,* **312**
Calliopsis (Liopoeum), 69*t,* 309, 309*k*-310*k,* 311, **312**
Calliopsis (Macronomadopsis), 663
Calliopsis (Micronomadopsis), 69*t,* 236*f,* 309, 310*k,* **312,** 660, 663
Calliopsis (Nomadopsis), 69*t,* 236*f,* 270*f,* 309, 310*k,* **312-313,** 663
Calliopsis (Perissander), 69*t,* 309, 310*k,* 312, **313,** 313*f,* 314
Calliopsis (Verbenapis), 20, 69*t,* 308*f,* 309, 309*k*-310*k,* **313-314**
Calliopsis abdominalis, 280
Calliopsis andreniformis, 306*f*-308*f,* 309, 311, 311*f*

Calliopsis anomoptera, 313, 313*f*
Calliopsis australior, 113*f*
Calliopsis edwardsii, 236*f*
Calliopsis hondurasicus, 309
Calliopsis laeta, 306*f*, 311
Calliopsis larreae, 309
Calliopsis parvus, 280
Calliopsis pugionis, 310
Calliopsis rozeni, 310
Calliopsis rudbeckiae, 281
Calliopsis s. str., 69*t*, 306*f*-307*f*, 308, 308*f*, 309, 309*k*-310*k*, 311, 311*f*, 312, 314, 655
Calliopsis scutellaris, 236*f*
Calliopsis subalpina, 309, 312*f*
Calliopsis trifasciatum, 113*f*
Calliopsis turnerae, 295
Calliopsis verbenae, 308*f*, 313-314
Calliopsis xenopous, 306*f*, 312
Calliopsis zonalis, 270*f*, 312
Callistochlora, 70*t*, 395, 396*f*, 397*k*, **406**, 406*k*
Callobombus, 799
Calloceratina, 73*t*, **614-615**, 614*k*, 616, 618
Callochile, 566
Callochlora, 406
Callocolletes, 67*t*, **166**, 166*k*
Callohesma, 68*t*, **215**, 220, 222, 224*k*, 226-227, 232
Callohesma calliopsiformis, 223*f*
Callomacrotera, 69*t*, 296*f*, 300, 301*k*, **302**
Callomegachile, 42, 72*t*, 551-553, 555, 555*f*, 557*t*, 558*k*, 560*k*-566*k*, 564*f*, 565, **567-569**, 568*f*, 572-578, 581, 586
Callomegachile (in *Chalicodoma*), 567
Callomegachile (in *Megachile*), 551, 553, 555, 555*f*, 557*t*, 558*k*, 560*k*-566*k*, 564*f*, 565, **567-569**, 568*f*, 572-578, 581, 586
Callomelecta, 775
Callomelecta pendleburyi, 775
Callomelitta, 67*t*, 133, 136, 138, 139*k*, **140-141**, 142, 171
Callomelitta antipodes, 140-141
Callomelitta picta, 133*f*, 140, 140*f*
Callonychium, 116, 307*k*, 309, **314**, 314*k*, 666
Callonychium (*Callonychium* s. str.), 69*t*, **314**, 314*k*
Callonychium (*Paranychium*), 69*t*, **314**, 314*k*
Callonychium argentinum, 314
Callonychium chilense, 314
Callonychium mandibulare, 314*f*
Callonychium minutum, 307*f*
Callonychium s. str., 69*t*, **314**, 314*k*
Calloprosopis, 67*t*, 189, 189*k*, **191**, 191*k*, 192
Calloprosopis magnifica, 191
Callorhiza, 68*t*, 190*f*, 196*f*, 215, **216**, 216*k*, 218
Callosphecodes, 364, **388-389**
Callosphecodes ralunensis, 388
Calloxylocopa, 600*k*, 609

Calohoplitis, 471
Calomelissa, 68*t*, 257*k*-258*k*, **260**
Calospiloma, 673
Calyptapis, fossil, 98*t*, 790
Calyptapis florissantensis, fossil, 98*t*
Camargoia, 827
Camargoia camargoi, 827
Campanularia, 211
Camptopaeum, 287
Camptopaeum hirsutulum, 312
Camptopoeum, 285*k*, **286-287**, 286*k*, 292, 655, 663
Camptopoeum (*Camptopoeum* s. str.), 69*t*, **286-287**
Camptopoeum (*Epimethea*), 69*t*, 236, 286, 286*f*, 286*k*, 287, 655
Camptopoeum (*Liopoeum*), 312
Camptopoeum chilense, 314
Camptopoeum friesei, 285*f*
Camptopoeum laetum, 311
Camptopoeum maculatum, 314
Camptopoeum nomadoides, 314
Camptopoeum nomioides, 314
Camptopoeum ochraceum, 309
Camptopoeum prinii, 308
Camptopoeum ruber, 285-287
Camptopoeum s. str., 69*t*, **286-287**
Camptopoeum subflava, 287
Camptopoeum submetallicum, 308
Camptopoeum trifasciatum, 308
Camptopoeum variegatum, 286*f*
Camptopoeumini (Tribe), 270
Campylogaster, 68*t*, 254*k*, 256*k*, **260**
Campylogaster fulvo-crustatus, 260
Canephora, 722
Canephorula, 33*t*, 74*t*, 588, 707-709, 710*k*, **722**, 729
Canephorula apiformis, 673, 709, 722, 722*f*
Canephorulina (Subtribe), 707-710
Canephorulini (Tribe), 81
Capanthidium, 71*t*, 498, 498*k*, **499-500**, 499*k*, 501-502, 512
Capicola, 70*t*, 417, 417*f*, 419, 419*k*, **420**, 422
Capicola (*Capicoloides*), 420
Capicola aliciae, 420
Capicola braunsiana, 420
Capicola cinctiventris, 420
Capicoloides, 70*t*, 419*k*, **420**
Caposmia, 71*t*, 489, **490**, 490*k*
Carandrena, 68*t*, 253*k*, 256*k*, **260**
Caranthophora, 74*t*, 744*f*, 745*k*, 747*k*, **748**
Carinandrena, 68*t*, 251, **260**
Carinanthidium, 72*t*, 510, **525**, 525*k*
Carinapis, 70*t*, 418, 419*k*-420*k*, **420**
Carinella, 567-568
Carinellum, 527
Carinorophites, 326
Carinula, 568
Carloticola, 72*t*, **516**, 516*k*
carpenter bees, 12
Cascapis, fossil, 98*t*, 831

Catoceratina, 73*t*, 613*k*, **615**
Caupolicana, 20, 80*t*-81*t*, 93, 105-106, 112, 174, **174**, 174*k*, 573, 648
Caupolicana (*Alayoapis*), 67*t*, **174**, 174*k*
Caupolicana (*Caupolicana* s. str.), 67*t*, **174-175**, 174*k*
Caupolicana (*Caupolicanoides*), 175
Caupolicana (*Willinkapis*), 67*t*, 174, 174*k*, **175**
Caupolicana (*Zikanapis*), 67*t*, 174, 174*k*, **175**
Caupolicana fulvicollis, 175
Caupolicana hirsuta, 125*f*, 174*f*
Caupolicana pubescens, 175
Caupolicana s. str., 67*t*, **174-175**, 174*k*
Caupolicana yarrowi, 173*f*, 175*f*
Caupolicania, 175
Caupolicanini (Tribe), 67*t*, 80, 80*t*, 129-130, 173, 173*k*, **174-176**, 174*f*, 174*k*, 175*f*
  biogeography of, 107*t*
Caupolicanoides, 175
Ceblurgus, 69*t*, 322-323, 324*k*, **325**, 328-329
Ceblurgus longipalpis, 325
Celetandrena, 68*t*, 244*k*, 248*k*, **260**
Celetrigona, 829
Cellaria, 346
Cellariella, 69*t*, **346**, 346*k*
Cellariella brooksi, 347*f*
Cemolobus, 74*t*, 717*k*, 719*k*, **722**
Cemolobus ipomoeae, 717*f*, 722
Centrias, 644
Centridini (Tribe), 9, 18, 33*t*, 74*t*, 81, 104, 116, 433, 590*k*, 668, 668*f*, 671, 687, 742, **753-761**, 753*f*, 754*k*, 759*f*, 762-763, 765*f*
  biogeography of, 108*t*, 109
Centris, 34, 52, 77*t*-80*t*, 105, 111-112, 433, 441, 545, 588, 652, 667, 691, 733, 742, 753, **754-759**, 754*f*, 754*k*-756*k*, 759*f*, 759-760, 766, 768, 784
Centris (*Acritocentris*), 74*t*, 754, **756**, 756*k*, 757
Centris (*Aphemisia*), 74*t*, 755*k*, **756**, 758
Centris (*Centris* s. str.), 74*t*, 753, 753*f*, 754, 754*f*, **755**, 756, 756*k*
Centris (*Cyanocentris*), 756
Centris (*Exallocentris*), 74*t*, 753-754, 755*k*-756*k*, **756-757**, 757
Centris (*Heterocentris*), 18*f*, 74*t*, 753-755, 755*k*, **757**, 758
Centris (*Melacentris*), 74*t*, 754-755, 755*k*, 756-757, **757**, 758-759, 759*f*
Centris (*Melanocentris*), 758
Centris (*Paracentris*), 18*f*, 74*t*, 754, 756-757, 756*k*, 757, **758-759**, 765*f*
Centris (*Penthemisia*), 757
Centris (*Poecilocentris*), 756
Centris (*Ptilocentris*), 74*t*, 754-755, 755*k*, **757-758**
Centris (*Ptilotopus*), 18*f*, 74*t*, 753-754, 754*f*, 755*k*, 757, **758**, 759, 764

Centris (*Rhodocentris*), 757
Centris (*Schisthemisia*), 74*t*, 755*k*, **758**
Centris (*Trachina*), 74*t*, 753-755, 755*k*, **758**, 759*f*
Centris (*Trichocentris*), 757
Centris (*Wagenknechtia*), 74*t*, 754, 755*k*-756*k*, 757, **758**
Centris (*Xanthemisia*), 74*t*, 753-755, 755*k*-756*k*, 757, **758**
Centris (*Xerocentris*), 74*t*, 753-754, 755*k*-756*k*, 756-757, **758-759**
Centris aenea, 753, 755
Centris aethiocesta, 754*f*
Centris agiloides, 759*f*
Centris americana, 755
Centris anomala, 756-757
Centris atra, 757-758
Centris atripes, 765*f*
Centris autrani, 756
Centris bicolor, 758
Centris bicornuta, 757
Centris caesalpiniae, 755
Centris caesalpiniae var. *rhodopus*, 757
Centris californica, 758
Centris cineraria, 758
Centris cingulata, 784
Centris clavipes, 827
Centris cornuta, 757
Centris difformis, 757
Centris dorsata, 757
Centris eisenii, 756
Centris eurypatana, 759*f*
Centris fasciatella, 756
Centris festiva, 758
Centris flavifrons, 755
Centris flavilabris, 758
Centris flavofasciata, 753
Centris labrosa, 757
Centris lineolata, 758
Centris longimana, 758
Centris mixta, 753, 758
Centris pallida, 9, 753, 755, 759
Centris plumipes, 756
Centris poecila, 753*f*
Centris rhodoleuca, 757
Centris ruthannae, 756
Centris s. str., 74*t*, 753, 753*f*, 754, 754*f*, 755, 756, 756*k*
Centris trigonoides, 18*f*
Centris umbraculata, 761
Centris vittata, 17
Centris zonata, 754*f*, 758
Centrosmia, 479, **482-483**
Cephalapis, **473-474**
Cephalictoides, 326
Cephalocolletes, 67*t*, 138*k*, 139, 144, 147*k*, 147*t*, **151**, 157-159
Cephalosmia, 71*t*, 476, 476*f*, 478*k*-479*k*, **480**, 484
Cephalotrigona, 75*t*, 812, 813*k*, 815, 820, **824-825**
Cephalotrigona capitata, 787*f*, 820*f*
Cephalurgus, 69*t*, 272, 273*k*, 280, 283, **283**, 283*k*
Cephalurgus anomalus, 283
Cephalylaeus, 67*t*, 198*t*, 199*k*, **203**
Cephen, 644
Cephylaeus, 67*t*, 198*t*, 199*k*-200*k*, **203**

Ceraheriades, 71*t*, 456, 456*k*, 457
Ceraplastes, 532
Ceratalictus, 70*t*, 396*f*, 400*k*, 402, 402*k*, **403**
Ceratias, 582
Ceratina, 8, 26, 28, 77, 77*t*, 78, 78*t*-80*t*, 81, 81*t*, 90, **613-619**, 613*k*-614*k*, 619-621, 621*f*, 623, 645, Plate 1
   of Eastern Hemisphere, 613*k*-614*k*
   of Western Hemisphere, 614*k*
Ceratina (*Calloceratina*), 73*t*, **614-615**, 614*k*, 616, 618
Ceratina (*Catoceratina*), 73*t*, 613*k*, **615**
Ceratina (*Ceratina* s. str.), 73*t*, 614*k*, **615**, 616-617
Ceratina (*Ceratinidia*), 73*t*, 613*k*-614*k*, **615**, 615
Ceratina (*Ceratinula*), 73*t*, 614*k*, **615-616**, 618
Ceratina (*Chloroceratina*), 73*t*, 613*k*, 615, **616**
Ceratina (*Copoceratina*), 73*t*, 614*k*, 616, **616**
Ceratina (*Crewella*), 73*t*, 614*k*, 615, **616**
Ceratina (*Ctenoceratina*), 73*t*, 612, 614*k*, **616**
Ceratina (*Euceratina*), 73*t*, 593, 612, 614*k*, **616**, 617
Ceratina (*Hirashima*), 73*t*, 614*k*, **616**
Ceratina (*Lioceratina*), 73*t*, 613*k*, 615, **616-617**, 618
Ceratina (*Malgatina*), 73*t*, 614*k*, 616, **617**
Ceratina (*Megaceratina*), 73*t*, 612, 613*k*, **617**
Ceratina (*Neoceratina*), 73*t*, 614*k*, 615, **617**
Ceratina (*Pithitis*), 73*t*, 612, 613*k*, **617**
Ceratina (*Protopithitis*), 73*t*, 612, 613*k*, **617**
Ceratina (*Rhysoceratina*), 73*t*, 614*k*, **617-618**, xiv
Ceratina (*Simiceratina*), 73*t*
Ceratina (*Simioceratina*), 614*k*, 616, **618**
Ceratina (*Xanthoceratina*), 73*t*, 613*k*, **618**
Ceratina (*Zadontomerus*), 73*t*, 611*f*, 613, 613*f*, 614*k*, 616, 618, **618**
Ceratina acantha, 612
Ceratina aereola, 617
Ceratina amabilis, 614
Ceratina armata, 616
Ceratina australensis, 612*f*
Ceratina azurae, 617
Ceratina bouyssoui, 617
Ceratina calcarata, 612*f*-613*f*
Ceratina capitosa, 614-615
Ceratina chalcites, 613*k*, 617
Ceratina cladura, 618
Ceratina cyanura, 616
Ceratina dalletorreana, 612, 616
Ceratina dupla, 612*f*, 622*f*
Ceratina ericia, 612
Ceratina exima, 614
Ceratina flavipes, 615
Ceratina flavopicta, 616

Ceratina hieroglyphica, 615
Ceratina japonica, 615
Ceratina laticeps, 613
Ceratina lucidula, 615
Ceratina madecassa, 616
Ceratina minuta, 616
Ceratina moerenhouti, 618
Ceratina montana, 617-618
Ceratina neomexicana, 593*f*
Ceratina nyassensis, 616
Ceratina okinawana, 615
Ceratina parvula, 615
Ceratina perforatrix, 615
Ceratina rupestris, 51*f*
Ceratina s. str., 73*t*, 614*k*, **615**, 616-617
Ceratina sculpturata, 617
Ceratina smaragdula, 617
Ceratina stilbonota, 618
Ceratina tejonensis, 618
Ceratina timberlakei, 611*f*
Ceratina titusi, 616
Ceratina volitans, 618
Ceratinae (Subfamily), 81*t*
Ceratinidae (Family), 79*t*
Ceratinidia, 73*t*, 613*k*-614*k*, 615, **615**
Ceratininae (Subfamily), 80*t*, 81
Ceratinini (Tribe), 13, 25, 28, 34, 73*t*, 79, 80*t*, 81, 90, 104, 112, 587, 588*k*, 593*k*-594*k*, 595, 597, **611-618**, 613*k*-614*k*, 619-620, 621*f*-622*f*, 623
   biogeography of, 107*t*
Ceratinoidea, 79, 79*t*
Ceratinula, 73*t*, 614*k*, **615-616**, 618
Ceratocolletes, 67*t*, 143-144, 147*t*, 149*k*-150*k*, **151**, 152
Ceratomonia, 70*t*, 426, **427**, 427*k*
Ceratomonia rozenorum, 426*f*, 427, 427*f*
Ceratopsithyrus, 800
Ceratorhiza, 68*t*, 190*f*, 214-215, 215*k*, **216**, 217
Ceratosmia, 484
Cercorhiza, 68*t*, 215, **216-217**, 216*k*
Cerianthidium, 503
Ceroliopoeum, 69*t*, 306*f*, 310*k*, **311**
Cestella, 72*t*, 557*t*, 564*k*, **569**
Ceylalictus, 346-347, 346*k*
Ceylalictus (*Atronomioides*), 70*t*, **346**, 346*k*
Ceylalictus (*Ceylalictus* s. str.), 70*t*, **346**, 346*k*
Ceylalictus (*Meganomioides*), 70*t*, **346-347**, 346*k*
Ceylalictus divisus, 113*f*
Ceylalictus s. str., 70*t*, **346**, 346*k*
Ceylalictus variegatus, 346, 346*f*
Ceylalictus warnckei, 346
Ceylonicola, 382
Ceylonicola atra, 382
Chacoana, 689
Chacoana melanoxantha, 689
Chaetalictus, 70*t*, 384*k*, **385**, 385*f*, 386
Chaetochile, 572-573
Chaetohesma, 68*t*, 224, **233**, 233*k*

Chaetohesma tuberculata, 233
Chaetostetha, 691
Chaeturginus, 69*t*, 126, 235, 272*k*-274*k*, **275-276**, 275*f*, 283
Chaeturginus alexanderi, 276
Chaeturginus testaceus, 270*f*, 275*f*, 276
Chalcidoidea (wasps), 65
Chalcobombus, fossil, 98*t*
Chalcobombus humilis, fossil, 98*t*
Chalcosmia, 481
Chalepogenoides, 692
Chalepogenus, **690**, 690*k*, 691
Chalepogenus (*Chalepogenus* s. str.), 73*t*, 689, 689*k*, 690, **690**, 690*k*
Chalepogenus (*Lanthanomelissa*), 73*t*, 674, 687, 689*k*, **690**, 690*k*
Chalepogenus betinae, 674
Chalepogenus caeruleus, 674, 688*f*, 690
Chalepogenus discrepans, 690
Chalepogenus goeldiana, 689*f*
Chalepogenus herbsti, 690
Chalepogenus incertus, 690
Chalepogenus leucostoma, 692
Chalepogenus s. str., 73*t*, 689, 689*k*, 690, **690**, 690*k*
Chalicodoma, 25-26, 72*t*, 77, 120, 536, 543, 551-556, 557*t*, 561*k*-564*k*, **569-570**, 577-578, 586
Chalicodoma (*Allochalicodoma*), 569
Chalicodoma (*Austrochile*), 567
Chalicodoma (*Callomegachile*), 567
Chalicodoma (*Carinella*), 567-568
Chalicodoma (*Cestella*), 569
Chalicodoma (*Chalicodoma* s. str.), 553
Chalicodoma (*Chalicodomoides*), 570
Chalicodoma (*Chelostomoda*), 570
Chalicodoma (*Chelostomoidella*), 570
Chalicodoma (*Cuspidella*), 572
Chalicodoma (*Dinavis*), 581
Chalicodoma (*Euchalicodoma*), 569
Chalicodoma (*Eumegachilana*), 567
Chalicodoma (*Largella*), 577
Chalicodoma (*Morphella*), 568
Chalicodoma (*Neglectella*), 581
Chalicodoma (*Neochalicodoma*), 586
Chalicodoma (*Parachalicodoma*), 569, 580
Chalicodoma (*Rhodomegachile*), 582
Chalicodoma (*Schizomegachile*), 583
Chalicodoma (*Stenomegachile*), 583
Chalicodoma (*Xenochalicodoma*), 569
Chalicodoma lefebvrei, 569
Chalicodoma mystaceana, 567
Chalicodoma pseudolaminata, 586
Chalicodoma quadraticauda, 572

Chalicodoma rufitarsis, 569
Chalicodoma s. str., 553
Chalicodoma semivestita, 577
Chalicodomoides, 72*t*, 555*f*, 557*t*, 565*k*-566*k*, 569, **570**, 577
Chalicodomopsis, fossil, 98*t*
Charitandrena, 68*t*, 243*k*, 248*k*, 252*k*, 254*k*, **260**
Chaulandrena, 261
Cheesmania, 68*t*, 216*k*, **217**
Chelostoma, 77, 78*t*, 126, 441-442, 445, 448-449, 449*t*, 450*k*, 452, 452*k*, **456-458**, 456*k*-457*k*, 461, 475, 486, 489-490, 533, 583
Chelostoma (*Cephalapis*), 473
Chelostoma (*Ceraheriades*), 71*t*, 456, 457, 457*k*
Chelostoma (*Chelostoma* s. str.), 71*t*, 456, 456*k*, 457
Chelostoma (*Eochelostoma*), 71*t*, **457**, 457*k*
Chelostoma (*Foveosmia*), 71*t*, 434*f*, 456*f*, 456*k*, 457, 458
Chelostoma (*Gyrodromella*), 71*t*, 456, 456*k*, **457-458**
Chelostoma (*Prochelostoma*), 71*t*, 448*f*, 456, 456*k*, 457, **458**
Chelostoma albifrons, 472
Chelostoma aureocinctum, 457
Chelostoma californicum, 434*f*, 456*f*
Chelostoma campanularum, 457
Chelostoma diodon, 457
Chelostoma distinctum, 457
Chelostoma florisomne, 457
Chelostoma foveolatum, 457-458
Chelostoma fuliginosum, 458
Chelostoma grande, 456
Chelostoma incisulum, 457
Chelostoma isabellinum, 456
Chelostoma jacintanum, 473
Chelostoma mocsaryi, 457
Chelostoma nasutum, 456-457
Chelostoma petersi, 456-457
Chelostoma philadelphi, 448*f*, 456-458
Chelostoma rapunculi, 458
Chelostoma rugifrons, 570
Chelostoma s. str., 71*t*, 456, 456*k*, 457
Chelostoma schmiedeknechti, 461
Chelostoma transversum, 457
Chelostoma ventrale, 457
Chelostomoides, 72*t*, 109, 453, 552-553, 554*f*, 555, 555*t*-556*t*, 557, 557*t*, 561*k*-562*k*, 565*k*-566*k*, 570, **570**, 576, 579
Chelostomoidella, 570
Chelostomoides, 72*t*, 109, 531, 543, 553, 553*f*, 555, 557, 557*t*, 558*k*, 560*f*, 560*k*, 567, **570-571**, 571*f*, 575-576, 582-583, 586
Chelostomopsis, 71*t*, 487*k*, **488**, 488*f*
Chelynia, 532
Chelynia labiata, 532
Chelynia rubifloris, 488
Chenosmia, 482-483

Chiasmognathus, 906
Chilalictus, 9, 10*f,* 70*t,* 297, 350, 371*t,* 372, 373*f,* 374*k,* 375*f,* 375*k,* **376**, 377, 407
Chilicola, 79, 79*t,* 80, 80*t,* 159, 180, **181-185**, 181*f,* 181*k*-182*k,* 182*f*-183*f,* 186, 319, 905
Chilicola (Anoediscelis), 67*t,* 181, 181*f,* **182**, 182*f,* 182*k,* 183*f,* 184
Chilicola (Chilicola s. str.*)*, 67*t,* 181, 181*k,* **183**
Chilicola (Chilioediscelis), 67*t,* 180, 181, 181*k,* 183, **183**
Chilicola (Heteroediscelis), 182
Chilicola (Hylaeosoma), 67*t,* 181, 181*f,* 181*k,* 182*f,* **183-184**, 183*f,* 184-185
Chilicola (Oediscelis), 67*t,* 181, 182, 182*k,* 183, **184**
Chilicola (Oreodiscelis), **184**
Chilicola (Oroediscelis), 67*t,* 181, 182*k,* **184**
Chilicola (Prosopoides), 67*t,* 181, 182*f,* 182*k,* **184**, 185
Chilicola (Pseudiscelis), 67*t,* 181, 182*f,* 182*k,* 184, **185**
Chilicola andina, 183
Chilicola ashmeadi, 180*f*-181*f,* 182, 182*f*-183*f*
Chilicola colliguey, 183
Chilicola friesei, 184
Chilicola gutierrezi, 184
Chilicola hahni, 184
Chilicola mantagua, 184
Chilicola megalostigma, 184
Chilicola mexicana, 181*f,* 183*f*
Chilicola minor, 182
Chilicola olmue, 182
Chilicola orophila, 182
Chilicola polita, 182*f,* 184
Chilicola prosopoides, 182*f*
Chilicola rostrata, 182*f*
Chilicola rubriventris, 183
Chilicola s. str., 67*t,* 181, 181*k,* **183**
Chilicolinae (Subfamily), 80*t*
Chilicolini (Tribe), 180*k*
Chilicolletes, 67*t,* 147*t,* 148*k,* **151**, 159
Chilimalopsis, 73*t,* 680, 681*k,* **682**, 684
Chilimalopsis parvula, 682
Chilimelissa, 67*t,* 180-181, 181*k,* **185**, 186, 905
Chilimelissa brevimaralis, 185
Chilimelissa luisa, 185, 185*f*
Chilimelissa rozeni, 185
Chilioediscelis, 180, 181, 181*k,* 183, **183**
Chilosima, 71*t,* 453, 453*k,* 454
Chlerogas, 70*t,* 116, 395*k,* 398*k,* **405**, 406, 895
Chlerogas hirsutepennis, 397*f*
Chlerogella, 398*k,* **405-406**, 406*k*
Chlerogella (Chlerogella s. str.*)*, 70*t,* 395*k,* **406**, 406*k*
Chlerogella (Ischnomelissa), 70*t,* 395*k,* 400*k,* 405, **406**, 406*k*
Chlerogella cyanea, 405
Chlerogella elongaticeps, 397*f,* 406

Chlerogella s. str., 70*t,* 395*k,* **406**, 406*k*
Chlerogella zonata, 405
Chlerogelloides, 70*t,* 394*k,* 398*k,* 405, **406**, 408
Chlerogelloides femoralis, 406, 407*f*
Chlerogus, 43
Chlidoplitis, 71*t,* 453, 464*k,* **468-469**
Chloralictus, 377-378
Chloralictus loureiroi, 407
Chlorandrena, 68*t,* 251*k,* 254*k,* 256*k,* 258*k,* **260**
Chloranthidiellum, 71*t,* 494, 502, 503*k,* **504**
Chloranthidium, 504
Chloroceratina, 73*t,* 613*k,* 615, **616**
Chlorosmia, 472
Chrisanthidium, 72*t,* 522*k,* **523**
Chromobombus, 802
Chrysandrena, 68*t,* 252*k,* 255*k,* **260**
Chrysantheda, 784
Chrysantheda nitida, 784
Chrysocolletes, 67*t,* 139*k,* **141-142**, 145, 156
Chrysocolletes moretonianus, 141, 141*f,* 145*f*
Chrysopheon, 540
Chrysopheon aurifuscus, 540
Chrysosarus, 72*t,* 551-552, 554, 556, 556*t*-557*t,* 558*k,* 560*k*-561*k,* **571**, 583, 585-586
Chrysosarus (Chrysosarus s. str.*)*, 586
Chrysosarus (Dactylomegachile), 571, 586
Chrysosarus (Zonomegachile), 585
Chrysosarus s. str., 586
Cilissa, 432
Cilissa erythrogaster, 266
Cilissa robusta, 432
cinctiventris group (of *Nomada*), 645
Cingulata, 212
Cirroxylocopa, 73*t,* 599*k,* 601*k,* **605**
Citrinopsithyrus, 800
Cladocerapis, 67*t,* 147*t,* 149*k,* 150, **151-152**, 153, 155-156
Claremontiella, 312
Clavicera, 77*t,* 615
Clavinomia, 69*t,* 334, 334*k,* 336, 336*k,* 337, **337**
Clavipanurgus, 287
Cleptommation, 32*t,* 70*t,* 352, 407*f,* 408, **409**, 409*k*
Cleptotrigona, 30, 75*t,* 805, 810-811, 814*k,* **815-816**, 817, 824
Cleptotrigona cubiceps, 811*f,* 816
Clisodon, 74*t,* 545, 742, 745, 745*k*-746*k,* 747-748, **748**, 749
Clistanthidium, 72*t,* 494, 514, 514*k,* **515**
Clypanthidium, 495, 502-503, 503*k,* 504, **504**
Cnemidandrena, 68*t,* 245*k,* 250*k,* 254*k,* 256*k,* 258*k,* **260**

Cnemidium, 783
Cnemidium viride, 783
Cnemidorhiza, 68*t,* 215, 216*k,* **217**
Coccineobombus, 75*t,* 793*k*-794*k,* 797, **797**
Cockerellia, 69*t,* 300, 301*k,* **302**, 302*f,* 305
Cockerellula, 69*t,* **297-298**, 297*k*
Coelioxinae (Subfamily), 79*t*
Coelioxita, 547
Coelioxoides, 33*t,* 56, 74*t,* 589*k,* 668, **694**, 694*k,* 695*f*
Coelioxoides exulans, 56*f,* 695*f*
Coelioxoides punctipennis, 694, 694*f*-695*f*
Coelioxoides waltheriae, 694*f*-695*f*
Coelioxula, 547
Coelioxyini (Tribe), 79*t*
Coelioxys, 32*t,* 33, 34-35, 42, 67, 78*t*-81*t,* 103, 112, 116, 538, **543-551**, 543*k,* 546*f,* 553, 557, 579, 586, 694, 906
 of Eastern Hemisphere, 546*k*-547*k*
 of Western Hemisphere, 545*k*-546*k*
Coelioxys (Acrocoelioxys), 72*t,* 546*k,* **547**
Coelioxys (Allocoelioxys), 72*t,* **547**, 547*k,* 549
Coelioxys (Argocoelioxys), 549
Coelioxys (Boreocoelioxys), 72*t,* 545*k*-546*k,* **547-548**, 547*k,* 548*f*
Coelioxys (Coelioxys s. str.*)*, 72*t,* 546*k,* 547, 547*k,* **548**
Coelioxys (Cyrtocoelioxys), 72*t,* 546*k,* **548**
Coelioxys (Dasycoelioxys), **548-549**
Coelioxys (Glyptocoelioxys), 72*t,* 546*k,* **548-549**
Coelioxys (Haplocoelioxys), 72*t,* 546*k,* **549**
Coelioxys (Hemicoelioxys), 549
Coelioxys (Intercoelioxys), 547
Coelioxys (Lepidocoelioxys), 547
Coelioxys (Liothgrapis), 549
Coelioxys (Liothyrapis), 72*t,* 543, 546*f,* 547*k,* **549**
Coelioxys (Melanocoelioxys), 547
Coelioxys (Mesocoelioxys), 72*t,* 547*k,* **549**
Coelioxys (Neocoelioxys), 72*t,* 546*k,* **549-550**
Coelioxys (Nigrocoelioxys), 547, 549
Coelioxys (Orientocoelioxys), 547, 549
Coelioxys (Paracoelioxys), 548
Coelioxys (Platycoelioxys), 72*t,* 546*k,* **550**
Coelioxys (Rhinocoelioxys), 72*t,* 546*k,* **550**
Coelioxys (Schizocoelioxys), 547
Coelioxys (Synocoelioxys), 72*t,* 546*k,* **550**
Coelioxys (Torridapis), 72*t,* 543, 546*f,* 547*k,* 549, **550**
Coelioxys (Tropicocoelioxys), 547, 549
Coelioxys (Xerocoelioxys), 72*t,* 545*f,* 546*k,* **550-551**

Coelioxys afra, 547
Coelioxys alata, 548
Coelioxys albiceps, 550
Coelioxys alternata, 544*f*
Coelioxys analis, 546*f,* 550
Coelioxys apicata, 549
Coelioxys argentea, 549
Coelioxys assumptionis, 549
Coelioxys costaricensis, 545, 548
Coelioxys decipiens, 546*f,* 549
Coelioxys dolichos, 547
Coelioxys ducalis, 550
Coelioxys echinata, 547
Coelioxys edita, 549
Coelioxys funeraria, 34-35, 547, 549*f*
Coelioxys fuscipennis, 547, 549
Coelioxys genoconcavitus, 547, 549
Coelioxys germana, 548-549
Coelioxys gracilis, 549
Coelioxys gracillima, 543, 549
Coelioxys mesae, 544*f*-545*f*
Coelioxys mexicana, 549
Coelioxys montandoni, 548
Coelioxys octodentata, 36*f,* 39*f,* 41*f,* 545*f*
Coelioxys otomita, 547
Coelioxys pergandei, 548
Coelioxys quadrifasciatus, 547, 549
Coelioxys ruficauda, 547
Coelioxys rufitarsus, 547, 548*f*
Coelioxys rufocaudata, 547
Coelioxys s. str., 72*t,* 546*k,* 547, 547*k,* **548**
Coelioxys scioensis, 546*f*
Coelioxys spatuliventer, 550
Coelioxys texana, 550
Coelioxys toltecа, 547
Coelioxys torrida, 550
Coelioxys tridentata, 540
Coelioxys vidua, 548
Coelioxys weinlandi, 550
Coelioxys zapoteca, 550
Colax, 762
Colletellus, 67*t,* 147*t,* 149*k,* **152**
Colletes, 21, 28, 33, 33*t,* 35, 42, 67*t,* 77*t,* 78, 78*t*-81*t,* 89, 93, 110, 131*k,* 132-134, 136, **167-169**, 167*k,* 168*f*-169*f,* 171, 193, 226, 353, 389, 645, 649-650, 673, 895, 905
Colletes (Albocolletes), 167
Colletes (Dentcolletes), 167
Colletes (Elecolletes), 167
Colletes (Nanocolletes), 167
Colletes (Pachycolletes), 167
Colletes (Ptilopoda), 167
Colletes (Puncticolletes), 167
Colletes (Rhinocolletes), 167-168
Colletes (Simcolletes), 167
Colletes albopilosus, 177
Colletes cercidii, 168*f*
Colletes chilensis, 177
Colletes cunicularius, 9
Colletes daviesanus, 169
Colletes elegans, 167
Colletes everaertae, 168*f*
Colletes fasciatus, 905
Colletes fulgidus, 84*f,* 133*f,* 168*f*
Colletes graeffei, 167
Colletes herbsti, 177
Colletes inaequalis, 125*f*
Colletes maculipennis, 167
Colletes nanus, 167

Colletes nasutus, 20, 167
Colletes occidentalis, 177
Colletes rubicola, 169
Colletes similis, 167, Plate 15
Colletes spiloptera, 167
Colletes zonalis, 163
Colletidae (Family), 20, 28, 32*t,* 34-35, 50, 52, 54, 61, 61*f,* 63, 65, 65*t,* 67*t*-68*t,* 79*t,* 80, 80*t,* 85, 87-88, 88*f,* 89-93, 95, 101, 111-112, 116-117, 123*k,* 125*f,* 127*k*-128*k,* 129, **132-135,** 134, 134*k*-135*k,* 173, 193, 235-236, 238, 284, 319, 345, 353, 389, 405, 438, 645, 648, 652, 673, 708, 905, Plate 1, Plate 4, Plate 15
 biogeography of, 107*t*
Colletides (Tribe), 78*t*
colletids. *See* Colletidae (Family)
Colletinae (Subfamily), 20, 34-35, 48, 60*f,* 67*t,* 80*t*-81*t,* 85, 87, 89, 93, 112, 116, 120, 126, 128*k,* 129, 132*f,* 133-134, 134*k,* 136, **136-172,** 137*k,* 138, 140, 141*f,* 156, 159, 171, 173, 179-180, 188, 220, 236-237, 323, 639-640, 664, 698, 895
Colletini (Tribe), 67*t,* 80*t,* 134, 136, 137*k,* 138, **167-170,** 167*k,* 905
 biogeography of, 107*t*
Colletoidea, 79*t*
Colletopsis, 67*t,* 147*t,* 149*k,* 152
Colocynthophila, 733
Coloplitis, 71*t,* 110, 463*k,* 467, **469**
Comeptila, 732
Compsoclepta, 675
Compsoclepta fasciata, 675
Compsomelissa, 623, 624*k,* 627, **627-628,** 627*k*
Compsomelissa (*Compsomelissa* s. str.), 73*t,* 620, 622*f,* **627-628,** 627*k,* 628, 631*f,* 632
Compsomelissa (*Halterapis*), 26, 73*t,* 620, 621*f*-622*f,* 623-624, 626*f,* 627*k,* **628-629,** 631*f*
Compsomelissa angustula, 629
Compsomelissa borneri, 627-628
Compsomelissa minuta, 629
Compsomelissa nigrinervis, 620, 621*f*-622*f,* 623-624, 626*f,* 629, 631*f*
Compsomelissa ocellata, 620*f*
Compsomelissa s. str., 73*t,* 620, 622*f,* **627-628,** 627*k,* 628, 631*f,* 632
Compsomelissa stigmoides, 622*f,* 631*f*
Compsomelissa zaxantha, 620*f,* 628
Conandrena, 68*t,* 245*k,* 248*k,* **260,** 261
Conanthalictus, 95, 322-324, 324*k,* **325-326,** 325*k,* 330
Conanthalictus (*Conanthalictus* s. str.), 69*t,* **325,** 325*k,* 326
Conanthalictus (*Phaceliapis*), 69*t,* 323*f,* **325-326,** 325*k*

Conanthalictus bakeri, 325, 326*f*
Conanthalictus caerulescens, 320*f*
Conanthalictus nigricans, 323*f*
Conanthalictus s. str., 69*t,* **325,** 325*k,* 326
Confusibombus, 75*t,* 792*k,* 795*k,* **798**
Confusobombus, 798
Congotrachusa, 72*t,* 533, 534*k,* **535**
Conohalictoides, 326-327
Conohalictoides lovelli, 326
Convolvulus, 20
Copoceratina, 73*t,* 614*k,* **616**
Copoxyla, 73*t,* 598, 602*k,* 604*k,* **605**
Coptepeolus, 651
Coptepeolus emarginatus, 651
Coptorthosoma, 606
Coquillettapis, 703
Coquillettapis melittoides, 703
Corbicula, 722
Corbicula apiformis, 722
corbiculate bees, 26, 28, 81-82, 97, 110, 117, 667-669, 668*f*-669*f,* 670-671, 686-689, 778, 780, 787*f,* 801, 816, 906
 fossil, 671, Plate 19-Plate 20
Cordandrena, 68*t,* 253*k,* 255*k,* **260**
Cornylaeus, 67*t,* 198*t,* **203-204,** 203*k*
Corynogaster, 406-407
Corynura, 83, 353, 393, 400*k,* **406-407,** 406*k,* 407, 410-411
Corynura (*Callistochlora*), 70*t,* 395, 396*f,* 397*k,* **406,** 406*k*
Corynura (*Corynura* s. str.), 70*t,* 394*k,* 395*f*-397*f,* 398*k,* **406-407,** 406*k,* 407
Corynura (*Corynuropsis*), 411
Corynura chilensis, 349*f,* 395*f*-397*f*
Corynura chloris, 396*f*
Corynura corynogaster, 396*f*
Corynura darwini, 411
Corynura flavofasciata, 406
Corynura gayi, 406-407
Corynura s. str., 70*t,* 394*k,* 395*f*-397*f,* 398*k,* **406-407,** 406*k,* 407
Corynurella, 410
Corynurella mourei, 410
Corynurina (Subtribe), 393
Corynuroides, 411
Corynuropsis, 411
Corythochila, 454
Crabronidae (Family) (wasps), 62-63, 63*f,* 64-65, 65*t,* 88, 90
Crabroninae (Subfamily) (wasps), 63
Crawfordapis, 67*t,* 173-174, 174*k,* **175-176**
Crawfordapis crawfordi, 176
Crawfordapis luctuosa, 176
Creightonella, 72*t,* 105, 120, 543, 551-556, 557*t,* 561*k,* 562, 562*k*-565*k,* **571-572,** 572*f,* 586
Creightoniella, 571

Cremnandrena, 68*t,* 246*k,* 248*k,* **260**
Cressoniella, 72*t,* 557*t,* 559*k*-560*k,* **572,** 584
Cressoniella (*Chaetochile*), 572
Cressoniella (*Neocressoniella*), 579
Cressoniella (*Orientocressoniella*), 568, 584-585
Cressoniella (*Ptilosaroides*), 581
Cressoniella (*Rhyssomegachile*), 582
Cressoniella (*Trichurochile*), 584
Cressoniella golbachi, 572
Cretotrigona, fossil, 98*t,* 100, 812, Plate 20
Cretotrigona prisca, fossil, 100-101, 812, Plate 20
Crewella, 73*t,* 614*k,* 615, **616**
Crinoglossa, 337
Crinoglossa natalensis, 337
Critotrigona [bee-like insect], 905
Crocisa, 775
Crocisa bicuspis, 772
Crocisa fulvohirta, 773
Crocisa lata, 766
Crocisaspidia, 69*t,* 339, 339*k,* **340**
Crocisaspidia chandleri, 340
Crocissa, 775
Crotchiibombus, 75*t,* 792*k,* 795*k,* 797, **798**
Crurilegidae, 78*t*
Cryptandrena, 68*t,* 251*k,* 253*k*-254*k,* **260,** 263
Cryptantha, 20
Cryptohalictoides, 326-327
Cryptohalictoides spiniferus, 326
Cryptosmia, 481
Ctenanthidium, 72*t,* 517-518, 518*k,* **519**
Ctenanthidium gracile, 519
Ctenaugochlora, 70*t,* 395*k,* 396*f,* **405,** 405*k*
Ctenioschelini (Tribe), 81, 763
Ctenioschelus, 74*t,* 80*t,* **764,** 764*k*
Ctenioschelus goryi, 768*f*
Ctenoapis, 437
Ctenoapis lutea, 437
Ctenoceratina, 73*t*
Ctenoceratina (*Hirashima*), 616
Ctenoceratina (in *Ceratina*), 612, 614*k,* **616**
Ctenoceratina (*Simioceratina*), 618
Ctenocolletes, 48, 67*t,* 93, 129, **130-131,** 130*k,* 131, 316
Ctenocolletes albomarginatus, 125*f*
Ctenocolletes fulvescens, 130
Ctenocolletes nicholsoni, 129*f*-130*f*
Ctenocolletes smaragdinus, 129, 129*f*
Ctenocorynura, 411
Ctenocorynura vernoniae, 411
Ctenonomia, 70*t,* 348, 361, 371*t,* 372-373, 373*k*-374*k,* **376-377,** 377, 380-381
Ctenonomia carinata, 376
Ctenoplectra, 18, 74*t,* 79, 79*t*-81*k,* 84, 97, 106, 116, 171, 697, **698-699,** 698*k,* 699*f*
Ctenoplectra (*Ctenoplectrina*), 699

Ctenoplectra albolimbata, 698*f*
Ctenoplectra armata, 699*f*
Ctenoplectra bequaerti, 698*f*
Ctenoplectra chalybea, 698
Ctenoplectra fuscipes, 697*f*
Ctenoplectra grimaldi, fossil, Plate 18
Ctenoplectra politula, 699
Ctenoplectra ussuriana, 431
Ctenoplectrella, fossil, 98*t,* 448
Ctenoplectrella viridiceps, fossil, 98*t*
Ctenoplectridae (Family), 697
Ctenoplectrina, 33*t,* 74*t,* 127, 698*k,* **699**
Ctenoplectrina politula, 698*f,* 699
Ctenoplectrinae (Subfamily), 80*t*
Ctenoplectrinellina (Subtribe), 448
Ctenoplectrini (Tribe), 18, 33*t,* 74*t,* 97, 104, 127, 413, 587, 590*k,* 667, 668*f,* 671, **697-699,** 698*f,* 698*k*
 biogeography of, 108*t,* 109
Ctenopoda, 605
Ctenosibyne, 67*t,* 158*f,* 162*k,* 163, **163**
Ctenosmia, 470
Ctenoxylocopa, 73*t,* 601*k,* 603*k,* **605,** 609
Cubiandrena, 68*t,* 252*k,* 254*k,* **261**
Cubitalia, 720*k,* **722-723,** 723*k*
Cubitalia (*Cubitalia* s. str.), 74*t,* **722,** 723*k*
Cubitalia (*Opacula*), 74*t,* **723,** 723*k*
Cubitalia (*Pseudeucera*), 74*t,* 723, **723,** 723*k*
Cubitalia donatica, 723
Cubitalia parvicornis, 20, 723
Cubitalia s. str., 74*t,* **723,** 723*k*
Cubitognatha, 71*t,* 453*k,* **454-455**
Cullumanibombus, 798
Cullumanobombus, 75*t,* 792*k,* 795, 795*k,* **798**
Curtisapis, 380
Curvinomia, 339, 341
Cuspidella, 72*t,* 557*t,* 563*k*-564*k,* **572**
Cyaneoderes, 606
Cyaneoderes fairchildi, 606
Cyanocentris, 756
Cyathocera, 344
Cyathocera nodicornis, 344
Cyphanthidium, 71*t,* 491*t,* 494, 494*k,* **512**
Cyphanthidium intermedium, 512
Cyphanthidium sheppardi, 512
Cyphepicharis, 74*t,* 760, 760*k*
Cyphepicharis borgmeieri, 760
Cyphomelissa, 767
Cyphomelissa pernigra, 767
Cyphopyga, 578
Cyphoxylocopa, 606
Cyprirophites, 326-327
Cyrtapis, fossil, 98*t*
Cyrtapis anomalus, fossil, 98*t*
Cyrtocoelioxys, 72*t,* 546*k,* **548**
Cyrtopasites, 655, 661
Cyrtosmia, 71*t,* 465, 465*k*-466*k,* **469,** 472

Dactylandrena, 68*t*, 245*k*, 247*k*, **261**
Dactylomegachile, 571, 586
Dactylurina, 75*t*, 806, 808, 810-812, 814*k*, **816**, 824, 828, 830
Dactylurina schmidti, 804*f*, 809*f*, 812*f*
Dalloapula, 73*t*, 621*f*-622*f*, 626, **626**, 626*f*, 626*k*
Dasiapis, 81*t*, 703-704, 706, **728**
Dasiapis ochracea, 703
Dasyandrena, 68*t*, 246*k*, 251*k*, **261**
Dasybombus, 75*t*, 792*k*, 796*k*, **798**, 801
Dasycoelioxys, 548-549
Dasycollettes, 155
Dasycollettes metallicus, 155
Dasycollettes ventralis, 154
Dasygastrae, 78*t*
Dasyglossa, 318
Dasyhalonia, 74*t*, 727, 727*k*, **728**
Dasyhalonia (*Pachyhalonia*), 729
Dasyhalonia (*Zonalonia*), 728
Dasyhalonia (*Zonohalonia*), 728
Dasyhalonia (*Zonolonia*), 728
Dasyhalonia justi, 728
Dasyhesma, 68*t*, **215-216**, 220, 223*k*
Dasyhesma abnormis, 221*f*, 226
Dasyhesma robusta, 215, 226
Dasymegachile, 72*t*, 557*f*, 558, 559*k*-560*k*, **572-573**
Dasymegilla, 74*t*, 746*k*, 747, 747*k*, **748**
Dasyosmia, 71*t*, 461*f*, 463, 465, 465*k*-466*k*, **469**
Dasypoda, 28, 70*t*, 77, 77*t*-81*t*, 117, 389, 413-417, **418**, 418*k*, 419
Dasypoda argentata, 418
Dasypoda braccata, 418
Dasypoda crassicornis, 418
Dasypoda hirtipes, 417*f*, 418
Dasypoda panzeri, 413*f*-414*f*
Dasypoda pyrotrichia, 418
Dasypodaidae (Family), 89, 95, 117
Dasypodainae (Subfamily), 35, 70*t*, 91, 95, 126, 413*f*, 414-415, 415*f*, 415*k*, **416**, 416*k*, 422, 426, 429, 659
Dasypodaini, 905
Dasypodaini (Tribe), 70*t*, 414, 416, 416*k*, **417-422**, 417*f*, 417*k*-418*k*, 422
biogeography of, 107*t*
Dasypodidae (Family), 117, 416
Dasypodinae (Subfamily), 80*t*
Dasypus, 416
Dasyscirtetica, 721
Dasystilbe, 783
Dasyxylocopa, 73*t*, 600*k*-601*k*, **605**, 609
Delomegachile, 584-585
Deltoptila, 74*t*, 743*k*, 745*f*, **750**
Deltoptila montezumia, 742*f*
Denticollettes, 167
Dentigera, 67*t*, 198*t*, 200*k*-201*k*, 202*f*, **204**
Dentirophites, 326-327
Deranchanthidium, 71*t*, 492*f*-493*f*, **513**, 513*k*, 520

Deranchylaeus, 67*t*, 198*t*, 203-204, 203*k*, **204**
Derandrena, 68*t*, 244*k*, 246*k*, 249*k*, **261**
Dermatohesma, 215
Derotropis, 578
Desmotetrapedia, 690
Diadasia, 20, 33*t*, 74*t*, 81*t*, 484, 648, 679, 701, 702*k*, **703-704**, 705-706, 728
Diadasia afflicta, 702*f*
Diadasia bituberculata, 704
Diadasia enavata, 704
Diadasia nemaglossa, 706
Diadasia rinconis, 700*f*
Diadasia sphaeralcearum, 701*f*
Diadasiana, 704
Diadasiella, 682
Diadasiella coquilletti, 682
Diadasina, 702, 702*k*, 704, **704**, 704*k*, 705
Diadasina (*Diadasina* s. str.), 74*t*, **704**, 704*k*
Diadasina (*Leptometriella*), 74*t*, 702-703, **704**, 704*k*
Diadasina distincta, 704
Diadasina s. str., 74*t*, **704**, 704*k*
Diagonozus, 70*t*, **390-391**, 390*f*, 390*k*
Diagonozus bicometes, 390
Dialictus, 18, 30-31, 31*t*-32*t*, 38*f*, 40*f*, 66, 70*t*, 103, 120, 332, 348-349, 351*f*, 352-354, 355*f*, 357, 358*f*, 358*k*, 371-372, 371*f*, 373, 373*k*-374*k*, 376, **376-377**, 378-383, 398, 407, 410, 587*f*, 623
Dialonia, 388
Diandrena, 68*t*, 244*k*, 246*k*-247*k*, **261**
Dianthidium, 491*t*, 496*k*, 505, 508, **512-513**, 513*k*, 521, 524, 531
Dianthidium (*Adanthidium*), 71*t*, 492*f*, 495*f*, 496, **513**, 513*k*, 525
Dianthidium (*Anthidulum*), 519
Dianthidium (*Bathanthidium*), 511
Dianthidium (*Deranchanthidium*), 71*t*, 492*f*-493*f*, **513**, 513*k*, 520
Dianthidium (*Dianthidium* s. str.), 25-26, 71*t*, 495*f*, 496, 502*f*, 512*f*, 513, **513**, 513*k*, 515, 524
Dianthidium (*Eoanthidium*), 515
Dianthidium (*Mecanthidium*), 71*t*, 512-513, **513-514**, 513*k*
Dianthidium (*Notanthidium*), 523
Dianthidium (*Spinanthidium*), 526
Dianthidium arizonicum, 495*f*
Dianthidium bifoveolatum, 511
Dianthidium chamela, 492*f*-493*f*, 513
Dianthidium concinnum, 23*f*
Dianthidium currani, 519
Dianthidium curvatum, 9, 495*f*, 502*f*, 513
Dianthidium discophorum, 492*f*
Dianthidium flavoclypeatum, 497

Dianthidium marshi, 496
Dianthidium paraguayense, 516
Dianthidium s. str., 25-26, 71*t*, 495*f*, 496, 502*f*, 512*f*, 513, **513**, 513*k*, 515, 524
Dianthidium subarenarium, 520
Dianthidium turnericum, 515
Dianthidium ulkei, 512*f*, 513
Diascia, 16, 18
Diaxylocopa, 73*t*, 599*k*, 601*k*, **605**, 607
Diceratosmia, 71*t*, 476-477, 478*k*, **480**, 481, 486
Dichanthidium, 72*t*, 510, 517-518, 518*k*, **519**, 521
Dichanthidium exile, 519
Dichroa, 388
Dicranthidium, 72*t*, **519**, 519*k*, 520
Dicromonia, 70*t*, **427-428**, 427*k*
Dictyohalictus, 70*t*, 384, 384*k*, **385-386**
Didonia, 68*t*, 252*k*-254*k*, 256*k*, 260, **261**
Didonia punica, 261
Diepeolus, 649
Dieunomia, 92, 332-333, **335**, 335*k*, 652
Dieunomia (*Dieunomia* s. str.), 69*t*, **335**, 335*k*
Dieunomia (*Epinomia*), 69*t*, **335**, 335*k*
Dieunomia nevadensis, 332*f*
Dieunomia s. str., 69*t*, **335**, 335*k*
Dieunomia triangulifera, 335
Digitella, 574
Digressobombus, 798
Digronoceras, 575
Dilobopeltis, 516
Dilobopeltis fuscipennis, 516
Dimorphides (Family), 78*t*
Dinagapostemon, 70*t*, 354, 356, 359*k*-360*k*, **364**, 383
Dinagapostemon gigas, 362*f*
Dinagapostemon orestes, 362*f*
Dinagapostemon sicheli, 359*f*
Dinavis, 581
Dinomia, 344
Dinoxylocopa, 610
Diomalopsis, 73*t*, **683**, 683*f*, 683*k*
Dioxoides, 540
Dioxyini (Tribe), 33*t*, 40, 43, 72*t*, 111, 116, 237, 435, 441, 442*k*, 442*t*, **538-542**, 539*k*
biogeography of, 107*t*
positions of spines and carinae on thoracic sclerites of, 539*t*
Dioxys, 72*t*, 78*t*-80*t*, 81, 81*t*, 538-539, 539*k*, **540**, 541-542
positions of spines and carinae on thoracic sclerites of, 539*t*
Dioxys cincta, 538-539
Dioxys formosa, 541
Dioxys longiventris, 542
Dioxys pannonica, 541
Dioxys pomonae, 539
Dioxys productus cismontanicus, 538*f*, 541*f*
Dioxys productus subruber, 539*f*
Dioxys quadrispinosa, 541

Dioxys schulthessi, 540
Dipedia, 703
Diphaglossa, 67*t*, 79*t*-80*t*, 177, 177*k*, **178**
Diphaglossa ecuadoria, 164
Diphaglossa gaullei, 179
Diphaglossa gayi, 61*f*, 177*f*, 178
Diphaglossinae (Subfamily), 35, 67*t*, 80, 80*t*, 93, 112, 128*k*, 129, 133-134, 134*k*, **173-179**, 173*f*, 173*k*, 317
biogeography of, 109
Diphaglossini (Tribe), 67*t*, 89, 133, 173, 173*k*, **177-178**, 177*f*, 177*k*
biogeography of, 107*t*
Diphysis, 536
Diphysis pyrenaica, 536
Disparapis, 70*t*, 419*k*-420*k*, **420**
Dissoglotta, 179
Dissoglotta stenoceratina, 179
Dissoglottini (Tribe), 67*t*, 89, 173, 173*k*, **179**
biogeography of, 107*t*
Distandrena, 68*t*, 252*k*, 254*k*, **261**
Dithygater, 74*t*, **740**, 740*k*
Dithygater seabrai, 740
Diversibombus, 798
Diversobombus, 75*t*, 793*k*, 796*k*, **798**, 799
Dizonamegilla, 743
Doeringiella, 33, 73*t*, 636, 647, 647*k*, **648**, 649*f*, 651
Doeringiella (*Doeringiella* s. str.), 648
Doeringiella (*Orfilana*), 648
Doeringiella (*Stenothisa*), 651
Doeringiella (*Triepeolus*), 639*f*
Doeringiella angustata, 651
Doeringiella arechavaletai, 649*f*
Doeringiella bizonata, 646*f*, 648
Doeringiella guttata, 649*f*
Doeringiella s. str., 648
Doeringiella singularis, 649*f*
Doeringiella thoracica, 641
Doeringiella variegata, 648
Dolichochile, 71*t*, 414-415, 429*f*, 431, **432**, 432*f*, 432*k*
Dolichochile melittoides, 432
Dolichosmia, 71*t*, 449, 487, 487*k*, **488**
Dolichostelis, 33*t*, 72*t*, 512*f*, 517, 530, 530*f*, **531**, 531*f*, 531*k*
Dolichotrigona, 829
Domanthidium, 71*t*, 498*k*, 499, **500**, 501, 512
Donatica, 723
Doxanthidium, 531
Drepanium, 388
Duckeanthidium, 71*t*, 491*t*, 496*k*, **514**
Duckeanthidium cibele, 514, 518
Duckeanthidium megachiliforme, 514, **514**
Duckeanthidium rondonicola, 514
Duckeanthidium thielei, 514
Duckeola, 75*t*, 820*f*, 824, **825**, 825*k*, 826-827

Dufourea, 69*t,* 77, 78*t,* 79, 79*t*-80*t,* 95, 322, 324*k,* 325, 325*k,* **326-327,** 328-330, 656-657
Dufourea (Afrodufourea), 326
Dufourea (Alpinodufourea), 326
Dufourea (Atrodufourea), 326
Dufourea (Dufourea s. str.), 327
Dufourea (Flavodufourea), 329
Dufourea (Glossadufourea), 327
Dufourea (Micralictoides), 328
Dufourea (Minutodufourea), 326
Dufourea (Trilia), 326
Dufourea alboclypeata, 422
Dufourea alpina, 326
Dufourea caeruleocephala cypria, 326
Dufourea calochorti, 320*f*
Dufourea cypria, 327
Dufourea flavicornis, 329
Dufourea gaullei, 326-327
Dufourea longiglossa, 326-327
Dufourea marginata, 322*f*-323*f*
Dufourea merceti, 326
Dufourea minuta, 326
Dufourea minutissima, 326
Dufourea monardae, 327
Dufourea muoti, 326
Dufourea novaeangliae, 327
Dufourea paradoxa, 327
Dufourea pectinipes, 327
Dufourea punica, 326
Dufourea rufiventris, 326
Dufourea s. str., 327
Dufourea spinifera, 327
Dufourea versatilis, 327
Dufoureidae (Family), 79, 79*t*
Dufoureinae (Subfamily), 80, 80*t.* See also Rophitinae

Ebmeria, 380
Ebmeriana, 663
Ecclitodes, 73*t,* 674, **676,** 676*k*
Echinanthidium, 506-508
Echthralictus, 32*t,* 40, 70*t,* 105, 352, 354, 362*k,* **364,** 370
Echthralictus extraordinarius, 38*f,* 40*f*
Eckfeldapis, fossil, 98*t,* 831
Eckfeldapis electrapoides, fossil, 98*t*
Ecplectica, 74*t,* 714*f,* 730*k*-731*k,* **731**
Ecplectica tintinnans, 731
Edriohylaeus, 67*t,* 93, 188, 197*f,* 197*k,* 198*t,* **204**
Edwynia, 157
Edwyniana, 151, 157
Effractapis, 73*t,* 110, 126, 623, 623*k,* 624, 627, **629,** 632
Effractapis furax, 31*t,* 629
Egapista, 179
Eickwortapis, fossil, 98*t*
Eickwortapis dominicana, fossil, 98*t*
Eickwortia, 70*t,* 372, 373*k,* **378**
Elandrena, 265
Elaphropoda, 74*t,* 743*k,* 750, **750,** 770
Elecolletes, 167
Electobombini (Tribe), fossil, 670
Electrapini (Tribe), fossil, 98*t*-99*t,* 670-671, Plate 19

Electrapis, fossil, 98*t,* 670-671, 831
Electrapis (Melikertes), fossil, 99*t,* 831
Electrapis (Protobombus), 831
Electrapis (Roussyana), fossil, 99*t,* 831
Electrapis stilbonota, fossil, 99*t*
Electroaugochlora, fossil, 98*t,* 393
Electrobombini (Tribe), fossil, 98*t,* 671
Electrobombus, fossil, 98*t,* 670
Electrobombus samlandensis, fossil, 98*t,* 671
Electrolictus, fossil, 98*t*
Electrolictus antiquus, fossil, 98*t*
Electrolictus antiquus, fossil, 357
Emphor, 79*t,* 705
Emphoridae (Family), 79*t,* 97
Emphorina (Subtribe), 700-701, 705
Emphorini (Tribe), 20-21, 58, 74*t,* 80*t,* 81, 116, 583, 590*k*-591*k,* 648, 668, 668*f,* 671, 680, 682, 687, **700-706,** 701*f,* 701*k*-702*k,* 702*f,* 709
biogeography of, 108*t*
Emphoropsis, 751
Emphoropsis murihirta murina, 751
Energoponus, 705
Energoponus strenuus, 705
Ensliniana, 72*t,* 442, 538, 539*k,* **540-541**
positions of spines and carinae on thoracic sclerites of, 539*t*
Ensliniana cuspidata, 540
Entechnia, 705
Eoanthidiellum, 515
Eoanthidium, 491*t,* 494, 494*k*-495*k,* 497, **514-515,** 514*k*-515*k*
Eoanthidium (Clistanthidium), 72*t,* 494, 514, 514*k,* 515
Eoanthidium (Eoanthidiellum), 515
Eoanthidium (Eoanthidium s. str.), 72*t,* **515,** 515*k*
Eoanthidium (Hemidiellum), 72*t,* 514, 514*k,* **515**
Eoanthidium (Salemanthidium), 72*t,* 494, 514*k,* **515**
Eoanthidium clypeare, 515
Eoanthidium insulare, 515
Eoanthidium nasicum, 515
Eoanthidium s. str., 72*t,* **515,** 515*k*
Eoanthidium semicarinatum, 515
Eochelostoma, 71*t,* **457,** 457*k*
Eomacropidini (Tribe), fossil, 98*t*
Eomacropis, fossil, 98*t*
Eomacropis glaesaria, fossil, 98*t*
Eomelipona, 817-818
Eopeponapis, 733
Eopsithyrus, 800
Eothrincostoma, 70*t,* 383, 385*f,* 390*k,* **391**
Eoxenoglossa, 74*t,* 722, **741,** 741*k*
Exylocopa, 606
Epanthidium, 491*t,* 496, 497*k,* 515-516, 515*k*-516*k,* 521

Epanthidium (Ananthidium), 72*t,* 515, **516,** 516*k*
Epanthidium (Carloticola), 72*t,* **516,** 516*k*
Epanthidium (Epanthidium s. str.), 72*t,* 515, 515*k*-516*k,* **516**
Epanthidium s. str., 72*t,* 515, 515*k*-516*k,* **516**
Epanthidium tigrinum, 515
Epeicharis, 690
Epeicharis mexicanus, 690
Epeolina (Subtribe), 647
Epeolini (Tribe), 73*t,* 80*t,* 81, 589, 590*k,* 636-637, 640, 643, **646-652,** 646*f*-647*f,* 647*k*-648*k,* 672, 678, 762
biogeography of, 108*t*
Epeoloides, 33*t,* 34, 73*t,* 80*t*-81*t,* 591*k,* 674, **675,** 675*k*
Epeoloides ambiguus, 675
Epeoloides coecutiens, 675, 675*f*
Epeoloides pilosula, 674*f,* 675
Epeoloidini (Tribe), 80*t,* 81, 674
Epeolus, 33, 35, 42, 73*t,* 77, 77*t*-81*t,* 646-647, 648*k,* **649-650,** 650, 650*k,* 651-652, 676, 678, Plate 2
Epeolus (Diepeolus), 649
Epeolus (Epeolus s. str.), 650, 650*k,* 652, **653,** 653*k*
Epeolus (Monoepeolus), 649
Epeolus (Trophocleptria), 647, 647*f,* 649-650, 650*k,* **653**
Epeolus aterrima, 676
Epeolus bifasciatus, 650
Epeolus compactus, 647*f*
Epeolus concavus, 651
Epeolus cruciger, 646*f*
Epeolus giannellii, 649
Epeolus glabratus, 649
Epeolus holmbergi, 648
Epeolus punctatus, 662
Epeolus rufiventris, 651
Epeolus s. str., 650, 650*k,* 652, **653,** 653*k*
Epeolus stuardi, 676
Epeolus viperinus, 673
Epicharana, 74*t,* 753, 753*f,* 760, **760-761,** 760*k,* 761, 765*f*
Epicharis, 33*t,* 56, 78*t,* 80*t,* 588, 671, 691, 753-754, 754*k,* **759-761,** 760*k,* 762, 768
Epicharis (Anepicharis), 74*t,* **760,** 760*k*
Epicharis (Cyphepicharis), 74*t,* **760,** 760*k*
Epicharis (Epicharana), 74*t,* 753, 753*f,* 760, **760-761,** 760*k,* 761, 765*f*
Epicharis (Epicharis s. str.), 74*t,* 759-761, 760*k,* **761**
Epicharis (Epicharitides), 74*t,* 759, 760*k,* **761,** 779*k*
Epicharis (Epicharoides), 74*t,* 759, 759*f,* 760*k,* **761,** 778*f*
Epicharis (Hoplepicharis), 74*t,* 760, 760*k,* **761**
Epicharis (Parepicharis), 74*t,* 760*k,* **761**

Epicharis (Triepicharis), 74*t,* 754, 760*k,* **761**
Epicharis albofasciata, 759*f*
Epicharis analis, 761
Epicharis bicolor, 761
Epicharis borgmeieri, 760
Epicharis cockerelli, 761
Epicharis dejeanii, 760
Epicharis elegans, 753, 753*f,* 761
Epicharis fasciata, 761
Epicharis maculatus, 761
Epicharis rustica, 753, 761, 765*f*
Epicharis s. str., 74*t,* 759-761, 760*k,* **761**
Epicharis zonata, 103, 753, 761
Epicharitides, 74*t,* 759, 760*k,* **761,** 779*k*
Epicharoides, 74*t,* 759, 760*k,* **761,** 778*f*
Epicharoides bipunctatus, 761
Epiclopus, 74*t,* 764*k,* **766,** 768
Epiclopus gayi, 763-764, 766
Epiclopus lendliana, 763
Epiclopus lendlianus, 766
Epiclopus wagenknechti, 763, 766
Epicolpus, 766
Epihalictoides, 326
Epimacrotera, 69*t,* 300*k,* **302-303,** 302*f,* 663
Epimelissodes, 74*t,* 715*f,* 734*k,* **735**
Epimethea, 69*t,* 236, 286, 286*f,* 286*k,* **287,** 655
Epimethea nana, 328
Epimethea variegata, 287
Epimonispractor, 684
Epimonispractor gratiosus, 684
Epinomia, 69*t,* **335,** 335*k,* 337
Epixylocopa, 609
Erandrena, 68*t,* 244*k,* 248*k,* **261**
Eremandrena, 265
Eremaphanta, 83
Eremaphanta, 91, 95, 109, 126, 345, 414-417, 417*k,* **418-419,** 418*k,* 905
Eremaphanta (Eremaphanta s. str.), 70*t,* 416, **418-419,** 418*k*
Eremaphanta (Popovapis), 70*t,* 418*k,* **419**
Eremaphanta s. str., 70*t,* 416, **418-419,** 418*k*
Eremaphanta dispar, 418*f*
Eremapis, 73*t,* 668*f,* 680-681, 681*k,* 682, **682**
Eremapis parvula, 682
Eremopasites, 659
Eremoplosmia, 490
Eremosmia, 71*t,* **455,** 455*k,* 476
Eriades, 460
Eriades fasciatus, 475
Eriades langenburgicus, 460
Eriades larvatus, 452
Eriades moricei, 489
Ericrocidini (Tribe), 33-34, 33*t,* 41, 74*t,* 104, 127, 589*k,* 651, 668, 668*f,* 671-673, 762, **763-769,** 763*f,* 763*k*-764*k,* 765*f*-768*f,* 770
biogeography of, 108*t,* 109
Ericrocini (Tribe), 80*t,* 81
Ericrocis, 41, 74*t,* 80*t,* 763, 764*k,* **766**

Ericrocis lata, 763*f*
Ericrocis pintada, 766*f*
Eriops, 289
Eryops, 289
Erythandrena, 69*t*, **268**, 268*k*
Erythandrena, 236
Erythronomioides, 347
Erythrosmia, 71*t*, 451, 475-476, 477*k*-478*k*, 480, **480**
Ethalonchopria, 67*t*, **142**, 142*k*
Euandrena, 68*t*, 243*k*, 246*k*-248*k*, 251*k*-252*k*, 255*k*, 257*k*-258*k*, **261**
Euaspis, 32*t*, 34, 72*t*, 491*t*, 494*k*, **516-517**, 529-530
Euaspis carbonaria, 114*f*
Eucara, 74*t*, 736, 736*f*, 736*k*, 737, **737**
Eucera, 42, 67, 77*t*-81*t*, 545, 645, 707-710, 720*k*, 722, **723-726**, 724*k*
Eucera (Agatheucera), 724
Eucera (Atopeucera), 724-725
Eucera (Eucera s. str.), 74*t*, 723-724, **724**, 724*k*
Eucera (Hemieucera), 724-725
Eucera (Hetereucera), 74*t*, 723-724, **724-725**, 724*k*
Eucera (Oligeucera), 74*t*, 724, 724*k*, **725**
Eucera (Pareucera), 724-725
Eucera (Pileteucera), 724
Eucera (Pteneucera), 74*t*, 724, 724*k*, **725**
Eucera (Rhyteucera), 724
Eucera (Stilbeucera), 724
Eucera (Synhalonia), 74*t*, 120, 711*f*, 713*f*, 715*f*, 718*k*, 720*k*, 723-724, 724*k*, **725-726**, 736-737, 739, 770
Eucera antennata, 737
Eucera atriventris, 711*f*, 713*f*, 715*f*
Eucera bidentata, 724
Eucera breviceps, 723
Eucera brevicornis, 483
Eucera caspica, 724
Eucera cearensis, 728
Eucera chrysopyga, 710*f*
Eucera cineraria, 724
Eucera cinerea, 724
Eucera clavicornis, 292
Eucera clypeata, 724
Eucera cressonii, 735
Eucera curvicornis, 330
Eucera euchnemidea, 725
Eucera euthyglossa, 676
Eucera floralia, 725
Eucera furfurea, 724
Eucera hamata, 726
Eucera herbsti, 721
Eucera hispana, 724
Eucera interrupta, 724
Eucera obscurior, 729
Eucera paraclypeata, 724
Eucera parvicornis, 723
Eucera parvula, 724
Eucera patellicornis, 728
Eucera popovi, 725
Eucera s. str., 74*t*, 723-724, **724**, 724*k*
Eucera seminuda, 724
Eucera venusta, 718
Euceratina, 73*t*, 593, 612, 614*k*, **616**, 617
Euceridae (Family), 79*t*, 97

Eucerina (Subtribe), 707-710
eucerine bees, 18, 23*f*, 667, 668*f*
Eucerini (Tribe), 20, 34-35, 74*t*, 80*t*, 97, 110, 112-113, 120, 165, 588, 590*k*, 648, 652, 662, 668, 668*f*, 671, 673, 680, 685, 697, 700, 701*f*, **707-741**, 708*f*-718*f*, 736*f*, 749, 770, 785, 895, Plate 11
  biogeography of, 108*t*
  of Eastern Hemisphere, 720*k*
  of North and Central America, 717*k*-720*k*
  of South America, 710*k*-717*k*
Eucerinoda, 74*t*, 590, 707-708, 710*k*, **726**
Eucerinoda gayi, 725*f*, 726
Eucerinodina (Subtribe), 707-708, 710
Eucerinodini (Tribe), 81
Euchalicodoma, 569-570
Eucharis, 761
Eucharis hirtipes, 761
Eucondylops, 30, 73*t*, 84, 619, 623, 624*k*, 626*k*, **629**
Eucondylops konowi, 31*t*, 622*f*, 629, 631*f*
Eucondylops reducta, 31*t*, 34*f*, 593, 628*f*, 629
Eudioxys, 72*t*, 538, 539*k*, **541**
  positions of spines and carinae on thoracic sclerites of, 539*t*
Euflorilegus, 74*t*, 712*f*, **726**, 726*k*
Eufriesea, 33*t*, 56, 75*t*, 779-780, 780*f*, 781-782, 782*k*, 783, **783**, 784
  fossil, 781
Eufriesea concava, 782*f*
Eufriesea pulchra, 518, 786*f*
Eufriesea purpurata, 779
Eufriesea surinamensis, 780*f*, 783*f*
Eufriesea violacea, 787*f*
Eufriesia, 783
Euglages, 653
Fuglages scripta, 653
Euglossa, 34, 75*t*, 77*t*-81*t*, 518, 545, 548, 778, 778*f*, 779, 779*f*, 780-782, 782*k*, **783-784**, Plate 15
  fossil, 781
Euglossa (Dasystilbe), 783
Euglossa (Euglossella), 783
Euglossa (Glossura), 778, 783
Euglossa (Glossurella), 783
Euglossa (Glossuropoda), 783
Euglossa annectans, 780
Euglossa atroveneta, 545, 780
Euglossa bursigera, 783
Euglossa cordata, 780*f*, 782*f*
Euglossa dodsoni, 778*f*
Euglossa gorgonensis, 778*f*
Euglossa hyacinthina, 781
Euglossa imperialis, 779*f*
Euglossa intersecta, 783
Euglossa nigropilosa, 783
Euglossa piliventris, 783
Euglossa pulchra, 783
Euglossa spinosa, 651
Euglossa villosa, 783
Euglossella, 783
Euglossidae (Family), 79*t*
euglossine bees. *See* Euglossini (Tribe)

Euglossini (Tribe), 4, 18-19, 21, 32*t*-33*t*, 34, 52, 75*t*, 80*t*, 81, 97, 104, 110-111, 117, 127, 492, 518, 588*k*, 667-668, 670-671, 754, **778-784**, 778*f*-780*f*, 781*k*, 782*f*, 785, 786*f*, 895, Plate 15
  biogeography of, 108*t*, 109
  fossil, 99*t*, 670, 670*f*, 671
Euherbstia, 69*t*, 94-95, 126, 131, 235-240, 240*k*, **267**, 268-269, 316-317
Euherbstia excellens, 267, 267*f*
Euherbstiinae (Subfamily), 80-81
Euherbstiini (Tribe), 239
Euhesma, 85, 224*k*, **226-227**, 227*k*, 231-233
Euhesma (Euhesma s. str.), 68*t*, **227**, 227*k*
Euhesma (Parahesma), 68*t*, **227**, 227*k*
Euhesma altitudinis, 226
Euhesma australis, 226
Euhesma crabronica, 226
Euhesma dolichocephala, 226
Euhesma fasciatella, 226, 230
Euhesma flavocuneata, 221*f*
Euhesma goodeniae, 221*f*, 226
Euhesma hemichlora, 226
Euhesma hemixantha, 221-222, 226
Euhesma hyphesmoides, 222-224, 226-227, 232-233
Euhesma maculifera, 226
Euhesma malaris, 226
Euhesma neglectula, 226
Euhesma palpalis, 226
Euhesma perditiformis, 226
Euhesma perkinsi, 226
Euhesma platyrhina, 226
Euhesma rainbowi, 226
Euhesma ridens, 226
Euhesma rufiventris, 226
Euhesma s. str., 68*t*, **227**, 227*k*
Euhesma serrata, 226
Euhesma tuberculipes, 227
Euhesma tubulifera, 226-227
Euhesma undulata, 226
Euhesma wahlenbergiae, 222, 226-227
Eulaema, 33*t*, 75*t*, 78*t*, 80*t*, 110, 112, 754, 780-781, 781*f*, 782, 782*k*, 783-784, **784**, Plate 15
Eulaema (Apeulaema), 784
Eulaema cingulata, 779*f*
Eulaema fasciata, 784
Eulaema meriana, 784
Eulaema meriana terminata, 781*f*
Eulaema nigrita, 780, 781*f*, 784
Eulaema polychroma, 778*f*, 784
Eulaema seabrai bennetti, 781*f*
Eulaenia, 784
Eulema, 784
Eulmites (Tribe), 78*t*
Eulonchopria, 116, 136, 138*k*, **142-143**, 145, 157, 164, 171
Eulonchopria (Ethalonchopria), 67*t*, **142**, 142*k*
Eulonchopria (Eulonchopria s. str.), 67*t*, **142-143**, 142*k*
Eulonchopria oaxacana, 143

Eulonchopria psaenythioides, **142-143**, 158
Eulonchopria punctatissima, 143
Eulonchopria s. str., 67*t*, **142-143**, 142*k*
Eumegachilana, 567, 569
Eumegachile, 72*t*, 551-554, 556*t*-557*t*, 561*k*-562*k*, 570, **573**, 579, 582-583
Eumegachile (Grosapis), 575
Eumegachile (Schrottkyapis), 583
Eumelissa, 74*t*, 768*k*, **769**
Eumelissodes, 74*t*, 730*k*-731*k*, 731, **731**
Eumorpha, 783
Eunomia, 335
Eunomia marginipennis, 335
Eunomioides, 346
Eupalaeorhiza, 68*t*, 216*k*, **217**
Eupalaeorhiza papuana, 217
Eupavlovskia, 75*t*, 770, 773, **773-774**, 773*k*
Eupetersia, 32*t*, 352-354, 361*k*, **364-365**, 364*k*, 382, 389
Eupetersia (Eupetersia s. str.), 70*t*, **364-365**, 364*k*
Eupetersia (Nesoeupetersia), 70*t*, 364*k*, **365**
Eupetersia neavei, 364
Eupetersia s. str., 70*t*, **364-365**, 364*k*
Eupetersia seyrigi, 364
Euphileremus, 73*t*, 661, **662**, 662*k*, 665-666
Euplusia, 783
Euprosopellus, 67*t*, 190*f*, 194*f*, 195*k*, 197*f*, 198*t*, 203, **204**
Euprosopis, 67*t*, 195*k*, 197*f*, 198*t*, **204-205**, 205
Euprosopoides, 67*t*, 188, 193, 195*k*, 197*f*, 198*t*, **205**, 208
Euryapis, 606
Euryglossa, 68*t*, 79*t*-80*t*, 133, 152, 213, 220, 224*k*, **227**, 232-233
Euryglossa (Callohesma), 215
Euryglossa (Dermatohesma), 215
Euryglossa (Euhesma), 227
Euryglossa (Euryglossa s. str.), 226
Euryglossa (Euryglossimorpha), 227
Euryglossa (Euryglossina), 229
Euryglossa (Parahesma), 227
Euryglossa (Xenohesma), 234
Euryglossa albocuneata, 213
Euryglossa calliopsiformis, 215
Euryglossa cupreochalybea, 227
Euryglossa flavicauda, 234
Euryglossa furcifera, 233
Euryglossa halictiformis, 159
Euryglossa laevigata, 223*f*, 227, 232
Euryglossa limata, 227
Euryglossa nigra, 227
Euryglossa paupercula, 229
Euryglossa s. str., 226
Euryglossa semipurpurea, 229
Euryglossa semirufa, 207
Euryglossa subsericea, 125*f*, 220*f*-221*f*
Euryglossa tuberculipes, 227
Euryglossa tubulifera, 227
Euryglossa wahlenbergiae, 227
Euryglossella, 68*t*, 93, 228, **228-229**, 228*k*, 229-230

Euryglossella minima, 228
Euryglossidia, 67t, 136, 139, 144, 147t, 149k, 150, **152-153**, 156
Euryglossidia rectangulata, 152
Euryglossimorpha, 227
Euryglossimorpha abnormis, 215
Euryglossina, 43, 220-222, 222k, 224, **227-230**, 228k, 231, 405
Euryglossina (Euryglossella), 68t, 93, 228, **228-229**, 228k, 229-230
Euryglossina (Euryglossina s. str.), 68t, 220, 223f, 227-228, 228f, 228k, 229, **229**, 230
Euryglossina (Microdontura), 68t, 222, 227, 228f, 228k, **229**
Euryglossina (Pachyprosopina), 68t, 228, 228k, **229**
Euryglossina (Quasihesma), 68t, 228-229, 228f, 228k, 229, **229-230**
Euryglossina aurantia, 227
Euryglossina chalcosoma, 230
Euryglossina gigantica, 230
Euryglossina hypochroma, 223f, 231
Euryglossina incompleta, 228-229
Euryglossina mellea, 227, 228f, **229**
Euryglossina moonbiensis, 228f
Euryglossina narifera, 231
Euryglossina nothula, 228f
Euryglossina paupercula, 228-229
Euryglossina proserpinensis, 229
Euryglossina pulchra, 220
Euryglossina s. str., 68t, 220, 223f, 227-228, 228f, 228k, 229, **229**, 230
Euryglossina sulphurella, 215
Euryglossinae (Subfamily), 18, 43, 45, 52, 61-62, 68t, 80t, 85, 88f, 90, 92-93, 114, 116, 128k, 133-134, 135k, 136, 142, 171, **220-234**, 221f-222f, 222k-224k, 223f, 228f, 237, 345, 418
 biogeography of, 107t, 109
Euryglossula, 61, 68t, 220-222, 222k, 228, **230**
Euryglossula chalcosoma, 222f-223f
Euryglossula fultoni, 231
Eurymella, 573-574, 581
Eurypariella, 71t, 462, 463k, **469**
Eurytis, 767
Eurytis funereus, 767
Euryvalvus, 289
Eusphecogastra, 68t, 215k, **217**
Eusynhalonia, 725
Eutechnia, 705
Euthosmia, 71t, 453, 476, 479k, **481**, 483-484
Euthyglossa, 676
Euthyglossa fasciata, 676
Eutricharaea, 72t, 554f-555f, 557t, 559k, 561k-563k, 565k-566k, **573-575**, 577, 579-581, 584

Eutrypetes, 460
Euxylocopa, 610
Eversmannibombus, 75t
Eversmannibombus (in *Agribombus*), 798
Eversmannibombus (in *Bombus*), 789, 794k, 796k, **798**
Evodia, 167
Evylaeus, 14, 32, 35f, 38f, 40f-41f, 66, 70t, 120, 348, 355f, 357f, 371, 371f, 372-373, 373k-374k, 375-376, **378-379**, 380, 382, Plate 16
Exaerete, 33t, 75t, 79t-80t, 780-782, 782k, **784**
Exaerete frontalis, 779f
Exaerete smaragdina, 778f-780f, 784, 787f
Exallocentris, 74t, 753-754, 755k-756k, **756-757**, 757
Exanthidium, 72t, 493, 526, **526-527**, 526k
Exanthocopa, 71t, 462-464, 464k, **469-470**, 472
Excolletes, 67t, 133, 144, 145f, 147t, 149k, **153**
Exilnobombus, 798
Exilobombus, 75t, 791, 796k, **798**
Exomalopsina (Subtribe), 681
Exomalopsinae (Subfamily), 81
Exomalopsini (Tribe), 35, 46, 73t, 80t, 81, 430, 437, 591k, 639-640, 642, 645, 667, 668f, 671, 674, **680-684**, 680f, 681k, 682f, 685, 687-688, 688f, 700
 biogeography of, 108t, 109
Exomalopsis, 28, 53, 79t-80t, 81, 81t, 640-641, 645, 667, 668f, 680-681, 680f, 681, 681k, 682, **682-684**, 683f, 683k, 684
Exomalopsis (Anthophorisca), 681
Exomalopsis (Diomalopsis), 73t, **683**, 683f, 683k
Exomalopsis (Exomalopsis s. str.), 73t, 680f, 683, 683f, 683k, **684**
Exomalopsis (Pachycerapis), 682
Exomalopsis (Panomalopsis), 681
Exomalopsis (Phanomalopsis), 73t, 683f, 683k, **684**
Exomalopsis (Stilbomalopsis), 73t, 683k, **684**
Exomalopsis analis, 684
Exomalopsis arcuata, 683
Exomalopsis aureosericea, 683f
Exomalopsis auropilosa, 683, 683f, 684
Exomalopsis bicellularis, 683, 683f
Exomalopsis bruesi, 683
Exomalopsis caerulea, 690
Exomalopsis chalybaea, 692
Exomalopsis cornigera, 682
Exomalopsis diversipes, 684
Exomalopsis fulvopilosa, 684
Exomalopsis jenseni, 684
Exomalopsis linsleyi, 681
Exomalopsis mellipes, 684-685
Exomalopsis nitens, 8
Exomalopsis pyropyga, 691
Exomalopsis s. str., 73t, 680f, 683, 683f, 683k, **684**

Exomalopsis similis, 680f
Exomalopsis solani, 680f, 684, 688f
Exomalopsis zexmeniae, 680f
Exoneura, 30, 31f, 79t-80t, 114, 620-621, 624k, 628, **629-630**, 629k-630k, 631-632
Exoneura (Brevineura), 73t, 622f, **630**, 630k, 631
Exoneura (Exoneura s. str.), 73t, 622f, 629, 629f, 630k, **631**
Exoneura (Exoneuridia), 632
Exoneura (Inquilina), 73t, 622f, 623, 629, 630k, **631**
Exoneura asimillima, 630
Exoneura bicolor, 13, 621, 623, 629f, 631
Exoneura concinnula, 622f, 630
Exoneura excavata, 622f, 631
Exoneura lawsoni, 629f, 631
Exoneura libanensis, 631
Exoneura obscuripes, 622f
Exoneura s. str., 73t, 622f, 629, 629f, 630k, **631**
Exoneura subbaculifera, 622f
Exoneura variabilis, 622f
Exonerella, 73t, 624k, **631-632**, 632
Exoneurella lawsoni, 622f
Exoneurella tridentata, 624, 631
Exoneuridae (Family), 79t
Exoneuridia, 623-624, 624k, 628, **632**, 632k
Exoneuridia (Alboneuridia), 73t, **632**, 632k
Exoneuridia (Exoneuridia s. str.), 73t, 632, **632**, 632k
Exoneuridia oriola, 632
Exoneuridia s. str., 73t
Exoneurula, 627
Exoneurula zavattarii, 627-628
Exosmia, 482

Fahrhalictus, 380
Fasciata, 212
Fascista, 212
Femorilegidae, 78t
Fernaldaepsithyrus, 800
Fertonella, 573
Fervidobombus, 75t, 788, 788f, 789-790, 793, 793k, 796k-797k, **798**, 800f
Fervidobombus (Rubicundobombus), 801
Festivobombus, 75t, 792k, 796k, **798**
Fidelia, 25, 80t, 81, 81t, 108, 121, 434, **438-440**, 438k-439k
Fidelia (Fidelia s. str.), 71t, **439**, 439k
Fidelia (Fideliana), 71t, **439**, 439k, xiv
Fidelia (Fideliopsis), **439**, 439k
Fidelia (Parafidelia), 71t, 434, 439, 439k, **440**
Fidelia braunsiana, 439
Fidelia kobrowi, 439
Fidelia major, 439
Fidelia paradoxa, 439
Fidelia s. str., 71t, **439**, 439k
Fidelia ulrikei, 439
Fidelia villosa, 438f
Fideliana, 71t, **439**, 439k, xiv

Fideliidae (Family), 89, 434, 436
Fideliinae (Subfamily), 21, 24, 71t, 80t, 81, 91, 434, 435k, **436**, 436f, 436k, 437-438, 441
Fideliini (Tribe), 61, 71t, 89, 96-97, 101, 122k, 126, 127k, 434-435, 435f, 436, 436k, 437, **438-440**, 438f, 438k, 510
 biogeography of, 107t, 108-109
Fideliopsis, **439**, 439k
Filiglossa, 67t, 85, 86f, 145f, 147t, 149k, **153**
Filiglossa filamentosa, 153
Fiorentinia, 690
Flavipanurgus, 69t, 272, 285, **288-289**, 288f, 288k, 289
Flavodufourea, 69t, 325, **329**, 329k
Flavomeliturgula, 69t, **291-292**, 291k
Flavomeliturgula tapana, 270f
Florentina, 690
Florilegus, 708-709, 713k, 715k, 719k, **726-727**, 726k, 729-730
Florilegus (Euflorilegus), 74t, 712f, **726**, 726k
Florilegus (Florilegus s. str.), 74t, **726**, 726k
Florilegus (Floriraptor), 74t, 713, **726-727**, 726k, 729
Florilegus barbiellinii, 712f
Florilegus condignus, 715f-716f, 726
Florilegus melectoides, 727
Florilegus riparius, 726
Florilegus s. str., 74t, **726**, 726k
Floriraptor, 74t, 713, **726-727**, 726k, 729
Foersterapis, 175
Formicapis, 71t, 450, 462-463, 464k-466k, 470, **470**, 472, 474
Formicapis clypeata, 470
fossil bees, 98, 98t-99t, 100-101, 353, 393, 448, 670-671, 781, 783, 790, 809, 812, 819, 831, 895, Plate 17-Plate 20
Foveosmia, 71t, 434f, 456f, **456**, **457**, 458
Fraternobombus, 75t, 791f, 792k, 795k, **798**
Friesea, 308
Friesea brasiliensis, 308
Friesella, 822
Frieseomelitta, 75t, 804, 807, 820f, 823-825, **825-826**, 825k, 826-828
Friesina, 282
Friesina carinulata, 282
Fumandrena, 68t, 252k, 254k, **261**
Funebribombus, 75t, 793k, 795k, **799**
Furcosmia, 468
Fuscandrena, 68t, 251, **261**

Gaesischia, 708, 713k-715k, 716, 716k-719k, 721, 726, **727-728**, 727k, 733-735, 895
Gaesischia (Agaesischia), 728

Gaesischia (Dasyhalonia), 74*t*, 727, 727*k*, **728**
Gaesischia (Gaesischia s. str.), 74*t*, 727, 727*k*, **728**
Gaesischia (Gaesischiana), 74*t*, 707*f*, 717*f*, 727, 727*k*, **728**
Gaesischia (Gaesischiopsis), 74*t*, 717*f*, 727, 727*k*, **728**
Gaesischia (Pachyhalonia), 74*t*, 727, 727*k*, **728**
Gaesischia (Prodasyhalonia), 74*t*, 727, 727*k*, **728**
Gaesischia exul, 707*f*, 713*f*, 717*f*, 728
Gaesischia flavoclypeata, 717*f*, 728
Gaesischia mexicana, 728
Gaesischia patellicornis, 727
Gaesischia s. str., 74*t*, 727, 727*k*, **728**
Gaesischiana, 74*t*, 707*f*, 717*f*, 727, 727*k*, **728**
Gaesischiopsis, 74*t*, 717*f*, 727, 727*k*, **728**
Gaesochira, 74*t*, 708, 713*k*, 716*k*, **729**, 734
Gaesochira complanata, 729
Gaesochira obscurior, 729
Gasparinahla, 69*t*, 291*k*, **292**
Gasparinahla megapalpae, 292
Gastrilegides (Family), 78*t*
Gastrohalictina (Tribe), 320, 354
Gastrohalictus, 377
Gastropsis, 131
Gastropsis victoriae, 131
Geandrena, 261
Geissandrena, 68*t*, 245*k*, 249*k*, **261**
Geniotrigona, 826
Genyandrena, 68*t*, 246*k*, 249*k*, **262**
Geodiscelis, 67*t*, 180, 181*k*, **185**
Geodiscelis megacephala, 185
Geoperdita, 303
Georgealictus, 382
Geotrigona, 75*t*, 803, 810-812, 821, 825*k*, 826, **826**, 829
Gephyrohylaeus, 67*t*, 192-194, 194*f*, 195*k*, 198*t*, 204, **205**, 206
Glaesosmia, fossil, 98*t*, 448
Glaesosmia genalis, fossil, 98*t*
Glazunovia, 74*t*, **738**, 738*k*
Glossadufourea, 326
Glossalictus, 70*t*, 371*t*, 373*f*, 374*k*, **380**
Glossamegilla, 743
Glossodialictus, 70*t*, 354, 361*k*, **365**
Glossodialictus wittei, 365
Glossopasiphae, 60*f*, 67*t*, 147*t*, 148*k*, **153**, 160, 163, 164*f*
Glossoperdita, 69*t*, 300*k*, 302, **303**
Glossoperdita pelargoides, 303
Glossosmia, 468
Glossura, 778, 783
Glossurella, 783
Glossurocolletes, 67*t*, 93, 139*k*, **143**, 145
Glossurocolletes bilobatus, 143, 143*f*
Glossurocolletes xenoceratus, 143, 143*f*, 145*f*
Glossuropoda, 783

Glyphandrena, 265
Glyptapina (Subtribe), 448
Glyptapis, fossil, 98*t*, 448
Glyptapis mirabilis, fossil, 98*t*, Plate 18
Glyptobasia, 399*f*, 404-405
Glyptobasis, 404-405
Glyptochlora, 405
Glyptocoelioxys, 72*t*, 546*k*, **548-549**
Gnathanthidium, 72*t*, 491*t*, 493, 493*k*, 495*k*, **517**, 527
Gnathanthidium (= Michanthidium), 520
Gnathanthidium prionognathum, 517
Gnathanthidium sakagamii, 520
Gnathias, 644
Gnathocera, 582
Gnathocera cephalica, 582
Gnathodon, 570
Gnathopasites, 657, 657*t*
Gnathopasites (Micropasites), 657
Gnathoprosopis, 68*t*, 194*f*, 196*k*, 198*t*, 204, **205**, 213
Gnathoprosopis (Sphaerhylaeus), 213
Gnathoprosopis globulifera, 213
Gnathoprosopoides, 68*t*, 188, 196, 196*k*, 198*t*, **205-206**, 213
Gnathosmia, 481
Gnathoxylocopa, 73*t*, 602*k*-603*k*, **605-606**
Gnathylaeus, 68*t*, 194, 198*t*, **206**
Gnathylaeus williamsi, 206
Goeletapis, 69*t*, 322, 324*k*, 325, **327-328**
Goeletapis peruensis, 327-328
Gonandrena, 68*t*, 245*k*, 249*k*, **262**, 264
Gongyloprosopis, 68*t*, 189*k*, 197*k*, 198*t*, 199*k*, **206**
Goniocolletes, 67*t*, 139, 144, 146*f*, 147*t*, 149*k*-150*k*, **153-154**, 156
Goniocolletes morsus, 153
Graecandrena, 68*t*, 252*k*, 254*k*, **262**
Grafanthidium, **514**
Grafanthidium amazonense, **514**
Grafella, 577
Greeleyella, 287
Greeleyella beardsleyi, 287
Gressittapis, 68*t*, 215*k*, **217**
Gronoceras, 72*t*, 552, 555, 557*t*, 558*k*, 560*k*, 563*k*-564*k*, 575, 575*f*, 580, 586
Gronoceras (Digronoceras), 575
Gronoceras wellmani, 575
Grosapis, 72*t*, 556*t*-557*t*, 558*k*, 560*k*, **575**
Gulanthidium, 71*t*, 506*k*, **507**
Gundlachia, 757
Gymnandrena, 263
Gymnus, 532
Gyrodroma (= Chelostoma), 457
Gyrodroma (= Stelis), 532
Gyrodroma ornatula, 533
Gyrodromella, 71*t*, 456, 456*k*, **457-458**

Habralictellus, 377-378
Habralictus, 345, 354, 356-357, 359*k*-360*k*, 363, **365-366**, 365*k*, 382-383, 418
Habralictus (Habralictus s. str.), 70*t*, 365*k*, **366**
Habralictus (Zikaniella), 70*t*, 365*k*, **366**
Habralictus crassiceps, 366
Habralictus flavopictus, 366
Habralictus s. str., 70*t*, 365*k*, **366**
Habralictus trinax, 355*f*
Habromelissa, 68*t*, 256*k*, 258*k*, **262**
Habrophora, 750-751
Habrophora ezonata, 750-751
Habrophorula, 74*t*, 743, 743*k*, 747*k*, **750**
Habropoda, 20, 74*t*, 742, 743*k*, 750, **750-751**, 770, 800
Habropoda depressa, 9
Habropoda impatiens, 750
Habropoda miserabilis, 745*f*, 751
Habropoda montezumia, 750
Habropoda nubilipennis, 750
Habropoda pallida, 745*f*
Hackeriapis, 72*t*, 109, 553, 557*t*, 565*k*, 566, 566*k*, 567, 570, 573, **576**, 584
Hadrorhiza, 68*t*, 216*k*, **217**
Haetosmia, 71*t*, 449*t*, 450*k*-451*k*, **458**, 462, 469
Halictanthrena, 67*t*, 146*f*, 147*k*, 147*t*, **154**
Halictanthrena malpighiacearum, 154
Halicti genuini, 354-355
Halicti hexagoni, 393
Halicti intermedii, 354-355
Halictidae (Family), 20, 25, 28, 31*t*-32*t*, 34-35, 45, 48, 52, 65*t*, 69*t*-70*t*, 79*t*, 80, 80*t*, 91, 93-95, 101, 103, 110, 112, 116-117, 122*f*, 123*k*, 124*f*-125*f*, 126, 127*k*-128*k*, 141, 157, 169, 236-237, **319-321**, 319*f*, 321*k*, 322-324, 345, 389, 418, 587*f*, 645, 652, 663, 698, Plate 4-Plate 5, Plate 16
  biogeography of, 107*t*
  fossil, Plate 17
  nest architecture of, 25, 25*f*
  phylogeny, 320, 321*f*
halictids. See Halictidae (Family)
Halictillus, 70*t*, 393, 398*k*, 400*k*, 406, **407**, 410
Halictina (Subtribe), 320, 354
Halictina (Tribe), 78*t*
Halictinae (Subfamily), 8-9, 12-15, 21, 23, 31, 31*t*-32*t*, 37, 43, 45, 48, 70*t*, 79, 79*t*-80*t*, 81, 81*t*, 83, 85, 101, 128*k*, 180, 271, 290, 296-297, 319-320, 321*k*, 322-323, 332-333, 345, **348-353**, 349*f*, 352*f*, 353*k*, 383, 389, 393, 407, 784
  fossil, 98*t*-99*t*
Halictini (Tribe), 18, 20, 24*f*, 28, 30-37, 31*t*-32*t*, 34,

35*f*, 36, 38*f*, 40*f*, 43, 70*t*, 80*t*, 81, 101, 105, 112, 116, 120, 320, 320*f*, 332, 345, 350, 352, 352*f*, 353, 353*k*, **354-392**, 354*f*-356*f*, 359*f*-360*f*, 362*f*, 390*f*, 393, 405, 407, 410
  biogeography of, 107*t*
  of Eastern Hemisphere, 361*k*-362*k*
  fossil, Plate 17
  of Western Hemisphere, 357*k*-361*k*
Halictoides, 77, 78*t*-79*t*, **326-327**
Halictoides (Amblyapis), 420
Halictoides (Cephalictoides), 326
Halictoides (Epihalictoides), 326
Halictoides (Parahalictoides), 326
Halictoides campanulae, 326
Halictoides dentiventris, 326
Halictoides ilicifoliae, 420
Halictoides paradoxus, 326
Halictoides ruficaudus, 328
Halictoidini (Subfamily), 80*t*
Halictomorpha, 377
Halictomorpha phaedra, 377
Halictonomia, 69*t*, 332, 334*k*, **335**
Halictonomia decemmaculatus, 335
Halictonomia minuta, 335
Halictus, 14, 31*t*, 42, 78, 78*t*-81*t*, 136, 167, **354-355**, 355*f*, 356-357, 356*t*, 357*f*, 358*k*-359*k*, 361*k*, 365*f*, **366-370**, 366*k*-367*k*, 371, 384, 393, 419, 645
Halictus (Acalcaripes), 368
Halictus (Argalictus), 70*t*, 365*f*, **367**, 367*k*, 368
Halictus (Calchalictus), 379
Halictus (Ceylalictus), 346
Halictus (Conanthalictus), 325
Halictus (Fahrhalictus), 380
Halictus (Gastrohalictus), 377
Halictus (Halictus s. str.), 70*t*, 352*f*, 365*f*, 366, **367-368**, 367*k*, 369
Halictus (Hexataenites), 70*t*, 367*k*, **368**
Halictus (Homalictus), 370
Halictus (Indohalictus), 370-371
Halictus (Inhalictus), 379
Halictus (Lampralictus), 70*t*, **368**
Halictus (Lampralicutus), 367*k*
Halictus (Leuchalictus), 380
Halictus (Lucasellus), 380
Halictus (Lucasiellus), 380
Halictus (Lucasius), 380
Halictus (Marghalictus), 377
Halictus (Microhalictus), 377
Halictus (Monilapis), 70*t*, 365*f*, 367*k*, **368**
Halictus (Nealictus), 70*t*, 367*k*, **368**
Halictus (Nesohalictus), 376
Halictus (Odontalictus), 70*t*, 365*f*, 367*k*, **368**
Halictus (Oxyhalictus), 376
Halictus (Pachyceble), 70*t*, 366, 367*k*, **368**, 369
Halictus (Pachyhalictus), 386
Halictus (Pallhalictus), 380

Halictus (Paraseladonia), 32*t*, 70*t*, 352, 361, 361*k*, 366, 367*k*, **368-369**
Halictus (Patellapis), 386
Halictus (Pauphalictus), 377
Halictus (Platyhalictus), 70*t*, 367*k*, **369**
Halictus (Prohalictus), 369
Halictus (Protohalictus), 70*t*, 365*f*, 367*k*, 368, **369**
Halictus (Puncthalictus), 377
Halictus (Pyghalictus), 377
Halictus (Ramalictus), 70*t*, 366*k*, **369**
Halictus (Rostrohalictus), 377-378
Halictus (Seladonia), 70*t*, 365*f*, 366, 367*k*, 368, **369**, 370
Halictus (Smeathhalictus), 377
Halictus (Thrincohalictus), 389
Halictus (Tytthalictus), 70*t*, 367*k*, **369**
Halictus (Vestitohalictus), 70*t*, 366, 367*k*, 369, **369-370**, 372
Halictus acuiferus, 376
Halictus alaris, 411
Halictus albofasciatus, 386
Halictus albomaculatus, 167
Halictus anomalus, 377
Halictus arcuatus, 378
Halictus auratus, 377
Halictus bellulus, 377
Halictus buccinus, 370
Halictus callopis, 408
Halictus cephalicus, 377
Halictus cephalotes, 377
Halictus chalybaeus, 368-369
Halictus chilensis, 406
Halictus chlerogas, 405
Halictus chloris, 406
Halictus clavipes, 380
Halictus cognatus, 376
Halictus conanthi, 325
Halictus confusus, 365*f*, 368
Halictus coriaceus, 380
Halictus cosmetor, 400
Halictus costulatus, 380
Halictus cressoni, 377
Halictus decemmaculatus, 335
Halictus divaricatus, 388
Halictus dybowskii, 375
Halictus dynastes, 380
Halictus etheridgei, 380
Halictus extraordinarius, 364
Halictus fahringeri, 380
Halictus farinosus, 55*f*, 84*f*, 91*f*, 96*f*, 122*f*, 368, Plate 4
Halictus gayi, 407, 595
Halictus hedini, 367
Halictus hesperus, 14*f*
Halictus horni, 346
Halictus inflaticeps, 411
Halictus insignis, 409
Halictus joffrei, 384
Halictus lasureus, 364
Halictus latesellatus, 381
Halictus latisignatus, 369
Halictus ligatus, 24*f*, 348, 365*f*, 368
Halictus longirostris, 377
Halictus lorentzi, 371
Halictus macrognathus, 391
Halictus maculatus, 30
Halictus malachurinus, 386
Halictus mediocris, 377
Halictus merescens, 386

Halictus micans, 377
Halictus minor, 369
Halictus modernus, 368
Halictus multiplex, 404
Halictus mutabilis, 388
Halictus nigromarginatus, 410
Halictus niveocinctulus, 369
Halictus nudatus, 376
Halictus nycteris, 378
Halictus ochrias, 408
Halictus osmioides, 377
Halictus pallens, 380
Halictus parallelus, 368
Halictus patellatus, 365*f*, 368
Halictus pauperatus, 377
Halictus pearstonensis, 385
Halictus peraustralis, 375
Halictus petrefactus, fossil, Plate 17
Halictus placidus, 369
Halictus podager, 383
Halictus prognathus, 389
Halictus proximus, 388
Halictus purus, 402
Halictus quadricinctus, 94*f*, 348, 352*f*, 365*f*, 368
Halictus repandirostris, 402
Halictus retigerus, 385
Halictus rhytis, 410
Halictus robbii, 376
Halictus rubellus, 406
Halictus rubicundus, 13, 349*f*, 351, 355*f*, 357*f*, 360*f*, 365*f*, 367, 369
Halictus rubricaudis, 376
Halictus s. str., 70*t*, 352*f*, 365*f*, 366, **367-368**, 367*k*, 369
Halictus scabiosae, 365*f*, 368
Halictus schultzei, 386
Halictus semiauratus, 403
Halictus sexcinctus, 351-352, 368
Halictus sicheli, 364
Halictus sordidus, 376
Halictus suarezensis, 384
Halictus subinclinans, 376
Halictus subopacus, 380
Halictus taclobanensis, 370
Halictus torridus, 391
Halictus vestitus, 369
Halictus virgatellus, 370
Halictus weenenicus, 385
Halterapis, 26, 73*t*, 620, 621*f*-622*f*, 623-624, 626*f*, 627*k*, **628-629**, 631*f*
Hamatothrix, 74*t*, 588, 713*k*, 715*k*, 722, 727, **729**, 733
Hamatothrix silvai, 729
Haplocoelioxys, 72*t*, 546*k*, **549**
Haplomelitta, 423, **424-425**, 424*k*
Haplomelitta (Atrosamba), 70*t*, 414, **424**, 424*k*
Haplomelitta (Haplomelitta s. str.), 70*t*, 423, **424-425**, 424*f*, 424*k*
Haplomelitta (Haplosamba), 70*t*, 424*k*, **425**
Haplomelitta (Metasamba), 70*t*, 423, 424*f*, 424*k*, **425**
Haplomelitta (Prosamba), 70*t*, 423, 424*k*, **425**
Haplomelitta atra, 424
Haplomelitta fasciata, 424*f*, 425
Haplomelitta griseonigra, 425

Haplomelitta ogilviei, 423*f*-424*f*, 425
Haplomelitta s. str., 70*t*, 423, **424-425**, 424*f*, 424*k*
Haplomelitta tridentata, 425
Haplosamba, 70*t*, 424*k*, **425**
Hauffapis, fossil, 99*t*, 831
Hauffapis scheuthlei, fossil, 99*t*, 831
Helianthus, 21
Helicosmia, 71*t*, 468, 471, 476, 477*k*-478*k*, 478, 479*k*, **481**, 482, 482*f*, 483-485, Plate 16
Heliomelissodes, 74*t*, 730*k*, **731**
Heliophila, 74*t*, 109, 667, 745*f*, 746*k*, 747, **748-749**, 749, 770, 777
Hemicoelioxys, 549
Hemicoelioxys gracillima, 549
Hemicotelles, 67*t*, **169**, 169*k*
Hemidiellum, 72*t*, 514, 514*k*, **515**
Hemieucera, 724-725
Hemihalictus, 21, 70*t*, 354*f*, 355, 371, 371*t*, 372-373, 373*k*, 376, 378, **380**
Heminomada, 644
Hemiosmia, 71*t*, 476, 477*k*, 478, 478*k*, 480, **482**
Hemiosmia (Hemiosmia s. str.), 482
Hemiosmia s. str., 482
Hemirhiza, 67*t*, 90-91, 93, 126, 132, 188, 189*k*, **191**, 191*k*
Hemirhiza melliceps, 187*f*, 190*f*, 191, 192*f*, 196*f*
Hemisia, 754, 756
Hemisia chilensis, 757
Hemisiella, 757
Hemisiini (Tribe), 80*t*, 81, 754
Henicohesma, 68*t*, **215**, 224, 224*k*
Herbstiella, 640
Herbstiella chilensis, 640
Heriades, 42, 53, 55, 77, 79*t*-80*t*, 105, 112, 116, 445-446, 448-449, 449*k*, 449*t*, 451*k*, 452, 456, **458-461**, 459*k*, 469, 472, 487-489, 502, 533
Heriades (Amboheriades), 71*t*, 458, **459**, 459*k*
Heriades (Eutrypetes), 460
Heriades (Heriades s. str.), 71*t*, 458, 459*k*, **460**
Heriades (Michenerella), 71*t*, 458, 459*k*, **460**, 487-488
Heriades (Micreriades), 467
Heriades (Neotrypetes), 71*t*, 458*f*, 459*k*, **460**
Heriades (Noteriades), 475
Heriades (Orientoheriades), 460
Heriades (Pachyheriades), 71*t*, 459*k*, **460**, 461
Heriades (Physostetha), 460
Heriades (Rhopaloheriades), 71*t*, 459*k*, **460-461**, 489
Heriades (Toxeriades), 71*t*, 459*k*, **461**
Heriades (Tyttheriades), 71*t*, 459*k*, **461**
Heriades apricula, 461
Heriades apriculus, 122*f*
Heriades aureocincta, 457
Heriades canaliculata, 459

Heriades carinata, 458*f*, 460
Heriades carinatum, 460
Heriades clavicornis, 460-461
Heriades crenulata, Plate 6
Heriades dalmatica, 460
Heriades floccifera, 458
Heriades foveolatus, 457
Heriades glaucum, 481
Heriades glutinosus, 488
Heriades mamillifera, 458
Heriades nigricornis, 457
Heriades opuntiae, 454
Heriades orientalis, 460
Heriades othonis, 460
Heriades paganensis, 460
Heriades parnesica, 467
Heriades philadelphi, 458
Heriades pogonura, 459
Heriades rapunculi, 457
Heriades rowlandi, 458
Heriades s. str., 71*t*, 458, 459*k*, **460**
Heriades sakaniensis, 490
Heriades schwarzi, 461
Heriades spiniscutis, 25, 26*f*, 460
Heriades testaceicornis, 487-488
Heriades turcomanica, 460
Heriades variolosa, 451*f*
Heriadina (Subtribe), 448, 449*t*
Heriadini (Tribe), 448
Heriadopsis, 72*t*, 442*t*, 553, 555, 556*t*, 557, 557*t*, 563*k*-564*k*, 576, **576-577**
Heriadopsis striatulus, 576
Hesperandrena, 68*t*, 244*k*, 246*k*, 248*k*, **262**
Hesperapis, 91, 95, 109, 111*f*, 414-417, 414*f*, 418, 418*k*, **419-421**, 419*k*-420*k*
Hesperapis (Amblyapis), 70*t*, **420**, 420*k*
Hesperapis (Capicola), 70*t*, 417, 417*f*, 419, 419*k*, **420**, 422
Hesperapis (Capicoloides), 70*t*, 419*k*, **420**
Hesperapis (Carinapis), 70*t*, 418, 419*k*-420*k*, **420**
Hesperapis (Disparapis), 70*t*, 419*k*-420*k*, **420**
Hesperapis (Hesperapis s. str.), 70*t*, 419*k*-420*k*, **421**
Hesperapis (Panurgomia), 70*t*, 419*k*, **421**, 659
Hesperapis (Xeralictoides), 70*t*, 419, 419*k*, **421**
Hesperapis (Zacesta), 70*t*, 419*k*-420*k*, **421**, 659
Hesperapis aliciae, 420
Hesperapis arenicola, 420
Hesperapis arida, 415*f*
Hesperapis braunsiana, 417*f*
Hesperapis carinata, 420
Hesperapis elegantula, 421
Hesperapis larreae, 659
Hesperapis laticeps, 421
Hesperapis pellucida, 413*f*
Hesperapis rufipes, 416-417, 656
Hesperapis s. str., 70*t*, 419*k*-420*k*, **421**
Hesperocolletes, 67*t*, 89, 133, 136, 139*k*, **144**, 145
Hesperocolletes douglasi, 143*f*, 144
Hesperonomada, 640
Hesperonomada melanantha, 640
Hesperoperdita, 69*t*, 300*k*, **303**
Heteranthidium, 72*t*, 509, 531,

533, 534f, 534k, **535**, 536-537, Plate 8
Heterapis, 206
Heterapis delicata, 206
Heterapis sandacanensis, 205
Heterapoides, 68t, 112, 115f, 192, 193f-194f, 195k, 198t, 204-205, **206**
Hetereucera, 74t, 723-724, **724-725**, 724k
Heterocentris, 18f, 74t, 753-755, 755k, **757**, 758
Heterocolletes, 155-156
Heterodasypoda, 418
Heteroediscelis, 182, 184
Heterohesma, 68t, 223k, 226, **230**, 232
Heteromegachile, 569-570
Heteroperdita, 69t, 300, 300k, **303**
Heterorhiza, 68t, 215, 215k, **217**, 218
Heterosarus, 69t, 94, 274-275, 275f, 278-279, 279k, **280**, 280k, 283-285, 655, 895
Heterosmia, 485
Heterostelis, 72t, 530f, 530k-531k, **531**, 532
Heterotrigona, 75t, 803-806, 807f, 812, 824-825, **826**, 826-829
Hexantheda, 67t, 111, 147t, 148k, **154**
Hexantheda missionica, 154
Hexaperdita, 69t, 300, 301k, 302f, **303**, 663
Hexataenites, 70t, 367k, **368**
Hexepeolini (Tribe), 73t, 590k, 636, **637-638**, 643, 658
biogeography of, 108t
Hexepeolus, 73t, **637-638**, 658
Hexepeolus mojavensis, 637
Hexepeolus rhodogyne, 634f, 637-638, 637f
Hexosmia, 71t, 448f, **455-456**, 455k, 476
Hirashima, 73t, 614k, **616**
Hofferia, 71t, 449, 449t, 450k, **461**, 488-489
Holandrena, 68t, 246k, 249k, 251, 253k-254k, 256k-258k, **262**
Holcomegachile, 567
Holcopasites, 43, 58, 73t, 112, 653, 653k, 654f, **655**, 657, 657t, 666
Holcopasites arizonicus, 654f-655f
Holcopasites bigibbosus, 654f
Holcopasites calliopsidis, 633f
Holcopasites heliopsis, 655f
Holcopasites illinoiensis, 634f
Holcopasitini (Tribe), 653
Holmbergeria, 67t, 146f, 147, 147t, 147t, **154**
Holmbergeria cristariae, 154
Holmbergiapis, 721
Holohesma, 232
Holonomada, 634f, 644-645
Homachthes, 663
Homalictus, 40, 92, 105, 296, 320, 332, 354, 356-357, 361, 362k, **370-371**, 376, 383-385
Homalictus (*Homalictus* s. str.), 70t, 354f, 364, **370-371**, 370k, 372

Homalictus (*Papualictus*), 70t, 370k, **371**
Homalictus (*Quasilictus*), 70t, 370k, **371**
Homalictus brevicornutus, 371
Homalictus dampieri, 354f
Homalictus eurhodopus, 370
Homalictus megalochilus, 371
Homalictus s. str., 70t, 354f, 364, **370-371**, 370k, 372
Homalictus silvestris, 369f
Homotrigona, 75t, 825k, 826, **826-827**, 827
Honanthidium, 499
honey bees, 1, 4, 7, 12, 21. See also *Apis mellifera*
Hoplandrena, 68t, 253k, 255k-258k, **262**
Hoplepicharis, 74t, 760, 760k, **761**
Hopliphora, 74t, 763, 764k, **767**
Hopliphora funerea, 765f
Hopliphora velutina, 767f
Hoplitella, 470
Hoplitella pentamera, 470
Hoplitina, 71t, 462, 465k-466k, **470**
Hoplitis, 32t, 120, 446, 449, 449f, 449t, 450k-452k, 453-455, 458, **461-474**, 476-477, 480-481, 486, 489, 533, 574, 895
of Eastern Hemisphere, 463k-465k
of Western Hemisphere, 465k-466k
Hoplitis (*Acrosmia*), 71t, 452, 462, 465k, **466**, 466k, 470, 476
Hoplitis (*Alcidamea*), 42, 71t, 448, 452, 453f, 456, 462, 464, 464k-465k, 466, **466-467**, 466k, 470-474, 502
Hoplitis (*Annosmia*), 71t, 462, 463k, **467**, 468-469, 471
Hoplitis (*Anthocopa*), 71t, 449, 451, 453, 461-464, 464k, 465, **467-468**, 468-471, 473, 475-476, 485, 490, 545
Hoplitis (*Bytinskia*), 32t, 71t, 451, 463k, 467, **468**
Hoplitis (*Calohoplitis*), 471
Hoplitis (*Chlidoplitis*), 71t, 453, 464k, **468-469**
Hoplitis (*Coloplitis*), 71t, 110, 463k, 467, **469**
Hoplitis (*Cyrtosmia*), 71t, 465, 465k-466k, **469**, 472
Hoplitis (*Dasyosmia*), 71t, 461f, 463, 465, 465k-466k, **469**
Hoplitis (*Eurypariella*), 71t, 462, 463k, **469**
Hoplitis (*Exanthocopa*), 71t, 462-464, 464k, **469-470**, 472
Hoplitis (*Formicapis*), 71t, 450, 462-463, 464k-466k, 470, **470**, 472, 474
Hoplitis (*Hoplitina*), 71t, 462, 465k-466k, **470**
Hoplitis (*Hoplitis* s. str.), 71t, 462-463, 463k, 466k, 467-469, **470**, 471-473, Plate 6
Hoplitis (*Jaxartinula*), 71t, 450, 464k-465k, **470-471**, 476

Hoplitis (*Kumobia*), 71t, 450, 451k, 463k, 467, **471**, 476
Hoplitis (*Megahoplitis*), 71t, 450k, 462, 463k, 465k, **471**
Hoplitis (*Megalosmia*), 71t, 464k-465k, **471**
Hoplitis (*Micreriades*), 472
Hoplitis (*Microhoplitis*), 71t, 463, 464k, 468, **471-472**
Hoplitis (*Monumetha*), 71t, 463, 464k-465k, 466, 466k, 467, 469, 471-472, **472**, 477
Hoplitis (*Nasutosmia*), 71t, 462, 464, 464k-465k, 470, 472, **472**, 473
Hoplitis (*Pentadentosmia*), 71t, 458, 462, 464, 464k-465k, 471, **472-473**
Hoplitis (*Penteriades*), 71t, 452, 462, 465, 465k, 466, 466k, 470, **473**, 476
Hoplitis (*Platosmia*), 71t, 464k-465k, 471, **473**
Hoplitis (*Prionohoplitis*), 71t, 464, 464k, 467, **473**
Hoplitis (*Proteriades*), 20, 71t, 448, 452, 461-463, 465, 465k, 466, 466k, 470, **473-474**, 476
Hoplitis (*Pseudosmia*), 468
Hoplitis (*Robertsonella*), 71t, 462, 465k-466k, **474**
Hoplitis acuticornis, 467
Hoplitis adunca, 463, 470, Plate 6
Hoplitis albifrons, 469, 472
Hoplitis anthocopoides, 463, 470
Hoplitis biscutellae, 461f, 463, 465, 469
Hoplitis bisulca, 464
Hoplitis brachypogon, 473
Hoplitis bunocephala, 470
Hoplitis carinata, 467
Hoplitis corniculata, 472
Hoplitis cristatula, 463, 468
Hoplitis curvipes, 464, 473
Hoplitis denudata, 469
Hoplitis erythrogastra, 468
Hoplitis fulgida, 463, 472
Hoplitis furcula, 465
Hoplitis heinrichi, 468-469
Hoplitis hypocrita, 469
Hoplitis illustris, 469
Hoplitis incanescens, 473
Hoplitis insolita, 473
Hoplitis jacintana, 473
Hoplitis karakalensis, 473
Hoplitis leucomelana, 467
Hoplitis marchali, 467
Hoplitis matheranensis, 451, 462, 464-465, 468
Hoplitis mitis, 465, 473
Hoplitis nasuta, 472
Hoplitis oxypyga, 470
Hoplitis papaveris, 463, 468
Hoplitis paroselae, 469
Hoplitis persica, 469
Hoplitis pici, 470
Hoplitis picicornis, 465
Hoplitis pilosifrons, 452, 470
Hoplitis plagiostoma, 466
Hoplitis praestans, 467
Hoplitis premordica, 469
Hoplitis producta, 453f, 467
Hoplitis pungens, 467

Hoplitis robusta, 470
Hoplitis rufopicta, 464
Hoplitis s. str., 71t, 462-463, 463k, 466k, 467-469, **470**, 471-473, Plate 6
Hoplitis serrilabris, 468
Hoplitis simplicicornis, 467
Hoplitis singularis, 464
Hoplitis sordida, 468
Hoplitis spoliata, 466, 472
Hoplitis tenuicornis, 471
Hoplitis tigrina, 471
Hoplitis tridentata, 464-465, 467, 471-472
Hoplitis tuberculata, 472
Hoplitis ursina, 468
Hoplitis xerophila, 465
Hoplitis zandeni, 472
Hoplitis zuni, 474
Hoplitocopa, 597f, 606
Hoplocolletes, 67t, 147, 147t, 148k, **154**
Hoplonomia, 69t, 333, 333f, 339, 339k, **340**, 341, 342f
Hoplonomia quadrifasciata, 340
Hoplopasites, 540
Hoploprosopis, 68t, 194, 198t, **206**, 207f, 209
Hoplosmia, 120, 442t, 449t, 450k-451k, 458, 462, **474-475**, 474f, 474k-475k, 476
Hoplosmia (*Hoplosmia* s. str.), 71t, 474f, 474k, **475**
Hoplosmia (*Odontanthocopa*), 71t, 474f, 474k-475k, **475**
Hoplosmia (*Paranthocopa*), 71t, 474k, **475**
Hoplosmia bidentata, 474f
Hoplosmia fallax, 474-475
Hoplosmia pinquis, 475
Hoplosmia s. str., 71t, 474f, 474k, **475**
Hoplosmia scutellaris, 474
Hoplosmia spinulosa, 474f
Hoplostelis, 31, 32t, 34, 491t, 496, 496k, 514, **517-518**, 517k, 531, 784
Hoplostelis (*Austrostelis*), 509, **510**
Hoplostelis (*Hoplostelis* s. str.), 31, 72t, 492, **517-518**, 517k
Hoplostelis (*Rhynostelis*), 72t, 496, 497k, 517k, **518**
Hoplostelis cornuta, 518
Hoplostelis multiplicata, 518
Hoplostelis s. str., 31, 34, 72t, 492, **517-518**, 517k
Hoploxylocopa, 606
Hortibombus, 799
Hortobombus, 799
Hylaeana, 68t, 198t, 198t, 199k, 203, **206-207**
Hylaeidae, 78t
Hylaeinae (Subfamily), 18, 25, 32t, 34, 45, 50, 52, 61-62, 67t-68t, 80t, 90-93, 110, 112, 116, 126, 128k, 132, 132f, 133-134, 135k, 136, 142, 171, 172f, 180-181, **187-219**, 187f-188f, 189k-191k, 190f, 220, 237, 905, Plate 1
Australian genera, 192f
biogeography of, 107t, 109

Hylaeoides, 214
Hylaeopsis, 68*t*, 198*k*, 198*t*, 199*k*, **207**, 212
Hylaeorhiza, 68*t*, 91, 132, 187*f*, 188, 188*f*, 189, 190*f*, 192, 194*f*, 195*k*, 197*f*, 198*t*, **207**
Hylaeosoma, 79*t*, 181, 181*f*, 181*k*, 182*f*, **183-184**, 183*f*, 184-185
Hylaeteron, 68*t*, 188, 193, 194*f*, 197*k*, 198*t*, **207**, 213
Hylaeus, 3, 10, 28, 42, 63, 77, 77*t*, 78, 78*t*-81*t*, 90, 93, 103, 105, 132-134, 169, 189, 189*k*-190*k*, 191, 191*k*, **192-214**, 194*f*, 197*f*, 201*f*, 207*f*, 211*f*, 214-215, 218, 643, 905
of Australia-New Guinea, 193*f*-197*f*, 194*k*-197*k*, 198*t*
Nearctic subgenera, 198*t*
Neotropical subgenera, 198*t*
Oriental subgenera, 198*t*
Palearctic subgenera, 198*t*, 200*k*-202*k*, 201*f*-202*f*
sub-Saharan subgenera, 198*t*, 202*k*-203*k*
of Western Hemisphere, 197*k*-200*k*
Hylaeus (Abrupta), 67*t*, 198*t*, 200*k*, 201*f*, 201*k*, **203**, 209
Hylaeus (Alfkenylaeus), 67*t*, 198*t*, 202*k*, **203**, 204, 210
Hylaeus (Analastoroides), 67*t*, 188, 190*f*, 193, 195*k*, 198*t*, **203**, 214
Hylaeus (Boreopsis), 208-209
Hylaeus (Cephalylaeus), 67*t*, 198*t*, 199*k*, **203**
Hylaeus (Cephylaeus), 67*t*, 198*t*, 199*k*-200*k*, **203**
Hylaeus (Cornylaeus), 67*t*, 198*t*, **203-204**, 203*k*
Hylaeus (Dentigera), 67*t*, 198*t*, 200*k*-201*k*, 202*f*, **204**
Hylaeus (Deranchylaeus), 67*t*, 198*t*, 203-204, 203*k*, **204**
Hylaeus (Edriohylaeus), 67*t*, 93, 188, 197*f*, 197*k*, 198*t*, 204, **204**
Hylaeus (Euprosopellus), 67*t*, 190*f*, 194*f*, 195*k*, 197*f*, 198*t*, 203, **204**
Hylaeus (Euprosopis), 67*t*, 195*k*, 197*f*, 198*t*, **204-205**, 205
Hylaeus (Euprosopoides), 67*t*, 188, 193, 195*k*, 197*f*, 198*t*, **205**, 208
Hylaeus (Gephyrohylaeus), 67*t*, 192-194, 194*f*, 195*k*, 198*t*, 204, **205**, 206
Hylaeus (Gnathoprosopis), 68*t*, 194*f*, 196*k*, 198*t*, 204, **205**, 213
Hylaeus (Gnathoprosopoides), 68*t*, 188, 196, 196*k*, 198*t*, **205-206**, 213
Hylaeus (Gnathylaeus), 68*t*, 194, 198*t*, **206**
Hylaeus (Gongyloprosopis), 68*t*, 197*k*, 198*t*, 199*k*, **206**
Hylaeus (Heterapoides), 68*t*, 112, 115*f*, 192-193, 193*f*-194*f*, 195*k*, 198*t*, 204-205, **206**

Hylaeus (Hoploprosopis), 68*t*, 194, 198*t*, **206**, 207*f*, 209-210
Hylaeus (Hylaeana), 68*t*, 198*k*, 198*t*, 199*k*, 203, **206-207**
Hylaeus (Hylaeopsis), 68*t*, 193, 198*k*, 198*t*, 199*k*, **207**, 212
Hylaeus (Hylaeorhiza), 68*t*, 91, 132, 187*f*, 188, 188*f*, 189, 190*f*, 192, 194*f*, 195*k*, 197*f*, 198*t*, **207**
Hylaeus (Hylaeteron), 68*t*, 188, 193, 194*f*, 197*k*, 198*t*, **207**, 213
Hylaeus (Hylaeus s. str.), 68*t*, 198*t*, 199*k*-200*k*, 201*f*-202*f*, 202*k*, **207-208**
Hylaeus (Imperfecta), 210
Hylaeus (Koptogaster), 68*t*, 188, 198*t*, 200*k*, 201*f*, 201*k*, **208**
Hylaeus (Laccohylaeus), 68*t*, 195*k*, 198*t*, **208**, 211
Hylaeus (Lambdopsis), 68*t*, 194, 198*t*, 200*k*, 201*f*, 201*k*, 202*f*, 202*k*, 207, **208-209**, 212
Hylaeus (Macrohylaeus), 68*t*, 188, 190*f*, 193, 195*f*, 196*k*, 198*t*, **209**
Hylaeus (Meghylaeus), 68*t*, 193, 195*f*, 196*k*, 198*t*, **209**, 214
Hylaeus (Mehelyana), 68*t*, 198*t*, 200*k*, 201*f*, 201*k*, 202*f*, 203, **209**
Hylaeus (Metylaeus), 68*t*, 198*t*, 202*k*, 206, 207*f*, **209**, 210
Hylaeus (Metziella), 68*t*, 93, 198*t*, 199*k*, **209-210**
Hylaeus (Nesohylaeus), 208
Hylaeus (Nesoprosopis), 32*t*, 34, 68*t*, 194, 198*t*, 200*k*, 202*f*, 202*k*, **210**, 212
Hylaeus (Nesylaeus), 68*t*, 194, 198*t*, **210**
Hylaeus (Noteopsis), 208-209
Hylaeus (Nothylaeus), 68*t*, 198*t*, 202*k*, 204, 206, 209, **210**, 211*f*
Hylaeus (Orohylaeus), 68*t*, **210-211**
Hylaeus (Paraprosopis), 68*t*, 194, 198*t*, 199*k*-200*k*, 201*f*, 201*k*, 202*f*, 210, **211**, 212
Hylaeus (Planihylaeus), 68*t*, 188, 195*f*, 197*k*, 198*t*, **211**, 213-214
Hylaeus (Prosopella), 68*t*, 198*t*, 199*k*, 207, 210-211, **211-212**
Hylaeus (Prosopis), 68*t*, 198*t*, 199*k*-201*k*, 202*f*, 202*k*, 210, **212**
Hylaeus (Prosopisteroides), 68*t*, 193, 196*k*, 198*t*, **212**, 213
Hylaeus (Prosopisteron), 68*t*, 188, 191, 193, 195, 197*f*, 197*k*, 198*t*, 202*k*, 203-205, 208-209, 211, **212-213**, 214
Hylaeus (Pseudhylaeus), 68*t*, 197*k*, 198*t*, 207, 211, **213**, 214

Hylaeus (Rhodohylaeus), 68*t*, 196*k*, 197*f*, 198*t*, **213**
Hylaeus (Spatulariella), 68*t*, 198*k*, 198*t*, 199*k*-200*k*, 201*f*, 201*k*, 210, **213**
Hylaeus (Sphaerhylaeus), 68*t*, 188, 195*f*, 196*k*, 198*t*, **213-214**
Hylaeus (Xenohylaeus), 68*t*, 195*k*, 197*f*, 198*t*, 213-214, **214**
Hylaeus albilabris, 615
Hylaeus albonitens, 205
Hylaeus albozebratus, 193
Hylaeus alcyoneus, 190*f*, 195*f*, 209
Hylaeus amiculus, 196
Hylaeus angustatus, 201*f*, Plate 1
Hylaeus annularis, 201*f*-202*f*
Hylaeus arnoldi, 203
Hylaeus asininus, 211
Hylaeus ater, 201*f*
Hylaeus aterrimus, 204
Hylaeus basalis, 94*f*
Hylaeus benoisti, 210-211
Hylaeus bicolorellus, 195*f*
Hylaeus bisinuatus, 193, 208
Hylaeus brachycephalus, 201, 204
Hylaeus brevicornis, 193, 201, 202*f*, 204
Hylaeus calvus, 211
Hylaeus cecidonastes, 207
Hylaeus ceniberus, 197*f*
Hylaeus chrysaspis, 194*f*
Hylaeus communis, 202*f*
Hylaeus comutus, 201*f*
Hylaeus confusus, 202*f*
Hylaeus cornutus, 203
Hylaeus crassanus, 209
Hylaeus cribratus, 207*f*
Hylaeus cruentus, 206
Hylaeus cyanophilus, 208
Hylaeus daviesiae, 185*f*
Hylaeus delicata, 115*f*
Hylaeus delicatus, 194*f*
Hylaeus douglasi, 213
Hylaeus dromedarius, 190*f*, 197*f*
Hylaeus elegans, 197*f*
Hylaeus ellipticus, 190*f*
Hylaeus episcopalis, 172*f*
Hylaeus euxanthus, 194*f*
Hylaeus extensus, 193*f*
Hylaeus fijiensis, 195*f*, 209
Hylaeus foveatus, 190*f*, 193, 203
Hylaeus friesei, 201*f*-202*f*, 209
Hylaeus globuliferus, 195*f*
Hylaeus grossus, 7
Hylaeus guamensis, 205
Hylaeus heraldicus, 211*f*
Hylaeus heteroclitus, 212
Hylaeus hurdi, 211-212
Hylaeus hyalinatus, 201*f*, 213
Hylaeus hyrcanius, 202*f*
Hylaeus ikedai, 208
Hylaeus insolitus, 212
Hylaeus interruptus, 379
Hylaeus larocai, 203
Hylaeus leptocephalus, 193, 202*f*, 208
Hylaeus lineolatus, 211
Hylaeus macilentus, 208
Hylaeus magnificus, 191
Hylaeus marginellus, 377
Hylaeus melba, 209
Hylaeus mucoreus, 370
Hylaeus namaquensis, 203

Hylaeus nanseiensis, 208-209
Hylaeus nesoprosopoides, 210
Hylaeus niger, 208
Hylaeus nippon, 210
Hylaeus nipponicus, 209
Hylaeus nubilosus, 187*f*-188*f*, 190*f*, 194*f*, 197*f*, 207
Hylaeus ofarrelli, 197*f*, 204
Hylaeus panamensis, 206-207
Hylaeus paulus, 201*f*
Hylaeus pectoralis, 193, 202*f*, 210
Hylaeus perconvergens, 197*f*
Hylaeus perhumilis, 197*f*, 213
Hylaeus philoleucus, 196
Hylaeus pictipes, 211
Hylaeus pilosulus, 202*f*
Hylaeus punctatissimus, 377
Hylaeus punctatus, 213
Hylaeus punctulatissimus, 201*f*
Hylaeus quadricornis, 206, 207*f*
Hylaeus rieki, 197*f*, 214
Hylaeus rubicola, 201*k*
Hylaeus s. str., 68*t*, 193*k*, 198*t*, 199*k*-200*k*, 201*f*-202*f*, 202*k*, **207-208**
Hylaeus sandacanensis, 205
Hylaeus sculptus, 194*f*, 205
Hylaeus semipersonatus, 195
Hylaeus semirufus, 194*f*
Hylaeus senilis, 367
Hylaeus serotinellus, 197*f*, 212
Hylaeus sinuatus, 202*f*
Hylaeus sparsus, 210
Hylaeus stevensi, 208
Hylaeus tomentosus, 368
Hylaeus tricolor, 193, 207
Hylaeus variegatus, 212, Plate 1
Hylaeus williamsi, 206
Hylaeus yapensis, 205
Hyleoides, 50, 68*t*, 132*f*, 187-189, 189*k*, 203, **214**, 218
Hyleoides concinna, 190*f*, 192*f*-193*f*, 196*f*, 214
Hyloeosoma, 183
Hymenoptera, 3, 6, 10, 13-15, 34, 43, 47, 51-52, 56, 59-61, 63, 65, 87-88, 101, 110-111, 116, 671, 787
Hypanthidiodes, 111, 491*t*, 496, 496*k*, 509, 513, 515, 517, 519-520
Hypanthidioides, **518-521**, 518*k*-519*k*
Hypanthidioides (Anthidulum), 72*t*, 495*f*-496*f*, **519**, 519*k*, 520-521
Hypanthidioides (Ctenanthidium), 72*t*, 517-518, 518*k*, **519**
Hypanthidioides (Dichanthidium), 72*t*, 510, 517-518, 518*k*, **519**
Hypanthidioides (Dicranthidium), 72*t*, **519**, 519*k*
Hypanthidioides (Hypanthidiodes s. str.), 72*t*
Hypanthidioides (Hypanthidioides s. str.), **519-520**, 519*k*, 521
Hypanthidioides (Larocanthidium), 72*t*, 518*k*, **520**
Hypanthidioides (Michanthidium), 72*t*, 517, 518*k*, **520**
Hypanthidioides (Mielkeanthidium), 72*t*, 518*k*, **520**
Hypanthidioides (Moureanthid-

ium), 72*t*, 518-519, 519*k*, 520, **520**
Hypanthidioides (Saranthidium), 72*t*, 517, 518*k*, 520, **521**
Hypanthidioides arenaria, 519
Hypanthidioides capixaba, 518
Hypanthidioides currani, 495*f*-496*f*, 502*f*, 521
Hypanthidioides exilis, 519
Hypanthidioides flavofasciata, 520
Hypanthidioides panamense, 521
Hypanthidioides s. str., 72*t*, **519-520**, 519*k*, 521
Hypanthidium, 491*t*, 497*k*, 509, 513, 518, 520, **521**, 521*k*, 895
Hypanthidium (Hypanthidium s. str.), 72*t*, **521**, 521*k*
Hypanthidium (Tylanthidium), 72*t*, **521**, 521*k*
Hypanthidium halophilum, 499
Hypanthidium s. str., 72*t*, **521**, 521*k*
Hypanthidium salemense, 515
Hypanthidium sheppardi, 512
Hypanthidium tigrinum, 516
Hypanthidium toboganum, 493*f*
Hypanthidium tuberigaster, 521
Hyperandrena, 68*t*, 252*k*, 255*k*, **262**
Hyphesma, 68*t*, 220, 222, 222*k*, 224, **230-231**
Hyphesma atromicans, 221*f*
Hypnorobombus, 801
Hypochrotaenia, 643-645
Hypochrotaenia (Alphelonomada), 644
Hypochrotaenia parvula, 643
Hypomacrotera, 69*t*, 309, 310*k*, 312, **312**, 312*f*, 655, 663
Hypomacrotera callops, 312
Hypotrigona, 75*t*, 805, 807-810, 814*k*, 816, **816**, 817, 821, 829, 906
Hypotrigona (Celetrigona), 829
Hypotrigona (Dolichotrigona), 829
Hypotrigona (Leurotrigona), 829
Hypotrigona (Trigonisca), 829
Hypotrigona braunsi, 811*f*

Icteranthidium, 72*t*, 491*k*, 494*k*, 515, **521**
Idiomelissodes, 74*t*, 709, 715*f*, 718*f*, 734*k*, **735**
Idioprosopis, 184
Idioprosopis chalcidiformis, 184
Immanthidium, 71*t*, 493, 497, 498*k*, **500**, 501, 535
Imperfecta, 204, 210
Indanthidium, 72*t*, 491*t*, 493*k*, **521-522**
Indanthidium crenulaticauda, 521-522
Indohalictus, 370-371
Inhalictus, 379
Inquilina, 30, 73*t*, 622*f*, 623, 629, 630*k*, **631**
Inquilina excavata, 31*t*
Inquilina schwarzi, 31*t*
Intercoelioxys, 547
Iomelissa, 68*t*, 243, 244*k*, 248*k*, 262, **262**
Ioxylocopa, 608
Ipomalictus, 376-377
Ischnocera, 764

Ischnomelissa, 70*t*, 395*k*, 400*k*, 405, **406**, 406*k*
Ischnomelissa zonata, 406
Isepeolini (Tribe), 33*t*, 34, 73*t*, 81, 116, 127, 587, 590, 591*k*, 667, 668*f*, 671, **672-673**, 672*f*, 672*k*-673*k*, 678, 763
biogeography of, 108*t*
Isepeolus, 73*t*, **673**, 673*k*
Isepeolus albopictus, 673
Isepeolus viperinus, 672*f*
Isomalopsis, 73*t*, 668*f*, 681, 681*k*, **682**, 682*f*
Isosmia, 71*t*, 452-453, 453*k*, **455**

Jaxartinula, 71*t*, 450, 464*k*-465*k*, **470-471**, 476
Jaxartinula malyshevi, 470

Kallobombus, 75*t*, 792*k*, 794, 795*k*, **799**
Katamegachile, 569-570
Kelita, 633*f*, 639, 639*k*, **640**, 640*k*
Kelita (Kelita s. str.), 73*t*, **640**, 640*k*
Kelita (Spinokelita), 73*t*, **640**, 640*k*
Kelita argentina, 640
Kelita s. str., 73*t*, **640**, 640*k*
Kelneriapis, fossil, 99*t*, 809, 812
Kelnermelia, fossil, 99*t*
Ketianthidium, 514
Ketianthidium zanolae, 514
Kirbya, 432
Koptobaster, 208
Koptogaster, 68*t*, 188, 198*t*, 200*f*, 201*f*, 201*k*, **208**
Koptorthosoma, 606
Koptortosoma, 73*t*, 597-598, 598*f*, 602*k*-604*k*, **606**, 608-609, Plate 10
Koptortosoma gabonica, 606
Kozlovibombus, 799
Kozlowibombus, 799
Krombeinictus nordenae (wasp), 60
Kronolictus, fossil, 99*t*
Kronolictus volacanus, fossil, 99*t*
Kumobia, 71*t*, 450, 451*k*, 463*k*, 467, **471**, 476
Kylopasiphae, 67*t*, 144, 146*f*, 147*t*, 148*k*, **155**

Laboriopsithyrus, 751, 800
Labrohalictus, 376
Laccohylaeus, 68*t*, 195*k*, 198*t*, **208**, 211
Laesibombus, 799
Laesobombus, 75*t*, 793*k*, 796*k*, **799**
Lagobata, 695-696
Lagobata diligens, 695
Lambdopsis, 68*t*, 194, 198*t*, 200*k*, 201*f*, 201*k*, 202*f*, 202*k*, 207, **208-209**, 212
Laminomada, 644
Lampralictus, 70*t*, **368**, 371
Lampralictus fossilis, 367*k*
Lamproapis, 644
Lamproapis maculipennis, 644
Lamprocolletes, 67*t*, 144, 147*t*, 149*k*-150*k*, **155**, 155-156, 159
Lamprocolletes (Cladocerapis), 155

Lamprocolletes antennatus, 151
Lamprocolletes bimaculatus, 155
Lamprocolletes bipectinatus, 151-152
Lamprocolletes cladocerus, 151
Lamprocolletes fulvescens, 156
Lamprocolletes venustus, 166
Lanthamelissa, 81*t*
Lanthanella, 690
Lanthanomelissa, 73*t*, 674, 687, 689*k*, **690**, 690*k*
Lanthanomelissa (Lanthanella), 690
Lanthanomelissa completa, 690
Lanthanomelissa discrepans, 690
Lapidariibombus, 799
Lapidariobombus, 799
Lapponicobombus, 801
Larandrena, 68*t*, 245*k*, 249*k*, 254*k*, 256*k*, **262**, 264
Largella, 72*t*, 557*f*, 561*k*-564*k*, 569, 575, **577**
Larinostelis, 32*t*, 34, 72*t*, 491*t*, 494*k*-495*k*, **522**, 532
Larinostelis scapulata, 522
Larocanthidium, 72*t*, 518*k*, 520
Larocanthidium emarginatum, 520
Larrea divaricata, 19-21
Larridae (Family) (wasps), 62-65
Larrinae (Subfamily) (wasps), 62-63
Larrini (Tribe) (wasps), 63
Lasanthidium, 511
Lasioglossum, 8, 66, 278, 320, 332, 354-355, 355*f*, 356-357, 356*t*, 357*f*, 358*k*-359*k*, 362*k*, 370, **371-382**, 371*t*, 373*f*, 383-384, 389, 392, 623, 645
of Australian region, 374*k*-375*k*
of Palearctic, Oriental, and African faunal regions, 372*k*-373*k*
of Western Hemisphere, 372*k*
Lasioglossum (Acanthalictus), 70*t*, 371*t*, 374*k*, **375**
Lasioglossum (Afrodialictus), 372, 377-378, 381
Lasioglossum (Australictus), 70*t*, 371*t*, 374*k*, **375**, 375*k*
Lasioglossum (Austrevylaeus), 70*t*, 361, 371*t*, 374*k*, **376**
Lasioglossum (Bluethgenia), 380
Lasioglossum (Callalictus), 70*t*, 371*t*, 374*k*-375*k*, **376**
Lasioglossum (Chilalictus), 9, 10*f*, 70*t*, 297, 350, 371*t*, 372, 373*f*, 374*k*, 375*f*, 375*k*, **376**, 377, 407
Lasioglossum (Ctenonomia), 70*t*, 348, 361, 371*t*, 372-373, 373*k*-374*k*, **376-377**, 377, 380-381
Lasioglossum (Dialictus), 18, 30-31, 31*t*-32*t*, 38*f*, 40*f*, 66, 70*t*, 103, 120, 332, 348-349, 351*f*, 352-354, 355*f*, 357, 358*f*, 358*k*, 371, 371*t*, 372-373, 373*k*-374*k*, 376, **376-377**, 378-383, 398, 407, 410, 587*f*

Lasioglossum (Ebmeria), 380
Lasioglossum (Eickwortia), 70*t*, 372, 373*k*, **378**
Lasioglossum (Evylaeus), 14, 32, 35*f*, 38*f*, 40*f*-41*f*, 66, 70*t*, 120, 348, 355*f*, 357*f*, 371-372, 371*t*, 373, 373*k*-374*k*, 375-376, **378-379**, 380, 382, Plate 16
Lasioglossum (Glossalictus), 70*t*, 371*t*, 373*f*, 374*k*, **380**
Lasioglossum (Hemihalictus), 21, 70*t*, 354*f*, 371, 371*t*, 372-373, 373*k*, 376, 378, **380**
Lasioglossum (Ipomalictus), 376-377
Lasioglossum (Labrohalictus), 376
Lasioglossum (Lasioglossum s. str.), 66, 70*t*, 350, 352, 354*f*, 357*f*, 371-372, 371*t*, 373*k*, 376, 379, 379*f*, **380**, 383*f*
Lasioglossum (Lophalictus), 380
Lasioglossum (Mediocralictus), 377-378
Lasioglossum (Paradialictus), 70*t*, 352, 371*t*, 374*k*, 380*f*, **381**
Lasioglossum (Parasphecodes), 70*t*, 371*t*, 373, 373*f*, 374*k*, 375, 375*k*, 380, **381**
Lasioglossum (Pseudochilalictus), 70*t*, 371*t*, 373*f*, 374*k*, **381**
Lasioglossum (Rubrihalictus), 376-377
Lasioglossum (Sellalictus), 70*t*, 371*t*, 373, 374*k*, **381-382**
Lasioglossum (Sericohalictus), 380
Lasioglossum (Sphecodogastra), 20, 70*t*, 320*f*, 348, 371*t*, 373*k*, **382**
Lasioglossum (Sudila), 70*t*, 371*t*, 373, 374*k*, 378, 381*f*, **382**
Lasioglossum acuticrista, 380
Lasioglossum aegyptiellum, 15, 352, 371
Lasioglossum albescens, 377
Lasioglossum allodalum, 348
Lasioglossum ankarantrense, 373
Lasioglossum asteris, 38*f*, 40*f*
Lasioglossum bidentatum, 381*f*
Lasioglossum breedi, 372
Lasioglossum calceatum, 124*f*, 380
Lasioglossum coriaceum, 355*f*
Lasioglossum costale, 379*f*
Lasioglossum crocoturum, 379*f*
Lasioglossum dybowskii, 375
Lasioglossum erythrurum, 376
Lasioglossum etheridgei, 373*f*, 380
Lasioglossum fulvicorne, 373, 379
Lasioglossum helichrysi, 373*f*, 376
Lasioglossum hemichalceum, 9, 10*f*, 376, 407
Lasioglossum idoneum, 375
Lasioglossum imitator, 373*f*, 381
Lasioglossum impavidum, 378
Lasioglossum leai, 373*f*
Lasioglossum leucozonium, 354*f*, 372

Lasioglossum lustrans, 21, 354f, 380
Lasioglossum malachurum, 32, 35f, 38f, 40f-41f, 351, 383f, Plate 16
Lasioglossum micante, 377
Lasioglossum microlepoides, 358f
Lasioglossum mirandum, 375f
Lasioglossum musicum, 373f
Lasioglossum mutilum, 352
Lasioglossum paralphenum, 382
Lasioglossum pauperatum, 378
Lasioglossum pavonotum, 373
Lasioglossum permetallicum, 381
Lasioglossum politum, 377
Lasioglossum polygoni, 376
Lasioglossum quebecense, 357f
Lasioglossum rhytidophorum, 351f, 587f
Lasioglossum s. str., 15, 66, 70t, 350, 352, 354f, 357f, 371-372, 371t, 373k, 376, 379, 379f, **380**, 383f
Lasioglossum saegeri, 376
Lasioglossum schubotzi, 379
Lasioglossum scitulum, 352
Lasioglossum seductum, 376
Lasioglossum sisymbrii, 357f
Lasioglossum synavei, 380f, 381
Lasioglossum texanum, 320f, 382
Lasioglossum tricingulum, 380
Lasioglossum umbripenne, 383
Lasioglossum villosulum, 349
Lasioglossum viride, 372
Lasioglossum wahlenbergiae, 373f
Lasioglossum zephyrum, 9, 373
Lasioglossum zonulum, 360f, 372
Lasius, 747-748
Lasius salviae, 747
leafcutter bee, 6f, 551-553
Leaf-cutter bees, 906
Legnanthidium, 72t, 534, 534k, **535**, 537-538
Leimelissa, 68t, 251, **262**
Leiopodus, 73t, 80t, **679**
Leiopodus abnormis, 679
Leiopodus lacertinus, 679, 679f
Leiopodus lecointei, 676
Leiopodus singularis, 678f, 679, 679f
Leiopoeodes, 306f
Leioproctus, 21, 28, 60f, 105, 112, 116, 120, 126, 133, 136, 138k, 139, 139k, 141-142, **144-161**, 161-166, 169, 171-172, 179, 237, 269, 278, 905
 of Australian region, 147t, 148k-150k
 of South America, 147k-148k, 147t
Leioproctus (Actenosigynes), 67t, 147t, **150**
Leioproctus (Albinapis), 67t, 143k, 147t, **150**
Leioproctus (Anacolletes), 155
Leioproctus (Andrenopsis), 67t, 144, 145f, 147t, 149k, **150**
Leioproctus (Baeocolletes), 67t, 145f, 147t, 148k, **150-151**
Leioproctus (Cephalocolletes), 67t, 138k, 139, 144, 147k, 147t, **151**, 157-159
Leioproctus (Ceratocolletes), 67t, 143-144, 147t, 149k-150k, **151**, 152
Leioproctus (Chilicolletes), 67t, 147t, 148k, **151**, 159
Leioproctus (Chrysocolletes), 141
Leioproctus (Cladocerapis), 67t, 147t, 149k, 150, **151-152**, 153, 156
Leioproctus (Colletellus), 67t, 147t, 149k, **152**
Leioproctus (Colletopsis), 67t, 147t, 149k, **152**
Leioproctus (Euryglossidia), 67t, 136, 139, 144, 147t, 149k, 150, **152-153**, 156
Leioproctus (Excolletes), 67t, 133, 144, 145f, 147t, 149k, **153**
Leioproctus (Filiglossa), 67t, 85, 86f, 145f, 147t, 149k, **153**
Leioproctus (Glossopasiphae), 60f, 67t, 147t, 148k, **153**, 160, 163, 164f
Leioproctus (Glossurocolletes), 143
Leioproctus (Goniocolletes), 67t, 139, 144, 146f, 147t, 149k-150k, **153-154**
Leioproctus (Halictanthrena), 67t, 146f, 147k, 147t, **154**
Leioproctus (Heterocolletes), 156
Leioproctus (Hexantheda), 67t, 111, 147t, 148k, **154**
Leioproctus (Holmbergeria), 67t, 146f, 147, 147k, 147t, **154**
Leioproctus (Hoplocolletes), 67t, 147, 147t, 148k, **154**
Leioproctus (Kylopasiphae), 67t, 144, 146f, 147t, 148k, **155**
Leioproctus (Lamprocolletes), 67t, 144, 147t, 149k-150k, 147t, **155**, 159
Leioproctus (Leioproctus s. str.), 60f, 67t, 141-142, 144-145, 145f, 147, 147k, 147t, 149k-150k, 151-153, **155-156**, 156-157, 159-160
 advenus group, 144, 151
 metallicus group, 156, 160
 platycephalus group, 144
Leioproctus (Microcolletes), 159
Leioproctus (Nesocolletes), 67t, 144, 145f, 147t, 149k, **156-157**, 160
Leioproctus (Nodocolletes), 145f
Leioproctus (Nomiocolletes), 67t, 142, 144, 147k, 147t, 151, **157**, 158, 164
Leioproctus (Odontocolletes), 67t, 143-144, 147t, 149k, **157**
Leioproctus (Perditomorpha), 60f-61f, 67t, 144, 146f, 147, 147f, 147t, 148k, 151-152, 154, **157-158**, 159-160, 640
Leioproctus (Protodiscelis), 60f, 67t, 147t, 148k, 153, **158-159**, 158f, 160
Leioproctus (Protomorpha), 143, 145f, 147t, 149, 149k, 150-151, 150k, 152-153, 157, **159**
Leioproctus (Pygopasiphae), 60f, 67t, 147t, 148k, 157, **159**

Leioproctus (Reedapis), 60f, 67t, 138k, 144, 147k, 147t, 151, 157, **159-160**
Leioproctus (Sarocolletes), 67t, 147t, 148k, 151, **160**
Leioproctus (Spinolapis), 67t, 144, 147f, 147t, 148k, 157, 158f, 159, **160**
Leioproctus (Tetraglossula), 67t, 147t, 148k, 153-154, 158, 158f, **160**, 163
Leioproctus (Torocolletes), 67t, 147k, 156, **160**
Leioproctus (Urocolletes), 67t, 111, 144, 145f, 147t, 149k, **160-161**
Leioproctus abnormis, 144, 148, 156
Leioproctus advenus, 150, 155-156, 189
Leioproctus antennatus, 151
Leioproctus anthracinus, 154
Leioproctus apicalis, 154
Leioproctus arnauellus, 157-158
Leioproctus bathycyaneus, 60f, 159-160
Leioproctus bicellularis, 157-158
Leioproctus bilobatus, 143
Leioproctus brunerii, 60f, 146f-147f, 157-158
Leioproctus caerulescens, 147f, 158f
Leioproctus calcaratus, 145f, 150
Leioproctus capillatus, 156
Leioproctus capito, 156
Leioproctus carinatus, 155
Leioproctus chrysostomus, 158
Leioproctus cinereus, 156
Leioproctus conospermi, 156
Leioproctus contrarius, 152
Leioproctus crenulatus, 141-142, 155-156
Leioproctus cyanescens, 136, 152-153
Leioproctus cyaneus, 160
Leioproctus delahozii, 151
Leioproctus dentiger, 145f
Leioproctus dolosus, 146f
Leioproctus douglasiellus, 150
Leioproctus duplex, 160
Leioproctus echinodori, 159
Leioproctus enneomera, 154
Leioproctus erithrogaster, 60f
Leioproctus eulonchopriodes, 146f, 147, 154, 158
Leioproctus excubitor, 156
Leioproctus fallax, 149
Leioproctus fiebrigi, 159
Leioproctus filamentosus, 86f, 145f
Leioproctus finkei, 153
Leioproctus flavorufus, 145f
Leioproctus fucosus, 158f
Leioproctus fulvescens, 145f
Leioproctus fulvoniger, 60f, 150
Leioproctus gracilis, 150
Leioproctus halictomimus, 159
Leioproctus herrerae, 158
Leioproctus heterodoxus, 152
Leioproctus iheringi, 154, 158
Leioproctus idiotropoptera, 905
Leioproctus imitatus, 155
Leioproctus impatellatus, 145f, 153
Leioproctus inconspicuus, 61f, 147f, 158
Leioproctus insularis, 155-156

Leioproctus laticeps, 139, 151
Leioproctus macmillani, 152, 156
Leioproctus maculatus, 156
Leioproctus malpighiacearum, 146f, 154
Leioproctus megachalcoides, 155-156
Leioproctus megachiloides, 155
Leioproctus microsomus, 153
Leioproctus missionica, 154
Leioproctus missionicus, 111
Leioproctus mourei, 158
Leioproctus mourellus, 159
Leioproctus nigrifrons, 150
Leioproctus nigriventris, 156
Leioproctus opaculus, 155
Leioproctus pacificus, 155
Leioproctus palpalis, 60f
Leioproctus platycephalus, 156
Leioproctus plaumanni, 60f, 153, 164f
Leioproctus plumosus, 155
Leioproctus pruinosus, 146f, 155
Leioproctus rhodopus, 156
Leioproctus rhodurus, 145f, 160-161
Leioproctus rubellus, 156
Leioproctus rubriventris, 146f
Leioproctus rudis, 155
Leioproctus ruficornis, 156
Leioproctus s. str., 60f, 67t, 141-142, 144-145, 145f, 147, 147k, 147t, 149k-150k, 151-153, **155-156**, 156-157, 159-160
Leioproctus semicyaneus, 133, 159-160
Leioproctus sexmaculatus, 155-156
Leioproctus simplicicrus, 142, 157
Leioproctus spathigerus, 158f, 159
Leioproctus striatulus, 152-153
Leioproctus subpunctatus, 150, 155-156
Leioproctus tarsalis, 145f, 149, 159
Leioproctus tomentosus, 134k
Leioproctus tristis, 158
Leioproctus tropicalis, 159
Leioproctus truncatulus, 156
Leioproctus tuberculatus, 155
Leioproctus unguidentatus, 145f, 156
Leioproctus velutinellus, 152
Leioproctus ventralis, 154
Leioproctus wagneri, 60f
Leioproctus worsfoldi, 155-156
Leioproctus zonatus, 158
Lepidandrena, 68t, 251k, 255k, **262-263**
Lepidocoelioxys, 547
Lepidorhopalomelissa, 337
Lepidotrigona, 75t, 812, 822, 824, 825k, 827, **827**
Leptergates, 703
Leptergatis, 703
Leptergatis halictoides, 703
Leptoglossa, 268
Leptoglossa paradoxa, 268
Leptometria, 703-704
Leptometria pereyrae, 703
Leptometria tucumana, 704
Leptometriella, 74t, 702-703, **704**, 704k

Leptophanthus, 69t, **277**, 277k
Leptorachina, 577
Leptorachis, 72t, 551, 557t, 558k-559k, 577, **577**, 581
Lestis, 73t, 596-597, 598f, 601k, 604k, **606**
Lestrimelitta, 30, 75t, 587, 803, 808, 810-811, 813k, **816-817**, 824
Lestrimelitta (Cleptotrigona), 815
Lestrimelitta cubiceps, 815
Lestrimelitta limao, 811f
Leucandrena, 68t, 243k, 245k, 248k-249k, 251k, 253k, 255k, **263**
Leuchalictus, 380
Leucobombus, 797
Leuconomia, 69t, 332, 339k, **340**
Leucosmia, 482
Leucostelis, 533
Leurotrigona, 829
Libellulapis, fossil, 99t
Libellulapis antiquorum, fossil, 99t
Lieftinckella, 606
Lioceratina, 73t, 613k, 615, **616-617**, 618
Liogastra, 762
Liogastra bicolor, 762
Liopodus, 679
Liopoeodes, 69t, 310k, **312**
Liopoeum, 69t, 309, 309k-310k, 311, **312**
Lioproctus, 155
Liosmia, 467
Liothgraphis, 549
Liothgrapis, 549
Liothyrapis, 72t, 543-544, 546f, 547k, **549**
Liotrigona, 75t, 632, 805, 808-810, 814k, 816, **817**
Liotrigona bottegoi, 817
Liotrigona mahafalya, 811f
Liotrigonopsis, fossil, 99t, 812
Liotrigonopsis rozeni, fossil, 99t
Liphanthus, 272, 272k, 273, 273k-274k, **276-278**, 276f, 276k-277k, 283, 640
Liphanthus (Leptophanthus), 69t, **277**, 277k
Liphanthus (Liphanthus s. str.), 69t, 276f, **277**, 277k
Liphanthus (Melaliphanthus), 69t, **277**, 277k
Liphanthus (Neoliphanthus), 69t, 276k, **277**, 277k
Liphanthus (Pseudoliphanthus), 69t, 276f, **277**, 277k
Liphanthus (Tricholiphanthus), 69t, 276f, **277-278**, 277k
Liphanthus (Xenoliphanthus), 69t, 276f, 276k-277k, **278**
Liphanthus atratus, 277
Liphanthus bicellularis, 277
Liphanthus leucostomus, 276f, 277
Liphanthus nitidus, 277
Liphanthus parvulus, 276f
Liphanthus rozeni, 277
Liphanthus s. str., 69t, 276f, **277**, 277k
Liphanthus sabulosus, 276f, 277
Liphanthus spiniventris, 276f

Lipotriches, 321, 333, **335-338**, 336k, 340, 645
Lipotriches (Afronomia), 69t, 335, **336**, 336k, 338
Lipotriches (Austronomia), 69t, 332-333, 332f, 333, 333f, 335, **336-337**, 336k, 338, 342f
Lipotriches (Clavinomia), 69t, 334, 334k, 336, 336k, **337**
Lipotriches (Lipotriches s. str.), 69t, 332, 333f, 334, 336, 336k, **337**, 344
Lipotriches (Macronomia), 69t, 335, 336k, **337-338**
Lipotriches (Maynenomia), 69t, 336k, **338**
Lipotriches (Melanomia), 69t, 336, 336k, **338**
Lipotriches (Nubenomia), 69t, 336, 336k, **338**
Lipotriches (Trinomia), 69t, 335, 336k, **338**
Lipotriches abdominalis, 337
Lipotriches australica, 332-333, 332f-333f, 342f
Lipotriches clavicornis, 337
Lipotriches halictella, 333f
Lipotriches maai, 333f, 337
Lipotriches notabilis, 334, 337
Lipotriches panganina, 337
Lipotriches reichardia, 334
Lipotriches s. str., 69t, 332, 333f, 334, 336, 336k, **337**, 344
Lipotriches testacea, 338
Lipotriches tuleareasis, 321
Lisotrigona, 75t, 810, 815k, **817**
Lissopedia, 692
Lithandrena, fossil, 99t
Lithandrena saxorum, fossil, 99t
Lithanthidium, fossil, 99t
Lithanthidium pertriste, fossil, 99t
Lithosmia, 468
Lithurge, 446
Lithurginae (Subfamily), 80, 80t, 444
Lithurgini (Tribe), 21, 71t, 79, 96, 434-435, 435f, 441, 442k, **444-447**, 444f-445f, 445k, 446f, 510, 538
  biogeography of, 107t
Lithurgomma, 447
Lithurgomma wagenknechti, 447
Lithurgopsis, 71t, **446**, 446k
Lithurgus, 6, 44, 78t-81t, 105, 434-435, 444-445, **445-446**, 445k-446k, 517, 576
Lithurgus (Lithurgopsis), 71t, **446**, 446k
Lithurgus (Lithurgus s. str.), 71t, 445k, **446**, 446k
Lithurgus apicalis, 444f-445f, **446**, 446f
Lithurgus atratus, 446
Lithurgus chrysurus, 446
Lithurgus huberi, 446
Lithurgus littoralis, 445f
Lithurgus pharcidontus, 446
Lithurgus rubricatus, 446
Lithurgus s. str., 71t, 445k, **446**, 446k
Lithurgus scabrosus, 446
Litocalliopsis, 69t, 307k, **314**
Litocalliopsis adesmiae, 314

Litomegachile, 72t, 557t, 558, 559k, 560f, 561k, 577, 578, 585
Liturgus, 446, 446f
Lobonomia, 341
Lomatalictus, 70t, 384, 384k, 385, **386**
Lonchoprella, 67t, 120, 138k, 145, 147f, 161, 161f, 162k, **163**
Lonchopria, 120, 133, 136, 138k, 140, 144-145, **161-164**, 162k, 166, 169, 172, 389
Lonchopria (Biglossa), 67t, 151, 158f, 161, 161f, **162-163**, 162k, 163-164
Lonchopria (Ctenosibyne), 67t, 158f, 162k, 163, **163**
Lonchopria (Lonchoprella), 67t, 120, 138k, 145, 147f, 161, 161f, 162k, **163**
Lonchopria (Lonchopria s. str.), 67t, 161, 162k, **163**, 164, 164f
Lonchopria (Porterapis), 67t, 158f, 162, 162k, 163, **163-164**
Lonchopria annectens, 147f, 161f, 163
Lonchopria bicellularis, 156
Lonchopria chalybaea, 161f, 163
Lonchopria cingulata, 158f, 163
Lonchopria fazii, 160
Lonchopria herbsti, 125f, 163
Lonchopria nivosa, 162
Lonchopria porteri, 158f, 163-164, 164f
Lonchopria robertsi, 158f, 163
Lonchopria rufipennis, 159
Lonchopria ruizii, 169
Lonchopria s. str., 67t, 161, 162k, **163**, 164, 164f
Lonchopria similis, 164f, 172f
Lonchopria thoracica, 161f, 162-163
Lonchopria zonalis, 162f
Lonchorhyncha, 67t, 138k, 145, **164**
Lonchorhyncha ecuadoria, 164, 164f
Longandrena, 68t, 251, **263**
Lophalictus, 380
Lophanthophora, 74t, 742, 744f, 745k, 747k, **749**
Lophopedia, 74t, 691f, 691k, **692**
Lophothygater, 74t, 711k, 715k, **729**
Lophothygater decorata, 729
Lophotrigona, 826
Loxoptilus, 74t, 711f, 718k, 720k, 737, **738**, 738k
Loxoptilus longifellator, 738
Loyolanthidium, 71t, **504**
Lucasellus, 380
Lucasiellus, 380
Lucasius, 380
Lutziella, 297
Lysicolletes, 152
Lysimachia, 17-18, 20

Maaiana, 73t, 602k-603k, **606**, 607
Machaeris, 388
Macrocera, 737
Macrocera analis, 739-740
Macrocera graja, 738

Macrocera mephistophelica, 722
Macrocera pruinosa, 733
Macrocera rustica, 732
Macrogalae antanosy, 906
Macrogalae maizina, 906
Macrogalea, 30, 31t, 73t, 114, 121, 593, 619-621, 623, 623k-624k, 631, **632**
Macrogalea candida, 593f, 622f, 630f, 632
Macrogalea mombasae, 623, 632
Macrogalea zanzibarica, 632
Macroglossa, 739-740
Macroglossa oribazi, 739-740
Macroglossapis, 739-740
Macrohylaeus, 68t, 188, 190f, 193, 195f, 196k, 198t, **209**
Macromegachile, 584
Macronomadopsis, 312-313, 663
Macronomia, 69t, 335, 336k, **337-338**
Macropididae (Family), 79t
Macropidinae (Subfamily), 80t, 429
Macropidini (Tribe), 429
Macropis, 17-18, 20, 28, 33t, 34, 56, 78t, 79, 79t-81t, 96, 414-416, 429, **430-431**, 430k-431k, 674-675, 695
Macropis (Macropis s. str.), 70t, 430, **431**, 431k
Macropis (Paramacropis), 70t, 422, 429-430, **431**, 431k
Macropis (Sinomacropis), 70t, 430, **431**, 431k
Macropis europaea, 86f, 413f
Macropis hedini, 415, 429, 431
Macropis omeiensis, 429
Macropis patellata, 415f
Macropis s. str., 70t, 430, **431**, 431k
Macropis ussuriana, 431
Macrotera, 9, 118, 296, **297-299**, 297k, 301f, 303-304, 376
Macrotera (Cockerellula), 69t, **297-298**, 297k
Macrotera (Macrotera s. str.), 69t, 296f, 297, 297k, **298**, 300f, 302f
Macrotera (Macroterella), 69t, 297, 297k, **298**
Macrotera (Macroteropsis), 69t, 297, 297k, 298, **298-299**, 298f-299f
Macrotera bicolor, 296f, 298
Macrotera cephalotes, 305
Macrotera mellea, 297-298
Macrotera opuntiae, 298
Macrotera portalis, 7, 297-298, 298f, 299, 299f, 301f
Macrotera s. str., 69t, 296f, 297, 297k, **298**, 300f, 302f
Macrotera texana, 14, 298, 300f, 302f
Macroterella, 69t, 297, 297k, **298**
Macroteropsis, 69t, 297, 297k, 298, **298-299**, 298f-299f
Maculonomia, 339-340
Madagalictus, 362, 384-385
Madrosoma, 179
Maiella, 606
Malanthidium, 72t, 497, 530, 530k, **532**

Malayapis, 68*t,* 251, 253*k,* **263**
Malgatina, 73*t,* 614*k,* 616, **617**
Manthidium, 71*t,* 511, **511**, 511*k*
Manuelia, 72*t,* 81, 81*t,* 592, 595, 597
Manuelia gayi, 593*f,* 595, 595*f*
Manueliini (Tribe), 72*t,* 588*k,* 592, 594*k,* **595**
  biogeography of, 107*t*
Margandrena, 68*t,* 253*k,* 256*k,* **263**
Marghalictus, 377
Martinapis, 106, 708, 717*k*-719*k,* **729-730,** 729*k*
Martinapis (*Martinapis* s. str.), 74*t,* **729-730,** 729*k,* 734
Martinapis (*Svastropsis*), 74*t,* 714*k,* 729*k,* **730**
Martinapis bipunctata, 730
Martinapis luteicornis, 713*f,* 718*f*
Martinapis s. str., 74*t,* **729-730,** 729*k,* 734
Martinella, 729
Massanthidium, 72*t,* 533, 534*k,* **535**
Mastrucatobombus, 797
Matangapis, 72*t,* 552-553, 555, 555*t,* 557*t,* 577
Maximegachile, 72*t,* 557*t,* 561*k*-564*k,* 568, **577-578,** 583
Maxschwarzia, 427
Maynenomia, 69*t,* 336*k,* **338**
Mecanthidium, 71*t,* 512-513, **513-514,** 513*k*
Mediocralictus, 377-378
Megabombus, 75*t,* 793*k,* 796*k,* **799**
Megabombus (*Exilnobombus*), 798
Megabombus digressus, 798
Megaceratina, 73*t,* 593, 612, 613*k,* 615, **617**
Megaceratina sculpturata, 617
Megachile, 9-10, 20-21, 28, 32*t*-33*t,* 35, 42, 44, 66-67, 77, 77*t*-81*t,* 105-106, 109, 112, 120, 435, 439, 444-445, 487, 517, 532-533, 543-545, 543*k,* 548-549, **551-586,** 554*f*-555*f,* 560*f,* 564*f,* Plate 7
  of Australian and Papuan regions, 565*k*-566*k*
  characters of, 556*t*-557*t*
  of Palearctic and Oriental regions, 561*k*-562*k*
  of sub-Saharan region, 562*k*-565*k*
  unplaced subgenera of, 586
  of Western Hemisphere, 558*k*-561*k*
Megachile (*Acentrina*), 578
Megachile (*Acentron*), 72*t,* 557*t,* 558*k*-559*k,* 566, 579
Megachile (*Addendella*), 584-585
Megachile (*Aethomegachile*), 906
Megachile (*Allomegachile*), 569-570
Megachile (*Amegachile*), 72*t,* 554*f*-555*f,* 557*t,* 562*k*-563*k,* 565*k,* **566-567,** 566*k,* 574, 576

Megachile (*Anodonteutricharaea*), 573
Megachile (*Archimegachile*), 581
Megachile (*Argyropile*), 72*t,* 557*t,* 558*k,* 561*k,* **567**
Megachile (*Austrochile*), 72*t,* 554*f,* 557*t,* 564*f,* 565*k,* **567**
Megachile (*Austromegachile*), 72*t,* 557*t,* 559*k*-560*k,* **567,** 582, 584
Megachile (*Berna*), 575
Megachile (*Callochile*), 566
Megachile (*Callomegachile*), 72*t,* 551-553, 555, 555*f,* 557*t,* 558*k,* 560*k*-566*k,* 564*f,* 565, **567-569,** 568*f,* 572-578, 581, 586
Megachile (*Carinula*), 568
Megachile (*Cestella*), 72*t,* 557*t,* 564*k,* **569**
Megachile (*Chalicodoma*), 25-26, 72*t,* 120, 536, 556, 557*t,* 561*k*-564*k,* **569-570,** 577
Megachile (*Chalicodomoides*), 72*t,* 557*t,* 565*k*-566*k,* 569, **570,** 577
Megachile (*Chalicodomopsis*), fossil, 98*t*
Megachile (*Chelostomoda*), 72*t,* 109, 453, 552-553, 554*f,* 555, 555*t*-556*t,* 557, 557*t,* 561*k*-562*k,* 565*k*-566*k,* 570, **570,** 576, 577
Megachile (*Chelostomoides*), 72*t,* 531, 553, 553*f,* 555, 557, 557*t,* 558*k,* 560*f,* 560*k,* 567, **570-571,** 571*f,* 575-576, 582-583, 586
Megachile (*Chrysosarus*), 72*t,* 551-552, 554, 556, 556*t*-557*t,* 558*k,* 560*k*-561*k,* **571,** 583, 585-586
Megachile (*Creightonella*), 72*t,* 105, 120, 556, 557*t,* 561*k,* 562, 562*k*-565*k,* **571-572,** 572*f,* 586
Megachile (*Cressoniella*), 72*t,* 557*t,* 559*k*-560*k,* **572,** 584
Megachile (*Cuspidella*), 72*t,* 557*t,* 563*k*-564*k,* **572**
Megachile (*Dactylomegachile*), 571, 586
Megachile (*Dasymegachile*), 72*t,* 557*t,* 558, 559*k*-560*k,* **572-573**
Megachile (*Delomegachile*), 584-585
Megachile (*Derotropis*), 578
Megachile (*Digitella*), 573-574
Megachile (*Eumegachilana*), 569
Megachile (*Eumegachile*), 72*t,* 551-554, 556*t*-557*t,* 561*k*-562*k,* 570, **573,** 579, 582-583
Megachile (*Eurymella*), 573-574
Megachile (*Eutricharaea*), 72*t,* 554*f*-555*f,* 557*t,* 559*k,* 561*k*-563*k,* 565*k*-566*k,* **573-575,** 577, 579-581, 584

Megachile (*Gronoceras*), 72*t,* 552, 555, 557*t,* 558*k,* 560*k,* 563*k*-564*k,* **575,** 575*f,* 580, 586
Megachile (*Grosapis*), 72*t,* 556*t*-557*t,* 558*k,* 560*k,* **575**
Megachile (*Hackeriapis*), 72*t,* 553, 557*t,* 565*k,* 566, 566*k,* 567, 570, 573, **576,** 584
Megachile (*Heriadopsis*), 72*t,* 442*t,* 553, 555, 556*t,* 557, 557*t,* 563*k*-564*k,* 576, **576-577**
Megachile (*Heteromegachile*), 569-570
Megachile (*Holcomegachile*), 567
Megachile (*Katamegachile*), 569-570
Megachile (*Largella*), 72*t,* 557*t,* 561*k*-564*k,* 569, 575, 577
Megachile (*Leptorachina*), 577
Megachile (*Leptorachis*), 72*t,* 551, 557*t,* 558*k*-559*k,* 577, **577,** 581
Megachile (*Litomegachile*), 72*t,* 557*t,* 558, 559*k,* 560*f,* 561*k,* **577,** 578, 585
Megachile (*Macromegachile*), 584
Megachile (*Matangapis*), 72*t,* 552-553, 555, 555*t,* 557*t,* **577**
Megachile (*Maximegachile*), 72*t,* 557*t,* 561*k*-564*k,* 568, **577-578,** 583
Megachile (*Megachile* s. str.), 72*t,* 551-552, 556*t*-557*t,* 558*k*-559*k,* 561*k*-562*k,* 573, 577, **578**
Megachile (*Megachiloides*), 72*t,* 551-552, 555*f,* 557*t,* 558, 558*k*-559*k,* 561*k,* 567, 574, **578**
Megachile (*Megella*), 72*t,* 553-555, 555*f,* 556*t*-557*t,* 561*k,* 563*k,* 564, 564*k,* **578-579**
Megachile (*Melaneutricharaea*), 573
Megachile (*Melanosarus*), 72*t,* 557*t,* 558*k*-559*k,* 578, **579**
Megachile (*Metamegachile*), 572
Megachile (*Mitchellapis*), 72*t,* 553, 555, 556*t*-557*t,* 565*k*-566*k,* 570, **579**
Megachile (*Moureapis*), 72*t,* 557*t,* 558*k*-559*k,* **579**
Megachile (*Neochalicodoma*), **586**
Megachile (*Neochelynia*), 72*t,* 557*t,* 559*k*-560*k,* 579, 584
Megachile (*Neocressoniella*), 72*t,* 555*f,* 562*k,* 574, **579-580**
Megachile (*Neoeutricharaea*), 573-574
Megachile (*Neomegachile*), 579
Megachile (*Paracella*), 72*t,* 555*f,* 557*t,* 562*k,* 564*k*-565*k,* 574, **580**
Megachile (*Parachalicodoma*), 72*t,* 555, 557*t,* 561*k*-562*k,* 569-570, **580**
Megachile (*Paramegachile*), 573

Megachile (*Paramegalochila*), 573
Megachile (*Phaenosarus*), 584-585
Megachile (*Platychile*), 566
Megachile (*Platysta*), 72*t,* 557*t,* 563*k,* 565*k,* 574, **580-581**
Megachile (*Pseudocentron*), 72*t,* 551-552, 557*t,* 558*k*-559*k,* 567, 579, **581**
Megachile (*Pseudomegachile*), 72*t,* 551, 555*f,* 557*t,* 558*k,* 560*k*-563*k,* 564*f,* 564*k,* 569, 572, 575, 577-578, **581,** 906
Megachile (*Pseudomegalochila*), 581
Megachile (*Ptilosaroides*), 72*t,* 557*t,* 559*k*-560*k,* 579, **581-582**
Megachile (*Ptilosarus*), 72*t,* 550, 557*t,* 559*k*-560*k,* 579, **582,** 582
Megachile (*Rhodomegachile*), 72*t,* 442*t,* 552-553, 557*t,* 565*k*-566*k,* 576, **582**
Megachile (*Rhyssomegachile*), 72*t,* 557*t,* 559*k*-560*k,* 567, 579, **582**
Megachile (*Sayapis*), 25, 72*t,* 439, 550-554, 555*f,* 556*t*-557*t,* 558*k,* 560*k,* 570, 573, 575, 579, **582-583,** Plate 7
Megachile (*Schizomegachile*), 72*t,* 557*t,* 565*k*-566*k,* 576, **583**
Megachile (*Schrottkyapis*), 72*t,* 551-552, 556*t*-557*t,* 558*k,* 560*k,* 570, 575, 582, **583**
Megachile (*Stellenigris*), **586**
Megachile (*Stelodides*), 72*t,* 551-553, 556*t*-557*t,* 558*k,* 560*k,* 570, 582, **583**
Megachile (*Stenomegachile*), 72*t,* 557*t,* 563*k*-564*k,* 568, 575, **583**
Megachile (*Thaumatosoma*), 72*t,* 552, 554*f,* 555, 557*t,* 565*k*-566*k,* **583-584**
Megachile (*Trichurochile*), 72*t,* 557, 557*t,* 559*k*-560*k,* **584**
Megachile (*Tylomegachile*), 72*t,* 551, 557, 557*t,* 559*k*-560*k,* **584**
Megachile (*Willinkella*), 578
Megachile (*Xanthosarus*), 72*t,* 551*f,* 555*f,* 557*t,* 559, 559*k,* 561*k*-562*k,* 580, **584-585,** 585*f*
Megachile (*Xenomegachile*), 581
Megachile (*Xeromegachile*), 578, 585
Megachile (*Zonomegachile*), 72*t,* 557*t,* 558, 560*k,* **585-586**
Megachile abdominalis, 582
Megachile addenda, 103, 555*f,* 559, 584-585, 585*f*
Megachile adeloptera, 554, 556
Megachile aethiops, 555*f,* **570**
Megachile africanibia, 575
Megachile albisecta, 553, 572
Megachile albitarsis, 566
Megachile albocincta, 581
Megachile aliceae, 574

Megachile alticola, 577
Megachile ambigua, 569
Megachile anthidioides, 579
Megachile anthracina, 555f, 580
Megachile apicalis, 30
Megachile apposita, 566
Megachile armaticeps, 571f
Megachile armatipes, 581
Megachile asiatica, 569
Megachile assumptionis, 575, 583
Megachile atrata, 105
Megachile atrella, 576
Megachile atropos, 514
Megachile auriculata, 574
Megachile azteca, 552
Megachile bertonii, 582
Megachile biseta, 568-569, 573, 575
Megachile bituberculata, 555f, 566
Megachile bombiformis, 575
Megachile bombycina, 573
Megachile breviceps, 568
Megachile brevis, 6f, 36f, 39f, 41f, 557, 577
Megachile campanulae, 558, 560f
Megachile carbonaria, 579
Megachile cariniventris, 525
Megachile centuncularis, 573, 578
Megachile cestifera, 569
Megachile chelostomoides, 583
Megachile chichimeca, 579
Megachile chilopsidis, 571f
Megachile chrysopyga, 543f, 554f
Megachile cincta, 575f
Megachile circumcincta, 584-585
Megachile cliffordi, 566
Megachile clotho, 567
Megachile cockerelli, 575
Megachile cognata, 571
Megachile combusta, 575
Megachile cornigera, 563
Megachile cristata, 467
Megachile crotalariae, 577
Megachile dawensis, 583
Megachile deanii, 582
Megachile dentipes, 583
Megachile detersa, 573
Megachile devexa, 569, 578
Megachile digiticauda, 573
Megachile discorrhina, 571
Megachile disjuncta, 569
Megachile dolichognatha, 563, 574
Megachile dolichosoma, 583
Megachile dubia, 447
Megachile duboulaii, 554f
Megachile edwardsi, 574
Megachile ericetorum, 551, 564f, 581
Megachile eurimera, 573
Megachile eurymera, 555f
Megachile euzona, 583
Megachile exilis, 571f
Megachile exsecta, 564, 579
Megachile fabricator, 579
Megachile felina, 575f
Megachile ferox, 576
Megachile fidelis, Plate 7
Megachile flavipes, 581
Megachile foliata, 566
Megachile fortis, 557, 584

Megachile frigida, 585
Megachile frontalis, 105, 572, 572f
Megachile frugalis, 575
Megachile fumipennis, 576
Megachile gemula, 585, 585f
Megachile gentilis, 577
Megachile georgica, 570
Megachile giraffa, 567
Megachile glaesaria, fossil, 98t
Megachile golbachi, 573
Megachile Group 1 *(Megachile),* 551-552, 557t
Megachile Group 2 *(Chalicodoma),* 551-553, 557t, 586
Megachile Group 3 *(Creightonella),* 551-555, 557t
Megachile guaranitica, 571
Megachile hohmanni, 573
Megachile incana, 580
Megachile indica, 570
Megachile infragilis, 578
Megachile ingenua, 585
Megachile integra, 552, 555f, 578
Megachile konowiana, 580-581
Megachile laeta, 577
Megachile lagopoda, 584-585
Megachile lanata, 551, 555f, 564f, 581, 906
Megachile lanigera, 573
Megachile larochei, 573
Megachile latimanus, 551f, 584, 585f
Megachile leachella, 573
Megachile louisae, 581
Megachile mackayensis, 576
Megachile macleayi, 576
Megachile malimbana, 555f, 578-579
Megachile manicata, 569
Megachile mariannae, 585-586
Megachile maritima, 584-585
Megachile maxillosa, 577
Megachile mcnamerae, 565
Megachile melanophaea, 114f, 555f, 585
Megachile mendica, 560f
Megachile michaelis, 563, 574
Megachile mitimia, 571
Megachile monstrosa, 583
Megachile montezuma, 567
Megachile monticola, 552, 569
Megachile montivaga, 551-552, 578
Megachile muansae, 581
Megachile mucida, 585
Megachile mystacea, 554f, 576
Megachile mystaceana, 553, 555f, 576
Megachile nasicornis, 566
Megachile neoxanthoptera, 581
Megachile nigrovittata, 576
Megachile occidentalis, 553f, 571f
Megachile oenotherae, 555f, 557t
Megachile orba, 584
Megachile parallela, 567
Megachile parsonsiae, 571
Megachile pascoensis, 578
Megachile paulista, 579
Megachile perihirta, 555f
Megachile petulans, 577
Megachile platystoma, 580
Megachile pluto, 568f, 569

Megachile policaris, 25
Megachile praetexta, 575
Megachile prosopidis, 571
Megachile pruina, 581
Megachile pseudoanthidioides, 552
Megachile pseudomonticola, 579
Megachile pugnata, 555f, 582
Megachile quadraticauda, 572
Megachile relata, 568, 584-585
Megachile relativa, 578
Megachile resinifera, 554f, 564f, 567
Megachile rhodura, 576
Megachile rotundata, 4, 7, 340, 574-575
Megachile rufimanus, 570
Megachile s. str., 72t, 551-552, 556t-557t, 558k-559k, 561k-562k, 573, 577, **578**
Megachile saulcyi, 572
Megachile sculpturalis, 551, 568-569
Megachile semierma, 574
Megachile semiluctuosa, 565, 576
Megachile semivenusta, 555f, 580
Megachile semivestita, 577
Megachile sericans, 553
Megachile simillima, 582
Megachile simplicipes, 584
Megachile sjoestedti, 566
Megachile spinotulata, 570
Megachile spissula, 554f, 570
Megachile spissula parvula, 570
Megachile steloides, 536
Megachile striatula, 577
Megachile subserricauda, 576
Megachile tapytensis, 552
Megachile thygaterella, 584
Megachile torrida, 564f, 568
Megachile tricarinata, 475
Megachile trichorhytisma, 906
Megachile uniformis, 579
Megachile ustulata, 565, 576
Megachile ustulatiformis, 566
Megachile variolosa, 460
Megachile vidua, 584
Megachile villosa, 468
Megachile willughbiella, 584
Megachile xylocopoides, 579
Megachile zambesica, 574
Megachile zapoteca, 572
Megachileoides, 536, 578
Megachilidae (Family), 20, 25, 30, 32t, 34, 37, 45, 52, 63f, 65t, 71t-72t, 77-78, 78t-79t, 80, 80t, 83, 89, 96-97, 111-112, 116-117, 121, 122f, 122k, 127, 127k-128k, **434-435**, 435f, 435k, 437, 441, 444, 449, 491-492, 510, 517, 587, 668f, 670, 753, 906, Plate 6-Plate 8, Plate 16
  biogeography of, 107t
  fossil, 101, Plate 18
Megachilidae (Subfamily), 78t
Megachilina (Tribe), 78t
Megachilinae (Subfamily), 25, 32t-33t, 34, 39f, 48, 55, 71t-72t, 79t-81t, 96-97, 110, 116-117, 121, 128k, 434-435, 435k, 436-437, **441-443**, 441k-443k,
442t, 444, 448, 491, 530, 538, 635
Megachilini (Tribe), 32t, 33, 36f, 55, 72t, 79, 79t-80t, 111-112, 117, 120, 435, 435f, 441, 442t, 443k, 510, 538, **543-586**, 543k, 906
  biogeography of, 107t
  fossil, 98t
Megachiloides, 72t, 551-552, 555f, 557t, 558, 558k-559k, 561k, 567, 574, **578**
Megachiloides (= *Trachusa*), 536
Megachiloides oenotherae, **578**
Megacilissa, 175
Megacilissa gloriosa, 318
Megacilissa luctuosa, 175
Megacilissa nigrescens, 174
Megacilissa superba, 175
Megacilissa tarsata, 176
Megadasypoda, 418
Megahoplitis, 71t, 450k, 462, 463k, 465k, **471**
Megalobombus, 799
Megalochila, 578
Megalocilissa, 175
Megaloheriades, 71t, 449, 486k, **487**
Megalopta, 322, 343, 348, 350, 388, 394, 395k, 399k, **407-408**, 408k, 410-412, Plate 5
Megalopta (*Megalopta* s. str.), 70t, 408, **408**, 408k
Megalopta (*Megaloptella*), 408
Megalopta (*Megaloptidia*), 408
Megalopta (*Noctoraptor*), 32t, 70t, 352, 407f, **408**, 408k
Megalopta bituberculata, 343
Megalopta byroni, 407f, 408
Megalopta contradicta, 408
Megalopta genalis, 393f, 396f-397f, 404f
Megalopta idalia, 408
Megalopta minuta, 409
Megalopta noctifurax, 407f
Megalopta s. str., 70t, 408, **408**, 408k
Megalopta sulciventris, 408
Megaloptella, 408
Megaloptera, 408
Megaloptidia, 70t, 394k, 398k, 408, **408**
Megaloptidia contradicta, 397f
Megaloptilla, 70t, 395k, 400k, **408**
Megaloptilla byronella, 408
Megaloptina, 70t, 409, **409**, 409k
Megaloptodes, 343
Megalosmia, 71t, 464k-465k, **471**
Megamegilla, 743
Megandrena, 129, 237, 239, 239k, 240, 240k, 263, **267-268**, 268k, 269
Megandrena (*Erythrandrena*), 69t, **268**, 268k
Megandrena (*Megandrena* s. str.), 69t, 236, **268**, 268k
Megandrena enceliae, 124f, 239f-240f, 268, 268f
Megandrena mentzeliae, 268
Megandrena s. str., 69t, 236, **268**, 268k

Meganomia, 70*t*, 414, 426, 426*k*, **427**
Meganomia binghami, 427, 427*f*
Meganomia gigas, 427, 427*f*
Meganomiidae (Family), 89, 95
Meganomiinae (Subfamily), 43, 70*t*, 95, 105, 414-415, 415*k*, 416, **426-428**, 426*f*, 426*k*-427*k*, 427*f*, 429
  biogeography of, 107*t*
Meganomioides, 70*t*, **346-347**, 346*k*
Meganthidium, 72*t*, 494, 528, 528*k*, **529**
Meganthidium (Asianthidium), 528
Meganthidium (Oxyanthidium), 528-529
Megapis, 831
Megascirtetica, 74*t*, 712*f*, 721, 721*k*, **722**, 727
Megaxylocopa, 600*k*, 607
Megella, 72*t*, 553-555, 555*f*, 556*t*-557*t*, 561*k*, 563*k*, 564, 564*k*, **578-579**
Meghylaeus, 68*t*, 193, 195*f*, 196*k*, 198*t*, **209**, 214
Megilla, 431, 747
Megilla fulvipes, 431
Megilla graminea, 410
Megilla labiata, 431
Megilla sesquicincta, 743
Megillina (Tribe), 78*t*
Megomalopsis, 684
Megommation, 393-394, 394*k*, 398*k*, 401, 403, 406, **408-409**, 409*k*, 411
Megommation (Cleptommation), 32*t*, 70*t*, 352, 407*f*, 408, **409**, 409*k*
Megommation (Megaloptilla), 408
Megommation (Megaloptina), 70*t*, 409, **409**, 409*k*
Megommation (Megommation s. str.), 70*t*, 408, **409**, 409*k*
Megommation (Stilbochlora), 70*t*, 408, **409**, 409*k*
Megommation eickworti, 409
Megommation insigne, 397*f*, 409
Megommation minutum, 407*f*, 409
Megommation s. str., 70*t*, 408, **409**, 409*k*
Mehelya, 209
Mehelyana, 68*t*, 198*t*, 200*k*, 201*f*, 201*k*, 202*f*, 203, **209**
Melacentris, 74*t*, 754-755, 755*k*, 756-757, **757**, 758-759, 759*f*, 778*f*
Melaliphanthus, 69*t*, **277**, 277*k*
Melanapis, 68*t*, 252*k*, 254*k*, **263**, 895
Melanapis violaceipennis, 263
Melandrena, 68*t*, 247*k*, 250*k*-251*k*, 253*k*, 255*k*, 257*k*-258*k*, **263**
Melanempis, 43, 73*t*, 110, 653, 661, 662*k*, **663**
Melaneutricharaea, 573-574
Melanoanthidium, 506-507
Melanobombus, 75*t*, 791*f*, 793*k*, 794, 794*k*, **799**, 801
Melanocentris, 758

Melanocoelioxys, 547
Melanomada, 73*t*, 634*f*, 639*f*, **640**, 640*k*
Melanomia, 69*t*, 336, 336*k*, 338, **338**
Melanosarus, 72*t*, 557*t*, 558*k*-559*k*, 578, **579**
Melanosmia, 71*t*, 472, 476-477, 477*k*, 478, 478*k*, 479, 479*k*, 481, **482-483**, 482*f*, 484, 486
Melanostelis, 532
Melanostelis betheli, 532
Melanthidium, 501, 506
Melanthidium carri, 506
Melea, 74*t*, 746*k*, 747, 747*k*, 748, **749**
Melecta, 77*t*, 78, 78*t*, 79, 79*t*-81*t*, 770-771, 771*k*, **773-774**, 773*k*, 775-776
Melecta (Eupavlovskia), 75*t*, 770, 773, **773-774**, 773*k*
Melecta (Melecta s. str.), 75*t*, 772*f*, 773*k*, **774**, 774-775, 775*f*, 776
Melecta (Melectomimus), 75*t*, 771, 773*k*, **774**, 776
Melecta (Melectomorpha), 776
Melecta (Nesomelecta), 776
Melecta (Paracrocisa), 75*t*, 772, 773, 773*k*, **774**, 774, 776
Melecta (Pseudomelecta), 75*t*, 773, 773*k*, **774**, 776
Melecta (Xeromelecta), 776
Melecta aegyptiaca, 773
Melecta albifrons, 765*f*, 773-774
Melecta bicolor, 767
Melecta californica, 776
Melecta chalybeia, 774
Melecta diacantha, 774
Melecta edwardsii, 774
Melecta emodi, 775
Melecta fumipennis, 772*f*
Melecta funeraria, 773
Melecta haitensis, 776
Melecta histrionica, 775
Melecta maculata, 777
Melecta megaera, 773
Melecta mucida, 773
Melecta oreina, 771
Melecta pacifica, 775*f*
Melecta s. str., 75*t*, 772*f*, 773*k*, **774**, 774-775, 775*f*, 776
Melecta thoracica, 774
Melectidae (Family), 79*t*
Melectidae (Subfamily), 78*t*
Melectini (Tribe), 8, 33-34, 33*t*, 74*t*-75*t*, 80*t*, 116, 116*f*, 127, 144, 588*k*, 589, 589*k*, 668, 668*f*, 671, 762-763, 765*f*, 768, **770-777**, 771*f*-772*f*, 775*f*
  biogeography of, 108*t*
  of Eastern Hemisphere, 771*k*-772*k*
  fossil, 99*t*
  of Western Hemisphere, 771*k*
Melectites (Tribe), 78*t*
Melectoides, 73*t*, 672*k*, **673**
Melectoides bellus, 673
Melectoides kiefferi, 672*f*
Melectoides politus, 672*f*
Melectoides senex, 673
Melectoides triseriatus, 672*f*

Melectoides tucumanus, 673
Melectomimus, 75*t*, 771, 773*k*, **774**, 776
Melectomorpha, 75*t*, 765*f*, 770, 770*f*-771*f*, 772, 775*f*, 776, **776**, 776*k*
Melikerria, 817-818
Melikertes, fossil, 99*t*, 670-671, 831
Melikertini (Tribe), fossil, 99*t*, 670-671, Plate 19
Meliphila, 705
Meliphila ipomoeae, 705
Meliphilopsis, 74*t*, 700-701, 701*k*, **704-705**
Meliphilopsis melanandra, 704-705
Meliphilopsis ochrandra, 705
Meliplebeia, 75*t*, 809*f*, 812, 812*f*, **818-819**, 818*k*, 819
Melipona, 7, 12, 20, 44, 75*t*, 77, 78*t*-81*t*, 111*f*, 807-809, 808*f*, 810, 814*k*, 817, **817-818**
Melipona (Eomelipona), 817-818
Melipona (Melikerria), 817-818
Melipona (Micheneria), 817
Melipona (Michmelia), 817
Melipona cacciae, 817
Melipona caerulea, 822
Melipona denoiti, 823
Melipona fasciata, 804*f*
Melipona favosa, 818
Melipona ferruginea, 818
Melipona fulva, 810*f*
Melipona impunctata, 820
Melipona marginata, 803, 817
Melipona prosopiformis, 820
Melipona quadrifasciata, 817-818
Melipona quadripunctata, 823
Melipona rufiventris, 787*f*, 809*f*
Melipona schrottkyi, 822
Melipona scutellaris, 817
Melipona testacea, 821
Melipona testaceicornis, 819
Meliponidae (Family), 81, 97
Meliponinae (Subfamily), 52
Meliponini (Subfamily), 81
Meliponini (Tribe), 12-13, 18-19, 21, 23, 26, 27*f*, 29-30, 43, 52, 56, 75*t*, 76, 80*t*, 81-82, 91, 97, 101-105, 111, 111*f*, 112, 117, 236-237, 538, 588*k*, 667- 671, 785, 787*f*, **803-829**, 804*f*, 807*f*, 809*f*-813*f*, 820*f*, 830-831, 906, Plate 9
  African, 814*k*-815*k*
  alternative cladograms for, 808*f*
  of Asia and Sunda Islands, 815*k*
  of Australia and New Guinea, 815*k*
  biogeography of, 108*t*, 109
  fossil, 98*t*-99*t*, 670, 670*f*, 671, Plate 20
  neotropical, 813*k*-814*k*
  recent genera of, 812-829
Meliponites (Tribe), 78*t*
Meliponorytes, fossil, 99*t*, 812
Meliponorytes succini, fossil, 99*t*
Meliponula, 669, 808-809,

811, 814*k*, **818-819**, 818*k*, 823-824
Meliponula (Axestotrigona), 75*t*, **818**, 818-819, 818*k*
Meliponula (Meliplebeia), 75*t*, 809*f*, 812, 812*f*, **818-819**, 818*k*, 819
Meliponula (Meliponula s. str.), 75*t*, 804*f*, 808, 809*f*, 810, 812*f*, 818, 818*k*, **819**, Plate 10
Meliponula beccarii, 809*f*, 812*f*, 819
Meliponula bocandei, 787*f*, 804*f*, 809*f*, 812*f*, 818-819, Plate 10
Meliponula lendliana, 819
Meliponula s. str., 75*t*, 804*f*, 808, 809*f*, 810, 812*f*, 818, 818*k*, **819**, Plate 10
Melissa, 769
Melissa decorata, 769
Melissa diabolica, 767
Melissa duckei, 764
Melissa lendliana, 766
Melissina, 738-739
Melissina viator, 738
Melissites, fossil, 99*t*, 670-671, Plate 19
Melissites trigona, fossil, 99*t*, Plate 19
Melissoda, 764
Melissoda latreillii, 764
Melissode, 732
Melissodes, 42, 80*t*, 105, 652, 709, 713*f*, 713*k*, 716*k*, 717, 717*k*, 719*k*, 721, 726-727, 729, **730-732**, 730*k*-731*k*, 733-735, Plate 11
Melissodes (Apomelissodes), 74*t*, 730*k*, **731**
Melissodes (Brachymelissodes), 735
Melissodes (Callimelissodes), 74*t*, 730*k*-731*k*, **731**
Melissodes (Ecplectica), 74*t*, 714*f*, 730*k*-731*k*, **731**
Melissodes (Eumelissodes), 74*t*, 730*k*-731*k*, 731, **731**
Melissodes (Heliomelissodes), 74*t*, 730*k*, **731**
Melissodes (Idiomelissodes), 735
Melissodes (Martinella), 729
Melissodes (Melissodes s. str.), 74*t*, 714*f*, 730*k*-731*k*, **732**
Melissodes (Psilomelissodes), 74*t*, 730*k*-731*k*, **732**
Melissodes (Tachymelissodes), 74*t*, 730*k*-731*k*, **732**
Melissodes agilis, 710*f*, 718*f*, 731
Melissodes albata, 738
Melissodes atripes, 735
Melissodes atropos, 726
Melissodes bimaculata, 708*f*
Melissodes bombiformis, 705
Melissodes callophila, 721
Melissodes compta, 734
Melissodes condigna, 726
Melissodes crassidentata, 733
Melissodes dagosa, 732
Melissodes desponsa, 715*f*, 731
Melissodes duplocincta, 735
Melissodes edwardsii, 725
Melissodes enavata, 703
Melissodes exquisita, 736
Melissodes fimbriata, 731

Melissodes fonscolombei, 732
Melissodes fulvitarsis, 725
Melissodes intorta, 732
Melissodes latreillei, 764
Melissodes leprieuri, 732
Melissodes lupina, 701*f*, 731
Melissodes luteicornis, 729
Melissodes nigrifrons, 703
Melissodes nigroaenea, 714*f*
Melissodes pallidisignata, 709
Melissodes pygmaea, 681
Melissodes s. str., 74*t*, 714*f*, 730*k*-731*k*, **732**
Melissodes stearnsi, 717
Melissodes tepaneca, 714*f*
Melissoptila, 74*t*, 648, 711*k*, 716*k*, 719*k*, 726-727, **732**, 735, 740, 895
Melissoptila (*Comeptila*), 732
Melissoptila (*Melissoptila* s. str.), 732
Melissoptila (*Ptilomelissa*), 709*f*-710*f*, 714*f*, 718*f*, 728, 732
Melissoptila bonaerensis, 714*f*, 732
Melissoptila s. str., 732
Melissoptila tandilensis, 732
Melitoma, 74*t*, 80*t*, 652, 679, 700-701, 701*k*, 704, **705**
Melitoma ameghinoi, 705
Melitoma euglossoides, 703, 705
Melitoma paraensis, 704
Melitoma taurea, 702*f*
Melitome, 705
Melitomella, 74*t*, 701, 702*k*, **705**
Melitomella grisescens, 704-705
Melitomella murihirta, 705
Melitomella schwarzi, 705
Melitomina (Subtribe), 700
Melitomini (Tribe), 81, 700
Melitribus, 131
Melitribus greavesi, 131
Melitta, 35, 77, 78*t*, 79, 79*t*-81*t*, 96, 131, 414-416, 422, 429, 429*f*, 430*k*, **431-432**, 431*k*-432*k*, 432*f*, 645
Melitta (*Dolichochile*), 71*t*, 414-415, 429*f*, 431, **432**, 432*f*, 432*k*
Melitta (*Melitta* s. str.), 71*t*, 429*f*, 431, **432**, 432*f*, 432*k*
Melitta (*Promelitta*), 422
Melitta americana, 432*f*
Melitta annularis, 208
Melitta barbilabris, 263
Melitta bimaculata, 265
Melitta californica, 432
Melitta dimidiata, 430*f*, 432
Melitta dorsata, 266
Melitta haemorrhoidalis, 429
Melitta labialis, 262
Melitta leporina, 124*f*, 413*f*, 429*f*, 432
Melitta melittoides, 429*f*, 431-432, 432*f*
Melitta minutissima, 377
Melitta minutula, 263
Melitta nigriceps, 260
Melitta ovatula, 266
Melitta s. str., 71*t*, 429*f*, 431, **432**, 432*f*, 432*k*
Melitta smeathmanella, 377
Melitta swammerdamiella, 418
Melitta tibialis, 265

Melitta tricincta, 414*f*, 432
Melitta trimmerana, 262
Melitta xanthopus, 380
Melittidae (Family), 16-18, 21, 24, 28, 34-35, 61, 65*t*, 70*t*-71*t*, 76, 80, 80*t*, 83, 87-89, 91-96, 101, 111*f*, 112, 117, 123*k*, 124*f*, 126-127, 127*k*, 129, 131, 133, 188, 236, 345, 389, **413-415**, 413*f*-415*f*, 415*k*, 418, 426, 434, 645, 674, 685, 695, 697, 905
biogeography of, 107*t*
Melittidae (Subfamily), 78*t*
Melittidia, 338
melittids. *See* Melittidae (Family)
Melittinae (Subfamily), 18, 20, 35, 70*t*-71*t*, 80*t*, 81, 81*t*, 413*f*, 414-416, 415*f*, 415*k*, 422, **429-433**, 430*f*, 430*k*
biogeography of, 107*t*
Melittini, 429
Melittoides, 68*t*-69*t*, 236-237, 240, 253*k*, 255*k*, **263**, 269*f*
Melittosmithia, 68*t*, 223*k*, **231**
Melittosmithia carinata, 223*f*
Melittoxena, 656-657
Melitturga, 69*t*, 77, 78*t*-79*t*, 80, 80*t*-81*t*, 83, 86-87, 114, 236-237, 291*k*, **292**, 316, 353, 389
Melitturga (*Australomelitturga*), 292
Melitturga (*Petrusianna*), 292, 895
Melitturga capensis, 292
Melitturga clavicornis, 124*f*, 291*f*, 293*f*
Melitturga spinosa, 292, 895
Melitturgini (Tribe), 69*t*, 80*t*, 105, 270-271, 272*k*, 290, **291-294**, 291*f*, 291*k*, 295, 895
biogeography of, 107*t*
Meliturga, 292, 655
Meliturgopsis, 751
Meliturgula, 69*t*, 270-272, 291-292, 291*k*, **292-293**, 655
Meliturgula (*Popovmeliturgula*), 292
Meliturgula arabica, 291
Meliturgula braunsi, 292, 293*f*
Meliturgula dzheddaensis, 293
Meliturgula ornata, 293
Meliturgula scriptifrons, 293
Meliturgulini (Tribe), 270
Meliwillea, 75*t*, 814*k*, **819**
Meliwillea bivea, 819
Mellinus bipunctatus, 212
Mellinus variegatus, 212
Mellitidia, 69*t*, 334*k*, **338**, 342-343, 538
Mellitidia gressitti, 333*f*, 343*f*
Mellitidia metallica, 338
Mellitidia simplicinotum, 334, 338
Mellitosphecidae (Family), 905
Mellitosphex, fossil, 905
Mellitosphex burmensis, fossil, 905
Mellturga, 292

Meloponinae (Subfamily), 79*t*
Mendacibombus, 75*t*, 76, 789, 791*k*, 794, 794*k*, **799**
Mentzelia decapetela, 16
Merilegides (Family), 78*t*
Mermiglossa, 69*t*, 271, 291*k*, **293-294**
Mermiglossa rufa, 293-294, 293*f*
Mermiglossini (Tribe), 270
Meroglossa, 68*t*, 90-93, 126, 132, 188-189, 189*k*, 191, 214, **214**
Meroglossa (*Meroglossula*), 214
Meroglossa canaliculata, 190*f*, 196*f*, 214
Meroglossa eucalypti, 214
Meroglossa impressifrons penetrata, 192*f*
Meroglossa lactifera, 218
Meroglossa sculptifrons, 191
Meroglossa torrida, 187*f*
Meroglossula, 214
Merrophites, 326
Mesanthidiellum, 71*t*, 499, 499*k*, 500, **500-501**, 502
Mesanthidiellum amoenum, 500
Mesanthidium, 71*t*, 498, 498*k*, 499-500, **501**
Mesocheira, 74*t*, 763*k*, 764, **767-768**
Mesocheira azurea, 769
Mesocheira bicolor, 765*f*-767*f*, 768, 768*f*
Mesocheira velutina, 767
M[eso]chira, 767
Mesocoelioxys, 72*t*, 547*k*
Mesonychium, 74*t*, 673, 764*k*, 766, **768**
Mesonychium coerulescens, 768
Mesonychium garleppi, 766*f*
Mesoplia, 78*t*, 80*t*, 651, 673, 764, 764*k*, 768, **768-769**, 768*k*-769*k*
Mesoplia (*Eumelissa*), 74*t*, 768*k*, **769**
Mesoplia (*Mesoplia* s. str.), 74*t*, 768*k*, **769**
Mesoplia azurea, 763*f*, 765*f*-767*f*
Mesoplia imperatrix, 768*f*
Mesoplia s. str., 74*t*, 768*k*, **769**
Mesotrichia, 73*t*, 597*f*, 598, 602*k*-604*k*, **606-607**, 608-609
Mesotrichia chiyakensis, 609
Mesotrichia torrida, 606
Mesoxaea, 69*t*, 317*f*, 317*k*, **318**
Metadioxys, 72*t*, 538, 539*k*, **541**
positions of spines and carinae on thoracic sclerites of, 539*f*
Metallinella, 25, 71*t*, 442*t*, 477*k*-478*k*, **483**, 509
Metamegachile, 572
Metapsaenythia, 69*t*, 236*f*, 274*k*, 278, 279*k*, **280-281**, 280*k*, 655
Metapsithyrus, 800
Metasamba, 70*t*, 423, 424*f*, 424*k*, **425**
Metatrachusa, 72*t*, 533, 534*k*, **535-536**, 536
Metylaeus, 68*t*, 198*t*, 202*k*, 206, 207*f*, **209**, 210
Metylaeus cribratus, 209
Metziella, 68*t*, 93, 198*t*, 199*k*, **209-210**

Mexalictus, 70*t*, 320, 354, 356, 358-359, 359*k*-360*k*, **382**, 386
Mexalictus (*Georgealictus*), 382
Mexalictus micheneri, 382
Mexalictus polybioides, 356, 382
Michanthidium, 72*t*, 517, 518*k*, **520**
Michenerapis, 68*t*, 215, 215*k*, **218**
Michenerella, 71*t*, 458, 459*k*, **460**, 487-488
Micheneria, 817
Michenerula, 69*t*, 328, **330**, 330*k*
Michenerula beameri, 330
Michmelia, 817
Micralictoides, 69*t*, 322, 324*k*, **328**
Micralictoides altadenae, 322*f*
Micramegilla, 743
Micrandrena, 68*t*, 246*k*, 250*k*, 252*k*, 253, 253*k*-256*k*, 258*k*, 261, **263**
Micrandrena pacifica, 263
Micranthidium, 72*t*, 526, 526*k*, **527**, 528
Micranthophora, 748
Micrapis, 831
Micraugochlora, 411
Micraugochlora sphaerocephala, 411
Micreriades, 467, 472
Microcolletes, 153, 159
Microdasypoda, 418
Microdontura, 68*t*, 222, 227, 228*f*, 228*k*, **229**
Microdontura mellea, 229
Microhalictus, 377
Microhesma, 68*t*, 215, 222*f*-223*f*, 224, 224*k*
Microhoplitis, 71*t*, 463, 464*k*, 468, **471-472**
Micromelecta, 777
Micrommation, 70*t*, 394*k*, 398*k*, 408, **409**
Micrommation larocai, 409
Micronomada, 644
Micronomadopsis, 69*t*, 236*f*, 309, 310*k*, **312**, 660, 663
Micronychapis, 74*t*, 713*k*, 716*k*, **733**
Micronychapis duckei, 733
Micropanurgus, 289
Micropasites, 73*t*, **657**, 657*k*, 663
Microrophites, 326
Microsphecodes, 30, 31*t*-32*t*, 70*t*, 352-354, 358*k*, 360*k*, **382-383**, 389
Microsphecodes kathleenae, 383
Microsphecodes truncaticaudus, 356*f*
Microstelis, 533
Microthurge, 71*t*, 444-445, 445*k*, 446, **446-447**
Microthurge corumbae, 7-8
Microthurge pharcidontus, 444*f*-446*f*
Mielkeanthidium, 72*t*, 518*k*, **520**
Mielkeanthidium nigripes, 520
Mimandrena, 266
Mimoxylocopa, 605
Mimulapis, 326-327
Mimulapis versatilis, 326
Minutodufourea, 326

Mirnapis, 74t, 710, 713, 713k, **733**
Mirnapis inca, 733, 895
Mischocyttarus (wasp), 193, 207
Mitchellapis, 72t, 553, 555, 556t-557t, 565k-566k, 570, **579**
Mitsukuriapis, 265
Mitsukuriella, 265
Momordica, 18
Monia, 167
Monia grisea, 167
Monidia, 167
Monilapis, 70t, 365f, 367k, **368**
Monilosmia, 482-483
Monoeca, 74t, 687-688, 688k, **690-691**, 693
Monoeca brasiliensis, 690
Monoeca haemorrhoidalis, 674
Monoeca lanei, 688f, 691f
Monoepeolus, 649
Monomorphides (Family), 78t
Monosmia, 71t, 475, 477k-478k, **483-484**, 484
Monoxylocopa, 73t, 599k, 601k, 605, **607**
Monumetha, 71t, 463, 464k-465k, 466, 466k, 467, 469, 471-472, **472**, 477
Morawitzella, 69t, 325k, **328**
Morawitzella nana, 328
Morawitzia, 69t, 83, 323, 325k, **328**, 329
Morawitzia panurgoides, 328
Morgania, 663-664
Morphanthidium, 506
Morphella, 568-569
Moureana, 578
Moureanthidium, 72t, 519k, 520, **520**, 518519
Moureapis, 72t, 557t, 558k-559k, **579**
Mourecotelles, 136, 167, 167k, 168, **169-170**, 169k
Mourecotelles (Hemicotelles), 67t, **169**, 169k
Mourecotelles (Mourecotelles s. str.), 67t, 169, 169k, **170**
Mourecotelles (Xanthocotelles), 67t, 169, 169k, **170**
Mourecotelles mixta, 170
Mourecotelles s. str., 67t, 169, 169k, **170**
Mourella, 822
Mucidobombus, 75t, 789, 793k, 796k, **799**
Mucidobombus (Exilobombus), 798
Mucidobombus eversmanniellus, 798
Mucidobombus exil, 798
Mucoreohalictus, 370
Mutillidae (Family) (wasps), 60
Mycterochlora, 402
Mydrosoma, 67t, **179**, 179k
Mydrosoma bohartorum, 179f
Mydrosoma brooksi, 173f
Mydrosoma metallicum, 179
Mydrosomella, 67t, 173, **179**, 179k
Mydrosomella gaullei, 179
Mydrosomini, 179
Mystacanthophora, 74t, 745f, 746k-747k, **749**
Mystacosmia, 71t, 476, 479k, 481, 483, **484**

Nananthidium, 509, 895
Nananthidium bettyae, 509
Nannotrigona, 75t, 811-812, 814k, 817, 819, **819**, 820, 822-823
Nanocolletes, 167
Nanorhathymus, 74t, **762**, 762k
Nanosmia, 71t, 469, 486, 487k, **488**
Nanoxylocopa, 73t, 598, 599k, 601k, **607**, 609
Nasutapis, 73t, 623, 624k, 627, **632**
Nasutapis straussorum, 31t, 34f, 622f, 631f, 632
Nasutitermes, 803, 823
Nasutosmia, 71t, 462, 464, 464k-465k, 470, 472, **472**, 473
Navicularia, 212
Neagapostemon, 70t, **387**, 387k
Nealictus, 70t, 367k, **368**
Neanthidium, 72t, 491f, 493, 493k, **522**
Neanthidium octodentatum, 522
Nectarodiaeta, 74t, 710, **739**, 739k, 740
Nectarodiaeta oliveirae, 739
Neffapis, 69t, 110, 271, 272k-274k, **278**
Neffapis longilingua, 8f, 86f, 278
Neglectella, 581
Nemandrena, 68t, 245k-246k, 248k, **263**
Neoashmeadiella, 453, 570
Neoceratina, 73t, 614k, 615, **617**
Neoceratina australensis, 617
Neochalicodoma, 72t, **586**
Neochelynia, 72t, 557t, 559k-560k, **579**, 584
Neochelynia paulista, 579
Neocoelioxys, 72t, 546k, **549-550**
Neocorynura, 70t, 348, 378, 397k, 400-401, 400k, 405, **409-410**, 905
fossil, 393
Neocorynura (Neocorynuroides), 410
Neocorynura colombiana, 410
Neocorynura perpectinata, 405
Neocorynura pubescens, 396f
Neocorynurella, 70t, 397k, 400k, **401**
Neocorynurella seeleyi, 401
Neocorynuroides, 410
Neocressoniella, 72t, 555f, 562k, 574, **579-580**
Neoeutricharaea, 573-574
Neofidelia, 71t, 438, 438k, **440**
Neofidelia longirostris, 438f, 440
Neofidelia profuga, 436f, 440
Neohalictoides, 326
Neolarra, **660**, 660k
Neolarra (Neolarra s. str.), 73t, **660**, 660f, 660k
Neolarra (Phileremulus), 73t, **660**, 660f, 660k
Neolarra pruinosa, 660
Neolarra s. str., 73t, **660**, 660f, 660k
Neolarra verbesinae, 634f, 660f
Neolarra vigilans, 634f, 660f
Neolarrini (Tribe), 73t, 589k, 656, 659, **660**
Neoliphanthus, 69t, 276k, **277**, 277k

Neomegachile, 579
Neopanurgus, 294
Neopanurgus richteri, 294
Neopasiphae, 67t, 139k, 145, **164**
Neopasiphae mirabilis, 164
Neopasites, 653, 655f, 656, 656k, **657**, 657k, 657t, 660
Neopasites (Micropasites), 73t, **657**, 657k
Neopasites (Neopasites s. str.), 73t, **657**, 657k
Neopasites (Odontopasites), 655
Neopasites (Trichopasites), 655
Neopasites arizonicus, 655
Neopasites cressoni, 657
Neopasites fulviventris, 634f, 656f
Neopasites insoletus, 655
Neopasites s. str., 73t, **657**, 657k
Neopasitini (Tribe), 653
Neoperdita, 303
Neoscirtetica, 721
Neosmia, 71t, 477k-478k, **484**
Neotrypetes, 71t, 458f, 459k, **460**
Neoxylocopa, 73t, 105, 598, 598f, 600k, 601, 605, **607**, 609
Neshylaeus, 210
Nesocolletes, 67t, 144, 145f, 147t, 149k, **156-157**, 160
Nesoeupetersia, 70t, 364k, **365**
Nesohalictus, 376-377
Nesohylaeus, 208
Nesomelecta, 75t, 770f, **776**, 776k
Nesomonia, 70t, 427, **428**, 428k
Nesoprosopis, 32t, 34, 68t, 194, 198t, 200k, 202f, 202k, **210**, 212
Nesosphecodes, 32t, 70t, 353-354, 359k, 361k, **383**, 389
Nesosphecodes anthracineus, 383
Nesothrincostoma, 391
Nesylaeus, 68t, 194, 198t, **210**
Nevadensibombus, 797
Nigranthidium, 71t, 498k, 500, **501**, 502, 507, 538
Nigranthidium tergofasciatum, 502
Nigrocoelioxys, 547, 549
Niltonia, 67t, 85, 138k, 145, **165-166**
Niltonia virgilii, 132f, 165-166
Nitocris, 340
Nivanthidium, 71t, 505, 506k, **508**, 529
Nobandrena, 68t, 253k, 255k, **263-264**
Nobilibombus, 797
Noctoraptor, 32t, 70t, 352, 407f, **408**, 408k
Nodocolletes, 145f, 155-156
Nodocolletes dentatus, 155
Nodula, 73t, 596, 602k, 603, 603k, 605-606, **607**, 608, 610
Nogueirapis, 75t, 808f, 812, 814k, **819-820**, 821, 829
fossil, 819
Nogueirapis silacea, fossil, 819, Plate 20
Nolanomelissa, 69t, **290**
Nolanomelissa toroi, 290
Nolanomelissini (Tribe), 69t, 271, 272k, **290**
biogeography of, 107t

Nomada, 3, 33, 35-36, 41-42, 73t, 77, 77t-78t, 79, 79t-81t, 108, 278, 533, 590, 636-637, 639, **643-645**, 643f, Plate 2
Nomada (Adamon), 644
Nomada (Asteronomada), 644
Nomada (Callinomada), 644
Nomada (Heminomada), 644
Nomada (Holonomada), 634f
Nomada (Laminomada), 644
Nomada (Melanomada), 640
Nomada (Micronomada), 644
Nomada (Nomada s. str.), 644
Nomada (Nomadula), 644
Nomada (Pachynomada), 644
Nomada (Phelonomada), 644
Nomada adducta, 644
Nomada albilabris, 388
Nomada americana, 644
Nomada annulata, 644f
Nomada antonita, 644
Nomada armata, 644, Plate 2
Nomada articulata, 644
Nomada basalis, 644
Nomada belfragei, 644
Nomada bella, 644
Nomada bifasciata, 644
Nomada cinctiventris, 645
Nomada coxalis, 643f
Nomada crotchii, 634f
Nomada cruralis, 644
Nomada erigeronis, 644
Nomada furva, 644-645
Nomada gibba, 388
Nomada gigas, 644-645
Nomada grindeliae, 640
Nomada hattorfiana, 260
Nomada hesperia, 644
Nomada histrio, 775
Nomada imbricata, 639f
Nomada integerrima, 644
Nomada integra, 644-645
Nomada japonica, 645
Nomada koikensis, 644
Nomada luteola, 644
Nomada modesta, 644
Nomada obliterata, 644
Nomada odontophora, 644
Nomada penangensis, 643f
Nomada pilosula, 675
Nomada roberjeotiana, 644-645
Nomada ruficornis, 644-645
Nomada s. str., 644
Nomada schottii, 656
Nomada scutellaris, 775
Nomada scutellata, 775
Nomada superba, 644-645
Nomada texana, 644
Nomada tibialis, 776
Nomada trispinosa, 644
Nomada truncata, 656
Nomada vegana, 644-645
Nomada vincta, 644-645
Nomadidae (Family), 79, 79t-80t
Nomadina (Tribe), 78t
Nomadinae, 7
Nomadinae (Subfamily), 33, 33t, 34-35, 37-40, 37f, 39f, 41, 41f, 54, 61, 73t, 77, 80, 80t, 81, 81t, 83, 95, 97, 116, 126, 128k, 317, 434, 533, 538, 587-588, 589k-590k, 592, **633-636**, 633f-634f, 636f, 639f, 643, 649,

657t, 659, 661, 666-667, 668f, 672, 674, 678, 770
  biogeography of, 108t
  tribal relationships, cladograms of, 635f
Nomadini (Tribe), 73t, 80t, 81, 590k, 636, 639, 641, 643-645, 646
  biogeography of, 108t
Nomadita, 643, 645-646
Nomadita montana, 643
Nomadopsis, 69t, 236f, 270f, 309, 310k, 312-313, 663
Nomadopsis (Macronomadopsis), 312-313
Nomadopsis (Micronomadopsis), 312
Nomadopsis fracta, 312
Nomadopsis micheneri, 312
Nomadosoma, 644
Nomadula, 644
Nomia, 66, 78t-81t, 94, 116, 157, 333, 334k, 336, **338-341**, 339k, 345, 383
Nomia (Acunomia), 69t, 332, 332f, 333, 333f, 339, **339-340**, 339k, 341, 343
Nomia (Austronomia), 333, 336
Nomia (Clavinomia), 337
Nomia (Crocisaspidia), 69t, 339, 339k, **340**
Nomia (Curvinomia), 339, 341
Nomia (Dinomia), 344
Nomia (Epinomia), 337
Nomia (Halictonomia), 335
Nomia (Hoplonomia), 69t, 333, 333f, 339, 339k, **340**, 341, 342f
Nomia (Leuconomia), 69t, 332, 339k, **340**
Nomia (Lobonomia), 341
Nomia (Macronomia), 337
Nomia (Maculonomia), 339
Nomia (Meganomia), 427
Nomia (Nomia s. str.), 69t, 338, 339k, **340-341**
Nomia (Nomiapis), 341
Nomia (Nubenomia), 338
Nomia (Pachynomia), 341
Nomia (Paranomia), 339
Nomia (Paranomina), 339
Nomia (Paulynomia), 69t, 339, 339k, **341**, 342f, xiv
Nomia (Ptilonomia), 342-343
Nomia (Reepenia), 342-343
Nomia (Spatunomia), 343
Nomia (Trinomia), 338
Nomia amboinensis, 339
Nomia amoenula, 341
Nomia andrei, 337
Nomia andrenoides, 337
Nomia angustitibialis, 333f
Nomia aurantifer, 341, 342f
Nomia australica, 333, 336
Nomia binghami, 427
Nomia burmica, 337
Nomia californiensis, 339
Nomia candida, 332, 340
Nomia chalybeata, 339
Nomia chlorosoma, 333f
Nomia clavicornis, 337
Nomia coelestina, 383
Nomia diversipes, 340-341
Nomia filifera, 343
Nomia flavilabris, 685
Nomia flavipennis, 339
Nomia flaviventris, 428

Nomia fuscipennis, 339
Nomia halictoides, 376
Nomia japonica, 265
Nomia joergenseni, 156
Nomia kirbii, 335
Nomia lobata, 341
Nomia lutea, 340
Nomia luteofasciata, 341
Nomia lyonsiae, 333f, 342f
Nomia maynei, 338
Nomia megasoma, 339
Nomia melanderi, 28, 319f, 332, 332f, 340, Plate 4
Nomia melanderi howardi, 333f
Nomia melanosoma, 338
Nomia nortoni, 332, 339
Nomia nubecula, 338
Nomia persimilis, 335
Nomia picardi, 336
Nomia platycephala, 337
Nomia plumosa, 342
Nomia pulawskii, 340
Nomia punctulata, 333, 340
Nomia robinsoni, 333f
Nomia ruficornis, 333
Nomia rugiventris, 341
Nomia s. str., 69t, 338, 339k, **340-341**
Nomia senticosa, 340
Nomia speciosa, 339
Nomia swainsoniae, 341
Nomia taiwana, 344
Nomia terminata, 339
Nomia testacea, 338
Nomia tetrazonata, 333, 333f, 339
Nomia theryi, 339
Nomia thoracica, 339
Nomia triangulifera, 335
Nomia tridentata, 338
Nomia tsavoensis, 428
Nomia variabilis, 343
Nomia viridicinctula, 339
Nomia yunnanensis, 339
Nomiapis, 341
Nomiidae (Family), 79t
Nomiinae (Subfamily), 34, 69t, 79-80, 80t, 85, 112, 116, 128k, 319, 319f, 321k, 323, **332-344**, 332f-333f, 342f-343f, 345, 348, 383, 664
  biogeography of, 107t
  of Eastern hemisphere, 334k
  of North America, 334k-335k
Nomiini (Tribe), 80t
  biogeography of, 109
Nomiocolletes, 67t, 142, 144, 147k, 147t, 151, **157**, 158, 164
Nomioides, 44, 70t, 80t-81t, 323, 345, 345k, **347**, 389, 418, 663
Nomioides (Cellaria), 346
Nomioides (Cellariella), 346
Nomioides (Erythronomioides), 347
Nomioides (Eunomioides), 346
Nomioides (Paranomioides), 347
Nomioides arnoldi, 346
Nomioides karachensis, 346
Nomioides minutissimus, 345f-346f
Nomioides socotranus, 347
Nomioides somalica, 346
Nomioides steinbergi, 347

Nomioidinae (Subfamily), 69t-70t, 74t, 128k, 320, 321k, **345-347**, 345f, 345k-346k, 346f-347f, 348, 353, 370, 906
  biogeography of, 107t
Nomioidini (Tribe), 80t, 81, 112, 348
  biogeography of, 109
Noonadania, 68t, 216k, **218**
Notagapostemon, 363
Notandrena, 68t, 245k, 249k, 253k, 256k-258k, 262, **264**
Notanthidium, 491t, 496, 497k, **522-523**, 522k-523k
Notanthidium (Allanthidium), 72t, 522, **523**, 523k
Notanthidium (Chrisanthidium), 72t, 522k, **523**
Notanthidium (Notanthidium s. str.), 72t, 522, 522k, **523**
Notanthidium rodolfi, 523
Notanthidium s. str., 72t, 522, 522k, **523**
Notanthidium steloides, 523
Noteopsis, 208-209
Noteriades, 71t, 449, 449k, 449t, 451k, **475**, 484, 490
Nothosmia, 479, 482-483
Nothylaeus, 68t, 198t, 202k, 204, 206, 209, **210**, 211f
Nothylaeus (Anylaeus), 210
Nothylaeus aberrans, 210
Notocolletes, 152
Notocolletes heterodoxus, 152
Notolonia, 74t, 720k, **733**
Notolonia astragali, 733
Notomegilla, 743
Notomelitta, 432
Notoxaea, 69t, 317k, **318**
Notoxaea ferruginea, 318
Notoxylocopa, 73t, 596f-597f, 600k, **607**, 607f
Nubenomia, 69t, 336, 336k, 338, **338**
Nyctomelitta, 73t, 601k, 603k, **607**
Nylaeus, 208
Nyssonidae (Family) (wasps), 63

Obertibombus, 801
Obertobombus, 801
Ochreriades, 71t, 442t, 449t, 450k, **475**
Ocymoromelitta, fossil, 99t
Ocymoromelitta sorella, fossil, 99t
Odontalictus, 70t, 365f, 367k, **368**
Odontanthocopa, 71t, 474f, 474k-475k, **475**
Odonterythrosmia, 475
Odontobombus, 799
Odontochlora, 402
Odontocolletes, 67t, 143-144, 147t, 149k, **157**
Odontopasites, 655
Odontostelis, 517
Odontotrigona, 826
Odyneropsina (Subtribe), 646
Odyneropsis, 589, 646, 646f, 647, 647k, 650, **650-651**, 650k, 762
Odyneropsis (Odyneropsis s. str.), 73t, 650, **650**, 650k

Odyneropsis (Parammobates), 650, **650-651**, 650k, 651
Odyneropsis holosericea, 650
Odyneropsis s. str., 73t, 650, **650**, 650k
Odyneropsis veseyi, 633f
Odynoeropsis (Parammobates), 73t
Oediscelis, 67t, 181-182, 182k, 183, **184**
Oediscelis (Pseudiscelis), 185
Oediscelis friesei, 184
Oediscelis herbsti, 182
Oediscelis inermis, 182
Oediscelis paradoxus, 184
Oediscelis plebeia, 184
Oediscelis prosopoides, 184
Oediscelis styliventris, 184
Oediscelis vernalis, 184
Oediscelisca, 184
Oestropsis, 131
Oestropsis pubescens, 131
Oligandrena, 68t, 243k, 249k, **264**
Oligeucera, 74t, 724, 724k, **725**
Oligochlora, fossil, 99t, 353, 393, Plate 17
Oligochlora (Soliapis), fossil, 99t
Oligochlora eickworti, fossil, 99t, Plate 17
Oligochlora rozeni, fossil, 99t
Oligotropus, 570
Oligotropus campanulae, 570
Olmecanthidium, 537
Omachtes, 664
Omachtes capensis, 664
Omachthes, 663-664
Omachthes capicola, 664
Omachthes carnifex, 663
Onagrandrena, 68t, 243, 245k, 247k, 250k, **264**
Opacula, 74t, **723**, 723k
Opacula donatica, 723
Opandrena, 262
Oragapostemon, 388
Orandrena, 68t, 251k, 255k, **264**
Oranthidium, 71t, 494, 498k, 499-501, **501**, 502, 505, 512
Orbitella, 606
Oreodiscelis, 184
Oreomelissa, 68t, 251, 254k, 258k, **264**
Oreopasites, 33, 73t, 634f, 661, 662k, **663**, 664
Oreopasites (Perditopasites), 663
Oreopasites arizonica, 663
Oreopasites barbarae, 661, 666
Oreopasites euphorbiae, 661f
Oreopasites linsleyi, 663
Oreopasites scituli, 663
Oreopasites vanduzeei, 634f
Orfilana, 648
Orientalibombus, 75t, 789, 791k, 794k, **799-800**
Orientalobombus, 799
Orientocoelioxys, 547, 549
Orientocressoniella, 568, 584-585
Orientoheriades, 460
Orientosmia, 71t, 476, 477k-478k, **484**, 486
Orientotrachusa, 527
Oroediscelis, 67t, 181, 182k, **184**
Orohylaeus, 68t, **210-211**

Orphana, 69*t,* 131, 236-239, 239*k*-240*k,* 267, **268-269**
Orphana inquirenda, 267*f,* 268
Orthanthidium, 72*t,* 533, 534*k,* 536, **536**
Osirini (Tribe), 33*t,* 34, 73*t,* 80*t,* 127, 587, 590*k*-591*k,* 667, 668*f,* 671, **674-677**, 674*f,* 674*k*-675*k,* 675*f*-676*f,* 678, 762
  biogeography of, 108*t*
Osirinus, 73*t,* **675-676**, 675*k,* 677, 762
Osirinus lemniscatus, 675
Osirinus rutilans, 676*f*
Osiris, 40, 73*t,* 79*t*-80*t,* 674, 674*f,* 675, 675*k,* **676**, 676*f,* 677-678
Osiris obtusus, 676
Osiris pallidus, 676
Osiris variegatus, 676*f*
Osmia, 20, 25, 77, 79*t*-81*t,* 120, 448-449, 449*f,* 449*k*-450*k,* 449*t,* 450*k*-451*k,* 452-455, 461-463, 471, 473, **475-486**, 482*f,* 533, 570, 895, Plate 6
  of Eastern Hemisphere, 477*k*-478*k*
  of Western Hemisphere, 478*k*-479*k*
Osmia (Acanthosmia), 466
Osmia (Acanthosmioides), 71*t,* 476, **479**, 479*k,* 482-483, 482*f,* 483
Osmia (Aceratosmia), 484-485
Osmia (Allosmia), 71*t,* 451, 475-476, 477*k*-478*k,* **480**, 485
Osmia (Annosmia), 467
Osmia (Arctosmia), 467
Osmia (Aspidosmia), 509
Osmia (Atoposmia), 455
Osmia (Caerulosmia), 485
Osmia (Caposmia), 490
Osmia (Centrosmia), 482-483
Osmia (Cephalosmia), 71*t,* 476, 476*f,* 478*k*-479*k,* **480**, 484
Osmia (Ceratosmia), 484
Osmia (Chalcosmia), 481
Osmia (Chenosmia), 482-483
Osmia (Cryptosmia), 481
Osmia (Ctenosmia), 470
Osmia (Diceratosmia), 71*t,* 476-477, 478*k,* **480**, 481, 486
Osmia (Erythrosmia), 71*t,* 451, 475-476, 477*k*-478*k,* 480, **480**
Osmia (Euthosmia), 71*t,* 453, 476, 479*k,* **481**, 483-484
Osmia (Exosmia), 482
Osmia (Foveosmia), 457
Osmia (Furcosmia), 468
Osmia (Glossosmia), 468
Osmia (Helicosmia), 71*t,* 468, 471, 476, 477*k*-478*k,* 478, 479*k,* **481**, 482, 482*f,* 483-485, Plate 16
Osmia (Hemiosmia), 71*t,* 476, 477*k,* 478, 478*k,* 480, **482**
Osmia (Heterosmia), 485
Osmia (Hoplosmia), 475
Osmia (Liosmia), 467
Osmia (Lithosmia), 468

Osmia (Megalosmia), 471
Osmia (Melanosmia), 71*t,* 472, 476-477, 477*k,* 478, 478*k,* 479, 479*k,* 481, **482-483**, 482*f,* 484, 486
Osmia (Metallinella), 71*t,* 442*t,* 477*k*-478*k,* **483**, 509
Osmia (Microhoplitis), 471
Osmia (Monilosmia), 483
Osmia (Monosmia), 71*t,* 475, 477*k*-478*k,* **483-484**, 484
Osmia (Mystacosmia), 71*t,* 476, 479*k,* 481, 483, **484**
Osmia (Neosmia), 71*t,* 477*k*-478*k,* **484**
Osmia (Nothosmia), 483
Osmia (Orientosmia), 71*t,* 476, 477*k*-478*k,* **484**, 486
Osmia (Osmia s. str.), 71*t,* 448*f,* 468, 471, 475*f,* 477, 477*k*-478*k,* 484, **484-485**, 509, Plate 16
Osmia (Ozbekosmia), 71*t,* 476, 477*k*-478*k,* 484, **485**, 486
Osmia (Pachyosmia), 484
Osmia (Pentadentosmia), 472
Osmia (Platosmia), 473
Osmia (Protosmia), 488
Osmia (Pyrosmia), 71*t,* 112, 448, 475-477, 477*k,* 478, 478*k,* 480-481, 485, **485-486**
Osmia (Tergosmia), 71*t,* 475-476, 477*k,* 478, 478*k,* 484-485, **486**
Osmia (Trichinosmia), 71*t,* 479*k,* 483, **486**
Osmia (Tridentosmia), 467
Osmia (Viridosmia), 485
Osmia agilis, 486
Osmia albiventris, 482
Osmia alticola, 482*f*
Osmia andrenoides, 480
Osmia annulata, 467
Osmia apicata, 483-484
Osmia argyropyga, 482
Osmia armaticeps, 480
Osmia arnoldi, 509
Osmia atriventris, 7
Osmia atrocaerulea, 483
Osmia aurulenta, 481, Plate 16
Osmia avosetta, 485-486
Osmia azteca, 477, 480
Osmia balearica, 482
Osmia bicolor, 484
Osmia bidentata, 475
Osmia bischoffi, 480
Osmia brachyura, 458
Osmia braunsi, 490
Osmia brevicornis, 483
Osmia bruneri, 483
Osmia bucephala, 479, 482-483
Osmia californica, 480
Osmia canadensis, 482
Osmia capensis, 420
Osmia cephalotes, 112, 448, 450, 475-476, 478, 486
Osmia claviventris, 467
Osmia coerulescens, 481-482
Osmia conjuncta, 480
Osmia copelandica, 455
Osmia cordata, 482
Osmia cornifrons, 4
Osmia curvipes, 473
Osmia cylindrica, 472
Osmia denudata, 469

Osmia difficilis, 482
Osmia dimidiata, 481
Osmia distincta, 482
Osmia emarginata, 484-485
Osmia enceliae, 455
Osmia eremoplana, 490
Osmia excavata, 485
Osmia fallax, 475
Osmia fedtschenkoi, 485
Osmia ferruginea, 477, 485-486
Osmia flavicornis, 489
Osmia fuciformis, 482
Osmia fulgida, 472
Osmia gallarum, 485
Osmia gaudiosa, 483
Osmia gemmea, 485
Osmia georgica, 481
Osmia glauca, 481
Osmia gracilicornis, 484
Osmia grandis, 471
Osmia hohmanni, 471
Osmia hypocrita, 469
Osmia indigotea, 481
Osmia inermis, 478, 483
Osmia integra, 479
Osmia laboriosa, 471
Osmia latipes, 458
Osmia latisulcata, 486
Osmia latreillei, 481
Osmia lignaria, 448*f,* 475*f,* 485
Osmia lunata, 486
Osmia maritima, 476, 482-483
Osmia maxillaris, 484
Osmia melanogaster, 481
Osmia mirhyi, 483
Osmia mongolica, 485
Osmia montana, 480
Osmia montivaga, 466-467
Osmia mustelina, 484-485
Osmia nana, 485
Osmia nasuta, 472
Osmia nemoris, 484
Osmia nigrifrons, 482*f*
Osmia nigriventris, 478, 483
Osmia nigrobarbata, 476
Osmia nigrohirta, 484
Osmia niveata, 482*f*
Osmia notata, 481
Osmia odontogaster, 479
Osmia oxypyga, 469
Osmia paradoxa, 488
Osmia penstemonis, 482
Osmia pinquis, 475
Osmia platalea, 473
Osmia platycera, 468
Osmia pumila, 483
Osmia 4-dentata, 480
Osmia quinquespinosa, 472
Osmia remotula, 473
Osmia rhodoensis, 486
Osmia ribifloris, 484-485
Osmia robusta, 470
Osmia robustula, 455
Osmia rufa, 485, Plate 16
Osmia rufigastra, 484
Osmia rufohirta, 480
Osmia rutila, 480
Osmia s. str., 71*t,* 448*f,* 468, 471, 475*f,* 477, 477*k*-478*k,* 484, **484-485**, 509, Plate 16
Osmia sanrafaelae, 483
Osmia satoi, 481
Osmia saxicola, 485
Osmia schultzei, 487
Osmia signata, 481
Osmia simillima, 482

Osmia singularis, 468
Osmia spoliata, 472
Osmia subaustralis, 476*f*
Osmia sybarita, 480
Osmia tanneri, 479, 483
Osmia tenuicornis, 471
Osmia tergestensis, 486
Osmia teunisseni, 485
Osmia texana, 481
Osmia tigrina, 471
Osmia tridentata, 467
Osmia triodonta, 455
Osmia versicolor, 477, 485-486
Osmia viridana, 485
Osmia xanthomelana, 482, 482*f,* 483
Osmia xerophila, 473
Osmia zandeni, 471
Osmiina (Subtribe), 448-449, 449*t*
Osmiinae (Subfamily), 79*t*-80*t,* 458, 471, 475
Osmiini (Tribe), 20, 32*t,* 34, 71*t,* 79, 79*t,* 110, 116, 120, 435, 435*f,* 441, 442*t,* 443*k,* 446, **448-490**, 448*f*-449*f,* 449*t,* 475, 502, 509-510, 538, 543, 545, 556, 574, 583, Plate 6
  biogeography of, 107*t*
  of Eastern Hemisphere, 449*k*-451*k*
  fossil, 98*t,* Plate 18
  of Western Hemisphere, 451*k*-452*k*
Osychnyukandrena, 68*t,* 251, **264**, xiv
Othinosmia, 448-449, 449*t,* 450*k*-451*k,* 474, **486-487**, 486*k*
Othinosmia (Afrosmia), 71*t,* 450, 451*k,* 486, 486*k,* **487**
Othinosmia (Megaloheriades), 71*t,* 449, 486*k,* **487**
Othinosmia (Othinosmia s. str.), 71*t,* 486*k,* **487**, 488
Othinosmia globicola, 487
Othinosmia s. str., 71*t,* 486*k,* **487**, 488
Othinosmia stupenda, 487
Oxaea, 69*t,* 79, 79*t,* 80, 80*t,* 81, 81*t,* 95, 110, 317*k,* **318**, 651, 905
Oxaea ferruginea, 318
Oxaea flavescens, 94*f,* 317*f,* 318
Oxaea nigerrima, 318
Oxaeidae (Family), 81
Oxaeinae (Subfamily), 69*t,* 80*t,* 91, 93-95, 111-112, 114, 123*k,* 126, 128*k,* 130-131, 173, 235-237, 237*k,* **316-318**, 316*f*-317*f,* 317*k,* 651-652, 905
  biogeography of, 107*t,* 109
Oxyandrena, 68*t,* 245*k,* 251*k,* **264**
Oxyanthidium, 528-529
Oxybelini (Tribe) (wasps), 63
Oxybiastes, 649
Oxybiastes bischoffi, 649
Oxyhalictus, 376-377
Oxynedis, 767
Oxynedys, 767
Oxynedys beroni, 767
Oxystoglossa, 402
Oxystoglossa decorata, 402

Oxystoglossa ephyra, 402
Oxystoglossa jocasta, 410
Oxystoglossa theia, 403
Oxystoglossella, 70*t,* 401, 401*k,* **402**
Oxystoglossidia, 402
Oxystoglossidia uraniella, 402
Oxytrigona, 75*t,* 812, 813*k,* **820,** 824, 827
Oxytrigona obscura, 820*f*
Oxyxylocopa, 606
Ozbekosmia, 71*t,* 476, 477*k-*478*k,* 484, **485,** 486

Pachyanthidium, 112, 491*t,* 492, 493*k-*495*k,* 508, 516-517, **523-524,** 523*k,* 535
Pachyanthidium (Ausanthidium), 72*t,* 523, **523-524,** 523*k*
Pachyanthidium (Micranthidium), 527
Pachyanthidium (Pachyanthidium s. str.), 72*t,* 523*k,* **524**
Pachyanthidium (Trichanthidiodes), 72*t,* 523*k,* **524**
Pachyanthidium (Trichanthidium), 72*t,* 523, 523*k,* **524,** 527
Pachyanthidium ausense, 524
Pachyanthidium bicolor, 524
Pachyanthidium micheneri, 523
Pachyanthidium occipitale, 524
Pachyanthidium prionognathum, 517
Pachyanthidium s. str., 72*t,* 523*k,* **524**
Pachyanthidium semiluteum, 524
Pachyceble, 70*t,* 366, 367*k,* **368,** 369
Pachyceble lanei, 368
Pachycentris, 690
Pachycentris schrottkyi, 690
Pachycephalopanurgus, 289
Pachycerapis, 682
Pachycolletes, 167
Pachyhalictus, 70*t,* 106, 370, 383-384, 383*f,* 384*k,* 385-386, 385*f,* **386**
Pachyhalictus (Dictyohalictus), 385
Pachyhalictus (Pachyhalictus s. str.), 384
Pachyhalictus (Rugalictus), 385
Pachyhalictus binghami, 384
Pachyhalictus s. str., 384
Pachyhalonia, 74*t,* 727, 727*k,* **728,** 729
Pachyheriades, 71*t,* 459*k,* **460,** 461
Pachymelopsis, 74*t,* 111, 751, 751*k,* 752
Pachymelus, 743*k,* **751-752,** 751*k*
Pachymelus (Pachymelopsis), 74*t,* 111, **751,** 751*k,* 752
Pachymelus (Pachymelus s. str.), 74*t,* **751-752,** 751*k*
Pachymelus bicolor, 752
Pachymelus conspicuus, 751
Pachymelus heydenii, 752
Pachymelus hova, 752
Pachymelus micrelephas, 751-752
Pachymelus ocularis, 751-752
Pachymelus peringueyi, 751-752
Pachymelus reichardti, 752

Pachymelus s. str., 74*t,* **751-752,** 751*k*
Pachymelus unicolor, 752
Pachynomada, 644-645
Pachynomia, 69*t,* **341,** 341*k,* 344
Pachyodonta, 140
Pachyosmia, 484
Pachyprosopina, 68*t,* 228-229, 228*k,* **229**
Pachyprosopis, 94, 134, 220-222, 222*k,* 224, 228-230, **231-232,** 231*k*
Pachyprosopis (Pachyprosopina), 229
Pachyprosopis (Pachyprosopis s. str.), 68*t,* 228, **231-232,** 231*k*
Pachyprosopis (Pachyprosopula), 68*t,* 230-231, 231*k,* **232**
Pachyprosopis (Parapachyprosopis), 68*t,* 231, 231*k,* **232**
Pachyprosopis angophorae, 231-232
Pachyprosopis atromicans, 230
Pachyprosopis cornuta, 228, 231
Pachyprosopis flavicauda, 231-232
Pachyprosopis haematostoma, 231
Pachyprosopis indicans, 231-232
Pachyprosopis kellyi, 232
Pachyprosopis mirabilis, 231
Pachyprosopis s. str., 68*t,* 228, **231-232,** 231*k*
Pachyprosopis xanthodonta, 231
Pachyprosopula, 68*t,* 230-231, 231*k,* **232**
Pachysvastra, 74*t,* 111, 710, 711*k,* 716*k,* 722, **733,** 735-736
Pachysvastra leucocephala, 733
Paedia, 653
Paidia, 653
Palaeapis, fossil, 98
Palaeorhiza, 90-91, 93, 126, 132, 188, 189*k-*190*k,* 191, 193, 207, **214-218,** 215*k-*216*k,* 219
Palaeorhiza (Anchirhiza), 68*t,* 190*f,* **216,** 216*k*
Palaeorhiza (Callorhiza), 68*t,* 190*f,* 196*f,* 215, **216,** 216*k*
Palaeorhiza (Ceratorhiza), 68*t,* 190*f,* 214-215, 215*k,* **216,** 217
Palaeorhiza (Cercorhiza), 68*t,* 215, **216-217,** 216*k*
Palaeorhiza (Cheesmania), 68*t,* 216*k,* **217**
Palaeorhiza (Cnemidorhiza), 68*t,* 215, 216*k,* **217**
Palaeorhiza (Eupalaeorhiza), 68*t,* 216*k,* **217**
Palaeorhiza (Eusphecogastra), 68*t,* 215*k,* **217**
Palaeorhiza (Gressittapis), 68*t,* 215*k,* **217**
Palaeorhiza (Hadrorhiza), 68*t,* 216*k,* **217**
Palaeorhiza (Heterorhiza), 68*t,* 215, 215*k,* **217**
Palaeorhiza (Michenerapis), 68*t,* 215, 215*k,* **218**
Palaeorhiza (Noonadania), 68*t,* 216*k,* **218**

Palaeorhiza (Palaeorhiza s. str.), 68*t,* 216*k,* **218**
Palaeorhiza (Paraheterorhiza), 68*t,* 215*k,* **218**
Palaeorhiza (Sphecogaster), 217
Palaeorhiza (Trachyrhiza), 68*t,* 216*k,* **218**
Palaeorhiza (Xenorhiza), 219
Palaeorhiza (Zarhiopalea), 68*t,* 216*k,* **218**
Palaeorhiza amabilis, 217
Palaeorhiza bicolor, 190, 215, 218
Palaeorhiza conica, 190*f,* 214-216
Palaeorhiza flavomellea, 215
Palaeorhiza gigantea, 209
Palaeorhiza gratiosa, 217
Palaeorhiza gressittorum, 216
Palaeorhiza hamada, 219
Palaeorhiza hilara, 218
Palaeorhiza mandibularis, 190*f,* 216
Palaeorhiza melanura, 217
Palaeorhiza melliceps, 191
Palaeorhiza miranda, 217
Palaeorhiza paradisea, 217
Palaeorhiza paradoxa, 217
Palaeorhiza parallela, 89*f*
Palaeorhiza rugosa, 218
Palaeorhiza s. str., 68*t,* 216*k,* **218**
Palaeorhiza sculpturalis, 218
Palaeorhiza stygica, 190*f,* 196*f*
Paleoeuglossa, fossil, 99*t,* 781, 783
Paleoeuglossa melissiflora, fossil, 99*t,* 895
Paleomelitta, fossil, 99*t*
Paleomelitta nigripennis, fossil, 99*t*
Paleomelittidae (Family), fossil, 99*t*
Palinepeolus, 673
Pallandrena, 68*t,* 252*k,* 255*k,* **264**
Pallhalictus, 380
Panomalopsis, 681
Panurgidae (Family), 77-78, 78*t,* 79, 79*t*
Panurgidae (Subfamily), 78*t*
Panurgillus, 69*t,* 280, 283*k,* **284,** 895
Panurginae, 24
Panurginae (Subfamily), 9, 14, 18, 20-21, 24, 28, 45, 58, 69*t,* 79*t,* 80, 80*t,* 83, 85-87, 91-95, 101, 110, 111*f,* 112, 114, 123*k,* 126, 127*k,* 235-238, 236*f,* 237, 237*k,* 239, **270-272,** 270*f,* 271*k-*272*k,* 295, 316, 323, 345, 418, 639-640, 655, 660, 663, Plate 5
fossil, 99*t*
Panurgini (Tribe), 69*t,* 80*t,* 270-271, 272*k,* **285-289,** 285*k,* 288*f*
biogeography of, 107*t*
Panurgini (Tribe), 270
Panurginus, 69*t,* 77, 78*t,* 285, 285*k,* 286, **287,** 288-289
Panurginus albopilosus, 287
Panurginus calcaratus, 113*f*
Panurginus clavatus, 285, 287, 288*f*
Panurginus niger, 287

Panurginus occidentalis, 285*f*
Panurginus polytrichus, 287, 288*f*
Panurginus vagabundus, 284, 895
Panurgites (Tribe), 78*t*
Panurgomia, 70*t,* 419*k,* 420, 421, 659
Panurgomia fuchsi, 421
Panurgus, 47, 77, 78*t-*81*t,* 271-272, 285, 285*k,* **287-289,** 288*k,* 645
Panurgus (Avpanurgus), 285, 285*k,* 286
Panurgus (Clavipanurgus), 287
Panurgus (Euryvalvus), 289
Panurgus (Flavipanurgus), 69*t,* 272, 285, **288-289,** 288*f,* 288*k,* 289
Panurgus (Micropanurgus), 289
Panurgus (Pachycephalopanurgus), 289
Panurgus (Panurgus s. str.), 69*t,* 288, 288*f,* 288*k,* **289**
Panurgus (Simpanurgus), 69*t,* 288*k,* **289**
Panurgus (Stenostylus), 289
Panurgus aethiops, 282
Panurgus anatolicus, 287
Panurgus andrenoides, 264
Panurgus annulatus, 287
Panurgus calcaratus, 124*f,* 285*f,* 288, 288*f*
Panurgus chalybaeus, 261
Panurgus cinerarius zizus, 294
Panurgus clavatus, 287
Panurgus farinosus, 287
Panurgus flavofasciatus, 286
Panurgus flavus, 288
Panurgus lustrans, 380
Panurgus manifestus, 259
Panurgus marginatus, 326
Panurgus maurus, 326
Panurgus nadigi, 286
Panurgus novaeangliae, 326
Panurgus ovatulus, 287, 289
Panurgus phyllopodus, 289
Panurgus rungsii, 289
Panurgus s. str., 69*t,* 288, 288*f,* 288*k,* **289**
Panurgus venustus, 288*f*
Papualictus, 70*t,* 371, **371,** 371*k*
Papuanorhiza, 21
Papuatrigona, 75*t,* 812, 822, 824, 825*k,* **827**
Paraanthidiellum, 527
Paraanthidium, 72*t,* 510, 524, 534*k,* 535, **536**
Paraanthidium (Orthanthidium), 536
Paracella, 72*t,* 555*f,* 557*t,* 562*k,* 564*k-*565*k,* 574, **580**
Paracentris, 18*f,* 74*t,* 754, 756-757, 756*k,* **757,** 758-759, 765*f*
Paracentris fulvohirta, 757
Parachalicodoma, 72*t,* 553, 555, 557*t,* 561*k-*562*k,* 569-570, 580, **580**
Paracoelioxys, 548, 586
Paracoelioxys barrei, 586
Paracolletes, 79*t-*80*t,* 139*k,* 144-145, 164, **165,** 165-166, 165*k,* 166
Paracolletes (Anthoglossa), 67*t,* 165, **165,** 165*k*
Paracolletes (Heterocolletes), 155

Paracolletes (Lysicolletes), 152
Paracolletes (*Paracolletes* s. str.), 67*t*, **165**, 165*k*
Paracolletes capillatus, 155
Paracolletes crassipes, 144, 165
Paracolletes cygni, 165
Paracolletes halictiformis, 159
Paracolletes minutus, 159
Paracolletes montanus, 139, 165
Paracolletes moretonianus, 141
Paracolletes pachyodontus, 157
Paracolletes plumatus, 165
Paracolletes s. str., 67*t*, **165**, 165*k*
Paracolletes singularis, 152
Paracolletinae, 905
Paracolletini (Tribe), 67*t*, 80*t*, 89, 93, 116-117, 133-134, 136, 137*k*, **138-166**, 164*f*, 171-172, 905
 of Australian region, 138*k*-139*k*
 biogeography of, 107*t*, 108-109
 of Western Hemisphere, 138*k*
Paracrocisa, 75*t*, 772, 773, 773*k*, 774, **774**, 776
Paracrocisa sinaitica, 774
Paradialictus, 32*t*, 70*t*, 352, 371*t*, 374*k*, 380*f*, **381**
Paradialictus synavei, 381
Paradioxys, 72*t*, 538, 539*k*, **541-542**
 positions of spines and carinae on thoracic sclerites of, 539*t*
Parafidelia, 71*t*, 434, 439, 439*k*, **440**
Parafidelia friesei, 440
Parafriesea, 308
Paragapostemon, 70*t*, 80*t*, 354, 356, 359, 359*k*, 364, **383**, 388
Paragapostemon coelestinus, 383
Parahalictoides, 326
Parahesma, 68*t*, 226, **227**, 227*k*
Paraheterorhiza, 68*t*, 215*k*, **218**
Paralictus, 31, 353, 357, 372, 377-378, 381, 623
Paramacropis, 70*t*, 422, 429-430, **431**, 431*k*
Paramegachile, 573
Paramegalochila, 573
Paramegilla, 74*t*, 746*k*-747*k*, **749**
Paramelittergini (Tribe), 270
Parammobates, 73*t*, 650, **650-651**, 650*k*, 651
Parammobates brasiliensis, 650
Parammobatoides, 43, 73*t*, 661, 661*k*, **663**
Parammobatoides gussakovskii, 906
Parammobatoides indicus, 663
Parandrena, 68*t*, 244*k*, 247*k*, 256*k*, 262, **264**
Parandrena atypica, 264
Parandrenella, 68*t*, 252*k*, 255*k*, **264**
Paranomada, 73*t*, 590, 639, 639*k*, **641**
Paranomada nitida, 641
Paranomada velutina, 641*f*
Paranomia, 79*t*, 339
Paranomina, 339
Paranomioides, 347

Paranthidiellum, 527
Paranthidium, 491, 491*t*, 496*k*-497*k*, 513, **524-525**, 524*k*, 536
Paranthidium (*Mecanthidium*), 513
Paranthidium (*Paranthidium* s. str.), 72*t*, 524, 524*k*, **525**
Paranthidium (*Rapanthidium*), 72*t*, 524, 524*k*, **525**
Paranthidium jugatorium, 513
Paranthidium jugatorium perpictum, 491*f*, 495*f*, 524
Paranthidium s. str., 72*t*, 524, 524*k*, **525**
Paranthidium sonorum, 513
Paranthidium vespoides, 525
Paranthocopa, 71*t*, 474*k*, **475**
Paranychium, 69*t*, **314**, 314*k*
Parapachyprosopis, 68*t*, 231, 231*k*, **232**
Parapartamona, 75*t*, 821, **821**, 821*k*
Parapolyglossa, 171
Paraprosopis, 68*t*, 194, 198*t*, 199*k*-200*k*, 201*f*, 201*k*, 202*f*, 210, **211**, 212
Parapsaenythia, 69*t*, 112, 273*k*, 274, **278**, 283
Parapsaenythia paspali, 278, 282*f*
Parapsaenythia serripes, 278
Pararhophites, 71*t*, 81, 81*t*, 92, 97, 121, 127, 434-435, **437**, 440, 510
Pararhophites orobinus, 436*f*-437*f*
Pararhophites quadratus, 437
Pararhophitinae (Subfamily), 434
Pararhophitini (Tribe), 71*t*, 81, 89, 96-97, 127*k*, 434, 435*f*, 436, 436*k*, **437**, 437*f*
 biogeography of, 107*t*, 108
Parasarus, 69*t*, 278, 279*k*-280*k*, **281**, 640
Parasarus atacamensis, 281
Paraseladina, 361
Paraseladonia, 32*t*, 70*t*, 352, 361, 361*k*, 366, 367*k*, **368-369**
Parasphecodes, 70*t*, 371*t*, 373, 373*f*, 374*k*, 375, 375*k*, 380, **381**
Parasphecodes (*Aphalictus*), 381
Parasphecodes bribiensis, 381
Parasphecodes hilactus, 381
Parasphecodes tooloomensis, 376
Paratetrapedia, 674, 687-688, 688*f*, 689, 689*k*, **691-692**, 691*k*, 693, 695
Paratetrapedia (*Amphipedia*), 74*t*, 688, 691, 691*k*, **692**
Paratetrapedia (*Lophopedia*), 74*t*, 691*f*, 691*k*, **692**
Paratetrapedia (*Paratetrapedia* s. str.), 74*t*, 687, 691*k*, **692**
Paratetrapedia (*Tropidopedia*), 74*t*, 691*k*, **692**
Paratetrapedia (*Xanthopedia*), 74*t*, 687, 691*k*, **692**
Paratetrapedia calcarata, 687*f*
Paratetrapedia haeckeli, 692
Paratetrapedia lugubris, 589*f*, 687*f*-688*f*

Paratetrapedia pygmaea, 691*f*
Paratetrapedia s. str., 74*t*, 687, 691*k*, **692**
Paratetrapedia seabrai, 692
Paratetrapedia swainsonae, 692
Paratetrapedia tricolor, 692
Parathrincostoma, 32*t*, 70*t*, 352, 354, 361*k*, **383**, 390
Parathrincostoma seyrigi, 383
Paratrigona, 75*t*, 803, 811, 814*k*, **820**, 821, 828
Paratrigona (*Aparatrigona*), 820-821
Paratrigona opaca, 820
Paratrigona prosopiformis, 820
Paratrigonoides, 75*t*, 814, 814*k*, **821**
Paratrigonoides mayri, 821
Paraugochlora, 404
Paraugochloropsis, 70*t*, 399*f*, **404-405**, 404*f*, 404*k*, 410
Paremisia, 758
Parepeolus, 674*k*, **676**, 676*k*
Parepeolus (*Ecclitodes*), 73*t*, 674, **676**, 676*k*
Parepeolus (*Parepeolus* s. str.), 73*t*, 674, **676**, 676*k*
Parepeolus minutus, 674*f*
Parepeolus niger, 674, 675*f*
Parepeolus s. str., 73*t*, 674, **676**, 676*k*
Parepeolus stuardi, 674, 676
Parepicharis, 74*t*, 760*k*, **761**
Pareucera, 724-725
Parevaspis, 516
Parevaspis basalis, 516
Pariotrigona, 75*t*, 810, 815*k*, 816, **821**
Paroxystoglossa, 70*t*, 394*k*, 399*k*, 400, 400*k*, 408, **410**
Paroxystoglossa jocasta, 395*f*
Paroxystoglossa transversa, 401*f*
Partamona, 42, 803, 808*f*, 812, 814*k*, 819, **821**, 821*k*, 822, 826
Partamona (*Nogueirapis*), 819
Partamona (*Parapartamona*), 75*t*, **821**, 821*k*
Partamona (*Partamona* s. str.), 75*t*, **821**, 821*k*
Partamona bilineata, 809*f*, 813*f*
Partamona grandipennis, 821
Partamona s. str., 75*t*, **821**, 821*k*
Partamona testacea, 27*f*
Partamona xanthogastra, 821
Pasiphae, 160
Pasiphae caerulescens, 160
Pasiphae flavicornis, 157
Pasites, 43, 73*t*, 641, 653, 661, 662*k*, **663-664**
Pasites (*Ebmeriana*), 663
Pasites (*Micropasites*), 663
Pasites (*Spinopasites*), 665
Pasites dichroa, 663
Pasites gnoma, 664
Pasites maculata, 663
Pasites maculatus, 110, 641, 664
Pasites punctatus, 656
Pasites pygmaeus, 664
Pasites spinotus, 665
 pasitines, 77, 78*t*-80*t*, 81, 81*t*
Pasitomachtes, 664
Pasitomachthes, 664
Pasitomachthes nigerrimus, 664
Passiflora lutea, 20

Patagiata, 208, 210
Patellapis, 320, 354, 356-357, 362*k*, 365, 370, **383-386**, 384*k*, 385*f*, 389
Patellapis (*Archihalictus*), 70*t*, 384, **384-385**, 384*k*, 386
Patellapis (*Chaetalictus*), 70*t*, 384*k*, **385**, 385*f*, 386
Patellapis (*Dictyohalictus*), 70*t*, 384, 384*k*, **385-386**
Patellapis (*Lomatalictus*), 70*t*, 384, 384*k*, 385, **386**
Patellapis (*Madagalictus*), 362
Patellapis (*Pachyhalictus*), 70*t*, 370, 383-384, 383*f*, 384*k*, 385-386, 385*f*, **386**
Patellapis (*Patellapis* s. str.), 70*t*, 356, 383, 384*k*, 385, 385*f*, **386**
Patellapis (*Zonalictus*), 70*t*, 361-362, 384, 384*k*, 385*f*, **386**
Patellapis albofasciata, 385*f*
Patellapis braunsella, 356, 365, 383, 386
Patellapis carinostriata, 384
Patellapis concinnula, 361-362
Patellapis intricata, 386
Patellapis joffrei, 384
Patellapis laevata, 384-385
Patellapis malachurina, 384
Patellapis merescens, 383*f*, 385*f*
Patellapis partita, 385*f*
Patellapis pearstonensis, 385*f*
Patellapis perineti, 384-385
Patellapis plicata, 384
Patellapis retigera, 384
Patellapis s. str., 70*t*, 356, 383, 384*k*, 385, 385*f*, **386**
Patellapis schultzei, 385*f*
Patellapis suarezensis, 362, 383
Patellapis zacephala, 385*f*
Patera, 821
Paulynomia, 69*t*, 332, 339, 339*k*, **341**, 342*f*, xiv
Pauphalictus, 377
Pavostelis, 533
Pectinapis, 74*t*, 711*f*, 718*f*, 718*k*, 720*k*, 737, **738**, 738*k*, 749
Pectinapis fasciata, 738
Pectinata, 208
Pelandrena, fossil, 99*t*
Pelandrena reducta, fossil, 99*t*
Pelicandrena, 68*t*, 246*k*-247*k*, **264**
Pemphredoninae (Subfamily) (wasps), 3, 61-63
Penapini (Tribe), 322-323
Penapis, 69*t*, 322-323, 324*k*, 325, **328**, **328**, 329
Penapis penai, 323*f*, 328
Peniella, 314
Pentadentosmia, 71*t*, 458, 462, 464, 464*k*-465*k*, 471, **472-473**
Pentaperdita, 69*t*, 300, 301*k*, **303**, 305
Penteriades, 71*t*, 452, 462, 465, 465*k*, 466, 466*k*, 470, **473**, 473-474, 476
Penthemisia, 757
Peponapis, 74*t*, 652, 708-709, 711, 711*f*, 713*k*, 715*k*, 717*k*, 719*k*-720*k*, 722, **733**, **733**, 736, 740-741

Peponapis (Austropeponapis), 733
Peponapis (Colocynthophila), 733
Peponapis (Eopeponapis), 733
Peponapis (Xenopeponapis), 733
Peponapis (Xeropeponapis), 733
Peponapis crassidentata, 733
Peponapis fervens, 23*f,* 722, 733
Peponapis pruinosa, 709*f,* 715*f,* 717*f*
Peponapis timberlakei, 719, 733
Perdita, 24, 28, 86-87, 110-112, 118, 235, 271, 289, 296, 296*f,* 297, 297*k,* 299-305, 300*f,* 300*k*-302*k,* 301*f,* 345, 353, 389, 418, 660, 663
Perdita (Allomacrotera), 69*t,* 300, 301*k,* **302**
Perdita (Alloperdita), 69*t,* 299, 301*k,* **302,** 305
Perdita (Callomacrotera), 69*t,* 296*f,* 300, 301*k,* **302**
Perdita (Cockerellia), 69*t,* 300, 301*k,* **302,** 302*f,* 305
Perdita (Epimacrotera), 69*t,* 300*k,* **302-303,** 302*f,* 663
Perdita (Geoperdita), 303
Perdita (Glossoperdita), 69*t,* 300*k,* 302, **303**
Perdita (Hesperoperdita), 69*t,* 300*k,* **303**
Perdita (Heteroperdita), 69*t,* 300, 300*k,* **303**
Perdita (Hexaperdita), 69*t,* 300, 301*k,* 302*f,* **303,** 663
Perdita (Lutziella), 297
Perdita (Macroterella), 298
Perdita (Pentaperdita), 69*t,* 300, 301*k,* **303,** 305
Perdita (Perdita s. str.), 69*t,* 296, 296*f,* 302, 302*f,* 302*k,* **303,** 304*f,* 305, 663, Plate 5
Perdita (Perditella), 69*t,* 301*k,* **304**
Perdita (Procockerellia), 69*t,* 300, 301*k,* 302, 302*f,* **304**
Perdita (Pseudomacrotera), 69*t,* 300*k,* **304**
Perdita (Pygoperdita), 69*t,* 300, 301*k,* **305**
Perdita (Tetraperdita), 303
Perdita (Xeromacrotera), 69*t,* 300, 301*k,* **305**
Perdita (Xerophasma), 69*t,* 299, 300*k,* **305,** 305*f*
Perdita acapulconis, 296*f*
Perdita ainsliei, 302
Perdita albipennis, 90*f,* 111*f,* 302
Perdita albonotata, 302*f,* 304
Perdita albovittata, 303
Perdita beata, 302
Perdita bequaertiana, 305, 305*f*
Perdita bishoppi, 302*f*
Perdita cephalotes, 305
Perdita chamaesarachae, 303
Perdita chihuahua, 296*f*
Perdita coreopsidis, 302*f*
Perdita euphorbiae, 302*f*
Perdita floridensis, 296
Perdita graenicheri, 271
Perdita halictoides, 296, 302*f,* 303
Perdita hyalina, 302
Perdita ignota, 303
Perdita interrupta, 305
Perdita kiowi, 16
Perdita laneae, 304
Perdita larreae, 304
Perdita latior, 298
Perdita maculigera maculipennis, 296, 296*f*
Perdita maritima, 302
Perdita maura, 303
Perdita mortuaria, 298
Perdita novaeangliae, 302
Perdita octomaculata, 300*f*-302*f*
Perdita opuntiae, 297
Perdita rhodogastra, 303
Perdita ruficauda, 303
Perdita s. str., 69*t,* 296, 296*f,* 302, 302*f,* 302*k,* **303,** 304*f,* 305, 663, Plate 5
Perdita sexmaculata, 303
Perdita sphaeralceae, 302*f*
Perdita stathamae, 304*f*
Perdita stephanomeriae, 302
Perdita tridentata, 114*f*
Perdita turgiceps, 304
Perdita zebrata, 303
Perdita zebrata zebrata, 114*f*
Perdita zonalis, 302*f,* 312
Perditella, 69*t,* 301*k,* **304**
Perditini (Tribe), 21, 58, 69*t,* 110, 271*k,* 287, **296-305,** 296*f,* 297*k,* 298*f*-302*f,* 304*f*-305*f*
biogeography of, 107*t*
Perditomorpha, 60*f*-61*f,* 67*t,* 144, 146*f,* 147, 147*f,* 147*t,* 148*k,* 151-152, 154, **157-158,** 159-160, 640
Perditomorpha brunerii, 157
Perditopasites, 663
Pereirapis, 70*t,* 396*f,* 402, 402*k,* **403**
Pereirapis rhizophila, 403
Perezia, 573
Perezia maura, 573
Perissander, 69*t,* 309, 310*k,* 312, **313,** 313*f,* 314
Perixylocopa, 610
Petalosternon, 74*t,* 746*k*-747*k,* 748, **750**
Petrusia, 292
Petrusianna, 292, 895
Phaceliapis, 69*t,* 323*f,* **325-326,** 325*k*
Phaenosarus, 555, 584-585
Phaeosmia, 455
Phanomalopsis, 73*t,* 683*f,* 683*k,* **684**
Pharohylaeus, 68*t,* 189*k,* 192, 215, **218**
Pharohylaeus lactiferus, 190*f,* 192*f,* 196*f,* 218
Pharohylaeus papuanus, 218
Phelonomada, 644
Phenacolletes, 67*t,* 139*k,* 141, 145, **165**
Phenacolletes mimus, 141*f,* 145*f,* 165
Phiarus, 653
Phiarus minutus, 663
Philanthinae (Subfamily) (wasps), 62-63
Philanthini (Tribe) (wasps), 63
Philanthus gibbosus, 63*f*
Phileremidae (Subfamily), 78*t*
Phileremides (Tribe), 78*t*
Phileremulus, 73*t,* **660,** 660*f,* 660*k*
Phileremulus vigilans, 660
Phileremus, 662
Phileremus (Melanempis), 663
Phileremus abdominalis, 653
Phileremus ater, 663
Phileremus emarginatus, 656
Phileremus fulviventris, 657
Phileremus illinoiensis, 655
Phileremus oraniensis, 662
Phileremus productus, 540
Philoxanthus, 302
Phor, 644
Phyllotoma, 467
Physostetha, 460
Pileteucera, 724
Pithitis, 73*t,* 612, 613*k,* **617**
Placidohalictus, 369
Planiandrena, 68*t,* 251, **265**
Planihylaeus, 68*t,* 188, 195*f,* 197*k,* 198*t,* **211,** 213-214
Plastandrena, 68*t,* 243*k,* 248*k,* 250*k,* 252*k,* 254*k,* 256*k*-258*k,* **265**
Platandrena, 266
Platinopoda, 606
Platosmia, 71*t,* 464*k*-465*k,* 471, **473**
Platychile, 566
Platycoelioxys, 72*t,* 546*k,* **550**
Platyhalictus, 70*t,* 367*k,* **369**
Platynopoda, 606-607
Platyspatulariella, 213
Platysta, 72*t,* 557*f,* 563*k,* 565*k,* 574, **580-581**
Platysvastra, 74*t,* 708, 710, 715*k,* 716, **734**
Platysvastra macraspis, 734
Platytrigona, 826
Plebeia, 103, 587, 807-809, 808*f,* 811-812, 814*k,* 817, 819, **822-823,** 822*k,* 823-825, 829
fossil, 819
Plebeia (Nogueirapis), fossil, 819
Plebeia (Plebeia s. str.), 75*t,* 669*f,* 804*f,* 805, 805*f,* 813*f,* 815, 818, **822,** 822-823, 822*k,* 823
Plebeia (Scaura), 75*t,* 805, 814, **822, 822-823,** 822*f,* 822*k,* 824
Plebeia (Schwarziana), 75*t,* 807, 822*k,* **823**
Plebeia caerulea, 805*f,* 822-823
Plebeia frontalis, 53*f,* 669*f,* 804*f,* 813*f*
Plebeia intermedia, 822
Plebeia latitarsis, 822-823, 822*f*
Plebeia s. str., 75*t,* 669*f,* 804*f,* 805, 805*f,* 813*f,* 815, 818, **822,** 822-823, 822*k,* 823
Plebeia schrottkyi, 804*f,* 822, 829
Plebeia silacea, fossil, 819
Plebeia tica, 822
Plebeia timida, 814, 822-823
Plebeiella, 818-819
Plebeina, 75*t,* 808-809, 811, 815, 815*k,* 818, **823**
Plebeina denoiti, 812*f,* 823
Plesianthidium, 491*t,* 494*k,* 510, **525-526,** 525*k,* 528
Plesianthidium (Carinanthidium), 72*t,* 510, 525, **525,** 525*k*
Plesianthidium (Plesianthidium s. str.), 72*t,* **525,** 525*k,* 526
Plesianthidium (Spinanthidiellum), 72*t,* 494, 525, 525*k,* 526, **526**
Plesianthidium (Spinanthidium), 72*t,* 494, 525, 525*k,* **526**
Plesianthidium calescens, 526
Plesianthidium cariniventre, 525
Plesianthidium fulvopilosum, 525
Plesianthidium rufocaudatum, 526
Plesianthidium s. str., 72*t,* **525,** 525*k,* 526
Plesianthidium volkmanni, 526
Plesiopanurgus, 69*t,* 271-272, 291*k,* 293, **294**
Plesiopanurgus (Zizopanurgus), 294
Plesiopanurgus cinerarius, 293*f,* 294
Plistotrichia, 685
Plusia, 783
Plusia superba, 783
Podalirius, 77*t,* 747
Podalirius (Amegilla), 743
Podalirius (Paramegilla), 749
Podalirius grisescens, 705
Podalirius vallorum, 748
Podasys, 418
Podilegidae, 78*t*
Podilegides (Family), 78*t*
Poecilandrena, 68*t,* 253*k,* 255*k,* 257*k*-258*k,* **265**
Poecilobombus, 801
Poecilocentris, 756
Poecilomelitta, 292-293
Poecilomelitta flavida, 292
Poecilomelitta lacrymosa, 291
Poecilomelitta ornata, 292
Poliandrena, 68*t,* 253*k,* 255*k*-256*k,* **265,** 266
Policana, 177
Polistes (wasps), 650, 663
Polybiapis, 644
Polybiapis mimus, 644
Polyglossa, 171
Polyglossa (Parapolyglossa), 171
Polyglossa capensis, 171
Polyglossa heterodoxa, 171
Pomibombus, 801
Pomobombus, 801
Pontanthidium, 506-507
Popovapis, 70*t,* 418*k,* **419**
Popovia, 292
Popovmeliturgula, 292-293
Porterapis, 67*t,* 158*f,* 162, 162*k,* **163-164,** 164*f*
Pratibombus, 801
Pratobombus, 801
Presbia, 595
Pressibombus, 75*t,* 793*k,* 794, 794*k,* 796*k,* **800**
Prionohoplitis, 71*t,* 464, 464*k,* 467, **473**
Prioanthidium, 71*t,* 493, 497, 505, 506*k,* 507, **508,** 527
Probombus, fossil, 99*t,* 790
Probombus hirsutus, fossil, 99*t*
Prochelostoma, 71*t,* 448*f,* 456, 456*k,* 457, **458**

Procockerellia, 69*t*, 300, 301*k*, 302, 302*f*, **304**
Prodasyhalonia, 74*t*, 727, 727*k*, **728**
Prodioxys, 72*t*, 442, 538, 539*k*, **542**
  positions of spines and carinae on thoracic sclerites of, 539*t*
Prodioxys cinnabarina, 542
Prohalictus, 369
  fossil, 99*t*
Prohalictus schemppi, fossil, 99*t*
Promelitta, 70*t*, 415-416, **422**
Promelitta alboclypeata, 417*f*, **422**
Promelitta plumipes, 422
Promelittini (Tribe), 70*t*, 416, 416*k*, 417*f*, **422**
  biogeography of, 107*t*
Pronomia, 340
Pronomia pulawskii, 340
Proplebeia, fossil, 99*t*, 812
Prosamba, 70*t*, 423, 424*k*, **425**
Prosapis, 212
Prosopalictus, 377-378
Prosopalictus micans, 377
Prosopella, 68*t*, 198*t*, 199*k*, 207, 210-211, **211-212**
Prosopidae (Family), 79*t*
Prosopidae (Subfamily), 78*t*
Prosopinae (Subfamily), 80*t*
Prosopis, 68*t*, 78*t*-79*t*, 192, 198*t*, 199*k*-201*k*, 202*f*, 202*k*, 210, **212**
Prosopis (*Abrupta*), 203
Prosopis (*Amblyspatulariella*), 213
Prosopis (*Auricularia*), 212
Prosopis (*Barbata*), 209
Prosopis (*Brachyspatulariella*), 213
Prosopis (*Campanularia*), 211
Prosopis (*Cingulata*), 212
Prosopis (*Dentigera*), 204
Prosopis (*Fasciata*), 212
Prosopis (*Fascista*), 212
Prosopis (*Hoploprosopis*), 206
Prosopis (*Imperfecta*), 204
Prosopis (*Koptobaster*), 208
Prosopis (*Koptogaster*), 208
Prosopis (*Lambdopsis*), 208
Prosopis (*Mehelya*), 209
Prosopis (*Navicularia*), 212
Prosopis (*Nylaeus*), 208
Prosopis (*Paraprosopis*), 211
Prosopis (*Patagiata*), 208, 210
Prosopis (*Pectinata*), 208
Prosopis (*Platyspatulariella*), 213
Prosopis (*Pseudobranchiata*), 208
Prosopis (*Spatularia*), 213
Prosopis (*Spatulariella*), 213
Prosopis (*Trichota*), 208
Prosopis alcyonea, 209
Prosopis annulata, 207
Prosopis apicatus, 216
Prosopis aterrima, 203
Prosopis basalis, 203
Prosopis bifasciata, 208
Prosopis cenibera, 213
Prosopis chalybaea, 209
Prosopis cruenta, 206
Prosopis curvicarinata, 204
Prosopis cyanophila, 208
Prosopis difformis, 208
Prosopis dromedaria, 204
Prosopis eburniella, 205

Prosopis elegans, 217
Prosopis elegantissima, 217
Prosopis euxantha, 205
Prosopis facialis, 212
Prosopis facilis, 210
Prosopis fijiensis, 209
Prosopis freisei, 209
Prosopis frontalis, 286
Prosopis fulvicornis, 205
Prosopis heraldica, 210
Prosopis husela, 204
Prosopis imperialis, 217
Prosopis mexicana, 207
Prosopis morosa, 191
Prosopis nubilosa, 207
Prosopis nubilosellus, 191
Prosopis perhumilis, 212
Prosopis perviridis, 218
Prosopis philoleucus, 205
Prosopis potens, 209
Prosopis pulchricrus, 207
Prosopis punctata, 213
Prosopis quadricornis, 206
Prosopis ruficeps, 205
Prosopis sparsa, 209
Prosopis sulphuripes, 213
Prosopis trilobata, 211
Prosopis trinotata, 212
Prosopis variegata, 212
Prosopis vidua, 209
Prosopis xanthopoda, 205
Prosopisteroides, 68*t*, 193, 196*k*, 198*t*, **212**, 213
Prosopisteron, 68*t*, 188, 191, 193, 195, 197*f*, 197*k*, 198*t*, 202*k*, 203-205, 208-209, 211, **212-213**, 214
Prosopisteron serotinellum, 212
Prosopites (Tribe), 78*t*
Prosopoides, 67*t*, 181, 182*f*, 182*k*, **184**, 185
Prosopoxylocopa, 73*t*, 601*k*, 604*k*, **608**, 610
Protandrena, 79*t*-80*t*, 112, 126, 235, 273, 273*k*-274*k*, **278-281**, 283, 655
Protandrena (*Austrandrena*), 281
Protandrena (*Austropanurgus*), 69*t*, 278-279, 279*k*, **280**, 280*k*
Protandrena (*Heterosarus*), 69*t*, 94, 274-275, 275*f*, 279, 279*k*, **280**, 280*k*, 284-285, 655, 895
Protandrena (*Metapsaenythia*), 69*t*, 236*f*, 274*k*, 279*k*, **280-281**, 280*k*, 655
Protandrena (*Parasarus*), 69*t*, 278, 279*k*-280*k*, **281**, 640
Protandrena (*Protandrena* s. str.), 69*t*, 272, 272*k*, 273, 273*k*, 274-275, 278-280, 279*f*, 279*k*, **281**, 283
Protandrena (*Pseudosarus*), 273
Protandrena (*Pterosarus*), 69*t*, 278, 279*f*, 279*k*, **281**, 283, 655
Protandrena abdominalis, 280-281
Protandrena abdominalis tricolor, 236*f*
Protandrena atacamensis, 281
Protandrena bakeri, 280
Protandrena bancrofti, 114*f*
Protandrena evansi, 279

Protandrena maculata, 274, 278, 281
Protandrena meridionalis, 275
Protandrena mexicanorum, 279*f*, 281
Protandrena neomexicana, 275*f*
Protandrena nigra, 280
Protandrena platycephala, 280
Protandrena punctata, 280
Protandrena readioi, 280
Protandrena rudbeckiae, 279*f*
Protandrena s. str., 69*t*, 272, 272*k*, 273, 273*k*, 274-275, 278-280, 279*f*, 279*k*, **281**, 283
Protandrena sonorana, 281
Protandrena virescens, 280
Protandreninae (Subfamily), 79*t*
Protandrenini (Tribe), 69*t*, 272*k*, **273-284**, 273*k*-274*k*, 282*f*, 285, 306, 895
  biogeography of, 107*t*
Protandrenopsis, 282
Protandrenopsis fuscipennis, 282
Protanthidium, 536
Protepeolini (Tribe), 33, 33*t*, 73*t*, 80*t*, 116, 127, 587, 591, 591*k*, 633, 636, 667, 668*f*, 671-672, **678-679**
  biogeography of, 108*t*
Protepeolus, 80*t*, 679
Protepeolus singularis, 679
Proteraner, 388
Proteriades, 20, 71*t*, 448, 452, 461-463, 465, 465*k*, 466, 466*k*, 469-470, **473-474**, 476
Proteriades (*Penteriades*), 473
Proteriades group (of *Hoplitis*), 20
Proteriades semirubra, 473
Protoanthidium, 536
Protoanthidium rufobalteatum, 536
Protoanthidium rufomaculatum, 504
Protobombus, fossil, 99*t*, 670-671, 831, Plate 19
Protobombus indecisus, fossil, 99*t*, Plate 19
Protodiscelis, 60*f*, 67*t*, 147*t*, 148*k*, 153, **158-159**, 158*f*, 160
Protodiscelis fiebrigi, 158
Protodufourea, 69*t*, 324, 324*k*, **328-329**, 656, 658
Protodufourea parca, 326*f*
Protodufourea wasbaueri, 328
Protohalictus, 70*t*, 365*f*, 367*k*, 368, **369**
Protolithurgini (Tribe), fossil, 99*t*
Protolithurgus, fossil, 99*t*
Protolithurgus ditomeus, fossil, 99*t*
Protomelecta, fossil, 99*t*
Protomelecta brevipennis, fossil, 99*t*
Protomelissa, 775
Protomelissa iridescens, 775
Protomeliturga, 69*t*, 83, 126, **295**, 295*f*, 323
Protomeliturga turnerae, 29, 270*f*, 295*f*

Protomeliturgini (Tribe), 69*t*, 271*k*, **295**
  biogeography of, 107*t*
Protomorpha, 67*t*, 143, 145*f*, 147*t*, 149, 149*k*, 150-151, 150*k*, 153, 157, **159**
Protomorpha tarsalis, 159
Protopithitis, 73*t*, 612, 613*k*, **617**
Protosiris, 73*t*, 675, 675*k*, **676-677**
Protosiris gigos, 674
Protosiris obtusus, 677
Protosmia, 43, 448-449, 449*k*, 449*t*, 451*k*-452*k*, 461, 472, 474, 476, **487-488**, 487*k*, 489
Protosmia (*Afrosmia*), 487
Protosmia (*Chelostomopsis*), 71*t*, 487*k*, **488**, 488*f*
Protosmia (*Dolichosmia*), 71*t*, 449, 487, 487*k*, **488**
Protosmia (*Nanosmia*), 71*t*, 469, 486, 487*k*, **488**
Protosmia (*Protosmia* s. str.), 71*t*, 486, 487*k*, **488**
Protosmia asensioi, 488
Protosmia burmanica, 487-488
Protosmia capitata, 488
Protosmia rubifloris, 488, 488*f*
Protosmia s. str., 71*t*, 486, 487*k*, **488**
Protostelis, 72*t*, 530, 530*f*, 530*k*, 531, **532**
Protoxaea, 69*t*, 106, 317*k*, **318**, 652
Protoxaea (*Mesoxaea*), 317*f*, **318**
Protoxaea (*Notoxaea*), **318**
Protoxaea (*Protoxaea* s. str.), 317*f*, **318**
Protoxaea gloriosa, 124*f*, 316*f*-317*f*
Protoxaea nigerrima, 317*f*
Protoxaea s. str., 317*f*, **318**
Proxylocopa, 73*t*, 101, 592, 596-597, 597*f*, 598, 601*k*, 604*k*, **608**
Proxylocopa (*Ancylocopa*), 608
Proxylocopa s. str., 608
Psaenythia, 69*t*, 272, 273*k*, 274-275, 281, **281-282**, 284, 640
Psaenythia (*Parapsaenythia*), 278
Psaenythia argentina, 278
Psaenythia bergi, 274*f*
Psaenythia nigra, 277
Psaenythia parvula, 278
Psaenythia philanthoides, 281, 282*f*
Psammandrena, 68*t*, 247*k*, 251*k*, **265**
Pseudagapostemon, 350, 352*f*, 354, 356, 358*k*-359*k*, **386-387**, 387*k*, 388
Pseudagapostemon (*Brasilagapostemon*), 70*t*, 387, **387**, 387*k*
Pseudagapostemon (*Neagapostemon*), 70*t*, **387**, 387*k*
Pseudagapostemon (*Pseudagapostemon* s. str.), 70*t*, **387**, 387*k*
Pseudagapostemon amabilis, 387
Pseudagapostemon brasiliensis, 387
Pseudagapostemon cyanomelas, 387

Pseudagapostemon fluminensis, 387
Pseudagapostemon jenseni, 356, 387
Pseudagapostemon perzonatus, 387
Pseudagapostemon puelchanus, 356
Pseudagapostemon s. str., 70*t*, 387, 387*k*
Pseudapis, 332-333, 334*k*, 337, 341-342, 341*k*, 344, 664
Pseudapis (Pachynomia), 69*t*, 341, 341*k*, 344
Pseudapis (Pseudapis s. str.), 69*t*, 333*f*, 341-342, 341*k*, 343*f*
Pseudapis anomala, 341
Pseudapis bispinosa, 333
Pseudapis diversipes, 333
Pseudapis s. str., 69*t*, 333*f*, 341-342, 341*k*, 343*f*
Pseudapis s. strrictissimo, 341
Pseudapis umesaoi, 333*f*
Pseudaugochlora, 44, 70*t*, 101, 393, 398*k*, 400*k*, 405, **410**
Pseudaugochlora graminea, 349*f*, 395*f*-396*f*, 410*f*
Pseudaugochlora sordicutis, 394*f*, 410
Pseudaugochloropsis, 404, 410
Pseudaugochloropsis costaricensis, 405
Pseudepeolus, 73*t*, 647-648, 648*k*, 651, **651**
Pseudepeolus angustata, 651
Pseudepeolus angustatus, 651
Pseudepeolus fasciatus, 651
Pseudepeolus willinki, 648
Pseudeucera, 74*t*, 723, **723**, 723*k*
Pseudhylaeus, 68*t*, 197*k*, 198*t*, 207, 211, **213**, 214
Pseudiscelis, 67*t*, 181, 182*f*, 182*k*, 184, **185**
Pseudiscelis rostrata, 185
Pseudoanthidium, 491*t*, 493*k*, 497, 499, 508, 517, 522, **526-528**, 526*k*
Pseudoanthidium (Carinellum), 527
Pseudoanthidium (Exanthidium), 72*t*, 493, 526, **526-527**, 526*k*
Pseudoanthidium (Micranthidium), 72*t*, 526, 526*k*, **527**, 528
Pseudoanthidium (Paraanthidiellum), 527
Pseudoanthidium (Pseudoanthidium s. str.), 72*t*, 526*k*, 527
Pseudoanthidium (Royanthidium), 72*t*, 493, 526, 526*k*, **527-528**, 528
Pseudoanthidium (Semicarinella), 72*t*, 526*k*, **528**
Pseudoanthidium (Tuberanthidium), 72*t*, 492*f*, 526, 526*k*, 527, **528**
Pseudoanthidium brachiatum, 492*f*, 526, 528
Pseudoanthidium eximium, 527
Pseudoanthidium latitarse, 528
Pseudoanthidium ochrognathum, 527
Pseudoanthidium reticulatum, 493

Pseudoanthidium s. str., 72*t*, 526*k*, **527**
Pseudoanthidium truncatum, 528
Pseudoanthidium wahrmanicum, 527
Pseudobranchiata, 208
Pseudocentron, 72*t*, 551-552, 557*t*, 558*k*-559*k*, 567, 577-579, **581**
Pseudocentron (Grafella), 577
Pseudocentron (Leptorachina), 577
Pseudocentron (Moureana), 578
Pseudocentron crotalariae, 577
Pseudochilalictus, 70*t*, 371*t*, 373*f*, 374*k*, **381**
Pseudocilissa, 432
Pseudocosmia, 468
Pseudodichroa, 73*t*, 663, **664**, 664*k*
Pseudoheriades, 71*t*, 442*t*, 449, 449*t*, 450*k*-451*k*, 453, 458, 461, 475, **489**
Pseudoheriades (Afroheriades), 452
Pseudoheriades (Stenoheriades), 489
Pseudoheriades hofferi, 489
Pseudoheriades primus, 452
Pseudoliphanthus, 69*t*, 276*f*, 277, 277*k*
Pseudomacrotera, 69*t*, 300*k*, **304**
Pseudomegachile, 72*t*, 551, 555*f*, 557*t*, 558*k*, 560*k*-563*k*, 564*f*, 564*k*, 569, 572, 575, 577-578, **581**, 906
Pseudomegalochila, 581
Pseudomelecta, 75*t*, 773, 773*k*, 774, 776
Pseudoosmia, 467
Pseudo-osmia, 467
Pseudopanurgus, 69*t*, 273*k*-274*k*, 278-281, **282-283**, 655
Pseudopanurgus aethiops, 124*f*, 275*f*, 279*f*, 282*f*
Pseudopanurgus carinulatus, 282
Pseudopanurgus crenulatus, 282*f*
Pseudopanurgus fasciatus, 282
Pseudopanurgus fulvicornis, 282
Pseudopanurgus trifasciatus, 282
Pseudopanurgus trimaculatus, 282
Pseudoparasitae, 78*t*
Pseudopasites, 664
Pseudopeolus, 651
Pseudophilanthus, 426*k*, **427-428**, 427*k*
Pseudophilanthus (Dicromonia), 70*t*, **427-428**, 427*k*
Pseudophilanthus (Pseudophilanthus s. str.), 70*t*, 427*k*, **428**
Pseudophilanthus s. str., 70*t*, 427*k*, **428**
Pseudophilanthus taeniatus, 428
Pseudophilanthus tsavoensis, 426*f*-427*f*
Pseudophilanthus wenzeli, 428
Pseudosarus, 273, 280
Pseudosarus virescens, 280
Pseudoscolia (wasp), 90, 92
Pseudosmia, 467-468
Pseudostelis, 72*t*, 531*k*, 532, **532**
Psilomelissodes, 74*t*, 730*k*-731*k*, **732**
Psilylaeus, 212

Psilylaeus sagiops, 212
Psithyridae (Subfamily), 78*t*-79*t*
Psithyrides (Family), 78*t*
Psithyrus, 30, 31*t*, 37, 41, 75*t*, 76-78, 78*t*-80*t*, 623, 699, 786*f*-787*f*, 788-791, 791*k*, 794*k*, **800**, 800*f*, 802
Psithyrus (Allopsithyrus), 800
Psithyrus (Ashtonipsithyrus), 800
Psithyrus (Ceratopsithyrus), 800
Psithyrus (Eopsithyrus), 800
Psithyrus (Fernaldaepsithyrus), 800
Psithyrus (Laboriopsithyrus), 751
Psithyrus (Metapsithyrus), 800
Psithyrus cornutus, 800
Psithyrus fernaldae, 800
Psithyrus klapperichi, 800
Pteneucera, 74*t*, 724, 724*t*, **725**
Pterandrena, 259
Pterandrena pallidifovea, 266
Pterandrena pallidiscopa, 265
Pterosarus, 69*t*, 278, 279*f*, 279*k*, **281**, 283, 655
Ptilandrena, 68*t*, 244*k*, 249*k*-250*k*, 253*k*, 256*k*, **265**
Ptilocentris, 74*t*, 754-755, 755*k*, **757-758**
Ptilocleptis, 32*t*, 70*t*, 352-354, 358*k*, 361*k*, 364, 382, **387-388**, 389
Ptilocleptis polybioides, 40*f*
Ptilocleptis tomentosa, 40*f*, 355*f*, 387
Ptiloglossa, 67*t*, 174*k*, **176**, 650, 652
Ptiloglossa (Ptiloglossodes), 176
Ptiloglossa chalybaea, 175
Ptiloglossa ducalis, 176
Ptiloglossa guinnae, 174*f*
Ptiloglossa mexicana, 175*f*
Ptiloglossa tarsata, 176
Ptiloglossa zikani, 175
Ptiloglossidia, 67*t*, **179**, 179*k*
Ptiloglossidia fallax, 179
Ptiloglossidiini, 179
Ptiloglossodes, 176
Ptilomelissa, 709*f*-710*f*, 714*f*, 718*f*, 728, 732
Ptilonomia, 69*t*, 334*k*, **342**, 343
Ptilonomia plumosa, 333, 333*f*, 343*f*
Ptilopoda, 167
Ptilosaroides, 72*t*, 557*t*, 559*k*-560*k*, 579, **581-582**
Ptilosarus, 72*t*, 550, 557*t*, 559*k*-560*k*, 579, 582, **582**
Ptilothrix, 21, 74*t*, 79*t*-80*t*, 106, 111, 679, 700-701, 702*k*, 704, **705-706**
Ptilothrix bombiformis, 90*f*, 706
Ptilothrix fructifer, 702*f*
Ptilothrix plumatus, 583, 701, 705
Ptilothrix relatus, 705
Ptilothryx, 705
Ptilotopus, 18*f*, 74*t*, 753-754, 754*f*, 755*k*, 757, **758**, 759, 764
Ptilotopus americanus, 758
Ptilotrigona, 812, 827
Ptoleglossa, 268
Punchalictus, 377-378
Puncticolletes, 167
Pycnanthidium, 71*t*, 503, 503*k*, **504**

Pycnanthidium solomonis, 504
Pyganthophora, 74*t*, 744*f*, 745*k*-747*k*, 749, **750**
Pyghalictus, 377
Pygidialia, 62, 79, 79*t*
Pygnanthidiellum, 504
Pygnanthidium, 504
Pygnanthidium (Pygnanthidiellum), 504
Pygomelissa, fossil, 99*t*
Pygomelissa lutetia, fossil, 99*t*
Pygopasiphae, 60*f*, 67*t*, 147*t*, 148*k*, 157, **159**
Pygoperdita, 69*t*, 300, 301*k*, **305**
Pyrobombus, 75*t*, 792*k*, 794-795, 796*k*, **800-801**, 800*f*
Pyrobombus (Festivobombus), 798
Pyrosmia, 71*t*, 112, 448, 475-477, 477*k*, 478, 478*k*, 480-481, 485, **485-486**
Pyrrhobombus, 800
Pyrrhomelecta, 649
Pyrrhopappus, 21

Quasihesma, 68*t*, 228-229, 228*f*, 228*k*, 229, **229-230**
Quasihesma moonbiensis, 229
Quasilictus, 70*t*, 370*k*, **371**

Radoszkowskiana, 32*t*, 34, 72*t*, 543*k*, **586**
Radoszkowskiana rufiventris, 586
Ramalictus, 70*t*, 366*k*, **369**
Ramphorhina, 454
Ranthidiellum, 71*t*, 503, 503*k*, **504-505**, 511
Rapanthidium, 72*t*, 524, 524*k*, **525**
Raphidostoma, 488
Raphidostoma ceanothi, 488
Rathymus, 762
Reanthidium, 527-528
Rediviva, 16, 16*f*, 18, 71*t*, 96, 415, 429-430, 430*k*, **432**, 433, 695
Rediviva colorata, 16*f*
Rediviva emdeorum, 16*f*-17*f*
Rediviva longimanus, 16*f*
Rediviva neliana, 432
Rediviva peringueyi, 16*f*, 430*f*
Rediviva rufocincta, 16*f*
Redivivini (Tribe), 430
Redivivoides, 71*t*, 430, 430*k*, **432-433**
Redivivoides simulans, 432
Reedapis, 60*f*, 67*t*, 138*f*, 144, 147*k*, 147*t*, 151, 157, **159**
Reepenia, 69*t*, 332, 334*k*, **342**, 343
Reepenia bituberculata, 333*f*, 343*f*
Rhacandrena, 68*t*, 246*k*-247*k*, 251*k*, **265**
Rhamphorhina, 454
Rhanthidiellum, 504
Rhaphandrena, 68*t*, 245*k*, 249*k*, **265**
Rhathymidae (Subfamily), 78*t*
Rhathymini (Tribe), 33*t*, 34, 74*t*, 80*t*, 104, 127, 589, 589*k*, 668, 668*f*, 671, **762**, 762*k*, 763, 768
 biogeography of, 108*t*
Rhathymites (Tribe), 78*t*

Rhathymodes, 762
Rhathymus, 74*t,* 77, 78*t,* 80*t,* 110, 674*f,* 762, 762*k*
Rhathymus acutiventris, 762
Rhathymus bicolor, 762
Rhectomia, 70*t,* 393, 396*k,* 400*k,* 407, **410-411,** 411
Rhectomia pumilla, 410
Rhinepeolus, 73*t,* 647, 647*k*-648*k,* 651, **651**
Rhinepeolus rufiventris, 651
Rhineta, 656
Rhinetula, 70*t,* 112, 354, 356, 359*k,* 363, **388**
Rhinetula denticrus, 359*f,* 388
Rhinochaetula, 420
Rhinochaetula ogilviei, 424
Rhinochaetula plumipes, 422
Rhinocoelioxys, 72*t,* 546*k,* **550**
Rhinocollettes, 167-168
Rhinocorynura, 70*t,* 393, 396*k,* 399*k,* 411, **411,** 412
Rhinocorynura briseis, 396*f*
Rhinocorynura inflaticeps, 397*f*
Rhinomegilla, 74*t,* 746*k,* 747, 747*k,* 748-749, **750**
Rhodanthidium, 491*t,* 495*k,* 522, **528-529,** 528*k*
Rhodanthidium (Asianthidium), 72*t,* 497, 528, **528-529,** 528*k*
Rhodanthidium (Meganthidium), 72*t,* 494, 528, 528*k,* **529**
Rhodanthidium (Rhodanthidium s. str.), 72*t,* 494, 528, 528*k,* 529, **529**
Rhodanthidium (Trianthidium), 528
Rhodanthidium aculeatum, 528-529
Rhodanthidium caturigense, 528
Rhodanthidium glasunovi, 528-529
Rhodanthidium infuscatum, 529
Rhodanthidium s. str., 72*t,* 494, 528, 528*k,* 529, **529**
Rhodanthidium septemdentatum, 529
Rhodanthidium siculum, 529, Plate 8
Rhodanthidium sticticum, 529
Rhodobombus, 75*t,* 793*k,* 796*k*-797*k,* **801**
Rhodocentris, 757
Rhodohylaeus, 68*t,* 196*k,* 197*f,* 198*t,* **213**
Rhodomegachile, 72*t,* 442*t,* 552-553, 557*t,* 565*k*-566*k,* 576, **582**
Rhodosmia, 488
Rhogepeolina (Subtribe), 646
Rhogepeolus, 73*t,* 646-647, 647*k*-648*k,* 650, **651**
Rhogepeolus bigibbosus, 651
Rhopalictus, 406
Rhopaloheriades, 71*t,* 459*k,* **460-461,** 489
Rhopalolemma, 41, 73*t,* 590, 656, 656*k,* **657-658**
Rhopalolemma robertsi, 657, 658*f*
Rhopalomelissa, 333, 336-337
Rhopalomelissa (Lepidorhopalomelissa), 337
Rhopalomelissa (Trichorhopalomelissa), 337
Rhopalomelissa (Tropirhopalomelissa), 337
Rhopalomelissa hainanensis, 337
Rhopalomelissa nigra, 337
Rhopalomelissa xanthogaster, 337
Rhophites. See Rophites
Rhophites (Pararhophites), 437
Rhophites dispar, 419
Rhophites s. str., 83
Rhophites vitellinus, 418
Rhophitidae (Subfamily), 78*t*
Rhophitoides, 69*t,* **329,** 329*k*
Rhophitoides distinguendus, 329
Rhophitulus, 274*k,* **283-284,** 283*k*
Rhophitulus (Cephalurgus), 69*t,* 272, 273*k,* 280, 283, **283,** 283*k*
Rhophitulus (Panurgillus), 69*t,* 280, 283*k,* **284**
Rhophitulus (Rhophitulus s. str.), 69*t,* 273*k*-274*k,* 275, 275*f,* 283*k,* **284**
Rhophitulus frisei, 275*f,* 284
Rhophitulus s. str., 69*t,* 273*k*-274*k,* 275, 275*f,* 283*k,* **284**
Rhophitulus testaceus, 275
Rhynchalictus, 377-378
Rhynchalictus rostratus, 377
Rhynchocolletes, 167-168
Rhynchocolletes albicinctus, 167
Rhynocorynura, 411
Rhynostelis, 72*t,* 496, 497*k,* 517*k,* **518**
Rhysoceratina, 73*t,* 614*k,* **617-618,** xiv
Rhysoxylocopa, 73*t,* 602*k*-603*k,* 605, 607, **608,** 610
Rhyssomegachile, 72*t,* 557*t,* 559*k*-560*k,* 567, 579, **582**
Rhyteucera, 724
Rivalisia, 404
Rivalisia metallica, 404
Robertsonella, 71*t,* 462, 465*k*-466*k,* **474**
Robertsonella gleasoni, 474
Robustobombus, 75*t,* 793*k,* 795*k,* **801**
Rophites, 77-78, 78*t*-81*t,* 323-325, 325*k,* 327-328, **329,** 329*k,* 656-657
Rophites (Carinorophites), 326
Rophites (Cyprirophites), 326
Rophites (Dentirophites), 326
Rophites (Flavodufourea), 69*t,* 325, **329,** 329*k*
Rophites (Merrophites), 326
Rophites (Microrophites), 326
Rophites (Rhophitoides), 69*t,* **329,** 329*k*
Rophites (Rophites s. str.), 20, 69*t,* 83, 323, 325, **329,** 329*k*
Rophites algirus trispinosus, 113*f*
Rophites atrata, 326
Rophites cana, 329
Rophites canus, 329
Rophites gusenleitneri, 329
Rophites quadridentatus, 326
Rophites quinquespinosus, 329
Rophites s. str., 20, 69*t,* 323, 325, **329,** 329*k*
Rophitidae (Family), 78
Rophitinae (Subfamily), 20-21, 43, 61, 69*t,* 79, 83, 85, 95, 112, 127*k,* 319, 319*f,* 320, 320*f,* 321*k,* **322-331,** 322*f*-323*f,* 330*f,* 335, 345, 633, 656-658
biogeography of, 107*t*
of Eastern Hemisphere, 324*k*-325*k*
of Western Hemisphere, 324*k*
Rophitini (Tribe), 323
Rostratilapis, 391
Rostrohalictus, 377-378
Roussyana, fossil, 99*t,* 670-671, 831
Royanthidium, 72*t,* 493, 526, 526*k,* **527-528,** 528
Rubicundobombus, 75*t,* 791*f,* 793*k,* 795*k,* 798, **801**
Rubrihalictus, 376-377
Ruderariobombus, 802
Rufandrena, 68*t,* 251*k,* 254*k,* **265**
Rufipedibombus, 75*t,* 792*k,* 794*k,* **801**
Rufipedobombus, 801
Rufocinctobombus, 798
Rugalictus, 385
Ruginomia, 341
Ruizantheda, 70*t,* 354, 356, 359*k*-360*k,* 387, **388**
Ruizantheda (Ruizanthedella), 388
Ruizantheda divaricata, 350*f,* 388
Ruizantheda divaricatus, 113*f*
Ruizantheda mutabilis, 388
Ruizantheda proxima, 388
Ruizanthedella, 388
Ruiziella, 309
Ruzapis, 309
Ruziapis, 309

Sabulapis, 291
Sabulicola, 388
Sabulicola cirsii, 388
Sakagamilla, 823
Sakagamilla affabra, 823
Salemanthidium, 72*t,* 494, 514*k,* **515**
Salix, 20
Salvia, 20
Samba, 70*t,* 414, 423, 424*k,* **425**
Samba calcarata, 423*f*-424*f,* 425
Sambini (Tribe), 70*t,* 95, 416, 416*k,* 422, **423-425,** 423*f*-424*f,* 424*k*
biogeography of, 107*t*
Santiago, 74*t,* 711, 713*k,* 715, 716*k,* 727, 733, **734,** 895
Santiago mourei, 734
Saranthidium, 72*t,* 517, 518*k,* 520, **521**
Sarocolletes, 67*t,* 147*t,* 148*k,* 151, **160**
Sarogaster, 570
Saropoda, 748
Saropoda bombiformis, 743
Sayapis, 25, 72*t,* 439, 550-554, 555*f,* 556*t*-557*t,* 558*k,* 560*k,* 570, 573, 575, 579, **582-583,** Plate 7
Scaphandrena, 69*t,* 246*k*-247*k,* 250*k*-251*k,* 253*k,* 255*k,* **265**
Scaptotrigona, 75*t,* 810, 812, 814*k,* 817, 819-822, **823,** 826

Scaptotrigona barrocoloradensis, 810*f*
Scaptotrigona mexicana, 809*f*-810*f,* 813*f*
Scaura, 75*t,* 805, 812, 814, 822, **822-823,** 822*f,* 822*k,* 824
Sceliphron (wasp), 463, 481
Schisthemisia, 74*t,* 755*k,* **758**
Schizandrena, 265
Schizocoelioxys, 547
Schizomegachile, 72*t,* 557*t,* 565*k*-566*k,* 576, **583**
Schmiedeknechtia, 73*t,* 653, 653*k,* 655, **655**
Schmiedeknechtia (Cyrtopasites), 655
Schmiedeknechtia oraniensis, 655
Schmiedeknechtia verhoeffi, 655
Schoenherria, 608
Schönherria, 608
Schonnherria, 73*t,* 598, 598*f,* 599*k*-601*k,* 605-607, **608,** 609
Schrottkya, 690
Schrottkyapis, 72*t,* 551-552, 556*t*-557*t,* 558*k,* 560*k,* 570, 575, 582, **583**
Schwarziana, 75*t,* 807, 822*k,* **823**
Schwarzula, 812, 822
Scirtetica, 721
Scirtetica antarctica, 721
Scirteticops, 721
Scitandrena, 69*t,* 252*k,* 254*k,* **265**
Scoliandrena, 69*t,* 244*k,* 250*k,* **265-266**
Scrapter, 67*t,* 112, 133, 136, 142, **171-172,** 171*f*-172*f,* 220, 287, 664, 699
Scrapter albitarsis, 171
Scrapter armatipes, 172
Scrapter bicolor, 171
Scrapter brullei, 287
Scrapter calx, 171
Scrapter carinata, 231
Scrapter heterodoxus, 171*f,* 172, Plate 4
Scrapter niger, 171
Scrapter nitidus, 171*f*
Scrapterinae, 905
Scrapteroides, 287
Scrapteroides difformis, 287
Scrapteropsis, 69*t,* 244*k,* 250*k,* 259, **266**
Scraptrinae (Subfamily), 136, 905
Scraptrini (Tribe), 67*t,* 90, 134, 136, 137*k,* 905
biogeography of, 107*t*
Seladonia, 70*t,* 365*f,* 366, 367*k,* 368, **369,** 370
Sellalictus, 70*t,* 371*t,* 373, 374*k,* **381-382**
Semicarinella, 72*t,* 526*k,* **528**
Senexibombus, 75*t,* 793, 793*k,* 796*k,* 799, **801**
Separatobombus, 75*t,* 792*k,* 795*k,* **801**
Serapis, 529
Serapis denticulatus, 529
Serapista, 72*t,* 491*t,* 493*k,* 495*k,* **529**
Sericogaster, 68*t,* 223*k,* **232**
Sericogaster fasciatus, 223*f,* 232

Sericohalictus, 380
Series Apiformes, 64-65
Series Spheciformes, 64-65
Severanthidium, 71*t*, 505, 506*k*, 507-508, **508**
Shornherria, 608
Sibericobombus, 75*t*
Sibiricobombus, 801
Sibiricobombus, 792*k*, 794, 794*k*, 796, 796*k*, **801**
Sigmatapis, 831
Silphium, 21
Simandrena, 69*t*, 246*k*-247*k*, 250*k*-252*k*, 255*k*, 257*k*-258*k*, **266**
Simanthedon, 74*t*, 708, 718*k*-719*k*, 729, **734**
Simanthedon linsleyi, 708*f*, 711*f*, 734
Simcolletes, 167
Simiceratina, 73*t*
Simioceratina, 614*k*, 616, **618**
Simpanurgus, 69*t*, 288*k*, **289**
Sinomacropis, 70*t*, 430, **431**, 431*k*
Sinomelecta, 75*t*, 771, 771*k*, 774-775
Sinomelecta oreina, 774
Smeathhalictus, 377
Smithia, 231
Solamegilla, 749
Solenopalpa, 261
Solenopalpa fertoni, 261
Soliapis, fossil, 99*t*, 393
Sophrobombus, fossil, 99*t*, 674
Sophrobombus fatalis, fossil, 99*t*
Soroeensibombus, 799
Spatularia, 213
Spatulariella, 68*t*, 198*k*, 198*t*, 199*k*-200*k*, 201*f*, 201*k*, 210, **213**
Spatulariella helenae, 213
Spatunomia, 69*t*, 334, 334*k*, **343**
Sphaerhylaeus, 68*t*, 188, 195*f*, 196*k*, 198*t*, **213-214**
Sphecidae (Family) (wasps), 62-65, 65*t*
Spheciformes (wasps), 3, 63, 65, 65*t*, 88*f*
  fossil, 100
Sphecinae (Subfamily) (wasps), 62-63
Sphecodes, 30-34, 31*t*-32*t*, 36, 70*t*, 77-78, 78*t*-81*t*, 141, 278, 319-320, 348, 352-354, 359*k*, 361*k*, 364, 382-383, 387-388, **388-389**, 784
Sphecodes (*Austrosphecodes*), 388-389
Sphecodes (*Callosphecodes*), 388-389
Sphecodes (*Microsphecodes*), 382
Sphecodes antennariae, 388
Sphecodes antipodes, 140
Sphecodes carolinus, 355*f*
Sphecodes chilensis, 40*f*, 388
Sphecodes confertus, 388
Sphecodes cribrosa, 337
Sphecodes falcifer, 388
Sphecodes gibbus, 320*f*, 356*f*
Sphecodes kathleenae, 382
Sphecodes monilicornis, 35*f*, 38*f*, 40*f*-41*f*, 360*f*
Sphecodes ranunculi, 388

Sphecodes scotti, 365
Sphecodes stygius, 388
Sphecodes texana, 382
Sphecodidae (Subfamily), 78*t*
Sphecodina (Subtribe), 320, 354
Sphecodinae (Subfamily), 79, 79*t*
Sphecodium, 388
Sphecodium cressonii, 388
Sphecodogastra, 20, 70*t*, 320*f*, 348, 371*t*, 373*k*, **382**
Sphecodopsis, 661*k*, 663, **664**, 664*k*
Sphecodopsis (*Pseudodichroa*), 73*t*, 663, **664**, 664*k*
Sphecodopsis (*Pseudopasites*), 664
Sphecodopsis (*Sphecodopsis* s. str.), 73*t*, **664**, 664*k*
Sphecodopsis s. str., 73*t*, **664**, 664*k*
Sphecodosoma, 322, 324*k*, 325, 329, **330**, 330*k*
Sphecodosoma (*Michenerula*), 69*t*, 328, **330**, 330*k*
Sphecodosoma (*Sphecodosoma* s. str.), 69*t*, **330**, 330*k*
Sphecodosoma dicksoni, 8
Sphecodosoma pratti, 320*f*, 323*f*, 330
Sphecodosoma s. str., 69*t*, **330**, 330*k*
Sphecogaster, 217
Sphecoidea (superfamily) (wasps), 65. See also Apoidea
Sphecophala, 344
Sphegocephala, 69*t*, 334*k*, 336, 338, **344**
Sphegocephala philanthoides, 344
Sphegodes, 388
Sphex (wasps), 65
Sphex gibba, 388
Sphex signata, 212
Spinanthidiellum, 72*t*, 494, 525, 525*k*, 526, **526**
Spinanthidium, 72*t*, 494, 525, 525*k*, **526**
Spinanthidium (*Spinanthidiellum*), 526
Spinanthidium volkmanni, 114*f*
Spinasternella, 489
Spinasternella mevatus, 489
Spinokelita, 73*t*, **640**, 640*k*
Spinolapis, 67*t*, 144, 147*f*, 147*t*, 148*k*, 157, 158*f*, 159, **160**
Spinoliella, 69*t*, 306, 307*k*, 309, 314, **314-315**
Spinoliella (*Claremontiella*), 312
Spinoliella (*Peniella*), 314
Spinoliella euxantha, 312
Spinoliella nomadoides, 314*f*
Spinopasites, 73*t*, 662*k*, 663, **665**
Spinopasites spinotus, 665
Steganomus, 69*t*, 332, 334*k*, **344**
Steganomus javanus, 344
Stegocephala, 344
Stelidae (Family), 79*t*
Stelidae (Subfamily), 78*t*
Stelidiella, 533
Stelidina, 533
Stelidinae (Subfamily), 79*t*
Stelidini (Tribe), 79*t*
Stelidium, 388, 533
Stelidium trypetinum, 533

Stelidomorpha, 72*t*, 495, 530, 530*k*, **532**
Stelis, 510
Stelis, 33-34, 33*t*, 78*t*-81*t*, 116, 442*t*, 491*t*, 494, 495*k*-496*k*, 502, 517, 522-524, **529-533**, 530*f*, 530*k*, 537-538
Stelis (*Dolichostelis*), 72*t*, 512*f*, 517, 530, 530*f*, **531**, 531*f*, 531*k*
Stelis (*Heterostelis*), 72*t*, 530*f*, 530*k*-531*k*, **531**, 532
Stelis (*Hoplostelis*), 784
Stelis (*Leucostelis*), 533
Stelis (*Malanthidium*), 72*t*, 497, 530, 530*k*, **532**
Stelis (*Pavostelis*), 533
Stelis (*Protostelis*), 72*t*, 530, 530*f*, 530*k*, 531, **532**
Stelis (*Pseudostelis*), 72*t*, 531*k*, 532, **532**
Stelis (*Stelidiella*), 533
Stelis (*Stelidina*), 533
Stelis (*Stelidomorpha*), 72*t*, 495, 530, 530*k*, **532**
Stelis (*Stelis* s. str.), 72*t*, 530*f*, 530*k*-531*k*, 531, **532-533**
Stelis abnormis, 517
Stelis annulata, 531
Stelis anthidioides, 531
Stelis australis, 530*f*
Stelis bidentata, 540
Stelis freygessneri, 532
Stelis hemirhoda, 533
Stelis hurdi, 531
Stelis labiata, 533
Stelis lateralis, 533
Stelis laticincta, 512*f*, 531
Stelis louisae, 530*f*
Stelis malacensis, 497, 532
Stelis malaisei, 511
Stelis montana, 533
Stelis rubi, 497*f*, 530*f*, 532
Stelis rudbeckiarum, 531*f*
Stelis s. str., 72*t*, 530*f*, 530*k*-531*k*, 531, **532-533**
Stelis signata, 532
Stelis simillima, 530
Stelis strandi, 532
Stellenigris, 72*t*, **586**
Stellenigris vandeveldii, 586
Stelodides, 72*t*, 551-553, 556*t*-557*t*, 558*k*, 560*k*, 570, 582, **583**
Stenandrena, 266
Stenanthidiellum, 71*t*, 511, **511-512**, 511*k*
Stenanthidium, 506-507
Stenocolletes, 69*t*; 138, 281, **284**
Stenocolletes pictus, 273, 284
Stenodiscelis, 182
Stenoheriades, 71*t*, 105, 449, 449*t*, 450*k*-451*k*, 453, 461, 487, **489**
Stenoheriades asiaticus, 451
Stenoheriades coelostoma, 451
Stenohesma, 68*t*, 223*k*, **232**
Stenohesma nomadiformis, 232
Stenomegachile, 72*t*, 557*t*, 563*k*-564*k*, 568, 575, **583**
Stenomelissa, 69*t*, 251, 256*k*, 258*k*, 264, **266**
Stenosmia, 71*t*, 449*t*, 450, 450*k*-451*k*, 462, **489**
Stenostylus, 289
Stenothisa, 651

Stenotritidae (Family), 18, 65*t*, 67*t*, 91-94, 123*k*, 125*f*, 126, 127*k*, **129-131**, 129*f*, 130, 130*k*, 133, 173, 235, 316
  biogeography of, 107*t*
  stenotritids. See Stenotritidae (Family)
Stenotritinae (Subfamily), 80*t*
Stenotritus, 67*t*, 79*t*-80*t*, 93, 129, 130*k*, 131, **131**, 133
Stenotritus (*Ctenocolletes*), 130
Stenotritus elegans, 131
Stenotritus nicholsoni, 130, 130*f*
Stenotritus pubescens, 129*f*
Stenoxylocopa, 73*t*, 598, 600*k*, **608-609**
Stictonomia, 341
Stictonomia punctata, 341
Stilbeucera, 724-725
Stilbochlora, 70*t*, 408, **409**, 409*k*
Stilbomalopsis, 73*t*, 683*k*, **684**
Stilpnosoma, 227
Stilpnosoma clypeata, 230
Stilpnosoma laevigatum, 227
Stilpnosoma semisericea, 232
stingless bees, 43. See also meliponine bees (Meliponini)
Strandiella, 171
Strandiella longula, 171
Suandrena, 69*t*, 252*k*, 255*k*, **266**
Subterraneibombus, 801
Subterraneobombus, 75*t*, 793*k*, 794, 796, 796*k*-797*k*, **801-802**
Succinapis, fossil, 99*t*, 670-671, Plate 19
Succinapis goeleti, fossil, Plate 19
Succinapis proboscidia, fossil, 99*t*
Sudila, 70*t*, 371*t*, 373, 374*k*, 378, 381*f*, **382**
Sudila bidentata, 382
Sulcobombus, 798
Sundatrigona, 826
Svastra, 652, 709, 713, 714*k*, 716, 716*k*, 719, 719*k*, 727-730, **734-735**, 734*k*
Svastra (*Anthedonia*), 20, 74*t*, 718*k*, 720*k*, **734-735**, 734*k*
Svastra (*Brachymelissodes*), 74*t*, 734*k*, **735**
Svastra (*Epimelissodes*), 74*t*, 715*f*, 734*k*, **735**
Svastra (*Idiomelissodes*), 74*t*, 715*f*, 718*f*, 734*k*, **735**
Svastra (*Svastra* s. str.), 74*t*, 734, 734*k*, **735**
Svastra bombilans, 735
Svastra duplocincta, 715*f*, 718*f*, 735
Svastra fulgurans, 728
Svastra obliqua, 19*f*, 37*f*, 41*f*, 709*f*, 715*f*
Svastra s. str., 74*t*, 734, 734*k*, **735**
Svastra sapucacensis, 728
Svastra subapicalis, 735
Svastrides, 74*t*, 648, 714*k*, 716*k*, 727, 734, **735**, 895
Svastrina, 74*t*, 111, 709, 713*k*, 716, 722, 733, **735-736**
Svastrina anaroliata, 735

Svastrina subapicalis, 736
Svastropsis, 74*t*, 714*k*, 729*k*, **730**
sweat bees, 12, 21, 319. *See also* Halictidae (Family); Halictinae
Symmorpha, 774
Symphata, 101
Synalonia, 725
Synapis, 831
  fossil, 99*t*
Synepeolus, 651
Synhalonia, 74*t*, 120, 652, 708-710, 711*f*, 713*f*, 715*f*, 718*k*, 720*k*, 723-724, 724*k*, **725-726**, 736-737, 739, 770
Synhalonia albicans, 682
Synocoelioxys, 72*t*, 546*k*, **550**
Syntrichalonia, 74*t*, 718*k*, 720*k*, **736**
Syntrichalonia exquisita, 715*f*
Systropha, 20, 43, 69*t*, 77, 78*t*-81*t*, 92, 322, 324, 325*k*, 327, **330-331**, 335, 656-657
Systropha (*Systrophidia*), 330
Systropha curvicornis, 125*f*, 319*f*, 322*f*
Systropha ogilviei, 330
Systropha planidens, 322
Systrophidia, 330

Tachymelissodes, 74*t*, 730*k*-731*k*, **732**
Tachysphex (wasps), 165
Tachytes (wasps), 63*f*
Taeniandrena, 69*t*, 247*k*, 250*k*, 253*k*-254*k*, 257*k*-258*k*, **266**
Tanguticobombus, 799
Tapinorhina, 690
Tapinorrhina, 690
Tapinotapis, 692
Tapinotaspidini (Tribe), 671
Tapinotaspidini, 33*t*
Tapinotaspidini (Tribe), 18, 35, 73*t*-74*t*, 104, 589*k*, 591*k*, 667, 668*f*, 674-675, 680, **687-693**, 687*f*-688*f*, 688*k*-689*k*, 689*f*, 691*f*, 693*f*, 700
  biogeography of, 108*t*
Tapinotaspis, 74*t*, 81*t*, 687, 688*k*, 690, **692**
Tapinotaspis (*Tapinorhina*), 690
Tapinotaspis chacabucensis, 692
Tapinotaspoides, 74*t*, 687-688, 689*k*, 690, **692-693**
Tapinotaspoides tucumana, 674, 693*f*
Tarsalia, 73*t*, 84, 685, 685*k*, **686**
Tarsalia ancyliformis mediterranea, 685*f*
Tarsalia hirtipes, 686
Tarsandrena, 69*t*, 253*k*, 255*k*-256*k*, **266**
Teleutemnesta, 705
Teleutemnesta fructifera, 705
Temnosoma, 32*t*, 70*t*, 79*t*-80*t*, 116, 348, 352, 394, 394*k*, 399*k*, **411**, 411*f*
Temnosoma (*Temnosomula*), 411
Temnosoma metallicum, 411
Temnosoma platensis, 411
Temnosoma smaragdinum, 349*f*, 395*f*-396*f*, 404*f*

Temnosoma sphaerocephala, 411
Temnosomula, 411
Teratognatha, 73*t*, 668*f*, 680, 681*k*, 682, **684**
Teratognatha modesta, 682*f*, 684
Teratognathina (Subtribe), 681
Teratognathini (Tribe), 680
Tergosmia, 71*t*, 475-476, 477*k*, 478, 478*k*, 484-485, **486**
Terrestribombus, 797
Tetrachlora, 404
Tetraglossula, 67*t*, 147*t*, 148*k*, 153-154, 158, 158*f*, **160**, 163
Tetraglossula deltivaga, 160
Tetragona, 75*t*, 808, 812, 820*f*, 824-826, 825*k*, 827, **827-828**
Tetragonilla, 826
Tetragonisca, 75*t*, 807-808, 825, 825*k*, 826-827, **828**, 829
Tetragonula, 803, 805, 824, 826
Tetralonia, 34, 165, 680, 708, 720*k*, 721, 723, 725-726, 736, **736-737**, 736*k*
Tetralonia (*Eucara*), 74*t*, 736, 736*f*, 736*k*, 737, **737**
Tetralonia (*Tetralonia* s. str.), 74*t*, 736*k*, 737, **737**
Tetralonia (*Thygatina*), 74*t*, 106, 736-737, 736*k*, **737**
Tetralonia bipunctata, 730
Tetralonia decorata, 729
Tetralonia duckei, 733
Tetralonia fervens, 733
Tetralonia gayi, 721
Tetralonia gilva, 721
Tetralonia hoozana, 775
Tetralonia leucocephala, 733
Tetralonia macrognatha, 736*f*
Tetralonia malvae, 20, 736-737
Tetralonia melanura, 735
Tetralonia melectoides, 726
Tetralonia melonis, 733
Tetralonia mimetica, 728
Tetralonia mirabilis, 740
Tetralonia nigriceps, 738
Tetralonia niveata, 682
Tetralonia s. str., 74*t*, 736*k*, 737, **737**
Tetralonia spiniventris, 728
Tetralonia tarsata, 750-751
Tetralonia terminata, 739
Tetraloniella, 20, 120, 545, 573, 652, 662, 708, 710, 719*k*-720*k*, 723-724, 726, 735-736, 737-739, 738*k*
Tetraloniella (*Glazunovia*), 74*t*, **738**, 738*k*
Tetraloniella (*Loxoptilus*), 74*t*, 711*f*, 718*k*, 720*k*, 737, **738**, 738*k*
Tetraloniella (*Pectinapis*), 74*t*, 711*f*, 718*k*, 718*k*, 720*k*, 737, **738**, 738*k*
Tetraloniella (*Tetraloniella* s. str.), 74*t*, 711*f*, 717*f*, 736*f*, 737, **738-739**, 738*k*
Tetraloniella albata, 711*f*, 716*f*-717*f*
Tetraloniella auricauda, 738
Tetraloniella dentata, 739
Tetraloniella fasciata, 738
Tetraloniella graja, 737
Tetraloniella intermedia, 737
Tetraloniella junodi, 736*f*

Tetraloniella longifellator, 711*f*
Tetraloniella mastrucata, 737
Tetraloniella nigriceps, 738
Tetraloniella ruficornis, 723
Tetraloniella s. str., 74*t*, 711*f*, 717*f*, 736*f*, 737, **738-739**, 738*k*
Tetralonioidella, 75*t*, 589, 770, 771*k*, 774, **775**, 776
Tetranthidium, 506-507
Tetrapaedia, 695
Tetrapedia, 33*t*, 74*t*, 80*t*-81*t*, 116, 589, 691, 694, 694*f*, 694*k*, **695-696**
Tetrapedia clypeata, 694*f*, 696
Tetrapedia diversipes, 695, 695*f*
Tetrapedia globulosa, 692
Tetrapedia goeldiana, 690
Tetrapedia haeckeli, 692
Tetrapedia iheringii, 692
Tetrapedia muelleri, 690
Tetrapedia ornata, 696
Tetrapedia peckoltii, 694*f*
Tetrapedia pygmaea, 692
Tetrapedia serraticornis, 692
Tetrapediini (Tribe), 18, 33*t*, 74*t*, 81, 104, 127, 589*k*, 667, 668*f*, 671, **694-696**, 694*f*, 694*k*, 695*f*
  biogeography of, 108*t*
Tetrapedium, 695
Tetraperdita, 303
Tetrigona, 826
Thalestria, 73*t*, 646-647, 647*k*, 648, 651, **651**
Thalestria smaragdina, 651
Thalestria spinosa, 651
Thalestriina (Subtribe), 647
Thaumastombus, fossil, 99*t*, 670-671
Thaumastombus andreniformis, fossil, 99*t*
Thaumatosoma, 72*t*, 552, 554*f*, 555, 557*t*, 565*k*-566*k*, **583-584**
Thaumatosoma burmanicum, 488
Thaumatosoma duboulaii, 583
Thaumatosoma moniliferum, 487
Thectochlora, 70*t*, 396*k*, 399*k*, **411**
Thectochlora alaris, 411
Thladiantha, 18
Thoracobombus, 30, 31*t*, 75*t*, 789, 793, 793*k*, 796*k*, 799, 802, **802**
Thrausmus, 388-389
Thrausmus grandidieri, 388
Thrinchostoma, 43, 106, 348, 354, 356, 361*k*, 383, 385*f*, **390-392**, 390*k*
Thrinchostoma (*Diagonozus*), 70*t*, **390-391**, 390*f*, 390*k*
Thrinchostoma (*Eothrincostoma*), 70*t*, 383, 385*f*, 390*k*, **391**
Thrinchostoma (*Thrinchostoma* s. str.), 70*t*, 385*f*, 390, 390*k*, 391, **391-392**
Thrinchostoma afasciatum, 385*f*
Thrinchostoma lettowvorbecki, 390*f*
Thrinchostoma productum, 385*f*, 391
Thrinchostoma renitantely, 391

Thrinchostoma s. str., 70*t*, 385*f*, 390, 390*k*, **391-392**
Thrincohalictus, 70*t*, 354, 361*k*, **389-390**
Thrincohalictus prognathus, 390
Thrincostoma, 391
Thrincostoma serricorne, 391
Thrincostomina (Subtribe), 354, 390
Thygater, 44, 708-710, 711*f*, 711*k*, 715*k*, 717*k*, 719*k*, 721, 729, **739-740**, 739*k*, 740
Thygater (*Nectarodiaeta*), 74*t*, 710, **739**, 739*k*, 740
Thygater (*Thygater* s. str.), 74*t*, **739-740**, 739*k*
Thygater analis, 709*f*-711*f*, 715
Thygater s. str., 74*t*, **739-740**, 739*k*
Thygatina, 74*t*, 106, 736-737, 736*k*, 737, **737**
Thygatina fumida, 737
Thynnus abdominalis, 516
Thyreomelecta, 75*t*, 772*k*, **775**
Thyreomelecta kirghisia, 775
Thyreothremma, 732
Thyreothremma paraguayensis, 732
Thyreothremma rhopalocera, 732
Thyreotremata, 732
Thyreus, 75*t*, 340, 770-771, 771*k*, 775, **775-776**, Plate 12
Thyreus histrionicus, Plate 12
Thyreus ramosus, 765*f*
Thysandrena, 69*t*, 247*k*, 251*k*, 253*k*, 255*k*, **266**
Tiphia brevicornis, 656
Titusella, 454
Titusella pronitens, 454
Tmetocoelia, 408
Torocolletes, 67*t*, 147*t*, 156, **160**
Toromelissa, 74*t*, 701*k*, 704, **706**
Toromelissa nemaglossa, 706
Torridapis, 72*t*, 543, 546*f*, 547*k*, 549, **550**
Townsendiella, 73*t*, **659**
Townsendiella (*Eremopasites*), 659
Townsendiella (*Xeropasites*), 659
Townsendiella californica, 634*f*, 658*f*, 659, 659*f*
Townsendiella pulchra, 636*f*, 659
Townsendiella rufiventris, 659
Townsendiellini (Tribe), 73*t*, 590*k*, 656, **659**, 660
  biogeography of, 108*t*
Toxeriades, 71*t*, 459*k*, **461**
Trachandrena, 69*t*, 244*k*, 250*k*, 252*k*, 254*k*, 256*k*, 258*k*, 259, 266, **266**
Trachina, 74*t*, 753-755, 755*k*, 758, 759*f*, 778*f*, 781
Trachusa, 20, 79*t*, 111, 491, 491*f*, 494*k*-495*k*, 497*k*, 507, 509-510, 515, 530-532, **533-537**, 533*k*-534*k*, 537, 573
Trachusa (*Archianthidium*), 72*t*, 533, 533*k*, **534-535**, 538
Trachusa (*Congotrachusa*), 72*t*, 533, 534*k*, **535**
Trachusa (*Heteranthidium*), 72*t*, 509, 531, 533, 534*f*, 534*k*, **535**, 536-537, Plate 8

Trachusa (Legnanthidium), 72*t,* 534, 534*k,* **535,** 537-538
Trachusa (Massanthidium), 72*t,* 533, 534*k,* **535**
Trachusa (Metatrachusa), 72*t,* 533, 534*k,* **535-536,** 536
Trachusa (Orientotrachusa), 527
Trachusa (Orthanthidium), 72*t,* 533, 534*k,* 536, **536**
Trachusa (Paraanthidium), 72*t,* 510, 524, 534*k,* 535, **536**
Trachusa (Trachusa s. str.), 72*t,* 533, 533*k,* **536,** 545, 578
Trachusa (Trachusomimus), 72*t,* 495*f,* 531, 533, 534*k,* **536-537**
Trachusa (Ulanthidium), 72*t,* 492*f,* 534*k,* 535, **537,** 749
Trachusa aquifilum, 536
Trachusa bequaerti, 534
Trachusa byssina, 536
Trachusa catinula, 531
Trachusa cincta, 540
Trachusa cordaticeps, Plate 8
Trachusa formosana, 536
Trachusa gummifera, 495*f*
Trachusa interdisciplinaris, 537
Trachusa interrupta, 536
Trachusa larreae, 509
Trachusa longicornis, 536
Trachusa occidentalis, 535
Trachusa orientalis, 535
Trachusa ovata, 536
Trachusa pendleburyi, 535
Trachusa perdita, 536
Trachusa pueblana, 492*f*
Trachusa ridingsii, 535
Trachusa s. str., 72*t,* 533, 533*k,* **536,** 545, 578
Trachusa schoutedeni, 535
Trachusa serratulae, 536
Trachusinae (Subfamily), 79*t*
Trachusoides, 72*t,* 442*t,* 443, 491*t,* 494*k,* 509, **537**
Trachusoides simplex, 537
Trachusomimus, 72*t,* 495*f,* 531, 533, 534*k,* **536-537**
Trachyrhiza, 68*t,* 216*k,* **218**
Trianthidiellum, 512
Trianthidium, 528
Trichanthidiodes, 72*t,* 523*k,* **524**
Trichanthidium, 72*t,* 523, 523*k,* **524,** 527
Trichchostoma, 391
Trichinosmia, 71*t,* 479*k,* 483, **486**
Trichocentris, 757
Trichocerapis, 711*k,* 715*k,* 727, 729, **740,** 740*k*
Trichocerapis (Dithygater), 74*t,* **740,** 740*k*
Trichocerapis (Trichocerapis s. str.), 74*t,* 712*f,* **740,** 740*k*
Trichocerapis mirabilis, 712*f*
Trichocerapis s. str., 74*t,* 712*f,* **740,** 740*k*
Trichocerapis seabrai, 740
Trichocolletes, 139*k,* 144-145, 161, 164-165, **165-166,** 166*k*
Trichocolletes (Callocolletes), 67*t,* **166,** 166*k*
Trichocolletes (Trichocolletes s. str.), 67*t,* **166,** 166*k*
Trichocolletes hackeri, 166

Trichocolletes pulcherrimus, 166
Trichocolletes s. str., 67*t,* **166,** 166*k*
Trichocolletes venustus, 147*f,* 162*f*
Tricholiphanthus, 69*t,* 276*f,* **277-278,** 277*k*
Trichonomada, 73*t,* 112, 639, 639*k,* 640, **641**
Trichonomada roigella, 635*f,* 641
Trichopasites, 655
Trichorhopalomelissa, 337
Trichostoma, 391
Trichota, 208
Trichothurgus, 71*t,* 444-445, 445*k,* 446, **447**
Trichothurgus dubius, 444*f*
Trichothurgus wagenknechti, 444*f*
Trichotosmia, 490
Trichotrigona, 75*t,* 112, 805, 812, 813*k,* **823-824,** 824
Trichotrigona extranea, 823-824
Trichurochile, 72*t,* 555, 557, 557*t,* 559*k*-560*k,* **584**
Tricornibombus, 75*t,* 793*k,* 796*k,* **802**
Tridentosmia, 467, 472
Triepeolus, 35, 42, 73*t,* 639*f,* 646, 646*f,* 647, 647*f,* 648, 648*k,* **651-652,** 653
Triepeolus (Synepeolus), 651
Triepeolus concavus, 37*f,* 39*f,* 41*f,* 647*f*
Triepeolus epeolurus, 652
Triepeolus grandis, 636*f*
Triepeolus insolitus, 651
Triepeolus roni, 647
Triepeolus minimus, 649
Triepeolus texanus, 647*f*
Triepeolus tristris, 652
Triepeolus ventralis, 652
Triepeolus verbesinae, 646*f*
Triepicharis, 74*t,* 754, 760*k,* **761**
Trigona, 3, 12, 21-22, 28-29, 42, 60, 103, 109, 503-504, 587, 671, 687, 691, 695, 803, 807-812, 813*k,*
Trigona (cont'd.)
 815, 815*k,* 816-817, 820, 824, **824-829,** 906
 fossil, 809
 Indo-Australian, 825*k*
 neotropical, 825*k*
Trigona (Cephalotrigona), 815
Trigona (Duckeola), 75*t,* 820*f,* 824, **825,** 825*k,* 826-827
Trigona (Frieseomelitta), 75*t,* 804, 807, 820*f,* 823-825, **825-826,** 825*k,* 826-828
Trigona (Geotrigona), 75*t,* 812, 821, 825*k,* 826, **826,** 829
Trigona (Heterotrigona), 75*t,* 804-806, 807*f,* 824-825, 825*k,* **826,** 826-829
Trigona (Homotrigona), 75*t,* 825*k,* 826, **826-827,** 827
Trigona (Hypotrigona), 816
Trigona (Lepidotrigona), 75*t,* 812, 822, 824, 825*k,* 827, **827**
Trigona (Lestrimelitta), 816
Trigona (Lophotrigona), 826
Trigona (Meliponula), 819

Trigona (Odontotrigona), 826
Trigona (Oxytrigona), 820
Trigona (Papuatrigona), 75*t,* 812, 822, 824, 825*k,* **827**
Trigona (Parapartamona), 821
Trigona (Paratrigona), 820
Trigona (Partamona), 821
Trigona (Patera), 821
Trigona (Platytrigona), 826
Trigona (Plebeia), 822
Trigona (Proplebeia), fossil, 99*t,* 812
Trigona (Ptilotrigona), 827
Trigona (Scaptotrigona), 823
Trigona (Scaura), 822
Trigona (Schwarziana), 823
Trigona (Sundatrigona), 826
Trigona (Tetragona), 75*t,* 820*f,* 824-826, 825*k,* 827, **827-828**
Trigona (Tetragonisca), 75*t,* 807, 825, 825*k,* 826-827, **828,** 829
Trigona (Tetragonula), 824, 826
Trigona (Trigona s. str.), 75*t,* 787*f,* 810-811, 813*f,* 820-821, 825*k,* 826, 828, **828-829,** Plate 9
Trigona amalthea, 813*f*
Trigona angustula, 828
Trigona apicalis, 826
Trigona atripes, 826
Trigona beccarii, 818
Trigona bottegoi, 817
Trigona butteli, 819
Trigona canifrons, 804, 824-826
Trigona capitata, 815
Trigona carbonaria, 100, 807*f,* 824
Trigona cassiae, 815
Trigona cilipes, 787*f,* 828-829
Trigona clavipes, 820*f*
Trigona corvina, 828
Trigona dominicana, fossil, 99*t*
Trigona duckei, 829
Trigona elongata, 827
Trigona eocenica, fossil, 99*t*
Trigona ferricauda, Plate 9
Trigona fimbriata, 826-827
Trigona flaveola mediorufa, 820
Trigona flaviventris, 824
Trigona fulviventris, 828
Trigona fuscobalteata, 806
Trigona genalis, 827
Trigona ghilianii, 820*f,* 825
Trigona gribodoi, 816
Trigona haematoptera, 826
Trigona heideri, 827
Trigona hockingsi, 824-825
Trigona huberi, 825
Trigona hypogea, 828-829
Trigona iridipennis, 824, 826
Trigona itama, 826-827
Trigona jaty, 828
Trigona keyensis, 824
Trigona latitarsis, 822
Trigona lendliana, 818
Trigona limao, 816
Trigona longicornis, 829
Trigona longitarsis, 829
Trigona lurida, 824, 827
Trigona madecassa, 817
Trigona mombuca, 826
Trigona moorei, 826
Trigona mosquito, 822
Trigona muelleri, 829
Trigona nebulata, 818

Trigona necrophaga, 828
Trigona nitidiventris, 827
Trigona paranigra, 824
Trigona pendleburyi, 821
Trigona planifrons, 824, 826
Trigona postica, 823
Trigona prisca, fossil, 98*t,* 100*t*
Trigona s. str., 75*t,* 787*f,* 810-811, 813*f,* 820-821, 825*k,* 826, 828, **828-829,** Plate 9
Trigona savannensis, 820*f*
Trigona silvestrii, 825
Trigona staudingeri, 816
Trigona terminata, 824
Trigona thoracica, 826
Trigona timida, 822
Trigona trochanterica, 827
Trigona varia, 807
Trigona williana, Plate 9
Trigona zonata, 821
Trigonella, 826
Trigonisca, 75*t,* 805, 807-810, 812, 813*k,* 817, 821, 825, **829**
Trigonisca buyssoni, 804*f,* 810*f*
Trigonisca longicornis, 809*f*
Trigonisca longitarsis, 813*f,* 829
Trigonisca muelleri, 829
Trigonopedia, 74*t,* 687, 688*f,* 689*k,* **693**
Trigonopedia oligotricha, 13*f,* 693, 693*f*
Trilia, 326-328
Trinchostoma, 391
Trinomia, 69*t,* 335, 336*k,* 338, **338**
Triopasites, 73*t,* 639, 639*k,* **641-642**
Triopasites penniger, 642*f*
Triopasites timberlakei, 641
Troandrena, 69*t,* 253*k*-254*k,* **266**
Tropandrena, 262
Trophocleptria, 647, 647*f,* 648-650, 650*k,* **653**
Trophocleptria variolosa, 649, 653
Tropidopedia, 74*t,* 691*k,* **692**
Tropirhopalomelissa, 337
Truncandrena, 265
Trypetes, 460
Trypetes productus, 460
Trypetini (Tribe), 79*t,* 448
Trypetoidea, 79*t*
Trypoxylon (wasp), 193, 207
Trypoxylonini (Tribe) (wasps), 63
Tuberanthidium, 72*t,* 492*f,* 526, 526*k,* 527, **528**
Tumidihesma, 68*t,* 222-224, 224*k,* 226-227, **232-233**
Tumidihesma tridentata, 232
Turkanthidium, 71*t,* 506*k,* 507, **508**
Turnerella, 228*f,* 229
Turnerella gilberti, 229
Tylandrena, 69*t,* 244*k,* 249*k,* **266**
Tylanthidium, 72*t,* **521,** 521*k*
Tylanthidium tuberigaster, 521
Tylomegachile, 72*t,* 551, 555, 557, 557*t,* 559*k*-560*k,* **584**
Tytthalictus, 70*t,* 367*k,* **369**
Tyttheriades, 71*t,* 459*k,* **461**

Ulandrena, 69*t,* 252*k,* 255*k,* **266**

Ulanthidium, 72*t*, 492*f*, 534*k*, 535, **537**, 749
Ulanthidium (Olmecanthidium), 537
Ulanthidium interdisciplinaris, 537
Ulanthidium mitchelli, 537
Ulugombakia, 74*t*, 720, **740**
Ulugombakia platytarsus, 740
Uncobombus, 801
Urocolletes, 67*t*, 111, 144, 145*f*, 147*t*, 149*k*, **160-161**
Urohalictus, 70*t*, 320-321, 348, 354, 361, 362*k*, **392**
Urohalictus lieftincki, 390*f*-391*f*, 392
Uromonia, 427*k*, **428**, 428*k*
Uromonia (Nesomonia), 70*t*, 427, **428**, 428*k*
Uromonia (Uromonia s. str.*)*, 43, 70*t*, **428**, 428*k*
Uromonia flaviventris, 428
Uromonia s. str., 43, 70*t*, **428**, 428*k*
Uromonia stagei, 427*f*, 428
Uruguay, fossil, 100*t*, 101

Vachalius, 400
Varroa (mites), 4
Verbena, 20
Verbenapis, 20, 69*t*, 308*f*, 309, 309*k*-310*k*, **313-314**
Vespa concinna, 214
Vespa pratensis, 212
Vestitohalictus, 70*t*, 366, 367*k*, 369, **369-370**, 372
Viereckella, 675
Viereckella obscura, 675
Viridosmia, 485
Volucellobombus, 801

Wagenknechtia, 74*t*, 754, 755*k*-756*k*, 757, **758**
Wainia, 449*k*, 449*t*, 451*k*, 487, **489-490**, 490*k*
Wainia (Caposmia), 71*t*, 489, **490**, 490*k*
Wainia (Trichotosmia), 490
Wainia (Wainia s. str.*)*, 71*t*, **490**, 490*k*
Wainia (Wainiella), 71*t*, 451*k*, 489, **490**, 490*k*
Wainia albobarbata, 490
Wainia algoensis, 451, 489-490
Wainia braunsi, 490
Wainia consimilis, 490
Wainia elizabethae, 451, 490
Wainia eremoplana, 490
Wainia lonavlae, 490
Wainia s. str., 71*t*, **490**, 490*k*
Wainia sakaniensis, 490
Wainiella, 71*t*, 451*k*, 489, **490**, 490*k*
Warnckeia, 501
Willinkapis, 67*t*, 174, 174*k*, **175**
Willinkella, 578

Xanthandrena, 261
Xanthemisia, 74*t*, 753-755, 755*k*-756*k*, 757, **758**
Xanthepicharis, 761
Xanthesma, 133, 220-221, 221*f*, 222, 224*k*, 226, 230, **233-234**, 233*k*
Xanthesma (Argohesma), 68*t*, 233, 233*k*

Xanthesma (Chaetohesma), 68*t*, **233**, 233*k*
Xanthesma (Xanthesma s. str.*)*, 68*t*, **233**, 233*k*, 234
Xanthesma (Xenohesma), 68*t*, 114, 233, 233*k*, **234**
Xanthesma baringa, 233
Xanthesma foveolata, 233
Xanthesma furcifera, 221*f*, 223*f*, 233
Xanthesma infuscata, 233
Xanthesma isae, 233
Xanthesma levis, 233
Xanthesma s. str., 68*t*, **233**, 233*k*, 234
Xanthesma striolata, 233
Xanthidium, 644
Xanthoceratina, 73*t*, 613*k*, **618**
Xanthocotelles, 67*t*, 169, 169*k*, **170**
Xanthocotelles adesmiae, 170
Xanthopedia, 74*t*, 687, 691*k*, **692**
Xanthosarus, 72*t*, 551*f*, 555, 555*f*, 557*t*, 559, 559*k*, 561*k*-562*k*, 580, **584-585**, 585*f*
Xanthosarus s. str., 585
Xanthosmia, 482
Xenanthidium, 71*t*, 499*k*, **501-502**
Xenanthidium biserratum, 501
Xenochalicodoma, 569-570
Xenochilicola, 67*t*, 181, 181*k*, 185, **185**
Xenochilicola mamigna, 185
Xenochlora, 70*t*, 395*k*, 398, 408, 410, **412**
Xenochlora ianthina, 412
Xenochlora ochrosterna, 412
Xenoglossa, 114, 652, 708-709, 717*k*, 719*k*, 722, 733, 736, 740-741, 741*k*
Xenoglossa (Eoxenoglossa), 74*t*, 722, **741**, 741*k*
Xenoglossa (Xenoglossa s. str.*)*, 74*t*, **741**, 741*k*
Xenoglossa angustior, Plate 11
Xenoglossa fulva, 741
Xenoglossa ipomoeae, 722
Xenoglossa kansensis, 713*f*
Xenoglossa s. str., 74*t*, **741**, 741*k*
Xenoglossa strenua, 717*f*
Xenoglossa utahensis, 733
Xenoglossodes, 708, 737-739
Xenohesma, 68*t*, 114, 221, 233, 233*k*, **234**
Xenohylaeus, 68*t*, 195*k*, 197*f*, 198*t*, 213-214, **214**
Xenoliphanthus, 69*t*, 276*f*, 276*k*-277*k*, **278**
Xenomegachile, 581
Xenopanurgus, 279-280
Xenopanurgus readioi, 280
Xenopeponapis, 733
Xenorhiza, 68*t*, 189*k*-190*k*, 215, **219**
Xenorhiza (Papuanorhiza), 219
Xenorhiza krombeini, 21
Xenostelis, 33*t*, 72*t*, 491*t*, 495*k*, **537**
Xenostelis polychroma, 537
Xenoxylocopa, 73*t*, 603*k*-604*k*, 604, 604*k*, **609**
Xeralictoides, 70*t*, 419, 419*k*, **421**
Xeralictus, 69*t*, 322-324, 324*k*, 328, **331**, 348, 421

Xeralictus timberlakei, 330*f*, 331
Xerammobates, 73*t*, **662-663**, 662*k*, 663
Xerocentris, 74*t*, 753-754, 755*k*-756*k*, 756-757, **758-759**
Xerocoelioxys, 72*t*, 545*f*, 546*k*, **550-551**
Xeroheriades, 71*t*, 448-449, 449*t*, 452*k*, **490**
Xeroheriades micheneri, 490
Xeromacrotera, 69*t*, 300, 301*k*, **305**
Xeromegachile, 578, 585
Xeromelecta, 8, 771*k*, **776**, 776*k*
Xeromelecta (Melectomorpha), 75*t*, 765*f*, 770, 770*f*-771*f*, 772, 775*f*, **776**, 776*k*
Xeromelecta (Nesomelecta), 75*t*, 770*f*, **776**, 776*k*
Xeromelecta (Xeromelecta s. str.*)*, 75*t*, **776**, **776**, 776*k*
Xeromelecta alayoi, 770*f*
Xeromelecta californica, 116*f*, 589*f*, 765*f*, 770*f*-771*f*, 775*f*
Xeromelecta larreae, 776
Xeromelecta pantalon, 776
Xeromelecta s. str., 75*t*, **776**, **776**, 776*k*
Xeromelecta tibialis, 776
Xeromelissa, 67*t*, 80*t*, 110, 181, 181*k*, 185, **186**, 905
Xeromelissa wilmattae, 110, 185*f*, 186
Xeromelissinae (Subfamily), 67*t*, 85, 92, 101, 116, 118, 127*k*, 133-134, 135*k*, 159, 171, **180-186**, 180*k*-181*k*, 220, 895, 905
Xeromelissini (Tribe), 110, 180, 185*f*
 biogeography of, 107*t*
Xeropasites, 659
Xeropeponapis, 733
Xerophasma, 69*t*, 299, 300*k*, **305**, 305*f*
Xerophasma bequaerti, 305
Xerosmia, 473-474
Xilocopa, 609
Xiphandrena, 69*t*, 245*k*, 249*k*, **266-267**
Xyelidae, 101
Xylocopa, 6, 16, 49, 52, 77, 77*t*-80*t*, 81, 81*t*, 97, 106, 112, 114, 545, 552-553, 569, 592, 595-597, 597*f*, **598-610**, 598*f*
 of Eastern Hemisphere, 601*k*-604*k*
 of Western Hemisphere, 599*k*-601*k*
Xylocopa (Acroxylocopa), 610
Xylocopa (Afroxylocopa), 606
Xylocopa (Alloxylocopa), 73*t*, 601, 602*k*-603*k*, **604**, 604*k*
Xylocopa (Apoxylocopa), 610
Xylocopa (Audinetia), 606
Xylocopa (Biluna), 73*t*, 598, 602*k*-603*k*, **604-605**, 608
Xylocopa (Bomboixylocopa), 73*t*, 601*k*-602*k*, 604*k*, **605**
Xylocopa (Calloxylocopa), 609
Xylocopa (Cirroxylocopa), 73*t*, 599*k*, 601*k*, **605**

Xylocopa (Copoxyla), 73*t*, 598, 602*k*, 604*k*, **605**
Xylocopa (Ctenopoda), 605
Xylocopa (Ctenoxylocopa), 73*t*, 601*k*, 603*k*, **605**, 609
Xylocopa (Cyphoxylocopa), 606
Xylocopa (Dasyxylocopa), 73*t*, 600*k*-601*k*, **605**, 609
Xylocopa (Diaxylocopa), 73*t*, 599*k*, 601*k*, **605**, 607
Xylocopa (Dinoxylocopa), 610
Xylocopa (Eoxylocopa), 606
Xylocopa (Epixylocopa), 609
Xylocopa (Euxylocopa), 610
Xylocopa (Gnathoxylocopa), 73*t*, 602*k*-603*k*, **605-606**
Xylocopa (Hoplitocopa), 597*f*, 606
Xylocopa (Hoploxylocopa), 606
Xylocopa (Ioxylocopa), 608
Xylocopa (Koptortosoma), 73*t*, 597-598, 598*f*, 602*k*-604*k*, **606**, 608-609, Plate 10
Xylocopa (Lestis), 73*t*, 598*f*, 601*k*, 604*k*, **606**
Xylocopa (Lieftinckella), 606
Xylocopa (Maaiana), 73*t*, 602*k*-603*k*, **606**, 607
Xylocopa (Maiella), 606
Xylocopa (Megaxylocopa), 607
Xylocopa (Mesotrichia), 73*t*, 597*f*, 598, 602*k*-604*k*, **606-607**, 608-609
Xylocopa (Mimoxylocopa), 605
Xylocopa (Monoxylocopa), 73*t*, 599*k*, 601*k*, 605*k*, **607**
Xylocopa (Nanoxylocopa), 73*t*, 598, 599*k*, 601*k*, **607**, 609
Xylocopa (Neoxylocopa), 73*t*, 598, 598*f*, 600*k*, 601, 605, **607**, 609
Xylocopa (Nodula), 73*t*, 596, 602*k*, 603, 603*k*, 605-606, **607**, 608, 610
Xylocopa (Notoxylocopa), 73*t*, 596*f*-597*f*, 600*k*, **607**, 607*f*
Xylocopa (Nyctomelitta), 73*t*, 601*k*, 603*k*, **607**
Xylocopa (Orbitella), 606
Xylocopa (Oxyxylocopa), 606
Xylocopa (Perixylocopa), 609
Xylocopa (Platynopoda), 606-607
Xylocopa (Prosopoxylocopa), 73*t*, 601*k*, 604*k*, **608**, 610
Xylocopa (Proxylocopa), 73*t*, 592, 597, 597*f*, 598, 601*k*, 604*k*, **608**
Xylocopa (Rhysoxylocopa), 73*t*, 602*k*-603*k*, 605, 607, **608**, 610
Xylocopa (Schoenherria), 608
Xylocopa (Schonnherria), 73*t*, 598, 598*f*, 599*k*-601*k*, 605-607
Xylocopa (Schönherria), 608
Xylocopa (Schonnherria), **608**, 609
Xylocopa (Stenoxylocopa), 73*t*, 598, 600*k*, **608-609**
Xylocopa (Xenoxylocopa), 73*t*, 603*k*-604*k*, 604, 604*k*, **609**
Xylocopa (Xylocopa s. str.*)*, 73*t*, 602*k*, 604*k*, **609**

Index of Taxa

Xylocopa (Xylocopina), 608-609
Xylocopa (Xylocopoda), 73t, 599k, 601k, **609**
Xylocopa (Xylocopoides), 73t, 592, 597f, 600k, **609**
Xylocopa (Xylocopsis), 73t, 600k, 607, **609**
Xylocopa (Xylocospila), 608
Xylocopa (Xylomelissa), 73t, 602k-604k, 604-605, 607-608, **609-610**
Xylocopa (Zonohirsuta), 73t, 602k, 604k, **610**
Xylocopa abbreviata, 607
Xylocopa absurdipes, 610
Xylocopa acutipennis, 606-607
Xylocopa amethystina, 596
Xylocopa appendiculata, 604
Xylocopa artifex, 608
Xylocopa assimilis, 597f, 606-607
Xylocopa augusti, 593f
Xylocopa bambusae, 598, 608
Xylocopa bentoni, 606
Xylocopa bimaculata, 605
Xylocopa bomboides, 605
Xylocopa bombylans, 598f
Xylocopa brasilianorum, 607
Xylocopa caffra, 598f, Plate 10
Xylocopa californica, 10, 597f
Xylocopa cantabrita, 608
Xylocopa capitata, 610
Xylocopa carinata, 609
Xylocopa chrysopoda, 608
Xylocopa ciliata, 598, 607
Xylocopa collaris, 610

Xylocopa confusa, 606
Xylocopa cyanescens, 598, 605
Xylocopa darwini, 105
Xylocopa dejeanii, 610
Xylocopa elegans, 609
Xylocopa erythrina, 610
Xylocopa flavorufa, 597f
Xylocopa fraudulenta, 610
Xylocopa funesta, 609
Xylocopa hottentotta, 609
Xylocopa io, 610
Xylocopa iris, 598
Xylocopa lugubris, 610
Xylocopa micans, 598f, 608
Xylocopa mirabilis, 608
Xylocopa muraria, 570, 578
Xylocopa nasalis, 604
Xylocopa nigrita, 598f
Xylocopa nitidiventris, 597f, 608
Xylocopa ocularis, 606
Xylocopa olivieri, 608
Xylocopa orpifex, 53f, 115f
Xylocopa pubescens, 598
Xylocopa ruficollis, 608-609
Xylocopa rufipes, 605
Xylocopa rufitarsis, 609
Xylocopa s. str., 73t, 602k, 604k, **609**
Xylocopa sicheli, 605-606
Xylocopa smithii, 606
Xylocopa tabaniformis, 596f-597f, 607, 607f
Xylocopa tabaniformis orpifex, 96f, 597f, 599f
Xylocopa tenuata, 609
Xylocopa truxali, 605

Xylocopa varipes, 606
Xylocopa varipuncta, 9, 598f
Xylocopa vestita, 605
Xylocopa virginica, 10-11, 599f
Xylocopa watmoughi, 598f
Xylocopidae (Family), 78, 78t-79t, 81, 97
Xylocopina, 600k, 608-609
Xylocopinae (Subfamily), 8, 12-13, 15, 24, 31t, 44, 72t-73t, 80t, 81, 90, 97, 110, 127, 434, 587-588, 588k, **592-594**, 592k-594k, 593f, 595-596, 667, 668f, 671, Plate 1
  biogeography of, 107t
Xylocopini (Tribe), 6, 28, 34, 61, 72t-73t, 80t, 81, 101, 104, 111, 587, 588k, 592, 592k, **596-610**, 611, 620, 623, Plate 10
  biogeography of, 107t
Xylocopites (Tribe), 78t
Xylocopoda, 73t, 599k, 601k, **609**
Xylocopoides, 73t, 592, 597f, 600k, **609**
Xylocopsis, 73t, 600k, 607, **609**
Xylocospila, 608
Xylomelissa, 73t, 602k-604k, 604-605, 607-608, **609-610**

Zacesta, 70t, 419k-420k, **421**, 659

Zacesta rufipes, 421
Zacosmia, 75t, 111, 770, 771k, 773, 777
Zacosmia maculata, 771f
Zadontomerus, 73t, 603f, 611f, 613, 613f, 614k, 616, 618, **618**
Zalygus, 228
Zalygus cornutus, 228
Zaodontomerus, 618
Zaperdita, 303
Zarhiopalea, 68t, 216k, **218**
Zebramegilla, 743
Zikanapis, 67t, 106, 174, 174k, **175**
Zikanapis (Crawfordapis), 175
Zikanapis (Foerstersapis), 175
Zikanapis foersteri, 175
Zikaniella, 70t, 365k, **366**
Zikaniella crassiceps, 366
Zizopanurgus, 294
Zonalictus, 70t, 361-362, 383-384, 384k, 385f, **386**
Zonalonia, 728
Zonamegilla, 743, Plate 13
Zonandrena, 69t, 253k, 255k, **267**
Zonohalonia, 728
Zonohirsuta, 73t, 602k, 604k, **610**
Zonolonia, 728
Zonomegachile, 72t, 557t, 558, 560k, **585-586**
Zosteranthidium, 71t, 492, 498k, 499, **502**